6-17   27-28

AUTOMOTIVE SERVICE

• Inspection • Maintenance • Repair

Third Edition

Tim Gilles

Professor
Santa Barbara City College
ASE Master Automotive Technician,
ASE Master Automotive Machinist

THOMSON

DELMAR LEARNING

Australia • Brazil • Canada • Mexico • Singapore • Spain • United Kingdom • United States

THOMSON
DELMAR LEARNING

Automotive Service: Inspection, Maintenance, Repair, Third Edition
Tim Gilles

Vice President, Technology and Trades ABU:
David Garza

Director of Learning Solutions:
Sandy Clark

Managing Editor:
Larry Main

Senior Acquisitions Editor:
David Boelio

Senior Product Manager:
Matthew Thouin

Marketing Director:
Deborah S. Yarnell

Marketing Manager:
Erin Coffin

Marketing Coordinator:
Patti Garrison

Director of Production:
Patty Stephan

Content Project Manager:
Barbara L. Diaz

Content Project Manager:
Cheri Plasse

Editorial Assistant:
Lauren Stone

Cover Image:
Rebecca C. Johnson/Alamy

Library of Congress Cataloging-in-Publication Data:

Gilles, Tim, 1951-
  Automotive service: inspection, maintenance, repair/Tim Gilles. —3rd ed.
    p. cm.
  ISBN 1-4180-3758-3
  1. Automobiles—Maintenance and repair—Text—books.   I. Title.
  TL152.G516 2008
  629.28'72—dc22

                                    2007022068

NOTICE TO THE READER

Publisher does not warrant or guarantee any of the products described herein or perform any independent analysis in connection with any of the product information contained herein. Publisher does not assume, and expressly disclaims, any obligation to obtain and include information other than that provided to it by the manufacturer.

The reader is expressly warned to consider and adopt all safety precautions that might be indicated by the activities herein and to avoid all potential hazards. By following the instructions contained herein, the reader willingly assumes all risks in connection with such instructions.

The publisher makes no representation or warranties of any kind, including but not limited to, the warranties of fitness for particular purpose or merchantability, nor are any such representations implied with respect to the material set forth herein, and the publisher takes no responsibility with respect to such material. The publisher shall not be liable for any special, consequential, or exemplary damages resulting, in whole or part, from the readers' use of, or reliance upon, this material.

TABLE OF CONTENTS

SECTION 1
■■ THE AUTOMOBILE INDUSTRY

SECTION 2
■■ SHOP PROCEDURES, SAFETY, TOOLS, AND EQUIPMENT

SECTION 3
■ VEHICLE INSPECTION (LUBRICATION/SAFETY CHECK)

SECTION 4
■ ENGINE OPERATION AND SERVICE

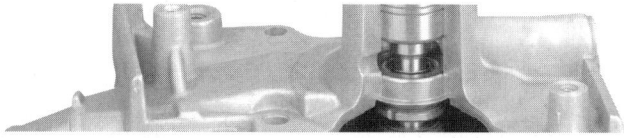

SECTION 5
■ COOLING SYSTEM, BELTS, HOSES, AND PLUMBING

SECTION 6
■ ELECTRICAL SYSTEM THEORY AND SERVICE

SECTION 7
■ HEATING AND AIR CONDITIONING

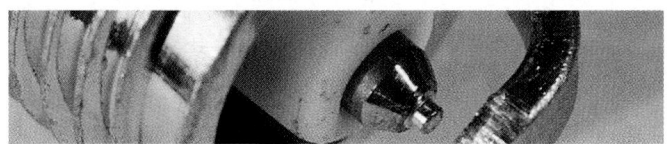

SECTION 8
■ ENGINE PERFORMANCE DIAGNOSIS: THEORY AND SERVICE

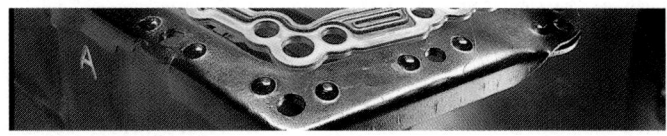

SECTION 9
■ AUTOMOTIVE ENGINE SERVICE AND REPAIR

SECTION 10
■ BRAKES AND TIRES

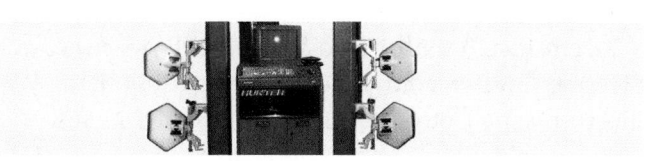

SECTION 11
■ SUSPENSION, STEERING, ALIGNMENT

SECTION 12
■ DRIVE TRAIN

From Tim Gilles:

Automotive Service: Inspection, Maintenance, and Repair evolved in part from my participation in a successful articulation program between local high schools and the community college where I have been a teacher for over 30 years. The text and art manuscripts of this best-selling book have once again been updated and improved in this third edition revision. An array of excellent technical reviews by a dedicated group of professional teachers and technicians ensures that this is the most technically accurate and up-to-date comprehensive automotive textbook available in the marketplace. The text has been written from a carefully detailed outline to allow each chapter to follow a logical, easy-to-understand path. Hundreds of new original color photographs and sketches have been added to update and supplement earlier material.

The transportation industry in North America is vast, with one in every six people contributing to it in some way. These include people of many levels of understanding and ability. With that in mind, the introductory fundamentals chapters are written at a lower level for all of the students, whereas the service chapters are for those who have mastered the introductory material. As a teacher, my philosophy is to challenge the best students in anticipation that the rest will be brought to a higher level: "a rising tide raises all boats."

The text can be used for a variety of educational purposes, including:

■ As a basic text in any automotive repair class
■ To train entry-level or apprentice technicians
■ To prepare more experienced technicians for ASE certification

Automotive Service: Inspection, Maintenance, and Repair, 3e is divided into 79 chapters that cover the NATEF Auto General Service Technician Program Standards and A1–A8 (the eight ASE automotive specialty areas). Advanced engine performance and emission controls are also covered in detail. In addition to coverage of the usual repairs performed in almost any automotive repair facility, the reader is introduced to the most frequently performed inspection and service procedures, from safety inspections to tire and wheel service.

The automotive repair industry of today is becoming more of a maintenance industry, as vehicles last longer and require fewer repairs. Long-term customer relationships, ethics, and professionalism have become more important to the success of a business. There are also environmental concerns today that were not a part of the industry of the past. The text includes chapters on safety, hand tools, and vehicle maintenance and lubrication that are more comprehensive than those found in most comparable texts. An accompanying lab manual emphasizes the NATEF Auto General Service Technician Program Standards, those jobs done in service stations, fast-lube outlets, or mass merchandisers (such as Sears, Goodyear, Firestone, K-Mart, or Montgomery Ward). An additional lab workbook, *Automotive Service Job Sheets for NATEF Task Mastery*, a new option with this edition, covers the eight ASE system areas.

Automobiles have become so complex in the last 30 years that to remain competent many of today's technicians specialize in one or more systems of the car. Basic theory of all automotive and light truck systems is covered so that service personnel will understand the function of the parts being serviced. When working in the industry, there is often no time for basic theory. Therefore, automotive class work could be a student's one and only chance to learn how systems operate so they can become better diagnosticians in the future.

This book is comprehensive in that it deals with the entire car and aims to teach theory of vehicle systems at an introductory student level, followed by service, diagnosis, and light repairs at a more advanced student level.

Most of the systems used in automobiles today are strikingly similar. Repair techniques universal to all automobiles are discussed and procedures or conditions unique to only one specific automobile make are purposely avoided. The reader is encouraged to refer to the service and repair information for the specific vehicle in question.

A major challenge for me as an automotive author is to keep the scope of the book from growing out of control. The emphasis must remain on preparation for *job-entry level*. Following the completion of this text, better students will be at a high job-entry level, at a skill level suitable for entry as an apprentice in one or more of the specialty areas of automotive repair in a new car dealership or an independent repair shop.

A primary objective of *Automotive Service: Inspection, Maintenance, and Repair* is to help the reader develop confidence in both thinking skills and problem-solving ability. One unique aspect of automotive education is that many automotive graduates venture into other professions and trades, such as engineering or construction. They will find much of the material learned in automotive classes to be very valuable and useful in their chosen fields. This aspect of the student's education is especially valuable when one considers how middle school and high school industrial arts programs have been scaled back in recent years. Dealing with such things as tools, soldering, basic electrical repairs, and repairing broken fasteners helps to provide some measure of practical education.

The tremendous decline in the number of corner gasoline service stations has resulted in a loss of those jobs formerly available in abundance to students. Successful service personnel who possess necessary basic automotive skills must continually learn new things in order to progress into other (higher-paying) specialty areas. The *Lab Manual to Accompany Automotive Service* contains service jobs that students should be able to perform before enrolling in an advanced automotive specialty area class.

NEW TO THIS EDITION

- Extensive new information on hybrid vehicles and alternative fuels.
- The latest information on engine oils and coolants.
- New engine performance and electronics coverage includes technology advances in on-board diagnostics, misfire detection, variable valve timing, displacement on demand, adaptive strategies, computer networks (including the latest CAN systems), drive-by-wire, and wide band oxygen sensors.
- Chassis and powertrain improvements include new technology in antilock brakes, stability control and traction control systems, electronic suspension systems, electric steering, and electronic transmission and all-wheel drive.
- Updated comfort and safety coverage includes the newest developments in air conditioning, supplemental restraints, lighting, entertainment systems, and adaptive cruise control.
- Vehicle electronics coverage has been improved and increased throughout, and electronics service has been expanded with more complete coverage of digital lab scopes, scan tools, and current probes.
- The table of contents has been improved to provide a more logical sequence of material, especially in the area of engine performance.
- End-of-chapter test questions have been rewritten to provide a better mix of question styles. New science, math, and history notes have also been added.
- An improved e.resource CD-ROM for instructors (see the Supplements page for more detail).
- A new back-of-book CD includes ASE review questions, and PowerPoint presentations with images, games and activities, and video clips for student self-review and assessment.

- In addition to the *Lab Manual to Accompany Automotive Service*, a new optional job sheet manual is available—*Automotive Service Job Sheets for NATEF Task Mastery*—that covers in depth the eight ASE areas.
- All automotive terms, abbreviations, and acronyms used in this text comply with the SAE Technical Standards Board Publication *SAE J1930*.

ACKNOWLEDGMENTS

I would like to extend special thanks to the following individuals, organizations, and companies:

- Thomson Delmar Learning Product Manager Matt Thouin for his helpful, positive attitude. Matt's dedication to excellence and his efficiency and organizational skills were very important to the overall improvement of this third edition.
- Denise Denisoff and Chris Shortt, developmental editors in earlier editions of the text.
- Cheri Plasse and Barbara Diaz, Content Project Managers, who managed the art, design, and production of the text. Cheri and Barbara have worked on several of my projects in the past and are always a pleasure to work with.
- Lori Hazzard and the staff at ICC Macmillan, Inc., who worked tirelessly behind the scenes. Their professional talent and effort with the text and art ensure a quality final product.
- Friends and colleagues Bob Stockero, Dave Brainerd, and Gary Semerdjian of Santa Barbara City College, and Chuck Rockwood of Ventura College for their continuous input and support.
- Members of the North American Council of Automotive Teachers (NACAT) and California Automotive Teachers (CAT), who provided a vast amount of input.
- *Students in the Automotive program at Santa Barbara City College, who provided continuous feedback and suggestions for improvement.*
- Physics professor Mike Young of Santa Barbara City College for his help with the Science Notes.
- Jack Rosebro of Perfect Sky, Inc., who helped with reviews of the hybrid vehicle material.
- Bernie Carr, Senior Engineer at Vetronix Corporation in Santa Barbara, who reviewed and helped with the new information on controller area networks (CAN) and other vehicle electronics material.
- Staff at all of the campuses of Universal Technical Institute, who provided helpful reviews of the manuscript.

ACKNOWLEDGMENTS

The contributions of the following reviewers are gratefully acknowledged:

Clayton Allen
Universal Technical Institute
Glendale Heights, IL

David Anderson
Universal Technical Institute
Orlando, FL

Jerry Andrews
Universal Technical Institute
Rancho Cucamonga, CA

John Archambault
NASCAR Technical Institute
Mooresville, NC

Wallace A. Armstrong
Universal Technical Institute
Rancho Cucamonga, CA

Jeffry Bankston
Universal Technical Institute
Houston, TX

David J. Bayley
Universal Technical Institute
Rancho Cucamonga, CA

William H. Beam
Universal Technical Institute
Exton, PA

Oliver Beckham
Universal Technical Institute
Houston, TX

David Bement
Universal Technical Institute
Glendale Heights, IL

W. Bogner
Universal Technical Institute
Houston, TX

James Bouwens
Paulding County Schools
Dallas, GA

Louis E. Brantmeyer
Universal Technical Institute
Rancho Cucamonga, CA

David R. Bruck
Universal Technical Institute
Rancho Cucamonga, CA

Bernard J. Carr
Vetronix Corporation
Santa Barbara, CA

Michael A. Carzoo
Universal Technical Institute
Rancho Cucamonga, CA

Mark Childers
Universal Technical Institute
Rancho Cucamonga, CA

Thomas M. Clark
Universal Technical Institute
Rancho Cucamonga, CA

David Cole
Universal Technical Institute
Rancho Cucamonga, CA

James Coll
Universal Technical Institute
Exton, PA

James Coombes
Universal Technical Institute
Exton, PA

Phillip Cooper
NASCAR Technical Institute
Mooresville, NC

William Crawford
Universal Technical Institute
Avondale, AZ

Matt Davis
Universal Technical Institute
Rancho Cucamonga, CA

Steve Davies
Universal Technical Institute
Exton, PA

Ryan Davis
Universal Technical Institute
Rancho Cucamonga, CA

Ed Deal
Universal Technical Institute
Glendale Heights, IL

Ralph DeMatteo
Universal Technical Institute
Phoenix, AZ

Rich Devito
Universal Technical Institute
Exton, PA

Michael Dommer
Indian Hills Community College
Ottumwa, IA

Jim Dorsten
Celina City School
Celina, OH

Tim Dwyer
Oklahoma State University
Okmulgee, OK

Robert Egli
Universal Technical Institute
Avondale, AZ

Russell Ferguson
Washtenaw Community College
Ann Arbor, MI

B. Fisher
Universal Technical Institute
Rancho Cucamonga, CA

Randall Glasser
NASCAR Technical Institute
Mooresville, NC

Randy Golding
Arizona Automotive Institute
Phoenix, AZ

K. Gordon
NASCAR Technical Institute
Mooresville, NC

Chuck Greene
York Tech College
Rock Hill, SC

David Haston
Universal Technical Institute
Rancho Cucamonga, CA

John S. Haston
Universal Technical Institute
Rancho Cucamonga, CA

Dennis W. Hayes
Universal Technical Institute
Rancho Cucamonga, CA

Anthony Herman
Universal Technical Institute
Avondale, AZ

Jamie C. Hixon
Universal Technical Institute
Rancho Cucamonga, CA

Lance Idecker
San Marcos High School
Santa Barbara, CA

Robert F. Jackson
Universal Technical Institute
Rancho Cucamonga, CA

Rodger D. Jerls
Universal Technical Institute
Rancho Cucamonga, CA

William R. Jones
Universal Technical Institute
Rancho Cucamonga, CA

Greg Jorgenson
Universal Technical Institute
Avondale, AZ

Roy Knight
Southern High School
Louisville, KY

Mike Lemke
Middle Bucks Institute of Technology
Jamison, PA

James Lewis
Universal Technical Institute
Exton, PA

Daniel Livingston
Jackson Community College
Jackson, MI

Joe Martinez
Universal Technical Institute
Rancho Cucamonga, CA

Bob Masterson
Universal Technical Institute
Houston, TX

Robert G. Mathers
Universal Technical Institute
Avondale, AZ

Carl McClain
Universal Technical Institute
Glendale Heights, IL

Harry McPheron
Celina City School
Celina, OH

David Mendoza
Universal Technical Institute
Rancho Cucamonga, CA

Marcos Meza
Universal Technical Institute
Rancho Cucamonga, CA

Mark J. Miller
Universal Technical Institute
Rancho Cucamonga, CA

Calvin Motley
ATI
Dallas, TX

Michael Murrill
Universal Technical Institute
Rancho Cucamonga, CA

Ken Nagel
Universal Technical Institute
Glendale Heights, IL

Mike F. Nitz
Universal Technical Institute
Rancho Cucamonga, CA

B. J. Nolek
Universal Technical Institute
Exton, PA

Robin Norris
Universal Technical Institute
Avondale, AZ

Dennis Palmer
Rock Valley College
Rockford, IL

Edward M. Peace
University of Alaska, Anchorage
Anchorage, AK

Rodney G. Perkins
University of Alaska, Anchorage
Anchorage, AK

Paul J. Petersen
Universal Technical Institute
Rancho Cucamonga, CA

Paul Peterson
Universal Technical Institute
Rancho Cucamonga, CA

Jeff Pittman
Universal Technical Institute
Orlando, FL

George Potter
Universal Technical Institute
Glendale Heights, IL

John J. Quiles
Universal Technical Institute
Rancho Cucamonga, CA

Brian Rasmussen
Universal Technical Institute
Rancho Cucamonga, CA

Scott M. Rittenhouse
Universal Technical Institute
Rancho Cucamonga, CA

Harold L. Robar
Universal Technical Institute
Rancho Cucamonga, CA

Chris Roberts
Universal Technical Institute
Glendale Heights, IL

Chuck Rockwood
Ventura College
Ventura, CA

Jack Rosebro
Perfect Sky, Inc.
Santa Barbara, CA

Brandon M. Sangster
Universal Technical Institute
Rancho Cucamonga, CA

Walter Shields
Des Moines Area Community College
Ankeny, IA

David Sitchler
Burlington County Institute of Technology
Westhampton, NJ

Peter Skoupas
Universal Technical Institute
Glendale Heights, IL

Skip Smith
Universal Technical Institute
Avondale, AZ

Raymond Brian Spray
Universal Technical Institute
Orlando, FL

Greg Stewart
San Bernardino High School
San Bernardino, CA

Bob Stockero
Santa Barbara City College
Santa Barbara, CA

Hunter S. Taylor
Halifax Community College
Weldon, NC

Philippe A. Toth
Universal Technical Institute
Rancho Cucamonga, CA

Bob T. Troxler
Universal Technical Institute
Rancho Cucamonga, CA

Joe Viveros
Universal Technical Institute
Rancho Cucamonga, CA

Michael K. Wallan
Universal Technical Institute
Rancho Cucamonga, CA

Michael Ward
Southern Westchester BOCES
Valhalla, NY

Steven Ware
Universal Technical Institute
Rancho Cucamonga, CA

Ken Welch
Mission Viejo High School
Mission Viejo, CA

Christopher R. Williams
Universal Technical Institute
Rancho Cucamonga, CA

Darryl Williams
Universal Technical Institute
Rancho Cucamonga, CA

Mike Witgen
Universal Technical Institute
Glendale Heights, IL

Marty J. Zuniga
Universal Technical Institute
Rancho Cucamonga, CA

The contributions of additional reviewers from the first and second editions of this text are also acknowledged:

Bill Routley, Ferris State University; Lorin Cuthbert, Santa Maria High School; Lee Haeberlein, Rio Hondo College; Michael McLaughlin, San Marcos High School; Dan Perrin, Trident Technical College; George Hritz, College of Marin; Quentin Swan, California Department of Education; and Patrick Brown Harrison, Southern Alberta Institute of Technology.

Robert Abbey, Santa Barbara City College, Santa Barbara, CA; Stu Bradholdt, Santa Barbara City College, Santa Barbara, CA; Joe Bradbury, Arroyo Grande High School, Arroyo Grande, CA; Frederick G. Heath, Heath and Associates, Louisville, KY; David Lannom, Lincoln Technical Institute, Grand Prairie, TX; Stephan Baldwin, Sullivan High School, Sullivan, IN; Richard Caperton, James L. Walker Vo-Tech, Naples, FL; James Pinto, Porter and Chester Institute, Wethersfield, CT; Charles Statz, Temple Jr. College, Temple, TX; Ronnie Bush, Tennessee Tech Center at Jackson, Jackson, TN; Patrick Devlin, Montgomery College, Rockville, MD; Brad Walsh, Mark Keppel High School, Alhambra, CA; Steve Ford, San Marcos High School, Santa Barbara, CA; Kim Edward Ray, Northwest Career Center,

Dublin OH; Robert Wenzlaff, Fullerton College, Fullerton, CA; Walter Bertotti, Porter and Chester Institute, Watertown, CT; Earl Friedell, DeKalb Technical Institute, Clarkston, GA; Paul Pate, Community College of Southern Nevada, Boulder City, NV; Johnny Beason, Murray County High School, Chatsworth, GA; Jim Zimmerman, Union Grove High School/Gateway Technical College, Union Grove, WI; Michael T. Maher, Fox Valley Career Center, DeKalb, IL; Tommy Wilson, Columbia Technical Institute, Fort Mitchell, AL; Daniel Wickware, Northeast Texas Community College, Mt. Pleasant, TX; Photographers: Kurt Rhody; Tim Gilles; Models: Bridget Berg; Gary Semerdjian; Janet Hill; Julio A. Limon; Darren George; Travis Alexander; Julianne Casey; Mark Kono; Kurt Reed; William Tomlin; and Cheryl Weikliem; Hsi Lin, Jeff Felske.

The author expresses sincere appreciation to the following companies and individuals who have supplied illustrations, technical information, or other assistance during the preparation of *Automotive Service: Inspection, Maintenance, and Repair:*

AERA; Tim Waters, All Pro/Bumper to Bumper; Guy Avellon, ASTM, Committee on Fasteners; Automotive Lift Institute; Bob Barkhouse, Yuba College; Tom Birch, Yuba College; Charlie Camp; Antonio Casillas, Santa Barbara High School; Joe Schuit, Channel City Engineering, Santa Barbara, CA; Clay Furtaw, DeWalt Machinery Co.; Boyce Dwiggins; Dema Elgin, Elgin Racing Cams, Redwood City, CA; EngineTech, Inc., Carrollton, TX; Jack Erjavec, Delmar Publishers; Rick Escalambre, Skyline College; Ray Fausel, Professor Emeritus, Cal-State Los Angeles; Jim Halderman, Sinclair Community College; Frederick G. Heath, Heath and Associates, Louisville, KY; Ed Hernandez, Santa Barbara City College; Barry Hollembeak, Technical Training, Inc., Rochester Hills, MI; Jim Hughes, Rio Hondo College; Don Knowles, Knowles Automotive Training, Moose Jaw, Saskatchewan; Mark Kreger, Ertel Products Co.; Dr. Norman Laws, Professor Emeritus, Chicago State University; Randy Ingle, General Motors Technical Instructor; Alan Nagel, General Motors Training Center, Calgary, Alberta, Al Santini, Professor Emeritus, College of DuPage; Toby's Engine Parts, Santa Barbara, CA; Phil Unander, Larry's Auto Parts, Santa Barbara, CA; Dr. Jay Webster, Professor Emeritus, Long Beach State, and Stan Crapo.

Portions of materials contained herein have been reprinted with permission of Alternative Fuel Systems, Inc., Calgary, Alberta, Canada; Deere & Company, John Deere Publishing, Moline, Illinois; General Motors Corporation, Service and Parts Operations, License No. 10713; and Tokico (USA), Inc., Performance Division, Rancho Dominguez, California, 1-800-548-2549.

DEDICATION

The completion of this book was made possible with help from a great many individuals. *Automotive Service: Inspection, Maintenance, and Repair, Third Edition* is dedicated to them and to my wife, Joy, and my daughter, Terri. Their organizational skills and able assistance were invaluable in preparing and organizing the art package. This was a giant task!

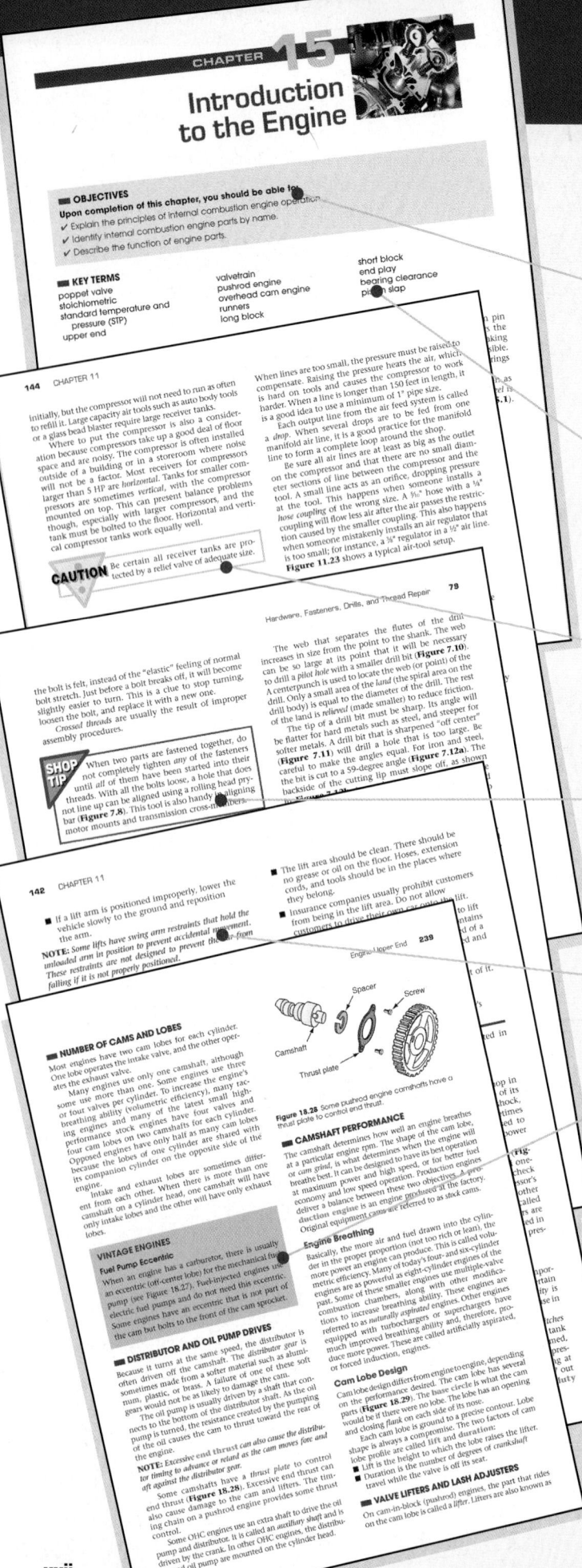

FEATURES
OF THE TEXT

■ OBJECTIVES

Each chapter begins with a list of the most important points discussed in the chapter. This list of objectives is intended to provide the student with a general idea of what he or she will be studying.

■ KEY TERMS

Each chapter contains a list of new terms to know. These terms are **highlighted in bold** in the text. Descriptions of these terms can be found in the glossary.

■ CAUTIONS

Cautions are urgent warnings that personal injury or property damage could occur if careful preventative steps are not taken.

■ SHOP TIPS

Appropriate shop tips are described throughout the text. These tips provide shortcuts and emphasize fine-tuning procedures to shop practices commonly performed by experienced technicians.

■ NOTES

Notes are used throughout the text to highlight especially important topics.

■ VINTAGE SYSTEM NOTES

Boxed information on vintage systems (and related parts) puts today's newer technologies in historical perspective and offers the reader insights into the development of the automobile.

SAFETY NOTES

Safety is the number-one priority in an automotive shop. There are numerous safety notes placed throughout the text. Most of the safety notes have been taken from real-world experiences and many are accompanied by actual case histories.

CASE HISTORIES

Case histories are presented throughout the text. These true stories recount actual automotive situations encountered by the author in over 30 years in automotive service. Case histories present the reader with examples of the pattern of critical thinking skills required to diagnose automotive problems.

SCIENCE, HISTORY, COMPUTER SYSTEM, AND MATH NOTES

These notes are included when interesting topics relating to them are covered in the text. The objective of these features is to pique the student's interest and show a correlation between his or her automotive studies and these areas of learning.

REVIEW QUESTIONS

These questions guide the student to the most important points in the chapter and act as a check for understanding of the material. Each chapter's review questions are presented in the same order in which they appear in the chapter. This provides an instructor with the flexibility to assign portions of the chapter to read and then follow up with a few of the study questions.

ASE-STYLE REVIEW QUESTIONS

These questions are designed to provide preparation for the certification examinations administered by the *National Institute for Automotive Service Excellence* (*ASE*). The ASE test is a task-oriented test (the test taker is supposed to be able to relate to shop-oriented questions), so the practice tests should help the student in becoming familiar with accepted trade procedures.

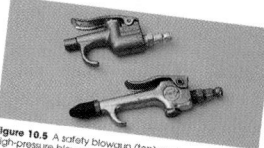

The following reproduce sample textbook pages shown at right:

---

124 CHAPTER 10

the soap suspends dirt so it can be washed away. Dirt is the only water-soluble material found on automotive parts.

Organic soils include petroleum by-products, gasket sealers and paints, carbon, and other by-products of combustion. These materials *cannot* be effectively cleaned with water. Chemicals are used to make these soluble so that water will be able to wash them off.

Cleaning with Bases

Alkaline materials cut grease very well and work best when heated. Most automotive soaps are alkaline. Using soap and hot enough water temperatures are both very important when cleaning grease. Soap lifts grease and makes cleaning easier and much more effective. Temperature is also very important. Remember how much faster hands clean up when using hot water instead of cold?

SCIENCE NOTE

A cleaning chemical will clean about 10 times as well at 150°F than at room temperature. From 140°F to 200°F cleaning time is cut in half by every increase of 20°F. Agitating, or moving the liquid, also shortens cleaning time.

Cleaning with Acids

Acids are useful only in removing rust and scale. Acid will not cut grease. Before rust can be removed, any oil or grease must first be removed with an alkaline material. When an acid is used to remove rust or scale, it

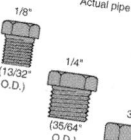

Figure 10.5 A safety blowgun (top) and a rubber-tipped high-pressure blowgun (bottom). (*Courtesy of Tim Gilles*)

Air Blowguns

Air blowguns can be used to blow off parts. There are two main types of blowguns. The *safety blowgun* (**Figure 10.5**) restricts pressure to around 35 psi for air-drying parts. When more pressure is required, a *rubber-tipped blowgun*, which allows full shop air pressure, is used.

SAFETY NOTE Be especially careful when using the high-pressure, rubber-tipped blowgun. Compressed air at full shop pressure can be dangerous and can actually penetrate your skin. Do not blow air against your skin. Wear eye protection and use care when blowing into holes. The liquid will come back out at you. Remember to blow down and away from yourself.

---

82 CHAPTER 7

CASE HISTORY An apprentice was chasing the threads in a cylinder block after removing it from the cleaning tank. He tapped out all of the holes, including those for the oil gallery plugs. Oil gallery plugs have tapered pipe threads. The apprentice turned the tapered pipe tap through the entire length of the thread until it turned easily. When the new oil gallery plugs were installed, they could not become tight because the threads in the block were now too big. The block had to be repaired.

Actual pipe thread sizes

1/8"
(13/32" O.D.)

1/4"
(35/64" O.D.)

3/8"

1/2"

---

124 CHAPTER 10

the soap suspends dirt so it can be washed away. Dirt is the only water-soluble material found on automotive parts.

Organic soils include petroleum by-products, gasket sealers and paints, carbon, and other by-products of combustion. These materials *cannot* be effectively cleaned with water. Chemicals are used to make these soluble so that water will be able to wash them off.

Cleaning with Bases

Alkaline materials cut grease very well and work best when heated. Most automotive soaps are alkaline. Using soap and hot enough water temperatures are both very important when cleaning grease. Soap lifts grease and makes cleaning easier and much more effective. Temperature is also very important. Remember how much faster hands clean up when using hot water instead of cold?

SCIENCE NOTE

A cleaning chemical will clean about 10 times as well at 150°F than at room temperature. From 140°F to 200°F cleaning time is cut in half by every increase of 20°F. Agitating, or moving the liquid, also shortens cleaning time.

Cleaning with Acids

Acids are useful only in removing rust and scale. Acid will not cut grease. Before rust can be removed, any oil or grease must first be removed with an alkaline material. When an acid is used to remove rust or scale, it

Figure 10.5 A safety blowgun (top) and a rubber-tipped high-pressure blowgun (bottom). (*Courtesy of Tim Gilles*)

Air Blowguns

Air blowguns can be used to blow off parts. There are two main types of blowguns. The *safety blowgun* (**Figure 10.5**) restricts pressure to around 35 psi for air-drying parts. When more pressure is required, a *rubber-tipped blowgun*, which allows full shop air pressure, is used.

SAFETY NOTE Be especially careful when using the high-pressure, rubber-tipped blowgun. Compressed air at full shop pressure can be dangerous and can actually penetrate your skin. Do not blow air against your skin. Wear eye protection and use care when blowing into holes. The liquid will come back out at you. Remember to blow down and away from yourself.

---

134 CHAPTER 10

REVIEW QUESTIONS

1. Wet parts can be carried from the solvent tank in a _____ _____ to prevent solvent from dripping on the floor.
2. Pure water has a pH rating of _____.
3. If the nozzle of a bead blaster is aimed at the window, it frosts the _____ _____.
4. Why should the nozzle not be aimed at the glove that is holding the part being blasted?
5. Be sure to rinse and _____ _____ to protect machined areas of ferrous metals after hot-tanking.
6. How can a metal be tested to see if it is ferrous or nonferrous?
7. An OHC head has oil galleries. This makes it impractical to clean these heads with a glass bead _____.
8. What is another name for the airless blaster?
9. _____ and open flame are two types of thermal cleaning ovens.
10. Rust and _____ are two types of materials that thermal cleaning removes that other cleaning methods do not.

ASE-STYLE REVIEW QUESTIONS

1. Which of the following could be used for removing rust or scale from the cooling surfaces of a cast iron engine?
   a. Base
   b. Acid
   c. Both A and B
   d. Neither A nor B
2. Technician A says that if the temperature of the hot tank liquid is too low, cleaning time will be longer. Technician B says that if the temperature of the hot tank liquid is too low a spray-type hot tank will foam over. Who is right?
   a. Technician A
   b. Technician B
   c. Both A and B
   d. Neither A nor B
3. Technician A says that larger shot hits a part with more impact. Technician B says that parts can be made stronger by shot peening. Who is right?
   a. Technician A
   b. Technician B
   c. Both A and B
   d. Neither A nor B
4. The temperature of the caustic cleaner in the hot tank should be about:
   a. 100°
   b. 150°
   c. 190°
   d. 250°
5. Which of the following could be used to clean an aluminum part?
   a. A shot blaster
   b. A soda blaster
   c. A caustic cleaner
   d. Any of the above

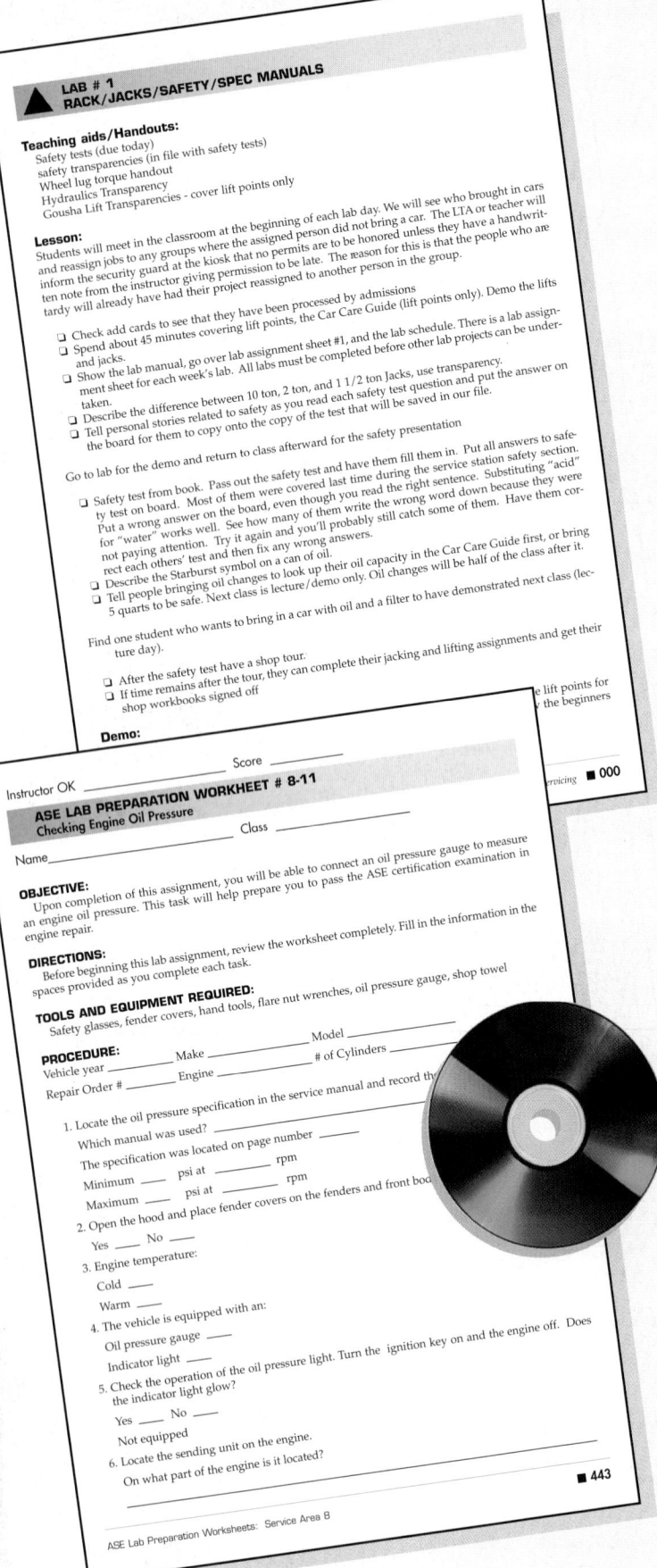

SUPPLEMENTS

■ **e.resource™**—The third edition e.resource™ CD-ROM includes the Instructor's Guide in Microsoft Word, a Computerized Test Bank with hundreds of modifiable questions (true/false, fill-in-the-blank, and ASE-style multiple choice), PowerPoint presentations with full-color images and video clips, a searchable Image Library of hundreds of full-color photos and line art from the core text, and Correlation Grids to the NATEF Automobile Program Standards. "Also, a new Distance Learning section on the e.resource provides the full Computerized Text Bank, ASE-style system-specific tests in ExamView, an complete PowerPoint™ presentations in a format ready for easy uploading into your Blackboard or WebCT online learning platforms."

■ **Instructor's Guide**—The Instructor's Guide contains answers to review questions, lesson plans (lecture outlines), lab exercises, and chapter and page number correlations to all eight topic areas of the NATEF Standards.

■ **Lab Manual**—The Lab Manual includes Worksheets that define each lab procedure presented in increasing levels of difficulty. Each project or lab assignment is built upon the next in a logical sequence in much the same manner as science instructional programs are constructed, and the reader completes one task before progressing to the next one. In addition, a variety of illustrations support the Worksheets and help visual learners better understand the jobs. The Worksheets are keyed to the NATEF Auto General Service Technician Programs Standards where applicable.

■ **Automotive Service Job Sheets for NATEF Task Mastery**—The Job Sheets in this manual cover all P-1, P-2, and most P-3 procedures as identified by the most recent NATEF Automobile Program Standards. Full-color illustrations offer visual support to the Job Sheets, and the sequence of topics follows that of the core text. Each Job Sheet includes a simple and clear rating rubric for instructor evaluation of student performance on the task, and a supplemental CD-ROM includes NATEF task-tracking software for easy recording of students' mastery of the procedures.

The Automobile Industry

THEN AND NOW: THE AUTOMOBILE INDUSTRY

On January 29, 1886, Karl Benz of Mannheim, Germany, patented the world's first automobile, the three-wheeled *Benz Motorwagen*. Later that same year, Gottlieb Daimler of Cannstadt, Germany, built a four-wheeled car. Its 1.5 hp engine had 50% more power than the Benz; the horsepower race had begun. In 1900, Benz's company became the biggest auto maker in the world, building 603 cars.

Long before Benz's patent, there were ingenious automotive inventors and tinkerers in the United States. But the 1896 *Duryea* of Massachusetts was the first car to be produced for sale in the United States, followed shortly by the *Haynes and Winton*. In 1900, Ransom E. Olds of Detroit became the first to mass-produce automobiles in America, the curved-dash "merry Oldsmobile" of the song "Come Away with Me, Lucille."

Henry Ford was the first to produce the automobile in mass quantities. His grand idea was to build a car that everyone could afford. In 1903, the current Ford Motor Company was founded. The first Model T was sold in 1908. In its 19-year run, 15 million copies of that rugged, simple automobile were produced. This record was not surpassed until the Volkswagen Beetle did so in the early 1970s.

Another early automotive giant was General Motors, which in 1908 bought Buick, Oldsmobile, Cadillac, and Oakland (which would become Pontiac). Within 2 years, 30 firms had been brought under the GM umbrella, including 11 auto makers.

Walter P. Chrysler merged Willys and Maxwell-Chalmers in 1920. The first car to bear the Chrysler name went on sale in 1924 and was a huge success.

A current trend among foreign car makers is to build assembly plants in the United States. This idea is far from new. In 1888, Steinway & Sons, the New York piano maker, obtained the rights to all Daimler patents in the United States. It produced engines and cars in this country between 1905 and 1907.

Today, one out of every nine working Americans builds, sells, or fixes and maintains motor vehicles. Automobile dealerships account for 29% of the retail business conducted in the United States. Automotive-related business represents about a third of a trillion dollars worth of our nation's economy in an average year. It is estimated that $150 billion of that total is spent on parts, repairs, and maintenance.

The 1886 Benz Patent Motorwagen. (© *Courtesy of Mercedes-Benz USA, LLC*)

A delivery to a modern dealership. (© *Courtesy of Mercedes-Benz USA, LLC*)

CHAPTER 1

Introduction to the Automobile

■ OBJECTIVES

Upon completion of this chapter, you should be able to:

✔ Describe the differences between the unibody design and frame and body design.

✔ Tell how the four-stroke cycle engine operates.

✔ Understand the purposes of the major engine support systems.

✔ Describe the parts of front- and rear-wheel drive powertrains.

✔ Explain major events in the history of the automobile.

■ KEY TERMS

horseless carriage	atomization	torque
chassis	butterfly valve	clutch
unibody design	ignition timing	transaxle
reciprocating engine	sensor	Otto-cycle
piston ring	actuators	Selden patent
four-stroke cycle	FWD	CAFE
stroke	RWD	

■ INTRODUCTION

Automobiles have been on the scene for over 100 years. Early vehicles were built on the principle of the horse and wagon and were called **horseless carriages** (**Figure 1.1**). Continuing developments in vehicle design since those early years have resulted in today's high-tech vehicle (**Figure 1.2**). Some of those developments are outlined later in this chapter.

Today, there are over 130 million passenger cars in the United States alone (1.5 per household). This is just over one-third of the cars in the world. The automobile is a direct or indirect source of employment for one of every nine workers in the country. Americans drive an average of 7,767 miles per year.

The automobile includes several systems. These systems include the body and suspension, the engine, the electrical system, the powertrain, emission controls, and accessory systems.

■ BODY AND CHASSIS

The **chassis** (pronounced "chassy") includes the suspension. It supports the engine and the car body. The chassis also includes the *frame*, the *brakes*, and *steering*

components. Older cars had heavy frames, but most cars are built today with what is known as a **unibody design** (**Figure 1.3**). This design has a floorpan and a small subframe section in the front and rear. The *body* includes the interior of the car, windows, door latches, the body electrical system, and the body accessories.

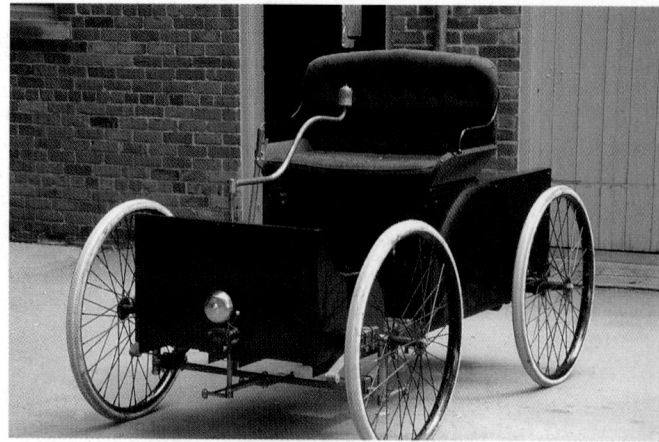

Figure 1.1 A horseless carriage. *(Courtesy of Tim Gilles)*

Figure 1.2 Today's automobile. *(Courtesy of DaimlerChrysler)*

Figure 1.4 A modern gasoline-powered engine. *(Courtesy of DaimlerChrysler Corporation)*

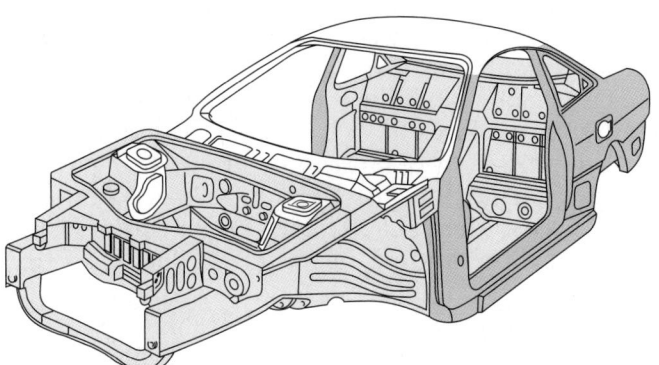

Figure 1.3 Unibody construction.

ENGINE PARTS AND OPERATION

Most of today's automobiles use spark-ignited *four-stroke reciprocating* gasoline engines. A limited number of cars use diesel engines. **Figure 1.4** shows a cutaway of a modern gasoline engine. In-depth coverage of engine operation is found in Chapter 15. A quick overview is given here.

A reciprocating gasoline engine has a round piston in a cylinder, a connecting rod, and a crankshaft. The principle of its operation is simple. The piston moves up in the cylinder, compressing a mixture of air and fuel in front of it. Compressing the air and fuel makes it very flammable. When the piston reaches the top of its travel, the *air-fuel mixture* is ignited. As the piston is pushed down in the cylinder by the expanding gases, it pushes on the rod, forcing the crankshaft to rotate.

Power from the rotation of the crankshaft turns the wheels. As the crankshaft turns, the piston is returned to the top of the cylinder to repeat the cycle again. The continuing up-and-down motion of the piston is why the engine is called a **reciprocating engine**.

The burning mixture is sealed into the cylinder on the top end by a *cylinder head* and a *head gasket* (**Figure 1.5**). The cylinder head has intake and

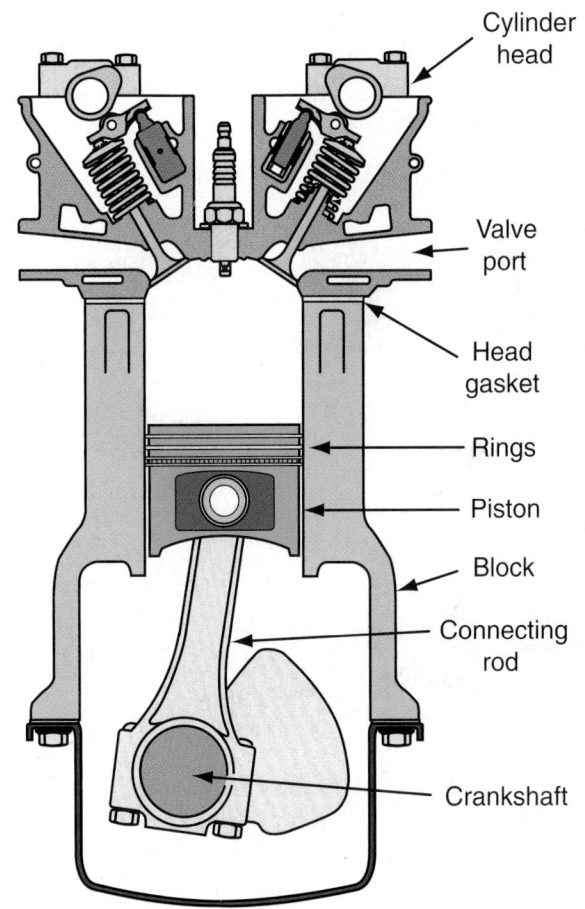

Figure 1.5 Engine parts.

exhaust *ports*. The intake port allows the flow of the air-fuel mixture into the cylinder. The exhaust port allows the escape of the exhaust gases after the mixture has been burned. Each port is sealed by a valve that is opened by a lobe on the camshaft and closed by a spring (**Figure 1.6**). The piston is sealed to the cylinder with **piston rings** that slide against the cylinder wall as the piston moves up and down.

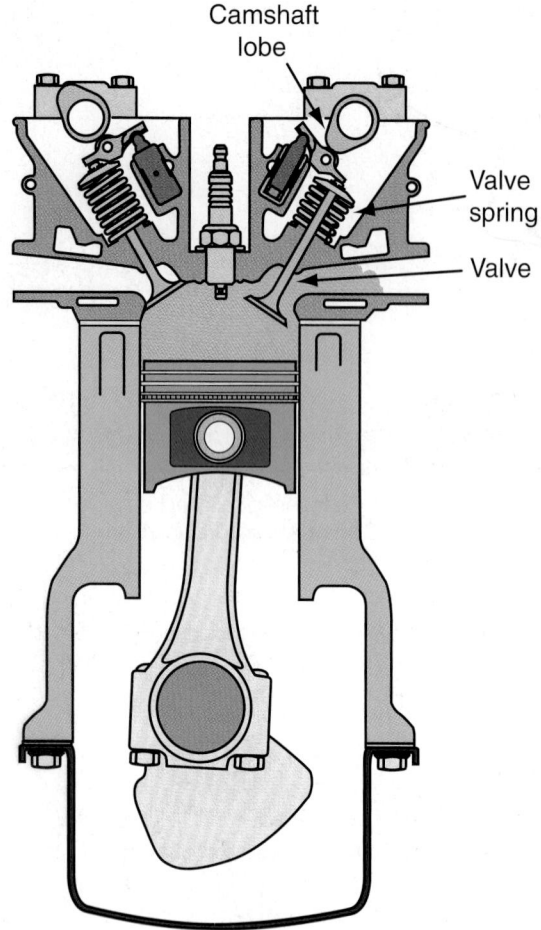

Figure 1.6 The valve is opened by a lobe on the camshaft and closed by a spring.

Four-Stroke Cycle

The **four-stroke cycle** is described here using a single cylinder engine (**Figure 1.7**). Automobile engines actually have multiple cylinders. The movement of the piston from the top of its travel to the bottom of its travel is called a **stroke**. Each cycle required to burn the air-fuel mixture has four strokes. Hence the name, four-stroke cycle.

During the *intake stroke*, the piston is pulled down by the turning crankshaft, creating a vacuum above it. Because the intake valve is open while the piston is moving down, the air-fuel mixture is drawn into the cylinder through the intake valve port. The mixture is supplied to the cylinder by the fuel system. In a spark-ignition engine, gasoline is especially combustible when one part of it is atomized with about 15 parts of air. **Atomization** makes the mixture like fog.

With both valves closed, the piston moves back up in the cylinder on the *compression stroke*, compressing the air-fuel mixture, making it far more combustible. As the piston approaches the top of its travel, a spark plug ignites the mixture.

During the *power stroke*, the burning fuel expands rapidly, forcing the piston to move back down in the cylinder. The exhaust valve opens as the piston approaches the bottom of its travel. This is so that burning gases can escape before the piston begins to move upward in the cylinder once again.

During the *exhaust stroke*, the piston moves back up, forcing any remaining exhaust gas from the cylinder through the open exhaust valve. As the crankshaft

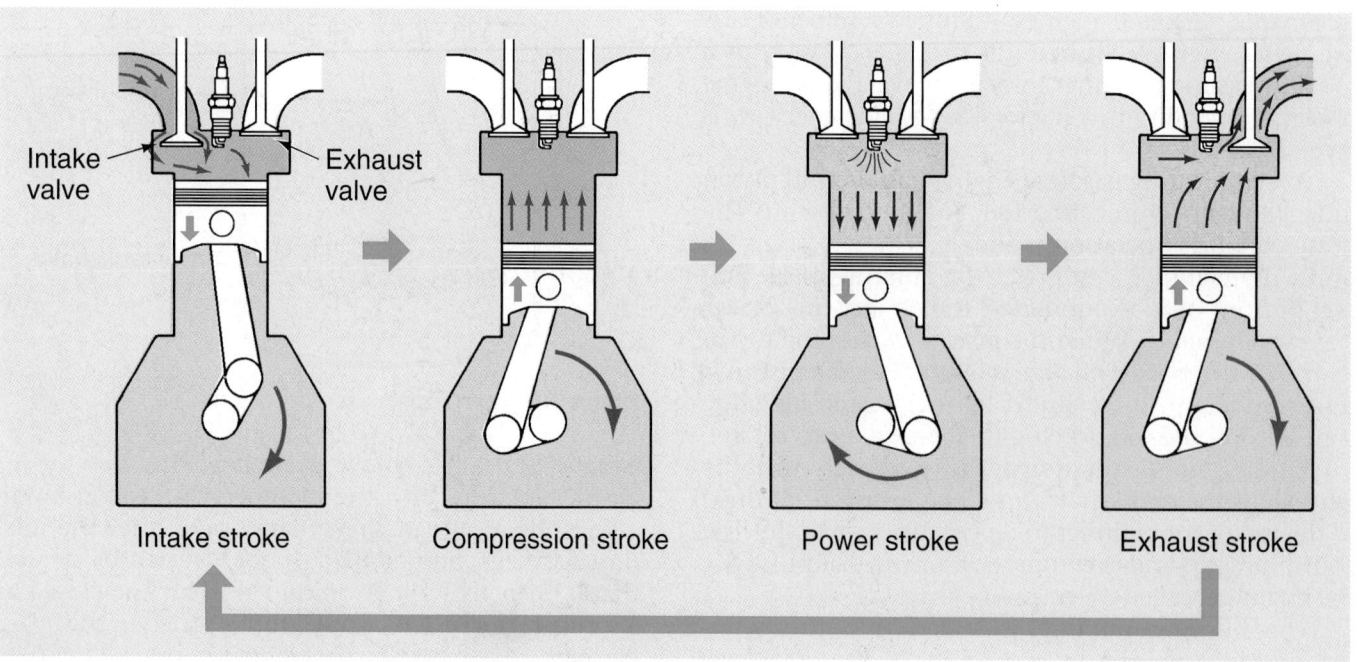

Intake valve   Exhaust valve

Intake stroke   Compression stroke   Power stroke   Exhaust stroke

Figure 1.7 The four-stroke cycle.

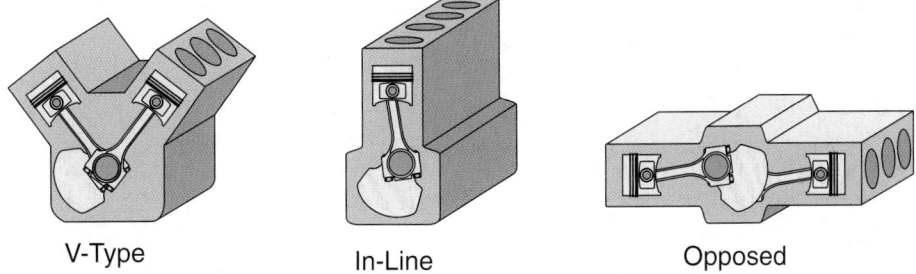

Figure 1.8 Common cylinder block arrangements.

V-Type   In-Line   Opposed

continues to rotate, the piston goes back down in the cylinder as the four-stroke cycle repeats itself.

Cylinder Arrangement

Car engines have multiple cylinders, commonly 4, 6, or 8 (**Figure 1.8**). Cylinder blocks have rows of cylinders that are arranged either *in-line,* or in a "V" or "W," or are *opposed* to each other.

■ ENGINE SUPPORT SYSTEMS

Several subsystems support engine operation. They are the *cooling system;* the *fuel system;* the *lubrication system;* the *electrical system,* including the *charging, starting, ignition,* and *computer systems;* and the *exhaust system.* These systems are covered in detail in later chapters. To provide improvements in fuel economy and exhaust emissions, new cars also have sophisticated computerized systems that operate fuel and emission control systems.

The Cooling System

As an engine operates, it creates a great deal of heat that is wasted and must be carried away by the cooling system (**Figure 1.9**) so the engine does not get too hot. *Coolant,* also known as *antifreeze,* is circulated by the *water pump* through *water jackets* in the engine's cylinder block. It carries heat to the radiator where it can be carried away by the outside air.

The cooling system's thermostat maintains the coolant at a constant temperature. It also speeds the warm-up of the engine so emission controls and the heater can operate. The engine experiences less wear once it is at operating temperature.

Fuel System

The fuel system is responsible for supplying the correct air-fuel mixture to the cylinder. Liquid gasoline does not burn. It must first be mixed with air in the correct proportion to form a vapor. The *air-fuel ratio* on gasoline engines ranges from about 12:1 (12 parts of air to 1 part of fuel) to 15:1.

There are three types of fuel delivery systems used on automobile four-stroke cycle engines. They are the *carburetor, gasoline fuel injection,* and *diesel fuel injection.* Carburetors are used on many cars built through about

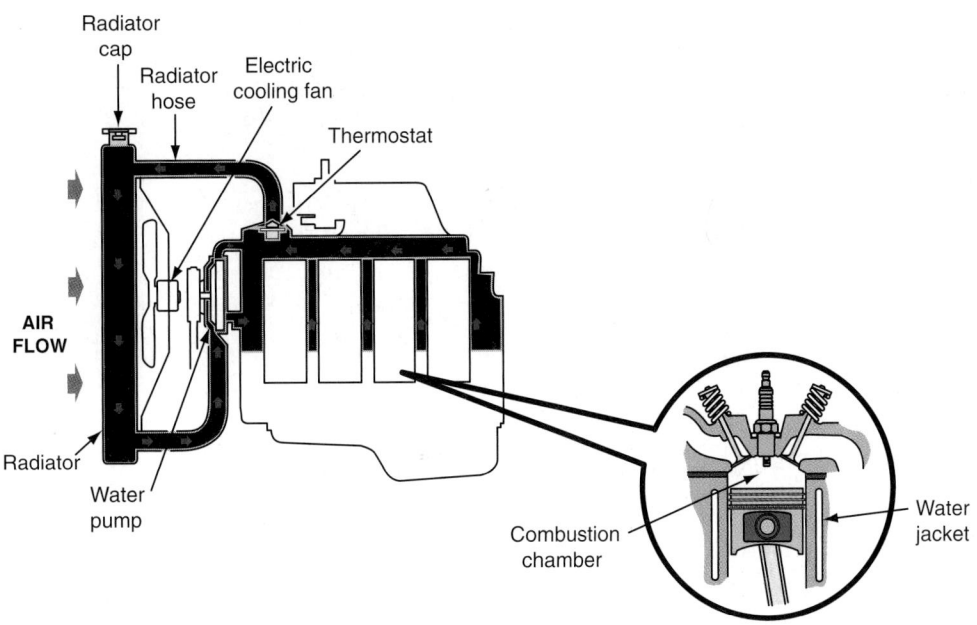

Figure 1.9 Parts of the cooling system.

the 1985 model year. Because of the need for more exact control of the fuel system for control of exhaust emissions, fuel injection systems have replaced carburetors.

The engine draws in a great deal of air as it runs. This is called engine vacuum. Air rushing through the carburetor draws fuel into it on its way to the cylinders. When the driver steps on the gas pedal (*throttle*), a **butterfly valve** opens to let in more air, and thus, more fuel (**Figure 1.10**). Opening the throttle valve allows in more air and raises engine speed.

Carburetor. On vintage engines, a carburetor mixes fuel and air in response to how much air is flowing through it. A mechanical fuel pump supplies the carburetor reservoir with fuel.

Gasoline Injection. Fuel injection systems use fuel injectors to spray fuel into the airstream flowing into the engine (**Figure 1.11**). Newer automobiles (since about 1985) use fuel injection fuel delivery systems operated by a computer.

An electric fuel pump provides fuel at a constant pressure to electronic fuel injectors (**Figure 1.12**). The injectors stay open a specified amount of time. This allows fuel to spray out of them in the exact amount called for by the operating condition at the moment. Open throttle allows the injectors to remain open longer; closed throttle leaves them open for only a short time.

Diesel Injection. In a diesel engine, only air is drawn into the cylinder through the air intake system. The air

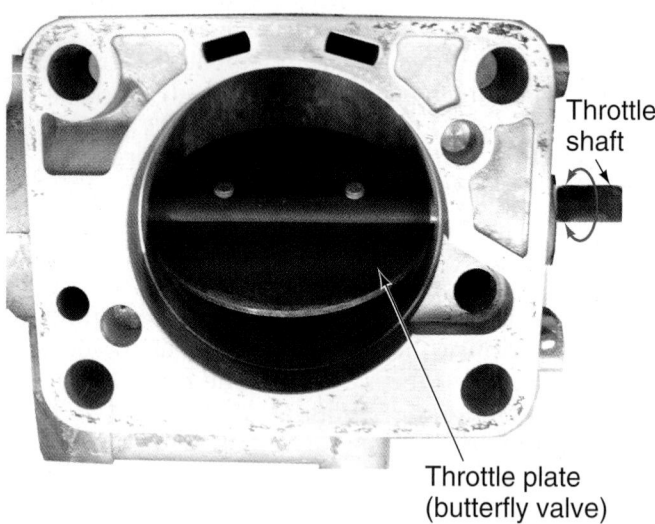

Figure 1.10 A butterfly valve controls airflow into the engine. *(Courtesy of Tim Gilles)*

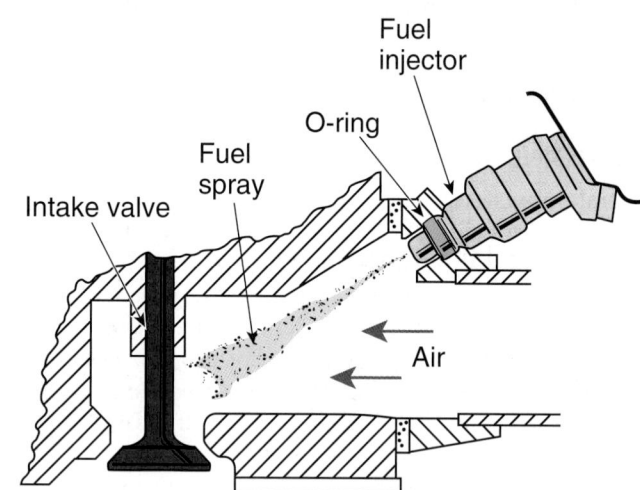

Figure 1.11 Fuel injection.

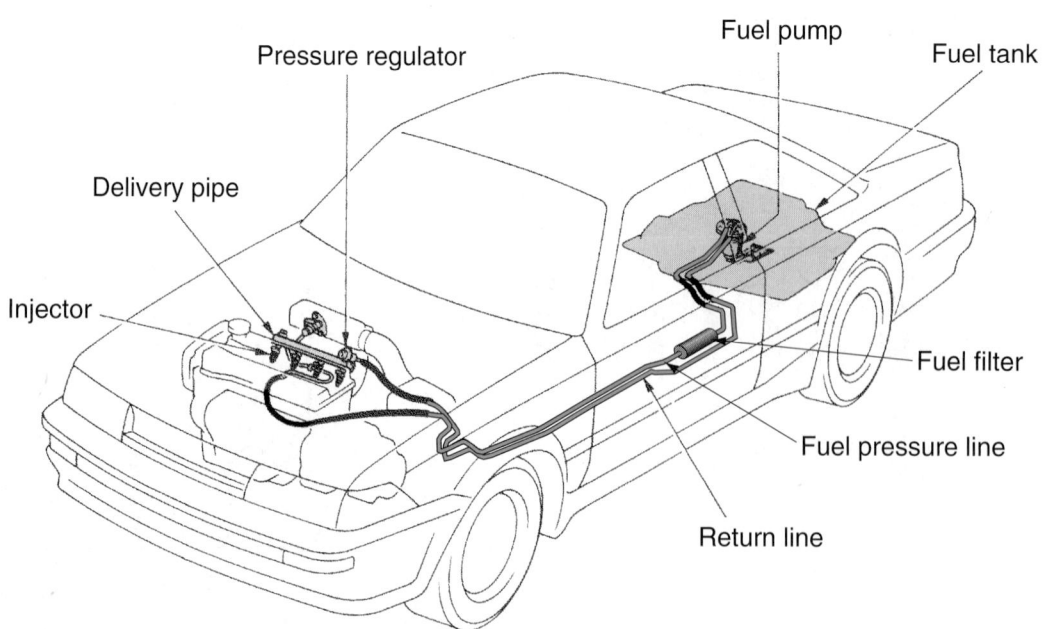

Figure 1.12 The fuel system.

is compressed to about one-half of the size that it would be compressed in a gasoline engine. This results in a great deal of heat. The heat would cause fuel to ignite during compression if it were already in the cylinder. Diesel engines do not have spark plugs or an ignition system, like gasoline engines. Instead, they rely on the injection system for ignition timing. When diesel fuel is injected by high-pressure injectors into the hot air in the cylinder, it ignites instantly (**Figure 1.13**). Because of the high pressure in the cylinder, the fuel system on a diesel must be under very high pressure.

The Lubrication System

The engine has a lubrication system that moves pressurized oil to all areas of the engine (**Figure 1.14**). A pump pulls oil out of the oil pan and forces it through a filter before it is distributed to the engine's parts. The oil prevents moving parts from touching each other.

NOTE: *In theory, during a 1,000-mile trip a properly operating lubrication system will allow about as much wear between parts as occurs during the first 15 seconds of engine operation in the morning before oil has reached all of the engine's parts.*

The Electrical System

The engine electrical system includes the *ignition system, starting system, charging system*, and fuel and emission systems controlled by computer. The body electrical system includes lighting and wiring systems.

The Ignition System. The ignition system has the job of creating and distributing a timed spark to the engine's cylinders (**Figure 1.15**). Through a process called

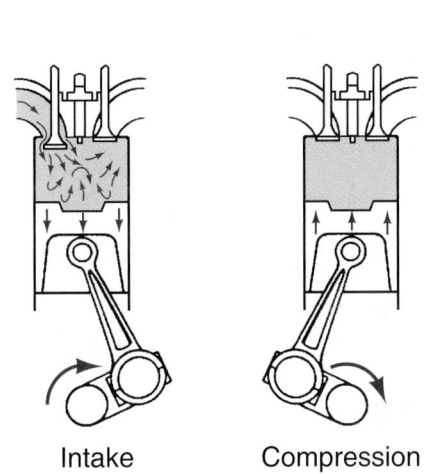

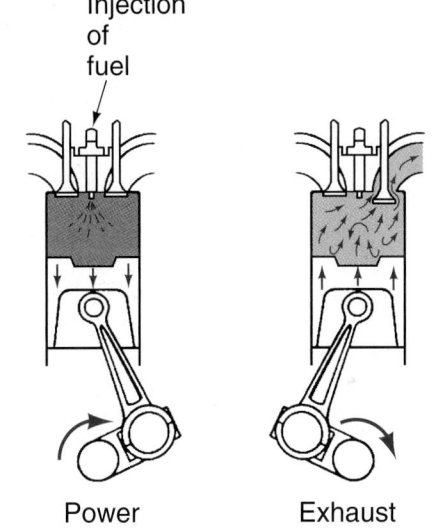

Intake   Compression   Power   Exhaust

Figure 1.13 Diesel four-stroke cycle.

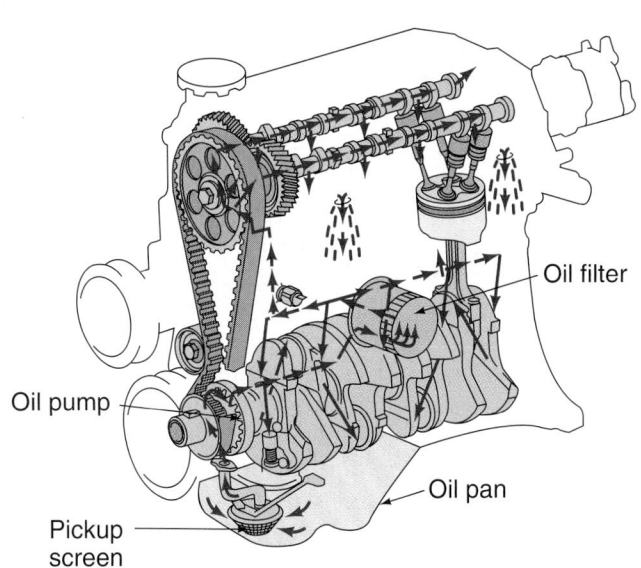

Figure 1.14 Engine lubrication system.

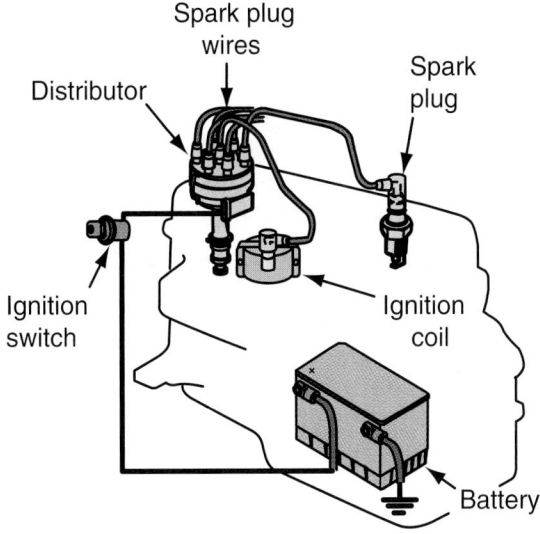

Figure 1.15 The ignition system provides a timed spark to each cylinder.

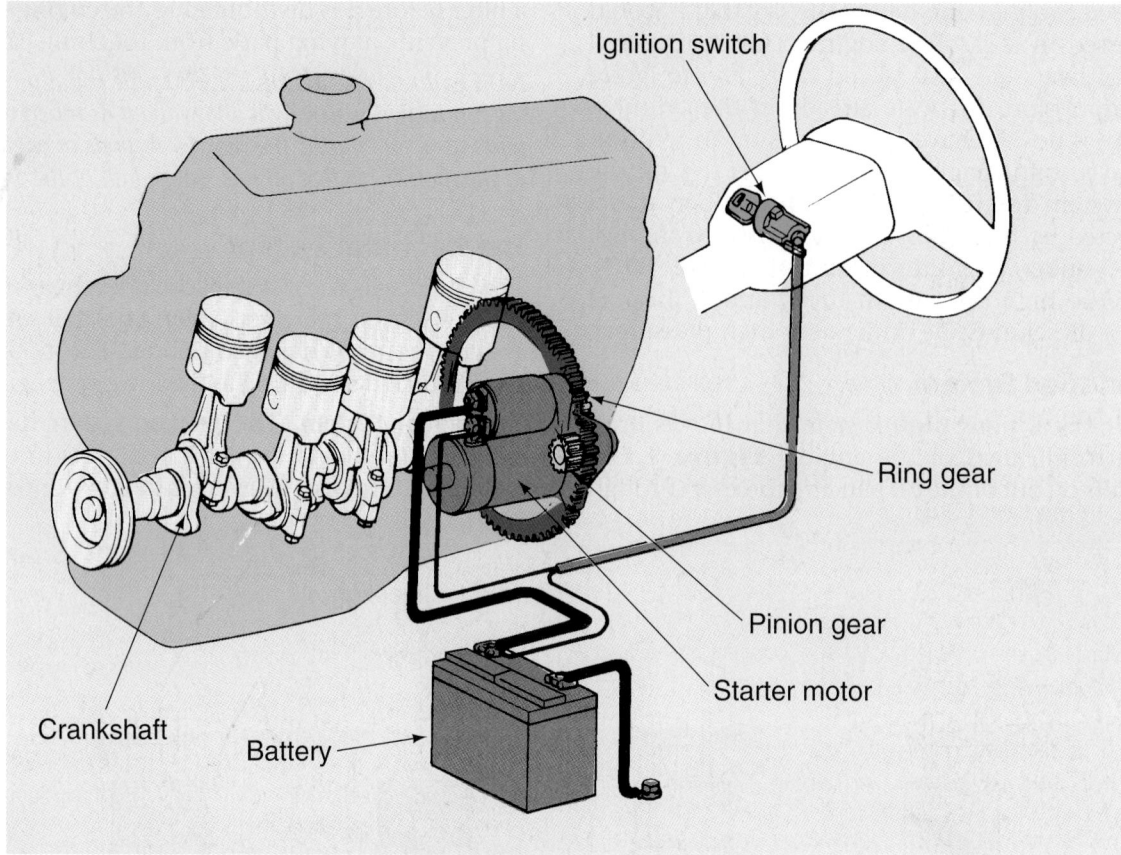

Figure 1.16 The starter motor turns the engine's crankshaft.

electromagnetic induction (see Chapter 25), a voltage of 5,000 to about 100,000 volts (on some of the newer systems) is created. The voltage causes a spark to jump a gap at the spark plugs to ignite the air-fuel mixture. The spark is timed to occur just before the top of the compression stroke. This is called *ignition timing*.

The Starting System. The starting system has an electric motor mounted low on the rear of the engine. It has a small *pinion gear* on it that meshes with a large *ring gear* on the engine's flywheel (**Figure 1.16**). The flywheel is bolted to the rear of the engine's crankshaft. The motor draws electrical current through a large cable from the car's battery. When the starter operates, the pinion gear turns the flywheel. This causes the crankshaft to rotate, drawing in air and fuel to start the engine.

The Charging System. As the vehicle is operated, electricity is drawn from the charging system to operate the ignition system, body electrical accessories, or lighting. The charging system (**Figure 1.17**) includes an *alternator,* driven by a belt on the engine's crankshaft pulley. The alternator produces electrical current and forces it into the battery to recharge it. Battery voltage is monitored, and the alternator is switched on or off depending on charging requirements.

The Computer System. Modern automobiles have a substantial amount of on-board electronics. Vehicle

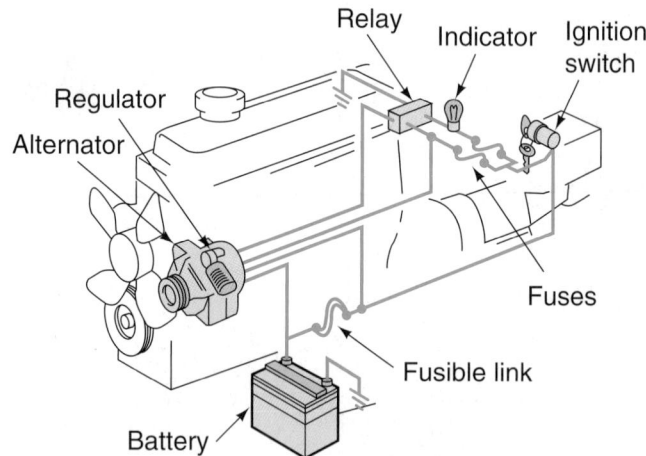

Figure 1.17 The charging system recharges the battery during engine operation.

electronics has become a specialty repair area with high earning potential. Qualified automotive electronics technicians command excellent pay. The computer system manages the operation of fuel injection, ignition, emission system components, automatic transmission shifting, antilock brakes, and body electrical accessories. Many of today's cars have several computers that manage these systems.

A main *computer*, called a *powertrain control module* (*PCM*) on late-model cars, controls the operation of all of the system components. **Sensors** react to temperature, airflow, engine load, road speed, and oxygen content in the exhaust stream. The various sensors send voltage signals to the computer. The computer processes the data and makes compensating adjustments using **actuators**. Operation of the computer system is covered in Chapter 39.

The Exhaust System

The exhaust system carries exhaust from the engine to the rear of the car. It also quiets sound. The exhaust manifold, pipes, a muffler, a catalytic converter, and sometimes a resonator make up the exhaust system components (**Figure 1.18**).

The Emission Control System

Since the mid-1960s, various emission devices have been installed on cars. Pollutants are produced during the combustion of air to fuel. The computer system provides the most complete combustion during all driving conditions. The emission system's purpose is to reduce or eliminate any remaining pollutants in the engine's exhaust. In addition to the control of exhaust pollutants, fuel vapors are controlled. Operation of these systems is covered in Chapter 45.

▬ THE POWERTRAIN

Engine power is transmitted to the wheels through the *powertrain*, which includes the *transmission* or transaxle, the clutch (used with manual transmissions) or *torque converter* (on automatic transmissions), and the *differential* and axles or half-shafts.

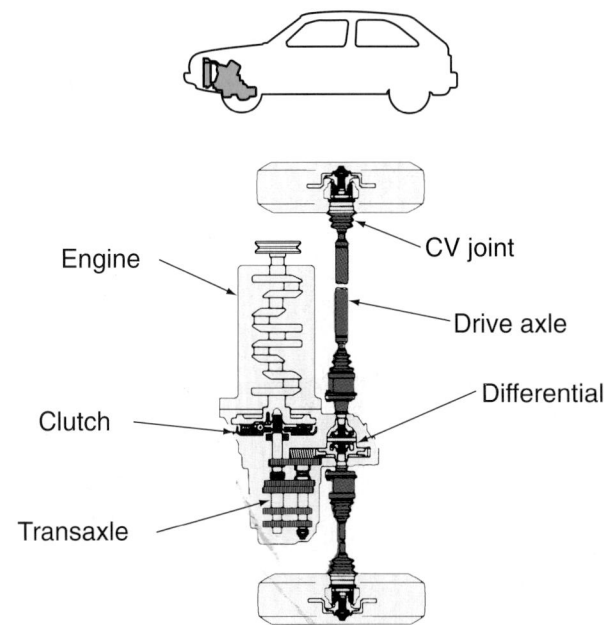

Figure 1.19 Front-wheel drive.

Vehicles have either *front-wheel drive* (**FWD**) or *rear-wheel drive* (**RWD**). Front-wheel-drive cars (**Figure 1.19**) use a transaxle and axle shafts, while rear-wheel-drive cars use a transmission and drive shaft coupled to a differential and rear axles (**Figure 1.20**). Transmissions can be either manually shifted using a clutch, or they can shift automatically.

Manual Transmission

The manual transmission (see Figure 1.20) provides for shifting of the gears by the driver. Changing gears results in a change in leverage or **torque**. The car's engine develops more pushing power when its

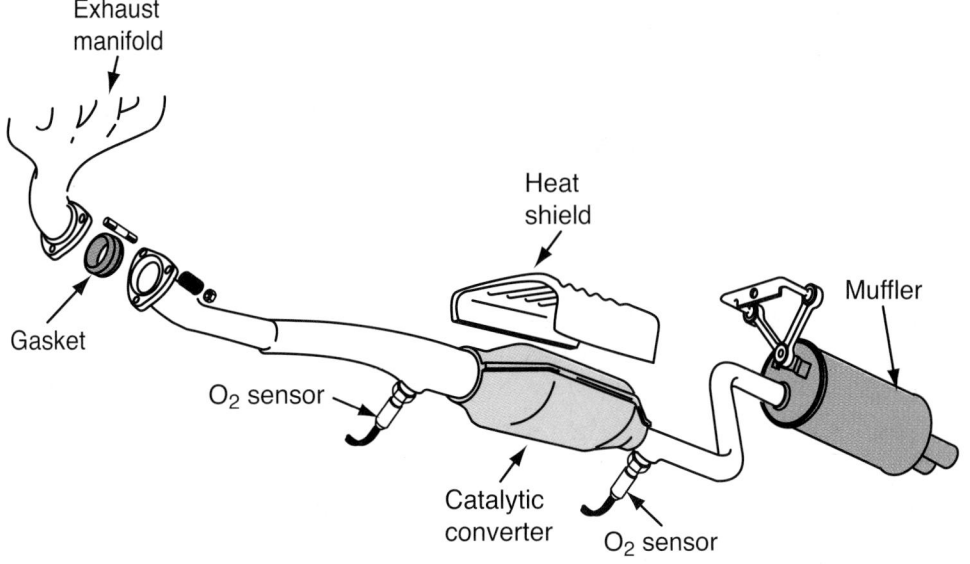

Figure 1.18 Parts of the exhaust system.

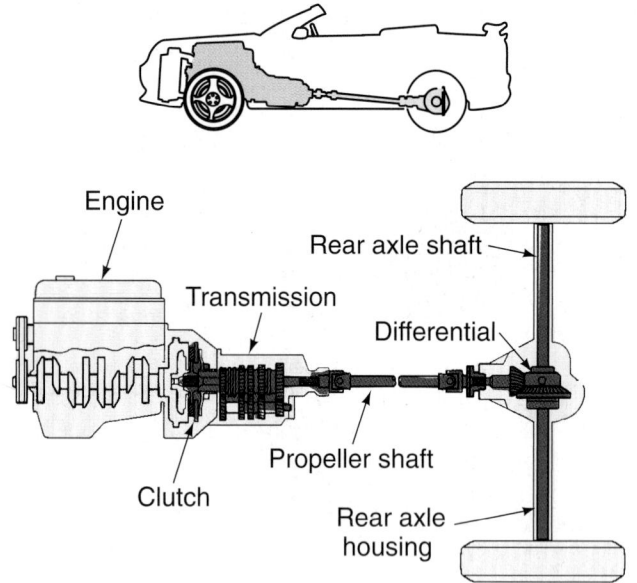

Figure 1.20 Rear-wheel drive.

crankshaft pins run at a certain number of revolutions per minute (rpm). Shifting the transmission into different gears allows the engine to be operated at the best rpm so that it can propel the car more efficiently. This is much the same as the gears on a ten-speed bicycle making it possible to climb a steep hill slowly or go fast downhill.

The Clutch. A **clutch** (see Figure 1.20) is used on cars with standard (manual shift) transmissions. Depressing the clutch pedal in the driver's compartment uncouples the powertrain from the engine.

Automatic Transmission

Automatic transmissions are found on most of the new cars manufactured today. Once an automatic transmission is shifted into drive, it does not require further shifting unless a direction change is desired. The torque converter is a fluid clutch that allows stopping and starting automatically. Gears are shifted automatically in response to changes in vehicle speed and engine load.

Drive Shaft

A *drive shaft*, or propeller shaft, is used on rear-wheel-drive cars to transfer power to the rear axle. It is a hollow metal tube that has a universal (swivel) joint at each end.

Rear Axle Assembly

The rear axle assembly has drive axles that go to each rear wheel and a *differential* assembly (see Figure 1.20). A differential allows the rear wheels to rotate at different speeds as the vehicle goes around corners.

Transaxle

A **transaxle** is used on front-wheel-drive vehicles (see Figure 1.19). It contains a transmission and differential within one housing. *Axle half-shafts* (*drive shafts*) with *constant velocity joints* (*CV joints*) are on each end to allow the axle to move up and down and operate at an angle. Their operation is covered in detail in Chapter 77.

■ ACCESSORY SYSTEMS

Accessory systems are also known as *comfort systems*. Included are *air conditioning* and *heating*, *power seats* and *power windows*, and *cruise control*. Today's cars have more "creature comforts", like navigation and sound systems, heated seats, rain-sensing wipers, and active cruise control.

■ HISTORY AND DEVELOPMENT OF THE AUTOMOBILE

Steam-Powered Vehicles

The first automobiles were powered by steam engines. The first steam engine was developed in 1698. In 1801, a steam-powered stage coach carried six people a distance of 6 miles. By 1825, a steam-powered bus was driven between Paris and Vienna at speeds of up to 22 miles per hour. Early vehicles had to operate at low speeds because of the poor quality of the roads. The early vehicles were also of large size and weight. Railroads were the solution to this problem.

The steam engine design is called an *external combustion engine* (**Figure 1.21**). This means that the fuel is burned outside of the engine. Locomotives used steam engines. Coal, wood, or oil is burned in a firebox under a boiler filled with water. The water is turned into steam, which pushes against the piston. The piston drives the wheels by way of a connecting rod (**Figure 1.22**).

In the early days of the automobile, over 60,000 Stanley Steamers were sold. Steam engines do not require a transmission, because they can be started, stopped, or reversed easily. Steam power did have its problems, however. Inventors of the time realized that an engine that did not use coal for fuel must be developed. The steam-powered automobile was gradually replaced with the gasoline-powered forerunner of today's power plant.

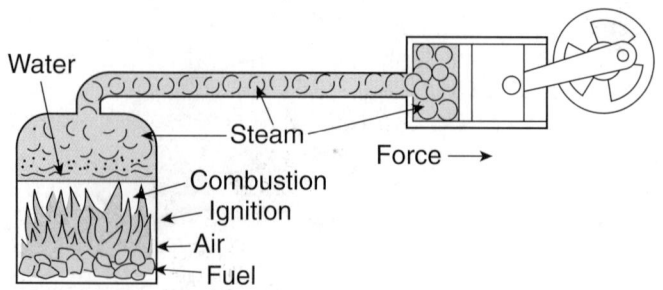

Figure 1.21 An external combustion steam engine.

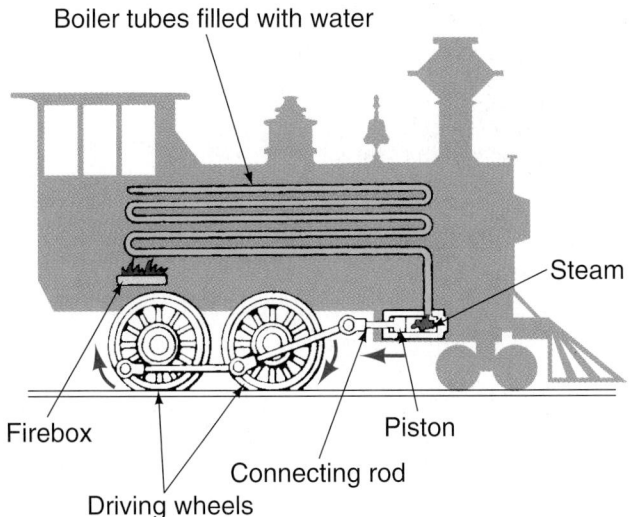

Boiler tubes filled with water

Steam

Firebox

Piston

Connecting rod

Driving wheels

Figure 1.22 A steam-powered locomotive.

Early Gasoline Engines

In 1876, Dr. Nicolas Otto patented the first slow-speed, four-stroke, internal combustion engine. Today, the gasoline-type four-stroke piston engine is still called the **Otto-cycle** engine. Compared to previous internal combustion engine designs using the same amount of fuel, Otto's four-stroke cycle engine weighed less, ran much faster, and required less cylinder displacement to produce the same horsepower. A few years later, an Otto-cycle engine would power a motorcycle and then a horseless carriage.

In 1885, Gottlieb Daimler patented a high-speed engine that ran on petroleum. Daimler is credited as the "Father of the Automobile." Today's engines are patterned after Daimler's invention. During the same year, Carl Benz of Mannheim, Germany, patented a gasoline engine tricycle. The following year, several engineers developed motor carriages. Benz developed a car with a one-cylinder, water-cooled engine capable of speeds of 10 miles an hour. Its use on streets was restricted to certain hours only.

The first car to be imported to the United States was a Benz, shown at the 1893 World's Fair in Chicago. You may recognize the names of Daimler and Benz from the corporate name of the company that manufactures Mercedes-Benz automobiles, known as Daimler-Benz until their 1998 merger with Chrysler.

Early cars were horseless carriages with single cylinder engines and chain drive (see Figure 1.1). They had tall wire wheels and solid tires. They were steered by a handle, rather than a steering wheel. Cars were so rare in 1900 that Barnum and Bailey Circus gave one top billing in its show. By 1919, 90% of cars still looked like carriages. Running boards were used until the late 1930s. Some cars, like the Volkswagen Bug still used them into modern times.

Early gasoline engines were underpowered, having few cylinders and low compression (about half of today's compression). Later engines of four or six cylinders replaced the early single cylinder engines. The cylinders were most often cast in pairs. The engines had very heavy connecting rods and pistons and ran at slow speeds.

As the automobile was developed, there came a need for better materials. Thus, the automobile contributed to scientific developments and provided the financial means to accomplish them. Present-day manufacturing methods including mass production of interchangeable parts and the assembly line are a result of early developments required to produce engines and automobiles.

Cars have resulted in great changes to our world. In less than 100 years, 3.8 million miles of roads in the United States have been developed. This is equivalent in size to the area of the entire state of Mississippi.

Auto Parts. There were no auto parts stores in the early days of the automobile. Parts had to be secured from the individual factories. Nothing came in the same size on any of the cars. Bolts had different threads, and wheels and tires were of different sizes. Engineers of the best factories met and organized the *Society of Automotive Engineers* (*SAE*).

Early Auto Racing

The first auto race was held in Chicago in 1895 as a promotion for the automobile. Cars improved at a rapid pace after about 1910. Speed and reliability contests were held, the most famous being the Indianapolis 500, which was started in 1913.

Early Transmissions

Early cars had the transmission on the rear axle. Later on, the transmission was attached to the rear of the engine.

Carburetors

Early carburetors had a wick saturated with gasoline that air was drawn through. Later carburetors had a bowl full of gasoline (**Figure 1.23**). The carburetor was usually in a difficult location for repair. This was because it had to be lower than the gasoline tank, located under the driver's seat. One advancement was to put the tank in the rear of the car and use a fuel pump to move the fuel to the carburetor. These cars could have a shorter intake manifold, which meant that atomized fuel could be more easily delivered to the cylinder.

Fuel Pumps

In 1915, Stewart Warner developed the vacuum tank, which was used on almost all cars to feed fuel from the tank. In 1928, the electric and mechanical fuel pumps, which were attached to the engine, improved the way fuel was moved.

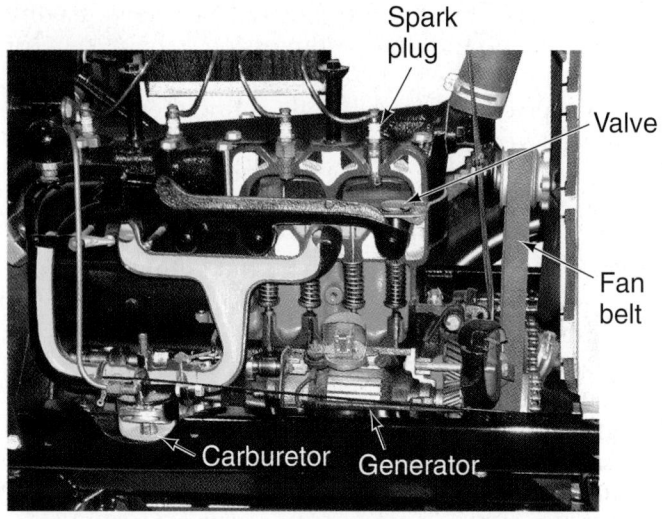

Figure 1.23 Cutaway of an early Ford Model T engine. It has an updraft carburetor, a gear-driven generator, and a flat fan belt. *(Courtesy of Tim Gilles)*

Lubrication System

Early engines used a drip oiler (**Figure 1.24**). Later cars had mechanical oiling with individual oil tubes to the bearings and cylinders. Later engines used a combination of splash oiling and pressure lubrication.

Tires

Early tires were solid (**Figure 1.25**). In 1900, a Frenchman named Michelin made the first pneumatic (air-filled) tires. Michelin's early tires had a single tube. In 1903, a casing and tube were developed. Tires ranged in diameter from 28 to 42 inches. Changing a tire was a big and dirty job. The tires had to be pumped full of air with a hand pump. By 1919, all cars were equipped with cord tires. In 1924, the balloon (low-pressure) tire was introduced.

Figure 1.24 This early engine had no valve cover or oil pan. Lubricant was provided by drip oil. *(Courtesy of Tim Gilles)*

Figure 1.25 Early solid rubber and pneumatic tires. *(Courtesy of Tim Gilles)*

Electrical Systems

Some of the early cars had 8-, 12-, or 24-volt electrical systems. Almost all of the early electrical systems combined the starter and generator in one unit. In 1915, separate units started to show up on cars and the 6-volt battery became standard. Battery ignition systems, which were combined with the generator, became popular. Today's cars use 12-volt batteries, which became standard in the middle 1950s.

Starter System

Early engines had to be hand cranked to get them started. In 1912, Charles F. Kettering developed the electric starter motor.

Early American Automobiles

Selden Patent. In 1879, a United States patent, known as the **Selden patent**, covering all gasoline-powered, self-propelled road vehicles was issued to George B. Selden. Selden did not actually invent the automobile, but all American companies had to pay licensing fees to him anyway. Henry Ford was a defendant in a case that was carried through the Michigan courts and on to the United States Supreme Court. In 1911, the patent was ruled to be void.

In 1892, Charles Duryea and his brother, Frank, built the first American car that would actually run. Alexander Winton sold the first automobile in the United States in 1898. Ransom Olds produced his Curved Dash Olds. A fire consumed all but the prototype model for the 425 that were produced that year.

Model T Ford. Henry Ford built his first car in 1893. Henry Ford produced the Model T Ford between 1908 and 1926. It was one single universal car. Most of them were all of the same color, black. One thousand of them were produced on the assembly line per day. Over 15 million Model Ts were produced in all. By 1926, over two-thirds of the cars on the road were Model Ts.

General Motors. General Motors' Billy Durant had the idea of producing a variety of cars, "one for every purse and purpose." Durant was a good promoter but a poor businessman. He was removed by GM when Dupont took over the company. Alfred P. Sloan became the new president of General Motors.

Chrysler. In 1919, Walter Chrysler left the Buick Division of GM to found the Chrysler Corporation. In 1924, he produced Chryslers and in 1928 acquired Dodge.

Later Developments

By the late 1950s, American cars began to have large, powerful engines. They had poor fuel economy, polluted the air badly, and had poor brakes compared to later cars. In the mid-sixties, large engines like the 427 (7 liter) produced by Ford and Chevrolet had over 400 horsepower. This was called the "muscle car" era.

Emission controls were added to cars starting in the early 1960s. These were brought about by clean air legislation. Horsepower output of engines was reduced as a result of the changes brought to these engines by emission controls.

Fuel Economy Standards

During the 1960s and early 1970s, a gallon of gasoline could be purchased in the United States for about 25 to 30 cents. An oil embargo by Middle Eastern countries in 1973 caused the price of gasoline to quadruple. In 1975, the United States Congress passed fuel economy standards. Corporate Average Fuel Economy (CAFE) is the name of these standards. Fuel economy for different models sold by each manufacturer are averaged to come up with the CAFE number. The CAFE standard at this writing is 27.5 miles per gallon. When a manufacturer makes cars that get poor fuel economy, they pay a heavy penalty to the government. When a consumer buys a "gas guzzler," a tax is included in the purchase price.

Modern Developments

Today's automobile has benefited greatly from technological developments made in the military and space programs. The automobile of 1965, while still having the same basic parts, is a dinosaur compared to today's high-tech vehicle. Today's car has a very good power-to-weight ratio, which means it gets good gas mileage and goes faster compared to comparably sized cars of the sixties. Computerized controls manage engine and body systems for comfort and efficiency. Yearly advancements are made that make vehicles safer and more reliable.

▄▄▄ REVIEW QUESTIONS

1. What is the name of the automotive chassis design that has a floorpan and a small subframe section in the front and rear?

2. When air and fuel are compressed, they become very_____.

3. What does reciprocating mean?

4. The valve is opened by a _____ on the camshaft.

5. What are the names of the parts that seal the combustion gases between the piston and the cylinder?

6. The movement of the piston from the top of its travel to the bottom of its travel is called a _____.

7. List the four strokes of the four-stroke cycle in order.

8. The air-fuel mixture is _____ as the piston moves up on the compression stroke.

9. What is another name for antifreeze?

10. The air-fuel ratio on gasoline engines ranges from about _____:1 to _____:1.

11. What is the name of the butterfly valve that controls the amount of air entering the engine?

12. The starter motor pinion gear drives the _____ gear on the crankshaft's flywheel.

13. What is the name of the part that switches the alternator on and off?

14. The parts of a computer control system that react to temperature, airflow, engine load, road speed, and oxygen content in the exhaust stream are called _____.

15. What was the name of the patent for the automobile?

■ ASE-STYLE REVIEW QUESTIONS

1. Technician A says that the intake valve is closed during the power stroke. Technician B says that the exhaust valve is closed during the compression stroke. Who is right?
   a. Technician A
   b. Technician B
   c. Both A and B
   d. Neither A nor B

2. Which of the following is/are true about fuel systems?
   a. Fuel injection has replaced the carburetor on modern engines.
   b. Fuel injection provides for more accurate control of exhaust emissions.
   c. To burn, gasoline must be mixed with air to form a vapor.
   d. All of the above.

3. Which of the following is/are true about diesel engines?
   a. Diesel engines do not compress the air in the cylinder as much as gasoline engines do.
   b. Diesel timing is controlled by when the fuel is injected into the cylinder.
   c. Diesel engines have spark plugs similar to the ones used in gasoline engines.
   d. All of the above.

4. Technician A says that an ignition spark is 12 volts. Technician B says that the spark happens just after the power stroke begins. Who is right?
   a. Technician A
   b. Technician B
   c. Both A and B
   d. Neither A nor B

5. Technician A says that most rear-wheel-drive cars have a transaxle. Technician B says that front-wheel-drive cars have two drive shafts. Who is right?
   a. Technician A
   b. Technician B
   c. Both A and B
   d. Neither A nor B

6. Technician A says that gears are changed to increase or decrease engine torque. Technician B says that the majority of new cars have automatic transmissions. Who is right?
   a. Technician A
   b. Technician B
   c. Both A and B
   d. Neither A nor B

7. Technician A says that a steam engine is an external combustion engine. Technician B says that a four-stroke piston engine is an internal combustion engine. Who is right?
   a. Technician A
   b. Technician B
   c. Both A and B
   d. Neither A nor B

8. All of the following are true about the four-stroke cycle gasoline engine *except*:
   a. It is called an Otto-cycle engine.
   b. During the compression stroke, both valves are closed.
   c. It is a compression-ignition engine.
   d. Piston movement from the top of the cylinder to the bottom is called a stroke.

9. Technician A says that Model Ts were all black. Technician B says that the 12-volt electrical system has been standard on American cars since 1940. Who is right?
   a. Technician A
   b. Technician B
   c. Both A and B
   d. Neither A nor B

10. Technician A says that CAFE standards govern vehicle emissions. Technician B says that CAFE standards govern fuel economy. Who is right?
   a. Technician A
   b. Technician B
   c. Both A and B
   d. Neither A nor B

CHAPTER 2

Automotive Careers and Technician Certification

■ OBJECTIVES

Upon completion of this chapter, you should be able to:

✔ Describe different automotive careers.

✔ Understand some of the skill requirements of an automotive technician.

✔ Know what kind of work is performed by technicians in the different automotive specialty areas.

✔ Understand the different types of certification and licensing.

■ KEY TERMS

PDI

service manager

service writer/service advisor

service dispatcher

shop foreman

fleet shop

ASE

master automobile
   technician

AERA

licensing

flat rate/commission

ASA

ASC

NADA

NATEF

■ INTRODUCTION

Demand for Technicians

There are over 240 million motor vehicles in the United States alone. Due partly to the increased complexity of vehicles, the demand for skilled technicians is high and will continue to climb.

Today's automotive technician must be able to perform a multitude of tasks requiring several different types of skills. It is said that a mechanic must be a "jack of all trades." The ability to repair automobiles includes skills in plumbing, metal working and welding, electrical, electronics/computer/radio, and air conditioning.

The repair and service of automobiles is a challenging and rewarding field. Most important of all is to enjoy the work. If you like the work and have the talent, you will never be bored and will continually learn new things. Following are some of the jobs that automotive training can prepare you for.

■ AUTOMOTIVE CAREER OPPORTUNITIES

A generation ago, most automotive technicians began their careers in service stations (also called gas stations or filling stations). Today's technicians usually start out in a technical education program in high school before continuing with postsecondary education and on-the-job training. Although some of today's gas stations still have automotive service facilities (**Figure 2.1**), the majority of these establishments combine high-volume fuel sales with a convenience store.

Figure 2.1 Service stations often perform automotive repairs. *(Courtesy of Tim Gilles)*

Another change has been that gas stations used to be full service. People would have underhood checks done on their cars at each fill-up. Today's gas stations are mostly self-service, and often consumers neglect their cars as a result. Lube outlets and mass merchandisers are filling the void in the periodic service business left by the closing of many former service stations.

Lubrication/Safety Shops

Many repair facilities have a "quick-lube" section to accommodate maintenance services. There are also specialty shops that only perform lubrication services (**Figure 2.2**). A *lubrication specialist* performs maintenance and safety services on the automobile. Changing engine oil and filters, fluids, belts and hoses, and inspecting and replacing safety items are included. A thorough description of lubrication and safety service is found in the early chapters in this book.

New Car Dealerships

New car dealers have a service department where warranty and customer-pay repairs can be made (**Figure 2.3**).

Figure 2.2 Some shops specialize in quick lubrication and safety service.

Figure 2.3 New car dealers have service and repair departments specializing in one or more makes of vehicles. *(Courtesy of Tim Gilles)*

Dealerships also perform pre-delivery inspections (**PDI**) and *warranty repairs* when the car is still new and a problem related to its manufacture develops. Warranty coverage is usually only given before a car has reached a certain age and a certain amount of mileage, 5 years and 50,000 miles, for instance. Dealerships often have body shops and used car departments also.

Automotive manufacturers all have training schools for updating the skills of their dealership technicians. A dealer technician has the advantage of being able to attend these schools. Many manufacturers also offer on-line or satellite training opportunities for their dealership technicians.

Independent Repair Shops

Independent repair shops can be one-person specialty shops or very large, full-coverage repair shops (**Figure 2.4**). They often specialize in one or more makes of vehicles. It is not uncommon for a dealer technician to start a competing business on one make of vehicle after leaving the employment of a dealership. One of the things that customers like about the smaller establishments is that they get to actually talk to the person who works on their car. In a dealership or large independent shop, the technicians do not normally have the opportunity to talk with their customers.

Independent technicians typically attend nighttime professional update classes that are periodically offered for working technicians. These classes, sometimes called clinics, are offered through parts companies like NAPA, Car Quest, and AC-Delco. Update classes are also offered by many community colleges and by private companies specializing in aftermarket training.

Operation of Large Shops

The **service manager** is responsible for the operation of the service department (**Figure 2.5**). This sometimes includes the management of the parts department,

Figure 2.4 Independent repair shops perform repairs on one or more makes of vehicles. *(Courtesy of Tim Gilles)*

Figure 2.5 In a large repair facility, the service manager is responsible for the operation of the service department. *(Courtesy of Tim Gilles)*

although dealerships often have parts managers, too. The service manager supervises the service writers, dispatcher, and technicians. He or she is also responsible for the management of warranty communications between the manufacturer, dealer, and consumer. Customer complaints and communication are important aspects of the job.

The link between the technician and the customer is the **service writer/service advisor**. A successful service advisor is a "people person" (someone who enjoys interacting with customers), is organized, and has good communication skills. He or she greets the customer, listens to the complaint, interprets it, and then writes the repair order for the job (**Figure 2.6**). The repair order includes the estimated cost of the repairs. When the cost of the repair goes over the estimate that the customer was informed of, the service writer will call the customer to explain the need for more repairs and get authorization to proceed. He or she also works with the technician to verify that the customer complaint is repaired.

In many states, consumer protection laws state that the customer must give consent for additional repairs before the additional work can be completed. The technician depends on the service writer to provide an accurate description of the problem so that the correct repair can be made at the estimated cost.

The **service dispatcher** is the person who organizes the repair orders and dispatches them to the technicians in the service bays. He or she keeps track of how work is progressing. This position requires a good deal of diplomacy. Because the technicians are paid on a commission basis, the jobs that they are assigned can make a big difference in earnings, even between technicians working in the same establishment.

Some large shops have a **shop foreman** who keeps repair work on track. He or she must be able to diagnose or troubleshoot difficult problems and must keep up on the latest repair procedures.

Fleet Shop

A fleet is a group of several vehicles owned by a company, utility, or municipality (government) (**Figure 2.7**). A taxi cab company, rental car company, or the telephone company are examples of fleets. Preventive maintenance is scheduled ahead of time. Public safety is the most important consideration, so safety inspections are done on a regular basis. Some **fleet shops** subcontract major repairs to another specialty shop. Others perform all of their own repairs. Fleet technicians often repair diesel and alternative fuel vehicles, too.

■ TECHNICIAN CERTIFICATION AND LICENSING

There is a tremendous and growing shortage of trained automotive technicians. Vehicles have become very complex, and the education and knowledge required

Figure 2.6 The service writer works with the customer. *(Courtesy of Tim Gilles)*

Figure 2.7 Fleet operations usually have their own repair facilities. *(Courtesy of Tim Gilles)*

of today's technician is much more than that required to repair and service vehicles produced 25 years ago.

ASE Certification

Due to the complexity of today's vehicles, technicians often choose one or more areas to specialize in. There are eight specialty areas under the automotive repair umbrella. Many technicians are certified through the voluntary certification program administered by the National Institute for Automotive Service Excellence (ASE). The eight automotive repair areas of specialization are:

■ Engine Repair
■ Engine Performance
■ Heating and Air Conditioning
■ Electrical Systems
■ Automatic Transmissions and Transaxles
■ Standard Transmissions and Axles
■ Brakes
■ Suspensions

Tests are given twice each year, usually at a school in the community. To become certified in one of the specialty areas, you must pass between 60% and 70% of the questions (depending on the difficulty of the particular test). **Figure 2.8** shows the number of test questions and the content of the specialty areas.

ASE sends test results that tell what percentage of test questions you passed in each area of the test. When you pass the test, a wall certificate and pocket protector with your certification dates are sent to you.

Certified technicians can be identified by a shoulder patch that says "Automotive Technician."

A **master automobile technician** is a journey-level professional who is certified in all eight of the ASE areas of specialization. A master technician has a different shoulder patch that says "Master Automobile Technician" (**Figure 2.9**).

Work Experience Requirement

You must have 2 years of automotive work experience (school training can count as one of the years). If you lack the experience requirement, you can still take the test. You will receive your test results, and ASE will certify you as soon as you meet the experience requirement.

Why Certify

Employers often complain that the technicians and machinists they hire are unqualified. Many employers now ask for ASE certification when they advertise a job opening. Although some employees continue to resist certification, it does provide a technician with a means of showing a prospective employer that a technician has work experience and has developed sufficient knowledge about automobiles to pass a test in an automotive specialty area. Certification also demonstrates that a technician can read—something all the best technicians of the future must be able to do.

Advanced Engine Performance Specialist

One of ASE's other certifications is called the *advanced engine performance specialist* (L1). To qualify to take this test, a technician must first be certified in the

TEST TITLE		TEST CONTENT
Engine Repair (60 Questions)	Test A1	Valve train, cylinder head, and block assemblies; lubricating, cooling, ignition, fuel, exhaust, and battery and starting systems
Automatic Transmissions/Transaxle (50 Questions)	Test A2	Automatic transmissions and transaxles; electronic and manual controls; hydraulic and mechanical systems
Manual Drive Train and Axles (40 Questions)	Test A3	Manual transmissions, clutches, front and rear drive axles, differentials, and four wheel systems
Suspension and Steering (40 Questions)	Test A4	Manual and power steering, suspension systems, alignment, and wheel and tires
Brakes (50 Questions)	Test A5	Drum, disc, combination and parking brake systems; power assist and hydraulic systems; ABS
Electrical/Electronic Systems (50 Questions)	Test A6	Batteries; starting, charging, lighting, and signaling systems; gauges, instrumentation, horns, wipers, and power accessories
Heating and Air conditioning (50 Questions)	Test A7	Refrigeration, heating and ventilating systems, A/C controls; refrigerant recovery and recycling
Engine Performance (65 Questions)	Test A8	Ignition, fuel, air induction, and exhaust systems; emission controls; computerized engine controls; OBDII and electrical systems

Figure 2.8 ASE tests cover the eight automotive specialty areas. Test content is listed here. *(Courtesy of ASE)*

Figure 2.9 An ASE certification patch. *(Courtesy of ASE)*

regular automotive categories of engine performance and electrical systems. This test covers several areas: general powertrain diagnosis, computerized engine controls diagnosis, ignition system diagnosis, fuel and air induction system diagnosis, emission control system diagnosis, and emission inspection and maintenance (IM) failure diagnosis.

Engine Machinist

Automotive machine shops also have an ASE certification program. Originally sponsored by the Automotive Engine Rebuilders' Association (**AERA**), there are three areas of specialization:

- Assembly Specialist (M1)
- Cylinder Block Specialist (M2)
- Cylinder Head Specialist (M3)

Machinists who qualify in all three categories are known as *Master Engine Machinists*. Engine Machinist Tests are administered as part of the regular ASE technician test series.

Automotive Service Consultant

Service consultants, also known as service advisors or service writers, can become ASE certified. The Automobile Service Consultant (C1) test includes questions on communications, customer and internal relations, vehicle systems knowledge, sales skills, and shop operations.

Parts Specialist

Parts specialists can become ASE certified in one or more of six parts specialty areas, including Automobile Parts Specialist and Medium- and Heavy-Truck specialties. Tests include questions on communication and sales skills, vehicle systems knowledge, vehicle identification, cataloguing skills, and inventory management. More information on automotive parts careers is offered later in this chapter.

Other ASE Test Areas

There are also certification categories in Heavy-Duty Truck, Auto Body, and Paint, as well as in Alternate Fuels.

To receive a bulletin of information about ASE certification, write to:

ASE
13505 Dulles Technology Drive
Herndon, VA 22071-3415

Complete information about ASE programs and services can be found on the Internet at http://www.ase.com.

Certification or License

Certification is a *voluntary process*. When state governments require a certain testing or skill level before allowing specialized work to be performed, this is called **licensing**. Emission testing and repair licensing is one area in which state governments have become heavily involved. Some state governments use the ASE advanced engine performance specialist certification as a prerequisite to licensing their emission technicians. Other states have their own testing programs.

■ TECHNICIAN SKILL LEVELS AND PAY

Flat Rate/Commission

Master technicians are often paid on a **flat rate/commission** basis. Each job has a listing in the *Parts and Time Guide* or *Flat-Rate Manual* (see Figure 5.16) that gives the amount of time it is supposed to take to complete. If the technician takes an hour and a half to complete a job that is listed as a 2-hour job, he or she is still paid for the 2 hours.

The opposite is also true. If it takes 3 hours to complete the job, it still only pays 2 hours. Technicians call this "flagging time." On a really good day, a technician might earn 12 hours of pay for an 8-hour day. On a poor day, 4 hours' pay might be all that is earned.

When a flat-rate work fails, the job is done again by the technician at no charge to the customer. A problem deep inside a recently rebuilt engine could require removal of the engine, a disaster for the technician who first performed the job.

For income protection, the best technicians sometimes have a "guarantee" in which the repair shop agrees to pay a minimum amount per pay period. In reality, the guarantee rarely comes into play with an experienced technician.

Automobiles have become so complex in recent years that technicians often earn more money than others who are in positions of management within the dealership. If a successful technician decides that a move into management is what he or she would like, management is often reluctant to have the technician make the move because he or she is too valuable in the shop.

An experienced master technician earning $50,000 a year is probably responsible for $110,000 to $120,000 per year of labor earnings for the business. This does not include the markup on the sale of required parts.

Job Shadowing

Job shadowing programs allow a student the opportunity to work behind an experienced and seasoned technician or other professionals in the automotive industry. The goal of job shadowing is to expose the student to the actual "real life" experience of being a professional in the automotive world.

Apprentice Technician

Some areas have unions that have formal apprenticeship programs. Apprentices work under journey-level or master technicians for a specified number of years before they work on their own. In areas where there is no formal apprenticeship, many repair shops and dealers use an apprenticeship system within their shops.

One scenario for a 1-year apprenticeship in a dealership works like this: A new technician is hired who has had some previous automotive work experience but is not a master technician. The apprentice will work for 2 months with each of the master technicians in the dealership. The apprentice is paid an hourly wage by the dealership during this first year.

As their incentive for mentoring the apprentice in the above system, the master technicians are paid for all of the hours that they earn in combination with the apprentice. In other words, they have free help for the 2-month period in exchange for giving instructions and sharing their expertise.

Another system used by some shops is the concept of a "team." Team members have several levels of expertise. They share pay for the jobs completed, with each team member being paid at a different level for the hours flagged. In this case, the apprentice at the bottom of the scale earns considerably less than the lead technician.

Mentoring

In a mentoring program, the students' mentor is in constant contact with them to answer any questions regarding the area of expertise that the students are interested in. The mentor also offers excellent "real world" experiences to the students.

■ OTHER AREAS OF SPECIALIZATION

Tire and Wheel (Chassis) Shops

Some repair shops specialize in tires and wheels (**Figure 2.10**). An ASE technician certified in brakes and suspension systems is usually employed by these shops and performs wheel alignment and suspension repairs related to tire wear. Sometimes brakes and

Figure 2.10 A tire and wheel specialty shop. *(Courtesy of Tim Gilles)*

alignment or other repair services are performed in the larger shops. A tire shop will have one or more *tire and wheel specialists* who mount and balance tires.

Muffler Shops

Muffler shops specialize in repairs and replacement of the vehicle's exhaust system (**Figure 2.11**). An *exhaust system specialist* fabricates new pipes and installs them. Muffler shops have large tubing benders for making new exhaust pipes from straight lengths of pipe. The new pipes and mufflers are welded together in place on the car. Muffler hangers are installed to keep the parts from vibrating and becoming damaged. Some muffler shops also do other things like quick lubes, tires, and brake and suspension work.

Transmission Shops

Some shops specialize in automatic transmission repair and rebuilding. Manual transmissions and differentials are repaired in some of these shops as well.

Merchandising Shops

Another specialty automotive area is audio system sales and installation. There are also shops specializing in the sport compact market, or tuners. Other companies deal with off-road vehicles and truck accessories.

Figure 2.11 Muffler shops specialize in repairs and replacement of the vehicle's exhaust system. *(Courtesy of Tim Gilles)*

■ OTHER AUTOMOTIVE CAREERS

Auto Parts

An *auto parts specialist* needs to have a basic knowledge of the operation of the vehicle. Some of the best parts people have repaired cars and have very good diagnostic abilities. In addition to automobile knowledge, a parts person must have good customer relations and telephone skills. When things get hectic (telephones are ringing and a line of customers is waiting), a parts person must be able to work without losing patience. Good reading skills and the ability to effectively use computer and parts books are all part of the skills required of the successful parts person (**Figure 2.12**).

Auto Body Technician

When auto body shops are associated with auto repair shops, the work area is located in a separate area. This is because of noise, dust, and vapors. An auto body technician fixes accident damage and paints vehicles. Specialized skills and training are required of a true professional in this area. Auto body shops are now highly regulated due to clean air laws.

Figure 2.12 An auto parts specialist identifies and orders replacement parts using (a) parts books or (b) a computer. *(Courtesy of Tim Gilles)*

Small Business Opportunities

Many independent repair shops are owned and operated by families. It is very common to find a husband and wife working together, managing and operating a business. Several kinds of abilities are required, including leadership and management talents. Public relations skills are very important to the success of a small business. The most talented auto technicians will quickly go out of business if they lack adequate people skills.

Hiring and firing employees are other aspects of running a business. The owner must assess the character of a potential employee, and if he or she does not work out, the owner must deal with terminating that person.

Other types of abilities required in managing a small business include payroll and bookkeeping. Some shops hire accounting firms for these duties.

Industry Professional Associations

Successful shop owners have leadership ability and are usually involved in automotive small business associations such as the Automotive Service Association (**ASA**) or Automotive Service Councils (**ASC**) (**Figure 2.13**). Dealers often belong to the National Automobile Dealers Association (**NADA**). These associations provide a way to meet other professionals with similar concerns. Local chapter monthly meeting agendas include legal, business, and training items. The association at the state and national level works to influence legislation that affects automotive businesses. Grouping together also provides increased benefits to the group members for items such as insurance for medical and workers' compensation.

■ ON THE JOB AS AN AUTOMOTIVE TECHNICIAN

Successful automotive technicians have several attributes besides the ability to repair automobiles. The success of a business depends on the performance of quality repairs. But a customer must also feel that he or she is being fairly treated.

Figure 2.13 Owners of automotive businesses often belong to professional associations. *(Courtesy of Automotive Service Association)*

Customer Relations

Technicians need to communicate effectively. In a large repair operation, this can be through a service writer. It is important that information be clearly written on a repair order. A customer's impression of the repair job and the business in general is affected by how professional the repair order appears. The repair order delivered to the customer must be neat and clean, without spelling or grammatical errors.

In a smaller business, technicians often communicate directly with customers (**Figure 2.14**). The success of the business can hinge on how considerate the technician is. Car maintenance is usually a planned and expected part of the customer's life. But sometimes a customer is unhappy because of a car repair that is unplanned and expensive. People skills are something that a good technician continually works to improve.

Appearance is another attribute that is important in customer relations. Many customers are bothered by dirty clothes and a lack of grooming. If you appear to lack pride in your appearance, customers feel that you do not treat their vehicle with respect either.

Every time a customer leaves a car at a repair shop, he or she is inconvenienced by having to arrange alternate transportation. Reliability is another attribute of a successful technician. A business depends on regular attendance by the technician so that promised repairs can be completed. Coming to work on time is also required. Repair shops sell the technician's time.

Customers often bring in their cars before going to work. In most shops, technicians begin working at 8 A.M. Under the flat-rate system, it is beneficial to show up early and be ready to start work. If a technician is late, the service dispatcher might give him or her a less desirable job. The other technicians will be able to begin their second jobs sooner and will probably complete more work for the day. If all of the shop's business is completed by 3 P.M., the technician who was late will have flagged fewer hours than the others.

Getting along with fellow employees and supervisors is also very important. One employee who always has something negative to say can poison the environment of an otherwise happy shop. A supervisor will usually favor the person with the good attitude.

NATEF School Certification

ASE also certifies schools with automotive programs. Schools certified in all eight ASE areas are *Master Certified* (**Figure 2.15**). For this certification, called National Automotive Technicians Education Foundation (**NATEF**), a team consisting of a certified team leader and technicians from local repair shops tours the campus. The team evaluates the facilities, tools and equipment, curriculum content and hours, and administrative support. School certifications such as the one provided by NATEF provide one means for a prospective student to evaluate a school he or she might like to attend for further automotive study.

Cooperative Work Experience Training

Some schools that are NATEF certified have cooperative work experience programs. Students attending school can earn school credit while working part-time. A contract is signed with the employer stating that the student will perform work in certain specialty areas. For instance, a student who wants to learn more about brakes will get to perform brake repairs while at work.

Figure 2.14 In a small business, the technician often communicates directly with the customer. *(Courtesy of Tim Gilles)*

Figure 2.15 The National Automotive Technicians Education Foundation (NATEF) evaluates school automotive programs for certification by ASE. *(Courtesy of ASE)*

▄▄ SUMMARY

Today's automotive technician must have a multitude of skills. There are many types of specialties under the automotive umbrella. Many skilled technicians are ASE certified. Technicians are often paid under the "flat-rate" commission system. There are large and small automotive businesses. Some are corporate and some are independently owned.

▄▄ REVIEW QUESTIONS

1. List five skills that an automotive technician must have.
2. What does PDI stand for?
3. What is the name of the person who greets the customer in a large automotive repair shop or dealership?
4. What is the name of the person who assigns the jobs to the technicians in a large shop?
5. What is the name of the kind of shop run by taxi cab companies or utilities?
6. List the eight ASE automobile specialty service areas that make up a master certification.
7. What is a technician called who is certified in all eight ASE areas?
8. What are the three ASE Engine Machinist areas of specialization?
9. Which is a voluntary process, certification or licensing?
10. What is the automotive system of pay that is based on the amount of work completed called?

Shop Procedures, Safety, Tools, and Equipment

THEN AND NOW:
TOOLS AND DIAGNOSTIC EQUIPMENT

The tools used to fix and maintain motor vehicles have changed almost as much as the cars themselves. In the early years, the only tool available to remove a nut or bolt was a crude "spanner" (an open-end wrench). Ownership of a set of spanners and an assortment of adjustable "monkey" wrenches, hammers, screwdrivers, and pliers were enough to qualify a person as a mechanic.

Diagnostic equipment was limited to intelligence and ingenuity, both of which are still essential today. Various devices for determining useful troubleshooting information started to show up in the early decades of the twentieth century. The vacuum gauge was considered essential, as was the compression tester. In the 1950s, the dwell meter (which told how long the ignition points were closed), the tachometer, the timing light, and the volt/ohmmeter became standard pieces of equipment in every service technician's toolbox. Possession of a secondary ignition oscilloscope and a distributor machine made a shop a high-tech diagnostic center.

Today's wrenches, whether the combination or socket type, are precisely made and finely finished in a fashion that reminds one of jewelry. Pneumatic tools, from impact guns to air ratchets and chisels, are of high quality and have become affordable. Hands and muscles are saved as these tools assist with the hard work.

Modern automobiles rely on complex electronic engine management systems, and diagnostic equipment has evolved into a computer-age level of sophistication. Large console-type engine analyzers check everything from compression to ignition performance. Extremely useful and convenient handheld scan tools plug into the vehicle's diagnostic connector (now standardized under OBD II regulations). These instruments instantly provide trouble codes and sensor voltages and values. Miniaturized lab scopes (also called "digital storage oscilloscopes" and sometimes referred to as a "visual voltmeter") catch electronic glitches that would be invisible to any other piece of equipment.

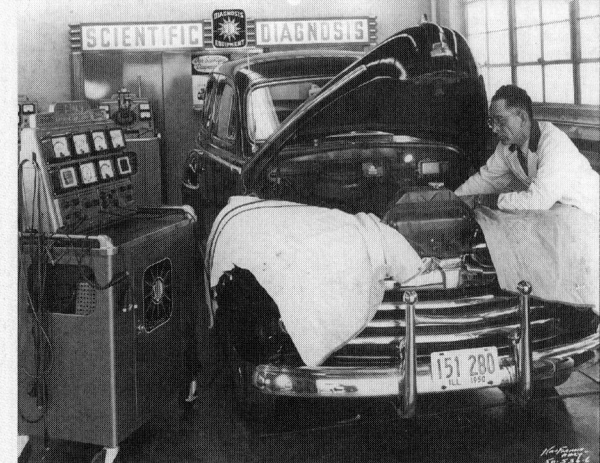

Automotive diagnostics circa 1950. *(Courtesy of Ogilvy, Adams, and Rinehart for Snap-On Tools Company)*

A modern computer system scan tool. *(Courtesy of Tim Gilles)*

CHAPTER 3

Shop Safety

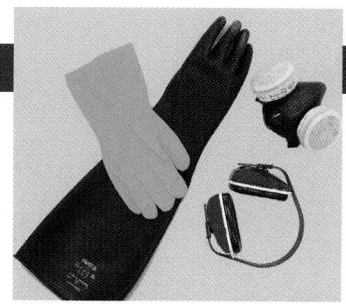

■■ **OBJECTIVES**

Upon completion of this chapter, you should be able to:

✔ Use shop tools and equipment safely.

✔ Understand safety rules.

■■ **KEY TERMS**

Class A fire	backfire	MSDS
Class B fire	popback	dermatitis
Class C fire	greasesweep	barrier creams
Class D fire	Hazard Communication	HEPA vacuum
flash point	Standards	electrolyte

■■ **INTRODUCTION**

The number one priority of any business is the health and safety of its employees. Shop safety issues are covered here such as:

■ A safety test is included at the end of the chapter.

■ As you read this chapter, realize that the situations described can and do occur, sometimes often.

■ Case histories presented throughout the book are true. Pay extra attention to the safety precautions described.

■ Safety precautions are presented for each piece of equipment described in this book.

The employer is responsible for safety training for each employee and for providing a safe working environment. According to federal law, each shop will have a safety training program. A shop owner in Los Angeles was prosecuted in the death of an employee for not providing proper safety training.

The employee is the one who is actually responsible for his or her own safety and the safety of others. Accidents that occur in an automotive shop often happen because safety considerations are not as obvious when repairing automobiles as they are in such trades as roofing or carpentry. Accidents are often caused by carelessness resulting from a lack of experience or knowledge or from being in a hurry and taking short cuts.

Injury accidents are often caused by someone other than the person injured. The person who caused the accident often suffers from guilt from the harm he or she has caused another. In the event of an accident, be sure to inform your instructor or supervisor, who knows what procedures to follow.

Injured persons often suffer from *shock* and should not be left unattended. When an injury does not appear to be serious enough to call an ambulance, a companion should go with the injured party to seek professional help.

The American Red Cross offers thorough first-aid training. Every shop should have someone trained to handle emergencies.

■■ **GENERAL PERSONAL SAFETY**

A first-aid kit (**Figure 3.1**) contains items for treating some of the small cuts and abrasions that often happen. Fires and accidents involving lifts and battery chargers happen occasionally in automotive shops. But the most common injuries are mostly preventable injuries involving the *back* or *eyes*.

Gloves, goggles, and respirators are called personal protective equipment, or PPE (**Figure 3.2**).

Wearing eye protection will prevent most eye injuries, so when using machinery eye protection is mandatory. Several types of eye protection are shown in **Figure 3.3**.

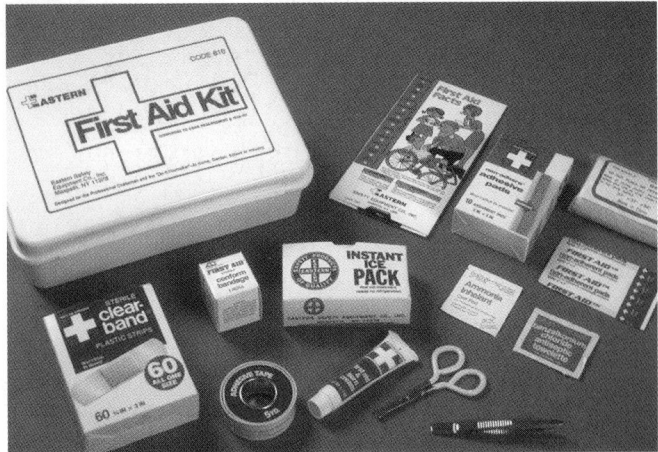

Figure 3.1 A first-aid kit.

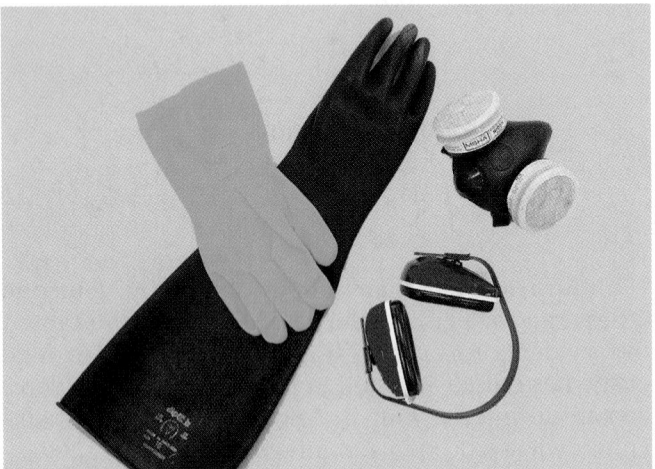

Figure 3.2 Personal protective equipment includes gloves, a respirator, and ear protection. *(Courtesy of Tim Gilles)*

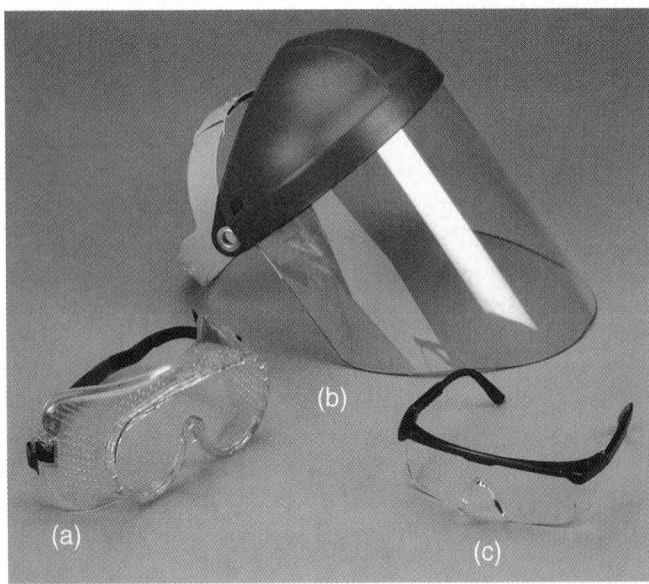

Figure 3.3 Eye protection. (a) Goggles. (b) Face shield. (c) Safety glasses.

Eye Protection Safety

Parts can explode when they are pressed or pounded on. Rotating tools can throw pieces of metal or grit, causing eye injuries. Safety goggles or a face shield should be worn when using shop tools and equipment. Because eye injuries are so common in an automotive shop, continual use of safety glasses or goggles is recommended. Prescription safety glasses are an advantage because the user always wears them. Eye protection is emphasized for your protection!

Face shields (see Figure 3.3b) are convenient because they can stay with the piece of equipment. They are also easily adjusted to fit your head.

The following list gives some hints for preventing eye injuries. Eye protection must be worn:

- Whenever working around moving parts and machinery (grinding, cutting, drilling, washing, etc.)
- When blowing off parts with compressed air
- When working under the vehicle
- When working on air conditioning
- When flame-cutting or welding

If your eye is accidentally contaminated with a dangerous liquid, flush thoroughly in an eyewash fountain (**Figure 3.4**) or other water source.

Back Protection Safety

Following safe lifting procedures will prevent most back injuries (**Figure 3.5**). The following are safety precautions to be used when lifting:

Figure 3.4 An eyewash fountain. *(Courtesy of Western Emergency Equipment)*

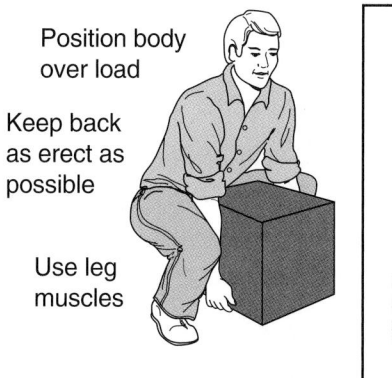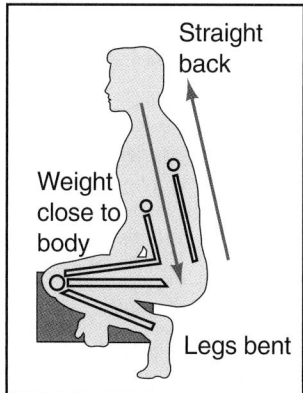

Position body over load

Keep back as erect as possible

Use leg muscles

Straight back

Weight close to body

Legs bent

Figure 3.5 Lifting precautions.

- Be sure to get help when moving heavy items. The normal tendency is to say that items are not that heavy and you hate to ask somebody for help. If something is in an awkward position for lifting, leverage and the position your back is in can make it easier for an injury to occur.
- If an item is too heavy to lift, use the appropriate equipment.
- Before moving a heavy item, plan the route that the item will be carried and how it is to be set back down when you get there.
- Lift slowly.
- Do not jerk or twist your back. Shift your feet instead.
- Bend your knees and *lift with your legs, not your back!* Also, keep your lower back straight when lifting. Think about thrusting your stomach out.

CASE HISTORY *A technician lifted a spare tire out of the trunk of a car. As he reached forward and tugged the tire up and out of the trunk, he felt a small pop in his lower back. The result was a herniated disk in his lower back, an injury that will affect him for the rest of his life.*

Ear Protection

Damage to ears happens due to exposure to loud noises over a period of time. When loud air tools are used, ear protection should be worn.

NOTE: *Once it has been damaged, your hearing will not recover.*

Clothing and Hair

Clothing (shirt tails) or hair that hangs out can get caught in moving machinery or under a creeper. Keep long hair tied or under a cap. Shirt tails should be tucked in, or shop clothing can be worn over the shirt.

Shoes or Boots

Leather shoes or boots offer much better protection than tennis shoes or sandals. Soles are available that resist slipping and are resistant to damage from petroleum products. Boots and shoes that have the toe reinforced with steel inserts are widely available.

Hand Protection

Cuts and scrapes are common when repairing automobiles. Be sure to keep your fingers away from moving machinery and hot surfaces.

Solvents used to clean parts can damage skin. Some types of solvents can actually penetrate your skin. Use the correct type of glove when handling chemicals (see Figure 3.2). Some inexpensive gloves will actually allow some chemicals to pass through them. The use of gloves is becoming more common in the automotive repair industry.

■ FIRE SAFETY

Two major items should be considered when dealing with fires. A little common sense is important. If the fire is burning so dangerously that your personal safety is jeopardized, leave the area immediately and call for help. If you can safely remove the source of fuel or heat to a fire, do so. This might include shutting off fuel to a fuel fire or disconnecting the electrical source from an electrical fire.

■ FIRE EXTINGUISHERS

A fire extinguisher is a portable tank that contains water or foam, a chemical, or a gas (**Figure 3.6**). There are four kinds of fires, each calling for a different type of fire extinguisher (**Figure 3.7**):

- A **Class A fire** is one that can be *put out* with water. Such items as paper and wood make up these kinds of fires.

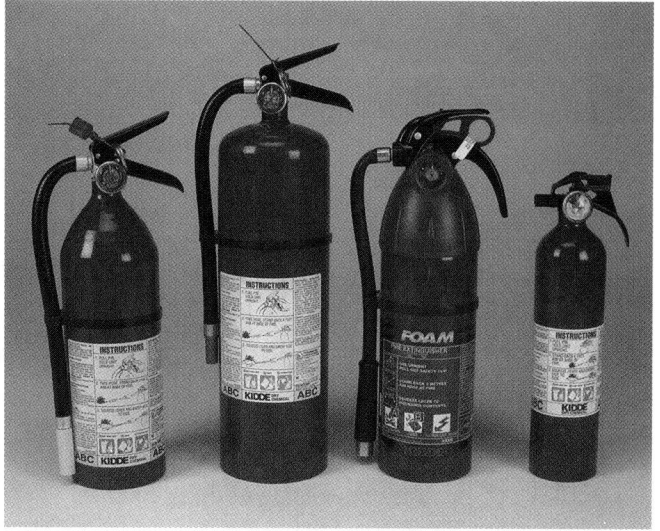

Figure 3.6 Fire extinguishers.

	Class of Fire	Typical Fuel Involved	Type of Extinguisher
Class A Fires (green)	**For Ordinary Combustibles** Put out a Class A fire by lowering its temperature or by coating the burning combustibles.	Wood Paper Cloth Rubber Plastics Rubbish Upholstery	Water*[1] Foam* Multipurpose dry chemical[4]
Class B Fires (red)	**For Flammable Liquids** Put out a Class B fire by smothering it. Use an extinguisher that gives a blanketing flame-interrupting effect; cover whole flaming liquid surface.	Gasoline Oil Grease Paint Lighter fluid	Foam* Carbon dioxide[5] Halogenated agent[6] Standard dry chemical[2] Purple K dry chemical[3] Multipurpose dry chemical[4]
Class C Fires (blue)	**For Electrical Equipment** Put out a Class C fire by shutting off power as quickly as possible and by always using a nonconducting extinguishing agent to prevent electric shock.	Motors Appliances Wiring Fuse boxes Switchboards	Carbon dioxide[5] Halogenated agent[6] Standard dry chemical[2] Purple K dry chemical[3] Multipurpose dry chemical[4]
Class D Fires (yellow)	**For Combustible Metals** Put out a Class D fire of metal chips, turnings, or shaving by smothering or coating with a specially designed extinguishing agent.	Aluminum Magnesium Potassium Sodium Titanium Zirconium	Dry powder extinguishers and agents only

*Cartridge-operated water, foam, and soda-acid types of extinguishers are no longer manufactured. These extinguishers should be removed from service when they become due for their next hydrostatic pressure test.
Notes:
(1) Freeze in low temperatures unless treated with antifreeze solution, usually weighs over 20 pounds, and is heavier than any other extinguisher mentioned.
(2) Also called ordinary or regular dry chemical. (solution bicarbonate)
(3) Has the greatest initial fire-stopping power of the extinguishers mentioned for class B fires. Be sure to clean residue immediately after using the extinguisher so sprayed surfaces will not be damaged. (potassium bicarbonate)
(4) The only extinguishers that fight A, B, and C class fires. However, they should not be used on fires in liquified fat or oil of appreciable depth. Be sure to clean residue immediately after using the extinguisher so sprayed surfaces will not be damaged. (ammonium phosphates)
(5) Use with caution in unventilated, confined spaces.
(6) May cause injury to the operator if the extinguishing agent (a gas) or the gases produced when the agent is applied to a fire is inhaled.

Figure 3.7 Guide to Fire Extinguisher Selection.

- A **Class B fire** is one in which there are *flammable liquids* such as grease, oil, gasoline, or paint.
- A **Class C fire** is *electrical*.
- A **Class D fire** is a *flammable metal* such as magnesium or potassium.
- Either *carbon dioxide* (CO_2) or a *dry chemical* fire extinguisher can be used on *Class B and C fires*.

A gauge on the top of the fire extinguisher tells whether it is fully charged or if the charge pressure has leaked off. Fire extinguishers in business establishments are periodically inspected by the local fire department or a fire extinguisher service company, but be sure to check the gauge on the extinguisher on a regular basis.

For auto repair shops, the size should be at least a 20B (this number designation is explained later). A popular fire extinguisher that people keep in their vehicles or boats is a 2A-10BC:

- The 2A means that the extinguisher can put out a 2 cubic foot fire of a material like paper.
- The 10B means that it can put out 10 square feet of burning liquid.

 **SAFETY NOTE** Be sure to have a large enough fire extinguisher on hand. An engine compartment fire often requires a larger extinguisher, depending on how far the fire has progressed.

- The C means that electrical fires can be extinguished.

Locate and check the type of fire extinguisher(s) in your shop. They should be located in a place *other* than where the most likely start of a fire would be. For instance, do not locate a fire extinguisher right over the welding bench or next to the solvent tank. If a fire began there, you would not be able to get to the extinguisher. Locating fire extinguishers on both sides of a common work area is a good idea.

◼ FLAMMABLE MATERIALS

Greasesweep and rags soaked in oil or gasoline should be stored in covered metal containers (**Figure 3.8**). Keeping oily materials separated from air prevents them from self-igniting, a process called *spontaneous combustion*. Flammable materials should be stored in a flammable storage cabinet (**Figure 3.9**).

Used greasesweep is kept in a flammable storage container because it may be reused until it becomes saturated (wet). Saturated greasesweep must be disposed of using a licensed waste hauler.

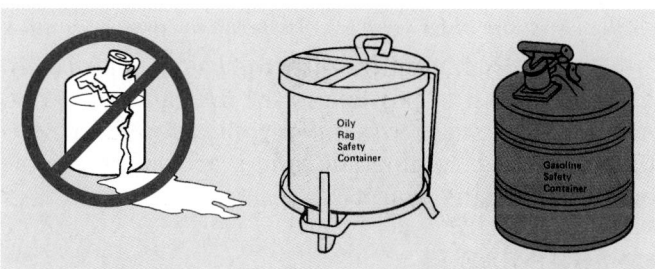

Figure 3.8 Keep combustibles in safety containers.

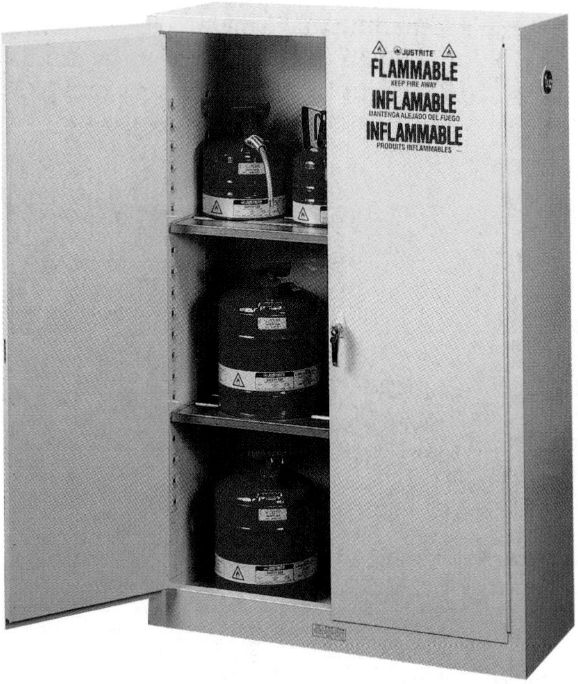

◼ FUEL FIRES

Gasoline is a major cause of automotive fires. Liquid gasoline is not what catches on fire. Rather, it is the *vapors* that are so dangerous. Gasoline vapors are heavier than air, so they can collect in low places in the shop. They can be ignited by a spark from a light switch, a motor, electrical wires that have been accidentally crossed, or a dropped shop light.

CAUTION Cold fuel spray can cause a bulb to shatter and the hot filament can start a very dangerous fuel fire.

There are two kinds of shop lights (**Figure 3.10**). One of them has a fluorescent bolt enclosed in a plastic tube. The other uses an incandescent lamp, or a light bulb. A fire hazard exists when using an incandescent lamp around a fuel system, especially high-pressure ones.

Rubber-coated bulbs, which are available, will reduce the chances of scattering glass if the bulb is broken, but the recommended style of shop light to be used around automobiles is the one enclosed in a plastic tube or an ordinary flashlight.

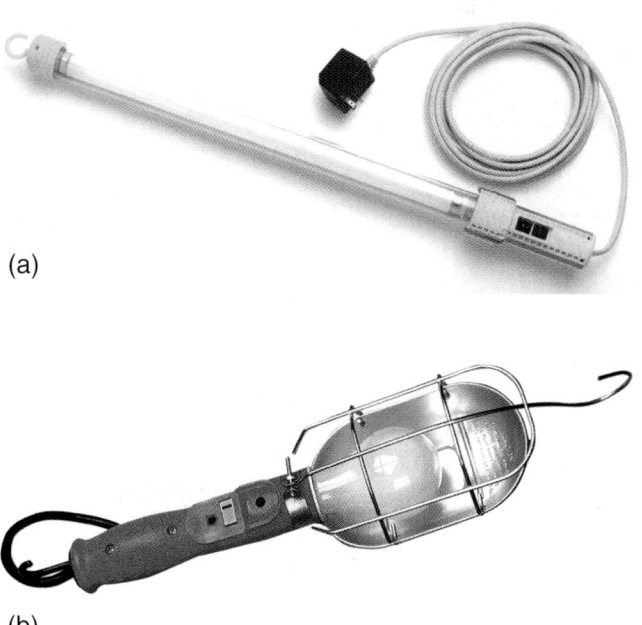

(a)

(b)

Figure 3.10 Two kinds of drop lights. (a) A fluorescent tube safety light. (b) The light shown here should be used with a shatterproof bulb. *(Courtesy of Woods Industries, Inc.)*

Figure 3.9 An approved flammable storage cabinet. *(Courtesy of Justrite Manufacturing)*

Gasoline Safety

1. Gasoline should be stored only in an approved safety container (see Figure 3.9) and never in a glass jar.

2. Never use gasoline to clean floors or parts. Parts cleaning solvent (Stoddard solvent is the industry term) has a higher **flash point** than gasoline. A flammable liquid's flash point is the temperature at which the vapors will ignite when brought into contact with an open flame.

3. Careless cigarette smoking or failing to immediately and thoroughly clean gasoline spills can contribute to a dangerous situation. People get used to working around gasoline and then begin to ignore how dangerous it can be.

4. Do not attempt to siphon gasoline with your mouth. Accidental breathing of gasoline into the lungs can be fatal.

A fire will go out if the source of oxygen is removed or if the temperature is lowered below the fuel's flash point. Firefighters can use a fog of water to cool a fuel fire and cut off its oxygen, but shop personnel should be knowledgeable about fire extinguishers and how to use them.

NOTE: Do **not** attempt to douse a fuel fire with water. Water will cause the fire to spread because fuel is lighter than water and floats on top of it.

 **CASE HISTORY** *A technician completed the installation of a rebuilt engine. When he started the engine, the float in the carburetor stuck. The results were gasoline pouring out of the carburetor, and an air-fuel mixture that was too rich. When the engine "popped back" through the carburetor, a fire began. The technician grabbed a nearby hose and attempted to douse the fire. The water only served to spread the flames. The result was a total loss of the vehicle.*

NOTES:

■ **Backfire** is a term for combustion that occurs in the vehicle's exhaust (after it leaves the cylinder).

■ **Popback** is a term for combustion occurring in the fuel induction system (before it enters the cylinder).

■ ELECTRICAL FIRES

Electrical fires are prevented by disconnecting the battery before working on the electrical system or around electrical components, such as the starter or alternator (**Figure 3.11**). Unbolt the ground cable first. This prevents the possibility of a spark occurring if a wrench accidentally completes a circuit between a "hot" cable and ground.

Figure 3.11 Remove the battery ground cable first.

NOTE: *The ground cable is the one that is bolted to the engine block. Do not assume that ground is the negative cable. On some older vehicles, the positive cable is ground.*

If there is an electrical fire, the *battery must be disconnected* as fast as possible so the fire can be put out. Another advantage to removing the ground cable is that an electric cooling fan cannot accidentally come on while working near it.

■ SHOP HABITS

The shop can be kept cleaner if a technician gets into the habit of using a shop towel when working. Greasy, oily tools and hands should be wiped clean, preventing the mess from being spread around the rest of the shop. Shop towels sometimes have metal burrs from machining operations. These are not removed by laundering (**Figure 3.12**). Do not rub your face or hands with a shop towel before carefully inspecting it.

Figure 3.12 Metal burrs are often found in laundered shop towels. *(Courtesy of Tim Gilles)*

An apprentice was removing a clutch from a car. After raising the vehicle on the lift, he began to remove the starter motor while the battery was still hooked to the circuit. He did not disconnect the wiring, which was in an inaccessible location on the top of the starter. When the unbolted starter came loose, bare electrical terminal ends became wedged against the frame, which resulted in a major electrical fire. The technician working next to him rushed over with a bolt cutter and cut the battery ground cable near where it was bolted to the block. Damage to the vehicle was limited to a wiring loom and the alternator. The apprentice went to the emergency room to be treated for burned hands and respiratory irritation from inhaling noxious smoke when the insulation on the wires caught fire.

Common sense around the shop dictates that spills be cleaned up as soon as they happen. Oil and solvent spills can be cleaned up with an absorbent material, such as rice hull ash or kitty litter **greasesweep**. Greasesweep is swept up and reused until it becomes too wet (**Figure 3.13**). It is better to use it when it is slightly wet, because the dust that results when using new greasesweep is avoided.

One of the results of environmental regulation is that greasesweep is classified as hazardous if it is used to soak up used engine oil. Waste disposal companies provide a service where superabsorbent cloths are used to soak up spills. The disposal companies then collect the cloths for proper treatment.

NOTE: *Because saturated greasesweep could be flammable, it must be stored in a metal can with a lid.*

Newer materials that are now being used instead of greasesweep are bioremedial oil absorbent products. These products have microbes that "eat" oil or fuel,

Figure 3.13 Greasesweep is used to absorb spills. *(Courtesy of Tim Gilles)*

converting it to harmless CO_2 and water. Concrete floors cleaned with this material are left clean and slip resistant. The biggest advantage to this method is that the need for hazardous disposal is reduced or eliminated.

There are also nontoxic water-based degreasers.

HAZARDOUS MATERIALS

Some materials routinely used in the shop may be dangerous to your health. In addition, many chemicals irritate skin. Cautions about skin and eye protection are covered throughout this chapter. A material is considered *hazardous* if it causes illness, injury, or death or pollutes water, air, or land.

Hazardous materials that automotive technicians commonly come into contact with include:
- Cleaning chemicals (carburetor cleaner, solvents, caustics, acids)
- Battery acid
- Fuels
- Paints and thinners
- Used oil and fluids
- Heavy metals
- Antifreeze/coolant
- Asbestos (brakes and clutches)
- Refrigerants

Many states publish pamphlets that specifically address the proper disposal of wastes generated in automotive repair shops. Many modern automotive repair shops use equipment specifically to reduce the generation of hazardous wastes. Some of these include coolant recyclers, air conditioning refrigerant recyclers, oil filter crushers, pyrolytic baking ovens, and waste oil furnaces for shop heating.

In the United States, materials that can harm workers and the environment are regulated by the following three agencies:
- The charge of the Environmental Protection Agency (EPA) is to protect the land, air, and water from hazardous materials.
- The *National Institute of Occupational Safety and Health* (*NIOSH*), formerly OSHA, manages the protection of workers from unsafe workplace hazards, including hazardous materials.
- The Department of Transportation (DOT) oversees the transportation of hazardous materials.

The EPA characterizes hazardous waste in two ways and has developed a list of over 500 specific hazardous wastes. A material is a hazardous waste if it is on this list or if it exhibits a characteristic from one of four categories: ignitability, corrosivity, reactivity, or toxicity. The following are examples of hazardous wastes from these four categories.
- *Ignitable wastes* burn readily. These materials have a flash point of less than 140°F. Automotive products such as cleaning solvents, paint solvents, degreasers, and gasoline are examples of ignitable materials.

■ *Corrosive wastes* burn the skin or corrode metals or other materials. They have a pH of lower than 2 or higher than 12.5. Examples include rust removers and acid or alkaline cleaning liquids.

■ *Reactive wastes* are unstable and react rapidly or violently when combined. A good example of reactive wastes include ammonia and bleach, which combine to form phosgene gas (nerve gas). There are hardly any reactive wastes used in the automotive industry. Acetylene (used in gas welding) is an example of a reactive substance.

■ *Toxic wastes* are often described as anything you cannot safely eat or drink. Examples of toxic wastes include aerosol spray paints and fuel system or brake cleaners that are commonly used in most automotive shops (**Figure 3.14**). Some products contain chlorinated compounds, and most contain various toxic organic chemicals, such as methyl ethyl ketone (MEK), acetone, methylene chloride, trichloroethane, toluene, and heptane. Some are hazardous air pollutants (HAPs), and chlorinated solvents are a threat to the earth's protective ozone shield. Other automotive pollutants on the EPA's air contaminant list include carbon monoxide from car exhaust and chlorofluorocarbon (CFC) from older air-conditioning refrigerants.

Petroleum-based solvents can cause nervous system damage, and some are carcinogens (potentially causing cancer). Wastes are also toxic if they contain certain heavy metals such as chromium, lead, or cadmium.

Bloodstream Hazards

Hazardous materials can enter the bloodstream through the lungs by breathing fumes. Some materials can pass right through a respirator. Asbestos is a good example. Asbestos fiber is so small that a special filter called a high-efficiency particulate air (HEPA) filter is required for breathing protection. Asbestos is known to cause

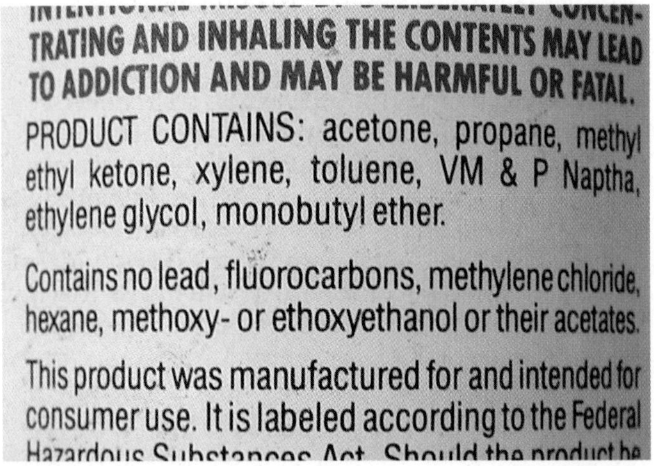

Figure 3.14 A contents label from a spray paint can. *(Courtesy of Tim Gilles)*

damage to lungs and has been linked to a specific cancer of the lung cavity.

Problems with Spray Cleaners. Aerosol spray cleaners tend to be more hazardous to your health because they enter the bloodstream when breathed. The base from which many spray brake cleaners is made is perchloroethylene (PERC). Other brake cleaners are volatile organic compound (VOC)-based solvents. These include xylene, toluene, and hexane.

■ PERC is a carcinogen.
■ Xylene causes birth defects.
■ Toluene causes central nervous system damage.
■ Hexane is a neurotoxin.

NOTE: *Water-based cleaners are preferable, especially in brake work. Some water-based cleaners contain solvents such as alcohol, terpenes, and glycol ethers, some of which may be toxic and damaging to hands. There are effective water-based cleaners that do not have solvent additives.*

Hazardous materials can also be ingested by eating or drinking. The skin protects the bloodstream from most liquids. However, the molecules of some toxic liquids are smaller than the membrane of human skin that allows sweat to escape. These liquids can, and do, penetrate the skin and enter the bloodstream. Always wear protective gloves to protect yourself. Be certain the gloves provide the necessary amount of protection. Some gloves can allow the same materials that can penetrate your skin to pass through. Obviously, materials can enter the bloodstream unimpeded by the skin if you have a cut or scrape.

NOTE: *Many technicians routinely wear blue nitrile gloves on their hands at all times. Nitrile gloves are not damaged by most solvents. They also provide some measure of protection from skin damage and penetration through the skin by solvents.*

Skin Damage

Automotive technicians often receive chronic exposure to cleaning materials. Using petroleum-based solvent to clean parts by hand without using gloves is one example. Years of this abuse can result in a painful condition in which hands are cracked, often bleed, and the skin peels off. This is called dermatitis. It can also be caused by using the wrong soap for the soil being cleaned. There are hand lotions and creams formulated for use to correct dermatitis. Be sure to wash your hands thoroughly before applying a lotion or cream.

NOTE: *If your employer provides you with personal protective equipment (PPE), you are required to use it. Eye protection is one example. NIOSH Regulations Section 29 CFR 1910.133 (a) (1) says, "each affected employee shall use eye or face protection when exposed to eye or face hazards from flying particles, molten metal, liquid chemicals, acids or caustic liquids, chemical gases or vapors, or potentially injurious light radiation" (such as arc welding).*

■ HAZARDOUS COMMUNICATION STANDARDS

In the United States, the EPA outlines regulations for the disposal of hazardous materials. These regulations are called the Hazard Communication Standards.

Companies must develop a hazardous waste policy, and each hazardous waste generator must have an EPA identification number. When waste is transported for disposal, a licensed waste hauler must transport and dispose of the waste. A copy of the written manifest—an EPA form—must be kept by the shop.

The Hazard Communication Program requires chemical manufacturers to publish information about the chemicals they produce and distribute. The employer then provides information by labeling materials and training employees in their proper uses. The label on a hazardous material must include the product manufacturer's name and address, its chemical name and trade name, and safety information about the chemical.

■ MATERIAL SAFETY DATA SHEETS

For any hazardous materials used, an employer must make available to employees a Material Safety Data Sheet (MSDS) (**Figure 3.15**). An MSDS provides details of the composition of a chemical, lists possible health and safety problems, and gives precautions for its safe use. A supplier of a chemical is required by law to provide an MSDS upon request.

```
HEXANE
=====================================================
MSDS Safety Information
=====================================================
Ingredients
=====================================================
Name: HEXANE (N_HEXANE)
% Wt: >97
OSHA PEL: 500 PPM
ACGIH TLV: 50 PPM
EPA Rpt Qty: 1 LB
DOT Rpt Qty: 1 LB
=====================================================
Health Hazards Data
=====================================================
LD50 LC50 Mixture: LD50:(ORAL,RAT) 28.7 KG/MG
Route Of Entry Inds _ Inhalation: YES
Skin: YES
Ingestion: YES
Carcinogenicity Inds _ NTP: NO
IARC: NO
OSHA: NO
Effects of Exposure: ACUTE:INHALATION AND INGESTION ARE HARMFUL AND MAY BE FATAL.
INHALATION AND INGESTION MAY CAUSE HEADACHE, NAUSEA, VOMITING, DIZZINESS, IRRITATION
OF RESPIRATORY TRACT, GASTROINTESTINAL IRRITATION AND UNCONSCIOUSNESS. CONTACT
W/SKIN AND EYES MAY CAUSE IRRITATION. PROLONGED SKIN CONTACT MAY RESULT IN
DERMATITIS (EFTS OF OVEREXP).
Signs And Symptons Of Overexposure: HLTH HAZ:CHRONIC:MAY INCLUDE CENTRAL
NERVOUS SYSTEM DEPRESSION.
Medical Cond Aggravated By Exposure: NONE IDENTIFIED.
First Aid: CALL A PHYSICIAN. INGEST:DO NOT INDUCE VOMITING. INHAL:REMOVE TO FRESH AIR. IF
NOT BREATHING, GIVE ARTIFICIAL RESPIRATION. IF BREATHING IS DIFFICULT, GIVE OXYGEN.
EYES:IMMED FLUSH W/PLENTY OF WATER FOR AT LEAST 15 MINS. SKIN:IMMED FLUSH W/PLENTY OF
WATER FOR AT LEAST 15 MINS WHILE REMOVING CONTAMD CLTHG & SHOES. WASH CLOTHING
BEFORE REUSE.
=====================================================
Handling and Disposal
=====================================================
Spill Release Procedures: WEAR NIOSH/MSHA SCBA & FULL PROT CLTHG. SHUT OFF
IGNIT SOURCES:NO FLAMES, SMKNG/FLAMES IN AREA. STOP LEAK IF YOU CAN DO SO W/OUT
HARM. USE WATER SPRAY TO REDUCE VAPS. TAKE UP W/SAND OR OTHER NON-COMBUST MATL &
PLACE INTO CNTNR FOR LATER DISP (SU PDAT)
Neutralizing Agent: NONE SPECIFIED BY MANUFACTURER.
Waste Disposal Methods: DISPOSE IN ACCORDANCE WITH ALL APPLICABLE FEDERAL, STATE AND
LOCAL ENVIRONMENTAL REGULATIONS. EPA HAZARDOUS WASTE NUMBER:D001 (IGNITABLE
WASTE).
Handling And Storage Precautions: BOND AND GROUND CONTAINERS WHEN TRANSFERRING LIQUID.
KEEP CONTAINER TIGHTLY CLOSED.
Other Precautions: USE GENERAL OR LOCAL EXHAUST VENTILATION TO MEET
TLV REQUIREMENTS. STORAGE COLOR CODE RED (FLAMMABLE).
=====================================================
Fire and Explosion Hazard Information
=====================================================
Flash Point Method: CC
Flash Point Text: _9F_23C
Lower Limits: 1.2%
Upper Limits: 77.7%
Extinguishing Media: USE ALCOHOL FOAM, DRY CHEMICAL OR CARBON DIOXIDE. (WATER MAY BE
INEFFECTIVE).
Fire Fighting Procedures: USE NIOSH/MSHA APPROVED SCBA & FULL PROTECTIVE
EQUIPMENT (FP N).
Unusual Fire/Explosion Hazard: VAP MAY FORM ALONG SURFS TO DIST IGNIT SOURCES & FLASH
BACK. CONT W/STRONG OXIDIZERS MAY CAUSE FIRE. TOX GASES PRDCED MAY INCL:CARBON
MONOXIDE, CARBON DIOXIDE.
=====================================================
```

Figure 3.15 A material safety data sheet (MSDS).

In addition to your local parts supplier, the Internet is a good source for accessing MSDSs. Typical shop practice is to obtain pertinent MSDSs and keep them in a binder where employees can have access to them. An MSDS must include any hazardous materials that are contained in a product. It lists the percentage of the product that is made up of each of the product's hazardous ingredients. For instance, a typical carburetor cleaner spray might be made up of 45% toluene, 20% acetone, and so on.

Interpreting an MSDS

The MSDS lists the *permissible exposure limit* (PEL) and the *threshold limit value* (TLV) of various vapors, dusts, or fumes. Exceeding the limit can result in illness or injury. Short-term exposure limits (STEL) are also included. The STEL is the maximum amount of exposure during a 15-minute period in one 8-hour workday.

NOTE: *The smaller the exposure limit on the MSDS, the more toxic the chemical is. NIOSH lists an exposure limit called immediately dangerous to life and health (IDLH).*

Be aware that exposure limits are simply estimates of danger for short-term exposure. There is no guarantee that you will not suffer long-term damage if you are regularly exposed to a chemical at a level that is less than the STEL.

Flammable limits for a material are also listed on an MSDS. LEL is the *lower explosive limit* of a material. This represents the lowest vapor concentration of a material that will ignite. Leaner mixtures than the LEL will not burn. UEL stands for the upper explosive limit. Concentrations richer than the UEL will not burn. Some chemicals list LFL and UFL standards. These are the same, except they mean lower flammable limit and upper flammable limit, respectively.

■ HAZARDOUS MATERIALS COMMON TO THE AUTOMOTIVE INDUSTRY

Solvents

Solvents are very popular for cleaning automotive parts. A solvent is a liquid organic chemical compound used to dissolve solid substances. Most are made from petroleum or synthetic materials. Solvents can dissolve plastics and resins such as those used in gaskets and sealers. They also evaporate quickly, leaving relatively little residue. Evaporation rate is important to the application. Gasket cements and paints must dry quickly. That is why solvents used in them have high evaporation rates.

Solvents are popular in the workplace because they work very well and make jobs go much faster. All solvents are toxic to a certain extent, however, and must be treated with respect if they are to be used. Used correctly, solvents can be very handy tools. Always read the MSDS!

Hazards Related to Solvents

In chemistry, classes are groups of chemicals with similar properties and molecular structures. There are several chemical classes of solvents, including alphatic, aromatic and chlorinated hydrocarbons, alcohols, esters, and ketones. Most solvents are flammable, and some can explode. Some can react with other substances or heat to create a different hazardous material.

NOTE: *Chlorinated solvents are not flammable but when exposed to flame can become phosgene gas. Phosgene gas is the nerve gas that killed so many soldiers in World War I.*

Symptoms of solvent exposure can show up immediately, acute symptoms, or they can arise later, chronic symptoms or effects. Acute symptoms include rashes and burns, nausea, or headaches. Solvents can cause irritation and damage to your skin, eyes, and respiratory tract, including the nose and throat. Continued irritation can result in chronic bronchitis and other lung diseases. These types of effects are called *"local effects"* because the harm is caused at its original point of contact.

All organic solvents can cause skin problems. Repeated contact causes the skin's protective fats and oils to dissolve. When the skin's protective barrier is removed, it becomes dry and cracked, leaving it vulnerable to absorption of toxic materials. Some solvents, such as MEK, can be absorbed through the skin. When a chemical enters the body in this way, it can travel to other organs, such as the liver, kidneys, muscles, heart, or brain. This type of exposure is called *"systemic."* Liver and kidney damage can result from systemic chemical exposure. The liver converts solvents to less toxic substances. The kidneys filter the results of that conversion and prepare them to leave the body. The liver and kidneys can be damaged by the substances they are attempting to eliminate.

Chronic effects are also called latent effects. They can be caused by repeated exposure over time or can result from a single exposure. Latent effects can take weeks or even years to show up. Chronic effects are often irreversible. These include cancer, heart disease, blood disease, and potential birth defects. Symptoms of chronic exposure to dangerous chemicals are documented over time. Heart damage was linked to solvents during a 5-year period when more than 100 people died from sniffing solvents.

Benzene, heptane, trichloroethylene, and methylene chloride are some of the chemicals that can cause arrhythmias. Arrhythmia is a term that applies to abnormal patterns in a heart's pumping cycle. Tetrachloroethylene, also called perchloroethylene or perc, and toluene have been implicated as spontaneous abortion risks in fathers who have been exposed to it in the workplace.

Some solvents can cause neurological symptoms such as narcosis. Central nervous system involvement can result in narcosis, or unconsciousness, dizziness, headaches, fatigue, and nausea. At high concentrations, symptoms can mimic drunkenness, followed by death.

The most common solvents in the automotive industry include hexane, mineral spirits, xylene, toluene, heptane, and petroleum napthas. Hexane is found in gasoline, rubber cements, and many spray products. Check the label on the container (**Figure 3.16**) and read the MSDS. Hexane can affect the central and peripheral nervous systems. Peripheral nervous system involvement causes a slowing of the speed of nerve impulses from the spine to the arms and legs accompanied by numbness, weakness, or paralysis. This is called peripheral neuropathy. Symptoms can be similar to multiple sclerosis. Another disorder of the peripheral nervous system is polyneuropathy. Symptoms include muscle spasms, leg pain and weakness, and tingling in the arms. Solvents implicated as possible causes include benzene, n-hexane, n-heptane, and toluene.

Sometimes two or more substances react with one another and can become more hazardous. This is called synergism. One example is the higher instance of health problems among asbestos workers who also smoke. In another example, when MEK is combined with n-hexane, severe nerve damage can result. Carbon tetrachloride causes serious liver damage, especially when combined with alcohol.

Heptane is a central nervous system depressant. Dizziness and loss of coordination are the result of short-term exposure, which is 2,000 parts per million for 4 minutes. Long-term exposure can result in minimal peripheral nerve damage.

Some solvents are known to cause cancer in humans and animals. Benzene can cause leukemia. Experts suspect that chlorinated solvents—those whose names include "chloride" or "chloro"—might be carcinogens

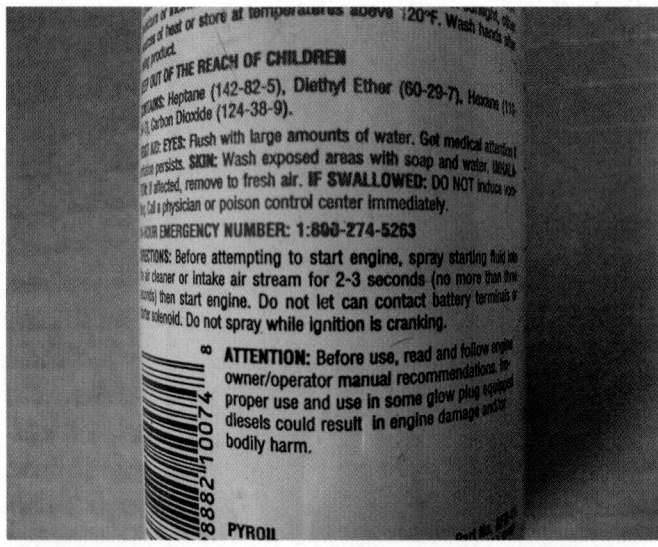

Figure 3.16 This contents label is from a can of starting fluid spray. *(Courtesy of Tim Gilles)*

and cause cancer. Twenty-one human carcinogens are regulated by NIOSH. There are hundreds of other suspected carcinogens under investigation.

Mutagens are chemicals that affect the genetic material in human cells. The result of exposure to a mutagen can include damage to sperm or egg cells, which can prevent conception. If conception does happen, a miscarriage or damage to the fetus can occur. A teratogen is a chemical that affects the development of a fetus. A fetus is often more sensitive to chemicals than its mother.

When dealing with hazardous materials, the most prudent course of action is to avoid contact altogether. If a part can be cleaned in a contained tank without dangerous contact to you, use this method whenever practical.

 Be aware of the materials you are dealing with. Remember, no one will be more concerned for your safety and health than **you!**

■ HOT TANK SAFETY PRECAUTIONS

Caustic or Acid Cautions

In addition to burns that can occur due to the high temperature of the solution, caustic solution from a hot tank can also eat your skin and cause blindness.

Hot Tank Safety

1. Always wear gloves and face protection when working around caustics.
2. If caustic contacts your skin, rinse immediately and seek medical attention.
3. If caustic gets into an eye, the caustic must be flushed from the eye immediately or blindness may result. Be sure to lift the eyelid and flush under it with water for 15 minutes. Continue repeated 15-minute washings while medical attention is sought.
4. Articles should be lifted from the tank with a lifting device. People have fallen into large hot tanks while trying to retrieve objects by hand.
5. Be sure that lifting slings are securely fastened to items in the tank. Failure to do so can cause a dangerous splash.

■ CLEANING SOLVENT SAFETY PRECAUTIONS

Cleaning Solvent Cautions

Cleaning solvents are used extensively in auto repair. Their use is generally safe provided the user has knowledge of hazards associated with them. A flammable liquid's flash point is the temperature at which it will catch fire. Stoddard solvent has a relatively high flash point, but fires can still result.

Cleaning Solvent Safety

1. Use eye protection.
2. Do not breathe vapors.
3. Clean only in areas with adequate ventilation.
4. Wear a respirator if using a blowgun to apply a chemical. (Nontoxic chemicals are available.)
5. Do not clean with gasoline. It is very explosive when it vaporizes.
6. Some solvents can catch fire on a hot engine. Be sure to use a nonflammable solvent.
7. Do not spray solvents if near any pilot lights.
8. Use a brass or nylon brush instead of a steel wire brush to avoid the possibility of an accidental spark.
9. The lid on the solvent tank should always be kept closed when not in use to prevent a fire hazard.
10. Do not smoke or weld anywhere near solvent.

■ SKIN CARE SAFETY PRECAUTIONS

Many solvents are damaging to skin. **Dermatitis** (irritated skin) can result from exposure. If hands are burned or swollen from solvent contact, seek medical attention.

■ Do not put hand creams on the skin as treatment. Hand cream can seal in the solvent irritant, causing even more damage.

■ Be sure the gloves you use do not allow penetration by the solvent being used. Neoprene, an oil-resistant artificial rubber commonly used for hoses and gloves, protects against some types of solvents, but not others. Latex, a material commonly used in making the surgical gloves popular with many mechanics, will also allow penetration by some chemicals.

■ Do not mix other chemicals with cleaning solvents.

To aid in cleaning hands, **barrier creams** are sometimes rubbed on the skin prior to working in a greasy environment. Although these creams are quite effective in making hand cleaning easier, they do not protect the skin against penetration by solvents and are not a substitute for gloves.

■ BREATHING SAFETY

There are many sources of breathing hazards. These can include paints, cleaning chemicals, burned wire insulation, grinding dust, asbestos, vehicle exhaust, and battery gas. Some of these hazards are only breathed by accident. Others are in the air and must be guarded against. Some are only dangerous after long-term exposure.

Figure 3.17 Use an exhaust ventilation system when working around a running engine. *(Courtesy of Tim Gilles)*

Sometimes you will need to work around a running engine. Be sure the parking brake is firmly applied and try to stand to the side of a running vehicle. Exhaust gas can contain large amounts of carbon monoxide. Protect yourself and others from exhaust gas by using an exhaust system (**Figure 3.17**).

Asbestos

Asbestos, found in some brakes, clutches, and cylinder head gaskets, is dangerous to breathe. In fact, the respirator shown in **Figure 3.18** will probably not completely protect your lungs from asbestos fibers. Dry brushing is expressly prohibited during brake cleaning. Using a drill-operated wire brush to clean fiber head gaskets can also release asbestos into the air. When working on brakes or clutches or when removing head gaskets from cylinder head surfaces, one of two approved methods can be used:

- A popular asbestos cleaning method is to use water with an organic solvent or wetting agent (**Figure 3.19**).

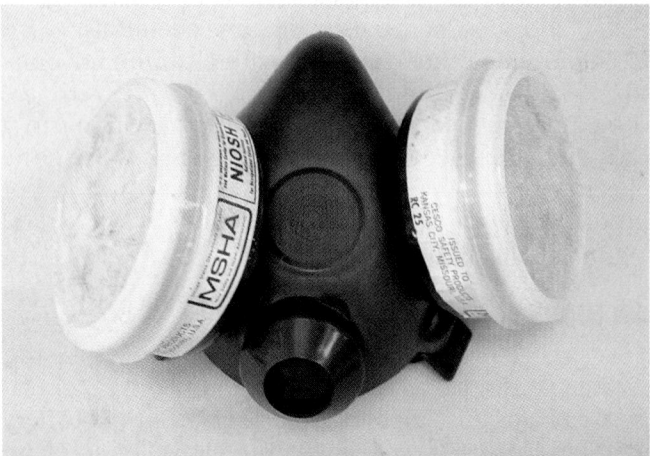

Figure 3.18 This respirator will not provide complete protection from asbestos. *(Courtesy of Tim Gilles)*

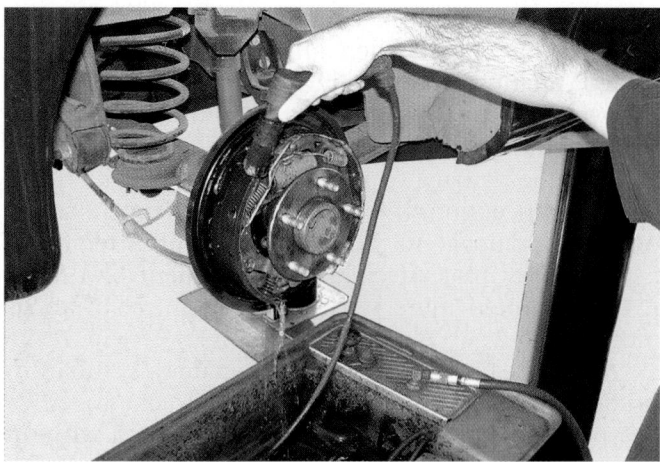

Figure 3.19 Cleaning brakes with a water-based cleaner. *(Courtesy of Tim Gilles)*

- Another asbestos cleaning method is to use a special apparatus (called a **HEPA vacuum**). It vacuums asbestos and captures the particles in a special filter. A regular vacuum will not do. Asbestos will go right through it.

HEPA stands for *high-efficiency particulate air*. A HEPA vacuum completely encloses the brake drum or disc assembly. Gloves and a window allow the technician to clean the brake or clutch assembly. After using the vacuum to suck up loose material, other particles are dislodged using compressed air and are sucked into the vacuum.

 **SAFETY NOTE** Be sure to inspect the enclosure to see that it is tight against the brake backing plate before beginning work.

Asbestos Disposal

A vacuum filter that is full must be wetted with a mist of water before removing it from the vacuum. Filters must be placed in impermeable containers, labelled, and disposed of in the manner prescribed by law.

When using the liquid method of asbestos protection, the wetting agent must flow through the brake drum *before* the drum is removed. When a drum is removed first, a quantity of asbestos can escape into the surrounding air. A catch basin is positioned under the brake assembly. After the drum is removed, the brake parts are thoroughly wetted before starting to clean. If the system has a filter, it must be disposed of in the same way as a HEPA filter.

An alternative method for shops that do fewer than five pairs of brakes or clutches in one week allows a low-pressure water system or fine-mist spray bottle to be used to wet the parts before beginning work. All parts are then wiped clean with a cloth, which must be disposed of in an approved, properly labeled container or laundered in a way that prevents the release of asbestos fibers. Dry brushing is prohibited in this method also.

▄▌ ELECTRICAL SAFETY PRECAUTIONS

Electric Shock

Twelve-volt direct current (DC) electrical systems, like the ones used in automobiles, do not cause serious electrical shock (unless the engine has a distributor-less ignition). Shop equipment, however, is powered by either 110V or 220V alternating current, which can be very dangerous.

Electrical Safety

1. Do not stand in water when using electric tools.
2. Be sure that a tool is turned off before plugging it in so that a spark does not jump from the outlet to the plug.
3. Electrical wall outlets should be properly grounded.
4. Three-wire electrical tools are the best choice for commercial work. The extra terminal is for ground (**Figure 3.20**). If you use a homeowner-type tool with a two-wire plug, it should be *double-insulated.*
5. Be sure to observe electrical wire color coding when repairing or replacing power cords on electric tools.

Hybrid electric vehicles (see Chapter 16) are also a potential source of dangerous high voltage. In these vehicles, high-voltage cables and connections are colored orange.

 **CASE HISTORY** *An electrical plug on a drill with a metal housing was damaged and needed to be replaced. A technician bought a new 3-wire plug, cut and stripped the wires, and installed it on the cord. He had been working on cars for many years and was used to the color code used around automotive batteries. Traditionally in automobile wiring, black is ground and red is positive, but in commercial wiring the* **green wire is ground**. *When he connected the black wire to the ground terminal, he was actually hooking one of the hot leads to ground (also the housing of the metal drill). When he plugged the drill into the wall, he was holding it in his hand and received a dangerous electrical shock. Luckily, he was not seriously injured.*

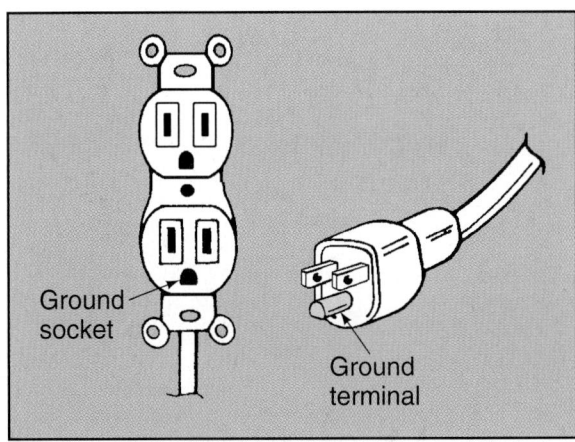

Ground socket

Ground terminal

Figure 3.20 The third wire terminal is for ground.

▄▌ COOLING FAN SAFETY

The fan that draws cool air across the radiator can be driven by either a belt or electricity. Rotating fans can be dangerous. Some are controlled by an automatic switch and can start unexpectedly.

1. Electric cooling fans should be disconnected when working around them (**Figure 3.21**).
2. Visually inspect fan blades for damage. A damaged fan blade will probably be out of balance, making it prone to fly apart.
3. Keep hair and clothing away from fans and fan drive belts.

▄▌ COOLANT BURNS

The most likely ways to be burned in an automotive shop are with engine coolant or by the exhaust system. Opening the radiator on a hot engine can be very dangerous. Disc brake rotors can also be hot after a car has recently been driven.

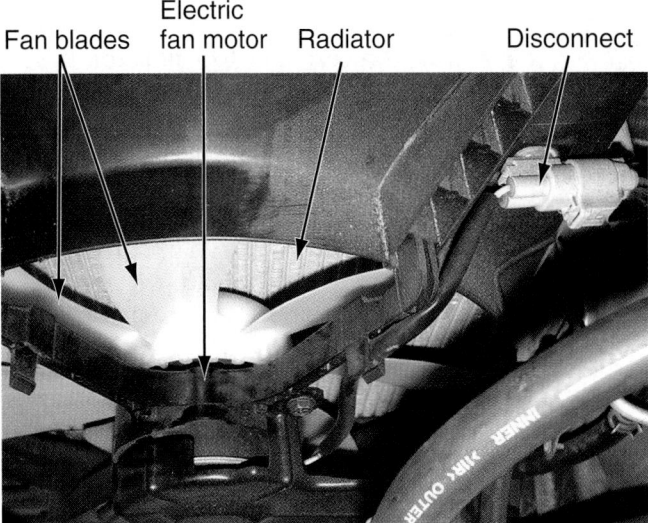

Fan blades   Electric fan motor   Radiator   Disconnect

Figure 3.21 Disconnect the power to an electric cooling fan before working on it. (*Courtesy of Tim Gilles*)

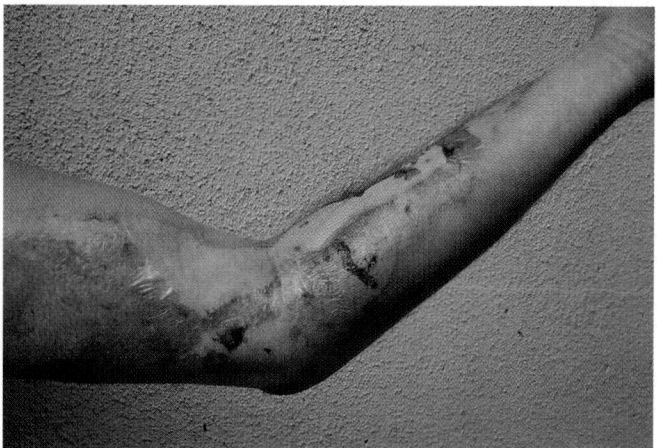

Figure 3.22 This arm was badly burned when the technician opened the radiator cap on a hot engine. *(Courtesy of Tim Gilles)*

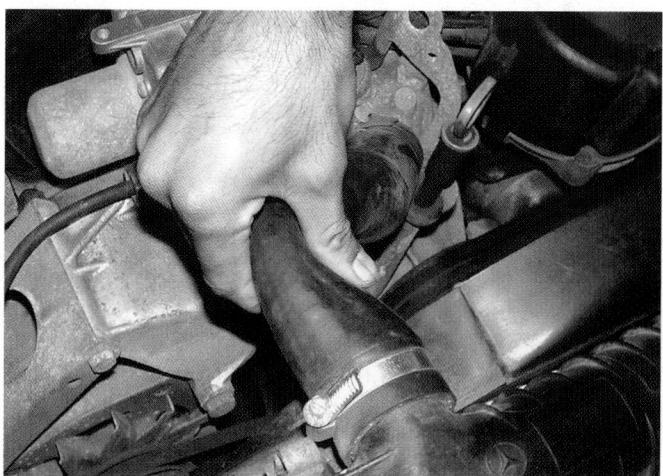

Figure 3.23 Squeeze the top hose before attempting to remove the radiator cap. *(Courtesy of Tim Gilles)*

If the hose is hard and feels like it is full of coolant, the coolant level is acceptable. If the hose collapses, the coolant is not under pressure but steam can still cause a burn. Exercise extreme caution when opening the cap.

CASE HISTORY *The radiator cap keeps pressure on the coolant when the engine heats up. When coolant is under pressure, its boiling point is higher. Loosening the cap removes the pressure and the coolant boils instantly. A student had just begun working in a full-service gas station. A customer asked him to check the coolant level in the radiator. When he opened the radiator cap, he turned it ½ turn. This defeated the cap's safety feature and the coolant boiled out, severely burning him (**Figure 3.22**).*

SAFETY NOTE Always squeeze the top radiator hose before opening a radiator cap (**Figure 3.23**). When the radiator cap is turned only ¼ turn, the safety catch will prevent it from coming all the way off, but superheated coolant can escape through the small hose that runs from beneath the radiator cap to the overflow reservoir. With the pressure removed, the coolant in the overflow reservoir will boil violently. When it comes out of the reservoir, it can cause serious burns.

■ GENERAL HAND TOOL SAFETY

Cuts and scrapes are common when working on cars. Experienced technicians cut themselves less often than beginners. This is because they use hand tools properly.

Refer to Chapters 8 and 9 for more information on the particular hand tools and equipment covered here. The following precautions should be used when handling hand tools:

1. Before using a tool, inspect it to see that it is not damaged or cracked.
2. Maintain tools in safe working condition.
3. Do not use worn or broken tools.
4. When using screwdrivers, do not hold small components in your hand in case the blade slips.
5. Do not put sharp tools in your pocket.
6. Be sure to always use eye protection when using hammers and chisels.
7. Mushroomed chisels and punches should be reground before use (**Figure 3.24**).
8. Be sure that a file has a handle installed on its tang before using it (**Figure 3.25**).

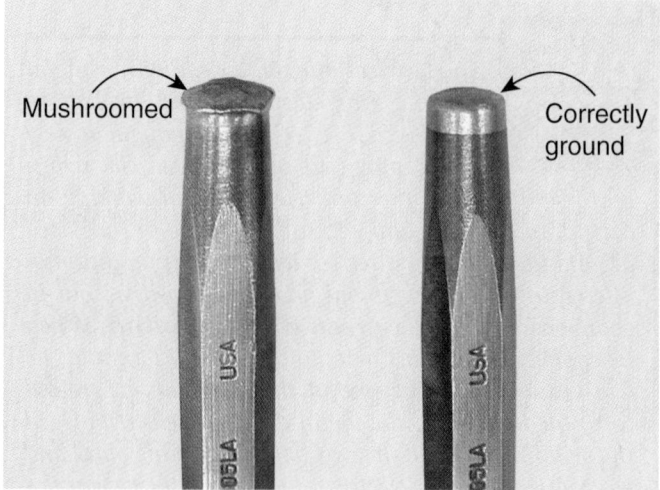

Mushroomed   Correctly ground

Figure 3.24 Regrind a chisel with a mushroomed head. *(Courtesy of Tim Gilles)*

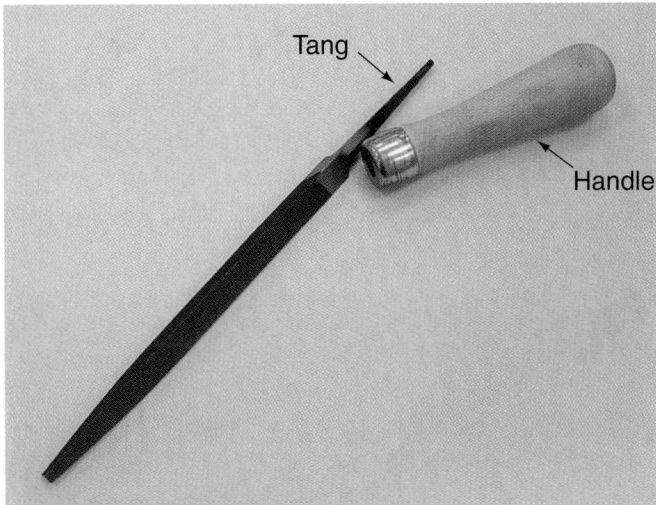

Figure 3.25 Install a handle on the tang of a file. *(Courtesy of Tim Gilles)*

▇ VISE SAFETY

Be careful not to tighten the vise too much at its wide-open position or the vise might break. Never use a piece of pipe on the handle of a vise; it can break the vise.

▇ PULLER SAFETY

Pullers can be dangerous, especially if used improperly. Parts that are being separated can be under a good deal of pressure. They can explode, especially when using an impact wrench or a hydraulic puller. The following are some general tips on safe use with pullers:
- Wear eye protection.
- Be sure the puller is aligned so it is perpendicular to the part being pulled.
- Do not use a puller with damaged or worn parts.
- Use the correct size puller so overloading is avoided.
- Use a 3-jaw puller instead of a 2-jaw puller when possible.

▇ MACHINERY SAFETY

There are many types of machinery used in automotive shops. Common sense will prevent most injuries and accidents.
- *Do not talk* to someone who is operating a machine.
- Do not talk to someone when you are operating a machine.
- Be cautious when working around a running engine or rotating machinery. Fingers can be cut off by a moving belt and pulley.

NOTE: *Sometimes, machinery rotating at high speed does not appear to be moving. One example of this is household lighting. House and shop light appear to be constant. However, they are powered by alternating current so they actually flicker at 60 times per second. This can produce a strobe effect on moving machinery.*

▇ ELECTRIC DRILL SAFETY

Drilling Safety Precautions

Drills are used often in repairing automobiles. As with other types of machinery, knowledge and common sense will prevent accidents from occurring. The following are some tips on correct use of electric drills:
- *Always* wear eye protection.
- *Release pressure occasionally* to allow chips to break off before they can become too long and dangerous.
- A drill bit may catch when it starts to break through the work being drilled. *Be sure that sheet metal is clamped to the worktable.* Let up the pressure on the bit as it starts to break through the bottom of the hole.
- If the drill grabs the work, *shut off the drill.* Never grab the moving work.
- *Be sure to remove the chuck key* from the chuck before drilling.
- *Never* stand in water when drilling. Standing in water increases the danger of electrical shock.

▇ GRINDER SAFETY

Grinder Safety Precautions

The grinder is a part of the machine inventory of all automotive shops. Serious accidents can occur if the grinder is not used properly. If you understand the following safety precautions about the grinder, it can be used safely and effectively:
- *Stand to the side when starting the motor.* The grinding wheel is more likely to explode during startup because of the inertia of the wheel.
- *Wear face protection.*
- *Position the tool rest as close to the wheel as possible,* so that nothing can get trapped between the wheel and the tool rest (**Figure 3.26**).

Figure 3.26 Position the tool rest as close to the grinding wheel as possible. *(Courtesy of Tim Gilles)*

- Use the shield installed on the grinder.
- *Do not grind on the side of the grinding wheel.*

■ WIRE WHEEL SAFETY

Wire Wheel Safety Precautions

Most grinders are set up with a wire wheel on one side and a grinding wheel on the other. The wire wheel is used to remove carbon, paint, or rust from parts and to deburr freshly machined parts. To avoid injury, use common sense when dealing with a wire wheel and follow the following precautions:

- Use face protection.
- Wear leather gloves if they are available.
- Do not push too hard against the wheel. Severe finger injuries can result if your hand slips.
- Wire wheels can damage soft aluminum surfaces.

■ COMPRESSED AIR SAFETY

Compressed Air Tool Safety Precautions

Compressed air is very useful to a technician but can be dangerous when used improperly. Horseplay has no place in a shop. A blast of air can result in a broken eardrum. Blowing compressed air into a body orifice can result in death. The following are compressed air tool safety precautions:

- Always wear eye protection when blowing off parts. Pieces of debris can be blown into eyes.
- Do not blow air against your skin; the high-pressure compressed air used in auto repair shops can penetrate skin.
- Compressed air is used to power chassis grease guns. Pressurized grease can penetrate skin.
- Hold onto the air hose when uncoupling an air line so it does not fly through the air.
- When possible, bleed off the air from an air line before uncoupling an air hose.
- Rubber-tipped blowguns are not safety regulated like other normal blowguns. When held against a part, full shop air pressure is available at the tip.

■ IMPACT WRENCH SAFETY

Air impact wrenches are used extensively in auto repair. They save time and make work easier when used with knowledge and common sense. The following are some safety tips when using impact wrenches:

- Be careful of loose clothing or hair that might become tangled in the tool.
- Use approved *impact* sockets, not ordinary chrome sockets, which can break (**Figure 3.27**).
- Be sure that the socket is secured to the air tool. The clip at the end of the tool's square drive can become worn so that it no longer holds the tool.

Figure 3.27 Regular and impact sockets. *(Courtesy of Tim Gilles)*

- Do not turn on the impact wrench unless the socket is installed on a nut or bolt, especially when using a wobble socket. The socket can fly off the impact wrench, possibly causing an injury.
- When the impact wrench fails to loosen a fastener, use a large breaker bar.

■ AIR CHISEL SAFETY

Examples of uses for air chisels include cutting metal, driving on valve guides, riveting, and replacing front suspension parts. An injury can occur if an air chisel is not used properly.

- Be sure to wear eye protection.
- Before pulling the trigger, be sure to have the tool bit against the workpiece (**Figure 3.28**). Otherwise, the tool might fly out of the air chisel.

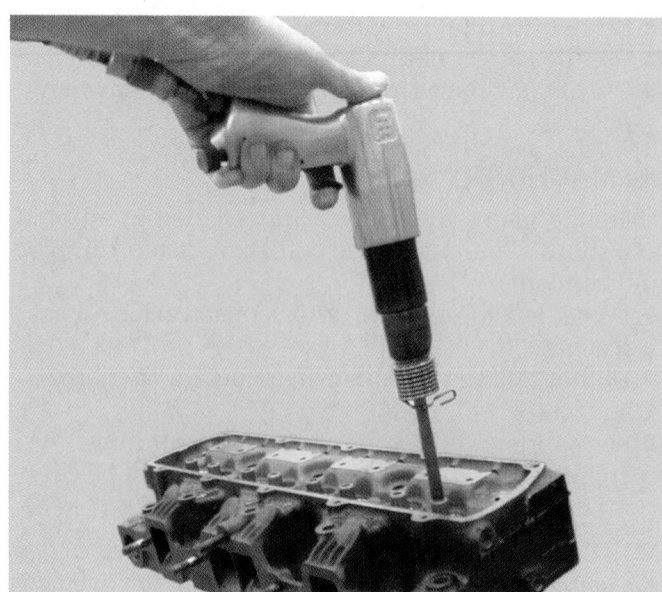

Figure 3.28 Hold the tool bit against the workpiece before pulling the trigger on the air chisel. *(Courtesy of Tim Gilles)*

DIE GRINDER/AIR DRILL SAFETY

Air drills and die grinders are used with small grinding wheels, burrs, and wire brushes. A die grinder is similar to an air drill but turns at a much higher speed. There are several safety precautions to use when using these tools:

- The tool is very loud. Use ear protection.
- Die grinders turn at speeds of up to 20,000 rpm. When using a grinding wheel or wire wheel at high speeds, be sure that the wheel is rated for such use.
- Some head gaskets contain asbestos. Do not remove head gaskets using a wire brush and die grinder or air drill.

PRESS SAFETY

A hydraulic press can be very dangerous, because parts being separated are under such pressure that they can explode. Common automotive presses develop hydraulic force in the range of 20 to 50 tons (40,000–100,000 pounds); see **Figure 3.29**. Bearings have been known to explode like a bomb, injuring the user with shrapnel. Be sure to use common sense.

- Use applicable safety guards.
- Wear face protection.
- When a press has fast and slow speeds, use the fast speed to press parts whenever possible. This position does not provide full hydraulic pressure.
- Use extreme caution as the pressure applied to the part becomes higher.

GENERAL LIFTING SAFETY

When lifting a vehicle, be sure that the lift or jack contacts the frame at the recommended lift point (**Figure 3.30**). To be sure that the vehicle is firmly supported, push on its fenders before going under it.

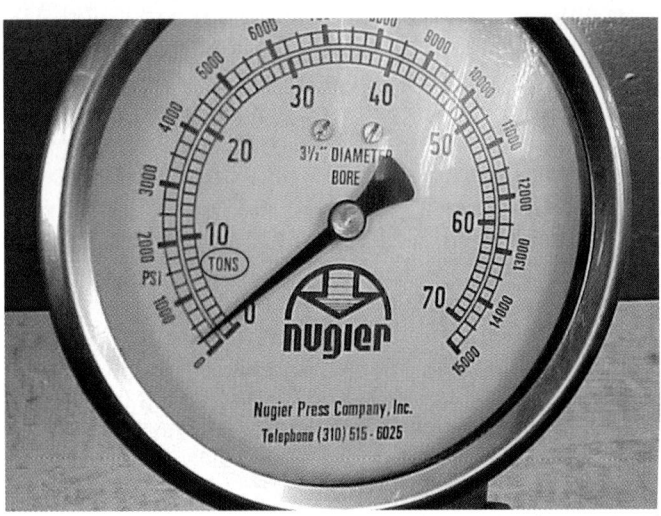

Figure 3.29 The gauge on a typical hydraulic press. How many pounds is 50 tons? (*Courtesy of Tim Gilles*)

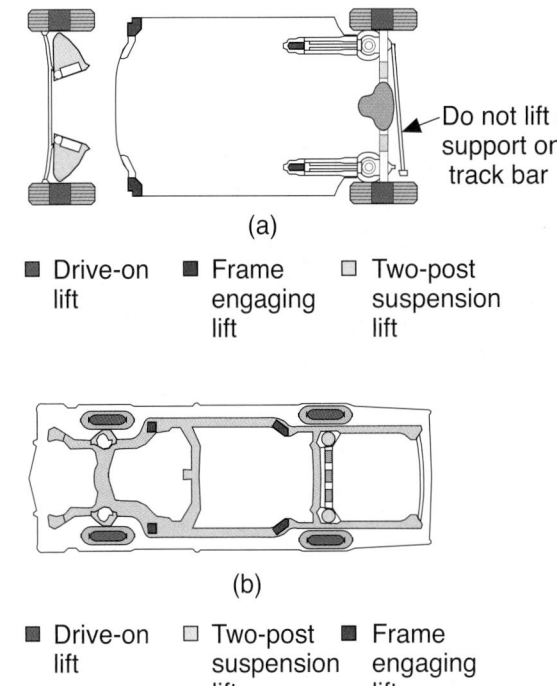

(a)

- ■ Drive-on lift
- ■ Frame engaging lift
- ☐ Two-post suspension lift

- ■ Drive-on lift
- ☐ Two-post suspension lift
- ■ Frame engaging lift

(b)

Figure 3.30 Lift a vehicle from one of the lift points shown in the shop service manual.

Figure 3.31 Raise the vehicle a few inches from the floor, then jounce it to be certain it is firmly positioned on the hoist pads. (*Courtesy of Tim Gilles*)

When lifting a vehicle on a frame-contact hoist, place the lift adapters at the specified locations and raise the vehicle about 6 inches (**Figure 3.31**). Then, shake it to see that it is firmly placed.

HYDRAULIC FLOOR JACK (SERVICE JACK) SAFETY

Hydraulic floor jacks (service jacks) are used extensively in automotive repair. Accidents do not occur often, but when they do they can be life threatening. Cars

Figure 3.32 Always use vehicle support stands when a vehicle is jacked up. *(Courtesy of Tim Gilles)*

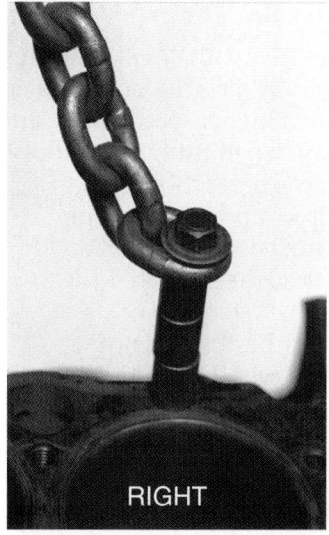 

RIGHT　　WRONG

Figure 3.33 The lifting fixture must be firmly tightened against the part to be lifted. The bolt on the right will be under extreme stress, which can cause it to break.

are often serviced while raised on a portable service jack, but this is a dangerous practice. Cars have fallen and crushed people who used jacks that failed. Follow these safety precautions when using a hydraulic floor jack:

- A hydraulic service jack should be used to raise and lower a vehicle only.
- Always use vehicle support stands (jackstands) (**Figure 3.32**).
- Use support stands in pairs.
- Be sure that the support stands are positioned in the recommended location on the frame (see Figure 3.30).
- Be sure to use support stands only on a level concrete surface. On a hot day, the legs can dig into asphalt and cause an accident.

SHOP TIP Place a piece of thick plywood or a steel plate under support stands that are used on asphalt.

It is foolish to crawl under a vehicle that is not resting *solidly* on vehicle support stands.

■ SHOP CRANE (ENGINE HOIST) SAFETY

When using a shop crane (cherry picker), several safety precautions are necessary. An engine is very heavy. It is important that the center of gravity be observed. When the engine is removed from the car, it is usually raised 3–4 feet off the ground to clear the radiator grill. *It is very dangerous to move the shop crane with the center of gravity this high because the crane can tip over easily.* Lower the engine close to the ground, and then roll the crane in as straight a line as possible to avoid tipping it over. This precaution is especially critical when the engine and transmission have been removed at the same time.

CASE HISTORY *An apprentice had just removed an engine and transmission from a rear-wheel-drive pickup. The shop crane was still in the high position that was required to clear the front grill and radiator mount. He attempted to move the engine and transmission without first lowering them close to the ground. As he tried to push the shop crane over a ledge at the entrance to the service bay, the balance of the crane was upset and the engine crashed to the ground. Luckily, the only damage was to the transmission oil pan and extension housing. One of the legs of the shop crane was bent so badly that it was taken out of service.*

- It is very important that the bolts that hold the engine sling to the engine block are not too short. The amount of thread on the bolt that enters the block should be of an amount that is at least 1½ times the diameter of the screw thread (usually about 6 turns).
- Bolts used to attach a lifting fixture to a heavy part must be firmly tightened against the device. Use spacers if necessary (**Figure 3.33**).
- Be sure to lower the engine as soon as possible before attempting to roll the crane to a different location.
- When an engine has been removed from a vehicle, be careful when raising the vehicle in the air while on a frame-contact lift. The vehicle will be rear-end heavy when the weight of the engine is removed from it. The result can be that the vehicle falls off the lift.

■ TRANSMISSION JACK SAFETY

A transmission jack is used when the vehicle is raised in the air on a lift. The safest use of this jack is when the vehicle is on a wheel-contact lift. When a vehicle is on a frame-contact lift, there is a danger that raising the transmission can result in the vehicle being lifted off the adapters on the lift. If its balance is upset, the vehicle could be knocked off of the lift.

■ BATTERY SAFETY

As a battery charges it gives off explosive hydrogen gas. Sparks must be avoided around batteries, which occasionally explode (**Figure 3.34**). The most common cause of battery explosions is when someone unhooks a battery charger without turning it off. A battery explosion can cause:

■ Damage to skin (from acid thrown by the explosion)
■ Permanent or temporary hearing loss
■ Blindness

■ HISTORY NOTE ■

Hydrogen gas is the same gas that was used in the Hindenburg, the passenger zeppelin that crashed and exploded in 1937. The space shuttle Challenger exploded in 1986 when its liquid hydrogen tank ignited.

Battery Safety

1. Always wear eye protection when working around batteries.
2. Unplug the charger before connecting or disconnecting any of the cables.
3. Be sure to shut off the battery charger before unplugging its 110V power cord, or a spark could happen when the electricity tries to jump from the outlet to the plug.

Jump-Starting a Car

When jump-starting a car, follow the proper procedures and wear eye protection (**Figure 3.35**). A battery that

Figure 3.34 Careless use of the battery charger caused this battery to explode. (*Courtesy of Tim Gilles*)

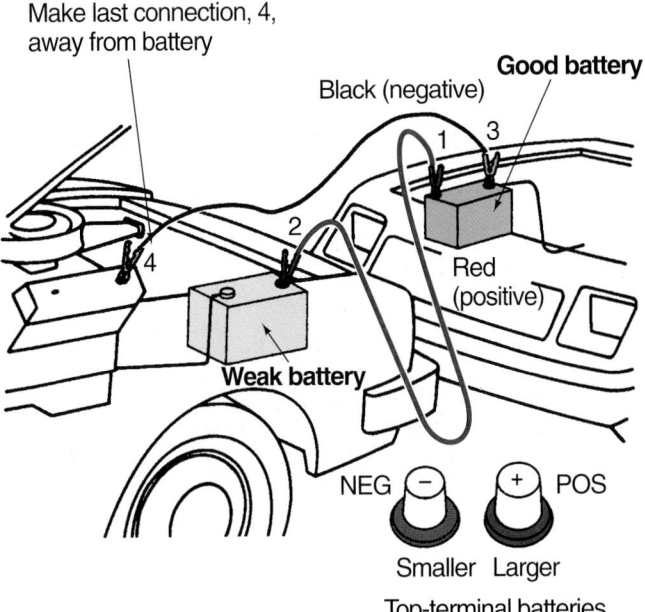

Figure 3.35 Correct connections for jump-starting a car.

has been charging normally (the host vehicle) could have trapped hydrogen gas under its cell caps or the battery terminal clamp. Be sure that no sparks occur near it. Always connect the dead battery last and hook up the ground cable to the engine block away from the battery.

Battery Acid Safety

Battery acid (**electrolyte**) is a chemical combination of *sulfuric acid* and *water*. Battery acid can harm the skin, eyes, and clothing.

 **SAFETY NOTE** If battery acid gets on your skin, immediately wash the area with water for at least 15 minutes.

Baking soda can be used to neutralize battery acid. After carrying or servicing a battery, keep your hands out of your pockets. The outside of a battery is usually covered with acid resulting from the mist from the vent caps during charging. The acid from your hands will cause holes to develop in your pockets after your clothes are washed.

General Hybrid Safety

If you ever work on a hybrid electrical system, your life will depend on knowing what and when something is safe to touch. Section 6 provides more detailed coverage of hybrid safety. Voltage can range up to 650 volts and 60 amps *instantly*. This can be deadly. All hybrid manufacturers provide ample information on the Internet to allow fire departments and other emergency personnel easy access for training. Toyota emergency information is available at the following Web site: http://techinfo. toyota.com

Figure 3.36 High-voltage cables on hybrid vehicles are orange. Accidental contact can kill you! *(Courtesy of American Honda Motor Co., Inc.)*

CAUTION Orange means high voltage. Do not forget this! Figure 3.36 shows typical orange-colored high-voltage connections in a hybrid vehicle.

SAFETY NOTE Liquid refrigerant contained in the air-conditioning system will instantly **freeze** anything that it comes into contact with.

■ REFRIGERANT SAFETY

Refrigerant is used in automotive air-conditioning systems. It is a vapor at room temperature and becomes liquid only when kept under pressure. When pressure is removed from it (such as when you disconnect one of the air conditioning lines), it boils vigorously as it turns into a vapor.

When working around refrigerants:

■ Wear gloves to protect your skin.
■ Wear goggles to protect your eyes. Refrigerant can cause instant blindness.

■ SCIENCE NOTE ■

It is against the law to allow R12 refrigerants found in most 1993 and earlier vehicles to escape into the atmosphere. R12 is a chlorinated fluorocarbon that is believed to damage the ozone layer. (See Chapter 35)

■ GENERAL SAFETY AROUND AUTOMOBILES

Do not leave things on the floor of the shop that people can trip on. Hydraulic service jacks (floor jacks) should be left with the handle up, rather than hanging down low where someone could trip on it. Floor creepers should be stood vertically against a wall or workbench when not in use (**Figure 3.37**).

Before a Test Drive

Before driving a customer's car, remember to check the operation of the brakes and look at the condition of the tires. Do not test drive a car with obvious safety hazards until they are corrected. It makes no sense to test drive a car with dangerous brakes.

Working Around Belts

One of the most common farm injuries is lost fingers. Farm machinery has many belts. Fingers are often cut off when they are caught between a belt and pulley.

Be sure that a helper understands what you are asking him or her to do. Assuming that they understand can result in an accident.

Before attempting a fan belt adjustment, be sure that the keys are out of the ignition. If someone cranks the engine over, fingers can be cut off.

Figure 3.37 Stand a creeper on end when not in use. *(Courtesy of Tim Gilles)*

▰▰ SAFETY TEST

Choose your answers to the Safety Test from this list:

6	charger	eye or face	ground	legs	talk
6	data sheet	fire	handle	manual	vapor
air	disconnected	flash	hydraulic	orange	water
asbestos	drill	freeze	hydraulic jack	reground	water
B	drilling	gasoline	impact	rest	water
back, eyes	electrical	gravity	instructor	side	wrench
belt	end	green	jack	spontaneous	
blowguns	explosions	grinders	jack	squeezing	

1. If you should become injured while working in the shop, inform your _____ immediately.

2. The most common injuries in an automotive shop involve the _____ and _____.

3. Lift with your _____, not your back.

4. If a _____ is burning out of control, leave the area immediately and call for help.

5. A flammable liquid fire is called a Class ____ fire.

6. Keep oily materials separated from air to prevent _____ combustion.

7. What form of gasoline is the most dangerous, liquid or vapor (circle one)?

8. Never use _____ to clean floors or parts.

9. The temperature at which a flammable liquid's vapors will ignite when brought into contact with an open flame is called the _____ point.

10. Disconnect the battery _____ cable before working on a vehicle electric system.

11. Before an _____ fire can be extinguished, the battery must be disconnected.

12. Do not attempt to put out a gasoline fire with _____.

13. The information sheet about a hazardous material is called a material safety _____ (MSDS).

14. If caustic Alkaline liquid (from the hot tank) gets into your eyes flush with _____, especially under the eyelids.

15. A HEPA vacuum is a special vacuum for _____.

16. The color of the wiring for ground on commercial and household electrical wiring is _____.

17. Electric cooling fans should be _____ before working around them.

18. Before opening a radiator, test for pressure in the system by _____ the top radiator hose.

19. Mushroomed punches or chisels should be _____ before use.

20. Always wear _____ protection when using hammers, chisels, pullers, batteries, air-conditioning machinery, compressed air, and other hazardous situations.

21. Always use a _____ on the tang of a file.

22. Compressed _____ can penetrate skin and must be used cautiously.

23. Do not _____ to someone who is operating a machine.

24. A _____ bit has a tendency to grab the work when it just begins to break through the metal.

25. When _____ sheet metal or large pieces of metal on the drill press, make sure they are clamped to the work table.

26. Stand to the _____ when starting the grinder.

27. Position the tool _____ as close to the grinding wheel as possible.

28. Rubber-tipped _____ work at full shop air pressure.

29. When using an air impact wrench, be sure to use an extra thick socket, called an _____ socket.

30. When an air impact _____ is used with a wobble socket, the socket can fly off if the tool is started when the socket is not on the bolt head.

31. Die _____ turn at speeds of up to 20,000 rpm. Be sure to use wheels that are rated for that high a speed.

32. Because they can develop force of from 20–40 tons, _____ presses can cause parts to explode.

33. Locate the correct lift points in the service _____ before using a hydraulic service jack or lift.

34. When lifting a vehicle on a frame-contact lift, place the lift adapters at the specified locations and raise the vehicle about _____". Then, shake it to see that it is firmly placed.

35. Use a _____ _____ to raise and lower a vehicle only.

36. A jacked-up vehicle should be placed firmly on _____ stands.

37. When moving an engine that is mounted on a shop crane, be sure the engine is lowered as far as possible to keep the center of _____ low.

38. Bolts that attach an engine sling or engine stand adapter to the block should enter the block at least _____ turns.

39. When using a transmission _____ on a car that is on a frame-contact lift, be careful not to raise the car off the lift adapters as you raise the transmission.

40. The most common cause of battery _____ is when someone unhooks the battery charger without turning it off.

41. Shut off and unplug the battery _____ before unhooking either of the cables.

42. If acid gets on the skin or eyes, immediately flush with _____ for at least 15 minutes.

43. If refrigerant from an air conditioning system contacts your skin or eyes it will _____ them.

44. The keys should be out of the ignition any time a _____ is being inspected or adjusted.

45. When a floor creeper is not in use, it should be stood on _____ so it is not accidentally stepped on.

46. Be careful around hybrid vehicle wiring that is colored _____.

CHAPTER 4

Shop Management, Service Records, and Parts

■ OBJECTIVES

Upon completion of this chapter, you should be able to:

✔ Understand the relationship between shop personnel and the customer.

✔ Know how to treat customers so that they will have confidence in your repair business.

✔ Interview a customer to find the true cause of an automotive problem.

✔ Correctly fill in a repair order.

✔ Understand how parts are priced by various establishments.

■ KEY TERMS

service record	jobber	net price
aftermarket	DIY	cost leader
OEM	TBA	OE
WD	list price	stock

■ INTRODUCTION

This chapter is about business practices. It describes such things as customer relations, filling out repair orders, and interviewing customers regarding problems with their cars. Also included is an explanation of how businesses buy parts and the discounts involved.

■ CUSTOMER RELATIONS

Automotive repair shops are in business to make a profit. If a shop does not make a profit, it goes out of business. Good relations with customers is the key to a successful business. It is important to make a good first impression with the customer so that he or she is confident in the shop (**Figure 4.1**). In other than large dealerships, most technicians will sometimes need to communicate directly with the customer.

Dealerships have professional salespersons who greet the customer and write up the repair order. These people are called service writers or service advisors. When a customer does not have a complaint about a car's performance but is simply seeking preventive maintenance, the customer experience is usually satisfactory. One major cause of customer dissatisfaction, however, is when the customer must return again in order to have the original problem corrected. In fact, in several surveys, the percentage of complaints not correctly diagnosed and repaired on the initial visit is nearly half. This problem can often be traced to faulty diagnosis by an undertrained service writer.

Service writing in smaller shops is usually done by the owner (who is also a technician). In this case, the customer

Figure 4.1 The customer's first impression is important.

gets to speak directly with the technician about a car's problem, which often results in a satisfied customer after only one trip to the shop. Part of the solution to the problem for larger shops is to hire service writers who understand the operation of the various systems of the car.

Telephone Service

Although the owner or manager of a business often answers the telephone, technicians in small shops usually share telephone responsibilities also. The phone should be answered promptly and courteously, stating the name of the business and the person who is speaking.

Most often there is a phone extension in the repair area. Although it is sometimes an annoyance to have to stop a job to answer a telephone call, the phone is an important source of business. The telephone often presents the customer with his or her first impression of the business. If the customer's first impression is that the business is professionally managed, things will go easier from that point on.

Remember: There is no second chance to make a first impression!

SERVICE RECORDS

A service record is written for every car that enters a repair facility. A multiple copy, numbered **service record**, repair order (R.O.), or work order (W.O.) is used for legal, tax, and general recordkeeping purposes. One of the copies of the R.O. is given to the technician. It lists the repairs needed and is used for making notations of repairs completed and items needing attention. One of the copies of the repair order includes the cost estimate and is for the customer. The remaining copy is for the shop's files.

The repair order (**Figure 4.2**) is important for several reasons. It is a legal document that

1. Fully identifies the customer and the vehicle
2. Gives the technician an idea of the reason the car is in the shop for repairs

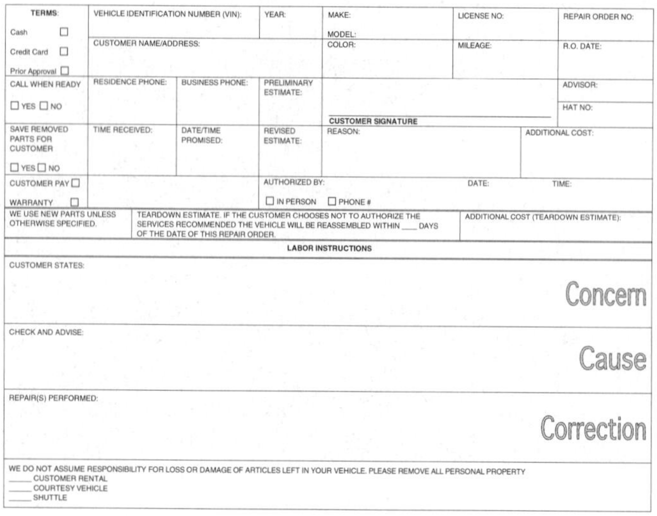
TERMS: Cash ☐ Credit Card ☐ Prior Approval ☐
VEHICLE IDENTIFICATION NUMBER (VIN):
YEAR:
MAKE: MODEL:
LICENSE NO:
REPAIR ORDER NO:
CUSTOMER NAME/ADDRESS:
COLOR:
MILEAGE:
R.O. DATE:
CALL WHEN READY ☐ YES ☐ NO
RESIDENCE PHONE:
BUSINESS PHONE:
PRELIMINARY ESTIMATE:
CUSTOMER SIGNATURE
ADVISOR:
HAT NO:
SAVE REMOVED PARTS FOR CUSTOMER ☐ YES ☐ NO
TIME RECEIVED:
DATE/TIME PROMISED:
REVISED ESTIMATE:
REASON:
ADDITIONAL COST:
CUSTOMER PAY ☐
WARRANTY ☐
AUTHORIZED BY: ☐ IN PERSON ☐ PHONE #
DATE:
TIME:
WE USE NEW PARTS UNLESS OTHERWISE SPECIFIED.
TEARDOWN ESTIMATE. IF THE CUSTOMER CHOOSES NOT TO AUTHORIZE THE SERVICES RECOMMENDED THE VEHICLE WILL BE REASSEMBLED WITHIN ___ DAYS OF THE DATE OF THIS REPAIR ORDER.
ADDITIONAL COST (TEARDOWN ESTIMATE):
LABOR INSTRUCTIONS
CUSTOMER STATES:
Concern
CHECK AND ADVISE:
Cause
REPAIR(S) PERFORMED:
Correction
WE DO NOT ASSUME RESPONSIBILITY FOR LOSS OR DAMAGE OF ARTICLES LEFT IN YOUR VEHICLE. PLEASE REMOVE ALL PERSONAL PROPERTY
_____ CUSTOMER RENTAL
_____ COURTESY VEHICLE
_____ SHUTTLE

Figure 4.2 A repair order. *(Courtesy of Tim Gilles)*

3. Tells the shop's hourly labor rate
4. Gives the customer an estimate of the cost of the repair
5. Gives the time the vehicle will be ready for the customer

The signature of the customer gives approval for the repair and agrees to pay for the shop's services when the job is completed.

Repair orders are numbered for future reference. Also entered on the repair order are the

1. Odometer mileage
2. Vehicle make (Ford)
3. Vehicle model (Mustang)
4. Vehicle model year (1996)
5. Vehicle license number (1VAL239)

Information that will identify the customer includes his or her name, address, and telephone numbers for home, cell, and business.

A space for labor instructions is located below the customer information area.

This area will be filled in with the customer's complaint, possible cause(s), and repairs to be performed.

> **SHOP TIP** The three items on the repair order are called the "3 Cs": concern, cause, and correction.

The customer will need to be questioned carefully about the symptoms. He or she often has a preconceived notion of what is wrong that is incorrect.

The following are examples of possible questions that could be asked if the customer complains that the vehicle runs hot:

1. Does it become hot right away after it is started?
2. Does it become hot only at freeway speeds?
3. Does the radiator consume coolant?
4. Does it become hot enough to boil over?
5. Have you heard any unusual noises from the engine?
6. Has the cooling system had recent service?
7. Has the ignition system been serviced recently?

There is a structured method of questioning that can be followed when asking questions of the customer. Many businesses use a structured format when questioning a customer and listing concerns.

1. First ask for a general description of the problem.
2. Ask whether it happens in the front, back, or under the hood.
3. Identify symptoms: Do you hear, feel, or smell something?
4. With the information you have learned, use your technical skills to identify the problem.

Figure 4.3 Repair shops use a personal computer for shop management activities. *(Courtesy of Tim Gilles)*

Figure 4.4 A carpet mat protects the car's carpet from grease. *(Courtesy of Tim Gilles)*

Computer Records

Service and repair facilities use personal computers for keeping records, maintaining a running inventory and ordering parts, and tracking employee productivity (**Figure 4.3**). Notes saved in memory can include personal notes regarding a particular customer and what occurred during his or her most recent visit to the shop.

> ### ▨ COMPUTER NOTE ▨
>
> *A computer locates a customer's records using the vehicle's license number, the owner's name, or the owner's address. The license number is the most popular means of access.*

Although a shop may keep computer records, a *hard copy* is still used by many shops for the technician, although some shops have a computer terminal in the service bay.

It is important that a shop keep written records of all repairs and recommendations. Litigation (court cases) often results in a losing case due to a lack of adequate records. A completed R.O. includes a written estimate of the cost of the repair and a record of phone conversations with an owner when an earlier approved estimate requires updating.

▬ KEEP THE CAR CLEAN

Be sure that your hands, shoes, and clothing are clean before getting into a customer's car. A dirty steering wheel or carpet will guarantee an unhappy customer, no matter how good of a repair job was done. Work shoes often have grease on their soles. Many shops use carpet mats made of paper to help keep a customer's carpet clean (**Figure 4.4**). Some shops use seat covers as well.

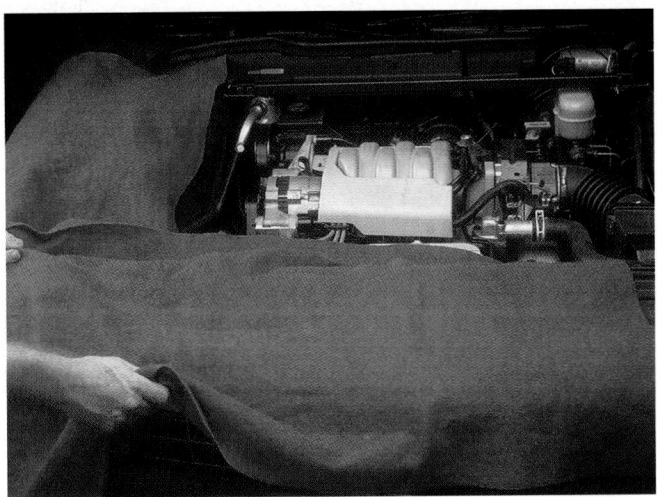

Figure 4.5 Fender covers are used when working under the hood.

Fender Covers

Fender covers are used whenever underhood work is done (**Figure 4.5**). Greasy hands or brake fluid can ruin the finish on a vehicle. Fender covers also protect against scratches from items such as belt buckles. Cars cost many thousands of dollars. When working on cars, be sure to respect them as much as the owner does.

▬ LINEN SERVICE

Most shops have weekly linen service for shop towels and uniforms. The shop is billed for the cost of any shop towels that show signs of acid exposure. Shop towels are usually dyed red. When they are exposed to battery acid or other acids, they turn blue. This alerts the linen service to the probability that the rags will disintegrate during laundry service (**Figure 4.6**).

Some shops purchase their own uniforms. Other times they are rented from the linen service. Some shops get shirt and pants service and others get shop coats or coveralls. Shop clothing is often made of materials resistant to battery acid.

Figure 4.6 This shop towel was damaged by battery acid. *(Courtesy of Tim Gilles)*

Figure 4.8 A wholesale and retail parts store. *(Courtesy of Tim Gilles)*

■■ WHOLESALE AND RETAIL DISTRIBUTION OF AUTO PARTS

The following section familiarizes the reader with some of the terms used in the automotive service business. Approximately one in every six jobs in North America is related to the automobile. The automotive **aftermarket** is supplied from the manufacturer to the customer through a large parts and service *distribution* system (**Figure 4.7**). The original equipment manufacturers (**OEMs**) and a large number of independent parts manufacturers sell parts to over 1,000 warehouse distributors (**WDs**) throughout the continent. Warehouse distributors are large distribution centers that sell to auto parts wholesalers, known in the industry as **jobbers**.

Jobbers

Jobbers (**Figure 4.8**) sell parts, accessories, and tools to several different markets. More than one-half of a typical jobber's customers are independent repair shops and service establishments. Fleet companies, farmers, trade and industrial accounts, and, occasionally, new car dealerships are some of a jobber's other customers Do-it-yourself (**DIY**) customers account for over

one-fourth of a typical jobber's business customers. Some parts, such as filters and ignition parts, sell in higher volume. In the do-it-yourself market, filters (oil, air, and fuel) account for over half of all parts sales. Parts manufacturers use codes to tell which parts average more units of sale than others (**Figure 4.9**). For instance, code 1 parts average 4 or more units of sale per year, while code 8 parts average less than 0.1 unit of sale per year. An average parts store will carry parts listed as code 1 through code 5. This will provide a parts inventory of about 75% of full coverage. Parts that are not often sold increase the cost of doing business. These parts will have to be marked up by a higher percentage in order to compensate.

Services provided to customers are often a prime element in gaining repeat business. Prompt delivery of phoned-in parts orders is provided to regular wholesale customers. Sometimes a parts jobber will have a machine shop, too. These machine shops perform such jobs as engine machining and rebuilding, flywheel surfacing, and hydraulic press work.

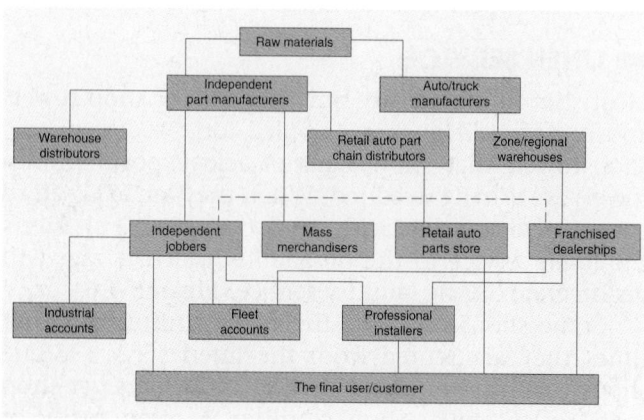

Figure 4.7 The auto parts supply network.

Class Code	Number of Parts	Probability of Sale	#of Parts Likely to Sell	#of Parts Likely Not to Sell
1	1,800	99.6%	1,793	7
2	700	97.0%	679	21
3	1,300	91.9%	1,195	105
4	3,400	77.8%	2,645	755
5	6,000	52.8%	3,170	2,830
6	7,000	31.3%	2,190	4,810
7	8,000	16.1%	1,290	6,870
8	50,000	4.9%	2,450	47,550

Figure 4.9 Probability of sale by part class. *(Courtesy of Aftermarket Auto Parts Alliance)*

Retail Chain Stores

Many parts retailers are managed by national or regional retail chains. *Chain stores* usually deal in faster moving items such as alternators, starters, tools, and car accessories. They often do not have as large an inventory of diverse parts as a jobber store. Some retail stores are owned by one or several owners, but they buy everything through the distribution system of the main company. Prices, especially on sale items, are sometimes lower than retail.

Mass merchandisers, such as large department store chains, sell fast-moving, popular items such as oil and filters, windshield wipers, and spark plugs to the DIY market. They often have a large service facility for doing maintenance and minor repairs on automobiles (**Figure 4.10**).

Dealership Parts Departments

Auto dealers have their own parts departments with a large organized inventory of original parts. Dealer parts departments usually have a separate counter for the technicians who work in the shop. Some dealerships have phone connections or an intercom in the service bay so the technician can call in a parts order before walking to the parts counter. Some dealerships have personnel who bring the cars to the service bay for the technician. This is one way that the technician can be more productive in a busy shop.

The operation of a dealership's parts business is often determined by its owner. Some dealers are very aggressive in pricing and services offered to the independent market and others are not. Discounts are offered in varying degrees to wholesale accounts. It is not uncommon for dealerships to deliver parts to independent repair shops within a 50-mile radius in some markets. Although they are competitors, independent parts establishments often have a give-and-take

arrangement with the dealers, supplying each other with parts when they are needed in a hurry.

Service Stations and Independent Shops

Service businesses are very concerned with the cost of its high-volume parts. A shop that sells five oil filters a day might be able to purchase them in quantity at half the cost and keep them in *stock* (on the premises). The same holds true with tires, batteries, and accessories (**TBA**). Service stations often buy TBA at very good discounts through the oil company whose products they sell. A higher volume shop will often have computerized inventory, where parts are automatically ordered over the Internet as the supply falls below a certain level.

Parts Pricing

Prices are usually determined by the price the jobber pays. The ability to find good prices on parts is an important part of a jobber's business. Jobbers are sometimes affiliated with cooperative groups. This gives them the ability to band together so that they can secure better prices by buying in huge quantities. They still continue to be independent owners, however, and may buy from whomever they choose.

Parts are priced to the customer according to list, net, or variable pricing (**Figure 4.11**). The **list price**, also called the retail price, is the suggested price of an item and is what the customer paying for car repair pays. In reality, very few stores charge list price to the walk-in DIY customer. Most parts are discounted, although in varying amounts. The **net price** is the price that the wholesale auto repair customer pays.

Some items like oil, oil filters, and coolant, are priced by most establishments with little or no markup, because most customers are aware of their prices. These parts, called loss leaders or **cost leaders**, are considered to be part of the cost of promoting a business (**Figure 4.12**). *Variably priced* items are those

Figure 4.10 Mass merchandisers often have large service facilities. *(Courtesy of Tim Gilles)*

Part Number	List	User Net	Dealer	Part Number	List	User Net	Dealer
5188	3.08	2.62	2.05	5282	20.94	17.80	13.94
5189	6.55	5.57	4.36	5283	22.40	19.04	14.92
5193B	13.48	11.46	8.98	5284	13.69	11.63	9.12
5195C	14.81	12.59	9.87	5287A	14.02	11.92	9.34
5197D	13.60	11.56	9.06	5288C	17.66	15.01	11.78
5197D	17.99	15.29	11.98	5289B	32.96	28.01	21.95
5198D	16.32	13.87	10.87	5290A	13.80	11.73	9.19
5200C	12.67	10.77	8.45	5291A	13.28	11.29	8.85
5201B	16.57	14.08	11.03	5292B	28.81	24.50	19.19
5202B	16.55	14.06	11.02	52393B	16.41	13.95	10.93
5204	18.73	15.92	12.47	5294	12.36	10.51	8.23
5208A	13.66	11.62	9.10	5296A	15.93	13.54	10.62
5210C	15.11	12.84	10.06	5297	10.94	9.30	7.29
5211E	20.19	17.16	13.45	5298	2.70	2.30	1.80

Figure 4.11 Example of list, net, and dealer prices.

Figure 4.12 Oil, oil filters, and coolant are loss leaders or cost leaders. (*Courtesy of Tim Gilles*)

that are slow movers and not readily available to the customer. These can be marked up a larger amount to help compensate for cost leaders.

When a service or repair business (installer) buys auto parts from a jobber, the price the installer is charged is sometimes dependent on the volume of business that the installer does with that parts establishment. More volume on a monthly basis can mean a higher percentage of discount to the shop. The part is marked up to the retail price for the customer who brings in a car for repair. This system can make a difference in the profitability of a service business.

Replacement Parts

Factory replacement parts are categorized as original equipment (**OE**). **Stock** means the part is the same as intended by the manufacturer. *Aftermarket* is a broad term that refers to parts that are sold by the non-OE market. Many OE parts are manufactured by the same manufacturer as aftermarket parts.

▬ REVIEW QUESTIONS

1. What are three names for the written record that is kept of a customer's service visit?

2. What is the most popular way of identifying a customer when entering information into the computer?

3. What are two things that fender covers protect the paint from?

4. What are wholesale parts sellers called?

5. What does DIY stand for?

6. What does TBA stand for?

7. What is another name for list price?

8. The price that a wholesale business pays for a part is called _____ pricing.

9. What does OE mean?

10. _____ is a broad term that refers to parts that are sold by the non-OE market.

▬ ASE-STYLE REVIEW QUESTIONS

1. Technician A says that using a computer to generate a repair order saves a good deal of paper. Technician B says that a hard copy of a repair order is needed for the customer and for the technician. Who is right?

   a. Technician A   **c.** Both A and B
   b. Technician B   **d.** Neither A nor B

2. Technician A says that a repair order contains a written estimate of the cost of a repair. Technician B says that a repair order lists a record of telephone conversations with the customer. Who is right?

   a. Technician A   **c.** Both A and B
   b. Technician B   **d.** Neither A nor B

3. Technician A says that red shop towels turn blue if they come into contact with battery acid. Technician B says that shop clothing is often

   made of materials that are resistant to battery acid. Who is right?

   a. Technician A   **c.** Both A and B
   b. Technician B   **d.** Neither A nor B

4. Which of the following is/are listed on a repair order?

   a. The concern   **c.** The correction
   b. The cause   **d.** All of the above

5. All of the following could be a source of irritation to a customer *except*:

   a. Grease on a fender

   b. Grease on the carpet or seat

   c. Grease on the steering wheel

   d. Grease on a wheel bearing

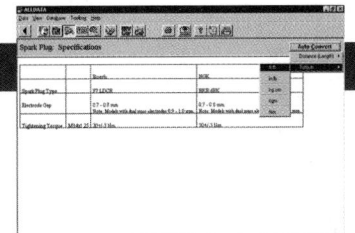

CHAPTER 5

Locating Service Information and Specifications

■ OBJECTIVES

Upon completion of this chapter, you should be able to:

✔ Describe the types of service information available.

✔ Locate repair information in shop manuals.

✔ Locate repair information on CD-ROM.

✔ Understand how microfiche works.

✔ Select the correct manual for estimating the time it will take to perform a task.

■ KEY TERMS

VIN
underhood emission
  control label

service manual
owner's manual

microfiche
TSB

■ INTRODUCTION

In 1965, the amount of material a mechanic needed to be able to read to be very knowledgeable about automobiles was about 5,000 pages. In 1974, the amount of published information for automobile repair was less than a quarter of a million pages. In 1993, that amount was 1,251,016, almost five times as many pages. Information is doubling two times a year. Today, a technician needs to have access to much, much more service information. Today, the amount of information a technician needs to have access to is approximately 1,601,300,400 (1.6 billion) pages and continuing to grow.

Repair manuals in 1965 were relatively small. Service information for only one of today's make of vehicle usually consists of more than one manual, or information stored on compact disks, DVDs, computer hard drives, networks, or on the Internet.

There are many procedures to be learned to successfully service and repair automobiles. This chapter deals with how to locate service information. Several types of service information are available, including traditional service manuals, computer libraries, and microfiche.

■ SERVICE LITERATURE

Service literature, whether in book form or digital, gives written instructions for repairs. Illustrations provide helpful hints.

Before working on any make of vehicle for the first time, it is a good idea to consult the service literature to find any helpful hints or precautions that can make your task more successful. Repair information is written for experienced technicians and makes the assumption that certain procedures are understood and will be followed.

Keep in mind that it is important to keep parts organized when you are disassembling a component or system. Be sure to label parts for reassembly. As you progress in your automobile repair skills, you will see why it is important to be methodical. When in doubt, always check the service literature. Even with the help of good information, it is very difficult to put back together something that you did not take apart yourself. It is especially helpful to note the location of accessory brackets, such as those for the alternator, smog pump, power steering pump, or air-conditioning compressor. After removing brackets from an engine, reassemble them to the unit they hold, so you do not forget where they belong.

It can be helpful to take a digital photograph of the engine compartment or of an assembled component before disassembly of something unfamiliar. Other helpful items to organize your project besides a camera include a permanent marking pen, masking tape, ID tags, and plastic baggies (**Figure 5.1**).

Identifying the Vehicle

Even though you may know the year and model of the vehicle, sometimes you may need to know more about the specific vehicle. Each vehicle comes with a vehicle identification number (**VIN**). Passenger vehicles sold in the United States (domestic and imported cars) have the VIN located on a plate on top of the dashboard, just under the windshield, on the driver's side of the vehicle (**Figure 5.2**). Until 1981, every manufacturer had its own sequence and meaning for the numbers of the VIN. In 1981, VIN codes were standardized. The VIN is required to include 17 digits. Each digit identifies a characteristic of the vehicle. Information identified by the VIN includes the country of origin, model year, body style, engine, manufacturer, vehicle serial number, and more (**Figure 5.3**). Each manufacturer can choose its own code for each of the VIN's digits, except for the first and tenth positions. These are saved for universal codes that all manufacturers must use. The first digit of the VIN identifies the country where the vehicle was manufactured, called the world manufacturer code. The tenth character identifies the model year of the vehicle. The model year was listed as a letter

Figure 5.1 Some helpful items for a first-time engine rebuild include a camera, masking tape, marking pen, and resealable plastic bags. *(Courtesy of Tim Gilles)*

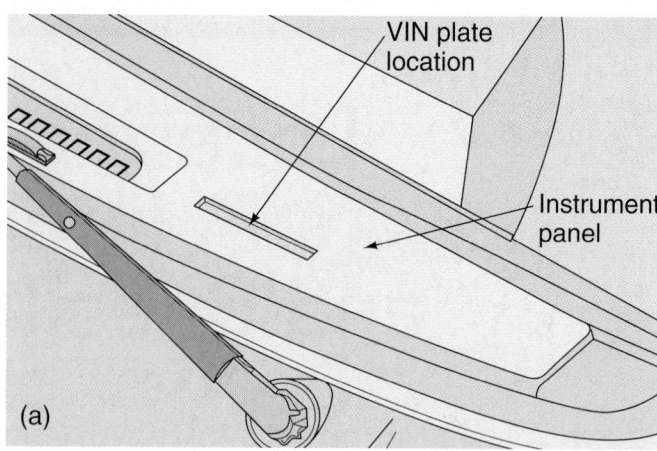

(a)

(b)

Figure 5.2 (a) The vehicle identification number (VIN) is viewed through the windshield. (b) Examples of VIN plates. *(a, Courtesy of DaimlerChrysler Corporation; b, Courtesy of Tim Gilles)*

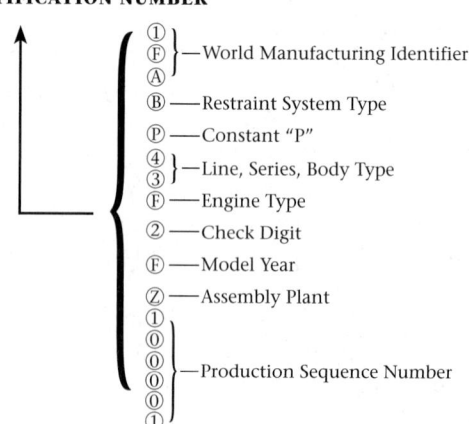

1FABP43F2FZ100001

VEHICLE IDENTIFICATION NUMBER

- ① ⓕ ⓐ — World Manufacturing Identifier
- ⓑ — Restraint System Type
- ⓟ — Constant "P"
- ④ ③ — Line, Series, Body Type
- ⓕ — Engine Type
- ② — Check Digit
- ⓕ — Model Year
- ⓩ — Assembly Plant
- ① ⓪ ⓪ ⓪ ⓪ ① — Production Sequence Number

Figure 5.3 Each digit in the VIN stands for something. *(Courtesy of Ford Motor Company)*

NOTE: *The letters I, O, and Q are never used.*

until 2001, when it changed to a number. A chart listing the first digit's countries of origin and the model year denoted by the tenth digit is located in the appendix at the back of the book.

Engine Identification

Identification numbers are also found on the engine. Some manufacturers use tags or stickers attached at various places, such as the valve cover or oil pan. Blocks often have a serial number stamped into them. **Figure 5.4** shows examples of several serial number locations. Service information will give the location of the code for a particular engine.

Underhood Label. Vehicles produced since 1972 are equipped with an underhood emission control label (**Figure 5.5**). This label gives useful information to the technician as well as to an emission control specialist. It contains such information as ignition timing specifications, emission control devices, engine size, vacuum hose routing, and valve adjustment specifications.

■ MANUFACTURERS' SERVICE MANUALS

Service Information

Three types of information are contained in shop manuals.
- Diagnostic or troubleshooting information
- Step-by-step repair information
- Specification charts (**Figure 5.6**)

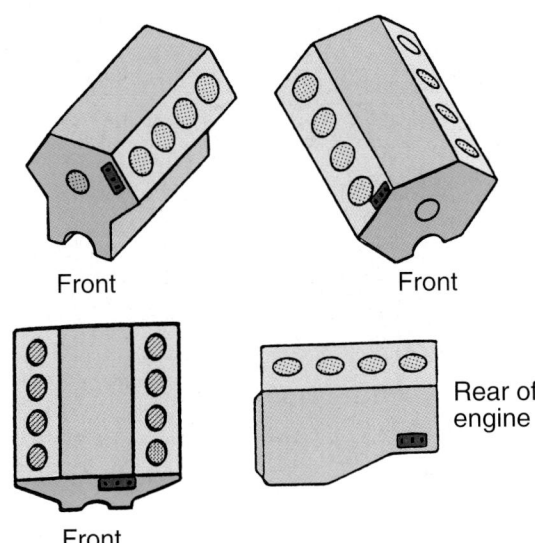

Front   Front

Rear of engine

Front

Figure 5.4 Engine serial number locations.

TIGHTENING SPECIFICATIONS

Application	Ft. Lbs. (N.m)
Camshaft Sprocket Bolts	20 (27)
Balance Shaft Retainer Bolts	27 (37)
Balance Shaft Gear Bolt	45 (61)
Connecting Rod Bolts	45 (61)
Cylinder Head Bolts	[1]60 (81)
Exhaust Manifold Bolts	37 (50)
Flywheel-to-Crankshaft Bolts	60 (81)
Front Engine Cover Bolts	22 (30)
Harmonic Balancer Bolt	219 (298)
Intake Manifold Bolts	[2]
Main Bearing Cap Bolts	100 (136)
Oil Pan Bolts	14 (19)
Outlet Exhaust	
Elbow-to-Turbo Housing	13 (17)
Pulley-to-Harmonic Balancer Bolts	20 (27)
Outlet Exhaust	
Right Side Exhaust	
Manifold-to-Turbo Housing	20 (27)
Rocker Arm Pedestal Bolts	37 (51)
Timing Chain Damper Bolt	14 (19)
Water Pump Bolts	13 (18)

[1]Maximum torque is given. Follow specified procedure and sequence.
[2]Tighten bolts to 80 INCH lbs. (9 N.m.)

Figure 5.6 Engine specifications.

■ GENERIC SERVICE MANUALS

Generic service manuals (**Figure 5.7**), dealing with all makes of either foreign or domestic cars, are produced by several companies. There are usually two types of generic manuals: those dealing with mechanical repairs, such as engines and transmissions, and those devoted to the areas of fuel/emission/ignition/air conditioning. Separate manuals are available for light trucks.

Using Service Manuals

Comprehensive generic repair manuals include sections for each make of car arranged in alphabetical order by maker. An index of cars is located at the front of the manual (**Figure 5.8**). On the first page of each section is an index of service operations that can be used to locate a particular service procedure. Many manuals have separate sections for component repair, such as carburetors, starters, alternators, transmissions, or differentials. A few pages of each section are devoted to specifications, including tune-up, electrical, wheel alignment, engine tightening, and valves (**Figure 5.9**).

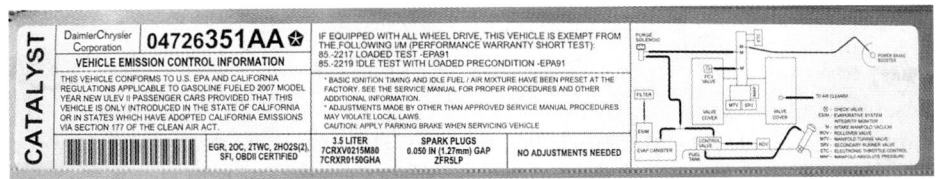

Figure 5.5 A typical underhood emission label. *(Courtesy of Tim Gilles)*

Figure 5.7 A generic service manual.

CONTENTS:

ENGINES
Section 5

CLUTCHES
Section 6

DRIVE AXLES
Section 7

TRANSMISSION
SERVICING
Section 12

GENERAL INFORMATION
Acura
Audi
BMW
Chrysler Mitsubishi
Daihatsu
Ford Motor Company
General Motors
Geo
Honda
Hyundai
Infiniti
Isuzu
Jaguar
Lexus
Mazda
Mercedes-Benz
Nissan
Saab
Subaru
Suzuki
Toyota
Volkswagen
Volvo

Figure 5.8 In the front of a repair manual is an index of cars by manufacturer.

■ LUBRICATION SERVICE MANUAL

The manual that is used most during lubrication/safety checks is a lubrication service manual (**Figure 5.10**). This manual has all of the most used specifications. It is arranged in three basic sections:

■ Lubrication and Maintenance Information
■ Capacities
■ Underhood Service Information

The lubrication and maintenance section includes lube and maintenance instructions, lubrication diagrams and specifications, vehicle lift points, and preventative maintenance intervals.

ENGINE SPECIFICATIONS

VALVE SPRINGS

Engine	Free Length In. (mm)	PRESSURE Lbs. @ In. (Kg @ mm)	
		Valve Closed	Valve Open
3.8L (VIN C)	2.03 51.6	100-110@1.73 (45-49@44)	214-136@1.30 (97-61@33)
3.0L & 3.8L (VIN 3)	2.03 51.6	85-95@1.73 (39-42@44)	175-195@1.34 (79.1-88.2@34.04)
3.8L (VIN 7)	2.03 (51.6)	74-82@173 (33-37@44)	175-195@1.34 (79.1-88.2@34.04)

CAMSHAFT

Engine	Journal Diam. In. (mm)	Clearance In. (mm)	Lobe Lift In. (mm)
3.0L	1.785-1.786 (45.34-45.36)	.0005-.0025 (.013-.064)	Int. .210 (5.334) Exh. .240 (6.096)
3.8L (VIN C)	1.785-1.786 (45.34-45.36)	.0005-.0025 (.013-.064)	¹ .272 (6.909)
3.8L (VIN 3)	1.785-1.786 (45.34-45.36)	.0005-.0025 (.013-.064)	¹ .245 (6.223)
3.8L (VIN 7)	1.785-1.786 (45.34-45.36)	.0005-.0025 (.013-.064)	.....

¹ – Specification applies to both intake and exhaust.

Figure 5.9 Some of the pages are devoted to specifications for different systems. *(Courtesy of Mitchell)*

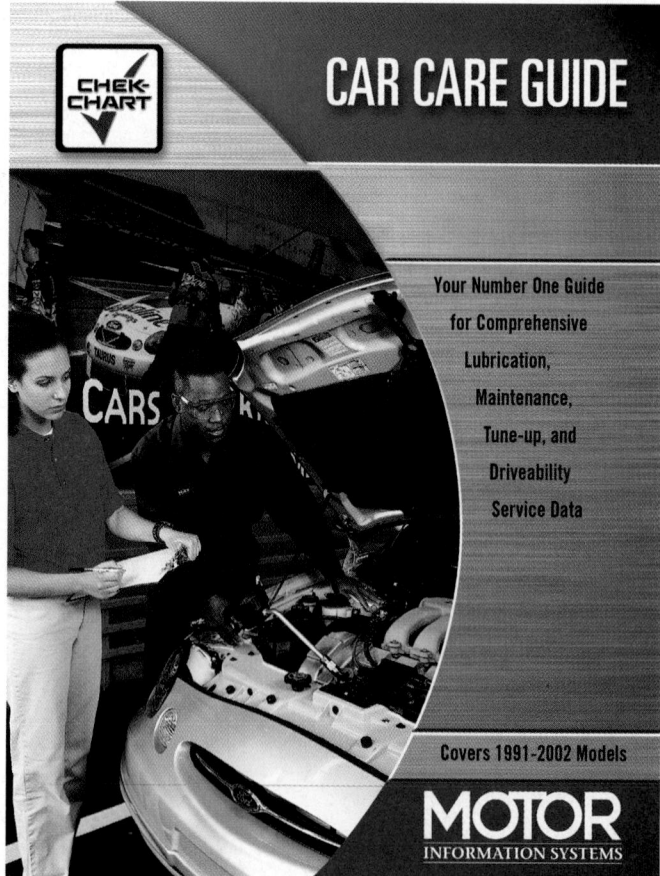

Figure 5.10 A lubrication service manual. *(Courtesy of Chek-Chart Publications)*

The capacities section includes cooling and air-conditioning system capacities, cooling system air bleed locations, wheel and tire specifications, and wheel lug torque specifications.

The underhood service information section includes underhood service instructions, specifications for engine tune-up; mechanical, electrical, and fuel systems; diagrams; and belt tensions.

■ OWNER'S MANUAL

An **owner's manual** is a booklet that comes with a new car (**Figure 5.11**). In addition to the information it contains about the operation of the vehicle's accessories, it also contains some information that can be useful to a technician. Features and safety are usually covered first. A maintenance schedule tells the owner about recommended service intervals. Sometimes an owner's manual contains instructions for maintenance for owners who want to do some of the maintenance themselves.

■ OWNERS' WORKSHOP MANUALS

Locating repair information for some cars can be more difficult. Owners' workshop manuals (**Figure 5.12**), often sold by parts stores, cover several years of one model of car. Workshop manuals are commonly available back to 1940.

■ ELECTRONIC SERVICE INFORMATION

Computers have become the predominant storage system for service information (**Figure 5.13**). Subscriptions to computer-based information libraries are available from publishing companies like AllData and Mitchell On Demand. The systems are easy to use, and information is easy to access from selections on the screen (**Figure 5.14**). The information can be read off the screen or printed by the computer printer and taken to the service bay or workbench.

AllData covers vehicles from the 1982 model year to the present. Mitchell On Demand goes back to 1984. They also sell a vintage system that dates back to 1960.

Figure 5.12 Owners' workshop manuals cover several years of one car model. *(Courtesy of Haynes Publishing Group)*

Figure 5.13 A computerized information library has a vast quantity of information. *(Courtesy of Tim Gilles)*

On single-user systems, information can be stored on CDs or DVDs that are updated quarterly. DVDs hold more information, so fewer disks are required than are needed with a CD system.

SHOP TIP One handy feature of shop information systems is that they are capable of converting specifications between the English and metric systems of measurement (**Figure 5.15**)

Many of today's repair shops have a computer in each service bay. All of the computers in the shop can

Figure 5.11 An owner's manual. *(Courtesy of Tim Gilles)*

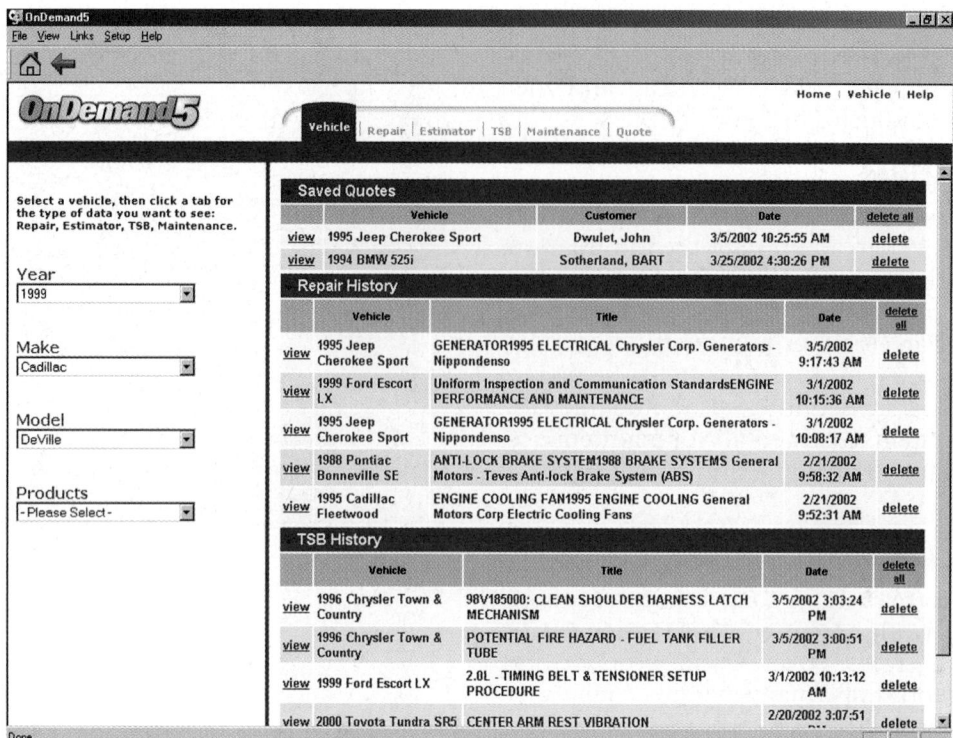

Figure 5.14 Selections are made from a menu on the screen. *(Courtesy of Mitchell)*

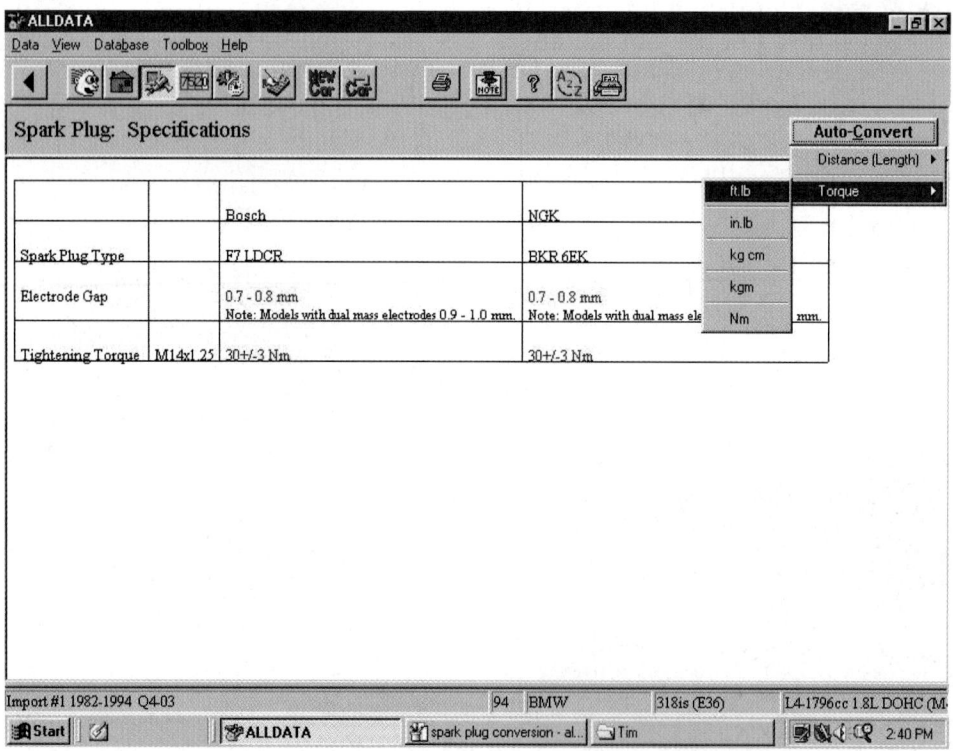

Figure 5.15 Shop information systems can convert between English and metric measurement systems.

be served by the shop's local area network. Some of the subscription service providers have information available directly from the Internet, where information is updated instantly and continuously:

■ Some hand-held automotive scan tools include access to subscription service information as well.
■ Parts and labor estimating guides are also included in some electronic information libraries.

Figure 5.16 A flat-rate manual.

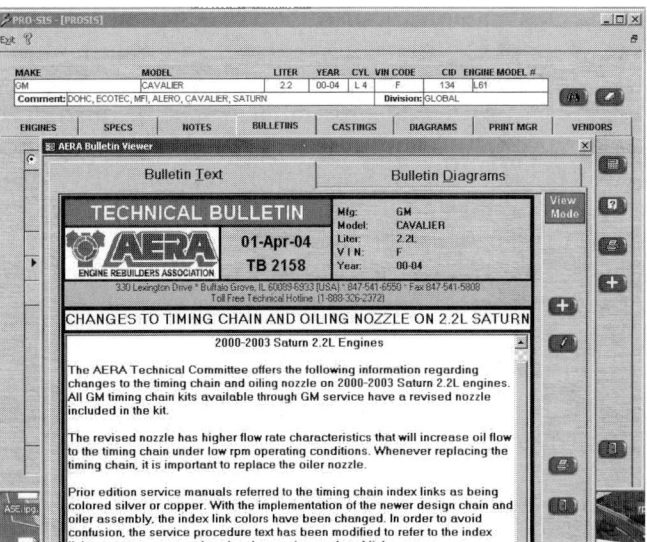

Figure 5.17 Typical technical service bulletin (TSB). *(Courtesy of AERA)*

Parts and Labor Estimating

When parts fail, they are replaced with either new or rebuilt parts. When the technician replaces the part, there is a time estimate available for this procedure. The time it should take a professional technician to perform a particular job is commonly called the *flat rate*. This information is available as part of the electronic subscription services or in a parts and time guide (**Figure 5.16**). Also included is an estimate cost of parts, so the business owner or service writer can provide the customer with a fairly accurate estimate.

■ TECHNICAL SERVICE BULLETINS

Manufacturers and industry professional associations such as the Automotive Engine Rebuilders Association (AERA) or Automatic Transmission Rebuilders Association (ATRA) publish monthly technical service bulletins (TSBs) for subscribers and members. In some of these bulletins, members report of experiences discovered through trial and error. Other times, information is contributed by manufacturers. The AERA technical committee writes technical bulletins from material contributed by the membership. An example of one of their TSBs is shown in **Figure 5.17**. Manufacturers make extensive

use of TSBs to inform technicians of important service changes to their vehicles.

Note: *Computer information systems like All-Data and Mitchell On-Demand, Shop Key, AERA's Prosis, and manufacturer-specific Web sites have technical bulletins. Always refer to these before attempting a large repair.*

VINTAGE ENGINES

Manufacturer Service Manuals and Microfiche

Until computers eventually made them obsolete, manufacturer service manuals were published each year for each make of vehicle. These were designed for use by the technicians in a dealership and covered only one year and model of vehicle. Every service operation was listed in detail. In the days of simple cars and trucks, when do-it-yourself was a popular service option, many new vehicle owners would purchase a dealer service manual to go with their vehicle. These are still available from some aftermarket publishers.

Another casualty of the computer revolution is microfiche. Although it is still available from some service literature providers, it has largely been replaced by computers and has become uncommon. Microfiche is a small plastic film card that is magnified by a microfiche reader. Many of these machines had copying capability so a hard copy of the information could be carried to the service bay.

■ HOT LINE SERVICES

Hot line services are those that provide answers to service information by telephone. Manufacturers provide help by telephone for the technicians in their dealerships. There are subscription services for independent repair shops to be able to get repair information by phone.

Some manufacturers also have phone modem systems that can transmit computer information from the vehicle to another location. The vehicle's diagnostic link is connected to the modem. The technician in the service bay runs a test sequence on the automobile. The system downloads the latest updated repair information on that particular model of vehicle. If that does not repair the problem, a technical specialist at the manufacturer's location will review the data and propose a repair.

Internet Service Information

Technicians make extensive use of the Internet. The International Automotive Technicians Network (IATN), made up of many thousands of technicians, is one example where technicians share diagnosis tips for solving tough problems. Tire, brake, and suspension part manufacturers have very detailed Web pages with much helpful information for technicians. The URL is http://members.iatn.net/.

Hollander Interchange Manuals

Sometimes a part is not available from a dealer. Salvage yards use an interchange manual (**Figure 5.18**) that lists parts that interchange from one vehicle to another. For instance, front spindles and brake calipers from one model of Chevrolet might fit a Cadillac of a different year. The interchange manual is updated with new material each year. Manuals are available for both domestic and imported vehicles.

■■ TRADE MAGAZINES

There are many professional magazines that technicians subscribe to. Examples are *Motor, Automotive Rebuilder, Motor Service, Motor Age, Import Service, Transmission Digest,* and *Precision Machine Shop.*

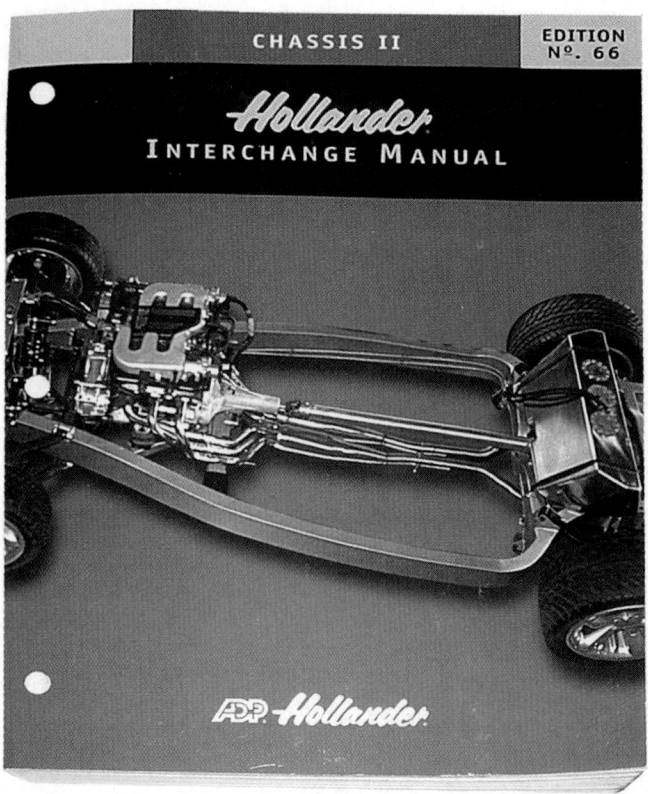

Figure 5.18 An interchange manual. *(Courtesy of Tim Gilles)*

These magazines usually include information from the latest technical service bulletins. Many also include a mail-in reader service postcard. The reader can circle numbers on this card that relate to items about which he or she would like to receive more information.

Many auto manufacturers offer their service information on a subscription or per access fee basis.

■■ REVIEW QUESTIONS

1. What number is used by manufacturers to identify a particular vehicle?
2. Which character of the VIN number on a domestic car tells the engine code?
3. What kind of information is included on an underhood label?
4. Which type of manual is the most comprehensive one for a particular vehicle?
5. What are the three types of information found in lubrication service manuals?
6. What is the name of the type of service information that requires a reader or magnifier to enlarge the information and put it on a screen?
7. What is the name of the kind of service information that is stored on compact disks?
8. What is the name of the manual that is used to determine the length of time to complete a job?
9. What are bulletins called that are produced by manufacturers and associations?
10. List six names of professional trade magazines.

▰ ASE-STYLE REVIEW QUESTIONS

1. Technician A says during disassembly, reassemble brackets to a component for easier reassembly. Technician B says to take a Polaroid photo of a complicated component before disassembling it. Who is right?

   a. Technician A   **c.** Both A and B

   b. Technician B   **d.** Neither A nor B

2. Which of the following can be found on an underhood label?

   a. Engine size/code

   b. Vacuum hose routing

   c. Spark plug gap

   d. All of the above

3. Step-by-step repair information is included in all of the following *except*:

   a. Mitchell On-Demand

   b. All-Data

   c. An owner's manual

   d. A Chilton's service manual

4. Which of the following is/are true about the vehicle identification number (VIN)?

   a. The eighth digit tells the year of the vehicle.

   b. The tenth digit tells the engine of the vehicle.

   c. Both A and B.

   d. Neither A nor B.

5. Technician A says that a flat-rate manual gives the amount of time it should take to complete a repair job. Technician B says that a flat-rate technician is paid for the time it actually takes to complete the job. Who is right?

   a. Technician A   **c.** Both A and B

   b. Technician B   **d.** Neither A nor B

Measuring Tools and Systems

■ **KEY TERMS**

British Imperial (U.S.) system	foot-pounds/Newton-meters	I.D.
metric system	plastigage	LCD
psi	O.D.	

■ INTRODUCTION

This chapter deals with the common measuring instruments and systems of measurement used in repairing automobiles. The two systems of measurement that an automotive technician must understand are the **British Imperial (U.S.) system** and the **metric system**.

The British system, using fractions and decimals, has been the basic measuring system in England, the United States, and Canada until recently. This system is based on inches, feet, and yards. In the twelfth century, 1 inch was decreed to be a length equal to three barley corns end to end. One yard was the distance from King Henry's nose to the end of his thumb. The basis for this system is somewhat ridiculous, but we have used it for so many years that it is hard to dismiss.

American manufacturers have changed their tooling to the metric system to be competitive in the rest of the world, which uses the metric system.

NOTE: *Beginning in 1975, Ford, General Motors, and Chrysler began to use metric fasteners in some locations on their vehicles. Today, all vehicles use metric fasteners.*

Technicians working on domestic automobiles will need to own tools of both metric and English sizes.

■ METRIC SYSTEM

The international system (S.I.) of measurement is known as the metric system. The basis of the metric system is the meter, which is 39.37" long, slightly longer than a yard.

Metric Engine Size Measurement

The metric measurement of volume is the liter. A liter is the quantity of liquid that will fill a cube 1/10 of a meter long on each edge. A liter is slightly larger than a quart.

NOTE: *There are 61.02 cubic inches in 1 liter.*

Because 1 *liter* is equal to 1,000 *cubic centimeters* (*cc*), a 2,000 cc engine is also called a 2-liter engine. Using the approximate measurement of 60 cu. in. to the liter, 2 liters are equal to about 120 cubic inches.

Weight Measurement

The metric unit of weight is the *gram*, which is the weight of the amount of water it takes to fill a cube that is 1/100 of a meter long on each side. A thumbtack weighs about 1 gram. The *kilogram* is the most common use of the gram. A kilogram equals about 2.2 pounds. All metric measurement units are related. For instance, 1,000 cc of water weighs 1 kilogram.

Pressure Measurement

The metric system equivalent to pounds per square inch (**psi**) is expressed as *kilograms per square centimeter*. Both of these measurements are used in measuring atmospheric pressure at sea level:

■ Atmospheric pressure is one *BAR* in the metric system (1 kilogram per square centimeter).

■ Atmospheric pressure in the English system is called one atmosphere (14.7 psi).

Temperature Measurement

On the Fahrenheit scale, water freezes at 32° and boils at 212°. In the metric system, temperature is measured in degrees *centigrade*, more often expressed as degrees *Celsius*. The temperature at which water freezes is zero degrees Celsius (0°C). Water boils at 100°C.

Torque Measurement

In a later chapter, torque wrenches are discussed. Torque readings are expressed in **foot-pounds** in the English system. The metric equivalent is expressed in **Newton-meters**.

A conversion chart from the English system to the metric system is included in the appendix at the back of the book.

Metric Measuring

The metric system is based on the number 10. It is a very easy system to use, because multiplying and dividing simply involves adding or subtracting zeros. For example:

- $\frac{1}{100}$ (0.01) of a meter is a centimeter (cm).
- 1,000 meters is a kilometer (km).
- $\frac{1}{1000}$ (0.001) of a meter is a millimeter (mm).

Because we know that 0.25 mm equals 0.010", we can make close approximations between the metric and U.S. systems when dealing with part sizes. For instance:

- A 0.75 mm oversize piston is 0.030" oversize.
- 0.50 mm undersize crankshaft bearings are 0.020" undersize in the inch system.

The appendix at the end of most repair manuals contains a metric conversion chart.

NOTE: *To convert to millimeters when the inch value is known, multiply by 25.4. To convert to inches when the metric value is known, divide by 25.4.*

▄▄ MEASURING TOOLS

The common *steel rule* is used to make approximate measurements. Common rulers used by technicians measure both in the metric system and in fractions of 1 inch (**Figure 6.1**). There are several common types of rulers. A *tape measure* is handy to have in a toolbox. The pocket-style 6" steel rule is also a popular tool. Rulers may be used with dividers or calipers to transfer measurements. Uses of the ruler are shown in subsequent chapters.

A *metric ruler* has lines that indicate hundredths and thousandths of a meter (see Figure 6.1b):

- The hundredths are called *centimeters*.
- The thousandths are called *millimeters*.
- The numbered marks are centimeters.
- Each of the 10 small lines between the numbers is 1 millimeter.

Thickness gauges are commonly called *feeler gauges*. They can be either the flat or wire type (**Figure 6.2**). Feeler gauges are commonly used to measure valve clearance, piston ring side clearance, crankshaft end

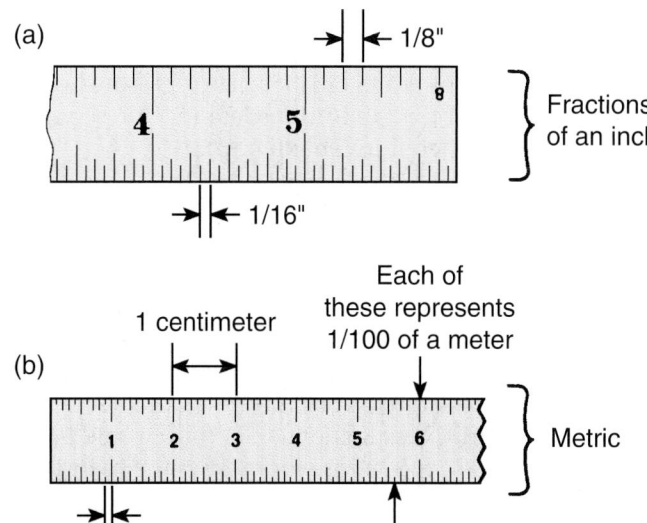

Figure 6.1 (a) Common rule measurements. (b) A metric ruler.

Figure 6.2 Flat and wire types of feeler gauges. *(Courtesy of Tim Gilles)*

play, and the gaps of spark plugs, piston rings, and ignition points.

When measuring clearance with a feeler gauge, there should be a slight drag when the feeler strip is moved through the area of clearance. If a feeler gauge that is 0.001" thicker will not fit into the gap, the correct measurement has been made. In industry, this is commonly called a "go/no-go" measurement.

Plastigage is a product used to measure oil clearance in bearings and oil pumps. It is a strip of plastic that deforms when crushed. The deformed plastic gives an idea of the amount of clearance present. Its use is covered in detail in Chapter 53.

■ PRECISION MEASURING TOOLS

The following tools are available in both English and metric types. Measurements in industry are commonly made in thousandths of an inch, or 0.025 mm. This is one of the advantages of the inch system. The metric system measurement of 0.1 mm is four-thousandths of an inch (0.004"), too large for precision measurement. The next closest metric measurement (0.01 mm) would be $\frac{1}{10}$ as large, or 0.0004", too precise for most of the industry.

Vernier Caliper

The vernier caliper was developed in the seventeenth century. A vernier caliper is one that has a movable scale that runs parallel to a fixed scale. The tool shown in **Figure 6.3** will measure outside diameter (**O.D.**) and inside diameter (**I.D.**) from 0 to 7 inches in thousandths of an inch, or from 1 to 180 millimeters. The inside measurement is somewhat limited by the length of the jaws, so it cannot be used deep in a hole. The vernier caliper shown here can also measure depth. When the cost of a vernier caliper is compared to the cost of an entire set of micrometers, which are covered later in this chapter, the reason for its popularity becomes apparent. A vernier caliper is the most versatile measuring tool. Every technician should own one.

How to Use a Vernier Caliper

The following explanation of the main and vernier scales deals with the inch-standard vernier caliper. The same principles apply to metric verniers, but measurements are made in decimal parts of a millimeter. The main parts of the scale are shown in Figure 6.3.

A fine-adjustment nut found on some vernier calipers is helpful in getting a proper feel.

Main Scale. The main scale of the caliper is divided into inches. Each inch is divided into 10 parts of $\frac{1}{10}$" (0.1") each (**Figure 6.4**). It is customary to read precision measurements in thousandths of an inch (0.001"). Therefore, each of these $\frac{1}{10}$" measurements is expressed as 0.100" (one hundred thousandths).

The area between the 0.100" increments is further divided into fourths ($\frac{1}{4}$ of 0.100"). Each of these lines is equal to 0.025".

Vernier Scale. The vernier scale divides each of the 0.025" sections on the *main scale* into 25 parts, each one being 0.001" (**Figure 6.5**). Measurements are determined by combining readings on the sliding vernier scale with those on the main scale. There are actually only 24 divisions on the main scale to the 25 on the sliding vernier, and only one line will line up exactly with another at one time. A magnifying glass is helpful for reading the vernier lines, but in the examples here they are highlighted.

Determining the Measurement. Locate the line on the main scale that the zero on the sliding vernier scale is lined up with, or is just beyond. In **Figure 6.6**, the reading on the main scale is 0.025". If the zero lined up *exactly* on the third line, then 0.075" would be the measurement (**Figure 6.7**). If the vernier scale zero did not line up exactly with a line on the main scale, then find the number on the vernier scale that lines up perfectly with *any* line on the main scale. Only one line on the vernier scale will line up perfectly. In **Figure 6.8a**, it is the line representing 0.005". The main scale reading is 0.100", so the reading is 0.105".

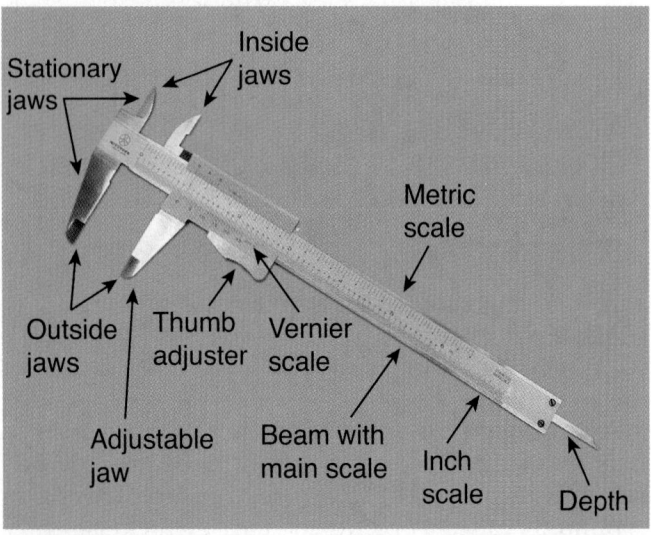

Figure 6.3 A vernier caliper. *(Courtesy of Tim Gilles)*

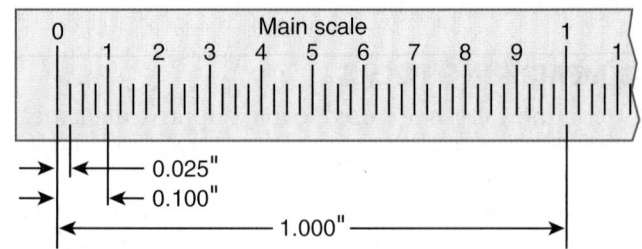

Figure 6.4 Measurements on the main scale were $\frac{1}{40}$ of an inch, or 0.025" (twenty-five thousandths).

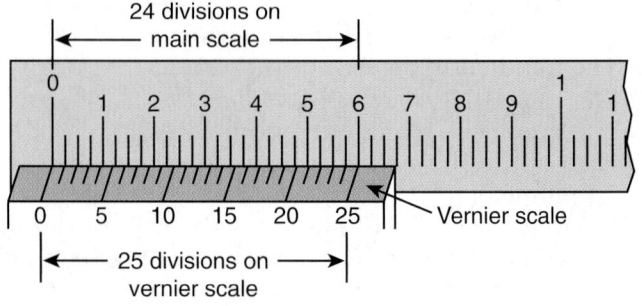

Figure 6.5 The vernier scale divides each of the main scale measurements into 25 parts.

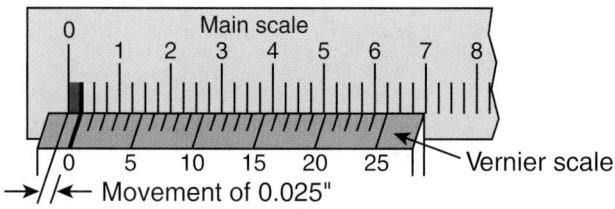

Figure 6.6 Measuring 0.025" on a vernier scale.

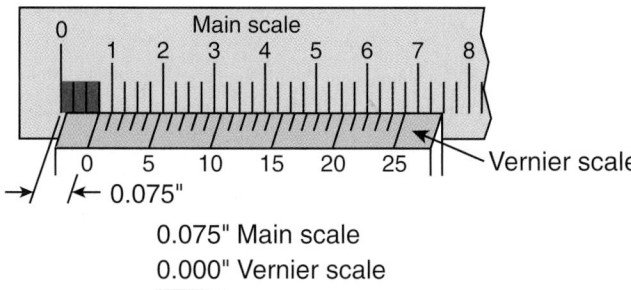

0.075" Main scale
0.000" Vernier scale
—————————
0.075" Total

Figure 6.7 A 0.075" reading on the vernier scale.

(a) Example: 0.100" Main scale
0.005" Vernier
—————
0.105" Overall

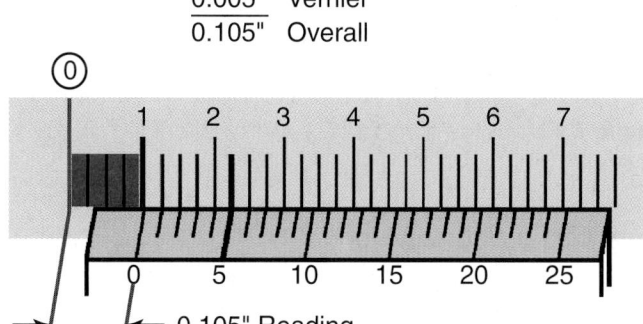

Figure 6.8 In examples (a) and (b), select the vernier scale line that lines up exactly with any line on the main scale.

(b) Example: 4.000" ⎫
0.275" ⎬ Main scale
0.012" Vernier
—————
4.287" Overall

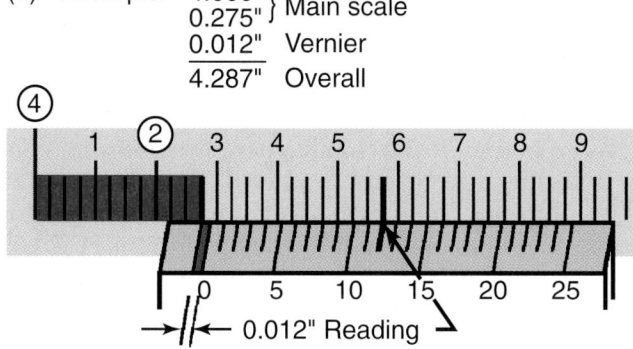

In **Figure 6.8b**, the main scale reading is 0.275" plus the vernier scale reading of 0.012" for a total of 0.287".

NOTE: *The main scale also shows 4", so the total reading is 4.287".*

An alternative to a standard vernier caliper is the *dial vernier* shown in **Figure 6.9**. It registers the 0.001"

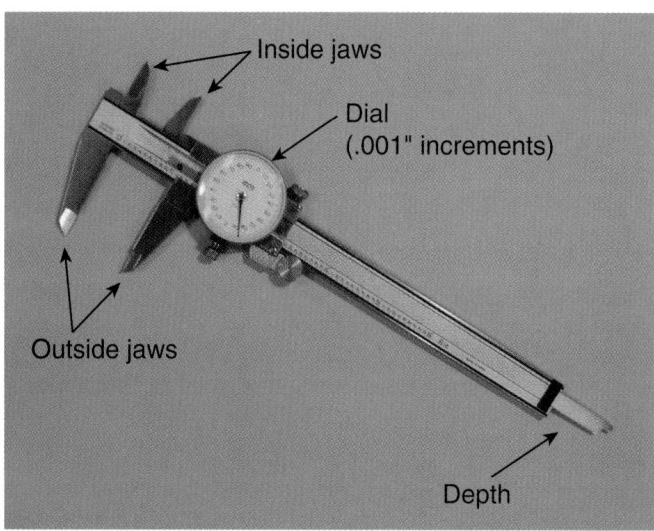

Figure 6.9 A dial vernier caliper. *(Courtesy of Tim Gilles)*

part of the measurement on its dial. This type of caliper is very popular, but more expensive.

Sometimes a vernier caliper main scale is divided into 50 sections instead of 25. There is a scale for reading inside measurements and a separate scale for reading outside measurements. There is only one division between the 0.100" line and the 0.200" line. Each of these divisions represents 0.050" instead of 0.025". Reading this vernier caliper is otherwise the same as reading the 25-section caliper. Be aware of which type you are reading.

An *electronic caliper* (**Figure 6.10**) works the same way as an ordinary caliper, but its measurement is shown on a liquid crystal display (**LCD**) to an accuracy of 0.0005. Measurements can be in either inch or metric.

Micrometer

Micrometers are often called "mikes" for short. **Figure 6.11** shows a set of *outside micrometers*. Each one has a range of only 1 inch. Because of the expense of a complete set of mikes, they are found mostly in machine shops.

Micrometers have several advantages over other types of measuring instruments. They are clear and easy to read. They measure consistently and accurately, and they have a built-in adjustment to compensate for wear. A cutaway view of an outside micrometer is illustrated in **Figure 6.12**.

Some micrometers have extra features:
- A ratchet, which sees that spindle pressure is always consistent.
- The locknut, which can hold a reading so it cannot be changed accidentally.
- The decimal equivalent chart stamped on the frame.
- Some micrometers have a digital readout. This feature may also give metric readings.

If a micrometer spindle must be moved a long way to adjust it, roll the thimble on your arm or hand (**Figure 6.13**).

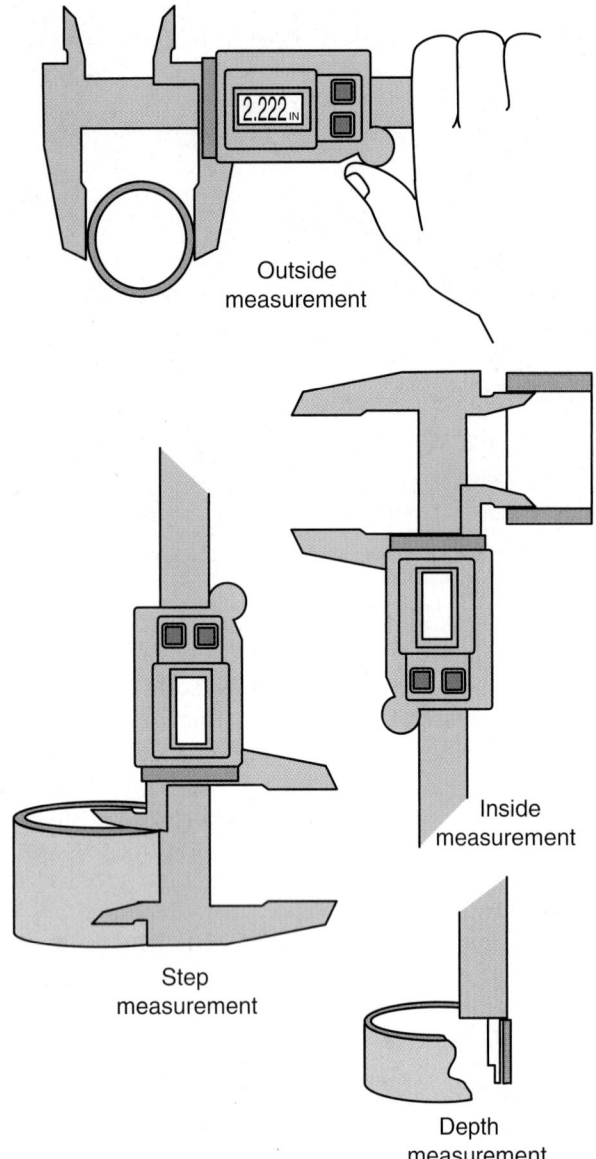

Outside measurement

Step measurement

Inside measurement

Depth measurement

Figure 6.10 An electronic caliper.

Figure 6.11 A set of outside micrometers with gauge blocks. *(Courtesy of Tim Gilles)*

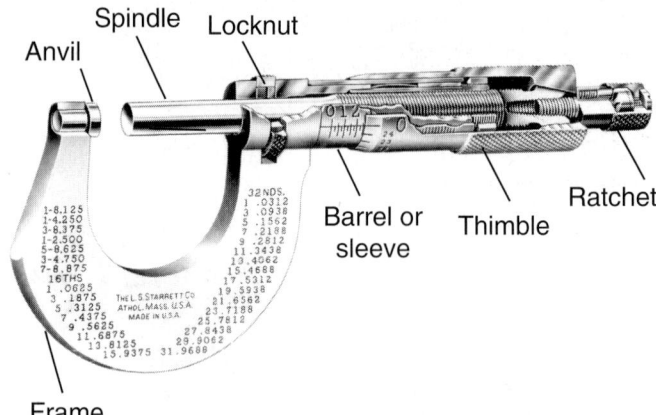

Anvil Spindle Locknut

Barrel or sleeve Thimble Ratchet

Frame

Figure 6.12 A cutaway view of an outside micrometer. *(Courtesy of L. S. Starrett)*

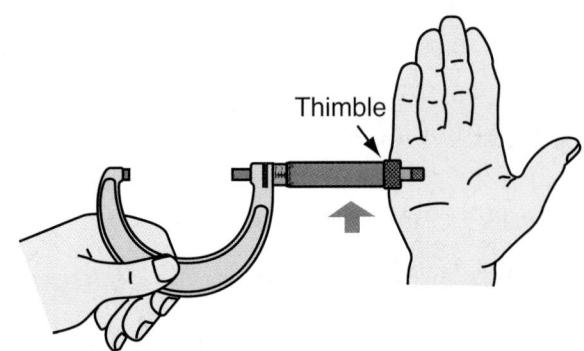

Thimble

Figure 6.13 Roll the thimble on your hand for long spindle movements.

Reading a Micrometer

The micrometer operates on a simple principle: The spindle (**Figure 6.14a**) has 40 threads per inch, so one revolution of the thimble will advance or retract the spindle $\frac{1}{40}$ of an inch (0.025") (**Figure 6.14b**). The sleeve, or barrel, is laid out in the same fashion as the frame on a vernier caliper. Each line on the barrel represents 0.025", and a new line is uncovered each time the thimble is turned one revolution. Measurement amounts are shown in **Figure 6.15**.

As each line on the thimble passes the zero line on the hub, the mike opens another 0.001". To read the exact number of thousandths, note the number on the thimble that lines up with the zero line. In **Figure 6.16**, this amount is 0.008". If you add 0.350 to 0.008, you get 0.358", which is the actual measurement in this reading.

NOTE: *The measurement reflects the actual size of the part.*

In **Figure 6.17**, the reading is 0.358". The reading is obtained by adding the barrel measurement to the thimble measurement. The reading is rounded off to the nearest $\frac{1}{1000}$ of an inch (0.001").

The micrometer is held as shown in **Figure 6.18**. An experienced technician will develop a feel for the micrometer. Compare your readings to those of

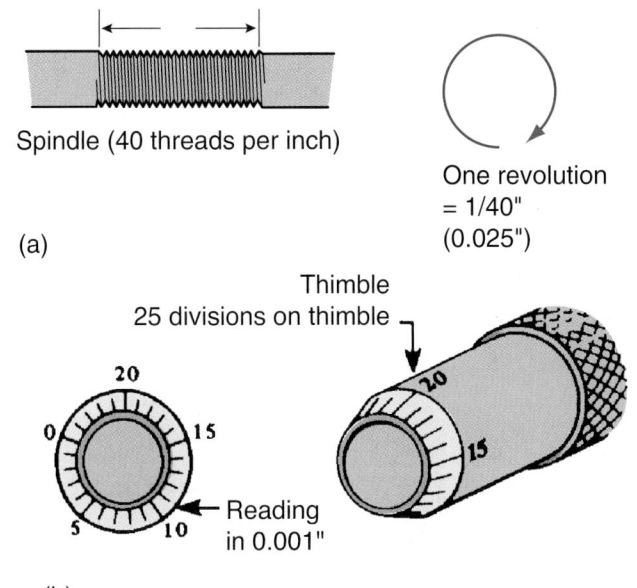

(a)

Spindle (40 threads per inch)

One revolution = 1/40" (0.025")

Thimble
25 divisions on thimble

Reading in 0.001"

(b)

Figure 6.14 Relationships between (a) micrometer spindle pitch (40 threads to the inch); spindle/thimble movement (¹⁄₄₀", or 0.025", per revolution); and (b) graduations on the thimble of an inch-standard micrometer.

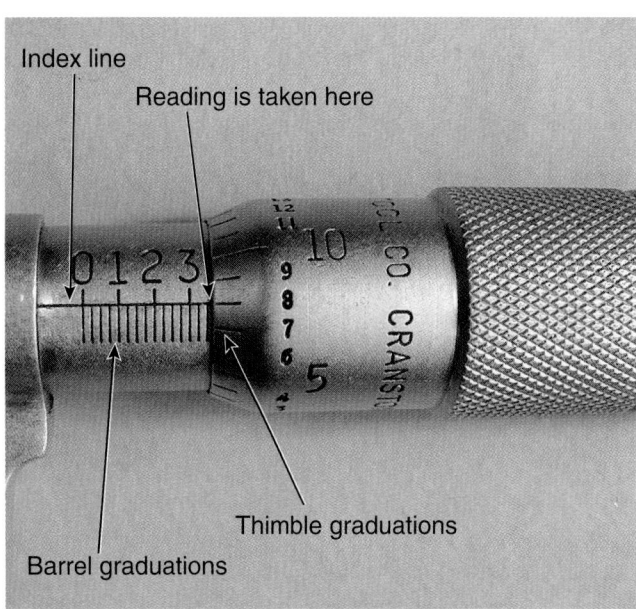

Index line

Reading is taken here

Thimble graduations

Barrel graduations

Figure 6.16 Reading a micrometer. *(Courtesy of Tim Gilles)*

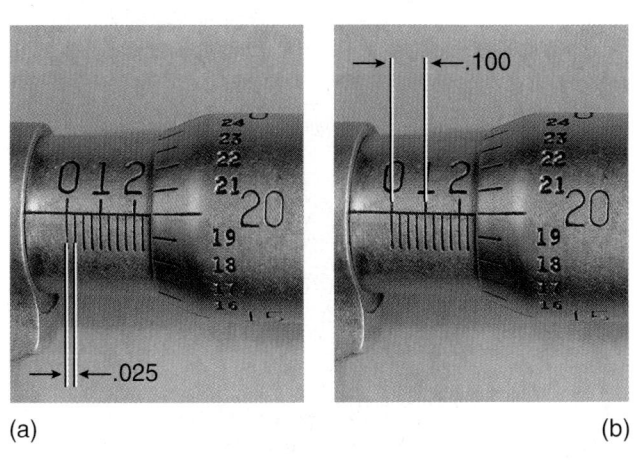

.100

.025

(a)   (b)

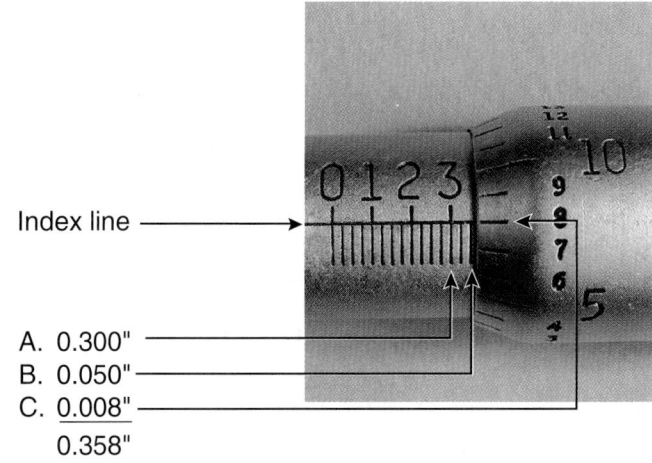

Index line

A. 0.300"
B. 0.050"
C. 0.008"
    0.358"

Figure 6.17 To read a micrometer to the nearest 0.001", add A, B, and C together. *(Courtesy of Tim Gilles)*

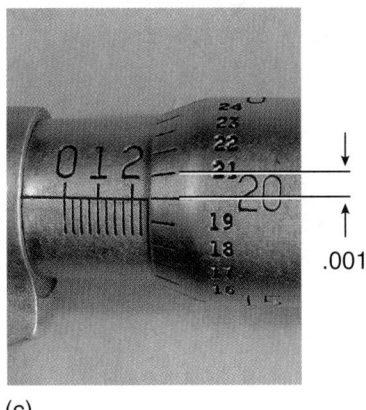

.001

(c)

Figure 6.15 Micrometer measurements. (a) Each line on the barrel is 0.025" (one revolution of the thimble). (b) Each number on the barrel is 0.100" (four revolutions of the thimble). (c) Readings on the thimble are 0.001" (¹⁄₂₅ revolution of the thimble). *(Courtesy of Tim Gilles)*

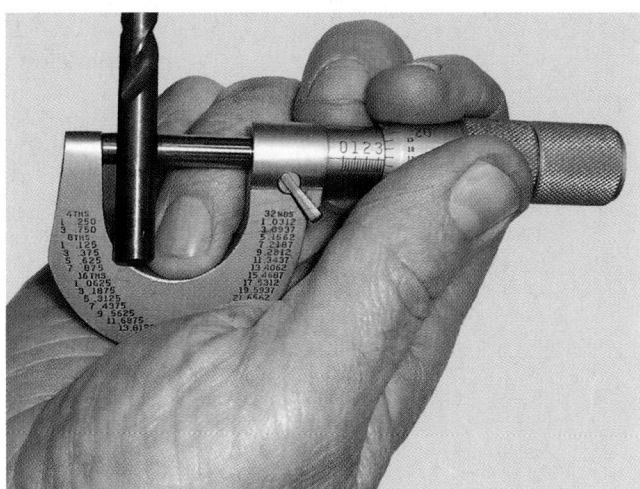

Figure 6.18 The proper way to hold a small micrometer. *(Courtesy of Tim Gilles)*

an experienced person until you are consistent. The readings might be off by as much as 0.001". When you have developed a consistent feel, you can calibrate your micrometers. Calibration is the process of setting a precision measuring device to read at zero. You can set your micrometers to read at zero by using a special *gauge block* that comes with the micrometer (see Figure 6.11).

Slide the micrometer over the part and roll it back and forth (**Figure 6.19**). It should be snug against the part but not feel tight. Use the ratchet until you develop a good touch.

Normally, a reading will be rounded off to the nearest $\frac{1}{1000}$ of an inch (0.001"). But sometimes an estimate can be made to the nearest ten-thousandths (0.0001") (**Figure 6.20**). This estimate is made by gauging the distance between the index line and the last number passed on the thimble.

Micrometer Vernier Scale

Some micrometers also have a *vernier scale* for making readings to 0.0001" (called *tenths*). Most measurements need not be more accurate than 0.001", but in case it is needed, the vernier scale provides the capability for more accuracy. The vernier scale divides a thousandth of an inch by $\frac{1}{10}$, into tenths of thousandths of an inch (**Figure 6.21**). There are 10 graduations on the top of the micrometer barrel that occupy the same space as 9 graduations on the thimble. The difference between the width of one of the 9 spaces on the thimble and one of the 10 spaces on the barrel is $\frac{1}{10}$ of one space.

Because each graduation on the thimble represents 0.001", the difference between graduations on the barrel and thimble is 0.0001". If the zero lines up with a line, the reading is accurate to three decimal places (**Figure 6.22**).

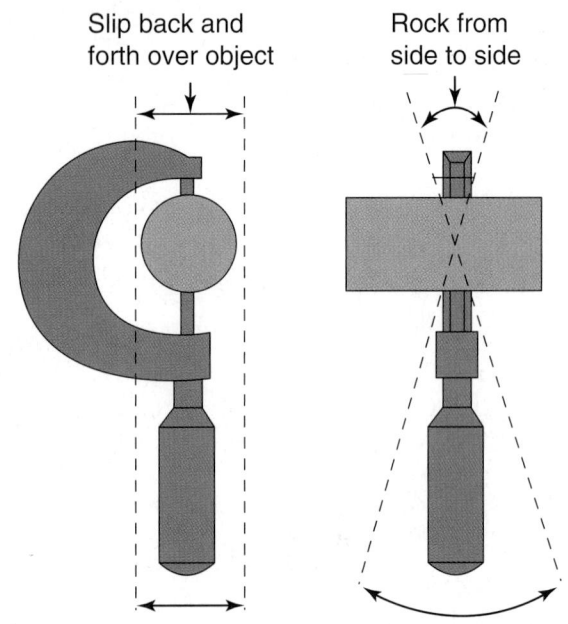

Figure 6.19 Adjust the fit of the micrometer to the part so that it is snug but not tight.

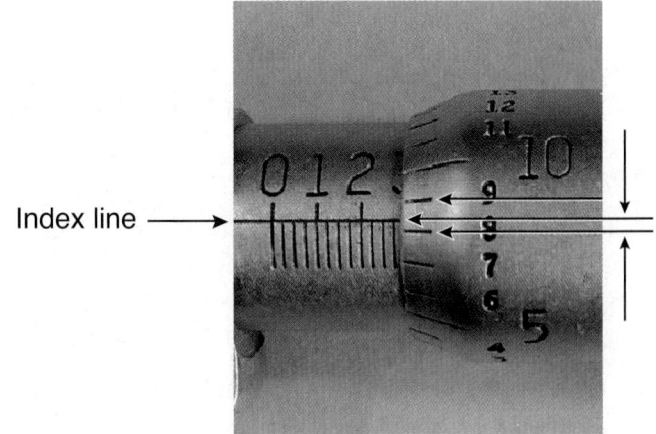

Figure 6.20 Make an approximate reading to four places (0.0003") by estimating the distance between the index line and the reading on the thimble. *(Courtesy of Tim Gilles)*

Vernier scale compared to thimble scale

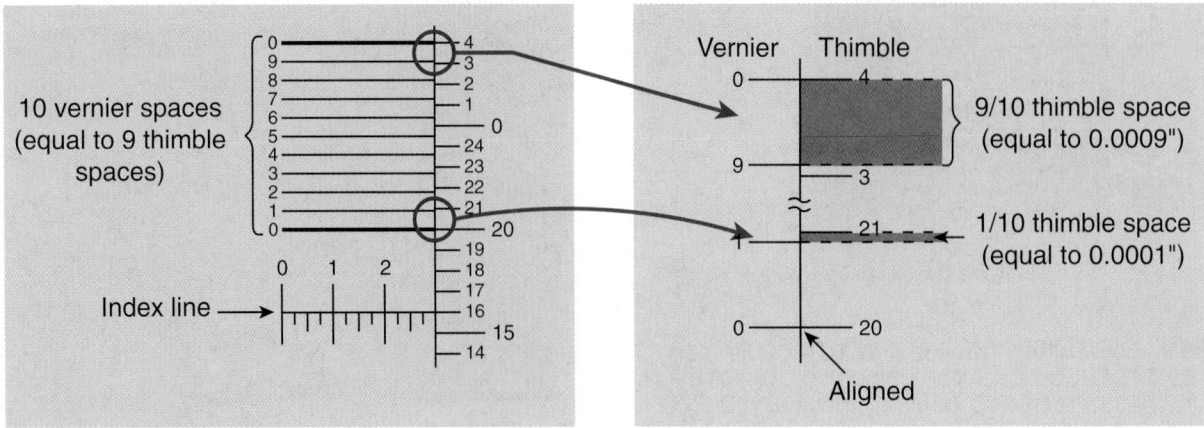

Figure 6.21 A micrometer vernier scale.

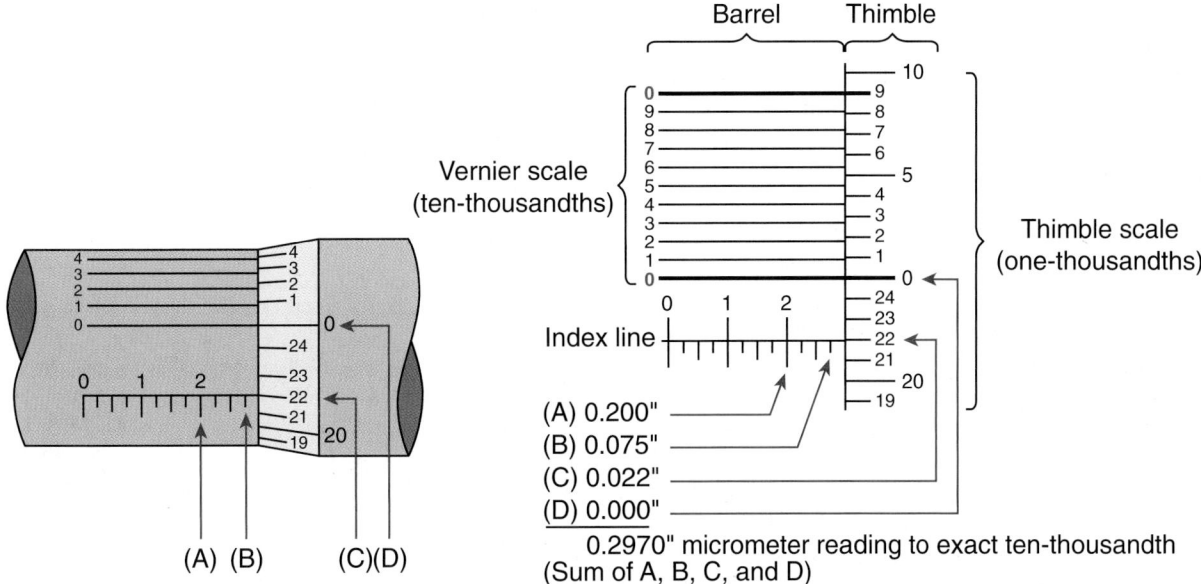

(A) 0.200"
(B) 0.075"
(C) 0.022"
(D) 0.000"

Index line

Vernier scale
(ten-thousandths)

Barrel   Thimble

Thimble scale
(one-thousandths)

(A) (B)   (C)(D)

0.2970" micrometer reading to exact ten-thousandth
(Sum of A, B, C, and D)

Figure 6.22 A reading of exactly 0.2970". Both zero lines are aligned with the thimble lines.

If no line on the thimble lines up exactly with the zero line, the vernier scale is used. To read to tenths, simply find the line on the vernier scale that lines up with any number on the thimble, and then add this number to your reading. The example in **Figure 6.23** reads 0.3812".

Metric Micrometer

Except for the graduations on the hub and thimble, metric micrometers look just the same as English mikes. One turn of the thimble turns the spindle 0.5 mm, as opposed to 0.025". The hub is graduated in millimeters and half millimeters (**Figure 6.24**). The lines below the index line are the half millimeters (0.5 mm). Two revolutions of the thimble equal 1.0 mm on the hub. The thimble usually has 50 divisions. Every fifth line has a number from 0 to 45. Each graduation on the thimble is equal to 0.01 mm (1/100 millimeter).

The metric micrometer is read in the same manner as its English counterpart. Just add the millimeter and half millimeter readings on the barrel to 1/100 (0.01) millimeters on the thimble.

Example

0.300"
0.075"
0.006"
0.381" sum of reading to nearest one-thousandth
0.0002" vernier scale reading
0.3812" sum of reading to nearest ten-thousandth

Figure 6.23 The vernier scale indicates 0.0002".

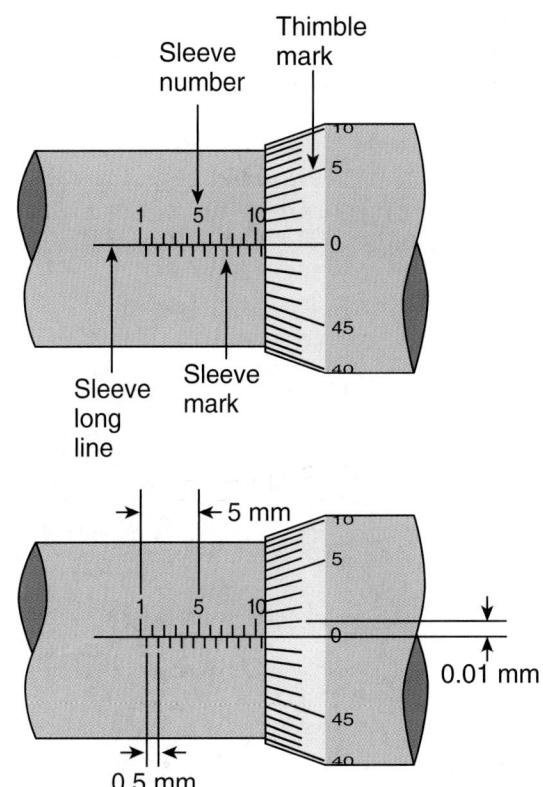

Sleeve
number

Thimble
mark

Sleeve
long
line

Sleeve
mark

← 5 mm

0.01 mm

0.5 mm

Figure 6.24 Metric micrometer markings and graduations.

The vernier scale feature gives the ability to read to two-thousandths of a millimeter (0.002 mm). There are only 5 divisions on this scale, compared to 10 on the English vernier scale. When one of the lines on the vernier scale lines up with *any* of the lines on the thimble, add that amount to the reading.

Combination Digital Mikes

There are combination micrometers that give both metric and English readings. There are two versions available. One of them gives inch readings in the conventional manner, while the other reads the metric system. LCD readings taken from the *digital readout window* (**Figure 6.25**) are accurate to 0.0001 and are convertible between inch and metric.

Inside Micrometers

An inside micrometer can be used to measure cylinder bores and main and rod bearing bores. The thimble of an inside micrometer does not move as freely on the barrel as the thimble of an outside micrometer. The added friction helps keep the reading from changing. Reading an inside micrometer accurately requires some practice.

Inside micrometers have extension rods to make them the proper size. They have handles for use in deep cylinders. **Figure 6.26** shows an example of an inside micrometer reading.

Telescoping and Split-Ball Gauges

An inside micrometer cannot work in cylinders smaller than about 2", so telescoping or small-hole gauges are used. These are known as *transfer gauges*, because the measurements they make are read with an outside micrometer.

The split-ball gauge is used on small holes such as valve guides. Be sure to take your reading at 90 degrees to the split in the ball (**Figure 6.27**).

Telescoping gauges can be used with a micrometer to measure cylinders (**Figure 6.28**), bearing bores, and so on.

Figure 6.25 A digital micrometer.

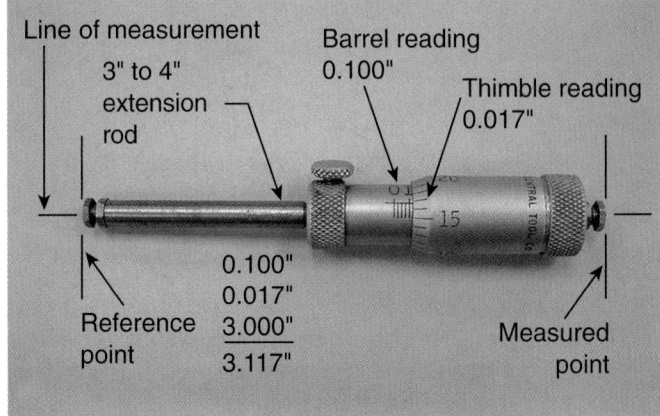

Figure 6.26 Taking an inside micrometer measurement. *(Courtesy of Tim Gilles)*

Dial Indicators

A dial indicator can measure movements such as end play of crankshafts and valve guide wear. It can be used to measure valve-in-head depth and cylinder and main bearing bores. These processes are covered in subsequent chapters.

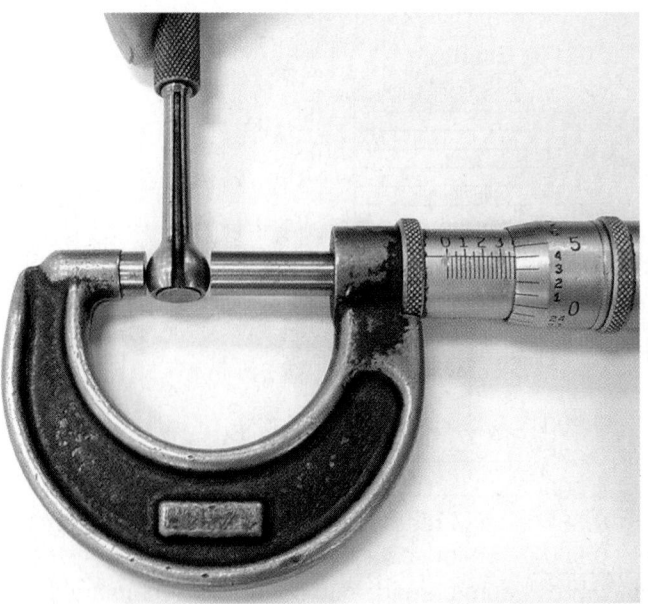

Figure 6.27 Small-hole gauges. After setting the gauge to size, read the size with an outside micrometer at 90 degrees to the split in the ball. *(Courtesy of Tim Gilles)*

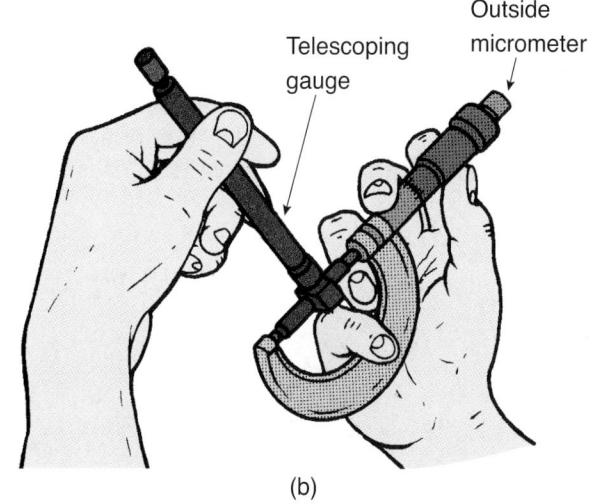

(a)   (b)

Figure 6.28 (a) A telescoping gauge in a cylinder bore. (b) The gauge is read with an outside micrometer.

An indicator (**Figure 6.29**) is a very sensitive instrument consisting of small gears activated by spindle movement. It can measure in inches or millimeters. Movement is transmitted through a small gear to an indicating hand on a dial. The dial can be either balanced or continuous (**Figure 6.30**). Long-range indicators are equipped with revolution counters.

Indicators are comparison instruments because an indicator reading must be compared to a known measurement (**Figure 6.31**). When measuring thrust (forward and backward movement), no comparison is necessary.

Figure 6.29 Dial indicator nomenclature.

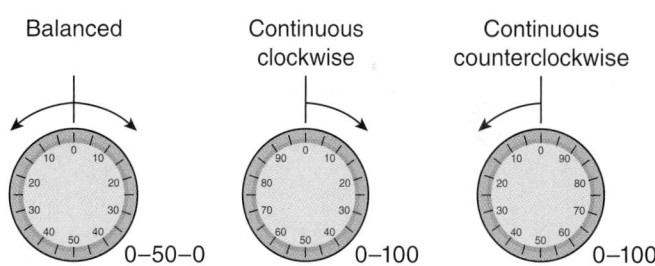

Figure 6.30 Types of dial indicator faces.

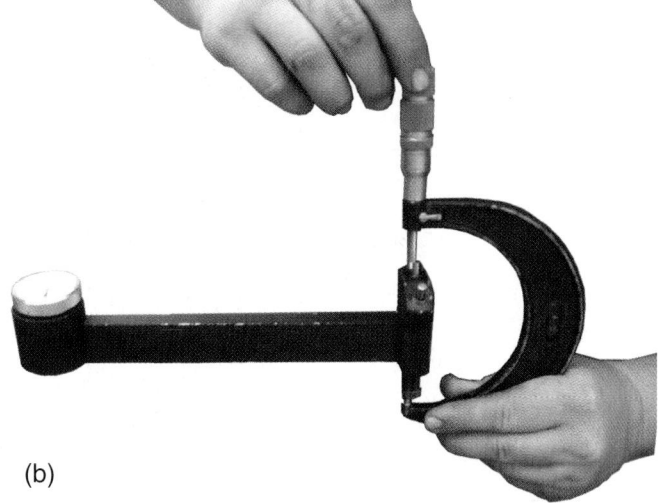

(a)   (b)

Figure 6.31 A dial indicator is measured for comparison. *(Courtesy of Tim Gilles)*

A special indicator and micrometer setting fixture used for measuring cylinder bores is shown in **Figure 6.32a**. The tool is a *comparator gauge*. It has a micrometer setting fixture (**Figure 6.32b**). If the setting fixture is set to read a cylinder bore of 4.375, then when the gauge on the indicator reads "0" it is actually reading 4.375".

One type of indicator has a magnetic base. Another type has various attachments for clamping it to the workpiece (**Figure 6.33**). The most popular and versatile dial indicator among technicians is the vise-grip indicator shown in **Figure 6.34**. It has many uses in the specialty areas of auto repair.

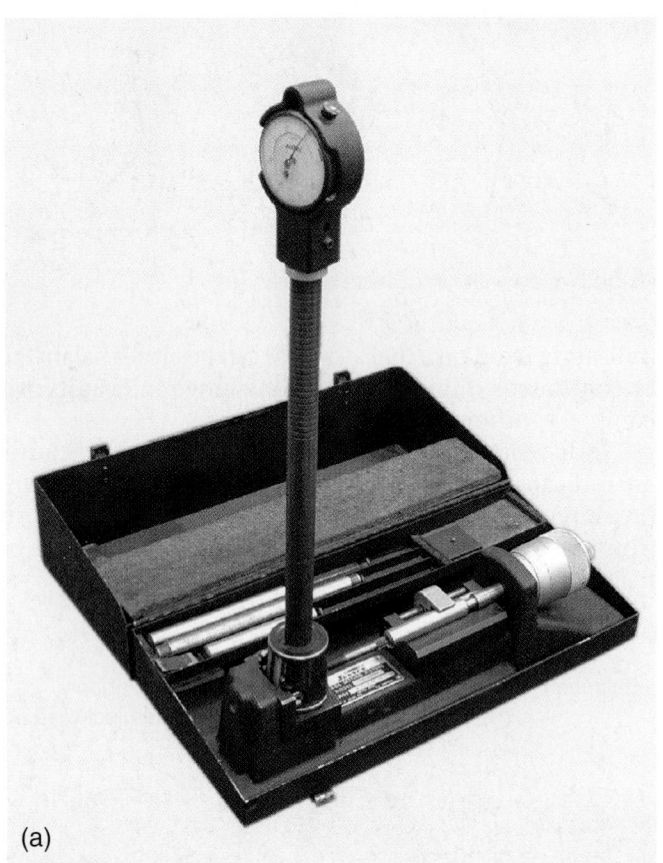

(a)

(b)

Figure 6.32 (a) A dial bore gauge is calibrated in a setting fixture. (b) A dial bore gauge measuring a cylinder bore. *(Courtesy of Tim Gilles)*

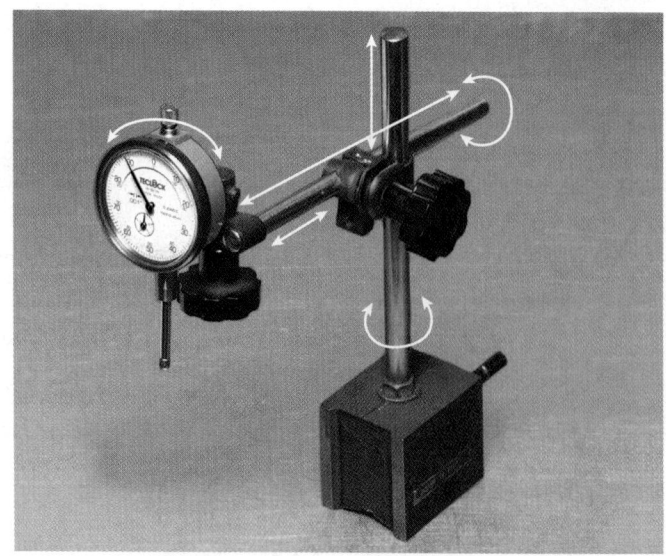

Figure 6.33 Various positions of a universal dial indicator set.

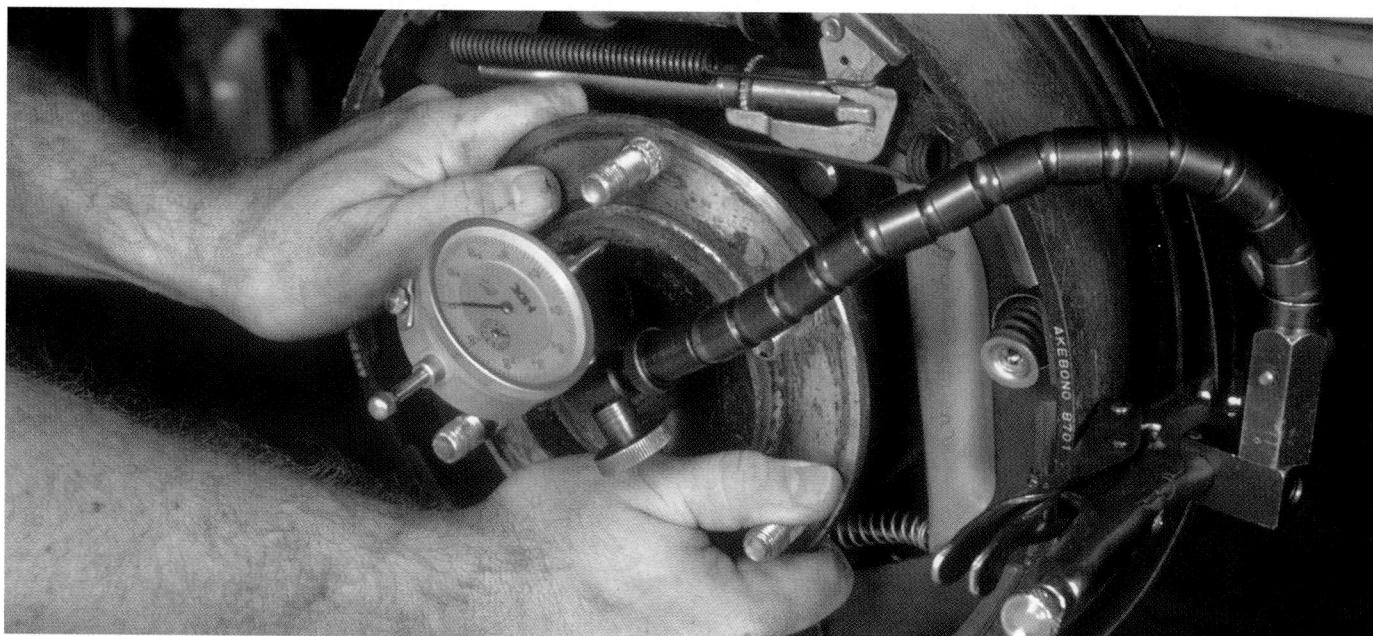

Figure 6.34 A vise-grip indicator in use.

▋▋ REVIEW QUESTIONS

1. On precision measuring instruments, how is ⅒" expressed with decimals?
2. Name the tool used to calibrate a micrometer.
3. Which scale on the micrometer is used to measure to 0.0001"?
4. When reading a small-hole gauge, how is the micrometer reading taken in relation to the split in the ball, on the split, or 90 degrees to it?
5. Name an instrument that would be used to measure end play of a shaft.

6. What is the extra gauge that long-range dial indicators have?
7. Convert 5 inches to millimeters.
8. Convert 127 millimeters to inches.
9. Approximately how many cubic inches are there in 1 liter?
10. Approximately how many cubic inches equals 2.7 liters?

▋▋ ASE-STYLE REVIEW QUESTIONS

1. One liter equals
   a. 1000 cubic centimeters
   b. 1 quart
   c. 1 gallon
   d. None of the above

2. Technician A says that one revolution of the micrometer thimble equals ¼₀ inch. Technician B says that one revolution of the micrometer thimble equals 0.025 inch. Who is right?
   a. Technician A   c. Both A and B
   b. Technician B   d. Neither A nor B

3. The metric measuring system is based on
   a. The gold standard
   b. The number 10

   c. Fractions of 1/64"
   d. The distance between the earth and the moon

4. A vernier caliper can measure
   a. Inside diameter
   b. Outside diameter
   c. Depth
   d. All of the above

5. Technician A says that a gauge block is used to measure valve lash (clearance). Technician B says that the vernier scale on a micrometer measures to 0.001". Who is right?
   a. Technician A   c. Both A and B
   b. Technician B   d. Neither A nor B

◼ MICROMETER PRACTICE

Enter the micrometer readings below. *(Courtesy of Tim Gilles)*

1.

2.
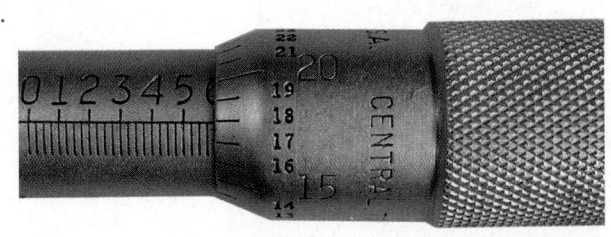

3.
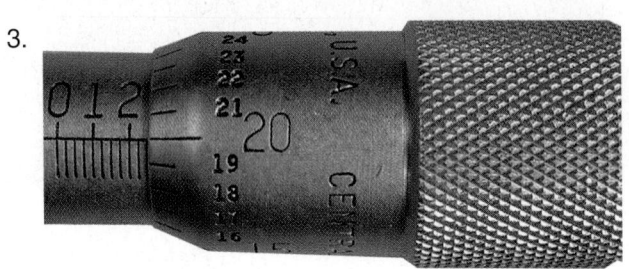

4.

5.

6.

7.
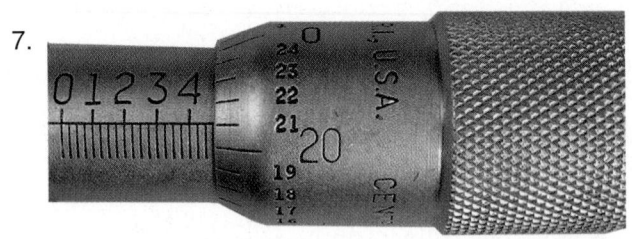

8.

CHAPTER 7

Hardware, Fasteners, Drills, and Thread Repair

■ OBJECTIVES

Upon completion of this chapter, you should be able to:

✔ List fasteners and the methods in which they are used for repair.

✔ Describe the various methods of thread repair.

✔ Describe characteristics of hand drills.

■ KEY TERMS

capscrew
temper
anneal
clamping force
elastic limit

tensile strength
bolt grade
ANSI
ISO
hydrostatic lock

through-hardened
  washer
case-hardened washer
tap drill
EDM

■ INTRODUCTION

A good automotive technician must know many things about fasteners (nuts and bolts), the different methods of repairing them when they break, and how not to break them. This chapter deals with these parts and procedures.

■ CHARACTERISTICS OF FASTENERS

A *bolt* is an externally threaded fastener that is used with a nut. When it is used without a nut in a threaded or blind hole application to hold components together, it is known as a **capscrew**. During manufacturing, screw threads are either rolled or cut. *Rolled screw threads* are larger than the shank of the bolt (the area between the thread and the head of the bolt). To determine the size of a fastener, measure the screw thread.

Standard threads come in both coarse and fine threads. The fine thread is called *national fine*; the coarse thread is *national coarse*. These are classified, respectively, as UNF and UNC. The "U" stands for *unified*; it means that the thread interchanges with British threads. A UNF thread callout for a ⅜" diameter thread would be ⅜ × 24. It would have 24 threads per inch. A UNC would be ⅜ × 16. It would have 16 threads per inch.

Metric screw threads are also coarse or fine. The last number in the callout represents the thread pitch in millimeters. It is different from the threads per inch

in the unified system. With metric screw threads, the thread pitch is the *distance* between threads in millimeters. This varies between 1.0 and 2.0 and depends on the diameter of the fastener. If the thread pitch is 1.0, this means there is 1 mm between each screw thread. An 8 mm fine thread could be either 8 × 1.0 or 8 × 1.25; a coarse thread would be 8 × 1.5. Notice that the lower pitch number denotes a finer thread.

Several different kinds of fasteners are used in automobiles. They are selected according to purpose and price. Fasteners are made of different types and grades of metals, and some are hardened.

■ To toughen, or **temper**, a metal, it is heated to a specific temperature and then quenched.

■ To soften, or **anneal**, a metal, it is heated and allowed to cool slowly.

Bolts can be broken or damaged by overtightening, bottoming out, or being forced into a thread that does not match.

Bolt Stretch

When a bolt is properly tensioned it will be "spring loaded" against the part it is fastening. Bolts are purposely stretched like this to provide **clamping force** or "spring tension" on the parts being held together. Usually a bolt is stretched to 70% of its **elastic limit** when properly tightened. The elastic limit of a fastener is reached when it will no longer return to its original shape when loosened. A torque wrench is used to tighten fasteners to a specified tension. Correct tightening methods and torque wrenches are covered in Chapter 51, Engine Sealing, Gaskets, Fastener Torque.

When a bolt is overtightened, it becomes overstretched. A stretched bolt can be identified when a nut can turn easily down the bolt threads until it encounters the stretched area, where it will become hard to turn (**Figure 7.1**).

Fastener Grades

Fasteners are of different quality grades, depending on their intended use. One fastener's use might call for flexibility, while another's might call for high strength. Some new vocabulary terms are necessary for understanding fastener quality:

■ **Tensile strength** is the maximum stress a material can withstand without breaking.
■ *Yield point* is the maximum stress a material can withstand and still be able to return to its original form without damage.
■ *Ultimate strength* is about 10% higher than the yield point (for high-quality steel). That is the point at which the fastener breaks.

The heads of capscrews or bolts have markings to identify their grades (**Figure 7.2**). On customary (inch) bolts, there are radial lines that indicate strength. Count the lines and add 2 to determine the strength of the bolt. These are known as SAE **bolt grades** and are evaluated by the American National Standards Institute (**ANSI**) standard, which is used to measure a fastener's tensile strength. Metric bolts are numbered. The higher the number, the greater the strength of the bolt.

In addition to the ANSI standard, there is another standard for fastener quality. The International Standards Organization (**ISO**) defines fastener quality in terms of tensile strength and yield strength. This standard is used for metric fasteners and is anticipated to eventually replace all other grading standards. When interpreting an ISO fastener grade number:

■ The first number is the tensile strength.
■ The second number is the yield strength.

For example, a metric bolt marked 8.8 has a tensile strength of 800 MPa (115,000 psi) and a yield strength

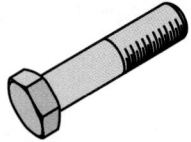

Customary (inch) bolts—Identification marks correspond to bolt strength—increasing numbers represent increasing strength.

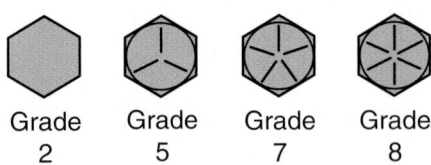

Grade 2   Grade 5   Grade 7   Grade 8

Metric bolts—Identification class numbers correspond to bolt strength—increasing numbers represent increasing strength.

4.6 4.8 5.8 8.8 9.8 10.9

Figure 7.2 Bolt grade markings.

of 80% of 800 MPa. A 10.9 bolt has a tensile strength of 1,000 MPa (145,000 psi) and a yield strength of 90% of 1,000 MPa. Markings are required only on bolts classed 8.8 or higher.

Metric fastener strength is called *property class* rather than *grade* (as with SAE fasteners). SAE grades and property classes are comparable:

SAE Grade	Property Class
Grade 2	5.8
Grade 5	8.8
Grade 7	9.8
Grade 8	10.9

In Appendix A at the end of the book, there are charts describing the tensile strengths and the torque recommendations for various sizes and grades of fasteners. These torque recommendations are *not* for gasketed joints or for soft materials.

NOTE: *The strongest bolts are not always the best choice. A high-grade fastener might be best for a front suspension bolt but may not be suitable for use in an engine. Using a higher grade of fastener allows a higher torque to be used. But if torque is not increased, there is no increase in the clamp load, whether a higher bolt strength is used or not. Also, higher strength bolts have less resistance to fatigue than softer bolts if they are undertightened.*

Automotive bolts are usually grade 5 or 6 (or 9.8 or 10.9 for metric use). Grade 6 bolts are usually used on

Reduced diameter

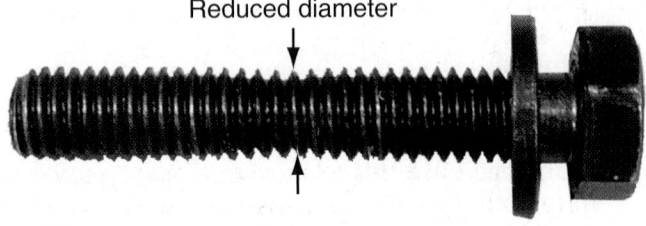

Figure 7.1 When a bolt is overstretched, the screw thread distorts. A nut will not turn easily on the bolt. *(Courtesy of Tim Gilles)*

engine main bearing caps. Grade 8 screws are used for flywheels and flexplates.

Thread Lubricants

An antiseize compound is used where a bolt might become difficult to remove after a period of time—for example, in an aluminum block or on exhaust manifold bolts (**Figure 7.3**).

Lubricants also introduce the possibility of a **hydrostatic lock**, where oil is trapped in a blind hole. When the bolt contacts the oil, it cannot compress it; therefore, the bolt cannot be properly tightened. A cracked part can result.

Nuts

The grade of nut used must match the grade of bolt that it is used with. Manufacturers use several different nut markings to denote grade identification. Some are marked on the top and others are marked on the sides (**Figure 7.4**). Grade 8 nuts are always marked, but grade 5 nuts sometimes are not.

Reuse of Nuts

A nut must be slightly softer than the bolt so that its threads can flow to fit the threads of the bolt (**Figure 7.5**). This thread distortion is permanent. Bolts other than torque-to-yield head bolts (see Chapter 51) are normally reused in the engine rebuilding process. Nuts, however, lose some effectiveness with repeated use. Reusing the nut does not allow as good a match as the first use. So there is more friction between the threads and an accurate torque reading cannot be achieved.

According to Bowman Distribution, a nut acts much like the coils of a spring when an assembly is tightened. Each thread in the nut will carry progressively less load:

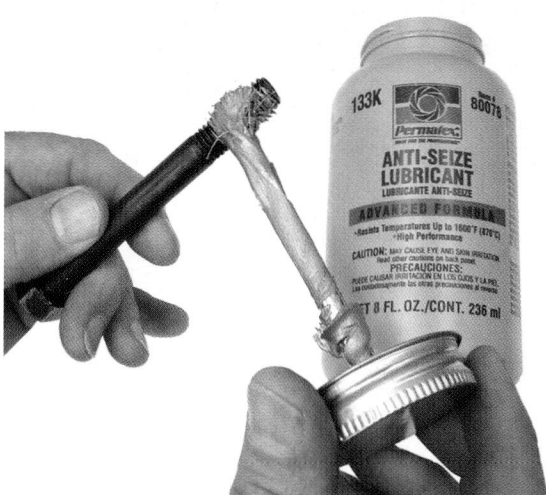

Figure 7.3 An antiseize compound is used where a bolt might be difficult to remove. (*Courtesy of Tim Gilles*)

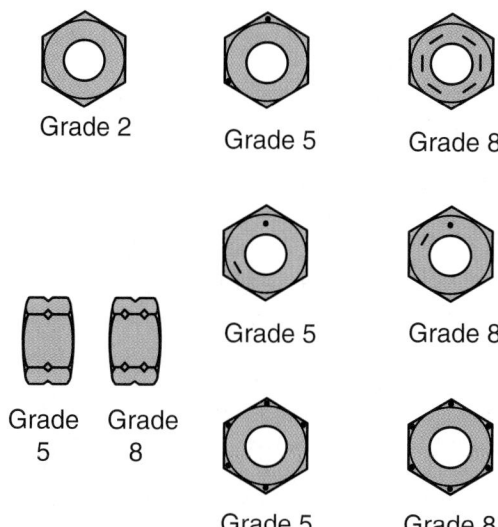

Grade 2   Grade 5   Grade 8

Grade 5   Grade 8

Grade 5   Grade 8

Grade 5   Grade 8

Figure 7.4 Nut grade markings.

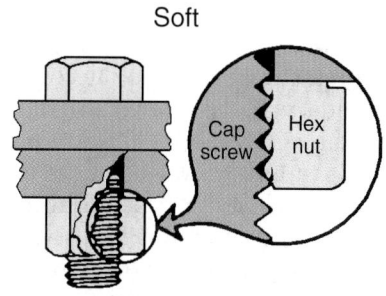

Soft

Cap screw   Hex nut

Times nut was reused	Percentage of tension lost	For maximum safety and performance, a nut should be discarded once it is removed.
1st	20%	
2nd	33%	
5th	47%	
9th	60%	

Figure 7.5 A nut conforms to the thread of a bolt, so it loses more holding ability each time it is reused. (*Courtesy of Lawson Products*)

- The first thread of a coarse fastener (the one against the work being clamped) will support about 38% of the total load on the bolt.
- The second supports about 25%.
- The third supports about 18%.

The first thread will be into the plastic range (yield), and the outer threads may not even touch the bolt at all.

Locknuts, like the type used on some rocker arm studs, are deformed at the top (**Figure 7.6**). These nuts lose some of their locking ability with repeated reuse. Do not remove and replace them any more than is necessary.

Another type of locknut is called a *castle nut*. It has slots at the top that align with a hole in the male screw thread. A cotter pin is inserted as a safety assurance that the nut cannot come loose. This kind of nut is commonly used to retain front wheel bearings and steering linkage parts.

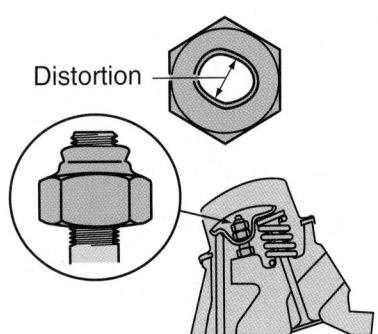

Figure 7.6 The locking nut used on rocker arm studs is distorted at the top.

A *wing nut* is a nut that is tightened by hand only. Only minor tightening torque is necessary with a wing nut. Examples of its use include holding an air cleaner housing together or fastening a battery holddown bracket.

Washers

Use of the proper washer is also necessary to achieve the correct load on a bolt. Use a **through-hardened**, heat-treated flat **washer**. These washers are hardened throughout. **Case-hardened washers** are hard on the surface only; the core of the washer is soft and will compress, allowing the connection to lose clamping force. Washers are also used to provide a hardened surface for a bolt to act against. This is especially important when the material to be clamped is not steel.

NOTE: *Every 0.001" of bolt relaxation will cause a loss of about 30,000 psi of clamping force.*

Removing a Stud

Studs have threads on both ends. They are often found on carburetor mounts and exhaust manifolds. To remove a stud, a special stud puller may be used (see Figure 8.43). When a puller is not available, use two nuts with a lock washer between them. Turn the nut that is farthest from the end of the stud to remove the stud.

▬ FASTENER FAILURES

Fatigue breaks account for about 75% of fastener problems. A bolt becomes fatigued from working back and forth when it is too loose. This problem occurs when the assembly is not properly torqued to stretch the bolt.

SHOP TIP Threaded holes in the block must be chased (cleaned) with a tap (**Figure 7.7a**) prior to reassembly, especially when the block has been hot tanked. The holes and the capscrew should receive a light coating of oil. Be sure also to clean the bolt thread (**Figure 7.7b**) and the underside of the bolt head (**Figure 7.7c**).

Figure 7.7 Preparing the threads. (a) Bolt holes in the block are chased with a tap to ensure correct clamping force after assembly. (b) Clean the bolt thread. (c) Clean *under* the bolt head. (*Courtesy of Tim Gilles*)

Shear or *torsion* breaks result from the use of a poor grade of fastener, too much friction, or an improper thread fit. Technicians sometimes fail to clean bolt threads in the block thoroughly.

Bolts can be broken when they are *bottomed out*. Normally, bolts should turn into the thread for four to six turns, or about 1½ times the diameter of the screw thread. If a bolt is bottomed out, the sudden stop of the bolt is felt, instead of the "elastic" feeling of normal bolt stretch. Just before a bolt breaks off, it will become slightly easier to turn. This is a clue to stop turning, loosen the bolt, and replace it with a new one.

Crossed threads are usually the result of improper assembly procedures.

> **SHOP TIP** When two parts are fastened together, do not completely tighten *any* of the fasteners until *all* of them have been started into their threads. With all the bolts loose, a hole that does not line up can be aligned using a rolling head prybar (**Figure 7.8**). This tool is also handy in aligning motor mounts and transmission cross-members.

■ DRILL BITS

Drill bits are used to make or enlarge holes in metal and to help remove broken fasteners. A drill bit has two *flutes*, which provide a channel for the cutting chips to escape into during drilling (**Figure 7.9**). They also allow cutting fluid to reach the cutting edge of the drill.

The web that separates the flutes of the drill increases in size from the point to the shank. The web can be so large at its point that it will be necessary to drill a *pilot hole* with a smaller drill bit (**Figure 7.10**). A centerpunch is used to locate the web (or point) of the drill. Only a small area of the *land* (the spiral area on the drill body) is equal to the diameter of the drill. The rest of the land is *relieved* (made smaller) to reduce friction.

The tip of a drill bit must be sharp. Its angle will be flatter for hard metals such as steel, and steeper for softer metals. A drill bit that is sharpened "off center" (**Figure 7.11**) will drill a hole that is too large. Be careful to make the angles equal. For iron and steel, the bit is cut to a 59-degree angle (**Figure 7.12a**). The backside of the cutting lip must slope off, as shown in **Figure 7.12b**, to produce the chisel edge on the drill. Too little clearance causes the drill to rub instead of cut. Too much clearance causes the bit to cut too rapidly and may break or chip.

Hand Sharpening a Drill Bit

Hand sharpening the drill bit requires some practice. Hold the drill against the wheel at a 59-degree angle, and rotate it in an arc movement (**Figure 7.13**). The angle and length of the lip can be checked with a gauge. A properly ground drill will give off even amounts of chips from both sides of the metal as it cuts (**Figure 7.14**).

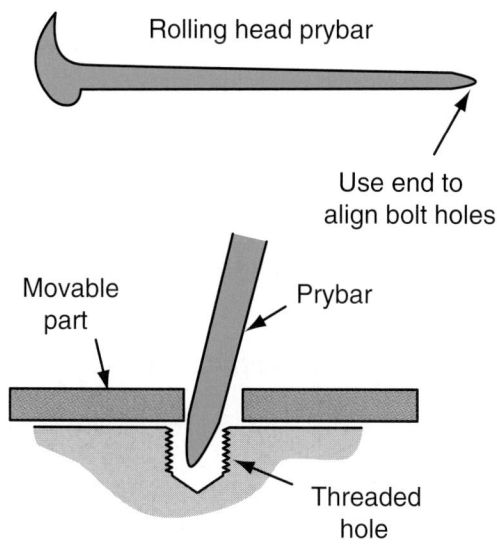

Figure 7.8 A rolling head prybar can be used to align parts.

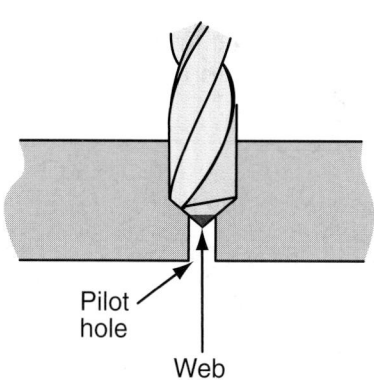

Figure 7.10 Before drilling with a large drill bit, drill a pilot hole.

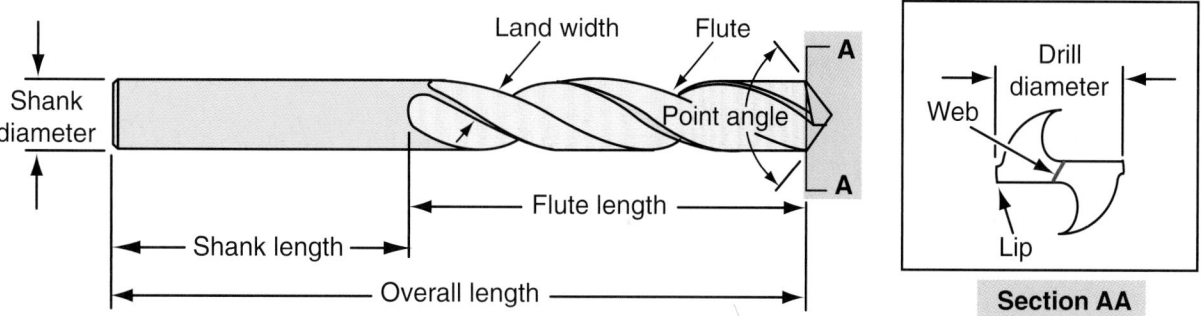

Figure 7.9 Parts of a drill bit.

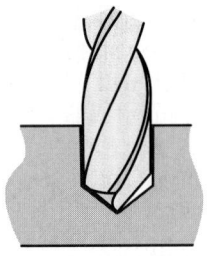 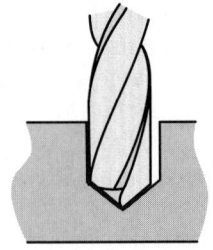

Lips of equal length but at unequal angles

Lips of unequal length but at equal angles

Figure 7.11 Results of incorrect drill sharpening.

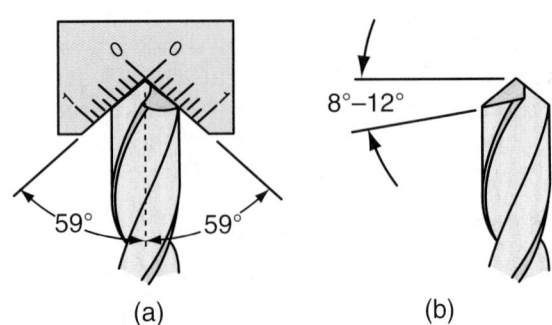

(a)   (b)

Figure 7.12 (a) Drill bits for iron or steel are ground to a 59-degree point angle. (b) The cutting lip must slope off as shown here.

SHOP TIP Be sure to quench the bit often to keep it cool. If it gets too hot, the temper can be drawn from it.

If much precision drilling is to be done, several manufacturers have drilling handbooks available.

Drill Speed and Lubricants

To prevent drill bit wear, keep as close as possible to the recommended drilling speed. Bits are usually made of *high-speed steel*; they will be marked "HS" or "HSS" on their shanks.

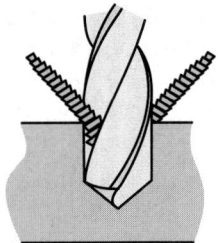

Figure 7.14 A properly ground drill bit will chip equally from both flutes.

SHOP TIP During grinding, different metals give off different color sparks:
- High-speed steel drill bits give off a dull red spark.
- Less expensive carbon steel drill bits give off white sparks.

Use carbon steel drill bits at half the cutting speed of high-speed steel because their resistance to heat is lower.

Approximate HS drill speeds when drilling mild steel are:
- 1,500 rpm for ¼"
- 1,000 rpm for ⅜"
- 750 rpm for ½"

Drill aluminum with the same size drill bit at an rpm twice as fast. Soft cast iron is generally drilled at the speed required for mild steel.
- When drilling steel, use a cutting oil.
- Cast iron and aluminum can be cut dry.

Drill Size

Drills are sized in a number of different ways. Most technicians own a *fractional set* of drills called a *drill index*. Sizes in the drill index increase by increments of 1⁄64". Less expensive drill indices are also available with drill sizes increasing in 1⁄32" or 1⁄16" increments. Decimal equivalents are also printed on the index. The size of a drill is printed on the shank of the drill

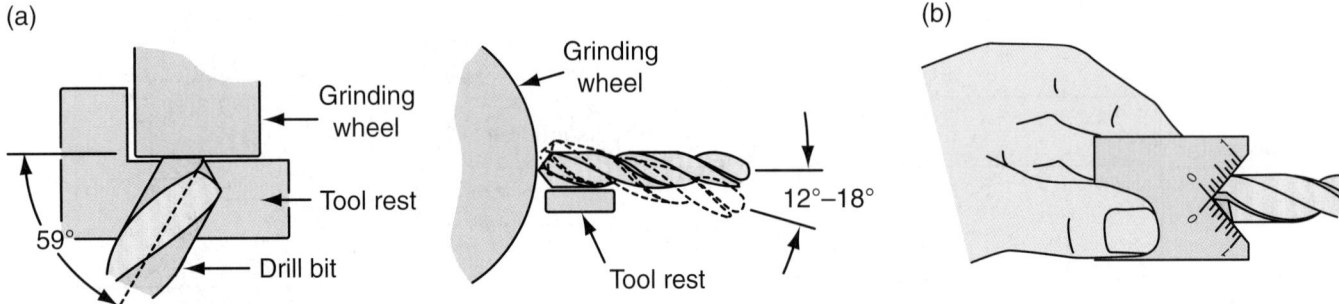

Figure 7.13 Sharpening a drill bit by hand. (a) Hold the drill on the workrest at about 59 degrees to the wheel, and rotate the front of the drill upward. (b) Check the angle on a drill gauge.

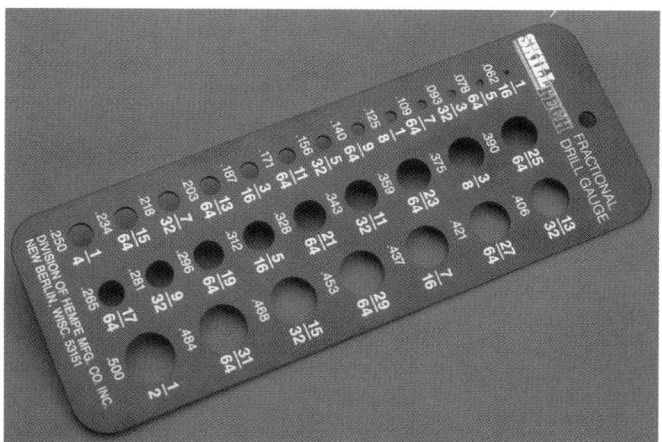

Figure 7.15 A drill gauge can be used to determine the size of a drill bit.

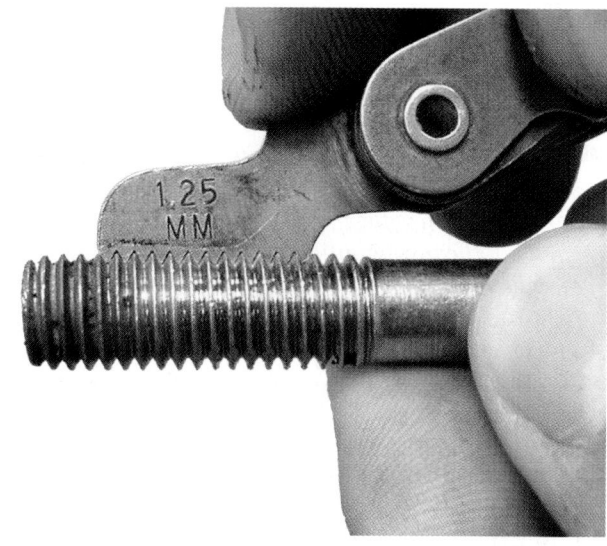

Figure 7.16 This screw has 1.25 threads per millimeter. (*Courtesy of Tim Gilles*)

bit. It is not unusual for a drill to slip in the chuck, erasing the size marking. The holes in the drill index correspond to the correct size of the drill. A micrometer or a special drill gauge with holes of the specified size can also be used to determine the size of a drill bit (**Figure 7.15**).

Drills also come in *letter* and *number* sizes. Letter-size drills increase in size from A (0.234") to Z (0.413"). They are larger than the largest number-sized drill.

Drills sized by number commonly range from 80 (0.0135") to 1, which is 0.228" (almost ¼"). The most widely used number-sized drills usually are numbers 1 to 60 (0.040"). Commonly used metric drills range from 0.20 mm to 16.00 mm in size.

■ TAPS AND THREADS

A tap is used to cut internal threads in a previously drilled hole or to *chase* (clean) existing threads. Standard and metric both have coarse and fine threads for each diameter of fastener. To determine the screw thread of a hole, screw in the proper tap and then read the size on the tap. A special *thread-pitch gauge* can be used, but this method is not always dependable; some screws have different diameters but the same number of threads per inch (**Figure 7.16**).

Types of Taps

There are three types of taps available for any given thread size. Two of the most common types are shown in **Figure 7.17**.
- The *tapered* tap helps pilot the tap into the hole.
- The *plug* tap is a standard tap.
- A *bottom* tap makes threads all the way to the bottom of a blind hole, and it chases threads to clean them.

A blind hole is one that does not go all the way through a part (**Figure 7.18**).

Taper tap

Plug tap

Bottoming tap

Figure 7.17 Three types of taps are available for each screw thread. The plug tap is found in most technicians tap sets. (*Courtesy of Tim Gilles*)

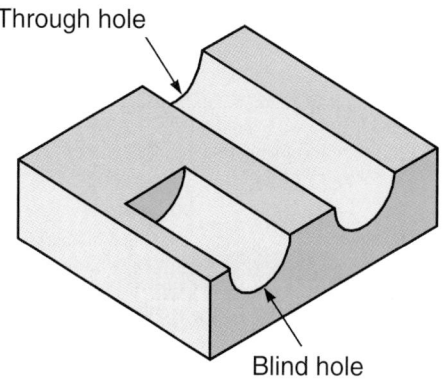

Through hole

Blind hole

Figure 7.18 A "blind" hole does not go all of the way through the part.

CASE HISTORY *An apprentice was chasing the threads in a cylinder block after removing it from the cleaning tank. He tapped out all of the holes, including those for the oil gallery plugs. Oil gallery plugs have tapered pipe threads. The apprentice turned the tapered pipe tap through the entire length of the thread until it turned easily. When the new oil gallery plugs were installed, they could not become tight because the threads in the block were now too big. The block had to be repaired.*

Pipe Threads

Pipe threads are used for heater outlets in the block and intake manifold, as well as for oil and coolant temperature sending units (**Figure 7.19**). Pipe taps have tapered threads (**Figure 7.20**) designed to wedge against each other. Therefore, when tapping pipe thread, do not continue turning the tap until it turns easily. Because the size of a pipe thread is determined by the inside diameter of

Figure 7.19 Pipe threads are found on threaded plugs, heater outlets, and oil and coolant temperature sending units. *(Courtesy of Tim Gilles)*

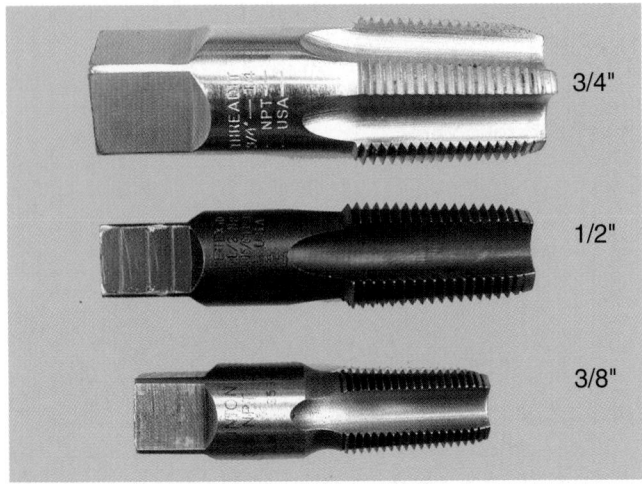

Figure 7.20 Pipe taps for NPT threads. *(Courtesy of Tim Gilles)*

Actual pipe thread sizes

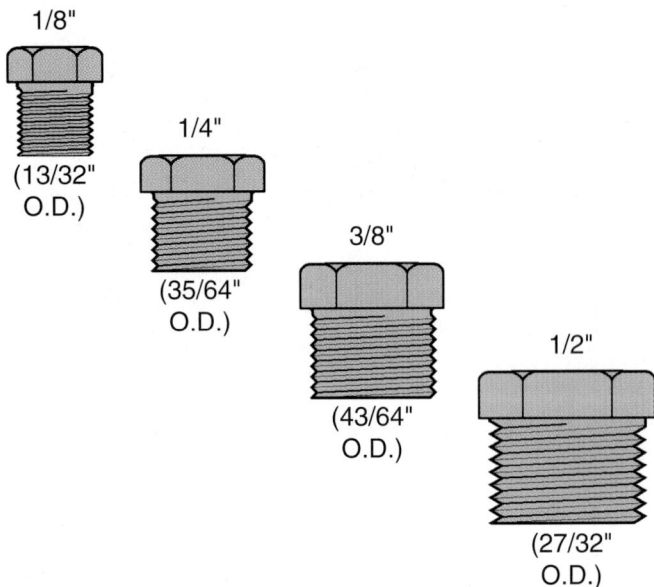

Figure 7.21 Pipe threads are sized according to the inside diameter (I.D.) of the pipe. The outside diameter (O.D.) can be considerably larger in diameter.

the piece of pipe, a ½" *National Pipe Taper* (*NPT*) tap will be quite a bit larger than ½" (**Figure 7.21**).

SHOP TIP There are special inside pipe wrenches for removing pipe nipples from cylinder heads and manifolds (**Figure 7.22**).

Determining Tap Drill Size

Before tapping a thread, a hole of the correct size must be drilled. A **tap drill** usually provides about 75% of a full thread. This allows some margin for error and keeps the tap from binding during the tapping operation. The tap drill chart (**Figure 7.23**) shows which drill to use.

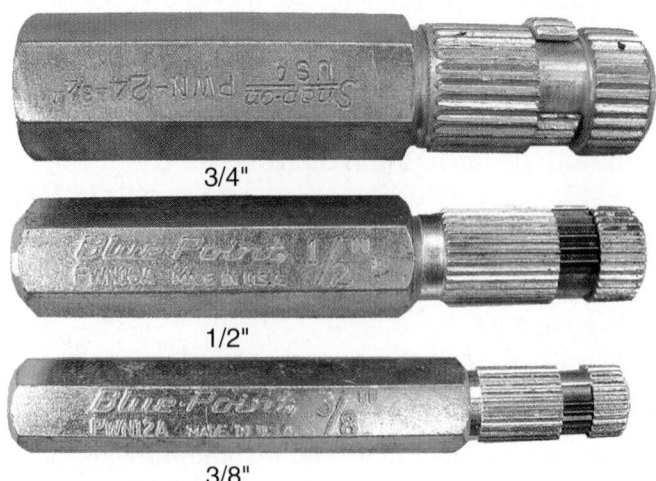

Figure 7.22 Inside pipe wrenches grip the I.D. of the pipe fitting. *(Courtesy of Tim Gilles)*

Size	Threads per inch			Outside Diameter Inches	Tap Drill Approx. 75% Full Thread	Decimal Equivalent of Tap Drill
	NC	NF	NS			
0	...	80	...	.0600	3/64	.0469
1	...	...	56	.0730	54	.0550
1	64	...	...	.0730	53	.0595
1	...	72	...	.0730	53	.0595
2	56	...	...	.0860	50	.0700
2	...	64	...	.0860	50	.0700
3	48	...	...	.0990	47	.0785
3	...	56	...	.0990	45	.0820
4	...	...	32	.1120	45	.0820
4	...	...	36	.1120	44	.0860
4	40	...	...	.1120	43	.0890
4	...	48	...	.1120	42	.0935
5	...	...	36	.1250	40	.0980
5	40	...	...	.1250	38	.1015
5	...	44	...	.1250	37	.1040
6	32	...	...	.1380	36	.1065
6	...	...	36	.1380	34	.1110
6	...	40	...	.1380	33	.1130
8	...	...	30	.1640	30	.1285
8	32	...	...	.1640	29	.1360
8	...	36	...	.1640	29	.1360
8	...	...	40	.1640	28	.1405
10	24	...	...	.1900	25	.1495
10	...	...	28	.1900	23	.1540
10	...	...	30	.1900	22	.1570
10	...	32	...	.1900	21	.1590
12	24	...	...	.2160	16	.1770
12	...	28	...	.2160	14	.1820
12	...	...	32	.2160	13	.1850
1/4	20	...	...	.2500	7	.2010
1/4	...	28	...	.2500	3	.2130
5/16	18	...	...	.3125	F	.2570
5/16	...	24	...	.3125	I	.2720
3/8	16	...	...	.3750	5/16	.3125
3/8	...	24	...	.3750	Q	.3320
7/16	14	...	...	.4375	U	.3680
7/16	...	20	...	.4375	25/64	.3906
1/2	13	...	...	.5000	27/64	.4219
1/2	...	20	...	.5000	29/64	.4531
9/16	12	...	...	.5625	31/64	.4844
9/16	...	18	...	.5625	33/64	.5156
5/8	11	...	...	.6250	17/32	.5312
5/8	...	18	...	.6250	37/64	.5781
3/4	10	...	...	.7500	21/32	.6562
3/4	...	16	...	.7500	11/16	.6875
7/8	9	...	...	.8750	49/64	.7656

Figure 7.23 A tap drill chart for inch-sized drills. A size conversion chart is located in the appendix. *(Courtesy of L.S. Starrett)*

A ⅜"-by-24 (24 threads per inch) tap would require a Q (0.332") drill. From a fractional-size drill index, select the next largest size available. In this case, the right choice is an ¹¹/₃₂" (0.343"), which is only 0.011" larger than a letter Q drill (see Appendix B for size conversion). If a tap drill chart is not available, select the largest drill bit that will fit into a threaded hole the same size as the desired thread (**Figure 7.24**).

NOTE: *Do not use a smaller drill; the tap could bind and break. A female screw thread that is only 50% of full depth has been shown to have so much strength that the bolt would strip before the thread would strip.*

Tapping a Hole

When tapping a hole, advance the tap in a clockwise direction. Then, back off about ¼ turn to break off any metal chips that might accumulate. These chips could ruin the new thread. The chips can be felt breaking as the tap is backed off. The broken chips gather in the flutes of the tap (the lower area next to the cutting teeth) (**Figure 7.25**). Thread lubricants are required when tapping steel and most nonferrous metals; cast iron can be tapped dry.

Removing a Broken Tap

When a tap is broken, try to drive it counterclockwise with a centerpunch or a special tap extractor to remove it (**Figure 7.26**). If the broken tap will not come-out of the hole, try to break it up with a centerpunch. If this does not work, a process called electrical discharge machining (**EDM**) can be performed by a machine shop that has the necessary machinery. EDM erodes the fastener, leaving the thread intact.

■ DIES

Dies are used to make external threads on a round rod. When making a thread, the die is advanced and then turned backward like when using a tap. The die can

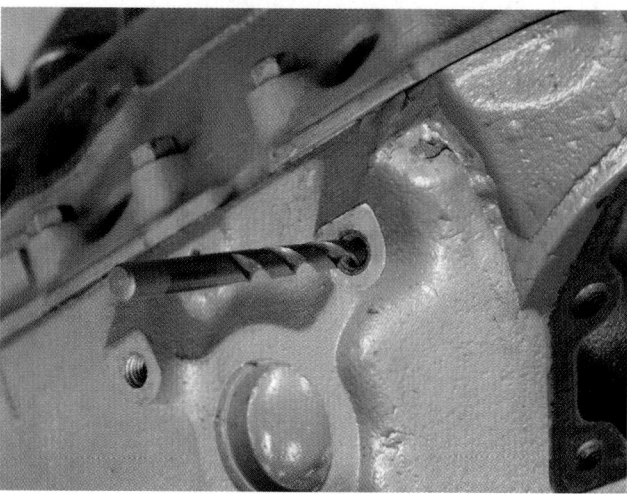

Figure 7.24 The largest drill that fits into a threaded hole of the same size can be used as a tap drill. *(Courtesy of Tim Gilles)*

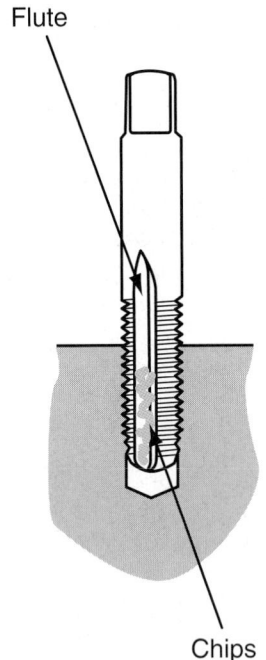

Figure 7.25 Chips are gathered in the flutes of the tap.

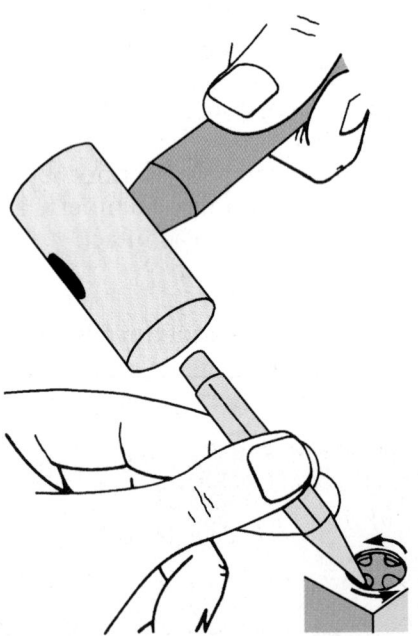

Figure 7.26 Try to remove a broken tap with a center-punch or a chisel.

be broken if it is forced. A die is especially useful for chasing *burred* threads. When shortening a bolt, first screw a die onto the bolt threads. After shortening and chamfering the end of the thread, removing the die will clean up the end of the thread. **Figure 7.27** shows a die being used to determine the size and pitch of a screw thread.

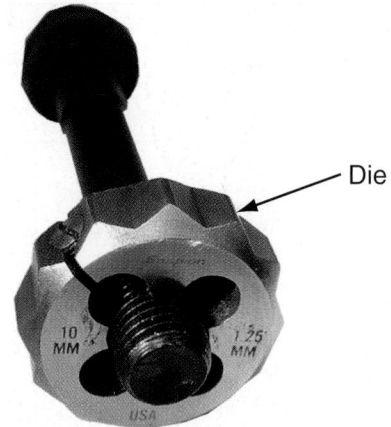

Figure 7.27 A die can be used to determine the size and pitch of a screw thread. *(Courtesy of Tim Gilles)*

■ REPAIRING BROKEN FASTENERS

Every technician has to remove a broken bolt occasionally. If a broken bolt has not bottomed out in the hole, it can sometimes be removed by driving it in a counterclockwise direction with a chisel, just as in removing a broken tap. A bolt that has broken off above the surface may be removable with pliers or a stud extractor (**Figure 7.28a**). Some types of stud pullers have interchangeable inserts for accommodating studs of different diameters. The center post is removed when exchanging inserts (**Figure 7.28b**).

> **SHOP TIP** Heat, applied with a torch, will often make removal of a broken stud or bolt easier. Applying wax (paraffin) to quench after heating lubricates the screw thread, making removal easier.

A left-hand drill bit can be used with a reversible drill motor to remove a broken bolt that is not bottomed out.

Screw Extractors

There are two common types of screw extractors. **Figure 7.29a** shows an extractor that is commonly known as an *easy out*. The other type of an extractor has flutes (**Figure 7.29b**). Both are used in the same manner.

The side of an easy out is stamped with the correct size of drill bit to use with it. A hole is drilled, and the easy out is pounded into it (**Figure 7.30**). Turning the easy out counterclockwise will remove the broken part (if the remaining screw thread is not bound up).

Extractors are hardened, which makes them difficult to remove when they break. A broken extractor can sometimes be removed by heating it with a torch and letting it cool slowly. This softens the metal for drilling. If an extractor cannot be removed by this

(a)

(b)

Figure 7.28 (a) A screw broken off above the work surface can sometimes be removed with vise grips or a stud puller. (b) Remove the center post on a stud puller to change inserts so the puller will fit the stud. (*Courtesy of Tim Gilles*)

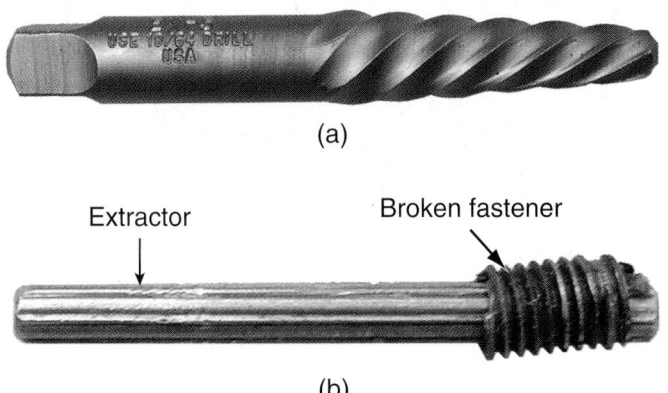

(a)

Extractor | Broken fastener

(b)

Figure 7.29 Two types of screw extractors. (a) An easy out. (b) A fluted extractor. (*Courtesy of Tim Gilles*)

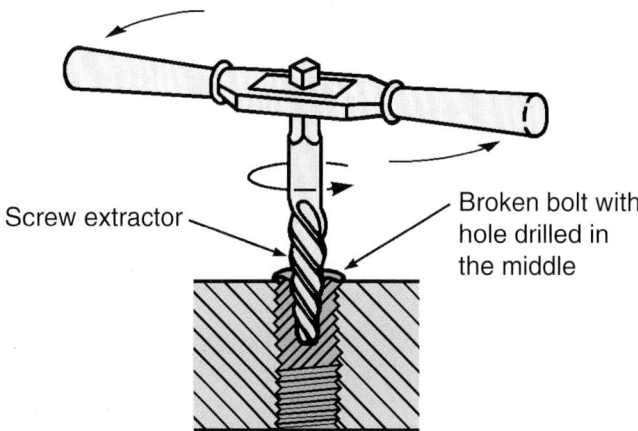

Screw extractor — Broken bolt with hole drilled in the middle

Figure 7.30 Using a screw extractor to remove a broken screw.

method, a machine shop can erode the screw with the EDM process discussed earlier.

A fluted extractor set includes several sizes of fluted extractors, the correct-sized drill bits, splined hex nuts, and drill guides (**Figure 7.31**). To remove a broken fastener with a fluted extractor set:

■ First, make a centerpunch mark *precisely* in the center of the fastener (**Figure 7.32a**).

■ Then, drill a hole of a precise size into the broken fastener (**Figure 7.32b**). The drilled hole should go all the way through the bottom of the fastener. The first hole should be drilled with a small drill bit. Then, drill with the largest drill bit in the set that is smaller than the fastener thread being removed.

NOTE: *If the fastener is broken off below the surface of the workpiece, a drill guide in the extractor set can be used to align the hole with the center of the broken screw (**Figure 7.33**).*

■ Drive the extractor into the hole with a brass hammer (**Figure 7.34a**).

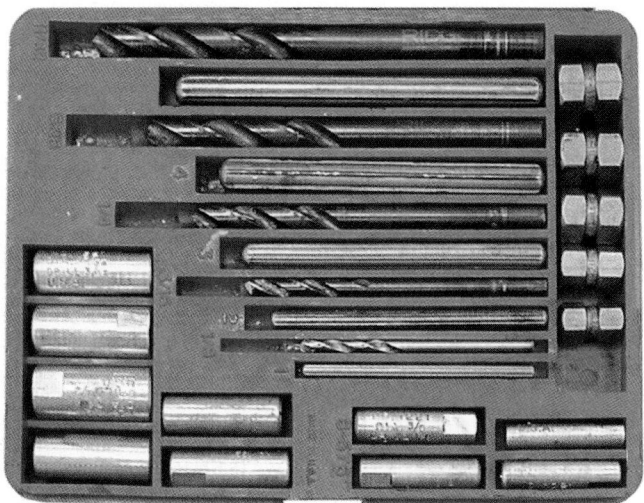

Figure 7.31 A fluted screw extractor set. (*Courtesy of Tim Gilles*)

Figure 7.32 Removing a broken stud or bolt. (a) The stud is punched *exactly* on center. (b) A hole is drilled precisely on center. *(Courtesy of Tim Gilles)*

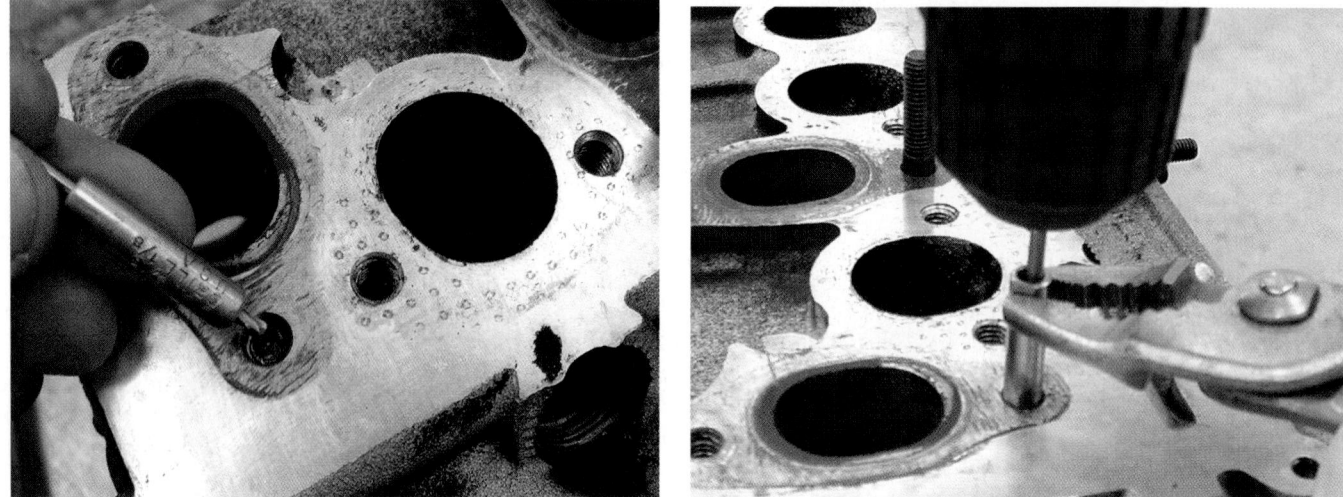

Figure 7.33 A drill guide centers the hole when a fastener is broken below the surface. *(Courtesy of Tim Gilles)*

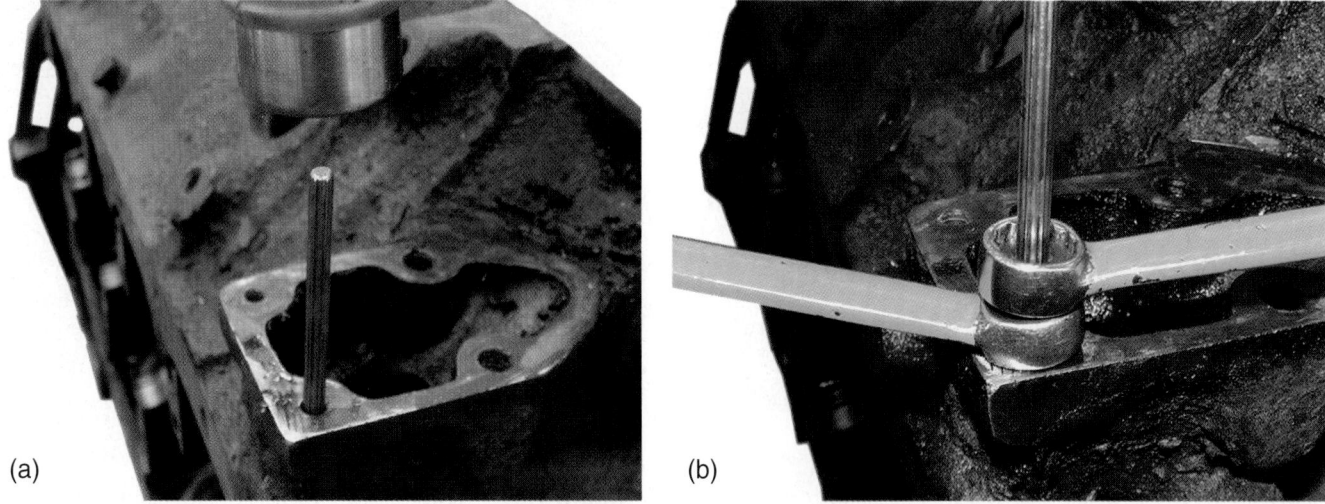

(a)   (b)

Figure 7.34 (a) The extractor is pounded into the hole. (b) The nut is slipped onto the extractor, and two wrenches are used to turn the nut counterclockwise. *(Courtesy of Tim Gilles)*

■ Then, unscrew the broken fastener (**Figure 7.34b**). Use as large an extractor as possible, so that the walls of the drilled stud will become as thin and flexible as possible. The extractor set includes a nut that fits over the extractor. Slide the nut down the flutes. Then grip the nut with a pair of wrenches (to exert equal force) and turn the extractor counterclockwise.

If the extractor begins to shear off, it should be removed from the drilled hole before it breaks off. The hole can then be heated with a torch (**Figure 7.35**), and the broken fastener cooled by dripping water onto it or applying wax (paraffin) to it. Then, try the extractor again.

Some machine shops will remove a broken fastener by placing a nut on top of it. Then, the inside of the nut is welded to the fastener with nickel welding rod (ni-rod). A wrench is then used on the nut to unscrew the fastener.

Drilling and Retapping

Sometimes a broken fastener can be repaired by drilling it out. The hole is then carefully tapped. If this fails, the next step would be to drill and then tap to the next larger thread size. When using an oversized screw thread, it is often necessary to drill an oversized hole in the part that the bolt fastens to (an exhaust manifold, for example).

Thread Inserts

When a screw thread is stripped, a better repair is to use a *thread insert* to preserve the original thread size. There are several varieties of thread inserts. They all require drilling and tapping a larger hole. The insert is then threaded into place.

A heli-coil is an oversized spring coil of stainless steel (**Figure 7.36**). The *heli-coil* method is shown in **Figure 7.37**.

■ First, drill an oversized hole.
■ Tap the hole.

Figure 7.35 If the fastener will not come out, heat the drilled hole with a torch. *(Courtesy of Tim Gilles)*

Figure 7.36 A Heli-Coil® thread insert. *(Courtesy of Emhart Fastening Teknologies)*

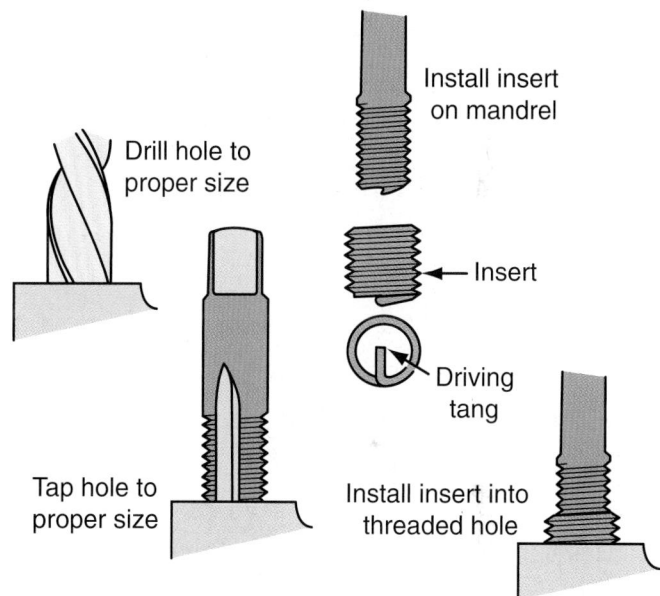

Figure 7.37 The heli-coil® repair method.

■ Apply Loctite® or a similar adhesive to the insert.
■ Install it in the hole with a special tool. There is a drive tang at the bottom of the heli-coil that locks to the tool during installation. It is notched for easy removal.
■ Break the driving tang off the bottom of the heli-coil with a hammer and punch.

Figure 7.38 shows the appearance of a repaired thread.

Locking Inserts

Solid-Threaded Inserts. Solid-threaded inserts are used in applications that require high strength, such as repairing stripped head bolt holes or main bolt holes in a block. The insert shown in **Figure 7.39** has bottom external threads that are cold rolled. The installation tool locks the insert into place by forcing the bottom threads to expand the mating external threads into the threads cut into the block.

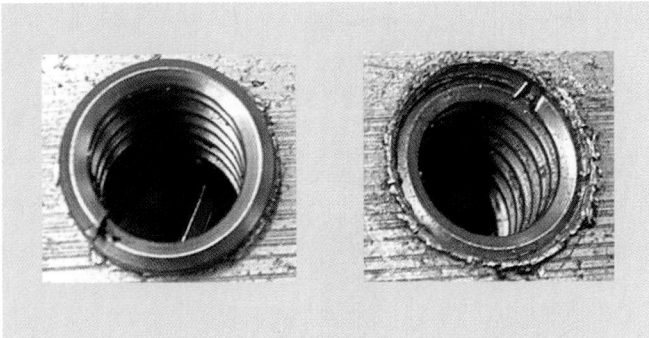

Figure 7.38 Repaired threads. The one on the left is not completely installed. *(Courtesy of Tim Gilles)*

Figure 7.40 A Keensert®. *(Courtesy of Tim Gilles)*

Figure 7.39 A solid threaded insert. *(Courtesy of Tim Gilles)*

Heli-coil also makes a screw lock insert for use with permanent studs. The center coil has a series of flats that hold the insert in place. One type of insert has a locking feature that is activated by pounding it into place. This method—the *Slimsert* method—requires a special, stepped drill:

- Tap the hole.
- Screw the insert into place; the top of the Slimsert should be approximately 0.015" below the work surface.
- Strike the top of the driver with a hammer until the nylon washer touches the work surface. This locks the slimsert in place.

Another type of locking insert, called the *Keensert*, has locking keys that are pounded into place (**Figure 7.40**).

Spark Plug Inserts

There are inserts available for repairing spark plug threaded holes in cylinder heads. Installation is the same as other inserts, but one caution must be observed. Some spark plugs have tapered seats, whereas others are flat and require a gasket. Be sure to install the correct insert that corresponds to the rest of the spark plugs in

the engine. Installing the insert from the combustion chamber side often works well for gasketed plugs.

Pop Rivets

Pop rivets are sometimes used to fasten sheet metal parts to other engine parts (**Figure 7.41**). A nail is pulled through inside of an aluminum rivet to expand it and fasten it securely to the hole. Common uses include attaching ID number plates, baffles, or heat deflectors. Pop rivets are inexpensive and easy to use. A pop rivet installation tool is shown in **Figure 7.42a**.

- Insert the nail end of the pop rivet into the installation tool (**Figure 7.42b**).
- Position the pop rivet in the hole (**Figure 7.42c**).
- An important step in the installation process is to push the installation tool firmly against the rivet head (**Figure 7.42d**).
- Squeeze the handle on the installation tool to pull the nail through the rivet and tighten it to the hole (**Figure 7.42e**).
- Increase the grip on the tool until the nail breaks off, leaving a finished pop rivet installation (**Figure 7.42f**).

Retaining Rings

Retaining rings, often called snaprings or lock rings, are used in some applications to prevent a pin, such as a piston pin, from sliding out of position. A retaining ring can be either internal or external, depending on whether it fits into a groove on the outside of a shaft or within a bore (**Figure 7.43**).

There are inexpensive wire lock rings as well as higher-quality retaining rings. Snaprings, sometimes called truarc® lock rings, have one square flat side and

Figure 7.41 A pop rivet. *(Courtesy of Tim Gilles)*

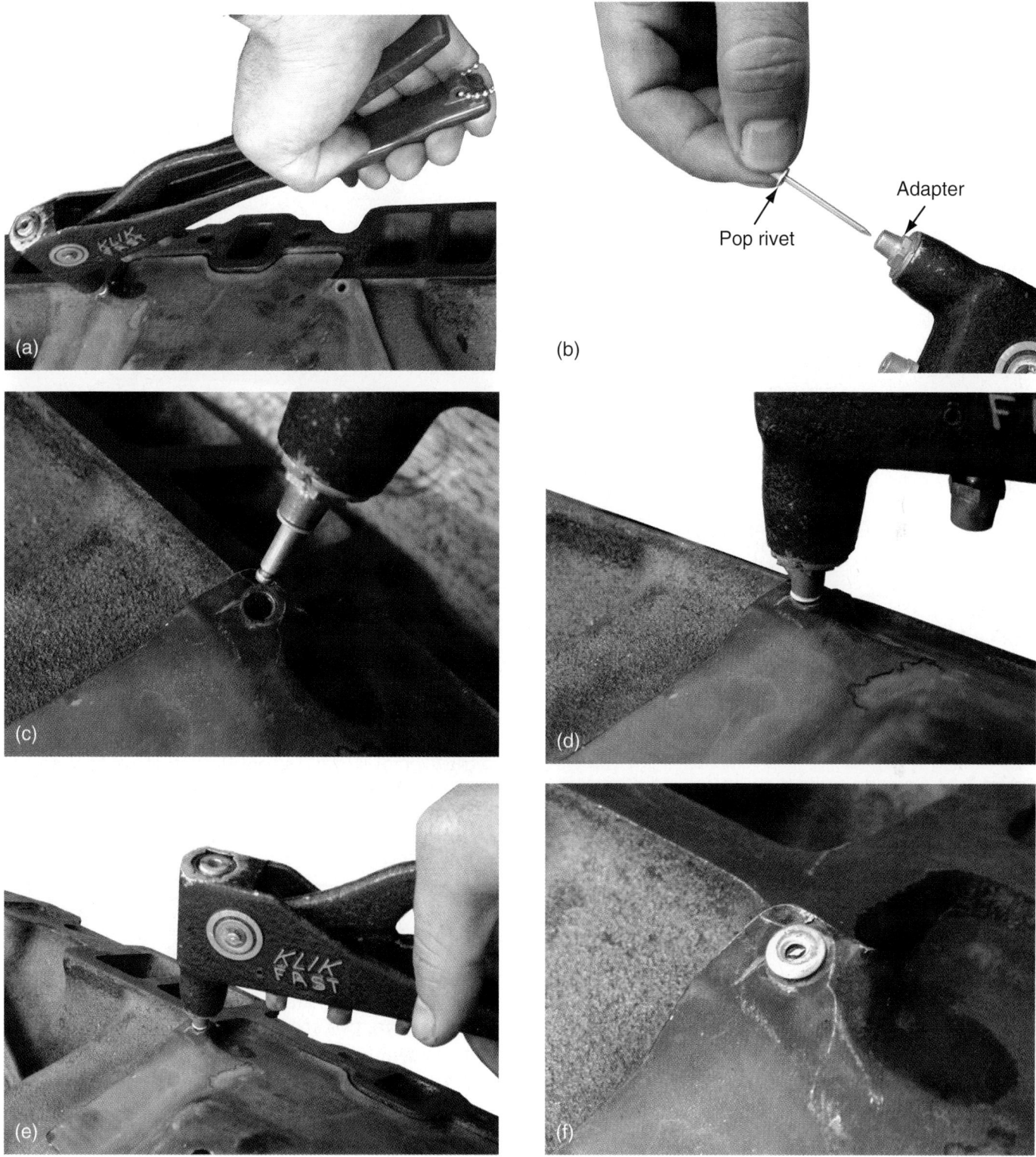

Figure 7.42 (a) A pop rivet installation tool. (b) Thread the correct size adapter into the tool and insert the pop rivet into the tool. (c) Position the head of the pop rivet in the hole. (d) Push down firmly on the pop rivet flange. (e) Squeeze the handle on the rivet tool to expand the rivet until the nail breaks off. (f) A finished pop rivet installation. *(Courtesy of Tim Gilles)*

one side that is slightly radiused (**Figure 7.44**). Install these retainers with their flat, sharp sides out.

Another type of retaining ring is the spiral type, sometimes called a spiral lock ring (**Figure 7.45a**), made of a coil of flat wire. It is wound into the groove while pressing against the ring until the entire ring is in the groove (**Figure 7.45b**). It can be removed with a screwdriver, starting at the gap.

Figure 7.43 An external and an internal snapring. *(Courtesy of Tim Gilles)*

(a)

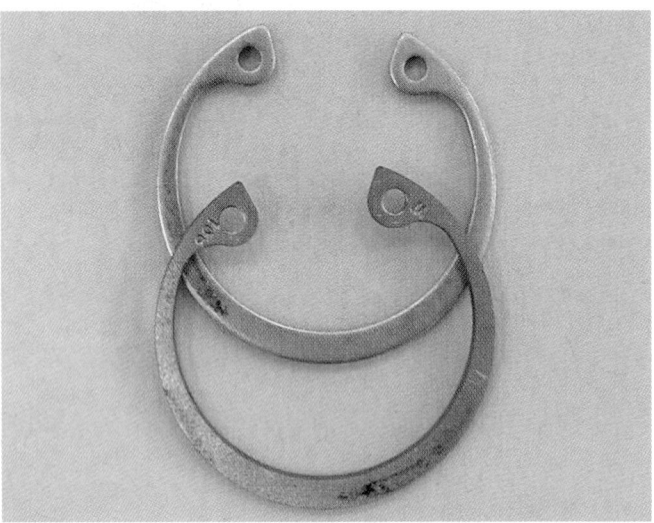

Figure 7.44 One side of the snapring is round. *(Courtesy of Tim Gilles)*

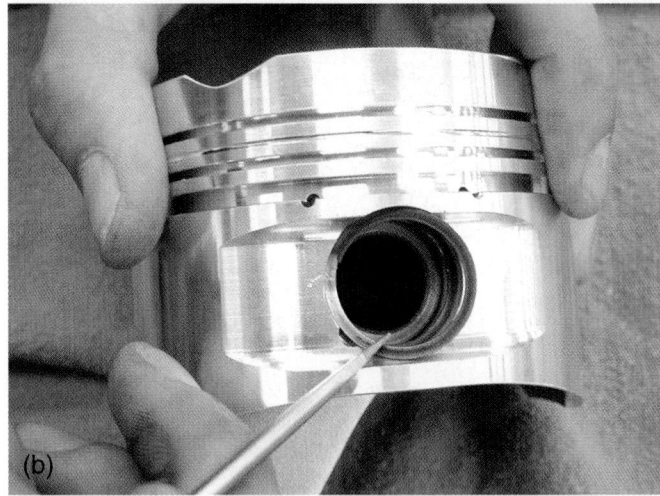

(b)

Figure 7.45 (a) A spiral retaining ring. The tab is provided to allow removal with a screwdriver or dental pick. (b) Installing a spiral piston retaining ring. *(Courtesy of Tim Gilles)*

■■ REVIEW QUESTIONS

1. Name four types of taps.
2. Refer to a tap drill size chart. What is the correct tap drill for a ¼ × 20 screw thread?
3. Why is a tap turned backward after turning forward when cutting a thread?
4. To what percent of its elastic limit is a bolt usually torqued?
5. If a nut turns easily for a few threads on a bolt and then begins to turn hard, what could the problem be?
6. What cutting lubricant, if any, is used when cutting cast iron?
7. What is one of the trade names for a replacement thread insert?
8. One way drills are classified is by fractions. What are two other methods?
9. Name one place where pipe threads are found in an engine.
10. What is the name of the machine shop process that erodes a broken fastener without damaging the screw thread?

▰▰ ASE-STYLE REVIEW QUESTIONS

1. Which of the following are methods by which twist drill sizes are classified?

   a. Number   **c.** Fraction

   b. Letter   **d.** All of the above

2. Technician A says that the "20" in ¼ × 20 signifies the total number of threads the fastener has. Technician B says that the "¼" signifies the socket size of the fastener head. Who is right?

   a. Technician A   **c.** Both A and B

   b. Technician B   **d.** Neither A nor B

3. Technician A says that a hole can be drilled in cast iron using no lubricant. Technician B says that a hole can be drilled in aluminum using no lubricant. Who is right?

   a. Technician A   **c.** Both A and B

   b. Technician B   **d.** Neither A nor B

4. How many radial lines does an SAE grade 8 bolt have on its head?

   a. 2

   b. 4

   c. 6

   d. 8

5. Technician A says that a nonhardened nut loses clamping ability each time it is retorqued. Technician B says that a capscrew loses clamping ability each time it is retorqued. Who is right?

   a. Technician A   **c.** Both A and B

   b. Technician B   **d.** Neither A nor B

CHAPTER 8

Shop Tools

■ OBJECTIVES

Upon completion of this chapter, you should be able to:

✔ Identify hand tools, air tools, and pullers used in an automotive repair shop.

✔ Understand the proper uses of hand tools, air tools, and pullers.

✔ Use shop tools safely.

■ KEY TERMS

hand tools	filings
flare nuts	collets

■ INTRODUCTION

An automotive service technician uses many tools and pieces of equipment. This chapter deals with the various types of **hand tools** and their proper uses and advantages. A technician who works on commission (flat rate) will purchase many of the smaller, personal tools to make a job go easier or faster.

■ TOOLS OF THE TRADE

Tools of the trade are classified either as hand tools, *portable power tools,* or *equipment.* Major pieces of equipment that are shared among all of the business's employees are typically owned by the employer. Hand tools and portable power tools that would be likely owned by an employee are covered in this section.

■ HAND TOOLS

The term *hand tools* refers to such tools as sockets, wrenches, and screwdrivers. Most automotive technicians own their own set of hand tools. A toolbox is a source of pride. A true professional keeps tools clean and well organized. It is not unusual to have several thousand dollars invested in tools (**Figure 8.1**). Loaning of tools between employees in a shop is generally not encouraged. If a technician finds the need to borrow a tool more than once, serious consideration should be given to the purchase of that tool. The master technician will often allow an apprentice to use his or her tools, but expects that the apprentice will start assembling his or her own set of tools.

One of the questions a prospective employer often asks during a job interview is, "Do you own a set of tools?" In fact, this is sometimes a condition of

Figure 8.1 A professional toolbox. *(Courtesy of Snap-on Tools Company, www.snapon.com)*

employment. The employer might want to inspect the tools to see whether the applicant has taken professional care of the tool set. A well-maintained toolbox indicates "pride in workmanship."

Neat and clean work habits show a professional attitude. Hand tools should be wiped clean of oil and grease following use. Experienced technicians usually stay considerably cleaner than beginners. Much of this cleanliness can be attributed to the use of shop towels in keeping tools clean during a job. Greasy tools tend to move grease around the shop (and into a customer's car).

Wrenches

Most screw or bolt heads are hexagonal (having six sides). The wrench size used on *hex heads* will almost always be either standard (British Imperial/U.S. Customary System) or metric. These measuring systems are covered in Chapter 6.

Wrenches are typically organized in racks so they can be easily located (**Figure 8.2**). There are several types of wrenches (**Figure 8.3**). The favorite among professional technicians is the *combination wrench*. Both ends of a combination wrench are the same size. One end is a *box wrench* and the other end is an *open-end wrench.*

The box end of the wrench is used for loosening or tightening a fastener, while the open end is used for turning it. The box end of a combination wrench is angled to provide clearance for hands (**Figure 8.4**). The open end of the wrench is angled 15 degrees so that it can be flipped over in tight quarters to provide access to the flats on the head of a nut or bolt (**Figure 8.5**).

Box wrenches are either *6-point* or *12-point* (**Figure 8.6**). The best choice for a box wrench is a 12-point because it is more versatile, fitting on the head of a bolt in more positions. Moving the wrench to a different position requires only 30 degrees of movement of the wrench head. A 6-point box-end is more difficult to

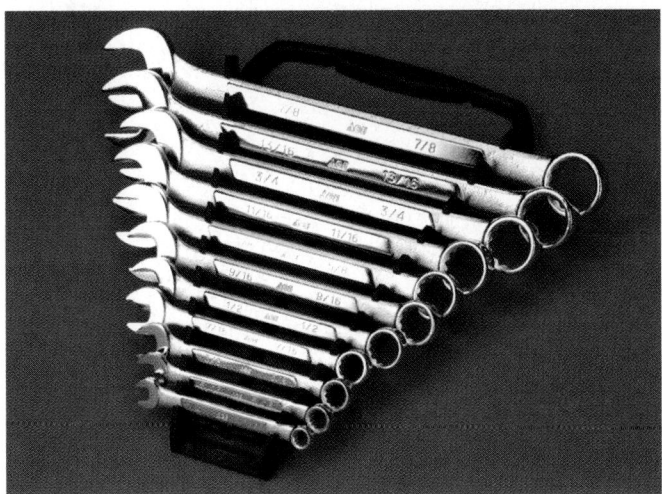

Figure 8.2 Wrenches are organized so they can be easily located.

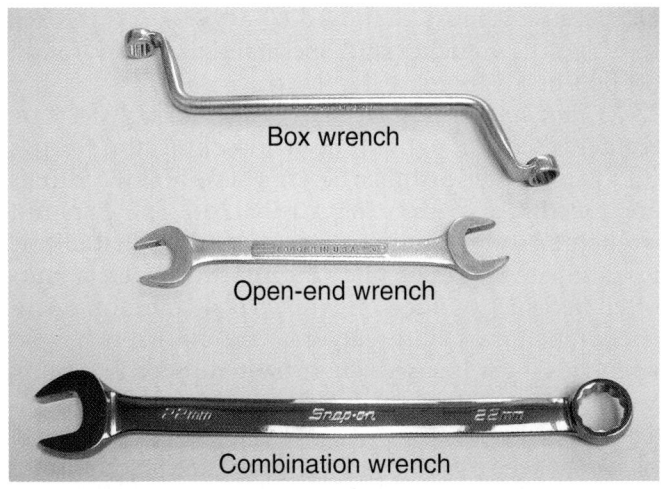

Figure 8.3 Common wrenches for bolts, screws, and nuts. *(Courtesy of Tim Gilles)*

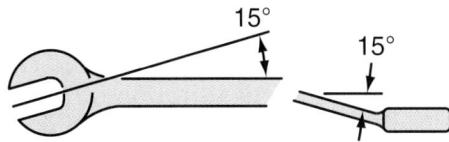

Figure 8.4 On a combination wrench, the open end is angled to the side and the box end is angled up.

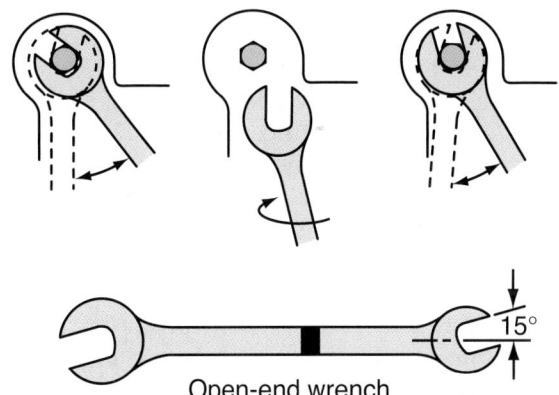

Figure 8.5 How to use an open-end wrench.

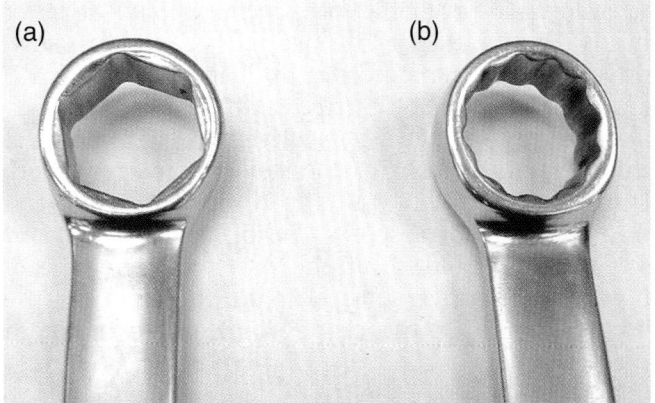

Figure 8.6 Box wrenches are either (a) 6-point or (b) 12-point. *(Courtesy of Tim Gilles)*

use because it must be moved 60 degrees. A drawback to 12-point wrenches and sockets is that they round off bolt heads more easily.

A *flare-nut wrench,* or *line wrench,* is a special box wrench used for tightening or loosening the fittings on "flared" fuel or brake lines. These hollow fittings are called **flare nuts** (**Figure 8.7a**). The flare-nut wrench has an opening on one of its sides that allows it to slip over a fuel line (**Figure 8.7b**). Flare nuts often become rounded off whenever a common open-end wrench is used instead of a flare-nut wrench. Two wrenches must be used when tightening or loosening a flare fitting so that the fuel line or a part is not damaged (**Figure 8.7c**).

A *ratcheting box wrench* (**Figure 8.8**) is sometimes used for loosening or tightening special applications such as muffler clamps or tie-rod end clamps on the steering linkage. This tool is especially handy when a nut is used on a very long bolt, where a deep socket would not be deep enough. The highest quality ratcheting box wrenches are very compact and can fit into tight spots. To change the direction of rotation, turn the wrench over and use it the opposite way.

An *adjustable wrench* (**Figure 8.9**), commonly called a *crescent wrench* after the original inventor of the tool, has one movable jaw. Different sizes are available. The medium-sized wrench is often used to adjust wheel bearings and perform other jobs on parts where substantial torque is not used. An adjustable wrench is a handy tool to keep in a vehicle for emergencies but is hardly ever the tool of choice when working on automobiles. The tool has a tendency to slip and round off the head of a bolt and is too bulky to fit easily into tight spaces. Figure 8.9 shows the right and wrong ways to use an adjustable end wrench.

An *Allen wrench* is a hex tool used to turn set screws and capscrews. An Allen wrench can be L-shaped, or it can be mounted on a socket driver. **Figure 8.10** shows several types of socket drivers.

Screwdrivers

Screwdrivers are available with several different styles of heads (**Figure 8.11**). The two most common screwdrivers are the standard *flat* tip blade (**Figure 8.12**) and the *Phillips* head (**Figure 8.13**).

Be sure to select a screwdriver that is the correct thickness and width for the slot in the screw head. Using a screwdriver with the wrong size blade can ruin the head of a screw. The blade on a standard screwdriver should fill up the slot in the screw head. A worn screwdriver blade can be reground on the grinder. Grind in the proper direction to maintain the squareness of the blade (**Figure 8.14**). Be sure to quench often.

NOTE: *It is important that the blade be quenched often during grinding so the hardness of the metal is not changed.*

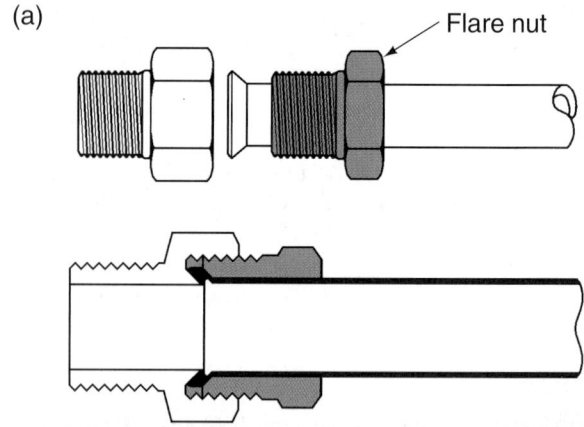

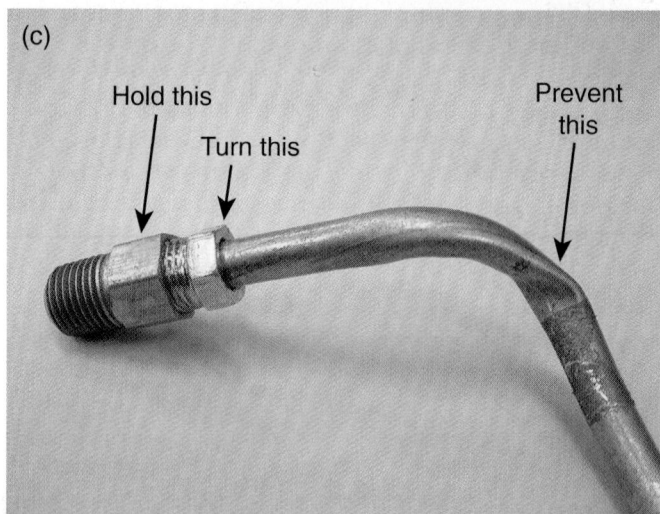

Figure 8.7 (a) Tubing fittings are called flare nuts. (b) A flare-nut wrench used with an open-end wrench to loosen a fuel line. (c) This fuel line was damaged when the apprentice tried to loosen the flare fitting without using two wrenches. (*a, Courtesy of Dana Corporation; c, Courtesy of Tim Gilles*)

Some tool manufacturers give a warranty for their screwdriver blades, so a worn or damaged one can be exchanged for a new one.

A Phillips head screwdriver pushes on four areas, rather than two like the standard flat tip. With a Phillips head, its number denotes its size. A #1 and a #2 are the

Figure 8.8 A ratcheting box wrench. *(Courtesy of Tim Gilles)*

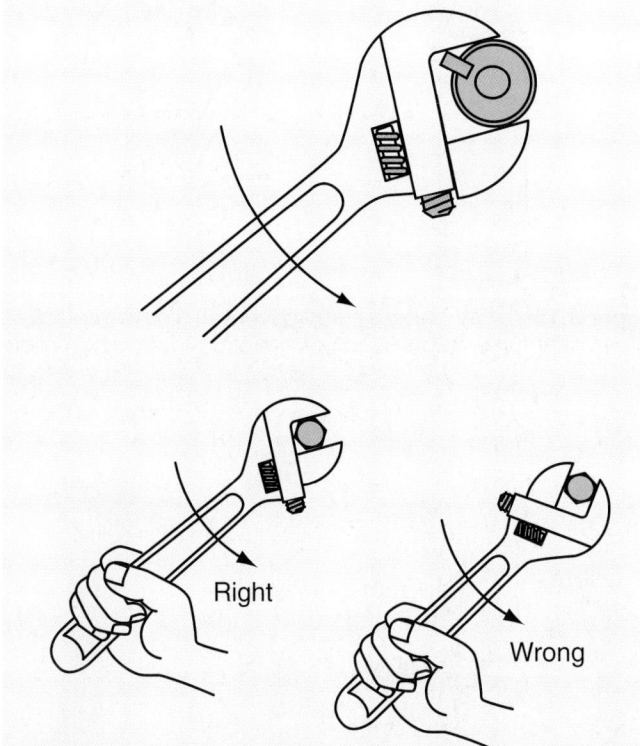

Figure 8.9 The right and wrong ways to use an adjustable end wrench. An adjustable wrench grasps the woodruff key to turn the crank. Is the wrench in this example being used properly?

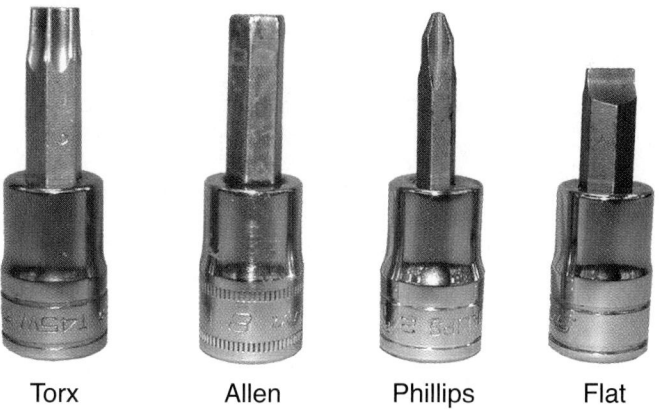

Torx   Allen   Phillips   Flat

Figure 8.10 A variety of socket drivers. *(Courtesy of Tim Gilles)*

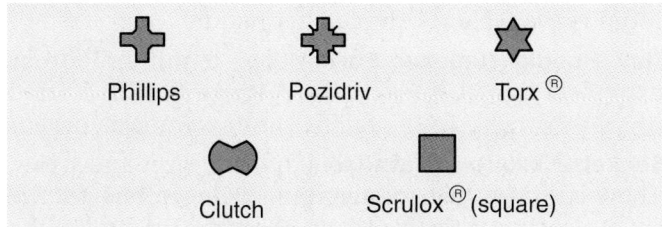

Phillips   Pozidriv   Torx ®

Clutch   Scrulox ® (square)

Figure 8.11 Miscellaneous screwdriver tips.

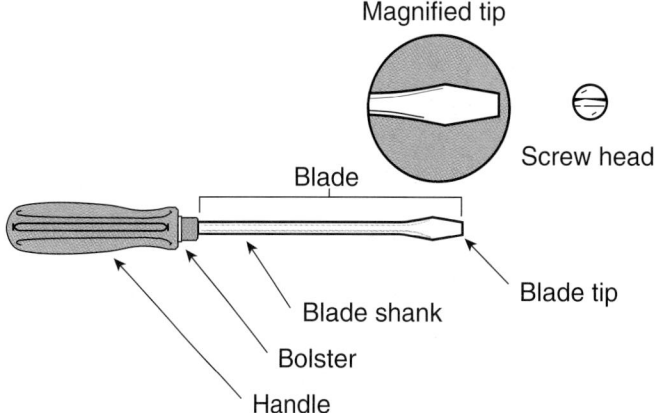

Magnified tip

Blade

Screw head

Blade shank

Blade tip

Bolster

Handle

Figure 8.12 A standard flat tip screwdriver.

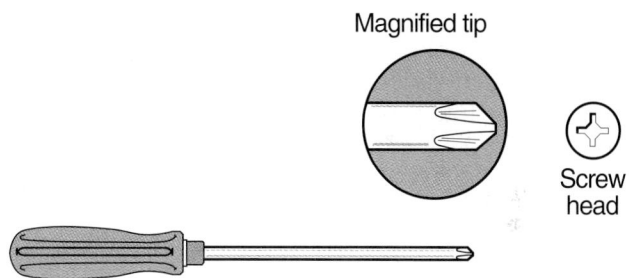

Magnified tip

Screw head

Figure 8.13 A Phillips head screwdriver.

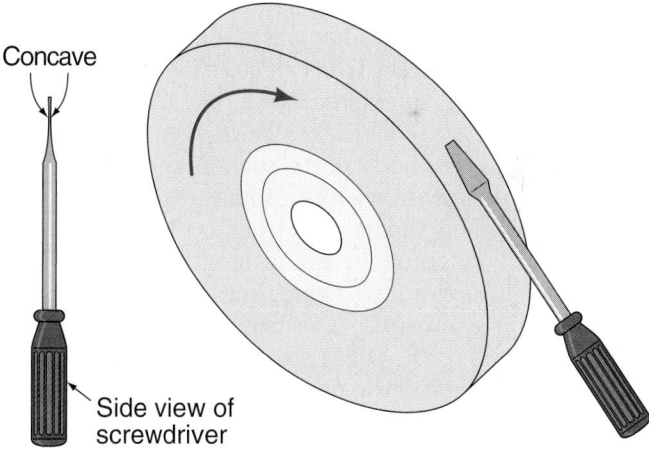

Concave

Side view of screwdriver

Figure 8.14 The proper way to grind a screwdriver.

most common sizes used in automobile repair. A #1 Phillips head is smaller than a #2.

SHOP TIP If a Phillips head is slipping in the screw slot, try putting some valve lapping compound on it.

A screwdriver that looks very similar to the Phillips is the *Reed and Prince*. Do not use a Reed and Prince on automotive screws. It has a deeper screw slot, and the walls that separate the slots are tapered. It was developed for the furniture industry because the standard flat tip

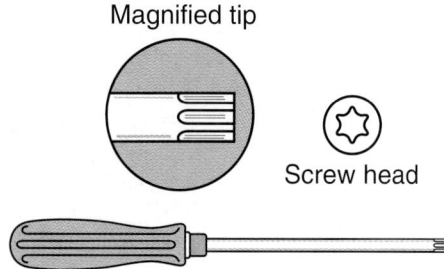

Figure 8.15 A Torx screwdriver.

screwdriver would tend to slide off and cut upholstery. The Reed and Prince is also used in IBM equipment.

A *Pozidriv* is also similar to a Phillips head, but it has little wings on the inside that provide a slightly better grip and hold onto fasteners for easier assembly work. Because Phillips and Pozidriv screws are different, usually a Pozidriv's handle is a different color so it is not confused with a Phillips. If two screwdrivers from the same manufacturer have different colored handles, this could be a clue that something is different about them.

In 1980, there were two billion Phillips screws used in automobile production. Because the Phillips only pushes on four areas, tool bit breakage was common. In the early eighties, the industry changed to *Torx* head fasteners (**Figure 8.15**). The Torx design provides greater contact with the head of the screwdriver and less chance of stripping the screw head. It was first used on *door strike plates* in the early 1960s. Torx can be either internal drive (like a screwdriver) or external drive (like a wrench).

A *clutch head* or *butterfly* screwdriver is found on some older vehicle body parts, such as door latches.

A *Scrulox* screwdriver has a square tip to fit screws found on campers, boats, and truck bodies.

Screwdrivers come in many lengths, from *stubby* (very short) to extra long (sometimes more than 2 feet long). A long screwdriver is usually easier to use, given sufficient access space. Offset screwdrivers are used when space does not permit the use of a stubby screwdriver.

NOTE: *Do not use a screwdriver as a prybar, unless it was designed for such use.*

The handle on a good screwdriver is usually large and comfortable. Handles on premium screwdrivers are made of hard plastic that will survive a lifetime of use. Some tool manufacturers provide replacement blades while reusing the old handle.

An *impact screwdriver* is very effective in removing very tight screws (**Figure 8.16**). The tool is held against the screw slot while twisting it in the direction desired. When the tool is struck with a hammer, the spring-loaded tool breaks the screw loose. The ⅜" socket drive end makes it possible to use the set's screwdriver bits with a ⅜" drive tool.

Sockets, Ratchets, and Breaker Bars

Whenever possible, a socket is used instead of a wrench because it is faster. A socket is used with either an *air*

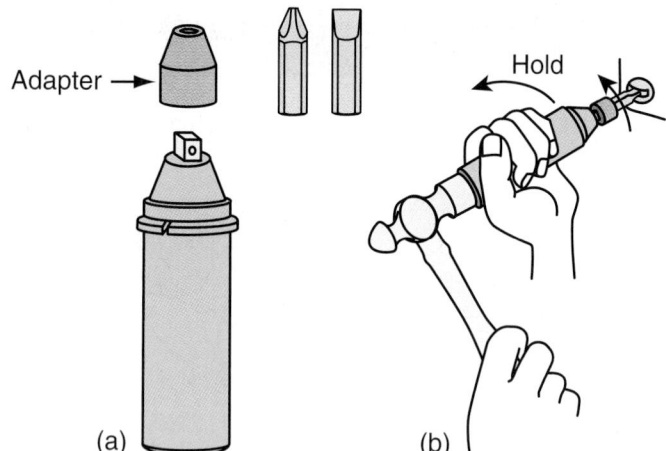

Figure 8.16 (a) An impact screwdriver set. (b) An impact screwdriver turns automatically when struck with a hammer.

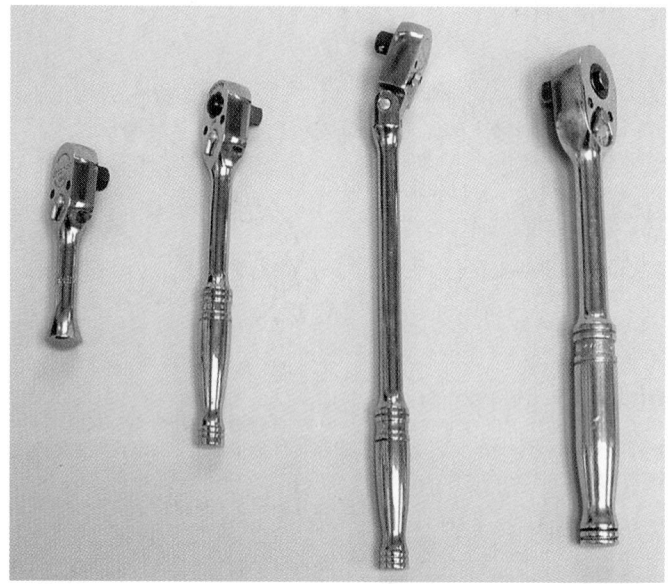

Figure 8.17 An assortment of ratchets. *(Courtesy of Tim Gilles)*

impact wrench (see Figure 8.45), a *ratchet* (**Figure 8.17**), or a *flex handle* (breaker bar). All of these tools allow for a change in direction so that a bolt can be either loosened or tightened.

Ratchet. A ratchet has teeth inside its head. **Figure 8.18a** shows the parts that make up the inside of a ratchet head. The teeth allow for a small amount of movement of the head of a bolt before the ratchet handle is moved back to its starting position. This allows for tightening when only a small amount of room is available for the ratchet handle to move.

Flex Handle (Breaker Bar). When a nut or bolt is extremely tight, a breaker bar (**Figure 8.18b**) is used so that the teeth in the ratchet head are not damaged.

Sockets. Sockets are available in many sizes and styles. There are two size designations: One refers to the square ratchet *drive end,* and the other refers to the *hex socket end* that fits the bolt head (**Figure 8.19**).

(a)

(b)

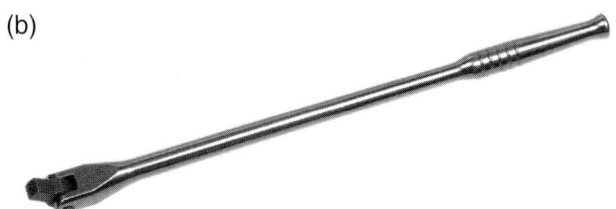

Figure 8.18 (a) Internal parts of a ratchet head make up a ratchet repair kit. (b) A breaker bar prevents the need for a ratchet repair kit. *(Courtesy of Tim Gilles)*

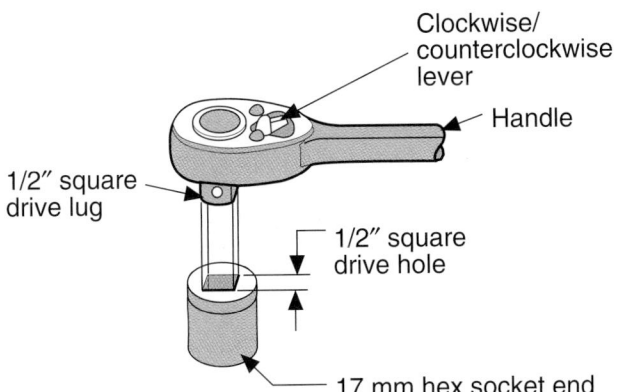

Figure 8.19 A 1/2" drive ratchet used with a 17-millimeter hex socket.

A typical toolbox will contain sets of both *standard* (fractional) and *metric* bolt head sockets. The drive end is always an inch size, whether a socket hex end is standard or metric. That way, different ratchets and extensions are not required when switching from a standard to a metric size.

Socket Drive Sizes. The most common drive sizes used on automobiles are the ¼", the ⅜", and the ½". The ⅜" drive is the most versatile for automobile work. A set of ⅜" drive sockets usually includes hex (bolt head) sizes ranging from ⅜" to ¾".

Socket Adapters. Occasionally, a ratchet of a different drive size is needed but is not available, so a *socket adapter* is used (**Figure 8.20**). For instance, a ⅜" to ¼" drive adapter might be used to drive a ⁵⁄₁₆" socket head because a ⁵⁄₁₆" socket would not normally be found in a ⅜" drive socket assortment (which usually includes ⅜" to ¾" only).

Figure 8.20 A 1/2" to 3/8" socket adapter and a male pipe plug driver.

CAUTION Adapters are often used with driving tools that are too large for the socket being driven (a ½" ratchet driving a ⅜" socket, for instance). Be certain to use the correct torque to avoid breaking a tool.

Plug Drivers. Do not confuse a socket adapter with a plug driver. The plug driver is solid. An adapter, which has a spring-loaded ball to hold the socket, will probably break if it is used to loosen a rusted gallery plug (which has a tapered locking thread). There are many fittings that use threaded tapered pipe plugs, especially in engines. Plug drivers (see Figure 8.20) are used here.

Extension Bar. When driving a socket with a ratchet or an air wrench, an *extension* is usually used (**Figure 8.21**). An extension allows the technician to have his or her hands clear of parts, preventing injury. Common sizes of extensions are 1", 3", 6", and 10". Very long extensions (2 or 3 feet) are used with a swivel socket in transmission *remove and replace* (*R&R*)

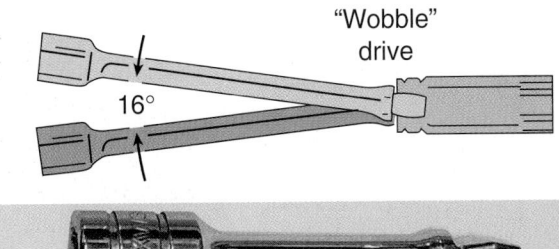

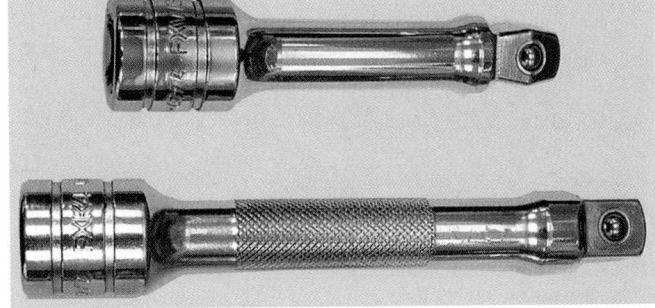

Figure 8.21 Standard and wobble extensions. The wobble allows the socket to pivot 16 degrees. *(Courtesy of Tim Gilles)*

jobs and other specialized applications. One of the extensions shown in **Figure 8.21** shows a *wobble drive* that allows the socket to pivot 16 degrees to allow easier access to bolts.

Sockets. Sockets are kept in order according to size in the same drawer or in the top of the toolbox so that they can be easily located when doing a job (**Figure 8.22**). Socket clip rails and trays are used to keep things in order. When the proper tools are used, a professional will rarely ruin a fastener. Like wrenches, the bolt end of most sockets is *6-* or *12-point*. A 6-point socket is the tool of choice because it is less likely to "round off" a fastener head. *Eight-point* sockets are available for driving square heads.

Deep Sockets. A set of *deep (bolt-clearance) sockets* is usually part of a complete tool set, too (**Figure 8.23**). A deep socket provides clearance when removing a nut from a long stud.

Special Use Sockets

A popular tool is a *swivel socket* that is an integral part of an individually sized socket. This tool is commonly known as a *wobble socket*. It is a much better choice than a universal swivel, which is used between a regular socket and a ratchet or extension so that it can pivot. Wobbles that are designed to be used with air-powered impact wrenches are very popular (see Figure 8.45)

Special, thicker, *impact sockets* are available for use with air tools (see Figure 8.47). Also, while standard sockets are chrome plated, impact and industrial sockets are not. A regular socket used with an air tool can result in a broken socket.

A *crowfoot socket* is a wrench head that has been adapted so that it can be used with a ratchet and extension. Some crowfoot wrenches are open-end and some are flare-nut box wrenches. They can be stored in an organized fashion using the holder shown in **Figure 8.24**.

Spark plug sockets for automobiles come in two sizes, ⅝" or ¹³⁄₁₆". They have a rubber or magnetic insert to hold the spark plug firmly inside the socket because spark plugs are ceramic and can easily be cracked if dropped.

Socket Drivers

Socket drivers are very popular and are used for a variety of applications (see Figure 8.10). These tools, which can be driven either by a ratchet or air impact wrench, are available with various heads such as *hex (Allen head), Phillips, Pozidrive, clutch head (butterfly), torx, and the standard flat tip.*

A popular driver for small work is one that resembles a screwdriver handle (**Figure 8.25a**). The driver, sometimes called a *nutdriver,* is commonly used with small ¼" drive sockets.

Another popular driver is a *speed handle* (**Figure 8.25b**). It is an excellent tool for bench work on engines and transmissions. While disassembling these parts with an air tool is popular, reassembly with an air tool is risky. A speed handle is almost as fast as using an air tool, but it minimizes the possibility of cross-threading or overtightening a fastener.

Figure 8.22 Six-point, 12-point, and 8-point sockets. *(Courtesy of Tim Gilles)*

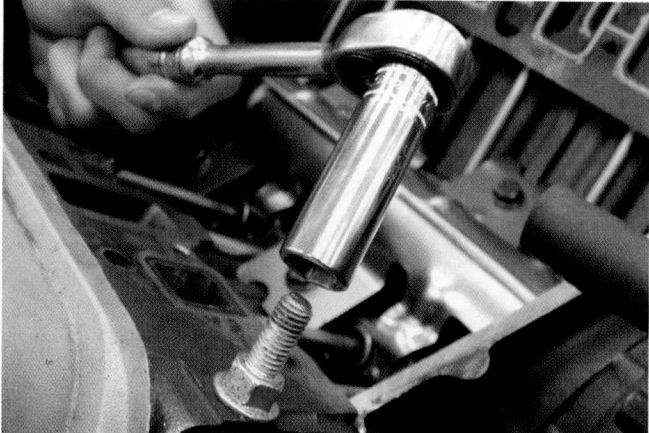

Figure 8.23 A deep socket used with a breaker bar to remove a nut from a long stud. *(Courtesy of Tim Gilles)*

Crowfoot socket wrenches

Figure 8.24 Crowfoot socket wrenches.

(a)

(b)

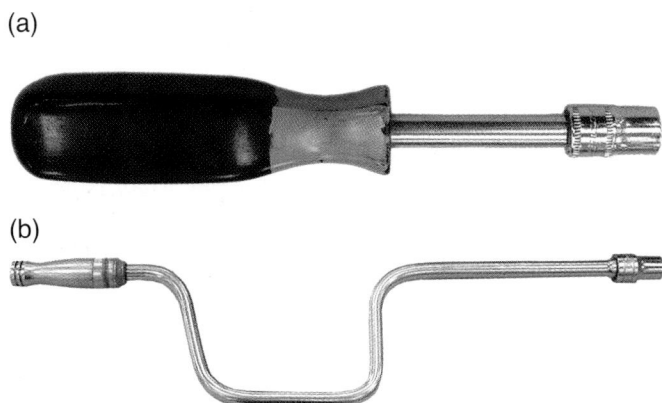

Figure 8.25 (a) A 1/4" socket set with a nutdriver handle. (b) A speed handle. (*Courtesy of Tim Gilles*)

Torque Wrench

Most fastener applications are designed to be tightened with a torque wrench. A torque wrench is used to tighten a fastener a prescribed amount. Various torque wrenches are illustrated and described in detail in Chapter 51.

Pliers

Pliers come in many types and sizes (**Figure 8.26**). It is not uncommon for the same tool to have more than one name, including the one from the tool company that originally developed the tool. Examples of this are the *rib joint pliers,* called *channel-locks,* and *locking pliers,* called *vise grips.* Other pliers include *cutting pliers, needle nose,* and *slip-joint utility* (*combination*) pliers.

Pliers are often misused for tightening and loosening fasteners. The result is most often a rounded-off fastener head. The proper use of a plier is dictated by its design. Most often a plier is for cutting, bending, or positioning a part (such as a cotter pin) or for stripping an electrical wire. The use of too small a plier for the size of the job can result in damage to the plier's jaws.

Channel-Locks. Channel-lock pliers, also called rib joint pliers, have tongue and groove rows that provide a means of jaw size adjustment. When the adjustment of the jaws is changed, they remain parallel for use.

Cutters. Cutters, such as diagonals (side cutters) are also popular in automotive work. They are used for cutting wire and removing and installing cotter pins. *Compound leverage cutters,* also called *bolt cutters,* have an extra fulcrum that increases the leverage on the jaws. Jaws on compound cutters move a smaller distance with handle movement than ordinary pliers.

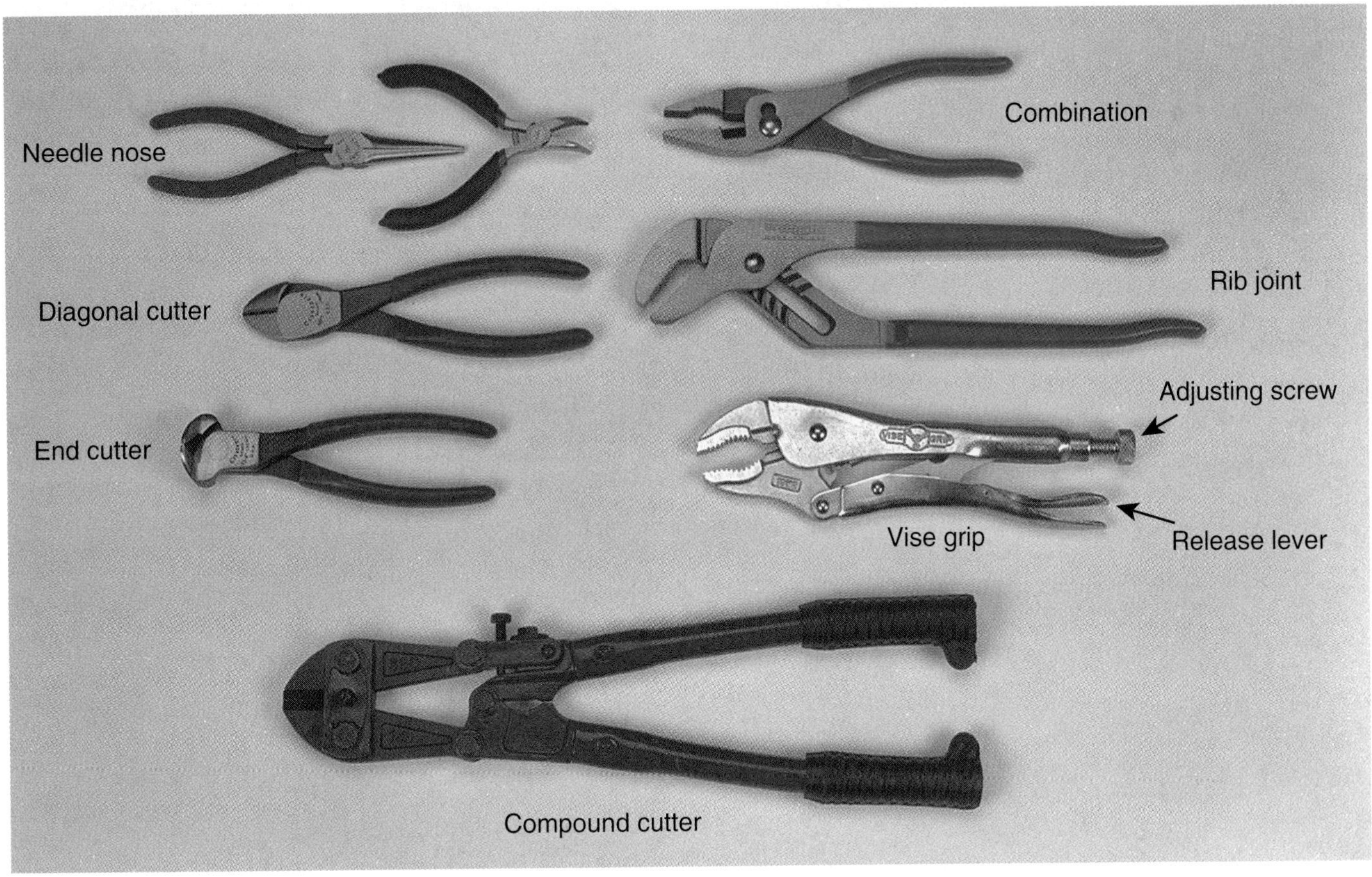

Needle nose

Diagonal cutter

End cutter

Combination

Rib joint

Adjusting screw

Vise grip

Release lever

Compound cutter

Figure 8.26 Different types of pliers.

Locking Pliers. Commonly called *vise grips*, locking pliers have locking jaws that are adjusted by turning a screw at the end of the plier handle (see Figure 8.26). The better quality vise grips also have a release lever. Tightly closing the jaws locks them, while pressing on the release lever releases them.

Needle Nose Pliers. *Needle nose* pliers are also called *long nose pliers* (see Figure 8.26). They are handy for positioning clips and very small parts.

NOTE: *Needle nose pliers are not designed to be used for bending. They can be damaged when overstressed.*

Special Use Pliers

Special use pliers include those used to remove and install spring-loaded hose clamps or pliers used to remove corroded battery bolts.

Lineman's Pliers. Lineman's pliers are used by electricians but are often seen in a technician's or machinist's toolbox. They have a cutting edge and gripping teeth. An area on the inside of the jaws is used to remove insulation from household electrical wiring (**Figure 8.27**).

High Leverage Combination Pliers. High leverage combination pliers (**Figure 8.28**) are also popular for cutting or gripping. They require only ¼ the normal

application of effort on the handle. Much movement of the handle results in little movement at the jaws. These are excellent pliers when a very strong grip is needed.

Snapring Pliers. A snapring is used to secure a bearing to a shaft or hold it inside of a housing. Snaprings are *external* when they retain a bearing on a shaft. They are *internal* when they hold a bearing in a housing. Snapring pliers come in several types to match the different types of snaprings available. **Figure 8.29** shows three types of snaprings and some of their uses.

Hammers

Hammers come in various sizes, weights, and materials (**Figure 8.30a**). A common hammer, the *ball-peen*

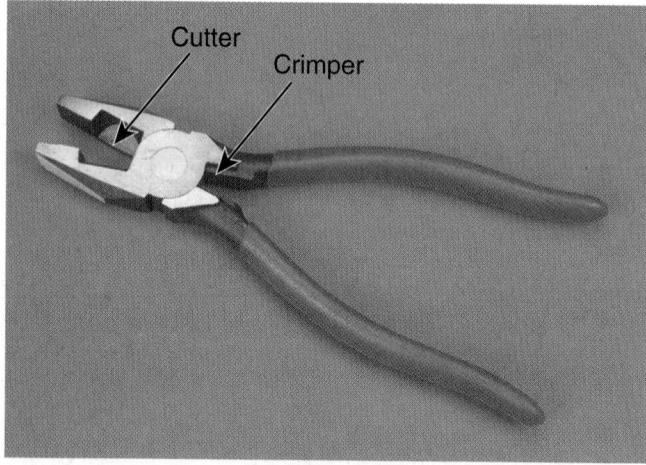

Figure 8.27 Linesman's pliers.

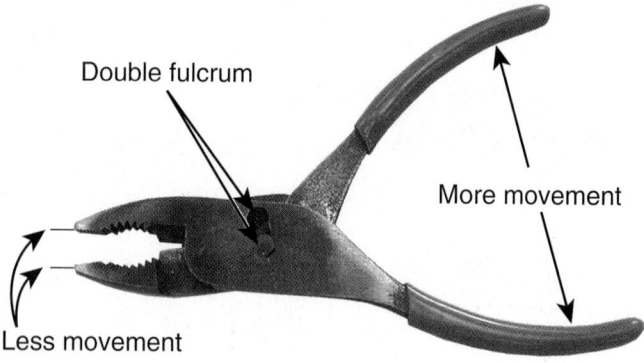

Figure 8.28 High leverage combination pliers. (*Courtesy of Tim Gilles*)

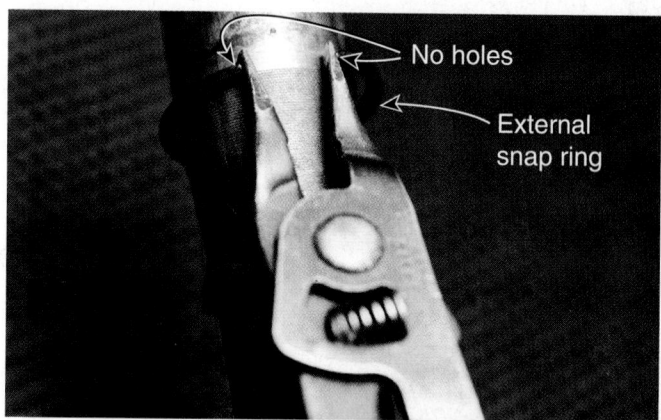

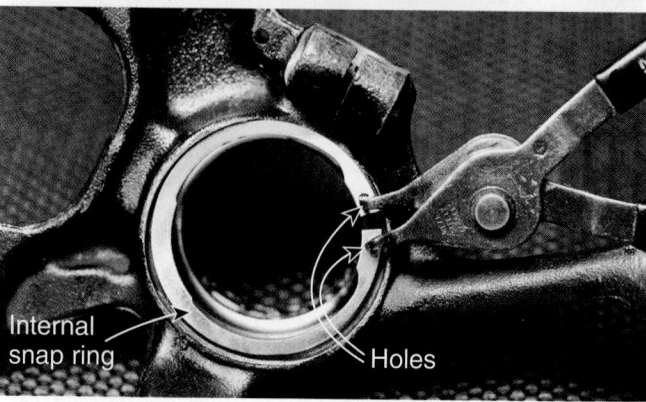

Figure 8.29 Types of snapring pliers. (*Courtesy of Snap-on Tools Company, www.snapon.com*)

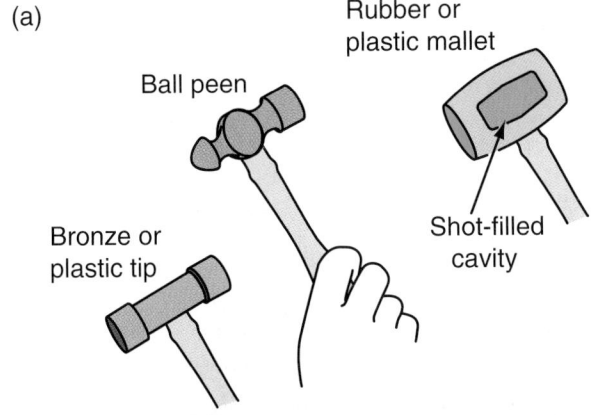

(a)

Ball peen

Rubber or plastic mallet

Bronze or plastic tip

Shot-filled cavity

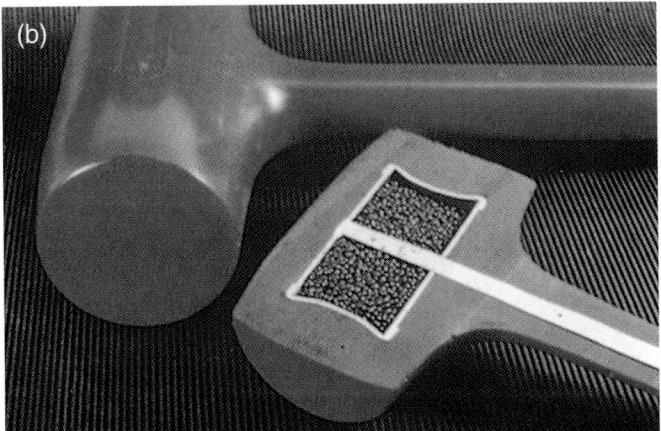

(b)

Figure 8.30 (a) Several types of hammers. (b) A dead blow hammer has a cavity full of metal shot. (b, Courtesy of Snap-on Tools Company, www.snapon.com)

hammer, comes in several different weights. The handle should be gripped near the end and the face of the hammer should squarely contact the object being struck.

Soft-faced brass and *soft plastic* hammers are also popular. One type of soft plastic hammer has a cavity in its head that is filled with metal shot (**Figure 8.30b**). This hammer, called a *dead-blow* hammer, will not bounce back upon impact.

Chisels and Punches

Chisels and punches, used with a hammer, have different applications. A chisel is used for cutting metal, while a punch is used to drive a pin or key, or to make an indentation in metal. Several shapes of chisels are shown in **Figure 8.31**.

 Be sure to always use eye protection when using hammers and chisels.

Various types of punches are shown in **Figure 8.32**. A starting punch is used to start driving

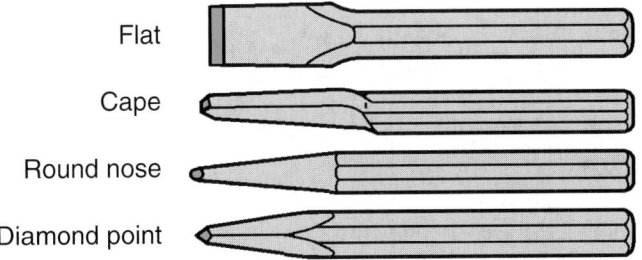

Flat

Cape

Round nose

Diamond point

Figure 8.31 Types of chisels. (Courtesy of Ford Motor Co.)

a pin or rivet from a hole. The pin punch is then used to finish driving it out. A centerpunch is used to make an indentation in metal before drilling a hole.

Punches and chisels often develop a "mushroomed" head from repeated pounding with a hammer (**Figure 8.33a**). Chips can break off a mushroomed tool head, causing injury. Regrind a mushroomed tool head to maintain the chamfer (**Figure 8.33b**), quenching often to avoid damage to the heat treatment of the tool.

NOTE: *Using a hammer that is too large for a punch or chisel can result in chipping of the punch or chisel head. According to Snap-on Tools, a hammer head should be no more than ⅜" larger in diameter than the head of the punch or chisel.*

The cutting edge of the chisel must be maintained in sharp condition (**Figure 8.33c**). The edge is ground to a slight radius (curve).

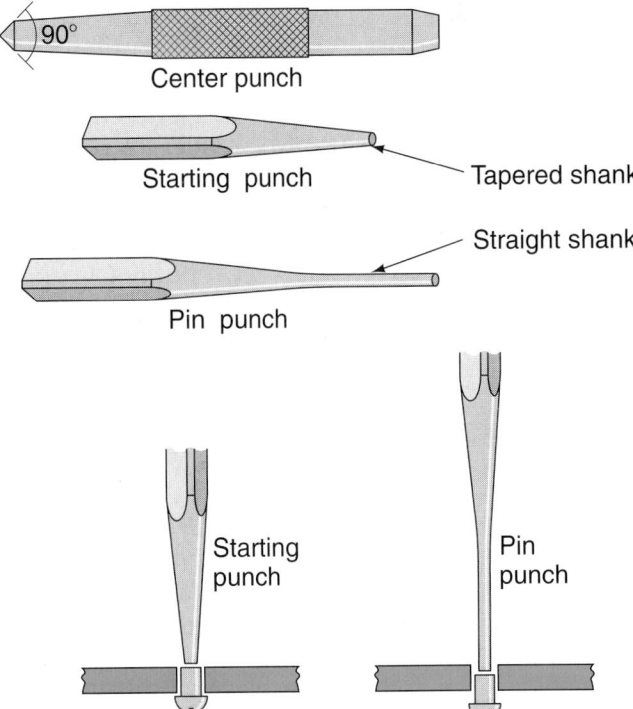

90°

Center punch

Starting punch

Tapered shank

Straight shank

Pin punch

Starting punch

Pin punch

Figure 8.32 Types of punches. (Courtesy of Snap-on Tools Company, www.snapon.com)

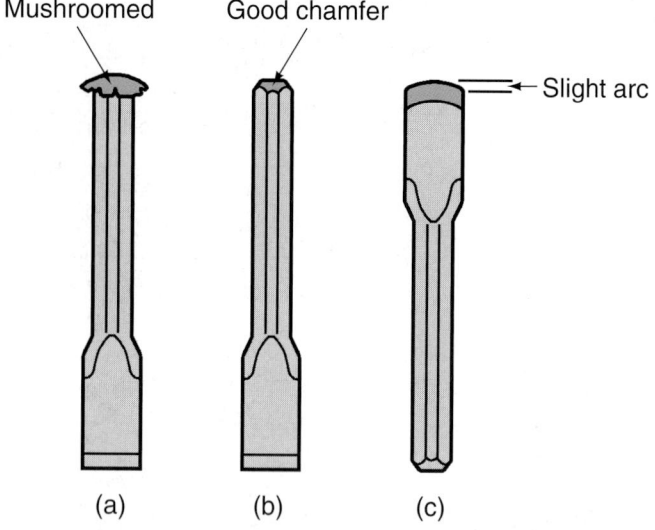

Mushroomed   Good chamfer   Slight arc

(a)   (b)   (c)

(d)

Figure 8.33 Chisel maintenance. (a) A mushroomed chisel head. (b) The head should have a chamfer. (c) The cutting edge of the chisel has a slight radius. (d) A chisel can be held with a chisel holder when cutting metal. (a, b, c, Courtesy of Snap-on Tools Company, www.snapon.com, d, Courtesy of Tim Gilles)

The chisel can be held with a holder when cutting metal (**Figure 8.33d**).

Hacksaws

A hacksaw is used to cut metal. The following are some tips on correct hacksaw usage:

■ The blade cuts on the forward stroke only. Applying pressure during the return stroke will dull the blade. Be sure the blade is properly installed with the teeth facing away from the handle (**Figure 8.34a**) so that they cut during the forward stroke.

■ Use the entire blade when cutting so that the entire blade gets a chance to wear evenly.

■ When cutting steel, cutting oil will make cutting easier and extend the life of the blade.

■ Some hacksaw blades have fewer teeth for faster cutting of thick metals.

■ For sheet metal, the hacksaw blade must have more teeth so that the sheet metal does not drop between the teeth, damaging them (**Figure 8.34b**). The

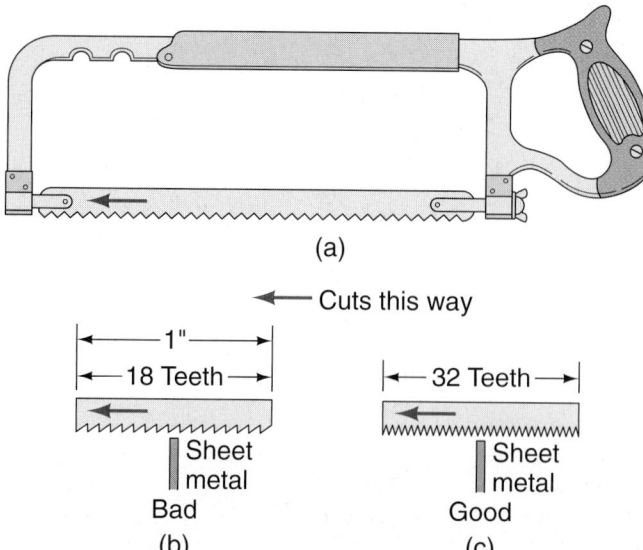

(a)

← Cuts this way

|← 1" →|
|← 18 Teeth →|

Sheet metal
Bad
(b)

|← 32 Teeth →|

Sheet metal
Good
(c)

Figure 8.34 Hacksaws. (a) The blade is installed with the teeth facing forward. (b) A hacksaw blade that is too coarse will not work well with sheet metal. (c) A fine-toothed blade is used with thin metal.

blade should have at least two teeth in contact with the metal at all times (**Figure 8.34c**).

Files

Files are used to shape, roughen, or smooth metal. They vary in size and roughness and can be of either a round, half-round, square, triangular, or flat shape (**Figure 8.35a**).

■ To avoid puncture wounds, a file should always be used with a handle on the tang.

■ A file cuts on the forward stroke. Push down gently on it when filing. Then, release pressure on the return stroke.

■ Filing softer metal, like aluminum or brass, calls for a coarser file.

■ For filing steel, use a fine file.

■ From roughest to finest, the names of file cuts are *bastard cut, second cut,* and *smooth.*

The diagonal rows of teeth on a file are either *single cut* or *double cut.* A double cut file has two rows of teeth that cross each other on an angle.

Filings are the pieces of metal that come off of the workpiece during filing. If they are allowed to remain trapped on the surface of the file, they can damage the filed surface and limit the file's effectiveness. Proper care of a file will prolong its life:

■ A *file card* is used to clean the teeth of a file (**Figure 8.35b**).

■ Putting chalk on a file's teeth will help avoid loading up the teeth when filing soft materials such as aluminum or brass.

■ Store a file carefully so that it will not rust or be nicked by other tools.

■ Keep files clean and free of oil or grease.

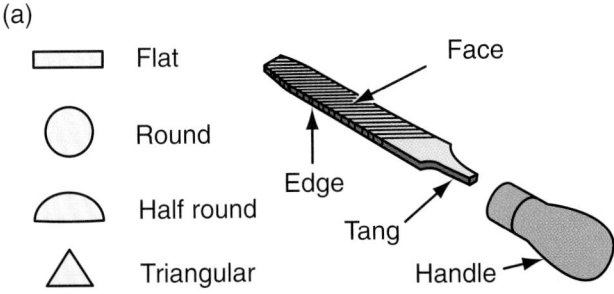

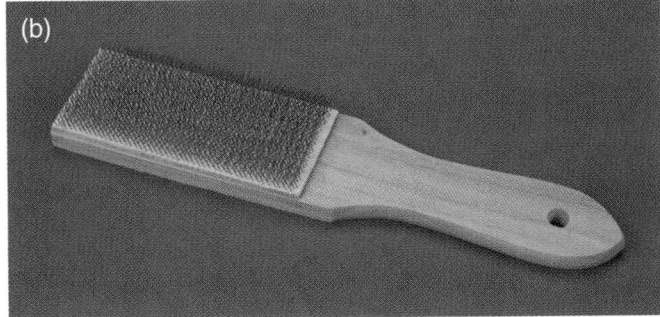

Figure 8.35 (a) Files come in several shapes. (b) A file card.

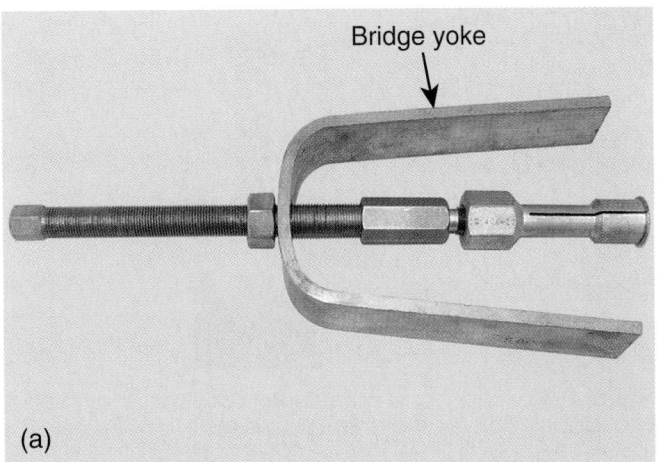

(a)

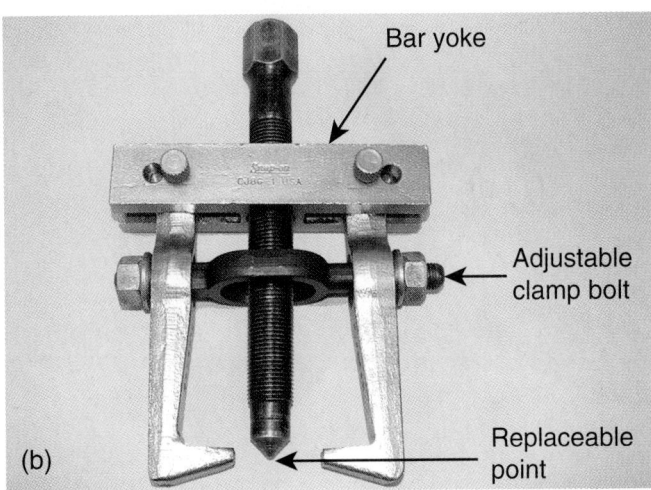

(b)

Figure 8.36 Pullers. (a) A bridge-yoke puller. (b) A bar-yoke puller. *(Courtesy of Tim Gilles)*

▬ PULLERS

There are many types of *pullers* used in automotive work, ranging from small to large. Pullers are used to remove or install pressed-fit gears, bushings, bearings, or other parts from shafts. Specific uses of many of the pullers discussed here are covered in later chapters.

Pullers are either *hydraulic* or *manual*. Manual pullers have a bridge yoke (**Figure 8.36a**) or a bar-type yoke (**Figure 8.36b**). A *slide hammer* or a *pressure screw* is threaded through the *yoke*.

- A *bridge-type yoke* pushes against the outside of the part being pulled.
- Some *bar-type yoke* pullers have jaws and others use bolts or **collets**.
- Jaw-type pullers sometimes have an adjustable clamp bolt that holds the jaws against the work for a more reliable pull.
- The end of the pressure screw has a replaceable hardened point.

On many pullers, the jaws can be turned around to accommodate inside and outside pulls (**Figure 8.37**). Jaws of different lengths and sizes can be installed to make the puller more versatile. The size of the jaw depends on the reach and spread required for the pulling application. Pressure screws of different lengths are installed to make it possible for the puller to use different length jaws.

Some pullers use bolts that are screwed into threaded holes in a gear or pulley (**Figure 8.38**). Others use bolts screwed into parts such as a vibration damper on the front of the crankshaft (**Figure 8.39**). The vibration damper puller is also called a *flange-type puller*. It can also be used for pulling

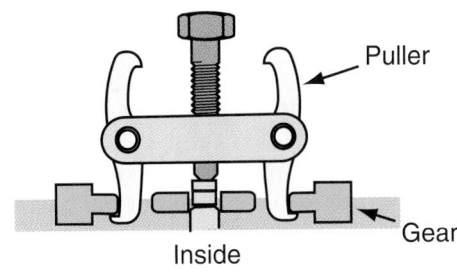

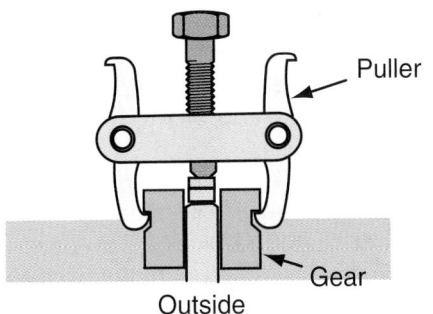

Figure 8.37 Puller jaws installed for inside and outside pulls.

Figure 8.38 A gear and pulley puller uses bolts that thread into tapped holes in the part. *(Courtesy of Tim Gilles)*

Figure 8.39 A flange-type puller can be used to pull a vibration damper or a steering wheel. *(Courtesy of Tim Gilles)*

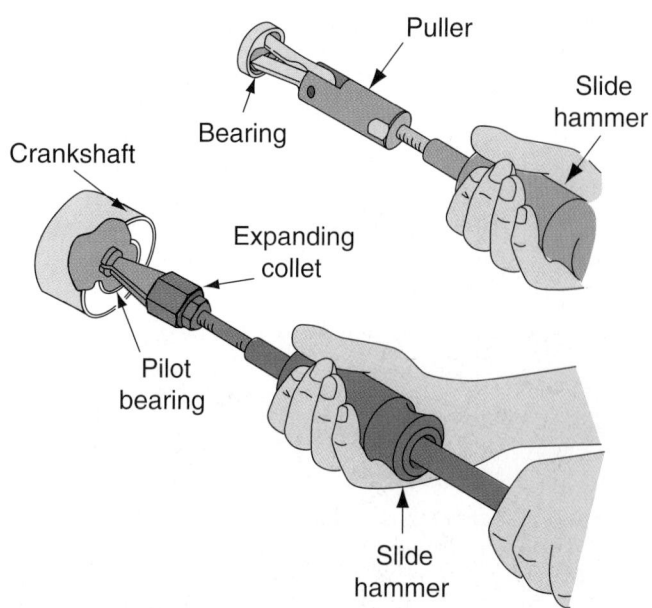

Figure 8.40 Two types of slide hammer pullers.

steering wheels. It works on parts that have either two or three threaded holes. Slotted holes in the puller body allow it to accommodate many different size parts.

A *slide hammer puller* is another common tool. The ones shown in **Figure 8.40** can be used to pull a bearing from a bored hole that it is pressed into. They are adjustable with either a collet or puller jaws. To remove the bearing, slide the weight back hard until it hits the flat area on the rear of the puller.

A larger puller is the hydraulic-type puller, often called a *porta-power* (**Figure 8.41**).

One of the most versatile pullers is the *bearing separator* and bar-type puller setup shown in **Figure 8.42**. The puller used in **Figure 8.43** is being used with a *step plate* to protect the end of the shaft from damage during pulling.

Stud pullers (**Figure 8.44**) are commonly used in engine repair. A stud is a fastener that is threaded on both ends. It therefore lacks a provision for using a wrench to turn it. The stud puller wedges against the stud (usually in an area in the center that has no threads) and loosens it. The tool works best if it is held all the way down against the work surface.

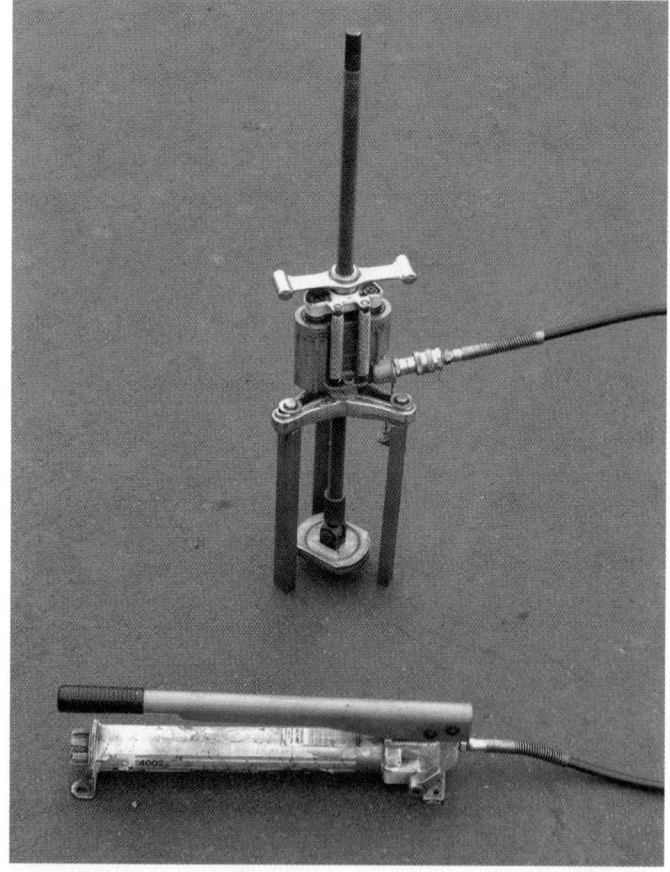

Figure 8.41 A hydraulic puller.

Puller Safety

1. Wear eye protection.
2. When heating a part to help free it, do not heat the jaws of the puller. This could change the temper of the metal.

Figure 8.42 A bearing separator plate and bar yoke puller.

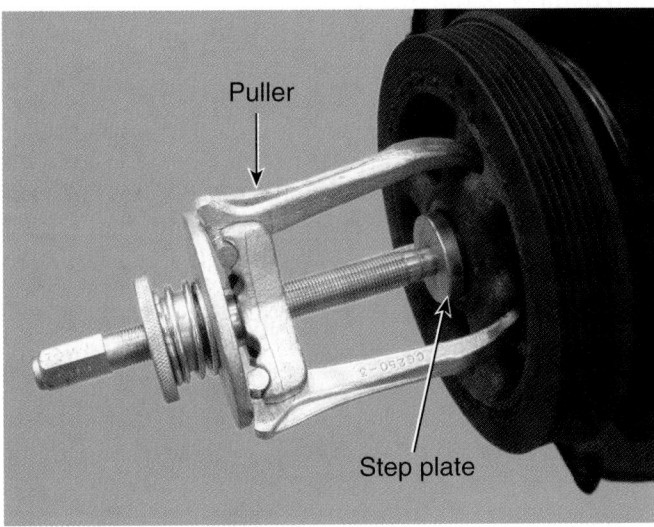

Puller

Step plate

Figure 8.43 The step plate protects the end of the shaft from damage. *(Courtesy of Tim Gilles)*

3. Be sure the pressure screw is clean and lubricated before using an impact wrench.

4. Be sure the removable point is installed on the puller shaft tip.

5. Be sure the puller is aligned so it is perpendicular to the part being pulled.

6. Do not use a puller with damaged or worn parts.

7. Use the correct size puller so overloading is avoided.

8. Use a 3-jaw puller instead of a 2-jaw puller when possible.

■ AIR TOOLS

Compressed air is used in practically all automotive shops. Air tools are great timesavers for technicians.

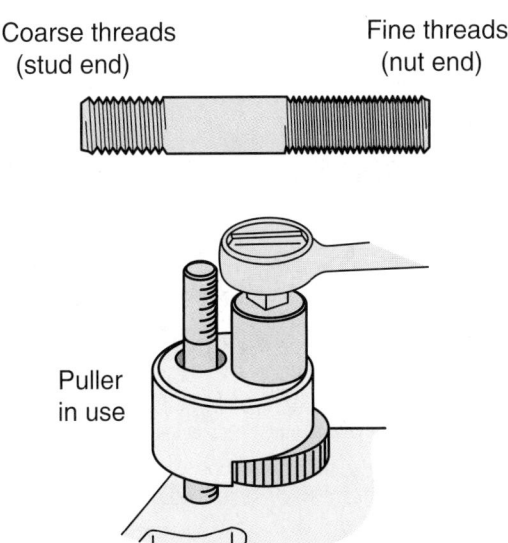

Coarse threads (stud end)   Fine threads (nut end)

Puller in use

Figure 8.44 A stud puller is used to loosen and tighten studs.

There are many air-operated tools, including air drills, air hydraulic jacks, and other specialized pieces of equipment that are explained in other chapters. Smaller, air-operated hand tools are explained here.

The *air compressor* provides air at a regulated pressure of 90 to 150 pounds per square inch (psi). Air tool manufacturers recommend regulating air pressure to 90 psi to get the best performance and reliability from air tools. Chapter 11 deals with air compressors.

SAFETY NOTE Blowguns for blowing off parts are regulated so that they do not produce more than 35 psi. Rubber-tipped blowguns, used to blow into brake or automatic transmission fluid passageways or engine oil galleries, do not have this safety feature. A worker should not use these tools until proper instructions on their safe use is understood.

There are many air tools. Only the impact wrench, air ratchet, air chisel, die grinder, and air drill are discussed here.

Impact Wrenches

The air impact wrench is one of the technician's most popular tools. Impact wrenches are available in various sizes with different size socket drives. A ⅜" drive impact wrench is very popular. It is generally used with a *wobble (universal) impact socket* (**Figure 8.45**).

A ½" drive impact wrench (**Figure 8.46**) is used to loosen large, very tight bolts. Special, extra-thick impact sockets must be used with the impact wrench (**Figure 8.47**). Regular sockets can crack or explode (see Figure 3.22).

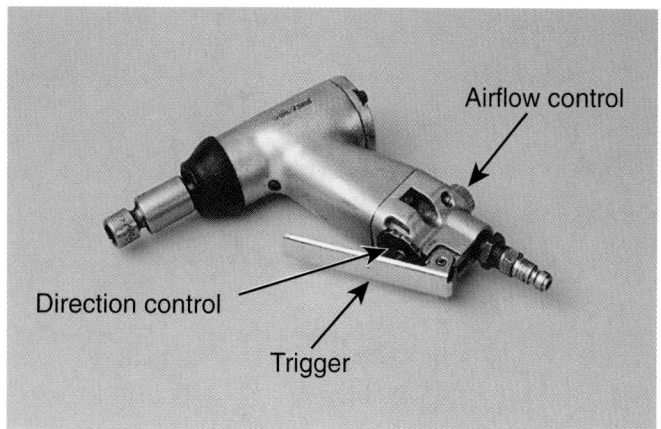

Figure 8.45 A 3/8" impact wrench with a universal impact socket. *(Courtesy of Tim Gilles)*

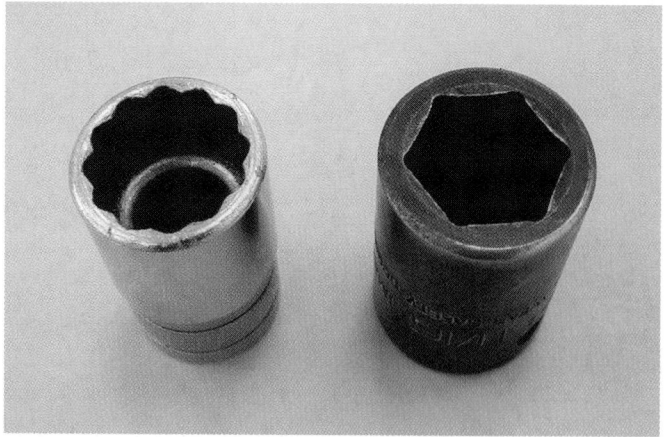

Figure 8.47 Regular and impact sockets. *(Courtesy of Tim Gilles)*

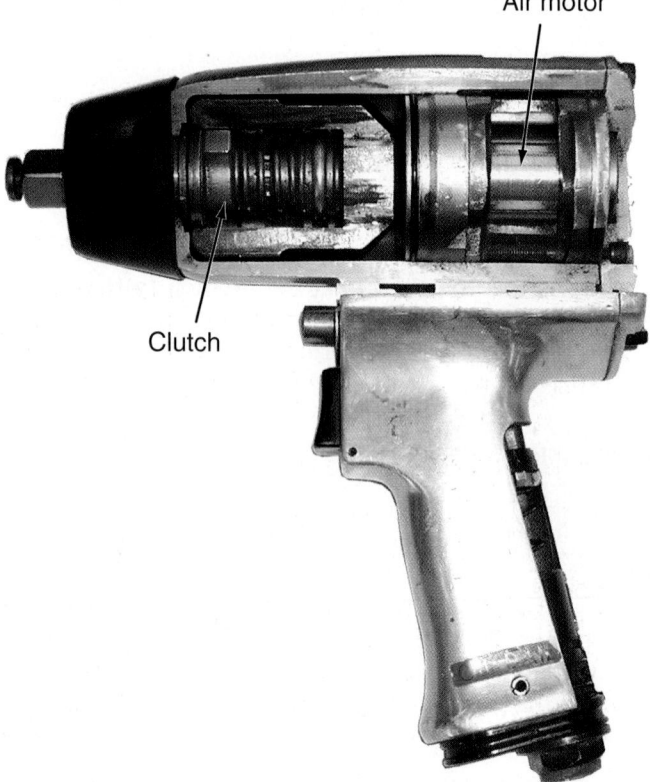

Figure 8.46 Cutaway of a heavy-duty 1/2" impact wrench. *(Courtesy of Tim Gilles)*

CAUTION When using a wobble socket, do not turn on the impact wrench unless it is installed on a nut or bolt. The socket can fly off the impact wrench, possibly causing an injury.

Impact wrenches are especially useful around exhaust bolts. The "impact" caused by the tool tends to shake the nuts and bolts loose instead of twisting them off. Impact wrenches are also handy on vibration damper and flywheel bolts because the crankshaft does not need to be restrained from turning. Wheel lug nuts can also be loosened on a raised car without holding the wheel from turning.

To allow the maximum amount of air assist, the air flow control knob is turned counterclockwise to open a passageway. When an especially tight bolt is encountered, it might be helpful to put an extra amount of air-motor oil into the wrench through the air inlet to help seal the vanes in the tool.

Some air systems have automatic oilers to provide lubrication for air tools. Electric impact wrenches are also available.

Always use approved impact sockets, not chrome sockets. Be sure that the socket is secured to the air tool. A clip at the end of the tool's square drive can become worn so that it no longer holds the tool. When the impact wrench fails to loosen a fastener, use a large breaker bar.

NOTE: *For larger, extremely tight fasteners, use a 1" impact wrench or a planetary gear head multiplier to gain additional leverage.*

Gear multipliers with a ½" input (female) end and a ¾" output (male) end can multiply input torque by from 3.3:1 up to 6:1 (**Figure 8.48a**). The multiplier uses *planetary gears* to achieve an incrase in torque. Planetary gears are used extensively in automatic transmissions. A simple planetary gear set includes four components: the *sun gear*, the *planetary carrier* or *cage*, *three pinion gears*, and the *ring gear* (**Figure 8.48b**). In order to operate, one component of the planetary gear set must be held while another part is turned (**Figure 8.48c**). The output is the remaining component. The handle of the gear multiplier is the ring gear that is held when turning the center of the tool.

Some gear multipliers can be stacked to increase torque even more. The need for this kind of torque would only be necessary for very heavy-duty applications,

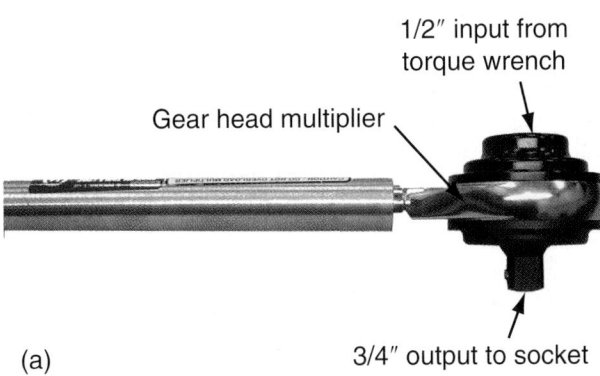

(a)

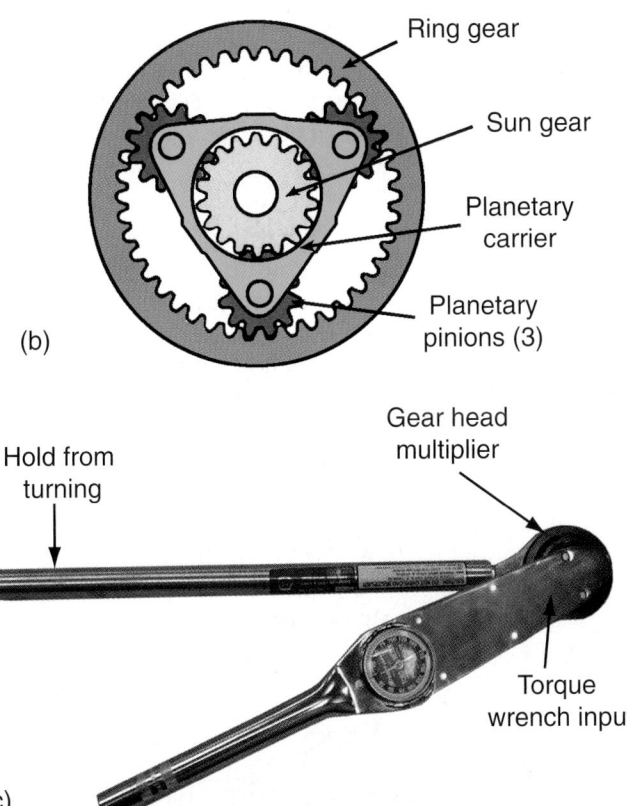

(b)

(c)

Figure 8.48 (a) A gear head multiplier. (b) Parts of a planetary gear set. (c) The multiplier is held from turning as torque is applied. (*a and c, Courtesy of Tim Gilles*)

such as truck and heavy equipment. Torque is discussed in detail in Chapter 51.

Air Ratchet

Another popular tool is the air ratchet shown in **Figure 8.49**. It is an excellent tool for use in confined places; ⅜" and ¼" drive sizes are common. A disadvantage about this tool is that it is noisy.

Air Chisel

The air chisel, or *air hammer*, is a miniature jackhammer. One is pictured in **Figure 8.50**. It is often used to drive valve guides out of cylinder heads, and it is useful

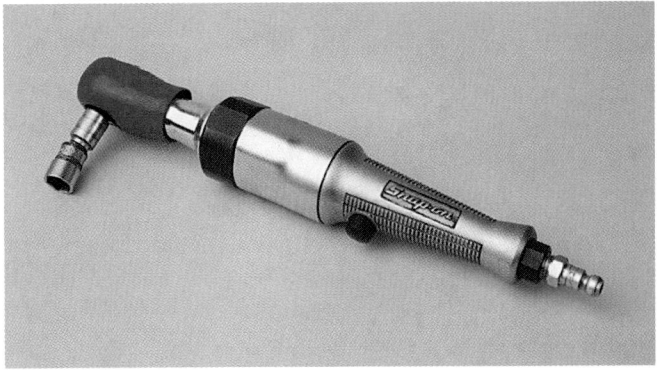

Figure 8.49 An air ratchet. (*Courtesy of Tim Gilles*)

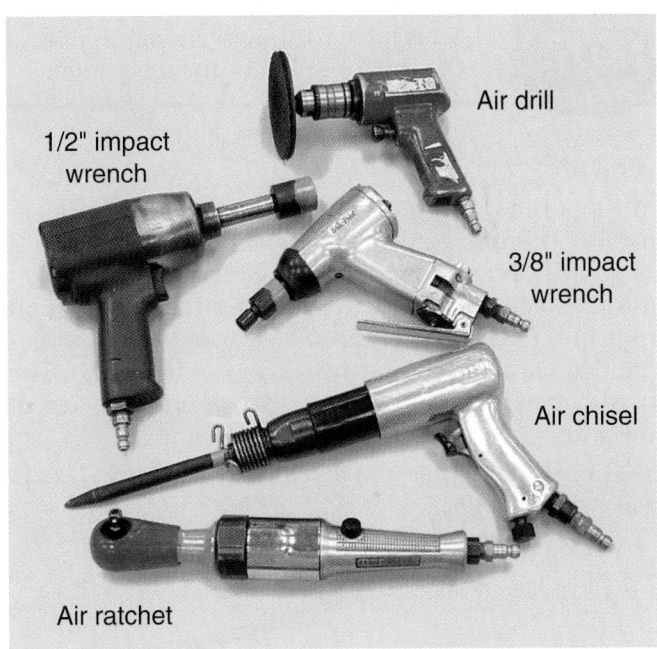

Figure 8.50 Miscellaneous air tools. (*Courtesy of Tim Gilles*)

in front-end repair and muffler work. There are many attachments available for a variety of uses.

 Before pulling the trigger, be sure to have the tool bit against the workpiece. Otherwise, the tool might fly out of the gun. Be sure to wear eye protection.

Die Grinder

Die grinders (**Figure 8.51**) turn at very high speeds (about 20,000 rpm). They are used to remove burrs using either small cutters or grinding wheels. Carbide-tipped cutters are also available. Use a cutter rather than a grinding wheel if the workpiece is a soft metal.

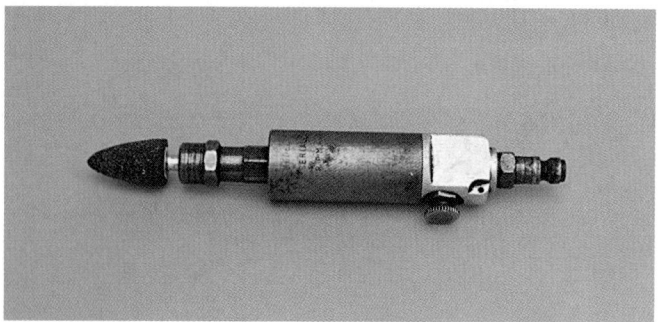

Figure 8.51 A die grinder. *(Courtesy of Tim Gilles)*

Be sure that the grinding wheel or wire wheel is rated for a high enough rpm to match the top speed of the die grinder.

Air Drill

An air drill is a popular tool with automotive technicians (**Figure 8.52**). Sometimes it is fitted with a small wire brush for removing gaskets or carbon on iron or steel. A special wire wheel is available for high rpm use. Called an "encapsulated wire wheel," it has molded plastic in between the wires of the brush. The plastic material wears away to expose only the tips of the wire.

NOTE: *Some air drills turn at too slow an rpm for effective wire brush cleaning.*

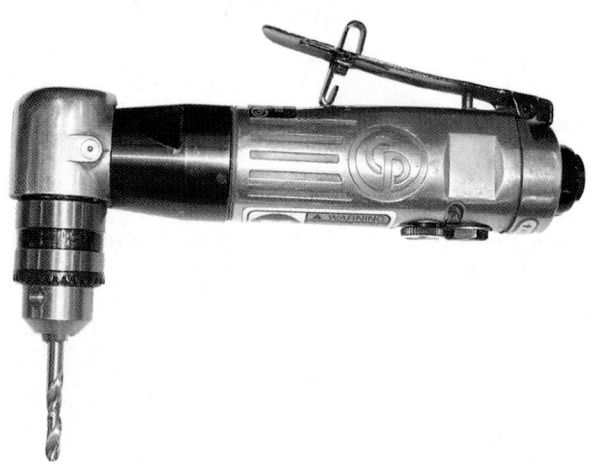

Figure 8.52 An air drill. *(Courtesy of Tim Gilles)*

For maintenance purposes, air tools that are not lubricated through an automatic lubricating system should have a few drops of air-motor oil put into the air inlet each time they are used.

■ Be sure to use face protection. The tool is very loud, so ear protection is also a good idea.
■ Some head gaskets contain asbestos. Do not remove head gaskets with the wire brush.

Angle Grinder

An angle grinder fitted with an abrasive wheel, or cutoff wheel, is used in many shops (**Figure 8.53**). The abrasive cutter is capable of cutting very hard materials, like piston pins and padlocks.

■ SPECIAL SERVICE TOOLS

There are many special tools required for some specialty area service and repair operations. Manufacturers identify special tools for the repair of specific components in their service manuals. Special tool usage is discussed in this text in those chapters dealing with the repair of specific components.

The hardware chapter, Chapter 7, contains information on specialty tools for extracting and repairing broken fasteners. Information in that chapter includes drill bits and taps and dies. The fuel system service chapter, Chapter 43, contains information on the flaring tool used to repair fuel lines. Check the index for the tool desired.

Figure 8.53 An air grinder fitted with an abrasive cutoff wheel. *(Courtesy of Tim Gilles)*

▬ REVIEW QUESTIONS

1. The head of an open-end wrench is angled at _____ degrees.

2. The name of the tool used to loosen and tighten fuel line fittings is a _____ nut wrench.

3. What is the name of a very short screwdriver?

4. An _____ screwdriver is pounded on with a hammer to loosen a screw.

5. For quickness and avoiding cross threading, a _____ handle is used when assembling parts.

6. An extra fulcrum provides some types of pliers or cutters with extra _____.

7. To give a dead blow (one that does not bounce) a soft-faced hammer will have metal shot in its _____.

8. What is the name of the type of puller that uses a heavy weight that is slid against its handle?

9. A _____ separator is a tool that is used with a puller to pull a bearing from a shaft.

10. A sun gear, _____ gear, and planetary pinions are the different gears in a gear head multiplier.

▬ ASE-STYLE REVIEW QUESTIONS

1. Two technicians are discussing hand tool use. Technician A says a 6-point wrench is easier to use in tight places than a 12-point. Technician B says that a ratchet is used to loosen fasteners that are very tight. Who is right?

   a. Technician A   **c.** Both A and B
   b. Technician B   **d.** Neither A nor B

2. Technician A says a special wire brush is needed if it will be rotated at high rpm. Technician B says some wire brushes have plastic molded into them. Who is right?

   a. Technician A   **c.** Both A and B
   b. Technician B   **d.** Neither A nor B

3. All of the following are true about hacksaws *except*:

   a. A hacksaw only cuts on the forward stroke.

   b. A course hacksaw blade (one with fewer teeth) is better for cutting thick steel than a fine blade.

   c. A fine hacksaw blade (one with many teeth) is better for cutting sheet metal.

   d. A hacksaw blade is hardened in the center, so it is best to saw only with the center portion of the blade.

4. Technician A says an 8-point socket can be used on square drive heads. Technician B says that a 12-point socket can be used on a 6-point head. Who is right?

   a. Technician A   **c.** Both A and B
   b. Technician B   **d.** Neither A nor B

5. Technician A says when using an impact wrench to remove a bolt from the front of an engine's crankshaft, the crankshaft must be held to keep it from turning. Technician B says when using an impact wrench to remove lug nuts on a raised vehicle, the brakes must be applied. Who is right?

   a. Technician A   **c.** Both A and B
   b. Technician B   **d.** Neither A nor B

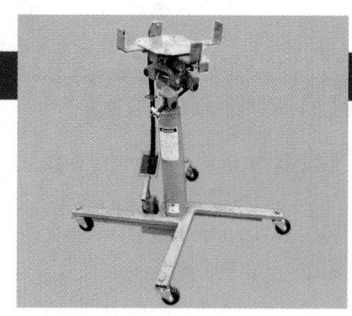

CHAPTER 9

General Shop Equipment

■ OBJECTIVES

Upon completion of this chapter, you should be able to:

✔ Identify shop equipment.

✔ Understand the proper uses of equipment.

✔ Use equipment safely.

■ KEY TERMS

ASME	chainfall	MIG welding
cherry picker	runout	inert gas shield
engine sling	arc welding	oxyacetylene welding

■ SHOP EQUIPMENT

Tools of the trade are classified either as *hand tools, portable power tools,* or *equipment.* Many types of equipment are owned by a typical automotive repair shop. Some are small and do not require power. Others are powered by electricity, hydraulic pressure, or compressed air.

Major pieces of equipment that are shared among all of the company's employees are typically owned by the employer. A well-equipped general shop would have most of the equipment required to assist the technician.

■ HYDRAULIC EQUIPMENT

Example of hydraulic equipment include jacks, shop cranes (engine hoists), presses, and lifts (covered in Chapter 11). It is possible to increase pressure by hydraulic means when a small piston is used to move a larger one. **Figure 9.1** shows hydraulic fluid pressure of 5 psi acting on a 2 square inch area to produce a force of 10 pounds. When that 5 psi is applied to a 4 square inch piston, the force that results is doubled to 20 pounds. The amount of movement required of the jack or press handle (input piston) will be more, but less effort will be needed.

There are many hydraulic lifting devices and supporting equipment in use in repair shops. Their operating principles are very similar. Only a few of the most popular ones have been selected for coverage in this chapter. Most of this equipment is described in an

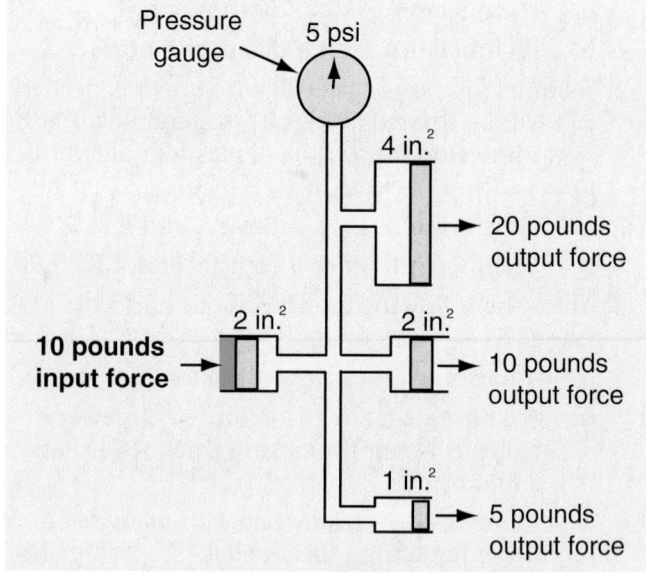

Figure 9.1 Using hydraulic pressure.

American National Standards Institute/American Society of Mechanical Engineers (ANSI/**ASME**) standard. This national standard lists the design, construction, and maintenance requirements for these devices.

Hydraulic Jacks

The hydraulic floor jack, called a *service jack* (**Figure 9.2a**), is an important piece of equipment owned by most shops. It is used to raise and lower the vehicle

(a)

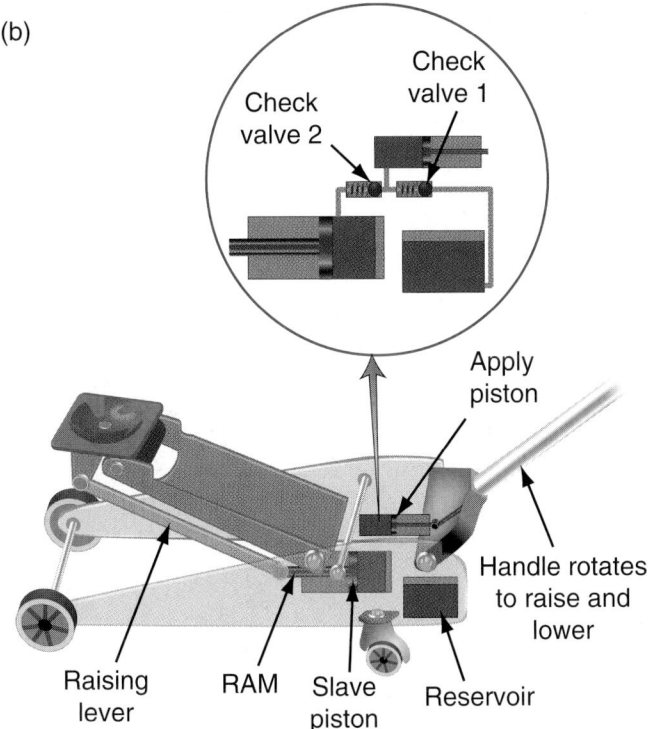
(b)

Figure 9.2 (a) A hydraulic floor service jack. (b) When the handle is raised, check valve #1 moves against its spring, drawing fluid from the reservoir. When the handle is pushed down, check valve #2 moves against its spring, allowing fluid to be forced against the slave piston. *(Courtesy of Tim Gilles)*

and to help position heavier components, such as engines and transmissions.

A hydraulic jack has a fluid reservoir and two one-way check valves (**Figure 9.2b**). Twisting the handle controls whether fluid pressure is allowed to build or escape. To raise the jack, the handle is turned clockwise to close off the return passage to the reservoir and allow pressure to build. When lowering the jack, turn the handle very *slowly* counterclockwise until fluid begins to bleed back to the reservoir and the vehicle slowly lowers.

Position the jack under an area of the vehicle frame, or at one of the correct lift points shown in a service

manual (see Chapter 3). Many vehicles do not have frames and may be damaged if lifted improperly. Do *not* jack under the vehicle floorpan, or under front-end linkages, which can be bent.

When lifting a vehicle from the rear, it can usually be lifted from the center (**Figure 9.3**). Service jack safety is covered in detail in Chapter 3.

SAFETY NOTE Position the jack so that its wheels can roll as the vehicle is lifted. Otherwise, the lifting plate may slip on the frame, or the jack may tip over.

Support Stands (Jackstands)

There are several different types of *vehicle support stands,* most often called jackstands. One type is shown in **Figure 9.4**.

Figure 9.3 The rear of the vehicle can usually be lifted at the center of the axle or frame. *(Courtesy of Tim Gilles)*

Figure 9.4 Jackstands.

The load should be transferred from the floor jack to the jackstand. The floor jack should be left in position as a secondary load support.

 **SAFETY NOTE** A car should always be positioned on jackstands when it is jacked up for service.

Another type of stand is used to stabilize a lifted vehicle when heavy components are being removed. This tall jackstand, called a *high reach supplemental stand*, has a screw adjustable top (**Figure 9.5**).

 **SAFETY NOTE** Be careful not to raise the vehicle off the lift when raising the adjustment on a high reach stand.

Creepers. A creeper has small wheels and saves wear and tear on the body when working in uncomfortable positions or working under a car that is resting on vehicle support stands. There are several special kinds of creepers. A floor creeper is shown in **Figure 9.6**. To avoid having someone accidentally step on it:
- Be sure to stand a floor creeper on end when not in use.
- When taking a break in the middle of a repair job, push the creeper under the raised car or stand if on end.

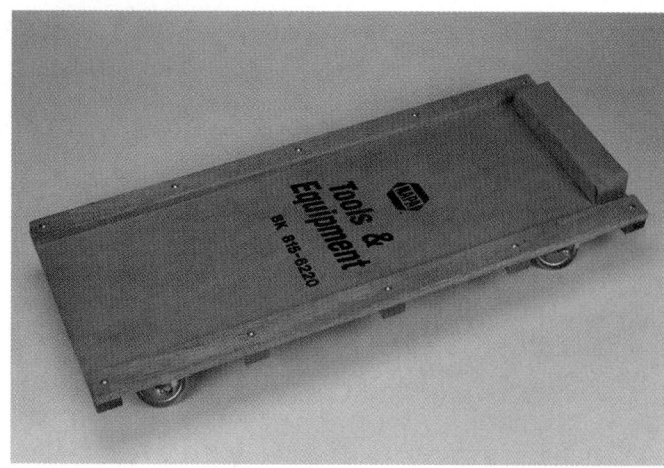

Figure 9.6 A floor creeper.

Presses

There are many sizes of presses. Some are mechanical, such as the arbor press (**Figure 9.7**), and some are hydraulic. The largest press is the most versatile, having both a fast and slow pumping speed, and the ability to separate axle bearings, which sometimes require 25 tons of pressure (**Figure 9.8**).

A 10-ton *electric/hydraulic* press is found in some shops. It is used for such things as pressing wrist pins into and out of connecting rods (**Figure 9.9**). This press does not require pumping of the handle. An electric motor provides pressure that is controlled by a control handle.

There are many special-use press fixtures available. For press work a *bearing separator plate* is often used

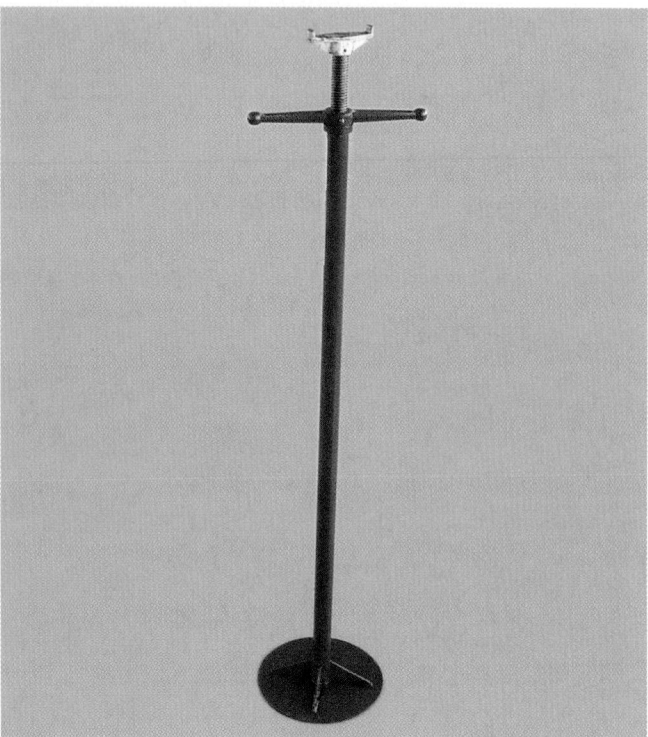

Figure 9.5 A tall jackstand is used to support components when the car is raised on a lift. *(Courtesy of Tim Gilles)*

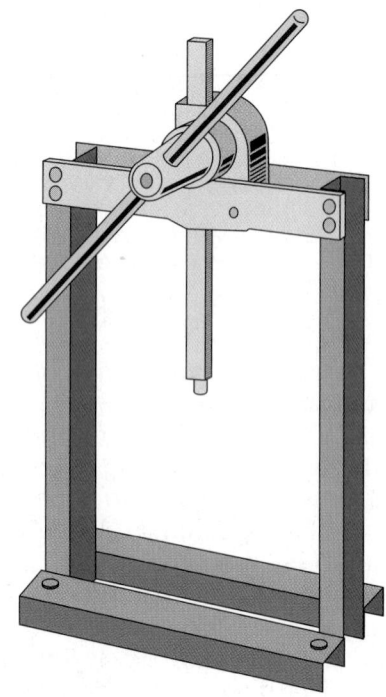

Figure 9.7 An arbor press.

Figure 9.8 A large 50-ton press. *(Courtesy of Tim Gilles)*

Figure 9.9 Pressing a piston pin with an electric/hydraulic press. *(Courtesy of Tim Gilles)*

(see Figure 8.42). Be sure to support it where the bolt holds the two halves of the tool together (**Figure 9.10**). If the separator is installed in the press 90 degrees to the correct position, the bolts will be bent and the tool can be damaged.

Shop Cranes (Hydraulic Engine Hoists)

A shop crane is commonly referred to as a **cherry picker** (**Figure 9.11**). It is used to remove the engine from the car.

Engine Lifting Sling. An **engine sling** (**Figure 9.12**) is used to hook the shop crane to the engine block.

Lifting Sling Safety

1. The sling must be tightened against the block, because the holding bolts can be overly stressed and break (see Figure 3.33).

2. The bolts should go into the block to a depth at least 1½ times the diameter of the fastener.

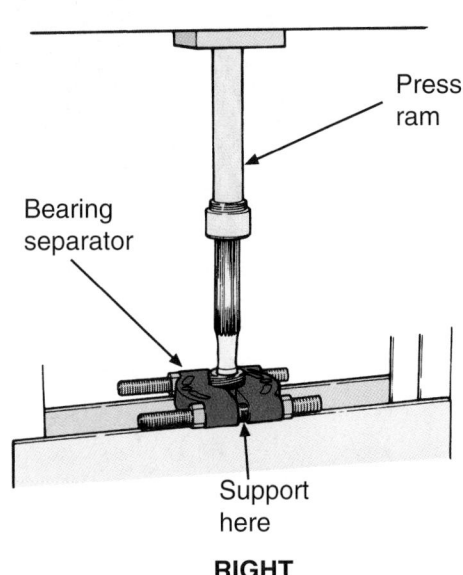

Press ram

Bearing separator

Support here

RIGHT

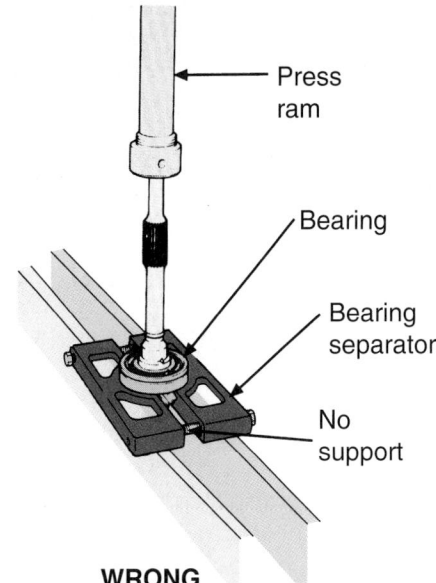

Press ram

Bearing

Bearing separator

No support

WRONG

Figure 9.10 Support the bearing separator under the bolts so the tool is not damaged.

Figure 9.11 A hydraulic engine hoist. *(Courtesy of Tim Gilles)*

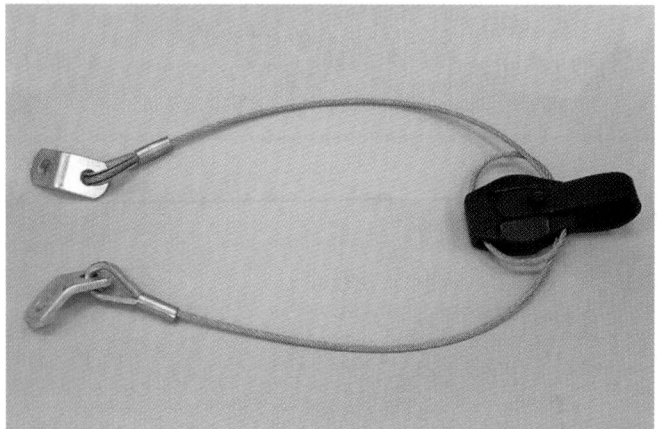

Figure 9.12 An engine sling. *(Courtesy of Tim Gilles)*

Be careful that the sling does not contact the distributor or the carburetor, which can be damaged during engine removal.

Chainfalls

A **chainfall**, or chain hoist, is also a common tool for removing the engine from the car. It is usually hung from an overhead structural beam. The beam can be part of the building structure or a portable I-beam structure with wheels. When the chainfall is used for engine removal, the car's hood must first be removed.

It is more difficult to maneuver an engine back into position during reinstallation using a chainfall.

Engine Stands. The engine stand provides a convenient means of turning the engine over for disassembly and reassembly (**Figure 9.13**). When an engine is removed from a car, it is a good idea to mount the engine on an engine stand immediately.

SAFETY NOTE Inexpensive engine stands are widely available. Be sure to purchase one that is designed to carry the maximum load that you might encounter. Do not use a light-duty stand for a heavy engine.

The stand's universal mounting adapter (**Figure 9.14**) is designed to fit almost any engine. It is bolted to the rear of the engine in place of the bell housing or automatic transmission converter housing.

First, mount the universal mounting adapter to the engine. Then, install the engine and mounting adapter on the engine stand. For more information on the engine stand, see Chapter 50.

SAFETY NOTE It is not safe to work on an engine that is hanging from a shop crane. An engine stand provides a margin for safety, as well as convenience.

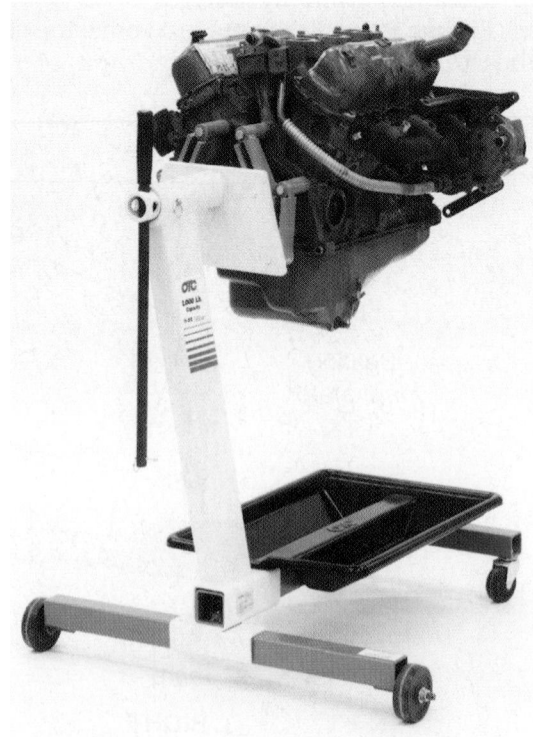

Figure 9.13 An engine stand. *(Courtesy of OTC/SPX Service Solutions)*

Figure 9.14 Mount the engine stand head to the engine. Use washers under the bolt heads. *(Courtesy of Tim Gilles)*

Transmission Jacks

The transmission jack is used to remove automatic and manual transmissions from vehicles. There are two major types of transmission jacks. One kind is used when the car is on support stands (**Figure 9.15**), and the other is used with the car raised on a lift (**Figure 9.16**).

> **SAFETY NOTE** When using a transmission jack with a car raised on a lift, be sure to use high reach stands to stabilize the vehicle. Also be careful not to raise the car off the lift pads as you raise the transmission. For safety reasons, it is preferable to use a wheel-contact lift when removing the engine or transmission.

■ SHOP ELECTRIC MACHINERY

Much of the machinery in a shop is powered by electricity. Electric machinery includes anything with a motor. Motors that power large machines usually run at a constant speed. Many pieces of machinery have different

Figure 9.15 A low lift transmission jack. *(Courtesy of OTC/ SPX Service Solutions)*

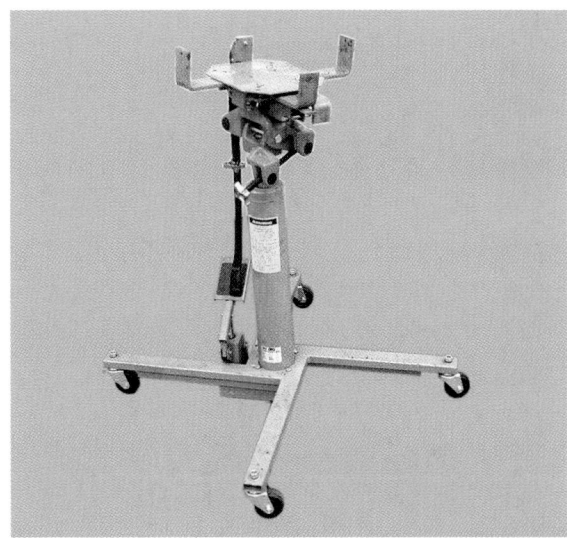

Figure 9.16 A tall hydraulic transmission jack. *(Courtesy of Tim Gilles)*

sized belt-driven pulleys or another means of adjusting the speed of the machine to fit the application.

Drill Press

The drill press (**Figure 9.17**) is a large, stationary machine used to drill parts that are clamped to the press table. There are various types of clamps available.

Drill presses have adjustable speeds. The speed is varied according to the type of metal and the size of hole being drilled. One type has optional pulley grooves for the v-belt drive. The other type (vari-speed) has a pulley that changes sizes by means of a crank. The speed on a vari-speed must be changed while the drill press motor is running.

A cutting lubricant is used to drill all metals except cast iron or aluminum. These metals can be drilled dry or cooled with a water based coolant or kerosene. For information on drilling speeds, drill bits, and lubricants, see Chapter 7.

Drill Motors/Portable Power Tools

There are a variety of sizes of hand-held drill motors (**Figure 9.18**). The part that grips the drill bit is called the *spindle chuck*, or chuck (**Figure 9.19**). A drill bit is tightened between the jaws of the chuck with a *chuck key*. A drill motor is classified according to the largest drill bit that its chuck can accommodate. The following are approximate speed ranges for different-sized drill motors. Notice that the larger drill bits turn at slower speeds (see Chapter 7).

- ¼" drill motor—1,500–2,000 revolutions per minute (rpm)
- ⅜" drill motor—1,000 rpm
- ½" drill motor—500 rpm

Some drills are variable speed; some are also reversible. Drills usually have a button located next to the trigger that can be used to hold the drill in the "on" position. To shut the drill off, just press the trigger again.

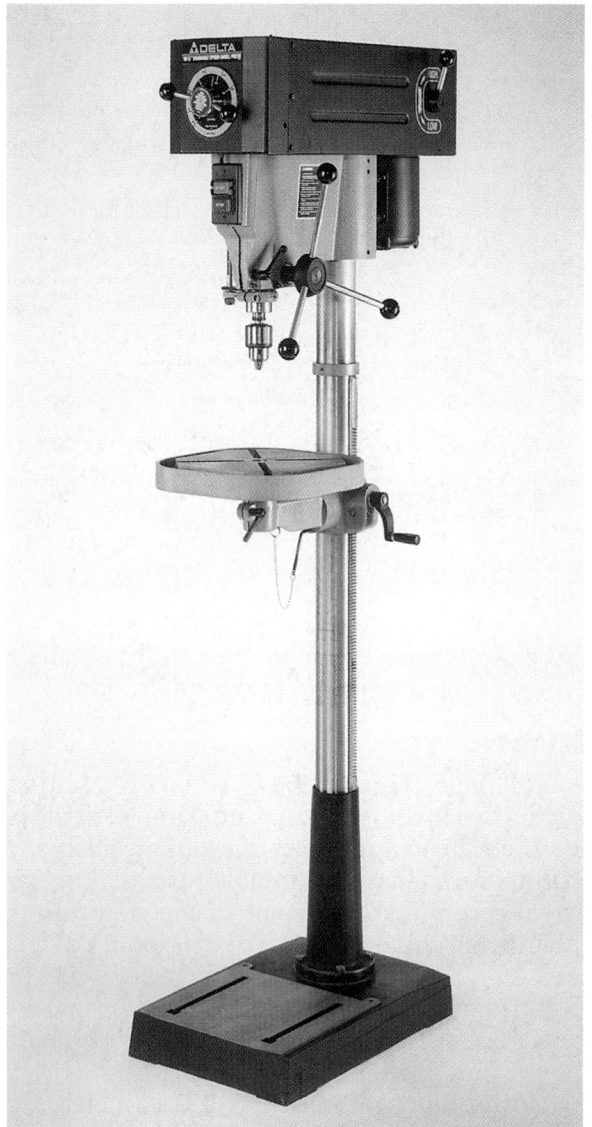

Figure 9.17 A drill press. *(Courtesy of Delta International Machinery Corp.)*

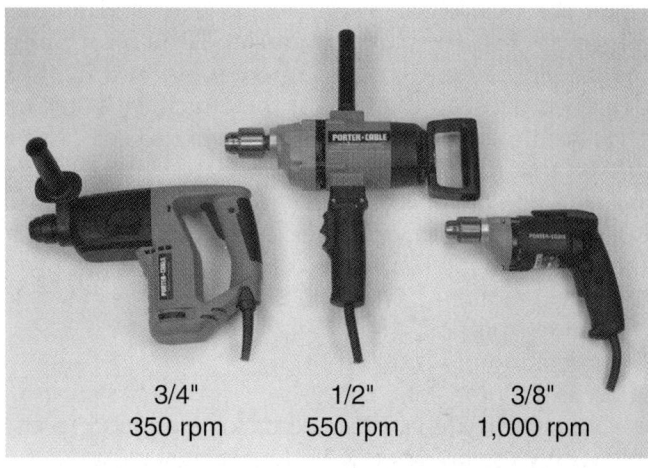

3/4"   1/2"   3/8"
350 rpm   550 rpm   1,000 rpm

Figure 9.18 Various hand-held drill motors.

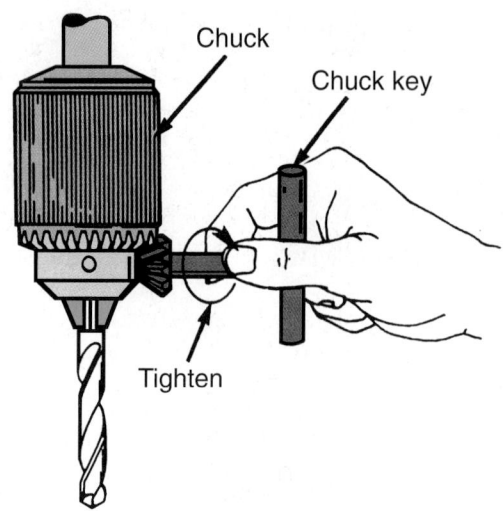

Figure 9.19 The drill bit is tightened into the chuck with a chuck key.

Heavy-Duty Power Tool Considerations

Power tools that are designed to be used by professionals are different than those that are used by the homeowner or hobbyist. That is why they cost more. A tool that will work well in a home environment will probably not survive the challenges of the workplace. **Figure 9.20** shows the parts of a drill motor.

Bearings. Bearings in a drill motor are located at the front and rear of the motor armature, and at the spindle chuck. Three types of bearings are used in power tools: ball, roller, and powdered metal sleeves. Motor armatures on heavy-duty tools use ball bearings, which have almost unmeasurable free play.

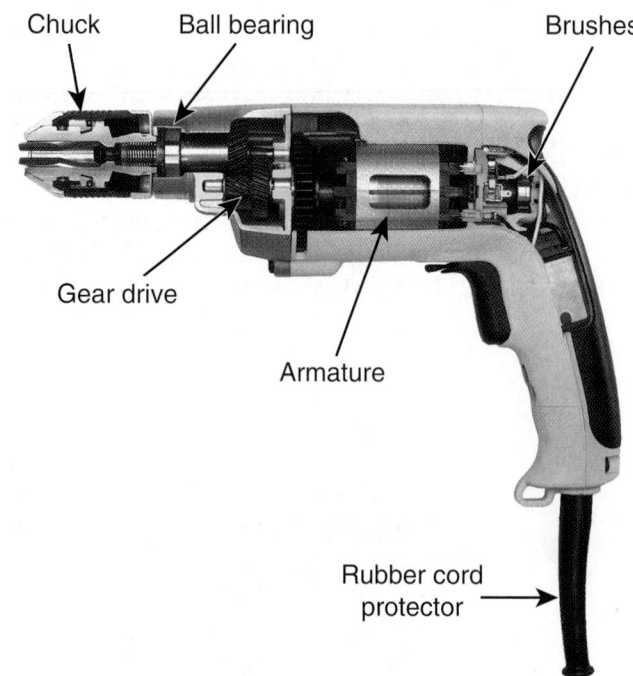

Figure 9.20 A cutaway of a drill motor. *(Courtesy of Black and Decker)*

The spindle chuck of a drill is often under high side-to-side and end loads. A quality commercial drill uses a *ball bearing chuck*, while consumer drills use sleeve bearings.

Chuck. The chuck on a professional drill is one of the most expensive parts of the drill. For durability, the jaws of the chuck, the parts that contact the drill or tool bit, are made of expensive case-hardened steel. More precise machining is done in a commercial drill chuck. Chuck **runout**, measured at a point 1" from the jaws, is the amount that the chuck wobbles. It is allowed to be up to 0.010" in a consumer drill, but only half that amount in a commercial drill.

Housing. Weight and durability are factors in the design of portable power tools. Plastic is usually used for the housings of modern tools. Plastic is an insulator, so the danger of electrical shock is minimized. The plastic used for professional tool bodies is often Super Tough™ nylon. It is practically indestructible and is resistant to oil, solvents, and corrosives, as well as heat.

Pilot pins align the handle, gear housing, and motor housing of a commercial power tool. Separate sets of screws hold one part to the next to ensure alignment of the parts, even after rough handling.

Power Cord. A consumer's power tool will probably have a short power cord. When the tool is used, an extension cord will be needed. A professional portable power tool usually has an 8–10' power cord, which almost always eliminates the need for an extension cord.

Battery-Operated Tools

Battery-operated tools, especially drill motors (**Figure 9.21**), are popular in shops. Some of the higher quality ones have a *clutch* that releases during overtightening or when a drill bit catches on the work being drilled. The clutch feature can prevent breakage of fasteners and drill bits. The versatility provided by not having to run an electrical cord is also a desirable feature.

A typical professional setup includes a separate battery charger and two batteries.

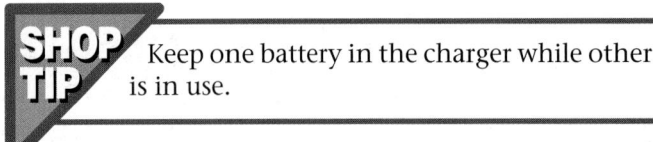

SHOP TIP Keep one battery in the charger while other is in use.

If a battery is fully discharged just before putting it in the charger, it will accept a charge more fully. It is not good for the battery to be allowed to remain in a state of discharge.

Grinder

Almost every shop has a grinder (**Figure 9.22**). Grinders can be either the bench type or the pedestal type (floor mounted). A typical grinder is equipped with a grinding wheel on one side and a wire brush for cleaning parts on the other side. Different-sized motors are available. The motor should be powerful enough to maintain speed under pressure. The grinding wheel is kept square by dressing it with the tool shown in **Figure 9.23**.

The grinder can be used to sharpen tools. A *water pot* is attached to the front of the grinder. The metal being ground must be constantly quenched. If the metal is allowed to become too hot, either of two things can occur:

1. Metal that is heated and then quenched will be made more brittle.
2. Metal that is heated and allowed to cool slowly will be softened (see Chapter 7).

Grinder safety precautions are outlined in Chapter 3.

Figure 9.21 A cutaway of a battery-operated drill. *(Courtesy of Black and Decker)*

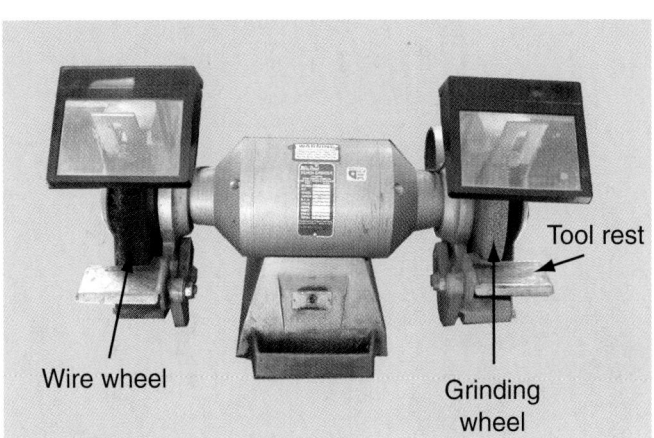

Tool rest

Wire wheel

Grinding wheel

Figure 9.22 A bench grinder with a wire wheel. *(Courtesy of Tim Gilles)*

Figure 9.23 Dressing a grinding wheel. *(Courtesy of Tim Gilles)*

Electric soldering gun

▓ OTHER ELECTRIC EQUIPMENT

Battery Charger

The *battery charger* (**Figure 9.24**) can be used to charge batteries as well as to help the car's battery start the engine. Some battery chargers have instructions that recommend against using the unit for starting the car. Be

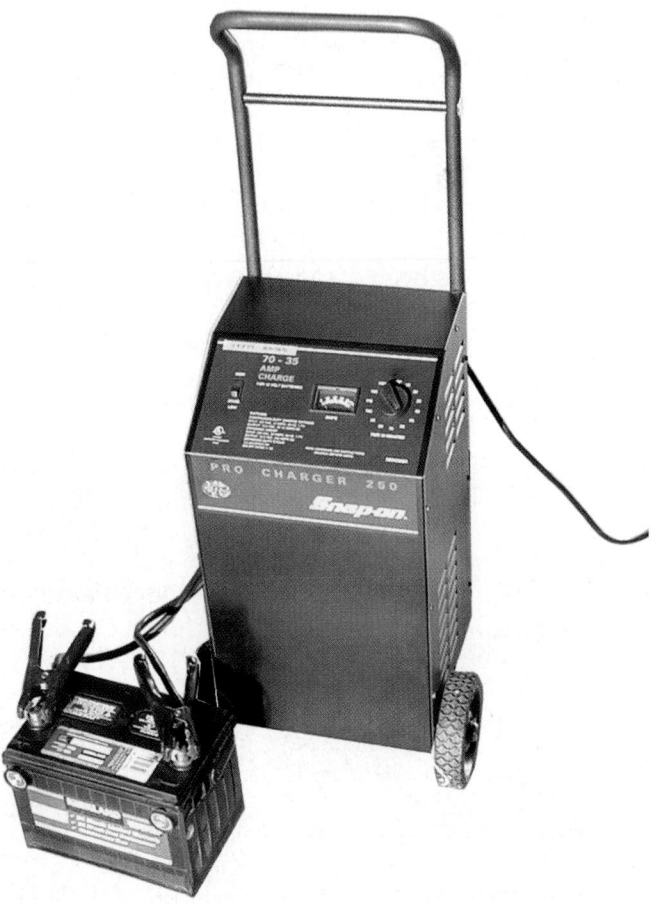

Figure 9.24 A battery charger. *(Courtesy of Tim Gilles)*

Propane soldering pencil

Figure 9.25 An electric soldering gun and a propane soldering pencil. *(Courtesy of Tim Gilles)*

sure to check the instructions for the unit you are using and be sure the cables are clamped to the proper battery terminal. Chapter 3 covers battery charger safety.

Soldering Tools

A soldering iron, soldering gun, or soldering pencil (**Figure 9.25**) is used to melt *solder* when joining electrical wires. Soldering is covered in Chapter 33.

▓ OTHER SHOP EQUIPMENT

The valve grinder, boring bar, cylinder hone, brake lathe, and wheel balancer are other pieces of equipment covered in greater detail in other chapters. The rest of this chapter deals with miscellaneous pieces of general shop equipment.

Vises

Every shop has a vise (**Figure 9.26**). Some vises have pivoting bases that allow them to be turned on the bench. The jaws of automotive vises generally have

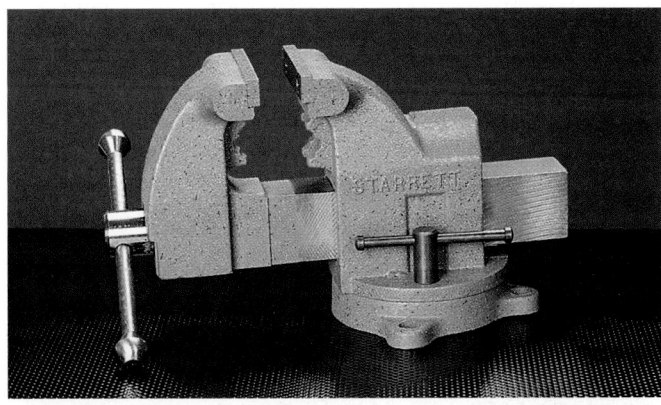

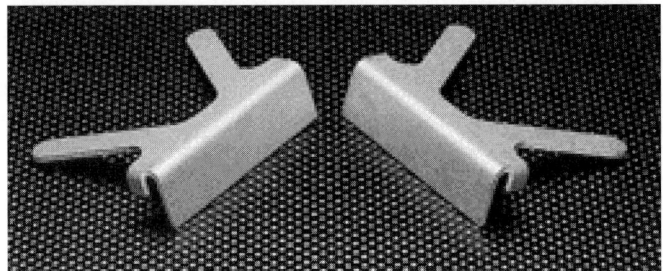

Figure 9.26 A bench vise with pivoting base and soft jaw caps. *(Courtesy of L. S. Starrett Co.)*

teeth, so a set of brass jaw caps is a good investment. They will prevent the vise from leaving a mark on the workpiece.

Vise Safety

1. Be careful not to tighten the vise too much at its wide-open position or the vise might break.
2. Never use a piece of pipe on the handle of a vise; it can break the vise.
3. Do not hammer on a vise, except on the anvil portion that some vises have.

Engine Analyzer

An engine analyzer, or scope, is useful in diagnosing engine problems as well as in tuning a car after an overhaul. The scope can help to diagnose compression problems, uneven balance between cylinders, worn timing chains, and air leaks. Be sure to keep the scope leads (wires) out of the fan.

Tire Changer

Shops that sell or repair tires have a *tire changer* (**Figure 9.27**). The wheel rim is mounted securely to the machine either by an air-operated or mechanical clamp. Some tire changers are totally mechanical, but most are electrical and air assisted. Air, under a high pressure, is used to unseat the bead of the tire from the wheel rim. A bar is inserted between the wheel rim and tire and the wheel is rotated by air pressure or electricity to remove the tire from the wheel.

Figure 9.27 A tire changer. *(Courtesy of Hunter Engineering Company)*

Welding Equipment

There are two main types of welding equipment: electric and gas. In both types, metal is melted into a *puddle*, which is moved by the operator to complete the weld. This is a mechanical skill and requires some practice.

Arc Welding. Electric welding is called **arc welding**. It is inexpensive and fast, but usually requires a 220-volt electrical hook-up (which some shops do not have). A ground (negative) clamp is hooked to the piece to be welded. A welding rod attached to the positive clamp is "struck" against the workpiece in a motion similar to striking a match. The electric arc melts the metal when electrons flow across the arc between the positive and negative electrodes.

MIG Welding. A very popular type of arc welding in automotive shops is **MIG welding** (**Figure 9.28**). MIG stands for *metal inert gas*. An **inert gas shield** (argon, helium, or CO_2) is applied around the arc area during welding to prevent oxidation of the metal. A MIG welder has a spool of wire that is automatically fed through the positive end of the torch. When aluminum wire is used, this kind of welding can be used on aluminum. Aluminum cannot be welded using ordinary arc welding. MIG welding is especially popular for welding sheet metal body panels.

Oxyacetylene Welding. Gas welding is called **oxyacetylene welding.** Two compressed gases, oxygen and acetylene, are stored in separate steel cylinders (**Figure 9.29**). Hoses connect the cylinders to the torch where the gases are mixed. When the gases are mixed in the correct proportion and ignited, intense heat is created at the tip of the torch. The heat is intense enough to melt steel for welding or cutting.

Welding Safety

1. Arc welding produces *ultraviolet rays*. A helmet with the appropriate shade of lens must be used to prevent damage to the eye.

2. Welding can result in splatter of molten metal. Be sure to wear appropriate protective gear.

3. Bystanders must have appropriate protective gear, too.

4. Cutting with a torch can result in a great deal of flying molten metal. Be sure that flammable materials are not in the area.

Figure 9.28 A MIG welder. *(Courtesy of Daytona MIG)*

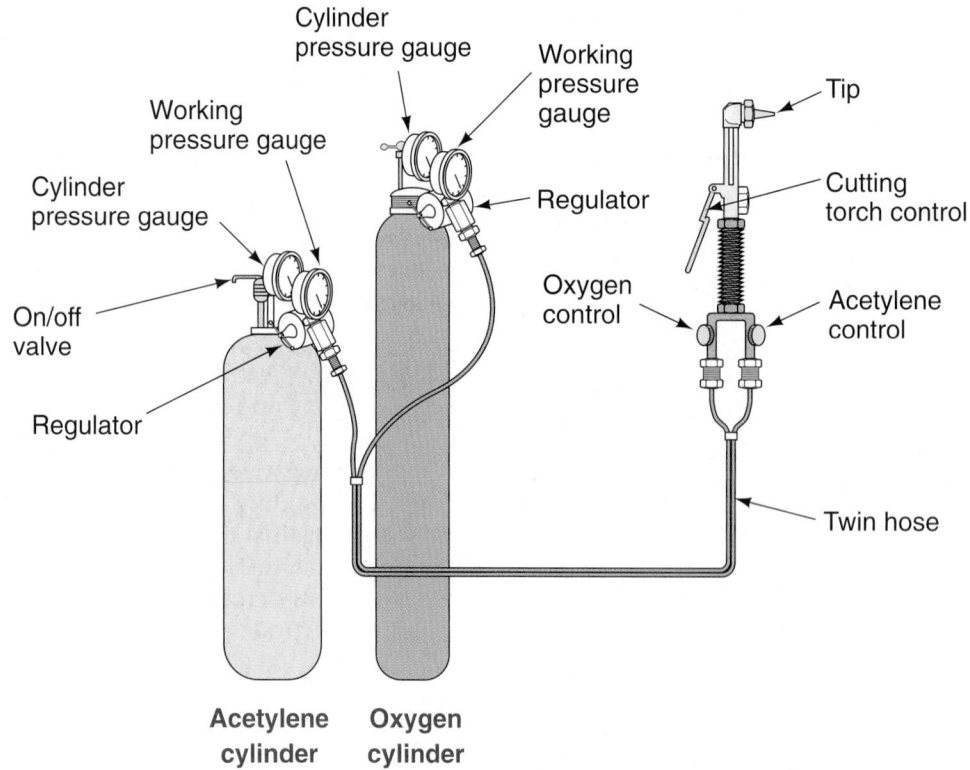

Figure 9.29 Oxyacetylene gas welding and cutting equipment.

Each cylinder is equipped with a regulator that controls the amount of pressure of the gas coming out of the cylinder. The regulator has two gauges: one shows the pressure inside of the cylinder and the other shows the pressure going to the torch.

Oxyacetylene welding has an advantage over arc welding in that it is more versatile. Gas welding equipment can be used to cut and heat metals. Technicians often use the torch to heat rusty or seized parts, such as exhaust components, for easier removal. Pressed fit parts can be heated for easier installation. With a *rosebud tip*, which puts out a great deal of heat, cylinder heads can even be straightened. An obvious disadvantage is that there is an open flame, which can result in burns and start fires.

◼◼ REVIEW QUESTIONS

1. What is a cherry picker?
2. A _____ is a chain or cable attached to the block to help hoist an engine out of a car.
3. The bolts holding an engine sling should go into the block to a depth at least _____ times the diameter of the fastener.
4. When mounting an engine on an engine stand, the mounting head adapter should be attached to the engine before placing it on the _____.
5. Name two metals that do not require a lubricant when they are drilled.
6. When metal being ground is not quenched and is allowed to cool slowly, what happens to it?
7. A battery _____ can be used to assist the battery in starting the car.
8. Jaw caps or soft jaws are used when clamping a part in a _____ to avoid marring the work surface.
9. What type of welding would you use for cutting or heating, oxyacetylene or arc?
10. What is the name of a kind of welding that uses an inert gas shield?

◼◼ ASE-STYLE REVIEW QUESTIONS

1. Technician A says that an arbor press is a type of hydraulic press. Technician B says to use a pipe on the handle of a vise to increase clamping force. Who is right?
   a. Technician A
   b. Technician B
   c. Both A and B
   d. Neither A nor B

2. All of the following are true about lifting an engine *except*:
   a. When a shop crane that is carrying an engine is being moved, the engine should be positioned as low to the ground as possible.
   b. A chainfall can sometimes damage vehicle paint.
   c. A shop crane is sometimes easier to use than a chainfall.
   d. Bolts fastened to an engine sling should be slightly loose against the engine block.

3. Which of the following is/are true about drilling holes?
   a. A single speed drill motor with a ½" chuck turns at a faster speed than one with a ⅜" chuck.
   b. Faster speeds are used with larger drill bits.
   c. A drill motor is classified according to the largest drill bit its chuck can accommodate.
   d. A chuck key is used to unlock a drill's chuck.

4. Technician A says that it is possible to increase pressure by hydraulic means when a large piston is used to move a smaller one. Technician B says that if a force of 10 pounds acts on a 2 square inch area, hydraulic fluid pressure of 5 psi is produced. Who is right?
   a. Technician A
   b. Technician B
   c. Both A and B
   d. Neither A nor B

5. Technician A says that metal that is heated and then quenched will be made more brittle. Technician B says that metal that is heated and allowed to cool slowly will be softened. Who is right?
   a. Technician A
   b. Technician B
   c. Both A and B
   d. Neither A nor B

10

Cleaning Equipment and Methods

■ KEY TERMS

blowby gases	acid	labor intensive
sludge	pH scale	ferrous
alkaline	base	scale

■ INTRODUCTION

This chapter deals with methods for cleaning various parts of vehicles. There are different types of materials that require cleaning. Some are hard deposits, and others are dirt, oils, greases, and contaminants.

The insides of engines and gear cases such as transmissions and differentials contain oils and contaminants that cannot be washed down drains. Cleaning methods differ, depending on whether materials to be cleaned are considered to be toxic or not.

The internal combustion engine in the modern automobile produces many by-products during operation. **Blowby gases**—gases that leak past the piston rings into the crankcase—contain acids, as well as burned and unburned fuel. Moisture also contributes to the accumulation of **sludge** inside the engine (**Figure 10.1**). In addition, carbon builds up in the combustion chambers because of oil that leaks past the rings from the crankcase (**Figure 10.2**). Oil that leaks out of the engine combines with dirt and road grime to make a greasy film on the outside of the engine. Also, mineral deposits can build up in the cooling system.

It is estimated that approximately ⅓ of a shop's expenses can be attributed to cleaning the engine. This chapter deals with the cleaning equipment used when preparing an engine for rebuilding.

■ GENERAL SHOP HOUSEKEEPING PRACTICES

Good housekeeping practices are essential when cleaning automotive parts. A clean, orderly shop is vital for impressing upon the public that a shop is thorough and competent and will do a professional job on its work. Of even more importance, however, is the health and safety of anyone in the shop area.

Slippery floors are dangerous. To avoid the possibility of a dangerous slip and fall, immediately clean up slippery spills. Examples of slippery spills include coolant, lubricants, solvents, glass beads, or steel shot. Preventing spills from occurring in the first place is best. When spills do occur, they must be dealt with

Figure 10.1 Water and by-products of combustion combine with engine oil to form sludge.

Figure 10.2 Carbon deposits sometimes form in the combustion chamber. *(Courtesy of Dana Corporation)*

Figure 10.4 To avoid dripping solvent on the floor, carry wet parts in a drain pan.

immediately. This will also keep the mess from spreading around the shop. Parts that are wet with solvent should be blown off into the solvent tank (**Figure 10.3**) or allowed to air dry before being moved.

> **SHOP TIP** Wet parts can be carried from the solvent tank in a drain pan to prevent solvent from dripping onto the floor (**Figure 10.4**).

When an engine is to be removed from a vehicle, oil and coolant should be drained from it first. The oil filter holds oil, so it should be removed, too.

A spill often occurs when an engine block mounted on a stand is turned upside-down during disassembly. There is usually some coolant still left in the block. Positioning a drain pan under the engine before turning it upside-down can control the spill.

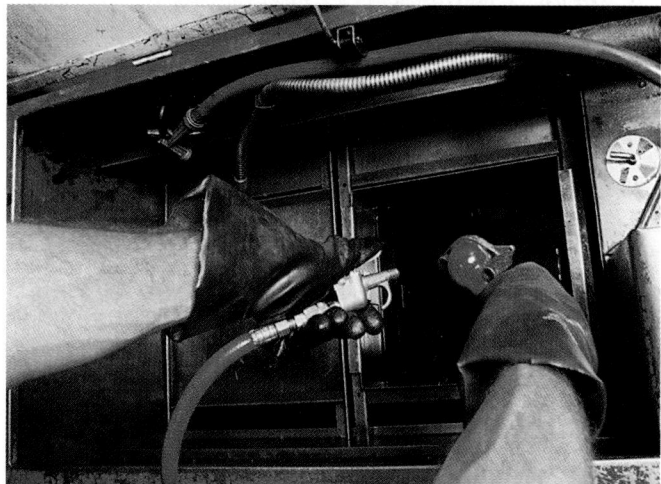

Figure 10.3 When drying parts with compressed air, blow solvent back into the solvent tank.

CLEANING METHODS

There are several categories of automotive materials that require cleaning. Water-soluble deposits (dirt), organic soils, scale, or rust all require different methods of cleaning. Cleaning methods include:
- Wet cleaning, which includes water and/or chemical solutions
- Abrasive cleaning
- Thermal cleaning

The most common of the three methods is wet cleaning. Methods and materials covered in this chapter are those that are the most popular ones used in repair and machine shops. Specific information about cleaning materials is best obtained from the supplier of the equipment.

Chemical Cleaning

Chemical cleaning in engine repair is of three main types:
- **Alkaline** (base)
- **Acid**
- Solvents

> **SCIENCE NOTE**
>
> *The acidity (or alkalinity) of a solution is measured on the* **pH scale***, which ranges from 1 through 14.*
> - *Pure water is rated at 7.*
> - *Solutions that are below 7 on the scale are acidic.*
> - *A pH rating of 1 is a strong acid.*
> - *A pH rating of above 7 is an alkaline cleaner, or* **base***.*
> - *A strong alkaline cleaner has a pH rating above 10.*

Soap and Chemicals. To clean *soils*, a chemical must wet the material. *Soap* is a wetting agent. After wetting, the soap suspends dirt so it can be washed away. Dirt is the only water-soluble material found on automotive parts.

Organic soils include petroleum by-products, gasket sealers and paints, carbon, and other by-products of combustion. These materials *cannot* be effectively cleaned with water. Chemicals are used to make these soluble so that water will be able to wash them off.

Cleaning with Bases

Alkaline materials cut grease very well and work best when heated. Most automotive soaps are alkaline. Using soap and hot enough water temperatures are both very important when cleaning grease. Soap lifts grease and makes cleaning easier and much more effective. Temperature is also very important. Remember how much faster hands clean up when using hot water instead of cold?

SCIENCE NOTE

A cleaning chemical will clean about 10 times as well at 150°F than at room temperature. From 140°F to 200°F cleaning time is cut in half by every increase of 20°F. Agitating, or moving the liquid, also shortens cleaning time.

Cleaning with Acids

Acids are useful only in removing rust and scale. Acid will not cut grease. Before rust can be removed, any oil or grease must first be removed with an alkaline material. When an acid is used to remove rust or scale, it also removes a small amount of the base metal. The chemical used to remove scale deposits from the vehicle's cooling system is an example of an acid used in automotive repair.

Cleaning with Solvents

Solvents are of three types:
■ Water-based
■ Mineral spirits (Stoddard solvent)
■ Chlorinated hydrocarbons (carburetor cleaner)

■ CLEANING THE OUTSIDE OF THE ENGINE

Before a component such as an engine or transmission is removed from a car, it can be cleaned by one of several methods. Pressure washing and steam cleaning are methods of grease removal whose popularity has declined recently due to environmental considerations. The sewer drain must be able to capture all of the hazardous contaminants that would otherwise go into the sewer. Also, these cleaning methods are **labor intensive**, which means that they require an operator. They can also cause damage to air conditioning and electrical components and connections.

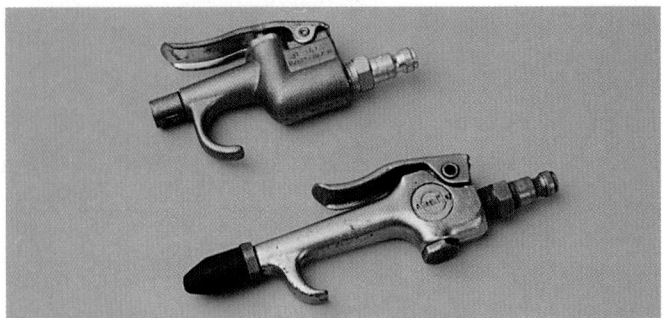

Figure 10.5 A safety blowgun (top) and a rubber-tipped high-pressure blowgun (bottom). *(Courtesy of Tim Gilles)*

Air Blowguns

Air blowguns can be used to blow off parts. There are two main types of blowguns. The *safety blowgun* (**Figure 10.5**) restricts pressure to around 35 psi for air-drying parts. When more pressure is required, a *rubber-tipped blowgun*, which allows full shop air pressure, is used.

SAFETY NOTE Be especially careful when using the high-pressure, rubber-tipped blowgun. Compressed air at full shop pressure can be dangerous and can actually penetrate your skin. Do not blow air against your skin. Wear eye protection and use care when blowing into holes. The liquid will come back out at you. Remember to blow down and away from yourself.

■ ENVIRONMENTAL CONCERNS WITH ENGINE CLEANING

Cleaning engines when they are installed in a vehicle that runs involves cleaning dirt from the outside of the engine. Cleaning oily engine parts that have been disassembled presents another problem. The inside of a dirty engine can harbor hazardous waste. When cleaning the inside of the engine, the steam cleaner and pressure washer are to be avoided. Engine bearings are made of lead and other metals. As bearings wear, the metal is deposited in the engine oil. Used engine oil also contains other contaminants that cannot be allowed to enter the sewer.

Some localities have environmental laws that require the liquid from cleaning operations to be contained to prevent contamination of sewage. Shops that perform these open cleaning operations have special waste water systems. Disposal of toxic waste is costly. Many methods of disposal have been developed by shop operators.

NOTE: *Even materials that are called biodegradable will become hazardous waste when they pick up hazardous materials during use.*

CLEANING INTERNAL PARTS

There are many methods for cleaning parts that have been removed from the engine. There are also additives for removing sludge from the inside of the crankcase of a running engine. The vehicle should *not* be driven while these additives are flushing the engine, because they do not have proper lubricating characteristics. Engine oil is changed immediately following the flushing procedure.

Methods of cleaning the inside of the engine include *chemical cleaning, abrasive cleaning,* and *thermal cleaning.* These cleaning methods must all keep contaminants contained for proper toxic waste handling.

CHEMICAL CLEANING

Chemical cleaning includes solvent tanks, small parts cleaners (carburetor cleaner), hot and cold tanks, spray washers, jet washers, ultrasonic cleaners, and salt baths.

Solvent Cleaning

Solvents are either mineral spirits, Stoddard solvent, carburetor cleaner, or water-based solvent. Once a solvent has been used, they all come under hazardous waste laws. Of these, water-based solvents are easiest to dispose of because they can be evaporated to reduce their volume prior to disposal. Stoddard solvent is commonly recycled, and although the use of gloves is recommended, it is not a major irritant to most people's skin. Carburetor cleaners are the most difficult of these three solvents to dispose of and fumes can be hazardous. Due to safety and environmental considerations, carburetor cleaners (chlorinated hydrocarbons) are not commonly used.

NOTE: *Chlorinated solvents are not legal to use in some communities. Consult the local authorities. Most newer solvents are nonchlorinated.*

Solvent Tanks

The *solvent tank* (**Figure 10.6**), or *cold tank,* is used for cleaning grease off smaller parts. Cold Stoddard solvent does a good job of cutting grease. Although it would clean better if it were hot, evaporation and fuming would result.

Water-based solvents have become increasingly popular. These solvents clean better when hot. They are usually heated to slightly above 100°F, which is comfortable to touch. Water-based chemicals are usually available in either liquid or dry forms. Powder weighs less, so it costs less to ship. If a spill occurs, powder can be swept up and used. Spilled liquid would be wasted.

SHOP TIP Scrape heavy deposits of grease from parts before cleaning them in the solvent tank.

Solvent is applied with a parts brush, which has stiff bristles that will not be softened by the solvent.

Figure 10.6 A water-based solvent cleaning tank. *(Courtesy of Tim Gilles)*

Some solvent tanks have agitators to keep the solvent moving. Agitation reduces cleaning time. Dirt settles in sediment trays in the bottom of the tank where it is periodically cleaned out.

When solvent becomes too oily, it is usually recycled. Companies that specialize in waste handling pump the old solvent from the tank and replace it with clean recycled solvent.

Caustic Cleaning

One cleaning solution for **ferrous** (iron and steel) materials is a mixture of water and caustic soda (lye) heated to about 190°. This strong alkaline mixture is higher than 10 on the pH scale. Brass, copper, or bronze can be cleaned in full-strength solutions because they do not react with the caustic.

NOTE: *If aluminum is cleaned in the solution designed for ferrous metals, it reacts and slowly dissolves.*

Special solutions that are not as strong are available for cleaning aluminum. Some of the aluminum solutions include brighteners. Shops that do a high volume of parts cleaning will have separate cleaning tanks for aluminum and ferrous materials.

CASE HISTORY *An apprentice mistakenly put an aluminum cylinder block into a full-strength caustic solution. When another employee removed it from the soak tank an hour later, it was a very dark grey.*

SHOP TIP If you cannot tell by a visual inspection whether a painted engine part is aluminum or ferrous, use a magnet. Ferrous materials are magnetic, aluminum is not.

Cleaning with a caustic solution leaves metals completely clean. Parts should be rinsed immediately to avoid continuing chemical reaction on the metal.

Preventing Rust. Immediately after cleaning iron or steel with water-based chemicals, it is important to lubricate parts to prevent rust. Because cleaning processes are so thorough, rust can begin to develop within minutes.

SHOP TIP Ferrous metal parts can be dipped in clean mineral spirits or WD40 following cleaning. Some shops use rust prevention products like Cosmoline (the original GI rust preventive).

CAUTION Caustic soda will eat your skin and can cause blindness. Chapter 3 contains a safety section about caustics.

Hot Tanks

The *hot tank*, or *soak tank*, is one of the oldest methods of cleaning engine parts. Materials are cleaned by soaking them for a period of time of from 1–8 hours, depending on the strength of the cleaning solution. Although hot tank cleaning takes a long time, it is popular in small shops because the equipment is inexpensive.

One of the disadvantages of a simple hot tank is that the user has to devise a method of finding items that are soaking in it. An *engine sling* or chain is bolted to the block. When a piece of wire fastened to the side of the hot tank somewhere above the water level is attached to the sling, it can be easily located to retrieve the part from the bottom of the tank. Some of the more expensive hot tanks have lifting tables or baskets built into them (**Figure 10.7**).

Spray Washers

A more expensive type of hot tank that is often found in automotive machine shops is the *hot spray washer*, often called a *jet washer* (**Figure 10.8**). A jet washer is like an automatic dishwasher. It sprays the cleaning solution, heated to 180°F or hotter, from spray heads mounted in a long pipe. The engine block is mounted on a rotating cleaning platform. The spray heads are positioned to spray the solution at the block from the sides, above, and below. Some two-stage spray washers include a soak tank, too.

One problem the spray washer has is that of foaming, which lengthens cleaning time. Foaming occurs if a cleaning cycle is attempted before the cleaning solution is sufficiently heated. A chemical is added to caustic solutions to prevent foaming. An *agitator*

Figure 10.7 A hot dip tank with a basket.

Figure 10.8 A jet washer.

diverts some of the pump water to the bottom of the tank where it stirs up any caustic that may have settled out of the solution. A control lever is positioned in the agitator position for the first 5 minutes of operation each day.

Advantages of a hot spray tank are:

1. Speed—cleaning is completed in 10–20 minutes.
2. Safety—chances of splashing are reduced. The machine will not operate unless the door is closed.
3. Drying—the temperature of the solution results in very quick drying of the parts.
4. Newer machines are more energy efficient.

A disadvantage is that spray washers do not clean oil galleries as well as the hot soak tank does.

Another type of washer has machinery that throws a large volume of cleaning solution at parts, rather than spraying it through nozzles. The advantage here is that the solution hits the dirty parts with a greater impact than the steady stream of a spray washer.

Cylinder heads can usually be hot-tanked before disassembly. This confines all oil and grease to the tank, helping keep the shop clean (**Figure 10.9**).

Be sure to rinse and lubricate all parts immediately after cleaning. If valve springs are allowed to rust, they will be ruined. New valve springs have a protective coating that is removed by the hot tank. Ideally, heads would be disassembled before hot-tanking, but this is not common trade practice.

NOTE: *Following cleaning in a caustic tank, oil galleries MUST be cleaned with a brush after the plugs sealing the ends of the galleries are removed. Over many miles of driving under various maintenance conditions, grime builds up in the oil galleries. This material is loosened during the cleaning process. If it is not physically removed, it will end up ruining new engine bearings and possibly more.*

Caustic Tank Hazardous Waste

Engine bearings must be removed before the block is put into caustic. Bearing lead that settles with other heavy materials in the sludge in the bottom of the tank causes it to become toxic waste.

Rebuilders have several methods of controlling or minimizing toxic waste. Skimming or overflowing the tank is done to remove oily contaminants from the surface of the solution. Some tanks have filtration systems. Others use evaporators to reduce the volume of the sludge. Sludge can be neutralized and solidified with chemicals so that they can be disposed of as non-hazardous waste.

NOTE: *Different communities have different laws governing the handling and disposal of waste materials. Be sure to check with the local agency in charge.*

Removing Scale

A chemical cooling system cleanser can be used to help remove **scale** from the radiator and water jackets *before* removing the engine. The engine shown in **Figure 10.10** has water jackets that are full of rust and scale. Acid cleaning will remove rust and scale, but alkaline cleaners will not.

Manual Cleaning Methods

A wire wheel (**Figure 10.11**) mounted on a bench grinder, on a pedestal grinder, or on a drill motor is the most common tool used for carbon removal.

Hand-held wire brushes are available for cleaning valve guides and oil galleries in the block, head, and crankshaft. Sandpaper is sometimes used to clean machined surfaces, but only very fine sandpaper should be used on bimetal engines that have super smooth head and block surfaces. An electric or air drill can be fitted with a small wire brush for removing

Figure 10.9 These heads are installed on a special head rack for cleaning before disassembly. *(Courtesy of Tim Gilles)*

Figure 10.10 The water jackets in this cylinder block have rust and scale. This is removable by acid cleaning, but not with base. *(Courtesy of Tim Gilles)*

Figure 10.11 A wire wheel mounted on a grinder. *(Courtesy of Tim Gilles)*

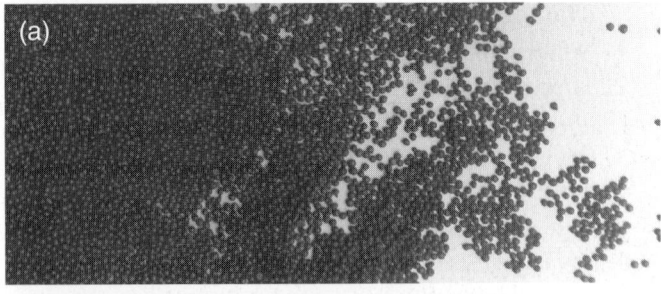

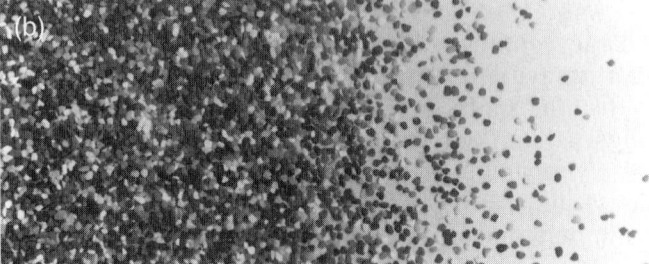

Figure 10.12 Blasting media. (a) Shot. (b) Grit.

gaskets or carbon from cast iron parts. These tools can damage aluminum, however. Plastic abrasive discs are also popular tools for cleaning aluminum surfaces (see Chapter 51). Some are flat mesh surface conditioning discs, and others are plastic dish-shaped bristle discs embedded with abrasive. As the brush wears away, fresh new abrasive is exposed. The grit is determined by the color of the disc: green is coarse, yellow is medium, and white is fine.

A special wire wheel called an encapsulated wire wheel is available for high-rpm use. It has molded plastic in between the wires of the brush. The plastic material wears away to expose only the tips of the wire.

Additional manual cleaning methods are described in Chapter 51.

Abrasive Cleaning

Materials to be cleaned by abrasive cleaning methods must be free of oil and grease, which can interfere with the proper operation of an abrasive cleaning machine. Following precleaning, two types of abrasive blasting are used for various cleaning applications. *Shot* is round and *grit* is sharp and angular (**Figure 10.12**). Several blast materials are used by engine rebuilders for cleaning parts. Steel shot and glass beads are used for automotive part cleaning when removal of the surface of the material being cleaned is not desired.

Beads and shot come in various sizes, depending on the application. Smaller beads are used where there are tight corners and crevices (like screw threads or gear teeth) to be cleaned. Large beads are better for cleaning flat surfaces or loosening heavy deposits. Other shot materials, like stainless steel, ceramic, plastic, walnut shells, or aluminum may be used for special applications. Plastic chips, for instance, are used on plastics and soft metals, or when there is a chance of trapped shot dislodging and destroying an engine or a transmission.

Beads or shot may be used to improve the strength of parts. While ordinary shot blast cleaning does

impart some strength to a part, it is not *peening*. Peening specifically for strength is done only in heavy-duty or high-performance instances. Peening for strength is a controlled process that uniformly compresses stressed areas of a part.

Steel, which is heavier than glass, is used more often for peening. It peens more intensely and lasts several times as long as glass beads before it is worn out.

Grit, used for heavy-duty cleaning, often uses the same blasting material, called *media*. The shape of the media is angular, rather than spherical, so it etches (removes) material from the part surface during cleaning. It provides excellent surface preparation prior to painting. Steel grit and aluminum oxide are the most common grit materials. Grit blasting also causes stress in the surface of machined parts. It is not widely used in the automotive industry.

Glass Bead Blasters

Glass bead blasting is a very effective means of removing carbon. *Bead blasters* (**Figure 10.13**) are found in most automotive machine shops. The finish left by the beads improves the surface of the material by removing flaws and stress spots. It is also said to provide an ideal bearing surface. Bead-blasted parts are shown in **Figure 10.14.**

A costly drawback to glass bead blasting is that the machine is *labor intensive* (requires an operator). Other popular cleaning methods, like shot blasting, do not require a machine operator. To operate the bead blaster, compressed air draws in glass beads and directs them at the parts to be cleaned. Most units have a foot pedal to control the flow of air and beads. The operator watches through a window and aims the nozzle at the parts.

Reclaimer

Figure 10.13 A glass bead blaster with a reclaimer. *(Courtesy of Tim Gilles)*

Figure 10.14 One-half of this cylinder head was bead blasted.

Figure 10.15 This bearing failed when glass beads became impregnated in it. *(Courtesy of Tim Gilles)*

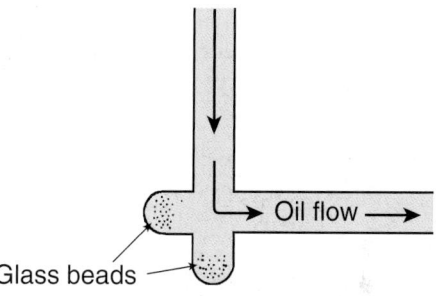

Oil flow →

Glass beads

Figure 10.16 Glass beads can become trapped in blind oil galleries during cleaning.

Beads become trapped

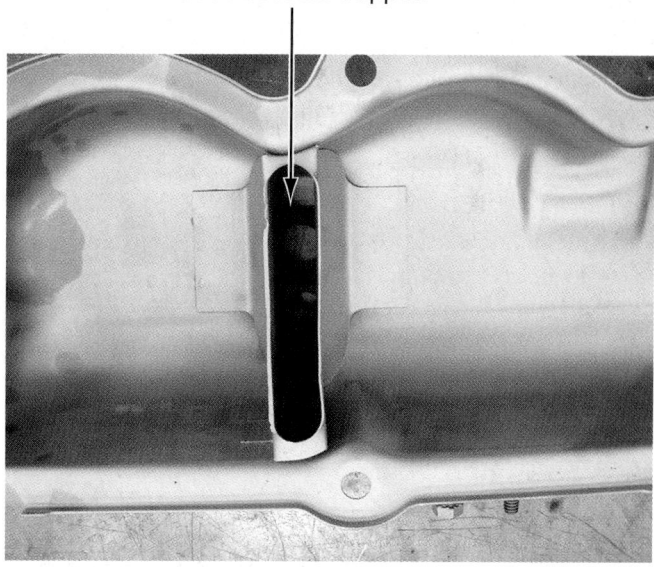

Figure 10.17 Glass beads can become trapped in inaccessible places during cleaning. *(Courtesy of Tim Gilles)*

Glass bead cleaning is often done improperly. Bead blasting a part that has oil galleries is not advisable. Oil galleries often have blind spots where beads can become lodged during cleaning. Trapped beads can escape into engine oil and embed in soft engine bearings (**Figure 10.15**). The bearings can swell, eliminating bearing oil clearance or causing abrasive wear:

■ Bead blasting a part that has oil galleries is not advised. Galleries are often drilled to intersect other oil passages. This process leaves blind spots where beads can become lodged, later to be pulled out into the oil stream by the flow of engine oil (**Figure 10.16**).

■ Sheet metal parts such as oil pans or valve covers often have spot-welded inserts that can trap beads during the cleaning process (**Figure 10.17**).

The cast aluminum valve cover shown in **Figure 10.18a** shows a hidden cavity that can trap contaminants. Do not use the bead blaster on valve covers or oil pans that have sheet metal baffles unless the baffle can be removed (**Figure 10.18b**).

Glass beads can cause abrasive wear, especially to aluminum. *Excessive* blasting can round out ring grooves, which must be square for proper ring sealing (**Figure 10.19**). Pistons should never be cleaned using a bead blaster unless the connecting rod has first been removed. Glass beads that get in between the piston

(a)

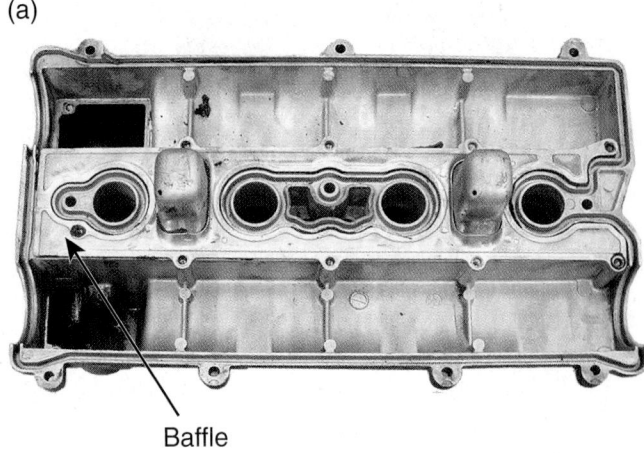

Baffle

(b)

Baffle

Dirt collects here

Figure 10.18 (a) Appearance of a cast valve cover with a steel baffle. (b) This hidden cavity can harbor contaminants. *(Courtesy of Tim Gilles)*

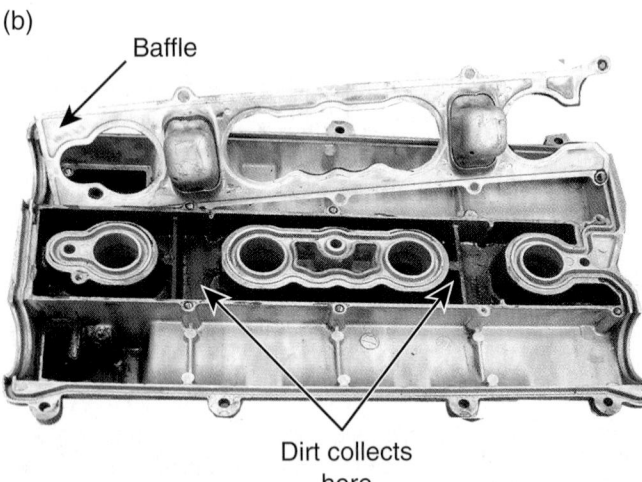

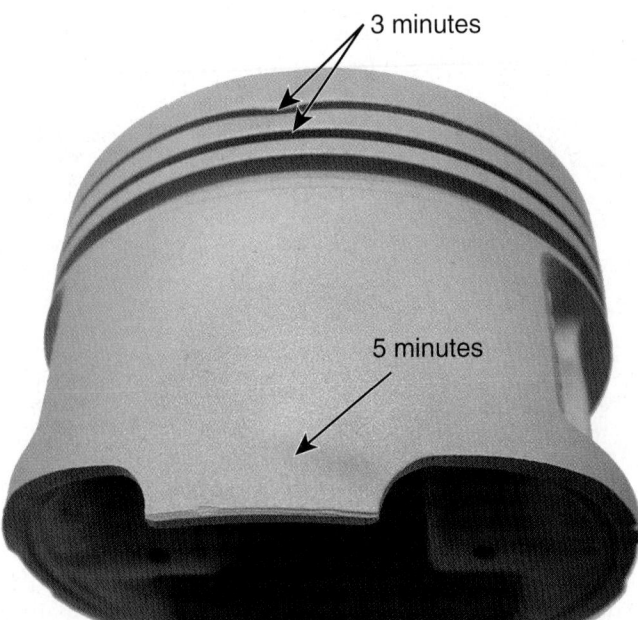

3 minutes

5 minutes

Figure 10.19 Appearance of surfaces of a piston that have been continuously bombarded with glass beads. *(Courtesy of Tim Gilles)*

CASE HISTORY *A machinist cleaned an aluminum cylinder head using a glass bead blaster. The engine was reassembled and installed in the car. The customer drove the car home (30 miles). The next morning the engine would not crank over. Upon disassembly, it was discovered that the engine bearings were embedded with glass beads. The beads swelled the bearings, taking up the normal bearing clearance and preventing the crankshaft from turning.*

pin and bore can cause wear to the aluminum piston pin bores. Beads left in piston ring grooves can interfere with proper ring function.

Following cleaning, glass beads must be thoroughly removed from all parts. It is important that the air supply to the bead blaster be dry, or beads will stick together, plugging up the machine. To clean parts after blasting, first use compressed air from the blowgun inside the blaster cabinet. Do not use solvent, which tends to stick the beads to each other. Parts can be dried in an oven to make removal of glass beads easier. A *tumbler*, used most often for removing steel shot following shot peening (covered later in this chapter), can also be used to remove glass beads (**Figure 10.20**). Machined areas on ferrous metals should be lubricated immediately to prevent rusting. The following precautions should be taken when using a bead blaster:

■ Do not accidentally blast the window in the blaster cabinet. The result is a "frosted" glass that must be replaced.

■ Do not blast parts unless the reclaim motor is on. The entire shop will be filled with glass dust.

Figure 10.20 A tumbler. Engine parts are placed in the center of the tires. As the tires roll, glass beads are shaken out of them. *(Courtesy of Tim Gilles)*

- Some bead blasters have separate reclaim cabinets. When glass beads become too small to be useful any longer, they are separated out by the reclaimer, which consists of several long, cloth tubes. Service to the reclaim unit includes moving a handle on the side of the cabinet in and out to shake spent beads off. They land in a tray beneath the tubes, which fills with worn-out beads and must be periodically emptied.
- Two rubber gloves extend into the cabinet for holding the blaster nozzle (**Figure 10.21**) and maneuvering parts. Try not to hold parts with the gloves. The fingers of the gloves get holes worn into them (**Figure 10.22**). A small spring clamp is effective for holding small parts. Do not use the bead blaster if there are holes in the gloves.

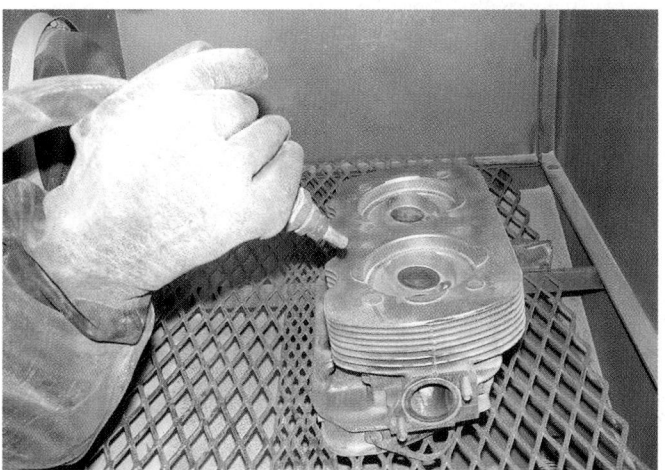

Figure 10.21 The blast nozzle held with a rubber glove. *(Courtesy of Tim Gilles)*

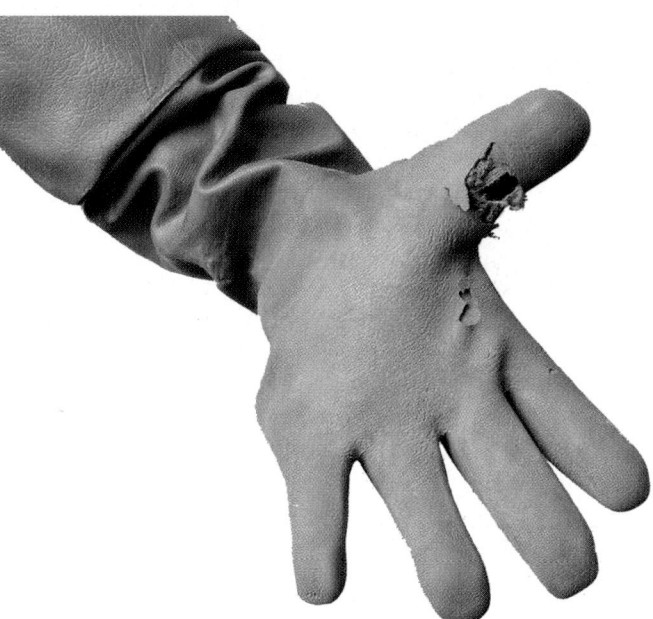

Figure 10.22 A worn left bead-blaster glove caused by holding the part while blasting. *(Courtesy of Tim Gilles)*

Soda Blaster

A popular, more recent method of cleaning is the *soda blaster*. Soda blasting is similar to glass bead blasting, but baking soda is used as the cleaning medium. The soda is used only once. There is no reclaiming as there is with glass beads. One advantage to soda blasting is that a greasy part can be put into the cabinet and cleaned without having been precleaned of oil or grease, as is required with glass bead cleaning (**Figure 10.23**). Another important advantage to soda blasting is that the removal of residual material is not as crucial as it is with glass bead cleaning. Soda dissolves in water and is easily washed out of oil galleries. Soda blasting is not as effective as bead blasting for removing rust, but it works very well for carbon removal.

Airless Blasters

An airless centrifugal blasting machine, also called a *shot blaster* (**Figure 10.24**), uses an impeller in a sealed cabinet to scatter steel shot at a part from above and below. Shot blasting has recently become more popular due to environmental concerns with other cleaning methods. While this cleaning process is not labor intensive, two extra operations are required: pre-cleaning of oil and grease, and removal of trapped shot from engine parts before beginning reassembly. A tumbler is used to remove any steel shot that remains in the part.

Shot blasting is widely used to clean ferrous parts such as the block, heads, and sometimes crankshafts and connecting rods. It is done before the parts are remachined because the process distorts machined surfaces. Shot blasting can damage valve springs; it changes their spring rate. Most shops use 100% steel

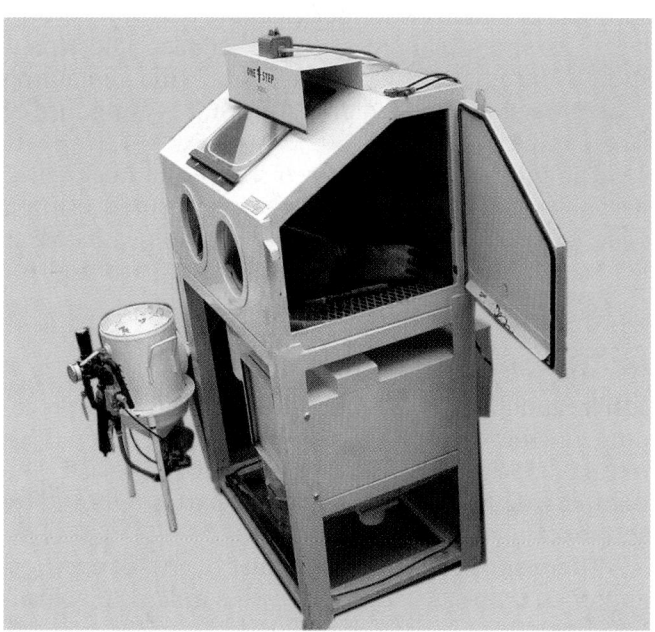

Figure 10.23 A soda blaster. *(Courtesy of Tim Gilles)*

Figure 10.24 A shot blaster.

Figure 10.25 A cleaning furnace. *(Courtesy of Pollution Control Products)*

shot, although some like to use a mixture of steel and aluminum shot to give a brighter looking surface.

Shot comes in different sizes and hardnesses. Popular sizes range from 110 (smaller) to 230 (larger). The most popular size is 170.

NOTE: *Changing the size of the shot affects the amount of impact on the part; larger shot has more impact.*

Using a 110 shot results in an impact that is more than four times less than using a 230 shot. However, a small shot works better for lighter cleaning and smaller surfaces. As shot wears, it becomes smaller. Old, spent shot is automatically separated from the remaining shot. New shot is added and the result is a mixture of larger and smaller sizes.

The most widely used steel shot has a Rockwell C hardness of 40–50 Rc. Hardness is measured using a scale called the Rockwell scale, with values listed as Rc. Harder shot is used for heat-treated parts and for peening.

■ THERMAL CLEANING

Many rebuilders use *thermal cleaning*; a cleaning procedure in which a *pyrolytic* (high-temperature) oven (**Figure 10.25**) cooks oil and grease, turning it to ash. The hard, dry deposit that is left on the part is removed by shot blasting or jet washing.

There are two types of thermal ovens: convection and open flame. A *convection oven* is a flameless, insulated oven that is heated by burners from the bottom. Parts are not exposed to flame and heat up gradually

as the oven heats up. This is said to be an advantage in that the gradual heating of parts gives less chance of warpage. Depending on the size of the oven and the quantity of parts being cleaned, the cleaning cycle lasts from 1–4 hours.

Temperatures for cleaning ferrous metals are about 700°F. Aluminum, which softens at about 650°F, is cleaned at about 450°F.

An *open flame oven* is like a rotisserie. Parts are mounted in a cage that avoids hot spots by slowly rotating the parts directly over a flame. The average temperature of the flame is about 1,100°F, but the temperature of the air inside the oven is only about 500°, allowing aluminum and ferrous metals to be cleaned together. After about 10 minutes of exposure to the flame, it goes out and a 20-minute baking cycle begins. The total cleaning cycle lasts about ½ hour.

With oven cleaning, three processes are actually used: precleaning, baking, and postcleaning. Shot blasting is the choice for postcleaning recommended by most oven companies because parts can be blasted without waiting for them to cool down as would be required with jet washing. Ash that remains is soft, and the time in the shot blaster is only a few minutes. Fifteen minutes in the tumbler finishes the process.

Advantages of thermal cleaning are:

1. Lower cost: not labor intensive, no operator is necessary except for transferring parts between the oven, blaster, and tumbler.

2. While lead or heavy metals still remain in the ash residue, there is a lower volume of hazardous waste to dispose of than with some of the other cleaning methods.

3. The inside of oil galleries in the block are thoroughly cleaned.

4. Heat turns rust and scale in the water jackets to powder. These materials would normally need to be removed with acid.

5. Carbon deposits in manifolds and combustion chambers are loosened. These deposits resist other cleaning methods.

6. Shot blasting removes stress raisers in the surface of parts. This can strengthen the part, lessening the tendency to crack.

7. Aluminum welding is easier after open flame cleaning because contaminants are so thoroughly cleaned.

8. Warped cylinder heads can be straightened in the oven (see Chapter 52).

Gallery plugs should be removed before cleaning ash with a shot blaster to prevent shot from being trapped in oil galleries. Studs in aluminum cylinder heads should be removed before oven cleaning. They will be difficult to remove afterwards.

Air pollution from vaporization of contaminants is dealt with by oxidizing the pollutants in the smokestack. Minimum temperature in the smokestack is from 1,400°–2,200°F so there is no visible soot or unburned hydrocarbons.

▬ VIBRATORY PARTS CLEANERS

A *vibratory parts cleaner* (**Figure 10.26**) is a vibrating tub that uses large beads of ceramic, aluminum, or plastic in a cleaning solvent (usually Stoddard solvent).

Figure 10.26 A vibratory parts cleaner. *(Courtesy of C&M Cleaning Systems)*

It is very effective on valves and valve springs and does not require a machine operator. It should be installed in a soundproof room. When the tub shakes, it is noisy.

▬ OTHER CLEANING METHODS

Other cleaning methods are used in larger and non-automotive applications. *Salt bath* cleaning systems are used mostly by large production engine rebuilders. Parts are soaked in a bath of 650°F molten oxidizing salt for 5–10 minutes and then rinsed.

Ultrasonic cleaning is used commonly by jewelers and dentists. It cleans by sound waves cycling through water and detergent that is heated to between 120°F and 140°F. As the sound waves cycle at about 40,000 times per second, bubbles open and collapse, tearing contaminants away from the metal. Some rebuilders use this method for small parts cleaning.

Marking Clean Parts

Clean parts can be marked with number or letter stamps or with a colored paint marker (**Figure 10.27**). Colored paint markers have white or yellow paint stored in a tube with a rolling ball point. A rubber bulb on the end is used to prime the ball point. The paint is resistant to oil and solvents. Parts like blocks, heads, or crankshafts can be readily labeled.

Figure 10.27 Labeling parts with a paint marker. *(Courtesy of Tim Gilles)*

■ REVIEW QUESTIONS

1. Wet parts can be carried from the solvent tank in a _____ _____ to prevent solvent from dripping on the floor.

2. Pure water has a pH rating of _____.

3. If the nozzle of a bead blaster is aimed at the window, it frosts the _____.

4. Why should the nozzle not be aimed at the glove that is holding the part being blasted?

5. Be sure to rinse and _____ to protect machined areas of ferrous metals after hot-tanking.

6. How can a metal be tested to see if it is ferrous or nonferrous?

7. An OHC head has oil galleries. This makes it impractical to clean these heads with a glass bead _____.

8. What is another name for the airless blaster?

9. _____ and open flame are two types of thermal cleaning ovens.

10. Rust and _____ are two types of materials that thermal cleaning removes that other cleaning methods do not.

■ ASE-STYLE REVIEW QUESTIONS

1. Which of the following could be used for removing rust or scale from the cooling surfaces of a cast iron engine?
   - **a.** Base
   - **b.** Acid
   - **c.** Both A and B
   - **d.** Neither A nor B

2. Technician A says that if the temperature of the hot tank liquid is too low, cleaning time will be longer. Technician B says that if the temperature of the hot tank liquid is too low a spray-type hot tank will foam over. Who is right?
   - **a.** Technician A
   - **b.** Technician B
   - **c.** Both A and B
   - **d.** Neither A nor B

3. Technician A says that larger shot hits a part with more impact. Technician B says that parts can be made stronger by shot peening. Who is right?
   - **a.** Technician A
   - **b.** Technician B
   - **c.** Both A and B
   - **d.** Neither A nor B

4. The temperature of the caustic cleaner in the hot tank should be about:
   - **a.** 100°
   - **b.** 150°
   - **c.** 190°
   - **d.** 250°

5. Which of the following could be used to clean an aluminum part?
   - **a.** A shot blaster
   - **b.** A soda blaster
   - **c.** A caustic cleaner
   - **d.** None of the above

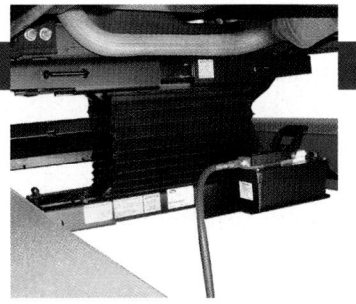

CHAPTER 11

Lifting Equipment and Air Compressors

■ OBJECTIVES

Upon completion of this chapter, you should be able to:

✔ Explain the differences between in-ground and surface mount lifts.

✔ Explain the differences between frame contact, wheel contact, axle engaging, and pad lifts.

✔ Understand hydraulic and semi-hydraulic lift operation.

✔ Understand lift inspection, maintenance, and safety requirements.

✔ Safely position and raise a vehicle on a lift.

✔ Understand the operation of air compressors.

✔ Understand the differences between types of air compressors.

✔ Properly maintain air compressors.

■ KEY TERMS

center of gravity
duty cycle

free air delivery
SCFM

■ INTRODUCTION

Hydraulic equipment is used for lifting vehicles and heavy parts. Portable equipment, like a hydraulic floor jack, is used in all automotive shops. That equipment is covered in Chapter 9. Most shops also have hydraulic or electric lifts for lifting vehicles high in the air for convenience when they are worked on. This chapter deals with lifts and the air compressors that operate some of them.

The Automotive Lift Institute (ALI) is a trade association that promotes the safe design, construction, installation, operation, maintenance, and repair of automotive lifts. ALI has several safety materials available.

Automotive *lifts* are often called *hoists* or *racks*. The two main categories are the *in-ground* lift and the *above ground* (surface mount) lift. When the lifting mechanism of a lift is located below the floor, the lift is an in-ground lift.

■ LIFT TYPES

Within the two categories of lifts (surface mount and in-ground) there are several styles, but two main types:

■ The *frame-contact* lift (**Figure 11.1**).
■ The *wheel-contact* (drive-on) lift (**Figure 11.2**).

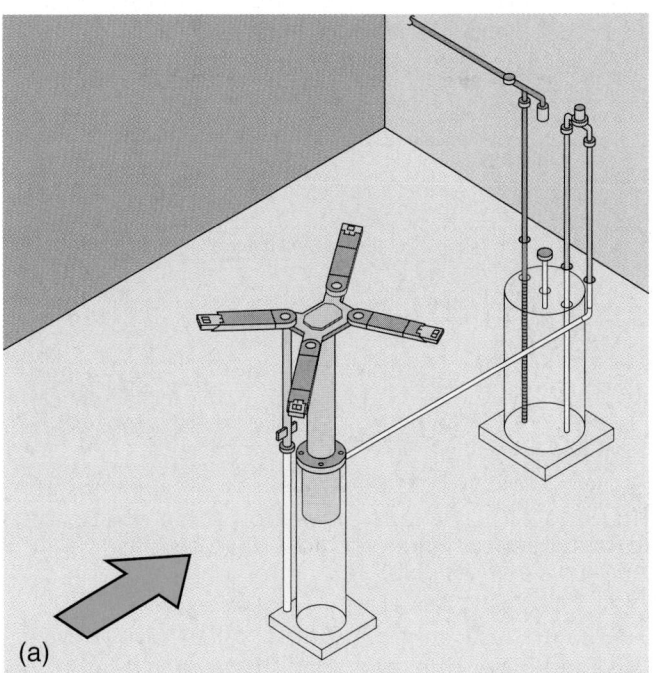

Figure 11.1 (a) Single-post frame-contact lift. (*Courtesy of Automotive Lift Institute*)

(continued)

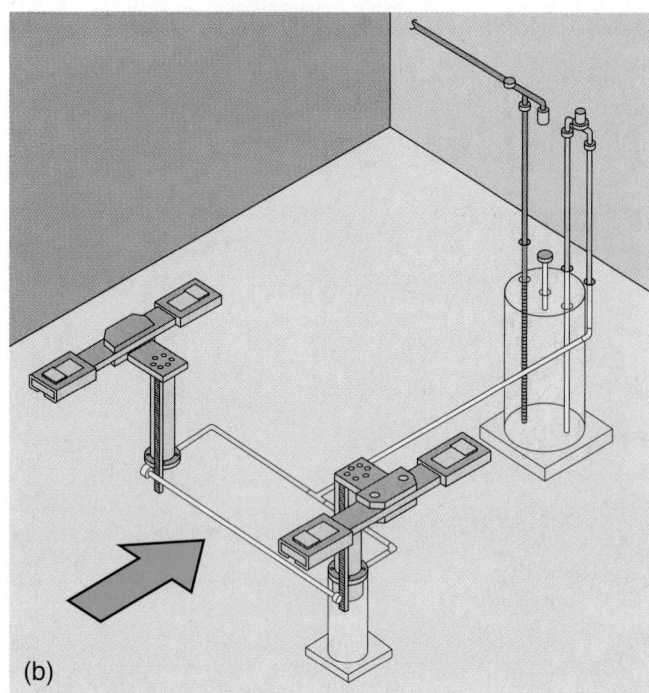

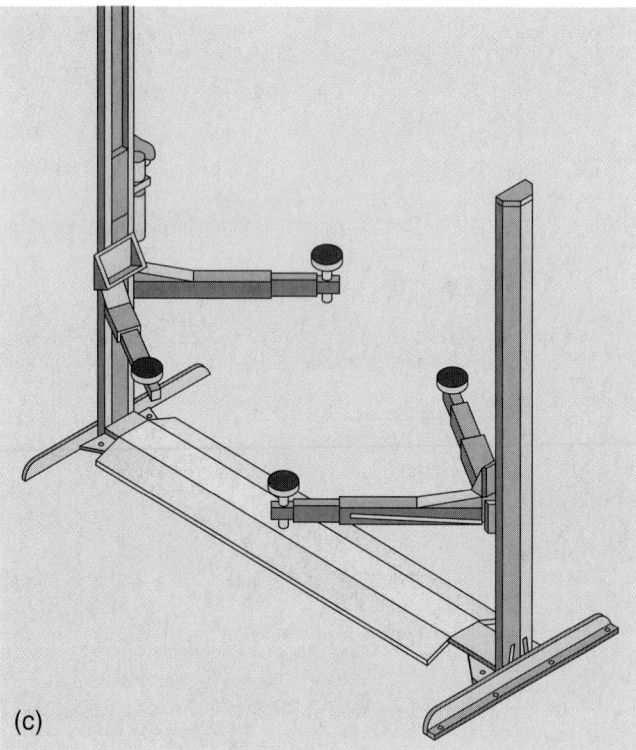

Figure 11.1 (continued) (b) Two-post frame-contact lift. (c) Surface mount frame-contact lift. *(Courtesy of Automotive Lift Institute)*

There are also in-ground lifts called *axle engaging* lifts, and other lifts called *pad lifts* (covered later). An advantage to frame-contact, axle engaging, or pad lifts is that the wheels hang free, which makes it easier to perform tire, brake, and suspension work. These lifts usually provide better access to the underside of the vehicle.

Figure 11.2 Wheel-contact lift. *(Courtesy of Rotary Lift)*

◼ FRAME-CONTACT LIFTS

Frame-contact lifts (see Figure 11.1a–c) have adapters at the end of adjustable *lift arms*. The lift adapters contact the vehicle frame at the manufacturer's specified lift points (at the rocker panels or on a section of the frame) (**Figure 11.3**). Be sure to check the service manual before trying to lift a car. Sometimes when a car is lifted improperly, body, suspension, or steering components can be bent. The windshield might even pop out or the vehicle might fall.

In 1992, the Society of Automotive Engineers adopted a standard (SAE J2184) for vehicle lift points. Some manufacturers label their vehicles with decals depicting the recommended lift points. These are located inside the passenger side front door. Permanent markings on the underside of the chassis also mark the lift points. These markings can include a hole, a boss, or a ¾" depression of a triangle.

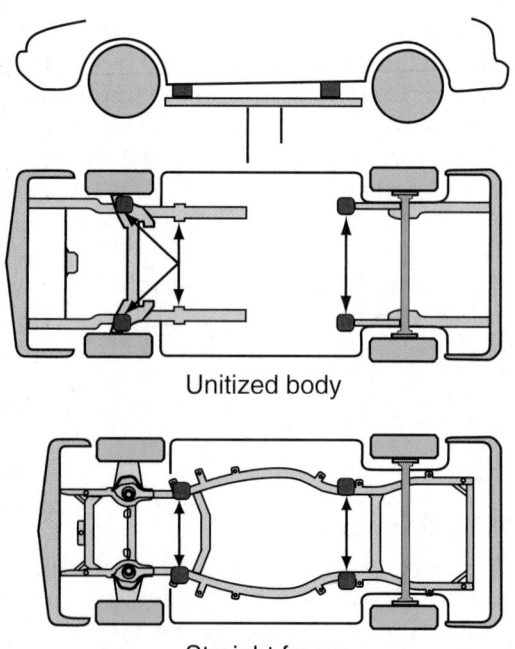

Unitized body

Straight frame

Figure 11.3 Check the service information for the correct lift points for a particular vehicle.

Some adapters (sometimes called *foot pads*) flip up. These adapters are adjustable to several positions to accommodate different heights between the lift points in the front and rear of the vehicle and to provide clearance between the rocker panel and the lift arm (**Figure 11.4**). Some foot pads have screw threads that allow them to be positioned higher or lower (**Figure 11.5**).

SAFETY NOTE

- Adapters must be placed at the manufacturer's recommended lifting points and must be set to raise the vehicle in a level plane or the vehicle will be unstable.
- Lift pads should face either toward each other or away from each other to form a "V" or an "A" (**Figure 11.6**).

Some adapters have rubber pads. Be sure that they are in good condition and that they are not covered with oil or grease that could make them slippery.

Passenger cars with unibody construction are lifted at points on the rocker panel (just under the bottom outside edge of the car body). There are special rocker panel lifts, called pad lifts (**Figure 11.7**), that can be used for these cars and cars with perimeter frames. Do not use pad lifts for trucks.

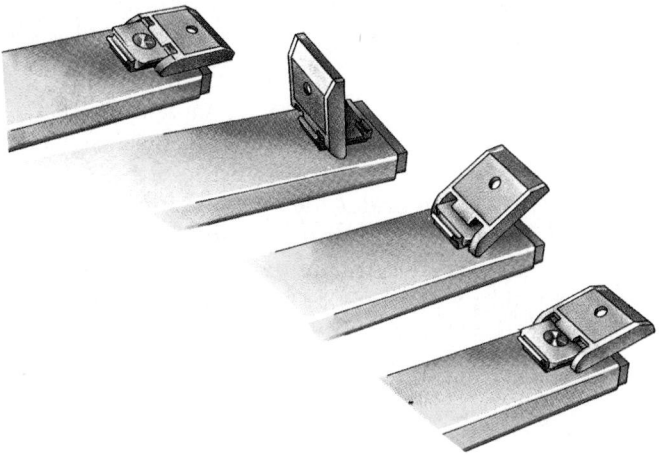

Figure 11.4 Adapters flip up to accommodate different frame heights. *(Courtesy of Automotive Lift Institute)*

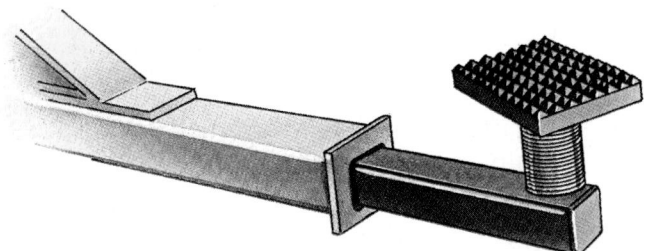

Figure 11.5 Some adapters have a screw adjustment. *(Courtesy of Automotive Lift Institute)*

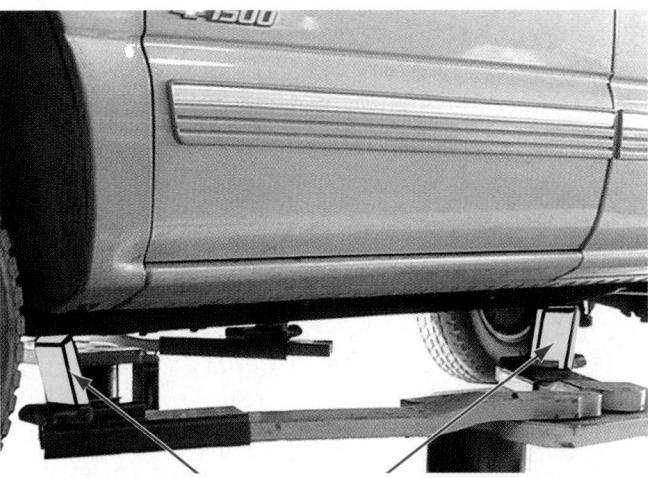

Figure 11.6 Lift pads should face away from each other like this ("V"), or toward each other ("A"). *(Courtesy of Tim Gilles)*

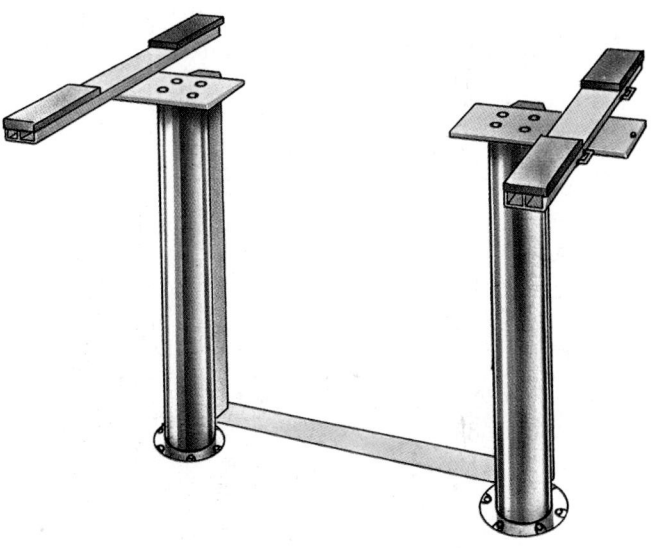

Figure 11.7 A rocker panel (pad) lift. *(Courtesy of Automotive Lift Institute)*

Special Auxiliary Adapters/Extenders

Sometimes extenders are used on the adapters (**Figure 11.8**). If the lift points on the vehicle are undercoated, a special adapter might be needed when using a lift that has steel adapters. Damaging the undercoating can void the vehicle's rust protection warranty. Some sport utility vehicles, light trucks, and vans require special adapters to provide clearance between the lift arm and the rocker panel. Most lift manufacturers make these available. Do not use makeshift extenders.

■ WHEEL-CONTACT LIFTS

A wheel-contact lift is one in which all four of the vehicle's wheels are supported (see Figure 11.2). Some of these "drive-on" lifts are of the in-ground type. Others are the four-post, surface mount type. The four-post type has the advantage of easier installation and removal.

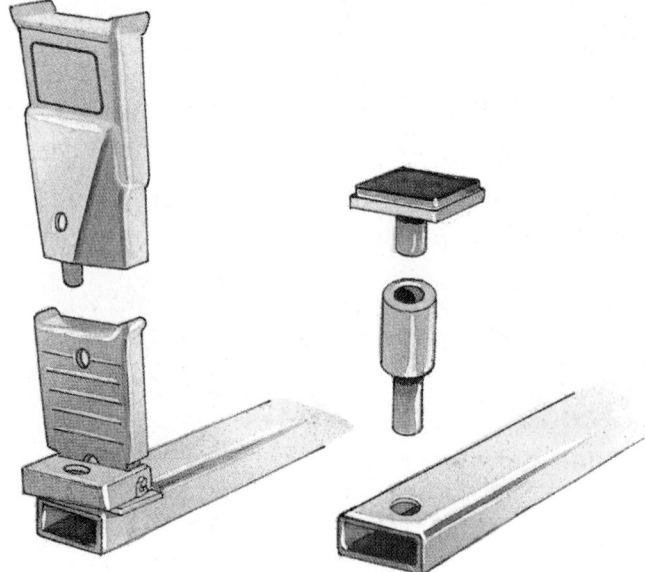

Figure 11.8 Extenders can be used with pickup trucks and vans. *(Courtesy of Automotive Lift Institute)*

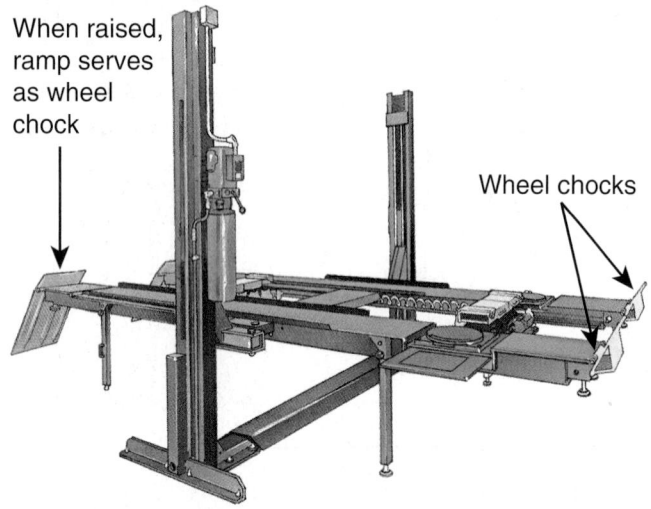

When raised, ramp serves as wheel chock

Wheel chocks

Figure 11.9 This lift has safety stops at each end. *(Courtesy of Automotive Lift Institute)*

Four-post wheel-engaging lifts are especially popular in service shops, engine and transmission repair, wheel alignment, and muffler shops.

When driving a car onto one of these lifts, be sure that the tires are an equal distance from the edges of the ramps. Wheel-contact lifts have secondary stops for roll-off protection at the front and rear (**Figure 11.9**). After spotting the vehicle, always use manual *wheel chocks* to prevent the car from rolling.

■ WHEEL-FREE JACKS

On a wheel-contact lift, the wheels are supported unless a *wheel-free* jack is used (**Figure 11.10**). These jacks, which are air powered and/or hydraulic, are used to raise either end of the car for wheels-free work.

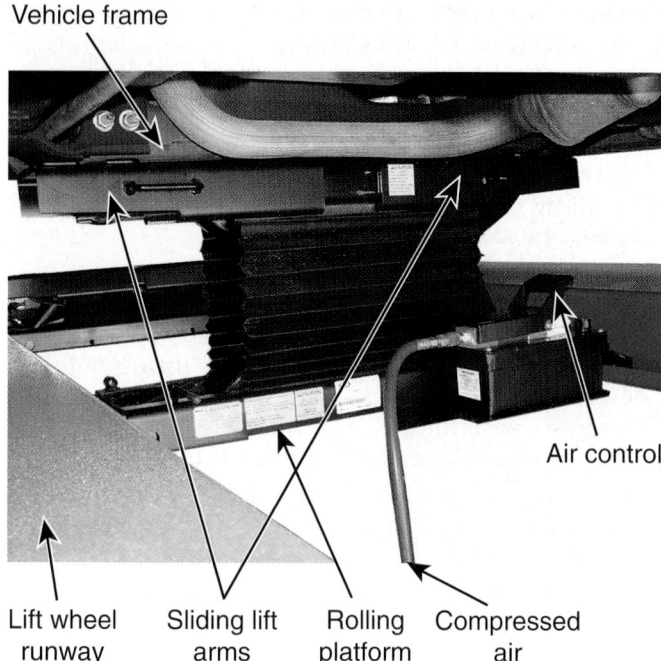

Vehicle frame

Air control

Lift wheel runway Sliding lift arms Rolling platform Compressed air

Figure 11.10 A wheels-free air-hydraulic jack. *(Courtesy of Tim Gilles)*

After the vehicle is raised, the jack is lowered onto a mechanical safety latch. When using one of these jacks, keep hands clear and be sure to extend each of the lift arms an equal amount to avoid uneven loading. Be sure that a wheels-free jack is lowered all of the way before driving into or out of the lift.

Some advantages of wheel contact lifts are:

■ The vehicle may be raised safely, even without the engine in it. Trying to lift such a vehicle can be dangerous on a frame-contact lift, because the unbalanced vehicle might tip off the lift.
■ The center of the underside of the vehicle is accessible on the twin-post hydraulic and four-post types. This is important for exhaust system and transmission work.
■ For quick, easy service, the vehicle requires no set-up of lift adapters.
■ Wheels are in contact with the wheel ramps. This is necessary for wheel alignment work.

■ IN-GROUND LIFTS

Most in-ground lifts used for raising automobiles have either one or two *pistons*, often called *posts*. In two-post lifts, the pistons are located in one of two ways:

■ One in front of the other
■ Side by side

Newer in-ground systems must be enclosed with double walls to prevent oil from leaking from them into the ground.

Frame-Engaging Lifts

In-ground, frame-contact lifts are either *single-post* or *two-post* (see Figure 11.1a and b). The two-post style is either *drive-through* (**Figure 11.11**) or *drive-over* (**Figure 11.12**). Drive-through lifts are more open in the center to allow easier access to the underside of the vehicle. The vehicle is driven between the lift arms.

The lift arms on a drive-over lift are located closer together so the vehicle can be driven over them without bumping into them. As vehicles have been downsized, drive-through lifts have become more popular. Using a drive-over lift to work on the underside of a small car is very difficult because access is extremely limited. Two-post lifts provide better access to the underside of the vehicle than single-post hoists do.

Above ground lifts also have the advantage of being relatively portable, in case the shop loses its lease. They also do not have the problem of pollution of the ground water in case of a hydraulic fluid leak like an in-ground hoist has.

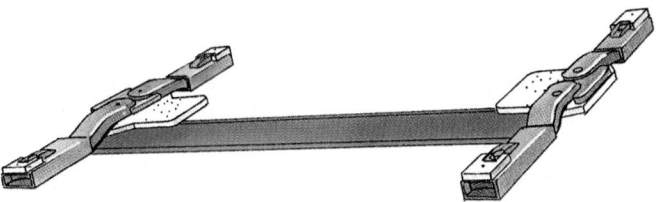

Figure 11.11 A drive-through lift. *(Courtesy of Automotive Lift Institute)*

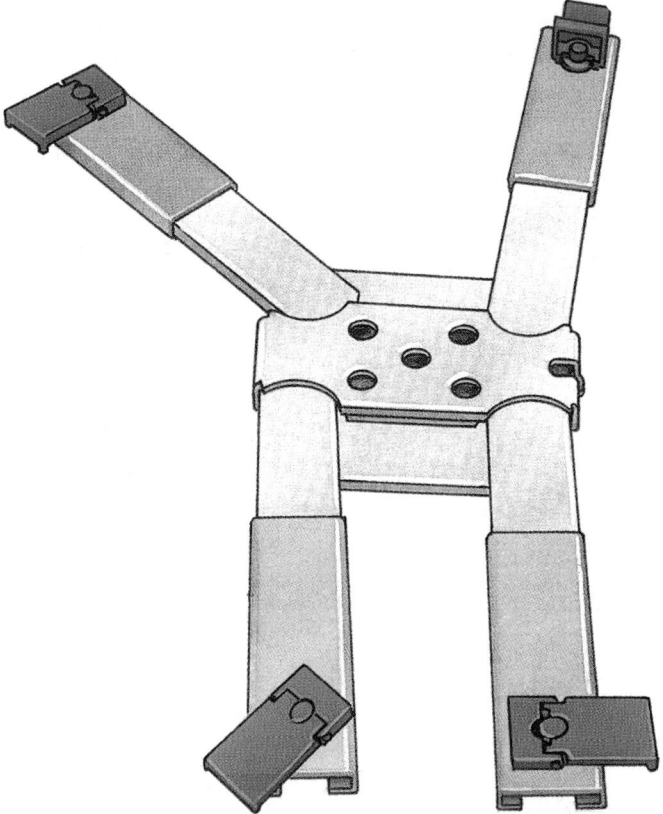

Figure 11.12 A drive-over lift. *(Courtesy of Automotive Lift Institute)*

Axle Engaging Lifts

An axle engaging lift, also called a *suspension contact* lift, has at least two posts (pistons) positioned front to rear (**Figure 11.13**). The front post is movable so that it can accommodate vehicles with different length wheelbases.

The front post of the lift has adjustable arms that are positioned just under the lower control arm, as far out toward the wheel as possible. The rear post engages the rear axle housing on rear-wheel-drive vehicles. Some front-wheel-drive cars have a *support rail* between the rear axles that allow the use of this type of lift, but this type of lift is not recommended for a majority of front-wheel-drive cars. Other lifts of this style provide the option of lifting the rear of the vehicle by the wheels (this option works for front-wheel-drive vehicles).

Semi-Hydraulic and Fully Hydraulic Lifts

In-ground lifts that are powered by compressed air are either semi-hydraulic or fully hydraulic. Other in-ground lifts use an electric pump to pressurize the hydraulic oil. Air pressure is not necessary in this system. Semi-hydraulic lifts have a self-contained air/oil reservoir. Fully hydraulic lifts have a separate air/oil reservoir. When compressed air is put into the tank containing the hydraulic oil, it pressurizes the oil to lift the vehicle. There are separate oil and air controls

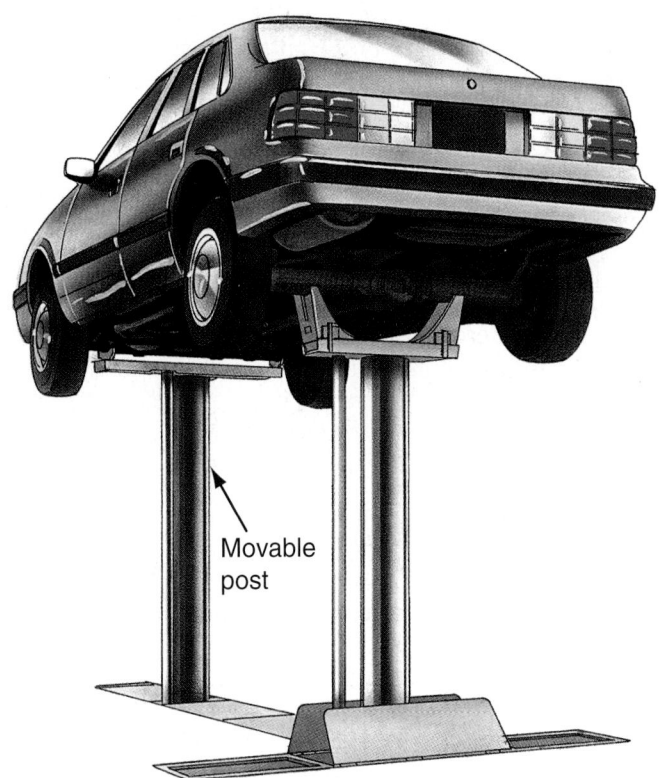

Movable post

Figure 11.13 An axle engaging, or suspension contact, lift. *(Courtesy of Automotive Lift Institute)*

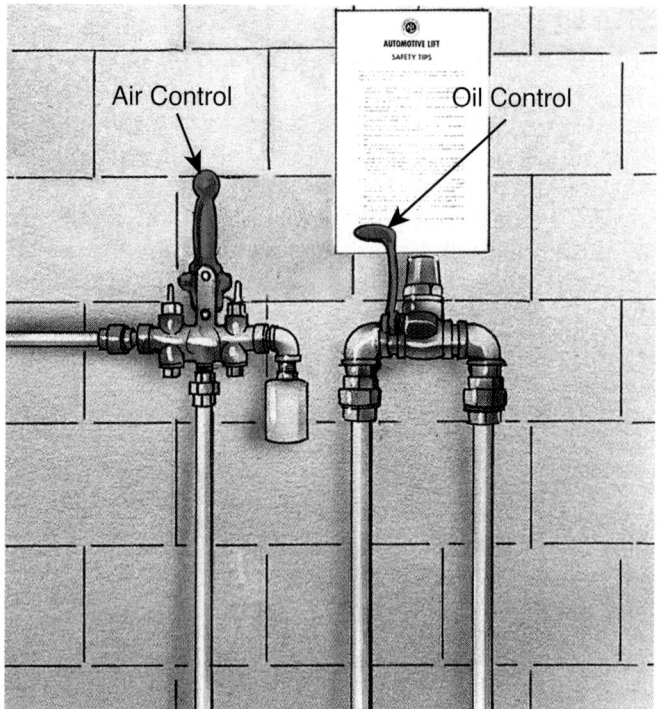

Figure 11.14 Fully hydraulic hoists have two controls. (*Courtesy of Automotive Lift Institute*)

Figure 11.15 A screw-drive hoist. (*Courtesy of Automotive Lift Institute*)

on fully hydraulic lifts (**Figure 11.14**). Semi-hydraulic lifts use only one control:

- The air control is opened first.
- When the air has had a chance to pressurize the air/oil tank, the second control handle is moved to allow the pressurized oil to lift the car.

When the shop's air compressor has not been on long enough to build adequate air pressure, when a compressor is too small, or when the shop has been using too much air, the lift will work very slowly.

In-Ground Lift Maintenance

If an in-ground lift vibrates while lifting the vehicle or will not raise to its full height, this could be due to an oil leak. If you hear air escaping or see or suspect an oil leak, stop using the lift immediately and release the air pressure.

Lifts are usually repaired only by qualified service personnel. Sometimes an experienced technician or business owner will perform maintenance to lifts, however. Always follow the lift manufacturer's maintenance instructions.

NOTE: *Be sure that all of the air pressure is exhausted from the lift before attempting to check or add oil.*

Remove the fill plug carefully by hand. Do not use an impact wrench. There could still be air pressure in the air/oil tank. Use the specified oil only.

■ SURFACE MOUNT LIFTS

Many newer lifts (since 1970), called surface mount lifts, are mounted completely above the floor (see Figures 11.1c and 11.2). An electric motor operates either a *screw drive* (**Figure 11.15**) or a *hydraulic pump and cylinders*.

Ease of installation is one advantage of a surface mount lift. They can be removed and reinstalled in another location if a business owner is renting a shop and moves. A surface mount lift is also easier to maintain or repair because it requires no excavation. Surface mount lifts are also less likely to leak oil into the ground (an environmental concern).

The most popular surface-mounted lift style is a two-post, drive-through, frame-engaging type. It has two lifting carriages that each support two swing arms. The carriages are synchronized so that they go up and down together. Either a steel chain, cables, synchronized motors, or hydraulic circuits are used to keep the two carriages moving together. **Figure 11.16** shows a chain synchronizer.

The synchronizing parts and the driving system are attached to the posts either overhead or across the floor. When they are across the floor, sometimes a wide groove is cut in the concrete. Parts that connect the two posts are submerged in the groove so

Figure 11.16 This lift has a chain-type synchronizer. (*Courtesy of Automotive Lift Institute*)

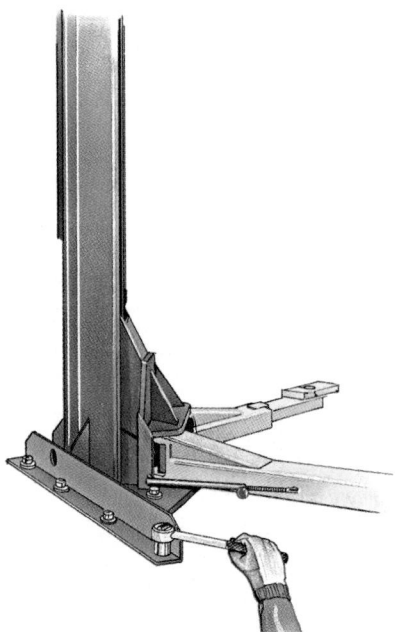

Figure 11.17 Periodically check the bolts of a surface mounted lift for tightness. *(Courtesy of Automotive Lift Institute)*

that the floor remains unobstructed. This is especially convenient when a transmission jack is used.

Surface Mount Lift Maintenance

To bolt a surface mount lift, the floor must have at least 5" of solid concrete. The bolts that hold the lift to the concrete floor should be inspected periodically for tightness (**Figure 11.17**). They can and do vibrate loose. If cracks develop in the concrete, use of the lift should be discontinued until it can be checked by a *professional*. Always follow the lift maintenance instructions.

■ LIFT SAFETY

Lifts have an excellent safety record, but, unfortunately, vehicles occasionally fall off of them. When a vehicle comes down by accident, this is usually due to carelessness, misuse, or neglected maintenance. Training in the use of the lift is mandatory before attempting to lift a car.

The American National Standards Institute (ANSI) and ALI have set the American National Standard (ALOIM-1994) for automotive lifts. This standard lists safety requirements for operation, inspection, and maintenance of lifts. It requires annual inspection of automotive lifts by a qualified lift inspector. This annual inspection is designed to keep automotive lifts in good operating condition.

Center of Gravity

Find the **center of gravity** of the vehicle and position it over the posts of the lift. The center of gravity is the point between the front and the rear wheels where the weight will be distributed evenly. Different positions are used for front-wheel-drive and rear-wheel-drive cars (**Figure 11.18**).

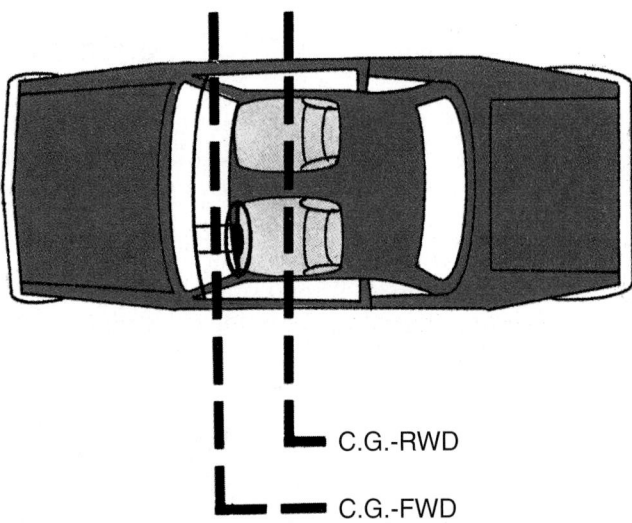

Figure 11.18 Center of gravity (C.G.) positions for front-wheel drive (FWD) and rear-wheel drive (RWD). *(Courtesy of Automotive Lift Institute)*

According to ALI:
- On rear-wheel-drive (RWD) cars, the center of gravity is usually below the driver's seat.
- On front-wheel-drive (FWD) cars, the center of gravity is usually slightly in front of the driver's seat, beneath the steering wheel.

 **SAFETY NOTE** You are lifting a vehicle that weighs several thousand pounds. The car must be correctly spotted on the lift so that the *center of gravity* is correct.

NOTE: *On a two-post above ground hoist, do not position the vehicle to the front or rear of the posts just so that the door can be opened easier. Positioning the center of gravity is very important.*

On one- or two-post lifts, position the vehicle's center of gravity over the posts. On four-post models, position the car equally between the front and rear.

Lift Operation Safety

- If there are any problems with the lift, do not use it. See your supervisor immediately. Do not take chances.
- When lifting, first raise the vehicle until its wheels are about 6" off the ground. Then, jounce the vehicle and double-check the contact between the adapters and the frame to be sure the vehicle is safely engaged.
- Be certain that all four lift pads are contacting their lift points and bearing a load. It is not unusual for three lift arms to be touching the car with the fourth one free to move. If a lift arm can be moved after the car is in the air, the car is unevenly loaded. Lower the car and reposition the arm.

- If a lift arm is positioned improperly, lower the vehicle slowly to the ground and reposition the arm.

NOTE: *Some lifts have swing arm restraints that hold the unloaded arm in position to prevent accidental movement. These restraints are not designed to prevent the car from falling if it is not properly positioned.*

- When performing vehicle repairs on a vehicle on a frame-contact lift, do not use a large prybar or do anything else that might knock the vehicle off of the adapters. When tight bolts are encountered, it is best to use an air impact wrench on them.

- Be sure that the lift contact points on the vehicle are in good condition and that there is no oil or grease on them.

- Some lifts have different length arms in the front than they do in the rear. These are called asymmetrical arms. Be sure to consult the manufacturer's instructions before using this type of lift.

- Some lifts have a safety locking device that holds the post should a hydraulic failure occur (**Figure 11.19**). Be sure it is engaged. If the lift is not raised high enough for it to engage or if the lift is not equipped with a safety device, use four high-reach supplementary stands (tall jackstands).

Figure 11.19 A locking device is engaged when the lift is at its highest travel. *(Courtesy of Automotive Lift Institute)*

- The lift area should be clean. There should be no grease or oil on the floor. Hoses, extension cords, and tools should be in the places where they belong.

- Insurance companies usually prohibit customers from being in the lift area. Do not allow customers to drive their own car onto the lift.

- Be sure that the lift has adequate capacity to lift the weight of the vehicle. If the vehicle contains any loads inside, in the trunk, or in the bed of a pickup, the center of gravity will be affected and the vehicle will be unsafe to lift.

- Be sure that the lift is all of the way down before attempting to drive a car into or out of it.

- Before lowering a vehicle, be sure to alert anyone nearby. Be certain that no tools or equipment are below the car. All of the car's doors should be closed.

Before lifting a vehicle consider the items listed in **Figure 11.20**.

■ AIR COMPRESSORS

Compressed air is used to power tools in the shop in much the same way that electricity is used. One of its advantages is that there is no danger of electrical shock, especially in auto body work where water is sometimes used when sanding. Compressed air can be used to blow off parts, to apply paint with a spray gun, power hand tools, or in an air jack for lifting vehicles.

Air compressors resemble small engines (**Figure 11.21**). They have one or more pistons and one-way check valves. When a piston moves down, a check valve allows air to be drawn into the compressor's cylinder. When the piston moves back up, another check valve directs the air into a holding tank, called a *receiver tank* (**Figure 11.22**). Larger compressors are often *two-stage*, which means that air is compressed in two stages. Two-stage compressors are used when pressures over 100 psi are required.

Air Compressor Size

When shopping for an air compressor, size is an important consideration. Each air tool consumes a certain amount of air during use. Air compressor *capacity* is dependent on the number of tools that will be in use in the shop at one time and their air requirements.

Air compressors are equipped with *pressure switches* to shut the compressor off when the receiver tank reaches a set pressure. As air in the tank is consumed, the pressure drops and the switch turns the compressor back on again. An air compressor is operating at its maximum practical capacity when it runs for 7 out of every 10 minutes. This is referred to as a 70% **duty cycle**.

AUTOMOTIVE LIFT
SAFETY TIPS

Post these safety tips where they will be a constant reminder to your lift operator. For information specific to the lift, always refer to the lift manufacturer's manual.

1. Inspect your lift daily. Never operate if it malfunctions or if it has broken or damaged parts. Repairs should be made with original equipment parts.

2. Operating controls are designed to close when released. Do not block open or override them.

3. Never overload your lift. Manufacturer's rated capacity is shown on nameplate affixed to the lift.

4. Positioning of vehicle and operation of the lift should be done only by trained and authorized personnel.

5. Never raise vehicle with anyone inside it. Customers or by-standers should not be in the lift area during operation.

6. Always keep lift area free of obstructions, grease, oil, trash, and other debris.

7. Before driving vehicle over lift, position arms and supports to provide unobstructed clearance. Do not hit or run over lift arms, adapters, or axle supports. This could damage lift or vehicle.

8. Load vehicle on lift carefully. Position lift supports to contact at the vehicle manufacturer's recommended lifting points. Raise lift until supports contact vehicle. Check supports for secure contact with vehicle. Raise lift to desired working height. CAUTION: If you are working under vehicle, lift should be raised high enough for locking device to be engaged.

9. Note that with some vehicles, the removal (or installation) of components may cause a critical shift in the center of gravity, and result in raised vehicle instability. Refer to the vehicle manufacturer's service manual for recommended procedures when vehicle components are removed.

10. Before lowering lift, be sure tool trays, stands, etc. are removed from under vehicle. Release locking devices before attempting to lower lift.

11. Before removing vehicle from lift area, position lift arms and supports to provide an unobstructed exit (See Item #7).

These "Safety Tips", along with general lift safety materials, are presented as an industry service by the Automotive Lift Institute. Visit our web site at <autolift.org> for more information on this material, or write to: ALI, PO Box 33116, Indialantic, FL 32903-3116.

Look For This Label on all Automotive Service Lifts.

FOUNDED 1945

AUTOMOTIVE LIFT INSTITUTE

MEMBER

This product was manufactured by a member of the Automotive Lift Institute, Inc., the North American trade association of lift manufacturers dedicated to promoting responsible design, construction, installation, use, and maintenance of automotive vehicle service lifts.

AUTOMOTIVE LIFT INSTITUTE, INC.

ALI-ST90(rev00)

Figure 11.20 Automotive lift safety. *(Courtesy of Automotive Lift Institute)*

Figure 11.21 An air compressor resembles a small gasoline engine. *(Courtesy of Campbell Hausfield)*

Air compressor

Receiver tank

Figure 11.22 The receiver tank acts as a reservoir for the air. *(Courtesy of Campbell Hausfield)*

There is more than one rating for air compressor capacity. The best measurement of useful capacity is **free air delivery**, measured in standard cubic feet per minute (**SCFM**). Most small air tools are designed to run at 90 psi, so it is necessary to know the air pressure at which the free air delivery was measured. According to the *Compressed Air and Gas Institute*, a good compressor choice for an automotive shop would be one with 15–22 SCFM of free air delivery at 150 psi. This amount of output is commonly achieved by a 5–10 horsepower (HP) two-stage, two-cylinder compressor. The air requirement depends on the number of service bays in the shop.

Air Compressor Tank

The correct receiver tank for an air compressor is an important consideration. A large tank takes longer to fill initially, but the compressor will not need to run as often to refill it. Large capacity air tools such as auto body tools or a glass bead blaster require large receiver tanks.

Where to put the compressor is also a consideration because compressors take up a good deal of floor space and are noisy. The compressor is often installed outside of a building or in a storeroom where noise will not be a factor. Most receivers for compressors larger than 5 HP are *horizontal*. Tanks for smaller compressors are sometimes *vertical*, with the compressor mounted on top. This can present balance problems though, especially with larger compressors, and the tank must be bolted to the floor. Horizontal and vertical compressor tanks work equally well.

 **CAUTION** Be certain all receiver tanks are protected by a relief valve of adequate size.

Compressor Motors

Permanently installed air compressors used in most shops are driven by 5 to 10 HP motors. Larger compressors usually require a 220-volt electrical supply. Portable compressors under 5 HP that run on 110-volt power are also available. Some compressors are driven by gasoline engines. These are usually used on portable compressors or permanent installations on trucks for use out in the field. Electric motors are less expensive, less noisy, cost less to operate, and they do not pollute the air in the shop.

Air Lines

It is important that air delivery lines be large enough so that enough air gets from the compressor to the tool.

When lines are too small, the pressure must be raised to compensate. Raising the pressure heats the air, which is hard on tools and causes the compressor to work harder. When a line is longer than 150 feet in length, it is a good idea to use a minimum of 1" pipe size.

Each output line from the air feed system is called a *drop*. When several drops are to be fed from one manifold air line, it is a good practice for the manifold line to form a complete loop around the shop.

Be sure all air lines are at least as big as the outlet on the compressor and that there are no small diameter sections of line between the compressor and the tool. A small line acts as an orifice, dropping pressure at the tool. This happens when someone installs a *hose coupling* of the wrong size. A $\frac{5}{16}$" hose with a $\frac{1}{4}$" coupling will flow less air after the air passes the restriction caused by the smaller coupling. This also happens when someone mistakenly installs an air regulator that is too small; for instance, a $\frac{3}{8}$" regulator in a $\frac{1}{2}$" air line. **Figure 11.23** shows a typical air-tool setup.

Air Compressor Maintenance

- Water is produced as outside air is condensed by the compressor into the receiver tank. The tank requires periodic *bleeding*, or *blow-down*, to drain the water that accumulates. Moisture in the air supply line is reduced if the receiver tank is drained daily. A valve or faucet is usually located at the bottom of the tank. Some compressed-air systems have water filters (traps) at the outlet of the compressor to remove the moisture in the air. Some shops have oilers in the line to provide lubrication for air tools. Other air systems have special dryers to provide high-quality air for spray painting.

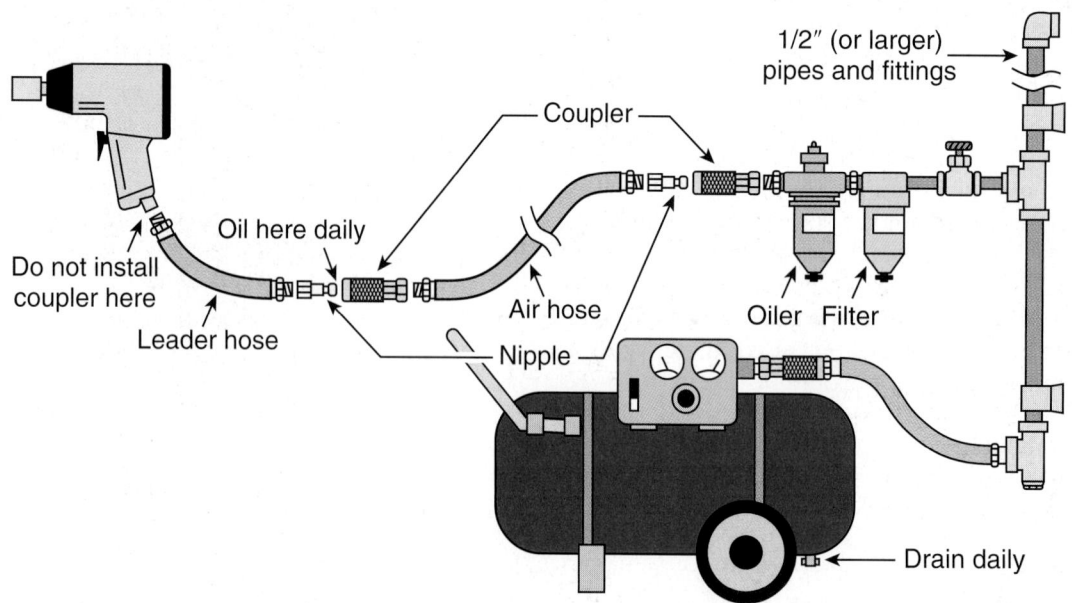

Figure 11.23 A typical setup for use of an air tool.

- The oil level in the compressor should be checked regularly, especially if it appears that there is oil leakage. The oil should be changed every 3 months.
- It is important that the compressor breathes clean air, much the same as with an internal combustion engine. The air filter must be cleaned regularly. If it becomes restricted, the compressor will overheat.
- Check the drive belts regularly for signs of wear or looseness. A loose belt will slip and wear the pulley.

REVIEW QUESTIONS

1. The two main categories of lifts are the in-ground lift and the _____ lift.

2. What kind of jack is used to raise the wheels off of a wheel-contact lift?

3. A single-post lift is handy for transmission removal. True or false?

4. What kind of lift has two controls, one for air and one for hydraulic pressure?

5. What kind of lift would you want to install if you were renting a building and had a short term lease, in-ground or surface mount?

6. Two ways that a surface-mounted lift is driven are with a hydraulic pump and cylinders and _____ drive.

7. If one of the lift arms can be moved after the vehicle is in the air, what would you do?

8. When would a two-stage air compressor be used?

9. If an air compressor runs for 6 out of every 10 minutes, what is its duty cycle?

10. The useful capacity of an air compressor is measured in standard _____ _____ per minute.

ASE-STYLE REVIEW QUESTIONS

1. Which of the following are true about lifting a vehicle using incorrect lift points?
   a. Damage to suspension or steering parts can result.
   b. The windshield can be loosened or damaged.
   c. The vehicle's center of gravity can be off.
   d. Any of the above

2. All of the following statements are true about lifting a vehicle *except*:
   a. Unibody cars be lifted from the edge of the car body at the rocker panel.
   b. A frame contact lift is the best choice for raising a vehicle with the engine removed.
   c. Raised lift pads should be positioned in a "V" or "A", but not facing the same direction.
   d. Always make sure all lift pads are touching the lift points.

3. Technician A says that on rear-wheel-drive cars, the center of gravity is usually just below the driver's seat. Technician B says that on front-wheel-drive cars, the center of gravity is usually just below the driver's seat. Who is right?
   a. Technician A   c. Both A and B
   b. Technician B   d. Neither A nor B

4. Technician A says to position a car on a two-post surface mount lift so that the driver's side door can be opened all of the way. Technician B says to raise the car until its wheels are about 6 feet off the ground and shake it to see if it is safely mounted. Who is right?
   a. Technician A   c. Both A and B
   b. Technician B   d. Neither A nor B

5. Technician A says that using a ⅜" regulator in the middle of a ½" air line will result in less airflow through the lines. Technician B says that water must be periodically bled from an air compressor. Who is right?
   a. Technician A   c. Both A and B
   b. Technician B   d. Neither A nor B

Vehicle Inspection (Lubrication/Safety Check)

THEN AND NOW: SAFETY

In 1769, a French army engineer named Nicolas Cugnot mounted a steam engine on a wooden wagon in an effort to create a better way to transport a cannon. In the process, he became the first person in history to produce a motor vehicle. During a test, his invention went out of control and ran into a wall, making him the first person in the world to have an automobile accident.

The survival of the driver and passengers has been looked upon as a serious matter since the beginning of the automotive era. In the 1920s, steering wheels replaced tillers. This safety development was followed by *safety glass*, a glass that crumbles instead of breaking into the knife-like fragments of ordinary broken glass. Four-wheel brakes were also developed. These helped stop a vehicle much sooner. All of these milestones saved many thousands of lives and prevented countless injuries. But a car was still an essentially hostile environment for a human being. There were 4,000 Americans killed in automobile accidents in 1913. By 1930, the death toll was 32,900. In some years since, over 50,000 have died—that is almost one fatality every 10 minutes!

In 1956, Ford Motor Company offered two safety options, seat belts and padded dashboards. Very few customers opted for them. In 1958, Volvos were sold with seat belts as standard equipment, followed the next year with shoulder harnesses. By 1966, the U.S. government introduced uniform vehicle safety regulations, one of which made dual circuit braking systems mandatory. This feature prevented total brake failures. Another safety regulation required seat belts on every car sold in America. Unfortunately, most people considered them a nuisance and never put them on. To remedy this, the seat belt interlock was mandated in 1974. It prevented the car from starting unless your seat belt was fastened. The public outcry against this inconvenience was so strong that the government rescinded the law.

Cars are much safer now than they were in the early days of motoring. The traffic death rate per 10,000 registered vehicles went down 94% between 1912 and 1995. This means where we once had 33 fatalities, we now have 2.

Antilock brakes, air bags, and electronic stability control are the biggest news on the safety front today. By pulsing the brakes, antilock brakes help avoid a skid that can result in an accident. If you are unfortunate enough to be involved in an accident, an air bag inflates instantly to cushion the impact and improve your chances of survival. Electronic stability control senses yaw and provides combinations of brake applications to prevent rollovers or loss of vehicle control.

Volvo introduced the shoulder harness in 1959. (*Courtesy of Volvo Cars of North America*)

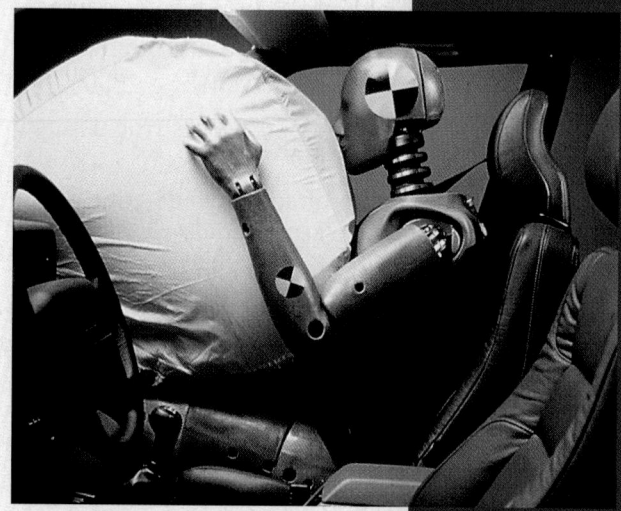

A passenger air bag. (*Courtesy of Volvo Cars of North America*)

CHAPTER 12

Engine Lubrication

■ **OBJECTIVES**

Upon completion of this chapter, you should be able to:

✔ Describe engine lubrication under different service conditions.

✔ Select the correct engine oil to use.

✔ Describe the operation of different types of oil filters.

■ **KEY TERMS**

dry start
crankcase
viscosity
multiple viscosity

polymer
viscosity index
API
boundary lubrication

full-flow oil filter
by-pass valve
anti-drainback valve
by-pass oil filter

■ **INTRODUCTION**

In theory, all moving parts are separated by a thin layer of oil (**Figure 12.1**). Oil is supplied to engine parts by an oil pump (**Figure 12.2**). If oil is properly maintained with no dirt allowed to accumulate in it, then very little wear should occur. Under normal conditions, the only time a breakdown in lubrication occurs is when the engine is first started in the morning. The crankshaft sits on the bearing until a wedge of oil is reestablished after pressurized oil reaches the bearing (**Figure 12.3**). A few seconds can pass before

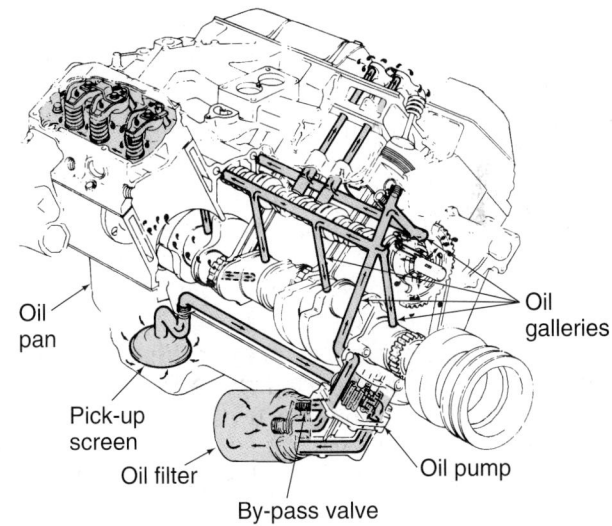

Figure 12.2 Oil is circulated through the system by an oil pump. *(Courtesy of DaimlerChrysler Corporation)*

the engine's oil pump can distribute oil to the entire engine (see Chapter 19). During this condition, called **dry start**, parts can rub and wear results.

NOTE: *The amount of wear that can happen during this short amount of time is equivalent to that caused by hundreds of miles of freeway driving. Today, most vehicle manufacturers require the use of thinner oils, which reach engine components more quickly after startup.*

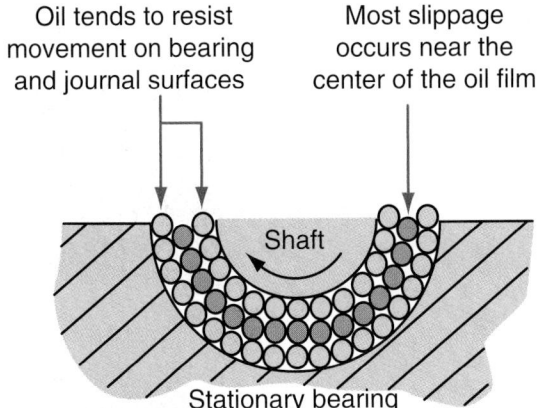

Figure 12.1 Moving parts are separated by a thin film of oil.

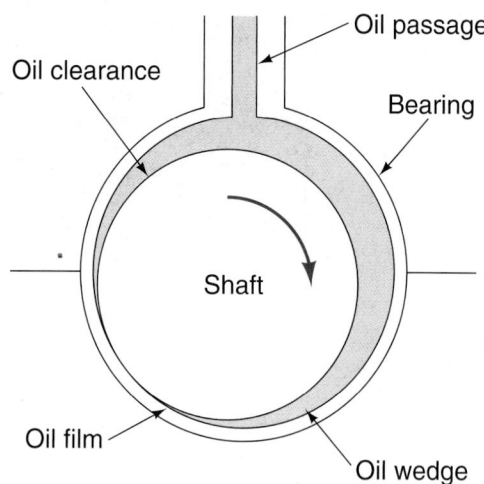

Figure 12.3 A wedge of oil lifts the shaft off the bearing.

ENGINE OIL

Engine oil is more than just basic crude. It contains a complicated additive package. Crude oil contains wax, most of which is removed during the refining process. The first oil additives were developed in the 1930s to deal with residual wax in the oil. More additives were developed during World War II. Over the years, this additive package has been improved. The result is that engines last longer and longer. Besides lubricating, engine oil cools, cleans, and prevents rust from forming inside the engine. Oil also fills hydraulic lifters and helps to seal the piston rings against the walls of the cylinder.

Oil Level

The correct oil level is designed to keep the oil pickup screen (see Figure 12.2) below the level of the oil under all operating conditions. If the oil level is allowed to drop too low, serious engine damage can result. The crankshaft bearings can be damaged (**Figure 12.4**) or the piston can become *scuffed*. **Figure 12.5** shows what can happen when an engine is run for an extended time while low on oil.

The **crankcase** is the area at the bottom of the engine enclosed by the oil pan. If there is *too much* oil

Figure 12.5 These pieces of engine parts were found in the oil pan of an engine that ran low on oil on a highway trip and blew up. *(Courtesy of Tim Gilles)*

in the crankcase, the crankshaft can dip into the oil as it spins. The oil is thrown onto the cylinder walls by the crankshaft in such an amount that the oil rings are overwhelmed. The result can be excessive smoke. Chapter 49 has more information on oil consumption. When the crankshaft hits the oil, it churns it up, aerating it (mixing it with air). Aerated oil does not provide for sufficient oil pressure and can result in collapsed hydraulic lifters, or even a broken crankshaft. The oil *dipstick* is located somewhere on the engine block. It has markings that indicate "full" and "add" (**Figure 12.6**). When the oil level is at the add line, the crankcase is one quart low.

NOTE: *Be sure that the engine dipstick is not being confused with the dipstick for an automatic transmission. The engine dipstick will be located somewhere on the engine block. The transmission dipstick often has instructions printed on it regarding fluid type and checking procedure. Also, the transmission fluid "add line" indicates 1 pint, instead of 1 quart.*

Figure 12.4 A worn crankshaft bearing. *(Courtesy of Tim Gilles)*

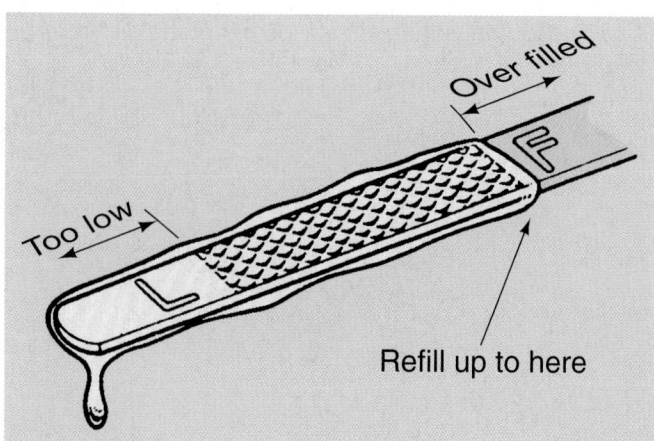

Figure 12.6 The dipstick indicates the oil level in the crankcase.

When checking the engine oil level:

- The vehicle should be on a level surface.
- The engine should be warm.
- The engine should have been off for at least 5 minutes, giving the oil a chance to drain back to the oil pan.
- Be sure the dipstick is pushed all of the way down against its seat.
- If the oil level reading on the dipstick is unclear, try looking at the backside of the dipstick, or dip it into the oil and attempt to read it again.
- If the oil level is low, check to see if the vehicle is due for service before adding a quart of new oil.

SHOP TIP Polish a dipstick with emery cloth to make it easier to read.

Oil Viscosity

Viscosity is the thickness or body of an oil. It is a measurement of its resistance to flow. The Society of Automotive Engineers (SAE) designates viscosity ratings. The SAE classification was adopted by the petroleum industry in 1911. An oil with a viscosity rating of 30 will have "SAE 30" displayed on the oil container. Viscosity, which is measured at 212°F (100°C), is higher if the oil is thicker. For instance, an SAE 10 oil is thinner than an SAE 30 oil. The higher the oil's viscosity is, the stickier it is and the more resistant to flow it is.

When a "W" accompanies the rating, it means that the oil's viscosity has been tested at 0°F; also, this oil is said to be a *winter-grade* oil.

SCIENCE NOTE

Viscosity is determined using a device called a viscometer. The viscosity of a liquid is determined by comparing it with the time required for a measured amount of reference fluid to pass through a calibrated opening at a specified temperature. Several methods are used for measuring viscosity. The common method in Europe is the Engler method: a ratio comparing the flow time for 200 mL of reference fluid to that of 200 mL of water at the same temperature.

In North America, oil viscosity is measured using a Saybolt viscometer. Saybolt viscosity increments are called Saybolt universal seconds. For higher viscosity measurements, like gear and engine oils, a different increment, Saybolt seconds furol (SSF), is used. The test liquid is 60 mL of furol, which is a shortened way of saying "fuel and road oil."

The numbers in the SAE oil viscosity rating are a reference to Saybolt seconds. For instance, 60 mL of SAE 10 oil will flow through the orifice in the viscometer in 10 seconds. The same amount of SAE 30 oil will take 30 seconds to flow through the same opening. Some modern winter oils can have a viscosity rating of 0W-30. Obviously, the measured volume of test oil cannot complete its flow in 0 seconds. Therefore, at this point the variation in the viscosity scale is based on a theoretical value.

Figure 12.7 compares SAE engine and gear oil viscosities with Saybolt viscosities and kinematic viscosities. You can see that engine oil and kinematic viscosity measurements have become the methods most commonly used for lubricants outside of the automotive industry. Kinematic viscosity is measured in stokes. Stokes are too small to work with, however. The number is multiplied by 100, converting the reading to centistokes. At 68.4°F, the kinematic viscosity of water is 1 centistoke (1 cSt). Figure 12.8 compares viscosities of different liquids in centistokes.

Multiple Viscosity Oils

Some oil has only one viscosity rating, but most new engine oils are **multiple viscosity** (*multi-vis*) (**Figure 12.9**). The following is an interpretation of a typical multi-vis oil (SAE 10W-30):

SAE = Society of Automotive Engineers

10W = The viscosity of the oil when measured at 0° (the "W" means winter grade).

30 = The viscosity of the oil when measured at 212°.

SAE Engine oil (cSt)	SAE Gear oil (cSt)	Kinematic viscosities (cSt)	Say Bolt viscosities (cSt)	
			[100°F]	[212°F]
		70	8000	300
	250	60	6000	
		50	4000	200
		40		
	140	30	2000	
				100
		20	1000	
50	90			80
40			600	
30	85W			60
	80W	10	400	
20		8		50
	75W	6	200	
10W				
5W		4	100	40

Figure 12.7 A comparison of different viscosity measurements at 212°F (100°C).

Liquid	Centistoke (cSt) Viscosity
Water	1
Milk	4
Vegetable oil	43.2
SAE 10 oil	110
Tomato juice	220
SAE 30 oil	440
Honey	2200

Figure 12.8 A comparison of viscosities of different liquids, in centistokes.

Figure 12.9 Various multi-viscosity oils. (*Courtesy of Tim Gilles*)

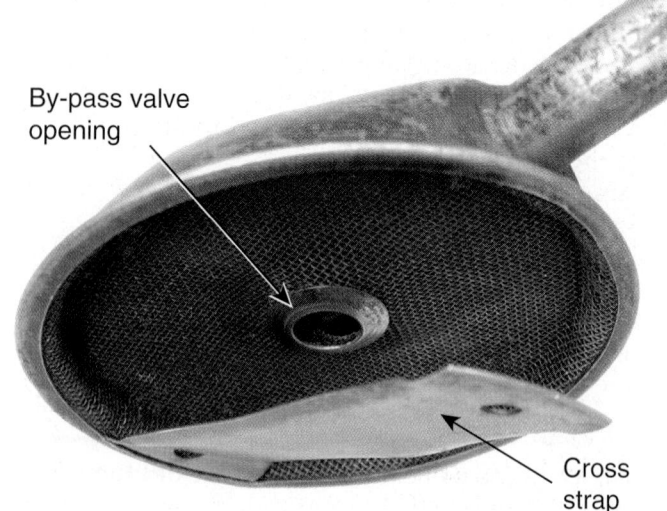

Figure 12.10 Pump screen by-pass valve and cross strap. (*Courtesy of Tim Gilles*)

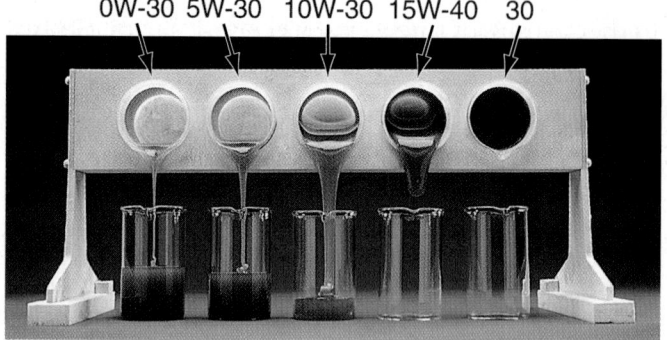

Figure 12.11 These oils were chilled to −35°C for 16 hours. The photo was taken 30 seconds after the caps were removed from the containers. (*Courtesy of Imperial Oil Limited*)

SAE 10W-30 has a base rating of 10 when cold. It will flow freely at temperatures as low as –20°F. Oil normally thins out when heated. Because it becomes thinner, its viscosity number becomes lower. An additive package containing **polymers** is blended into the oil. Polymers expand when heated, so the oil actually maintains its viscosity to the point where it is equal to what a hot SAE 30 oil would be. This ability to resist change in viscosity as an oil heats up is called its **viscosity index**.

A multi-vis oil helps combat engine wear because it flows more quickly to the bearings when the engine is first started—especially important during cold weather. The oil pickup has a screen with a relief opening that opens, allowing oil that is too thick or dirty to pass through to engine parts (**Figure 12.10**). The theory is that thick or dirty oil is better than no oil at all. Thick oil can also bypass the oil filter (see "Oil Filter" later in this chapter). In the past, some manufacturer's recommendations included SAE 30 in temperate climates (above 40°) because it flows acceptably at that temperature. Due to advances in lubrication technology, this is no longer the case. **Figure 12.11** shows how oils of different viscosity ratings flow. In some overhead

cam engines, using an oil with a viscosity higher than 5W-20 can result in failure of the cam and cam followers or problems with variable valve timing.

NOTE: *Always follow the manufacturer's recommendations for oil selection (**Figure 12.12**).*

Oil viscosity is also critical on diesel engines with electronically controlled hydraulic fuel injection. The oil needs to flow at the correct rate if it is to operate the injector as designed.

Oil Pressure

Pressure cannot develop unless there is a resistance to flow. As an engine wears, clearance between its crankshaft and bearings increases. The oil pump can no longer pump sufficient oil to fill this extra clearance. The result is that oil pressure is low whenever the engine rpm drops to idle speed. As the engine is accelerated, the oil pump (which is driven by the camshaft or the crankshaft) turns faster, so oil pressure will rise back to normal. Because a multi-vis SAE 20W-50

Figure 12.12 The oil filler cap often has the manufacturer's oil recommendation. *(Courtesy of Tim Gilles)*

oil is thicker both when hot and cold, it can provide higher oil pressure to an older, idling engine. But this same oil will flow more slowly (than an SAE 10W-40) to the bearings of a new engine when the engine is first started in the morning.

■ ENGINE OIL LICENSING AND CERTIFICATION

The SAE decides when new oil specifications are needed. The American Society for Testing and Materials (ASTM) sets the performance specifications according to the SAE's needs. The American Petroleum Institute (**API**) administers the licensing and certification of the oils. Prior to 1992, these three organizations were the only ones responsible for the specifications and classifications of engine oils. In 1992, the International Lubricant Standardization and Approval Committee (ILSAC) was formed to provide manufacturers minimum lubrication performance standards for gasoline-fueled (GF) passenger car and light truck engine oils. ILSAC includes the American Automobile Manufacturers Association (AAMA), the Japanese Automobile Standards Organization (JASO), and the Engine Manufacturers Association. Today, these six organizations comprise the Engine Oil Licensing and Certification System (EOLCS), administered by the API.

Starburst Symbol

The API licenses engine oil marketers to display its starburst certification mark (**Figure 12.13**) on their containers provided they meet ILSAC GF-4 requirements. Since 2005, no oils can display the starburst symbol unless they meet GF-4 standards. GF indicates "gasoline fuel," and GF-4 is the latest ILSAC gasoline-fueled engine oil standard. The starburst signifies that the oil meets the requirements necessary for it to be recommended by vehicle manufacturers. Many manufacturers recommend only oils displaying the starburst symbol. The manufacturers are particularly concerned

Figure 12.13 The starburst symbol.

that oils not damage catalytic converters, which now carry a warranty of 150,000 miles. Phosphorus, a long-time antiwear additive, is implicated in shortening catalytic converter life, so it has been replaced with other additives.

Oil Service Ratings

The API sets the service ratings of oil, which progress from SA through SM. Straight mineral oil, with no additives, is classified "SA." SA and SB graded oils are good only for very light-duty applications and are obsolete for automotive engines, although they are sometimes found on market shelves. Customers should be advised against their use. SM oil has many high-quality additives and will work well in any engine.

Gasoline and diesel engines have different rating systems:
- The "S" means the oil is for engines with spark ignition.
- Diesel engine oils are rated from CA through CI. The "C" means commercial, or that the oil is rated for engines with compression ignition. The latest diesel engine rating is CI-4 PLUS.

The API has a service symbol "donut" (**Figure 12.14**) made up of three sections. The top half lists the API performance, SM or CI, for instance. The center of the donut lists the SAE oil viscosity. The bottom half identifies if the oil is energy conserving when compared to a reference oil.

Figure 12.14 The viscosity and API rating of the oil are printed on the oil container using this standard industry symbol.

European ACEA Oils

The ACEA engine oil standards were developed by European auto manufacturers. ACEA stands for European Automobile Manufacturers Association (when translated from French to English). The test sequence has 11 laboratory categories and 9 engine categories. Some of the categories exceed the standards of ILSAC oils, but many of the standards are the same as the ASTM tests for American oils. Very few oils sold in the United States list ACEA on the label, and it is considered safe to use ILSAC GF-4 oil with the starburst label in the engines of most European cars sold in North America but always check the manufacturer's recommendation.

Energy-Efficient Oils

The ASTM certifies an oil as *energy conserving* (*EC*) if it passes certain tests. Energy conservation ratings began in 1994 and are restricted to multigrade oils with a hot viscosity of less than SAE 30, such as SAE 5W-30 or SAE10W-30. These oils can provide a 1% to 4% improvement in miles per gallon over regular oil. While improvements of this smaller amount would be practically impossible for a car owner to observe, these oils help the manufacturers to meet government fuel economy standards. An energy conserving oil that states "ECII" on its label will have better than a 2.7% fuel economy increase over regular oil. Be sure to use a high-quality oil. The use of an SA oil can cause serious damage to a modern engine (**Figure 12.15**).

■ OIL ADDITIVES

Oils (other than SA) contain an additive package that can make up as much as ⅓ of the volume of the oil. These additives have very important functions. Some of the additives found in engine oil are:

- *Pour-point depressants* that allow the oil to flow in very cold weather
- *Corrosion and rust inhibitors* that help the oil to stick to metal surfaces
- *Anti-foam additives*
- *Friction modifiers*
- *Oxidation inhibitors*
- *Anti-wear additives*

When the oil film becomes too thin or starts to break down under load, this is known as **boundary lubrication**. The primary job of anti-wear additives is to protect the cam and the lifters, which are under a severe load in today's small high-compression engines. During boundary lubrication, the additives combine chemically with engine metals to provide a lubricating layer.

The use of phosphate anti-wear additives has been reduced in SM oils because of their negative effect on catalytic converter life.

NOTE: *The reduced amount of anti-wear additive in the latest oils can result in an increase in camshaft and lifter wear in older engines that do not have roller lifters. These engines should use racing oil or truck oil.*

Detergents and *dispersants* are added to oil to keep small particle contaminants suspended in the oil (**Figure 12.16**). If they are large enough, they will be trapped by the oil filter. But if they are too small, they will not be removed from the engine until the oil is replaced.

As oil decomposes at high engine temperatures, a chemical reaction between oil and oxygen causes a gummy mixture to form that can plug oil passages. Detergents make these deposits oil soluble so that they can be suspended in the oil. Detergents in oils are similar to those found in household cleaning products. The difference is that they are oil soluble rather than water soluble.

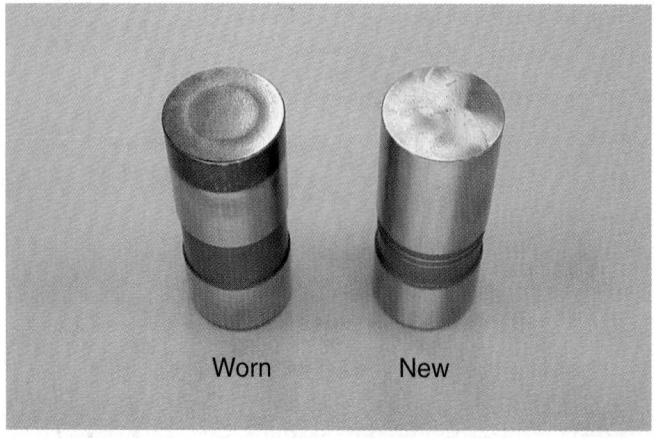

Figure 12.15 Worn and new lifters. *(Courtesy of Tim Gilles)*

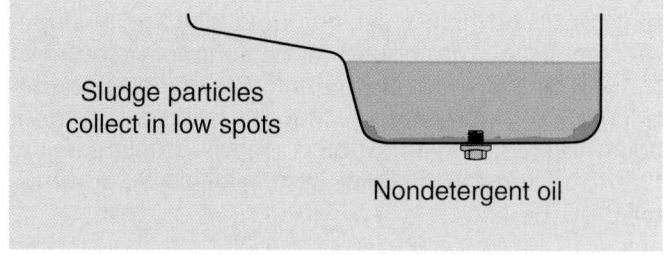

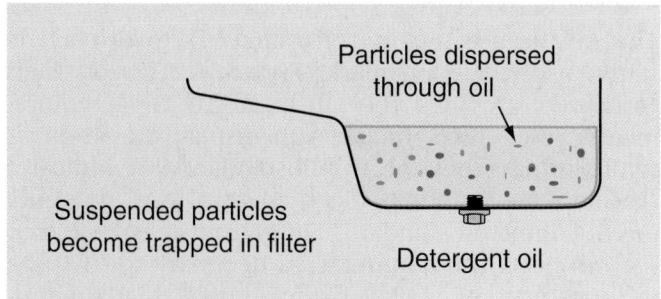

Figure 12.16 Detergent suspends particles so they can be trapped by the oil filter.

Figure 12.17 The sludge in this engine is the result of a lack of maintenance. Many by-products result from the burning of fuel. Engine overheating can also result in thermal breakdown of the oil, and sludge formation. *(Courtesy of Tim Gilles)*

Sludge is a mixture of moisture, oil, and contaminants from combustion (**Figure 12.17**). It can clog the oil screen and oil lines if allowed to accumulate. Moisture accumulates from condensation and is also a product of combustion. When an engine gets warm enough, the positive crankcase ventilation (PCV) system removes the water vapors. Short trips allow the accumulation of sludge. Accumulated fuel and moisture do not get a chance to evaporate in an engine that never fully warms up.

NOTE: *It takes 10 to 15 minutes for the cylinder walls to warm to the point where evaporation begins to take place.*

The oil level on the dipstick can be artificially high due to a high water or gasoline content.

Dispersants cause the oil to become discolored because they keep sludge-forming particles in suspension in the oil. Excessively rich air-fuel mixtures can cause soot contamination, which turns oil black.

Oil without additives (SA oil) is referred to as *nondetergent* oil. Sometimes mechanics used to suggest using nondetergent oil until the first oil change. The use of lower quality oil was recommended in the late 1950s to help the older-style, plated piston rings to seat. Today, the recommendation is to use the warranty-approved product for the entire life of the engine. Most service technicians use SM oil exclusively.

Oil oxidizes at temperatures in excess of 250°F, becoming thicker and forming varnish deposits. The oxidation rate of oil actually doubles with every 20°F rise in temperature above 140°F until about 800°F, when carbon forms. *Antioxidants* combat the effects of heat on the oil. If the oxidation inhibitor becomes depleted, the oil becomes thicker and thicker. Detergents make the varnish oil soluble so that they remain suspended in the oil. Excessive oil temperatures can be caused by:

- Lean air-fuel mixtures
- Retarded ignition timing
- Using the air conditioner
- An automatic transmission-equipped vehicle pulling a trailer

When possible, oil temperature should be kept below 220°F. The cooling system temperature required for water to vaporize out of the oil is 185°F–195°F. The ideal cooling system temperature for oil is therefore about 195°F.

NOTE: *Oil temperature is usually 10°F to 25°F hotter than coolant temperature. Fuel and moisture are effectively removed at oil temperatures of between 215°F and 220°F.*

Synthetic Oils

Synthetic lubricants have been available for many years, but their use has not become widespread until recently, as synthetics have become the factory fill in more and more vehicles (**Figure 12.18**). Compared to conventional mineral oil molecules, which are of different sizes and shapes, synthetic oil molecules are nearly the same. Less friction and heat are created as synthetic oil molecules move against each other. They can withstand higher loads without breaking down and losing their structure. They exhibit exceptional low and high temperature characteristics, have a better viscosity index, and provide increased fuel economy, power, and lower oil consumption.

HISTORY NOTE

The first synthetic hydrocarbon was created in the late 1800s. Germany was the leader in synthetic oil development. In 1913, a German scientist named Friedrich Bergius developed a process for producing synthetic oil from coal. In the late 1920s, Standard Oil of Indiana developed and produced commercial synthetic oils through the polymerization of olefins. During the 1930s, Union Carbide developed polyalkylene glycol (PAG), a synthetic lubricant still used today in air-conditioning systems.

Synthetic lubricants became well known during World War II, when they were used successfully in German tanks and jets. As World War II approached, Germany lacked access to sufficient fuel and oil supplies. In 1936, Adolph Hitler's government embarked on a major synthetic fuel and lubricant program. By 1939, synthetic hydrocarbons were commercially produced from carbon monoxide and hydrogen. In the 1940s, the United States Naval Research Laboratory developed ester base stocks. These oils were superior at the high and low temperatures encountered by jet aircraft engines.

Figure 12.18 Synthetic oils. *(Courtesy of Tim Gilles)*

Conventional oils have wax, which prevents easy oil flow at low temperatures. Most of this wax is removed during the refining process. Synthetics do not have mineral oil's wax or impurities, so they are especially suited to very low temperature uses, where they retain their ability to flow. Some of the better conventional oils are rated to flow at −38°F, while synthetics are rated to flow at −65°F. Synthetic oils can also withstand higher temperatures. This makes them especially suited for racing engines or engines in passenger vehicles with high underhood temperatures or turbochargers.

Engines using synthetic oils experience lower oil consumption, due to the uniform size of the synthetic oil molecule. Conventional oil molecules range from large to small, and the smaller molecules can sometimes leak past the piston rings into the combustion chamber. There they are burned off, leaving thicker oil and ash from the combustion process. Ester-based synthetic oils do not leave ash deposits when burned in the combustion chamber.

Synthetic oil's primary drawback is its price. Conventional oils are refined from crude, whereas synthetic oil production requires a more complex manufacturing process. Chemists continue to improve synthetic oils. The first synthetic motor oils were esters. Polyalphaolefin (PAO)-based synthetics have displaced them because of their lower cost and formulating similarities to conventional mineral oil. The better fully synthetic oils are usually a combination of esters and PAOs, with the ester content varying from 5% to 25%.

SCIENCE NOTE

Synthetic oils start as simple and pure materials that are synthesized to produce custom, predetermined molecular structures. Less expensive semi-synthetic oils are a mixture of conventional and synthetic oils that share some of a synthetic's improved properties.

Lubrication is provided by the oil's base stock, which is pure oil before it is supplemented with additives. Synthetic engine lubricants can be made from two primary base stocks: esters and PAOs. Both are branched hydrocarbons. PAOs are less expensive than esters. PAOs are made from hydrocarbons and are similar to the petroleum molecule. Their high temperature performance is the same as petroleum oil, and they produce ash when they burn, contaminating the oil.

Another class of synthetic lubricant is ester, which is more expensive. Ester has been the exclusive lubricant of jet engines since they were first developed. A jet engine breathes high-altitude air, which is 50°F to 60°F below zero. It must also deal with internal temperatures approaching 600°F. Piston air compressors also live in a hot environment, so an ester-based synthetic is their preferred lubricant.

Acid esters are a product of reactions between acids and alcohols. There are many different ester molecule possibilities because of the high numbers of acids and alcohols available, so custom lubricants can be designed to fit a particular application. The primary difference in structure between an ester and PAO is the ester's multiple linkages, which cause its molecules to have polarity. One end of the oil molecule is negatively charged, and the metal in engine parts is positively charged. This causes oil and engine parts to attract, resulting in more lubricity and protection during startup. Polarity also causes ester molecules to be attracted to one another. Thus, more energy is required to cause them to change from a liquid to a gas. This means they have a lower vapor pressure that results in a lower flash point and evaporation rate. Generally, when an ester has more ester linkages, its volatility is lower so less oil is burned in the combustion chamber.

Esters can cause swelling in some elastomer seals. In mixed PAO and ester oils, this is controlled in the formulation of the base stocks. PAOs sometimes have the opposite effect on elastomer seals, shrinking and hardening them. When synthetic oil was used in some engines made prior to 1980, seal leakage sometimes occurred. Modern synthetic rubber seals are better suited for use with synthetic lubricants.

One advantage to synthetic oil is that it can be changed less frequently. ExxonMobil says Mobil 1 synthetic oil is good for 25,000 miles or 1 year, but they recommend replacing the filter at regular intervals to get rid of combustion by-products and contaminants. One of the important reasons for changing oil is to remove dilution and contamination. When a vehicle is relatively new, be certain to follow the manufacturer's oil change recommendations so you do not void the warranty.

The API defines five groups of lubricating oils. Groups I through III are conventional petroleum-based oils. Group IV is PAO synthetic, and group V includes the rest of the synthetics, such as ester, PAG, and others.

■ CHANGING ENGINE OIL

A factor in the life of an engine is whether the engine oil is drained and refilled at regular intervals. Oil does not wear out, but it becomes diluted and contaminated. Contaminated oil will wear out engine bearings and seals permanently. In a cold engine, contaminants condense on the cylinder walls and migrate to the crankcase where they are mixed with the engine oil. Oil also becomes contaminated when an engine is run with a missing or dirty air filter. If an air cleaner is carelessly changed or serviced, dirt can enter the intake manifold. Leaking vacuum accessories and lines can also result in dirty air being drawn into the engine's cylinders. Following are some benefits to changing the oil:

- An oil's additives are depleted over time and changing the oil replenishes them.
- Because the oil filter takes out only particles larger than a certain size, periodic oil changes help to clean the smaller contaminants from the oil.
- Unburned contaminants like fuel and acids are removed with the oil.
- Sludge is removed.

Oil Change Intervals

Internal combustion engines produce many by-products during operation. *Blowby gases* (gases that leak past the piston rings into the crankcase) contain acids as well as unburned and burned fuel. When moisture combines with these materials, sludge forms inside the engine. If the engine is never fully warmed up or if oil is not changed often enough, more sludge develops. A vehicle driven primarily on the freeway—where it runs more efficiently—can have less frequent oil change intervals (7,500 miles or more) than one that is driven primarily on short trips. One thing that has become even more important with newer engines with variable valve timing (see Chapter 18) is engine maintenance. Regular oil changes with high-quality, low-viscosity oil are mandatory. Lack of maintenance can result in sludging of the oil galleries and failure of the VVT mechanism to operate.

NOTE: *For city driving, it is a good practice to change the oil every 90 days or 3,000 miles. Oil is inexpensive insurance against engine damage. On late-model vehicles, this old recommendation might not be best, however. There is concern that anti-wear additives in the oil can damage the catalytic converter (CAT). More frequent oil changes can result in more contamination to the CAT.*

Changing Brands of Oil

Oils with the same API service rating are usually compatible. But today's oils can be made up of ⅓ additives, so a chemical reaction is possible. Although the chemical reactions have *not* been found to be serious:

- It is a good idea to avoid mixing brands of oils *between* oil changes.
- Changing brands of oil is best done when the oil is being changed.
- When there is not a choice of brands available, it is better to add any brand of high-grade oil than to operate the engine with a low crankcase oil level.

Changing Oil

It is best to change the oil while it is still hot because the detergents in the oil will suspend contaminants, allowing them to be removed with the oil. While the oil is draining, the filter can be changed. Check the lubrication manual for the capacity of oil that the crankcase holds and the API and viscosity classifications of the oil.

SHOP TIP If an engine's oil is to be changed, remove the oil filler cap and place it somewhere obvious. Make a habit of not replacing it until all of the new oil has been added after the oil change.

When changing the oil:
- Check the condition of the drain plug gasket (**Figure 12.19**). The gasket is usually made of a synthetic material like nylon or a soft metal such as aluminum or copper. Drain plug gaskets are usually reusable, but at the first sign of wear or damage they should be replaced.
- Be careful not to strip the threads in the oil pan.

SHOP TIP When starting a thread into a hole, first turn counterclockwise (direction of loosening) until you feel the thread drop into place. Then, tighten the plug.

- Always turn the drain plug all of the way into the thread *by hand*. Only use a wrench to torque the plug when its gasket has begun to be compressed against the oil pan. This will avoid the possibility of accidentally stripping a thread.

Figure 12.19 Check the condition of the gasket on the drain plug.

- If a stripped thread is encountered, *self-tapping drain plugs* are available (**Figure 12.20**).

  Sometimes a service technician can be distracted by a phone call or a customer with a problem and accidentally forget to refill the crankcase with oil. This can result in serious engine damage.
- Be sure to check the oil level, run the engine, and check for leaks following completion of the oil change.
- Close the hood and check to see that the latch is secure.
- Wipe off any grease or handprints that might have accidentally gotten on the hood or steering wheel.

Mileage Service Record. When oil is changed, it is customary to put a reminder of the mileage somewhere where the customer will be able to refer to it. The mileage can also be noted in the customer's service record book, usually found in the glove box. This is usually a door record sticker, placed on the driver's side door jamb (**Figure 12.21**) or a transparent sticker placed on the upper inside of the windshield.

Oil Monitor System Reset

Many late-model vehicles have an oil monitor system. It uses a computer-based software algorithm to compute the necessary oil change interval and illuminates a message on the instrument panel when an oil change is required (**Figure 12.22**). The oil change interval is based on the mileage since the last oil change and the type of driving experienced by the engine. Following an

Figure 12.20 A self-tapping drain plug and gasket. *(Courtesy of Tim Gilles)*

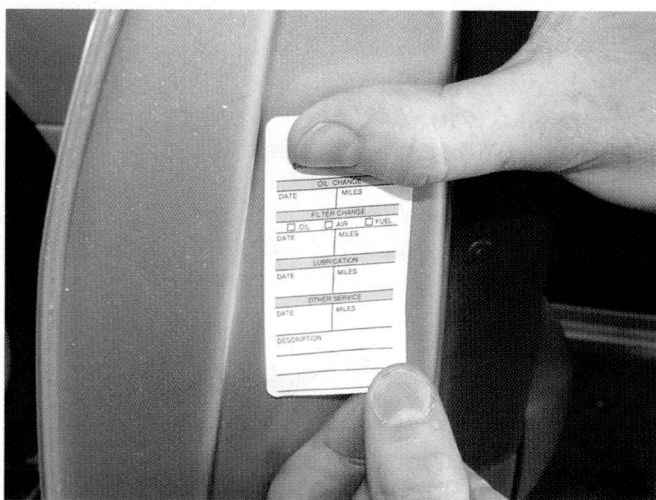

Figure 12.21 A door record sticker. *(Courtesy of Tim Gilles)*

Figure 12.22 Instrument panel reminder message when oil change is required. *(Courtesy of Tim Gilles)*

oil change, a procedure must be followed to shut off the light and reset the monitor so it can begin measuring conditions for the next oil change reminder. Procedures vary between vehicles and are listed in service information libraries and in various aftermarket booklets. One typical reset procedure requires the throttle pedal to be floored three times within 5 seconds while the ignition key is in the "on" position. The indicator on the instrument panel will flash a message saying that the monitor has been reset (**Figure 12.23**).

Figure 12.23 A new message that appears when the reminder has been reset. *(Courtesy of Tim Gilles)*

■ OIL FILTER

An engine's oil filter prevents harmful abrasive particles in the oil from damaging internal parts of the engine. Most oil filters are the *spin-on* type (**Figure 12.24**). The sheet metal shell of the filter contains a filtering element, usually made of pleated paper. Oil flows through the paper from the outside to the inside, taking advantage of the entire surface area of the paper for maximum flow (**Figure 12.25**). A metal tube in the center of the filter keeps the paper element from collapsing inward when the oil is thick or contaminated.

Full-Flow Oil Filter

Today's passenger cars use the **full-flow oil filter** system. In full-flow systems, all of the oil supplied by the pump is *supposed* to flow through the oil filter on its way to lubricate the engine bearings (**Figure 12.26**). The filter used is made of tough, treated paper that has a minimum resistance to flow. It is called a *surface-type* filter.

Because all of the oil supplied by the pump must first flow through the filter before it can reach the bearings, a full-flow filter must have a **by-pass valve** (**Figure 12.27**). The by-pass valve opens to let the unfiltered oil flow to the engine when:

■ The oil is cold and thick.
■ The filtering material is plugged due to a lack of proper maintenance.

Some engines have a built-in by-pass valve. Otherwise, it is incorporated into the filter.

■ The by-pass valve on most passenger car oil filters will open when resistance to flow reaches about 8 psi. At normal oil operating temperature, oil flowing at 4 gallons per minute through a paper filter will have a pressure drop that is less than 1 psi.

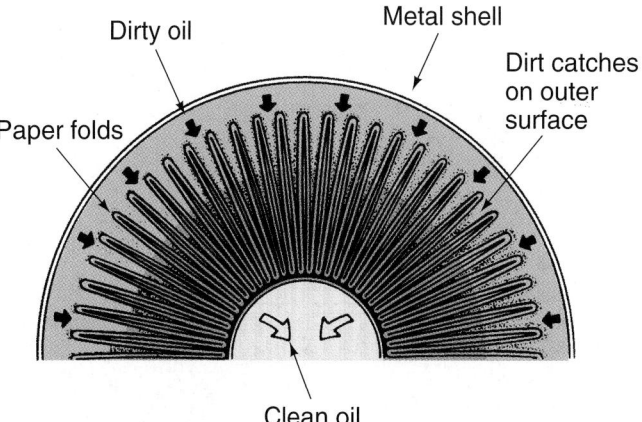

Figure 12.25 Oil flows through the paper in the filter. *(Courtesy of Dana Corporation)*

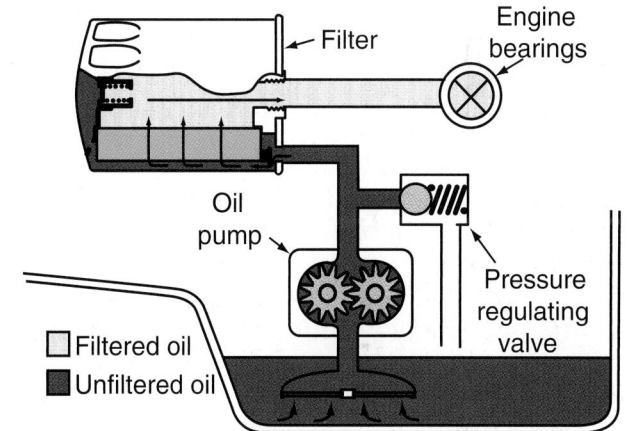

Figure 12.26 In a full-flow system, all oil is pumped to the filter.

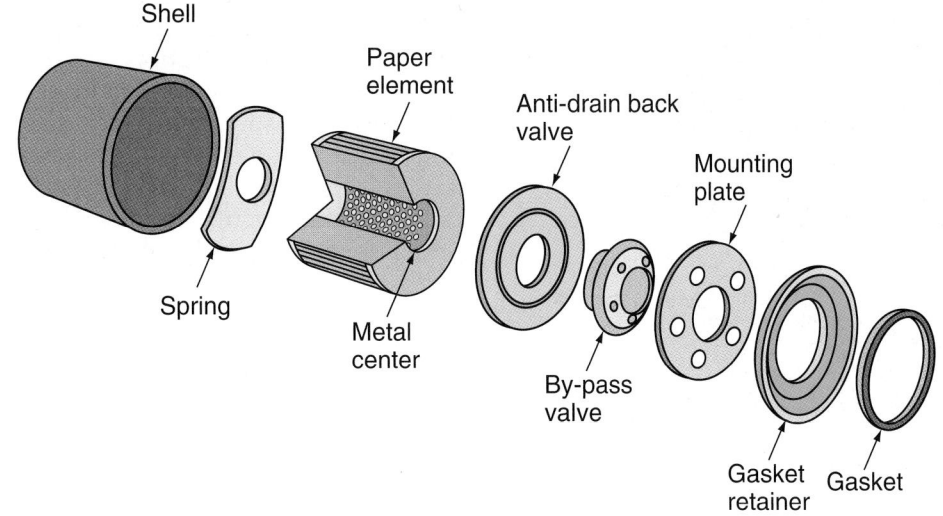

Figure 12.24 Parts of a pleated oil filter.

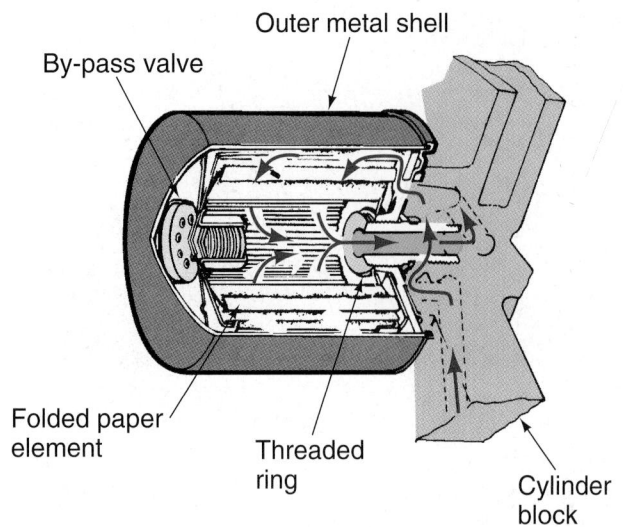

Figure 12.27 An oil filter has a by-pass valve. *(Courtesy of DaimlerChrysler Corporation)*

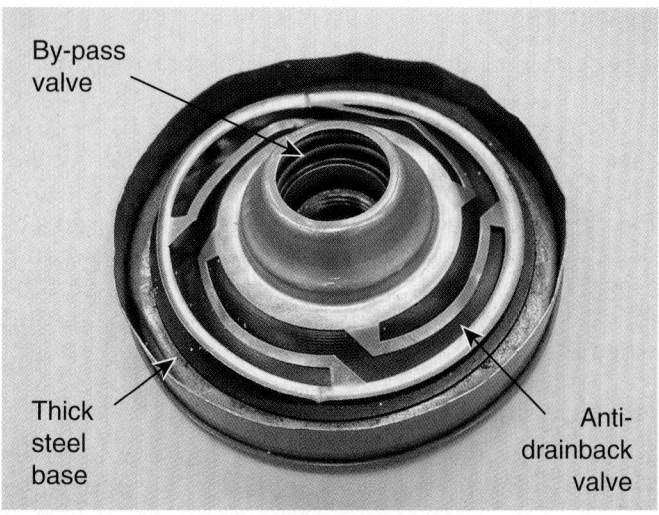

Figure 12.29 Cutaway of an oil filter showing the by-pass and anti-drainback valves. *(Courtesy of Tim Gilles)*

■ Full-flow filters filter out only relatively large particles. According to General Motors, their AC full-flow paper oil filter elements will trap and hold any particle that is larger than 40 microns (100 microns is equal to 0.004"). A human hair is usually about 60 to 80 microns in diameter, so the size of filtered particles would be visible to the eye. Particles smaller than the filter can trap cause only minor wear when proper service intervals are maintained. They remain in the oil, held in suspension by the detergents and dispersants. If the filtering material was dense enough to filter out all of the smaller particles too, it would not be able to allow enough oilflow to be able to function as a full-flow filter.

Oil flows into the filter through the small holes that are in a circle in the filter base. It flows out through the large center hole (**Figure 12.28**). Many oil filters are

Figure 12.28 Oil flows into the filter through the small holes and out through the large center one.

mounted horizontally on the engine block. When the engine is shut off, about half of the oil in these filters can empty out through the filter inlet holes.

Horizontally mounted filters must have an **anti-drainback valve** to prevent this from happening (**Figure 12.29**). If the drainback valve fails or if the wrong filter is used, the lubrication system will be starved of oil until the filter is filled back up. On engines with hydraulic lifters or overhead cam engines that have hydraulic timing chain tensioners, this will result in a noisy condition in the morning.

By-pass Filters

Supplemental add-on filters and oil filters used on heavy trucks are often of the **by-pass** or partial-flow **oil filter** design (**Figure 12.30**). A by-pass filter traps very fine materials and allows very little oil to flow through it. It only filters about 10% of the oil at one time. Oil is picked up at any pressurized point and allowed to "leak" through the parallel path provided by the filter on its way back to the oil pan. Little, if any, pressure drop occurs from this leak because the amount of flow is so little.

By-pass filter elements are made of such materials as cotton fiber, shredded paper, stacked high-density filter papers, and wood chips, which can remove particles down to 1 micron in size (one thirty-nine millionth of an inch). They are called *depth-type* filters. These filters are capable of removing sludge and were used in older vehicles that used straight mineral (SA) oil (which produced sludge rapidly). Today's oils do not allow the formation of sludge unless detergent/dispersants have been depleted.

Variations in Filters

Oil filters are identified by a number that is printed on the metal shell of the oil filter and on the filter box. Usually, printed on the box will be an application chart

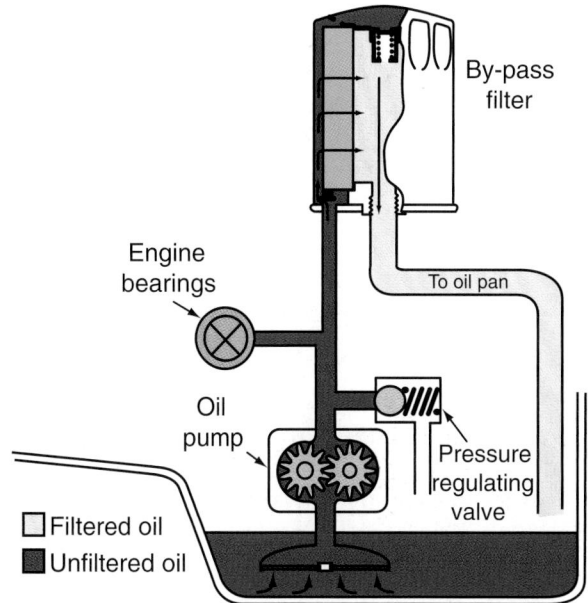

Figure 12.30 Heavy-duty filtering systems use a by-pass filter that filters only a small amount of the oil at any one time.

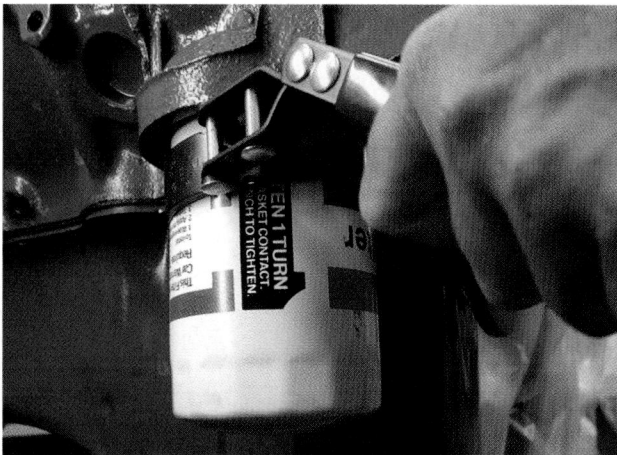

Figure 12.31 This type of oil filter wrench should be placed as close to the filter base as possible. (*Courtesy of Tim Gilles*)

that tells which cars a filter will fit. A cross reference is often included that tells which other manufacturers' filters interchange with a particular filter.

Be sure to follow the filter manufacturer's catalog recommendations when selecting an oil filter. There are metric and U.S. threads found in a mix in both import and domestic filters, so a filter might fit on the mounting base and yet not have the correct thread. A loose thread fit could result in serious engine damage if the filter becomes loose. A tight thread fit can result in a stripped thread. Some filters have anti-drainback valves and by-pass valves; others do not.

■ CHANGING THE OIL FILTER

The sheet metal shell on an oil filter is very thin. It can easily become collapsed with improper handling, making it difficult to remove. Many technicians prefer to use a professional-quality *oil filter wrench* with a pivoting handle (**Figure 12.31**). This tool allows for easy positioning at the *base* of the filter. The sheet metal filter case is supported at the base by the heavy mounting plate (see Figure 12.29), so it will not collapse. There are also specialized oil filter wrenches available to fit the many different filter applications.

There is a rubber O-ring on the bottom of the filter that seals it against the engine block. Inspect it to see that it is in place on the filter and is undamaged (**Figure 12.32**), and be certain the old seal is not stuck to the engine block. The Automotive Filter Manufacturer's Council recommends that the filter's O-ring be lubricated with oil before installing the filter (**Figure 12.33**). They recommend against using grease.

Figure 12.32 The O-ring gasket is out of its groove on the base of this oil filter. (*Courtesy of Tim Gilles*)

Figure 12.33 Lubricate the filter O-ring before installing the filter.

Most filters have instructions printed on the outside stating the amount that the filter should be tightened. An example of such instructions is: "tighten the filter ¾ turn after the gasket contacts the block." If the filter cannot be tightened a sufficient amount with bare hands, a filter wrench can be used. Overtightening a filter will make future removal difficult.

NOTE: *When refilling the crankcase with oil, an extra amount of oil equivalent to the oil filter's capacity must be added. Service information sometimes specifies an amount that includes the filter capacity. Some technicians fill the filter with oil before installation to prevent the engine from running "dry." This is especially recommended on engines with turbochargers.*

■ REVIEW QUESTIONS

1. Under normal conditions, when is the only time when an oil film does not separate engine parts?

2. Give two possible problems that can occur if the oil level in the crankcase is too high.

3. When the oil level is below the "add" line on the dipstick, how much oil should be added?

4. How will an engine dipstick be different than an automatic transmission dipstick?

5. What does SAE stand for?

6. What is the name of the part of the oil additive package that swells, causing the oil to maintain its viscosity as it heats up?

7. What causes oil oxidation?

8. What is the name of the oil filter part that allows the oil to get to the engine bearings even if the filter is plugged?

9. If sludge forms in a modern engine, what additive is probably depleted?

10. When figuring the amount of oil to add to the crankcase, add the amount specified plus some extra to compensate for what?

■ ASE-STYLE REVIEW QUESTIONS

1. Which of the following oils has the highest viscosity at engine operating temperature?
   a. SAE 10W-40
   b. SAE 30
   c. SAE 20W-50
   d. SAE 20W-20

2. Which of the following is *not* a reason for changing engine oil?
   a. To remove small particles from the oil
   b. To remove sludge from the oil
   c. To replenish additives that have been depleted
   d. To replace worn-out oil with new oil

3. Which of the following situations could allow oil to bypass a full-flow oil filter and go directly to the engine bearings?
   a. When the oil is cold and thick
   b. When the oil filter is dirty and restricted
   c. Either A or B
   d. Oil never bypasses a full-flow oil filter.

4. Technician A says that the type of oil filtering system used on modern cars is called a full-flow system. Technician B says that the type of filter that filters out sludge and very small particles in engine oil is called a full-flow filter. Who is right?
   a. Technician A
   b. Technician B
   c. Both A and B
   d. Neither A nor B

5. Which of the following statements is *not* correct?
   a. Thick oil has a lower viscosity than thin oil.
   b. Multi-viscosity oil can reach bearings faster than single viscosity oil when the engine is first started.
   c. Detergents keep small particles suspended in the oil so the filter can remove them.
   d. Detergents clean the surfaces within the engine.

6. Which type of oil filter uses an anti-drainback valve to prevent it from emptying of oil when the engine is shut off?

a. A vertically mounted full-flow filter

b. A horizontally mounted full-flow oil filter

c. Both a and b

d. Neither a nor b

7. Technician A says that when removing an oil filter, position the oil filter wrench at the inner end of the filter. Technician B says that when loosening an oil pan drain plug, turn it clockwise. Who is right?

a. Technician A   **c.** Both A and B

b. Technician B   **d.** Neither A nor B

8. The most recent API service classification for gasoline automotive engine oil is:

a. SA   **c.** CE

b. SM   **d.** None of the above

9. Which of the following is *not* true when installing an oil filter?

a. Tighten the filter an additional 2/3 to 3/4 turn after the filter O-ring touches the mount surface.

b. Compare the old filter mount surface to the new filter.

c. Be sure that the rubber O-ring is dry.

d. Start the engine after adding oil and check for leaks.

10. Each of the following statements about oil is true *except*:

a. When oil oxidizes it becomes heavier and forms varnish deposits.

b. Driving on the freeway is generally easier on the engine than in-town driving.

c. The "W" in SAE 10W-40 stands for "weight".

d. Oil does not wear out, but its additives do.

CHAPTER 13

Underhood and Body Inspection (Car on Ground)

■ OBJECTIVES

Upon completion of this chapter, you should be able to:

✔ Pinpoint safety items in need of attention.

✔ Perform on-ground brake system checks.

✔ Perform on-ground steering and suspension checks.

✔ Perform an exhaust system inspection.

✔ Be knowledgeable about car body service.

✔ Perform fuel systems checks.

✔ Perform cooling system checks.

✔ Perform electrical systems checks.

✔ Safely and competently perform a complete underhood service.

■ KEY TERMS

DOT	ATF	hydrometer
dry park check	PCV system	motor mounts

■ INTRODUCTION

During a vehicle inspection, items in need of repairs are located and documented. A lubrication/safety service includes underhood and underbody inspections. This chapter deals with the *underhood and body inspection*, which is done while the car is on the ground. The next chapter describes an underbody inspection done with the vehicle on a lift.

New car dealers often have a technician in charge of new vehicle pre-delivery inspections (*PDI*), lube/safety service, and tire and wheel service. Non-dealer shops perform the same jobs, with the exception of PDIs. Some shops have senior personnel or the owner of a small business working the "lube rack" because of their experience in spotting needed repairs.

In a full-service facility (where all systems of the car are repaired), lube personnel sometimes earn a commission. Job responsibilities include inspection, diagnosis, and documenting problems with a vehicle. It is usually an inconvenience to the customer to arrange to leave a car for repairs or service. When contacted, the customer often decides to have needed repairs performed while the car is still in the shop. Thus, the lube person who highlights needed repairs can be a source of additional business. The customer is better served because needed repairs may be spotted before a road failure occurs or a potential safety hazard becomes dangerous.

■ BRAKE SYSTEM INSPECTION

Brakes are probably the most obvious safety item on a customer's mind, and the customer expects the service specialist to give a complete appraisal regarding their condition. Only obvious conditions are inspected. A complete brake inspection is usually *not* included in a lube service unless an inspection warrants further investigation. If the wheels need to be removed, there is usually an extra charge, which is sometimes applied to the cost of a future brake job done by the same shop. Also, when a wheel is removed, the shop assumes a liability.

As part of the lubrication and safety service, the following items in the brake system are inspected. There is also a good deal of related information that a service specialist should know in order to give the vehicle more than just a cursory brake inspection. That information is covered in Chapter 58, which deals with brakes.

Brake Pedal Travel and Feel

Brake pedal travel should be checked to see that it is not too low (**Figure 13.1**) and that the pedal feels firm when applied. Proper operation and condition of the master cylinder can be checked by holding light pressure on the pedal. If the pedal slowly slips toward the floor, an internal seal leak is indicated.

NOTE: *Prior to taking a test drive to check out a complaint involving the brakes, be sure to check the master cylinder fluid level and the brake pedal height. A vehicle with faulty brakes should never be driven.*

Brake adjustment can be checked by pumping the pedal rapidly. If the height of the pedal rises significantly on the second application, the drum brakes need an adjustment. If the pedal feels soft and "spongy"

there could be air in the system, which can be removed by bleeding the brakes (see Chapter 58). Other things like a loose disc brake caliper can cause a mushy pedals as well. Further diagnosis will be needed.

Parking Brake Travel and Adjustment

The brake system is required by law to include a parking brake (also called *emergency brake*) that operates independently of the hydraulic *service brakes*. Apply the parking brake to see that it is fully applied at about ½ of the full travel of the handle or pedal. The parking brake pedal or hand brake should be firm. When fully applied, it should work well enough to hold the car when stopped on a hill. If the service brakes require adjustment, the parking brake travel will also be excessive. After a brake adjustment, the emergency brake should have proper travel.

NOTE: *Drum brakes should always be adjusted before attempting to adjust the emergency brake.*

Inspecting Brake and Clutch Fluid

Nearly all brake systems and many clutch systems use hydraulic fluid. **Figure 13.2** shows the location of typical master cylinders for both of these systems. A vehicle with an automatic transmission will not have the clutch master cylinder shown in the photo. The plastic caps shown in the photo are flexible and are easily removed. Other master cylinders have plastic screw-on caps or a metal bail that is pulled over the reservoir cover.

Clean around the top of the cover before removing it so the fluid does not become contaminated.

Some reservoirs have a diaphragm attached to the inside of the cover that keeps air, dirt, and moisture from contact with the brake fluid (**Figure 13.3**). Be sure the diaphragm is not damaged. The cover should not be left off the master cylinder for any period of

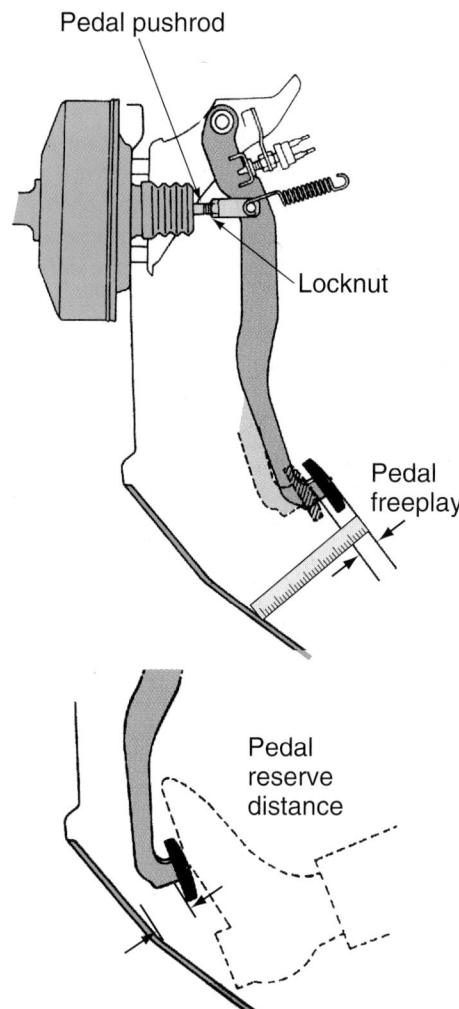

Figure 13.1 Check brake pedal travel.

Figure 13.2 Typical hydraulic master cylinders and reservoirs for brake and clutch systems. *(Courtesy of Tim Gilles)*

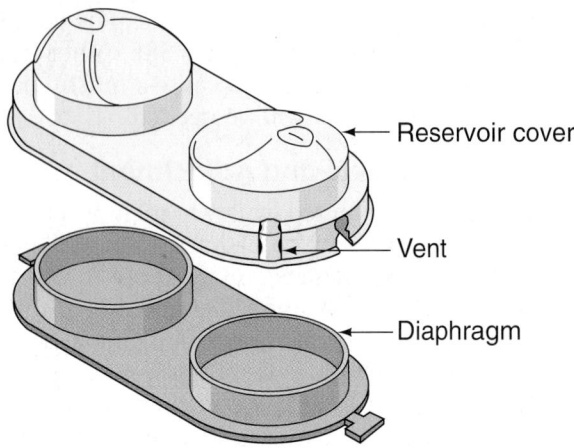

Figure 13.3 A diaphragm keeps moisture out of the brake fluid. Check the reservoir vent to see that it is not plugged.

time, because brake fluid absorbs moisture rapidly. Exposure to air can result in a lowering of its boiling point, which can cause brake failure. Brake fluid checks are covered in Chapter 58.

When fluid level is low, a more complete brake inspection is required. Check for obvious signs of leakage, which might help identify the problem. See if the master cylinder is wet, indicating a leak. The backing plates on drum brakes can be wet with brake fluid, and tires will often be streaked with fluid and dirt. Sometimes the outside of the tire can show signs of leakage.

There are two chambers in master cylinder reservoirs of all passenger vehicles built since 1967. Some cars have a disc and drum combination system, called hybrid, with disc brakes on the front and drum brakes on the rear (**Figure 13.4**). Hybrid master cylinders usually have two chambers of different sizes. The disc brake chamber is the larger of the two reservoirs. When inspecting brake fluid level, a service specialist can often tell if the brake system will need a more thorough inspection. When disc brake pads wear, the fluid level in the master cylinder drops.

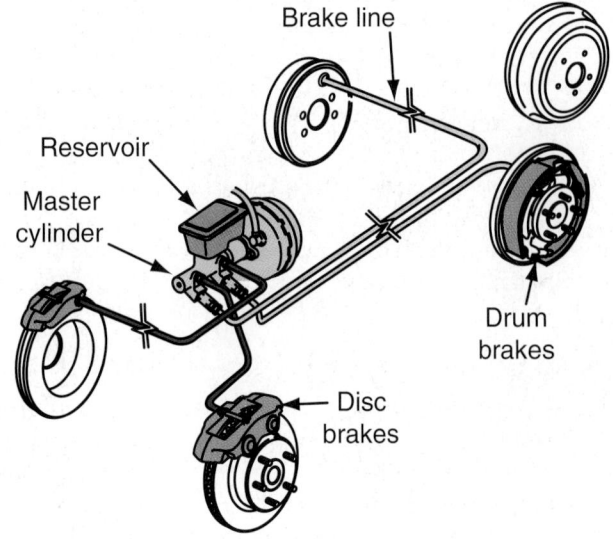

Figure 13.4 A disc and drum combination brake system.

When brake fluid is dirty, further inspection of the system is warranted. Dirty fluid indicates a system that has not been serviced for some time. See Chapter 57 for a more complete explanation of brakes and fluids.

NOTE: *Be careful when checking or adding brake fluid. Brake fluid can damage a car's paint. Always use a fender cover when working around the master cylinder. Be sure to wash off any accidental brake fluid spills immediately with water or denatured alcohol. Some technicians keep a spray bottle of water handy for such spills.*

Use only an approved brake fluid with a high boiling temperature. Brake fluid is regulated by the Department of Transportation (**DOT**) and will have "DOT" on the label. When adding brake fluid, the new fluid must be DOT 3 or DOT 4.

NOTE: *Accidental contamination of the brake hydraulic system with oil or transmission fluid can cause serious damage to the rubber parts in the brake system, including swelling of the rubber in the reservoir diaphragm.*

Some cars with manual transmissions have a clutch master cylinder located next to the brake master cylinder. It uses brake fluid also. Check the fluid level and inspect the clutch master cylinder the same way as for a brake master cylinder.

Disc Lining Inspection

If the master cylinder fluid level in the large reservoir of a master cylinder is low, inspect the front brake linings for wear. The lining should be as thick as its metal back. There are several systems to warn the driver that brake pads are worn. One system uses an electronic sensor that causes a dash light to illuminate. Another warning system uses an electronic sensor that causes the pedal to pulsate when the pads become too worn. The oldest warning system is called the audible sensor system. A metal tab on the disc brake lining causes noise to warn the driver of lining wear (**Figure 13.5**).

The linings can usually be seen through an opening in the outside of the brake caliper (**Figure 13.6**). It is usually necessary to remove the wheel to perform the inspection. When brakes are worn, be sure to inspect the rear brakes for wear also.

■ ON-GROUND STEERING AND SUSPENSION CHECKS

Another part of a lubrication/safety service is inspection of the steering and suspension systems. Look for obvious defects, such as loose steering parts and worn tires. More detailed inspections are covered in the next Chapter, "Undercar Inspection and Service," and also in Section 11.

Steering Looseness

Steering wheel *freeplay* is the first item to check. With the wheels straight ahead, there should only be a minimal amount of freeplay (less than 30 mm). Excessive

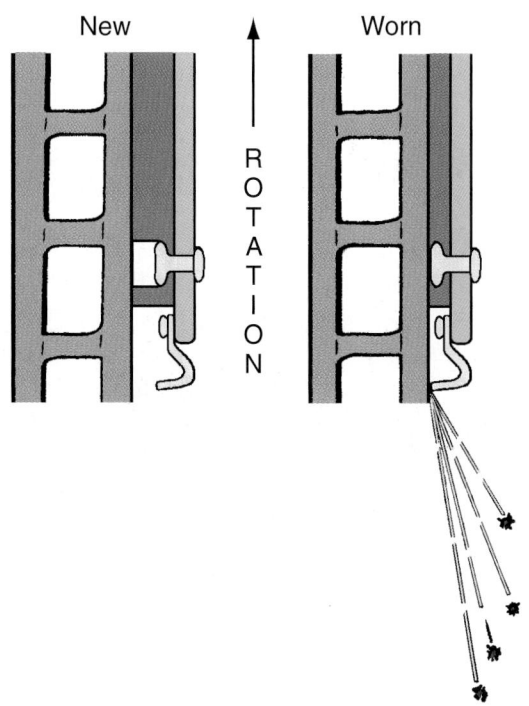

Figure 13.5 When the lining is worn too much, the metal warning sensor contacts the brake rotor.

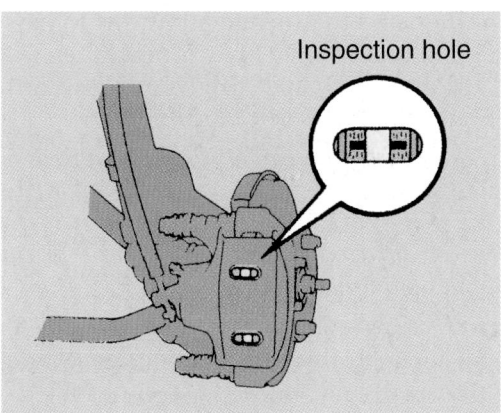

Figure 13.6 Some calipers have an inspection hole to see if the lining is worn excessively.

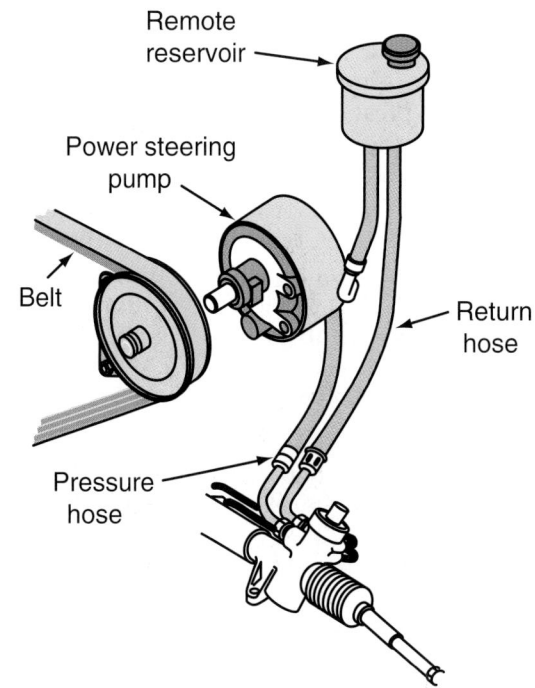

Figure 13.7 Parts of a power steering system include the pump and fluid reservoir.

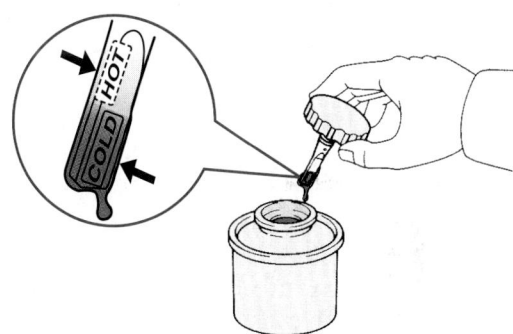

Figure 13.8 The dipstick for the power steering has hot and cold levels.

looseness can be checked by performing a **dry park check**. With the tires on the ground, have a helper repeatedly turn the steering wheel back and forth (about 3" or so). Look under the car for any signs of looseness between steering linkage parts. Good linkage parts will allow pivoting only. During this test, slack between parts will show up because of the resistance to the tires as they try to turn against the ground.

Steering Fluid Level

The power steering fluid reservoir is often part of the pump, which is mounted on the front of the engine and is driven by a belt (**Figure 13.7**). Fluid level is checked with the engine off. The filler cap often has a dipstick, which includes "hot" and "cold" fill levels (**Figure 13.8**). Fluid level is checked when hot with

the engine off. The level will be lower when the car has not been driven and the fluid is cold.

When adding fluid, be sure to use the one specified by the manufacturer. Some power steering systems use automatic transmission fluid (**ATF**). Others use a specially formulated fluid. Check the power steering system for fluid leaks at the hoses, the pump, or the steering gear.

Manual steering gears use different fluids. Consult the service information for the lubrication requirements for the gearbox and the location of the inspection and fill plug.

Check Shock Absorbers (Car on Ground)

Shock absorbers, commonly called *shocks*, are designed to control the initial movement and subsequent oscillations of the spring during jounce (downward movement of the car) and rebound (movement of the car back up). Shock absorbers are checked twice: once while the car is on the ground and again when it is in the air.

First check to see that the vehicle sits level and does not sag to one side. Then, push down on each of the vehicle's bumpers to check resistance of the shocks. Do not press on the fenders. They can become greasy and might get dented when you press on them. Normal shocks will allow the spring to rebound only once. This is only a cursory check of shocks. A more complete check can be done to shocks while the car is in the air. That information is covered in the section of the text on undercar service. See Section 11 for more information on shock absorbers.

FUEL SYSTEM INSPECTION

Flexible hoses are used to connect gaps between sections of steel tubing that run along the frame from the fuel tank to the engine. Check for leaks and inspect all fuel hoses to see that they are not cracked, hard, or leaking. Inspect the fuel filter to see if it needs to be replaced.

 **SAFETY NOTE** Hard or cracked fuel hoses must be replaced! A dangerous fuel fire can result.

Check to see that the gas cap fits properly and that its gasket (if it has one) is in good condition.

Air Filter

An air filter cleans the air entering the engine (**Figure 13.9**). It is important that air be unable to leak past the filter into the engine.

Leaking vacuum lines or vacuum accessories can allow dirt to be drawn into the engine as it operates. Dirt is the number one cause of engine failure. Hold a

Figure 13.9 An air filter cleans the air entering the engine. Lift it out of the housing to inspect it. Trapped dirt will be on the side of the filter that is underneath. (*Courtesy of Tim Gilles*)

Dirt particles   Outside air enters here

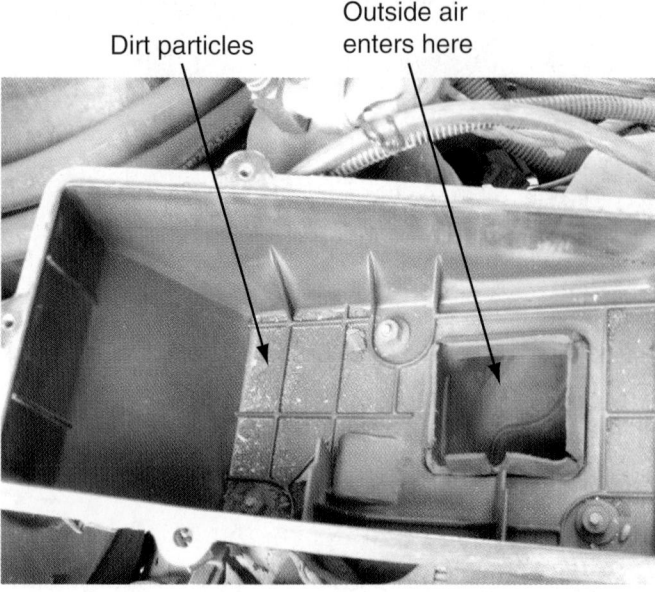

Figure 13.10 This dirt fell out of the filter when it was removed. Be careful that no dirt is allowed into the other half of the air filter housing where it could enter the engine. (*Courtesy of Tim Gilles*)

shop light near the filter and observe how much light comes through. Look to see that there are no small holes in the paper that would allow dirt to pass. Compare the amount of light that a new filter will allow to pass through the paper to give yourself an idea of whether an old filter is dirty or not.

NOTE: *Sometimes, when an air filter is changed, dirt from the outside of the dirty filter is carelessly allowed to enter the engine (**Figure 13.10**). The equivalent of two aspirin tablets of dirt causes more wear than that caused by 75,000 miles of normal driving.*

Figure 13.11 shows a cylinder with scratches caused by a leak in the air intake system.

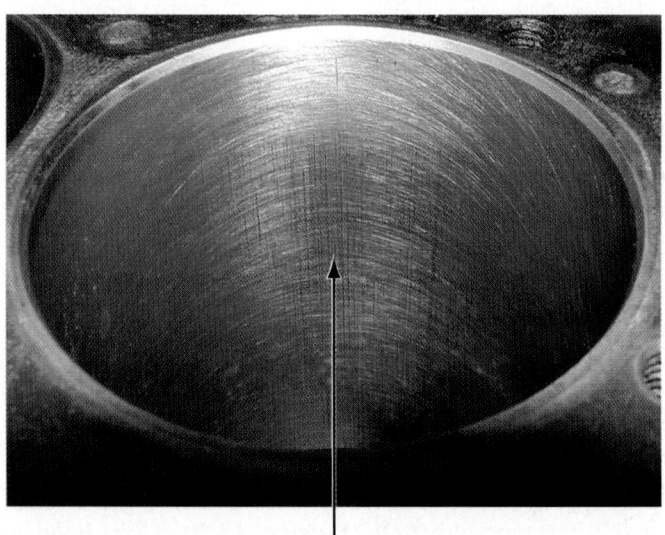

Figure 13.11 A cylinder with scratches caused by a leak in the air intake system. (*Courtesy of Tim Gilles*)

Air filters are covered in more detail in Chapter 44.

Crankcase Ventilation System

Engines have a positive crankcase ventilation system, called a **PCV system**. The PCV valve is usually serviced and replaced as part of a tune-up. Its service is covered in detail in Chapter 46. The system must have a means of drawing in filtered air. This sometimes includes an auxiliary air filter that can require service also.

If the crankcase ventilation system is operating properly, there should be a slight vacuum at the oil filler opening when the engine is running at idle. There are testers available for checking this.

▉ COOLING SYSTEM INSPECTION

The radiator has a pressure cap and an overflow tank (**Figure 13.12**). The radiator pressure cap increases the pressure on the coolant so that its boiling point will be higher than normal. The boiling point increases approximately 3 degrees for each pound of pressure. The pressure rating of the cap is stamped on its top. When opening the radiator cap, a safety catch stops it from turning after ¼ turn. To remove it the rest of the way, it must be pushed down and turned.

SAFETY NOTE A gallon of 50/50 coolant and water mixture will expand by about ⅓ pint when heated to engine operating temperature (about 195°F). Before opening the radiator cap on a hot engine, squeeze the top radiator hose to be sure that the coolant is not under pressure. It is not a good idea to check coolant level on a hot engine. The coolant is best checked when cold.

Remove the radiator cap for a visual inspection. Check its pressure relief valve, upper sealing gasket, and vacuum valve (**Figure 13.13**) The operation of these parts is discussed in Chapter 20. The lower seal is made of rubber and can become brittle or indented over a period of use. Dirty coolant can interfere with the operation of the vacuum valve and rubber seal. Replace the cap if any of these conditions exist. Clean these areas of the cap before reinstalling it if it is in good condition (**Figure 13.14**).

Coolant Inspection

Visually inspect the coolant for chemical breakdown. Coolant can be several different colors depending on the manufacturer. See that the level is correct and that the coolant appears clean. Coolant strength can be checked with a coolant **hydrometer** (**Figure 13.15**). This is similar to a battery acid hydrometer (used to measure a battery's state of charge).

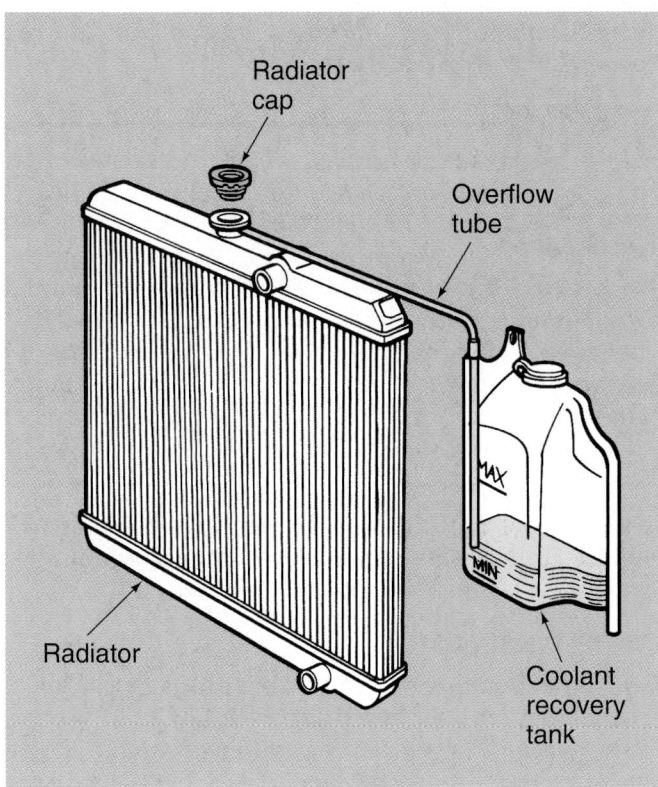

Figure 13.12 A radiator and recovery tank. *(Courtesy of DaimlerChrysler Corporation)*

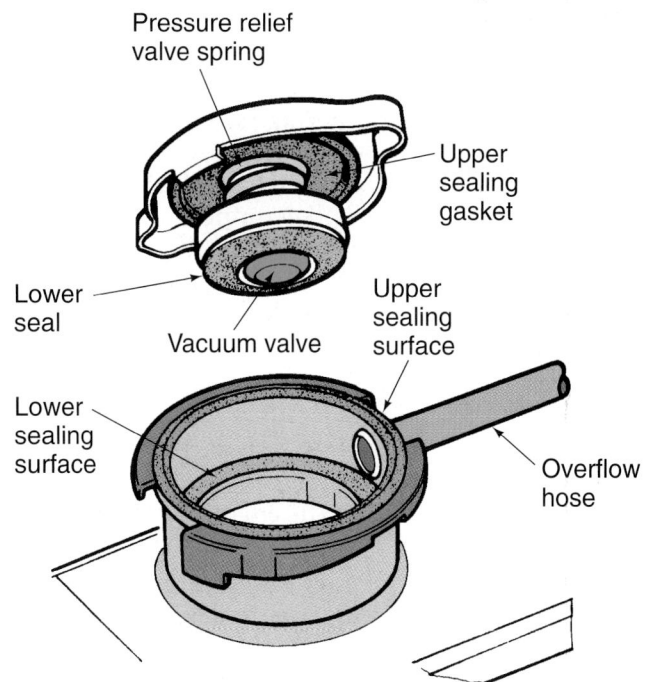

Figure 13.13 Remove and inspect the parts of the radiator cap. *(Courtesy of DaimlerChrysler Corporation)*

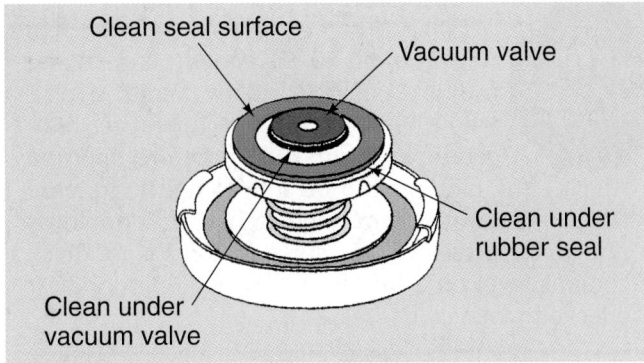

Figure 13.14 If the cap is in good condition, clean the vacuum valve and rubber seal. *(Courtesy of DaimlerChrysler Corporation)*

Figure 13.15 A coolant hydrometer can be used to check the strength of the coolant. *(Courtesy of Tim Gilles)*

Coolant is drawn out of the radiator and sampled to compare its weight to that of pure water (see Chapter 21). One kind of coolant hydrometer has several different weight balls in a small chamber. The strength of the coolant is determined by how many of the balls in the chamber float in the coolant. The more that float, the stronger the coolant concentration. A mixture of water and coolant provides better protection than pure coolant.

Cooling System Leaks

Look for leaks in the cooling system. When the coolant level is low, a pressure test can determine whether the leak is internal (within the engine) or external (see Chapter 21).

NOTE: *When looking for leaks, be aware that the vehicle's air conditioner collects moisture and allows it to drain from the passenger compartment to the ground through a hose. It is not unusual for a customer to complain of a coolant leak when the only problem is that the air-conditioning compressor is on and operating normally.*

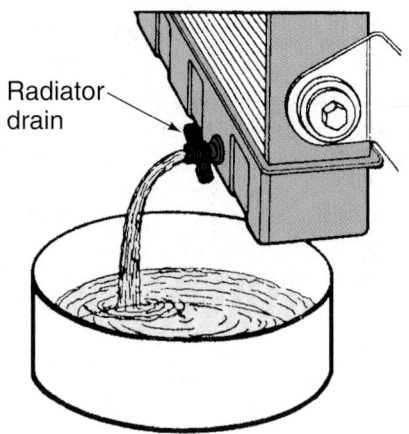

Figure 13.16 Before removing a coolant hose, drain some coolant from the radiator. *(Courtesy of DaimlerChrysler Corporation)*

Inspect Cooling System Hoses

Inspect all coolant hoses. They should not be swollen or cracked.

- Oil damages rubber cooling system hoses. Leaking oil usually does the most damage to lower hoses.
- Hoses at the top of the engine get the hottest coolant, so they are most prone to failure from heat.
- When a hose is collapsed, look for a defective radiator cap.

Check both heater hoses, too. Be aware that aging hoses can also fail from the inside.

NOTE: *When a hose is removed, coolant will probably escape. Drain some coolant from the radiator into a clean container first (**Figure 13.16**).*

> **CASE HISTORY** *A customer with 60,000 miles on her car had a coolant flush done while she was getting an oil change. The next time the car was in for service, the shop replaced a heater hose on her vehicle. New coolant was added to the system and charged to the customer's bill. The customer realized she had recently paid for new coolant during the earlier coolant flush, but she did not protest the bill. She never returned to this shop again, however, and she told several of her friends.*

Coolant that has leaked will create a mess that will require unnecessary cleanup. In some communities coolant is classified as toxic.

■ BELT INSPECTION

Check the condition of accessory drive belts by looking to see that they are not cracked or glazed. A quick inspection of the tension of a belt can be performed with the engine off. Try to turn a pulley by hand to see if the belt holds or slips. If the alternator belt tension is too loose, the alternator will not be able to keep the battery charged. Belts can be either V-belts or V-ribbed belts. More complete information on belts is included in Chapter 22.

ELECTRICAL SYSTEM INSPECTION

Give the entire electrical system a visual inspection, looking for obvious problems. Check for bare or burned wires that could possibly complete a circuit to ground. This would probably result in a blown fuse, but could also cause a spark, resulting in an explosion of fuel or the battery. Inspect the connections on the alternator and ignition system wiring. Look for loose or corroded connections, which can cause annoying intermittent electrical problems (**Figure 13.17**).

Battery Check

Visually inspect the battery (**Figure 13.18**). Look for damaged or corroded cable clamps. Cleaning the cable clamps is not usually done during a lubrication/safety

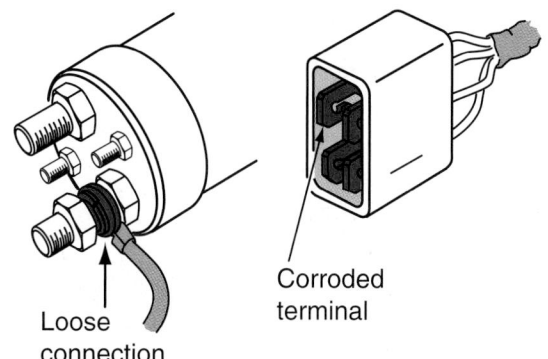

Figure 13.17 Look for loose or corroded connections.

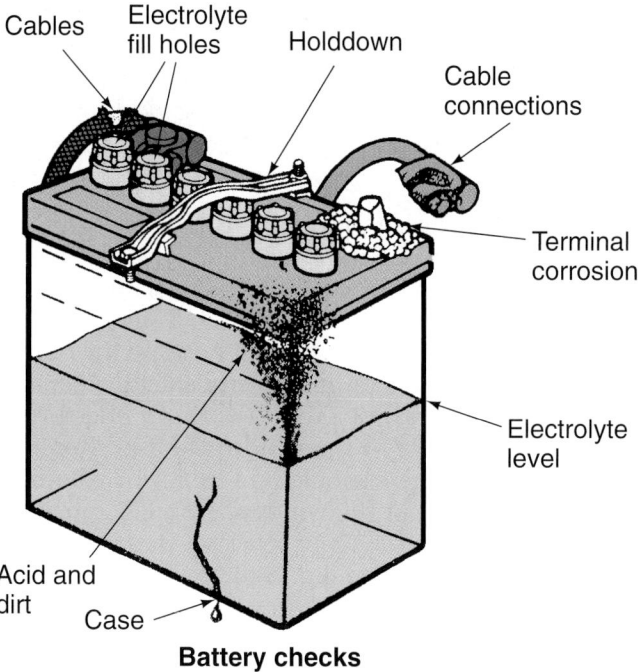

Figure 13.18 Inspect the battery, cable connections, and holddown.

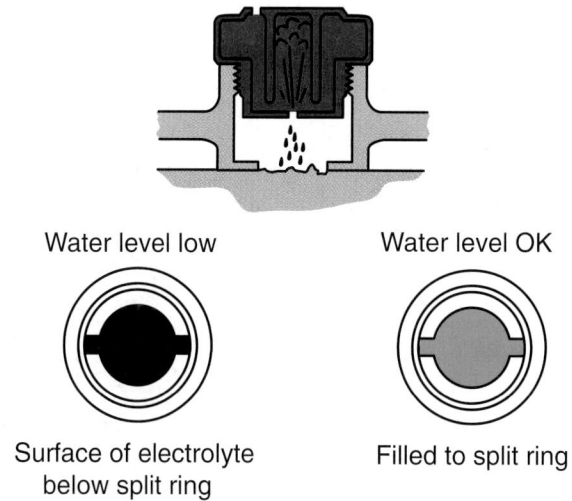

Water level low — Surface of electrolyte below split ring

Water level OK — Filled to split ring

Figure 13.19 The battery electrolyte level is correct when the water level reaches the split ring.

service but is part of a tune-up or maintenance service. Check the condition of the battery holddown and holddown clamps.

NOTE: *If the terminal clamps are to be removed for any reason, be sure to remove the ground cable first. This will eliminate the chance of a spark that could cause a battery to explode. Computer memories can be preserved using a small battery connected through the cigarette lighter.*

Battery Electrolyte Level

Battery *electrolyte* is sulfuric acid and water. When a battery is recharged, some water may evaporate from the electrolyte mixture. On many batteries, clean water can be added to replenish the electrolyte. Twelve-volt batteries used on modern cars have six cells. Some batteries have fill holes (see Figure 13.18). If the battery has removable caps, inspect the electrolyte level. The level should be up to the split ring seen inside the fill hole (**Figure 13.19**).

■ Battery acid can be neutralized with baking soda. A mixture of baking soda and water may be used to clean the top of a battery.
■ In addition to being dangerous to skin and eyes, battery acid can damage a car's paint. Be sure to install fender covers on the vehicle when working on the battery. Wash off any spilled electrolyte immediately with water.

CHECK OPERATION OF LIGHTS

Inspect the operation of all of the lights on the vehicle. For some of the lamps to operate on some cars, the ignition switch might have to be turned on. Check all lenses for cracks. A crack in a lens will allow moisture to get in, causing corrosion and failure of the circuit. A bulb is usually removed by pushing it in and

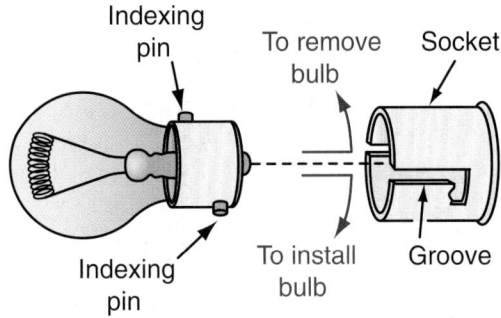

Figure 13.20 Some bulbs are removed by pushing and turning counterclockwise.

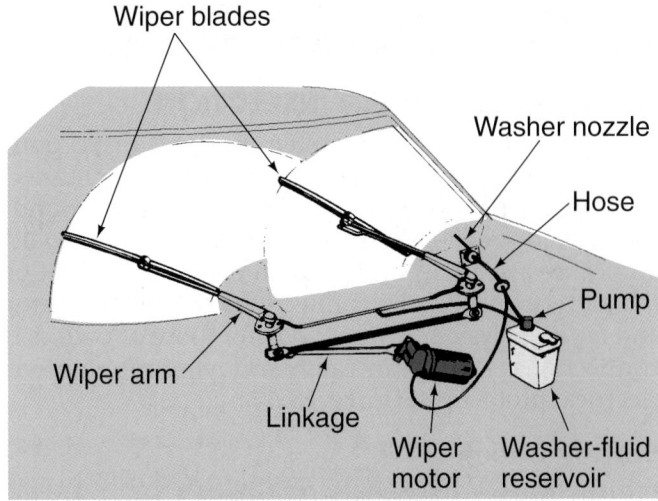

Figure 13.21 Parts of the windshield wiper system. (*Courtesy of Trico Products Corporation*)

turning it counterclockwise in its socket (**Figure 13.20**). Chapter 33 covers the various methods of removing bulbs and lenses.

After checking the operation of external and internal bulbs, check the dash indicator lights. Dash lights that require the key to be on to test them include the oil pressure indicator, water temperature indicator, safety belt reminder, emission reminder, malfunction indicator light, and brake warning lights.

NOTE: *Burned-out bulbs can be caused by excessive voltage, so a charging system check may be called for.*

To check obvious problems with the aim of the headlights, shine them against a wall. Check low and high beam operation and that they appear to be level.

■ VISIBILITY CHECKS

Visibility items include the windshield and windows, mirrors, wipers, and defroster. Windows are checked for damage, condition of weatherstripping rubber, and operation. Operate power windows to see if they operate too slowly, indicating a problem needing attention.

Mirrors should not be cracked or fogged and should be fastened securely to the vehicle.

Check the operation of the defroster. The blower motor is tested to see that it works. Operate the heater/blower selector levers. Duct hoses should also be inspected when possible to see that they are in good condition.

Windshield Wipers

Parts of the windshield wiper system are shown in **Figure 13.21**. Check the condition of the wiper blades. Wiper arms should be installed so that they do not hit the trim around the window. They should have sufficient tension to contact the entire glass surface.

Check the level in the *windshield washer reservoir* (**Figure 13.22**). Washer fluid used to fill the reservoir is usually blue and is one of two types. Check the label for the level of freeze protection. Some summer washer fluids have the same freeze point as water. Winter washer fluid, colored blue, yellow, or orange, contains methanol or another type of alcohol to lower the freeze point of the liquid. In winter climates, some

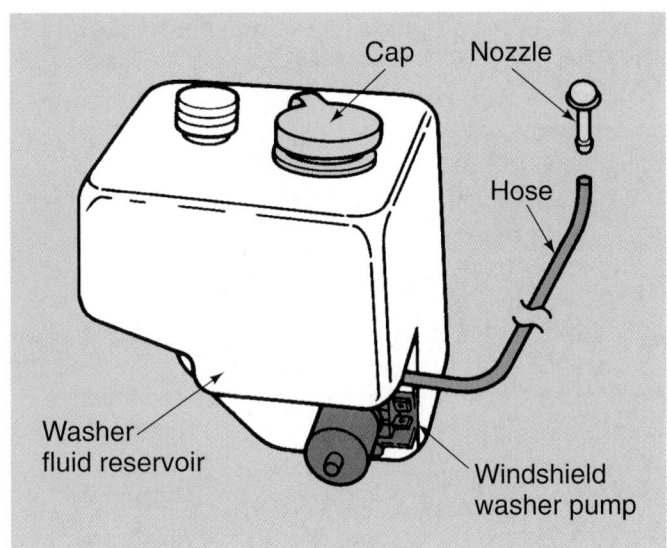

Figure 13.22 A windshield cleaning fluid reservoir. (*Courtesy of Volkswagen of America, Inc.*)

washer fluids are also promoted as a de-icer. There is a higher cost for this type of fluid. The windshield wiper reservoir on most new cars has a small rotary pump. When the switch is activated, a small impeller pushes washer fluid through small hoses to the nozzles below the windshield or on the windshield wiper arm. The windshield washer should work on the driver's and passenger's side of the windshield and should be aimed correctly. There should be a sufficient volume of washer fluid to clean the window. Nozzles sometimes become plugged. They can be cleaned with a small piece of wire or very small drill bit.

Wiper Blades

Windshield wipers have a blade, an arm, and a spring (**Figure 13.23**). Blades can become dull from sliding on dirt or ice, or they can become torn or hard and

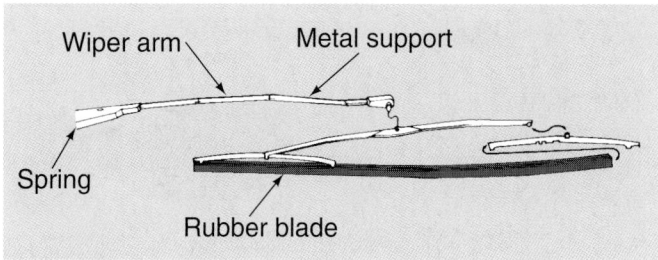

Figure 13.23 The wiper assembly has an arm, a blade, and a spring. *(Courtesy of DaimlerChrysler Corporation)*

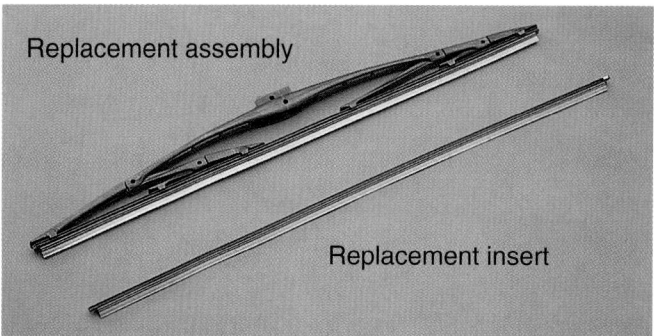

Figure 13.24 Replaceable wiper blades.

inflexible. They are made of rubber, which deteriorates as it ages or is exposed to the sun, heat radiated from hot windshield glass, wind, cold weather, and road oils.

Sometimes, wiper blade wear is evident to the eye, and other times the wiper just stops doing a good job. When a blade no longer cleans the window effectively, try cleaning it with soap and water. Wet the window and try it again. If it still leaves a smear, it should be replaced.

> **SHOP TIP** Some technicians clean wiper blades with very fine wet or dry sandpaper during a service job.

Wiper blades must contact the window evenly across the entire blade. Occasionally the spring on the wiper arm breaks, requiring replacement of the entire arm. If the metal on the blade assembly is bent or a pivot point is corroded, the blade will not lie evenly on the window. In this case, the entire blade assembly is replaced.

Removing and Replacing Wiper Blades

Wiper blades, sometimes called refills, squeegees, or cartridges, are usually replaced once or twice a year, depending on conditions. Many wiper blade assemblies have replaceable inserts (**Figure 13.24**), usually purchased in pairs. Sometimes a wiper arm is not serviceable and must be replaced as an assembly.

For the DIY market it is often easier and nearly as economical to replace the entire wiper arm, but in service establishments replaceable inserts are often used. Many shops replace the entire arm because worn arm pivots can cause poor wiper operation, like chattering or streaking. There are three designs of wiper blade refill assemblies (**Figure 13.25**). One uses a metal clip on the end, and another has a plastic release button. Import cars often use a system with a special clip, sometimes only available at the dealer. These types can be replaced with one of the more common types by simply buying the entire blade assembly the first time wipers are replaced. Subsequent replacements can be done with a refill.

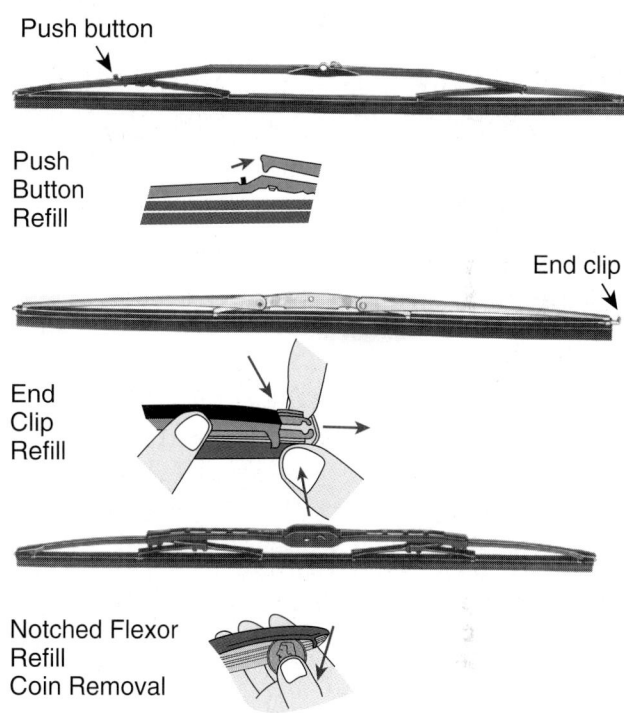

Figure 13.25 Three designs of wiper blade refills.

NOTE: *It is easy to accidentally scratch a window when replacing a wiper blade. To avoid this, remove the entire blade assembly before attempting to replace a refill. When the refilled blade assembly is reinstalled, pull on it to see that it is firmly in place.*

Several types of connector designs are used to fasten the blade assembly to the wiper arm (**Figure 13.26**). The *bayonet connector* has a small round dimple on its top that fits into a hole to help hold the blade assembly on the arm. A spring catch or a lever on the underside of the arm is released to disengage the dimple from the hole. Another style uses a *pin connector*. Other styles include the *inner lock connector*, the *screw-type connector*, the *shallow hook connector*, and the *shepherd's crook connector*.

Removing and Replacing Wiper Arms

Wiper *arms* are mounted on a splined post or a threaded screw shaft (**Figure 13.27**). The splined posts have either a release lever called a clip tab or a slide latch

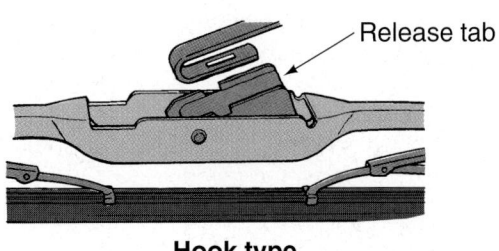

Hook type

Bayonet type

Pin type

Inner lock type

Screw type

Center hinge types   **Side latch types**

Figure 13.26 Several connector designs fasten the blade assembly to the wiper arm. *(Courtesy of DaimlerChrysler Corporation)*

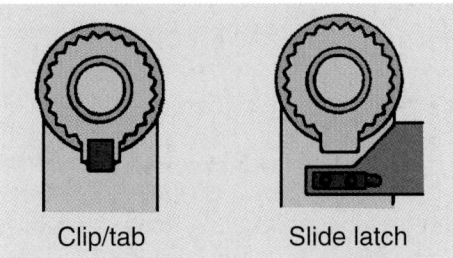

Clip/tab   Slide latch

Figure 13.28 Splined posts use a clip tab or a slide latch. *(Courtesy of Federal-Mogul Corporation)*

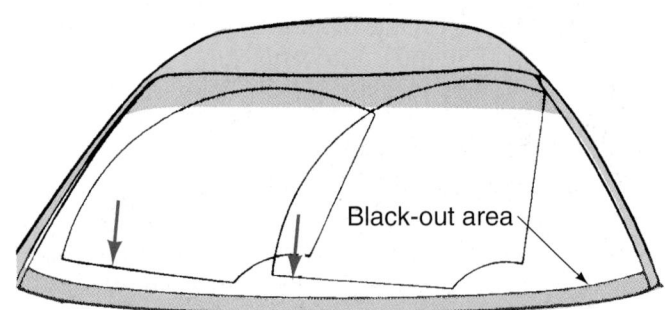

Black-out area

Figure 13.29 Position the wiper arm so that it cannot hit the window frame. *(Courtesy of DaimlerChrysler Corporation)*

electrical systems allow the arm to continue to move until it reaches full travel at the bottom. A relay then shuts off the motor. Position the arm so the blade is about an inch from the windshield frame. Then, cycle the wipers to see that the blade assembly does not hit the side frame at the other end of its travel.

■ OTHER SAFETY CHECKS AND SERVICE

Check the operation of the horn. Then, check the operation of both hood latches. There is a *primary hood latch* and also a *safety catch* in case the hood is not all the way latched. The safety catch should hold when tested. If it lets loose when pulled against, it is not working properly and will require attention.

NOTE: *Always document items in need of attention on the repair order.*

Its pivot points can be lubricated.

The door striker plate can be lubricated with wax lubricant (called *Door Ease*) that comes in stick form. Silicone spray lube can be applied to the door latch rotor.

Door locks are lubricated with dry graphite. Because it is not sticky, it does not attract dust like oily products do. The dust would interfere with operation of the tumblers in the lock.

Check the seat belts to see that they are in place and operating correctly.

In addition to maintenance for the sake of vehicle upkeep, the owner will be especially interested in items related to the safety of his or her family. The shop owner or service manager is often in a position of trust.

Figure 13.27 Wiper arms are mounted on a splined post or a metric screw shaft. *(Courtesy of Federal-Mogul Corporation)*

(Figure 13.28). To remove the wiper arm, raise it as high as it goes and release the clip tab or tap on the slide latch with a screwdriver. Carefully pry the arm off using two screwdrivers, prying a little on one side and then on the other. A special tool is available for prying off splined wiper arms. Threaded screw shafts use a locknut.

SHOP TIP The paint on the car can easily be chipped during this operation. Put some cardboard down under the wiper post when prying in case something accidentally slips.

Most wiper arms can be installed at any point on the spline. When reinstalling a wiper arm, position it where it cannot hit the frame around the windshield **(Figure 13.29)**. When the wipers are turned off, most

Recommendations of the shop will be what the owner relies on for his or her decision making. Safety is the highest priority in vehicle repair. Never fail to document (write down) safety-related items on the repair order.

General Safety Check

The safety checklist in **Figure 13.30** is a simple version of that done by a professional facility. Its purpose is to provide an apprentice with a basic familiarity with those parts that will be inspected during a professional lubrication and safety service. Locating and repairing these items can please customers and increase their loyalty.

Automatic Transmission Fluid Check

Automatic transmissions do not normally consume fluid, but leaks sometimes develop. Follow these precautions when checking ATF:

- When the car first arrives at the shop and the fluid is still hot, lift the hood and check the ATF level. Fluid expands as the transmission warms up (**Figure 13.31**).
- Be sure the parking brake is firmly set and the vehicle is on level ground. Automatic transmission fluid level is checked with the engine running. The transmission must be in a specified gear position.

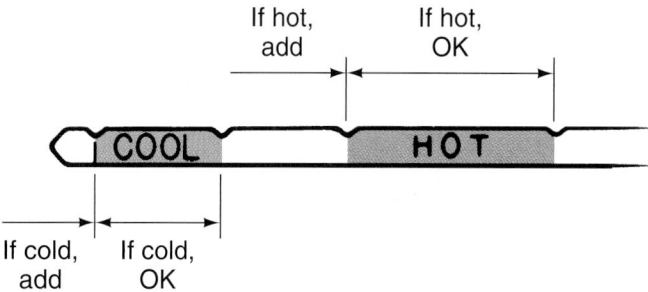

Figure 13.31 Automatic transmission fluid expands as it warms up.

Most transmission fluid levels are checked with the transmission in park. Some transmissions are checked in neutral and will give a different dipstick reading when checked in park. If fluid is overfilled, it can become foamy as it gets whipped up, filling with air. Foamed fluid is compressible, which can interfere with the operation of the transmission. Aerated fluid also results in a higher fluid level, which can cause fluid to leak out of the dipstick tube or transmission vent.

CAUTION If a transmission (older Chrysler product) is moved into neutral for a fluid level check, be sure someone is in the driver's seat with a foot firmly on the brake.

The dipstick is located behind the engine. It fits into a large round tube.

NOTE: *When reading the ATF level on the dipstick, be aware that the add mark represents 1 pint, and not 1 quart as the engine oil dipstick does.*

Some automatic transaxles do not have dipsticks and must be checked following the manufacturer's recommended procedures.

Fluid Condition

Under normal use, fluids can last for a very long time. Under severe service conditions such as hot weather driving, trailer towing, or stop-and-go city driving, the heat that develops can damage the fluid. When checking the fluid level, also inspect its condition. A burnt smell could indicate possible transmission damage. Some fluids (synthetics or partial synthetics) have an unusual smell to begin with, so check with a transmission specialist if in doubt.

When fluid has a milky appearance, water or coolant may have gotten into it. Check to see that the dipstick has been properly seated at the top of the dipstick tube. Wipe the dipstick to see if it feels sticky. This could indicate excessive fluid oxidation due to heat. Changing the fluid might solve the problem, but a complete overhaul will probably be needed.

Safety Checklist (Car on Ground)		
Safety Check:	**OK**	**Needs Attention**
Lights		
Brake pedal travel		
Wiper blade condition		
V-belt tension and condition		
Tire pressure check (cold)		
Tire condition (excessive wear)		
Spare tire pressure and condition		
Hood latch		
Fuel hose condition		
Exhaust leaks (listen)		
Horn		
Brake fluid level		
Emergency brake adjustment		
Battery water check and fill		
Lug nut torque check (ft.-lb.)		
Door latches (lube)		
Electrical wiring		
Hose condition (radiator, heater)		
Power steering hoses		
Defroster		
Mirrors (condition and tightness)		
Interior and dash lights		
Seat belts (not loose or damaged)		

Figure 13.30 Safety checklist.

Always check the manufacturer's minimum service interval requirement and use this as a rule of thumb for normal transmission fluid service. Modern electronic transmissions require more frequent fluid change intervals than earlier transmissions. Also, today's vehicles have a life expectancy that is 50,000 to 100,000 miles longer than those of 30 years ago. Most fast-lube and automotive shops have equipment for quickly exchanging transmission fluid.

Automatic Transmission Fluid

The oil used in automatic transmissions is called "ATF" (automatic transmission fluid). It is usually colored with a red dye to prevent confusion with engine oils. The red color also helps in determining the source of leaks. Transmission fluid serves several purposes. It must transmit engine torque to the wheels, act as a coolant, provide lubrication to moving parts, and seal and transmit hydraulic pressure. All of this is done in a hot, hostile environment. Just as engine oil deteriorates with excess heat, transmission fluids are affected also.

Transmission fluid is made up of a high-quality mineral oil with an additive package. The additive package contains foam suppressors, oxidation inhibitors, viscosity improvers, corrosion and rust inhibitors, detergents, friction modifiers, and seal swelling agents.

Transmission parts are made of steel, aluminum, bronze, and also some nonmetal materials. Manufacturers use different friction materials for their transmission clutches, so there are different types of fluids available that have different friction characteristics.

NOTE: *Be sure to check the service information for the type of ATF to use. Using a friction-modified fluid in a transmission designed for nonmodified fluids can cause slippage and transmission damage.*

Dexron-III Mercon fluid is a fluid that can be used in most of today's transmissions. Always check manufacturers' recommendations. Some manufacturers recommend a different fluid.

Tires

Tires can be checked for obvious wear when the car is on the ground. It is easier to check for nails and obvious tire damage when the car is on the lift. Lug nut torque is checked while the car is on the ground. Check for the correct specifications.

NOTE: *If tires are rotated, on some vehicles you must reset each tire's pressure monitor (See Chapter 62).*

Check Tire Pressure

Tires develop heat as the car is driven. Driving on tires that are low on air pressure is very damaging to the tire. A tire typically loses about 2 pounds of air pressure a month, so periodic refilling is necessary. By the time a

tire appears to be underinflated, it will be well below specifications and tire damage may already have occurred. Tire pressures should be checked monthly, when the tire is cold. Because the tire heats up during operation, the air in the tire gets hot, which causes more pressure in the tire.

NOTE: *Every 10°F change in temperature results in a change in tire pressure of about 1 psi.*

Correct pressure is listed in the owner's manual or on a sticker on one of the door jambs or in the glove compartment (**Figure 13.32**). Also, check the tire sidewall for the maximum safe operating pressure. Gauges that are part of the air hose filler are frequently inaccurate due to abuse. Use a dial- or pocket-type pencil gauge. Press the gauge firmly onto the valve stem. The scale will pop out the end of the gauge, showing the amount of pressure in the tire (**Figure 13.33**). When

Figure 13.32 The correct tire pressure is listed on the door placard. (*Courtesy of Tim Gilles*)

Figure 13.33 The scale pops out of the gauge to show the tire pressure. (*Courtesy of Tim Gilles*)

checking the air pressure in the tires, inspect each valve stem carefully to see that it is not cracked or leaking. Be sure to check the spare tire.

Engine and Transmission Mounts

The engine and transmission are supported on the vehicle frame or unibody with vulcanized rubber and steel mounts (**Figure 13.34**). Engine mounts,

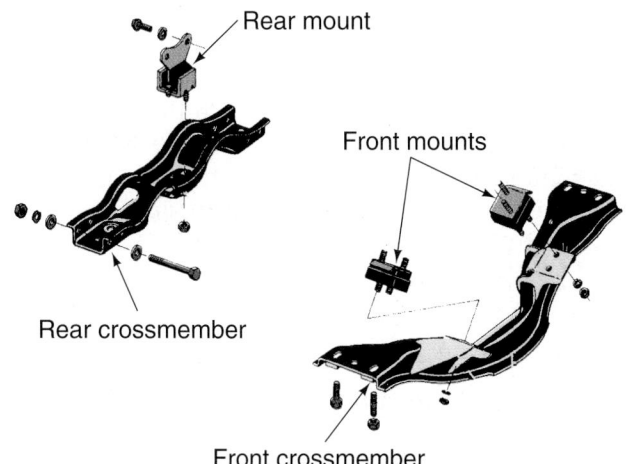

commonly called **motor mounts**, can fail due to unusual stress and strain or because they have been contaminated by leaking oil.

A loose or broken motor mount can cause vibration during acceleration. This is especially true during engagement of the clutch on a vehicle with a manually shifted transmission.

To test for broken motor mounts, two people are needed:

■ First, set the emergency brake firmly and double-check to see that it holds the vehicle against the torque of the engine.

■ While a helper puts the car in gear with the engine running, observe the mount on one side of the engine to see if the engine lifts away from it.

 **SAFETY NOTE** Be sure that you are at one side of the vehicle, and not in front of it.

■ Then, have the helper shift the transmission into reverse and check the mount on the other side of the engine.

Figure 13.34 Typical motor mounts for a rear-wheel-drive engine. (*Courtesy of DaimlerChrysler Corporation*)

■ REVIEW QUESTIONS

1. What is the minimum standard brake fluid that can be used in automobiles?

2. What is the job of a shock absorber?

3. What checks are done to a gas cap?

4. What can be done to the top radiator hose to see if the radiator is under pressure before removing its cap?

5. When a stream of water leaks out from under the passenger compartment of the car, what could be the cause?

6. Which radiator hose is most likely to suffer heat damage, the top or bottom?

7. When a technician attempts to turn a fan pulley, what is being tested?

8. When installing a new wiper arm on a splined post, if it is installed in the wrong position, what can happen?

9. How many hood latches should a car have?

10. List a common type of ATF.

■ ASE-STYLE REVIEW QUESTIONS

1. When pumping the brake pedal, it rises to a higher level on the second pump. This is likely due to:
   a. A defective master cylinder
   b. Drum brakes that need adjusting
   c. Air in the hydraulic system
   d. All of the above

2. Technician A says that brake fluid that is allowed to remain uncovered absorbs water. Technician B says that if brake fluid is accidentally spilled on a fender of a car, it can damage the paint. Who is right?
   a. Technician A   c. Both A and B
   b. Technician B   d. Neither A nor B

3. Technician A says that the strength of coolant can be checked with a coolant hydrometer. Technician B says that before removing a radiator cap, you should always squeeze the top radiator hose to see if it is hard due to hot coolant under pressure. Who is right?

   a. Technician A   **c.** Both A and B

   b. Technician B   **d.** Neither A nor B

4. A master cylinder has two chambers, one large and the other small. If the fluid level is low in the larger chamber, this is most likely due to:

   a. A leak in the hydraulic system

   b. Worn rear drum brakes

   c. Worn front disc brakes

   d. None of the above

5. All of the following are true *except* that:

   a. A loose alternator drive belt can cause a dead battery.

   b. The positive battery cable should be removed first.

   c. Battery electrolyte level is full when the electrolyte is up to the split ring inside the fill hole.

   d. Baking soda neutralizes battery acid.

6. Technician A says that tire pressure should be checked when the tire is hot. Technician B says that a typical tire loses 2 pounds of air pressure in a month. Who is right?

   a. Technician A   **c.** Both A and B

   b. Technician B   **d.** Neither A nor B

7. Technician A says that ATF is normally dyed red in color. Technician B says that it is normal for an automatic transmission to consume a small amount of oil under normal use. Who is right?

   a. Technician A   **c.** Both A and B

   b. Technician B   **d.** Neither A nor B

8. The brake pedal is applied during a safety service. Technician A says that brake adjustment is being checked. Technician B says that master cylinder function is being checked. Who is right?

   a. Technician A   **c.** Both A and B

   b. Technician B   **d.** Neither A nor B

9. Which of the following is *not* true when checking fluid level in all automatic transmissions?

   a. The vehicle is parked on level ground.

   b. The fluid is hot.

   c. The transmission is in neutral.

   d. The engine is idling.

10. Which of the following statements about power steering fluid is *not* true?

    a. Fluid level is checked with the engine off.

    b. Automatic transmission fluid can be used in all power steering systems.

    c. Some power steering pumps have a dipstick.

    d. Fluid level is higher when hot.

CHAPTER 14

Undercar Inspection and Service

■ **OBJECTIVES**

Upon completion of this chapter, you should be able to:

✔ Inspect tires.

✔ Inspect the brake system.

✔ Be knowledgeable about chassis lubricants.

✔ Perform chassis lubrication.

✔ Perform suspension and steering checks.

✔ Perform driveline checks.

✔ Safely and competently perform a complete undercar service.

■ **KEY TERMS**

tread wear indicator	chassis lubricant	hypoid gear
zerk fitting	rebound bumper	E.P.
thickening agent		

■ **INTRODUCTION**

After completing inspection and maintenance under the hood, the vehicle is raised in the air to perform an undercar inspection. The oil and filter are usually changed while the vehicle is in the air. After reading instructions in Chapter 11 for lifting the vehicle, raise it in the air.

When performing undercar service, practice developing an efficient routine. After raising the car on the hoist, start at the front on the passenger side and work around the car, finishing up at the front again. Undercar service, including draining the oil, is done prior to underhood service. Then, the technician can refill the crankcase from the top while completing underhood services.

■ **TIRE VISUAL INSPECTION**

Visually inspect the tires for problems. Inspect the tire tread for signs of wear. A properly inflated tire will contact the road evenly. If the center of the tread is worn more than the outside edges, the tire has probably been run with too much air pressure (**Figure 14.1**). A more typical situation is to find wear on both outside edges of a tire's tread (**Figure 14.2**). This indicates an underinflated tire. When a car has been driven through corners at excessive speeds, wear to the outside edges of the tread can also result.

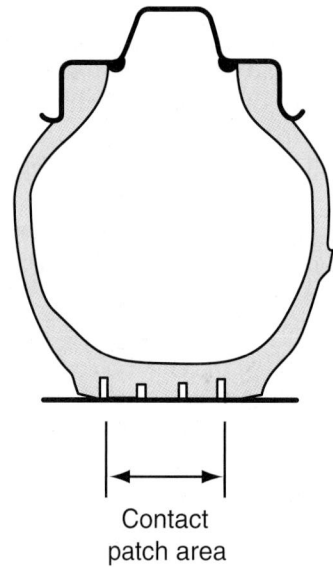

Contact
patch area

Figure 14.1 Overinflation wears the center of the tread.

177

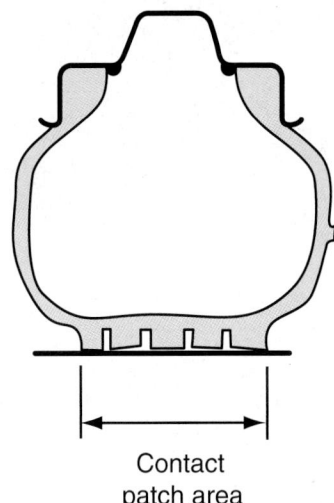

Figure 14.2 Underinflation results in excessive outer tread wear.

With the car in the air, rotate each tire slowly around while inspecting the tread area for nails, glass, screws, tears, and bulges. **Tread wear indicators** (*wear bars*) are molded into the tire's tread (**Figure 14.3**). Tires should have at least 1/16" of tread remaining. The wear bars are at this depth. When the wear bar shows across the entire tread, the tire is worn beyond legal limits and should be replaced. When tread depth is shallow through part of the circumference (usually on one side of the tread), this indicates that a belt has slipped and the tire should be replaced.

Look for *scalloped* wear or bald spots that would indicate the need for a wheel balance or a suspension system repair (**Figure 14.4**). Uneven wear on the tire tread or a *feathered edge* across the entire tread indicates the need for a wheel alignment (see Chapter 67). When a car is aligned, front and rear suspension and steering parts are adjusted to minimize tire wear by allowing the tread of the tires to sit as flat as possible on the road during most driving conditions. Correct adjustment will also allow the car to *track* properly

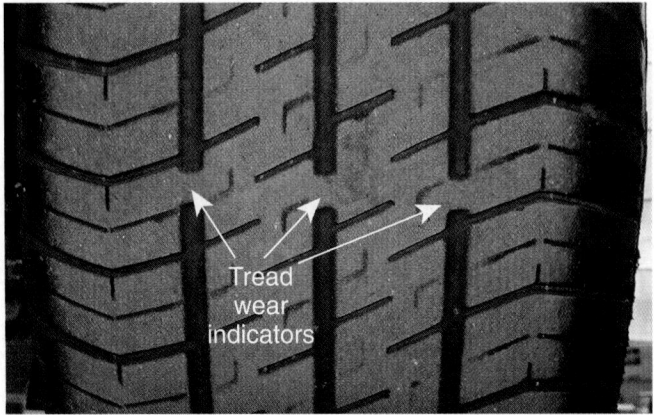

Figure 14.3 Tread wear indicators. *(Courtesy of Tim Gilles)*

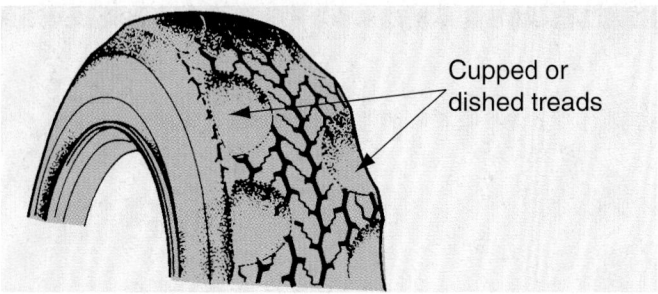

Figure 14.4 Bald spots and scalloped wear indicate loose parts or a need for a wheel balance. *(Courtesy of Ford Motor Company)*

(go straight, with no pull to one side or the other). Tires are rotated so that they will wear evenly (see Chapter 62).

Inspect the tire sidewall for cracks and signs of tread separation (bubbles). With the car in the air, spin the tires to see if they appear to be "true." *Runout* is the term that defines the up-and-down or side-to-side movement of a rotating tire. Check all of the lug bolts to see that none are damaged, missing, or broken and that they are torqued (tight). See that the wheels all spin with the same amount of effort. If one is tight, further investigation for a problem will be necessary.

■ UNDERCAR BRAKE CHECKS

Steel brake tubing runs the length of the vehicle's frame. Rubber hoses on both front wheels and above the rear axle provide a flexible connection from the steel tubing to the brake system components (**Figure 14.5**). Flexible rubber connections are required, because the front wheels pivot during steering and the rear axle moves up and down as the springs deflect. When the vehicle has independent rear suspension or disc brakes, each rear wheel is served by its own flexible hose.

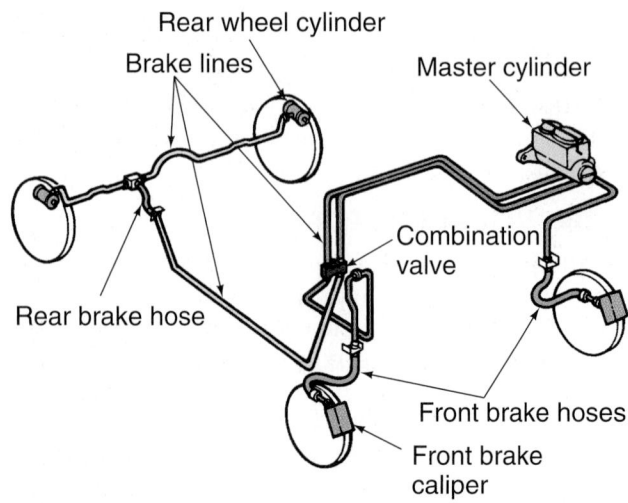

Figure 14.5 Hydraulic lines and hoses connect the master cylinder to the wheel brake units.

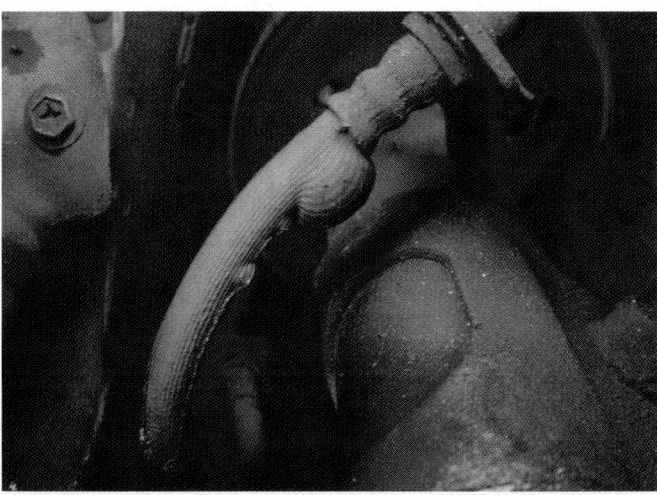

Figure 14.6 Look for obvious signs of damage to a brake hose. (Courtesy of Tim Gilles)

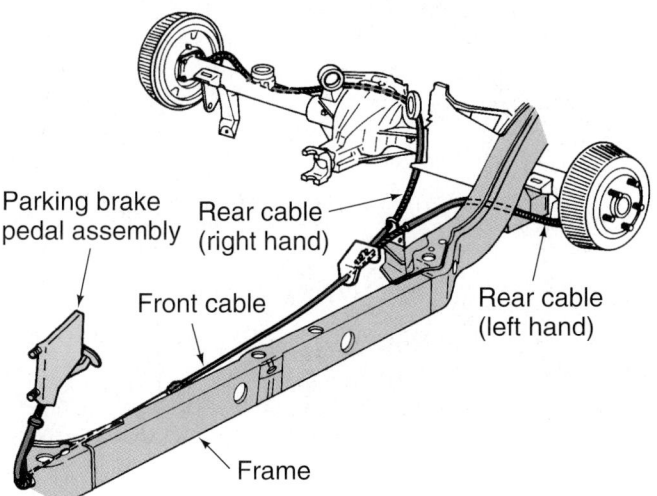

Figure 14.7 Typical parking brake cable.

Inspect brake hoses for cracks, signs of leakage, or other obvious damage (**Figure 14.6**). Check to see that the front hoses have not been rubbing on tires. Check also to see that the rear hose is not too close to an exhaust pipe where it can be burned (or rubbed against).

Metal brake lines are checked for leaks at all threaded connections. Look for any signs of leakage along the fluid lines or on brake backing plates. Metal lines are usually fastened to the frame or axle. Check to be sure that they are properly secured. Look for excessive rust on metal brake lines, which can eventually burst, causing a loss of breaks pressure.

Look at the emergency brake cable (**Figure 14.7**) to see if there are signs of rust or damage to a cable sheath.

■ EXHAUST SYSTEM INSPECTION

Part of a safety inspection includes checking the condition of the exhaust system. Exhaust leaks can result in carbon monoxide poisoning of the passengers riding in the vehicle. With the engine running, listen for leaks at

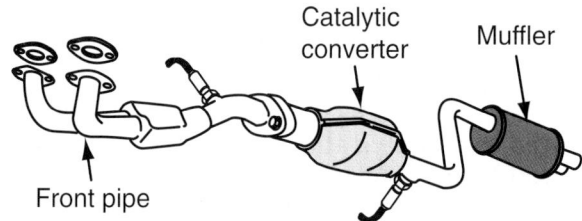

Figure 14.8 Listen for leaks in the exhaust system.

the exhaust manifold, muffler, and pipes (**Figure 14.8**). A stethoscope with the metal end removed can help to pinpoint the location of the leak. With the car in the air, check for broken muffler hangers that could allow a muffler to break loose. Exhaust components falling from a vehicle could cause a serious accident. Look to see that a muffler that was replaced has not been positioned near a fuel or brake line, where it might rub or cause excessive heat.

■ CHASSIS LUBRICATION

When two parts move against each other, they pivot on a bushing or bearing. Pivot points such as *tie-rod ends*, *ball joints*, and *constant velocity* (CV) *joints*, are sealed with rubber boots (**Figure 14.9**). Joints with boots are usually permanently sealed, but there are many older cars and newer light trucks that still have fittings provided for lubrication. To be certain, always check the lubrication chart showing lube points for the vehicle.

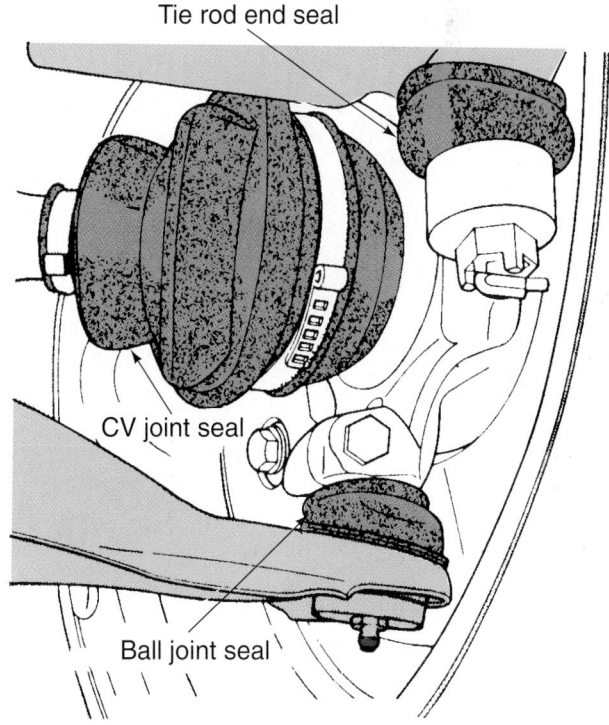

Figure 14.9 Pivot points are sealed with rubber boots. *Check that boots are not cracked or torn. (Courtesy of DaimlerChrysler Corporation)*

Grease fittings are called **zerk fittings** or *standard nipple* type fittings. They are threaded to fit into holes in the part to be lubricated (**Figure 14.10**). The thread on a fitting is usually tapered pipe thread, which forms a wedged tight seal as the fitting is tightened into its hole. Zerk fittings have a one-way spring-loaded check valve. The valve allows grease to enter the joint, but not to leak back out. Be sure to wipe off the end of the zerk fitting before applying grease to it. Dirt can plug up the check valve, allowing grease to leak out and water and dirt to leak in.

Some vehicles have threaded plugs instead of zerk fittings (**Figure 14.11a**). To apply grease, an adapter can be threaded into the hole. The plug is replaced after grease is applied. Metal plugs can be reused, but rubber or plastic plugs should be replaced. Plugs can also be replaced with new zerk fittings (**Figure 14.11b**). Rubber-tipped adapters are also available for lubricating through threaded holes.

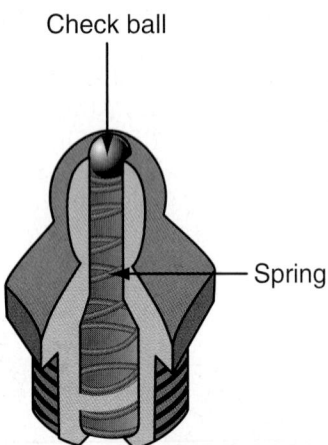

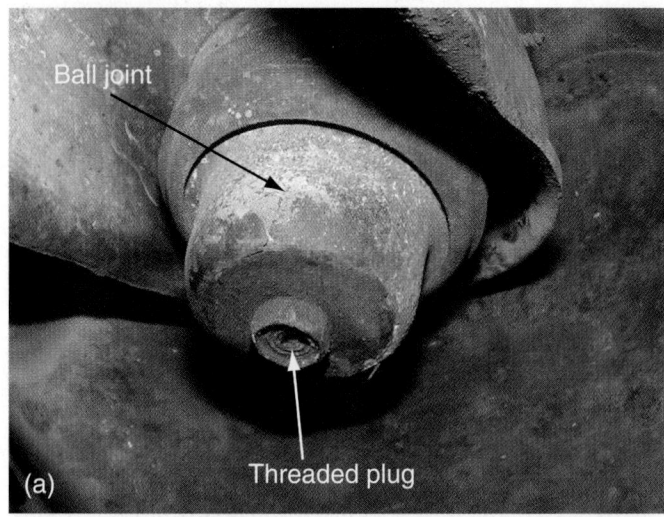

Figure 14.11 (a) A threaded plug. (b) A zerk fitting. (*Courtesy of Tim Gilles*)

Grease Guns

Hand-held grease guns (**Figure 14.12**) are available that can be used safely by anyone. Grease cartridges are available to provide a convenient grease gun refill. Cartridges are an advantage when special lubricants or multiple greases are required for the same car. Professional lubrication equipment uses a high-pressure grease gun, powered by compressed air (**Figure 14.13**).

 Be careful not to apply the grease gun against skin, or an infection can result.

Figure 14.10 Zerk fittings. (*Photo courtesy of Tim Gilles*)

Grease gun nozzles have teeth that grip the fitting tightly when grease pressure is applied (**Figure 14.14**). A grease fitting is often more easily accessible when the wheels are turned to the side. If the ignition switch

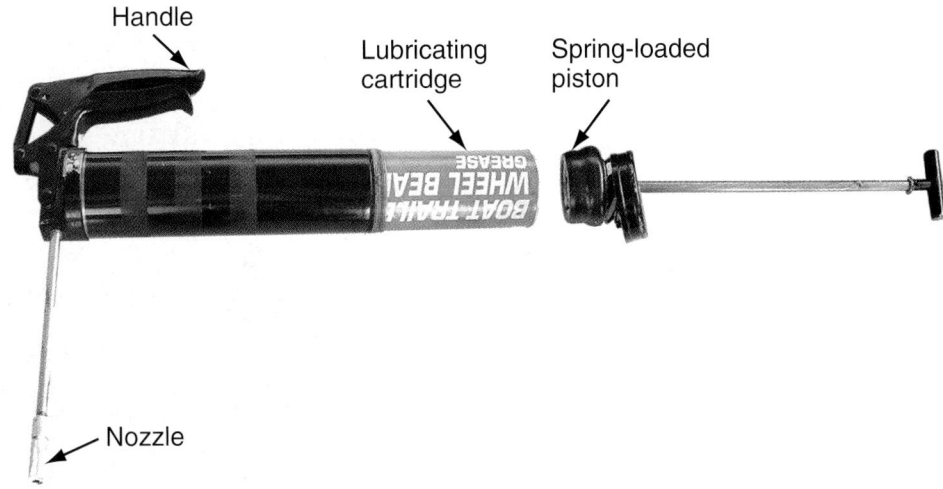

Figure 14.12 A hand-held grease gun. *(Courtesy of Tim Gilles)*

Figure 14.13 The lube gun nozzle grips the fitting.

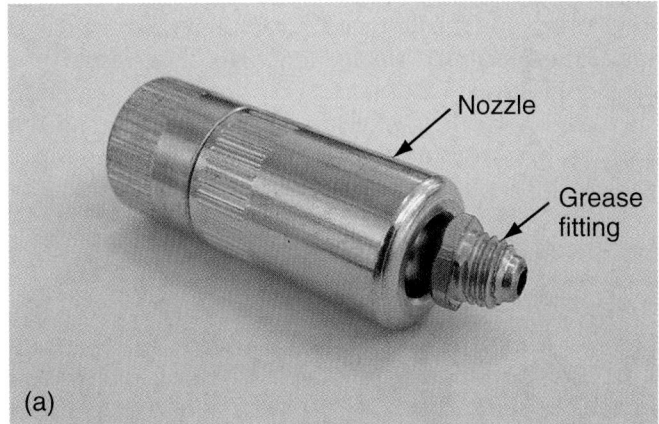

(a)

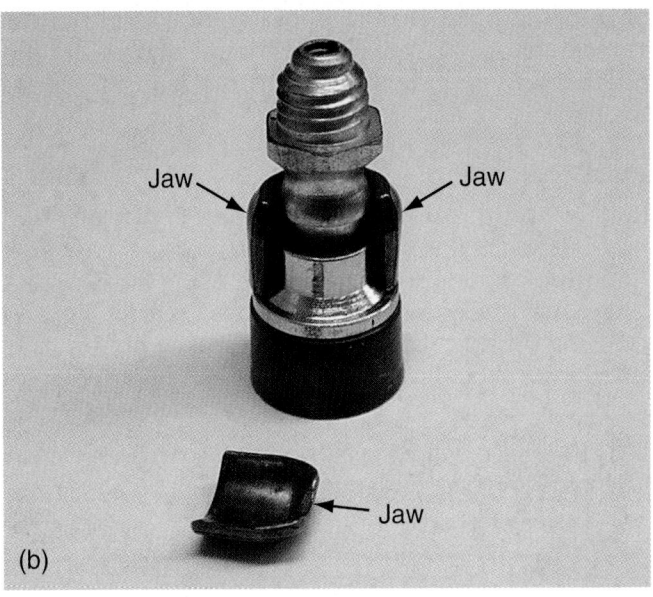

(b)

Figure 14.14 (a) The grease gun nozzle grips the fitting as pressurized grease is applied. (b) Parts of a grease gun nozzle disassembled to show how the jaws grip the grease fitting. *(Courtesy of Tim Gilles)*

is left unlocked, it is easy to turn the wheels when the car is in the air and the weight of the vehicle is off them. Unless the joint being greased is unloaded on a wheel-contact hoist, a pressurized grease gun may be necessary.

Add grease until fresh grease is seen at the vent hole in the rubber boot. Rubber boots sometimes have a small hole that allows air or excessive grease to bleed off. The boot can still be damaged, however, if too much pressure is allowed to enter the joint at once.

SHOP TIP If lubricant will not go into a fitting, the joint may be frozen up. Try turning the wheels from side to side while applying pressurized grease. If this fails, put the vehicle back on the ground and push on the bumper before trying to add grease again.

Sometimes the zerk fitting is frozen or plugged and must be replaced.

Grease

Grease used for automobile chassis lubrication is liquid petroleum oil made semi-fluid by adding a metallic soap **thickening agent**, such as aluminum or lithium. Unlike oil, grease does not leak or flow out of a bearing. This is what makes it desirable as a lubricant.

Chassis lubricant is grease with a consistency that allows it to be applied through a zerk fitting with a grease gun. It is highly resistant to being diluted or washed away with water.

A *multipurpose grease*, which is the most common type used in automotive service, satisfies the requirements of chassis, wheel bearing, and universal joint lubricants. Multipurpose does not mean all-purpose; it only meets certain requirements. There are several types of greases, which are discussed in greater detail in Chapter 60. The lubricant most often used in chassis lubrication is a *multipurpose lithium-based grease*.

Some greases contain solid lubricant materials such as *molybdenum (moly)* or *graphite*. These are often used to lubricate speedometer cables, emergency brake cables, splines, and leaf springs.

■ SUSPENSION AND STEERING CHECKS

Inspect Ride Height

Before raising the vehicle, see if it appears to be riding low or leaning to one side. If anything appears out of the ordinary, check for tire wear and then perform a *ride height check* as described in Chapter 68. Wheel alignment angles change when the ride height of the vehicle changes. Lower ride height due to worn springs often results in front tire wear on the inside of the tread.

Rubber Bushings

Rubber bushings separate many pivoting suspension parts. The bushings stretch but do not move, so friction that would exist between moving parts is not there. A bushing can be visually checked for signs of cracking (**Figure 14.15**). Use a prybar to pry on parts that have rubber bushings to see if there is any sign of looseness.

NOTE: *Do not lubricate rubber parts with oil; it will rapidly deteriorate a rubber bushing.*

 **SAFETY NOTE** Bushing replacement is done frequently by tire and wheel specialists. Proper training is required. This job should not be attempted without trained assistance.

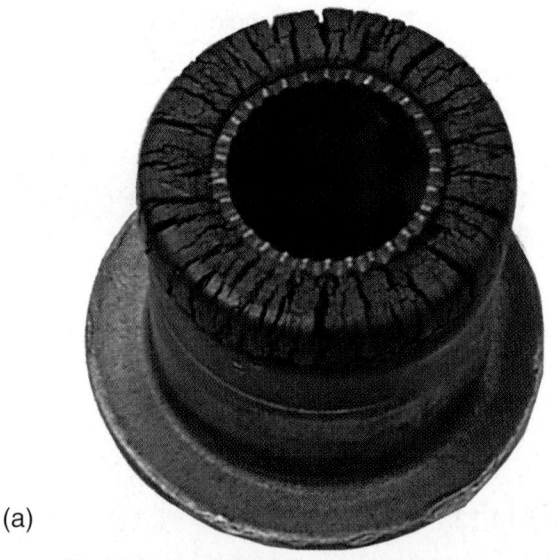
(a)

(b)

Figure 14.15 (a) This suspension bushing was replaced because it dried out and cracked. (b) These sway bar bushings need to be replaced. *(Courtesy of Tim Gilles)*

Steering Linkage Checks

Check for looseness in front-end parts with the car in the air. Twist the tie-rods and other steering linkages (**Figure 14.16**), trying to move them up and down. Parts should be free to move when twisted by hand but should not move up and down, which indicates excessive looseness. Check the condition of rubber lubrication boots. A permanently sealed joint with a torn boot should be replaced, especially if dirty rain water has gotten into the joint.

Ball Joints

Check ball joints for excessive wear and for torn grease seals. Look for unusual tire wear. Worn ball joints or other front suspension parts that cause looseness can result in cupped wear of the tire tread. If the tires are cupped, check the ball joints as described in

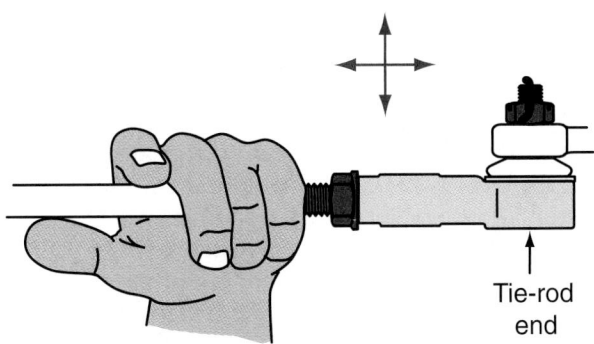

Figure 14.16 Twist and shake steering linkages to test for looseness.

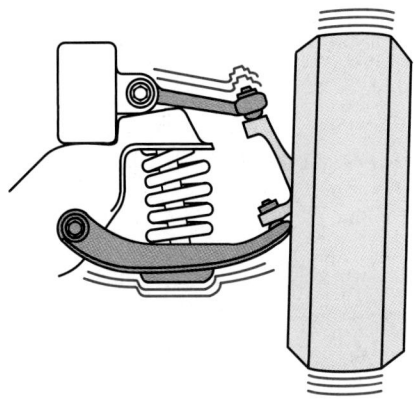

Figure 14.18 A worn shock absorber will allow wheel tramp.

Chapter 64. A torn or missing ball joint seal will require replacement of the joint by a front suspension specialist. Bad shock absorbers or a tire out of balance can also result in cupped wear.

Shock and Strut Inspection—Undercar

Shock absorbers are located at each corner of the vehicle, near the spring (**Figure 14.17**). The one shown in Figure 14.17 is in a MacPherson strut. Their purpose is to dampen spring oscillations. A worn shock absorber will allow the wheel to *tramp*, or hop up and down on the road (**Figure 14.18**).

Rubber **rebound bumpers** are bolted to the lower control arm or chassis. These rubber bumpers come into use only when the suspension has reached the full limit of its travel. They do not normally function unless the car has been driven in extremely rough conditions or if the shock absorbers are worn out.

SHOP TIP Worn or damaged rebound bumpers often indicate a need for shock absorber replacement. Check to see if they have come into contact with the frame of the vehicle.

Inspect the shock absorbers for leaks or signs of physical damage. Inspect the rubber bushings and mounting bolts for signs of wear or looseness. The appearance of a slight amount of moisture at the shock absorber seal is normal. But the outside of a shock should not show any drips or be entirely wet. Physical damage to a shock is cause for replacement of *both* shocks on that end of the vehicle. Shocks are always replaced in pairs. If shocks need to be replaced, follow the procedure in Chapter 64. If the vehicle has MacPherson strut shock absorbers (see Chapter 64), inspect them for leakage or damage.

If there is a reason to question the condition of a shock, unbolt it from the bottom. Then, push it and pull it through its regular travel (**Figure 14.19**). Resistance should be felt in both directions, through the entire travel of the shock. Shock absorbers have multiple stages of operation, so this test does not determine that they are in perfect operating condition, but it does pinpoint problems in the first stage of operation. See Chapter 63 for more information on shock absorber operation.

NOTE: *The wheels must be supported when removing shocks from rear-wheel-drive cars with coil springs.*

■ DRIVELINE CHECKS

Inspect Clutch and Transmission Linkage

Vehicles with manual transmissions use various methods to shift gears and disengage the clutch. Some use cables; others use linkage or a hydraulic master and slave cylinder. Cars with automatic transmissions must also have some form of linkage or cables for shifting. Visually inspect the ends of linkages to see that rubber, plastic, or brass bushings are not worn or broken. Grasp

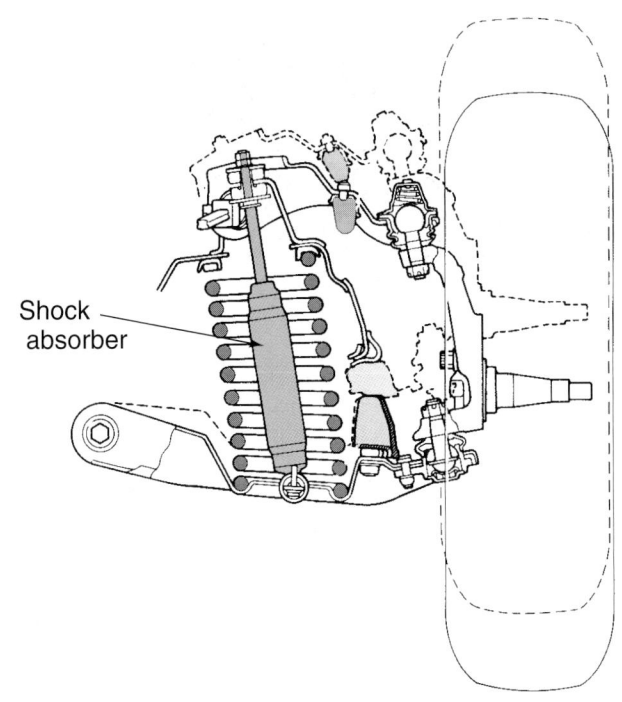

Shock absorber

Figure 14.17 A shock absorber.

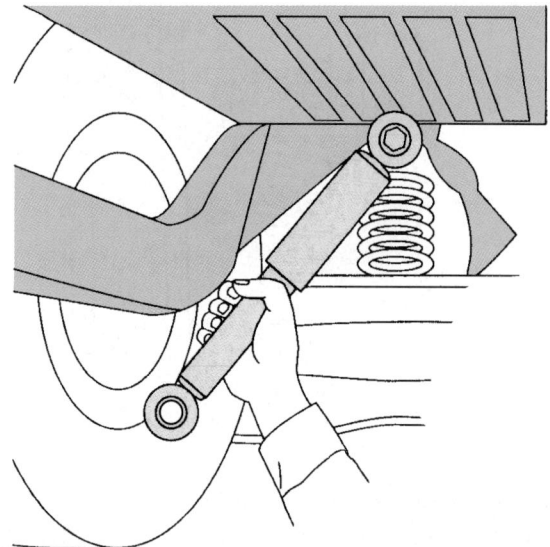

Figure 14.19 Check a shock by loosening one end and pulling it both ways through its full travel. *(Courtesy of Tim Gilles)*

linkages and try to move them to see if they are loose or worn.

Look for signs of leakage on a clutch master and slave cylinder.

Inspect for Transmission and Axle Leaks

Inspect the transmission or transaxle for leaks, especially if it has a low oil level. **Figure 14.20** shows the location of the filler plug and drain plugs on a manual transmission.

NOTE: *Some manual transmissions have external bolt heads that hold internal parts. Do not mistake these for a fill or drain plug.*

Check the service information for the correct type of lubricant. Transmissions and transaxles can use automatic transmission fluid (ATF), engine oil, gear oil, or synthetic oil.

On rear-wheel-drive cars, a *differential* is located between the rear wheels. It splits the engine's torque equally to the drive wheels (see Chapter 75). There is a

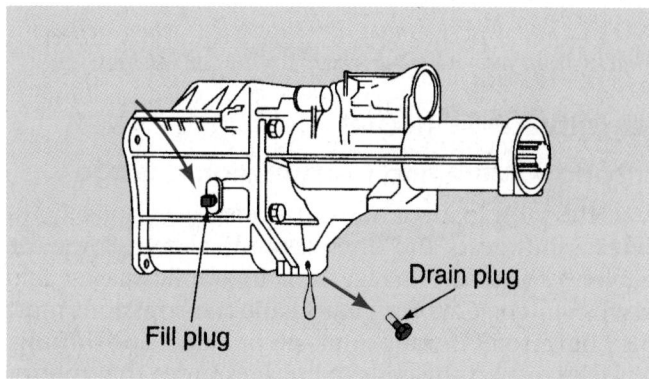

Figure 14.20 Location of manual transmission fill and drain plugs.

vent on the top of the axle housing. A hose is usually installed from the vent into the trunk or frame to protect against water entry. Check to see that the hose is not damaged.

Check the differential and both rear axles for oil leaks. **Figure 14.21** shows the location of typical differential fill plugs. The oil level should be at the bottom of the threaded fill plug hole (**Figure 14.22**). If there is a leak, check to see that the vent is not plugged.

When gear oil leaks onto a rear brake assembly, the brakes will need to be replaced after repairing the axle leak.

SHOP TIP To tell the difference between gear oil and brake fluid, apply water to the wet area near the leak. Brake fluid is water soluble; gear oil is not.

Many shops have air- or pump-operated gear oil dispensers. A gauge tells how much lubricant has been

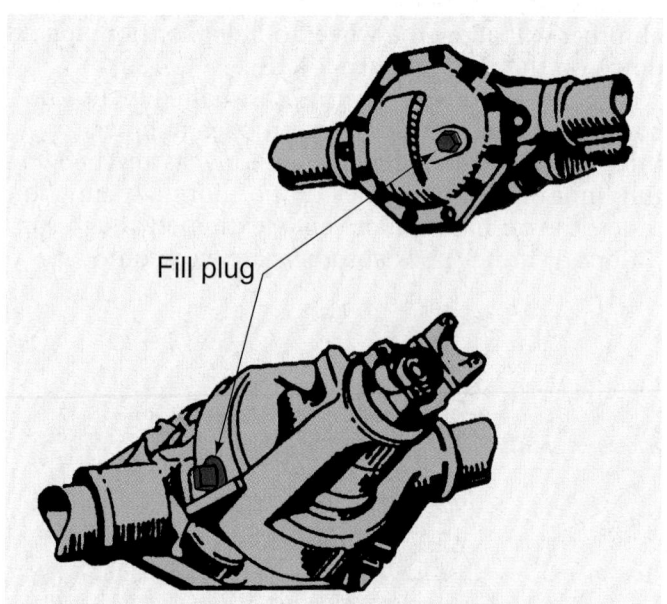

Fill plug

Figure 14.21 Typical locations of differential fill plugs.

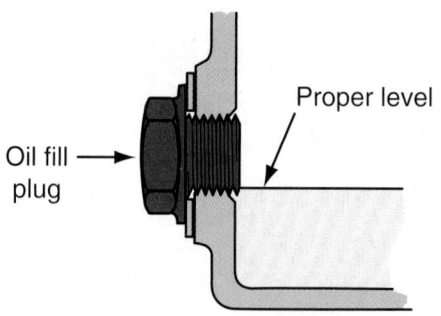

Proper level

Oil fill plug

Figure 14.22 The fluid level should be at the bottom of the fill plug hole.

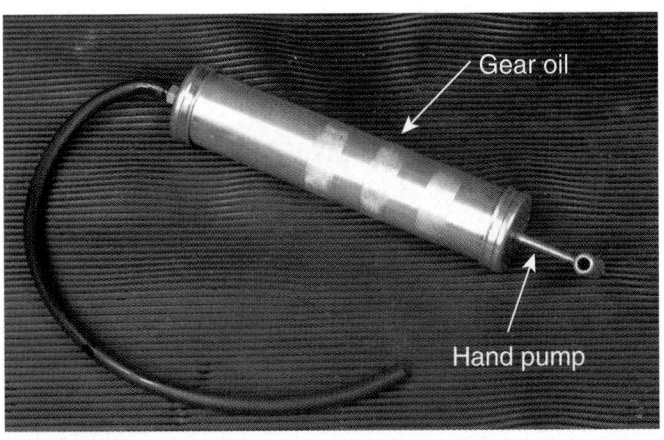

Figure 14.23 A tool used for pumping transmission lubricant.

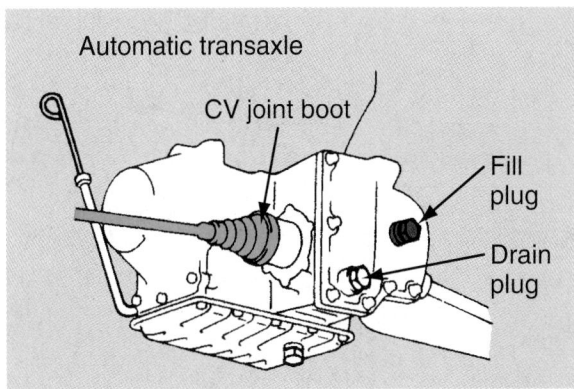

Figure 14.24 Typical location of a transaxle fill plug.

dispensed. Some shops have dispensers that are rolled around on the floor. Other shops have bulk oil supplied to the dispenser handle through hoses mounted on overhead reels. There are also portable hand pumps available (**Figure 14.23**).

Gear Oil

In rear-wheel-drive cars, differentials have curved **hypoid gears** and use SAE 90 E.P. gear oil. The **E.P.** means extreme pressure. Hypoid gears have a ring gear and a pinion gear that intersect below the centerline of the ring gear (see Chapter 75). This is so that the hump in the floor of the vehicle can be lower. The teeth on hypoid gears are spiraled and curved. They "wipe" across each other on a very small area of the gear tooth. The tremendous load that results calls for the use of E.P. gear oil.

NOTE: *E.P. gear oils contain zinc. They are dark in color and have a strong odor.*

Some differentials are of the *limited-slip* or *positraction* type. These axles, which usually have an I.D. tag, require a special *limited-slip lubricant* containing a *friction modifier*.

Transaxles

Front-wheel-drive (FWD) cars have a transaxle at the front of the vehicle. Low-viscosity engine oil or automatic transmission fluid can be used to lubricate the differential section because hypoid gears are not necessary. **Figure 14.24** shows the location of a typical transaxle's drain and fill plugs. Be sure to check the service information for the proper lubricant to use.

CV Joints

Front-wheel-drive cars have CV joints at each end of their two drive axles (see Figure 14.24). They are flexible joints requiring special grease sealed permanently into the joint with a flexible air-tight rubber seal, commonly called a boot. Sometimes CV joint boots tear or rub on something and become damaged. Inspect the boots to see if they show signs of leakage. If so, the boot must be replaced *immediately*. If the leak has been allowed to continue for a period of time, the axle has probably

been subjected to water (especially when driving in the rain). Dirt is carried into the joint by water and damage quickly results. Replacement of CV joint seals usually requires removal of the axle and disassembly of the entire joint (see Chapter 78).

U-Joints

Rear-wheel-drive cars have a *drive shaft* that carries power between the transmission and the differential (**Figure 14.25**). The drive shaft has a *universal joint*, also called a *U-joint*, at each end. Two-piece drive shafts often have a U-joint in the center, too. To test a U-joint for damage, grasp the drive shaft near one end and work it up and down while watching each U-joint for signs of looseness (**Figure 14.26**). Inspect the seals on the bearing cups at each end of the U-joint for signs of rust.

Replacement aftermarket universal joints usually have a zerk fitting to provide lubricant after installation and for periodic lubrication thereafter. Universal joint grease or multipurpose grease is added to a U-joint with a lube gun adapter with a special pressure-control attachment to prevent damage to the seal. Squeeze the handle using short, quick, grips on the handle to supply grease in small spurts until evidence of the grease appears at the seals. Further service of the U-joint is covered in Chapter 76.

Sometimes, long drive shafts have two pieces with a center support bearing surrounded by rubber (**Figure 14.27**). If the bearing seizes up, the rubber will be torn away from the bearing. The rubber can also deteriorate due to age or contamination with oil. The drive shaft will operate at the wrong angle and vibration or noise will result.

Exhaust Inspection

Exhaust systems rust due to acids and moisture in the engine's exhaust. An engine that never fully warms up will cause rust from moisture that pools in the lowest spots in the exhaust system. This is where holes usually develop.

Listen for exhaust leaks while the engine is running. Check for leaks at all pipe connections and

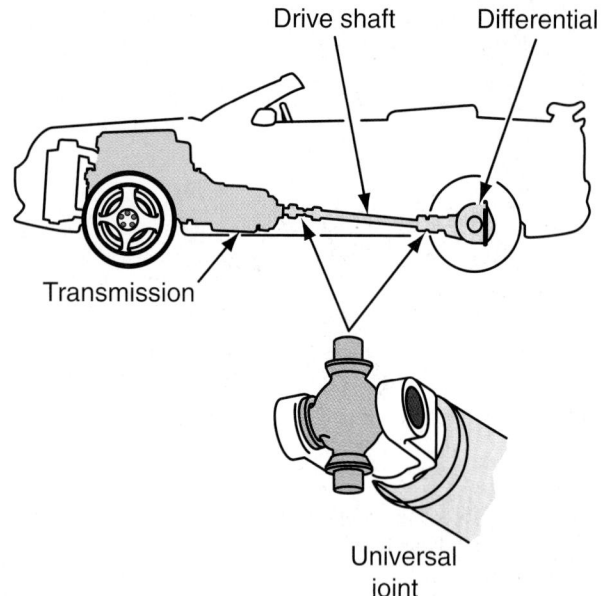

Drive shaft   Differential

Transmission

Universal joint

Figure 14.25a Rear-wheel-drive cars have a drive shaft carrying power between the transmission and the differential.

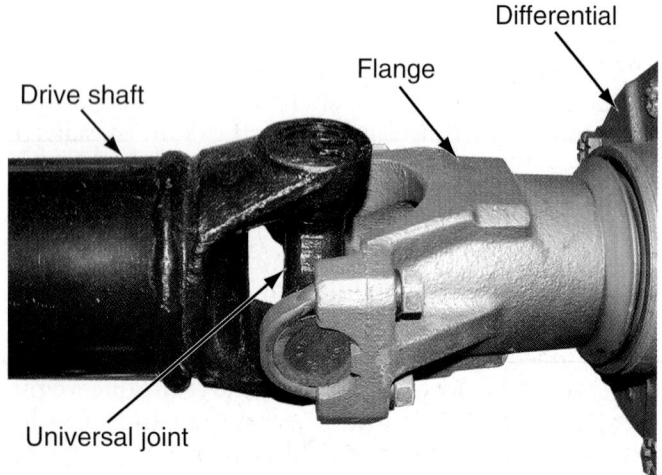

Differential

Flange

Drive shaft

Universal joint

Figure 14.25b A universal joint. *(Courtesy of Tim Gilles)*

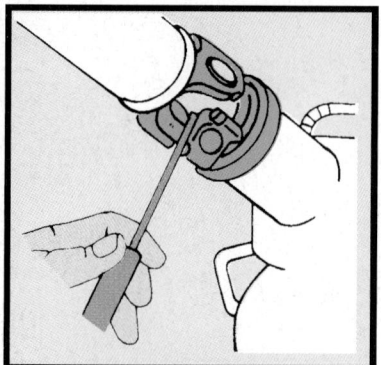

Check for rust

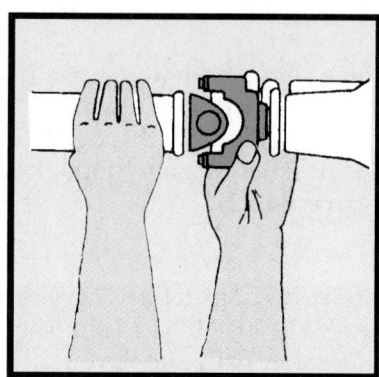

Use hand force to check for looseness

Figure 14.26 Testing a U-joint for signs of obvious damage. *(Courtesy of Federal-Mogul Corporation)*

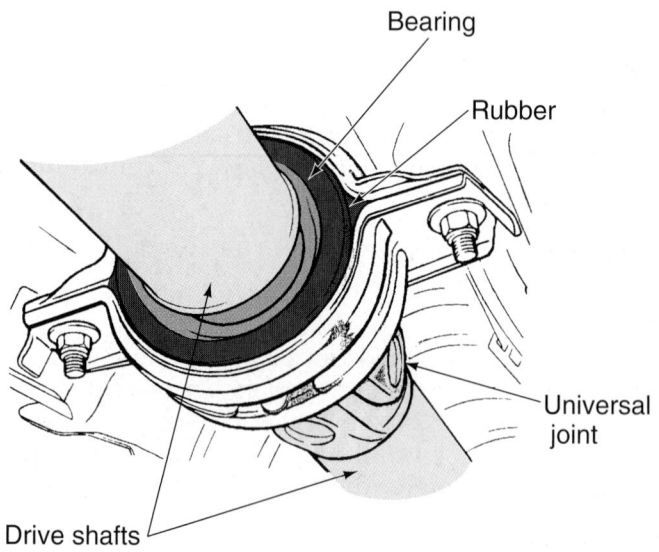

Bearing

Rubber

Universal joint

Drive shafts

Figure 14.27 A center support bearing. *(Courtesy of DaimlerChrysler Corporation)*

at low spots in the system. Remove the metal end from a stethoscope and use just the hose to listen for leaks.

NOTE: *Sometimes there is a small hole at the bottom rear of the muffler that allows moisture to drain out.*

Additional information on exhaust system service is found in Chapter 44.

Inspect the condition of all clamps and hangers. When the lubrication and safety service has been completed, fill out a door record sticker and put it on the door jamb. A door sticker has a place for listing the mileage when this service was performed. Also fill in the kind of oil that was used.

Items on the worksheet in **Figure 14.28** are representative of those found on a typical commercial lubrication service record. A space on the worksheet asks for "items needing attention" to note any items requiring further service or repair.

CERTIFIED CAR CARE SERVICE

Customer Name _____

Address _____ City _____ Zip Code _____ Phone _____

Date _____ Time _____

Vehicle _____ Year _____ Model _____ License Number _____ Odometer Reading _____

ELECTRICAL SYSTEM CHECKS
_____ Wiring Visual Inspection
_____ Battery
_____ Top Off Water Level
_____ Posts and Cables
_____ Clean _____ Corroded
_____ Damaged
Battery Condition
_____ Good _____ Replace
_____ Recharge

LIGHTS
_____ Park _____ Brake
_____ Signal _____ Emergency
_____ Dash Lights Back-up
Headlight Operation
_____ High Left
_____ Low Left
_____ High Right
_____ Low Right
_____ Horn Operation

FUEL SYSTEM CHECKS
_____ Condition of Hoses
_____ Gas Cap Condition
_____ Air Cleaner
_____ Crankcase Vent Filter
_____ Fuel Filter (miles until change suggested) _____

COOLING SYSTEM CHECKS
_____ Level
_____ Strength of Coolant
_____ (Protection to _____ °)
_____ No Leaks
Condition of Hoses
_____ Radiator
_____ Heater
_____ Thermostat Bypass
_____ Hose (if so equipped)
Pressure Test
_____ Radiator
_____ Cap
_____ Condition of Coolant
_____ Pump Belt

BRAKE INSPECTION
_____ Pedal Travel
_____ Emergency Brake
_____ Brake Hoses and Lines
_____ Master Brake Cylinder-
_____ Fluid Level and Condition

ON-GROUND STEERING, SUSPENSION, DRIVELINE CHECKS
_____ Steering Wheel Freeplay
_____ Power Steering Fluid Level
_____ Shock Absorber Bounce Test
_____ Good _____ Unsafe
_____ Front
_____ Rear
_____ No Squeaks
_____ Ride Height Check
_____ Check ATF Level

VISIBILITY
_____ Mirrors
_____ Wiper Blades
Wiper Operation
_____ fast _____ slow
_____ Washer Fluid and Pump
_____ Clean and Inspect all Glass

UNDERCAR SERVICE

_____ Drain Crankcase (if ordered)
_____ Remove and replace oil filter
_____ Inspect Under Car for
_____ Oil, Gasoline, and Coolant Leaks
_____ Check Crankcase Oil Level
_____ Check Oil Filter for Leaks

INFLATE AND CHECK TIRES
Inflate to _____ lb.

Tire Condition:
_____ Good _____ Fair _____ Unsafe

RF	_____	_____	_____
LF	_____	_____	_____
RR	_____	_____	_____
LR	_____	_____	_____

Inflate and Check Spare
_____ Good _____ Fair _____ Unsafe

SUSPENSION AND STEERING
_____ Inspect Steering Linkage
_____ Inspect Shock Absorbers
_____ Inspect Suspension
_____ Bushings
_____ Clean Lubrication Fittings
_____ Lubricate Fittings
_____ Ball Joints
_____ Inspect Ball Joint Seals
_____ Ball Joint Wear
_____ Inspection
_____ Inspect Ride Height

UNDERCAR FUEL SYSTEM CHECKS
_____ Condition of Fuel Hoses
_____ Condition of Fuel Tank

DRIVELINE CHECKS
_____ Check Universal
_____ or CV Joints
_____ Inspect Gear Cases
_____ Transmission
_____ Differential
_____ Replace Drain Plugs
_____ Inspect Motor Mounts
_____ Lubricate Door and Hood
_____ Hinge and Latches

EXHAUST SYSTEM CHECKS
_____ Mufflers and Pipes
_____ Pipe Hangers
_____ Exhaust Leaks
_____ Heat Riser

FINAL VEHICLE PREPARATION
_____ Replace Crankcase Oil
_____ Clean Windows, Vacuum
_____ Interior
_____ Fill Out and Affix Door
_____ Jamb Record to Door Post
_____ Complete a Repair Order

Figure 14.28 A complete lubrication, maintenance, and inspection worksheet.

■■■ REVIEW QUESTIONS

1. Which tire tread wear location is more typical: outside edges or center?

2. When the wear bars on a tire's tread are even with the remaining tread, how much tread depth remains?

3. What type of tubing connects the brake master cylinder to the brake hoses at either end of the suspension?

4. Lubrication fittings are commonly called _____ fittings.

5. Before squirting grease into a fitting, _____ off the fitting.

6. Three possible uses for a multipurpose grease include lubricating the chassis, wheel bearing, and _____ joint.

7. Loose parts, such as worn ball joints, can cause _____ tire wear.

8. When a rubber rebound bumper has been damaged because the suspension reached its limit of travel, what could be the cause?

9. Do holes in the exhaust system happen at the highest or lowest parts of the system?

10. What does the "E.P." on a gear oil label stand for?

■■■ ASE-STYLE REVIEW QUESTIONS

1. Technician A says that most cars today have several lubrication fittings. Technician B says that if a ball joint will not accept grease, try turning the wheel from side to side while applying grease. Who is right?
   a. Technician A
   b. Technician B
   c. Both A and B
   d. Neither A nor B

2. Technician A says that one advantage to a hand-held grease gun is that special-use grease cartridges can be used. Technician B says that grease is made of soap and oil. Who is right?
   a. Technician A
   b. Technician B
   c. Both A and B
   d. Neither A nor B

3. When a tire has been run for a long time with too much air pressure, where does wear often occur?
   a. In the center of the tread
   b. On its inside edges
   c. Near the wheel rim
   d. On the sidewall

4. When a rubber boot on a steering or suspension joint becomes torn:
   a. The boot should be replaced.
   b. The entire joint or axle assembly will have to be replaced.
   c. No service is required.
   d. It can be fixed using a special epoxy.

5. All of the following are true about hypoid differentials *except*:
   a. They use 90W gear lube.
   b. Some of them use engine oil.
   c. They use oil with E.P. additives.
   d. Some of them require special limited-slip gear oil.

6. Rubber bushings should be lubricated:
   a. With oil
   b. With rubber lubricant
   c. Bushings are designed to remain dry.
   d. None of the above

7. Technician A says that the part at each end of a rear-wheel-drive drive shaft is called a CV joint. Technician B says that the joints at the ends of front-wheel-drive axles are called universal joints. Who is right?
   a. Technician A
   b. Technician B
   c. Both A and B
   d. Neither A nor B

8. A car has a bad shock absorber. Technician A says to replace the shock absorber. Technician B says to replace both shock absorbers on that end of the vehicle. Who is right?
   a. Technician A
   b. Technician B
   c. Both A and B
   d. Neither A nor B

9. When checking steering linkage for wear with the vehicle off the ground:
   a. Steering linkage parts should not be able to move up and down.
   b. Steering linkage parts should have no movement.
   c. Steering linkages do not wear out.
   d. Steering linkage is allowed to move no more than ¼" from side to side.

10. A car with an independent rear suspension has at least _____ flexible brake hoses.
    a. One
    b. Two
    c. Three
    d. Four

Engine Operation and Service

THEN AND NOW: ENGINE EVOLUTION

A Frenchman named Etienne Lenoir built the first internal combustion engine in 1860. It ran on *town gas* that was used for lighting in those days. In general, this engine was very different from the engines of today. It was not until 1876 that German inventor Nikolas Otto produced the original four-stroke cycle engine of the type used today.

Early engines did not put out much power for their size. The original single-cylinder Benz of 1886, for instance, displaced 58 cubic inches (954 cc—almost a liter). Yet it produced only three-quarters of a horsepower and ran at a mere 400 rpm.

There are numerous ways of increasing the power and fuel efficiency of an engine. Raising the compression ratio can result in a substantial increase. But when this was tried with ordinary gasoline, severe detonation occurred. In 1922, Thomas Midgley, Jr., discovered that adding *tetraethyl lead* to gasoline quieted the explosions in the cylinders. This was the birth of high-octane gasoline.

The design of the combustion chamber and valve train are both extremely important in determining the power an engine produces. Flathead engines, with their valves alongside the cylinders, were the design of choice for many decades. They were simple and dependable. The later development of *overhead valves* (OHV) increased power immensely. Overhead valve engines even appeared on some early cars, such as the 1910 Buick.

Another means of increasing power is for the engine to spin faster without damage. Early design engines had heavy cast iron pistons. The change to much lighter aluminum helped allow higher engine rpm. Higher rpm is also made possible by the *overhead camshaft* (OHC), which eliminates trouble-prone pushrods and, in many cases, rocker arms. The OHC advance is credited to French car maker Adolph Clement in 1902.

Force-feeding pressurized air into the engine's cylinders results in performance equal or greater to that of an engine that is much larger. A supercharger was used with some success on sports cars of the 1920s, and turbochargers began to be used on cars beginning in the 1960s. Detonation and added complications are drawbacks to these systems.

Today, multi-valve, computer-managed, fuel-injected, high-rpm engines produce amazing amounts of power for their displacement. For example, the Acura Type R with its V-Tech valve system pumps 195 hp at 8,000 rpm out of a mere 110 cubic inches (1.8 liters).

The 1910 Buick had overhead valves. *(Courtesy of Bob Freudenberger)*

A V6 engine with dual overhead camshafts and variable valve timing. *(Courtesy of Toyota Motor Sales, U.S.A., Inc.)*

CHAPTER 15

Introduction to the Engine

■ OBJECTIVES

Upon completion of this chapter, you should be able to:

✔ Explain the principles of internal combustion engine operation.

✔ Identify internal combustion engine parts by name.

✔ Describe the function of engine parts.

■ KEY TERMS

poppet valve
stoichiometric
standard temperature and
   pressure (STP)
upper end

valvetrain
pushrod engine
overhead cam engine
runners
long block

short block
end play
bearing clearance
piston slap

■ INTRODUCTION

The first chapter in this book gives a basic description of engine operation. More in-depth information is given in this chapter. Such things as firing orders, valve adjustment, oil pressure testing, and engine diagnosis are covered in later chapters.

■ BASIC ENGINE OPERATION

Most modern automobiles use a *four-stroke reciprocating* gasoline engine. The *four-stroke cycle* is described here using a single-cylinder engine. Automobile engines actually have multiple cylinders.

A simple reciprocating engine has a cylinder, a piston, a connecting rod, and a crankshaft. If the cylinder is compared to a cannon, the piston is a round plug that can be compared to a cannonball. In a spark-ignited internal combustion engine, a mixture of air and fuel is compressed in the cylinder.

Fuel that is burned in a "gasoline engine" must be a liquid that vaporizes easily (such as gasoline, methanol, or ethanol), or a flammable gas (such as propane or natural gas). When the air-fuel mixture is compressed and then burned, it pushes a piston down in a cylinder. This action turns a crankshaft that powers the car (see Figure 1.5).

A cylinder head bolted to the top of the block closes off the end of each cylinder. A piston is connected to

the crankshaft by a connecting rod and a piston pin (also called a *wrist* pin). This arrangement allows the piston to return to the top of the cylinder, making continuous rotary motion of the crankshaft possible. The piston is sealed from the crank case by piston rings that slide against the cylinder wall.

Because of the powerful pulses on the piston as the fuel is burned in the cylinder, a heavy *flywheel* is bolted to the rear of the crankshaft (**Figure 15.1**).

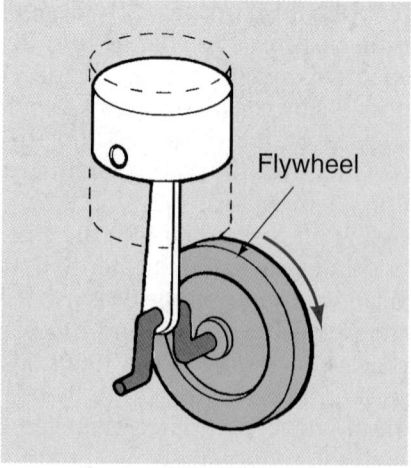
Flywheel

Figure 15.1 A flywheel is installed at the rear of the crankshaft. (*Courtesy of Ford Motor Company*)

The weight of the flywheel blends together the power pulses into one continuous motion of the crankshaft.

The cylinder head has an intake valve port for each cylinder that allows an air and fuel mixture to flow into the cylinder. An exhaust valve port in the head allows the burned gases to flow out. Each port is sealed off by a **poppet valve** (**Figure 15.2**). The head is sealed to the cylinder block with a *head gasket* (**Figure 15.3**). The opening of the valves is controlled by the *camshaft* (see Figure 1.6). The camshaft is turned by the crankshaft with gears or sprockets driven by a belt or chain.

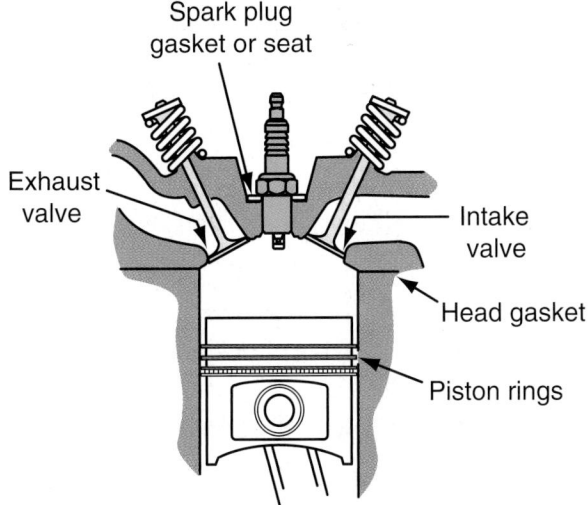

Figure 15.2 Poppet valves seal off the valve ports.

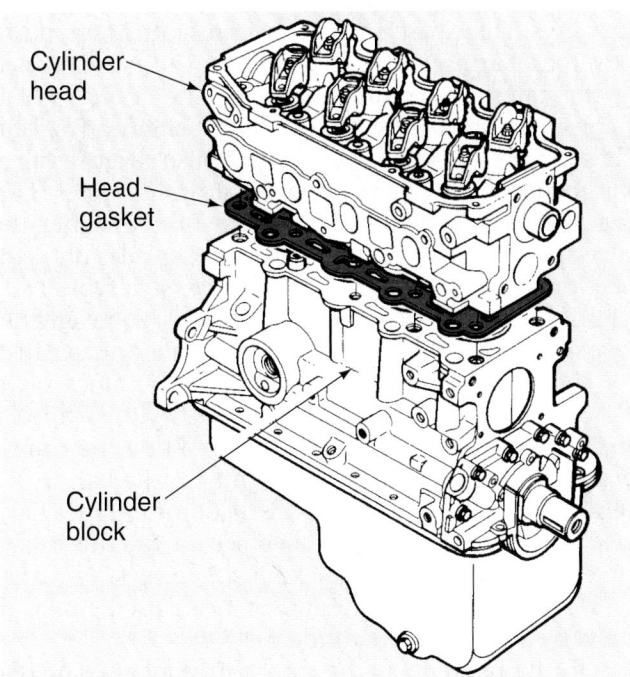

Figure 15.3 A head gasket seals the head to the block. (*Courtesy of Federal-Mogul Corporation*)

■ FOUR-STROKE ENGINE OPERATION

Piston Travel

The piston travel in the cylinder is limited in both directions. The upper limit of piston travel is called *top dead center* (*TDC*), and the lower limit is called *bottom dead center* (*BDC*).

Piston Stroke

A stroke is the movement of the piston from TDC to BDC, or from BDC to TDC. There are four strokes in the engine's *four-stroke cycle*: the intake stroke, compression stroke, power stroke, and exhaust stroke.

Intake Stroke.

- Gasoline will not burn unless it is mixed with the correct amount of air. It is very explosive when 1 part is mixed with about 15 parts of air by weight.
- As the crankshaft turns, it pulls the rod and piston down in the cylinder. This action creates a lower pressure suction called *engine vacuum*. Atmospheric pressure pushes the lower pressure mixture of air and fuel through the open intake valve (see Figure 1.7).
- The air-fuel mixture is supplied by the *fuel* system. The ideal mixture for the combined purposes of engine performance, emission control, and fuel economy is about 15 to 1 by mass. This is called a **stoichiometric** mixture.

> ### ▓ SCIENCE NOTE ▓
> *Air-fuel ratio is measured by weight. A 15:1 air-fuel ratio has 15 pounds of air to 1 pound of gasoline. When the air-fuel mixture is measured by volume, there are 10,000 gallons of air for each gallon of gasoline.*

Volume of Engine Air. Approximately 10,000 gallons of air would fill a single-car garage. At sea level, a box with dimensions of 28 inches on all sides will hold 1 pound of air with a volume of about 22,000 cubic inches.

Air volume is measured at a **standard temperature and pressure (STP)** of 25°C at sea level. Volume is affected more by altitude than by temperature, changing about 3.5% with each 1,000-foot increase or decrease.

To determine the volume of each individual cylinder, divide the displacement of the engine by the number of cylinders. The answer can be converted to weight.

Weight of Engine Air. The weight of air displacing the volume of a cylinder can be calculated. One cubic foot of air weighs 0.081 pound at 32°F (0.075 pound at 70°F). One pound of air equals 12.4 cubic feet. Atmospheric pressure at sea level is 14.7 pounds per square inch. This means a 1-inch column of air that reaches from the outer edge of space to the Earth weighs 14.7 pounds.

Something's molecular weight is 1 mole. One mole of air weighs approximately 1 ounce. Air is a mixture of oxygen and nitrogen, but it is not a pure molecular compound, because other things are part of air as well. One mole of nitrogen (N_2) weighs 28 grams (slightly less than 1 ounce). One mole of oxygen (O_2) weighs 32 grams (slightly more than 1 ounce).

Comparing Volume and Weight. One mole of STP gas takes up 23.4 liters (6.18 gallons) in volume at standard temperature and pressure. If an ounce of air takes up 6⅙ gallons, then a pound of air takes up 98 gallons (13 cubic feet). Therefore, a pound of air takes up an area slightly less than two 55-gallon drums.

Metric Conversion. Engine displacement is measured in cubic centimeters (cc) or liters (1,000 cc). To convert cubic centimeters to cubic inches, divide by 16.4. Then, convert to cubic feet by dividing the result by 1,728.

NOTE: *Most vehicles since the early 1980s have fuel injection systems with computer controls. The computer monitors the oxygen content in the vehicle's exhaust and then adjusts the fuel supply to provide the correct amount of fuel and air for each intake stroke.*

Compression Stroke.
- The compression stroke begins at BDC after the intake stroke is completed. The intake valve closes during the compression stroke as the piston moves up in the cylinder, compressing the air-fuel mixture (see Figure 1.7).
- Compressing the mixture of air and fuel into a smaller area heats it and makes it easier to burn.

Power Stroke.
- As the piston approaches TDC on its compression stroke, the compressed air-fuel mixture becomes very flammable (see Figure 1.7). When the ignition system produces a spark at the spark plug, the air-fuel mixture ignites.
- As the air-fuel mixture burns, it expands, forcing the piston to move down in the cylinder until it reaches BDC. The action of the piston turns the crankshaft to power the car.
- Some leakage of gases past the rings occurs during the power stroke. This leakage, called *blowby*, causes pressure in the area around the crankshaft (**Figure 15.4**).

Exhaust Stroke.
- As the piston nears BDC on the power stroke, the exhaust valve opens, allowing the burned gases to escape. Because the burning gases are still expanding, they are forced out through the open valve.
- As the crankshaft continues to turn past BDC, the piston moves up in the cylinder, helping to force the remaining exhaust gases out through the open exhaust valve (see Figure 1.7).
- A few degrees after the piston passes TDC, the exhaust valve closes.

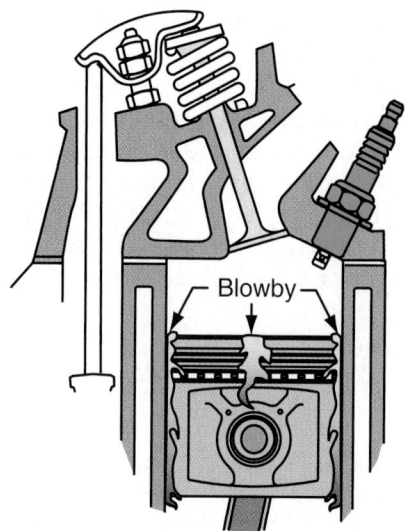

Figure 15.4 Leakage of gases past the ring is known as blowby.

The entire four-stroke cycle repeats itself, starting again as the piston moves down on the intake stroke.
- One four-stroke cycle requires two 360-degree revolutions (720 degrees) of the crankshaft.
- While the crankshaft turns twice, every cylinder's intake and exhaust valve opens once.
- Ignition occurs once.

ENGINE UPPER END

Parts of the **upper end** of the engine include the cylinder head(s) and valvetrain.

VALVETRAIN

The **valvetrain** includes the parts that open and close the valves. These parts include the camshaft, lifters, pushrods, rocker arms, valves, and springs. Valves are opened by the camshaft, commonly called the *"cam,"* which is considered to be the "heart" of the engine. The cam has eccentric (off-center) *lobes* that push against valvetrain parts, causing valves to open at precise times (**Figure 15.5**). When the lobe turns away from the lifter, the valve spring closes the valve. The camshaft has one cam lobe for each intake and exhaust valve, so a typical six-cylinder engine would have 12 cam lobes. Some engines have multiple valves per cylinder for intake and exhaust.

The cam can be located either in the block or in the cylinder head. The cam-in-block style is called a **pushrod engine** (see Figure 15.5). When the cam is located on the top of the head, this is called an **overhead cam engine** (**Figure 15.6**). Some engines have *dual overhead cams*. These have a cam for the intake valves and another for the exhaust valves.

Depending on the design of head used, some of the valvetrain parts are different.

On many engines, the camshaft may also include another gear that drives the ignition distributor and oil pump. The distributor is usually meshed to the oil pump so the one gear operates them both.

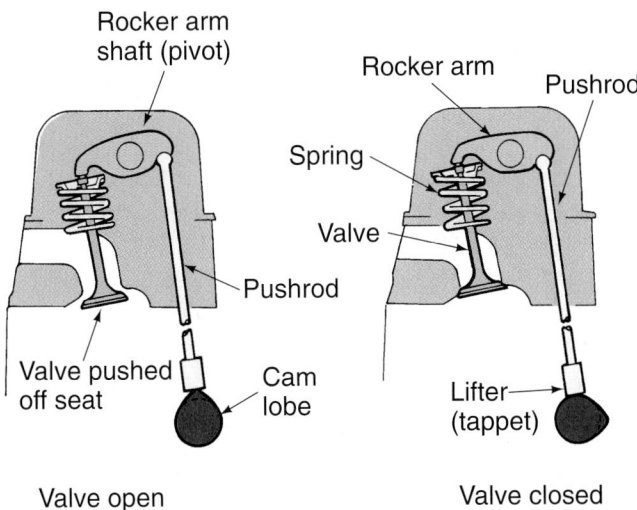

Figure 15.5 The cam lobe opens the valve. *(Courtesy of Ford Motor Company)*

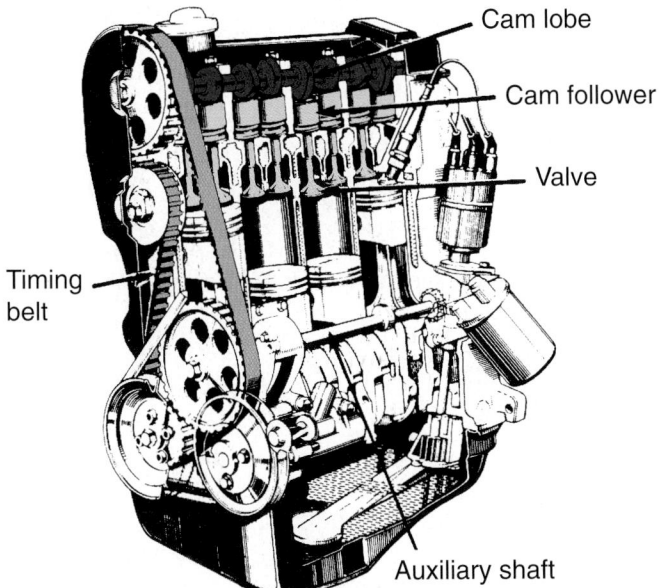

Figure 15.6 Cutaway of a four-cylinder OHC engine with the cam positioned over the valve. *(Courtesy of Daimler Chrysler Corporation)*

VINTAGE ENGINES

On older engines, which used a carburetor, the camshaft also includes another eccentric lobe. This lobe drives its mechanical fuel pump. The fuel pump eccentric has a rounder shape than an ordinary cam lobe. The fuel pump rocker arm is pushed up against spring pressure by the eccentric.

On a pushrod engine, *valve lifters*, sometimes called *tappets*, fit into bores above the cam lobes. The lifters act on *pushrods* and *rocker arms* to open the valves.

Rocker arms are mounted on top of the cylinder head. They take the up-and-down motion of the pushrod

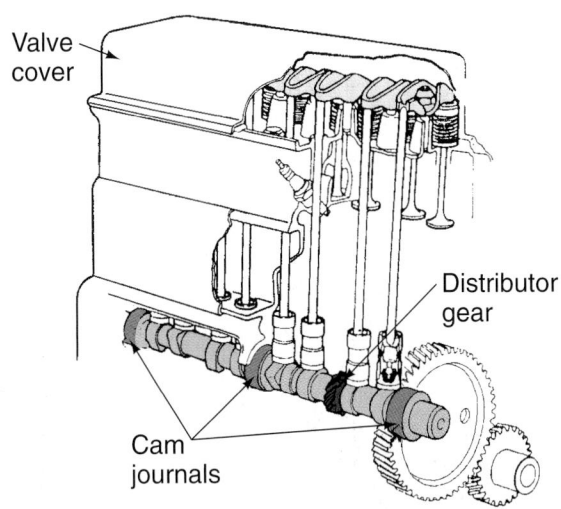

Figure 15.7 Cam journals ride inside of cam bearings to support the camshaft.

and transfer it to the valve. As the rocker arm rocks up, it pushes the valve down in the opposite direction.

Pushrods are usually hollow tubes. One end of the pushrod fits into a socket in the lifter, and the other fits into one end of the rocker arm. Overhead cam engines do not use pushrods.

Camshaft journals are machined surfaces that ride inside of cam bearings that are pressed into bores in the block on a pushrod engine (**Figure 15.7**). On an overhead cam engine, the cam journals ride in bores in the cylinder head.

■ CYLINDER HEAD

The cylinder head(s) bolts to the top of the engine block, sealing off the cylinders (see Figure 15.3). The term *combustion chamber* refers to the small chambers in the cylinder heads that are located just above the pistons (**Figure 15.8**). The spark plugs are threaded into holes in the combustion chambers.

Valve Parts

Refer to **Figure 15.9** for the locations of valve parts that are explained here. There are usually two valves per cylinder located in the combustion chambers. Some engines have three or four valves per cylinder.

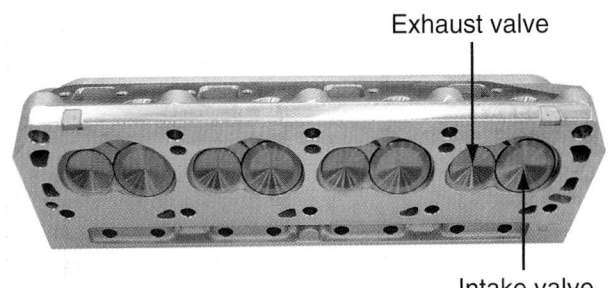

Figure 15.8 Combustion chambers on the bottom side of a cylinder head. *(Courtesy of Tim Gilles)*

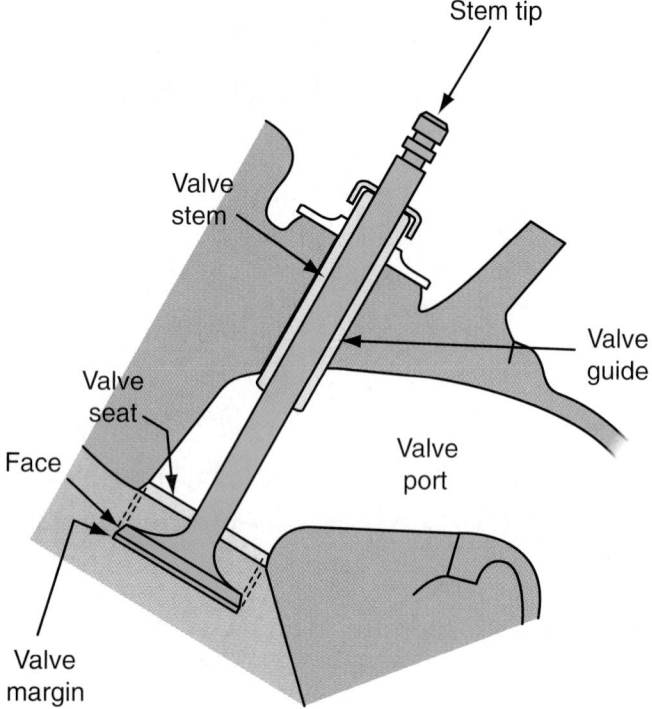

Figure 15.9 The valve face seals against the valve seat.

The *intake valve* is the larger of the two valves because the incoming air-fuel mixture takes up a lot of space and is not under higher pressure like the exhaust (see Figure 15.8).

NOTE: *Intake and exhaust valves are made of different materials.*

■■■■ SCIENCE NOTE ■■■■

Intake valves are usually magnetic, but the exhaust valves, which are made of stainless steel, are not. Check an intake and exhaust valve with a magnet to verify this. You might also find that some exhaust valves are magnetic on the stem, but not on the valve head, because the valve is made of two different metals.

- *Valve seats* are the machined areas under the valves. The valve and its seat fit against each other to provide a tight seal.
- The *valve port* is the area under the seat.
- The *valve face* is the machined area that the valve seat rides against.
- The *valve head* is the part of the valve that is exposed to combustion.
- The *valve margin* is located between the valve head and the valve face. If the margin of the valve is too thin (after excessive grinding), the valve can

become too hot when exposed to combustion and it will burn.

- The *valve stem* is smooth and polished and fits into the valve guide.
- *Keeper grooves* are at the top of the stem. *Keepers* are the small, half-moon pieces that keep the valve spring attached to the top of the valve.
- The *valve stem tip* is usually hardened on exhaust valves to protect against wear.

NOTE: *The exhaust valve has to open against the pressure of combustion. Because of this, exhaust lobes and related parts usually fail before intake parts.*

Other Valve Parts

- A *valve spring* closes the valve after the camshaft opens it.
- A *valve spring retainer* fits on top of the spring and is held to the stem tip by the keepers.
- *Valve guides* are bores in the bottom of the valve port that the valves fit into. The valves slide up and down in the valve guides (**Figure 15.10**).
- A *valve guide seal* is positioned on each of the valves. It keeps excessive oil from entering the valve guide. Defective valve guide seals can result in a serious oil consumption problem (see Figure 15.10).
- The *valve cover* encloses the valve spring area of the cylinder head (see Figure 15.7). It is made of plastic or sheet metal and has a *valve cover gasket* sealing it to the head. Some valve covers are sealed with silicone sealant instead of a gasket.

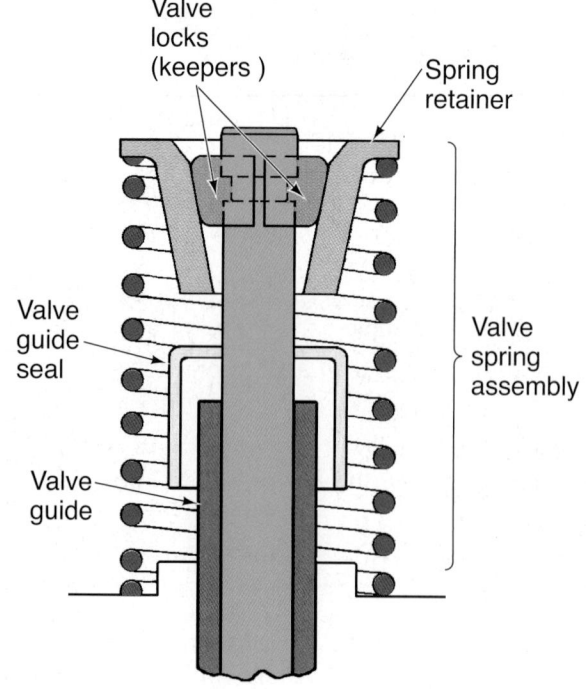

Figure 15.10 A valve guide and seal. (*Courtesy of Federal-Mogul Corporation*)

Manifolds

The *intake manifold* (**Figure 15.11**) is bolted to the side of a head or in between cylinder heads. The manifold provides a passage to the intake valve ports in the head. A carburetor or fuel injection system supplies fuel to mix with the air that enters the **runners** of the manifold.

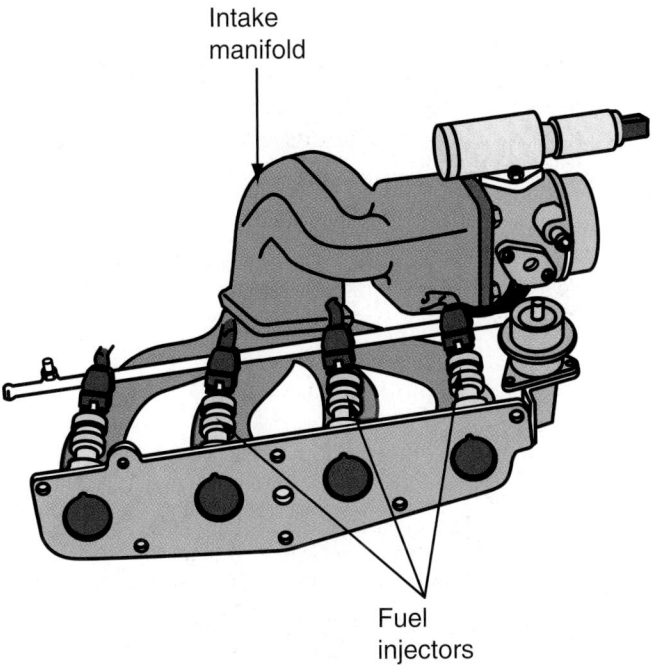

Figure 15.11 An Intake manifold provides a passage to the intake valve ports in the head.

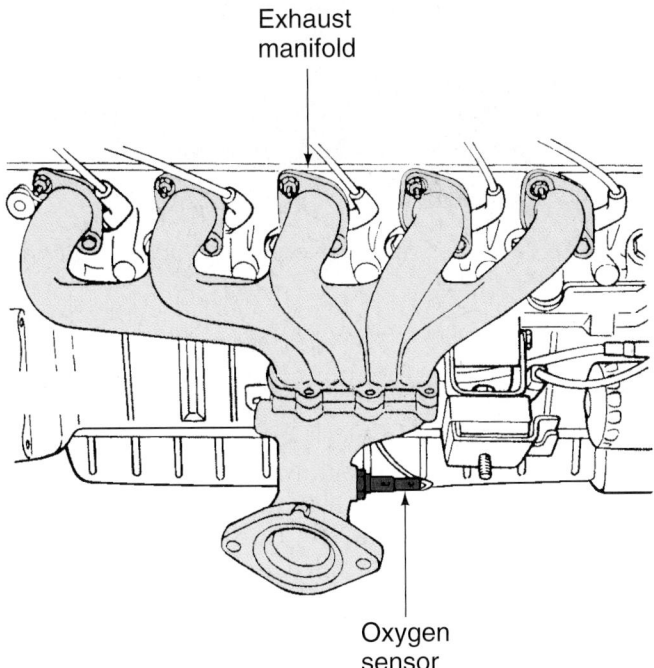

Figure 15.12 Exhaust manifold. *(Courtesy of DaimlerChrysler Corporation)*

Exhaust manifolds are bolted to the cylinder head (**Figure 15.12**). They provide a channel to carry exhaust gases away from the engine. Exhaust manifolds must carry hot exhaust gases, so they are usually made of cast iron. Some manifolds are constructed of steel tubing.

■ ENGINE FRONT

The front of the engine has a camshaft drive, a timing cover or front cover, and the crankshaft vibration damper or pulley. The front of the engine is the side opposite the transmission.

Cam Drive

The camshaft is driven by the crankshaft by two *timing gears* or by *sprockets* used with a *timing chain* or *timing belt* (**Figure 15.13**).

NOTE: *The crankshaft must turn twice for every one turn of the cam. For that reason, there are half as many teeth on the crank drive as there are on the cam drive.*

> **▬ MATH NOTE ▬**
>
> *An engine running at 3,000 rpm has to open and close a valve 1,500 times per minute (25 times per second)! Each spark plug must also fire this frequently.*
>
> *3,000 rpm = 3,000 crankshaft revolutions per minute*
>
> *3,000 rpm = 1,500 four-stroke cycles per minute*
>
> $\dfrac{1,500}{60} = 25$ *four-stroke cycles per second*

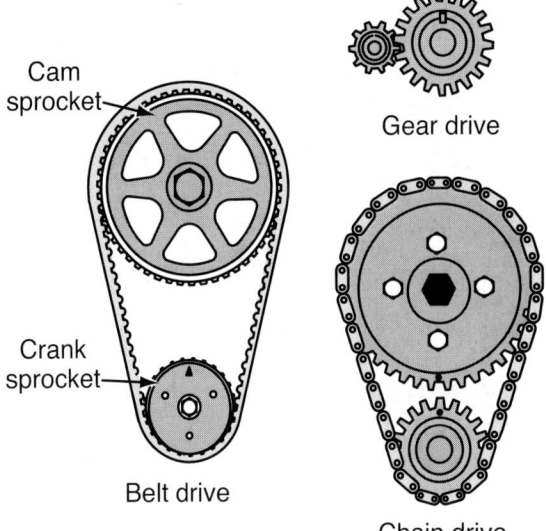

Figure 15.13 There are half as many teeth on the crank drive as there are on the cam drive.

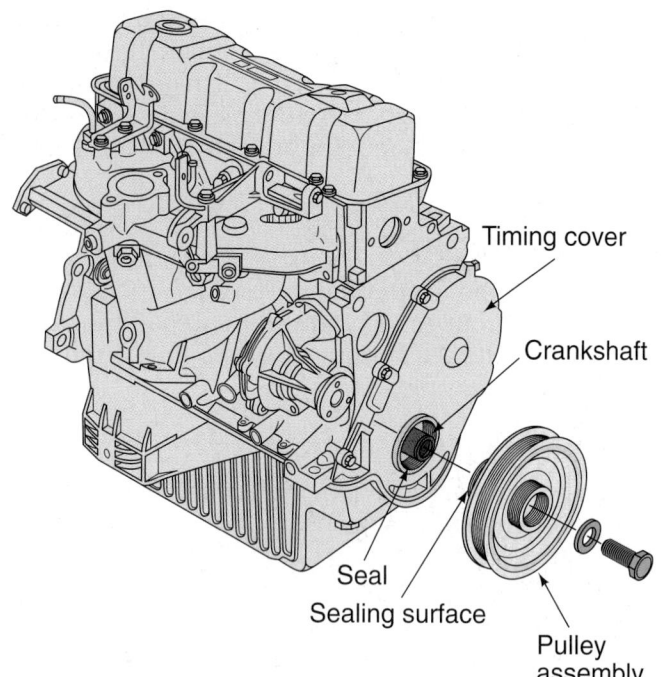

Figure 15.14 Timing cover and pulley. (*Courtesy of Ford Motor Company*)

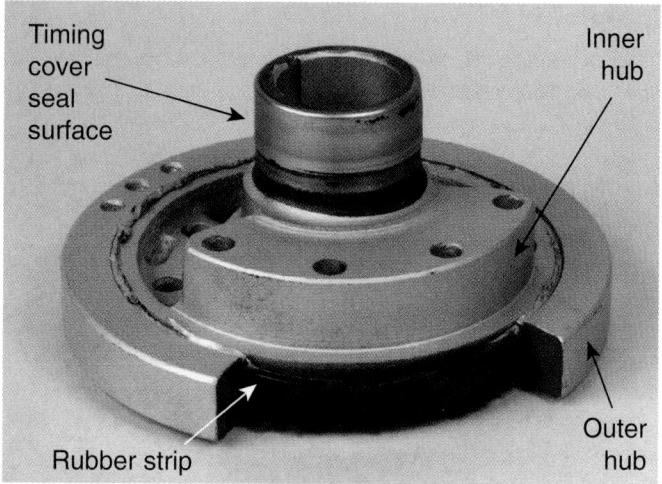

Figure 15.15 This cutaway shows the three pieces of a damper—the inner and outer hubs separated by a rubber strip—along with the timing cover seal surface. (*Courtesy of Tim Gilles*)

Timing Cover

When an engine has a chain or gears, they are lubricated with engine oil. The timing cover on these engines must seal against oil leakage. In this case, it will have a full-round seal that rides on the sealing surface of the vibration damper or pulley (**Figure 15.14**). When the engine has a timing belt, the seal will be on the crankshaft, behind the belt sprockets. A timing cover for a belt does not need to seal in lubricants; it just needs to keep the elements out.

Vibration Damper or Pulley

Some engines use a *vibration damper* to minimize vibrations in the crankshaft, preventing damage to it. A damper has three pieces: an outer and an inner ring separated by a thin strip of rubber (**Figure 15.15**). The damper is also called a *harmonic balancer*.

Some engines do not use a damper. They just have a one-piece *pulley* to which the belt pulleys can be bolted.

■ CYLINDER BLOCK ASSEMBLY (LOWER END)

The cylinder block is cast from iron or aluminum (**Figure 15.16**). The casting has oil galleries as well as cooling passages called water jackets. Precision machining of the block's cylinder bores allows the

Figure 15.16 A cylinder block casting. (*Courtesy of Tim Gilles*)

pistons to slide against the cylinder walls with very little clearance.

Figure 15.17 shows the parts of the *lower end* of the cylinder block assembly. The area of the block where the crankshaft and bearings are located is called the *crankcase* (**Figure 15.18**). Horizontal holes in the lower end of the block called *main bearing bores* are precisely bored to accommodate the crankshaft. The bottom half of the main bearing bores include removable *main bearing caps* that allow for installation and removal of the crankshaft.

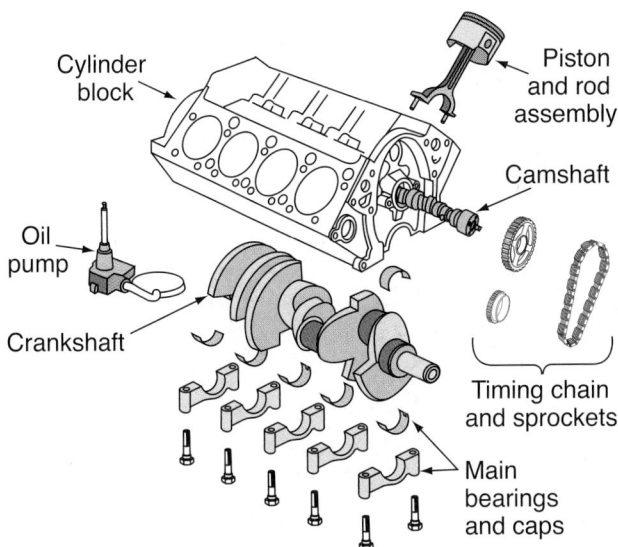

Figure 15.17 Lower end parts.

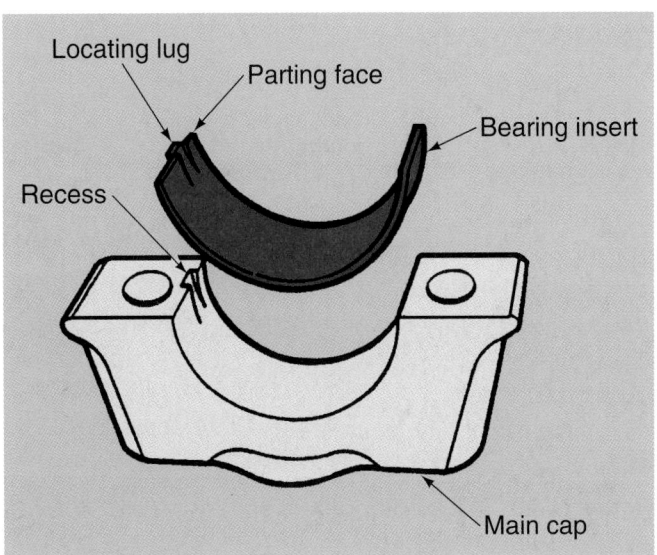

Figure 15.19 Bearing insert. *(Courtesy of Federal-Mogul Corporation)*

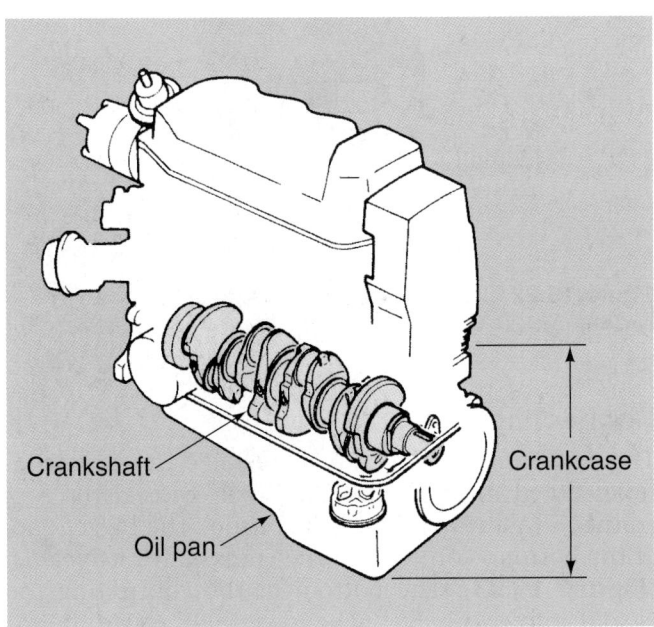

Figure 15.18 The oil pan encloses the crankcase. *(Courtesy of Federal-Mogul Corporation)*

Replaceable *bearing inserts* are installed in the top and bottom halves of the inside diameters of the bearing bores (**Figure 15.19**).

The block *deck* is the top surface of the block that supports the cylinder head. A *cylinder head gasket* fits between the head and the deck.

At the factory, the camshaft holes, cylinder bores, cylinder-head mounting surface, all threaded holes, and all gasket surfaces are machined automatically and in perfect alignment to each other.

■■■ SHORT BLOCK AND LONG BLOCK

A complete block assembly with the entire *valvetrain* (cylinder heads and related parts) included is called a **long block**.

The cylinder block assembly (without the heads installed) is called a **short block** (see Figure 15.17). The short block for a pushrod engine includes the crankshaft, piston and rod assembly, camshaft and cam drive components, and all bearings. Short blocks are not usually available for OHC engines.

Crankshaft

The *crankshaft* changes the *reciprocating* (up and down) motion of the pistons to rotation. Its polished, machined bearing surfaces for the main and rod bearings are called *journals* (**Figure 15.20**).

The *main bearing journals* are in a row down the center of the crankshaft. The front and rear crankshaft journals and all journals in between are main journals.

One of the main bearings has *flanges* (sides) that ride on corresponding surfaces on the crankshaft to limit the fore and aft movement (**end play**) of the crankshaft (**Figure 15.21**). This bearing is called the *thrust main*.

Rod journals are offset from the main journals. Connecting rods are bolted to the rod journals.

Oil comes to the main bearings through *oil galleries* in the block. There are oil holes in the upper bearing inserts to allow the oil to enter the bearing so that metal-to-metal contact between the bearing and the crankshaft is avoided.

Lubrication oil holes are drilled from the main bearing journals to the rod bearing journals to provide pressurized lubrication.

Bearing clearance is the name of the space between the bearing insert and the crankshaft.

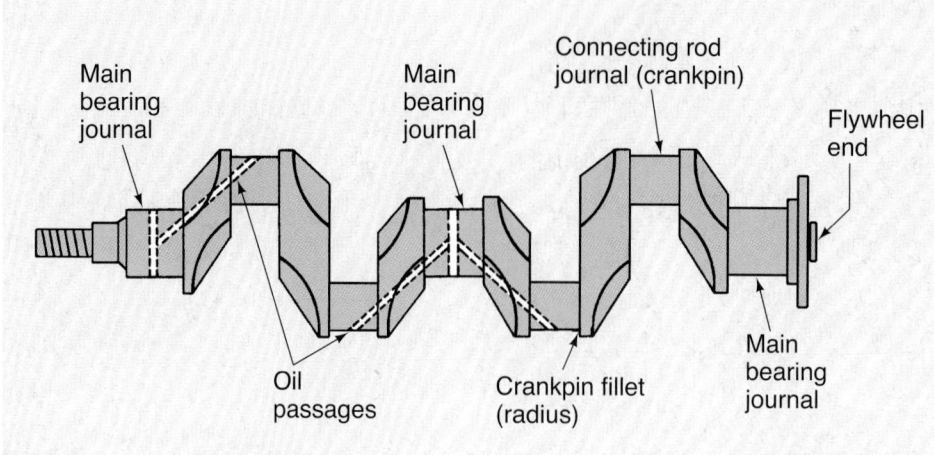

Figure 15.20 Crankshaft parts. *(Courtesy of Ford Motor Company)*

Figure 15.21 The Thrust surfaces control fore and aft movement of the crankshaft. *(Courtesy of Tim Gilles)*

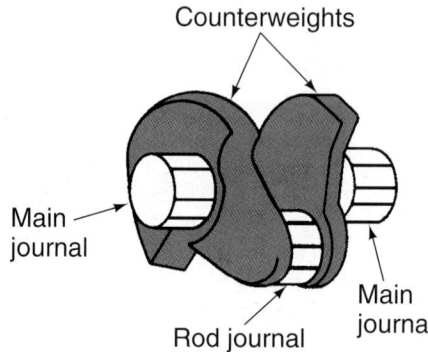

Figure 15.22 Crankshaft counterweights. *(Courtesy of Federal-Mogul Corporation)*

Counterweights are part of the crankshaft. When the crankshaft rotates, their weight offsets the weight of the connecting rods on the opposite side (**Figure 15.22**).

The *snout* is the front of the crankshaft. Mounted on the snout are the crankshaft gear or sprocket and the damper or pulley.

The back of the crankshaft has a *flange* to which the flywheel or flexplate is bolted. On the rear of the crankshaft, the center of the flange has a precision hole. This hole is used to align the front of the transmission with the engine. It will either locate the hub of an automatic transmission torque converter or it will hold a pilot bushing or bearing that a standard transmission input shaft fits into.

Connecting Rod

The *connecting rod*, which is shaped like an I-beam, has holes bored in each end. The small end, at the top, connects to a pivot pin in the piston. The bigger end, at the bottom, connects to the crankshaft's *rod journal* (**Figure 15.23**). The bottom of the connecting rod is split into two halves. The lower half, called the *rod cap*, bolts to the bottom of the rod. A bearing insert fits between the rod journal and the big bore of the rod.

Piston

The piston skirt (**Figure 15.24**) fits closely to its cylinder bore, where it rides on a thin layer of oil as it moves up and down in the cylinder. During combustion, pressure from the expanding gases pushes on the piston, which pushes against the connecting rod to turn the crankshaft.

Piston rings seal the piston to the cylinder wall. A typical piston has three grooves machined just above the piston skirt. *Ring lands* separate the rings from each other. The two grooves on top are for compression rings, whereas the wider, bottom groove is for oil control. Behind the oil ring, an opening in the

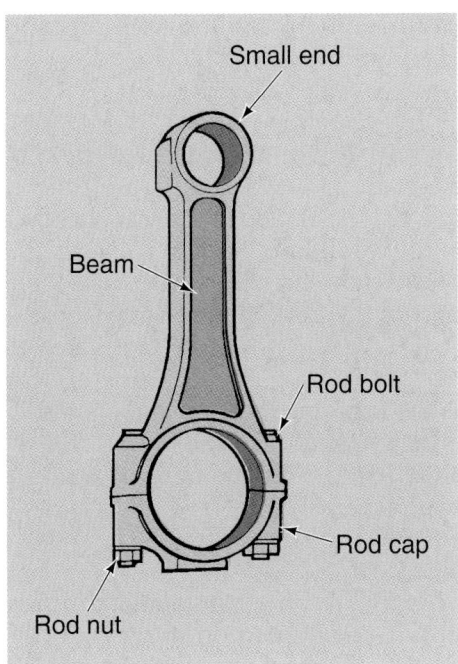

Figure 15.23 Connecting rod. *(Courtesy of Ford Motor Company)*

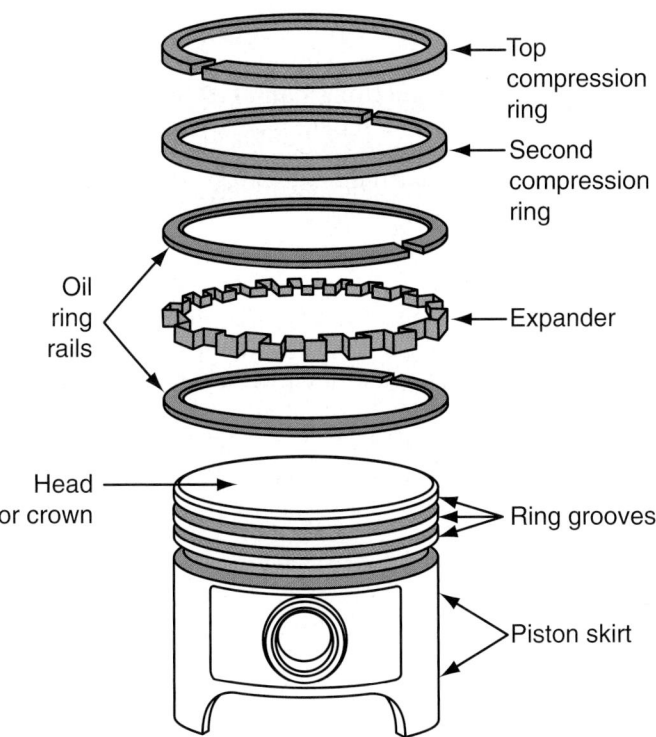

Figure 15.24 Piston and rings.

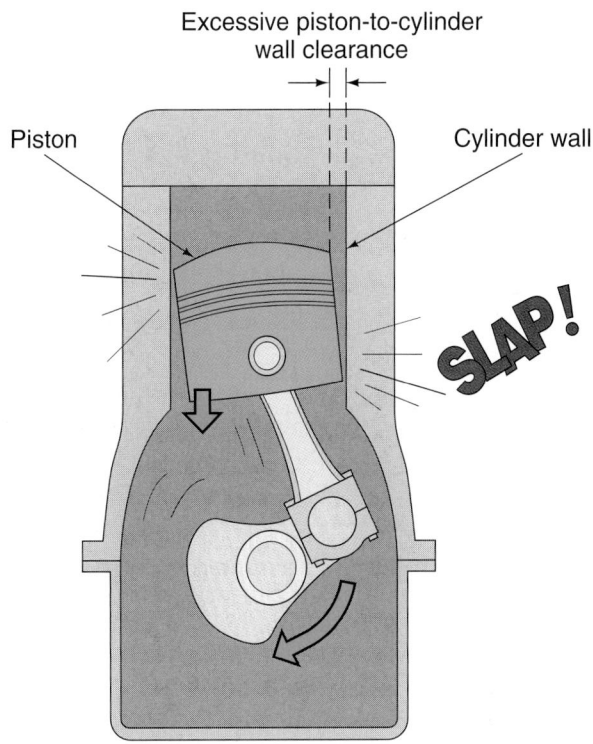

Figure 15.25 Piston slap. *(Courtesy of Ford Motor Company)*

piston ring groove allows oil to flow to the inside of the piston and back into the crankcase.

Pistons are made of aluminum, which has a melting point of about 1,225°F. During combustion, temperatures are well in excess of this. The piston must be thick enough to withstand the intermittent higher temperature without melting. There is only a small amount of clearance between the piston skirt and the cylinder wall. Excessive clearance can cause **piston slap**, resulting in noise (**Figure 15.25**). The piston expands in a controlled manner as the engine warms up. Some piston skirts are coated with an antistick material that holds oil and provides extra protection against damage to the piston skirt or cylinder wall.

Piston Pin

The small end of the connecting rod connects to the piston pin, sometimes called a wrist pin. Machined into each side of the piston are large holes, called pin bores. They provide a pivot point for the piston pin. The surrounding reinforced area is called the *pin boss*. Most piston pins are pressed-fit in the small end of the connecting rod, but some have a bushing that allows the pin to rotate freely in both the piston and the connecting rod. The piston pin moves freely in the piston pin bore.

Piston Rings

Piston rings seal between the piston ring grooves and cylinder wall, keeping combustion pressure from entering the crankcase. Leakage past the piston rings is called blowby (see Figure 15.4). Piston rings are lightly spring loaded to keep them in position against the cylinder wall until combustion pressure from the ignited fuel charge forces them more tightly against the cylinder wall (**Figure 15.26**). This pressure, combined with the

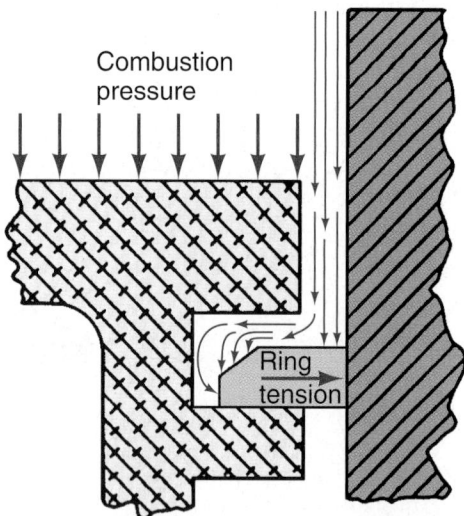

Figure 15.26 Combustion forces the ring against the cylinder wall. *(Courtesy of Federal-Mogul Corporation)*

sealing action of the small amount of oil that is in the pores of the cylinder wall, ensures a very good seal.

The *ring gap* is the space where the ends of the piston ring come together after installation on the piston. The ring is constructed in an arc that is larger than the cylinder bore. When the ring is compressed and installed, it will form a spring-loaded circle that matches the cylinder bore.

The lower ring is the *oil control ring* (see Figure 15.24). It scrapes oil that lubricates between the cylinder wall and piston skirt. Oil must be kept from entering the combustion chamber, where it would be burned. There are slots in the oil ring that allow oil to reach the rear of the ring groove where it can return to the crankcase through the slot or holes (**Figure 15.27**). Excessive

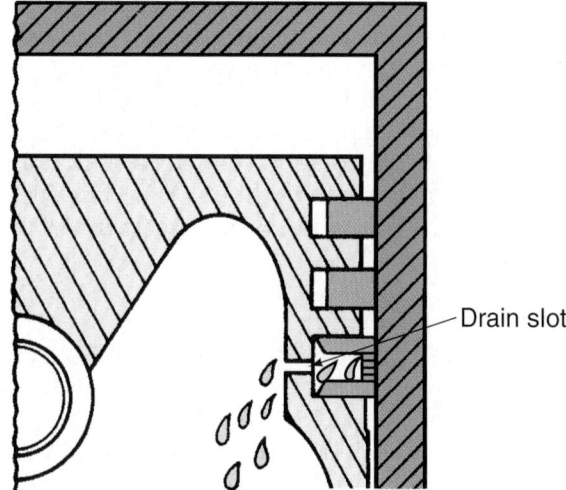

Figure 15.27 Oil is scraped off the cylinder wall by the oil ring and returned to the crankcase through holes in the piston.

leakage past the oil ring results in oil consumption and blue smoke from the exhaust.

Oil Seals

Oil seals are installed on the front and rear of the crankshaft to keep oil in the crankcase. Seals are either one piece or two halves. The lip of the seal rides on a smooth surface on the crankshaft. The *rear main seal* is the seal at the rear of the crankshaft. The *front seal* is located in the timing cover or the front of the engine block.

Oil Pan

The oil pan is a stamped sheet metal or plastic part that encloses the crankcase (see Figure 15.18). It provides a reservoir where the engine oil is cooled as air passes across its surface.

Flywheel

A *flywheel* is used with a standard transmission (**Figure 15.28**). It is mounted on the rear of the engine's crankshaft. The flywheel does three things:
- The weight of the flywheel helps in carrying the crankshaft beyond BDC after the power stroke and smoothes out the power impulses of multiple cylinders.
- A ring gear mounted on the circumference of the flywheel provides a gear drive for the starter motor. Ring gears are sometimes damaged by faulty starter motors; they can be replaced easily while the engine is out of the car.
- A flywheel also provides a surface for the clutch to work upon.

When an automatic transmission is used, a torque converter and *flexplate* take the place of the flywheel and clutch (see Figure 15.28). The torque converter provides the necessary weight. The flexplate is simply a flat piece of steel, usually with a starter ring gear on its outside diameter.

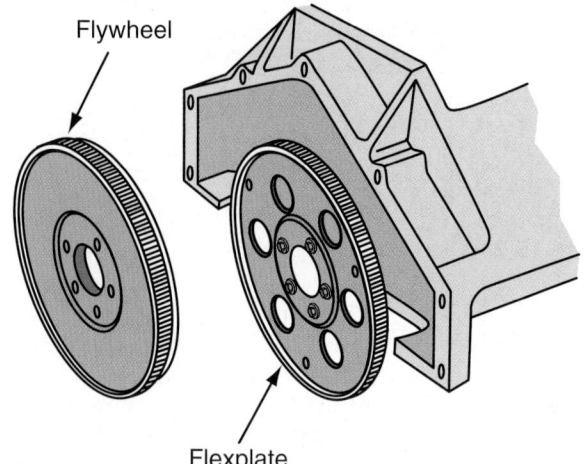

Figure 15.28 A flywheel and flexplate.

■ SUMMARY

■ Most of today's cars and light trucks are powered by Otto-cycle engines. During one four-stroke cycle, the intake, compression, power, and exhaust strokes are completed. This action takes 720 degrees or two crankshaft revolutions.

■ The camshaft turns once and the crankshaft turns twice during one four-stroke cycle.

■ The up-and-down motion of the piston is changed to usable rotary motion by the connecting rod and crankshaft.

■ A flywheel gives momentum to the crankshaft and smoothes the impulses between power strokes.

■ The camshaft and valve train control the engine's intake and exhaust flow in and out of the cylinders.

■ REVIEW QUESTIONS

1. How many times does each valve open during 720 degrees of crankshaft rotation?

2. What does OHC mean?

3. Which valve is usually not magnetic, intake or exhaust?

4. When an engine is running at 6,000 rpm, how many times does each valve have to open and close every second?

5. What is another name for the harmonic balancer?

6. What is the term for an engine block assembly that is sold without cylinder heads?

7. Beside spring pressure, what forces the piston ring against the cylinder wall?

8. If there is too much clearance between the piston and the cylinder wall, what happens?

9. What do the top two piston rings do?

10. On the sketch in **Figure 15.29**, list which valve is open during each of the strokes. Only list a valve as open if it is open for the entire stroke. Draw an arrow next to the piston to show in which direction it is moving.

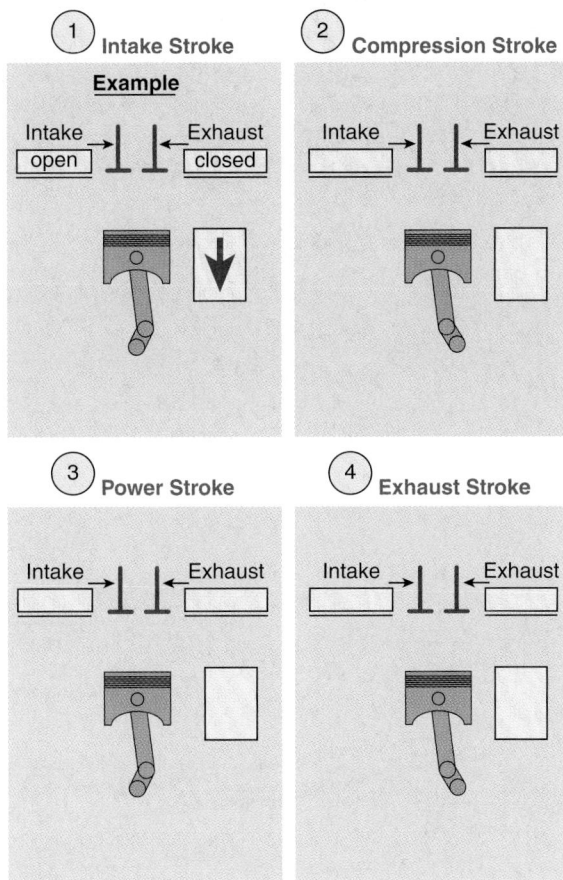

Figure 15.29 On a separate piece of paper, list which valve is open during each of the strokes, and tell which direction the piston is moving. Follow the example shown in the intake stroke above.

◼◼◼ ASE-STYLE REVIEW QUESTIONS

1. Technician A says that the crankshaft turns 720 degrees during one four-stroke cycle. Technician B says that the camshaft turns 360 degrees during one four-stroke cycle. Who is right?
   - **a.** Technician A
   - **b.** Technician B
   - **c.** Both A and B
   - **d.** Neither A nor B

2. Which of the following is/are true about camshafts?
   - **a.** The camshaft on a typical six cylinder engine has 12 cam lobes.
   - **b.** The crankshaft sprocket has half as many teeth as the camshaft sprocket.
   - **c.** Older engines had an extra eccentric lobe to operate a mechanical fuel pump.
   - **d.** All of the above

3. Technician A says that the oil pump is sometimes driven by the camshaft. Technician B says that the distributor is sometimes driven by the camshaft. Who is right?
   - **a.** Technician A
   - **b.** Technician B
   - **c.** Both A and B
   - **d.** Neither A nor B

4. Which of the following is/are true about intake and exhaust valves?
   - **a.** Exhaust valves are larger than intake valves.
   - **b.** Intake valves open against higher pressure than exhaust valves.
   - **c.** Both A and B
   - **d.** Neither A nor B

5. Technician A says that the intake valve is open during part of the compression stroke. Technician B says that the exhaust valve is open during part of the compression stroke. Who is right?
   - **a.** Technician A
   - **b.** Technician B
   - **c.** Both A and B
   - **d.** Neither A nor B

CHAPTER 16

Engine Classifications and Advanced Transportation Technologies

■ OBJECTIVES

Upon completion of this chapter, you should be able to:

✔ Explain various engine classifications and systems.

✔ Know the various differences in cylinder heads.

✔ Describe differences in operation between gasoline and diesel four-stroke piston engines.

✔ Explain the operation of two-stroke and Wankel rotary engines.

■ KEY TERMS

cylinder bank	I-Head	compression ratio
firing order	OHC	Wankel engine
companion cylinders	SOHC	two-stroke engine
oxides of nitrogen (NO$_x$)	DOHC	ZEV
bimetal engine	cross-flow head	regenerative braking
electrolysis	stratified charge	hybrid electric vehicles (HEVs)
L-head	diesel-cycle	

■ INTRODUCTION

A service technician needs an understanding of the basic designs and configurations of automobile engines to be able to intelligently use service manuals or communicate with customers or peers. Examples of engine-related terms found in a service manual might include reference to a 1.6L OHC V6 engine, or the term "crankcase capacity." This chapter deals with the terms related to engines. After reading it, you should be able to look under the hood of an automobile and describe what type of engine it has.

■ ENGINE CLASSIFICATIONS

Piston engines all have the same basic parts, as we learned in the last chapter. But differences in design can affect how you will go about repairing them. Understand the following engine classifications as you read this chapter:

■ Cylinder arrangement

■ Cooling system: air or liquid

■ Valve location: head or block

■ Combustion chamber design

■ Cam location: head or block

■ Fuel used: gasoline or alternative fuel

■ Ignition system: spark or compression

■ Number of strokes per cycle: two or four

■ CYLINDER ARRANGEMENT

An automobile engine has three, four, five, six, or more cylinders. Cylinders are arranged in several ways including: *in-line*, in a "V" arrangement, or *opposed* to each other (**Figure 16.1**). Modern engines come with 3, 4, 5, 6, 8, 10, or 12 cylinders.

An in-line engine has all of its cylinders lined up in one row. Some in-line engines have the cylinders vertically arranged and some of them are slanted at various angles. This is so that they can fit into an

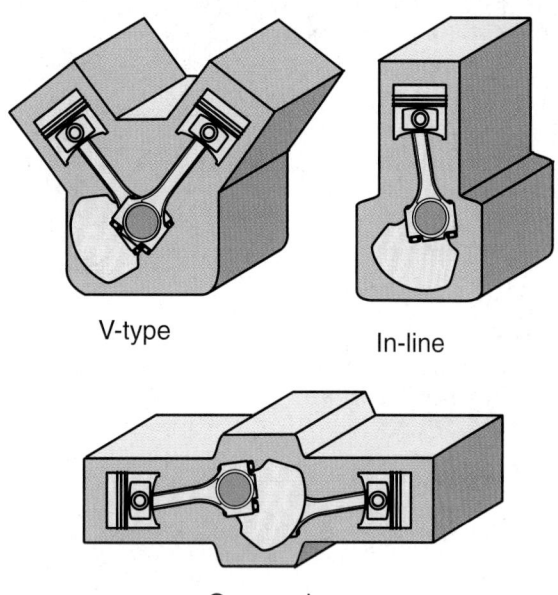

V-type

In-line

Opposed

Figure 16.1 Cylinder arrangements.

Figure 16.3 A crankpin for a V8 engine is double wide so it can fit two connecting rods. *(Courtesy of Tim Gilles)*

engine compartment with a lower hood line for less wind resistance.

A V-type engine looks like the letter "V" when looked at from the end. V-type cylinders are cast in two rows, called the left and right **cylinder banks**. Left and right banks are identified when viewed from the flywheel end of the engine. V8 blocks are cast with the cylinder banks separated by a 90-degree angle. V6 blocks have either 60 degrees or 90 degrees between banks. The V6 shown in **Figure 16.2** is a 60-degree V.

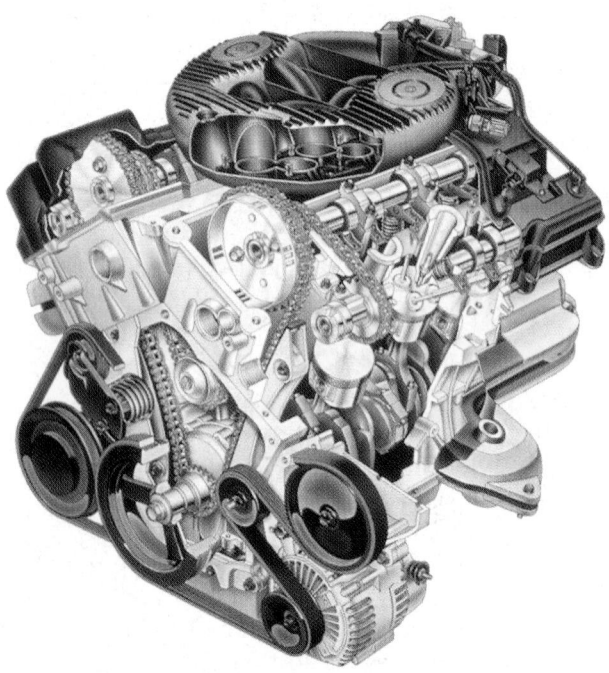

Figure 16.2 A 60-degree V6 engine. *(Courtesy of DaimlerChrysler)*

There are *big block* and *small block* V8 engine designs. Smaller, lighter blocks are more popular in passenger cars because of their fuel efficiency. An *intake manifold* covers the area between the heads known as the *valley*.

The V cylinder arrangement has an advantage over the in-line arrangement when an engine has more than four cylinders. The V-type engine is considerably shorter in length, and a completely assembled V-type engine can typically be lighter in weight than an in-line engine. Connecting rods from two cylinders on opposite sides of the engine will share one crankpin (**Figure 16.3**). Thus, the engine block requires fewer main bearing supports.

Opposed engines, sometimes called "pancake" engines, have their cylinders arranged so they face each other from opposite sides of the crankshaft. Opposed engines are especially suited for smaller underhood areas. They have been found on such cars as Porsche, Volkswagen, and Subaru.

▬ FIRING ORDER

To make a smooth-running engine, multiple-cylinder engines have their power strokes spaced at specified intervals. In a four-cylinder engine, one cylinder starts a power stroke at every 180 degrees of crankshaft rotation (**Figure 16.4**). This interval between power strokes is known as the *ignition interval*.

Within two turns of the crankshaft, all of the engine's cylinders will have fired once. The order in

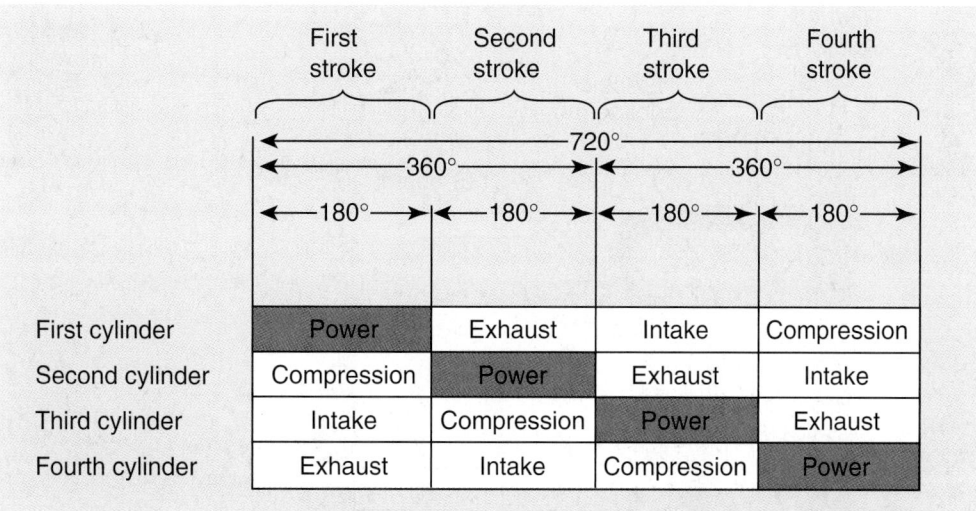

	First stroke	Second stroke	Third stroke	Fourth stroke
First cylinder	Power	Exhaust	Intake	Compression
Second cylinder	Compression	Power	Exhaust	Intake
Third cylinder	Intake	Compression	Power	Exhaust
Fourth cylinder	Exhaust	Intake	Compression	Power

Figure 16.4 A four-cylinder engine has one cylinder on a power stroke every 180 degrees of crankshaft rotation.

which the cylinders fire is known as the **firing order**. The firing order does not usually follow the order of cylinder numbering. It is not unusual for two V6 or V8 engines to have different firing orders.

Companion Cylinders

Any engine with an even number of cylinders will have pairs of cylinders called **companion cylinders**, or *running mates*. The pistons go up and down in pairs. The companions in the in-line six-cylinder engine shown in **Figure 16.5** are 1 and 6, 2 and 5, and 3 and 4. When one piston is starting its power stroke, its companion piston will be at the start of its intake stroke. To find out which cylinders are companions, take the first half of the engine's firing order and place it above the second half. The firing order for the engine shown in Figure 16.5 is 1-5-3-6-2-4. To determine the companions, put numbers 1, 5, and 3 above 6, 2, and 4.

first revolution,   1   5   3
           | | |
second revolution,   6   2   4

For a V8 with a 1-8-4-3-6-5-7-2 firing order, put numbers 1, 8, 4, and 3 above numbers 6, 5, 7, and 2.

first revolution,   1   8   4   3
           | | | |
second revolution,   6   5   7   2

Remember, the crankshaft turns two revolutions (720 degrees) to complete one four-stroke cycle. The first half of the firing order represents one crankshaft revolution (360 degrees). The second half of the firing order represents the second revolution of the crankshaft (360 degrees).

In the above example, when cylinder number 7 is beginning its intake stroke, cylinder number 4 is beginning its power stroke. This eight-cylinder engine would have one power stroke every 90 degrees of its 720 degrees four-stroke cycle. Therefore, a V8 should have banks 90 degrees from each other for even firing; a V6 should have 120 degrees or 60 degrees between its banks.

$$\frac{720}{8 \text{ (cyl.)}} = 90 \text{ degrees}$$

■ ENGINE COOLING

Engines use either of two types of cooling systems: *air cooling* or *liquid cooling*. Air-cooled engines are found on lawnmowers, motorcycles, and some automobiles. Cooling happens when air is circulated over cooling fins that are cast on the outside of engine parts. Because of higher running temperatures that cause increased **oxides of nitrogen (NO_x)** emissions (a major component in photochemical smog), production of these engines has been curtailed in recent years.

Liquid-cooled engines have *water jackets* to cool the areas around all cylinders and valve-seat areas.

Figure 16.5 This vintage L-head in-line six-cylinder engine shows companion cylinder pairings. The pistons in cylinders 1 and 6, 2 and 5, and 3 and 4 go up and down together. *(Courtesy of Tim Gilles)*

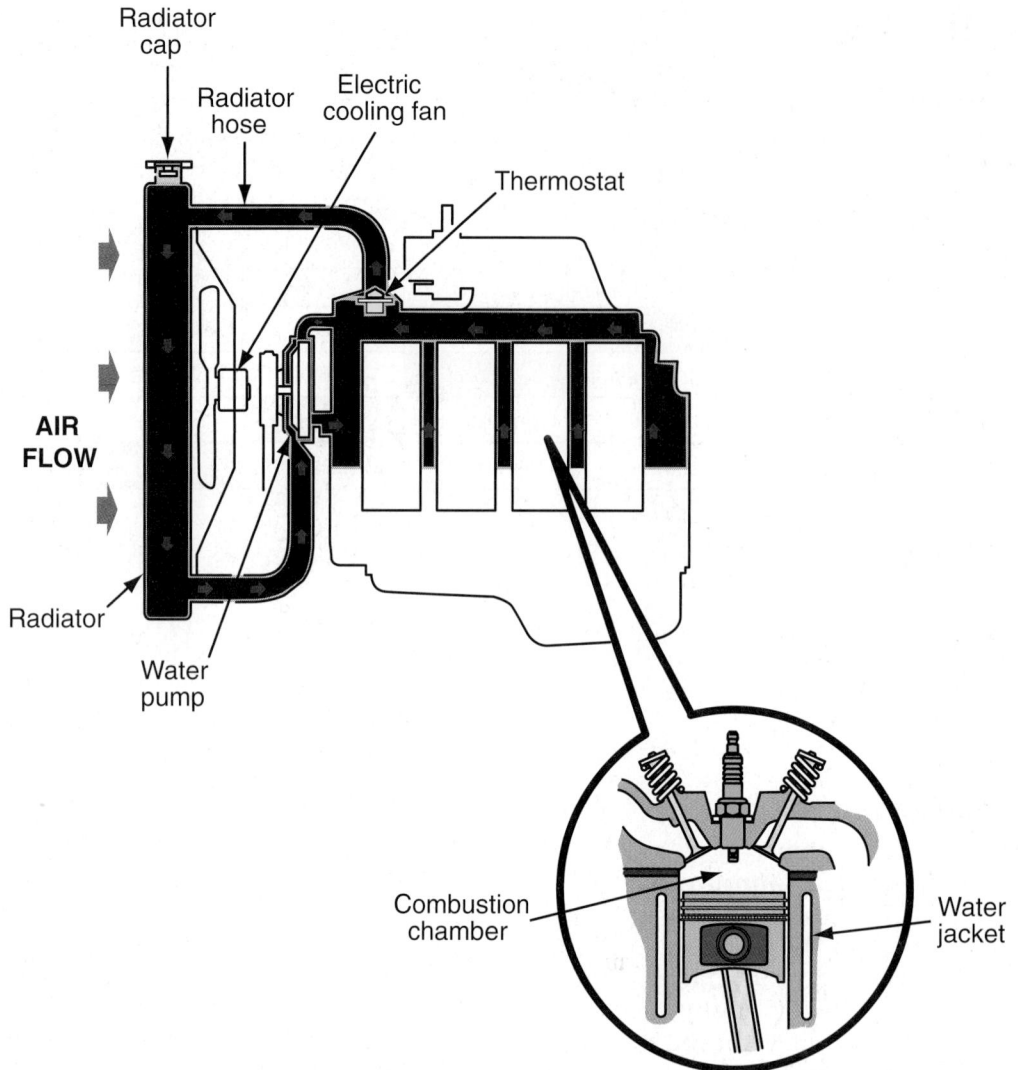

Figure 16.6 Cooling system parts and coolant flow.

Coolant is pumped through the system by a coolant pump, commonly called a *water pump*. A *thermostat* controls the flow of coolant between the engine and radiator to maintain temperature at a specified level. **Figure 16.6** shows parts of a cooling system.

A *coolant* mixture with a concentration of about 50% water and 50% anti freeze provides freezing and boiling protection. The coolant is designed to prevent rust and electrolysis, which causes corrosion. In the **bimetal engine** found in most of today's cars, the combination of iron cylinder blocks and aluminum cylinder heads is found. These two dissimilar metals promote **electrolysis**, or the creation of an electrical current. Electrolysis causes much faster deterioration of the metals.

▬ VALVE LOCATION

Engines are also classified by where valves are located. There are two common valve arrangements for internal combustion four-stroke engines: **L-head** and **I-head** (**Figure 16.7**). The I-head is used in today's automobiles.

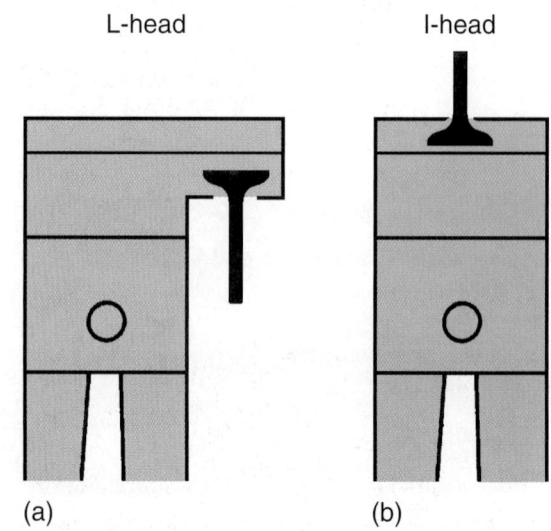

Figure 16.7 (a) An L-head, or flathead, engine has the valves in the block. (b) When the valves are in the cylinder head, the engine is known as an I-head engine.

L-Head (Vintage)

Many automobiles until the early 1950s had L-head engines (see Figure 16.5). The valve configuration resembles an upside-down "L" (see Figure 16.7). These engines are also called *flatheads* or *sidevalves* and are still used in lawnmowers, generators, and other industrial engines. The advantage of the L-head is that it is less expensive to manufacture. Its main disadvantage is that it produces more exhaust emissions than other engine designs. The flathead is also limited in its compression ratio and valve lift (the height of the valve opening). Increased valve lift would require more clearance in the combustion chamber, which would lower compression.

I-Head

The overhead valve (OHV) engine found in today's cars is known as an I-head or *valve-in-head* engine (see Figure 16.7). It breathes better than the flathead because it has a more direct path of air-fuel flow. I-heads also create less smog because they have less surface area in the combustion chamber; more surface area causes more quenched, unburned fuel. Unlike the L-head, an I-head valve job can be performed with the head on the workbench. The head may be cleaned before it is reinstalled so that grinding grit will not enter the engine.

Higher compression is possible with the I-head because, unlike the flathead, the combustion chamber does not need extra volume to accommodate the valves. Increased valve lift is possible because, as the valve opens, the piston is moving down in the cylinder (on the intake stroke) or is already at BDC (on the exhaust stroke). Increasing the valve opening to a certain point is necessary to allow enough air-fuel mixture into the cylinder to develop maximum power. The more air-fuel mixture that is packed into the cylinder, the more power will be developed. This is called volumetric efficiency, and it is the reason that *turbocharging* or *supercharging* is so effective in producing extra power from relatively small engines. In these engines, an air pump forces more air-fuel mixture into the cylinder at higher engine speeds (see Chapter 42).

■ CAMSHAFT LOCATION

The camshaft on I-head engines is located in either the cylinder head or in the cylinder block. The cam-in-block engine is called a *pushrod engine,* and the cam-in-head design is called an *overhead cam engine* (OHC).

In a pushrod engine, the camshaft acts on pushrods that operate rocker arms to open the valves (see Figure 15.12). In late-model vehicles, pushrods are found most often on V-type engines.

A more popular type of valve operating arrangement for in-line engines is the *overhead cam* design, or OHC (see Figure 15.6). This type of engine has the camshaft mounted on top of the cylinder head, just above the valve. The OHC is popular for high-speed operation. It has the advantage of having fewer parts and less weight.

Some OHC engines have a single overhead cam (**SOHC**). Each cylinder is provided with two separate lobes to operate the intake and exhaust valves.

High-performance OHC engines often have two cams per head. This engine design is called dual overhead cam (**DOHC**). One cam operates the intake valves; the other operates the exhaust valves.

To drive the cam, the OHC engine uses a long chain or belt from the crankshaft sprocket to the camshaft sprocket (**Figure 16.8**). Some OHC engines use an *auxiliary shaft* to drive the ignition distributor; others use the crankshaft to drive it. OHC was limited in the past to smaller, in-line engines, except for its use in luxury or racing automobiles. In recent years, belt-driven OHC engines have become commonplace.

(a)

(b)

Figure 16.8 V-type overhead cam engines. (a) Belt-driven overhead cam V6. (b) Chain-driven overhead cam V8. *(Courtesy of Tim Gilles)*

OTHER CYLINDER HEAD VARIATIONS

When intake and exhaust manifolds are on opposite sides on an in-line engine, the head is called a **cross-flow head** (**Figure 16.9**). This design improves engine breathing (how efficiently intake and exhaust mixtures can flow through it). Cross-flow heads have a coolant passage that provides the intake manifold with heat to help vaporize the fuel.

High-Performance Breathing Arrangements

Multiple Valve Heads. High-performance late-model engines commonly use three or four valves per cylinder (**Figure 16.10**). Some exotic engines even have five or six valves per cylinder. The use of multiple valves has become popular due to its improved higher rpm breathing and reduced valve weight.

- A greater amount of flow area for a given amount of valve lift is possible compared to two valve heads.
- When an engine will be operated at high engine rpm, the light weight of valves becomes more important.
- Combining smaller combustion chambers (because of multivalves) with a more central spark plug location has decreased the chances for an engine to knock. This allows the use of higher compression ratios, and delivers more power.

To burn very lean air-fuel mixtures, the fuel must be mixed well. At high speeds, there is plenty of turbulence, so this is not a problem. But multivalve heads tend to allow fuel to fall out of the mixture at low speeds. Some multivalve heads use control valves that cause only one intake valve to open at low rpm and another to open at higher rpm. This helps maintain velocity and swirl at low speed and high flow at high speed (**Figure 16.11**).

Other multivalve heads use two intake manifold runners per cylinder. These manifolds are variably tuned using a butterfly control valve to control airflow.

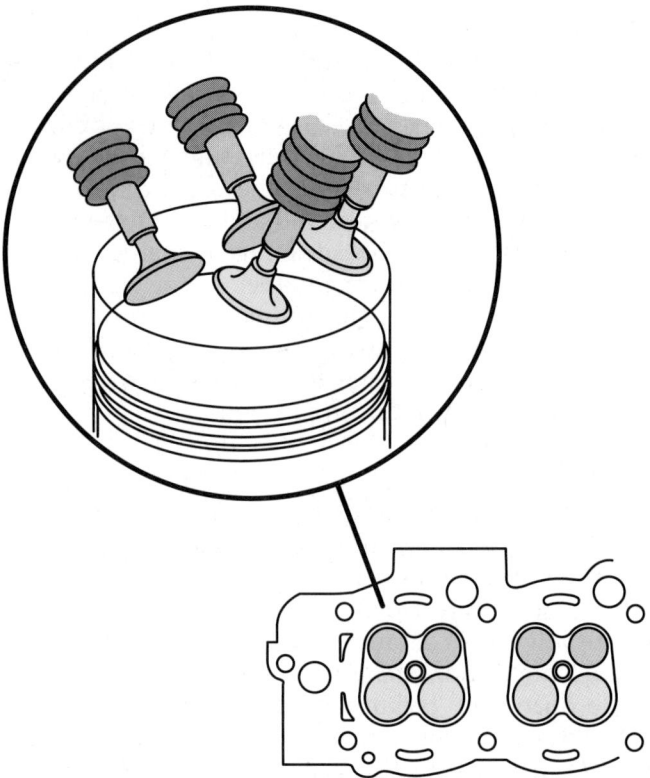

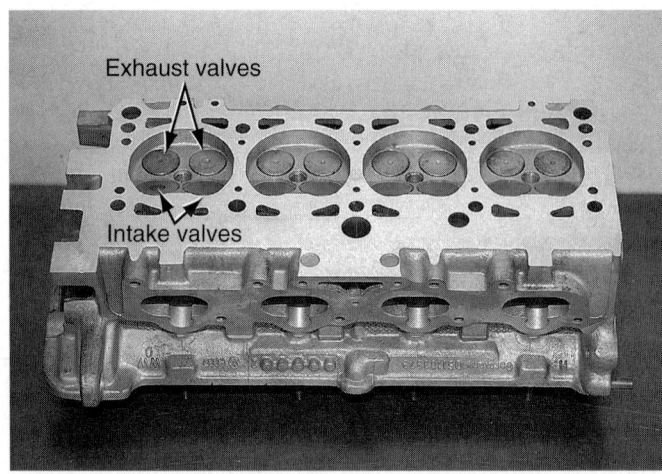

Figure 16.10 Four-valve combustion chamber.

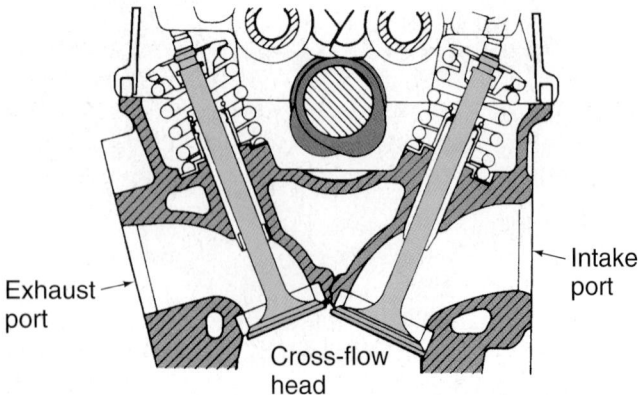

Figure 16.9 Cross-flow head. (*Courtesy of DaimlerChrysler Corporation*)

COMBUSTION CHAMBER DESIGNS

Common combustion chamber designs include the *hemi* and the *wedge* (**Figure 16.12**). Other chamber designs include the *pent-roof* and chambers shaped like a "D" or a heart. The wedge chamber is mostly used in pushrod engines, with the camshaft located in the block. It has a squish/quench area that causes movement (*turbulence*) of the air-fuel mixture and cooling of the gases to prevent abnormal combustion. This movement causes more complete burning at lower speeds with less chance of detonation.

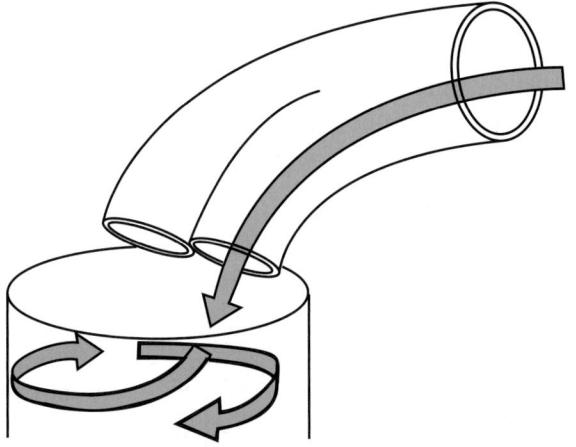

Below 2,500 rpm

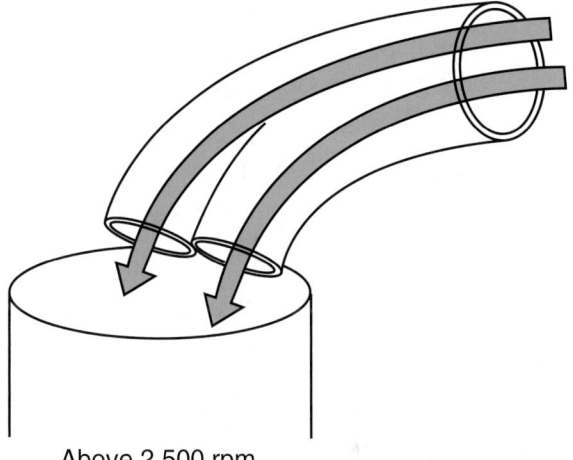

Above 2,500 rpm

Figure 16.11 At low rpm, velocity and swirl are maintained. At high rpm, there is high flow.

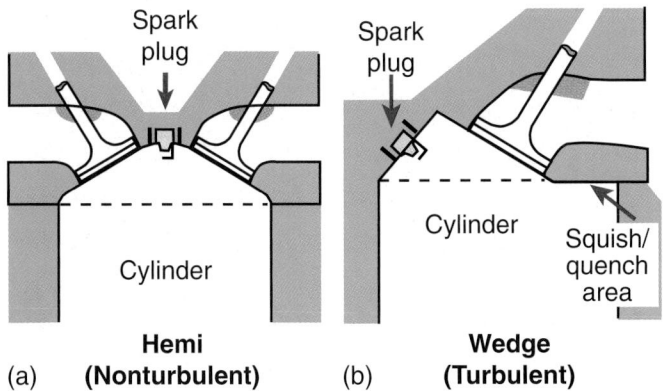

Figure 16.12 Hemi and wedge combustion chambers.

There are *turbulent* and *nonturbulent* combustion chambers. Turbulent combustion chambers, like the wedge, can cause air and fuel to separate from each other at high speeds. A nonturbulent combustion chamber, the hemispherical (hemi) design, is more

efficient for high-speed use. Because the mixture is centered near the spark plug, the flame spreads evenly. A hemi chamber also allows the use of bigger valves. Sometimes hemis have a tendency to "spark knock" when using lower octane fuels (see Chapter 39).

Diesel engines have no chamber in the cylinder head itself. The combustion chamber side of the head is virtually flat. Turbulence and squish in the cylinder are controlled by the shape of the piston head.

A *pent-roof* combustion chamber is shaped like a "V." This design is popular for use with *four-valve per cylinder* designs. The pent-roof and other newer designs are designed for more efficient combustion and better emission control. In a *high swirl chamber*, like in the wedge chamber, areas on the head surface are raised to cause a planned turbulence of the air-fuel mixture.

A stratified charge design was pioneered by Honda in the 1970s. The name comes from the stratification, or layering, of different densities of air-fuel mixtures, where a very small amount of rich mixture ignites a very lean (normally unburnable) mixture in a small precombustion chamber. When it is ignited by the spark plug, the advancing flame front from this small, rich mixture ignites the leaner mixture in the main cylinder. This makes it possible for the engine to run on an air-fuel mixture that is leaner than normal. Newer direct-injected gasoline engines use sophisticated computer controls to cycle high-pressure fuel injectors on and off during the combustion event, providing a very controlled stratified charge.

■ SPARK AND COMPRESSION IGNITION

Although this text does not deal specifically with diesel engines, most of the automobile engine information included here applies to light-duty diesel engines found in some passenger cars and light trucks. **Diesel-cycle** and four-cycle gasoline engines share the same basic principles of operation. The difference is in the way the fuels are ignited.

The gasoline engine is called a *spark ignition (S.I.)* engine. A spark is created in the ignition system. A distributor, geared to the camshaft, times and distributes the spark to the spark plug at exactly the correct instant. Late-model engines have computer-controlled spark ignition. Many new engines have *distributorless ignition systems (D.I.S.)* with ignition coils that are triggered by the computer in response to a signal from a camshaft or crankshaft sensor.

Diesel Engine

The diesel engine was invented by Rudolph Diesel in 1892 in Germany. Diesel engines, which can be either two- or four-stroke cycle, are used extensively in heavy equipment and were not used in automobiles until the 1930s. The operation and appearance of the diesel engine is very similar to the gasoline engine.

A diesel is a *compression ignition* (C.I.) engine. It does *not* use a spark to ignite the fuel. Basically, when air is compressed and fuel is injected into it, the fuel ignites.

Compression ratio is the comparison between the volume of the cylinder and combustion chamber when the piston is at TDC and BDC (see Figure 17.4). Diesel compression ratios can be in the neighborhood of 20:1. Gasoline engine compression ratios are usually from 8:1 to 10:1.

When air is compressed, it heats up. Because an air-fuel mixture explodes if it is compressed too much, the diesel engine compresses only air. Diesel fuel does not burn at room temperature. But when it is injected into the cylinder at the exact moment ignition is desired, it burns easily in the hot environment of the compressed air (approximately 1,000°F).

Instead of the ignition system used with gasoline engines, a diesel uses either mechanical injectors operated by a camshaft, a precision fuel distributor and individual injectors, or electronic injectors (**Figure 16.13**). With either type of injector the pressure of the fuel must be very high in order to overcome the pressures in the cylinders that are reached during the compression stroke. Electronic diesel direct injection is covered in Chapter 42.

The diesel can run at very lean mixtures at idle and is generally about one-third more efficient on fuel, although it produces less power than a gas engine. In gasoline engines, the amount of air is changed to control speed and power. In the diesel, the amount of air remains the same while the fuel mixture is changed to control speed and power. The mixture can be as rich as 20:1 under load and as lean as about 80:1 at idle.

Problems with the diesel are its high particulate emissions (soot) and the high temperature of combustion, which produce high levels of NO_x emissions. Diesels also have starting problems in cold weather, and they require more frequent oil changes and other owner maintenance.

ALTERNATE ENGINES

Almost all automotive and truck engines use internal combustion four-stroke piston engines. Over the years, several other engine types have been developed by designers, but only the Wankel rotary and the two-stroke piston engines are found in today's vehicles. The gasoline internal combustion four-stroke piston engine has proven to be the best engine choice to date.

The Wankel Rotary Engine

The **Wankel engine** is also known as the rotary engine. It operates on the four-stroke cycle, although there are actually no strokes. Automotive rotary engines have two rotors that rotate inside of a chamber that looks like a modified figure eight (**Figure 16.14**). The rotor has three sides that act as pistons. While one of the chambers is experiencing intake, the others will be doing other parts of the cycle. Thus, one revolution of the crankshaft produces the equivalent of three power strokes.

As the rotor turns, the end of one of the lobes moves past the intake port, drawing in fuel and air. Turning further, the mixture is compressed as it nears the spark plug. The spark plug ignites the air-fuel mixture, and the rotor continues revolving until the exhaust port is uncovered. When the exhaust has escaped, the rotor is in position above the intake port to begin the cycle again.

Rotary engines do not have pistons that have to start and stop moving hundreds of times per second at high rpm like reciprocating engines do. There are no poppet valves to open and close either. This means that these engines run very smoothly at higher rpm.

Rotary engines require complicated emission control systems. The result is that they are not as fuel efficient as they could be. The engine has been in

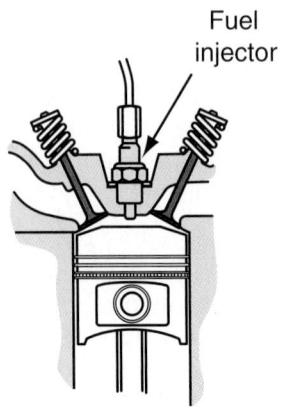

Fuel injector

Figure 16.13 A diesel engine has a timed high-pressure fuel injector to control the point of ignition.

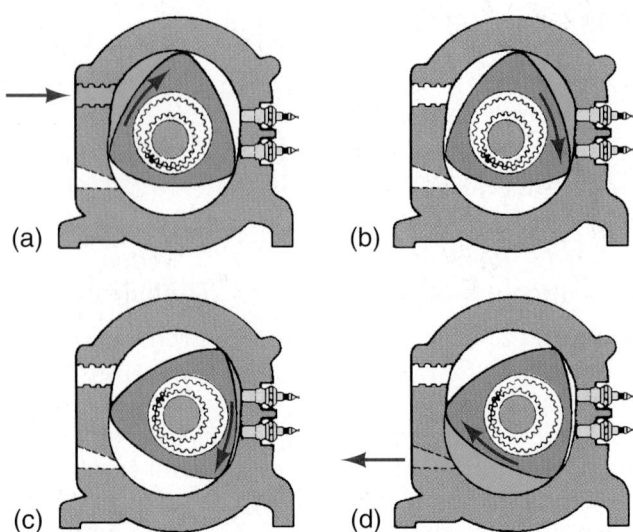

(a)　　　　(b)

(c)　　　　(d)

Figure 16.14 Rotary engine cycle.

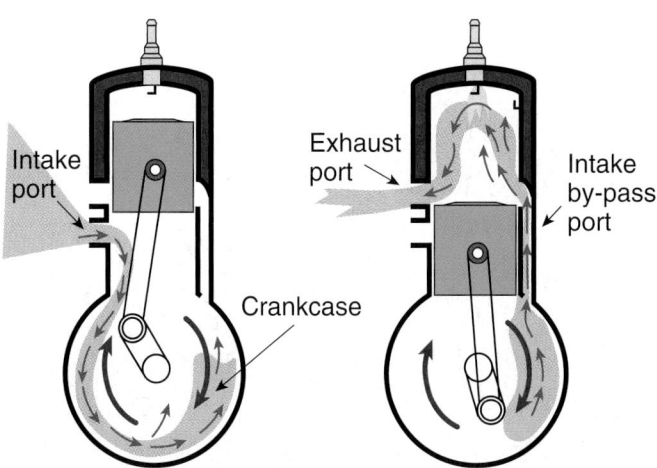

Figure 16.15 A two-stroke cycle engine.

limited use in Mazdas. If readily available alternative clean-burning fuels become a reality, the rotary engine could become a popular choice.

Two-Stroke Cycle

A **two-stroke engine** can be made smaller and lighter than a four-stroke engine of comparable size. Two-strokes, used for years in outboards, chainsaws, and motorcycles, use a mixture of oil and gasoline for lubrication of the crank, rod, and piston. Some of the new designs for automobiles have crankcases lubricated with pumped oil.

A two-stroke engine has a power stroke every crankshaft revolution. The two-stroke cycle begins with the piston at TDC on the power stroke. The cylinder has intake and exhaust ports, which are openings in the side of the cylinder (**Figure 16.15**). As the piston reaches the bottom of the power stroke, the exhaust port is opened to release exhaust gases. Shortly after the exhaust port opens, the intake port opens and the air-fuel mixture is pushed into the cylinder. This action also helps to push the exhaust out. As the piston moves up on its compression stroke, both the intake and exhaust ports are covered.

Older two-stroke engines used a mixture of oil and fuel. This mixture lubricated the *lower end* (crankshaft and bearings) as it flowed through the crankcase on its way to the cylinder. New direct-injection two-strokes use fuel injectors to put fuel into the combustion chamber. Air is pushed into the cylinder using a super-charger (see Chapter 44). The crankcase is pressure lubricated in these engines just like in four-stroke engines. Some older diesel truck engine designs also use a two-stroke cycle, but these engines have high exhaust emissions.

■ NEW GENERATION VEHICLES

The United States Environmental Protection Agency is very interested in clean air. California and other states have regulations requiring manufacturers to sell a certain percentage of ultra-low emission vehicles (ULEVs) like hybrids, and zero emission vehicles (**ZEVs**), like hydrogen fuel cell electrics.

Electric vehicles (EVs) lack exhaust emissions and are therefore ZEVs (other than the power sources used to charge their batteries). Advantages include reduced noise and excellent acceleration torque. If you have ever driven an electric golf cart, you probably have noticed that electric motors have a tremendous torque advantage over gasoline engines during acceleration from a stop.

There are some concerns regarding battery EVs and hybrid EVs. One disadvantage is that they must carry many heavy nickel metal hydride or lithium-ion batteries. Also, in the event of an accident, electrolyte in the batteries and their dangerous high voltage require specialized hazard and safety training for emergency service personnel.

Most manufacturers developed EVs on an experimental basis in the 1990s, often with research subsidies from the government. Due to practical considerations, the manufacture of EVs powered solely by an electric motor has been discontinued, however. A key problem was their somewhat limited range between recharges (often less than 100 miles). Also, recharging the battery pack required several hours at best. This limited their practicality to short-trip commuter driving. Electric-powered heating and air conditioning were also concerns.

The technology that was developed for EVs was not wasted, however. Currently, hybrid vehicles couple existing internal combustion engine (ICE) technology with EV technology to produce a viable alternative to battery EVs. Fuel cell EVs that have been produced and are undergoing further development are also based on technology developed for EVs.

Regenerative Braking

Electric and hybrid vehicles use a large, high-efficiency electric motor and a very large battery pack, controlled electronically. During deceleration, the motor is used as a generator, producing electricity to recharge the batteries as it slows the vehicle down. This feature, called **regenerative braking**, also increases the longevity of the vehicle's conventional friction brake linings.

■ HYBRID VEHICLES

Hybrid electric vehicles (HEVs) (**Figure 16.16**) overcome some of the shortcomings of EVs by combining the electric motor(s) with one or more additional power sources. Hybrids offer improved fuel economy, increased performance, and a reduction in exhaust pollutants and greenhouse gas emissions. A moped (**Figure 16.17**) is one example of a hybrid, with power provided by either the pedaling effort of the rider or by a gasoline engine, or a combination of the two. Submarines are hybrids as well, combining electric power with

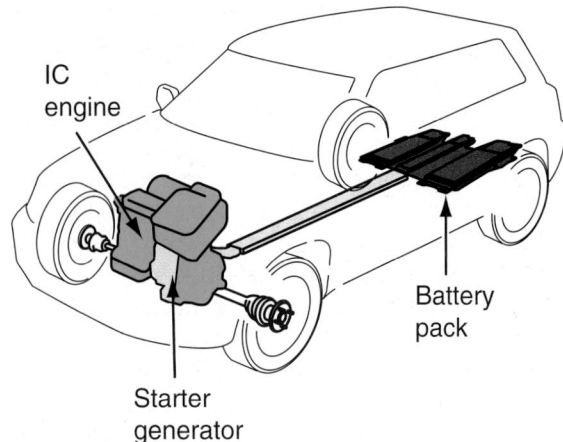

Figure 16.16 Most hybrid automobiles are powered with an internal combustion engine or a battery-powered electric motor, or both at the same time.

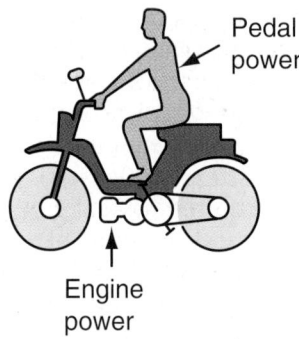

Figure 16.17 A moped is one example of a hybrid.

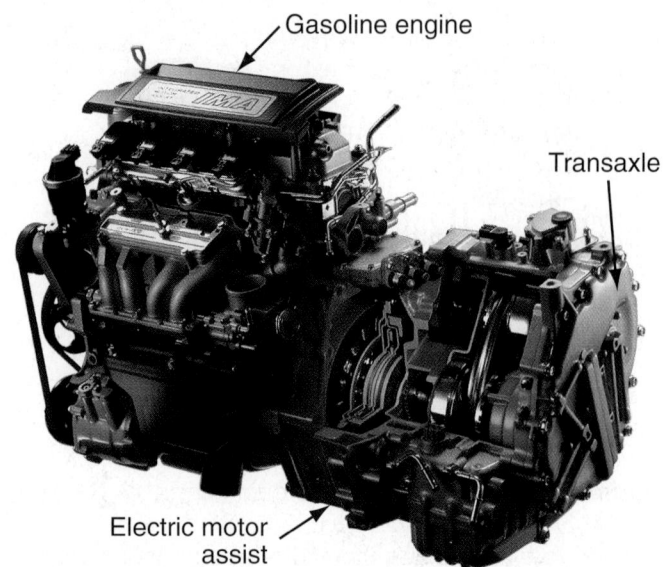

Figure 16.18 A Honda Civic hybrid has an electric assist motor/generator between the engine and transaxle. *(Courtesy of America Honda Motor Co., Inc.)*

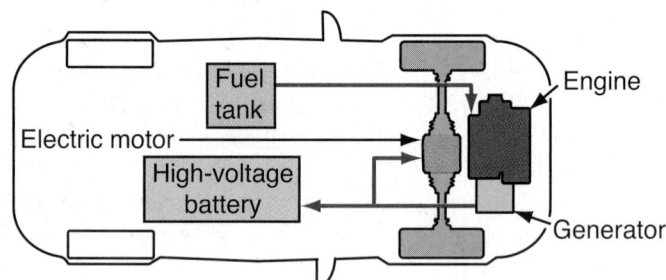

Figure 16.19 A series hybrid.

either nuclear or diesel power. Buses, trucks, and locomotives can also be diesel-electric hybrids. Another type of hybrid is the hydraulic hybrid, used mostly in trucks. The energy that would normally be used in braking is captured and stored in hydraulic fluid storage tanks.

Most hybrid automobiles are powered with an internal combustion engine or a battery-powered electric motor, or both at the same time.

NOTE: *A motor and an engine are two different things. The following illustrates the difference in HEV terminology:*

- *Engine: piston powered by burning fuel*
- *Motor: armature powered by electricity*

Figure 16.18 shows the relationship between the engine, motor generator, and transaxle in a Honda hybrid.

Series and Parallel Hybrids

Most automotive hybrids have a four-stroke cycle gasoline- or diesel-powered engine, as well as a battery pack and electric motor. The engine and motor can be coupled in series or parallel, or a combination of series and parallel.

Series Hybrid. A *series hybrid* is closer to a true EV because it is powered only by the electric motor (**Figure 16.19**).

A gasoline- or diesel-powered engine turns a generator to recharge the battery pack, but the generator does not have direct input to the transmission. The generator can either charge the batteries or power the electric motor that drives the transmission. A series hybrid requires significant battery size and output. No production series hybirds are sold in North America.

Another type of series hybrid is the fuel cell hybrid (covered later in this chapter), which replaces the engine and motor/generator with a fuel cell. A fuel cell produces only electricity, so fuel cell hybrids are configured in series.

▥ HISTORY NOTE ▥

A Santa Fe locomotive used diesel engines to create electricity to run motors and move the train. It was a series hybrid, not used in automobiles because the size of the battery would be extremely large. Trains do not have storage batteries. This technology could have future use in hydrogen fuel cells.

Parallel Hybrid. A *parallel hybrid* uses the engine and motor together to provide power to the transaxle (**Figure 16.20**). The motor is typically mounted in line with the engine, either between the engine and transaxle or inside the transaxle itself. Power is sent by the transaxle to the wheels.

Series/Parallel Hybrid. Many automotive hybrids use a *series/parallel* configuration. Power can be provided by the motor or the engine individually, or the motor and engine together. Some hybrids have two motor generators in the front and another one in the rear (**Figure 16.21**). These provide all-wheel drive and a substantial boost in torque output.

Automotive parallel and series/parallel hybrids have their batteries charged by the engine or by *regenerative braking*. Instead of using conventional braking, the electric motor used in a parallel hybrid acts like a generator as it slows the car down. **Figure 16.22** shows an all-wheel-drive hybrid during regenerative braking.

When the batteries are sufficiently charged, power to move the vehicle is provided by the electric motor. To allow the regenerative brakes to work more often and to prevent overcharging, the target charge is usually around 60%. The battery can be allowed to charge to as much as 80% to 90% and discharge down to about 30% to 40% of charge while waiting for opportunities for regenerative braking to recharge it without using fuel.

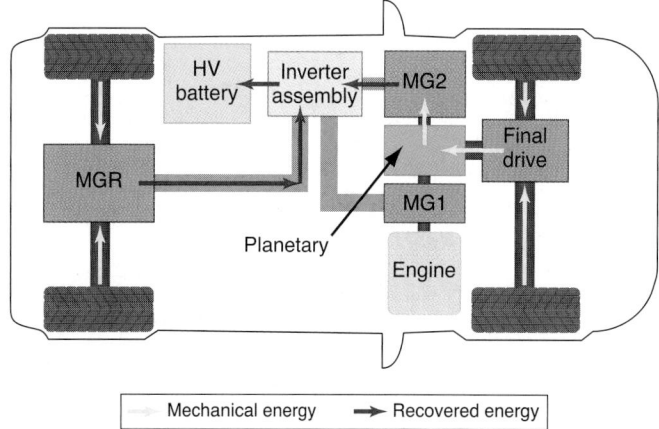

| Mechanical energy | Recovered energy |

Figure 16.22 An all-wheel-drive hybrid during regenerative braking.

Differences between Conventional Power and Hybrid Power

The major operating difference between a hybrid and a conventional vehicle powered only by an engine is that the engine in a hybrid vehicle stops running at idle (when certain operating conditions are met). When the hybrid's computer program calls for the engine to be restarted, the motor/generator reverses to provide a powerful starting motor. When starting a conventional internal combustion engine, a starter motor turns the crankshaft at about 250 rpm. On a hybrid engine, the motor/generator turns the crankshaft at 500–1,000 rpm, much faster than normal. This feature allows the engine to stop running when it would normally be idling, when the vehicle is at rest or decelerating. The engine can seamlessly restart, without the driver or passengers being aware.

One objective of the hybrid design is to improve fuel economy. An engine in a hybrid car is often smaller than what would normally be required to run a gasoline-powered vehicle of similar weight. Used in conjunction with an electric motor, a reasonable amount of power is available to accelerate the vehicle. When extra power is needed for climbing a hill, the electric motor assists the small engine's power.

Disadvantages to hybrid vehicles include their high initial cost and concerns for technician safety from electrocution. Hybrids are more expensive than ordinary ICE vehicles due to the additional expenses of the battery pack, motor controller, and the expensive starter generator. An additional concern is that when freeway driving is the primary mode of operation, regenerative braking is not used. Without regenerative braking, fuel economy suffers to a considerable degree.

Types of Hybrids

There are several classifications of hybrids, depending on the manufacturer. Hybrid classifications include mild, medium, full, and plug-in.

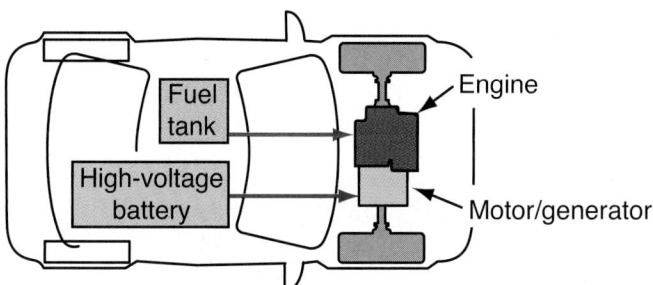

Figure 16.20 A parallel hybrid.

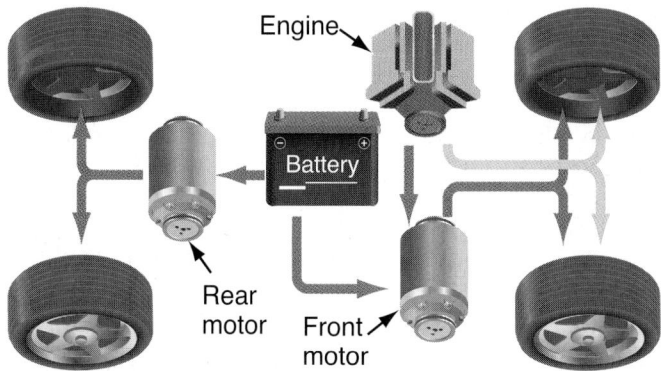

Figure 16.21 An all-wheel-drive hybrid with front and rear electric motors.

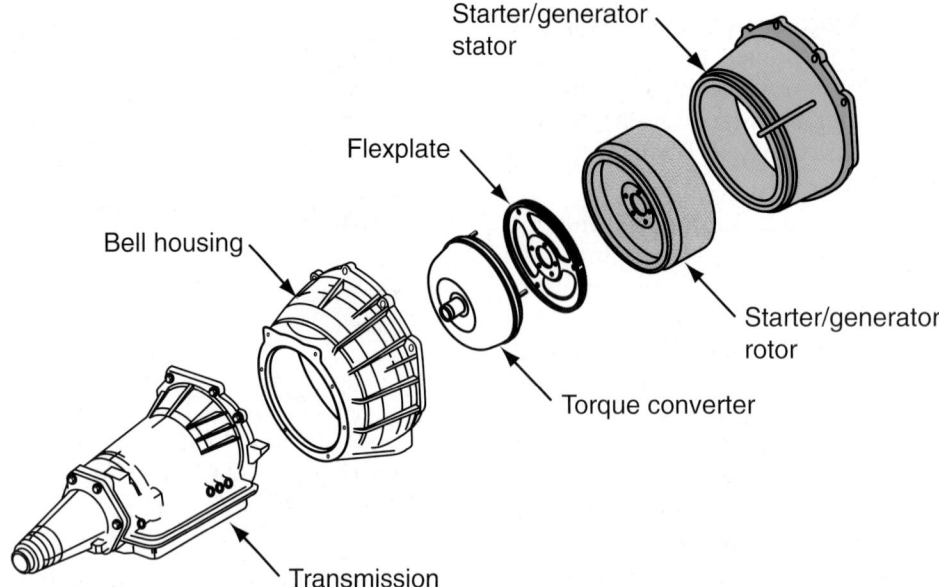

Figure 16.23 A mild hybrid with a starter/generator attached to the crankshaft-driven flexplate.

Mild Hybrid. A *mild hybrid* cannot move the vehicle without assistance from the ICE. A mild parallel hybrid like the General Motors or Dodge light truck provides a gain of about 15% in fuel economy because the engine shuts off at idle. Its key component is a starter/generator attached to its crankshaft by way of the flexplate (**Figure 16.23**). It has a standard 12-volt battery with a 14-volt charging system. Three additional absorbed glass mat 12-volt batteries and a 42-volt charging system are included as well. Regenerative braking recharges the batteries, except when the antilock brake system is operating. This system cannot power the vehicle with its electric motor, and there is no fuel economy improvement at freeway speeds.

A belt alternator starter (BAS) hybrid is a variation of the mild hybrid system. In this system, a belt-driven starter/generator is located in place of a conventional AC generator (**Figure 16.24**).

Medium Hybrid. A *medium hybrid* has high-voltage (144V) idle stop capability. It can propel the car using its electric motor, but only after the ICE has started the vehicle in motion. The high-voltage electrical system allows for electric power steering.

Full Hybrid. *Full hybrids* operate with voltages ranging from 300V to 650V. They do everything that medium hybrids do, but they can also power the vehicle from a standing stop. A typical full hybrid idles with the ICE off, using the electric motor for initial acceleration before restarting the engine. The electric motor also provides an additional power boost to the engine when needed. In larger-sized hybrid vehicles, this provides a high-performance feel. Unlike ICEs, electric motors provide full torque from a standing start. Torque is what gives you the feeling of being pushed

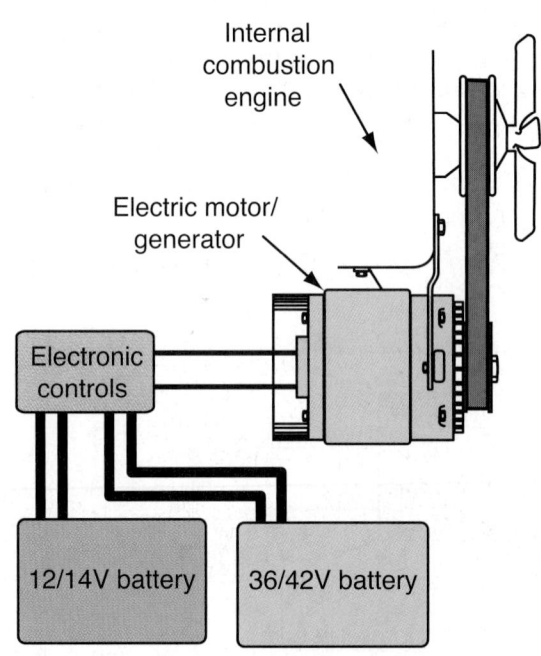

Figure 16.24 A belt alternator starter hybrid system.

back into the seat. "Customers buy horsepower, but they drive torque." Electric motors have 100% torque all of the time. This is why an electric go-cart with only 8 HP feels like it is "fast." In a hybrid system, as the motor speeds up but its torque remains the same, the engine provides supplementation.

From a fuel economy standpoint, the high-performance aspect of hybrid vehicles can result in worse mileage when coupled with an overly enthusiastic driver. Wide-open throttle kills the battery pack and ruins fuel economy. Driving in bursts, accelerating and decelerating, is the most fuel efficient.

NOTE: *Toyota, Ford, and Chrysler use the Toyota hybrid system, which has a series/parallel full hybrid configuration. Honda and GM have regenerative braking and idle stop. Honda uses the engine primarily, with the electric motor providing additional assist. This is the opposite of the Toyota system, which uses the motor for primary power and the engine for recharging and assist. The GM and Dodge truck systems cannot power the vehicle using the electric motor alone.*

Plug-In Hybrid. *Plug-in hybrids* are a newer development that will become more popular as battery technology improves. A power socket allows their larger batteries to be recharged by an external source of electricity. Conventional hybrids cannot run for more than a very short distance on battery power alone.

NOTE: *The hybrid was not designed to be driven as an electric vehicle. If the engine fails or runs out of fuel, the electric motor can still move the vehicle for 1–2 miles.*

Plug-in hybrids can run for longer distances, although, like EVs, their range is limited to about 70 miles between charges. From an environmental standpoint, driving on electricity alone reduces greenhouse gases and exhaust pollutants, and can provide very high fuel economy (about twice that of a conventional hybrid). They are more expensive than ordinary hybrids due to their higher battery cost.

Hybrid Vehicle Service and Safety

For your personal safety, you will need to be aware of some important considerations when you service a hybrid vehicle. This book is written with the objective of preparing a student for an entry-level job. Most experienced automotive technicians will not work on the electrical systems of hybrid vehicles. Manufacturers are very concerned about the possibility of electrocution (**Figure 16.25**), so they provide training and certification only to highly skilled master technicians. Therefore, the emphasis to this point has been on hybrid vehicle theory.

Become familiar with the three color designations of electrical conduit: Black conduit signifies ordinary low 12 volts. Yellow conduit indicates 42 volts. This provides enough voltage to jump a gap and arc weld when a wire is disconnected. Orange (**Figure 16.26**) is the electric/hybrid color (144 to 650 volts). **This can kill you!**

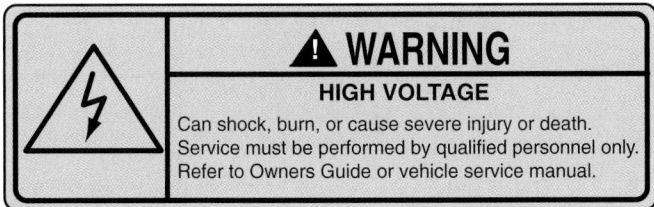

⚠ **WARNING**

HIGH VOLTAGE

Can shock, burn, or cause severe injury or death.
Service must be performed by qualified personnel only.
Refer to Owners Guide or vehicle service manual.

Figure 16.25 A high-voltage warning label from a hybrid vehicle.

Figure 16.26 On hybrid vehicles, orange cables are high-voltage cables and are dangerous! *(Courtesy of Tim Gilles)*

NOTE: *Some hybrids use a "smart key." If a hybrid is left in the "ready" position, even though the engine is off, it can restart if the transponder is within a few feet of the vehicle. This is especially unnerving to a technician when the vehicle is in the air with the oil drained out of the engine.*

Other Hybrid Vehicle Operation, Safety, and Service

More information on hybrid vehicles can be found in other chapters in this book. Operation, safety, and service of hybrid vehicle electrical systems are described in Chapters 30 and 31 ("Charging System Fundamentals" and "Charging System Service", respectively). Battery operation, safety, and service are described in Chapters 26 and 27. Hybrid CVT and planetary operation and power flow are described in Chapter 73, Automatic Transmissions Fundamentals. A good Web site with in-depth information on hybrids and fuel cells is http://www.world.honda.com.

■ FUEL CELL ELECTRIC VEHICLES (FCEVs)

When an EV is powered by a fuel cell, it is called a fuel cell electric vehicle (FCEV or FCX). **Figure 16.27** shows an FCEV's design layout. An FCEV does not store electricity but converts it when needed to power the electric motor. The only exhaust by-products are water and heat. Different types of fuel cells exist, but the proton exchange membrane (PEM) fuel cell (covered later in this chapter) is thought by many to be the power source most likely to replace the ICE in the next generation.

FCEVs are technically hybrid vehicles. They use an electric engine rather than a heat engine to power an electric motor. An FCEV has a battery module to provide backup or supplemental power, like an electric

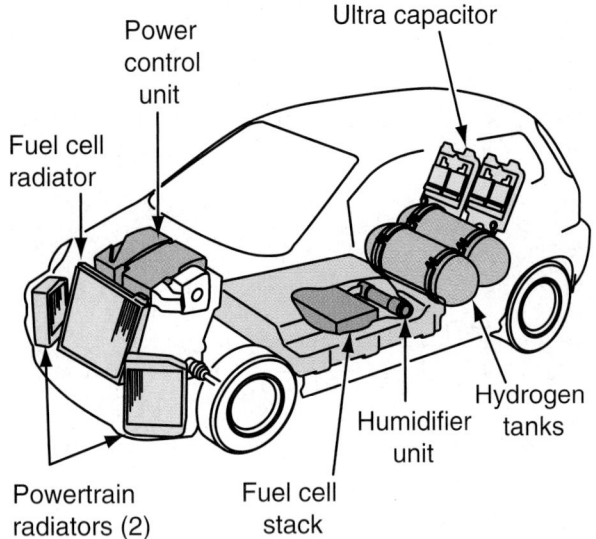

Figure 16.27 Layout of a typical fuel cell vehicle.

supercharger. The battery module is recharged by the fuel cell as well.

Fuel cells, invented in the late 1800s, began to gain popularity when they were used by NASA for electrical power generation and water production during manned space flights. They are currently viable for power generation in stationary applications but have always been large in size. Recent advances in technology have resulted in their size being diminished to the point of practicality for installation in automobiles, trucks, and buses. Small fuel cells are now available for powering portable electronic devices.

Fuel cells use an electrochemical reaction to produce electricity, much the same as an ordinary car battery. But, unlike the car battery, fuel cells do not wear out or require recharging. They feed on a fresh supply of hydrogen and oxygen. Pure hydrogen is the fuel of choice for a fuel cell, but it must be kept chilled to –253°C to remain a liquid. This presents serious refueling and storage problems.

Hydrogen can be "reformed," using an onboard reformer, from hydrocarbons such as methanol, gasoline, or natural gas, to extract their hydrogen content. There is a big loss of efficiency in the reforming process, and it is only a bridging strategy until a future time when pure hydrogen fuel is readily available. The advantage to gasoline reforming is that the refining and distribution system is already in place.

Fuel Cell Operation

Fuel cells use hydrogen for fuel, with oxygen from air as the oxidant. Combining hydrogen and oxygen produces electricity. The fuel cell operates similar to a conventional car battery, which produces direct current using an electrochemical process (see Chapter 26). But unlike car batteries, fuel cells never run dead.

■■■ **SCIENCE NOTE** ■■■

A fuel cell can be described as a sandwich (Figure 16.28). Its two electrodes, the anode and the cathode, are the "bread." They are separated by a polymer membrane electrolyte, the "filling." Each electrode has a platinum catalyst bonded to one side. Oxygen passes over one electrode, and hydrogen passes over the other.

■ *At the anode, the hydrogen reacts to the catalyst and separates into positively charged protons (H+ = positive hydrogen ions) and negatively charged electrons (e-). The protons and electrons take different return paths to the cathode. The free electrons produce usable electrical current, which is conducted through the electrical circuit.*

■ *The oxygen enters the fuel cell through the cathode. The protons move through the electrolyte membrane to the cathode. The catalyst layer on the cathode causes the protons to combine with oxygen from the air and leftover electrons from the electrical circuit. The result is heat and pure water.*

Fuel Cell Construction

Individual cells generate a relatively small voltage, on the order of 0.7 to 1.0 volt each, after accounting for resistance losses. To develop higher voltages, cells are "stacked" and connected in series. Fuel cells are arranged in stacks of several hundred cells that generate direct current (**Figure 16.29**). This is converted into alternating current to power the electric motor.

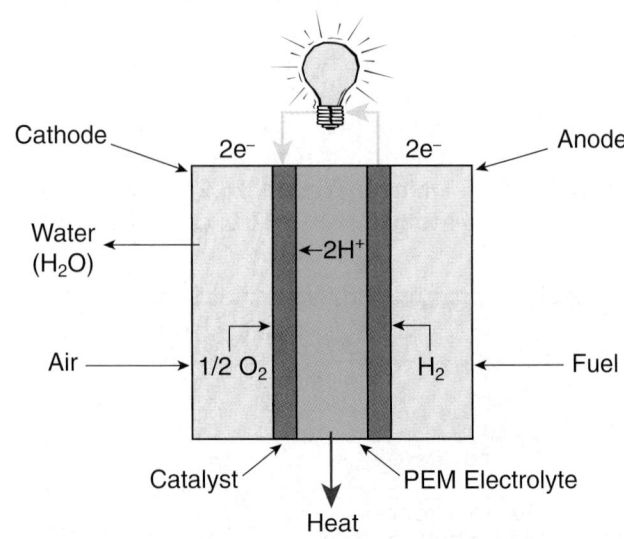

Figure 16.28 Operation of a fuel cell. (*Courtesy of Ballard Power Systems*)

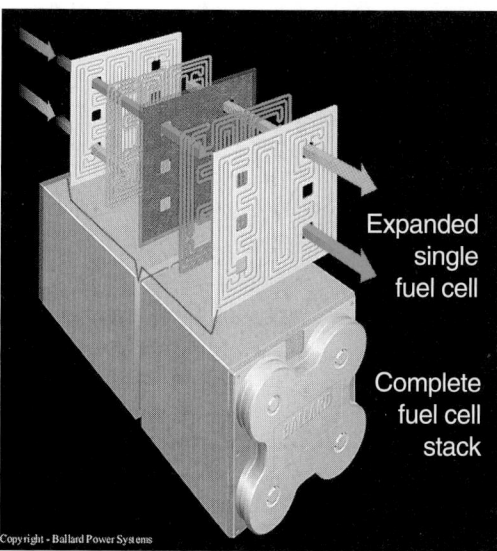

Figure 16.29 A fuel cell stack. *(Courtesy of Ballard Power Systems)*

Ultracapacitors

Instead of a battery, some fuel cell systems use an ultracapacitor (**Figure 16.30**). It operates in place of a battery, supplying supplemental power to the electric motor during startup and when the fuel cell is under load. It also supplies power for electrical accessories when the vehicle is stopped.

Like a battery, a capacitor is an electrical storage device. But unlike a battery, a capacitor does not use a chemical reaction when charging and discharging. It has lower internal resistance and is capable of nearly twice the output of a battery of the same weight. It discharges and stores electricity as the output of the fuel cell stack changes and does not require a converter for voltage regulation as in a battery system. Further information on capacitor operation can be found in Chapter 25, and battery theory is described in Chapter 26.

The Future of Fuel Cells

Since the late 1990s, most vehicle manufacturers have built prototype FCEVs, and some have been put in use by the public on an experimental basis (**Figure 16.31**). Fuel cell vehicles are very responsive and have been well received by the public. However, there are some serious concerns that will need to be addressed before fuel cell automobiles become commonplace. A production and distribution system for hydrogen fuel is just one of the problems. Another serious concern is that fuel cells are expensive and have not proven to have sufficient longevity. Replacing a very expensive fuel cell at 60,000 miles is not a viable option. The idea of a fuel cell vehicle is highly appealing, and in some form it will become the way in which future vehicles are powered. However, a commercially viable fuel cell vehicle is not yet on the horizon in the near future.

New Vehicle Development

Many types of vehicles have been invented. But whether they come to production depends on many things. The availability and distribution of fuel, cost of the vehicle, customer satisfaction, drivability and performance, and fuel economy are all considerations.

Alternative fuel vehicles are used extensively in fleet operations such as utility companies. Specialized training in these technologies is available. Alternative fuels are discussed in Chapter 41.

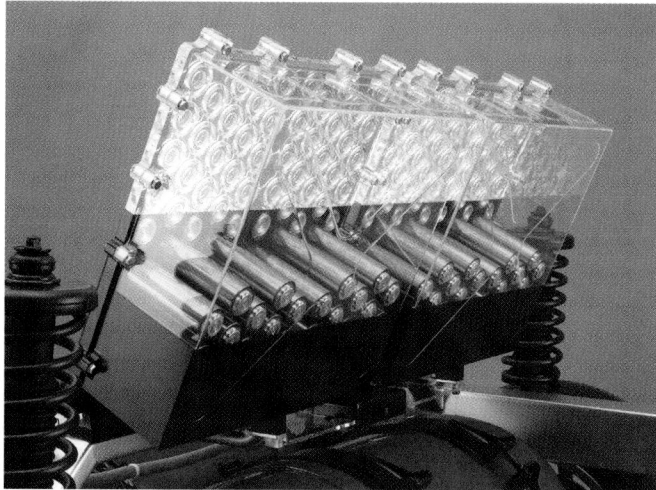

Figure 16.30 An ultracapacitor that supplements a fuel cell to provide a burst of power to the electric motor and run accessories when needed. *(Courtesy of American Honda Motor Co., Inc.)*

Figure 16.31 A fuel cell installed under the hood of an experimental vehicle. *(Courtesy of DaimlerChrysler)*

SUMMARY

- Piston engines share common parts but there are many different design variations.
- Cylinders are arranged in-line, in a V, or opposed to each other. The most popular automotive engines have either four, six, or eight cylinders.
- Cylinders rows, called banks, are determined from the flywheel end of the engine.
- The crankshaft turns two revolutions (720 degrees) to complete one four-stroke cycle. The first half of the firing order represents one crankshaft revolution (360 degrees). The second half of the firing order represents the second revolution of the crankshaft (360 degrees).
- Engines use either liquid or air cooling.
- The two camshaft designs are pushrod and overhead.
- An engine operating at 3,000 rpm on the freeway has to open and close each valve 25 times per second.
- Alternatives to the four-stroke piston engine are not yet viable.

REVIEW QUESTIONS

1. What are three ways that cylinders are arranged in a block?
2. What are the rows of cylinders in a V-type block called?
3. What is the name of the area between the heads on a V-type block?
4. List two firing orders for a V8 and for a V6 (see Chapter 37).
5. If a four-cylinder engine has a firing order that is 1-3-4-2, which cylinder is on its exhaust stroke when cylinder #2 is on its intake stroke?
6. What is the name of the design that uses a small amount of rich air-fuel mixture to ignite a leaner mixture?
7. What kind of engine has a combustion chamber that is flat?
8. What is the name of the engine design that is called compression ignition?
9. What kind of engine is a Wankel, reciprocating or rotary?
10. What kind of engine is often found in chainsaws and outboard motors?

ASE-STYLE REVIEW QUESTIONS

1. Technician A says that it takes two revolutions of the crankshaft to fire all eight cylinders on a V8. Technician B says that it takes two revolutions of the crankshaft to fire all six cylinders on a V6. Who is right?
   - **a.** Technician A
   - **b.** Technician B
   - **c.** Both A and B
   - **d.** Neither A nor B

2. An eight-cylinder engine has a firing order of 1-8-4-3-6-5-7-2. Technician A says that if cylinder #2 is at the beginning of its power stroke, cylinder #3 is starting its intake stroke. Technician B says that if cylinder #3 is at the beginning of its power stroke, cylinder #2 is starting its intake stroke. Who is right?
   - **a.** Technician A
   - **b.** Technician B
   - **c.** Both A and B
   - **d.** Neither A nor B

3. What valve arrangement is found in modern passenger cars?
   - **a.** The L-head
   - **b.** The I-head
   - **c.** Both A and B
   - **d.** Neither A nor B

4. All of the following statements are true *except*:
   - **a.** The camshaft in an OHC engine is located in the cylinder head.
   - **b.** Some OHC engines have one camshaft only.
   - **c.** Some OHC engines have two camshafts.
   - **d.** The camshaft in a pushrod engine is located on the cylinder head.

5. Which of the following is known as a turbulent combustion chamber?
   - **a.** The hemispherical combustion chamber
   - **b.** The wedge combustion chamber
   - **c.** Both A and B
   - **d.** Neither A nor B

6. Which of the following engines is likely to be the heaviest?
   - **a.** An in-line six cylinder
   - **b.** A V6
   - **c.** An in-line four cylinder
   - **d.** An opposed four cylinder

7. How many camshafts does a V8 DOHC engine have?

   a. One   **c.** Three

   b. Two   **d.** Four

8. Which of the following describes pistons for companion cylinders?

   a. Two pistons that are next to each other in the block

   b. Two pistons that are both at TDC or BDC at the same time

   c. Two pistons that are the same size

   d. All pistons in one bank of a V-type engine

9. Most automotive hybrids are _____ hybrids.

   a. Series   **c.** Fuel cell

   b. Parallel   **d.** Plug-in electric

10. All of the following are true about hybrid vehicles *except*:

   a. The starter cranks at a higher speed than a normal starter.

   b. The engine is shut off at higher speeds.

   c. The battery pack is recharged as the vehicle slows down.

   d. Hybrid vehicles are more expensive than ordinary internal combustion engine vehicles.

CHAPTER 17

Engine Size and Measurements

■ OBJECTIVES

Upon completion of this chapter, you should be able to:

✔ Describe various ways of measuring engine size.

✔ Understand the effects of engine compression ratio.

✔ Explain the principles of engine power and efficiency.

✔ Relate torque to horsepower.

✔ Understand the operation of various types of dynamometers.

■ KEY TERMS

bore	force	Btu
piston stroke	work	horsepower
crank throw	foot-pound	dynamometer
oversquare	watt	prony brake
undersquare	joule	road horsepower
displacement	energy	mechanical efficiency
engine displacement	inertia	thermal efficiency
swept volume	momentum	BMEP
compression pressure	power	

■ INTRODUCTION

This chapter provides an understanding of various engine size and performance measurements. Methods of understanding and measuring engine power output are also discussed.

■ ENGINE SIZE MEASUREMENTS

Engines have different numbers of cylinders and different sizes of pistons and cylinders. An engine's size is determined by the volume of air that its pistons displace in the cylinders. You will need to know an engine's size to be able to order parts for it. You will also need to know an engine's size to be able to look up wear specifications for it. A sample general engine specifications chart is shown in **Figure 17.1.**

Cylinder Bore and Stroke

Cylinder **bores** in automobile engines usually range from about 3.5" to 4". The bore size is the diameter of the cylinder and is measured across the cylinder (**Figure 17.2**).

The **piston stroke** is the distance that the piston moves from TDC to BDC (see Figure 17.2). The stroke is controlled by the length of the throw of the rod journal (**Figure 17.3**). The length of the **crank throw** is one-half of the total stroke.

The average engine has a stroke of from 3.5" to 4". In the service manual, bore and stroke are listed together. The bore is the first measurement and the stroke comes last. For instance, 4.000" × 3.500" would mean a 4" bore and 3½" stroke (see Figure 17.1).

When an engine has a cylinder bore that is larger than its stroke, this is called an **oversquare** engine. When the bore is less than the stroke, it is an **undersquare** engine.

Characteristics of oversquare engines (short stroke) include:

■ Faster revving

■ Lack low-speed torque

Specifications

GENERAL ENGINE SPECIFICATIONS

| Year | Engine | | Fuel System | Bore & Stroke, Inches | Comp. Ratio | Net HP @ RPM | Maximum Torque, Ft. Lbs. @ RPM | Normal Oil Pressure, psi @ 3000 RPM |
	Liter	VIN Code①						
1998–2000	2.0L	C	SMPI	3.45 x 3.27	9.8	132 @ 6000	129 @ 5000	25–80
	2.4L	X	SMPI	3.44 x 3.98	9.4	150 @ 5200	167 @ 4000	25–80
	2.5L	H	SMPI	3.29 x 2.99	9.4	168 @ 5800	170 @ 4350	35–75
2001–02	2.4L	X	SMPI	3.44 x 3.98	9.5	150 @ 5200	167 @ 4000	25–80
	2.7L	U	SMPI	3.386 x 3.091	9.7	200 @ 5800	190 @ 4850	45–105

SMPI — Sequential Multi-Port Injection

① — Eighth digit of Vehicle Identification Number (VIN) denotes engine code.

Figure 17.1 A service manual specifications chart. *(Published by Motor Information Systems, a division of Hearst Business Publishing, Inc. A unit of the Hearst Corporation)*

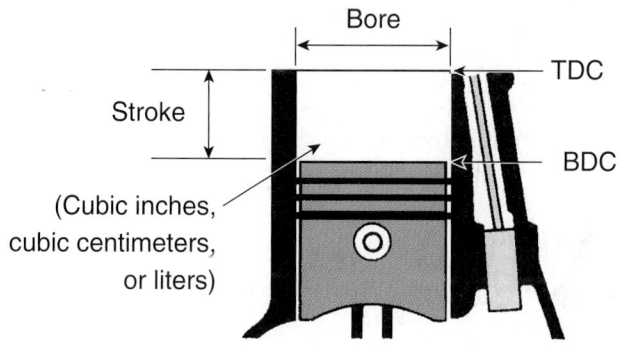

Bore² x stroke x 0.7854 x number of cylinders

Figure 17.2 Cylinder terms.

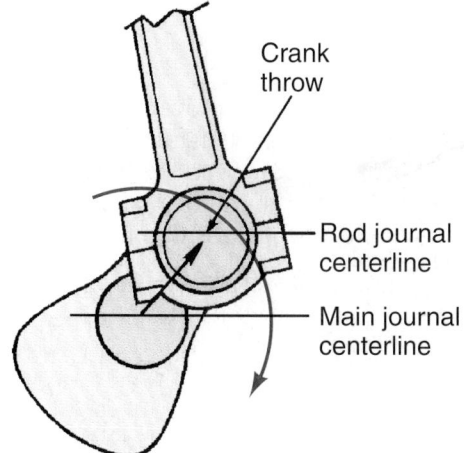

Figure 17.3 Crank throw.

Characteristics of undersquare engines (long stroke):
- Slower revving
- Good low-speed torque (make good engines for work trucks)

Displacement

Piston or cylinder **displacement** refers to the volume that the piston displaces in the cylinder as it moves from TDC to BDC (see Figure 17.2). It is measured in cubic inches, cubic centimeters, or liters.

The easiest formula to use to determine a cylinder's displacement is

$$\text{Bore}^2 \times \text{Stroke} \times 0.7854$$

Engine displacement is determined by multiplying the cylinder displacement by the number of cylinders.

The engine's total displacement is determined by multiplying

$$\text{Bore}^2 \times \text{Stroke} \times 0.7854 \times \text{number of cylinders}$$

The larger the displacement, the larger the engine.

SHOP TIP Appendix D is a chart showing the displacements of various engines. As an experiment, add 0.062" to the bore size of your engine. This is approximately the amount that the displacement of your engine would increase by if the cylinders were rebored to maximum oversize for new pistons.

In late-model vehicles, volume is described in liters or cubic centimeters (converting between metric system measurements and English system measurements is covered in Chapter 6).

Another means of determining displacement is to multiply the **swept volume** of the cylinder by the number of cylinders. Swept volume is determined by multiplying

$$\frac{\text{Bore}}{2} \times \frac{\text{Bore}}{2} \times \text{Stroke} \times \pi = \text{Swept Volume}$$

or

$$\frac{(\pi \times B^2 \times \text{stroke}) \times \#\text{cyl}}{4}$$

■ COMPRESSION RATIO

The compression ratio determines how much the air and fuel are compressed on the compression stroke. As the piston moves from BDC to TDC in a gasoline engine, the mixture is compressed to about ⅛ of the volume it occupied when the piston was at BDC. In this case, the *compression ratio* is said to be 8:1 (**Figure 17.4**). This means the maximum volume of the cylinder and combustion chamber is eight times the combustion chamber's volume. If the mixture was compressed to 1/12 its original volume, the compression ratio would then be 12:1. In a diesel engine, the compression ratio is much higher (from about 17:1 to over 20:1).

Gasoline engines of the "muscle car era" (approximately 1958–1970) had higher compression ratios, sometimes as high as 12:1. A higher compression ratio can increase an engine's power and fuel economy. Older cars with high-compression engines used leaded gasoline that had a higher octane rating. Higher compression engines also produce more exhaust emissions, so compression ratios have been lower for many years (about 8:1 to 9:1). The compression ratio on newer engines is trending higher due to improved computer programming.

Compression Ratio and Engine Power

During combustion the potential energy of the air-fuel mixture is turned into thermal (heat) energy and kinetic energy. Compression ratio has an effect on the amount of power an engine can produce by increasing the thermal efficiency of the engine. Squeezing the mixture into a smaller space results in higher combustion pressure and more expansion of the mixture throughout the power stroke. More of the heat energy of the fuel is converted to work, as less heat is allowed to escape from the engine.

Compression Pressure

Compression pressure is the amount of pressure made by the piston moving up in the cylinder in pounds per square inch (psi) or kilopascals (kPa). Gasoline engines typically produce compression pressure of

> **MATH NOTE**
>
> *Calculating an Engine's Compression Ratio (assuming piston deck height is zero)*
>
> *To calculate an engine's compression ratio, add together:*
> - *The volume of the cylinder*
> - *Piston volume*
> - *Compressed head gasket volume*
> - *Combustion chamber volume (clearance volume)*
>   *Total #1_____*
>
> *The sum of these is divided by the sum of:*
> - *Cylinder volume*
> - *Compressed head gasket volume*
> - *Combustion chamber volume*
>   *Total #2_____*
>
> *Total #1 divided by Total #2 = Compression Ratio*

125–175 psi. The pressure is measured by installing a compression tester into the spark plug hole and cranking the engine through several revolutions (see Chapter 48). If compression pressure is low, a burned valve, a broken piston ring, or a blown head gasket could be causing a leak.

■ PHYSICAL PRINCIPLES OF WORK

Force

Force is measured in pounds or Newtons. A force can be applied in the form of a push, pull, or lift. It is defined as any action that changes, or tends to change, the position of something.

Work

Work is when an object is moved against a resistance or opposing force. The movement can be either lifting or sliding (**Figure 17.5**). Work is measured in **foot-pounds** or **watts**. In the metric system, work is measured in *Newton-meters*, or **joules**. The formula for work is

Force × Distance = Work

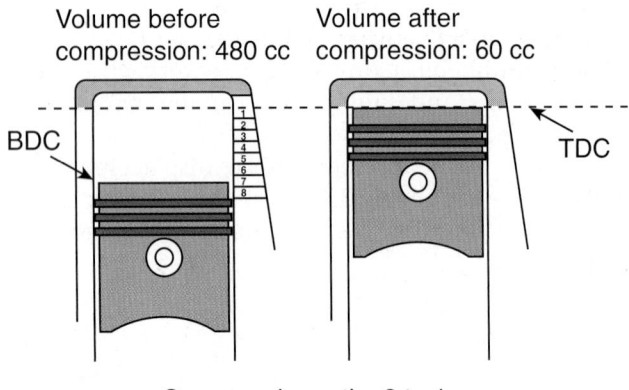

Compression ratio: 8 to 1

Figure 17.4 Compression ratio is a comparison of the volume of the air space above the piston at BDC and TDC. In this example, the compression ratio is 8:1.

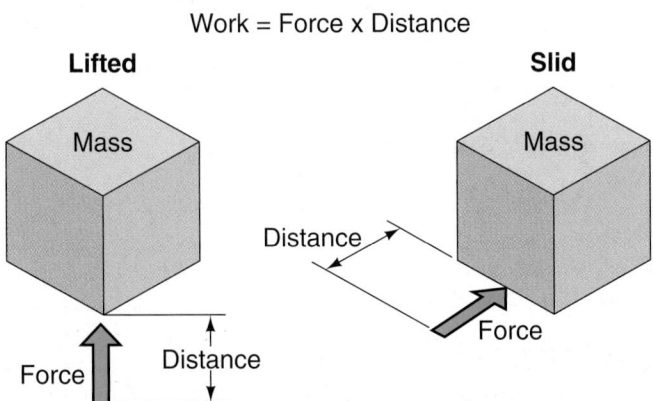

Figure 17.5 When work is performed, a mass is slid or lifted a certain distance.

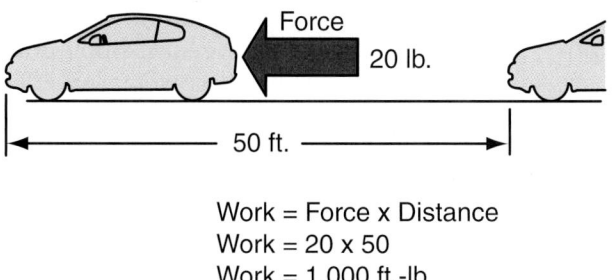

Work = Force x Distance
Work = 20 x 50
Work = 1,000 ft.-lb

Figure 17.6 Work measured in foot-pounds.

One *foot-pound* is when 1 pound is moved for a distance of 1 foot. For example:
■ Moving a 20-pound weight 50 feet results in 1,000 foot-pounds of work (**Figure 17.6**).
■ Lifting a 10-Newton weight a distance of 1 meter results in 10 Newton-meters of work.

The four-stroke cycle can be used to illustrate work:
■ Intake stroke—the air-fuel mixture has work done on it to get it into the cylinder.
■ Compression stroke—work is done as the mixture is compressed.
■ Power stroke—the expanding air-fuel mixture works on the piston and crankshaft.
■ Exhaust stroke—work is performed as the exhaust gas is expelled from the engine.

Energy

Energy is the ability to do work, or the ability to produce a motion against a resistance. When a weight is lifted, energy is stored in it. Dropping it performs work.

Inertia

Inertia is the tendency of a body to keep its state of rest or motion. The larger the mass, the more it is affected by inertia. Inertia and energy are stored in a flywheel.

Momentum

When a body is in motion, it has **momentum**. Momentum is a product of a body's mass and speed. A body going in a straight line will keep going the same direction at the same speed if no other forces act on it.

Power

Power is how fast work is done or how fast motion is produced against a resistance.

■■ TORQUE

Torque is a turning motion. It is the ability to make power, and it is defined as the tendency of force to rotate a body on which it acts. Tightening a bolt is a use of torque (**Figure 17.7**). Torque in an engine is the amount of turning force exerted by the crankshaft (**Figure 17.8**).

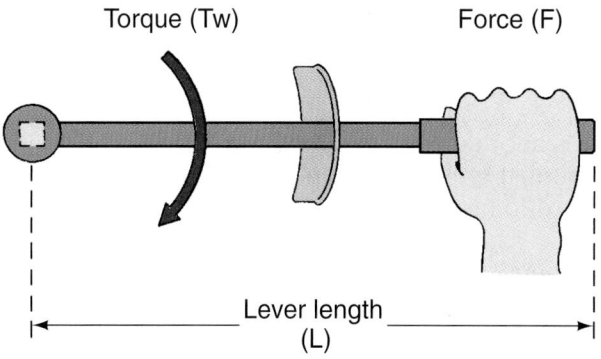

Torque (Tw) = L x F

Figure 17.7 Tightening a bolt is a use of torque. *(Reproduced by permission of Deere & Company, John Deere Publishing, Moline, IL. All rights reserved.)*

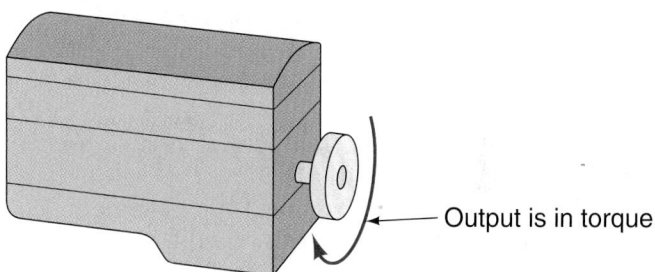

Output is in torque

Figure 17.8 Torque is the amount of turning force exerted by the crankshaft.

NOTE: *Although the measurement of torque is commonly referred to in foot-pounds, it should actually be expressed as pounds-feet to distinguish it from work.*

> **■■ SCIENCE NOTE ■■**
>
> *The definitions for torque and work are similar; both involve a force being multiplied by a distance. However, torque and work are very different quantities. Work is force times the distance moved (see Figure 17.6), and torque is force times leverage, the distance from a pivot point to the applied force (see Figure 17.7). When distinguishing work from torque, the metric system (SI) preferred unit of torque is the Newton-meter (N•m) and the metric work unit is the joule (J). The English imperial system expresses torque as pounds-feet, whereas work is expressed as foot-pounds.*

Engine torque varies with rpm. A force of 1 pound exerted at a distance of 1 foot from the center of a crankshaft results in 1 foot-pound of torque. The pulling ability of a car from a standing start depends on its engine's torque. This means that torque should be high at lower speeds.

To convert a torque reading to Newton-meters: multiply ft.-lb × 1.356.

To convert a torque reading to ft.-lb: multiply Newton-meters × 0.737.

Heat

Heat is another form of energy. It is measured in British thermal units (**Btu**). One Btu is the amount of heat required to heat 1 pound of water by 1°F.

One *joule* is an equivalent value that compares heat energy (Btu) to mechanical energy (ft.-lb.).

$$1 \text{ Btu} = 778 \text{ ft.-lb}$$

■ HORSEPOWER

Horsepower is the measurement of an engine's ability to perform work in a specified time. One horsepower equals 33,000 foot-pounds of work per minute. When measuring 1 horsepower of work, it is the amount of work required to lift 550 pounds 1 foot in 1 second (**Figure 17.9**). In Figure 17.1 notice that the torque and horsepower readings are given at a specific rpm.

In the metric system, horsepower is measured in *watts*. One watt is the power to move 1 Newton-meter per second. Because this is so small a measurement, *kilowatts* (kw) are used. One horsepower equals 0.746 kw.

Horsepower is a measure of work performed in a straight line in a specified time. Torque measures force in a rotating direction.

■ SCIENCE NOTE ■

In 1984, the United States used 84 × 10^{15} kilojoules of energy. The average person in the United States used 12 kilojoules of energy per second that year. Since 1 kilojoule per second is equivalent to 1 kilowatt, the energy consumption per person in the United States was equivalent to keeping 120 100-watt light bulbs burning continuously for the entire year.

Power produced at the crankshaft is called *gross horsepower*. Accessories that rob power include the alternator (charging system), air conditioning, coolant pump, cooling fan, power steering, and smog pump. These absorb about 25% of the power available at the crankshaft. The power that remains for use is called *net horsepower*. Power is also lost through friction in the driveline (transmission and differential), and due to wind resistance, vehicle weight, tires, and weather.

There are several measurements of engine horsepower:

■ *Brake horsepower* (*bhp*) is the usable horsepower at the crankshaft.

■ *Indicated horsepower* (*ihp*) is the amount of pressure made in the combustion chambers. It is measured with special instruments and varies throughout the four-stroke cycle. This is a theoretical measurement, which does not consider friction losses.

■ *Frictional horsepower* (*fhp*) is the power lost due to friction. It is the difference between brake horsepower and indicated horsepower (**Figure 17.10**). It considers the power needed to compress the air-fuel mixture and friction between engine parts such as the piston rings and cylinder walls.

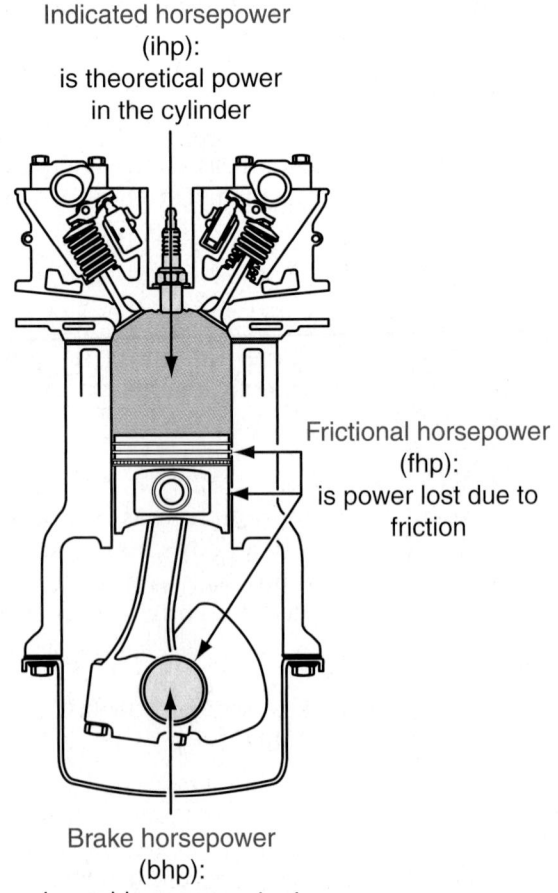

Indicated horsepower (ihp): is theoretical power in the cylinder

Frictional horsepower (fhp): is power lost due to friction

Brake horsepower (bhp): is usable power output

Figure 17.10 Frictional horsepower is the difference between brake horsepower and indicated horsepower.

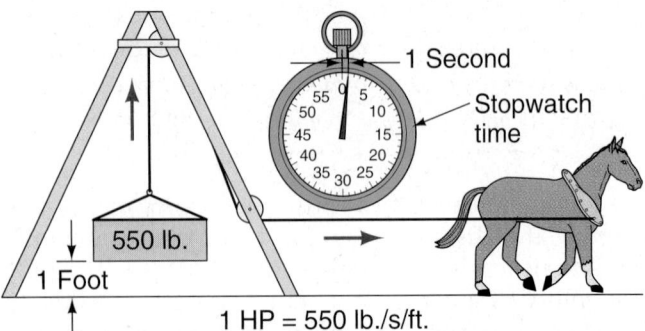

1 Second

Stopwatch time

550 lb.

1 Foot

1 HP = 550 lb./s/ft.

Figure 17.9 One horsepower is the amount of work required to lift 550 pounds 1 foot in 1 second.

- *Net horsepower* is the maximum power available from the engine when all the accessories are turned on.
- *Gross horsepower* (*ghp*) is the power available with only the water pump and alternator using power.

■ DYNAMOMETER

An engine's output can be measured using a **dynamometer**, commonly called a *dyno*. The engine must be loaded (braked) to measure the torque it can produce. Depending on the type of dyno, braking can be done by electricity, hydraulics, or friction.

A simple dynamometer that uses friction is called a **prony brake** (**Figure 17.11**). An arm pushes on a scale to provide a reading in pounds. When the length of the arm is known, the measurement can be converted to foot-pounds or Newton-meters.

Engine Dynamometer

An engine dyno measures horsepower coming out of the engine. The horsepower measured is called brake horsepower because the dyno acts as a brake on the engine's crankshaft.

Chassis Dynamometer

A chassis dyno (**Figure 17.12**) measures horsepower available at the car's drive wheels. This is called **road horsepower** (**Figure 17.13**). It is always less than brake horsepower because of friction losses through the driveline.

NOTE: *Chassis dynamometers are used in emission control testing programs because testing of oxides of nitrogen can only be done when the engine is under load. This is called loaded mode testing. Emission testing is covered in Chapter 44.*

On a chassis dynamometer, one roller is attached to the power absorption unit (covered later) and the other is an idle roller. The idle roller has the speedometer pickup hooked to it.

Measuring Torque and Horsepower

The engine is a big air pump. After the engine starts to run, both the intake air and exhaust air have momentum.

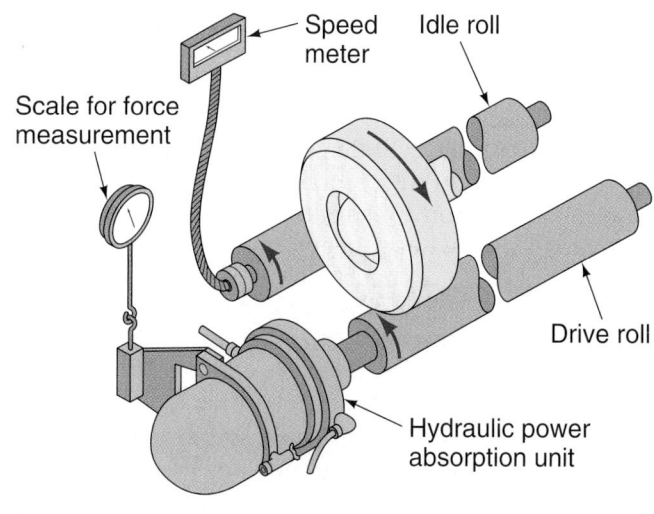

(a)

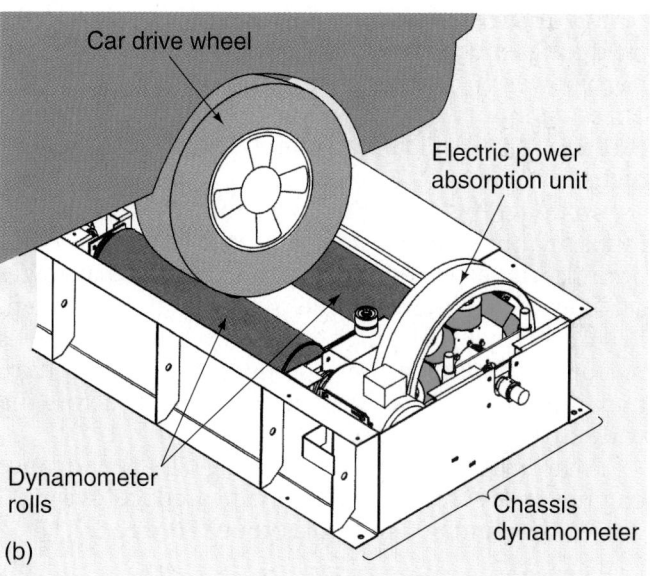

(b)

Figure 17.12 The rollers on a chassis dynamometer run off the car's drive wheels a) a dyno with a hydraulic absorption unit, b) a dyno with an electric absorption unit. (*Courtesy of Clayton Industries*)

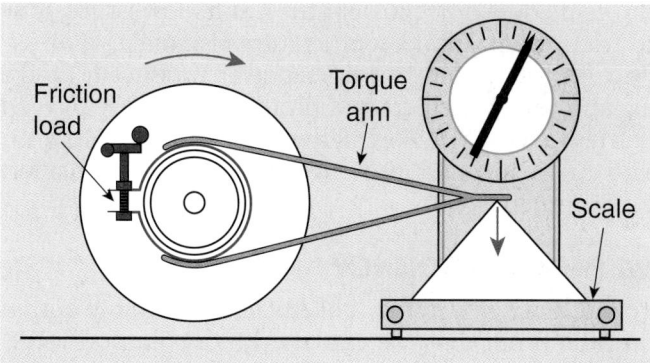

Figure 17.11 A prony brake.

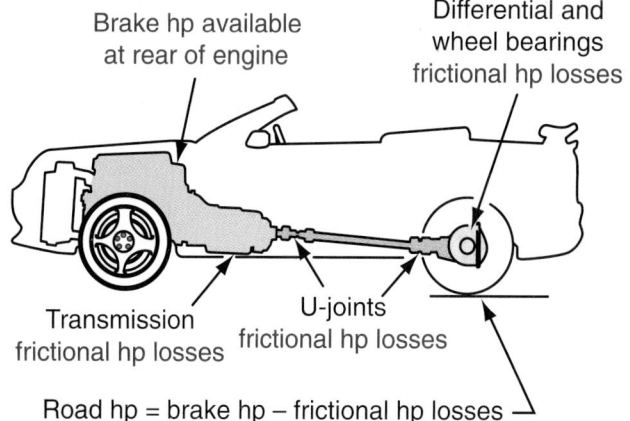

Road hp = brake hp − frictional hp losses

Figure 17.13 Road horsepower is always less than brake horsepower because of friction losses through the driveline.

After the piston completes its downward movement on the intake stroke, the intake valve remains open for a period of time into the compression stroke (see Figure 18.47). Momentum of the incoming air-fuel mixture continues to fill the cylinder, even after the piston starts moving upward on the compression stroke.

After the piston finishes its upward movement on the exhaust stroke and begins moving down on the intake stroke, the exhaust valve remains open for a period of time. This period, when both the intake and exhaust valves are open, is called valve overlap (see Figure 18.48).

At higher speeds, the momentum of the exhaust gas moving through the exhaust system continues to pull exhaust from the cylinder. The vacuum created behind the moving exhaust stream also helps draw in fresh air and fuel through the open intake valve. This is called the *scavenging effect*.

An engine that does not breathe well will not produce as much power. Engines produce torque and horsepower in different amounts as rpm changes. These amounts are listed on a performance chart (**Figure 17.14**). Fuel consumption can also be predicted.

In designing an engine, selection of the camshaft, intake manifold, exhaust system, and valves is made to optimize horsepower and torque curves. A passenger car engine has more torque available at lower rpm. Racing engines make high horsepower at high speeds but operate poorly at low speeds. The idea is to have maximum torque and horsepower available across the widest rpm band of use.

Many manufacturers are producing high-performance engines today, some of which can produce over 500 horsepower and 400 lb-ft. of torque (**Figure 17.15**).

Dynamometer Operation

Some dynos have computers that plot horsepower and torque curves. To make an engine perform work

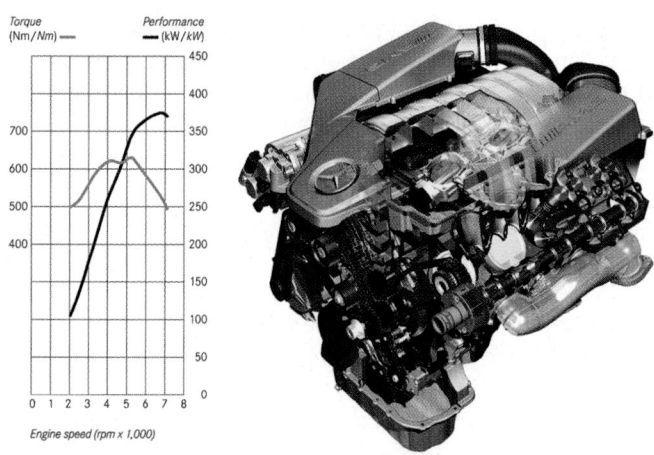

Figure 17.15 One of the world's fastest naturally aspirated engines is the Mercedes amg 6.3L (385 cu. in.) V8. It has 510 HP (375 kw) at 6,750 rpm and 461 lb.-ft (625 N•m) of torque at 5,200 rpm. (© *Courtesy of Mercedes-Benz USA, LLC*)

during a dyno test, the engine is put under load using a power absorption device. Most automotive dynamometer absorption units are fluid types controlled by the amount of water that enters the device.

The power absorption unit is a fluid coupling consisting of two members: a turbine and a stator (see Figure 73.5). The turbine tries to move the water, but the stator prevents it from moving. The load unit is like a torque wrench that measures the load applied. The amount of load put on the engine and the amount of torque it produces are used to calibrate horsepower.

The formula for horsepower is

$$\frac{\text{Torque} \times \text{rpm}}{5,250} = \text{HP}$$

An engine tested at 2,625 rpm that develops 500 foot-pounds of torque produces 250 horsepower.

$$\text{Torque } (500) \times \text{rpm } (2,625) = 1,312,500$$

$$\frac{1,312,500}{5,250} = \text{HP}(250)$$

Torque readings are made at every 500 rpm. An engine that is warmed to operating temperature produces its best horsepower and has its lowest loss due to friction. Air intake temperature also makes a difference in the amount of horsepower produced; intake air at a lower temperature produces a higher oxygen content, which makes better combustion. More air and fuel can enter the engine in a colder (smaller) environment.

■ ENGINE EFFICIENCY

To rate an engine in terms of efficiency, both the output and the input must be expressed in a common value. There are three types of engine efficiency measurements: **mechanical efficiency**, *volumetric efficiency*,

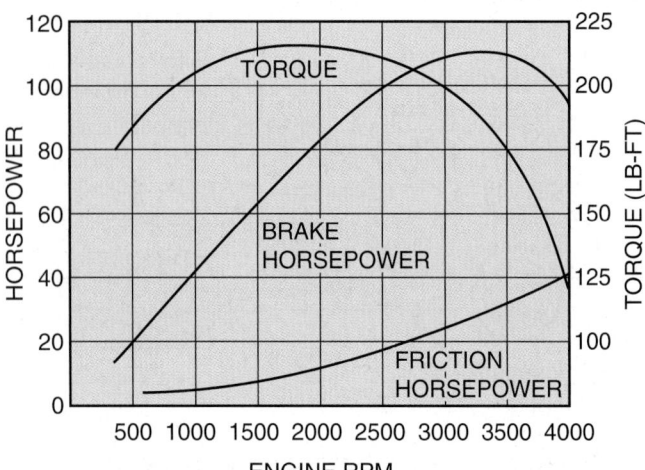

Figure 17.14 A performance chart lists torque and horsepower for an engine at a given rpm.

and **thermal efficiency**. An efficiency measurement is a value less than 100%. The difference between the efficiency measurement and 100% is the amount of loss.

Mechanical efficiency describes all of the ways friction is lost in an engine (**Figure 17.16**). Horsepower is a value that can be used to compare the mechanical efficiency of two engines. Brake horsepower (bhp) divided by the indicated horsepower (ihp) gives the mechanical efficiency of the engine.

The formula is

$$\text{Mechanical efficiency} = \frac{\text{bhp (engine output)}}{\text{ihp (engine input)}}$$

If an engine had 100 indicated horsepower and 80 brake horsepower, it would have a mechanical efficiency of 80%.

$$\text{Mechanical efficiency} = \frac{80 \text{ bhp}}{100 \text{ ihp}} = 80\%$$

Volumetric Efficiency

The measurement comparing the volume of airflow actually entering the engine with the maximum that theoretically could enter it (this is the same as the displacement) is called volumetric efficiency (V.E.) (**Figure 17.17**). Volumetric efficiency determines the engine's maximum torque output.

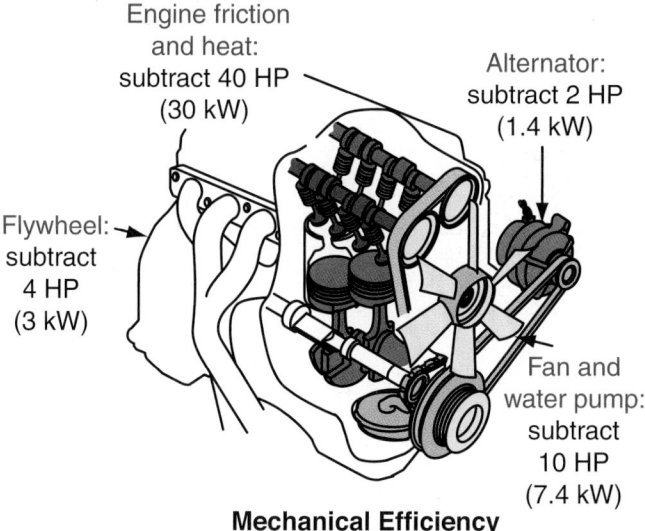

Engine friction and heat: subtract 40 HP (30 kW)

Alternator: subtract 2 HP (1.4 kW)

Flywheel: subtract 4 HP (3 kW)

Fan and water pump: subtract 10 HP (7.4 kW)

Mechanical Efficiency

1. Indicated horsepower (ihp) = 200 HP (150 kW)

2. Flywheel or brake horsepower (bhp) =
$$\begin{array}{r} 40 \\ 10 \\ 4 \\ +2 \\ \hline 56 \end{array}$$
$$200 - 56 = 144$$

3. Mechanical efficiency: 144/200 x 100 = 72%

Figure 17.16 Mechanical efficiency.

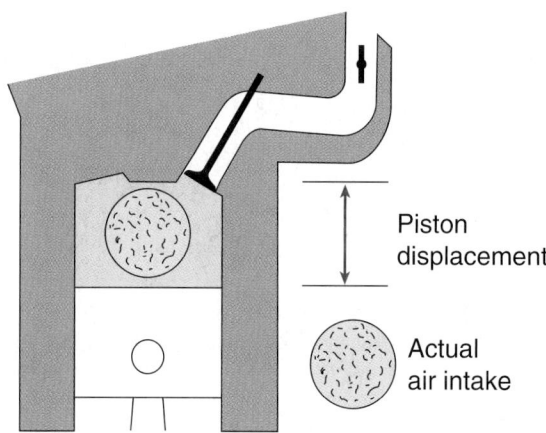

Piston displacement

Actual air intake

Figure 17.17 Volumetric efficiency.

The rpm at which the engine does its best breathing usually determines its maximum torque. Remember that the engine is like a big air pump. Actually, a pump uses energy to compress air; an internal combustion engine gets its power from expanding air. Heat is what makes the air expand.

V.E. changes with temperature, engine speed, load, and throttle opening. For instance, at 2,000 rpm V.E. may be 85%, while at 4,000 rpm it may be only 60%. At lower engine speeds, the engine has enough time to fill with air at atmospheric pressure. With increases in speed, there is less time for the air to move through the intake and exhaust systems. This results in decreased volumetric efficiency. Closing the throttle also causes a restriction resulting in lowered V.E.

Thermal Efficiency

Thermal efficiency is the ratio of how effectively an engine converts a fuel's heat energy into usable work. Each fuel has a certain amount of heat or thermal energy. Gasoline's thermal energy varies between fuels, but its average is about 19,000 Btu per pound. The energy of the fuel has the potential to produce a certain amount of work when burned in the engine. This thermal efficiency is a theoretical value.

A more useful measurement of thermal efficiency is called *brake thermal efficiency*. This is the brake horsepower converted to Btu, divided by the fuel's heat input in Btu with the result multiplied by 100. **Figure 17.18** shows how the formula works in a typical engine.

In a spark ignition engine, only about ¼ of the energy from the burning of the fuel is converted to work at the crankshaft. The remainder of the energy is wasted as heat; part of it goes out the exhaust or is lost to the air. The other part is carried off by the cooling system (**Figure 17.19**). If the thermal efficiency of a gasoline engine was doubled, its fuel economy would also double.

Diesel fuel has more heat energy than gasoline (19,000–20,000 Btu/lb.). Diesel engines also have higher compression ratios. Both of these are reasons why

1. ENGINE DEVELOPS 100 FLYWHEEL OR BRAKE H.P. (75 kW) PER HOUR
2. 2545 × 100 B.H.P. = 254,500 B.T.U. (75 kW)
3. FUEL BURNED PER HOUR = 800,000 B.T.U. (234 kW)
4. BRAKE THERMAL EFFICIENCY =

$$\frac{254,500}{800,000} \times 100 = 31.8\%$$

Figure 17.18 Brake thermal efficiency for a typical engine. *(Reproduced by permission of Deere and Company, John Deere Publishing, Moline, IL. All rights reserved)*

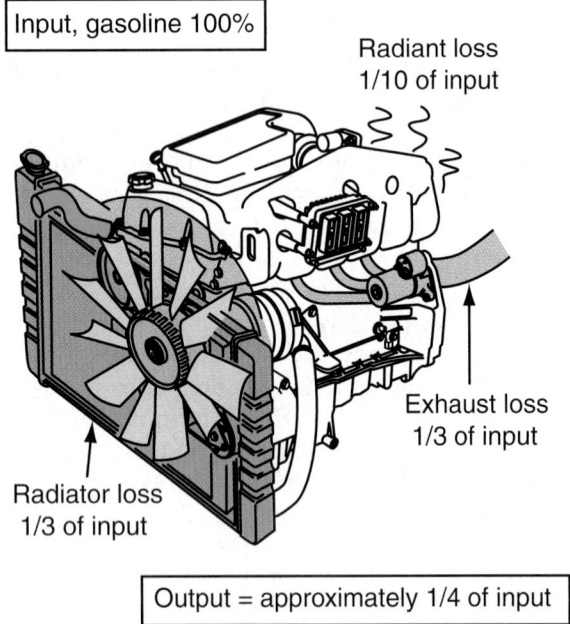

Input, gasoline 100%

Radiant loss 1/10 of input

Exhaust loss 1/3 of input

Radiator loss 1/3 of input

Output = approximately 1/4 of input

Figure 17.19 A gasoline engine loses the majority of its heat energy.

diesels get better fuel economy than gasoline engines. Passenger car diesel engines have lower power and weigh more than gasoline engines. Therefore, they do not perform as well. The overall efficiencies of different types of engines are shown in **Figure 17.20**.

EFFICIENCIES OF DIFFERENT ENGINES	
Gasoline Engine	25–28%
Diesel Engine	35–38%
Aircraft Gas Turbine	33–35%
Liquid Fuel Rocket	46–47%
Rotary Engine	20–22%
Steam Locomotive	10–12%

Figure 17.20 Overall efficiencies of different types of engines.

■ MEAN EFFECTIVE PRESSURE

The pressure within the cylinder increases during the compression stroke and becomes highest after ignition. Peak cylinder pressure should occur between 10 and 20 degrees past TDC. Compare this to the point a bicycle rider puts the most leverage on the pedal. As combustion pressure moves the piston down in the cylinder, the pressure drops as the cycle continues (**Figure 17.21**).

■ SCIENCE NOTE ■

There are two kinds of mean effective pressure. The average pressure within the cylinder is called indicated mean effective pressure (IMEP). *This calculation requires equipment usually found only in laboratories.*

Brake mean effective pressure (BMEP) is a term commonly heard in automobile racing. It is calculated from the horsepower reading on a dynamometer.

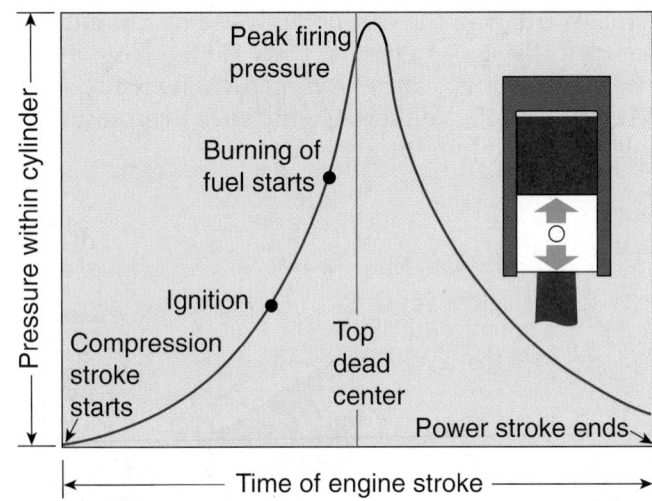

Pressure within cylinder

Peak firing pressure

Burning of fuel starts

Ignition

Compression stroke starts

Top dead center

Power stroke ends

Time of engine stroke

Figure 17.21 Pressure in the cylinder is called mean effective pressure.

REVIEW QUESTIONS

1. When an engine has a cylinder bore that is larger than its stroke, what is this called?

2. What is the name for the volume that the piston displaces in the cylinder as it moves between TDC and BDC?

3. If the mixture is compressed to about 1/10 of the volume it occupied when the piston was at BDC, what is the compression ratio?

4. About how much change in horsepower does each point of change in compression ratio result in?

5. What is typical gasoline engine compression pressure?

6. Moving a 10-pound weight 50 feet results in _____ foot-pounds of work.

7. What is the name for the amount of turning force exerted by the crankshaft?

8. What is the term for work, pounds-feet or foot-pounds?

9. What is the name of the piece of equipment used to measure an engine's output?

10. What is the name of the measurement comparing the volume of airflow actually entering the engine with the maximum that theoretically could enter it?

ASE-STYLE REVIEW QUESTIONS

1. All of the following is/are true about internal combustion engine torque output *except*:

   a. It is effected by engine rpm.

   b. It is used in the equation for horsepower.

   c. Short stroke engines have more torque than long stroke engines.

   d. It is a measurement of the amount of turning force exerted by the crankshaft.

2. Which of the following is/are true about the length of the crankshaft throw?

   a. It is half the total stroke.

   b. It is twice the total stroke.

   c. It is the same as the total stroke.

   d. None of the above.

3. Technician A says that a dynamometer absorption unit is like a torque wrench for measuring engine output. Technician B says that more air and fuel can enter the engine in warmer weather. Who is right?

   a. Technician A   **c.** Both A and B

   b. Technician B   **d.** Neither A nor B

4. Technician A says that a spark ignition engine converts about 75% of the energy from the burning of the fuel to work at the crankshaft. Technician B says that diesel fuel has more heat energy than gasoline. Who is right?

   a. Technician A   **c.** Both A and B

   b. Technician B   **d.** Neither A nor B

5. The amount of usable horsepower at the crankshaft is called:

   a. Indicated horsepower

   b. Frictional horsepower

   c. Brake horsepower

   d. None of the above

Engine Upper End

■ **OBJECTIVES**

Upon completion of this chapter, you should be able to:

✔ Identify all of the parts of the engine's upper end.

✔ Understand the difference between cylinder head designs.

✔ Understand the variations in camshaft design.

✔ Describe the different camshaft grinds and the uses they are best suited for.

✔ Identify the different cam drive arrangements.

✔ Describe the difference between freewheeling and non-freewheeling engines.

■ KEY TERMS

integral seat	production engine	backlash
induction hardened valve seat	base circle	freewheeling engine
	lift	interference engine
positive stop	duration	valve overlap
end thrust	zero-lash	

■ INTRODUCTION

In earlier chapters, you became familiar with basic parts of the engine and the different types of engines. This chapter goes into more detail about the *upper end* of the engine (**Figure 18.1**). The upper end includes:

■ Cylinder head or heads

■ Valvetrain (camshaft and cam drive)

Intake and exhaust manifolds, which are sometimes considered to be a part of the upper end, are covered in another chapter. In later chapters, you will learn to diagnose and repair engines. This chapter helps prepare you for these areas.

■ CYLINDER HEAD CONSTRUCTION

Cylinder heads are made of either cast iron or aluminum. A *bare head* is a head without any of its loose parts installed. The rest of the loose parts include the intake and exhaust valves, the keepers and retainers, the valve guide seals, the valve springs, and rocker arms. Parts of the cylinder head are shown in **Figure 18.2**.

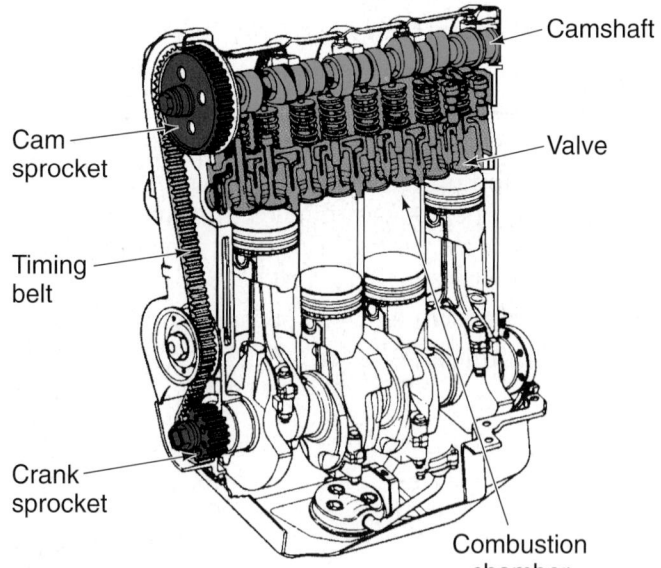

Figure 18.1 The upper end. *(Courtesy of DaimlerChrysler Corporation)*

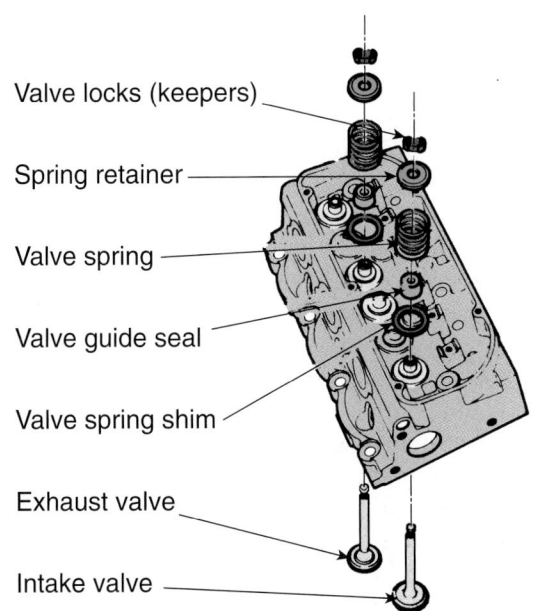

Figure 18.2 Parts of a cylinder head. *(Courtesy of Ford Motor Company)*

Valve locks (keepers)
Spring retainer
Valve spring
Valve guide seal
Valve spring shim
Exhaust valve
Intake valve

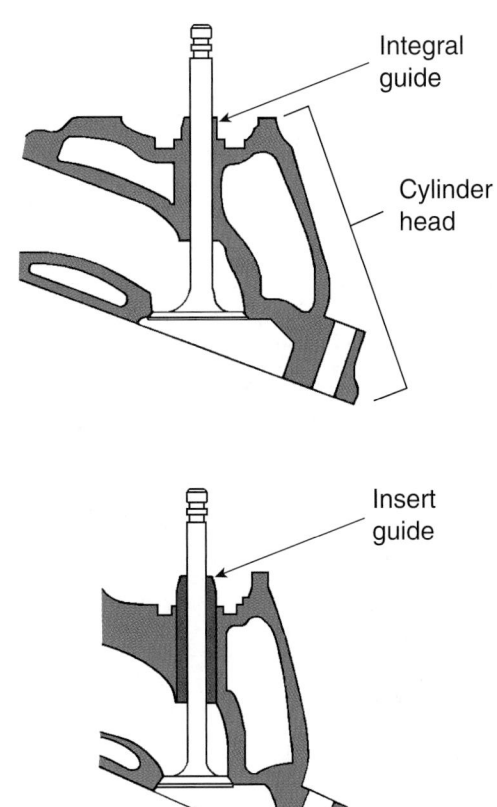

Integral guide
Cylinder head
Insert guide

Figure 18.3 Valve guides are integral or replaceable inserts. *(Courtesy of Federal-Mogul Corporation)*

Sometimes a head is cracked or unrepairable and the rest of the parts are good. A bare head would be purchased in this instance.

VALVE GUIDES

The valve guide is a hole in the cylinder head that the valve slides up and down in. There are two types of guides: *integral* and *replaceable insert* (**Figure 18.3**). An integral valve guide is a hole bored into an iron cylinder head. Integral guides lower production costs but cannot be used in aluminum heads.

Some iron heads and all aluminum heads have pressed-in guides, called replaceable inserts. Insert guides are made of iron or bronze alloy.

Valve guides become worn, which usually results in oil consumption when combined with a bad valve guide seal (**Figure 18.4**).

VALVE GUIDE SEALS

Leaking valve guide seals account for nearly half of oil consumption complaints. A faulty intake seal allows oil to be drawn into the cylinder during the intake stroke (**Figure 18.5**). This results in burnt oil exhaust smoke when the engine decelerates because the oil is *not* being diluted with gasoline during that time. Under normal driving conditions, the air-fuel mixture would mix with the oil that leaked in, making it less visible. Next time you are driving down a long grade, notice how many cars in front of you have exhaust smoke. These are the ones that are using oil through the valve guides.

Besides causing oil consumption, a leaking intake seal can result in carbon buildup on the stem of the valve (which interferes with engine breathing) (**Figure 18.6**). The carbon buildup interrupts the flow of incoming air

and fuel as it passes the irregular carbon surface. This causes turbulence in the airflow, which reduces volumetric efficiency (breathing ability) and decreases power.

Even though the exhaust guide is not exposed to engine vacuum on the intake stroke, it can also cause oil consumption. **Figure 18.7** shows how the exhaust

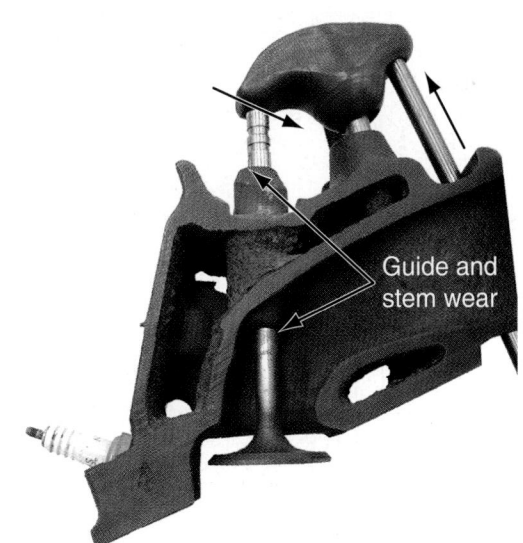

Guide and stem wear

Figure 18.4 The valve stem and guide wear on opposite sides at the top and bottom due to rocker arm movement. *(Courtesy of Tim Gilles)*

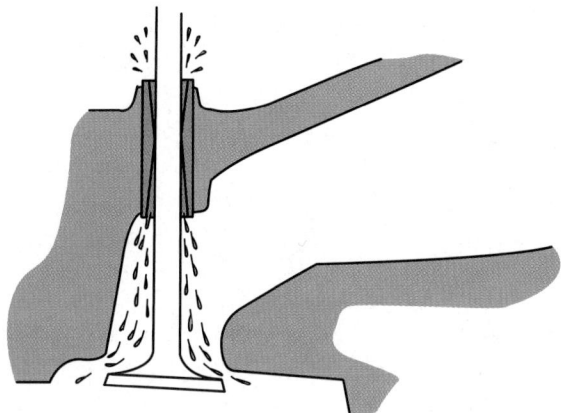

Figure 18.5 Bellmouth wear of the guide causes oil consumption.

Figure 18.6 Carbon buildup on the neck of the valve. *(Courtesy of Tim Gilles)*

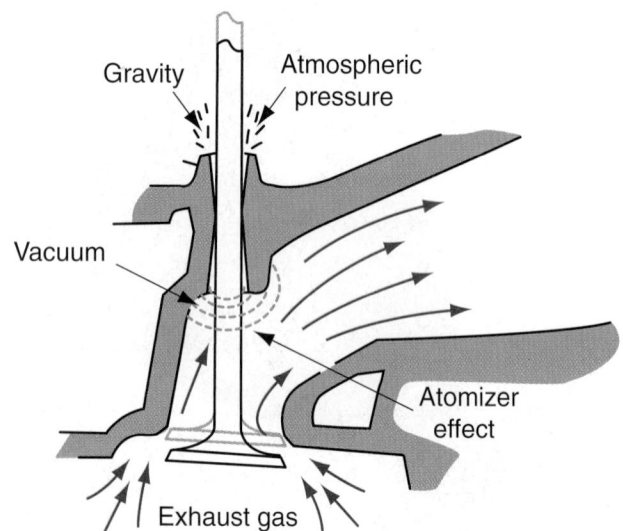

Gravity

Atmospheric pressure

Vacuum

Atomizer effect

Exhaust gas

Figure 18.7 The atomizer effect results in oil being drawn down the exhaust guide as exhaust gas moving past the bottom of the valve guide creates a vacuum.

gas moving past the exhaust valve causes a vacuum that pulls oil down the guide and into the hot exhaust. This is called the *atomizer effect.*

Valve seals come in three types: *umbrella, positive,* and *O-ring* (**Figure 18.8**). An umbrella seal fits inside the valve spring (see Figure 15.10). It fits the valve snugly and rides up and down with it, keeping oil from entering the valve guide. If an umbrella seal becomes brittle with age, it can become loose on the valve stem, which allows oil to leak down the valve guide.

Sometimes umbrella seals break up, floating in the oil until the pieces are picked up by the oil pump. This can result in engine damage when the oil pump locks up.

A positive valve seal fits on the top of the valve guide. The valve moves up and down inside of a positive valve guide seal. Positive seals are used on overhead cam engines because the camshaft is heavily lubricated right above the valves. Occasionally, positive seals are also used on pushrod engines.

An O-ring seal is the type used by General Motors. It fits under the keepers, inside of the spring retainer (**Figure 18.9**). Oil is pumped through the pushrod to a hole in the rocker arm and spills onto the retainer.

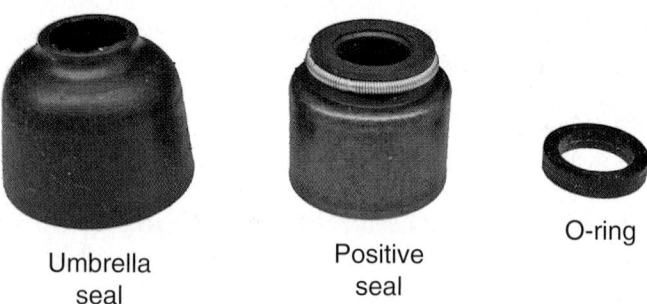

Umbrella seal

Positive seal

O-ring

Figure 18.8 Three types of valve guide seals. *(Courtesy of Tim Gilles)*

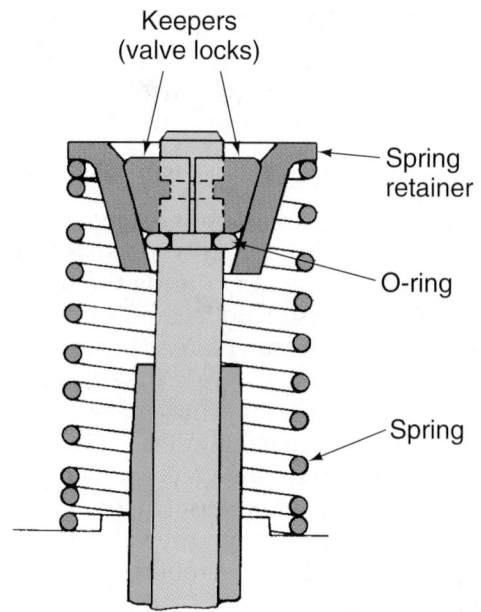

Keepers (valve locks)

Spring retainer

O-ring

Spring

Figure 18.9 An O-ring seal fits beneath the keepers. *(Courtesy of Federal-Mogul Corporation)*

Without the O-ring seal, oil would be able to leak down the stem and into the guide.

Valve Guide Seal Materials

Valve guide seals are made of several materials having different resistances to high temperatures. Although most kinds of artificial rubber look and feel alike, they are not. The premium seals resist heat up until about 440°F. Lower quality seals resist heat only to 250°F. The hottest area, at the combustion chamber end of the valve stem, runs at a temperature in excess of 1,300°F, so you can see that resistance to temperature is important.

Premium Viton and Teflon seals are used for the highest temperature applications. These materials are found only in positive valve guide seals. Umbrella seals are made of nitrile (used only on the intake side) or silicone, lower temperature materials.

Valve Seats

Valve seats are usually ground to a 45-degree angle, although some older heads have 30-degree seats. Like valve guides, seats can either be part of the head (integral) or replaceable inserts (**Figure 18.10**). **Integral seats** are only found in cast iron heads. *Replaceable seats* are found in original equipment aluminum heads and in iron heads that have been repaired by a machine shop. The inserts are made from iron, *stellite* (hard seats), or bronze.

Induction Hardening of Seats

Integral seats on heads manufactured since about 1970 are **induction hardened** to reduce **valve seat** wear. Induction hardening hardens only the seat area of the head. One advantage to integral seats is that the temperature they operate at is approximately 150°F cooler than replaceable seat inserts.

▄ VALVES

Automotive valves are *poppet valves*, which were first used in steam engines. They were designed to require little or no lubrication, and they operate well in high temperatures. To match the valve seat, the valve face angle is usually 45 degrees, although some valves (mostly intakes) have a 30-degree angle.

Parts of the valve are shown in **Figure 18.11**. The valve is machined on the face surface, the tip, and the stem. There is a groove, or multiple grooves, in the valve stem near the tip that *keepers* (**Figure 18.12**) fit into.

Valve Size

The intake valve is usually 35%–40% larger than the exhaust valve (**Figure 18.13**). Intake air and fuel take up quite a bit of space, requiring a larger valve opening.

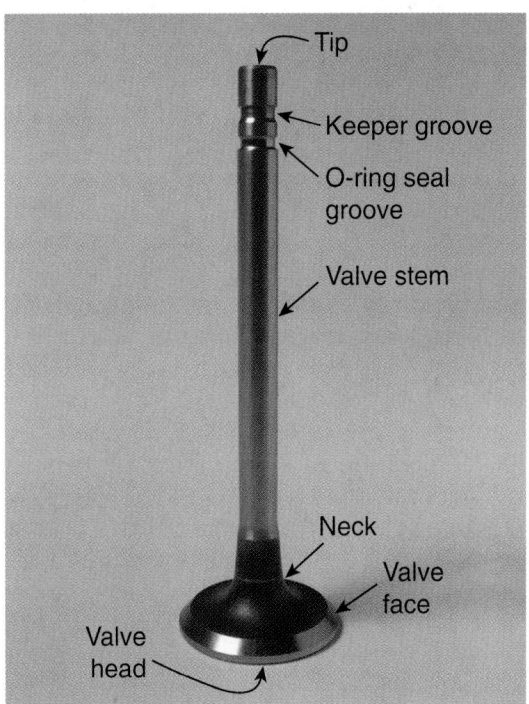

Figure 18.11 Parts of a poppet valve. *(Courtesy of Tim Gilles)*

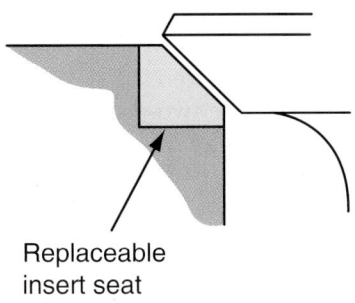

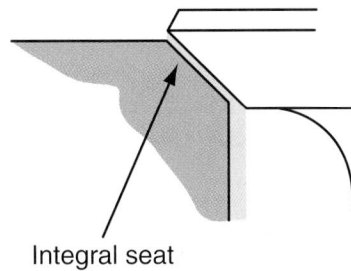

Figure 18.10 Integral and replaceable valve seats.

Figure 18.12 Valve locks are also called "keepers." *(Courtesy of Tim Gilles)*

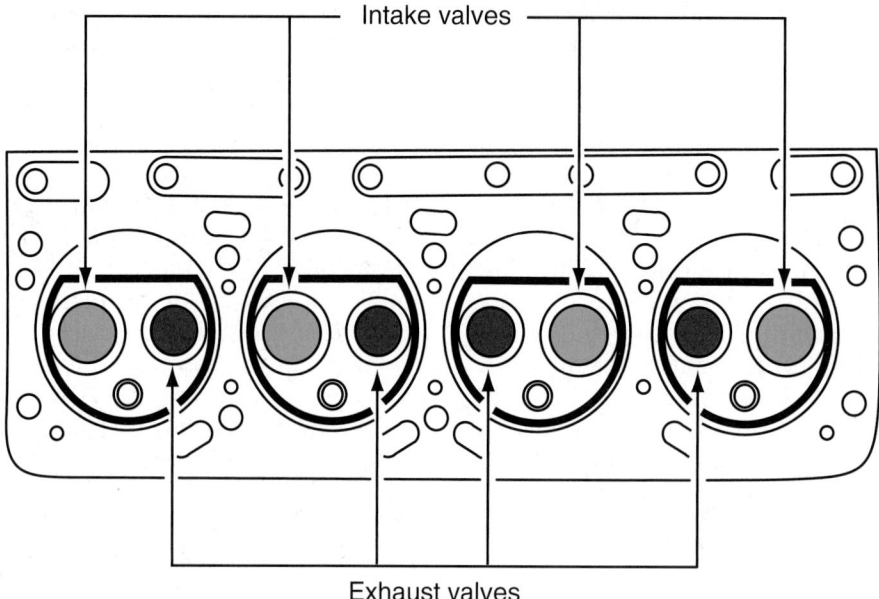

Figure 18.13 Intake valves are larger than exhaust valves.

Exhaust is under a great deal of pressure, so it does not require as large an opening. Exhaust gas can actually go through a smaller opening because it takes up less space after it is burned. Because it is under pressure, it requires an even smaller opening.

Valve Temperature

The *combustion chamber* (see Figure 18.1) is subjected to severe conditions during engine operation. Valves run at high temperatures, opening and closing at half engine speed. Exhaust seats run at about 800°F. The neck area of the exhaust valve sometimes runs at temperatures in excess of 1,300°F. This is above the melting temperature of aluminum, and it is red hot for these steel valves. **Figure 18.14** shows typical temperatures of a passenger car exhaust valve.

The intake valve runs at a temperature that is usually below 1,000°F. It is cooler than the exhaust valve because it is bathed in cooler intake air and fuel during every intake stroke. Water jackets are in the head behind the valve seat and around the valve guide (**Figure 18.15**). The heat in the valves is transmitted from the valve stem to the valve guide and then into the engine coolant. Valves depend on good contact between the guide and seat for proper cooling.

Valve Materials

Intake and exhaust valves are usually made of different materials. Exhaust valves are constructed of higher quality steel. Even though they are smaller than intake valves, they cost more money.

- The *austenitic stainless steel* that exhaust valves are constructed of is usually nonmagnetic.
- The carbon or alloy steel that intake valves are made of is magnetic.

▧ SCIENCE NOTE ▧

Iron is magnetic because it has unpaired bonding electrons. The interatomic distance between iron atoms is just right for these unpaired electrons to line up and reinforce each other. Stainless steel, an iron/chromium/nickel alloy is not magnetic even though both chromium and iron have unpaired electrons. The reason stainless steel is not magnetic is because the iron/chromium interatomic distances are not right for the unpaired electrons to line up and reinforce each other.

The valve stem is often of a different material than the valve head. The two materials are *spin-welded* together. Two-piece welded valves typically have a *forged stainless steel* head welded to a *high carbon steel* stem about ⅔ down from the stem tip.

Valve stems are often chrome plated, so the weld is not visible. The weld is so strong that failure of valves usually occurs at the neck area of the valve, not at the weld.

Valve Stem and Face Coatings

Valve stems have two common types of protective coating to prevent stem wear. *Chrome plating*, which reduces stem wear by up to 80% over 50,000 miles, is commonly done in North America. In Europe, the popular stem coating is *bath nitriding*, which leaves a black finish that is as effective as chrome.

Valve Burning

The *margin* of the valve gives an idea of how much of the valve remains after valve grinding. If the margin is too thin, it will heat up and the valve will burn

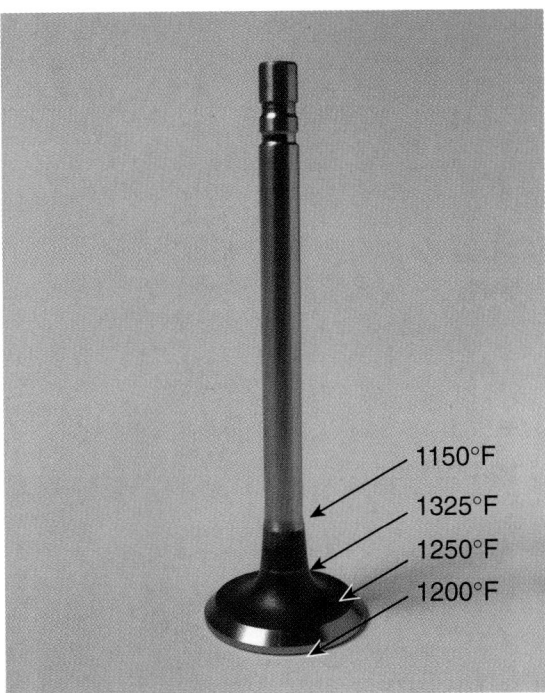

Figure 18.14 Typical temperatures of an automotive exhaust valve. (*Courtesy of Tim Gilles*)

Figure 18.15 Water jackets are behind the valve seat and around the valve stem. (*Courtesy of Tim Gilles*)

(**Figure 18.16**). A burned valve will cause a misfire at idle speed. Because it runs so much hotter, the exhaust valve is usually the one that burns (**Figure 18.17**).

SHOP TIP To test for a burned exhaust valve, hold a piece of paper over the exhaust pipe as the engine idles. You will see it suck into the pipe every time there is an intake stroke. This is because the air and fuel filling the cylinder draws on the exhaust stream through the opening (burned area) in the exhaust valve.

SCIENCE NOTE

Steel is either carbon steel or alloy steel. Ninety percent of all steel produced is carbon steel, which contains no other metals. Alloy steels contain carbon but also different metals used to give the steel different properties. Stainless steels contain chromium and nickel. Chromium improves hardness and resistance to corrosion, while nickel adds toughness. Other mineral additives such as vanadium and manganese provide the properties of springiness and wear resistance. Different rates of quenching also change the properties of steel. For flexibility, steel is cooled slowly. This results in large smooth steel crystals. If a good cutting edge was the desired property, the steel would be hardened by quenching rapidly, which causes small, jagged steel crystals.

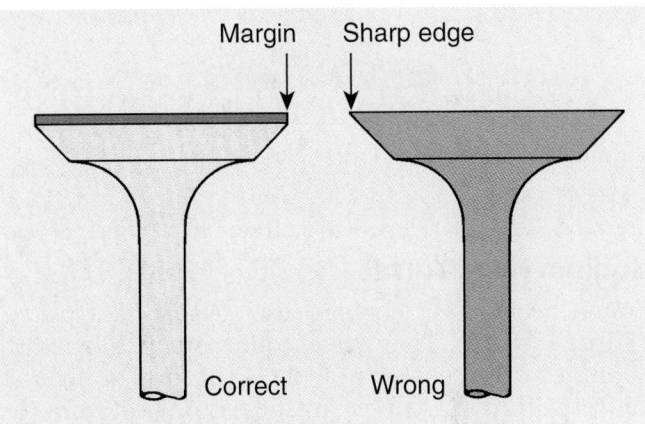

Figure 18.16 If the margin is too thin, the valve can burn. (*Courtesy of Federal-Mogul Corporation*)

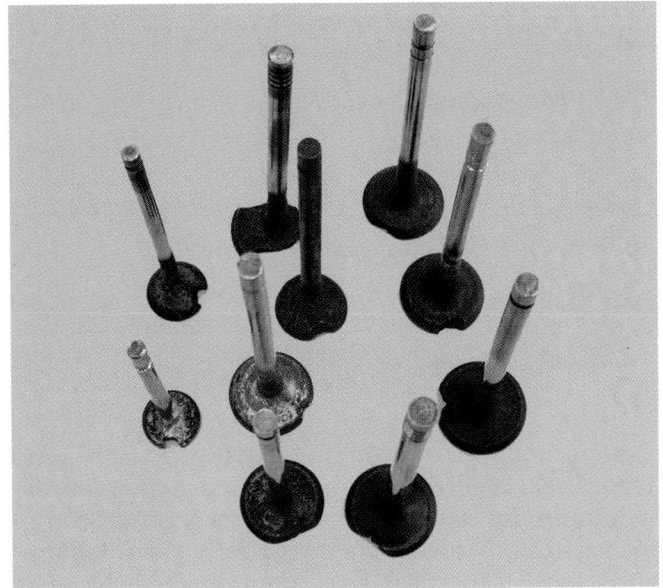

Figure 18.17 Severely burned valves. (*Courtesy of Tim Gilles*)

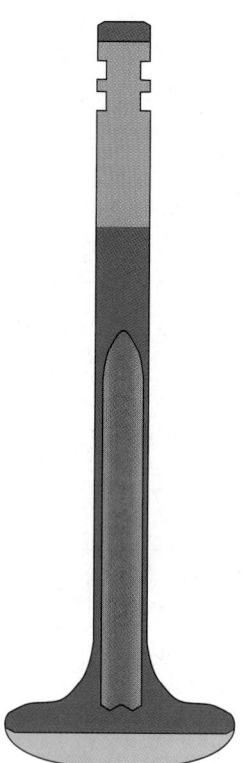

Figure 18.18 A sodium-filled valve.

Sodium-Filled Valves

Some heavy-duty engines use *sodium-filled valves* (**Figure 18.18**). They have hollow stems filled with sodium, with the weld near the stem tip. The sodium moves within the valve stem and carries heat from the neck of the valve to the cooler valve stem area that is surrounded by the water jacket. Maximum valve temperatures are actually cut by about 350°F.

■ SCIENCE NOTE ■

Metallic sodium is a solid powder in the valve stem until it reaches 208°F. It is a liquid at engine operating temperature. Materials that are good conductors of electricity are often also excellent heat conductors. Thermal energy is transported rapidly from one part of the metal to another by conduction through the valence electrons, which roam freely throughout the material when solid or liquid.

The hollow cavity in the valve is actually only about half full of sodium. The movement of the sodium in the valve stem helps to carry off heat from the neck of the valve to the cooler valve stem. Sodium-filled valves can fail if valve stem-to-guide clearance becomes excessive because the heat cannot dissipate from the valve stem.

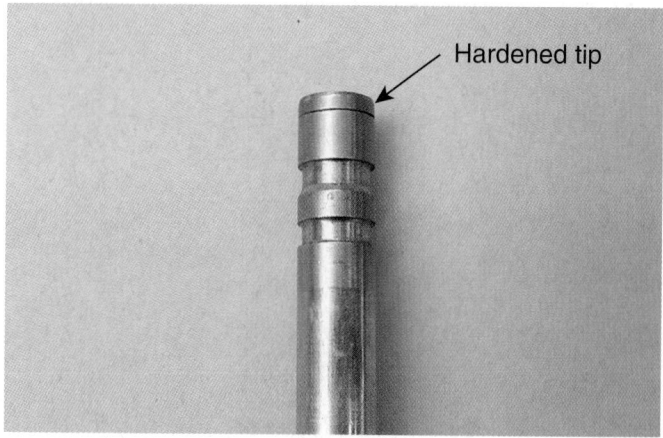

Figure 18.19 An exhaust valve stem with a welded-on tip. (*Courtesy of Tim Gilles*)

CAUTION Metallic sodium will burst into flame if it comes into contact with moisture, so it is dangerous to grind or break these valves. Sodium is safe as long as it is contained within the valve.

Valve Tip

The tip of the valve is hardened. The exhaust valve must open against very high pressure in the cylinder during the power stroke, so many exhaust valve stems have a welded-on hardened tip (**Figure 18.19**). An engine that has mechanical valve adjustment would tend to *mushroom* the end of the stem if it is not hardened or if the valve clearance becomes excessive.

■ RETAINERS AND KEEPERS

The retainer and valve keepers, or *locks*, hold the spring and valve together. Keepers fit into the groove in the valve stem. The retainer, which resembles a large washer, holds the spring against the spring seat on the top of the cylinder head. Some retainers have a feature that causes the valve to rotate to aid in heat dissipation. These retainers are called *valve rotators* (**Figure 18.20**).

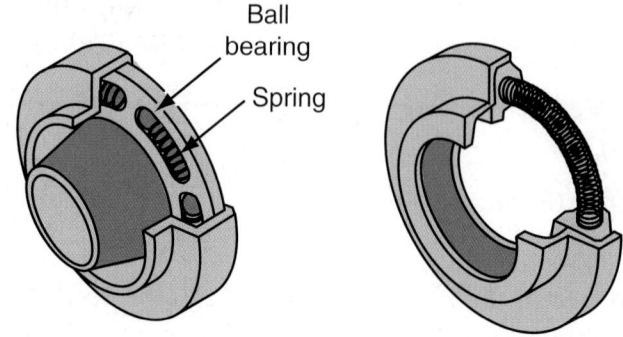

Figure 18.20 Valve rotators.

■ VALVE SPRINGS

After the valve is opened by the valvetrain, a valve spring is needed to close it. Some engines have heavier valve springs than others. On camshafts with higher valve lift (covered later in this chapter), more spring pressure is often required. The thickness of the spring is limited by how close together its coils are. When the valve opens, the spring could stack against itself causing damage. This is called coil bind.

If a stronger spring is required, two springs are used (one inside the other). The spring on the outside is wound in a direction opposite the inside spring. When a spring is compressed, it tends to bend inward. Think about how a "slinky" bends to go down stairs. Having the two springs wound in different directions tends to minimize this tendency.

Some springs have a flat wound *dampener* on the inside to dampen spring vibrations. Springs vibrate at a certain frequency that can result in engine damage if allowed to continue. Think about how an opera singer can cause a glass to break when a note of the right pitch is sung. The dampening coil is wound in a direction opposite to the valve spring also.

■ ROCKER ARMS

Rocker arms either pivot on a rocker shaft or on rocker studs. *Shaft-mounted rocker arms* are usually made of iron. Older engines used more expensive cast iron rocker arms. Most modern engines use the less expensive stamped-steel rocker arms.

Stud-mounted, or *pedestal-mounted,* rocker arms are made of steel stamped in a "canoe" shape. They are less expensive than the shaft-mounted type, and are commonly replaced if they are worn. The rocker arm shown in **Figure 18.21** has a ball pivot and self-locking adjusting nut mounted above it.

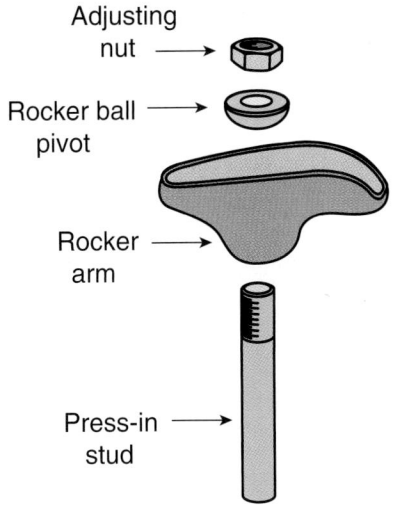

Figure 18.21 A stamped-steel, stud-mounted rocker arm.

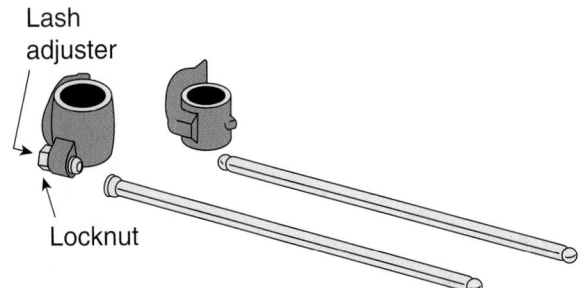

Figure 18.22 Adjustable and nonadjustable shaft-mounted rocker arms.

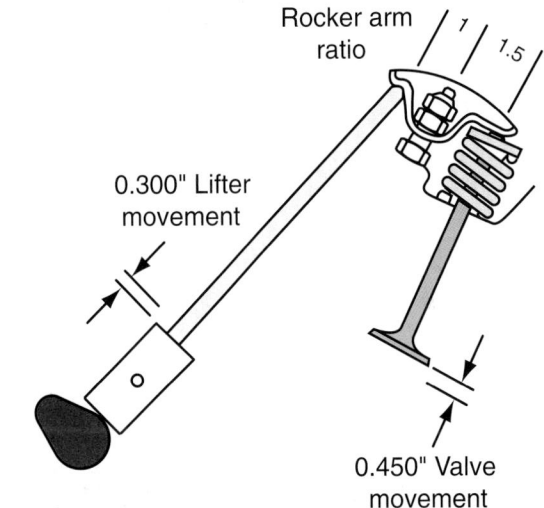

Figure 18.23 Rocker arm ratio.

Stud-mounted rocker arms are mounted on a stud that is pressed or threaded into the head. One kind of stud has a shoulder that the adjusting pivot stops against when the rocker stud is torqued. This is called a **positive stop** arrangement and requires no valve adjustment. **Figure 18.22** compares adjustable and nonadjustable rocker arms. Nonadjustable rockers can only be used in positive stop engines.

Rocker Arm Ratio

Rocker arms sometimes have different *rocker arm ratios* (**Figure 18.23**). Rocker ratio is usually 1.4:1 to 1.75:1, depending on the manufacturer. Changing the ratio can cause movement of the lifter at the cam lobe to be *increased* at the valve, making it possible for an engine to be designed with smaller cam lobes.

Pushrods

Cam-in-block engines use pushrods to transmit motion from the lifters to the rocker arms (**Figure 18.24**). Most pushrods are hollow. Some have oil holes to provide lubrication to the rocker arms and valves (**Figure 18.25**). The ends of pushrods have many shapes, including concave (indented) and convex (protruding).

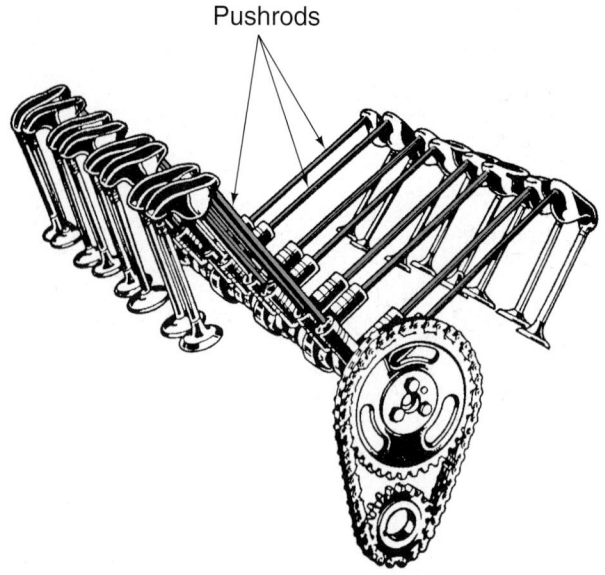

Pushrods

Figure 18.24 Cam-in-block engines have pushrods that transmit motion from the lifters to the rocker arms. *(Courtesy of Crane Cams, Inc.)*

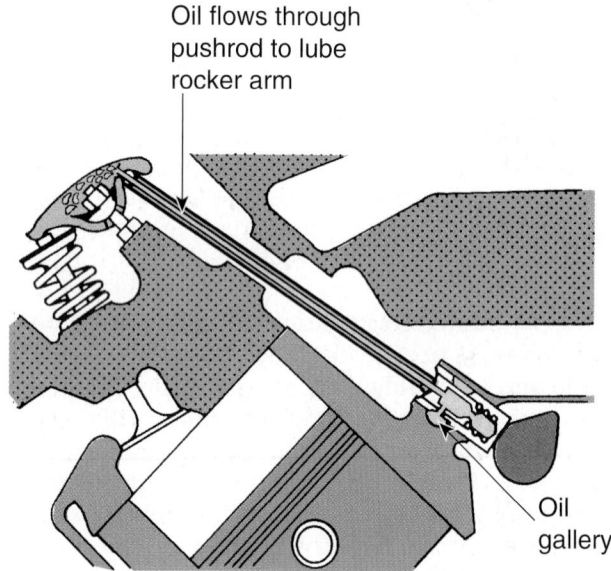

Oil flows through pushrod to lube rocker arm

Oil gallery

Figure 18.25 The stud-mounted rocker arm is lubricated through hollow pushrods.

Some engines use a *guide plate* to limit side movement of the pushrods. Other engines have small holes in the cylinder head or intake manifold that act as guides. Another design variation has rails on the rocker tip that keep the pushrod "captured" on the top of the valve stem.

Overhead Cam Valve Operation

Some overhead cam (OHC) engines use rocker arms. One end of the rocker arm rests against the valve stem, and the other end sits on a ball pivot or hydraulic lash adjuster (**Figure 18.26**). The cam lobe pushes on a pad in the middle of the rocker arm. The pivot is

Figure 18.26 Hydraulic lash adjusters/compensators used with OHC rocker arms. *(Courtesy of Tim Gilles)*

sometimes equipped with threads for adjusting valve clearance, or *lash*. Other OHC engines have cam lobes that act directly on the valves. These cam lobes are larger than lobes on engines with rocker arms. This is because there is no rocker arm ratio to multiply movement caused by the lobe. Lash is sometimes adjusted on this type of engine by adding or subtracting shims (called *lash pad adjusters*). Most new engines use hydraulic adjusters to automatically maintain correct valve clearance. Their operation is covered later in this chapter.

■ CAMSHAFT

Camshaft Construction

The camshaft (**Figure 18.27**), commonly called "the cam," controls the opening and closing of the valves. It is usually made of hardened cast iron, although cams used with roller lifters are made of steel. *Cam journals* support the camshaft at several places. The number of *lobes* a camshaft has depends on how many valves the engine has.

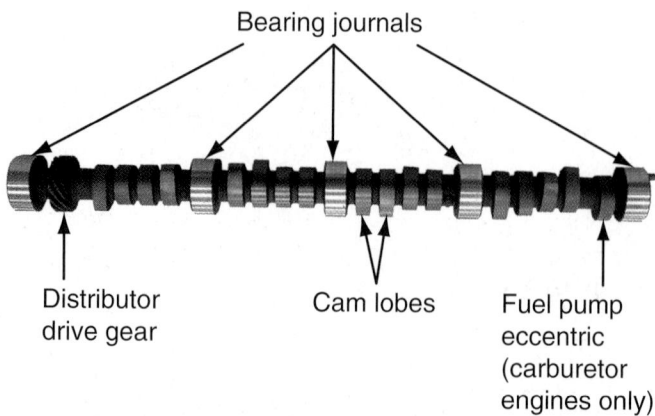

Bearing journals

Distributor drive gear

Cam lobes

Fuel pump eccentric (carburetor engines only)

Figure 18.27 Camshaft parts. *(Courtesy of Tim Gilles)*

NUMBER OF CAMS AND LOBES

Most engines have two cam lobes for each cylinder. One lobe operates the intake valve, and the other operates the exhaust valve.

Many engines use only one camshaft, although some use more than one. Some engines use three or four valves per cylinder. To increase the engine's breathing ability (volumetric efficiency), many racing engines and many of the latest small high-performance stock engines have four valves and four cam lobes on two camshafts for each cylinder. Opposed engines have only half as many cam lobes because the lobes of one cylinder are shared with its companion cylinder on the opposite side of the engine.

Intake and exhaust lobes are sometimes different from each other. When there is more than one camshaft on a cylinder head, one camshaft will have only intake lobes and the other will have only exhaust lobes.

VINTAGE ENGINES

Fuel Pump Eccentric

When an engine has a carburetor, there is usually an eccentric (off-center lobe) for the mechanical fuel pump (see Figure 18.27). Fuel-injected engines use electric fuel pumps and do not need this eccentric. Some engines have an eccentric that is not part of the cam but bolts to the front of the cam sprocket.

DISTRIBUTOR AND OIL PUMP DRIVES

Because it turns at the same speed, the distributor is often driven off the camshaft. The *distributor gear* is sometimes made from a softer material such as aluminum, plastic, or brass. A failure of one of these soft gears would not be as likely to damage the cam.

The oil pump is usually driven by a shaft that connects to the bottom of the distributor shaft. As the oil pump is turned, the resistance created by the pumping of the oil causes the cam to thrust toward the rear of the engine.

NOTE: *Excessive* **end thrust** *can also cause the distributor timing to advance or retard as the cam moves fore and aft against the distributor gear.*

Some camshafts have a *thrust plate* to control end thrust (**Figure 18.28**). Excessive end thrust can also cause damage to the cam and lifters. The timing chain on a pushrod engine provides some thrust control.

Some OHC engines use an extra shaft to drive the oil pump and distributor. It is called an *auxiliary shaft* and is driven by the crank. In other OHC engines, the distributor and oil pump are mounted on the cylinder head.

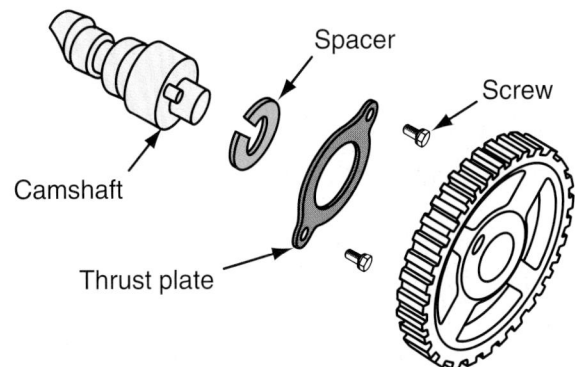

Figure 18.28 Some pushrod engine camshafts have a thrust plate to control end thrust.

CAMSHAFT PERFORMANCE

The camshaft determines how well an engine breathes at a particular engine rpm. The shape of the cam lobe, or *cam grind*, is what determines when the engine will breathe best. It can be designed to have its best operation at maximum power and high speed, or for better fuel economy and low speed operation. Production engines deliver a balance between these two objectives. A **production engine** is an engine produced at the factory. Original equipment cams are referred to as *stock* cams.

Engine Breathing

Basically, the more air and fuel drawn into the cylinder in the proper proportion (not too rich or lean), the more power an engine can produce. This is called volumetric efficiency. Many of today's four- and six-cylinder engines are as powerful as eight-cylinder engines of the past. Some of these smaller engines use multiple-valve combustion chambers, along with other modifications to increase breathing ability. These engines are referred to as *naturally aspirated* engines. Other engines equipped with turbochargers or superchargers have much improved breathing ability and, therefore, produce more power. These are called artificially aspirated, or forced induction, engines.

Cam Lobe Design

Cam lobe design differs from engine to engine, depending on the performance desired. The cam lobe has several parts (**Figure 18.29**). The **base circle** is what the cam would be if there were no lobe. The lobe has an opening and closing *flank* on each side of its nose.

Each cam lobe is ground to a precise contour. Lobe shape is always a compromise. The two factors of cam lobe profile are called **lift** and **duration**:
- Lift is the height to which the lobe raises the lifter.
- Duration is the number of degrees of *crankshaft* travel while the valve is off its seat.

VALVE LIFTERS AND LASH ADJUSTERS

On cam-in-block (pushrod) engines, the part that rides on the cam lobe is called a *lifter*. Lifters are also known as

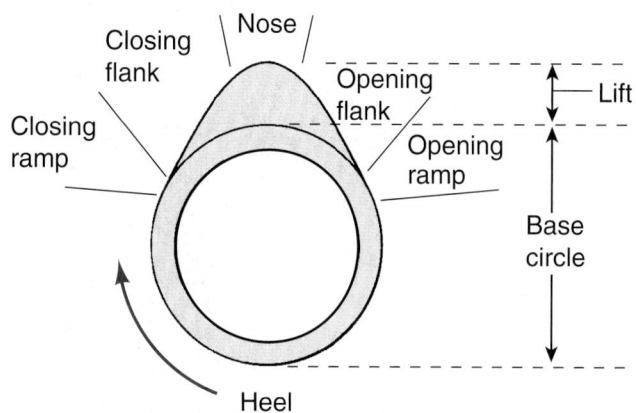

Figure 18.29 Parts of a cam lobe.

tappets. Valves require a clearance adjustment to compensate for wear and changes in engine temperature. Lifters that are used in an engine requiring a specified clearance are called mechanical lifters. Those with automatic clearance adjustment are called hydraulic lifters.

On overhead cam engines, the part that rides on the cam lobe is called a *cam follower*. Hydraulically adjusted cam followers are often called hydraulic lash adjusters.

Lifter and Cam Lobe Relationship

On cam-in-block engines, the lifter and cam lobe have a unique relationship. The lifter face has a convex shape (protruded face), and it is offset from the center of the cam lobe (**Figure 18.30**). In addition, the cam lobe is tapered from approximately 0.0007" to 0.002" across its face. These factors are all introduced to cause the lifter to spin. Spinning is desirable because it helps to dissipate the tremendous load—up to 100,000 psi—that the lifter is subjected to. **Figure 18.31** shows examples of two lifters that failed to spin.

The lifter's convex shape also helps to prevent *edge loading*, which can destroy a cam lobe. Because only the center of the lifter contacts the face of the cam lobe,

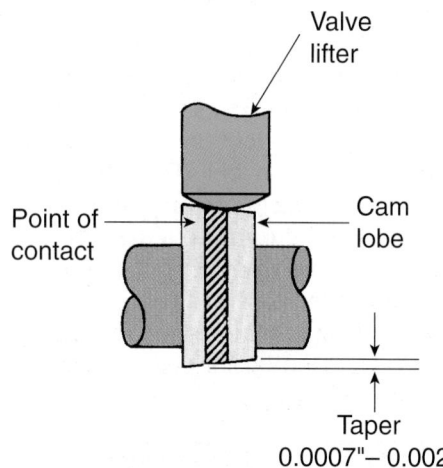

Figure 18.30 The lifter face is convex and is offset from the center of the cam lobe.

Figure 18.31 Two examples of wear that results when a lifter fails to spin. *(Courtesy of TRW, Inc.)*

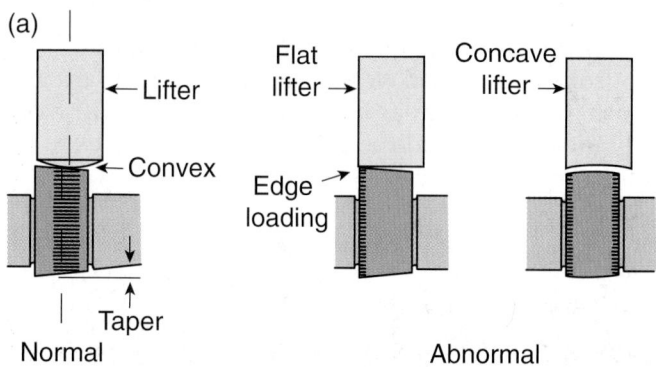

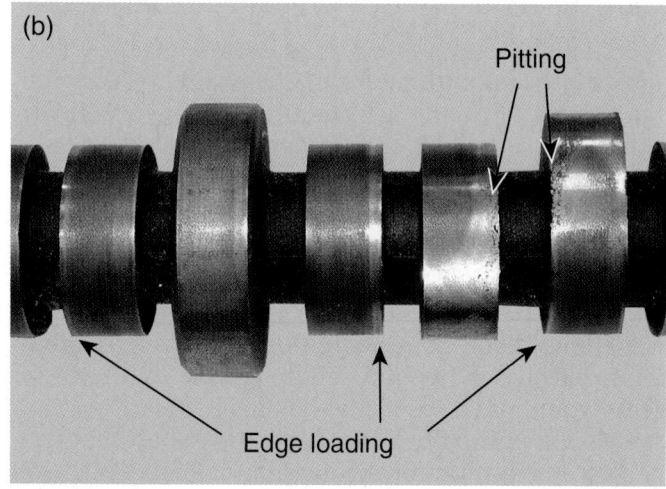

Figure 18.32 (a) A convex lifter prevents edge loading. (b) Notice the bright area on the bottom of the lobes. This indicates edge loading. *(b, Courtesy of Tim Gilles)*

normal wear occurs near the center of the cam lobe. If the lifter face contacts the outside edge of the cam lobe, severe loading causes rapid wear of the cam lobe, making the lifter face become concave. A concave lifter will wear the edges of a cam lobe (**Figure 18.32**).

■ HYDRAULIC LIFTERS

Hydraulic lifters automatically maintain a condition called **zero-lash**. This is a point where there is no clearance and no interference between parts of the valve opening mechanism. Hydraulic lifters eliminate

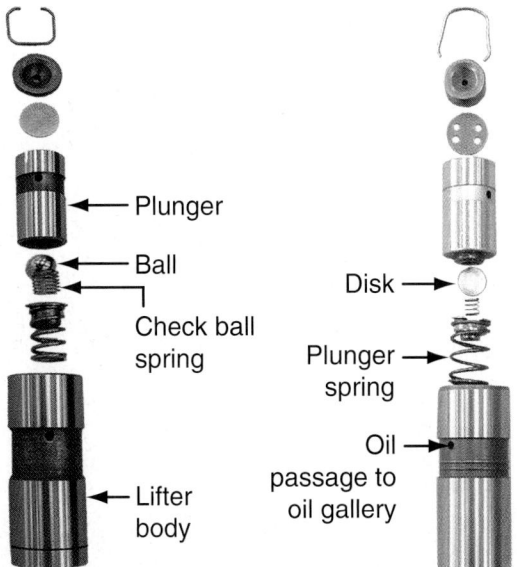

Figure 18.33 Parts of a hydraulic valve lifter. The lifter on the left has a ball check valve. The lifter on the right uses a disk check valve. *(Courtesy of Tim Gilles)*

the need for periodic valve adjustments, are quiet, and reduce unnecessary valve train wear.

Hydraulic Lifter Construction

The lifter consists of a plunger, a plunger return spring, and a check valve assembly (**Figure 18.33**). The plunger is made of steel that is chrome plated to resist wear and corrosion. Lifters from different manufacturers sometimes have different outward appearances, even though they are for the same engine.

Hydraulic Lifter Operation

Whenever clearance occurs in the valvetrain, a spring between the plunger and lifter body causes the lifter to expand (**Figure 18.34**). Oil under pressure fills the cavity created under the plunger. A very small amount of leakage of oil between the plunger and the lifter body allows a lifter with too much oil in it to leak down. The lifter *"leaks down"* as the other valve train parts expand when the engine warms up.

Lifter leak-down is very small, because there is only about 0.0002" (count the zeros) clearance between the plunger and the lifter body.

This small clearance is beneficial because dirt particles are usually larger than 0.0002" and cannot easily lodge between the lifter body and plunger.

▬ MATH NOTE ▬	
0.200	*two hundred thousandths*
0.020	*twenty thousandths*
0.002	*two thousandths*
0.0002	*two tenths of a thousandth*

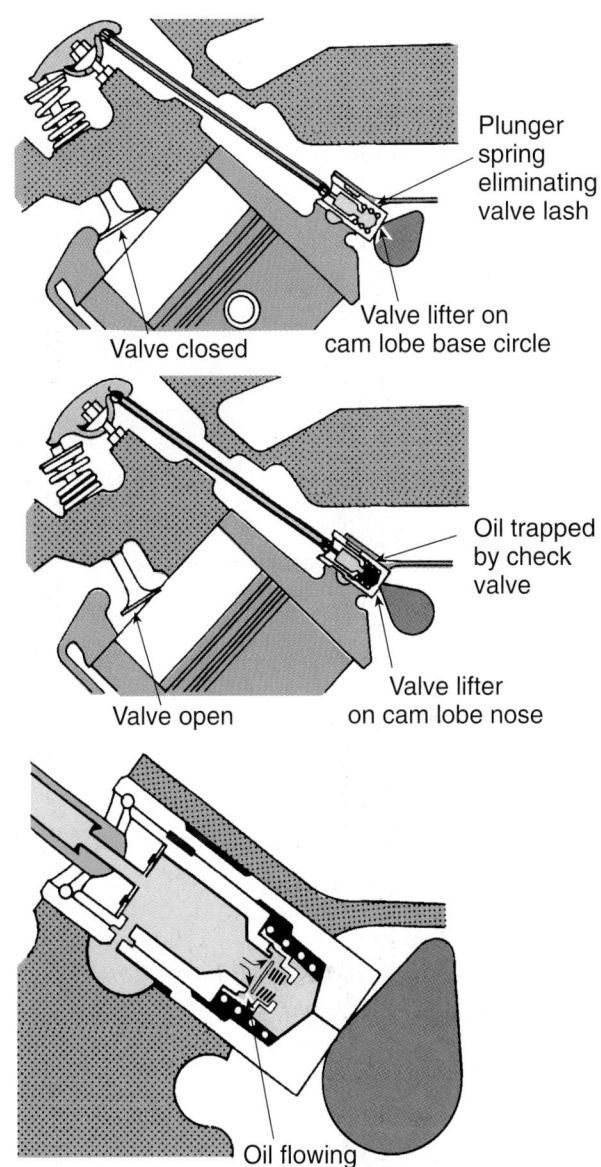

Figure 18.34 Hydraulic lifter operation.

Overhead Cam Hydraulic Lash Adjusters. An OHC with hydraulic lash adjusters is shown in **Figure 18.35.** It works in the same manner that an ordinary hydraulic lifter works (**Figure 18.36**).

Overhead Camshaft Lubrication

An OHC camshaft does not have oil thrown off from the valve lifters like a cam-in-block engine does. Some OHC engines have hollow camshafts with oil holes on each cam lobe to provide pressure lubrication. These camshafts have long lives as long as the oil holes are kept clean. Periodic oil changes are especially important in these engines.

▬ ROLLER CAM AND LIFTERS

In the mid-1930s, the *roller lifter* was developed (**Figure 18.37**). In earlier years, roller lifters have been

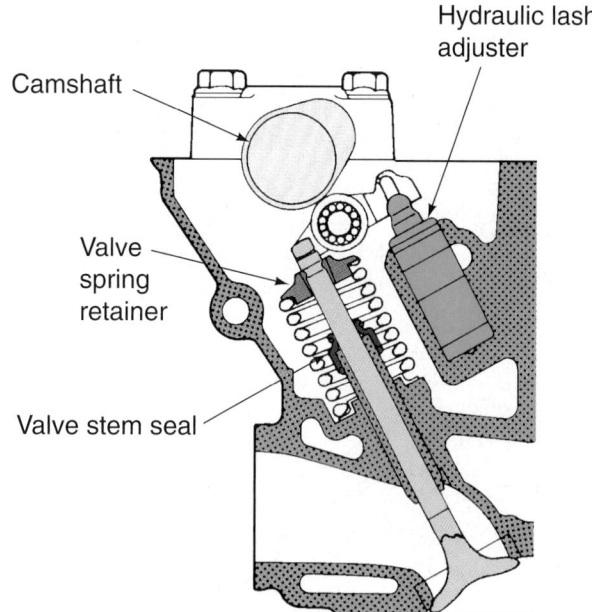

Figure 18.35 An overhead cam head with hydraulic lash adjusters. *(Courtesy of Hyundai Motor America)*

Figure 18.36 A bucket-type hydraulic lash adjuster, assembled and disassembled. *(Courtesy of Tim Gilles)*

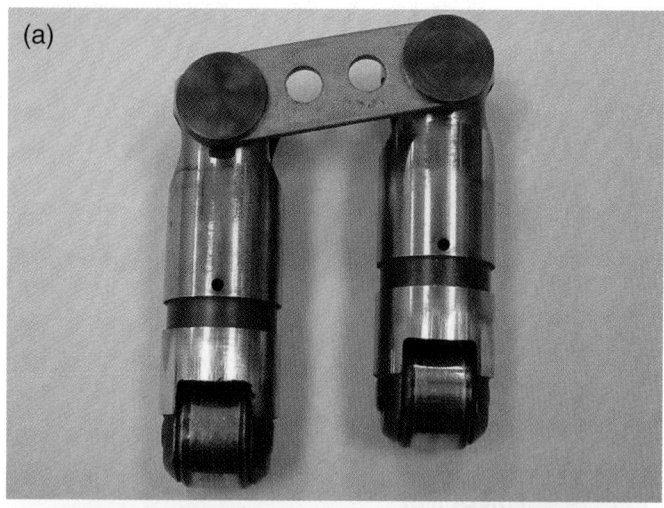

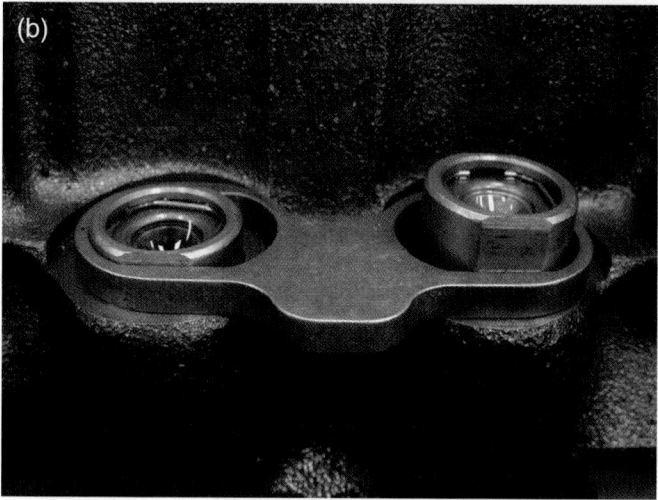

Figure 18.37 (a) A bracket keeps roller lifters from turning in their bores. (b) Another style of roller lifter bracket. *(Courtesy of Tim Gilles)*

Figure 18.38 A roller camshaft. *(Courtesy of Tim Gilles)*

used in diesels and race cars. Today, they are used in production gasoline engines. A roller lifter can accept a much higher rate of movement than a flat tappet without wear to the lifter or cam lobe. The roller lifter cuts valvetrain friction almost in half, resulting in an increase in horsepower and fuel economy.

Roller lifters need a means to keep them from turning. If the lifter could turn sideways, the roller would not work and damage would result. Either a pin or a bracket holds two lifters in the proper plane.

A *roller camshaft* is made of steel. The lobes have a different shape than standard cam lobes, which allows faster acceleration rates (the valve can be opened faster). **Figure 18.38** shows a roller camshaft.

"Flat tappet" cam-in-block camshafts, also known as "non-roller" camshafts, are made of cast iron coated with a dark material that provides resistance to scuffing, especially during the break-in period between the cam and lifter (**Figure 18.39**).

Figure 18.39 A camshaft for a pushrod engine has a black coating to help protect the lobes. Note there is no coating on the bearing journals. *(Courtesy of Tim Gilles)*

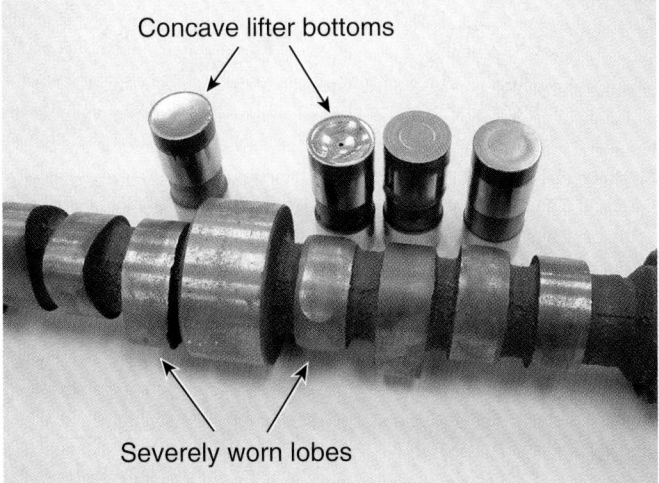

Figure 18.40 This cam has several worn lobes. Notice that the corresponding lifters are concave. *(Courtesy of Tim Gilles)*

Sometimes cam lobes wear severely (**Figure 18.40**). This usually happens on the exhaust lobe because it has to open against combustion pressure.

■ CAM DRIVES

The cam is driven by the crankshaft using one of three methods:
■ Gear drive
■ Sprockets and timing chain
■ Sprockets and timing belt

Chain and belt drives are shown in Figure 16.8.

Gear drives are more positive than a chain (less chance of excessive **backlash** or clearance between the parts) but are limited by gear size and cost.

To reduce noise, the cam gear is usually made of a soft material, either fiber or aluminum. Gear-drive engines use helical gears (see Chapter 71); the crank drives the cam in the direction opposite to the direction of crank rotation.

NOTES:
■ *A conversion kit that changes a chain drive to a gear drive is available from racing parts distributors. It uses idler gears to make the cam and crank rotate in the same direction. A disadvantage of these gear drives is that they have spur gears, so they are noisy.*

■ *In some V-type engines used for racing or in expensive automobiles, a multiple gear arrangement drives dual overhead cams. This is impractical for use in conventional passenger cars.*

Sprockets and Gears

Sprockets and gears are made of steel, iron, aluminum, or fiber. Some aluminum sprockets have nylon teeth. These sprockets can be injection molded, which makes them less expensive to produce. When a chain used with nylon gear teeth develops slack, the nylon teeth sometimes shear off of the sprocket (**Figure 18.41**). The broken teeth can stick in the oil-pump relief valve. Nylon gears should not be used in heavy-duty vehicles, such as trucks or high-performance cars with standard transmissions. A worn sprocket (**Figure 18.42**) should not be used with a new chain; it must be replaced.

Figure 18.41 Some of the nylon teeth are missing from the camshaft sprocket of this timing set. *(Courtesy of Tim Gilles)*

Worn sprocket teeth

Figure 18.42 The teeth on this timing sprocket show wear. It should not be reused. *(Courtesy of Tim Gilles)*

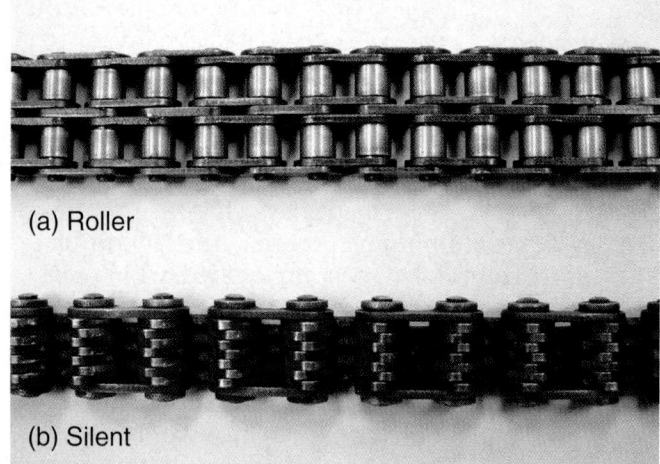

Figure 18.43 Timing chain types. (a) Roller. (b) Silent. *(Courtesy of Tim Gilles)*

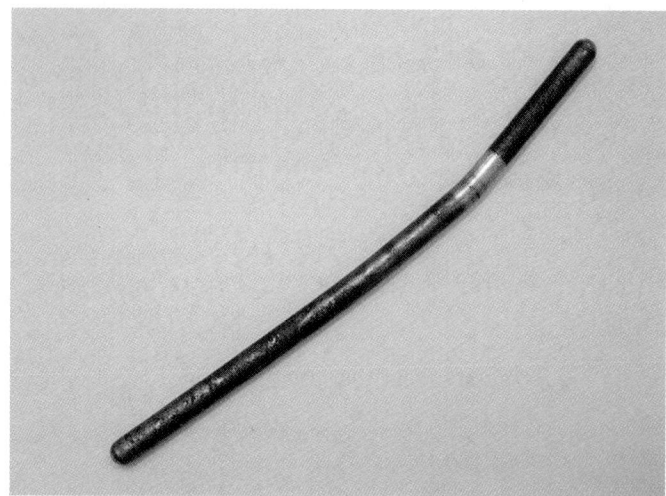

Figure 18.44 This pushrod was bent when a valve contacted a piston. *(Courtesy of Tim Gilles)*

Timing Chains

The two types of chains are the *roller chain* and the Morse-type *silent chain* (**Figure 18.43**). *Double roller chains* are usually used on overhead cam engines.

Pushrod engines usually use a silent chain. New silent chains are of the *large-pin* design. These "floppy" chains are twice as strong as older silent chains. They are less susceptible to misalignment and overload problems than roller chains, due to greater flexibility. The large-pin silent chain eliminates chordal action; a problem that roller chains have. Chordal action means the chain operates on a constantly changing diameter, which causes varying camshaft speeds.

Noninterference and Interference Engines

As an engine ages, its timing chain stretches. In fact, this is really the main obstacle to long engine life. If the valve timing is wrong or the cam sprocket keyway becomes stripped in a noninterference, or **freewheeling engine**, the pistons and valves cannot come into contact with each other (as piston-to-valve clearance is always maintained).

Some engines can experience piston-to-valve interference if the timing chain or belt skips or breaks. These engines are called non-freewheeling or **interference engines**. When this happens on a pushrod engine, pushrods will usually bend (**Figure 18.44**). The valves on a pushrod engine might survive, but on OHC engines, the valves must bend (since there are no pushrods to bend under the stress).

Timing chains have a drive, or tension, side and an opposite side where slack accumulates. Long chains, such as those used on OHC engines, use chain tensioners to take up slack as the chain stretches (**Figure 18.45**). Some chain tensioners are spring loaded and hydraulic. Timing chains in these engines can make noise until the

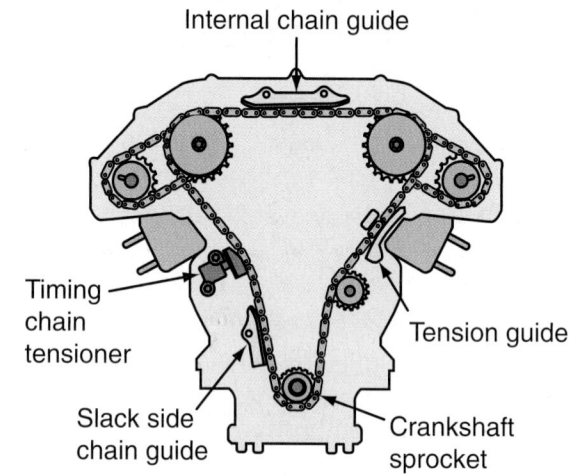

Figure 18.45 Parts of an OHC chain drive.

engine runs for a few seconds and the tensioner gets oil pressure (making it move against the chain freeplay).

If sprocket teeth become damaged or chain slack becomes excessive to the point where the crankshaft sprocket can skip forward a link or more on the timing chain, all of the engine's exhaust valves can be bent to the same angle (**Figure 18.46**). Normally, as a piston moves up in the cylinder on its exhaust stroke, the exhaust valve is closing just in front of it, narrowly avoiding a collision. When the crankshaft sprocket skips forward, the camshaft timing is late and all valves open and close later than normal. As the pistons move up in the cylinders toward TDC, they collide with the valves, which are too far open to allow sufficient clearance.

Timing Belts

The camshafts on some OHC engines are driven by a cogged belt drive, called a *timing belt*. Oil pumps and

Figure 18.46 These exhaust valves from a four-cylinder OHC engine were all bent when the timing chain skipped. *(Courtesy of Tim Gilles)*

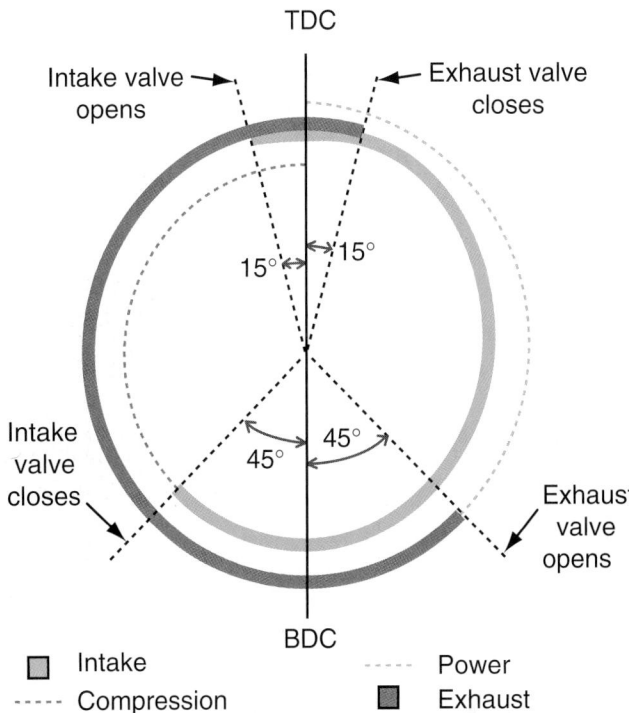

Figure 18.47 A valve timing chart.

water pumps can also be driven by the timing belt. Compared to chain drives, belt drives are:
- Quieter
- Less costly
- Lighter in weight
- Easier to replace
- Lubrication-free
- Stretch-free

NOTE: *A non-freewheeling engine often has a decal on the valve cover stating that the belt must be changed at least every 60,000 miles. The mileage amount varies. Many late-model belts are designed to go even longer.* More information about timing belts is covered in Chapter 22.

■ VALVE TIMING

Figure 18.47 shows an example of a valve timing chart. Several points should be noted:
- The intake valve must open *before* TDC, when the piston starts down on the intake stroke.
- The intake valve stays open long after BDC (into the compression stroke) to take advantage of the velocity of the incoming gases, which helps to "ram" additional air-fuel mixture into the cylinder.
- The exhaust valve must open long before the end of the power stroke in order to bleed off the pressure of expanding gases in the cylinder before the piston tries to move up against that pressure on its exhaust stroke. It does no harm to open the exhaust valve at this point because most of the power of the exploding gases has been delivered to the piston by about halfway through the power stroke.
- The exhaust valve is allowed to remain open past TDC. At higher speeds, the inertia of exhaust moving out of the cylinder can create a vacuum, sucking more exhaust gas from the cylinder. Removing as much exhaust as possible

makes room for more fresh air-fuel mixture and, therefore, increased power.

Valve Overlap

Valve overlap is the number of degrees of crankshaft rotation that occurs during the time that both the intake and exhaust valves are open at the same time. This happens at the top of the intake stroke when the intake valve has begun to open and the exhaust valve has not yet finished closing (**Figure 18.48**). In the valve timing chart in Figure 18.47, the amount of overlap is 20 degrees.
- Additional overlap provides more efficient cylinder filling at high rpm but causes lower engine vacuum, poorer performance, and less fuel economy at idle and low speed. A high overlap cam may have high valve lift. It is designed for high-performance, high rpm use.
- Less overlap results in more pressure in the cylinder at lower rpm. This results in greater torque at low rpm. The "RV" or "high gas mileage" cams fit this profile. It is designed for low rpm use.

NOTE: *RV means uses such as off-road and does not apply to motorhomes that would be travelling mostly at highway speeds.*

NOTE: *Overlap is what gives high-performance engines their distinctive "lope" at idle. A cam grind that will benefit performance at high rpm will hurt performance at low rpm. The reason for this is the inertia (momentum) of the gases. The low vacuum associated with lope at idle can lead to problems such as inconsistent vacuum power*

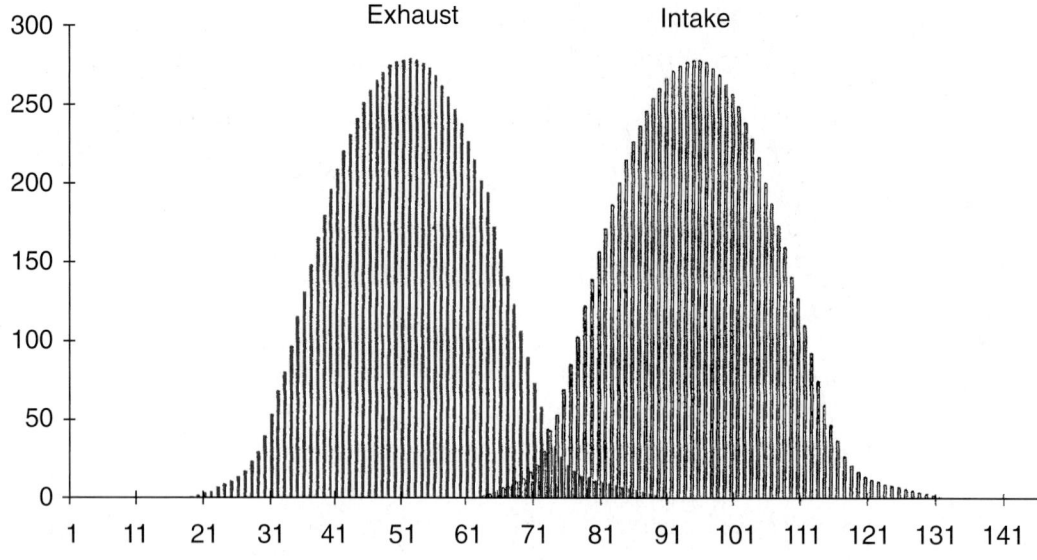

Figure 18.48 *Valve overlap. (Courtesy of Dimitri Elgin and D. Elgin Cams)*

brake applications. Overlap is also a factor that can lead to clearance problems between the valves and pistons.

Before changing to a non-stock camshaft, you will need to know information that is not included in this text. Be aware that a high-performance camshaft will probably have low engine vacuum at idle. Power will not come on until higher engine rpm, and fuel economy will suffer. This is why newer engines use variable valve timing; they can have a smooth idle and good low-speed performance as well as high performance at higher engine rpm.

Changes in Cam Timing

Most cams are ground with 4- to 6-degree advance for better low-end torque and to compensate for timing chain stretch. The results of changes in cam timing are as follows:

■ Advanced cam timing improves low and mid-range torque.
■ Retarded cam timing improves high-rpm power.

Mechanical Valve Lash

With solid lifter cams, valve adjustment is very important. Solid cams have a clearance ramp at the beginning and end of each lobe. This clearance ramp can be compared to a wedge. Its purpose is to cushion the opening and closing of the valve (**Figure 18.49**). A change of 0.001" in valve lash will result in approximately 3 degrees of change in duration. Too much clearance will cause the lifter to miss the clearance ramps altogether. This, in turn, can cause extra shock loads that can seriously damage valvetrain parts.

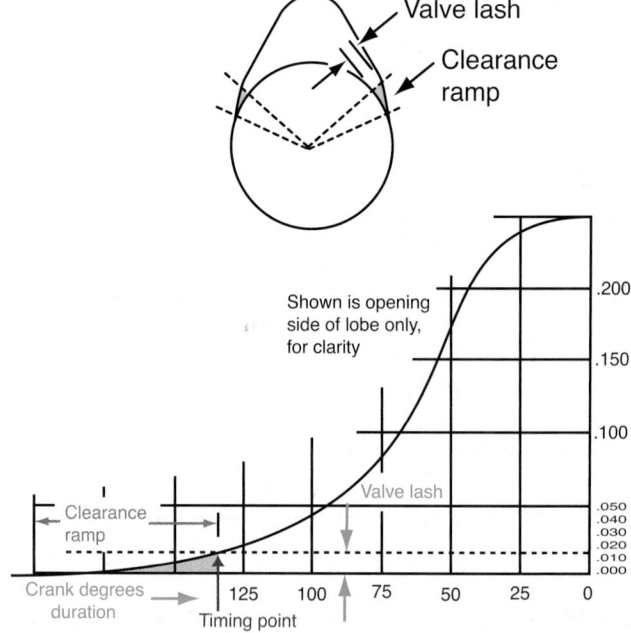

Figure 18.49 A clearance ramp cushions the opening and closing of the valves.

■ VARIABLE VALVE TIMING

Most manufacturers now use variable valve timing (VVT). The camshaft profile in conventional engines has always been compromised between low and high rpm demands (**Figure 18.50**). High-speed, high-performance engines call for very different cam lobe profiles than low-rpm four-wheel-drive off-road vehicles. VVT has become commonplace in recent years. With computer controls, it is a relatively inexpensive way to significantly increase horsepower and control emissions. Power and torque can be improved across a wider rpm range. Increased torque at low rpm and increased power at higher rpm are the result. Typical

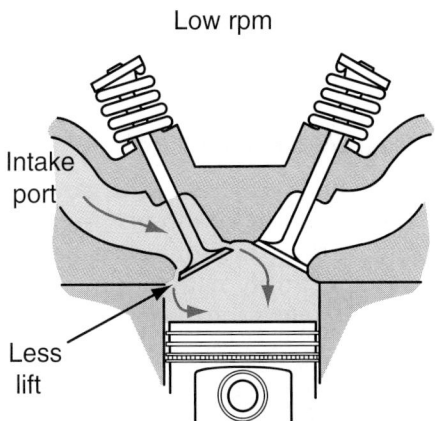

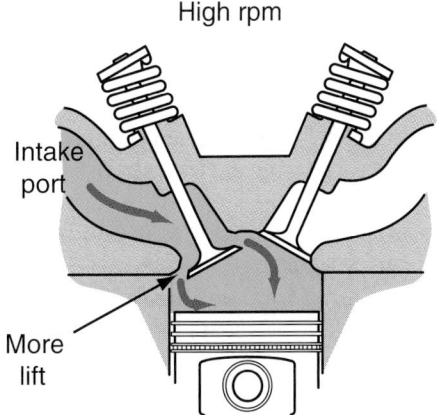

Figure 18.50 Variable valve timing changes duration between low- and high-rpm operation.

power increase can be 25%. Horsepower has climbed above 100 horsepower per liter and is still climbing.

There are two basic types of VVT systems. One type varies the lift and duration. The other system varies only valve timing by changing the cam-to-crank phasing (**Figure 18.51**). A third type of VVT is a combination of the other two.

Variable Valve Timing and Lift

The *variable timing and lift* system, pioneered by Honda in 1988, is a simple and effective system in which each pair of valves has three cam lobes. Two of the lobes operate the valves at low rpm, and the third comes into play at higher rpm (approximately 4,500–6,500). The low-rpm lobes are mild, giving good low-end driveability. When the high-rpm lobe takes over, its more aggressive profile locks onto the two low rpm rocker arms and overrides their operation (**Figure 18.52**). Mitsubishi, Nissan, and Toyota have similar systems of two-stage variable valve lift and duration.

Variable Camshaft Phasing

Some VVT systems vary only the valve timing but not the lift and duration as described previously. A sophisticated attachment on the cam sprocket advances and retards the cam using either a vane or piston and gear

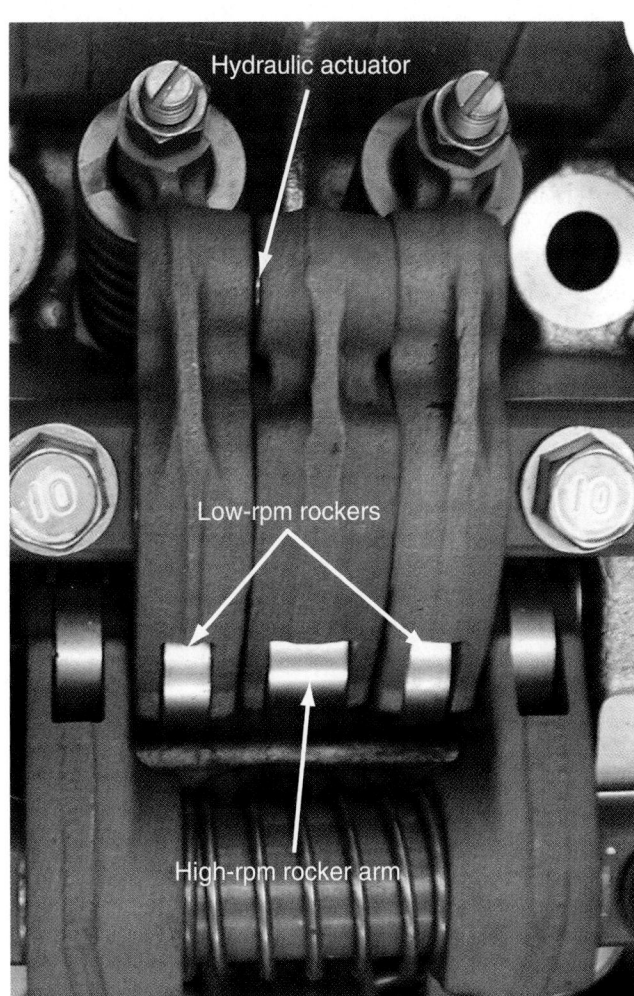

Figure 18.52 On Honda's V-tec system, a hydraulic actuator disconnects the outer rocker arms at high rpm. The center rocker arm rides on a more aggressive cam lobe. (*Courtesy of Tim Gilles*)

Figure 18.51 Variable valve timing. (*Courtesy of Toyota Motor Sales, U.S.A., Inc.*)

assembly operated by oil pressure and controlled by the vehicle's computer. This can be accomplished by moving either the intake or the exhaust camshaft, or both:

- For performance enhancement, changing the intake valve timing affects the torque characteristics of the engine, whereas changing the exhaust valve timing does not.
- Variable timing of the exhaust lobe is used for the control of emissions and fuel efficiency.

Most systems control the intake valve only. Simpler systems have only two or three fixed phasing angles. More complex systems are continuously variable and have a position sensor.

Two methods of hydraulically varying the cam timing are shown in **Figure 18.53**. Figure 18.53a shows rotary vane actuation. Figure 18.53b shows how a piston in the center of the cam sprocket moves, changing the relationship between the camshaft and timing sprocket.

Variable Timing and Lift plus Cam Phasing

A more complex VVT system combines VVT and lift with variable camshaft phasing (**Figure 18.54**). These systems are used on both intake and exhaust camshafts, with continuously variable cam phasing and two-stage variable valve lift and duration. This system is more costly but has the advantage of improved torque through a wider rpm band.

In addition to providing a substitute for the EGR valve, VVT can also be used to make an engine easier to crank during starting. Some of the more advanced systems that use computer-controlled valve actuation vary the valve lift and duration throughout the entire

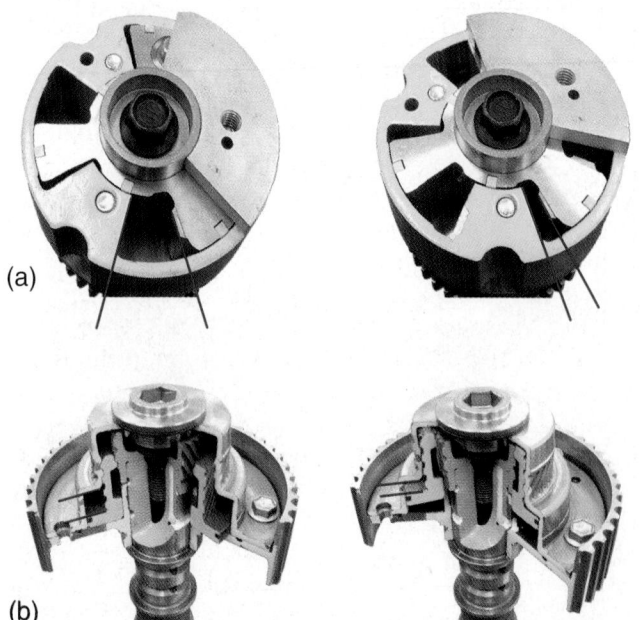

Figure 18.53 (a) Rotary vane actuation. (b) How a piston in the center of the cam sprocket moves. *(Courtesy of Tim Gilles)*

rpm range. This allows for the elimination of the throttle valve, because airflow into the engine is varied by the amount of valve lift.

Atkinson Cycle Engine

Some hybrid vehicles use a gasoline engine to assist an electric motor, or motors, to power the wheels. The engine is the Atkinson cycle design, which operates within a relatively narrow rpm band (4,500 maximum rpm). VVT allows the intake valve to close later, lowering the engine's operating compression ratio (dynamic compression ratio). Lower compression during idling and startup allows the engine to restart more easily because it does not have to overcome normal compression.

The intake valve opens late, remaining open until the piston is about ⅔ of the way back up in the cylinder. In a conventional engine this would begin to push the air charge back out the open intake valve, but this is prevented by restrictive intake manifold in the Atkinson cycle engine.

The "shorter" compression stroke is followed by a "longer" power stroke to maximize the energy of the burning fuel. Under heavy loads, the valve timing changes to allow operation like that of a conventional engine. When the Atkinson engine is supercharged, this is called the Miller cycle.

Displacement on Demand

With displacement on demand, when power is not needed the engine runs on only four cylinders. Special valve lifters can be activated or inactivated to enable or disable the opening of the valves and allow normal cylinder operation or operation with the contribution of fewer cylinders (**Figure 18.55**). Half of the lifters are normal, and the others are operated by solenoids. Cylinders are disabled when the valves remain closed. For smooth operation, the engine runs on all of its cylinders when idling.

Shutting down cylinders became more common beginning in 2005. Unlike earlier displacement-on-demand engines of the early 1980s, modern computer-controlled systems allow "seamless" operation that cannot be felt by the driver. Today's sophisticated computers control throttle opening, fuel injection, and spark timing.

Controlling Duration with Solenoids

Some manufacturers have developed systems that control cam timing using electric solenoid actuators to directly control the valves. This is a technology that manufacturers hope to perfect. Solenoid actuators replace the cam and cam drive. Rockers or buckets allow valve control completely independent of cylinder position. A computer controls valve opening in the same manner that fuel injectors now control fuel flow, allowing more open time with increased engine rpm.

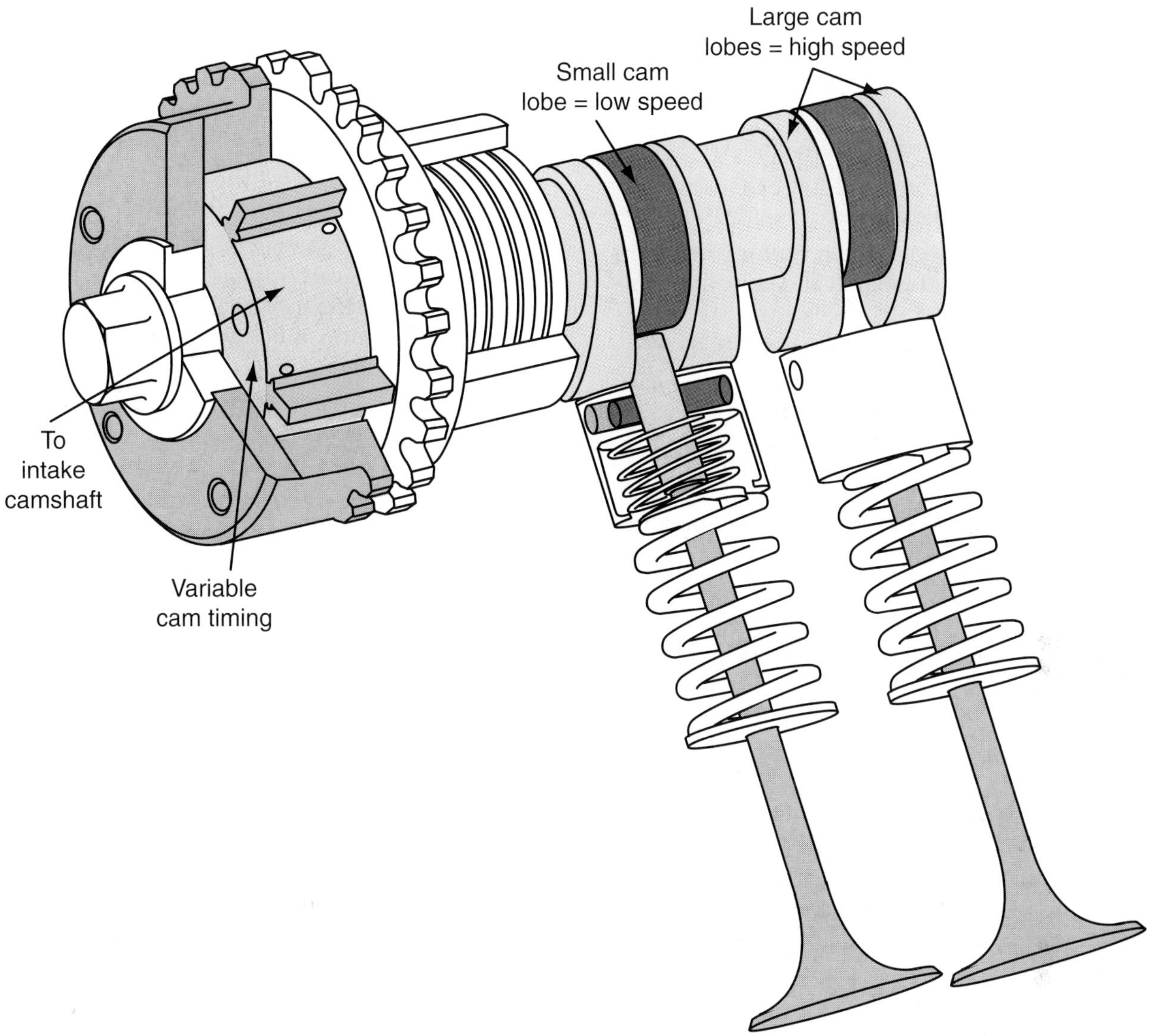

Small cam
lobe = low speed

Large cam
lobes = high speed

To
intake
camshaft

Variable
cam timing

Figure 18.54 Porsche's VarioCam Plus varies the cam phasing and also provides a high lift and duration performance position.

Figure 18.55 A displacement-on-demand engine uses special valve lifters to disable cylinders in response to computer commands. *(Courtesy of DaimlerChrysler)*

This system also allows for interesting possibilities, such as dynamic braking and shutting down cylinders for better fuel economy (sometimes called displacement on demand). These engines are also capable of higher rpm than are conventional engines.

On the downside, larger electrical systems are required for these modifications. There is also the problem of how to close the valves *gently* 25 times per second at 3,000 rpm, or even 50 times per second at 6,000 rpm. On an engine with a camshaft, the shape of the cam lobe takes care of that. The valve shuts very fast but slows down just before it seats. Electronic controls are used to solve this problem. Maximum rpm is also a concern. High-current solenoids have a difficult time working fast. It is hoped that the valves will be able to open so fast that mid-range performance will be what high-rpm performance is now. Vehicles equipped with this system will have a 36-volt battery, 42-volt charging system.

■■■ REVIEW QUESTIONS

Cylinder Head Questions

1. What two materials can cylinder heads be made of?

2. What are the two types of valve guides?

3. If a car decelerating on a hill has exhaust smoke, what are two possible causes?

4. The _____ effect is the term for the reason that a vacuum can happen at the bottom of the exhaust valve guide.

5. What are the three types of valve guide seals?

6. What temperature is a low-quality valve guide seal good until? A high-quality seal?

7. What is the name of the process that hardens integral valve seats?

8. What can happen to a valve that has a margin that is too thin?

9. Why would a spring have two coils?

10. What is the name of the type of rocker arm stud that does not require a valve adjustment?

Camshaft/Cam Drive Questions

1. What are three things that the camshaft can drive besides opening the valves?

2. What causes the camshaft to thrust toward the rear of the engine?

3. An engine without a turbocharger or supercharger is called a naturally _____ engine.

4. What is the term for the time when both the intake and the exhaust valves are open at the same time?

5. If the intake valve opens 15 degrees before TDC and the exhaust valve closes 10 degrees after TDC, how much valve overlap does the camshaft have?

6. What is the name for when there is no clearance between parts of the valve opening mechanism?

7. What is the name for the kind of lifter that must be held from turning?

8. List three ways that a camshaft can be driven.

9. The name for a part that is returned to replace a rebuilt or remachined part is a _____.

10. When an engine breaks a timing belt but cannot possibly have interference between the valves and pistons, this type of engine is called a _____ engine.

■■■ ASE-STYLE REVIEW QUESTIONS

Cylinder Head Questions

1. All of the following are true about valve guides *except*:

   a. Integral valve guides are found in aluminum heads.

   b. Integral valve guides are found in iron heads.

   c. Insert guides are found in iron heads.

   d. Insert guides are found in aluminum heads.

2. Technician A says that a leaking valve guide seal can result in oil consumption. Technician B says that a leaking valve guide seal can result in engine breathing difficulties. Who is right?

   a. Technician A   c. Both A and B

   b. Technician B   d. Neither A nor B

3. Technician A says that an umbrella seal moves up and down with the valve stem. Technician B says that a positive valve guide seal is attached to the top of the valve guide. Who is right?

   a. Technician A   c. Both A and B

   b. Technician B   d. Neither A nor B

4. Which of the following is/are true about intake and exhaust valves?

   a. Exhaust valves are larger than intake valves.

   b. Intake valves run at cooler temperatures than exhaust valves.

   c. Both A and B.

   d. Neither A nor B.

5. Which of the following is/are true statements about engine valves?

   a. Some valves are magnetic and some are not.

   b. Valves are sometimes constructed of more than one type of metal.

   c. Valve stems are either chrome plated or nitrided to provide a hard coating.

   d. All of the above.

Camshaft/Cam Drive Questions

1. Which of the following is/are true about valve timing?

   a. Duration is the number of crankshaft degrees that the camshaft turns while the valve is open.

   b. The exhaust valves opens well before the end of the power stroke.

   c. The exhaust valve closes during the intake stroke.

   d. All of the above

2. All of the following statements about variable valve timing are true *except*:

   a. Advanced cam timing improves low and mid-range torque.

   b. Retarded cam timing improves high-rpm power.

   c. Some variable valve timing systems change both lift and duration.

   d. Changing the exhaust valve timing changes the engine's torque output.

3. Technician A says that the lifter on a pushrod engine is supposed to be concave on its bottom. Technician B says that the cam lobe on a pushrod engine is tapered. Who is right?

   a. Technician A   **c.** Both A and B

   b. Technician B   **d.** Neither A nor B

4. Which of the following is/are names for valve lifters?

   a. Cam followers   **c.** Tappets

   b. Lash adjusters   **d.** All of the above

5. Technician A says that a flat tappet lifter is supposed to spin in its bore. Technician B says that a roller lifter is supposed to spin in its bore. Who is right?

   a. Technician A   **c.** Both A and B

   b. Technician B   **d.** Neither A nor B

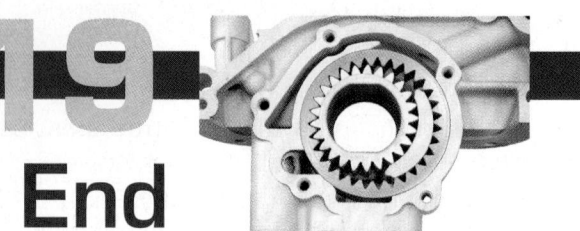

Engine Lower End and Lubrication System Theory

▪▪ OBJECTIVES

Upon completion of this chapter, you should be able to:

✔ Describe the related theory of all of the parts that make up the lower end.

✔ Tell how a cylinder block is made and understand the functions of its parts.

✔ Understand how pistons are constructed and the reasons for their various design considerations.

✔ Discuss the various types of piston rings and be able to make the correct choice when selecting rings for an engine.

✔ Know about the various types of engine bearings.

✔ Describe the parts of the crankshaft and their functions.

▪▪ KEY TERMS

lower end	torsional vibration	bearing crush
cylinder bore taper wear	bearing spread	cam ground

▪▪ INTRODUCTION

The crankshaft assembly, the piston, and the rod are referred to as the **lower end**. This chapter deals with all lower end parts, various service procedures performed on them, and diagnosis of part failures. Also included are coverage of the engine block and lubrication system.

▪▪ CYLINDER BLOCK CONSTRUCTION

The cylinder block, which is made of cast iron or aluminum, is cast around a sand mold called a *core* (**Figure 19.1**). The core is suspended in a container with a liner that will provide the shape for the outside surface of the engine block. The core is supported at several points around the outside of the core box, which leaves core holes in the finished block. When molten iron is poured into the core box, the heat of the casting process cooks the sand. When the casting cools, the sand breaks up. The casting is shaken out. Any remaining sand is washed away through the core holes leaving the finished casting, complete with water jackets. The core holes are closed off with core plugs (**Figure 19.2**).

▪▪ CORE PLUGS

Core plugs are usually made of steel or brass, although rubber and copper expandable plugs are available, too. Brass core plugs are superior because they do not rust. Brass plugs are not used in new cars because of their extra cost, and because new engines are filled with coolant, which prevents rust. Core plugs are also known as expansion plugs, welsh plugs, freeze plugs, or soft plugs.

▪▪▪ HISTORY NOTE ▪▪▪

Many years ago, core plugs were sometimes called "freeze-out plugs," or "freeze plugs."

There is a misconception that core plugs will push out if the coolant freezes, preventing the block from being damaged by expanding ice, but this is not usually the case.

A welsh plug is an older-style core plug, different from the standard cup-type core plug of today. It is a slightly convex circular piece of metal that is installed against a recess in the core opening. It is tightened by striking its convex center to expand it against the block.

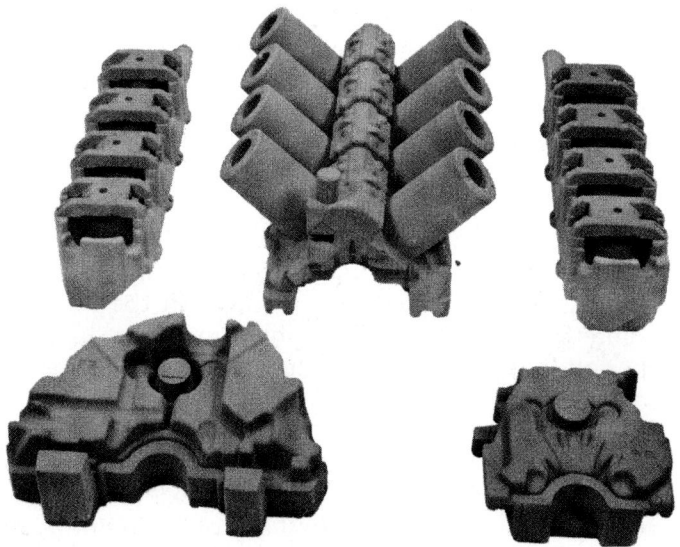

Figure 19.1 Casting cores.

Figure 19.2 Core plugs.

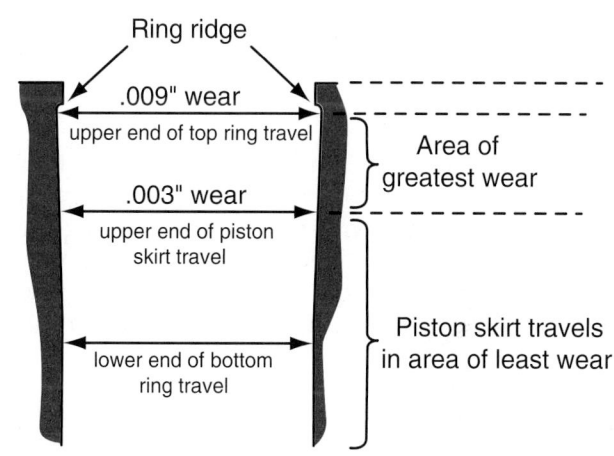

Figure 19.3 A ring ridge.

CYLINDER BORE

Cylinders are bored in the block. They wear in a *taper* and *out-of-round* fashion. The top of the cylinder receives less lubrication, and it is subjected to the high pressure of the piston rings against the cylinder wall as the air-fuel mixture is ignited. This causes **cylinder bore taper wear** (**Figure 19.3**), which forms the ring ridge at the top of the ring travel. When piston-to-cylinder-wall clearance is within specifications, out-of-round wear is minimal. When there is excessive piston skirt clearance, out-of-round wear can result when the piston tilts from one side to the other as it stops and changes direction at the top of the cylinder.

CYLINDER SLEEVES

Some engines use replaceable cylinder bores, called wet or dry *sleeves*. Aluminum blocks usually have permanently installed iron cylinder sleeves. Sometimes a cylinder wall is cracked or rusted, or the block has a serious defect (see Chapter 53). If all the other cylinders are in good shape, the damaged cylinder can be bored oversize to accept a *dry* sleeve installed with an interference fit. Sleeves usually come in ⅛" and ³⁄₃₂" wall thicknesses. Some heavy-duty engines use removable *wet* sleeves. These differ from conventional sleeves

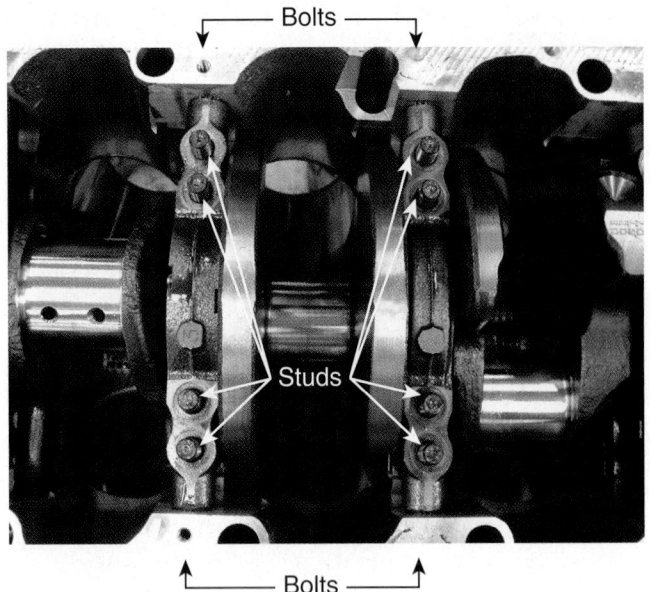

Figure 19.4 These main bearing caps have four studs and two more from the side. *(Courtesy of Tim Gilles)*

in that wet sleeves only contact the block at the upper and lower ends. The middle portion of the sleeves are exposed to coolant in the water jackets, so the sleeves must be sealed at each end.

MAIN BEARING CAPS

Main bearing bores are bored at the factory with the bearing caps in place. Main caps are not interchangeable and must be returned to the same bearing bores from which they were removed. Some heavy-duty engines use four-bolt and six-bolt main caps instead of two-bolt main caps for extra strength (**Figure 19.4**).

LIFTER BORES

Lifter holes are bored in the block on engines with camshafts in the block (pushrod engines). They have very little clearance to the lifters, and the lifters spin in the lifter bores.

CRANKSHAFT DESIGN

The crankshaft converts the up-and-down (reciprocating) motion of the pistons to rotary motion. Its polished bearing surfaces are called *journals*. The journals that support the crankshaft as it turns in the block are called *main bearing journals*. Those bearing journals that are in line on the same axis as the front and rear journals are all mains.

Journals that are offset from the main bearing journal centerline are called *rod journals* or *crankpins*. Connecting rod journals transfer up-and-down motion between the crankshaft and connecting rod. Rod journals are also known as crankpins. As described in Chapter 15, connecting rod journals are offset at 90-degree angles for eight cylinders, 120-degree angles for even-fire six cylinders, and 180-degree angles for four cylinders.

The crankshaft has oil passages drilled from the main journals to the rod journals. This allows oil under pressure to reach the connecting rod journal. With proper maintenance, crankshafts often last until a complete engine overhaul with very little wear.

Counterweights

Opposite each rod journal on the crankshaft is a counterweight that precisely balances the combined rotating mass of the offset rod journals and the rod (see Figure 15.22). The counterweight is much heavier than the rod journal to compensate for the opposite reciprocating weight of the connecting rods, bearings, and piston assembly.

An eight-cylinder, V-type crank sometimes has fewer main bearing journals than an in-line six cylinder. **Figure 19.5** compares the crankshafts of an in-line 4, an in line 6, and a V8. V8 engine rod journals are wide enough for two connecting rods (see Figure 16.3).

A V6 that has 60 degrees between the banks, like the smaller V6, is *even-fired* (a cylinder fires every 120 degrees of crank rotation). Some 90-degree V6 engines are also even-fired, and some are not. To achieve 120-degree even-firing with a 90-degree V6 block requires the crankpins to be offset **Figure 19.6**. The amount of offset varies between manufacturers.

Cast or Forged Cranks

Crankshafts are either cast or forged. Forged cranks are stronger but more costly. Cast cranks are of high enough quality to do an adequate job.

NOTE: *The cast crank has a very narrow parting line where the casting mold halves were joined. The forged crank has a wider, ground-off area. The difference between the two types can be seen in* **Figure 19.7**.

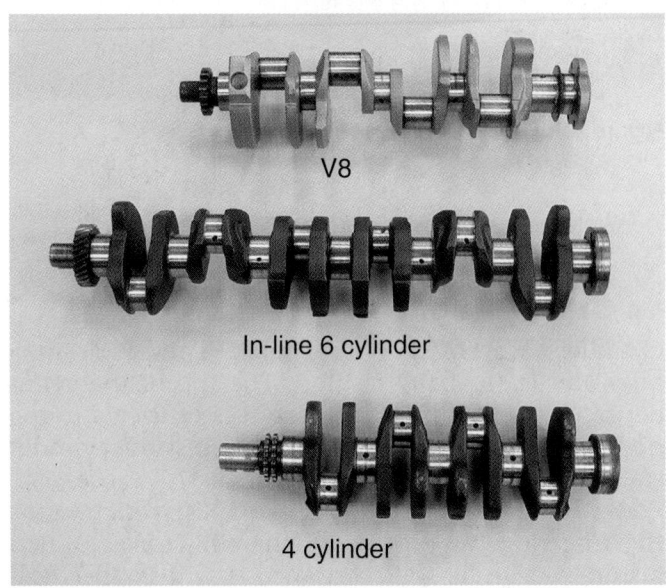

Figure 19.5 Crankshaft design differences. From top to bottom: A V8 crank, an in-line 6 crank, and an in-line 4 crank. *(Courtesy of Tim Gilles)*

Figure 19.6 Splayed crankshaft journals on a 90-degree V6 are used to minimize engine imbalance. *(Courtesy of Tim Gilles)*

(a)

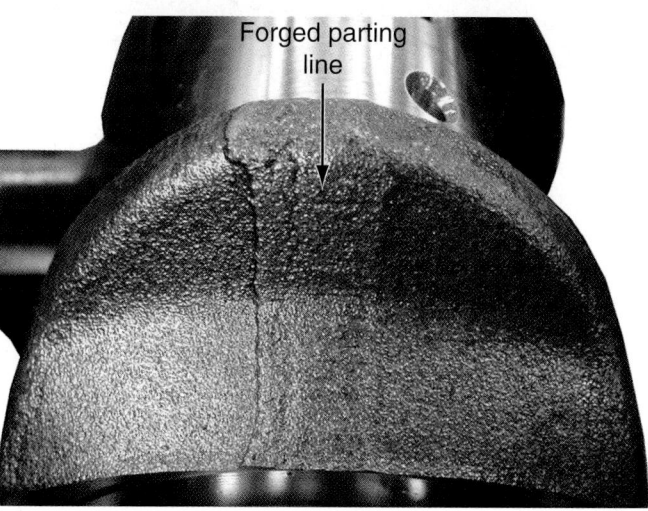

(b)

Figure 19.7 (a) A cast crank. (b) A forged crank. *(Courtesy of Tim Gilles)*

Cast cranks must have larger counterweights because cast metal is not as dense as forged metal, so therefore it is lighter.

CRANKSHAFT END THRUST

The crankshaft is pushed forward by pressure in the torque converter or by the release spring pressure of the clutch. This is called *end thrust*. One of the crankshaft main bearings has a precision bearing surface ground on its sides, called thrust surfaces (see Figure 15.21). A *flanged thrust bearing* fits between the crankshaft thrust surfaces, (**Figure 19.8**) controlling back-and-forth movement (end thrust). Sometimes thrust insert half-washers are used instead of a flanged bearing (**Figure 19.9**).

DIRECTION OF CRANKSHAFT ROTATION

The Society of Automotive Engineers (SAE) has a standard for engine rotation. Most automotive engines (other

Thrust faces

Figure 19.8 A main bearing thrust insert. *(Courtesy of Tim Gilles)*

Thrust inserts

Figure 19.9 Thrust inserts. *(Courtesy of Tim Gilles)*

than Hondas and Hyudais) rotate counterclockwise when viewed from the flywheel end of the engine. *Transverse*-mounted engines also follow this standard. A *longitudinally* mounted engine (called front-to-rear or north-to-south) would turn clockwise when viewed from the front.

VIBRATION DAMPER

During combustion, the force on the piston twists the crankshaft and the crankshaft tries to straighten itself out again. It overcorrects, twisting back in the other direction. Like a tuning fork, these *oscillations* occur for several cycles before fading out. When oscillations from several cylinders are occurring at the same time, a **torsional vibration** can actually cause the crankshaft to break. Timing chain and sprocket wear can also result. Most of the vibration occurs at the front of the crankshaft because the flywheel is resistant to oscillations.

The vibration damper or *harmonic balancer* at the front of the crankshaft dampens torsional vibration. It has a heavy outer *inertia ring* and an *inner hub* separated by a synthetic rubber strip (see Figure 15.15). The two parts stretching against the rubber strip absorb vibrations. Four-cylinder engines usually have only a pulley and do not use a damper. Six- and eight-cylinder engines usually have dampers.

CRANKSHAFT HARDNESS

Crankshafts used by some manufacturers (mostly imports and heavy duty) are specially hardened. If these cranks are reground, the manufacturers recommend that they be rehardened. Crankshafts that have *not* been previously hardened can suffer misalignment if they are subjected to hardening.

NOTE: *Most crankshafts are not specially hardened, but they tend to work-harden with use. Nickel in the crankshaft comes to the surface. A used, polished crankshaft will have a yellow tint because of this nickel.*

BEARINGS

Bearing Inserts

Bearing inserts are made from many different materials. Dissimilar metals slide against each other with lower friction than similar metals, so bearings usually have a steel back with a soft alloy surface. Bearings have been made with surfaces of babbit, copper-lead, or aluminum alloy.

- The copper-lined bearings are premium bearings.
- Some bearings are made of solid aluminum alloy with no steel backs.
- Multilayered bearings have steel backs with one or more layers of lining material (**Figure 19.10**).
  Some bearings have pores filled with an antistick coating like PTFE (Teflon). Others are manufactured with a porous surface that provides pockets for lubricant storage.

Figure 19.10 A multilayered steel-backed bearing. *(Courtesy of Dana Corporation)*

Bearing Properties

The bearing surface has three primary properties that make it suitable for use in an engine:

- *Embeddability* (**Figure 19.11**) helps prevent crankshaft wear. When foreign particles are forced between the bearing and journal, they are absorbed by the softer bearing material. When the particle is too large to be completely absorbed, it creates a high spot on the bearing.
- *Conformability* is when the crankshaft journal surface is damaged and the bearing metal "flows" to conform to it (**Figure 19.12**).
- *Fatigue strength* is the bearing's ability to withstand intermittent loads without deteriorating.

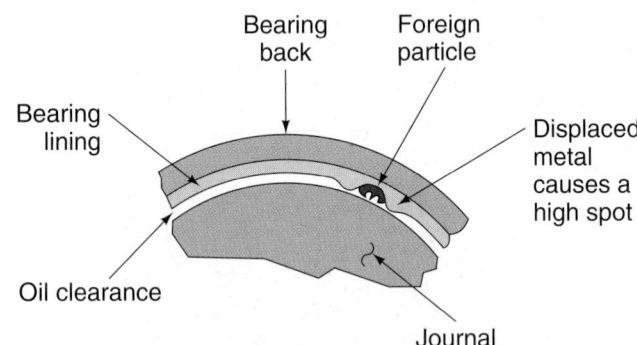

Figure 19.11 Embeddability.

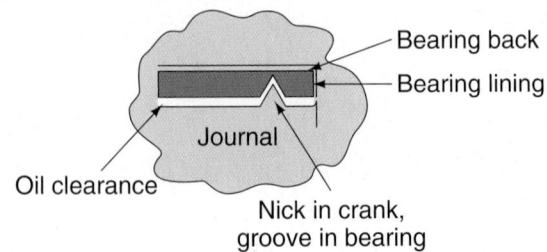

Figure 19.12 Conformability.

Different bearing materials are best for specific uses. Babbitt bearings, which have the best embeddability, were widely used in the past, but the loads of modern engines call for more load-carrying capacity.

Bearings are subjected to normal loads from three sources:

- Pressure from the flame front against the piston
- Centrifugal force from the rotating weight of the rod and crankpin
- Inertia from the up-and-down motion of the piston and rod assembly

Most bearing manufacturers prefer trimetal bearings, which can carry at least three times the load of babbitt bearings and are acceptable in the other categories. Aluminum bearings have the best load-carrying capacity but are poor in embeddability.

Locating Lug and Oil Groove

Bearing inserts are positioned in the bearing bore by a locating lug (tang) or by a *dowel* (**Figure 19.13**).

NOTE: *Locating lugs are used to locate or position the bearing laterally, not to keep it from spinning.*

The lug is located on the parting face. It fits into a recess in the bearing bore. When a dowel is used, it fits into a hole in the bearing back.

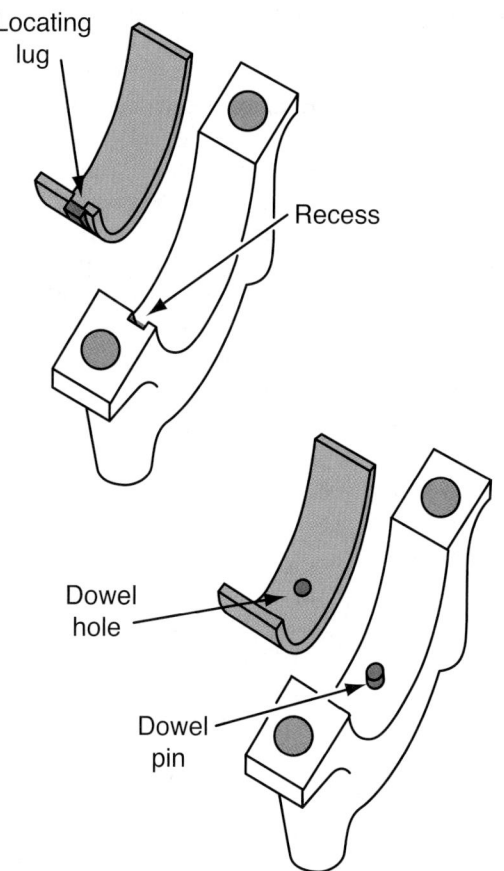

Figure 19.13 Bearings are located in the bearing bore by a locating lug or a dowel.

Bearing Spread and Crush

Bearing spread (**Figure 19.14**) is where the measurement across the parting face is slightly larger than the diameter of the bearing bore. Spread allows the bearing to snap into place and remain in place during assembly of the engine.

Bearing crush is when the bearing extends slightly above the parting line of the bearing bore half by about 0.0005"–0.0015" (**Figure 19.15**). Crush promotes better heat transfer and prevents the bearing from spinning in the bore. Excessive crush would distort the bearing at its parting lines. Insufficient crush would allow the bearing to move back and forth in the housing because there will not be enough radial pressure on the bearing.

Bearing Undersizes

Bearings come in standard sizes and in 0.010", 0.020", and 0.030" undersizes. The size is usually marked on the back of the bearing. Undersized bearings are used when a crankshaft is reground. They are also used on some new engines to compensate for errors in production. The assembled inside diameter of undersized bearings is smaller to compensate for the smaller, reground crankshaft bearing journals. (**Figure 19.16a**).

NOTE: *A crankshaft that is ground to a 0.030" undersize has had 0.015" of metal removed from its surface so both of its new bearing inserts will be 0.015" thicker (*Figure 19.16b*).*

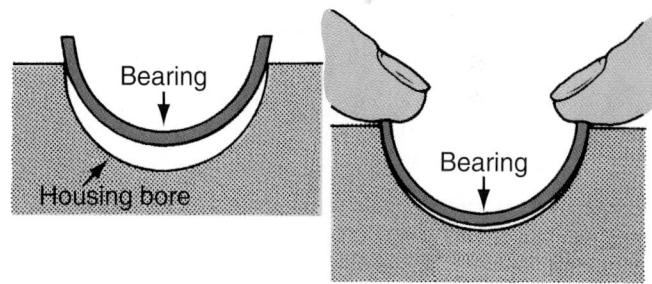

Figure 19.14 Bearing spread permits the bearing to be snapped into place.

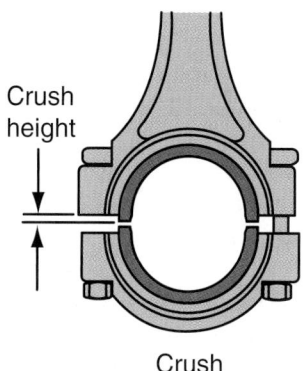

Figure 19.15 Bearing crush prevents bearing movement in the bore.

Precision insert for a standard size crankshaft

Precision insert for a 0.030" undersize crankshaft

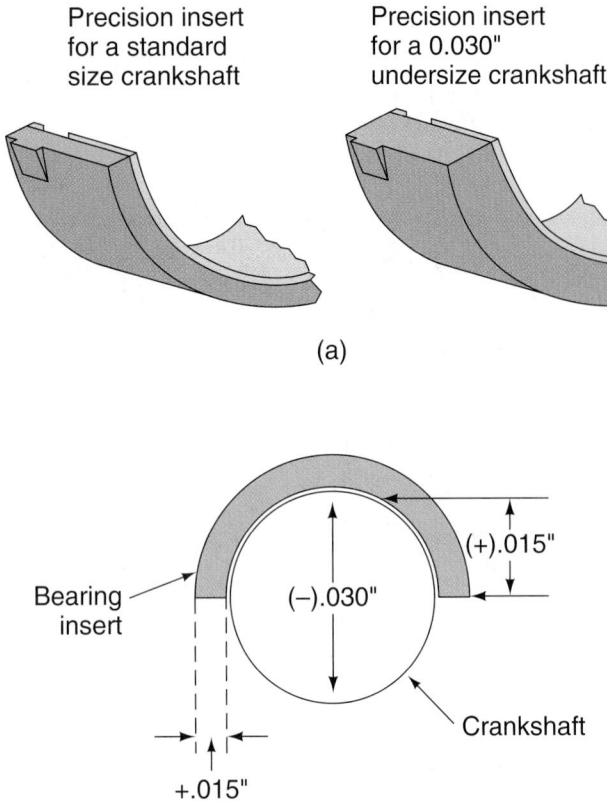

(a)

Bearing insert

(+).015"

(−).030"

Crankshaft

+.015"

(b)

Figure 19.16 (a) An undersized bearing insert is thicker. (b) A 0.030" undersized crankshaft will require bearing inserts that are 0.015" thicker (on the bearing back).

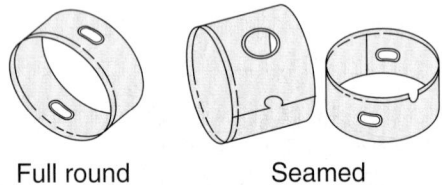

Full round   Seamed

Figure 19.17 Cam bearings are either seamless or they have an interlocking or straight seam. (*Courtesy of Federal-Mogul Corporation*)

Bearings are also available in 0.001" and 0.002" undersizes, but not 0.011" and 0.012".

Cam Bearings

Most cam bearings are made from seamless steel tubing with lining material bonded to the inside. Heavy-duty cam bearings are made in flat strips. The lining material is bonded to the strip, and the strip is rolled to form the bearing. The seam will be either interlocking or straight (**Figure 19.17**).

■ CONNECTING RODS

Connecting rods are usually made of forged or cast steel formed in an I-beam shape for strength. Some racing rods are made of forged aluminum (**Figure 19.18**). Forged rods are stronger than cast rods, but casting techniques

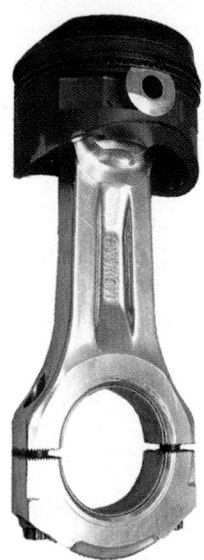

Figure 19.18 This is a racing rod made of aluminum. Connecting rods are usually made of forged or cast steel. (*Courtesy of Tim Gilles*)

have been improved to the point that many late-model passenger car engines use cast rods because they cost less. Some newer connecting rods are cast in one piece before being scribed and fractured to create the rod cap. This creates a more precise fit between the rod and the cap.

The big end of each rod is precisely ground to achieve perfect oil clearance when the rod is installed with its bearing inserts on the crank journal with its bearing inserts. Rod caps are not interchangeable. If they are interchanged, the oil clearances of the bearings can vary greatly and the crank might not even be able to turn.

■ Rod bolts usually have slightly enlarged shanks to hold them tight in the rod (**Figure 19.19**). The cap has precise holes that line up with this enlarged area of the bolt to prevent misalignment of the rod and cap when they are assembled.

■ The notches cut in the rod for the bearing-locating lugs face each other when correctly installed. Incorrect installation of the rod cap can cause uneven bearing wear.

Figure 19.19 The shanks of rod bolts fit tightly in the rod cap for precise alignment of the machined halves.

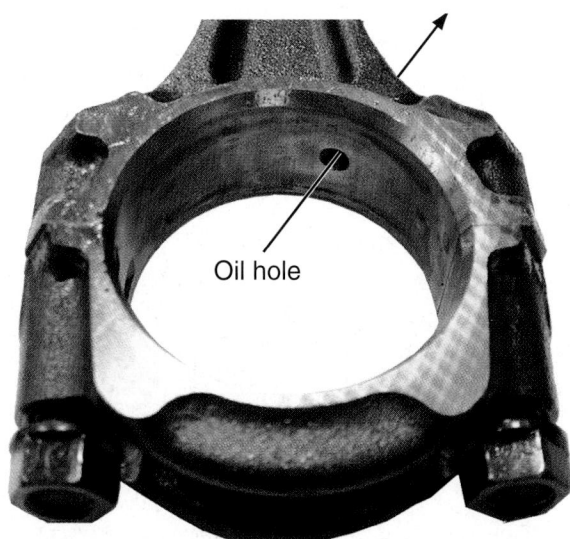

Figure 19.20 Some connecting rods, usually on older engines, have a hole to help lubricate the cylinder walls at low rpm. *(Courtesy of Tim Gilles)*

Rod Oil Holes

Some rods have a "spit" hole to provide cylinder wall lubrication at low engine speeds. In-line engines often have a hole drilled in the rod above the rod bolt so that each rod can lubricate its own cylinder. Some V-type engines have a groove between the rod and cap to lubricate the cylinders on the opposite bank (**Figure 19.20**).

Piston Pin Offset and Spit Hole Orientation. Following are some handy rules to refer to when dealing with piston pin offset and rod spit hole orientation. Nearly all engines are left-hand engines. A left-hand engine is one that turns counterclockwise when viewing the engine from the flywheel side. On a rear-wheel-drive vehicle, this is the orientation from the driver's seat. When a left-hand engine has oil spit holes on the connecting rods, the spit holes are to the right when the piston notches are facing forward. This provides lubrication to the working surface of the piston against the cylinder wall. When viewing the piston from the right-hand side, the piston pin will be offset to the right side of the piston (the major thrust surface).

NOTE: *Most late-model engines do not have oil spit holes but depend on increased rod side clearance to throw oil onto the cylinder walls at low speed.*

■ PISTONS

Piston Parts

Older pistons were made of cast iron, but today's pistons are cast or forged aluminum. Aluminum is an excellent material for pistons because of its strength, light weight, and ability to dissipate heat. The piston undergoes remarkable stresses. It starts and stops twice in each revolution of the crank. At a highway speed of 3,000 rpm, the piston must change direction (start and stop) 6,000 times per minute, or 100 times per second.

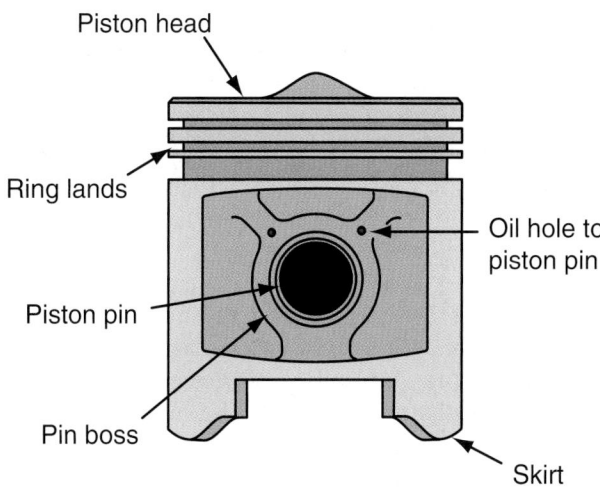

Figure 19.21 Parts of a piston.

Obviously it must be as light as possible. The top of the piston is exposed to the heat of combustion and must be able to get rid of heat in a hurry or it will melt. **Figure 19.21** shows the parts of a piston.

■ PISTON HEAD

The piston head (or *crown*) is round; the skirt is usually oval. The diameter of the piston head is smaller than the diameter of the skirt by about 0.019" to 0.048" (most are about 0.022" less) (**Figure 19.22**).

■ PISTON RING GROOVES

The top piston ring is positioned as high as possible on the piston to lessen piston slap, help ring durability, and lower emissions. The top ring groove suffers the most abuse.

The holes in the *oil ring groove* allow excess cylinder wall oil to return to the crankcase by way of the inside of the piston (see Figure 15.27).

Heat Transfer

Crown heat is transferred through the piston rings to the water jackets behind the cylinder walls. Some pistons have a slot below the oil groove, and some have a heat dam above the top groove to keep crown heat

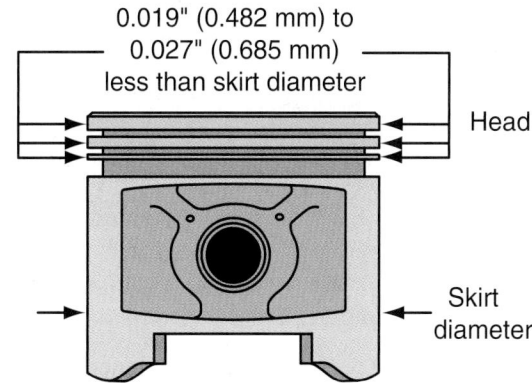

Figure 19.22 The piston head is smaller than the skirt.

from traveling down to the top ring groove. Sometimes oil return holes are long slots that also help to prevent the transfer of heat from the piston crown to the skirt. Some piston heat is also transferred to the oil beneath the crown. **Figure 19.23** shows carbon that has formed beneath the crown from excessive piston heat. Carbon like this sometimes clogs the oil pump inlet screen.

Piston Head Shapes

Some manufacturers use many different piston head shapes available for the same engine to allow variations in compression ratios (**Figure 19.24**). High-compression pistons can only be installed in one

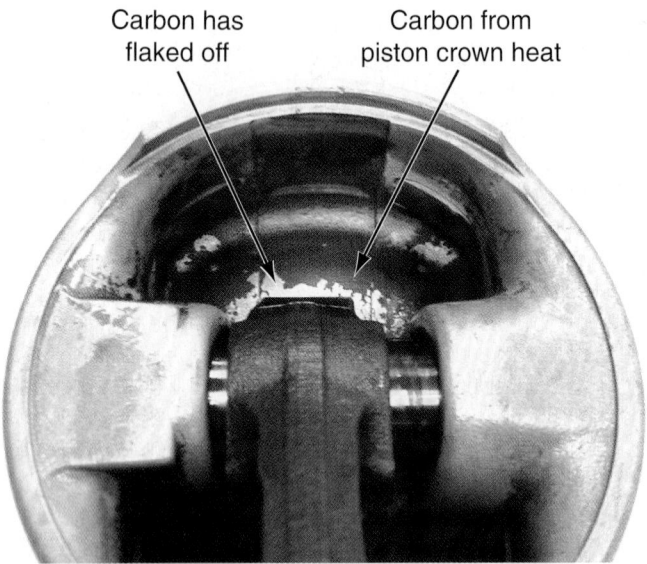

Carbon has flaked off   Carbon from piston crown heat

Figure 19.23 Excessive temperature in the combustion chamber heats the center of the piston, resulting in the formation of carbon and soot. (*Courtesy of Tim Gilles*)

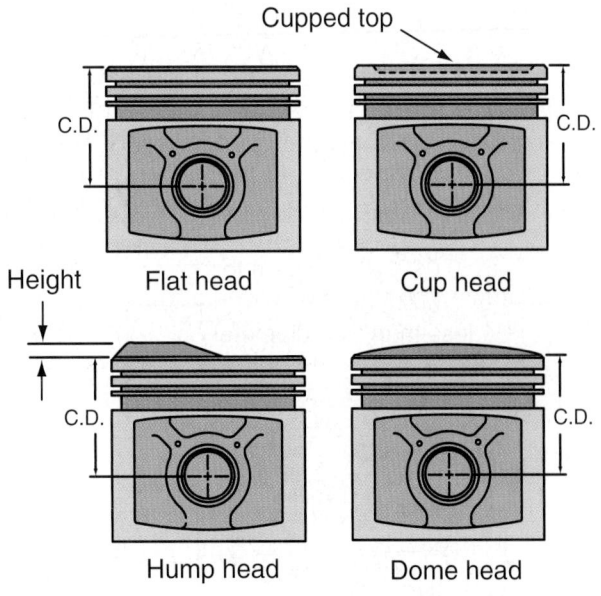

Cupped top

Height   Flat head   Cup head

C.D.   C.D.

Hump head   Dome head

C.D. = Compression distance

Figure 19.24 Common piston head shapes.

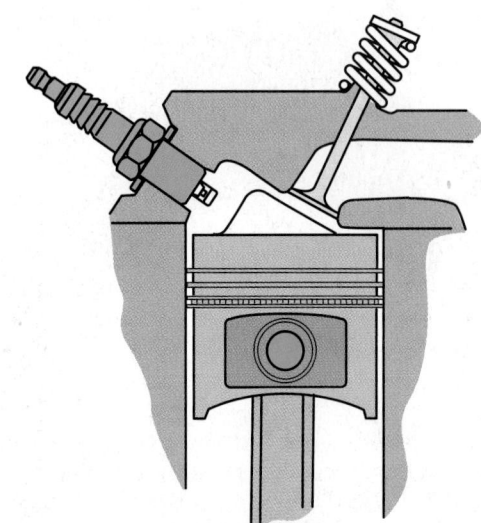

Figure 19.25 A section view of a combustion chamber and cylinder showing the relationship between a high-compression piston and the valve and cylinder head.

direction in the cylinder (**Figure 19.25**). If they are installed backward, the piston crown will interfere with the combustion chamber or valves.

■ CAST AND FORGED PISTONS

Cast pistons are the most common. Molten aluminum is poured into a mold to form a piston casting (**Figure 19.26**). Forged pistons are available for heavy-duty or high-performance use. Forged pistons have a *dense* grain structure (**Figure 19.27**) and are said to be 70% stronger than cast pistons. Forged pistons also dissipate heat better (**Figure 19.28**). Cast pistons are not strong enough for prolonged use above 5,000 rpm. Cast pistons have a *porous* grain structure (see Figure 19.31).

Pistons in newer vehicles are often *hypereutectic*, which refers to the silicon content of the aluminum. This allows them to be made lighter. A hypereutectic piston cannot withstand the tensile loads that a forged

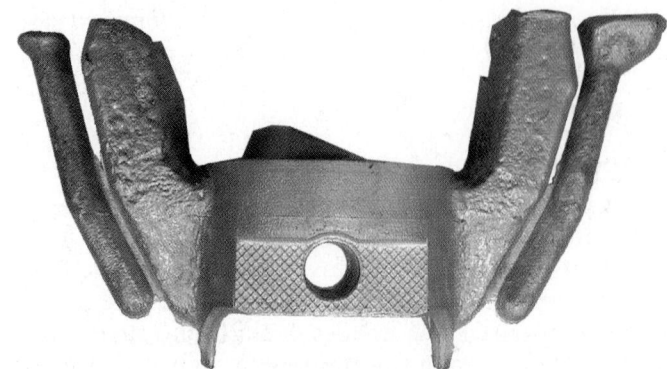

Figure 19.26 This is what a piston looks like when it comes out of the foundry. (*Courtesy of Tim Gilles*)

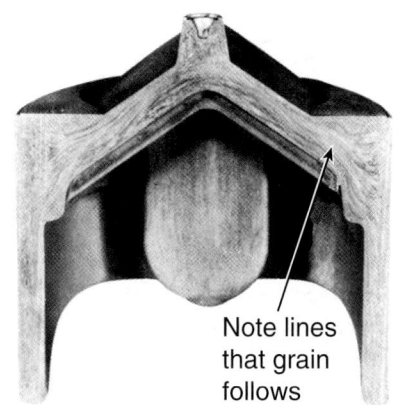

Figure 19.27 A forged piston has a dense grain structure. Note that it has a grain structure similar to wood.

piston can, but it has better wear characteristics, such as in the ring groove and pin bore areas. They also have lower rates of thermal expansion, and they can run with lower clearances.

■ PISTON SKIRT

The surface of the piston skirt is purposely left slightly rough, with small machine marks less than 0.001" deep. This somewhat rough surface is superior for retaining lubrication.

An aluminum piston expands at about twice the rate of the cast-iron block. Three things can be done to control this expansion:

■ Most piston skirts are tapered (**Figure 19.29**). They are larger in diameter at the bottom because the bottom of the piston runs cooler than the top. At operating temperature, the piston will fit the cylinder wall properly. Many newer pistons are *barrel shaped*; smaller at both the top and bottom.

■ The piston skirt is **cam ground** (**Figure 19.30**). This allows it to fit the cylinder with only 0.0005" to 0.0025" of clearance. The smaller the clearance, the less chance of piston "slap" and the oval cylinder wear that it causes. Expansion takes place perpendicular (at a 90-degree angle) to the piston skirt.

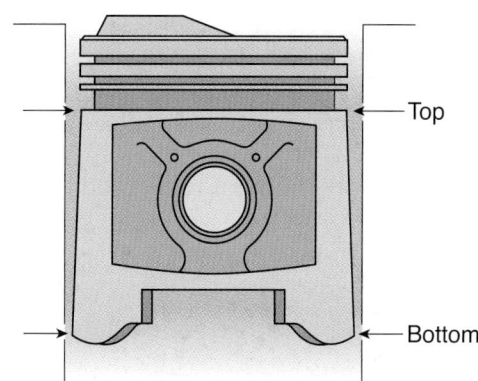

Figure 19.29 Piston skirt taper.

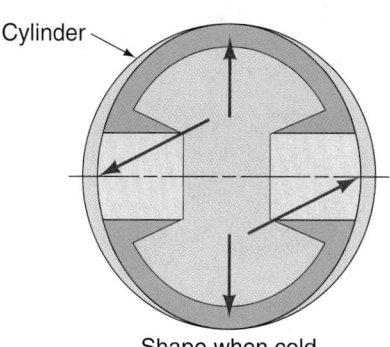

Shape when cold

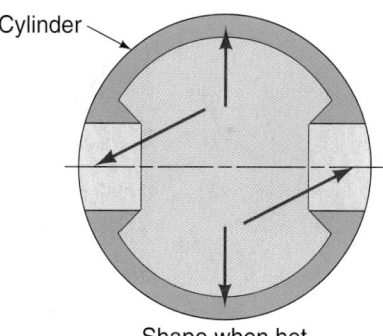

Shape when hot

Figure 19.30 Cam ground piston skirt. (*Courtesy of Sunnen Products Company, St. Louis, Missouri*)

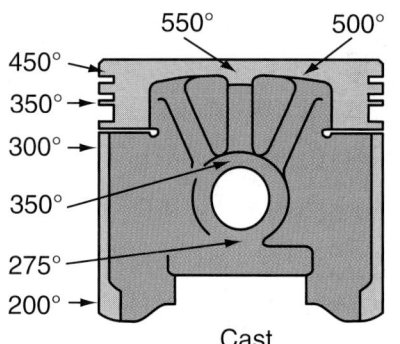

Cast

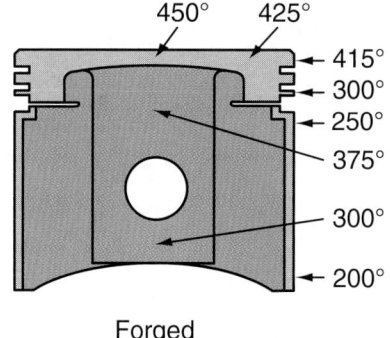

Forged

Temperature distribution

Figure 19.28 A forged piston dissipates heat better than a cast piston.

■ Cast pistons often have struts of spring-loaded steel cast into them (**Figure 19.31**). Struts help the piston resist expansion as it warms up. Newer cast pistons do not have struts and require about 0.005" extra clearance. The broken piston in Figure 19.31 is the result of overtightening of a rod bolt during engine assembly.

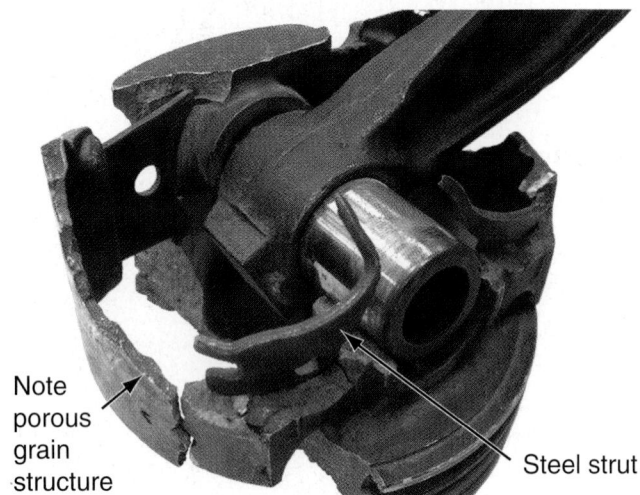

Figure 19.31 This piston was destroyed when an overtightened connecting rod bolt gave way. Notice the steel expansion strut around the wrist pin. (*Courtesy of Tim Gilles*)

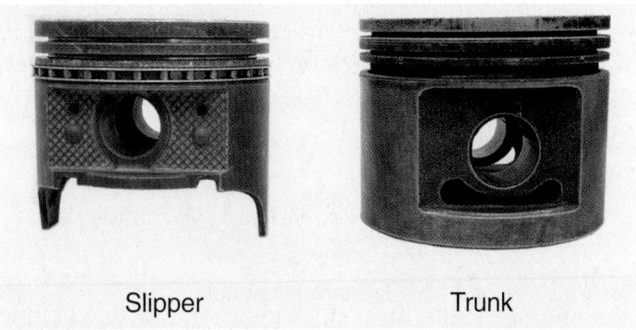

Figure 19.32 Slipper and trunk-skirt pistons. (*Courtesy of Tim Gilles*)

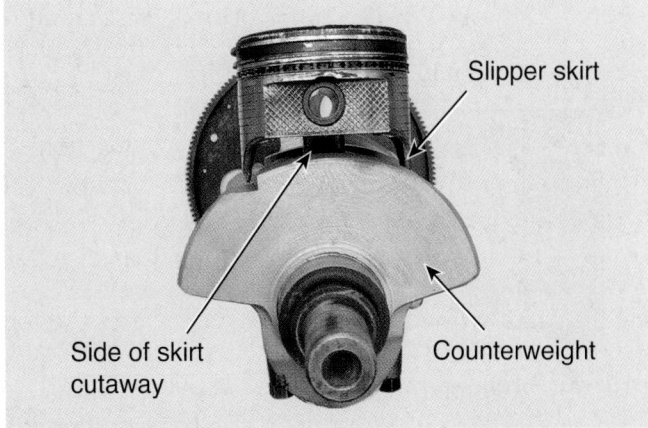

Figure 19.33 The slipper-skirt piston is cut away to clear the crankshaft counterweights. (*Courtesy of Tim Gilles*)

Skirt Types

Figure 19.32 shows slipper-skirt and trunk-skirt pistons. A full-skirt piston used on longer stroke engines is known as a *trunk piston*. Most of today's automotive pistons are slipper-skirt pistons. A *slipper-skirt* is designed to clear the crankshaft counterweights on engines with short strokes (**Figure 19.33**). The surfaces of the piston skirts that are 90 degrees to the wrist pin are called *thrust surfaces*.

■ PISTON PIN OFFSET

Piston pins are often offset approximately ¹⁄₁₆" (0.062") from the piston centerline (**Figure 19.34**). The connecting rod changes to the other side between the compression and power strokes. When the piston rocks from one skirt to the other at TDC, the side of the piston that is "thrust" against the cylinder wall changes.

A piston usually has a direction-identifying notch that faces the front of the engine. On V-type engines, the pins on one bank of pistons will be offset to the intake manifold side of the cylinder, and the pins in the opposite bank will be offset toward the lower side of the cylinder (exhaust manifold side of the cylinder).

Some *crowned head* (*pop-up*) pistons used in high-compression engines (see Figure 19.25) are not offset at all, so that all of the pistons for a V-type engine can be identical. The engine will not turn over if these pistons are accidentally installed backward.

Piston Pin Height

The piston pin can be placed at different heights also (see Figure 19.24). This dimension, called *compression height*, is the distance from the center of the pin bore to the flat area of the top of the piston.

NOTE: *When pistons from engines of the same bore have different displacements (because of different crankshaft strokes), do not interchange them without checking this critical dimension.*

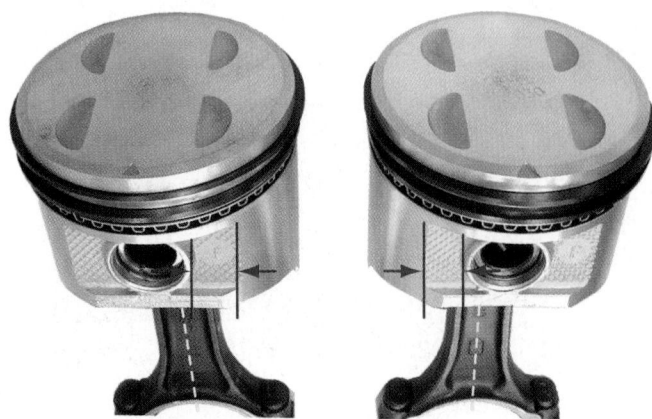

Figure 19.34 The piston pin is often offset. (*Courtesy of Tim Gilles*)

To calculate compression height, measure the distance to the top of the piston from the top of the pin hole. Add one-half of the pin diameter to this amount.

WRIST PINS

The piston is attached to the connecting rod with a wrist pin (also called a *piston pin*). The wrist pin is lubricated from either a hole in the top of the pin boss (see Figure 19.22) or through an angle-drilled hole that runs from the oil ring groove to the pin boss. The pin boss is the metal area of the piston that surrounds the wrist pin. Modern automobile piston pins are either press-fit in the connecting rod or full floating (**Figure 19.35**).

Pressed-Fit in Rod

The most common method of attaching the wrist pin is to have it *float* (pivot) in the piston with a very close clearance (about 0.0005"). A pin with this fit feels almost tight when cold but will move freely when hot. The pin is pressed into the rod with a force of about 2 tons (4,000 lb.). The *interference* fit is about 0.001".

Full-Floating Pins

A full-floating pin floats in the "eyes" of both the rod and piston. Retainers (lock rings) are required to prevent the pin from sliding out and damaging the cylinder wall (**Figure 19.36**). When lock rings come out

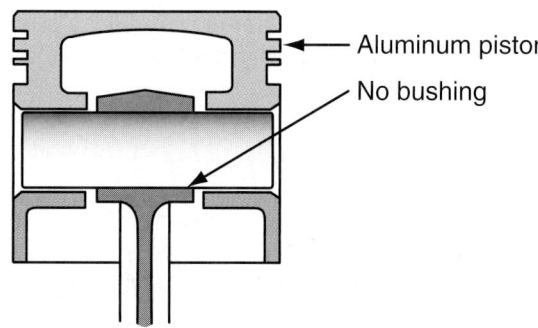

Press fit in connecting rod

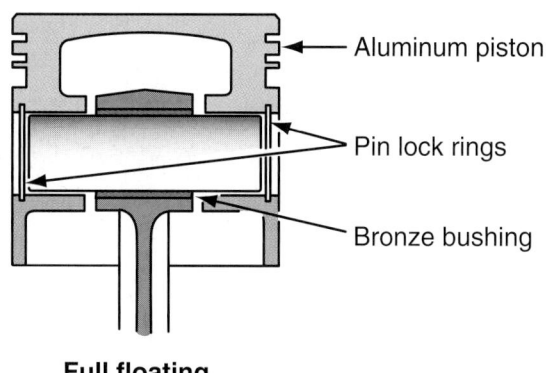

Full floating

Figure 19.35 Piston pins are either pressed into the rod or they pivot in both the piston and rod.

Lock ring

Figure 19.36 Lock rings are installed on full-floating pins with the open end facing down. *(Courtesy of Tim Gilles)*

of their grooves during service, the cause is usually excessive crankshaft end thrust, a tapered rod journal, or a misaligned connecting rod.

PISTON RINGS

Piston rings do a truly remarkable job in the engine. Most engines today use two compression rings and one oil ring (see Figure 15.24). The top ring is exposed to the flame of combustion during every power stroke. In addition to sealing the tremendous combustion pressures, piston rings also help cool the piston and control oil consumption.

COMPRESSION RINGS

Compression rings are forced against the cylinder wall by combustion pressure at the top and back of the ring (see Figure 15.26). The *top ring* controls most of the sealing of combustion. It operates under a load of 110,000 psi and a temperature of 600°F. As long as the top ring seals, the piston stays in good condition. When excessive blowby and particles of combustion begin to leak by along with the flame, the oil rings will become stuck and plugged up, and the piston and rings will rapidly deteriorate.

NOTE: *A square ring groove is a must. A scratch in the bottom of a ring groove can leak as much as a more obvious scratch in a cylinder wall.*

The *second ring* runs at 35,000 psi and a temperature of 300°F. It performs two functions, serving as a backup for the compression ring above it and the oil ring below it. It captures what little pressure escapes past the top ring. It also scrapes oil back down the cylinder wall, where it can be returned to the crankcase.

Compression rings are cast in groups. They are installed on a mandrel and machined out of round on a lathe. A slot is milled, which becomes the ring gap.

When the ring is compressed, closing the large gap, the ring forms a spring-loaded circle. Rings are made as narrow as possible in order to keep them light in weight so they will not flutter. Flutter occurs when inertia from high speed causes the ring to stay against the top of the groove, causing leakage, blowby, and, ultimately, a broken ring. Increased crankcase pressure can also help to unseat the ring from the bottom of the groove causing flutter.

Low-Tension Rings

Most of the friction within an engine is between the valve lifter and cam lobe or between the piston rings and cylinder walls. To improve fuel economy, *low-tension piston rings* were introduced in 1985. Also called *low-friction* or *shallow groove rings*, these have reduced cylinder wear rates considerably.

Compression rings account for about 20% of the total drag on a cylinder wall, and the oil ring accounts for 60%. Low-friction ring sets have reduced ring pressures by 60% for the oil ring and 10% for the compression rings.

■ COMPRESSION RING DESIGN

Several ring designs are shown in **Figure 19.37**. The simplest cast-iron rings are square. Other rings are *taper faced* or have *chamfers* or *counterbores* that cause the ring to twist when compressed. The intent of these features is to cause the ring to contact the cylinder wall with only a narrow part of its face (the part of the ring that slides against the cylinder wall). This creates higher sealing pressure on the power stroke while maintaining lower friction on the other three strokes. These rings will seat more quickly, and their downward scraping action will help to control oil. There are several designs of piston ring. Their theory of operation is beyond the scope of this text.

■ COMPRESSION RING MATERIALS AND COATINGS

Piston rings are either plain cast iron, moly, or chrome.

Cast-Iron Rings

Rings are often made of plain cast iron. Iron rings are used in re-ring jobs when the cylinder walls have not been rebored. Iron rings are very forgiving of cylinder wall imperfections and are very popular with technicians because they seat easily. When a plain cast iron ring is machined on the surface that rides on the cylinder wall, some manufacturers leave a threaded finish on the ring. The peaks of thread help the ring to seat rapidly to the bore. The threads also hold oil for break-in.

NOTE: *The threaded finish can often be seen on the upper part of the face of a taper-faced second ring when an engine is disassembled for a rebuild.*

Chrome-plated and moly rings do not have the threaded finish, because they are lapped at the factory.

NOTE: *Where premium rings are used in the top groove, iron rings are often used in the second groove.*

Iron rings can become loose and their spring tension is reduced at higher temperatures.

Moly Rings

Some premium cast-iron rings, called *moly rings*, have a groove machined on their faces. A plasma spray gun is used to fill the groove with molten molybdenum (**Figure 19.38**), which solidifies as soon as it contacts the ring. After machining, the moly is about 0.004"–0.006" thick. The groove is cut into the ring because the moly does not stick to the ring's base material as well as chrome does. The groove keeps the moly from coming off the ring.

Moly top rings have become popular on a majority of today's new cars. Their high melting temperature makes them extremely resistant to scuffing.

■ SCIENCE NOTE ■

Molybdenum has a melting temperature of 4,750°F; the melting temperature of iron is 2,250°F. When molten molybdenum solidifies on the ring surface, it becomes molybdenum oxide. This oxide is a ceramic and is very hard, unlike the metal molybdenum.

Moly rings are called self-lubricating; that is, they have a porous surface that helps them retain lubricant

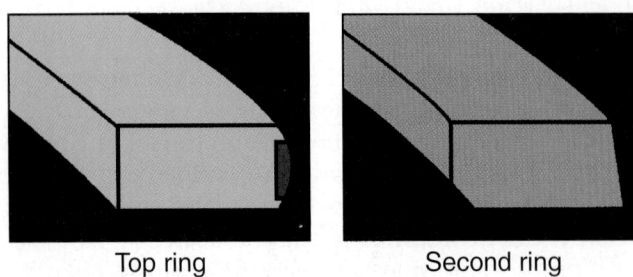

Figure 19.37 Typical compression ring designs. *(Courtesy of Federal-Mogul Corporation)*

Top ring　　　　Second ring

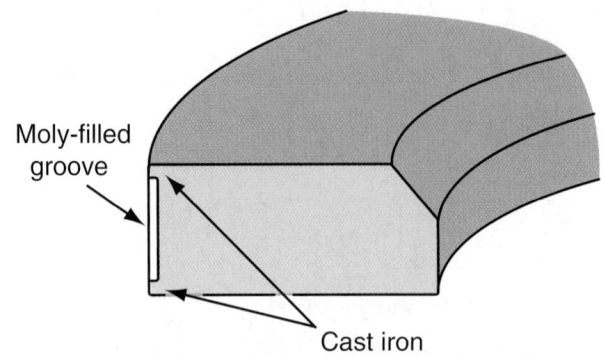

Moly-filled groove

Cast iron

Figure 19.38 A moly (molybdenum) piston ring.

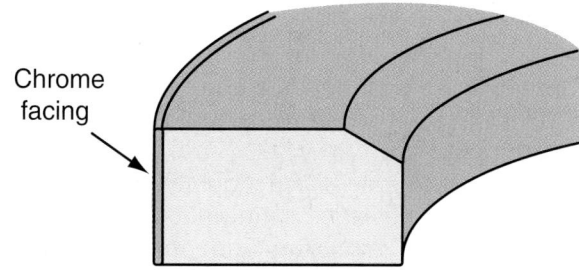

Figure 19.39 A chrome piston ring.

for less wear. They are also said to seat instantly because they are preground to make them perfectly round. Moly rings are ground rather than lapped; this prevents the porous surface from filling with lapping abrasive. As development of these rings has progressed, the porosity (5%) of the surface has become less and less. Older moly rings, with more porosity (20%), trapped dirt to become abrasive.

Chrome Rings

Some rings are chrome plated (**Figure 19.39**). These rings have about twice the resistance to abrasive wear as moly rings but are difficult, or impossible, to seat in worn (oval) cylinders. They are lapped at the factory when new so they can seat in true cylinder bores. Chrome-plated rings are especially recommended for engines used where excessive dirt will be encountered. They should *not* be used with either propane or natural gas fuels. They also have a problem that moly does not have—moly will not weld, but chrome can overheat and weld to the cylinder bore (scuffing). Chrome rings typically last for the life of the engine with virtually no wear.

Premium Ring Combination

Moly has become more popular than chrome for passenger car use because it wears very well under normal conditions. A common piston ring combination found on passenger cars uses a moly barrel-faced top ring, a reverse-torsion second ring, and a three-piece chrome oil ring (**Figure 19.40**).

High-Strength Rings

Ductile Iron Rings. Special ductile iron rings are often used in the top groove to withstand the higher temperatures in some of today's high-heat engines. Ductile iron, also called *nodular iron*, is about twice as strong as plain cast iron. It is always coated with moly or chrome.

Steel Rings. Steel rings are made from steel wire. They are coated with chrome or plasma moly because they would be prone to scuffing on cast-iron cylinder walls. The width of these rings is so narrow (most are 1.2 mm, or 0.047") that a groove for moly is not easily machined; so these rings are usually chrome plated.

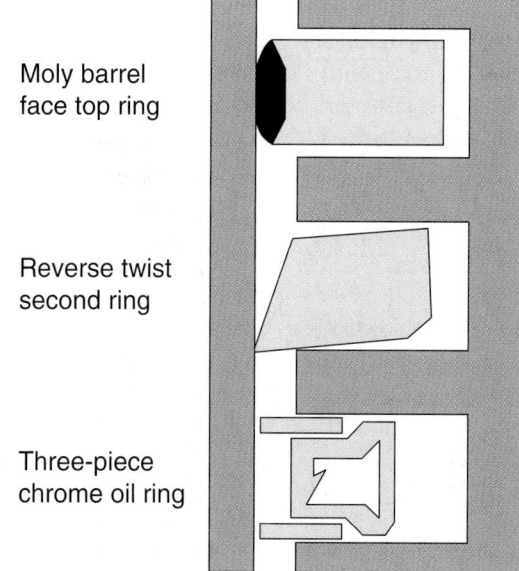

Figure 19.40 A typical piston ring set.

These rings cost less to manufacture than ductile iron rings and provide the same advantage in ductility.

SHOP TIP To test a moly or chrome ring to see if it is ductile iron, simply twist it. If it is regular iron, it will break in two.

Plasma Ring Coatings

Plasma Ceramic. A newer ring face coating is plasma ceramic.

> **SCIENCE NOTE**
>
> *A plasma ceramic ring facing is a composition of titanium oxide and aluminum oxide. It is applied to nodular iron rings with a plasma torch at 30,000°F to 40,000°F. With this process, materials that are not normally sprayable can be sprayed. This provides a combination of metal and ceramic that has even higher resistance to wear than chrome.*

According to TRW, plasma ceramic rings are five times as strong as a stock ring. The rings are said to resist detonation damage, cause less cylinder wear, and have excellent break-in characteristics. Cylinder preparation is the same as for moly rings.

■ OIL CONTROL RINGS

Oil control rings do a remarkable job. The oil ring runs at a temperature of 250°F. Older engines often used three or four compression rings and oil rings at the top and bottom of the piston skirt.

- According to Sealed Power Corporation, if $\frac{1}{10}$ of a drop of oil were consumed during each power stroke in a vehicle driven at 60 mph, an eight-cylinder engine could consume about 90 quarts of oil in one 600-mile trip. Actual oil consumption per power stroke in an average engine is from $\frac{1}{1000}$ to $\frac{2}{1000}$ of a drop.
- An engine that uses a single drop of oil with every power stroke would use 1 quart of oil every 2 miles.
- Oil consumption increases with engine speed. A typical engine uses about seven times as much oil at 70 mph as it does at 40 mph, depending on the rear axle ratio.
- Vacuum during deceleration increases with compression ratio. Earlier, lower compression smog control engines developed only about 20 inches of vacuum, but today's higher compression engines can develop up to 25 inches. This makes the oil ring's job that much more difficult.

Oil rings fail if they become plugged because of improper maintenance, temperatures too high or too low, fuel with too much lead or impurities, or improper air-fuel mixtures.

Most engines have only one oil ring located below the two compression rings. There are several oil ring designs. Some are single-piece cast types, but most oil rings in passenger cars are the three-piece type. These consist of a stainless steel expander and two chrome rails (**Figure 19.41**). Oil rings rarely have problems because they operate in a relatively cool, well-lubricated environment.

The three-piece oil ring prevents oil from being drawn around the sides or back of the ring. Most of the oil is actually scraped off by the lower rail, so barely any oil must actually pass through the expander to the inside of the piston.

The oil ring contributes the most friction to the engine. The chief advantage of the three-piece oil ring is that it can operate with less tension and, therefore, less friction. Low-tension rings have rails that are not as deep and a weaker expander is used.

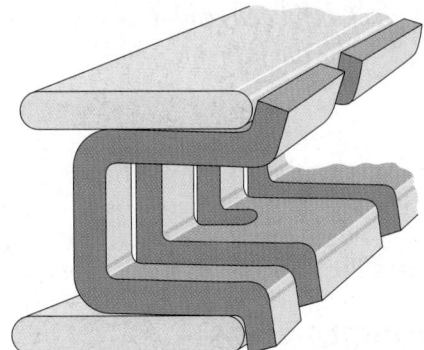

Figure 19.41 A three-piece oil control ring. *(Courtesy of Dana Corporation)*

ENGINE BALANCING

When moving parts are not in balance, the result can be compared to a washing machine with all the clothes on the wrong side of the tub. The result is vibration and worn parts.

Every time engine speed doubles, the force that results from the imbalance is multiplied by 4. An imbalance of ¼ ounce (7 grams, or about one sheet of paper) at a 4" radius on a rotating part creates forces of

- 7 lb. at 2,000 rpm
- 28 lb. at 4,000 rpm
- 63 lb. at 6,000 rpm
- 112 lb. at 8,000 rpm

Counterweights are added to the crankshaft to counteract the force of up or down vibration. They are positioned so they arrive at exactly TDC when the piston is arriving at BDC and vice versa.

An engine can be balanced to prevent vibration. Pistons and rods are weighed on an accurate scale. Material is removed from the heavier parts until they weigh the same as the lightest parts. Balancing theory is covered in detail in Chapter 54.

Balance Shafts

Some of today's engines have used one or two counterweighted balance shafts that are driven by the crankshaft in opposite directions at twice the crankshaft speed (**Figure 19.42**). These shafts, also called *silent shafts*, have counterweights that are timed to cancel out the engine's imbalance.

THE LUBRICATION SYSTEM

The lubrication system is perhaps the least understood area of engine repair. Too often, when engines are repaired, important aspects of this system are neglected. This neglect can result in a very expensive failure.

In a typical lubrication system (**Figure 19.43**), oil is drawn in through the pump screen, and then forced by the pump through an oil filter to the engine's *main oil gallery*. Oil from the main gallery provides lubrication to the camshaft and cam bearings. In a typical pushrod engine with stud-mounted rocker arms, oil is also pumped to the hydraulic lifters. From the lifters, the oil is pumped through hollow pushrods to the rocker arms and valve guides.

- The crankshaft receives oil at its main bearings.
- Oil is pumped through drilled oil galleries from the main journals to the rod journals.
- The cylinder wall and piston are lubricated by oil thrown or sprayed from the rod journals.
- The timing chain, the distributor, and the oil pump drive gear are all splash lubricated.

OIL PUMPS

On cam-in-block engines and some OHC engines, a gear on the camshaft usually drives the oil pump.

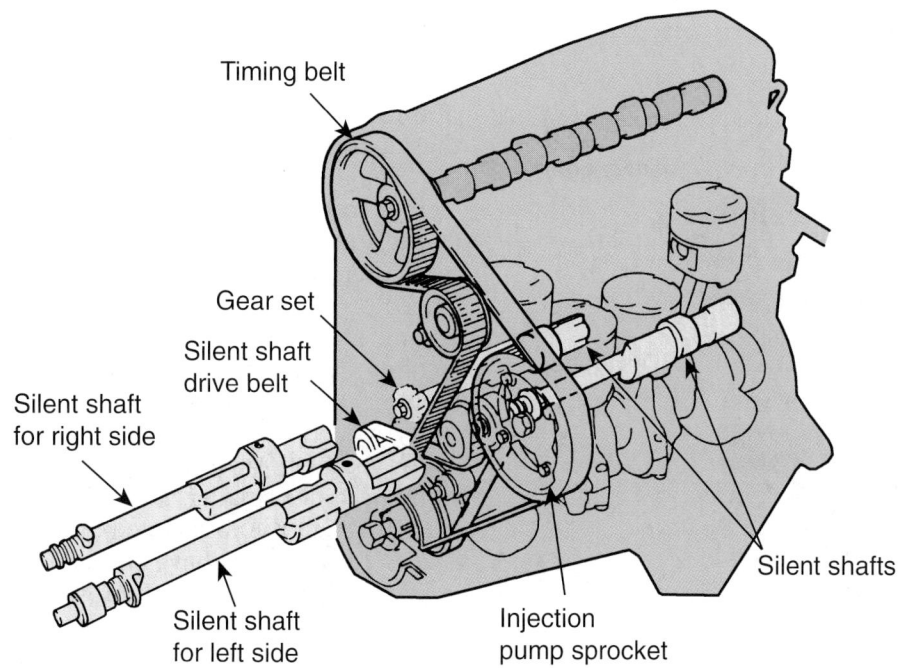

Timing belt

Gear set

Silent shaft drive belt

Silent shaft for right side

Silent shaft for left side

Injection pump sprocket

Silent shafts

Figure 19.42 Balance shafts, also called silent shafts. *(Courtesy of Ford Motor Company)*

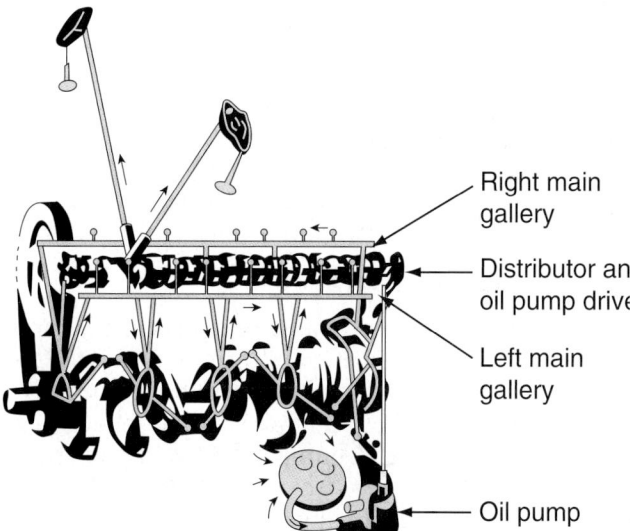

Right main gallery

Distributor an oil pump drive

Left main gallery

Oil pump

Figure 19.43 A typical lubrication system.

Oil pump

Figure 19.44 A crankshaft-driven oil pump. *(Courtesy of Tim Gilles)*

Camshaft-driven pumps are usually attached to the bottom of the distributor by a tang or hex drive shaft (**Figure 19.44**). On the bottom of most distributors is a gear that meshes with a drive gear on the camshaft. Some OHC oil pumps are crankshaft driven (**Figure 19.45**).

NOTE: *Automotive oil pumps are positive displacement pumps; they deliver the same amount of oil with each rotation. Without a pressure relief valve, if the outlet were plugged, the pump would explode.*

Three types of oil pumps are used in automotive engines: the *external gear* (**Figure 19.46a**); the rotor, or *gerotor* (**Figure 19.46b**); and the *internal gear*, or

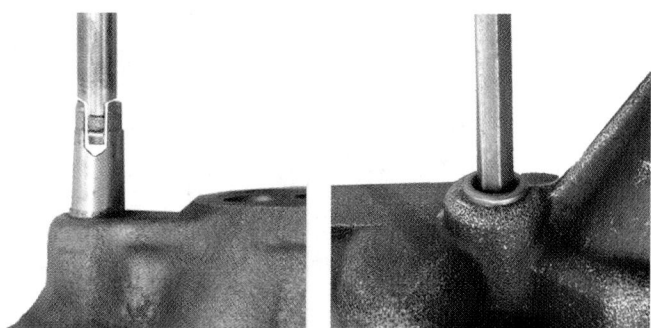

Figure 19.45 Two common distributor oil pump drives. *(Courtesy of Tim Gilles)*

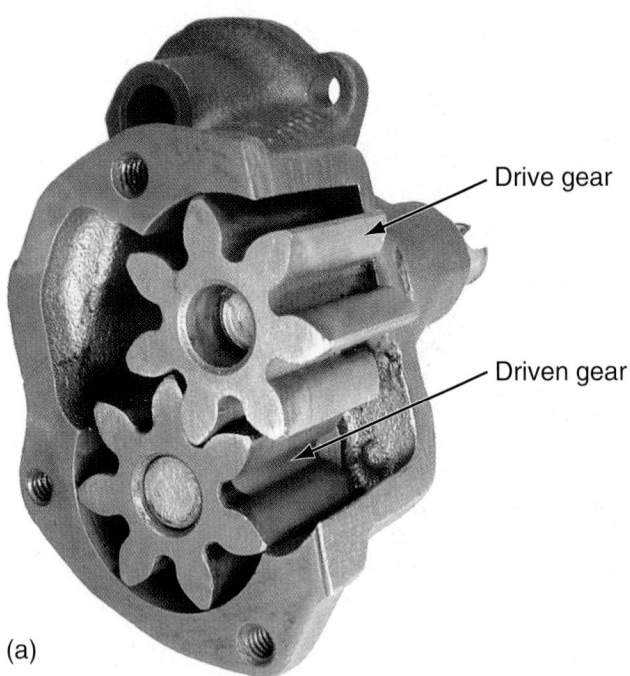

(a)

(b)

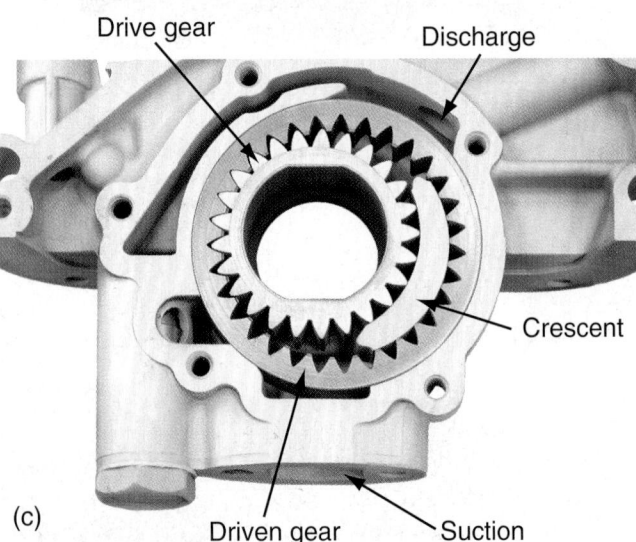

Drive gear   Discharge

Crescent

(c)   Driven gear   Suction

Figure 19.46 (a) An external gear oil pump.
(b) A crankshaft-driven gerotor pump. (c) A crankshaft-driven internal gear pump. (*Courtesy of Tim Gilles*)

crescent, pump (**Figure 19.46c**). External gear pumps and rotor pumps are the most common types on cam-in-block engines. Internal gear and gerotor pumps are found on OHC engines, especially when the engine has distributorless ignition.

The outer rotor of a gerotor pump is mounted on center to the crankshaft or camshaft drive. The inner rotor is mounted off-center and has one less lobe than the outer rotor, so the outer rotor rotates at 80% of the speed of the inner rotor. As the inner rotor turns, the chamber of the pump enlarges, creating a vacuum and pulling in oil. The pump chamber reaches its largest volume when the lobes and tips of the rotors seal both the inlet (low-pressure) and outlet (high-pressure) sides. As the rotor continues to turn, the chamber volume becomes smaller, and oil is pumped under pressure through the discharge port.

Gerotor pumps have smooth pumping action and less aeration of the oil than gear-type pumps, which change their volume at about each 7 degrees of rotation. With each rotor revolving on its own center, pressure remains more constant, instead of constantly raising and lowering. Because there are fewer moving parts and a lower rotational speed of 600 rpm at 3,000 shaft rpm, less wear is the result.

Crankshaft-Driven Oil Pumps

Crankshaft-driven gerotor and internal gear pumps turn at twice the speed as camshaft-driven pumps, so they provide more oil flow at idle. They also save space, having thinner gears, and the pump turns twice as fast.

▬ PRESSURE RELIEF VALVE

The oil pump must be capable of supplying oil under sufficient pressure to all areas of the engine *at idle speed*. The faster the pump turns, the more oil it pumps. For this reason, it must have a relief valve to relieve excessive pressure when the pump turns at higher speeds (**Figure 19.47**).

- Most oil pump relief valves limit maximum oil pressure by diverting excess oil back to the inlet side of the pump.
- The maximum oil pressure in an engine is controlled by the tension of the relief valve spring. Too much pressure can cause an oil filter to burst.

▬ OIL PUMP SCREEN BY-PASS VALVE

Most screens have a *by-pass valve* that can open when the screen becomes plugged, or when the oil is cold and too thick to flow freely. The by-pass valve is normally seated against the *cross strap* on the screen (**Figure 19.48**). If a screen by-pass valve has opened and remains open, the screen will have to be replaced. When the screen is restricted and the bypass opens, a large (about ½") opening results. This opening allows large particles to enter the pump because oil goes to the pump before it goes to the filter (**Figure 19.49**).

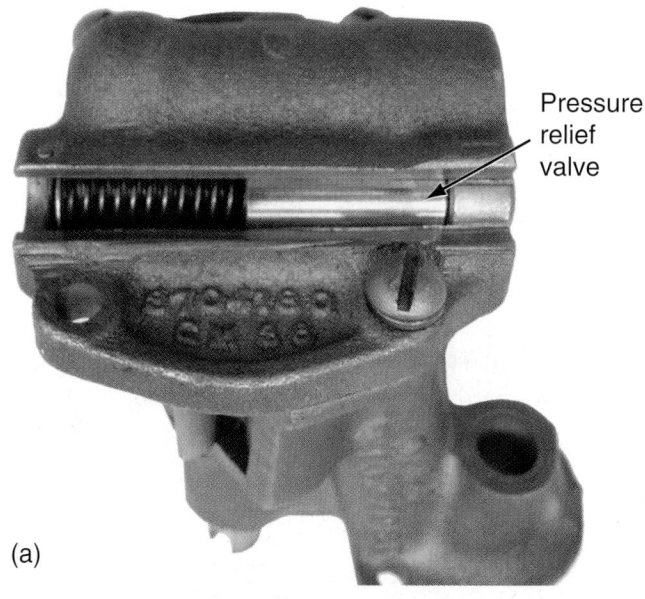

(a)

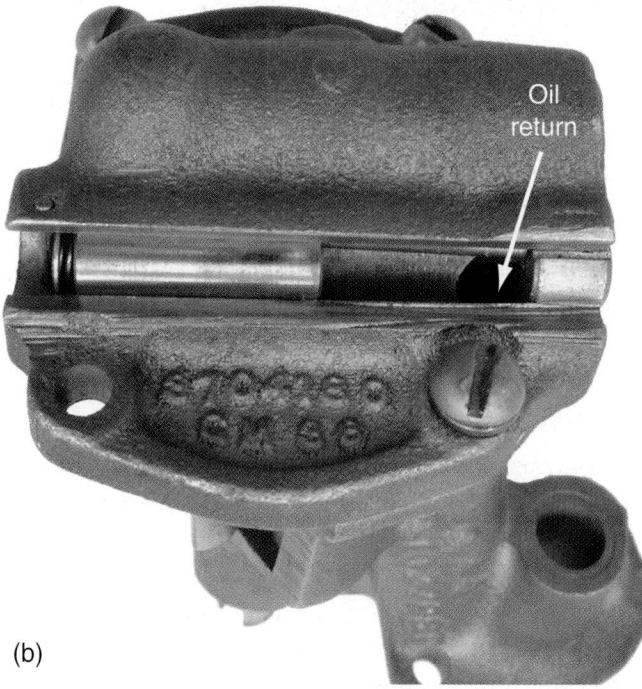

(b)

Figure 19.47 Oil pump pressure relief valve operation. (a) Pressure relief valve closed. (b) When the valve is open, the oil return to the crankcase is uncovered. *(Courtesy of Tim Gilles)*

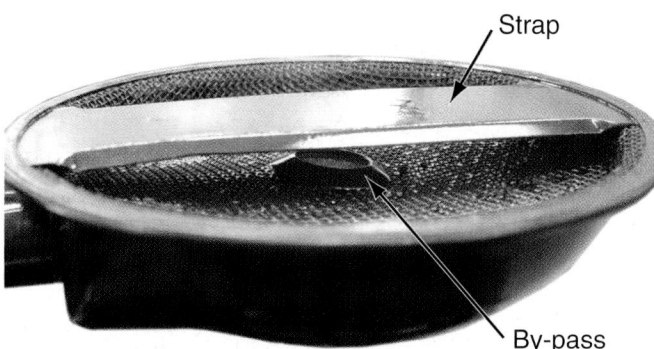

Figure 19.48 This pump screen is defective; the bypass should be seated against the strap. *(Courtesy of Tim Gilles)*

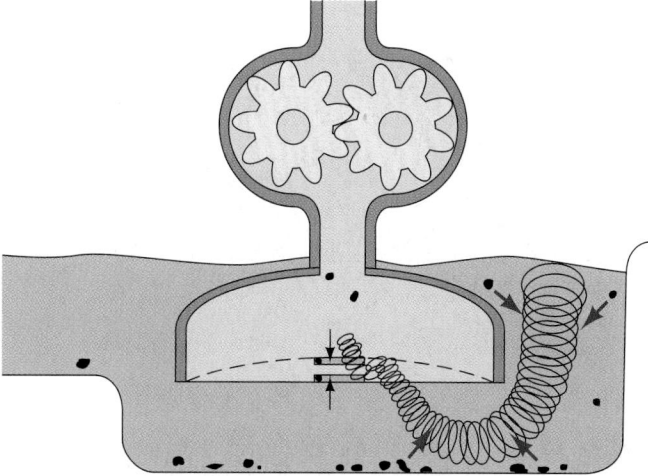

Figure 19.49 When the by-pass valve opens on a cold morning, the foreign material will be sucked into the oil pump. *(Courtesy of Federal-Mogul Corporation)*

■ OIL PRESSURE

Proper lubrication in an engine is achieved only by the distribution of clean oil under pressure. Oil *pressure* can only be created when there is *resistance to the flow* of oil from the oil pump (see Pascal's law, Chapter 57).

The correct amount of bearing clearance is very important. If there is too much clearance, the engine will have low oil pressure at idle and will probably *knock*. If there is too little oil clearance, the crankshaft could seize or a bearing could be burned.

A *positive displacement pump*, like those used in automobiles, has a fixed output per revolution. The faster it turns, the more oil it pumps. If correct bearing clearances are not maintained, oil will not reach all areas of the engine when the engine is idling. At idle speed, the pump is turning slowly and cannot fill the system if clearances are excessive.

Excessive oil *clearance* near the oil pump—at a main bearing, for instance—will result in insufficient oil pressure at the rocker arms or valve lifters, which are farther away from the pump. Noisy hydraulic lifters are often an indication of an oil pressure problem.

NOTE: *Most manufacturers recommend a minimum of 10 psi of oil pressure for every 1,000 rpm. Some engines have different requirements, so check manufacturer's specifications.*

Satisfactory oil pressure at idle is normally in the neighborhood of 25 psi, but pressure indicator lights do not normally come on until pressure drops below 10 psi.

■ HIGH-VOLUME OIL PUMPS

The output per revolution of an oil pump depends on the diameter and thickness of the rotors or gears. Oil pumps with about 20%–25% more volume are available. High-volume pumps deliver more oil per revolution due to the increased size of the pump cavity (**Figure 19.50**). These pumps are able to provide more

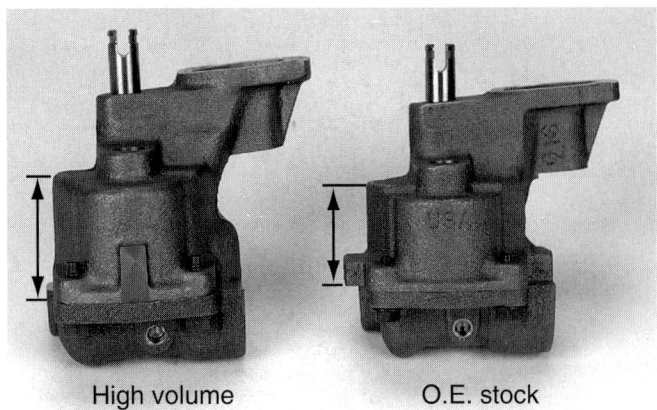

Figure 19.50 A high-volume oil pump is larger than a stock pump.

Figure 19.51 An oil pan with a built-in windage tray. *(Courtesy of Tim Gilles)*

oil to a worn engine at idle. Because they pump a larger volume of oil per revolution, this results in an increase in idling oil pressure in such an engine. However, they may not necessarily provide any other added advantages to a passenger car engine. Consider the application carefully before replacing a standard pump with a high-volume one.

NOTE: *Some racing engines use a dry sump system in which the oil is held in a large holding tank, which uses a high-volume external oil pump, usually mounted on the front of the crankshaft. An oil cooler might also be used in this system. Oil is pumped from the pan to the holding tank. A system of this type holds more oil. The problem of foaming and aeration that would occur if the oil was stored in the crankcase is eliminated.*

▬ WINDAGE TRAY AND BAFFLES

At high speeds, the revolving crankshaft creates wind. This can cause air pockets around the oil pump screen, which can cause the pump to lose its prime. The objective is to keep the pump screen continuously immersed in oil. A windage tray helps to prevent these air pockets (**Figure 19.51**). Some windage trays bolt onto the main bearing caps; others are built into the oil pan.

Some oil pans also have built-in baffles that help to keep the oil from sloshing back and forth when driving over hills or bumps or during hard cornering or braking. Be sure to check for pieces of foreign material trapped under a windage tray or baffle.

▬ REVIEW QUESTIONS

1. What are two ways that cylinder bores wear?
2. When a cylinder wall is damaged, what is the name of the part that is installed to correct it?
3. What is the force that causes the crankshaft to move to the front or rear called?
4. The SAE standard for crankshaft rotation says that crankshafts will rotate (circle one) clockwise, counterclockwise when viewed from the flywheel end.
5. What are two names for the part that is on the front of the crankshaft to control vibrations?
6. What is the name for the bearing design feature that allows it to snap in place and remain there during engine assembly?
7. What keeps the bearing from spinning in its bore?
8. At 6,000 rpm, how many times does the piston have to start and stop in 1 second?
9. Name three types of ring facings.
10. What part usually drives the oil pump drive shaft?

◾ ASE-STYLE REVIEW QUESTIONS

1. Technician A says that when a cylinder wears in a tapered fashion, a ridge is left at the top of the cylinder. Technician B says that a ridge forms at the top of the cylinder because the rings slowly push the metal up from the cylinder wall below it. Who is right?

   a. Technician A   **c.** Both A and B

   b. Technician B   **d.** Neither A nor B

2. All of the following statements about crankshafts and bearings are true *except*:

   a. A replacement bearing insert for a 0.020" undersized crankshaft is 0.010" thicker than a standard one.

   b. A 0.010" undersized crankshaft that is worn by 0.001" can be refitted with 0.011" undersized bearings.

   c. Copper lined bearings are premium bearings.

   d. A 0.020" undersized crankshaft journal has had 0.010" of metal removed from its original surface.

3. Which of the following statements is/are true about pistons and piston rings?

   a. The piston pin is usually offset from the centerline of the piston to prevent piston slap.

   b. The head of the piston is round and the skirt is oval.

   c. Piston rings often have chamfers or counter-bores to cause them to twist.

   d. All of the above.

4. Technician A says that the oil pump screen has a relief valve in it in case the screen is plugged or the oil is too thick. Technician B says that excessive main bearing clearance can result in low oil pressure that happens only at high speeds. Who is right?

   a. Technician A   **c.** Both A and B

   b. Technician B   **d.** Neither A nor B

5. Technician A says that a longitudinally mounted engine would turn counterclockwise when viewed from the front. Technician B says that transverse engines turn clockwise when viewed from the flywheel end. Who is right?

   a. Technician A   **c.** Both A and B

   b. Technician B   **d.** Neither A nor B

Cooling System, Belts, Hoses, and Plumbing

THEN AND NOW: WATER PUMPS

Henry Ford's Model T moved North America into the age of the automobile, selling 15 million between 1909 and 1927. But in spite of the clever engineering that made it so practical, the "Tin Lizzie," as the Model T was nicknamed, did not have a water pump. The liquid coolant in those days was plain water, sometimes combined with a little alcohol to prevent ice from forming. It was circulated by natural *convection*—just like hot air, hot liquids rise. The water absorbed heat from the block, causing it to rise and enter the top of the radiator. As it cooled, it sank to the bottom entering the lower part of the engine. While this thermo-siphon principle worked well enough for normal service, heavier hauling jobs required spending a few dollars for an aftermarket add-on water pump.

Water pumps became standard on almost all other cars after the Model T. They move a huge volume of coolant—from a couple of hundred gallons per hour at idle speed, to several thousand in a big V8 cruising on the freeway.

A number of engines today have a water pump powered by an overhead cam's timing belt. An advantage is that the engine cannot run without the pump spinning, preventing one possibility of overheating. Also, the engine can be made a little shorter because the pump pulley is eliminated—an important consideration with today's crowded engine compartments. Another recent advance in driving the pump is the serpentine accessory belt, most examples of which are self-tensioning. Serpentine belts take up less space, last longer, run more quietly, and are less likely to slip than the traditional "V" belt.

The biggest change in water pump design occurred when the ceramic seal was adopted many years ago. This seal increased the life expectancy of the pump, adding far more resistance to wear and abrasion than the rubber or leather types used previously.

Premature pump failure can also be prevented by using quality antifreeze, which contains additives that lubricate and protect the water pump shaft seal. Changing the coolant at reasonable intervals will also add to the life of the pump. Water pump replacement is a fact of life, however. Sooner or later almost every car will start to leak at its water pump seal.

An aftermarket water pump on a Ford Model T. (*Courtesy of Bob Freudenberger*)

A modern water pump. (*Courtesy of Tim Gilles*)

CHAPTER 20

Cooling System Theory

■■ OBJECTIVES

Upon completion of this chapter, you should be able to:

✔ Describe the operation of a cooling system.

✔ Understand the construction and theory of operation of cooling system parts.

✔ Explain the importance of coolant.

✔ Describe why regulating engine temperature is important.

■■ KEY TERMS

down-flow radiator
cross-flow radiator
heat exchanger
transmission oil cooler
detonation

preignition
open loop
thermostat bypass
sending unit
cavitation

fan clutch
bimetal coil spring
heater core

■■ INTRODUCTION

Combustion temperatures sometimes reach as high as 4,500°F; average temperatures of combustion are close to 2,000°F. Aluminum pistons melt at about 1,225°F, so this heat must be carried off rapidly to prevent engine damage (**Figure 20.1**). Most of this heat is not used to propel the piston and must be carried off by the cooling system.

Theory of the cooling system is covered here because it is an essential but often neglected area of vehicle maintenance. A thorough knowledge of the operation of the system provides the service technician with the ability to serve the customer's needs (and sell profitable service).

■■ LIQUID AND AIR COOLING

Some smaller engines are only cooled by air, but most automotive cooling systems use liquid coolant to carry off engine heat. The engine block and cylinder head have cast-in water jackets surrounding its cylinders and combustion chambers (**Figure 20.2**).

■■ LIQUID COOLING SYSTEM PARTS

A liquid cooling system is made up of the following parts (**Figure 20.3**):

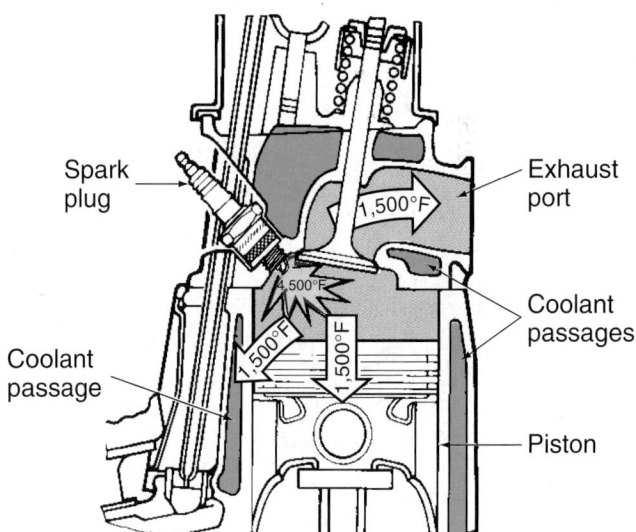

Figure 20.1 Heat must be removed from the engine by the cooling system.

■ The water pump, or coolant pump, circulates water throughout the system.

■ The radiator conducts heat away from the coolant.

■ Radiator hoses connect the parts of the cooling system.

Figure 20.2 Coolant passages surround the cylinders. *(Courtesy of Tim Gilles)*

- The fan pulls air through the radiator when the car is not moving fast enough to move the air.
- The thermostat controls the temperature of the coolant.
- Coolant is circulated to the heater in the passenger compartment.

■ COOLING SYSTEM CIRCULATION

Coolant exits from the bottom of the radiator after it is cooled. It is drawn into the coolant pump (commonly called a *water pump*) (**Figure 20.4**), which pumps it into the block and head where it picks up heat. The coolant is then returned to the top of the radiator where it is cooled again.

Coolant Pump

The coolant pump is a centrifugal impeller-type pump. It is mounted on the front of the engine and is usually belt driven by the crankshaft.

An *impeller* fits in the circular pump housing. As it turns, centrifugal force draws coolant into the pump inlet near the center of the pump (**Figure 20.5**). It exits from the outlet at the outside of the pump. Centrifugal pumps are *variable displacement*, which means that their output changes with speed. The pump is sometimes called upon to circulate as much as 7,500 gallons of coolant in 1 hour.

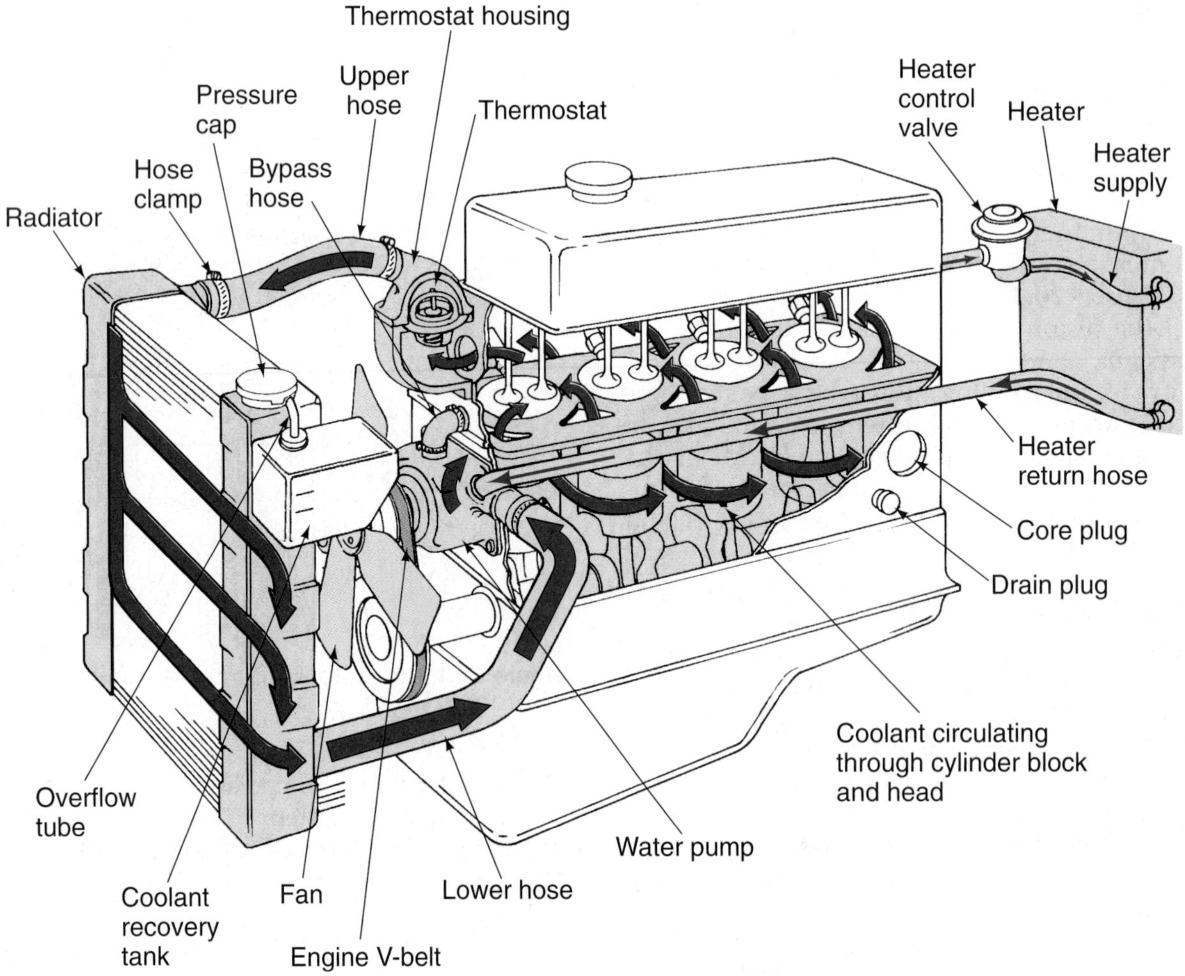

Figure 20.3 Parts of a liquid cooling system. *(Courtesy of the Gates Rubber Company)*

Thermostat housing
Upper hose
Pressure cap
Thermostat
Heater control valve
Heater
Hose clamp
Bypass hose
Heater supply
Radiator
Coolant circulating through cylinder block and head
Heater return hose
Core plug
Drain plug
Overflow tube
Coolant recovery tank
Fan
Lower hose
Engine V-belt
Water pump

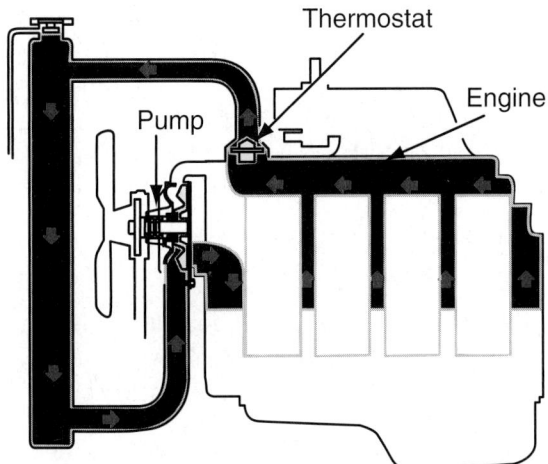

Figure 20.4 Coolant exits the bottom of the radiator and is drawn into the coolant pump.

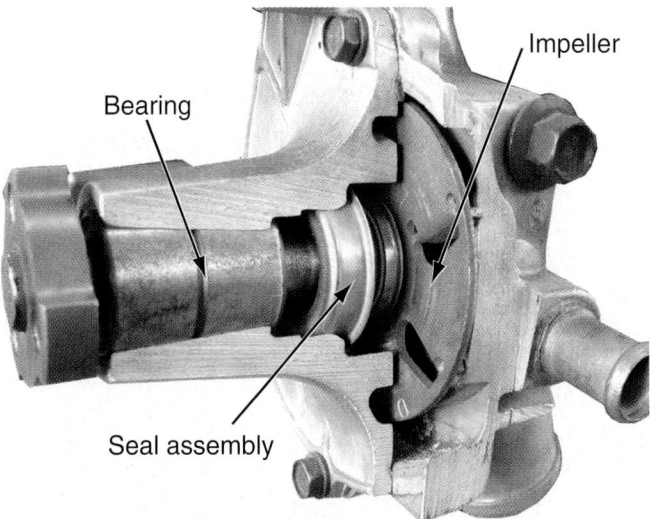

Figure 20.6 A cutaway coolant pump showing the relationship between the bearing and its spring-loaded mechanical seal. *(Courtesy of Tim Gilles)*

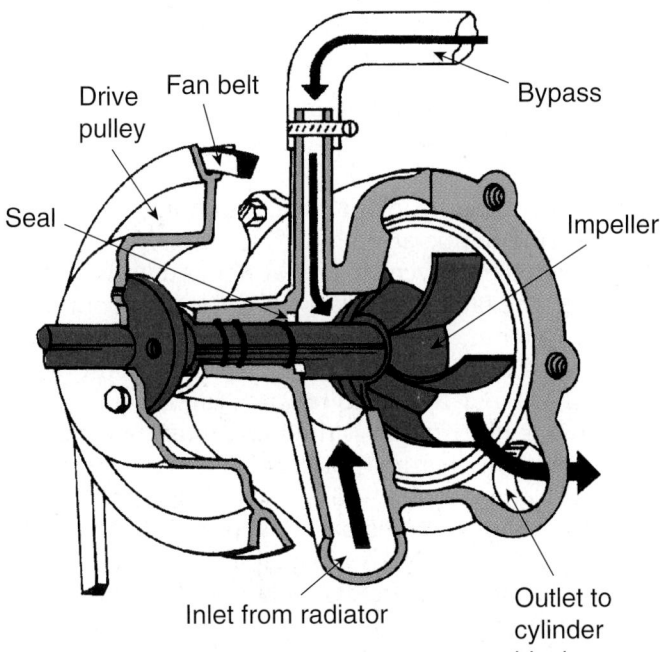

Figure 20.5 Parts of a coolant pump. *(Courtesy of the Gates Rubber Company)*

The coolant pump has a spring-loaded, permanent seal to keep coolant from leaking out. The pump cutaway in **Figure 20.6** shows the relationship between the bearing and its spring-loaded *mechanical seal*.

The coolant pump housing is either aluminum or cast iron. Aluminum is subject to erosion. As the aluminum erodes, the pump cavity becomes larger. The reduced pump efficiency can cause air pockets. Severe erosion will also result in serious internal or external leaks.

Thermostat

The temperature of the coolant must be regulated to be neither too hot nor too cold. A thermostat (see Figure 20.4) regulates the temperature of coolant

leaving the block. When the coolant temperature is too low, it is trapped in the block until it heats up sufficiently to cause the thermostat to open.

Water Jackets

Hollow water jackets are cast into all of the areas of the block and heads that are subjected to high temperatures. Coolant is circulated through these passages to carry away excess heat. When the block is cast at the foundry, metal is poured around sand cores that are supported through core openings. The holes left by these core supports are machined smooth and closed with interference-fit *core plugs*.

■ COOLANT

All automotive coolants are glycol based. The most common automotive coolants consist of a mixture of ethylene glycol with other additives and some water. Concentrated coolant is about 95% ethylene glycol, about 2.5% water, and the rest are additives. It is called permanent antifreeze/coolant because it does not evaporate.

NOTE: *A common misconception is that the term* coolant *refers to a 50/50 mixture of antifreeze and water. The terms* coolant *and* antifreeze, *however, are used interchangeably by their manufacturers. In the industry, antifreeze/coolant is often simply called coolant.*

The other glycol-based coolant is propylene glycol. The major difference between the two coolants is that propylene glycol is not as toxic.

Pure water has 1.4 times the heat-carrying ability of pure ethylene glycol, so water would be the best coolant to use if the only consideration in selection of a coolant was its ability to carry off heat. But water has other limitations. It forms *rust* on iron engine parts. Cylinder walls vibrate during combustion,

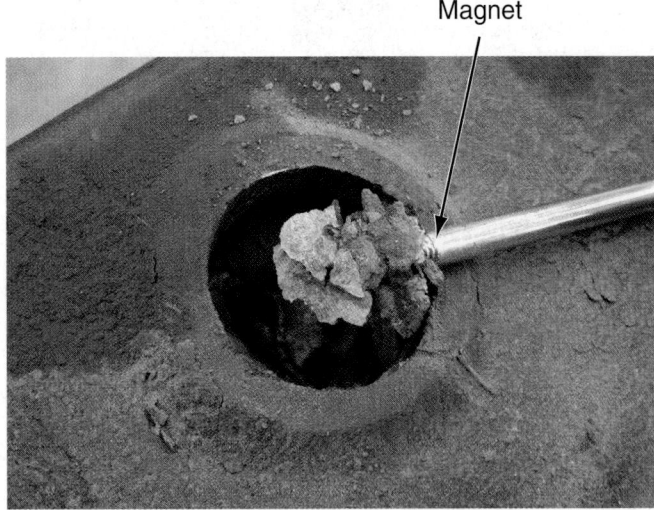

Figure 20.7 This sediment behind a core plug in the water jacket of an iron block is magnetic. *(Courtesy of Tim Gilles)*

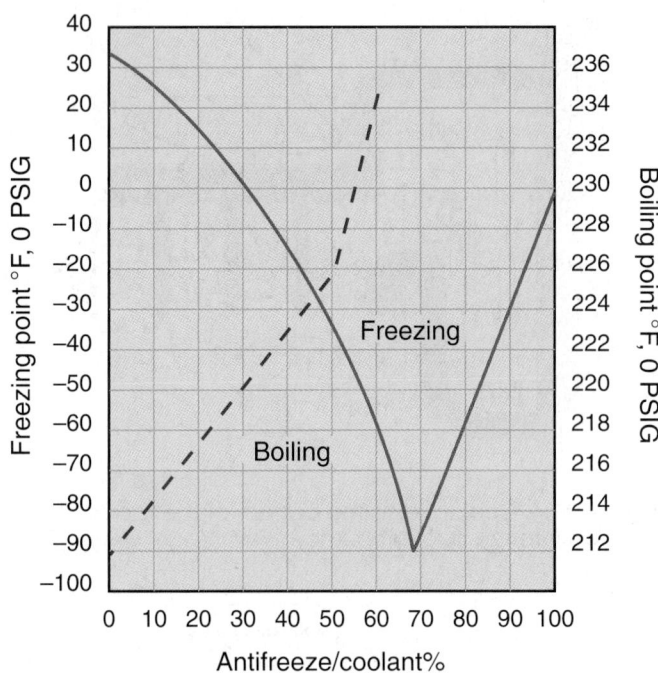

Figure 20.8 The percentage of coolant changes its freezing and boiling points.

breaking off the rust. The coolant carries the rust to other cooling system areas. Hard water forms mineral deposits. The resulting *corrosion* interferes with heat transfer even before the results of it plug the radiator and fill the waterjackets with sediment.

On older vehicles, the cooling system was often neglected, especially in frost-free areas of the country where people often used ordinary tap water to fill their radiators. This resulted in plugged radiators and thermostats, rusted radiator cap springs, rusted core plugs, plugged heater cores, rusted radiator hose reinforcement springs, metallic sediment in the coolant jackets (**Figure 20.7**), and a host of other problems. Most of these problems would have been avoided had the owner simply used a mixture of coolant and water instead of 100% tap water.

Coolant Boiling and Freezing Points

The freezing point for coolant is that point at which ice crystals begin to form. As coolant gets slushy, it will plug the radiator of a cold engine. When slushy coolant trapped in the block gets hot, boilover can occur.

The freezing point of pure water is higher than some winter temperatures, and its boiling point is lower than the coolant temperature at which a hot (but not overheated) engine can operate. Coolant has a higher boiling point than water. This is especially an advantage on today's hotter running vehicles with multiple accessories and air conditioning.

Water absorbs more heat than pure ethylene glycol coolant. Therefore, a coolant/water mixture is a better choice than concentrated coolant. Also, the freezing point of ethylene glycol concentrate is higher if it is diluted with water. For motor vehicle use, concentrated coolant is mixed with water until it is the correct mixture. The chart in **Figure 20.8** shows the

effects of various coolant concentrations on boiling and freezing temperatures.

- Pure ethylene glycol freezes at 8°F and boils at 330°F. The factory radiator fill is usually about 50% water and 50% coolant, providing protection against freezing to −34°F and a boiling point of 265°F when a 15-pound radiator cap is used.
- A 70% concentration is the maximum that should be used. The freezing point of a 70% concentration of ethylene glycol is –85°F. Its boiling point is 11°F higher than would be provided by a 50/50 mixture.

Antifreeze/coolant is also available in a ready-to-use 50/50 mixture of antifreeze/coolant and purified water (**Figure 20.9**).

Figure 20.9 Coolant is available in a 50% mixture of antifreeze and purified water. *(Courtesy of Tim Gilles)*

NOTE: *Some older automobiles did not have a temperature gauge. If the engine reached a certain temperature, a switch in a sending unit closed and a high-temperature warning lamp on the instrument panel illuminated. The triggering temperature of the sending unit was usually well above the boiling point of water (212°F at sea level). With pure water, the warning light might never get a chance to work, as the water would have already boiled out before the switch reached a temperature high enough to close it.*

Coolant Additive Package

There are many different types of ethylene glycol coolant using various additive packages. During the warranty period, it is especially important that the manufacturer's recommendations be followed when replacing coolant. After the warranty period is up, you will probably be able to use one of three coolants to cover most vehicle makes. Those choices are discussed later.

NOTE: *Color cannot be used to identify coolants. There are many different colors, but there is no guarantee that a certain color is a particular type of coolant. Conventional coolant is usually green in color. Extended-life coolant is usually red or orange in color, although conventional coolant can also be red, orange, or gold.*

Coolants have an additive package called supplemental coolant additive (SCA). Conventional coolant has its own SCA package. Extended-life coolants use a different additive package that includes organic acids. These are called OAT organic acid technology (OAT) coolants.

A variety of additives are included in automotive coolants, including inhibitors for aluminum protection, absorbers and buffers, cavitation retarders, electrical galvanic activity preventers, antifoam additives, and dyes. Cavitation happens in the cooling system when liquid forms air pockets as it boils due to a drop in pressure. When the bubbles collapse, they create potentially damaging shock waves in the coolant. Cavitation is covered in more detail later in this chapter. A supplemental coolant additive package can include:

- **Borax** (sodium tetra borate), a buffer and neutralizer of acid.
- **Phosphate** (dipotassium phosphate). American and Japanese manufacturers use phosphate as a cavitation retarder and acid buffer. Phosphates protect cast iron but when used in very hard water (as found in Europe), scale forms.
- **Sodium molybdate**, an expensive corrosion inhibitor for heat-rejecting aluminum (when aluminum, such as in the coolant pump, is heating the coolant). Used as a cavitation retarder in nonphosphate coolants.
- **Sodium silicate** for aluminum protection.
- **Sodium nitrate**, which resists solder corrosion and pitting of aluminum radiators.
- **Triazoles**, chelants that grab and form a film over metal ions like brass and copper to prevent formation of a galvanic cell.
- **Sodium hydroxide**, used to keep the pH stable.
- **Phosphorescent dye**, which shows up under ultraviolet light. The colors vary and are for consumer identification only.
- **Benzoate**, which protects ferrous metals and high-lead solder. It is found in most coolants.
- **Water**, used to make the additives soluble.

Coolant Life

Like oil, coolant does not wear out, but the SCA package in conventional coolants requires replenishment on a regular basis, usually by exchanging the old coolant for new. Extended-life coolant has a long shelf life and can remain in storage for many years without deteriorating. Conventional coolants with silicate additives have a shorter shelf life.

NOTE: *One important consideration with long-life coolants is that they can turn to gel when air has entered the cooling system. To prevent this, be certain the radiator cap is replaced on a regular basis and that all air is bled from the system during cooling system service or engine replacement. Thorough flushing must be done before adding new coolant.*

Differences in Coolant

Coolants are not the same for European, Asian, and North American vehicles. There is some controversy regarding potential damage in using the wrong coolant in some manufacturers' vehicles.

- Most North American coolants contain phosphates and aluminum-protecting silicate additives.
- European coolants contain silicates but in lower concentrations than North American coolants.
- Japanese coolants do not contain silicates but use phosphates instead. Some say silicates are hard on coolant pump seals. Others say there is no evidence that silicates can harm today's carbide coolant pump seals.

SCIENCE NOTE

Phosphate is often used as a corrosion inhibitor in North American and Asian coolants. The water in Europe is very hard. As phosphate softens water, it forms solids of calcium or magnesium salts. These can drop out of solution, blocking the cooling system. Using purified water would prevent this problem from occurring.

Electrolysis

Electrolysis, which can be of three types, is very destructive to the engine's cooling system. The first type, chemical electrolysis, results when coolant has become too acidic. Automotive cylinder heads are most often made of aluminum, and they are bolted to cylinder blocks made of cast iron. Chemical electrolysis

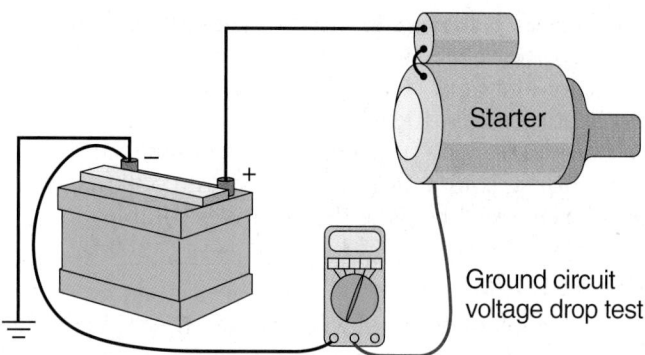

Figure 20.10 Voltage drop test connections for the ground circuit. A bad ground circuit can cause extreme cooling system corrosion.

occurs when two dissimilar metals in a liquid carry an electrical current. Just like in a battery, one of the metals acts like a positive plate, and the other acts like a negative plate. During electrolysis, metal leaves one of the "plates" (the aluminum head) and is deposited on the other "plate" (the iron block). A corrosion inhibitor additive is required to protect the aluminum. To protect aluminum from corrosion, use a coolant that specifically states it can be used with aluminum.

A second type of electrolysis can be caused by a faulty electrical ground circuit, which can result in extreme corrosion. Perform a voltage drop test on the engine's ground circuit during engine cranking. To perform a voltage drop test, connect a voltmeter with one lead on the negative battery post and the other lead on the starter housing at the engine block. **Figure 20.10** shows how a voltage drop test is conducted on the ground side of the electrical circuit. No voltage will be displayed on the meter until the engine cranks and current is flowing in the circuit. Any voltage displayed on the meter is what is dropping across the circuit due to resistance. A reading of 0.5 volt is excessive. Less than 0.3 volt is desirable.

The third type of electrolysis results from static electricity. This is the condition that results in heater core failures, as described earlier.

 When community water has an abundance of minerals, distilled water should be used.

■ COOLING SYSTEM PRESSURE CAP

A pressure cap is used to seal the opening on top of the radiator (see Figure 13.14) or coolant reservoir. Its functions are to pressurize the coolant and keep it from surging out as it is circulated through the radiator.

Cooling systems are pressurized for two reasons:
■ The efficiency of the water pump is increased.
■ Putting pressure on the coolant raises the boiling point of the liquid in the cooling system.

When the system is not under pressure, the water pump does not always remain full of coolant. *Aeration* is the result, which causes a loss of pump efficiency of about 15%.

Normal engine operating temperatures are between 180° and 212°F. The typical coolant/water mixture boils at somewhere between 220° and 230°F without pressure. When the cooling system is overworked, temperatures can rise as high as 270°F without the coolant boiling. Pure water boils at 212°F at sea level, but its boiling point is less than that at higher altitudes because atmospheric pressure on the coolant is less.

NOTE: *Each pound of pressure on the coolant raises its boiling point by about 3°F.*

A 15-pound cap will raise the boiling point of the water by about 45°F to 257°F.

> 212°F (boiling point of water at sea level)
> + 45°F (3° × 15 lb)
> 257°F

Also, if coolant is mixed with water, the boiling point will be higher.

Pressure Cap Valves

The two valves in the pressure cap are the *pressure* valve and the *vacuum* valve (**Figure 20.11**)

Pressure Valve. Most pressure caps pressurize the system from 13 to 17 pounds. (Metric caps are usually 0.98 AE.) The large spring in the cap is what maintains this pressure by forcing the rubber *pressure seal* against the seat in the *radiator filler neck* (**Figure 20.12**) or coolant reservoir. The pressure valve maintains pressure on the coolant even if the engine is shut off.

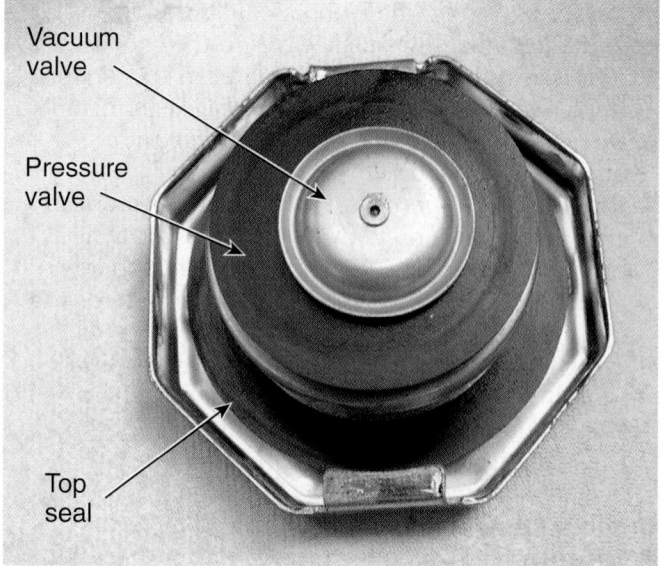

Figure 20.11 The two valves in the radiator cap are the pressure valve and the vacuum valve. (*Courtesy of Tim Gilles*)

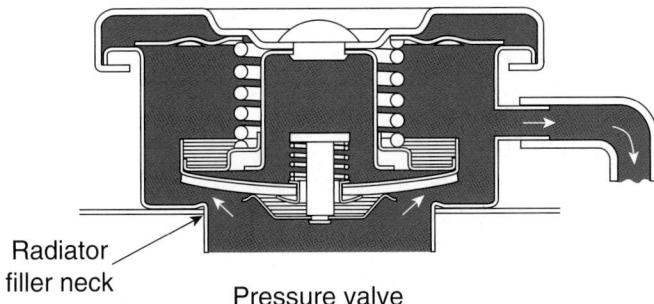

Figure 20.12 Pressure valve operation.

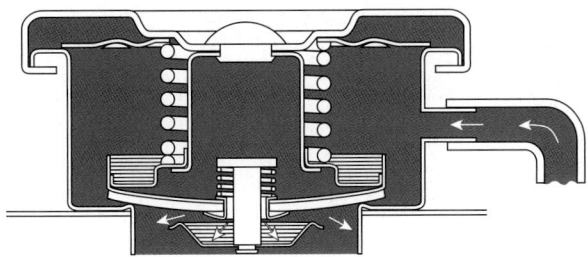

Vacuum valve

Figure 20.13 Vacuum valve operation.

Vacuum Valve. The vacuum valve (**Figure 20.13**) protects the radiator as the temperature of the coolant drops. Hot liquid shrinks to occupy less space. This could create a vacuum on the inside of the radiator if the spring-loaded vacuum valve did not open to prevent it. Excess coolant (which went into the recovery tank during warmup) is drawn back into the radiator.

Some vacuum valves are spring-loaded and some are loose in the cap:

■ A *constant pressure* cap has a spring-loaded vacuum valve. It allows the system to begin building pressure as soon as the coolant starts to expand.

■ An *atmospheric pressure* cap has no spring on the vacuum valve. It does not build up pressure until the system gets hot.

■ COOLANT RECOVERY SYSTEM

Coolant recovery systems have been included on most new cars since 1969. **Figure 20.14** shows a typical system. Coolant expands by over 1/10 of its original volume as it reaches engine operating temperature. Expanding coolant goes into the recovery tank instead of onto the ground. When the coolant temperature drops, the vacuum valve opens, drawing coolant back into the system from the recovery tank. Because air never enters the cooling system, a properly operating recovery tank system increases overall cooling efficiency by 10%.

The top radiator seal (see Figure 20.11) prevents steam from escaping when the cap is turned a ½ turn to the pressure release position. It also seals when the coolant temperature drops in the radiator, causing coolant to be drawn in from the coolant recovery tank.

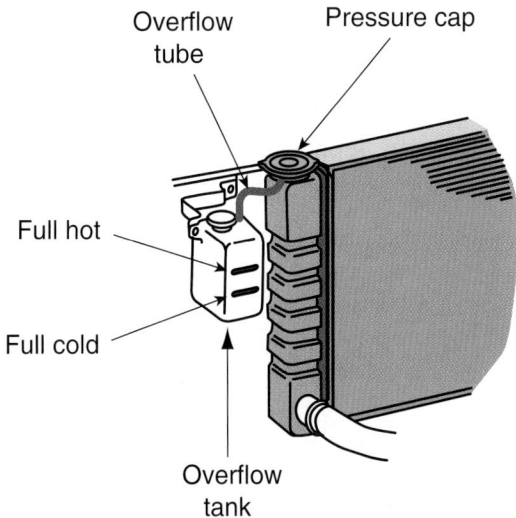

Figure 20.14 A coolant recovery system.

■ RADIATOR

Radiators are designed with either of two flow patterns; down flow or cross flow (**Figure 20.15**). Both designs work equally well, but most automobiles use the cross-flow design. It can be made wider with a lower profile, allowing for a lower hoodline (less wind resistance).

A radiator has top and bottom, or side *tanks* that are attached to a *core* that contains *tubes* with air *fins* attached to them (**Figure 20.16**). Coolant circulates through the radiator tubes. When air travels across the fins, it carries off the heat that is transferred from the coolant to the tubes.

Radiator cores are made of copper or aluminum. Older radiator *tanks* were brass and were soldered to the copper radiator core. The radiators on most newer cars have plastic tanks and vacuum-brazed aluminum cores. The vacuum-brazed aluminum core/plastic tank radiator has a good life expectancy and has proven to be durable. But these radiators are not as tolerant of poor-quality coolant, neglected maintenance, and low coolant level as are copper/brass radiators.

Transmission Heat Exchanger

A car with an automatic transmission usually has a heat exchanger in the radiator (**Figure 20.17**). This part is also called a transmission oil cooler.

■ THERMOSTAT

The thermostat is usually located where the coolant leaves the top of the block, just below the upper radiator hose (see Figure 20.3). When the engine is cold, the thermostat remains closed, trapping the coolant in the block. The coolant is circulated in the block by the coolant pump until it warms up to a predetermined point and the thermostat opens (**Figure 20.18**). Hot coolant can then circulate to the radiator, where it is cooled

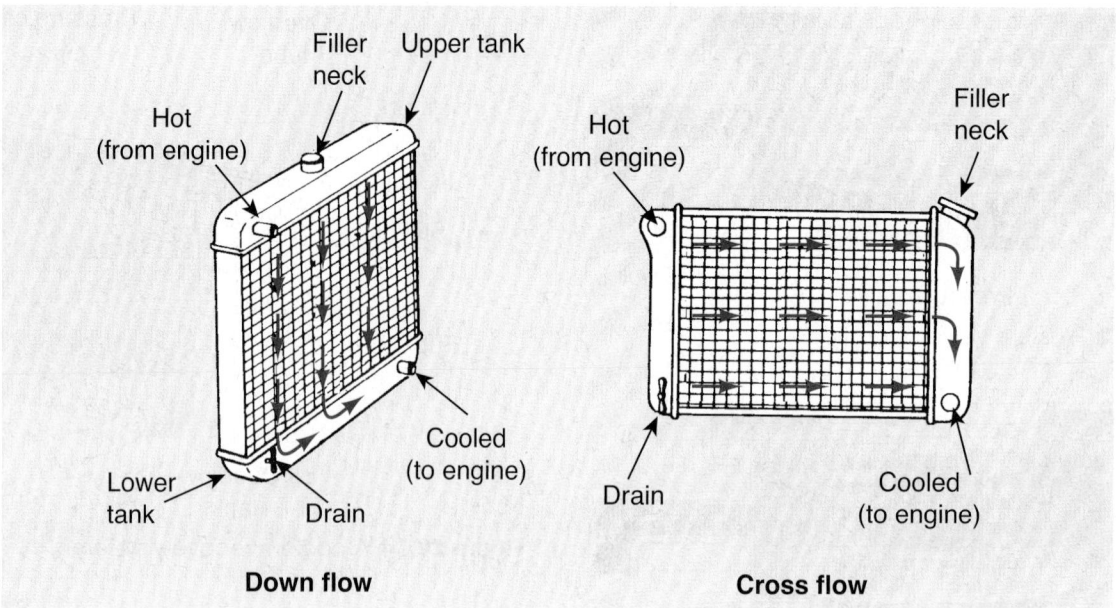

Figure 20.15 Radiator coolant flow.

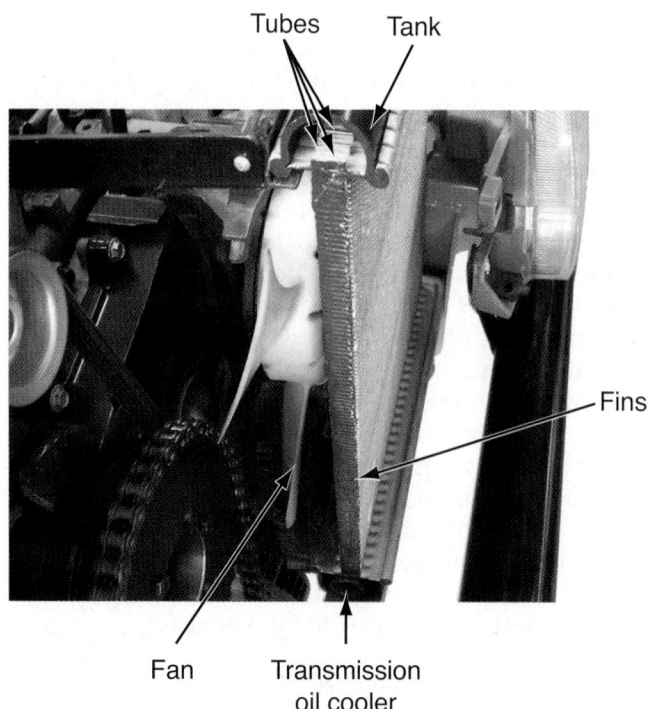

Figure 20.16 A cutaway showing the parts of a typical radiator. *(Courtesy of Tim Gilles)*

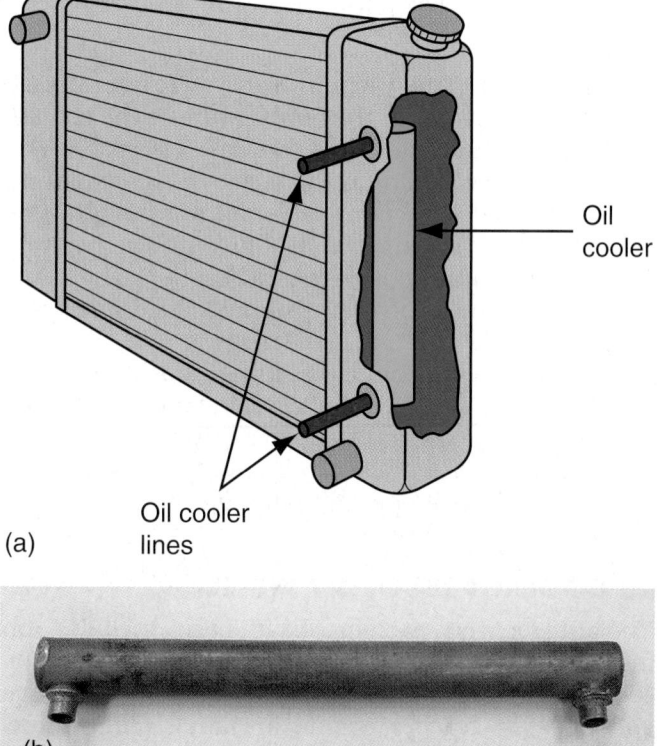

Figure 20.17 (a) An automatic transmission cooler in the radiator. (b) A transmission oil cooler. *(b, Courtesy of Tim Gilles)*

and returned to the block. As soon as the temperature of the coolant in the block falls below a predetermined point, the thermostat closes once again and the cycle repeats.

The *wax pellet* thermostats used in automobiles have an expansive wax compound that expands at a predictable rate to open the valve in the thermostat (**Figure 20.19**). A thermostat is rated according to the engine operating temperature that it is supposed to

start to open at. The temperature of the thermostat is stamped on the bulb on its bottom in either Celsius or Fahrenheit (**Figure 20.20**). Thermostats of several different ratings are available. The temperature rating of a thermostat is somewhere above 180°F on modern cars.

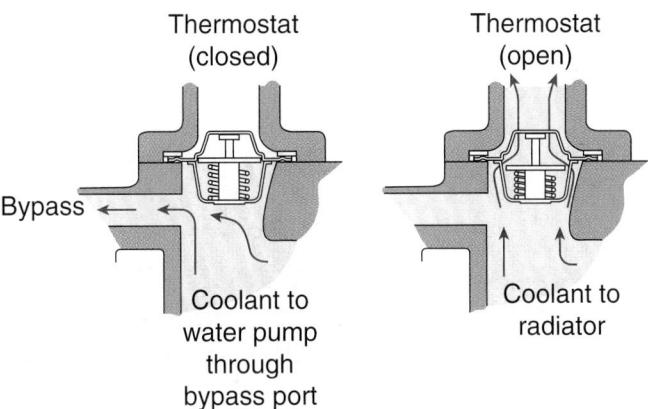

Figure 20.18 Operation of a "positive actuator" thermostat.

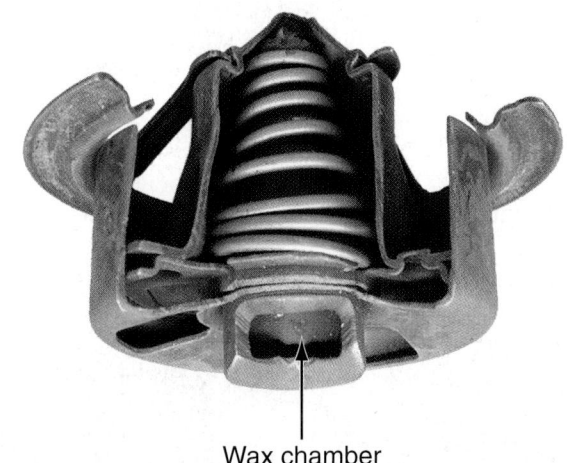

Wax chamber

Figure 20.19 A cutaway of a thermostat. *(Courtesy of Tim Gilles)*

180°F   87°C

Figure 20.20 Thermostat rating. The operating temperature of the thermostat is stamped on the bottom in Fahrenheit (left), or in Celsius (right). *(Courtesy of Tim Gilles)*

When an Engine Runs Too Hot

It is important for several reasons that the thermostat allow the coolant to quickly reach and maintain a high enough temperature. The following are results of an engine that runs too hot:

- Its oil becomes thinner. Thinner oil does not provide the same level of lubrication and can escape more easily past the piston rings and into the combustion chamber.
- Oil oxidizes more rapidly at temperatures above 240°F. Oxidation causes the oil to thicken and can also result in varnish buildup. Varnish can cause hydraulic lifters to stick and fail and can also plug the oil return channels in the oil rings.
- Excessive engine temperatures can result in engine damage. A hot engine can cause pistons to expand excessively. When a piston scuffs, it can actually weld to the cylinder wall momentarily.
- **Detonation** or **preignition** occurs when fuel explodes due to high engine temperatures. Engine damage can result. It is important that the fuel in the combustion chamber be able to burn evenly.

Cold Engine Operation

When the engine runs too cold, it experiences poor fuel economy and produces considerably more emissions. Metal parts, such as pistons and rings, conform to the cylinder walls when warm. When they are cold, they do not fit as well. Blowby, which is exhaust gas that escapes past the rings during combustion (see Figure 15.4), increases when the engine is cold. The engine will experience far less wear if it is brought quickly to operating temperature. If a car has a *computer feedback* fuel system, that system will not operate until its sensors have all reached a predetermined operating temperature. This will cause the fuel management (computer) system to continue to operate in the **open loop** mode (this information is covered in a later chapter). High emissions and poor fuel economy result when an engine does not reach operating temperature.

Sludge in the oil will increase if the engine is running at too cold a temperature for too long. Sludge is a mixture of moisture, by-products of combustion, and water. Considering that about 1 gallon of water is produced for each gallon of fuel burned, it is especially important that the engine be run at a temperature that results in the evaporation of condensation.

An engine will run without a thermostat in an emergency. But in addition to the problems already mentioned, leaving it out can cause an engine to run hot. Although the coolant temperature is relatively low, the coolant moves too quickly through the water jackets to absorb heat from the engine's parts.

Thermostat Operating Characteristics

When an engine is warm, the thermostat can be open either partially or fully, fluctuating to maintain a constant temperature. Should the cooling requirements increase in hot weather or under a load, the thermostat will open all the way.

No matter how hot the coolant is, a properly operating thermostat remains closed until its predetermined opening point. A common misconception is that installing thermostat with a lower temperature rating will result in a cooler running engine. This practice will simply result in the engine operating temperature being lower during cooler weather. Another practice based on faulty logic is to install a thermostat with a higher opening temperature in an attempt to cause the engine to warm up faster. The engine will warm up at the same rate. Its operating temperature will simply be higher.

Types of Thermostats

There are two types of *wax pellet* thermostats commonly used in pressurized cooling systems: the *diaphragm* and the *positive piston* actuator. Figure 20.18 shows how a positive piston actuator thermostat operates. There is sometimes a small check valve in the thermostat that allows air to escape during filling of the cooling system (**Figure 20.21**). It also allows a small amount of coolant circulation around the wax sensor for more accurate temperature control.

Thermostat Bypass

When the thermostat is closed, the coolant is circulated in the block by the coolant pump. There is a passage (either a hose or a designed-in passage) that allows the coolant to circulate when the coolant is cold and the thermostat is closed. This is called a **thermostat bypass** (**Figure 20.22**). By-pass hoses are usually molded hose (**Figure 20.23**).

A few engines, called reverse flow, have their thermostats located at the top of the bottom radiator hose at the inlet to the coolant pump (**Figure 20.24**). The purpose of this is to minimize "thermal cycling." The

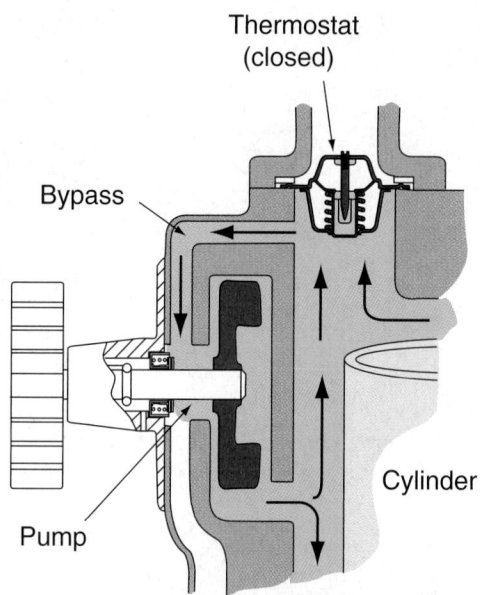

Figure 20.22 The cooling system includes a bypass that functions when the thermostat is closed.

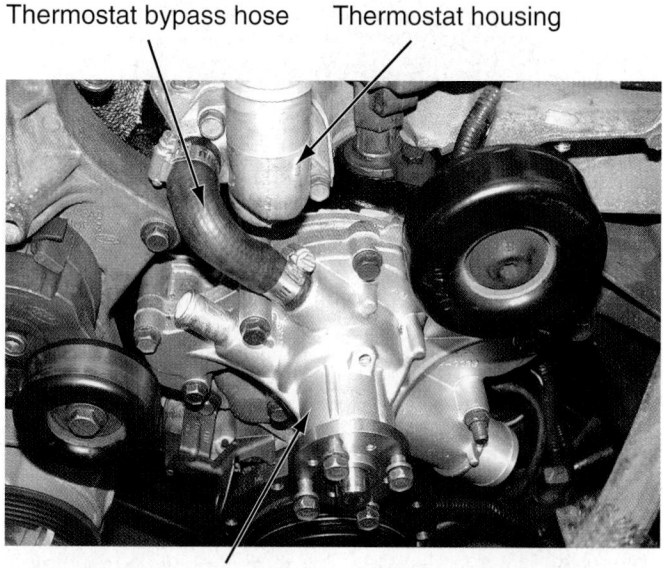

Figure 20.23 A typical thermostat bypass hose is a molded hose. *(Courtesy of Tim Gilles)*

thermostat begins to open in response to increased temperature of the coolant flowing through the coolant pump. The cold coolant entering the engine from the radiator closes the thermostat. This results in the engine warming up more gradually. The temperature of the heater is also more consistent, and the temperature gauge does not fluctuate. Cylinder heads are cooled first, which prevents detonation.

■ TEMPERATURE WARNING LIGHT OR GAUGE

A *warning light* or *temperature gauge* is located on the instrument panel. When the temperature level of the coolant reaches unsafe levels, the light or gauge warns

Figure 20.21 A thermostat with a ball check valve to help bleed trapped air. *(Courtesy of Tim Gilles)*

Thermostat housing

Figure 20.24 This thermostat is located at the top of the *lower* radiator hose. *(Courtesy of Tim Gilles)*

Figure 20.25 Cavitation on and near a coolant pump impeller. *(Courtesy of Tim Gilles)*

the driver of the problem. A sending unit for the gauge, also called an *engine coolant temperature sensor*, is screwed into one of the coolant passageways in the head or manifold. It is bathed in coolant and sends the temperature signal to the gauge.

A sending unit can signal a slow rise in temperature or a fast rise due to a stuck-closed thermostat, but it only works effectively when submerged in coolant. Sometimes the coolant boils out or is lost due to a fast leak. With no coolant in the cooling system, the sensor might not get hot enough to register the excessive temperature.

SCIENCE NOTE

Cavitation happens when air bubbles form in the coolant. It can occur when coolant boils, when air gets into the coolant, or when suction is restricted and over-pumping occurs. If the water pump tries to move more coolant than is possible, a pressure drop at the suction side of the pump results in cavitation on the pump impeller (Figure 20.25). Bubbles can also form from vibration of the cylinder walls during combustion, especially during detonation. Bursting bubbles can create pressures of up to 60,000 psi as they burst. As boiling coolant travels through the engine, the collapsing bubbles chip off metal.

Figure 20.26 shows an example of cavitation erosion in an aluminum cylinder head. The flaking aluminum can plug a radiator. Phosphate additives in the coolant help to control cavitation. Good cooling system maintenance, including good antifreeze, a good pressure cap, and tight connections, can help control cavitation.

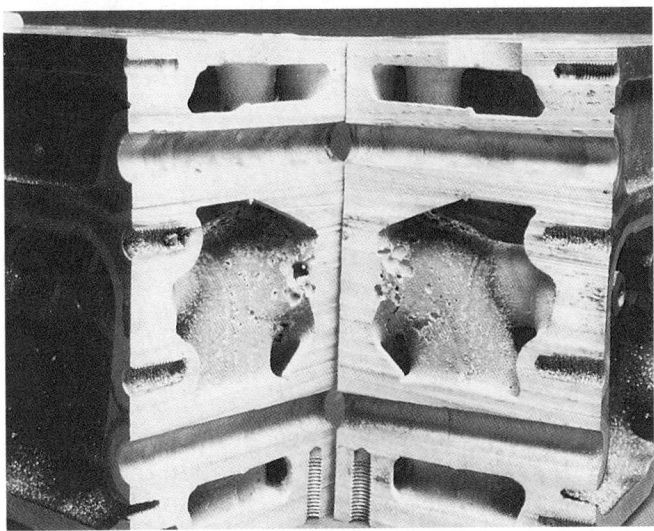

Figure 20.26 Cavitation erosion in an aluminum cylinder head. *(Courtesy of Prestone Products Corporation)*

■ FANS

The fan's purpose is to draw air through the radiator when the vehicle is not moving fast enough to provide enough air circulation. It is really only necessary at idle and low speeds.

Belt-Driven Fans

Belt-driven fans are usually mounted on the front of the coolant pump, which is driven by a belt that rides in a pulley groove on the crankshaft. They are found on most rear-wheel-drive vehicles. Fans powered by electric motors and controlled by engine temperature are found on most front-wheel-drive vehicles (**Figure 20.27**).

Electric Fans

With an electric fan, the fan motor is switched on and off as the engine temperature rises and falls. The fan might also be switched on to cool the air-conditioning

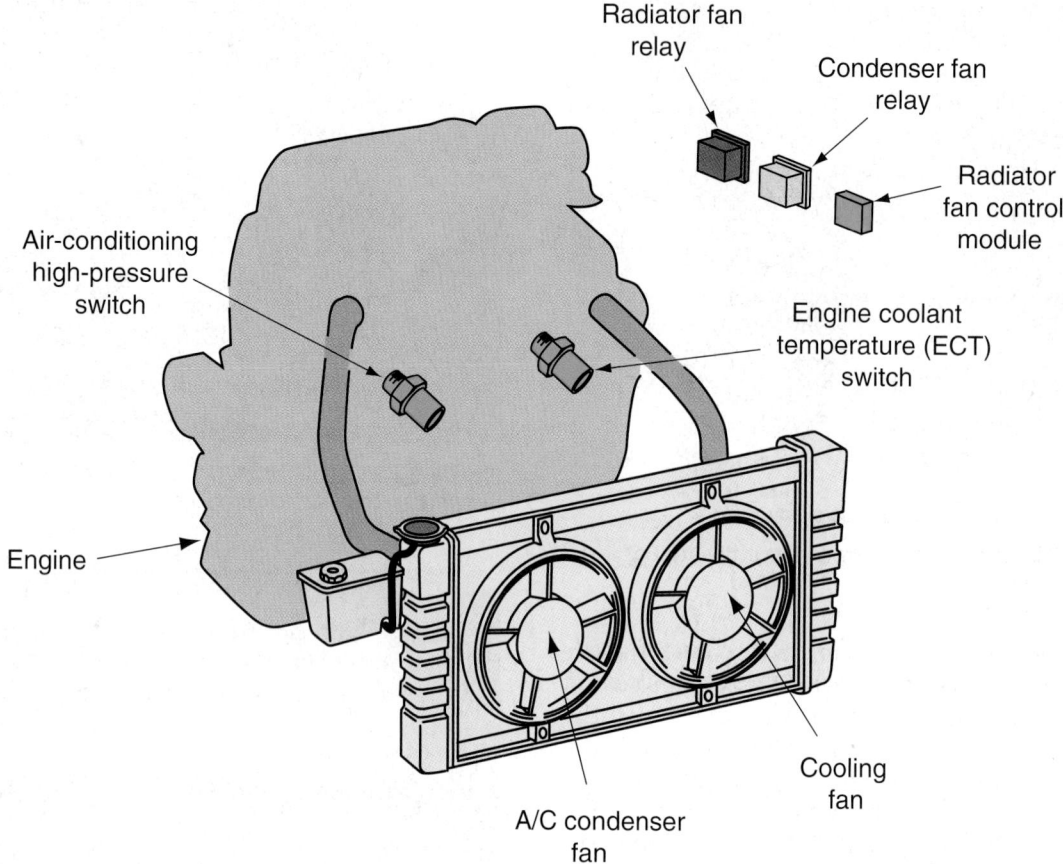

Figure 20.27 An electric cooling fan system.

condenser. The fan can also operate after the engine is shut off. A timer controls this function so that the battery will not run dead. Aftermarket add-on fan kits are also available.

The disadvantage of a belt-driven fan is that at higher speeds, when the fan is no longer required, it can rob the engine of horsepower. Manufacturers have developed several ways to minimize this. Fan blades are made of aluminum, flexible plastic, or steel.

Flex Fan

A flex fan has blades with a flexible trailing edge and a rigid leading edge. Flex fan blades have a high angle at low speeds. At high speeds, the blades flatten out, reducing the horsepower required to turn them.

Besides being a safety hazard, a fan blade can cut a power steering or radiator hose, a brake line, or other part. Any time a fan blade loses a piece, the fan will become unbalanced. This can cause coolant pump failure.

CAUTION A cracked or bent flex fan should be replaced. Flex fans can break up. There have been several recalls on them.

■ FAN CLUTCH

A **fan clutch** reduces horsepower requirements (**Figure 20.28**). They are found especially on vehicles with air conditioning. There are two kinds of fan clutches: the temperature regulated and the speed sensitive (viscous).

Figure 20.28 A fan clutch. *(Courtesy of Tim Gilles)*

The *temperature-controlled (thermal) fan clutch*, generally used with a heavy-duty fan blade, is the type most used in original equipment. It is usually controlled by a **bimetal coil spring** (**Figure 20.29**), a thermostatic coil consisting of two types of metal wound together. When the coil is heated, it expands. When it cools, it shrinks. The fan only works when the engine is hot. When the engine is cold, the fan freewheels.

When the air coming through the radiator is hot, the bimetal spring causes an internal valve to open; silicone fluid from a reservoir expands to fill the working chamber and engages the clutch. When the air coming through the radiator is cool enough, the bimetal spring cools, the valve opens, and the silicone fluid moves back into the reservoir to disengage the clutch.

Thermal fan clutches also respond to engine rpm through slippage in the fluid chamber. Slippage occurs when ram air through the radiator from vehicle movement is sufficient for cooling.

A *speed-sensitive (viscous) fan clutch* slips when the resistance of air coming through the fan becomes higher. This fan is similar in operation to the flex fan in that it uses some horsepower at all times. A viscous fan clutch is not as efficient as a thermal fan clutch. A thermal fan clutch reacts not only to engine temperature, but also to rpm. So it would be the clutch of choice for high-performance or heavy-duty cooling applications.

▄▄ RADIATOR SHROUD

The radiator shroud makes the fan far more effective at pulling air through the radiator. If the fan shroud is damaged or missing, the vehicle could overheat in traffic.

▄▄ HEATER CORE

The **heater core** is a small heat exchanger that engine coolant is circulated through (**Figure 20.30**). It is usually located inside the driver's compartment (see Figure 20.3). A blower motor passes air across the fins of the heater core, transferring heat from the engine to the passenger compartment.

The heater core is supplied with engine coolant through two *heater hoses*. One hose carries coolant leaving the engine. A valve installed in that hose controls the flow of coolant to the heater core when the heater dash controls are set for "heat." A return hose carries the coolant back to the engine for reheating.

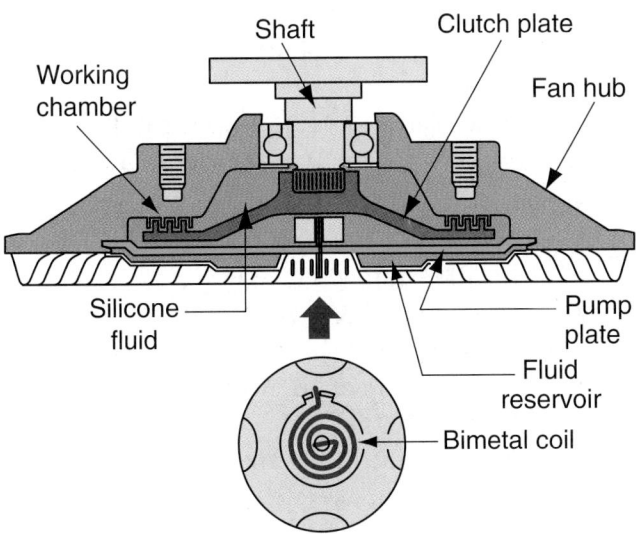

Figure 20.29 A cutaway of a fan clutch.

Shaft · Clutch plate · Working chamber · Fan hub · Silicone fluid · Pump plate · Fluid reservoir · Bimetal coil

Figure 20.30 A heater core. (*Courtesy of Tim Gilles*)

▄▄ REVIEW QUESTIONS

1. Automotive coolant is made of _____ and water.

2. Cylinder blocks are made of either iron or _____.

3. When an electrical current develops between two dissimilar types of metal in the cooling system, this is called _____.

4. _____ is the name of a coolant additive that protects aluminum.

5. What is the normal operating temperature of an engine?

6. For each pound of pressure on the coolant, approximately how much will its boiling point increase?

7. What is the name of the small valve in the center of a radiator pressure cap?

8. Two names for the part of the radiator that automatic transmission fluid flows through are the transmission oil cooler and the heat _____.

9. _____ in the thermostat expands to cause it to open.

10. What passage allows coolant to circulate within the block when the thermostat is closed?

■ ASE-STYLE REVIEW QUESTIONS

1. Which of the following statements is/are true about the purpose of a radiator cooling fan?
- **a.** It pulls air through the radiator when the vehicle is traveling at freeway speed.
- **b.** It pushes air through the radiator.
- **c.** Both A and B.
- **d.** Neither A nor B.

2. Which of the following has the best cooling ability?
- **a.** Water
- **b.** Copper
- **c.** Ethylene glycol coolant
- **d.** A miniature cooling fan powered by political hot air

3. Which of the following has the highest boiling point?
- **a.** Pure water
- **b.** Ethylene glycol
- **c.** Air-conditioning refrigerant
- **d.** Alcohol

4. Which of the following is not a function of a cooling system pressure cap?
- **a.** It raises the boiling point of the coolant.
- **b.** It prevents coolant from surging out of the top of the radiator.
- **c.** It prevents air from entering the system.
- **d.** It prevents the coolant from freezing.

5. What is the name of the small metal valve in the center of a pressure cap?
- **a.** Pressure valve
- **b.** Top seal
- **c.** Vacuum valve
- **d.** None of the above

6. All of the following are true about thermostats *except*:
- **a.** Installing a thermostat with a lower temperature rating will allow an engine to run cooler in hot weather.
- **b.** Running an engine without a thermostat will result in lower fuel economy.
- **c.** A correctly operating thermostat will prevent undue engine wear.
- **d.** The thermostat allows for quicker heater operation.

7. Which of the following could cause a computer-controlled fuel system to run rich?
- **a.** A stuck-closed thermostat
- **b.** A stuck-open thermostat
- **c.** Both A and B
- **d.** Neither A nor B

8. Technician A says that radiators are made of copper and brass. Technician B says that radiators are made of aluminum and plastic. Who is right?
- **a.** Technician A
- **b.** Technician B
- **c.** Both A and B
- **d.** Neither A nor B

9. Which of the following is the best practice when changing coolant?
- **a.** Use 100% ethylene glycol.
- **b.** Use a mixture of 70% water and 30% ethylene glycol.
- **c.** Use a mixture of 30% water and 70% ethylene glycol.
- **d.** Use 100% distilled water.

10. Which of the following is/are true about extended-life coolants?
- **a.** They are usually red or orange in color.
- **b.** They use an additive package that includes organic acids.
- **c.** Both A and B.
- **d.** Neither A nor B.

Cooling System Service

■ **KEY TERMS**

rod out
coolant hydrometer
hydrolocked engine

thermoplastic seizure
block check test

■ **INTRODUCTION**

This chapter deals with cooling system problem diagnosis, maintenance, and repairs. Cooling systems are generally quite dependable. However, they do require periodic maintenance. Cooling system service is one of the best values for the customer in terms of preventative maintenance. Working on the cooling system is usually not very difficult and is sometimes very profitable. This chapter describes those repairs and services.

■ **DIAGNOSING COOLING SYSTEM PROBLEMS**

Maintaining Correct Engine Temperature

The engine is designed to run at a predetermined temperature. There are several possible causes of incorrect engine temperature, including leaks, overheating, and overcooling. If an engine does not get warm enough, emissions rise, fuel economy suffers, the heater does not work, and the engine can suffer excessive wear. Overheating will cause serious damage to an engine. The following are some of the cooling system problems that can affect engine temperature (**Figure 21.1**).

Coolant Level. This is the first thing to check.

Restricted Radiator. Rust and scale form in the cooling system when it has been neglected.

NOTE: *A plugged radiator will usually cause overheating on the highway. The restricted circulation can cause the coolant to be pumped out the radiator overflow.*

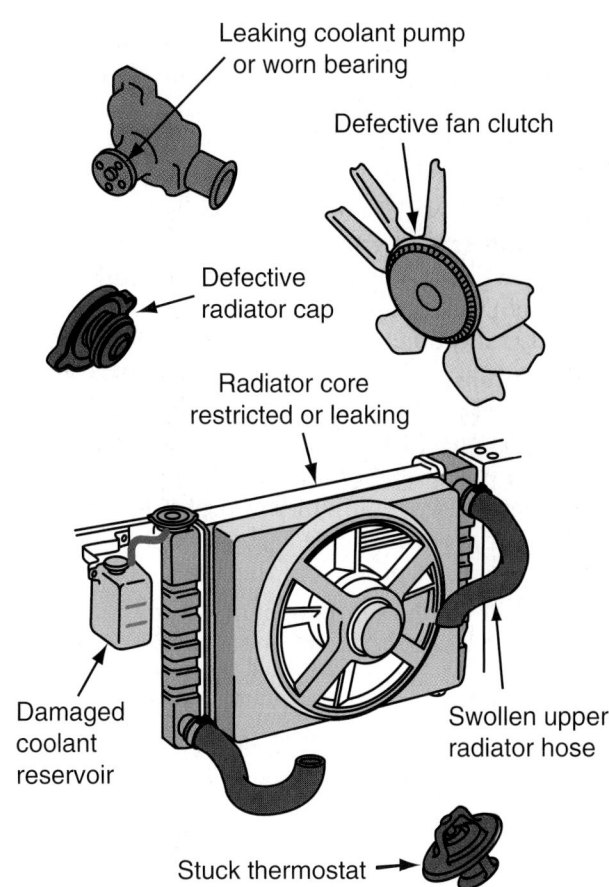

Figure 21.1 Common problems with a cooling system.

Leaking coolant pump or worn bearing

Defective fan clutch

Defective radiator cap

Radiator core restricted or leaking

Damaged coolant reservoir

Swollen upper radiator hose

Stuck thermostat

Stuck Thermostat. A stuck thermostat will cause an engine to overheat or to not heat enough. If it is stuck partially open, the engine can overheat on highway trips and not warm up as soon as it should in town. A thermostat that sticks closed will cause the engine to overheat very quickly (in the time that it normally takes for your heater to start to work in the morning).

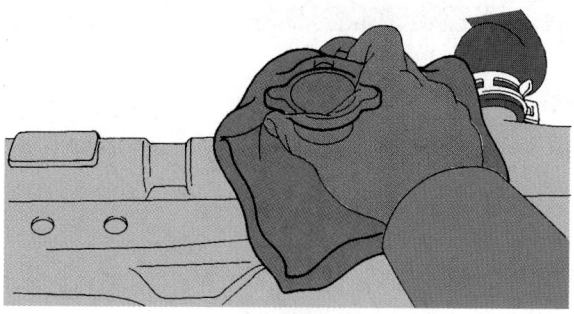
Cover cap with shop towel

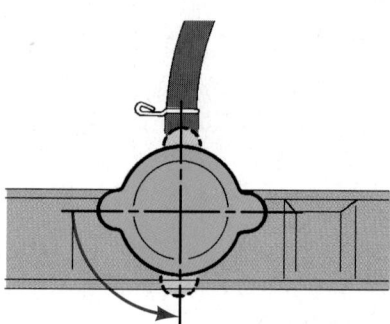
Turn slowly counterclockwise

Figure 21.2 Cover the radiator cap with a shop towel and turn it slowly counterclockwise while holding down against spring pressure.

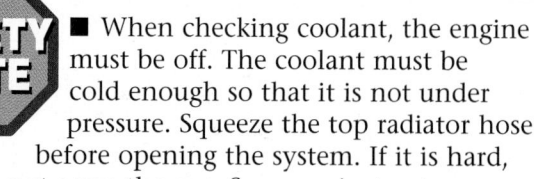

> **SAFETY NOTE**
>
> ■ When checking coolant, the engine must be off. The coolant must be cold enough so that it is not under pressure. Squeeze the top radiator hose before opening the system. If it is hard, do not open the cap. See your instructor.
>
> ■ If you let the pressure off of the coolant, it will boil. This is a dangerous situation! Yet, circumstances sometimes call for a hot system to be opened. When opening the radiator on a system that is hot, fold a shop towel and place it over the radiator cap (**Figure 21.2**). Holding pressure down on the shop towel, use it to turn the radiator cap ¼ turn until its first stop. Letting pressure off of the shop towel will allow any remaining pressure on the coolant to escape. Remember, the maximum pressure on any radiator is only about 15 to 17 pounds. This is not so much that you cannot hold pressure against it.
>
> ■ Check to see that the cap is loose before turning it further and removing it.

Late Ignition Timing. On older engines, without computer controls, ignition timing that is retarded more than 2° or 3° from specifications can cause engine overheating.

Loose Belt Tension. If the belt that drives the coolant pump is too loose, overheating can result. Two things will be apparent if this is the case:
■ The belt will probably squeal as it slips.
■ The charging system light will probably come on or the battery will go dead. The same belt usually drives the water pump and the alternator.

Bad Coolant Pump. Coolant pumps can suffer a failure in the bearing, seal, or impeller. The pump has a *static seal* and *bearing* that can fail. The bearing is permanently sealed and can be damaged by excessive belt tension. When a bearing fails, the static seal is also damaged, which can result in a coolant leak. The gasket to the back plate can also leak.

Fan Shroud. When a fan shroud is loose, broken, or missing, the engine can overheat. On today's high-temperature engines, the fan is inefficient without the shroud.

Frozen Coolant. When coolant has not been properly maintained, it can freeze in cold temperatures. When an attempt is made to start the engine in the morning,

the engine is seized because the coolant in the water pump has frozen solid. Removing the pump drive belt and starting the engine will confirm your diagnosis.

As coolant begins to freeze, it gets slushy as ice crystals start to form. Slushy coolant can plug the radiator in a cold engine. As the coolant trapped in the block begins to heat up, boilover can occur.

Cooling Fan. When a cooling fan does not work properly, the engine can overheat. This problem will be more acute when idling or driving in town.

Exhaust Blockage. A partially blocked catalytic converter or exhaust system can contribute to overheating and loss of power.

Inoperative EGR Valve. Detonation can result from an inoperative EGR valve (see Chapter 44). This can contribute to engine overheating.

■ RADIATOR CAP

One inexpensive item that a vehicle often needs is a new radiator cap.

Radiator Cap Inspection

The rubber seal on the pressure cap can become worn or damaged with age, the pressure spring can rust, or the radiator filler neck seat can be damaged. A damaged cap can allow pressure to escape from the radiator. If pressure is not maintained on the coolant, its boiling point can drop 40°F to 50°F. This causes the

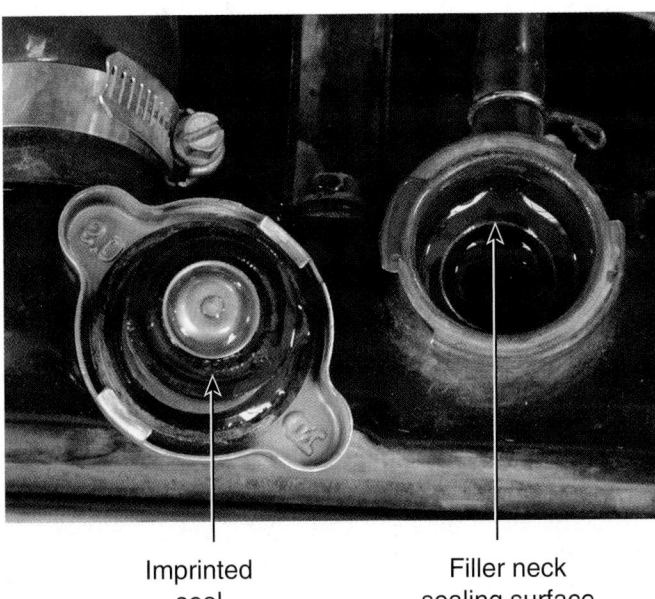

Imprinted
seal

Filler neck
sealing surface

Figure 21.3 Check the condition of the radiator cap pressure seal. *(Courtesy of Tim Gilles)*

Figure 21.4 A pressure tester is used to test the pressure cap. *(Courtesy of Tim Gilles)*

radiator to boil over in hot weather or when driving at high altitude.

Several things are checked on a radiator cap:

- Check to see that the rubber seal is not torn or imprinted so that it no longer provides an effective seal (**Figure 21.3**).
- Inspect the pressure valve spring for rust damage and freedom of movement.
- Inspect the vacuum valve to see that it is not stuck, broken, or plugged, and that the seal is not damaged. Vacuum valves can be of two types: spring loaded and ones with no spring.
- The cap should be of the correct pressure as specified by the manufacturer.

Radiator Pressure Tester. A pressure tester (**Figure 21.4**) can be used to test the cap's pressure valve.

- First, visually inspect the sealing surfaces and the vacuum valve.

SHOP TIP Moisten the sealing surface with water to help it seal during testing.

- Install the radiator cap on the adapter. Attach the pressure tester to the other end of the adapter.
- Pump the handle until the gauge reaches its highest point. The needle should reach a high point and remain constant. The pressure rating of the cap is stamped or printed on the top of the radiator cap (**Figure 21.5**). This is the point where the pressure cap should relieve pressure.

16 lbs

0.9 bar

Figure 21.5 Cooling system pressure is printed or stamped on the cap. Imported cars often have a barometric pressure rating. *(Courtesy of Tim Gilles)*

- Before removing the cap, release pressure from the system by pushing the tester hose sideways at the cap.

Radiator Inspection

Inspect the radiator for leaks, flaking, crushed or bent fins (**Figure 21.6**), or damage to the filler neck seat. Look for obstructions to airflow such as bugs and so

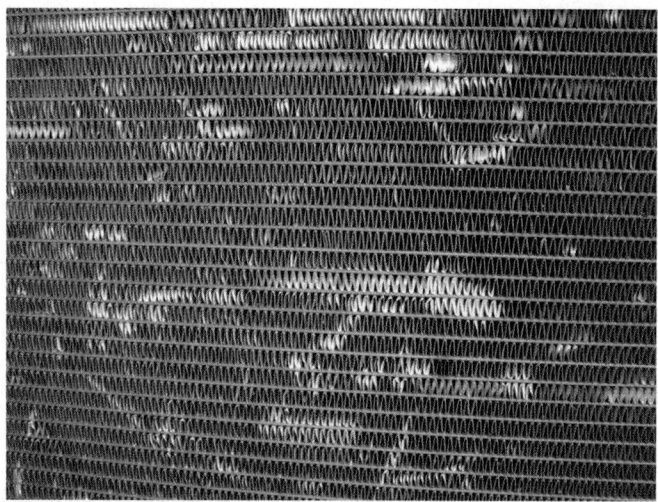

Figure 21.6 These radiator fins, bent during careless handling, can probably be straightened to allow airflow. *(Courtesy of Tim Gilles)*

forth, which can be washed out with water from the engine side of the radiator. Inspect the overflow hose and passage, if possible.

> **CASE HISTORY**
>
> *A technician was performing an underhood inspection on a car. He noticed that the upper radiator hose was collapsed. When he removed the radiator cap, the hose swelled up to its normal size. An inspection of the radiator cap's vacuum valve showed that it was in good condition. He connected a hand vacuum pump to the hose going to the overflow tank and discovered that there was an obstruction. Further investigation of the hose showed that it had been pinched under the battery box when it was removed during a recent collision repair.*

Radiator Flow

With the radiator full, watch the flow of coolant when the lower hose is removed from the radiator. If flow is sufficient, coolant should fill the entire radiator opening as it flows out.

Another test for radiator flow involves feeling the outside temperature of a radiator in various areas from top to bottom. If the radiator is allowing a sufficient volume of coolant to flow, the fins nearest the radiator inlet (the hose that goes to the thermostat) will feel hot to the touch. The areas near the radiator outlet will feel cooler. Cold spots indicate a restriction at the cold area of the radiator.

Cleaning a Radiator Core

Radiator cores are made of copper or aluminum. If the fins of the core are in good condition, a restricted copper/brass radiator can be disassembled and **rodded**

out by a radiator shop. Rods are forced through the tubes to clean them out.

Replacing a Radiator Core

Cooling efficiency suffers if the tubes have become plugged or if the fins are corroded or broken loose from the tubes. Check the condition of the fins by rubbing something gently against them (be careful not to bend them). If they are in good condition, they will be rigid and will not flake away.

When the cooling fin metal rots, the radiator will require replacement. Older radiator *tanks* are brass and are soldered to the copper radiator core. On copper/brass radiators, the top and bottom tanks are reused by soldering them to the new core. The filler neck is also soldered to the brass tank. This is done in a radiator shop.

The radiators on newer cars have plastic tanks and vacuum-brazed aluminum cores. It is especially important that these radiators be periodically maintained as they are more prone to corrosion/electrolysis than copper/brass radiators. Aluminum radiators are relatively inexpensive to replace, so they are usually not repaired when damaged or restricted.

Radiator Damage

A radiator can be damaged if a fan blade collides with it. This can happen when a fan clutch or coolant pump bearing fails. If the engine has a broken mount, the engine can move excessively, which can tear radiator hoses or allow contact between the fan and the radiator. A copper/brass radiator that is not too badly damaged can be soldered by a radiator shop.

Most automatic transmissions have heat exchangers (oil coolers) built into the radiator. Look for damage to the transmission oil cooler lines and fittings.

Transmission Heat Exchanger Leaks

If a leak develops in a heat exchanger, transmission fluid can be pumped into the radiator when the engine is running. The engine's crankshaft drives the transmission pump. With the engine running, pressure inside of the radiator heat exchanger is about 35–55 psi. Radiator cap pressure is only about 15 psi so transmission fluid migrates to the radiator.

After the engine is shut off, the pressure in the radiator is still 15 psi. Because the transmission has no pressure when the engine is not running, coolant can enter the transmission. The coolant gums up the transmission, meaning it will probably have to be rebuilt.

■ COOLANT SERVICE

After it has been in use for a period of time, coolant loses some of its protective ability and can become corrosive from contaminants it has picked up.

Coolant Inspection

Inspect coolant while it is cold. Open the radiator cap and check inside the filler neck with your finger. Look for deposits of grease, dirt, or rust. Look for *corrosion bloom*, a white deposit that attaches to the tops of the tubes. Deposits like these indicate a need for a coolant change because the coolant additives have been depleted. If coolant appears to be rusty or contaminated, a radiator flush will be required.

Checking Coolant Conductivity

Electrolysis in the cooling system makes small holes in cooling system parts as metal transfers from one electrically charged part to another. Electrolysis can produce a voltage in the cooling system.

The coolant's conductivity can be checked with a voltmeter.
- Ground the positive probe by attaching it to the radiator.
- Insert the negative probe in the coolant.
- A reading of 0.2 volt or less is good.

If the voltmeter reads 0.5 volt or more, the system should be flushed and refilled with new coolant to prevent corrosion of the metal parts in the system. Besides the internal problems this condition can cause, an inaccurate coolant temperature sensor reading can result when the cooling system charge is above 0.4 volt. The coolant solution should be close to 50% strength to provide adequate protection against electrolysis.

Coolant Change Interval

Most car manufacturers recommend that coolant be changed *at least* every 3 years or 30,000 miles or more. By-products of combustion contaminants can get into the coolant by leaking past the head gasket. Additives in the coolant package also become depleted and can be replenished by changing the coolant. There are also coolant recycling machines that clean and treat old coolant before returning it to use.

When mixing coolant and water, very hard water should not be used, especially with aluminum heads. Phosphate corrosion inhibitors can drop out of coolant with very hard water. Use distilled water instead.

Older vehicles require more frequent coolant changes. Inhibitors in the coolant wear out. Silicate additives that protect aluminum have a shorter life span than other additives because as they react with the metals, they are consumed.

Draining Coolant

Most modern radiators have drain valves made of plastic (**Figure 21.7**). Loosen the drain plug and drain the coolant. If the coolant is to be used again, be sure to drain it into a clean pan. When the radiator has no drain plug, remove the lower radiator hose to drain it. Twist the hose before trying to pull it off its

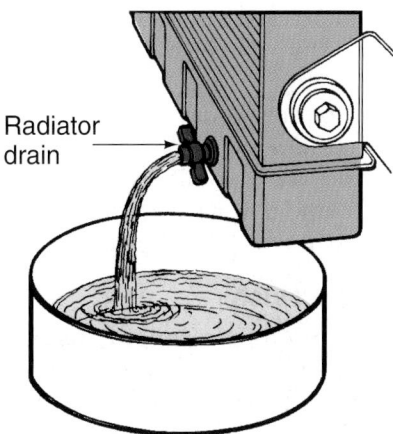

Figure 21.7 A radiator drain valve. *(Courtesy of DaimlerChrysler Corporation)*

connection, being careful not to damage the radiator. The hose connection is soldered to the lower tank on copper/brass radiators. This seal can become broken or the connection (which is soft brass) can be easily deformed. Special tools are available for separating a stuck hose from its connection (**Figure 21.8**). If it does not separate easily, it will be necessary to cut the hose (**Figure 21.9**). A special hose-cutting knife is available

Figure 21.8 A special tool for loosening radiator hoses. Wet the tool first by dipping it in the radiator coolant. *(Courtesy of Tim Gilles)*

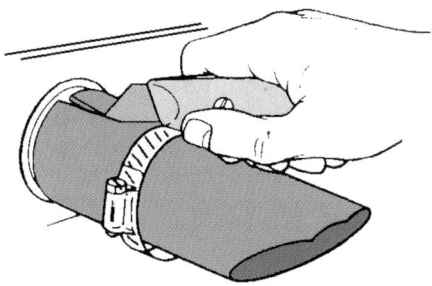

Figure 21.9 If a radiator hose will not come off without forcing it, cutting the hose might be necessary.

from tool manufacturers. Be especially careful not to cut through the thin brass on the inlet or outlet of a heater core.

There are also drain plugs on the side of the block. They are tapered pipe plugs and often become rusted to the block and are difficult to remove. If they are accessible and come out easily, remove them to drain the block. Use sealer on them on reinstallation.

Cooling System Flush

When coolant is not professionally maintained, mineral deposits and dirt can build up in the water jackets (**Figure 21.10**). Without removing the radiator from the car, the cooling system can be flushed, either with a cooling system flusher or by back flushing. A back flush is when water is run through the system backwards from its normal direction of flow. The coolant escaping the top of the radiator should be captured and disposed of in an environmentally safe manner.

NOTE: *Many communities have regulations governing the disposal of coolant. Be sure to follow regulations in your area. Ethylene glycol is biodegradable when new, but when it has been used, corrosion and heavy metals from within the engine can be present. In addition, coolant is poisonous to humans and animals.*

Coolant Exchanger

A popular piece of shop equipment is a coolant exchanger (**Figure 21.11**). When used correctly, little or no spillage of coolant occurs. The machine is easy to use. The upper radiator hose is disconnected and an adapter is attached with one end on the radiator and the other end on the upper hose (**Figure 21.12**). Before removing a hose, the machine is used to lower the radiator level.

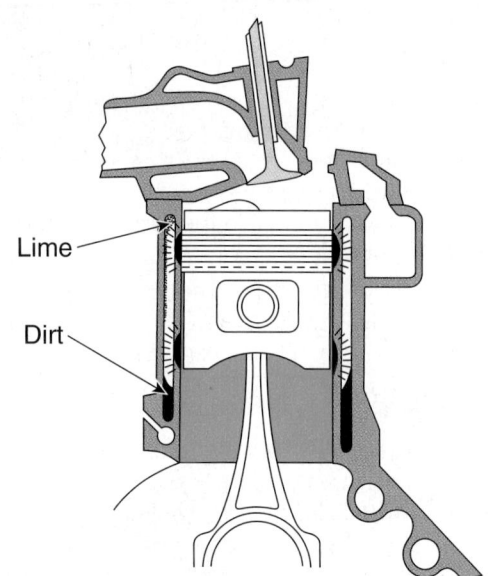

Lime

Dirt

Figure 21.10 Using tap water results in dirt and mineral buildup in the water jackets.

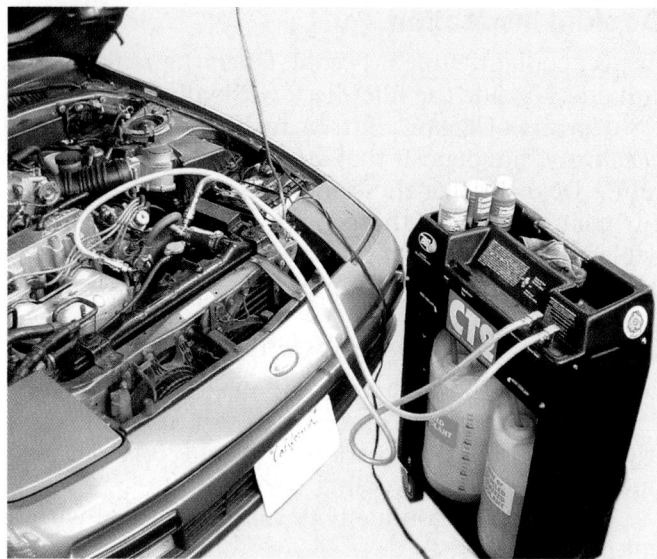

Figure 21.11 A coolant exchanger. (*Courtesy of Tim Gilles*)

Figure 21.12 Attaching one end of the adapter in place of the radiator hose. (*Courtesy of Tim Gilles*)

No cutting of hoses is needed. The machine has adapters for several sizes of radiator hose. **Figure 21.13** shows the coolant flush adapters installed in series with the upper radiator hose.

Many manufacturers recommend periodic replacement of the thermostat, which can be done as part of a flushing procedure. Thermostat removal and replacement is covered later in this chapter. When the thermostat is not being replaced, the engine is run during the coolant exchange. It must be warmed up first, however, until the thermostat opens.

Premixed coolant is used, or coolant and distilled water are mixed in the correct proportion and amount in one of this machine's two containers (**Figure 21.14**). During the coolant exchange, one of the machine's containers collects the old coolant while new coolant is pumped into the cooling system from the other

Figure 21.13 Several sizes of adapters. *(Courtesy of Tim Gilles)*

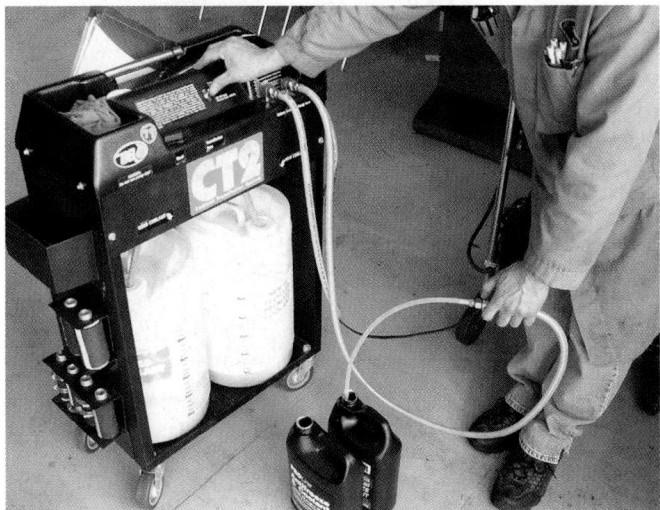

Figure 21.14 One container collects old coolant and the other supplies new coolant. *(Courtesy of Tim Gilles)*

Figure 21.15 The position of this switch determines whether old coolant or new coolant is pumped. *(Courtesy of Tim Gilles)*

the dashboard heater levers to the "heat" position to be sure that the heater core will be flushed, too.

Some radiator cleaners are acids, which are effective for removing rust and scale. They must be neutralized with a base following use, or damage to the cooling system can result.

 **SAFETY NOTE** Be sure that you wear safety goggles when using the radiator flush chemical. It can cause blindness.

Run the car for the required period of time. If the cleaning chemical requires a neutralizer, add it to the remaining water in the radiator after flushing is complete. The liquid remaining in the block will be 100% water.

Aluminum Oxide Contamination

Aluminum oxide is very abrasive; it is what sandpaper is made of. It does not form unless the coolant has become too diluted with water. If there is aluminum oxide in the cooling system, it looks like black sand beads. When flushing a cooling system, heat the engine to operating temperature and reflush the system at least three times, heating to full temperature each time. After flushing, run the water through a coffee filter to see if any black sand particles remain.

Testing Coolant Condition and Strength

Test strips (**Figure 21.16**) can be used to test the condition of a coolant. Different strips are used for conventional coolants and organic acid (OAT) coolants. Some single test strips can check pH, cavitation additive protection, and coolant concentration. There are also test strips that can tell if different types of coolants have been mixed.

container. Following the exchange of coolant, the overflow reservoir is pumped out and cleaned, using some of the used coolant pumped from the exchanger. The reservoir is filled to the correct level to complete the job. **Figure 21.15** shows the different switch positions used during the flush procedure. The machine is also handy for draining and refilling original coolant during an engine repair.

The coolant exchange process sometimes includes a chemical flush when a radiator becomes partially plugged with soft sludge. Using a commercial chemical cleaner, the radiator can be flushed without removing it from the car. With a hose running into the radiator, the engine runs at idle while the system is flushed. Move

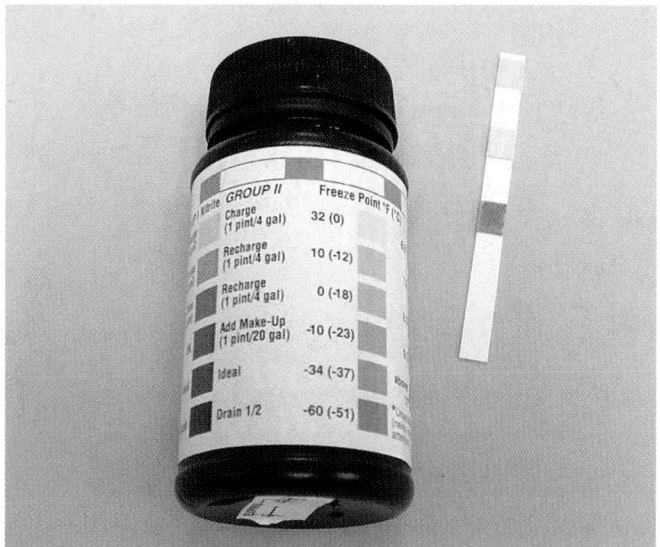

Figure 21.16 Different coolant test strips are used depending on the type of coolant. *(Courtesy of Tim Gilles)*

Conventional coolant has a higher pH than extended-life coolants. The additives in the coolant give it a pH level of about 10.5 when new. As coolant ages, acids form. Used coolant must continue to contain a sufficient amount of corrosion inhibitor to neutralize these acids. This neutralizing ability is called reserve alkalinity. Preserving an engine's cooling system depends on changing the coolant before its reserve alkalinity is depleted. Used conventional coolant should test at a pH level of at least 9.0. Extended-life coolant, which is more acidic due to its organic acid package, should test at a pH level of at least 7.5. When the additives become depleted, the acid level rises (pH level drops) and corrosion begins. The first things to fail are usually the radiator and the heater core because these are the thinnest cooling system parts.

Coolant Density Testers

Two types of testers can be used to measure a coolant's freeze point. Both determine the density of the coolant. The technician needs to know what kind of coolant is being tested. A problem that arises when testing coolant is how to identify it. One cannot tell simply by the color if the coolant is the factory-recommended coolant or an aftermarket replacement like propylene glycol.

A Coolant *hydrometer* is one way to test the strength of coolant (see Figure 13.15). Draw some coolant into the hydrometer and read the gauge. It compares the weight of ethylene glycol to the weight of pure water. Check the instructions on the tester.

Another tool used to measure coolant concentration is the *refractometer*, an optical tester that measures how much light can be refracted (bent) by a liquid.

Starting with a cool engine, use an eyedropper to remove coolant from the radiator (**Figure 21.17a**). Place a drop or two of coolant on the viewing surface under the cover of the tester (**Figure 21.17b**). Close the cover and hold the tester up to a light source while looking into the eyepiece (**Figure 21.17c**). The freeze protection is viewed in degrees.

Coolant Concentration

Ethylene glycol coolant is mixed with purified water until its concentration is correct. Coolant in too high a concentration can be further diluted. If coolant strength has been allowed to become very weak, a coolant flush and change is recommended. Draining off a quart of coolant and adding straight coolant can strengthen a slightly weak concentration. After running the engine, the strength is checked again.

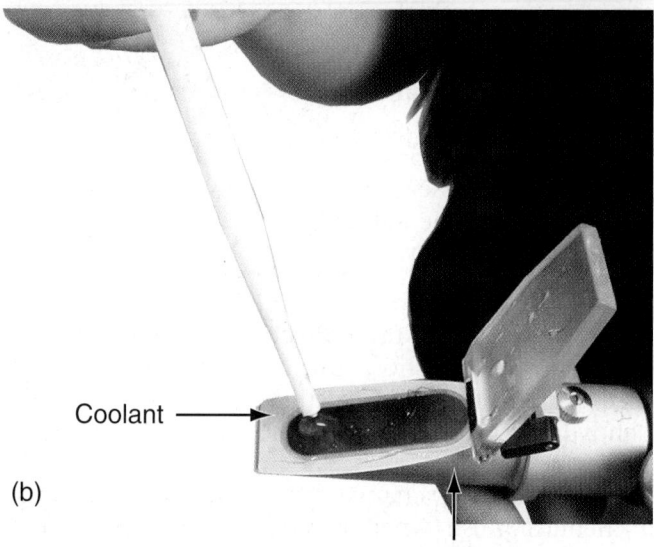

Figure 21.17 Using a refractometer. (a) Remove a coolant sample from the radiator. (b) Place a drop of coolant on the viewing surface. *(Courtesy of Tim Gilles)*

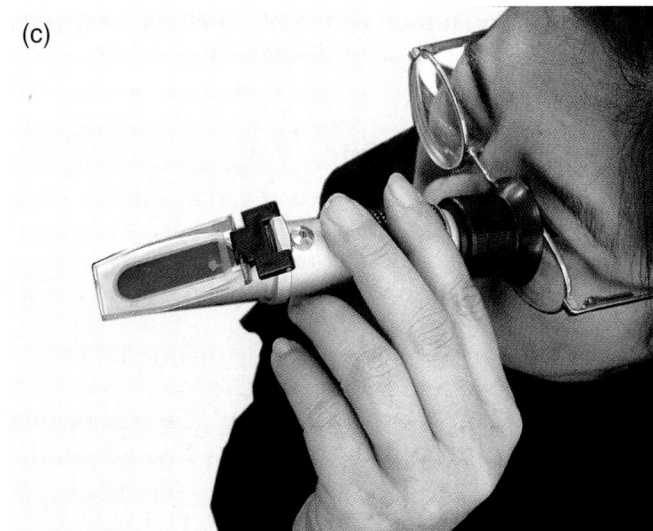

Figure 21.17 (c) Read the freeze protection through the eyepiece. *(Courtesy of Tim Gilles)*

FREEZE/BOIL PROTECTION CHART *Using a 15 psi Pressure Cap	% of Cooling System Capacity	PROTECTS FROM	
		Freezing down to	Boiling up to*
	50	-34°F	265°F
	60	-62°F	270°F
	70	-84°F	276°F

Cooling System Capacity in Quarts	Quarts of Antifreeze Required for Protection to Temperatures (°F) Shown								
	3	4	5	6	7	8	9	10	11
8	-7	-34	-69						
9	0	-21	-50	-70					
10	4	-12	-34	-62					
11	8	-6	-23	-47	-65				
12	10	0	-15	-34	-57				
13		3	-9	-25	-45	-64			
14		6	-5	-18	-34	-54	-68		
15		8	0	-12	-26	-43	-62		
16		10	2	-8	-19	-34	-52	-64	
17			5	-4	-14	-27	-42	-58	-69
18			7	0	-10	-21	-34	-50	-62
19			9	2	-7	-16	-28	-42	-56
20			10	4	-3	-12	-22	-34	-48

Figure 21.18 Typical coolant concentration chart found on a coolant container.

The maximum concentration that should be used is 70% coolant and 30% water. Look up the cooling system capacity in the service literature so that you can add the correct mixture of coolant and water (approximately 50% of cooling system capacity). If an engine has a cooling system capacity of 16 quarts, use 2 gallons of coolant to get an approximate 50% concentration.

Coolant is sold in gallon containers. It is available in quarts but is more expensive that way. If a cooling system holds 13 quarts, 1½ gallons (6 quarts) will provide a mixture that fits into the ⁴⁰‰ range. Unless the winter weather in the area is especially harsh, this will provide a good mixture. **Figure 21.18** shows a chart commonly found on the back of a coolant container.

Deciding Which Coolant to Use

Sodium silicate is a very good aluminum protection additive, but silicate additives can cause problems when used in too high a concentration. A majority of coolants contain this additive, but many modern coolants contain a lesser amount. Heavy-duty truck manufacturers and some automotive manufacturers specify the use of coolants without silicates. Be sure to use the specified coolant.

VINTAGE COOLANT

Older engines had cast iron heads and cylinder blocks. When an engine's cooling system does not contain any aluminum parts, a good rule of thumb is to use a coolant without silicate additives.

As the ratio of coolant to water is increased, the solubility of the silicates decreases. Silicates in a heavy concentration can gel on heat transfer surfaces. Engine overheating and poor heater operation can result. Abrasive sand granules can also form, resulting in coolant pump leaks.

NOTE: *Coolant should be used before the shelf life date printed on the container expires. If coolant with silicate additives is stored for too long, it will get gummy.*

The corrosion inhibitor used in DexCool extended-life coolant is based on two organic acids. These replace the additives found in conventional coolants. Because there are no phosphates in the additive package, hard water deposits are virtually eliminated. One of the organic acid additives, 2-EHA, has been associated with softening of some plastic parts, including coolant gaskets and coolant pump impellers. Be sure to use DexCool in vehicles if it is specified by the manufacturer.

G-05 is a popular coolant supplied as the factory fill in Mercedes and Chrysler vehicles. It is called hybrid organic acid technology (HOAT) coolant. It is like OAT coolant except that it contains silicates, although at a lower level than conventional American coolant. Like most coolants other than OAT coolant, it also contains benzoate.

Following the warranty period, some repair shops recommend a thorough flush and refill with a coolant containing phosphates and/or silicates like conventional green coolant or G-05. These provide extended head gasket life and protect the coolant pump from cavitation damage. Be sure the manufacturer does not recommend against this practice. Some original equipment coolant pump seals have failed from silicate abrasion.

NOTE: *All manufacturers recommend that conventional and extended-life coolant not be mixed. If a vehicle comes with one type of coolant, it is prudent to continue to use that type of coolant.*

There are many different coolants, but you will probably be able to satisfy most coolant needs with three coolants. If the engine originally used DexCool

---

SCIENCE NOTE

Silicon is the second most abundant material in the Earth's crust. It is the most important industrial semi-metal, its major use being in electronic components such as transistors. Glass is another material made from silicon. When silicon is dispersed in automotive coolant it protects aluminum, too.

■ Ethylene glycol coolant is poisonous. Ingestion of 4 ounces is sufficient to kill a human being. It has a sweet taste, so it is especially attractive to animals.

■ Be careful when handling extremely cold ethylene glycol. It can freeze-burn skin when it has been stored in the trunk of a car during very cold weather.

■ Ethylene glycol can be ignited by a flame at 474°F. Research by General Motors has shown that an explosion can actually occur if a mist of pressurized ethylene glycol coolant mixture is sprayed on an open flame.

SCIENCE NOTE

A hydrocarbon in which one of the hydrogen atoms has been replaced with a hydroxyl group (OH) is an alcohol. The hydroxyl group bestows waterlike properties to the hydrocarbon, so alcohols are soluble in water. Glycols are dihydroxy alcohols (they contain 2 OH groups instead of one). Coolant (ethylene glycol) is made from ethane (CH_3 CH_3) by replacing a hydrogen on each carbon with a hydroxyl group. When ethylene glycol coolant freezes, it forms a slushy mass rather than a solid block like water. In the absence of antifreeze, the 9-volume expansion that takes place when water freezes would generate a force of 30,000 psi at –22°C. This is sufficient force to damage an engine block.

Brake fluid is a chemical relative to coolant. It is a polyglycol, a hydrocarbon containing two or more hydroxyl groups, making it also soluble in water.

(GM, VW/Audi), top off with that. Otherwise, use G-05 or conventional green North American coolant.

■ THERMOSTAT

Testing the Thermostat

Testing the thermostat in the car can be done using a thermometer attachment with a hand-held multimeter or by putting a thermometer into the coolant while the engine runs and observing it while the engine warms up. With the engine running, the coolant will begin to show signs of movement when the thermostat opens. Note the temperature at that point.

If the coolant begins to circulate at a temperature that is different than the rating of the thermostat, the thermostat is defective and must be replaced. If the temperature does not rise to the proper point, the thermostat could be stuck open or missing completely.

Thermostat Check after Removal

After a thermostat is removed, it can be tested. To check its operation, suspend it in a pan of hot water. Use a thermometer while heating the water (**Figure 21.19**). If the thermostat is fully closed when cold and opens all of the way within a few degrees of its rating, it is good. Most thermostats start to open at between 188°F and 195°F. They are fully open at 212°F (water's boiling point).

The thermostat should open at least ¼" or more when immersed in boiling water. The thermostat should be fully closed at room temperature. Hold it up to the light to see if it is. If light comes through the thermostat at the sealing surface, replace it.

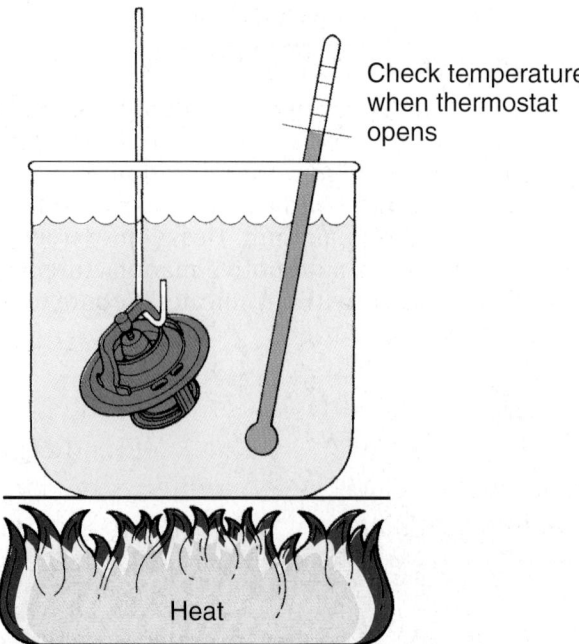

Check temperature when thermostat opens

Heat

Figure 21.19 Checking thermostat operation. *(Courtesy of Ford Motor Company)*

Remove the Thermostat

Drain some coolant into a clean container until the coolant level is below the thermostat housing. Remove the upper radiator hose connection from the thermostat housing. Loosen the housing bolts and remove the housing. Remove the gasket or seal and carefully clear the surface of the housing and the mounting surface on the engine. If any gasket remains on either of the surfaces, there will probably be a coolant leak after reassembly. Some engines use a rubber O-ring to seal the thermostat housing.

Replacing the Thermostat

Compare the new thermostat to the old one. They are of different sizes, types, and temperature ratings. The temperature rating is stamped on the sensing bulb on the bottom of the thermostat (see Figure 20.19). The temperature bulb faces into the block (**Figure 21.20**). When replacing a thermostat, be sure that the thermostat fits into the groove in the block or outlet housing (**Figure 21.21a and b**). Be certain to carefully scrape the gasket surfaces.

SHOP TIP When a paper gasket is used and the recess is in the thermostat housing, it is a good practice to position the thermostat into the recess and glue the gasket to hold it in place. If it falls out of its groove during installation, the outlet housing can be cracked or a coolant leak will result. Before tightening the water outlet housing, try to rock it back and forth to be sure it is flush. Housings are often cracked during this step.

Reinstall the thermostat housing. Refill the system and run the engine or pressure test to check for leaks.

Figure 21.20 The thermostat must face the correct direction. *(Courtesy of Tim Gilles)*

Sensing bulb faces into block

(a)

Recess

(b)

Figure 21.21 (a) The thermostat fits into a recess; (b) Correctly installed thermostat. *(Courtesy of Tim Gilles)*

When the engine has reached operating temperature, make sure the thermostat opens. You should be able to see coolant circulating within the radiator.

CAUTION If the radiator is filled to the top with coolant and the engine is run without the radiator cap in place, the coolant will expand and spill over as the engine warms up.

Another way of checking thermostat operation is to feel the top radiator hose or use a thermometer or multimeter with a temperature probe to confirm that the coolant is warming up. If the engine is overheating but the top hose is still cool to the touch, the thermostat is stuck closed and must be replaced.

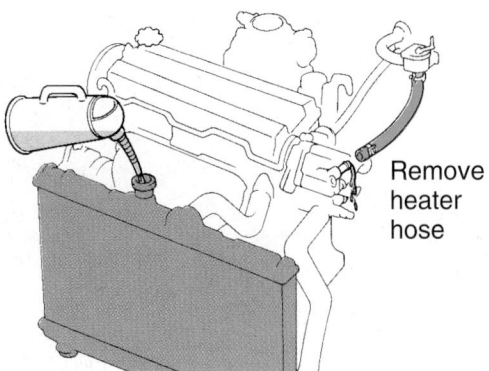

Figure 21.22 When refilling the system, bleed air off by removing a hose from the highest place in the system.

Bleeding Air from the System

There is sometimes a small check valve in the thermostat that allows air to escape during filling of the cooling system (see Figure 20.21). Some cooling systems are difficult to fill without trapping air. Overheating can result unless the air is bled off. Because of aerodynamics, many of the new cars have radiators that are lower than some of the other cooling system parts. These other parts trap air that cannot be bled out from the radiator cap as it normally would.

Check the service literature for instructions on bleeding the system. Some procedures call for removing a temperature sending unit.

Always recheck to see that the system is full and all air has been purged from the system before releasing the vehicle to a customer.

SHOP TIP To bleed a cooling system, remove a heater hose at the highest place in the system while filling the radiator (**Figure 21.22**). Removing a heater hose bleeds air from the heater core, too. If a flushing T has been installed in the top hose, unscrew its cap and raise the hose until it is higher than the heater core. Then top off the coolant until it comes out the hole. Some engines are equipped with a bleed valve located at a high point in the cooling system **Figure 21.23**.

■ LOCATING LEAKS

Whenever possible, locate a leak before starting a repair procedure. Leaks can be external or internal. External leaks are usually obvious and easily observable. Internal leaks can be through a leaking gasket or a crack. Tests for leaks are covered in this section.

Pressure Tester

When an internal or external leak is suspected, a pressure tester is helpful in locating it (**Figure 21.24**). Sometimes when trying to determine whether there is

Coolant bleed screw

Figure 21.23 Some engines have a bleed screw for removing trapped air from the cooling system. *(Courtesy of Tim Gilles)*

Figure 21.24 A pressure tester installed on a radiator filler neck. *(Courtesy of Tim Gilles)*

an internal leak, the engine must be tested at different temperatures. Cracks often leak when the engine is cold but not after it warms up. Also, although it is easier to pressure test a cold engine, leaks sometimes will not show up when the engine is cold. Test stubborn leaks with the engine both hot and cold.

NOTE: *Perform the pressure test with the engine off. The pressure tester does not blow off like a radiator cap does. The cooling system can be damaged if pressure is allowed to rise above normal system pressure.*

Pump on the handle of the pressure tester to pressurize the cooling system to the pressure marked on cap. After 5 minutes, the pressure on the gauge should remain steady, which indicates no leakage. If the gauge pressure drops and an external leak is not apparent, an internal leak is indicated.

NOTE: *A very small leak may not be evident in this short a time. Also, some pressure drop can occur as the coolant shrinks if the cooling system temperature drops.*

■ EXTERNAL LEAKS

If the pressure reading drops during a pressure test, first look for signs of external leakage.

SHOP TIP When a leak is in a position on the engine where it is not easy to see, an inspection mirror can usually be used with a flashlight to help to locate it. When the flashlight is shined on the mirror to bounce the light onto the area of the suspected leak, the leak will be visible in the inspection mirror.

Check for external leaks at the heater core, heater hoses, radiator hoses, thermostat housing, core plugs, radiator, or the hole on a coolant pump (see Figure 21.30). It is not uncommon for a leak that appears to be coming from the back of the engine (between the engine and transmission) to be running down the side of the block and along the side of the oil pan. There are core plugs behind the flywheel on some (but not all) engines. A black light tester (see Chapter 49) can be helpful in determining the location of difficult leaks.

CASE HISTORY *A student wanted to replace leaking core plugs on the back of the block between the engine and the transmission. He removed the automatic transmission from the car and removed the flexplate from the back of the crankshaft. Unfortunately, the block he was working on did not have core plugs on the back. When he pressurized the cooling system, he found that a core plug behind one of the side motor mounts was leaking. Coolant was running down the edge of the oil pan to the rear of the block, where it appeared to be coming from behind the flywheel. Using a pressure tester and flashlight to locate the leak before attempting to repair it would have saved a good deal of needless work.*

Coolant Outlet Housing Inspection

Inspect the coolant outlet (thermostat) housing for leaks or damage. Aluminum housings often suffer electrolysis damage. When possible, replace the housing with one made of the same material as the rest of the engine block or the cylinder head that it bolts to. For instance, install an iron housing on an iron head and an aluminum housing on an aluminum head. Be sure to position the hose clamp so as not to allow corrosion between the hose and the housing (**Figure 21.25**).

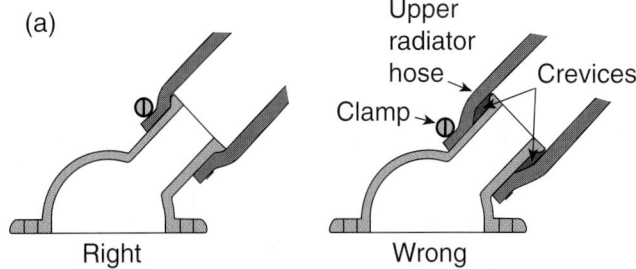

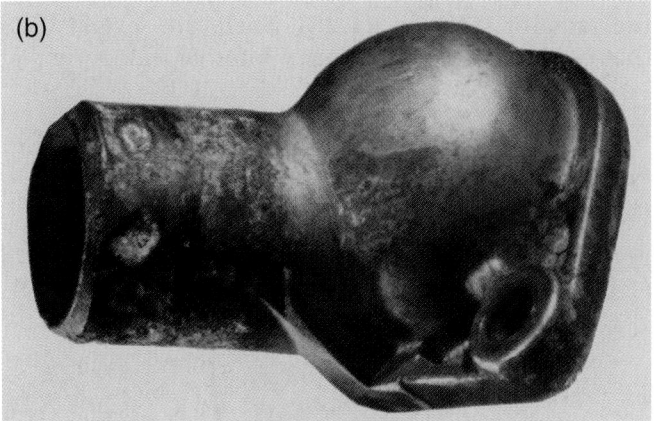

Figure 21.25 (a) Position the hose clamp near the ridge on the outlet housing. (b) This aluminum water outlet became corroded when the clamp was positioned improperly. *(Courtesy of Prestone Products Corporation)*

Core Plug Inspection

When the cooling system has not been serviced regularly or when someone fills the system with water after a leak has been repaired, corrosion inside the system can occur. Core plugs are made of steel (unless they have been replaced with brass or stainless steel). They commonly rust out and begin to leak (**Figure 21.26**).

Core plugs are usually located on the sides of the block. Sometimes they are found on the front and rear of the block, too. Use a pressure tester, mirror,

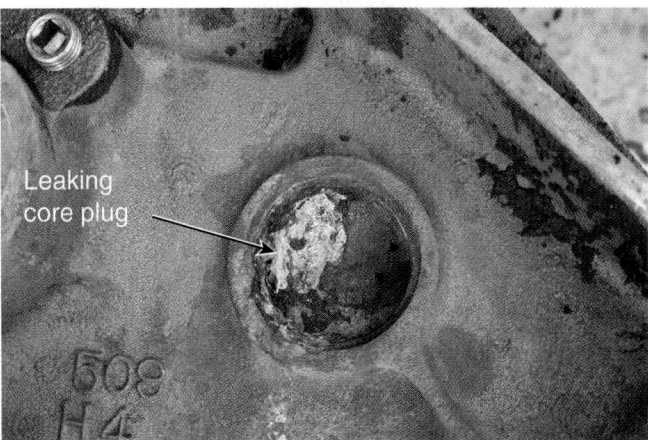

Figure 21.26 This core plug behind the flywheel was leaking. *(Courtesy of Tim Gilles)*

and flashlight to inspect core plugs for signs of rust or leakage.

■ INTERNAL LEAKS

A leaking head gasket or a crack in a cylinder head or bore can result in an internal leak (**Figure 21.27**). When there is an internal leak, coolant will flow into the cylinder during the intake stroke and when the engine is off. During combustion, exhaust gas is forced into the cooling system and can appear as bubbles in the radiator (**Figure 21.28**). There are several tests that can be done to confirm an internal leak.

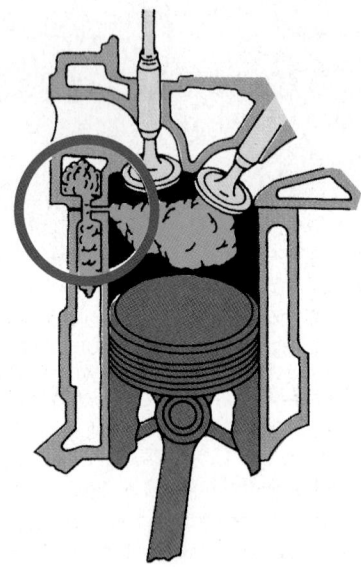

Figure 21.27 When there is an internal leak, coolant will flow into the cylinder when the engine is off and during the intake stroke. During combustion, exhaust gas migrates into the cooling system. *(Reproduced by permission of Deere & Company, John Deere Publishing, Moline, IL. All rights reserved)*

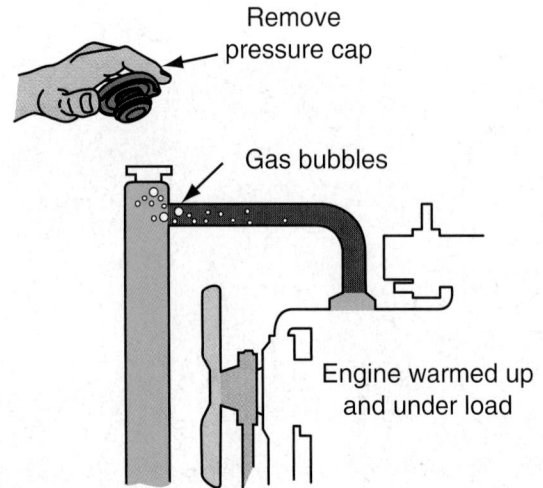

Figure 21.28 Exhaust gas leaking into the cooling system can result in bubbles in the radiator.

Bubble Test

Look for bubbles in the radiator when the engine is warm and under a load. Rapidly accelerating the engine is usually enough of a load to produce the bubbles.

NOTE: *Cracks tend to leak more when the engine is cold. After warmup, the crack closes.*

The radiator cap will blow off when the pressure from the combustion leak exceeds radiator cap pressure. To see if combustion pressure is indicated, put the radiator overflow hose into a container of water while the engine runs (**Figure 21.29**). If bubbles are evident, combustion pressure is getting in.

Bubbles could also be present because the cooling system is drawing in air. To eliminate this possibility, shut off the engine, loosen the drive belt to the coolant pump, and repeat the test. If the bubbles disappear, air was getting into the pump.

When there is air in the system, corrosion occurs at about three times the normal rate. Air can leak into the cooling system through a leak in the lower radiator hose. The lower hose is the suction hose, where coolant is drawn into the pump. Air can leak in even though water may not leak out. Air can also leak past a worn coolant pump seal. To test a cooling system for air leakage, tape the filler neck of the radiator closed. Put a hose from the radiator overflow pipe into a jar of water. With the engine running, look for bubbles in the jar.

SHOP TIP Here is a very effective test that can be done for final confirmation of an internal combustion leak before removing the cylinder heads:
- Loosen or remove the water pump belt and remove the thermostat. Reinstall the thermostat housing.
- Unhook the top radiator hose from the radiator and fill the hose with water and put a thermometer in it.
- Run the engine and look for bubbles in the coolant before the water's boiling point is reached. On a V-type engine, remove the radiator hose and look into the thermostat housing to see which side of the engine the bubbles are coming from to pinpoint the bank that has the leak.

Hydrostatic Lock

Sometimes an internal leak can result in one or more cylinders filling up with coolant after the engine is shut off. This happens because the radiator cap continues to exert pressure on the coolant, even though the engine is off. If the engine stops with a piston down in the cylinder while both of its valves are closed, the engine will be **hydrolocked** and the crankshaft will not be able to turn.

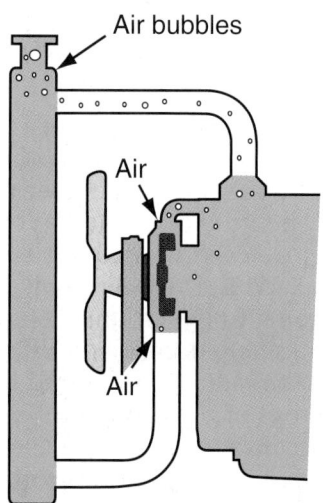

Figure 21.29 Air leaking into the system results in air bubbles coming out of the overflow tube.

Sometimes a leaking head gasket or a cracked cylinder allows a concentrated mixture of ethylene glycol coolant to leak into the crankcase (**Figure 21.30**). The

SHOP TIP Remove spark plugs on a hydrolocked engine. Crank the engine to let the coolant out so that the car can be driven to the repair shop.

CAUTION If you smell gasoline after the spark plugs are removed, do not crank the engine. A malfunction in a fuel system could have resulted in gasoline locking up the engine.

result is varnish-like oil that can plug oil rings and ruin valve guide seals. This sticky substance can actually seize the crankshaft (called **thermoplastic seizure**). The problem will happen again if the source of the leak is not found and repaired. Then the engine and cooling system must be flushed.

CASE HISTORY *A student's car would not crank over after he had been working on it in the shop. The instructor told him to remove the spark plugs to see if the engine turned over. The student turned the engine over and gasoline came out of one of the cylinders. It was ignited by one of the unhooked spark plug wires that was lying near the spark plug hole. The burning gasoline was sprayed onto the back of one of the students who was helping with the job. The student panicked and started to run across the shop. He was tackled by the instructor and the fire was extinguished from the back of his jacket and his hair. He was okay but he was lucky. Do not run if you are on fire. Drop and roll on the ground to extinguish the flames.*

Block Check Test

A leaking head gasket will not always show up on a pressure test. Another means of testing for leakage of exhaust gas into the cooling system is the **block check test**, also called *combustion leak test* (**Figure 21.31**). The tester samples air in the filler neck of the radiator. Unlike the pressure tester, the block check is used with the engine running. If there is carbon monoxide (CO) exhaust gas in the radiator, the color of the tester fluid will change. Carbon monoxide is a by-product of combustion, so the tester will not work if the leaking

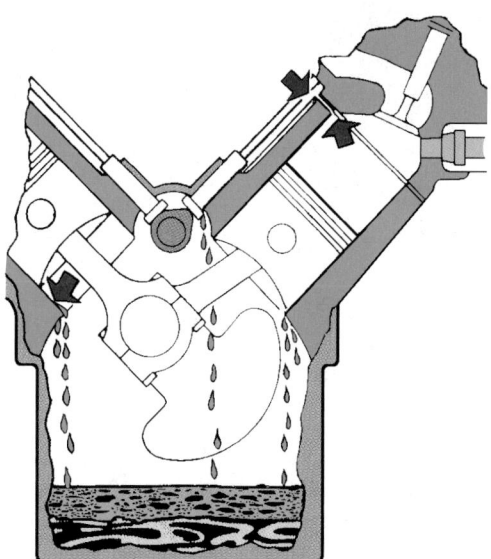

Figure 21.30 An internal leak can result in coolant entering the crankcase. *(Courtesy of John Deere)*

Figure 21.31 A block check test samples air above the coolant. *(Courtesy of Tim Gilles)*

cylinder's spark plug is not firing. Also, if compression is too low or if coolant entering the cylinder causes the plug not to fire, the tester will not give a reading.

A small amount of a special blue fluid is used to fill the tester to its fill line. If there is exhaust gas in the coolant, it will react with the tester fluid. First, the fluid turns from blue to green. Then it turns yellow.

This test is done on a warm engine. Be sure that the thermostat is open. First, the level in the radiator must be lowered if necessary until it is 2" below the top of the filler neck. This is done with the engine off. With the engine idling, place the tester on the radiator filler neck and pump the bulb several times to suck *air* from above the coolant.

NOTE: *As the coolant gets hotter, its level will rise. Letting coolant into the test fluid will ruin the fluid and void the test.*

If the results show no exhaust gas in the coolant after performing the following test, try the tester at the opening of the car's exhaust to see how the fluid reacts. When the fluid is left exposed to air, it will return to its original blue color.

Infrared Analyzer

An infrared exhaust analyzer can also be used to check if there is exhaust gas in the coolant. The analyzer probe is inserted into the neck of the radiator, in the air space above the coolant. The analyzer samples air, drawing it in through the probe. Be careful that the coolant is not accidentally sucked into the probe.

- This method works whether the mixture has been burned or not.
- Load the engine by accelerating it in gear with the brakes on for 3 seconds or less.
- Hold the probe over the radiator filler neck to check for hydrocarbons in the coolant. Be sure not to suck coolant into the tester probe. It can be quite expensive to replace a filter.
- Testing coolant with an infrared analyzer is most reliable when both HC and CO are present. The presence of HC only is not always a reliable indicator, because other chemicals can trigger an HC reading. If there is CO in the cooling system, exhaust gases must be getting in during combustion.

■ RECOVERY TANK SERVICE

An overflow (recovery) tank (see Figure 20.14) is a good feature to add to an older car if it does not already have one. A replacement tank can also be installed on newer cars with a damaged or leaking tank. Some of the less-expensive tanks have a single molded plastic mounting hole in the tank. Original systems that have these holes have several of them for support. Be sure to buy an original equipment one or one with a high-quality bracket.

NOTE: *One gallon of water weighs about 8 pounds. When the tank vibrates on rough roads, it can be torn from its mount.*

If the radiator cap needs to be replaced on a recovery tank system, be sure the new cap includes a seal that works against the top of the filler neck, too (see Figure 20.11a). The hose from the top of the radiator allows coolant to escape when pressure exceeds the rating of the cap. With a recovery tank, it is recycled instead of just going to the ground.

Recovery tanks help decrease corrosion. When the coolant level in the radiator is allowed to drop below the level of the tops of the cooling tubes, *solder bloom* corrosion can occur. This corrosion is lead oxide that forms when oxygen in the air reacts with the solder at the top of the tubes.

■ COOLING SYSTEM REPAIRS

Replacing Core Plugs

Cup-type core plugs are the most common type. To replace a core plug, pound it sideways with a *blunt* drift punch (**Figure 21.32a**). A sharp punch will penetrate the core plug. The objective is to turn the plug sideways inside of its bore so that it can be easily removed. When it is sideways, it can be pulled with pliers (**Figure 21.32b**). Try not to let it go into the water jacket.

NOTE: *Do not leave an old core plug inside of the block. This is an unprofessional practice that can result in further cooling system damage.*

Sometimes a slide hammer with a hook can also be used to remove a core plug. Sometimes the plug is so rusty that removal is difficult.

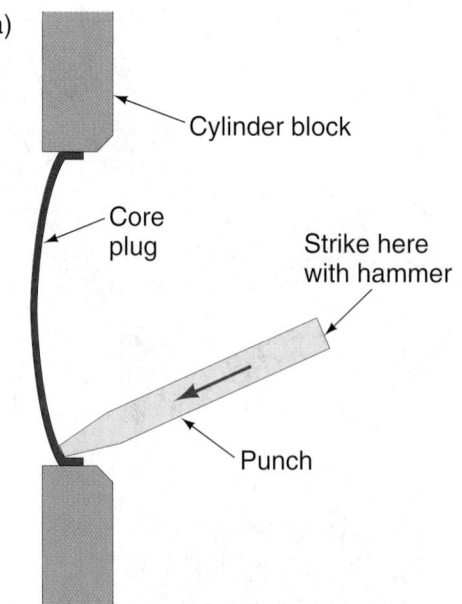

Figure 21.32 (a) Removing a core plug. Use a blunt punch to knock the core plug sideways.

Figure 21.32 (b) Remove the core plug with pliers. *(Courtesy of Tim Gilles)*

CAUTION Sometimes a core plug is positioned on the block with very little clearance to the back of wall of a cylinder **Figure 21.33**. Be careful not to pound the core plug against the cylinder wall.

Core Plug Installation

Before installing a core plug, clean the opening in the block with emery cloth. Put some sealer on the sides of a new plug and also on the side that will be exposed to the coolant. Pound the core plug in with a driver or a socket that fits *loosely* into the inside diameter of the plug (**Figure 21.34**). Check to see that the driver contacts the core plug on its inside surface (not on its

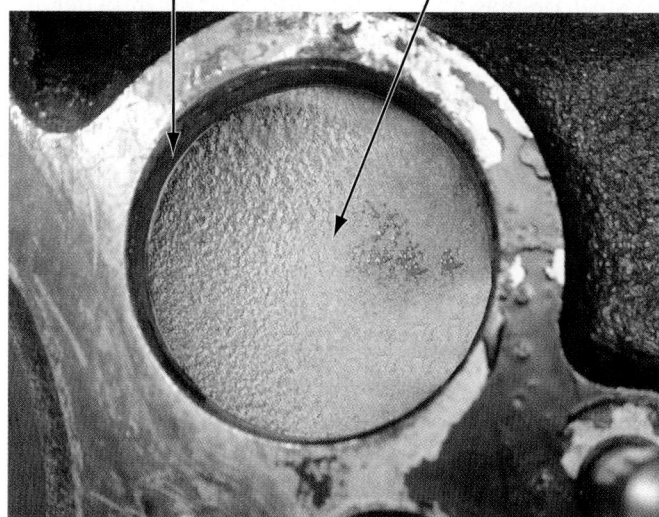

Core plug opening   Back side of cylinder

Figure 21.33 This cylinder wall casting is directly behind the core plug opening, making core plug removal more difficult. *(Courtesy of Tim Gilles)*

Figure 21.34 Drive in the core plug. Be sure to use a brass hammer to avoid damaging the tool. *(Courtesy of Tim Gilles)*

outer sealing edge). Be sure the core plug goes straight into the hole.

CASE HISTORY *An apprentice technician was overhauling an engine and removed the core plugs. The wall of one of the cylinders was positioned directly behind a core plug. The core plug became wedged between the block and the cylinder wall, so he forced it out. A cylinder wall is not very rigid and can easily become distorted when something such as a core plug is forced against it.*

During reassembly, the apprentice attempted to reinstall the piston and rings, but the piston would not go into the cylinder. When the shop master technician measured the cylinder, he found that it was out of round by 0.005". The piston clearance specification was 0.002". The block had to be sent out to a machine shop for boring and honing.

A core plug seals on its outer lip. It is correctly installed when the lip is against the bore. Pound it in until the outside sealing edge is just below the chamfer that is on the outside of the core plug bore (**Figure 21.35**).

NOTE: *Some technicians prefer to replace steel core plugs with brass ones that will not corrode.*

■ COOLANT PUMP SERVICE

Coolant pumps are often replaced after many years and miles of service when they begin to leak or make noise.

Leaking Pump

Some pumps have a vent hole or weep hole in the bottom of the pump (**Figure 21.36**). Leakage from

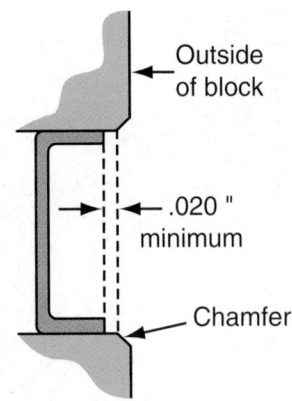

Figure 21.35 Installing a core plug. Install the core plug until it is all the way into the hole.

Look here for leakage

Figure 21.36 The pump seal has failed when leakage is evident at the vent hole. *(Courtesy of Tim Gilles)*

a coolant pump will usually be visible from the hole. A coolant pump leak can also appear to be from the lower radiator hose. Use a mirror to look at the weep hole. Pressurize the system if necessary with a pressure tester.

- A pump seal can fail because of bearing failure, corrosion of the shaft, or dirt.
- The seal can become red hot if it is run without coolant when the radiator boils over. Adding cold water to an overheated system causes the hot seal to crack.
- When a vehicle has been allowed to sit with a dry or dirty cooling system, the water pump seal will sometimes stick to the shaft. When the engine is started, it breaks loose, resulting in a leak.

Worn Bearing

Sometimes a worn bearing will result from the failure of the seal. With the engine running, a stethoscope can

be used to listen to a bad bearing. Before replacing a pump, loosen the drive belt and feel for roughness in the bearing. It should turn freely and smoothly, without end play.

Worn or Broken Impeller

Sometimes a pump impeller can be loose or broken, but this is rare. Look for water pump action in the radiator with the engine warm and running.

Coolant Pump Replacement

There are various types of pump installations on new vehicles. When in doubt about the procedure to follow for removal of the pump, consult the service manual for the vehicle. Sometimes, other accessory belts must be removed so that the belt that drives the pump can be removed or installed. Other times, they will not have to be removed. Simply loosen the bolts *slightly* so that the belts can be loosened. Unbolt the brackets to the air conditioning compressor or power steering pump if necessary.

SHOP TIP Drain the coolant into a clean drain pan (especially if it is to be reused). Sometimes a customer has recently had the coolant changed. He or she will not be happy to pay for new coolant again. Coolant is also costly to dispose of.

Remove all of the bolts and the pump. Clean all remnants of the old gasket off of the engine. Be careful not to gouge aluminum parts.

Inspect the Old Pump

Inspect the old pump. Pump impellers are made either of steel or plastic. Carefully inspect the impeller for erosion, looseness, or breakage. If pieces break off the impeller they will go into the block and, possibly, the radiator. It may be necessary to disassemble the radiator to be sure that all pieces are removed.

Selecting a Replacement Pump

Replacement water pumps are usually new, although sometimes rebuilt pumps are available at reduced cost.

SHOP TIP In some applications that use serpentine belts, the water pump is driven from the bottom side, using the back of the belt. The water pump **turns the opposite direction** of a V-belt–driven pump. Be sure the correct pump is installed or the impeller will be rotating backward, resulting in engine overheating.

When buying a rebuilt pump, the old one must be turned in as a core, which is an old rebuildable part.

Compare the new pump with the old one (before leaving the parts store, if possible).

Installing the New Pump

A coolant pump often has a steel plate bolted to its back. The plate gasket sometimes dries out during shipping. If this happens, the bolts will be loose, causing a leak. Before installing a new pump, it is a good idea to remove the bolts and cement both sides of the plate gasket. Be sure that the screws that hold the cover on the back of the pump are tight.

When replacing a pump, be sure that all gasket material is thoroughly removed from the block mating surface and that any O-rings, hoses, or gaskets are not damaged or forced during assembly. Use sealer to glue the gasket to the water pump. Sometimes a chemical gasket is used. Be sure that the surfaces of the pump and block have been cleaned of all oil and coolant so that the chemical can stick. Be sure the pump is perfectly flat against the block before tightening any fasteners.

Refilling the System

It is a good idea to fill the cooling system with water and pressurize it with a pressure tester before completing the reassembly of the belts and radiator. If there is a leak, it can be easily fixed at that point without wasting newly added coolant.

After the job is completed, the engine needs to be run. When the thermostat opens, the water level will probably drop as the coolant leaves the radiator to fill up the empty cylinder block.

Refill the Radiator. Inspect the hose from the radiator filler neck to the recovery tank. Be sure the hose is in good condition and tight on its fittings. Fill the recovery tank to about ½ full.

■ FAN INSPECTION

There are different types of fans. Procedures for checking them vary.
- An out-of-balance fan assembly can lead to coolant pump shaft and bearing failure.
- A leaking fan clutch, bent or broken fan, or a cocked or cracked aluminum fan spacer are all possible causes of pump failure. Be sure to clean all mating surfaces and tighten the fan bolts *evenly* to avoid causing a cocked assembly.

■ FAN CLUTCH INSPECTION

Inspect the fan clutch for fluid leaks and to see if it is loose or frozen. There are several ways to test the temperature-controlled clutch.

Figure 21.37 Inspect the fan clutch for leakage. *(Courtesy of Tim Gilles)*

Physical Tests

- First, with the engine off, turn the fan by hand. There should be a slight resistance, but the fan should turn without roughness, which would indicate a bad bearing.
- Rock the fan up and down to see if it is too loose.
- If there is a buildup of greasy dirt in the bearing areas of the clutch, the silicone fluid has probably leaked out (**Figure 21.37**). If the bearing in the clutch fails, replace the clutch; otherwise, the resulting imbalance will ruin the coolant pump.

Engine Running Tests

Block off the radiator and run the engine with the air conditioner operating to help warm the coolant.
- When the engine is cool, the fan will not pull much air.
- As the engine warms, there should be a noticeable increase in the noise level from the fan.
- If the fan clutch does not engage before the temperature gauge shows hot, it must be replaced.
- When a warm engine is shut off, the fan can turn a small amount, but it should not continue to freewheel. If it turns more than four or five turns, it is probably defective.
- When the engine cools down after the radiator is unblocked, the fan should disengage.

SHOP TIP A handy way to check fan clutch engagement is to write a number on the engine side of each fan blade with a marking crayon, and then point a timing light at the fan. If the drive pulleys are of equal size, the numbers will be stationary if the fan is locked up. If the clutch is slipping, the numbers will run backward.

■ ELECTRIC COOLING FAN SERVICE

Electric fans are turned on and off in response to a signal from a coolant temperature switch threaded into the radiator, water outlet housing, or a water jacket in the engine block or head. When the engine temperature goes above a predetermined temperature, the fan comes on to provide extra cooling.

If the fan is not working, look for an obvious cause such as a disconnected wire. Then, check the fuse panel to see if the fuse is burned out. Next, check the switch to see if it is operating properly using the following procedure:

■ When the engine is cold, disconnect the electrical connector to the coolant temperature switch. Use an ohmmeter to read across the two terminals of the switch. It should show an open switch (infinite resistance).

■ With the wires to the switch connected, run the engine until it is warm. The fan should come on, indicating a switch that is good.

■ If the fan does not come on, disconnect the wires to the switch. With the ohmmeter, the switch should now show continuity (low resistance), indicating a closed switch. If not, replace the switch.

This will involve checking the coolant temperature sensor and any relays that apply.

 Be sure to disconnect an electric fan motor whenever working near it.

■ HEATER CORE SERVICE

A heater core can leak or become plugged. Removing it to repair or replace it in the event of a leak is sometimes a big job requiring the removal of many parts under the dash.

NOTE: *If the windshield becomes more fogged when the defroster is turned on, this can be a symptom of a leaking heater core. But momentary fogging of the windshield is common in combination heater/air-conditioning systems when the heater is first turned on (even when the heater core is good). Operate the air conditioning with the heater control in the defrost position to dry the inside of the windshield.*

The heater core is supplied with engine coolant through two *heater hoses*. One hose carries coolant leaving the engine (**Figure 21.38**). A return hose carries the coolant back to the engine for reheating. If both heater hoses are removed during service or repair, mark one of them so that they can be returned to their proper positions.

On older cars, a valve installed in that hose controls the flow of coolant to the heater core when the heater dash controls are set for "heat." Newer cars have coolant flowing through the heater core at all

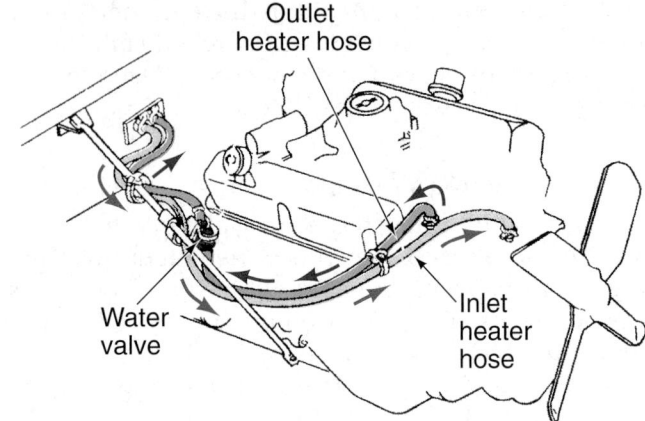

Figure 21.38 Direction of coolant flow to the heater core. *(Courtesy of DaimlerChrysler Corporation)*

times. Heat demands are controlled by controlling the doors to the ducts around the heater core.

Hybrid Cooling Systems

Hybrid vehicles have unique cooling considerations. With the vehicle power provided by one or more electric motors, the engine shuts off and does not produce heat. Hybrid high-voltage batteries produce heat, but they are air cooled. The high-voltage electrical system requires a special cooling system for its inverter (**Figure 21.39**). **Figure 21.40** shows a typical inverter and its cooling system reservoir.

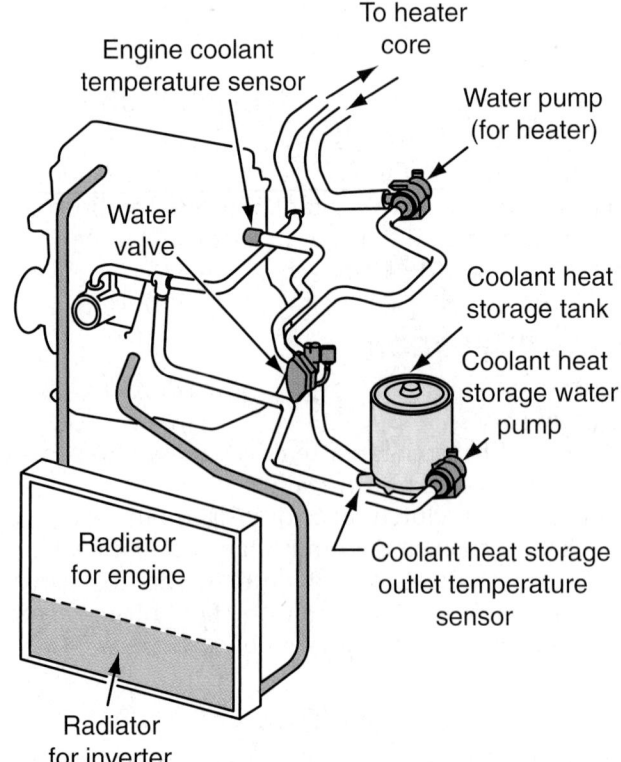

Figure 21.39 The high-voltage inverter for a hybrid automobile requires its own cooling system.

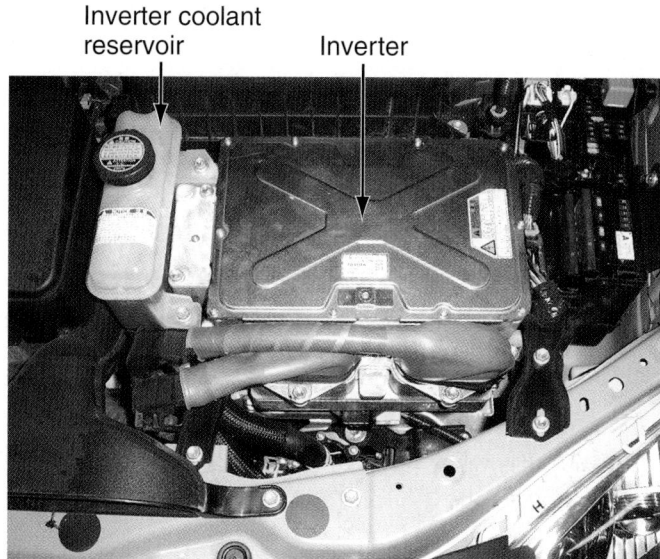

Figure 21.40 Coolant reservoir for a hybrid high-voltage system. *(Courtesy of Tim Gilles)*

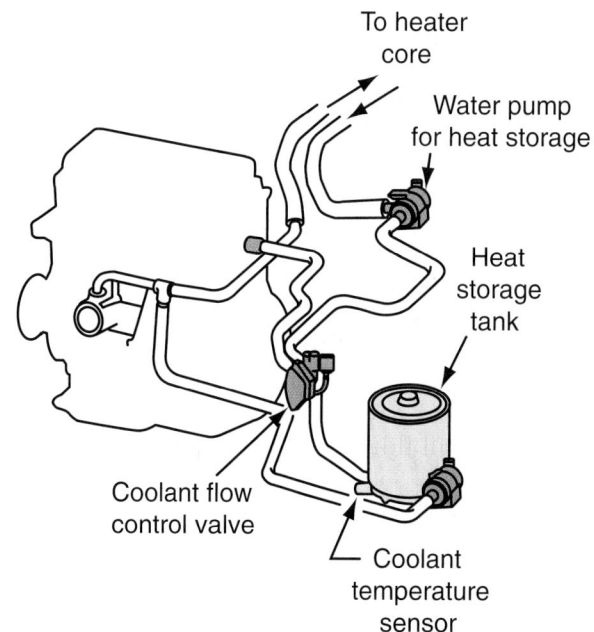

Figure 21.41 A coolant heat storage system provides heated coolant to a cold engine.

Coolant Heat Storage. Toyota uses an insulated tank to store heated coolant at 176°F for up to 3 days. When the engine is cold, an auxiliary pump supplies hot coolant to the engine. This reduces hydrocarbon emissions because internal combustion engines produce most of their pollution at low engine temperatures

(**Figure 21.41**). When changing the coolant in these systems, disconnect the electrical connector to the coolant heat storage coolant pump to prevent its operation, and refer to the service literature for instructions.

▇▇ REVIEW QUESTIONS

1. When the bottom of a radiator core feels considerably colder to the touch than its top, it is probably plugged. True or False?

2. When a radiator is disassembled and cleaned out, this is called _____ out.

3. When measuring the voltage of coolant, what is the limit before the system should be flushed out?

4. What is the name of the tool that compares the density of water to that of coolant?

5. What type of cleaner removes rust and scale?

6. Ethylene glycol coolant is dangerous to animals because it is _____.

7. Why is there a small hole in some thermostats?

8. What is the name of an old, rebuildable part that is turned in to the parts house when a rebuilt part is bought?

9. What is the name of the test that uses colored liquid to find exhaust gas in the cooling system?

10. When an engine will not turn over because a cylinder has become full of coolant, this is called _____ lock.

ASE-STYLE REVIEW QUESTIONS

1. A car that overheats on the freeway, but not in town, could have
   a. A stuck closed thermostat
   b. A partially restricted radiator
   c. Both A and B
   d. Neither A nor B

2. A transmission heat exchanger has failed. Technician A says if the engine is running, oil will flow from the transmission to the radiator. Technician B says when the engine is off, coolant will flow into the transmission. Who is right?
   a. Technician A   c. Both A and B
   b. Technician B   d. Neither A nor B

3. Two technicians are attempting to locate a coolant leak using a pressure tester attached to a radiator. Technician A says if the location of the leak cannot be seen, use a mirror and flashlight. Technician B says to run the engine when performing a pressure test. Who is right?
   a. Technician A   c. Both A and B
   b. Technician B   d. Neither A nor B

4. Which of the following is/are true when replacing a thermostat?
   a. The temperature sensing bulb must face out of the block.
   b. The thermostat often fits into a recess in the engine or water outlet housing.
   c. Both A and B
   d. Neither A nor B

5. Technician A says some radiator cleaners are acids, which are effective for removing rust and scale. Technician B says acid cleaners must be neutralized with base. Who is right?
   a. Technician A   c. Both A and B
   b. Technician B   d. Neither A nor B

6. Which of the following is/are true about coolants?
   a. Engine with cast iron heads should use anti-freeze with silicate additives.
   b. HOAT coolant contains silicates.
   c. Both A and B
   d. Neither A nor B

7. Which of the following is/are true about aluminum oxide?
   a. It is hard and abrasive
   b. It is what sandpaper is made of
   c. It is black
   d. All of the above

8. Technician A says electrolysis can produce voltage in the coolant. Technician B says electrolysis can cause holes in engine parts. Who is right?
   a. Technician A   c. Both A and B
   b. Technician B   d. Neither A nor B

9. Which of the following can be determined using coolant test strips?
   a. pH
   b. Cavitation additive protection
   c. Coolant concentration
   d. All of the above

10. Which of the following is/are true about coolants?
    a. Conventional coolant has a pH level of about 10.5 when new.
    b. Extended life OAT coolant has a pH level of about 7.5 when new.
    c. Extended life coolant is more acidic than conventional coolant.
    d. All of the above.

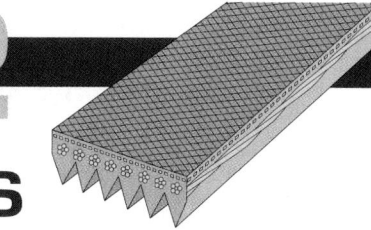

Automotive Belts

KEY TERMS

tensile cords

neoprene

high cordline belt

V-ribbed belt

serpentine belt

jackscrew

INTRODUCTION

The topic of this chapter is the theory and service of all types of belts. Accessories are usually driven with a belt from the crankshaft (**Figure 22.1**). Pumps and air-conditioning compressors are driven either by a V-belt or a V-ribbed serpentine belt. On some engines, the camshaft is also driven by a belt called a timing belt.

BELT MATERIAL

Belts are very strong and flexible with **tensile cords** to provide strength (**Figure 22.2**). The overcord material on the top of the belt is made of **neoprene** or another kind of oil-resistant artificial rubber. The undercord of the belt is the area beneath the tensile cords. It supports the cord and transfers loads to the pulleys. Sometimes, the undercords have a cord support platform with textile cords running perpendicular to the tensile cords (see Figure 22.2). Tensile cords are used to prevent the belt from sagging in the middle, which results in uneven load distribution and early belt failure.

V-BELT

Early cars used a flat drive belt. The V-belt was invented in 1917. It has more surface area in contact with the pulley groove than a flat belt of the same width would have.

V-belts must be the correct size:

■ A belt should extend slightly out of the pulley groove (**Figure 22.3**). The belt has a cord member to provide strength. If the belt rides too high in the groove, the belt will wear below the cord member.

■ When a belt rides below the edge of the groove this indicates either a worn belt, a worn pulley groove, or a belt that is too small.

■ A belt that is too small for the pulley will bottom out in the pulley groove, and the sides of the belt will not grab.

■ Some small diameter pulleys would cause severe bending stress on a belt. In this case, a notched belt is used.

Belt Cords

Most V-belts have polyester tensile cords. The strength of a belt is determined by the placement of the belt's tensile cords.

■ **High cordline belts** (**Figure 22.4**) are stronger but require more material to manufacture.

■ *Center cord* belts are cheaper but do not last as long.

High cord belts have about 40% more cords because the cord is at a wider part of the belt. In SAE tests, high cord belts lasted about four times longer than center cord belts.

In past years, the edges of premium belts were covered with a fabric cover to protect them from the elements. Today's belts have no cover and do not show wear as easily. Sometimes they can appear good, even though they are ready to fail.

Some engines use dual belts to drive accessories; these must be replaced in pairs when they are worn.

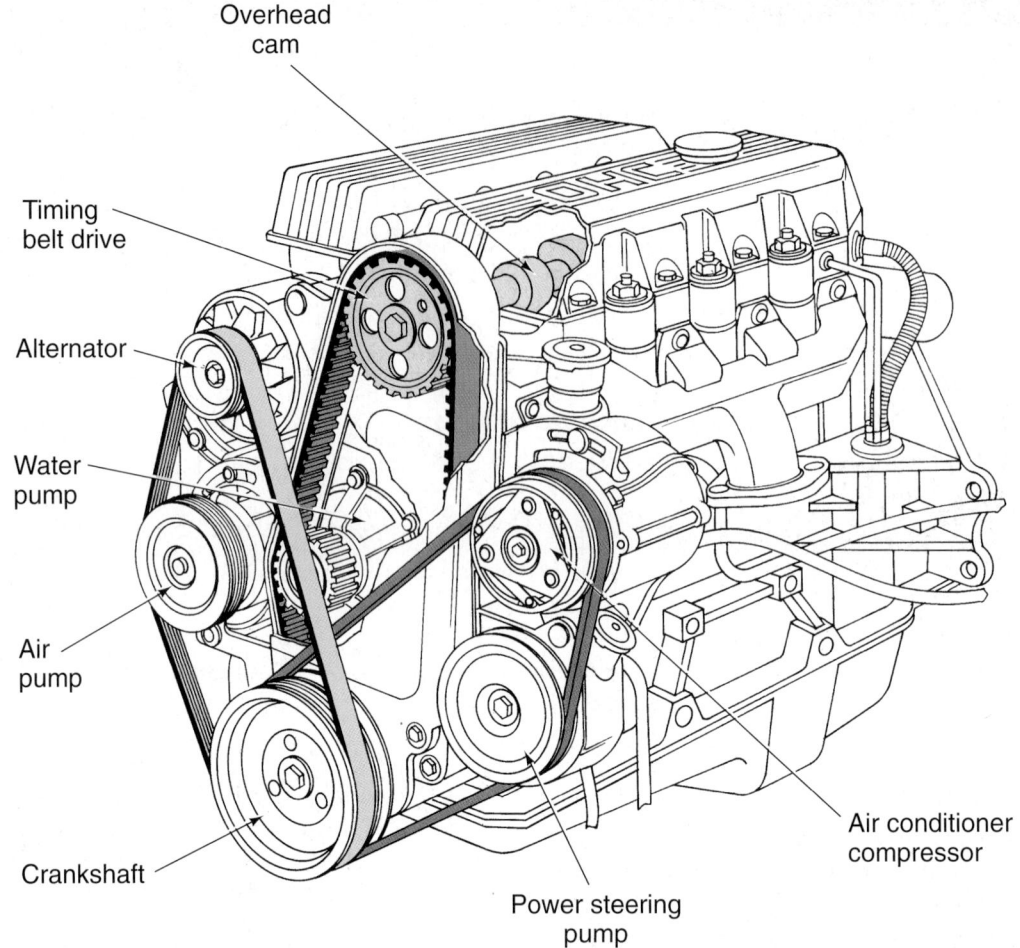

Overhead
cam

Timing
belt drive

Alternator

Water
pump

Air
pump

Crankshaft

Power steering
pump

Air conditioner
compressor

Figure 22.1 Belt-driven accessories and camshaft. *(Courtesy of The Gates Rubber Company)*

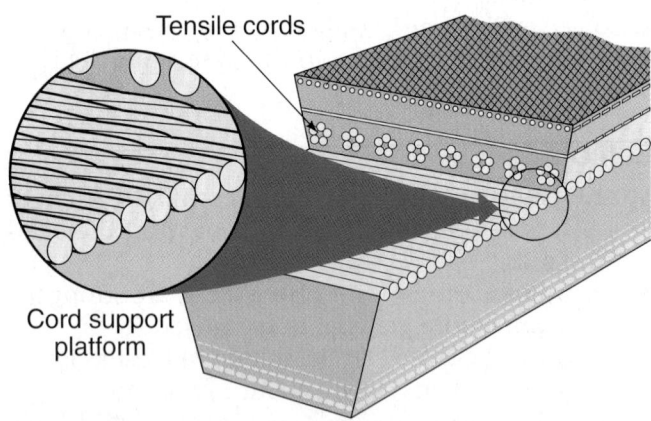

Tensile cords

Cord support
platform

Figure 22.2 Tensile cords and the cord support platform.
(Courtesy of The Gates Rubber Company)

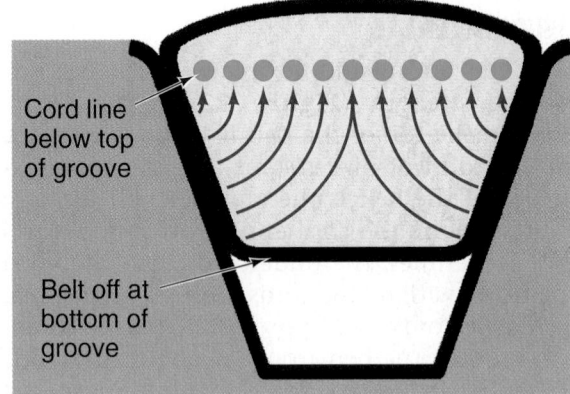

Cord line
below top
of groove

Belt off at
bottom of
groove

Figure 22.3 This V-belt fits well in the pulley groove.
(Courtesy of The Gates Rubber Company)

■ V-RIBBED BELTS

V-ribbed belts are ribbed on one side (**Figure 22.5**) and flat on the other. **Figure 22.6** compares the pulleys for V-belts and V-ribbed belts. The thinness of the belt makes it more flexible so it can bend around smaller pulleys and also be bent backward so both sides can be used to transmit power. Usually the ribbed side matches the pulley grooves of accessories and the flat

side goes against a spring-loaded tensioning roller. But the flat side is also capable of transmitting power.

■ SERPENTINE BELT DRIVE

V-ribbed belts are used in newer cars in a conventional manner or in a **serpentine belt** drive. Serpentine V-ribbed belts, which first appeared in the late 1970s, are used on many new engines today. One belt is used

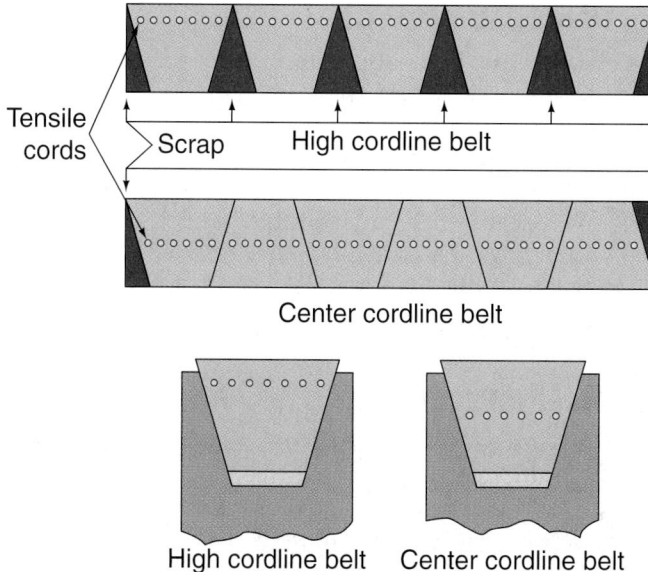

Figure 22.4 High cordline belts are stronger but require more material to manufacture. The higher cord position is evident when viewing the belt from the side. *(Courtesy of The Gates Rubber Company)*

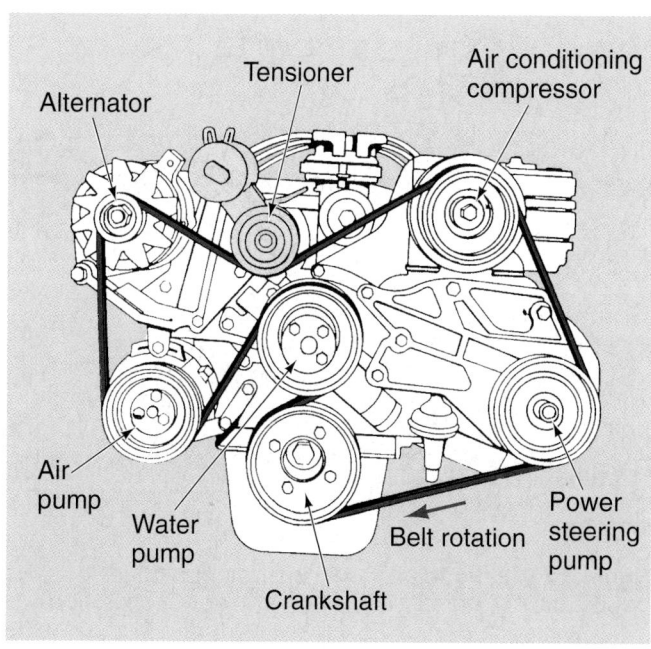

Figure 22.7 A serpentine V-ribbed belt drive. *(Courtesy of The Gates Rubber Company)*

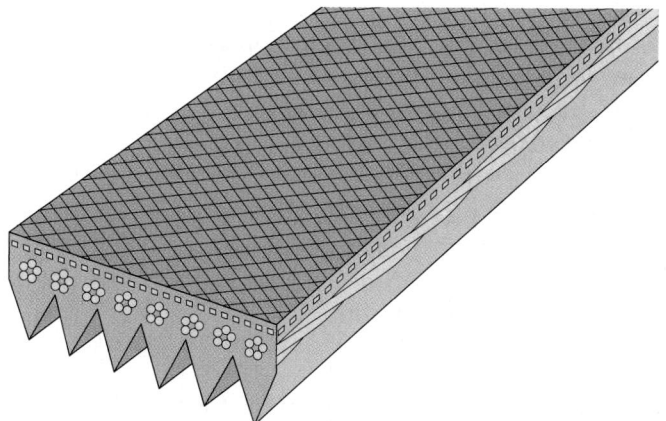

Figure 22.5 Construction of a V-ribbed belt. *(Courtesy of The Gates Rubber Company)*

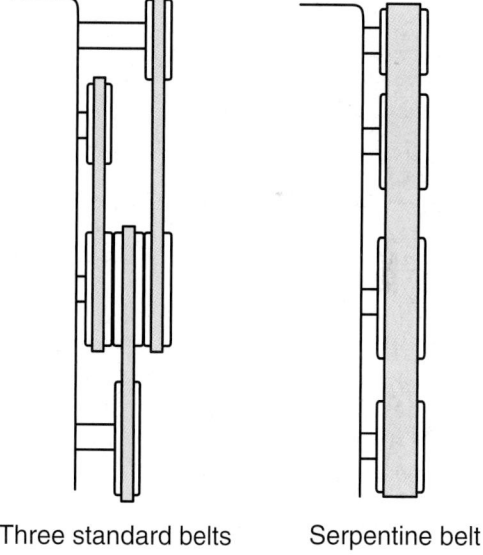

Three standard belts   Serpentine belt

Figure 22.8 Serpentine belts save space.

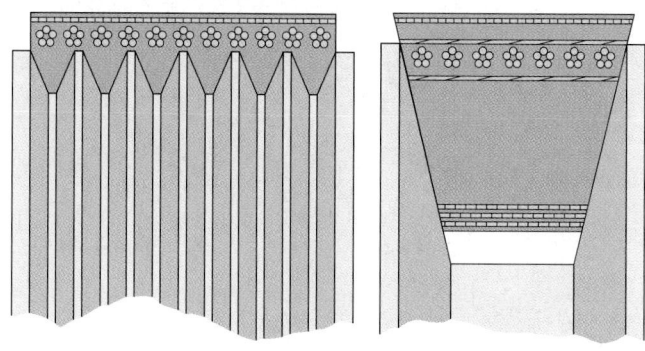

V-ribbed belt   V-belt

Figure 22.6 Comparison of a V-ribbed and a V-belt in their pulleys. *(Courtesy of The Gates Rubber Company)*

to operate all accessories (**Figure 22.7**). The belts are called serpentine because they follow a snake-like path, weaving around the various pulleys. Compared to V-belts, serpentine belts are easier to install, take up less space (**Figure 22.8**), transmit power more efficiently, and last longer.

■ TIMING BELTS

On some overhead cam engines, a timing belt drives the camshaft. Timing belts were introduced in the 1960s. Compared to a timing chain, they are quieter, do not require lubrication, are more efficient, and resist

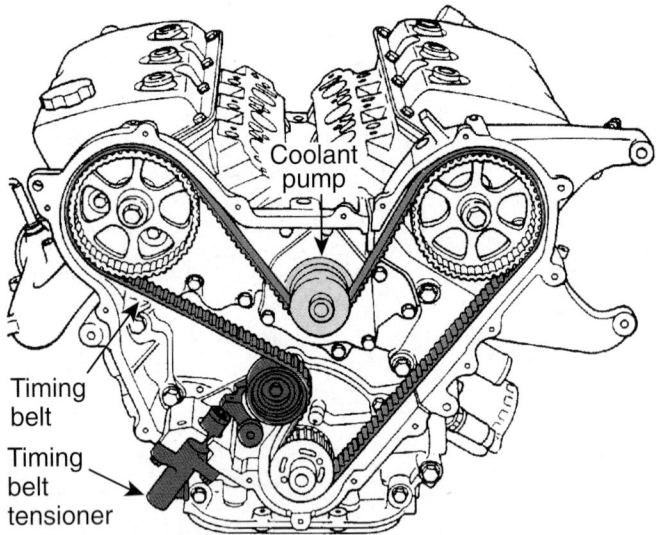

Figure 22.9 Both sides of a timing belt can drive accessories. *(Courtesy of DaimlerChrysler Corporation)*

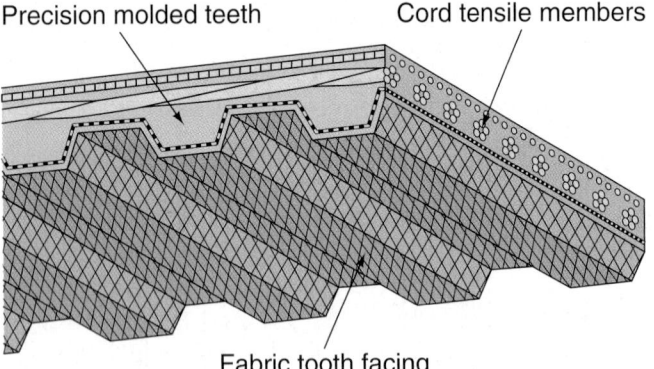

Figure 22.10 Parts of a timing belt. *(Courtesy of The Gates Rubber Company)*

stretching. Both the inner cogged surface and the outer flat surface can be used to drive accessories. The timing belt sometimes drives a coolant pump (**Figure 22.9**), an oil pump, and/or a balance shaft.

Timing belts have a very strong fiberglass cordline and rubber-impregnated molded teeth (**Figure 22.10**). The fiberglass cords are stronger than the polyester cords found in conventional belts, but they are more fragile. Bending them in any direction can cause them to break.

The rubber material in timing belts can be either high-temperature neoprene or highly saturated nitrile (HSN). HSN is a superior material for high-performance applications. A comparison of belt materials by the Gates Rubber Company gives the following estimation of belt life at sustained high temperatures:

Neoprene: 200 hours
High-temperature neoprene: 500 hours
HSN: 1,500 hours

More information on timing belts can be found in Chapter 18.

DRIVE BELT SERVICE

Accessory drive belts and timing belts are very strong and dependable, when they are replaced at reasonable intervals. A failed water pump drive belt can cause engine failure if the owner ignores the problem and continues to drive the car until the engine overheats severely. Because belts are so important, it is advisable to change them periodically *before* they fail, regardless of appearance. Belt failures rise significantly after four years of use, so that is the recommended interval for replacement.

BELT INSPECTION AND ADJUSTMENT

One serpentine belt usually drives all of the accessories. The belts are relatively trouble-free, but if one fails, the cooling and electrical systems and power steering will all cease to operate. The engine will still run, but driving the car will result in serious engine damage.

CASE HISTORY *A motorist was driving in the fast lane at 75 mph in heavy freeway traffic. The air-conditioning compressor clutch failed, causing the serpentine belt to be destroyed (**Figure 22.11**). The motorist was concerned for her safety and continued driving the vehicle until she could get off the freeway and locate a telephone. By the time she was able to leave the freeway, the engine was seriously damaged by overheating.*

Check all drive belts for wear, cracks, or damage (**Figure 22.12**). A belt that has any of these conditions should be replaced. Be certain that there is sufficient tension on the belts.

Figure 22.11 This serpentine belt was destroyed when the air-conditioning clutch failed. *(Courtesy of Tim Gilles)*

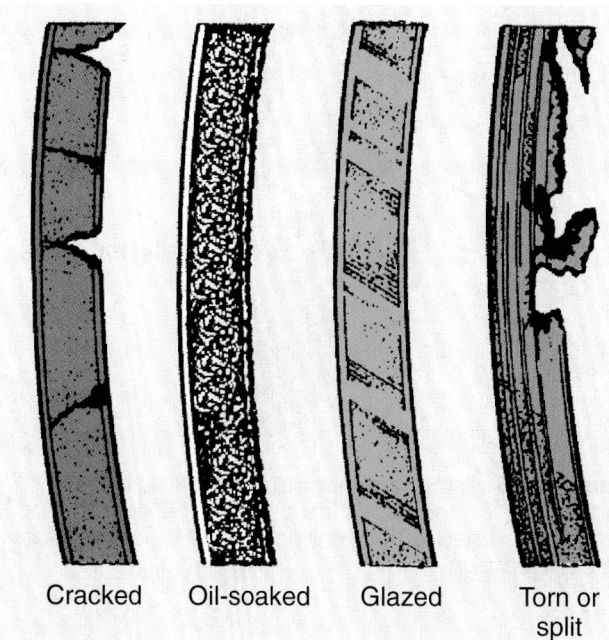

Cracked · Oil-soaked · Glazed · Torn or split

Figure 22.12 A worn belt can result in serious engine damage. *(Courtesy of DaimlerChrysler Corporation)*

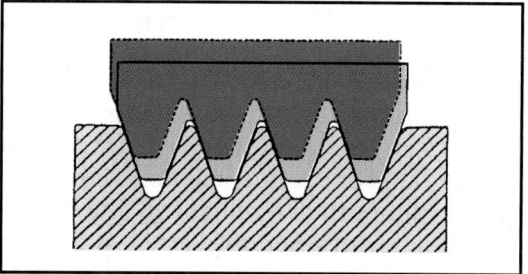

Figure 22.13 Noise happens when a misaligned belt vibrates in the pulley groove. *(Courtesy of The Gates Rubber Company)*

can be due to the pulley shafts being unparallel or due to an accessory being located improperly. Misalignment can cause rapid belt and pulley wear and thrown belts. When pulleys are out of alignment, the belt will chirp. Noise is a result of vibration. In this case, the vibration results because the belt must continually slide into the groove as it moves against the pulley (**Figure 22.13**).

> **SHOP TIP** With the engine off, check water pump/alternator belt tension by trying to turn the pulley on the alternator by hand. If it slips, the belt is too loose or glazed and must be replaced.

When a belt slips on the pulley, it becomes glazed. A glazed belt must be replaced because it will not grip the pulley tightly, even if tension is at specification.

Inspecting Belts

Normally, a V-belt contacts only the sides of the pulley groove. When a V-belt is the wrong size or has become worn from slipping, the bottom of the belt can contact the pulley groove bottom (see Figure 22.3). Cracks on the face of an old V-ribbed belt are normal, but replacement is suggested.

Inspect the pulley grooves for oil, rust, or wear. Smaller pulleys usually show wear first.

NOTE: *Surface finish of the pulley is more important with a V-ribbed pulley than with a V-type pulley.*

Some shops use belt dressings to help a belt to grip. The problem with these substances is that they are sticky and attract dirt, which wears the pulley groove. Some belt dressings actually attack and damage the belt material.

Belt Alignment

Inspect belt alignment before disassembly. V-belt pulleys must be in alignment within $\frac{1}{16}$" for each foot of the distance between the pulleys. Misalignment

---

> **SCIENCE NOTE**
>
> *When an object vibrates, it causes the molecules of air surrounding it to compress and then expand. This cycle moves through the air much like a ripple through water. If the compression/expansion wave vibration is at or near the natural frequency of the part, damage can occur. This is why the entire Tacoma Narrows Bridge collapsed in 1940.*
>
> *The movement of the air can be compared to a wave, with the crests corresponding to forward movement (expansion) and the troughs corresponding to backward movement (contraction). The distance between the crests is called a wavelength. The number of crests that bypass a given point in one second is called the frequency of the vibration. The shorter the wavelength, the smaller the frequency, and vice versa.*

A misaligned V-ribbed belt can walk off the pulley (**Figure 22.14**) or can tear off a rib. The accessory can sometimes be realigned using 0.030" front end alignment shims.

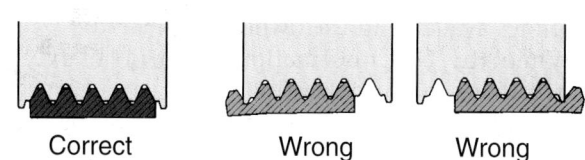

Correct · Wrong · Wrong

Figure 22.14 A V-ribbed belt can walk off the pulley if it is misaligned.

▪ REPLACING BELTS

SAFETY NOTE Before replacing a belt, disconnect the battery to prevent an electric cooling fan from accidentally coming on. Also, the output terminal on the alternator will be hot when the battery is connected. Prying against the outside of the alternator can result in an accidental short circuit.

Loosen bolts on any accessories that act as drive belt adjusters and remove the belt. Compare the new belt to the old belt. If no belt is available, use a piece of string in the pulley groove to estimate the replacement belt size. Belts usually change in size in ½" increments, which is reflected in the part number of the belt.

SAFETY NOTE Be sure the new belt is the right size. A belt that is too long can rub on a radiator hose or fuel line after tension is adjusted.

SHOP TIP If a belt seems too small to install, try installing it in a different order. For instance, if the belt will not go over the coolant pump pulley last, try installing it over the alternator pulley last.

V-Ribbed Belt Replacement

Most V-ribbed belts have constantly spring-loaded tensioning idler pulleys that contact the smooth backside of the belt (**Figure 22.15**). A few others have "locked center" drives. In these, tension is controlled in one of the following manners:

- A tensioner pulley with an off-center bolt (**Figure 22.16**)
- An adjustable **jackscrew** (**Figure 22.17**)
- An accessory such as the alternator (**Figure 22.18**)
  Before removing a serpentine belt, make a sketch of how it is installed. It can be complicated figuring out the path of one of these belts. Most vehicles have an underhood diagram label. Belt manufacturers produce catalogs with belt routing diagrams for the various makes of vehicles. These are available from parts stores and are usually free of charge to customers. Belt routing information is sometimes available in the owner's manual, too.
  When the belt information is not available, the following tips might be useful:
- Remember that both sides of the belt can be used to drive accessories. The V-grooved side of the belt will mesh with pulley grooves. The flat side of the belt will go against a flat pulley.

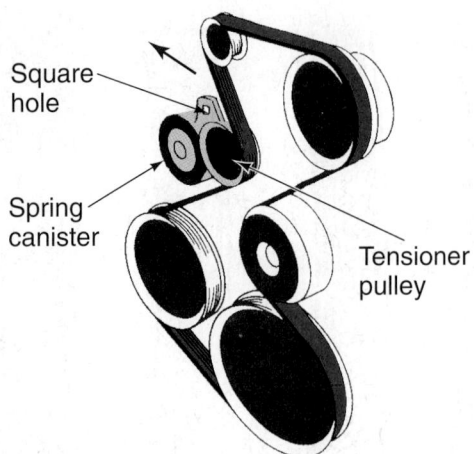

Figure 22.15 To install a V-ribbed belt with a spring-loaded belt tensioner, pull the pulley in the direction of the arrow to release belt tension. *(Courtesy of The Gates Rubber Company)*

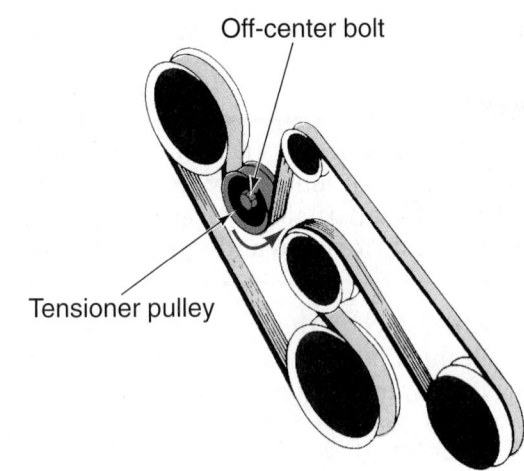

Figure 22.16 To loosen tension on this belt, turn the off-center bolt in the direction of the arrow. *(Courtesy of The Gates Rubber Company)*

- The belt is routed around the outside of the pulleys and is drawn in and looped around the smooth pulleys (**Figure 22.19**).
- Only one pulley can be threaded incorrectly. It is always near the center and is a smooth pulley. If the belt does not seem to fit properly one way, try another way (**Figure 22.20**).

SHOP TIP When a belt is replaced, the old belt can be stored in the trunk for emergencies. If a belt fails, power to all accessories and the coolant pump will be lost.

Figure 22.17 This belt is tensioned by an adjustable screw. *(Courtesy of Ford Motor Company)*

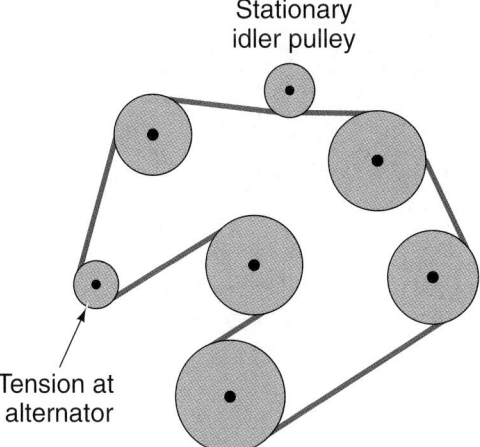

Figure 22.18 This belt is tensioned by the alternator. *(Courtesy of The Gates Rubber Company)*

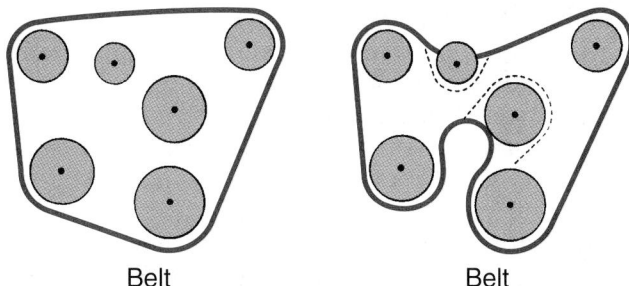

Figure 22.19 The belt is drawn in from the outside and looped around smooth pulleys. *(Courtesy of The Gates Rubber Company)*

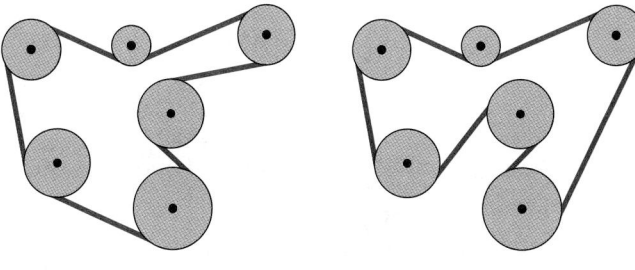

If this routing doesn't work ...   try this

Figure 22.20 Different ways to position a serpentine belt. *(Courtesy of The Gates Rubber Company)*

■■■ BELT TENSION

Belt tension is important for long belt life. Belts stretch slightly in the first few minutes of operation, and then remain constant in length. If they are overtightened, parts can be overloaded.

■ Too much belt tension can cause failure of the coolant pump bearing, the alternator, or the front main bearing (**Figure 22.21**).

■ Improperly tightened belts can also cause coolant pump and alternator bearing noise.

■ Loose belts can cause overheating and abnormal combustion, noise, or a dead battery.

Tightening a Belt

There are several ways that belts are tensioned:

■ Some accessories are provided with a place to pry against (**Figure 22.22**).

■ Brackets sometimes have a hole that a prybar can be inserted into.

■ Manufacturers often provide some provision for tightening, such as a ½" square hole in the mounting bracket for a ½" breaker bar to be inserted into (**Figure 22.23**).

■ Sometimes a jackscrew adjustment is provided (see Figure 22.17).

Figure 22.21 This front upper main bearing shows wear caused by too tight a belt. *(Courtesy of Tim Gilles)*

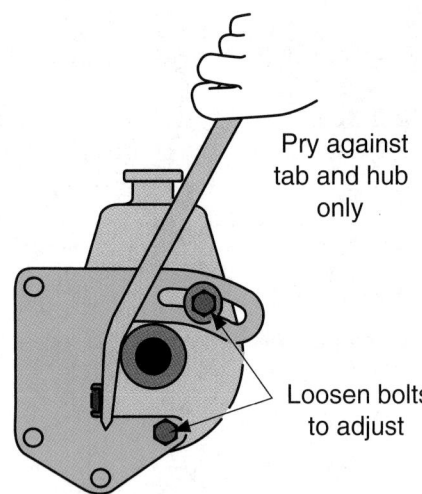

Figure 22.22 When tightening the power steering belt, pry only in the designated areas.

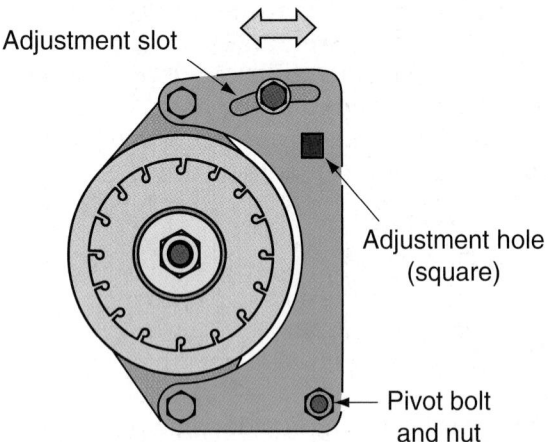

Figure 22.23 Use a breaker bar or ratchet in the square hole to tighten the belt.

Figure 22.24 Notice how thin the housing is on this cutaway power steering pump. Do not pry on it. (*Courtesy of Tim Gilles*)

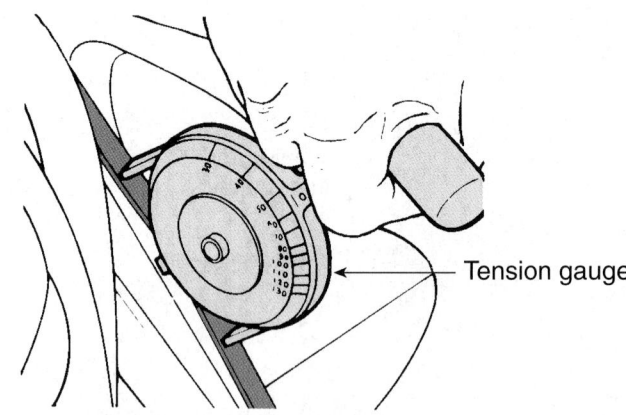

Figure 22.25 Use a belt tension gauge when adjusting the belt. (*Courtesy of DaimlerChrysler Corporation*)

When there is no provision for tightening the belt, pry against a strong area of the accessory. Be sure that the outside of an alternator or power steering pump is not accidentally damaged. Power steering reservoirs are made of sheet metal or plastic. Prying against unreinforced sheet metal will cause a dent, which can damage internal components of the pump. **Figure 22.24** shows a cutaway power steering pump. Notice how thin the housing is. Do not pry on the power steering reservoir or any other delicate part.

V-Belt Tension

New belts stretch slightly during the first few minutes of operation. After that, their length will remain constant. When adjusting a new V-belt, set the tension about 15 pounds higher than the recommended specification. After running the engine for about 15 or 20 minutes, recheck and adjust the belt tension using a belt tension gauge (**Figure 22.25**).

V-Ribbed Belt Tension

V-ribbed belts usually use more tension than V-belts. Maximum tension should be limited to 30 pounds per rib, checked at a splice-free area.

■ Use a *click-type tension gauge* for V-ribbed and timing belts (**Figure 22.26**).
■ After a new belt is installed, run the engine. Then, loosen the tensioner bolt and retighten it.

After some initial tension is lost, V-ribbed belts will maintain 20 pounds of tension per rib for a long time. Used belts should have 15 to 20 pounds of tension per rib.

Serpentine belts with spring-loaded tensioners have a belt length variation gauge (**Figure 22.27**). If a new belt is too short, it can damage the tensioner. If a belt is too long, it will have low spring tension and might possibly rub against something.

Figure 22.26 A click-type belt tension gauge is used for measuring tension on V-ribbed and timing belts. *(Courtesy of Tim Gilles)*

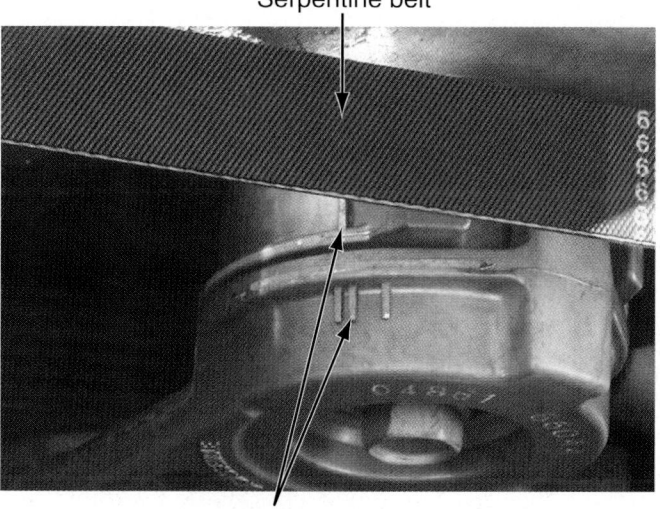

Serpentine belt

Belt length variation gauge

Figure 22.27 A spring-loaded serpentine belt tensioner has a tension gauge. *(Courtesy of Tim Gilles)*

Timing belts are regularly replaced. Manufacturers list various replacement intervals for them. Neglecting this important service can result in catastrophic engine damage when the belt breaks or strips (**Figure 22.28**). It can also result in an unsafe condition on the roadway when an engine stops dead.

The procedure for replacement of a timing belt is covered in Chapter 52.

Stripped cogs

Figure 22.28 Teeth are missing from this timing belt. *(Courtesy of Tim Gilles)*

REVIEW QUESTIONS

1. What are the two types of accessory drive belts?
2. Which belt has more surface area in contact with the pulley groove, a V-belt or a flat belt of the same width?
3. Which cord design is used to make a better belt, high cordline or center cordline?
4. What is the name of the belt drive that is snake-like?
5. If there is a ½" square hole in the alternator bracket, what tool would you use to tighten the drive belt?
6. After a new V-belt is installed and the engine has been run for 15 minutes, what should be done?
7. Before removing a serpentine belt, what should be done?
8. What is the maximum tension per rib of a new V-ribbed belt?
9. What kind of belt tension gauge is used for V-ribbed belts and timing belts?
10. After a new belt is installed and the engine has been started and shut off, what do you do to the belt?

ASE-STYLE REVIEW QUESTIONS

1. Which of the following belts can drive from both sides?

   a. A V-belt   **c.** Both A and B

   b. A V-ribbed belt   **d.** Neither A nor B

2. Which of the following is/are more sensitive to surface finish imperfections?

   a. A V-ribbed belt   **c.** Both A and B

   b. A V-belt   **d.** Neither A nor B

3. Technician A says that adjusting a belt to too high a tension can result in a worn upper front main engine bearing. Technician B says that too much belt tension can ruin a coolant pump bearing. Who is right?

   a. Technician A   **c.** Both A and B

   b. Technician B   **d.** Neither A nor B

4. Technician A says that belt failures rise significantly after 2 years of age. Technician B says to pry on a power steering reservoir when adjusting belt tension. Who is right?

   a. Technician A   **c.** Both A and B

   b. Technician B   **d.** Neither A nor B

5. Which of the following conditions can result from a loose belt?

   a. Engine overheating

   b. Low battery charge

   c. Squealing and pulley wear

   d. All of the above

Automotive Hoses

■ **KEY TERMS**

vulcanizing   worm gear clamp
abrasion   rolled edge clamp
banjo fitting

■ INTRODUCTION

There are more than ten different types of hoses found on modern automobiles (**Figure 23.1**). Hoses are found on the radiator, heater, automatic transmission, fuel and emission systems, brake system, lubrication system, and air-conditioning system. This chapter deals with the theory and service of all automotive hoses.

■ HOSE THEORY

Hoses consist of an inner rubber tube, reinforcement, and the outer rubber cover that are bonded together with adhesives. They have different kinds of tubes and covers and are reinforced differently to withstand different amounts of pressure.

SAFETY NOTE Using the wrong hose can result in a fire or damage to a major component. Also, use caution around refrigerant lines. (Refrigerant service is discussed in Chapter 36.)

■ HOSE SIZE

Hoses are sized according to their inside diameter (called I.D.), the same way that pipes are.

■ Common heater hose sizes are ⅝" and ¾".

■ Common fuel hose and line sizes are 5⁄16" and ⅜" for pressure hoses and ¼" for return lines.

None of these sizes correspond to the outside diameter (O.D.) of the hose. A chart in Appendix E at the end of the book shows actual sizes of flare fittings and pipe thread.

■ UNREINFORCED HOSE

Hoses used for vacuum, windshield washer, and drains are made of unreinforced rubber. These hoses are not under much strain but can present safety problems if they fail (when an engine stalls in an intersection, for instance). They harden with age and are routinely replaced.

■ RADIATOR HOSE

Radiator hoses are designed to have a burst strength of between five and six times the working pressure of the cooling system.

■ HOSE TYPES

There are three kinds of hoses: the *straight*, the *curved*, and the *universal* (**Figure 23.2**). Examples of uses of the straight hose include fuel or vacuum lines. It kinks when it is bent too much, so it is not used when sharp bends are encountered.

Curved Hose

Curved hose, also called *molded* hose, is preformed with the required bends. The hose is heated on a mandrel of the desired shape to vulcanize it. **Vulcanizing** is a process that cures rubber and causes it to set in a predetermined shape.

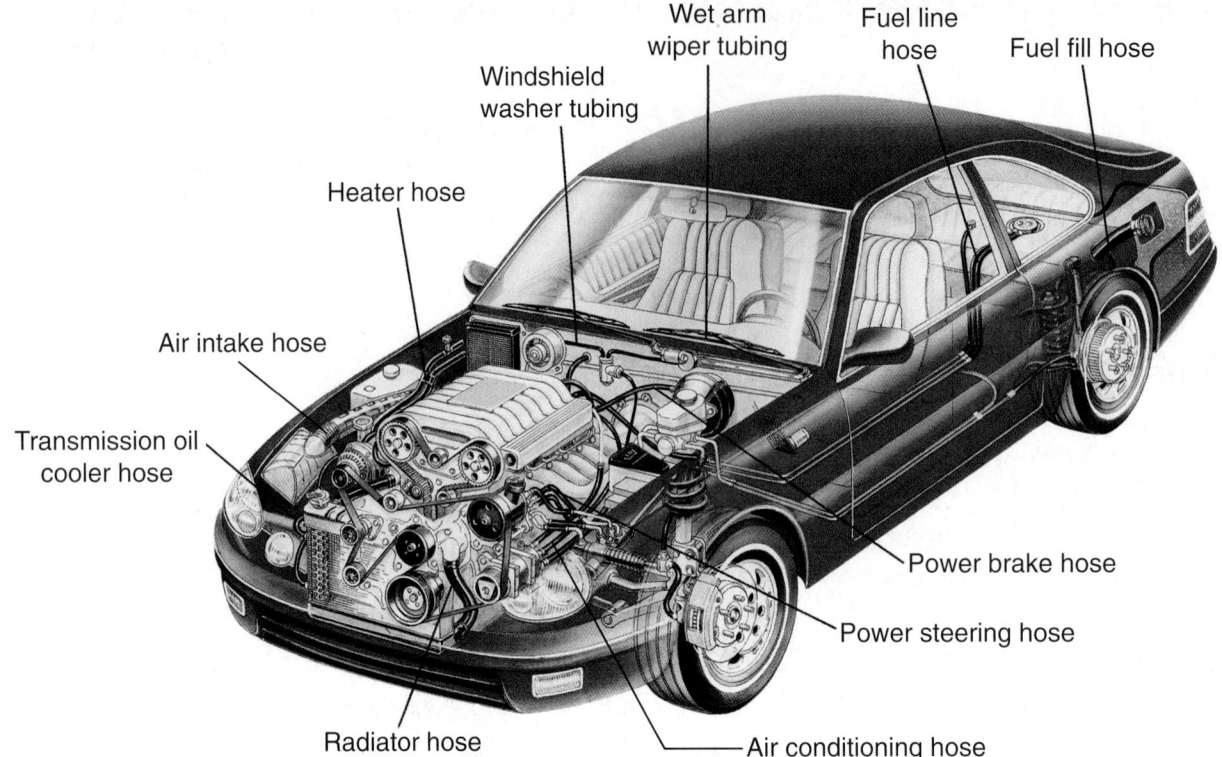

Figure 23.1 Many types of hoses and tubings are found on today's vehicles. *(Courtesy of The Gates Rubber Company)*

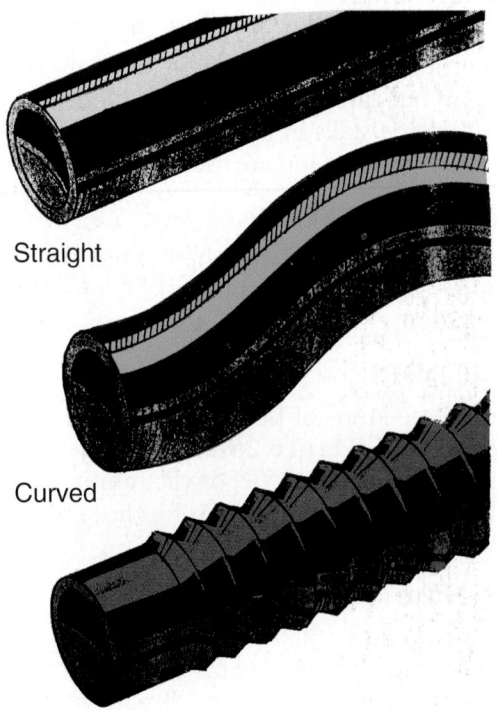

Straight

Curved

Flex or universal

Figure 23.2 Three kinds of hoses. *(Courtesy of The Gates Rubber Company)*

Molded hoses are used in over 90% of replacement radiator hoses. They often come in longer lengths than the desired replacement. Some of them are printed with cutoff line marks so that they can be cut for use in several different applications. The hose manufacturer's catalog tells which line to cut on.

NOTE: *Always exercise care when trimming a hose to length.*

This type of inventory control is done whenever possible so that a business can keep fewer part numbers in stock. Otherwise, prices would have to be higher to compensate for the different cataloging and packaging.

Universal Hose

Universal or flex hose can be clamped on one end and then bent until it assumes the desired shape. It is reinforced with wire so that it will not collapse. Its disadvantage is that the ridges in it resist the flow of water, so pump efficiency is hampered.

Formable Hose

Bending conventional hose will cause it to kink, restricting water flow. Formable hose has a metal wire inside of it (**Figure 23.3**). This allows formable hose to be bent without kinking it. This type of hose is popular

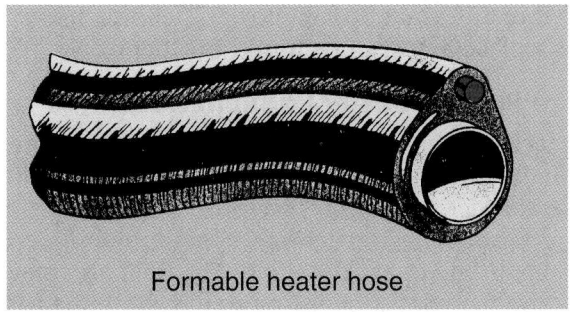

Figure 23.3 This hose can be bent without distorting. *(Courtesy of The Gates Rubber Company)*

when replacing formed hoses, such as those found in the cooling and emission control systems.

By-pass Hoses

Hoses for the thermostat bypass (**Figure 23.4**) are produced in various lengths. They are molded to the proper curve but must be cut to length for the application.

Fuel Hose

The fuel delivery system from the fuel tank to the engine uses metal lines (tubing). These are connected to the tank and engine by rubber hoses. One line delivers fuel from the tank, and the other is a return line. Fuel delivery lines range in size from 5⁄16" to 3⁄8". Return or vapor hoses are usually ¼".

For safety reasons, fuel hose must also have toughness to resist abrasion (rubbing). The inside of fuel hoses and vapor hoses must be resistant to attack from ethanol and methanol fuels. The outside cover must be oil and temperature resistant.

The outside of a fuel hose is usually temperature rated for about 300°F. Because fuel is relatively cool,

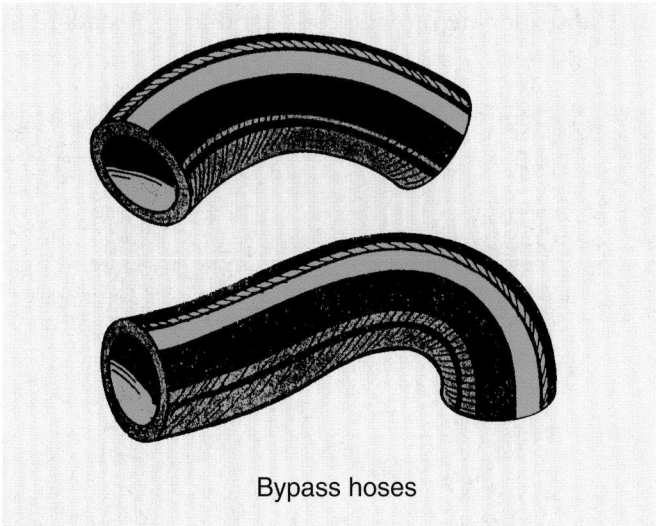

Bypass hoses

Figure 23.4 Thermostat bypass hoses. *(Courtesy of The Gates Rubber Company)*

the inside of a fuel hose does not require a high temperature rating. The inside of the hose should not be subjected to temperatures in excess of 257°F.

NOTE: *Do not use fuel hose for an application in which hot oil will be used, such as for a transmission cooler line.*

Carburetor Fuel Hose

Carburetor fuel systems are under low pressure. The hose consists of a braided reinforcement with a synthetic rubber cover and inner tube. Fuel hose is only rated at 50–75 psi. It is made of a special compound that does not react to gasoline or alcohols.

Fuel Injection Hose

Fuel injection systems usually run under much higher pressure than carburetor systems. Pressure in some of these systems sometimes reaches 175 psi. Special hose and special hose clamps (covered later) must be used with these systems. Fuel injection hose has a burst strength in excess of 900 psi. On many systems, the hoses are crimped to the tubing, like high-pressure power steering hoses.

Transmission Oil Cooler Hose

Hot oil is circulated from the automatic transmission to its cooler (located inside or outside of the radiator). When hoses are used to connect metal tubing in these systems, they are subjected to two of rubber's biggest enemies: oil and heat. Any rubber product is broken down by oil, shortening its life. Rubber products suffer a shortened life of about 50% for every 18°F increase in temperature. Transmission oil cooler (TOC) hose can withstand a constant temperature of up to 300°F. Its pressure rating is 450 psi.

Compare fuel hose's ratings for temperature and pressure, and you can see why fuel hose should not be used in place of TOC hose. Both of these types of hose look identical from the outside. **Figure 23.5** shows the differences between these hoses internally.

CASE HISTORY *A technician installed an auxiliary transmission oil cooler on a motor home. Because of the heavy loads and high temperatures encountered by an automatic transmission in a motor home, this is a common installation. He installed the cooler in series with the existing cooler in the radiator. The connections were made with fuel hose. After using the motor home for one season, a leak developed in one of the hoses. The transmission fluid leaked onto a hot exhaust manifold, and in the resulting fire the motor home suffered serious damage. Luckily, no one was hurt.*

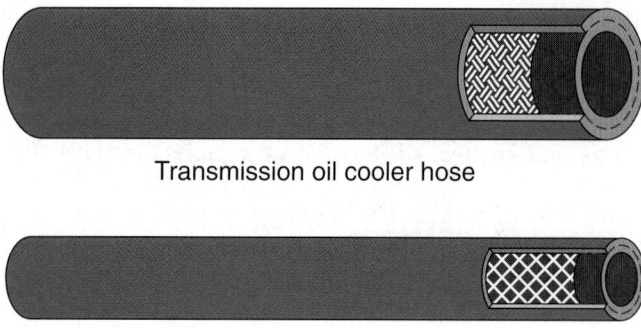
Transmission oil cooler hose

Fuel line hose

Figure 23.5 Transmission oil cooler hose and fuel hose look identical from the outside. *(Courtesy of The Gates Rubber Company)*

Power Steering Hose

There are usually two power steering hoses. One is a pressure line and the other is a return line (**Figure 23.6**). The return hose is not under the high pressure that the pressure side is. The hose on the pressure side is made of reinforced, oil-resistant synthetic rubber that is crimped to metal tubing. It must be able to withstand temperatures of 300°F and handle pressures up to 1,500 psi. Some pressure hoses also have a flow reducer, an orifice to help control flow. The return line is usually connected to the power steering pump with a hose clamp. It must be able to withstand high temperatures but operate under low pressure. It must also be resistant to damage by power steering fluid.

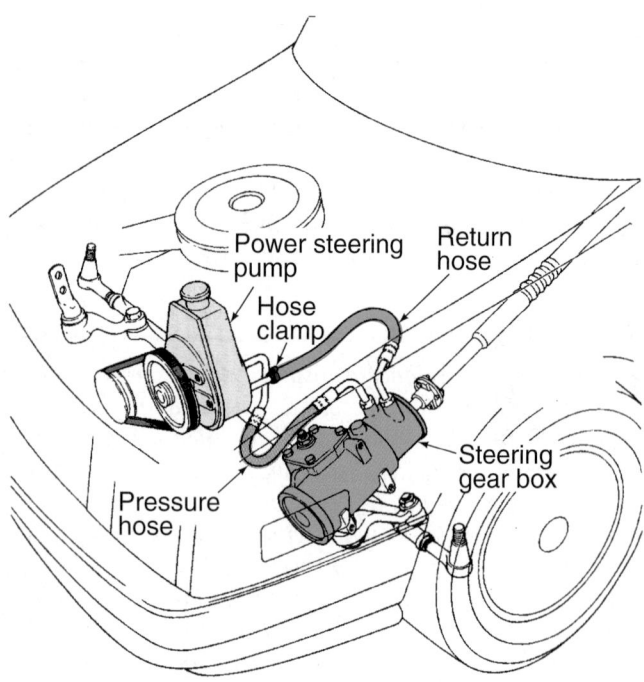

Figure 23.6 Power steering hoses. Pressure hoses are crimped. Return hoses, under less pressure, usually have hose clamps. *(Courtesy of Federal Mogul Corporation)*

More information on power steering hoses and their replacement can be found in Chapter 66.

Brake Power Booster Vacuum Hose

A replacement hose to a power booster must be reinforced and fuel resistant. Fuel will damage the diaphragm in a power booster. Some hoses have an inline filter in them to prevent fuel from entering the power booster during a misfire in the intake manifold. The filter also prevents evaporating hydrocarbons from leaking into the passenger compartment.

Brake Hose

Brake hoses provide a flexible connection to the wheels, which must be able to turn back and forth during steering. They are called upon to carry very high pressures. Because brake hoses are a safety item, federal motor vehicle standards require that they have a burst strength of at least 5,000 psi.

Brake hoses come in different lengths and have fittings swaged or crimped on the ends (**Figure 23.7**). The hose is made of rubber reinforced with woven fabric, usually rayon. The fabric's job is to prevent the hose from expanding under pressure. The federal safety standard requires that the outside of a brake hose be labeled with either *HR*, which means that the hose expands normally, or *HL*, which means that the hose has a lower amount of expansion.

The hose has an inner layer of rubber that comes into contact with the brake fluid. Most brake hoses have two *plies* that make up the middle fabric layers of the hose.

The *jacket* of the hose is the outside layer that protects the plies. The outside of a hose is subjected to damage from ozone in the atmosphere, which can cause the rubber to crack.

Federal safety standards also call for the hose to have either raised ribs or two 1/16" stripes painted on it. The stripes are provided so that during installation,

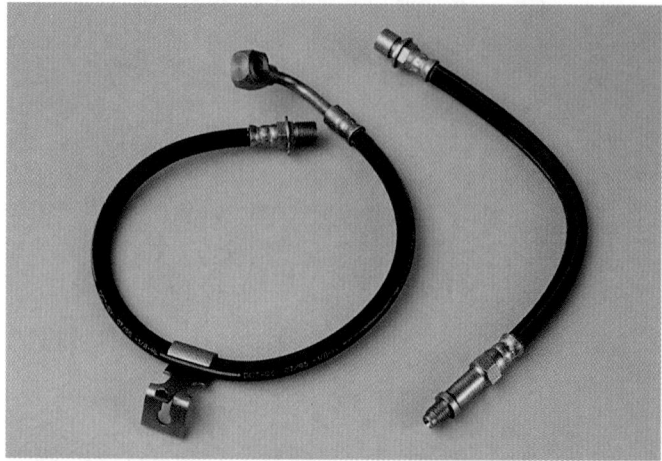

Figure 23.7 Brake hoses have fittings crimped on their ends.

the technician can tell whether the hose is twisted excessively. Excessive twist in a hose can cause it to be damaged internally.

Internal damage is difficult to diagnose. Depending on which way the internal rubber flap is torn, the hose could cause delayed engagement of the brake or act as a one-way valve, preventing fluid pressure from releasing from the brake after application. Internal damage also often results when a brake caliper is allowed to hang on a hose during a brake job. These hoses often become damaged when a technician clamps one off using vise grips.

Brake *tubing* always has male fittings on both ends. Brake *hose* fittings have either female threads, male threads, or banjo fittings. **Banjo fittings**, named because they resemble a banjo, allow brake fluid to make a 90-degree turn in a tight space (**Figure 23.8**). Brake fluid goes into the banjo fitting and exits through a hollow bolt. A copper gasket is used to seal both sides of the fitting.

NOTE: *The copper gasket, which is often reusable, is easily lost during a brake service. The best practice is to keep a supply of these gaskets on hand and replace them any time a brake hose is removed.*

Tapered pipe threads are not usually found in brake systems. This is because the female part of the fitting can split when overtightened due to the taper of the thread. Male compression fittings are used instead, usually with copper washers that are compressed when the fittings are torqued. Other times, the end of the male thread has a tapered *seat* that seals at the bottom of a female fitting.

SHOP TIP Once a compression fitting with a copper washer is tightened, the hose is no longer free to rotate. Therefore, the compression fitting must be installed before the other end of the hose is installed.

Air-Conditioning Hose

Air-conditioning systems carry liquid and vapor under high pressure. As with power steering systems, thes high-pressure hoses are reinforced and have crimped connectors. High-quality coatings are also used on the hoses. Oil circulates along with the refrigerant in the system, and temperatures are high. Newer air-conditioning systems use R134A refrigerant. R134A consists of a considerably smaller molecule than the older R-12 (Freon) refrigerant does. Consequently, R134A refrigerant can leak through the older air-conditioning hoses. As a result, today's new car refrigerant hoses incorporate a barrier of nylon or similar material to prevent leakage.

■ HOSE CLAMPS

Clamps and hoses are used to connect two metal lines together. There are several styles of clamps (**Figure 23.9**). One of the most popular hose clamps for all applications (except fuel injection) is the **worm gear clamp**. It is usually reusable and is easy to install. A worm gear clamp is shown on the left-hand side of the Figure 23.9.

Twin wire clamps are strong hose clamps but can cut into the hose if tightened too far. **Rolled edge clamps** (**Figure 23.10**) are designed so that they will not cut into a hose. They are a good hose clamp for fuel injection, where fuel pressures are higher than those found in the cooling system.

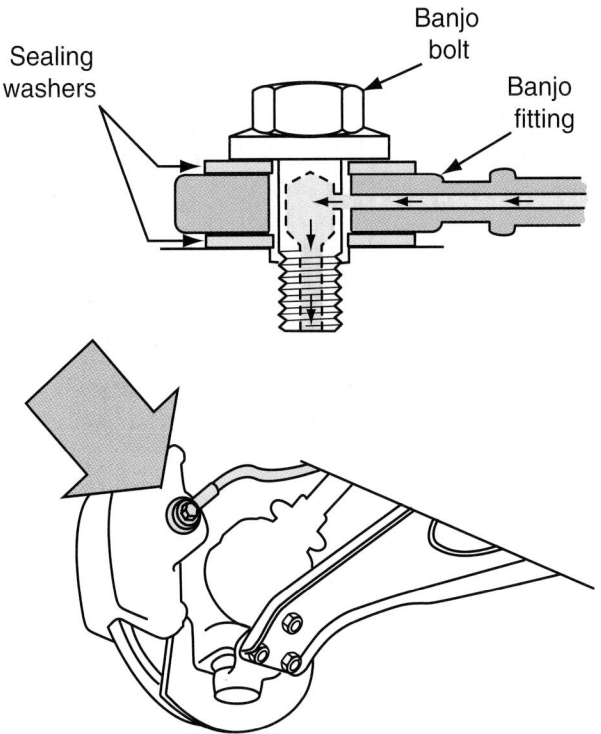

Figure 23.8 A banjo fitting makes tight turns possible.

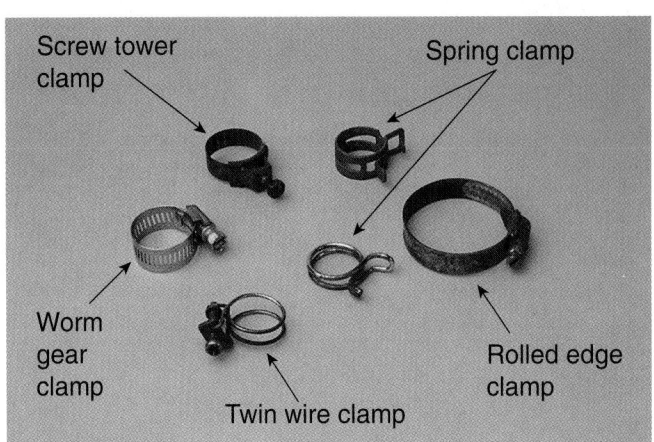

Figure 23.9 Several types of hose clamps.

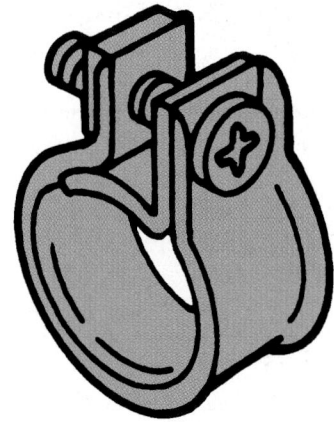

Figure 23.10 A rolled edge hose clamp.

Screw tower clamps are strong clamps too, but they are difficult to remove. Technicians usually cut these off with cutters. *Spring clamps*, used as original equipment by some manufacturers for cooling system hoses, are not used for most fuel system applications, but they come with replacement in-line fuel filters on engines with carburetors. They do not provide a strong enough clamping force for fuel injection applications and can cut into a hose, causing a fuel leak.

▄▄ HOSE INSPECTION

Check the condition of all radiator, heater, and fuel hoses. Hoses are normally firm, yet flexible. Look for leaks, stiffness, sponginess, hidden rot, rubbed or burned areas, and oil soaking. Hoses that appear to be cracked, swollen, hard, soft, or broken should be replaced.

Hard Hoses

When checking hoses for hardness, try to bend them, especially at the ends.

 Be careful when bending a hose at the end. A bad hose might break, allowing dangerous fuel or hot, pressurized coolant to escape.

The heater core, a small radiator, is located inside the driver's compartment. It has two hoses to carry engine coolant: one coming from the engine and one for return (**Figure 23.11**). Heater hoses often become brittle due to heat and aging.

 Sometimes cutting off the first inch of hose will eliminate a hardened section.

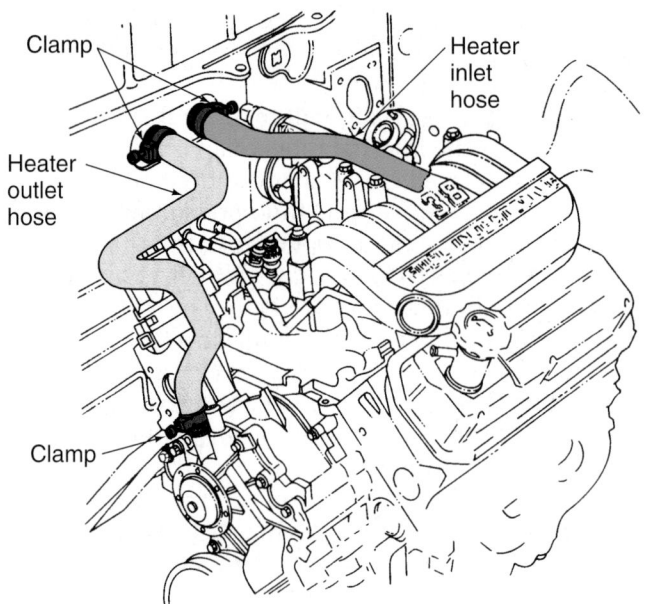

Figure 23.11 The heater has an inlet hose and an outlet hose. (*Courtesy of Ford Motor Company*)

Oil Damage

Rubber hoses that are not used in fuel- or oil-related environments can deteriorate rapidly when they come into contact with oil. Oil, an enemy of rubber hoses, can cause them to swell. Be sure to fix any oil leaks. Also, water pump lubricants found in some cooling system additives contain petroleum, which can damage the inside of the hose.

NOTE: *Defective hoses do not always outwardly appear so. Deterioration on the inside of the hose can cause small particles to flake off and fall into the coolant (**Figure 23.12**).*

Check the rubber on the inside of the hose to see that it is not deteriorating. Occasionally, a relatively

Figure 23.12 A hose that is defective on the inside. (*Courtesy of The Gates Rubber Company*)

low-mileage vehicle will experience the failure of a hose that appears to be good on the outside.

Electrochemical Damage and Radiator Hoses

A newly discovered condition discovered independently by both Gates and Goodyear has been found to lead to hose failure. It is called *electrochemical degradation,* or *ECD.* ECD occurs when the hose, engine coolant, radiator, engine, and fittings form a galvanic cell, or battery. This causes small cracks in the inside of the hose, allowing coolant to penetrate into the hose reinforcement. New coolant hose that is electrochemically resistant is made of *ethylene propylene rubber* (EPDM).

To check for ECD damage, squeeze the hose in several places to see if the rubber has the same consistent feel throughout. Be sure to thoroughly check both hoses. The upper radiator hose (engine outlet end) suffers the most abuse because it is exposed to fully heated coolant before it reaches the radiator. Lower hoses, on the other hand, are more difficult to inspect and are often overlooked.

Some lower radiator hoses have a coil of wire in them. The coil is installed at the factory so that the hose does not collapse during the factory coolant fill procedure. A rusty cooling system will often result in pieces breaking off and circulating in the system. If the wire is missing, coolant circulation, especially at highway speeds, can be hampered. Whenever one of these hoses is removed during service, check to see that the wire is not rusted.

 **SAFETY NOTE** Any questionable hose should be replaced. If a hose fails, an engine can be ruined.

■ REPLACING HOSES

When a coolant hose is to be replaced, first drain off coolant to a level below the height of the hose. A hose clamp is used to fasten a hose to its connection. Hoses are usually easily removed by loosening the clamp and twisting gently. If the clamp is rusted in place, cut it off with side cutters or tinsnips.

The worm gear clamp is the most popular clamp used in the aftermarket. Most worm gear clamps have screw heads with both a screwdriver slot and a hex head (**Figure 23.13**). They can be removed with either a screwdriver or a socket.

Some car makers use a spring-type clamp for radiator connections. Special pliers with grooves in the jaws accommodates the clamp for removal (**Figure 23.14**).

When a hose does not come loose from its fitting easily, do not force it. This can damage the fitting,

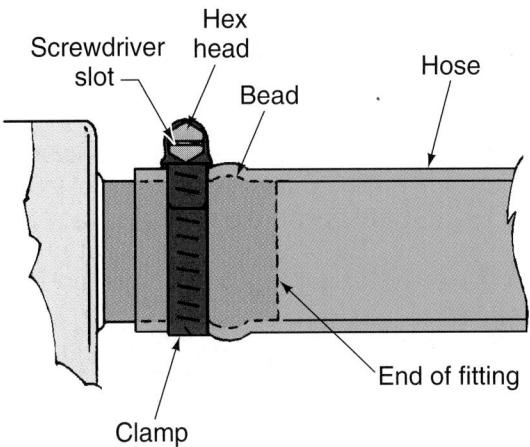

Figure 23.13 Worm gear clamps usually have a hex head and a screwdriver head. *(Courtesy of The Gates Rubber Company)*

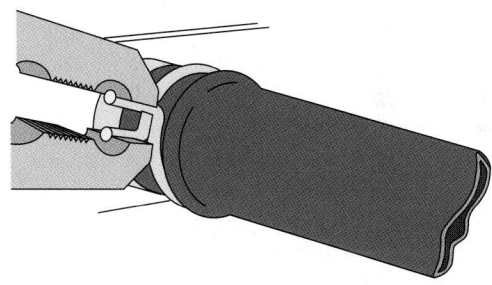

Figure 23.14 Spring clamps are removed using pliers with a groove in their jaws. *(Courtesy of The Gates Rubber Company)*

radiator, or heater core outlet. Cut it off carefully with a knife (see Figure 21.9). Most EPDM hoses tend to bond to metals, and during removal they *must* be cut off.

Clean the fitting of corrosion or pieces of old hose (**Figure 23.15**). Hose connections on the radiator and engine have a small ridge that helps seal the hose. Install the hose and position the clamp as close to the ridge on the connection as possible to prevent the possibility of corrosion (see Figure 21.25). Positioning the clamp on top of the ridge can cut the hose.

Be sure that all hose clamps are in good condition and are tight.

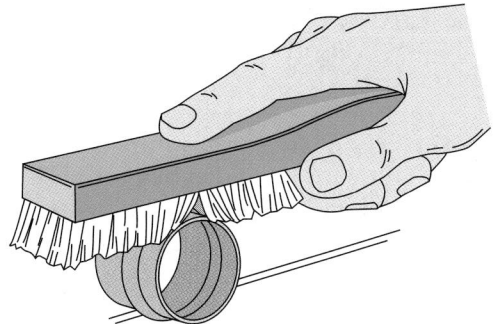

Figure 23.15 Clean the hose fitting before installing the new hose. *(Courtesy of The Gates Rubber Company)*

NOTE: *When new radiator and heater hoses are installed, they take a set following the engine's first heating and cooling cycle. This sometimes leaves the clamps loose and they must be retightened.*

Because a molecule of air is smaller than a molecule of coolant, air can be drawn in through a loose connection (even if the opening is not large enough to let coolant seep out). This is especially true with the lower radiator hose and heater return hose because they are *suction hoses*.

 Check to be sure that the hose does not contact fan belts, fuel lines, or the fan.

Replacing a Heater Hose

Sometimes the heater hoses on a car are two different diameters. The size is determined by the inside diameter (I.D.) of the hose. Hose comes in long rolls. Be sure that you select the correct size hose before cutting it off of the main hose roll.

Twist the old heater hose to loosen it, and cut it carefully with a sharp knife if it does not remove easily. When installing the new section of hose, check the following:

■ Be certain that the hose does not interfere with manifolds, belts, or spark plug wiring.
■ Check to see that the hose will not be damaged by movement of the engine or accessories.

Position the screw side of the clamp for easy access. On lower radiator hoses, an extra-long screwdriver can be used from the top of the engine compartment. Be sure the screw on the hose clamp is positioned so it cannot accidentally come into contact with the cooling fan.

SHOP TIP If the hose is difficult to install, apply a small amount of soap to the connection.

By-pass Hose Replacement

Obtain the correct replacement hose and compare it to the existing hose (**Figure 23.16**).

CAUTION Do not try to substitute an unmolded hose for a molded by-pass hose. It can fold when bent, restricting coolant flow.

Power Steering Hose Service

When checking a power steering hose, look for signs of leakage or dampness at the connections. Also, look for

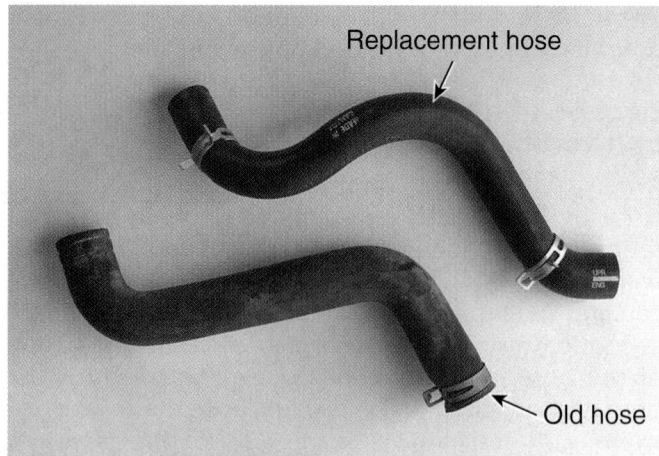

Figure 23.16 Compare the old hose to the new hose before installing it.

signs of deterioration, such as cracks, signs of rubbing, or swelling. A failed pressure hose can leak, blowing oil onto an exhaust manifold where it can cause a fire.

Get the correct replacement hose and replace the old one. Parts stores have catalogs that have extensive illustrated listings for power steering hoses. It is also a good idea to compare the new hose to the old hose to be sure that the length and fittings are identical.

CAUTION Power steering fluid, which is not flammable at engine operating temperature, is flammable at temperatures a little above 300°F (lower than the temperature of an exhaust manifold).

SHOP TIP Sometimes a replacement hose is not available. Some businesses have special equipment for making hydraulic hoses. Check with a business that deals with heavy truck, industrial, or farm implement accounts.

Refilling the Power Steering System

Be sure to use the correct fluid when refilling the system. Earlier systems traditionally used automatic transmission fluid, which is red in color. Many late-model vehicles use power steering fluid, which has a lower viscosity. It is colored either yellow or yellow/green.

After refilling the system, run the engine and check for leaks. On hoses that use O-rings, new O-rings should be used.

SHOP TIP If a flared fitting leaks, try loosening the fitting and twisting the tubing to seat it against the flare seat. Then retighten the fitting.

Air in the power steering system will cause erratic operation or a groaning sound when the steering wheel is turned. Bleed air from the system by turning the steering wheel back and forth a few times. This should be done with the tires in the air. Otherwise a flat spot can be worn on the tires. If the fluid appears to be foamy after this, let it sit for several minutes and allow the foam to disappear. In extreme cases, a vacuum pump can be applied to the top of the power steering reservoir filler neck to remove stubborn air.

NOTE: *Some cars have hydroboost power brakes, which are operated off of the power steering pump. On these systems, depress the brakes several times while turning the steering wheel to help rid the system of air.*

Fuel Injection Hose Replacement

For safety reasons, it is important to replace fuel injection hoses before a failure occurs.

An original equipment manufacturer (OEM) requirement is that the outside of fuel injection hose be stamped with *"Fluoroelastomer"* or *EFI* in yellow print. Always use the specified hose and do not connect high-pressure lines with hoses and clamps. Instead, use hoses with crimped ends.

 Be sure to use the correct fuel hose designed to withstand the pressure of the fuel injection system.

Fuel injection lines maintain pressure even when the engine is off. This is so the engine can easily restart without the system having to build pressure first.

 ■ Be sure to carefully relieve the system of fuel pressure before replacing a fuel line or fuel filter (see applicable service manual or a manufacturer's hose and tubing catalog).
■ Fuel hose must always be routed so that it is at least 4" (100 mm) away from any part of the exhaust system. Catalytic converters (see Chapter 43) run hotter than the rest of the exhaust system. Fuel hoses should not be any closer than 10" (284 mm) from a catalytic converter.

 When hose clamps and hose are used, be sure to position the hose clamp behind the ridge on the fitting, not on it (**Figure 23.17**). Failure to do this can result in a cut hose and dangerous fuel leak.

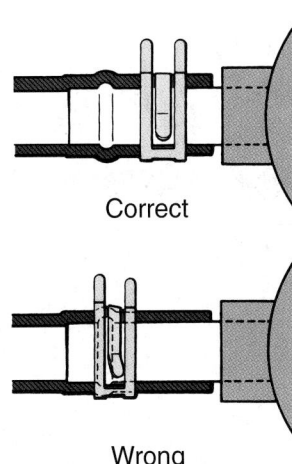

Figure 23.17 Installing a fuel hose clamp on the ridge can damage the hose.

Repairing damaged tubing with a section of hose is covered in Chapter 24. Nylon tubing used in fuel injection systems is covered in that chapter also.

Air-Conditioning Hose Service

Factory-installed air-conditioning systems come with a very wide variety of hose styles and configurations. Air-conditioning shops commonly make up their own lines from bulk hose and a selection of fittings kept in stock. Parts stores keep a supply of the more common fittings on hand. Usually, the old fitting can be reused. The crimp connection must be cut off carefully to avoid damage to the fitting (**Figure 23.18**). The hoses used are identified using a chart like the one in **Figure 23.19**, which compares the actual sizes of the hoses.

In the past, worm gear hose clamps were used with *3-barbed fittings* when making repairs (**Figure 23.20**). This

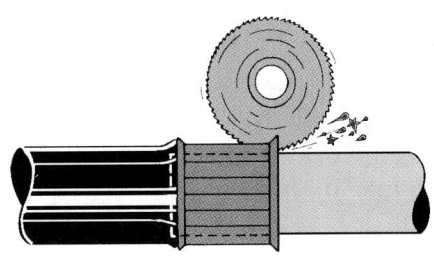

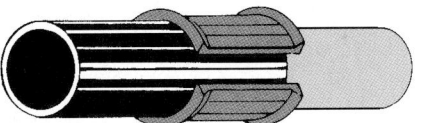

Figure 23.18 The crimp connection is carefully cut off the old hose. *(Courtesy of The Gates Rubber Company)*

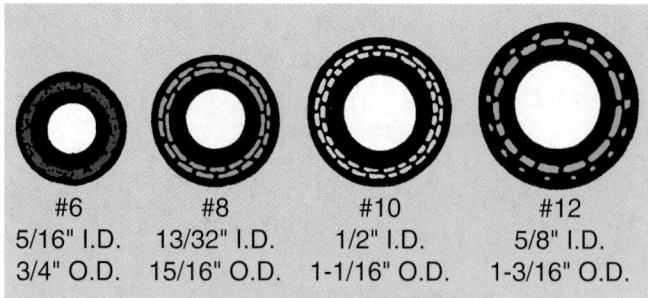

Figure 23.19 A refrigerant hose size chart. *(Courtesy of CARQUEST)*

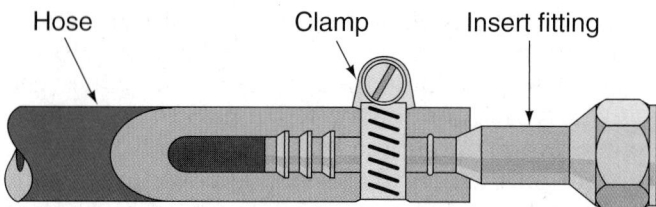

Figure 23.20 A worm gear hose clamp used with a 3-barbed fitting.

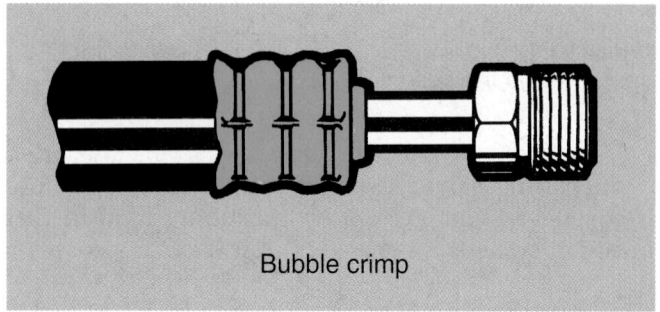

Bubble crimp

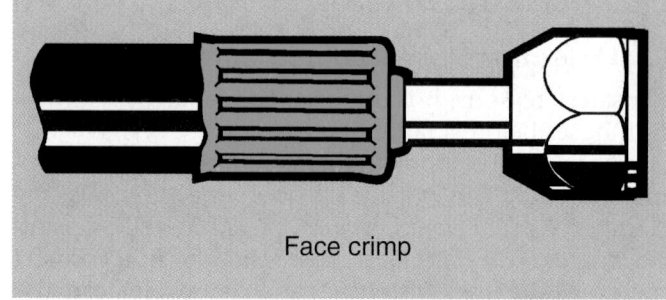

Face crimp

Figure 23.21 Two styles of crimps. *(Courtesy of CARQUEST)*

method of repair is no longer recommended and must be absolutely avoided with R134A beadlock fittings.

There are two styles of crimp used with air-conditioning hoses (**Figure 23.21**). When crimping a fitting onto an R134A line, a *bubble crimp* is used. A *face crimp* is recommended for use with 3-barbed fittings. Hydraulic press crimping or less-expensive hand crank crimping tools are available.

▬▬ REVIEW QUESTIONS

1. Hoses are sized by their inside/outside (circle one) diameter.

2. What special feature allows formable hose to bend without kinking?

3. The inside of transmission oil cooler hose can withstand constant temperatures of up to ____°.

4. When there are two hoses in a power steering system, what are they for?

5. What is the name of the type of hose clamp to use for fuel injection system hoses?

6. What is a drawback to screw tower hose clamps?

7. Which radiator hose suffers the most abuse, the upper or the lower? Why?

8. If a hose does not come off a radiator fitting easily, what should be done?

9. What can happen if a regular hose is substituted for a molded bypass hose?

10. Can power steering fluid catch fire?

▬▬ ASE-STYLE REVIEW QUESTIONS

1. Which of the following statements is/are true about a fuel hose?

   a. It can be used to replace a damaged section of transmission oil cooler tubing.

   b. Fuel injection hoses have pressure in them even when the engine is off.

   c. Both A and B.

   d. Neither A nor B.

2. Technician A says that a worm gear clamp is the best style to use in a fuel injection system. Technician B says that for a hose clamp to work properly, it should be positioned on top of the sealing ridge on the connection. Who is right?

   a. Technician A   **c.** Both A and B

   b. Technician B   **d.** Neither A nor B

3. All of the following are true statements about hoses *except*:

   a. A lower radiator hose is more likely to leak air than an upper hose after a new hose shrinks.

   b. Brake hose has either raised ribs or stripes to help guard against twisting it during installation.

   c. Clamps used on neoprene radiator hoses do not have to be retightened because this kind of hose does not shrink like EPDM hose.

   d. A power steering return line often has a hose clamp, rather than a crimped fitting.

4. Which of the following can cause damage to a rubber hose?

   a. Heat   **c.** Coolant

   b. Oil   **d.** A and B only

5. Technician A says that a defective brake hose can cause delayed engagement of the brake. Technician B says that a defective brake hose can prevent brakes from releasing. Who is right?

   a. Technician A   **c.** Both A and B

   b. Technician B   **d.** Neither A nor B

CHAPTER 24

Automotive Plumbing: Tubing and Pipe

■ OBJECTIVES

Upon completion of this chapter, you should be able to:

✔ Describe the different types of tubing used on automobiles.

✔ Understand the different types of tubing connections.

✔ Repair damaged tubing.

■ KEY TERMS

work harden
double-walled tubing
furnace brazing
double flare
ISO flare

inverted flare nut
compression fittings
ferrule
union
street elbow

close nipple
long nipple
pipe dies

■ INTRODUCTION

Tubing and pipe are found on automobiles and on shop equipment. This chapter covers how to service various types of tubing and pipe and what types of plumbing parts are available. Also covered are the different types of connectors used with tubing and pipe. Understanding these topics is just one of the considerations that separates good technicians from average ones.

■ TUBING

Tubing found on automobiles, often called *line*, does not have threads at its ends like pipe. Tubing can be made of either copper, steel, or plastic. Different tubing materials are selected depending on the intended use. Manufacturers choose the least expensive alternative that will do the job safely and dependably. The engineering rule of thumb for tubing selection is that the maximum pressure that the material will be exposed to should be less than ⅕ the rated pressure of the tubing.

NOTES:

■ *Tubing's size is determined by its outside diameter (O.D.).*

■ *Pipe and hose are sized by their inside diameter (I.D.).*

Compare three pieces of ½" diameter pipe: one plastic, another steel, and the other copper. All of them have the same inside diameter, but they are very different in appearance (**Figure 24.1**).

Copper Tubing

Copper tubing can be either soft or rigid. It is softer and easier to bend than steel. Soft copper with a 0.020" wall thickness can withstand 1,000 psi of pressure. This is sufficient strength for *low-pressure* applications, like carburetor fuel systems, vacuum lines, oil pressure gauge lines, and other lines that carry lubricants. Copper **work hardens**, which means that it gets brittle with repeated flexing. Therefore, it should not be used where flexing will be encountered.

1/2"
plastic

1/2"
copper

1/2"
steel

Figure 24.1 Comparison of sizes of plastic, copper, and steel pipes. *(Courtesy of Tim Gilles)*

Copper tubing should not be used for brake lines. It is *single walled* and does not have sufficient burst strength. It also work hardens from vibration, which can cause it to crack. Copper also corrodes easily and holes develop from electrolysis.

Plastic Tubing

Plastic tubing is used for vacuum lines or oil pressure lines. Polyethylene tubing has a rated pressure of only 200 psi. Nylon tubing is used on some fuel injected vehicles and for air brake lines on trucks (not hydraulic lines). Nylon has a rating of 300 psi. Plastic tubing is not used where it will be exposed to heat. Plastic can burn through if it comes into contact with an exhaust manifold, for instance. Fittings used with plastic tubing are covered later in this chapter.

 **SAFETY NOTE** Brakes and power steering systems develop very high pressures, sometimes in excess of 1,000 psi. Do not use copper for these applications.

Steel Tubing

Steel tubing, galvanized to prevent rust, can be used almost anywhere on an automobile. A flaring tool (covered later) is used to flare the ends of the tubing. Tubing diameter ranges from ⅛" to ⅜".

Steel tubing used for brake lines is **double-walled tubing** so that it can bend easily. To make the tubing, copper-plated sheet steel is used. One type of tubing called *seamless tubing* is made by rolling the steel sheet until it has two layers and one seam, providing a double wall. Then, the copper is fused to this seamed tube using a **furnace brazing** process. When the copper melts, fusing to the tubing, the seam disappears.

Another type of brake tubing is made in two different layers with the seams separated by at least 120 degrees. Furnace brazing is used to bind the copper into the seams of this tubing also.

Most brake tubing used on automobiles has an O.D. of 3/16". This is the size used for disc brakes. Drum brake systems sometimes use ¼" O.D. tubing.

An SAE standard for brake tubing dictates that it must be able to be bent in a full circle around a mandrel five times its diameter and not kink or be damaged in any way.

Sometimes there is a steel coil around the outside of a brake tube (**Figure 24.2**). The coil, called "armor," is there to protect the line from damage where it might be prone to rub against something. It also allows the tubing to be bent without kinking.

When tubing is subjected to vibration or flexing, it is usually coiled (**Figure 24.3**). This is especially common on lines below the brake master cylinder. Because

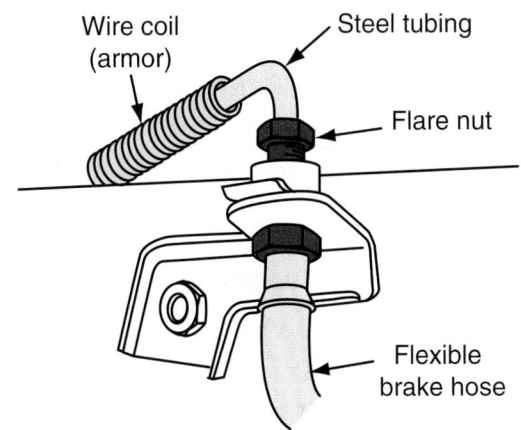

Figure 24.2 Armor protects the brake tubing from abrasion.

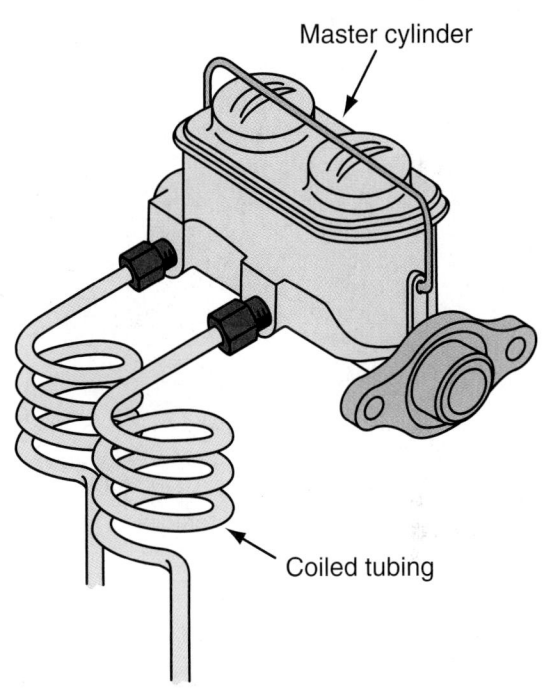

Figure 24.3 Tubing is coiled to prevent damage from vibration.

it is mounted on the car body and the brake lines are secured to the frame, flexing must occur there.

Tubing Fittings

There are many types of fittings that are used to join tubing to components. Be sure when using fittings that they are the correct ones for the application. Flare fittings, compression fittings, and pipe fittings all have different types of threads and seats within them.

NOTE: *Threads can be either male or female. Female threads are those that are internal. Male threads are external.*

Sometimes manufacturers use connectors with oversized threads to prevent assembly errors. When

Figure 24.4 Adapter fittings allow different sizes and types of fittings to be connected together. *(Courtesy of Tim Gilles)*

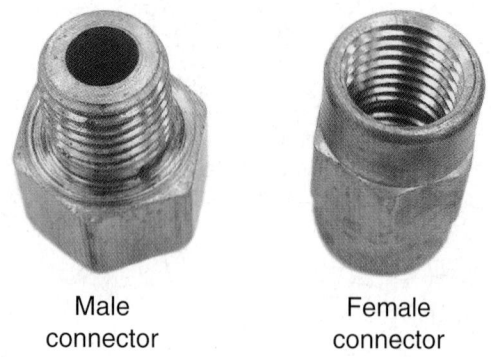

Male
connector

Female
connector

Figure 24.5 Connectors are used for joining tubing to parts. *(Courtesy of Tim Gilles)*

replacing or adapting a part, this sometimes causes problems. *Step-up* or *step-down adapters* (**Figure 24.4**) are available for making these connections.

Connectors (**Figure 24.5**) are used between tubing and a part like the oil pump, carburetor, fuel pump, or brake parts. They can also serve as adapters between different types of fittings. An example of this would be if a pipe thread and flare fitting were to be joined.

CAUTION When two parts with different types of threads are accidentally joined, damaged threads and a leak can result.

When nylon lines are connected to steel lines, one of two types of *push connectors* is used. These are covered later in this chapter.

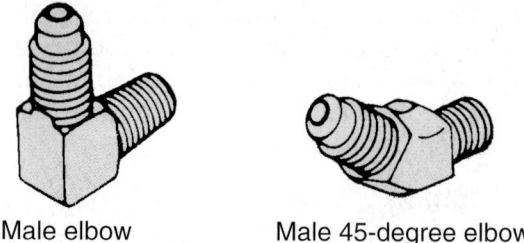

Male elbow

Male 45-degree elbow

Figure 24.6 Elbows are used for making sharp bends.

An *elbow* is used when a sharp turn is made. Bending the tubing takes too much space to do. Elbows can be either 30 degrees, 45 degrees, or 90 degrees (**Figure 24.6**).

▬ FLARED CONNECTIONS

Flare fittings are sometimes used when two steel fuel lines are connected. A flare is suited for high-pressure applications and *must* be used for brakes or power steering. Any type of tubing that can be formed with a flaring tool (copper, aluminum, or steel) can have a flare installed on it.

Figure 24.7 shows the relationship between the parts of a flared connection. The end of the line is tapered outward. This is called a *flare*. The flare on the end of the line is wedged between the flare fitting and the flare nut on the line. Long and short flare nut designs are available. The long nut is used when the part is subject to vibrations because it supports the tubing further away from the connection.

There are two kinds of flares used in automobiles, the *SAE type 45-degree* **double flare** or the International Standards Organization (**ISO**) **flare**, also called a *bubble flare* (**Figure 24.8**). SAE automotive fittings are flared at 45 degrees. There are also 37-degree SAE flares. Be sure you are using SAE 45-degree fittings.

A 45-degree SAE-type double flare is usually used with an **inverted flare nut** (**Figure 24.9**). The inverted-type flare nut is more common on

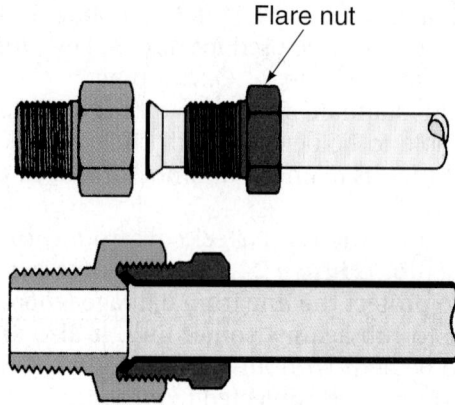

Flare nut

Figure 24.7 Relationship between parts of a flared coupling. *(Courtesy of Dana Corporation)*

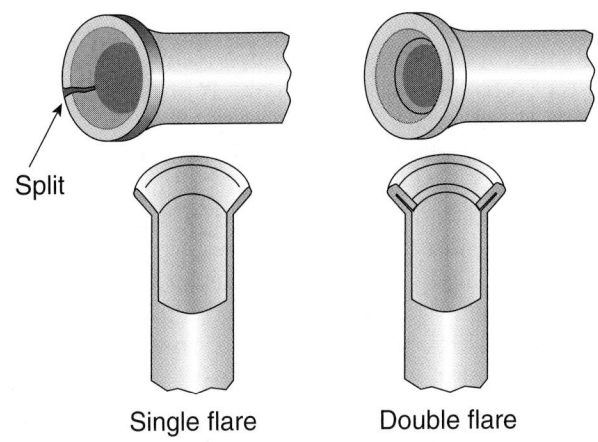

Figure 24.10 A single flare will result in a split in the tubing.

automobiles. The standard flare is found in such applications as household natural gas lines.

The flare can be either a single or double flare. A single flare is not used on small automotive tubing, because it will split the tubing (**Figure 24.10**). A double flare is a two-step process (described later).

ISO flares have been found on automobiles since the early 1980s. A bubble or ridge is formed in the line a short way back from its end. A single flaring operation can be done without danger of splitting the line (like with a single 45-degree flare).

■ COMPRESSION FITTINGS

Flareless fittings, called **compression fittings**, are also found on automobiles (**Figure 24.11**). One kind uses a brass *sleeve*, called a **ferrule**. The sleeve is installed loosely on the tubing. When the compression nut is drawn against the fitting with a wrench, the sleeve is pinched tightly into the wall of the tubing to provide the seal.

To install a compression fitting:
- First, slide the nut onto the tubing.
- Next, slide a ferrule over the line to be compressed.
- Then, slide the sleeve onto the tubing.

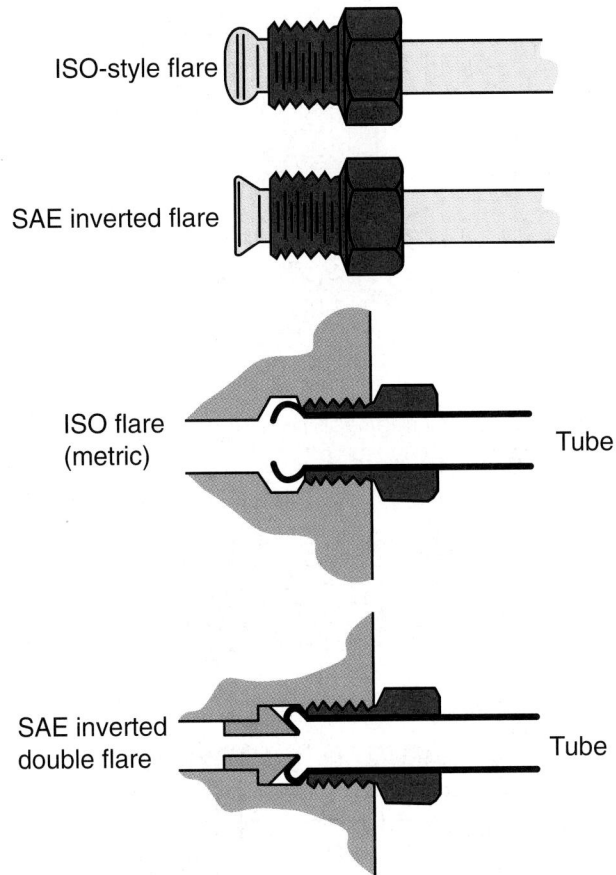

Figure 24.8 SAE and ISO flares.

Figure 24.9 Comparison between inverted and SAE flares.

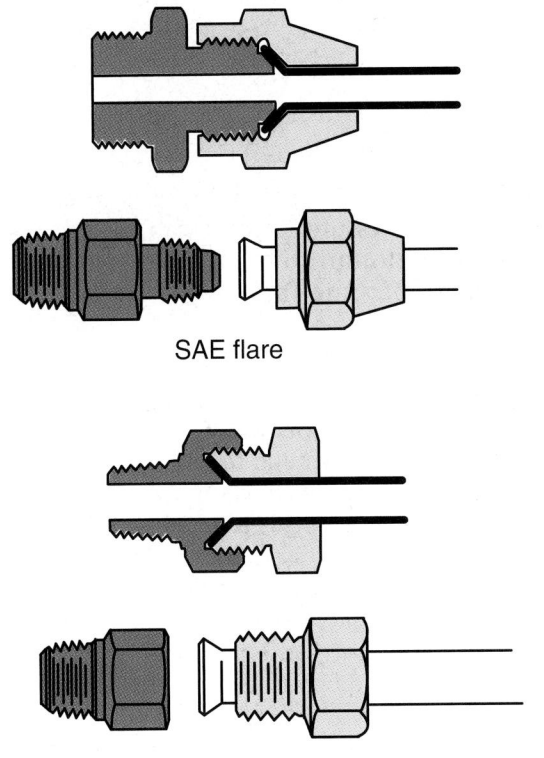

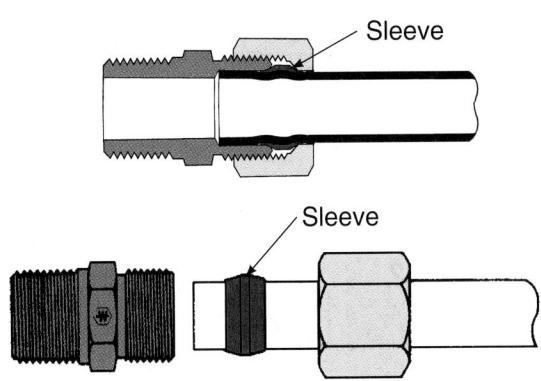

Figure 24.11 When a compression fitting is tightened, the sleeve compresses into the tubing to make the seal. *(Courtesy of Dana Corporation)*

■ Finally, insert the tubing as far into the fitting as possible. Hold it in that position as you use one wrench to tighten the nut and another wrench to hold the fitting.

■ After both halves of the fittings contact, they are tightened further 1¼ turns. This compresses the sleeve into the tubing to form the seal.

Compression fittings should not be used on high-pressure applications such as brakes or power steering systems. There are specialized compression fittings available for high-pressure applications.

Compression fittings can also be used with rigid plastic tubing. With softer plastic tubing, an insert is put inside the end of the tubing so it does not get crushed when the sleeve is compressed (**Figure 24.12**).

Another kind of compression fitting, called a *double compression fitting*, does not have a separate sleeve but compresses the front part of the nut against the tubing (**Figure 24.13**). After both halves of the fittings contact, they are tightened further 1½ turns.

O-ring connections are sometimes used to seal fittings. An O-ring is an artificial rubber donut that is used with straight threads to seal when compressed (**Figure 24.14**).

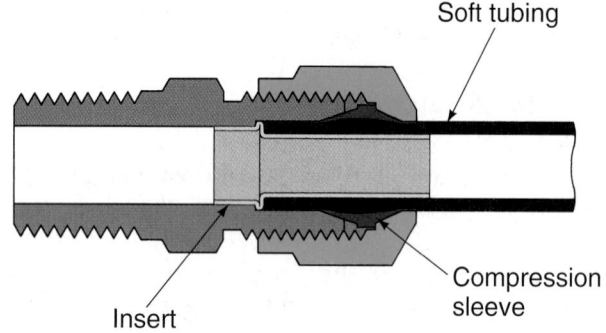

Figure 24.12 An insert is used when a compression fitting is used with plastic tubing. (*Courtesy of Dana Corporation*)

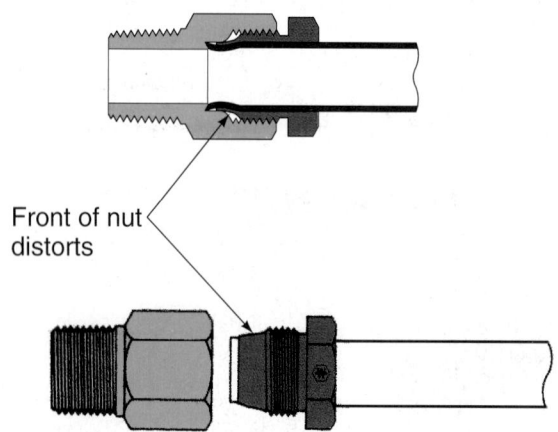

Figure 24.13 When a double compression fitting is tightened, the front part of the nut compresses into the tubing. (*Courtesy of Dana Corporation*)

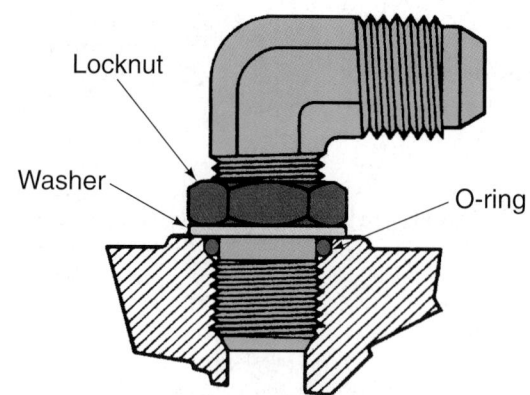

Figure 24.14 An O-ring seals between parts. (*Courtesy of Dana Corporation*)

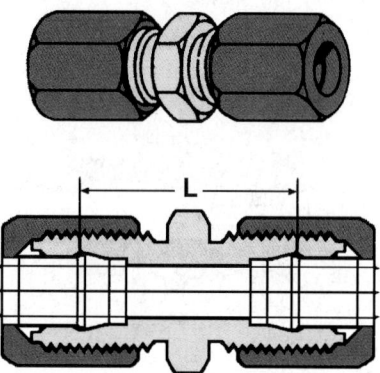

Figure 24.15 A union. The length that is removed from the tubing should equal **L**. (*Courtesy of Dana Corporation*)

Unions

A **union** joins two pieces of tubing together (**Figure 24.15**). This type of fitting is often used on vacuum or air pressure lines. It can be disassembled without having to turn the tubing. It is a good repair for the higher pressure that runs in automatic transmission cooler lines.

■ PIPE FITTINGS

On automobiles, pipe threads are used for heater outlets in the block and intake manifold, oil gallery and coolant drain plugs, as well as for oil and coolant temperature sending units (**Figure 24.16**). Pipe fittings are also used for compressed air lines in the shop.

Pipe fittings on copper, brass, or iron pipe use tapered threads that wedge together as they are tightened. An airtight seal is formed between the two tapered threads. The pressure capability of the joint is rated at 1,000 psi.

The size of a pipe thread is determined by the I.D. of the piece of pipe. The O.D. of a ½" *National Pipe Taper* (*NPT*) thread will be quite a bit larger than ½" (**Figure 24.17**).

Figure 24.16 Pipe threads are found on threaded plugs, heater outlets, and oil and coolant temperature sending units. *(Courtesy of Tim Gilles)*

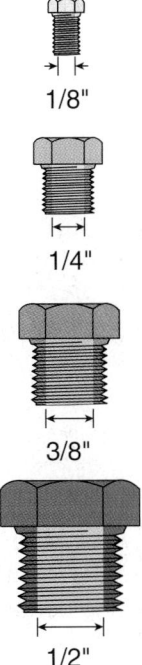

Figure 24.17 Pipe threads are sized according to the inside diameter (I.D.) of the pipe *so the thread diameter is larger than the callout size.*

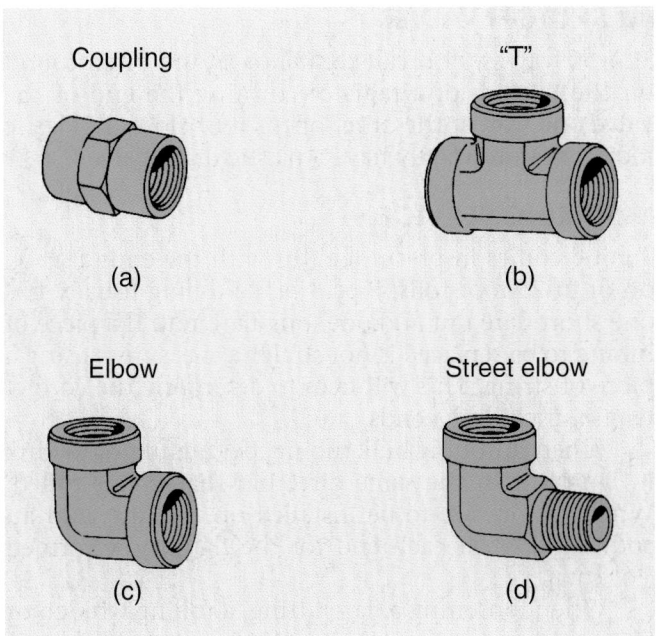

Figure 24.18 Different types of pipe connections. *(Courtesy of Dana Corporation)*

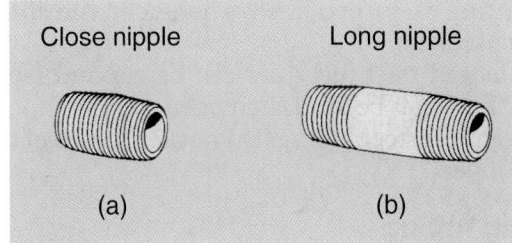

Figure 24.19 Pipe nipples. *(Courtesy of Dana Corporation)*

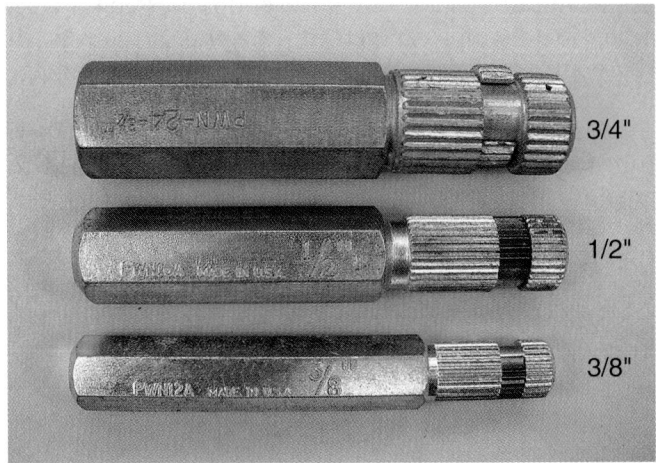

Figure 24.20 Internal pipe wrenches. *(Courtesy of Tim Gilles)*

Two pipes can be joined together with a pipe coupling (**Figure 24.18a**). When a pipe is joined to two other sections of pipe, the fitting is called a *"T"* (**Figure 24.18b**). A female pipe coupling that makes a turn is called an *elbow* (**Figure 24.18c**). When there is a male thread on one end, the coupling is called a **street elbow** (**Figure 24.18d**).

A small section of pipe is called a nipple. A **close nipple** has tapered threads on each end that join in the middle (**Figure 24.19a**). A **long nipple** has a section of plain pipe separating the threads (**Figure 24.19b**).

An internal pipe wrench is used to remove a nipple (**Figure 24.20**). Attempting to remove a nipple with an ordinary pipe wrench will collapse the nipple and/or damage its threads.

■ SHUTOFF VALVES

Shutoff valves, also called *draincocks*, are often found on the bottom of a radiator. Turning the end of the valve one way or the other opens it or closes it. Plastic radiator tanks usually have a plastic drain valve.

■ TUBING SERVICE

Tubing comes in precut lengths with the ends already on or in 25-foot rolls. Precut brake tubing usually has one short flare nut and one long flare nut. If a piece of tubing to be replaced is not straight, measure it using a piece of string. This will help to determine the correct length, including bends.

When unrolling bulk tubing, be careful not to kink it. Unroll it in the same direction that it was rolled. When fittings are to be installed on the line, add an additional ⅛" at each end for the flare to be formed (¼" total).

When loosening a flare fitting, a tubing wrench or flare nut wrench is used. Flared lines are held against a seat in the fitting. *Always* use two wrenches. The second wrench holds the female part of the fitting. If the female fitting is allowed to turn when the male flare fitting is turned with a wrench, the line will become kinked.

Damaged steel fuel lines can be cut and repaired, or new lines can be fabricated using a flaring tool. Use only seamless steel tubing. Do not replace steel tubing with copper.

Cutting Tubing

When cutting tubing to length, be sure to cut it so it is square on the end. It is best to use a tubing cutter (**Figure 24.21**). A hacksaw will leave a rough edge that might not be square. When the end is dressed off with a file, metal chips can get into the end of the tube. If these are not thoroughly cleaned out, serious damage to a component can result.

The tubing cutter is first tightened against then rolled around the tubing. The handle is tightened to advance the cutter as the tubing is cut. Do not overtighten it or the cutter will cut through too soon and the tubing could be damaged.

A burr usually remains on the end of the tubing after cutting it. This should be removed with the reamer blade that most tubing cutters are equipped with. Be sure to remove any chips from the end of the tubing after completing the cut.

Some tubing cutters are very small. In tight quarters, these can be used for repairing damaged tubing on the vehicle. The tubing cutter shown in **Figure 24.22** is especially handy for this.

Bending Tubing

When bending tubing, remember that too sharp of a bend will result in a kink or restriction. Tubing can be bent with a tubing bender (**Figure 24.23**). It can also be bent by holding it over a large piece of pipe and slowly forming it using the tubing that is being replaced as a guide when possible. Be careful not to bend the tubing too sharply.

It is better to install fittings and flare both ends before bending tubing. Otherwise, if the bend is too close to the flare, the flaring tool will not be able to clamp to the line. It is best if the bend is not too close to the flare. Leave at least a couple of inches when possible.

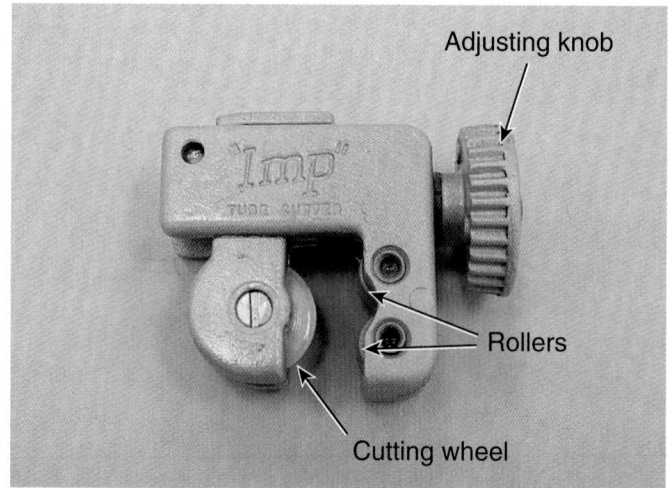

Figure 24.22 A small tubing cutter is handy for working in tight spaces on the vehicle. *(Courtesy of Tim Gilles)*

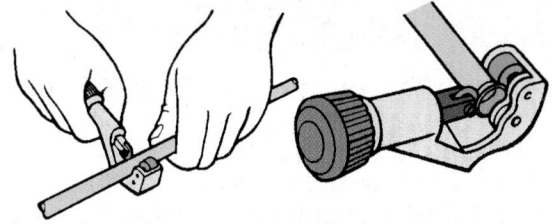

Figure 24.21 Using a tubing cutter to cut tubing. *(Courtesy of DaimlerChrysler Corporation)*

Figure 24.23 A tubing bender. *(Courtesy of Tim Gilles)*

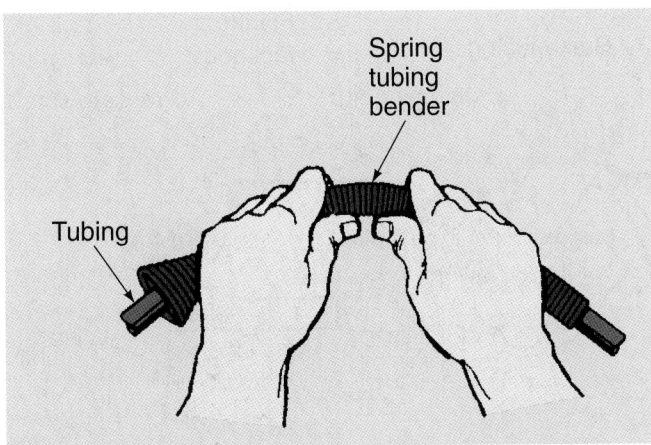

Figure 24.24 A bending spring helps prevent the tubing from becoming kinked.

A bending spring (**Figure 24.24**) can be installed over the tubing. It will help keep the tubing from becoming kinked during the bend. When using a bending spring, one end of the line must not have its fitting installed yet. This is so that the spring can be installed and removed from the line.

Steel lines should not have long, straight runs. They are difficult to remove and replace, and they can fatigue at the connections. When a short section of tubing is replaced, at least one end should have a bend so that the connection can flex (**Figure 24.25**).

Long runs should be supported with clamps. Heavy connections such as distribution blocks must be securely mounted. The ends of the tubing should align with the fitting. Be sure that the threads can be turned all the way into the fitting easily.

Flaring the Ends of Tubing

Tubing is flared with either a double flare or an ISO (metric double) flare. Different tools are required to perform these jobs. Be sure to use the same kind of flare as was used on the original line. The different types of flares are not interchangeable.

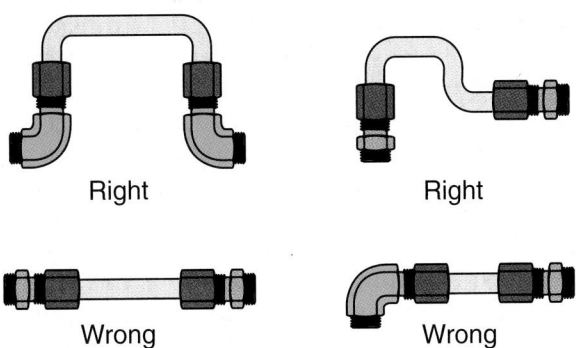

Figure 24.25 A bend in a rigid length of tubing can prevent vibration damage.

Double flaring is a two-step procedure. The fuel line is clamped in a special tool while its end is formed. The tube is then folded over itself to complete the double flare.

- First, slip the fitting onto the line.
- Select the correct size hole in the flaring tool bar.
- Clamp the line in the flaring tool bar. It should extend out of the bar by the width of the flaring tool adapter (**Figure 24.26**).
- A threaded flaring cone and clamp are used to form the end of the tubing (**Figure 24.27a**). **Figure 24.27b** shows the appearance of the end of the tubing following the forming operation.
- Insert the adapter in the end of the tubing and tighten down on it with the threaded flaring tool until it bottoms out (**Figure 24.28a**).

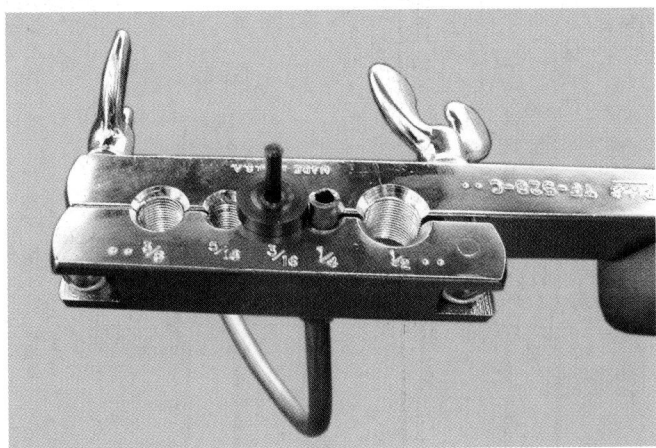

Figure 24.26 Clamp the tubing in the flaring bar with it protruding about the thickness of the adapter. (*Courtesy of Tim Gilles*)

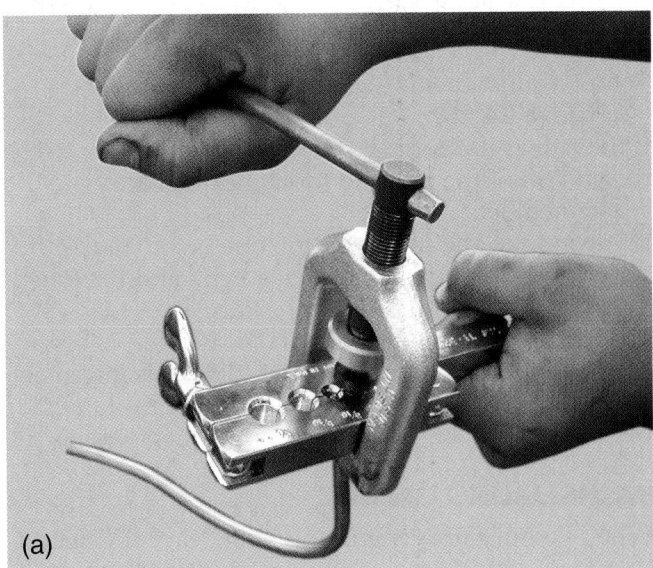

(a)

Figure 24.27 (a) A threaded flaring cone and clamp are used to form the end of the tubing. (*Courtesy of Tim Gilles*)

(*continued*)

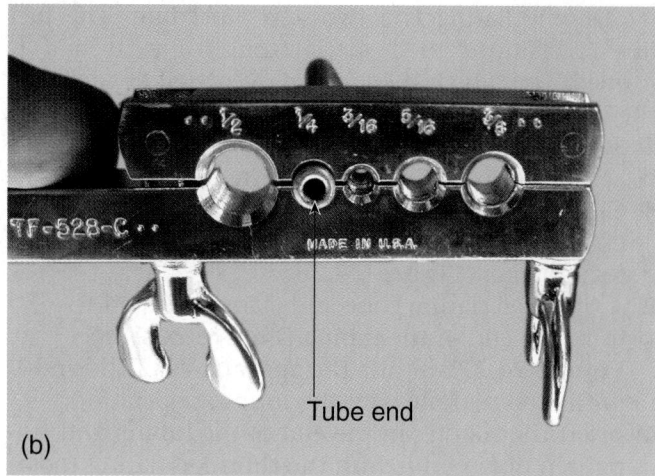

(b)

Figure 24.27 (b) The end of the tube following the forming operation. *(Courtesy of Tim Gilles)*

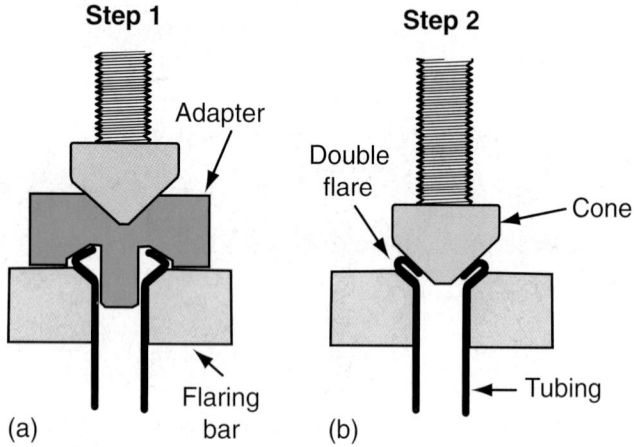

Step 1   **Step 2**

Adapter

Double flare

Cone

Flaring bar

Tubing

(a)   (b)

Figure 24.28 (a) The adapter is forced into the end of the tube. (b) The flaring cone completes the double flare.

■ Remove the adapter and tighten the flaring tool against the line again to complete the flare (**Figure 24.28b**).

If the flare is not formed properly, you will have to cut the end off of the line and form another flare.

Remember:

■ Always put the fitting on the line before flaring it.
■ Leave enough space between a bend and the flared fitting so that the fitting can slide.

Thread sealers or Teflon tape are not necessary when using flare fittings. A flare fitting seals internally, and the threads should not be exposed to liquids. If the threads get wet, the flare is not a quality joint.

■ INSTALLING TUBING

When you install a length of tubing, leave the first fitting loose after threading it into the fitting. After starting the fitting on the other end of the line, tighten them both. Hold the fitting with a wrench while tightening the flare nut. Do not overtighten the fittings.

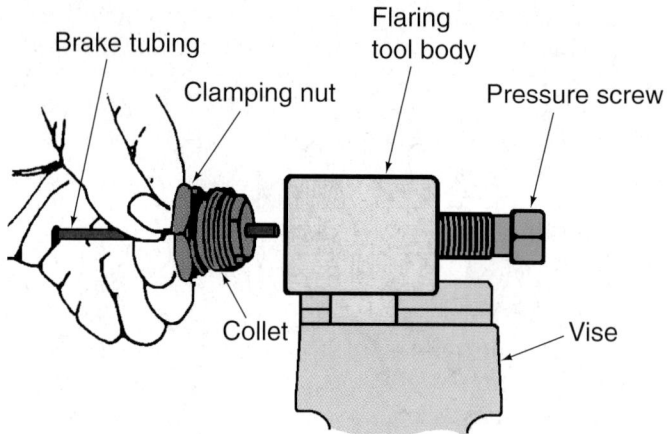

Figure 24.29 An ISO flaring tool. *(Courtesy of Dana Corporation)*

After the flare is brought into contact with the fitting, turn it an additional ⅙ turn only.

ISO Flaring

To perform a bubble flare (ISO flare), a different flaring tool is used (**Figure 24.29**). The tubing is inserted through the end of the clamping nut. When the clamping nut is tightened into threads in the flaring tool body, a collet is compressed to grip the outside of the tubing. Tightening the pressure screw forms the bubble on the tubing.

Union Repairs

Sometimes a union is used to repair a damaged line (see Figure 24.15). A section of the line slightly less than the length of the fitting is cut from the old line. Some unions require the ends of the line to be flared first. Others use compression fittings. With compression fittings, flaring is not required.

A union is a better repair than hose for higher pressure (30–35 psi) lines. When there is enough straight line remaining behind a flare, cut the kink from the line and install a union to couple the two pieces of line together.

Using Hoses to Repair Tubing

When a section of steel fuel line is damaged, it is best to replace it with a new section of line with flared ends and fittings. Besides being stronger and more reliable, steel lines allow better cooling of the fuel.

Sometimes a small section of damaged fuel tubing is replaced with hose. When using fuel hose to repair a damaged section of metal fuel line, use a section of hose that is 4" longer than the section of tubing removed. A raised bead is formed on each end of the tubing with a flaring tool. The hose is installed with 2" of hose overlapping each end of the tubing. The clamp is positioned ⅛" from each end of the tubing.

If a section of tubing is removed that is over 6", the recommended safe practice is to use two sections of hose separated by a piece of steel tubing. Be sure to secure the tubing so that it cannot rub against body parts. Following completion of the repair, start the engine to check for leaks. After 1 week, the clamp should be retightened because hoses take a set and the clamp can become loose.

■ TRANSMISSION OIL COOLER LINE REPAIRS

The best repair for a transmission cooler line is a union. But these lines are often repaired with hose. For repairs to transmission cooler lines, be sure to use hose that is rated for high temperature, pressure, and oil. Do not use ordinary fuel hose, which deteriorates in the hostile environment of the hot oil. When a transmission cooler line is repaired using a hose, power steering hose can be used if transmission cooler hose is not available.

If a rubber hose is used to repair transmission lines, be sure to flare both ends of the metal tubing that is being joined. Otherwise, the high pressure will blow the hose off the lines. Over time, the edges of a double flare can cut the rubber hose as it works against the flare. Performing the first part of a double flare on the ends of the lines (see Figure 24.27b) will present a smoother edge for the hose to squeeze against.

 *A technician repaired a damaged transmission cooler line by cutting out the bad section, flaring both ends of it, and installing a piece of hose with worm gear hose clamps. After a few thousand miles, the sharp edge of the single flared line cut through the rubber hose, resulting in a serious fluid leak. Luckily, the owner spotted smoke and stopped the car before a fire or damaged transmission could occur.*

■ NYLON FUEL INJECTION TUBING

Some vehicles use nylon fuel line, which must first be soaked in boiling water to allow it to stretch during installation. When nylon line is connected to steel line, a push connector is used.

■ To remove a "hairpin" push connector (**Figure 24.30**), separate the clip legs by about ⅛" and pull the triangular tab to remove the clip. Then, firmly and gently pull the nylon line from the steel line.

■ To remove a "duckbill" connector (**Figure 24.31**), narrow-jawed pliers are used to compress both retaining clips on the side of the push connector at once. The retaining clip remains on the metal tube when the connector is removed.

Some nylon lines are coupled together with a "spring lock" connector (**Figure 24.32**). The flared end on the female fitting of this coupling lodges

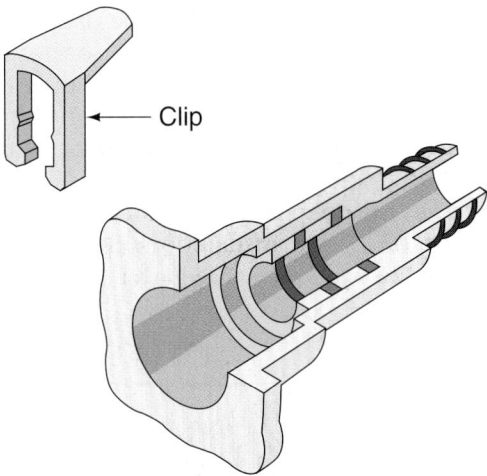

Figure 24.30 A hairpin push connector. Separate the clip legs by about 1/8" and pull the triangular tab to remove the clip.

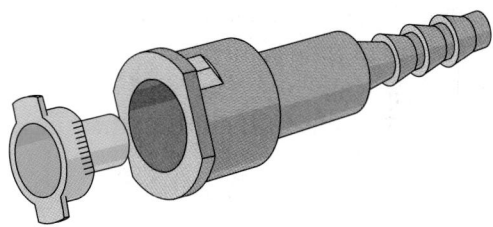

Figure 24.31 A duckbill connector. Use narrow-jawed pliers to compress both retaining clips on the side of the push connector at once.

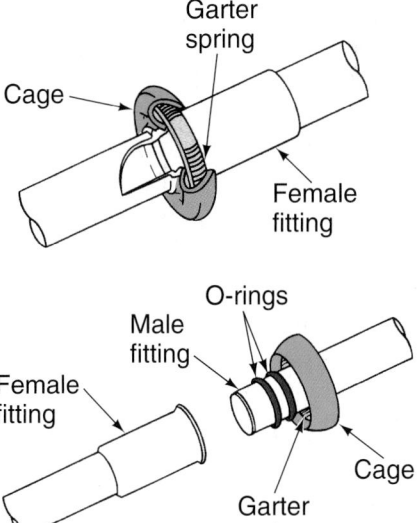

Details of spring lock (garter) connector

Figure 24.32 Some nylon lines are coupled together with a "spring lock" (garter) connector.

behind the garter spring inside the cage on the male fitting. This holds the two couplings together.

There are two O-rings on the male coupling. They are made of fluoroelastomer. When replaced, it is important that the correct O-rings be used. A special tool is used to release the garter spring so that the two halves of the coupling can be released. Sliding the tool toward the garter spring causes a ledge on the tool to go under the spring, raising it so the flare can clear it.

If the garter spring is damaged or missing, it can be replaced in the coupling. Lubricate the new O-rings with engine oil before installation. Be sure that the flare is all of the way under the garter spring.

■ PIPE SERVICE

Pipe is cut with a pipe cutter, which resembles a large tubing cutter. **Pipe dies** are used to form the threads on the outside of the pipe. A thread sealer is used between the threads. Teflon tape is often used (see Chapter 51). After hand-tightening the pieces of pipe, tighten further a minimum of 2½ turns.

■ REVIEW QUESTIONS

1. List three materials that tubing can be made of.
2. What happens to copper when it is repeatedly flexed?
3. Why is tubing sometimes coiled when installed?
4. What are two types of flares used on automobiles?
5. What kind of flare is most often found on cars, inverted or standard?
6. What is the name of the fitting that uses a sleeve?

7. How much further should a compression fitting be tightened after the parts of the fitting are snug?
8. What is the name of a 90-degree pipe fitting that has a male thread on one end and a female thread on the other?
9. How much should be added to the length of a new section of tubing to compensate for the fittings?
10. How much is a flared connection tightened after the nut becomes snug?

■ ASE-STYLE REVIEW QUESTIONS

1. Which of the following materials is/are used for brake lines?
   a. Copper
   b. Steel
   c. Both A and B
   d. Neither A nor B

2. Technician A says that a single flare is not used because it will split the tubing. Technician B says that Teflon® tape or thread sealer is used with flared connections. Who is correct?
   a. Technician A
   b. Technician B
   c. Both A and B
   d. Neither A nor B

3. Technician A says that tubing's size is determined by its outside diameter. Technician B says that pipes and hoses are sized by their outside diameter. Who is right?
   a. Technician A
   b. Technician B
   c. Both A and B
   d. Neither A nor B

4. Which of the following is/are true about steel tubing?
   a. It is galvanized to provide strength.
   b. It is threaded at its ends.
   c. Both A and B.
   d. Neither A nor B.

5. Technician A says to install a flare fitting on tubing after it has been bent. Technician B says pipe fittings have tapered threads. Who is correct?
   a. Technician A
   b. Technician B
   c. Both A and B
   d. Neither A nor B

Electrical System Theory and Service

THEN AND NOW: ELECTRICAL SYSTEMS

Many early cars had no electrical system whatsoever—not one wire in the entire vehicle! The engine was cranked by hand. Ignition might have been supplied by a platinum tube heated by a flame. If a driver was adventurous enough to attempt driving at night, acetylene lamps were used to light up the road.

Hand-cranking limited the use of the automobile to those with strong arms and backs. The hand crank also presented a dangerous situation when it kicked back. These were big factors in the electrification of the automobile.

In 1910, Bryon T. Carter, whose Cartercar Company had been bought by GM, stopped to help a woman whose car had stalled on a bridge. While attempting to start the engine, the hand crank kicked back, breaking his arm and jaw. He was admitted to the hospital, where he contracted pneumonia and died.

Henry Leland, head of Cadillac at that time, knew Carter. He was horrified to hear of Carter's death and directed his executives to find a way to eliminate the hand crank. Charles "Boss" Kettering was commissioned by Cadillac for this job. This ingenious inventor (one of the founders of Delco) came back with a design for the first integrated electrical system. It included not only an electric starter and battery, but also a generator. The job of the generator was to keep the battery charged and to power the system's accessories, which included headlights and a dependable point-type electrical ignition system. This landmark in automotive history appeared on the 1912 Cadillac.

As more and more electrically powered accessories appeared on cars, electrical load requirements began to exceed the capacity of the direct current (DC) generator used until the 1960s. In 1960, Chrysler introduced the *alternator,* which is actually an alternating current (AC) generator. Diodes within the alternator convert its AC output to the DC required by the automobile electrical system. Unlike a DC generator, an alternator produces plenty of current at low engine rpm.

The dangerous hand crank. *(Courtesy of Bob Freudenberger)*

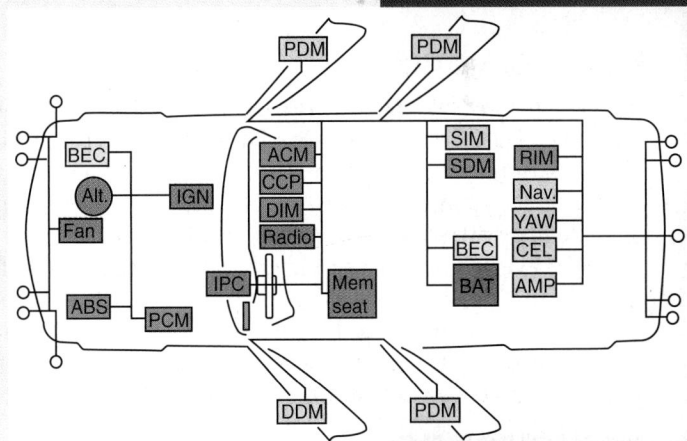

A modern electrical system.

The ever-increasing list of electrical accessories also resulted in larger bundles of wires under the dash. Locating a short circuit or a bad ground requires an accurate wiring diagram and can be time-consuming.

Today, electronic components have been added to vehicle electrical systems. Electronic engine management systems provide increased performance, better fuel economy, and lower exhaust emissions. Many cars have a sophisticated on-board computer that controls several specialized computers. The use of fiber optics and multiplexing has made it possible to eliminate the large bundles of wiring found on older vehicles. Troubleshooting and repair of these electronic systems requires a good deal of training and experience.

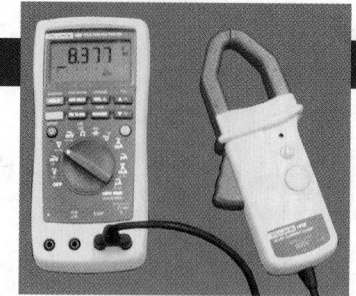

Basic Electrical System Theory and Repairs

■ KEY TERMS

conductor
insulator
EMF
volt
amp
DC
AC
Ohm's law

ohm
resistor
series circuit
parallel circuit
electromagnetic
   induction
capacitor
semiconductor

diode
transistor
analog meter
DMM
DVOM
impedance
voltage drop

■ INTRODUCTION

Almost every system of the car uses electricity or electronics in some manner. These systems include antilock brakes, engine emission control devices, dash warning lights and gauges, electronic fuel injection, electrically shifted transmissions, and others. Technicians in every specialty service area must have a basic understanding of electricity to be successful. This chapter covers the basic items needed to begin learning about more advanced topics. Learning these topics now will make mastering more advanced topics easier.

■ ELECTRON FLOW

Electricity cannot be seen. It is made up of very small objects flowing at very high speed. All matter is composed of *atoms*. Atoms are composed of *protons, neutrons*, and *electrons* (**Figure 25.1**). Electrons have a negative charge and orbit around the positively charged protons. Protons and neutrons (which have no charge) are located in the *nucleus* (center) of the atom. The number of protons and electrons in an atom is what determines what the atom is. **Figure 25.2** compares aluminum and silver atoms.

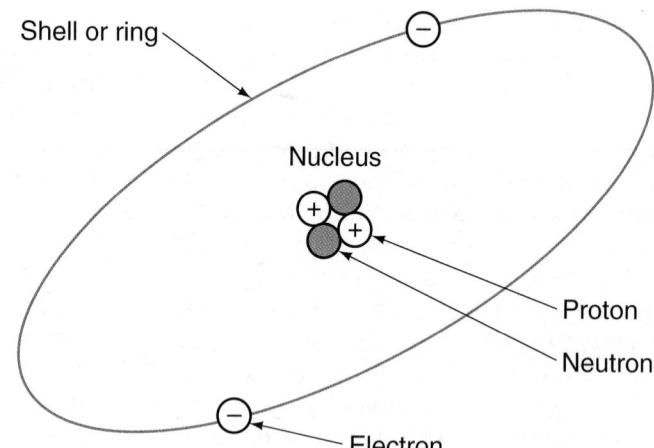

Figure 25.1 Atoms are composed of protons, neutrons, and electrons.

The electrons remain in orbit around the nucleus because they have an electrical attraction to its protons. This is because opposite charges attract (**Figure 25.3**).

Each atom tries to remain electrically neutral. When it is neutral, it has no charge; the number of protons and electrons is equal.

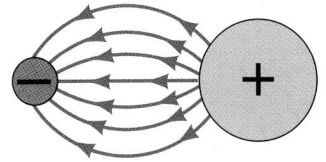

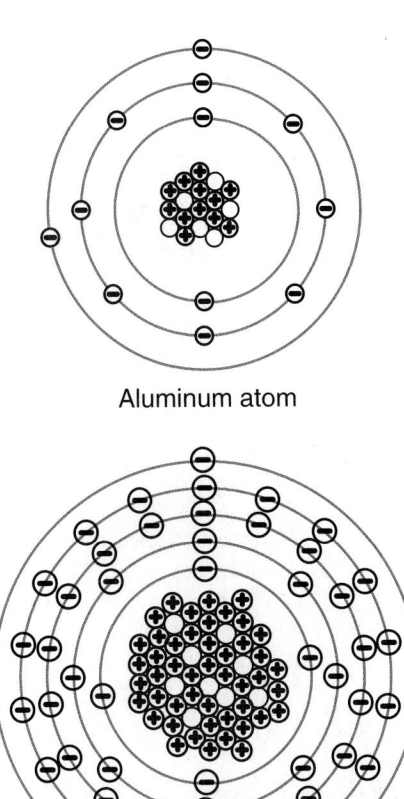

Aluminum atom

Silver atom

Figure 25.2 The number of protons and electrons in an atom determines its identity.

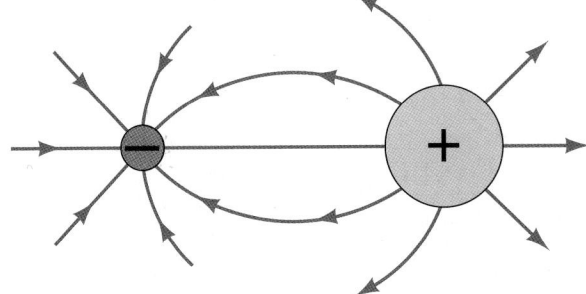

Close together, all lines interlink

Far apart, there is little field interlinkage

Figure 25.3 Opposite charges attract. *(Courtesy of Ford Motor Company)*

Electricity is the flow of electrons from one atom to another (**Figure 25.4**). Protons are tightly bound within the nucleus of the atom, but electrons are free to move in their orbits at a fixed distance from the nucleus. The electrons are in orbit and stay a fixed distance from each other because they are of like charge and repel one another.

Conductors

A **conductor** is metal or another material such as carbon from which wires and electrical parts are made. Conductors have atoms that allow electricity to flow freely. Atoms attempt to remain electrically neutral. To remain in electrical balance, an atom will shed or attract electrons from neighboring atoms. When an

atom is lacking electrons and another atom has too many, electrons will flow between atoms to equalize the charges. This is what makes up electricity.

Conductors are atoms with *free electrons* (**Figure 25.5**). These are extra electrons not bound to the protons in the nucleus. The attraction of a negatively charged electron for the positive charge in the atom's nucleus is less when the electron is in an outer orbit.

An atom is identified by its number of protons and its number of electrons in their orbits about the nucleus. Hydrogen is the simplest atom, with only one electron in orbit. A complex atom is compared to a hydrogen atom in **Figure 25.6**.

Copper is a more complex atom, with 29 electrons in various orbits. But it has only one electron in its valance ring (**Figure 25.7**). The electron can move easily from one atom to the next, which makes it a good conductor. Some good conductors used in wiring are silver, copper, and aluminum.

Insulators

An **insulator** is the opposite of a conductor. It is a material with no, or few, free electrons. Because their electrons are tightly bound in their orbits, an insulator prevents the flow of electrons between two conductors. Glass, rubber, and porcelain are good insulators.

Conductor

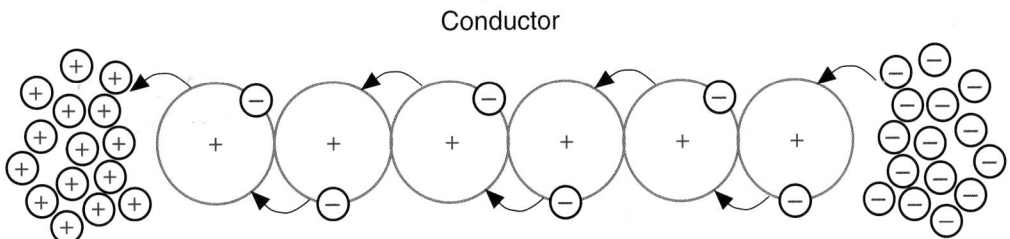

Figure 25.4 Electricity is the flow of electrons from one atom to another.

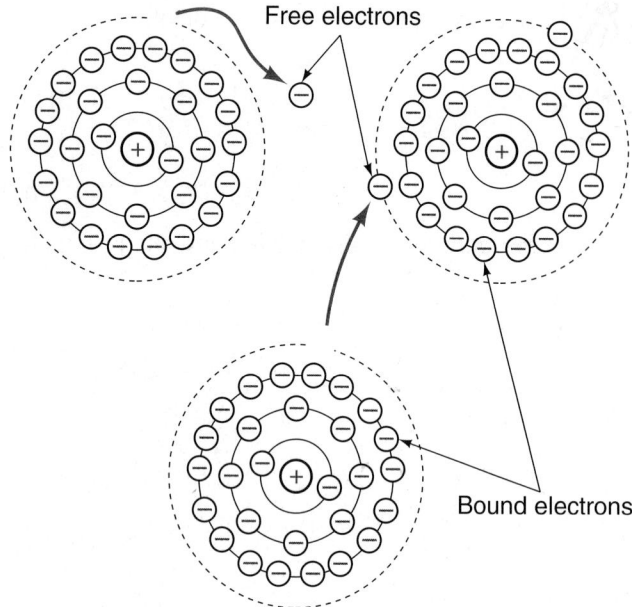

Figure 25.5 Conductors are atoms with free electrons.

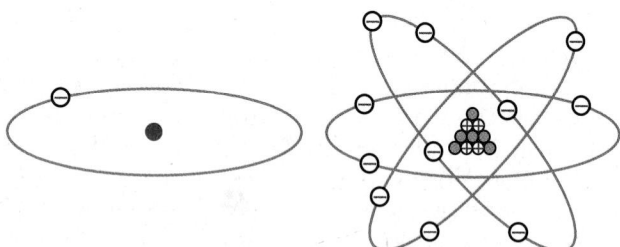

Figure 25.6 A complex atom compared to a hydrogen atom. *(Courtesy of Federal-Mogul Corporation)*

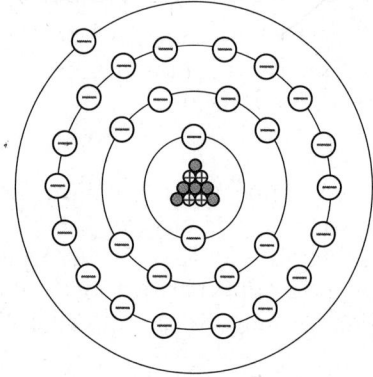

Figure 25.7 Copper has only one electron in its valence ring. *(Courtesy of Federal-Mogul Corporation)*

Circuit

A complete *circuit* is when a full circle is provided for electrical flow. A light filament hooked across a battery, between its positive and negative posts, is an example of a circuit (**Figure 25.8**). On automobiles, the ground

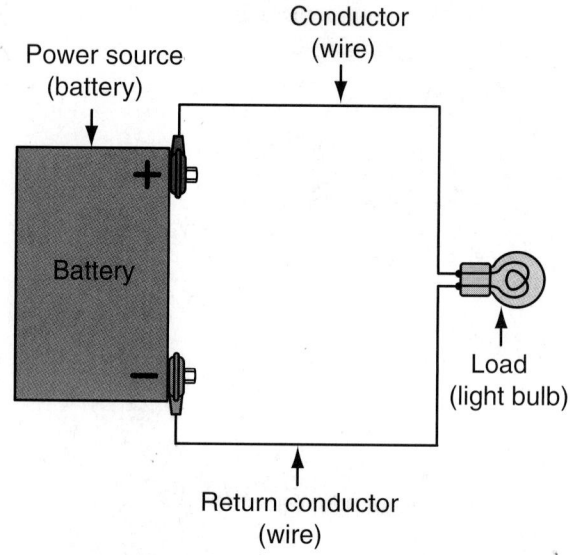

Figure 25.8 A circuit.

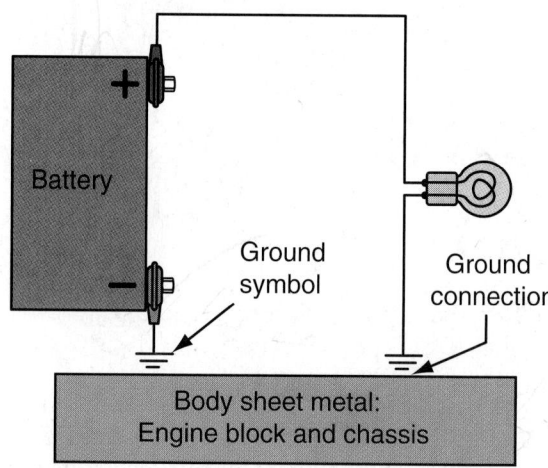

Figure 25.9 The ground part of the circuit is provided by the frame and body metal.

part of the circuit is provided by the frame and body metal (**Figure 25.9**).

Wiring

Electricity always takes the path of least resistance to ground. Without insulation, electricity can take undesired paths. With enough voltage present, electricity can jump air gaps too. Wiring from the battery's positive terminal is insulated so it cannot find ground.

Most automotive wiring is insulated with *polyvinyl chloride* (PVC). This is a durable plastic that is designed to insulate, yet remain flexible and resistant to heat and attack from fuels and oils. The ground side of a circuit usually does not require insulation. More information on wiring is found in Chapter 32. Electrical service technicians rely on wiring diagrams to solve problems in much the same way a motorist relies on a map to reach a destination. Wiring diagrams are covered in greater detail in Chapter 33.

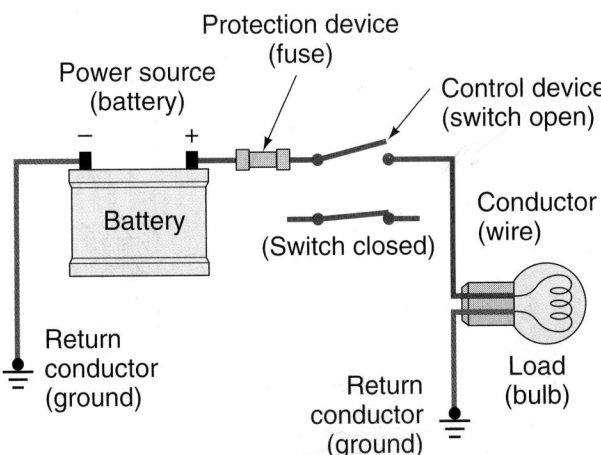

Figure 25.10 A switch provides a means of turning the circuit on and off. *(Courtesy of Ford Motor Company)*

Control and Protection Devices

- A *switch* can be used to open a circuit, providing a means of turning it on and off (**Figure 25.10**).
- Protection devices, *fuses* or *circuit breakers*, are installed in the circuit. These are designed to open (fail) before damage can occur elsewhere in an electrical circuit. Protection devices are covered in detail in Chapter 32.

■ BASIC AUTOMOTIVE ELECTRICAL SYSTEM

Electricity powers the *starting, ignition, lighting,* and *accessory systems* of the car. When parts of these systems operate, there is an electrical *load* on the vehicle's charging system. When a load is connected to the positive and negative terminals of an electrical power supply, electricity flows through the circuit. The electricity is supplied by a *storage battery*, replenished by the charging system. These systems are covered in later chapters.

When pressure is applied to one end of a circuit, electrons repel each other, causing movement in the circuit (**Figure 25.11**). Electrons do not actually move more than a few inches a second. But the energy applied to one end of the circuit is transmitted to the

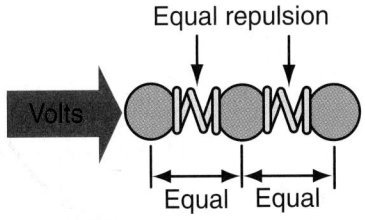

Figure 25.11 Electrons move in response to electrical pressure (volts).

other end at the speed of light (186,000 miles per second). The energy transmission can be compared to the action on a row of billiard balls (**Figure 25.12**).

■ ELECTRICAL TERMS

Electricity is invisible. To be able to see what is happening in an electrical system, you will need to learn electrical terms and how to use a meter. Volts, ohms, and amps, terms for electrical pressure, resistance, and flow, are the common electrical terms you will need to know before using electrical meters.

The flow of electricity will be compared to the flow of water in the following discussions on voltage, current, and resistance.

Voltage

The force that is needed to push or pull an electron out of its orbit is called electromotive force (**EMF**), which means electron moving force. EMF can be compared to the action of a pump on a water supply. It is a measurement of electrical *pressure* and is measured in **volts** using a voltmeter (**Figure 25.13**).

HISTORY NOTE

Alessandro Volta *(1745–1827) discovered the* voltaic pile *(battery), the first source of direct electrical current. The French named the unit of electrical pressure, the volt, in his honor in 1881 after he had been dead for many years.*

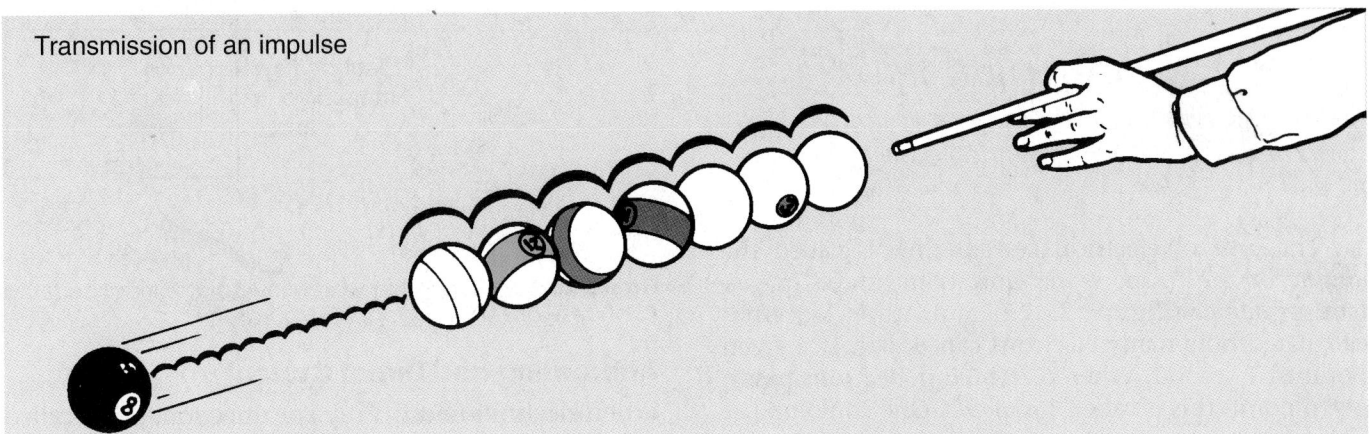

Transmission of an impulse

Figure 25.12 Energy is transmitted almost instantaneously. *(Courtesy of Ford Motor Company)*

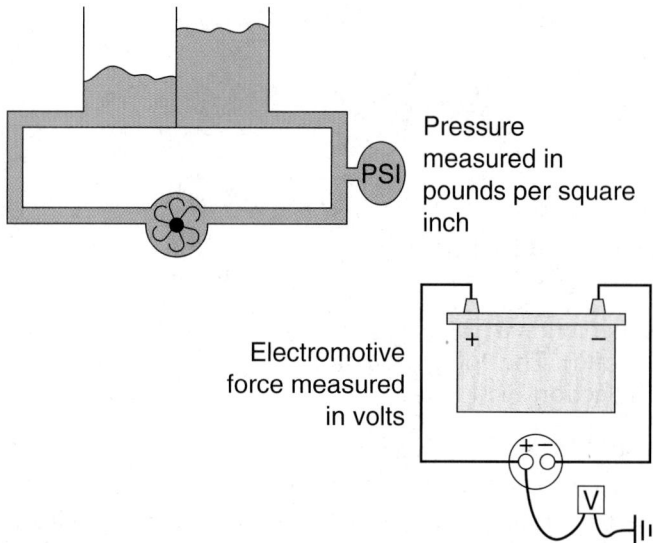

Figure 25.13 Electrical pressure is measured in volts. *(Courtesy of Federal-Mogul Corporation)*

The lead acid storage battery used in automobiles produces approximately 2.1 volts per cell. There are six cells in a 12-volt battery, actually producing 12.6 volts.

Voltage controls how strongly a load in a circuit is operated. If voltage is low, a light will not be as bright or a heater defroster motor will not spin as fast. Voltage drops as various loads are placed on a battery's circuit. This means that electrical flow is directly proportional to the amount of voltage pushing it (**Figure 25.14**).

Voltage Differential. Electrons flow because of the difference between battery voltage at one end of a circuit and zero voltage on the ground side. There must be a difference in voltage to create a current flow.

Current

The correct term for the flow of electricity is *current*. Current is the number of electrons flowing per second in a circuit. It can be compared to how much water flows through a pipe.

HISTORY NOTE

Andre Ampere *(1775–1836) was a French physicist who discovered the unit of electrical current flow.*

The unit of electrical current flow is called the *ampere* or the **amp**. Water flow is measured in *gallons per minute* (**Figure 25.15**). Similarly, science has calculated how many electrons can flow past a given point in 1 second. When 6.28 billion electrons pass a given point, this is called 1 *coulomb*. One coulomb per second is equal to 1 amp or ampere of current. Current flow is measured with an ammeter.

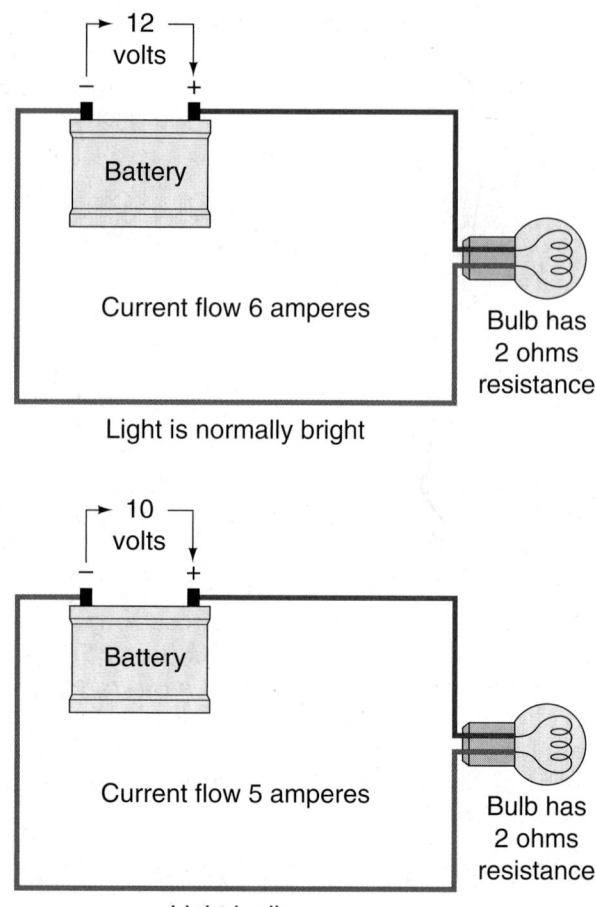

Figure 25.14 Electrical flow is directly proportional to the amount of voltage pushing it. *(Courtesy of Ford Motor Company)*

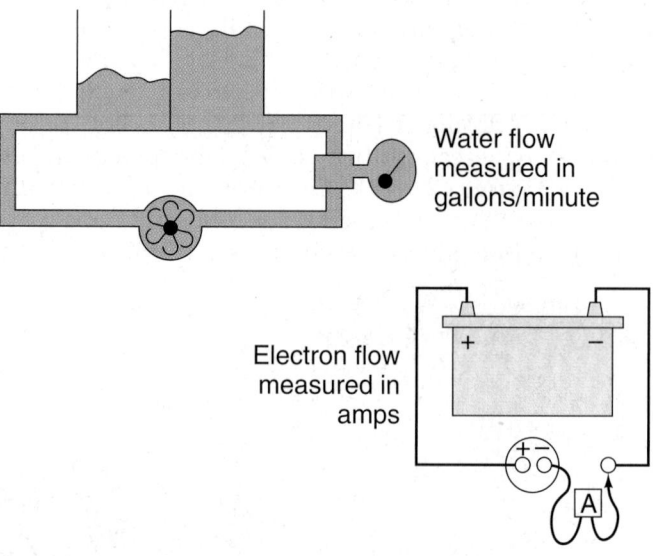

Figure 25.15 Comparing water flow to electrical current. *(Courtesy of Federal-Mogul Corporation)*

Alternating and Direct Current

When electrons flow in only one direction, this is called direct current (**DC**). The automotive electrical system uses direct current because it can be stored in a battery.

When electrical current surges from positive to negative and back again, this oscillation is called alternating current (**AC**). Alternating current is used for household electricity. AC current cannot be stored in a battery. An alternator driven by a belt from the engine's crankshaft provides electrical power to the vehicle. The alternator makes alternating current, but it is rectified (changed to direct current) before it is used to recharge the battery.

Current Flow and Electron Flow

Current flows from positive to negative. Electrons flow from negative to positive. This sometimes causes confusion. In the automotive industry, electrical flow has traditionally been described as moving from positive to negative. This is known as the *conventional theory* and is the way electrical schematics are written in service literature. The *electron theory* is the way electrical flow is described in the electronics industry. Under the electron theory, electrons will flow from a point that is less positive to one that is more positive, or from negative to positive. You probably will not need to use electron theory unless you are studying to become an electrical engineer.

Resistance

Resistance acts as an obstruction to electrical flow (similar to when a smaller water pipe is used) (**Figure 25.16**). *Georg Ohm* (1787–1854), a German physicist, discovered **Ohm's law,** which governs the relationship between volts, ohms, and amps. An **ohm** is the unit of electrical resistance measurement expressed with the Greek letter omega (Ω). One ohm is the resistance that will allow 1 ampere to flow when pushed by 1 volt.

Changes in Current Flow

You learned earlier that the amount of current flow in a circuit is directly proportional to circuit *voltage*. It is

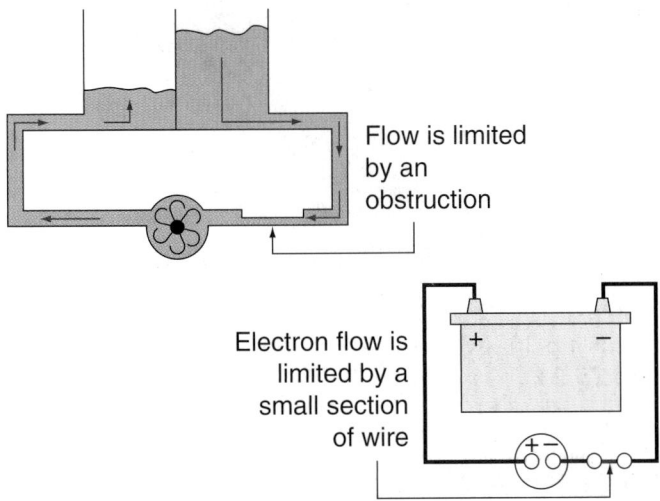

Flow is limited by an obstruction

Electron flow is limited by a small section of wire

Figure 25.16 Resistance. *(Courtesy of Federal-Mogul Corporation)*

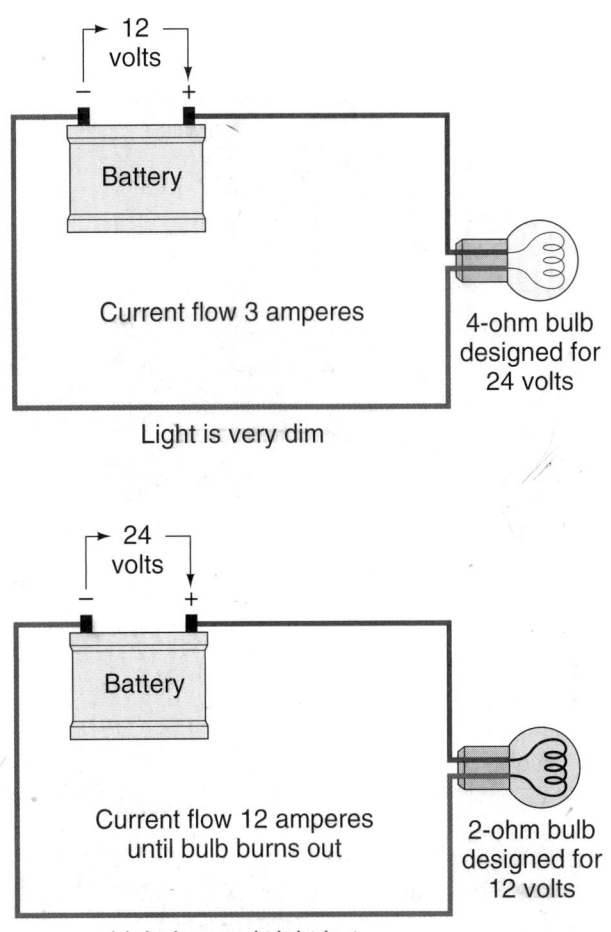

12 volts

Battery

Current flow 3 amperes

4-ohm bulb designed for 24 volts

Light is very dim

24 volts

Battery

Current flow 12 amperes until bulb burns out

2-ohm bulb designed for 12 volts

Light is very bright but bulb burns out quickly

Figure 25.17 Current draw. Using the incorrect bulb can result in electrical problems. *(Courtesy of Ford Motor Company)*

inversely proportional to the *resistance* in the circuit. This means that the more resistance there is, the less current flow there will be. Current can be increased either by increasing voltage or decreasing resistance.

The term *current draw* refers to the amount of current required to operate a load. In **Figure 25.17** current draw is compared with different voltages and resistances (light bulbs). *Loads* refer to lamps, motors, solenoids, or other electrical accessories on an automobile.

Light, heat, or motion energy results when resistance opposes the flow of current, and electrical energy changes from one form to another. **Resistors** are used to make heat or to control the intensity of a load. On a typical heater blower motor, two resistors control the motor's speed. Either none, one, or both resistors can be used to provide a stepped measure of control for low, medium, or high speeds (**Figure 25.18**).

Resistors are of different *values* (resistances). Larger resistors often have their listed values printed on them. Smaller resistors are color coded with from three to six color bands.

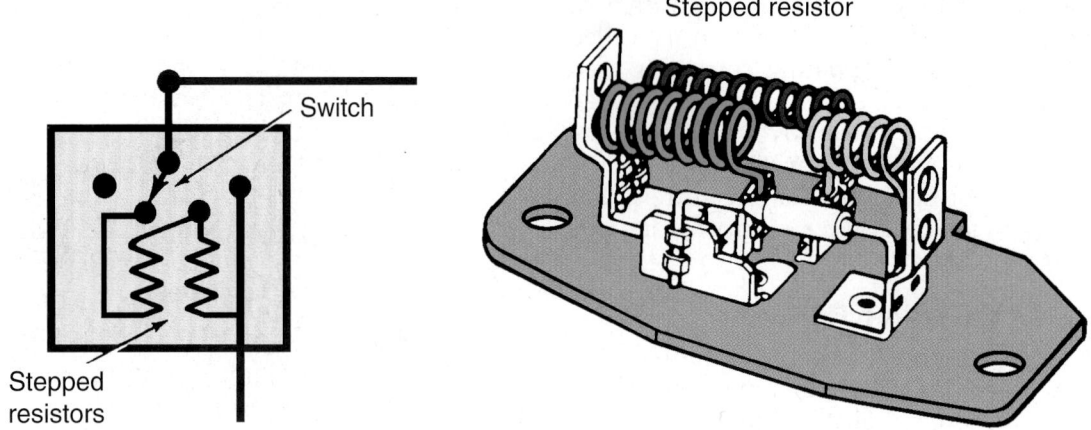

Figure 25.18 Stepped resistance controls a heater motor. *(Courtesy of Ford Motor Company)*

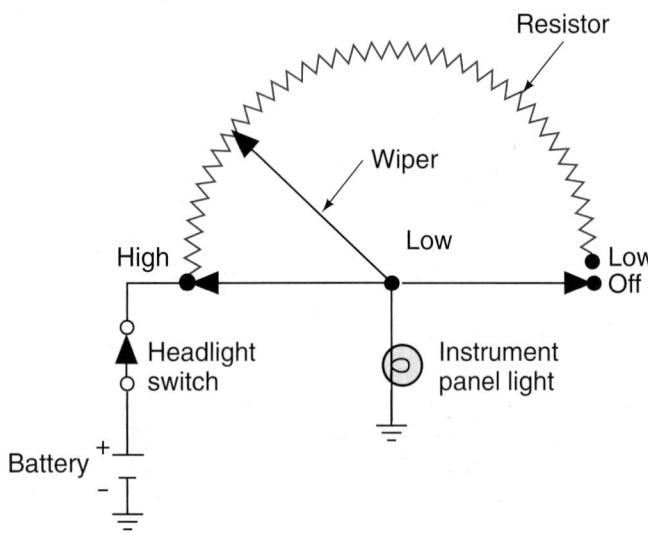

Figure 25.19 A rheostat varies current flow through the circuit.

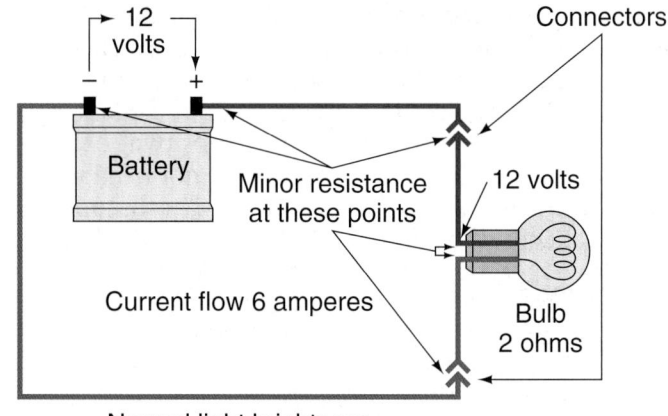

Figure 25.20 Current flow is affected by resistance. *(Courtesy of Ford Motor Company)*

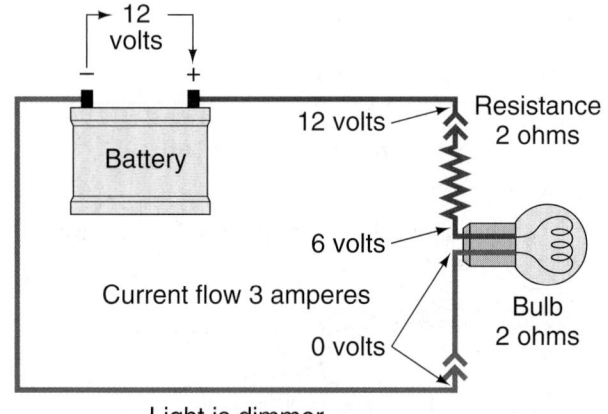

Figure 25.21 Higher resistance drops electrical flow. *(Courtesy of Ford Motor Company)*

Variable resistors, called rheostats or potentiometers, are also used to control speed and intensity of electrical loads. A *rheostat* varies *current* flow through the circuit as a movable wiper runs along a resistor (**Figure 25.19**). A *potentiometer* varies the *voltage* in a circuit. A variable resistor is commonly called a "*pot*," whether it is a potentiometer or a rheostat.

Many of these devices are found in computer circuits in modern automobiles. This information is covered in Chapter 45.

Conductors offer some resistance in a circuit, but the circuit is designed so that this is not a governing factor in current flow. The longer a wire must be, the larger its diameter will be.

■ The longer a conductor is, the more resistance it has.
■ The larger its diameter, the less resistance it has.

A circuit is designed to have a certain amount of resistance in it. This determines the amount of current flowing in it (**Figure 25.20**). Problems arise when resistance becomes higher in a circuit. This happens when corrosion, grease, dirt, or another contaminant builds up in switches or on loose connections (**Figure 25.21**).

■ CIRCUITRY AND OHM'S LAW

Types of Circuits

Circuits can be arranged in three ways: series, parallel, and series-parallel.

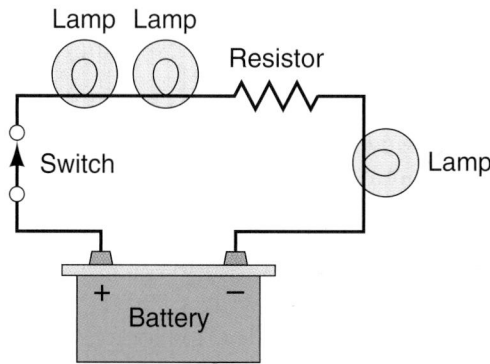

Figure 25.22 A series circuit.

A **series circuit** is when current flows equally through all parts of a circuit, first to one load and then on to the next one (**Figure 25.22**). When current flows in the circuit, both bulbs light. If one of the bulbs fails, this acts like a switch and both bulbs go out.

In a series circuit, the resistances of all of the loads add up (**Figure 25.23**). Adding loads reduces current flow.

A **parallel circuit** is when a circuit starting from a common point has different branches through which electricity can flow (**Figure 25.24**). Each branch of the circuit has a load and a separate ground. A bulb failure in one branch will not affect the other branches of the circuit.

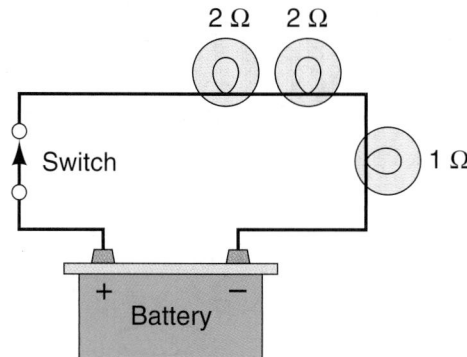

Figure 25.23 In a series circuit, the resistances of all of the loads add up.

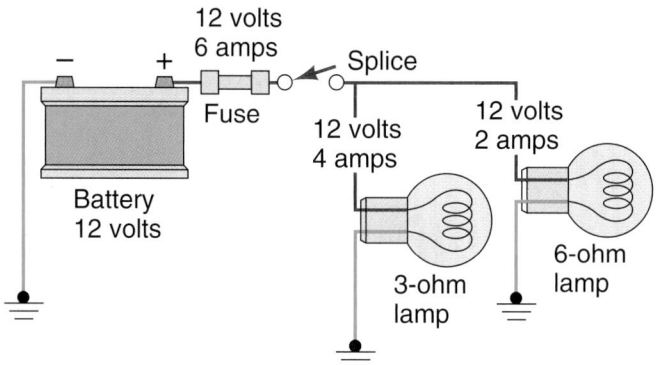

Figure 25.24 A parallel circuit. *(Courtesy of Ford Motor Company)*

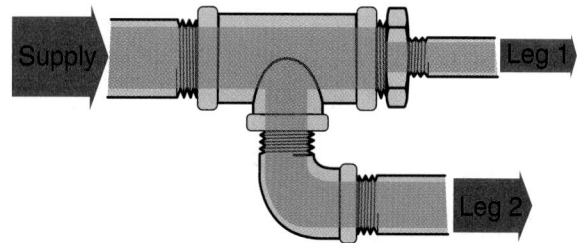

Figure 25.25 Comparing electron flow to water flow in a parallel circuit.

In a parallel circuit, the total resistance is less than the sum of the individual resistances. This is because the current has two paths it can take. This can be compared to water flow as in **Figure 25.25**. When the same voltage is applied to the different branches, the resistance within each branch will determine the current flow. Voltage is constant in all branches until it is dropped by the load.

Remember that in a series circuit, adding loads reduced the current. In a parallel circuit, adding loads will overload the circuit.

Series-parallel circuits combine the two types of circuits. Actually, the circuit shown previously in Figure 25.24 is the simplest kind of series-parallel circuit. The bulbs are the parallel part of the circuit, and the switch is in series with both of them.

Ohm's Law

The relationship among voltage, amperage, and resistance can be predicted using Ohm's law. **Figure 25.26** shows what happens as various parts of the equation are changed.

To use Ohm's law to determine what will happen when any two values are known, use the formulas in **Figure 25.27**. The sketch shows a typical pie chart that electronics technicians use. In the chart, *E* stands for volts. It is *E* instead of *V* because voltage is really electromotive force. Current is expressed as *I* (intensity).

OHM'S LAW RELATIONSHIP TABLE		
Voltage	**Resistance**	**Amperage**
Up	Down	Up
Up	Same	Up
Up	Up	Same
Same	Down	Up
Same	Same	Same
Same	Up	Down
Down	Down	Same
Down	Up	Down
Down	Same	Down

Figure 25.26 The relationship among voltage, amperage, and resistance can be predicted using Ohm's law.

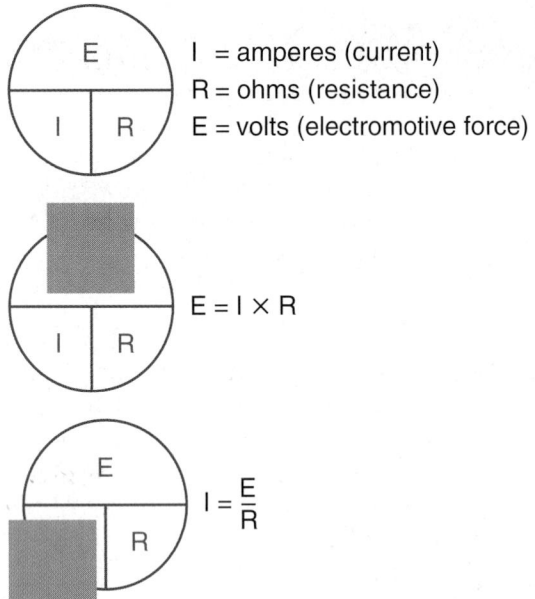

Figure 25.27 Ohm's law formulas.

I = amperes (current)
R = ohms (resistance)
E = volts (electromotive force)

$E = I \times R$

$I = \dfrac{E}{R}$

▰▰ MAGNETIC FIELDS

Magnetism is used in the automobile to produce electrical current in the charging system. It is also used to make the starter motor turn. An electronic fuel injection system or an electronic automatic transmission both use magnetism to operate injectors and solenoids.

A magnet is a material that will attract iron, steel, and a few other materials. Many of the laws that govern electricity also govern magnetism.

Magnets have polarity and if allowed to hang free will align themselves with north and south. The north-facing end is called the *north pole* and the south-facing end is called the *south pole*. Like poles repel each other and unlike poles attract (**Figure 25.28**).

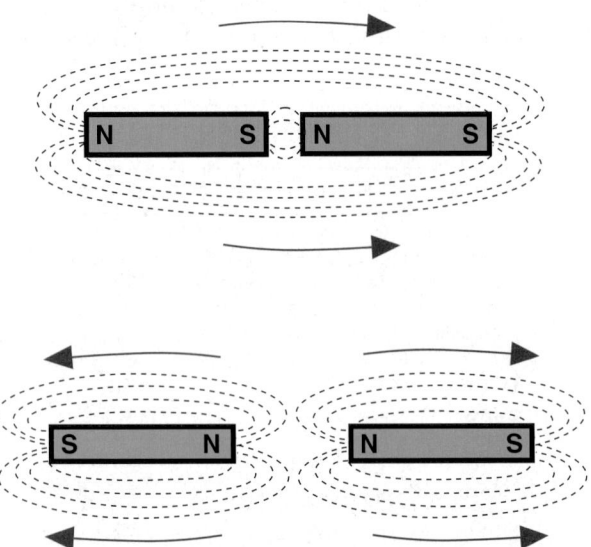

Figure 25.28 Like poles repel each other. Unlike poles attract.

A magnetic field is the space around the outside of the magnet that contains its lines of magnetic force. Magnetic lines of force can penetrate all objects. There is no known insulator of magnetic fields. The lines of force are detected only by other magnetic materials or another magnetic field.

Electromagnetism

A magnetic field is created around the outside of a conductor with electricity flowing in it (**Figure 25.29**). When the wire is coiled, the magnetic field is made stronger (**Figure 25.30**). Adding an iron core to the inside of the coil of wire will make a still stronger magnetic field (**Figure 25.31**). Increasing the number of coils increases the strength of the field. The strength of the field is also related to the amount of current flowing in the conductor (**Figure 25.32**).

Electromagnetic Induction. Electricity can be produced by moving a magnetic field over a conductor (**Figure 25.33**). This is the basic principle behind the operation of the alternator that charges the car's battery. Alternators are covered in a later chapter. **Electromagnetic induction** is also used to produce the spark used in the ignition system.

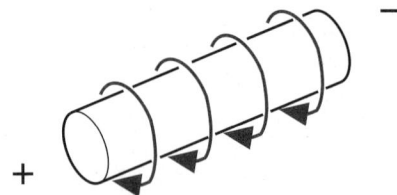

Figure 25.29 A magnetic field surrounds a conductor with current flowing in it.

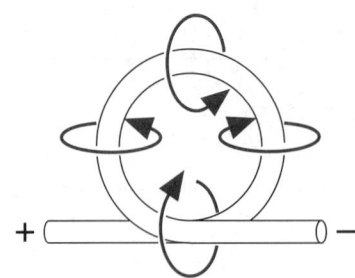

Figure 25.30 Coiling the conductor makes a stronger magnetic field.

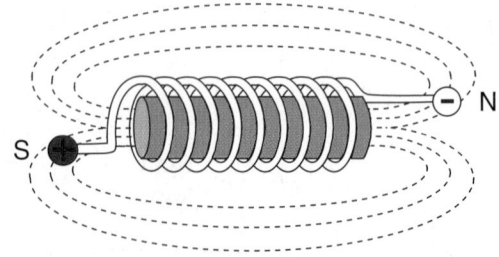

Figure 25.31 Adding an iron core concentrates the magnetic field.

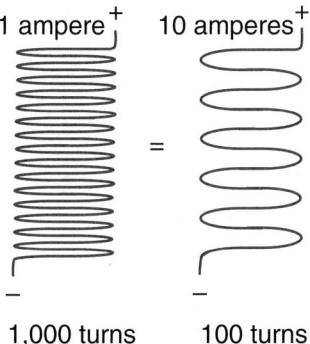

Figure 25.32 Magnetic field strength is determined by the amount of current flow and the number of coils.

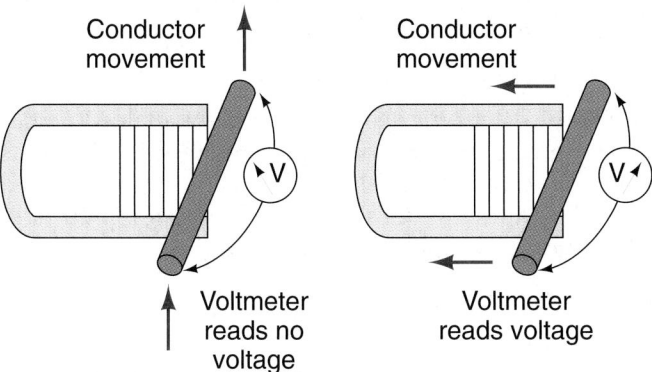

Figure 25.33 Electricity can be produced by moving a magnetic field over a conductor.

Relays

One application of magnetism is in a *relay.* A relay is a magnetically controlled switch. It is common to use a relay when a large load must be controlled by a small wire. Examples of circuits where relays are used include the starter motor and the horn. The small wire provides power to a coil of wire to create a magnetic field. The field pulls on an armature to close a switch (**Figure 25.34**).

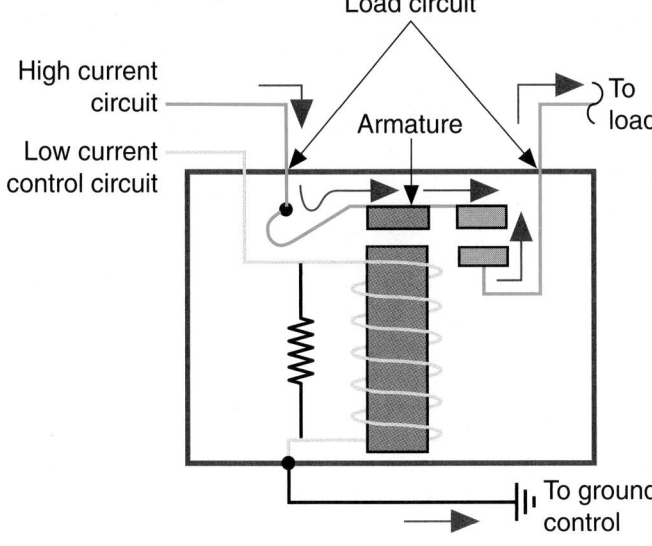

Figure 25.34 Operation of a relay.

■ CAPACITORS

A **capacitor** (condenser) stores electricity. Capacitors do not use energy but return all voltage to the circuit when they discharge. They are used to absorb voltage changes, called *voltage spikes,* in the system. This prevents damage to electronic components. In a DC circuit, electrons cannot flow through a capacitor.

Capacitors were used sparsely in the past in the ignition distributor or for silencing radio noise. Today's electronic cars have capacitors in most circuits. A capacitor is connected in parallel in a DC circuit (**Figure 25.35**). In an AC circuit, electricity will flow through a capacitor as if it were part of the wiring.

A capacitor is made up of two pieces of foil separated by an insulator (**Figure 25.36**). The ground side of the capacitor (case) is connected to one piece of foil.

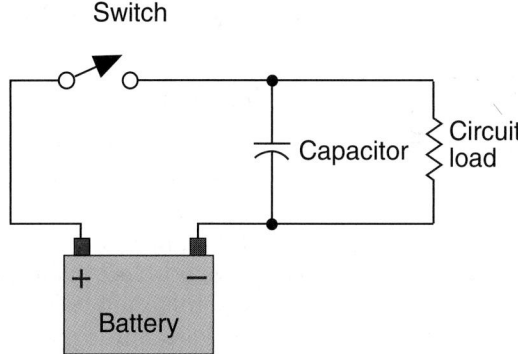

Figure 25.35 A capacitor connected in parallel in a circuit.

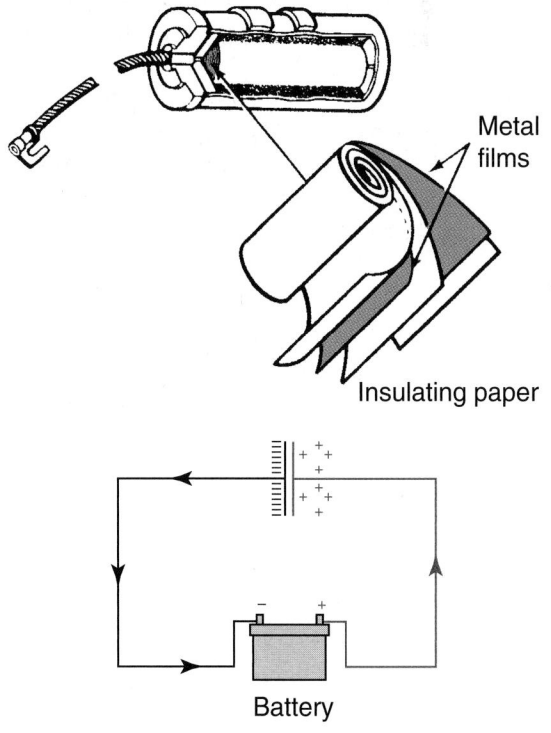

Figure 25.36 A capacitor has two layers of foil separated by an insulator.

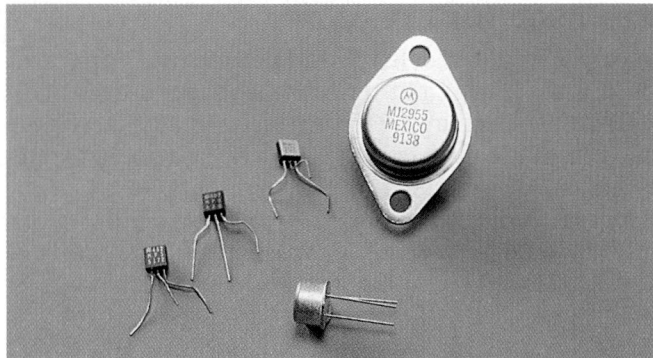

Figure 25.37 Transistors used in automotive circuits.

The positive side is connected to the other piece. Electricity always looks for the shortest path to ground. The capacitor's job is to fool the electricity so that it loads one of the foil pieces with electrons. They enter the capacitor because of the attraction of the other piece of foil's opposite charge.

■ AUTOMOTIVE ELECTRONICS

Electrical components use moving or mechanical parts. These can wear, burn, or pit and are slow when compared to electronic parts. Electronic systems use *solid state* parts, which have no moving pieces.

- A **semiconductor** can act as both an insulator and a conductor. Common semiconductor materials are silicon (Si) and germanium (Ge).
- A **diode** is an electronic one-way check valve. It allows electricity to flow in only one direction.
- A **transistor** (**Figure 25.37**) is an electronic relay. The word is a combination of the two words *transfer* and *resist*. It is a very fast switch with no movable parts or contacts that can burn. It can resist electrical flow or allow a predetermined amount of current to flow.

    Chapter 45 deals with automotive electronics in detail.

■ BASIC ELECTRICAL TESTS

Meters

A meter can be used to measure volts, ohms, and amps. Meters used for testing automotive electricity are either permanent magnet or digital. A permanent magnet meter is called an **analog meter**. This means it has a dial with a spring-loaded needle that is moved by a magnet (**Figure 25.38**). When three magnets are placed in a line, the center one will try to pivot into alignment by turning 180 degrees.

    The center magnet in the meter is an electromagnet. When its coil is energized, it operates on a second magnetic field produced by the permanent magnet. The electromagnet has a positive and a negative pole like the permanent magnet. Because like poles repel, the meter dial moves in proportion to the electricity flowing through it.

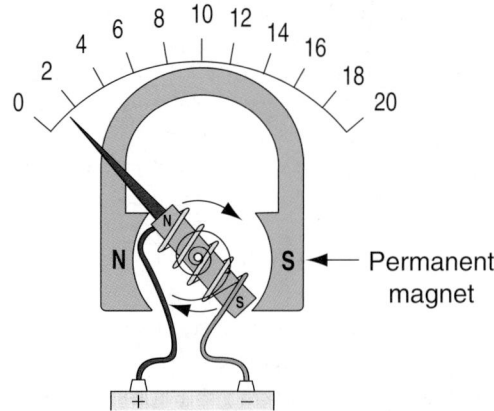

Figure 25.38 An analog meter has a dial with a needle attached to an electromagnet.

Vintage Electrical

Before the electronic age all meters were analog, including voltmeters, ammeters, and ohmmeters. These meters were found on vehicle dash panels and were also used in test equipment. Some types of test equipment still use analog meters, but most are digital.

Digital Volt Ohmmeter

A common technicians's tool is a digital multimeter (**DMM**) (**Figure 25.39**). When the meter has only

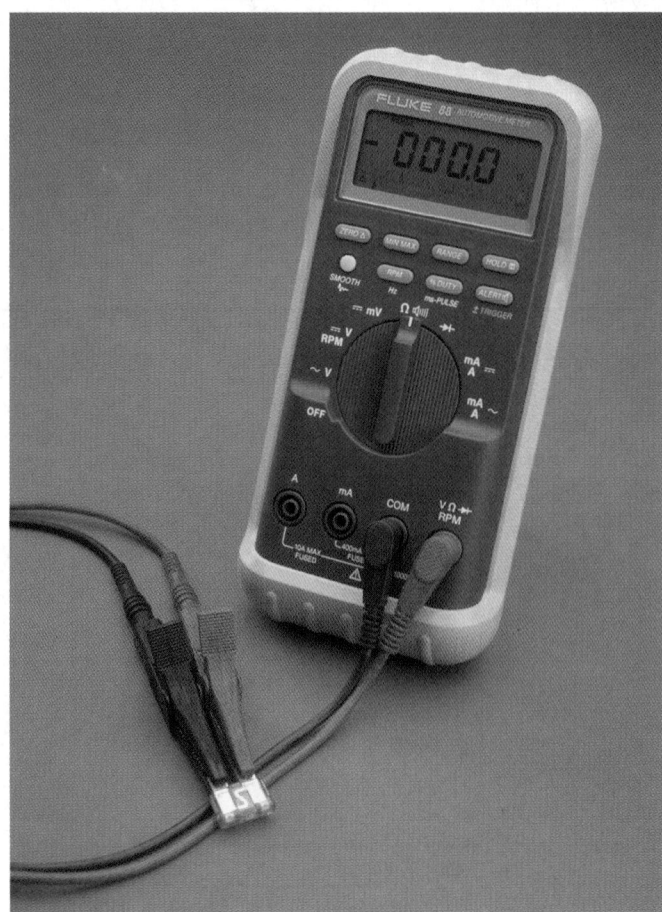

Figure 25.39 A digital multimeter.

a voltmeter and ohmmeter, it is called a digital volt-ohmmeter (**DVOM**). Digital multimeters also have an ammeter (usually only up to 10 amps maximum). Because of the low amount of amperage that can be measured, a technician will need another meter for measuring amperage. Starting and charging systems can have amperage flow in excess of 100 amps. A larger volt amp tester is needed for these systems (see Figure 27.37).

The digital multimeter is more popular today than the analog meter because of its sensitivity to electronic (very small amounts of electricity) circuits and its automatic features. Because computer systems run on very small amounts of electricity, most magnetic meters will load the circuit and change the test results. Digital meters have high input resistance and have a digital display like a watch. Having high input resistance prevents the meter from drawing current when connected to the circuit. This maintains the accuracy of the meter reading and prevents damage to the circuit.

NOTE: *An analog meter usually cannot be used to measure computer circuits unless it is one of the more expensive meters having more than 10 megohms (10 million ohms)* impedance *(internal resistance). Most analog meters have 20,000–30,000 ohms of impedance. They can increase amperage in the circuit and damage the component or the circuit. Using a meter with resistance this low can destroy an oxygen sensor, for instance.*

Digital meters are not polarity sensitive. This means they can be hooked up in either direction. The meter will simply give a positive (+) or negative (−) reading on the dial. Most newer digital meters are self-scaling, too. This means that the dial does not have to be adjusted to read a certain level of volts, ohms, or amps like on an analog meter.

Voltmeter

A voltmeter is used for several tests:

■ To measure available system voltage at the battery or alternator
■ For checking the difference in voltage between two points
■ For checking for excessive voltage drop due to resistance in a circuit

To read system voltage, the meter is connected to the circuit in parallel (**Figure 25.40**). The test leads (wires) are usually color coded, with red being positive and black negative. Connect the red lead to the positive side of the circuit and the black lead to ground. If an analog meter is connected backwards, the needle will deflect in the opposite direction. A digital meter will simply read (−) instead of (+).

A voltmeter can be used to check for voltage at several loads in a simple electrical circuit. In **Figure 25.41**, with the negative wire connected to a good ground, the voltmeter should display full battery

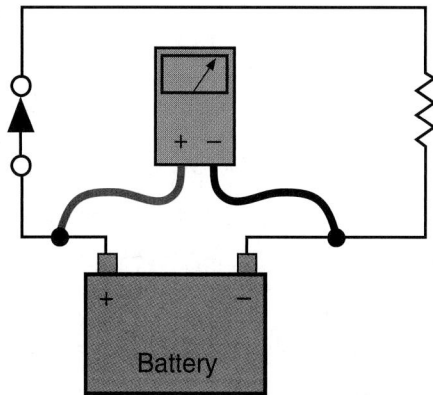

Figure 25.40 Voltmeter connection.

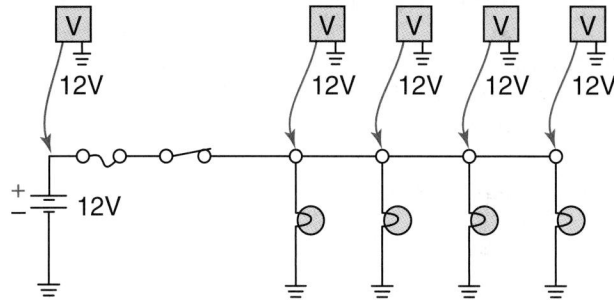

Figure 25.41 Using a voltmeter to check voltage at different loads in a circuit.

voltage when the positive wire is touched to various positive connections in the circuit. If there is a resistance in part of a lighting circuit, one or more of the lights might be out or fail to illuminate as brightly as the others and lower voltage would be evident on the meter.

Voltage Drop. A voltmeter can also be used to find resistance in a circuit. **Voltage drop** is a loss of voltage caused by current flow through a resistance. This measurement is widely used and is very popular with electrical service technicians. Voltage drop must be measured when the circuit is under load.

Kirchoff's law states that the total voltage drop in an electrical circuit will always be equal to the available voltage at the source. All of the voltage at the source must be dropped in the circuit before it returns to the source. When resistance is excessive somewhere in a circuit, it will result in a voltage drop across that portion of the circuit. Using Ohm's law to calculate a voltage drop is done by multiplying resistance by current.

To measure voltage drop, the voltmeter is connected to two places on the same side of the circuit (**Figure 25.42**). For instance, to test a battery terminal connection, one lead of a voltmeter is connected to the battery terminal and the other lead is connected to the battery post. When the engine is cranked, any reading on the voltmeter indicates that voltage is being used up, driving current through the

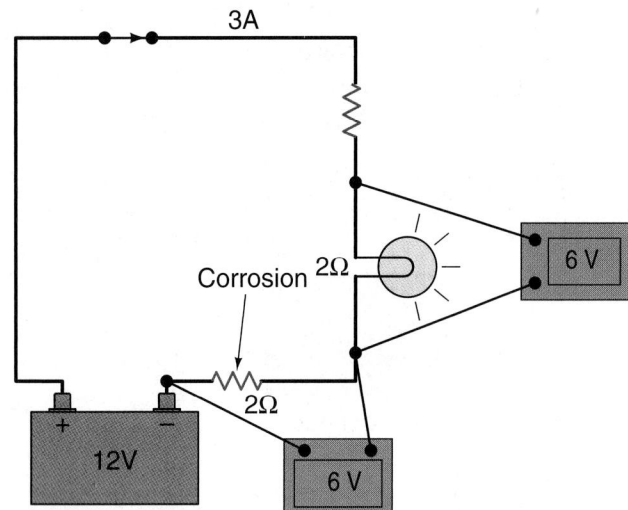

Figure 25.42 To measure voltage drops with current flowing, the voltmeter is connected to two places on the same side of the circuit. This circuit shows 2 Ω resistance from corrosion.

resistance. More than a specified amount means too much resistance. This test is handy because the system does not have to be disassembled to find a problem.

NOTE: *Results of a voltage drop test appear to be valid only in a circuit that carries higher amounts of current. In low-voltage computer circuits, there does not seem to be enough amperage to use this test. In computer circuits, significant voltage drops are measured in the .001V (mV) range.*

Ammeter

An ammeter measures amperage or current flow in a circuit. To measure water flow, a flowmeter would have to be installed in series in the water line. The same thing is true to measure electrical current flow. The ammeter must be hooked in *series* with the electrical load (**Figure 25.43**) or an *inductive pickup* must be used (**Figure 25.44**). An inductive pickup encircles the outside of a wire and measures the amount of electrical flow using the principles of electromagnetism.

Testing amp draws is done with the system under load. Using a meter that does not have enough capacity for the load will result in a blown fuse.

Current Probe. Many technicians use a digital lab scope or graphing meter for electrical diagnosis. Current probes are available for low-amp and high-amp uses. Some have their own integral meters. The tool shown in **Figure 25.45** is an inductive clamp that converts current flow in a wire or cable so it can be read by another digital device. More information on lab scopes and graphing multimeters is found in Chapter 46.

Typical Amp Draw. The following list from the *Battery Council International (BCI)* gives the current (amps) loads that a typical passenger car might have.

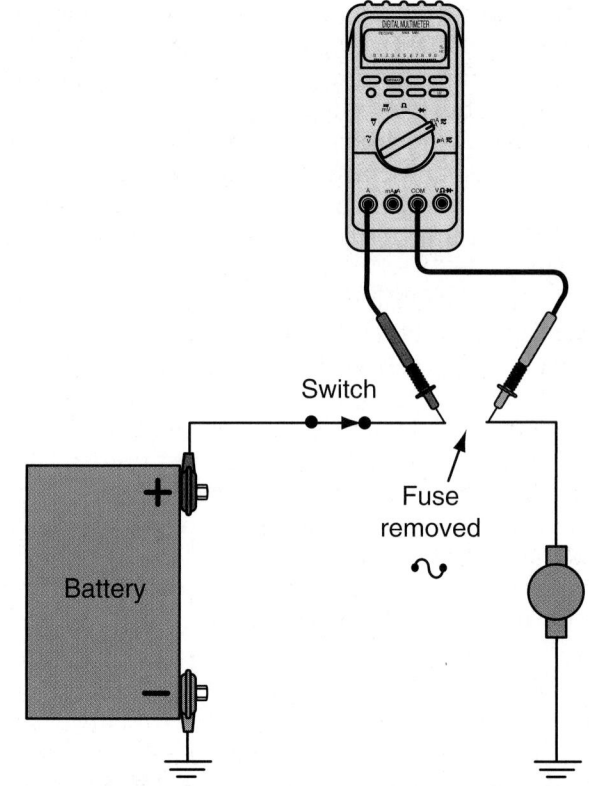

Figure 25.43 An ammeter is connected in series so electrical current can flow through it.

Typical Current Loads	
Load	**Amp Draw**
Stoplights	8
Interior Lights	2
Blower (low)	6
(medium A/C)	16
(high A/C)	35
Heated Back Window	22
Ignition	6
Radio	0.5
Windshield Wipers	7.5
Headlights (low beam)	9
(high beam)	13
Parking Lights	7
Base Load with A/C	
(summer)	50
(winter)	45
Summer Starting	
(gas)	150–250
(diesel)	450–550
Winter Starting (very low temperatures)	
(gas)	250–350
(diesel)	700–800

Figure 25.44 An ammeter with an inductive pickup.

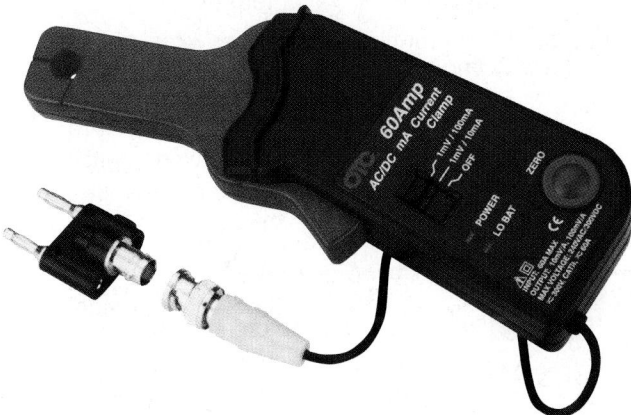

Figure 25.45 A current probe is an inductive clamp that converts current flow in a wire or cable so it can be read by another digital device. *(Courtesy of OTC/SPX Service Solutions)*

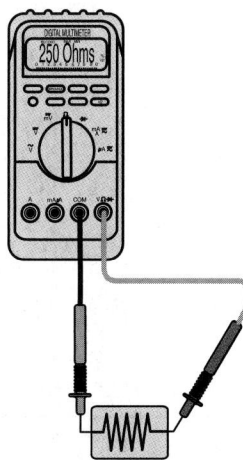

Figure 25.46 An ohmmeter measures resistance by measuring the amount of current that can be forced by a small voltage through a circuit or device.

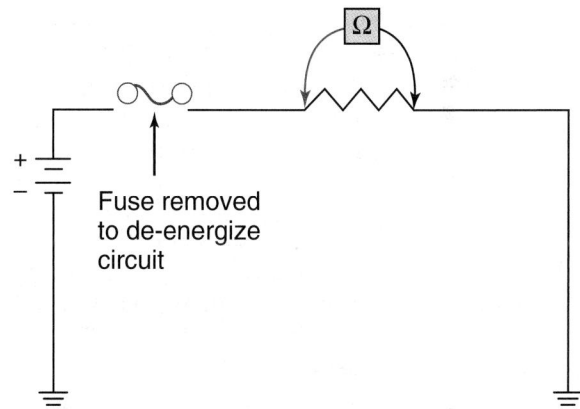

Figure 25.47 An ohmmeter is used with the power off.

Ohmmeter

An ohmmeter measures the amount of resistance in an electrical circuit by measuring the amount of current (amperage) that can be forced to go through a circuit by a small voltage (**Figure 25.46**). To do this, it must have its own battery.

Because batteries do not always have the same state of charge, the ohmmeter must either be a self-calibrated type or it must be hand calibrated to adjust its output voltage before it is used. An ohmmeter can be hand calibrated by connecting its wires together to read 0 resistance. Some meters have an indicator that tells when the battery is low.

NOTE: *Never connect an ohmmeter across an energized circuit. No electricity can be flowing in it (**Figure 25.47**).*

Circuit Problems

- An *open circuit* is when there is a break in the path for electrical flow, an incomplete circle.
- A *short circuit* is when the electrical path has been shortened, when wires accidentally touch each other or grounded metal, for instance (**Figure 25.48**).

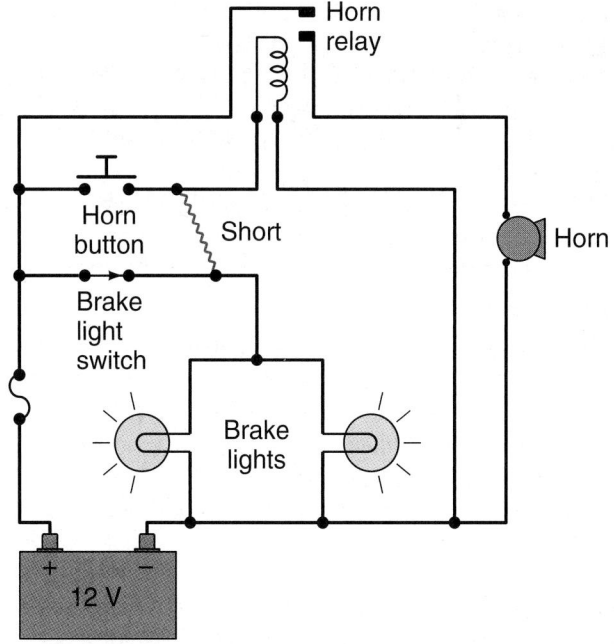

Figure 25.48 A short circuit.

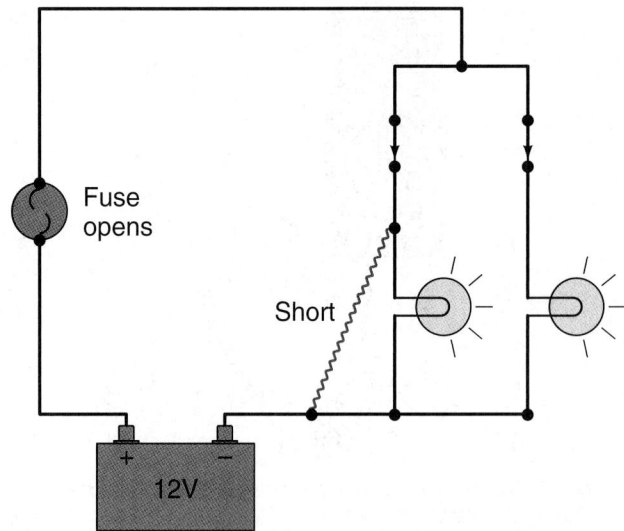

Figure 25.49 A grounded circuit.

- A *grounded circuit* is like a short circuit, but the current goes directly to ground (**Figure 25.49**). Notice the ground symbol in the sketch. A short to ground lowers circuit resistance and allows very high current flow that can damage wiring and circuit loads.

Jumper Wires

A jumper wire is a simple wire with alligator clips attached to its ends (**Figure 25.50**). It is used primarily for finding **open circuits**, or "opens". **Figure 25.51** shows how to eliminate different places in the circuit as possible causes of an open circuit. An open is when there is a break in a circuit's continuity.

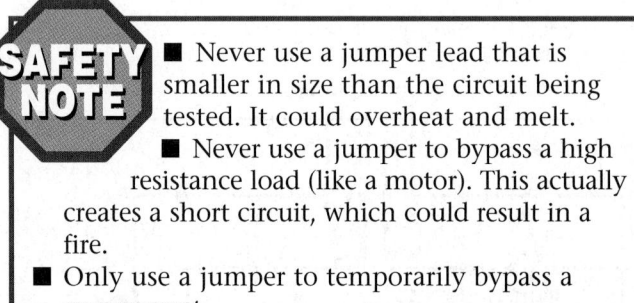

SAFETY NOTE
- Never use a jumper lead that is smaller in size than the circuit being tested. It could overheat and melt.
- Never use a jumper to bypass a high resistance load (like a motor). This actually creates a short circuit, which could result in a fire.
- Only use a jumper to temporarily bypass a component.

Test Lights

A 12-volt test light is a handy technician's tool (**Figure 25.52**). To use a test light:
- Test its operation by connecting it across the battery.
- Connect the alligator clip to a good ground.
- Probe where necessary with the pick end (see Chapter 46 for more information on backprobing electronic connectors).

Testing for opens can be done with a test light. With a jumper lead, the process was to try to bypass the open to operate the circuit. With a test light, the process is to locate where voltage is in the circuit (**Figure 25.53**).

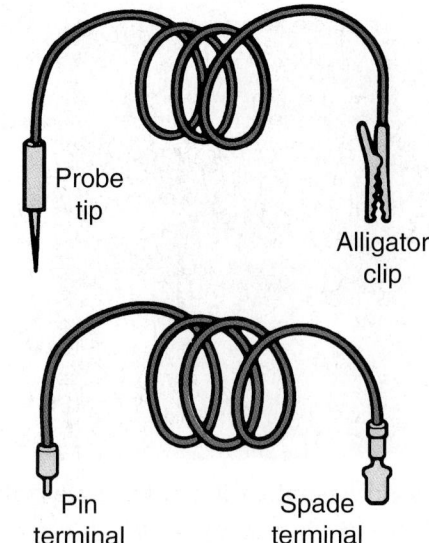

Figure 25.50 Different kinds of jumper wires. (*Courtesy of Ford Motor Company*)

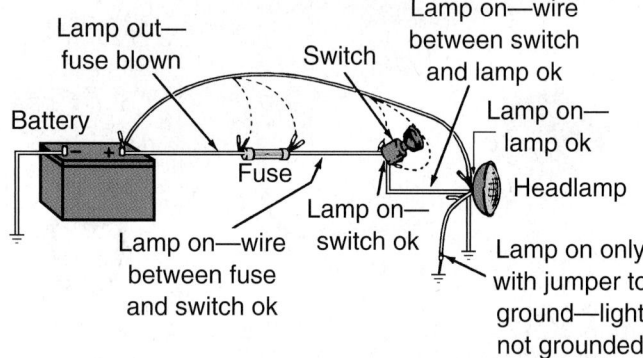

Figure 25.51 Using a jumper wire to find an open circuit. (*Courtesy of Ford Motor Company*)

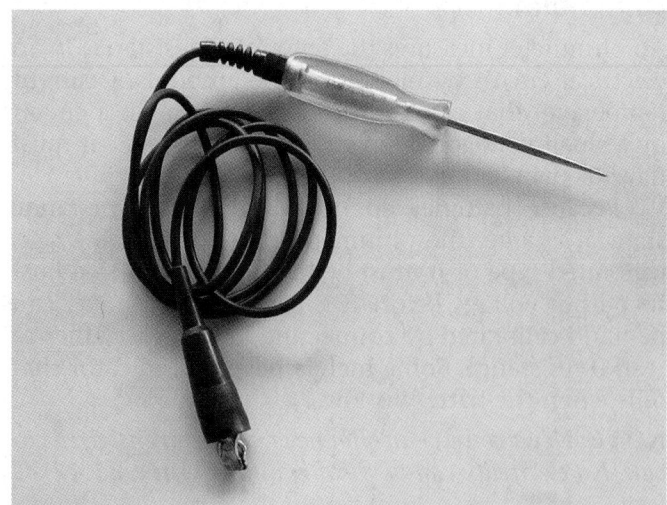

Figure 25.52 A 12-volt test light. (*Courtesy of Tim Gilles*)

Self-Powered Test Light. A self-powered test light is like a flashlight. It has its own battery and is used to test for continuity much like an ohmmeter. The tester can be used for testing for open or short circuits when

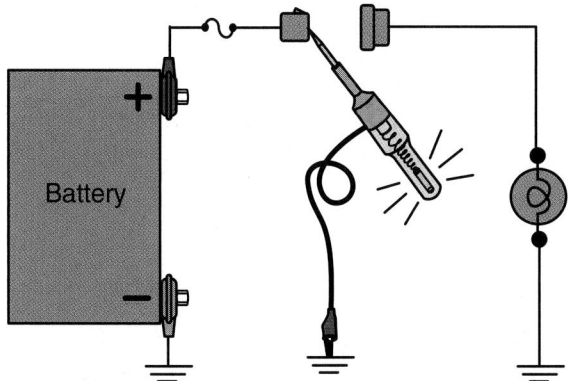

Figure 25.53 Checking for opens with a test light.

power is disconnected. If the light glows, the circuit or part has *continuity*. **Figure 25.54** shows how a self-powered test light can be used to test a light filament. Testing for a complete ground circuit is done as shown in **Figure 25.55**. Other tests and repairs to electrical wiring are covered in Chapter 33.

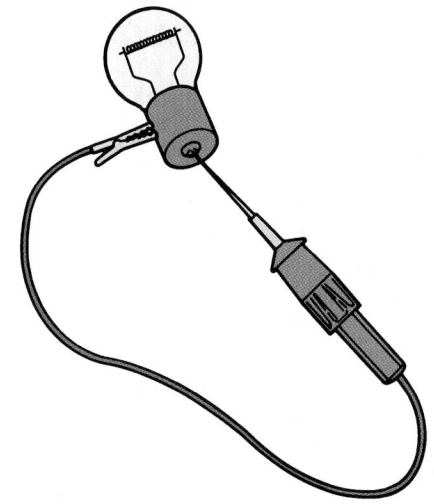

Figure 25.54 Using a self-powered test light to check a light bulb.

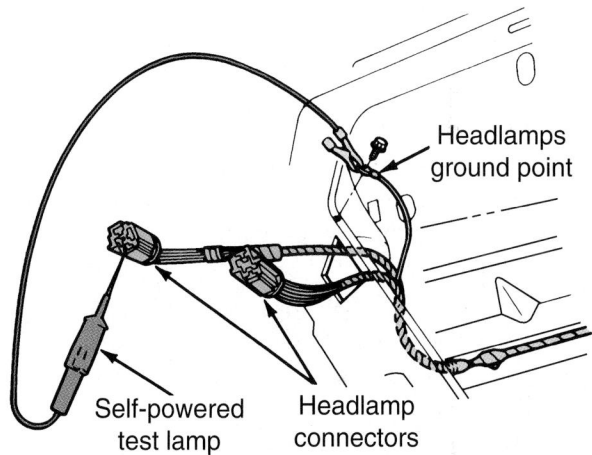

Figure 25.55 If the ground circuit is complete, the self-powered test light will light up. *(Courtesy of Ford Motor Company)*

■■ REVIEW QUESTIONS

1. What is the name for the materials that are made of atoms that allow electricity to flow freely?

2. An _____ prevents the flow of electrons between two conductors.

3. Name three good insulators.

4. What is the name for a complete circle provided for electrical flow?

5. What are the names of two circuit protection devices?

6. How fast does light travel?

7. What kind of current is used for houses, AC or DC?

8. What is the unit of electrical resistance measurement called?

9. If there is more resistance in a circuit, what happens to current flow?

10. What is the name of the electrical part that is used to change the amount of current that flows to a heater motor?

11. What kind of circuit has current flowing in a line first to one load and then on to the next one?

12. What kind of circuit has different branches that current can flow through starting from a common point?

13. What happens when a magnetic field is moved over a conductor?

14. When a large load must be controlled by a small wire, what device is used?

15. Which kind of meter is used only on a circuit that has no electrical power: voltmeter, ohmmeter, or ammeter?

■ ASE-STYLE REVIEW QUESTIONS

1. Technician A says that voltage is electrical pressure. Technician B says that amperage is electrical flow. Who is right?

   a. Technician A **c.** Both A and B

   b. Technician B **d.** Neither A nor B

2. Which of the following can result from voltage that is too low?

   a. A light that is dim

   b. A heater defroster motor that spins slower

   c. Both A and B

   d. Neither A nor B

3. All of the following are true about electrical systems *except*:

   a. Direct current (DC) is used in automobile electrical systems.

   b. Magnetic poles attract each other and unlike poles repel.

   c. Current flow is measured with an ammeter.

   d. A diode is an electrical one-way check valve.

4. Technician A says that a rheostat varies voltage through a circuit. Technician B says that a potentiometer varies the current flow in a circuit. Who is right?

   a. Technician A **c.** Both A and B

   b. Technician B **d.** Neither A nor B

5. Which of the following is/are true about resistance in an electrical circuit?

   a. The longer a conductor is, the more resistance it has.

   b. The larger a wire's diameter, the less resistance it has.

   c. Resistance determines the amount of current that can flow in a circuit.

   d. All of the above.

6. Loads are being added to a circuit. Technician A says that in a series circuit, current flow will be reduced. Technician B says that additional loads can overload a parallel circuit. Who is right?

   a. Technician A **c.** Both A and B

   b. Technician B **d.** Neither A nor B

7. Which of the following is/are true about magnetic fields?

   a. Lead is a good insulator for magnetic fields.

   b. A magnetic field can be made stronger by coiling a current-carrying wire.

   c. Both A and B.

   d. Neither A nor B.

8. Which of the following is an electronic relay?

   a. A diode

   b. A transistor

   c. Both A and B

   d. Neither A nor B

9. Which of the following is/are true about digital meter?

   a. It has a dial readout.

   b. It is the best choice for electronic circuit testing.

   c. Both A and B.

   d. Neither A nor B.

10. Technician A says that a voltmeter can be used to check resistance in a circuit. Technician B says that an ohmmeter can be used to check resistance in a circuit. Who is right?

    a. Technician A **c.** Both A and B

    b. Technician B **d.** Neither A nor B

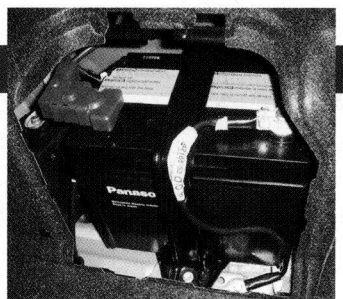

Battery Fundamentals

KEY TERMS

element	reserve capacity	low-maintenance
battery terminal	deep-cycle	battery
CCA	maintenance-free battery	sulfation

INTRODUCTION

The car's battery is the heart of the vehicle's electrical system. It converts electrical energy received from the charging system into chemical energy that it stores. Then, it converts the chemical energy back into electrical energy.

- When the engine is off, the battery provides power for electrical accessories that are turned on.

- During cranking, the battery supplies current to the starter. When the engine is running and electrical demands are more than the charging system is capable of contributing, the battery serves as a reservoir, supplying electrical energy for the extra loads.

- The battery acts like a capacitor, absorbing high voltages that happen in the electrical system. These high voltages would damage other electrical system components if the battery did not serve this function.

When a *load*, such as a light bulb, is connected to the positive and negative terminals of the battery, the bulb will light up because electrons produced by a chemical reaction in the battery are flowing through its filament (**Figure 26.1**). Chemical energy is being converted to electrical energy.

The largest load is the starter motor, which draws a good deal of current from the battery. The starter is the

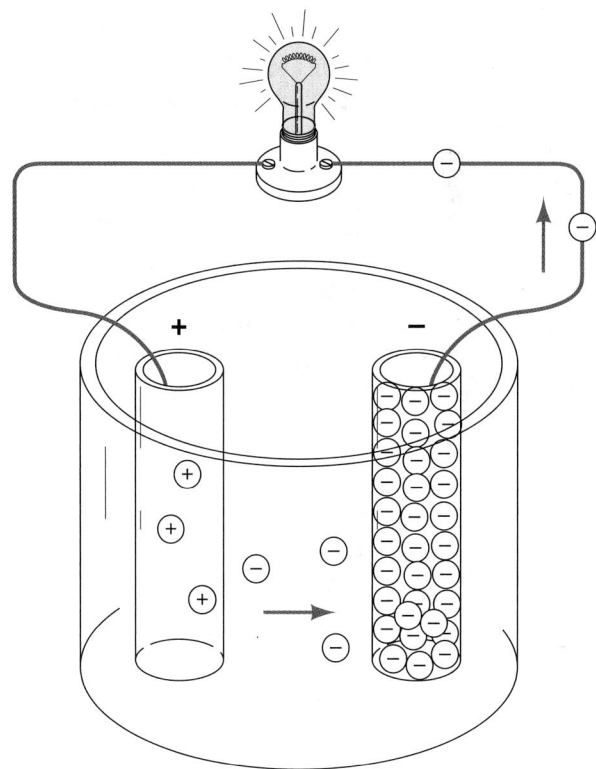

Figure 26.1 When a load is placed between the positive and negative battery terminals, electrons flow.

reason that the battery cables are as large as they are. If they were smaller, they would get hot as the electrons tried to flow through them to the starter. The starter draws more than 100 amperes, but the charging system is only capable of generating 35 to 100 amps, depending on its alternator's capacity.

After the starter is used, the battery voltage is momentarily low. After the engine starts, the battery needs to be recharged. It is recharged by the charging system. This is a conversion of electrical energy to chemical energy; the opposite of what took place when the battery was discharged. With all accessories running, the charging system may not be able to provide all the electricity needed. In this case, the battery will supply the extra energy needed until the load is reduced or the battery fails.

■■ BATTERY PARTS AND OPERATION

When two different metals are immersed in an electrolyte solution, a direct current (DC) voltage is produced. You can do this using a potato, an orange, or a soft drink as the electrolyte. Poke two types of metal into the potato and use a voltmeter to read the voltage between them.

Different metals used to construct other types of batteries provide various voltages. For instance, a flashlight battery's 1.5 volts comes from the reaction of the electrolyte and carbon (the positive carbon center rod) and zinc (the negative battery case). A standard flashlight has two batteries stacked together in *series* to provide the 3 volts required to light its bulb. Notice how the positive and the negative flashlight battery terminals touch each other.

In automotive batteries, 2.1 volts is the amount that the two dissimilar materials of the plates provide. The typical 12-volt automotive battery has six 2.1-volt *cells* hooked to each other in series (from positive to negative to positive). This provides about 12.6 volts when the battery is fully charged (**Figure 26.2**).

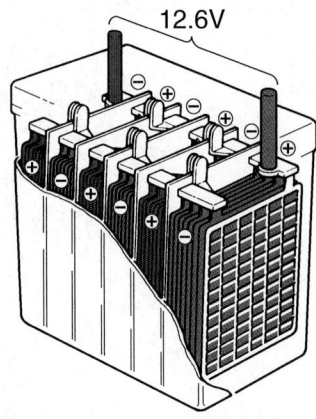

12.6V

Figure 26.2 An automotive battery has six cells connected together in series.

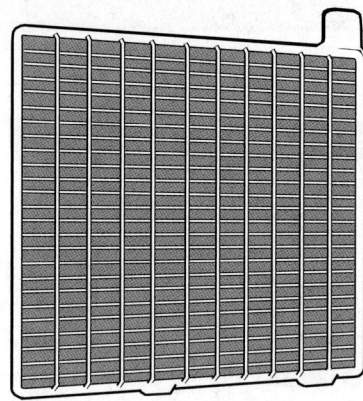

Figure 26.3 A battery grid. *(Courtesy of Ford Motor Company)*

Plates

Battery plates are constructed of grids with horizontal and vertical bars (**Figure 26.3**). A paste is pressed into the grid. The paste is made up of lead oxide, acid, and material expanders. After a forming charge is given to the positive plate, its lead oxide turns into lead peroxide. When the negative plate is given its forming charge, the same paste turns into sponge lead. This provides the two dissimilar metals needed to make a voltage when they are immersed in electrolyte.

Electrolyte

The battery's case is filled with an electrolyte mixture of *sulfuric acid* and *water*. Electrical current in the battery is produced by a chemical reaction between the *battery acid* (another name for electrolyte) and two different types of lead material on the battery's positive and negative *plates*.

Battery Cells

Each battery cell consists of a packet of several positive and negative plates (**Figure 26.4**) insulated from each other and connected in *parallel* (positive to positive and negative to negative). The positive and negative plates in the cell are held together in an **element** by plate straps (**Figure 26.5**). The element is submerged in electrolyte. Each cell is then hooked to its neighboring cell in series (the positive plates of one cell are hooked to the negative plates of the next cell) (see Figure 26.2).

Battery Terminals

The battery has positive and negative terminal connections made of lead. Automotive **battery terminals** will be either on the top or on the side (**Figure 26.6**). When battery terminals are on top, they are called *battery posts* (Figure 26.6a). The positive post is larger in diameter than the negative post. The top of the battery case is usually molded with labels that say "positive" or "negative." Sometimes, the top of the larger positive

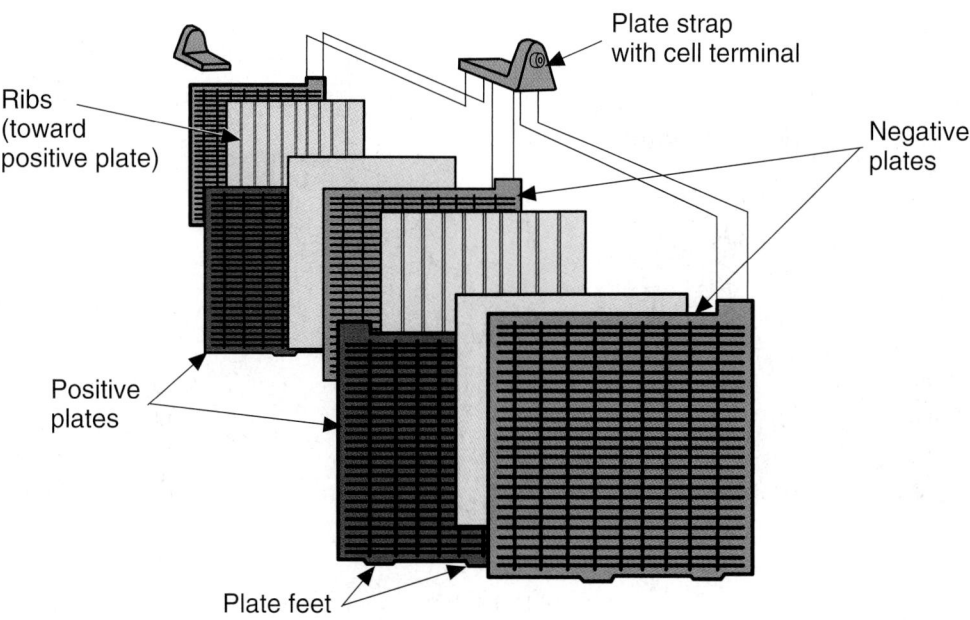

Figure 26.4 A battery cell has alternating positive and negative plates. *(Courtesy of Ford Motor Company)*

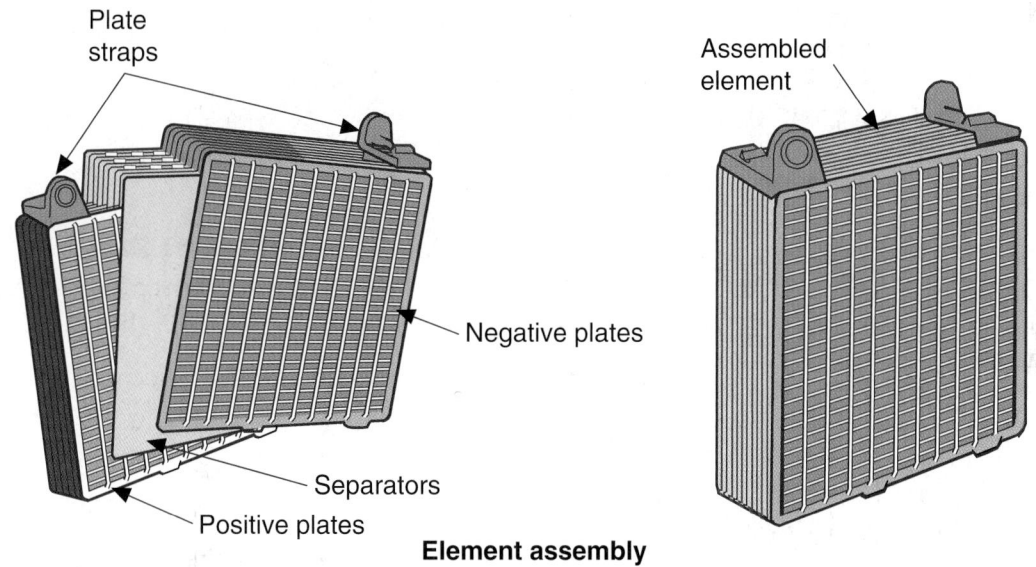

Element assembly

Figure 26.5 The cell is made up of a battery element. *(Courtesy of Ford Motor Company)*

post is painted red and can have a "+" stamped into it. The top of the smaller negative post can be painted black and have a "−" stamped into it.

Some batteries have *side terminals* that have reinforced internal threads (Figure 26.6b). A special terminal adapter bolt threads into the terminal.

Batteries with *L terminals* are found on marine applications, motorcycles, and some imported cars.

Battery Case

Most battery cases are constructed of lightweight plastic (polypropylene) with a one-piece cover. Cases can also be made of hard rubber and other plastics. The case must be able to resist acid absorption and

withstand extreme changes in temperature. It must be durable enough to resist vibration. The bottom of the case often has raised supports (**Figure 26.7**). The cell elements sit on top of these.

As a battery oxidizes, the plates shed (slough) material. Positive plates shed more material than negative plates. The loose plate material could cause a short between the cell positive and negative plates. Sloughed-off plate material collects in the bottom of the case between the raised cell supports. This area is called the *sediment chamber*. One problem with having the plates rest on the cell supports is that vibration has more of an effect on the plates. Another disadvantage is that the electrolyte that surrounds the loose material in

Figure 26.6 Different types of terminal connections. *(Courtesy of Tim Gilles)*

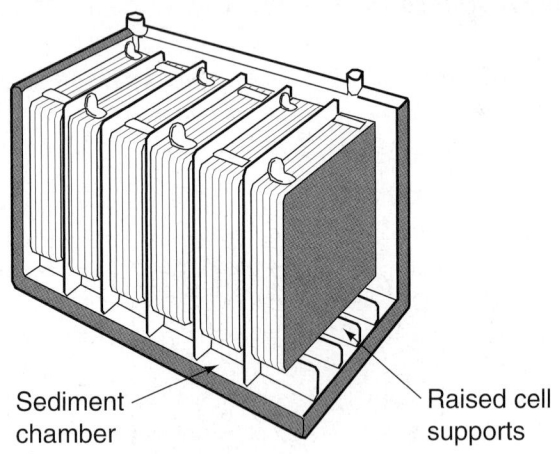

Sediment chamber

Raised cell supports

Figure 26.7 Cells in some batteries rest on raised supports in the bottom of the battery case. *(Courtesy of Ford Motor Company)*

Figure 26.8 This battery has removable cell caps.

the bottom of the battery is inactive. A bigger case is required to hold the desired amount of electrolyte.

Some cell elements have plastic envelopes that fit around the cell plates. The envelope, collects material that falls off the plates. Electrolyte can flow through the plastic envelope, but plate material that is shed is contained by it. The envelope prevents shed material from causing a short between the positive and negative plates. With envelopes around the plates, adding water is required less often because there is more reserve water above the plates. The envelopes also allow the battery case to be made smaller because a large sediment catch basin is no longer necessary.

Cell Caps

As a battery charges, hydrogen gas is released, so the case cover must have vents. Although many batteries today are maintenance-free (without cell caps), most modern battery tops have removable cell caps (**Figure 26.8**). Maintenance-free batteries are covered later in this chapter. In batteries with removable cell caps, the vents are in the caps. The caps also act as spark arrestors.

Charging and Discharging

As a battery is charged, current flows into it from the vehicle's charging system. The battery stores the energy until it is needed. When a battery is charging, the alternator puts electrons (negative charges) on the negative plate. The result is that the positive and negative plates have a difference in voltage (called electrical pressure or potential).

SAFETY NOTE Hydrogen gas is very explosive. It is lighter than air, so it gets trapped under the battery terminal connections and cell caps as the battery charges. Battery explosions put over 15,000 people a year in the hospital.

When the battery is discharging, current flows out of it. When this happens, it is releasing stored energy. The positive and negative plates in the cell have separators between them. These allow electrolyte to flow but insulate between the positive and negative plates and prevent them from touching.

The voltage imbalance between the positive and negative plates can only be equalized through a current path outside of the battery. When an electrical load is placed between the positive and negative terminals on the battery, a chemical reaction takes place. Current flows to the load as the voltage difference between the plates tries to neutralize. Chemical action of the battery allows electrons to build up on the negative plate. Electrons pass from negative to positive through the electrical load (**Figure 26.9**).

A fully charged positive plate is a combination of lead and oxygen called lead dioxide (PbO_2). The negative plate is sponge lead (Pb). The electrolyte is sulfuric acid (H_2SO_4) and water.

Discharge.   During discharge (**Figure 26.10**) the lead (Pb) from the positive plate combines with the sulfate (SO_4) from the battery acid. This forms lead sulfate ($PbSO_4$) on the plate. Oxygen in the positive plate combines with hydrogen in the electrolyte to form water

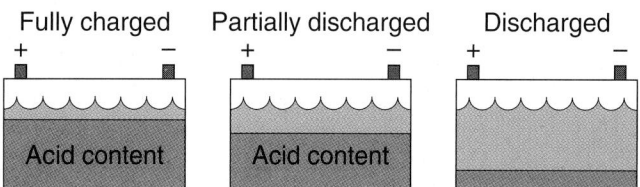

Figure 26.11 As a battery discharges, the strength of the electrolyte becomes weaker. *(Courtesy of Interstate Batteries)*

(H_2O). This results in a diluted concentration of electrolyte (**Figure 26.11**).

Positive and negative plates are becoming alike as the battery discharges. The negative plate (Pb) combines with sulfate (SO_4) to form lead sulfate ($PbSO_4$) on the plate. When the cell is dead, the positive and negative plates are identical. As the battery discharges, additional water forms in the electrolyte to dilute it.

Charge.   During charging (**Figure 26.12**), the process is reversed. The lead sulfate in both the positive and negative plates is changed back to lead and sulfate as it was originally. The battery "gasses"; the water in the electrolyte is split into hydrogen and oxygen. The negative plates give off hydrogen gas, and the positive plates give off oxygen, which is a very explosive combination. The hydrogen (H_2) combines with the sulfate (SO_4) to become sulfuric acid (H_2SO_4) once again. The leftover oxygen as the water separates combines with the positive plate to form lead dioxide (PbO_2). The battery's plates and electrolyte are now back in their original form and the battery is charged.

NOTE: *The battery gasses the most as it reaches full charge.*

As a battery is repeatedly charged and discharged, the active material on the cell plates is slowly worn away. The plates oxidize, and more plate material is sloughed off. Finally, the battery must be replaced.

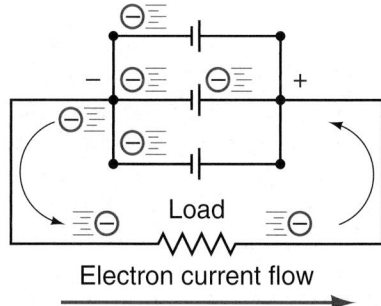

Figure 26.9 Electrons pass through the electrolyte solution as they move from the negative plates to the positive plates. *(Courtesy of Interstate Batteries)*

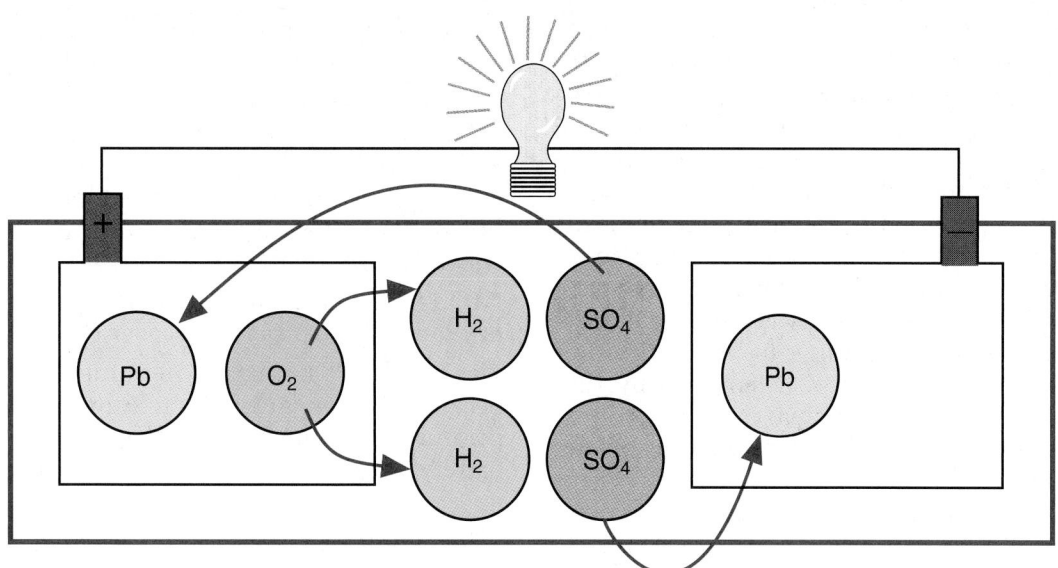

Figure 26.10 Chemical action during discharge of a battery.

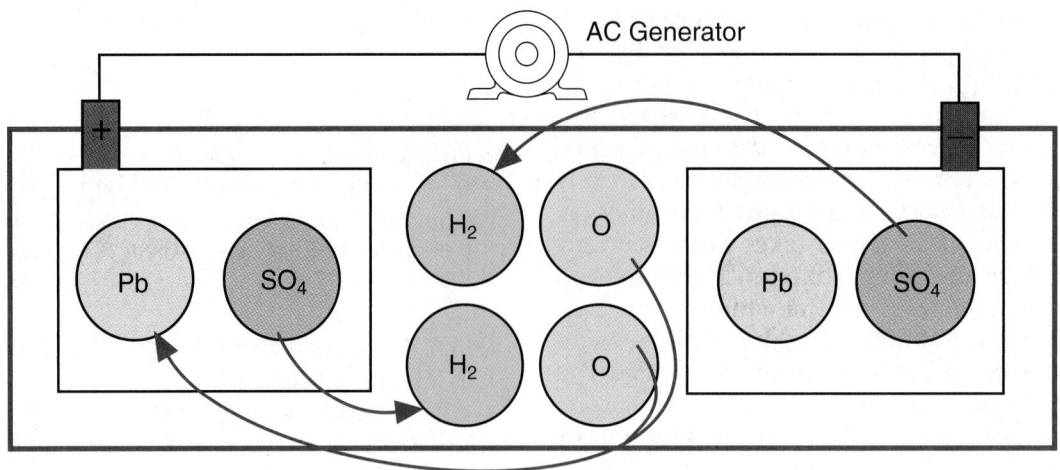

Figure 26.12 Chemical action during recharge of a battery.

With the engine off, most good batteries can provide about 25 amperes for 2 hours before discharging to the point where the engine will not start. The amount of energy that the battery is capable of providing depends on the amount of plate surface material and the concentration of acid in the electrolyte.

■ BATTERY CAPACITY RATINGS

Battery capacity ratings are established by the Battery Council International (BCI) and the Society of Automotive Engineers (SAE). Battery capacity is related to the following factors:

■ The surface area of the plates
■ The weight of the active materials on the plates
■ The strength of the electrolyte solution

A battery's current capacity rating reflects its ability to deliver cranking power to the starter motor and provide reserve capacity to the electrical system.

Manufacturers in the United States primarily use the CCA rating. Other auto manufacturers use other rating systems. In Europe one battery rating is called European Norm (EN). Another, used in Germany, is called (DIN). European and Asian manufacturers use amp hour (AH) and reserve capacity (RC) ratings as well.

Cold Cranking Amp (CCA) Rating

The most common method of rating automotive batteries in the United States is cold cranking amps (**CCA**). CCA is the common standard for low-maintenance batteries. The previous standard, called the *amp-hour* rating, is no longer widely used, except in some Asian cars.

For a 12-volt battery, CCA is determined by the number of amperes a battery can deliver for 30 seconds at 0°F (−17.7°C) without terminal voltage falling below 7.2 volts.

A battery should have at least 1 CCA for each cubic inch (cu. in.) of engine displacement (61 cu. in. = 1,000 cc or 1 liter). The formula for metric is 1 CCA for each 16 cubic centimeters (cc) of displacement. A metric engine with 1,600 cc of displacement will require a battery of 100 CCA. Engines with more accessories require more CCA. The usual range for passenger cars and light trucks is between 300 and 600 CCA. Some batteries have a rating as high as 1,100 CCA.

Reserve Capacity

Another test of a battery's capacity is **reserve capacity**, also called staying power. This is a measurement of the battery's ability to provide current when there is no electricity from the charging system. Reserve capacity gives an indication of how long a vehicle can be driven after a charging system failure. It is also an important rating when selecting **deep-cycle** batteries for recreational vehicles. This test is done under the most strenuous conditions and is usually not applied to car batteries.

Reserve capacity is determined by the length of time in minutes that a battery can be discharged at 25 amps without its individual cell voltage dropping below 1.75 volts. A battery with a reserve capacity of 100 would be able to deliver 25 amps for 100 minutes before the voltage would drop below 10.5 volts (6 cells × 1.75 volts).

Watt-Hour Rating

Some battery manufacturers rate their batteries in watt-hours. This is because the battery's electrical energy is converted into mechanical energy (watts).

NOTE: *A watt is the unit of electrical power that is equivalent to horsepower. One horsepower is equal to 746 watts.*

The watt-hour rating is determined at 0°F (−17.7°C) because a battery's capacity in watts changes with temperature. The rating is determined by multiplying a battery's amp-hour rating by the battery voltage.

■ BATTERY TYPES

Batteries are designed for specific uses. The state of charge of an automotive battery is kept about 90% to

100% most of the time. If the battery's state of charge falls below 75%, the battery needs to be recharged. Allowing a battery to be discharged until dead or near dead will shorten its life.

Batteries in automobiles are of three main types: maintenance-free (low water loss), low-maintenance and deep-cycle (R.V.). A recreational vehicle or a boat's electric trolling motor would need to supply a relatively small amount of current for long periods of time. A passenger car, on the other hand, would require a large amount of current draw during cranking, followed by a period of recharge.

Battery grids are made of lead. But lead by itself is too soft, so it must be combined with another metal to make the grid material harder. Conventional deep-cycle batteries have grids constructed of lead and antimony. Low water loss batteries have grids constructed of lead combined with either calcium, cadmium, strontium, or lower concentrations of antimony.

Conventional Deep-Cycle Batteries

Conventional lead antimony batteries are the original battery design that was used in cars for many years. They are very good for deep cycles, have high CCA ratings, and are easily recycled. A disadvantage is that they can continue to accept up to 5 amps of current during a recharge, even though they are fully charged. Because of this, they are no longer used in passenger cars. They use considerably more water, and they gas heavily during overcharging. The gas results from electrolysis, which is when water is separated into hydrogen and oxygen. The hydrogen gas, which is dangerous, must be vented outside of the battery.

When a battery gases, corrosion of the terminals, holddowns, and battery tray occur from the sulfuric acid fumes in the escaping gas. Lead antimony batteries also have considerable terminal corrosion and require periodic charging when stored.

These batteries can go through many deep cycles. A deep cycle is when the battery is allowed to run almost completely dead and then is recharged. Deep-cycle batteries have a thicker layer of paste on their plate grids and can stand hundreds of repeated discharges with very little damage. The thick plates are porous and can give off more electrons, but they do this more slowly. Thicker plates results in less surface area in these batteries than would be the case if there were more, thinner plates.

Antimony, used in these batteries, is a stiffening material for the grids. It is not as resistant to overcharging as the other grid materials. It is temperature sensitive; as it gets hot, its resistance drops. This results in an increase in the breakdown of water into hydrogen and oxygen. Antimony becomes very hot if the charging voltage is allowed to climb above 14.5 volts. As more of the water is gassed from the battery, the plates get even hotter.

Maintenance-Free Batteries

Maintenance-free batteries (**Figure 26.13**) have cell plates made of a slightly different material. Calcium or strontium is used to strengthen the plate grids instead of the antimony used in conventional batteries (**Figure 26.14**). The electrical resistance of calcium does not drop when hot nearly as much as antimony. When overcharged, calcium will only use ⅓ of the water that antimony does. Corrosion on the terminals and surface of the battery is minimized also.

Maintenance-free batteries are sealed except for small breather holes (see Figure 26.13). This is because the lead and calcium combination is susceptible to damage from even a small amount of dirt or grease. The battery does not require water to be added *if the charging system is operating properly*. If the voltage regulator allows the alternator to overcharge the battery, the water level will drop. If it drops too low, the battery must be replaced.

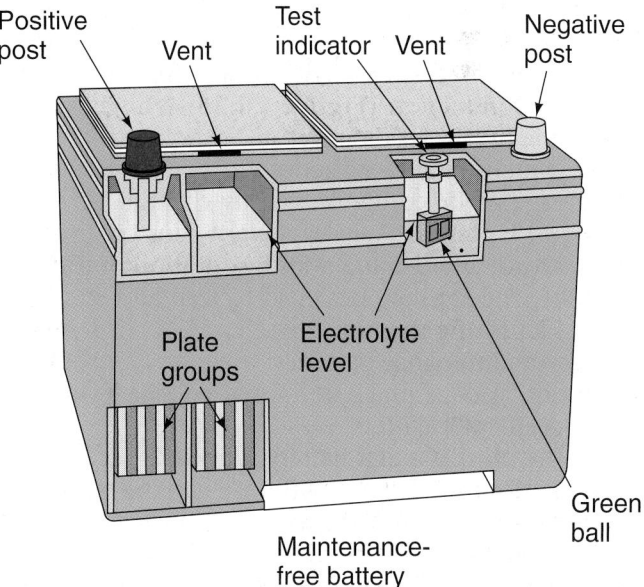

Figure 26.13 Parts of a maintenance-free battery. *(Courtesy of DaimlerChrysler Corporation)*

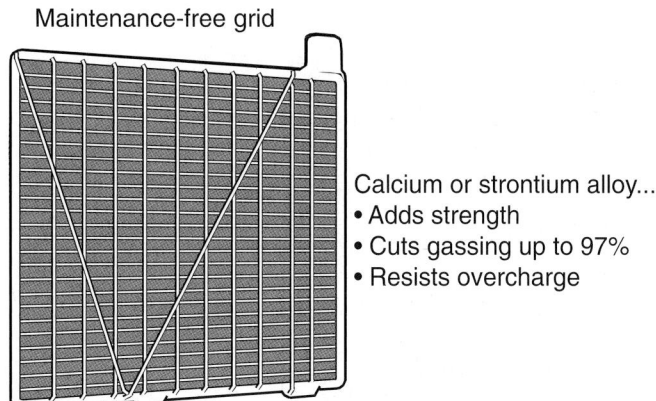

Figure 26.14 Maintenance-free battery grids are strong and thin. *(Courtesy of Ford Motor Company)*

Additional advantages to the maintenance-free battery are:

- It can be stored for long periods of time without experiencing self-discharge (three times as long as lead-antimony). Eighteen months on the shelf without self-discharge is common.
- It normally does not require an activation or boost charge when new.

Disadvantages of the maintenance-free battery are:

- It can be damaged with repeated deep discharges, reducing its plate material significantly.
- Because there are no cell caps, hydrometer testing of the battery's state of charge is not possible (see Chapter 27).
- It has a lower reserve capacity.
- Its lead-calcium material is very brittle and breaks easily.

Low-Maintenance Batteries

A **low-maintenance battery** is a revision of the maintenance-free battery. Its grids contain about 3.4% antimony, and it is called a dual alloy battery. The grids intersect at a different angle for strength and better electrical conduction (**Figure 26.15**). This is known as a *hybrid* battery grid design.

Low-maintenance batteries are not as susceptible to damage from contaminants as calcium or strontium grid batteries. They have vent caps on top that allow hydrometer testing and the addition of water to the cells. The caps are designed to trap and condense vapors, returning them to the cell.

Low-maintenance batteries:

- Do not "gas" as much at normal voltage levels as the deep-cycle battery
- Are the most popular battery for passenger car use
- Have a longer shelf life without charging than deep-cycle batteries (about 1 year)
- Have a higher CCA rating
- Can be supplied in a dry state, ensuring that they are fresh

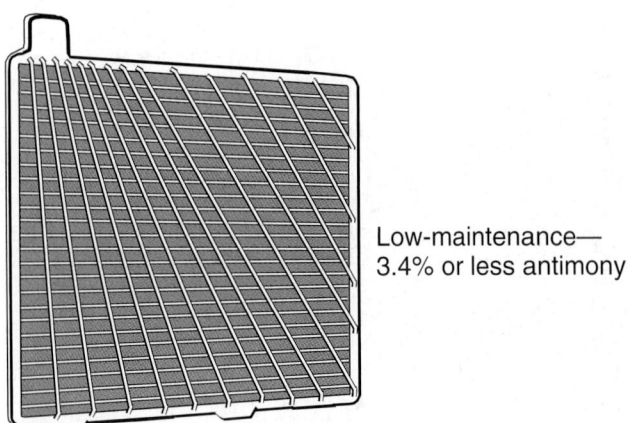

Low-maintenance—
3.4% or less antimony

Figure 26.15 A low-maintenance battery grid has a reduced antimony content. *(Courtesy of Ford Motor Company)*

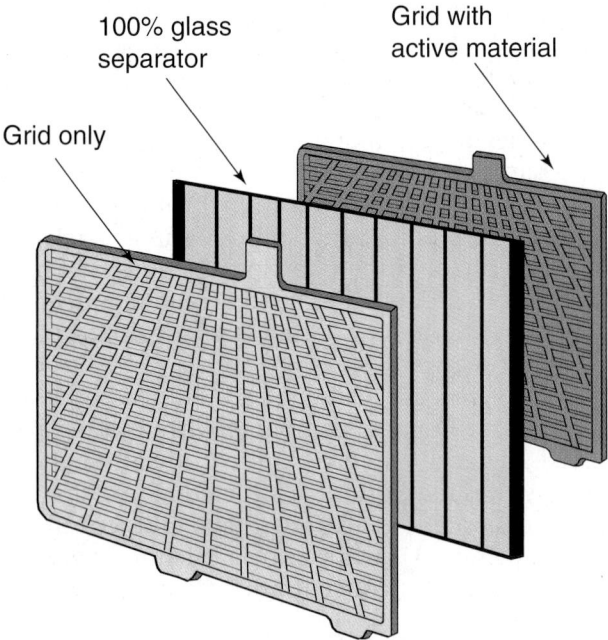

100% glass separator

Grid with active material

Grid only

Figure 26.16 A hybrid battery grid design. *(Courtesy of Ford Motor Company)*

- Are more resistant to damage from deep discharge

Whereas a conventional battery has grid bars arranged vertically and horizontally, a hybrid battery grid design uses bars arranged in a manner that provides a shorter current path (**Figure 26.16**). The result is a battery that produces current more rapidly for better starter motor operation. These batteries can undergo more deep cycles (from fully charged to fully dead) while maintaining their original reserve capacity.

▇ BATTERY PLATE SIZE

The size of a battery is usually related to its reserve capacity. The amount of active plate material is what actually determines the capacity of the battery. The thicker the plates in the cells are, the more reserve capacity.

Cars with larger engines and more electrical accessories require batteries with more plate surface area (higher CCA rating). A battery with thin plates can supply a high current for a short period (such as when operating a cranking motor). A deep-cycle R.V. battery, which must supply a smaller amount of current for a longer period of time, might have fewer plates but thicker plates.

A *series* hookup is when the positive and negative plates are hooked to each other, adding up the voltages of the cells (see Figure 26.2). Two 12-volt batteries can actually be hooked in *parallel* to provide more plate capacity. In fact, R.V. often have this arrangement to provide sufficient storage capacity to supply all of their electrical needs for a night. With a parallel hookup, voltage stays the same, but the amount of plate material doubles.

Recombination Batteries

Recombination batteries are sealed batteries that use electrolyte gel or microporous nonwoven glass instead of liquid electrolyte. Because they do not give off hydrogen gas, vents are not necessary. These batteries are so named because oxygen given off by the positive plates during charging recombines hydrogen in the negative plates to form water within the battery. They have pressure relief valves that do not allow free-flowing acid to escape but do allow pressure to be relieved during overcharge or under extreme temperatures.

Absorbed glass mat (AGM) batteries do not use free-liquid or gel electrolyte. Instead, electrolyte is absorbed into fiberglass mats sandwiched between the plates. AGM battery construction is sometimes the same as a conventional battery but with flat plates. Most of these batteries have cells that are spiral wound, which means they are tightly wound in a circle and enclosed within a round cylinder (**Figure 26.17**). AGM batteries cannot leak if they become damaged, and they will work when tilted in any direction without losing electrolyte. They are especially popular for off-road vehicles and for trunk installations, where battery leakage would be problematic. Because the cell plates are tightly held together and insulated by the fiberglass mats, the lead/tin alloy used in production can be softer than the lead used in conventional battery plates. They can be recharged more quickly than conventional batteries due to their inherent low resistance.

Figure 26.17 An AGM battery with spiral-wound cells. *(Courtesy of Exide Technologies)*

■ BATTERY SELECTION

The BCI lists group numbers to indicate the physical size of batteries (**Figure 26.18**). Cold cranking power is also listed with the number of months warranty on the battery. The following are some considerations when selecting a battery:

Grp. Size	Vlt.	Cold Cranking power-amps for 30 secs. at 0°F*	No. of mo. warranted	Size of battery container in inches (incl. terminals)		
				Lgth.	Wd.	Ht.
17HF	6	400	24	7¼	6¾	9
21	12	450	60	8	6¾	8½
22F	12	430	60	9	6⅞	8⅛
	12	380	55	9	6⅞	8⅛
	12	330	40	9	6⅞	8⅛
22NF	12	330	24	9½	5½	8⅞
24	12	525	60	10¼	6⅞	8⅝
	12	450	55	10¼	6⅞	8⅝
	12	410	48	10¼	6⅞	8⅝
	12	380	40	10¼	6⅞	8⅝
	12	325	36	10¼	6⅞	8⅝
	12	290	30	10¼	6⅞	8⅝
24F	12	525	60	10¼	6⅞	8⅝
	12	450	55	10¼	6⅞	8⅝
	12	410	48	10¼	6⅞	8⅝
	12	380	40	10¼	6⅞	8⅝
	12	325	36	10¼	6⅞	8⅝
	12	290	30	10¼	6⅞	8⅝
27	12	560	60	12	6⅞	8⅝
27F	12	560	60	12	6⅞	9
41	12	525	60	11⁹⁄₁₆	6¹³⁄₁₆	6¹⁵⁄₁₆
42	12	450	60	9⅝	6⅞	6¾
	12	340	40	9⅝	6⅞	6¾
45	12	420	60	9½	5½	8⅞
46	12	460	60	10¼	6⅞	8⅝
48	12	440	60	12	6⅞	7½
49	12	600	60	14½	6⅞	7½
56	12	450	60	10	6	8¾
	12	380	48	10	6	8¾
58	12	425	60	9¼	7¼	6⅞
71	12	450	60	8	7¼	8½
	12	395	55	8	7¼	8½
	12	330	36	8	7¼	8½
72	12	490	60	9	7¼	8¼
	12	380	48	9	7¼	8¼
74	12	585	60	10¼	7¼	8¾
	12	525	60	10¼	7¼	8¾
	12	505	60	10¼	7¼	8¾
	12	450	55	10¼	7¼	8¾
	12	410	48	10¼	7¼	8¾
	12	380	40	10¼	7¼	8¾
	12	325	36	10¼	7¼	8¾

*Meets or exceeds Battery Council International rating standards.

Figure 26.18 BCI battery group numbers tell the size and power of the battery. *(Courtesy of Battery Council International)*

- The battery must fit the battery box.
- The posts must be on the correct side of the battery.
- The battery holddown must fit the battery. Sometimes, there is a holddown bracket cast into the bottom side of the case of the battery.
- The battery cannot be so high that it shorts out on the hood.

■ BATTERY SERVICE LIFE

The average battery has a service life of about 3 to 5 years. The negative plates become soft during the repeated charge and discharge cycles. A battery's life is shortened by improper (too high or low) charging system amperage or voltage. Too high a charging rate is the greatest cause of shortened battery life.

The actual length of a battery's life is determined by the amount of material that has been shed from the surface of its positive plates. Plates shed material during overcharging or vibration.

When a battery discharges, a neutral coating of $PbSO_4$ coats the plates. The effects of this **sulfation** are usually reversible, but when it becomes too extensive, the battery is scrapped (see Chapter 27).

Effect of Temperature on Batteries

Batteries do not work as well in colder weather. The following chart shows the effect that cold temperatures actually have on a battery's cranking power.

Temperature	% of Cranking Power
80°F (26.7°C)	100
32°F (0°C)	65
0°F (−17.8°C)	40

Other battery problems caused by cold weather include:
- The engine becomes harder to crank because oil becomes thicker in colder temperatures.
- A battery that is not fully charged will freeze easier.

NOTE: *Parts stores in high-elevation resorts sell many batteries when tourists from areas with milder climates have batteries that are marginal. The cold weather causes battery failure.*

■ BATTERY CABLES

Battery cables must be large enough to carry all of the current demanded by the starter and vehicle's electrical system. The large cable goes directly to the starter because it requires a great many more electrons to operate than any of the vehicle's other electrical loads. If a cable is too small, it will become hot as the engine is cranked. Battery cables for 12-volt systems are usually 4 or 6 gauge. As in the selection of other electrical

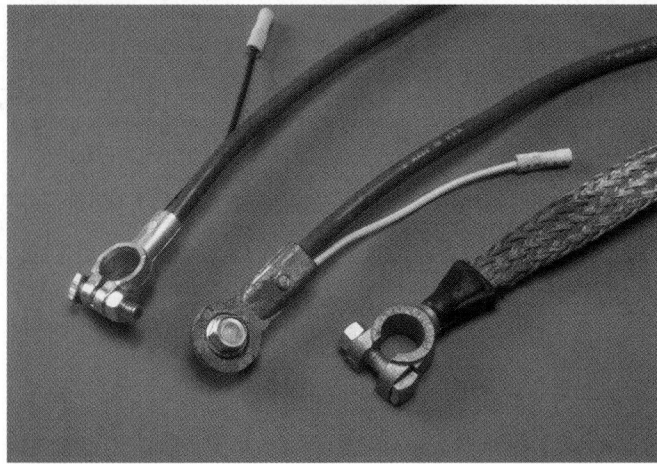

Figure 26.19 Different battery cable designs.

Figure 26.20 Battery cable clamps. *(Courtesy of Tim Gilles)*

wires, a larger cable will not be a problem. A cable that is too small will not deliver enough electrons to the starter motor and could possibly start a fire.

Figure 26.19 shows several designs of battery cable. The uninsulated woven cable is a ground cable. If it were the hot cable and touched against a ground, a spark would result. The positive cable is usually red. Do not take this for granted!

Battery Cable Clamps. Battery cable clamps are made of steel or lead, depending on manufacturer preference (**Figure 26.20**). The cable is usually crimped to the clamp, although some marine and outdoor equipment uses a bolt and wing nut to attach a crimped cable end to the clamp.

■ BATTERY HOLDDOWNS

A battery must be held in its tray because it can fall out as the car travels over bumps. Also, excessive vibration can harm the battery, causing the plates to shed more of their material. Battery holddowns (**Figure 26.21**) are

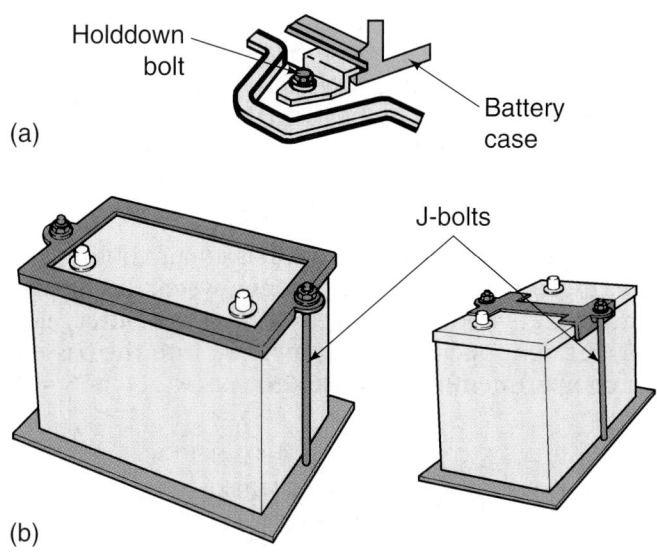

(a)

(b)

Figure 26.21 Different types of battery holddowns. (a, Courtesy of DaimlerChrysler Corporation; b, Courtesy of Ford Motor Company)

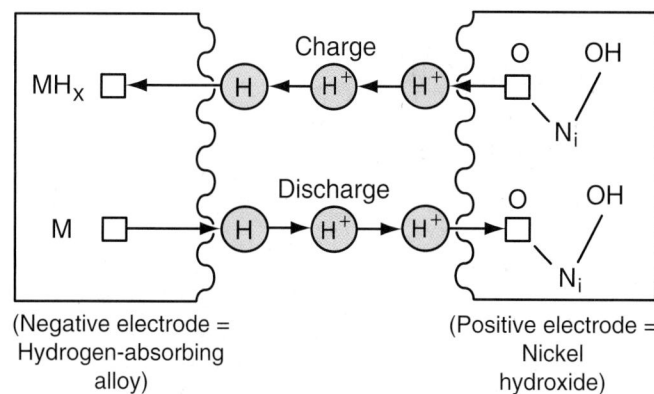

(Negative electrode = Hydrogen-absorbing alloy)

(Positive electrode = Nickel hydroxide)

Figure 26.22 During NiMH discharge, hydrogen moves from the negative plate to the positive plate.

SCIENCE NOTE

Ions are groups of atoms with a positive or negative charge. Most rechargeable batteries move ions from the anode to the cathode during discharge and back during recharge.

---

made of steel or plastic. Some batteries have a molded piece near the bottom of the case that a holddown bracket is clamped to. Others have a bracket that fits around or across the top of the battery. Long threaded rods (called *J-hooks*) are hooked into holes in the battery tray.

■ REASONS FOR BATTERY FAILURE

- A damaged battery case can leak electrolyte.
- Undercharging can result in sulfation of the battery plates, which limits electron flow.
- Overcharging can result in buckled and warped plates from excess heat. Material will be shed from the plates.
- Vibration can cause more material to be shed from the plates.
- A short between the plates can cause a dead cell.

■ HYBRID ELECTRIC VEHICLE AND OTHER BATTERY TYPES

Other types of batteries include nickel-based batteries and lithium-based batteries. There are two common nickel-based batteries: nickel metal hydride (NiMH) and nickel-cadmium (NiCd). They are similar to each other, except for the material used in the anode. Both have 1.2 volts per cell. NiMH batteries are found in most hybrid vehicles because they have more energy available in a smaller space and do not contain cadmium, which is an environmental danger. NiCd batteries can tolerate deep cycling about three times as much as NiMH, however, so they may have a future use in battery electric vehicles.

Nickel Metal Hydride Batteries

The positive plate in a NiMH battery is made of nickel hydroxide. The negative plate contains metal alloys that absorb hydrogen. The separator material between the plates is a fiber sheet with an alkaline electrolyte of potassium hydroxide. A typical cell is housed in a sealed metal housing with a high-pressure safety vent.

When fully charged, a NiMH cell measures 1.2 volts.

NOTE: *Rechargeable NiMH batteries are used to replace conventional AA alkaline batteries in cameras and other 1.5-volt applications. A conventional alkaline battery has 1.5 volts but only when it is new. In contrast, the average voltage during the life of a NiMH battery is said to be 1.2 volts, and it stays at this level for about 80% of its discharge cycle.*

During NiMH discharge, hydrogen moves from the negative plate to the positive plate (**Figure 26.22**). As the cell recharges, hydrogen moves from the positive plate to the negative plate. The electrolyte level does not change during charging and discharging, hence no service is required.

The Memory Effect. The memory effect is something that occurs in some rechargeable batteries. Ni-Cd batteries are particularly susceptible. The memory effect causes a battery to hold less charge and lose its maximum capacity if it is repeatedly discharged only partway before recharging. NiMH and lithium-ion batteries do not suffer from the memory effect and can be "topped off" repeatedly with very little negative effect.

Lithium-Ion Batteries

Lithium-ion (Li-ion) is a term used for all batteries that use lithium. Lithium-ion batteries are like nickel-based

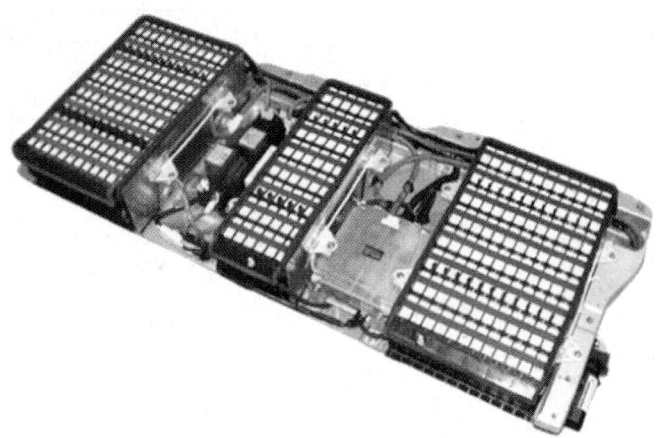

Figure 26.23 The battery pack assembly includes modules, a controller, relays, and solenoids. *(Courtesy of Toyota Motor Sales, U.S.A., Inc.)*

batteries, but they have several advantages, including no memory effect and less damage to the environment. They can also be made smaller than nickel-based batteries.

However, lithium metal is not only corrosive but highly flammable, oxidizing very fast when exposed to air and water. Therefore, battery cells use lithium compounds rather than lithium metal. Lithium-ion cell output is around 3.6 volts. There is an abundance of research on lithium-ion technology for future use in electric vehicles.

Typical Hybrid Battery

Most current hybrid automobiles use NiMH batteries. Honda uses 120 individual NiMH cells that resemble standard D cell flashlight batteries. The cells are positioned end to end in groups of six called modules. A 144-volt Honda battery pack is smaller than a typical Toyota 273-volt battery. The battery pack in some Toyota hybrids has 228 1.2-volt cells in 38 six-cell modules that produce 273.6 volts of DC current. The current generation Prius uses a 201.5-volt battery with 168 1.2-volt batteries in 28 six-cell modules.

A Toyota battery module is slightly larger than an 18-volt cordless electric drill battery. The battery pack includes modules, relays, and solenoids in a metal case (**Figure 26.23**) similar to a suitcase. In some hybrids, battery packs are split into smaller units due to space considerations and because they have lower voltages when separated for service. A Honda battery pack weighs about 48 pounds, compared with a Toyota battery pack, which weighs about 110–150 pounds, depending on its voltage.

Hybrid Battery Electrolyte. The cells in the hybrid NiMH battery packs each have their own sealed case. Each cell contains thin paper membranes that absorb a potassium hydroxide (KOH) electrolyte. The KOH electrolyte solution in the high-voltage batteries was originally liquid during battery manufacture, but the

batteries are called dry cells because the liquid electrolyte is almost totally absorbed in the battery membranes. Unlike ordinary lead-acid batteries, hybrid battery cell electrolyte gel is highly alkaline, with a pH of 13.5.

Hybrid Battery Cooling System

The battery pack gets hot, so it has an air cooling system (**Figure 26.24**). The air cooling system has several sensors and various fan speeds. When the battery gets hot, the vent opens to bring more air into the passenger compartment (**Figure 26.25**).

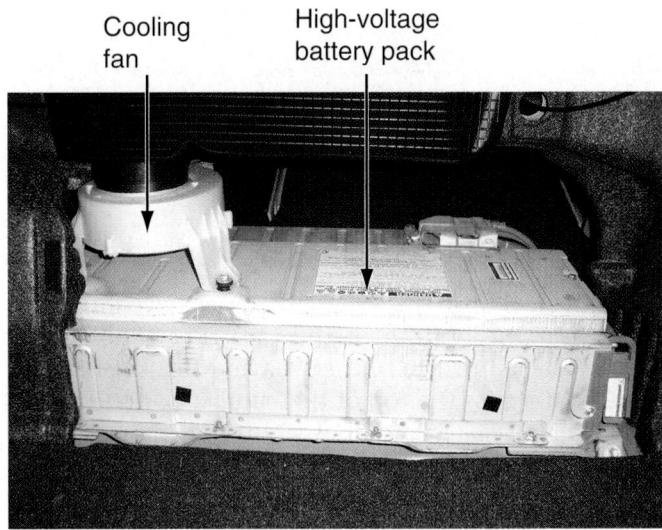

Figure 26.24 The high-voltage battery pack has a cooling system. *(Courtesy of Tim Gilles)*

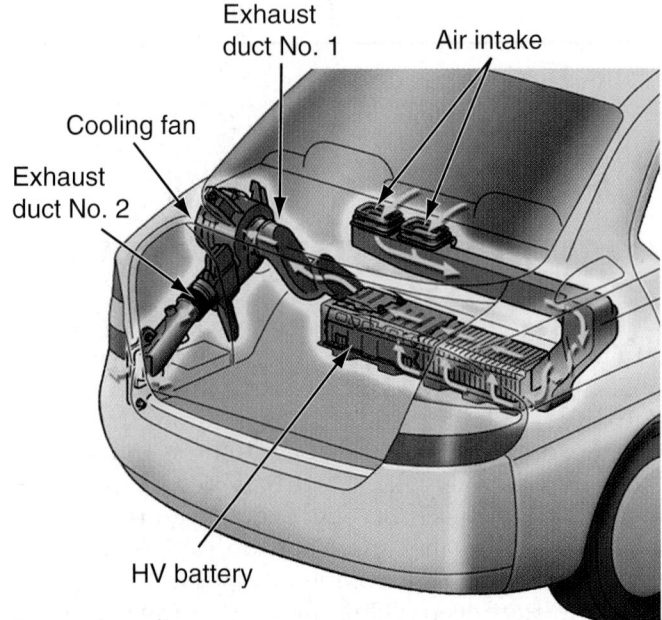

Figure 26.25 Air is circulated from the passenger compartment to the high-voltage battery. *(Courtesy of Toyota Motor Sales, U.S.A., Inc.)*

■ REVIEW QUESTIONS

1. Is the electricity produced by a battery AC or DC?

2. How many volts will a fully charged battery cell have?

3. What kind of gas is given off as a battery is charged?

4. When does a battery give off the most gas, when charging has just begun or when the battery is almost fully charged?

5. What is the most common method of rating battery capacity for automobiles?

6. What is the battery test for deep-cycle batteries called?

7. What is the electrical unit of measurement that is equivalent to horsepower?

8. What percentage of a battery's cranking power will be available at 0°F?

9. What is the name of the battery type used in hybrid vehicles that does not have a memory effect and is less damaging to the environment than nickel-based batteries?

10. Which type of battery do most current hybrid electric vehicles use?

■ ASE-STYLE REVIEW QUESTIONS

1. Technician A says that a deep-cycle battery uses more water than a battery that uses calcium as a plate-strengthening material. Technician B says that a maintenance-free battery will not discharge as much on the shelf as a deep-cycle battery. Who is right?
   - **a.** Technician A
   - **b.** Technician B
   - **c.** Both A and B
   - **d.** Neither A nor B

2. When an automobile battery is fully charged, a voltmeter connected to it while the key is off will register _____.
   - **a.** 12 volts
   - **b.** 12.6 volts
   - **c.** 13.8 to 14.2 volts
   - **d.** None of the above

3. All of the following are true about battery posts *except*:
   - **a.** They are often labeled ⊕ or ⊖.
   - **b.** The negative battery post is larger than the positive post.
   - **c.** They are made of lead, which oxidizes when exposed to air.
   - **d.** They can be on the side or top of the battery.

4. Technician A says that in a completely discharged battery, the positive and negative plates are the same material. Technician B says that in a fully charged battery, the positive and negative plates are the same material. Who is right?
   - **a.** Technician A
   - **b.** Technician B
   - **c.** Both A and B
   - **d.** Neither A nor B

5. Connecting two batteries _____ will provide more plate capacity for use as a recreational vehicle power supply.
   - **a.** In parallel
   - **b.** In series
   - **c.** Both A and B
   - **d.** Neither A nor B

6. Technician A says that batteries are rated by cold cranking amperage. Technician B says that a battery can be rated according to its reserve capacity. Who is right?
   - **a.** Technician A
   - **b.** Technician B
   - **c.** Both A and B
   - **d.** Neither A nor B

7. Which of the following could cause a battery cable to become hot when the engine is cranked?
   - **a.** A cable too large
   - **b.** A cable too small
   - **c.** Using the wrong battery for the vehicle
   - **d.** None of the above

8. Two technicians are discussing the memory effect. Technician A says that NiCd batteries suffer from this. Technician B says that the memory effect becomes worse if a battery is not discharged completely before it is recharged. Who is correct?
   - **a.** Technician A
   - **b.** Technician B
   - **c.** Both A and B
   - **d.** Neither A nor B

9. The vehicle's largest load on the battery is the _____.
   - **a.** Air-conditioning blower
   - **b.** Horn
   - **c.** Electric seats
   - **d.** None of the above

10. Technician A says that a battery gives off the most hydrogen when it is almost fully charged. Technician B says that the most hydrogen given off is when the battery is dead and beginning to charge. Who is right?
    - **a.** Technician A
    - **b.** Technician B
    - **c.** Both A and B
    - **d.** Neither A nor B

CHAPTER 27

Battery Service

■ **OBJECTIVES**

Upon completion of this chapter, you should be able to:

✔ Inspect a battery and recommend the correct service for it.

✔ Service a battery.

✔ Perform a variety of tests on a battery and make a diagnosis from the results.

✔ Select the best charge rate and charge a battery.

✔ Perform battery service safely.

✔ Safely and correctly jump start a car.

■ **KEY TERMS**

lead oxidation	surface charge	inductive pickup
specific gravity	carbon pile	parasitic load

■ **INTRODUCTION**

The battery is the heart of the electrical system. The battery must be in good condition and fully charged if tests on other electrical system components are to be performed. Batteries last an average of 3 years and are routinely replaced when they fail. Tests will show a battery's state of charge and output voltage. Results will tell if the battery is good, needs a recharge, or must be replaced. The following case history illustrates the importance of having a good battery.

> **CASE HISTORY**
>
> *A customer complained that his car's idle would change sometimes when he stepped on the brake pedal. The technician tested the power brake booster (a typical cause of this problem) and determined that it was good. After further testing, the technician determined that the problem was that the alternator had a bad diode. When the battery is not fully charged, the computer on modern automobiles will not work properly. On some systems, if the voltage drops below 11.6 volts at idle, the idle speed-control motor will raise the idle speed to increase the charging rate. When the voltage is borderline, even the current draw requirements of the brake lights can cause the idle to raise when the brake pedal is depressed.*

Date of Manufacture

Manufacturers provide a code on the battery to tell when and where it was made. The number is hot stamped into the case, on or near the cover (**Figure 27.1**). For aftermarket battery sales, a round sticker

↑ Manufacturer code

Figure 27.1 A manufacturer code stamped into the battery case tells when it was made. *(Courtesy of Tim Gilles)*

Figure 27.2 Inspect the condition of the cables and terminals.

Figure 27.3 A battery water filler is handy for filling batteries.

with a letter and a number is often installed as well. For most manufacturers, the letters *A* through *M* represent the month of manufacture (with *A* being January and *M* being December) and the number represents the year. A sticker with "B8" would denote a battery manufactured in February 2008.

■ BATTERY INSPECTION

Several things are looked at when determining a battery's condition. Inspect the following items:

- Date code on the battery label. This tells the date that the battery was installed in the vehicle.
- Battery case condition. Electrolyte deposits on the surface of the case can conduct electricity between the positive and negative terminals, discharging the battery. Check for damage to the battery case.
- Level of the electrolyte. Add distilled water as needed to fill the battery to the split ring under the cell cap. If one of the cells is low, there could be a short in that cell. Low electrolyte throughout the battery indicates the possibility of overcharging. Charging system tests are covered in Chapter 31.
- Condition of the cables and terminals. Clean them as needed (**Figure 27.2**)
- Battery holddown and tray.
- Built-in hydrometer (some batteries).

■ BATTERY SERVICE

Battery service includes replenishing water (some batteries) and service to the terminals and clamps.

Refilling a Battery with Water

Batteries lose electrolyte over a period of time due to electrolysis, which is using electricity to break down water into hydrogen and oxygen. If the water level is allowed to drop below the level of the plates, they will rapidly become sulfated. Sulfation is discussed later in this chapter. In most batteries, water that has

evaporated can be replenished through holes in the battery top.

Only water is added, not electrolyte, because the acid does not evaporate. When a battery is low on water, it has more internal resistance. This causes more heat, which is also hard on the battery.

Use clean filtered water when filling a battery. Distilled water is best. A *battery water filler* is a handy tool for filling batteries (**Figure 27.3**). The water filler has a nozzle that allows water to flow when it is pushed down against the split ring indicator in the cell opening (see Figure 13.19). When the cell is full, the flow of water stops automatically. A battery with an electrolyte level that is filled to the split ring will have the correct specific gravity when fully charged.

Cleaning Battery Terminals and Clamps

In a survey of auto repair shops by the *Car Care Council*, a public information agency, technicians listed battery cable maintenance as one of the most overlooked items. Battery terminal connections that have become corroded are the *most common cause of hard-starting complaints*. Terminal posts are made of lead, which forms an insulating oxide coating when exposed to air. Sometimes an oxidized terminal is black, but not always.

Lead oxidation is most often found on the positive post. This coating, which does not conduct electricity, will reduce or stop the flow of current between the battery and the vehicle's electrical system. Removing the oxide layer exposes fresh lead underneath. Use an emery cloth or a terminal cleaner to clean both posts and the inside of the terminal clamps (**Figure 27.4**). **Figure 27.5** shows several popular styles of terminal cleaners.

Corrosion, which looks like white powder on and around battery posts, sometimes happens when battery gas escapes up the terminal post from inside the case. Corrosion on both battery terminals is sometimes

 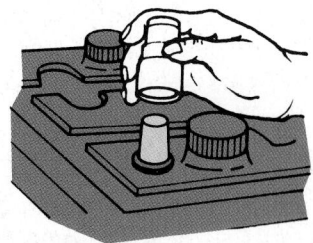

Figure 27.4 A battery terminal cleaner that cleans both the inside of the clamp and the outside of the post.

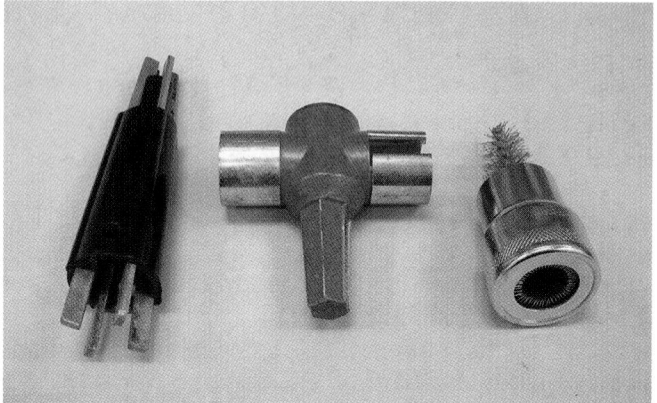

Figure 27.5 Several popular styles of battery terminal cleaners. *(Courtesy of Tim Gilles)*

due to battery caps that are not installed tightly all of the way. Today's maintenance-free batteries are less prone to post corrosion because they do not gas as much. When they have *case vents,* the vents are aimed away from the posts. Side-terminal batteries are also less prone to this problem.

Battery Acid Is Corrosive. Battery electrolyte, which is sulfuric acid and water, will make holes in many types of clothing. It is not unusual to find holes in coveralls or pants after carrying a battery. The outside of some batteries is covered with a mist of acid as a battery gasses. When this gets on hands, it can be accidentally transferred to clothing.

NOTE: *To avoid the possibility of staining a car's paint, get in the habit of always using fender covers.*

Sometimes the battery box will need to be painted after it has been thoroughly cleaned.

> **SHOP TIP** *Acids* can be *neutralized* with a *base,* such as baking soda. A mixture of water and baking soda or a strong detergent can be used to wash a battery (**Figure 27.6**). Be sure not to get any of the mixture into the cells. The battery box is cleaned with the same solution.

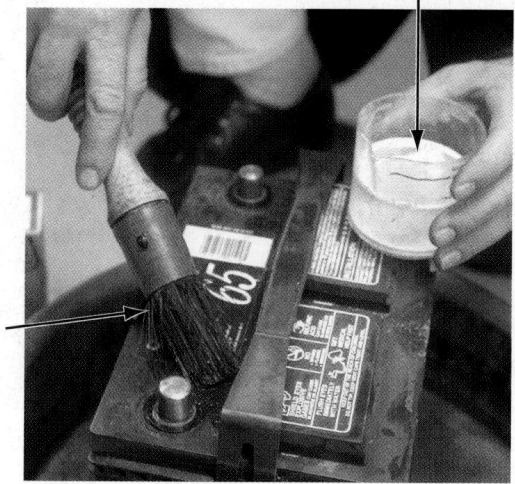

Soda/water solution

Brush

Figure 27.6 A mixture of water and baking soda or a strong detergent can be used to wash a battery.

Dry Batteries

One of the advantages of maintenance-free batteries is that they have a long shelf life. This means that they can be stored and shipped full of electrolyte. New batteries with cell caps are sometimes shipped dry, requiring the addition of electrolyte. New electrolyte, which is approximately 25% acid and 75% water (by volume), is contained in a large plastic bag inside a sturdy cardboard box. The plastic bag has a tube on it and a pinch device to shut off the flow of liquid when the plates of the cell are covered. The tube is sealed for shipping, and the end needs to be cut off before the liquid can be poured.

After filling a dry battery, allow it to sit for at least 15 minutes so that the plates can soak up the electrolyte. Then, charge the battery at 30 amps until the battery is charged. This is checked with a hydrometer (covered later in this chapter). Then, check the electrolyte level and add more electrolyte to fill the cells to the correct level.

> **SAFETY NOTE** Be sure to wear safety goggles when performing this task. Concentrated acid will burn skin and destroy clothes. Spilled electrolyte must be neutralized immediately or it will etch a concrete floor, causing a bright white stain.

> **SHOP TIP** Fill the battery until the electrolyte just covers the plates. Do not fill the cells to the normal fill line until *after* the battery has been charged. The electrolyte might spill over when it becomes heated during charging.

■ REPLACING A BATTERY

When servicing a battery, the vehicle should be outdoors so that the battery box can be washed out. Be sure to use a fender cover on the fender nearest the battery because battery acid can ruin a paint job.

Remove the Terminal Clamps

When a battery is removed, the ground cable is always disconnected first (**Figure 27.7**). The ground cable is the one that is connected to the block or chassis. It is usually, but not always, the negative post. Double-check to be sure, especially on older cars (1955 and earlier and British cars before 1970) because some of them are *positive ground*.

 Be certain that all electrical circuits are shut off before disconnecting the battery. Otherwise a spark could cause a battery explosion (see Figure 3.34).

Check the bolt and nut used to clamp a post-type terminal clamp. Due to corrosion, they will probably be smaller than they once were. This means that a wrench will not fit them. Special battery bolt pliers are available for removing these bolts (**Figure 27.8**).

Terminal clamps are usually easy to remove. Do not twist the battery posts. They are made of lead, which is a soft metal. The terminal clamp or the battery itself can be easily damaged. The terminal connection inside a battery is heated during manufacture to "weld" it to the inside of the battery terminal post. Twisting the terminal clamp can damage the fragile lead post inside the battery (**Figure 27.9a**). Use a battery terminal puller when a stubborn clamp is encountered (**Figure 27.9b**).

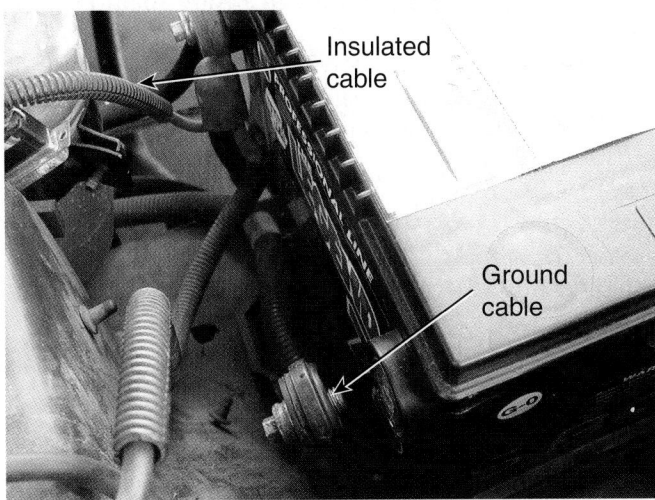

Figure 27.7 Always disconnect the battery ground cable first. *(Courtesy of Tim Gilles)*

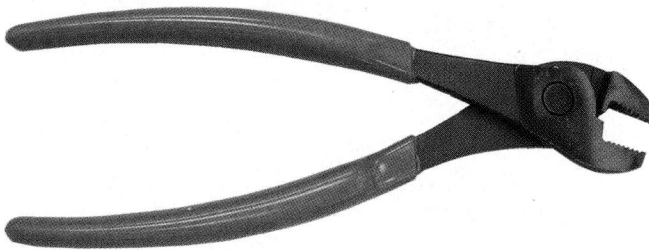

Figure 27.8 Battery terminal nut pliers. *(Courtesy of Tim Gilles)*

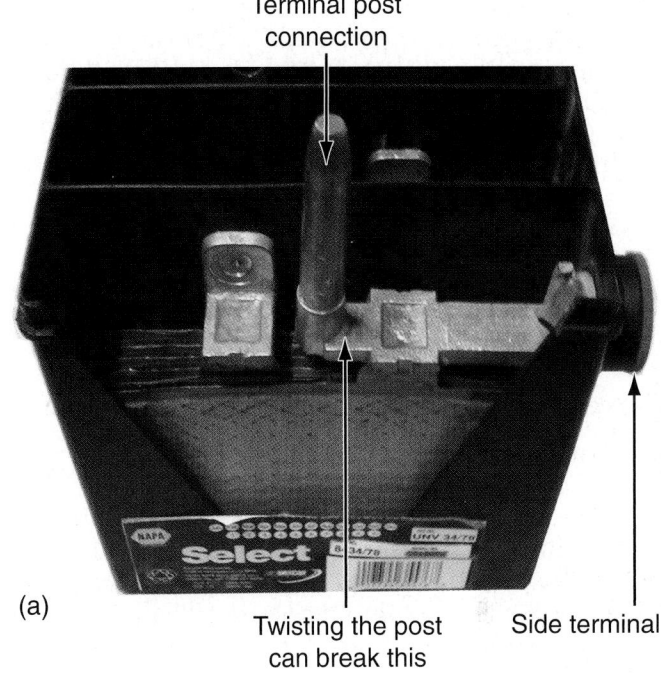

(a)

Terminal post connection

Twisting the post can break this

Side terminal

(b)

Figure 27.9 (a) Battery cutaway of the terminal stud that will be melted to join it to the inside of the top battery terminal post. Twisting the terminal clamp can damage this fragile part. (b) A battery terminal puller. *(a, Courtesy of Tim Gilles)*

NOTE: *When replacing a battery on computer-equipped cars, memory can be lost for certain settings such as radio stations, clock, block-learned driveability, seat settings, fuel consumption rate, and temperature control. Diagnostic codes in the computer's memory will be lost, too. After the battery is reinstalled, the computer will relearn the lost information in about 50–100 miles. The owner might notice a slight difference in the driveability of the vehicle until that time.*

SHOP TIP To avoid the need to reset these things, a small transistor radio battery (9 volt) can be connected to the battery cables before removing them. A special tool is available that can be connected into the cigarette lighter. The other end of it is connected to another battery to keep the circuits alive (**Figure 27.10**). This method also prevents battery terminal arcing when the underhood battery is disconnected. Be sure to remove the underhood light (if applicable) and keep the doors closed so the dome light will not discharge the 9-volt battery.

- The positive battery cable under the hood has power at all times while this tool is connected. Do not let the positive battery cable touch ground or the cigarette lighter fuse will blow. This will electrically disconnect the spare battery from the vehicle. All computer memory will be lost.
- Be sure to observe correct polarity when connecting the alligator clips to the spare battery.
- Verify that the cigarette lighter works before inserting the tool into the cigarette lighter receptacle. This is to verify that voltage from the spare battery will actually be able to get into the electrical system to maintain memory circuits.

Remove the Battery

Lubricate the holddown clamp threads; they are probably rusty. Remove the holddown clamp and lift the battery out of its tray. Remember to keep the battery away from clothing. Clean the battery and its tray with baking soda and water. Take care that soda is not accidentally spilled into any of the cells. Check and refill the battery with water as needed.

Battery Terminal Clamps and Cables

Resistance at the cable connections is a major cause of starting system problems. There are two causes of increased resistance: battery acid vapors and air. The easiest way to prevent the resistance is to make sure that neither of these contact the critical areas. One fix is to install a felt washer under the terminal clamp to prevent acid from gassing out between the terminal post and battery case (**Figure 27.11**). Another method that can be used to prevent corrosion buildup is to cover the terminal and post with silicone gasket cement following assembly.

When a terminal clamp becomes corroded beyond repair, it can be replaced. After the old clamp is cut off and the insulation is removed from the end of the cable, a new clamp can be soldered to the old cable (**Figure 27.12**). When the lug on the other end of a cable is replaced, it can be crimped on (**Figure 27.13**).

One popular type of repair for worn cable ends is a *bolt-on type terminal clamp* (**Figure 27.14**). These are handy but are not recommended by manufacturers, who call them *emergency repair battery terminal clamps*. New replacement battery cables are available and should be installed as soon as possible after the emergency repair. They are better than bolt-on terminal clamps because their terminations are more

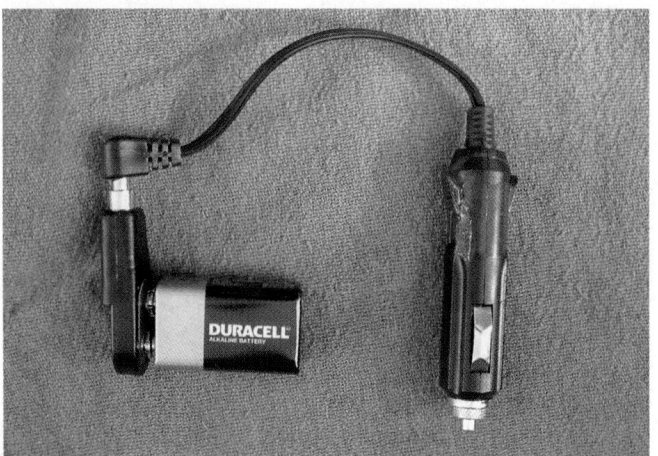

Figure 27.10 A tool for keeping computer memory alive while a battery is removed for service. *(Courtesy of Tim Gilles)*

Figure 27.11 A felt washer under the terminal clamp helps keep the battery from gassing through the terminal post.

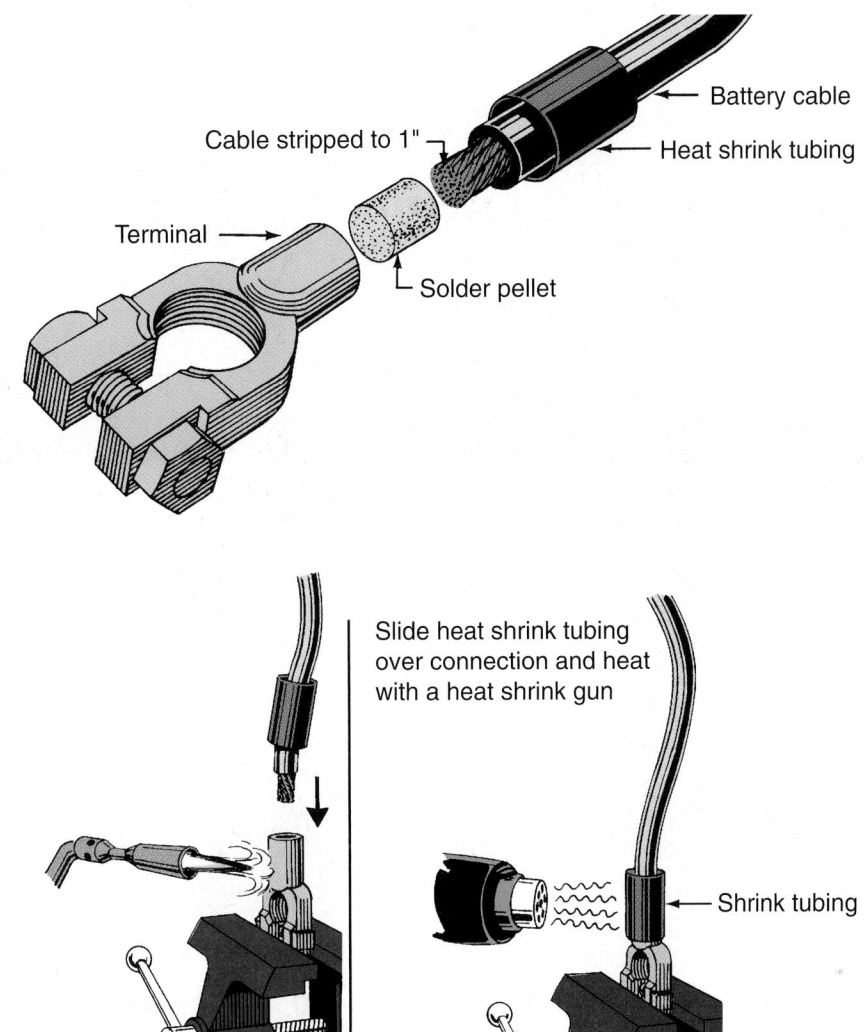

Figure 27.12 Battery cable repair. Soldering a terminal. *(Courtesy of Federal-Mogul Corporation)*

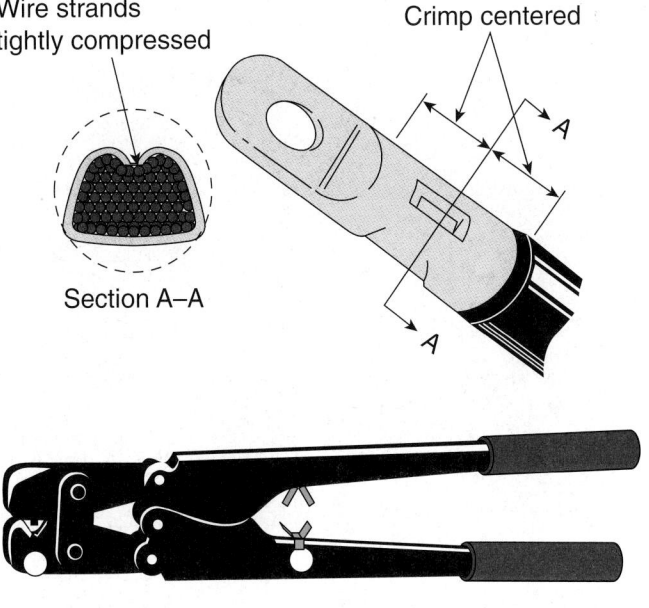

Figure 27.13 A new battery cable lug can be crimped to the cable. *(Courtesy of Ford Motor Company)*

thoroughly crimped or soldered to the cable. Before either type of terminal clamp is used, the end of the cable must be bright and clean. Use a sharp knife to strip back about ½" of insulation from the end of the cable. If a bolt-on type clamp is used, be sure it is sufficiently tightened to the cable so that it cannot come loose.

NOTE: *Whether a cable is to be repaired or replaced, it must remain long enough after installation to permit enough slack for the engine to be able to freely rock on its mounts.*

A cable that is too short can also rub against something until its insulation wears through, causing a short to ground (**Figure 27.15**). A cable that is too long and not restrained properly can burn through on an exhaust manifold. Always perform a visual check for possible electrical system problems before attempting a diagnosis.

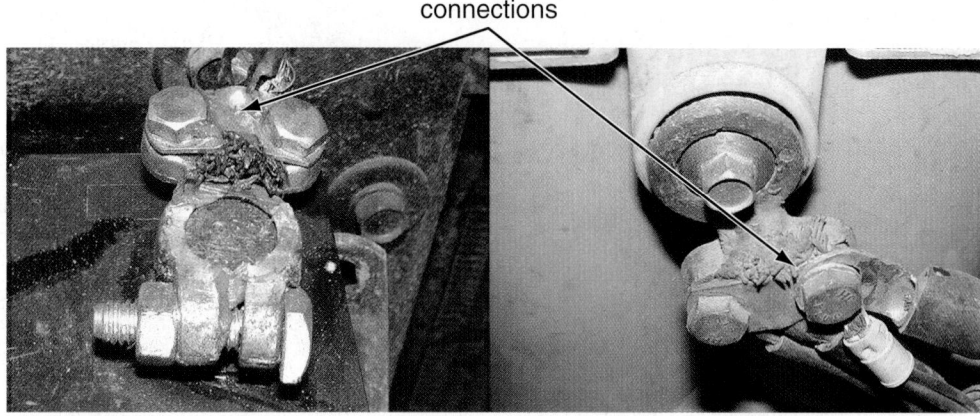

Temporary connections

Figure 27.14 Temporary terminal connections are subject to corrosion. The one shown on the right has several connections that are potential failure points as well. *(Courtesy of Tim Gilles)*

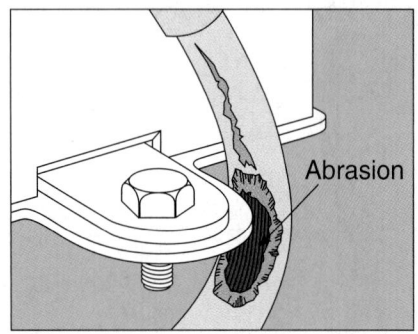

Abrasion

Figure 27.15 The insulation can be worn away on a cable that is too short. *(Courtesy of Federal-Mogul Corporation)*

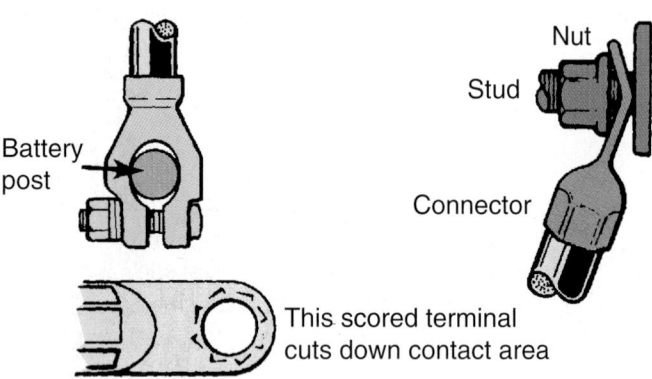

Battery post

Nut

Stud

Connector

This scored terminal cuts down contact area

Figure 27.16 These cables do not have full contact with ground. *(Courtesy of Ford Motor Company)*

CASE HISTORY *A customer complained that his battery kept going dead. He had already replaced his battery. His neighbor had replaced his alternator and voltage regulator, but the battery continued to go dead. The customer took his car to a qualified technician in a repair shop. The technician performed a visual inspection first. She found a battery cable that had been resting against the exhaust manifold. Its insulation had burned, allowing a direct short to ground. Replacing and relocating the cable solved the problem.*

Sometimes, a battery terminal does not have good contact with the terminal post or the cylinder block (in the case of the ground cable) (**Figure 27.16**).

Battery Holddowns

The life of the battery will be shortened if it is not held firmly in place (**Figure 27.17**). If it is allowed to bounce and vibrate against the battery box as the car travels over bumps in the road, material will be shed from the battery's plates. A battery can also fall out of its battery tray and cause damage or a fire.

Figure 27.17 A battery installation in an off-road vehicle. The battery is securely held from moving. *(Courtesy of Tim Gilles)*

CASE HISTORY *An auto shop student put a new battery in his Jeep. The battery tray is high on the bulkhead under the hood. He did not have a battery holddown but planned to buy one soon. After a week or so, he forgot about the holddown. While driving to school, he heard a loud pop and oil began to leak from the bottom of his Jeep. He had recently rebuilt the engine and was very concerned. Upon opening the hood, he discovered that the battery had fallen down against the engine. Its positive post had grounded out against the oil filter and burned a hole in it (**Figure 27.18**). He was lucky that he stopped and did not run the engine long enough without oil to cause any damage.*

Reinstall the Battery in the Vehicle

Reinstall the battery and holddown clamps.

NOTE: *Tighten the holddown clamp only until it is snug. DO NOT OVERTIGHTEN the clamp (**Figure 27.19**). The battery can be damaged.*

Before attempting to reinstall the old terminal clamp, clean the terminal clamps and enlarge their holes using expanding pliers (**Figure 27.20**). Reinstall the hot—usually (+)—lead. Reinstall the ground—usually (−)—lead. The ground is always the one connected last.

Corrosion can be prevented by not allowing air to contact the terminal posts. Spread silicone RTV (see Chapter 51) around the base of the terminal or use battery spray or treated felt washers to prevent oxidation (needed on non-maintenance-free batteries only). Some technicians like to use grease but, it liquifies at high underhood temperatures and coats the battery, providing a current path from positive to negative. After washing your hands, check to see that the battery starts the car.

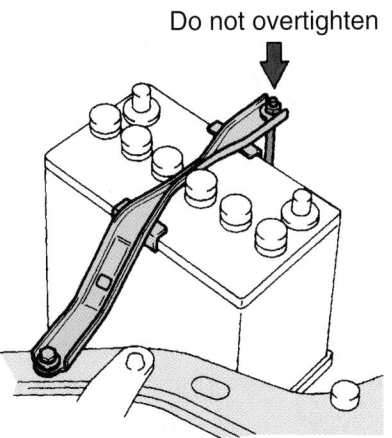

Figure 27.19 Do not overtighten the holddown.

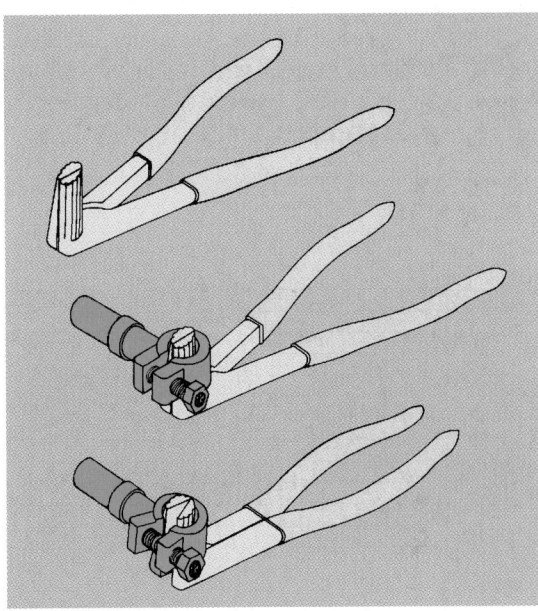

Figure 27.20 Enlarge the hole in the terminal clamp with the expanding pliers. *(Courtesy of Battery Council International's Battery Service Manual)*

■ BATTERY TESTING: MEASURING A BATTERY'S STATE OF CHARGE

Battery Hydrometer Testing

There are several ways of testing batteries. The most popular way to determine a battery's state of charge is by checking the strength of the electrolyte with a hydrometer (**Figure 27.21**). When a battery discharges, its **specific gravity** lessers. This is because as the battery discharges, water dilutes the electrolyte as oxygen from the positive plate joins with hydrogen in the electrolyte.

A hydrometer compares the weight of a liquid that is drawn into it to the weight of pure water. Pure water is given the reading of 1.000. Acid weighs more than pure water (pure acid has a specific gravity of 1.840) so the more sulfuric acid in the electrolyte, the higher the specific gravity reading on the hydrometer. A fully

Figure 27.18 The hole in this oil filter happened when the battery fell against it. *(Courtesy of Tim Gilles)*

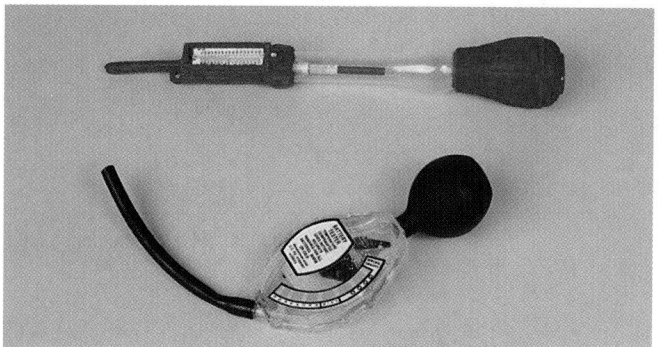

Figure 27.21 Two types of battery hydrometers.

charged battery sold in the United States usually has an electrolyte reading of 1.260–1.270 (this is the specific gravity of new electrolyte) measured at 80°F.

Batteries can freeze in cold weather. A dead battery (1.110 specific gravity) will freeze at about 19°F, whereas a fully charged battery will never get cold enough to freeze (**Figure 27.22**).

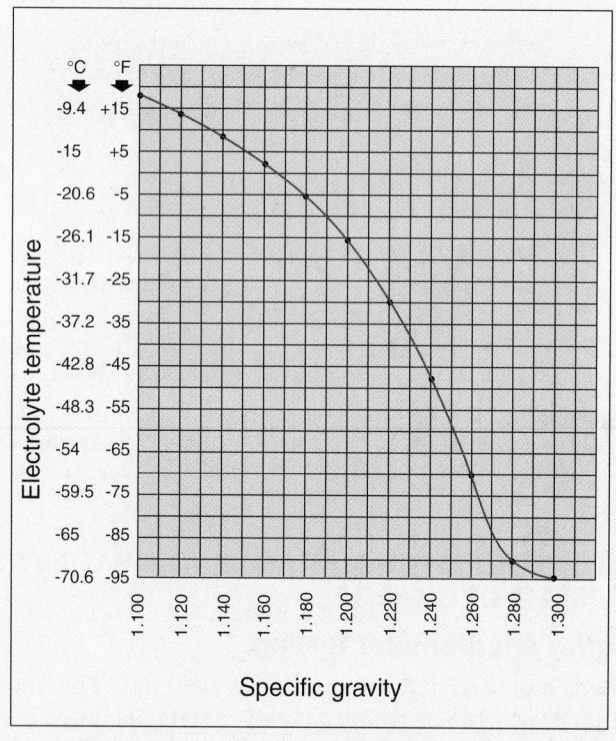

Specific gravity corrected to 80°F (26.7°C)	Freezing temperature	
1.280	-92°F	(-69°C)
1.265	-71.3°F	(-57.4°C)
1.250	-62°F	(-52.2°C)
1.200	-16°F	(-26.7°C)
1.150	+5°F	(-15°C)
1.100	+19°F	(-7.2°C)

Figure 27.22 A fully charged battery will never get cold enough to freeze.

NOTE: *A battery that is at least ¾ charged (1.225 specific gravity) is in no danger of freezing.*

Electrolytes are formulated to correspond to the temperature where the battery will be used. A battery made for use in extremely cold weather has more concentrated electrolyte. In addition to preventing the electrolyte from freezing, this increases the *cold cranking power* of the battery. Higher specific gravities decrease the service life of the battery, however.

Electrolyte in tropical climates (where it never freezes) ranges from 1.210 to 1.230. The battery is more efficient at these temperatures, and the loss of cold cranking power is not important.

Reading the Hydrometer

Remove the vent caps and lay them on the top of the battery.

NOTE: *Before taking a hydrometer reading, be sure that the battery water level is up to the bottom of the filler tube of each cell. If the water level is low, the specific gravity will read higher than it actually is. If the electrolyte level is too low, refill and recharge the battery.*

- Draw electrolyte into the hydrometer (**Figure 27.23**) to the line on the tester bulb (if so equipped).
- Hold the hydrometer vertically so the float can rise to its proper level. The gauge must float freely. Read the specific gravity on the float and record it on your sheet (**Figure 27.24**). Make a temperature correction, if necessary.

NOTE: *A hydrometer is not accurate at temperatures above or below 80°F. Most of the popular hydrometers are temperature compensated. If not, a correction of +0.004 is made for each 10°F change above 80°F. Subtract for temperatures below 80°F. The compensation factor can be important at temperature extremes.*

Figure 27.23 Draw electrolyte into the hydrometer. *(Courtesy of Tim Gilles)*

Figure 27.24 Read the specific gravity on the hydrometer float. *(Courtesy of Tim Gilles)*

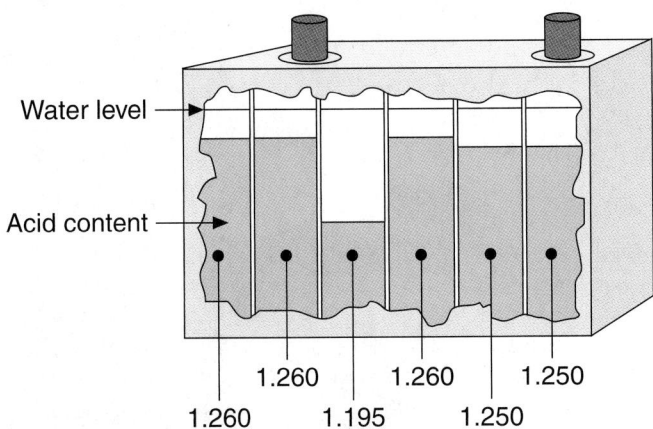

Water level

Acid content

1.260   1.260   1.250

1.260   1.195   1.250

Figure 27.25 Hydrometer readings should vary no more than 0.050. This battery has a defective cell.

■ Return the electrolyte to the cell and repeat the process for the other cells.
■ Differences between the cell readings should be less than 0.050 (**Figure 27.25**). All of the battery's cells are electrically connected to one another. If

one cell is low, it will pull down the voltages of the remaining cells. While the battery might function effectively immediately after a recharge, it will discharge overnight.

NOTE: *Shorted cells often are low on electrolyte while the other cells are all full.*

Built-In Hydrometer Batteries

Some maintenance-free batteries have a built-in hydrometer eye (**Figure 27.26**) that tests their state of charge. These batteries have no provision for adding water. A maintenance-free battery that does not have a built-in hydrometer cannot have its specific gravity tested, because it is sealed. The following are interpretations of hydrometer eye readings for one manufacturer's battery that uses a green ball.

■ If the eye is yellow or clear, the electrolyte is low and the battery should be discarded.
■ If the eye is green with a black spot in the center, recharge it (**Figure 27.27a**).
■ If the eye is green and the spot shows, the battery is at least 60% charged and is fine.

    Some manufacturer's use different hydrometer eye interpretations (**Figure 27.27b**)

NOTE: *The hydrometer checks the electrolyte strength of only the cell in which it is installed, so other cells could still have a problem.*

Refractometer

A refractometer is another way of measuring a battery's state of charge. Refractometers are more expensive than hydrometers but are more accurate and more versatile. The same tool can be used to measure coolant concentration. It automatically corrects for changes in temperature. An electrolyte sample is taken from the battery cell, and a drop or two is placed on the measuring prism; the reading is then viewed (**Figure 27.28**).

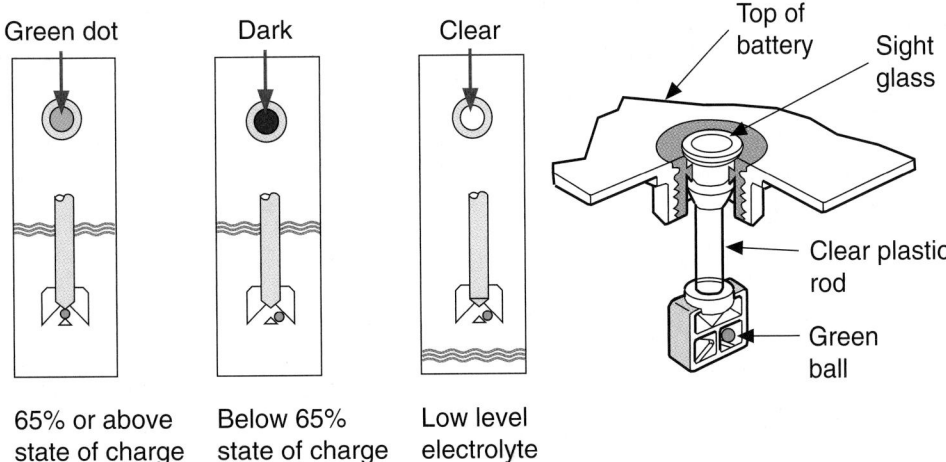

Green dot   Dark   Clear

Top of battery   Sight glass

Clear plastic rod

Green ball

65% or above state of charge   Below 65% state of charge   Low level electrolyte

Figure 27.26 A built-in hydrometer in a sealed battery indicates the condition of the battery.

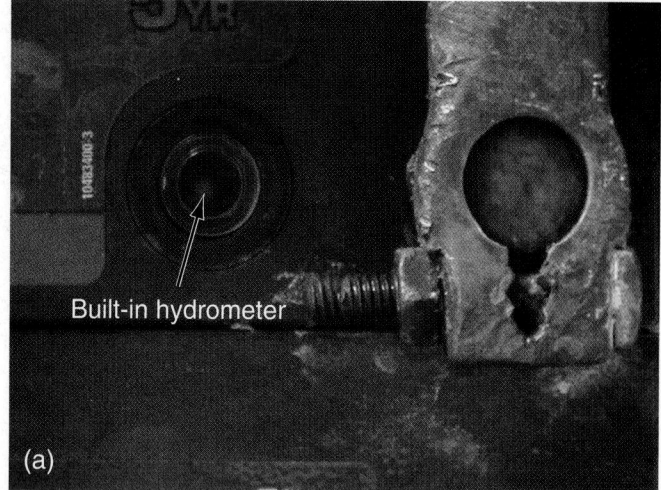

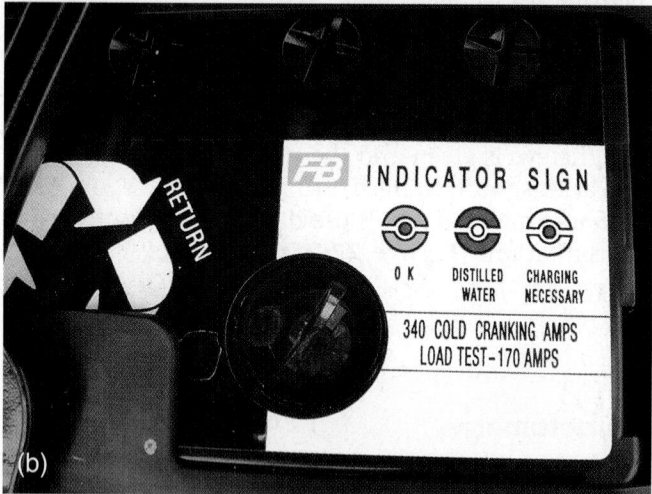

Figure 27.27 Two batteries with built-in hydrometers showing sufficient charge: (a) The eye on this battery is green. (b) This eye has a red dot surrounded by a blue ring. *(Courtesy of Tim Gilles)*

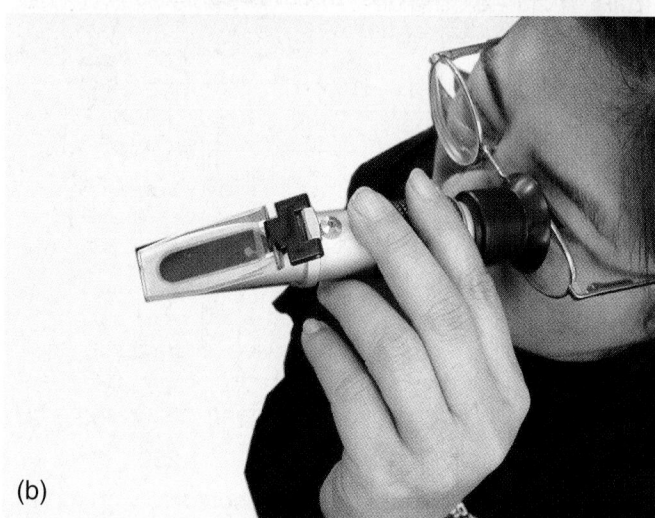

Figure 27.28 (a) Removing an electrolyte sample from the battery. (b) Reading the strength of the sample with the refractometer. *(Courtesy of Tim Gilles)*

SCIENCE NOTE

A refractometer uses the bending of light to determine the concentration of a substance. The angle at which light is refracted depends on the density of the sample. Differing materials will refract (bend) light in different manners.

Some testers are for multiple types of fluids, coolant and battery electrolyte, for instance. A different tester is required for brake fluid.

Open Circuit Voltage

Open circuit voltage, or *constant battery voltage*, can be checked to see if the battery has a sufficient state of charge. A battery is considered charged if its state of charge is 75% or more. The chart in **Figure 27.29** shows the relationship between open circuit voltage, specific gravity, and state of charge.

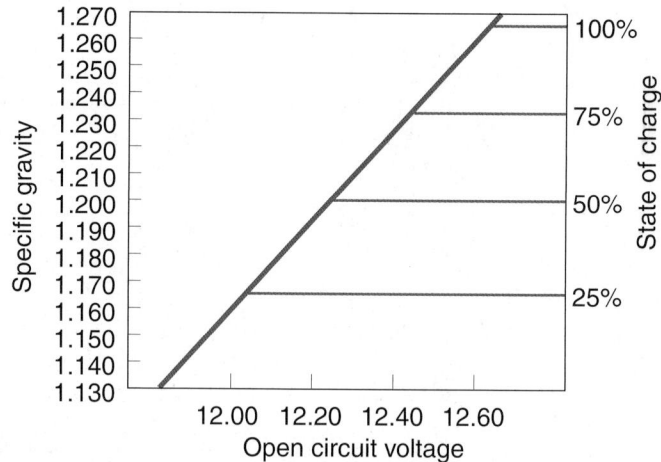

Figure 27.29 A chart comparing open circuit voltage to specific gravity. *(Courtesy of Ford Motor Company)*

Figure 27.30 Reading open circuit voltage with a digital voltmeter. *(Courtesy of Tim Gilles)*

Before making the test, remove the **surface charge** from the battery plates by connecting a 300-amp load across the battery for 15 seconds with a carbon pile. A battery capacity test (covered later) can also be performed to remove the surface charge.

> **SHOP TIP** The charge on the surface of the plates can actually be enough to start a car that has been recently run. But after the car sits for the night, the charge might be gone and the car will not be able to start.

Allow at least 10 minutes after a load test for the battery's voltage to stabilize.

NOTE: *Before reading open circuit battery voltage after charging or boosting, the battery voltage must be stabilized. Wait for at least 4 hours, or overnight.*

With the terminals removed from the battery, check the voltage across the positive and negative terminals (**Figure 27.30**). A fully charged battery will have an open circuit voltage of 12.6 volts or higher after the surface charge has been removed. Open circuit battery voltage must be above 12.4 volts or the battery must be recharged.

■ BATTERY CHARGING

Most battery chargers are the *constant voltage* type. A constant voltage charger reduces the amount of charging current while maintaining the output voltage at the same level as the battery reaches full charge. This type of battery charger keeps gassing to a minimum and is easiest on the battery. Other types of battery chargers can be dangerous. When they are left on all night, voltage can climb dangerously high. Trickle chargers are often of this type.

Be sure to check the electrolyte level in the battery before attempting to charge it. Dead batteries are sometimes low on water. Fill a battery only halfway prior to fast-charging. During a fast charge, the electrolyte will become heated and expand. If the battery is full, it could overflow onto the floor. Acid etches concrete. If the battery is a sealed type and electrolyte is low, do *not* attempt to recharge it.

> **SAFETY NOTE** A battery being charged gives off hydrogen. A spark can cause a dangerous explosion! Exploding batteries are the cause of over 15,000 injuries per year that require hospitalization.

> **SAFETY NOTE**
> ■ Sparks, flames, and cigarettes must be kept away from batteries.
> ■ Always unhook the ground cable first. If you remove the hot lead first, you might accidentally touch the pliers or wrench to a grounded part. This can cause a spark.
> ■ Wear eye protection when working around batteries.
> ■ Do not disconnect the battery charger while it is turned on.
> ■ Let the battery chemically stabilize before removing the charging cables.
> ■ Do not attempt to recharge a frozen battery. Forcing current through it can cause it to explode.
> ■ Keep the vent caps in place on the battery. The vent caps act as spark arrestors. Put a wet cloth over the top for additional protection.
> ■ Check that the cable polarity is correct—(+) to (+) and (−) to (−).

Some manufacturers require that the ground cable be disconnected from the battery if it is to be charged while it is in the vehicle. This is to protect the computer system.

Connect the battery charger cables to the terminals on the battery. The positive cable is usually red, and the ground cable is usually black. Do not assume this. See if the negative cable is connected to the block and the positive cable is connected to the starter motor.

On top-post batteries, twist the clamps back and forth so that they dig into the lead on the terminals. On side-terminal batteries, a special adapter is required for charging batteries (**Figure 27.31**).

The ignition switch must be off during charging on computer-controlled cars. Be sure to remove the ignition key as an extra precaution.

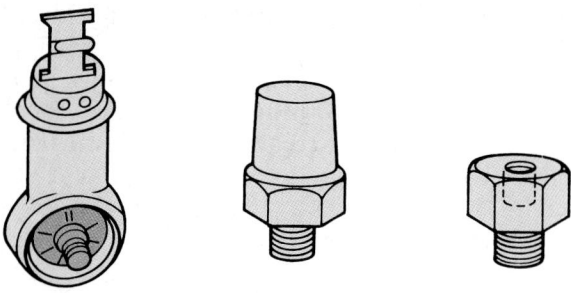

Figure 27.31 Adapters for testing and charging side terminal batteries.

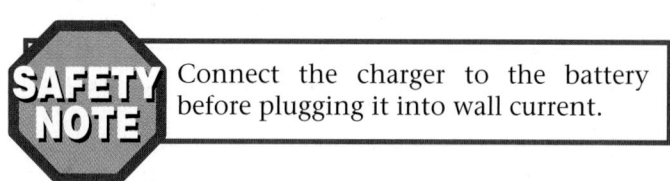

Figure 27.32 Battery charging rates and times.

Charging Amperage	5 Amps	10 Amps	20 Amps	30 Amps
Open Circuit Voltage	Hours Charging at 21°C (70°F)			
12.25 to 12.39	6 Hrs.	3 Hrs.	1.5 Hrs.	1 Hr.
12.00 to 12.24	8 Hrs.	4 Hrs.	2 Hrs.	1.5 Hrs.
11.95 to 12.09	12 Hrs.	6 Hrs.	3 Hrs.	2 Hrs.
10.00 to 11.95	14 Hrs.	7 Hrs.	3.5 Hrs.	2.5 Hrs.

SAFETY NOTE Connect the charger to the battery before plugging it into wall current.

Rate of Charge

Batteries are charged either fast or slow. A *fast charge* is when there is a high rate of current for a short period of time. Start charging the battery. If the battery is taking a charge that is over 30 amps, adjust its output to a lower setting.

NOTE: *If the battery will not take a charge, first check to see that the polarity of the connections is correct. Automotive battery chargers must sense some voltage in the battery or they will not work. Some battery chargers have a jump start button that will force electrons into a dead battery to give its plates a small surface charge. Depress this button for 1 minute and then release it. The battery should then take a charge.*

During a fast charge, the temperature of the electrolyte should not be allowed to climb higher than 125°F. If the charging voltage is kept below 15 volts, the temperature should remain low.

According to *Interstate Batteries*, 58% of battery failures are caused by overcharging. A sulfuric acid smell and/or boiling electrolyte indicates that a battery is being charged too fast. The electrolyte should never boil. It should just show a small amount of bubbling.

Check across the battery with a voltmeter to see that voltage is less than 15 volts. If it is higher, reduce the charging amperage until the voltage on the battery is less than 15 volts. Charge rates are listed in **Figure 27.32**. Do not fast-charge a battery for longer than 2 hours.

Sulfation

When a battery will not accept a charge, it is probably sulfated. This happens when the battery is allowed to remain in a discharged state. The lead sulfate in the battery plates becomes hard and resistant to recharging (**Figure 27.33**).

A dead battery that is not sulfated or damaged should readily accept a charge. If the charger is equipped with an ammeter, it should register that some amperage is flowing into the battery. If the battery still will not accept a charge, replace it. Some chargers apply a higher voltage initially to a battery. This can help a sulfated battery accept a charge.

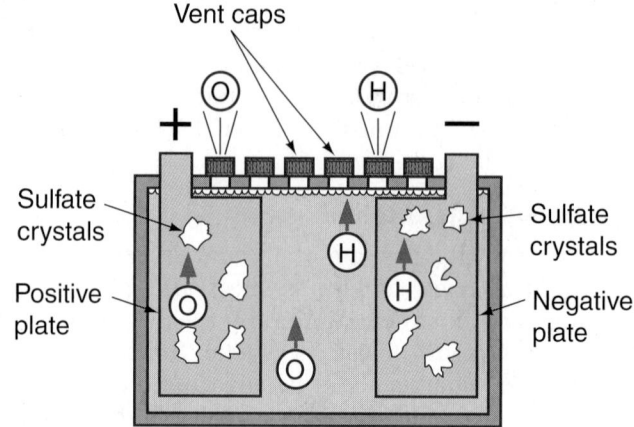

Figure 27.33 Sulfate crystals penetrate the battery plates when a battery becomes sulfated. They prevent the battery from accepting a charge.

A sulfated battery can sometimes be saved. Hook up a load, such as a headlight bulb, between the positive and negative posts. Let the bulb totally drain the battery. Then recharge it slowly. Repeating the cycle of charging and discharging the battery will sometimes knock enough of the sulfation off of the plates so that the battery can charge. If after two cycles the battery still will not accept a charge, replace it.

SCIENCE NOTE

During discharge, insoluble lead sulfate forms at each electrode. The lead sulfate actually in contact with the electrode can chemically adhere to the electrode. During charging, the lead sulfate is supposed to decompose into PH, H_2SO_4, and PbO_2. However, this decomposition is never 100% due to lead sulfate's insolubility. This means that over time less and less surface area of the electrode plates is available for recharging the battery.

Slow Charge

A slow charge is easiest on the battery. It is also the only way a battery can be restored to a fully recharged state. Slow-charging causes the lead sulfate on the battery plates to convert to lead peroxide and sponge lead throughout the plate's total thickness. Fast-charging only converts the lead sulfate on the outside of the plates.

Unfortunately, slow-charging is usually impractical unless the battery has been removed from a vehicle and can be left by the customer. If a customer needs to have a fast charge to get on the road as soon as possible, he or she should be encouraged to return the vehicle for a slow charge when time permits.

On a slow charge, the rate of charging is between 3 amps and 15 amps. A guideline for slow-charging is to allow 1 amp for each positive plate in one cell. The time required for a slow charge is related to a battery's reserve capacity. The chart in **Figure 27.34** shows the amount of time required for a slow charge at different rates of charge.

On a slow charge, a battery is fully charged when its specific gravity does not climb higher during two checks done 1 hour apart. The chart in **Figure 27.35** gives an approximate specific gravity relationship between a fully charged battery and one that is dead.

NOTE: *Following a recharge, if the difference between readings taken from different cells is more than 0.050, the battery should be replaced.*

Battery Capacity (Reserve Minutes)	Slow Charge
80 minutes or less	10 hours @ 5 amperes
	5 hours @ 10 amperes
Above 80 to 125 minutes	15 hours @ 5 amperes
	7.5 hours @ 10 amperes
Above 125 to 170 minutes	20 hours @ 5 amperes
	10 hours @ 10 amperes
Above 170 to 250 minutes	30 hours @ 5 amperes
	15 hours @ 10 amperes

Figure 27.34 Table showing the time and amperage for charging a battery according to its reserve capacity. *(Courtesy of Battery Council International)*

Specific Gravity (Cold and Temperate)	State of Charge	Specific Gravity (Tropical)
1.265	100%	1.225
1.225	75%	1.185
1.190	50%	1.150
1.155	25%	1.115
1.120	Dead	1.080

Figure 27.35 Table showing the relationship between a battery's state of charge and its specific gravity.

■ STORING A VEHICLE

When a vehicle is left for a month at a time without being started, its battery is still being discharged by loads that are present with the key off. A battery that is only half charged can often start an engine during warm weather. This is not only hard on the battery, but it presents extra challenges for the alternator that can cause it to overheat. To prevent this condition, disconnecting the battery any time it is not to be used for 10 days or more is a good recommendation.

■ BATTERY CAPACITY TEST

A battery capacity test, also called a heavy load test, is the first test to make when testing a battery with at least a 75% state of charge. Using the starter cranking amps will generally provide too light an amp load to be of value for this test. Determine the proper discharge amount to use during the battery capacity test in one of the following ways:

a. Find the CCA of the battery (**Figure 27.36**) and divide by 2.

NOTE: *Some batteries list cranking amps (CA). This is a higher value than CCA and should not be used for this test.*

b. Find the cubic inch displacement of the engine.

    1) Multiply by 2 for a four-cylinder engine.

    2) Multiply by 1.5 for a six-cylinder engine.

    3) Multiply by 1 for an eight-cylinder engine.

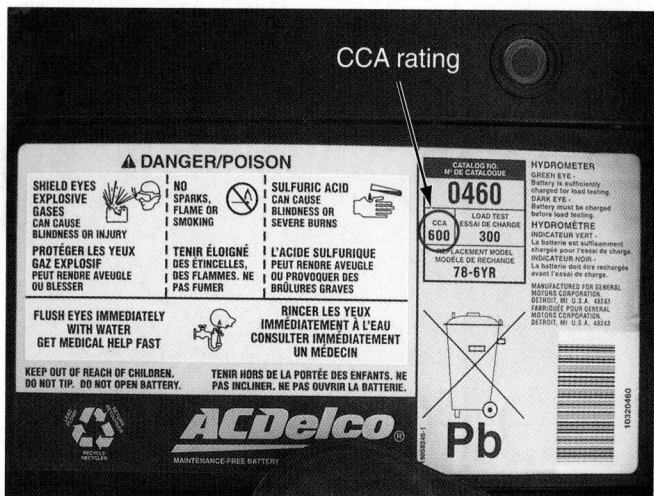

Figure 27.36 The cold cranking amperage is listed on the top of this battery. Divide by 2 to determine the discharge amount during the battery capacity test. *(Courtesy of Tim Gilles)*

NOTE: *This will give you a cold cranking amperage (CCA) value of a suitable battery for an average vehicle. For heavy electrical loads or extra-heavy starter draws, this may have to be increased by 50 amps.*

    c. Multiply the amp-hour rating by 3.

    To test the battery's capacity (load test), discharge the battery for 15 seconds at one of the above discharge amounts. Battery voltage should remain above 9.6 volts at the end of 15 seconds.

Volt-Amp Tester (VAT)

The tester most often used in repair shops for the battery capacity test is a volt-amp tester, or VAT (**Figure 27.37**). It has a voltmeter, an ammeter, and a variable **carbon pile** rheostat. The large tester leads are used for current flow (amperage). When the carbon pile is compressed, current flows through them. Voltage can

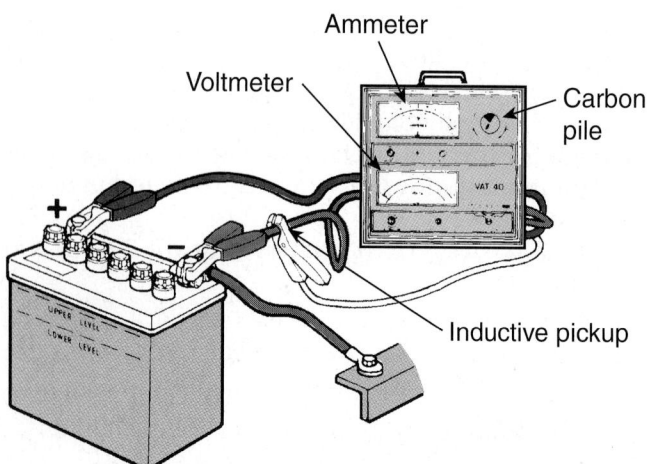

Figure 27.37 Parts of a volt-amp tester.

be read through a small wire because voltage is only potential, not current flow.

    Amperage is usually read by inserting the gauge in series, but modern ammeters have an **inductive pickup** that wraps around one of the big leads. It uses an inductive pickup to sense the amount of current flowing through the wire.

SCIENCE NOTE

Electricity can be produced by magnetic induction. Whenever a conductor is moved through magnetic lines of force, or vice versa, a potential difference or voltage is created between the ends of the conductor. As soon as the conductor or magnetic field lines stop, the voltage ceases to exist. The induced voltage can be increased by increasing the speed of movement of the lines of force or by increasing the number of conductors that are cut.

NOTE: *Side-terminal batteries require a special adapter if the cables are to be disconnected (see Figure 27.31).*

    The carbon pile is made up of alternating positive and negative layers of carbon that are compressed against each other when a large knob on the face of the tester is tightened. Turning the large knob clockwise will cause a rapid discharge of the battery. As the layers of carbon under the knob are compressed, the battery is discharged at a higher rate as more current travels between positive and negative.

SAFETY NOTE Before connecting the cables to the battery, be sure that the control knob on the carbon pile (large knob) is off (counterclockwise).

    Watch the ammeter to determine the amount of current flow. Turn the control knob clockwise. The ammeter needle should move. If not, determine the cause before proceeding. Maintain the desired amperage for 15 seconds while watching the voltmeter. Read the voltmeter. Then, turn off the carbon pile control knob (counterclockwise). The voltmeter reading while the carbon pile is on should be above 9.6 volts at the end of the 15-second test. For every 10 degrees below 70 degrees, the voltage may be 0.1 less than 9.6.

    If a battery's state of charge is 75% or higher on an open circuit voltage test (12.4 volts) and the battery fails the load test, replace the battery.

NOTE: *On a battery with a hydrometer eye:*

■ *If it is yellow, the electrolyte level is low. If this is a sealed battery, it must be replaced regardless of the test results.*

■ *A eye "dot" shows sufficient charge (see Figure 27.27).*

Battery capacity can be tested *without* a carbon pile volt-amp tester in the following manner:

- Disable the engine's ignition system.
- Remove the surface charge from the battery's plates by turning on the high beam headlights for 30 seconds.
- Connect a voltmeter to the positive and negative terminals of the battery.
- Crank the engine with the starter motor for 15 seconds.
- If battery voltage is 9.5 volts or higher, the battery is good.
- If battery voltage is below 9.5 volts, recharge the battery and repeat the test. If the result is still below 9.5 volts, check for excessive starter motor draw. If the starter is good, replace the battery.

Battery Conductance Testing

Battery conductance, or capacitance, testing is a means of quickly testing a battery within seconds to see if it can conduct current. Conductance testing gives an indication of the amount of battery plate surface available to react chemically with the electrolyte and produce current. The test is recommended by many vehicle and battery manufacturers. The tester (**Figure 27.38**) sends a small current through the battery and measures the amount of alternating response. A new battery will conduct at a high rate, from 110% to 140% its CCA rating. When a battery ages, it can shed active material from the surfaces of its plates, impacting its efficiency and lowering the test results. This can give an indication of remaining battery life.

The conductance test can also detect shorts and open circuits in cells. Related battery information, such as CCA, is entered into the tester. Following the short, automated test, the tester displays whether the battery is good or not, or if it requires recharging before performing the test again. While this test is relatively reliable, it will occasionally give a satisfactory test result on a fully charged defective battery that has failed a hydrometer or heavy load carbon pile test.

▆ BATTERY DRAIN TEST

If a battery that is in good condition continually goes dead, there may be a circuit that is causing the drain when everything is supposed to be off. This is called a **parasitic load** or drain. Electronic components draw small amounts of current at all times. If a voltmeter is used to check for drains, it may show battery voltage due to the parasitic loads of the vehicle's electronics.

The most correct way to test for drains is to use an ammeter that can read in tenths of amps (**Figure 27.39**). Less than 0.050 A (50 mA) is an acceptable amount of draw. Newer vehicles may require up to 0.100 A (100 mA).

NOTE: *Many electronic circuits will drain when battery power is lost, such as when the key is shut off for about 5 minutes or the battery is disconnected. They may require several minutes before they "time out" and stop pulling a higher amp draw.*

If a sensitive ammeter is not available, use a test light (**Figure 27.40**). If a 1-candlepower or smaller bulb does not light, the drain is too small to be of concern.

- With the key off, check all lights and accessories to see that they are off. Check all courtesy lights, such as those that come on when a car door is open, to see that they do not remain on at all times.
- Disconnect the battery ground cable. Connect a test light between the battery cable end and the battery post. It should not light.

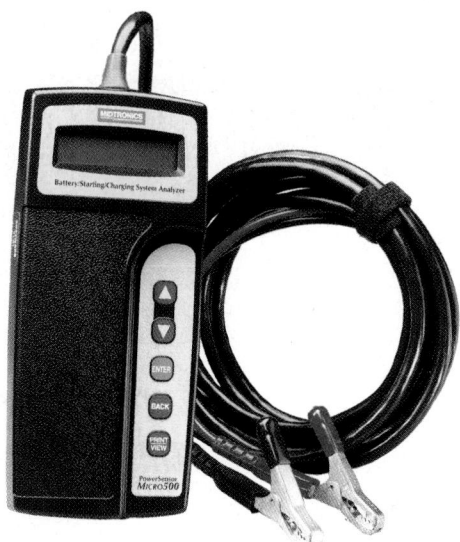

Figure 27.38 A battery conductance tester capable of testing the battery and the starting and charging systems.

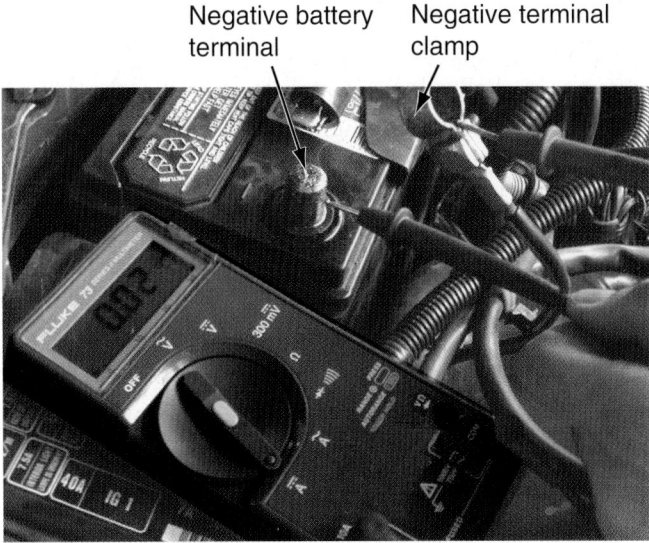

Negative battery terminal   Negative terminal clamp

Figure 27.39 Check for parasitic drains by inserting an ammeter in series with the negative battery cable. *(Courtesy of Tim Gilles)*

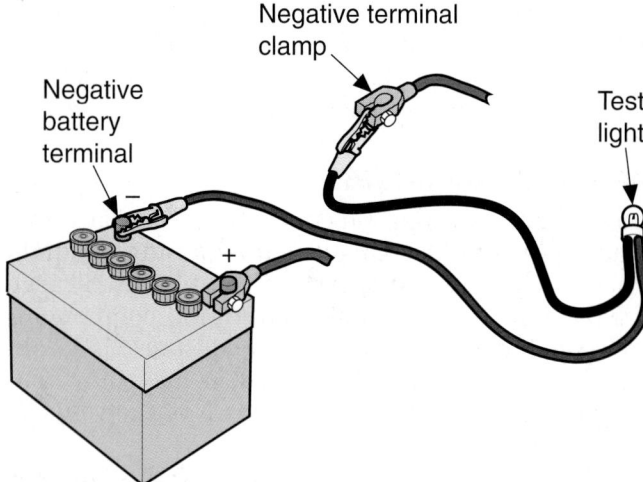

Figure 27.40 Checking for parasitic drains using a test light. *(Courtesy of Ford Motor Company)*

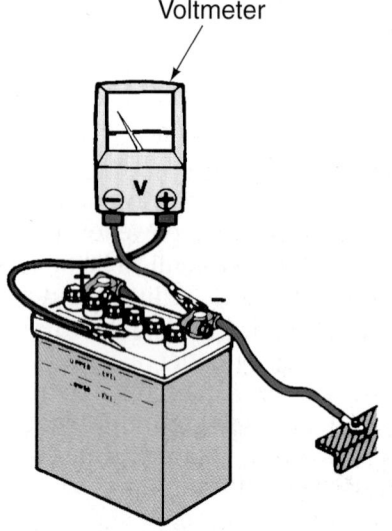

Figure 27.41 Use a voltmeter to check if dirt on the battery top is causing a drain.

■ If the test light comes on during the last part of the test, check the lights in the trunk, ashtray, engine compartment, and glove box to see that they are not the cause of the problem. Pulling fuses one at a time can isolate the problem. Remove a fuse and see if the test light goes out. A brake light switch that stays on could also cause the problem.

A dirty battery case can also cause a drain. Test with a voltmeter between the negative battery post and the top of the case (**Figure 27.41**).

■ BATTERY JUMP-STARTING

Dead batteries that are in good condition are common. Customers often leave lights on or crank an engine until the battery loses its charge. Jump-starting from another vehicle can help to get the vehicle running again.

Jumper Cables

Use good-quality braided-copper jumper cables. They are flexible and will not overheat if they are large enough. Using stiff aluminum jumper cables that are too small will result in hot cables and possibly a burn to the technician.

CASE HISTORY

A technician had completed the installation of a rebuilt engine in a motor home. When he tried to crank the engine over, it would not turn. He was concerned that maybe something in the engine was too tight. The battery was located in the rear, and the cable run was quite long. The headlights were dim, so he checked the battery's specific gravity. It was about 1.200 (low).

While the battery charged, he connected the shop's good jumper battery directly to the starter motor and engine block. This eliminated all of the other possible problems related to the wiring in the motor home. The engine still would not crank. He attached a ratchet and socket to the bolt on the vibration damper on the front of the crankshaft. The crankshaft turned with a reasonable amount of resistance.

Next, he connected the ammeter connection of the volt-amp tester around one of the jumper cables. Attempting once again to crank the engine, he found that the starter motor was not drawing enough current. He pulled back the insulation on one of the clamps on the jumper cable. Someone had switched jumper cables with the shop's original high-quality ones. The substituted cables were of too high a gauge number (too thin), although they appeared to be good because the insulation on them was thick. When he substituted another pair of jumper cables, the engine cranked normally.

The chart in **Figure 27.42** shows the size of jumper cables needed for different size engines. The

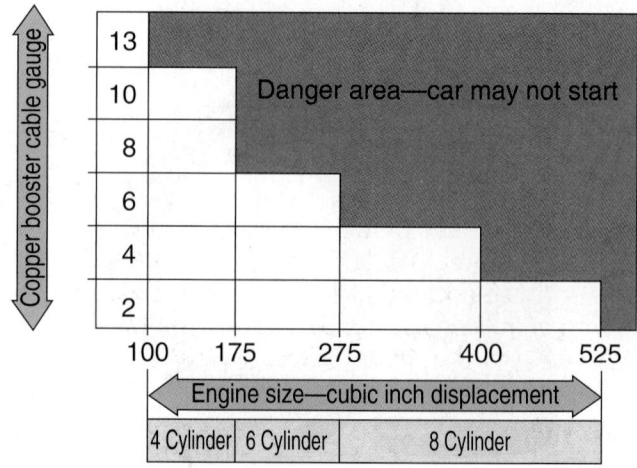

Figure 27.42 This chart shows recommended minimum jumper cable sizes. *(Courtesy of Federal-Mogul Corporation)*

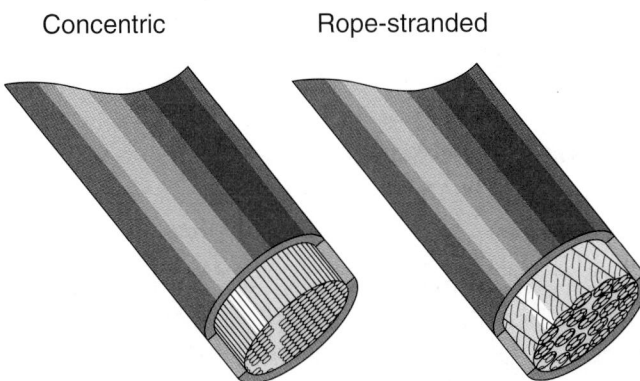

Concentric   Rope-stranded

Figure 27.43 Rope-stranded cables are flexible.
(Courtesy of Ford Motor Company)

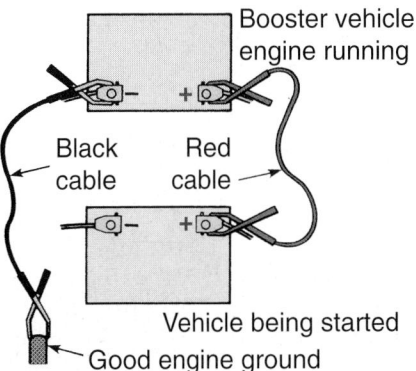

Booster vehicle engine running

Black cable   Red cable

Vehicle being started
Good engine ground

Figure 27.44 Jump-start procedure.

best jumper cables are rope stranded (**Figure 27.43**). These cables are made up of wires that are much smaller. This allows the cable to be very flexible.

CAUTION
- Wear eye protection.
- Do not try to jump-start a vehicle in freezing weather. Jump-starting a frozen battery can cause it to explode.
- Serious damage to a car's electrical system can result if the cables are connected backward (with the wrong polarity).
- When possible, computer-controlled vehicles should have a dead battery removed and charged rather than jump-starting. All connections must be made with the key off on these cars.
- If a computer-equipped car is to be jump-started, it is mandatory that both cars have their ignitions in the "off" position when the jumper cables are connected.
- Check to see that the battery has electrolyte above its plates. If not, do not attempt to jump start the car.
- The two vehicles should not be touching each other. This could provide an unwanted ground path.

Jump-Start Procedure

Turn on the heater blower motor in the vehicle with the dead battery. Turning on the blower motor will allow it to help absorb any damaging voltage spikes. Turn off all other switches and lights. Connect the hot cable first and the ground cable last. When attaching the ground cable from the booster battery to the dead battery, do *not* connect the other end of the negative cable to the dead battery. Connect it instead to a ground on the engine (**Figure 27.44**). A metal bracket or the end of the negative battery cable that is attached to the block will do.

SAFETY NOTE There will be an arc (spark) when the last cable is connected as the dead battery tries to equalize itself with the booster battery. Connecting cables away from the battery avoids the possibility of a spark near a battery that can cause it to explode.

As soon as the dead vehicle starts, disconnect the jumper cable immediately from the block. Remove the cables in the reverse order that they were installed.

Low-maintenance batteries have a higher internal resistance than conventional lead-antimony batteries. The jumper cables may need to remain in place for a minute or so before attempting to start the disabled vehicle so that the dead battery can take on a charge.

Run the host vehicle at 2,000 rpm to allow its charging system to recharge the battery.

■ HYBRID BATTERY SERVICE

Hybrids have both low-voltage and high-voltage batteries. Service procedures are unique to each.

Hybrid Low-Voltage Battery Service

In a hybrid, the low-voltage electrical system supplies power to control the high-voltage modules. It also supplies power for the engine and accessories. Some hybrids use a conventional battery that is mounted under the hood. Other hybrids use a small 12-volt auxiliary battery (**Figure 27.45**) located in the trunk, where a conventional battery, which vents hydrogen gas, would be dangerous. When the auxiliary battery is located in the trunk, it will be an absorbed glass mat (AGM) battery. The auxiliary battery compartment is vented to outside air.

The 12-volt battery is charged through the inverter from the high-voltage charging system. A hybrid battery is matched to the charging system. If it requires replacement, it must be replaced with the recommended battery.

AGM batteries cannot be fast-charged. They are trickle-charged at a maximum charge of 3.5 amps for

Figure 27.45 A hybrid absorbed glass mat (AGM) auxiliary battery located in a compartment in the trunk. *(Courtesy of Tim Gilles)*

8 hours. A conventional 1.5-volt trickle charger will work, but it will take longer. Special pulse chargers are available in some dealerships, but they are not widely used due to their expense.

Most hybrids cannot be push-started. When jump-starting, connections are made in the same manner as a conventional battery. When the 12-volt battery is in the trunk, there is often a dedicated connector under the hood for jump-starting.

Hybrid High-Voltage Battery Service

Manufacturer warranties on hybrid batteries are relatively long; many are guaranteed for 8 years or 100,000 miles. If a 30% to 70% battery charge is maintained, the battery should last 150,000 miles. In fact, some warranties are for 10 years or 150,000 miles. Battery packs can cost from $3,000 to $8,000 to replace, depending on the manufacturer and battery type. The battery pack is made up of modules. A typical module contains up to six 1.2-volt cells. The on-board computer can check each battery module. If one is bad, it can tell which one to replace.

Hybrid Battery Cooling System. Intake air for the high-voltage battery cooling system is often on the shelf above the back seat. Hybrid vehicle owners need to know that if they block the intake vents with personal items, the battery temperature will rise. Temperature sensors will cause the vehicle to run in fail safe mode and a diagnostic trouble code will be set. The vehicle will run only on the gasoline engine with limited speed and acceleration.

Hybrid High-Voltage Battery Charging. The state of charge of the high-voltage hybrid battery pack does not usually fall below the point at which the vehicle would no longer operate. If this happens, service at a dealership will be required. The battery charger for hybrid batteries can raise their charge to 40% to 50% in

about 3 hours. These battery chargers are uncommon and sometimes must be shipped to the dealership. Only specially trained technicians may operate them.

General Hybrid Safety

If you ever work on a hybrid electrical system, your life will depend on knowing what and when something is safe to touch. The Ford Hybrid has the highest voltage of the hybrid vehicles, with a 330-volt system, although it does not boost its voltage. The Toyota Highlander, Lexus RX400h, and Camry hybrids all boost their voltages in amounts that can range up to 650 volts and 60 amps *instantly!*

Before working on anything to do with the electrical system, you will need to know how to eliminate system power. On a typical hybrid electrical system, turning the key off opens the orange high-voltage cable contacts. When the key is on, the system is dangerous. Never leave the key in the ignition.

> **SAFETY NOTE** Standard industry practice has always been to leave the keys with the vehicle during service. With a hybrid, this can be dangerous. Besides the danger of electrocution, some hybrids can start when the key is within 15 feet of the vehicle.

Depowering the high-voltage system is sometimes as easy as turning it off and removing the key, but you cannot afford to assume this. All hybrid manufacturers provide ample information on the Internet to allow fire departments and other emergency personnel easy access for training. Emergency information is available at the following URLs:

http://techinfo.toyota.com/public/main/erg.html
http://www.firehouse.com/extrication

Remember:
- **Orange means high voltage. Do not forget this!**
- The high-voltage battery pack will still contain high voltage even after you have depowered the system. Every manufacturer says to wait at least 5 minutes after disconnect, due to the storage capacitor. Capacitors, which are about the size of a soda can, are rated as high as 750 volts. Always use a test instrument to check for high voltage before touching a part. Your multimeter must be able to read at least 600 volts and be CATIII or CATIV approved.
- When working on a high-voltage system, keep one hand in your pocket and do not lean against the car. Do not wear rings or watches. Wear insulated gloves rated at a minimum of 1,000 volts. Gloves need to be recertified every year at about the same cost as buying new ones. If there is an orange

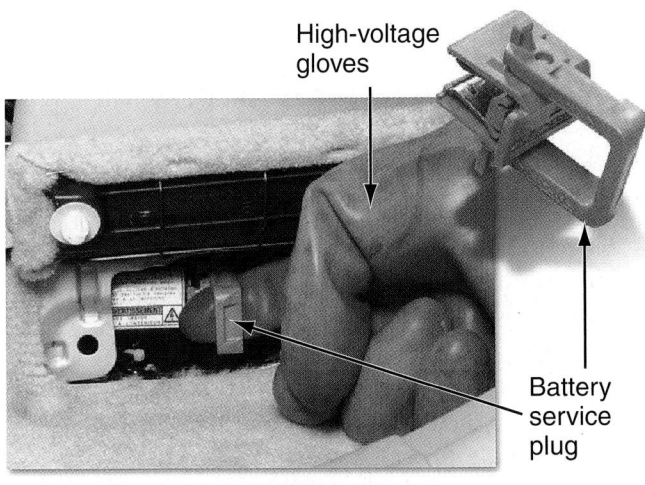

High-voltage gloves

Battery service plug

Figure 27.46 Use high-voltage gloves to pull the battery source plug. *(Courtesy of Toyota Motor Sales, U.S.A., Inc.)*

cable, get your meter out and put on your gloves just in case!

■ Use high-voltage gloves to pull the battery service plug (**Figure 27.46**). Keep the battery service plug in your pocket so nobody can put it back in but you.

REVIEW QUESTIONS

1. If a battery is filled to the split ring, will it have the correct specific gravity?

2. When battery terminal posts are exposed to air, they oxidize and change color. True or False?

3. Battery corrosion can be prevented by covering terminals with grease or installing washers under the terminals. True or False?

4. What is the name of the tester that compares the weight of pure water to the weight of electrolyte?

5. What is the approximate specific gravity of new electrolyte?

6. What kind of gas does a battery give off as it charges?

7. What kind of battery charge penetrates the entire plate, fast or slow?

8. What does VAT stand for?

9. A current draw on the battery when the key is off, is called a _____ load.

10. What color is used to identify electrical wiring on dangerous high voltage hybrid electrical systems?

ASE-STYLE REVIEW QUESTIONS

1. A battery has one cell low on electrolyte. What would be the most probable cause?
   a. A voltage regulator allowing too high a charge rate
   b. A shorted cell
   c. A leak in the case
   d. All of the above

2. Two technicians are discussing volt and amp testing. Technician A says that large test leads are for load testing a battery. Technician B says that small test leads are for reading voltage. Who is right?
   a. Technician A   c. Both A and B
   b. Technician B   d. Neither A nor B

3. Technician A says that storing materials on the shelf above the back seat can cause a hybrid vehicle to set a code and run in fail safe mode. Technician B says that green electrical wiring signifies high voltage. Who is right?
   a. Technician A
   b. Technician B
   c. Both A and B
   d. Neither A nor B

4. What can be used to neutralize battery acid?
   a. A base
   b. Baking soda
   c. Both A and B
   d. Neither A nor B

5. Technician A says that disconnecting a battery on a computer-controlled car can result in a change in a car's driveability. Technician B says that a small battery is all that is needed to maintain computer memory. Who is right?

   a. Technician A   **c.** Both A and B

   b. Technician B   **d.** Neither A nor B

6. Before charging a deep-cycle battery, the electrolyte level should be:

   a. Just above the top of the plates.

   b. Just below the top of the plates.

   c. Even with the split ring in the fill opening.

   d. There should be no electrolyte in the battery.

7. All of the following are true statements about jumper cables *except*:

   a. Using jumper cables that are too small in diameter can prevent the engine from starting.

   b. When the last connection is made there will be an arc.

   c. On negative ground systems always connect the positive cable last.

   d. It is usually a better choice to charge a battery first rather than to jump it.

8. Technician A says that the battery ground cable is often black. Technician B says that black is negative and red is positive on all vehicles. Who is right?

   a. Technician A   **c.** Both A and B

   b. Technician B   **d.** Neither A nor B

9. Technician A says that battery terminal posts are made of lead. Technician B says that battery posts turn green when they oxidize. Who is right?

   a. Technician A   **c.** Both A and B

   b. Technician B   **d.** Neither A nor B

10. All of the following are true about battery conductance testing *except*:

   a. It is a fast way to test a battery to see if it can conduct current.

   b. It is better than hydrometer and carbon pile testing at determining battery condition.

   c. The test sends a small current through the battery and the amount of alternating response is measured.

   d. It can detect shorts and open circuits in cells.

Starting System Fundamentals

■ OBJECTIVES

Upon completion of this chapter, you should be able to:

✔ Explain electric motor principles.

✔ Describe starter parts.

✔ Understand the operation of a solenoid.

✔ Discuss starter drive operation.

■ KEY TERMS

solenoid

pole shoe

field coil

overrunning clutch

inertia starter drive

■ INTRODUCTION

The starting system is an important part of the automotive electrical system. Without a starter, the car would have to be push-started. Most of today's cars have automatic transmissions, which prevent push-starting. Henry Ford's Model T had a hand crank for the engine. This chapter deals with the operation of the starting system. If you are unclear about any of the basic principles described here, refer to Chapter 25.

■ STARTER MOTOR

The *starter circuit* includes a powerful *starter motor* and *starter drive, battery, ignition switch,* and *solenoid* (relay) (**Figure 28.1**). A starter operates at a high rpm. It has a gear (*drive pinion gear*) on the end of its *starter drive* that meshes with a large gear (*ring gear*) on the engine's flywheel (**Figure 28.2**). A gear ratio provides the starter with the leverage (torque) necessary to be able to turn the crankshaft against engine compression. The gear ratio between the two gears is about 18:1. To crank the engine at normal cranking speed (200 rpm), the starter motor must be turned at 3,600 rpm.

■ STARTER MOTOR FUNDAMENTALS

Starters use electromagnetism to convert electrical energy stored in the battery to mechanical power to crank the engine. Other electric motors on the car work in the same manner. Because like charges repel each other and unlike charges attract, magnetic fields can be used to cause motion. Chapter 25 gives an explanation of magnetic fields.

Figure 28.3 shows the magnetic field surrounding a conductor positioned between the north and south poles of a horseshoe magnet. There are two separate magnetic fields. One is produced by the horseshoe magnet, and the other results from the current flowing through the conductor, which represents one loop in the armature.

There is a push-pull effect on the armature (**Figure 28.4**) that causes the conductor to want to move from a stronger magnetic field to a weaker field (from left to right in the illustration). This action is stronger if the magnetic field is stronger or current flow in the conductor is higher.

In a motor, the conductor is formed into a loop. When electrons flow through the loop, the rotating force will be pushing in different directions on opposite sides of the loop (**Figure 28.5**). The two magnetic fields work together on one side of the armature loop to make one strong field. The other side of the loop has a weaker field. This causes the loop to turn a small amount.

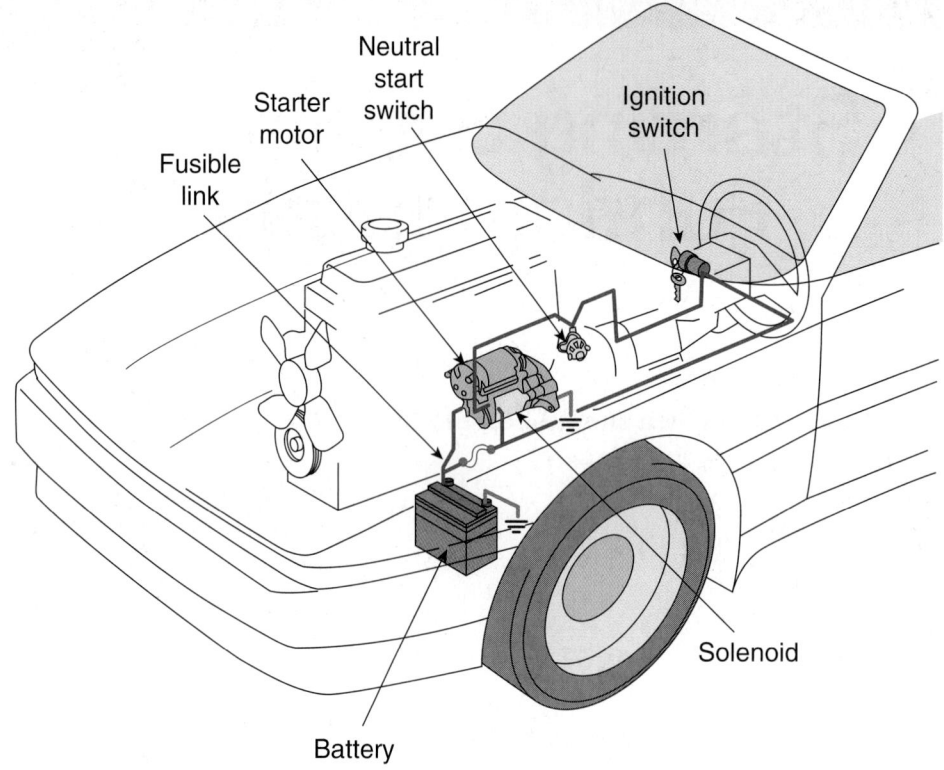

Figure 28.1 Parts of a starter circuit.

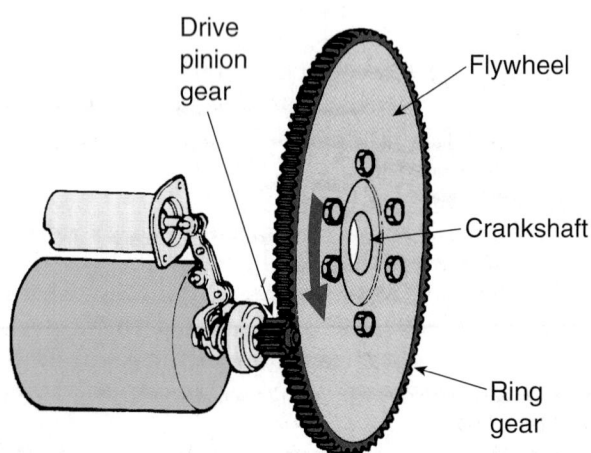

Figure 28.2 Starter drive parts. (*Courtesy of Ford Motor Company*)

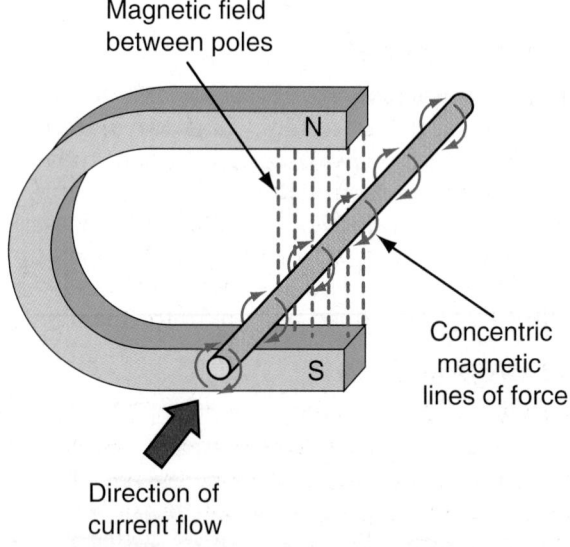

Figure 28.3 Two magnetic fields are produced when current runs in the wire.

Parts of a basic motor are illustrated in **Figure 28.6.** A loop of wire is placed between two electromagnetic pole shoes. The ends of the wires have *commutator* bars that electrically conducting *brushes* ride upon (**Figure 28.7**).

Armature

Multiple loops make up an armature (**Figure 28.8**). As each loop rotates slightly, a new magnetic field from the next loop reacts with the magnetic pole shoe. This causes the armature to continue to spin.

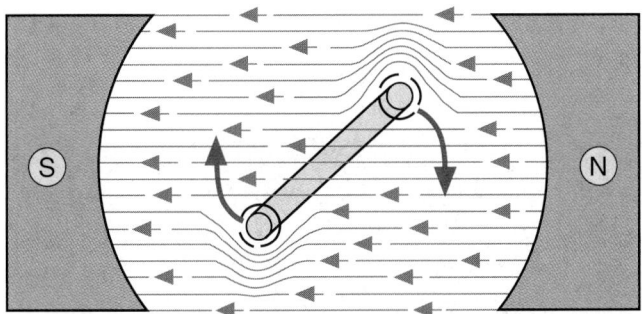

Figure 28.4 The conductor attempts to move from the strong field to the weak field.

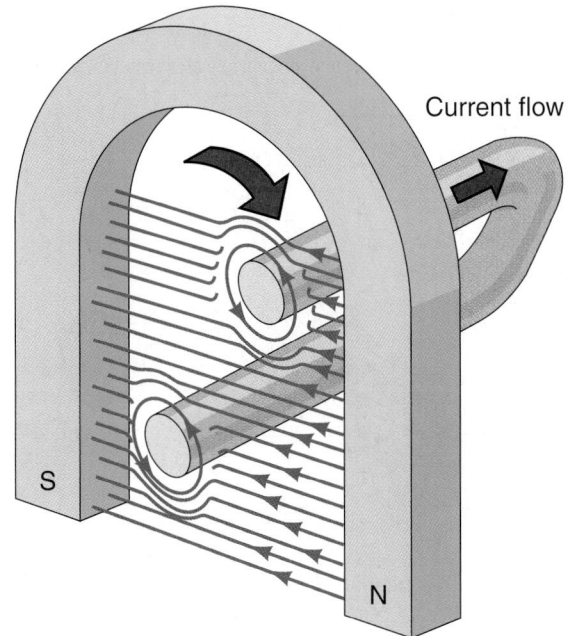

Figure 28.5 The rotating force pushes in different directions on opposite sides of the loop. (*Courtesy of Ford Motor Company*)

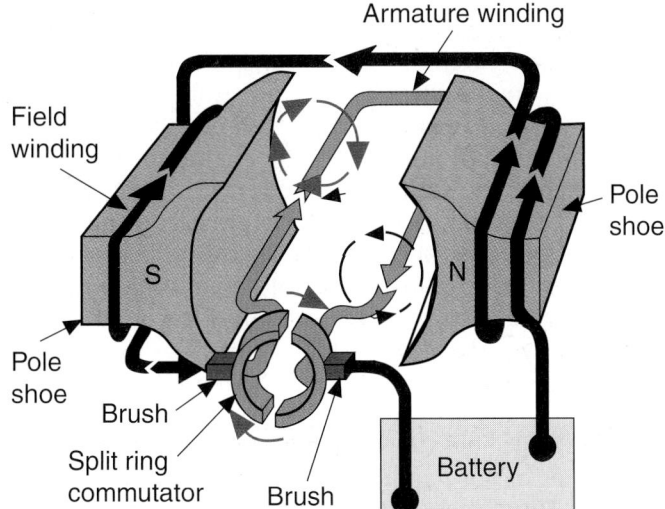

Figure 28.6 Basic parts of a motor.

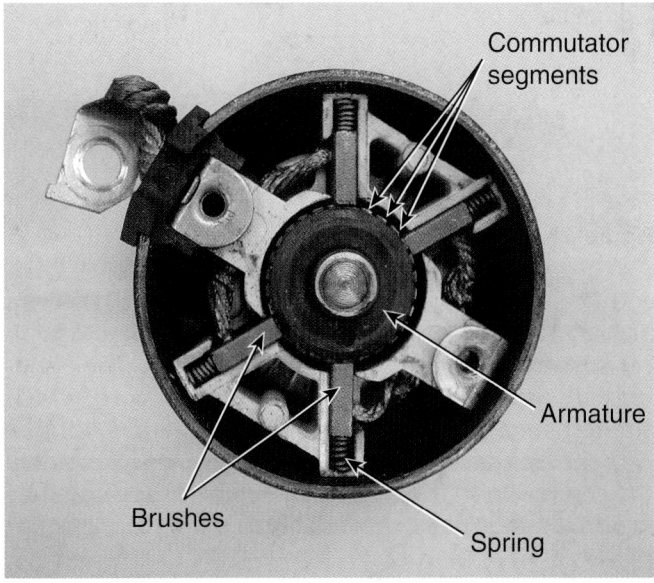

Figure 28.7 Starter armature and brushes. (*Courtesy of Tim Gilles*)

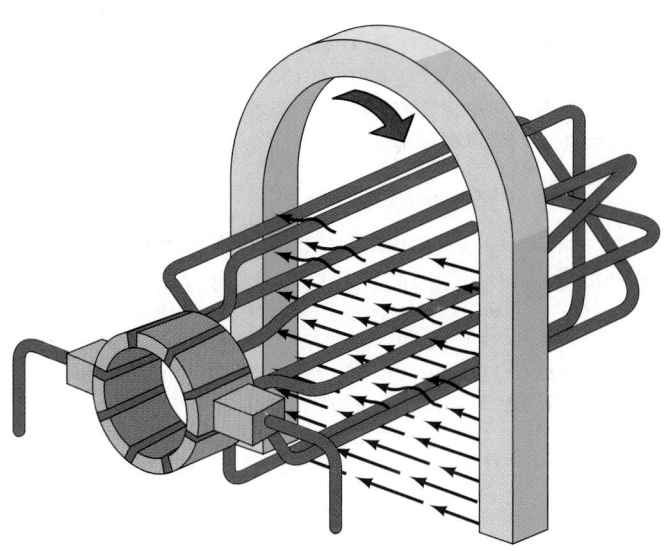

Figure 28.8 Multiple loops make up an armature. (*Courtesy of Ford Motor Company*)

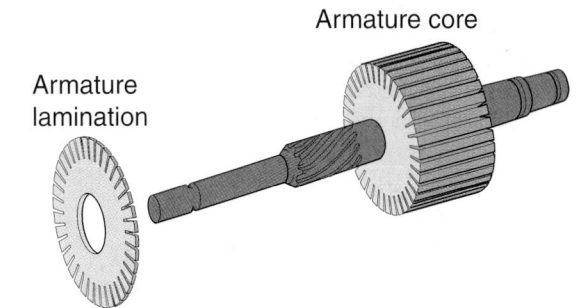

Figure 28.9 The core is made up of laminations, each insulated from the other.

The armature has a soft iron *core* (**Figure 28.9**). The core is made up of soft iron laminations (thin layers of iron, each insulated from the other). If the core were solid iron (without the laminations), it would generate

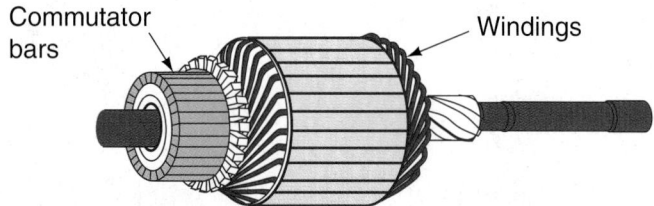

Figure 28.10 Parts of an armature.

eddy currents. The core is wrapped with many loops of heavy insulated copper wire called *windings*. A typical armature will have about 30 windings. The windings are arranged lengthwise on the iron core, which strengthens and concentrates the magnetic field. The electrically conducting end of the armature is called the commutator. It has multiple segments separated by insulation strips called *mica* (**Figure 28.10**).

Field Coils

The starter must be a very powerful motor in order to spin an engine fast enough to start it. It requires very strong electromagnets to make a magnetic field strong enough to move the armature. The field coils are made of heavy copper ribbons wound around soft iron cores called pole shoes (**Figure 28.11**).

Permanent Magnet Starter

Some starter motors used since the mid-1980s have no field coils but have permanent magnets instead (**Figure 28.12**). A special alloy magnet was developed that is ten times as strong as earlier permanent magnets. A permanent magnet starter is simpler, weighs less, and creates less heat than a conventional field coil starter. With no field coils, current goes directly to the armature through the commutator and brushes.

Brushes

Brushes, usually made of carbon, are lightly held against the commutator by springs. There are usually four brushes, which are together in pairs (see Figure 28.7). They supply electricity to the armature windings. One brush is in contact with one commutator bar, and the other is in contact with the commutator bar on the other side of the same winding (loop). One of the brushes is positive and the other is negative.

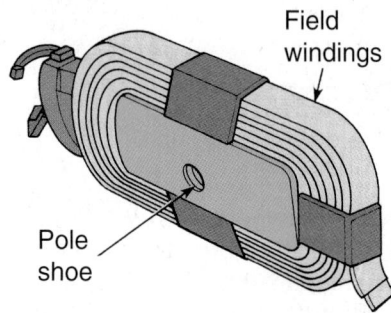

Figure 28.11 Field windings and pole shoe.

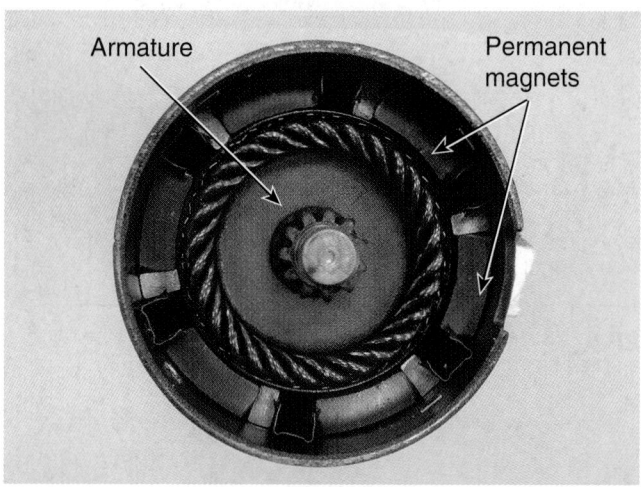

Figure 28.12 A permanent magnet starter. *(Courtesy of Tim Gilles)*

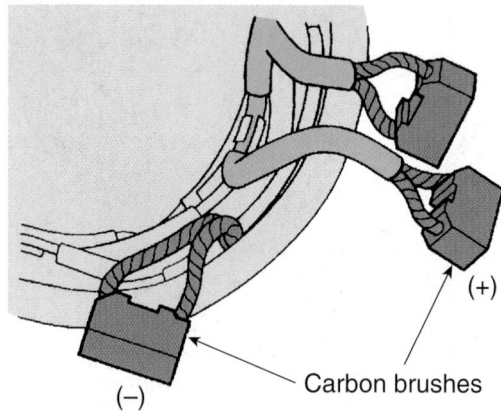

Figure 28.13 The positive brush is insulated from the frame. *(Courtesy of Ford Motor Company)*

The positive brush is insulated; the negative brush is connected to the starter housing (**Figure 28.13**).

■ STARTER DRIVES

Overrunning Clutch Drive

Starter drives have an overrunning, or one-way, clutch. The **overrunning clutch** transmits motion from the starter to the flywheel but not from the flywheel to the starter (**Figure 28.14**). It disengages from the engine at startup, preventing the engine from driving the starter. If the starter were to remain engaged with the flywheel after startup and engine rpm increased to 2,000 rpm (fast idle), the starter would be forced to turn at 36,000 rpm. At speeds of about 10,000 rpm, a starter can be destroyed by centrifugal force. The teeth on the starter drive gear are tapered to allow for smooth engagement with the flywheel (**Figure 28.15**).

■ STARTER ELECTRICAL CIRCUIT

The starter motor requires a large amount of current to operate. The battery must be in good enough condition to be able to provide substantial current for at least 15 seconds. The starter or its relay is connected directly

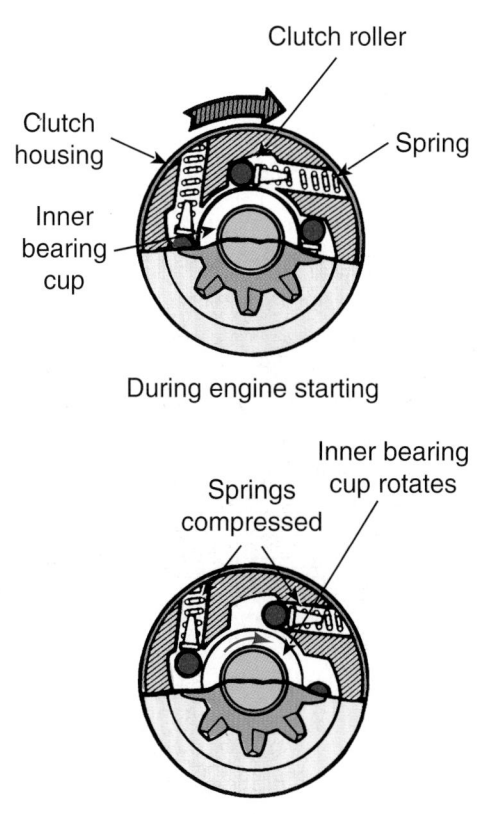

Figure 28.14 Overrunning clutch operation.

Figure 28.15 The starter drive pinion gear has tapered teeth.

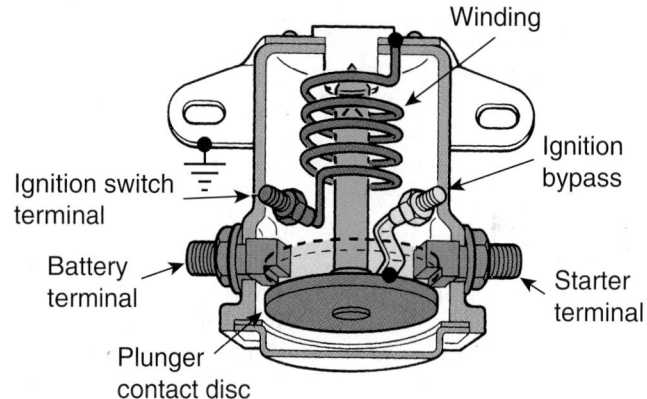

Figure 28.16 A magnetic switch. (*Courtesy of Ford Motor Company*)

to the top of the battery by a heavy cable. The return for the electrical circuit is through another cable (usually the negative) after flowing through the engine block.

The starter is switched on by the ignition switch. A relay is required so that a high amount of current can be controlled by a small wire to the ignition switch. If the same amount of electricity went through the ignition switch and the starter, the switch would have to be very large and expensive. A starter uses a relay called a solenoid, or *magnetic switch* (**Figure 28.16**).

Solenoid

Most cars use a solenoid, a combination magnetic switch and mechanical device that engages the starter drive pinion with the flywheel ring gear. A relay has contact points, whereas a solenoid uses a movable iron piston attached to a *contact disc*. Most solenoids are mounted on the top of the starter motor (**Figure 28.17**). Some solenoids are mounted in a remote location.

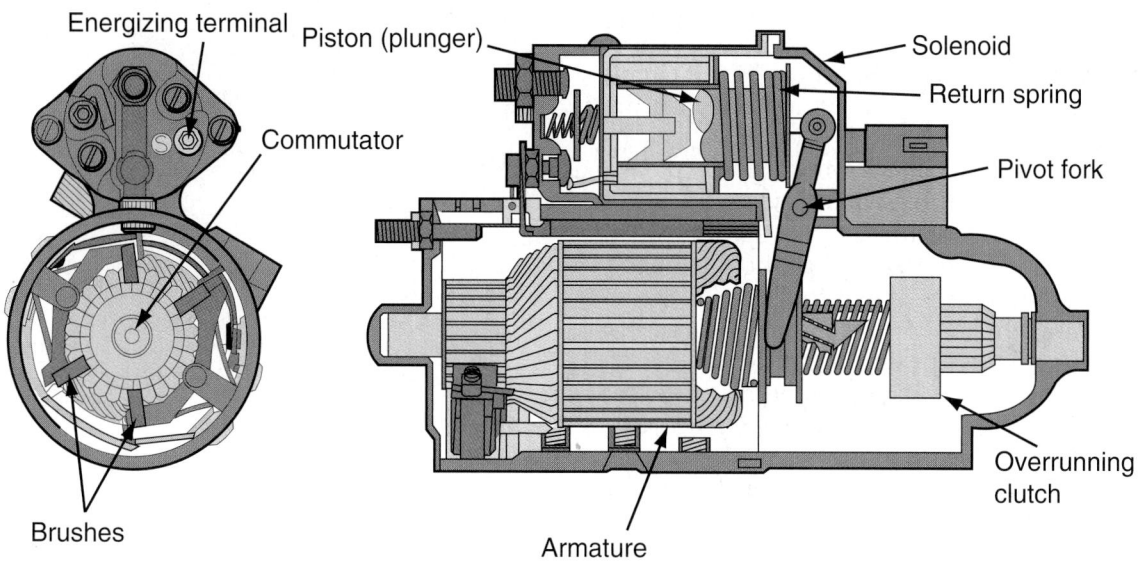

Figure 28.17 A starter motor with a solenoid.

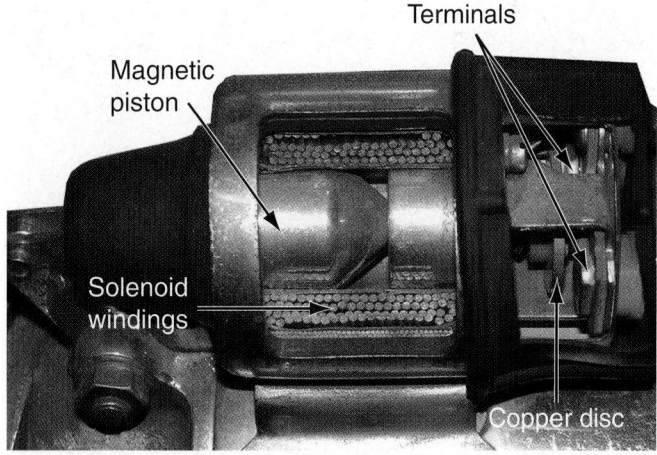

Figure 28.18 A cutaway solenoid. The copper disc is forced into contact with the terminals by the magnetic piston. *(Courtesy of Tim Gilles)*

When the ignition switch is turned to the "start" position and the safety switch is closed, electricity flows from the battery to the solenoid. When a coil of wire in the solenoid is energized, a magnetic field draws the piston into the coil (**Figure 28.18**). On one end of the piston is a copper disc that can carry a high amount of current without being damaged. It bridges the gap between two terminals in the relay to allow current to travel from the battery cable to the starter. A spring returns the piston to its original position, disengaging the starter drive from the flywheel when the key is released.

It requires more current to pull the solenoid piston against the spring than it does to keep it in engagement. There are two windings in the solenoid, a *pull-in winding* and a *hold-in winding*. Both windings are energized when the piston is pulled into the solenoid housing (**Figure 28.19**). When the contact disc contacts the battery electrical contacts, it shorts out the pull-in winding, leaving only the hold-in winding energized. This frees up some electrical current to operate the starter.

Ignition Switch

The ignition switch (**Figure 28.20**) closes the circuit that runs to the starter. It is usually mounted on the steering column and is operated by a linkage rod. As the key is turned, sliding contacts move to complete a circuit as desired. The ignition switch powers the starter relay whenever the key is turned to the spring-loaded "start" position. The starter relay or solenoid connects the starter to battery power. When the key is released, it returns to the "run" position, where it continues to provide battery power to the ignition system.

There are two paths in the starter electrical circuit that electricity can take. One of them is the large cable that goes directly from the battery to the starter. The other is the small wire that goes from the battery to the ignition switch, safety switch, and solenoid (**Figure 28.21**). Although the battery cable remains

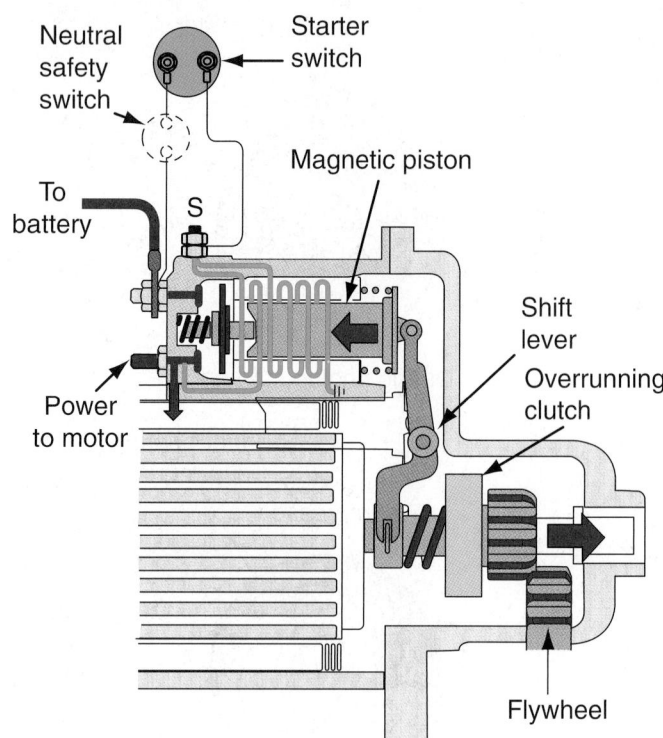

Figure 28.19 Solenoid electrical wiring.

connected to solenoid, the solenoid breaks electrical contact to the starter motor.

Safety Switches

The circuit on newer cars with automatic transmissions has a *neutral safety*, or *neutral start switch*, installed

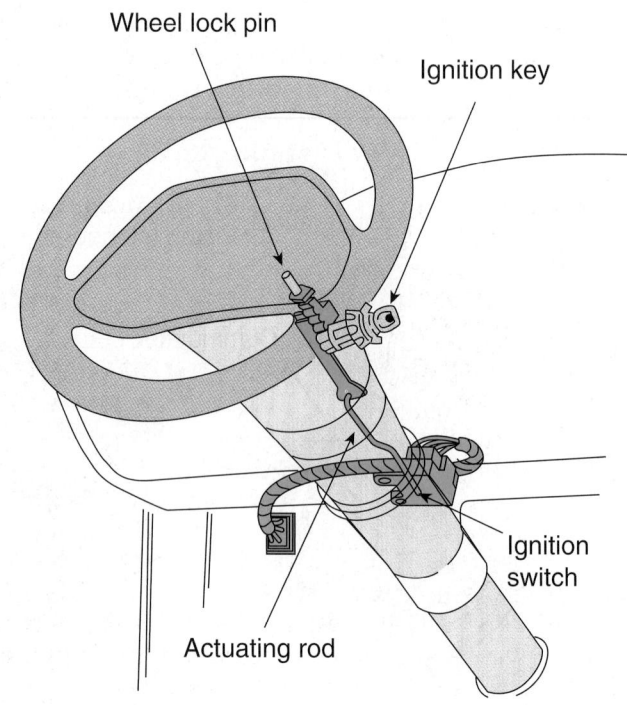

Figure 28.20 The ignition switch is operated by the key switch.

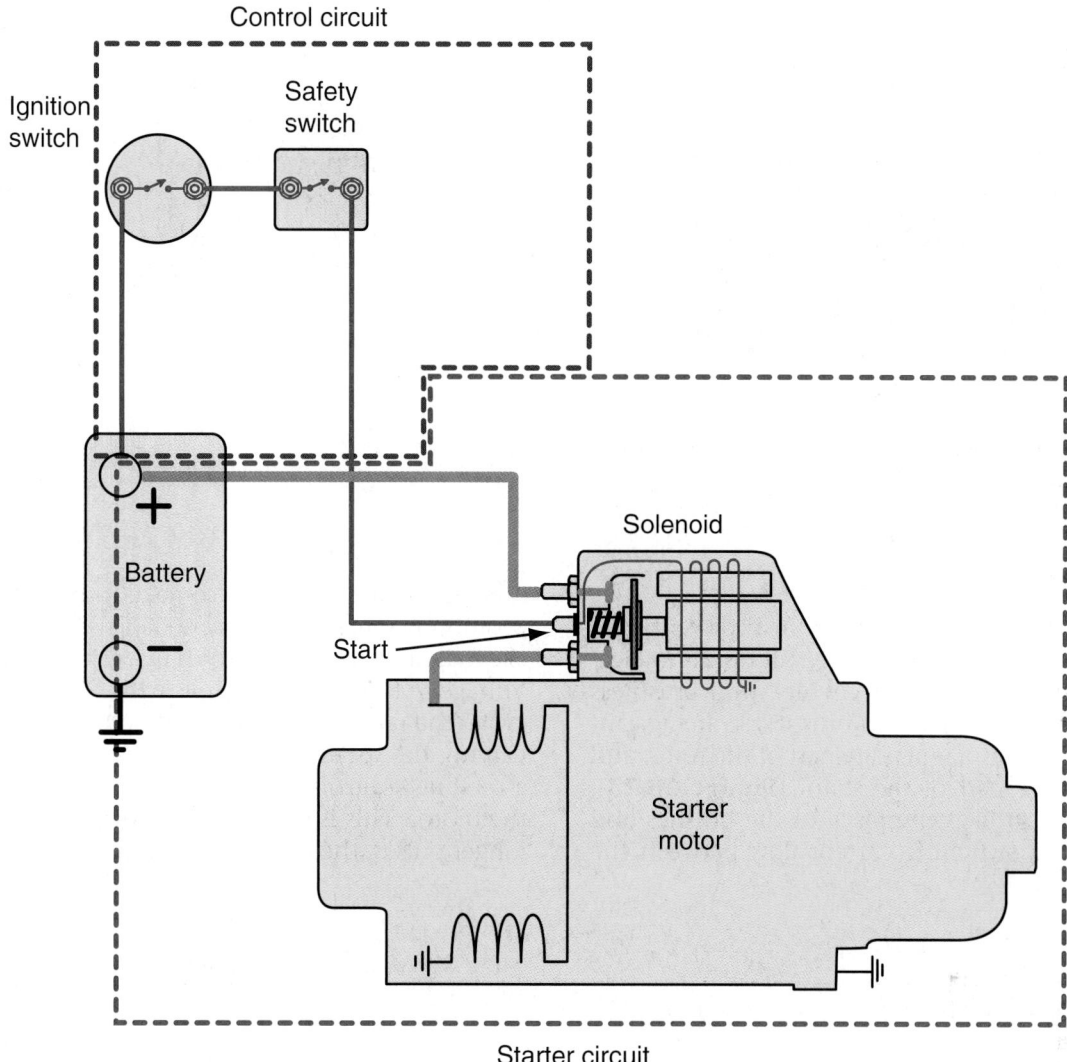

Figure 28.21 A parallel electrical path powers the solenoid.

between the ignition switch and the starter. The ignition switch cannot operate the starter motor unless the transmission is in park or neutral.

Late-model vehicles with manual transmissions have a *starter/clutch interlock switch* that requires the clutch to be depressed to the floor.

Starter Relay

Some cars (Fords, in particular) have a separate magnetic switch called a starter relay (see Figure 28.16). The starter relay is mounted on the fender well near the battery. Some cars use the relay by itself; others use a relay to reduce the load on the neutral safety and ignition switches.

Cars built before the early 1960s that were equipped with a remote starter relay had **inertia starter drive** units. These drives, also called *Bendix* drives, operate like a heavy nut on a screw. When the starter motor is turned on, the weight of the unit keeps it from turning while the armature shaft threads into it. Some British cars still had Bendix drives until the late 1960s.

Later cars with a starter relay use a *positive engagement starter* that uses the magnetic force of one of the field coils to engage the pinion with the flywheel ring gear. One of the field windings has a hollow coil called a drive coil. A movable pole shoe (the part that fits within the field coil) is drawn into the drive coil when the starter is energized by the relay. The other end of the movable pole shoe has a lever that pushes the starter drive into engagement. These starters are used mostly on Fords.

▄▄▄ GEAR REDUCTION STARTERS

Some manufacturers use gear reduction starters. A gear reduction of from 2:1 to 4:1 is developed by ordinary gears or planetary gears (**Figure 28.22**). These starters are gaining popularity because they are lighter and require less current to operate. Although they are small, their lower gear ratio gives them enough torque to turn the engine. Smaller battery cables can also be used because these starters draw less current.

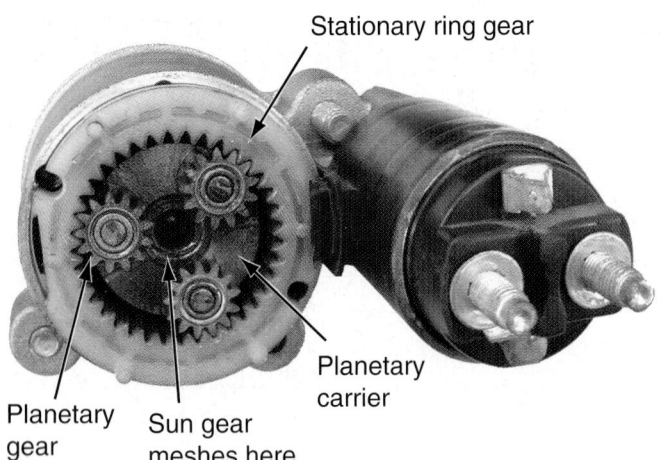

Figure 28.22 A gear reduction starter. *(Courtesy of Tim Gilles)*

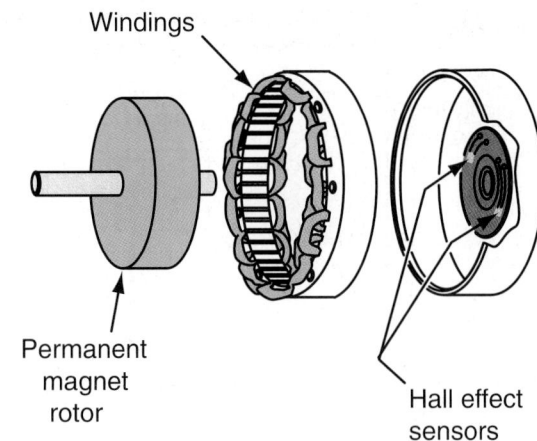

Figure 28.23 Parts of a brushless DC motor.

BRUSHLESS DC MOTORS

Hybrid vehicles use brushless motors. In a conventional DC motor, arcing not only causes wear, it results in electrical interference in computer circuits. In a brushless motor there is no commutator or brushes, so arcing cannot occur. Permanent magnets are part of the rotor, and electromagnets are part of the stator (**Figure 28.23**). Electronic circuitry takes the place of the brushes and commutator bars, switching current flow between the different stator windings to make the rotor turn. Rotor position is sensed either by a Hall switch or by sensing the magnetic field strength in unexcited field windings. Voltage in the windings varies with changes in the duty cycle (the ratio or percentage of on time to off time). By varying the speed at which power is turned on and off, a module controls current flow to change the speed of the motor. This is called pulse-width modulation. With longer pulses, the motor turns faster.

REVIEW QUESTIONS

1. To crank an engine at 200 rpm, how fast must an average starter motor turn?
2. _____ fields can be used to cause motion.
3. What are the names of the parts at the ends of the armature loop that the brushes slide against?
4. The armature core is made up of soft iron _____ insulated from each other.
5. What are the loops of heavy insulated copper wire in the armature called?
6. What is another name for the one-way clutch in the starter drive?
7. If an average starter remains engaged and is turned at 2,000 engine rpm, how fast would it turn?
8. How does the electrical current return from the battery to the starter?
9. What is the name of the electrical device that uses a movable iron core attached to a contact disc?
10. Which solenoid winding is shorted out when the starter motor turns?

ASE-STYLE REVIEW QUESTIONS

1. Which of the following would make the action of a starter motor stronger?
   a. A stronger magnetic field
   b. Higher current flow in the conductor
   c. Both A and B
   d. Neither A nor B

2. Technician A says that unlike charges attract and like magnetic charges repel. Technician B says that there is a stronger magnetic field on one side of a conductor in a motor. Who is right?
   a. Technician A   **c.** Both A and B
   b. Technician B   **d.** Neither A nor B

3. All of the following are true about gear reduction starters *except*:

    a. They have a reduction of from 2:1 to 4:1.

    b. They take less current to operate.

    c. They use larger battery cables.

    d. They tend to be smaller and lighter than other starter designs.

4. Which of the following is/are true about starter brushes in a negative ground system?

    a. Positive starter motor brushes must be insulated to keep them from touching the starter housing.

    b. Negative starter motor brushes must be insulated to keep them from touching the starter housing.

    c. Both A and B.

    d. Neither A nor B.

5. Two technicians are discussing starter solenoids. Technician A says that they have two separate windings. Technician B says that a neutral start switch prevents current from flowing to a solenoid unless the car is in park or neutral. Who is right?

    a. Technician A   **c.** Both A and B

    b. Technician B   **d.** Neither A nor B

6. Which of the following is another name for a solenoid-type starter?

    a. Bendix

    b. Overrunning clutch

    c. Both A and B

    d. Neither A nor B

7. What holds a Bendix starter drive from turning while the armature shaft is threading into it or out of it?

    a. Another clutch

    b. A brake band

    c. Its own weight

    d. None of the above

8. What is the name of the part of the starter that the brushes ride against?

    a. The armature

    b. The commutator

    c. The solenoid

    d. The overrunning clutch

9. Which of the following devices completes the circuit to a starter motor from the battery?

    a. The magnetic starter switch or relay

    b. The starter solenoid

    c. Both A and B

    d. Neither A nor B

10. Technician A says that the electrically conducting ends of the armature make up the area called the commutator. Technician B says that multiple loops make up an armature. Who is right?

    a. Technician A   **c.** Both A and B

    b. Technician B   **d.** Neither A nor B

CHAPTER 29

Starting System Service

■ **OBJECTIVES**

Upon completion of this chapter, you should be able to:

✔ Measure amperage draw on a starting system.

✔ Measure voltage drops on both the positive and ground sides of the starting circuit.

✔ Diagnose no-crank conditions with a test light.

✔ Replace a solenoid and starter drive.

■ **KEY TERMS**

voltage drop testing

■ **INTRODUCTION**

This chapter deals with the process for testing and repairing common problems with the starting and charging systems. Principles of operation and electrical fundamentals learned in earlier chapters will be important here. Diagnosis of failures is very important before parts are replaced. Most parts stores will not accept returns of electrical items.

NOTE: *Better techs often refer to less skilled mechanics as "parts replacers." In the long run, parts replacers cost the motoring public a great deal of unnecessary expense. In some cases, service manuals do call for testing by replacing with a known good part.*

■ **STARTING SYSTEM SERVICE**

Starting System Diagnosis

Examples of starting system problems include a no-crank condition and slow cranking. Methods of diagnosing those problems are covered in this chapter. Also included are on-the-vehicle tests of the starting system when the starter does not crank, and preventive maintenance and diagnosis. Replacement of relays, solenoids, and entire starters is stressed. Local labor rates dictate whether a starter can be rebuilt economically by the shop or a rebuilt unit is purchased.

When testing a starter, always follow a logical procedure and do not skip steps. Some starter tests are done on an engine that still cranks over. Others must be done on a starter that will not crank at all. Be sure that the starter is the cause of the problem before removing it from the engine. Over half of the starters returned on warranty claims are not defective. Sometimes, the starter cranks normally but the engine does not start. This can be unrelated to the starter itself.

Problems can be of two types: mechanical and electrical. Mechanical problems are often identified by the noise that results. Noises that can be heard when a starter is bad include multiple clicking and single clicking, humming after cranking, and metallic noises related to the gear drive and flywheel teeth.

Electrical problems can be identified during visual and electrical tests.

Visual Check. Problems are often found during a visual check. Check the wiring connections to see that they are clean and tight. Loose or dirty connections can cause excessive voltage drop. Cables should be of the correct diameter and length. Too long a cable can cause too much resistance. The cable should not get hot during cranking, which would indicate excessive resistance.

Volt-Amp Tester. The volt-amp tester (VAT) is a fundamental piece of electrical test equipment that will be used to test the starting and charging systems. The VAT's ammeter is capable of measuring 500 amps, which could be reached in the test of a faulty starter. There is also a voltmeter and a carbon pile that is used to load the battery.

STARTING SYSTEM TESTS

Test the Battery First

The biggest cause of starter motor failure is low battery voltage. Remember that a battery can cause starting problems and a starter can cause battery problems. Be sure that a battery has at least 75% of full charge before attempting to test the starter. The battery must be able to keep voltage above 9.6 volts while delivering needed current to the starter.

One quick check of battery condition is done with the headlights on while cranking the engine:

■ If the engine does not crank and the headlights stay off, the battery is dead.

■ If the lights go out while cranking, the battery is weak.

■ If the lights stay bright but the engine does not crank over, there could be high resistance or an open circuit in the ignition switch or the solenoid. If the solenoid will not click in the starter, brushes are probably worn out, not grounding the solenoid.

When the battery is weak, a solenoid can make a series of rapid clicks. This happens as the winding in the solenoid pulls the plunger and disc into contact with the contacts that operate the starter motor. When the starter begins to turn, battery voltage decreases rapidly to a point where the windings in the solenoid can no longer hold the solenoid disc against the contacts. This releases the load made by

the starter, so the solenoid gets enough voltage to operate again (click).

Electrical power is measured in watts (amps × volts). When battery voltage drops, amperage decreases, cranking rpm is lower, and ignition voltage drops. The net result is poor or no starting.

NOTE: *A starter will draw twice the current (amps) if the battery voltage drops to half.*

Increased amperage results in more heat in the starter and cables. Perform battery tests as outlined in Chapter 27.

Disable the Fuel or Ignition System. During starter tests, the engine must be prevented from starting. Be sure the transmission is out of gear and the parking brake is set. The fuel or ignition system must be disabled. On most cars, there is an easy way to trigger the starter motor from under the hood with a *remote starter switch* (**Figure 29.1**).

If there is a fuel pump or ignition fuse, remove it. If not, disable a distributor ignition system using one of the following procedures:

■ Remove the center lead from the distributor cap and attach a jumper lead from it to ground (**Figure 29.2**).

■ Disconnect the wiring harness connector from the distributor or coil (**Figure 29.3**).

Do not crank the engine with the coil to distributor wire simply disconnected. This could result in damage to the coil and ignition module.

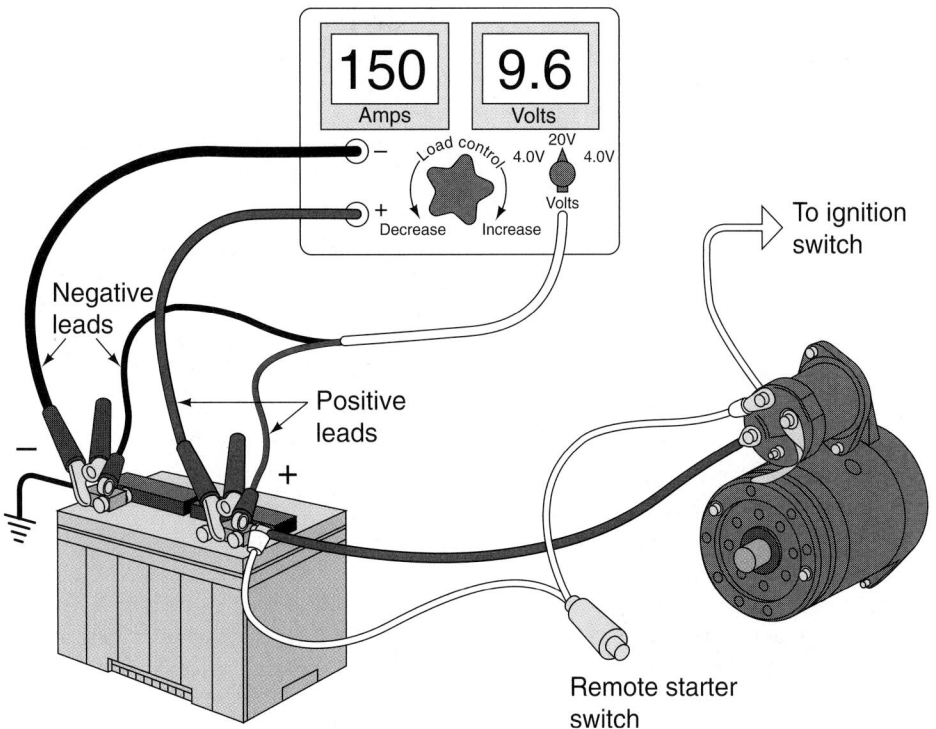

Figure 29.1 With the ignition system disabled, a remote starter is used to power the starter.

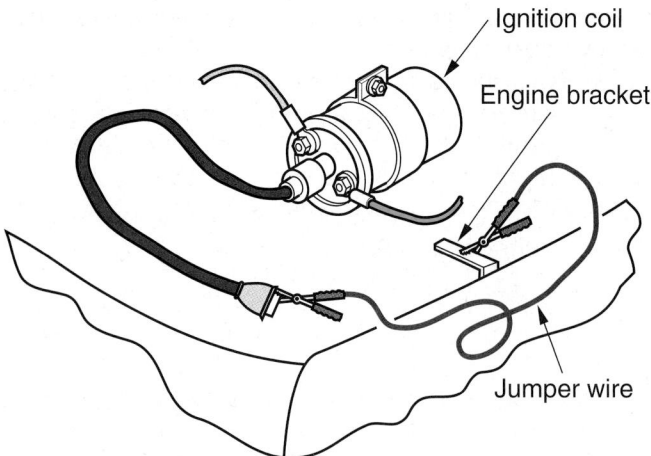

Figure 29.2 Disabling a distributor ignition system.

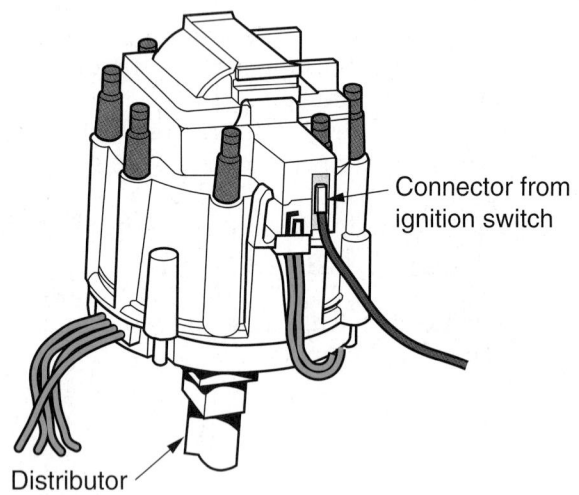

Figure 29.3 Disconnect the connector at the distributor to disable a distributor ignition.

■ CRANKING VOLTAGE AND AMPERAGE TEST

The inductive cable on the VAT is installed around the battery cable (**Figure 29.4**). Be sure the carbon pile is off. The voltmeter should read battery voltage and the ammeter should read zero unless there is a load, such as a dome light or the ignition switch is on. When you turn on the headlights, the ammeter should move toward negative. Turn off the headlights and adjust the zero adjust knob if the machine has one.

Move the meter to where you can observe it while cranking the engine for about 5 seconds. Most V8s will draw about 200 amps, six cylinders about 150 amps, and four cylinders about 125 amps. These are just estimates. The service manual will list actual specifications.

NOTE: *The starter motor is designed to be used for short periods of time only. In normal use, the starter should never need to be operated longer than 15 seconds at a time. When testing, do not use the starter for more than 30 seconds. After 30 seconds of operation, allow it to cool for at least 2 minutes.*

Cranking Test Results

Normally, when the voltage drops, the amperage draw goes up. If the reading was low on voltage and high on amperage, you know that more tests are required.

When there is resistance, there is a voltage drop. Cable connections or resistance in relay contacts can cause voltage to drop and amperage to become higher.

The next tests will show how to look at individual components.

■ CRANKING SPEED

Generally, 250 engine rpm is the speed for a standard starter. A gear reduction starter will turn at about 200 rpm (sometimes they turn faster). This can be

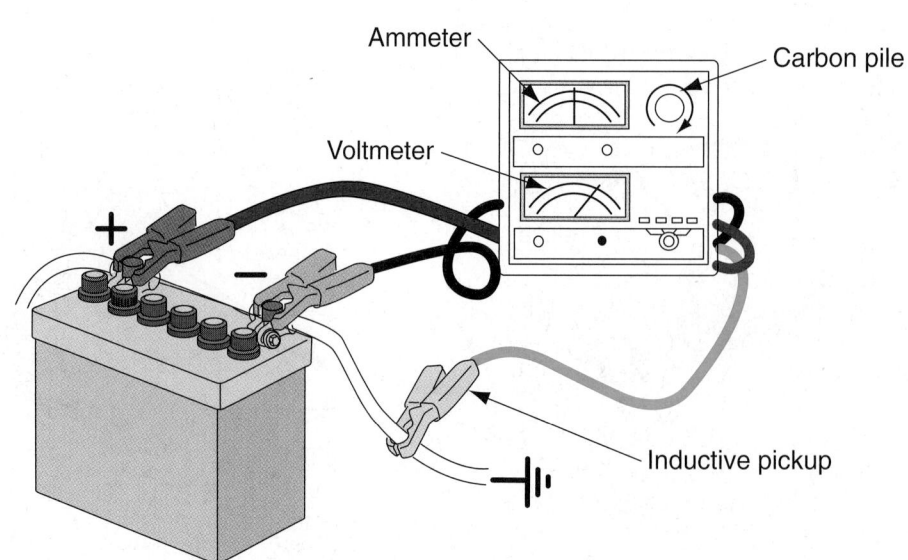

Figure 29.4 Volt-amp tester connections.

measured with a tachometer, but a trained ear can tell if the speed is correct.

When there is high resistance in a battery cable or its connections, they will become hot to the touch. When resistance is within acceptable limits, cables and connections will not become hot.

NOTE: *When there is a combination of slow cranking speed and lower than normal cranking amperage, the starter will need to be removed. This happens only when there is too much resistance internally in the starter.*

■ CIRCUIT RESISTANCE TEST

Voltage drop testing is a very popular way of checking for resistance in a circuit. It is a fast and simple test because it does not require the disassembly of any connections.

A connection that appears to be corroded will most often be found to be adding unwanted resistance to the circuit. In this case, testing the connection is not necessary. Disassemble the terminal connection, clean it, install the proper washers, and reassemble it. A connection without visible corrosion can still have high resistance. A voltage drop test is a fast way of determining the integrity of the connection.

Voltage drop is the amount of voltage used trying to force current through a resistance. This test measures the voltage lost when current flows from the source (battery) to the load (starter) and back to the source (battery). Voltage drop can only be measured when there is current flowing in the circuit. Excessive voltage drop indicates excessive resistance. This will be accompanied by current flow that is lower than it should be. The result is that the starter turns the engine at a slower cranking speed and the engine does not start as rapidly.

The starting circuit is designed to have very little resistance. Voltage drop should be less than 0.1 volt per connection. The battery ground side is allowed a 0.2-volt drop because there are two connections—one at the battery terminal and the other where the ground cable is bolted to the block.

Many vehicles also have a ground cable from the negative side of the battery to the vehicle body. This wire serves as a ground for the lights and accessories and should be clean and tight.

The positive side of the starting circuit is allowed a 0.4-volt drop because on most vehicles there are four connections at each end of the positive cable and two connections on the solenoid.

When there is less than 0.6 volt of total voltage drop in the starting circuit, no service work is required. More than a 0.6-volt drop in the total starting circuit calls for pinpoint tests to determine the exact location of the excessive resistance. Before making this test, disable the ignition system so the engine will not start. Connect a voltmeter across the battery and observe the voltmeter reading while the engine is cranked. If the voltmeter reading wavers, write down an average reading.

Next, remove the voltmeter connections from the battery. Connect the red voltmeter lead to the starter positive terminal, and the black lead to one of the screws that secures the starter to the block (**Figure 29.5**). Crank the engine once again while observing the voltmeter. If the second voltmeter reading is within 0.6 volt of the first reading, the circuit resistance is within limits. If voltage readings exceed 0.6 volt difference, further pinpoint tests will be needed to locate the excessive resistance.

Excessive voltage drop can also mean that the battery cables are of too little diameter. A typical indication that a cable is too small is when a battery cable gets warm to the touch.

For the voltage drop test, adjust the voltmeter to read on its lowest scale. This will probably be about 2 to 4 volts.

NOTE: *An analog voltmeter can be damaged if it attempts to read full battery voltage when it is set on the low scale. To avoid this problem when testing voltage drop on a starter solenoid, use the high scale first, then switch to the low scale or use a digital multimeter.*

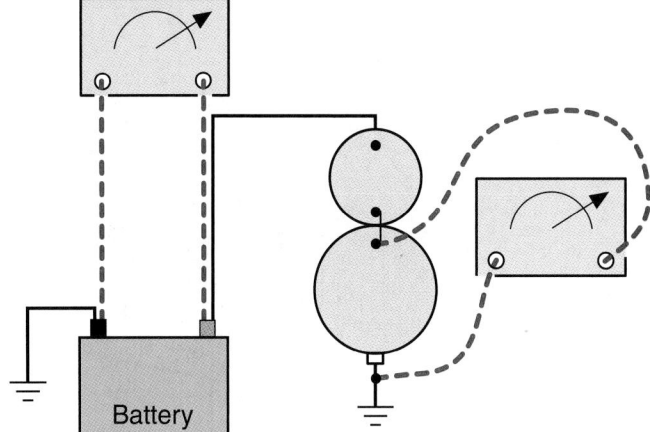

Figure 29.5 A voltage reading taken between the starter input and the starter case should be within 0.4 volt of battery voltage.

Connect the positive tester lead to the most positive side of the circuit. If a voltmeter is connected backwards, the meter will read backwards.

The voltage drop test is done with the engine cranking. With current flowing, a resistance will have a different voltage (pressure) on each of its sides. First check the entire positive circuit (**Figure 29.6**), then the entire negative circuit (**Figure 29.7**) to isolate on which side the excessive resistance is located. When the high-voltage drop side has been isolated to either the positive or negative circuit, the circuit can be broken into smaller pieces during further voltage drop tests. When checking for voltage drop in the positive insulated circuit:

■ Normal voltage drop would be about 0.2 volt when normal current is flowing. A rule of thumb

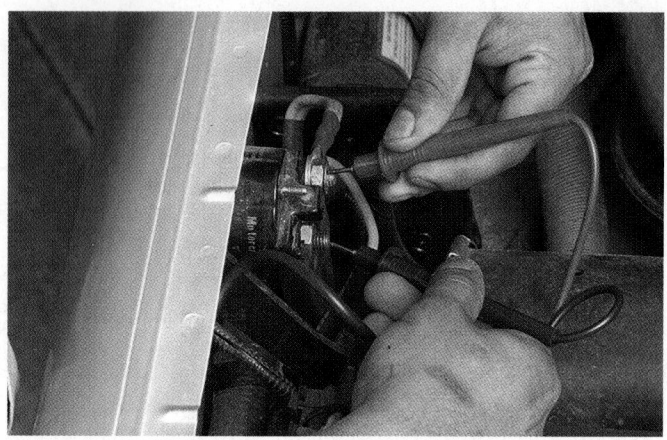

Figure 29.8 Voltage drop across a starter relay or solenoid should be less than 0.2–0.3 volt.

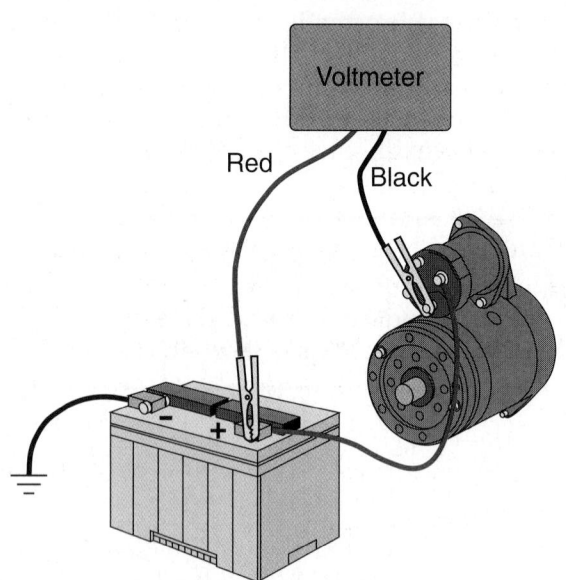

Figure 29.6 Insulated circuit voltage drop check.

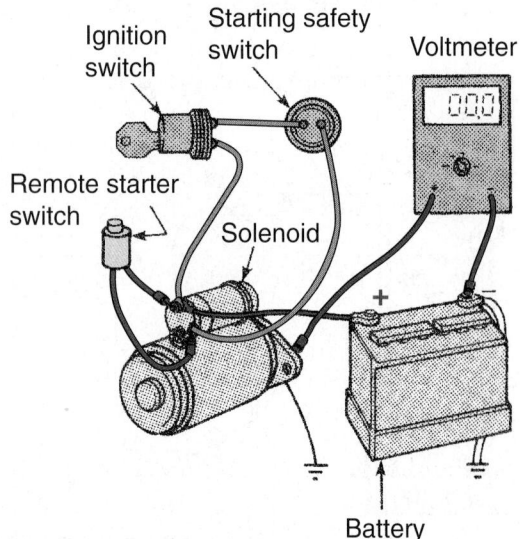

Figure 29.7 Isolate which side of the circuit the resistance is on by checking the voltage drop.

is 0.1 volt for each connection and 0.2–0.3 volt for the solenoid (**Figure 29.8**).

■ If 100% more current is flowing, adjust the specification up by 100% to 0.4 volt. Voltage drop will increase as current increases.

If the voltage drop reading is high, check each connection in the circuit. If a connection has voltage drop that is above normal, inspect it for corrosion or looseness. The solenoid is part of the overall 0.2-volt drop. The only part of the circuit that is expected to have more than 0.2-volt drop is the starter motor. The starter motor should drop almost all of the voltage during cranking. For instance, if 10 volts are measured across the battery during cranking, at least 9 volts should be measured across the starter motor.

After voltage drop is restored to a normal level, cranking amperage can be used to diagnose a starter. A higher than normal amperage draw combined with no excessive resistance in the circuit points to an engine that is too tight or a starter that is bad. Remove the starter and disassemble it. Look for obvious causes of excessive current (amp) draw like worn bushings, a rubbing armature, or shorted windings. When there are two field windings, a shorted winding results in less resistance, half the magnetic field, and twice the current flow.

■ NO-CRANK TESTS USING A TEST LIGHT

Voltage drop testing works while the engine is still able to crank. When the engine does *not* crank, this is usually because of an open circuit. A 12-volt test light is used for this diagnosis.

NOTE: *Always test voltage first with a meter (see Chapter 31).*

Start at one end of the circuit with the key in the start position. Use the 12-volt test light for a quick check to see if there is power at the outlet of the solenoid (**Figure 29.9**) (where power enters the starter). Because this is the end of the circuit, power here will

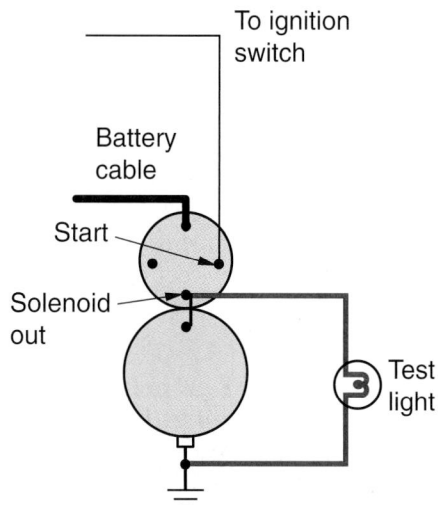

Figure 29.9 Use the 12-volt test light to check if there is power at the outlet of the solenoid.

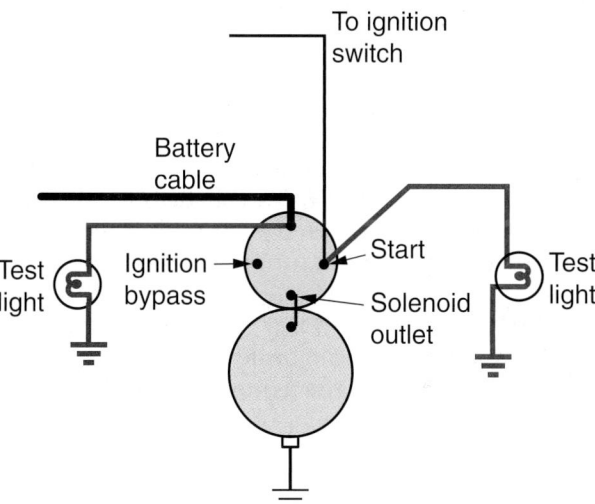

Figure 29.10 Test light connections.

tell you that the positive side of the circuit is complete. It will not tell you whether there is sufficient available voltage, however. A meter must be used for this. If there is no light, work your way back through the system until you find power (see Figure 25.52). Check to ensure that the ignition switch and safety switches are functioning. Place the test light on the start terminal of the solenoid (**Figure 29.10**). There should be power when the key is rotated to the start position. Connect the test light to the end of the battery cable at the solenoid (see Figure 29.10). The light should be on and stay on when the key is turned to the start position. If it goes out, repair the cable or connection.

The next test is to the starter's ground path. Clean off an area of the starter body and connect the test light's ground connection to it. Connect the positive probe of the test light to the known good terminal at the outlet to the solenoid. If the light comes on, the

ground path is good. This means that the starter must be the problem. Always check the power and ground circuits before pulling and replacing the starter.

▬ SOLENOID PROBLEMS

A rapidly clicking solenoid can be caused by several things, including:
■ A weak battery
■ A corroded or loose battery cable connection
■ An open circuit in a hold-in winding

Before replacing a solenoid, check the condition of the battery. A single click when the battery is in good condition can often be traced to burned contacts in the solenoid. Sometimes these contacts can be replaced or repaired. On other solenoids, the entire solenoid is replaced. Be certain to correctly diagnose the problem before replacing the solenoid. Sometimes, they are as expensive as an entire rebuilt starter.

> **SHOP TIP**
> A typical pull-in winding will draw 72 amps. This will drop to 22 amps during the hold-in phase. Sometimes the hold-in winding connection becomes disconnected (open circuit). The hold-in winding has about 2,500 pounds of hold-in strength. With the starter off the car and the solenoid energized, a technician should not be able to physically pull the piston out of the solenoid.

When the contact disc in a starter solenoid has become burned, high resistance develops in the positive side of the circuit (across the solenoid terminals). Using the voltage drop test, the resistance can be located. The symptom of resistance here is that the starter solenoid will click instead of starting the car, or the starter will turn slower than normal.

> **SAFETY NOTE**
> The test described above should only be done on a solenoid that is removed from a starter. When the solenoid is energized, its piston will move very fast and with great force. With the solenoid assembled to the starter, solenoid current is best tested with an ammeter to avoid the chance of personal injury.

Neutral Start Switch

The neutral start switch is attached to either the steering column or the side of the transmission. Because it is in the relay circuit, it uses small wires. It is usually adjustable and sometimes controls the back-up lights also. Occasionally, moving the shift lever while turning the ignition key to the start position will allow the engine to crank. This means that the start switch needs to be adjusted.

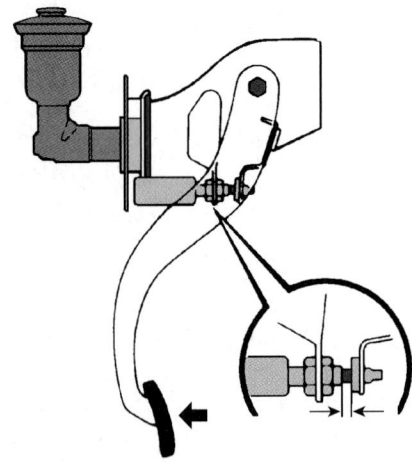

Figure 29.11 A clutch start switch closes to allow current flow when the clutch pedal is depressed.

Clutch Start Switch

Later-model cars with manual shift transmissions have a clutch start switch (**Figure 29.11**). This ensures that the clutch pedal is depressed before the starter motor gets electricity. Using an ohmmeter, there should be no continuity when the clutch pedal is up. Continuity when the clutch is depressed allows current to flow through the switch to the starter.

■ STARTER REPAIR

Starters are not always economical to rebuild. This depends on the availability of competitively priced parts, wages, and whether the shop's work load is low on a particular day when a bad starter requires repair. Many shops replace starter drives and solenoids. Procedures covered here apply to replacements of those parts within the starter. The procedure for disassembly and reassembly is also covered here so that you can disassemble a starter and inspect its working parts for a better understanding of its operation.

■ STARTER DISASSEMBLY

Get in the habit of marking parts that are disassembled so they can be reassembled in their original positions. Some starters have pins that align the parts; others do not.

The following procedure applies to General Motors starters. Other starters are similar. To remove the solenoid, disconnect the solenoid's electrical terminal(s). Remove the two screws that hold the solenoid on the starter housing. Twist the solenoid until the locking flange is free. The spring will push the solenoid away from the housing. There are two long bolts that go through the starter frame. Remove the bolts, the end frame, and the starter body from the drive end housing. Remove the armature from the housing.

NOTE: *Solenoids and starter relays used with vehicles with computers have an internal diode to provide protection from*

voltage spikes that are produced when the magnetic field in the relay collapses (see electromagnetic induction in Chapter 25). A replacement solenoid with a diode may be used on applications that do not have computers. Using an early model solenoid on a computer-equipped car can result in problems.

Bearing or Bushing Service

Inspect the bearings or bushings at both ends of the housing. Inspect bearings by feeling them.

NOTE: *Commercial rebuilders have listening devices that spin a bearing and tell whether it makes too much noise.*

If a bearing is tight or feels rough, replace it. Most of today's starters use bushings because they do not rotate for long periods of time. Bushings are visually inspected. When they wear, it is usually on one side. A worn bushing is pounded or pressed out.

■ STARTER DRIVE SERVICE

Starter drives often do not last the life of the starter motor. Replacing a starter drive is commonly done by technicians. Noise that can be heard coming from a starter is usually from the starter drive. The starter whines or grinds and does not turn the engine over reliably. Sometimes, the overrunning clutch will slip in one direction when it wears internally. It might only fail to work occasionally. Test it by trying to turn it in both directions (**Figure 29.12**). Sometimes, it is necessary to wiggle the gear and lightly attempt to turn it to get it to slip.

> **SHOP TIP** A starter drive can be bench tested to see if it will slip. Clamp the removed starter drive in a suitable holding fixture. Locate a 12-point socket that will fit snugly over the pinion gear teeth. Using a torque wrench, see that the drive clutch will not slip when a 50 foot-pound load is applied to it.

Replacing a Starter Drive

On the General Motors starter, the drive unit is held in place on the armature shaft by a split metal ring covered by a full round ring.

- Tap down on the outer ring to remove it from covering the inner ring (**Figure 29.13**).
- Remove the snapring (**Figure 29.14**).
- Slide the starter drive from the armature.

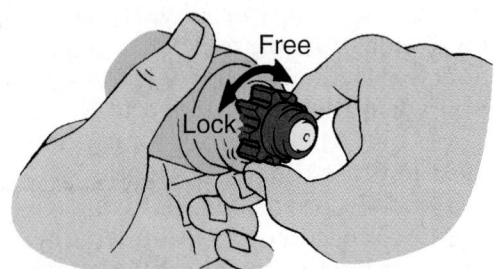

Figure 29.12 Test the starter drive one-way clutch.

Figure 29.13 Remove the outer ring from the retainer using a socket and a soft hammer.

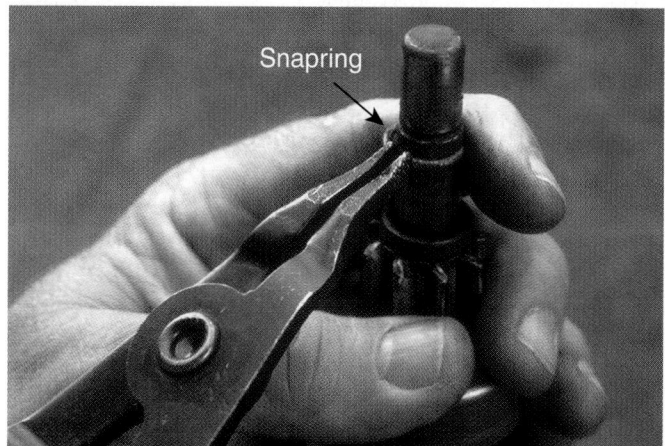

Figure 29.14 Remove the snapring.

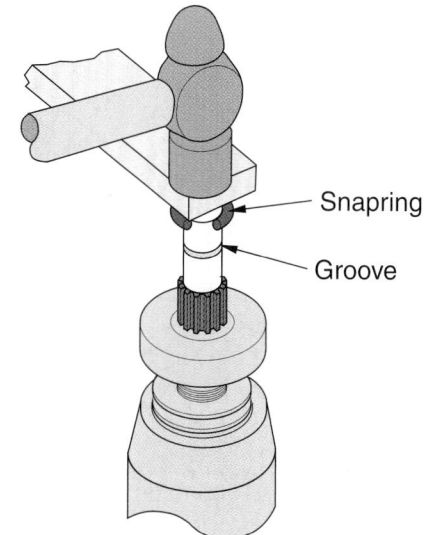

Figure 29.15 Reinstalling the snapring.

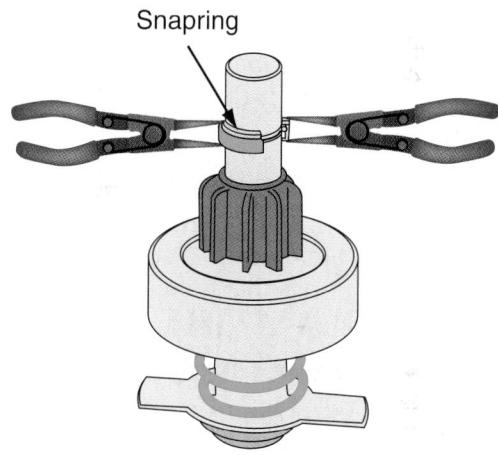

Figure 29.16 Compress the retainer assembly together.

NOTE: *If the starter drive is to be reused, do not immerse it in cleaning solvent. The bearing is permanently lubricated and can be damaged by solvent.*

- Before replacing the starter drive with a new one, count the number of teeth on the drive pinion and match up the old and the new ones to see that they are the same.
- When replacing a starter drive, be sure to inspect the starter ring gear for damage, especially if the starter drive pinion shows signs of wear or damage.
- To reinstall the snapring, use a hammer and a block of wood, or a rubber mallet (**Figure 29.15**).
- Use two pairs of pliers to compress the ring assembly together (**Figure 29.16**).

Brush Service

Brushes occasionally wear too thin and have to be replaced. Some must be soldered in, and others have connectors and can be installed with screws (**Figure 29.17**). Whenever a starter is taken apart, new brushes should

Figure 29.17 This cutaway of a starter motor shows brushes that are attached with screws. (*Courtesy of Tim Gilles*)

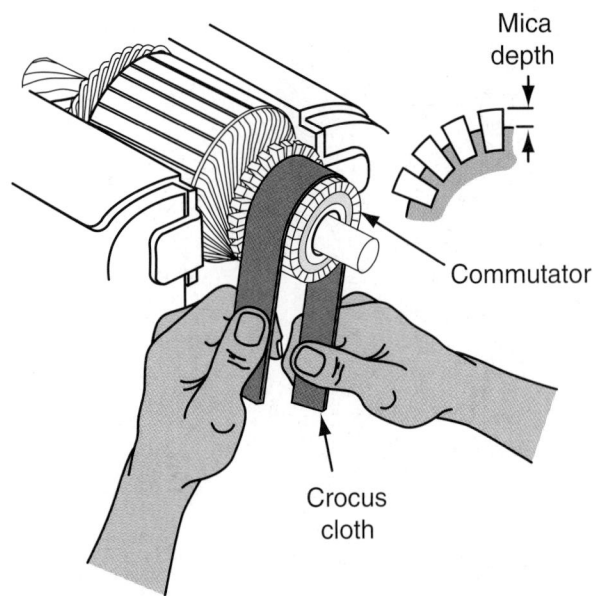

Figure 29.18 Cleaning the commutator with crocus cloth. Check the mica depth.

be installed. They are very inexpensive. If the commutator is in good condition, it can be cleaned with crocus cloth (**Figure 29.18**).

■ STARTER REASSEMBLY

Reassemble the starter. Some starters have brushes that are on pivots and are easy to install. Others require pulling up on the springs that hold the brushes against the commutator (**Figure 29.19**).

■ PINION CLEARANCE TESTS

There are two tests that can be made. One is with the starter off the car. With the solenoid energized, push the pinion back toward the armature to remove any slack. Check the clearance with a feeler gauge (**Figure 29.20**). The General Motors specification is 0.010"– 0.140".

With the starter on the car, the pinion to flywheel ring gear clearance is checked (**Figure 29.21a**). If the pinion clearance to the ring gear is excessive, the

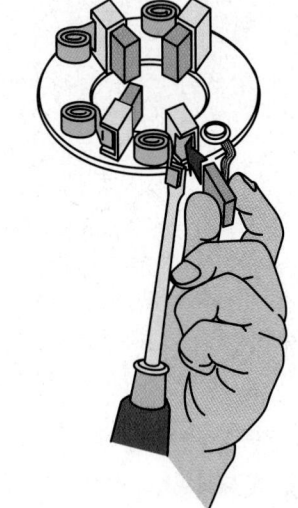

Figure 29.19 Installing brushes during starter reassembly.

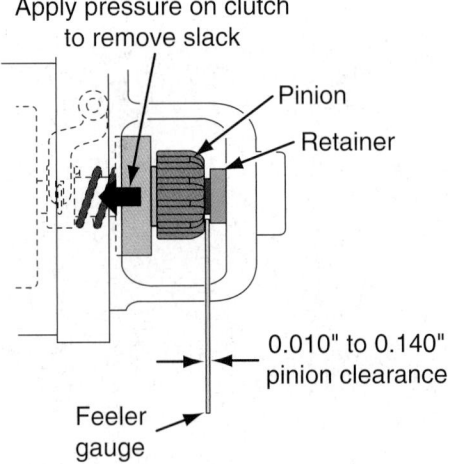

Figure 29.20 Measuring pinion gear to drive housing clearance.

starter can be loud and the teeth can be damaged. With too little clearance, the starter could bind and the amp draw could be higher. The clearance is adjusted using a shim or shims (**Figure 29.21b**). Each 0.015" shim changes the clearance by about 0.005".

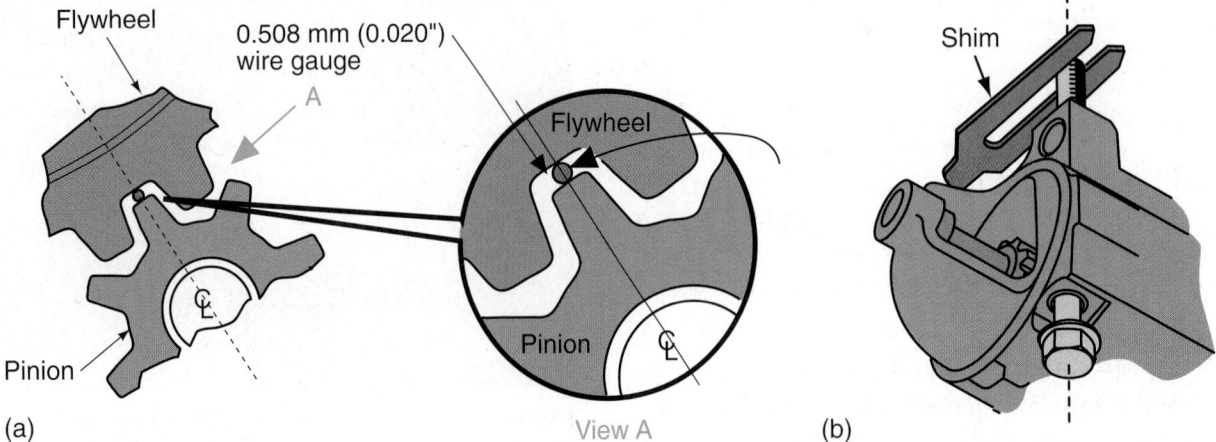

Figure 29.21 (a). Checking starter pinion to flywheel ring gear clearance. (b). Adjusting pinion depth with shims.

■ REVIEW QUESTIONS

1. Problems in a starter can be either electrical or _____.

2. What is the biggest cause of starter motor failure?

3. Be sure that a battery has at least _____ of full charge before attempting to test the starter.

4. What part makes a series of clicking sounds when the battery is low or there is a bad connection?

5. What tool is used to energize the starter motor with the ignition system disabled?

6. When testing, do not use the starter for more than _____ seconds.

7. What resistance test can be made with a voltmeter?

8. Worn bushings, a rubbing armature, or shorted windings could cause excessive _____ draw.

9. What part of the starter does not always last the life of the starter?

10. Before replacing a starter drive, what do you compare between the old gear teeth and the new gear teeth?

■ ASE-STYLE REVIEW QUESTIONS

1. Two technicians are attempting to diagnose an engine being cranked with the headlights on. Technician A says that if the lights go out while cranking, the battery is weak. Technician B says that if the lights stay bright but the engine does not crank, there could be high resistance or an open circuit. Who is right?
   - **a.** Technician A
   - **b.** Technician B
   - **c.** Both A and B
   - **d.** Neither A nor B

2. Which of the following is NOT a method of preventing the engine from starting during starter motor testing?
   - **a.** Cranking the engine with the wire from the coil to the distributor disconnected
   - **b.** Removing the fuel pump fuse
   - **c.** Removing the ignition fuse
   - **d.** Disconnecting the wiring harness connector to the coil

3. Which of the following would NOT cause a solenoid to click repeatedly?
   - **a.** A weak battery
   - **b.** An open circuit in the solenoid hold-in winding
   - **c.** A faulty neutral start switch
   - **d.** None of the above

4. Technician A says that a starter will draw half the amperage if the battery voltage drops to half. Technician B says that most V8 engines will draw about 50 amps during cranking. Who is right?
   - **a.** Technician A
   - **b.** Technician B
   - **c.** Both A and B
   - **d.** Neither A nor B

5. Each of the following is a true statement about cranking tests *except*:
   - **a.** During engine cranking the normal meter readings will be low voltage and high amperage.
   - **b.** Voltage drop indicates resistance in a circuit.
   - **c.** With a standard starter motor, cranking speed is typically about 250 engine rpm.
   - **d.** Low resistance in a battery connection can result in a hot battery cable.

6. Technician A says that current must not be flowing in a circuit during a voltage drop test. Technician B says that when voltage drop through a circuit is low, resistance is high. Who is right?
   - **a.** Technician A
   - **b.** Technician B
   - **c.** Both A and B
   - **d.** Neither A nor B

7. Which of the following is/are true about an engine that cranks slowly with cranking amperage lower than normal?
   - **a.** There is too much resistance internally in the starter.
   - **b.** The starter will have to be removed from the vehicle for repair or replacement.
   - **c.** Both A and B.
   - **d.** Neither A nor B.

8. Technician A says that when current is flowing through a resistance, the resistance will have the same voltage at each of its ends. Technician B says that if there is excessive resistance in a circuit, a test light will be brighter. Who is right?

   a. Technician A   c. Both A and B

   b. Technician B   d. Neither A nor B

9. Which of the following is/are true about voltage drop testing?

   a. The starter motor is allowed more of a drop in voltage than the rest of the starting circuit.

   b. Voltage drop testing is used to diagnose a no-crank condition.

   c. Both A and B.

   d. Neither A nor B.

10. When attempting to start an engine, a starter with good circuit wiring makes a strong, *single* click. Technician A says that the solenoid electrical contacts are probably burned. Technician B says that a low battery could be the cause. Who is right?

   a. Technician A   c. Both A and B

   b. Technician B   d. Neither A nor B

Charging System Fundamentals

OBJECTIVES

Upon completion of this chapter, you should be able to:

✔ Explain electrical generation principles.

✔ Describe alternator parts.

✔ Explain the operation of a voltage regulator.

KEY TERMS

generator
alternator
sine wave voltage

heat sink
three-phase electrical output
voltage regulator

zener diode
pulse width modulation

INTRODUCTION

The charging system is an important part of the automotive electrical system. The charging system allows a battery to maintain a charge and operate electrical accessories. This chapter deals with the operation of the charging system. If you are unclear about any of the basic principles described in the chapter, refer to Chapter 25.

CHARGING SYSTEM

The charging system consists of the alternator, the voltage regulator, the dash light or gauge, and related wiring (**Figure 30.1**). A battery alone can supply a vehicle's electrical needs for a period of time, but a charging system is needed to replenish the battery. The charging system output is increased whenever the load of various components causes battery voltage to drop below a certain point. The starter motor is a particularly large load on the battery. The charging system works hardest after the engine is first started, as well as when all electrical options are operating.

The alternator is driven by a belt from the engine crankshaft pulley (**Figure 30.2**). From the time a car is first started until the alternator is rotating fast enough to begin charging, the battery supplies the electrical needs of the vehicle. When the alternator starts to work, it recharges the battery and supplies all electrical power for all the needs of the vehicle.

DIRECT CURRENT (DC) GENERATORS

Older cars used direct current (DC) generators, driven by a belt on the crankshaft (**Figure 30.3**). A **generator** resembles an electric motor, having a stationary magnetic field with an armature output winding spinning inside. Although the generator produces alternating current, its output is DC because its commutator has brushes on its north and south poles.

SCIENCE NOTE

Anytime a rotating member carrying current passes a stationary conductor, the current flow in the conductor changes direction as it enters the magnetic field and leaves it. The generator actually produces alternating current, which is rectified by the commutator on the generator armature.

HISTORY NOTE

Generators were used on automobiles until the early 1960s, when they were replaced by alternators.

There are some drawbacks to the generator when compared to the alternator, including the following:

■ For a generator to have high output, more current must flow through the brushes. Current flow causes

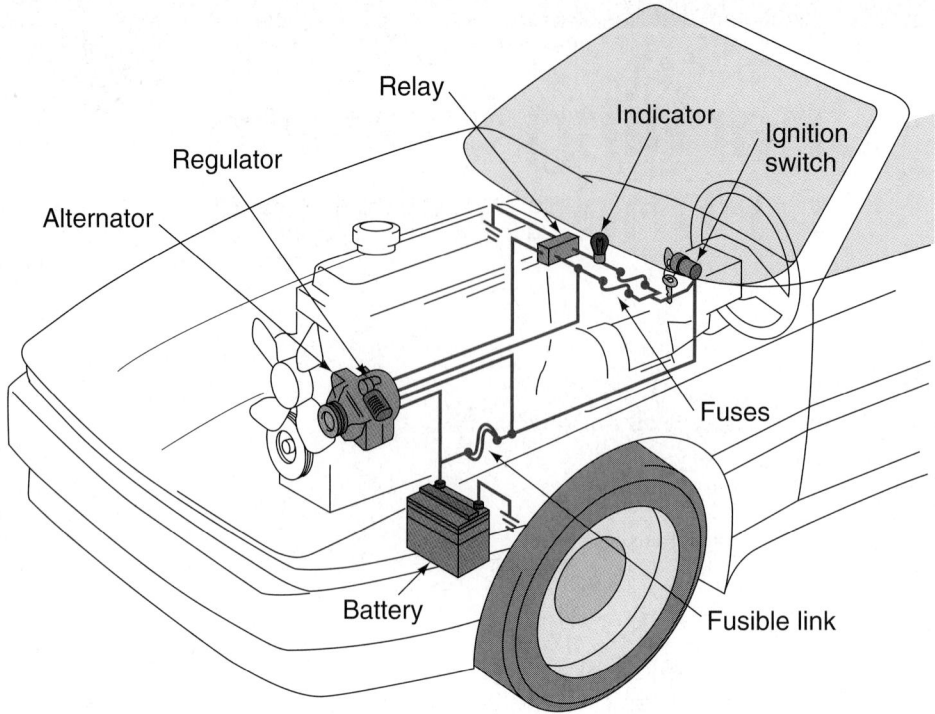

Figure 30.1 Parts of the charging system.

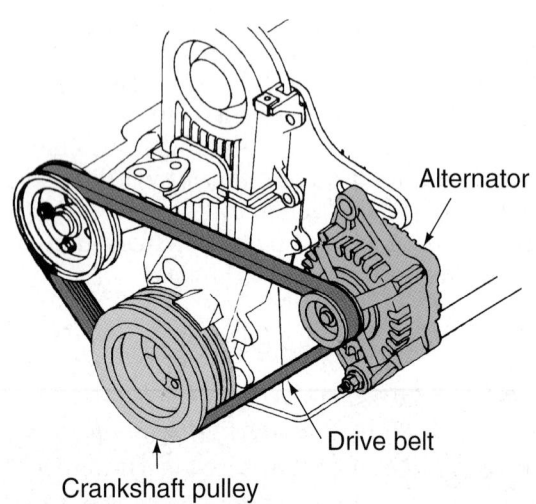

Figure 30.2 The alternator is driven by a belt from the engine crankshaft pulley.

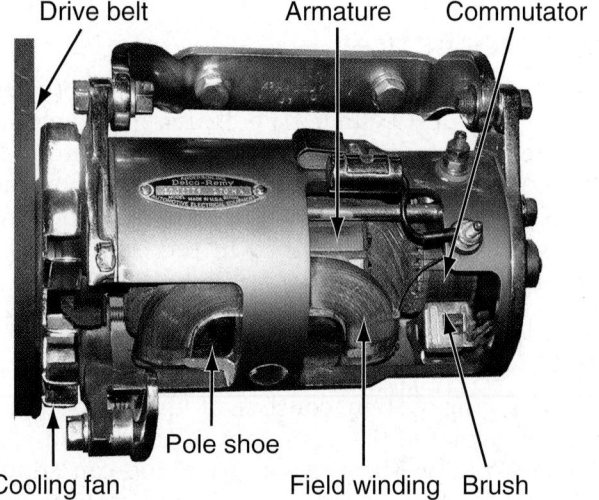

Figure 30.3 A cutaway of a DC generator. Notice how closely it resembles a starter motor. *(Courtesy of Tim Gilles)*

brushes to wear out. If a generator were designed to put out 60–80 amps, it would have to have from six to eight brushes.

- Generator speed is limited to about 10,000 rpm.
- Generators do not produce enough output at low speeds (city traffic) to satisfy the demands of electrical accessories.

■ ALTERNATORS

An **alternator** (**Figure 30.4**) is an *alternating current (AC) generator*. It is lighter, more efficient, and more dependable than a generator. Because of its design, an alternator uses brushes that require only 3–4 amps of

current flow. Brushes on alternators commonly last for the life of the original engine.

The alternator works on two basic electrical principles (see Chapter 25). When current passes through a coiled wire, a magnetic field builds up around the wire (see Figure 25.29). This is called an *electromagnet*. When a magnetic field is passed over a wire, voltage is induced. When a complete circuit is added, current flow results in the wire (see Figure 25.33). This is called *electromagnetic induction*.

In an alternator, an electromagnet passes across coils of wire to induce voltage in the coils. It is different from a generator in that the magnetic field spins and the output coils are stationary.

Figure 30.4 Cutaway of an AC Generator. *(Reprinted with permission from Bosch)*

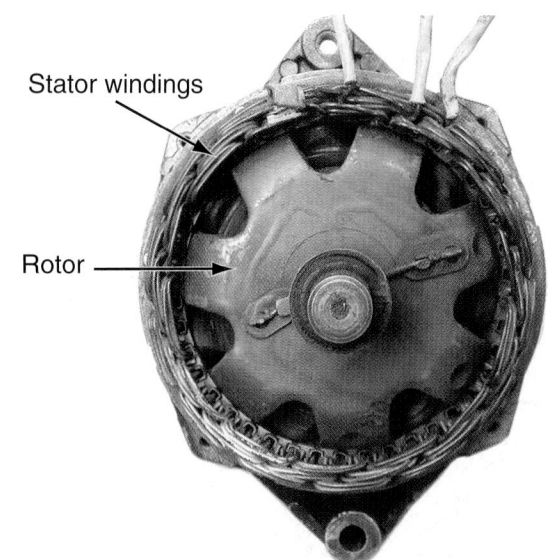

Figure 30.6 The rotor is very close to the stator windings. *(Courtesy of Tim Gilles)*

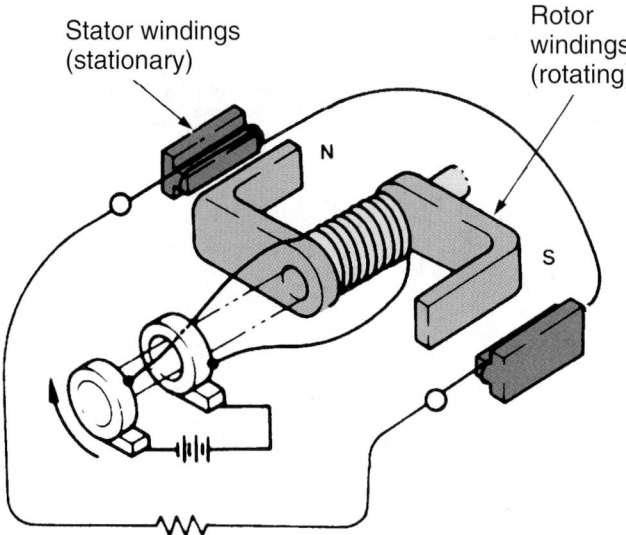

Figure 30.5 The two windings in an alternator.

An alternator has two different electrical windings, each having a separate function (**Figure 30.5**).
■ The *stator* is a stationary conductor.
■ The *rotor* is a rotating electromagnetic field.
As the rotor spins, its magnetic field cuts across the windings of the stator. This causes current to flow in the stator windings.

Rotor Construction

The rotor is the magnetic field that rotates within the stator's wire windings. There is very little clearance (about 0.015") between the two so that the effect of the magnetic field will remain strong (**Figure 30.6**). The spinning rotor has a magnetic field that cuts across the windings in the stator producing voltage in them.

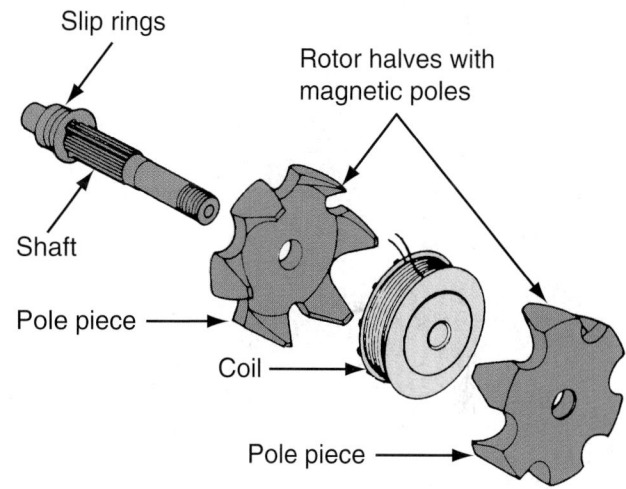

Figure 30.7 Parts of an alternator rotor.

The rotor's field coil has electrical wire wound around a shaft to make a powerful electromagnet (**Figure 30.7**). *Slip rings* connect to both ends of the field coil. Carbon *brushes* ride on the slip rings to provide electricity to the field coil (**Figure 30.8**). Slip rings are smooth so they do not wear the brushes.

One brush is insulated. Power comes in through that brush, goes through the field coil winding, and leaves through the ground brush.

Poles, which are shaped like claws, fit into each other. The pole pieces of the rotor make several pairs of north and south poles. This increases the magnetic flux and provides smoother electrical output. A magnetic field is formed between the pairs of north and south pole pieces (**Figure 30.9**). When the rotor spins, an alternating north/south polarity is created (AC current) (**Figure 30.10**). This is called **sine wave voltage**. Current flow follows the same path.

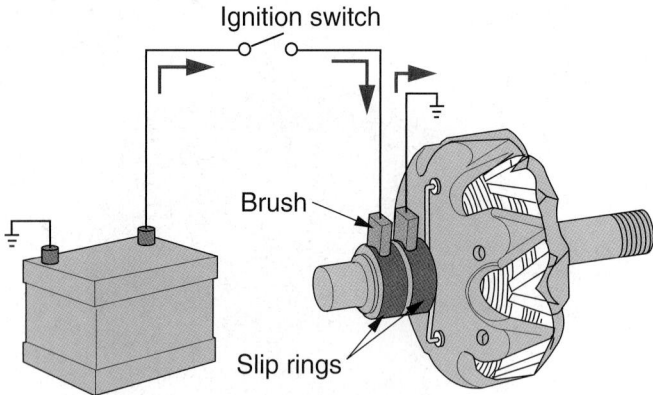

Figure 30.8 Brushes provide electricity to the field coil.

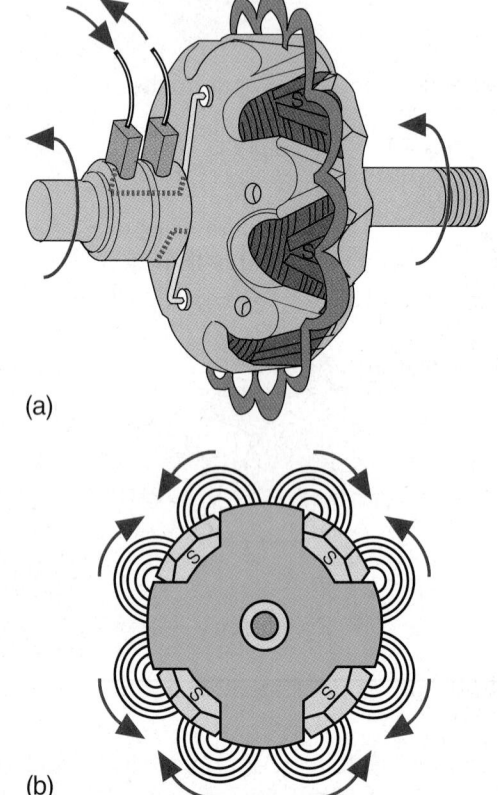

(a)

(b)

Figure 30.9 A magnetic field is formed between the pairs of north and south pole pieces. *(b, Courtesy of Ford Motor Company)*

An average rotor can spin at about 13,500 rpm. This is not engine rpm, but alternator rpm. The differences in size between the pulleys on the crankshaft and alternator determine the speed at which the alternator spins. The engine usually turns at about one-third the speed of the alternator.

Stator Windings

The stator fits in the frame of the alternator, between the front and rear halves of the cast aluminum housing (**Figure 30.11**). A stator has three sets of windings wrapped around slots in a laminated round iron frame

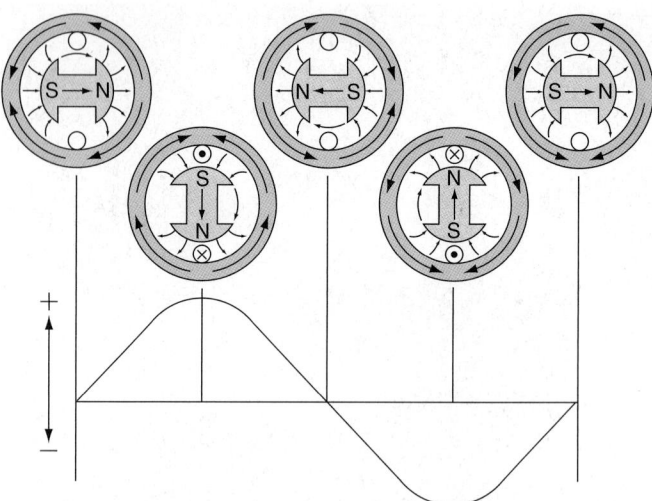

Figure 30.10 Alternating current is produced as the poles rotate.

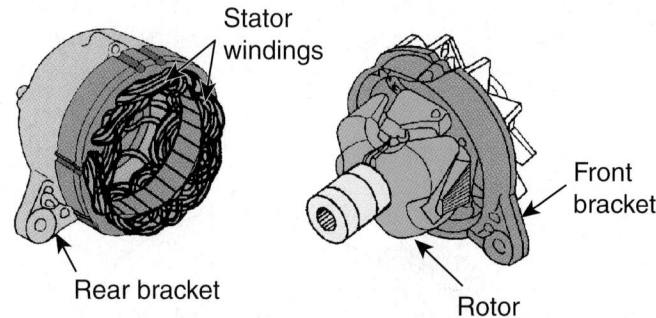

Figure 30.11 The stator fits between the front and rear halves of the alternator. *(Courtesy of DaimlerChrysler Corporation)*

called a core (**Figure 30.12**). Each winding has the same number of coils as the rotor has pairs of north and south poles. The windings are overlapped and positioned at designed places on the core.

Each winding has two leads, one for current to enter and the other for it to exit. There are two ways of connecting the leads, the wye (Y) winding and the delta winding. The wye (Y) is the most common in later vehicles. It provides higher output at lower engine speeds and is more efficient throughout the entire range of speed of alternator operation. Its magnetic field can excite (begin to work) at less than idle speed.

In the wye (Y) winding, one lead from each winding (three altogether) is connected to a common connection in the middle. The remaining three leads are branched out in a "Y" pattern (**Figure 30.13**). Each one connects to a diode (covered later).

When a stator is wound with a delta connection, the leads of the windings are connected together in series (**Figure 30.14**). The form resembles the Greek letter delta (Δ). A delta winding is used in high-output alternators. It can put out about one-third more current at highway speeds than a Y-wound stator. Because it is about one-third less efficient at idle speed, there

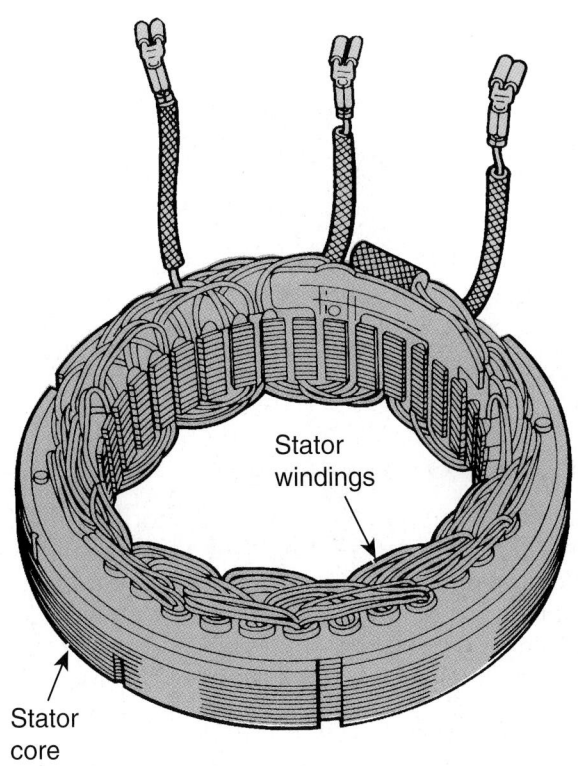

Stator windings

Stator core

Figure 30.12 A stator has three windings. *(Courtesy of Ford Motor Company)*

is a problem with delta-wound alternators. One complaint is the dash light will not go out until the engine is revved up enough to excite its field.

NOTE: *By cutting and changing the connections from a wye (Y) to a delta, the amp output of the alternator will go from 63 amps to 80 amps.*

Rectifier Construction

The current that an alternator produces is *alternating current*. This means that the current flows first in one direction and then in the other. The battery cannot

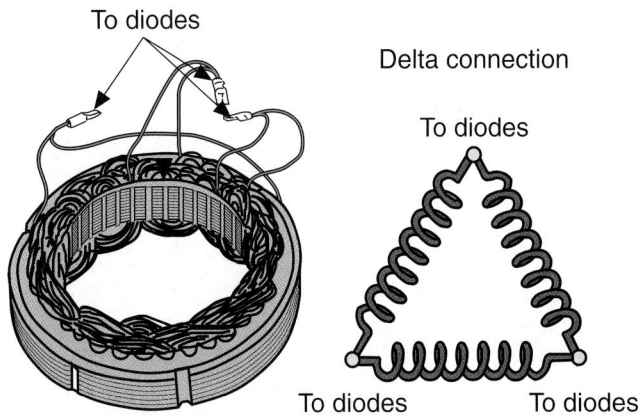

To diodes

Delta connection

To diodes

To diodes   To diodes

Figure 30.14 A delta stator winding.

use AC, so alternator output must be converted to DC. A diode rectifier bridge is used for this purpose.

A *diode* is a one-way check valve for electricity. There are positive and negative diodes to control the flow in either direction. When the AC current reverses itself, the diode *blocks* and no current flows. **Figure 30.15** shows the pulsing DC current that results when a positive diode blocks negative current.

A pair of diodes is used for each stator winding (a total of six in all) (**Figure 30.16**). Three positive diodes are mounted in a heat sink (**Figure 30.17**). A **heat sink** helps to dissipate the heat that occurs as electrons try to pass the diode. The three negative diodes are mounted in the alternator frame.

Using a pair of diodes that are reverse biased to each other, both sides of the AC sine wave can be *rectified* to positive DC. The three windings of the stator produce their currents in phases. The output is called **three phase**. When the three phases of AC are rectified, the result is almost uniform DC voltage (**Figure 30.18**). There is very little pulsation (about 42,000 pulses per minute at 500 engine rpm).

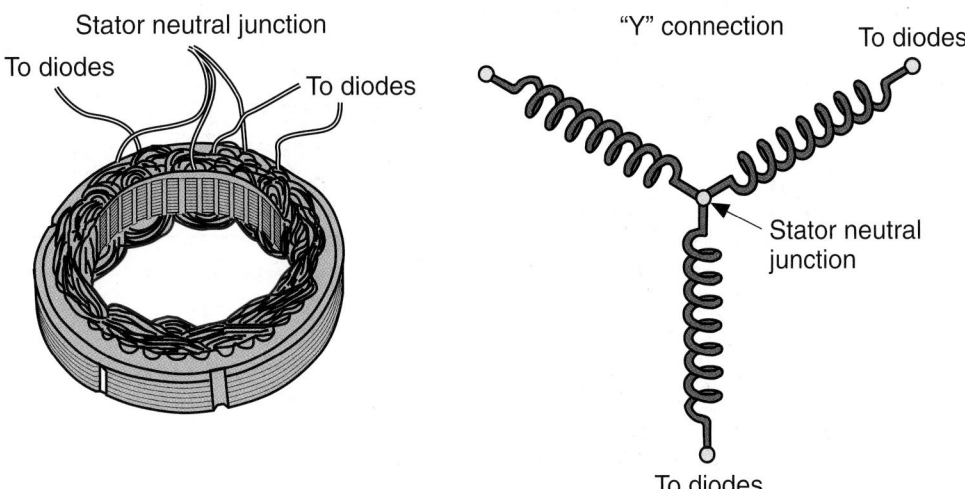

Stator neutral junction

To diodes   To diodes

"Y" connection   To diodes

Stator neutral junction

To diodes

Figure 30.13 A wye (Y) stator winding.

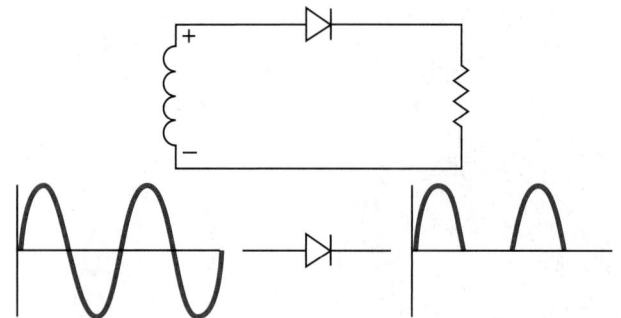

Figure 30.15 A positive diode blocks negative current flow.

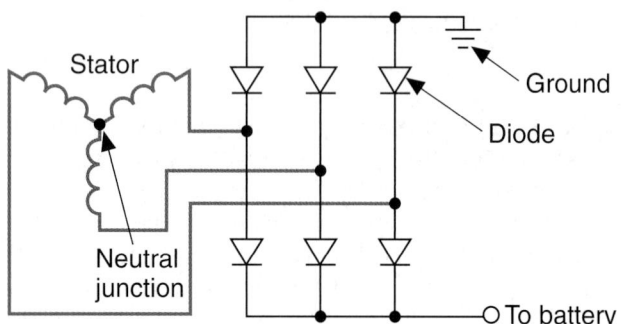

Figure 30.16 A pair of diodes is used for each stator winding.

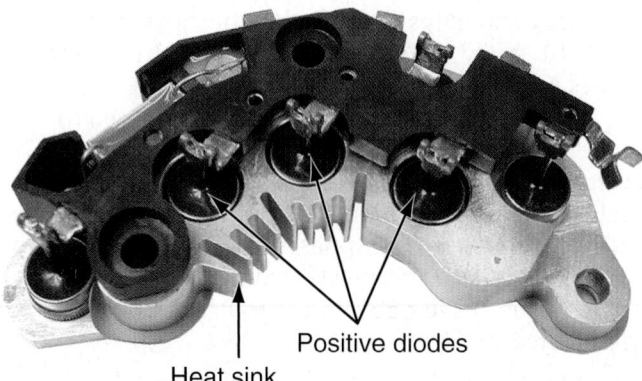

Figure 30.17 Positive diodes are mounted in a heat sink. (*Courtesy of Tim Gilles*)

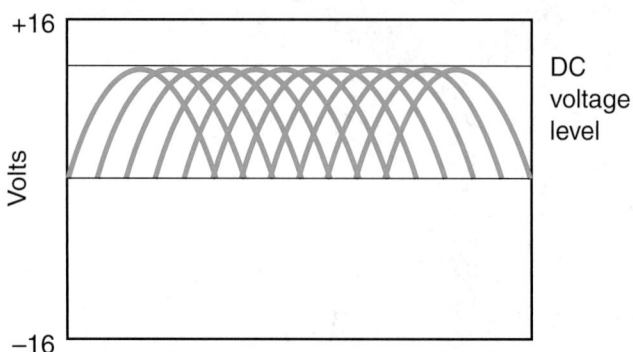

Figure 30.18 Three-phase output results in very little alternator pulsation.

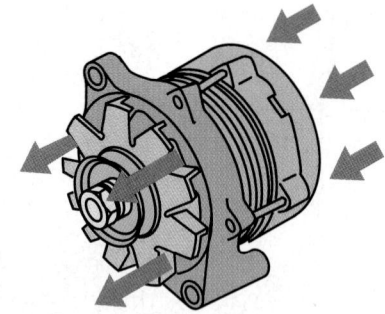

Figure 30.19 The fan draws air from the rear of the alternator for cooling.

■ ALTERNATOR BEARINGS

The rotor is supported in the alternator housing using either ball or roller bearings. The bearings are usually sealed and packed with grease. The front bearing fits into an indent in the case and is retained with a small plate and screws. The rear bearing is usually press-fit into the case. The rotor shaft slides into the rear bearing.

Alternator Fan

The cooling fan draws air into the alternator through openings at the rear of the alternator. The air leaves the alternator through openings at the front of the alternator behind the cooling fan (**Figure 30.19**). Some later model alternators also have an internal cooling fan.

■ VOLTAGE REGULATOR

The **voltage regulator** controls the current passing through the windings of the electromagnetic field in the rotor. This determines the amount of current produced in the stator (**Figure 30.20**). It senses battery voltage and decides how much current to put into the

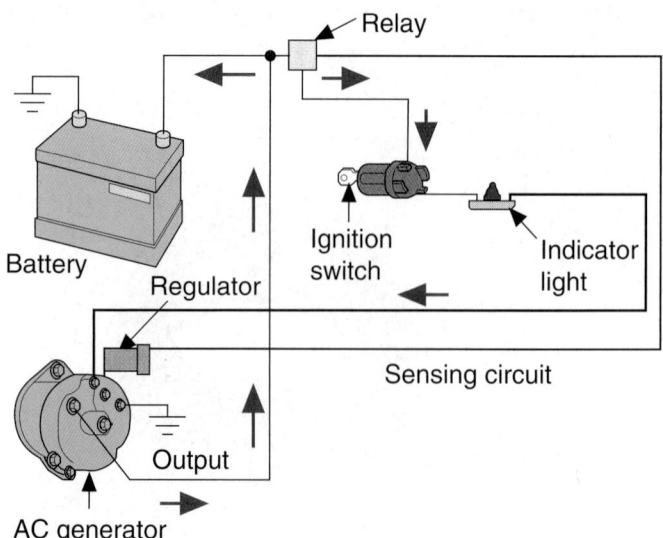

Figure 30.20 The voltage regulator controls the strength of the electromagnetic field in the rotor.

rotor (electromagnet). This determines the amount of alternator output.

When charging system voltage is low and accessories are loading the system, the regulator will increase the alternator current output. If system voltage is normal and there is a heavy accessory load, the alternator will run the accessories with little drain on the battery. When current demands are low and the battery is charged, alternator current output is low. To reduce the amount of output, the regulator puts more resistance between the battery and field coil in the rotor.

Electronic voltage regulators were phased in during the late 1970s. They have no moving parts or contacts to wear or burn out, so they are very reliable. Electronic regulators are either external or integral.

Integral regulators are mounted on or in the alternator housing (**Figure 30.21**). They work the same way as external regulators, but are smaller with integrated circuits. Some of them are inside the alternator. Others are on the outside for easier replacement (**Figure 30.22**).

Voltage regulation on many late-model vehicles is done by the on-board computer. This has eliminated the need for an integral voltage regulator.

Figure 30.21 This alternator has an integral voltage regulator. *(Reprinted with permission from Bosch)*

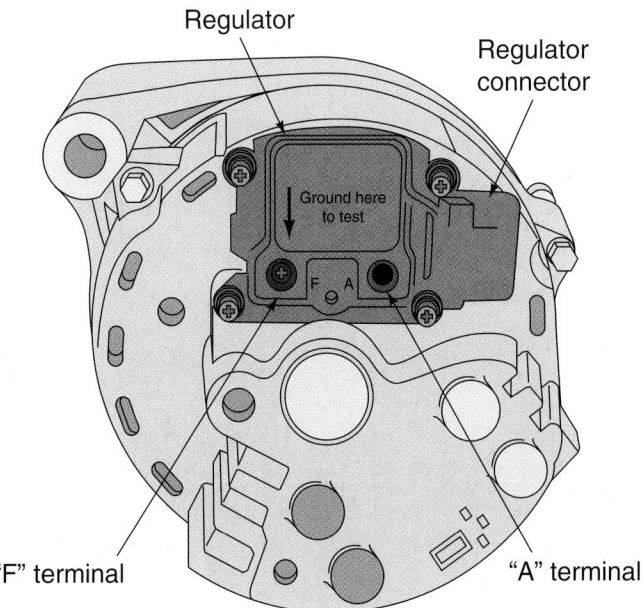

Figure 30.22 An integral regulator. *(Courtesy of Ford Motor Company)*

Electronic Regulator Operation

A **zener diode** is an electronic component in the voltage regulator that only conducts electricity when a certain voltage is reached. This capability allows an electronic regulator to keep battery voltage between two specified points. Electronic regulators are not usually adjustable.

The voltage regulator controls current in the field circuit. It can be located on either the ground side or the positive side of the field circuit. An electronic regulator has no moving parts, so it can cycle electronically between 10 and 7,000 times per second. This quickness means that the regulator can control the output of the alternator more accurately. Turning the alternator rapidly on and off to achieve a precise output is called **pulse width modulation**.

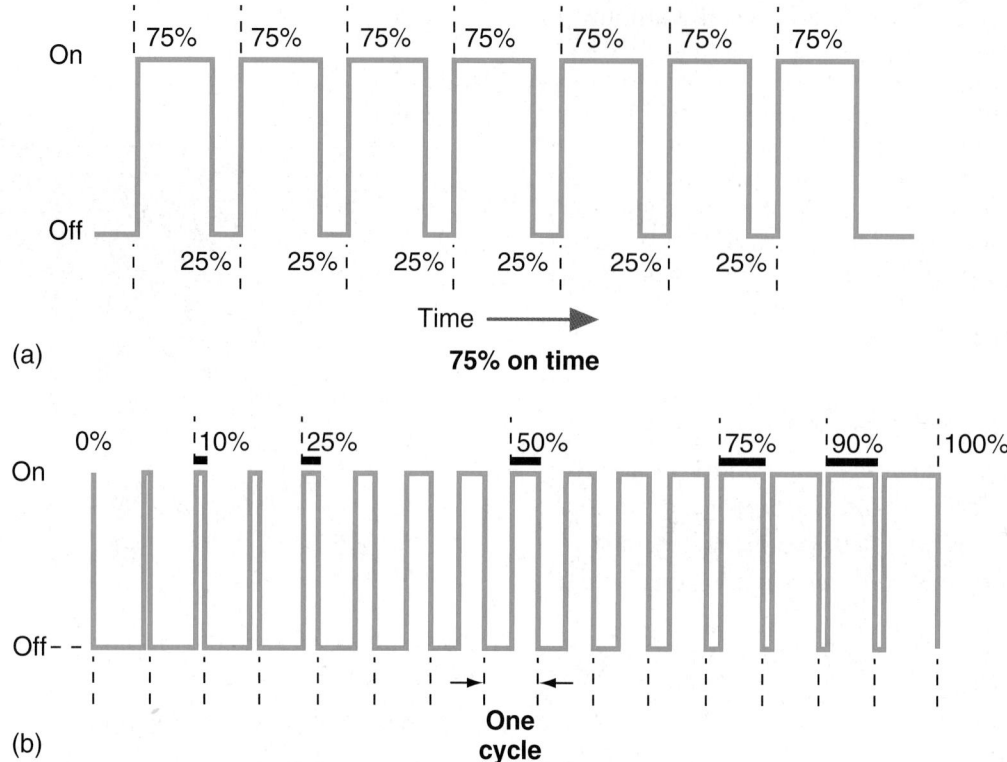

Figure 30.23 (a) Duty cycle with 75% "On" time. (b) Pulse width modulation is when the duty cycle varies.

(Figure 30.23b). Pulse width modulated output, averaged over time, provides an exact voltage input to an electrical device. For example, if the duty cycle output spends exactly 50% of the time with an output of 14 volts and 50% of the time at 0 volts, then the average output would be exactly 7 volts.

CHARGING SYSTEM INDICATORS

When there is a problem with the charging system, the driver is warned with one of three types of charge indicators: a warning light, a voltmeter, or an ammeter. The indicator is on the car's dashboard. Most cars have an indicator light. A voltmeter or ammeter is an option found on some vehicles.

The *alternator warning light* is wired into the charging circuit and will come on when system output drops below a certain level. On some older systems, the light was wired in *series* with the field current. If the light burned out, the charging system would not charge. A resistor was wired parallel to the bulb to provide a path for current flow if the bulb burns out. **Figure 30.24** shows a wiring diagram for a typical Y-wound alternator circuit with an indicator light. Testing the bulb is done by turning on the ignition switch without the engine running.

A *voltmeter indicator* shows system voltage when the engine is running. At 80°F, a fully charged battery is about 12.6 volts. The voltage regulator must allow higher voltage to come from the alternator to force electrons into the battery. A typical electronic voltage regulator maintains voltage at between 13.5 and 15.5 volts (check manufacturer's specifications). Voltage is closer to 15.5 volts whenever the system is very cold, and closer to 13.5 volts when the system is hot. The voltmeter indicator can give the driver a good idea of how well the charging system is working.

An advantage to a voltmeter is that it is a parallel circuit (**Figure 30.25**). No current runs through the wire or the indicator, so there is no danger of a fire if the wire becomes pinched or grounded. A problem with a voltmeter is that it shows what the charging system is doing at any one moment. An uninformed driver might visit the dealership to question a nonexistent problem.

An *ammeter indicator* gives the amount of current flowing to or from the battery. The needle in the center represents zero. Usually, if the needle moves to the left side, the alternator is discharging the battery. When it moves to the right, the system is charging the battery. It is normal for the gauge to read a high amount of current after the engine is first started. The amount of charge will drop off as the battery is recharged.

One disadvantage to the ammeter is that its wire carries high current. A fire under the dash can result if it grounds. An ammeter is also too expensive because the meter must be of the correct size to carry current for the entire charging load. Charging system output

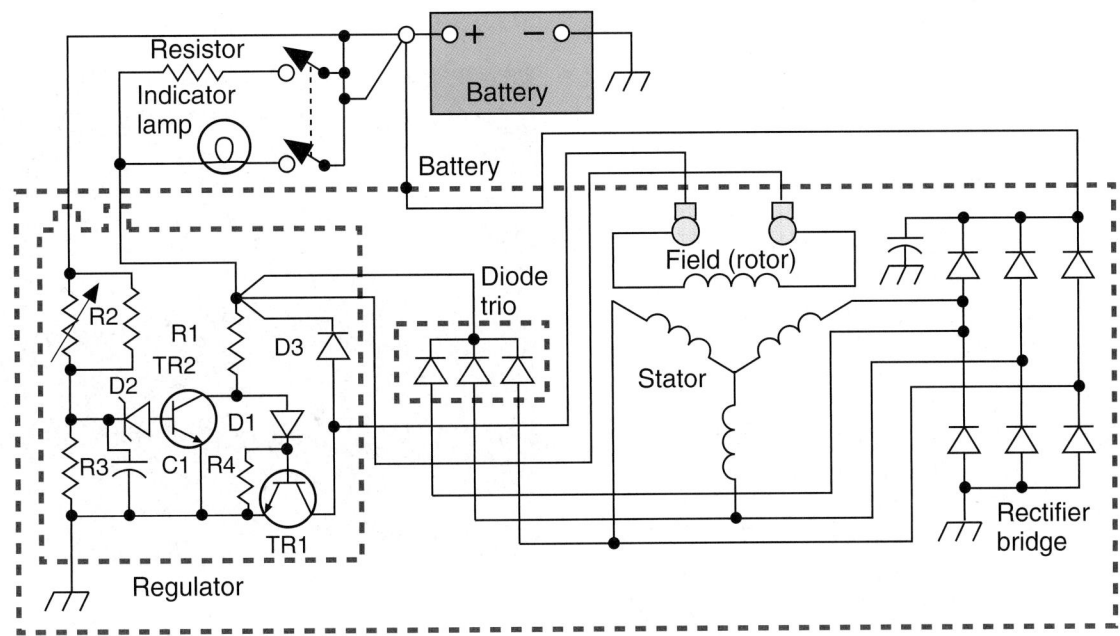

Figure 30.24 A wiring diagram for an alternator circuit with an indicator light.

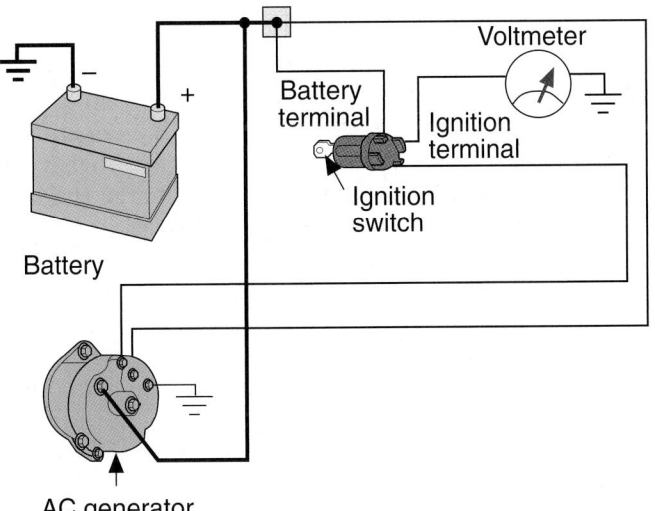

Figure 30.25 A voltmeter is wired in parallel.

has become very high in comparison to cars built in the sixties when they had ammeters.

■ HIGH-VOLTAGE CHARGING SYSTEMS

The typical electrical requirements of a 1970 automobile amounted to about 500 watts. In 2007, a typical vehicle's electrical demand was about 4,000 watts, or eight times as much. Alternators (AC generators) on 1970s vehicles typically put out from 35 to 50 amps. Today, it is not unusual to find 150-amp alternators on passenger vehicles. As vehicles evolve, electrical demand is projected to increase to more than double today's level, that is, to 10,000 watts or more. Additional electrical accessories will include items like electric water pumps, electric AC compressors, electric

power steering, electromechanical intake and exhaust valves, electric oil pumps, electric brakes ("brake by wire"), and electrically heated catalytic converters. Automotive charging systems have traditionally been called 12-volt systems, although they are really closer to 14 volts when operating. For today's low 14-volt charging system to produce 10,000 watts, alternator output would have to be over 700 amps. This would require much larger wiring.

Before hybrid vehicles became popular, 42-volt electrical systems were being designed for conventional automobiles. Because the voltage is tripled from the conventional 14-volt level, wiring size can be reduced by two-thirds. However, connectors, fuses, relays, switches, and motors all must be redesigned to be used with 42-volt and higher voltage systems like those used in hybrids.

Several manufacturers have moved away from 42-volt system development. Higher voltage systems are more prone to corrosion, but this is not the most serious problem. In addition to the danger of electrical shock, connectors are much more sensitive to high voltage if they are disconnected when under power. This is because a much greater arc is produced, especially when compared to a 14-volt system (a 42-volt arc can reach 6,000°F). A very important advantage to the conventional 14-volt electrical system is that when disconnected while under load, it does not sustain an electrical arc.

Hybrid Charging System Operation

Hybrid vehicles usually have more electrically powered items than conventional automobiles. This is partly because the internal combustion engine is usually off at idle, so it cannot provide power for the electrical accessories. Electrically powered devices can include an electric water pump, electric AC compressor, and electric power steering.

As described in Chapter 16, the combination starter-generator used in a hybrid automobile is usually located between the engine and transmission. Its rotor is connected to the crankshaft, and its stator is attached to the engine block. The combination starter-generator is often called an integrated starter generator (ISG), but it also can be called an integrated starter alternator (ISA) or an integrated motor generator (IMG). The ISG uses a three-phase AC motor "integrated" with the internal combustion (IC) engine.

A generator creates AC voltage, which is rectified by an AC to DC converter so it can be stored in a battery and used in the automobile's DC electrical system.

NOTE: *When AC to DC conversion is done within the generator, the device is customarily called an alternator rather than a generator.*

Hybrid vehicles use a generator with an *inverter/converter* (**Figure 30.26**). It is the "high-voltage train station" that controls electrical flow between the ISG and the vehicle's high- and low-voltage electrical systems. The inverter can invert in both directions, AC to DC and DC to AC:

- It inverts the generator's AC output to DC so it can be stored in the batteries.
- It inverts the high-voltage (HV) battery pack's DC to AC so it can be used by the motor.

The inverter/converter also converts hybrid battery pack voltage to low voltage so it can power the computer and electrical accessories.

The ISG cannot power and charge simultaneously, although it can change very quickly and seamlessly between its duties as a generator and a motor. ISG high-voltage systems can operate at voltages varying

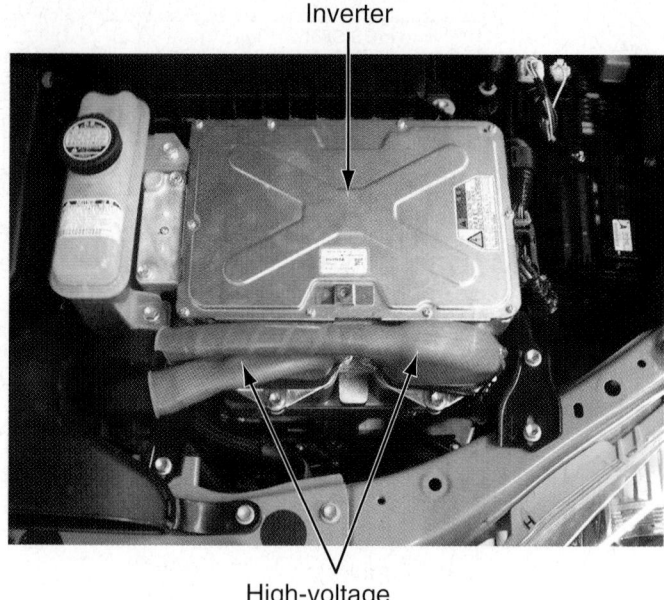

Inverter

High-voltage (orange) cables

Figure 30.26 An inverter located under the hood of a hybrid vehicle. *(Courtesy of Tim Gilles)*

between 42 and 158 volts using a sophisticated energy management system. Hybrid systems vary depending on the manufacturer. A typical hybrid uses high voltage from the three-phase generator to charge the battery pack, some of which store over 200 volts. When the ISG is acting as a motor, it can use 500 volts or more to power the vehicle. The internal combustion engine and the rest of the vehicle electrical accessories/demands use a conventional 12- to 14-volt electrical system. The one exception is an electric air-conditioning compressor (if equipped), which is the only high-voltage accessory. It typically runs on HV battery voltage inverted to AC.

▦ REVIEW QUESTIONS

1. What kind of electrical output results from the generator commutator and brushes, AC or DC?

2. What is another name for an AC generator?

3. How many amps of current flow through alternator brushes?

4. What is the name of the stationary conductor in an alternator?

5. What is the name of the rotating electromagnet in an alternator?

6. What is the name of a one-way check valve for electricity?

7. What is it called when the AC sine wave is changed to all positive pulses of current?

8. Which way does air move through an alternator, front to back or back to front?

9. What is the name of the device that controls the amount of electricity entering the rotor?

10. What is the name of the electronic component in the voltage regulator that only conducts electricity when a certain voltage is reached?

■ ASE-STYLE REVIEW QUESTIONS

1. What is the name of the part that allows a generator to produce direct current output?

   a. Armature   **c.** Commutator

   b. Stator   **d.** None of the above

2. Technician A says that all brushes in an alternator must be insulated. Technician B says that all diodes in an alternator must be insulated. Who is right?

   a. Technician A   **c.** Both A and B

   b. Technician B   **d.** Neither A nor B

3. At what speed does an alternator rotate?

   a. Crankshaft rpm   **c.** Distributor rpm

   b. Camshaft rpm   **d.** None of the above

4. The stator winding that has three wires connected to each other at a common place in the middle is called a _____ winding.

   a. Delta   **c.** Primary

   b. Wye (Y)   **d.** Secondary

5. Which of the following is/are true about alternator diodes?

   a. Three positive diodes will be mounted in a heat sink.

   b. Three negative diodes will be pressed into the alternator frame.

   c. Both A and B.

   d. Neither A nor B.

6. Alternator output is _____ phase.

   a. Single   **c.** Three

   b. Two   **d.** Four

7. Turning the alternator rapidly on and off is called:

   a. Rapid cycling

   b. Pulse width modulation

   c. Ground side switching

   d. Alternating current

8. Which wire to a charging system component can start a fire if it grounds out?

   a. The voltmeter

   b. The ammeter

   c. Both A and B

   d. Neither A nor B

9. Alternator cooling fans are:

   a. On the outside of the alternator

   b. Enclosed within the alternator housing

   c. Both A and B

   d. Neither A nor B

10. What allows an alternator to put out direct current?

   a. A rectifier.

   b. Diodes.

   c. Both A and B.

   d. Alternators put out alternating current only.

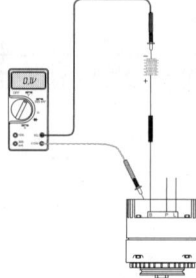

Charging System Service

■ KEY TERMS

charging system output test	full-field test	B-circuit
regulator maximum voltage test	circuit resistance test	isolated field
	A-circuit	

■ INTRODUCTION

This chapter deals with the process of testing and repairing common problems with the charging system. Principles of operation and electrical fundamentals learned in earlier chapters will be important here. Diagnosis of failures is very important before parts are replaced. Most parts stores will not accept returns of electrical items.

■ CHARGING SYSTEM SERVICE

Charging system problems usually become evident when the battery goes dead or a customer notices a noise.

The following are four common charging system complaints:
- Dead battery
- Battery water low (overcharging)
- Indicator light glows or incorrect voltage is indicated
- Noises

Test Battery First

Just as with the starting system, the battery must be in good condition if its state of charge is to be maintained by the charging system. A complete charging system diagnosis must be performed before replacing any charging system parts. If you replace a battery just because it is old, that will not solve a charging problem caused by a loose drive belt. The car will leave the shop with a new, freshly charged battery, but the customer will be upset when the battery goes dead once again and more work is required on the vehicle. Odds are that the customer will go to another shop for the work. Always test and retest your work.

Battery Condition. First, check the battery's state of charge. A battery will not be fully charged unless the charging system is putting some current into it. If the battery state of charge is low, the problem can be due to a parasitic drain when the engine is off. If the battery is at least 80% charged, perform a 15-second load test to find out if it has the capacity to stay above 9.6 volts.

If a battery fails the load test, perform the 3-minute charge test to see if the battery is sulfated. Sulfation occurs when the battery has been allowed to remain in a discharged state (see Chapter 27). Simply replacing this battery will not solve the problem. Testing the charging system is always a part of battery replacement.

■ TESTING THE CHARGING SYSTEM

First, perform a visual inspection, looking for obvious problems. If the battery fails when tested, substitute a known good battery before testing the charging system.
- Check for corroded or broken wire connections.
- Wiggle wires while the engine runs. Have an assistant sit inside the car to see if an indicator light goes out or a gauge begins to read correctly.

- Listen for noises as the engine runs. Be careful not-to use a stethoscope close to alternator wiring or belt.
- Look for a loose or damaged alternator drive belt.

SHOP TIP With the engine off, try to turn the alternator fan and pulley. If the pulley slips on the belt, it is too loose or is glazed and must be replaced.

A loose drive belt can cause two problems. The obvious problem is a low charge rate. When the regulator energizes the field in the alternator rotor, a load is put on the drive belt. If it is loose, it will slip and the charging system will not work to capacity. Today's synthetic belt materials do not squeak as badly when they slip as did older belts, which contained cotton fibers. Another problem caused by a slipping belt is that it can heat up the rotor shaft and cause failure of the drive end bearing.

CAUTION Use a belt tension gauge on belts that do not have a spring-loaded tensioner. If a belt is too tight, bearing wear to the water pump, alternator, and engine crankshaft can result.

SHOP TIP There is a simple test to see if an alternator field is strong, and it is done with the key on and engine off. Check the rotor shaft at the outside of the housing of the rear bearing for magnetism. If there is enough magnetism to lightly hold a screwdriver, the rotor is energized. If the rear bearing is not magnetized, the problem could be related to the regulator, the brushes, or a defective rotor.

Check Alternator Rating

Check to see what the alternator's maximum rated output is. This can be found in the specification listing in the service information library. Some manufacturers identify their alternators by stamping the output on the alternator frame or by color coding. The more accessories the vehicle has, the larger the alternator will be. **Figure 31.1** shows the typical amp draw of accessories that the charging system might have to supply.

Connect the volt-amp tester to the battery (see Figure 29.4). The inductive pickup should be as close to the battery as possible to avoid picking up a stray magnetic signal from the alternator or another electrical device under the hood. Set the meter range to the highest scales during starting and then adjust them to lower ranges as needed.

There are four charging system tests to perform:
- Charging system output test
- Regulator voltage settings

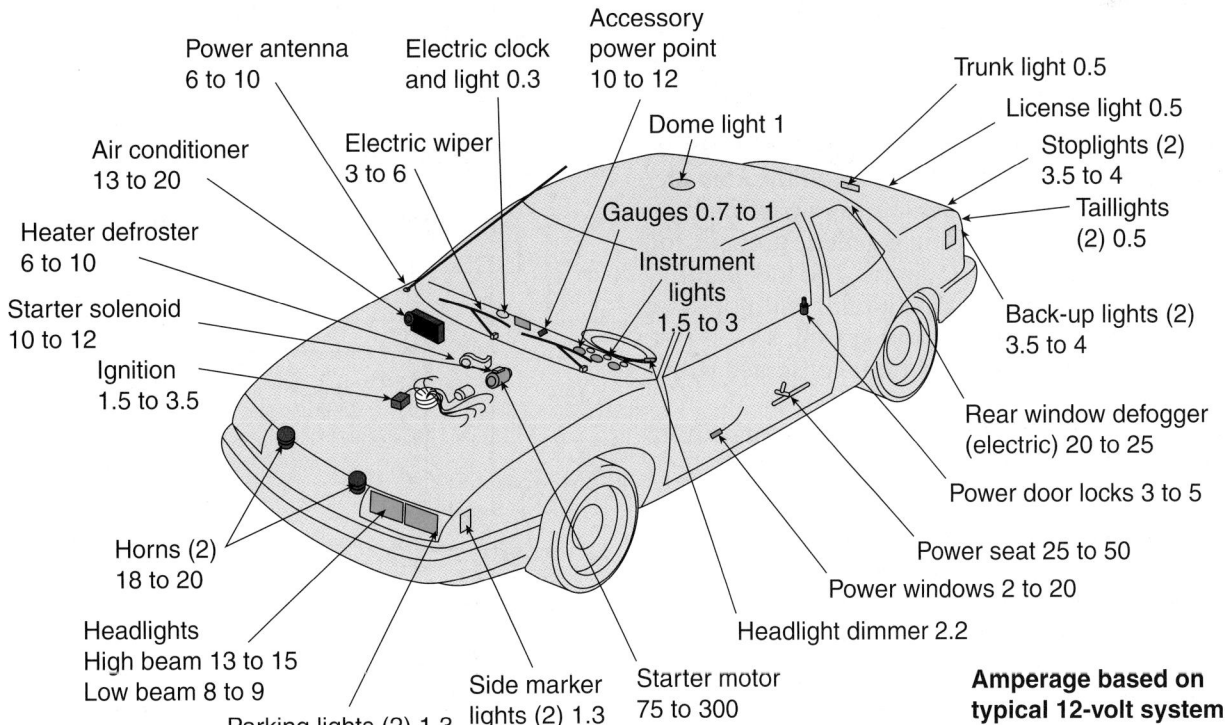

Figure 31.1 Typical amp draw for accessories.

- Alternator full-field test (when available)
- Charging circuit resistance tests

The **charging system output test** loads the charging system and measures its current output.

The **regulator maximum voltage test** checks to see that the voltage regulator is energizing the rotor field and is not allowing the alternator to overcharge.

The **full-field test** of the alternator takes the regulator out of the circuit and causes the alternator to give full output.

Circuit resistance tests check the resistance in the ground and insulated circuits by measuring voltage drop.

NOTES:

- *Alternator ground wires are usually braided ground straps. The condition of these wires is as important as the insulated wires because they carry the same current flow.*
- *Some alternators are insulated in rubber bushings. These must have a separate ground strap.*

▨ CHARGING SYSTEM OUTPUT TEST

The following outlines the test of charging system output with a volt-amp tester:

- Start the engine. The speed of the alternator determines its output. At idle, the alternator is not capable of making maximum output.
- Raise the idle speed to about 2,000 rpm as you turn the knob on the carbon pile to lower battery voltage. As the voltage is lowered, the voltage regulator will energize the alternator's rotor.
- Read the amount of current being put into the battery by the alternator.
- The maximum output of the alternator will be more than the amount that shows on the gauge because the vehicle consumes some current while the engine is running. This amperage amount is never delivered to the battery where it could be read on the ammeter.
- A quick estimate of the alternator's maximum output can be made by using a generic amount for this calculation, such as 15 amps. The fuel pump, injectors, and many ignition systems will not draw current with the engine off and key on, so it is not a practical way to estimate current consumed with the engine running.
- For a more accurate estimation of actual maximum alternator output, move the amp probe to the B+ wire of the alternator.

NOTE: *A hot alternator loses 25% of its rated output. Alternators are rated at 160°, so they could have higher than the rated output when cold. This is because the electrical resistance of most conductors goes up with heat.*

Regulator Voltage Setting Check

The regulator voltage test checks to see the high and low settings of the regulator. A regulator must be able

to full-field an alternator immediately. It must also be able to keep the system at a predetermined voltage at normal operating temperature (usually about 13.5–14.5 volts). With lower temperatures, the regulator will raise the voltage.

With the engine running, the meter should show a gradual decrease in the charging current. The voltage should remain within normal operating limits. A defective electronic voltage regulator is not a typical cause of a low charge problem, but a bad regulator can result in higher than normal charging voltage.

NOTE: *A voltage drop (resistance) on the ground side of the regulator will raise system voltage by the amount of the voltage drop.*

Test the regulator ground with a voltmeter (**Figure 31.2**). If the ground side has no resistance, check the sensing voltage input to the regulator to see that it is the same as the battery voltage. If it is, the regulator should be able to keep voltage within specifications. It must be replaced if it does not.

Full-Field Test

When the amount of output of an alternator is too low, this could be due to the alternator or the regulator. Full-fielding the alternator is a test that eliminates the regulator from the circuit and energizes the rotor fully. This will cause the alternator to produce full output. If the output is too low with the alternator full-fielded, the alternator must be the source of the problem.

There are several procedures for full-fielding alternators depending on the design, although some manufacturers recommend against this procedure. Be sure to follow manufacturers' directions.

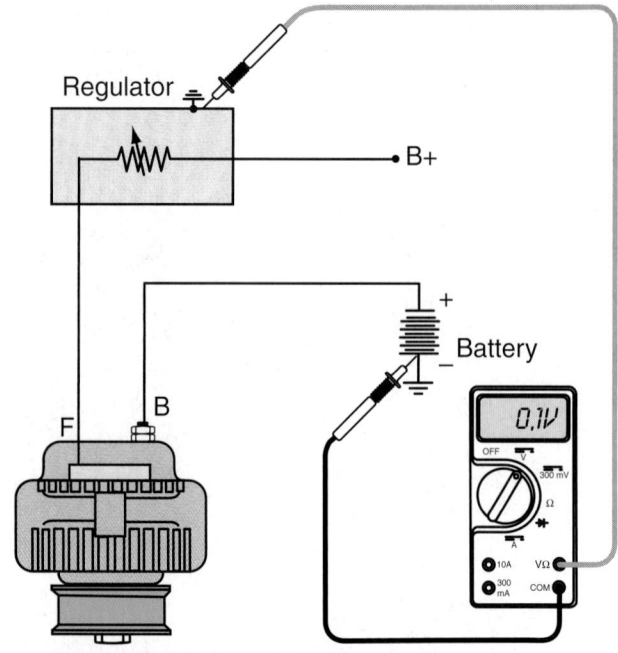

Figure 31.2 Performing a voltage drop test on the voltage regulator ground circuit with the engine running.

NOTE: *Some late-model alternators cannot be full-fielded.*

Regulators are connected to the alternator in one of two ways called, A- and B-circuits. The **A-circuit** has the regulator in the ground circuit. This type is used most often with solid-state (electronic) systems. One brush is connected to the battery and the other is grounded through the regulator. A **B-circuit** regulator has its regulator between the positive feed (insulated side) and the field coil, which is grounded inside of the alternator. This type is usually found on the older electromechanical voltage regulators.

Directions for full-fielding alternators vary according to the type of circuit. They can be found in service manuals and information sheets provided by electrical test equipment manufacturers.

- To full-field an A-circuit, supply a ground to the field terminal.
- To full-field a B-circuit, supply battery voltage (B+) to the field.

NOTE: *If you are not sure whether you are working with an A type or a B type, try both power and ground to the field terminal. You will not do any damage. The alternator will just not put out current if hooked up incorrectly.*

- Some regulators are of the **isolated field** type. They have two field leads and must have one lead jumped to battery positive (B+) and the other to ground.
- Some late-model alternators with integral regulators (like some Robert Bosch models) do not provide a method of full-fielding. They must be disassembled and the components individually tested, or you can use the charging system output test.

One way of full-fielding an alternator with an integral voltage regulator is shown on the General Motors (GM) alternator in **Figure 31.3**. A screwdriver is used to short a tab on the frame to ground. On some alternators with an external regulator, full-fielding is done by unplugging the connector to the regulator. A jumper

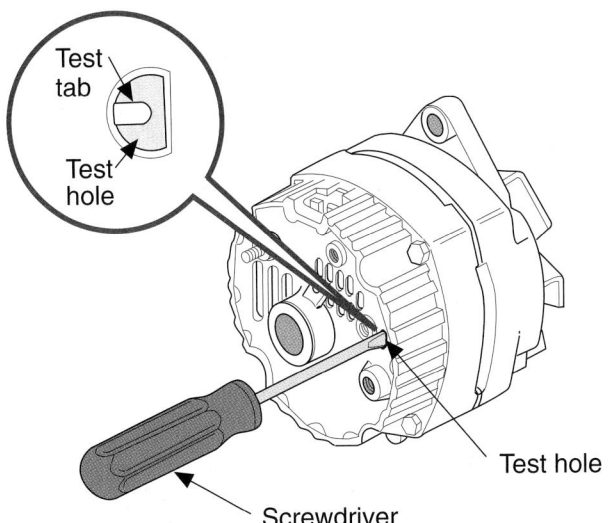

Figure 31.3 Full-fielding a GM alternator.

Alternator Testing Precautions

- Connecting the wrong wire to a ground or hot lead can ruin diodes or a regulator.
- Full-fielding should not be done for more than 10 seconds.
- Do not connect or disconnect electrical components with the key on. They could be damaged.
- Do not connect the regulator with the engine on.
- Do not operate the alternator without an external load connected to it. This can cause extremely high voltage and ruin the alternator.
- Do not run the engine with the battery disconnected. The battery helps stabilize voltage and acts like a shock absorber for damaging voltage spikes.

Figure 31.4 Alternator testing precautions.

wire is used to bypass the regulator, connecting the field directly to the battery.

Many computers have taken over the regulator function by controlling the alternator duty cycle. These are tested in the same manner as other alternators, usually by grounding the field terminal. Follow the manufacturer's recommendations. If the regulator has failed, these vehicles will require a new computer. Alternator testing precautions are listed in **Figure 31.4**.

CAUTION Be careful not to let the voltage in the system rise above regular regulated voltage. This can be controlled with the carbon pile. If you let the voltage continue to climb without the voltage regulator in control, an electronic component might be accidentally damaged. A safe maximum is about 16 volts.

Full-Field Test Results. When you bypass the regulator, the voltage should be controlled to the same amount it reached during maximum output with the regulator in control. These figures will be compared. If the alternator puts out more at the same voltage with the regulator bypassed, the regulator is faulty. When the current output remains below specification, a bad alternator or bad drive belt is the probable cause. Before condemning the regulator, full-field the alternator at the field terminal of the alternator. This will eliminate excessive resistance in the wiring as a possible cause.

Differences between A- and B-Circuit Faults. If an A-circuit regulator loses battery (B+) voltage, the alternator will overcharge if the field coils still have power. If it loses its ground, the alternator will stop working.

If a B-circuit regulator loses ground, it will overcharge. If it loses its battery (B+) voltage, the alternator will not work.

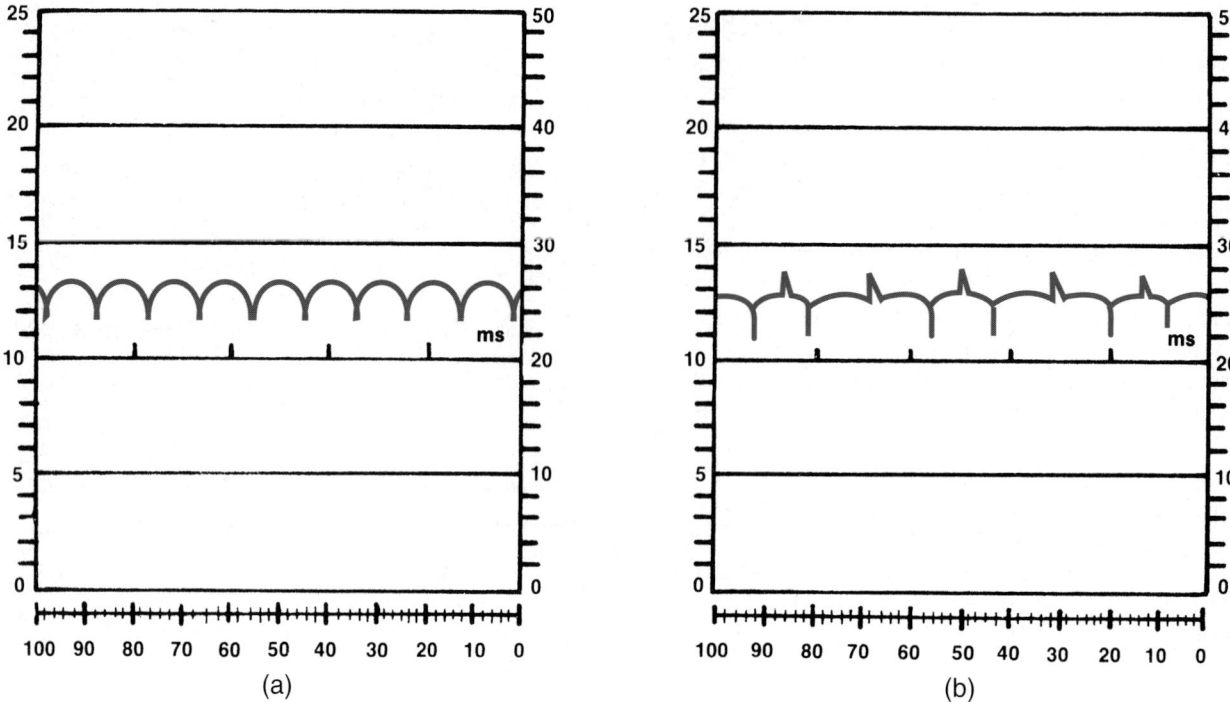

Figure 31.5 Normal alternator patterns under (a) full load and (b) no load.

■ DIODE TESTS

Diodes must be in good condition if the alternator is to function correctly. One or more leaking diodes will decrease an alternator's output. In addition to rectifying the alternator's AC output to DC, diodes prevent AC from leaving the alternator in the B+ output wire. If more than 0.5 AC volt is found in the B+ output, defective diodes are indicated. When testing, use a DMM on the AC setting and connect the positive to the alternator output and negative to ground. A low amp current probe can also be used to test diodes while the alternator is still assembled. With the engine off, there should be less than 0.5 milliamp of current flow in the B+ output wire. Diode service with the alternator disassembled is covered later in the chapter.

Oscilloscope Testing

A lab scope can be used to look at alternator patterns. **Figure 31.5** shows normal scope patterns. Notice how the patterns are equal. When there is a regular dropout in the pattern, a problem is indicated. This can be an open, a short or ground in a stator winding, or a shorted diode (**Figure 31.6**). The pattern for an open diode is shown in **Figure 31.7**. Whenever an electronic part has failed, it is a good idea to check the alternator diode pattern. DC electronic components cannot be exposed to AC. If you simply replace an electronic component and do not fix the cause of the problem, the new part might fail, too.

Diode Noises

When there is excessive AC coming from the alternator, a car radio will sometimes make a whining sound like a siren. The whine changes with engine rpm. When this sound is loud enough for a customer complaint, a bad diode is probably the cause. Newer car radios are less likely to be susceptible to such noise.

A bad diode can also cause the alternator to growl or whine. The pitch of the noise changes with engine rpm.

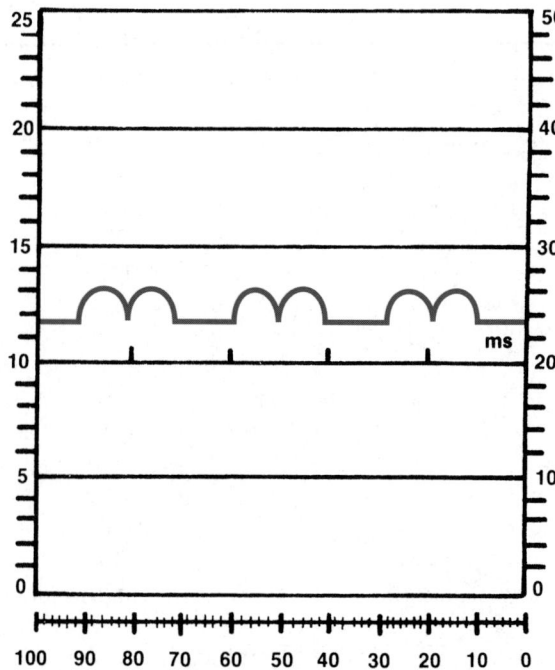

Figure 31.6 Scope pattern for a bad stator winding or shorted diode with the alternator under full load.

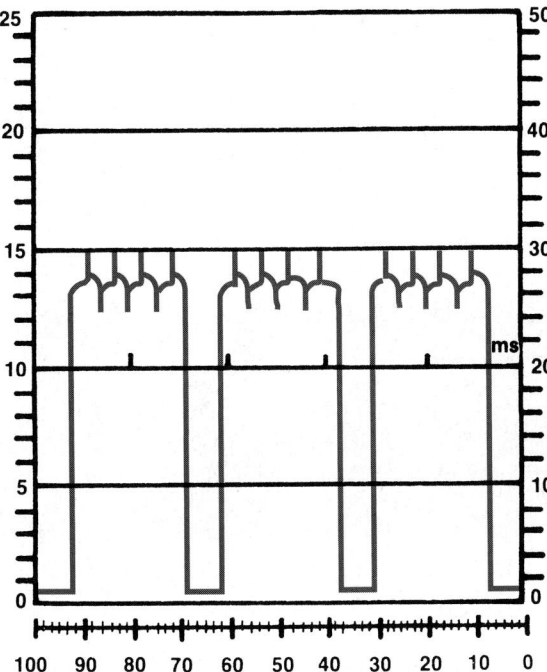

Figure 31.7 Scope pattern when there is an open diode.

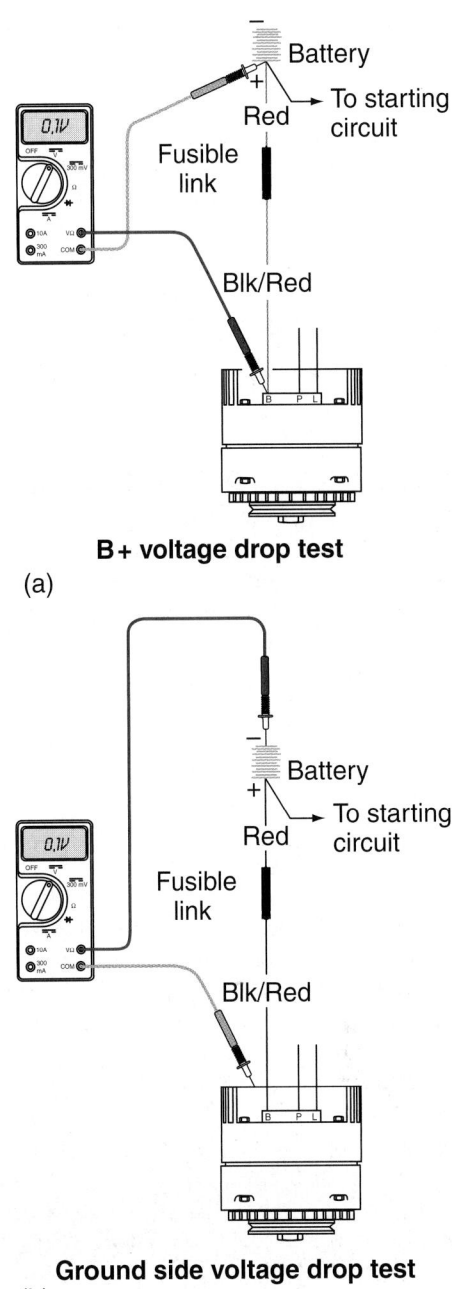

Figure 31.8 Charging system voltage drop tests. (a) The insulated (positive) side test. (b) The ground side test.

CHARGING SYSTEM VOLTAGE DROPS

Checking the resistances in the charging circuit is done when locating a stubborn problem that does not appear to be related to either the alternator or regulator. Resistance due to corrosion usually takes a long time to get bad enough to result in a failure. Checking voltage drops as a preventive maintenance measure will catch potential problems before they fully develop. The tests tell if the battery, regulator, and alternator are each working at the same potential.

NOTE: *Extra resistance in the regulator circuit will cause the battery to have a higher than normal charging voltage.*

Perform a voltage drop test in the same way you did on the starter circuit. **Figure 31.8** shows voltage drop test connections for the insulated (B+) and ground sides of the circuit. Current must be running through the system to be able to perform the test. Turn on the headlights or use the carbon pile on the volt-amp tester to load the system. The combined voltage, both insulated and ground sides, should not be more than 0.5 volt maximum.

- If the system has an indicator lamp, when checking the positive circuit with 10 amps flowing, no connection should have more than 0.3-volt drop.
- With an ammeter, 0.7-volt drop is the limit.
- A small amount of voltage drop past the fuse link, which is a smaller wire than the rest of the circuit, is normal. The amount of voltage drop is very small because the fuse link is so short.

There are no fuse links in the ground circuit, so lower voltage drop (0.1 volt) is desired.

ALTERNATOR SERVICE AND REPAIR

Many general automotive repair shops do not repair alternators. Most communities have specialty shops that rebuild and repair alternators, generators, and starters. There are also large remanufacturing companies that specialize in electrical system component rebuilding. After testing has determined an alternator to be defective, a new or rebuilt alternator is ordered. The defective unit is traded for a remanufactured one.

When the alternator has been determined to be the cause of a charging system problem, remove it from the vehicle.

NOTE: *Always disconnect the battery ground cable first. Label the connections to the alternator before disconnecting them.*

Loosen the drive belt and remove it. Then unbolt the alternator. Be careful that any spacers that are used in mounting are kept with their original bolts. This will make correct reinstallation much easier.

A bad alternator could be the result of several problems. A shop that specializes in alternator rebuilding will have a test bench for alternators. This tester provides for fast, easy mounting of the alternator. The alternator can be tested for rectifier and stator grounds and for opens or shorts in the field winding.

Alternator Disassembly

The alternator is easy to disassemble. Although alternators are commonly replaced with a professionally rebuilt unit, the procedure for disassembly is covered here so that you can become familiar with the parts. Some shops will also replace brushes or diodes. Procedures for the electrical testing of rotors and stators are described in applicable shop literature.

As you did when disassembling a starter, mark the side of the case for easy realignment of parts on reassembly (**Figure 31.9**). After removing the bolts, the rotor is separated from its housing (**Figure 31.10**).

Brush Service

If only the brushes are to be serviced, the rest of the alternator will not have to be disassembled. Some alternators have brushes that can be replaced from the outside without splitting the alternator housing

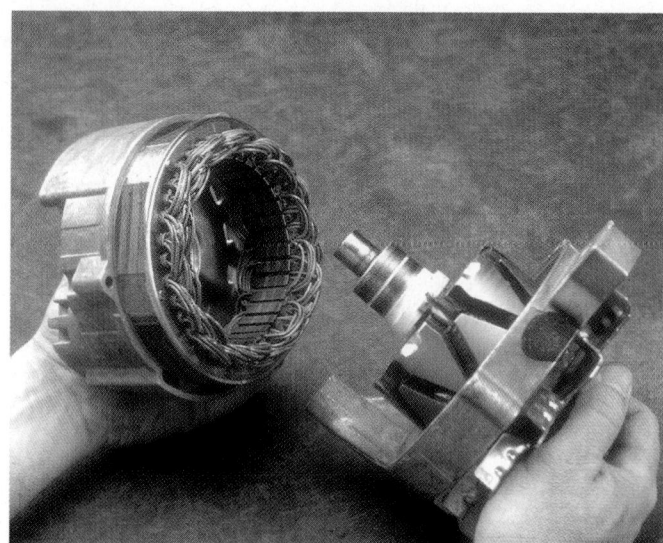

Figure 31.10 Separating the halves of the alternator.

(**Figure 31.11**). Brushes are spring loaded. When they wear, spring tension becomes less. Brushes that are too short can cause an alternator to stop working intermittently (once in a while) before failing completely.

Stator Service

After removing the stator leads from the rectifier bridge (the part that holds the diodes), the stator can be removed from the alternator housing. The stator is discarded if it has any signs of burned or overheated insulation. It can be checked for opens (**Figure 31.12**) and shorts to ground (**Figure 31.13**).

Rotor Service

To remove the rotor from the front frame requires removal of the pulley and fan. Use a small impact wrench to loosen the nut so you do not have to restrain the rotor

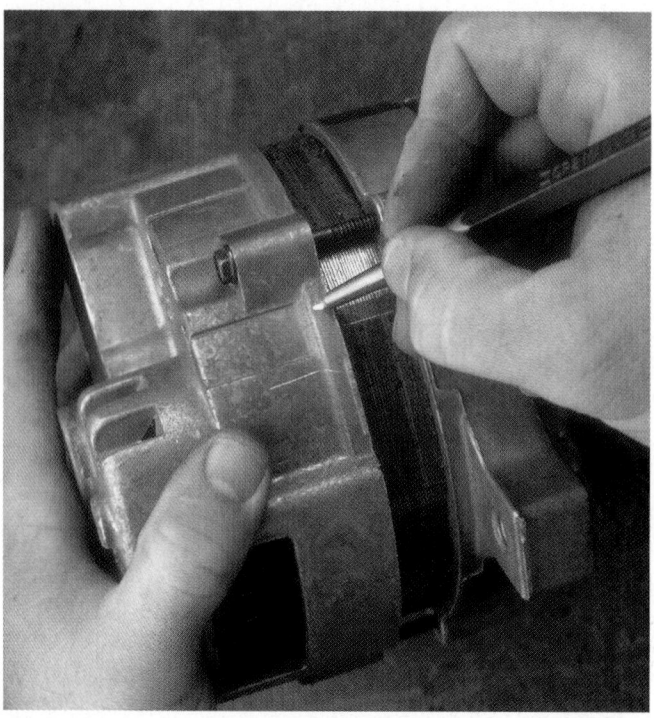

Figure 31.9 Mark the parts of the alternator before disassembly.

Figure 31.11 Some alternators have brushes that can be replaced without disassembling the alternator.

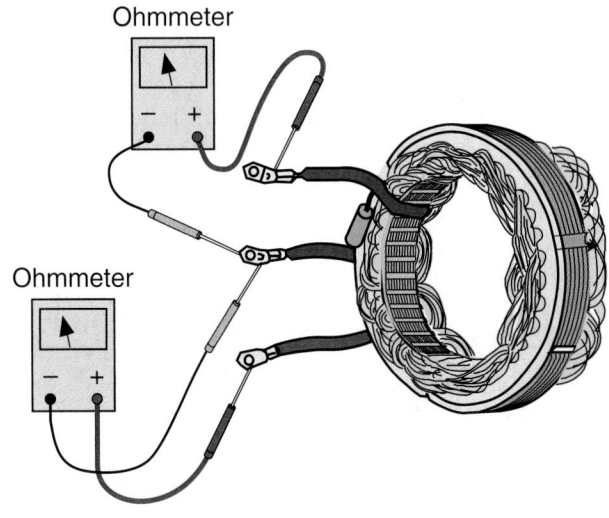

Checks for opens

Figure 31.12 Checking a stator for an open circuit.

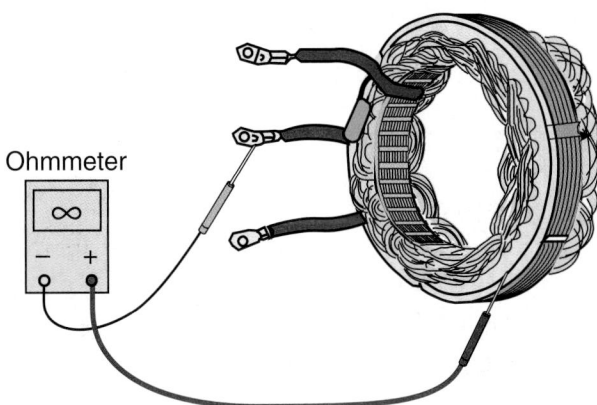

Checks for shorts to ground

Figure 31.13 Checking a stator for a grounded wire.

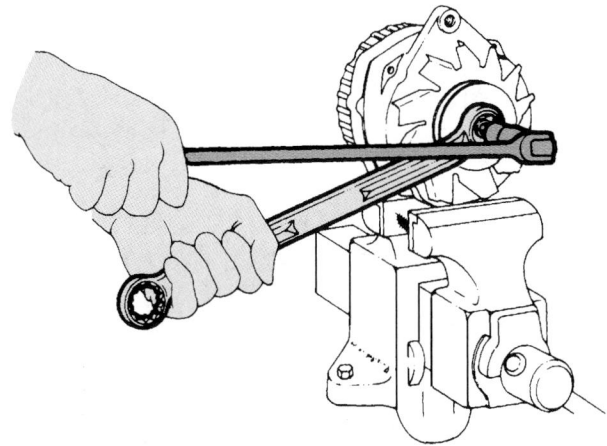

Figure 31.14 Holding the rotor shaft while tightening the pulley nut. *(Courtesy of DaimlerChrysler Corporation)*

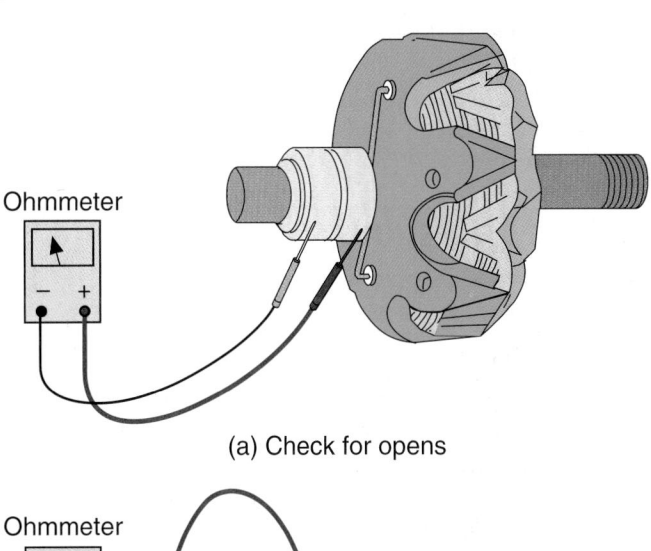

(a) Check for opens

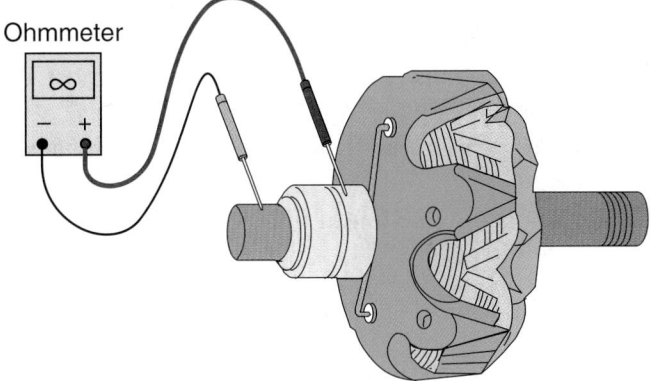

(b) Check for shorts to ground

Figure 31.15 Testing a rotor for (a) opens and (b) grounds.

from turning. Do not put the rotor in a vise to hold it. It can be easily damaged. Some rotor shafts have a means of holding the rotor from turning (**Figure 31.14**). This is important for reinstallation torque.

Test the continuity of the rotor winding to see that it does not have any opens (**Figure 31.15a**), shorts, or grounds (**Figure 31.15b**). Resistance is checked with an ohmmeter. Manufacturers' specifications are different and range from 2.4 to 6.0 ohms.

Any of the following are cause for replacement of the rotor:
- A reading below specs points to a shorted rotor.
- High resistances mean connections are corroded.
- Infinite resistance means the rotor is open.

Diode Service

Bad diodes are a common cause of alternator charging problems. A diode must be tested under a load. There are several ways of testing diodes. A diode tester is a common way. Using an ohmmeter is another way. A diode should allow current flow in one direction only.

Figure 31.16 shows a diode trio. Connect the test leads of the ohmmeter to one of the diode trio leads and the other to the case.
- A good diode will show high resistance in one direction and low resistance in the opposite direction.
- A shorted diode will have low readings in both directions.
- An open diode will show high resistance in both directions.

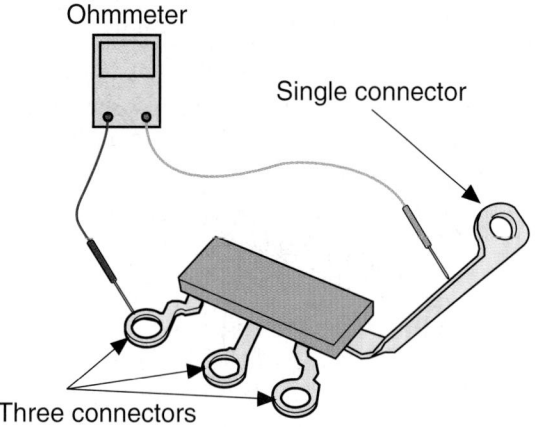

Figure 31.16 Using an ohmmeter to test a diode trio. It should show continuity in one direction only.

Individual diodes are tested in the same manner.

NOTE: *A digital volt-ohmmeter (DVOM) may not be able to test a diode using the ohm scale. Use the digital meter's diode test feature. It measures the voltage drop of the diode instead of its resistance in ohms.*

Sometimes, diodes are soldered together and must be unsoldered to test them. They will have to be resoldered after the test. Be sure to apply the solder as soon as the wire is hot enough. Overheating a diode can ruin it. Pliers can be clamped onto the diode lead and used as a heat sink during soldering to avoid overheating the diode.

ALTERNATOR REASSEMBLY

Reassembling the alternator is usually not difficult. When the brushes are inside of the alternator, they must be held up against their springs so the slip rings of the rotor can be positioned under them. There is usually a hole that a pin, paper clip, or drill bit can be installed in while doing this. The drill bit is removed from a hole in the back of the alternator after the rotor is installed in the frame. Brushes that mount from the outside are simply compressed against their springs as they are installed.

REVIEW QUESTIONS

1. What is the name of the test that takes the regulator out of the charging circuit and causes full alternator output?

2. What test with a voltmeter measures the resistance while current flows in the alternator circuits?

3. As the voltage is lowered during the charging system output test, what part of the alternator does the regulator give current to?

4. What happens to electrical resistance of most conductors when heated?

5. What is the normally expected voltage range of a regulator?

6. What is the maximum time that full-fielding should be done?

7. Two components that should not be disconnected with the engine running are the alternator and the _____.

8. What is a safe voltage maximum to keep under when full-fielding an alternator?

9. Performing _____ _____ tests to a charging system on a preventive maintenance basis will catch resistance problems before they become too serious.

10. When an electronic component has failed, what type of part failure should you look for on an oscilloscope pattern?

ASE-STYLE REVIEW QUESTIONS

1. Where should the inductive pickup on a volt amp tester be clamped on the battery cable?
   a. As close as possible to the battery.
   b. As far away as possible from the battery.
   c. Anywhere on the cable.
   d. It should not be clamped on the battery cable.

2. Two technicians are discussing testing alternator diodes with the alternator on the vehicle. Technician A says to use a DMM. Technician B says to use a low amp current probe. Who is right?
   a. Technician A   c. Both A and B
   b. Technician B   d. Neither A nor B

3. If voltage output is higher than specified, this could be due to a regulator with a bad _____ connection.
   a. Insulated   c. Crimp
   b. Ground   d. All of the above

4. Where is the voltage regulator located in an alternator A-circuit?

   a. In the ground side   **c.** Under the dash

   b. In the insulated side   **d.** None of the above

5. Two technicians are discussing ways to full-field an A-circuit alternator to test its output. Technician A says to supply a ground to the field terminal. Technician B says to apply battery voltage to the B+ terminal. Who is right?

   a. Technician A   **c.** Both A and B

   b. Technician B   **d.** Neither A nor B

6. Technician A says that applying the wrong polarity (+ or −) to the field terminal can ruin diodes. Technician B says that connecting the wrong wire to a ground or hot lead can ruin diodes or a regulator. Who is right?

   a. Technician A   **c.** Both A and B

   b. Technician B   **d.** Neither A nor B

7. Extra resistance in a voltage regulator circuit will cause the battery to have _____ normal charging voltage

   a. Higher than

   b. Lower than

   c. Half

   d. None of the above

8. During a full field test, the current output remains below the specification. Technician A says that a bad alternator could be the cause. Technician B says that a bad drive belt could be the cause. Who is right?

   a. Technician A   **c.** Both A and B

   b. Technician B   **d.** Neither A nor B

9. Technician A says that a good diode will allow electricity to flow freely in both directions. Technician B says that a diode can be ruined when soldering it to a wire. Who is right?

   a. Technician A   **c.** Both A and B

   b. Technician B   **d.** Neither A nor B

10. Which of the following is/are true about alternator diodes?

   a. They rectify AC output to DC.

   b. They prevent alternating current from leaving the alternator in the B+ output wire.

   c. If more than 0.5 AC volt is found in the B+ output, defective diodes are indicated.

   d. All of the above.

CHAPTER 32

Lighting and Wiring Fundamentals

■ OBJECTIVES

Upon completion of this chapter, you should be able to:

✔ Describe differences between wire and cable.

✔ Explain the fundamentals of operation of automotive lighting and wiring.

✔ Relate when different circuit protection devices would be used.

■ KEY TERMS

fusible link	filament	high intensity discharge (HID)
primary wiring	candlepower	head lamps
secondary wiring	type I headlamp	xenon head lamps
cable	type II headlamp	bulb trade number
AWG	halogen headlamp	NA
SFE	composite headlamp	signal flasher
bimetal strip		

■ INTRODUCTION

Lights, and the wiring to power them, make up a sometimes complicated system. The lights and wiring on older cars were simple, consisting of dash lights, the dome light, headlights, taillights, and the license plate light. Today, there are many lights for all kinds of conveniences. This chapter deals with lighting, wiring, and circuit protection devices.

Electricity, for lights and other systems and accessories, is provided through wiring, cables, or fibers (**Figure 32.1**). Circuit protection devices include fuses, circuit breakers, and **fusible links**. They prevent fires and damage in the event a circuit becomes shorted or grounded.

■ WIRE AND CABLE

Chemicals, corrosion, vibration, and heat can damage wiring. Also, aftermarket electrical accessories may be added to cars, and sometimes this work is done carelessly. A service technician should know the basics of wiring to be able to perform professional service and minor repairs.

Low-voltage wire and cable are made up of one or more copper conductors encased in an insulator. In most cases, this insulator is plastic and is color coded so it can be traced and/or located using a wiring diagram.

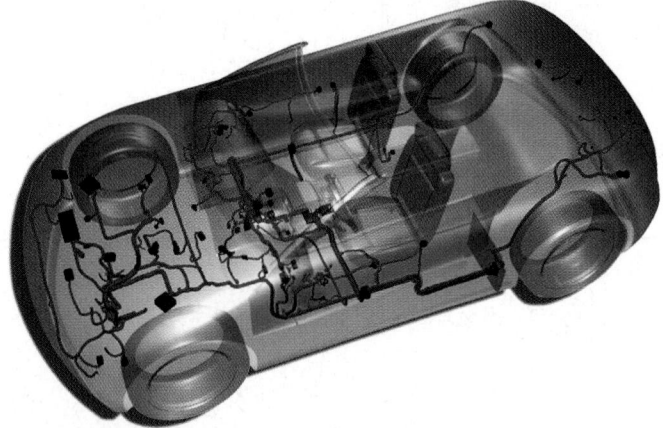

Figure 32.1 Power in the electrical system is provided through wiring, cables, or optical fibers. *(Courtesy of Delphi)*

Primary and Secondary Wiring

Low-voltage wiring on a car is called **primary wiring**. Ignition wiring (high voltage) is called **secondary wiring**. The terms *primary* and *secondary wiring* are used in describing transformer wiring. The ignition coil is an example of a transformer. The low-voltage wiring supplying current is termed primary wiring, and the output wiring is called secondary. Secondary wiring differs from primary in that it has very thick insulation so it can carry voltages of up to 100,000 volts.

Cables are large wires that allow more electrical current to flow. The size of the wire or cable used is described using either the American Wire Gauge (**AWG**) numbers (larger wire equals smaller number) or metric size designations. A chart of wire gauge sizes is included in Appendix D at the back of the book.

■ CIRCUIT PROTECTION DEVICES

Fuses

A fuse is a protective device designed to melt when the flow of current becomes too high for the wires or loads in the circuit (**Figure 32.2**). Without the fuse, the circuit would be damaged. Each fuse has an amp rating at which it is designed to fail. Each vehicle's electrical system has a fuse panel (**Figure 32.3**).

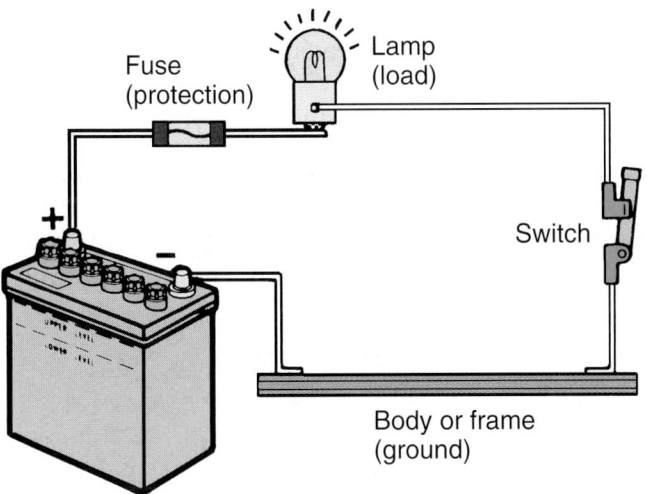

Figure 32.2 A fused circuit.

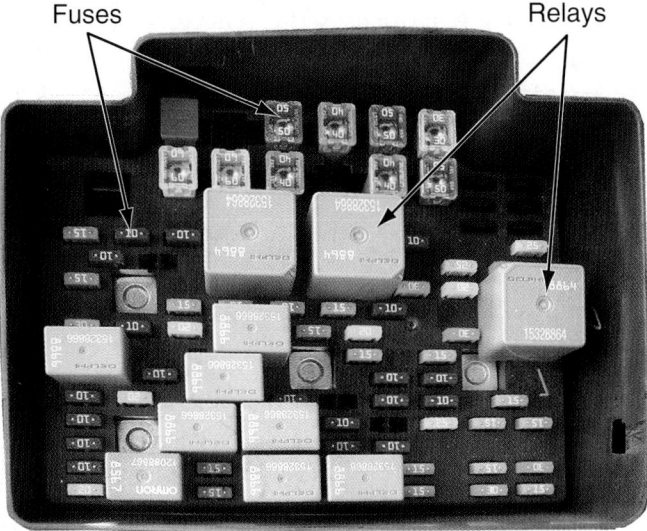

Figure 32.3 A fuse panel. *(Courtesy of Tim Gilles)*

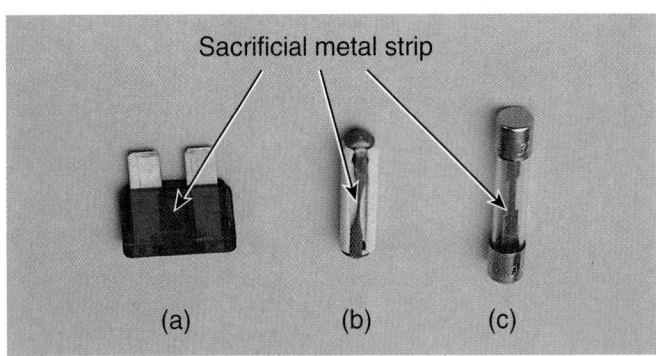

Figure 32.4 Three types of fuses: (a) blade type, (b) ceramic, and (c) glass cartridge. *(Courtesy of Tim Gilles)*

There are three types of replaceable fuses: the *cartridge* type (*glass tube* type), the *ceramic* type, and the *blade* type (**Figure 32.4**).

Glass Tube Fuses. Glass tube fuses have a small wire sacrificial strip inside of a glass tube. They are usually ¼" in diameter and come in lengths increasing in ⅛" increments from ⅝" to 1⅞". There are three common lengths used in automobiles. Usually, the longer the fuse, the higher the amperage rating.

There are two classifications for glass tube fuses:

■ SFE fuses are designed to meet the standards of the Society of Fuse Engineers (**SFE**). These fuses will be longer if their amperage rating is higher.

■ Automotive glass (AG) fuses, also called *Bussman*, or *Buss*, are manufactured by McGraw Edison Company. They have a number code to determine their application and have identification letters, such as AGA or AGC or AGW. These are simply Bussman designations, with no other meaning. Fuses within each of these classifications are all the same length no matter what the amperage. For instance, AGA fuses are ⅝", AGC fuses are 1¼", and AGW fuses are ⅞". A 5-amp and a 15-amp AGC fuse would each be the same length.

Some fuses are "slow-blow" to allow for a heavier draw when wiper or blower motors first start to operate.

Ceramic Fuses. Ceramic fuses are used in some imported cars. The fuse strip is exposed to the air and is mounted over the outside of its ceramic backing (see Figure 32.4b). Ceramic fuses are color coded, with the amperage rating cast into the ceramic back of the fuse.

Blade-Type Fuses. The blade type has been popular since the late 1970s in both import and domestic vehicles. The fuse element is cast into a clear plastic outer body (see Figure 32.4c). Blade-type fuses are color coded according to the SAE color code.

There are two blade fuse sizes. The smaller of the two is called a mini®fuse. The original fuse was called the ATO®, or Autofuse®. The smaller fuse size allows the installation of more fuses in the same space. These are used on some later-model cars so that more circuits can be individually fused.

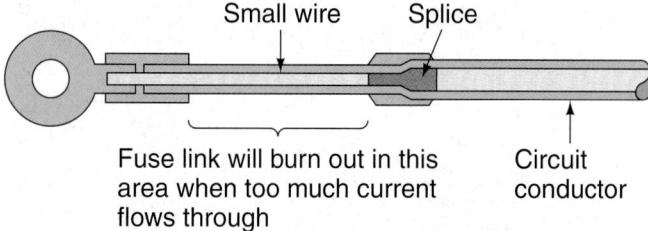

Small wire   Splice

Fuse link will burn out in this area when too much current flows through

Circuit conductor

Figure 32.5 A fuse link will burn through if the wire is overloaded. *(Courtesy of Ford Motor Company)*

Fuse Links

In some circuits in which current control is not as critical, a fuse link will be used. A fuse link is a small length of wire that is smaller in diameter than the wire it is connected to. It is usually installed close to the power source. When the wire overheats, it will melt, opening the circuit (**Figure 32.5**).

Circuit Breakers

Sometimes a circuit only has temporary overloads or must have power restored if it goes out. This is true with the headlight circuit. If the dimmer switch develops a short or ground, you would not want the headlights to go out and remain out.

A circuit breaker used in automobiles is usually self-resetting. In other words, when there is an overload in the circuit, the breaker "trips" and then resets. It has two metal strips with different expansion rates (**Figure 32.6**). This is called a **bimetal strip**. During an overload, the high-expansion metal will become longer, breaking the switch contact. When it cools, contact is reestablished. Unlike the fuse and fusible link, the circuit breaker is not destroyed by overloads. Different styles of circuit breakers are shown in **Figure 32.7**.

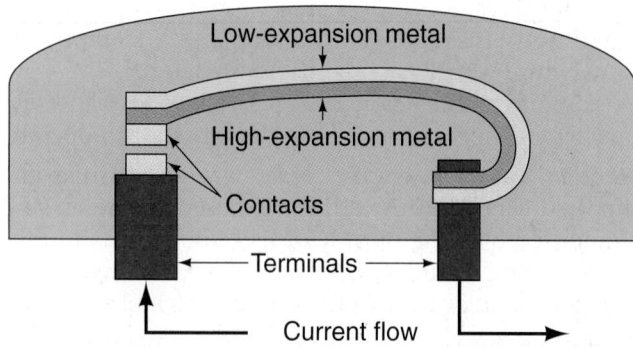

Low-expansion metal

High-expansion metal

Contacts

Terminals

Current flow

Figure 32.6 Circuit breaker operation. *(Courtesy of Ford Motor Company)*

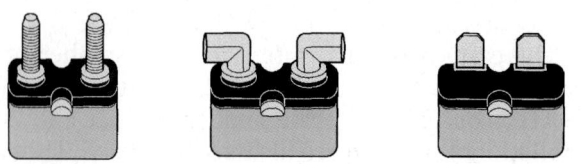

Figure 32.7 Different kinds of circuit breakers. *(Courtesy of Cooper Bussmann, St. Louis, MO)*

When a circuit breaker is manually resettable, a button pops out on the case. It must be reset by depressing the button.

■ LIGHTING

Light bulbs, often called *lamps*, have **filaments** that electricity flows through. The filament provides a resistance to electron flow so it heats up, causing light.

■ HEADLAMPS

The terms *headlight* and *headlamp* are used interchangeably. Automobiles in the United States have been equipped with *sealed-beam* headlights since 1940. The *four-light* system, with separate high-beam lamps, has been used in some cars since 1958. Until 1975, when the *rectangular headlamp* was introduced, all headlamps were round. There are two sizes of both the round and the rectangular lamps. *High-beam* lamps are used at highway speeds with no vehicles oncoming or in front of the car. *Low beams*, used for city driving, are used much more often.

The intensity of a headlamp is rated in **candlepower** (cp). Maximum candlepower of headlights is limited by law. Older lamps were limited in intensity by the Department of Transportation (DOT) to 75,000 cp. European lamps were considerably brighter (300,000 cp). Since 1979, allowable light intensity has been increased to 150,000 cp. This standard was originally allowed for high-beam lamps only but now applies to all lamps.

Sealed-Beam Headlamp Construction

Sealed-beam headlights have an inner glass or plastic *reflector surface* that is sprayed with an aluminum reflective material (**Figure 32.8**). An outer glass or plastic *lens* is fused to the reflector, and the lamp is then filled with *argon* gas. The beam of light from a 50-cp bulb's *tungsten* bar filament is aimed at the focal point of the reflector, where it is intensified to 20,000 cp. The outer lens focuses the reflected light into a beam pattern (**Figure 32.9**).

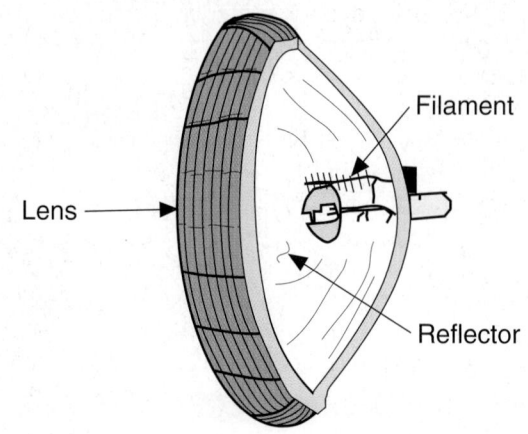

Filament

Lens

Reflector

Figure 32.8 Parts of a sealed beam headlight.

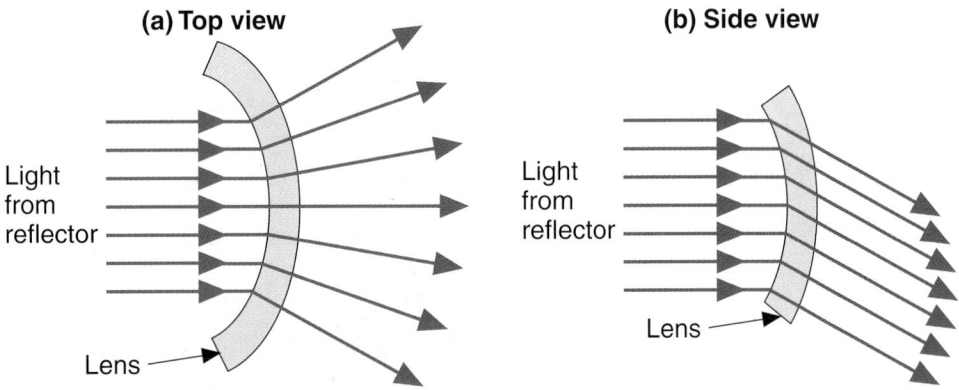

(a) Top view

Light from reflector

Lens

(b) Side view

Light from reflector

Lens

Figure 32.9 The outer glass lens focuses the reflected light into a beam pattern.

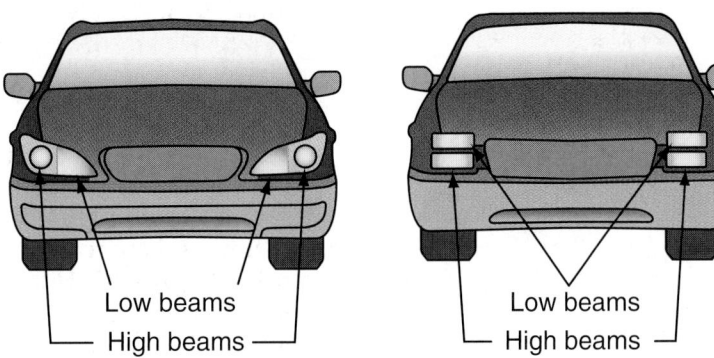

Low beams

High beams

Low beams

High beams

Figure 32.10 Four-light headlight system placement.

In a *dual beam headlight*, which has filaments for both low and high beam, the upper (high-beam) filament is offset so it aims higher. A **type I lamp** has high beam only, with two lugs on its electrical connector. A **type II lamp** has both low and high beams and a three-lug connector. When a type II headlamp is on low beam, only one of its filaments is lit. When the high beam is turned on, the second filament lights too. There are only three wires on the connector because both filaments use the same ground wire.

When there are four headlamps, the single filament of the type I lamp lights on high beam, along with the two filaments in the type II lamp. A type II lamp will be located above or outside of the type I lamp on a four-lamp system (**Figure 32.10**).

Halogen Lamps

Halogen headlamps say "halogen" on the center of the lens. These lamps, used in many new cars since 1979, produce a 25% higher output of "whiter" light on the same amount of power as a conventional sealed-beam headlamp. The tungsten filaments are enclosed in pressurized halogen vapor that allows them to be heated to a higher temperature, making a higher output. Halogen lamps are so bright that, even in a four-light system, no more than two filaments will be on at once.

Halogen lamps are either of the conventional sealed-beam type or they are **composite headlamps**,

with a fixed lens cover and a replaceable halogen bulb (**Figure 32.11**). **Figure 32.12** shows a dual-filament halogen bulb. A composite headlamp has a glass balloon that the halogen lamp fits inside of. There are both two-light and four-light halogen replacement bulb systems (**Figure 32.13**). On dual-filament lamps, the

Replaceable halogen bulb

Figure 32.11 A headlight with a replaceable halogen bulb. *(Courtesy of Tim Gilles)*

Figure 32.12 A dual-filament halogen replacement bulb. *(Courtesy of Tim Gilles)*

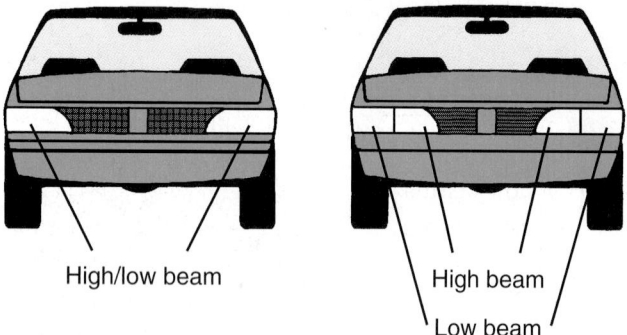

High/low beam

High beam

Low beam

Figure 32.13 Two- and four-light halogen systems. *(Courtesy of GE Lighting)*

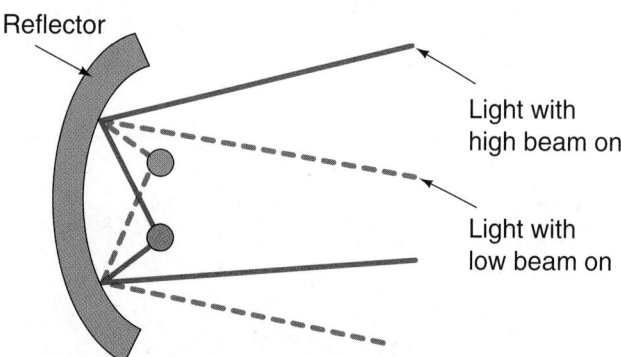

Reflector

Light with high beam on

Light with low beam on

Figure 32.14 The location of the filament and the reflector determines how the light is projected.

location of the filament and the reflector determines how the light is projected (**Figure 32.14**).

Some composite headlamp designs develop condensation inside of the lens. The moisture dissipates from the heat of the lamps once the lights are turned on. Other composites are sealed and are defective if condensation develops in them.

High-Intensity Discharge (HID) Lamps

High-intensity discharge (HID) headlamps, often called **xenon headlamps,** have been available

Figure 32.15 A high-intensity discharge (xenon) headlight emits blue light. (© *Courtesy of Mercedes-Benz USA, LLC*)

on some cars since the mid-1990s. The blue-white light put out by these lamps (**Figure 32.15**) provides three times more light than a conventional halogen headlamp, and twice as much light is spread out on the road (**Figure 32.16**). HID lamps produce light in ultraviolet and visible wavelengths, which cause reflective materials like highway signs to glow. Visibility is also better during rainy or foggy conditions because the light reflected back is better light.

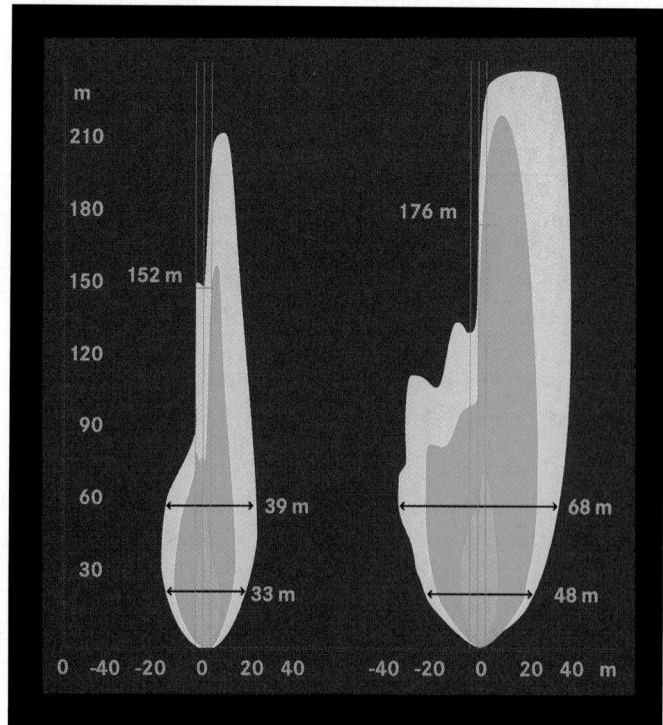

Halogen   Xenon HID

Figure 32.16 A xenon headlight shines light over a wider area and with greater intensity than a halogen headlight. *(Courtesy of DaimlerChrysler)*

Normal headlamps produce light from a glowing filament. HID headlamps produce light when a controlled high voltage arcs across an air gap between two electrodes in a small quartz capsule, called an arc tube, surrounded by an inert gas, xenon—a gas similar to helium (**Figure 32.17**). The inert gas amplifies the effect of the light produced by the arcing. The quartz capsule also contains mercury and metal halide salts.

When the headlamp switch is turned on, battery voltage powers the input side of the ballast, which allows it to charge high-voltage output to the headlamp arc tube. Eight hundred volts (800 V) fire up the striker to operate the ballast. An arc generates between the electrodes in the arc tube. The ballast provides the high-level voltage required to keep the lamp on until steady state operation occurs. Full brilliance is achieved in 3–5 seconds. The result is about twice the light output while using about 25% less power than is required for halogen lamps.

Although HID's continuous electrical draw is less than a sealed-beam headlight, the battery must be able to provide a minimum of 175 amps to start operating the lamps.

Figure 32.17 A xenon headlight. *(Courtesy of DaimlerChrysler)*

Figure 32.18 An adaptive headlight assembly. *(Courtesy of DaimlerChrysler)*

Adaptive Headlights

Some luxury vehicles have adaptive headlights (**Figure 32.18**) that can swivel up to 15 degrees right or left in response to signals received from a steering wheel angle sensor (see Chapter 65). When cornering, this illuminates part of the road that would not normally be seen.

Headlight Switch

Some headlight switches are part of a multifunction switch on the steering column, whereas others are mounted on the dash panel. Older vehicles used a pull-on dash switch that could be rotated to adjust the intensity of instrument panel lights and to turn on the dome light. Newer dash-mount switches often have separate rheostat controls for the instrument panel and dome lights (**Figure 32.19**).

A typical headlight switch has two or three positions besides the *off* position. Battery voltage is usually applied to two terminals so the lights can be operated without the key turned on. In the *park* position, all other lights but the headlights and dome light(s) are on. The next switch position is for *headlights*. Some vehicles are also equipped with an *automatic* position on the headlight switch.

Dimmer Switch

A dimmer switch is used for changing headlights from low beam to high beam. It is usually attached to the multifunction switch or turn signal lever (**Figure 32.20**). Older vehicles used a separate dimmer switch that was installed in the floor. One position turns on the high beams (brights), and the other switches on the low beams for city driving. When the dimmer switch is activated by raising the turn signal switch, the switch is called a *multifunction switch*.

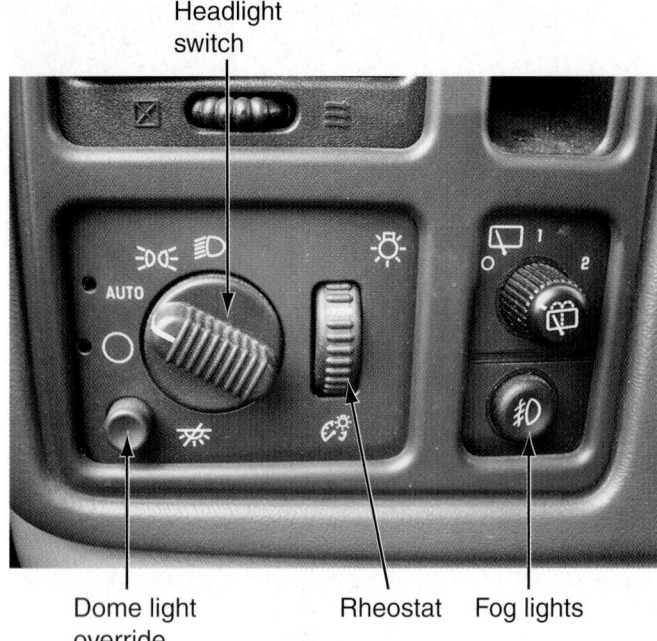

Headlight switch

Dome light override

Rheostat   Fog lights

Figure 32.19 Lighting controls. *(Courtesy of Tim Gilles)*

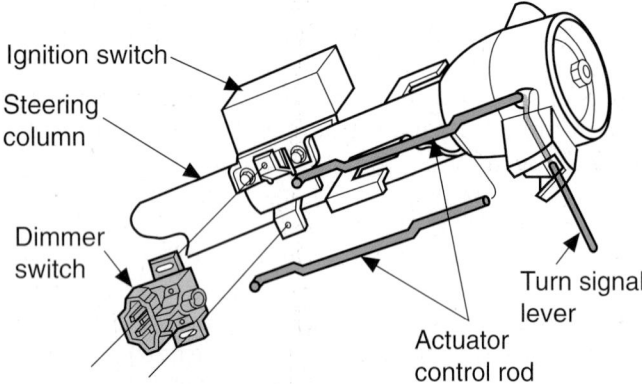

Ignition switch

Steering column

Dimmer switch

Turn signal lever

Actuator control rod

Figure 32.20 A column-type dimmer switch. *(Courtesy of Ford Motor Company)*

Automatic Headlight Dimmer

Some cars have an automatic system for dimming the headlights. This system has a light sensor that detects oncoming headlights. On some systems, it switches the high beams back on when there is no oncoming light. The sensor can be located in the grille or on the top of the dash.

Courtesy Lights

Interior lights are called security lights or courtesy lights. These include the dome light and underdash lights. They are usually operated by turning a potentiometer knob on the instrument panel. When any door is opened, its switch turns the lights on. Most cars use the switch to provide a ground for the light circuit. Door switches on Ford vehicles open and close the power side of the circuit. Many newer vehicles use a computer to control operation of the courtesy lights.

Headlight Wiring

The headlight circuit shown in **Figure 32.21** has battery power to the headlight switch with power supplied to the headlights only when the switch is closed. Many manufacturers use electrical systems that are ground-side controlled with the headlight switch completing the circuit to ground.

A resetting-type circuit breaker is installed in the headlight feed. If there is a problem that trips the circuit breaker, the lights will go out and come back on repeatedly as the breaker resets and breaks the circuit again.

Daytime Running Lights. Many new vehicles have daytime running lights for safer driving. The lights are powered whenever the ignition is turned on and the engine is running. The lights do not operate when the parking brake is applied. Daytime running lights are illuminated at a lower intensity than normal headlights. Some vehicles use special lights for this function; others use the high-beam headlights with their voltage cut in half.

Automatic Headlights

Automatic headlight systems use a light-sensitive photocell sensor to determine the intensity of outside light. The sensor is located under the windshield at the top of the dash panel or in the defroster grille. Some manufacturers use two sensors for system input: one to monitor the sky above the vehicle and the other to monitor light in front of the vehicle. When the sun goes down and outside light levels become lower, the resistance in the sensor increases until it reaches a set point. Then the module directs current to the headlight relay to illuminate the headlights, dash lights, and running lamps. The lights stay on until outside light becomes bright enough or when the key is shut off or the driver overrides the system.

A delay system on most vehicles keeps the lights on for a short period after the key is shut off. Sometimes there is a switch so the driver can adjust the time period of the delay or shut it off altogether.

Flash to Pass

Moving a turn signal mounted dimmer lever activates the high beams on most vehicles. This feature, called flash to pass, works when the headlights are off as well as when the dimmer switch closes the circuit to the high beams.

Driving Lights and Fog Lamps

Some customers want to add auxiliary lighting to their vehicle. Driving lights and fog lamps are occasionally installed as original equipment, but most of them are added in the aftermarket. Each requires the addition of a separate relay and switch and should be installed on a vehicle with a charging system of sufficient output.

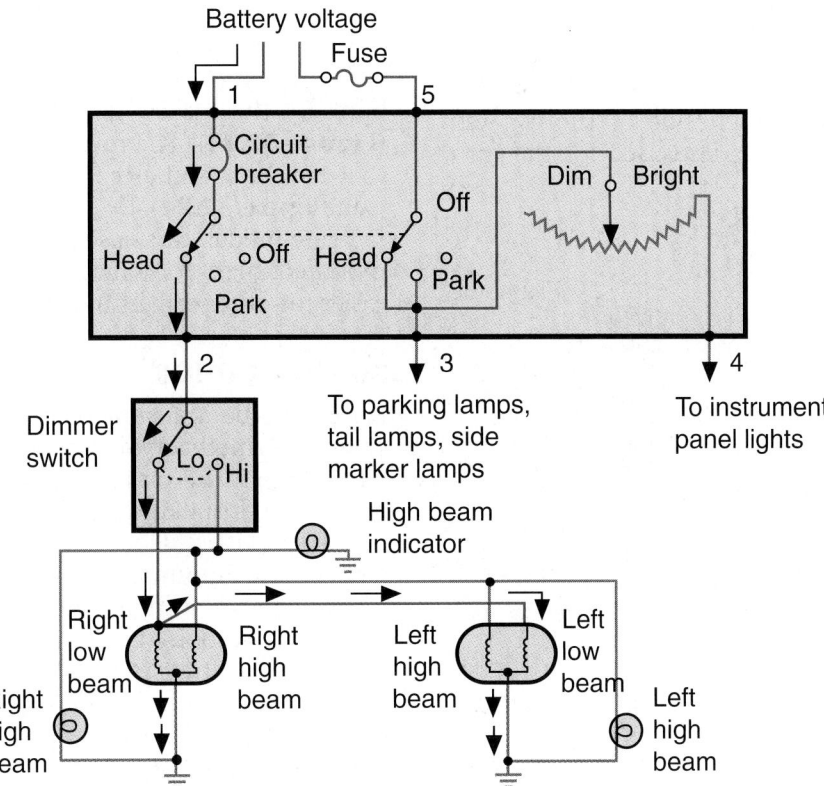

Figure 32.21 A headlight circuit with battery positive power feed and the dimmer switch selecting low beams.

Driving lights add to the illumination of normal high-beam headlights. Typical driving lights use an H3 or H4 quartz halogen bulb with a reflector and lens to lengthen the distance of light projection. These lights are connected so they can only come on when the high beams are on, and it is important that they be accurately aimed because they have a narrow light beam.

Normal lights do not work well in fog because the fog reflects the light back at the driver. Fog lamps are installed low on the vehicle and aimed parallel to the road surface in an attempt to project light beneath the fog.

■ TAILLIGHTS

Taillight bulbs come in many sizes and types. Bulbs have different numbers of filaments and different methods of grounding. Some bulbs have two terminals, and others have only one. Bulbs usually ground through the case to the light socket. The light socket usually will be grounded to the frame through a wire or directly. A bulb used on a ceiling dome light or a fiberglass vehicle, such as a boat or motor home, will have two terminals for only one filament. The second terminal is to provide a ground path for the filament.

Many newer vehicles use bulbs that have a plastic socket mounted in lightweight plastic light housings. Many of these lights use two wires, one of which attaches to a ground connection. Two filament bulbs

of this type will have three wires: one for ground and the other two to power each filament.

■ IDENTIFICATION OF LIGHT BULBS

It is important that the correct bulb be used or damage to a circuit could result. Automotive bulbs are numbered by the American National Standards Institute (ANSI). No matter who the manufacturer is, the bulb will still have the same number, called the **bulb trade number**.

■ When there is an *A* or an *NA* after the bulb number, that means that the bulb is an amber (yellow) color.

■ The natural amber (**NA**) bulb is colored glass, whereas the less expensive A is painted glass.

Bulbs that are smaller than headlamps are also known as *miniature lights*.

Stoplight Switch

Brake lights (stoplights) are activated by a mechanical switch on the brake pedal (see Figure 57.68). Some older cars had hydraulically activated stoplight switches. The wire that supplies current to the brake lights is interrupted at the stoplight switch. When the brakes are applied, the two ends of the wire are connected electrically to allow current to travel to the light filaments. On some cars, the brake lights operate through the turn signal switch.

Taillights and Brake Lights

Taillight bulbs have one or two filaments (**Figure 32.22**). When the brakes are applied, the switch activates one filament on each side of the car. On domestic cars, this filament is also used for the turn

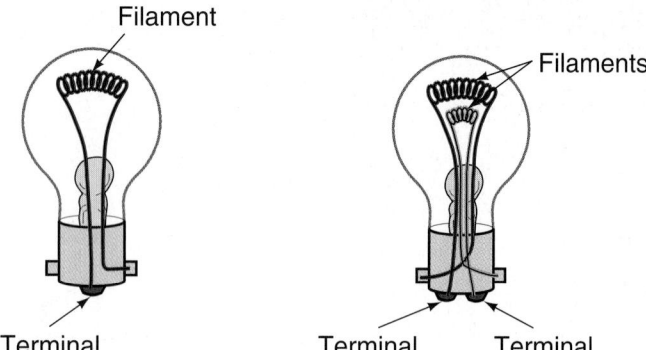

Figure 32.22 Light bulbs with one and two filaments.

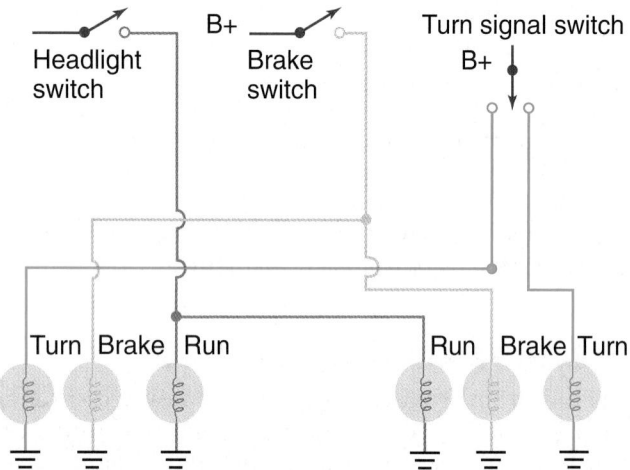

Figure 32.23 A three-bulb taillight circuit with turn signals, brake, and running lights.

Figure 32.24 This taillight has a yellow turn signal light and red brake and running lights.

signals and emergency flashers. The circuit is controlled through the turn signal switch. Imported cars and some later model domestics often use three bulbs for the brakes, turn signals, and running lights (**Figure 32.23**). The turn signal bulb has a lens cover that is yellow and a brake light lens cover that is red (**see Figure 32.24**).

Late-model cars also have a center brake light, mounted high for safety purposes. These lights are easier for other drivers to see in a panic stop situation.

▆ FIBER OPTICS

Light can be transmitted through a special plastic called polymethacrylate. The light rays remain parallel, although the plastic bends and weaves to accommodate its installation in the car body. The term for this is *fiber optics*. Fiber-optic strands are used in some vehicles to illuminate an indicator on the top of a fender when a turn signal is flashing. The fiber-optic wires usually look like ordinary electric wires. Outside door locks, ashtrays, and other very low light needs are other common uses for fiber optics. The light source is a normal light bulb with a special bulb clip to hold the fiber-optic strand under the focus of the bulb.

LED and Neon Lights

Headlights, brake lights, turn signals, and taillights are sometimes light-emitting diode (LED) lights or neon lights. LEDs are energy efficient, long lasting, and take up little space, so several of them are installed behind one lens (**Figure 32.25**). They do not need to heat a filament to provide illumination, so they are able to come on very fast. An LED can illuminate in less than 1 millisecond (0.001 second), compared to the 200 milliseconds required to fully illuminate a conventional light bulb. This provides more safety for the driver behind during an emergency stop. The development of high-brightness (HB) LEDs has resulted in their use in about half of the center high-mount brake lights produced for new vehicles. High-power white LEDs have also been used for some headlights.

Neon lights, used in some cars, can also be illuminated faster than a conventional bulb (approximately 3 milliseconds). A neon light lasts longer than an ordinary light bulb because it does not have a filament. There are also a few automotive applications that use laser lighting, which is similar to LED technology and is very energy efficient.

Adaptive Brake Lights

The size and intensity of adaptive brake lights increases as a driver pushes harder on the brake pedal and during an antilock brake stop. During a normal stop, the brake lights and center high-mount lamp will illuminate normally. **Figure 32.26** compares normal and emergency stop brake lighting with adaptive brake lights.

Figure 32.25 LED taillights are used on many new cars.

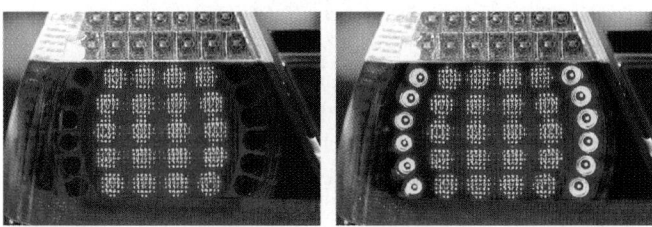

Figure 32.26 Adaptive brake lights during normal and hard stopping. (*Courtesy of BMW NA, LLC*)

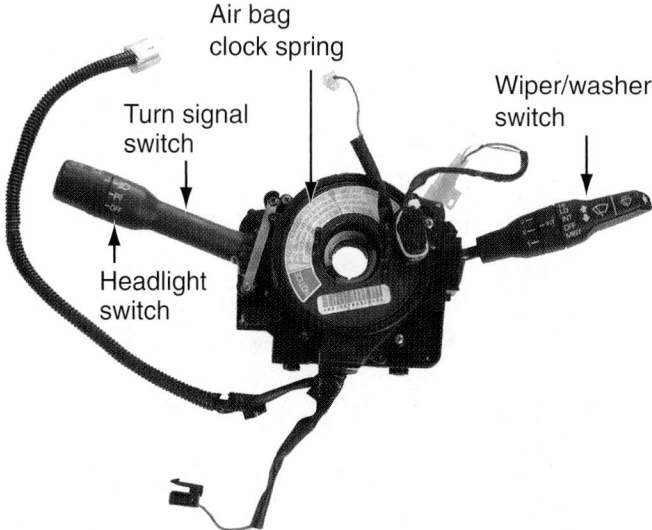

Figure 32.27 This multifunction switch includes the turn signal switch, wiper and washer switches, and air bag clock spring. (*Courtesy of Tim Gilles*)

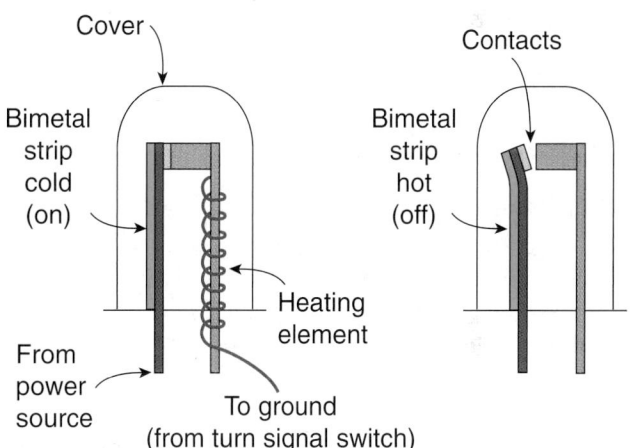

Figure 32.28 Mechanical turn signal flasher operation.

■ TURN SIGNALS

The circuit for the turn signals has a switch, a **signal flasher**, two indicator bulbs in the dashboard, the stoplight filaments of the taillights or rear stoplight bulbs, and the two bulbs in the front of the car. The turn signal switch is usually a part of a multifunction switch (**Figure 32.27**) located in the steering column beneath the steering wheel.

■ SIGNAL FLASHER OPERATION

Turn signal flashers are either mechanical or electronic. A mechanical flasher is activated by the heat of the electricity traveling through it to the signal bulbs. It normally flashes at between 60 and 120 cycles a minute. It has a bimetal strip (**Figure 32.28**) and works the same way that a circuit breaker does. There are usually two flashers, one for the turn signals and one for safety hazards.

Electronic turn signal flashers operate when a transistor is triggered on and off. A transistor is an electronic switch with no moving parts (see Chapter 37). Electronic signal flashers often operate both the turn signals and the hazard warning lights. **Figure 32.29** shows a schematic of an electronic signal flasher.

■ HAZARD FLASHERS

Hazard flashers are required on all cars manufactured in 1967 and later. This system is powered directly from the battery, independent of the turn signals, and turns on all of the turn signals and the dash indicators at once. The emergency flasher switch is usually located on the side of the steering column, on the dashboard, or in the glove box. The flasher used in emergency warning systems is a *variable load* one so that it will flash whether or not bulbs are burned out.

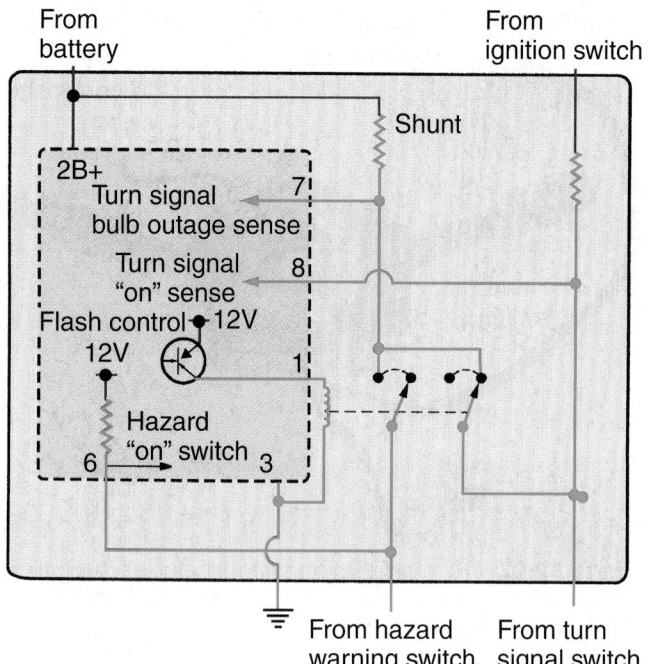

Figure 32.29 A schematic of an electronic turn signal flasher circuit. *(Courtesy of DaimlerChrysler)*

■■ WIPERS AND HORN

Windshield Wiper Motor

The windshield wiper circuit has a fuse or circuit breaker and a wiper switch. A relay is used to complete the circuit because of the relatively large amp draw of the motor. The relay is often located on the fuse panel (see Figure 32.3).

The motor assembly includes a transmission (plastic gears), a housing, a drive crank, and a park switch (**Figure 32.30**). Linkage arms connect the wipers to the crank. The parts of the entire wiper assembly are shown in **Figure 32.31**.

A typical wiper motor has either two or three speeds. Like starter motors, wiper motors can have fields that are either electromagnets or permanent magnets.

Electromagnet Wiper Motors. Electromagnet wiper motors have two brushes on the armature: one positive and the other negative. Three-speed motors (and some two-speed motors) have two field windings wound in opposite directions. One is wired in series with the armature, and the other is shunted through a resistor to ground. **Figure 32.32** shows a schematic of a two-speed electromagnet wiper motor circuit with two windings. The speed of the motor is determined by the strength of the magnetic fields. In the low-speed position, current in the shunt field is strong, creating a strong magnetic field that works against the magnetic field in the series field to slow the motor down.

Permanent Magnet Wiper Motors. A typical permanent magnet motor has three brushes, one of which is a common ground. The other two are low- and high-speed brushes that determine the motor speed by controlling which armature windings are energized. Energizing more windings serves to slow the motor down.

Intermittent Wipers

Intermittent wipers allow variations in the time interval a wiper sweeps across the windshield during foggy or misty conditions. The switch on most new vehicles is a rheostat located on the wiper control arm. The rheostat

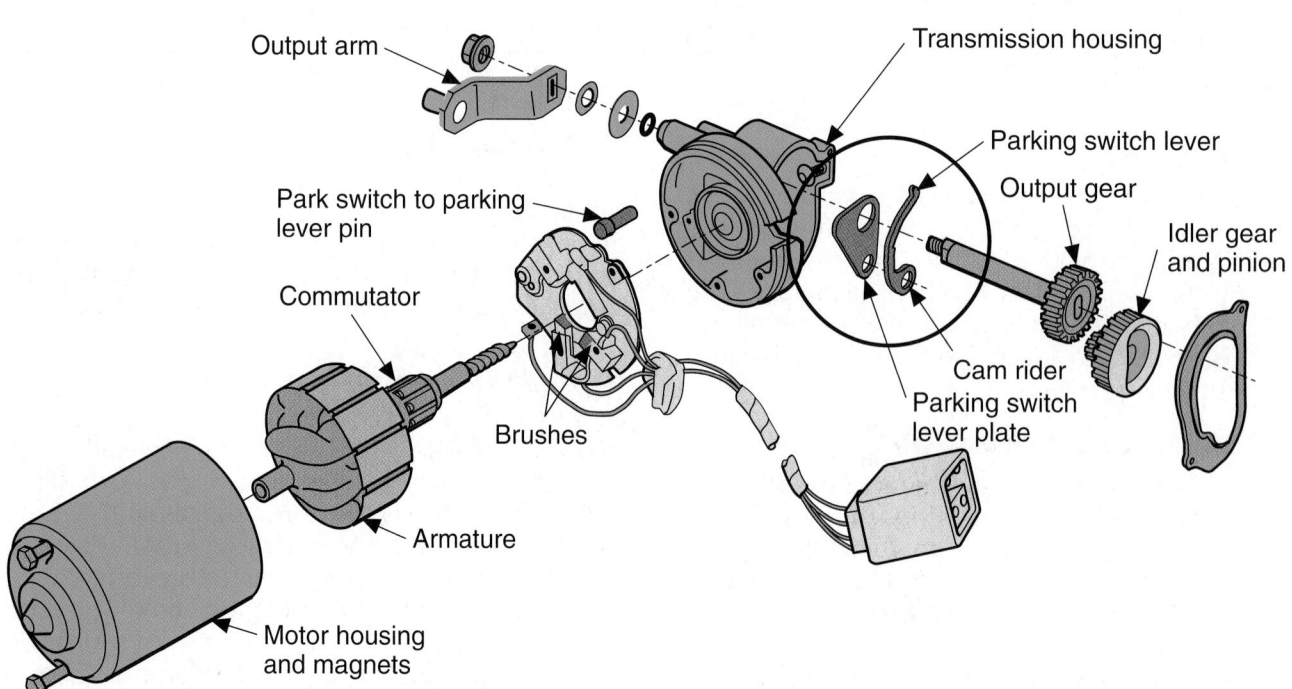

Figure 32.30 Parts of the windshield wiper motor.

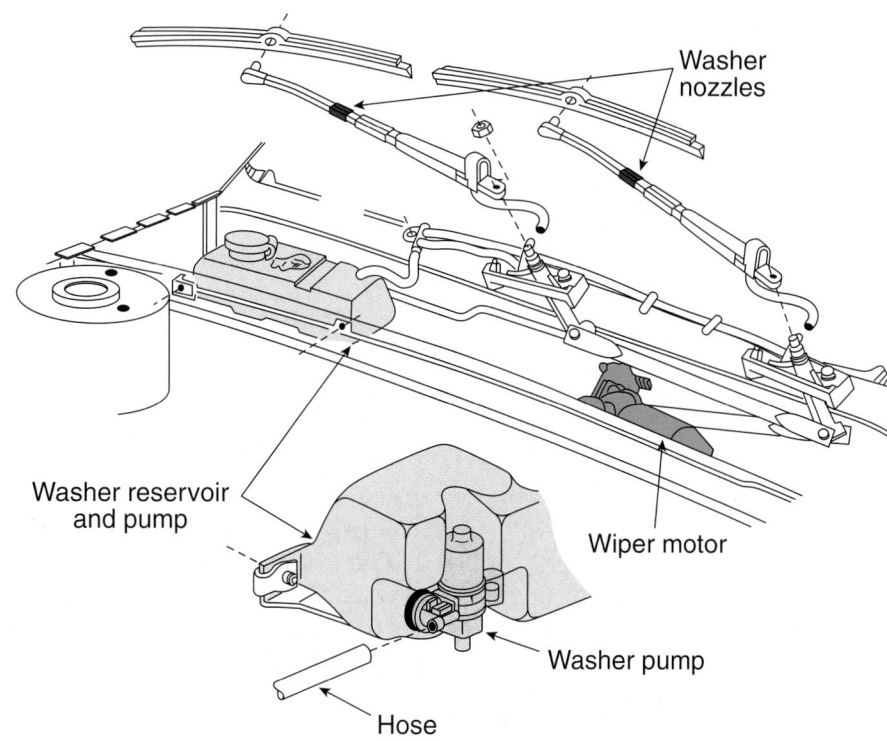

Figure 32.31 Parts of the windshield wiper assembly.

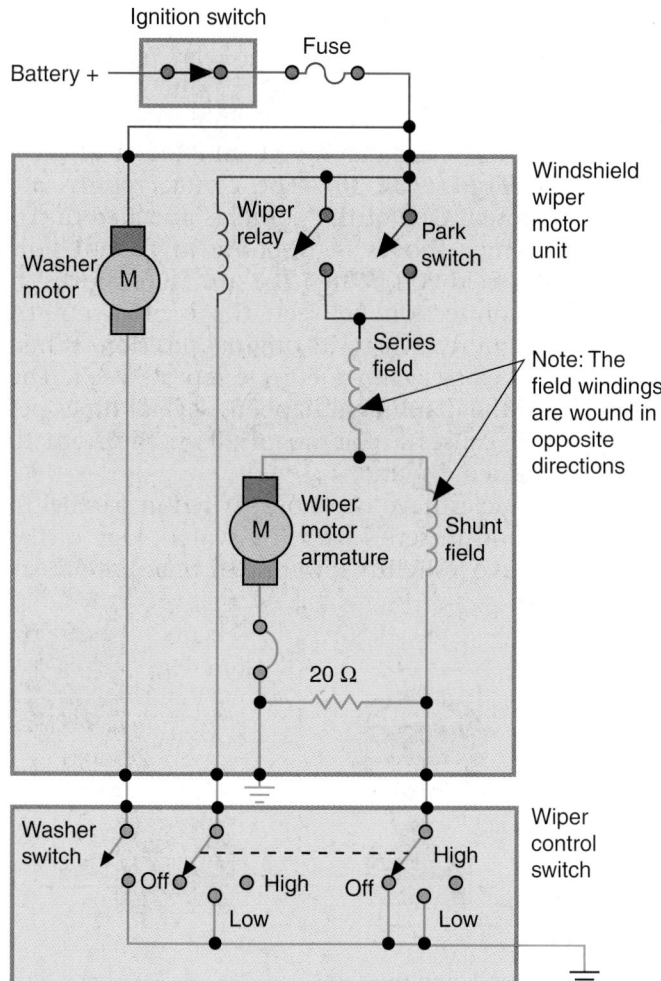

Figure 32.32 A schematic of a two-speed electromagnet wiper motor circuit with two windings.

allows the wiper interval to be adjusted. Older systems have a delay function that works like a mechanical turn signal flasher. Most of today's vehicles use a governor module to operate the intermittent wipers. It is often located at the base of the steering column (**Figure 32.33**). A capacitor within the module is charged through the control switch on the wiper lever. The rheostat varies the amount of time it takes to recharge the capacitor, which triggers an electronic switch to begin wiper operation. When the wiper completes a sweep on the windshield, a "park" switch in the motor shuts the wiper off until it is once again triggered by a full capacitor.

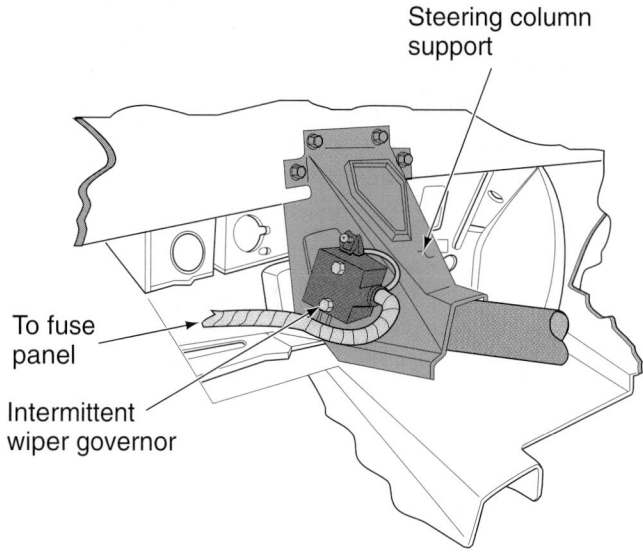

Figure 32.33 An intermittent wiper governor module located at the base of the steering column. (*Courtesy of Ford Motor Company*)

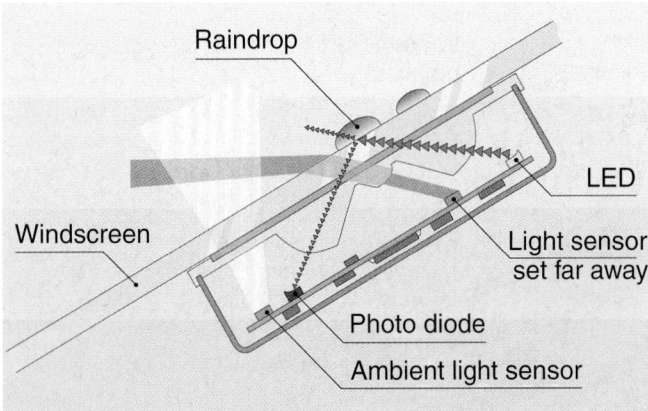

Figure 32.34 Rain-sensing wipers operate when more light passes through a wet windshield. *(Reprinted with permission from Bosch)*

Rain-Sensing Wipers. Some wiper systems can begin operation automatically when water is sensed on the windshield. The sensor, located behind the rear view mirror (**Figure 32.34**), uses an infrared sensor to detect moisture. It transmits an infrared light beam through the windshield. All of the light is reflected back through the windshield to the sensor whenever the windshield is dry. Moisture on the windshield allows more infrared light to pass without reflecting back. The sensor determines how fast and how often to sweep the windshield wipers. The driver can also adjust the sensitivity of the system.

Speed-Sensitive Wipers. A few vehicles have wiper systems controlled by a control module with input from the vehicle speed sensor. These systems change the speed and interval of wiper operation according to how fast the vehicle is traveling.

HISTORY NOTE

The first windshield wipers had to be operated by hand. Early powered wipers operated on vacuum from the engine. This presented a unique problem. When a vehicle was under a load, such as when climbing up a hill, the wipers would lose vacuum and stop moving. When the accelerator was released, an increase in vacuum caused the wipers to move rapidly again. Electric wipers solved this problem. A 1968 federal law called for all vehicles to have wipers with at least two speeds and a washer system.

Windshield Washer

The windshield washer has a reservoir, a switch, a pump, washer nozzles below the windshield, and hoses and connections. Inexpensive antifreeze solvent is used in the reservoir to clean the windows.

There are two types of pumps used, a rotary type and a bellows (diaphragm) type. Most new cars use a rotary pump. It is located in the solvent reservoir. A small motor spins an impeller. The bellows-type pump is usually mounted on the wiper motor.

Rear Window Wiper-Washers. Hatchbacks, station wagons, and SUVs typically have a rear windshield wiper motor and washer controlled by a separate switch. The motor "parks" the same as a front wiper motor. If there is an intermittent function, it operates in the same manner as well.

Headlight Wiper-Washers. A few vehicles use headlight washers and fog lights. Some of them have their own switch; others work with the regular windshield wipers.

Windshield Washer Low-fluid Indicator. Many vehicles have a low-fluid level indicator within the washer fluid reservoir. When the fluid level drops below ¼ full, a switch closes to illuminate the indicator on the instrument panel.

Horn

The horn circuit includes the horn, a fuse and wiring, the horn button in the steering wheel, and a relay (**Figure 32.35**). Pressing on the horn button closes a circuit to activate the horn relay. The relay provides the current to the horn. The horn takes quite a bit of current, so most horns use a relay like motors do.

Most horns have a diaphragm vibrated by an electromagnet (**Figure 32.36**). The contact points are normally closed. One of the points is attached to the armature, which moves in response to current flow through the field coil. When the armature moves, it opens the connection between the contact points. The diaphragm retracts to its original position. When the contacts close again, the cycle repeats itself. The vibration of the diaphragm happens several times per second. This causes a column of air in the horn to vibrate, producing sound.

Most vehicles have two horns wired in parallel to each other and in series with the switch. One of the horns will have a slightly lower pitch than the other.

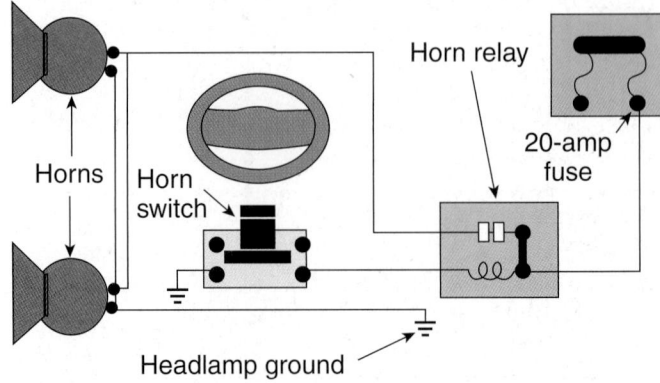

Figure 32.35 The horn circuit.

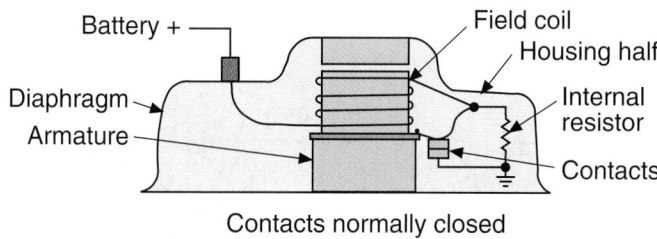

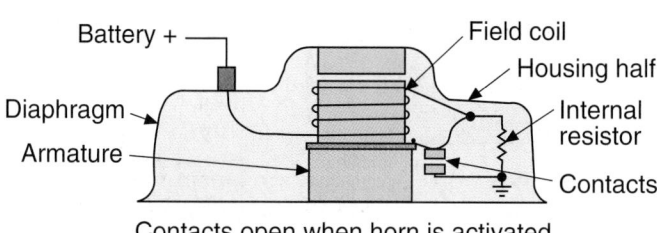

Figure 32.36 Operation of a horn.

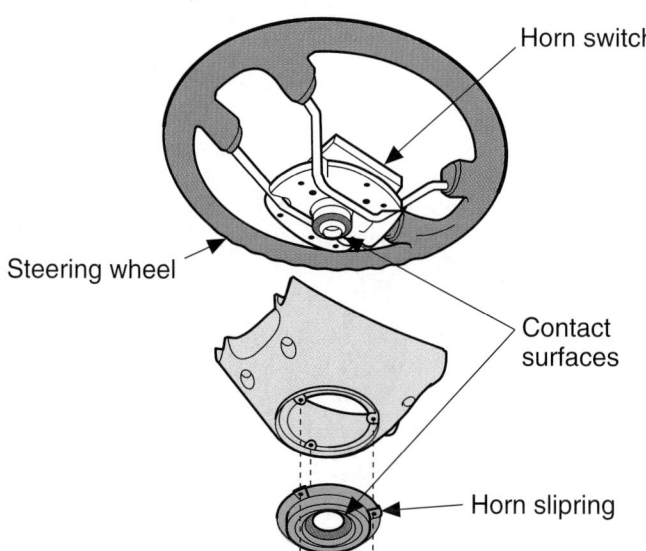

Figure 32.37 The horn has a sliding electrical contact beneath the steering wheel. *(Courtesy of Ford Motor Company)*

The horn switch is either a single or multiple button switch in the steering wheel. Sliding contacts must be used so that the steering wheel can turn while maintaining contact with the switch (**Figure 32.37**). There is a circular contact in the steering wheel that slides against a spring-loaded contact in the steering column.

Electrical Instrumentation/Instrument Panel

Older vehicles had a dashboard with a mechanical speedometer and gauges. Today's dashboard is called an instrument panel with combined analog and digital displays. Instruments that use analog displays include the speedometer; tachometer; and fuel, temperature, and electrical gauges (**Figure 32.38**). Analog displays

Figure 32.38 An analog instrument display. *(Courtesy of Tim Gilles)*

Figure 32.39 A digital instrument display. *(Courtesy of Siemans VDO Automotive)*

are those that show a moving needle or graph. They are popular because they are often easier to see and can show relative movement such as climbing or falling engine rpm or temperature or vehicle speed. Digital electronic displays (**Figure 32.39**) are of three different types (see Chapter 45).

Electric Gauges

Gauges operate on input received from a sensor or sending unit. Older mechanical gauges received their input directly, but modern vehicle electronic systems receive information at one or more computers before sending it to a display. Electrical gauges use *instrument voltage regulation* (IVR) to stabilize the signal. As an example, if a fuel gauge did not have a stabilizing influence, it would respond with a change in fuel level when the vehicle went over a bump or climbed a hill. An IVR is shown in **Figure 32.40**.

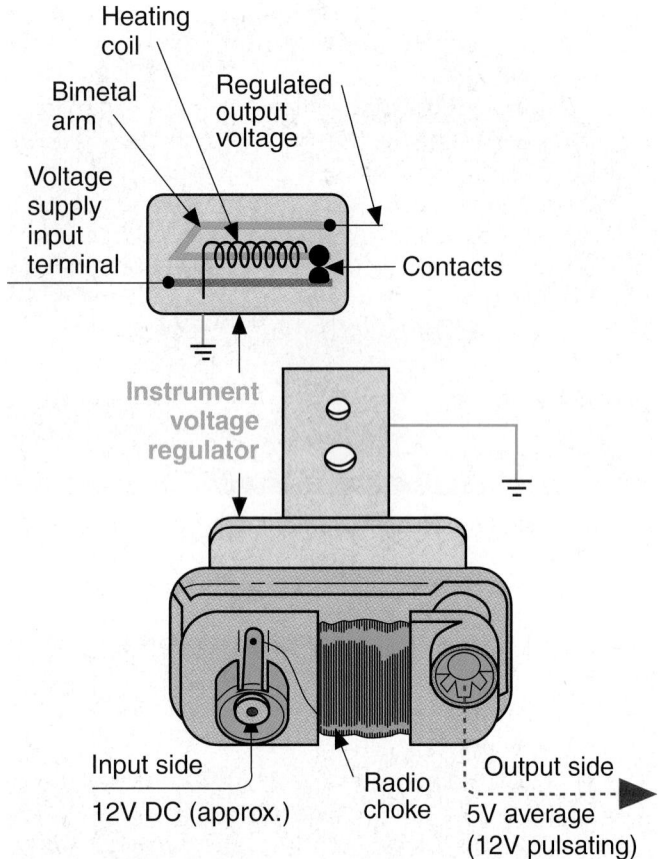

Figure 32.40 An instrument voltage regulator (IVR) is used to stabilize a fuel gauge.

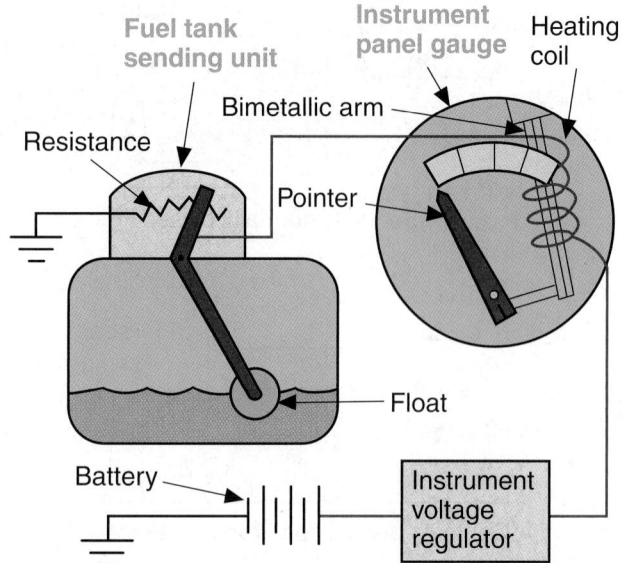

Figure 32.41 A thermal/bimetal gauge moves less with decreased current when the sending unit float is low in the tank. (*Courtesy of DaimlerChrysler Corporation*)

Electrical analog gauges are either thermal or magnetic.

Thermal/Bimetal Gauges. Thermal gauges are also called bimetal gauges. The sending unit varies the amount of current to the gauge, which has a *bimetal spring* that acts to move a needle. A bimetal spring is one made of two different types of metal with different rates of expansion that are welded together. As the temperature changes, a bimetal spring will bend. **Figure 32.41** shows a thermal fuel gauge circuit. Increased current from a higher float level in the fuel tank causes heat, which bends a bimetal arm and moves the needle on the gauge.

Magnetic Gauges. Magnetic gauges are of different types. The simplest is the ammeter type of gauge discussed in Chapter 25. It uses a permanent magnet that attracts a pivoting magnetic pointer, causing it to move. A coil of wire, or armature, is wound around the base of the pointer, creating an electromagnet (**Figure 32.42**). When current from the sending unit flows through the armature, a magnetic field is created, which opposes the force of the permanent magnet and moves the needle. The direction the needle moves is dependent on the direction of current flow in the armature. This gauge is often called a bobbin gauge, because the armature is also known as a bobbin.

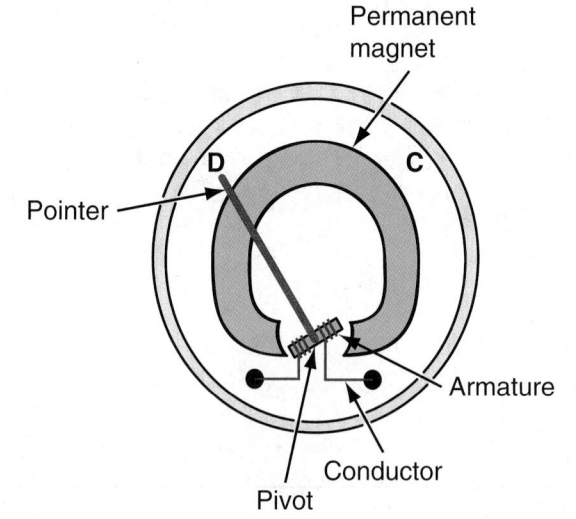

Figure 32.42 In a magnetic gauge, a coil of wire wound around the base of the pointer creates an electromagnet.

A balancing coil gauge is another type of gauge. It does not have a permanent magnet but uses two electromagnets (**Figure 32.43**). If resistance in the sending unit is high, the coil on the left has more current, creating a stronger magnetic field to pull the needle to the left. Low resistance in the sending unit results in more current in the right-hand coil and needle movement to the right.

Digital Panel Gauges. Digital panel gauges consist of a number of segments, or bars. When the sending unit allows more current to flow to the gauge, more bars are illuminated.

■ MISCELLANEOUS GAUGES

Several gauges are found on the instrument panel, including the speedometer; tachometer; and gauges

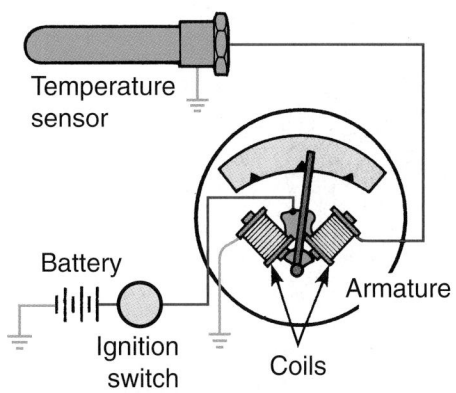

Figure 32.43 A balancing coil gauge has two electro-magnet coils. *(Courtesy of Ford Motor Company)*

for fuel level, oil pressure, engine coolant temperature, and other warning gauges. Some of these gauges are covered in other chapters.

Speedometer/Odometer

Older speedometers were nonelectric mechanical gauges. A speedometer-driven cable attaches to the back of the dashboard speedometer. The cable rotates a magnet within a cup (**Figure 32.44**). The cup is attached to the speedometer needle and is held against zero by a small spring called a hairspring. As the cable rotates faster, the magnet forces the cup to turn, moving the needle in response to increasing vehicle speed.

Mechanical odometers use a row of rotating drums geared together. When the drum on the far right completes one revolution, it turns the neighboring drum 1/10 of a turn. Most mechanical odometers have six drums only, so when 99,999 miles is reached, the odometer returns to zero on the next revolution. Some odometers have a trip odometer as well that can be reset to zero by the driver.

Digital Electric Speedometers. The most common electronic speedometer uses the transmission's vehicle speed

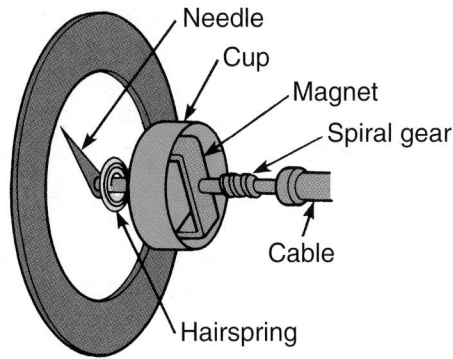

Figure 32.44 A mechanical speedometer has a rotating magnet within a cup. The speedometer cable rotates the magnet.

sensor (VSS) to send a signal to the instrument panel through the computer (**Figure 32.45**). Every time the VSS pulses 40,000 times, 1 mile is logged. The instrument panel reading is an analog display positioned by the computer in response to the digital frequency of the VSS signal. Digital speedometers are calibrated according to the tire size and final drive ratio.

An electric odometer records mileage to seven places, so it can display up to 1 million miles. Mileage is stored in nonvolatile memory so it is not lost when the battery dies or is disconnected. When a digital speedometer is replaced its odometer is programmed to display the existing vehicle mileage using a scan tool.

Tachometer

A tachometer displays engine rpm in response to an ignition pulse signal relayed from the computer. The display is on an analog display, a balanced coil gauge, or a digital readout.

Engine Coolant Temperature Gauge

Older coolant temperature gauges had a mercury tube threaded into a cooling passage in the engine. The other end was attached to a gauge in the passenger compartment. When the coolant heated up, the mercury inside the tube would expand, moving the needle on the gauge.

An electric engine coolant temperature (ECT) sensor is a thermistor, a variable resistor that regulates current flow through a winding in a gauge or sends a varying signal to a computer that controls a digital display. Different resistances reflect changes in coolant temperature. Chapter 46 covers ECT sensors in more detail.

Engine Oil Pressure Gauge

Older mechanical oil pressure gauges had a tube running from an engine oil gallery to the back of the gauge in the passenger compartment. The gauge consisted of a flat, hollow tube like a party curler that expanded with increasing pressure to advance a needle on the gauge face.

Modern oil pressure gauges are electric and receive a signal from a sending unit in an engine oil gallery. Pressure gauges do not typically display a pressure reading because that can cause confusion with some customers. A typical oil pressure gauge is simply labeled with an area that denotes "good," and the needle is programmed to be centered when oil pressure is normal. A minimum oil pressure standard is 10 psi for every 1,000 engine rpm, so a warning light is not triggered for an engine idling at 600 rpm until oil pressure drops below 6 psi.

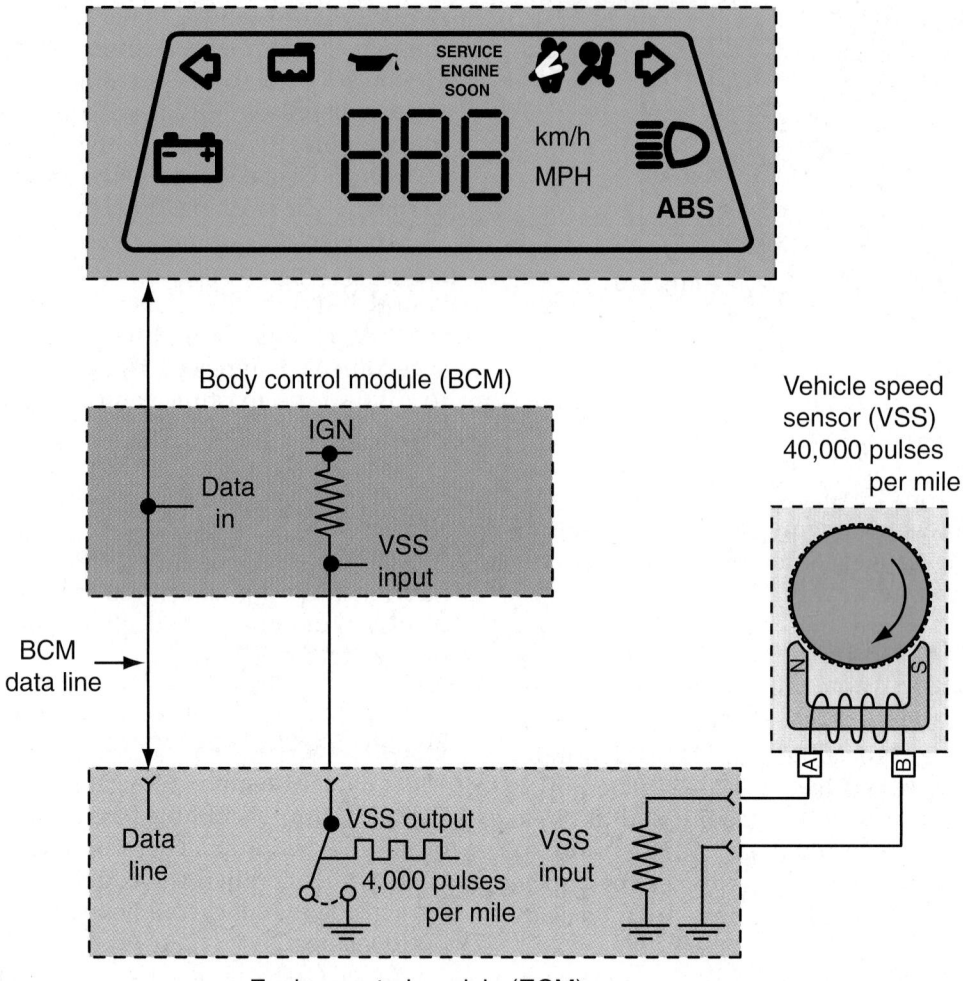

Figure 32.45 An electronic speedometer using input from the vehicle speed sensor (VSS).

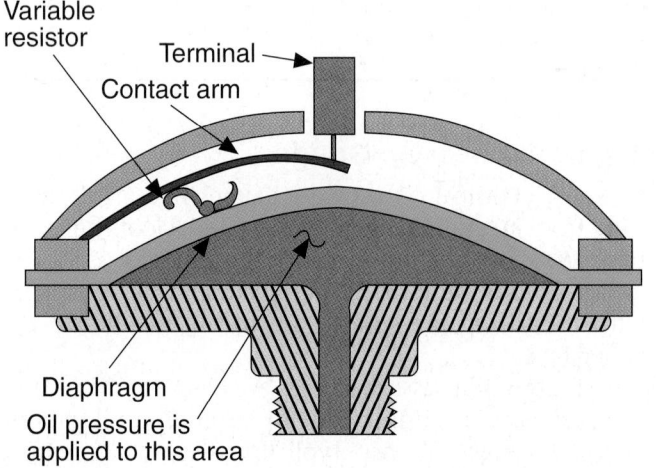

Figure 32.46 Oil pressure is measured by a piezoresistive sensor.

The oil pressure sensor is a piezoresistive sensor (**Figure 32.46**). A flexible diaphragm within the sensor moves in response to changes in oil pressure. A sliding contact arm moves against the resistor, changing the amount of current flow to the gauge.

Fuel Level Gauge

The fuel gauge receives signals from a sending unit in the fuel tank. The sending unit is part of a fuel pump cluster. The float moves a wiper arm across a variable resistor to change resistance as fuel level changes. This results in increased or decreased current flow to the gauge or digital display. Most fuel systems include a low fuel warning (LFW) module to warn the driver when the fuel level lowers to within 1/8 to 1/16 of full. The module also reduces fuel gauge movement from sloshing fuel.

Lamp-out Warning Light

When a light burns out, an electronic warning module completes a circuit to illuminate an instrument panel warning light. When all bulbs are operating in a circuit, 0.5 volt is provided to the module through resistive wiring. The module monitors small voltage level changes in the wires. If one bulb fails and does not draw current, voltage input on the resistive wire drops by about 0.25 volt to trigger the warning light.

Other Warning Lights

There are many other warning lights in vehicle electric systems. Some of these include:

- Air bag readiness light
- Antilock brake fault light
- Brake indicator lights for pad wear, fluid level, and hydraulic failure
- Cruise control light
- Door ajar indicator
- Four-wheel-drive lamp
- Rear defrost warning light
- Seat belt fasten indicator
- Tire pressure monitors
- Traction and stability control system problem light
- Transmission fail-safe mode light

Other Electrical Accessories

Operation of other electrical accessories is found in Chapter 34.

REVIEW QUESTIONS

1. What is the name for the large wires that allow more electrical current to flow?

2. With American Wire Gauge (AWG) numbers, does a larger wire mean a smaller or larger number?

3. What is another name for a colored plastic fuse?

4. Which kind of glass fuses are all the same length no matter what the amp rating?

5. What is the name of the type of fuse that allows for a heavier draw when wiper or blower motors first start to operate?

6. What is the name for two metal strips with different expansion rates that are fastened to each other?

7. When the dimmer switch is on the turn signal lever, what is the switch called?

8. In what year were rectangular headlamps introduced?

9. What measurement is the intensity of a headlamp rated in?

10. When the high beam is turned on and the second filament lights are too, is this a type I or a type II headlamp?

11. What is the name of the type of headlamp that has a small bulb that can be removed and replaced in its large lens housing?

12. What color is a bulb that has an *A* or an *NA* in its ANSI number?

13. Besides the regular turn signal flasher, what other flasher is located on vehicles since 1967?

14. How often do the contacts in the horn open and close?

15. How are two horns wired, in series or parallel?

ASE-STYLE REVIEW QUESTIONS

1. Each of the following is true about fuses *except*:
   a. Each fuse has an amp rating at which it is designed to fail.
   b. A fuse is designed to melt when the flow of voltage in the circuit becomes too high.
   c. Ceramic fuses are color coded.
   d. Mini-fuses are color coded.

2. Which of the following wire sizes is smaller?
   a. AWG size 0   c. Metric size 2.0
   b. AWG size 20   d. Metric size 8.0

3. Technician A says that some circuit breakers must be reset after they are tripped by an overload. Technician B says that some circuit breakers are self-resetting. Who is right?
   a. Technician A   c. Both A and B
   b. Technician B   d. Neither A nor B

4. All of the following statements are true about newer-style windshield wipers *except*:
   a. Wiper interval on most new systems is controlled by a rheostat.
   b. Intermittent operation is usually controlled by a governor module.
   c. The delay function works like a mechanical turn signal flasher.
   d. A capacitor triggers an electronic switch to begin wiper operation.

5. Technician A says that a type II headlamp has two wires on the connector. Technician B says that a type I headlamp has three wires on the connector. Who is right?
   a. Technician A   c. Both A and B
   b. Technician B   d. Neither A nor B

6. None of the following is true *except*:

   a. Any halogen headlamp with moisture in it is defective.

   b. A halogen headlamp burns brighter than a standard sealed beam headlamp because the tungsten filament gets hotter.

   c. If an ANSI light bulb has two terminals, one must be for ground.

   d. A bulb with an ANSI designation of NA is painted glass.

7. Technician A says that on many imported cars the turn signal lens is red. Technician B says that on many imported cars, the brake light lens is yellow. Who is right?

   a. Technician A   **c.** Both A and B

   b. Technician B   **d.** Neither A nor B

8. A _____ in the wiper switch allows the wiper speed to be changed.

   a. Thermostat   **c.** Time switch

   b. Potentiometer   **d.** Rheostat

9. Technician A says that high-voltage wiring on a car is called primary wiring. Technician B says that low-voltage wiring is called secondary wiring. Who is right?

   a. Technician A   **c.** Both A and B

   b. Technician B   **d.** Neither A nor B

10. A fuse link is a section of _____ diameter wire.

   a. A smaller   **c.** The same size

   b. A larger   **d.** None of the above

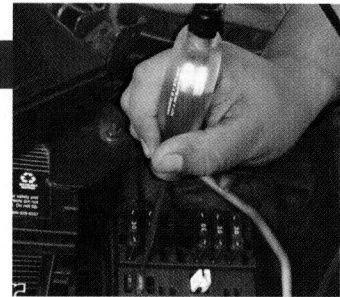

Lighting and Wiring Service

◾ OBJECTIVES

Upon completion of this chapter, you should be able to:

✔ Diagnose problems in lighting and wiring systems.

✔ Adjust headlamp aim.

✔ Make repairs to automotive wiring, lamps and bulbs, and protection devices.

◾ KEY TERMS

open circuit	flux	heat-shrink tubing
short circuit	tinning a wire	Gauss meter
grounded circuit	cold solder connection	dielectric
resistance		

◾ ANALYZING ELECTRICAL PROBLEMS

In this chapter, you will build on your knowledge of electrical system service. Wiring serves all of the electrical components in the automobile, and a circuit is sometimes the cause of a problem. Before attempting to repair electrical wiring and lighting problems, you must understand how the system works and how to use electrical system tools. You need to be methodical in going about your repair. Do not just replace a starter or alternator until you have thoroughly diagnosed the cause of the problem.

- The first step is to verify the complaint to make sure what the problem is.
- Check to see if there are related symptoms. A related symptom might be that the lights dim when the engine is cranked. Check a wiring diagram to see if there are common problems that could be tied together by a bad ground or power feed.
- Take time to consider the symptoms and their possible causes.
- Check for a quick fix. Sometimes a loose or corroded connection could be the cause of a problem.

Types of Electrical Problems

There are three categories of electrical problems: opens, shorts, and excessive resistance:

- When a circuit has a break in its continuity, this is called an **open circuit**.

- A **short circuit** is a current path that is unwanted. It can cause increased current flow that can result in burned wires or parts (see Figure 25.48). Damaged wiring insulation is a major cause.
- Shorts can also be directly to ground. This is called a **grounded circuit** (see Figure 25.49).
- Excessive **resistance** in a circuit causes reduced current flow. A corroded connector is a common cause of electrical problems.

Testing procedures for electrical problems are covered in Chapter 25.

SAFETY NOTE When disassembling a number of wires or vacuum hoses that are not clearly color coded, use numbered marker tape to keep them in order.

◾ WIRING SERVICE

Wiring Diagrams

Manufacturers' service information includes detailed wiring diagrams, sometimes called schematics, of all of a vehicle's electrical circuits. A wiring diagram is like a road map. Wires are colored and/or numbered. To trace a circuit, follow the colored wires.

Diagrams show wiring for the power side of the circuits, including electrical parts, splices, connections,

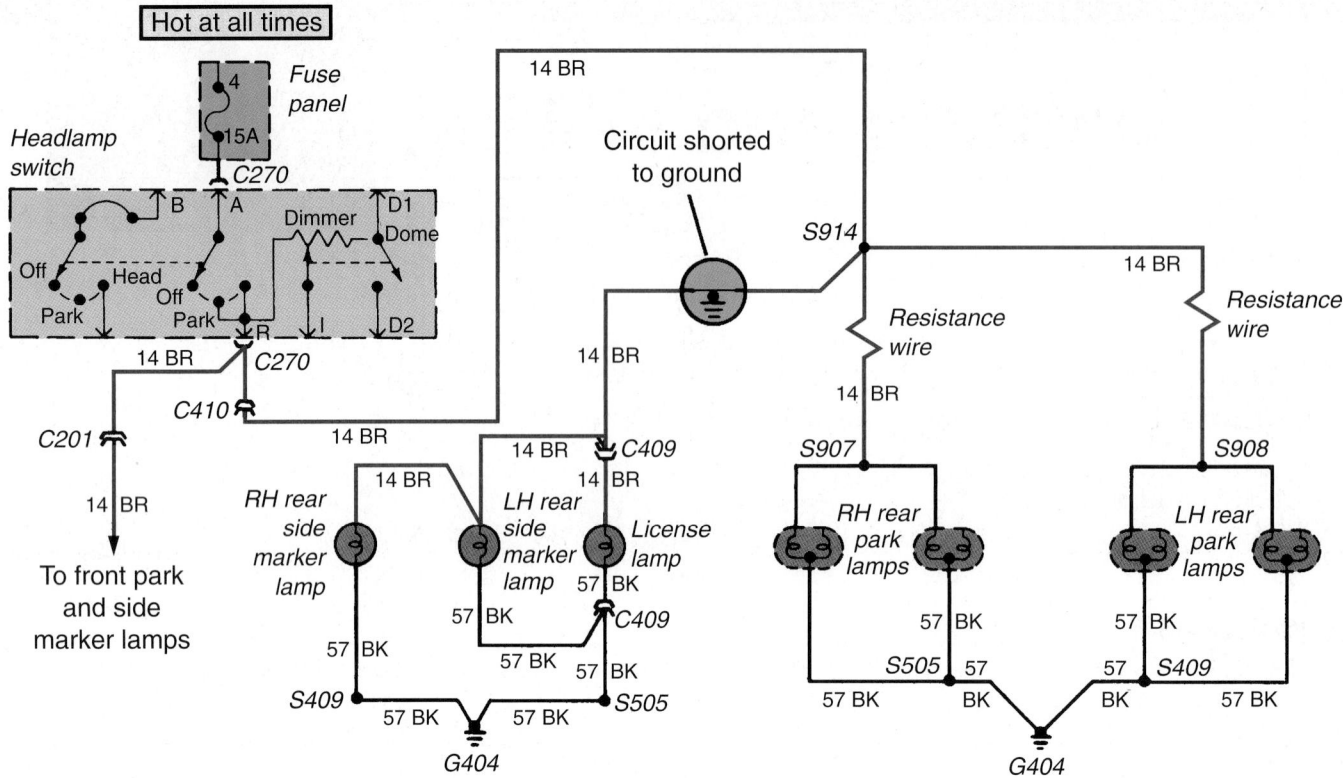

Figure 33.1 Part of a wiring diagram. *(Courtesy of Ford Motor Company)*

and return paths to ground. Letters or numbers located near parts identify a wire's colors. When there is more than one letter, the first is the color of the insulation and the second is the color of the stripe. If there is more than one stripe, or tracer, a slash (/) will separate them. Wire colors are abbreviated and are obvious to the eye. For instance, BRN indicates brown and DK BLU is dark blue.

Numbers can refer to the wire gauge size or locations. A typical scenario would be for numbers 100–199 to be underhood locations, 200–299 to be underdash locations, 300–399 to be locations in the passenger compartment, and 400–499 to be located in the trunk. Letters preceding the number are G (grounds), S (splices), and C (connectors) (**Figure 33.1**). Following the map, you will sometimes encounter a change in wire color after a splice.

Vehicle electrical systems are much more complex than they used to be. A complete wiring diagram in the past might have been 4 pages long, whereas today's typical wiring diagram approaches 12 pages. Wiring diagrams seem complex when you look at the entire diagram. The trick to analyzing a diagram is to study the individual subsystem or circuit you are concerned with. SAE wiring diagrams are arranged so that the upper right side of the page is the electrical power feed and the lower right is ground. Typical wiring diagram symbols are shown in **Figure 33.2**.

An electronic information library, such as All-Data or Mitchell On Demand, has an electrical component locator section. Pictorial wiring diagrams in the locator show the location of the components, grounds, and splices on the automobile.

Control modules are a newer feature on wiring diagrams. A control module is a micro-computer that monitors information and acts on it. These units control functions like the lighting system, antilock brakes, automatic transmissions, climate control, and others. The module is shown in the wiring diagram in block form with its electrical feeds and grounds. Sometimes a pictorial sketch is shown that depicts how the device would work if it were mechanical rather than electronic—a switch instead of a transistor, for instance.

■ CRIMP TERMINALS

Crimp (solderless) terminals are a popular way of repairing wire ends. There are many varieties of terminals and connectors available (**Figure 33.3**). Terminals are made with different size crimping tabs to accommodate wire sizes from 22 gauge (small) to 10 gauge (larger). They can be soldered or crimped with a special wire stripper/crimper (**Figure 33.4**).

To install the terminal:

■ About ¼" of the insulation is stripped from the end of the wire. If the wire is not clean and

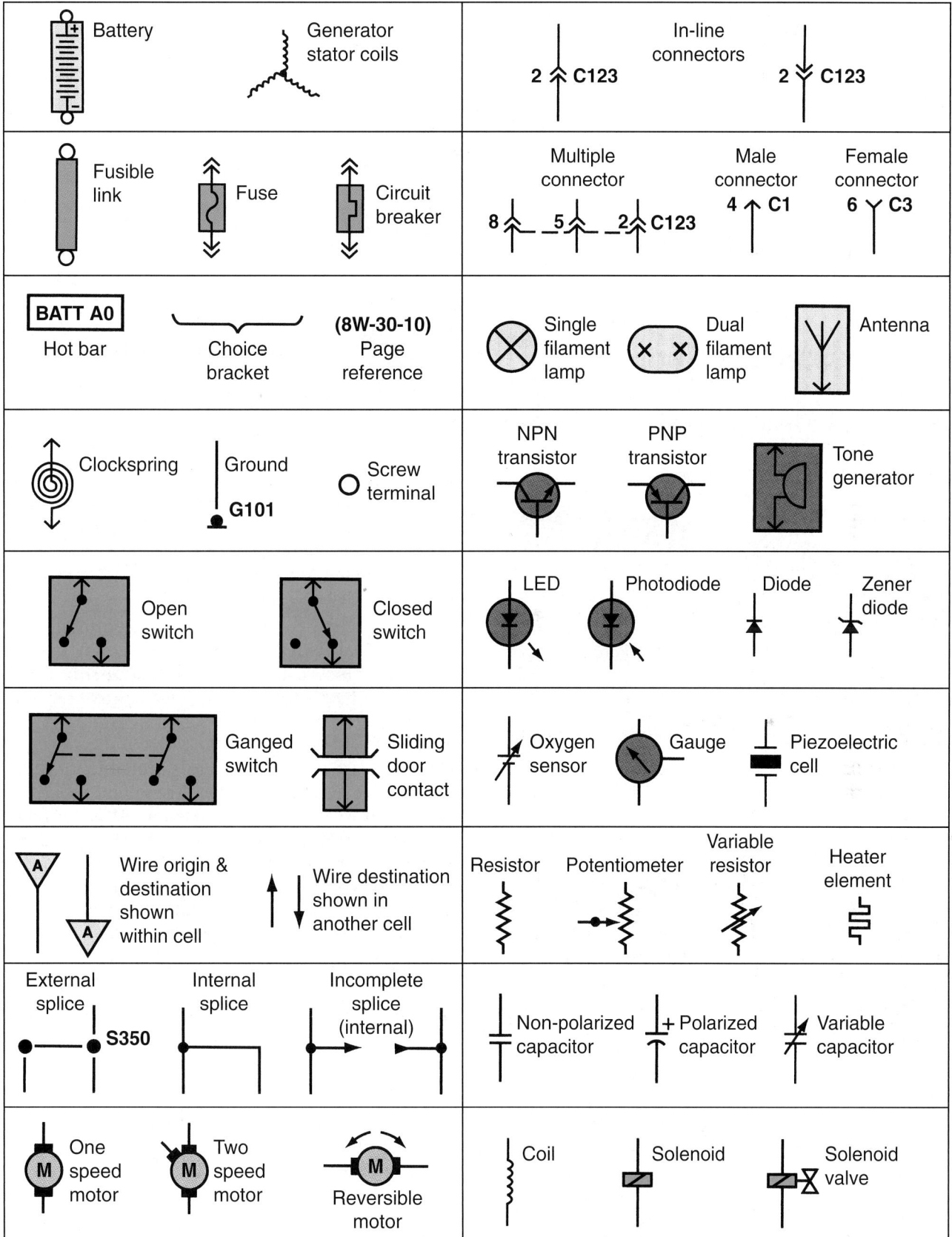

Figure 33.2 Common electrical symbols used on wiring diagrams. *(Courtesy of DaimlerChrysler Corporation)*

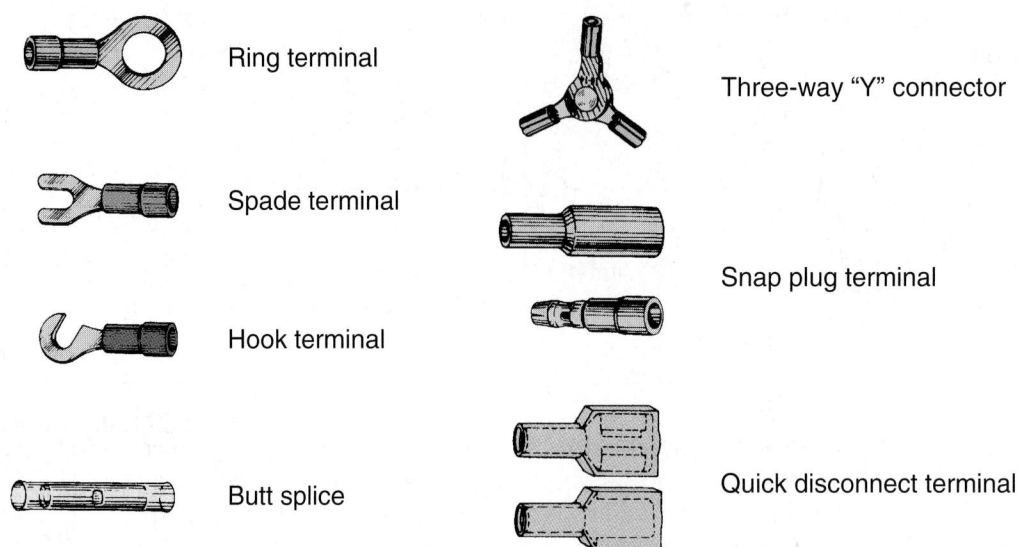

Figure 33.3 Different terminals and connectors. (*Courtesy of Federal-Mogul Corporation*)

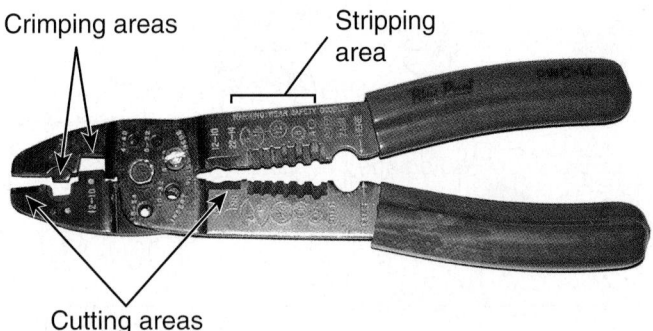

Figure 33.4 A wire crimping tool. (*Courtesy of Tim Gilles*)

shiny, cut the insulation and wire back further until it is.

- Insert the end of the wire into the terminal and crimp it using a crimping tool. The dimple from the crimping tool should be opposite to the seam in the connector.
- The crimping tool will usually have labeled grooves of the proper size to crimp most sizes of wires.

Figure 33.5 shows how to crimp an open type of connector end.

A poor crimp will create excessive resistance, intermittent operation, and early failure of the circuit.

NOTE: *Crimp connectors are not used in computer circuits. Special weatherproof connectors are used to prevent corrosion. This type of connector prevents air from entering and corroding the connection. Corrosion causes resistance that will change the computer's sensor readings. These connectors are covered later in this chapter.*

Terminals are often held in a terminal block to keep several wires organized. Removing terminals from the block usually requires depressing a locking tang. **Figure 33.6** shows a typical connector release.

Crimp Connectors

There are also crimp connectors available for splicing a wire together (**Figure 33.7**). These *butt connectors* are a fast and relatively inexpensive method of repairing wires, provided the crimp is solid and is done carefully. Insert both ends of the wire into the connector so that both wires are side by side and then crimp.

■ SELECTING REPLACEMENT WIRE

Replacement wire and cable comes wound in spools. Be sure to use wire of adequate size for the load. A chart in Appendix E shows the wire size needed for different length runs. Replacement wire should always be at least

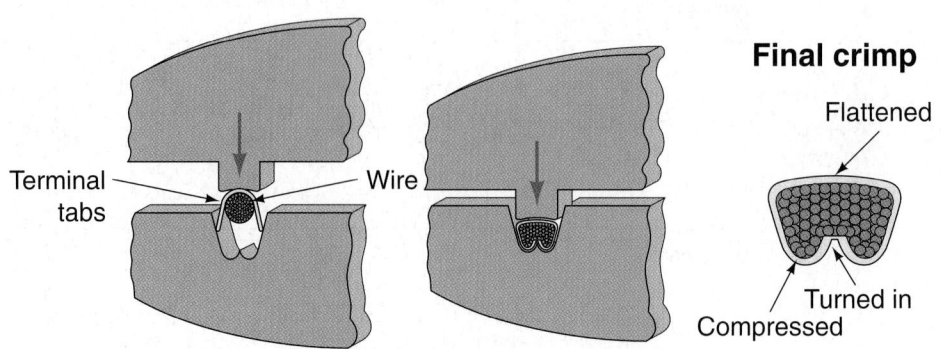

Figure 33.5 Crimping an open connector. (*Courtesy of Ford Motor Company*)

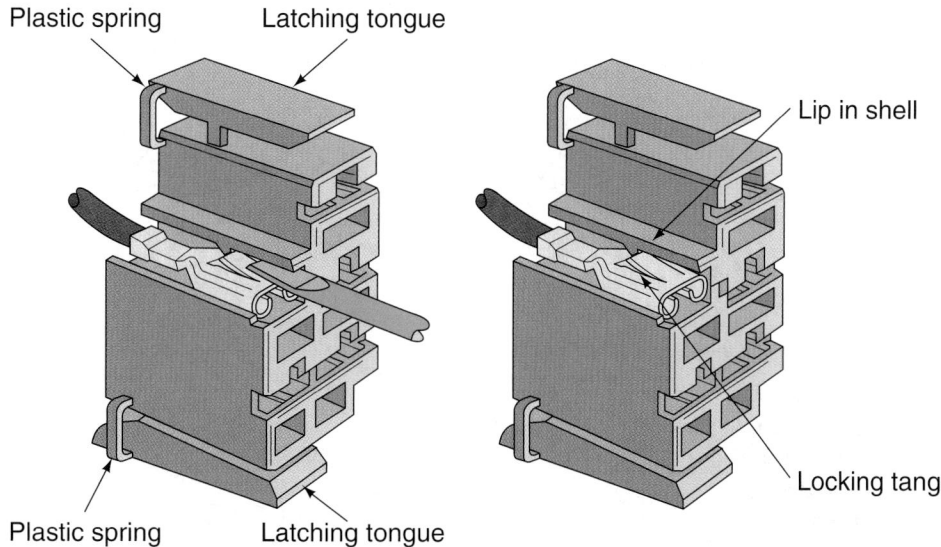

Figure 33.6 Depress the lip to remove the connector from the junction. *(Courtesy of Ford Motor Company)*

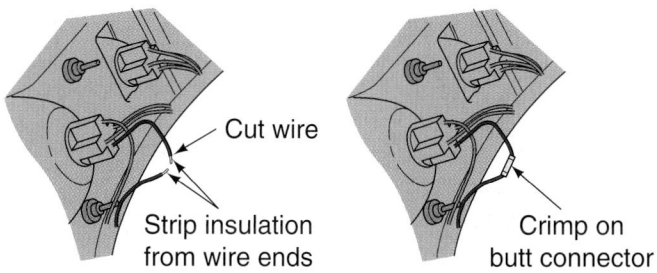

Figure 33.7 A butt connector is used to join wire ends. *(Courtesy of Ford Motor Company)*

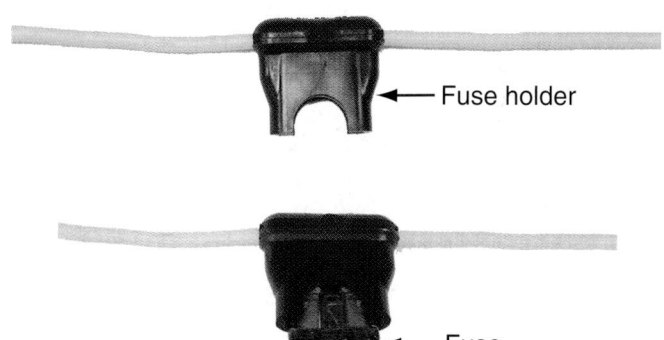

Figure 33.8 A fuse holder is installed in series in a line. *(Courtesy of Tim Gilles)*

as big as the original wire. Wire that is too small or too long for a circuit can cause excessive resistance, resulting in a loss of lighting or accessory efficiency. Wire of double the original length has twice the resistance.

■ ADDING ELECTRICAL ACCESSORIES

Be sure that proper techniques are followed when adding items to a vehicle's wiring system. When an aftermarket accessory is added to the vehicle, it is usually necessary to add a separate fused circuit. Trailer lights are commonly added with an aftermarket trailer hitch. The color code for the existing lights should be followed when possible. A variety of multiple wire harness adapters are available.

NOTE: *Adding extra loads, such as stereos, CB radios, or fog lights can tax a vehicle's charging system so that it may no longer be adequate. Be sure to investigate this first before adding any extra load.*

Fuse Holders

When an original equipment accessory is added to a vehicle's electrical system, power can usually be taken from a fuse on the fuse panel.

If a fuse is not available, a fuse holder can be connected to an existing adequate power source. A radio will usually come with a fuse holder and instructions

for its installation. A handy *blade-type* fuse holder can be easily installed into the wire (**Figure 33.8**) to provide a safe source of power.

A *tap splice* connector can be used to tap into a power wire without the need to strip or solder. This connector is used for tapping another wire into an existing wire without cutting it or stripping its insulation. **Figure 33.9** shows how it is used. This kind of connector is used in electrical circuits only, not in electronic circuits.

A tap splice, however, is not a high enough quality connection for such use. In regular electrical circuits, these connections sometimes fail after a few years of service. Give careful consideration to the use of the connector. This connector is most common for trailer connections. An underdash installation of an electrical component would be ill advised. A soldered connection (covered later) would be a better choice.

■ SOLDERING

Soldering is a professional method of repairing a connection. For *electronic* connections, such as those in computer circuits, soldering is especially preferred

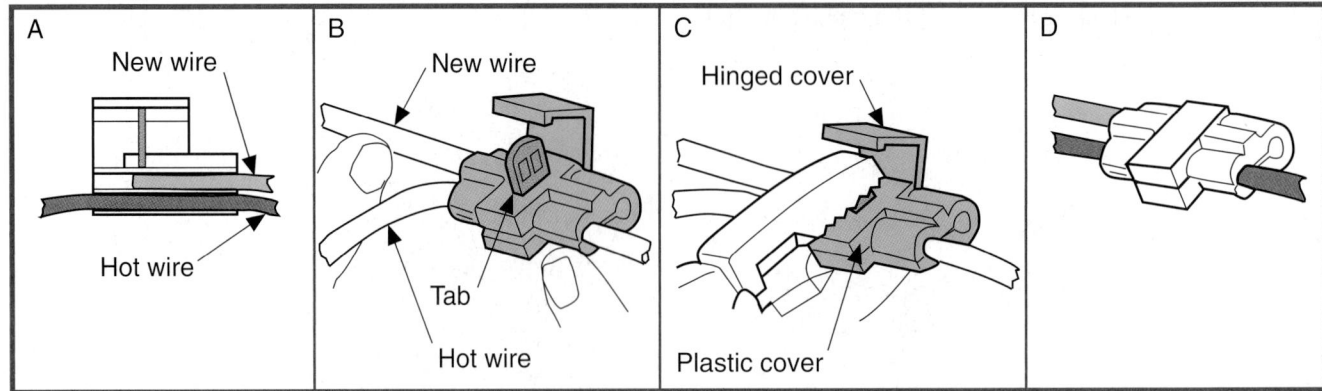

Figure 33.9 Installing a tap splice.

because of the small amount of electrical current that runs through them. A properly soldered connection will not suffer increased resistance due to oxidation with the passage of time.

A *soldering iron* or a *soldering gun* (see Figure 9.25) is used for connections of wiring. A small *propane flameless soldering pencil* can also be used in tight places (**Figure 33.10**). For radiators or larger cable terminal connections, a propane, acetylene, or Mapp gas torch is used.

Sometimes, soldering is not convenient to do, because of the unavailability of electricity, because of the need to avoid heat in the circuit, or because a crimp connector is simply faster and more convenient. A crimp connector is used to join stainless steel wire because stainless is not solderable. Stainless steel wire is used in boat trailers instead of copper because of its resistance to corrosion.

Solder will not stick to metals unless they are very clean and not corroded. Corroded sections of wire are cut out and replaced. When making a repair to an old wire, a new section at the end of a wire is stripped of insulation to provide a clean area to solder.

Flux is used to clean metal so that solder will stick to it. Solder usually has a hollow core full of flux. Solder for *electrical repairs* is commonly a 40/60 tin/lead mix with a *rosin* flux core (**Figure 33.11**). *Acid-core* solder is also available, but it should not be used on electrical connections, because it will corrode them.

Heat the soldering gun until the tip is hot. Clean the hot tip with a wet towel. Flux and renew the solder coat.

A wire terminal can be soldered to the wire to make a professional connection (**Figure 33.12**). After the wire is stripped of insulation, **tin a wire** by heating it and applying solder (**Figure 33.13**) before attempting to solder the terminal to it.

Figure 33.11 This rosin core solder is an alloy of tin and lead.

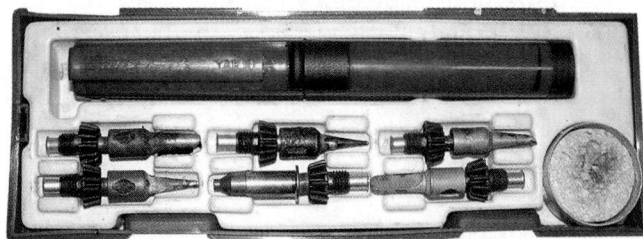

Figure 33.10 A propane soldering pencil. (*Courtesy of Tim Gilles*)

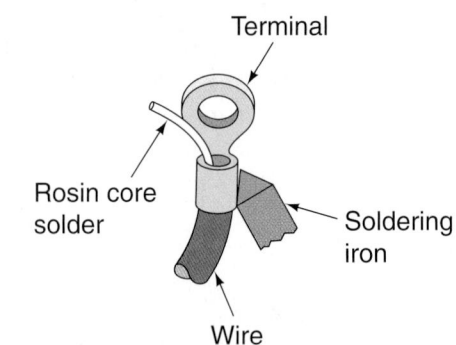

Figure 33.12 Soldering a terminal to a wire.

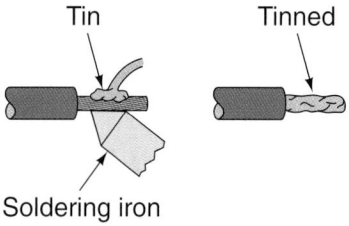

Figure 33.13 Tinning a wire.

To join two wires, splice their stripped ends together. **Figure 33.14** shows several methods of splicing. The wires are soldered after completing the splice.

When using a soldering iron or gun, apply solder to the opposite side of the connection from the tip of the iron (**Figure 33.15a**). Heat the wire until the solder flows into the strands of wire (**Figure 33.15b**). This ensures that a **cold solder connection** does not happen. A cold connection happens when the solder is melted by the iron but the wire is too cold to bond to it. Also, be careful not to heat the wire too much. Overheating a wire can ruin its insulation.

Heat-Shrink Tubing

Vinyl **heat-shrink tubing** can be used to insulate a solder joint and make it airtight. Cut the tubing to the correct length and slip it over one of the wires *before* they are spliced (**Figure 33.16a**). After the soldering job is completed, the shrink tubing is slid over the soldered connection (**Figure 33.16b**). When the tubing is heated with a flame or a heat gun, it shrinks (**Figure 33.16c**) to provide an airtight connection (**Figure 33.16d**).

NOTE: *Heat-shrink tubing is thin. It will not withstand abrasion.*

If heat-shrink tubing is not used, insulate the connection with vinyl electrical tape.

Ordinary shrink tubing is available in bulk rolls (**Figure 33.16e**). It is usually made of PVC, which shrinks to half its original size.

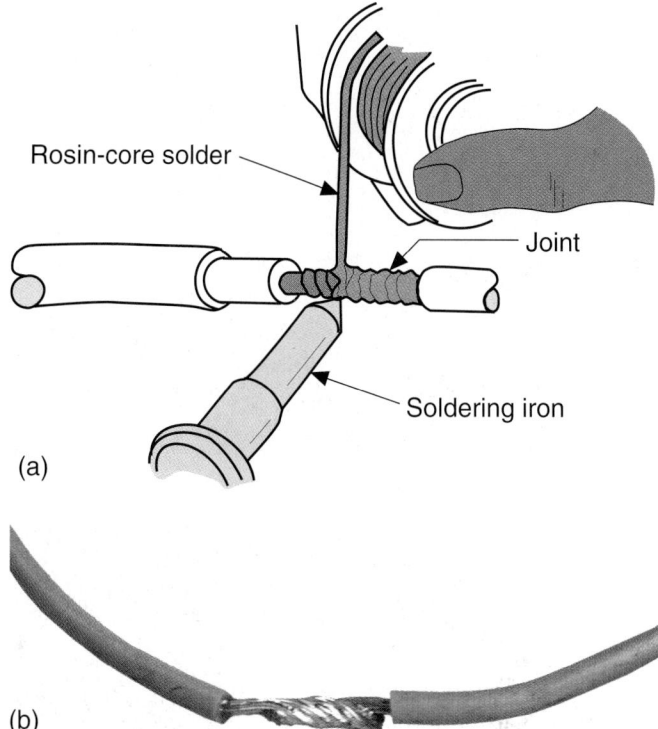

(a)

(b)

Figure 33.15 (a) Apply the solder to the side of the wire opposite to the tip of the soldering iron or gun. (b) Heat the wire until the solder flows into the strands of the wire like this. (*b, Courtesy of Tim Gilles*)

Adhesive-Lined Shrink Tubing and Connectors. Due to the very low amounts of current flow in electronic circuits, a corrosion-free environment is crucial. Manufacturers recommend variations of shrink tubing, which include adhesives, to seal the ends of the connections from air. DaimlerChrysler Corporation recommends a shrink tubing lined with a melting adhesive to seal the ends of the tubing when heated. General Motors uses *crimp-and-seal* connectors that resemble ordinary butt connectors. They have an adhesive sealant also.

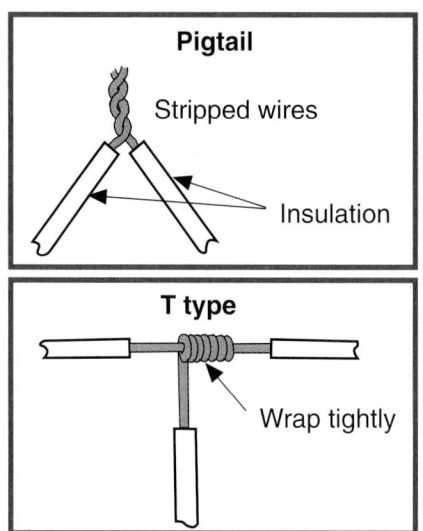

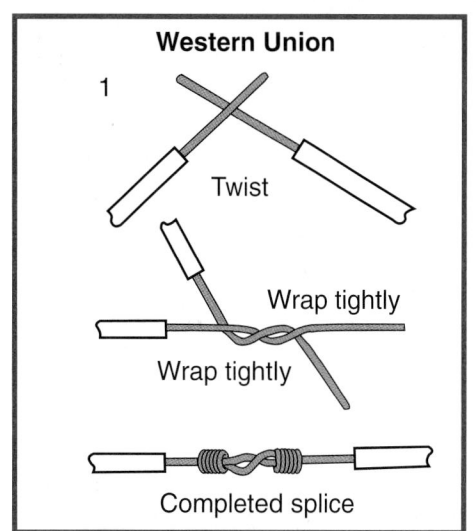

Figure 33.14 Several methods of splicing wires.

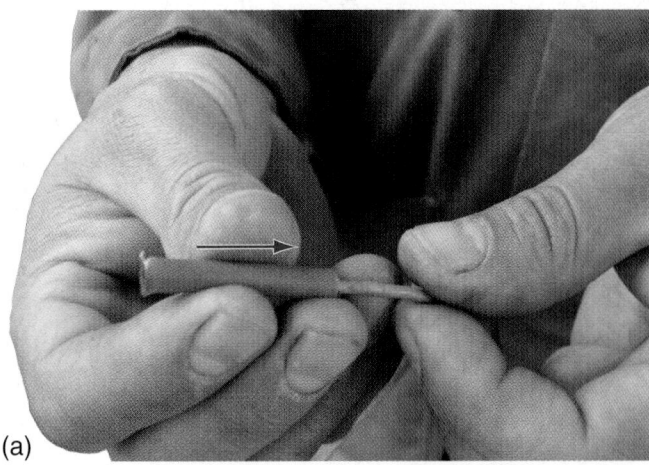

(a)

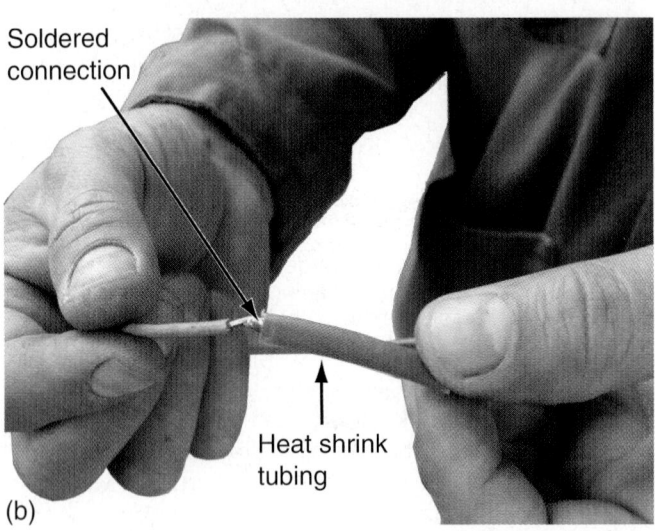

Soldered connection

Heat shrink tubing

(b)

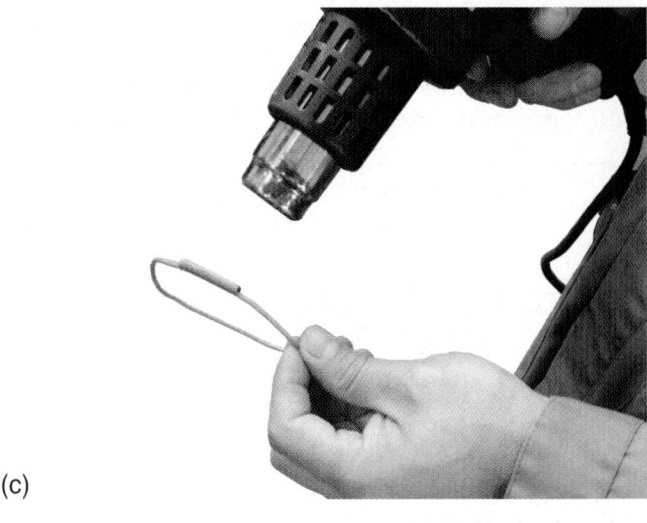

(c)

(d)

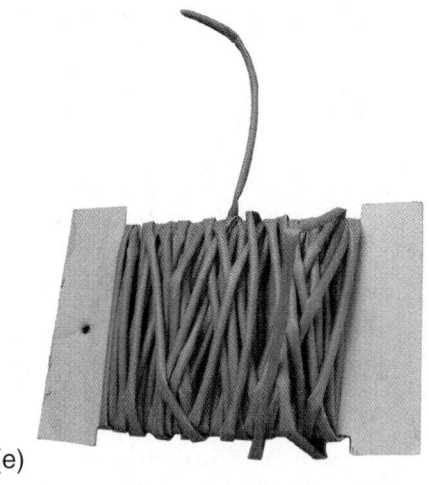

(e)

Figure 33.16 Using a heat-shrink tubing. (a) Slip the heat-shrink tubing over the wire before splicing. (b) After soldering, slip the tubing over the soldered joint. (c) Heat the shrink tubing with a heat gun. (d) The appearance of a completed job. (e) Shrink tubing is available in bulk rolls. *(Courtesy of Tim Gilles)*

Adhesive-lined connectors and tubing are made of polyolefin, which shrinks to one-third its original size for a tighter connection.

■ BROKEN OR DAMAGED GROUND STRAPS

The engine and the car body are isolated from the frame with rubber mounts. Rubber is an insulator, so ground straps are installed between the engine and chassis (see Figure 50.6). This provides electricity with a complete path back to the battery through the frame.

A broken ground strap can cause electricity to hunt for a different path to ground. When this happens, strange part failures can occur. Some of the possible results are

■ A burned transmission bushing and drive shaft yolk
■ Burned emergency brake cables
■ A burned carburetor return spring
■ Flickering headlights
■ Burned front-wheel bearings or CV joints (on front-wheel-drive cars)

■ CIRCUIT TESTING AND SERVICE

Fuse Failure

A fuse failure could be a once only occurrence, or there could be an electrical problem that must be fixed. Replacing a fuse or fuse link or resetting a circuit breaker does not fix the problem that caused the overload. The overload device will probably just fail again. Always fix the problem before restoring circuit protection.

Occasional fuse failure can be due to a defective fuse. Corrosion on the end of the fuse can cause failure, too.

Automotive fuses are rated according to current capacity, not voltage. A 12-volt fuse will also work in a

110-volt application. Voltage does not cause a fuse to blow; current does. Fuses have a 10% overload factor to guard against minor power surges.

▄▄▄ FINDING GROUNDS

Locating the cause of a grounded circuit with a test light is not usually possible, because the fuse blows as soon as the circuit is energized. A circuit breaker can be installed temporarily in place of a fuse for diagnosing grounded wires. Install a test light in series with the circuit breaker while making the test (**Figure 33.17**). Disconnect individual circuits until the light goes out. The circuit that was disconnected when the light went out is the one at fault.

A compass or **Gauss meter** (a gauge that detects magnetism) can be used to locate the exact location of a ground when a wire has been accidentally pinched (**Figure 33.18**).

Another method of detecting a grounded circuit is to use an ohmmeter. With the fuse removed, connect it to the circuit side of the fuse holder and ground.

NOTE: *Be careful not to allow the ohmmeter to contact the power side of the fuse holder. When there is continuity between the load and ground, the circuit is grounded.*

▄▄▄ FUSE TESTING AND SERVICE

The fuse panel is usually under the hood, under the instrument panel, or in a kick panel in the driver's compartment.

A visual check of the fuses in question can often show a burned fuse (**Figure 33.19**). Sometimes a fuse can blow where it cannot be seen (**Figure 33.20**). Remove and test a questionable fuse with an ohmmeter.

With the fuse in the socket, a test light can be used to test the fuse (**Figure 33.21**). With one end of the test light clipped to ground, the tester should light up when probing both sides of the fuse.

■ If the tester glows when touched to either side of an installed fuse, the fuse is good and the voltage is on.

■ If the tester glows only when touched to one end of the fuse, the fuse is defective. The side that does not light is the ground side of the circuit.

■ The end of a blade-type fuse has a small hole that exposes its fuse metal to the outside of the fuse so it can be probed with a test light.

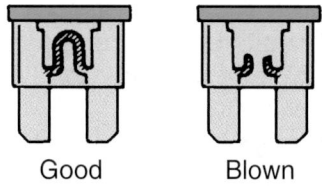

Good   Blown

Figure 33.19 Good and blown fuses.

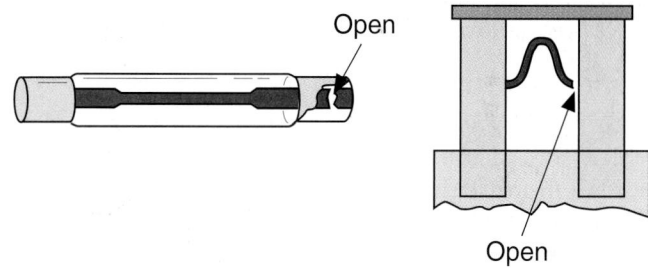

Figure 33.20 Sometimes a fuse can blow where it is difficult to see. Test a questionable fuse with an ohmmeter.

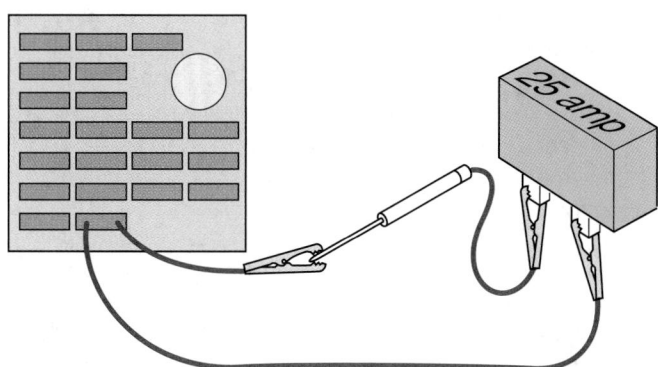

Figure 33.17 A test light installed with a circuit breaker for finding a faulty circuit.

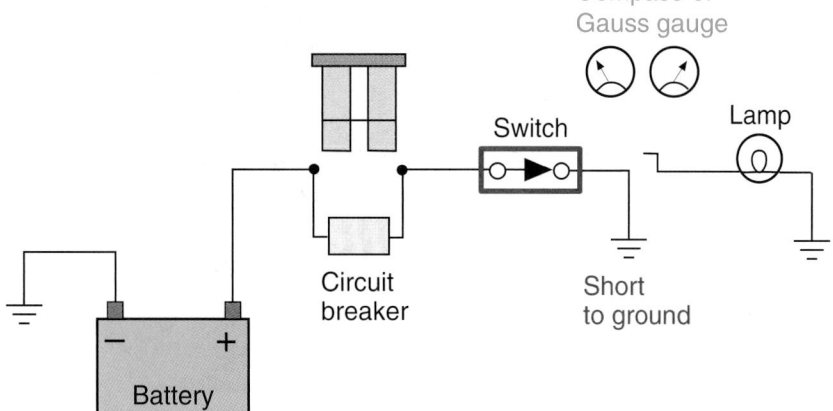

Figure 33.18 Use a compass or Gauss gauge to check for changes in the magnetism in the circuit at the location of the pinched wire.

Figure 33.21 Using a test light to test a fuse in a fuse panel. *(Courtesy of Tim Gilles)*

> **SAFETY NOTE** If you use a screwdriver or a needle nose pliers to remove a glass cartridge or ceramic fuse, wear eye protection. The glass tube or the ceramic body could shatter.

Removing Fuses

With glass cartridge or blade-type fuses, a special fuse removing tool can be used to remove a bad fuse. Blade-type fuses are easy to remove by hand or with a needle nose pliers.

Testing the Fuse Socket

With the fuse removed, use the test light to determine which side of the fuse socket is the ground side. The tester will not light when attached to that side. If the tester does not glow on either side of the circuit:

- The circuit is shut off.
- The circuit is broken.
- The tester does not have a good ground connection.

Replacing the Fuse

Be sure to replace a fuse with the correct one for the application. SFE and Bussman glass cartridge fuses will interchange with one another, so care is required. Be sure that the fuse is the correct rating. Too low an amp rating will result in burned fuses. Too high a rating can cause a fire.

Check the end of the fuse for its amp rating. The correct amp rating of the fuse to be used is listed on the fuse block.

■ FUSE LINK SERVICE

Fuse link wire is covered with insulation that bubbles if a fuse link melts. The blisters on the insulation indicate the failure in the fuse link. To replace a burned fuse link, cut out the damaged part of the wire. Splice a new fuse link in with a butt connector. Be sure to use a fuse link of the correct size.

■ HEADLAMP SERVICE

Headlamp Replacement

The back of a headlamp has tabs that align to holes in the headlight brackets. Halogen and conventional sealed beams are both of the *standard SAE design* and will fit into the same brackets. Round type I and type II lamps will *not* interchange, because they are of different sizes.

NOTE: *While standard and halogen lamps can be interchanged, this should only be done in an emergency. Conventional lamps will not illuminate equally with halogen. Also, there can be a false signal generated to the headlight warning light on the dashboard due to different electrical resistances.*

The headlight is held in a bracket that has adjusting screws on it.

NOTE: *If the headlight is being replaced because it is burned out, the new bulb will not normally require readjustment.*

Before attempting to remove a headlight, locate and identify the aiming screws (**Figure 33.22a**). If

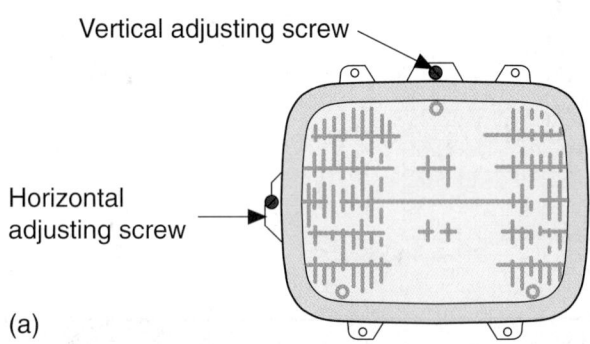

(a)

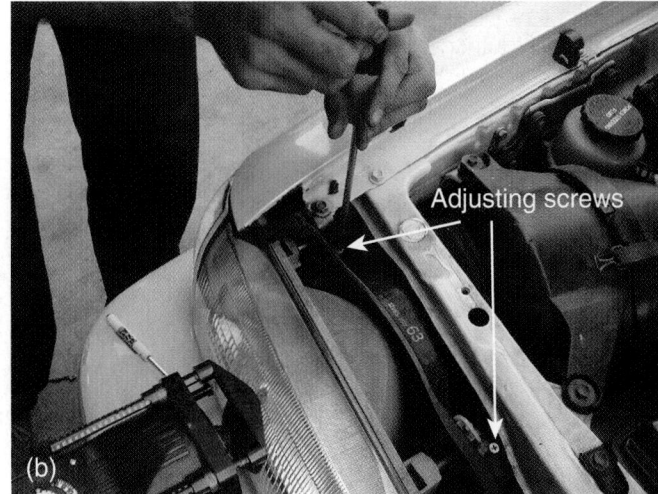

(b)

Figure 33.22 (a) Headlight aiming screws. (b) Vertical and horizontal adjusting screws on the back of the headlight mounting fixture. *(b, Courtesy of Tim Gilles)*

these screws are accidentally turned, the adjustment of the headlights will be changed. Remove only the retaining screws for the trim piece around the headlight. Some headlights have screws located on the rear of the headlight mount that are used for horizontal and vertical headlight adjustment (**Figure 33.22b**).

Use a test light to check both the high and the low beams of the electrical plug of a suspected burned-out headlight. If there is no power to the connection, consult the service information. Use a nonconductive (**dielectric**) grease on the connection when installing the new headlight.

NOTE: *The headlamp only fits properly one way. Locate the alignment tabs on the rear of the lamp when installing the lamp in the bracket.*

Halogen Lamp Replacement

Sealed-beam halogen lamps are replaced as a unit and are more costly than conventional sealed-beam lamps.

- When the outer housing breaks on a halogen bulb, the inner housing, which could still be intact, will still light up. The quality of the light will be poor, however.
- A halogen lamp contains high-pressure gas. If it is dropped, it may shatter.
- Halogen lamps burn at very high temperatures. When a lamp has been recently lit, do not attempt to replace it until it has cooled.
- When removing a composite halogen replaceable lamp, touch it only on its plastic base (**Figure 33.23**). If oil from skin gets on the lamp, it will shatter when it lights.
- Remove the old bulb only when ready to install the new one so dirt and moisture do not get into the lamp housing.

Figure 33.23 Removing or installing a halogen bulb. Do not touch it with your fingers. *(Courtesy of Tim Gilles)*

HID Lamp Service

HID lamps are expensive but reliable. They are two to three times less likely to fail than a conventional lamp. However, in the event service is required, one of the following conditions could exist:

- One or both lamps do not come on.
- Lamps do not restrike when they are hot.
- The lamp is wearing out. This can result in a change in color of the light to a reddish color, accompanied by a blinking or flickering of gradually decreasing frequency. A variation in color when the lights are first turned on is normal. The color should stabilize quickly.
- Current supply to the ballast is too low. This results in a flicker also. The ballast provides enough current for a short time before current decreases to an unworkable level.

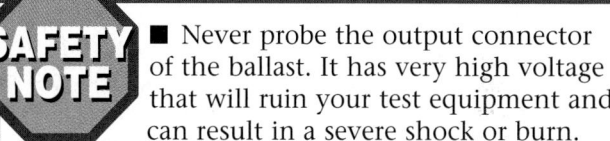

SAFETY NOTE
- Never probe the output connector of the ballast. It has very high voltage that will ruin your test equipment and can result in a severe shock or burn. Never probe either power or ground between the arc and tube assembly.
- Each headlight assembly has a **high-voltage** generator. Be sure to disconnect the car's battery before attempting to service HID lights and follow the manufacturer's service instructions.
- Do not touch the red or amber insulation on the wiring harness to the lamps while the headlights are on.
- Never turn on the headlight when the bulb is out of the socket.
- Never service an HID system while your hands are wet.

▬ HEADLIGHT AIMING

When properly aimed, low-beam headlights face down and to the right. This is so they do not shine in the eyes of oncoming drivers. Headlights are adjusted to a point that is 25 feet in front of the car. At that distance, the allowable margin of error is 4". From 300 feet, a driver is supposed to be able to see obstacles. But if a headlight is out of adjustment by the allowable amount of 4" at 25', the amount of the error is 4' at 300'. This can be dangerous, especially at high speeds.

Mechanical headlight aiming equipment is lightweight and is usually stored in a small suitcase. It is accurate and simple to use. Vertical and horizontal adjustments are made by turning two screws on the headlight bracket (see Figure 33.22). A headlight aiming set consists of two headlight aimers, mounting adapters, and level compensating adapters. The two adapters are

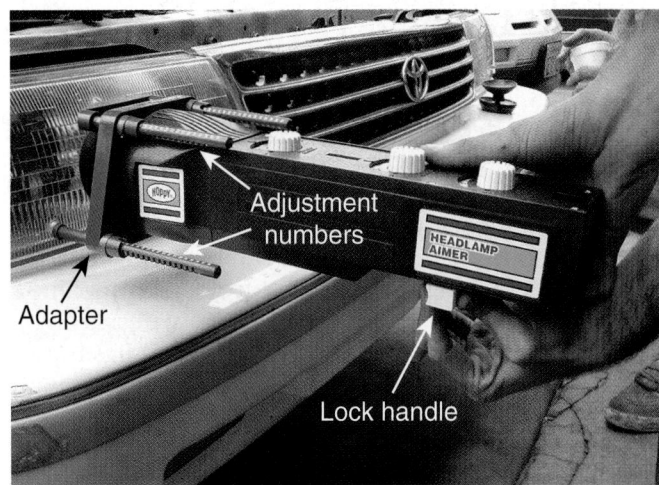

Figure 33.24 Attaching a headlight aimer to the headlight. *(Courtesy of Tim Gilles)*

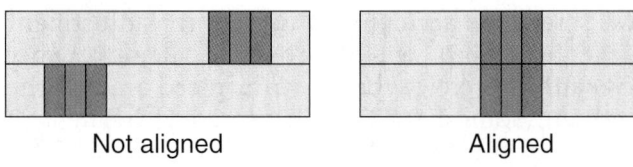

Not aligned   Aligned

Figure 33.26 When the headlights are correctly aimed, both halves of the screen align. *(Courtesy of Hopkins Mfg. Co.)*

mounted on the headlights using suction cups. When attaching the aimer to the headlight, first push the handle located on the bottom of the aimer toward the headlight (**Figure 33.24**). Check to see that the rubber suction cup is securely attached to the lens on the light. Then pull the handle away from the light to lock it.

Before adjusting headlights, several things must be checked. The vehicle should be carrying its usual load. The trunk should contain its typical amount of material. Ideally, the gas tank would be half-full and the driver's weight in the front seat would be taken into account. Tire pressures should be adjusted.

Headlight aiming equipment uses a *bubble level* to calibrate vertical alignment. Horizontal alignment is made by comparing one headlight to the other to see if they are parallel. A target (sight bar) on one aimer is viewed through mirrors inside a window on the other aimer (**Figure 33.25**). The target on the aimer being viewed will appear as two lines when the two aimers are not parallel. As adjustment is corrected, the two lines merge to become one (**Figure 33.26**). Access to the adjusting screws is often possible without having to remove trim pieces or headlight doors.

The aimers must be adjusted to compensate for the level of the floor that the vehicle is on. When one area of the shop is always used for adjusting headlights, the compensation step can be skipped.

Mounting the Aimers

The aimers are attached to the headlights using their suction cups. The headlight must be clean for the suction cup to stick to it. Some lamps require a suction cup extension that is screwed into threads on the inside of the aimer's suction cup. Otherwise the suction cup will not reach the lamp. Different styles of headlights require different adapters to allow the aimer to fit the light.

Some new vehicles have composite lamps with a curved surface. Mounting the aimers to these lamps requires different length rods that are part of the adapter kit (see Figure 33.24). American-style headlamps have three guide pins on the front outside of the lens that locate the adapter. Each guide pin has a number next to it. Each rod is adjusted to the number on its corresponding headlight guide pin (**Figure 33.27**).

Prior to adjusting headlamps, each aimer can be adjusted using either its vertical or horizontal knob (see Figure 33.25). These knobs can also be used to determine the amount that a headlight is out of adjustment before aiming it. With the aimer mounted on the headlight, the amount of out-of-adjustment is determined by reading the gauge dial. Prior to reading the gauge dial, center the level bubble using the vertical knob and turn the horizontal knob until the split image is aligned.

NOTE: *If the gauge reads 6", the aim is off calibration 6" at a distance of 25' in front of the headlight.*

Some states have requirements other than zero for proper headlight adjusting.

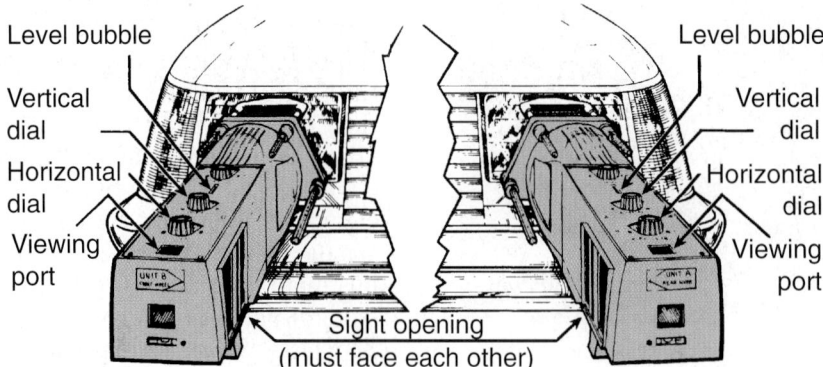

Level bubble

Vertical dial

Horizontal dial

Viewing port

Level bubble

Vertical dial

Horizontal dial

Viewing port

Sight opening
(must face each other)

Figure 33.25 Headlight aimers read off of each other. *(Courtesy of Hopkins Mfg. Co.)*

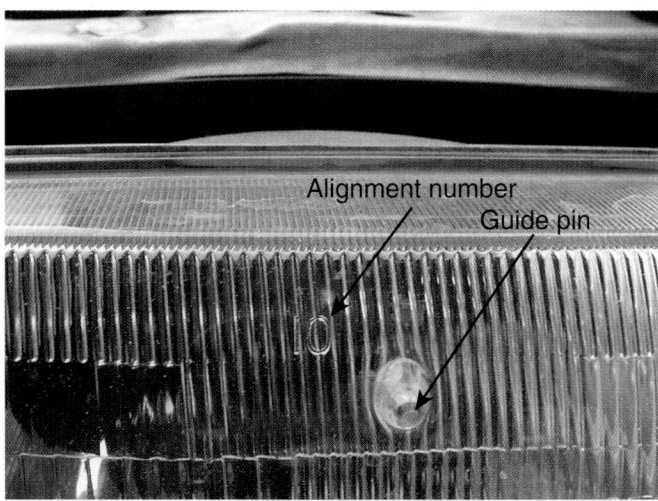

Figure 33.27 The aimer adapter fits over the guide pin. It will be adjusted to the number 10. *(Courtesy of Tim Gilles)*

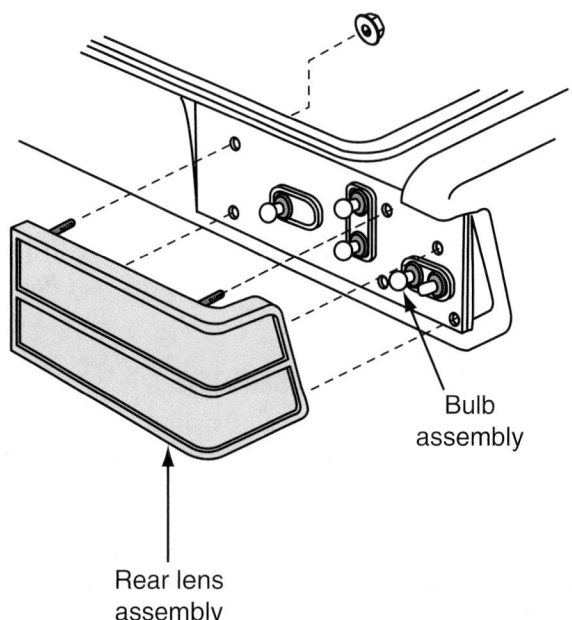

Figure 33.29 Access to this light requires removal of the lens.

▬ TAILLIGHT AND PARK LIGHT BULB SERVICE

Bulbs in some taillights are easy to remove and can be accessed from inside the trunk. The light socket is turned slightly to remove it from the taillight housing (**Figure 33.28**).

Other lights require removal of the light lens (**Figure 33.29**). Be careful when removing a light lens. The lens sometimes gets stuck to the rubber molding around it. If it is forced, an expensive lens can be broken. Be careful when reinstalling the lens that the gasket is in place in its groove.

To remove a bulb from its socket:

■ Push it in and turn it approximately ⅛ turn counterclockwise.

■ It will spring out of the socket when it is released. Excessive voltage shortens the life of a light bulb. Voltage as little as 5% higher can reduce a bulb's life by half. The only cause of excessive voltage is a malfunctioning charging system. Be sure to test the voltage regulator setting if a vehicle requires frequent bulb replacement. Be sure the bulb socket is clean of oxidation, which would prevent electron transfer.

Check the part number (called the bulb trade number) printed on the base of the bulb to be sure it is the correct one. A wrong bulb could burn with a different intensity compared to the one on the other side of the vehicle. The bulb trade number is the same for a specified bulb, no matter who the manufacturer is.

You should be able to verify the condition of most bulbs by visually checking the filaments. A bad filament is usually obvious. If a problem with the lights is difficult to diagnose, either replace the bulb or use a self-powered test light to test it as shown in Figure 25.54.

An ohmmeter can be used to check the continuity of the bulb filament (**Figure 33.30**). If the reading between

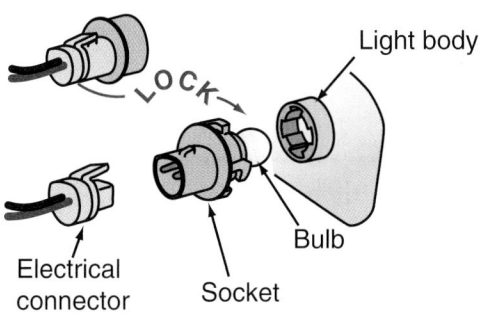

Figure 33.28 A removable light socket.

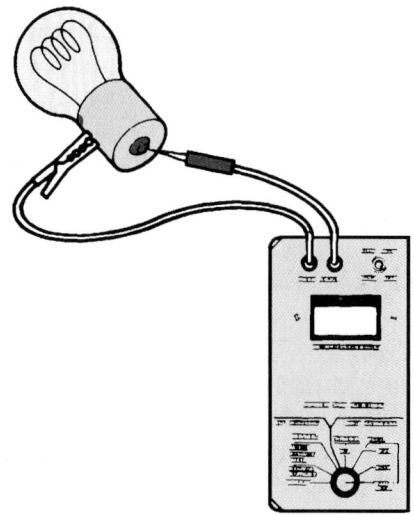

Figure 33.30 Checking a bulb with an ohmmeter. *(Courtesy of Ford Motor Company)*

the case (ground) and the terminal is infinity, the bulb is open and must be replaced. A relatively low reading on the ohmmeter gauge means the bulb is good.

If the bulb is a combination brake and taillight, it will have two filaments and two terminals. To check which of the wires is hot during a particular application, use a test light.

NOTE: *The socket is usually the ground for most bulbs. If you short the tester between the hot wire and the light socket housing, a fuse will blow.*

■ STOPLIGHT SWITCH SERVICE

To test a suspected bad stoplight switch, bypass it with a jumper wire. The stoplights should light when the wires are connected. Use the test light to see if there is power at one of the wires.

- If the lights come on when the wires are connected but not when the brakes are applied, replace the switch.
- If the stoplights operate without the key on, remove the stoplight fuse before removing the switch.
- If the new switch is adjustable, adjust it so it is open when the pedal is released. The lights should come on about ¼" after the pedal is applied.

■ BACK-UP LIGHTS

The back-up light circuit includes a fuse, a switch mounted on the shift lever or transmission, wiring, and lights. Sometimes the shift lever switch is adjustable. It is possible that the back-up lights could come on in a gear range other than reverse. Check the service manual for adjustment procedures.

■ TURN SIGNAL SWITCH

When a turn signal switch is operating properly, as the steering wheel returns to straight ahead after a turn, the signals should cancel. Following a minor turn, such as a lane change, the turn signals will have to be manually canceled. If the switch is defective, it may not cancel the signals following a sharp turn.

Brake light problems can sometimes be traced to a defective turn signal switch. The steering wheel must be removed to service the switch. Follow instructions in the service manual for steering wheel removal. Be especially careful if there is an air bag in the steering wheel (see Chapter 66).

■ SIGNAL FLASHER

Electronic signal flashers can sense when a bulb has failed and causes the remaining good bulb and the dash indicator to flash *faster* than normal. The opposite is true with a mechanical flasher.

If a bulb burns out, there is less current traveling through a mechanical flasher and it will flash *slower* or not at all. The speed of some flashers is dependent upon the current flowing through the circuit. The use

Figure 33.31 *Standard and heavy-duty mechanical flashers. (Courtesy of Tim Gilles)*

of the wrong bulb can cause the flasher to flash too fast or too slow. Turn signals on American cars use either an 1157 or a 1034 bulb, which are interchangeable. An 1157 bulb draws more current than a 1034, however, and if it is used in place of a 1034, the flasher will flash faster. If a 1034 bulb is used in place of an 1157 bulb, it will flash slower.

There are different flashers available to match the load of the bulbs in the circuit (**Figure 33.31**). For instance, if a trailer is being towed and its lights are wired into the turn signal circuit, the extra current of the trailer lights will cause the flasher to flash too fast. A heavy-duty (variable load) flasher will solve the problem. If the heavy-duty flasher is used when the trailer is disconnected, the signals would normally flash more slowly. But a heavy-duty flasher always flashes at the same speed, no matter what the load (even if a bulb burns out). The disadvantage to a heavy-duty mechanical signal flasher is that it does not flash slower to alert the driver to a burned-out light.

To check the filaments of all of the turn signal bulbs and the brake lights, turn on the hazard flashers and walk around the car.

■ LOCATING A SIGNAL FLASHER

A flasher can be located in the fuse panel (see Figure 32.3), under the dash in a wiring loom, in a glove box, under the hood, or somewhere else. Most flashers are located under the dashboard on the driver's side. When working properly, a flasher is easy to locate because it makes a ticking sound. When it is not functioning, a service manual may be needed to help locate it. Lighting company catalogs generally have pages listing the locations of flashers.

SHOP TIP When there are two flashers and you do not know which is which, turn on the hazard flashers and the turn signal lever and see which one goes out when you pull one of the flashers.

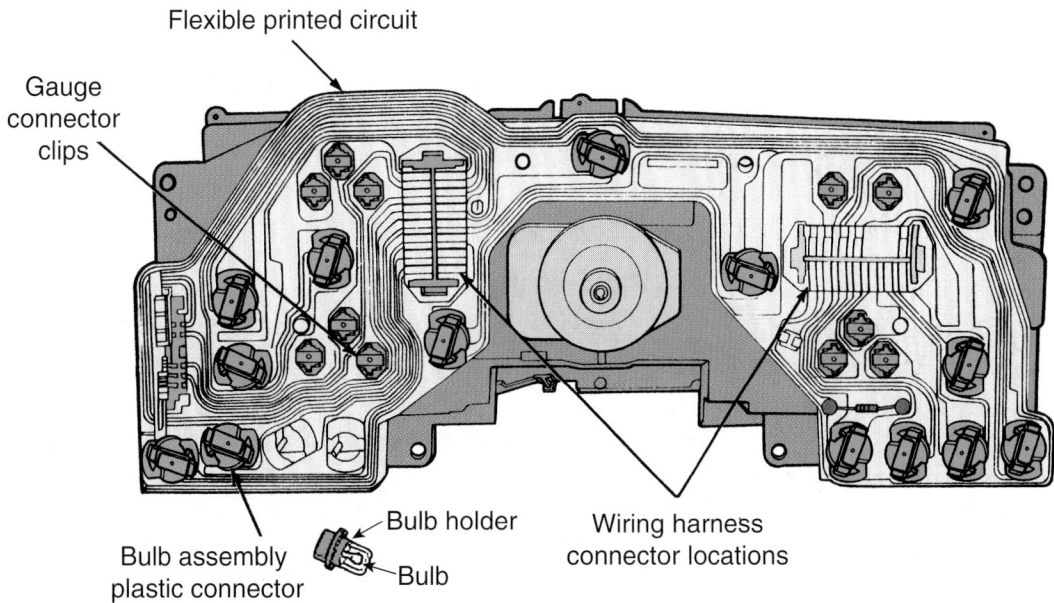

Flexible printed circuit

Gauge connector clips

Bulb holder

Bulb assembly plastic connector

Bulb

Wiring harness connector locations

Figure 33.32 Typical printed circuit instrument panel.

■ DASH LIGHT BULBS

Dash lights are usually very small bulbs that resemble small photographic flash bulbs. In printed circuit dashboards, they are housed in a plastic connector (**Figure 33.32**). Turning the connector ¼ turn counterclockwise removes it from the dash. The bulb is removed by pulling it straight out of the connector.

■ WINDSHIELD WASHER SERVICE

The primary reason for windshield washer problems is restrictions in the washer nozzles. This happens when people add fluid other than windshield washer fluid or purified water. To determine if the problem is due to the pump or a restriction, pull one of the hoses off a nozzle and operate the washer. If fluid is pumped from the end of the hose, clean the nozzle with a pin or replace it. If no fluid is pumped, check the fuse and determine if there is power to the pump. If the problem is with the pump, replace it (**Figure 33.33**).

■ HORN SERVICE

The horn pitch can be adjusted by changing the spring tension on the armature. Turning a screw on the horn housing changes the rate of vibration in the horn.

Coverage of the service and repair of other electrical accessories is found in Chapter 34.

■ GAUGE TESTING

When a gauge does not operate, first check its fuse. If the fuse is intact but there is no power, check the wiring diagram and see if there are other gauges on the

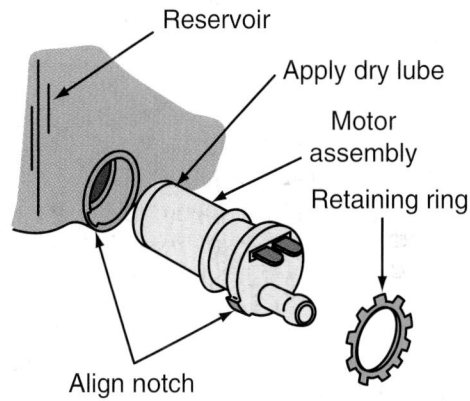

Reservoir

Apply dry lube

Motor assembly

Retaining ring

Align notch

Figure 33.33 A windshield washer pump in the bottom of the reservoir.

circuit that are not operating either. If so, work your way toward the battery, checking for voltage at each point in the circuit that is common to the problem gauge.

Piezoresistive Oil Pressure Sending Unit Test. Use an ohmmeter to check resistance between the terminal of the sending unit and ground. If specifications do not match, install a pressure tester and verify that the engine is producing oil pressure.

Coolant Temperature Sending Unit Test. Measure the resistance between the sending unit terminal and ground using an ohmmeter. The sending unit is a variable resistor, so the resistance values should change with changes in coolant temperature.

REVIEW QUESTIONS

1. What is the name of the road map that shows all of the car's wiring and loads?

2. Besides color coding, striping, or hash marks, how might a wire be marked from the factory?

3. What is the name of the type of splice that can be used to join one wire to another without having to strip the first wire?

4. What kind of wire must be joined with a connector because it cannot be soldered?

5. What is the name of the material used to clean wire prior to soldering?

6. What two metals is solder made of?

7. What is it called when a wire is coated with solder before joining it to a connection?

8. What is the name of the vinyl material that is used with a heat gun to insulate an electrical connection?

9. What can be used to locate the exact location of a ground when a wire has been accidentally pinched?

10. What kind of fuse has a small hole that exposes its fuse metal to the outside of the fuse so it can be probed with a test light?

11. For which kind of fuse should you wear eye protection when removing?

12. What happens to the insulation on a fuse link when it fails?

13. What is nonconductive grease called?

14. Where is the target located that the mirror on a headlight aimer "sees"?

15. What do you do with the two wires on a stoplight switch to make the rear brake lights illuminate?

ASE-STYLE REVIEW QUESTIONS

1. What kind of solder can be used for electrical connections?
   - **a.** Acid core solder
   - **b.** Rosin core solder
   - **c.** Both A and B
   - **d.** Neither A nor B

2. Technician A says that a wire of double the length has twice the resistance. Technician B says that a wire that is too small for a circuit can cause excessive resistance. Who is right?
   - **a.** Technician A
   - **b.** Technician B
   - **c.** Both A and B
   - **d.** Neither A nor B

3. Which of the following is/are true about headlights?
   - **a.** Headlight aim must be readjusted after replacing a headlight.
   - **b.** Some halogen headlamps can be ruined if they are handled by bare hands.
   - **c.** Both A and B.
   - **d.** Neither A nor B.

4. Technician A says that a burned transmission bushing and drive shaft yoke could be caused by a broken ground strap. Technician B says that burned emergency brake cables could be caused by a broken ground strap. Who is right?
   - **a.** Technician A
   - **b.** Technician B
   - **c.** Both A and B
   - **d.** Neither A nor B

5. Which of the following is/are true about fuses?
   - **a.** A 12-volt fuse will also work in a 110-volt application.
   - **b.** Voltage does not cause a fuse to blow.
   - **c.** Both A and B.
   - **d.** Neither A nor B.

6. Technician A says that installing a fuse with too high an amp rating will result in a burned fuse. Technician B says that installing a fuse with too low an amp rating can cause a fire. Who is right?
   - **a.** Technician A
   - **b.** Technician B
   - **c.** Both A and B
   - **d.** Neither A nor B

7. Which of the following is/are true about ohmmeter use?
   - **a.** Use an ohmmeter on a circuit that is powered.
   - **b.** Remove and test a questionable fuse with an ohmmeter.
   - **c.** Both A and B.
   - **d.** Neither A nor B.

8. Excessive _____ shortens bulb life.
   - **a.** Voltage
   - **b.** Amperage
   - **c.** Resistance
   - **d.** Amplitude

9. Two technicians are using ohmmeters to check a light bulb whose ground is its case. Technician A says that if the reading between the case and the terminal is infinity, the bulb is good. Technician B says that a high reading means the bulb is good. Who is right?

a. Technician A   **c.** Both A and B

b. Technician B   **d.** Neither A nor B

10. Each of the following statements about signal flashers is true *except:*

a. Using the wrong tail lamp can cause a mechanical flasher to flash at a different speed.

b. A failed bulb can cause an electronic flasher to flash faster than normal.

c. A failed bulb can cause a mechanical flasher to flash faster than normal.

d. A heavy-duty mechanical signal flasher with a failed bulb will flash at normal speed.

CHAPTER 34

Safety, Security, Comfort Systems, and Electrical Accessories

■ KEY TERMS

active restraint
passive restraint
inertia wheel
air bag module
safing sensors
squibs

multistage ABS
OSS
shorting bar
resistance key
transponder key
passive alarm system

active alarm system
amplitude modulation
frequency modulation
crossover

■ INTRODUCTION

Each year brings more advanced features to automobiles. What was only found on expensive luxury vehicles a few short years ago is now commonplace on many cars. Safety, security, and navigation systems are covered in this chapter, as well as other systems like sound systems, which are common features on today's vehicles. Some comfort systems are added on to a new vehicle, while others are provided as original equipment. **Figure 34.1** shows some typical safety and comfort enhancements on a modern car.

■ SUPPLEMENTAL RESTRAINT SYSTEMS

It is estimated that somewhere in the world, a person dies in an automobile accident every second. During a collision, the force of a 2-ton vehicle must come to a stop very fast. The force on the vehicle is great because its momentum changes instantly, but the passenger's momentum has not changed—yet. Seat belts and air bags are examples of supplemental restraint systems. They are designed to stop the passenger's momentum with as little damage as possible. This is a huge objective. The driver and passengers are located very close to the steering wheel and dash, and a fraction of a second is all that is available to accomplish the task. That small space in time is enough of a lifesaving opportunity to allow the system to slow the passenger, instead of allowing an instant halt to his or her motion.

Automobile manufacturers realize that safety ratings are very important to the public. Their vehicles are designed to be able to absorb the force of an accident. Intensive crash tests are performed on vehicles from the front, side, and rear.

One in four accidents is a side impact, in which the passengers are especially vulnerable. Side impact

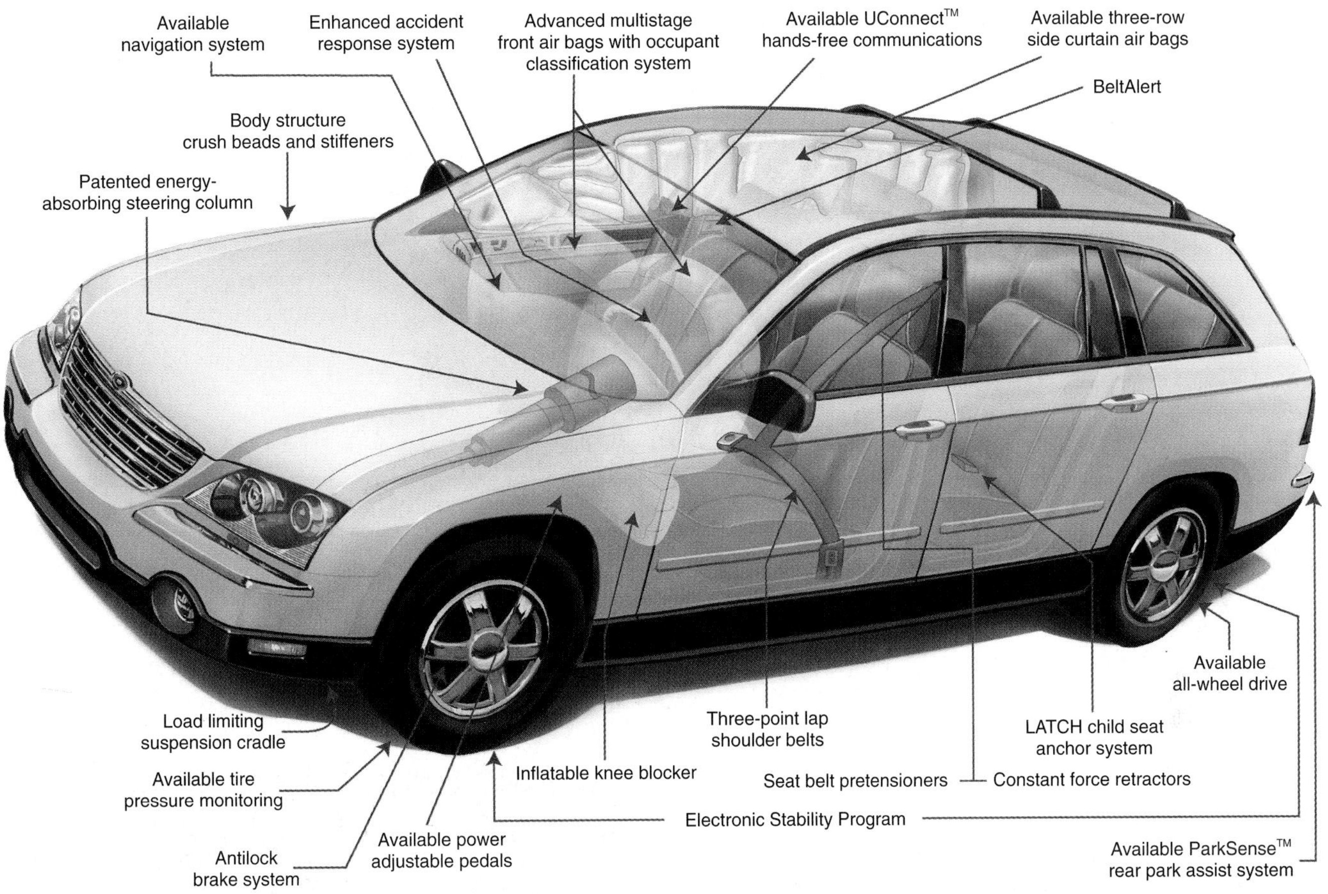

Available navigation system

Enhanced accident response system

Advanced multistage front air bags with occupant classification system

Available UConnect™ hands-free communications

Available three-row side curtain air bags

BeltAlert

Body structure crush beads and stiffeners

Patented energy-absorbing steering column

Load limiting suspension cradle

Available tire pressure monitoring

Antilock brake system

Available power adjustable pedals

Inflatable knee blocker

Three-point lap shoulder belts

Seat belt pretensioners

Constant force retractors

Electronic Stability Program

LATCH child seat anchor system

Available all-wheel drive

Available ParkSense™ rear park assist system

Figure 34.1 Modern vehicles have many safety and comfort enhancements. *(Courtesy of DaimlerChrysler)*

beams are installed during manufacturing. They are designed to minimize injury to the passenger from collisions from the side. Additionally, pillars are installed to protect the passengers in case of a rollover.

■ ACTIVE AND PASSIVE RESTRAINTS

Passenger protection includes active and passive restraints:

- An **active restraint** is one that the passenger must activate. An example of an active restraint is a manually buckled seat belt.
- A **passive restraint** is something that takes place automatically to protect the occupant of a vehicle. Automatic seat belts or air bag systems are examples of passive restraints (**Figure 34.2**).

Seat Belts

The first seat belt in an American car was installed on a 1956 Ford. By 1964, seat belts were available as standard equipment. In 1967, they were mandatory equipment. Today seat belts and air bags are required on all cars and light trucks.

A seat belt is often a combination of a lap belt and a shoulder belt. This is called a three-point belt. The belt that goes across a person's lap is called a lap belt. A shoulder belt is the part of the belt that extends

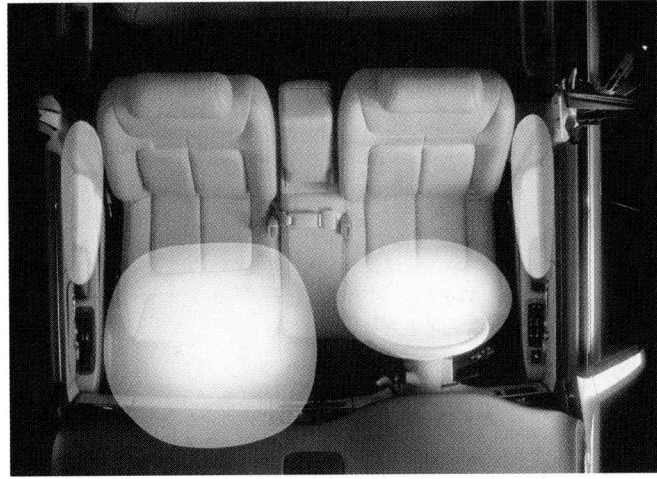

Figure 34.2 An example of passive restraint: an air bag system as seen from above. *(Courtesy of Delphi)*

across a person's shoulders and chest. One end of the belt has a buckle, and the other end is attached to an anchor that attaches it to the floor or seat (**Figure 34.3**). The buckle must be able to withstand a 2-ton load without releasing.

Five-point safety harnesses are used in children's seats and in racing (**Figure 34.4**). The lap part of

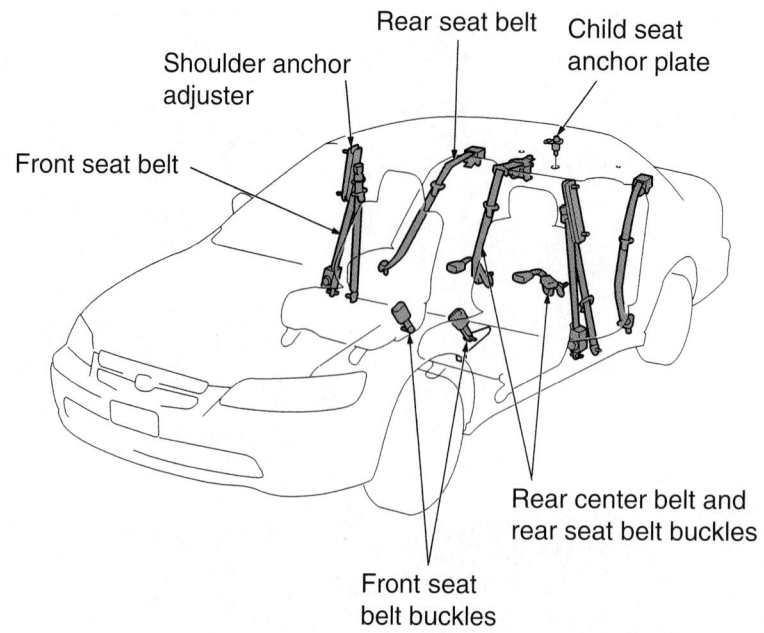

Figure 34.3 Parts of a seat belt system. *(Courtesy of Nissan North America, Inc.)*

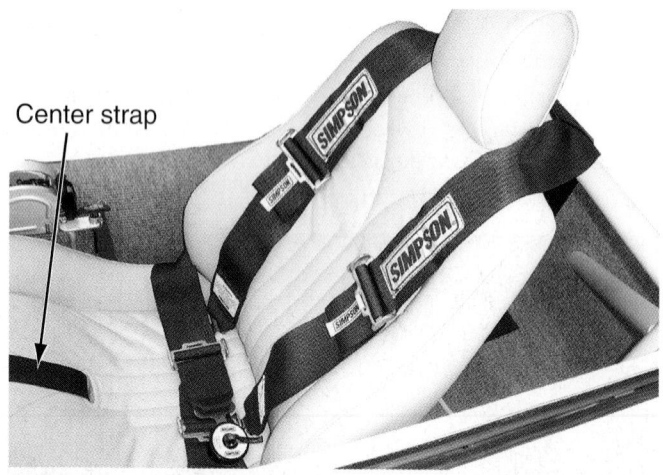

Figure 34.4 A five-point harness used for racing has a center strap. *(Courtesy of Tim Gilles)*

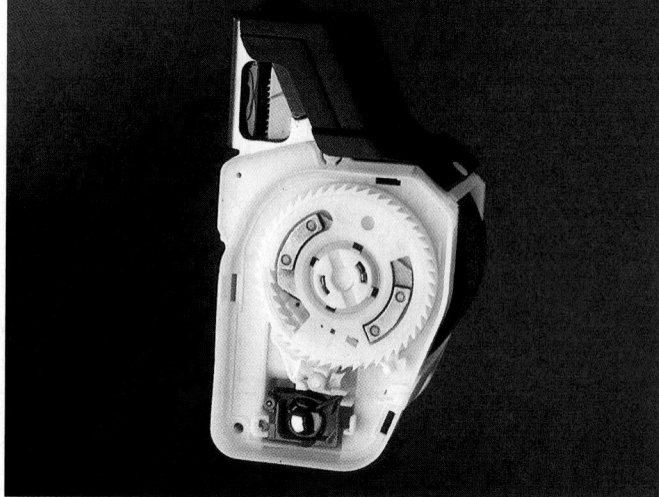

Figure 34.5 A seat belt retractor. *(Courtesy of Autoliv, Inc.)*

the belt is connected to a harness between the legs. Six point harnesses used in racing have two leg harnesses because this is the area most likely to fail in a high-speed collision.

Current seat belt use is estimated at only 70%.

Early seat belts were sometimes difficult to buckle and/or adjust. In 1968, Volvo introduced the seat belt **inertia wheel**. It allowed the belt to be more comfortable for the passengers and eliminated the need to adjust the belt for each passenger. Today's modern vehicles have belt pretensioners or retractors (**Figure 34.5**). Late-model, smart seat belts control the slack in seat belts during a collision. These are covered later in this chapter.

▪ HISTORY NOTE ▪

The three-point safety belt was invented by Nils Bohlin and patented by the Swedish auto maker Volvo in 1959. Volvo was a pioneer in the field of automotive safety. By 1963, all Volvos came equipped with three-point belts. Three years later, in 1966, Volvo released a report called "the 28,000 accident report" showing that thousands of lives had been saved already by three-point belts. This report proved to be the basis for the 1977 legislation requiring the installation of passive seat belts in all cars sold in the United States by the 1984 model year.

There are three different styles of pretensioners:

- Mechanical pretensioners use an inertia wheel, a pendulum-like device that locks during sudden deceleration (**Figure 34.6**).
- Electric pretensioners are activated by a sensor. They can be a part of the air bag circuitry.
- Pyrotechnic tensioners work with air bags. They are also activated by a sensor. An explosive charge removes slack and locks the belt. They must be replaced once they have been used. These devices are covered later in this chapter.

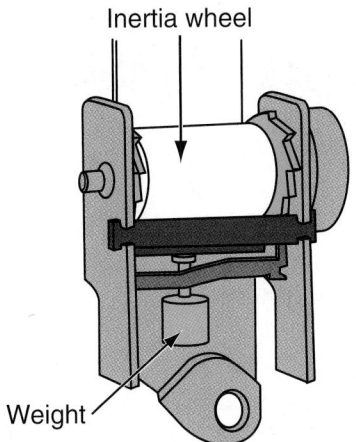

Figure 34.6 An inertia wheel locks during sudden deceleration.

Seat belt systems also include a sensor and a warning system consisting of a dash light with a bell or buzzer. With a passive system, the warning light is illuminated for a few seconds. If the belt is not buckled, an audible warning sounds. The audible system can also sound when a door is open.

In a passive seat belt system, electric motors operated by the ignition switch move shoulder belts across the driver and front seat passenger (**Figure 34.7**). The shoulder and lap belts are separate, so there are two retractors.

Air Bags

Air bag systems are called air bags, supplemental inflatable restraints (SIR), supplemental air restraints (SAR), or supplemental restraint systems (SRS). They have been required on all new cars sold in the United States since 1996. It is estimated that front air bags reduce head-on fatalities by 25% when drivers are wearing seat belts and 30% when they are not. When the seat belt is combined with air bags, serious head and chest injuries are reduced by 65% to 75%.

An air bag is a flexible nylon bag that inflates almost instantaneously in the event of a serious collision. The inflating bag protects the passenger from a serious upper body or head injury. The air bag inflates at about 200 mph. The entire inflation and deflation cycle occurs in less than one-half second. The average speed of all manufacturers' air bag deployment is 33 mS.

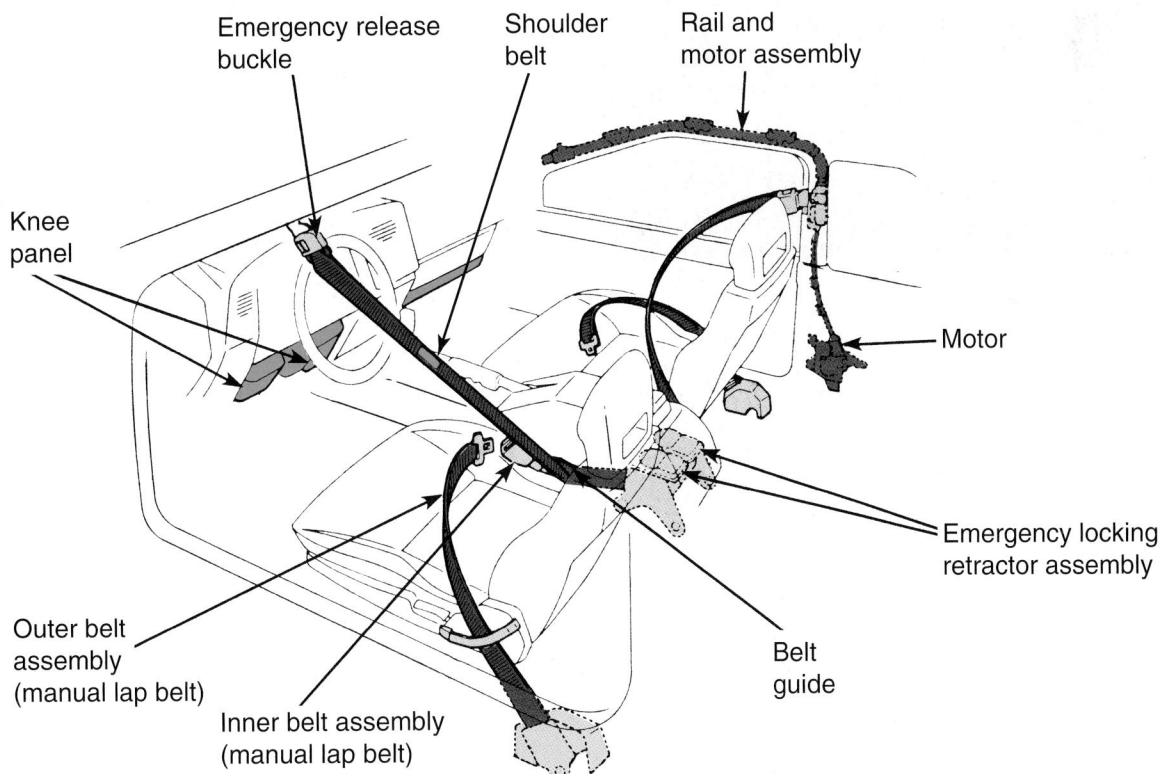

Figure 34.7 A passive seat belt system. *(Courtesy of Ford Motor Company)*

In the early air bag systems, the force of a collision sufficient to deploy the air bag needed to be equivalent to that of a vehicle running into a wall at approximately 14 miles per hour (mph). Hitting a solid wall at this speed is equivalent to a moving vehicle colliding with a stationary vehicle at about 28 mph. This threshold prevented accidental deployment due to hitting a curb or during a panic stop. With the car parked and the ignition off, deployment was unlikely because there was no power to the circuits to deploy the air bag. Newer systems have far more sophisticated deployment criteria; they are discussed later in the chapter.

Driver-side air bags are mounted in the steering wheel. The air bag is folded with talcum powder or cornstarch in a cover that is designed to break open when the bag is deployed (**Figure 34.8**). The steering wheel cover in a car with an air bag usually is labeled (**Figure 34.9**) with SRS or SIR.

An early, simple air bag system included remote mechanical impact sensors with an additional sensor in the control module used to deploy a single driver-side

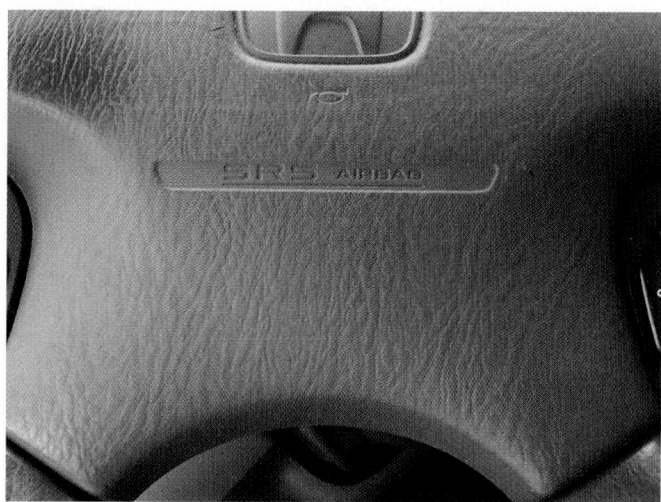

Figure 34.9 A driver-side air bag mounted in the center of the steering wheel. *(Courtesy of Tim Gilles)*

air bag (**Figure 34.10**). These systems used a pyrotechnic device for inflation like today's electronic systems. Today's systems, however, have become considerably more sophisticated and complicated.

Following are some of the additional types of air bags available:

- Passenger-side air bags
- Side air bags in seat backs
- Side curtain air bags above the headliner
- Knee air bags
- Rear seat air bags

These are all used with seat belt pretensioners, which remove the slack from seat belts.

Passenger Air Bags. A passenger-side air bag is located in the top of the dash on the passenger side. It is contained under a small door in the instrument panel (**Figure 34.11**). The windshield is an integral part of the passenger-side air bag system, and it must be correctly glued in place. In some passenger air bag systems, the air bag deployment bounces off the windshield and is very violent. A passenger-side air bag deployment will often break the windshield. When an air bag inflates and the windows are all up, doors can be forced open and damage to the vehicle body can occur due to the increase in passenger compartment pressure.

Most often both front air bags are deployed together. It is rare that only one goes off. For that reason, the undeployed bag is often replaced as well.

Side Air Bags. The side area of the vehicle is very close to the driver. One-quarter of all injuries from automobile accidents are from side impacts. Yet these accidents account for over a third of the serious injuries and fatalities. At least half of the injuries resulting from side impacts are head injuries. The United States Department of Transportation estimates that one-quarter of all automobile-related deaths could be prevented by side air bags designed to protect the head.

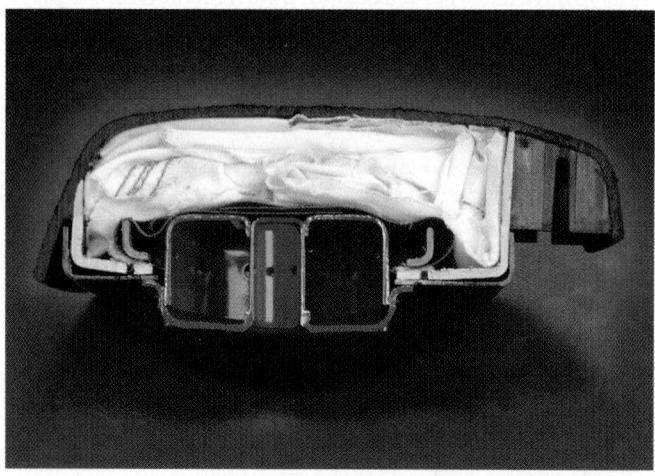

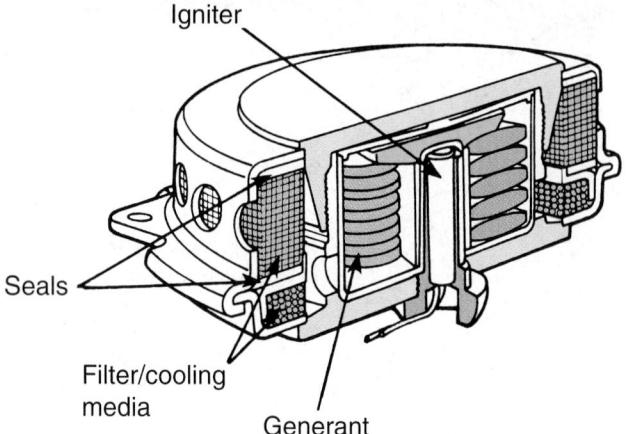

Figure 34.8 (top) Cutaway of a driver-side air bag and inflator; (bottom) An igniter assembly is beneath the bag. *(top, Courtesy of Breed Technologies)*

Igniter

Seals

Filter/cooling media

Generant

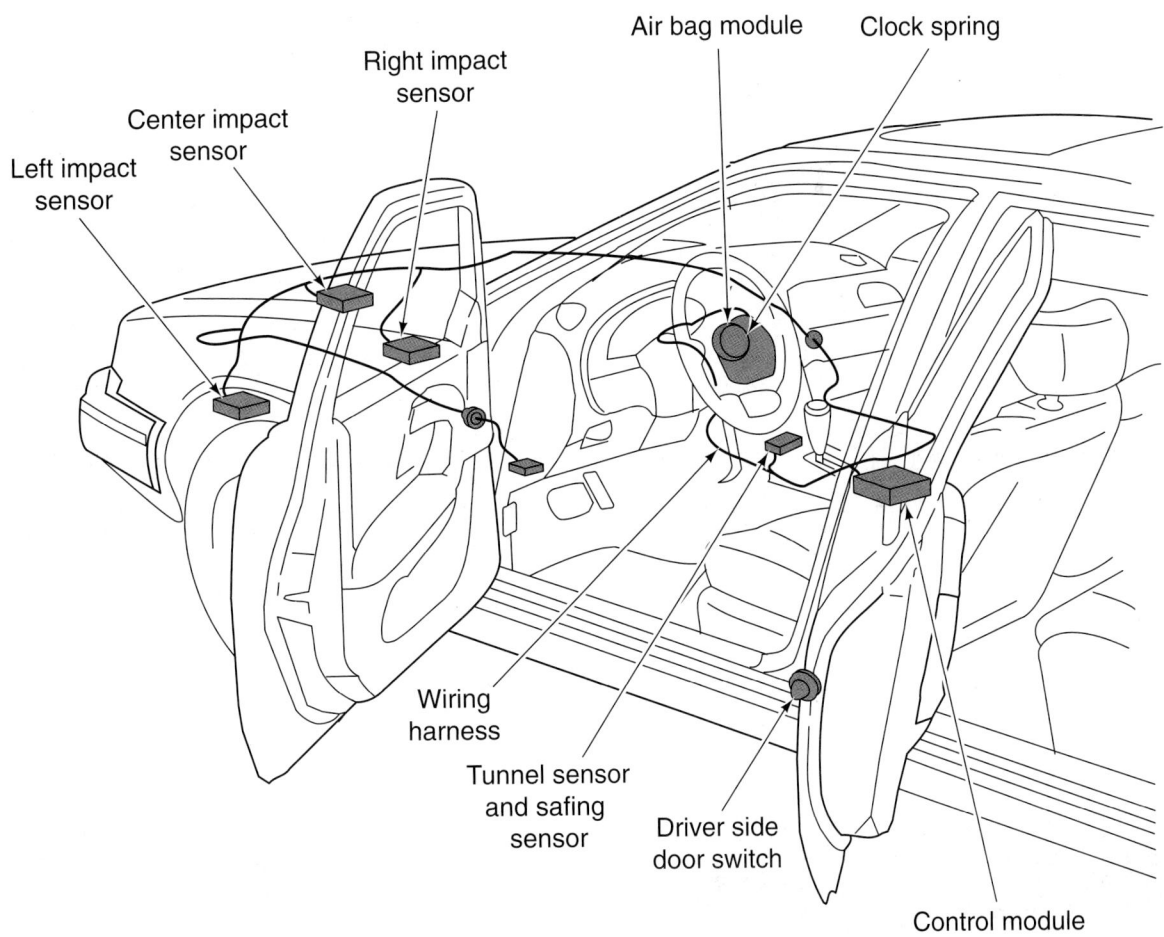

Figure 34.10 An earlier, simpler air bag system.

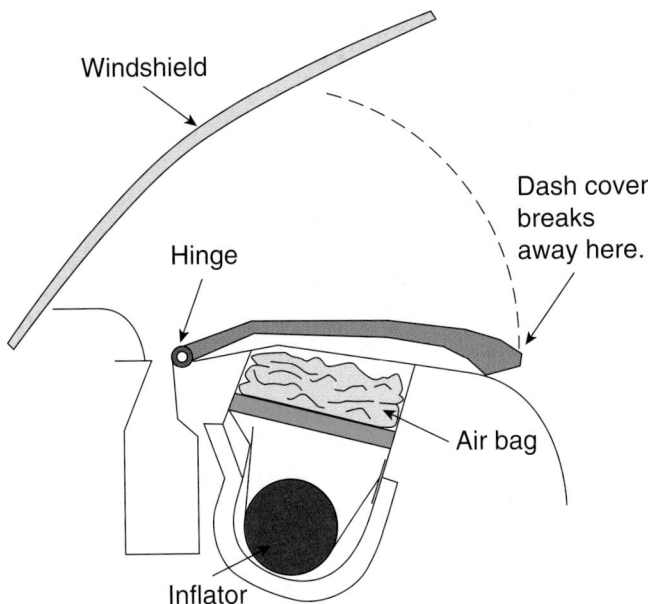

Figure 34.11 A passenger-side air bag.

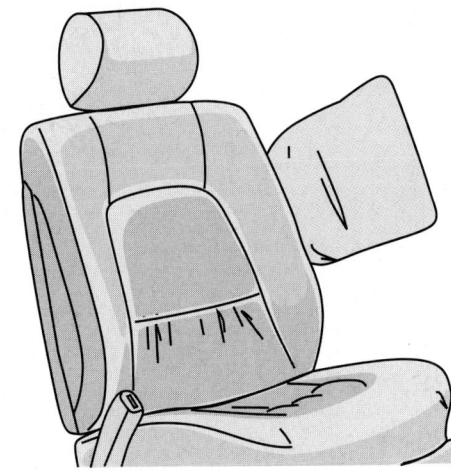

Figure 34.12 A side air bag in the seat.

less than 10 mS. Side curtain air bags are discussed later.

Seat Belt Pretensioners. Seat belt pretensioners are used with the air bag system. They are located on the seat belt buckles. A small explosive charge is designed to take up to 6" of slack out of the lap and shoulder harnesses during a front collision (**Figure 34.13**). They are usually deployed at the same time as the driver- and

Side air bags have inflator modules in either the front door or front seat (**Figure 34.12**) on either side of the vehicle. There is very little time to inflate the bag in a side collision. The bag must inflate in

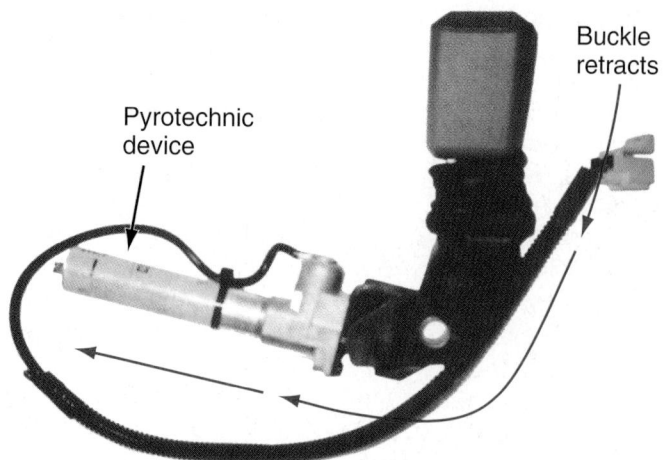

Figure 34.13 This seat belt tensioner has an explosive charge that pulls the belt tighter when the air bag is deployed. *(Courtesy of Tim Gilles)*

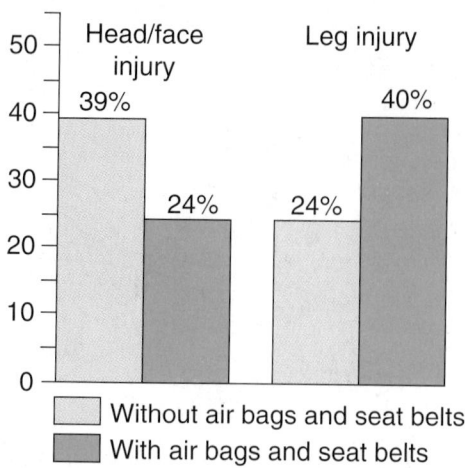

Figure 34.15 Leg injuries account for a majority of injuries where an air bag is deployed. *(Courtesy of Autoliv, Inc.)*

passenger-side air bags. The belt tensioner can only deploy if the seat belt is buckled.

When a seat belt tensioner deploys, the system has more time to inflate the air bag. Modern systems can inflate the bag in stages, depending on the severity

Figure 34.14 A knee air bag. *(Courtesy of Delphi)*

of the crash. More time allows for gentler inflation, which results in fewer injuries caused by the air bag.

Knee Protection Systems. The part of the dash designed to protect the driver's and passenger's knees is called a knee bolster. Some air bags are for the knees (**Figure 34.14**). General Motors' Delphi Automotive calls theirs an active knee bolster. When a passenger wearing a seatbelt survives an accident during which an air bag deploys, leg injuries account for most of the injuries (**Figure 34.15**). Sixty percent of the leg injuries are below the knee.

The knee air bag reduces the risk of the person sliding free under the seat belt during an accident. It also keeps the person in a better position for front air bag protection.

Rear-Seat Air Bags. Rear-seat air bags are relatively uncommon. They are installed in the rear cushion of the front seats in some luxury automobiles to protect passengers in the rear in the event of a front impact.

Air Bag System Parts

A modern air bag system has an electronic control unit (ECU). The air bag and its inflator are called an **air bag module**. There are two modules if there is a passenger-side air bag. The air bag system also includes sensors, an ignition device, explosive propellant, and other parts.

An air bag warning lamp is located in the instrument panel. It lights when a problem is detected in the system.

A clock spring makes the connection between the steering wheel and column (**Figure 34.16**).

A control module is the computer that monitors the system. It also has an electrical storage capacitor as a safety backup in the event that the vehicle's battery is damaged in the collision.

The earliest air bags, called first-generation air bags, were first used on vehicles in the late 1980s. GM's

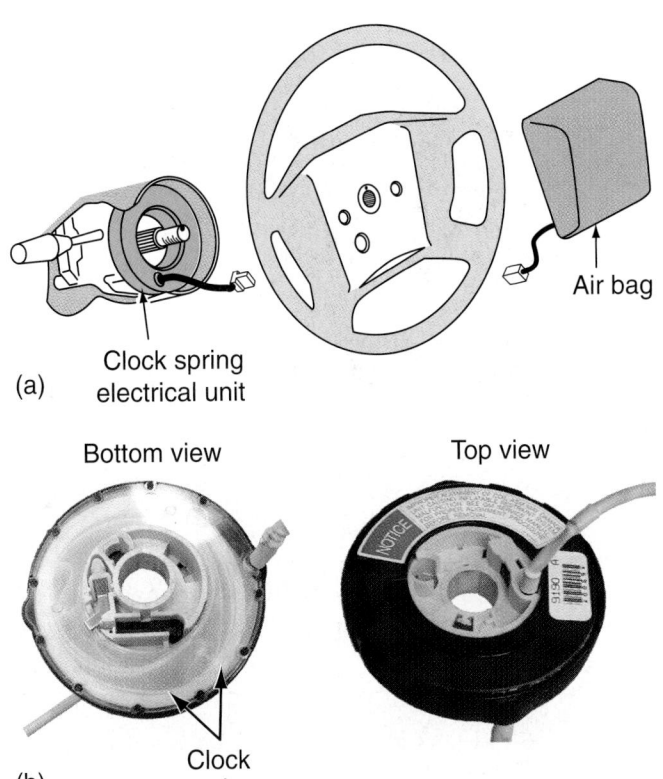

Figure 34.16 (a) A clock spring allows electrical continuity when the steering wheel turns. (b) Bottom and top views of a clock spring. (b, Courtesy of Tim Gilles)

computer was called a diagnostic energy reserve module (DERM). The DERM was not required for the deployment of the air bag. Arming and discriminating sensors completed the electrical path for the deployment. Early and late systems have an electrical energy reserve in case the battery becomes disconnected during a crash.

Later air bag systems, called second generation, have an sensing diagnostic module (SDM) that controls multiple-stage front air bag deployment. These advanced systems, which first appeared in 1994, include side air bags, electronic frontal sensors, and occupant detection and position sensors. The SDM also has energy reserve in case the battery is disconnected, but unlike the first-generation system, the SDM completes the electrical circuit and deploys the bag(s). A dual-stage system has twin deployment circuits, so it has a special clock spring.

Collision Sensors. The first-generation systems are covered here first. These systems used electromechanical crash sensors. Later systems have taken advantage of advances in microelectromechanical technology. Most use a single, very sophisticated electronic sensor. These systems are covered later.

The amount of physical damage to the vehicle is not what determines whether or not an air bag will be deployed. In first-generation systems, there are usually three or more crash sensors that work with the control module to determine the difference between a crash and a noncrash by measuring the severity of the impact. Sensors are either on the outside or the inside of the vehicle, depending on their function. Front impact sensors, sometimes called discriminating sensors, are usually found in the engine compartment near the radiator.

Inside sensors are also called **safing**, or *arming*, **sensors**. Their function is to determine that there has actually been a crash. Safing sensors are most often located in the center console. They can also be located in different locations in the passenger compartment, depending on the manufacturer. Sometimes they are part of the control module.

During sudden deceleration of the vehicle, a sensor's contacts close electrically to send a signal to the control module. The module must receive a signal from the rear (safing) sensor first.

NOTE: *At least two sensors must say there is a crash before an air bag will be deployed. The safing sensor is on the power side of the circuit. One of the front discriminating sensors must provide the ground to the squib or no deployment will take place.*

In first-generation systems, most impact sensors operate when a sufficient impact occurs within a 60-degree window, 30 degrees to either side of the vehicle centerline (**Figure 34.17**). When sensing is only in this one direction, the system is called single-point sensing. Some systems have remote sensors in other vehicle locations and are called multipoint sensing. Second-generation systems can sense impacts at 90 degrees to either side of the vehicle (180 degrees).

Sensor operation is similar. The location of a sensor determines what it is called. A sensor can be electronic or mechanical. One type of mechanical sensor is the ball-and-magnet type (**Figure 34.18**). A gold-plated ball is the sensing mass, which is normally held in place by a magnet. The force of a crash breaks the ball away from the magnet, where it completes an electrical

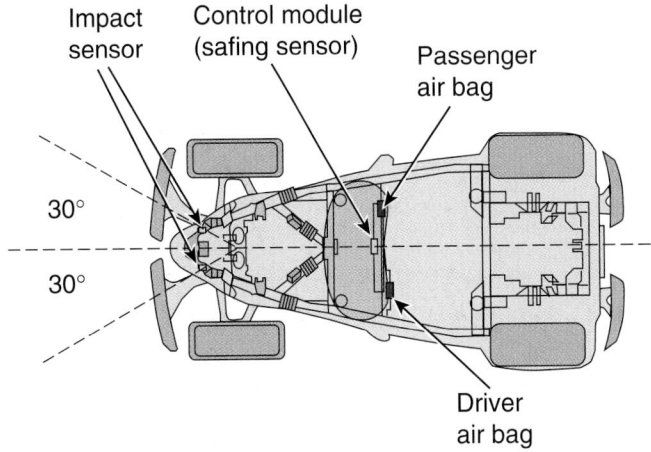

Figure 34.17 A modern air bag system. (Courtesy of DaimlerChrysler Corporation)

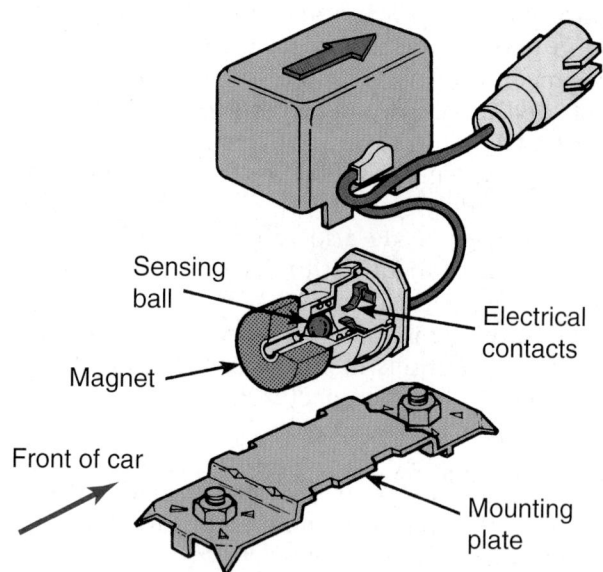

Sensing ball

Magnet

Electrical contacts

Front of car

Mounting plate

Figure 34.18 A ball-and-magnet sensor. *(Courtesy of DaimlerChrysler Corporation)*

circuit between two contacts. The ball only makes electrical contact for a short time before it is drawn back to the magnet once again. The strength of the magnet and the weight of the ball determine the calibration of the sensor. The correct installation of the sensor in the vehicle is also critical to its calibration.

Another type of sensor has a stainless steel ribbon that is rolled up inside the sensor. The sensor is airtight and is filled with nitrogen gas to prevent corrosion. During an impact the ribbon unrolls and closes an electrical circuit. It rolls back when the force stops.

An electronic accelerometer, sometimes called a strain gauge, is a recent addition to automotive safety systems. This technology, called microelectromechanical systems (MEMS), advanced in the 1990s to the point where it could survive the harsh environment of the automobile. An accelerometer is an electronic sensor that can sense impact (**Figure 34.19**). The MEMS accelerometer senses a net change in forward velocity and can sense up to ±90 degrees.

SCIENCE NOTE

MEMS integrates mechanical parts (sensors/actuators) and electronics on a common silicon substrate. This miniaturization is called microfabrication technology. A human hair is 100 microns. The following are sizes of other typical MEMS devices:

- *miniature gear* — *300 microns diameter*
- *pressure sensor* — *200 microns wide*
- *resonant gyroscope accelerometer* — *100 microns wide*
- *angular rate yaw sensor* — *50 microns wide*

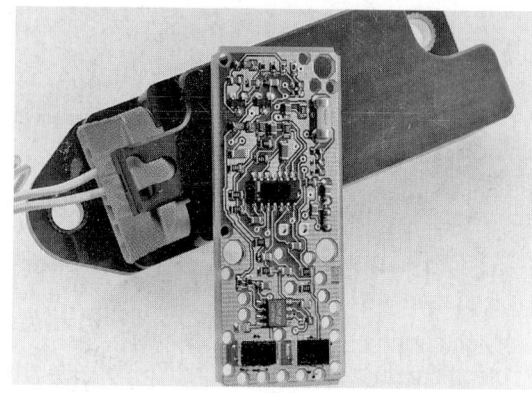

Figure 34.19 An accelerometer registers the force of a collision electronically. *(Courtesy of Siemens VDO Automotive)*

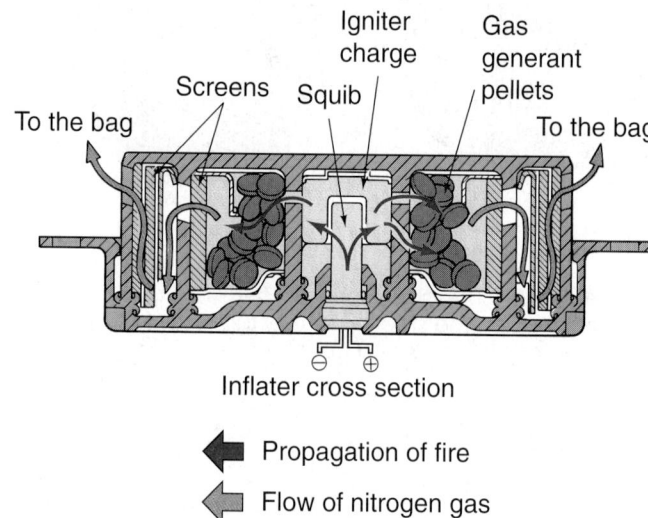

Igniter charge | Gas generant pellets | Screens | Squib | To the bag | To the bag

Inflater cross section

⬅ Propagation of fire

⬅ Flow of nitrogen gas

Figure 34.20 A squib fires the igniter.

Air Bag Electronic Control Module. The electronic control module used with today's air bags has become very sophisticated. Early modules only had to deal with the driver-side air bag. New ones must deal with passenger-side air bags and seat belt pretensioners as well. The module can be mounted in the steering wheel only if there is a driver-side air bag.

Squibs. An air bag control module receives input from the impact sensor and matches deployment to the conditions of impact. One or more fuse-like ignition devices called **squibs** are used in each air bag (**Figure 34.20**). A squib can also be called an initiator or an igniter. It is the heating element that ignites the gas-generating material to inflate the bag. It has an electrical igniter and a small explosive charge.

New-Generation Air Bags

The latest air bag systems are called smart restraint systems or third-generation air bags. Smart air bag systems match the speed of deployment, the crash characteristics of the vehicle, and the size and weight of the

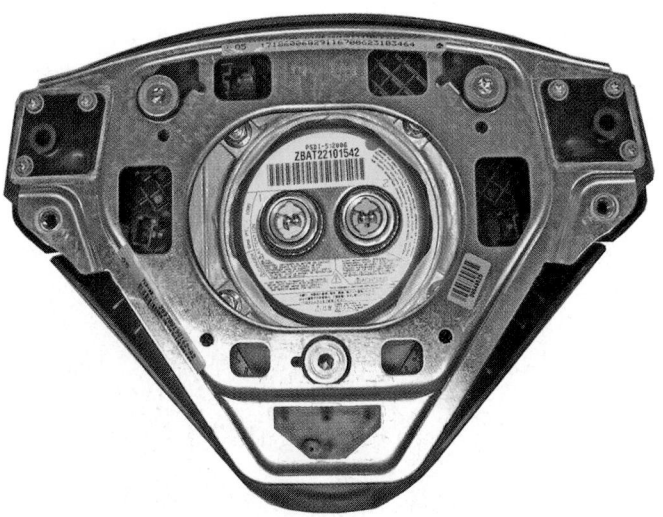

Figure 34.21 A multistage air bag with two squibs. (*Courtesy of Tim Gilles*)

Figure 34.22 Cutaway of a hybrid inflator. (*Courtesy of Autoliv, Inc.*)

passenger. Multistage systems can have approximately 18 deployment loops, with up to six side air bags and seat belt tensioners. Ultrasonic sensors monitor the front passengers for seat position and weight, the position of the driver in relation to the steering wheel, seat belt usage, and how severe a crash is. A **multistage ABS**, in which air bags can have more than one squib, can inflate an air bag in stages (**Figure 34.21**). In a less severe crash, only one of the squibs will be fired by the computer. In an intermediate collision, one squib will fire, followed by the other a few milliseconds later. In severe impacts, both squibs are fired at once.

Air Bag Deployment

The deployment process takes about 0.1 second from start to finish from the moment a crash is detected. First, the sensors must confirm that a collision has occurred. On systems with driver-side-only air bags, a current (about 2 amps) is directed to an igniter. On multiple air bag systems, the sensor signal is evaluated by the electronic control unit, which directs electrical current to the correct air bag's squib. The current heats a filament, which ignites a capsule. The heat from the capsule ignites solid pellets, which generate harmless nitrogen gas when they burn. The rapidly expanding gas inflates the air bag.

The air bag breaks its way through the steering wheel cover or the plastic or fabric dashboard on the passenger side. Following deployment, the air bag begins to deflate immediately. The harmless gas vents through holes in the back of the bag or through the fabric.

Most systems use a pyrotechnic inflator that uses sodium azide pellets (rocket fuel) to produce nitrogen gas to fill the air bag. Some newer systems use heated gas inflators (HGI). They use a nontoxic gas (hydrogen) and air, which turn to water vapor when released after the bag deploys. Other systems use high-pressure stored argon gas to help inflate the bag. These systems are more common for use in passenger-side air bags. A cross-section cutaway of a *hybrid inflator* is shown in **Figure 34.22**.

A two-stage bag has two chambers with cartridges full of argon gas compressed to 3,000 psi and a solid propellant. When the squib fires, it ignites the solid propellant in one of the chambers. The rapidly burning propellant heats the cool argon gas, which helps to inflate the bag. Firing multiple squibs results in even more expansion.

NOTE: *The air bag sensing and diagnostic modules on later vehicles record data during deployment that can be accessed by accident investigators when determining the cause of a collision. Crash event recordings can tell the speed of the vehicle, whether the brakes were applied and ABS was engaged, how far the throttle was open and the rpm of the engine, whether the seat belts were in use, and the amount of change in velocity of the vehicle during the crash.*

Side Air Bag Deployment. Side air bags (see Figure 34.12) deploy individually when there is an impact on either side of the vehicle but not during a front or rear impact or a rollover. Both side air bags cannot deploy at the same time unless the vehicle is hit from both sides. Side air bags have a single-point sensing system controlled by a side impact module. An accelerometer (see Figure 34.19) is located in the side pillar behind the front door. When an impact of sufficient magnitude is sensed, the control module supplies the squib with enough current to light it off. When the air bag deploys, it splits the seat back fabric or trim. Unlike front air bags, side air bags are unvented so they deploy three times as fast. Side curtain air bags use the same kind of sensor/controller but are located in the headliner and stretch all along the side of the car (**Figure 34.23**). They protect passengers from head injuries in a side collision.

Smart Air Bags. The smart air bag system uses various inputs, including speed at impact and passenger body

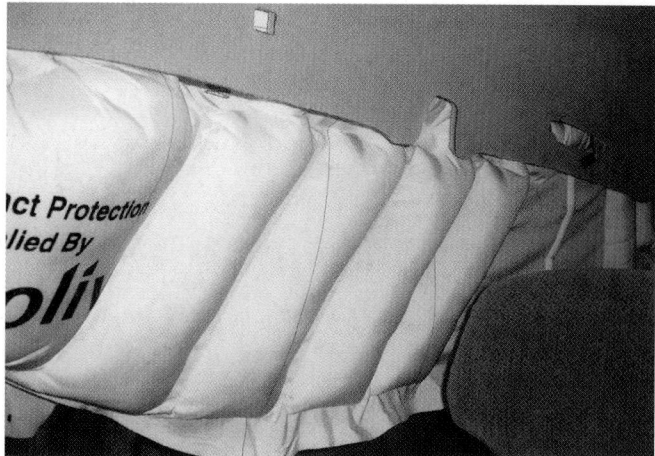

Figure 34.23 A side curtain air bag protects the driver's or passenger's head. *(Courtesy of Tim Gilles)*

weight and location, to automatically adjust the force of the air bag deployment. It can sense when to step down the force of the deployment. It can also decide that a deployment is not needed.

Occupant Sensing System. One smart air bag system feature is the occupant sensor system (**OSS**), which determines where passengers are located—leaning forward, back, or to the side. It also learns if there is a large or small passenger, or a baby seat, sitting in the front seat on the passenger side. Air bag deployment can be violent, especially for smaller passengers and infants. The Federal Motor Vehicle Safety Standard 208 has been amended to protect passengers weighing less than 100 pounds (45 kg) by using a less violent deployment based on the severity of the impact. The safety amendment also specifies that the air bag be disabled when the front seat is occupied by an infant in a rear-facing baby seat or a child weighing less than the average 6-year-old. A warning light must illuminate when the passenger seat is occupied and the air bag is disabled (**Figure 34.24**).

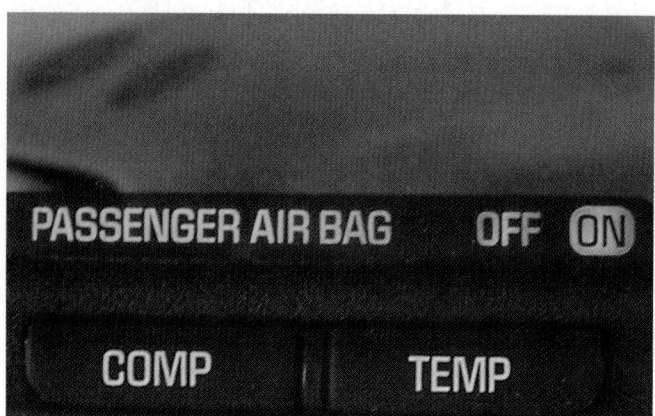

Figure 34.24 A warning light illuminates when the passenger seat is occupied and the air bag is disabled. *(Courtesy of Tim Gilles)*

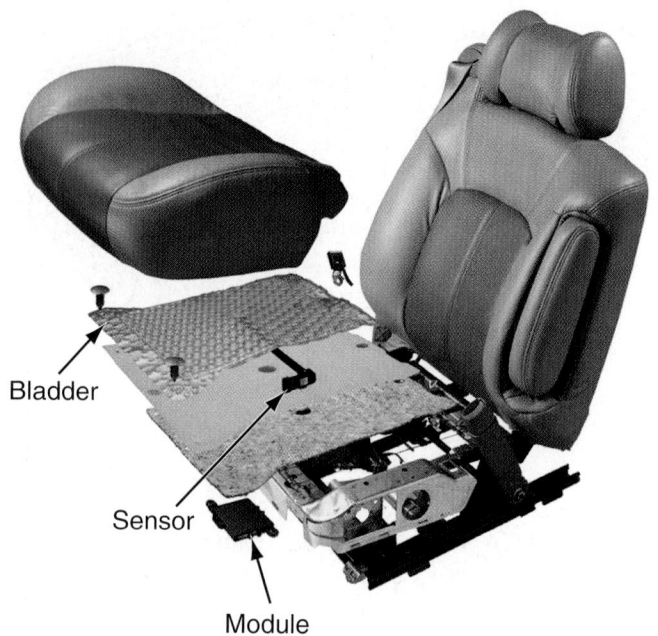

Figure 34.25 Parts of an occupant-sensing seat. *(Courtesy of Delphi)*

Occupant sensing can be done using a silicone-filled bladder under the seat foam (**Figure 34.25**). A pressure sensor is connected to the bladder by a hose. The pressure reading is analyzed by the occupant-sensing module. A belt tension sensor, located at the bottom of the seat belt, indicates how tightly the belt has been fastened because a tightly cinched belt around a baby seat can change the weight reading from the bladder and pressure sensor.

Another method of weight sensing is to use a strain gauge at each corner of the seat frame (**Figure 34.26**) where they support the seat. A circuit board bonded to each gauge has a metallic foil grid that changes resistance under strain. When the weight on it changes, the sensor's voltage output changes, providing information to the occupant-sensing module, sometimes called the occupant classification module. Driver- and passenger-side seat track position sensors are also part of some systems. This information is analyzed by the computer when it deploys the air bags.

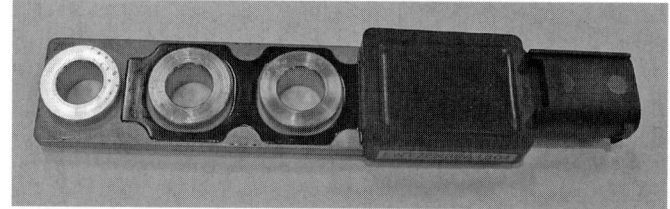

Figure 34.26 A strain gauge used at the corners of a passenger seat frame for occupant sensing.

The computer monitors the accelerometer value repeatedly during fractions of a second. In GM's system, if an average of four consecutive samples exceed a force of 1–2 g, the algorithm is enabled and determines whether a deployment should occur. This decision is made very fast (less than 20 mS). An air bag is usually fully inflated within 50 mS of impact.

Although the mathematical calculations regarding crashes are becoming more sophisticated with experience, accidents like collisions with a tree or pole are difficult for the algorithm to resolve because it provides a slower event. Imagine driving into a saw blade that cuts the car in half but does not provide the accelerometer with a "g" force that it can measure.

The newest designs have the ability to consider vehicular forces during a rollover. The sensor, called an angular rate sensor, tells how quickly the vehicle is rolling over and how many times it rolls. It includes automatic fuel cutoff and automatic battery disconnect. Current systems allow the air bag to stay inflated for up to 7 seconds. Chances of a rollover lasting longer than this are remote.

Communication between computers in the air bag system is on the vehicle network so the vehicle's satellite communication system can sense that there has been an accident and dispense emergency vehicles to the site.

Radar-based systems are able to sense that a vehicle will not be able to stop in time to avoid an accident. The computer will begin air bag deployment. This same system is used for adjustable distance cruise control and is discussed later.

■ RESTRAINT SYSTEM SERVICE

After an accident, all of the parts of the passive restraint system are checked. The seat belts are inspected for fraying and ease of operation. Pull the web all the way out of the retractor to give it a thorough inspection. Connect the ends of the buckle to be sure it clicks and releases correctly. If a problem is found, replace the belt assembly.

Check a centrifugal belt retractor by pulling it hard. Centrifugal force should make it catch. The belt should be able to move out of the retractor when pulling it slowly. A pendulum belt retractor depends on the vehicle stopping quickly to provide inertia to set the belt. During a road test, brake suddenly at a speed over 10 mph to determine if this retractor is working.

When seat belts are equipped with pretensioners, you can tell they were deployed if it is difficult buckling up the belt. A pretensioner that has been deployed must be replaced.

Air Bag Service

It is necessary to use the manufacturer's service and diagnostic information when servicing air bag systems. A typical information library will include the system's components and how they operate. Repair procedures typically provide a chart listing the procedure to follow when making a repair. Detailed wiring diagrams and component locators are also provided. Air bags are explosive devices, so safety precautions are described. Cautionary labels can be found in the engine compartment, on the back of sun visors, on the back of the air bag (**Figure 34.27**), and on the wiring harness (**Figure 34.28**). Also note that air bag connectors are colored cautionary yellow.

Diagnostic trouble codes can be used to determine the cause of active codes, which are those that illuminate the dash light. They can also tell you which squibs

Sun visor label

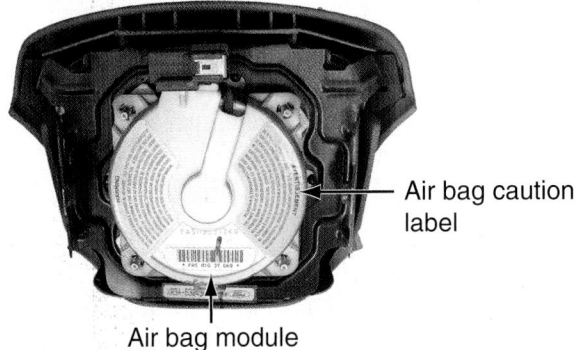

Air bag caution label

Air bag module

Figure 34.27 Air bag caution labels from the back of a sun visor and on the back of an air bag module. (*Courtesy of Tim Gilles*)

Figure 34.28 Air bag wiring harness connectors are colored cautionary yellow. (*Courtesy of Tim Gilles*)

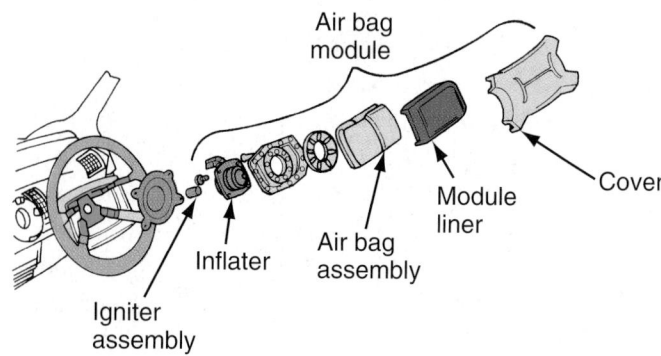

Figure 34.29 Parts of a typical driver-side air bag assembly. (*Courtesy of DaimlerChrysler Corporation*)

in a multistage system have fired. Stored codes can help you determine the cause of an intermittent problem.

Newer systems have special tools that can be substituted for parts of the system to simulate a resistance during testing. A problem can sometimes be isolated by substituting a resistance that is known to be good into the circuit. Be certain that all connectors are in good condition and that none of them are bent.

One test is to follow the manufacturer's instructions and remove the air bag. The load tool is installed in place of the air bag. If the problem exists when the battery is reconnected and the ignition key is turned on, it is not because of the air bag module. The battery should be disconnected once again and the clock spring disconnected. The load tester is installed in place of the clock spring while the battery is reconnected and the key is energized. If the code is not active any longer, the problem has been isolated to the clock spring.

Many manufacturers have a safety device called a **shorting bar** that automatically shorts the circuit when an air bag module connector is disconnected. A problem can occur when the connector is reconnected if the shorting bar does not release as designed. The air bag will remain short-circuited, and the diagnostic module will set a code and illuminate the air bag light.

When a new computer is installed in a vehicle, it must be programmed with the correct VIN or the system will not operate. The VIN tells the system what type of system the vehicle has, such as whether there are side air bags, for instance.

Replacing an Air Bag

Air bags are expensive and cannot be reused once they have been deployed. Many manufacturers stipulate additional parts to be replaced when an air bag has deployed. If all of the vehicle's air bags have been deployed, this can be enough to cause an insurance company to declare a vehicle to be a total loss.

When one air bag deploys but the other does not, both air bags are often replaced because the bag that did not go off could be defective. Some shops explode the replaced undeployed air bag inside a cage.

An air bag explosion is a violent event. A driver who survives a crash during which an air bag has been deployed often has burn marks on his or her arms from holding the steering wheel when the bag went off. The surface of a deployed bag will be covered with powder, which is mostly cornstarch or talc, and by-products of the chemical reaction. One of the by-products is sodium hydroxide, or lye. This caustic mixture reacts very quickly with the air to form sodium carbonate and sodium bicarbonate (baking soda). Although it is unlikely that the dust on a deployed air bag is caustic, use gloves and goggles to protect yourself just in case.

Figure 34.29 shows the parts of a typical driver-side air bag assembly. Details on air bag care, removal, and disarming are given in Chapter 66. That chapter deals with steering wheel service. Because all new vehicles are now equipped with air bags, it is necessary to present that material in that chapter.

Air Bag Sensor Service

Air bag sensors are sensitive to damage from mishandling. Do not use an impact wrench around a sensor, and be careful not to drop it. When installing a sensor, it must be positioned so that the arrow on its housing is aimed to the front of the car.

Use caution when adjusting a door strike plate. Hitting the pillar with a hammer while the key is on could deploy a side air bag.

■ SECURITY, NAVIGATION, AND ELECTRICAL ACCESSORIES

Automobile theft is a serious problem. Security systems have been on automobiles for many years in an effort to make automobile theft more difficult. Antitheft devices include locking systems, alarms, and disabling systems.

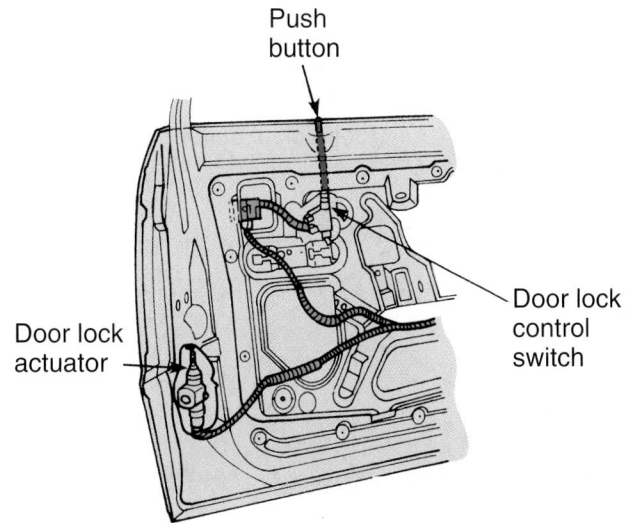

Figure 34.30 An electric door lock.

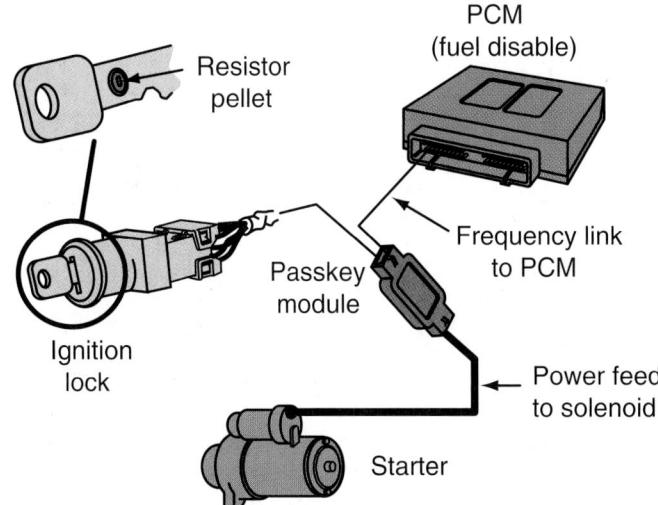

Figure 34.31 A resistance key unlocks this security system.

Door Locks

Locks have always been provided for doors and the trunk. Newer systems include electronically coded resistance keys, keyless entry systems, and light delay systems. Internal and external hood locks prevent a thief from gaining access to the underhood area. This makes it difficult to electrically disconnect an alarm system's siren or horn. Most cars have a release cable or electrical switch accessed from inside the vehicle to open the fuel filler door. This helps prevent the theft of gasoline.

Electric Door Locks. A popular feature on many vehicles is electric door locks (**Figure 34.30**). They can be operated by a switch. Electric door locks allow manual operation in case of electrical failure. Some cars do not have them on the rear doors, and some only have them on the driver's door. When they are on all doors, for child safety, the driver's side has a switch position that makes unlocking of the other doors possible from that seat only. Some doors automatically lock when the shift selector is moved out of the park position, and some engage automatically at speeds over 8 mph.

A solenoid or permanent magnet motor operates the locks. Large solenoids were used on old cars. Since the mid-1970s, permanent magnet reversible motors have been used to operate the rod that moves the lock. They do not require a ground, because the motor's polarity switches with the direction of current flow through its two wires. The ground connection for the entire system is at the master door lock switch.

▤ SECURITY SYSTEMS

Pass Key Systems

Security keys can have a resistor or a transponder to protect against unauthorized copying of the key. The key copy might fit the lock, but the engine will be prevented from starting if the computer does not sense the correct key being used. The ignition system cannot be bypassed by "hot-wiring" when this system is engaged.

Electronic Valet

Some systems have an electronic valet feature. It allows access through the car door and lets the engine be started but prevents unlocking the trunk, glove compartment, or console.

Resistance Key

A **resistance key** is a normal key with a resistance pellet imbedded in it (**Figure 34.31**). There is a small resistance that the computer recognizes as the one that should go with the vehicle. There are several different resistance values that can be assigned to the vehicle, and both the resistor and tumblers must match. If the resistance is incorrect, the computer opens the starter circuit so the engine will not crank. Depending on the vehicle, it might also disable other systems.

Transponder Key

A **transponder key** is used by some late-model systems as a theft deterrent. The transponder in the key receives a radio signal from the computer each time the engine is started. The transponder alters it and sends it back to the computer. The transponder's signal is different each time the engine starts, but the computer knows what the signal should be. If the signal is not as expected, the starting, fuel, and ignition systems are disabled.

Keyless Entry

A keyless entry system, common on most new cars, lets the driver open the doors or trunk (**Figure 34.32**). A key fob transmitter (**Figure 34.33**) or a keypad on

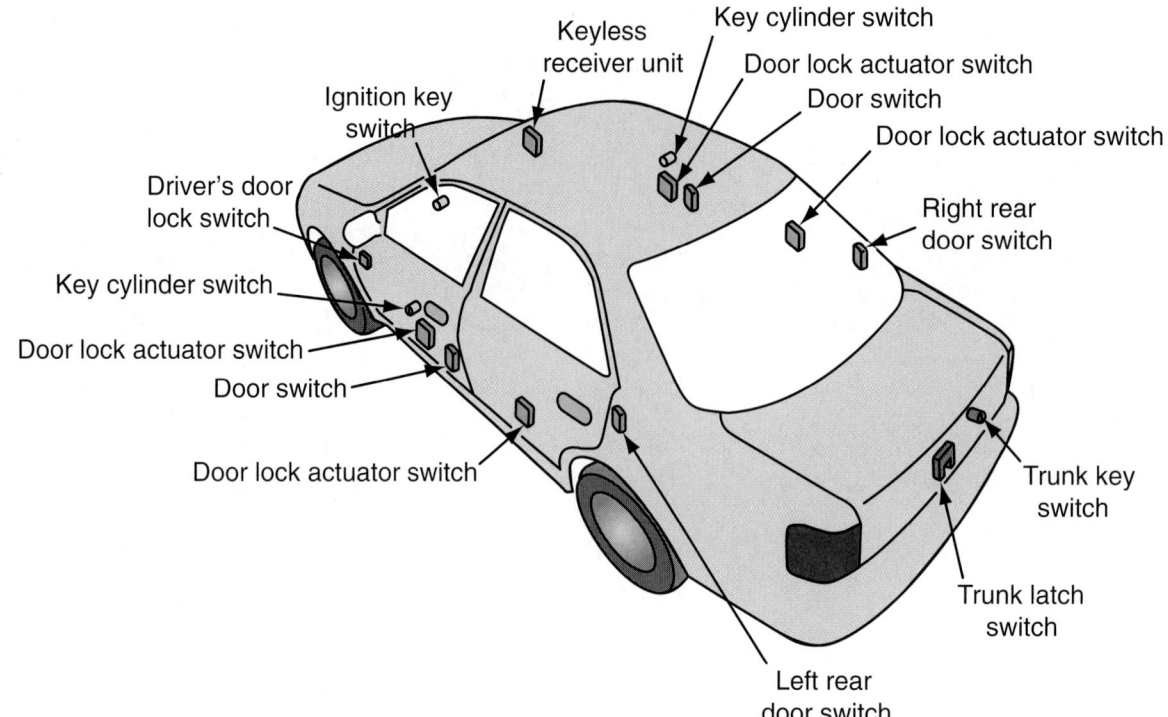

Figure 34.32 A keyless entry system.

Figure 34.33 Examples of keyless entry transmitters. *(Courtesy of Tim Gilles)*

An *intelligent key system* allows control of engine starting, door unlocking and locking, and seat position adjustment without using a mechanical key. The intelligent key (**Figure 34.34**) communicates with the intelligent key module in the vehicle when the transponder is within approximately 31.5 inches (80 cm) of the receiving antenna for the driver and passenger side doors and the back door (**Figure 34.35**). There are two sensitive outside antennae located in the mirrors or door handles and three inside antennae. His and hers keys are identified by the module and respond by setting the car seat position and mirrors.

A mechanical key within the transponder can be used to mechanically operate the system. If the battery in the key transponder goes dead while the car is being driven, the engine will not shut off, but it will not restart

the driver's door is used to operate the system. Interior lights and the security alarm are also operated by the transmitter. Some systems can open the windows and sunroof, too. Holding down the door unlock key is the usual method of activating this feature.

Electronic Key Systems

Some vehicles use an electronic key that transmits a radio frequency to open and close door locks and the trunk. Some have a metal key that slips out of the key fob to operate an ignition switch without conventional tumblers. The "key" uses infrared technology to unlock the steering column and start the engine.

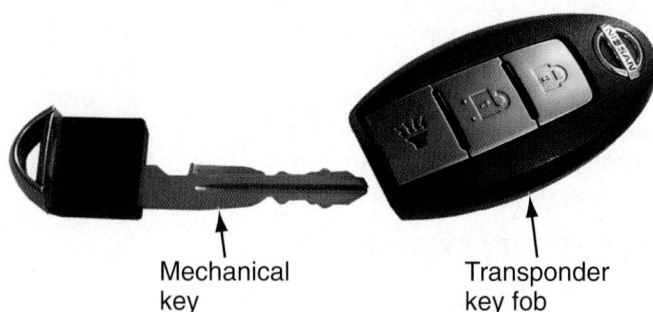

Mechanical key   Transponder key fob

Figure 34.34 An intelligent key can automatically open doors or start the engine when it comes within range of the system's antennae. *(Courtesy of Tim Gilles)*

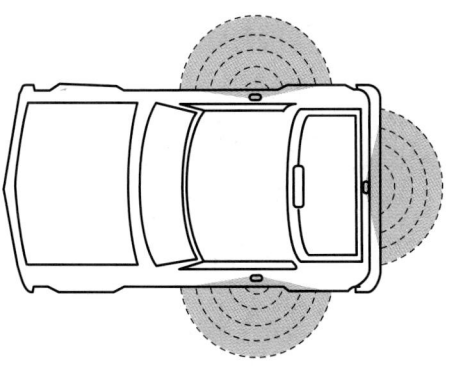

Figure 34.35 An intelligent key system has antennae at each front door and at the back for the trunk or hatchback.

once the car is shut off. Programming the key is done with a scan tool used with a dedicated program card.

Remote Engine Starting. Intelligent key systems can be set for remote engine starting without installing the key. Remote engine starting has long been a popular aftermarket addition to vehicles, especially in cold climates where it is advantageous to have the engine warmed up with the heater working before entering the vehicle. However, as sometimes also occurs with aftermarket car alarms, add-on features sometimes cause problems when they are not compatible with sophisticated vehicle electronic systems. Original equipment engine starting systems prevent this problem. **Figure 34.36** shows a schematic for an engine starting system.

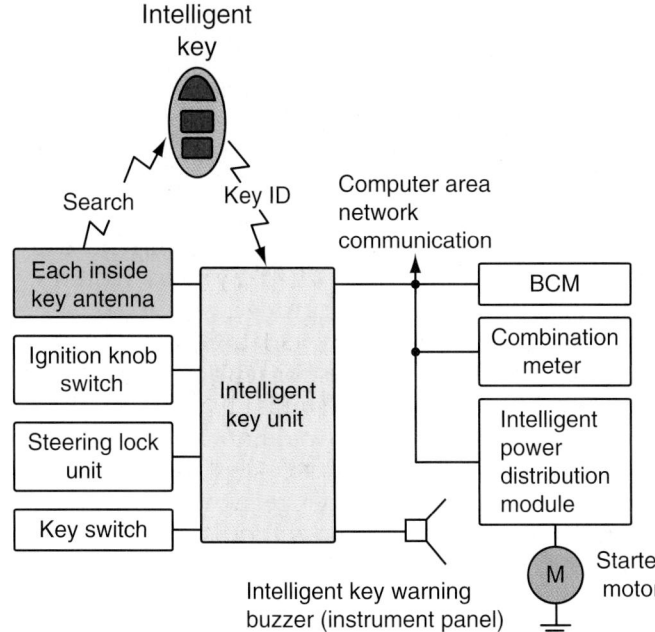

Figure 34.36 An intelligent key schematic with engine starting capability.

Illuminated Entry

Some vehicles are equipped with illuminated entry. This system turns on the interior lights when the outside door handle is moved and the doors are locked. The keyhole in the door also illuminates on many cars.

Theft-Warning Systems

Vehicle alarm systems are either passive or active. A **passive alarm system** is "armed" when the key is removed from the ignition, the hood and trunk are closed, and all the doors are locked. An **active alarm system** must be turned on. Older systems used the key, a keypad, or a switch. Newer systems are activated by the keyless entry transmitter.

Doors can be locked using the key, the button, or the remote control transmitter. The system can be disarmed only by using the key or remote control. An alarm is triggered if the door, trunk lid, or hood is opened without using the key or remote control transmitter. The alarm system uses several simple mechanical switches to provide signals to a computer. **Figure 34.37** shows a typical system. The chart lists pin numbers that would be used with the electrical wiring diagram when troubleshooting problems in the system. The switches provide an "on" or "off" ground signal to the computer. Once the system is armed, an alarm is triggered if the computer loses a ground signal from any of the switches.

Some systems use ultrasonic motion sensors. A drawback to these systems is that they generate annoying false alarms. Someone leaning on the car or bumping against it can set off the alarm.

The computer determines if the vehicle is being tampered with while the alarm is armed. The computer can disable the starter and lock the car. Typically, the horn sounds and/or the headlights flash for 2 or 3 minutes. The system then resets provided that the intruder is no longer attempting to enter the car. This is to prevent draining the battery. Some alarm systems are equipped with a "panic" feature. Holding the button down on the remote control will activate the alarm.

NOTE: *If the alarm system is accidentally triggered, a typical system can be shut off by turning a key in the door lock.*

The front door key cylinder switches are an important part of the system. A pair of magnetic reed switches are mounted on the body of each of the lock cylinders (**Figure 34.38**). A magnet on the key cylinder sweeps past one of the reed switches when the key is turned to the unlock position. It sweeps across the other one when turned to the lock position. The trunk lock uses a similar system. It puts the system on standby when the trunk is opened with the key while the system is armed. Closing the trunk rearms the system.

Switch	Pin #s	Location	Function
Door switches	14, 15, 34	Door jamb of each door	Reports if door is open or closed (shared with dome light)
Door key cylinder switches	29, 30	Attached to key cylinders on the left and right front doors	Reports if door lock is turned with the key
Hood switch	28	Top of inner fender	Reports if hood is open or closed
Trunk switch	25	Trunk latch	Reports if trunk is open or closed (shared with trunk light)
Trunk key cylinder switch	26	Attached to trunk key cylinder	Reports if trunk lock is turned with the key
Door lock actuator switches	11, 12, 13	Part of door lock actuators in each door	Reports if door is locked or unlocked
Ignition key "in" switch	23	Part of ignition cylinder assembly	Reports if key is in or out of ignition switch

Figure 34.37 An example of a theft-warning system using a series of simple switches as sensors.

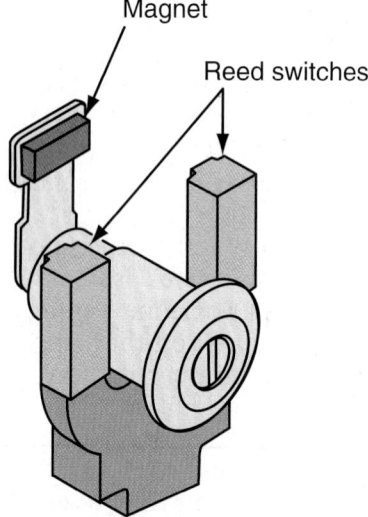

Figure 34.38 A door lock switch for an alarm system. When the key is turned, a magnet sweeps across a reed switch to signal the computer that the door is unlocked.

If an alarm triggers too easily, check the adjustment and switch position for the door, trunk, or hood. Some systems have a key cylinder tampering switch. This prevents the door or trunk locks from being punched out during a burglary. The switch can also be triggered by unlocking a door with a Slim Jim.

Transponder Passive Antitheft Systems

Many vehicles produced since the mid-1990s use similar passive antitheft systems with an encoded transponder ignition key. The following is a generic description of one of the simplest systems. A transponder, which is an electronic device installed permanently in the key, has its own unique identification code. Each key must be programmed to match the vehicle's computer before it will start the engine. Also, if a PCM is replaced, the engine will not start unless the PCM is reset to match the transponder.

Some manufacturers' antitheft systems disable the starter motor, whereas others' do not. Most systems will start the engine for 1 second. A transceiver module located near the steering column verifies the signal from the key during the start attempt by comparing it to information stored in nonvolatile memory. If a problem is sensed, the fuel pump and fuel injectors are disabled and the engine dies. Once a car starts and runs for 1 second, the antitheft system is disabled; therefore, if the engine runs for longer than 1 second, the antitheft system is not the cause of engine stalling.

A theft indicator light flashes every 2 seconds when the ignition is off to verify system operation. The light illuminates for 3 seconds when the key is turned to start or run. If there is a problem in the system, the light will remain on or will flash repeatedly. Diagnosing problems with these systems requires a scan tool with a dedicated cartridge.

Manufacturers use many variations of this system and have different processes for replacing keys. All of them are expensive; for example, if you lose all but one key on a Lincoln, you would not be able to make another because you need two keys to program the new one. The car learns the frequency that the key sends. To program a new key with this system, put the first key in the ignition and cycle it on and off. Cycle the second key and then cycle the new key to program it. Additional keys can be programmed by cycling them at this time.

CASE HISTORY

A vehicle was towed into a shop following a no-start. In the service bay, the vehicle started and the problem would not repeat. The owner's wife picked up the car and drove it home without a problem. The next morning, the man attempted to start the car. It would not start, so it was towed once again to the shop. Further investigation by the service advisor determined that there was a gasoline speed-pass on the man's key ring, next to the transponder key. A speed-pass is a transponder as well, and it transmits on the same frequency as the key transponder. When two transponders are on the same key ring, they interfere with each other and the key is disabled. The wife's key ring did not have a speed-pass, so she was able to drive the car without a problem. A tollway pass can also cause this problem.

■ VEHICLE TRACKING/NAVIGATION SYSTEMS

In 1993, the United States Air Force launched the 24th Navstar satellite into orbit to complete a network of satellites called the Global Positioning System (GPS). GPS is a worldwide radio-navigation system formed from a constellation of 24 satellites and their ground stations. The satellites are "man-made stars." GPS uses these as reference points to calculate positions to closer than 1 centimeter, giving every square meter of the Earth its own address.

Developed by the United States Department of Defense for use at sea during the Cold War, the objective of GPS was to be able to pinpoint the exact positions of nuclear submarines on the surface of the ocean in minutes. Cost of development for the system was $12 billion.

Some newer vehicles are equipped with navigation systems tracked by the GPS (**Figure 34.39**). Many navigation systems are incorporated into the radio and use a GPS antenna to provide the exact location of the vehicle. A gyroscope determines when the vehicle is turning. Navigation road maps are shown on a thin film transistor (TFT) LCD color display. Newer systems can respond to voice prompts using the hands-free phone system. Roadside assistance is available through the satellite connection when needed. If keys are left in the vehicle, the driver can also call roadside assistance and have the vehicle unlocked. When an air bag is deployed, information is relayed by satellite so police and ambulance agencies can be notified. If a vehicle is stolen, the vehicle-tracking system can pinpoint its exact location.

Night Vision

Infrared technology is used to assist a driver in avoiding hazards when driving at night. The system can "see" five times farther than the driver can with the headlights on low beam. A camera-like sensor located in the grille reacts to an object's infrared energy. The computer processes the information and projects it as a video image on the windshield, left of center, through a heads-up display. Warmer objects (people, animals, and moving vehicles) appear light in color or white. Colder objects (sky, signs, and parked vehicles) are darker in appearance.

The night vision feature requires little service. Removal of mud and snow from the camera lens is essential. If the camera requires aiming, refer to the service information for the procedure.

Ultrasonic Rear Parking Assist

Behind the rear bumper is a blind spot in the driver's field of rear vision. The blind spot is from about 10 inches above the pavement to the top of the trunk to a depth of up to 5 feet. Rear assist systems operate whenever a vehicle is in reverse to help a driver determine the distance to a still or moving object when backing. It can also help the driver into a tight parking spot, although the system does not detect small objects that are close to the ground. The sensor is hidden in a hole behind the bumper fascia to protect it from low-speed impacts.

Rear assist systems are either ultrasonic or radar based. Radar-based systems can detect objects at further distances (**Figure 34.40**), providing a driver with more time to brake. A typical radar-based system uses dual-beam continuous wave (non-Doppler) radar to monitor an area behind the vehicle approximately 5 meters in depth and 2.1 meters wide. A speed-sensitive algorithm triggers an audible *imminent collision warning*. Some systems have optional visual alerts, and some smart CAN-based systems can give optional braking commands, too. The distance-based alert changes frequency depending on how close the vehicle is to an object.

Figure 34.39 A GPS navigation system can display a vehicle's exact location. *(Courtesy of Delphi)*

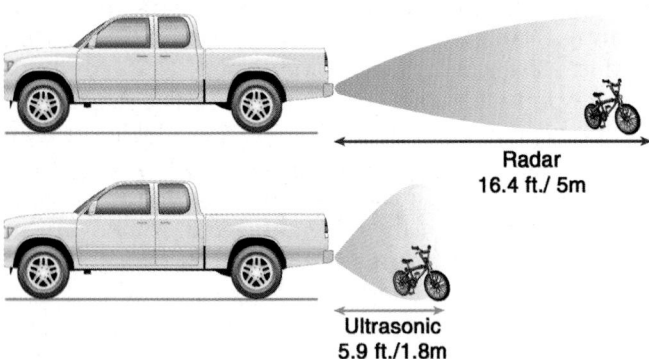

Radar vs. Ultrasonic Detection Zones

Radar
16.4 ft./ 5m

Ultrasonic
5.9 ft./1.8m

Figure 34.40 A rear assist system calculates the distance to an object. *(Courtesy of Delphi)*

■ OTHER COMFORT SYSTEMS

Some of the following systems are installed on vehicles at the factory, whereas others are installed in the aftermarket. Aftermarket accessories are not covered by the manufacturer's warranty. When installed incorrectly, some of these devices can interfere with the vehicle's electrical system, causing serious hard faults and intermittent electrical problems that can void the warranty. Grounds must be provided in sufficient quantity and quality. For most devices, the power feed should be fused to an ignition-controlled circuit so it only comes on with the engine on.

Audio Systems

The first sound system for a car was the tube-type AM radio in 1929. A hybrid tube-type AM radio with a transistor-type power supply was installed on the 1957 Chevrolet. In the late 1960s, FM began to be the frequency of choice. Today's audio systems are of very high quality due to the advances in electronics but are still made up of the same components as the earlier systems: receiver, antenna, amplifier, and speakers. Today's systems use electronically tuned receivers (ETRs).

AM Radio. AM stands for **amplitude modulation**, which varies the strength, or amplitude, of the broadcast signal. Frequencies range from 550 to 1,600 kilohertz. AM signals bounce off the ionosphere (part of the atmosphere), so they can be received a long distance from where they are transmitted. When the antenna is not of sufficient quality or when the station's signal is weak, AM is subject to a higher amount of radio frequency interference (RFI) than is FM.

> **HISTORY NOTE**
>
> *The AM car radio was invented by William Lear, of Lear Jet fame, in the early 1920s. He sold his patent rights to Paul Galvin in 1924. In 1929, Galvin produced the Motorola car radio. The name was developed by combining "motor" and "Victrola." Philco was another popular make of early radios. By the end of the 1930s, one in five automobiles came equipped with factory-installed AM car radios.*

FM Radio. FM stands for **frequency modulation**. Whereas AM varies the *strength* of the broadcast signal, FM varies the *frequency* of the signal. FM radio signals range between 88 megahertz and 108 megahertz. This frequency is very high compared to AM and is not reflected off the atmosphere. The range of the FM broadcast is restricted to "line-of-sight" distances, typically 35 miles, or more if the station's signal strength is higher. The signal can be blocked by a hill or a group of buildings. In large cities where the radio signal is strong, the FM waves can bounce off other tall buildings to provide reception that ordinarily would not be possible.

FM broadcasts are free of RFI noise compared to AM broadcasts. Changes in amplitude (AM) are more susceptible to noise interference than changes in frequency (FM).

> **HISTORY NOTE**
>
> *The FM theory was first envisioned by an American named John Carson in 1922. Another man, Howard Armstrong, patented the technology in 1933. Few radio companies were interested in the idea because AM was already well entrenched.*
>
> *The German company Blaupunkt, the mobile electronics division of Robert Bosch, claims to have produced the first FM car radio. FM stereo came of age in the 1960s and was soon being installed in vehicles because of the superior sound quality that it produced.*

Antenna. A typical electromagnetic radio signal received from the broadcast antenna only has a strength of about 25 microvolts (0.000025 V). Such a small signal requires an antenna to increase its strength. AM radios work best with the longest antenna possible, but FM antennas should be exactly 31 inches (79 centimeters) long. FM windshield antennas are exactly this long. A power antenna that is not achieving its full height of 31 inches will not provide a strong signal to an FM radio. When an antenna is defective, AM reception will suffer most.

Listening to the radio can help determine whether an antenna is good or bad. A bad cable to the antenna will tend to give very poor AM reception, while FM reception will be weaker than normal. Check the cable with an ohmmeter. It should read infinite resistance between the center antenna case and its center lead. The antenna case must also have a good ground to the vehicle body.

Interference. An antenna picks up different voltages as signal strength varies. Voltage variations can cause radio noise called "whine." Any device that contains a coil can cause radio interference. The ignition system or the alternator are possible noise sources. Either of these noises changes with engine rpm. Installing a capacitor, or a coil-like device called a choke, on the power side connection can reduce or eliminate radio noise. Alternator whine is controlled by installing a choke in series with the power feed to the radio. Ignition noise is controlled by installing a capacitor on the positive side of the ignition coil. The capacitor must be connected to a good ground. Two or more capacitors can be installed in parallel if more capacitance is needed.

Most noise complaints come after the installation of an aftermarket amplifier, graphic equalizer, or other radio accessory. Poor ground connections are a probable cause. Antenna amplifiers are another possible cause of radio noise.

Power Antenna. An electrically powered antenna (**Figure 34.41**) has an antenna mast with a nylon cord attached to it. It is powered by a reversible electric motor through a relay, most often with a circuit breaker. Most power antennas operate automatically when the radio is turned on. The antenna assembly includes upper and lower limit switches.

An antenna mast is cleaned by wiping it with a cloth and then oiling it. Testing is the same as with an ordinary antenna. An infinite ohmmeter reading should be the result when testing between the center antenna terminal and the grounded housing. Some power antennas are replaced as a unit. Others have a replaceable cord and mast assembly (**Figure 34.42**).

Speakers. Amplified electrical energy from the radio, tape, or CD player is turned into acoustical energy by a speaker (**Figure 34.43**). The sound a speaker makes comes from the vibration of its diaphragm. The speaker has a permanent magnet surrounded by a coil

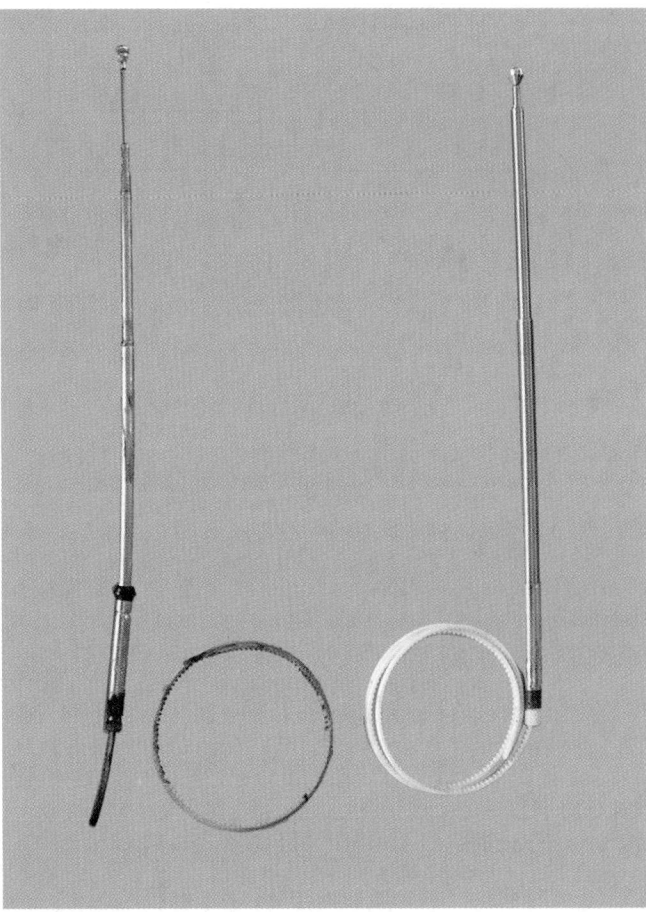

Figure 34.42 A broken electric antenna insert and a new replacement cartridge. *(Courtesy of Tim Gilles)*

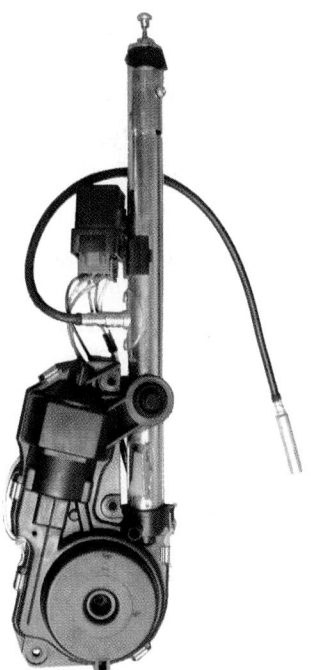

Figure 34.41 A power antenna. *(Courtesy of Tim Gilles)*

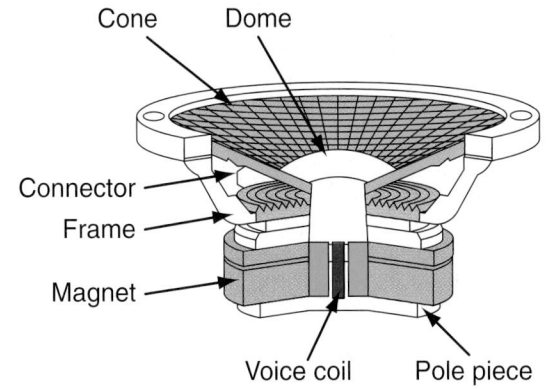

Figure 34.43 Amplified electrical energy is turned into acoustical energy by a speaker.

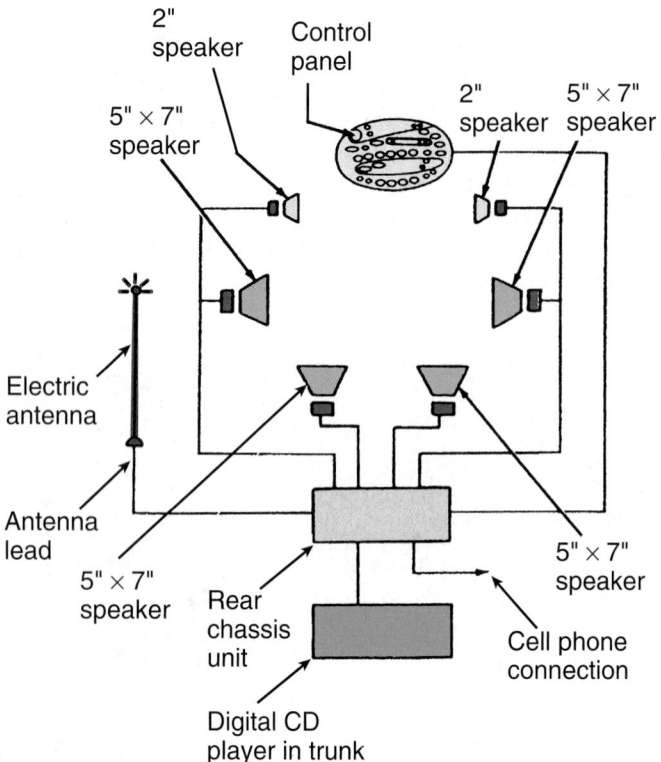

Figure 34.44 Chart showing the layout of a typical audio system. (*Courtesy of Ford Motor Company*)

of wire that causes the speaker to vibrate in response to the acoustical energy, which moves air. The vibration produces pressure waves to make sound.

One speaker is not capable of reproducing sound in the entire range that can be heard by the human ear. The ear can hear sounds ranging from 20 hertz (cycles per second) at the low end to as high as 20 kilohertz (20,000 Hz). Different speakers are needed to produce different ranges of sound. **Figure 34.44** shows the layout for a complete sound system.

Tweeter. A tweeter produces high-frequency sound (4 to 10 kilohertz). Tweeters produce directional sound. This is why they are usually mounted higher on a door panel or on the top of the instrument panel. You will be more able to locate a tweeter by listening than you will be to find speakers with more of a bass sound. Midrange speakers produce sound from 400 Hz to 5 KHz. These are relatively easy to locate by listening as well.

Woofer. A woofer produces bass, or low-frequency, sound in the range of 125 Hz or less. It is sometimes called a subwoofer. Bass sounds are not directional, so you cannot tell where they are coming from. This means they can be located almost anywhere in the vehicle. They are usually in the rear, where there is more room for them because they are large. There are also midbass speakers that produce frequencies between 100 Hz and 500 Hz.

Coaxial Speakers. A coaxial speaker has more than one speaker contained within its frame. These speakers have a separate midrange and tweeter to provide a wider frequency range than an ordinary single cone speaker. Triaxial speakers have three speakers within the frame.

Speaker Impedance. Speakers should have the same impedance (total resistance in the circuit). If two rear speakers have 6 ohms of resistance each, wiring them in parallel will provide 3 ohms of resistance. The front speakers should also have an equal load. Rear speakers are typically bass speakers with more load. The front speakers can be connected in series to raise and balance the load. Connecting the rear speakers in parallel will lower the load.

Speaker Wiring. Wires must be large enough to provide the specified power to the speakers. Speaker wire is sometimes as small as 22 gauge. A more practical size to use is 14 gauge, especially if larger, higher-impedance speakers might someday be installed on the same wiring. Remember, a smaller number (14) is a larger wire size. Soldering the connections ensures a quality connection.

Amplifying Devices

A **crossover** is a device that blocks certain frequencies to a speaker. For instance, to make tweeter circuits more efficient at producing mid- and high-range sound, a bass blocker (capacitor and coil) is sometimes used to keep low-frequency sound from reaching the smaller speakers. This is called a high-pass filter. A low-pass filter is one that passes (transfers) only low-frequency sound. Higher frequencies are blocked.

A passive crossover does not use an external power source. An electronic crossover, also called an active crossover, is a more expensive device with filters that are powered (amplified). At least two amplifiers are needed to make an effective active crossover. One amplifier covers high and midrange; the other is for the woofers.

Making deep bass sounds consumes a large amount of electrical current. A large capacitor is installed in the power line to the amplifier. Battery power is often too slow to respond to the current demand. Remember that a capacitor is a device that stores an electrical charge. The capacitor, in this case called a power line capacitor, attempts to maintain a consistent voltage level at the amplifier by discharging current when it senses a demand.

A capacitor must be precharged before installing it. Otherwise it will draw too much current and blow the in-line fuse. To precharge the capacitor, follow the instructions that come with it. A typical procedure would be to connect the capacitor's negative terminal to ground and then connect its positive terminal to a battery with a 12-volt lamp in series between the two. While the capacitor is charging, the 12-volt lamp will be lit. It will go out when the capacitor is fully charged. The capacitor can then be connected to the circuit without blowing the fuse.

Satellite Radio

Satellite radio receivers can be original equipment (**Figure 34.45**) or they can be added to a vehicle in the aftermarket (**Figure 34.46**). Satellite subscription service purchased by the vehicle owner is received as a digital signal from an orbiting satellite. The signal is received by a satellite digital audio receiver (SDAR) and is routed by the regular radio amplifier to the speakers. If the signal is lost, no interference noise is heard. The regular radio will still work, but that mode must be selected first.

A satellite antenna is often located at the centerline of the roof. When communication between the satellite and the antenna is blocked by an overpass or a tall building, a buffer prevents brief disruptions in the service.

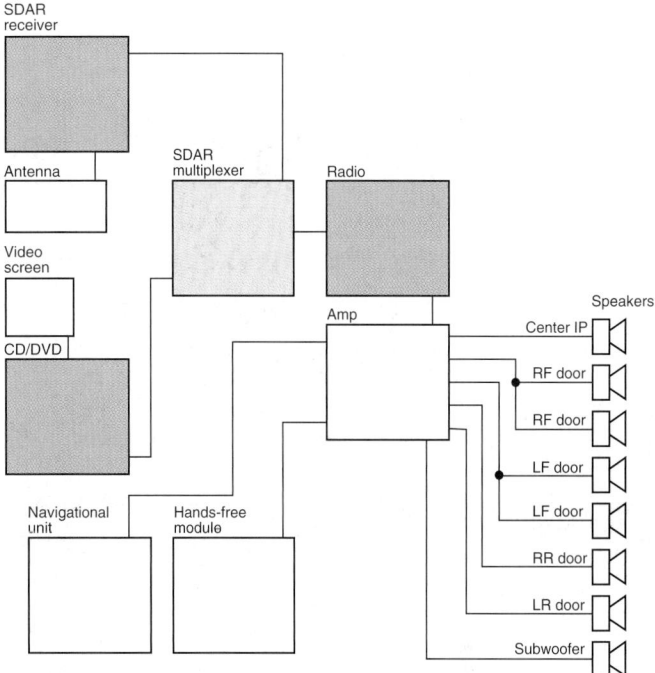

Figure 34.47 A multiplexer acts as an electronic switch, sending the selected signal to the radio.

Figure 34.45 An original equipment radio with satellite capability. *(Courtesy of Delphi)*

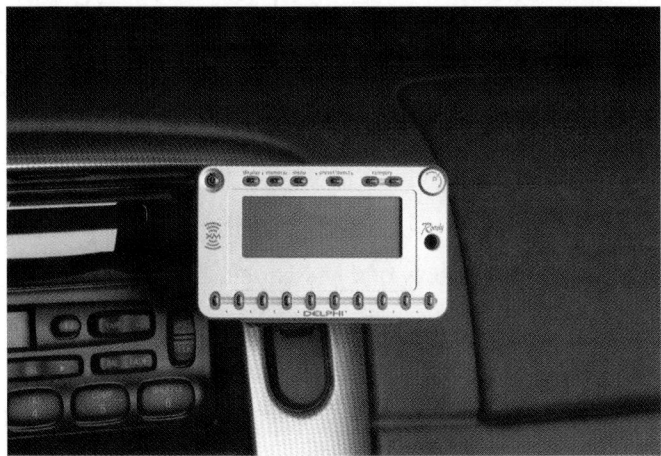

Figure 34.46 An aftermarket satellite radio installation. *(Courtesy of Delphi)*

Multiplexer

When different audio sources are to be connected to a radio that does not have multiple accessory plugs, a multiplexer is used. This device allows the addition of inputs like a CD, DVD, satellite radio service, or hands-free cell phone. The multiplexer acts as an electronic switch between the items, sending the signal output to the radio. **Figure 34.47** shows a schematic for a multiplexer.

DVD Systems

Digital video disc (DVD) systems are available on many vehicles. Some DVD monitors are installed in the rear of a front headrest or seat, but most flip down from the ceiling (**Figure 34.48**). Audio plays through the regular audio system using surround sound when a DVD is playing. This sends more of the sound to the rear of the vehicle.

Jacks often permit connection to video games or video cameras. Wireless headphones are available as well. With headphones, the speakers can play the radio or CDs while the children in the back seat watch and listen to DVDs through the headphones.

Hands-Free Cellular Phones

There are different types of wireless cellular phone technologies. One type is Bluetooth™, which allows different modules to communicate, such as a cell phone and the on-board receiver in the vehicle. The system is capable of recognizing several cell phones from within the vehicle once a phone is paired to the

Figure 34.48 A DVD system with a flip-down monitor. *(Courtesy of Delphi)*

begun, the radio stores its volume level and tuning information so it can return to those settings when the call is over. The speaker volume fades and the call is broadcast on the speakers.

The hands-free system is operated using voice recognition software. A microphone module for the system can be located in the mirror, in the center console, or on the dash panel. The module includes the microphone, a preamplifier, and electronics. The module is adaptive and can learn a voice as it improves system operation.

Complaints of loss of signal are most often related to the service provider. A cell phone's transmitting power is quite low, usually less than 3 watts. This is possible because of the close proximity of the cellular phone network's antennas. One "cell," as they are called, transmits to another, and so on. When the distance from the transmitter (cell phone) to the receiver becomes too great, the signal becomes weak or is lost. When the distance doubles, the signal strength becomes one-quarter of what it was.

Electrical interference from the engine and power accessories can cause cell phone problems. Defective spark plug wires will cause a pop heard during conversations. Test the operation of the phone with all of the vehicle's power off to see if a problem goes away.

Rear Window Defoggers

Most cars have an electric grid baked onto the rear window. When activated, it warms the glass to above 80°F to melt frost or clear the window of dew. A switch timer relay controls its operation. A window grid draws a relatively high amount of current, up to 30 amps. When cold, resistance is less so current flow is higher. The timer limits on time to about 10 minutes. If needed again, the driver can reset the switch to allow an extra 5 minutes of operation. **Figure 34.49** shows a rear defogger circuit. If the vehicle has heated mirrors, they are also activated by the timer switch through a separate fused circuit.

system during a setup process. The hands-free module stores the cell phone's IP address, and the cell phone learns the module's IP address.

The radio broadcasts cell phone calls through the vehicle's audio system. If the radio is already playing, the cell phone has priority. When a cell phone call is

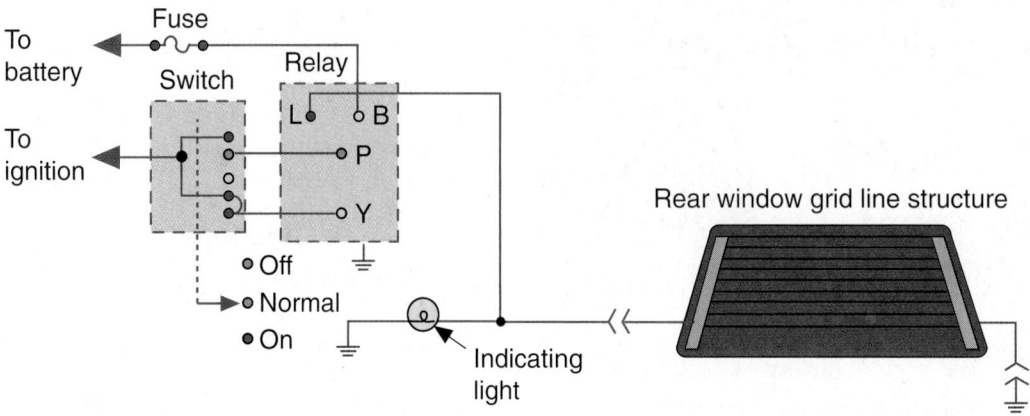

Figure 34.49 A rear defogger circuit.

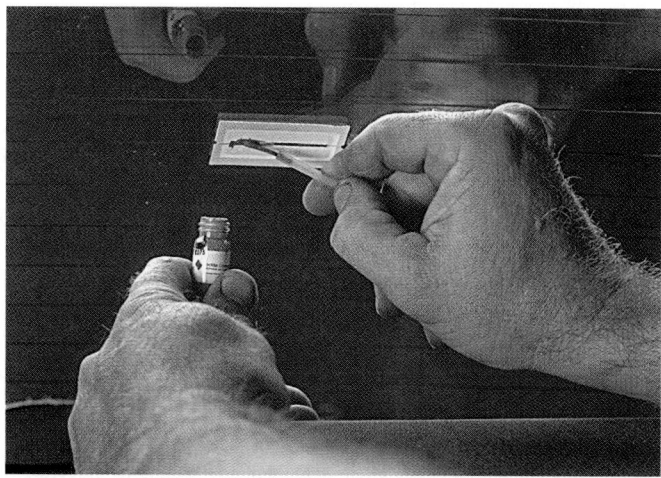

Figure 34.50 Repairing a short section of electric grid on the rear window: Apply two rows of tape to the window to construct a clean channel for the epoxy.

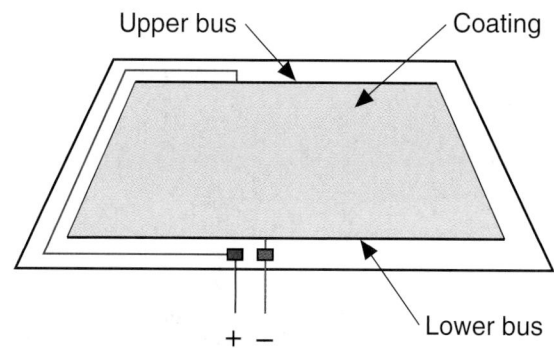

Figure 34.51 A self-defrosting windshield has silver bus bars fused to the coating at the top and bottom of the windshield to provide power and ground connections. *(Courtesy of Ford Motor Company)*

The grid is easily damaged. Some vehicles use the rear window grid as the radio antenna, too. Operation of the radio can be affected when there is damage to the grid. Damage to the grid often occurs when tape or a decal is removed using a razor or knife blade. The horizontal wires of the grid are connected in parallel, so if one row fails the rest will continue to work.

A short broken section of grid can be repaired using an epoxy repair kit. The kit resembles a bottle of fingernail polish and applicator. An epoxy-based liquid that conducts electricity is applied to the break in the grid wire and allowed to dry. Clean the window with alcohol and apply two rows of tape to the window to construct a clean channel for the epoxy to flow into (**Figure 34.50**). Allow 24 hours for the epoxy to cure.

If no part of the grid works, check for power at the grid with a test light or meter. If there is no power, check at the relay timer and switch. Most systems have a lighted switch, which is a pretty good indication of power availability. A bad ground on the side of the grid opposite the power side can also cause the outage.

Heated Windshield

Some newer vehicles are equipped with self-defrosting windshields that melt ice and frost faster than a conventional system that uses air supplied by a blower motor. A wire grid, as in the rear systems, would interfere with the driver's vision. Improved production techniques in glassmaking allow an extremely thin, practically invisible metallic coating to be part of the inside of the windshield. Silver bus bars fused to the coating at the top and bottom of the windshield provide power and ground connections (**Figure 34.51**). A sensor that monitors voltage drop within the windshield will shut off power to the circuit if the window has been damaged.

Heated Mirrors. Heated mirrors are found on some cars. They are sometimes tied into the same circuit as a heated windshield or rear window.

Intelligent Windshield Wipers. Intelligent windshield wiper systems that are sensitive to moisture are found on some cars. These systems are covered in Chapter 32.

Power Mirrors

Power mirrors are common on many cars. A "joystick" controls both the driver and passenger side mirrors. Each mirror has a dual motor drive, which allows up/down and side/side movement of the mirror (**Figure 34.52**). The joystick works for either side. The right or left side is selected by rotating the joystick knob clockwise or counterclockwise or with a separate switch.

Automatic Rear View Mirrors. Some interior rear view mirrors have a directional compass displayed on the lower left corner. Mirrors on some newer vehicles have automatic, power-free glare reduction. These electrochromic or photochromatic mirrors operate like the newer eyeglasses that automatically darken in response to sunlight.

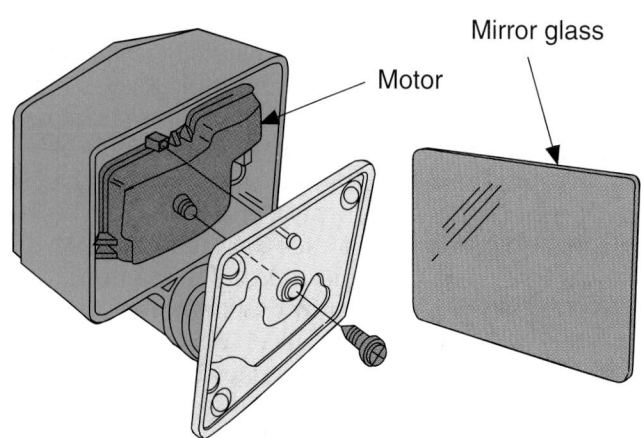

Figure 34.52 Parts of a power mirror.

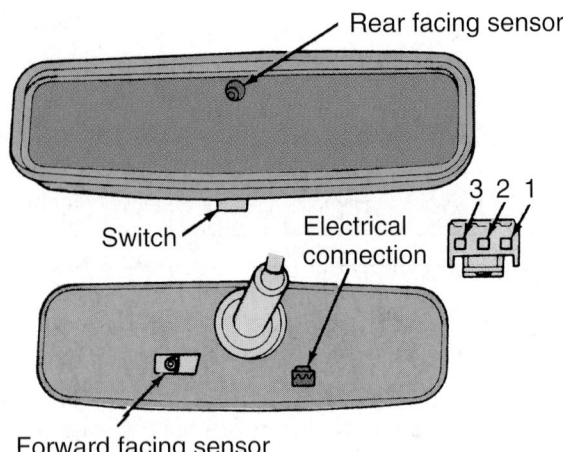

Figure 34.53 An automatic day and night mirror. *(Courtesy of DaimlerChrysler Corporation)*

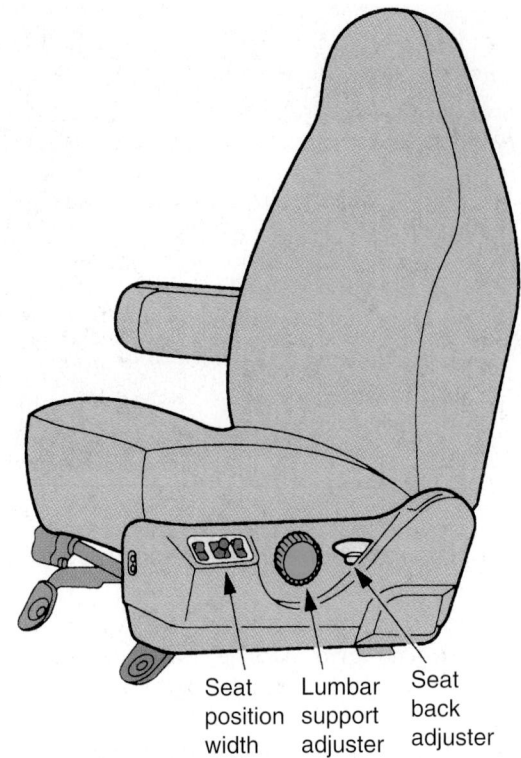

Figure 34.54 Seat position is controlled by switches or a joystick. *(Courtesy of Ford Motor Company)*

Electrochromic Mirrors. Electrochromic mirrors can be found on inside and outside mirrors. Automatic day/night inside rear view mirrors have two layers of conductive glass with an electrochromic material sandwiched between to provide self-dimming in response to glare. The driver can shut off this feature using a switch, except when the transmission is shifted into reverse. Photosensors facing the front and rear adjust the reflection level of the mirror (**Figure 34.53**). A faulty mirror cannot be serviced and must be replaced.

Automatic Tilt Rear View Mirrors. Some vehicles have automatic inside rear view mirrors that tilt automatically when too much light is reflected into the driver's eyes. The mirror housing has two photocells: one measures the light within the vehicle (ambient light), and the other measures the intensity of light on the mirror. When the light on the mirror reaches a set point in relation to ambient light, a solenoid moves the mirror to a different position.

Power Seats

Power seats allow easy adjustment of seat position (**Figure 34.54**). Seat position is controlled by switches or a joystick. Motors and gear drives are located under or in the seat. A typical seat uses multiple reversible DC motors. The rotating motion of the motor is changed to linear motion to move the seat. The motion of the motor is transmitted to the seat tracks by a steel cable that is like a speedometer cable with a very thick steel housing. Some drives are rack and pinion; most are screw drive. Electrically, they operate the same.

Seats are typically four way or six way. In a four-way seat, the entire seat moves up or down and forward or backward. A six-way seat operates in the same way, but it also allows independent adjustment of the front or rear height of the seat. Up and down adjustment is a desirable feature for taller or shorter individuals.

Figure 34.55 shows a schematic of a six-way seat system. Power is supplied to the seats at all times. The ignition key does not need to be turned on.

Four-way systems are used on bench seats, and six-way systems are used on bucket or split bench seats. Two motors are usually used for four-way seats. Three are used on six-way seats (**Figure 34.56**), although some seats use a single motor and a transmission to provide all six movements. Some seats have as many as eight motors to control the six-way functions as well as headrest height, seat length, and side bolsters.

Troubleshooting Power Seats. When troubleshooting electric seats, use common sense. If only the up and down positioning has failed, check that motor and circuit first. If both seats are not working, look to the power source. The most common problems with power seats are with the switches and motors. Power is supplied to the seats at all times. If the seats do not operate, first check the fuse or circuit breaker. Most vehicles use circuit breakers due to the possibility of something physically becoming lodged between the seat and floor. Sometimes the relay clicks when the switch is activated but the seat does not move. In this case, the breaker is operating but the relay or motor could be defective. A test light should light up on both sides of the circuit breaker, which is usually located on

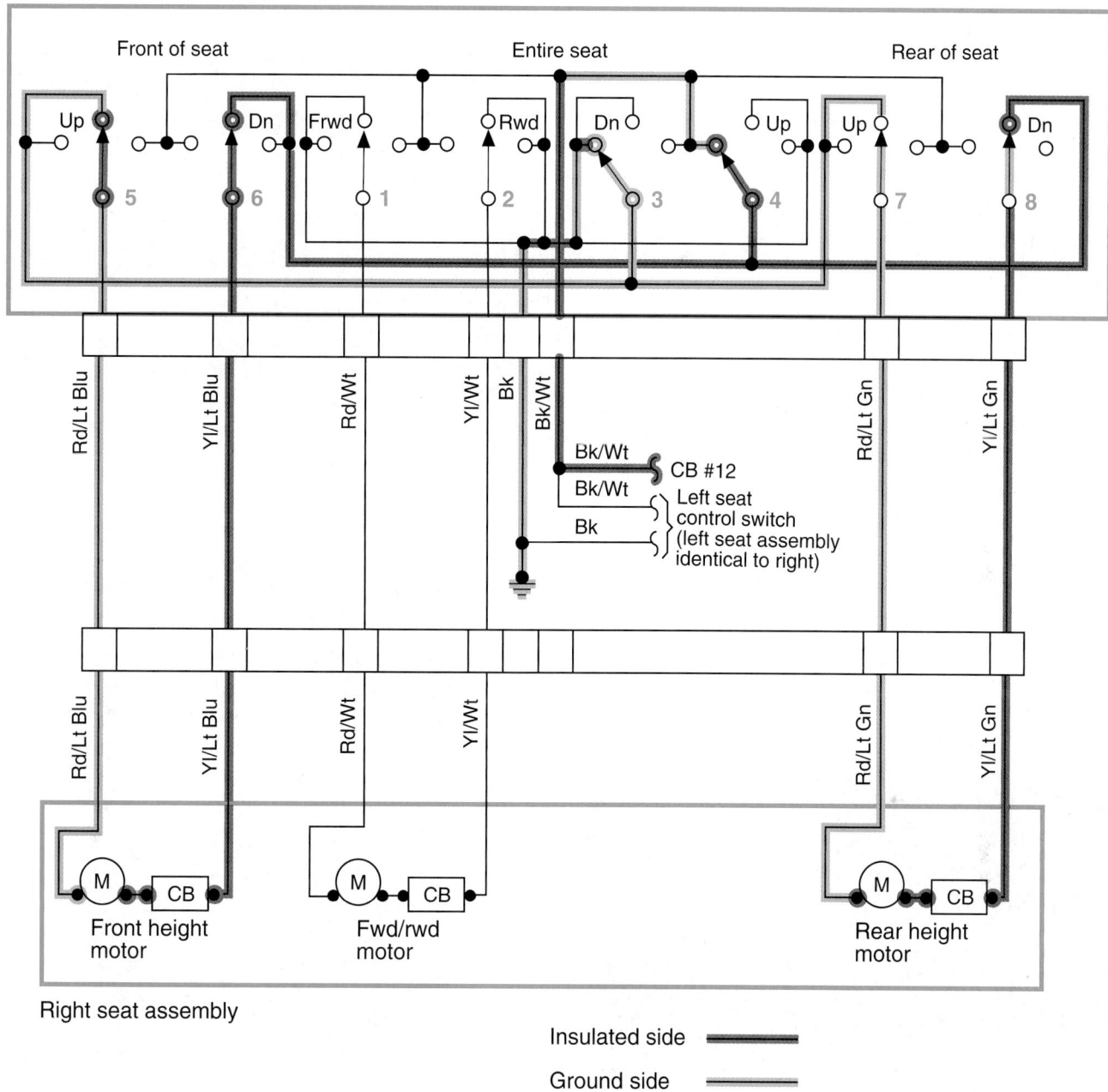

Figure 34.55 A schematic for a power seat.

the fuse panel. Remove the seat control and check for voltage and ground connections.

Sometimes a power seat problem is not electrical. The causes of the problems vary based on the design of the system. Check the manufacturer's service information to learn the details of the repair procedure.

Memory Seats

Some cars have memory seats. A computer remembers the seat position, and some computers remember the steering wheel position and rear view mirror position as well. Typically, switches allow programming for at least two driver's seat positions (**Figure 34.57**). The module

is programmed by putting the seat in the position desired. Then a button is depressed and held to enter the position into memory. For safety reasons, the power seat memory in many cars only works if the transmission selector is in park or neutral. During service, be sure the selector is in the correct position.

Adaptive Seats

Adaptive seats use memory seat positions but move the adjustment to fit the driver as she or he shifts in the seat. Some luxury vehicles have massaging seats that move rows of rollers up and down 2" for a few seconds when the driver presses a button.

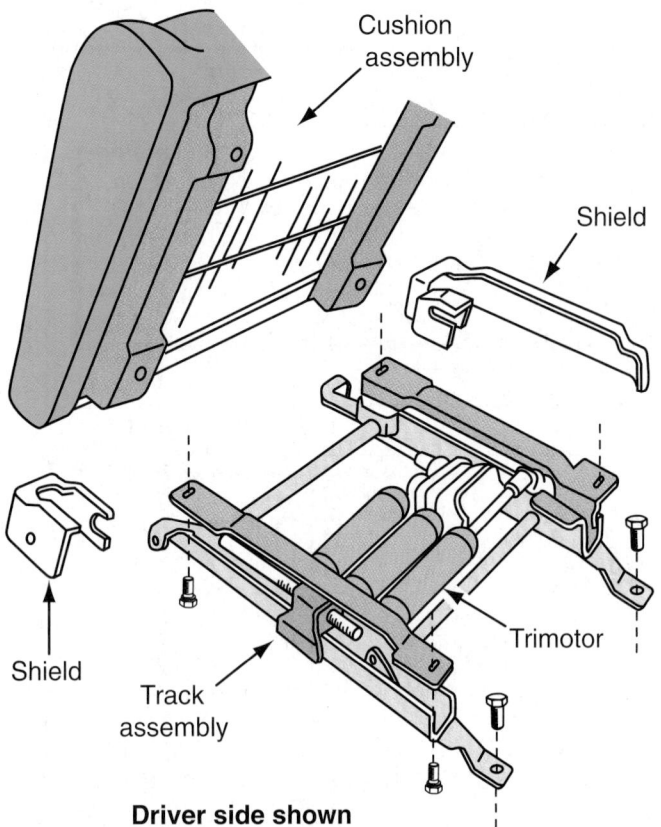

Driver side shown

Figure 34.56 A six-way seat with three motors in one unit.

Figure 34.57 Seat memory and adjustment switches. This one has three memory settings. *(Courtesy of DaimlerChrysler Corporation)*

Adjustable Pedal Height

To accommodate shorter drivers, some cars have an electric motor–operated assembly that can raise the brake and accelerator pedals up to 3" higher (**Figure 34.58**). This system can be part of the seat memory as well.

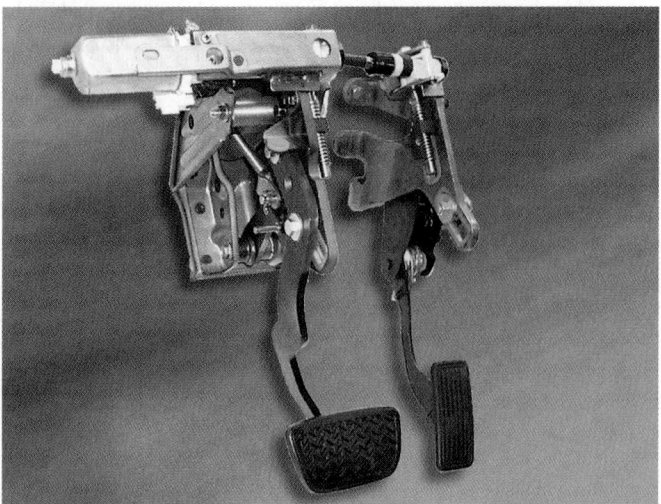

Figure 34.58 Some vehicles have pedal height adjustment operated by a motor. *(Reprinted with permission).*

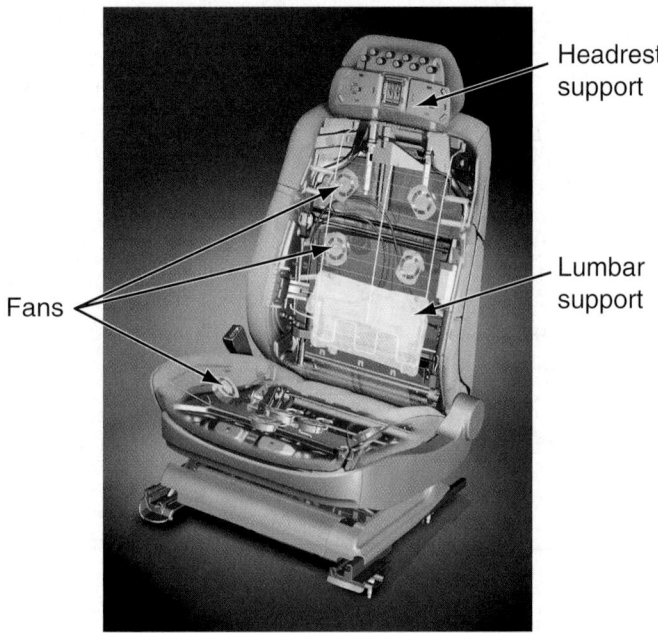

Figure 34.59 A seat with adjustable lumbar and headrest support. The fans are part of the seat's heating and cooling system. *(Courtesy of BMW NA, LLC)*

Heated Seats

Some luxury cars have heated seats that warm seat cushions in cold weather (**Figure 34.59**). A pair of large resistance wires or more than one multiple heating element is routed through the seat cushions. Activating the seat warmer switch energizes its holding relay, which stays on when energized. The relay routes current that heats the seats. The heat stays on until the key or switch is shut off.

Check for a lack of heating in the normal way. Look for power and ground problems, and check the switch and relay. If power reaches the heating element but it does not heat, it has an opening in its circuit and needs to be replaced.

Climate-Controlled Seats

Climate-controlled seats have a seat control unit under the seat. A thermoelectric device not only heats but cools the seat (**Figure 34.60**). The unit does not heat as well as a conventional heating unit, and the circuit has only a 15-amp fuse. A blower motor beneath the seat cushion pushes heated or cooled air through perforations in the leather seat, producing a light feeling of air movement (**Figure 34.61**). The heating and cooling operation of a climate-controlled seat is made possible by the Peltier effect.

> ### ▨▨▨ SCIENCE NOTE ▨▨▨
>
> *The Seebeck effect converts differences in heat into electricity. Also called the thermoelectric effect, it is named after the physicist Thomas Johann Seebeck, who discovered in 1821 that a voltage developed between two ends of a piece of metal when there was a difference in temperature in the metal. Thermocouple thermometers are based on this principle.*

> *In 1834, Jean Peltier discovered the opposite of the Seebeck effect. The Peltier effect is when a difference in heat is created when current is applied to a thermocouple. Current passed through the two dissimilar metals of a thermocouple connected at two junctions creates a temperature difference as the current pushes heat from one of the junctions to the other. As one of the junctions cools, the other heats up. In modern applications of the Peltier effect, n-type and p-type semiconductors, used in place of a thermocouple, are connected to each other at the two junctions. When several semiconductors are connected to the circuit, the Peltier effect is increased.*

Although the Peltier effect is not as efficient as other heating and cooling methods, Peltier devices are easily adjustable and accurate. In climate-controlled seats, the Peltier effect creates a small heat pump using the same unit to both heat and cool the seat. When current flows through the circuit (**Figure 34.62**), heat is produced at the upper junction and absorbed at the lower junction. Reversing the current flow causes the upper junction to absorb heat while the lower junction produces heat.

Power Lumbar Supports

Power lumbar supports are a feature of some seats, usually on the driver side. The lumbar area describes the lower area of the back. A bladder in the lower seat back is inflated or deflated according to the desires of the person in the seat (see Figure 34.59).

Power Windows

Electric windows are opened and closed by drive motors. The motors drive a window regulator similar

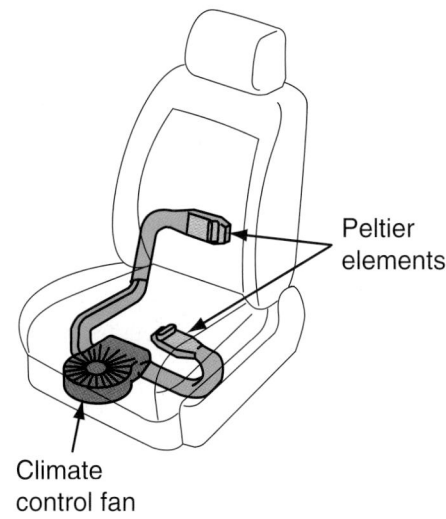

Figure 34.60 Seats can be heated and cooled using the Peltier effect.

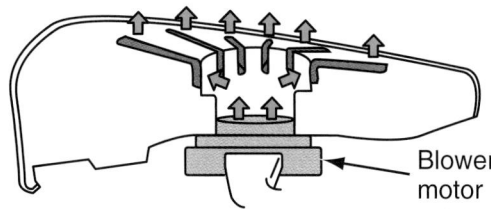

Figure 34.61 A blower motor beneath the seat cushion pushes heated or cooled air through perforations in the seat.

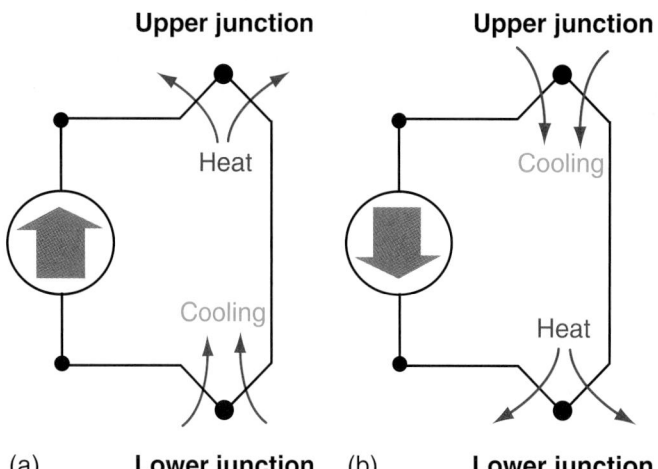

Figure 34.62 The Peltier effect. (a) When current flows toward the upper junction it gives off heat as the lower junction absorbs heat. (b) When current is reversed the upper junction absorbs heat as the lower junction gives off heat.

Window   Regulator   Window
          assembly   motor

Figure 34.63 The door skin has been removed to reveal the motor and regulator for an electric window. *(Courtesy of Tim Gilles)*

to the ones on hand-crank windows (**Figure 34.63**). The window regulator is attached to the window. It is a gear drive that opens and closes the window. Window adjustments for upper and lower stops and the tilt of the glass are the same as for manual windows.

The motors are controlled by individual switches at each window or by a master switch, usually located on the driver's door (**Figure 34.64**). Power for the individual switches comes from the master switch. The two switches are wired in series, so a problem with either of them will prevent the window from operating.

Permanent-magnet DC motors are used in most systems. These motors can be driven in either direction, depending on the polarity of the wires to the motor. Battery power is supplied to the center of a double-pole, double-throw switch. To operate the window, battery positive (B+) power is directed through the motor from the center terminal and flows back to ground through the other side of the switch (**Figure 34.65**). Moving the switch in the other direction reverses the polarity of the motor to move the window in the opposite direction. The motor is not grounded to the door but receives its ground through a connection at the master switch.

Each motor is protected by its own internal circuit breaker. A window can jam up due to ice formation, be out of adjustment in its track, or something can be binding in a mechanism or track.

On some cars, a window lock on the master switch activated by the driver prevents the individual switches from operating the windows. This feature is a safety device designed to protect children. Power windows will only run when the ignition switch is in the run or accessory position. This is a safety feature as well. Some systems have a time delay that allows windows to be powered for a short time after the engine is shut off. This allows passengers to close windows prior to leaving the car. Opening a door defeats this feature.

Some vehicles have a feature that opens the driver's side window all the way when the window switch is held down for more than a third of a second then released. The window can be stopped anywhere in its travel by depressing the switch once again. When using this feature, an electronic module energizes a

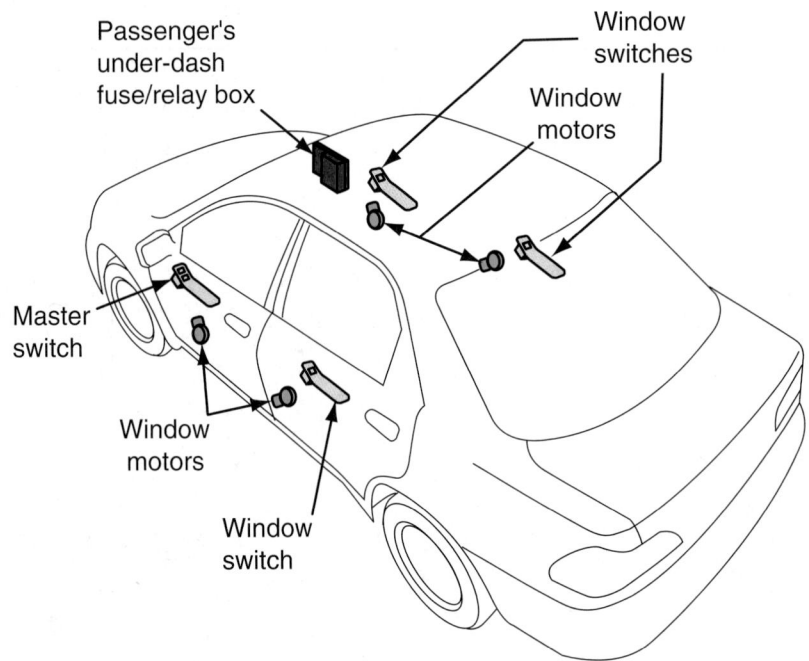

Passenger's under-dash fuse/relay box

Window switches

Window motors

Master switch

Window motors

Window switch

Figure 34.64 Parts of a power window system.

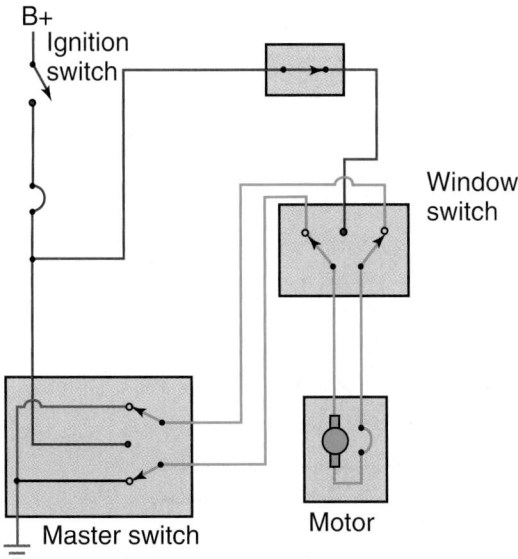

Figure 34.65 Schematic of a simple power window circuit. *(Courtesy of Ford Motor Company)*

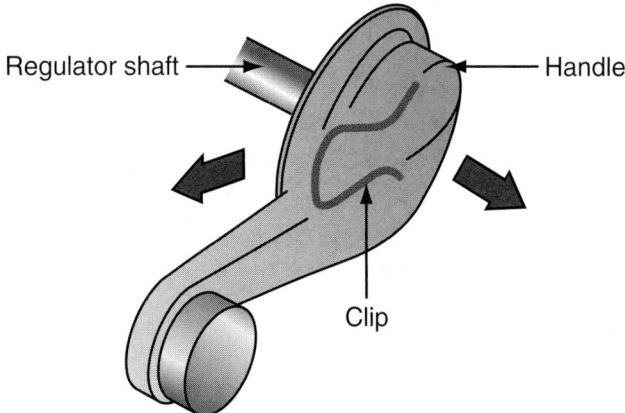

Figure 34.66 A hook used to remove a window crank.

relay to power the motor. When the window reaches its bottom stop, current increases and the module responds by opening the circuit.

Some cars have more sophisticated power window systems that can sense when something is in the way. This is a safety feature, especially for children, who tend to get their fingers stuck in windows. Some systems sense when the window comes into contact with something, which causes it to reverse its operation. Others use an infrared light sensor. When the infrared light beam is interrupted by an obstruction, the window reverses direction.

Troubleshooting Electric Window Problems. When troubleshooting electric window problems, try all of the windows to see if a problem is something they have in common. An open circuit in one of the control wires from the master switch to a motor can result in a window that operates in only one direction. If one of the directional wires from the window switch to the master switch has an open circuit, the window will not operate at all.

When both rear windows fail to operate, check the circuit for the window lockout. If all windows are not working, check the fuse or circuit breaker first. Then check the ground wire connection at the master switch in the driver's door panel or below the instrument panel.

When a window does not operate in either direction, a defective motor could be the cause. The circuit breaker could also be opening the circuit if a window is binding in the track. Grab the edge of the window and try to pull it to see if it is free to move. If it is, the window should be able to move the glass.

To test a circuit, verify that both the master switch and the individual switch have voltage. Moving the switch from the up position to the down position should result in a switch of B+ from one side of the switch to the other.

After testing the switches, the motor can be tested by jumping its terminals to B+ and ground. The window should operate. Reversing the jumper leads should operate the window in the opposite direction.

Power door locks are covered earlier in this chapter as part of the security system.

Removing a Door Panel

When repairing an electric window motor or electric door lock, the inner door panel is removed. This is accomplished by removing the armrest, window crank and door handles, and any other pieces of trim. Window cranks and door handles are retained by either a screw or a clip. A hooked tool or bent wire can be used to pull the clip from its groove (**Figure 34.66**). To finish removing the door panel, pop loose the plastic or metal clips attaching the outside of the door panel to door. There are special tools available for this purpose.

■ CRUISE CONTROL

Most cars are equipped with cruise control capable of maintaining the vehicle at a constant speed above 30 mph without the driver pushing on the accelerator. The driver controls vehicle speed from a switch on the steering wheel or at the end of the turn signal lever (**Figure 34.67**). A light on the instrument panel tells the driver when cruise control is active. Stepping on the brake or clutch pedal will deactivate cruise control, allowing the vehicle to decelerate. A "resume" function reinstates cruise control to the originally selected vehicle speed.

Earlier cruise control systems consisted of a transducer vacuum-operated servo and a transducer. The transducer sensed vehicle speed by monitoring the speed of the drive shaft. When vacuum was applied or released from the servo, it pulled on the accelerator or released it to increase or decrease vehicle speed.

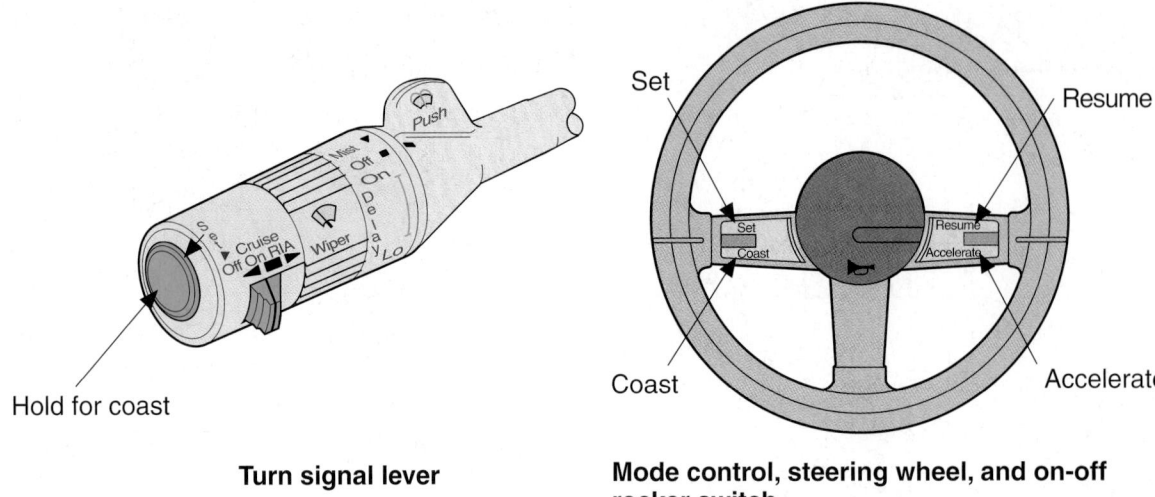

Turn signal lever

Mode control, steering wheel, and on-off rocker switch

Figure 34.67 The cruise control switch on the turn signal lever or steering wheel is used to set speed, accelerate, or resume after braking. The coast position does not apply the brake lights in case a law enforcement vehicle is behind you. *(Courtesy of Ford Motor Company)*

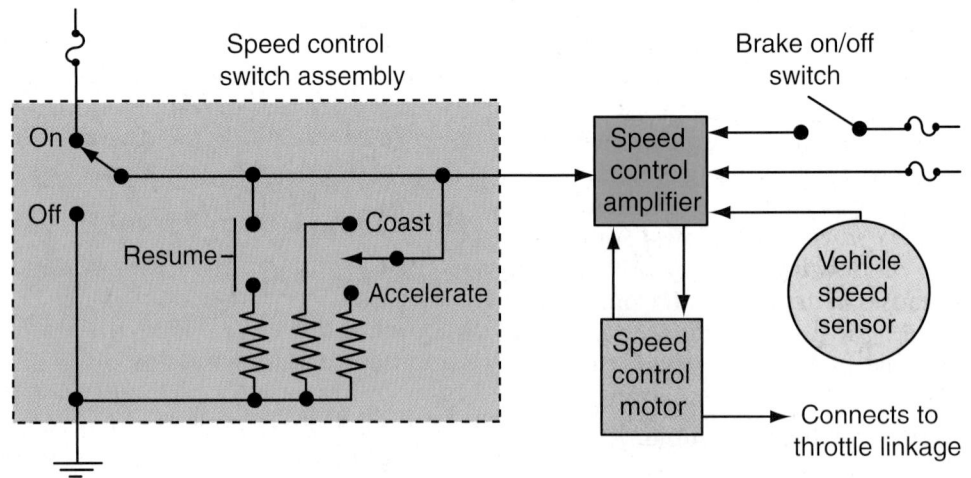

Figure 34.68 Electronic cruise control inputs include a vehicle speed sensor (VSS) and a brake on/off switch.

Electronic Cruise Control

The vehicle computer for electronic cruise control analyzes and acts on inputs of vehicle speed and engine load. Most of today's vehicles use a stepper motor that pulls on a strap to move the throttle. Many vehicles use "drive by wire," in which the throttle is controlled by a computer. These vehicles are basically on full-time cruise control, which is continuously selected by throttle position as the driver moves the gas pedal. Electronic cruise control inputs include a vehicle speed sensor to sense vehicle speed and switches to release the clutch and/or brake (**Figure 34.68**).

Intelligent Cruise Control

Some newer vehicles have an optional *intelligent cruise control*, also called adaptive cruise control, that maintains a selected distance from the vehicle ahead dependent on its speed (**Figure 34.69**). A radar or infrared sensor in

Figure 34.69 Adaptive cruise control maintains a selected distance from the vehicle in front. *(Courtesy of Delphi)*

Figure 34.70 A sensor located in the grille senses the distance to a vehicle in front and provides information to the control module. *(Courtesy of Delphi)*

the grille of the vehicle (**Figure 34.70**) detects objects in front and sends a signal to a module. **Figure 34.71** shows an electrical diagram of the system. The computer can detect a vehicle at up to 390' (120 m), determining its speed, and applies either the brakes or accelerator to maintain the designed distance. Typical distance settings at 60 mph are 195', 130', and 90'. The brakes can apply

at 25% braking power, which is a substantial amount of braking. When the road ahead is clear, the system accelerates once again to the preset speed. During slowing, the system illuminates the brake lights. A driver has the option of using either the intelligent system or conventional cruise control.

The distance sensor has a narrow range of operation so it does not see oncoming vehicles on narrow roads. However, system performance can be poor on winding roads. The sensor can only sense hard objects like a motorcycle but cannot detect soft objects like animals or humans. The system disengages during wet weather to prevent contamination of the sensor and confusion of the computer.

Cruise Control Service

When servicing all types of cruise control systems, begin by checking the fuse. Check linkage to the throttle control to see that it is not broken or binding up. On older cruise control systems look for damaged vacuum hoses and bad electrical connections. Check the drive shaft sensor or vehicle speed sensor on the transmission. Service literature provides instructions for cruise control calibration so the vehicle does not accelerate or decelerate too roughly.

Always check the brake lights to see that they operate. The brake on and off switch is an essential part

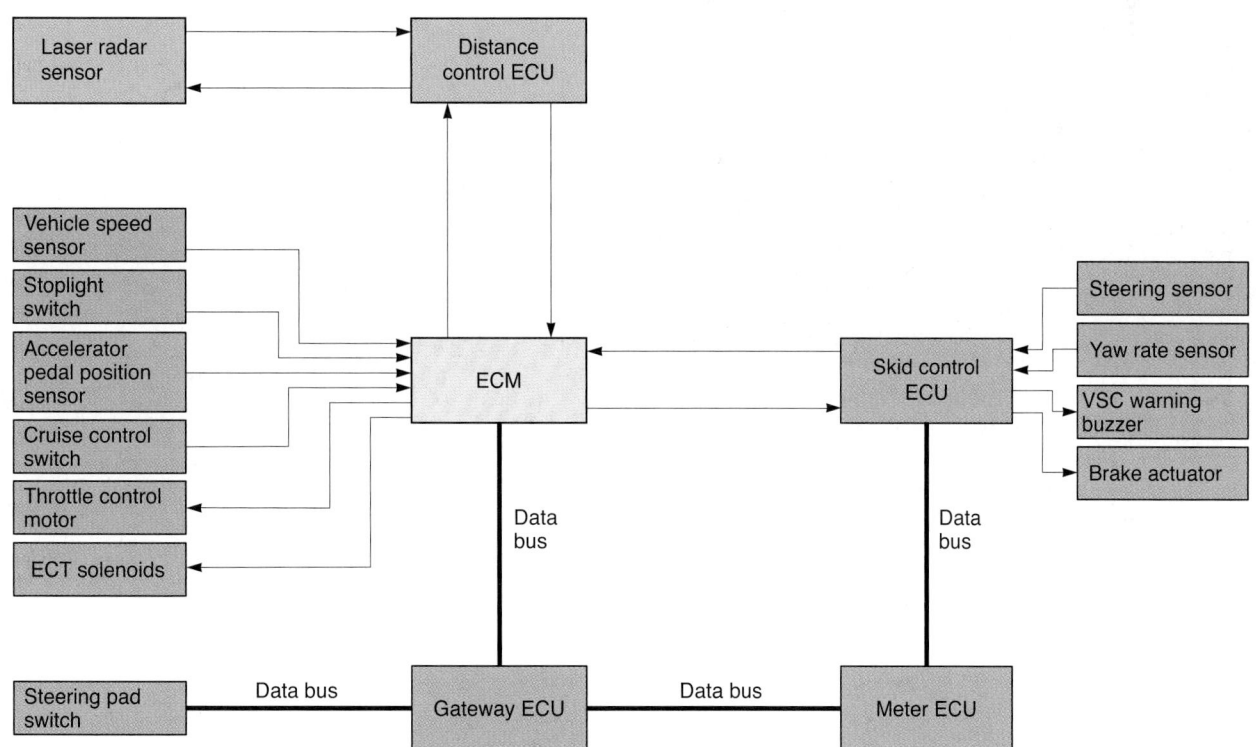

Figure 34.71 An electrical schematic of a laser radar system.

of cruise control, and if the brake lights do not work, cruise control cannot operate.

Computer-controlled cruise control diagnosis begins with a look at the service information. This is followed by testing with a scan tool to find defective sensors and switches or to locate a problem in an electrical circuit.

■ LANE DEPARTURE WARNING SYSTEM

A lane departure warning (LDW) system uses a camera that sees the lines on the road (**Figure 34.72**). A module flashes a light and sounds a chime to alert the driver if the vehicle moves too close to the paint stripe marking either edge of the lane. The system does not operate below 45 mph or when the turn signal has been activated on the side of the vehicle that is approaching the paint stripe.

A typical LDW system has a camera lens with a control unit beneath the headliner. LDW is on the CAN system, so it uses very few wires. **Figure 34.73** shows a CAN schematic for an LDW system.

Service to the system is minimal. The calibration process for the camera is done electronically following service instructions that aim the camera at a target a specified distance to a wall. A scan tool does a self-calibration.

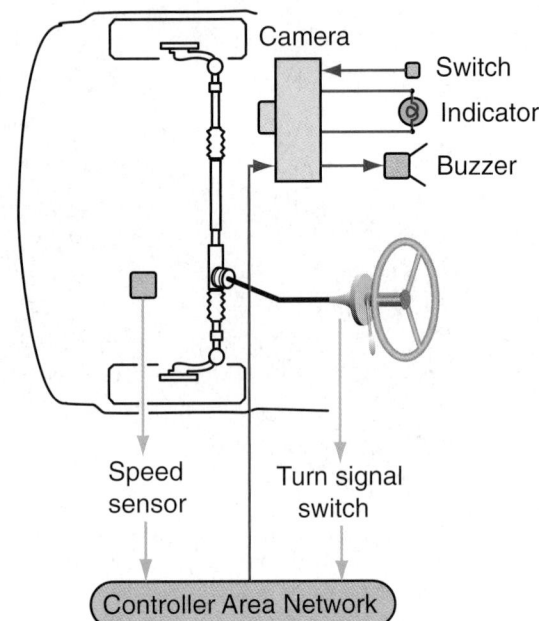

Figure 34.73 A CAN schematic for a lane departure warning system.

Figure 34.72 A lane departure warning system uses a camera that sees the lines on the road.

■ REVIEW QUESTIONS

1. Automatic seat belts or air bag systems are examples of _____ restraints.

2. What is a combination lap belt and shoulder belt called?

3. Since the year _____, supplemental restraint systems have been required on all new cars sold in the United States.

4. The _____ air bag reduces the risk of the person sliding free under the seat belt and keeps him or her in a better position for protection by a front air bag.

5. The air bag control module has an electrical storage capacitor as a safety backup in the event that the vehicle's _____ is damaged in the collision.

6. What is the name of the sensor used to determine that there has actually been a crash?

7. What are three names for the fuse-like ignition device used in an air bag?

8. What is the name of a restraint system with ultrasonic sensors that monitor passenger position and seat belt usage and adjust the force of the air bag deployment?

9. A sodium azide pyrotechnic inflator produces _____ gas to fill the air bag.

10. Where is the ground connection for the entire system located on electric window systems?

11. What is the name of the worldwide radio-navigation system formed from a constellation of 24 satellites and their ground stations?

12. How much farther can infrared night vision "see" than the driver can with the headlights on low beam?

13. Which kind of radio signal varies the *strength* of the broadcast signal? What kind varies the *frequency* of the signal?

14. What kind of radio speaker produces nondirectional sound (you have trouble telling where it comes from)?

15. Which radio speaker is more likely to be located high on a door panel, a woofer or a tweeter?

■ ASE-STYLE REVIEW QUESTIONS

1. All of the following are true about cruise control *except*:

   a. Stepping on the brake or clutch pedal will deactivate cruise control.

   b. A driver-controlled switch can reinstate cruise control to the originally selected vehicle speed.

   c. Adaptive cruise control uses radar or a laser sensor mounted near the rear bumper.

   d. On some systems if the brake lights do not work, cruise control cannot operate.

2. Technician A says that FM radio is subject to a higher amount of interference than AM radio. Technician B says that the range of the AM broadcast is restricted to "line-of-sight" distances. Who is right?

   a. Technician A   **c.** Both A and B

   b. Technician B   **d.** Neither A nor B

3. Technician A says that it is rare for both front air bags to deploy together. Technician B says that at least two sensors must say there is a crash before an air bag will be deployed. Who is right?

   a. Technician A   **c.** Both A and B

   b. Technician B   **d.** Neither A nor B

4. Each of the following statements about antitheft systems is true *except*:

   a. Each key must be programmed to match the vehicle's computer before it will start the engine.

   b. If a PCM is replaced the engine will not start unless the PCM is reset to match the transponder.

   c. All manufacturers' antitheft systems disable the starter motor.

   d. If an engine starts and runs for longer than 1 second, the antitheft system is not the cause of engine stalling.

5. Which of the following is/are true about electric seats?

   a. Four-way seats usually have two motors.

   b. Some six-way seats have one motor.

   c. Some seats have eight motors.

   d. All of the above.

6. Technician A says that window motors must be properly grounded in the door. Technician B says that the polarity of a window motor changes with the direction of current flow through its two wires. Who is right?

   a. Technician A   c. Both A and B

   b. Technician B   d. Neither A nor B

7. Technician A says that FM radios need the longest antenna possible. Technician B says that AM antennas should be exactly 31 inches (79 centimeters) long. Who is right?

   a. Technician A   c. Both A and B

   b. Technician B   d. Neither A nor B

8. Each of the following statements is true about adaptive cruise control *except*:

   a. A brake switch signal shuts off the cruise control.

   b. Adaptive cruise control can apply the brakes.

   c. The distance sensor can see oncoming vehicles on narrow roads.

   d. The distance sensor cannot detect animals.

9. Two technicians are discussing power seats. Technician A says that most vehicles use a fuse in the power feed to the seat. Technician B says that a four-way seat allows independent adjustment of the front or rear height of the seat. Who is right?

   a. Technician A   c. Both A and B

   b. Technician B   d. Neither A nor B

10. Technician A says that electric windows are powered by two switches wired in parallel. Technician B says that a problem with either the window switch or the master switch will prevent the window from operating. Who is right?

    a. Technician A   c. Both A and B

    b. Technician B   d. Neither A nor B

Heating and Air Conditioning

THEN AND NOW: CLIMATE CONTROL

The dangers and discomforts in the early days of motoring would probably discourage even the toughest motocross racer today. Headlights were so dim you could barely see a locomotive in your path. Primitive suspensions were rough enough to chip your teeth. Climate control systems (air conditioning and heating) were simply nonexistent.

The first heater simply used hot bricks lined up on the floorboards. The next step was a footwarmer that burned blocks of charcoal. Those ideas were obviously inefficient since internal combustion engines create a good amount of waste heat anyway. Early car makers developed ducts that directed air over the exhaust manifold into the passenger compartment, but this was never very satisfactory.

Using the liquid from water-cooled engines to provide heat made good sense. This idea was tried as early as 1897 on the Canstatt-Daimler. It was not until 1931, however, that Lincoln produced a heater of the type we know today. It had finned tubes in a housing (heater core) coupled with an electric fan to move air. Flaps were used to control the volume and direction of heat.

The *gasoline heater* was made popular by Ford in the 1930s. A blowtorch-like flame and a fuel line inside the passenger compartment would seem dangerous today, but this system was very effective in dealing with frost.

A car's interior is very successful in collecting solar energy. There were attempts to chill the passenger compartment long before refrigeration—or even the automobile—was invented. In 1884, William Whiteley placed blocks of ice in trays under horse carriages. A fan attached to a wheel forced air to the interior of the carriage. A bucket of ice in front of a floor vent was the automotive equivalent to this early comfort control system.

The first car with real air conditioning was the 1939 Packard. A huge evaporator was mounted in the trunk, leaving very little space for luggage. Cadillac was the next car with air conditioning. In 1941, 300 Cadillacs were built with the air-conditioning option.

A recent change in modern air-conditioning systems has been the switch from R-12 Freon® (which damages the Earth's ozone layer) to a more environmentally friendly refrigerant, R-134A. Freon production has been outlawed, and older cars are being converted to the new refrigerant.

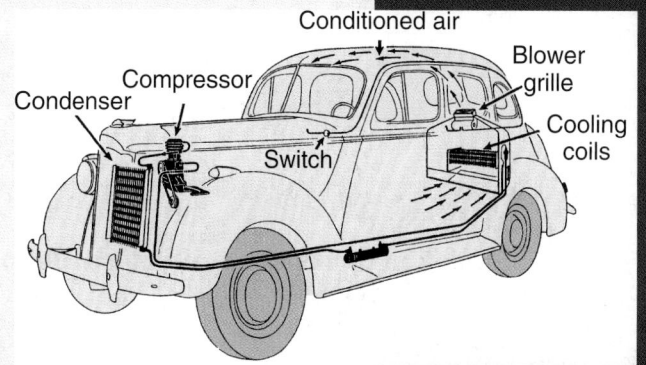

A 1939 Packard with A/C.

A gauge set for servicing A/C systems with the more environmentally friendly R-134A refrigerant. *(Courtesy of Bob Freudenberger)*

 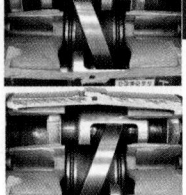

Heating and Air-Conditioning Fundamentals

■ **OBJECTIVES**

Upon completion of this chapter, you should be able to:

✔ Explain refrigeration fundamentals.

✔ Describe the difference between the high and low sides of the system.

✔ List the major heating and air-conditioning parts and describe their operation.

■ **KEY TERMS**

squirrel cage fan	sensible heat	ambient air
blend air door	latent heat	high side
air conditioning	latent heat of vaporization	low side
convection	condensation	desiccant
radiation	latent heat of condensation	CFC
evaporation	tons rating	inches of mercury (in. Hg)
humidity	saturated vapor	

■ **INTRODUCTION**

In this chapter, the basic operation and service of heating and air-conditioning systems are covered. As in previous chapters, the emphasis is on learning how the system operates and how to perform minor service and diagnostic procedures.

■ **SOURCES OF HEAT**

Like systems for buildings, the automobile heating and air-conditioning system is called a heating, ventilation, and air-conditioning system (HVAC). In the winter, the system adds heat to the inside of the vehicle. In the summer, it removes interior heat.

Heat comes from several sources (**Figure 35.1**).

■ Each passenger adds heat to the inside of the car, mostly from their breath. Each passenger's temperature is approximately 98.6°F. The more passengers, the more heat.

■ Heat from the outside air contributes to 15% of the total heat.

■ Heat that comes off the road, the engine, and the catalytic converter accounts for 20%.

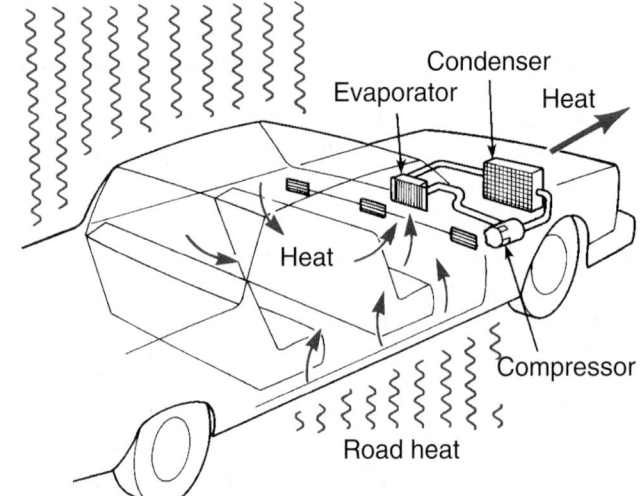

Figure 35.1 Heat from the sun, air, and road is removed from the vehicle's interior by the air-conditioning system. *(Courtesy of Four Seasons, Division of Standard Motor Products, Inc.)*

- Most of the heat comes from sunlight, which radiates through glass or shines on the painted surfaces of the car. Insulation of the interior helps to keep the heat inside the car.

VENTILATION

Fresh air is introduced into the vehicle interior to replace stale air and to prevent the possibility of carbon monoxide entering from the vehicle's exhaust. Venti-

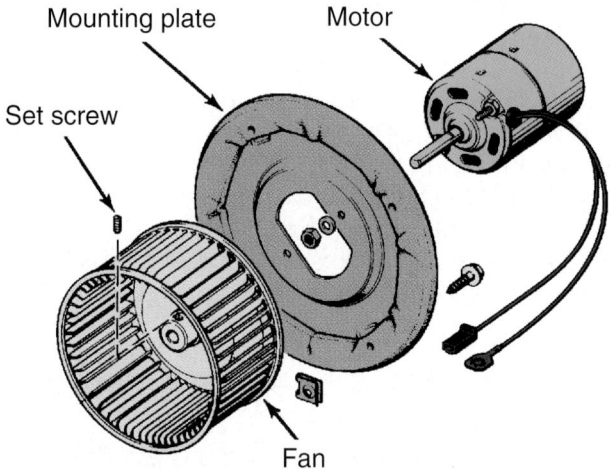

Figure 35.2 A blower motor and squirrel cage fan. *(Courtesy of DaimlerChrysler Corporation)*

lation can be provided by blowing or ducting outside air into the vehicle's interior. Air ducts, or vents, allow outside air to ventilate the inside of the car. Ventilation from outside air does not work when the car is moving at slow speeds or is stopped in traffic. The ventilation system's electrically driven blower motor and **squirrel cage fan** (**Figure 35.2**) take care of these needs.

HVACs share the same blower motor, which has multiple speed settings. On many cars, instead of an "off" setting, the blower runs at low speed when the ignition key is on. This maintains a fresh flow of air into the interior of the car and also creates positive cabin pressure to keep out exhaust when the vehicle is stopped.

HEATING

Hot air is provided by routing engine coolant to a heater core in the passenger compartment (**Figure 35.3**). When the heater operates, outside air or air from the car's interior passes over the fins of the heater core, radiating heat to the passenger compartment.

NOTE: *An engine cooling system operating at the correct temperature with a properly operating thermostat is necessary for the heater to be effective.*

Air Distribution

Some systems have a heater control valve (**Figure 35.4**) operated by a cable, vacuum, or electricity. Other

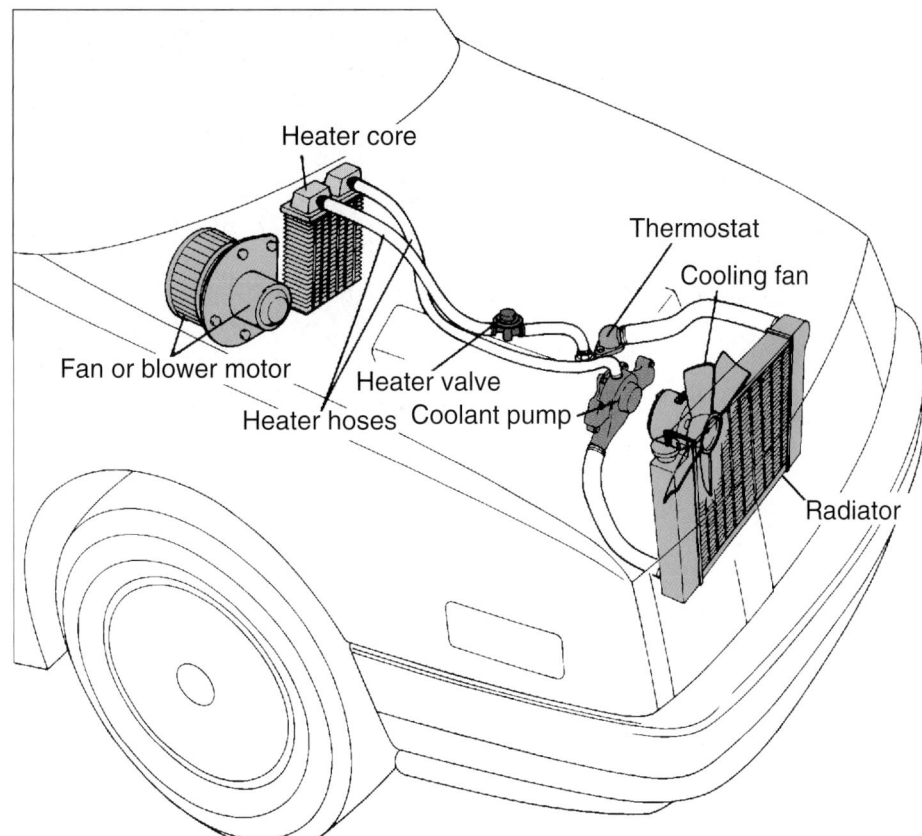

Figure 35.3 Coolant flows through the heater core to warm the passenger compartment. *(Courtesy of Four Seasons, Division of Standard Motor Products, Inc.)*

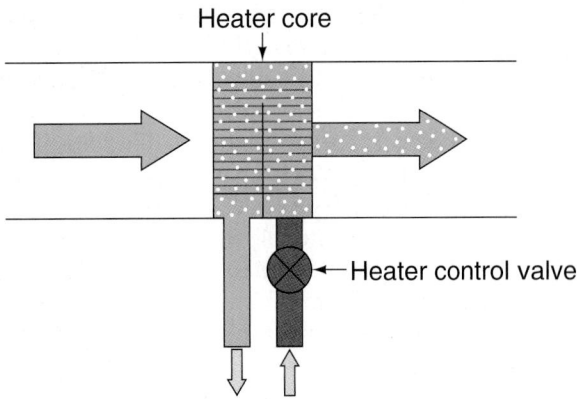

Heater core

Heater control valve

Figure 35.4 This heater uses a control valve.

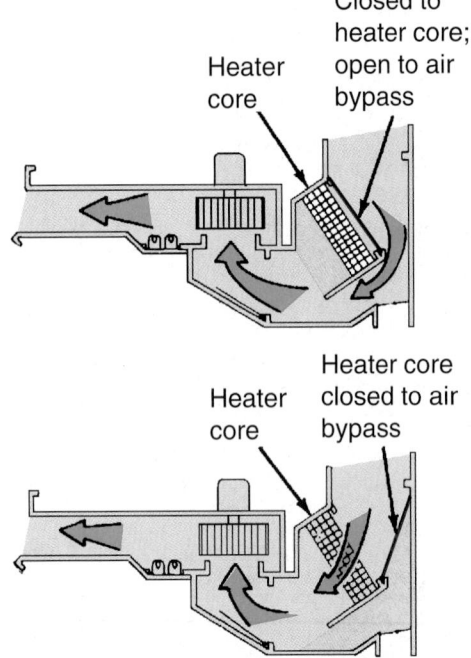

Closed to heater core; open to air bypass

Heater core

Heater core closed to air bypass

Heater core

Figure 35.5 The blend, or mixture control, door is closed to control the flow of air across the heater core. *(Courtesy of DaimlerChrysler Corporation)*

systems allow coolant to flow constantly through the heater core. Heat to the passenger compartment in these systems is controlled by whether or not air is passed over the heater core. This system is more responsive to automatic controls and is found on most new cars. The *air mix damper,* or **blend air door**, can be opened and closed to provide hot air immediately (**Figure 35.5**). If this type of system has a heater control valve, it is used during periods of maximum-cooling mode only.

There are doors for three purposes: defrost, vent, and floor. The airflow doors control whether air from the heater or A/C goes to the windshield defroster (on the top of the dashboard) or to the ventilation outlets (**Figure 35.6**). Sometimes, ducts are provided under the front seats for heating or cooling the rear seat passengers. Another door controls how much air is allowed to pass across the heater core. A third door may regulate the proportion of outside air to recirculated interior air that is ducted to the blower. Doors are controlled by cables, by vacuum motors, or by electric motors (**Figure 35.7**).

Figure 35.7 An electric air door actuator motor.

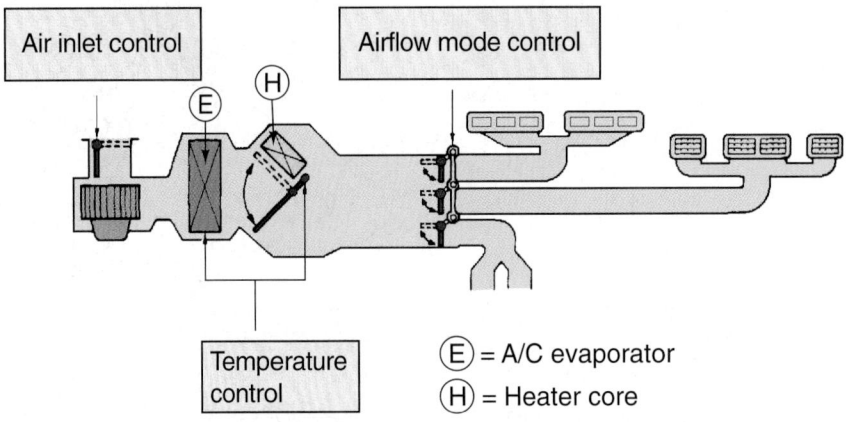

Air inlet control

Airflow mode control

Temperature control

E = A/C evaporator
H = Heater core

Figure 35.6 Doors control whether the air goes to defrost, the floor, or the dash.

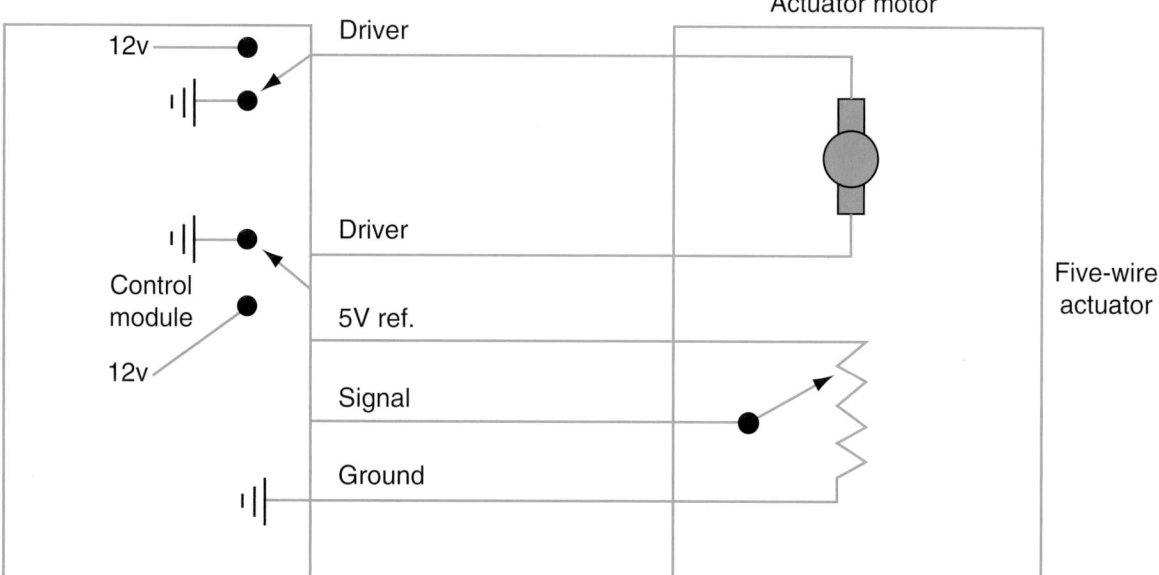

Figure 35.8 A feedback potentiometer door circuit that uses a variable resistor to provide door position input to the control module.

Cables are simple and reliable, but with time they tend to stretch. Vacuum doors require a reservoir so they can work when the vehicle is going up a hill or under acceleration, especially at wide-open throttle.

Electric motor–controlled doors are usually found on late-model vehicles because they are reliable and can be controlled very effectively by a computer. A simple two-wire door motor circuit moves either open or closed. Many newer vehicles use a feedback door with a five-wire actuator. The door uses a feedback potentiometer, a variable resistor that provides input to the control module as the door moves (**Figure 35.8**). The computer can command the door into many different positions between open and closed. Some actuators use a logic module to control blend door position.

■ AIR CONDITIONING

Air conditioning is the process in which air inside of the passenger compartment is cooled, dried, and circulated. Comfort depends on temperature, humidity, and air movement. Heat is removed from inside the vehicle and transferred to the outside air.

■ HISTORY NOTE

In the 1920s, household refrigerators were first introduced. The first refrigeration unit on an automobile was available as a luxury extra on the Packard automobile in 1939, but air conditioning did not become popular with the motoring public until the 1960s. Today, over 80% of cars sold have air conditioning. Refrigeration systems are also used on off-road and farm machinery.

Air-conditioning systems originally were manually controlled by the passengers. Many of today's systems are automatic. At one time, it was thought that air conditioning was not energy efficient because it put an additional load on the engine. But today's aerodynamic cars are designed to be run on freeways with the windows up for the least wind resistance. A car gets better freeway fuel economy with the air conditioning on and the windows up than it does with the windows down and the air conditioning off. In fact, at speeds above 40 mph more gas is used with the windows down.

■ AIR-CONDITIONING PRINCIPLES

Automotive air conditioning works on the same principles on which household refrigerators and air conditioners work. A liquid refrigerant is changed to a gas and then back to a liquid again (**Figure 35.9**). If a change of state of the refrigerant is to take place, there must be a transfer of heat. Two principles apply here:

■ For a liquid to change to a gas, it must absorb heat.
■ For a vapor to change to a liquid, it must release heat.

■ HEAT TRANSFER

Whenever there is a difference in the temperatures of two objects, heat can be transferred. Heat will flow to anything that has less heat. Heat transfer occurs by **convection**, **radiation**, or **evaporation**.

Convection

When a body gives off heat, the surrounding air becomes warmer and moves upward. Air that contains less heat takes its place. This is convection, one

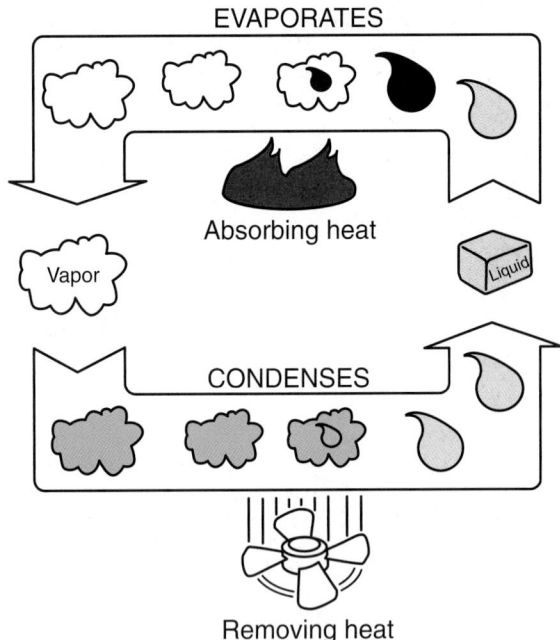

Figure 35.9 The refrigeration process.

of the processes of removing heat. Two principles of convection are

■ Heat rises (notice how steam or smoke rises).
■ Heat always flows from hot to colder.

In a car air-conditioning system, air is not actually cooled; heat is taken away. Have you ever walked past the outside of a house air conditioner installed in a window? Did you feel the hot air coming off the back of it? Heat is being removed from the house.

Radiation

Another way heat is transferred is by radiation. The effects of the sun's radiation are felt by the human body when moving from the shade into bright sunlight. Dark colors absorb and radiate heat better than light colors.

Evaporation

Heat is also transferred by evaporation. As moisture is vaporized it absorbs heat, cooling the surface. Your body feels the effect of evaporation as you perspire. The cooling effect is more noticeable with wind, as it speeds up the vaporization process.

■ HUMIDITY

Besides cooling the air, air-conditioning systems provide **humidity** control too. When humidity is 100%, the air is totally saturated with moisture. When humidity is 50%, the air is holding half the amount of moisture that it is capable of holding at a given temperature. Low humidity (dry air) permits heat to be taken away from the human body by evaporation of perspiration. Notice how after a shower if a breeze hits you, you feel cool from evaporation.

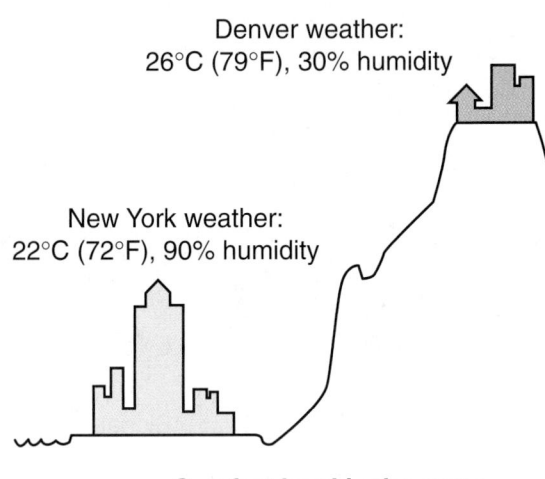

Denver weather:
26°C (79°F), 30% humidity

New York weather:
22°C (72°F), 90% humidity

Comfort level is the same

Figure 35.10 Humidity affects the level of comfort.

High humidity makes evaporation more difficult. The human body is most comfortable at between 72°F and 80°F (22.2°C to 26.6°C) with humidity at 45% to 50% (in street clothes while riding in an enclosed vehicle). People feel just as cool at 79°F with 30% humidity as they do at 72°F and 90% humidity (**Figure 35.10**).

■ STATES OF MATTER

All common matter can exist in three different states depending on its temperature. The three states of matter are *solid*, *liquid*, and *gas*. When a solid is heated above its freezing point, it begins to melt, becoming a liquid.

■ LATENT HEAT

Heat that goes into matter and results in a temperature increase is called **sensible heat** because it is easy to understand (sensible). Before matter can actually change its state, extra heat is required. This additional heat is called **latent heat** (hidden heat). Automotive air conditioning operates using this principle.

NOTE: *After the refrigerant reaches its boiling temperature (−22°F at atmospheric pressure) it absorbs more and more heat, yet its temperature does not increase. It is simply absorbing heat as it attempts to change its state from a liquid to a gas.*

Latent heat cannot be recorded on a thermometer.

Quantity of Heat

One candle's temperature is measured at 500°F, but if you add another candle, the temperature of the two candles combined is still 500°F. This is called quantity of heat, which can be measured in British thermal units (Btu), or calories. One Btu is the amount of energy required to raise the temperature of 1 pound of water by 1°F.

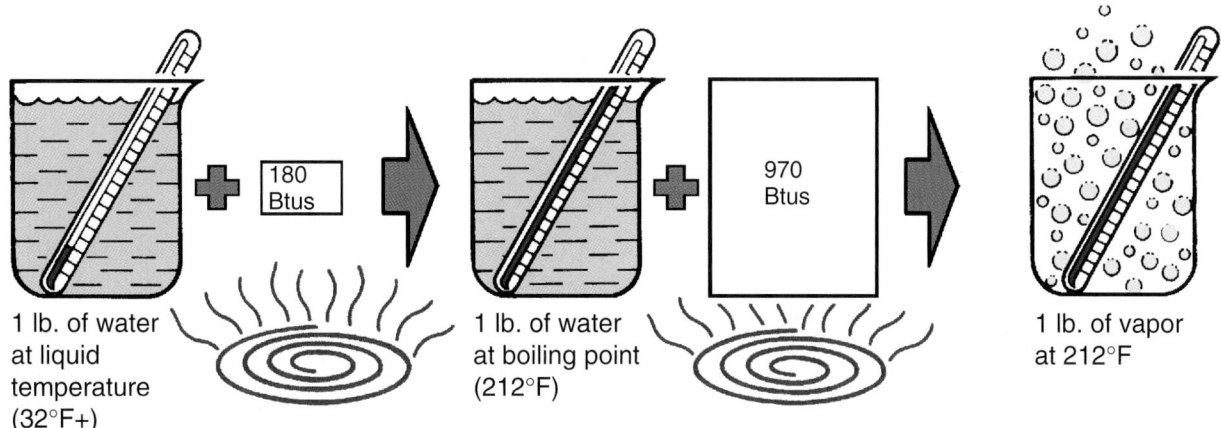

Figure 35.11 It takes 970 Btus to make water boil without raising its temperature. *(Courtesy of Ford Motor Company)*

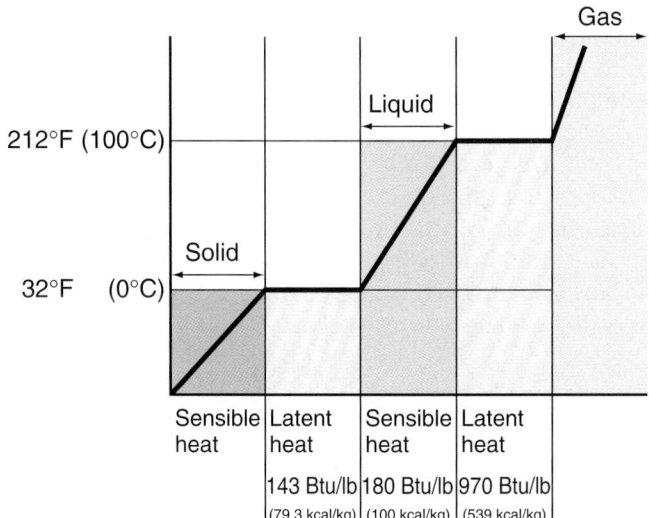

Figure 35.12 The relationship between latent heat and sensible heat.

Vaporization

The boiling point of water is 212°F at sea level. One pound of water that has already been heated to 212°F will require an additional 970 Btu of heat to make it boil. None of this heat can be recorded on a thermometer, so the temperature of the boiling water and steam will remain at 212°F. This is called the **latent heat of vaporization** (**Figure 35.11**). **Figure 35.12** shows the relationship between latent heat and sensible heat.

NOTE: *The amount of energy it takes to raise water's temperature from freezing to boiling is much less than the amount of latent heat of vaporization (the amount of heat needed to make the water boil).*

Condensation

When moisture from steam condenses on a cool bathroom mirror, a vapor changes to a liquid. This is called **condensation**. As a vapor is condensing, it releases its latent heat. When steam condenses back to water, it releases 970 Btu of heat per pound. This heat released during condensation is called **latent heat of condensation**.

Cooling means "taking away heat." During evaporation heat is absorbed. The lower the boiling temperature of a liquid, the easier it evaporates. Alcohol feels cooler on your hands than water does. This is because alcohol has a lower boiling point. As it evaporates it pulls heat from your hands.

A household window air-conditioning unit is rated in how many British thermal units it removes (20k Btu, for instance). A typical GM vehicle air conditioner is rated at about 20,000 Btu. It needs to be so much greater because of the heat load mentioned before. The way that the heat load is figured is with the **tons rating**, or tonnage. The tons rating is how much heat needs to be added in 24 hours to turn 1 ton of ice at 32° into water; 1 ton = 12,000 Btu/hr.

■ AIR-CONDITIONING SYSTEM OPERATION

An air-conditioning system consists of four major devices in a closed system: the *compressor*, the *condenser*, the *evaporator*, and a *metering device*. Refrigerant circulates among these devices. A flow control device regulates the flow of refrigerant among them (**Figure 35.13**). For an air-conditioning system to operate, there must be large differences in pressure within the system. Changing pressure and the gas or liquid state of the refrigerant regulates the operation of the cooling cycle. The air-conditioning cycle has four stages: *compression*, condensation, *expansion*, and vaporization.

■ ABSORBING HEAT

Located inside of the car's interior is a small, radiator-like device called the evaporator. *Liquid* refrigerant is circulated to the evaporator. As the refrigerant exits the metering device, it loses pressure. There is a slight amount of flash gas at the metering device. Then liquid flows into the evaporator.

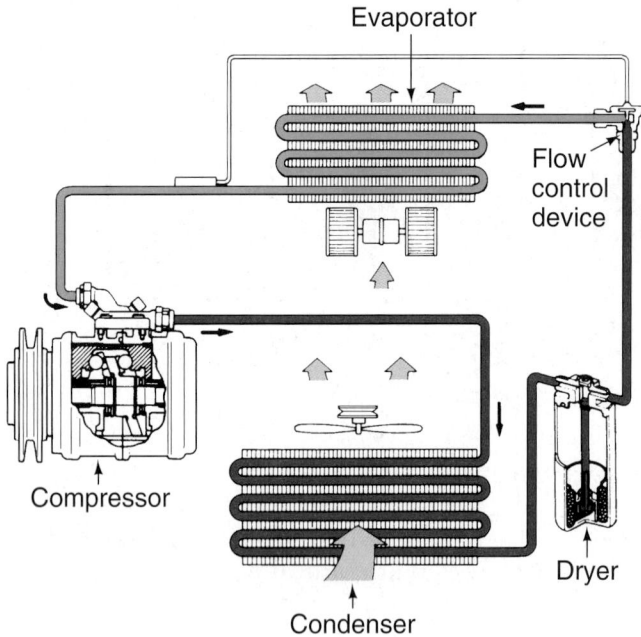

Figure 35.13 A flow control device regulates the flow of refrigerant.

The refrigerant is a **saturated vapor**. A saturated vapor is a liquid that is in contact with its vapor within an enclosed space.

The liquid refrigerant absorbs heat from the inside of the car. This is the same as when alcohol or another liquid evaporates from skin, leaving it cool. When the refrigerant absorbs enough heat from the air flowing across the evaporator, it boils and changes to a vapor.

When the refrigerant is pressurized again in the engine compartment, it gives off the heat to the surrounding outside air. When the air-conditioning system is first turned on, each cycle through the evaporator absorbs at least 25°F of heat from the air blowing across it. Typical temperature readings would be 90°F air entering the evaporator and 65°F air exiting it. As the temperature in

the car goes down, there is not as much heat to remove, so the temperature difference will become less.

■ REDUCING HUMIDITY

Humidity can enter the car's interior as moisture from the outside air or from the breathing of passengers. When moisture condenses on the cool window surfaces, visibility problems can result. One of the air-conditioning system's functions includes dehumidifying the air inside the car. Warm air holds more moisture than cold air. When warm air passes across the cool fins of the evaporator, it loses heat. Moisture in the air tends to condense on the evaporator fins, like it does on the cool glass of a mirror.

The moisture that deposits on the cold evaporator core collects in a drain pan beneath it and is drained off through the floor as water. This accounts for why a car recently parked after the air conditioning has been operating will often have a stream of water draining onto the ground from the bottom of the front/right side of the passenger compartment.

Because of the principle of latent heat, as the vapor turns to a liquid more heat is released into the surrounding air. Therefore, when the humidity is high, the air-conditioning system does not cool the air as much. You will get more comfortable, however, because once the humidity is lowered your body can lose its heat more effectively.

Defroster Operation

When a driver selects the defrost mode, outside air is pulled in by the blower and pushed across the fins of the evaporator. Any moisture in the air condenses onto the cold evaporator, which dries the air. The dried cool air moves through the heater core, where it absorbs heat before it is blown onto the windshield (**Figure 35.14**).

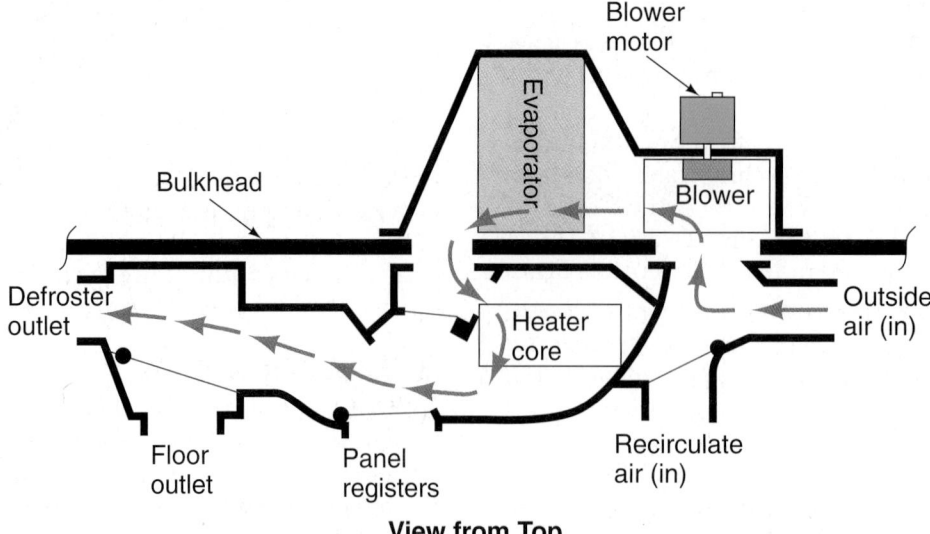

View from Top

Figure 35.14 In defrost mode the evaporator dries the air before it is heated by the heater core.

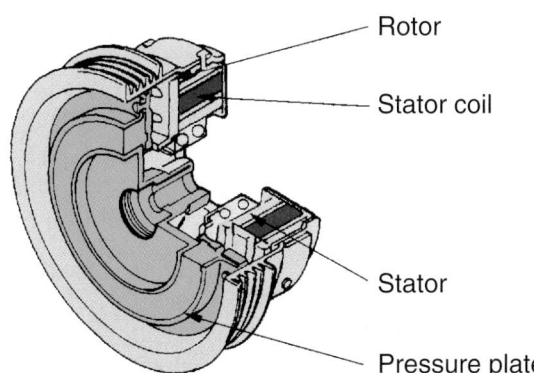

Figure 35.15 Cutaway of a magnetic air conditioner clutch.

NOTE: *Many vehicles sold in Alaska are ordered without air conditioning, but areas of the state can be very moist. A defroster without an air conditioner does not work as well under these circumstances.*

■ COMPRESSING THE REFRIGERANT

The vaporized refrigerant is pulled through the suction line from the evaporator to the *compressor* in the engine compartment. Most compressors are driven by a belt from the engine crankshaft, although many hybrid cars use electrical compressors. The compressor pressurizes the heated refrigerant, further increasing its temperature. Raising the temperature of the refrigerant before it goes to the condenser makes the condenser more efficient at removing heat.

Compressor Clutch

The compressor has an electromagnetic clutch to connect and disconnect it from the crankshaft pulley (**Figure 35.15**). In some systems, clutches are only released during periods of heavy loads. Other systems use the clutch to cycle the compressor on and off. Many clutch assemblies are controlled through the computer.

■ TRANSFERRING REFRIGERANT HEAT TO OUTSIDE AIR

From the compressor, the refrigerant is pumped through the dischargeline to the condenser, located in front of the engine's radiator (**Figure 35.16**). The condenser has metal tubes/fins. The condenser is a radiator for refrigerant. Its job is to transfer the heat that was absorbed in the passenger compartment to the cooler air, or ram air, blowing through it. The refrigerant's boiling temperature becomes higher when it is pressurized. This is the same idea as pressurizing radiator coolant with a radiator cap to raise its boiling point. Cooling the pressurized refrigerant in the condenser causes it to change from a gas to a slightly cooled but still warm liquid.

Pressure in the condenser must be high enough to operate the flow control device and raise the refrigerant temperature well above that of the **ambient** (surrounding) **air**. The refrigerant becomes concentrated

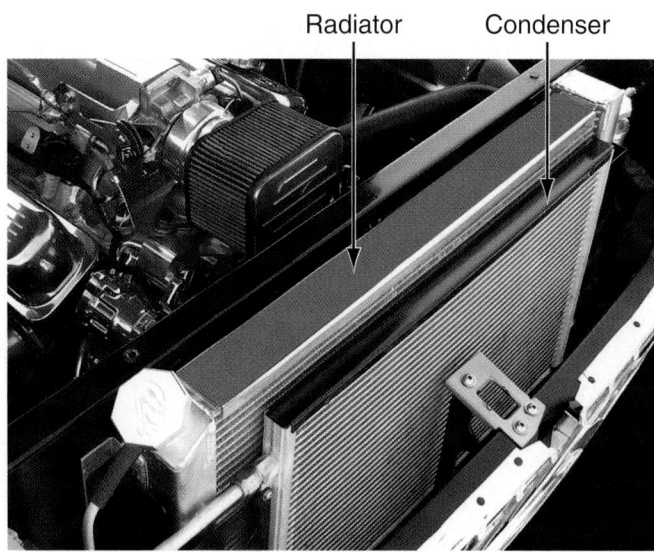

Figure 35.16 The condenser is located in front of the radiator. *(Courtesy of Tim Gilles)*

and very hot when it is compressed. It must be hotter than the air coming across the condenser or heat transfer cannot take place. This allows rejection of the heat that was absorbed as the refrigerant was flowing through the evaporator.

■ FLOW CONTROL DEVICES

Air-conditioning systems require a device to control refrigerant flow. There are two basic systems in use: one using a thermostatic expansion valve, and the other using an orifice tube. An expansion valve system is used with a receiver/dryer, and an orifice tube system is used with an accumulator dryer.

To raise pressure, there must be a restriction in the air-conditioning system. The restriction divides the system into the **high side** and the **low side**. The terms *high side* and *low side* refer to high pressure and low pressure within the system (**Figure 35.17**). The flow control device, located in the liquid line between the condensor and the evaporator, lets the high pressure off the refrigerant as it "trickles" into the evaporator. When the pressure on the refrigerant is lowered, it can evaporate at a lower temperature. During evaporation, heat is absorbed from the passenger compartment. The low-pressure side is located after the flow control device.

The refrigerant flowing through the evaporator has boiled, and yet it continues to absorb more heat and increase in temperature. Even though the boiling point is cold (−22°F), the refrigerant is still superheated past the evaporator.

Superheat refers to temperatures that are above a liquid's boiling point. Superheat in the air-conditioning system is the temperature difference of the refrigerant between the inlet and outlet of the evaporator. It is the difference between the refrigerant's boiling point at system pressure at that moment and the outlet temperature of the refrigerant.

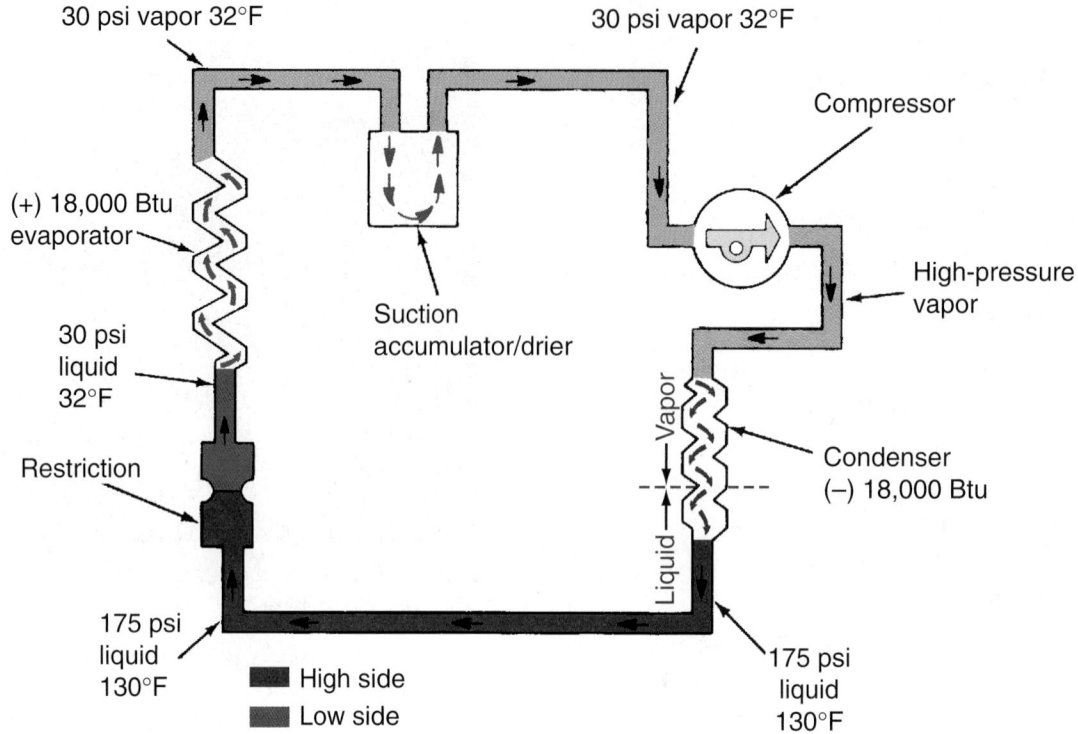

Figure 35.17 The high-pressure side and the low-pressure side of the air-conditioning system.

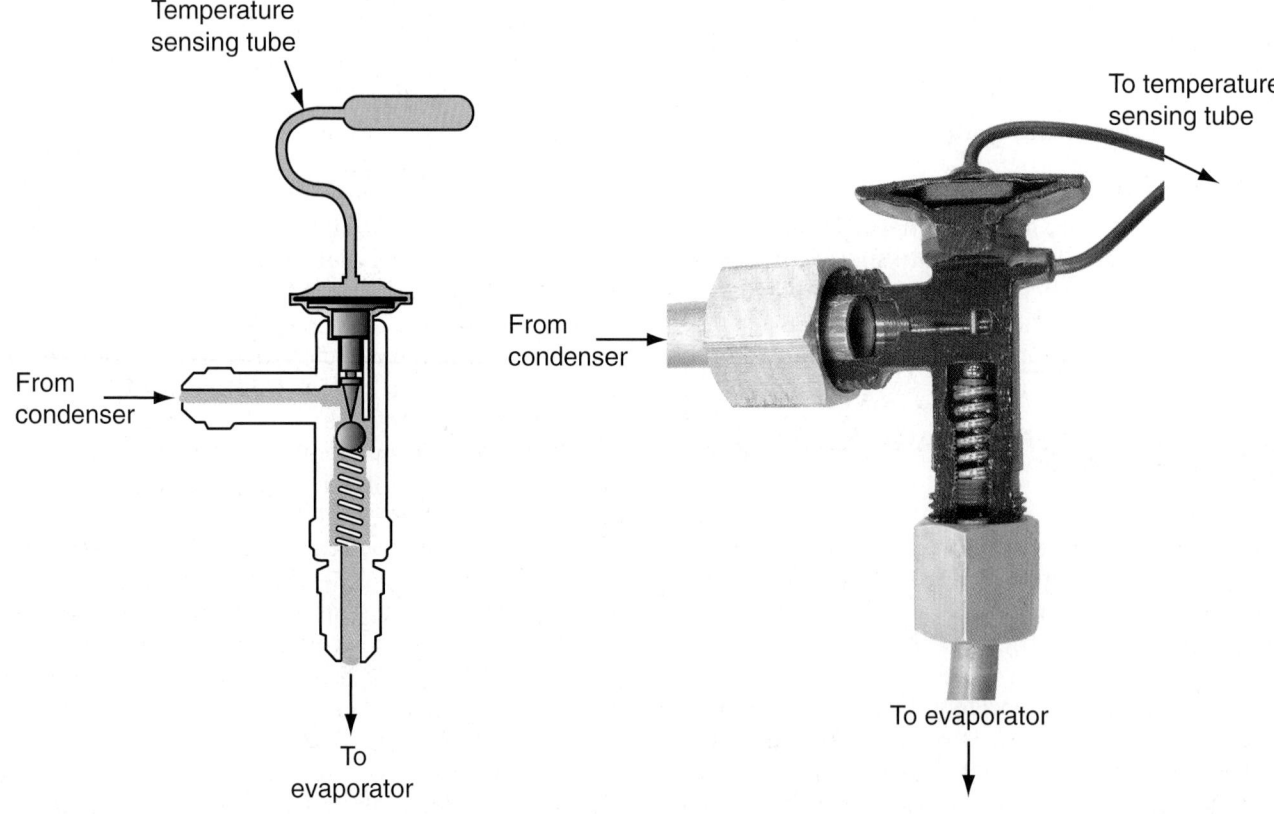

Figure 35.18 Cutaway of an expansion valve. *(right, Courtesy of Tim Gilles)*

Expansion Valve

One type of flow control device is a metering valve called an expansion valve (**Figure 35.18**). A *thermostatic expansion valve* (*TXV*) controls the amount of refrigerant allowed to flow to the evaporator (**Figure 35.19**).

A temperature sensing bulb on the evaporator outlet controls the expansion valve.

A primary purpose of the flow control device is to control the amount of refrigerant flowing into the evaporator. A colder evaporator removes more

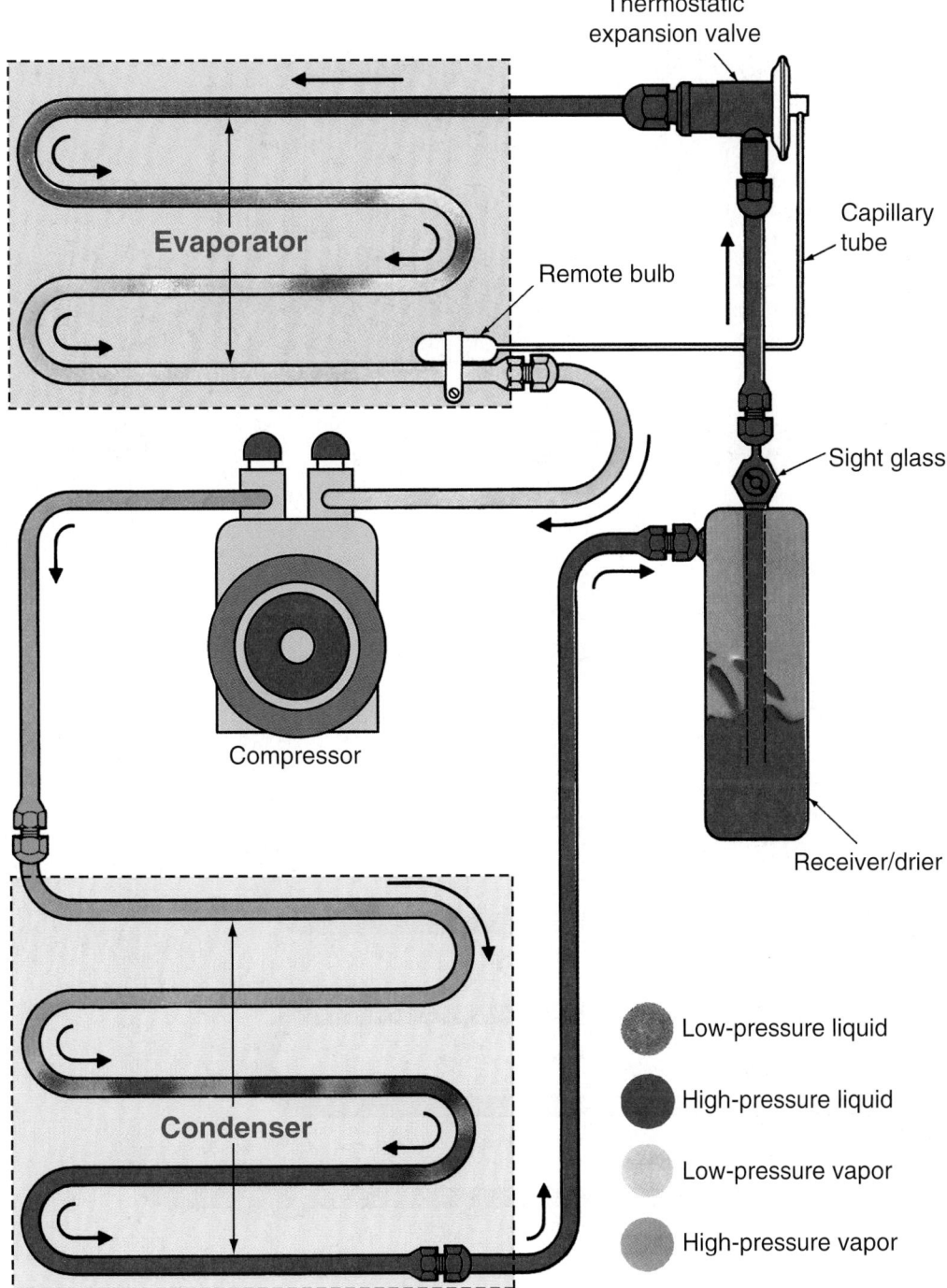

Figure 35.19 This system uses an expansion valve as its flow control device.

humidity, but if the expansion valve allows too much refrigerant to flow, flooding of the evaporator can occur. When there is too much liquid refrigerant, boiling cannot occur. The amount of heat absorbed by the evaporator is directly related to how much of the liquid refrigerant inside the evaporator boils.

If enough refrigerant does not flow into the evaporator, a starving condition occurs. This causes the evaporator to increase in temperature as vaporized refrigerant absorbs more heat (called "superheating"). Once again, little cooling takes place. This can be due

to a low charge of refrigerant or a restriction in the expansion valve or the line to it.

When the evaporator becomes too cold, moisture that has accumulated on the fins freezes, preventing air-flow and hampering heat dissipation. The evaporator can also freeze up if its drain is plugged, allowing water to build up in the evaporator case.

Orifice Tube

An orifice tube system (**Figure 35.20**) cycles the compressor clutch instead of using an expansion valve.

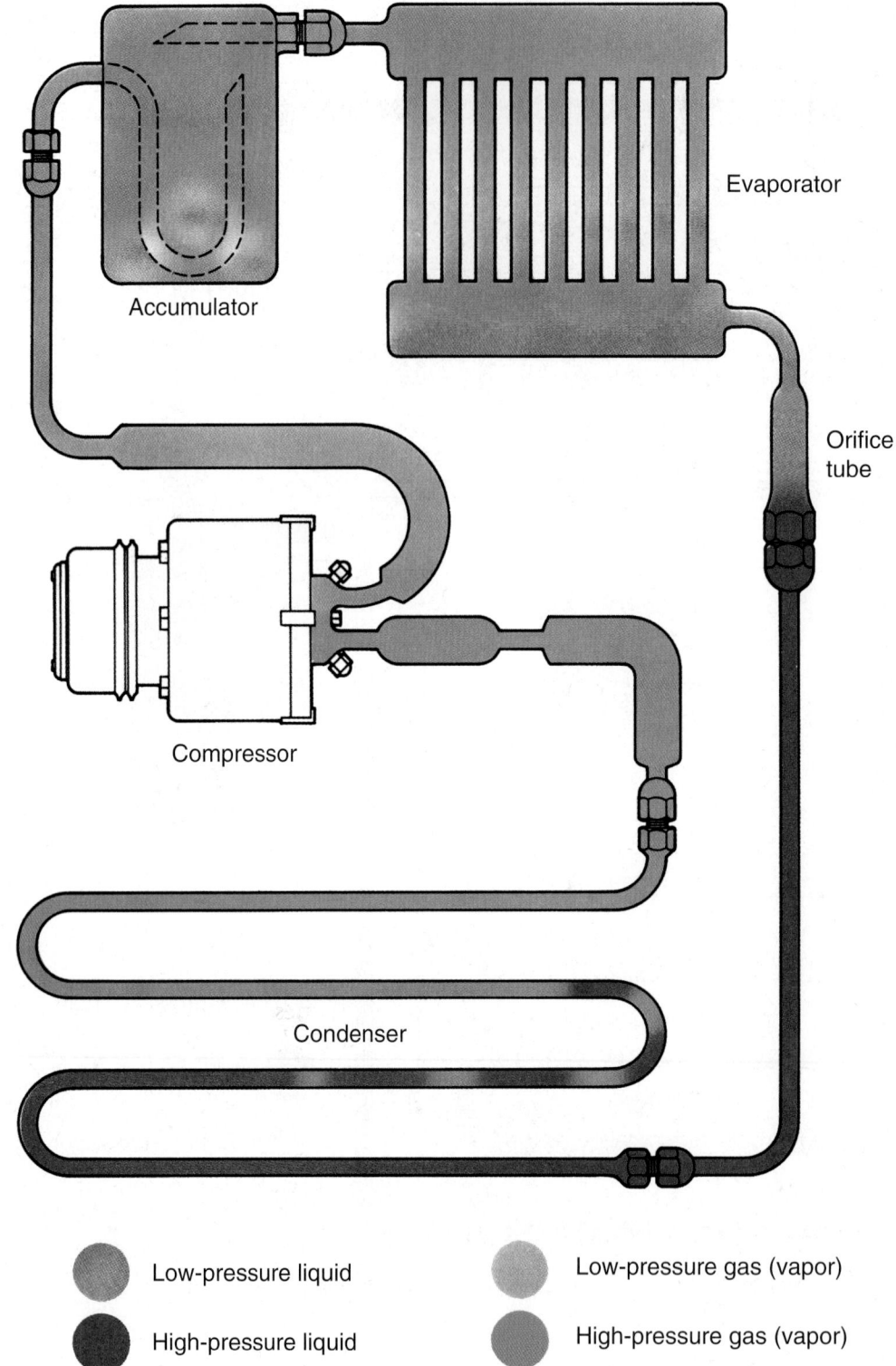

Evaporator

Accumulator

Orifice
tube

Compressor

Condenser

Low-pressure liquid

Low-pressure gas (vapor)

High-pressure liquid

High-pressure gas (vapor)

Figure 35.20 This system uses an orifice tube as its flow control device.

It has a *fixed orifice* between the condensor outlet and the inlet to the evaporator, which is simply a hole with screens on either side that cannot change size like the expansion valve does (**Figure 35.21**). This system is called a *cycling clutch orifice tube* (*CCOT*) design. The magnetic clutch on the compressor cycles the compressor on and off to control temperatures in the evaporator.

There are several thermal expansion valve designs (**Figure 35.22**). Most have a temperature-sensing bulb that is inserted into the evaporator (**Figure 35.23**). The H-valve design (see Figure 35.22), which is used

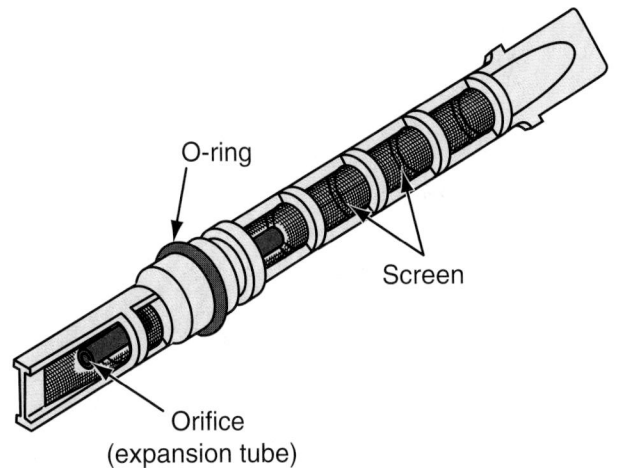

Figure 35.21 Appearance of a typical orifice tube.

O-ring

Screen

Orifice
(expansion tube)

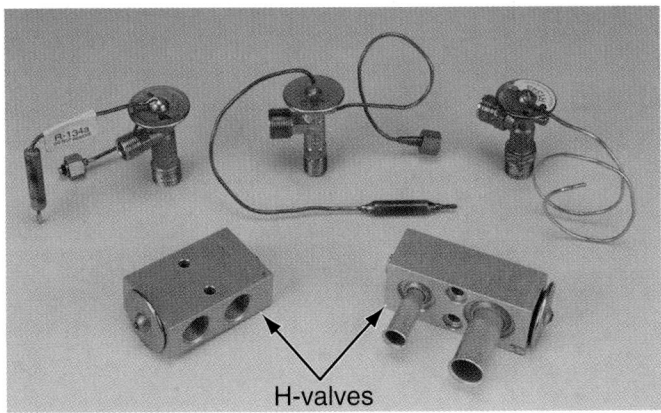

H-valves

Figure 35.22 Various thermostatic expansion valve designs.

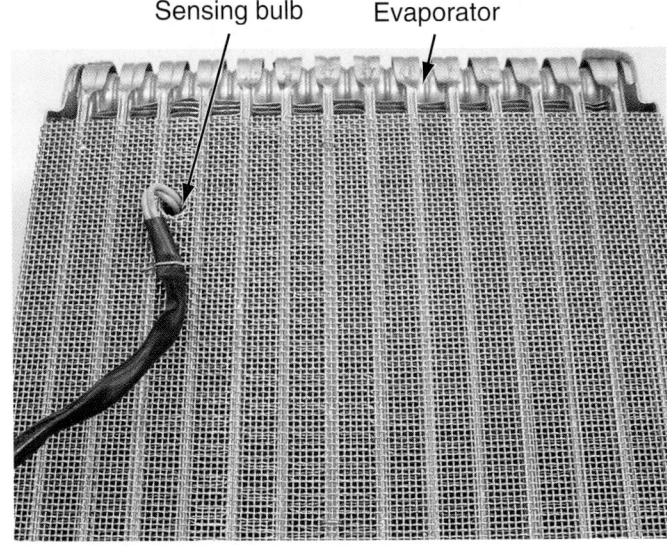

Sensing bulb   Evaporator

Figure 35.23 Most expansion valves have a temperature sensing bulb that is inserted between the fins of the evaporator. *(Courtesy of Tim Gilles)*

in many Chrysler vehicles and a few others, has no capillary tube. It is mounted to both the inlet and outlet of the evaporator. Like a traditional TXV, it has a sealed gas-controlled diaphragm. This device, called a *power dome*, is internal, with its control mechanism exposed to the refrigerant vapor leaving the evaporator.

There are other orifice tube system designs that do not cycle the clutch but have variable displacement or variable output compressors. The clutch remains engaged except during high load conditions, like wide-open throttle (WOT) or during high power steering pressure needs. This system is used with some smaller displacement engines because it allows steadier engine operation without the feeling of the clutch cycling on and off.

■ AIR-CONDITIONING COMPRESSORS

Earlier compressors were large and heavy. Today's compressors are small and light. Compressors have become much smaller. The larger cast iron A-6 compressor used by GM up until the early 1980s weighed nearly 35 pounds. Its aluminum replacements weigh much less.

There are several types of compressor designs. Three of the designs use pistons and reed valves. Earlier *crankshaft-type compressors* have two cylinders and resemble a gas engine. Crank compressors can be either in-line or V-type. They have cast iron piston rings.

Later compressors use Teflon® rings. These have low friction so they can be used in an aluminum bore while causing very little wear.

The compressor works just like an air compressor (see Chapter 11). The *reed valve* is a thin piece of sheet metal that works as a one-way check valve. It is opened and closed to allow the compressor to draw in refrigerant and compress it (**Figure 35.24**). Reed valves are like the ones used in some two-stroke engines.

An *axial compressor* has four or more cylinders. The pistons move lengthwise in the compressor body. Connected to the drive shaft is an axial plate called a *swash plate*. It is mounted at an angle and wobbles when it rotates. This pivots the pistons back and forth in their bores (**Figure 35.25**). The pistons are double ended so

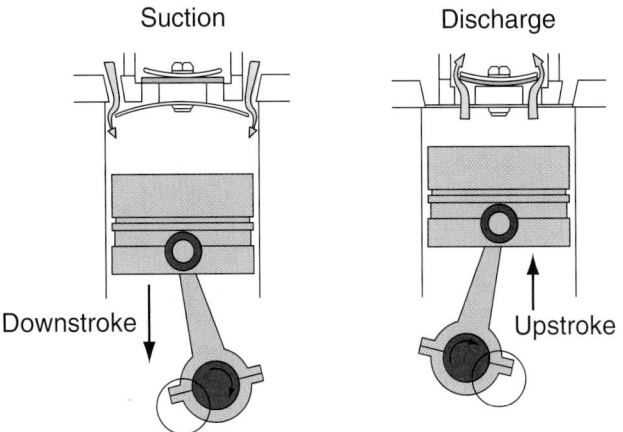

Suction   Discharge

Downstroke   Upstroke

Figure 35.24 Operation of a reed valve.

Figure 35.25 As the swash plate rotates it pivots the pistons back and forth in their bores. (*Courtesy of Tim Gilles*)

both sides can pump. Six- and 10-cylinder axial compressors are the most common. They have three and five pistons, respectively. **Figure 35.26** shows a six-cylinder axial compressor body and reed valve plate.

Axial compressors sometimes have a *wobble plate* instead of a swash plate (**Figure 35.27**). Where a swash plate rotates with the drive shaft, the wobble plate does not rotate. It simply wobbles in place. The pistons are only one-sided and have connecting rods. Swash plate compressors always have an even number of cylinders, but wobble plate compressors tend to have an odd number of cylinders, usually five or seven.

Several manufacturers make newer wobble plate compressors that are *variable displacement compressors* (**Figure 35.28**). A wobble plate moves to shorten the stroke. This reduces the load on the compressor when it is not needed or wanted. These dependable compressors work well because you do not feel the compressor kicking in and out when it is used on a low-power engine. With cycling clutch compressors, the clutch shuts off the compressor to prevent icing

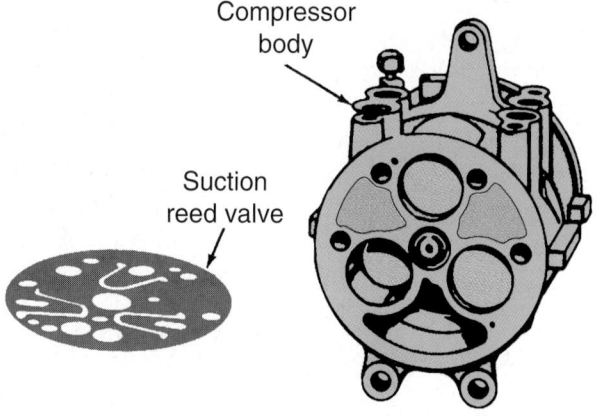

Figure 35.26 An axial compressor body and reed valve. (*Courtesy of DaimlerChrysler Corporation*)

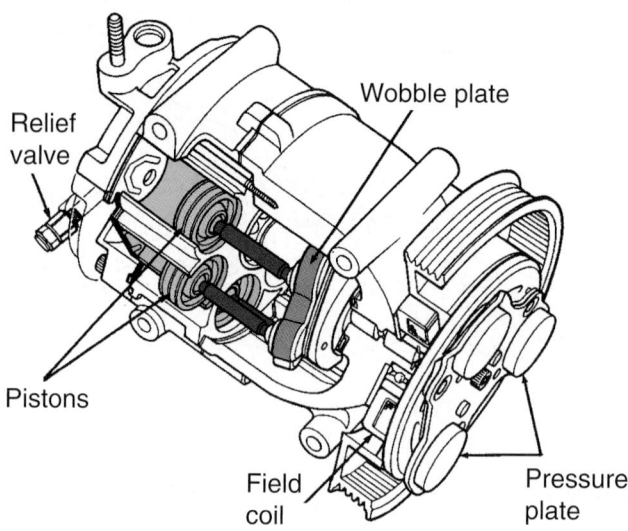

Figure 35.27 A wobble plate compressor.

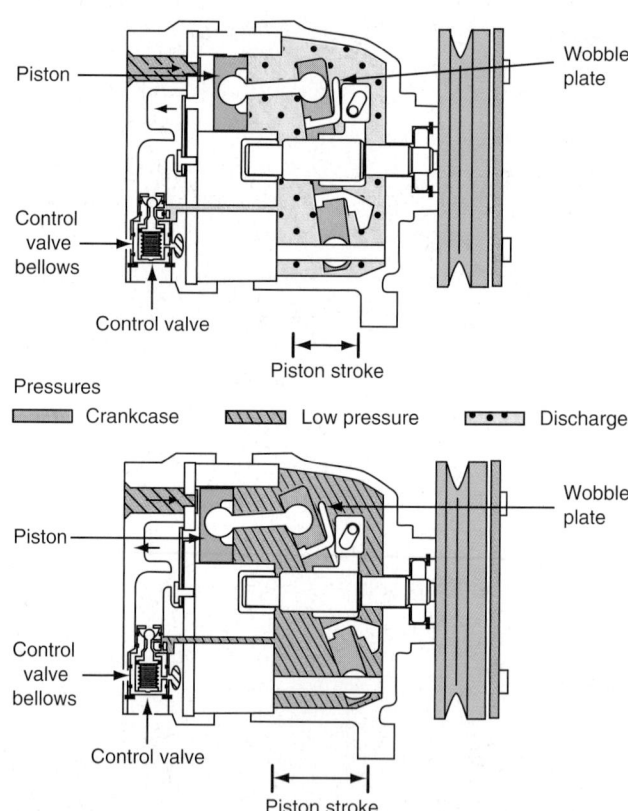

Figure 35.28 A variable displacement compressor.

of the evaporator. With a variable displacement compressor, no clutch cycling is used. The compressor just changes its displacement. **Figure 35.29** shows a variable displacement compressor without its housing.

A *radial compressor* (**Figure 35.30**) is like a radial aircraft engine. It has multiple cylinders with pistons and one eccentric crankshaft throw called a *Scotch yoke*. It uses reed valves, too.

Another compressor type is the *scroll compressor*, which has a fixed and a movable scroll (**Figure 35.31**). As the moveable scroll oscillates around the fixed scroll,

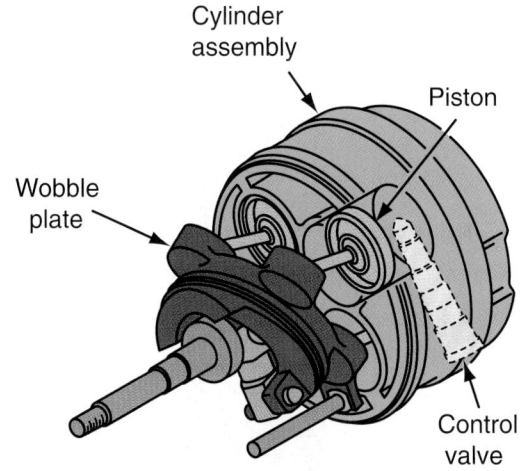

Figure 35.29 Parts of a variable displacement compressor.

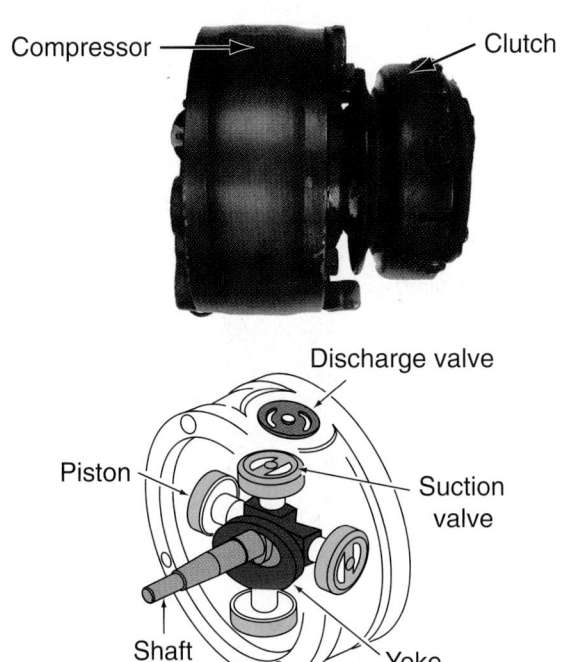

Figure 35.30 An R4 radial scotch yoke compressor. *(top, courtesy of Tim Gilles)*

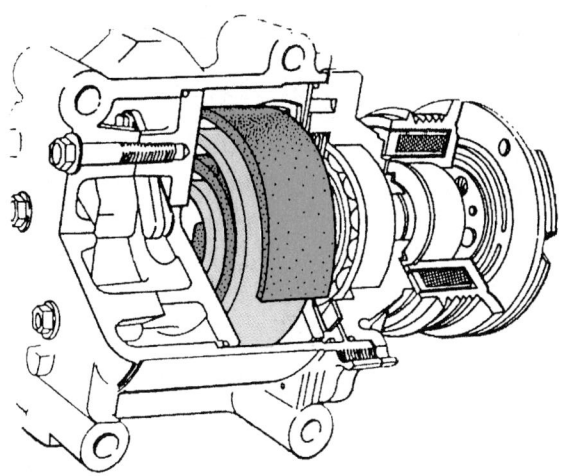

Figure 35.31 A scroll-type compressor.

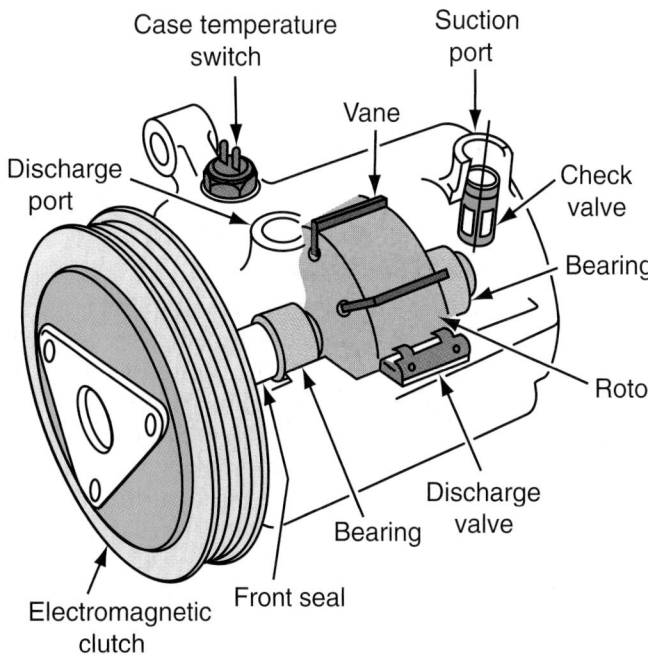

Figure 35.32 A rotary vane compressor.

a pumping chamber forms that is open at the outer end. The chamber becomes smaller as the scroll rotates. Advantages to this compressor include its smooth operation and lower power consumption, an important consideration especially with small engines.

One other compressor style, the *rotary vane*, has blades like a power steering pump or smog pump (**Figure 35.32**).

Electric Compressors

Hybrid vehicles often operate with the internal combustion off, so a belt-driven compressor would not always have power when needed. Some hybrid vehicles use a conventional air-conditioning system but operate the internal combustion engine any time the air conditioning is on. Many hybrid vehicles use an electric compressor (**Figure 35.33**) to pressurize the

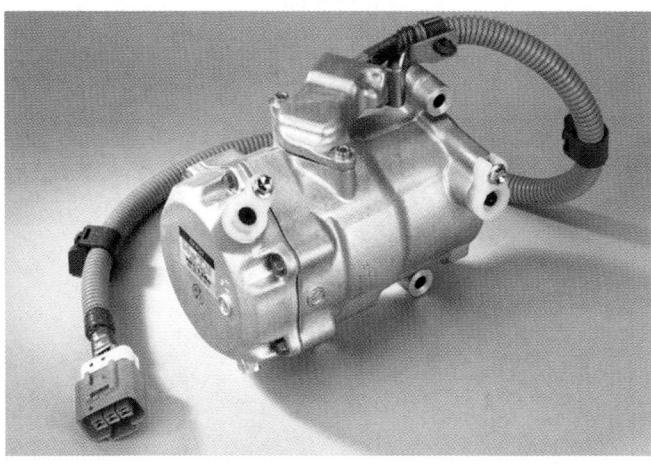

Figure 35.33 An electric air-conditioning compressor is often used on hybrid vehicles. *(Courtesy of Toyota Motor Sales, U.S.A., Inc.)*

air-conditioning system so the engine does not have to be run. It is a scroll compressor powered by a DC motor. It does not need to be mounted on the engine, so it can be located anywhere on the vehicle the engineer desires. Other than the electric compressor, a typical hybrid air-conditioning system operates in the same way as a conventional system.

■ COMPRESSOR LUBRICATION

Compressors are similar to two-stroke gasoline engines in the way they are lubricated. Most compressors are lubricated by oil that is carried in the refrigerant. Oil also seals and cools the compressor. Heat transfer is diminished when oil is flowing along with the refrigerant. Newer air-conditioning compressors are designed to keep oil within the compressor. More information on air-conditioning lubrication is covered in Chapter 36.

■ MUFFLER

Some air-conditioning systems have a muffler installed on the outlet of the compressor (**Figure 35.34**). This is because some systems make pumping noises due to high- or low-side pressure vibrations that result from the use of multiple pistons. Mufflers are occasionally wrapped with insulation to further reduce noise.

■ ACCUMULATOR OR RECEIVER/DRYER

Air conditioners use either a *receiver/dryer* (**Figure 35.35**) or an *accumulator* (**Figure 35.36**). The two share two basic functions:

■ Both devices have a **desiccant** that removes moisture from the system. Moisture is an enemy to air-conditioning systems because it can react with refrigerant and corrode the inside of the system (see Chapter 36). Refrigerant 134A is more prone

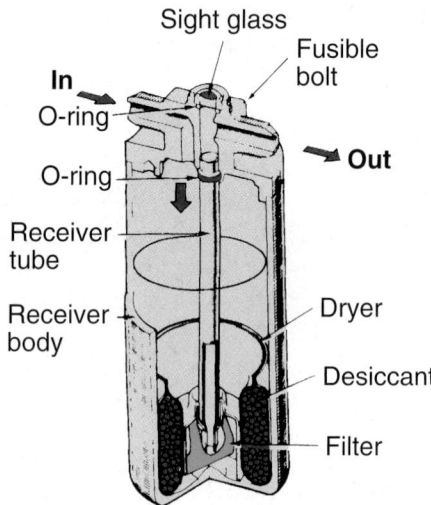

Figure 35.35 A receiver/dryer fills and siphons liquid refrigerant from the bottom.

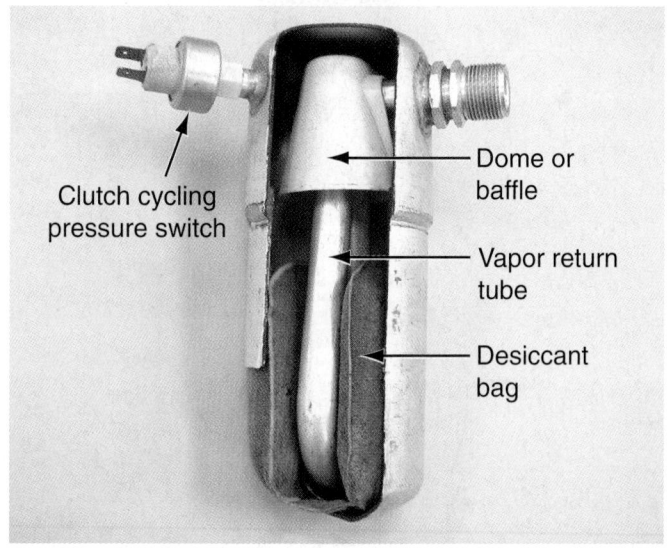

Figure 35.36 Cutaway of an accumulator. (*Courtesy of Tim Gilles*)

to moisture absorption than R-12 so the receiver/dryer or accumulator is sometimes slightly larger so it can hold more desiccant.

■ SCIENCE NOTE ■

The desiccant contains a chemical drying agent. It is a molecular sieve, which means it has uniform small pores through which refrigerant can pass. Water is a larger molecule that is trapped by the desiccant, which can absorb up to 22% of its weight in water.

■ Providing a reservoir for extra system capacity on a cool day is also a function of these devices. The amount of refrigerant needed by the system varies according to heat load and temperature.

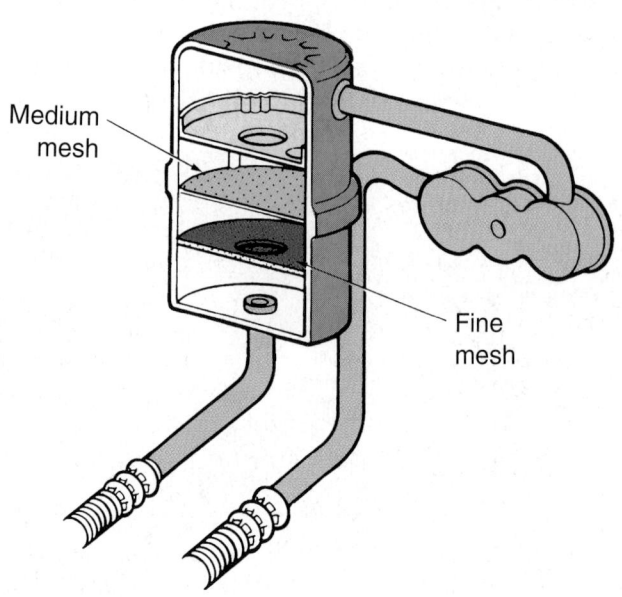

Figure 35.34 Some systems have a muffler for the compressor.

These devices also vary slightly as to function and differ as to where they are installed in the system. A receiver/dryer is located in the high-pressure side of the system, whereas an accumulator is located on the low side:

When a receiver/dryer is used, it is installed in the *high-pressure side* of the system commonly referred to as the *high side*. One of its functions is to ensure that pure liquid refrigerant is supplied to the expansion valve.

When an accumulator is used, it is installed in the *low side*. An accumulator accumulates liquid and lets it turn back to a vapor before it goes back into the system (**Figure 35.37**). One of its functions is to ensure that pure refrigerant vapor is supplied to the compressor. In an orifice tube/accumulator system, if everything is working as designed, nearly all of the refrigerant will remain as a liquid in the evaporator. The refrigerant leaves the evaporator as a liquid, carrying lubricant along with it. The refrigerant leaving the evaporator enters the top of the accumulator. The vapor return tube leading to the accumulator outlet is curved through the bottom of the accumulator housing. The accumulator might be half full of liquid refrigerant, but the vapor return tube inlet is above the liquid level so it only brings vapor to the compressor. A small orifice beneath the filter in the bottom of the vapor tube allows some oil to enter the vapor stream so it can lubricate the compressor.

When a receiver/dryer is used, it is installed after the condenser where it provides a storage place for

Sight glass

Figure 35.38 A receiver/dryer with a sight glass. *(Courtesy of Tim Gilles)*

excess liquid refrigerant until it is needed again by the evaporator. Remember, refrigerant is a high-pressure liquid after the condenser.

NOTE: *A sight glass located on the top of a receiver/dryer (**Figure 35.38**) was used on older air-conditioning systems to provide a quick check to see if there was a problem in the system like low refrigerant level or air in the refrigerant. It was a common practice to add refrigerant until the bubbles disappeared from the sight glass. On newer systems, the sight glass is often covered with paint. Today's systems have much less refrigerant and are susceptible to compressor damage if the system is incorrectly charged. Refrigerant charging procedures and concerns are provided in Chapter 36.*

If the system has an accumulator instead of a receiver/dryer, it will be found in the low side at the outlet to the evaporator. All of the refrigerant does not vaporize in the evaporator core, so the accumulator must capture any liquid to keep it from going to the compressor. Liquid refrigerant would "slug" the compressor, ruining it. The compressor is only supposed to pump vapor. A sight glass would not work on the low side, because it is full of vapor, not liquid.

An accumulator is used with an orifice tube. The hole in the orifice tube is often only about 0.070", depending on the vehicle. This sounds small, but 200 psi will blow a substantial amount of refrigerant through that size on an opening.

■ EVAPORATOR ICING CONTROL

There are several methods of shutting off the clutch to keep the evaporator from freezing on an orifice tube system. A *thermostatic switch* on the evaporator is one method (**Figure 35.39**). It has a tube usually containing mercury or CO_2 that senses when the temperature

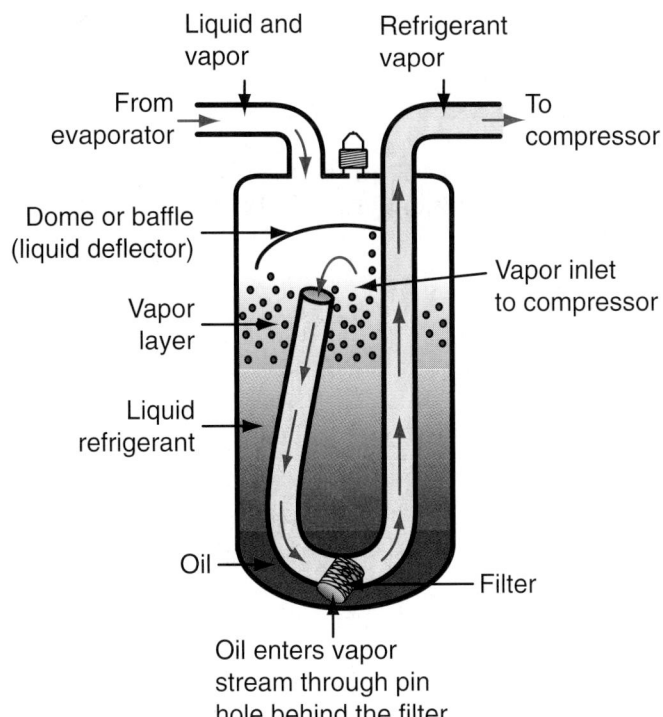

Oil enters vapor stream through pin hole behind the filter

Figure 35.37 The accumulator accumulates liquid and lets it turn back to a vapor before it goes to the compressor. The vapor is pulled off from the top.

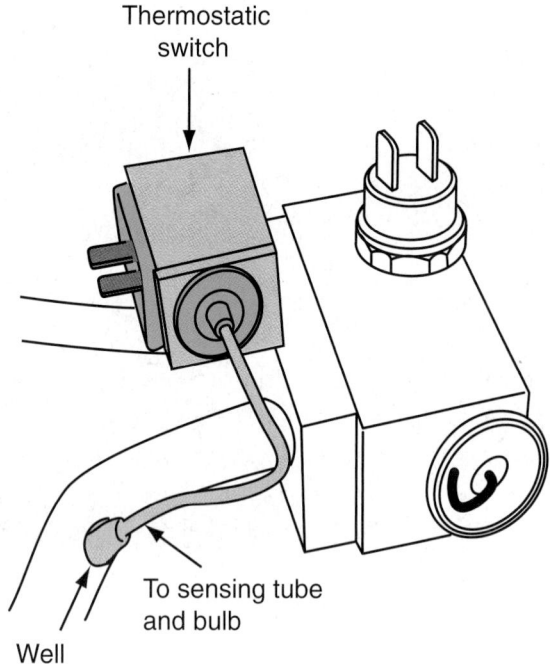

Thermostatic switch

To sensing tube and bulb

Well

Figure 35.39 A thermostatic switch on the evaporator shuts off the clutch when the evaporator starts to freeze.

of the evaporator falls to about 32°F. The switch turns off the compressor clutch. Remember, the outer surface of the evaporator accumulates moisture as it dehumidifies the air in the passenger compartment. This water would freeze if the air-conditioning system were to continue to pump a fresh supply of refrigerant to the evaporator core. Air from the blower would not be able to carry heat away from the passenger compartment and the system would cease operation.

Sometimes a *pressure cycling switch* is used instead of a thermostatic switch (**Figure 35.40**). The temperature

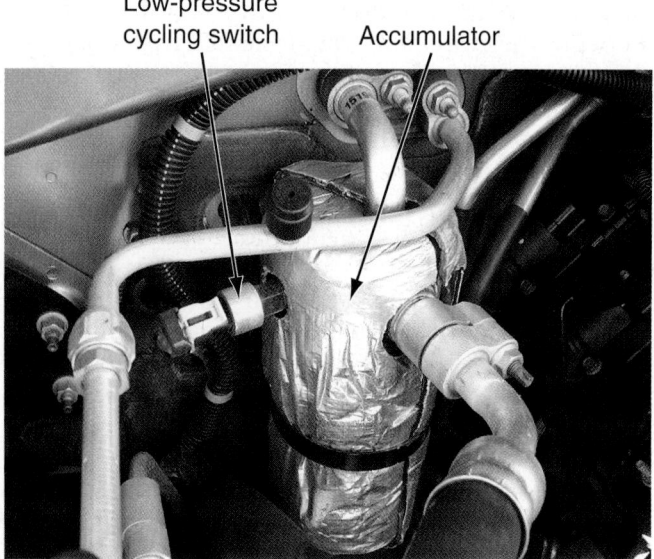

Low-pressure cycling switch   Accumulator

Figure 35.40 A pressure cycling switch on the accumulator. (*Courtesy of Tim Gilles*)

in the evaporator can be predicted by the pressure of the refrigerant inside. The pressure cycling switch is mounted on the accumulator at the evaporator outlet.

VINTAGE AIR CONDITIONING

Some older systems in 1960s and 1970s vehicles use a *suction throttling valve* (*STV*) installed between the evaporator and compressor. These systems did not cycle the clutch. Lower pressure means that the system is getting colder. An STV monitors pressure in the evaporator and shuts off refrigerant flow when the pressure drops below 30 psi with R-12 refrigerant. An STV can also be called a *pilot-operated absolute* (*POA*) valve or an *evaporator pressure regulator* (*EPR*) valve.

■ SYSTEM SWITCHES

There are many different switches and controls used to protect the system. Some of them are listed here:
- Some air-conditioning systems run in the defrost position to dry the air going to the windshield. An *ambient temperature switch* keeps the compressor from working when outside temperatures are cold (about 35°F to 42°F). This protects the compressor from damage. Oil does not flow properly in very cold weather.
- The system must not be allowed to operate when there is little or no refrigerant, because there will be no lubrication for the compressor. When pressure in the system drops too low, the *low-pressure cutout switch* will open the circuit to the magnetic clutch (**Figure 35.41**). This switch is most commonly mounted in the low-pressure side.

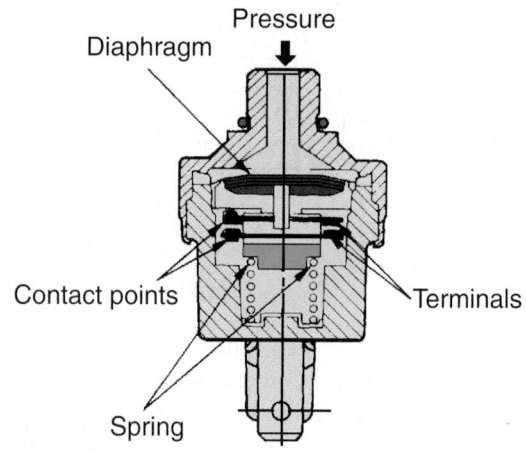

Pressure

Diaphragm

Contact points   Terminals

Spring

Figure 35.41 A low-pressure cutout switch will shut off the clutch if the system becomes too low on refrigerant.

- A *high-pressure cutout switch* mounted on the high side shuts off the compressor if discharge pressure becomes too high. This prevents compressor damage.
- A pressure relief valve bleeds off excess pressure. It can be located on the compressor, receiver/dryer, or anywhere else on the high side of the system.
- A *cutoff switch* can be used to shut off the clutch during WOT operation on small cars. It can also shut off operation during high power steering pressure.
- An air-conditioning control switch can be turned on or off from the passenger compartment.

■ HEATING AND AIR-CONDITIONING CONTROLS

The heating and air-conditioning system is controlled either manually or automatically. The driver has control over whether to turn the system on or off.

In a manual control system, the driver controls the blower speed, position of air doors, and temperature. A typical manual dash control is shown in **Figure 35.42**. Electric switches, cables, or vacuum motors are used to activate the selected positions.

The normal position brings in fresh air. When the selector switch on the dashboard is in the "max" (recirculating) position, only about 7% outside air is brought into the vehicle.

In semiautomatic temperature control (SATC) systems, the driver controls the blower speed and modes that control where the air comes out. Lowering the

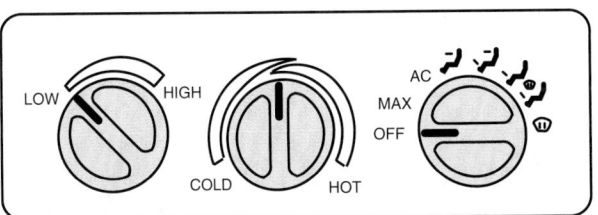

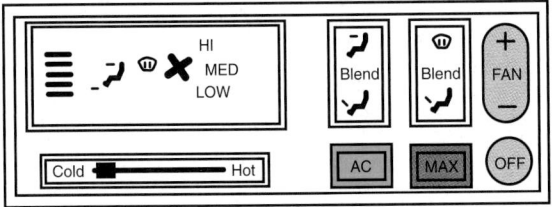

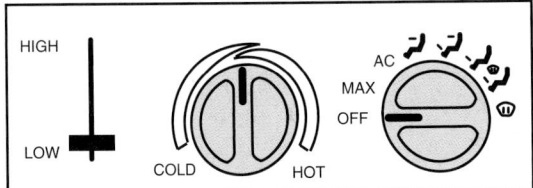

Figure 35.42 Typical manual dash controls for heating and air conditioning.

Figure 35.43 A typical automatic air temperature control. (*Courtesy of Tim Gilles*)

speed results in less noise and lower air movement, but the air coming out of the duct will be cooler.

Automatic temperature control (ATC) systems are also called automatic climate control (ACC) systems. Setting the temperature on the dash control is all that the driver has to do (**Figure 35.43**). A control module and sensors control the temperature setting in the passenger compartment. Although these systems have many inputs, they have few problems.

Automatic Blower Control

The speed of the blower motor on many ATC HVAC systems is controlled by a module responding to an input from the HVAC control module. The blower is pulse-width modulated, which means it is repeatedly cycled on and off. Unlike mechanical systems, the interval between on and off is changed by the controller by commanding the module to vary the speed of the blower across the entire rpm range. The blower can run at whatever speed the programmer determines is best for customer comfort.

■ AUTOMATIC AIR-CONDITIONING SENSORS

Newer air-conditioning systems use sensor inputs to maximize automatic air-conditioning system performance. Some sensors tell the control module the outside temperature and the intensity of the sun. Others tell the control module what the passengers are feeling in the *cabin* (passenger compartment). Inputs include the discharge temperature at the ducts, the temperature in the cabin, and the position of the air doors.

Outside Air Temperature (OAT) Sensor

The outside air temperature (OAT) sensor, sometimes called an *ambient sensor*, is usually located in front of the condenser in the grille (**Figure 35.44**) and sends input to a temperature display on the mirror. It requires airflow to be accurate, so an algorithm delays the display

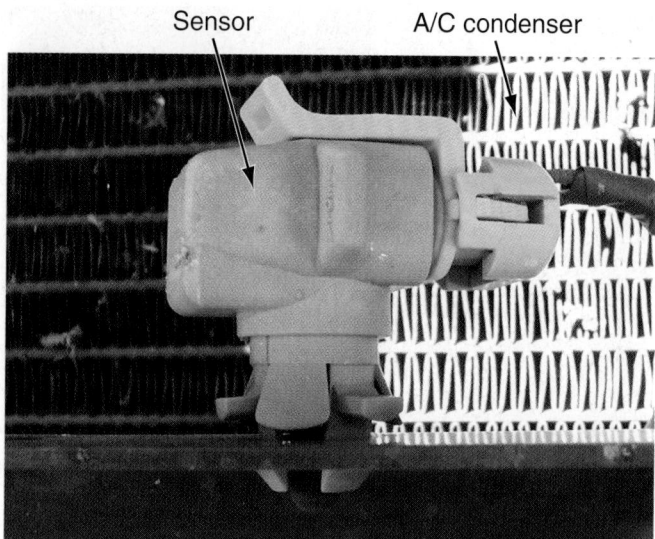

Figure 35.44 An outside air temperature sensor, or ambient sensor. *(Courtesy of Tim Gilles)*

and updates it with vehicle speed. When outside air temperature drops below somewhere between 35°F and 42°F, the system denies air-conditioning compressor operation. Some systems require a vehicle speed sensor signal before the temperature will update. An OAT sensor is a negative coefficient thermistor (NCT), so its resistance goes down as outside temperature increases.

Sunload Sensor

The sunload sensor is located in the defroster outlet grille or on the dash (**Figure 35.45**). It is a photodiode, cadmium dioxide sensor (CDS) that blocks current flow during darkness. The sunload sensor reacts to incandescent light or sunlight but not to fluorescent light. As light intensity increases, its resistance drops.

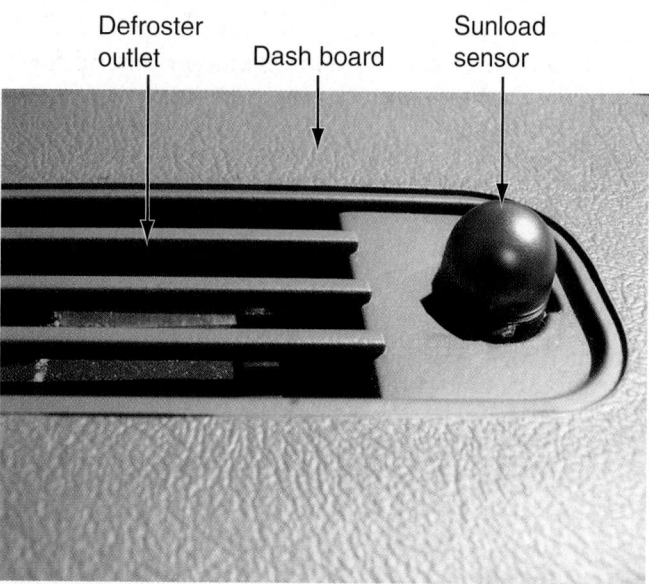

Figure 35.45 A sunload sensor located in the defroster outlet grille. *(Courtesy of Tim Gilles)*

Bright sunlight signals the control module as sunlight intensity increases so the air-conditioning system can compensate before the heat instead of after the passenger compartment becomes too hot. More intense sunlight will cause the control module to open the blend door and spin the blower at higher speed. Some systems use more than one sensor. If it is hotter on one side of the vehicle, the air discharge temperature on that side of the vehicle will be lowered to compensate.

Output Duct Temperature Sensor

The discharge duct temperature sensor is an NCT that provides the control module with the temperature of the air leaving the duct. A typical system uses one sensor in the center air-conditioning discharge duct and another in the heater duct. Dual-control air-conditioning systems with driver and passenger temperature control have four sensors, two on each side of the vehicle (**Figure 35.46**).

Interior Temperature Sensor

An interior temperature sensor is usually mounted under the dashboard. In a typical ATC system, a port in the dash runs to a low-pressure area of the blower housing to pull air into the tube from the passenger compartment and pass it across the sensor (**Figure 35.47**). It is an NCT, so its resistance goes down with heat. Some manufacturers use this sensor for the first few minutes of vehicle operation only. The discharge duct temperature sensors provide input after that.

NOTE: *Some vehicles have a smog sensor. When hydrocarbon emissions get too high, the control module closes the recirculation door.*

Figure 35.46 A discharge duct temperature sensor is located on each side of the vehicle with dual-control systems controlled by the driver and passenger.

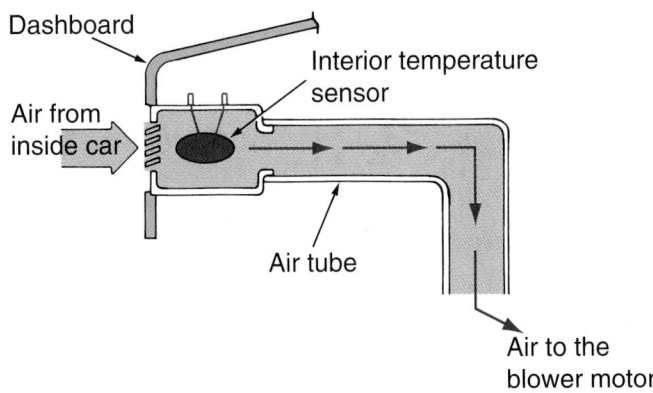

Figure 35.47 The blower motor pulls air across the interior temperature sensor.

REFRIGERANTS AND THE ENVIRONMENT

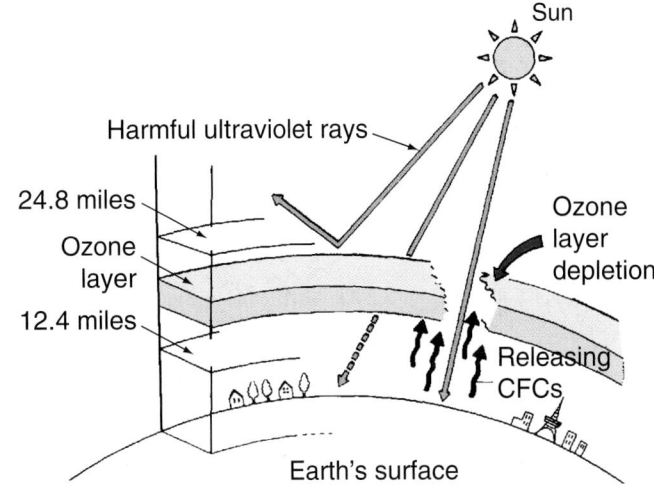

Figure 35.48 The ozone layer filters out the sun's harmful rays.

SCIENCE NOTE

The refrigerant that was widely used in the past was called R-12 or Freon. It had been around for over 60 years. It is nonpoisonous, stable, and inexpensive to manufacture. Due to environmental concerns, it has been phased out in new vehicles and a substitute refrigerant has been retrofitted to many older vehicles.

*Located in the Earth's stratosphere (10–30 miles above the surface of the Earth) is its protective ozone layer, which filters out most of the sun's harmful ultraviolet rays (**Figure 35.48**). The amount of ozone, which is what makes the sky appear blue, has remained about the same in the stratosphere for centuries. Freon (R-12) is a chlorofluorocarbon (CFC). CFCs are depleting this protective ozone layer through a chemical reaction. When CFCs are released into the atmosphere, they slowly travel into the stratosphere where they can remain for 100 years or more (**Figure 35.49**).*

At ground level, ozone is a part of photochemical smog. A lack of ozone in the stratosphere actually increases ozone at ground level. Because of centrifugal force, the ozone layer is thickest at the equator and thinnest at the north and south poles.

CFCs are made of chlorine, fluorine, and carbon. They have been used for such items as fire extinguishers, dry cleaning fluid, solvents, and aerosol can propellants. In 1978, due to concerns about CFCs contributing to global warming, they were banned by the United States and other countries for use as aerosol propellants and dry cleaning solvents.

The concentration of chlorine in the stratosphere was measured at 80 parts per trillion (ppt) in 1978.

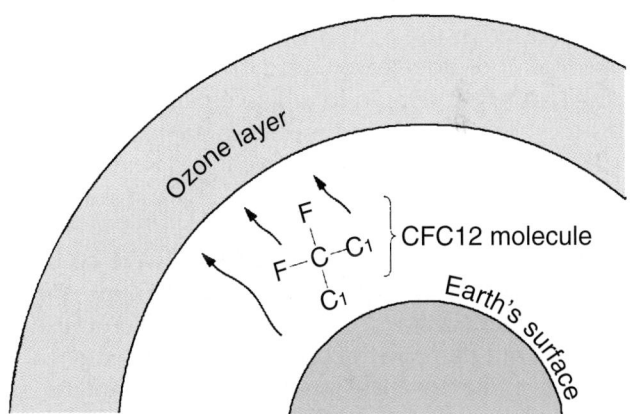

Figure 35.49 CFC migration to the stratosphere.

*In 1985, scientists confirmed that the ozone layer was depleted to the point that it had developed a large hole over the south pole. This was attributed to CFCs. Chlorine in the stratosphere was measured in 1990 to be 499 ppt. Chlorine attacks the ozone in the stratosphere and destroys it. Without ozone in the stratosphere, ultraviolet light reaches the Earth's surface (**Figure 35.50**).*

In 1987, the United States and 22 other countries signed an agreement known as the Montreal Protocol. The agreement set limits on the production of ozone-depleting chemicals, which were totally phased out by the year 2000. In Europe and the United States, they have been banned since 1996.

It is estimated that about 30% of released CFCs are from mobile air-conditioning sources. Some leaks out naturally, but most of the problem comes during repair and service. By 1993, service technicians were required to be licensed to work on air-conditioning systems or be able to purchase CFC refrigerants.

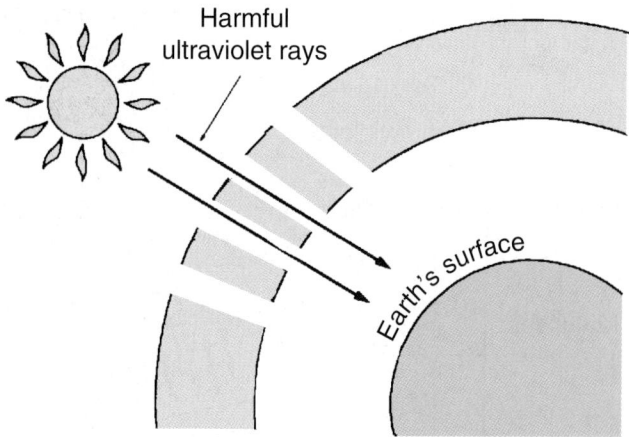

Figure 35.50 Without the ozone in the stratosphere, ultraviolet light reaches the Earth's surface.

Two primary refrigerants used in cars are R-12 and R-134A. R-22 is used in stationary air conditioners and refrigerators. In the past, inexpensive R-12 refrigerants were commonly released into the air. Today, those refrigerants that remain in older vehicles are recovered and recycled. Recovering means to capture the refrigerant and store it for reuse. Recycling is when a special machine is used to remove water, air, and oil from it so it can be reused.

Beginning in 1994, almost all cars were equipped with R-134A refrigerant. This refrigerant is a *hydrofluorocarbon*, rather than a CFC. It is simply a greenhouse gas and does not damage the ozone layer. There have been changes made to the systems to accommodate the different characteristics of the refrigerants. The most notable is that the condenser is larger. The system will cool effectively with the new changes.

Future Refrigerants

Although R-134A does not damage the ozone layer, it is still a greenhouse gas. A typical vehicle loses 2.3 pounds of refrigerant during its lifetime, and an estimated 40 million pounds leak into the atmosphere each year. Several refrigerants are being investigated for their potential as a replacement for R-134A. Two of them are R-152A and CO_2 (R-744). R-152A has about one-tenth the global warming potential of R-134A and allows an increase in fuel economy. It is a class 2a flammable, however, which is not very flammable but is still a concern. Currently it is used as an aerosol propellant. CO_2 is a very high-pressure system, with operating pressures of +2,000 psi on the high side. European systems use metal woven hoses. Although service safety and component durability are concerns, CO_2 provides an advantage in the areas of flammability and toxicity.

▮ TEMPERATURE AND PRESSURE

When dealing with refrigerants and diagnosing problems with the air-conditioning system, you will need an understanding of the temperature and pressure and their relationship to one another.

The atmosphere consists primarily of the gases oxygen (21%) and nitrogen (78%). It extends about 600 miles above the Earth's surface. The pressure exerted by the weight of atmospheric gases on the Earth's surface is about 14.7 pounds per square inch.

On a pressure gauge, pressures above atmospheric are called pressure and those below atmospheric are called *vacuum*. This is because the gauge is designed to read 0 psi and 0 vacuum regardless of actual atmospheric pressure. Pressure above atmospheric is actually *PSIG*, which means *pounds per square inch gauge*. With this system 0 on the gauge represents 0 at sea

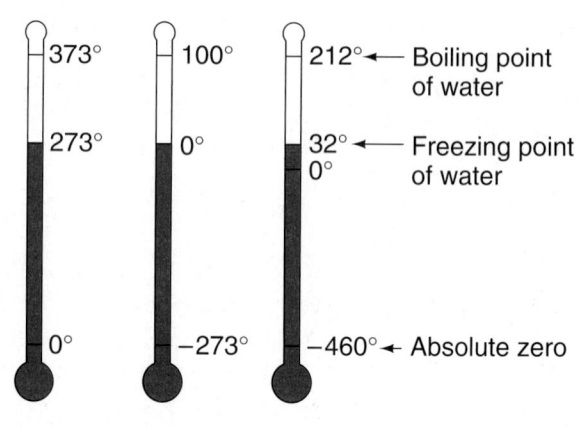

Figure 35.51 Comparison of different temperature scales.

Pressure vs. Vacuum			
PSIG	PSI (abs)	BAR	IN. Hg
2	16	1.14	-
1	15	1.07	-
0	14.2	1.00	0
−1	13	0.94	2
−2	12	0.87	4
−3	11	0.80	6
−4	10	0.72	9
−5	9	0.64	11
−6	8	0.57	13
−7	7	0.50	15
−8	6	0.44	18
−9	5	0.37	20
−10	4	0.30	22
−11	3	0.22	24
−12	2	0.14	26
−13	1	0.07	28
−14	0	0.00	30

Figure 35.52 Relationship between vacuum (in. Hg) and various pressure measurement scales.

level. Air-conditioning pressures are measured in this way.

Pressure below atmospheric is called vacuum. It is measured with a gauge that lifts a column of the liquid heavy metal mercury (Hg) up a small tube. Vacuum is measured in **inches of mercury (in. Hg)**. Hg is the scientific abbreviation for mercury. The strongest vacuum that can be achieved in the Earth's gravity is 30 in. Hg. A chart in **Figure 35.52** shows the relationship between pressure measured in PSIG, barometric (BAR), and in. Hg.

The relationship between boiling point and pressure is different for every liquid. Because water is a liquid everyone is familiar with, examples are given here using *water*. Water boils at 212°F at sea level. Its boiling point is lower above sea level because atmospheric pressure is lower. Every 1,000 foot increase in altitude results in a boiling point 1.1°F lower than 212°F.

Example:
at 8900' altitude:
8.9 (8900 divided by 1000) × 1.1° = 9.8°F
212°F − 9.8°F = 202.2°F

Increasing the pressure on water by 1 pound will raise its boiling point by about 3°F. Cooking food in a pressure cooker raises the boiling point of water so that the temperature of the water can be above 212°F. The radiator cap on the engine's cooling system raises the boiling point of coolant.

■■■ REVIEW QUESTIONS

1. What is the name of the device that engine coolant flows through to heat the car's interior?

2. Comfort depends on temperature, _____, and air movement.

3. Heat transfer occurs by _____, radiation, or evaporation.

4. When the air is totally saturated with moisture, this is called 100% _____.

5. What is the name of the type of heat that is required before matter can actually change its state?

6. How is latent heat measured?

7. What is the name of the small, radiator-like device that absorbs heat from the car's interior?

8. Each cycle through the evaporator absorbs at least ____° of heat from the air blowing across it.

9. What is the name of the part in front of the radiator that carries heat away from the refrigerant?

10. The flow control device can be either an orifice _____ or an expansion valve.

11. What is the name of the one-way check valve used in an air-conditioning compressor?

12. The _____ compressor with a swash plate uses pistons that are double ended so they can pump in each direction.

13. Which compressor type resembles an aircraft engine?

14. Which refrigerant is a CFC, R-12 or R-134A?

15. What is the name of the substance that is a pollutant at ground level, but when in the stratosphere protects the Earth's surface from ultraviolet rays?

▰▰ ASE-STYLE REVIEW QUESTIONS

1. Which of the following driving conditions uses more fuel at highway speeds?

   a. Windows up and air conditioner on

   b. Windows down and air conditioner off

   c. Windows up and air conditioner off

   d. Going down hill with a tailwind

2. Technician A says that a receiver/dryer is installed in the high side of the system. Technician B says that a receiver/dryer supplies refrigerant vapor to the expansion valve. Who is right?

   a. Technician A   c. Both A and B

   b. Technician B   d. Neither A nor B

3. Which of the following is/are true about humidity?

   a. In hot weather, the body is more comfortable when humidity is high.

   b. The "MAX" air conditioner control lever position is better for controlling humidity.

   c. Both A and B.

   d. Neither A nor B.

4. Which of the following is/are true?

   a. At sea level both water at 212°F and steam are the same temperature.

   b. Refrigerant cools off when it is compressed.

   c. Too little refrigerant flow in the evaporator will result in a frozen evaporator core.

   d. All of the above.

5. Technician A says that the expansion valve keeps the evaporator from freezing. Technician B says that cycling the clutch keeps the evaporator from freezing. Who is right?

   a. Technician A   c. Both A and B

   b. Technician B   d. Neither A nor B

6. An air-conditioning compressor is designed to pump:

   a. Liquid   c. Both A and B

   b. Vapor   d. Neither A nor B

7. Which of the following is/are true about cycling clutch/orifice tube air-conditioning systems?

   a. An accumulator is installed in the low-pressure side of the system.

   b. The accumulator supplies refrigerant vapor to the compressor.

   c. Both A and B.

   d. Neither A nor B.

8. Refrigerant is a high-pressure liquid when it leaves the:

   a. Condenser   c. Both A and B

   b. Evaporator   d. Neither A nor B

9. Technician A says that a low-pressure cutout switch will open the circuit to the magnetic clutch if the refrigerant level is low. Technician B says that a high-pressure cutout switch shuts off the compressor if discharge pressure drops too low. Who is right?

   a. Technician A   c. Both A and B

   b. Technician B   d. Neither A nor B

10. Technician A says that when refrigerant is pumped from the compressor it is a high-pressure gas. Technician B says that when refrigerant enters the compressor it is a liquid. Who is right?

    a. Technician A   c. Both A and B

    b. Technician B   d. Neither A nor B

CHAPTER 36

Heating and Air-Conditioning Service

OBJECTIVES

Upon completion of this chapter, you should be able to:

✔ Locate obvious problems in heating and air-conditioning systems with a visual inspection.

✔ Test air conditioner efficiency and pressures.

✔ Locate leaks in the refrigeration system.

✔ Diagnose and repair problem components.

✔ Evacuate and recharge a refrigeration system in a safe and legal manner.

KEY TERMS

discharge service valve
suction service valve
compound gauge

static pressure
PAG
ester

dual pass machine
single pass machine

HEATER SERVICE

Complaints in the heating system are usually related to coolant leaks or inappropriate temperatures for the season. With modern automobiles and the necessary practice of maintaining the coolant concentration, many of the old problems related to coolant flow have disappeared. If a heater is not often used, rust can accumulate in it.

The heater flow can be checked by feeling the hoses at the inlet and outlet to the heater core. If the heater has a control valve and it is open, both hoses should be warm. This indicates that coolant is flowing through the heater core. If the outlet hose is cold, there must be an obstruction in the heater core or control valve.

NOTE: *Be sure to check the coolant level first. If the coolant level is low, the heater might not have sufficient coolant to operate.*

A plugged heater core can often be flushed with pressurized water and chemical compounds designed for flushing a cooling system. Unhook both hoses at the most accessible locations and force water through one hose while it comes out the other.

Heater Core Replacement

When a heater core leaks, a new heater core is installed or the old one is repaired. The heater housing is usually under the dash and must be removed to gain access to the heater core (**Figure 36.1**). Once the housing is removed, it can be dismantled to gain access to the heater core. Some housings have two halves, which are held together by clips (**Figure 36.2**).

AIR-CONDITIONING SERVICE

Air-conditioning service is a challenging and rewarding specialty area. Due to concern about the Earth's ozone layer (see Chapter 35), technicians must be licensed to service all air-conditioning systems. This has resulted in the work that was formerly done by do-it-yourself people at their homes now being done by repair shops.

There is an abundance of air-conditioning work, but there are also several different types of refrigerants. A technician must have a good understanding of the entire air-conditioning system before attempting to work on it.

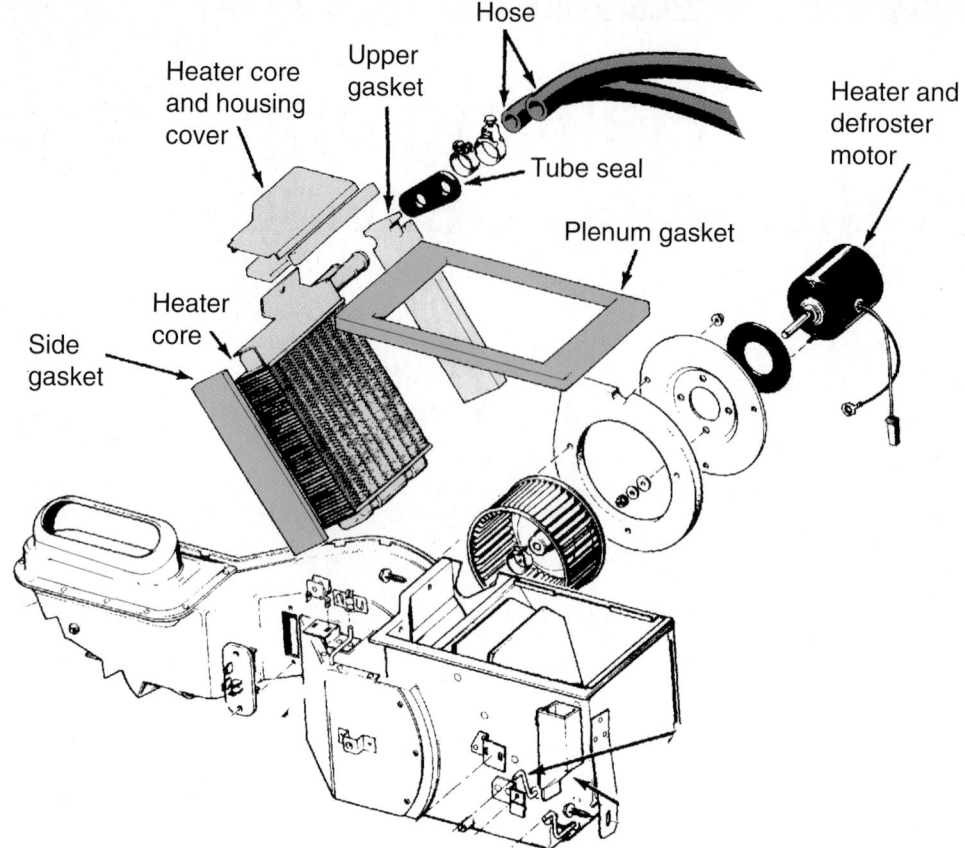

Figure 36.1 The heater housing is disassembled to get to the heater core. *(Courtesy of DaimlerChrysler Corporation)*

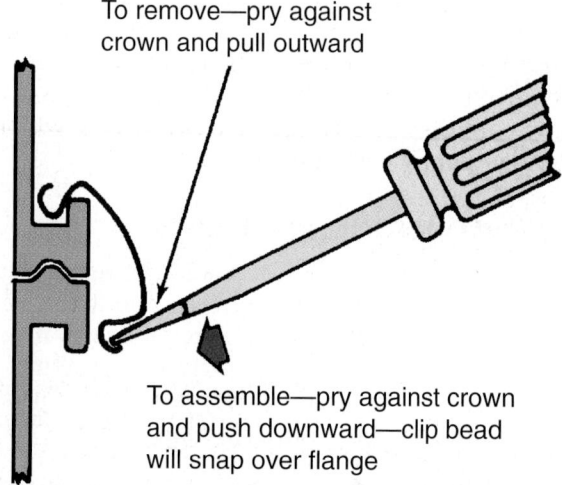

Figure 36.2 To separate the halves of the heater housing, use a screwdriver on the clip. *(Courtesy of DaimlerChrysler Corporation)*

Air-Conditioning Safety

Safety is always a top priority. When working on air conditioning, be careful around fans and hot coolant. Refrigerant requires other safety precautions.

CAUTION ■ Be sure to wear eye protection when attaching pressure gauges or servicing the system. Have you ever seen a demonstration in a science class where a rubber band is frozen with liquid nitrogen? When it is dropped on the table it shatters. Refrigerant will instantly freeze anything it comes into contact with. If it touches your eyes, it will cause blindness.
■ R-12 refrigerants produce nerve gas (phosgene) when exposed to an open flame.

NOTE: *Be sure to use the correct refrigerant. Blend refrigerants are not approved by automotive manufacturers. Some of them are flammable and can cause swollen seals.*

■ AIR-CONDITIONING SYSTEM SERVICE AND DIAGNOSIS

When diagnosing the air-conditioning system you will use the basic understanding you gained of system theory while reading Chapter 35. Check simple things first to try to form a good idea of where a problem might be. To review the main parts of the system see Figure 35.19 and 35.20.

Foam insulation

Figure 36.3 A decayed foam seal can interfere with the operation of an air door. *(Courtesy of Tim Gilles)*

Figure 36.4 An underhood cabin air filter with the access door removed.

Air Distribution System Diagnosis

Temperature doors are controlled by cables, vacuum, or electricity. Cables can stretch, freeze up, or break, usually at the end.

Vacuum doors require a reservoir so they can work when the vehicle is accelerating, for instance, going up a hill or at wide-open throttle. During a failure, they are required by law to default to the defroster. Cold air coming up the defroster indicates a problem.

Electric doors typically have a five-wire actuator. Three of the wires send a reduced voltage from the 5V input back to the PCM to shut off the door.

Sometimes blend air doors fail, resulting in full heat. This can be due to decayed grey foam seals interfering with an air door (**Figure 36.3**). To test for this problem, pinch a heater hose to see if this changes the amount of heat coming from the duct. If ice is on the air-conditioning line, there could be a problem in the air distribution system. Chilling takes place very quickly when there is no air movement.

Cabin Filters

A cabin filter, which is installed on many newer vehicles, cleans circulated air. It is usually an air filter resembling the one used to filter air entering the engine's intake manifold. Some cabin filters have a charcoal section for controlling odors. The filter element can be located on either side of the blower motor. Filters that clean outside air are usually accessible from under the hood (**Figure 36.4**). Plenum air filters are found inside the cabin and clean all of the air that runs through the system, whether fresh or recirculated. They are sometimes replaced through a door in the back of the glove compartment (**Figure 36.5**). Sometimes the filter is located in the center console. Insufficient airflow through the ducts can be due to a plugged filter or an oily evaporator when dust sticks to it.

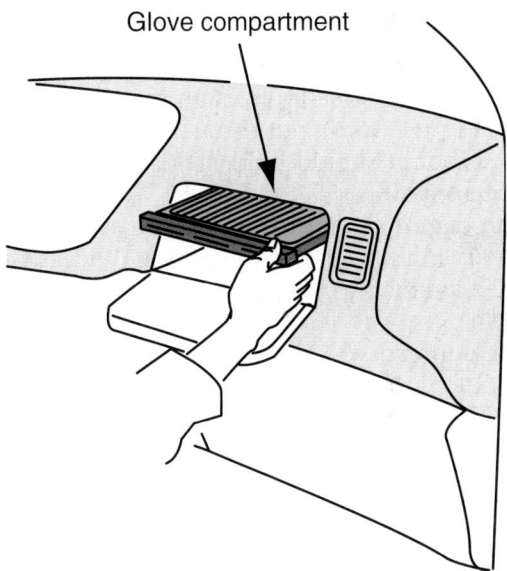

Glove compartment

Figure 36.5 A cabin air filter that is replaced through a door in the back of the glove compartment.

Many drivers leave their air conditioning in the MAX position. When the system is shut down, it continues to sweat. With warm air, the combination results in the growth of mildew and fungus. If you can see the area, it is dark in appearance.

SHOP TIP To assist customers in preventing the formation of mildew and fungus in their vehicles, you can suggest that they shut off the air conditioning before arriving at home and use the vent position instead of MAX.

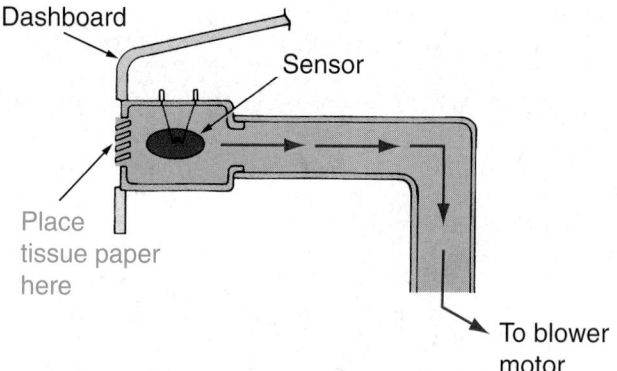

Figure 36.6 To verify operation of an interior air temperature sensor, put tissue paper across the sensor air inlet. It should hold and then drop off when it shuts off.

To dry the evaporator, some manufacturers use an afterblow module that turns on the blower for about 10 seconds after shutoff and cycles again every few minutes.

Sensor Testing

Automatic temperature control systems have several sensors. To verify operation of an interior temperature sensor (**Figure 36.6**), put tissue paper across the input to the sensor. It should hold the tissue and drop it off when it goes off.

The engine coolant temperature sensor will shut off the air conditioning if the engine is hot, so a defective ECT can be the cause of a compressor that will not come on.

The sunload sensor located in the defrost grille is a photodiode. As temperature increases, the resistance should drop. With bright light, the sensor will provide 4 ohms of resistance, and 6 ohms with dim light.

> ### ▰ HISTORY NOTE ▰
>
> *Before the damage to the Earth's ozone layer was discovered, R-12 refrigerant cost less than $1 a pound and was used wastefully. During an air-conditioning service, it was a common practice to let the refrigerant escape into the atmosphere. Not only is this now against the law, but it can be expensive. Some large systems with rear air can take up to 9 pounds of refrigerant. Making a mistake can be quite costly.*

It is a mistake to allow an air-conditioning system to remain empty. This allows moisture to enter the system, ruining a receiver/dryer or an accumulator. Replacing these parts is costly, especially when combined with the cost of lost refrigerant.

Visual Inspection

Always perform a quick visual inspection first, looking for obvious causes of problems. Check the coolant level in the radiator. The engine's cooling system has an effect on the operation of both the heating and refrigeration systems. Look for blockage of the condenser fins with either mud or leaves. Clean mud and dirt away with water.

Check the tension and condition of the air-conditioning drive belt. If the belt is not a spring-loaded serpentine belt, it probably requires higher tension than the other V-belts. Air-conditioning compressors use a good deal of power when they run. A belt that has been allowed to slip will be glazed and will no longer hold the pulley even if properly tensioned.

Look for wet stains around connections of hoses and parts of the system. These would be from the oil that circulates in the refrigerant.

> **SHOP TIP** Old R-12 systems used mineral oil, which showed up when it leaked. New systems use PAG oil, which is water soluble so it washes away and does not stick to the hose or part when it leaks.

Check the operation of the clutch. You should be able to hear the clutch click as it engages periodically. The electric cooling fan should begin to run at low speed when the clutch comes on. Check wiring and hoses for obvious problems.

Air Conditioning Noises

Some air conditioning noises are normal, while others can indicate a potential problem. A hissing or whistling sound after the engine is shut off is a normal sound that can result when high side pressure moves through the metering device as the system pressures equalize. Although a clicking noise when the clutch engages and disengages is normal, a worn clutch will tend to make a louder click. Compressor noises can be the result of a loose mounting bracket. A compressor that is failing can rattle or knock, with the noise becoming worse as engine rpm increases. An incorrectly routed hose that is rubbing against a body part can also cause noise.

System Operation Quick Check

Run the engine for 5 to 10 minutes with the air-conditioning system operating. Check for hot and cold as you trace your way through the system. Starting at the compressor, feel the tubes on the high and low sides of the system. If a line does not feel hot or cold where it is supposed to, look for a restriction at that point in the line. A *frost ring* often forms at the restriction.

The high side should be warm or *hot* as it compresses vapor. This is because the refrigerant is under pressure and wants to lose heat to the condenser.

Be careful, hoses may burn you!

There is low-pressure liquid after the orifice tube.

On the low-pressure side, evaporator lines should be *cold*. On accumulator systems, the accumulator's job is to separate liquid from vapor. The accumulator should feel as though ice water were running through it because the refrigerant coming to it from the evaporator is mostly vapor. The boiling point of the refrigerant is lower at low pressure, so the line feels cold because the vapor within is absorbing heat.

If frost accumulates on the receiver/dryer, it could be restricted.

When the low side is cool but the system does not cool the passenger compartment, the problem could be in an air door or panel controls. Check their operation next. If the high side is not hot and the low side is not cool, you will know the problem is in the refrigeration section of the system.

Testing Air-Conditioning Efficiency

Before temperature and pressure tests, the following list of items should be met:

- Set the temperature control to maximum cool and normal AC. Some manufacturers call for MAX AC so that recirculating air is used. Be sure to check the specifications.
- Set the blower position to high.
- The temperature inside of the car must be stabilized (not getting cooler).
- Engine speed must be at least 1,500 rpm to ensure adequate refrigerant flow.
- The compressor clutch must be engaged.

Air Discharge Temperatures

Most air-conditioning work deals with temperatures on the Fahrenheit scale. Conversions between Fahrenheit and the metric (Celsius) system may be necessary. A conversion chart can be used, or use the following equations:

- To change C to F—multiply C × 1.8 and add 32°
- To change F to C—subtract 32° and multiply by 0.556

Several types of thermometers are available. The *stem/dial thermometer* is the most popular one for air-conditioning service. There are two common ranges: from 0°F to 220°F and from −40°F to +160°F (this one is the most popular for air-conditioning work). Both ranges can work on either the high side (in the condenser) or the low side (in the outlets in the car's interior) (**Figure 36.7**).

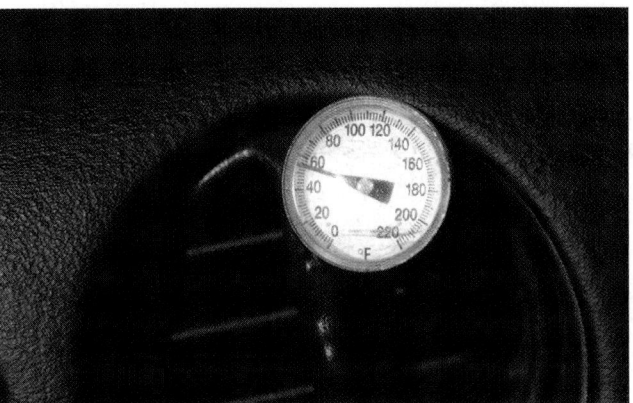

Figure 36.7 A thermometer used for checking condenser and evaporator temperatures.

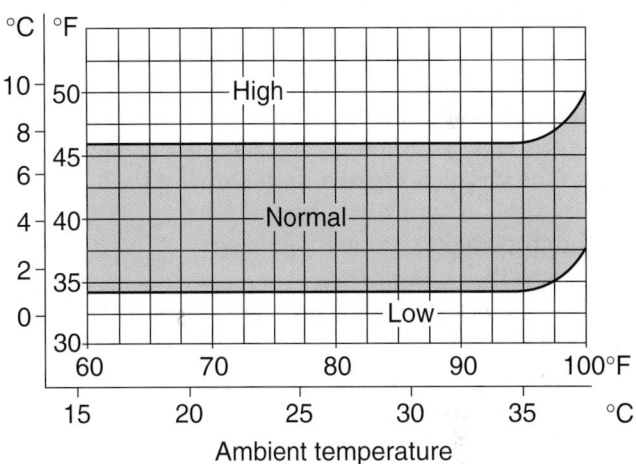

Figure 36.8 A chart showing expected air discharge temperatures at different outside temperatures. (*Courtesy of Ford Motor Company*)

Most discharge temperatures specified by manufacturers run from 35°F to 55°F, depending on temperature and humidity. **Figure 36.8** shows a chart of expected air discharge temperatures at different outside temperatures. The vertical temperature readings refer to actual duct readings. The horizontal readings refer to the day's outside temperature. The drop in the air's temperature as it passes through the evaporator should be at least 20°F. It is very difficult to tell how well air-conditioning works on a cool day.

Rules of thumb:

- With 70°F air entering the evaporator, expect about a 20°F drop in air temperature (50°F).
- With 80°F air entering the evaporator, expect about a 25°F drop in air temperature (55°F).
- With 90°F air entering the evaporator, expect about a 30°F drop in air temperature (60°F).

It is important that the hood be closed during this test. The hardest work an air conditioner does is on "normal" because it is cooling outside air.

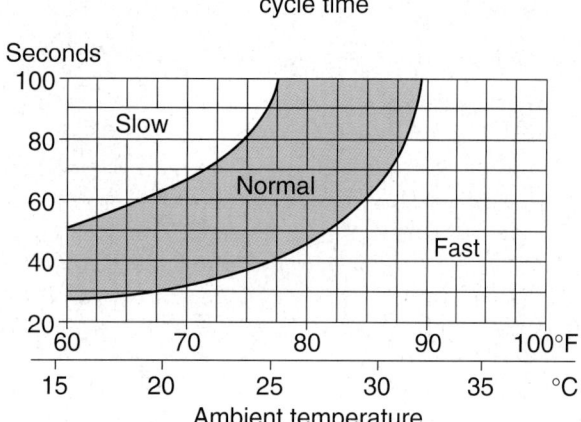

Total clutch
cycle time

Figure 36.9 A chart showing typical clutch cycling times. (*Courtesy of Ford Motor Company*)

Clutch Cycling Times

The majority of vehicles built after 1980 have an orifice tube system. Total cycling time is the length of time between clutch engagement and re-engagement after going through a complete engagement/disengagement cycle. The chart in **Figure 36.9** shows a typical clutch cycling time. A typical cycle would be:
- Clutch on—60 seconds
- Clutch off—15 seconds

In this example, the total clutch cycle would be 75 seconds.

▬ PRESSURE TESTING

A pressure gauge set is used to measure pressures in the system. One of the gauges is for high-side pressure and the other is for low-side pressure. The gauges and hoses are of a standard color code (**Figure 36.10**).

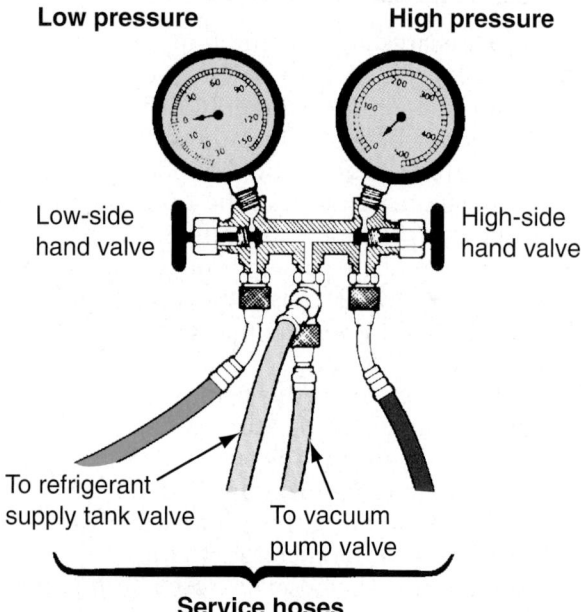

Figure 36.10 A pressure gauge set is used to measure pressures in the system. (*Courtesy of Ford Motor Company*)

- *Red is the high side.*
- *Blue is the low side.*
- *Yellow is for charging and discharging equipment.*

NOTE: *This is the same color code that plumbing follows for hot and cold water.*

The gauges are connected to one manifold that has an additional hose fitting for performing service procedures. Evacuation and recharging of the system can be done through this port.

Service Valves

Service valves for refrigerant installation are located in the high and low sides of the system.
- The **discharge service valve** is on the output side of the compressor on the *high side*.
- The **suction service valve** is on the inlet to the compressor on the *low side*.

The service port on the low side tends to be positioned in the easiest place for the manufacturer to put the vehicle together.

CAUTION Be careful that the connections are made to the correct sides of the system. Older R-12 systems (before 1976) can have fittings that are the same size. On later systems, the high-side fitting is always in a smaller diameter line than the low-side fitting.

Most service valves are *Schrader-valves* (**Figure 36.11**). When the gauge connection is installed over a Schrader valve, it depresses the center pin in the valve. This opens the port to allow refrigerant pressure to act on the gauge.

R-134A service ports use a different type of connector so they cannot accidentally be mixed with R-12 service equipment. A quick coupler is used for connecting and disconnecting hoses from the service ports on R-134A systems. When the hoses are disconnected, they seal off to keep refrigerant trapped in the hoses. The high-side fitting is larger than the low side.

Sealed caps are installed over the ports to prevent refrigerant leakage when they are not attached to the gauges. The caps have an O-ring and require finger-tightening only. Adapters are often required to attach the gauges to R-12 high-side fittings.

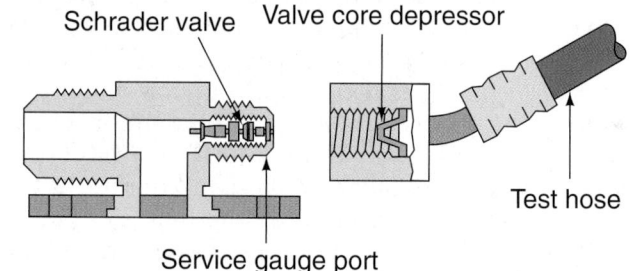

Figure 36.11 A Schrader valve service port.

Start the threads onto the service port threads a couple of turns. Hold the knurled knob against the fitting while you tighten it to avoid refrigerant escaping before the valve is tight. Only finger tighten; do not use pliers.

Connecting the Gauges

Before connecting the gauges to the service ports, close the manifold hand valves and any valves on the service hoses at a refrigerant tank or vacuum pump. Be careful that the hoses do not hang where the fan or an exhaust manifold can damage them when you connect the gauges (**Figure 36.12**).

Before the damage to the ozone layer was discovered, the recommendation was that hoses be purged of air by leaving one of the hoses off while allowing some refrigerant to leak out. This process is no longer allowed. Newer hoses have a check valve that keeps refrigerant in the line.

The low-pressure gauge is a **compound gauge** that reads in either low pressure or vacuum. This is because it is in the suction side of the system. A hand-operated valve opens and closes the suction line to the service hose that charges the system. The gauge is always open to the suction line and registers low-side pressure.

The high-side gauge registers *pressure* only. The gauge usually reads to 500 psi. Most high-side gauges have a small orifice in the inlet that keeps the needle from fluctuating rapidly as the system pulses. Many gauges have a zero calibration adjustment screw on the gauge face also. Sometimes gauges have a scale to convert pressure readings to relative temperature.

The hand control valve for the high side opens and closes the line to the high-pressure service hose used for discharging the system. The gauge always measures pressure in the high-pressure side.

Modern service hoses have a check valve, called an antiblowback valve, in the end of the line to prevent refrigerant from escaping when the line is disconnected. A small amount can still get out. When removing a hose from a connection, put a shop rag around it to prevent oil and refrigerant from spraying. Disconnecting the low side while the system is running results in the lowest pressure at that connection.

▄▄ STATIC PRESSURE READING

The pressure reading in the system when it is not operating is called the **static pressure**. This will be at least 50 psi with a normal refrigerant charge. High- and low-side pressures will be equal when there is no movement through the system. Remember Pascal's law? *Pressure is equal when in a confined area.* The pressure should be the same as the temperature/pressure relationship for the refrigerant (**Figure 36.13**). At 70°, system pressure should be about 70 psi.

High pressure = high temperature
Low pressure = low temperature

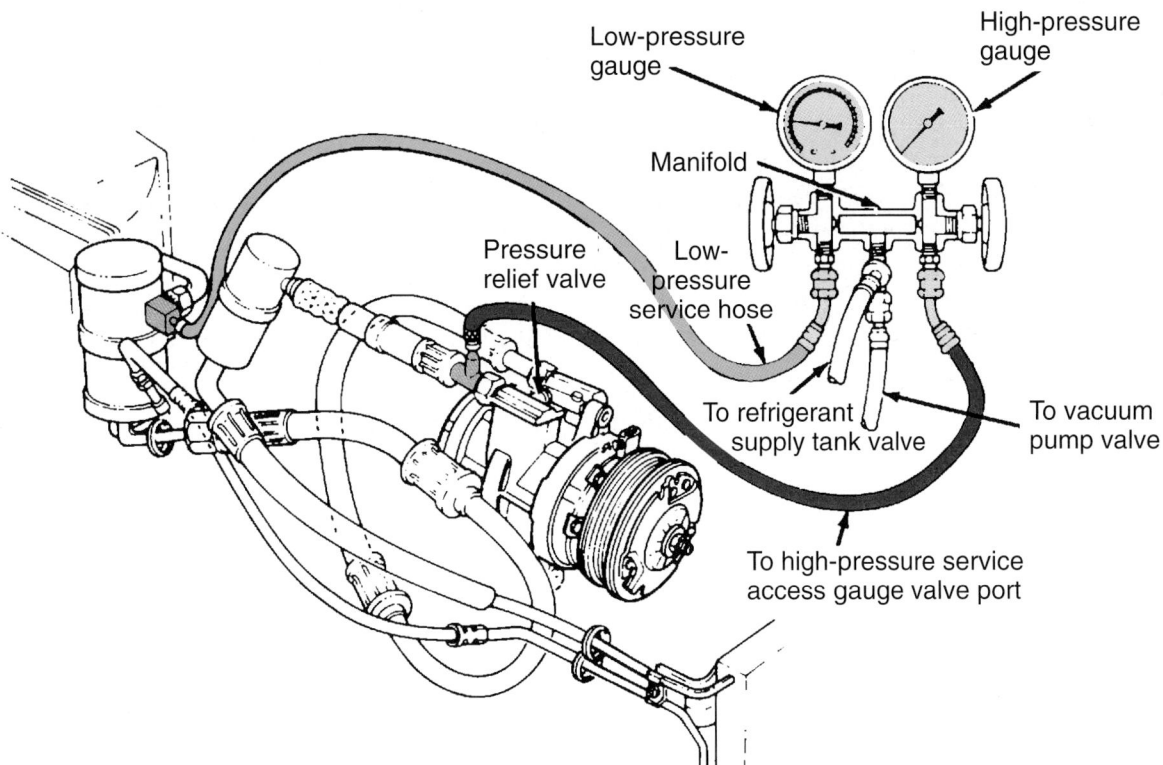

Figure 36.12 Connect the gauges to the service ports. (*Courtesy of Ford Motor Company*)

Pressure Temperature Chart

Temperature		Pressure		Temperature		Pressure	
°F	°C	HFC-134a	CFC-12	°F	°C	HFC-134a	CFC-12
−60	−51.1	21.8	19.0	55	12.8	51.1	52.0
−55	−48.3	20.4	17.3	60	15.6	57.3	57.7
−50	−48.6	18.7	15.4	65	18.3	63.9	63.8
−45	−42.8	16.9	13.3	70	21.1	70.9	70.2
−40	−40.0	14.8	11.0	75	23.9	78.4	77.0
−35	−37.2	12.5	8.4	80	26.7	86.4	84.2
−30	−34.4	9.8	5.5	85	29.4	94.9	91.8
−25	−31.7	6.9	2.3	90	32.2	103.9	99.8
−20	−28.9	3.7	0.6	95	35.0	113.5	108.3
−15	−26.1	0.0	2.4	100	37.8	123.6	117.2
−10	−23.3	1.9	4.5	105	40.6	134.3	126.6
−5	−20.6	4.1	6.7	110	43.3	145.6	136.4
0	−17.8	6.5	9.2	115	46.1	157.6	146.8
5	−15.0	9.1	11.8	120*	48.9	170.3	157.7
10	−12.2	12.0	14.6	125	51.7	183.6	169.1
15	−9.4	15.0	17.7	130	54.4	197.6	181.0
20	−6.7	18.4	21.0	135	57.2	212.4	193.5
25	−3.9	22.1	24.8	140	60.0	227.9	206.8
30	−1.1	26.1	28.5	145	62.8	244.3	220.3
35	1.7	30.4	32.6	150	65.6	261.4	234.6
40	4.4	35.0	37.0	155	68.3	279.5	249.5
45	7.2	40.0	41.7	160	71.1	298.4	265.1
50	10.0	45.3	46.7	165	73.9	318.3	261.4

Red figures — in. Hg Vacuum   * Do not heat can above 120°F

Black figures — PSIG

Figure 36.13 A temperature/pressure relationship chart for refrigerant. At 70°F pressure should be about 70 psi.

SCIENCE NOTE

A 1787 gas law is named for Jacques Charles, a French physicist and inventor who is also credited with being the first one to use hydrogen in balloons (in 1783). Charles' law states that the volume of an amount of a hypothetical dry ideal gas is directly proportional to its temperature provided the amount of gas and the pressure on it remain constant. As temperature is increased the volume of the gas increases.

A related gas law is Boyle's law, named after Robert Boyle in the mid-1600s. Boyle studied the relationship between pressure and volume on a confined ideal gas, proving that when temperature remains constant, pressure and volume are also constant.

■ OPERATING PERFORMANCE TEST

When you start the engine and turn on the air-conditioning system, the low-side pressure will drop and the high side will increase. Pressure on the low side will change according to the temperature in the evaporator. The high side will reflect the temperature of the liquid refrigerant leaving the condenser. The pressures will continue to change until the system stabilizes (reaches its normal operating temperature). The system pressures are stabilized when the system cycles off and maximum high-side pressure does not change from cycle to cycle, and duct temperatures have stabilized.

Check the Sight Glass

Inspect the sight glass if the system has one. This is done in the shop out of the sun. The sight glass will be located on top of the receiver/dryer or in a high-side refrigerant line. When the sight glass is on the receiver/dryer, you are looking to see if the bottom of the pickup tube is uncovered. This would tell you there is not enough reserve refrigerant in the system.

NOTE: *Normal appearance in the sight glass is for the refrigerant to be clear after a burst of bubbles flowing past just after the clutch comes on. After the clutch cycles off, the burst of bubbles will appear again within half a minute or so. Excessive bubbles can indicate a low system. A low system can also appear clear. Bubbles can also indicate air in the system. In R-12 systems, a cloudy appearance means that desiccant has escaped from its bag or there is moisture in the system. A few bubbles or cloudy appearance is normal in R-134A systems.*

■ SYSTEM OPERATING PRESSURES

The pressures will vary with humidity because high humidity puts a higher load on the system. Adequate airflow over the condenser is also required. A portable fan might be necessary to blow additional air through the grill on a car with a belt-driven fan.

SHOP TIP A rule of thumb for normal high-side pressure is with the temperatures of the condenser and compressor at 130°F, the high-side pressure should equal the outside (ambient) air temperature + 100 (+ or −20°F).

Expected low-side pressures differ between expansion valve and orifice tube systems (see chart in Appendix K). Troubleshooting charts in manufacturers' service manuals will help pinpoint the causes of pressure problems.

The following are rules of thumb for the pressure test:

■ The higher the high-side pressure, the more heat there is.

■ The lower the low-side pressure is, the lower the temperature will be at the outlet.

■ High-side pressure affects low side. When you squirt water on the condenser, the high and low pressures *both* drop at once.

■ LEAK DETECTION

Today's air-conditioning systems have a smaller refrigerant charge than those of the past. A 3-ounce leak represents about 20% of the capacity of today's typical system. Some original equipment R-134A systems are designed to operate with 2.5% to 3.5% air in the refrigerant; however, as little as 2% air can cause problems. The amount of excess air resulting from a 1% refrigerant leak drops the cooling efficiency of an air-conditioning system by 1°F. For instance, a 2% leak can cause an orifice tube system to freeze up.

NOTE: *According to the EPA, service port valves are responsible for half of all air-conditioning refrigerant leaks. A missing Schrader valve cap can result in a leak of 1 pound of refrigerant in a year.*

Leaks can be found using several means, including soapy water, colored or fluorescent internal dyes, and several types of electronic leak detectors.

Soapy Water

Soapy water is made by mixing water and soap to a thick solution that can be applied with a small brush. If the leak is large enough, soap bubbles will appear (**Figure 36.14**).

Soap bubbles are only effective for checking leakage that exceeds 40 ounces in a year. Electronic leak detectors are much more effective and can check to less than ½ ounce per year.

UV Dye Leak Detection

One method of finding difficult leaks is with ultraviolet (UV) leak detection in which fluorescent dye is injected into the refrigerant on the low side of the system. A black light is used to locate the leak (**Figure 36.15**).

> ### ▰▰▰ SCIENCE NOTE ▰▰▰
>
> *Pure UV black light is 400–700 nanometers, most of which is not visible to the naked eye. Pure UV light will illuminate a leak better than blue light. Battery-powered light is blue light that appears bright. When the battery starts to discharge, ultraviolet light will still show. Goggles must be used when viewing white light/blue light.*

Figure 36.14 Soap bubbles showing a leak in a hose.

Figure 36.15 A fluorescent dye is added to the air conditioning system before checking for leaks with a UV/Blue light inspection lamp. *(Courtesy of Tracer Products)*

This method is particularly effective because it can pinpoint difficult leaks that only happen during the vibration that results from driving the vehicle. For testing, the system is charged to at least 50 psi static pressure. Static pressure is when the engine is off.

Technicians like UV leak detection because it is something they can show the customer. One problem with UV leak detection is that evaporator leaks are difficult to find using dye.

There are many types of leak detection dyes. Some kits allow dye to be injected with the system fully charged and use an injector similar to a caulking gun. Be sure the one you use meets SA standards for compatibility. Some dyes are installed at the factory. GM and Ford use dye in all of their air-conditioning fluids. During air-conditioning service, however, OE dye can lose its density because it is removed and commingled with other recycled refrigerant during recovery.

When using dye, follow the manufacturer's instructions. Too much dye makes a UV test less effective. Also, too much dye can dilute the oil, resulting in compressor noise or failure. The recommended amount is usually only ¼ ounce, although dual systems may require more. You can put a small amount of leaked refrigerant on a shop towel and check it with a UV light. If it glows, there is enough dye.

Some manufacturers do recommend dyes in their systems; however, a compressor failure might not be covered by the warranty if there is dye in the system.

NOTE: *Sealants, the wrong dye, and flushing agents are all bad for air-conditioning systems. Sometimes the small refrigerant cans used in the do-it-yourself market include sealants with the refrigerant.*

Colored Dye

Another internal charge leak detection method includes injecting a coloring agent into the low side of the system. Some red dyes are available in pound-size refrigerant containers. The dye is harmless to the refrigerant and will be left in the system after the test. If dye leaks out, it will be visible at the source of the leak.

Problems with Dye. Some OEMs do not want dye used in their air-conditioning systems. A problem with red dyes is that they can leak onto interior carpet and cause a damaging stain. If there is a leak, the refrigerant that is recovered from the system for reuse will be full of dye. This means all of the refrigerant in the charging station tank will become dyed to some extent, although the oil separator on the refrigerant recycler will remove most of the dye.

Electronic Leak Detection

Electronic leak detectors are the most popular ones used today. Many of them work on both R-12 and R-134A refrigerants. There are three types of electronic leak detectors. Two of the types are the corona discharge and the infrared. Infrared sensors are said to be more sensitive. Ultrasonic detectors listen for the sound of gas leaking; therefore, the user wears headphones. The third type of electronic tester, which is a heated diode or heated triode, is considered by many to be the best (**Figure 36.16**).

The tester is moved around the various joints in the system. When a leak is found, the tester lights up and its slowly pulsing, high-pitched horn or clicking sound becomes more rapid. The tester has a sensitivity adjustment that you can adjust until it just stops ticking. When it starts to tick again, you might have found a leak. The "gross leak" position can find leaks of 4 ounces a year. The sensitivity in this position is less so you can locate the exact source of the leak. Leak detection works for finding big leaks. When it detects things besides leaking refrigerant, it is called false triggering.

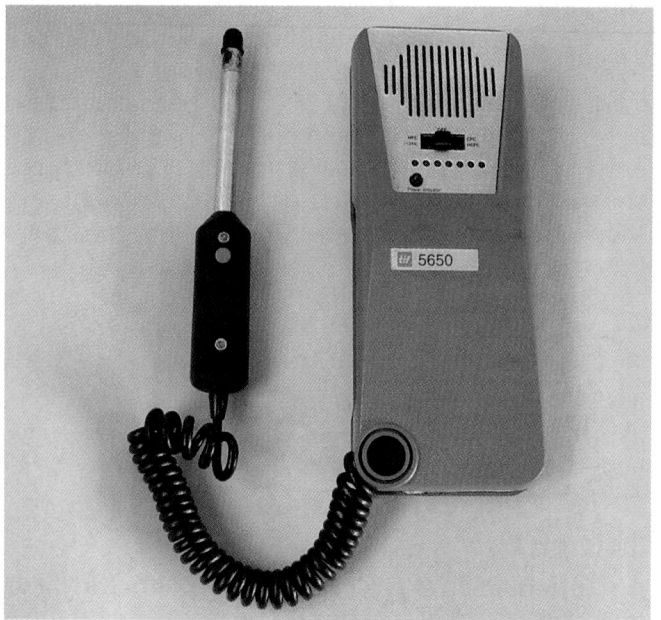

Figure 36.16 An electronic leak detector that works on both R-12 and R-134A refrigerants. (*Courtesy of Tim Gilles*)

There is a correct way to use a leak detector. Do not move the probe too fast. It should be moved at about 1 to 2 inches per second. Keep the end of the probe closer than ¼ inch to the point being checked. Refrigerant is heavier than air, so pass the probe under the area in question. Be aware that if the engine is hot, however, air will rise and pick up the leaking refrigerant.

Leaks are most often found at connections in the system. Sometimes, simply tightening a fitting a small amount will eliminate the leak. Be careful not to overtighten a fitting. If an O-ring is damaged, tightening the fitting will not correct the problem.

With a small leak of ½ ounce of refrigerant per year, it will take 32 years to lose 1 pound. Automobiles normally leak that much from the front compressor seal. Refrigerant that leaks from that seal can become pocketed around the compressor. Blow off the area with compressed air and test again before condemning the seal. If it really is a leak, it will still be there. The tester is very sensitive. Check the evaporator core at its lowest point where the water from condensation comes out.

SHOP TIP Use a clear tube on the end of the tester so you do not accidentally suck water or oil into the tester.

Test the bottom of the accumulator during an air-conditioning leak test in case its metal housing leaks. The accumulator is spin welded during manufacture, so this is a possibility. Test pressure switches for leakage also.

O-rings and Seals

Threaded fittings and block fittings are often sealed with O-rings. O-rings are used because the system must be able to tolerate vibrations. A rigid pipe joint would fail over time. O-rings used since the late 1980s are oval shaped (**Figure 36.17**). The newer O-rings do not fill the O-ring groove completely. They are said to last longer. The new-style O-ring can also be used on older cars.

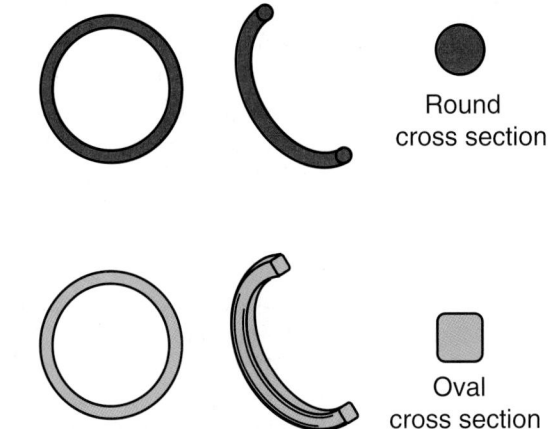

Figure 36.17 Different types of O-rings.

Round
cross section

Oval
cross section

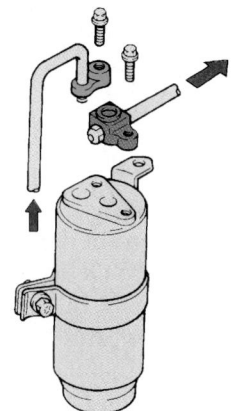

Figure 36.18 Block fittings are used at connections to parts because they allow sealing surfaces to be more accurately positioned.

The oil used with R-134A will damage conventional rubber seals. Be sure to use compatible O-rings. Newer O-rings are usually color coded. R-134A O-rings are also thicker.

Block fittings (**Figure 36.18**) are used at connections to parts because they allow sealing surfaces to be more accurately positioned.

Always coat O-rings with mineral oil before installing them. When they are moistened, they seal better. Newer style sealing washers do not require lubrication and, in fact, will leak if lubricated. Be sure to use mineral oil. PAG oil will pull moisture into the fittings and cause corrosion.

■ COMMON AIR-CONDITIONING COMPONENT PROBLEMS

Prior to component removal and replacement, be sure to capture all of the refrigerant in the system. Remember that the system does not like being exposed to moisture even for a short time. Be ready to cap off all components and seal the system while you work on

a part. Do not leave the system open or it will quickly take on excess moisture, causing the receiver/dryer to require replacement!

Accumulator or Receiver/Dryer Service

Because it can react with refrigerant and corrode the inside of the system, moisture is an enemy to air-conditioning systems.

> ### ■ SCIENCE NOTE ■
>
> *When moisture combines with R-134A, acids are not formed. The only result is corrosion. When moisture gets into an R-12 system, three types of acids are formed: carbonic, hydrofluoric, and hydrochloric. These acids destroy aluminum. When they mix with the refrigerant oil, sludge forms. The majority of the sludge accumulates in the evaporator, causing corrosion and leaks.*

To remove moisture from the system, air conditioners use either an accumulator or a receiver/dryer (see Chapter 35).
- A receiver/dryer is installed in the high side.
- An accumulator is installed in the low side.

NOTE: *Both systems contain a desiccant that requires periodic replacement of the receiver or accumulator.*

Desiccants are pellets held in place by a screen or a bag. A desiccant removes moisture from refrigerant. It is a chemical that forms a molecular bond to water (H_2O). Refrigerant can pass through the desiccant, but moisture bonds permanently to the pellets. The desiccant changes into a different molecule so there is no way water can be separated from the pellets.

NOTE: *A desiccant becomes saturated when it becomes full of water. On a day with 80% humidity, a desiccant can become completely saturated if left fully open to the atmosphere for as little as 10 minutes.*

A receiver/dryer or accumulator should be replaced whenever its system has been leaking, left empty, or left open for a long time, or when a compressor or its reed valve has failed.

The material used for desiccants since the early eighties is a *molecular sieve* that will not crumble when wet. The sieve material can absorb 45 grams of water and is said to be good for 10 years of normal use when the system has never been without pressure.

When retrofitting an R-12 system with R-134A, a new accumulator or receiver/dryer is installed. The desiccant used in R-12 systems was XH5, which is not compatible with R-134A. XH7 is the desiccant used with R-134A systems, and it can be used with R-12 as well.

▣ REFRIGERANT OIL

Air-conditioning systems use a special nonfoaming refrigeration oil to lubricate the compressor, gaskets and seals, and expansion valve.

NOTE: *Oil is to an air-conditioning system as blood is to a human: you must have the correct type. Be sure to use the correct oil. It must be the correct amount and viscosity.*

When the air-conditioning system is operating, there is always a small amount of oil located in each of the system's components. The amount of oil in most systems can vary from less than 6 ounces to 9 ounces. The older GM A-6 compressor was unusual in that it used 12 ounces of oil and had its own oil sump.

All of the system's components contain oil. When a part is replaced, a certain amount of oil is added to the part before replacing it. Oil requirements for an orifice tube (expansion tube) system are shown in **Figure 36.19**.

Oil collects in the bottom of the compressor during use. When the old compressor is replaced, drain out the oil and measure it so a like amount of oil can be added to the new compressor.

NOTE: *Like brake fluid, refrigerant oil should be stored with the cap on. It can absorb moisture, which is damaging to the air-conditioning system.*

Some replacement compressors are shipped with no oil, whereas others come with a full 5- to 9-ounce charge in the crankcase. This oil must be drained or an overcharge will result. Every system is different, so be sure to check manufacturer's recommendations for the correct amount of oil. Too much oil in the system

takes up space that would normally be used for the refrigerant. This reduces cooling system capacity. If an oil overcharge is suspected, try to drain the accumulator. A two-bag accumulator should hold 3 ounces of oil. Some aftermarket accumulators may have only one bag. The desiccant will hold the oil, so there should not be very much oil that pours out.

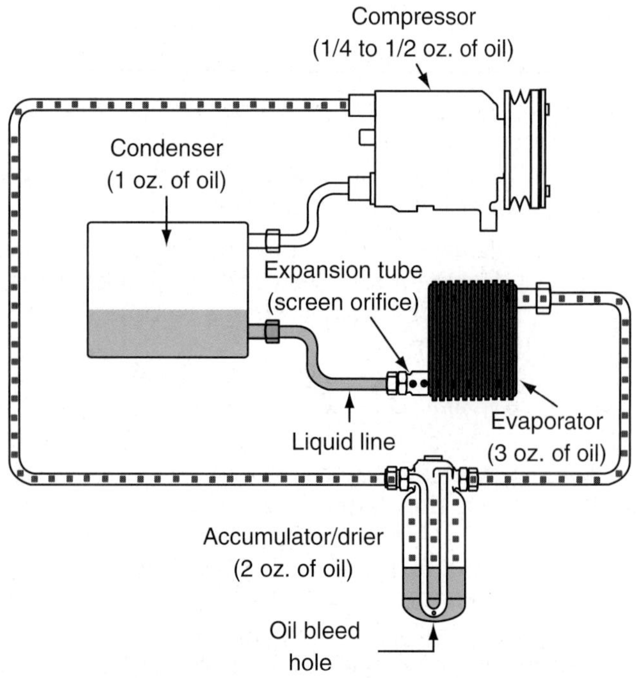

Choosing the Right Oil

The oil must be compatible with the refrigerant. R-12 uses *mineral oil* and R-134A uses *polyalkylene glycol* (**PAG**) or polyalester (**ester**) oil. The correct oil is very important in R-134A systems. OEM compressor manufacturers recommend their own lubricants. An underhood air-conditioning label is shown in **Figure 36.20**. Oils used in different systems are not the same. Check the compressor label or the service information to see what the manufacturer recommends. The mineral oil used with R-12 systems is not soluble in R-134A or PAG oil, so it will separate and form sludge in recesses in the system.

> **CAUTION** ■ PAG oil is nontoxic. R-134A is also nontoxic. But when R-134A and PAG oil are mixed together, both are toxic.
> ■ PAG oil can damage some paints.
> ■ Use regular mineral oil as a lubricant. Do *not* use PAG oil as a lubricant, because it absorbs water more easily than mineral oil. If PAG oil leaks, it will corrode fittings. One-half of every connection is exposed to air. This can cause corrosion on the exposed area of an R-134A system.

NOTE: *If mineral oil is mistakenly installed in an R-134A system, it will not lubricate the compressor properly and will separate in the accumulator*

PAG oil does not tolerate chlorine, which is part of CFC gas. It will turn to a gelatin-like substance if it is accidentally put into an R-12 system. PAG oil is used in most vehicles, except for Volvo, Saab, Jaguar, and Land Rover, which use ester oil. A new universal PAG oil is used on some GM vehicles. PAG oils are hygroscopic, which is a good trait because moisture is absorbed into the oil and does not settle out in the system.

Figure 36.19 Oil requirements for different parts of an orifice tube system. Oil in this system constantly circulates.

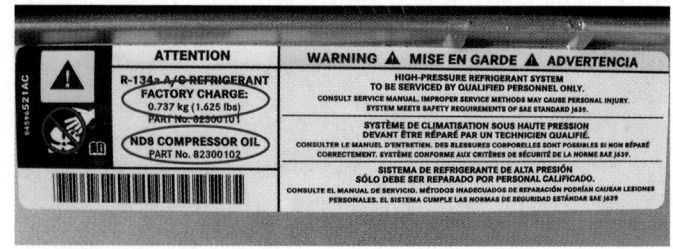

Figure 36.20 An underhood air-conditioning label listing the compressor oil type and the amount of refrigerant installed at the factory. (*Courtesy of Tim Gilles*)

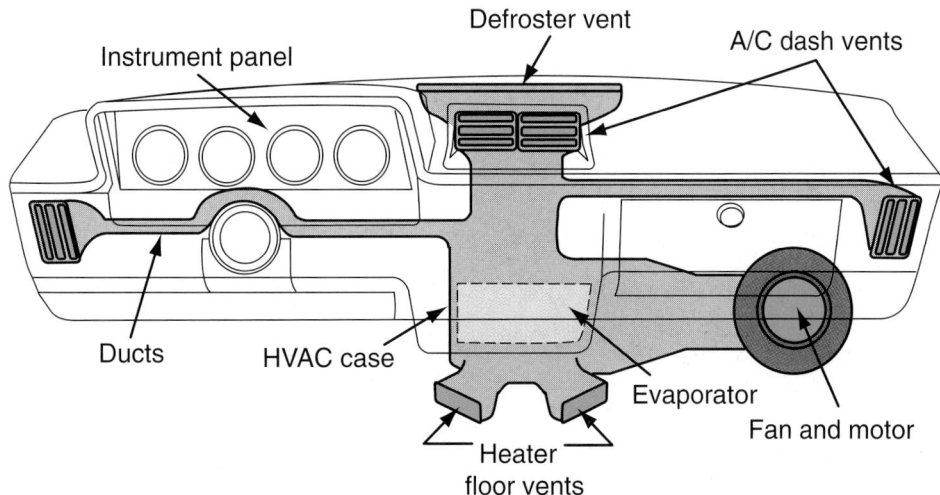

Figure 36.21 The evaporator is often located within the dashboard inside the HVAC case.

Because of the moisture it contains, air is the primary contaminant in an air-conditioning system. Moisture mixed with oil will not boil. Moisture control in R-134A is provided by the desiccant in the receiver/dryer or the evaporator.

NOTE: *Moisture causes fewer problems in R-134A systems because it mixes with the PAG oil. Esters, which are oils used in commercial refrigeration, are not hygroscopic.*

Refrigerant Viscosity. Be sure to use the correct type and viscosity of oil recommended by the manufacturer because there are many different ones. There are three main viscosity R-134A oils: 46, 100, and 150. These oils are all thinner than SAE 15 oil because they are rated in Saybolt Universal Seconds (SUS). Mineral oils used in R-12 systems can have Saybolt viscosities ranging from 300 to 1,000, about the same as SAE 30–SAE 50 engine oil.

Oil in the Evaporator. In an orifice tube/accumulator system, if everything is correct, nearly all of the refrigerant will remain as a liquid in the evaporator. The refrigerant leaves the evaporator as a liquid, carrying lubricant along with it.

■ EVAPORATOR PROBLEMS

Some manufacturers put a layer of foam on the outside of the passenger side of an evaporator. This is to prevent water from being blown on the passengers. Mildew is a problem with air-conditioning systems. Check the evaporator drain through the floorboard to be sure it is unrestricted.

SHOP TIP Some technicians spray Lysol® on the foam if it mildews. There are also commercial evaporator cleaning kits available.

If an evaporator leaks, it must be replaced. The evaporator is often located within the dashboard, (**Figure 36.21**) and gaining access to it can be time consuming. Consult the service information for instructions before attempting evaporator removal. The system is evacuated before removing the evaporator and recharged after the new one is in place to test for leaks before reinstalling the parts of the dashboard.

■ THERMAL EXPANSION VALVE OR ORIFICE TUBE PROBLEMS

An orifice tube system cycles the clutch to control refrigerant flow through the system. In a thermal expansion valve (TXV) system, the size of the TXV opening varies based on the temperature and pressure in the evaporator. When more cooling is required, it opens as far as it can, but it must still allow all refrigerant in the evaporator to boil before it exits. Too much refrigerant will flood the evaporator.

In either a TXV or an orifice tube system debris can cause a restriction to refrigerant flow. If the TXV is not operating as it should further tests can be done before attempting a repair.

Expansion Valve

A bad expansion valve will cause low readings on both the low and high side. Failure of the sensing bulb is the most common problem. Sometimes the valve can become plugged with debris.

NOTE: *If the expansion valve requires replacement for any reason, replace the receiver/dryer. It is right in front of the expansion valve and should have prevented the corrosion that ruined the expansion valve. The evaporator would fail next if the expansion valve was open.*

If the sensing capillary tube is warm, there is high pressure that pushes on the diaphragm. The sensing bulb senses the temperature of the outlet line of the evaporator. If it is warm, it opens the valve. Pressure on top of the diaphragm tries to push the valve open. The operation of the expansion valve can be checked.

Some expansion valves are wrapped with insulation. When the sensing bulb is in the evaporator case, it does not need to be wrapped, because it stays cold. When an expansion valve is packed in that type of system, it is only to silence the noise that the evaporator makes.

Some vans have a rear evaporator that has an expansion valve. There is always some flow in the rear unit to keep the expansion valve bulb cold. If you can pinch the hose to the rear unit while the rear unit is off and this changes pressure drastically, then the expansion valve is stuck open.

Orifice Tube Service

With an orifice tube system, also called an expansion tube system, the accumulator should feel cold and sweaty. The same thing occurs in it as in the evaporator. Like a bad expansion valve, a plugged orifice tube will cause low system pressures (**Figure 36.22**). There is a filter screen in the inlet to the orifice tube that can become plugged. A saturated desiccant can rupture, sending its material throughout the system. This can

Figure 36.22 A plugged orifice tube. *(Courtesy of Dave Brainerd)*

result in a plugged orifice tube screen. In a TXV system, this can plug the expansion valve. In an orifice tube system, the accumulator is after the orifice tube, so the desiccant has a long way to travel to get to the orifice tube. A more common cause of orifice tube plugging is pieces from the compressor.

The orifice tube is easily checked and replaced once the refrigerant has been recovered. An orifice tube can be installed in either direction, but there is one right way. It has an arrow that points in the direction of refrigerant flow. If you put it in backwards, the filter screen will be on the wrong end. It might plug up and be difficult to remove. When it has removal tabs on one end, simply grasp it by that end when you reinstall it. A special tool makes installation the correct way easier (**Figure 36.23**).

Occasionally, an orifice tube gets stuck in its pipe. A special removal tool can help to remove it. Sometimes, they break up inside the pipe. When the tube is stuck in the line and cannot be removed, there is a special kit that is used. The old section of line is cut off and is replaced with a new section of line that holds the new orifice tube (**Figure 36.24**).

SHOP TIP To assist in removing a stuck orifice tube, use a heat gun. It will soften the plastic at the ends of the orifice tube, helping to loosen it from its housing.

■ COMPRESSOR SERVICE

Air-conditioning compressors are relatively reliable and usually fail as the result of a leak. Compressors can leak

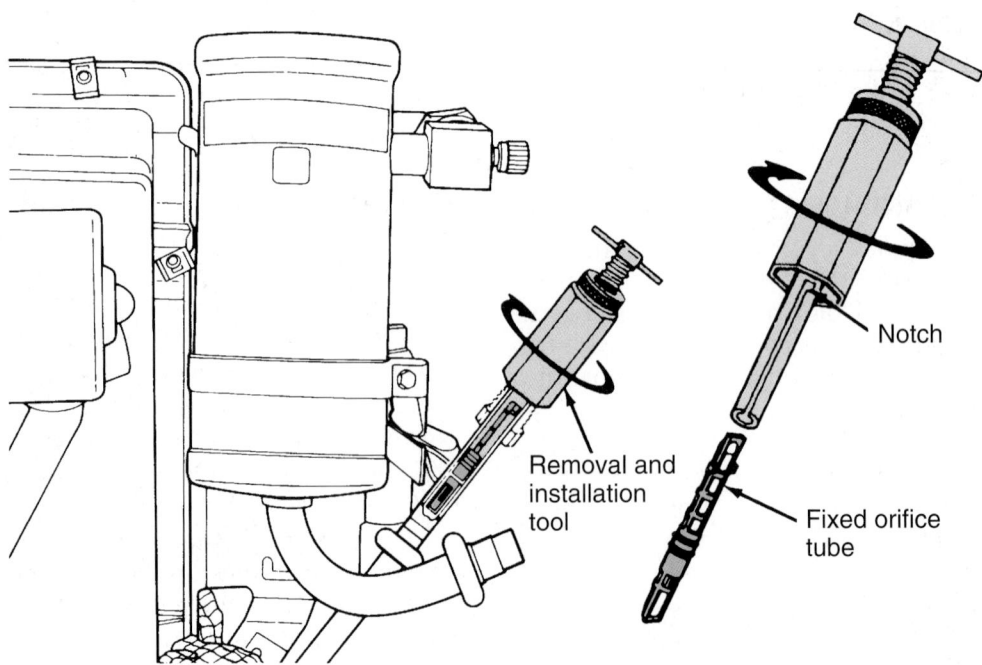

Notch

Removal and installation tool

Fixed orifice tube

Figure 36.23 A special tool for removing and replacing orifice tubes. *(Courtesy of Ford Motor Company)*

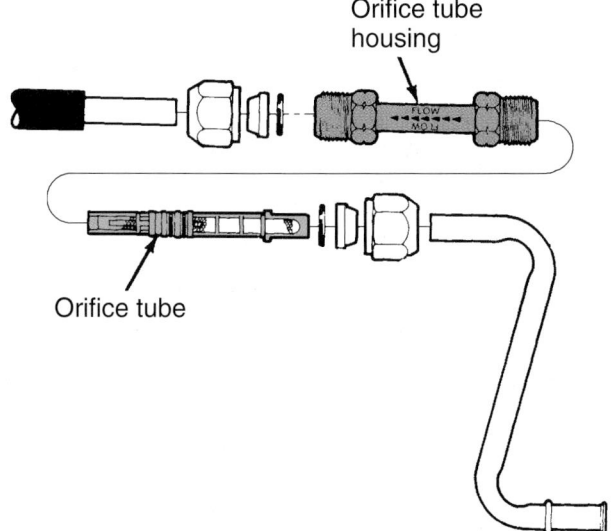

Figure 36.24 An orifice tube kit for replacing a damaged orifice tube. *(Courtesy of Ford Motor Company)*

from the front shaft, from seals on the case, or from O-rings on the back (**Figure 36.25**). Sometimes, the aluminum tubing that supplies the back of the compressor develops a leak (Figure 36.25). There are hundreds of configurations of these. Aftermarket companies produce them and have catalogs listing them by number. Other common causes of failures include excessive clutch cycling or lack of lubricant in the refrigerant. When a compressor has previously been replaced,

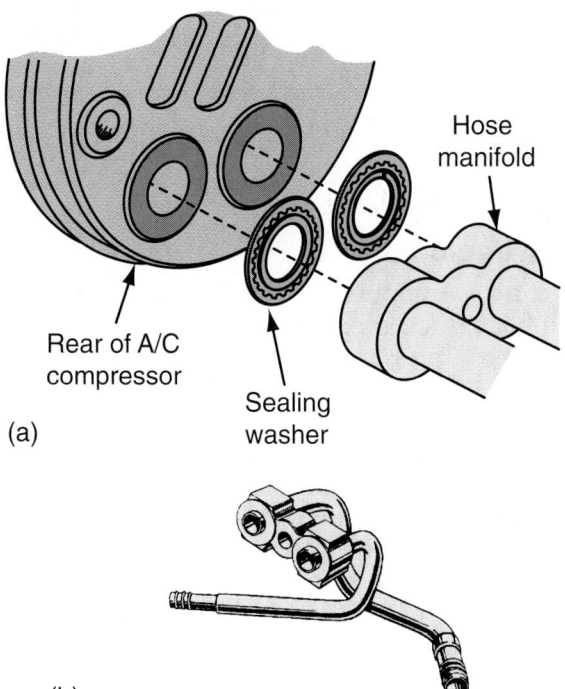

Figure 36.25 (a) Sealing washers on the back of a compressor. (b) An aluminum tubing manifold. *(b, Courtesy of Four Seasons, Division of Standard Motor Products, Inc.)*

debris left in the system from the previous compressor failure is often the cause of the repeat failure.

Piston compressors are the most popular. Vane and scroll compressors are particularly susceptible to failure from insufficient refrigerant. The refrigerant is needed to circulate the oil throughout the system and to provide heat transfer.

Replacing a Compressor

When you replace one compressor with another, it is important that the correct one be used. Be sure you have the correct part number. On the GM radial six-cylinder compressor, there are many different part numbers. The difference is in where the compressor is mounted. It is important that the correct mounting be used so the oil passage is in the right place.

When installing a new compressor, the Mobile Air Conditioning Society (MACS) recommends you rotate it by hand a dozen times with a spanner wrench or a ratchet and socket to expel the shipping oil and distribute the new oil throughout the compressor.

Replacement parts installed with a new compressor include the receiver/dryer or accumulator and an orifice tube or expansion valve. A clogged condenser or muffler must also be replaced. The following important considerations are also related to compressor replacement.

NOTES:

■ *Some mufflers have a filter that must be replaced in the event of a compressor failure.*

■ *It is a good idea to install a filter in the liquid line to trap debris from the previous compressor failure. It is better that the filter become plugged and restrict flow than to allow the debris to take out the new compressor or cause a restriction in the condenser. Flow tubes in the condenser are very, very small.*

■ *If the orifice tube is plugged, the condenser probably is too. After the system has been recharged and is operating again, test the condenser inlet and outlet temperatures. The average difference in temperature between the inlet and outlet ranges between 20°F and 50°F. If there is this much difference when the system is under a significant load, the condenser is probably not plugged.*

■ *When a compressor is replaced on an orifice tube system, the accumulator is also replaced because it has a pinhole at its bottom that pulls in oil. There is a high probability that the accumulator orifice will be plugged, which could result in failure of the new compressor.*

Refrigerant Filters

A filter can be added to an air-conditioning system (**Figure 36.26**). When a compressor has failed, be sure to put a filter after the outlet of the compressor. Filters are available either with or without an orifice tube. Put the new filter in the line out of the condenser where it is most convenient to install it.

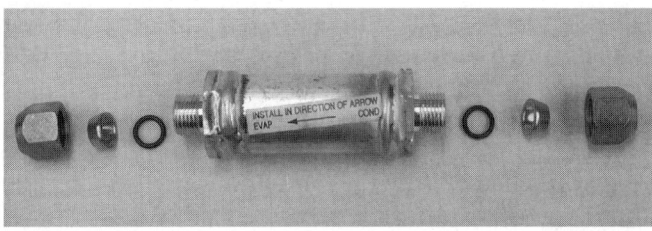

(a)

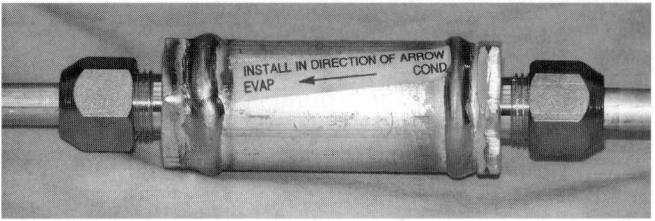

(b)

Figure 36.26 An air-conditioning liquid line filter.

Flushing an Air-conditioning System

Aftermarket compressor manufacturers require solvent flushing for the warranty to be honored in the event of a failure. Flushing must be done with a liquid, so grit is carried away.

> **SAFETY NOTE** Do not attempt to use shop air. Shop air and R-134A can cause an explosion.

Use only approved solvents for the type of system you are working on and completely remove them when finished. If the system has a plugged muffler, it cannot be flushed and, therefore, must be replaced.

There are still some older R-12 systems on the road. They are flushed with mineral spirits, which are chemically related to mineral oil. These cannot be used on R-134A systems.

A *manifolded condenser* came out in the late 1980s. It cannot be flushed like the *dual parallel condenser*. It has its inlet and outlet on the same tube but has seams that separate about every five tubes down. Some manufacturers will not pay warranty time on a system flush but will pay for a filter.

Compressor Shaft Seal

Compressor shaft seals occasionally require replacement. On rear-wheel-drive cars, this can often be done using special tools after the system has been discharged while the compressor remains on the engine. There are special tools available for removing and installing the clutch seal (**Figure 36.27**).

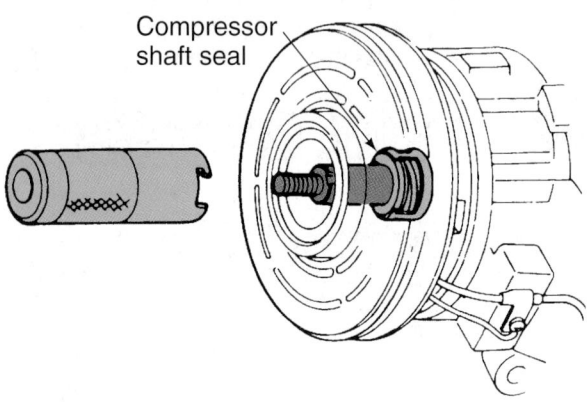

Figure 36.27 Installing a compressor shaft seal.

Clutch Problems

When clutch engagement problems occur, check for voltage from the source. Then use a fused jumper wire to apply power and ground to the terminals on the clutch. This will bypass the protection devices and determine whether the problem is the clutch or the circuitry. Parts of the clutch and compressor are shown in **Figure 36.28**.

Clutch Failures

There are several reasons why a clutch might fail. One reason is an excessive air gap. Always check to be certain the gap is not too wide (**Figure 36.29**). Another possible cause of clutch failure is excessive system pressure, which will cause the clutch to work too hard. This can happen if the system is charged with too much refrigerant or if the high side has a restriction. Check the available voltage and ground because insufficient current flow to the clutch can cause it to slip. Always perform voltage drop tests on both power and ground sides whenever you replace a clutch.

Clutch Replacement

New clutch assemblies are commonly available. The clutch plate is removed first. A special spanner wrench keeps the compressor shaft from turning while the nut is loosened (**Figure 36.30**). The clutch plate is removed by threading a special removal tool into the threaded hole in its center (**Figure 36.31**). Sometimes a retaining ring holds the pulley in place (**Figure 36.32**). **Figure 36.33** shows a puller being used to remove a pulley. There are many different possibilities for clutch removal, so refer to the manufacturer's instructions. There are also special tools for installing a clutch coil on the compressor.

NOTE: *Before installing a new clutch assembly, be certain to count the number of grooves on the pulley. Some compressors use pulleys with differing numbers of grooves.*

Some clutch assemblies use shims to obtain the correct air gap.

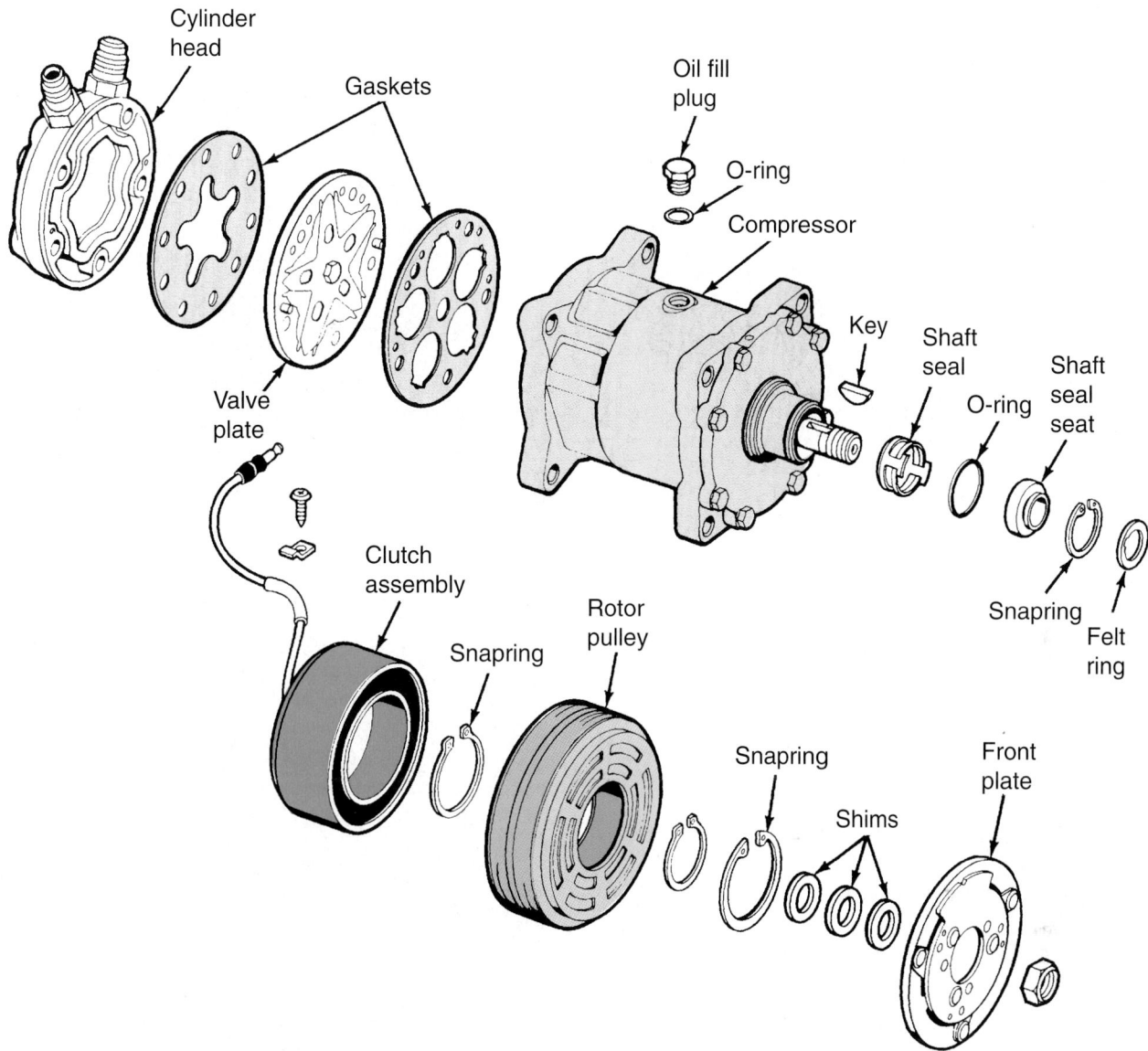

Figure 36.28 Parts of the clutch and compressor. *(Courtesy of DaimlerChrysler Corporation)*

Clutch Break-In. New clutches must be broken in. Run the engine at high idle and cycle the clutch on and off for a second 25–50 times at a time.

◼ EVACUATING AND RECHARGING

Two additional hoses can be attached to the center connection on the manifold set. These are for a refrigerant supply tank hookup and for a remote vacuum pump. Yellow or white are the preferred hose colors for R-12, and solid yellow with a black stripe is the R-134A color. Service ports use color-coded quick couplers that automatically close the hose end when disconnected.

Refrigerant Identification

Before evacuating refrigerant into a charging station, you must be certain it is the same refrigerant as what is in the tank reservoir. A refrigerant identifier can do this for you (**Figure 36.34**). Some refrigerants are blends,

and the identifier will tell if the refrigerant is 100% R-12 or R-134A. If it is not 100%, you will not want to contaminate your refrigerant tank. Some refrigerant identifiers will also tell air content. Air is considered to be a noncondensable gas.

◼ VACUUMING A SYSTEM

When a system is empty or has been repaired and drained, all of the air and moisture must be removed. They are removed with a vacuum pump. The vacuum pump removes the air right away. To remove the moisture, it must be boiled out. The system does not have to be heated to make moisture boil. It simply has the pressure removed from it to at least 28 in. Hg. Pressure this low will boil water in a glass at room temperature (**Figure 36.35**).

Vacuuming a system is done not only to dry out moisture but to help the efficiency of the system. Air builds up in the top of the condenser. Cooling can not

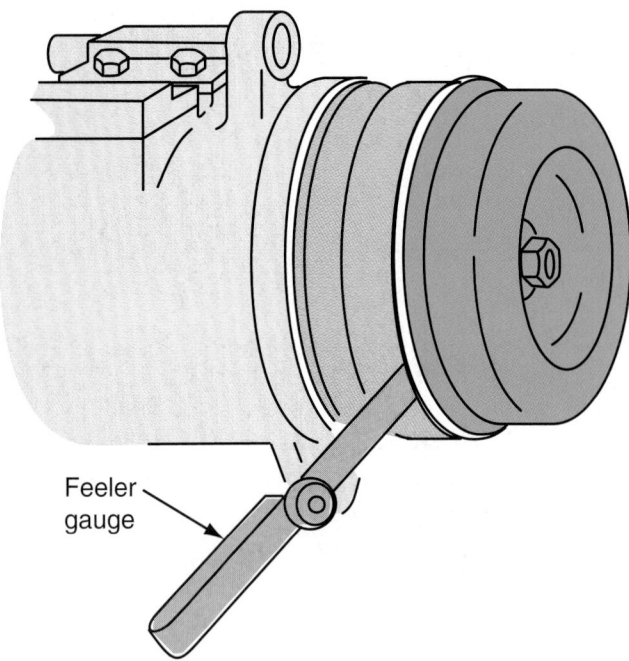

Figure 36.29 A compressor clutch can fail if the clutch plate air gap clearance is excessive.

Figure 36.30 The compressor shaft is held from turning while loosening the shaft nut.

Figure 36.31 A special puller is installed into the threads in the clutch plate. *(Courtesy of Tim Gilles)*

Figure 36.32 A retaining ring on the pulley assembly.

take place until the refrigerant gets below the air level, so you lose condenser efficiency. The result is that when the refrigerant comes into the expansion valve or tube, it carries more heat. Vacuuming can drop the outlet temperature by 6°F to 8°F.

Vacuum Pump Rating

Although the primary purpose of system evacuation is to remove air and moisture, you cannot actually get all of it out. You must have a high-quality vacuum pump if you expect to remove most of the moisture from the system. Vacuum pumps have different capabilities, depending on design and maintenance.

The micron rating of the vacuum pump tells how deep a vacuum it will pull. Too small a pump will take a long time to evacuate the system. A really good vacuum pump will pull to about 50 microns. This is

Figure 36.33 Removing the pulley with a special puller. *(Courtesy of Tim Gilles)*

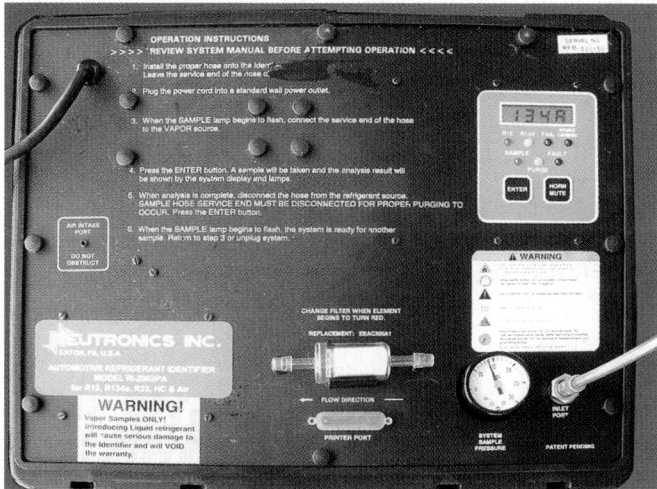

Figure 36.34 A refrigerant identifier. *(Courtesy of Tim Gilles)*

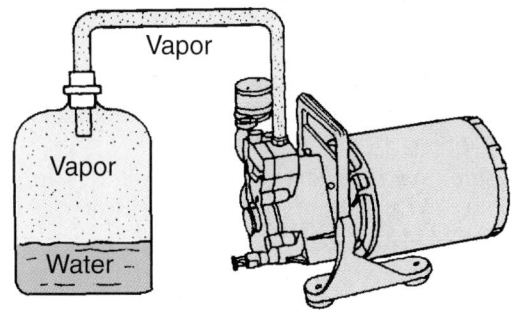

Figure 36.35 A vacuum pump reduces pressure within the system to cause water to boil at room temperature.

more than 29 in. Hg and drops water's boiling point to below zero. With practice and a good pump, you can pull the air out of a system in just 3 minutes. For best results, try to achieve at least 29.5" of vacuum. To verify the quality of your evacuation procedure,

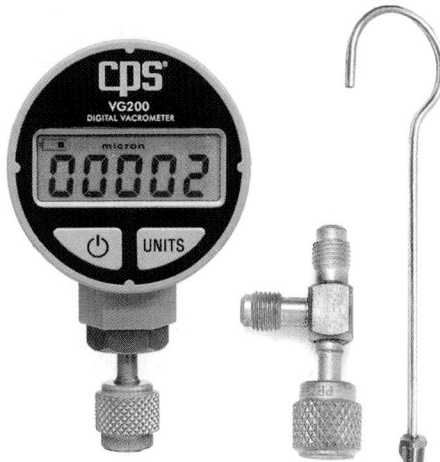

Figure 36.36 A thermistor vacuum gauge evaluates the quality of vacuum placed on the system. *(Courtesy of CPS Products, Inc.)*

install a thermistor vacuum gauge that measures in microns (**Figure 36.36**) in series. This will evaluate the strength of your vacuum pump and the quality of the vacuum being placed on the system.

NOTE: *The required time for factory evacuations is only about 45 seconds, extremely fast compared to service in the aftermarket.*

A drop of only 1 inch of applied vacuum results in a tremendous difference in negative pressure. Compare the following vacuum results from the Mobil Air Conditioning Society (MACS):

28.92" = 25,400 microns (this will not boil)
29.92" =   2.54 microns (this will boil
                            extremely fast)

SHOP TIP To be assured of no leaks and no air, an acceptable pump should be able to pull a system down to 100 microns; but if you can "pull the system down" to 500 microns, this will provide a satisfactory result.

Vacuum works better with R-12 because oil and water do not mix and so moisture is more easily removed. During the evacuation procedure, you need to get below the boiling temperature of the refrigerant. The refrigerant boiling point of −24 is 254 microns.

NOTE: *Protect a vacuum pump by changing its oil every 10 hours or less.*

■ EVACUATE THE SYSTEM

Evacuation and recharging of the system is done with the engine off by connecting the center hose to both service ports by opening both of the valves (**Figure 36.37**). To evacuate the system, connect the center service hose to the vacuum pump. Open both

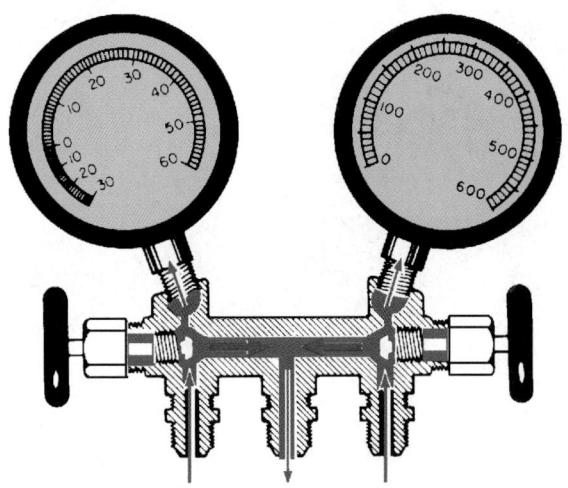

Figure 36.37 To evacuate and recharge the system, connect the center hose to both service ports and open both of the valves.

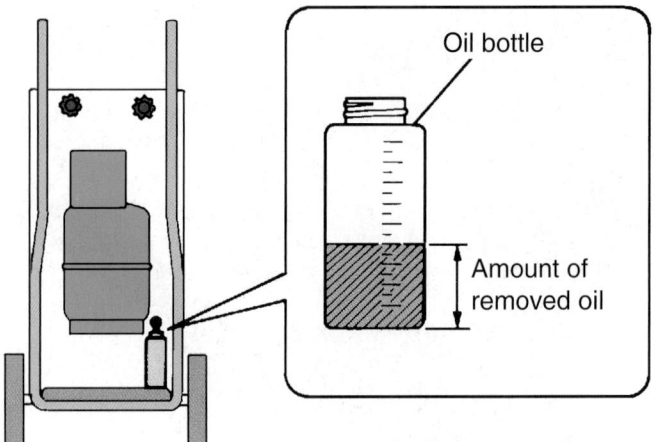

Figure 36.38 Measure the amount of oil that is removed during recovery and add that amount to the system during recharge.

valves on the manifold and the one on the vacuum pump (if it has one) and start the vacuum pump.

When the vacuum pump starts, you should see system pressure drop. When you do not have a thermistor vacuum gauge, vacuum is typically applied to the system for ½ hour or so. Keep a vacuum on it for another 15 minutes after it reaches its lowest point. As water boils in the system, it takes up more space. Watching the low-side pressure will give an indication of the amount of moisture to be boiled away. If the system drops right away to 28 in. Hg, it is fairly dry. If the system drops to 26 in. Hg and holds for a while, there is probably a good deal of moisture in the system.

NOTE: *Some commercial desiccants will give up moisture under vacuum, but not those used in automobiles.*

If vacuum does not drop below 20 in/Hg within 5 minutes, a leak is likely. Close all valves and shut off the vacuum pump. If vacuum drops and pressure rises, there must be a leak. The leak must be located and repaired before the system can be evacuated.

Following evacuation, close all valves and shut off the pump. Check the low-side pressure and recheck after 5 minutes for comparison. Pressure should remain steady if there is no leak.

NOTE: *Vacuum leak checking is not a dependable way of locating tough leaks. An O-ring could leak under pressure but not under vacuum.*

Replacing Oil. If you are 10% low on refrigerant on today's low-charge systems, the pressures may be too low to properly circulate the oil through the system. During recovery, oil is removed with the refrigerant. It collects in a bottle in the recovery station. Remember to empty the oil reservoir first. After recovery, measure the amount that is removed and add new oil in that amount when you recharge the system (**Figure 36.38**). During recovery of R-134A, you will usually lose 1 ounce of oil at least. This is much more than is lost with R-12.

When recharging a system, 1 or 2 ounces of extra oil is usually added with the new refrigerant. Do not add too much, or cooling efficiency will be inhibited.

The following are approximate oil amounts that are added to compensate when air-conditioning parts have been replaced:

   Accumulator: 2–3 oz
   Condenser: 1 oz
   Evaporator: 2–3 oz
   Receiver/dryer: 1 oz

When a part is replaced due to a refrigerant leak, extra oil is also added to compensate for the oil that escaped with the leaking refrigerant.

 **SHOP TIP**   To determine how low a system charge was, recover the refrigerant and weigh it.

■ MODERN AIR-CONDITIONING SERVICE EQUIPMENT

A modern charging station (**Figure 36.39**) will usually have a switch to control the process. Electric solenoids open the valves.

Separate systems are needed for dealing with R-134A and R-12 refrigerants. They cannot be mixed, or contamination results.

With a **dual pass machine**, first recover then recycle. With a **single pass machine** (one motion), there is no need to ever charge with vapor (from the low side). The machine removes the refrigerant and recycles it during the 15-minute minimum vacuum cycle. The refrigerant bottle will be heated sufficiently by that time so there is enough heat to blow the refrigerant into the high side without starting the vehicle. The machine has a switch that allows changing between recycled and new refrigerants.

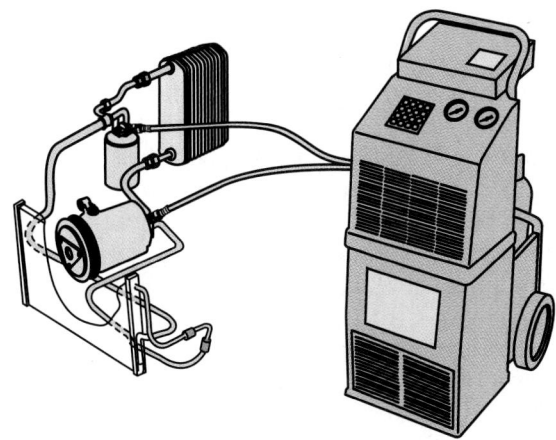

Figure 36.39 A modern charging station.

▣ REFRIGERANT CHARGING

After the evacuation process is complete, refill the system with the correct amount of refrigerant. The refrigerant charge amount is critical, and each system is different. Installing too much or too little refrigerant will hamper cooling efficiency. Overcharging can overload the accumulator, allowing liquid refrigerant to go to the compressor.

At the assembly plant, air-conditioning systems undergo a very fast refrigerant installation procedure that takes 15 seconds to vacuum, test, and fill. There is only one correct aftermarket fill procedure. It uses a calibrated scale and precision fill system with a positive displacement pump and the refrigerant amount corrected for differences in temperature.

Evaporator and Accumulator Temperature Tests

In addition to analyzing pressure readings, you can determine whether a refrigerant is liquid or vapor by its temperature.

Evaporator Inlet and Outlet Temperature Test. For a quick test of an orifice tube system to see if it is charged correctly, the inlet and outlet temperatures of the evaporator should be very close to the same (within about 5°F of each other, **Figure 36.40**). This test is done while the system is loaded (high blower with the windows open). Run the engine for 5 minutes or more. If the outlet temperature is higher than the inlet temperature by 2°F to 5°F, the refrigerant charge could be low.

Accumulator Inlet and Outlet Temperature Test. If there is less than a 2°F difference between the outlet and inlet temperatures of the accumulator, the refrigerant leaving it is still liquid. A 5°F difference between the accumulator inlet and outlet temperatures or frost on the accumulator can indicate two things: either the system is overcharged or the clutch is not cycling.

An accumulator system is also called a "flooded evaporator" system. In an expansion valve system, a flooded evaporator is bad. Remember that the accu-

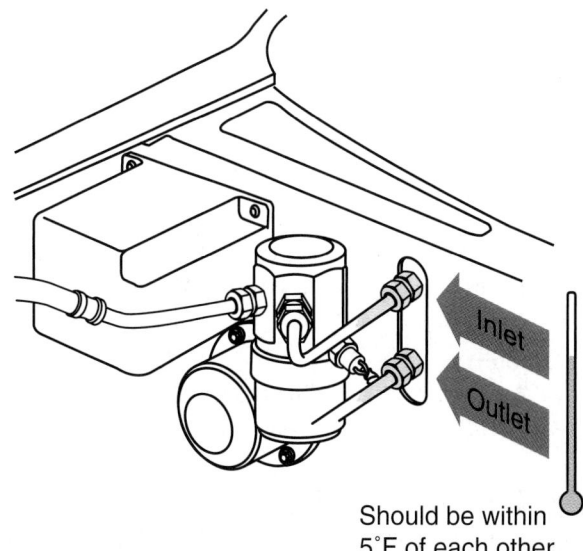

Should be within 5°F of each other

Figure 36.40 The inlet and outlet temperature of the evaporator should be within 5°F of each other.

mulator accumulates liquid and lets it turn back to a vapor before it goes back into the system. In an orifice tube/accumulator system, if everything is correct, nearly all of the refrigerant will remain as a liquid in the evaporator. The refrigerant leaves the evaporator as a liquid, carrying lubricant along with it. If the system has a low charge, the refrigerant will turn to a vapor in the evaporator and leave the oil behind, preventing the compressor from being lubricated.

Incorrect Refrigerant Fill

Pressure gauge readings, especially on the low side, are not a dependable way of testing the fill. There are several reasons why a fill might not be correct.

Sight Glass. On modern air-conditioning systems, you cannot rely on what you see in a sight glass because the system will still be undercharged when the bubbles disappear. Because of this, some sight glasses are painted or have a sticker over them.

Small Refrigerant Cans. Do not attempt to recharge using 12-ounce refrigerant cans while watching gauge pressures. A few shops still like to use small cans because it is an easy way to control inventory. Today's systems have small capacities, however, and the margin of error no longer allows for the use of small cans like it did in the old days of R-12 and larger capacity systems. For instance, with GM's charge tolerance of 10% and a typical new system charge of between 13 and 16 ounces, how could you use a 14-ounce can and hope to meter the correct amount of refrigerant? Today's condensers use less refrigerant. In one GM model, an older system with a charge of 26 ounces was replaced with one holding only 13 ounces with only 2 ounces of tolerance before the evaporator is starved. BMW's Mini Cooper lists a charge tolerance of only +/− 0.33 ounce.

The correct way to use a can is to rotate it slowly. Still, when you disconnect you can lose up to 20% of the contents of the can. Also, you cannot go by weight when using small cans because there will always be some refrigerant left in the can.

Residual refrigerant in the hoses. A typical service hose is 6 feet long and holds about 2 ounces of refrigerant. When hoses have been emptied, you need to add 6 ounces for the hoses. Some machines do this for you, but you need to know so you can empty the hoses or not.

Scale Accuracy. Your scale may not be accurate if it has not been calibrated. The typical tolerance is +/− 1 ounce.

SHOP TIP Place 27 pennies on the scale and see if the gauge reads 0.1 pound (2 ounces). Be certain you are reading specifications in the correct range. You can use an on-line service to convert between ounces and grams, or use the chart in Figure 35.52.

Determine the Correct Refrigerant Amount

The amount of refrigerant an air-conditioning system requires is normally found on an underhood label. However, there have been many TSBs updating incorrect labels. To be safe, the best practice is to refer to an electronic service library for the latest specifications. Older systems had larger receiver/dryers or accumulators. These systems are more forgiving if the recharge amount is excessive.

Refilling through the Low Side

Recharging the system from a large bulk tank installed in a charging station is pretty much the rule now that refrigerant has become so expensive. You will have to weigh the can during the refill. You can approximate the amount left in the can by feeling the can. Liquid refrigerant at the bottom of the can will be colder than the vapor in the top.

During the fill, the refrigerant boils as it is leaving the container. The first part goes in quickly because the system is under vacuum. The boiling refrigerant generates pressure and the process slows down. Larger stations have heaters to heat and raise pressure in the refrigerant container. Be sure to keep the can upright when filling the system so only gas enters. Liquid refrigerant entering the low side can damage the compressor.

SHOP TIP A good rule of thumb is if the system calls for 3½ pounds of refrigerant and you install the whole 4 pounds for every 5 psi of unwanted high side pressure it will cost ¹⁄₁₀ mpg of fuel economy on a 4-cylinder vehicle.

Completing the Refill

After the refrigerant starts to go in more slowly, the air-conditioning system can be run so that refrigerant can be sucked into the low side. Be sure that the high-side valve is always closed when the system is working. If the system has a low-pressure cutout, it will not operate. These can be jumpered to bypass them temporarily.

The system should begin to operate normally when the fill level is about ½ pound less than full. The remaining part is to fill the accumulator or receiver/dryer. When the system is being refilled with a charging station, the amount of the fill is entered into the machine and it takes care of the rest.

■■■ RECYCLING AND RETROFIT

Since CFC refrigerants have been phased out, their cost has skyrocketed. R-134A has been the refrigerant that new vehicles are equipped with since 1994. Look under the hood for a label that states what kind of refrigerant the vehicle has (**Figure 36.41**).

Retrofitting an R-12 system is possible, although there are some drawbacks. These are different refrigerants with different properties. Many things must be considered. With the cost of R-12 refrigerant increasing rapidly, do not make the mistake of substituting the wrong refrigerant. R-134A is the only refrigerant any manufacturer is recommending.

Manufacturers have kits that generally go back ten model years for coverage. Follow manufacturers recommendations for retrofitting.

A retrofit might require some of the following items:
- A compressor front seal
- A condenser replacement
- An extra cooling fan
- Hoses

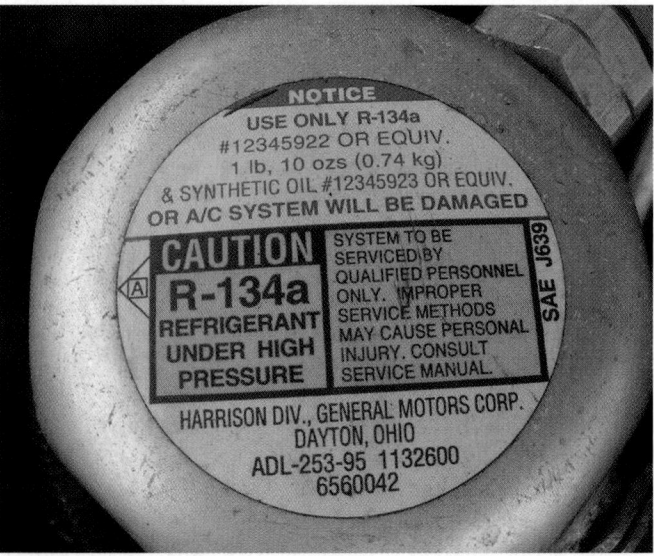

Figure 36.41 Look for an underhood label like this one on the accumulator that identifes the type of refrigerant used. (*Courtesy of Tim Gilles*)

The larger condenser required for an R-134A system is the biggest hardware change.

There are many differences between the way R-134A and R-12 react in the air-conditioning system:

- R-134A has a 30–60 psi increase on the high side and it is 6°–7° warmer.
- The low-side pressure with R-134A is typically 3 pounds less than with R-12. This means the pressure sensing switch will probably have to be changed during a retrofit.
- R-134A requires less weight of refrigerant during a refill because R-134A is a different density. At 79°F, R-12 weighs 81.5 pounds per cubic foot and R-134A weighs 75 pounds per cubic foot. That is why an R-134A 30-pound container is larger than an R-12 30-pound container.
- R-134A is about 15% less efficient than R-12. That is why the condenser is larger on cars that come equipped with it.

Refrigerant Storage

Refrigerant storage is the same for R12 and R134A. Commercial refrigerants are typically sold in 30 lb containers that resemble the propane bottles commonly found on backyard BBQs. The containers are disposable and have blow-off protection in case internal pressure becomes too high due to extreme heat. When a container is empty, it should be completely evacuated prior to disposal.

Different refrigerants should not be mixed because this will form a new chemical compound that can cause higher pressures resulting in damage to the compressor and other refrigerant system components. To prevent accidental cross-contamination, containers for different types of refrigerants have dispensing valves that are incompatible with one another. They also have labels with a different color background. R134A labels are sky blue and R12 labels are white. Other types of refrigerants use labels of other colors.

▬ DIAGNOSING HVAC ELECTRONIC PROBLEMS

Most automatic temperature control systems have self-diagnostic capability and can set a diagnostic trouble code when an electric signal is out of the expected range (**Figure 36.42**). In these systems, after your visual inspection for obvious problems, verify that the problem is present by connecting a scan tool and checking for codes. A scan tool designed for OBDII emission systems will not necessarily have all of the information for a particular manufacturer's air-conditioning system. A dedicated manufacturer scan tool or a special cartridge for an aftermarket scan tool might be necessary.

Compressor Clutch Control

The HVAC module controls the compressor clutch circuit, so you need to understand system operation in order to diagnose a problem. Some clutches will not engage when the temperature of the refrigerant is above 240°F.

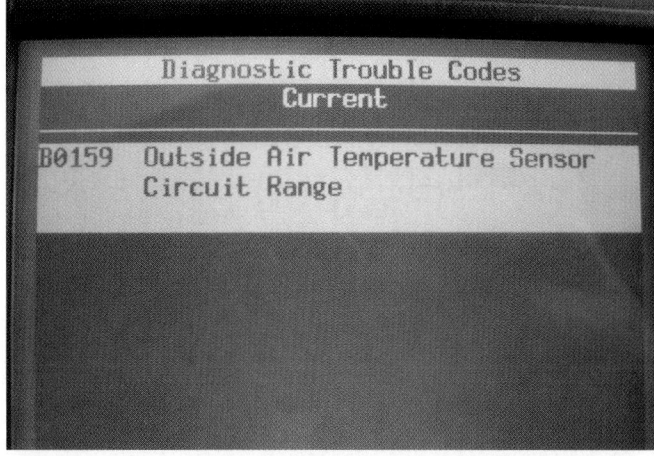

Figure 36.42 An air-conditioning DTC displayed on a scan tool.

Automatic temperature control systems monitor system pressure and do not allow the compressor clutch to engage when the refrigerant charge is too low. The computer will set a code that must be cleared before the system can operate after the problem has been corrected.

Engine vacuum and throttle opening are monitored to shut off the clutch when a heavy load is sensed. The amount of engine vacuum that triggers clutch shutoff varies between vehicles.

PCM idle air control seeks a desired idle and if it cannot be achieved, the compressor clutch shuts off. With some systems, you can step on the gas pedal to raise rpm and see if the clutch turns back on. Power steering pressure above 500 psi can also shut off the compressor clutch because most power steering occurs at idle or when there is very little throttle opening. This prevents the engine from stalling during parking maneuvers.

Electronic Control Problems

Electronic control problems with an HVAC system are often due to an electromechanical cause. Testing electronic control module problems is done by eliminating other possible problems as the cause. If a problem with a computer is suspected, be certain to test all power and ground circuits. Sometimes a controller will have more than one power and ground circuit. If there is a problem with the electrical circuit, the computer cannot function as designed.

When evaluating the performance of the system, operation of the HVAC sensors can be viewed on the scan tool (**Figure 36.43**). When testing any of the HVAC temperature sensors, measure the resistance of the sensor and compare it with a chart that specifies resistance at different temperatures (**Figure 36.44**). If the sensor is out of range, replace it. If its resistance is within specs, check for the correct reference signal voltage while the sensor connector is removed. If voltage is low, check for bad circuit connections.

If the HVAC electrical system checks out, be certain all mechanical parts of the system are operating correctly.

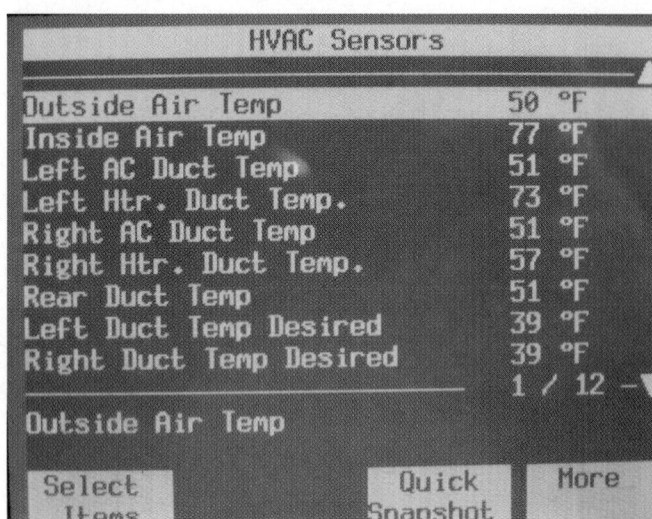

Figure 36.43 Operation of HVAC sensors displayed on a scan tool.

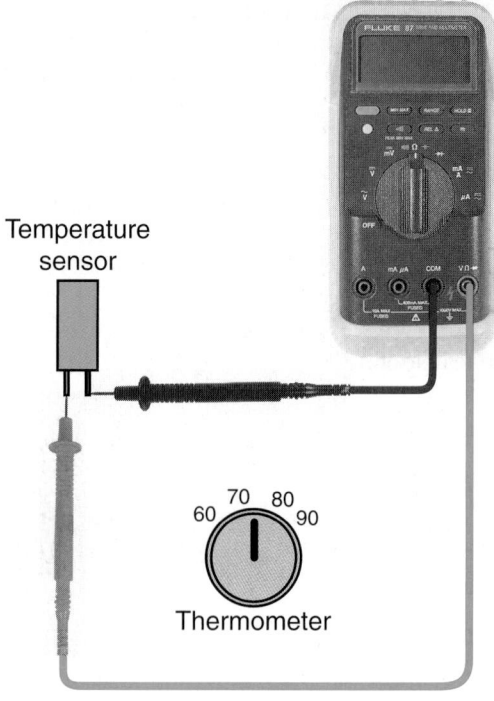

Figure 36.44 Sensor resistance is compared against a chart in the service information that lists expected resistances at different temperatures.

■ REVIEW QUESTIONS

1. If the heater has a control valve and it is open, the hoses should both be _____.

2. What gas is produced when R-12 is exposed to flame?

3. When there is a restriction in a line or part, what will commonly appear downstream of that spot?

4. Most air discharge specifications are between 35°F and ____°F.

5. If the high- and low-side fittings are not the same size, which one will be bigger?

6. What is the name of the material that absorbs water in the air-conditioning system?

7. When a part of the system is replaced, a measured quantity of _____ is added.

8. After a compressor failure, a _____ is usually installed in a line after the condenser.

9. What is used for flushing a system?

10. When refilling from the low side, should the can be upside down or right side up?

■ ASE-STYLE REVIEW QUESTIONS

1. Which of the following air-conditioning lines should feel cold to the touch?

   a. The high side   **c.** Both A and B

   b. The low side   **d.** Neither A nor B.

2. Technician A says that air temperature should drop by at least 20°F as it passes through the evaporator. Technician B says that the higher the temperature of ambient air, the more it cools as it passes through the evaporator. Who is right?

   a. Technician A   **c.** Both A and B

   b. Technician B   **d.** Neither A nor B

3. Which of the following is/are true about the air distribution system?

   a. Vacuum motors require a reservoir so they can work when the vehicle is going downhill.

   b. If a vacuum door fails, it is required by law to default to the floor duct.

   c. Cold air coming from the defroster ducts when the engine is warm indicates a problem.

   d. All of the above.

4. Technician A says that PAG oil absorbs moisture when exposed to air. Technician B says that too much oil in the refrigerant reduces cooling efficiency. Who is right?

  a. Technician A   **c.** Both A and B

  b. Technician B   **d.** Neither A nor B

5. Which of the following is/are true about the relationship between refrigerant pressure and temperature?

  a. High pressure means high temperature.

  b. Low pressure means low temperature.

  c. Both A and B.

  d. Neither A nor B.

6. Technician A says that a bad expansion valve can cause low pressures on both sides of the system. Technician B says that the accumulator must be replaced when an expansion valve is replaced. Who is right?

  a. Technician A   **c.** Both A and B

  b. Technician B   **d.** Neither A nor B

7. Technician A says that evacuating a system will result in more efficient cooling. Technician B says that water can be boiled at room temperature by pressurizing it. Who is right?

  a. Technician A   **c.** Both A and B

  b. Technician B   **d.** Neither A nor B

8. All of the following could be reasons for an air-conditioning clutch to fail *except*:

  a. Too wide an air gap

  b. Too much refrigerant charge

  c. A low-side restriction

  d. A bad clutch circuit ground

9. Which of the following is/are true about AC system pressure?

  a. High-side pressure affects the low side.

  b. High pressure means low temperature.

  c. Both A and B.

  d. Neither A nor B.

10. Technician A says that an accumulator should feel cold and sweaty. Technician B says that when there is ice on an air-conditioning line, the problem is in the air distribution system. Who is right?

  a. Technician A   **c.** Both A and B

  b. Technician B   **d.** Neither A nor B

Engine Performance Diagnosis: Theory and Service

THEN AND NOW: IGNITION SYSTEMS

Believe it or not, some of the earliest cars used *hot-tube ignition*—a platinum tube extended through the cylinder wall into the combustion chamber. A flame from an alcohol lamp outside kept the tube red hot. Even more unbelievable was the *flint and steel* idea: a rough piece of steel mounted on the top of the piston rubbed against a spring-loaded flint to make a spark.

A Frenchman named Etienne Lenoir invented the spark plug in 1860, 26 years before the first automobile appeared. But most early cars did not use spark plugs. Instead, they had *low-tension magnetos,* which produced electricity from magnets and coils of wire. A spark resulted when two contacts inside the cylinder were pulled apart by a complex mechanism.

The *high-tension magneto* appeared in 1903. It added an induction coil to produce enough voltage to jump the gap of a spark plug. This ignition system was popular in Europe into the 1930s. The *trembler coil* was first used on the 1896 Benz. It made a steady stream of sparks from a vibrating blade that controlled current through the coil. The trembler coil was very successful in the United States, sparking the 15 million Ford Model Ts built between 1909 and 1927.

Charles "Boss" Kettering, one of the founders of Delco, developed the dependable *breaker point and coil ignition system,* which was in use for over half a century. A drawback to this type of ignition was that the rubbing block of the movable point slowly wore down against the distributor cam. This resulted in spark timing occurring later and later. Also, efficiency fell off dramatically at high rpm.

Electronic ignition solved this problem. Chrysler used it on all models in 1973. By 1975, most cars sold in the United States had electronic ignition systems. With nothing to wear out, spark timing remained constant throughout the life of the car. The coil also got its current so fast there was no danger of running out of spark during hard acceleration.

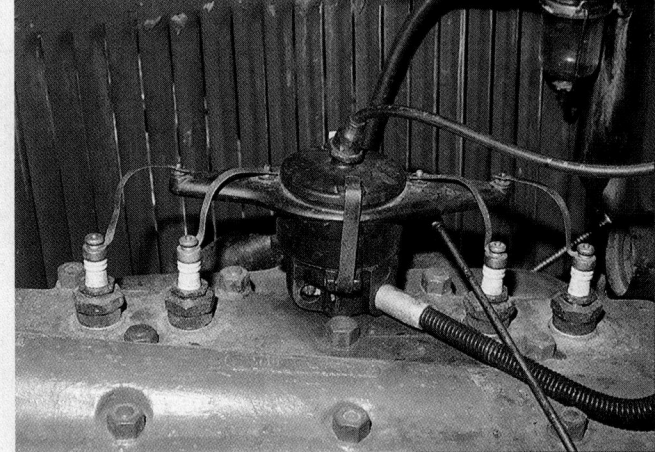

A Ford Model A ignition system. *(Courtesy of Bob Freudenberger)*

A modern platinum-tipped spark plug. *(Courtesy of Bob Freudenberger)*

Modern *distributorless ignition systems* (DIS) came on the scene with the 1984 Buick. Camshaft and crankshaft position sensors inform an electronic ignition module of the position of each piston in its cylinder in terms of the combustion cycle. The ignition module on this system fires multiple coils, which have spark plug wires running directly to the spark plugs. This design eliminated the need for the potentially troublesome distributor rotor and cap of the earlier distributor systems. Another DIS version, called *direct ignition,* eliminates the need for spark plug wires by mounting the coils in an assembly that snaps onto the spark plugs.

Ignition System Fundamentals

■ INTRODUCTION

The ignition system's job is to create a timed spark and distribute it to the engine's cylinders. The ignition system also serves the function of turning the engine on and off. The spark is distributed to the spark plugs, where it jumps the gap and ignites the air-fuel mixture just before the piston reaches the top of its compression stroke. The timing of the spark varies with engine speed because the amount of time it takes for the fuel to burn in the cylinder is relatively constant.

■ BASIC IGNITION SYSTEM

Modern vehicles have computer-controlled ignition systems. There are two main categories: **distributor ignition (DI)** and **electronic ignition (EI)**, also called **distributorless or direct ignition system (DIS)** or coil over plug (COP). The DI system uses a distributor and spark plug cables (**Figure 37.1**). All ignition system types use a battery, an ignition switch, a coil, a switching device, and spark plugs.

Ignition systems have two circuits, called primary and secondary. The primary is the low-voltage (battery) part of the system, and the secondary is the high-voltage (spark) side of the circuit.

■ PRIMARY CIRCUIT

The primary circuit carries current from the battery to the coil. When the flow of current in the primary circuit in a coil is switched on, a magnetic field is energized. When the current suddenly stops flowing, a high-voltage spark from the ignition coil to the spark plug is created.

The primary ignition system includes the following parts:

■ Battery
■ Charging system
■ Ignition switch
■ Ignition coil primary windings
■ Switching device
■ Distributor cam lobes or crank/cam sensor
■ Ground return path

Battery voltage is converted to high voltage by an ignition coil. An electrical spark jumps across a gap at the end of a spark plug. Electricity for the ignition system is supplied with battery voltage when the engine is cranking. Once the engine is running, the alternator

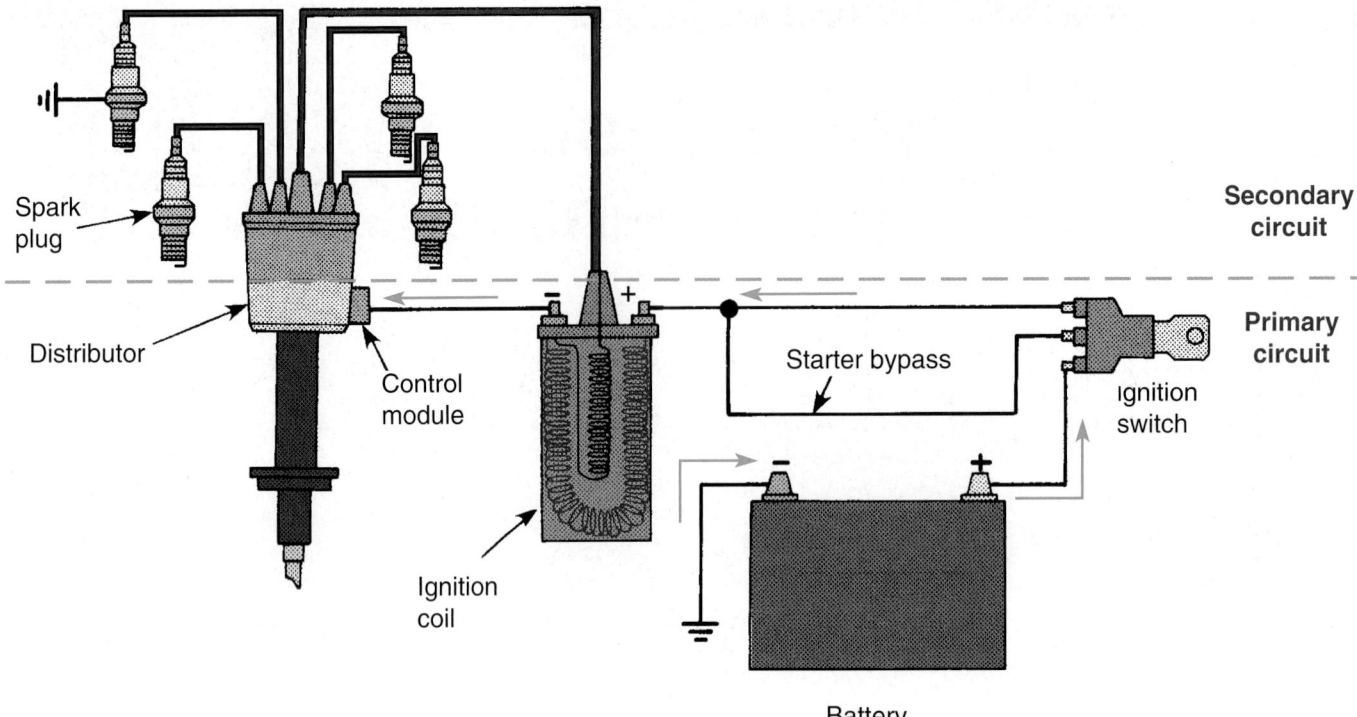

Figure 37.1 Parts of a distributor ignition system.

supplies the necessary electrical flow. Spark timing is critical to power output. It cannot be too early or too late. An ignition system varies the spark timing to control this.

Ignition Switch

A multiposition switch turns power to the ignition circuit on and off. In the *start* position, it triggers the starter motor and provides a parallel path for electrical flow to the ignition system. After starting, in the *run* position it continues to allow current to flow to the ignition system and the entire electrical system. The *accessory* position powers electrical accessories when the engine is off. This key position is sometimes when the key is turned counterclockwise. Other times, the key is turned to the first position in clockwise rotation.

The ignition switch operates the steering wheel lock (**Figure 37.2**) and a buzzer or light. If the key is in the ignition when the engine is off, a door is ajar or open, or seat belts are not buckled, the buzzer and/or light will operate. More information on the ignition switch is found in Chapter 28.

Ignition Coil

The coil is the heart of the ignition system. Battery voltage (12.6 V) does not provide enough electrical pressure to push a spark across the spark plug gap. The coil is a transformer that converts battery voltage into many thousands of volts.

A coil has a low-voltage *primary winding* and a high-voltage *secondary winding* (**Figure 37.3**). Because it has both low- and high-voltage windings, part of

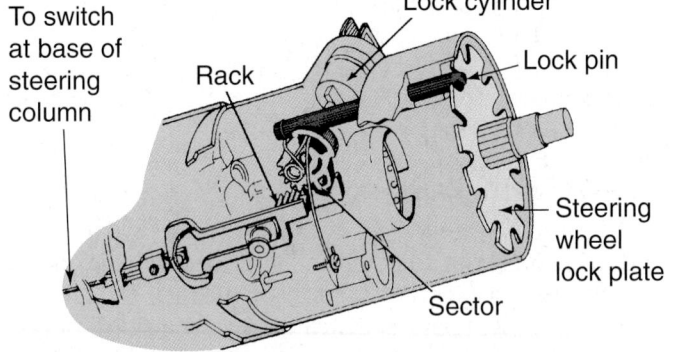

Figure 37.2 The key switch also operates the steering wheel lock. *(Courtesy of General Motors Corporation, Service Technology Group)*

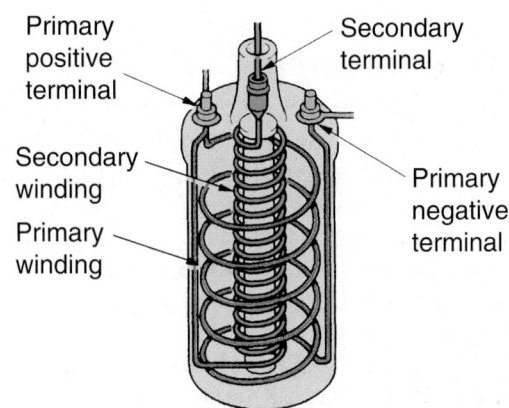

Figure 37.3 A coil has a low-voltage primary winding and a high-voltage secondary winding.

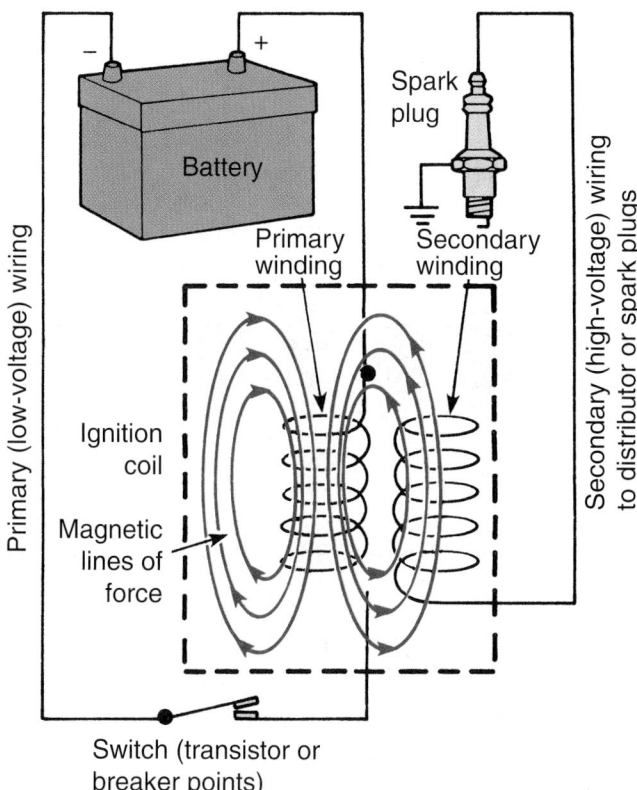

Figure 37.4 Current flowing in the primary coil winding creates a magnetic field.

the coil is in the primary system and part of it is in the secondary system.

In Chapter 25, you learned that a magnetic field surrounds a conductor when electrical current flows through it (**Figure 37.4**). Coiling the conductor around an iron core creates a stronger field. The high-voltage spark is created by the ignition coil when the electrical current supply in the primary winding is discontinued and the magnetic field surrounding the coil breaks down and crosses the secondary winding.

The more windings there are in a coil, the stronger the magnetic field. When there are fewer coils in the primary part of a transformer, it is a *step-up transformer*

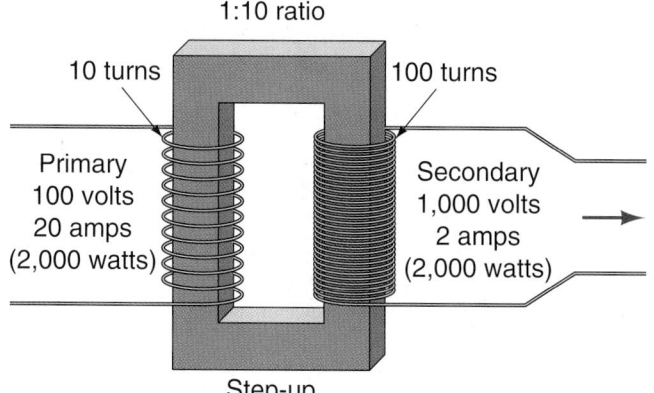

Figure 37.5 A step-up transformer has fewer windings in its primary winding. *(Courtesy of Ford Motor Company)*

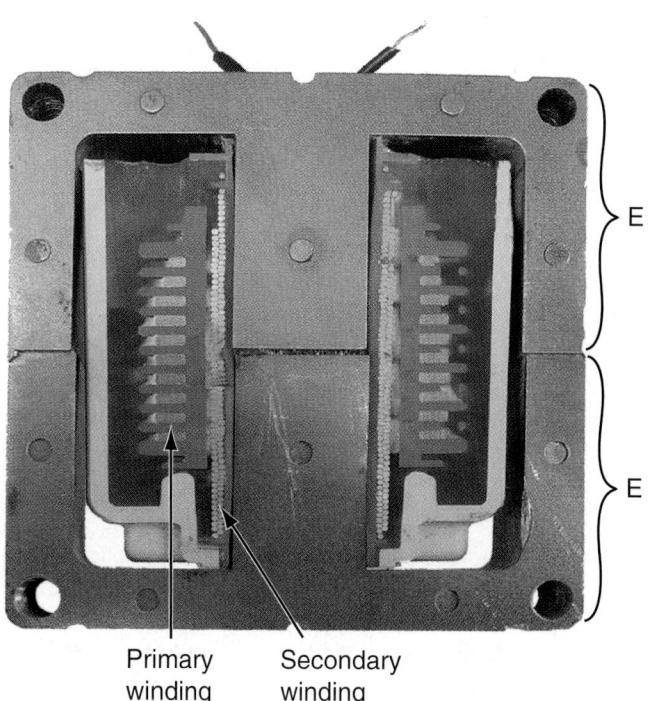

Figure 37.6 This cutaway of an e-coil shows laminations in the shape of two letter *E*s facing each other. *(Courtesy of Tim Gilles)*

(**Figure 37.5**). Fewer coils in the secondary would make a *step-down transformer*. An ignition coil is a step-up transformer.

The coil's primary winding is relatively large wire coiled 150 to 250 times. It carries several amps of current flow at battery voltage. Within the center of the primary winding is a secondary winding. It is made up of about 20,000 turns of very small, hairlike insulated wire. The secondary winding is wrapped around strips of iron layered together to make up a laminated iron core. Older coils were cooled with oil, whereas newer coils are usually air cooled. They have a core with laminations in the shape of the letter *E*, with the primary and secondary windings wound around it (**Figure 37.6**).

One end of the secondary winding goes to a terminal where the high-voltage pulse will leave the coil on its way to the spark plug gap. The other end of the secondary winding is connected to one side of the primary winding. This gives it a path to ground for a complete circuit.

Remember that moving a magnetic field across a coil of wire creates electricity. When current flow is interrupted in the primary winding, the magnetic field surrounding the primary and secondary windings collapses. The magnetic lines of force cut across the secondary windings. This creates high voltage and low amperage in the winding.

High voltage means high pressure. The pressure at the coil's high tension terminal is high and the electricity is looking for a path to ground through the spark plug. The spark jumps the gap between the spark plug's positive and negative electrodes.

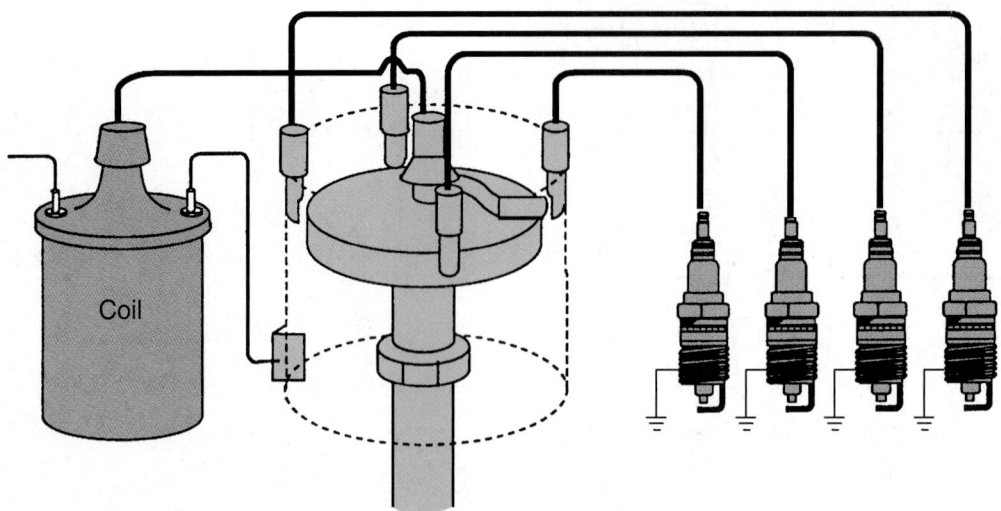

Figure 37.7 The secondary circuit is the high-voltage side of the ignition system. *(Courtesy of OTC/SPX Service Solutions)*

When did the spark happen? When the switching device broke the circuit in the primary system, allowing the magnetic field to collapse. Switching the primary current on and off was done mechanically in older cars using points. Today it is done electronically.

Saturation and Dwell. When the magnetic field has finished its buildup inside of the coil, the coil is said to be *saturated*. **Coil saturation** time depends on the amount of current flowing in the primary winding. Older contact point systems had slower saturation times because they were limited to about 2–3 amps of current flow. Some of the newer electronic ignition systems can flow current of up to 20 amps. This allows much faster saturation time.

Dwell is the length of time that primary current is flowing in the primary winding. It is measured in degrees of distributor rotation (½ crankshaft speed). In a contact point system, dwell is the length of time that the points are closed. In an electronic system, it is the length of time that the ignition module allows current to flow through the coil primary winding. A coil takes a certain amount of time to fill or saturate. This is why dwell is important.

> ### ▨ SCIENCE NOTE ▨
> *When current first begins to flow into the coil, it flows rapidly until resistance is encountered. Part of the resistance is due to the size of the wire, but another type of resistance results from the counterelectromotive force that happens as the magnetic field is building around the coil winding. The magnetic field in one loop of the coil winding passes over the next loop of the winding and so on, building a barrier to current flow. The term for this temporary resistance, which delays coil saturation, is* **reactance***.*

Most newer ignition systems have longer dwell periods.

▨ SECONDARY IGNITION PARTS

The secondary circuit is the high-voltage side of the ignition system. It delivers high voltage from the coil to the spark plugs (**Figure 37.7**).

Distributor ignition (DI) systems have a distributor that includes a *cam* that triggers an electronic *module*. The *distributor cap and rotor* are also part of the distributor. Many engines have no distributor. In a distributorless system, the crankshaft has a pickup that takes the place of the distributor. Distributorless systems have multiple ignition coils (usually one for every one or two cylinders). Spark plugs are fired based on a signal processed by the computer, sometimes through a plug wire and sometimes directly (with no plug wire).

Distributor Cap and Rotor

In DI systems, electricity flows from the coil to a distributor cap and rotor (**Figure 37.8**). Wires plugged into holes in the top of the distributor cap connect the distributor cap and spark plugs. The ends of the plug wires fit snugly into the distributor cap.

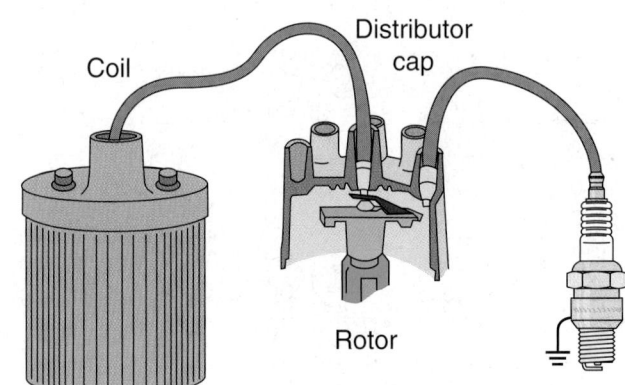

Figure 37.8 Electricity flows from the coil to a distributor cap and rotor.

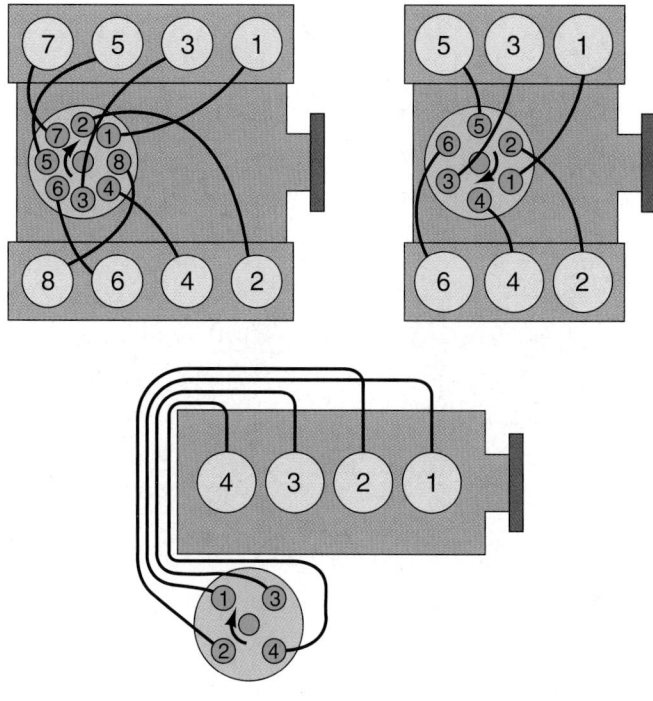

Common firing orders

Figure 37.9 The spark plug cables are arranged around the cap following the engine's firing order.

The distributor shaft is driven by the engine's camshaft. Remember that the camshaft turns once for every two turns of the crankshaft, so a distributor rotates at one-half crankshaft speed. A spark occurs toward the end of the compression stroke, then the crankshaft rotates through four more strokes (two turns) before another spark is required in that cylinder.

The rotor is located on top of the distributor shaft. As it turns, it points to a different terminal inside of the distributor cap. The center of the distributor cap has a button made of carbon. Electricity is fed into the center of the cap through a *high-tension lead* from the coil. From there, it goes to a tab on the center of the rotor. Electricity flows through the rotor to a different terminal on the cap each time the primary circuit magnetic field breaks down. The spark plug cables are arranged around the distributor cap in the engine's firing order (**Figure 37.9**). For more information on an engine's firing order, refer to Chapter 16.

■ SPARK PLUGS

In gasoline engines, spark plugs provide a gap for the spark to jump. The spark ignites the compressed air-fuel mixture in the cylinders. **Figure 37.10** shows the parts of a spark plug. A spark plug has a metal shell with threads that mesh with threads in the cylinder head. A metal conductor runs down the center of the plug, surrounded by a ceramic insulator. The end of the conductor forms the *positive electrode*. On the end of the metal case is the **ground spark plug electrode**.

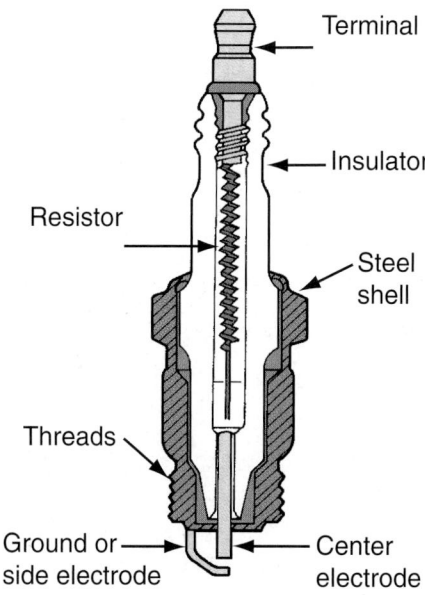

Figure 37.10 Parts of a spark plug.

The ground electrode can be bent toward or away from the center electrode to make the correct spark plug gap. The gap varies depending on the kind of ignition system used. Spark plugs in older point-type ignition systems specified small gaps of around 0.035". The gaps on newer electronic ignition cars can vary up to 0.080". A gap this large produces a very hot spark.

Spark Plug Threads
Spark plugs have different thread diameters (10, 12, 14, and 18 mm). The length of the threaded area of a spark plug is called its **reach**. A plug with too long a reach will extend into the combustion chamber. If the reach is too short, the end of the plug will not reach the ends of the threads in the head (see Chapter 38).

Heat Range
A spark plug's **heat range** is an indication of how fast heat can travel away from the spark plug center electrode to the cooling system's water jackets in the cylinder head. There are several different spark plug heat ranges available. The length of the ceramic insulator into the combustion chamber is what determines heat range. **Figure 37.11** shows how the path of heat transfer determines the temperature at which the plug will run. According to Champion Spark Plugs, over 90% of a spark plug's heat leaves it at the edge of the threads. The length of the ceramic insulator above the metal makes very little difference.

The use of spark plugs of the wrong heat range can cause major engine damage. If the heat path is long, the plug will run hot. A short path means a cold plug. If the plug is too hot, it will burn the plug electrode. It can also cause pre-ignition and can burn a piston. If a plug is too cold, deposits will build up on it and it

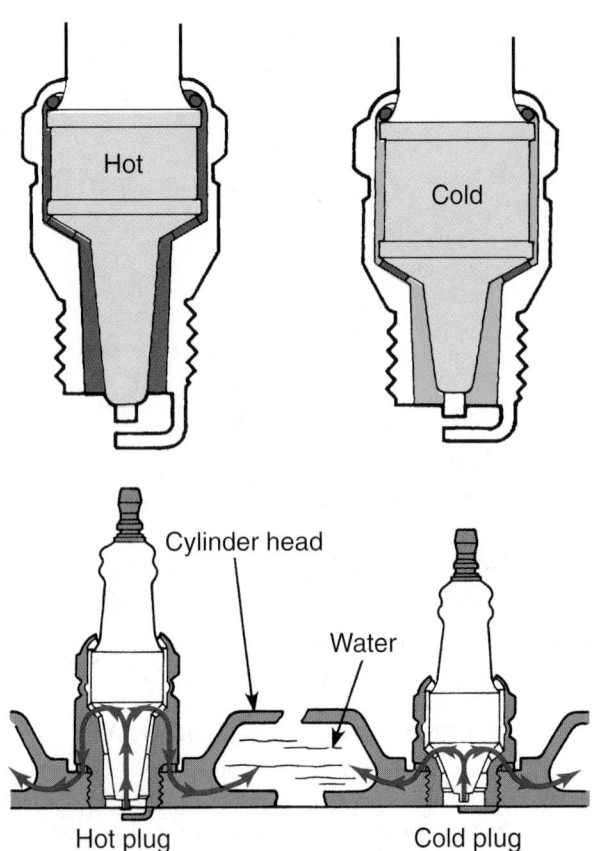

Figure 37.11 The length of a spark plug's ceramic insulator determines its heat range.

Figure 37.12 Tapered seat and gasketed spark plugs. (*Courtesy of Tim Gilles*)

will foul. Spark plug fouling means that it gets covered with material and a spark will not jump its gap.

The amount that the plug projects out below the threads has very little to do with the heat range. The reason for the projection is to get the spark deeper into the combustion chamber. A flush tip plug is more likely to foul when used under low-speed, light-load conditions. The extended tip acts like a hotter plug during those conditions. At higher speeds there is a charge cooling effect, which lets it perform like a cold plug. A projected tip maintains a more stable temperature range. The only reason a flush tip plug is used is to avoid contact with a piston. Most modern pistons have enough clearance to be able to use an extended-tip plug.

Taper- and Gasket-Type Plugs

To seal it against the cylinder head, a spark plug has either a tapered seat or a flat seat with a gasket (**Figure 37.12**). Tapered seats are used in all sizes of spark plugs, but especially in the smaller space-saving spark plugs. Spark plugs with gaskets are still popular also. The two types are not interchangeable.

Long-Life Spark Plugs

Many new vehicles come equipped with spark plugs advertised to have a 100,000-mile life. There is some controversy regarding this claim, especially when

normal vehicle use includes stop and go driving in heavy traffic. Modern fuels and fuel management systems help keep spark plugs clean, but difficult driving conditions allow deposits to form on spark plugs. In most vehicles, spark plugs will probably require replacement before 100,000 miles.

Precious metals are used for electrodes in pure or alloy form. An alloy is a combination of metals. Silver is one of the metals used for center electrodes. As you can imagine, the electrode diameter is reduced in size. Platinum and platinum alloys are popular electrode materials because platinum is resistant to chemical erosion and corrosion (**Figure 37.13**). There are single and double platinum designs. In a single

Figure 37.13 A spark plug with a platinum tip.

platinum plug, only the center electrode is platinum plated. In waste-spark ignition systems (covered later), sometimes the side electrode is plated as well. Platinum is expensive, so other precious metals like iridium are sometimes used to prolong electrode life.

Resistor Plugs and Wires

Resistance is added to the secondary ignition system, either in the spark plugs or spark plug cables. This increases the required firing voltage, which raises the spark plug firing frequency out of the range of radio interference.

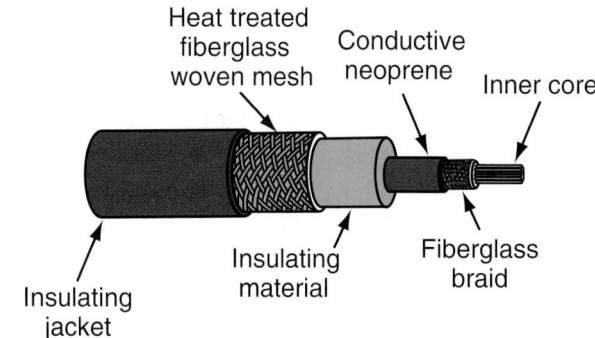

Figure 37.14 Static suppression ignition cable.

SCIENCE NOTE

As the plug is fired, the plug voltage rises and falls rapidly. The plug is actually refired several times during about ¹⁄₁₀₀₀ of a second. Dropping the voltage with a resistance changes the intensity of the magnetic field that happens along the length of the wire, which also lowers radio noise.

HISTORY NOTE

The wireless telegraph, invented by Guglielmo Marconi in 1895, was a spark transmitter similar to the ignition system on a car. Pulsed messages could be sent using the long and short dashes and dots of the Morse code. The transmitter had a wire antenna attached to its gap. It could be tuned by changing the width of the gap or the shape of the electrodes. This allowed each transmitter to have its own characteristic sound that could be identified by radio operators. This was important if signals were being received by more than one transmitter at a time.

When radios began to be installed in automobiles, it became apparent that ignition systems with metal spark plug wires were actually transmitting signals similar to the telegraph. Adding resistance to the circuit changed the frequency of the ignition pulses so that they would not interfere with radio reception.

Spark plugs often have resistors. A resistor inside the spark plug raises the firing voltage required by the coil (see Figure 37.30). Some spark plugs have carbon resistors. Others have a semiconductor suppressor that is not energized until 1,000 volts pass through it. This kind of resistor plug cannot be checked with an ohmmeter.

Spark Plug Cables

Most spark plug cables (wires) provide a resistance, too. Resistor cables, called television/radio suppression (**TVRS**) cables, are made of braided aramid fiber impregnated with graphite and latex (**Figure 37.14**).

These cables are very fragile. Some European spark plug cables are metal with a resistor boot at the end. These are used with a nonresistor spark plug.

Secondary wiring must be well insulated. If there is a leak in the insulation on the secondary spark plug cables, a spark will leak to ground, preventing it from reaching the spark plug. Older plug wires had 7 mm (0.282") Hypalon or CPE rubber insulation. Cars equipped with newer electronic ignition systems use 8 mm (0.312") silicone for insulation. Electronic ignition creates higher voltages, requiring thicker insulation. Silicone has better high temperature resistance than Hypalon. It is rated for 400°F continuous use.

ELECTRONIC IGNITIONS

EI systems are similar to the electromechanical contact point systems they replaced, although electronic systems are far superior and require very little maintenance. The key difference between the two systems is in the **trigger** mechanism that controls the flow of current in the primary winding. Most ignition systems control primary current flow on the ground side of the coil. Refer to **Figure 37.15** and compare the primary circuits of electronic- and point-type ignitions. Although ignition contact points are hardly ever encountered any longer, they are used here to help you more easily visualize how a transistor makes and breaks the primary circuit, causing the spark. Perhaps a more important reason is that antique and collector cars are very popular and a technician will often encounter restoration work during his or her career.

Nonelectronic ignition systems used mechanical contact points to alternatively energize and then open the primary ignition circuit. The points are opened by lobes on a distributor cam with as many lobes as the number of engine cylinders (**Figure 37.16**). Separating the contact points opens the primary circuit, causing the magnetic field in the coil to break down. This results in a spark from the secondary winding of the coil.

When an eight-cylinder engine is running at 3,000 rpm, which is not really that fast, its ignition coil must do remarkable things. Its magnetic field must be able to build up and collapse 200 times in 1 second.

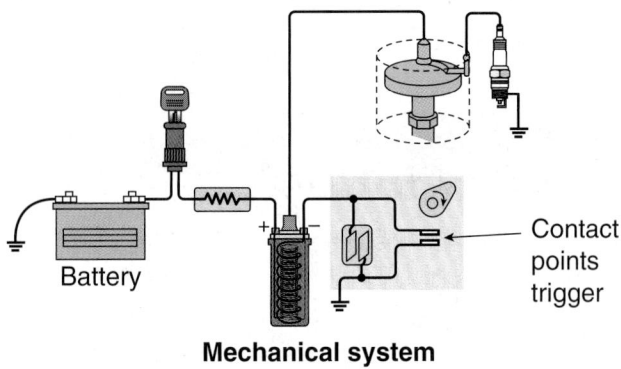

Mechanical system

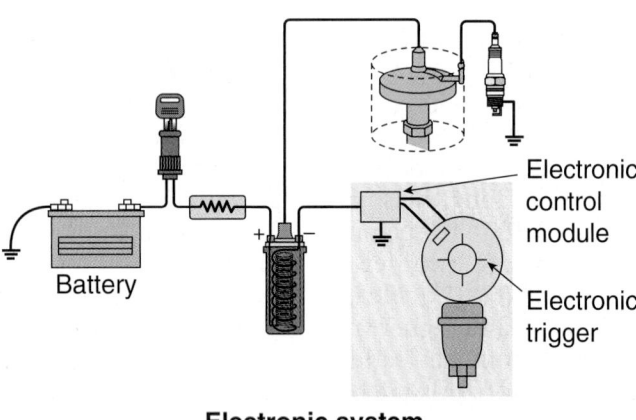

Electronic system

Figure 37.15 A comparison of primary triggers. *(Courtesy of OTC/SPX Service Solutions)*

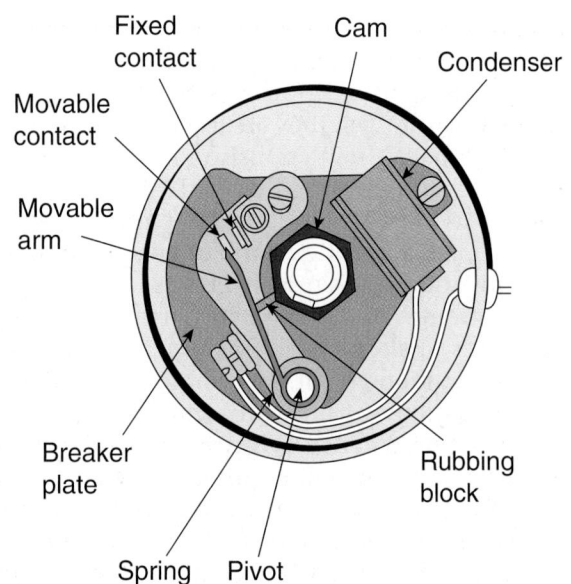

Figure 37.16 A distributor cam has as many lobes as the engine has cylinders. This is a vintage ignition point distributor.

This means that at ordinary highway speeds, that vintage set of contact points had to open and close 200 times per second. Imagine what an eight-cylinder Indy car's ignition system must do at 15,000 rpm. It has 1,000 spark plug firings per second.

Contact points require periodic replacement because they suffer from wear and pitting. Points suffer from wear to the rubbing block that rides against the distributor cam, retarding ignition timing.

Electronic Ignition Operation

Electronic ignition systems are reliable and provide improved emissions, performance, and fuel economy over electromechanical ignition systems. A transistor triggers the buildup and collapse of the magnetic field. The transistor is housed in an ignition module (**Figure 37.17**) or in the powertrain control module (PCM).

A transistor is an electronic switch or relay. A transistor is a semiconductor like a diode. The operation of semiconductors is covered in Chapter 45. Most ignition systems have one or more *power transistors* that switch the coil on and off (**Figure 37.18**). Power transistors can regularly carry 10 amps of current; far more than contact points could. The power transistor(s) are controlled by a *driver transistor* that gets its signal from a triggering device.

A transistor has an emitter, a collector, and a base. The emitter is the input, and the collector is the output. The transistor switches when a small current is applied to its base. This allows a larger amount of current to flow from the emitter to the collector and through the coil primary winding (**Figure 37.19**). When the switching device (trigger) interrupts the current to the base of the transistor, current flow between the collector and emitter is halted. This results in a spark at the spark plug. More information on transistors can be found in Chapter 45.

Because there is no mechanical switch or electrical arcing with electronic ignition systems, regular maintenance is not a normal short mileage occurrence as it was with breaker points. Later EIs have ignition advance controlled electronically. In earlier EI systems, the only parts that suffered wear were part of the advance mechanisms and bushings in the distributor. In DI systems, the distributor simply consists of a switching device, rotor, and shaft. Sometimes the ignition module is located in, or on, the distributor.

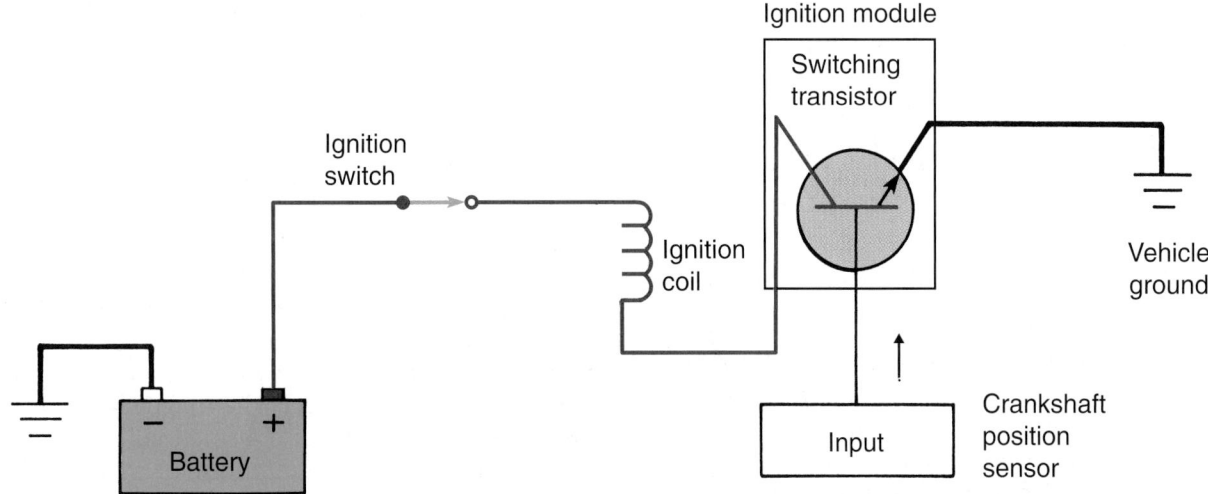

Figure 37.17 The switching transistor is in the ignition module or PCM.

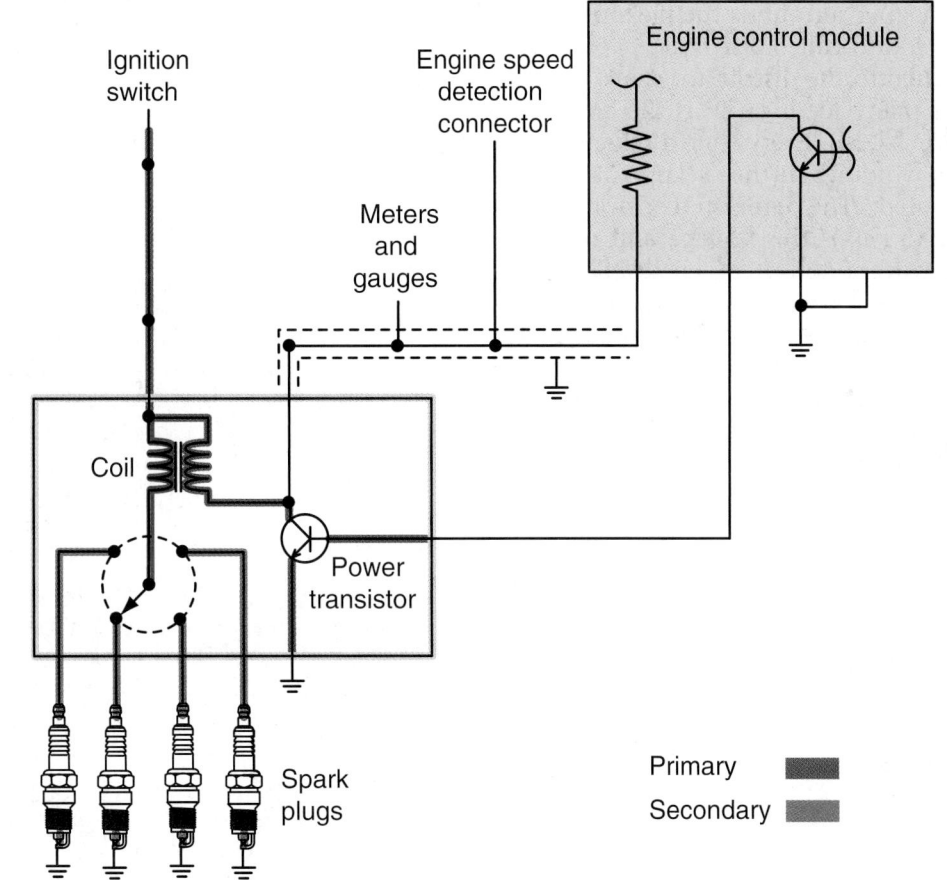

Figure 37.18 A power transistor switches the coil on and off.

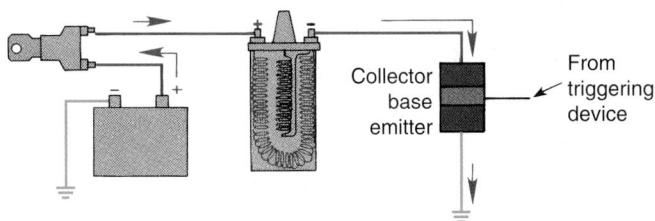

Figure 37.19 Current supplied to the base of the transistor from the trigger causes current to flow through the circuit.

■ ELECTRONIC IGNITION VARIATIONS

There are several variations of EIs. To solve engine performance problems, you will need to understand how these systems operate. The objective in this text is to provide a simple explanation of each. If you understand the basics, you will be able to understand the in-depth instructions that comprise the service literature on a particular product. Mastering the basics in this book will prepare you for job entry level, where you will learn product-specific information.

The sensor that triggers spark timing differs between manufacturers. It is located either in a distributor or on the crankshaft. The four types of sensors are the permanent magnet AC generator, the Hall switch, the magnetoresistive sensor, and the optical sensor.

AC Generator Systems

A common electronic trigger is the permanent magnet (PM) AC generator pickup in a distributor. A PM pickup works like an alternator. Different manufacturers call the parts by different names, but they all work the same way. A stationary sensor called a *pickup coil* sends signals to a transistor located in the *ignition module.*

The pickup coil is wrapped around an iron *pole piece* near a permanent magnet. The magnetic field of the permanent magnet surrounds the pickup coil.

A *trigger wheel*, also called a *reluctor, pulse ring,* or *armature,* is attached to the distributor shaft. The trigger wheel has as many teeth on it as the engine has cylinders (**Figure 37.20**). Every time a reluctor tooth passes across the windings in the pickup coil, a small voltage is generated. The ignition control module (ICM) or the PCM senses this voltage and uses it to control primary current "on-time," or dwell. Current flow in the coil primary winding causes a magnetic field to build up and surround the primary and secondary coil windings. When the module switches primary current flow off, the magnetic field breaks down. This creates the high-voltage surge that results in secondary ignition at the spark plug.

NOTE: *Energy is saved when the time interval of current flowing into the coil is reduced. The length of time required for ignition coil primary winding saturation varies with engine speed. At low engine rpm, there is more time between cylinder firings than at high speeds, so dwell time can be shorter.*

The trigger wheel has low magnetic reluctance. Low reluctance can be compared to low resistance. Magnetism is easily absorbed by a material with low reluctance. As the tooth of the trigger wheel is rotated into the magnetic field around the pole piece and pickup coil, magnetic flux becomes concentrated in the tooth (**Figure 37.21**). The magnetic field around the pole piece changes and a small voltage is induced in the pickup coil. The amount of voltage changes with the rate of the magnetic flux. As the reluctor tooth moves away from the pole piece, the magnetic field becomes weaker (**Figure 37.22**).

The output voltage signal varies between positive and negative (alternating current). As the reluctor approaches the pole piece, the voltage increases

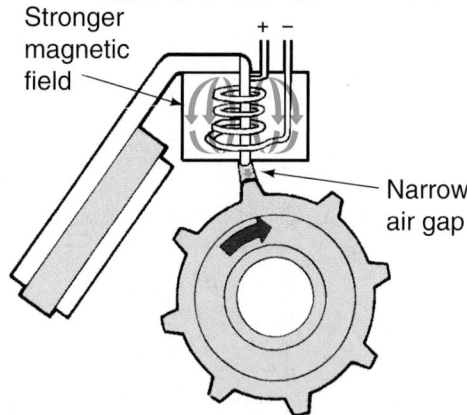

Figure 37.21 As the tooth of the trigger wheel lines up, magnetic flux concentrates in the tooth. *(Courtesy of DaimlerChrysler Corporation)*

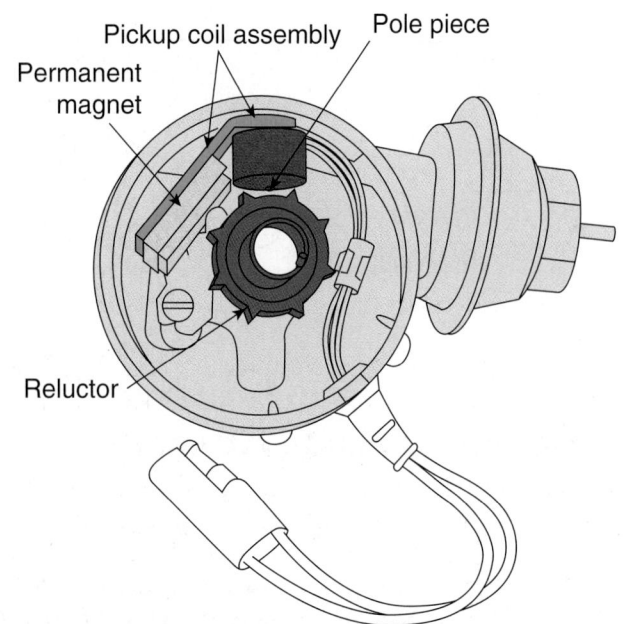

Figure 37.20 A magnetic-type AC generator pickup. *(Courtesy of DaimlerChrysler Corporation)*

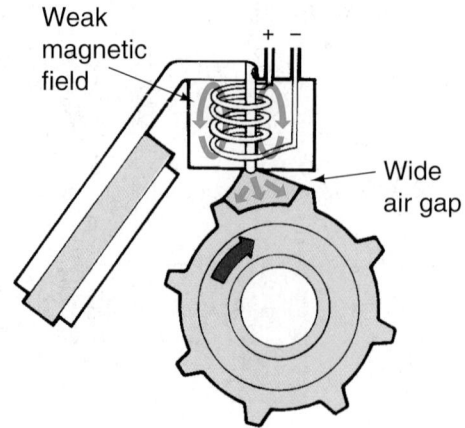

Figure 37.22 As the reluctor tooth moves away from the piece, the magnetic field becomes weaker. *(Courtesy of DaimlerChrysler Corporation)*

toward positive. When the reluctor teeth align with the pole piece, voltage is at zero. The voltage increases in a negative direction when the reluctor moves further away from the pole piece.

> ### ▦ SCIENCE NOTE ▦
>
> *When the reluctor tooth is aligned with the pole piece of the pickup coil, the voltage is zero even though the magnetic field is strongest. This is because there is no change in the rate of change in the density of the flux at this point.*

Most EI systems open the circuit as the polarity changes from positive to negative. The outcome is the same as when mechanical contact points open, resulting in a spark at the plug. As the tooth moves out of the magnetic field of the pole piece, the next tooth on the trigger wheel comes into the field and the cycle repeats itself. Current once again flows to create another magnetic field.

In the ignition module, the alternating signal from the PM generator is converted to direct current (DC) by an AC-DC converter. The converted DC signal is sent to the dwell control section of the module, which triggers the power transistor on and off to create a timed spark at the spark plug.

Hall-Effect Pickups

In the mid-seventies, the **Hall-effect switch** was introduced. The Hall switch, named for its inventor, has become the most popular electronic ignition triggering device. Hall-effect devices are found in distributors and as crankshaft position sensors (CKP). Like the AC generator, a Hall switch has a stationary sensor and rotating trigger wheel. It requires an input voltage to operate. Because the input voltage it controls is consistent, it works well at any speed, including low rpm. This is an advantage over the AC generator type of trigger, which must turn faster to generate sufficient voltage.

A Hall-effect signal does not generate an AC signal. Its signal is a rise in voltage, followed by a drop in voltage. Parts of a Hall switch include a permanent magnet, a *semiconductor Hall element*, and a cupped metal ring with vanes and windows. Parts of a Hall-effect distributor are shown in **Figure 37.23**. The permanent magnet is mounted with a small gap separating it from the Hall element. As the trigger wheel rotates, the cupped metal ring passes through the space between the element and the magnet. There are as many windows and vanes as the engine has cylinders.

When the ring rotates, a window opens (**Figure 37.24**). This allows the magnetic field to surround

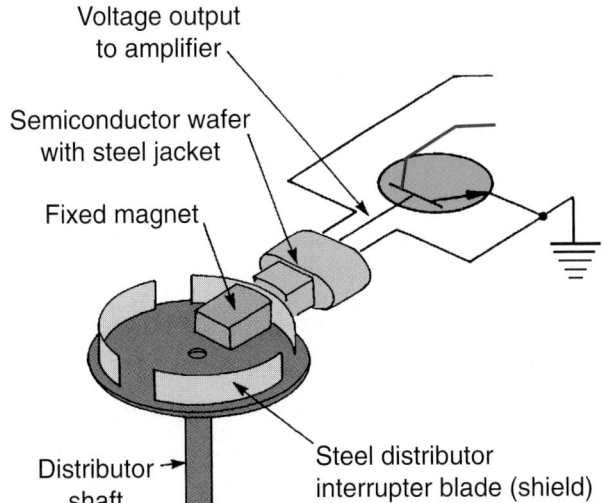

Figure 37.23 Parts of a Hall-effect distributor. *(Courtesy of OTC/SPX Service Solutions)*

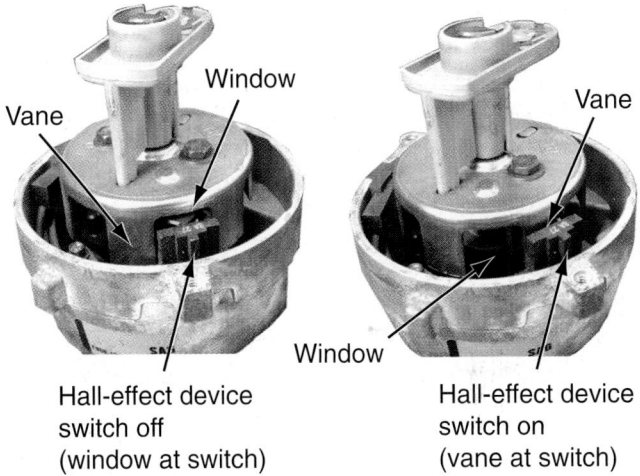

Figure 37.24 The windows and vanes move in and out of the magnetic field that acts on the Hall-effect device. *(Courtesy of Tim Gilles)*

the Hall element. In the **Hall-effect** process, current is passed through a thin semiconductor material while a magnetic field passes through it. This produces a small voltage (about 0.4 V) in the semiconductor. With the window open, the magnetic field causes a small voltage to be produced by the Hall element (**Figure 37.25**).

When the next metal tab on the shutter wheel enters the gap between the magnet and Hall element, the magnetic field is shielded from the Hall element. This stops the Hall voltage from being produced by the element. The module reacts to this by switching on the primary circuit, allowing the coil primary winding to saturate.

A square wave signal is more compatible with computer systems than an analog wave. A Hall switch creates a small analog voltage signal that is strengthened by an amplifier. Next, a device called a Schmitt

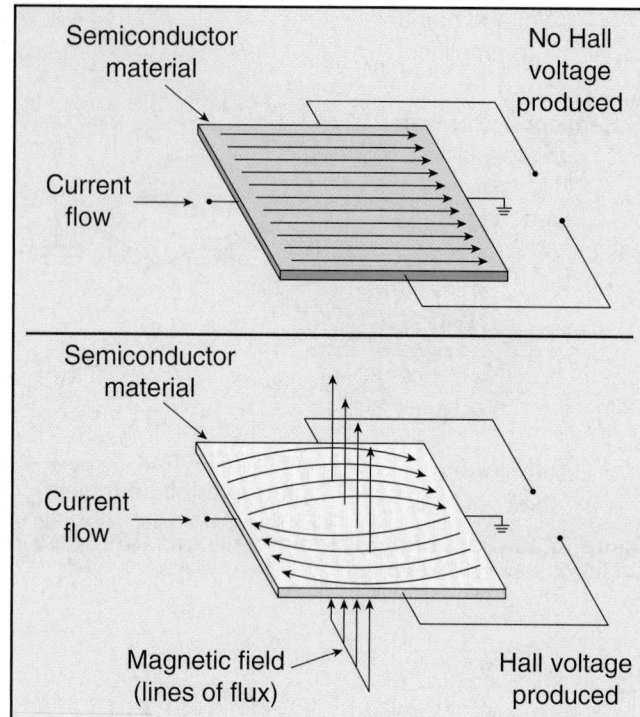

Figure 37.25 The Hall effect. When current is passed through a thin semiconductor material while a magnetic field passes through it, a small Hall voltage is produced. (*Courtesy of Ford Motor Company*)

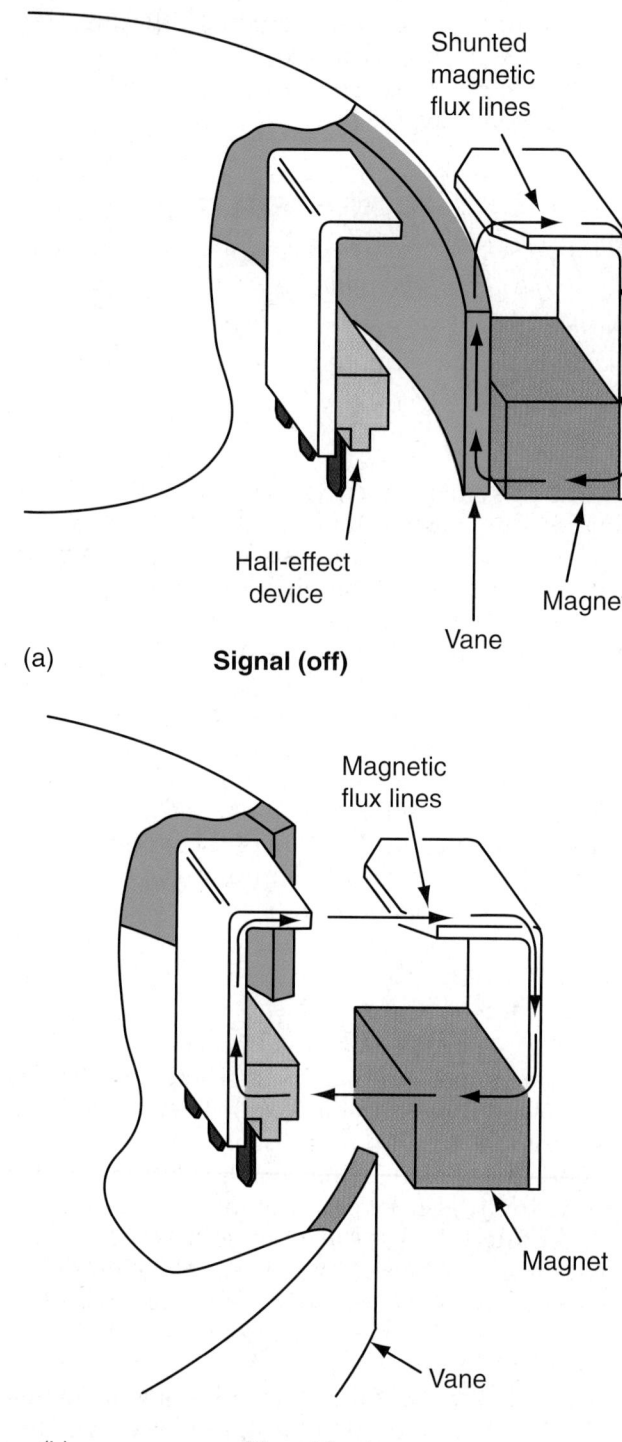

Figure 37.26 Hall-effect operation. (a) Output voltage rises when the vane interrupts the magnetic field. (b) When the vane moves past the magnet output voltage drops.

trigger converts it to square wave (pulsating DC), and the signal is sent to a switching transistor.

When the shutter blade blocks the magnetic field, a control module will either turn the primary on or off, depending on system design. As the leading edge of the shutter vane enters the air gap between the Hall switch and the permanent magnet, the magnetic field is deflected away from the Hall switch, decreasing Hall voltage (**Figure 37.26**) and increasing the modified signal from the amplifier. This turns on a transistor that controls current flow in the coil primary winding. As long as the vane blocks the air gap, current flows in the primary winding.

When the vane begins to move out of the gap, the modified signal from the amplifier drops once again. This shuts off the transistor, stopping current flow in the primary winding. The module determines the correct spark timing and opens the primary circuit at precisely the right time. This causes the magnetic field surrounding the coil windings to collapse, resulting in a spark in the cylinder at the plug gap.

Hall switches are also used to generate rpm signals and cause sequential fuel injection systems to pulse. Hall-effect devices are very accurate to within ± ¼ degrees of distributor rotation. Hall switches are also found on the crankshaft (**Figure 37.27**). The computer monitors the rate at which the voltage level from the switch rises and falls. From this information, it determines the position of the piston in the cylinder. The Hall switch is called a *crankshaft position sensor* (*CPS*) when it is used this way. Not all Hall devices use vanes. Some, like crankshaft position sensors, use notches or slots.

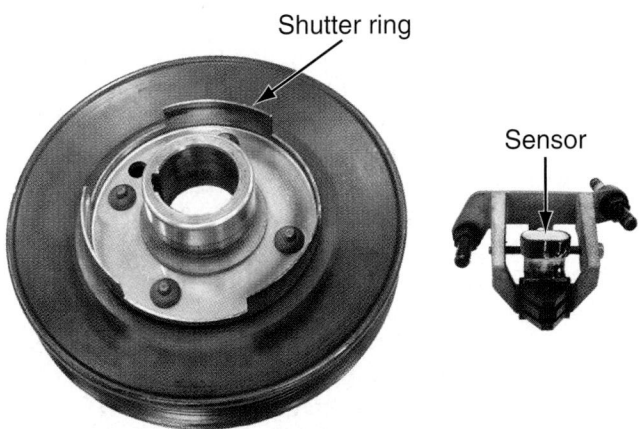

Figure 37.27 This crankshaft damper has a shutter ring. A Hall-effect sensor is mounted opposite the ring. The sensor shown here is for a damper with two shutter rings. (*Courtesy of Tim Gilles*)

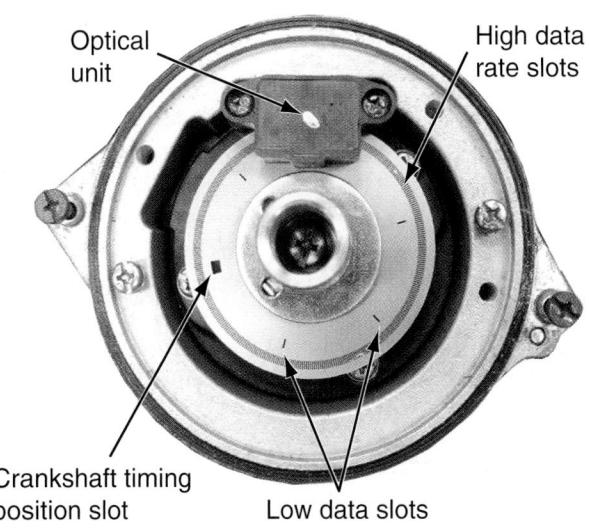

Figure 37.28 A photoelectric sensor. (*Courtesy of Tim Gilles*)

Magnetoresistive Sensors

Like the Hall-effect sensor, magnetoresistive (MR) sensors create a square wave digital signal. An MR sensor operates in a very similar way to a Hall sensor, and they both have three wires. The only real difference is in the reference voltage. Whereas the Hall-effect sensor receives its 5-volt reference signal from the computer or from the ignition module, the magnetoresistive sensor creates its own 5-volt signal. A permanent magnet at the end of the MR sensor is sandwiched between two sideways magnetic reluctance pickups. When the reluctor wheel passes the sensor one of the pickups gets the signal sooner than the other one. Signals from the two sensors are the same but occur at slightly different times. The differential signal switches a Schmitt trigger.

In one manufacturer's system, a 24-notch reluctor wheel has notches of two different widths positioned 15 degrees from each other. For faster engine starting during engine cranking, the computer can determine a cylinder's location within the first 45 degrees of crank rotation.

Optical Sensors

Optical sensors, also called *photoelectric sensors*, use a beam of light to control the primary circuit. The sensor shines a light beam on one side of a slotted disc (**Figure 37.28**). A photo cell on the other side of the slotted disc changes the light beam into voltage pulses. When there is voltage present, the control unit provides current to the ignition coil primary winding. When the disc interrupts the light beam, the voltage stops. The control unit shuts off current flow, collapsing the magnetic field and causing a spark at the plug.

Photoelectric sensors used for automotive engines are called **crank angle sensors** because they actually sense two things. One signal is generated by the sensor every degree of rotation by the 360 slots in the disc. These are called high data rate slots, or high resolution slots. There are also low data rate slots: four slots spaced every 90 degrees for four-cylinder engines and six slots at 60-degree intervals for six-cylinder engines. The computer determines crankshaft position, cylinder identification, and rpm from these openings, and information is also gathered from a section of the high data rate ring that does not have any slots, or a low data slot that is wider. The low data rate signal also tells the computer to provide spark and fuel.

An optical sensor makes an AC (analog) sine wave signal. Remember, an analog signal is not square like a digital computer needs, so it is converted to a square on and off signal by a processor.

Optical distributors operate very well at both low and high engine speeds. They tend to have problems in low temperatures, however, and can suffer from moisture, dirt, and corrosion.

▬ IGNITION MODULES

There are different types of electronic ignition modules. Most of the newer ones do four things: turn primary current on, turn primary current off, limit current, and vary dwell. These functions can all be seen on an oscilloscope.

Most ignition systems have a resistance somewhere in the primary circuit. It can be within the coil or outside of it in the form of a resistor wire or ballast resistor. The most common form of resistance is a variable resistance within an ignition module. This is called a **current limiting system**. The ignition module shuts back on current flow as soon as the coil primary winding is saturated. This increases the life of the coil and yet allows voltage to be higher when needed.

Newer ignition systems have **variable dwell**. At low engine rpm, dwell is shorter; at higher rpm, it will be longer.

IGNITION TIMING

The primary ignition circuit controls the point at which the secondary ignition circuit fires the spark plugs. The spark must ignite the air-fuel mixture at just the right moment so the flame front expands without exploding and pushes the piston toward the bottom of the cylinder. Ignition is timed to occur just before the piston reaches the top of the compression stroke, so peak combustion pressures are reached soon after the piston starts down on the power stroke. All of the air-fuel mixture should be ignited before the piston reaches 23 degrees ATDC. The initial spark timing setting, which is adjustable on some engines, is called base spark timing. This is the timing setting before the computer or mechanical advance mechanism has had a chance to make changes.

Ignition Timing Variation

A computer engine management system receives input that includes engine load and speed signals.

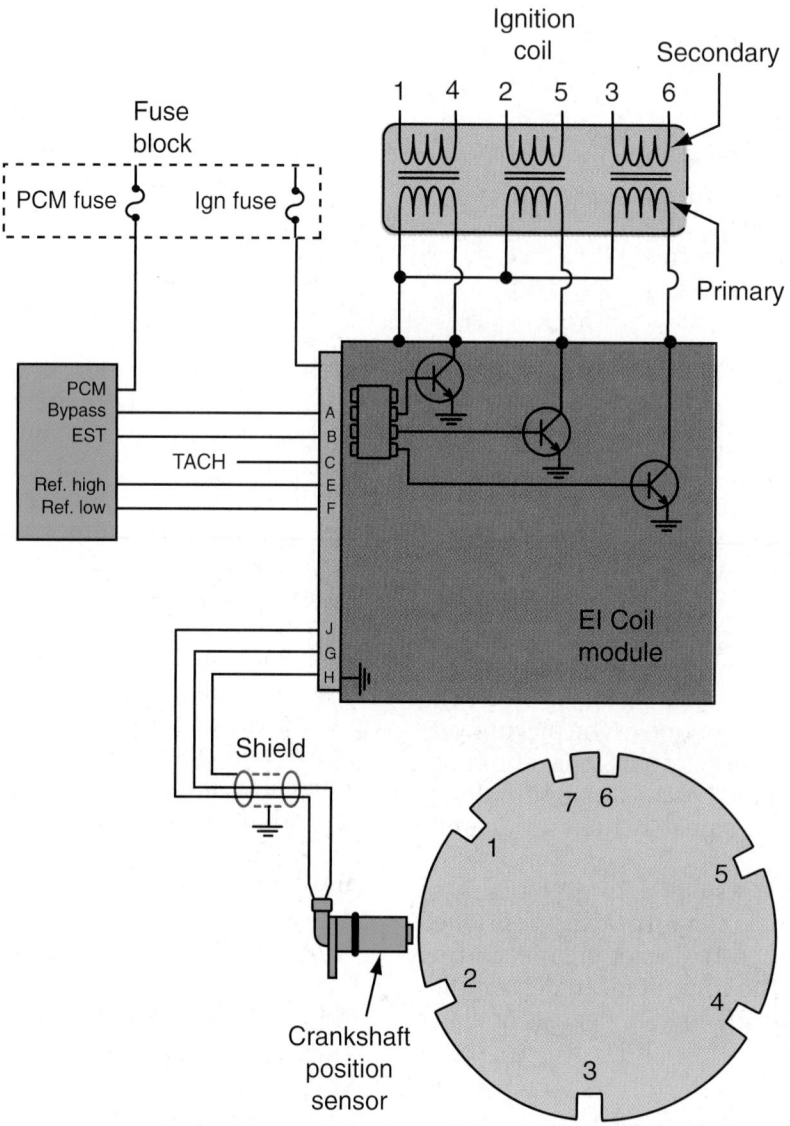

Figure 37.29 An electronic ignition system. This one uses a crankshaft position sensor instead of a distributor.

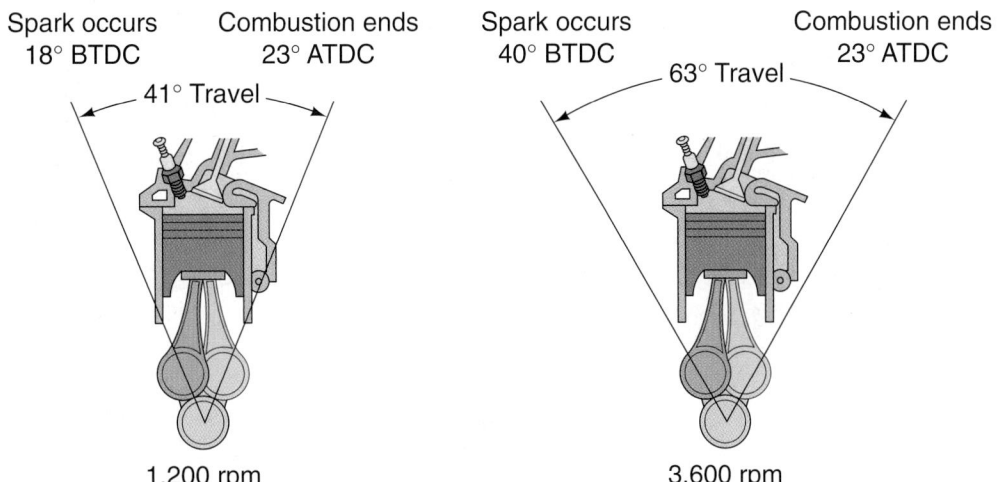

Figure 37.30 The spark must occur sooner at faster engine speeds. *(Courtesy of Federal-Mogul Corporation)*

From this information, it determines the best ignition timing setting, making changes in a fraction of a second. Consideration is given to piston speed and the amount of air and fuel in the cylinders. Ignition timing is advanced or retarded in response to engine speed and load changes, altitude, and engine temperature.

Older engines used two different spark advance mechanisms to accomplish the same thing. Centrifugal or mechanical advance sensed engine speed, and vacuum advance sensed engine load. Electronic systems perform the same functions as mechanical systems but do them more accurately and consistently using computer-controlled spark timing (**Figure 37.29**). Distributor vacuum and mechanical advance are no longer needed in these cars.

Whether the engine is running at low speed or at high speed, the same amount of time is required to burn the air-fuel mixture (0.003/second). Spark timing must be advanced at higher engine speeds if the mixture is to be ignited soon enough to complete its burn (**Figure 37.30**). As engine speed increases, the timing must advance. Engines with newer combustion chamber designs do not require as much spark advance as older systems.

Older engines used spring-loaded weights on the distributor shaft that moved outward in response to centrifugal force, triggering the spark earlier. Electronic systems use the engine rpm input and respond accordingly with changes in ignition timing.

Sensing Load with Engine Vacuum

Intake manifold vacuum is used to sense the load on the engine. When the throttle plate is all the way open, the engine fills with more air than the pistons can keep up with. This results in low engine vacuum. When the throttle is released, the pistons draw harder against the throttle plate, resulting in high engine vacuum. Under light load, vacuum is high. The mixture is lean, and there is more room between its fuel molecules. This means it takes longer for the flame front to move across the cylinder. A lean mixture needs to be ignited sooner so that the mixture has time to complete its burn by 23 degrees ATDC. Engine management systems analyze inputs from the manifold absolute pressure (MAP) sensor or mass airflow sensor (MAF) to determine engine load and adjust ignition timing accordingly. Some EI modules also allow for retarded timing during engine cranking. Retarding the timing allows the engine to crank faster for easier starting.

Older systems used a vacuum advance diaphragm to advance timing by pulling a point plate or pickup coil against the direction of distributor rotation. As vacuum decreased, the plate was returned to its original position by a spring. Modern distributors house only the primary switching trigger and sometimes an ignition control module. Secondary high voltage is still distributed through the rotor and distributor cap to the spark plug wires, but the distributor shaft is simply that—a shaft with a gear drive on the bottom and a rotor on the top.

Computer systems continuously adjust spark timing to provide the best engine power and lowest emissions throughout all operating conditions. The powertrain control module (PCM), also called a central processing unit (CPU), receives input signals on throttle position, manifold and barometric pressures, engine rpm, crankshaft position, and coolant temperature. Some of these extra functions were not possible with mechanical distributors.

■ A throttle position sensor (TPS) sends a voltage signal to the computer that tells it what position the throttle is in.

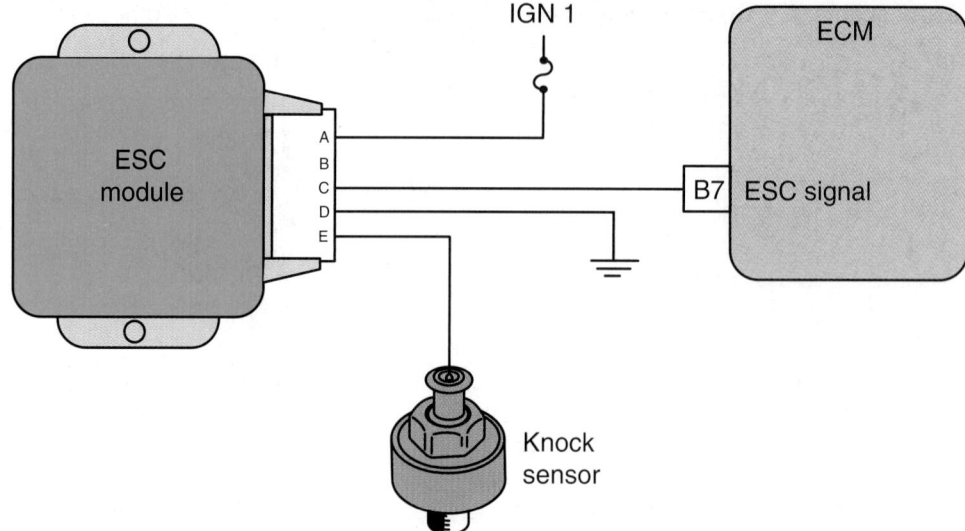

Figure 37.31 When a knock sensor detects detonation, the electronic spark control module retards timing.

■ An MAP sensor sends the computer data on the pressure in the intake manifold. This replaces the vacuum advance unit and also gives an indication of altitude.

■ The primary trigger in the distributor or on the crankshaft gives the computer the ability to interpret engine speed.

■ A coolant temperature sensor lets the computer make adjustments for changes in engine temperature.

Based on information from engine sensors, the PCM or CPU decides what the timing should be and signals the module to fire the spark plugs.

▄▄ DETONATION SENSOR

Some modern EI systems use a detonation sensor, or knock sensor, to control maximum spark advance (**Figure 37.31**). The sensor, a piezoelectric crystal (see Chapter 45), detects the frequency of spark knock and is capable of retarding the timing up to 12 or more degrees. When the air-fuel mixture in any of the cylinders detonates, the PCM retards the timing in steps until the vibration stops. Then it advances the timing until knocking occurs once again. It readjusts until it settles on the maximum spark advance possible without detonating. The PCM also checks to see if outside air temperature is high so it can prevent detonation from occurring.

▄▄ DISTRIBUTORLESS IGNITION

In SAE terminology, ignition systems without distributors are known as EI. They are also commonly known as DISs. Advantages of EI include reduced cost and lower maintenance. These ignition systems perform all of the same functions as a DIS but without

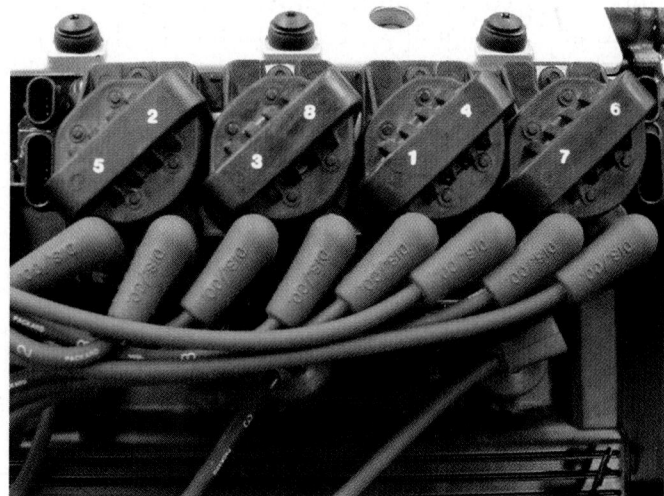

Figure 37.32 A coil pack for an eight-cylinder DIS. *(Courtesy of Tim Gilles)*

a distributor. EI has no rotor or distributor cap and, sometimes, no spark plug cables. Multiple coils, called a coil pack (**Figure 37.32**), a control unit, sensors, and a computer make up the rest of the DIS. An ignition module tells the coils when to fire with input provided by a crankshaft position sensor. A typical crank or cam sensor (**Figure 37.33a**) has a spaced ring with one or more unevenly spaced teeth that the computer can use to determine crankshaft position. Sometimes multiple sensors are used. By counting the time between impulses, the ignition module can tell when the unevenly spaced slot passes (**Figure 37.33b**). This information is used to tell the coils the correct time to spark. **Figure 37.34** shows how this works with different numbers of cylinders.

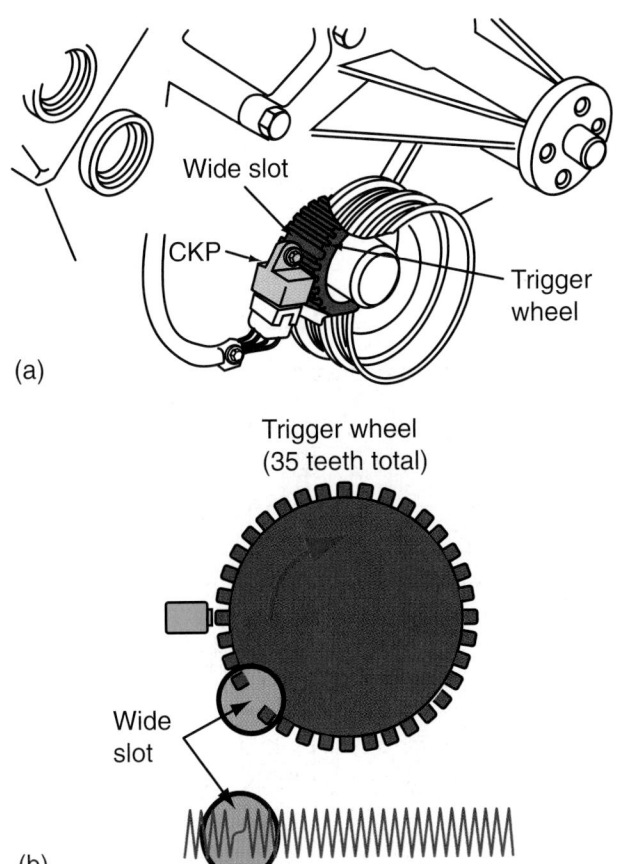

(a)

(b)

Figure 37.33 (a) A typical crankshaft position sensor (CPK). (b) By counting the time between impulses, the ignition module can tell when the unequally spaced slot passes the sensor.

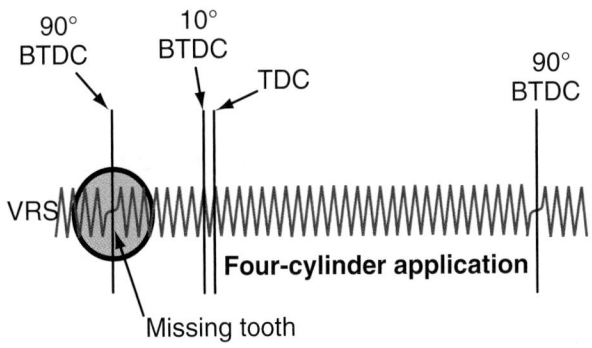

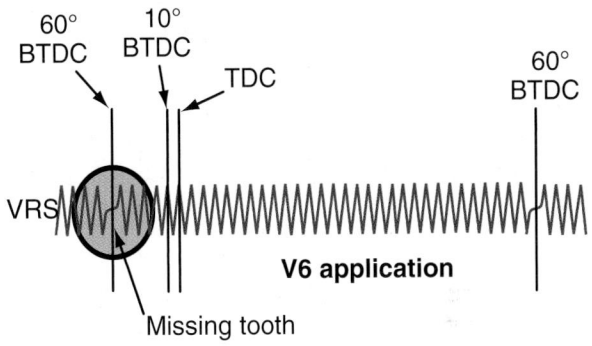

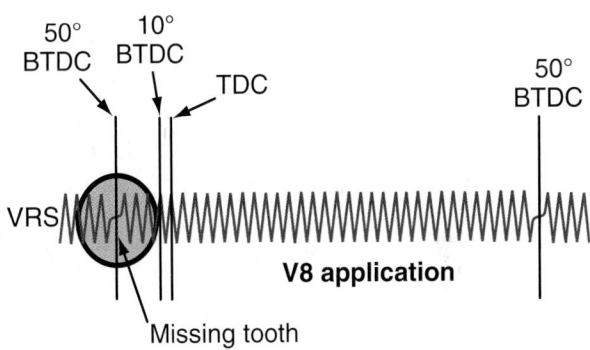

Figure 37.34 Crankshaft position sensor (missing tooth) signals for different numbers of cylinders.

Crankshaft Position Sensor

The CKP sensor, sometimes called the crank angle sensor, determines the speed of the engine and the position of the crankshaft (called crankshaft angle). Cars produced since 1996 come under federal emission on-board diagnostic II (OBD II) standards that require misfire detection capability. When a misfire occurs, the CKP sensor provides this capability by detecting very small changes in crankshaft velocity.

Some CKPs have more than one trigger. The sensor ring with more trigger notches provides a higher resolution signal used for lower engine rpm. The lower resolution signal is for higher engine speeds.

Camshaft Position Sensor

Many ignition systems also have a camshaft position sensor (CMP) that tells the computer when the #1 cylinder is on the compression stroke (**Figure 37.35**). This information is used for sequencing the fuel injection system and to sequence coil firing with coil-on-plug and coil-near-plug ignition systems (covered later in this chapter). Other distributorless systems fire the spark plug according to the CKP sensor, so camshaft position is not necessary. Many engines also have variable valve timing (VVT), which uses CMP sensor information to verify system operation. Some engine management systems use the CMP sensor to tell the sequential fuel injection system when to pulse.

Waste Spark

Some DIS systems use one coil per plug with the coil's secondary output attached directly to the spark plug. Other systems have one coil for every two spark plugs (**Figure 37.36**), often referred to as the **waste spark method** (**Figure 37.37**). In this system, both ends of each coil's secondary winding are attached to one of a pair of companion cylinders' spark plugs. This is a series system where both cylinders' spark plugs fire every revolution of the crankshaft. One cylinder will be at the top of its compression stroke, and the

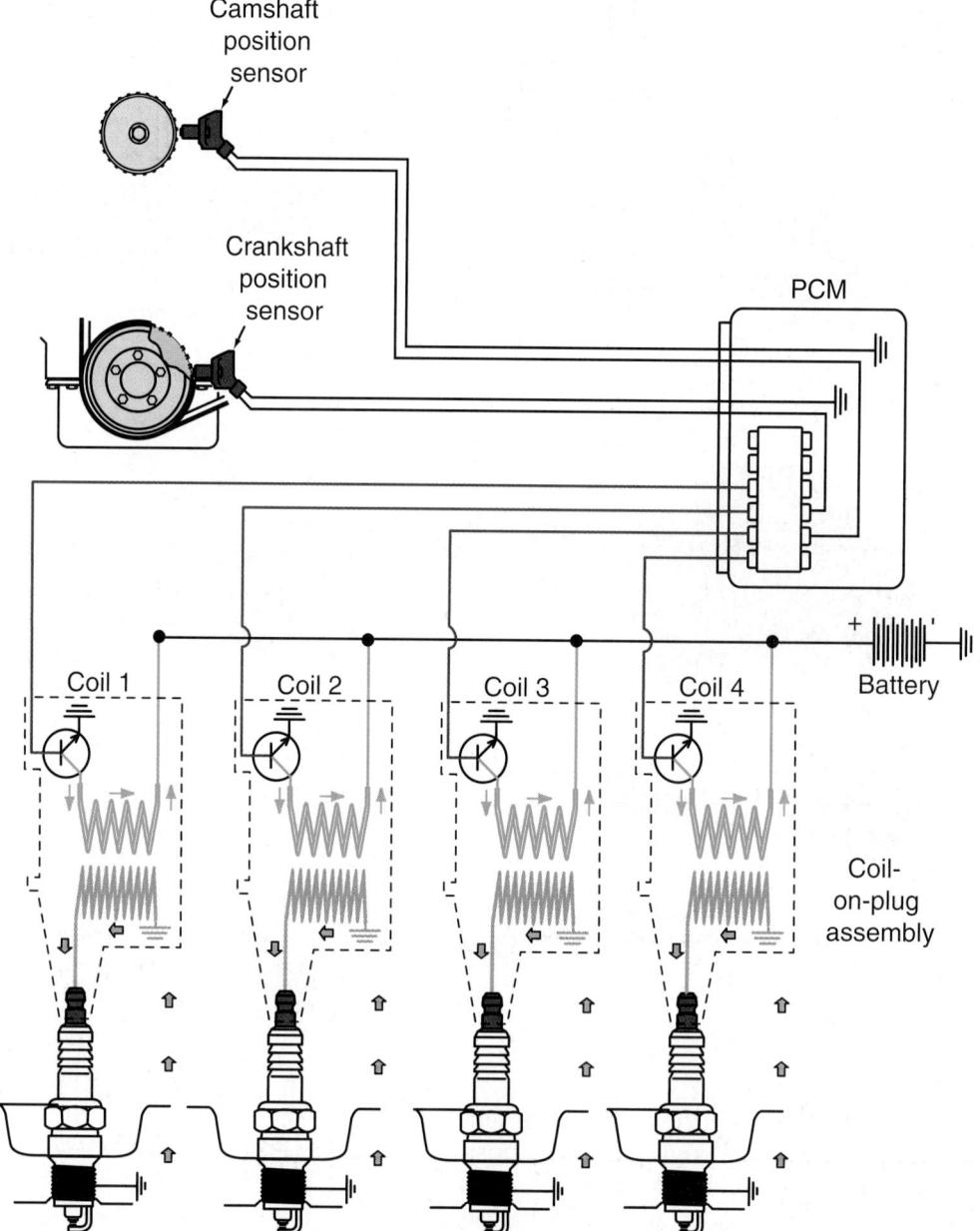

Figure 37.35 A coil-on-plug ignition system with camshaft and crankshaft position sensors.

companion will be on its exhaust stroke. The one on the exhaust stroke is the "wasted" spark because it does not do anything.

One spark plug fires from the center electrode to the side electrode. The other has reversed polarity and fires from the side electrode to the center electrode (**Figure 37.38**). In a conventional ignition system, reversed polarity would result in less available voltage. But these systems can produce up to 100,000 volts. The plug that fires on the exhaust stroke requires very little voltage because there is no compression in the cylinder (**Figure 37.39**). Compression makes it harder for a spark to jump a gap.

One DIS variation has individual coils mounted near each cylinder and is called *coil-near-plug* (CNP). Ignition coils/modules are mounted above each cylinder (**Figure 37.40**). A short length of cable connects the coil and spark plug. Each ignition coil module has its own control circuit activated sequentially by the PCM. This system provides quick response to signals from the computer.

Some EI systems have individual coils mounted directly on the spark plugs. These are called coil-on-plug (COP) or coil-over-plug (**Figure 37.41**). One variation of coil-on-plug uses waste firing and half of the cylinders have spark plug wires. A six-cylinder

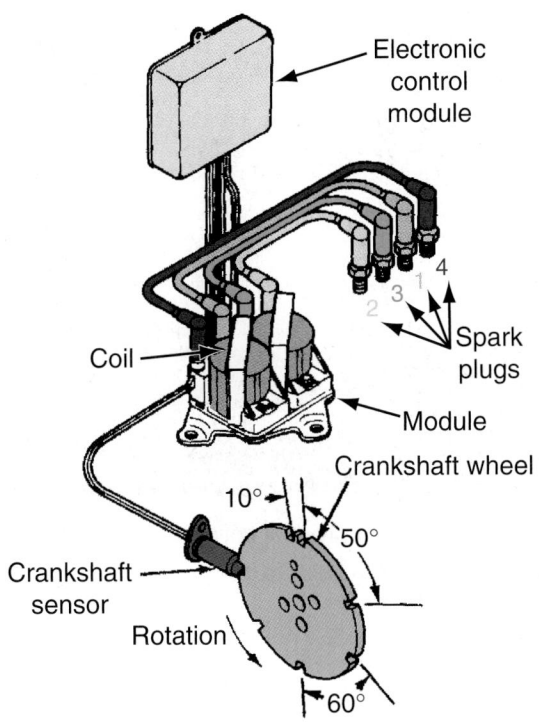

Shown at TDC cylinders 1 and 4

Figure 37.36 Parts of a distributorless ignition system. Note that each coil is supplying two cylinders. (*Courtesy of Federal-Mogul Corporation*)

version would have three coils mounted on spark plugs, with three spark plug cables coming from the coils and routed to companion cylinders.

Two Spark Plugs per Cylinder

Some engines use two spark plugs per cylinder (**Figure 37.42**). One is located on the combustion chamber's intake side and the other is on the exhaust side. The mixture is ignited at two places in the cylinder to help keep the flame burning. This ignition system,

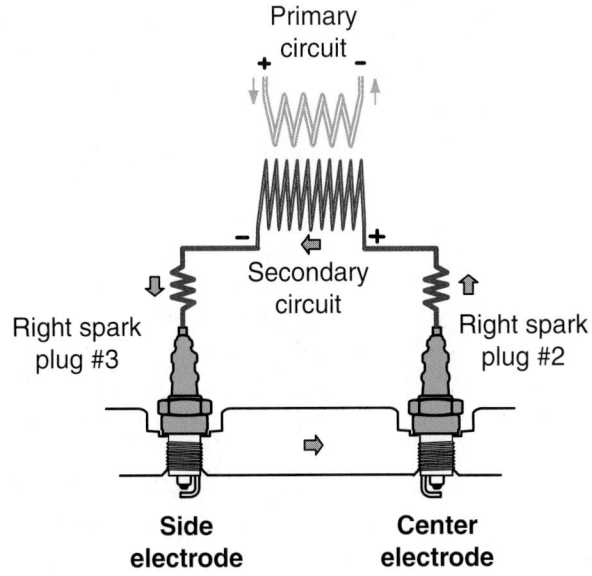

Figure 37.37 The waste spark method. Both ends of each coil's secondary winding are attached to one of a pair of companion cylinders' spark plugs.

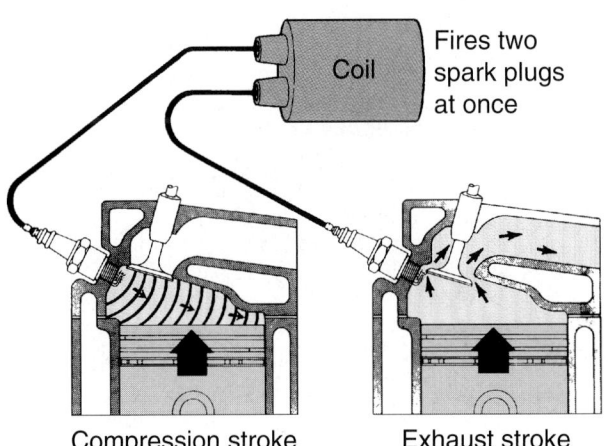

Figure 37.38 The spark in the exhaust cylinder is "wasted." (*Courtesy of Federal-Mogul Corporation*)

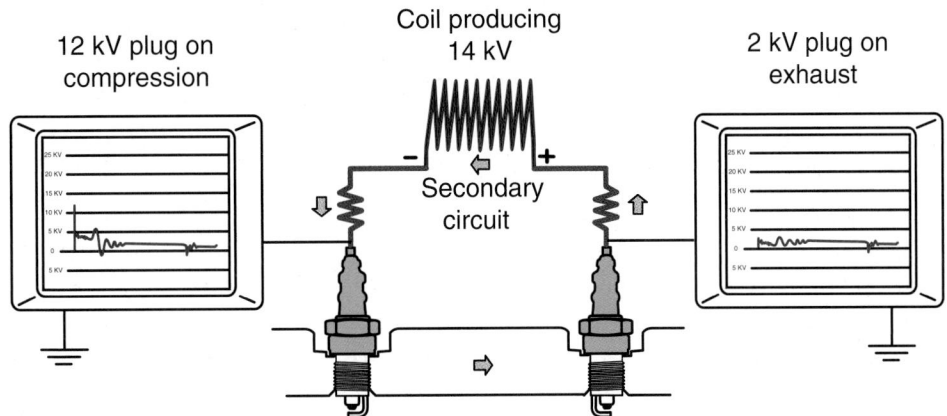

Figure 37.39 Coil voltage divided between two spark plugs. The plug that is firing on the exhaust stroke requires much less voltage due to low pressure in the cylinder.

Figure 37.40 A coil-near-plug system. (*Courtesy of Tim Gilles*)

Individual coils   Valve cover

Figure 37.41 A coil-over-plug or coil-on-plug system. (*Courtesy of Tim Gilles*)

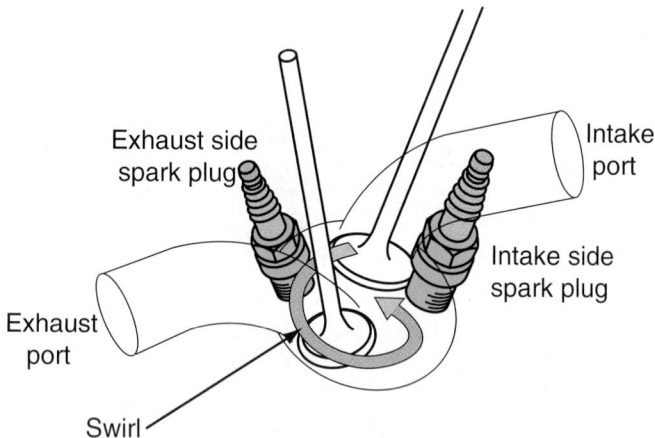

Figure 37.42 Some engines use two spark plugs per cylinder: one on the intake side and the other on the exhaust side.

which Ford calls *multi-strike*, provides some engines with smoother idle and lower emissions.

During cranking, some of these engines use only one of the cylinder's spark plugs. One coil pack is used for the intake side and another coil pack is used for the exhaust side. Multiple plug firing is used from idle to 1,500 rpm but shuts off after that to keep the coil from overheating.

▬ REVIEW QUESTIONS

1. The ignition system's job is to create and distribute a _____ spark.

2. An ignition system has two circuits called _____ and _____.

3. What is the name that describes the length of time in degrees that primary current is flowing in the primary winding?

4. In a distributorless system, the _____ has a pickup that takes the place of the distributor.

5. After a spark plug fires, the crankshaft rotates ___ turns before another spark is required in that cylinder.

6. Over 90% of a spark plug's heat leaves it at the edge of the _____

7. _____ is used for the electrode in long-life plugs because it is resistant to chemical erosion and corrosion.

8. When an engine is running at 3,000 rpm, its magnetic field must be able to build up and collapse _____ times in 1 second.

9. A primary trigger can be either a magnetic AC generator, a _____ switch, or an optical (photoelectric) sensor.

10. The spark must ignite all of the air-fuel mixture before the piston has reached _____ ATDC.

◼◼◼ ASE-STYLE REVIEW QUESTIONS

1. Which of the following is/are true about the primary ignition system?
   a. The primary side of the ignition system is the low-voltage side.
   b. The ignition spark comes from the primary part of the system.
   c. Both A and B.
   d. Neither A nor B.

2. Technician A says that the high-voltage surge to the spark plug is created in the coil when electricity stops flowing in the coil primary winding. Technician B says that the spark impulse comes from the primary winding when the magnetic field in the secondary winding breaks down. Who is right?
   a. Technician A   c. Both A and B
   b. Technician B   d. Neither A nor B

3. Technician A says that using a spark plug with too hot a heat range can cause preignition. Technician B says that using a spark plug with too hot a heat range can result in a burned piston. Who is right?
   a. Technician A   c. Both A and B
   b. Technician B   d. Neither A nor B

4. Which of the following is/are true about resistance spark plug cables?
   a. Resistance in the cables raises the firing voltage level.
   b. Raising the resistance lowers radio interference.
   c. Both A and B.
   d. Neither A nor B.

5. Which of the following is/are true about resistor spark plugs?
   a. Some resistor spark plugs have carbon resistors.
   b. Some spark plugs have semiconductor resistors.
   c. Both A and B.
   d. Neither A nor B.

6. Electronic ignition systems control the flow of current on the _____ side of the coil.
   a. Ground   c. Voltage
   b. Positive   d. Magnetic

7. On an eight-cylinder engine running at 3,000 rpm, the ignition system produces a spark _____ times every second.
   a. 25   c. 200
   b. 100   d. 10

8. Which of the following ignition triggers works best at any speed?
   a. A magnetic pickup trigger
   b. A Hall-effect switch
   c. Both A and B
   d. Neither A nor B

9. All of the following are true about spark timing control *except*:
   a. Timing adjusts to compensate for engine speed.
   b. Timing adjusts to compensate for engine load.
   c. Timing advances when the mixture is leaner.
   d. Timing is varied so combustion can be completed by 23 degrees BTDC.

10. Which of the following is/are true about a waste spark system?
    a. Spark plugs for companion cylinders fire every revolution of the crankshaft.
    b. One spark fires from its center and the other fires in reverse.
    c. Both A and B.
    d. Neither A nor B.

Ignition System Service

Upon completion of this chapter, you should be able to:

✔ Diagnose common ignition system problems.

✔ Service ignition systems and distributors properly.

✔ Install a distributor and adjust ignition timing.

✔ Operate an oscilloscope and interpret scope patterns.

■■ **KEY TERMS**

fouled spark plug	hard fault	superimposed pattern
crossfire induction	scope pattern	display pattern
carbon trail	firing line	stress test
timing light	spark line	
static timing	raster pattern	

■■ **IGNITION SYSTEM SERVICE AND REPAIRS**

Tune-up is a term that originated during the days when cars had ignition points and required periodic service.

Today, a tune-up is usually referred to as a *30,000- or 60,000-mile service*. This service can include replacement of all filters, belts, hoses, and fluids. The thermostat is replaced and the cooling system flushed, too. A few engines require mechanical valve lash measurement and adjustment. A timing belt replacement can also be part of a 60,000- or 90,000-mile service.

■■ **SPARK PLUG SERVICE**

Some of today's engines use expensive spark plugs designed to last for up to 100,000 miles. Others need to be changed periodically. When a spark plug is dirty or has worn electrodes, a misfire can result. Symptoms of a misfire can include a rough idle, uneven power on acceleration, and increased exhaust emissions.

NOTE: *At freeway speeds, a single cylinder that is misfiring part of the time might not be noticeable. At 3,000 rpm each spark plug fires 25 times every second.*

Engines built since 1996 come under federal OBDII guidelines, which call for misfire detection (see Chapter 47). These engines will illuminate a malfunction indicator light if a misfire becomes regular enough that damage to a catalytic converter could result.

Replacement Plugs

Be sure to use the correct replacement plugs. The code number on the spark plug tells about its heat range, thread size, type of seat, whether its tip is extended, and whether or not it has a resistor (**Figure 38.1**).

If you use a cross-reference chart on spark plugs, you will probably put in the wrong plug. Many spark plug–related problems are caused by misapplication. As an example, the Champion Spark Plug Company has 267 different OE spark plug applications at the time of this writing.

Be sure to use the correct spark plug heat range. Too hot a plug can cause serious engine damage (**Figure 38.2**). Also, be sure to replace a resistor plug with a resistor plug and non-resistor with non-resistor.

Some spark plugs have carbon resistors. Others have a semiconductor suppressor that is not energized until 1,000 volts. This kind of resistor plug *cannot* be checked with an ohmmeter. Check the size of the spark plug threads; 14 mm and 10 mm are the most popular. Check also to see whether the old plug has a gasket or a tapered seat without a gasket. Some square seat plugs are used without the gasket. Installing a spark plug without a gasket when one is required will allow the threads of the plug to hang into the combustion chamber. This can cause it to overheat, resulting in preignition.

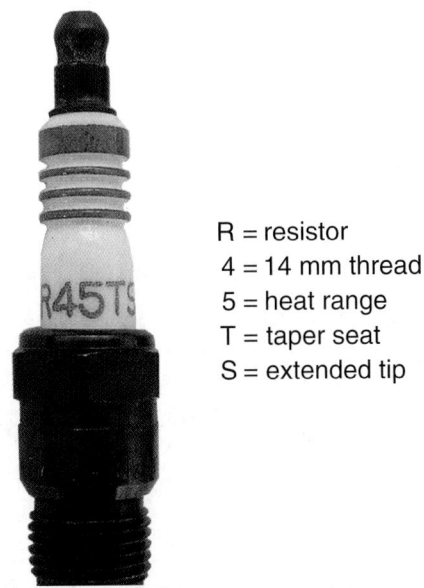

Figure 38.1 Spark plug code numbers tell about the plug. *(Courtesy of Tim Gilles)*

R = resistor
4 = 14 mm thread
5 = heat range
T = taper seat
S = extended tip

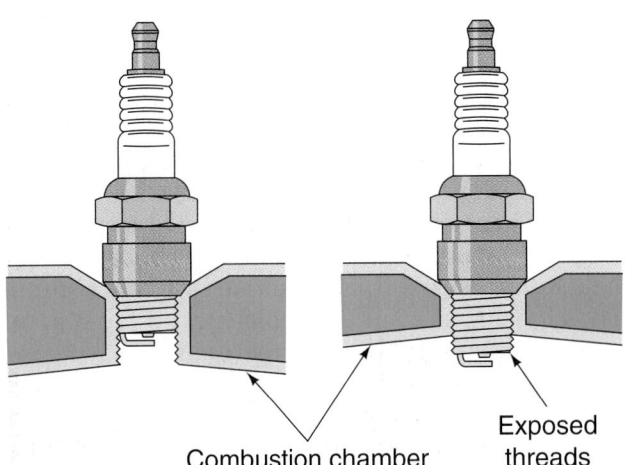

Combustion chamber

Exposed threads

Figure 38.3 Incorrect reaches.

Figure 38.2 Too hot a plug can cause serious engine damage like the hole on the top of this piston. *(Courtesy of Tim Gilles)*

Check the reach or thread length (**Figure 38.3**). The use of a plug with too short a reach can result in carbon buildup in the end of the spark plug hole. Too long a plug might possibly interfere with the piston. In either case, carbon on the threads can result in a ruined spark plug hole.

Removing Spark Plugs

Tapered seat plugs in particular are often very tight and can be difficult to remove. As you remove the spark plugs, inspect each one. Keep them in order and compare them.

■ SPARK PLUG DIAGNOSIS

The life of a spark plug is determined by the temperature of combustion, the quality of the electrical system, the correct air-fuel ratio, and good oil control in the cylinder. The temperature of a spark plug relates directly to combustion chamber temperature. There are many other factors that govern the temperature of a spark plug. These include ignition timing, workload, compression ratio, and air-fuel ratio. With a gasoline engine, 1,600°F is the maximum temperature that should ever occur.

The minimum spark plug temperature of 850°F is where carbon forms. If a conventional spark plug can operate at 1,200°F, it will have very little wear, and a life expectancy of 30,000 miles or more can be expected. The plug does need to get hot enough to burn off normal deposits, however. In normal driving, the load on a spark plug is very light until about 30 mph. Plugs are more apt to foul if the car is never driven over 30 mph. Spark plug fouling does not occur when cruising at 50 mph.

Reading Spark Plugs

Reading the condition of used spark plugs will give an idea of the condition of the engine and fuel system. Compare the plugs to see if there is a difference between cylinders. This can tell you if the engine can be tuned or will require repair.

When engines had leaded gas, plugs that were operating properly were tan to tan/grey in color. With today's fuels, the color of correctly operating plugs has changed. They should look nearly new with very little discoloration.

Blistering of the insulator tip or specks that resemble pepper indicate that a plug has been too hot. When a plug has been overheated, the center electrode's color resembles the burnt head of a match.

Heat dissipation is critical. When heat results, a hole gets burned in a piston and/or the spark plug center electrode melts. Figure 39.25 shows the results of preignition and detonation damage.

An electrode that has pointed wear often indicates a fuel mixture that is too lean. Lean mixtures burn hotter. In fact, when a drag racer runs out of fuel during a quarter mile run, a burned piston is a common result. This kind of wear can also be due to ignition timing that is too advanced or an inefficient cooling system.

Spark plugs must be properly torqued to avoid misfires. Look for evidence of etching on all threads that could indicate incorrect tightening. An overheated plug can result if it was not tightened correctly or due to a tapered plug seat that is dirty. The thread and thread seat are where over 90% of the plug's heat dissipation occurs. The threads must be clean and undamaged. If the spark plug does not unscrew easily with finger pressure only, clean the threads with a thread chaser. If all of the plugs appear to be overheated and the threads and plug seats are clean, try a colder heat range plug.

DIS Spark Plug Wear

On distributorless ignition systems (DIS), every other plug fires toward the ground electrode. The rest of the plugs fire toward the center electrode. The positively fired plug wears on its ground electrode, and the negatively fired plug wears on its center electrode. After 40,000 miles or so, the positive electrodes are the ones that tend to show wear first. Some manufacturers have recommended rotating the plugs for this reason or have a platinum pad on a different electrode depending on whether it is positively or negatively fired. Although they are expensive, today many manufacturers and technicians use double platinum plugs. Often called *double plat* plugs, they have platinum-plated ground and center electrodes, eliminating most wear and erosion of the gap. Double plat spark plug life is estimated to be 100,000 miles.

Fouled Plugs

Oil and ash deposits can be found on some plugs. A plug that is wet with fuel or oil usually indicates a misfiring cylinder.

Dirt on the spark plug's insulator can also cause misfiring by providing an electrical path for the spark to jump to ground.

A spark plug that has a buildup of carbon that shorts it out is called a **fouled spark plug**. The spark plug insulator looks smooth, but under a microscope it is porous. When a spark plug has fouled from running too rich, it fills with tiny metallic additives from the fuel. Because the insulator now conducts electricity, it will not work any longer (**Figure 38.4**). Once a plug has fouled once, it makes sense to replace it rather than to clean it. If an engine problem is suspected, perform a spark plug deposit test as described in Chapter 49.

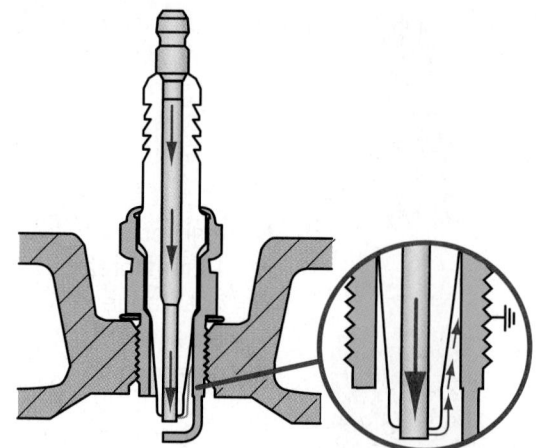

Figure 38.4 A fouled plug shorts to ground rather than firing normally.

SHOP TIP After new plugs are installed, the car is taken on a test drive. Hard acceleration can loosen carbon deposits and they can stick to the new plugs. Some technicians like to use a combustion chamber cleaner and allow a 20- to 30-minute soak period. Another technique is to do a hard acceleration test drive with old plugs to clean out the combustion chamber deposits before replacing the plugs. The problem with this is that the spark plugs and exhaust manifold will be hot when you want to remove the spark plugs.

Spark Plug Wear

The spark occurs at the outside edges of the electrode (**Figure 38.5**). A worn rounded electrode has more surface area that will need to be charged up with electrons during sparking, and electrons can jump easier from a sharp edge than a round one. In the distant past, spark plugs were cleaned, filed, and regapped.

NOTE: *Many years ago, it was common to clean spark plugs. On modern engines, this practice is not cost effective.*

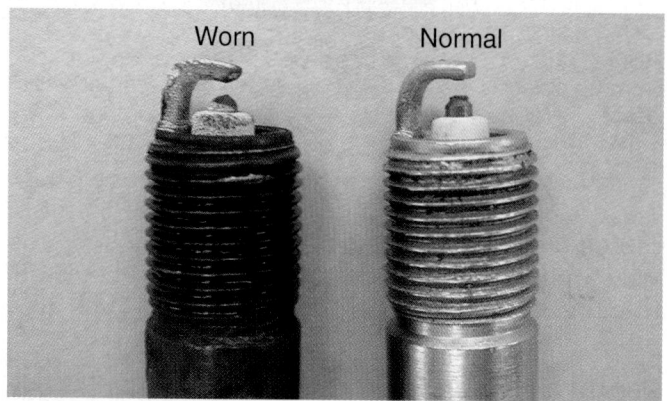

Figure 38.5 Worn and normal spark plugs. (*Courtesy of Tim Gilles*)

Spark plugs are inexpensive, and on some engines labor times to remove and replace them can be substantial.

Checking and Adjusting the Spark Plug Gap

The spark plug gap is usually set from the factory. Always double-check the gaps before installation. It is not unusual to find closed gaps or gaps that have changed due to rough handling. Also, spark plug companies often market the same plug for several different applications. It would be too costly to number them differently just because their gaps are of different sizes.

Spark plugs in older point-type ignition systems specified small gaps of around 0.035". The gaps on newer electronic ignition cars can vary up to 0.080". Be certain to look up the specification for the gap.

A properly gapped spark plug will have its ground electrode surface parallel with the center electrode. A wire gauge can be used to check the gap (**Figure 38.6**). It has an arm that is designed to be used to reposition the bendable electrode on the plug (**Figure 38.7**). The body of the spark plug is ceramic. It is fragile and can easily be cracked (**Figure 38.8**). A cracked insulator will allow the spark to go to ground rather than jumping the gap (**Figure 38.9**).

Spark plugs are removed and installed with a special socket. Spark plug sockets have either a rubber insert or a magnet inside to hold the plug in the

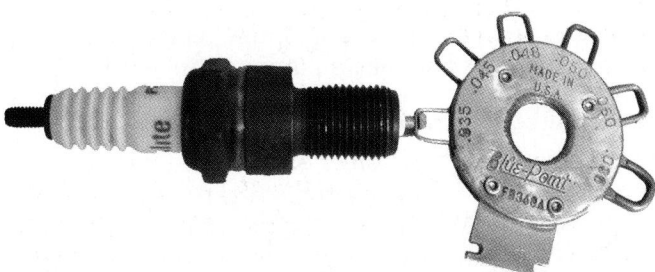

Figure 38.6 A wire gauge used to measure the spark plug gap. *(Courtesy of Tim Gilles)*

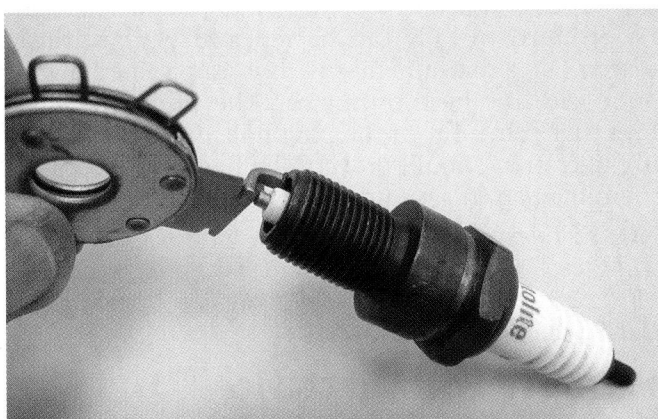

Figure 38.7 This tool is used to bend the ground electrode to set the gap. *(Courtesy of Tim Gilles)*

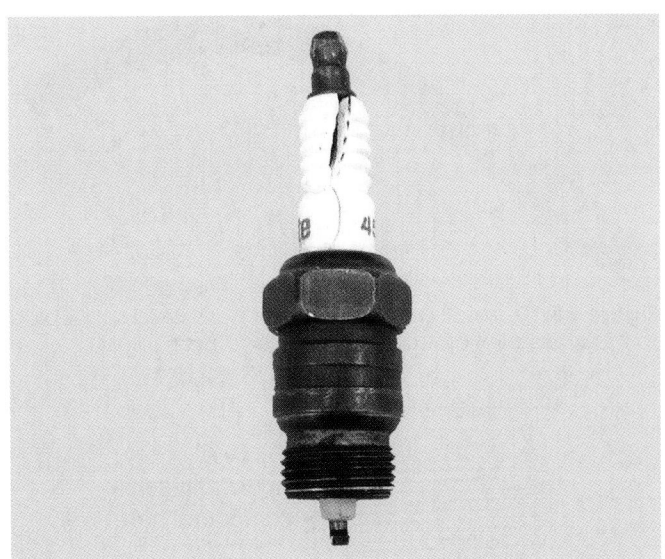

Figure 38.8 The body of the spark plug is ceramic, which can crack with rough handling. *(Courtesy of Tim Gilles)*

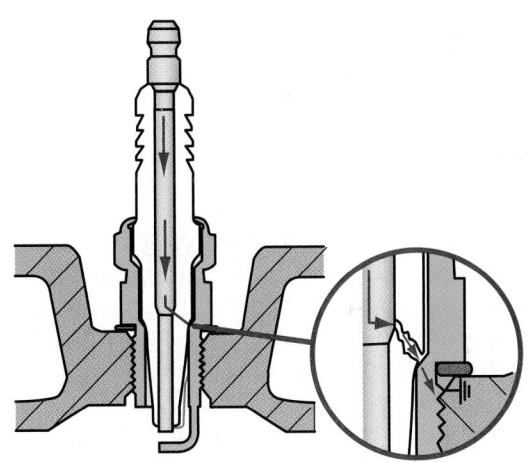

Figure 38.9 A cracked insulator will allow the plug to short to ground.

socket so it does not fall out during installation. When removing and installing spark plugs, be careful not to allow the spark plug socket to cock to the side.

NOTE: *Be careful not to drop a good plug on the ground. A plug that has been dropped on the ground could be cracked and should be replaced.*

■ INSTALLING SPARK PLUGS

SHOP TIP Install a short length of vacuum hose on the end of a spark plug prior to threading it into its hole in the cylinder head (**Figure 38.10**). Installing spark plugs in this manner is easy and prevents the possibility of accidentally cross-threading the spark plug.

Figure 38.10 Install a short length of hose on a spark plug to make installation easier. *(Courtesy of Tim Gilles)*

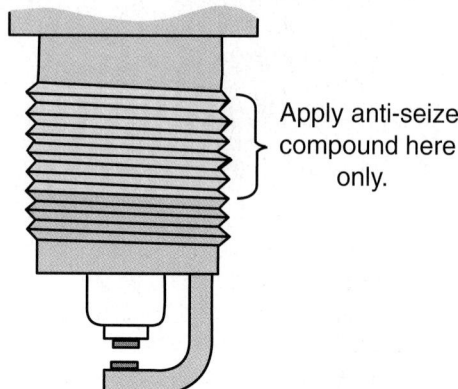

Apply anti-seize compound here only.

Figure 38.11 A small amount of anti-seize compound can be applied to the upper area of the spark plug threads.

Put a little bit of anti-seize compound on the outer two threads of the plug before installation (**Figure 38.11**). Anti-seize is especially needed on aluminum heads to prevent electrolytic action between the steel spark plug body and aluminum head. Do not use too much. Many newer spark plugs have coatings that do not require the use of anti-seize.

When tightening the plug, the seal of the plug can be ruined by overtightening. Using anti-seize on spark plug threads reduces the torque requirement. Some makers of spark plugs recommend tightening by angle or degrees rather than by torque wrench. Tighten a gasketed plug finger tight, then ¼ turn further. A taper seat plug is installed finger tight, then ¹⁄₁₆ turn further. Small "peanut" plugs will twist off at 50 to 60 pounds of torque. They are usually only torqued to about 10 or 15 foot-pounds. Check the specifications until you are used to the type of car you are working on. Some newer plugs are specified for a ½ turn after gasket contact. Most newer cylinder heads are made of aluminum. An overly tightened spark plug in an aluminum head can result in stripped threads when attempting to remove the plug at its next replacement interval.

Today's spark plugs have a longer lifetime than plugs in older vehicles. Oxygenated fuels tend to form more carbon. These two changes result in more deposits on spark plug threads.

NOTE: *Spark plugs that are not tight will run hot and can cause serious engine damage. If you cannot seat the spark plug all the way with your fingers, there is probably carbon in the threaded hole in the cylinder head. Chase the threads*

with a thread chaser before installing the spark plug. When removing plugs, examine the seat area closely for evidence that the plug was seated all the way in its previous installation. On plugs with gaskets, the gasket should be compressed. With tapered spark plugs, the seat area of the plug should show a "witness mark" (proving that it was seated). Spark plugs that fail because they were loose can have a blown-out center electrode or insulator, or the outer shell can be blue.

Indexing a Spark Plug

Indexing a plug is a high-performance procedure usually used only in high compression racing engines. The idea is to have the back side of the ground electrode face the exhaust valve so the fuel coming in through the intake valve can blow directly into the gap. Copper shims are used to get the indexing right. There are 0.060", 0.080", and 0.135" shims.

Platinum-Tipped Spark Plugs. Platinum plating on spark plug electrodes reduces wear, maintaining a more consistent air gap. Platinum spark plugs come in two designs. "Single plat" plugs have platinum-plated center electrodes. A double platinum plug has platinum plating on both the positive (center) and the ground electrodes. Single platinum plugs can be used on engines with coil-on-plug (COP) or distributor ignition systems because electron flow to the spark plug is always in the same direction, which results in the positive electrode experiencing the most wear. Waste spark design distributorless ignition systems will cause wear on half of the positive electrodes and half of the ground electrodes. To reduce wear on all of the spark plugs, replacement plugs in these engines are double platinum tipped.

■ REPAIRING DAMAGED SPARK PLUG THREADS

Spark plug threads are sometimes stripped during a careless installation. Stripping threads can be avoided if a ratchet is never used to install a plug, only to tighten it. A piece of vacuum hose can be installed on the end of the spark plug, or a special rubber spark plug tool can be used. The socket can be used without the ratchet also. If the threads are aligned properly, the plug should go all the way in to touch its seat without the aid of a ratchet.

Stripped threads can be replaced with a thread insert. On aluminum heads, this can sometimes be done with the head on the car. The aluminum shavings that come out of the plug hole during the reaming operation should prove to be harmless if some of them happen to get into the combustion chamber.

NOTE: *Tapered seat spark plug ports cannot be repaired with thread inserts. The tap used to prepare the hole for the thread insert will ruin the taper in the hole, resulting in a leaking spark plug.*

■ SPARK PLUG CABLE SERVICE

Replacing Spark Plug Cables

Removing spark plug cables must be a careful operation if the cables are to be reused. The rubber boots usually

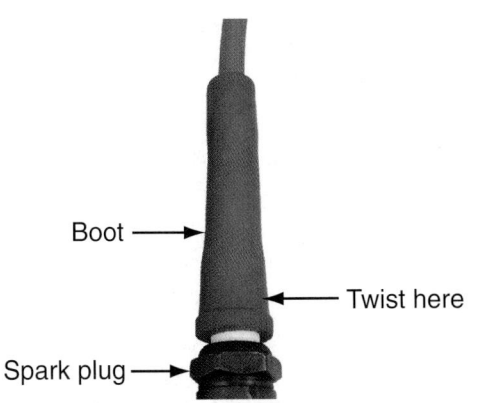

Figure 38.12 Twist the boot to loosen it from the spark plug insulator. *(Courtesy of Tim Gilles)*

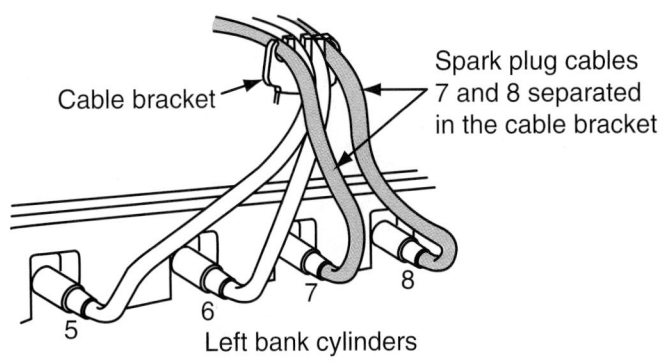

Firing order 1-5-4-2-6-3-7-8

Figure 38.14 Avoiding induced sparking.

Figure 38.13 The inside of a TVRS spark plug cable is fragile. *(Courtesy of Tim Gilles)*

become formed to the ridges on the spark plugs. They must be twisted to loosen them (**Figure 38.12**). Television/radio suppression (TVRS) cables are made of braided fiber and graphite and are very fragile (**Figure 38.13**). Handle them only by the plug boots so they do not suffer internal breaks.

NOTES:

■ *Cables should be held in looms away from the exhaust manifold and engine drive belts.*

■ *Do not allow any cables to run over the computer wiring loom where it could set up radio interference for the computer signals. This can cause an engine to run very poorly.*

■ *When installing the cables on the engine, do not allow cables from cylinders that fire after one another in the firing order to run parallel to each other (**Figure 38.14**). The first one to fire can induce a spark in the next one, causing it to fire before its time.*

SHOP TIP To test spark plug wire insulation for leaks, use a jumper wire to connect the blade of a large screwdriver—one with a large, well-insulated handle—to ground. With the engine idling, pass the screwdriver blade over the spark plug wires. A spark will jump to the screwdriver if the insulation is damaged.

Another test is to spray soapy water on the wires to see if the engine misfires.

SHOP TIP Changing cables one at a time will avoid accidentally mixing them up in the firing order. You will also be able to put them in their correct places in their existing loom.

Induced spark is called **crossfire induction**. It usually happens on V8s. In GM V8s, cylinders 5 and 7 fire after each other (1-8-4-3-6-5-7-2). In Fords, cylinders 7 and 8 fire after each other (1-5-4-2-6-3-7-8). These cylinders are next to each other in the block.

Be sure that the cable ends fit snugly into the distributor cap. Some manufacturers recommend putting some silicone dielectric compound on the inside of each boot. You should feel the metal clip on the end of the plug cable snap into the distributor cap socket.

Testing Cable Resistance

To determine that the internal structure of a cable is sound, check its resistance with an ohmmeter (**Figure 38.15**). The normal range of resistance in a new cable is about 5,000 to 10,000 ohms per foot. A defective cable can have 50,000 (50K) or 100,000 (100K) ohms of resistance, or it can be totally open (*infinite resistance*).

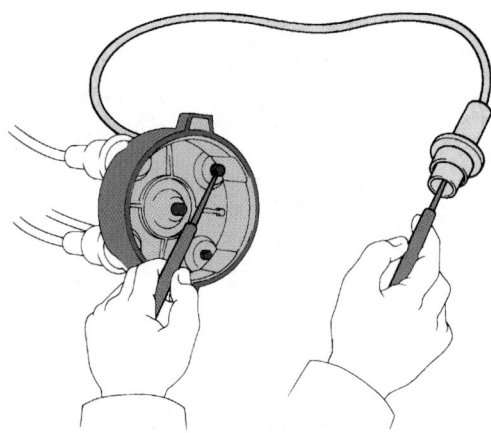

Figure 38.15 Checking the resistance of a spark plug wire and distributor cap.

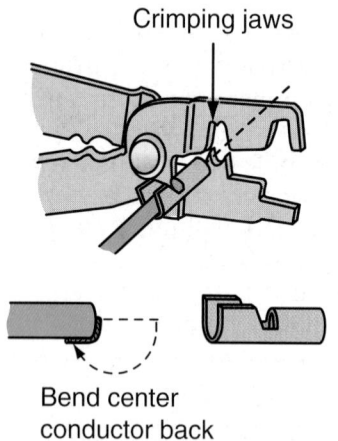

Crimping jaws

Bend center
conductor back

Figure 38.16 Replacing a spark plug terminal.

Repairing Spark Plug Cable Ends

The ends of the spark plug cables that connect to the inner fiber material are made of metal. The most common kind is crimped to the end of the plug cable. They can be replaced individually (**Figure 38.16**). The insulation is stripped off of about ⅜" of the end of the wire. The conductive fiber core is folded back over the outside of the insulation. The terminal is slipped over the outside of the conductor and insulation, then it is crimped tightly to the cable. A rubber boot is installed on the outside of the cable end to provide insulation. The boot is slipped over the metal end before it is installed. Special crimping pliers are used to crimp the terminal to the cable.

Sometimes, there is a resistor installed in the end of a spark plug cable. In this case, the cable will be solid cable. The spark plug might also *not* be a resistor plug.

Spark Plug Cable Installation

Spark plug cables are installed in the holes around the outside of the distributor cap in the engine's firing order (**Figure 38.17**).

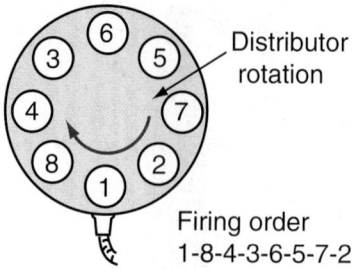

Distributor rotation

Firing order
1-8-4-3-6-5-7-2

Clockwise rotation distributor

Figure 38.17 Install spark plug cables in the engine's firing order.

CASE HISTORY *A customer took her Bronco into a service station for a tune-up. When she returned to get it, she was informed that the cam timing had skipped and the engine would no longer run. The service station did not do that kind of work, so she had to have the vehicle towed to another repair facility. The technician at the new facility performed a diagnosis on the engine before removing the timing cover. He discovered that the spark plug cables had been installed in the distributor cap in the wrong firing order. The engine ran fine after the order of the cables was corrected.*

■ DISTRIBUTOR IGNITION SERVICE

Although most engines use DIS, some still have distributors. Distributor ignition systems have some unique maintenance requirements. Service to the ignition trigger components is similar, whether the ignition system has a distributor or not, so those items are discussed later.

Distributor Cap and Rotor Service

Some caps are held on with screws; others have clips that snap into place. A distributor cap or rotor sometimes wears, cracks, or becomes corroded, requiring replacement (**Figure 38.18**). When a distributor cap is cracked, a **carbon trail** often forms along the crack. The carbon will conduct electricity, usually to a wrong cylinder or to ground.

When there is excessive resistance in the secondary system, the tip of the rotor can sometimes develop a hole. This *puncture* causes unusual symptoms. Replacement of the rotor is usually a part of a maintenance tuneup. The cap is always inspected. Some manufacturers recommend smearing some silicone dielectric compound on the tip of the rotor. This cuts down on arcing and radio interference, extending the life of the rotor and cap.

Rotors are attached to the distributor shaft in different ways. All of them have a feature that correctly

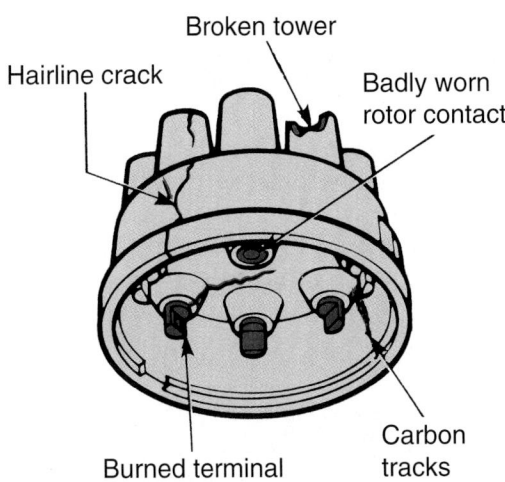

Figure 38.18 Places where a distributor cap can be damaged or worn.

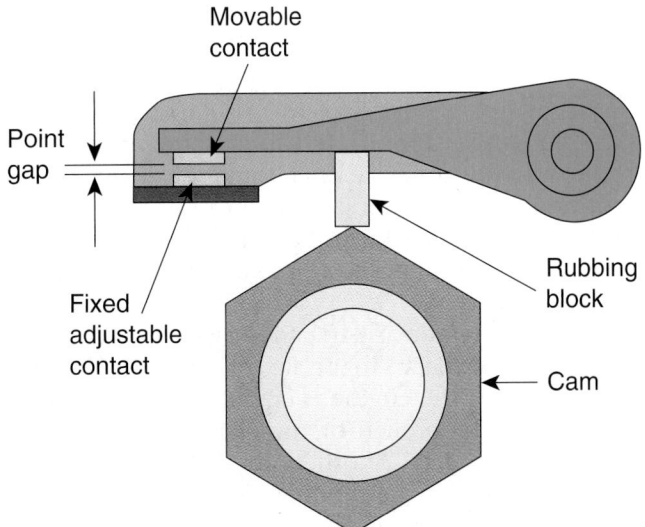

Figure 38.20 When adjusting the point gap on a vintage vehicle, the rubbing block is against the lobe on the cam.

aligns them to the shaft. Some fit snugly over the shaft and are aligned by a plastic tab that fits into a notch in the shaft. Others screw on and align with a square or round peg that fits into a corresponding notch. Be sure the aligning feature goes the correct way. Do not force it!

The distributor cap also has an aligning feature. There is a tab or notch somewhere on the distributor body that must align with a corresponding tab or notch in the cap. Once again, do not force something.

■ IGNITION TIMING

Ignition timing on DIS is controlled by the PCM. Some distributor ignition systems have adjustable base ignition timing that is adjusted with the engine running using a **timing light** (**Figure 38.19**). A timing light is a strobe light. The light turns on in response to an inductive trigger signal picked up from the number 1

spark plug cable. When the spark plug fires, the timing light flashes and "freezes" the crankshaft's position.

CAUTION The strobe makes the crankshaft look as though it is standing still. The parts are still moving and they are dangerous.

The ignition timing is changed by loosening the distributor body and rotating it in one direction or the other. On most cars, the distributor is turned until a timing mark on the front of the engine crankshaft pulley lines up with another mark on a tab on the timing cover.

Cars produced since 1972 have an underhood emission label that lists the timing specification. In order to be able to check base timing, or initial timing, on computer-controlled cars, a wire that gives information to the computer is usually disconnected. The wire is called a *by-pass wire*. Ford calls it a *spout* wire. The required procedure is described on the underhood label.

On old cars with points, the point gap can be adjusted with a feeler gauge. Position the distributor cam so that its lobe is against the rubbing block that holds the points open (**Figure 38.20**). When the cam turns to the position between the lobes, the points will be closed. Changing the point gap changes the ignition timing because it changes the dwell. If the new points are reinstalled with the correct gap, the timing should remain the same.

■ ELECTRONIC IGNITION DISTRIBUTOR SERVICE

Normal electronic ignition service on DI systems is limited to replacement of the rotor. Often, for preventive

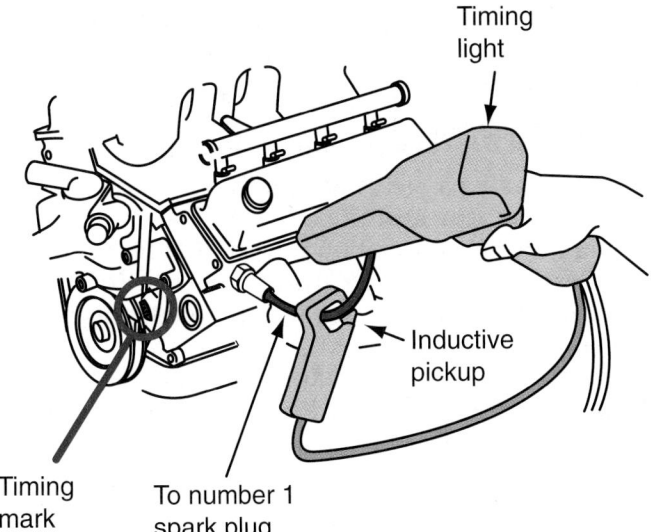

Figure 38.19 Checking ignition timing with a timing light.

maintenance purposes, both the cap and rotor are replaced.

No Spark

The term *ignition trigger* refers to either contact points or the electronic device that signals the module to make and break the primary circuit in the coil. A triggering mechanism and module take the place of the breaker points. Diagnosis of problems in those systems are covered here.

When there is no spark, the trigger might not be opening the primary circuit. Be sure that the distributor turns when the engine is cranked. A timing chain or belt could be broken or the distributor drive gear could be damaged. Another possibility is that the trigger is good but the module is defective. Although unlikely, a bad ignition coil could also cause a no-spark condition.

Module Tests

A bad module could also be a possible cause of a no-start, no-spark condition. The best test for a bad module is to test the other parts of the system first. If they are within specifications and are all functioning properly, the module is replaced. There are expensive testers available for modules, but they only work on specific systems.

First, check for power in the primary circuit. Check at the coil negative terminal with a digital logic probe or an electronic circuit tester. As the engine cranks, the test light should flash.

CAUTION If you use a regular 12V test light on a Hall circuit, too much current will flow through the parallel circuit provided by the light. Too much current flowing throughout the system can damage the electronic circuitry. A regular test light can be used with a magnetic trigger, but the clamp is connected to battery positive instead of ground. A logic probe (see Chapter 46) is the best way to test a Hall switch. Photoelectric sensors can be tested with a labscope.

Use a test light to see if there is power in the run and crank positions. A wiring diagram will tell you the correct wires to check. To test for power in the run position, with the key off disconnect the connections at the module. If there is no power in any of the following conditions, there is an open circuit that must be repaired.

CAUTION Do not disconnect an ignition module with the ignition switch on. A voltage spike can ruin the module.

Turn the key on and probe the power connection. The test light should light brightly. The (+) (battery) side of the coil should also light the test light when the key is in the run position.

Next, test for power in the start position. First, remove the connector from the module. Probe the connector that the wiring diagram says is for the start position while holding the key in the start position. The engine will crank unless you disconnect the wire to the S terminal on the starter solenoid. There should also be power at the battery side of the coil.

NOTE: *A defective coil can cause module failure. A shorted primary winding or open secondary winding can be the cause. Check the resistance in each coil winding (**Figure 38.21**). When primary winding resistance is less than specified, there is probably a short, allowing too much current flow. Current is what damages a module.*

SHOP TIP A coil can be tested while it is removed from the circuit using a spark tester. As shown in (**Figure 38.22**), connect one jumper wire from the coil (+) terminal to the battery. Connect one end of another jumper wire to the coil (−) terminal and the other end to a 0.22 microfarad capacitor. Another wire connects the end of the capacitor to ground. Take another wire and momentarily jump across both ends of the capacitor. When the wire is disconnected, the coil should create a spark across the gap of the spark tester.

Ignition modules are usually dependable. High underhood temperatures contribute to electronic failure, so modules are more prone to failure during very hot weather. On distributor-mounted modules, failing to correctly mount a module can result in its premature failure. They must be mounted securely with a layer of silicone dielectric grease under them to help keep them cool.

■ OTHER DISTRIBUTOR REPAIR SERVICE

Distributors have gears and bushings that can wear out or break. The cost of parts and labor has become too high for custom rebuilding of distributors in the service industry. Rebuilt distributors are commonly available from large rebuilders through local parts stores.

Sometimes the distributor drive gear can become stripped. It can be replaced by driving out a roll pin that is pressed fit into the bottom of the distributor shaft. The shaft bushing can wear out. Bushing wear will show up when checking ignition timing, causing the timing mark to be erratic. On a scope pattern, the start of the dwell section will vary.

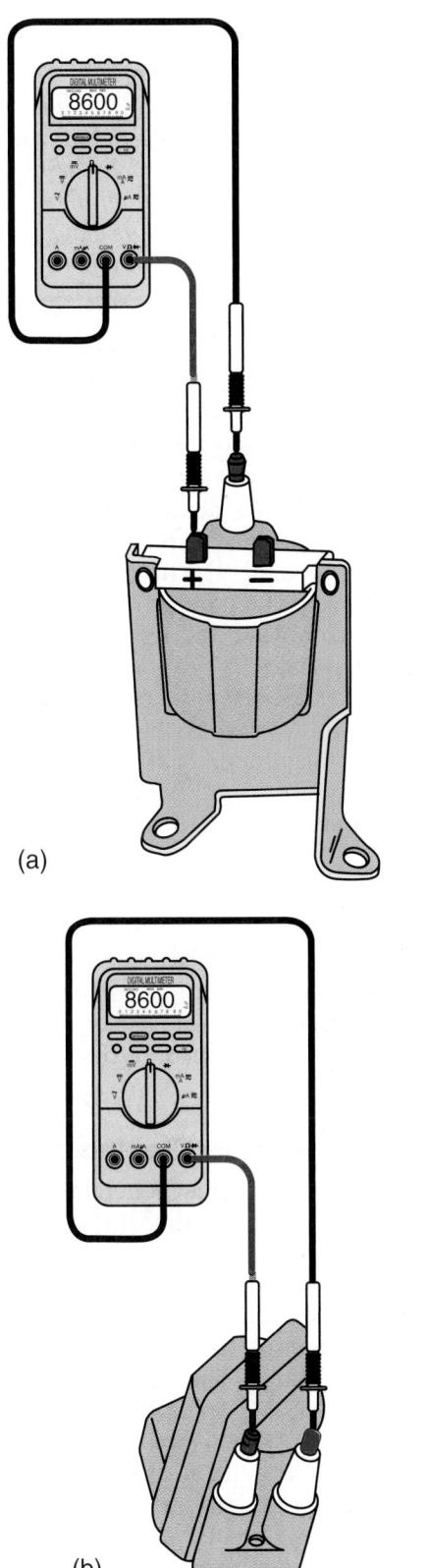

(a)

(b)

Figure 38.21 (a) Checking the resistance in coil windings. (b) Checking a waste spark coil.

■ DISTRIBUTOR INSTALLATION

Before installing a distributor, align the timing mark on the damper with the pointer on the timing cover.

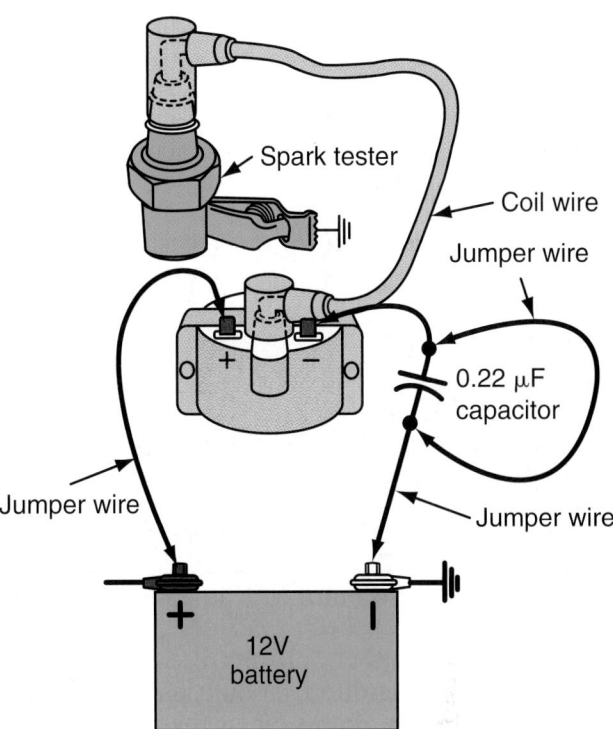

Figure 38.22 Test firing a coil using a spark tester.

When the marks are aligned, either the number 1 cylinder or its companion cylinder is at the top of its compression stroke. The spark plug should be ready to fire. If the distributor is installed 180 degrees off, backfiring will occur and the engine will not run. To determine whether the number 1 cylinder is on its compression stroke, remove its spark plug and slowly "bump" the engine over using the starter motor while feeling for escaping air, then align the timing mark.

Often on pushrod engines the distributor will not drop down all the way into the block because its bottom is not aligning with the drive lug on the oil pump. With the distributor in place, while holding gentle downward pressure, the engine can be cranked or turned by hand until the distributor and oil pump are aligned. Then the distributor can drop in the last ¼" or so until it is flush with the block.

SHOP TIP When a distributor is being installed during an engine reassembly, cylinder firing can be determined prior to installing the valve cover(s). To be sure that the number 1 cylinder is on its compression stroke, observe the rocker arms to see that they are both moving. Then, rotate the crankshaft one revolution until the rocker arms for the *companion* to cylinder number 1 are both moving at TDC. Align the timing mark and install the distributor with its rotor pointing to the number 1 spark plug cable in the distributor cap.

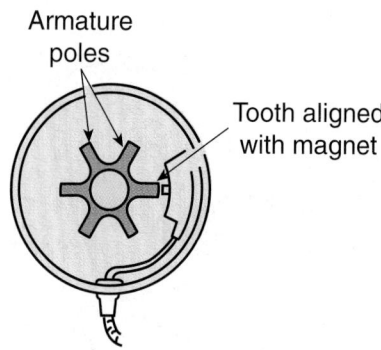

Figure 38.23 Timing an electronic ignition distributor. This one is from a six-cylinder vehicle.

▄▄▄ STATIC TIMING

Timing a distributor with the engine off is referred to as **static timing**. The following describes the procedure for timing a distributor with a permanent magnet generator.

Turn the distributor against its direction of normal rotation until the ignition module pulses. The electronic ignition pulse occurs when the armature pole, or trigger, lines up with the magnetic pickup in the distributor (**Figure 38.23**). The tooth on the armature must align perfectly with the pickup, or an error in timing will result.

The ignition system can be static timed by causing the ignition system to make a spark.

- With the number 1 cylinder on its compression stroke, align the timing indicator at the crank pulley to the desired timing specification.
- Install the distributor with the rotor pointing to the number 1 plug cable.
- Rotate the distributor body until the armature pole piece aligns with the permanent magnet.
- Tighten the distributor holddown.

The timing should now be set closely enough so that the engine will start right up.

▄▄▄ ELECTRONIC IGNITION SYSTEM SERVICE

Replacing a defective electronic component without determining the cause of the problem can result in a repeat failure of an expensive part. Determine the cause of the problem before replacing the part!

Ignition Problem Diagnosis

There is a logical way to go about determining the cause of a problem. If you start with the easiest things, you might find the cause of the problem quickly. First perform a visual inspection, looking for obvious problems like loose wires, broken vacuum hoses, or corroded terminals. Be sure to determine that battery voltage is sufficient.

Always check for technical service bulletins (TSBs). Others might have encountered the problem before you, and the problem might not be related to the ignition system.

CASE HISTORY *A vehicle had a misfire. The technician tested all of the ignition and fuel system components but could not find the cause. Finally, he checked for technical service bulletins in the service information library and found a bulletin saying that other misfire diagnoses had been caused by a faulty serpentine belt and tensioner.*

Anything attached to the crankshaft, like a loose damper or a loose torque converter bolt, could cause the PCM to inaccurately detect a misfire.

Sponsoring members of the International Automotive Technicians Network (IATN) can check their archives for the possible source of a problem.

If no obvious problems are found during the visual inspection, check the service library for TSBs. Then connect a scan tool to the data link connector and check for diagnostic trouble codes. Try to isolate the problem to one cylinder or to all cylinders. Electronic ignition system problems are usually common to only one or two cylinders. In a waste spark system, if a coil is malfunctioning, two cylinders will be affected. A problem that is currently occurring is known as a **hard fault**. Hard faults can be diagnosed and repaired. *Intermittent faults* often require guesswork. A scan tool can help locate an intermittent fault (see Chapter 46).

Signal Generator Testing

Different types of ignition system signal generators include crankshaft position sensors and distributor pickups. The different sensor types are common to both types of systems.

Pickup Coil Testing and Replacement

Permanent magnet sensors generate an AC voltage when the trigger wheel (reluctor) rotates past the permanent magnet sensor. Sensor signals have very low current. The sensor must be the correct distance from the trigger and wiring, and connections must be in excellent condition or the circuit will fail. Most sensors are not adjustable. The frequency of the signal is important to the PCM. The amount of voltage generated changes with speed, ranging from 0.2 volt during cranking to 100 volts at higher rpm. Interference in the signal is a possibility, so check the condition of the wiring harness. Twisted pairs of wires should not be unwound during a repair, and alternator wiring or spark plug cables should not pass close to the wires. Be careful not to switch wires during a repair, or a driveability problem could result.

Use an ohmmeter to test the continuity of the pickup coil (**Figure 38.24**). There are only two leads, so this is a simple procedure. Typical resistance readings range from 150 to 1,500 ohms. You can also test the

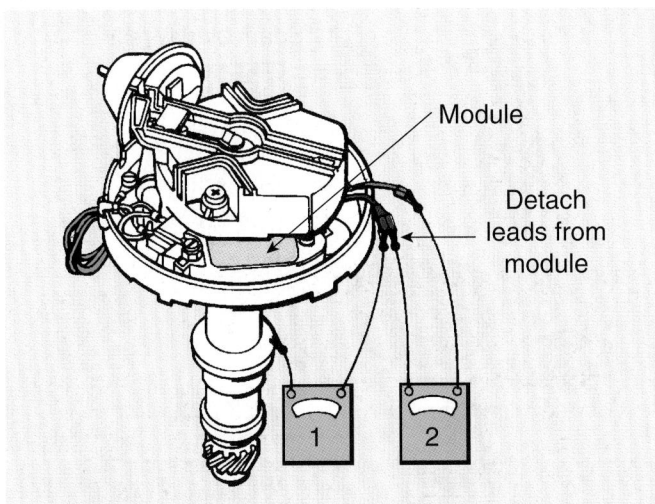

Figure 38.24 Checking a pickup coil with an ohmmeter.

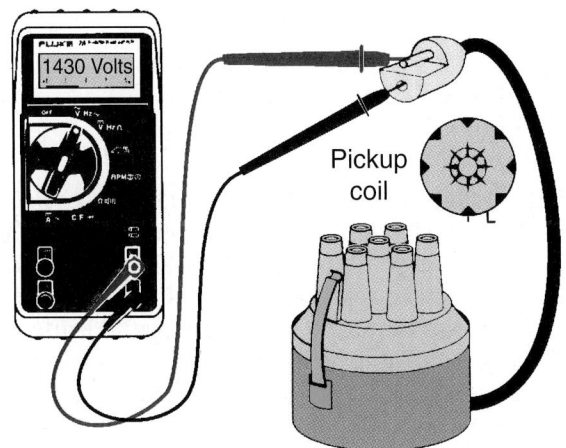

Figure 38.25 Checking a pickup coil with a voltmeter.

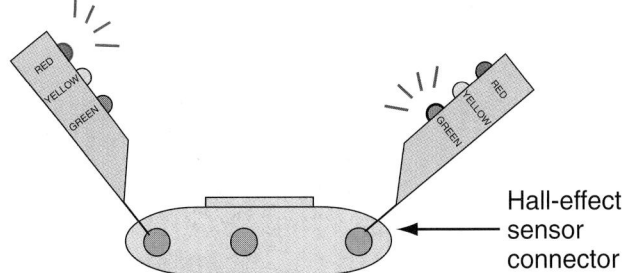

Figure 38.26 Testing a Hall-effect switch with a logic probe. On one side, the red light should come on. The green light should illuminate when the probe is touched to the other side.

pickup coil by cranking the engine with a multimeter connected (**Figure 38.25**). Look for an AC signal of at least 0.1 volt. If there is an AC signal, it will trigger the module. A lab scope will show a sine wave.

Hall Switch Testing

A Hall-effect switch has three wires: battery voltage, an output signal, and ground. It produces a voltage signal that varies according to whether or not there is a magnetic field in the circuit. When the interrupter ring tooth moves between the magnet and the semiconductor, Hall voltage drops off, and vice versa.

To generate a signal, a Hall switch requires voltage and ground. The Hall switch can be tested using a logic probe (**Figure 38.26**). There are three wires on a Hall switch. With the key on, connect the logic probe to each of the end wires. The red light should illuminate on one of the outside connections, indicating battery power. The green light should illuminate on the other outside connection, indicating ground. Backprobe the center connection and test it with the engine cranking.

All three lights should flash. The yellow light is flashing as the voltage switches back and forth between low and high. When voltage is high, the red light comes on; and when voltage is low, the green light comes on.

Hall-effect failures include broken or damaged teeth and dirt buildup. Poor connections can also cause problems. The problem can be intermittent, which makes it more difficult to diagnose. Experienced technicians like to watch the lab scope during a "wiggle test," where wires are manipulated while observing the pattern on the scope.

Transistor Problems. The transistor in the Hall sensor must be able to pull the reference voltage to ground. A bad sensor might not give a strong enough signal for the transistor to be able to complete the circuit to ground. Transistor problems tend to happen when the temperature is high.

Circuit Testing. Check to be sure the system had power. The ignition switch is a common cause of power supply problems. The sensor makes a 5-volt reference signal. With the key on, turn the crankshaft and watch when the sensor switches high. It should be near 5 volts.

Check the voltage drop in the ground side of the circuit between the battery negative terminal and the ground side of the switch. It should be less than 50 millivolts, although the sensor might still work when there is higher resistance in the circuit.

A graphing multimeter is good for testing Hall-effect output because it can display the sensor's frequency digitally over time so intermittent signal drops are visible. The transistor can stick low or high, or it can flutter between power and ground. Graphing multimeters are discussed in Chapter 46.

Air Gap Measurement

A few crankshaft position sensors have a provision for adjustment. Many sensors are not adjustable, but the air gap must still be correct or the sensor is replaced. Some manufacturers have special tools to check air gap, and others use a feeler gauge (**Figure 38.27**). Check at three locations 120 degrees apart to determine if there is excessive runout in the sensor ring.

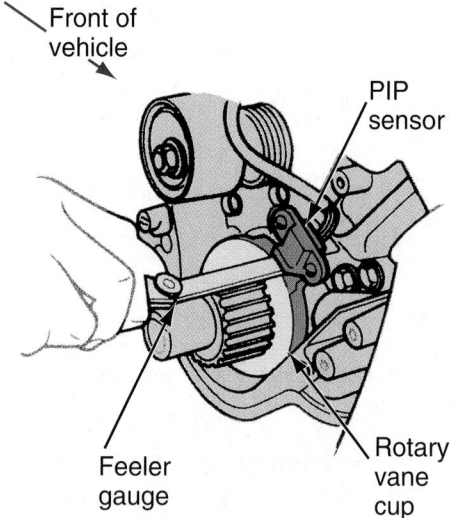

Figure 38.27 Checking crankshaft position sensor air gap. *(Courtesy of Ford Motor Company)*

Magnetoresistive sensor and optical sensor signals are the same as Hall sensor signals. Use the same procedures to test for battery voltage and ground, average voltage output, and signal frequency. Optical sensors do not have the same problems of transistor sticking and fluttering that plague Hall sensors. A big problem with these sensors is oil, coolant, or dirt filling or covering the slots.

Scopes

Oscilloscopes are helpful tools for pinpointing ignition and engine problems. Older analog scopes were large, but modern digital storage oscilloscopes (DSOs), commonly called lab scopes, are much smaller and have different capabilities (**Figure 38.28**). DSOs show a simulated pattern on a display screen. Input wires to the scope are attached to the battery, the number 1 spark plug, the ignition coil, and ground (**Figure 38.29**).

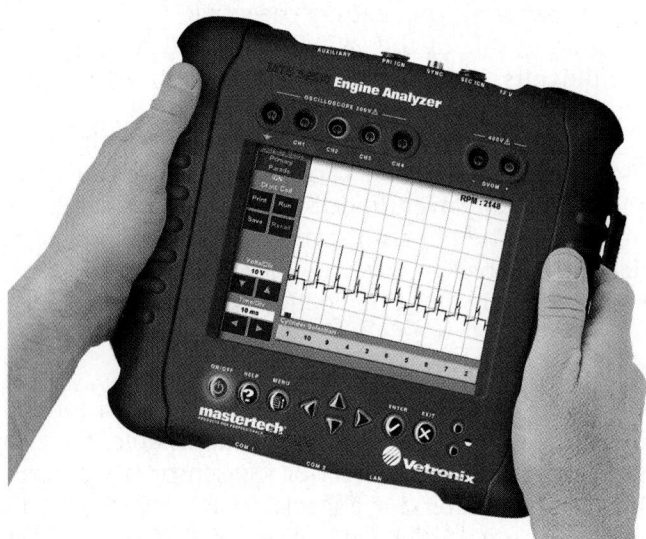

Figure 38.28 A digital storage oscilloscope. *(Reprinted with permission from Bosch Diagnostics)*

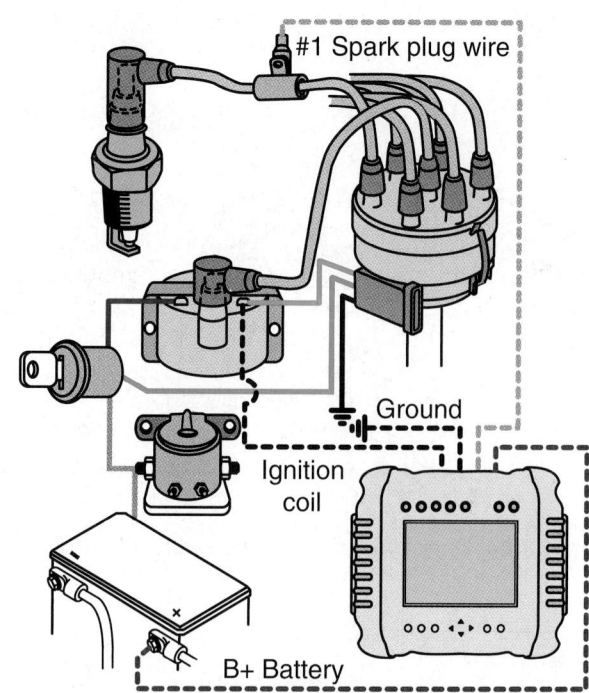

Figure 38.29 Typical scope connectors for a distributor ignition system.

The image on the screen is called a **scope pattern**. A scope is a high-speed voltmeter with a built-in time frame. Vertical spikes indicate voltage and the horizontal pattern represents time (**Figure 38.30**).

An ignition pattern displays the initial firing of the spark plug followed by the rest of the ignition cycle,

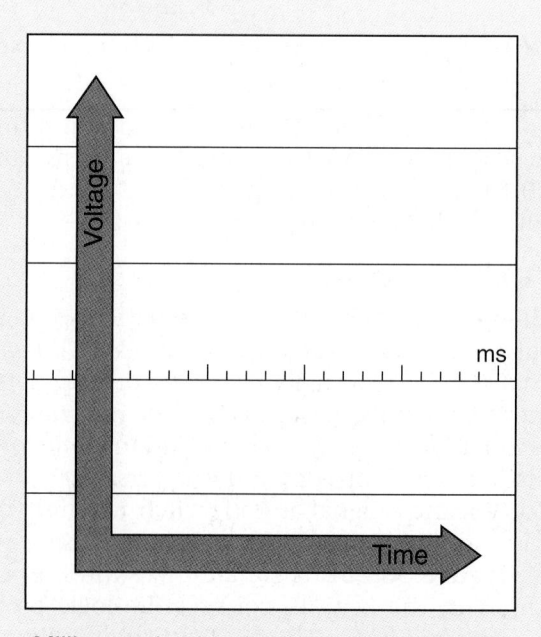

Figure 38.30 A scope is a voltmeter that also measures time.

leading up to the next firing. A technician can analyze the pattern, making comparisons between cylinders to determine which cylinder(s) are causing a problem. Electrical problems, combustion problems, and some engine problems can be diagnosed.

Ignition Patterns

Different types of scope patterns can be selected. All ignition systems use a primary winding to induce ignition in the secondary winding, but with many manufacturers primary voltage is not accessible. When it is accessible, the secondary pattern is the one most often used for ignition system diagnosis. It also indicates what is occurring in the primary side of the system. The secondary voltage is not induced by a voltage but by an electromagnetic force resulting from primary voltage.

Distributor ignitions usually have external coils, which make secondary testing easy. Systems with an internal coil, like the older GM HEI system, had a coil incorporated into the distributor cap, which required a special scope pickup that fit over the distributor cap.

Waste spark systems need special adapters. Their waveform is not too stable, because the ground moves back and forth due to system design. Some waste spark systems have coils hidden within cavities in a valve cover and require an inductive wand or other special testing.

Traditionally, analog oscilloscopes have always displayed the pattern with the *secondary firing section* as the first part of the pattern (**Figure 38.31**), with the firing line displayed at the end of the pattern. DSO patterns can be displayed with the firing section anywhere on the screen, and the firing section is often displayed in the center of the pattern (**Figure 38.32**).

The upward line that starts the firing section is called the **firing line**. Voltage builds up in the ignition coil until it is greater than the secondary resistance against it. This is the voltage required to start the spark across the spark gap, called ionizing the gap. The firing line will usually be somewhere between 5 kV and 15 kV.

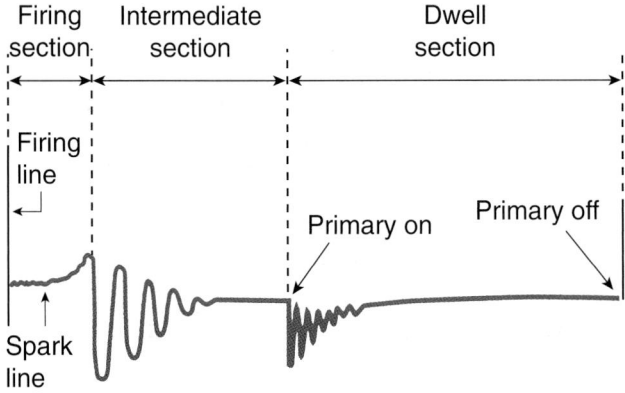

Figure 38.31 The dwell section of the pattern represents the time that the primary current is switched on.

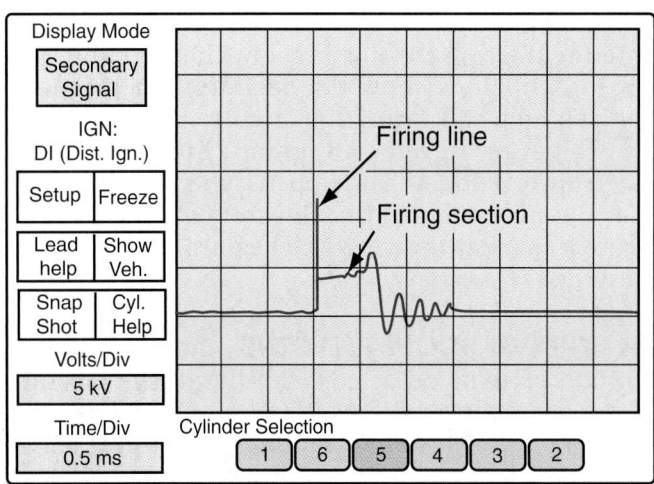

Figure 38.32 A single cylinder secondary ignition pattern on a DSO.

NOTES:

- *Higher spikes indicate more resistance to spark. This can be due to a lean air-fuel mixture, a wide spark plug gap, an open circuit in a spark plug cable, or higher compression because of carbon buildup.*
- *Lower spikes indicate lower resistance to spark. This can be due to a spark plug gap that is too narrow, a rich air-fuel mixture, or compression that is too low.*

Spark Line

The **spark line** is a horizontal line that begins at the voltage level where electrons start to flow across the spark plug gap. This horizontal pattern, made during the combustion process, indicates the continuous firing voltage requirement of the spark plug. A technician can use this to determine conditions in the combustion chamber.

The length of the spark line shows how long the spark actually lasts. The spark drains the voltage off of the coil secondary winding. When the voltage lowers to a certain level where it can no longer flow, this part of the pattern stops. The duration of the spark should be somewhere between 1 and 2 milliseconds on the scope pattern.

Intermediate Section. The *intermediate section* is the next part of the pattern. It has oscillations of the remaining voltage in the coil as it dissipates back to zero. Notice how the pattern's oscillations become lower and lower. Distributor ignitions should have at least three oscillations. Distributorless ignitions may have fewer, and sometimes no oscillations could be normal.

Dwell Section

The dwell section of the pattern represents the time that the primary current is switched on (see Figure 38.31). A short downward line shows where current starts to flow into the coil. As current is entering the coil, the line is below the zero line. The small oscillations show the beginning of magnetic field buildup. As saturation occurs, the horizontal line comes up to zero. An upward line is the point where dwell ends. What happens when dwell ends is that the magnetic field in the coil collapses, resulting in a spark at the plug (the beginning of the next pattern).

▄▄ TYPES OF SCOPE PATTERNS

Ignition patterns on a scope are the raster, superimposed, and display patterns. The secondary parade pattern is not available on COP and coil-near-plug (CNP) systems. CNP has a short section of plug wire exposed so individual secondary testing is possible.

Raster Pattern

A **raster**, or *stacked*, **pattern** displays all of the cylinders, one above the next (**Figure 38.33**). The voltage firing line is difficult to see when using this pattern. Patterns for each individual cylinder are arranged in the firing order and are sometimes displayed from bottom to top. The pattern shown is for an eight-cylinder engine with the firing order 1-8-4-3-6-5-7-2. The scale on the bottom represents percent of dwell. Some scopes measure this in degrees.

Superimposed Pattern

A **superimposed pattern** compares all of the cylinders while their patterns are displayed one on top of

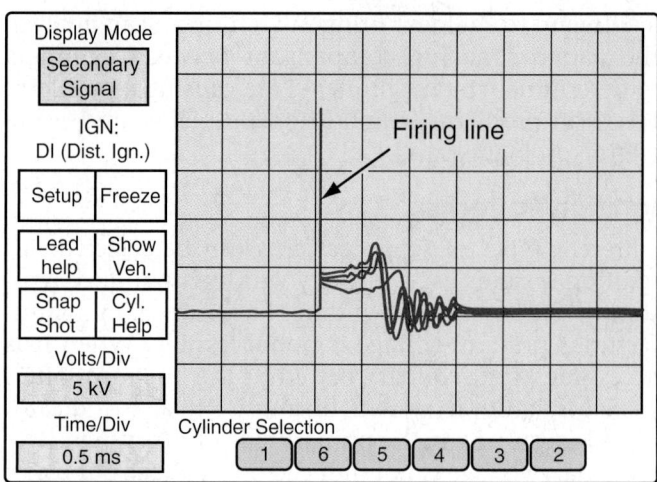

Figure 38.34 A superimposed pattern shows all of the cylinders' patterns on top of each other.

the other (**Figure 38.34**). If one cylinder is different from the others, it will be easy to see.

Display or Parade Pattern

A *parade* or **display pattern** displays all of the cylinders next to each other side by side so that the heights of the firing lines (voltage spikes) can be compared. Patterns from each cylinder are displayed in the order of the engine's firing order from left to right (**Figure 38.35**).

NOTE: *In a display pattern, many scopes display the number 1 spark plug firing spike at the far right, with the rest*

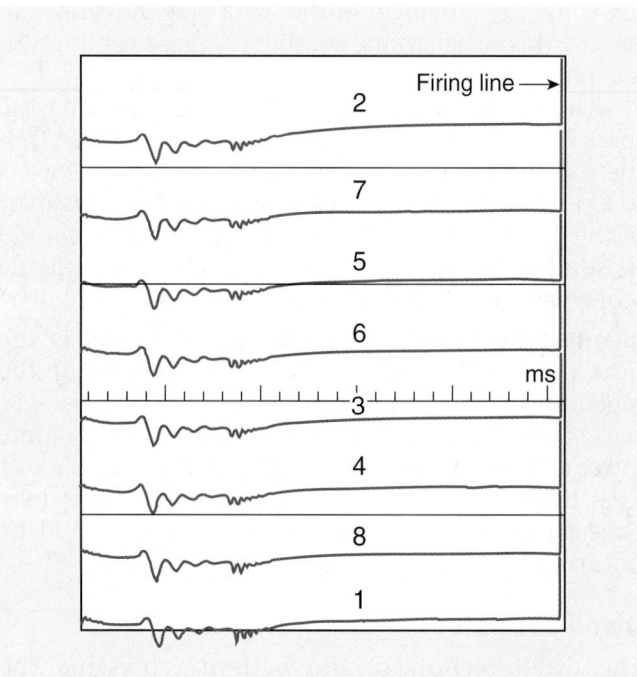

Figure 38.33 A raster pattern in the engine's firing order.

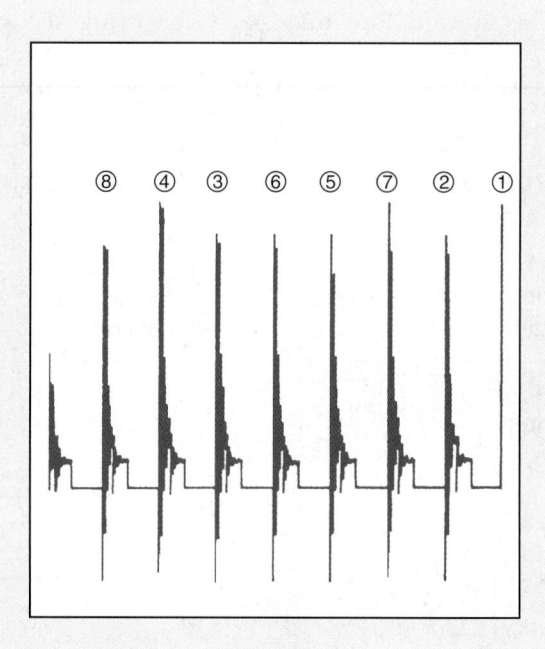

Figure 38.35 Uneven firing voltages of more than 3 kV in a parade pattern indicate a problem.

of the number 1 pattern at the left. The firing order shown here is 1-8-4-3-6-5-7-2.

■ OSCILLOSCOPE TESTS

With older contact point systems, a popular test was the available voltage test. A plug cable was removed using insulated pliers while the engine idled. A coil could put out somewhere around 20,000 volts. Ignition system energy requirements have increased significantly. Today's engines have high swirl combustion chambers and use exhaust gas recirculation (EGR) to dilute the air-fuel mixture with inert gas. Today's systems can produce in the neighborhood of 100,000 volts. On modern electronic systems, the open circuit test is no longer advisable. It can result in a breakdown of the insulation of the coil and overload the ignition coil module. Today, the special spark plug tester is used (**Figure 38.36**). It requires about 35 kV to fire this special plug.

Connect the spark plug tester to a spark plug cable. Then, crank the engine and watch the scope pattern. If the spark jumps the gap, the system has enough available voltage. If it does not, the voltage can be read as less on the scope. Battery voltage can be too low, the module could have excessive resistance, or the coil could have a problem.

Reading a Parade Pattern

Run the engine at 1,200–1,500 rpm for oscilloscope tests. Select the parade pattern and look at all the firing lines to see that they are within 3 kV of the height of each other (see Figure 38.35). With new spark plugs, the firing voltage will be lower. As the gap becomes larger with wear, the firing voltage requirement will increase.

A leaner air-fuel mixture causes a higher resistance to spark, requiring more voltage to jump the plug gap. Watch the height of the firing lines. A poorly performing cylinder with a high firing line can indicate a lean air-fuel mixture; a low firing line can indicate low compression.

Acceleration Load Test. Rapidly snap the throttle open and closed. Engine rpm does not need to climb above 2,000 for this test. This puts a load on the ignition system so you can see if it misfires. The firing lines will all rise equally as more load is put on the system. Again, the difference between the cylinders should be no more than 3 kV.

Reading a Raster Pattern

The *spark section* of the pattern is viewed from the raster or superimposed pattern selection. The spark sections should all be the same length and have a slight downward slope. The length of this line gives an indication of the amount of voltage in reserve. This part of the pattern gives an indication of what is going on in the combustion chamber. Further investigation is called for when not all cylinders are the same (**Figure 38.37**). For example:

- When the slope starts at the top of the firing line, a fouled plug is indicated.
- An upward slope in this section can indicate that the combustion chamber is receiving less fuel or there is high plug resistance.
- Normal turbulence in the air-fuel mixture can show oscillations in this line.

The intermediate section is the next part of the pattern. There should be five to seven gradually diminishing oscillations. If there are fewer, there could be shorted windings in the coil.

The *dwell section* is the next part of the pattern. The length of the dwell section will vary with design.

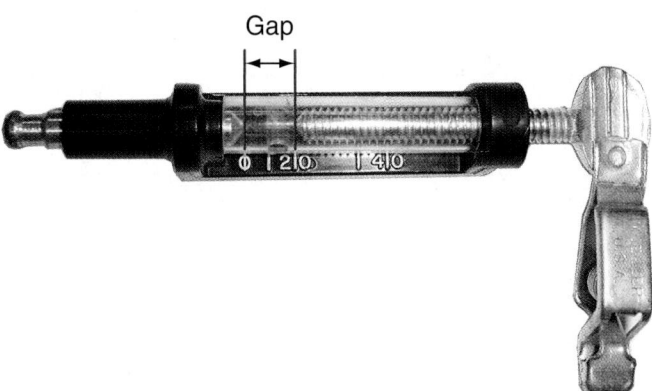

Figure 38.36 An adjustable spark tester. *(Courtesy of Tim Gilles)*

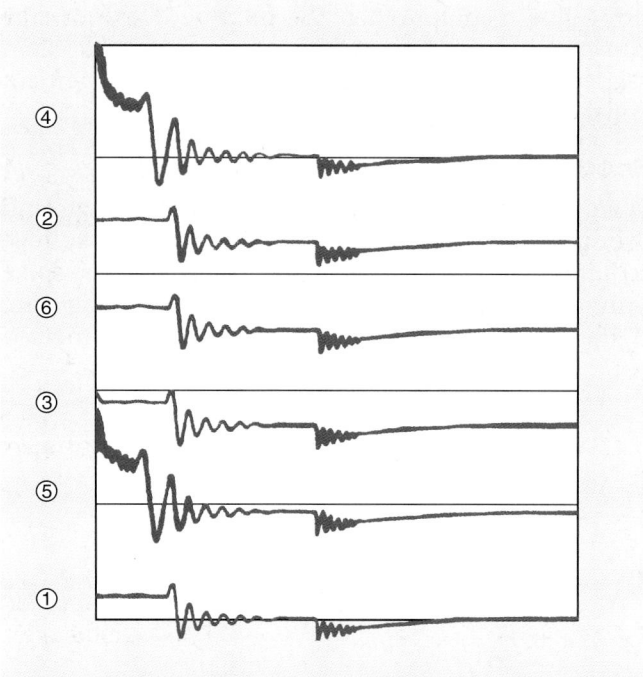

Figure 38.37 Two cylinders have problems. If the firing order is 153624, then cylinders number 5 and 4 are the ones to check.

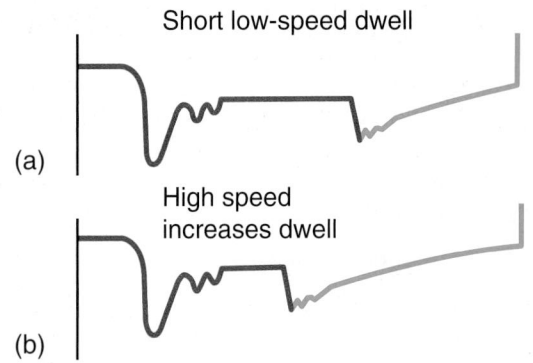

Short low-speed dwell

(a)

High speed
increases dwell

(b)

Figure 38.38 Dwell varies with engine speed.

To observe dwell variation, look at the dwell section of the pattern (**Figure 38.38**). A worn distributor shaft bushing can cause dwell to jump around. When the throttle is snapped, the dwell will often stabilize.

Dwell varies with engine speed. If the engine is running faster, the coil needs more time to saturate and so dwell is longer. Raising engine rpm will lengthen the dwell section of the scope pattern. When an ignition system does not have variable dwell, a faulty module is usually the cause.

When a system does not have a primary resistor, it has a current limiting function. There will be an additional hump in the dwell section of these patterns (**Figure 38.39**).

Scope Diagnosis of Engine Problems

Engine problems can show up in a scope pattern, too. Sticking valves and air leaks can show up in the scope's firing line section. When the mixture does not stay constant, the problem will be intermittent. A problem related to electrical resistance would tend to remain constant.

Scope Diagnosis of Ignition Problems

When the engine is off, use a ground probe to ground the spark plug end of a plug cable. The firing voltage requirement will drop, leaving the height of the spike only as the amount remaining for the rotor air gap. If the pattern is upside down, the coil is connected

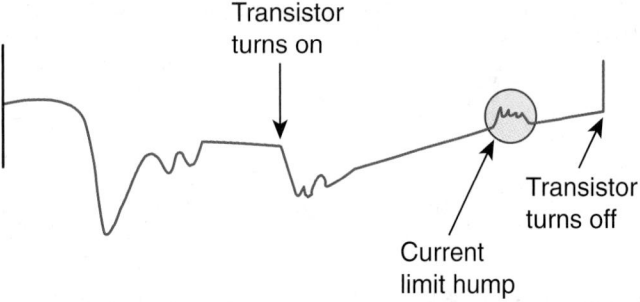

Transistor
turns on

Current
limit hump

Transistor
turns off

Figure 38.39 A pattern for an ignition system with current limiting.

backwards (reversed polarity). Reversed wires on waste spark systems can cause misfire detection.

Stress Test

Temperature can cause an electronic ignition to fail. To **stress test** parts like a module or pickup coil, parts can be cooled or heated. Direct compressed air at the base of the distributor to cool it. With the scope set on raster pattern, watch for changes in the dwell zone.

Watch the firing line and spark line while spraying water on the distributor cap and plug cables. Spark plug cable insulation can break down under moisture or load. To create a simulated load, you can put the engine under a light load with the transmission in gear and the brakes applied. Then bring the rpm up to approximately 2,000.

Problems related to heat will usually be found in the dwell section of the pattern. You can use a heat gun to warm up a module, trigger pickup, and electronic ignition connectors.

Primary Voltage Ignition Pattern

The secondary pattern is the best pattern for displaying an ignition system because it shows primary as well as secondary current flow. It can do this using induced voltage in the secondary because primary current only flows at times when the secondary part of the ignition system is not active. When the secondary pattern is not accessible (COP systems and combination COP system and DIS), primary voltage testing is sometimes used. On a CNP system and DIS primary voltage is not available or is tough to access, but secondary voltage is accessible.

The beginning of a primary voltage pattern shows when the switching transistor turns off, ungrounding the coil to stop current flow through the primary winding (**Figure 38.40**). The scope's voltage setting has been set to a maximum of 200 volts, so we do not see the total peak firing voltage. Primary peak firing voltage is much less than secondary, over 350 volts instead of many thousands, but primary voltage can

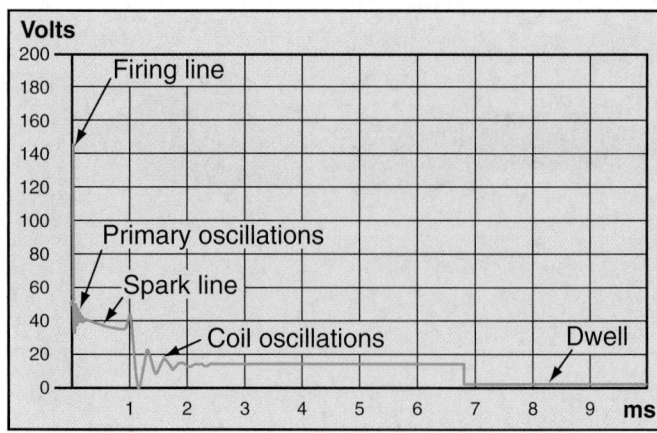

Figure 38.40 Parts of a primary ignition waveform.

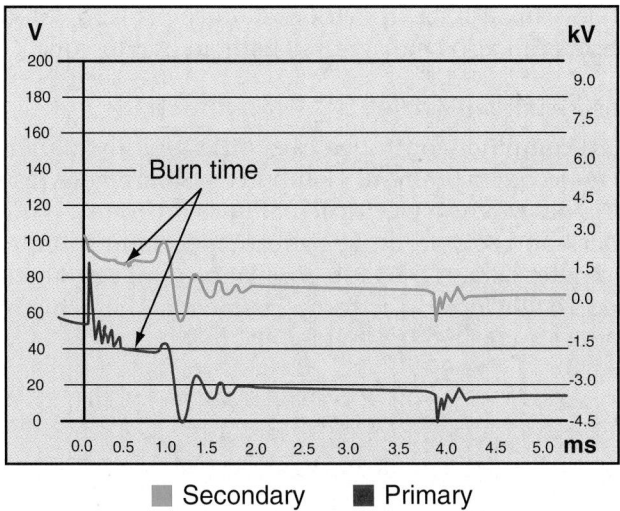

Figure 38.41 Comparison of primary and secondary ignition patterns.

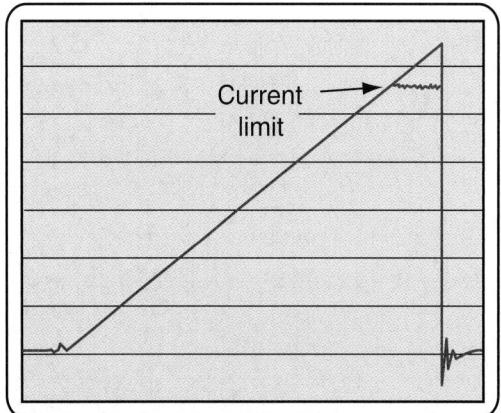

Figure 38.42 A coil current ramp using a low-amp current probe. Some patterns ramp all the way up, and others are current limited.

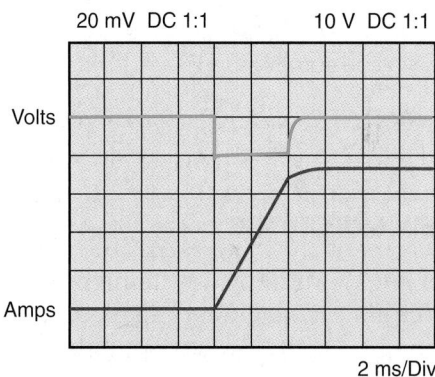

Figure 38.43 A dual trace DSO reading comparing current and voltage.

be correlated with secondary voltage and they are similar in appearance.

Compare different cylinders to see that they have the same firing voltage requirements. In **Figure 38.41**, you can see the spark line running at about 40 volts for about 1 millisecond. This *burn time* is normal. You can also see primary ignition oscillations at the start of the spark line that are not visible in a secondary pattern. This is because the secondary pattern only shows the primary pattern when nothing is happening in the secondary circuit. Primary and secondary waveforms of the same ignition system are compared in this figure.

Primary Current Ramp Test

Testing the resistance of coil windings is the traditional way of determining whether the coil has the ability to make a spark. But this test does show if the coil will have the correct amount of current flow when the system is under load. A low-amp current probe can be clamped around a coil's power or ground. This will allow you to watch the current, both as it ramps up and at its peak. Specifications are not usually available, but you can make comparisons with the system's other coils. All of the patterns should look the same. Some coils are called ramp and fire. The pattern is a straight ramp to the top (**Figure 38.42**). When the ramp levels off at the top, the coil is current limited.

The DSO can be set to show a dual trace, one for amps and the other for volts (**Figure 38.43**). Current can be tested anywhere in the circuit. To connect for voltage, use the negative terminal of the coil.

Vacuum Waveforms

A vacuum transducer can be connected to a DSO to test for ignition problems. It measures vacuum in each cylinder and is triggered using the firing order by clipping an inductive pickup over coil number 1 or the number 1 fuel injector. Problem cylinders will have less vacuum than others. This test can be a time-saver when attempting to diagnose a misfire on COP ignitions. During a brake torque test, which is a stall test of less than 5 seconds, a defective COP coil will arc beneath the coil, causing vacuum to drop on that cylinder.

When connecting a vacuum transducer on a V-type engine, look for a vacuum source that is not oriented to one bank or another. If you tap into one plane of the manifold vacuum, readings will not be accurate. Be sure to connect with a short hose, as close as possible to the vacuum source, or "ringing" in the hose will affect the vacuum reading.

▄ OTHER SCOPE TESTS

Tests of individual components can be done with the scope. When checking a trigger on an oscilloscope, a Hall switch, contact points, and an LED all give a square wave. The pickup coil gives an AC sine wave that increases with speed (**Figure 38.44**). An alternator

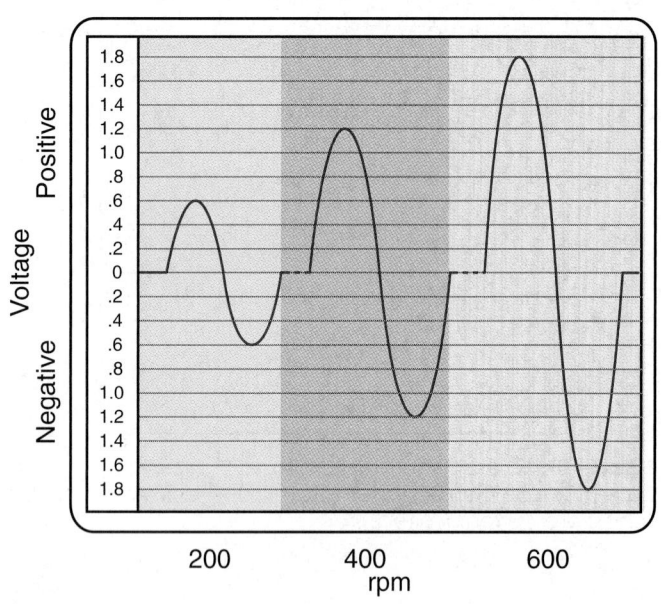

Figure 38.44 The pickup coil gives an AC sine wave that increases with speed.

diode can also be tested. Fuel injectors and knock sensors can also be diagnosed with an oscilloscope.

■ OTHER DIAGNOSTIC INSTRUMENTS

With computer-controlled cars, other test instruments have become common. Computer systems store trouble codes when an electrical malfunction occurs. Hand-held scan tools are used to read codes and interpret computer data. A DSO is a popular piece of equipment used to diagnose electronic problems. These instruments are covered in Chapter 40.

■ REVIEW QUESTIONS

1. In an engine running on the highway at 3,000 rpm, each cylinder's spark plug will be firing _____ times every second.
2. The spark plug thread and thread seat is where over _____ of the plug's heat dissipation occurs.
3. A spark plug that has a buildup of carbon that shorts it out is called _____.
4. Where should anti-seize compound be installed on the spark plug?
5. Tighten a gasketed plug finger tight, then _____ turn further.
6. The normal range of resistance in a new spark plug cable is about _____ to _____ ohms per foot.
7. Intake manifold _____ is used to sense the load on the engine.
8. Vertical spikes on the screen indicate _____ and the horizontal pattern represents time.
9. A _____ or parade pattern displays all of the cylinders next to each other side by side.
10. If the scope pattern is upside down, what is wrong?

■ ASE-STYLE REVIEW QUESTIONS

1. Which of the following could cause engine damage?
   a. Using a spark plug with too hot a heat range
   b. Using a spark plug with too cold a heat range
   c. Both A and B
   d. Neither A nor B
2. Two technicians are discussing DIS waste spark ignition systems. Technician A says that half of the plugs fire toward the ground electrode. Technician B says that only half of the negative electrodes will exhibit wear. Who is right?
   a. Technician A   c. Both A and B
   b. Technician B   d. Neither A nor B
3. An engine has one overheated spark plug. Technician A says that this can be due to incorrect spark plug torque. Technician B says that the fuel injection system could be running too lean. Who is right?
   a. Technician A   c. Both A and B
   b. Technician B   d. Neither A nor B

4. Which of the following can be repaired using a thread insert?

  a. Gasketed seat spark plug threads

  b. Tapered seat spark plug threads

  c. Both A and B

  d. Neither A nor B

5. Technician A says that some manufacturers recommend smearing some silicone dielectric compound on the tip of the distributor rotor. Technician B says to put a very small amount of anti-seize compound on spark plug threads used in some aluminum heads. Who is right?

  a. Technician A   **c.** Both A and B

  b. Technician B   **d.** Neither A nor B

6. Technician A says that a distributor rotor must be aligned in the correct position on the distributor shaft. Technician B says that a distributor cap must be installed on the distributor in only one position. Who is right?

  a. Technician A   **c.** Both A and B

  b. Technician B   **d.** Neither A nor B

7. All of the following signals can be used to sense the load on an engine *except*:

  a. Engine intake manifold vacuum

  b. Engine rpm

  c. Mass airflow

  d. MAP

8. A four-cylinder engine's ignition trigger must switch the coil primary on and off _____ times per second at 3,000 rpm.

  a. 25   **c.** 100

  b. 50   **d.** 200

9. Which of the following could be caused by a shorted winding in an ignition coil?

  a. A bad spark plug

  b. A bad spark plug cable

  c. A failed ignition module

  d. None of the above

10. Which of the following patterns do not show voltage spikes?

  a. Raster   **c.** Parade

  b. Display   **d.** All of the above

CHAPTER 39

Fuels, Alternative Fuels, and Advanced Transportation Technologies

■ OBJECTIVES

Upon completion of this chapter, you should be able to:

✔ Understand how petroleum is refined.

✔ Describe the different characteristics of various blends of gasolines.

✔ Know the effects of the different types of abnormal combustion.

✔ Decide on the best choice of gasoline or diesel fuel for a vehicle.

✔ Diagnose rich and lean fuel mixture problems.

✔ Describe the advantages and disadvantages of various types of alternative fuels.

■ KEY TERMS

cloud point	dieseling	carbonaceous deposits
cetane number	flame front	octane
volatility	abnormal combustion	gasohol
vapor lock	detonation sensor	LPG
RVP	knock sensor	CNG
surge	oil-based carbon deposits	

■ INTRODUCTION

Motorists often have questions about the fuel used in their cars. There are several kinds of fuels used in motor vehicles. A service technician should have a basic understanding of these fuels and their characteristics. This chapter deals with gasoline, diesel, and alternative fuels. Also included are discussions of rich and lean air-fuel mixtures and abnormal combustion. These conditions can result in engine damage, poor fuel economy, and poor performance (driveability problems).

■ CRUDE OIL

Raw petroleum, also called crude oil, is used to make such products as gasoline, diesel fuel, motor oil, solvents, and liquified petroleum gas (LPG), among others. **Figure 39.1** lists many of the products made from oil.

Crude oil is pumped from wells in the ground (**Figure 39.2**). It is a variable combination of large and small hydrocarbons together with petroleum gas, hydrogen sulfide, and water. The surplus petroleum gas can be used directly for fuel for the refinery.

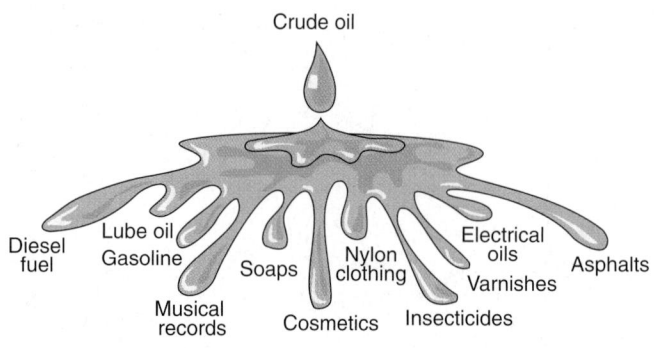

Figure 39.1 Some of the products made from oil.

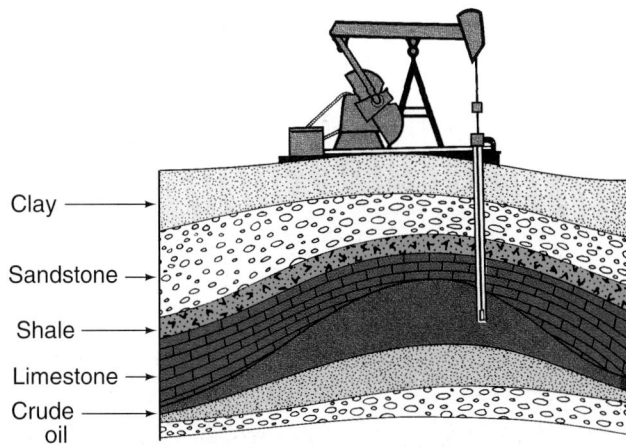

Figure 39.2 Oil is pumped from deep in the ground. *(Courtesy of Toyota Motor Sales, U.S.A., Inc.)*

Clay
Sandstone
Shale
Limestone
Crude oil

Figure 39.3 Fractionating or distillation towers are the tall columns that highlight the skyline of oil refineries. *(Courtesy of Chevron Texaco)*

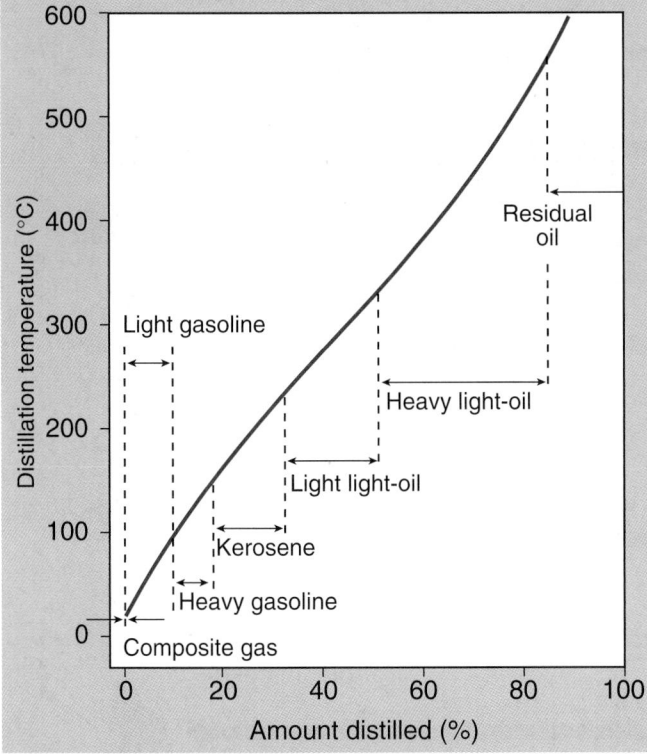

Figure 39.4 Crude oil fractions have different boiling points ranging from about 100°F to 700°F.

The liquid *hydrocarbons*, consisting of approximately 12% hydrogen and 82% carbon, are of many kinds. They are like a keg of nails of many different sizes all mixed together. They must be sorted before they can be used. Hydrogen is a light gas vapor. Carbon is a heavy black solid. These materials as well as hydrogen sulfide and water are removed during refining.

Oil is separated into many useful products at the oil refinery. During refining, the crude is first heated by pumping it through pipes in hot furnaces into a fractionating column. Light hydrocarbon molecules are separated from heavier ones in these tall narrow columns sometimes called distillation towers. They are what make up the distinct skyline of oil refineries (**Figure 39.3**).

The refining process breaks the crude down into different parts called *fractions*. Crude is not a single substance, like water. Water has one single boiling point (212°F). Crude oil fractions have different boiling points, ranging from about 100° to 700°F. (**Figure 39.4**).

The fractioning tower has draw pipes at different heights for pulling the desired petroleum materials out of the tower (**Figure 39.5**). The lightest products are a

Figure 39.5 Draw pipes at different heights pull lighter and heavier materials from the tower. *(Courtesy of Chevron Texaco)*

gas at room temperature. They are taken from the top of the fractioning tower as the crude is boiled in the refining process (**Figure 39.6**). After that comes gasoline. The remaining fractions have higher and higher boiling points. The heaviest products boil last and are taken from the bottom where the temperature is the highest. They are tarlike and are at "the bottom of the barrel."

Some of the fractions are used as raw materials to be blended into gasoline to correct octane, emissions, volatility, and storage life. The next fractions, such as kerosene or diesel, can sometimes be used directly. The last fraction, which has a very high boiling point, can be used as fuel oil or as a basic stock for lubricants. They can also be reformulated by heat or catalysts to produce more of the lighter stocks. The distillation process is repeated in other plants as the oil is further refined for its most efficient use (**Figure 39.7**).

Fractions from different geographical locations vary widely. Some crudes are light in color and flow easily. Other heavier crudes must be heated to make them flow. Some fractions are only useful for a particular product, such as gasoline or diesel. Pennsylvania-grade

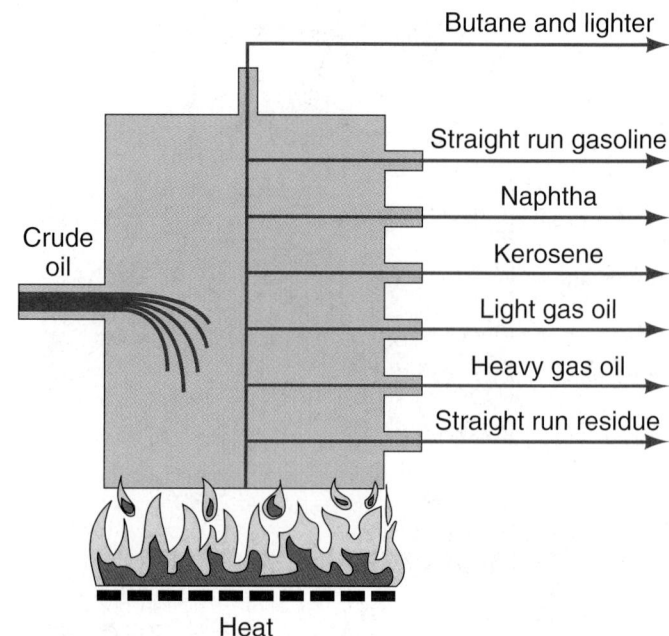

Figure 39.6 Lighter products come off the tower closer to the top. *(Courtesy of Toyota Motor Sales, U.S.A., Inc.)*

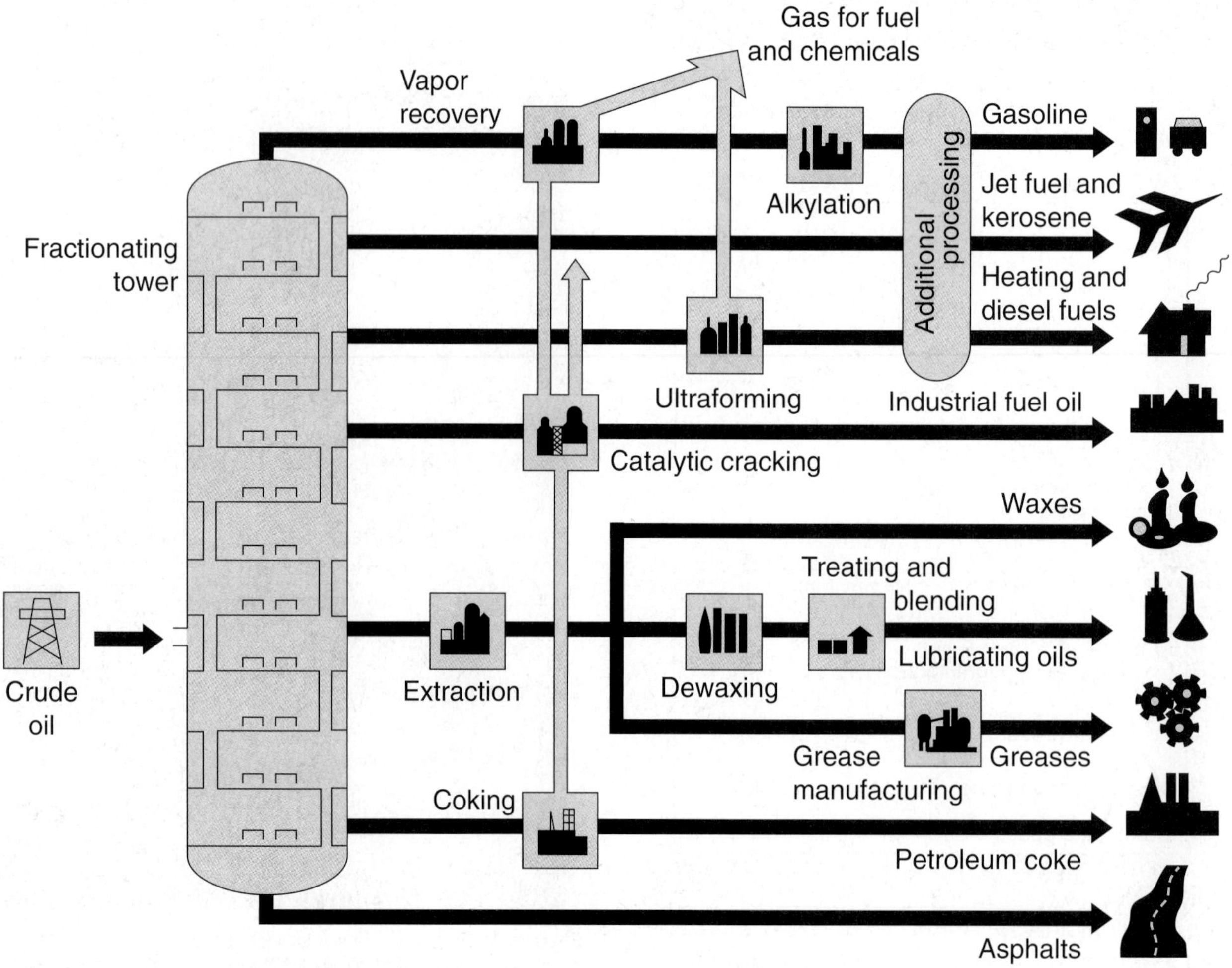

Figure 39.7 Oil refining process. *(Courtesy of American Petroleum Institute)*

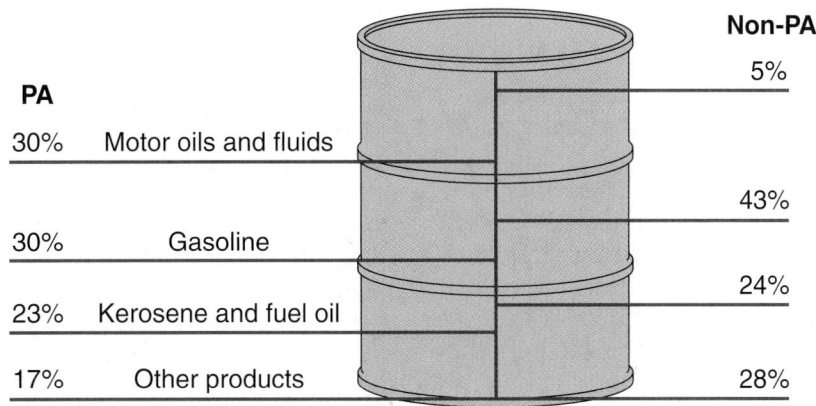

Figure 39.8 Average yields from a U.S. barrel of crude oil. The left side (PA) is Pennsylvania grade crude. The right side is an average of all others. *(Courtesy of Toyota Motor Sales, U.S.A., Inc.)*

crude, found only in Pennsylvania, New York, West Virginia, and Ohio, yields more high-quality lubricating oil. Crude oil is measured and sold by the barrel, which is 42 gallons. **Figure 39.8** shows the percentage of various products coming from an average barrel of petroleum.

▄▄ DIESEL ENGINES AND DIESEL FUEL

Diesel engines burn diesel oil as fuel. A diesel engine is called a compression ignition engine. A four-stroke diesel engine operates in a similar manner to a gasoline engine. The difference is that ignition in a diesel is controlled by the injection of fuel into the cylinder.

A diesel has a very high compression ratio (about 16:1 to 20:1) (see Figure 17.4). If fuel were drawn in and compressed as it is in a gasoline engine, it would self-ignite before the piston could reach TDC. In a diesel, the mixture is injected by a high-pressure injector at the instant that ignition is desired. The compressed, hot air in the cylinder causes the fuel to vaporize and burn.

Diesel fuel is light oil that is refined as part of the same process that makes gasoline. It has several properties that make it useful as a fuel.

Diesel Volatility

While gasoline has high volatility, diesel fuel has low volatility. It is safe at room temperature and only gives off vapors if heated. Besides their reliability and fuel economy, one major reason that diesel engines are so heavily used in boats is that they are much safer than gasoline in case of a fuel leak.

Diesel Grades

There are two grades for automotive diesel fuel.
- Grade number 1-D, known as *number 1 diesel,* is more volatile, is thinner, and is used in very low temperatures.
- Grade number 2-D, known as *number 2 diesel,* is of a lower volatility and is used for most automotive driving conditions.

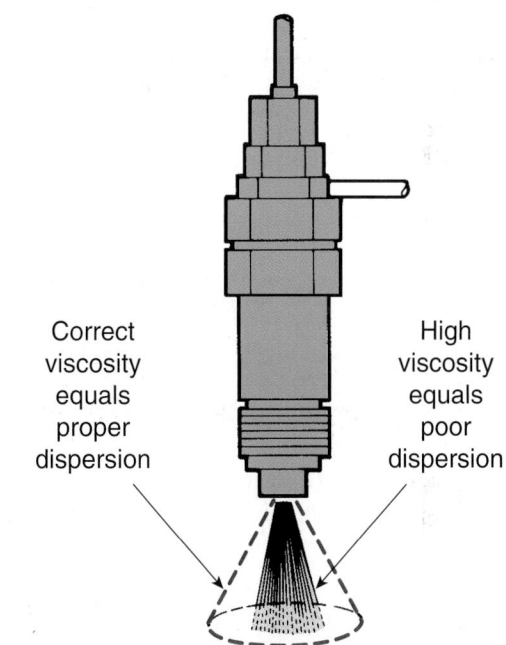

Correct viscosity equals proper dispersion

High viscosity equals poor dispersion

Figure 39.9 Diesel fuel must be of the correct viscosity. *(Courtesy of Ford Motor Company)*

Diesel fuel must be able to flow easily through the fuel system and be sprayed by the injectors, so it must be of low viscosity (**Figure 39.9**). Viscosity must be just high enough to provide lubrication to the fuel injectors and fuel pump. Number 1 diesel is only used in very cold weather because number two tends to become too thick as it gets colder. But number 1 might not provide sufficient lubrication.

Cloud Point

Diesel fuels contain some paraffin (wax). In very cold weather, the paraffin can separate from the fuel. This is called the **cloud point** because the fuel will appear cloudy when the wax separates out. The paraffin can clog fuel filters, causing the engine not to run.

Cetane Rating

The temperature at which a diesel fuel will ignite is called its ignition point. Diesel has good ignition quality, which means that it will burn without detonation soon after it is injected into the cylinder.

The **cetane number** of a diesel fuel describes how easily the fuel will ignite. The test number is derived by comparing how well a fuel burns with a sample of cetane. Cetane is a colorless, liquid hydrocarbon with excellent ignition quality. Its cetane rating is 100.

With gasoline octane ratings, the higher the octane number, the more resistant the fuel is to knocking. Diesel cetane ratings work in an opposite fashion (**Figure 39.10**). The higher the cetane rating, the easier it ignites (**Figure 39.11**). Fuels with a high cetane number burn as soon as they are injected, so no spark knock occurs. Number 45 is an average cetane

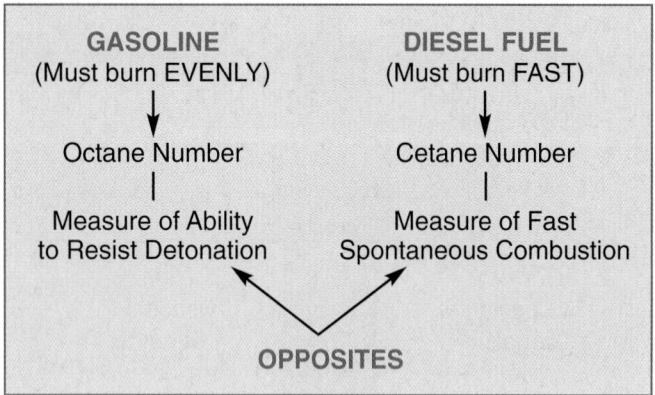

Figure 39.10 Octane and cetane numbers are opposites. *(Reproduced by permission of Deere & Company, John Deere Publishing, Moline, IL. All rights reserved)*

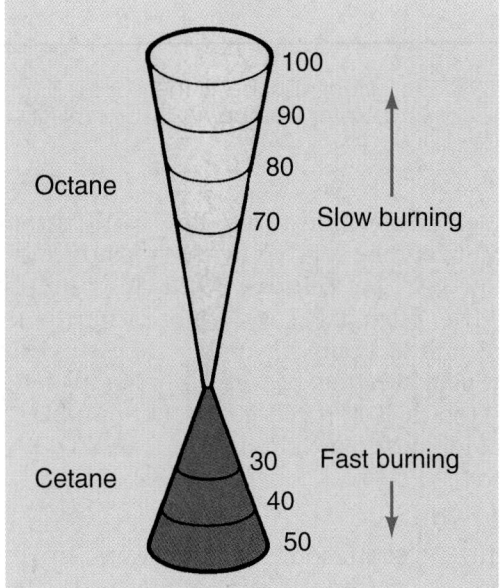

Figure 39.11 Gasoline with a higher octane rating burns slower. Diesel with a higher cetane rating burns more quickly. *(Courtesy of Ford Motor Company)*

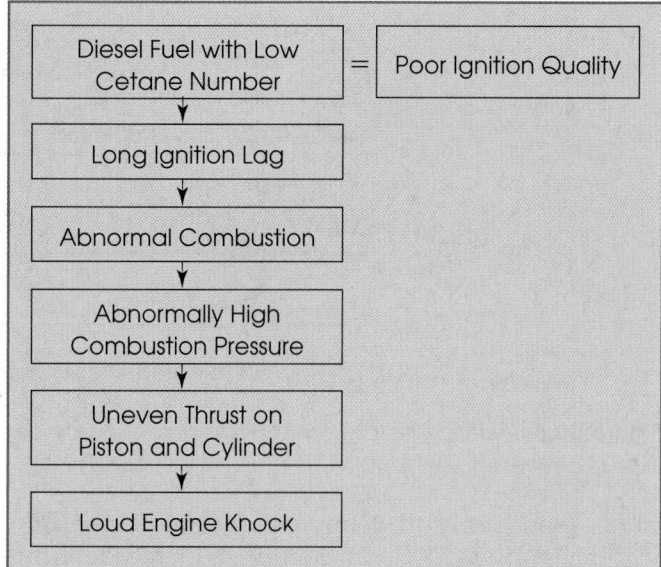

Figure 39.12 Problems caused by a fuel with too low a cetane rating. *(Courtesy of Ford Motor Company)*

value for #2 diesel and is what most auto manufacturers recommend.

The lower the cetane number, the higher the temperature required to ignite the fuel. If the cetane number is too low for the conditions, the fuel will take too long to light and will build up in the cylinder. Engine knocking can occur under this condition when the fuel is burned in too much quantity and out of control. **Figure 39.12** lists some of the problems caused by a fuel with too low a cetane rating.

Diesel Maintenance

Water is an enemy to diesels and must not be allowed to accumulate in the fuel system. It can rust the internal parts of the fuel injection system and can plug filters. Some diesel cars have a water detector in the fuel tank that completes a circuit to a dash light when the water level in the tank reaches a certain point. Other cars have fuel and water separators, which must be periodically drained.

More frequent oil changes are required with diesels. They generally have a larger capacity oil pan too. It is not uncommon to find oil change intervals of 3000 miles. If a car is driven in severe service conditions, the oil will need to be changed more often. Under a heavy load, a diesel makes soot, which is abrasive and is hard on engine parts. While heavy trucks have transmissions with many gears and are designed to pull heavy loads, passenger cars have only the normal number of gear ranges.

Biodiesel

Biodiesel is a renewable fuel (fatty acid alkyl esters) that can be made from recycled restaurant grease and oil, vegetable oil, or animal fat (**Figure 39.13**). Biodiesel is physically similar to petroleum diesel, and a blend of petroleum diesel with 20% biodiesel (B20) is usable in

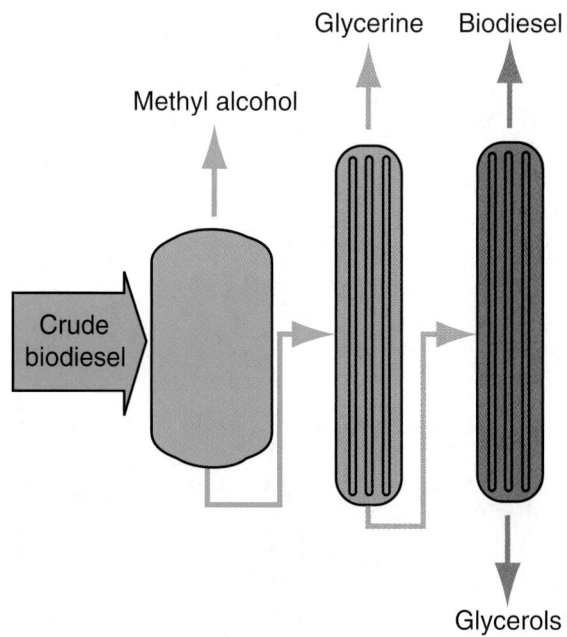

Figure 39.13 The biodiesel refining process.

During combustion, the hydrogen and carbon combine with oxygen. Combining hydrogen and oxygen produces water (H_2O), and combining carbon and oxygen produces carbon dioxide (CO_2). These harmless gases are what would result if all of the hydrogen and carbon in the fuel combined with oxygen during combustion. But incomplete combustion results in some HC and some carbon monoxide (CO instead of CO_2) (see Chapter 44). Emission controls are installed on the car to deal with these by-products of combustion.

CAUTION Gasoline is very dangerous because it is so flammable. Gasoline *vapor* mixed with air is actually what is so flammable. Be careful around electrical equipment, hot metal, and open flames such as are found with natural gas and propane appliances and equipment. Gasoline should never be used as a cleaning solvent.

■ SCIENCE NOTE ■

*Automobile gasoline HCs contain anywhere from 6 to 12 carbon atoms. The more carbon atoms, the higher the boiling point (**Figure 39.14**). These carbon atoms usually form branching-style molecules that reduce engine knocking. For instance, iso-octane, a common gasoline constituent with an octane rating of 100, contains three branching carbons (**Figure 39.15**).*

unmodified diesel engines. Although biodiesel is cleaner burning than petroleum diesel, pure biodiesel (B100) requires fuel system modifications on pre-1994 diesel engines. There is some concern regarding engine durability, and B100 is not suitable for cold weather operation.

■ GASOLINE

Gasoline makes up a substantial percentage of the crude oil. The amount depends on the location of the well and oil field. Gasoline is a very flammable hydrocarbon (HC).

Usage	Number of Carbon Atoms	Boiling Point (°C)
Natural gases/Solvents	C_1 to C_6	−160 to 60
Automobile/Motor gasoline	C_6 to C_{12}	60 to 80
Kerosene/Diesel fuel	C_{12} to C_{18}	180 to 140

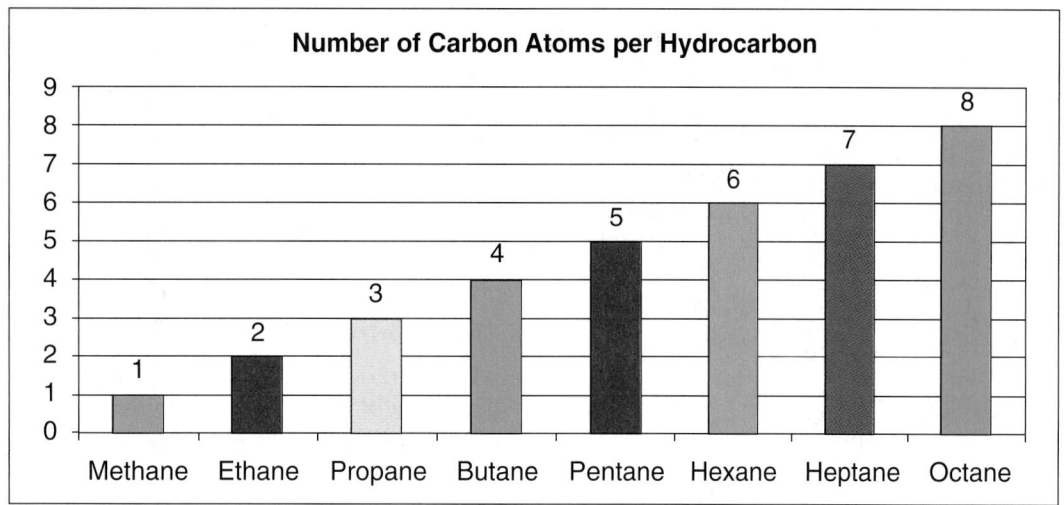

Figure 39.14 The relationship between petroleum products, the number of their carbon atoms, and their boiling points.

Chemical name	Structure	Relative knocking	Octane number
n-Heptane	H₃C—CH₂—CH₂—CH₂—CH₂—CH₂—CH₃ Straight	Extremely high	0
n-Hexane	H₃C—CH₂—CH₂—CH₂—CH₂—CH₃ Straight	High	42
2-Methyl Butane	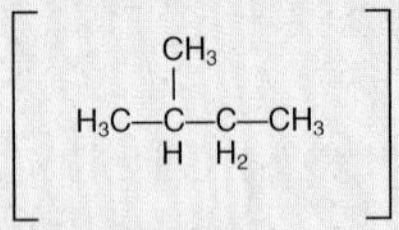 CH₃ / H₃C—C—C—CH₃ / H H₂ Branched	Low	93
Iso-octane	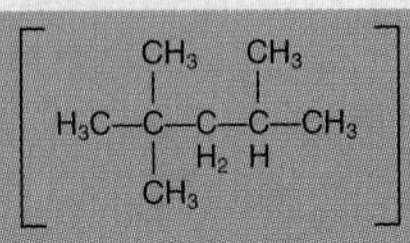 CH₃ CH₃ / H₃C—C—C—C—CH₃ / H₂ H / CH₃ Branched	Very low	100
Methyl Tertiary-Butyl Ether (MTBE)	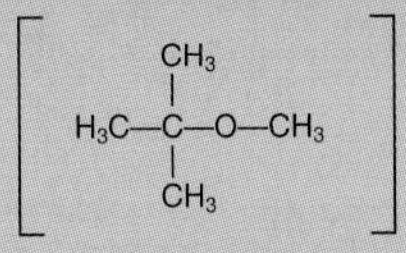 CH₃ / H₃C—C—O—CH₃ / CH₃ Branched ether	Extremely low	111

Figure 39.15 Various constituents of gasoline showing their structure and octane rating.

Gasoline Characteristics

Two characteristics of gasoline that are important to engine operation are volatility and resistance to spark knock.

Volatility is a measure of how easy a fuel evaporates. For fuel to burn properly in the cylinder, its vaporization point should be near the temperature in the intake manifold. If liquid fuel that is not atomized enters the cylinders, it will not burn. This wastes gasoline, which produces extra hydrocarbon and carbon monoxide emissions in the exhaust. Unburned gasoline also washes lubricating oil off of the cylinder walls, increasing wear.

When fuel does not vaporize easily enough, this is called *low volatility*. Low volatility results in the following:
■ Difficulty in cold starting
■ Poor driveability in cool weather
■ Unequal fuel distribution
■ Increased spark plug and combustion chamber deposits

If a fuel vaporizes too easily this is called *high volatility*. The resulting **vapor lock** causes the engine to stall because liquid fuel will not reach the carburetor. Other problems that can result are
■ High evaporative emissions
■ Poor driveability when hot
■ Lower fuel economy

Port fuel injection reduces volatility problems because the fuel is injected on or near the intake valves. The valves are hot and aid vaporization.

Refiners blend gasolines for the season and for the geographic area in which they will be used. These gasolines have different *vapor pressures*. Gasoline blended for summer use is less volatile (does not burn as easily). In higher altitude areas, fuels must have higher volatility because they can boil at lower temperatures. During unseasonable weather, cars can experience problems related to their fuel. Fuel that was refined for use in the summer that is used during the winter can cause hard starting. If winter fuel is used in the summer, vapor lock and carburetor icing can result.

Vapor lock happens to a carbureted car when fuel boils in the fuel line. Unlike liquid fuel, vapor is compressible. This means that the fuel cannot be pumped to the carburetor, so the engine stalls. After the fuel line cools sufficiently, the engine will run again. Fuel injected engines use electric fuel pumps that keep the fuel under higher pressure. Higher pressure raises the fuel's boiling point so vapor lock does not occur.

Measuring Volatility

The American Society for Testing and Materials (ASTM) has six volatility classes for gasoline: AA, B, C, D, and E. AA is the least volatile (**Figure 39.16**). In Figure 39.16 the 10% standard for AA gasoline means that 10% of the fuel would be evaporated before it reaches a temperature of 158°F, and so on. For all volatility classes, gasoline will have evaporated by 437°F.

ASTM D 4814 Gasoline Volatility Requirements							
Vapor Pressure/ Distillation Class	Distillation Temperatures				Vapor Pressure psi/Max.	Vapor Lock Protection Class	Vapor-Liquid Ratio of 20 °F Min.
	10% Evap. Maximum °F	50% Evap. °F	90% Evap. Maximum °F	End Point Maximum °F			
AA	158	170–250	374	437	7.8	1	140
A	158	170–250	374	437	9.0	2	133
B	149	170–245	374	437	10.0	3	124
C	140	170–240	365	437	11.5	4	116
D	131	150–235	365	437	13.5	5	105
E	122	150–230	365	437	15.0	6	95

Figure 39.16 Gasoline volatility requirements. *(Courtesy of Downstream Alternatives Inc.)*

Volatility is measured in one of three ways. The best-known standard is called Reid vapor pressure (**RVP**). During the RVP test, a sample of gasoline sealed in a metal chamber with a pressure measuring device is submerged in 100°F water. More volatile fuels vaporize more easily, creating more pressure. The vapor pressure is measured in pounds per square inch (psi). Because other ways of measuring volatility are becoming more popular, RVP is becoming known as VP.

In the United States, volatility standards for fuel are required by the Environmental Protection Agency (EPA) from June 1 to September 15 at retail gasoline stations. Gasoline vapor pressure must be below 9.0 psi. In southern high exhaust emission areas, called *non-attainment areas*, VP must be lower than 7.8 psi. VP for gasohol can be up to 1 psi higher.

■ AIR-FUEL MIXTURE

For an engine to start and run with good driveability and no internal damage to engine parts, it must have the correct air-fuel mixture. The air-fuel ratio of the engine is measured by weight in pounds. The desirable air-fuel mixture is about 15 pounds of air to 1 pound of fuel. A 15:1 air-fuel ratio is 9,000 gallons of air to 1 gallon of fuel.

A normal air-fuel mixture for high power is about 12 parts of air to 1 part of fuel (**Figure 39.17**). This is a 12:1 mixture and is called *rich*. Maximum power occurs from 12:1 to 12.5:1 air-fuel ratio. The ratio for maximum economy is 15:1–16:1, called *lean*.

Computer fuel management systems tailor the air-fuel mixture to adapt to various engine conditions. Stoichiometric (14.7:1) is the best air-fuel ratio for the most complete combustion for emission purposes (see Chapter 43).

Figure 39.18 shows air-fuel ratios for different operating conditions.

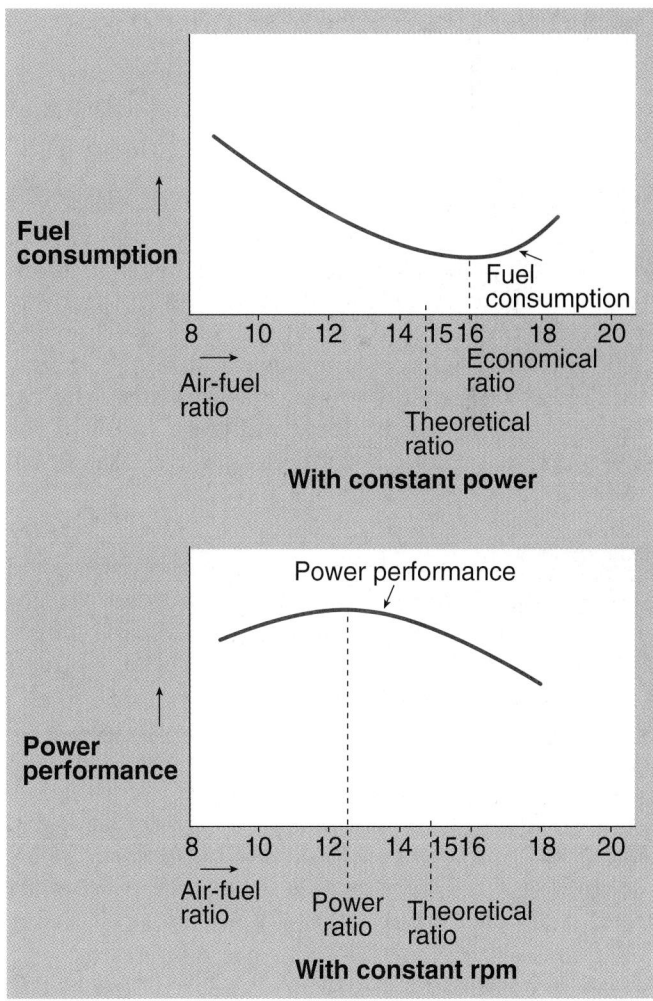

Figure 39.17 Engine operation with various air-fuel mixtures.

Rich Air-Fuel Mixture

A rich mixture results when there is too much fuel for the amount of air. 9:1 would be a rich mixture. A slightly rich mixture (12:1 for instance) will improve power but

ENGINE OPERATING CONDITION	AIR-FUEL RATIO (AIR:FUEL)
Starting (Air temperature approx. 0°C)	Approx. 1 : 1
Starting (Air temperature approx. 20°C)	Approx. 5 : 1
Idling	Approx. 11 : 1
Running slow	12 – 13 : 1
Accelerating	Approx. 8 : 1
Max. output (full load)	12 – 13 : 1
Running at medium (economical) speed	16 – 18 : 1

Figure 39.18 Air-fuel ratios for different operating conditions.

will result in poor fuel economy and increased exhaust emissions.

Lean Air-Fuel Mixture

An excessively lean mixture has a larger amount of air, 20:1 for example. With mixtures leaner than about 16:1 driveability suffers. A car's **surge** on the highway is a symptom of a lean condition. Surge is when the car does not keep constant power. Although it may not feel as though the engine has a misfire, the car will feel as though it is a boat riding on swells in the ocean. As the mixture becomes leaner, the engine will not run.

> **SHOP TIP** Lean mixtures can result in burned parts, such as head gaskets. Pointed spark plug electrodes are often the first indication of the problem (**Figure 39.19**).

NOTE: *A leaner mixture results in a higher idle speed. Listen to a chain saw or a lawn mower as it runs out of gas. The speed of the engine will increase momentarily before the engine quits. This is because the mixture is becoming leaner as the engine runs out of gas.*

Gasoline Engine Run-on

Fuel-injected cars shut off the fuel to the injectors when the key is off so those engines do not run on. When an engine continues to run even after the ignition key is turned off, this is called **dieseling** or *run-on*. It is called dieseling because a diesel engine's fuel ignites without a spark plug simply because of heat caused by pressure

Figure 39.19 This spark plug was overheated by operating with a lean air-fuel mixture. (*Courtesy of Federal-Mogul Corporation*)

in the cylinder. In a gasoline engine with a carburetor, if the engine runs on, the cause is usually too high an idle speed or too lean an air-fuel mixture. Run-on can damage an engine. It can result in a broken crankshaft if allowed to continue. Putting the transmission in gear with the brakes applied when shutting off the engine can stop run-on.

■ SPARK KNOCKS, CARBON NOISE, AND ABNORMAL COMBUSTION

When the air-fuel mixture is ignited during normal combustion, a **flame front** travels across the combustion chamber and pushes the piston down in the cylinder (**Figure 39.20**). The flame front travels at from 50 to 250 meters per second (depending on rpm and load) to *push* the piston down in the cylinder. This is very fast but it is *not an explosion*.

- During cranking, pressures in the cylinder are usually around 140–170 psi.
- Following ignition, cylinder pressures reach about 400 psi at TDC and peak at around 600 psi around 15° after TDC.
- Normal burning of the fuel takes about $\frac{1}{300}$ of a second.
- By the time the piston has traveled halfway down its bore, the flame has consumed most of the fuel and oxygen.

During normal combustion, the air-fuel mixture burns in a controlled manner to steadily force the piston down in the cylinder. Abnormal combustion can cause noise, shock damage, and burning of parts. It can be caused by the following:

- Cylinder temperatures that are too high
- Too lean an air-fuel mixture
- Engine overheating
- A driver that lugs an engine

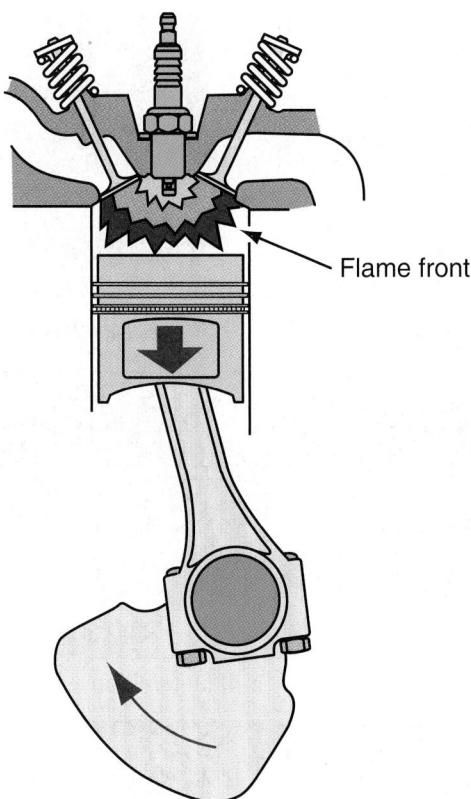

Figure 39.20 A flame front pushes the piston down in the cylinder.

Sounds caused by spark knocks can be mistaken for worn parts. Damage can result if these sounds are ignored.

ABNORMAL COMBUSTION

Two common **abnormal combustion** conditions can cause *spark knock* and engine damage: *preignition* and *detonation*. Sometimes, engine knock is inaudible due to the absorption of the sound by the piston and cylinder head. Other times, the rattling that results from detonation can be very loud.

NOTE: *The noise from spark knock is caused by the vibration of the combustion chamber walls.*

Occasional minor spark knock does not usually result in engine damage. During spark knock, there is a measurable loss of power on a dynamometer.

Variations in the flame front during normal combustion, preignition, and detonation are shown in **Figure 39.21**. Preignition and detonation are slightly different.

Preignition

Preignition allows the high heat of combustion to remain for too long on engine parts. The rise in combustion chamber temperature can cause parts such as valves, pistons, and head gaskets to be burned.

Preignition is commonly called *ping*. It occurs when the air-fuel mixture ignites before the regular spark occurs. Causes of preignition include:
- Spark plugs of too high a heat range
- Hot spots in the combustion chamber (from sharp edges on valves, head gaskets, or hot carbon particles)
- A loose spark plug

Notice the burning that results from preignition shown in **Figure 39.22**. Passenger car engines can usually survive detonation for a period of time, but preignition can damage an engine very quickly.

SCIENCE NOTE

The air-fuel mixture is more easily ignited before pressure begins to build in the cylinder when the piston is at BDC as it begins its compression stroke. Lighting the mixture becomes more difficult as the piston moves up, so preignition is more likely to occur earlier in the compression stroke (provided fuel is present). When the air-fuel mixture is preignited by a hot spot while the piston is at BDC, the heat from combustion is exposed to the piston, cylinder walls, combustion chamber, and valves for a much longer time than normal. The additional burn time can be well over 100 degrees of crankshaft rotation before the normal spark ignition point, heating the engine abnormally. The pressure in the cylinder intensifies further as the piston moves up, with the burning flame front continuing its expansion against it.

An engine can quickly suffer damage from the extra heat and pressure that occur during preignition.

NOTE: *Unlike detonation, preignition does not produce a knocking sound, so the impending damage is not apparent until it is too late.*

A typical symptom of preignition is a hole burned in a piston or melted spark plug electrodes. The center of the piston burns because it is relatively thin and aluminum absorbs heat more rapidly than the cylinder wall or cylinder head.

Sometimes preignition can be caused by detonation. When a heavily loaded engine has been detonating for an extended period of time, the spark plug can overheat and become a source of preignition as well. This is known as *detonation induced preignition*.

Detonation

Detonation, also called ping, spark knock, or engine knock, is when the air-fuel mixture self-ignites due to increased pressure in the cylinder (**Figure 39.23**).

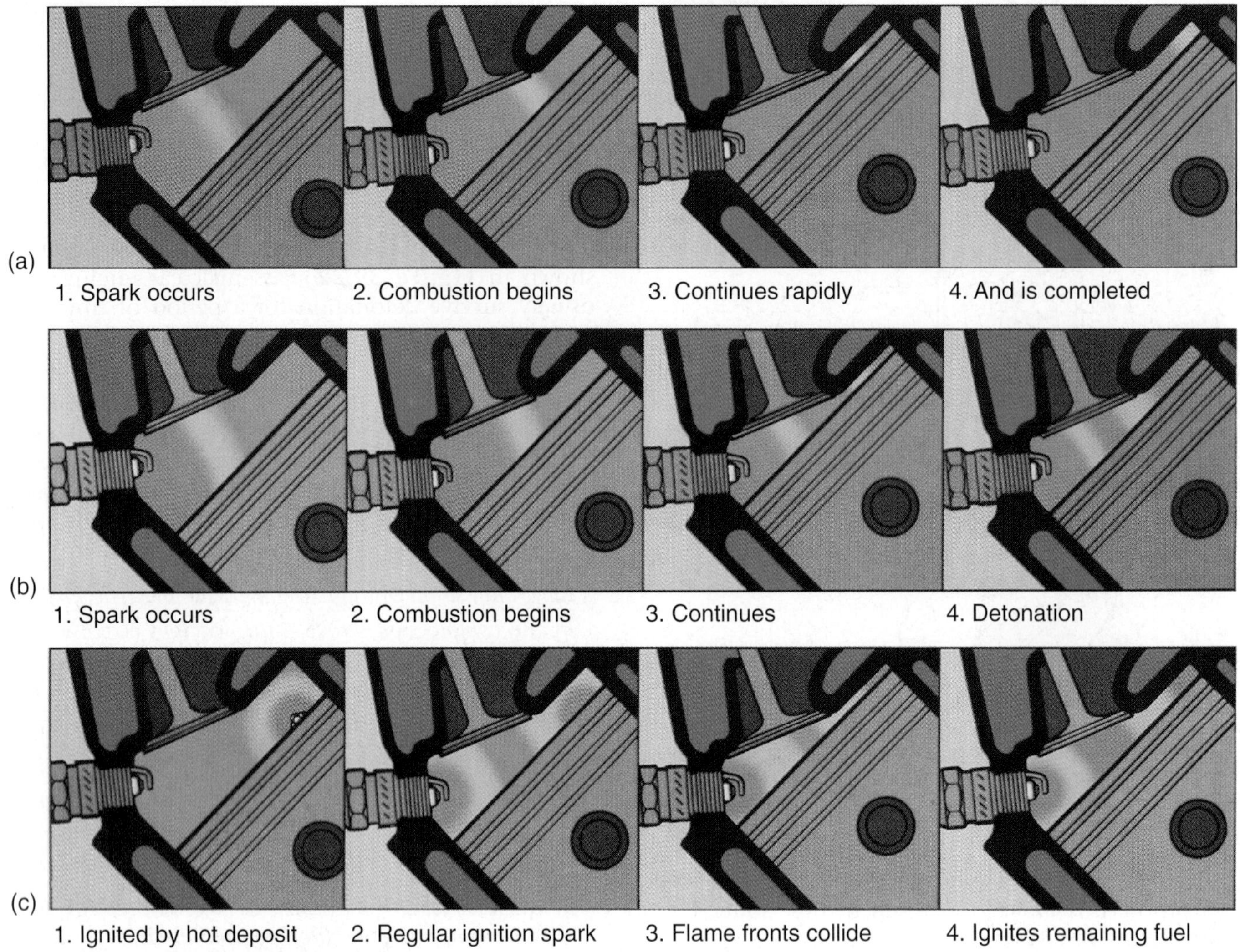

(a)

1. Spark occurs   2. Combustion begins   3. Continues rapidly   4. And is completed

(b)

1. Spark occurs   2. Combustion begins   3. Continues   4. Detonation

(c)

1. Ignited by hot deposit   2. Regular ignition spark   3. Flame fronts collide   4. Ignites remaining fuel

Figure 39.21 (a) Normal combustion, (b) detonation, and (c) preignition. *(Courtesy of Federal-Mogul Corporation)*

Figure 39.22 This piston burned due to preignition when a spark plug with too hot a heat range was mistakenly installed in the cylinder head.

It occurs when the unburned part of the air-fuel mixture explodes violently *after* regular spark ignition has occurred. The normal burn is already in progress when detonation happens. After the air-fuel mixture is ignited it expands, moving the piston down in the cylinder. Suddenly the mixture explodes, producing a brief, extreme spike in pressure as the rest of the combustible energy is expended. The explosion causes parts to break. The explosive result of detonation is shown in **Figure 39.24**.

Detonation can also cause the piston to expand too much, resulting in cylinder wall scuffing at the four corners of the piston skirts. The aluminum melts and runs into the piston ring grooves, causing them to stick. This causes a compression loss, and sometimes the flame front leaks past the rings, burning out an area on the ring lands. Damage of this type is often confused with preignition because of the burned ring lands. The original cause, however, was detonation.

NOTE: *This type of detonation failure is especially common in the air-cooled engines found in snowmobiles, jet skis, quads, and motorcycles.*

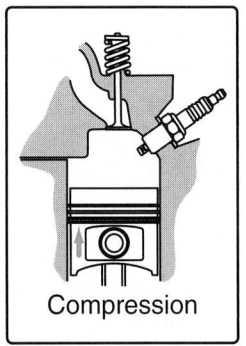
Compression

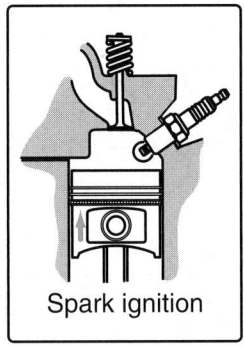
Spark ignition

Combustion

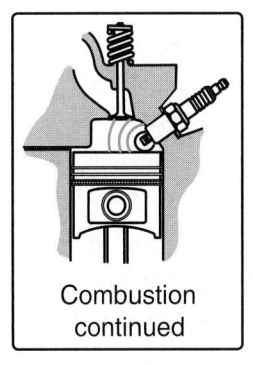
Combustion continued

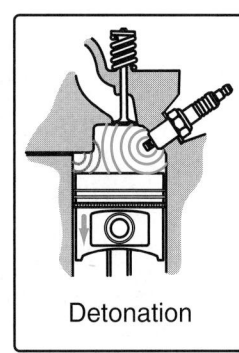
Detonation

Detonation

Figure 39.23 Detonation.

Detonation damage

Figure 39.24 This ring land was broken by detonation. *(Courtesy of Tim Gilles)*

Detonation is a race between the flame front and heat buildup during combustion. The faster the burn is completed, the more immune the engine is to detonation. If the air-fuel burn is completed before the temperature and pressure reach the point of detonation, abnormal combustion will *not* occur. Newer engine designs avoid detonation by speeding up the burn time using a specially designed pent-roof combustion chamber and centrally located spark plug or multiple spark plugs. Firing the spark closer to TDC means that the piston is not working against compression for as long a time. Older engines required more spark advance to complete combustion by the targeted point.

Detonation Noise. Sometimes, the pistons and cylinder head absorb the sound of engine knock, making it inaudible. This condition is called cold knock or inaudible ping. Other times, the rattling that results from detonation can be very loud. The noise is caused by vibration of the combustion chamber walls. When the air-fuel mixture detonates, the explosion is instantaneous and the loss of power can result in serious engine damage, including broken pistons, piston rings, and head gasket failure.

Before sophisticated computer-controlled spark timing, a common cause of detonation was over-advanced ignition timing.

- Excessively advanced ignition timing can cause a burned piston but will not cause an increase in engine temperature.
- Excessively retarded ignition timing will cause an engine to overheat but will not burn a piston.

SHOP TIP
- Spark plugs are a window to combustion chamber action. **Figure 39.25** shows spark plug conditions that resulted during preignition and detonation.
- Dynamometers (see Chapter 16) are used to test engines under load. Experienced dyno operators know that the temperature of exhaust gas in a detonating engine will drop.

Abnormal combustion can also be caused by cylinder temperatures that are too high, too lean an air-fuel mixture, lugging, and overheating.

A leading source of detonation is an inoperative exhaust gas recirculation (EGR) valve (see Chapter 43). This can be confirmed while running the engine at high idle (1,800–2,000 rpm) while the knock occurs. Manually open the valve (if possible). If the noise disappears, the valve is at fault.

SAFETY NOTE  Be careful of hot exhaust components when working around the EGR valve.

On older engines, a temperature-controlled air cleaner stuck in the heat-on position or a malfunctioning early fuel evaporation system can also cause detonation (see Chapter 44).

Preignition damage

Detonation damage

Figure 39.25 Spark plug conditions that resulted during preignition and detonation. *(Courtesy of Champion Spark Plugs)*

EXCESSIVE CARBON BUILDUP

In a car with relatively low mileage, carbon buildup in the combustion chambers can cause an increase in the compression ratio. This problem is not as common since the introduction of unleaded gas, which leaves fewer deposits, but some late-model engines develop carbon problems in less than 10,000 miles.

There are two kinds of deposits: oil-based and carbonaceous:

- **Oil-based carbon deposits** are the traditional gummy, black ones like those sometimes found on intake valves. They are caused when oil and heat come together.
- **Carbonaceous deposits** are from fuel. They are called *cauliflower deposits* because of their resemblance to the vegetable. These deposits are not as thick as oil deposits and are hard, dry, and tougher to remove. Driveability problems can result from them. Oil companies are developing fuels to minimize this problem.

Service procedures for removing carbon deposits are covered in Chapter 41.

REGULAR VERSUS PREMIUM FUELS

Increasing pressure in the combustion chamber can increase engine power. But increasing the pressure in the engine raises the likelihood of engine knocking. Gasoline can be refined so it will not knock as easily. This is called high **octane**.

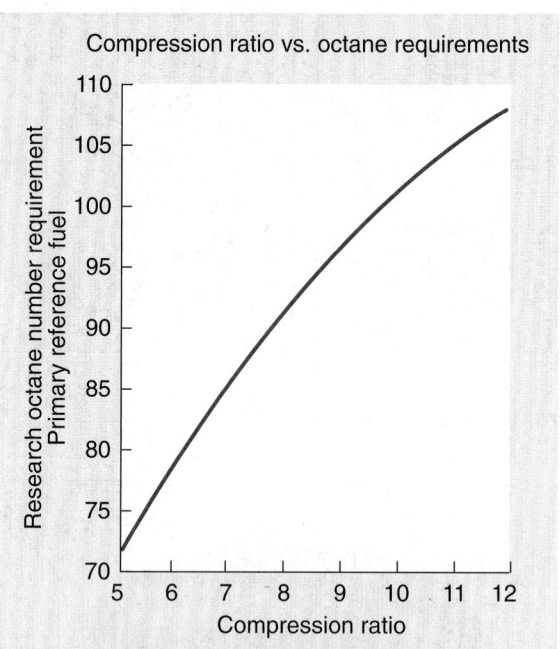

Figure 39.26 Octane requirement goes up with increased compression ratio. *(Courtesy of Downstream Alternatives, Inc)*

Most gasoline stations offer three octane grades of unleaded gasoline: *regular* (87 octane), *midgrade* (89 octane), and *premium* (93 octane). Higher compression engines require a higher octane fuel. The amount of octane needed is called the engine *octane number requirement* (ONR) (**Figure 39.26**).

Octane, a measurement of a fuel's ability to resist explosion during combustion, compares the *antiknock* qualities of different fuels. Most cars manufactured since the early 1970s can use regular gas. Check the manufacturer's recommendation.

NOTE: *Using premium gasoline in a car designed to run on regular will not harm it, but the premium costs considerably more and no advantage is gained.*

Design engineers test an engine on a dynamometer using fuels of different octanes. Different spark advances are tested at various speeds and loads to determine the engine's highest torque output. Ignition timing is best when it is advanced to the furthest point without causing detonation. Modern computer-controlled engines use a **detonation sensor**, sometimes called a **knock sensor**. When it senses the vibration caused by detonation, the computer retards the ignition timing until the spark knock goes away. The computer continually relearns the best timing for the fuel by advancing the timing until detonation occurs, then retarding the spark slightly until detonation stops. All modern automotive engines have this capability.

NOTE: *Manufacturers recommend fuel of a particular octane rating for each of their vehicles. When fuel of a lower octane rating is used, the timing will be adjusted to a lower amount of advance. The price is a slight penalty in fuel economy and performance.*

When fuel with too low an octane rating is used in an engine with a high-compression ratio, detonation often occurs at low or moderate speed.

SCIENCE NOTE

Newer engines have combustion chamber designs that can lower the octane requirement. The engines, which have fast-burn combustion chambers, need less spark advance than older ones with conventional combustion chambers. This provides a mechanical advantage resulting in improved fuel economy. When the spark is more advanced, the pistons have to push against pressure for a longer amount of crankshaft rotation. With a fast-burn combustion chamber, less spark advance is required so the pistons do not work against pressure as long. The engine becomes more efficient and pumping losses are reduced.

Air-fuel mixtures of 14.7:1 require the highest octane fuel. Anything richer or leaner will have a lower ONR.

Higher combustion temperatures will increase ONR. Combustion temperatures are affected by cooling system temperature, inlet air temperature, exhaust gas recirculation, and intake manifold heating.

Changes in outside air temperature, humidity, and pressure all have an effect on the octane number requirement. If you drive in the high altitude of the Colorado Rockies, you will notice that premium fuel has a lower octane number. This is because barometric pressure and the oxygen content of the air is lower.

SCIENCE NOTE

Octane requirements are a measure of how easily gasoline can be ignited. The higher the octane, the smoother and slower the fuel burns. Generally, the higher the octane, the greater the number of hydrocarbons containing larger numbers of carbon atoms. More carbon atoms per hydrocarbon means that more oxygen and more heat are needed to burn the fuel. As air temperatures fall, lower octane fuels can be used. This is also true at higher altitudes where there is less oxygen. With increases in pressure come increases in temperature, requiring higher octane fuels.

The octane rating is based on a "0" rating for n-heptane, which has the greatest tendency to cause knocking, and a "100" rating for iso-octane, which has a very low tendency to knock. For example, a fuel with an octane rating of 87 produces knocking equivalent to a mixture of 87% iso-octane and 13% n-heptane (see Figure 39.15).

Many of the variables that affect the octane number requirement are taken into consideration in the design of the engine's fuel and ignition system. In addition to having a knock sensor, some engines use a barometric pressure sensor that sends signals to a computer. Changes are then made in the ignition timing and air-fuel mixture to compensate.

■ OCTANE STANDARDS

The *ASTM* sets gasoline quality standards. ASTM standards are voluntary, but because of their effects on air quality, the EPA and some state standards now require all or part of these standards. The measurement of gasoline octane quality most often used is the *antiknock index (AKI)*. It is an average of the two ways that octane can be measured; the *research octane number (RON)* and the *motor octane number (MON)*. Both the RON and MON measure the same things but do it at different speeds, temperatures, and spark advances.

The research method (RON) gives a higher reading for the same fuel as the motor method (typically 8 to 10 octane numbers higher). It affects low-speed knock and engine run-on (dieseling). The motor method (MON) gives a measurement of how much engine knock will be present under heavy loads, such as when passing or climbing hills.

The antiknock index is stated as *(R+M)/2*. This is the number required by law to be listed on the octane decal on the gasoline pump. A third method of rating octane is the *road method*, which is not used often.

■ GASOLINE ADDITIVES

Gasoline additives are expensive and are added in minute quantities to fuel. A small amount is all that is necessary. A large effect on the quality of gasoline comes from the use of detergents and deposit control additives. These are used to keep port fuel injectors from becoming fouled. Deposit control additives have been required by law since 1995 in all 50 states.

HISTORY NOTE

Antiknock compounds such as lead were used in the past. Lead is a poison and also causes problems with catalytic converters used in the car's emission system. Lead was eliminated from the fuel of cars with catalytic converters in the mid-1970s. Government regulations resulted in the elimination of all lead from automobile fuel in 1996. Unleaded fuel can be refined in a more costly manner that results in a higher octane.

■ REFORMULATED GASOLINES

The way gasolines are refined can have a large effect on air pollution. Reformulated gasolines clean the air by providing more complete combustion. This kind of gasoline is required by the EPA in U.S. cities with the worst air pollution. The use of reformulated gasoline in older cars can result in damage to older rubber fuel lines. Reformulated gasoline has less energy content, so fuel economy will be less also.

OXYGENATED FUELS/ALCOHOLS

Oxygenated fuels are gasolines blended with ethers or alcohols. Ethyl alcohol (ethanol) at a 10% concentration (E10) and MTBE (methyl-tertiary butyl ether) at a 15% concentration are the most common oxygenates. These fuels are blended to enhance octane. They also provide wintertime control of carbon monoxide because of more complete combustion.

Ethanol Blends

The most used alcohol/gasoline mixture usually contains about 10% ethyl alcohol (ethanol), which can be made from grain. This used to be called **gasohol** in the late 1970s. It was first used to extend gasoline supplies during shortages. Ethanol is about 35% oxygen, so a 10% concentration adds about 3.5% oxygen to the mixture. There is a slight difference in odor noticed during a fill-up. A 10% ethanol mixture will raise an 87 octane fuel by at least 2.5 octane numbers. **Figure 39.27** shows the octane values of gasoline and different oxygenates.

Gasohol with less than 10% alcohol does not require any changes to the fuel system, although a fuel filter or two might need to be changed due to the cleaning effect the alcohol has on the fuel tank.

Oxygenates suspend water in the fuel and tend to keep it from accumulating in the gas tank. Gasoline cannot hold much water, so it separates and accumulates at the bottom of the tank. Alcohol and ether attract and hold water. In fact, the water removers that are on the market that are added to fuel tanks contain alcohol. The following are the amounts of water that 1 gallon of each fuel can hold in suspension:

- Gasoline = 0.5 teaspoon
- 15% MTBE/gasoline = 1.5 teaspoons
- 10% ethanol = 12 teaspoons

Gasoline Fuel Economy

Several variables can change a vehicle's fuel economy (**Figure 39.28**). Driving conditions such as air temperature, hill climbing, head winds, overloaded vehicles, and poor driving habits can contribute to a drop in economy. Engine condition and tire air pressure can also be factors. During the winter, fuel economy can be 10% to 20% lower. Fuel economy can be influenced by the energy

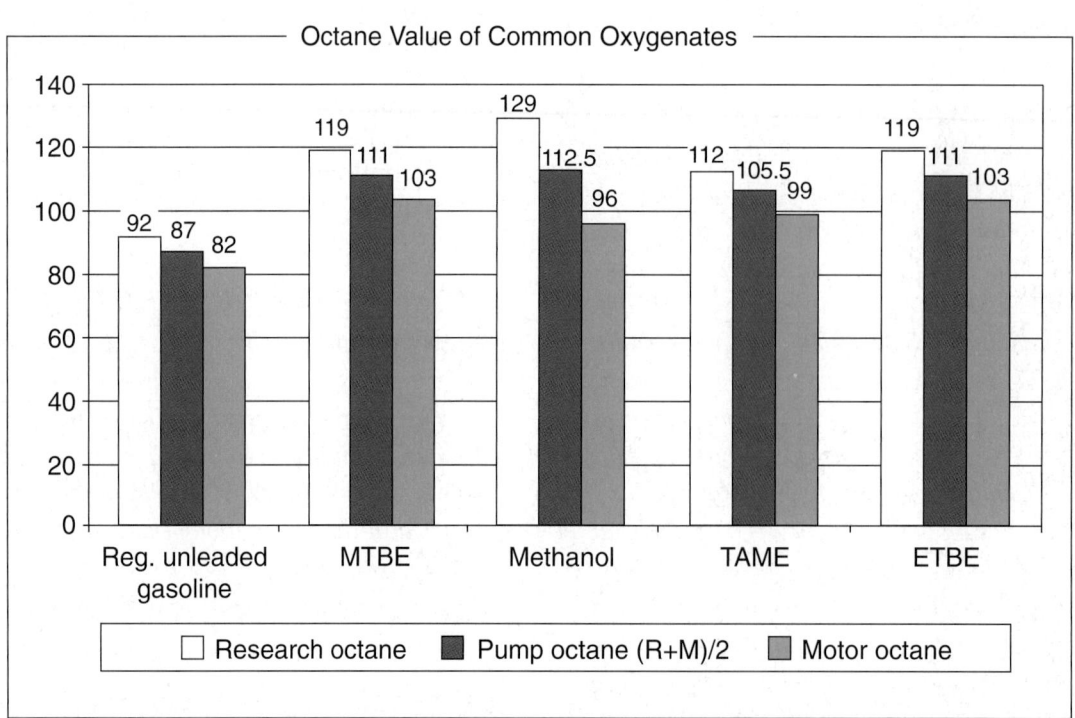

Figure 39.27 Octane values of gasoline and different oxygenates.

Factors That Influence Fuel Economy of Individual Vehicles		
Factor	Fuel Economy Impact	
	Average	Maximum
Ambient temperature drop from 77°F to 20°F	−5.3%	−13.0%
20 mph head wind	−2.3%	−6.0%
7% road grade	−1.9%	−25.0%
27 mph vs. 20 mph stop and go driving pattern	−10.6%	−15.0%
Aggressive vs. easy acceleration	−11.8%	−20.0%
Tire pressure of 15 psi vs. 26 psi	−3.3%	−6.0%

Figure 39.28 Fuel economy variables. *(Courtesy of Downstream Alternatives, Inc.)*

	Summer-Grade Btu	Winter-Grade Btu
Maximum	117,000	114,000
Minimum	113,000	108,500
%	3.4	5.0

Difference between summer maximum and winter minimum: 7.26%

Figure 39.29 Gasoline energy content. *(Courtesy of Downstream Alternatives, Inc.)*

content of the gasoline. During the summer, gasoline energy content is higher (**Figure 39.29**). A 10% ethanol mixture contains 3.4% less energy than pure gasoline.

SCIENCE NOTE

During the combustion of gasoline, the hydrocarbon fuel is converted by oxygen to CO_2 and H_2O. Chemically, this process is called oxidation (the addition of oxygen to a chemical compound). When gasoline is burned, oxygen is added to the hydrocarbon molecule (a molecule contains only hydrogen and carbon) to form CO_2 and H_2O. The heat produced from the combustion process comes from the addition of oxygen. This is the reason using gasohol results in fewer miles to the gallon. Gasohol contains 10% ethyl alcohol (CH_3CH_2OH). Since ethyl alcohol already contains oxygen, it has been partially oxidized already. Therefore, it produces less heat during combustion. In fact, running your car on pure alcohol would require 75% more fuel than running it on gasoline.

■ ALTERNATIVE FUELS

Alternative fuels are those other than gasoline and diesel fuel. Liquified petroleum gas and alcohol are currently in use in automobiles and trucks. Hydrogen may be used in the future.

According to the Federal Energy Policy Act of 1992, alternative fuel vehicles (AFVs) include any flexible fuel or dual-fuel vehicle designed to operate on at least one alternative fuel. A flexible fuel vehicle (FFV) is one that can use ethanol, gasoline, or any combination of the two fuels.

Ethanol and Biomass

Although gasoline engines often use gasoline blended with alcohol, oxygenated fuels are not classified as alternative fuels. To qualify for tax incentives, ethanol vehicles must be designed to run on a blend of up to 85% denatured ethanol and 15% gasoline (*E85*). During cold weather operation, the upper limit of 85% ethanol is lowered and the percentage of gasoline is raised.

An FFV uses the same fuel system, whether its engine is running on gasoline or ethanol. FFVs are different from dual-fuel vehicles that use natural gas or propane as a fuel, because those vehicles must have two distinctly different fuel systems. The vehicle owner's manual lists whether the vehicle can run on E85. E85 fuel is more prevalent in the midwestern United States due to its proximity to corn and other *biomass* sources of agricultural waste like corn stover (the stalks that remain after corn harvest), pulpwood, rice straw, switchgrass, and garbage.

P-Series Fuel

P-series fuel is a liquid blend that includes ethanol, a biomass co-solvent, and natural gas liquids. It is designed to be used in FFVs but is not currently produced in large enough quantities for widespread use.

Methanol

Methanol is methyl alcohol that can be burned in an internal combustion engine. But it produces only about half the energy that gasoline does. So the fuel system needs to be adjusted to provide an air-fuel ratio of about 6.5:1. Methanol has been used for years in race cars and will probably become more widely used in passenger cars. It can be made from coal, natural gas, oil shale, wood, or garbage.

Characteristics of Alcohol Fuels

- A major disadvantage to alcohol fuels when compared to gasoline is that they are invisible when burning. That is why race car drivers wear fire suits and emergency crews douse them quickly with fire extinguishers after an accident, even though the racing fans cannot actually see the fire.
- Methanol is very corrosive and is poisonous. It attacks aluminum, some plastics, and other materials. Gas tanks must be lined with resistant materials and fuel systems must be totally made of noncorrosive materials.

LP Gas

Liquified petroleum gas (**LPG**) is similar to gasoline chemically and is a product of gasoline refining. It can also be obtained from natural gas. LPG is mostly propane but contains a small percentage of butane (up to 8%). Because of this, many people simply refer to LPG as propane. LPG is a vapor above –40°F rather than a liquid. It is called "liquified" because it is stored as a liquid in a bottle under pressure. The pressure increases its boiling point and turns it into a liquid (**Figure 39.30**). It takes about 250 gallons of LP gas to make 1 gallon of compressed liquid. Propane's boiling point is about −44°F (7°C). At temperatures lower than that, propane is not under pressure. Increasing the temperature rapidly increases the pressure. From 40°F to 65°F, the pressure will increase from about 65 psi to about 100 psi. Butane's pressure will increase from about 3 psi to about 15 psi in the same range.

LPG requires a different fuel system to meter the gas vapor into the engine. Because the fuel is already under pressure, no fuel pump is required. When a carburetor is used, it is similar to a gasoline carburetor. The main difference is that there is no float bowl (fuel reservoir).

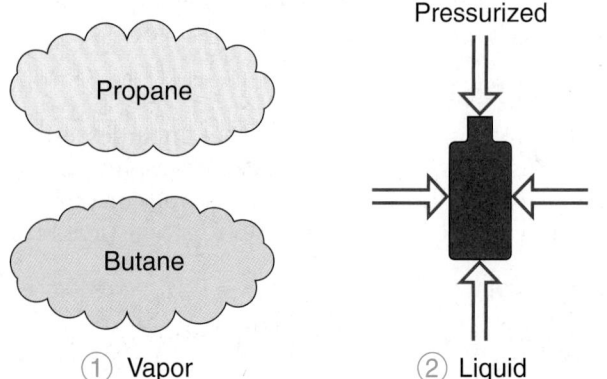

Figure 39.30 Pressure increases the boiling point of a gas and turns it into a liquid. *(Reproduced by permission of Deere & Company, John Deere Publishing, Moline, IL. All rights reserved)*

The following are advantages and disadvantages of LPG when compared to gasoline.

Advantages of LPG:

- LPG burns cleanly, produces fewer emissions, and lacks the objectionable exhaust odor of incompletely burned gasoline.
- Because it vaporizes easily, LPG does not tend to settle out as a liquid in the intake manifold. The engine can start easily in very cold weather and cold driveability is better.
- Less frequent oil change intervals are required because LPG is a dry gas and so does not create carbon in the combustion chamber. Also, there is no fuel wash to dilute the engine oil.
- LPG has a higher octane; therefore, it can be used in engines with higher compression ratios. Pure butane has a 93 octane rating, and pure propane is 100 octane. The octane of the LPG will depend on the ratio of the mixture of these two fuels.

Disadvantages of LPG:

- The heat energy per volume of LPG is less than gasoline.
- LPG must be stored under high pressure, so its storage tank must be strong.
- It is also more difficult to locate refueling stations for LPG.
- Refilling an LPG tank is time consuming.

LPG is often used in fleets, such as utility companies, taxis, buses, and for indoor machinery such as forklifts.

Compressed Natural Gas

Compressed natural gas (**CNG**), long used as a home energy source, is an excellent alternative fuel for internal combustion engines. It has long been used in fleets, buses, and taxicabs, and it is beginning to be found in more passenger cars, especially in states that provide tax incentives to encourage its use in motor vehicles. In some places, a natural gas vehicle (NGV) is allowed to use the carpool lane with only one passenger in the car.

The scarcity of public refueling stations is one problem with CNG. Refueling stations make it possible to refill the tank in a similar amount of time as refilling a gasoline vehicle. There are home refilling units (**Figure 39.31**), but a full refill using these devices requires several hours.

CNG is stored under pressure (3,000 psi) in a large cylinder. The cylinders are very strong, and proponents of natural gas fuels have done many tests to prove their safety. The size of the cylinder is one of the reasons why it is not in common use in passenger cars (**Figure 39.32**). CNG usually comes to the filling station in regular distribution lines as a low-pressure gas. At the facility, it is compressed and dried to remove any moisture from it. Some refueling stations are equipped with low-pressure (time-

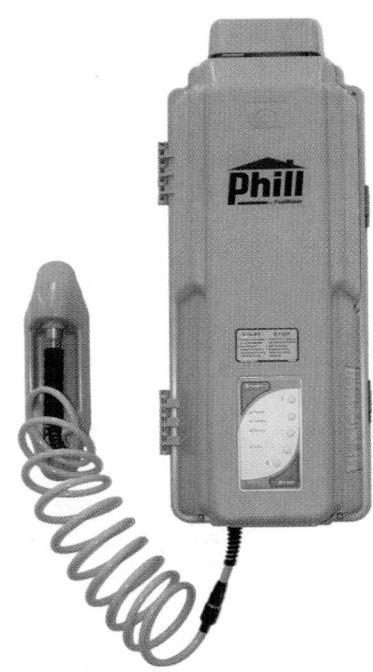

Figure 39.31 A home refueling unit for CNG.

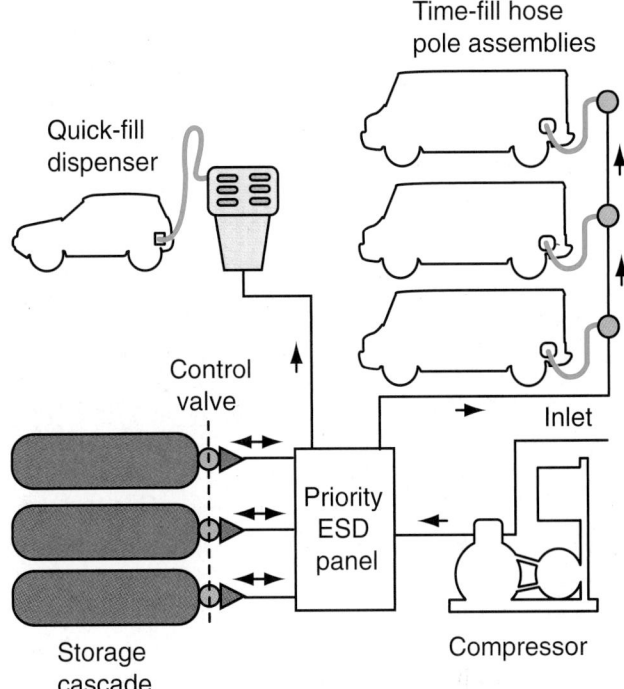

Figure 39.33 Some CNG refueling stations have quick-fill and time-fill capabilities.

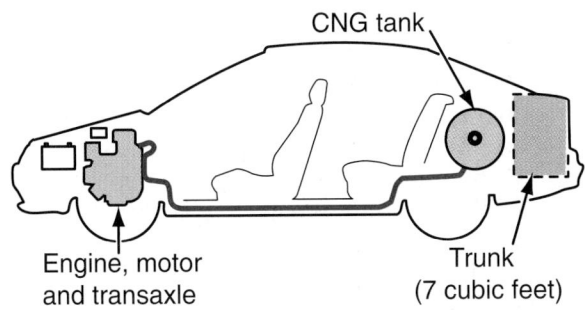

Figure 39.32 A CNG storage tank takes up much of the trunk space in this small vehicle.

Figure 39.34 When filling the fuel tank at a high-pressure refueling station, the nozzle is pushed into the filler neck and turned to seal it.

fill) and high-pressure (quick-fill) filling capabilities (**Figure 39.33**). When filling the fuel tank at a high-pressure refueling station, the nozzle is pushed into the filler neck (**Figure 39.34**) and turned ¼ turn to seal it against the CNG compressor's pressure of 3,600 psig. Turning the nozzle another ¼ turn causes filling to begin. When the tank is full, gas flow stops. Natural gas is the cleanest burning of the alternative fuels, with 99% less carbon monoxide and 85% less hydrocarbon emissions than gasoline. Because it burns cleanly, the oil change interval can be increased and spark plugs last longer. Natural gas is available in vast quantities in North America.

Modern technology is being used to run large equipment, automobiles, and trucks on compressed natural gas.

HISTORY NOTE

A drawback to older carbureted CNG systems was that they lacked an accelerator pump circuit and were therefore prone to lean hesitation on acceleration when used in motor vehicles. This is no longer a problem on newer CNG systems with special CNG fuel injectors.

Fuel flow is controlled by the ECM (computer), which provides the necessary enrichment during acceleration. The system shown in **Figure 39.35** uses

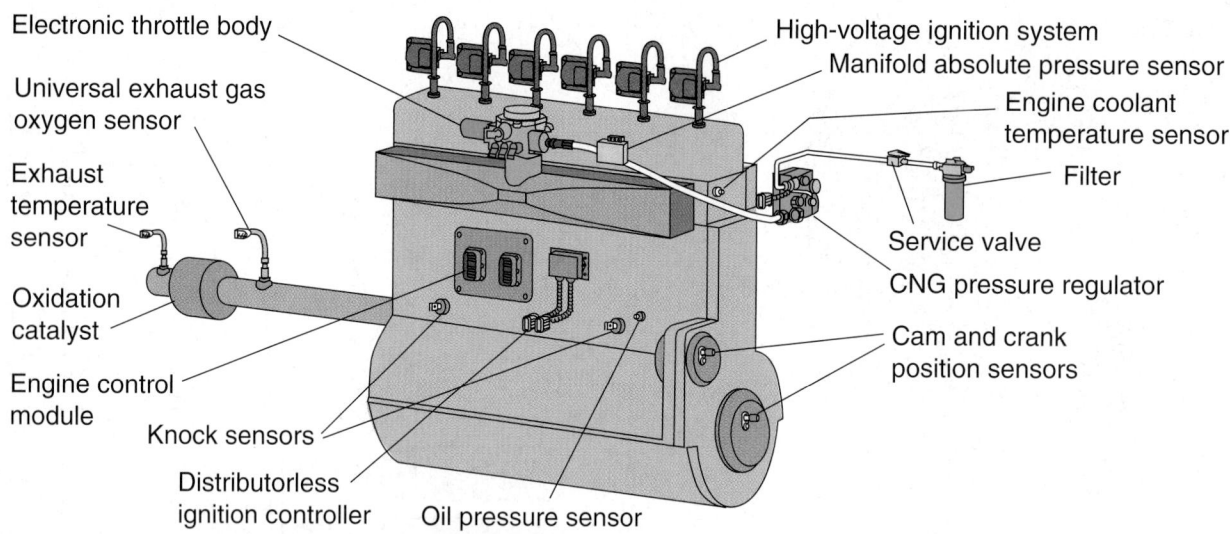

Figure 39.35 A compressed natural gas (CNG) system in a re-engineered diesel engine.

CNG in a re-engineered diesel engine with a lower compression head and spark plugs. These engines emit very low emissions, compared to the diesel engines they replace.

Gasoline Engine CNG Retrofits. Gasoline engines can be modified relatively easily for use with CNG. Retrofit systems are certified by the U.S. EPA. In some systems, the fill valve is located within the engine compartment. Gas under pressure of about 3,200 psi is routed through a high-pressure pipe from the tank to a reducer, which drops the pressure in stages. To prevent the reducer from freezing, it is warmed by engine coolant. From the reducer, the gas goes to the engine at atmospheric pressure.

Bi-fuel engines **Figure 39.36** can run on either gasoline or CNG and can switch seamlessly from CNG to gasoline. In a bi-fuel engine, natural gas and air flow through the intake manifold to the cylinders. When the gasoline switch is activated, the natural gas

system shuts off. Dedicated NGVs use special natural gas injectors to inject the gas directly into the cylinders. Electronic engine controls adjust ignition timing, which can be more advanced due to the higher octane level of CNG. These systems are relatively expensive.

Dual-fuel systems that use diesel and CNG are more difficult to deal with than dedicated (single fuel) systems or gasoline/CNG systems.

Dual-fuel operation, called fumigation, is when gaseous fuel and diesel are burned at the same time in a diesel engine. Due to the premixed air and fuel, the engine is a hybrid compression ignition/spark ignition engine with diesel fuel injectors replacing spark plugs as the source of ignition.

Liquefied Natural Gas

Liquefied natural gas (LNG) is another form of natural gas that is becoming more popular in heavy-duty transportation. Both LNG and CNG are delivered to the engine as a vapor under low pressure (1–300 psi). When cooled to −163°C (−260°F) at atmospheric pressure, LNG becomes a liquid weighing less than half as much as water. It is stored at about 1/600 the volume of natural gas regulated to the burner of a home appliance.

LNG will only burn when mixed with air in a ratio of 5% to 15%. It is colorless, odorless, nontoxic, and noncorrosive. LNG has an ignition temperature approximately 500°F above that of gasoline. It is not explosive in either liquid or vapor form. If there is a leak and a spill, LNG will evaporate very quickly.

Pipeline natural gas is about 92% methane but is purified during liquefaction to make LNG (not to be confused with LPG, which is 90% propane). Vehicle storage tanks are heavily insulated and pressurized at relatively low pressures. In order to remain a liquid, LNG must be kept below −117°F.

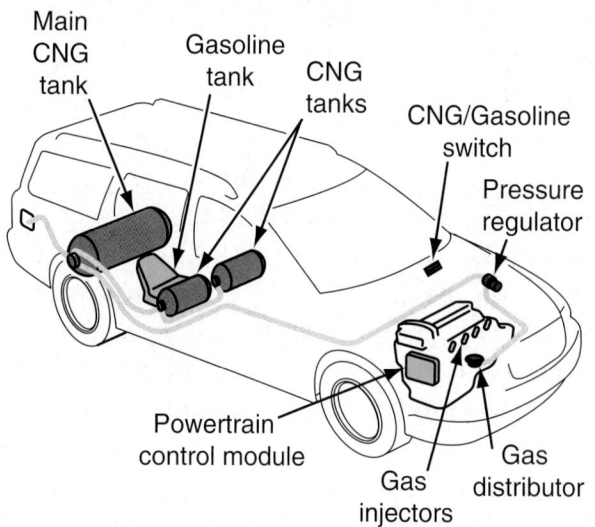

Figure 39.36 A bi-fuel CNG system.

The temperature of LNG will be about −200°F at 100 psig. LNG is kept cold using the principle of autorefrigeration. LNG boils at a very low temperature, 472°F colder than the boiling point of water. It remains at the same temperature as long as its pressure does not change. Like boiling water, which maintains its temperature as it is cooled by the evaporation of steam, the temperature of LNG remains the same as long as any steam that boils off is evacuated from the storage tank. If the vapor is allowed to remain inside the enclosure, the temperature and pressure will rise.

More in-depth information on alternative fuels can be found at http://www.naftc.wvu.edu/.

Hydrogen

Hydrogen is one of the most abundant elements in the universe. It is highly flammable and is a promising fuel for the future. It is an ideal fuel, producing no emissions other than water and carbon dioxide, which is a nonpolluting and nonpoisonous gas. The sun is an example of burning hydrogen.

Hydrogen is difficult to store due to the small size of its molecule. It can be produced from water (H_2O) by electrolysis. Electrolysis is a process where electrical current is applied to water. Producing hydrogen in this manner is expensive. Hydrogen is not very reactive unless it is subjected to high temperatures or pressures. Hydrogen and oxygen mixtures are very explosive when sparked or heated to high temperatures.

Hydrogen fuel can be used in an internal combustion engine, but it produces only about two-thirds of the normal power of a conventional gasoline-powered engine. The future of hydrogen as a fuel is widely believed to be in powering fuel cells (see Chapter 16). Pure hydrogen does not occur in nature. Producing it and distributing it economically are technologies of the future.

REVIEW QUESTIONS

1. What is the term for how easily a fuel evaporates?

2. What are the two types of abnormal combustion?

3. What is the term that describes a gasoline's ability to resist explosion during combustion?

4. What kind of ignition timing can cause an engine to overheat, advanced or retarded?

5. What is it called when an engine continues to run after its ignition is shut off?

6. Which kind of fuel system prevents engine run-on, carburetor or fuel injection?

7. What is the term that describes how easily a diesel fuel will ignite?

8. Which engine requires more frequent oil changes, diesel or gasoline?

9. What happens to propane or natural gas when they are pressurized?

10. What kind of fuel can natural gas be used to make?

ASE-STYLE REVIEW QUESTIONS

1. Which of the following would be more likely to experience vapor lock in hot weather?
   a. An engine with fuel injection.
   b. An engine with a carburetor.
   c. Both A and B.
   d. Neither A nor B.

2. Technician A says that exploding fuel pushes the piston down in the cylinder when the ignition timing is correct. Technician B says that exploding fuel pushes the piston down in the cylinder during normal combustion. Who is right?

   a. Technician A   c. Both A and B
   b. Technician B   d. Neither A nor B

3. Which of the following is most likely to result in a broken, unburned piston?
   a. Detonation   c. Both A and B
   b. Preignition   d. Neither A nor B

4. Technician A says that a car with a rich air-fuel mixture will surge on the highway. Technician B says that a lean mixture can result in burned parts. Who is right?
   a. Technician A   c. Both A and B
   b. Technician B   d. Neither A nor B

5. Which of the following is/are true about alcohol and gasoline?

   a. Alcohol fuels can carry more water than straight gasoline.

   b. Alcohol fuels have a higher octane number than gasoline.

   c. Both A and B.

   d. Neither A nor B.

6. Air and fuel mixed at a ratio of 9:1 is considered to be _____ in a gasoline internal combustion engine.

   a. Rich   **c.** Perfect

   b. Lean   **d.** Stoichiometric

7. Which of the following is/are true about B20 biodiesel fuel?

   a. It can be used in unmodified diesel engines.

   b. It is 20% nonpetroleum.

   c. It lowers exhaust emissions.

   d. All of the above.

8. Which of the following is/are true about alternative fuel vehicles?

   a. A flexible fuel vehicle (FFV) uses the same fuel system, whether its engine is running on gasoline or ethanol.

   b. In a bi-fuel engine, natural gas and air flow through the intake manifold to the cylinders.

   c. Dedicated NGVs use special natural gas injectors to inject the gas directly into the cylinders.

   d. All of the above.

9. Technician A says that a dual-fuel engine burns an air-fuel mixture combination of natural gas and diesel. Technician B says that a bi-fuel engine is one that can be switched to burn on one of two different fuels. Who is right?

   a. Technician A   **c.** Both A and B

   b. Technician B   **d.** Neither A nor B

10. Which of the following fuels does a flexible fuel vehicle (FFV) use?

   a. Ethanol   **c.** Both A and B

   b. Gasoline   **d.** Neither A nor B

CHAPTER 40

Fuel System Fundamentals

■ OBJECTIVES

Upon completion of this chapter, you should be able to:

✔ Explain the operation of the various carburetor systems.

✔ Compare fuel injection to carburetion.

✔ Identify the different types of fuel injection.

✔ Describe the design and function of electronic fuel injection components.

✔ Understand how a computer feedback system works.

■ KEY TERMS

filter sock
intake manifold vacuum
vaporization
venturi
accelerator pump
power valve
choke
feedback carburetor
mechanical fuel injection
continuous

fuel distributor
electronic fuel injection
pulse width
throttle body injection (TBI)
central fuel injection (CFI)
port injection system
MPI
SFI
ground side switching
sequential fuel injection system

multiport fuel injection (MFI)
central multiport fuel
injection (CMFI)
returnless fuel system
speed density system
air density system
actuators
closed loop

■ INTRODUCTION

The fuel system must be able to deliver a proper mixture of air and fuel to the engine so that it can be burned efficiently. It must also be able to store enough fuel so that the car can complete a trip of a few hundred miles. This chapter provides an overview of the operation, uses, and advantages of the different types of fuel delivery systems.

■ FUEL SYSTEM

The entire fuel system from the tank to the engine's intake valve is called the fuel delivery system. It includes a storage tank, a pump, a pressure regulator, one or more filters, fuel lines, and hoses used to deliver fuel to the engine's carburetor or fuel injection system (**Figure 40.1**). The job of the fuel induction system, whether it be fuel injection or carburetion, is to provide the engine with the correct mixture of burnable air-fuel mixture in all driving ranges from

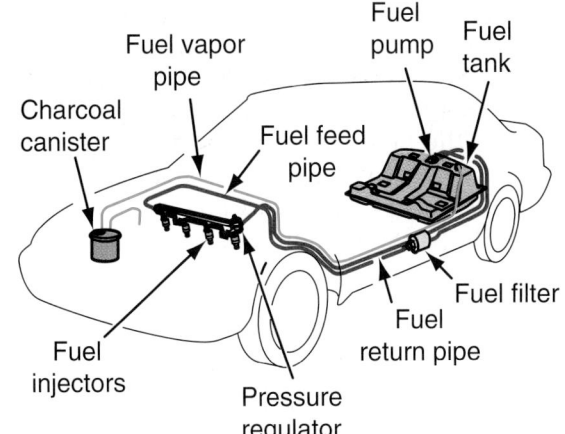

Figure 40.1 Parts of a typical fuel system.

idle to wide-open throttle (WOT). The combustion of gasoline and other fuels is covered in Chapter 39. The operation of carburetors and the major types of fuel injection systems are covered in this chapter, following coverage of their fuel supply systems.

FUEL TANKS

Fuel tanks on most cars hold between 12 and 20 gallons of fuel. Tanks are made of corrosion-resistant *galvanized* steel or plastic. Galvanized metal has zinc applied to it by electrolysis in much the same manner that chrome is applied to bumpers. This makes it resistant to rust.

The tank is made in two sections. Before the two halves are welded together, a vertical sheet metal *baffle* is installed in the tank (**Figure 40.2**). Fuel has a tendency to slosh back and forth in the tank when the car is driven. The baffle prevents this.

The fuel tank is usually hung in the back of the vehicle. On older vehicles, it was located under the trunk. In an effort to make vehicles safer in the event of a rear-end collision, gas tanks are now usually in front of the rear axle.

Hanger straps hold the tank in place. The fuel *pickup tube* is installed through a hole in the top of the tank. It is usually positioned about ½" from the bottom of the tank so that dirt and water will not be drawn into the fuel system. Water and gasoline do not mix. Because water is heavier than gasoline, it accumulates at the bottom of the tank where it can cause problems in below freezing temperatures.

A fuel tank *cluster* assembly includes a fuel pickup tube, the fuel gauge sending unit, and in fuel injected engines, a fuel pump. Most cars have an *in-tank filter* installed at the end of the pickup tube (**Figure 40.3**). It is usually made of woven plastic, sometimes called a filter strainer. Sometimes, over a period of years these filters plug up, requiring replacement. When an older car has a cloth-type filter, this is called a **filter sock**. The term is often used to describe a strainer.

The fuel gauge has a float attached to a sliding contact. As the float moves up or down, the contact position changes, varying the signal that the fuel gauge receives. An electric fuel pump that runs while immersed in gasoline is included in most late-model fuel tank clusters.

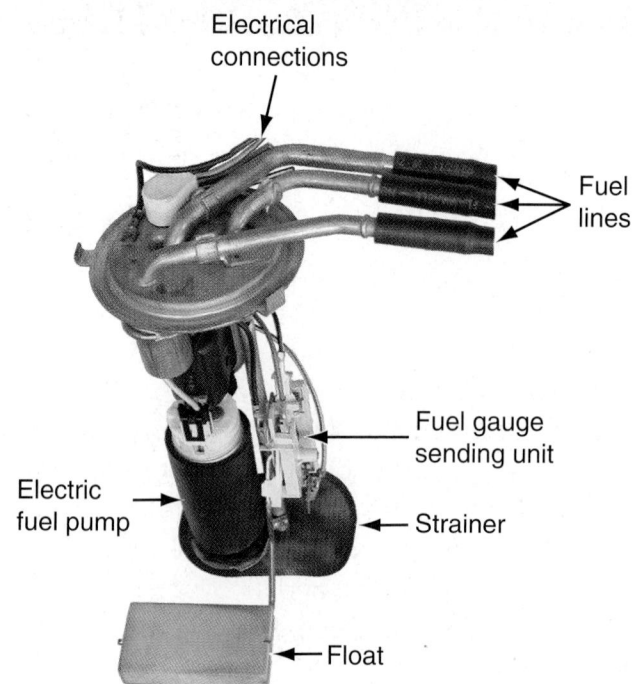

Figure 40.3 Combination electric fuel pump and sending unit. *(Courtesy of Tim Gilles)*

The tank also includes expansion and overfill protection, covered in Chapter 43.

FUEL LINES, HOSES, AND FITTINGS

To transport fuel from the tank to the engine, steel lines made of seamless tubing run the length of the frame. Wherever there is a flexible connection, a hose is used. Fuel tubing, fittings, hoses, and clamps are covered in other chapters.

FUEL PUMPS

Earlier carbureted cars used a low-pressure mechanical fuel pump with a diaphragm driven by an eccentric on the engine's camshaft. Two one-way check valves allow fuel to flow into and out of the pump.

Fuel for both mechanical and electronic fuel injection systems is supplied by an electric fuel pump (**Figure 40.4**). Electric fuel pumps are of either the

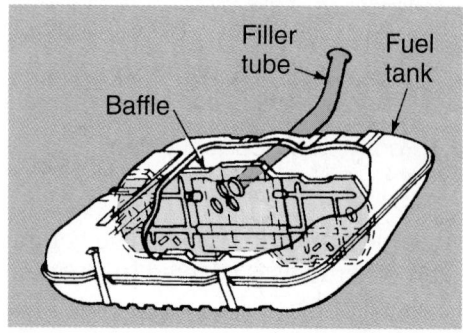

Figure 40.2 A gasoline tank with a baffle.

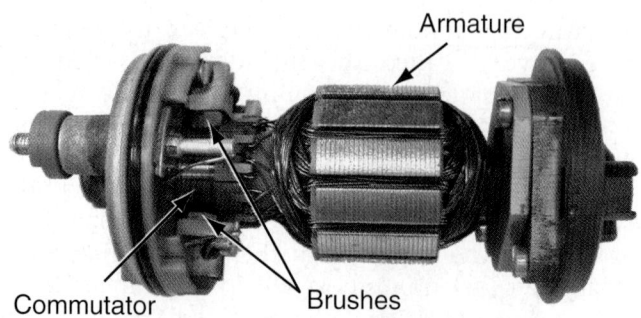

Figure 40.4 Parts of an electric fuel pump. *(Courtesy of Tim Gilles)*

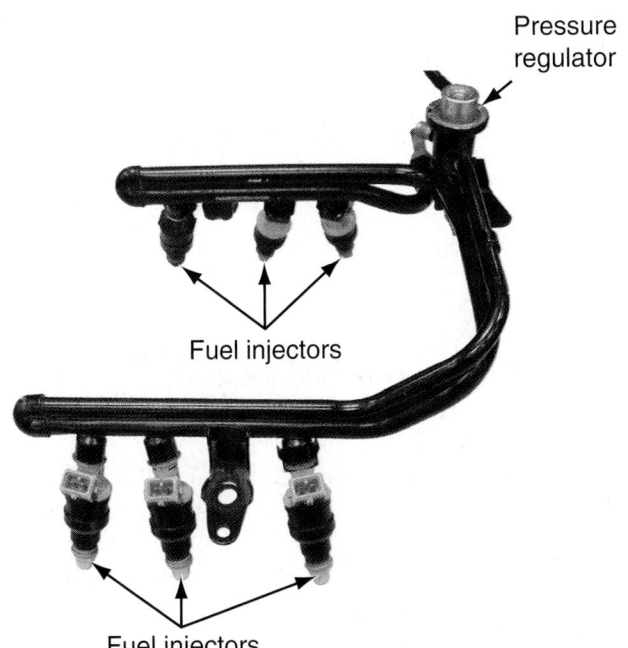

Figure 40.5 A fuel rail for a V6 engine. *(Courtesy of Tim Gilles)*

external or internal (inside the fuel tank) types. Fuel from the pump flows in a fuel rail loop between the engine and the fuel tank (**Figure 40.5**). A pressure regulator controls fuel pressure within the system so an exact, predetermined amount of fuel will be injected each time an injector is opened.

The electric fuel pump has a one-way check valve that maintains pressure in the fuel loop when the engine is off. This residual pressure allows for fast starting, as the fuel is already at the correct pressure for the fuel injectors to operate during engine cranking.

Electrical current is supplied to the fuel pump armature through brushes to commutator strips so current can continue to flow as the armature spins. Fuel

pumps with more commutator segments tend to last longer than those with fewer.

At first glance, it would seem that having the electric fuel pump in the fuel tank would be dangerous. The fuel pump is at the lowest portion of the fuel tank and fits into a well, or reservoir, of fuel. Because it is always immersed in fuel, it is always cooled and the brushes cannot spark because there is no air.

Fuel Pump Electrical Circuit

Electric fuel pumps on modern vehicles are computer controlled. In a typical circuit (**Figure 40.6**) when the ignition switch is in the crank or run position, the computer provides a ground path to the fuel pump relay. With the relay energized, current flows to the fuel pump, which remains on when the engine is cranking or running. When the ignition is turned on and the engine is not cranked, the pump only remains on for 0.5 to 2 seconds. On some engines, an oil pressure switch must detect oil pressure for the pump to remain on. Others use an engine intake airflow signal or a tachometer signal to ensure that the engine is running and allowing electrical power to continue to the fuel pump.

■ FUEL FILTERS

Fuel filters can be located in a fuel line or in the tank (**Figure 40.7a**). Fuel filters that are installed on the outlet side of a fuel pump are called *outlet filters* (**Figure 40.7b**).

Fuel injection systems require large, heavy-duty filters. They filter out smaller particles of dirt while still allowing the pump to supply a substantial amount of fuel.

■ FUEL INJECTION AND CARBURETION

There is a difference in pressure between outside air (atmospheric pressure) and the pressure inside of the intake manifold (**intake manifold vacuum**). As

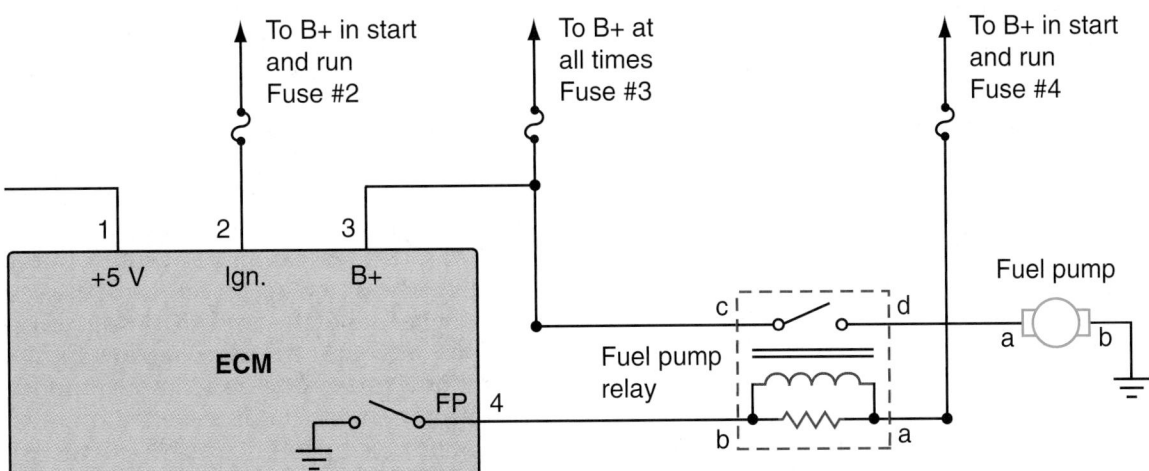

Figure 40.6 A wiring diagram for a typical fuel pump circuit. *(Courtesy of ASE)*

(a)

Inlet

Fuel enters here

FLOW

Outlet

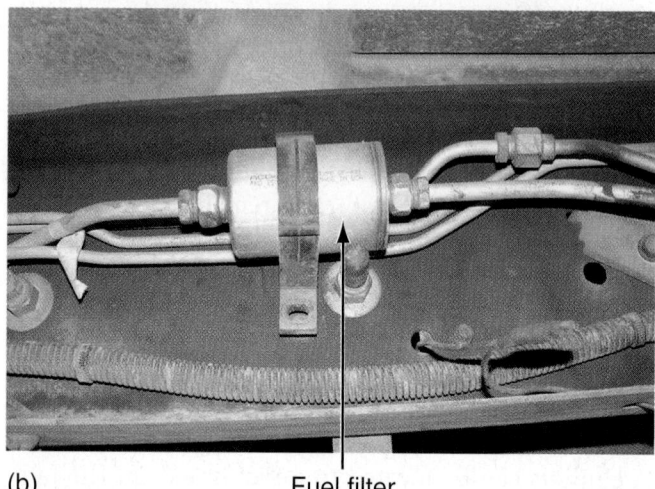

(b)

Fuel filter

Figure 40.7 (a) Cutaway of a typical fuel filter. Note the direction of fuel flow. (b) An external fuel filter. *(Courtesy of Tim Gilles)*

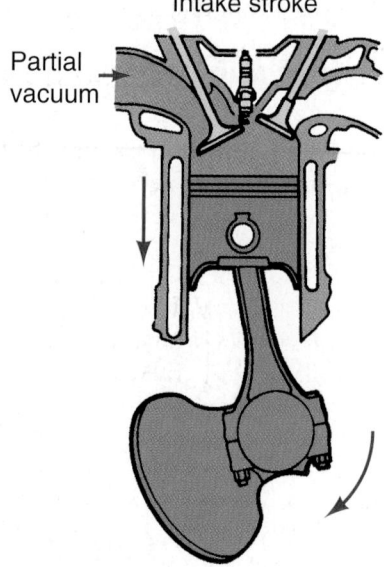

Intake stroke

Partial vacuum

Figure 40.8 A lower pressure area develops in the intake manifold as the piston moves down. *(Courtesy of Ford Motor Company)*

the engine's pistons move down in their cylinders, a lower pressure area develops in the intake manifold (**Figure 40.8**). Outside air moves into the intake manifold because it is under higher pressure.

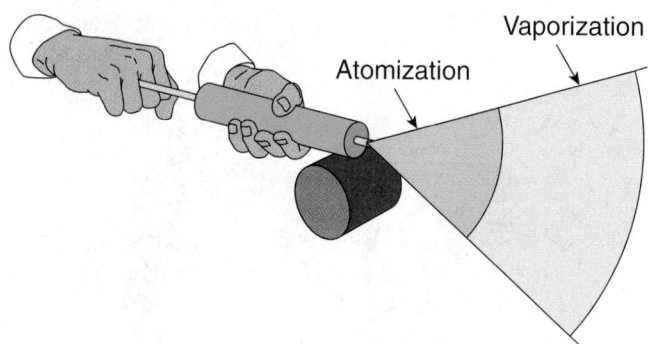

Atomization

Vaporization

Figure 40.9 Atomization and vaporization.

Fuel and air need to arrive at the cylinders as atomized mist, like fog, so it can burn easily. *Atomization* is the name of the process by which fuel is suspended in air in a mist of tiny droplets (**Figure 40.9**). Atomized fuel is more exposed to air, which helps it to vaporize. Liquid fuel must become a vapor before it can burn in the cylinders. **Vaporization** occurs when atomized fuel turns into a gas (vapor). In carbureted systems, vaporization occurs in the carburetor and intake manifold.

Most vehicles built since the mid-1980s have been equipped with fuel injection, but carburetors are still found on small engines and motorcycles. Carburetors are old technology, but there are still many collector cars or hot rods on the road that have a carburetor. As government exhaust emission and fuel economy standards became more stringent, the increasing complexity of carburetors caused them to become more and more costly. Fuel injection became competitive with them and is the prevalent system used today (**Figure 40.10**).

The same basic principles apply to both carburetors and fuel injection, so we will start with carburetors and then build on that theory through fuel injection.

The carburetor's job is to atomize air and fuel in the proper proportion. The carburetor is mounted on top of an intake manifold that vaporizes and distributes the mixture to the intake valve ports.

The inside of the carburetor barrel has a **venturi** that restricts airflow (**Figure 40.11**). The smallest area of the venturi has the least amount of pressure. This is where the main *nozzle* is positioned. The lower end of the passage to this nozzle goes to the *float bowl*, which serves as a reservoir for the fuel.

Fuel in the float bowl is under atmospheric pressure. When the engine is running or cranking, air rushes through the carburetor venturi into the intake manifold. The difference in pressure between the float bowl (higher) and the venturi (lower) causes fuel to be drawn into the stream of air flowing through the carburetor (**Figure 40.12**). As more air flows through the venturi, more vacuum is exerted on the fuel in the fuel nozzle. The reduced pressure in the venturi also helps the fuel to atomize.

If the engine ran at the same speed all of the time, no provision for changing the amount of air and fuel

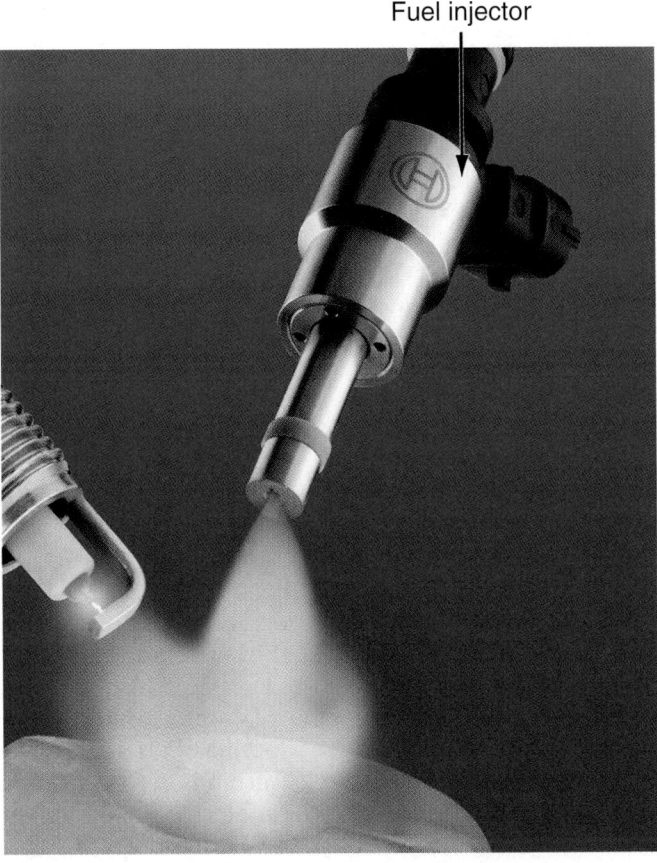

Figure 40.10 Fuel injectors spray fuel into the intake manifold or valve port. *(Reprinted with permission from Bosch)*

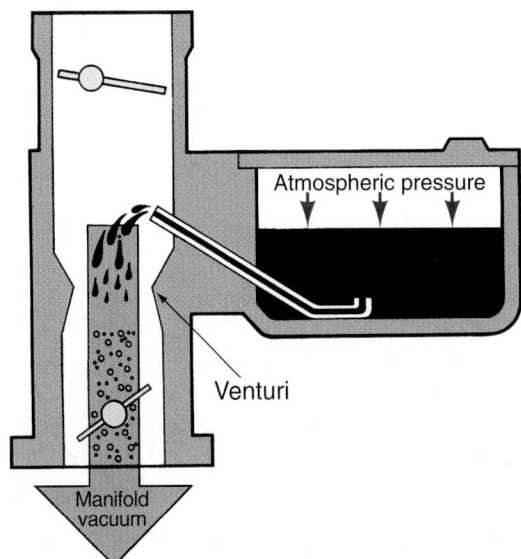

Figure 40.12 Higher pressure in the float bowl and lower pressure in the venturi result in fuel being drawn into the stream of air flowing through the carburetor.

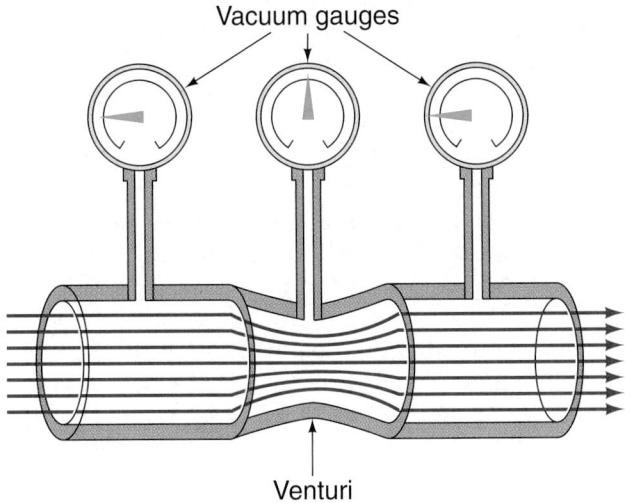

Figure 40.11 The center gauge shows more vacuum (less pressure) in the venturi.

intake would be needed. But the flow of air and fuel needs to be changed to allow for differences in engine load and speed. **Figure 40.13** shows the parts of a basic carburetor.

Throttle Plate

Airflow through the engine is changed by opening the *throttle plate*. The throttle plate is a butterfly valve positioned in the bottom section of a carburetor or at the inlet to a fuel injection system (see Figure 1.11).

When the driver depresses the accelerator pedal, the throttle plate opens. This allows more air-fuel mixture to enter the engine, raising engine rpm. Closing the throttle causes engine speed to drop.

Carburetors have several circuits: the float circuit, the idle and low-speed circuits, the main metering circuits, the power enrichment circuit, the accelerator pump circuit, and the choke circuit.

- The *float circuit* works the same way that a toilet does; when the level of fuel rises to a predetermined level, the float closes a *needle valve* against its *seat*, stopping further flow from the fuel pump. As the engine uses fuel the float level drops, opening the needle valve and allowing fuel to flow in again from the pump. The *float level* is important. If it is too high, the air-fuel mixture will be too rich; if too low, the mixture will be too lean.

- A replaceable *main jet* provides a specified opening that meters the amount of fuel that can enter the fuel nozzle from the float bowl. Flow through the jet determines how rich the air-fuel mixture will be.

- An idle port located just below the throttle plate allows a small amount of air and fuel to be metered into the intake manifold when the throttle is closed.

- An **accelerator pump** with a small, spring-operated piston or diaphragm provides an extra squirt of fuel when the vehicle is accelerated quickly, such as when passing another car.

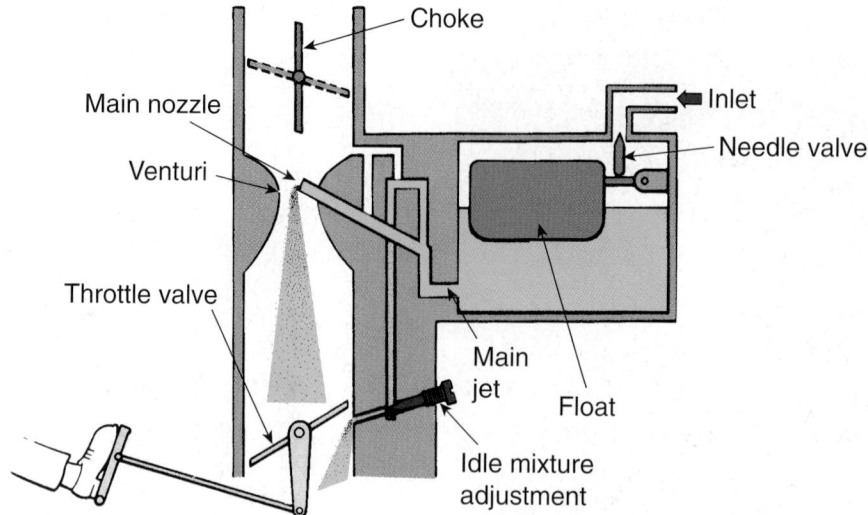

Figure 40.13 Basic parts of a carburetor.

- When vacuum is low during heavy acceleration, a **power valve** allows extra fuel to bypass the main jet.
- The **choke** is a butterfly valve like the throttle valve located in the top of the carburetor. When it is "closed," incoming air is restricted, causing a rich air-fuel mixture to help the engine run better when cold.

Feedback Carburetors

As fuel injection was coming into widespread use, a new carburetor was developed. These **feedback carburetors** meter the fuel according to how much oxygen is sensed by an oxygen sensor in the engine's exhaust. The oxygen sensor sends a signal to the computer, which adjusts the air-fuel ratio accordingly.

Fuel Injection Operation

Due to concern about excessive use of foreign oil and a desire for cleaner air, the federal government established standards for fuel efficiency and pollution control in the early 1970s. Because it provided a means of meeting these standards, fuel injection became the standard fuel system used in automobiles.

New cars and trucks are equipped with fuel injection systems that serve the same function as the carburetor that they replaced, but fuel injection provides a better means of controlling exhaust emissions and fuel economy.

There have been many different fuel injection system designs prior to the sequential fuel injection systems used on modern automobiles.

The **mechanical fuel injection** systems found on some imported cars are called **continuous** (**Figure 40.14**). A high-pressure pump routes fuel to a **fuel distributor**, where tubes carry it to an injector at each intake valve. Mechanical fuel injection systems

are trouble-free but expensive. They are found mainly on older European cars.

Electronic fuel injection systems are found on most modern cars. Earlier domestic engines used throttle-body injection. Port injection is used on newer engines. Both systems use similar electromagnetic fuel injectors, pulsed electrically by the computer to turn them on and off (**Figure 40.15**). The computer controls the amount of fuel injected during each pulse, determining the length of time each fuel injector should remain open. Because the fuel pressure in the supply rail to the injectors is closely controlled, the amount of fuel passing through an injector can be predicted. The length of time that an injector remains open is called its **pulse width**.

Throttle-body injection (TBI) systems (**Figure 40.16**) were a bridging technology between carburetors and fuel injection and used an intake manifold similar to a carbureted system. The throttle body has one fuel injector per barrel precisely controlled by the computer, but, like a carburetor, it is difficult to provide equal delivery of atomized fuel to all of the engine's cylinders. **Central fuel injection (CFI)** systems are similar to TBI systems in that they supply fuel from an injector in a central location.

A **port injection system** uses a fuel rail with individual fuel injectors at each intake port. **Sequential fuel injection systems** are those in which fuel is injected into the intake port at a precise point in relation to crankshaft and piston position. **Multiport fuel injection (MFI)** describes systems where injectors are pulsed in groups.

Central multiport fuel injection (CMFI) is a system with a central injector in the intake manifold. Individual nylon fuel tubes and poppet nozzles deliver metered fuel to the cylinders (**Figure 40.17**).The poppet nozzles have ball check valves that are forced open

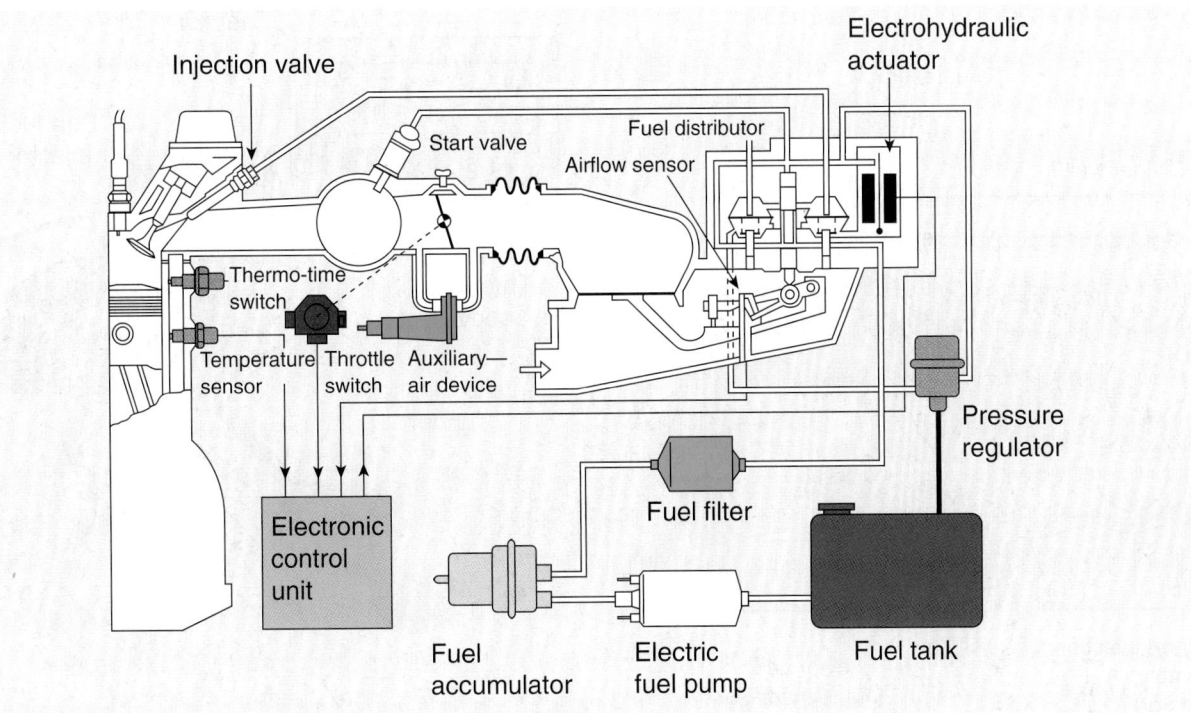

Figure 40.14 A mechanical injection system uses a fuel distributor.

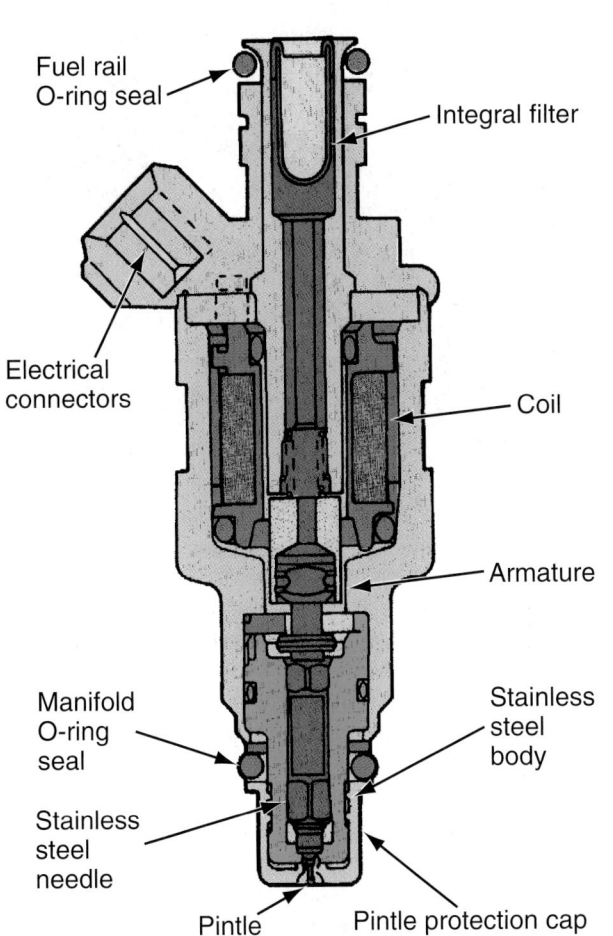

Figure 40.15 A typical fuel injector used in multiport fuel injection systems. *(Courtesy of Ford Motor Company)*

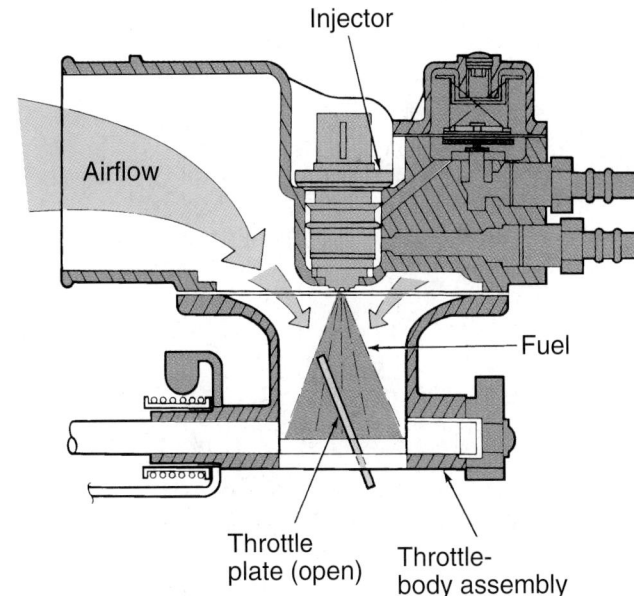

Figure 40.16 Throttle-body injection.

against spring pressure (**Figure 40.18**). When the fuel in the tube reaches a specified pressure, fuel is sprayed into a cylinder's intake port.

Another version of CMFI is *central sequential multiport fuel injection*. Instead of a central injector, this system has one sequentially fired injector for each poppet nozzle. Due to problems with sticking poppet nozzles, newer systems have been redesigned with the injectors moved out to the ports, although the fuel injection system is still housed within the intake manifold.

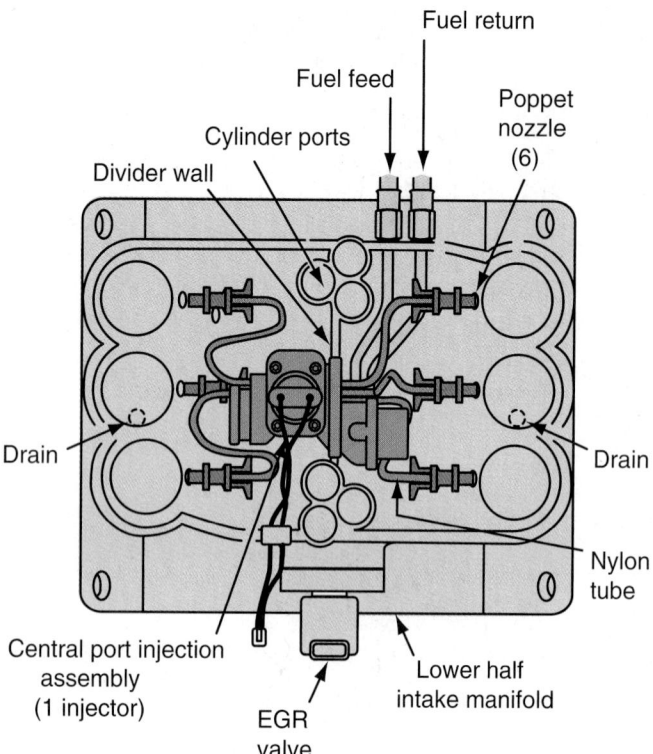

Figure 40.17 In central multiport fuel injection, the central injector is in the intake manifold.

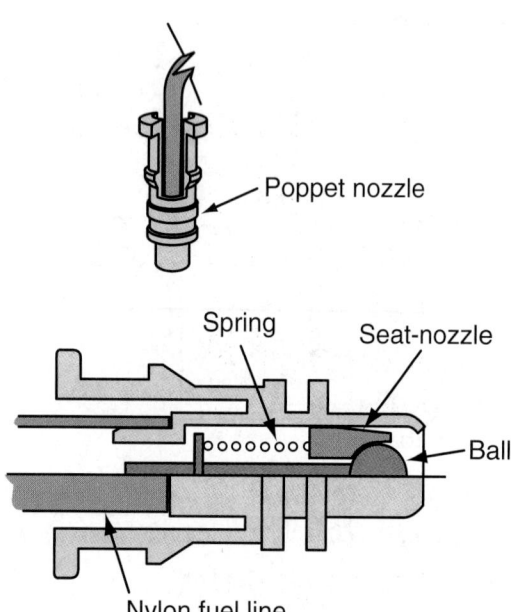

Figure 40.18 Poppet nozzles have ball check valves that are forced open against spring pressure.

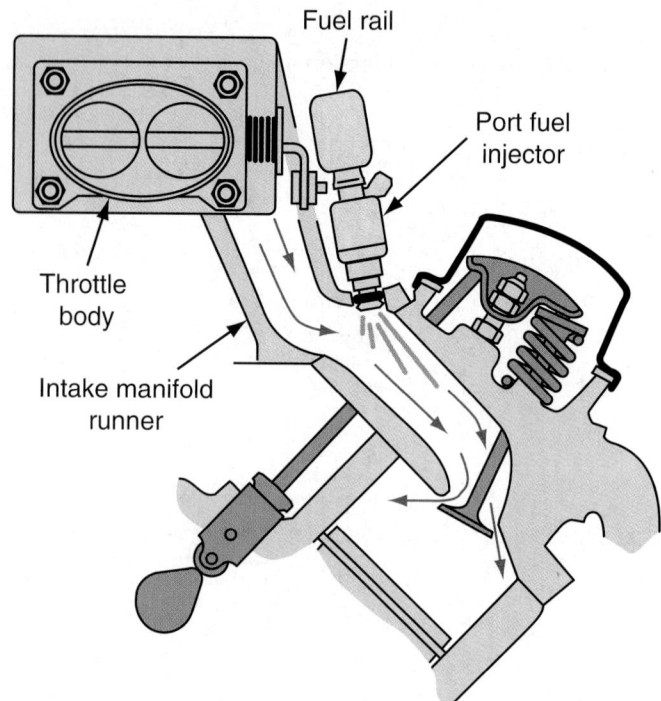

Figure 40.19 In a port fuel injection system, air and fuel are mixed right outside the combustion chamber.

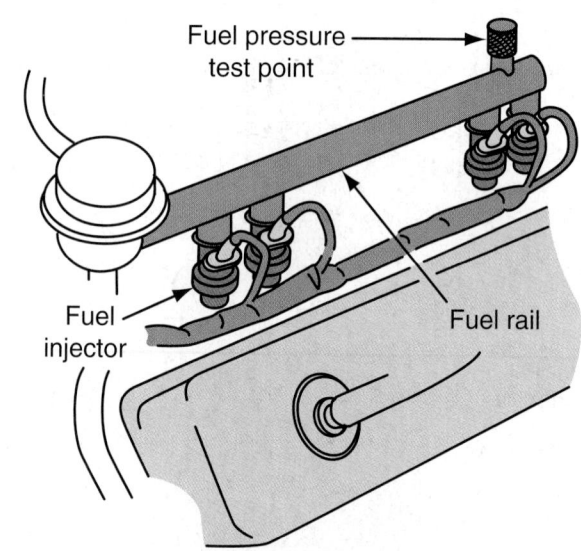

Figure 40.20 A port injection system.

Typical port fuel injectors spray fuel into the intake port about 3" above the valve (**Figure 40.19**). A fuel rail connects the individual injectors (**Figure 40.20**).

A major advantage of port fuel injection over TBI is that engineers can design intake manifolds with long, tuned runners. This results in higher engine torque at lower engine speeds. The intake manifold can be tuned this way because it only carries air, not air and fuel like TBI manifolds. Fuel injectors are fed either from the top or from the bottom. Bottom-feed injectors are used in TBI systems.

Port Injection Firing

Port fuel injection systems fire their injectors in different ways. Older MFI systems fire their injectors in pairs or groups (**Figure 40.21**). The computer fires the injectors a sufficient amount of time prior to intake valve opening so the intake port is filled with fuel. When there are two groups, one group fires during

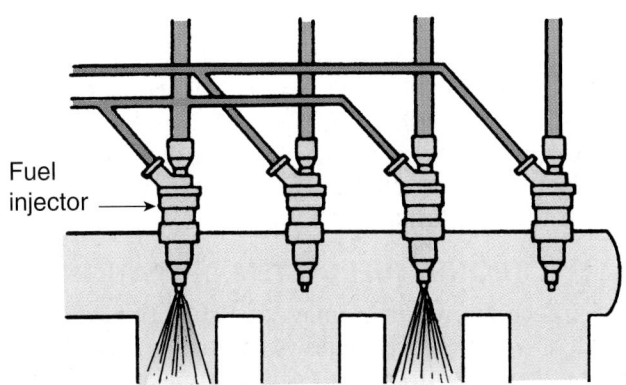

Figure 40.21 Grouped single-fire port injection.

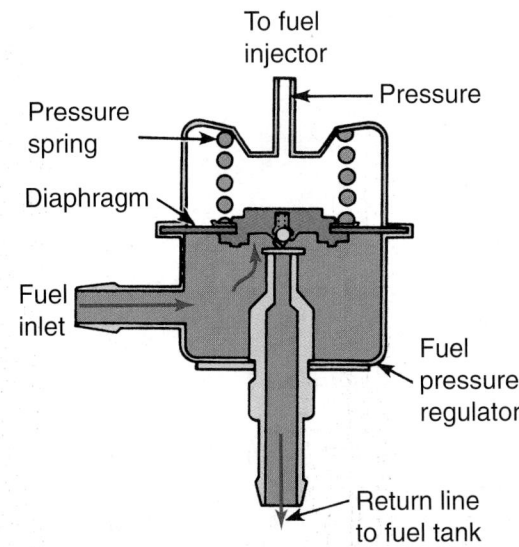

Figure 40.22 A fuel pressure regulator.

each alternate revolution of the crankshaft. Some MFI systems fire their injectors every crankshaft revolution. They inject only half as much fuel, because they do it with every turn of the crankshaft. In these systems, fuel must "wait its turn" in its port until the intake valve opens before it is drawn into the cylinder. Because ignition cycles are occurring at such a fast pace, even at idle, this is not a problem.

Sequential Fuel Injection

Advanced computer controls provide more accurate injection of the fuel. A sequential fuel injection (SFI) system opens each injector just before its intake valve opens. The mixture does not have to "wait" in the intake port, and adjustments to the fuel mixture can be made very quickly. Inputs from the camshaft and crankshaft position sensors tell the computer when to fire individual injectors.

Each injector has its own connection to the computer. With **ground side switching**, the computer completes the ground for each injector in sequence, causing it to fire. MFI systems use a common ground wire to the computer for their injector groups. European injection systems use the computer to power the injectors, rather than ground them.

NOTE: *Some port fuel-injected engines that have four valves per cylinder use two fuel injectors per cylinder. The primary injector is used all of the time, and the secondary one is used for high power, injecting fuel into the high-speed ports in the intake manifold.*

■ PRESSURE REGULATOR OPERATION

Fuel is pushed from the fuel tank by the fuel pump. The pressurized fuel is sent to the fuel filter and to the injector rail. A *fuel pressure regulator* (**Figure 40.22**) controls the system's maximum amount of pressure. In a loopfuel system, excess fuel is directed by the regulator to the return line to the fuel tank. Some systems have a damper with a diaphragm that controls slight variations in fuel pressure from the pump, and some have adjustable fuel pressure regulators.

Figure 40.23 A typical fuel pressure regulator for a port injection system.

Port injectors are located in the intake port so they are exposed to intake manifold vacuum. Because changes in vacuum can change the amount of fuel injected, port injection systems use a pressure regulator sensitive to manifold vacuum. It changes fuel pressure so pressure across the tips of the fuel injectors remains constant.

Under load, when intake manifold vacuum drops, the regulator maintains higher fuel pressure to compensate. At idle where intake manifold vacuum is higher and fuel can enter the cylinder more easily, the regulator compensates by lowering fuel pressure. A pressure regulator for a port injection system is shown in **Figure 40.23**. Notice the vacuum line connected to it.

Returnless Fuel Pressure Regulators

Many newer vehicles use **returnless fuel systems**. Typical fuel pressure regulators have a return line to the

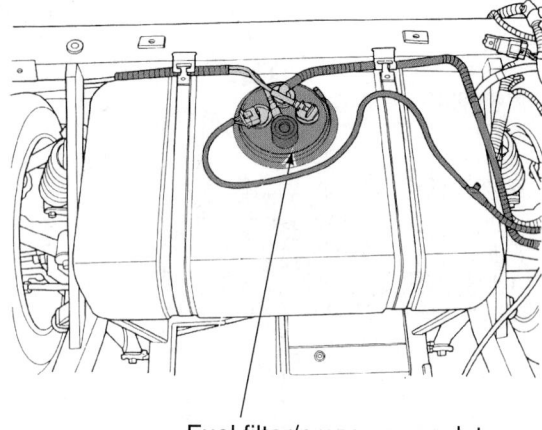

Fuel filter/pressure regulator

Figure 40.24 Returnless fuel system with the pressure regulator and filter mounted in the fuel tank with the fuel pump and fuel gauge sending unit. *(Courtesy of DaimlerChrysler Corporation)*

tank. Returnless systems have the fuel pressure regulator and filter mounted in the gauge and pump cluster mounted in the top of the fuel tank (**Figure 40.24**). The regulator exhausts excess fuel pressure directly to the tank.

Returnless systems have only one fuel line between the fuel pump and the fuel rail to the injectors. Unlike looped injection systems, no return line runs from the fuel rail back to the fuel tank. Having no return line lessens the chance of heat from the engine compartment raising the temperature of the fuel, resulting in increased evaporative emissions. Besides the lack of a return line, a returnless system can be identified by the lack of a pressure regulator on the rail.

Returnless means that the fuel does not move through the fuel rail, although the fuel pump provides more fuel than is needed and still returns the rest to the tank. Typical fuel travel is as follows: Fuel is pumped from the tank, through a filter to a "T"connector (**Figure 40.25**). From there, any excess

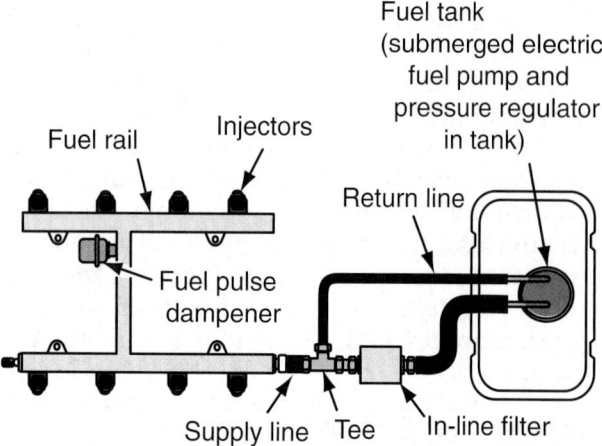

Figure 40.25 Returnless system with fuel pumped from the tank through a filter to a "T" connection.

fuel not required by the injectors returns to the fuel tank by way of the regulator in the fuel gauge sending unit at the top of the fuel tank (see Figure 40.24). The fuel either leaves the tank and returns after the filter or it circulates within the tank, depending on whether the fuel filter is externally mounted or is housed within the fuel tank.

▪▪ ELECTRONIC FUEL SYSTEM OPERATION

Fuel injectors are electromagnetic solenoid controlled nozzles. When electricity flows through a coil of wire, a magnetic field is produced. Each fuel injector is supplied with power when the ignition is on. The computer controls the ground or power to the injector to complete the circuit. In most systems when the injector is grounded, the solenoid coil of the injector is energized, creating a magnetic field. The plunger of the injector is pulled against spring tension by the magnetic field. This pulls the valve from its seat (see Figure 40.26).

Figure 40.26 shows the components of a typical electronic fuel injection system. On older cars, when the coolant temperature is low, a *cold start injector* operates during engine cranking. To prevent accidental flooding, a *thermal time switch* limits the maximum time this injector can operate.

Port fuel injectors are installed in a hole in the manifold or cylinder head. An insulating O-ring seals vacuum and keeps fuel in the injector from becoming heated by surrounding engine heat (**Figure 40.27**).

▪▪ AIRFLOW MEASUREMENT

The fuel injection system requires a means of determining how much air is flowing into the engine. There are different ways of accomplishing this.

Speed Density (MAP) Systems

Speed density systems are used on some engines. The computer uses a *manifold absolute pressure* (MAP) *sensor* and engine rpm to calculate the amount of air entering the engine. The MAP sensor (**Figure 40.28**) indicates the pressure in the intake manifold. Manifold pressure, density of the air, and engine speed are all considered by the computer before it calculates the correct amount of fuel to inject into the cylinders.

Other inputs are used to fine-tune the air-fuel ratio in a speed density system. If the throttle position sensor indicates sudden acceleration, for instance, the computer provides a momentarily richer air-fuel mixture.

Air Density Systems

Another system used by EFI systems is the **air density system**. An *airflow sensor* measures the volume of air entering the engine. Types of airflow sensors include the vane, the grid, and the hot wire.

Vane-Type Airflow Sensor. The vane-type mass airflow (MAF) sensor is found on many import and domestic

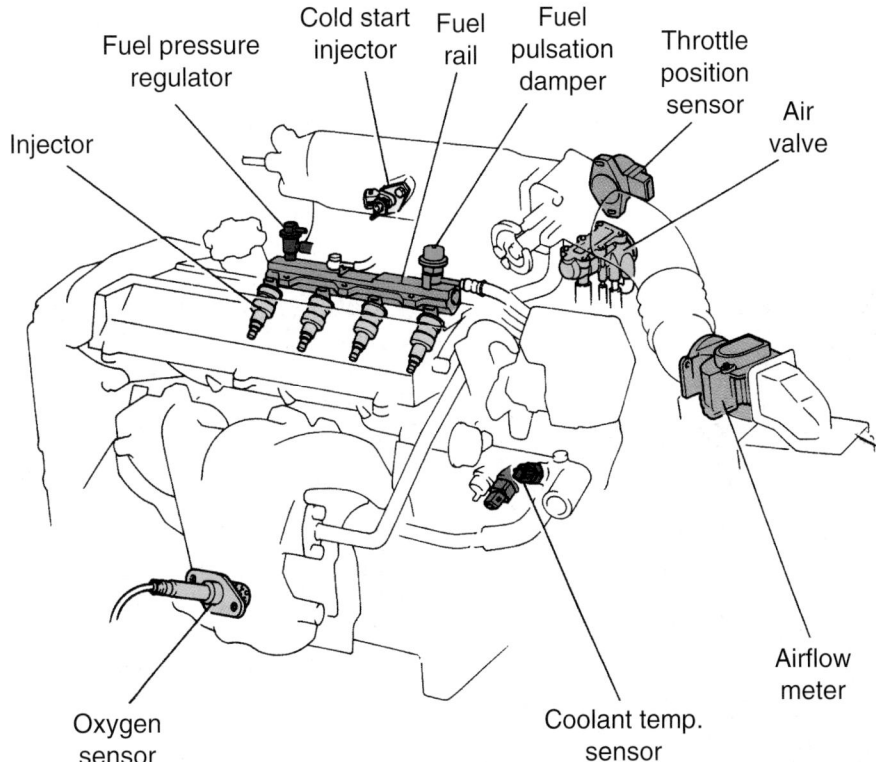

Figure 40.26 Components of a typical electronic fuel injection system.

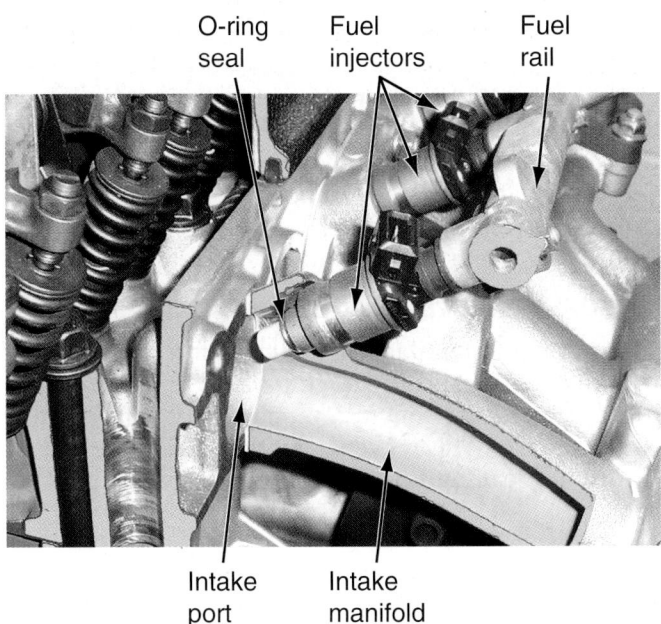

Figure 40.27 An insulating O-ring seals vacuum and keeps heat from the injector. (Courtesy of Tim Gilles)

vehicles with EFI. All intake air must flow through the sensor. Some MAF sensors are called *volume airflow meters*. A pivoted air-measuring plate is held closed with a light spring. Air movement across the plate moves it to the open position (**Figure 40.29**). A movable pointer attached to the plate wipes across a

Figure 40.28 A MAP sensor.

potentiometer. Movement of the plate sends a variable signal to the computer.

Heated Resistor MAF Sensor. A heated resistor airflow sensor has a resistor or electric grid in the air intake that is heated to a certain temperature whenever the key is on. As the engine is accelerated, more air flows across the resistor to cool it. More current is required to keep the resistor at the same temperature. The computer uses this signal to determine the amount of fuel to inject into the cylinder.

Hot Wire MAF Sensor. In a hot wire MAF sensor system, a sensor located next to a hot wire tells the computer

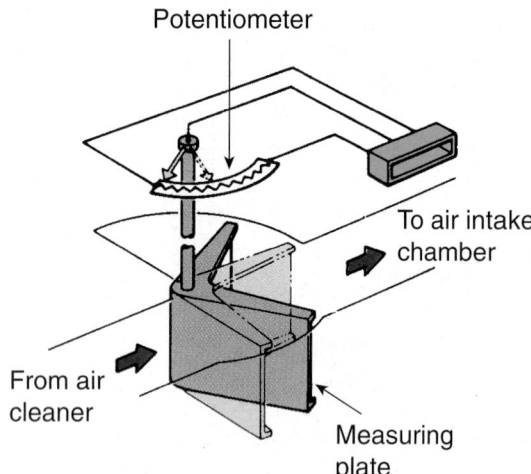

Figure 40.29 Movement of the air measuring plate changes the output from the potentiometer to the computer.

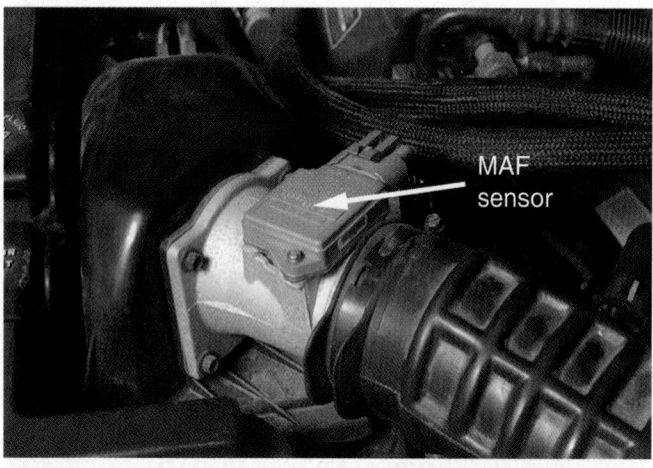

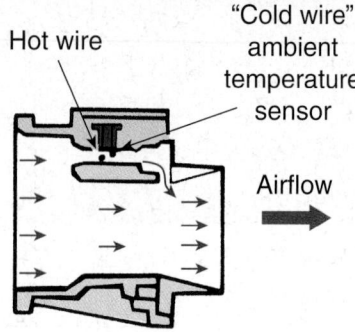

Figure 40.30 A hot wire mass airflow sensor. *(bottom, Courtesy of Ford Motor Company)*

the temperature in the intake system (**Figure 40.30**). When the ignition is on, the computer heats the wire to a specified temperature above the temperature of the air in the intake manifold. When the engine is accelerated, fresh air cools the wire and more electrical current is provided to heat it up. The computer calculates how much air is flowing into the engine by interpreting the signal generated by this action.

NOTE: *The Robert Bosch Corporation is a major fuel injection manufacturer. The three main types of fuel injection system are characterized by Bosch as follows:*

- *D-Jetronic – this early electronic fuel injection (EFI) system is a speed density system that uses a MAP sensor.*
- *L-Jetronic – this EFI system uses a mechanical (air door) or electronic (mass airflow) sensor and is an air density system.*
- *K-Jetronic – this is the mechanical fuel injection system.*

■ IDLE SPEED CONTROL

When an engine is cold or when there are extra loads on it (air conditioning, charging, and so on), the engine's idle speed needs to be raised to compensate. Idle speed is raised by allowing more air to bypass the throttle plate. Some systems use an *auxiliary air valve*. Others use an *air by-pass valve* or an *idle speed control* *(ISC) motor*.

A *throttle position sensor (TPS)* attached to the throttle body on the air inlet senses how many degrees the throttle plate is open.

A *coolant temperature sensor (CTS)* installed in the cooling system tells the computer the coolant's temperature. The computer responds by increasing injector duration, improving cold engine driveability.

An intake *air charge temperature sensor* senses the temperature of the incoming air (**Figure 40.31**). As temperature changes, volume and density of air change as well. Although the volume of air measured entering the engine is the same, the amount of fuel that is injected will be varied with temperature. When the temperature is above about 70°F, the computer decreases the volume of fuel coming from the injector. Below 70°F, the injector volume is increased.

■ FUEL PUMP RELAY

During engine cranking, the fuel pump is energized. If the key is on and the engine has not been cranked for 2 seconds, the fuel pump relay shuts off power to the fuel pump. This prevents flooding or engine lockup (in case of leaking injectors). It also prevents the fuel

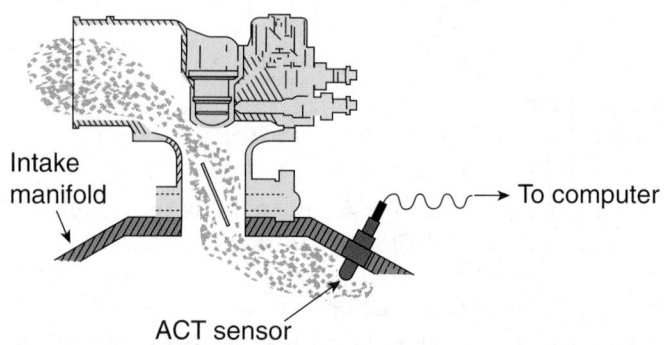

Figure 40.31 An air charge temperature sensor.

pump from running with the ignition switch on, in case of an accident or broken fuel line.

Some engines are equipped with an oil pressure switch hooked in parallel to the fuel pump relay. If the fuel pump relay is defective, the *oil pressure switch* will allow the fuel pump to keep operating.

■ COMPUTER-CONTROLLED FUEL SYSTEMS

Precise fuel metering to the engine is provided by onboard computers. These systems first became available on some cars in the late 1960s to control the fuel system. New vehicles today have computers controlling many systems, from transmission shift points to accessories. The *powertrain control module* (PCM) is the one that controls engine performance, including the fuel system.

Automotive ignition and electronics are complex specialty areas. Discussion here deals with electronic fuel system controls only. The intent is to provide a general idea of the operation of the system.

■ FEEDBACK FUEL SYSTEMS

Three types of devices make up an automobile computer system: the computer, *sensors,* and **actuators**. A computer monitors input from various sensors. Sensors relay information to the computer about throttle position, air and coolant temperature, airflow, manifold

pressure, and barometric pressure, among others. Actuators carry out the assigned change from the computer (**Figure 40.32**). In fuel injection systems, the actuators are the fuel injectors. Electronic systems have an advantage over older carburetor systems, which did not compensate for changes in altitude and temperature. Electronic systems can vary the amount of fuel according to information received from sensors and acted upon by actuators.

Engines with computer feedback have a sensor located in the exhaust manifold called an *oxygen sensor* (**Figure 40.33**). Other names for the oxygen sensor are O_2 *sensor*, *lambda sensor*, or *heated exhaust gas oxygen* (HEGO) *sensor*. O_2 sensors send a voltage signal to the computer that is generated in an amount that varies according to the amount of oxygen in the engine's exhaust.

In a feedback fuel system, the computer acts to make corrective changes to the air-fuel mixture entering the engine's cylinders (**Figure 40.34**). The air-fuel mixture is varied by increasing or decreasing the fuel injector pulse width.

Feedback carburetors were used on some older cars. These electronic carburetors changed the air-fuel mixture by varying the amount of pulsations a solenoid

Figure 40.32 The computer receives signals from sensors and sends commands to actuators.

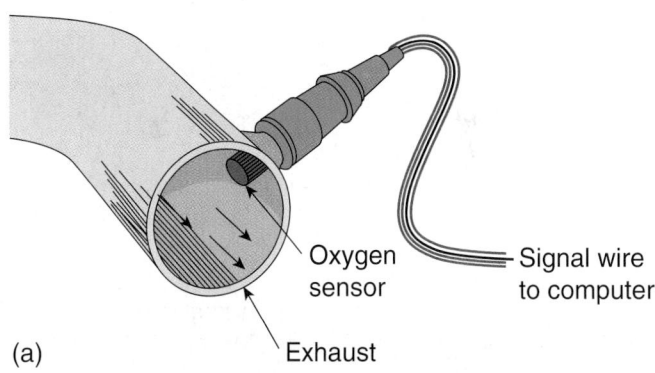

(a)

(b)

Figure 40.33 An oxygen sensor in an exhaust manifold. (*b, Courtesy of Tim Gilles*)

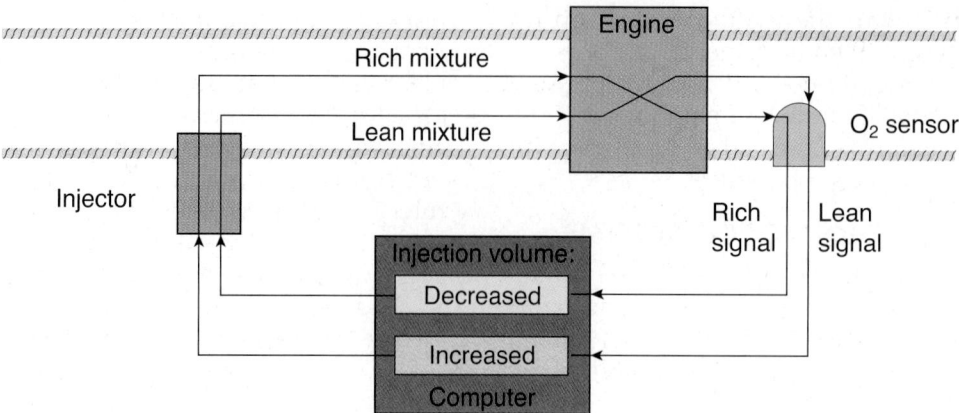

Figure 40.34 The computer changes the air-fuel mixture, in response to feedback from the O_2 sensor.

made. A fuel specialist can interpret meter readings of feedback fuel systems.

Open and Closed Loop

In *open loop,* the O_2 sensor does not send signals to the computer. The O_2 sensor cannot operate until it reaches about 600°F. **Closed loop** fuel control occurs when the engine reaches operating temperature and the computer starts acting upon information received from the O_2 sensor (**Figure 40.35**). The oxygen sensor reads and adjusts the air-fuel mixture about 10 times/second.

O_2 Sensor Operation

There is more than one kind of O_2 sensor used on cars. Also, some cars use more than one sensor. The most popular one, the *zirconium oxide* (ZrO_2) works like a

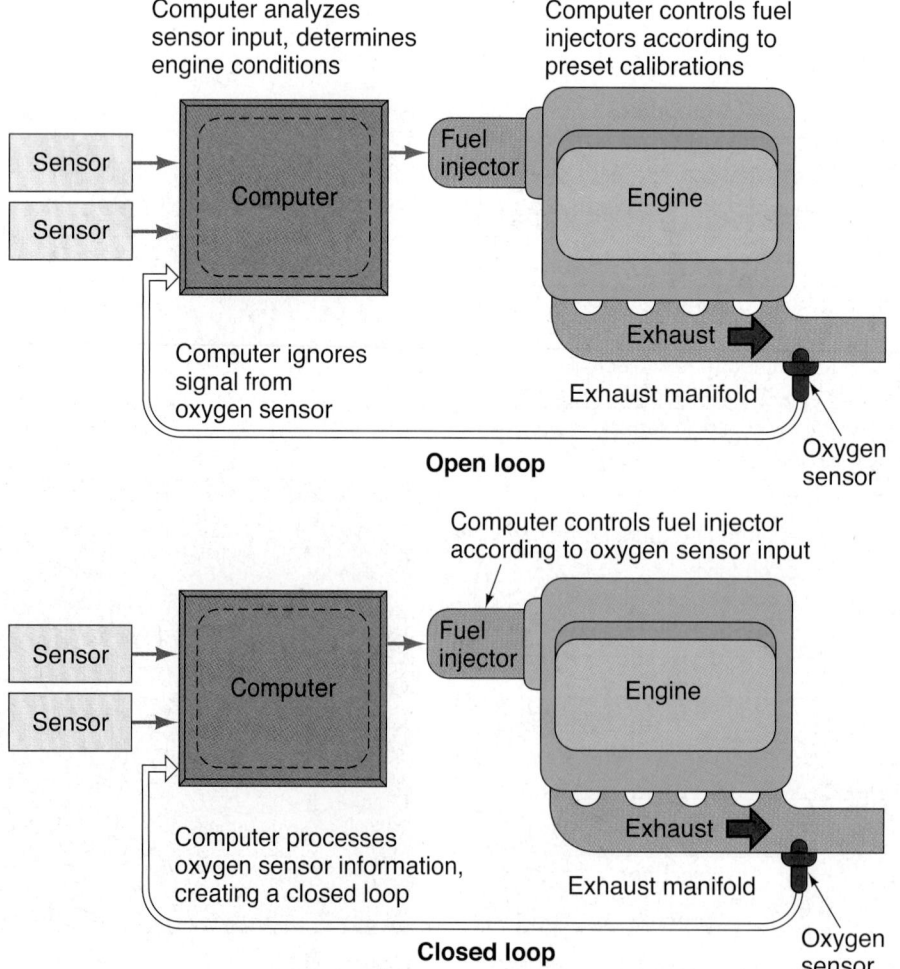

Figure 40.35 Open and closed loop. *(Courtesy of Ford Motor Company)*

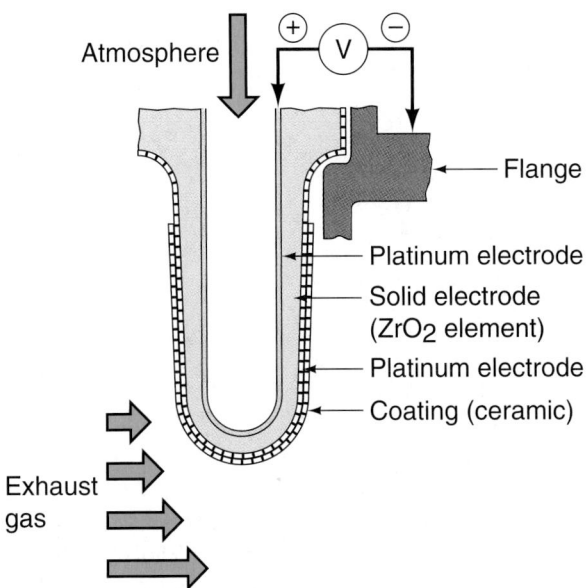

Figure 40.36 In an oxygen sensor, one platinum electrode is exposed to outside air and the other to the exhaust stream.

small battery, generating a variable DC voltage of between 0.1 volt (100 millivolts) and 0.9 volt (900 millivolts). The computer can use the voltage to tell whether the exhaust stream is rich or lean. When the mixture is rich, a signal of over 0.45 volt is generated. When the mixture is lean, the voltage signal from the sensor will be less than 0.4 volt.

When heated to about 600°F, a voltage is generated on two platinum electrode plates inside the sensor (**Figure 40.36**). One of these plates is exposed to outside air and the other to the exhaust stream (**Figure 40.37**). The two plates develop different voltages in response to the amount of O_2 ions they attract. The difference in these voltages is what makes the sensor act like a battery.

When there is a rich air-fuel mixture, there is a shortage of oxygen in the exhaust gas. The outside air side of the sensor has an abundance of oxygen. Oxygen ions, which are negatively charged atoms,

move from the atmospheric side platinum plate to the exhaust side plate. This transfers a negative charge to the exhaust side of the sensor, which results in a signal generation of 0.45 volt or higher. The opposite occurs when the air-fuel mixture is lean. The sensor can change from rich to lean almost instantaneously (about $1/10$ of a second). When there is more oxygen in the exhaust stream, there is a lean air-fuel mixture. The computer sees a voltage of less than 0.4 volt and compensates by driving the fuel system rich (more injector pulse width).

Lambda. Lambda, the eleventh letter in the Greek alphabet, is a name sometimes used for an O_2 sensor (especially in European cars) because its output is proportional to the lambda of the exhaust. Lambda is the ratio of the engine's actual air-fuel mixture in the exhaust to what would be an ideal, or stoichiometric, mixture. When lambda is 1.000, the mixture is stoichiometric. When lambda is less than 1, the mixture is rich; more than 1, and the mixture is lean.

Heated Sensor. Some O_2 sensors have three or four wires coming out of them (see Figure 40.33). These are heated sensors so that the computer can go into closed loop sooner. The OBD term for these sensors is HO_2S.

Wide Range Oxygen Sensors

In the mid-1990s, some manufacturers began using a different type of zirconia O_2 sensor capable of accurately detecting air-fuel ratios over a wider range than an HO_2S. An ordinary HO_2S can only determine whether the air-fuel ratio is leaner or richer than lambda (1.000) but not how much richer or leaner. This sensing capability is important for better computer control of the exhaust entering the catalytic converter. These sensors also allow much lower exhaust emissions on the federal test procedure (see Chapter 47). Later designs called planar sensors go into closed loop faster than earlier ones. They need to operate at a higher temperature than regular O_2 sensors, 1,200°F (650°C) instead of 400°F (750°C).

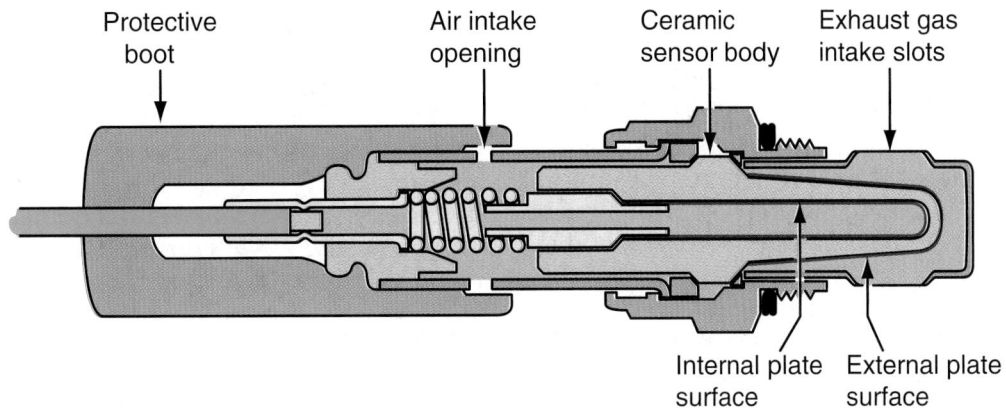

Figure 40.37 Outside air enters near the top of the O_2 sensor. Exhaust gas enters the sensor at the tip.

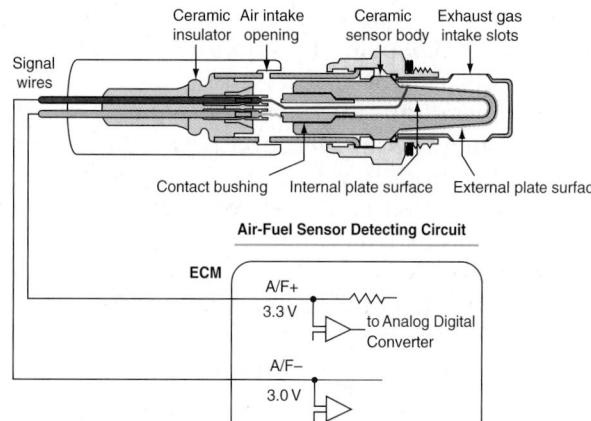

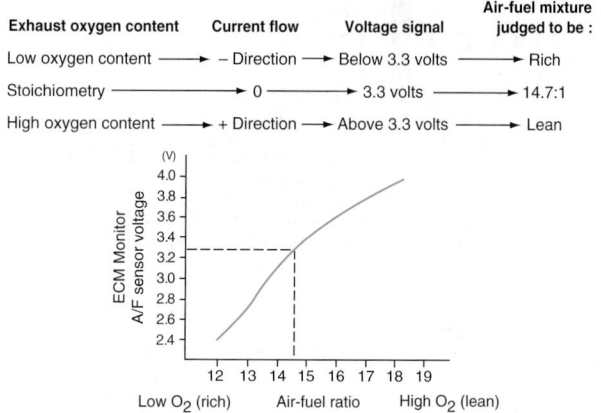

Figure 40.38 Operation of a wide range air-fuel sensor.

Wide range O_2 sensors are sometimes called wide band sensors, AFR (air-fuel ratio) sensors, lambda sensors, or lean air-fuel (LAF) ratio sensors. They look like ordinary four- or five-wire O_2 sensors.

Wide Range O_2 Sensor Operation

A wide range O_2 sensor is actually two zirconia sensors, one within the other. The amount of energy difference between the two sensors determines the air-fuel ratio (**Figure 40.38**). Like an ordinary HO_2S, a wide range sensor measures the amount of oxygen in the exhaust. With a wide range sensor, however, the computer wants to prevent it from cycling like a normal O_2 sensor. The PCM applies current to a pump cell to move oxygen in and out of the sensor. This changes its output from the normal operating range of an O_2 sensor. By applying positive and negative current, the PCM maintains O_2 sensor output at a constant voltage.

Remember that a wide range O_2 sensor is basically two sensors, one combined within the other. The outside of the combined sensor measures exhaust oxygen content, while the outside of the inner sensor samples outside air. A sealed diffusion chamber separates the two sensors. Instead of the outside sensor comparing exhaust stream oxygen content to the outside air, it now compares to the oxygen in the diffusion chamber, which is controlled by the PCM. Information on wide range O_2 sensor service is discussed in Chapter 46.

Diesel Direct Injection

Modern diesel engines have been mandated to have exhaust emissions that are nearly free of particulates. Engine manufacturers have been able to accomplish this using computerized engine controls with altered engine designs. These *common rail direct injection* diesels first appeared in the mid-1990s.

A tube called a *common rail* (**Figure 40.39**) connects the fuel injectors with diesel fuel under very high pressure of nearly 20,000 psi (1,360 BAR). The fuel system injects a small amount of high-pressure fuel before and after the main fuel charge. High pressure in the common

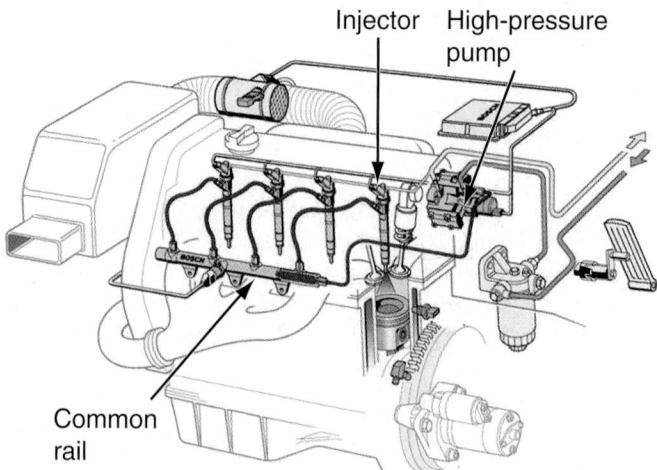

Figure 40.39 A common rail diesel injection system. *(Reprinted with permission from Bosch)*

rail thoroughly atomizes the diesel fuel, mixing it with air. This results in less unburned fuel and cleaner exhaust gas. Electronic piezoelectric injectors precisely control the fuel (**Figure 40.40**). Direct injection engines have lower emissions, are very responsive, and get better fuel economy than the older diesel engines.

Gasoline Direct Injection Systems

Direct injection into the combustion chamber is not new, but until recently it was only done with diesel engines. **Figure 40.41** compares conventional and direct injection. Gasoline direct injection (GDI) allows an engine to run under very lean conditions when cruising. With GDI, gasoline is injected directly into the combustion chamber rather than in the intake port just above the valve. **Figure 40.42** shows the components in a GDI system.

GDI has the ability to run the engine with a variable air-fuel mixture that can be extremely lean with an overall average ratio of around 40:1. This can increase fuel economy by as much as 30%, and exhaust emissions are reduced substantially.

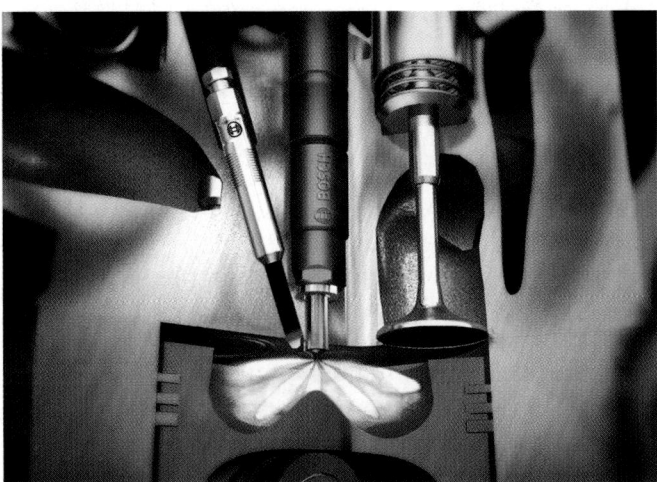

Figure 40.40 A common rail piezo inline diesel fuel injector. *(Reprinted with permission from Bosch)*

Indirect injection into intake manifold

Direct injection into combustion chamber

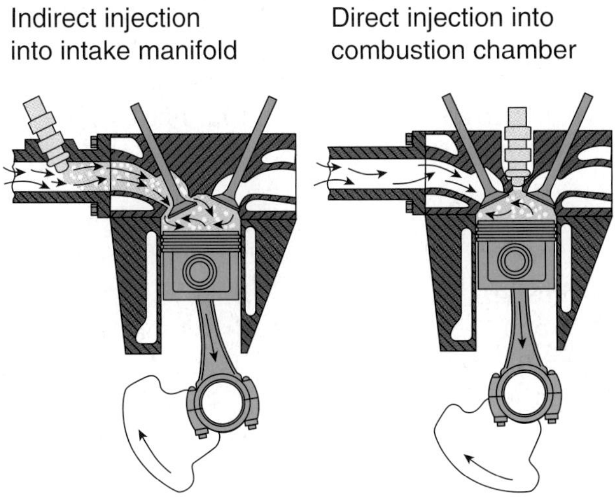

Figure 40.41 Indirect and direct injection.

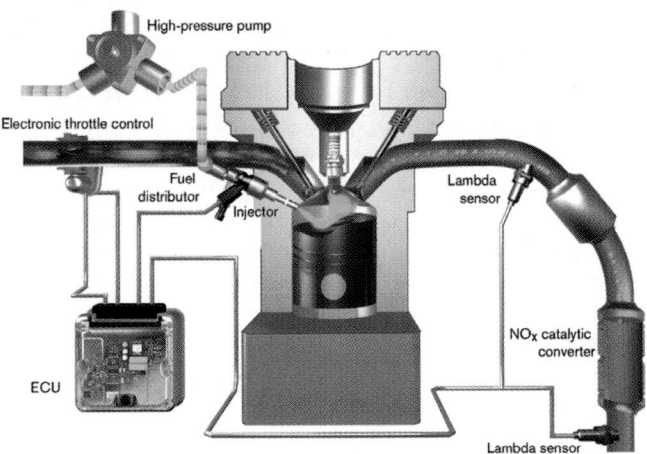

High-pressure pump

Electronic throttle control

Fuel distributor

Injector

Lambda sensor

ECU

NOₓ catalytic converter

Lambda sensor

Figure 40.42 A gasoline direct injection system. *(Reprinted with permission from Bosch)*

In GDI systems, fuel is under very high pressure. It does not boil due to the pressure, and it vaporizes as it is injected into the cylinder. Special fuel injectors

Figure 40.43 A special fuel injector designed to close against high pressure. *(Courtesy of Delphi Corporation)*

are designed to close against this high pressure. (**Figure 40.43**). Direct injectors are exposed to the high pressure of combustion, so they need to be able to inject fuel at even higher pressure. They must also close completely after spraying their fuel charge to prevent combustion pressure from entering the fuel system. The engine-driven fuel pump is supplied with fuel by an in-tank electric fuel pump. A computer controls timing of the ignition and injection for each cylinder.

With normal SFI, fuel is injected just before the intake stroke begins. With high-pressure direct injection, computer control of fuel timing means that the fuel can be injected at any time. Injectors can be pulsed more than once, even during the power stroke, to help maintain combustion. **Figure 40.44** shows how this provides a stratified fuel charge that concentrates around the spark plug and insulates the rest of the cylinder with a layer of air, resulting in lower exhaust emissions. Under conditions of light load, the intake stroke brings in air only, and fuel is injected near the end of the compression stroke just before the spark causes ignition. During part load, vaporization of the fuel helps cool the cylinder. Under heavier load, the computer calls for more fuel to be

Stratified mode

Homogeneous mode

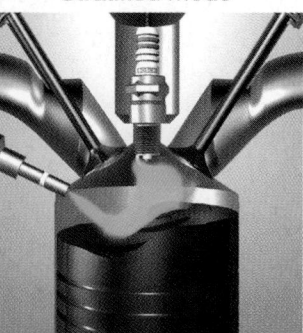
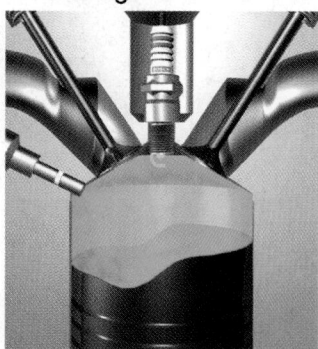

Figure 40.44 Computer control of the direct injector makes it possible to stratify the fuel charge. *(Reprinted with permission from Bosch)*

injected during the intake stroke. The resulting homogeneous fuel charge is not efficient at light loads and engine idle. GDIs precise control of fuel timing cools combustion under heavy load, however, lowering the fuel's tendency to knock, so lower octane fuel can be used. Engines can be designed with higher compression ratios to increase torque and power output while maintaining fuel economy.

In later chapters, you will learn about exhaust emissions. Oxides of nitrogen (NO_x) are a major component of smog. With GDI, NO_x emissions tend to increase due to the higher temperatures that result from lean air-fuel mixtures. Current GDI engine designs require an EGR valve to control NO_x.

■ REVIEW QUESTIONS

1. What is another name for the filter that is in the fuel tank at the bottom of the fuel tank cluster?

2. What is the term for the pressure inside the engine's intake manifold when the engine is running?

3. What is the name of the process where fuel is suspended in the air in a mist?

4. Fuel injection systems can be either mechanical or _____.

5. What is the length of time that a fuel injector is open called?

6. The two main categories of fuel injection airflow sensors are the air density and the _____ density systems.

7. If the key is on and the engine has not been cranked for _____ seconds, the fuel pump relay shuts off power to the fuel pump.

8. The three types of devices that make up a computer system are the computer, sensors, and _____.

9. What is the name of the sensor in the vehicle's exhaust that tells what the air-fuel mixture is?

10. When the computer is receiving signals and acting on them, this is called _____ loop.

■ ASE-STYLE REVIEW QUESTIONS

1. Which type of fuel injection do most new vehicles use?
   a. Continuous
   b. Sequential port
   c. Throttle body
   d. None of the above

2. All of the following are electronic fuel injection systems *execpt*:
   a. Throttle-body injection
   b. Port fuel injection
   c. Central fuel injection
   d. Continuous fuel injection

3. Technician A says that when there is a rich air-fuel mixture, the oxygen sensor will give a signal that is higher than 0.5 volt. Technician B says that if there is a low voltage signal (0.4 volt or less) coming from the oxygen sensor, the computer will drive the system rich. Who is right?
   a. Technician A
   b. Technician B
   c. Both A and B
   d. Neither A nor B

4. Technician A says that the computer on some fuel injection systems pulses several injectors at the same time. Technician B says that the computer on some fuel injection systems pulses each fuel injector individually. Who is right?
   a. Technician A
   b. Technician B
   c. Both A and B
   d. Neither A nor B

5. Technician A says that a speed density fuel injection system uses an airflow sensor. Technician B says that an air density system uses a manifold pressure sensor. Who is right?
   a. Technician A
   b. Technician B
   c. Both A and B
   d. Neither A nor B

Fuel System Service

■ KEY TERMS

flat spot	glitch	O_2 sensor safe
accelerator pump	schematic	min/max
flooding	cross counts	PROM
scan tool	RTV	carbon blaster
unmetered air		

■ INTRODUCTION

This chapter includes diagnosis and service of a number of the fuel system's parts: fuel pumps, filters, tanks, and fuel injection systems and their computer controls. Diagnosis and repair of problems in electronic systems can be complicated, and parts are expensive, especially when they are not necessary. Information in this chapter is provided for you to gain a basic, generic understanding of related problems. Successful specialists in these systems are highly educated, experienced, and able to locate service information. Check Chapter 40 for information on how fuel system components operate.

■ FUEL SUPPLY SYSTEM SERVICE

Fuel Tank Service

It is sometimes necessary to lower or remove a fuel tank from a car when the tank is corroded or to do the following:

■ replace fuel or vapor hoses

■ repair a fuel tank sending unit

■ replace an in-tank electric fuel pump

Today's engines typically last longer than 150,000 miles, but electric fuel pumps often have shorter lives. Most new vehicles have the fuel pump located in the fuel tank. Sometimes the fuel pump cluster can be removed through an access hole, but often the fuel tank must be removed from the vehicle.

The fuel pump can be tested using a lab scope and a low amp current probe (see Chapter 46). Fuel tank repair is commonly done by radiator shops. Because of the danger associated with gasoline, many repair shops choose not to make tank repairs. The information provided here is included as a precaution for those who choose to remove fuel tanks.

Before removing a fuel tank, fuel should be pumped from the tank using commercial fuel handling equipment (**Figure 41.1**). The fuel is pumped out either

Figure 41.1 Equipment for the safe handling of fuel. *(Courtesy of Tim Gilles)*

Figure 41.2 When a fuel hose is flexible, pinch pliers can be used to prevent a leak. *(Courtesy of Tim Gilles)*

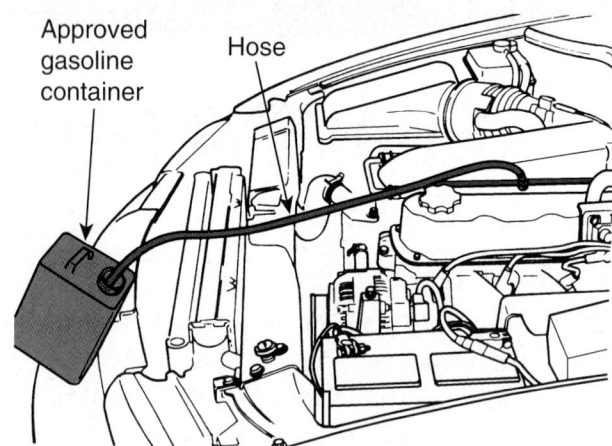

Figure 41.3 Bleed pressure from the fuel system before working on it. *(Courtesy of DaimlerChrysler Corporation)*

through the fill opening or through the outlet line to the fuel pump.

Bleed Fuel System Pressure

Fuel injection systems are designed to remain pressurized after the engine is shut off so the engine can start quickly. If one of the fuel line hose clamps is loosened, pressurized fuel can escape. To prevent gasoline from escaping, pinch pliers or long nose vise grips can be installed on the hose on the fuel tank side of filter (if the fuel hose is flexible) (**Figure 41.2**).

Several methods can be used to bleed pressure from the system before working on it. Check the manufacturer's service information for the recommended procedure.

CAUTION ■ Before removing a fuel tank, remove the battery ground cable to avoid an accidental spark.
■ Gasoline weighs about 7 pounds per gallon. A full tank could easily weigh in excess of 100 pounds. Attempting to remove it while full could result in a dangerous gasoline spill or an injury. Remember: gasoline vapor is *extremely* flammable.
■ Wear safety glasses and do **not** use a drop light. Liquid falling on the hot glass can cause the bulb to explode. If it is accidentally dropped or has a bad switch or loose connections, it can cause a fire.
■ Work in a well-ventilated area with no possible source of ignition, such as running cars or electrical equipment.
■ After draining a fuel tank, seal all openings.
■ Even if a tank has been thoroughly flushed, it will still contain flammable vapors. Attempting to weld it can result in an explosion. Before welding, the tank is filled with an inert gas such as CO_2.

The following is a very simple procedure for bleeding pressure from a fuel system:
■ Disconnect the fuel pump electrical connector or remove the fuel pump relay or fuse.
■ Crank the engine briefly to drop the fuel pressure or run the engine until it stalls.

Other methods can be used to bleed pressure from a fuel system. One way is to energize the fuel injector by applying positive voltage to one of the injector terminals and a ground to the other.

NOTE: *12 volts should not be applied to an injector for longer than 5 seconds.*

Another pressure bleeding procedure:
■ Disconnect the battery ground cable.
■ Remove the filler cap from the fuel tank.
■ Some systems have a *Schrader* valve that can be used to bleed off pressure from the system before disassembly. (A Schrader valve is the kind found on tire valve stems—see Chapter 61.) Remove the threaded cap from the fuel pressure test port on the fuel rail to find the Schrader valve.
■ Use a special hose that has a pressure relieving tool on the fuel rail end to drain fuel to a gas can (**Figure 41.3**).

Fuel Gauge Sending Unit Removal

Prior to removing the tank, the fuel gauge sending unit and float must be disconnected from it (**Figure 41.4**). The fuel lines, which usually enter the tank at this point, are disconnected too. The tank sending unit must be removed to get to the pickup strainer.

 Use a brass drift to loosen the lock ring to prevent the danger of a spark. When a spanner wrench is available for this purpose, this is the tool of choice.

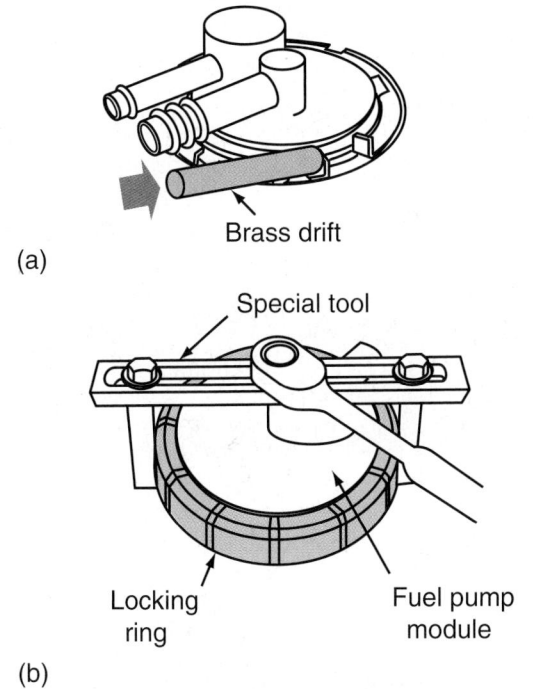

(a)

(b)

Brass drift

Special tool

Locking ring

Fuel pump module

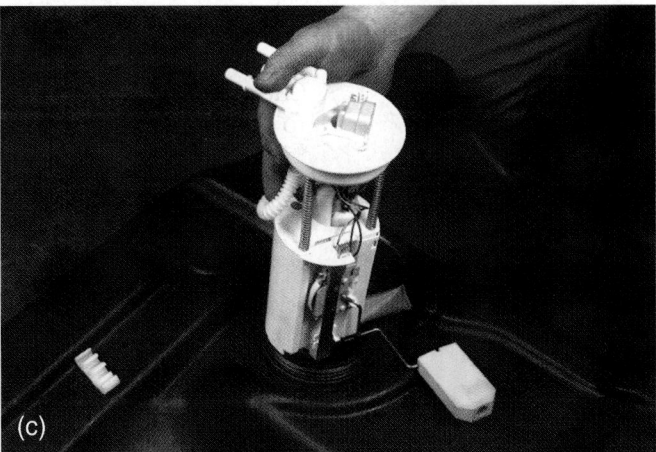

(c)

Figure 41.4 (a, b) Remove the fuel sending cluster from the fuel tank before removing the tank. (c) A fuel cluster includes the fuel pump and fuel gauge sending unit.

Vapor Recovery

Fuel vapors need someplace to go when filling a fuel tank or when fuel expands or contracts in the tank due to temperature changes. The fuel tank is connected to a *vapor recovery device*. Operation of these devices is covered in Chapter 43. A service technician will often need to replace the fuel hoses that connect these devices. Replace hoses one at a time so they are not accidentally reinstalled in the wrong place.

Replacing Hoses

Hoses should be inspected often, as they deteriorate over time and can fail from the inside out, plugging things up. Most industry authorities recommend that hoses be replaced every 3 years. If a fuel hose fails, a fire can result. If a hose needs to be replaced, be sure to use one of the proper type (see Chapter 23).

 Be sure that a hose is not positioned near any part of the exhaust system or the catalytic converter.

To inspect a hose, flex it back and forth to see if it breaks or shows cracks. If the end of a vapor hose has gotten hard and the hose is long enough, sometimes a small section can be cut off of the end. The remaining section of the hose is reinstalled on its fitting. This is done only on hoses that do not have a hose clamp. Clamped hoses usually contain pressurized fuel.

SAFETY NOTE Any suspected bad line should be replaced immediately so a dangerous situation does not develop. After getting the customer's approval and replacing the line, always start the engine and check for leaks.

Disconnecting fuel hose connections sometimes requires the use of a special tool (**Figure 41.5**). A tool kit with several tools is shown in **Figure 41.6a**. Removal of a fuel filter is shown using one of the tools in **Figure 41.6b**.

There are two types of fuel connections that do not require a special tool (**Figure 41.7**). Removal of one type calls for turning the fitting, while the other type has squeeze release tabs.

Some fuel hoses are retained with crimp clamps. Special pliers are used to squeeze a new clamp against a hose (**Figure 41.8**).

■ FUEL FILTER SERVICE

Fuel filters have specified intervals for replacement. They can be located in a fuel line, in the tank (see Chapter 40), in a carburetor, or in any combination of these. Outlet fuel filters can be found on the outlet side of a fuel pump.

In-line Fuel Filter

Fuel filters in the fuel line are replaced at specified intervals. If the fuel filter becomes plugged in a fuel injection system, fuel to the rail will be reduced, which results in hard starting, lean running conditions, misfiring, and so on. When a filter starts to plug, the engine can feel like it is being slowed under heavy load on hills.

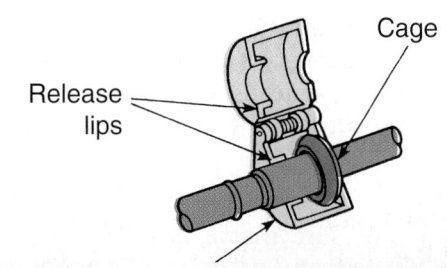

(a)

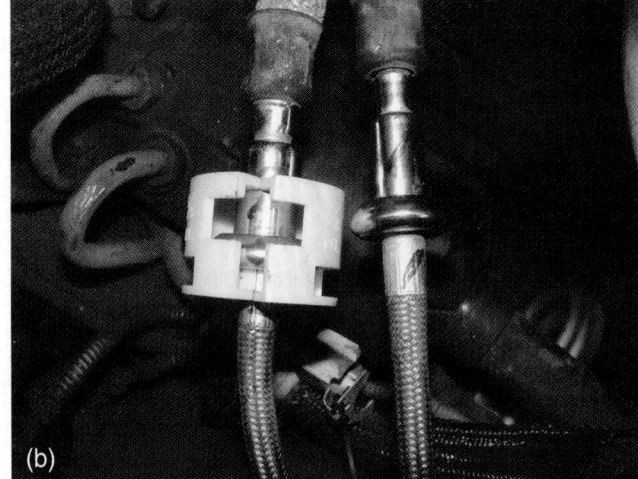

(b)

(c)

Figure 41.5 (a) Slide the tool into the connector. (b) This color tool was the wrong size. (c) This tool was the correct size. *(b and c Courtesy of Tim Gilles)*

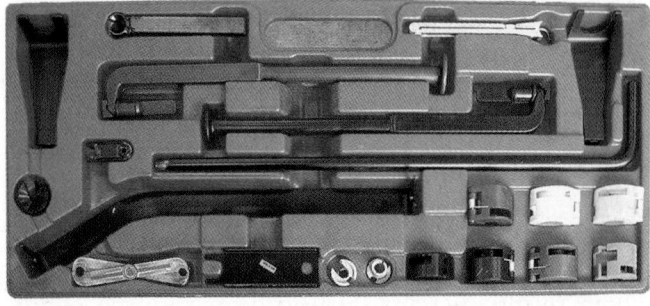

(a)

Figure 41.6 (a) A fuel filter removal tool kit. *(Courtesy of Tim Gilles)*

(b)

Figure 41.6 (b) One common tool for fuel filter removal. *(Courtesy of Tim Gilles)*

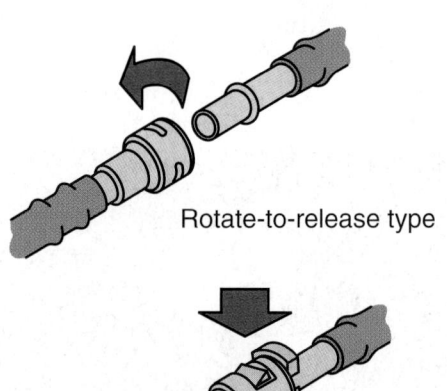

Rotate-to-release type

Squeeze-to-release type

Figure 41.7 Quick-disconnect hand-releasable fuel line fittings.

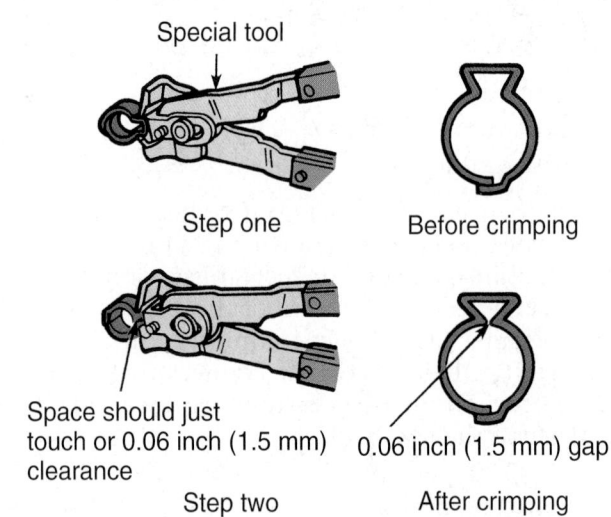

Special tool

Step one

Before crimping

Space should just touch or 0.06 inch (1.5 mm) clearance

Step two

0.06 inch (1.5 mm) gap

After crimping

Figure 41.8 A special tool is used to tighten crimp clamps.

CAUTION Bleed off residual pressure from a fuel injection system before removing any fuel lines. To do this, disable the fuel pump. Then start the engine and let it run until it runs out of fuel.

NOTE: *One unfortunate result of a restricted filter is that the fuel pump will draw more electrical current. This is hard on the fuel pump brushes and can result in premature failure of the pump.*

Before disconnecting a fuel line, place a shop towel (or drain pan, if possible) under the filter to catch fuel that spills.

SHOP TIP When installing a fuel line, jiggling the line back and forth makes it easier to turn the nut by hand.

When there is a flared connection, be sure to hold the nut with an open-end wrench while tightening the flare nut with a flare nut wrench. Some filters use special hose connectors. Refer to Chapter 24 for information about these.

A filter will usually have an arrow to tell the proper direction of installation.

▬ THROTTLE LINKAGE

Throttle control to carburetors on older cars is often by linkages and pivots. On fuel injection systems and newer carburetors, there is usually a cable. Throttle linkages have clips of various types (**Figure 41.9**). One kind of clip fits around a 90-degree bend on the end of the linkage and then rotates to clip around the linkage. This clip can be either right or left hand. Be sure to use the correct one.

▬ CARBURETOR PROBLEMS AND SERVICE

Carburetors are found on vehicles that are at least 20 years old, so in-depth carburetor service is not covered here. However, some of the easier troubleshooting

items are discussed. Carburetor operating principles and diagnosis can often be applied to fuel injection diagnosis as well.

Before attempting a fuel system repair or service, perform a visual inspection. Look for obvious problems, missing parts, and torn or damaged hoses. With the engine running, listen for hissing, which could indicate an air (vacuum) leak. With carburetors and early fuel-injected engines, the engine idle is controlled by an adjustment of the throttle plate opening, using a screw. An air leak is usually accompanied by a rough, higher idle and hesitation on acceleration. Problems caused by rich and lean air-fuel mixtures are covered in detail in Chapter 39.

A carburetor's float level is important. Too low a float level will cause a lean air-fuel mixture, and too high a float level will cause the engine to run rich. Urethane plastic floats become saturated with fuel over a period of years, making them heavier than normal and resulting in a higher fuel level and a richer air-fuel mixture.

A lean air-fuel mixture can cause engine surging and burned parts as well as a **flat spot** during initial acceleration. A flat spot is when the car stumbles when first accelerated. Flat spots were a common problem with carburetors, occuring when the **accelerator pump** wore out. Other causes of flat spots include retarded ignition timing or a bad vacuum advance unit. All of these problems have become obsolete with modern computer controls.

To test the accelerator pump, remove the air cleaner. With the engine off and the choke plate open, look down the carburetor while opening the throttle (**Figure 41.10**). A strong squirt of gasoline should be visible in the venturi. If there is no fuel, or only a dribble, the accelerator pump is not working correctly.

Figure 41.10 Checking the operation of the accelerator pump. *(Courtesy of Tim Gilles)*

Figure 41.9 Various types of throttle linkage clips. *(Courtesy of Tim Gilles)*

Accelerator pumps are part of a carburetor rebuild kit, but sometimes they can be purchased separately.

Sometimes carburetor linkages get gummed up, causing problems like too high an idle speed, a sticking throttle, or a stuck choke.

Carburetor power valves are made to open at a specified engine vacuum, allowing extra fuel to enter the air stream. The opening point can have a big effect on the vehicle's fuel economy. This is especially true with non-aerodynamic vehicles such as trucks and motorhomes.

NOTE: *Driving at 55 mph into a slight wind might cause an engine to run with 5" of intake manifold vacuum. If the power valve in this example opens at 6" of vacuum and driving at 50 mph results in 7" of vacuum, a good deal of fuel might be saved by driving at the slower speed.*

When carburetors were found on most vehicles, carburetor rebuilding was commonly done in most repair shops. Today, carburetors are no longer commonly rebuilt in repair shops. Some large companies specialize in rebuilding carburetors, and replacement or rebuilt carburetors are available for some of the more common engines. When a rebuilt carburetor is installed, the old carburetor is returned to the rebuilder as a core for future rebuilding.

Starting the Engine

Carbureted and fuel-injected cars require different treatment during starting. When a carbureted car was started in the morning, the driver depressed the accelerator pedal once to release the choke butterfly, allowing it to spring closed. Stepping on the accelerator pedal also results in a squirt of fuel from the accelerator pump. Both of these actions result in a rich mixture needed for starting. Cars with fuel injection sense when a cold engine is starting. They do *not* require the pedal to be depressed as carbureted cars do. If there is an engine misfire, it is possible that EFI parts in some systems can be damaged if the throttle is depressed during startup.

Clearing a Flooded Engine

Flooding occurs when an engine gets too much fuel. During flooding, spark plugs are wet and a spark cannot be generated across the plug gap. The engine will not start. This was common on carbureted engines when the choke was misadjusted or stuck, or when the driver pumped the pedal too many times while attempting to start a car with a low battery or in need of a tune-up. Carbureted engines were especially prone to flooding because whenever the pedal was depressed, the accelerator pump squirted a fresh charge of fuel into the intake manifold.

When a carbureted engine is flooded, hold the pedal all the way to the floor during cranking, raising the pedal very slowly until the engine starts. The air rushing through the engine carries the excess fuel out the exhaust with it. Some fuel-injected cars react to a floored pedal by shutting off the flow of fuel to the fuel injectors. This is called a "clear flood" condition.

■ EXHAUST GAS ANALYSIS

The exhaust stream from a running engine can be tested using an *infrared exhaust analyzer*. Infrared light, light that is invisible to our eyes, is used to measure these emissions. Modern exhaust gas analyzers test five gases: *hydrocarbons (HC), carbon monoxide (CO), carbon dioxide (CO_2), oxides of nitrogen (NO_x), and oxygen (O_2).* Information from an exhaust analysis can be used to diagnose incorrect air-fuel mixtures, engine and ignition system conditions, and operation of emission system components.

The air-fuel mixture can be checked under cruise conditions. Hold the accelerator at 2,500 rpm and observe the CO reading. CO that is lower than 0.5% will result in a flat spot on acceleration. This excessively lean mixture can cause burned spark plug electrodes. Premature exhaust valve burning can also result. First, check the float level to see if it is too low. If not, carburetor jets of a larger size can be purchased.

NOTE: *An operating catalytic converter will clean the exhaust. Significant readings for these gases need to be made in front of the catalytic converter.*

Additional information on exhaust analysis is given in Chapter 44.

■ FUEL INJECTION DIAGNOSIS AND SERVICE

Modern automobiles have electronic fuel injection systems. Diagnosis and repairs are similar among the various different EFI designs. Defects in other engine systems can be mistaken for fuel injection problems. Other areas to be checked include emission controls, ignition system operation, engine compression, and battery state of charge.

Visual Check

A visual check can often locate an obvious problem, such as a disconnected or damaged hose or wire. Electrical connections sometimes become corroded. Take the connections apart and look for corrosion (**Figure 41.11**). Taking the connections apart and putting them back

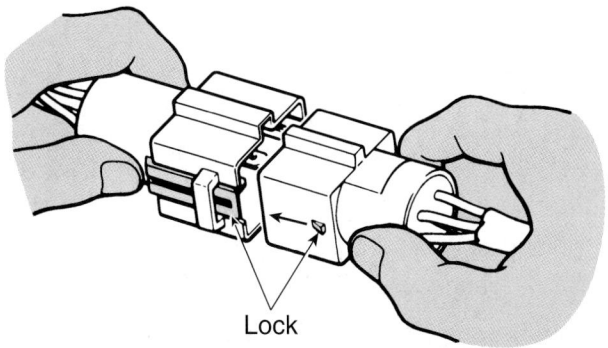

Figure 41.11 Disconnecting and reconnecting electrical connections sometimes solves the problem.

Figure 41.12 A small amount of dielectric grease is put on the connectors to prevent corrosion. *(Courtesy of Tim Gilles)*

together is sometimes enough to improve the electrical connection and solve the problem. Ford calls this a *wiggle test*.

NOTE: *Plastic electrical connections often become brittle with age. Handle them carefully.*

Put a small amount of dielectric grease on the connectors to prevent corrosion before reconnecting them (**Figure 41.12**).

■ EFI COMPUTER SELF-DIAGNOSTICS

Late-model cars have self-diagnostic provisions. There are codes for many malfunctions that the computer can diagnose. The easy way to read these codes is to use a scan tool, a hand-held diagnostic tool. A high impedance voltmeter can also be used (see Chapter 46).

■ AIR-FUEL MIXTURE PROBLEMS

If a computer feedback car has a driveability problem that only occurs when the engine is cold, check for external causes such as an intake manifold leak or incorrect fuel system pressure.

Intake Leaks

A leak in an intake manifold gasket can allow air into the engine that is not measured by an airflow sensor. This is called **unmetered air**. The computer pulses the injectors for less time than the amount of air warrants. This results in a lean air-fuel mixture when the engine is cold and the system is in open loop. Remember, open loop is when the engine is cold before the oxygen sensor starts working.

After the engine is warm and the computer is receiving feedback from the oxygen sensor, the computer can correct the injector pulse width to compensate for air-fuel mixture problems unless the leak is too large.

Pressure Testing

When fuel system pressure is not within specifications, the fuel injectors will not inject the correct amount of fuel. If the pressure is not too far off, the computer will adjust the air-fuel mixture after the system begins closed-loop operation. The engine can have a cold driveability problem that will disappear when it is warm.

To test fuel pressure, the pump must be operating. If it is not, check the fuel pump fuse. To hear if a fuel pump is running, put a long funnel in the fuel tank opening and listen. Check vehicle specific service information. For instance, on Fords an inertia switch in the circuit that powers the fuel pump is located in the trunk. Push the reset button on it first to see if the problem goes away.

It is possible that a fuel pump produces the specified pressure when the key is turned on or at engine idle but does not produce enough pressure under load. If the complaint is that the engine cuts out at higher speeds, the car should be road tested with a pressure gauge installed. A pressure gauge with a long hose can be taped to the windshield so it can be viewed from the passenger compartment.

NOTE: *Before a pressure gauge is installed, relieve the pressure in the fuel system.*

Pressure specifications vary among manufacturers and fuel injection systems. Mechanical injection systems (those with a fuel distributor) have pressures in excess of 50 psi. Throttle-body injection system pressure is typically low (9–13 psi), although there are high-pressure TBI systems, too. High- and low-pressure systems are found on port fuel injection systems as well. Typical specifications for high-pressure systems are 35–45 psi.

Installing a Pressure Gauge

Testing pressure in one type of throttle-body fuel injection requires that the line entering the throttle body be disconnected. A pressure gauge is installed in series with the line. Another testing method recommended by some manufacturers is to install a gauge in series with the fuel filter inlet (**Figure 41.13**). After the gauge is disconnected, install new gaskets on the banjo fitting. Copper gaskets are reusable if not damaged or imprinted.

In port fuel injection systems, the gauge is installed at the Schrader valve on the fuel rail (**Figure 41.14**), or on the fuel line to the cold start injector or to the connection of a fuel damper to the fuel rail (**Figure 41.15**).

To perform the pressure test, the fuel pump is energized. One method is to run the engine at idle speed. Another way is to cycle (turn on and off) the ignition key several times. Sometimes the engine will not run and it may be necessary to energize the pump electrically. Procedures vary between manufacturers. Consult the applicable service information.

Fuel pressure must be at least equal to manufacturer's specifications. Causes of low pressure could be a plugged fuel tank inlet sock, a kinked inlet line, or a bad auxiliary fuel pump. Some mechanical systems use

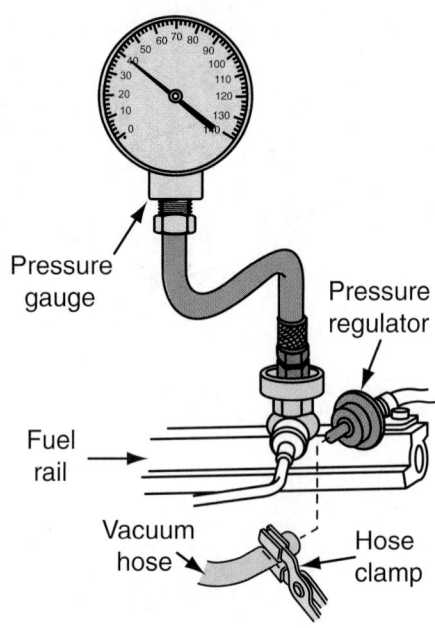

Figure 41.15 A gauge installed into the fuel damper on the fuel rail.

an auxiliary pump in the fuel tank to push fuel to a higher pressure pump outside of the tank.

When pressure is too low, a fuel pressure regulator could also be the cause. Fuel pressure regulators usually fail by causing pressure that is too low rather than too high.

Testing a Pressure Regulator

A pressure regulator can become defective due to foreign material or a ruined diaphragm. The result will be hard starting, poor idle quality, and lack of power. Burned spark plugs can result from operating with a lean air-fuel mixture.

High pressure is usually caused by a bad pressure regulator or on looped systems when a pressure return line to the fuel tank is kinked. When there is a vacuum line to the pressure regulator, part of the procedure is to pull the vacuum hose off while the engine idles. The pressure should rise about 5–10 psi in most systems when the hose is pulled. Fuel pressure should drop about 5–10 psi when the hose is reattached. Another test is to remove the vacuum hose from the regulator and apply 20 inches of vacuum to the regulator diaphragm (**Figure 41.16**). It should not leak, and the pressure should change in response to the pressure.

If gasoline drips from the vacuum hose when it is removed from a fuel pressure regulator, it is defective. The engine will run rich under light load because the fuel regulator "sees" low engine vacuum and increases pressure about 5–10 pounds. If the leak is big enough, fuel will be drawn into the intake air stream through the vacuum hose. In extreme cases, fuel can enter a cylinder when the engine is off and hydro-lock the engine.

On nonadjustable systems, the regulator must be replaced if pressure is not correct. If a pressure

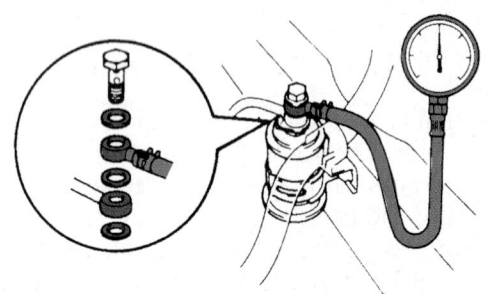

Figure 41.13 A fuel pressure gauge in series with the fuel filter inlet.

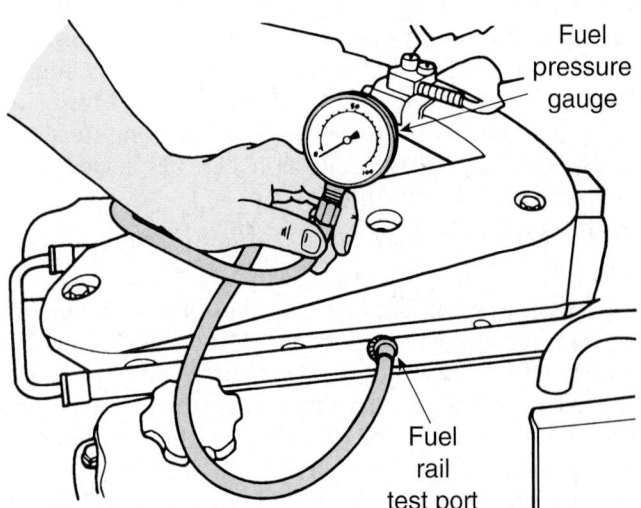

Figure 41.14 A gauge installed in the fuel pressure test port on the fuel rail. (*Courtesy of DaimlerChrysler Corporation*)

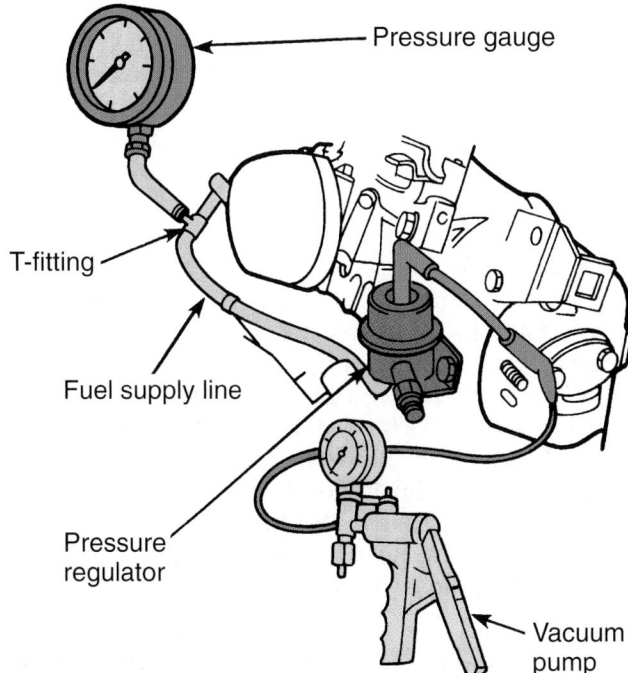

Figure 41.16 In one type of test the vacuum hose is removed from the regulator and vacuum is applied to the regulator diaphragm.

regulator requires adjustment, something else might be wrong with the system.

Fuel Pump Volume

Fuel pump volume is usually a better indicator of problems than pressure. A pump might build pressure but does not maintain sufficient flow. Minimum pump flow should be 1 pint in 15–20 seconds. If a restriction is found and it turns out to be the filter, cut the filter open and see what is in it. Any debris in the filter had to flow through the pump to get there, so the pump could have suffered damage as well. There is a strainer on the pump inlet that must be properly installed to prevent damage to the replacement pump.

When a pump is replaced, the tank should be removed and cleaned. New fuel should be installed. The very fine silt that is in the old fuel can damage the new pump, too.

Here is some good advice for your customers: It is better for your fuel system if you use the first half of a tank of fuel and then refill. The fuel pump is cooled by fuel. If there is more fuel, it will be able to cool better. Also, the silt that floats around in an almost empty tank is far more concentrated than it would be in a full tank.

■ INJECTOR PROBLEMS

Fuel injectors can be bad, leaking or dirty, shorted or open. On *port fuel-injected* cars, individual injectors for each cylinder are located in the intake valve port in the cylinder head. A stethoscope can be used to listen to the opening and closing of the injector as the engine operates (**Figure 41.17**).

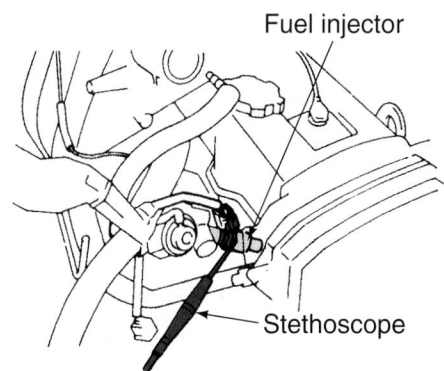

Figure 41.17 Listen to the fuel injector with a stethoscope.

Injector electrical test lights (called *noid lights*) provide an easy way to test to see if there is power to the injectors (**Figure 41.18**). If there is power to the injector, the light will flash on and off as the computer attempts to cycle the injector. If the light does not light, there is no power to the injector. A wiring harness or the computer could be the cause. Sensor input to the computer could also be the cause. On many systems, if there is no tach signal, the fuel injection system will not operate.

When there is power but there is no spray from an injector, the injector is probably bad. Use an ohmmeter to check the resistance and for shorts to ground (**Figure 41.19**). If the injector is open or shorted, replace the injector.

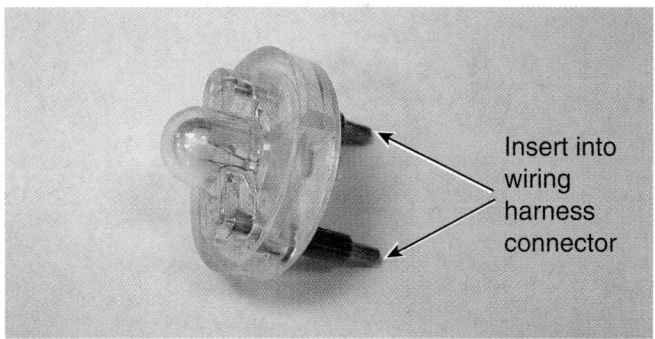

Figure 41.18 A special test light, called a noid light, is used to see if there is power at the injector. (*Courtesy of Tim Gilles*)

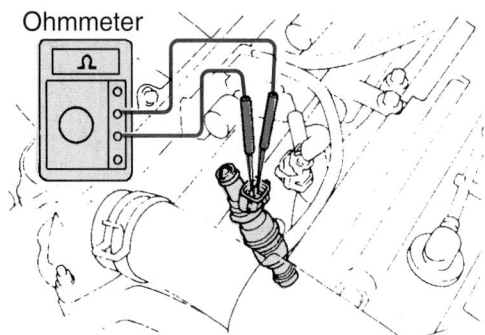

Figure 41.19 An ohmmeter is used to check for shorts and grounds.

Sometimes a failed injector can be the result of a pinched wire that runs from the injector to the ECM (computer). The ECM grounds the injectors to make them turn on, so pinching the wire to ground will cause the injector to energize and remain on. The injector will overheat and fail. Installing a new injector without repairing the electrical wiring problem will result in a repeat failure.

Checking Injector Pulse Width

Injector pulse width can be checked using some types of digital multimeters or a scan tool. An oscilloscope can also be used. A modified square wave pattern would be the result. The pattern on an oscilloscope shows voltage vertically (**Figure 41.20**). Above the zero line is positive voltage. Horizontal movement represents time.

Most injectors are powered on one side and controlled through the ground circuit. Attaching the meter feed to the ground side of one of the injectors will show the pulse width of the rest of the injectors.

Fuel injectors have two terminals. To connect the meter or scope, backprobe one of the terminals of an easily accessible injector with a paper clip. Check either terminal on the injector for voltage. When the key is turned on, there will be power to one side of the injector. The other side is the control (ground) side. In most cases, this is the side the pattern will be able to be viewed from (ground side controlled).

Computer fuel injection systems use adaptive fuel trim to continually adjust the air-fuel ratio in response to both immediate (short-term) and long-term conditions. Further explanation of the theory of computerized fine-tuning of the air-fuel ratio is given in Chapter 45.

Testing Fuel Injector Flow

When low-quality fuel is used, individual port fuel injectors are prone to plugging up from fuel deposits.

An electronic fuel injector tester is available for testing injector balance (**Figure 41.21**) while reading system pressure on a gauge. With the system at the specified pressure, each injector is activated for an equal period of time. This allows fuel in the loop to escape from the injector. The drop in pressure is recorded (**Figure 41.22**). The system is repressurized after each injector is bled, and the next injector is operated with the tester. There should be an equal amount of pressure drop

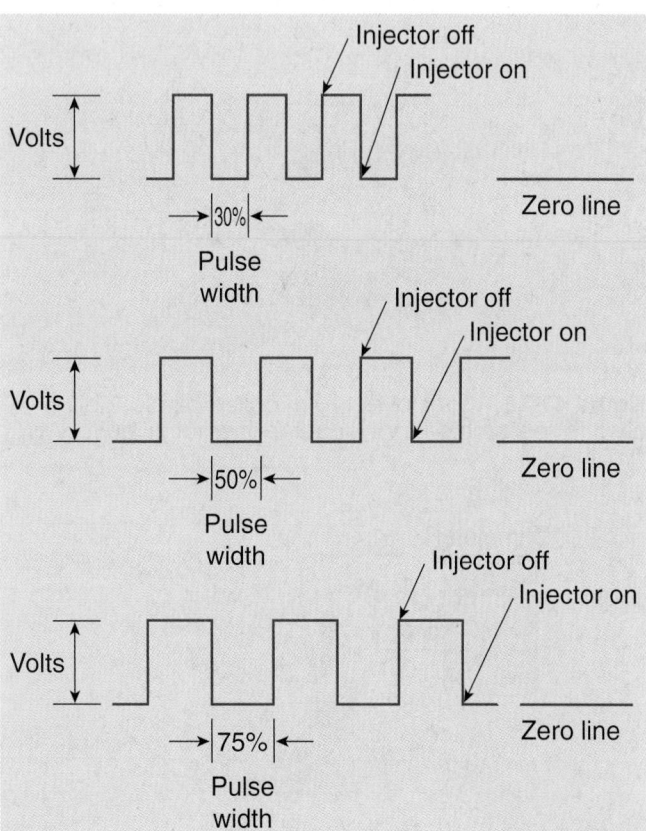

Figure 41.20 A square wave pattern of the pulse width.

Figure 41.21 A fuel injector tester.

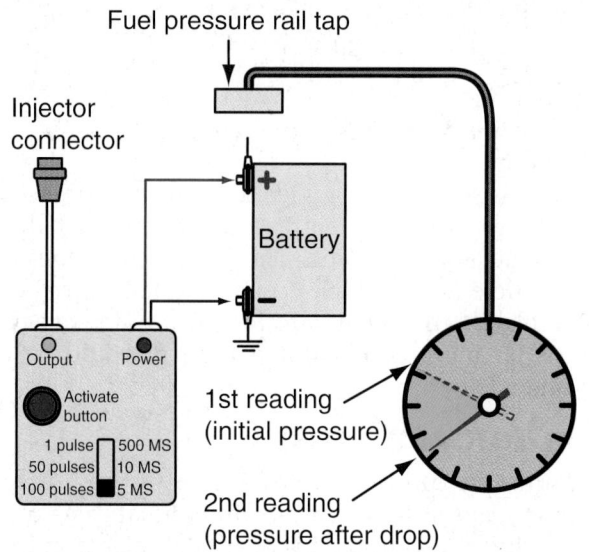

Figure 41.22 Check to see how much the pressure drops after the tester is activated.

from each injector. If the pressure drop is above or below average by 1.4 psi (10 kPa), the injector is defective.

Removing and Replacing Injectors

To replace an injector, bleed pressure from the system first. Be careful that no dirt gets into the system while the fuel rail is removed. Do not soak an injector in cleaning solvent. This can damage or contaminate the injector. Be sure to replace the rubber O-rings that insulate the injectors. Failing to do this can result in a vacuum leak and consequently, rough idle.

Some manufacturers use an injector flow test with the injectors and fuel rail removed from the engine. Each injector is energized for 15 seconds while holding it over a graduated container (**Figure 41.23**). The volumes of fuel flowing from each injector are compared. Variations of more than 5 cc calls for replacement of the injector.

Some older vehicles have cold start injectors. When foreign material gets into a cold start injector, it can leak. This will result in rough idle and backfire. Residual fuel pressure—pressure remaining in the line after the pump is off—can cause an injector to leak. This can cause hard starting and an overly rich mixture during starting.

Remove the cold start injector and hold it in a container. When the engine coolant is below a certain temperature, the injector should spray, but only when the engine is cranked. It is powered off of the starter solenoid circuit. If the injector does not work, check to see that it has power to it. If not, check the thermal time switch.

Leakage Test

During a system pressure test when the fuel pump is turned off, pressure in the system should remain constant for the next startup. After shutting off the engine, pressure should not drop more than 20 psi in 20 minutes. With a pressure gauge attached to the fuel rail at the Schrader valve, turn the key to the "on" position or start the engine and watch the pressure build on the gauge to normal, probably 35–45 psi, depending on the system specifications.

Shut off the engine and wait 20 minutes to see that the pressure does not drop more than 20 psi. If

pressure drops more than that, it could be due to a leaking check valve in the fuel pump, leaking fuel injectors, or a defective pressure regulator. To determine which of these is the cause:

- Energize the fuel pump and clamp the fuel supply hose from the pump.

NOTE: *Only pinch rubber fuel hoses. Do not clamp plastic lines.*

- Wait 10 minutes. If the pressure *does not* drop, the fuel pump check valve is the cause. Replace the fuel pump to correct the problem.
- If the pressure *does* drop, re-energize the pump and clamp the fuel return hose to the tank. If the pressure drop is now acceptable, replace the fuel pressure regulator.
- If pressure still drops, one or more injectors are leaking. If possible, remove the fuel rail while the engine is off. Repressurize it and observe which injector(s) is leaking.

If pressure leaks from the system, cycling of the ignition switch more than once is often required before starting the engine. This is because more than 2 seconds of pump operation might be required to fill the fuel rail.

Cleaning Injectors

Injector cleaning fluid is a mixture of the cleaner and gasoline. There are a variety of injector cleaning machines and processes. One type uses a canister, pressurized by shop air (**Figure 41.24**). A hose is connected from it to the Schrader valve on the fuel rail. The engine burns the pressurized fuel and cleaning solution as it

Figure 41.24 An injector cleaner connected to the Schrader valve on the fuel rail is pressurized by shop air. *(Courtesy of OTC/SPX Service Solutions)*

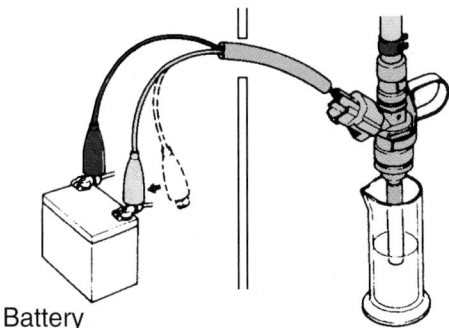

Battery

Figure 41.23 Each injector is energized for 15 seconds while holding it over a graduated container.

runs. The fuel pump is disabled and the pressure regulator return line is blocked to prevent the solution from returning to the fuel tank.

THROTTLE PLATE SERVICE

Occasionally, gum and carbon can accumulate around the throttle plate on throttle-body and port injection cars. The result is a rough or low idle. A spray can of throttle plate cleaner and a brush can be used to clean this area. Be sure the spray cleaner used is safe for oxygen sensors. If spray cleaner does not work, the throttle assembly will have to be removed and soaked in a cleaner.

> **CAUTION** Some throttle plates are coated with a special material, and cleaning the throttle plate can remove this material.

EFI ADJUSTMENTS

Common adjustments to EFI systems include idle speed, throttle stop, idle air-fuel mixture, and throttle cable. Raising idle speed means allowing more air to pass the throttle plate. Sometimes there is a screw that opens or closes a passageway. An idle air control (IAC) motor (**Figure 41.25**) is used to raise the idle speed when extra loads such as air conditioning are placed on the engine. Occasionally carbon blocks all or part of the air passage, resulting in engine stalling or erratic idle speed.

On later model vehicles, the idle speed is controlled by the computer. These systems have an idle speed control (ISC) motor (**Figure 41.26**). As with any new part, when one of these motors is replaced be sure to compare the old part to the new one. On one kind of motor, the pintle on the end of it must be pushed into the motor until a specified distance is reached. If the pintle is too far out, it could be damaged during installation.

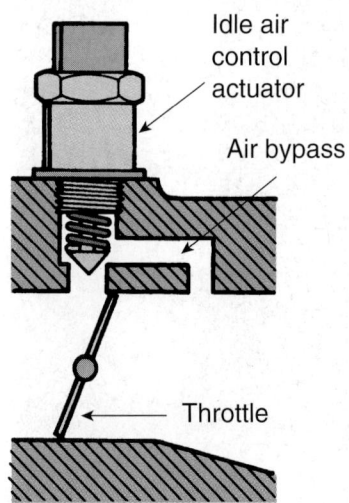

Figure 41.25 Allowing air to bypass the throttle plate raises the engine idle.

Idle air control actuator

Air bypass

Throttle

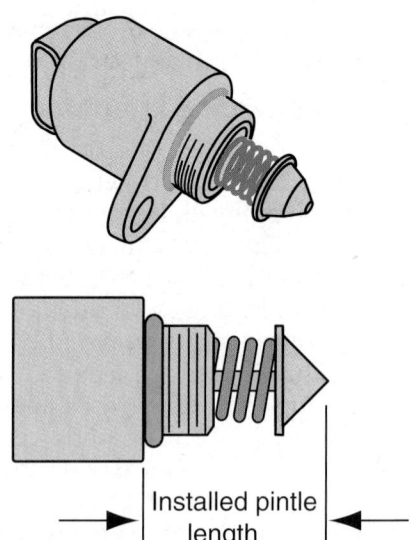

Installed pintle length

Figure 41.26 An idle speed control motor.

When the engine is shut off, most IACs move to the open position in readiness for the next engine start. On restart, engine idle will be high again when the engine is cold. As the engine warms up, the computer will adjust the idle speed by changing the number of steps on the IAC motor.

When a warm engine is restarted, it is normal for the engine idle to increase for a short time and then decrease to normal. This is called engine flare. If an engine does not flare, the IAC motor could be stuck.

SENSOR SERVICE

Testing procedures vary for the various fuel injection system sensors. Before disconnecting a computer system component, be certain that the ignition key is turned off. Use a scan tool, a digital volt-ohmmeter, or a test light. Always follow manufacturer's service information procedures. A scan tool is a device that reads computer self-diagnosis signals (see Chapter 46).

THROTTLE POSITION SENSOR

A bad throttle position sensor (TPS) can cause a change in idle speed, a stumble on acceleration, or engine stalling. Follow manufacturer's procedures for testing. The TPS has a metal wiper arm that rubs against another metal strip (**Figure 41.27**). If the strip wears away, momentary interruptions of the electrical signal can occur. These interruptions are called **glitches**.

A sensor can be tested with a voltmeter while its electrical wiring is still *connected* (this is called *backprobing* a connector). With the wiring *disconnected*, the TPS can be tested with an ohmmeter. Move the throttle slowly from closed to open. At different throttle openings, varying resistances are specified. Watch for any glitches.

Some throttle position sensors have three wires and some have four. With a four-wire sensor, the fourth wire is for the idle switch. Procedures for testing are given in the service information.

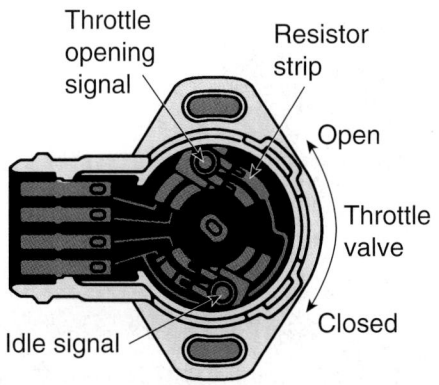

Figure 41.27 A throttle position sensor has a wiper arm that rubs against a resistor strip.

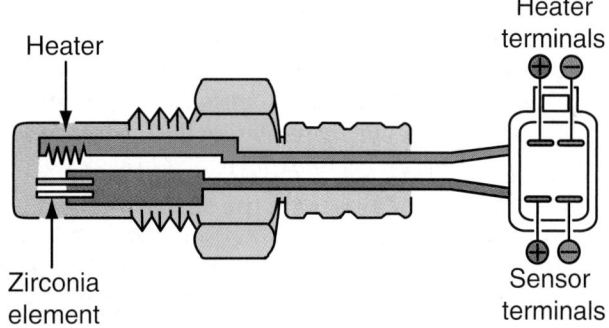

Figure 41.29 A four-wire heated O_2 sensor.

Most throttle position sensors are made so that their adjustments are tamperproof. They have screws that are either soldered or staked. To remove the switch, these might need to be drilled or filed off. After the new switch is installed and adjusted, the new mounting screws are restaked.

■ OXYGEN SENSOR SERVICE

The oxygen sensor is tested with the engine running at operating temperature. Check the manufacturer's service information for the correct procedure. Use a scan tool, a DSO, or a digital voltmeter. An analog voltmeter (one with a needle dial) can damage the sensor.

The voltmeter is connected to the O_2 sensor wire and grounded to perform the test. Some older sensors have only one wire coming from them. These sensors ground through the body of the sensor. To test this kind of sensor, ground one of the voltmeter wires. Attach the other lead to the wire from an older sensor. If there are two wires coming from the sensor, one is for ground. **Figure 41.28** shows a schematic of typical oxygen sensor wiring.

Heated O_2 Sensors

If there are three or four wires coming from the sensor, it is a heated sensor. These are found on any vehicle manufactured since 1996 with OBD II. The sensor is heated so the computer can go into closed loop sooner

and stay in closed loop during long periods of idle. Two of the wires are for the heater. With four wires, one is for the ground signal to the computer (**Figure 41.29**). Probe the other wire to get the computer signal (this wire is usually a different color). Check with a voltmeter at each wire to see which one reads between 0 and 1 volt. With the key turned on (not starting), the reading will be about 0.4 volt on most cars.

The two extra wires are for the heater and will read battery voltage across their connector. The O_2 sensor heater can be tested with an ohmmeter. If it does not have the specified resistance, the heater coil is bad and the sensor is replaced.

Testing the O_2 Sensor

When the engine is running, O_2 sensor voltage fluctuates rapidly and repeatedly from 0.2 volt to 0.8 volt or so. This means that the sensor is working properly. Sometimes an oxygen sensor is "lazy." This means that it does not sweep back and forth fast enough between a rich and a lean signal.

NOTE: *The speed at which this signal fluctuates is called* ***cross counts***. *On OBD II vehicles, the cross count rate is measured using a scan tool.*

When the O_2 sensor voltage remains low, the air-fuel ratio could be too lean, the sensor could be defective, or the wire between the sensor and computer could have high resistance.

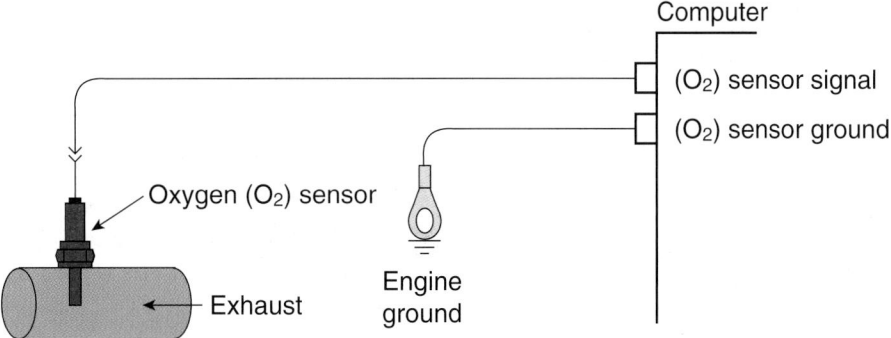

Figure 41.28 Wiring schematic for a simple oxygen sensor circuit (with no heater).

NOTE: *The O$_2$ sensor only senses O$_2$. A misfiring cylinder can give a false signal to the computer. Because of the abundance of O$_2$ in the unburned mixture, the reading sent to the computer from the O$_2$ sensor will be lean. The computer will compensate by increasing the injectors' pulse width, richening the air-fuel mixture.*

If the voltage remains high, the air-fuel ratio might be too rich or the sensor could be contaminated. When the platinum strip becomes insulated with a buildup of foreign material, it cannot react to oxygen ions to provide a signal to the computer.

Contaminated O$_2$ Sensor

An O$_2$ sensor can become contaminated in several ways. If the pores of the sensor become plugged, its response time and output voltage will drop. Zirconium sensors can be poisoned from either side. One side is exposed to outside air around the exhaust manifold, and the other side is exposed on the inside to the engine's exhaust stream.

Fumes from some types of silcone **RTV** sealants used on the engine can contaminate O$_2$ sensors. Be sure that the RTV that you use lists **O$_2$ sensor safe** on its label.

Lead in fuel can also contaminate an O$_2$ sensor. Lead is in some gasoline octane boosters.

Carbon buildup on the sensor can be from an overly rich air-fuel mixture, engine oil consumption, or a bad turbocharger turbine seal (see Chapter 42). Coolant leaking from a bad head gasket or cracked cylinder head can also contaminate a sensor. Check for the source of the contamination and correct the problem before replacing the sensor. Diagnosis of these problems is as follows:

■ Silicone contamination causes smooth, chalky white deposits on the tip of the sensor. Sensor output voltage can also go negative as a result of silicone contamination.
■ Engine oil leaves a brown residue that causes sensor speed to slow down.
■ A rich air-fuel mixture leaves a black coating that can be burned off by running the engine lean at fast idle for a few minutes. To make the engine run lean, pull a vacuum hose and unhook the O$_2$ sensor. This will result in a hotter exhaust.
■ Coolant leaves a white flaky deposit that sometimes has the sweet smell of ethylene glycol.
■ Contamination from the outside of the sensor can include brake fluid, power steering fluid, oil leaking from a valve cover, and dirt. These can block the air entrance to the sensor and slow down response time.

Sometimes, an O$_2$ sensor can suffer physical damage to itself or its electrical connections. Wires can

Figure 41.30 A special socket for removing an oxygen sensor has a slot to provide clearance for the wiring. *(Courtesy of Tim Gilles)*

be pinched or burned by a hot exhaust manifold. Connections can become corroded. The sensor housing is a ceramic. Therefore, it is brittle and can be broken like glass if bumped. If it rattles when shaken, it has been broken.

Some engine performance specialists routinely test new oxygen sensors after installation to see that they are operating correctly and within calibration.

Replacing the O$_2$ Sensor

The O$_2$ sensor is threaded into the exhaust manifold. Because it can be difficult to remove, use anti-seize compound on its threads. Torque the sensor to 30 foot-pounds using a special socket (**Figure 41.30**). A sensor that is too loose, or a cracked exhaust manifold, can result in a lean signal to the computer.

■ COOLANT TEMPERATURE SENSOR

The resistance of the coolant temperature sensor (ECT) varies with changes in temperature. The service information gives the resistance values at different temperatures (**Figure 41.31**). The sensor can be tested in hot water using a thermometer and an ohmmeter (**Figure 41.32**). Because of the time involved in removing an ECT sensor to test it in hot water, a technician will usually test it on the car using a scan tool (see Chapter 45).

CAUTION Do not test an ECT sensor with an open flame. The sensor will be damaged.

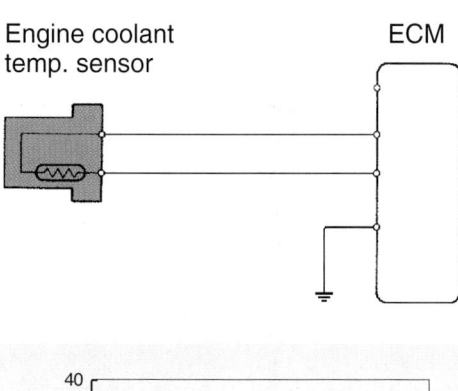

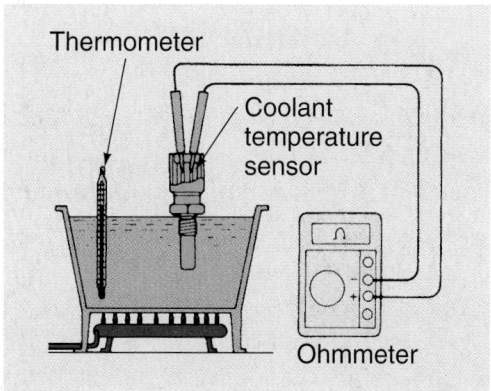

Figure 41.32 Testing a coolant temperature sensor in hot water using a thermometer and an ohmmeter.

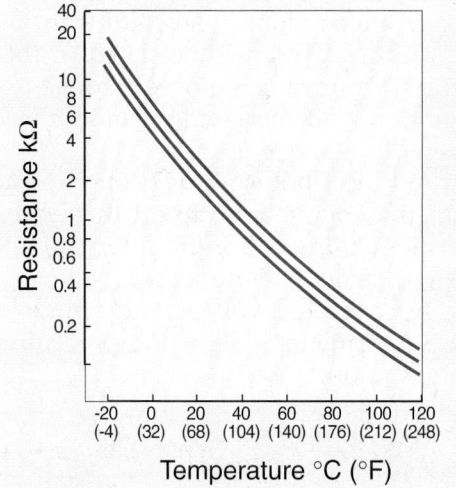

Figure 41.31 A coolant temperature sensor gives different resistance readings as its temperature changes.

■ INLET AIR TEMPERATURE SENSOR

The sensor that measures the temperature of incoming air is called an air charge temperature (ACT) sensor. It can be removed and tested in hot water like an ECT sensor (see Figure 41.32).

■ MAP SENSOR DIAGNOSIS

A defective manifold absolute pressure (MAP) sensor can cause the engine to run rich or lean. When the engine is off and the key is on, MAP sensors are supposed to tell the computer what the barometric pressure is. The service information provides the voltage specification for this test, which will vary with altitude and the weather. Voltage readings at various barometric pressure readings are provided. **Figure 41.33** is an example of a chart used by one manufacturer. You will have to check with a source about the weather or have your own barometer to find out what the present barometric pressure is.

Absolute Baro Reading	Lowest Allowable Voltage at –40°F	Lowest Allowable Voltage at 257°F	Lowest Allowable Voltage at 77°F	TBI MAP Sensor Designed Output Voltage	Highest Allowable Voltage at 77°F	Highest Allowable Voltage at 257°F	Highest Allowable Voltage at –40°F
31.0"	4.548 V	4.632 V	4.716 V	4.800 V	4.884 V	4.968 V	5.052 V
30.9"	4.531 V	4.615 V	4.699 V	4.783 V	4.867 V	4.951 V	5.035 V
30.8"	4.514 V	4.598 V	4.682 V	4.766 V	4.850 V	4.934 V	5.018 V
30.7"	4.497 V	4.581 V	4.665 V	4.749 V	4.833 V	4.917 V	5.001 V
30.6"	4.480 V	4.564 V	4.648 V	4.732 V	4.816 V	4.900 V	4.984 V
30.5"	4.463 V	4.547 V	4.631 V	4.715 V	4.799 V	4.883 V	4.967 V
30.4"	4.446 V	4.530 V	4.614 V	4.698 V	4.782 V	4.866 V	4.950 V
30.3"	4.430 V	4.514 V	4.598 V	4.682 V	4.766 V	4.850 V	4.934 V
30.2"	4.413 V	4.497 V	4.581 V	4.665 V	4.749 V	4.833 V	4.917 V
30.1"	4.396 V	4.480 V	4.564 V	4.648 V	4.732 V	4.816 V	4.900 V
30.0"	4.379 V	4.463 V	4.547 V	4.631 V	4.715 V	4.799 V	4.883 V

Figure 41.33 MAP sensor voltage signals at various barometric pressures. *(Courtesy of DaimlerChrysler Corporation)*

The atmosphere extends from the surface of the Earth to a height of over 600 miles, where it gradually merges with the solar wind. Immediately surrounding the Earth's surface is a blanket of air approximately 6 to 8 miles thick known as the troposphere. The air within the troposphere exerts the majority of the pressure felt at sea level. We can show the existence of this pressure by filling a long tube with mercury and inverting it in a dish of mercury to create a barometer. When the tube is inverted some, but not all, of the mercury runs out of it. The fact that not all of the mercury runs out of the tube shows that there must be a pressure (barometric pressure) exerted on the surface of the mercury in the dish, which is sufficient to support the amount of mercury remaining in the tube. If no pressure were exerted, there would be nothing to stop all of the mercury from running out of the upside-down tube.

Continue the test by applying vacuum to the MAP sensor with a hand vacuum pump. A typical test procedure is to apply 5 inches of vacuum to the MAP sensor and watch for the voltage reading to drop by a specified amount. Next, the pressure is lowered (vacuum is raised) to 10 inches. The voltage should change once again. The test continues by applying vacuum in 5-inch intervals until 25 inches is reached. If the readings are out of range, the sensor is replaced.

Some MAP sensors produce a signal that varies in frequency. This is called a voltage frequency signal. A tester is used that changes the frequency voltage to analog voltage so the voltmeter can read it. The tester is shown in **Figure 41.34**.

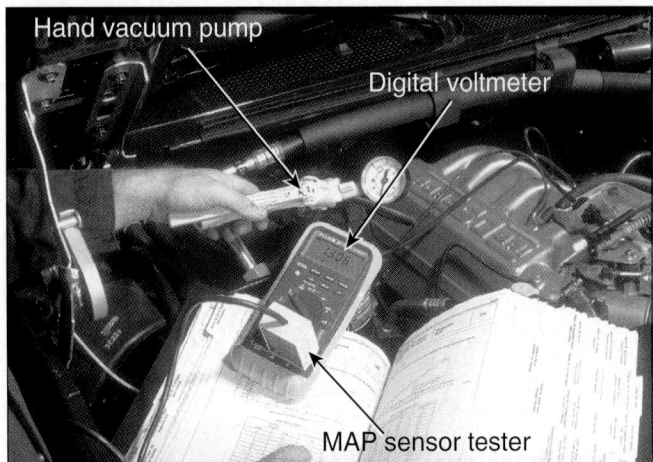

Figure 41.34 A tester that changes the frequency voltage to analog voltage is used so the voltmeter can read it.

■ MASS AIRFLOW (MAF) SENSOR DIAGNOSIS

Testing a mass airflow (MAF) sensor requires a voltmeter with a **min/max** feature. With the key on, the min/max button is pressed. With a vane-type MAF sensor, the air vane is moved from closed to wide open and back to closed (**Figure 41.35**). Pressing the min/max button will give the maximum voltage obtained. Pressing it again will give the lowest voltage (**Figure 41.36**). If minimum voltage is zero, there might be an open circuit in the sensor's variable resistor.

With some vane-type MAF sensors, manufacturers provide ohmmeter specifications. At some terminals, a thermistor might allow temperature to affect the resistance readings. This will be stated in the service information. The air vane is moved through its normal range of motion while observing a smooth reading on the voltmeter.

When testing a hot wire or heated resistor MAF sensor, run the engine and tap on the sensor. If the engine misfires, there is a loose internal connection requiring replacement of the sensor. There is a simple procedure for checking voltage and frequency readings with a multimeter. Follow the procedure in the service information.

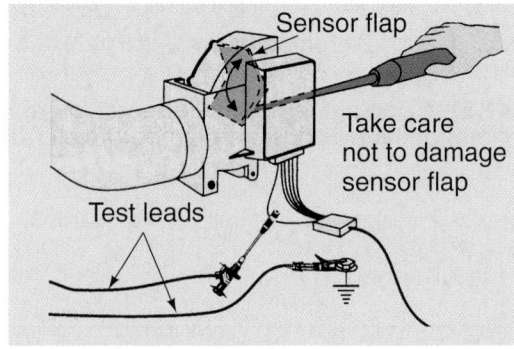

Figure 41.35 Move the sensor air vane from open to closed to test it. (*Reproduced with permission from Fluke Corporation*)

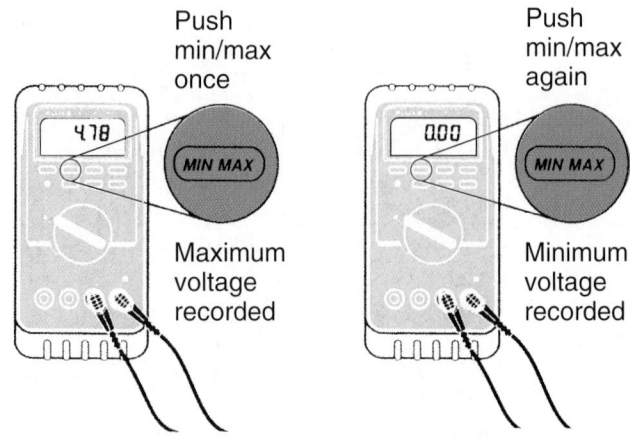

Figure 41.36 Use the min/max feature to read minimum and maximum sensor voltage signals. (*Reproduced with permission from Fluke Corporation*)

Figure 41.37 In this fuel system airflow is sensed before the hose to the manifold. A leak in the hose allows in air that the computer cannot "see." *(Courtesy of Tim Gilles)*

NOTE: *MAF systems can experience driveability problems due to "false air" entering the intake system. There is a hose between the MAF sensor and the intake manifold (**Figure 41.37**). When these hoses get older they sometimes crack. This results in engine performance problems as the engine moves on its engine mounts, opening and closing the crack and allowing unmetered air to enter the engine.*

TESTING OTHER SENSORS

Other EFI sensors are tested in ways similar to those discussed previously. Check resistance values using a scan tool and the vehicle's computer self-diagnostic system or use an ohmmeter or voltmeter. Refer to the service information for the procedures and values of each type of sensor.

COMPUTER SERVICE

The computer is rarely the cause of problems in the fuel system. If it is, be sure to locate any problem in the system that might have caused it to fail. When a computer is faulty, it is replaced. Remanufactured computers are widely available at a reduced cost for popular makes of cars. Some older computers have a replaceable element called a **PROM**. This stands for programmable read-only memory. The old PROM is installed in the replacement computer. Later-model computers have *EE Flash PROMs*. These are programmed electronically, using a scan tool with up-to-date software or by download over the Internet. See Chapter 46 for more about computer systems.

MECHANICAL INJECTION

Servicing Continuous Injectors

Compared to electronic systems, mechanical fuel injection systems run under very high pressure. They have

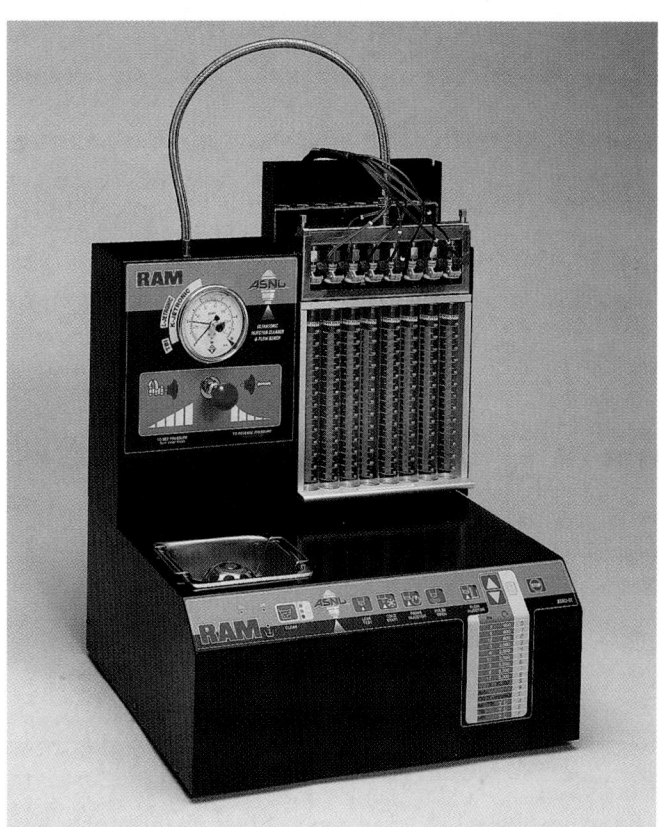

Figure 41.38 A mechanical fuel injection tester. *(Courtesy of Dana Corporation/Echlin)*

a spring-loaded valve that requires at least 50 psi to open. When they spray, there is an even fuel pattern. To check the operation of the injectors, remove each of them using two wrenches. Pull each injector out of the head and put them in a container to catch the fuel. Activate the fuel pump and move the airflow sensor to cause fuel flow to change. Check the spray pattern. A tester is available that keeps the fuel contained for a safer and more thorough check (**Figure 41.38**).

 ■ There is a danger of fire from spilled gasoline when it is sprayed at higher than 50 psi from the injectors.
■ Use a plastic container. Never use glass. If it breaks, you could be badly burned and the car could be destroyed in a fire.
■ Get help when performing this test. Use two hands to hold the container.

Volume of flow can be checked. Each injector should flow an equal amount. If not, there is a problem with either the fuel distributor or fuel injectors.

Switch injectors between cylinders. If the problem remains at the original cylinder, the fuel distributor is at fault. Fuel distributors are replaced as a unit. Refer to a service information for instructions.

CARBON DEPOSIT SERVICE

Carbon deposits cause driveability problems because fuel vapors can be absorbed into them. This results in rough idling when cold, as well as loss of power, surging, and high emissions. Carbon deposits on valves sometimes cause problems in as little as 5,000 miles.

One method of removing carbon deposits is a tool that has been around for many years. A **carbon blaster** uses crushed walnut shells blasted by compressed air to remove the carbon deposits. If any pieces of the shells remain in the engine after cleaning, they will be burned up during combustion. The process is as follows:

- The intake manifold is removed.
- The intake hose is attached to the intake port or fuel injector hole.
- The outlet hose is attached to the spark plug hole.
- The blaster is operated for a couple of minutes.
- Last, air only is used for about a minute to blow out the shells.

Sometimes carbon that has accumulated on a valve or in the combustion chamber can drop off and be crushed against the cylinder head. If this occurs during cranking, the piston can stop on its upstroke. When this happens at low rpm with the engine running, the noise resembles the sound of a bad rod bearing.

NOTE: *When carbon increases the engine's compression ratio, detonation can result. "Carbon knock" can be very noticeable when the engine is cold. Typically, engine knock occurs under load, but carbon knock can occur even when not under load.*

Carbon can be removed by using an additive such as GM Top Engine Cleaner.

NOTE: *Carbon removers can damage catalytic converters. Follow the manufacturer's instructions for their use.*

REVIEW QUESTIONS

1. Approximately how much does a gallon of gasoline weigh?
2. When the car stumbles when accelerated, this is called a _____ spot.
3. What is the kind of light that is invisible to your eyes called?
4. An engine is _____ when too much raw fuel has entered it and it will not run.
5. When the engine is cold, a computer feedback system is in open/closed (circle one) loop.
6. When a fuel pressure regulator fails, is the pressure usually too high or too low?
7. Most fuel injectors are powered on one side and controlled by the computer through the _____ circuit.
8. The O-ring around the base of a fuel injector does two things, insulate heat and prevent _____ leaks.
9. When testing an oxygen sensor, the voltage should fluctuate between 0.2 and _____ volt(s).
10. Some types of _____ sealer can damage oxygen sensors.

ASE-STYLE REVIEW QUESTIONS

1. When a fuel-injected engine is shut off:
   a. No fuel pressure remains in the supply line to the fuel injectors.
   b. The fuel supply line remains pressurized.
   c. Both A and B.
   d. Neither A nor B.
2. Which of the following can result from an intake manifold air leak (vacuum leak)?
   a. A rough idle
   b. A higher idle speed
   c. A flat spot on acceleration
   d. All of the above
3. Which of the following is a possible location for a fuel filter?
   a. In the fuel line to the fuel injectors
   b. In the fuel tank
   c. In a carburetor inlet
   d. All of the above

4. Remove a flared connection using:

   a. A box wrench

   b. Vise grips

   c. A flare-nut wrench

   d. An adjustable end wrench

5. A fuel pressure gauge can be installed at the Schrader valve on:

   a. Throttle-body injection

   b. Looped port fuel injection

   c. Both A and B

   d. Neither A nor B

6. Burned spark plugs can result from operating with a fuel system that is:

   a. Too rich   **c.** Either A or B

   b. Too lean   **d.** Neither A nor B

7. Which test is the best indicator of a fuel pump problem?

   a. Pressure test   **c.** Compression test

   b. Volume test   **d.** Aspiration test

8. The throttle plate can be held open by carbon and gum deposits on:

   a. Throttle-body injected engines

   b. Port injected engines

   c. Both A and B

   d. Neither A nor B

9. Technician A says that to start a fuel-injected car, the accelerator should be depressed once. Technician B says that to start a car with a carburetor, do not touch the accelerator pedal until the engine is running. Who is right?

   a. Technician A   **c.** Both A and B

   b. Technician B   **d.** Neither A nor B

10. A computer feedback fuel system has poor driveability when cold. Technician A says that this could be due to a bad oxygen sensor. Technician B says to check for external causes such as an intake manifold leak or incorrect fuel system pressure. Who is right?

    a. Technician A   **c.** Both A and B

    b. Technician B   **d.** Neither A nor B

CHAPTER 42

Intake and Exhaust Systems/ Turbochargers and Superchargers

◼ OBJECTIVES

Upon completion of this chapter, you should be able to:

✔ Explain the operation of the air intake system.

✔ Describe the parts and operation of exhaust system components.

✔ Understand the operation of a muffler.

✔ Explain the differences between turbochargers and superchargers.

✔ Diagnose problems with turbochargers and superchargers.

◼ KEY TERMS

pleated	catalytic converter	turbo lag
plenum	draw-through	intercooler
siamese runner	blow-through	blower
dual-plane manifold	turbocharger	Roots-type supercharger
single-plane manifold	normally aspirated	plain bearings
header	waste gate	NCFR
resonator	boost pressure	

◼ INTRODUCTION

This chapter deals with the parts, operation, and service of intake and exhaust systems, turbochargers, and superchargers. The intake system is covered first, followed by the exhaust system. Turbochargers and superchargers are covered last.

◼ INTAKE SYSTEM FUNDAMENTALS

There are two sources of engine contaminants: internal contaminants generated by heat and friction within the engine, and dirt that enters through the air intake system. Every gallon of gas burned requires about 9,000 gallons of air. The air that enters the engine must be filtered, but the filter must allow sufficient airflow for good engine operation. The air filter also muffles the sound of the air rushing into

the engine. Another job of the air cleaner is to act as a flame arrestor in case of a popback in the intake manifold.

NOTE: *A popback is an explosion that occurs in the intake system. A backfire is an explosion that occurs in the exhaust system.*

The common types of filter in use today are the dry paper type, made of pleated paper, and the oil wetted polyurethane type. Air filters are rated for *efficiency, flow,* and *capacity.*

- ◼ If 100 grams of dirt enter the filter housing and 99 grams are filtered out, the filter is said to be 99% efficient.
- ◼ The air filter must allow enough air to pass at the engine's maximum speed to not interfere with flow.
- ◼ Capacity is the amount of dirt an air filter can hold before it becomes restricted.

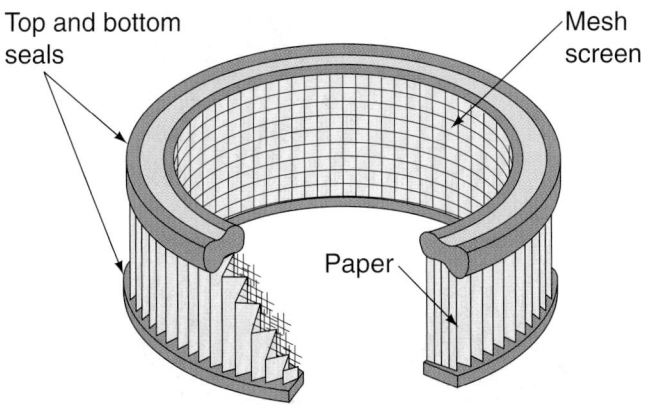

Figure 42.1 Parts of a throttle body injection or carburetor air filter element.

There must be a balance of efficiency, flow, and capacity for a filter to be effective.

Most paper air filters have an outer screen that protects the paper and holds it in place. The paper is **pleated** so it can offer a greater surface area in a small package. Seals at the ends of the filter keep it from leaking at the filter housing.

Air filters used with throttle body injection (TBI) or carburetors are usually round (**Figure 42.1**). An inner screen supports the paper and absorbs heat if there is a popback in the intake manifold. The filter housing is metal and has provisions for heating the incoming air during cold operation. Some air cleaners have fresh air ducted in from the front of the car to improve combustion.

Filters used with fuel injection systems are usually flat and fit into a plastic housing (**Figure 42.2**). Large rubber hose or ductwork carries the filtered air to the air

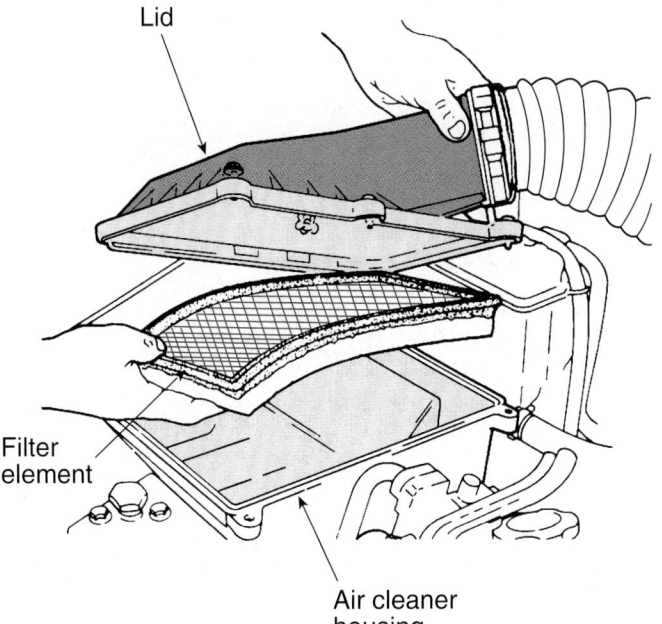

Figure 42.2 A typical fuel injection air cleaner. (*Courtesy of DaimlerChrysler Corporation*)

inlet. A hole in this hose can allow dirt into the engine or cause an unmetered air leak, depending on the fuel injection system design. An air leak from a bad hose can cause intermittent driveability problems resulting from the engine moving on its mounts. Information on air filter service is found in Chapter 13.

■ MANIFOLDS

There are two kinds of manifolds that are part of an engine's breathing system: the intake manifold and the exhaust manifold. Manifolds are carefully designed to provide a uniform air-fuel mixture to all cylinders. If they are the wrong size or design, the engine will not be able to breathe properly.

■ INTAKE MANIFOLDS

The passages in an intake manifold are known as runners. In engines that have TBI or a carburetor, the intake manifold is designed to provide optimum flow for the *air-fuel mixture*. Engines with port fuel injection (PFI) inject the fuel directly above the intake valve, so the manifold is designed for *airflow only*.

Port injection manifolds look different than other manifolds. They can be designed with larger runners than air-fuel manifolds. When the manifold flows air only, the runners can also have sharper bends because these manifolds do not have to keep fuel suspended in air.

Carburetors have not been found on engines since the early 1990s, but these cars are still required to pass emission control inspections in some states. Carburetors and intake manifold combinations are still sold in the aftermarket and are used in the auto racing world. Intake manifold design is crucial to engine operation in much the same manner as camshaft design. The purchase of a high-performance manifold without the addition of matching components will probably hurt performance. Each design is a compromise. Breathing parts must be properly matched.

NOTE: *Better high-rpm performance means worse low-rpm performance.*

Throttle-body injection uses the same type of manifold as carburetors. Intake manifolds that flow air and fuel are designed to keep fuel suspended in the air in fine droplets. By the time the mixture reaches the combustion chamber, most of the fuel should be evaporated so it will burn easily. If the speed of the mixture drops too low, droplets of fuel can fall out of the mixture.

Whether a manifold carries fuel or not, it will usually have small diameter runners that are longer, while larger diameter runners are shorter. Manifold runner sizes must be a compromise. Large-diameter runners flow well at high speeds, but the fuel will separate from the air at lower speeds. Smaller diameter manifolds provide enough flow and will keep the fuel

Intake
runners

Figure 42.3 These intake manifold runners are short, large, and relatively straight. *(Courtesy of Tim Gilles)*

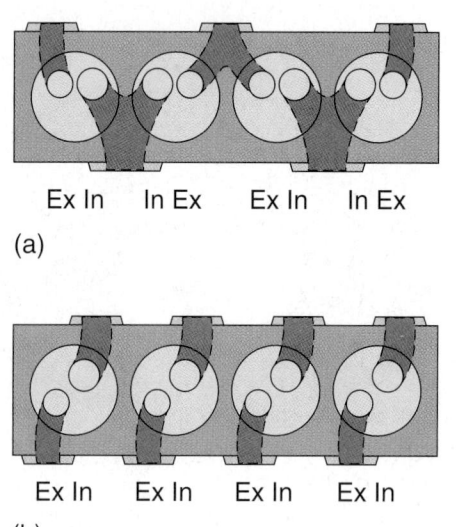

Ex In   In Ex   Ex In   In Ex
(a)

Ex In   Ex In   Ex In   Ex In
(b)

Figure 42.4 The top sketch (a) shows "siamese" valve ports that share a manifold runner. The bottom sketch (b) shows individual ports.

in suspension throughout the average rpm range of a passenger car. Runners have as few bends as possible in order to reduce the chances of the vaporized fuel turning back into liquid fuel (**Figure 42.3**).

The air space below a throttle body or carburetor is known as the **plenum**. The floor of the plenum is flat and often has ridges cast into it to catch fuel that drops out of the mixture. This makes it easier for the fuel to evaporate or to rejoin the moving air-fuel mixture as it flows through the manifold.

Manifold Runner Arrangement

Manifolds used in in-line engines are simple. Sometimes one runner will feed two neighboring cylinders. These are known as **siamese runners** (**Figure 42.4**). Manifolds for six- or eight-cylinder V-type carbureted engines have two barrels, or feed openings. In a **dual-plane manifold**, each barrel is independent of the other, although they share the same float bowl. On a V8 with a dual-plane two-barrel manifold, each barrel supplies fuel to four cylinders.

The dual-plane manifold has smaller runners and is better suited to lower rpm use. To keep the runners as much the same length as possible, one barrel serves both the inner two cylinders on the opposite side of the engine and the outer two cylinders on its own side (**Figure 42.5**). Knowledge of this is important when troubleshooting vacuum leaks, ignition, or fuel problems because sometimes the problem is only in those cylinders served by one barrel.

A **single-plane manifold** is one in which both barrels serve all eight cylinders. It is more suited for high-speed use.

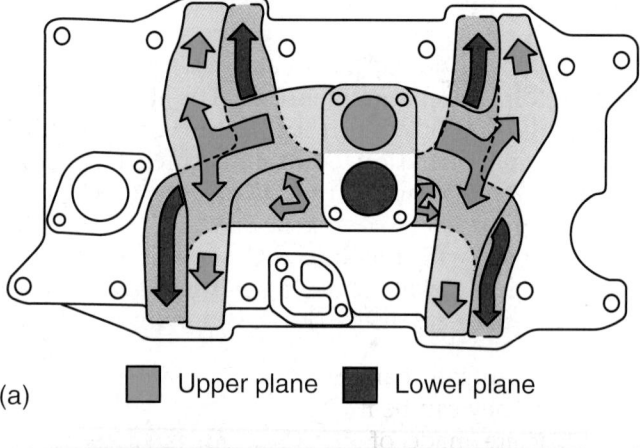

Upper plane   Lower plane
(a)

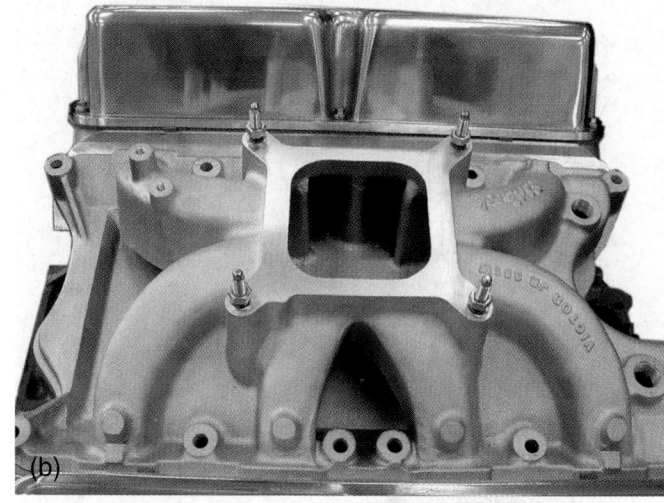

(b)

Figure 42.5 (a) A closed-type dual-plane V8 intake manifold. The arrows show that each carburetor barrel supplies fuel to four cylinders. (b) A single-plane manifold. *(b, Courtesy of Tim Gilles)*

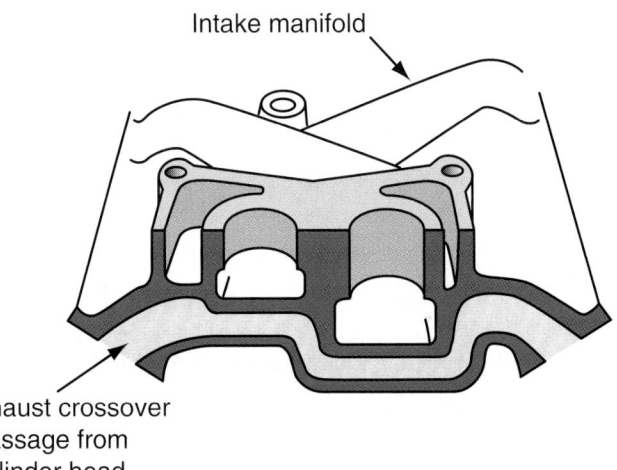

Intake manifold

Exhaust crossover passage from cylinder head

Figure 42.6 An exhaust gas passage in an intake manifold.

Manifold Heat Passage. Intake manifold heating is done on TBI and carbureted cars. An exhaust gas passage in the bottom of the intake manifold helps to vaporize the air-fuel mixture when the engine is cold (**Figure 42.6**). Newer engines use port fuel injection, which does not require heating.

Coolant Passage. An intake manifold on a V-type engine with a TBI or a carburetor usually has a coolant crossover passage. This passage connects the heads and provides the water outlet where the thermostat is found. A crack in this passage can cause a leak that can be difficult to diagnose. Manifolds on port-injected engines often do not have a coolant passage. These are called dry manifolds.

Port Injection Intake Manifolds

There is no intake manifold heating on port-injected engines, so they can be made of space age plastic materials. Most are made of plastic (**Figure 42.7**) or cast

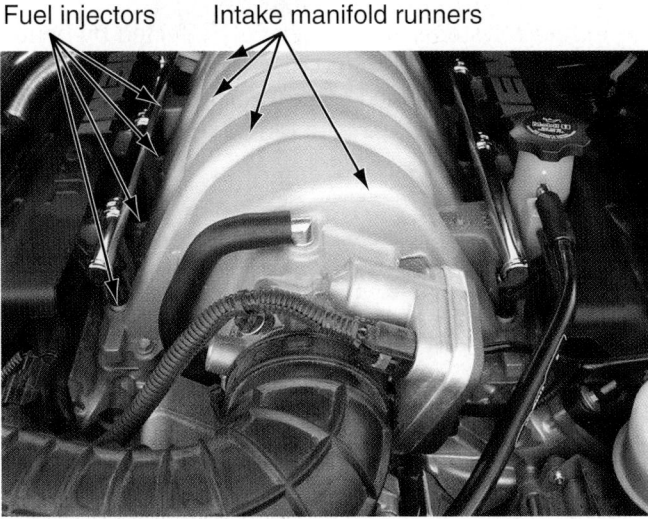

Fuel injectors   Intake manifold runners

Figure 42.7 An intake manifold on a late-model fuel-injected engine. *(Courtesy of Tim Gilles)*

Figure 42.8 This port-injected intake manifold has long runners of varying length. *(Courtesy of BMW NA, LLC)*

aluminum. There are other advantages to port injection. Cooler air is more oxygen dense than heated air, and manifolds can be made with longer curved air passages to increase airflow and power. **Figure 42.8** shows an intake manifold with variable length runners that would never be possible if fuel had to be carried with the air.

Multiple Valve Heads

Some high-performance late-model engines use three, four, or even five valves per cylinder (**Figure 42.9**). The

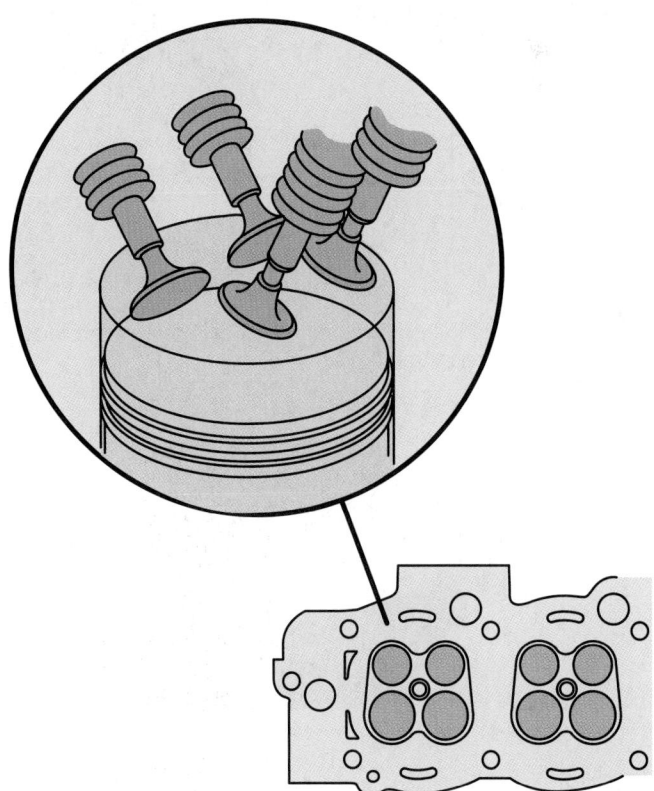

Figure 42.9 A four-valve combustion chamber.

use of multiple valves has become popular because this design helps higher rpm breathing. A greater amount of flow area for a given amount of valve lift is possible compared to two valve heads. A discussion of volumetric efficiency is covered in Chapter 17.

Variable Intake Manifolds

As was discussed previously, intake manifolds are designed for either low-speed or high-speed use. Engines with four valves per cylinder and variable valve timing are more capable of breathing and operating with excellent performance across a broader rpm range. The speed of air movement, and therefore pressure, is important to effective engine breathing. For comparison purposes, imagine trying to suck a drink into your mouth through a very small straw and a very large straw. If you suck softly through the small straw, it works very well. But if you try to suck too hard, no more liquid will flow through the straw.

Varying the size of the intake manifold can be done by changing back and forth between one and two intake manifold runners using a butterfly control valve to maintain velocity and swirl at low speed and high flow at high speed (**Figure 42.10**). Runners of differing lengths, short and long, can also be part of the design (**Figure 42.11**).

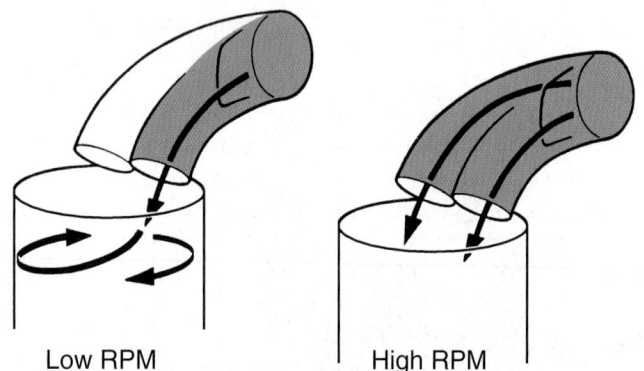

Low RPM   High RPM

Figure 42.10 At low rpm, velocity and swirl are maintained. At high rpm, there is high flow.

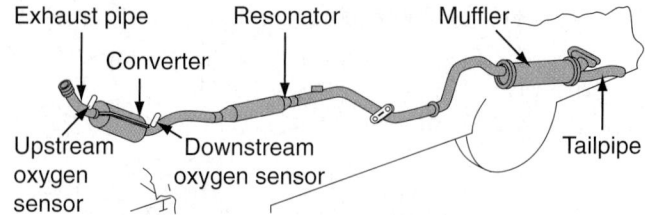

Figure 42.12 Parts of an exhaust system. *(Courtesy of DaimlerChrysler Corporation)*

■ EXHAUST SYSTEM FUNDAMENTALS

Parts of the exhaust system include the exhaust manifold, exhaust pipe, the catalytic converter, the muffler (and sometimes a resonator), the tailpipe, and all of the hangers (**Figure 42.12**). The exhaust system serves three functions:

■ It carries burned exhaust gases away from the passenger compartment of the car.
■ It quiets the engine.
■ Most new cars have one or more catalytic converters to control exhaust emissions.

Backpressure

It would do no good to have a large intake system, with large valves and ports, if the exhaust system were restricted. In four-stroke engines, restrictions in the exhaust can cause backpressure especially at higher speeds. Excessive backpressure reduces performance and fuel economy. The small amount of backpressure caused by the exhaust system is desirable so that the combustion chambers stay hot enough for more complete combustion. A small amount of backpressure (about 2¾ pounds) keeps the air-fuel mixture from going out the exhaust during valve overlap and results in a denser air-fuel charge in cylinder. The restrictions that the exhaust system provide also keep the exhaust system hot after the engine is shut off so that cold air cannot enter and warp an exhaust valve.

Exhaust systems are not severely affected by bends in the pipe as long as the cross-sectional area of the pipe is not diminished. There are special bending machines that provide a true radius to exhaust pipe.

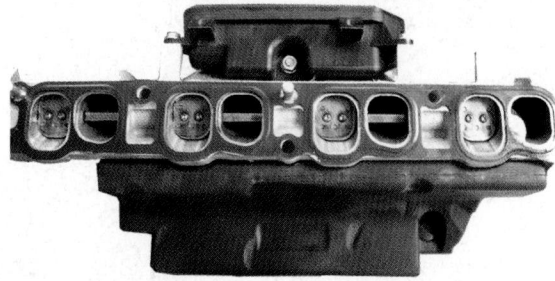
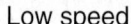

Low speed   High speed

Figure 42.11 Butterfly valves control airflow between the short and long manifold runners based on engine requirements. *(Courtesy of Tim Gilles)*

■ EXHAUST MANIFOLDS

Exhaust manifolds are mounted to the cylinder head's exhaust ports (**Figure 42.13**). V-type engines have two exhaust manifolds, and in-line engines usually have one. When intake and exhaust manifolds are on opposite sides of an in-line engine, the head is called a *cross-flow head* (**Figure 42.14**). This design improves breathing.

Exhaust manifolds are typically made of cast iron or steel, although some late-model cars use stainless steel manifolds. Cast iron is a good material for exhaust manifolds. Like the frying pan on your stove, it can tolerate fast, severe temperature changes.

Exhaust gas temperature is related to the amount of load on the engine. When the engine works hard, or when it has a lean air-fuel mixture, the exhaust

Figure 42.15 A header. *(Courtesy of Tim Gilles)*

manifold can run almost red hot. Usually, the manifold runs cooler, especially at idle.

NOTE: *At the factory, exhaust manifolds are sometimes bolted to the heads with no gaskets because the machined surfaces are perfectly flat. In service, replacement gaskets are usually used.*

Headers

Aftermarket manifolds made of tube steel are called **headers** (**Figure 42.15**). Headers are used when more airflow at higher speeds than stock is desired. Headers are short and have a large cross section to provide maximum high-rpm power. The greater pressure from longer pipes tends to favor low-rpm performance. Headers do not improve performance at low rpm unless they are specifically tuned for low-rpm use. They also have a relatively short service life because they can rust out.

Headers sometimes require modification to install. Emission requirements must be met or modified parts for street use cannot be installed. Any component that is substituted for an original emission part must have a Bureau of Automotive Repair (BAR) number, or federal tampering laws will be violated.

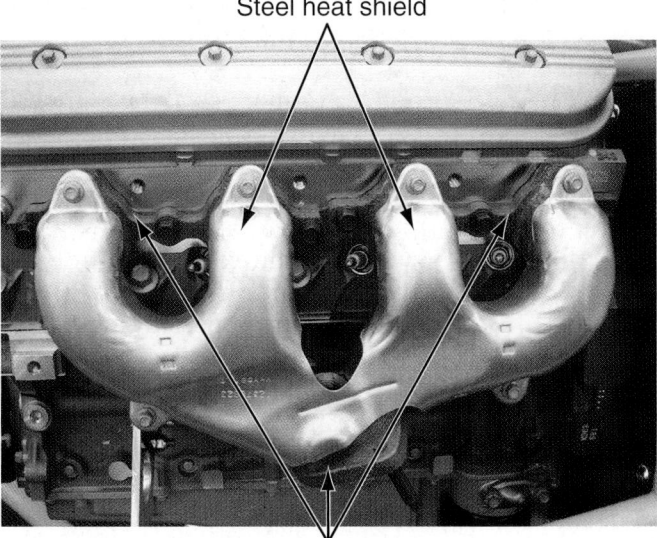

Steel heat shield

Cast iron manifold

Figure 42.13 An exhaust manifold. *(Courtesy of Tim Gilles)*

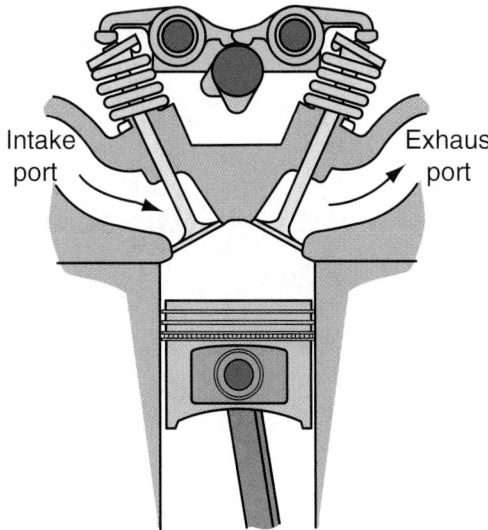

Intake port

Exhaust port

Figure 42.14 A cross-flow head.

■ EXHAUST PIPES

Exhaust pipes are made of steel. There are usually three exhaust pipes: the *header pipe* or *exhaust pipe*, an *intermediate pipe* between the muffler and catalytic converter, and the *tailpipe*.

Exhaust systems are of three styles:

■ In-line engines have a simple design that runs down one side of the engine. Called a *single exhaust system*, there is only one pipe to the rear of the car. Single exhausts can be found on all engine sizes.

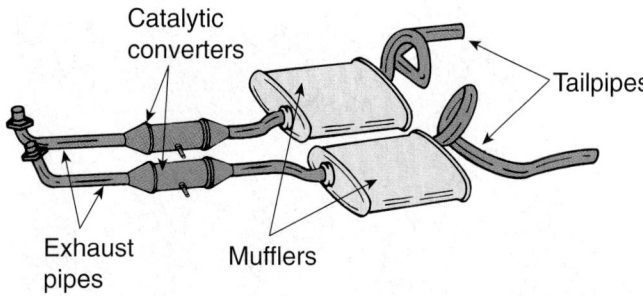

Figure 42.16 A dual exhaust system.

- V-type engines with *dual exhaust* have two pipes, two mufflers, and two catalytic converters (**Figure 42.16**). Large V-type engines use dual exhausts, especially when they are high performance. A dual exhaust allows better breathing when the engine is under load.
- A *crossover pipe* joins the two manifolds into one pipe as they leave the engine on a V-type engine's single exhaust system.

Muffler

Sound is vibration in the air. Each of the engine's exhaust valves releases a burst of pressurized exhaust every two turns of the crankshaft. The resulting noise from all of the cylinders blending together results in a loud roar. The roar is really many pressure bursts or vibrations.

The muffler has tubes and chambers that smooth vibrations out by letting the gas expand and cool (**Figure 42.17**). When the exhaust leaves the muffler, the pressure on it is more even, so it vibrates less. The result is less noise.

Mufflers are constructed of special materials and have different sizes and backpressures available to suit the particular need. There are two muffler designs: *straight through* and *reverse flow*. The straight through has a perforated inner pipe enclosed by the muffler housing, which is usually about three times the inner pipe's diameter. The reverse flow has short pipes and baffles that make the exhaust gas move back and forth as it travels through.

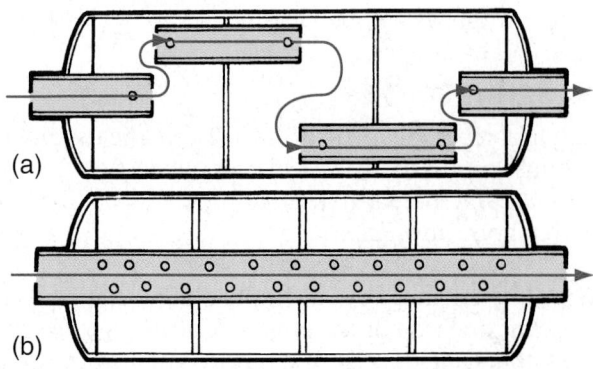

Figure 42.17 Two muffler designs.

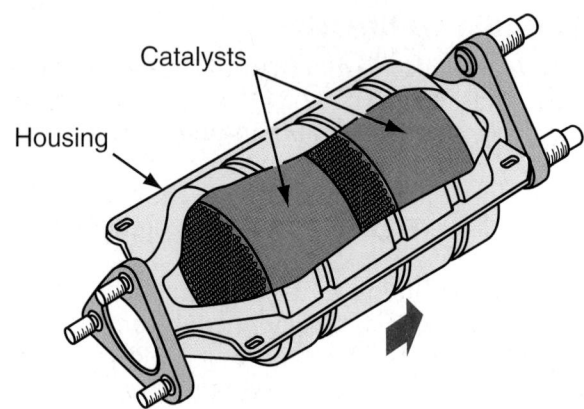

Figure 42.18 A catalytic converter.

Resonator

A **resonator** is a second muffler in line with the other muffler. It further reduces the noise level.

Catalytic Converter

Catalytic converters have been installed on many cars since the mid-1970s. They resemble a muffler, but they contain catalysts to clean up an engine's emissions before they leave the end of the exhaust pipe (**Figure 42.18**). Some catalytic converters have plumbing to them from the smog pump on the engine. There is often a heat shield on catalytic converters because they get very hot when the engine has a misfire (**Figure 42.19**). A misfire allows raw gas to enter the converter because it is not burned in the cylinder. This can cause the converter to overheat and melt. Operation of the catalytic converter is covered in detail in Chapter 43.

Muffler Hangers

Muffler hangers support the muffler and pipes. Some hangers are a piece of fabric and rubber that resembles

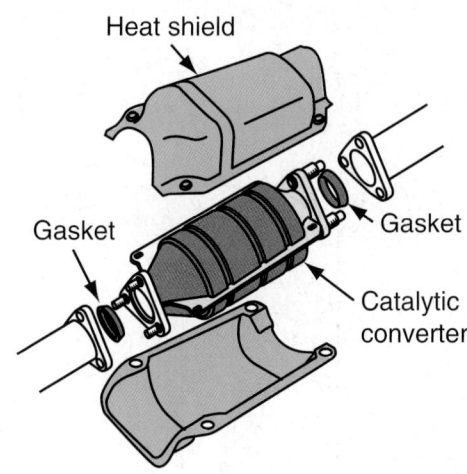

Figure 42.19 Some catalytic converters have heat shields.

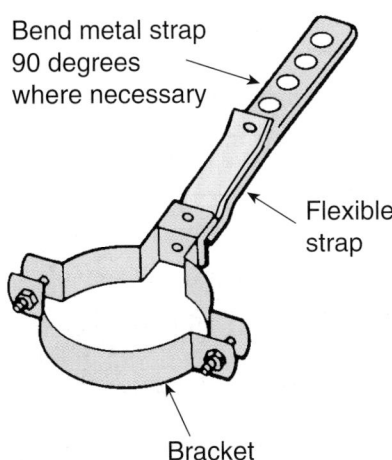

Figure 42.20 A muffler hanger.

a tire (**Figure 42.20**). Each end is riveted to a piece of metal that is fastened on one end to the car body or frame. The other end is clamped to the pipe or muffler. The hanger allows the exhaust to be positioned away from other parts of the car and allows some flexibility from engine torque and vibration. Using a piece of rubber isolates the exhaust noise from the car body. The vibrations would also be felt in the passenger compartment.

Another type of muffler hanger used on many vehicles consists of a pair of steel rods welded to the frame and exhaust pipe. A piece of rubber connects the two steel rods to support the pipe (**Figure 42.21**).

U-bolts of the same size as the exhaust pipe's outside diameter are used to clamp together sections of pipe. One end of one of the pipes is expanded to fit over the other piece of pipe (**Figure 42.22**).

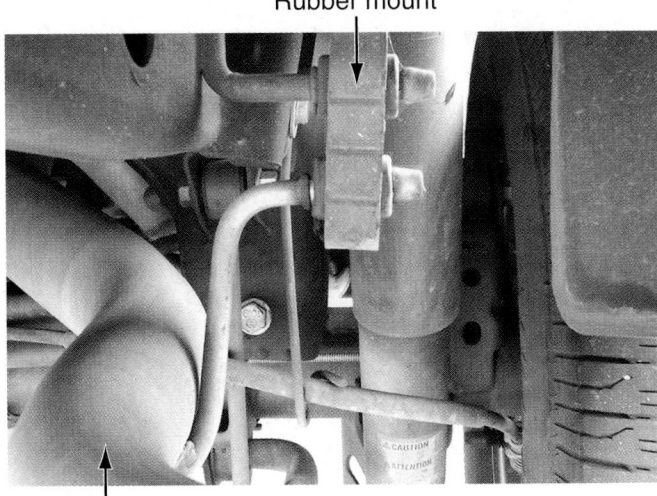

Figure 42.21 A piece of rubber connecting two steel rods supports this exhaust pipe. *(Courtesy of Tim Gilles)*

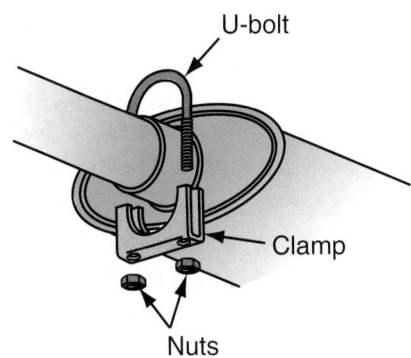

Figure 42.22 A muffler clamp.

■ SUPERCHARGING

Supercharging uses an air pump to increase the density of the air in the cylinder. Each cylinder of a four-cylinder, 2-liter (2,000 cc) engine has a displacement of 500 cc. If the piston is at BDC and the intake valve is open the cylinder will fill with 500 cc of air. This is 100% volumetric efficiency (see Chapter 17). However, if the engine is running, atmospheric pressure is not a sufficient force to fill the cylinder completely with air and its volumetric efficiency, will be less than 100%. An engine's power output is directly related to its volumetric efficiency, and supercharging provides a means of filling the cylinder more completely. Racers call supercharging "a replacement for displacement."

The two primary categories of automotive supercharging are the exhaust-driven turbocharger and the belt-driven supercharger. There are also electric superchargers available in the aftermarket.

Draw-Through or Blow-Through

Superchargers are either **draw-through** or **blow-through**. On carbureted engines, a draw-through system pressurizes the intake manifold after the carburetor and air cleaner (**Figure 42.23**). This is the only practical way to install a Roots-type blower (discussed later in this chapter). Fuel-injected engines have air pumped directly into the intake manifold.

Another type of installation, called blow-through, pressurizes the air cleaner above the carburetor or fuel injection system. These make easier aftermarket installations and will fit more easily under a low hood line (**Figure 42.24**).

Turbochargers

Some engines have a turbocharger installed in the exhaust manifold (**Figure 42.25**). A **turbocharger**, often called a turbo, is a small radial fan pump driven by the energy of the exhaust flow (**Figure 42.26**). The turbocharger helps a smaller engine provide more power than it would otherwise be capable of making, providing approximately a 40% increase in torque and horsepower over a regular **normally aspirated** engine. When the engine is not under load and the

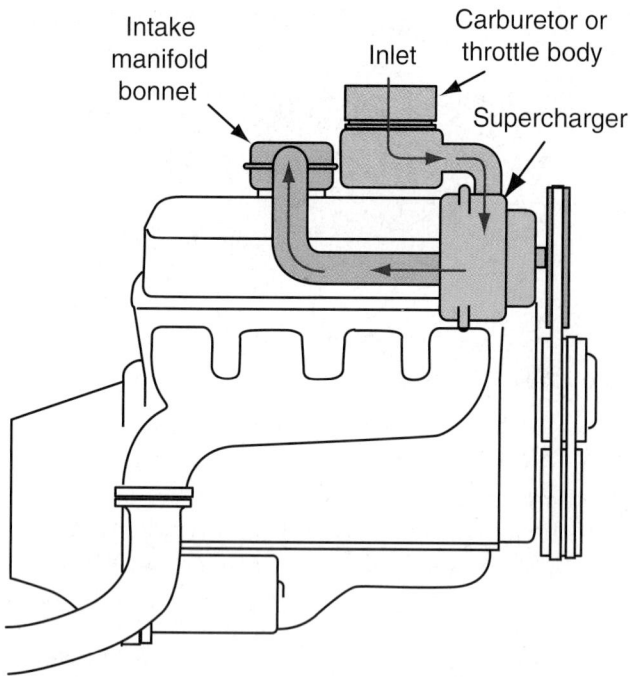

Figure 42.23 A draw-through supercharger.

Figure 42.24 An aftermarket centrifugal belt-driven supercharger. It is a blow-through, fuel-injected design. (*Courtesy of Tim Gilles*)

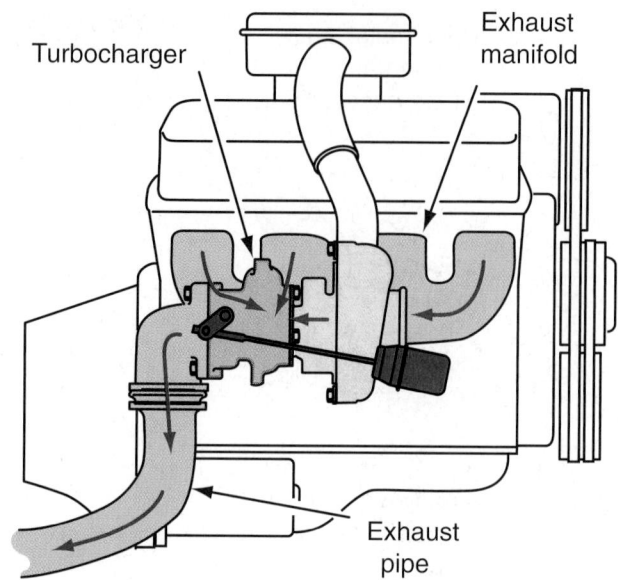

Figure 42.25 A turbocharger on a four-cylinder engine.

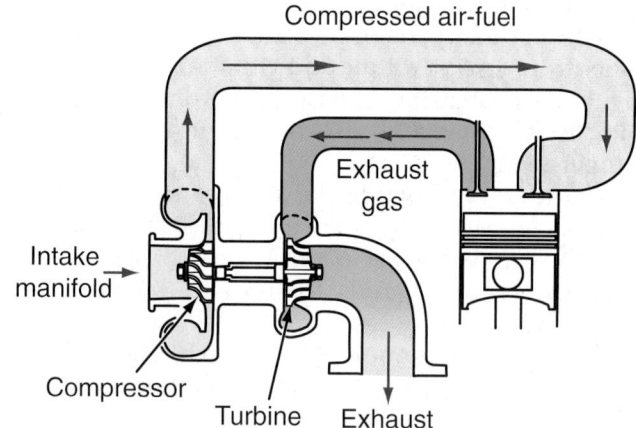

Figure 42.26 A turbocharger uses the energy of exhaust gas to force more air-fuel mixture into the cylinder to increase engine power.

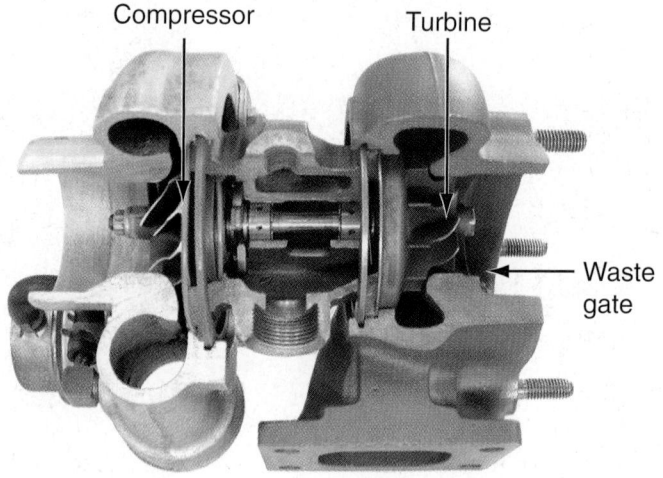

Figure 42.27 Turbocharger cutaway. (*Courtesy of Tim Gilles*)

turbo is not in use, the smaller engine gets better fuel economy than a larger, nonturbocharged engine of comparable power. A turbo can spin in excess of 100,000 rpm. In fact, turbos are balanced to run in excess of 150,000 rpm, 25 times the maximum rpm of most engines. For purposes of comparison, a typical alternator will spin up to about 20,000 rpm.

A turbocharger is a *centrifugal* pump. Centrifugal force takes the incoming exhaust and throws it out of a snail-shaped outlet. The pump has two impellers on one shaft. The one in the exhaust side is called the turbine; the one that forces the air into the engine is called the compressor (**Figure 42.27**). As the exhaust forces one

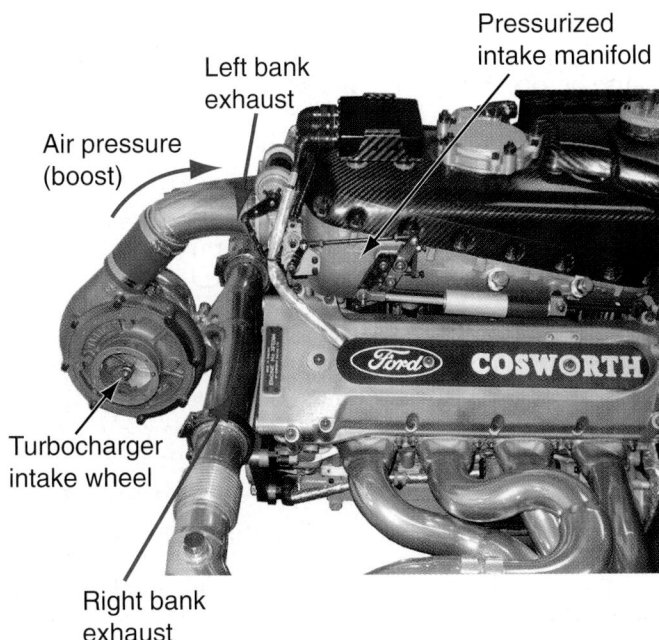

Air pressure (boost)

Left bank exhaust

Pressurized intake manifold

Turbocharger intake wheel

Right bank exhaust

Figure 42.28 A blow-through turbocharger on a racing engine. (*Courtesy of Tim Gilles*)

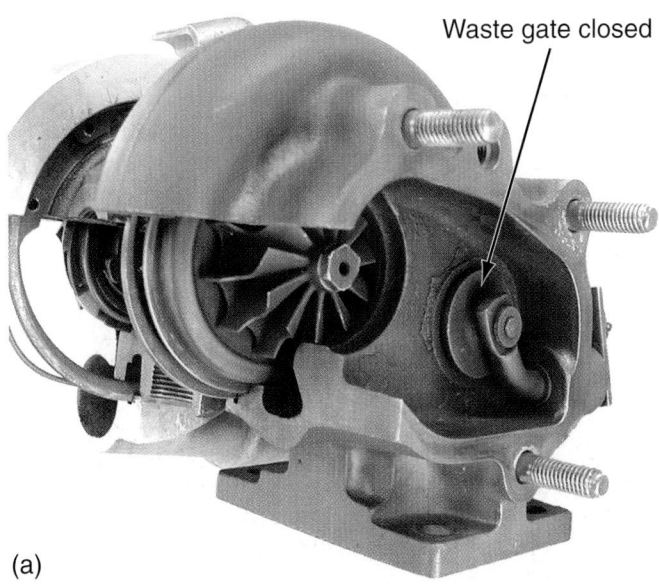

Waste gate closed

(a)

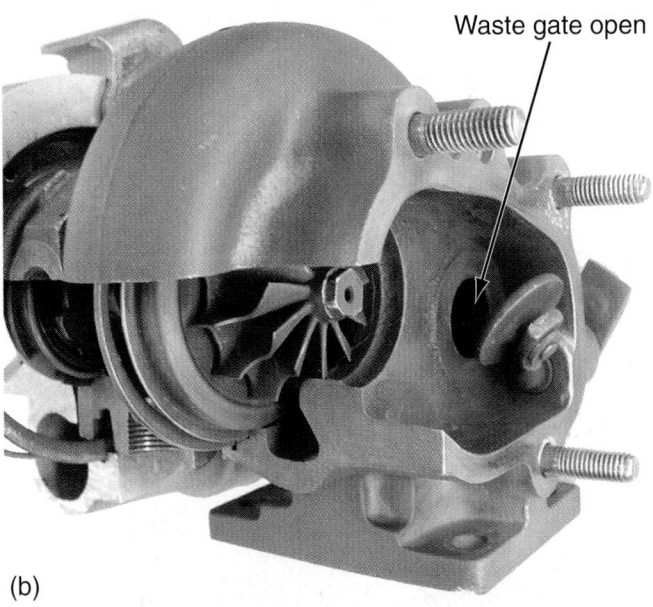

Waste gate open

(b)

Figure 42.29 (a) Waste gate in the closed position. (b) Waste gate in the open position. (*Courtesy of Tim Gilles*)

impeller to turn, the other impeller forces more air-fuel mixture past the intake valve and into the cylinder. This increases the efficiency of the engine.

Another turbo design, called the variable nozzle or variable displacement turbo, changes its nozzle opening in response to changes in engine load.

A drawback to the turbocharger is decreased engine life. The smaller the engine, the larger the percentage of time the turbo is used for accelerating and climbing hills. This results in a hotter running engine.

NOTE: *A normally aspirated engine will lose 3% of its horsepower with every 1,000-foot increase in altitude. A turbocharged vehicle will not lose power as it is driven into higher altitude.*

Figure 42.28 shows a blow-through turbocharger on a racing engine.

Waste Gate. Because the system would only work well at very high rpm when the exhaust is moving fast, a higher efficiency turbo is used. The turbo works better at a lower rpm; at higher speeds, it builds excessive pressure, so a relief valve is needed. This relief valve is called a **waste gate**. **Figure 42.29** shows a waste gate in the open and closed positions.

Boost. The amount of air density a turbo provides is known as **boost pressure**. A waste gate is set to open at a specific number of *pounds of boost*. Without a waste gate, the turbo could provide the engine with so much power that it could literally destroy itself. There is a relief valve to protect the system in case the waste gate becomes stuck.

NOTE: *The waste gate is the part that allows the driver of an Indy car to adjust the boost during a race. Of course, raising the boost also results in the use of more fuel, so this is not done without a great deal of thought.*

The point at which boost first starts (1800 rpm, for instance) is called the *boost threshold*.

Turbo Lag. When the turbo is spinning at low speed, little or no boost is produced. The time required to bring the turbo up to a speed where it can function effectively is called **turbo lag**. This is a hesitation in throttle response that is noticed when coming off idle. Turbo lag, which is actually a lag in the air-fuel ratio, is more pronounced when a carburetor or throttle body is too far away from the intake valve. The problem is less noticeable with port fuel injection (when the fuel injector is located directly above the intake valve).

Intercoolers. An **intercooler**, known also as a *charge air cooler* or an *aftercooler*, is an option that may be installed with a turbo (**Figure 42.30**). Theoretically, a turbocharged engine operating at sea level with 7 psi boost should have about 1½ times the power of a normally aspirated engine. Unfortunately, air that is compressed becomes hotter, so fewer air molecules can enter the cylinders on each intake stroke.

An intercooler is simply a very sophisticated and expensive air cooler installed after the turbo (**Figure 42.31**). For every 10°F that the air-fuel mixture is cooled, a power gain of about 1% is achieved. Intercoolers cool the turbocharged air by about 100° before it enters the engine. With that amount of cooling, an intercooler can provide about 10% more power.

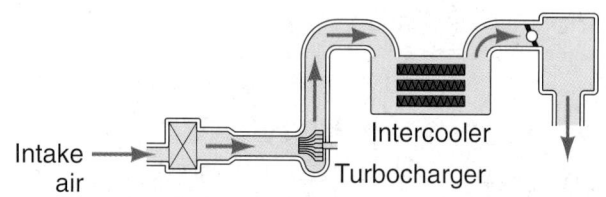

Figure 42.30 An intercooler cools the compressed air after it leaves the turbocharger.

Figure 42.31 An intercooler increases power by about 1% for every 10°F cooling of incoming air. *(Courtesy of Tim Gilles)*

Intercoolers use either air or water cooling. They are relatively trouble free. Pressure from the turbo can cause detonation of the fuel in the cylinder. When the temperature of the incoming air is 90°F, it will become 276°F if it is compressed at 14 pounds of boost. Most production cars run 6 or 7 pounds of boost so that they can live a long life on the low-octane fuels that are available. This is about half of atmospheric pressure.

Belt-driven Superchargers/Blowers

Superchargers are air pumps, commonly called **blowers**. They can easily produce 50% more power than a normally aspirated engine of the same size. The crankshaft usually drives the supercharger with a belt, but it is sometimes driven by a chain or gears. **Figure 42.32** shows a cutaway of an original equipment passenger car supercharger.

Belt-driven superchargers spin at 10,000 to 15,000 rpm, much slower than turbochargers. Although boost is limited by engine speed, blowers still have a pressure relief valve (called pop-off valves) (**Figure 42.33**). This valve opens when there is excess pressure due to a fuel explosion. Because it is belt driven, a supercharger provides more torque at lower speeds than a turbocharger. It also has a quicker response (no turbo lag).

There are several kinds of supercharger pumps, including centrifugal, Roots, vane (Judson), Lysholm twin screw, rotary (Wankel), and axial flow fan (like a jet turbine). The most common ones in automotive use are the centrifugal and the Roots types (**Figure 42.34**). A belt-driven Roots blower is the one commonly seen on dragster engines (**Figure 42.35**).

There are two groups of superchargers: the positive displacement and the dynamic.

Positive Displacement Superchargers. A positive displacement pump delivers the same amount of air

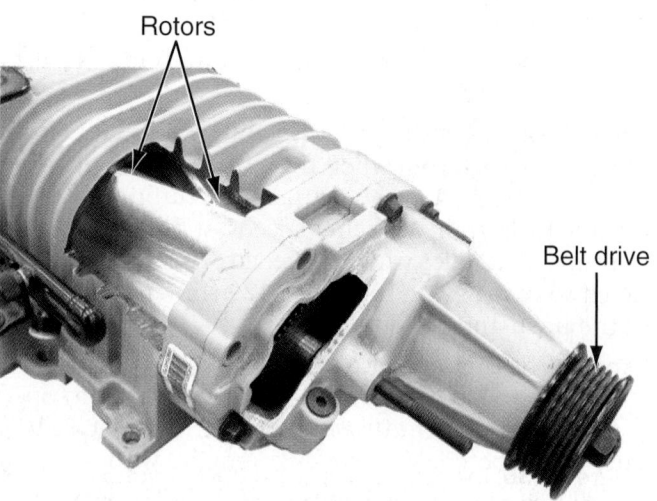

Figure 42.32 Cutaway of an original equipment supercharger. *(Courtesy of Tim Gilles)*

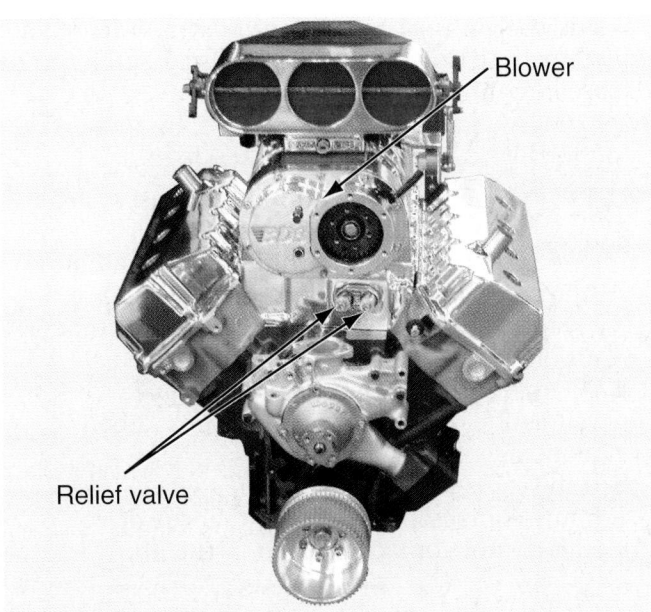

Figure 42.33 A pressure relief (pop-off) valve on a blower. (*Courtesy of Tim Gilles*)

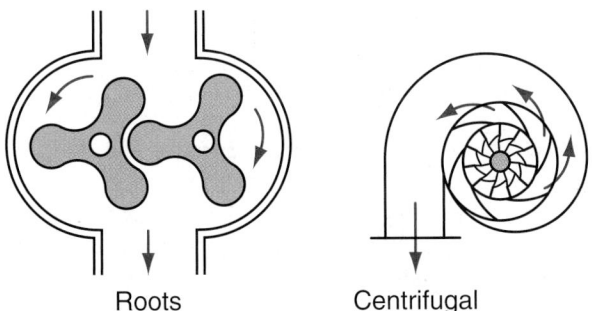

Figure 42.34 The most common automotive supercharging designs are the Roots-type and the centrifugal-type pumps.

Figure 42.35 This funny car engine has a Roots-type blower. The belt arrangement over the top provides protection if it explodes (as they sometimes do). (*Courtesy of Tim Gilles*)

with each revolution regardless of the speed. The faster it turns, the more air it pumps. The most popular positive displacement supercharger is the **Roots type**, called a lobe-type supercharger. Roots-type blowers are used on top-fuel engines because they are the only type of forced induction permitted in that class of drag racing. Meshing rotors pump and compress air when they are rotated. Sometimes lobes are *helical* (twisted), and other times they are *straight cut*. Lobe superchargers tend to pulse at low speeds, but a helical rotor tends to smooth these pulses out. A three-lobe rotor tends to pulse less than a two-lobe rotor. When the engine is not under a heavy load, intake manifold vacuum turns the rotors like a windmill, thereby using less power. Unlike turbos, however, blowers consume horsepower as they are driven. A Ford blower uses 60 horsepower at 5,000 rpm but only uses 1/2 horsepower when cruising on the highway.

Other Supercharger Types. Volkswagen's scroll-type G-lader is so called because *lader* means "charger," and the shape of its chambers resembles the letter "G." The displacer inside its housing moves in an off-center circular motion like a hula-hoop.

Another positive displacement blower is the vane type, which compresses air inside its housing before forcing it through its outlet. Electrical axial flow fan-type superchargers draw high amperage and are used most often on small engines with high boost. When installed correctly, they operate only at full throttle so they do not wear out the brushes in the electric motor.

Dynamic Superchargers. The three types of dynamic superchargers are the centrifugal, the axial flow, and the pressure wave. A centrifugal supercharger is similar in design to a turbocharger but is crankshaft driven rather than exhaust driven. The axial flow pump is not often found, because of its expense to manufacture. The pressure wave pump is used mostly for two-stroke diesels.

A dynamic supercharger is like a turbocharger in that its output increases as the square of engine speed. If the engine turns twice as fast, boost output is quadrupled, so this pump operates best at high speeds but has less boost off idle.

■ EXHAUST SYSTEM SERVICE

Exhaust systems can rust out because there are acids and a good deal of moisture in the engine's exhaust. An engine that is never fully warmed up will experience rapid rusting in the exhaust system because the moisture never gets a chance to dry out. It will pool in the lowest spot in the exhaust system.

NOTE: *Mufflers often rust out on the lowest end.*

Usually the first clue that there is a leak in the exhaust system is the sound of leaking exhaust. Listen for leaks while the engine is running. Check for leaks at all pipe connections and at low spots in the system. A stethoscope can be used. Remove the metal end from the stethoscope and use just the hose to listen.

Exhaust Gaskets

Exhaust gaskets suffer a good deal of abuse. Parts expand when heated. Sometimes an aluminum part will be clamped to an iron or steel exhaust part. These metals have different expansion rates. Gaskets between an exhaust manifold and cylinder head are assembled with the shiny side out toward the exhaust manifold. This allows the hotter manifold to be able to slide against the gasket as it expands.

A round sealing ring, often called a *donut*, fits between the manifold and the exhaust pipe (**Figure 42.36**). Donuts can be made of aluminum, steel, fiber, or ceramic. They allow a transverse engine to be able to move on its mounts without exhaust leaking from the flange.

Muffler shops specialize in this type of work. They have a large inventory of many sizes of straight pipe. Hydraulic tubing benders are used to make new pipes. Preformed pipes are also available from aftermarket suppliers for technicians who want to perform the work themselves.

It is not unusual for the entire system to require replacement. If only the muffler is rusted out, it can be replaced, reusing the old pipes.

CAUTION ■ Be careful around exhaust system parts. They are very hot.
■ Wear eye protection. When pounding on exhaust system parts to get them apart or put them together, rust can come off old pipe parts and get in your eye.
■ Do not run the engine without adequate ventilation. Engine exhaust contains carbon monoxide.
■ If the car is to be run while on a wheel-free, frame-contact lift, be certain that the drive tires do not contact any part of the lift.

Header Pipe

Sometimes the section of pipe that is attached to the exhaust manifold (header pipe) is stainless steel. This is to comply with the federal law stating that any emission related component must be warranted for 5 years or 50,000 miles or longer, depending on the year of the vehicle. A header pipe is often two laminated layers. The inside of these pipes can collapse, resulting in a restriction in the exhaust (**Figure 42.37**). A discolored stripe on the outside of the burned section of pipe often gives a clue to where the restriction is. Mufflers that have collapsed internally will often exhibit a stripe around the outside of the restricted area also. Refer to Chapter 48 for the procedure for diagnosing a collapsed exhaust.

Exhaust Part Replacement

Penetrating oil can be applied to rusted fasteners and joints in the pipe before disassembly, but parts

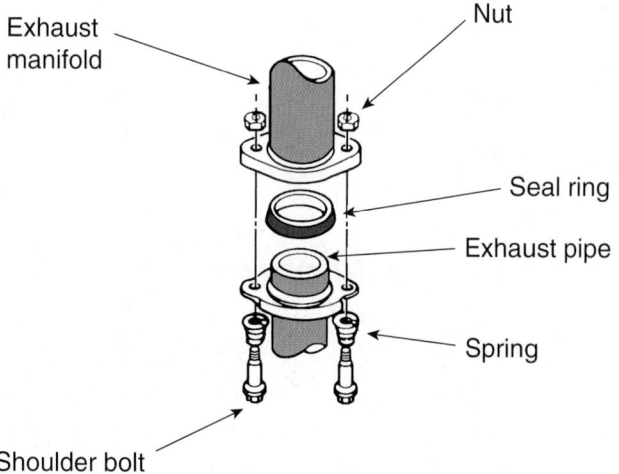

Figure 42.36 An exhaust seal, called a donut. (*Courtesy of DaimlerChrysler Corporation*)

Exhaust manifold
Nut
Seal ring
Exhaust pipe
Spring
Shoulder bolt

Figure 42.37 This collapsed laminated exhaust pipe caused breathing problems. (*Courtesy of Tim Gilles*)

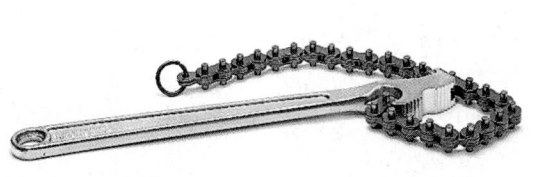

Figure 42.38 A chain-type exhaust tubing cutter. *(Courtesy of Snap-on Tools Company, www.snapon.com)*

Figure 42.39 A hydraulic pipe expander. *(Courtesy of Snap-on Tools Company, www.snapon.com)*

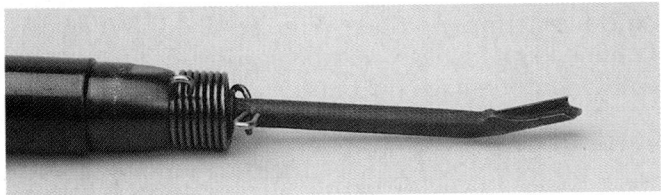

Figure 42.40 This tool is used to slice a rusty exhaust pipe before removal. *(Courtesy of Tim Gilles)*

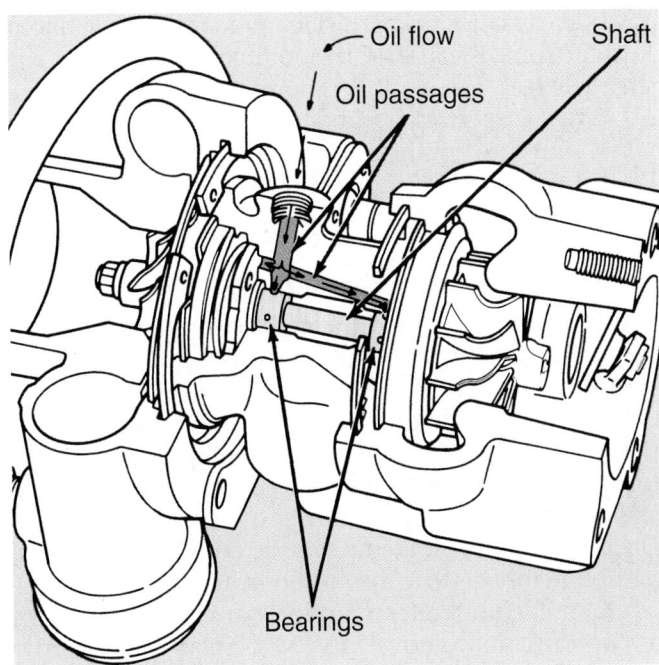

Figure 42.41 Turbocharger oil passages. *(Courtesy of Ford Motor Company)*

often break when they are rusted. Some of the tools used for exhaust work include a chain-type pipe cutter (**Figure 42.38**), a roller-type exhaust cutter; a pipe expander (**Figure 42.39**), deep sockets to clear the long, threaded studs; and a chisel cutter (**Figure 42.40**). The chisel cutter is often used with an air chisel. An oxyacetylene torch is the tool of choice in muffler shops.

A pipe shaper that resembles a tapered cone can be used with an air chisel, too. It is for straightening out ends of pipe that are bent.

■ TURBOCHARGER SERVICE

Turbo Lubrication

Lubrication is crucial to turbochargers (**Figure 42.41**). The rear bearing in a turbo can get extremely hot if the engine is shut off immediately after the turbo is used (*hot shutdown*). Motor oil burns at 430°F. The rear bearing, on the exhaust side of the turbo, can reach temperatures in excess of that when the engine is shut off after a period of acceleration. Some turbochargers have rear bearings cooled with engine coolant to help combat this problem. Synthetic oil burns at a higher temperature, so it is the oil of choice for use with a supercharger.

> **SHOP TIP** When oil is changed, disable the ignition system and crank the engine for 30 seconds to supply oil to the turbo. Failure to do this allows the turbo to run without oil when the engine is started. This can ruin an expensive turbo.

Turbocharger Replacement

Rebuilding a turbo is not commonly done in the trade. There are too many different part numbers available, and some turbos are balanced by grinding on the shaft nut. When this nut is removed, balance, which is critical, is upset. Unless many units are rebuilt, it is not economical for a repair shop to rebuild their own units.

A new or rebuilt unit, called a *cartridge*, is obtained and installed in the old compressor housing. The compressor housing must usually be removed from the intake manifold before the turbo cartridge can be removed from it. Failure to do this can result in a bent impeller blade. The turbo will howl and bearing failure will usually result.

Turbos, other than German ones, use full-floating **plain bearings** (see Figure 42.41). Full-floating bearings have oil clearance on both sides and they spin at about ⅓ shaft rpm. Because of this double oil clearance, the shaft will feel "loose." This is normal.

When inspecting a turbo, a rule of thumb is "if the wheel rubs the housing, the turbo needs to be replaced." Do not try to make it rub; simply see if it has been rubbing while in service.

Turbocharger Troubleshooting

The turbo runs in a difficult environment of high rpm and high heat. Exhaust gas is red hot (1,800°F to 1,950°F). This heat is constant, compared to that of the exhaust valve (which is cooled three-quarters of the time). The primary cause of turbo failure can be overlooked, resulting in failure of a second turbo. Blockages, leaks, and foreign object damage are common failures. Diagnose engine problems as if the engine did *not* have a turbo before getting into turbo troubleshooting.

Manufacturers refer to unnecessary warranty returns as "no cause for removal" (**NCFR**). About half of turbo warranty returns are NCFR. The leading cause of NCFRs is blowby, which can cause oil in a turbo housing.

The rings that seal the turbo shaft are for sealing exhaust pressure out of the center housing. A plugged PCV system causes pressure to flow *into* the oil drain from the turbo. Restrictions in the oil return (**Figure 42.42**) will cause pressure that results in oil

Figure 42.43 Damaged compressor wheels.

consumption and smoke. If a plugged drain line is not repaired, a new turbo will leak too.

A defective compressor shaft seal can also cause oil consumption and an intake manifold air leak. The PCV valve must close when the intake side of the turbo is under boost or the crankcase can be pressurized through the PCV system. This can cause oil to enter the intake side of the turbo from the now pressurized center housing. Be sure the proper PCV valve is used and is in good operating condition.

When the turbo has suffered foreign object damage, look at the wheel to see which direction the damage came from. Foreign object damage to the turbine (outlet) can be due to broken piston rings, valves, pistons, or any other engine failure that enters the exhaust. Damage to the compressor wheel is due to objects entering the intake (**Figure 42.43**). When contaminated oil damages a turbo bearing, heavy particles will tend to damage the outside of the bearing.

Turbocharger Care

Heat can be an enemy to the turbo. Oil flow to a turbo is only about 8% of an engine's total oil flow, but the turbo heats the oil 40% more than engine bearings do. Following high-output use, allow the engine to idle for 1 minute or drive slowly for 1,000 yards (about a kilometer) or so to let the turbo get resupplied with cool oil before shutting the engine down.

Early turbos in the late 1970s were mounted in the rear of the engine compartment and therefore ran hotter. *After-lube* was especially important in these units. Some of today's turbos have special lubrication and cooling features that function even after shutdown.

Figure 42.42 A turbo oil return can become plugged with carbon.

> **CASE HISTORY**
>
> *A driver of a turbo car was testing out his engine in a remote area when he was stopped by the police for speeding. He pulled over and shut off his engine. While the engine was off, the oil solidified in the oil return, resulting in excessive smoke.*

After a period of storage, a turbo can be prelubed by disabling the ignition and cranking the engine for

15 seconds or so until oil pressure is registered on the dash gauge or light. There are special turbo oils available. Multiviscosity oils are also recommended because cold 30W can bypass the filter and the unfiltered oil will go directly to the turbo (see Chapter 12).

SUPERCHARGER SERVICE

Because superchargers are not powered by exhaust gas energy, they do not get hot. Lubrication is not a big problem like it is with turbos. In fact, the Roots-type units are lubricated with their own supply of SAE 90 gear oil, which has no specified oil change interval. Blowers are very dependable, but dirt is a prime enemy to them.

Vacuum leaks (intake side) suck in dust that can ruin the unit. Exhaust-side air leaks (boost side) hurt performance. Vacuum leaks can also fool the computer, causing the engine to run lean. A leak on the boost side, on the other hand, will cause a rich condition. The oxygen sensor in these systems (the device that tailors the air-fuel mixture by what is coming out the vehicle's exhaust) will only make relatively minor corrections to the mixture. It cannot compensate for major leaks.

Leaks are usually accompanied by a whistling sound that can be easily located by listening for its source. On modern OBD II engines and later, a code will be set and the malfunction indicator light will illuminate if a leak is excessive.

REVIEW QUESTIONS

1. Which manifold runners are smaller in diameter, long ones or short ones?

2. What is the name of the manifold air space below the carburetor?

3. When one runner feeds two neighboring cylinders, what is this called?

4. Draw a sketch showing how a dual plane manifold distributes air and fuel to a V8 engine.

5. What is the name of the turbocharger wheel that is on the intake side?

6. If a normally aspirated engine climbs to 3,000 feet in altitude, how much power will it lose?

7. What is the name of the relief valve used to limit turbo boost pressure?

8. The time required to bring the turbo up to a speed where it can function effectively is called turbo _____.

9. Which uses power when cruising, a supercharger or a turbocharger?

10. Another name for a lobe-type supercharger is _____ type.

ASE-STYLE REVIEW QUESTIONS

1. Which of the following is/are duties of an air cleaner?

   a. It silences incoming air.

   b. It acts as a flame arrestor.

   c. Both A and B.

   d. Neither A nor B.

2. Technician A says that manifolds for throttle-body fuel injection (TBI) are designed to flow air only. Technician B says that manifolds designed for carburetors are made to flow air and fuel. Who is right?

   a. Technician A   c. Both A and B

   b. Technician B   d. Neither A nor B

3. Technician A says that bends in the exhaust system do not always cause a restriction. Technician B says that headers improve engine performance at low rpm. Who is right?

   a. Technician A   c. Both A and B

   b. Technician B   d. Neither A nor B

4. What is a turbocharger that pressurizes the intake manifold after the throttle plate called?

   a. A draw-through turbo

   b. A blow-through turbo

   c. A normally aspirated turbo

   d. None of the above

5. Which of the following is/are true about turbochargers?

   a. Turbochargers are usually called blowers.

   b. A car with a turbo will not operate as well in high altitude.

   c. Cooling the compressed air from the turbocharger by 100°F can increase engine power by 10%.

   d. All of the above.

6. All of the following are true about superchargers *except*:

   a. Superchargers are louder than turbochargers.

   b. Superchargers run cooler than turbochargers.

   c. Turbochargers spin slower than superchargers.

   d. Lubrication is not as big a problem with superchargers as turbochargers.

7. Exhaust systems usually rust out:

   a. At the highest point.

   b. At the lowest point.

   c. In the center.

   d. There is no location that rusts out more often than another.

8. Technician A says to take the metal end off a stethoscope to listen for exhaust leaks. Technician B says that engines that are not often fully warmed up tend to experience more rusting in the exhaust system. Who is right?

   a. Technician A   **c.** Both A and B

   b. Technician B   **d.** Neither A nor B

9. All of the following are likely symptoms of a restricted exhaust system *except*:

   a. A stripe that runs around the exhaust pipe

   b. A hissing sound out the end of the pipe under heavy acceleration

   c. Air rushing back out of the intake during acceleration

   d. Poor engine performance at idle and off idle

10. Technician A says that if a turbocharger wheel has been rubbing against the housing the turbocharger must be replaced. Technician B says that a bad turbocharger can cause engine oil consumption. Who is right?

   a. Technician A   **c.** Both A and B

   b. Technician B   **d.** Neither A nor B

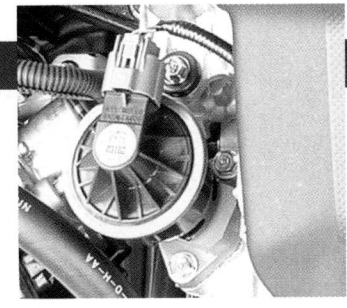

CHAPTER 43

Emission Control System Fundamentals

■ OBJECTIVES

Upon completion of this chapter, you should be able to:

✔ Describe the different types of air pollution caused by motor vehicles.

✔ Explain the fundamentals of the major emission control systems.

✔ Label the parts of emission control systems.

✔ Explain the operation of computer-controlled emission systems.

■ KEY TERMS

photochemical smog
inversion layer
EPA
skin effect
particulate
closed ventilation system
air injection system

smog pump
air switching valve
inert gas
catalyst
monolithic catalyst
reduction catalyst
oxidation catalyst

charcoal canister
hot soak
liquid/vapor separator
purge
malfunction
   indicator light
surface-to-volume ratio

■ INTRODUCTION

Since emission controls began to be included on cars in the 1960s, this has become a complicated specialty area. In most states, emission specialists are required to be licensed to perform repairs. The emphasis in this chapter is to cover the basic theory of emission devices. Understanding this information is necessary for you to learn to diagnose and repair emission systems.

■ AIR POLLUTION

Air is polluted by many different things: oil exploration, industry, nature, home fireplaces, furnaces, paints, and transportation (cars, trucks, trains, boats, and airplanes). **Photochemical smog** is produced when hydrocarbons and oxides of nitrogen react with sunlight to form a dirty yellow-brown haze in the air. Warm air close to the ground rises, where it is cooled by cooler air at higher altitudes. This process usually rids the air of smog. But when there is a warm air **inversion layer**, the warm air becomes trapped within 1,000 feet of the ground. This trapped air causes high concentrations of smog, resulting in health hazards.

Beginning in the early 1960s, the United States Congress began to pass laws related to air pollution.

These laws are administered by the Environmental Protection Agency (**EPA**). Vehicles have been designed to lessen the amount of pollutants produced when they burn fuel. Engine controls and emission devices have reduced vehicle emissions to the point that a car manufactured today will produce less than 5% of the air pollution that a 1960 model did.

■ AUTOMOTIVE EMISSIONS

There are three sources of emissions: crank-case, evaporation, and exhaust. The exhaust emission control system is designed to minimize by-products of combustion leaving the engine in its exhaust. The emission control system is designed to control or eliminate unburned *hydrocarbons (HC)*, *carbon monoxide (CO)*, and *oxides of nitrogen (NO$_x$)*.

Automotive fuels are hydrocarbons. When they are not completely burned in the engine and enter the air during daylight, smog can result. Vehicle emissions can come from several places (**Figure 43.1**):

■ The exhaust pipe—all three emissions

■ The crankcase—mostly HCs

■ Vapors—HCs that evaporate from the fuel tank or vents in the fuel system

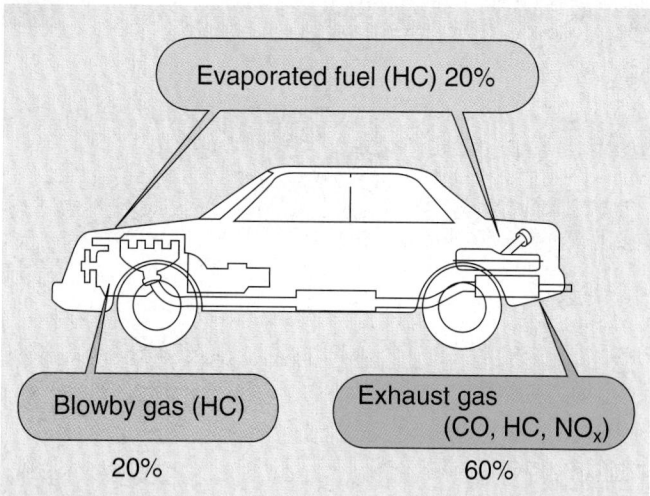

Figure 43.1 Sources of automotive air pollution.

Composition Type of Gas	CO	HC	NOₓ
Exhaust gas	100%	55%	100%
Blowby gas	—	25%	—
Evaporated fuel	—	20%	—

Figure 43.2 Composition of various emissions.

The percentages of various emissions in each of these categories are listed in **Figure 43.2**. All of these emissions are defined together as *exhaust emissions*.

Hydrocarbons

HC forms from several sources:

■ Blowby gases are those that leak past the piston rings. Crankcase ventilation systems, explained later, remove these vapors from the crankcase.

■ After each four-stroke cycle is completed, raw gasoline remains on all areas of the combustion chamber. This happens because the surface area of the combustion chamber is cold so fuel burning is *quenched* (**Figure 43.3**). This is called the **skin effect**. HC emissions are worse when there is more surface area.

■ When combustion is incomplete because of too rich an air-fuel mixture, raw gas results in the exhaust.

■ If a cylinder does not produce sufficient compression to burn the mixture, unburned hydrocarbons will exit the cylinder.

■ When the ignition system does not provide adequate spark.

Carbon Monoxide

Carbon monoxide (CO) emissions result when gasoline is not completely burned. The amount of CO in the

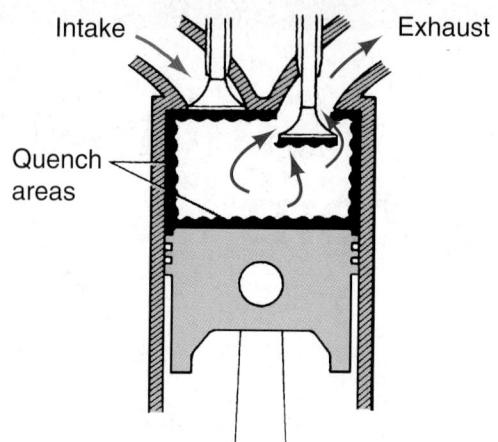

Figure 43.3 Unburned fuel is left on all surface areas.

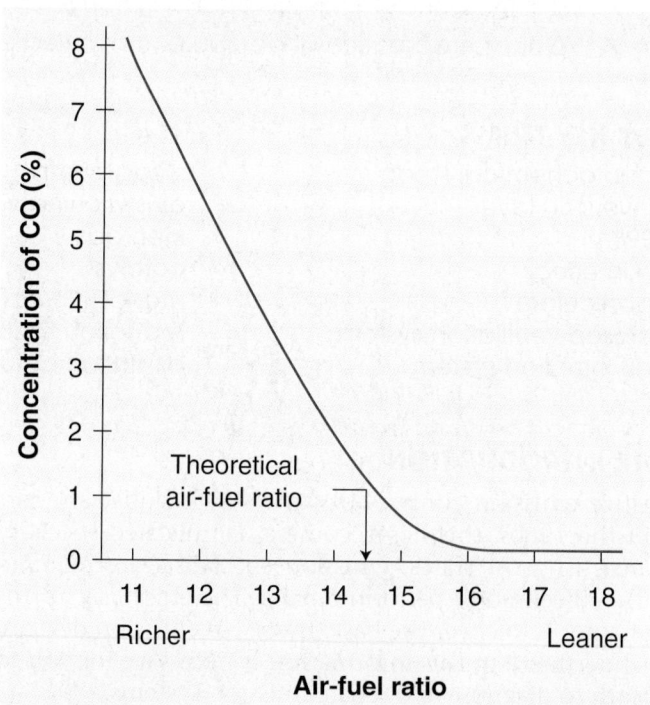

Figure 43.4 The amount of carbon monoxide (CO) in the exhaust varies with the air-fuel ratio.

exhaust varies with the air-fuel ratio (**Figure 43.4**). When an air-fuel mixture is too rich, there will be more CO in the exhaust. Uneven air-fuel distribution in the combustion chamber and too much quench from low cylinder wall temperatures can also cause CO to be high.

Oxides of Nitrogen

Oxides of nitrogen (NO_x) are produced when combustion temperatures are too high (above 2,500°F), causing nitrogen to react with oxygen. High compression and excess cooling system temperatures can be causes.

Particulates

Particulates are airborne microscopic particles like dust and soot. These are a source of secondary air pollution that contains carbon. Particulates can be harmful to your health. They are especially prevalent in diesel engine exhaust.

CO_2 and O_2

There are two other gases that are measured in vehicle exhaust. These two, *carbon dioxide* (CO_2) and *oxygen* (O_2), are not harmful but can be used to diagnose combustion problems. A four- or five-gas emission analyzer, covered in Chapter 44, reads these two gases as well as other polluting gases.

Greenhouse Gases

CO_2 is a greenhouse gas that traps heat in the atmosphere. Some of the sun's energy is absorbed by the surface of the Earth, warming it. Energy emitted from the Earth's surface radiates back out to space but at a slower rate than energy from the sun. Greenhouse gases radiate some of this energy toward space, but the rest of it is radiated back into the atmosphere, resulting in the *greenhouse effect*.

▨ SCIENCE NOTE ▨

Most scientists agree that gradual global warming is occurring, although they are not all in agreement as to how fast it is occurring. Regardless of the rate, scientists do agree that if it is allowed to continue for a sufficient amount of time, even a 1°F to 3°F change in temperature could dramatically affect coastlines and crop growing areas.

Greenhouse gases include water vapor, CO_2, methane, nitrous oxide, ozone, and others. Some of the atmosphere's greenhouse gases are naturally occurring, whereas others are a result of human industry.

▬ POLLUTION CONTROL

In the early 1960s, California became the leader in emission control legislation. This was due to the smog problem of the Los Angeles basin, caused in part by the large vehicle population. Studies at that time showed that 25% of HC emissions came from the open automotive crankcases of that day.

Beginning with the 1961 model year, all new cars sold in California were required to have a crankcase emission system. In the 6 years following the introduction of this system, 1 million more cars were registered in California, but emission levels remained the same. In 1963, this system became a requirement on all cars sold in the United States.

By 1966, new California cars were required to have exhaust emission systems. Some early systems included an air pump for reducing HC and CO emissions. Adding air to the exhaust ports helped the air-fuel mixture to continue to burn in the vehicle's exhaust system. Burning pollutants more completely lowers air pollution.

In 1970, the United States Congress passed the *Clean Air Act*. Maximum levels of pollution were set nationwide for HC, CO, and NO_X for automobiles and industry. The Arab oil embargo of 1973 caused oil prices to rise significantly. The United States Congress passed a law that required improved fuel economy on vehicles. This is the Corporate Average Fuel Economy (CAFE) standard. In response, manufacturers designed cleaner, more fuel-efficient engines. Engines since 1973 have had many improvements made to them, some of which are discussed in this chapter.

▨ SCIENCE NOTE ▨

Burning a pollutant more completely reduces pollution by one of the following methods:
- *Providing sufficient oxygen so that all carbon in the compound is oxidized to CO_2*
- *Providing enough heat so that large, tightly packed complex compounds can relax and stretch out enough so that oxygen can penetrate the inside of the compound for efficient burning*
- *Breaking toxic compounds down into nontoxic ones*

One method of disposal for toxic materials is incineration in furnaces on ships 200 miles off shore. In the automotive industry, parts cleaned by pyrolytic cleaning have very little toxic waste residue.

▬ AUTOMOBILE EMISSION CONTROL SYSTEMS

In addition to modified fuel blending techniques, a combination of things has been done to lower exhaust emissions. These include:
- Engine design
- Fuel and ignition system controls
- Devices designed specifically for the control of emissions

Computers manage emission devices on today's cars. A computer uses an engine load (vacuum) signal, a road speed (speedometer) signal, temperature signals, and the O_2 sensor reading from the vehicle's exhaust. It then controls all emission devices and engine management functions.

Older cars had many vacuum hoses to control or supply various emission devices. Vacuum spark timing, controlled by temperature-sensitive switches, was also popular. New vehicles have electronic controls for these items.

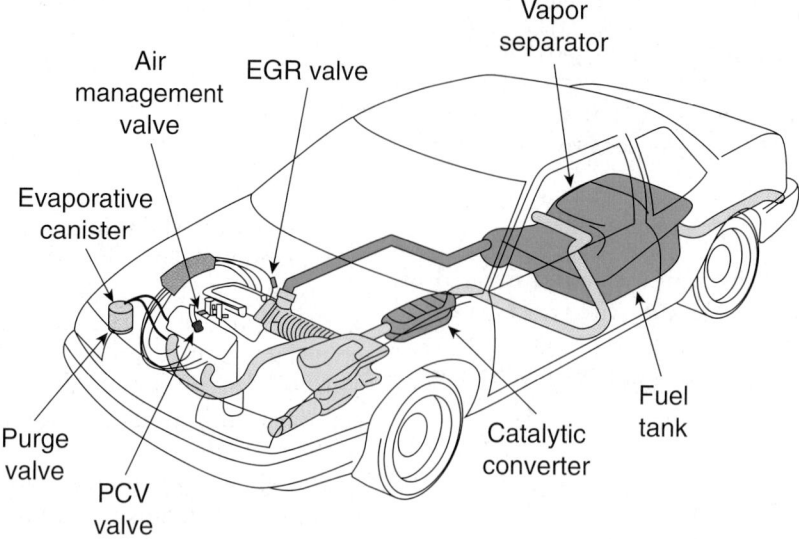

Figure 43.5 Typical emission control system.

Most emission control devices used on each car since 1972 are listed on the underhood emission label. Some of those devices are shown in **Figure 43.5**.

■ CRANKCASE VENTILATION

The first emission control device, installed on all new cars since the early 1960s, was the PCV system. PCV stands for positive crankcase ventilation. It prevents the emission of hydrocarbons (blowby gases) from the crankcase by scavenging them and reintroducing them into the combustion chambers.

NOTE: *Some people mistakenly call this system a PVC system. PVC stands for polyvinyl chloride, a common plastic used for wire insulation and plastic pipe.*

Early engines were equipped with a *road draft tube,* the end of which was positioned below the bottom of the engine to help purge the crankcase of harmful vapors. Since the early 1960s, cars have had crankcase ventilation systems instead of road draft tubes (**Figure 43.6**). Most PCV systems allow a small amount of intake manifold vacuum to leak to the crankcase (**Figure 43.7**).

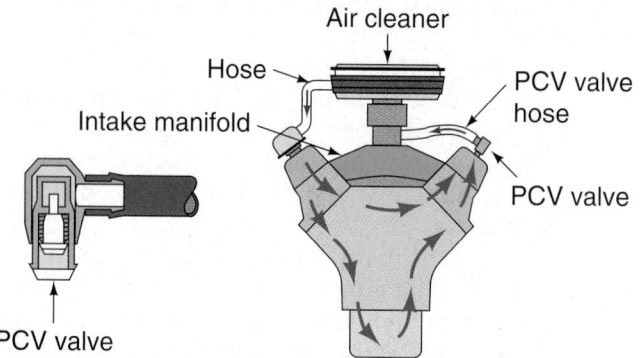

Figure 43.6 A closed crankcase ventilation system. *(Courtesy of Federal-Mogul Corporation)*

Idle or deceleration

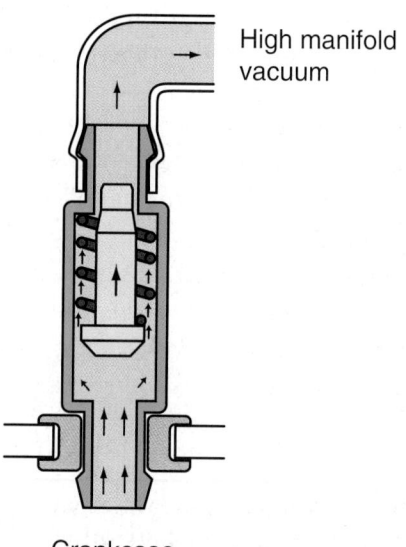

Figure 43.7 A PCV valve allows a small amount of intake manifold vacuum to leak to the crankcase.

A PCV valve has a major advantage over a road draft tube. The road draft tube depended on air movement to operate; it did not work when the vehicle was not moving. The PCV system works at idle and part throttle.

The modern crankcase ventilation system (since the mid-1960s) is a **closed ventilation system** (see Figure 43.6). Atmospheric pressure must enter the crankcase for the ventilation system to operate. Filtered intake air is supplied through a hose from the air cleaner. The fresh air in the hose (under atmospheric pressure) flows toward the negative pressure created by the PCV valve, which is an engineered vacuum leak.

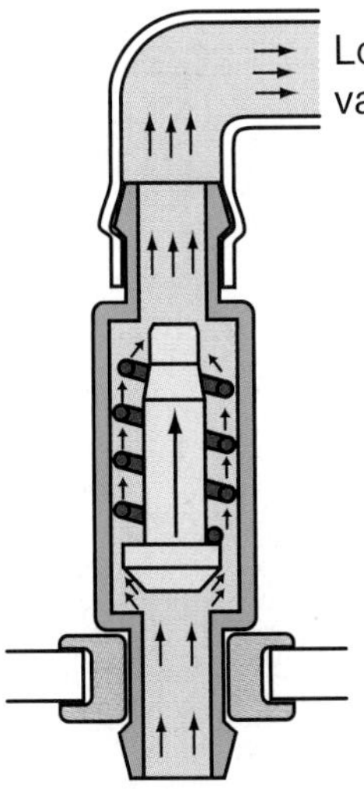

Heavy load

Lower manifold vacuum

Crankcase vapors

Figure 43.8 When the engine is under load, the spring moves the PCV valve to allow more airflow.

The fresh-air hose also provides an escape route for excessive crankcase pressure when blowby is more than the PCV valve can handle or the PCV valve is plugged. Excess pressure escapes up the hose into the air cleaner, where it will be drawn once again into the intake system with incoming fresh air. On the older open systems, excess blowby could escape through a breather into the outside air.

The orifice in the crankcase ventilation system (usually part of the PCV valve) is a specific size to match the size of the engine. Therefore, it is very important that the correct valve be used. Performance problems, as well as oil leaks, can be traced to the use of a wrong-sized PCV valve.

Because it operates on intake manifold vacuum, the PCV valve does not function as well under a load (when there is less vacuum). Therefore, it opens up further to allow more blowby to pass through (**Figure 43.8**).

Besides eliminating one source of air pollution, the PCV system has other benefits.

- It has cleaned up the insides of engines enormously. Acids are not formed in stagnant air like with a road draft tube.
- It reduces sludge by removing moisture from the crankcase.
- Oil leakage caused by excessive crankcase pressure is reduced.

When the engine is turned off, or if there is an intake manifold pop-back, the PCV valve is pushed against its seat by the spring, closing it off (**Figure 43.9**). If the valve was not against its seat during a pop-back, the flames from the explosion would be able to enter the crankcase.

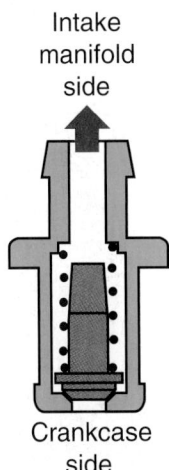

Intake manifold side

Crankcase side

Figure 43.9 When the engine is off, or if there is an explosion in the intake manifold, the PCV valve is closed to the crankcase.

NOTE: *A few PCV systems do not use PCV valves. Some use a metered orifice, and others use a breather hose from the crankcase to the air cleaner.*

■ AIR INJECTION SYSTEM

An **air injection system** has two basic functions. It provides a low-pressure air supply needed to lower HC and CO in the exhaust. It can also provide air to the catalytic converter to help heat it up when it is cold. The first air injection systems appeared in California in 1966 (**Figure 43.10**).

Some unburned gases remain in the exhaust after combustion in the cylinder. In early emission controlled engines, engine modifications were used

Check valve   Diverter valve   Air pump

Muffler   Injection tube to exhaust manifold

Figure 43.10 Parts of a vintage air injection system. Modern systems are similar. *(Courtesy of Tim Gilles)*

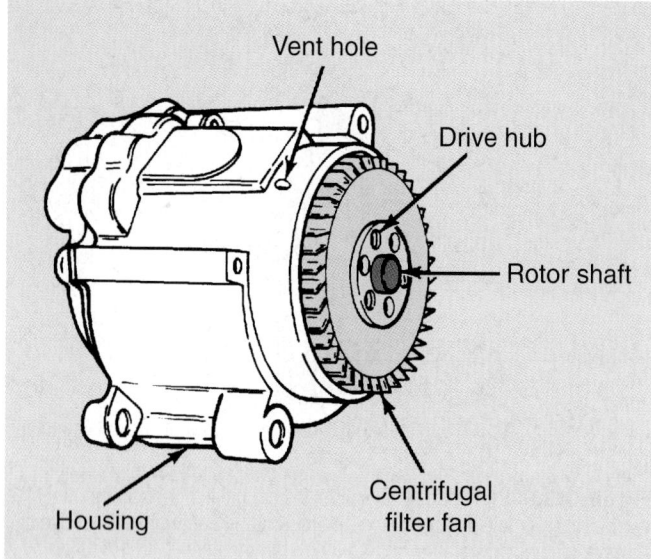

Figure 43.11 Parts of a smog pump. *(Courtesy of DaimlerChrysler Corporation)*

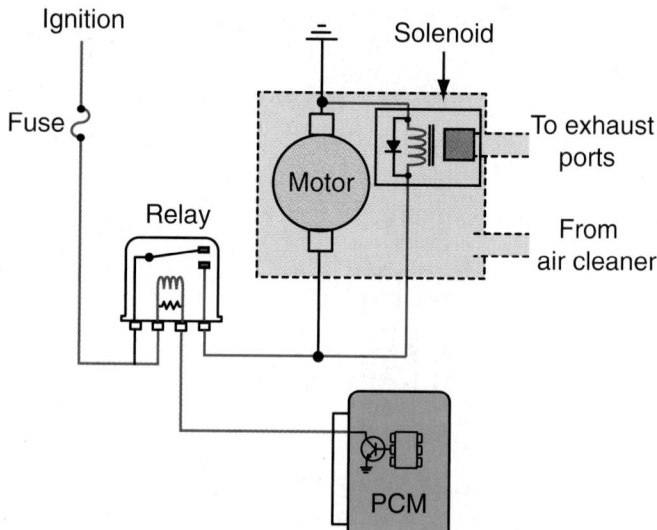

Figure 43.12 A circuit for a typical electric air pump.

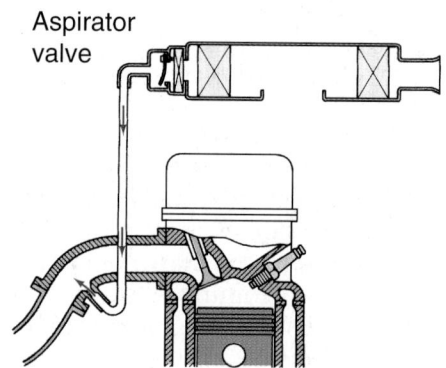

Figure 43.13 An aspirator valve injecting air into the exhaust manifold.

to control these emissions. Another method for reducing these emissions is to keep them burning in the exhaust pipe. An air injection system is used to feed these hot gases and keep them burning. Air is provided by either a belt-driven air pump, an electric motor–driven pump, or by a non-pump pulse air system.

The air injection system uses an air pump (commonly called a **smog pump**), control valves, and lines to the manifolds.

- A low-pressure *vane-type pump* driven by a crankshaft belt provides air (**Figure 43.11**).
- A *diverter valve,* or a similar type of device, prevents a backfire during deceleration. When a rich mixture might be ignited by the fresh air from the pump contacts, the diverter sends the air outside through a muffler or the air cleaner.
- A one-way *check valve* keeps the exhaust from backing up into the hoses and pump.
- An air distribution manifold or cast in passageways in the head distribute the air to the exhaust ports.
- Computer-controlled three-way catalyst systems (covered later in this chapter) include an **air switching valve** that moves the injected air into the exhaust manifold or into part of the catalytic converter.

Electric Air Pumps

Some vehicles have electric air pumps. The powertrain control module processes engine input signals and commands the air pump to inject air into the exhaust until the engine warms up. Injected air heats the catalysts so they can begin to work sooner. **Figure 43.12** shows a circuit for a typical electric air pump.

■ ASPIRATOR VALVE OR PULSE AIR SYSTEM

At the end of the exhaust stroke when the intake valve opens, a momentary low-pressure condition (pulse) occurs. The *aspirator valve* or *pulse air system* uses these pulses to blow fresh air into the exhaust (**Figure 43.13**). This system saves the cost of the air pump. It works well at idle and low speed but is not efficient at high speeds. It is mainly found on small engines. One type is used mainly to heat a three-way catalytic converter (**Figure 43.14**).

The main part of the aspirator system is a one-way check valve called an aspirator valve. It allows fresh air from the air cleaner to flow through it when there is a vacuum created by the pulse of the exhaust. It closes when exhaust pressure begins to build.

■ EXHAUST GAS RECIRCULATION

Scientific studies of the late 1960s discovered that NO_x was a part of smog, so the EPA added NO_x to its list of regulated items. In the early 1970s, manufacturers were designing their engines for lower HC and CO emissions. Unfortunately, the resulting lean air-fuel mixtures and

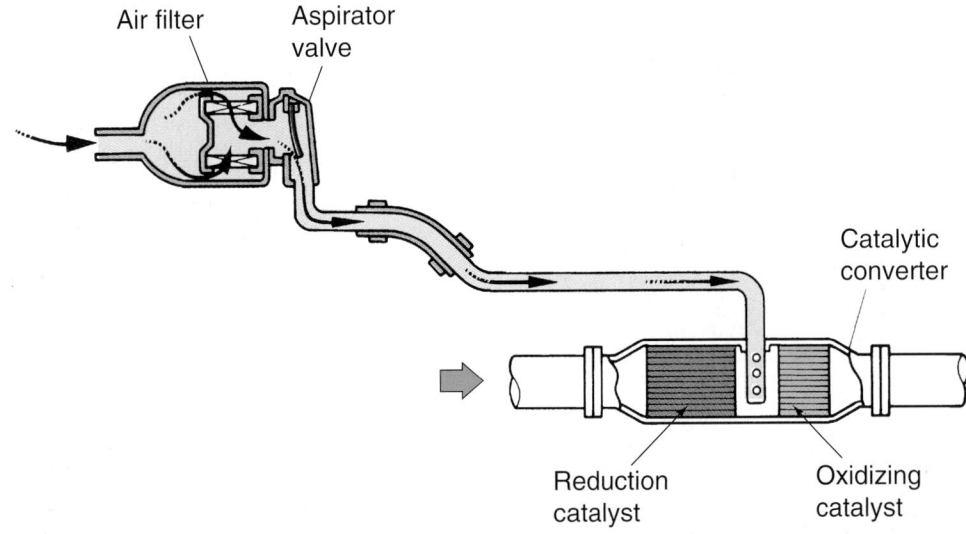

Figure 43.14 An aspirator valve system providing air to a catalytic converter.

higher engine operating temperatures raised NO_x in the exhaust. To lower NO_x formation, an **inert gas**—one that does nothing but take up space—was added into the incoming air and fuel. Because the inert gas will not burn, it does not contribute to emissions at the tailpipe.

The inert gas that is used is exhaust gas. The *exhaust gas recirculation* (*EGR*) system allows a small amount of exhaust gas (less than 10% of the total) to be routed into the incoming air-fuel mixture. Diluting the air-fuel mixture with this already-burned gas lowers combustion temperatures by about 300°F because there is very little free oxygen left in exhaust to support combustion. Remember from the earlier discussion that NO_x are formed when combustion temperatures are above 2,500°F. The purpose of EGR is to reduce NO_x emissions at specified times.

EGR is also used to improve fuel economy. Metering the valve provides a means of varying the engine's compression ratio. Under light load conditions, the compression ratio can be lowered so the engine can operate without having to compress mixtures to needlessly high levels. The PCM also uses EGR selectively to control engine knock.

By 1973, most cars used an EGR system, but many of the early systems developed a bad reputation. They were often disabled to improve fuel economy and driveability. Later systems are vastly improved.

■ EGR SYSTEM OPERATION

A simple EGR system has an EGR valve operated by engine vacuum. The valve is located on the intake manifold (**Figure 43.15**). There is either an internal passageway in the manifold or cylinder head, or a small pipe from the exhaust manifold that carries exhaust to the valve (**Figure 43.16**).

Very little NO_x is formed at idle or when the engine is cold, so the EGR valve is closed at those times. EGR is needed to reduce combustion temperatures when

Figure 43.15 An EGR valve mounted on an intake manifold.

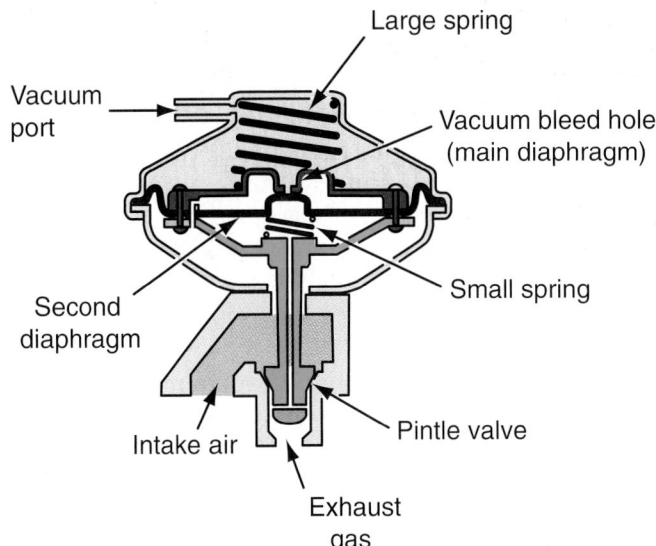

Figure 43.16 The EGR system allows a small amount of exhaust gas to leak into the intake stream. This EGR valve is in the *closed* position.

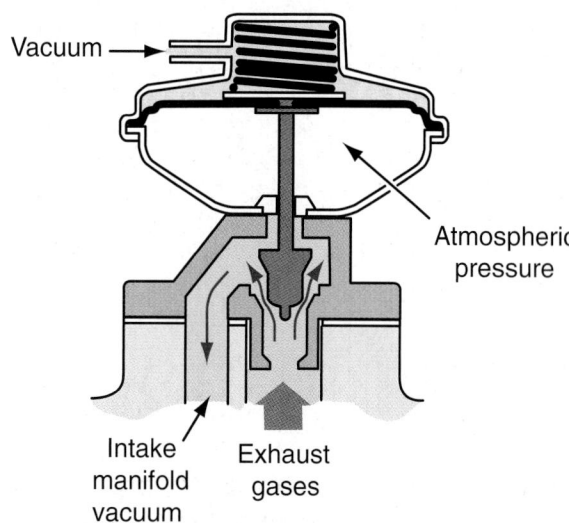

Figure 43.17 A simple EGR valve allows exhaust to flow into the intake when it is opened by vacuum.

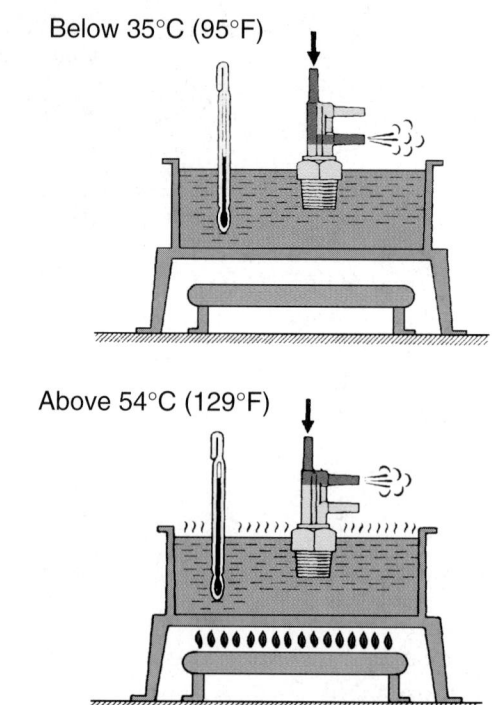

Figure 43.18 Testing a thermal vacuum valve.

accelerating under load or during part throttle driving. As the throttle is opened, vacuum from a port above the throttle plate is applied to the valve. The valve opens to allow the passage of exhaust gas into the intake manifold (**Figure 43.17**).

Having the valve open is like having an air leak (vacuum leak). It is only open when the throttle is open so it will not have an effect on engine idle quality. It cannot be open during engine cranking either. This would act like an air leak and make the engine hard to start.

Backpressure EGR Valve. Exhaust backpressure is a good indicator of engine load. When an engine is under heavy load, more NO_x is formed. So more EGR flow is needed. A *backpressure transducer valve* can be either inside the valve or within its own housing. A diaphragm in the valve senses exhaust pressure and closes a vacuum bleed hole in response to it. Opening the bleed hole reduces the vacuum available to the EGR valve so it does not open as far. When the engine is under load and backpressure is high, the bleed hole is covered. This allows full exhaust gas recirculation.

There are two types of backpressure transducers: positive and negative. The one described above is the positive type. The negative type does the same thing but reacts to vacuum in the exhaust (decreasing back pressure) to regulate valve flow.

Thermal Vacuum Valve. A thermal vacuum valve (TVV), also called a ported vacuum switch (PVS), is found in a cooling passage on precomputer cars. This prevents vacuum from operating the EGR valve before the engine is warmed up. The bottom of the TVV has a wax element heated by the surrounding coolant. The wax expands, moving a plunger to open a vacuum passage (**Figure 43.18**).

■ COMPUTER-CONTROLLED EGR SYSTEMS

Today cars have computer-controlled EGR systems, which do not need many of the mechanical controls used previously. Input signals of engine temperature and load are used to control EGR valve operation. The computer needs to see a vehicle speed signal or a signal from the PRNDL switch indicating that the vehicle is in gear before the EGR valve operates.

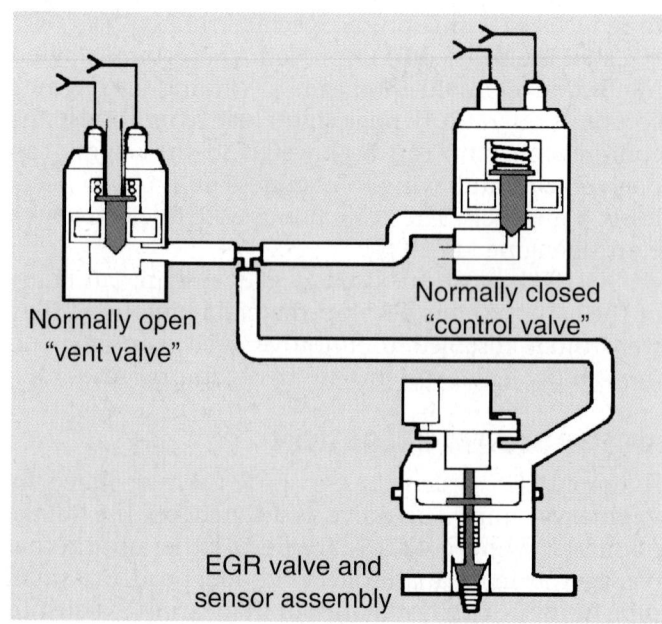

Figure 43.19 Computer-activated solenoids control the vacuum to the EGR valve. *(Courtesy of Ford Motor Company)*

Many EGR valves on late-model engines have *position sensors*. Some manufacturers have used these for many years. Different manufacturers have different ways of sensing the position. Some use temperature, and others use actual physical position of the pintel.

Some systems have an EGR valve with computer-activated solenoids that control vacuum (**Figure 43.19**). These valves control vacuum coming to the solenoid and bleed off vacuum when it is not wanted in the same manner as the backpressure transducer.

To regulate EGR more accurately, some EGR systems use pulse width modulation, like on fuel injection systems, to rapidly cycle an EGR vacuum control solenoid on and off.

Digital EGR valves are used on some cars. EGR flow is regulated by the computer by controlling a series of solenoids. There are three metered orifices opened and closed by solenoids (**Figure 43.20**).

Some newer EGR systems use a *linear EGR valve* (**Figure 43.21**). Instead of being opened by engine vacuum, this valve has a stepper motor. This means that the valve does not depend on engine vacuum and can be partly or fully opened in increments. More information on EGR systems is found in Chapter 47.

■ CATALYTIC CONVERTER

A **catalyst** is a substance that causes a chemical reaction to occur without undergoing any change to itself. The chemical reaction is one that normally would not occur at all, or one that occurs at a much faster rate than normal due to the catalyst. Catalytic converters have been installed on most cars since 1975, the same time unleaded fuel came into use. The *catalytic converter* is often called a *cat*. It is located in front of the muffler in the exhaust system and looks like a heavy muffler.

Before it can begin to operate, a catalytic converter must be hot. It is mounted closer to the engine than a muffler so it is quickly heated by exhaust. The point where the converter begins to work is called its *light off temperature*, which is about 500°F.

Prior to the invention of the catalytic converter, engine emissions were controlled by engine systems that included leaner air-fuel mixtures, which resulted in lower performance and fuel economy. The catalytic converter allowed automotive manufacturers to improve engine performance and fuel economy and let the converter take care of the emissions.

Catalytic Converter Construction

The catalyst is either **monolithic** or it has *pellets*. A monolithic catalyst is like a big honeycomb. It has a thin coating of *platinum* and *palladium* applied to either a ceramic or metal *monolith* coated with alumina. Alumina is an oxide of aluminum that is very porous. The catalyst's metals (platinum and palladium) fill the holes in the alumina. Later catalysts also included the rare silvery white metal rhodium.

Monolithic catalysts have become the design of choice because they warm up more quickly and produce less backpressure than the older pellet design, although they are more costly because they require more platinum and palladium to manufacture.

Both types of converters are contained within a stainless steel shell. In order to provide a long life to the expensive components within the converter, stainless steel is required. It resists corrosion better than the galvanized steel used in mufflers.

Late-model vehicles also have pre-catalysts, or warm-up converters (**Figure 43.22**). These smaller converters are located near the engine's exhaust manifold, where they are heated more quickly so they can begin operating sooner, before the downstream converters warm up.

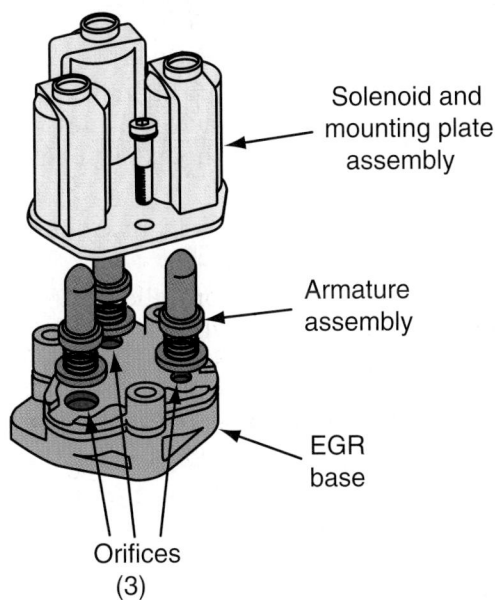

Solenoid and mounting plate assembly

Armature assembly

EGR base

Orifices (3)

Figure 43.20 A digital EGR valve with three electric solenoids.

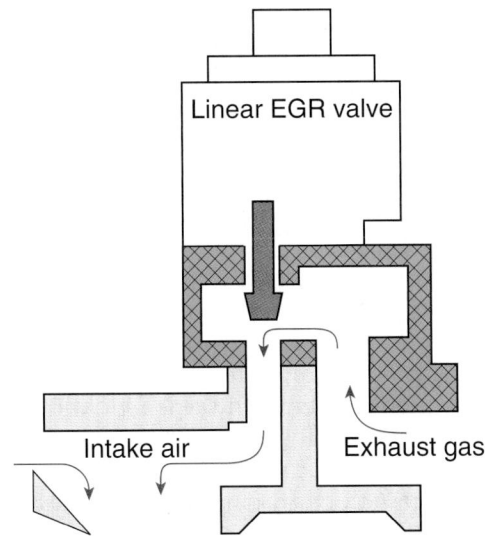

Linear EGR valve

Intake air   Exhaust gas

Figure 43.21 A linear EGR valve has a stepper motor.

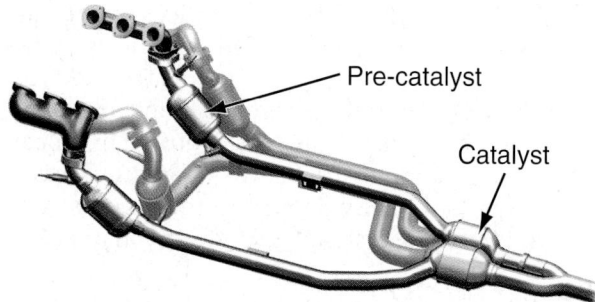

Figure 43.22 Pre-catalysts located near the cylinder head warm up quickly. This drawing shows two design variations. *(Courtesy of DaimlerChrysler Corporation)*

■ TYPES OF CATALYTIC CONVERTERS

There are three types of catalytic converters:

■ The earliest catalytic converter, called a *two-way converter*, was designed to oxidize HC and CO, converting them into H_2O and CO_2.

■ The *three-way single bed converter* oxidizes HC and CO and also reduces harmful NO_X into harmless nitrogen and O_2. The NO_X portion of the converter is called a **reduction catalyst**. The O_2 produced in the NO_X portion of the converter aids in oxidizing HC and CO.

■ The *three-way dual bed converter* has two chambers. It has a tube between the two chambers to provide oxygen from the emission system's air pump to its rear oxidation chamber (**Figure 43.23**).

Two-Way Oxidation Converters

A *two-way catalyst*, also called an **oxidation catalyst**, changes harmful HC and CO into harmless carbon dioxide (CO_2) and water vapor (H_2O). It is found mostly on cars built prior to 1980 without O_2 sensors, and it is often used with air injection to help it operate more efficiently.

An oxidizing catalyst begins working at temperatures between 500°F and 600°F, but works most effectively when heated to 750°F (400°C) or higher (**Figure 43.24**). Converters often operate at temperatures

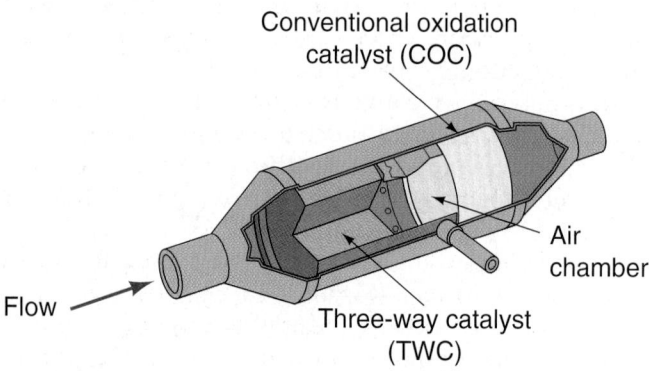

Figure 43.23 A three-way oxidation catalyst. *(Courtesy of Ford Motor Company)*

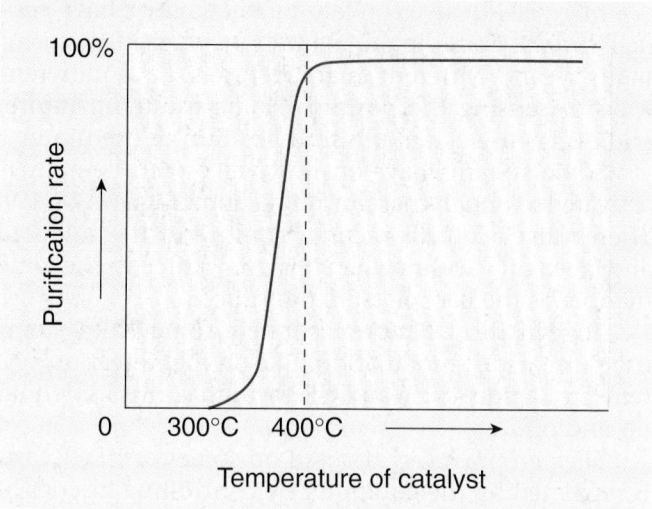

Figure 43.24 An oxidation catalyst does not work until it is hot.

near those in the combustion chamber. The graph shows a purification rate, which is a measure of the proportion of pollutants in the exhaust that can be converted to nonpollutants.

NOTE: *When a catalytic converter works, heat is a by-product. The gas leaving the converter can be from 50°F to 200°F higher than the exhaust coming into it. Insulating or shielding material keeps the heat inside of the converter and away from the floor of the car.*

Three-Way Catalytic Converters

A three-way catalytic converter, used with oxygen sensor feedback systems, reduces NO_X and oxidizes HC and CO. Three-way converters are either a *single bed* or *dual bed design*. A dual bed converter has its reduction and oxidation parts in separate chambers within a single housing. It has a combined three-way oxidizing and reduction catalyst in front of a two-way catalyst that only oxidizes. Sometimes, there is a separate smaller three-way converter in front of an oxidation converter. A single bed three-way converter is a single unit that reduces and oxidizes all three emissions at once.

Three-way catalytic converters use *rhodium* (or sometimes palladium) as a catalyst for their reduction portion (the part that controls NO_X). They only work properly when operated at the correct air-fuel mixture. Extra O_2 hinders the conversion of NO_X. This is why the tube from an air pump–equipped engine enters between the chambers in a dual bed cat. When the front part of the converter catalyzes NO_X, it creates more O_2, which assists in the rear oxidation catalytic reaction. In fact, a substantial amount of CO is oxidized to CO_2 before leaving the reduction part of the converter. The reduction part of the catalyst is therefore always located in front of its oxidation counterpart. The reduction process does not produce enough

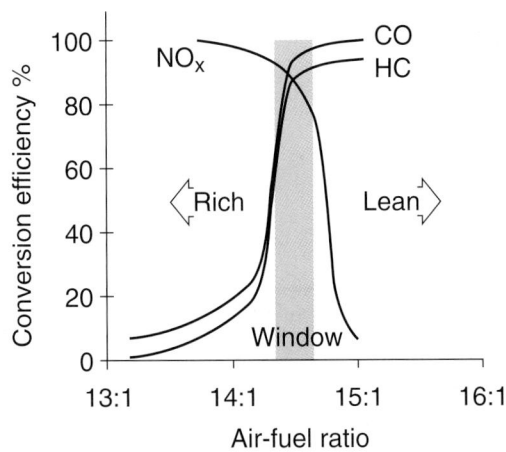

Figure 43.25 The air-fuel mixture needs to be close to 14.7:1 for the three-way catalytic converter to be effective. *(Courtesy of DaimlerChrysler Corporation)*

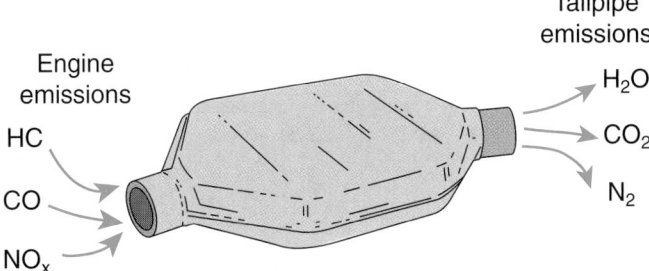

Figure 43.26 The catalytic converter converts harmful emissions to environmentally safe gases.

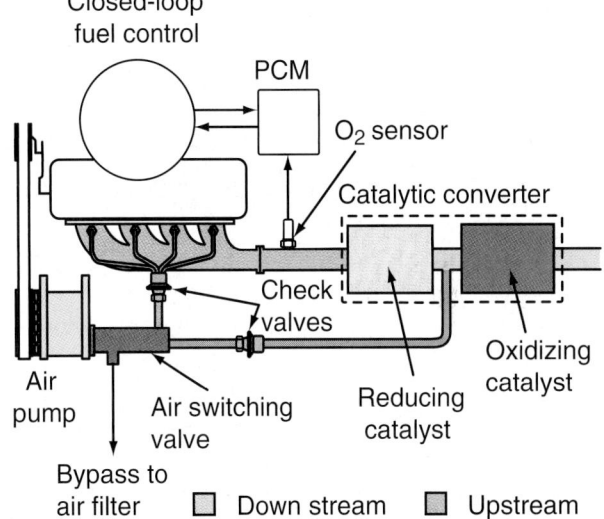

Figure 43.27 An air switching valve directs the air from the pump to go upstream, downstream, or bypass to the air filter.

O_2 for the oxidation catalyst to work at its best, however. Air added at the center of the converter assists in the continuation of the oxidation process.

A three-way converter needs heat and a regulated air-fuel mixture to work effectively. It works far less efficiently when the mixture is too rich or too lean. Fuel mixtures controlled by conventional O_2 sensors are tailored to a ratio of 14.7 to 1 to help the cat to function efficiently (**Figure 43.25**). The oxidizing part of the catalyst does not remove CO and HC very well at mixtures *richer* than 14.7 to 1. NO_x reduction in the catalyst does not work well at mixtures *leaner* than 14.7 to 1. The air-fuel mixture in an O_2 feedback system constantly fluctuates between rich and lean. This allows the oxidation part to operate when the mixture is lean and the reduction part to operate when the mixture is rich. The reduction part needs the extra CO to scavenge the O_2 it produces.

SCIENCE NOTE

*The catalytic converter produces both chemical oxidations and reduction reactions as it converts engine emissions to environmentally safe gases (**Figure 43.26**). During the oxidation process, oxygen (O_2) molecules combine with hydrocarbon (HC) and carbon monoxide (CO) molecules. During the reduction phase, diatomic oxygen molecules are removed from oxides of nitrogen (NO_x). This happens during reactions with water (H_2O), CO, and HC.*

HC and CO oxidize more efficiently when the exhaust is lean because it contains more O_2. NO_x reduction occurs more efficiently when the mixture is rich because it contains less O_2.

This is the reasoning behind closed-loop fuel control, where the air-fuel mixture alternates rapidly between rich and lean.

Since the 1990s, many cats contain cerium. Cerium is an element that stores O_2 when the mixture is lean and releases O_2 when the mixture is rich.

Air Switching Valve. Dual bed catalytic converters include an air switching valve and a diverter that directs the *upstream* air *downstream*, or to the air filter (*bypassing*) (**Figure 43.27**). The valve sends the air upstream to the exhaust manifold when the engine is cold. This helps heat the converter and the O_2 sensor (**Figure 43.28**). If air was injected into the manifold when the engine was in closed loop, the O_2 sensor would think the air-fuel mixture was too lean and that the computer would cause the air-fuel mixture to become too rich.

Injecting air in front of the oxidation (HC/CO) part of the catalytic converter helps that part of the cat to work more efficiently. When the engine is warm, the air is directed downstream in the exhaust to the oxidation chamber (**Figure 43.29**). *Preventing* the injection of air into the reduction (NO_x) chamber of the cat helps reduce NO_x. A small amount of CO (rich mixture) is required for the *reduction* part of the catalyst to function.

When the engine is cold, the air pump can blow upstream to the manifold and flow through the entire converter. At lower engine temperatures, not as much

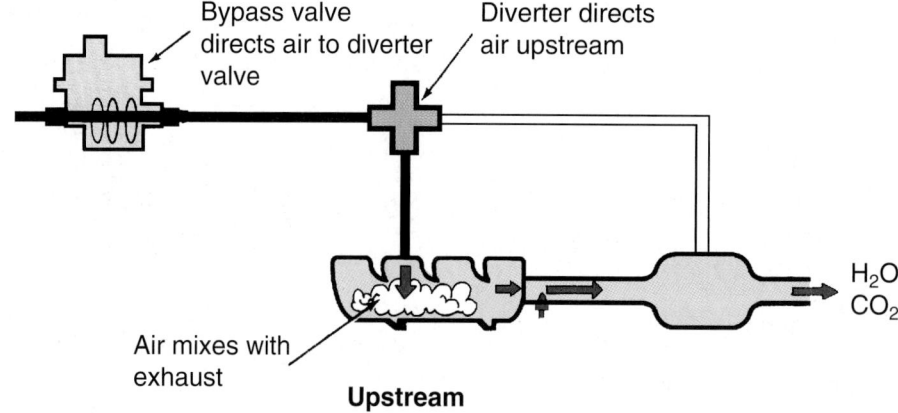

Figure 43.28 The valve sends the air upstream to the exhaust manifold when the engine is cold.

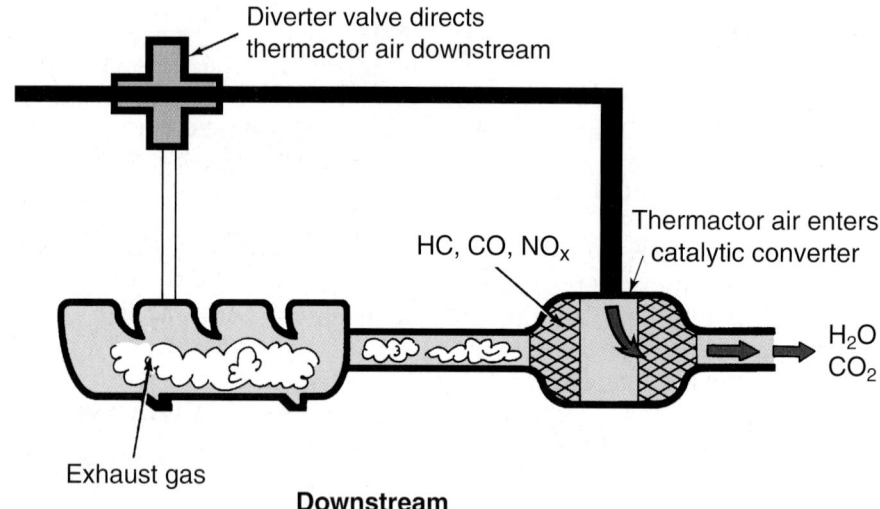

Figure 43.29 When the engine is warm, the air is directed downstream in the exhaust to the catalytic converter to help it operate.

NO_x is produced and the computer is not reacting to the O_2 sensor. The air provided by the air pump helps to heat the O_2 sensor and catalytic converter more quickly to operating temperature. The switching valve also sends air to the atmosphere on some cars during deceleration to prevent backfire during open-loop operation, when the engine is cold. It also can divert air from the catalytic converter during periods of extended idle.

Some cars use electric motor–driven air pumps (**Figure 43.30**). For more complete information on these pumps and their computer controls, see Chapter 47.

■ EVAPORATIVE CONTROLS

Since the early 1970s, evaporative emission controls have been required on all vehicles sold in the United States. Evaporative control systems reduce the emission of gasoline vapors from the fuel tank (**Figure 43.31**). In warm weather, gasoline evaporates more easily even though its vapor pressure is more closely controlled. Evaporative emissions account for about 20% of vehicle emissions. A parked car can emit substantial emissions if the system is not working properly.

The heart of the evaporative emission system is a **charcoal canister**. Activated charcoal can store gasoline vapors until they can drawn into the engine and burned. When the engine is not running, vapors are routed through lines and hoses to a charcoal canister where they are stored. When the engine is running under the correct operating conditions, the vapors are purged from the canister and again drawn to the intake manifold before burning in the cylinders.

Sealing the System

Emission of fuel vapors from the fuel tank can be controlled by sealing the fuel system from the atmosphere. Air must be allowed into the fuel tank to take the place of fuel as it is consumed. Otherwise, the fuel tank would collapse as the fuel pump continued to create a vacuum within it. Flow of fuel would also be restricted to the engine. The fuel tank can be vented by the gas cap or by a vent solenoid (**Figure 43.32**).

On a hot day, gasoline in the tank expands. A tank filled on a cool morning must have sufficient reserve capacity to hold the expanded fuel.

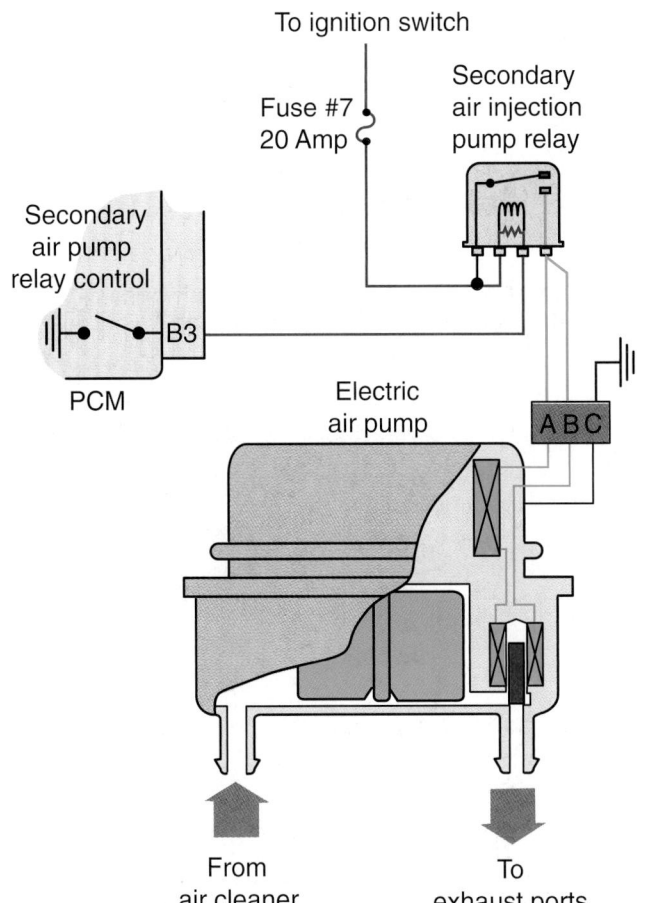

Figure 43.30 An electric motor-driven air pump.

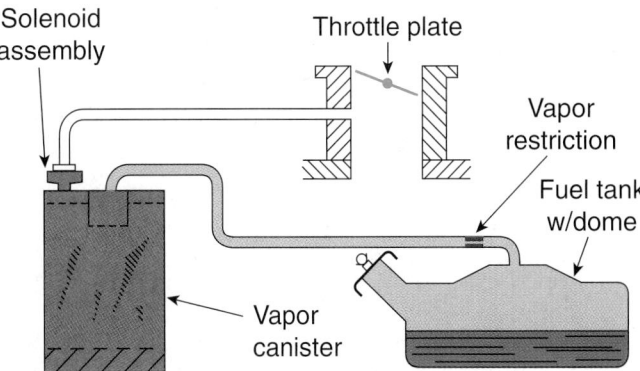

Figure 43.31 A simplified evaporative emission control system.

then stop running when driven for a few miles. Loosening the gas cap results in a large rush of air, and then the car will run again. The same effect would happen if the float bowl were not vented.

As the temperature of the engine and air under the hood increases, the evaporation rate of the fuel increases. With the engine running, fresh fuel is always coming into the float bowl. This keeps it cool. Once the engine is shut off, a condition known as **hot soak** occurs. On hot days, gasoline can boil out of the float bowl. The evaporative emission system must be able to store fumes for a long time.

Fuel Tank

Gasoline tanks in today's cars are designed to allow for expansion of the fuel by around 10% of the total volume. Major parts of the different designs of fuel tank systems include a *pressure-vacuum filler cap*, an *expansion area*, and a **liquid/vapor separator**. An expansion area's function is to prevent the overfilling

The carburetor float bowl must have means of venting also. Without a vent, fuel could not flow and the air-fuel mixture would become lean. In fact, if a gas cap check valve becomes restricted, a car will surge and

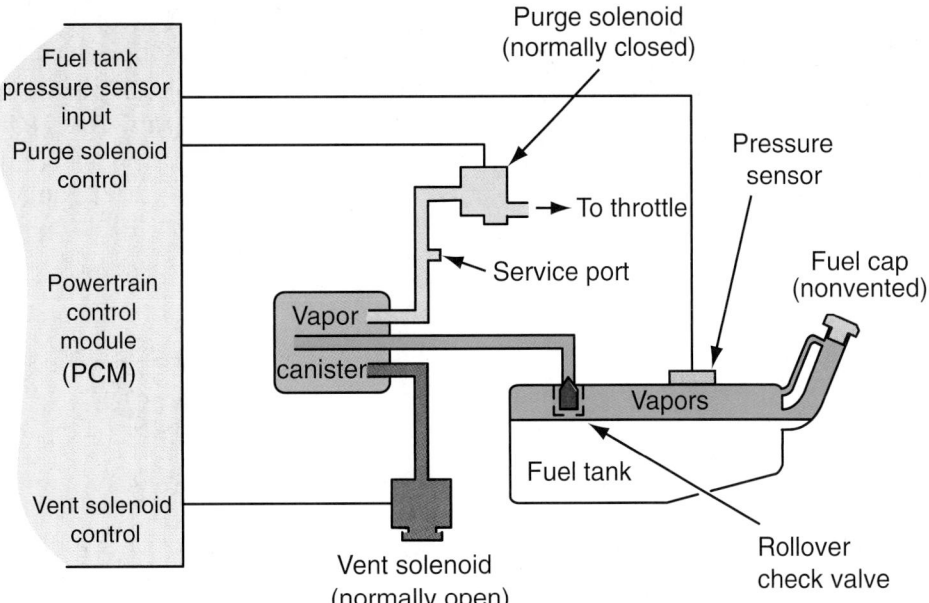

Figure 43.32 This evaporative system has a nonvented fuel cap with vent and purge solenoids controlled by the PCM.

of fuel, which could allow liquid into the vapor recovery system. During a tank fill, the filler nozzle will shut off before the expansion area can be filled. There are several ways of providing a reservoir for expansion of excess fuel.

Expansion Dome and Liquid/Vapor Separator. Some cars have a dome above the fuel tank for excess fuel. A liquid/vapor separator keeps liquid fuel or bubbles out of the evaporated system (**Figure 43.33**).

Expansion Tank. An expansion tank is sometimes located inside of the main gas tank (**Figure 43.34**). When the gasoline in the main tank expands on a hot day, it can slowly fill the expansion tank. There are small holes drilled in the expansion tank that allow the expanding fuel to enter it.

Some manufacturers install an extra tank that holds 1 or 2 gallons of fuel to cover expansion.

Filler Neck Design. The filler neck can be designed to prevent overfilling. On one type, when the gasoline reaches a certain level during a tank fill it flows through a tube back into the filler neck (**Figure 43.35**). This shuts off the gas nozzle. If the customer does not try

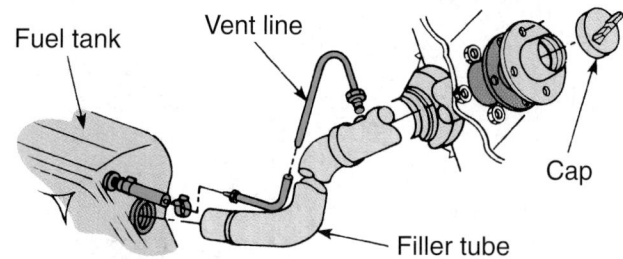

Figure 43.35 When gasoline reaches its maximum designed level in the tank, it flows through a vent line back into the filler neck.

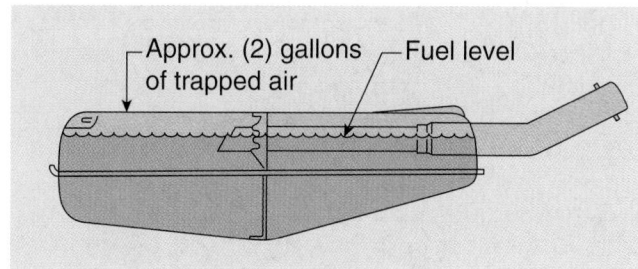

Figure 43.36 Overflow protection on this fuel tank design has the end of its filler tube positioned lower in the tank.

to further top off the tank, it will not be overfilled. On another type, the lower end of the tank filler tube is positioned so its end is low enough to prevent the tank from overfilling (**Figure 43.36**).

Gas Cap

Gas caps have either a *pressure vacuum valve* or are *sealed*. If pressure in the tank exceeds 1 psi, the pressure vacuum cap will vent. It works like a radiator cap with pressure and vacuum valves (**Figure 43.37**). With the sealed gas caps, tank venting can be done by a three-way valve in the vapor line to the charcoal canister. Be sure to replace a cap with the correct one, sealed or vented.

NOTE: *A loose gas cap can turn on the malfunction indicator lamp on OBD II cars. This information is covered in Chapter 47.*

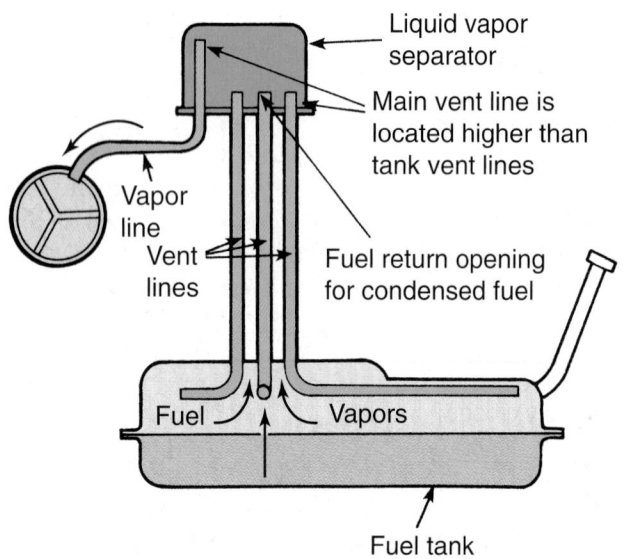

Figure 43.33 A liquid/vapor separator prevents liquid fuel from entering the evaporative system.

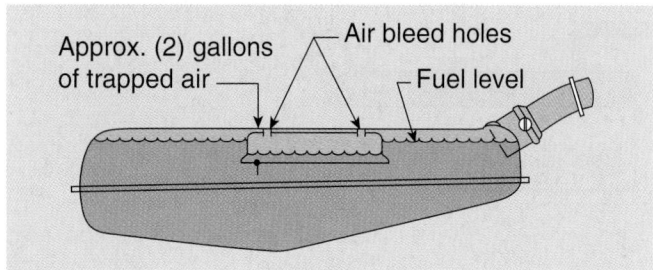

Figure 43.34 Some fuel tanks include an internal expansion tank.

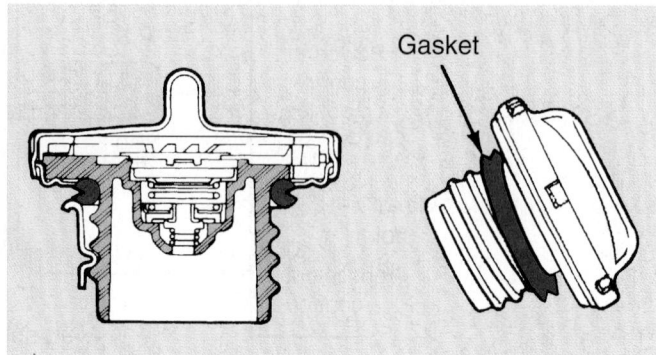

Figure 43.37 A pressure vacuum gas cap. *(Courtesy of DaimlerChrysler Corporation)*

■ OTHER PARTS OF THE FUEL TANK SYSTEM

A liquid vapor separator is part of the fuel tank or the expansion tank. Its function is to keep liquid fuel from being drawn into the charcoal canister. There are several types of liquid/vapor separators.

Some cars have a *rollover valve* in the vent line from the fuel tank. The valve has a ball or plunger that blocks the line if the car is in an accident and rolls over. This prevents liquid fuel from being able to leak out.

Charcoal Canister

The charcoal canister is a small plastic or steel container found in the engine compartment or in a front fender-well. The canister stores vapors from the gas tank, until they can be safely burned in the engine, rather than allowing them to escape to the atmosphere.

The canister has about 1.5 pounds of activated charcoal that can soak up twice its weight in fuel. It stores vapors until the engine is running, then the vapors are drawn into the intake system and burned in the cylinder. Air on some systems enters the canister through a filter in the bottom. On other systems, fresh air enters through the top of the canister. Air is drawn through the charcoal and leaves through the purge line.

A vapor canister is only allowed to **purge** (empty) itself of vapor storage when the engine is running (**Figure 43.38**). It collects vapors when the engine is off. Control is usually done by either a purge control valve or an electric solenoid (when there is computer control).

Heated Air Inlet. Throttle-body injection and carbureted engines had heated air controls in a *thermostatic air cleaner* (TAC). The density of air changes with temperature and fuel remains atomized better in warm air. Consistent temperatures of intake air made it possible to maintain a more consistent air-fuel mixture.

Heated Intake Air and Early Fuel Evaporation

Older vehicles had throttle-body injection or carburetion. Because air and fuel were mixed as they entered the intake manifold instead of at the valve, special design considerations were needed for cold driveability and emissions. In these engines, incoming air was heated. The intake manifold was heated by exhaust diverted beneath its floor.

Manifold Heat Control Valve. Cars with PFI do not require preheat to vaporize the fuel. A manifold heat control valve was used on many cars with carburetors or throttle-body fuel injection to reduce exhaust emissions and improve *driveability* during engine warmup. The manifold heat control valve routed exhaust gas under the floor of the intake manifold when the engine was cold to improve the vaporization of the cold fuel.

Some heat control valves were operated by engine vacuum; others had a bimetal spring that expands when heated to open the valve. This type of valve was known as a heat riser.

When the valve is controlled by engine vacuum, the vacuum signal can be controlled either by a computer or by a thermal vacuum switch (TVS). This kind of valve is called an *early fuel evaporation* (EFE) valve. Some EFE systems use an *electrically heated grid* installed under the base of the carburetor or throttle body to improve vaporization of the fuel during the first few minutes of engine operation.

■ ON-BOARD DIAGNOSTICS

The first on-board diagnostics regulations (OBDI) were developed by the California Air Resources Board (CARB) in 1985 and became law in 1988. The regulations require the computer to monitor the engine's O_2 sensor, EGR valve, and charcoal canister purge solenoid to see that all of these systems continue operating properly. A **malfunction indicator light** (MIL)

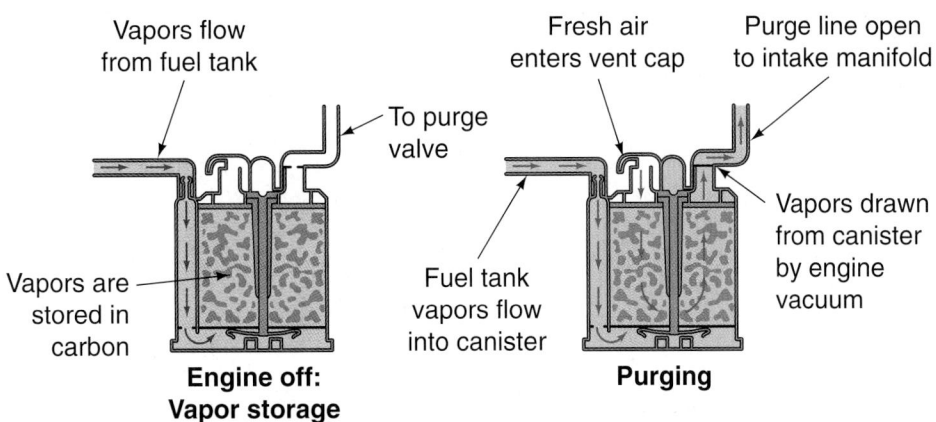

Figure 43.38 Vapors are stored in the charcoal canister until they can be purged with the engine running. *(Courtesy of Ford Motor Company)*

was also required by the regulations. Newer regulations were later developed that were more thorough on emission problem detection.

There have been many types of emission controls and also different names for the same part, depending on the manufacturer. On late-model cars, many of the major systems that have been covered in this chapter have become somewhat standardized. Names of emission parts and connections for test equipment were also standardized by the Society of Automotive Engineers. These changes are part of the OBD II regulations.

On-board diagnostics are covered in detail in Chapter 47.

ENGINE EMISSION MODIFICATIONS

Manufacturers use engine, ignition, and fuel modifications to achieve federally mandated standards. Listed here are some of the more common modifications:

■ A cool area of fuel remains unburned along all cool surface areas. The term for the amount of surface area is called **surface-to-volume ratio** (**Figure 43.39**). Engines have as little exposed surface area as possible on the combustion chamber and the top of the piston. The top piston ring has been moved as close as possible to the top of the piston so there is not much surface area of the piston exposed to flame.

■ To improve thermal efficiency, engines run with *higher cooling system temperatures*. Newer cars operate at 190°F to 218°F, whereas older cars ran at 160°F to 180°F.

■ *Advancing ignition timing* increases HC and NO_x but if done properly increases fuel economy. More efficient use of fuel helps reduce emissions overall. Using the knock sensor, the computer provides the best spark advance for the condition at hand.

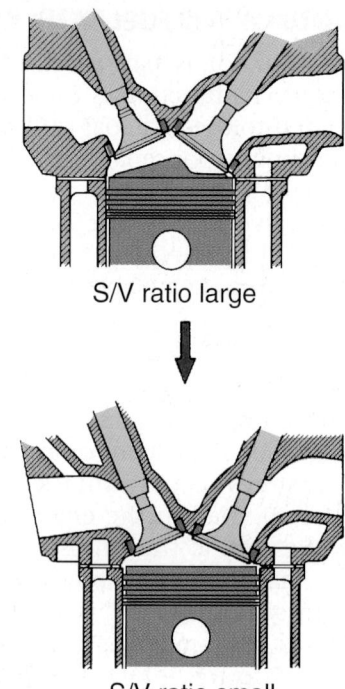

S/V ratio large

S/V ratio small

Figure 43.39 Surface-to-volume ratio is kept small for control of hydrocarbons.

■ *Changing the cam design specifications* can result in different emissions, especially at idle. This is done with variable valve timing, often allowing elimination of the EGR valve. Increasing valve overlap increases HC emissions but lowers NO_x emissions at low engine rpm.

Today's engines have many refinements that have made their engine design, emission, and fuel systems operate very well with each other.

REVIEW QUESTIONS

1. Smog is produced when hydrocarbons and oxides of nitrogen react with _____ to form a dirty yellow-brown haze in the air.

2. A warm air _____ layer traps air pollution within 1,000 feet of the ground to produce smog.

3. What are the three main by-products of combustion that parts of the emission control system are designed to control or eliminate?

4. _____ gases are those that leak past the piston rings.

5. What are two other nontoxic gases included in the vehicle exhaust?

6. What does PCV stand for?

7. What air injection part prevents a backfire during deceleration?

8. A one-way _____ _____ keeps the exhaust from backing up into the hoses and pump.

9. What type of emission does the EGR valve diminish?

10. When the engine is under load, is the exhaust backpressure high or low?

11. A _____ is a substance that causes a chemical reaction to occur faster without any measurable change to itself.

12. What is the name of the catalytic converter type that is a honeycomb style?

13. When is the evaporation rate higher in the fuel system, when the engine is running or after it is shut off?

14. How much extra tank capacity do fuel tanks have after filling?

15. When does the purge valve allow the canister to get rid of vapors in storage, when the engine is running or when it is off?

▬ ASE-STYLE REVIEW QUESTIONS

1. When an air-fuel mixture is too rich, which of the following gases will increase?
   a. Hydrocarbons
   b. Carbon monoxide
   c. Both A and B
   d. Neither A nor B

2. Which of the following is/are true about crankcase ventilation systems?
   a. The system does not work unless the vehicle is moving.
   b. The system is called a PVC system.
   c. During an explosion in the crankcase, the valve opens all the way.
   d. There is a metered vacuum leak to the crankcase.

3. Technician A says that air injection lowers HC and NO_x in the exhaust. Technician B says that air injection heats the catalytic converter when it is cold. Who is right?
   a. Technician A   c. Both A and B
   b. Technician B   d. Neither A nor B

4. Which of the following are results of high combustion temperatures?
   a. An increase in hydrocarbons
   b. An increase in oxides of nitrogen
   c. Cooler spark plug temperatures
   d. An increase in global warming

5. Technician A says that NO_x forms when the engine is under load. Technician B says that NO_x forms when the engine is hot. Who is right?
   a. Technician A   c. Both A and B
   b. Technician B   d. Neither A nor B

6. When is the EGR valve open?
   a. At idle
   b. During engine cranking
   c. Both A and B
   d. Neither A nor B

7. Each of the following is a true statement about catalytic converters *except*:
   a. Some catalytic converters have a tube in the center attached to the air injection system.
   b. The reduction catalyst reduces oxides of nitrogen.
   c. An oxidation catalyst reduces HC and CO.
   d. A monolithic converter has platinum-coated beads.

8. What is a catalytic converter with an air tube called?
   a. A single bed catalyst
   b. A dual bed catalyst
   c. Both A and B
   d. Neither A nor B

9. Technician A says that if air is directed downstream during closed loop, the computer will drive the fuel system lean. Technician B says that if air is directed upstream during closed loop, the computer will drive the fuel system rich. Who is right?
   a. Technician A   c. Both A and B
   b. Technician B   d. Neither A nor B

10. Which of the following emissions is lowered using the EGR valve?
    a. Hydrocarbons
    b. Carbon monoxide
    c. Oxides of nitrogen (NO_x)
    d. Carbon dioxide

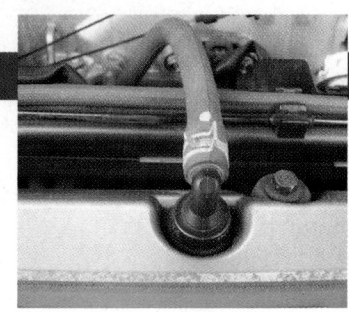

CHAPTER 44

Emission Control System Service

▬ OBJECTIVES

Upon completion of this chapter, you should be able to:

✔ Determine the causes of various emission system problems.

✔ Perform service on the emission control system.

✔ Repair or replace defective emission control parts.

✔ Diagnose emission problems with an exhaust gas analyzer.

▬ KEY TERMS

converter light off
high-temperature
   pyrometer

infrared thermometer
oxidizing
two-gas analyzer

four-gas analyzer
five-gas analyzer

▬ INTRODUCTION

In the United States, the federal Environmental Protection Agency (EPA) is involved in the regulation of air quality. Emission specialists are required to be licensed to perform repairs to the emission system.

The aim of this chapter is to provide you with sufficient understanding to be able to service emission systems, diagnose problems, or to remove and replace components when performing other types of repairs. Diagnosing emission problems using the five-gas exhaust emission analyzer is also stressed in this chapter.

Basic, generic service information is presented here. Learning the specifics of a particular vehicle will require use of manufacturer or aftermarket service information. Shop practice is a necessity if you are to become familiar with the different vehicle systems.

▬ INSPECTING EMISSION CONTROL SYSTEMS

Many states require periodic emission inspections. Vehicles produced since 1972 have an underhood label (see Figure 5.5) listing necessary emission information for the engine.

A vacuum hose routing label is usually found under the hood also (**Figure 44.1**).

Aftermarket *Emission Control Applications service information* provides EGR testing procedures, reminder

Underside of hood

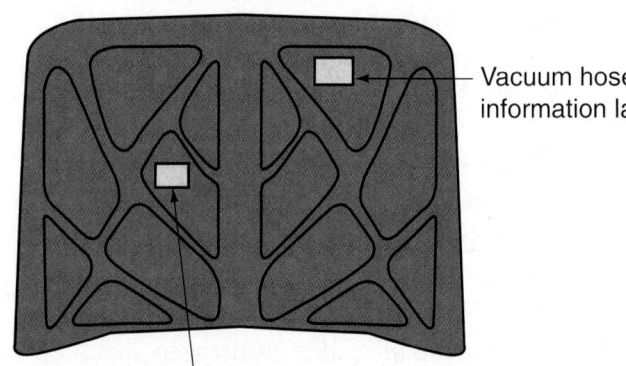

Figure 44.1 Sometimes vacuum and emission labels are located on the underside of the hood.

light instructions, ignition timing procedures, emission equipment requirements, and more.

As you have learned to do in other diagnostic work, a visual inspection is performed first. Look at hoses and wires and check that all parts are in place. Run the engine and listen for air leaks with a length of hose or stethoscope with the metal probe removed from the end. Check the condition and tension of the pump drive belt. Inspect the air cleaner to see that it is clean and unrestricted.

■ COMPUTER-CONTROLLED EMISSION SERVICE

Today's new cars have sophisticated emission systems that are controlled by a computer, sensors, and actuators. When problems occur, many of them are recorded in the computer's memory. A fault code is set and a warning light on the dash will come on. A technician will use a scan tool to read the codes in the computer. This information is covered in Chapter 45.

Some cars have emission maintenance reminder lights that indicate service needed to emission control devices based on time or mileage. There are many ways of testing the devices and resetting the light once it has been illuminated. Refer to an emission control applications manual or a manufacturer's service information for instructions.

■ CRANKCASE VENTILATION SYSTEM SERVICE

Crankcase ventilation system maintenance includes inspecting the PCV valve and checking hoses and passages for condition and internal deposits.

Checking Crankcase Ventilation System Operation

The crankcase system must be airtight (**Figure 44.2**). A leaking or misplaced gasket to a valve cover or intake manifold will cause air leakage sufficient to render the PCV system inoperable. Instead of creating a vacuum in the crankcase, the PCV valve will not be suction sufficient because the engine is not airtight.

NOTES:

■ *Gasket failures that result in failure of the ventilation system can cause oil leaks. With sufficient suction in the crankcase, a slight leakage past a seal should result in outside air leaking **into** the crankcase, rather than oil leaking out.*

■ *Leaking vacuum hoses, vacuum control units, vacuum accessories, or manifold leaks allow dirt to enter the engine unfiltered. This can cause engine wear.*

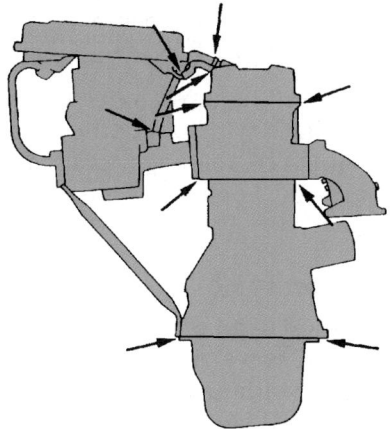

Figure 44.2 The crankcase system must be airtight for the ventilation system to operate efficiently.

Figure 44.3 A PCV tester installed at the oil filler hole to check for vacuum in the crankcase at idle. *(Courtesy of Tim Gilles)*

■ *A system that is not airtight can also cause driveability problems on computerized fuel systems that measure air density.*

Testing the PCV System

With the engine running and the PCV valve installed, there should be a vacuum at the oil filler opening. The tester shown in **Figure 44.3** can be used to see if there is a vacuum in the crankcase. Another simple test to see if there are any leaks in the crankcase is to pull one end of the hose from the valve cover to the air cleaner (at the air cleaner side). Put your thumb over the end of the hose and wait for a couple of seconds. You should feel vacuum if the system is working properly.

Locating a PCV Leak. To locate a leak in the crankcase ventilation system:

■ Seal the breather and PCV valve.
■ Blow *lightly* into the dipstick tube using a rubber-tipped blowgun.
■ Listen for leaks using a piece of hose or a stethoscope with the metal end pulled off.

A leak is not always readily apparent, especially at the top side of a valve cover gasket or where the intake manifold meets the block at the front or back.

Testing the PCV Valve

To test the PCV valve:

■ Pull the PCV valve from its mounting. There are several places the PCV valve can be located. It is usually in a grommet in the valve cover (**Figure 44.4**) intake manifold, or a hose from the crankcase.
■ With the engine stopped, remove and shake the PCV valve. It should rattle, although the one used on some smaller engines valves will not.

PCV valve

Grommet

Figure 44.4 A PCV valve sealed to the valve cover with a synthetic rubber grommet. *(Courtesy of Tim Gilles)*

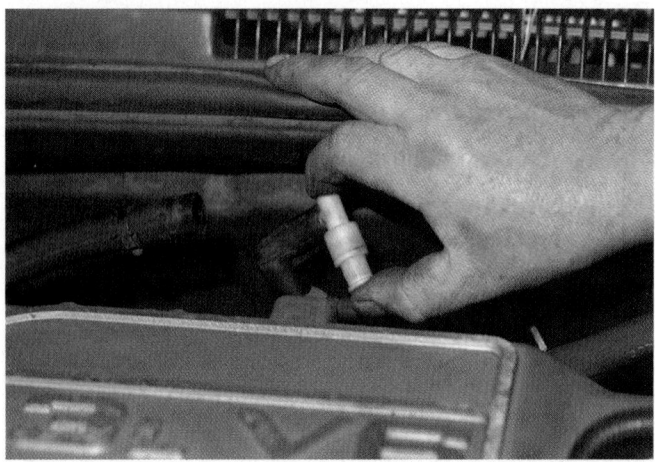

Figure 44.5 Checking a PCV valve; with the computer control of the ignition timing disabled, the engine's idle should drop at least 50 rpm when the valve is plugged.

■ With the engine running, cover the end of the PCV valve with your thumb or finger (**Figure 44.5**) or pinch the line leading to it. Engine rpm should drop 50 to 80 rpm, because the slight air leak from normal PCV valve operation has been closed off.

NOTE: *This test will not work on computer-controlled vehicles with idle speed management unless the system is disabled.*

■ Pull the fresh-air hose from the air cleaner. With the engine idling, a piece of paper placed over the end of the hose should remain suspended there.

Blocking the flow of air to the PCV valve results in an enriched air-fuel mixture because less air is entering the cylinder.

If a PCV valve is replaced, be sure to get the correct one for the vehicle. Valves with different rates of flow are available. Installing the wrong one can result in

poor engine operation, especially in open loop when the engine is cold.

Check the PCV Breather Hose

Make sure the breather hose to the air cleaner is not kinked or restricted. During heavy engine loads, this hose must allow blowby to escape from the crankcase; otherwise, excessive pressure can build up and cause oil leakage and oil consumption.

NOTES:

■ *Any blowby that can cause crankcase pressure **must** be able to escape through the hose to the air cleaner. An oily air cleaner is often a clue to a crankcase pressure problem. Excess pressure in the crankcase causes blowby to escape up the hose to the air cleaner.*

■ *On distributor ignition systems, crankcase pressure can cause oil to migrate up the distributor shaft and into the distributor.*

■ EVAPORATIVE CONTROL SYSTEM SERVICE

One of the major causes of air pollution from motor vehicles is fuel evaporation. Evaporative emission controls are installed on gasoline filler nozzles at filling stations (**Figure 44.6**). These pump nozzles will not pump unless the pleated rubber end is compressed against the vehicle fill opening or manually held in the compressed position during pumping. This device is responsible for preventing massive amounts of air pollution.

Vehicles have their own evaporative systems. These systems are often neglected until a driveability problem occurs. Evaporative fuel controls are included in an emission inspection.

■ EVAPORATIVE SYSTEM MAINTENANCE

Filter Replacement

Fresh air is drawn into the canister during purging. Because the charcoal canister breathes fresh air, some canisters have a replaceable filter (**Figure 44.7**). The filter is either foam or fiberglass. Check the manufacturer's specifications for the recommended service interval, which is usually every 2 years. In dusty conditions, the filter will require service more often. The filter is held in place with a retaining ring or a bar. If the canister must be removed to service the filter, be sure to mark the hoses so they can be correctly reinstalled.

Check Condition of Hoses

Check the condition of all system hoses. When replacing old or damaged hoses, use fuel hose or hose that is designed to be used with fuel vapors. Ordinary vacuum hoses will be damaged by fuel vapors.

Care should be used when disconnecting a vacuum hose from a plastic fitting on a vacuum "T" connection

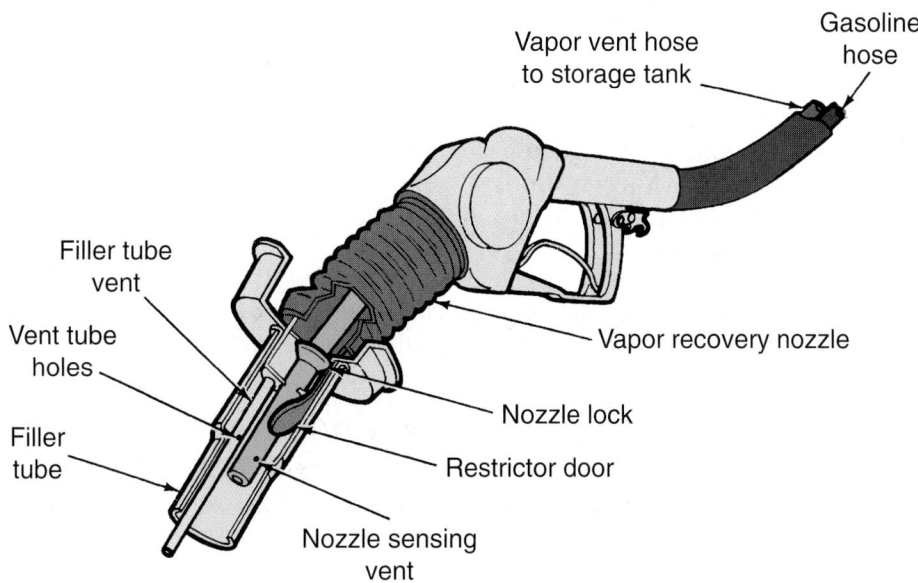

Figure 44.6 An evaporative control filling nozzle prevents HC emissions at gasoline stations. *(Courtesy of DaimlerChrysler Corporation)*

Figure 44.7 Some vapor canisters have a replaceable filter on the bottom. *(Courtesy of DaimlerChrysler Corporation)*

or a thermal vacuum switch. These can be easily broken, causing unnecessary replacement of parts.

SHOP TIP A small pick or screwdriver can be inserted between the hose and fitting before attempting to remove the hose.

DIAGNOSIS OF EVAPORATIVE SYSTEM PROBLEMS

A leaking evaporative (EVAP) system can allow gasoline vapors to escape into the air. **Figure 44.8** shows a typical EVAP system.

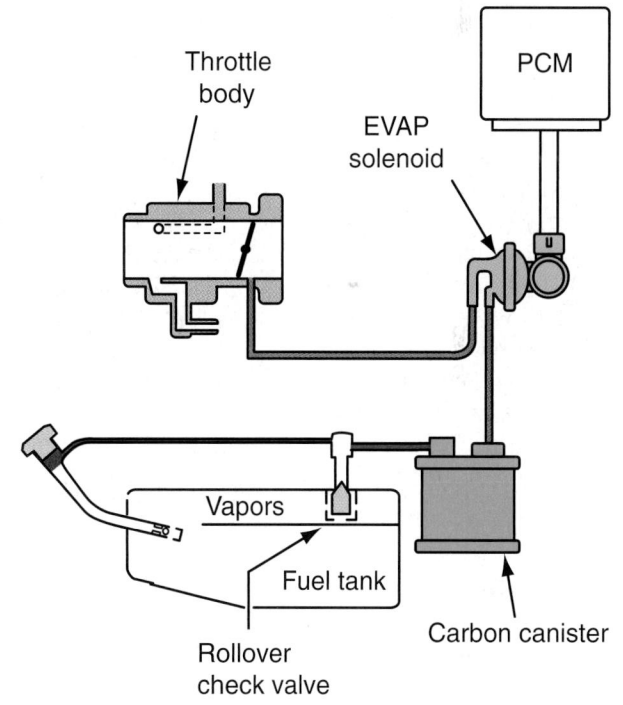

Figure 44.8 A typical evaporative system.

CAUTION If you notice the smell of gasoline, check the fuel system for leaks and check the evaporative system and its hoses. Gasoline vapor is very dangerous. If you can smell it, then it is a vapor.

Perform a visual inspection of all hoses for damage or loose connections. Check all electrical connections to see that they are clean and tight. If the malfunction indicator lamp is illuminated, use a scan tool and check for diagnostic trouble codes (DTCs) to help locate the problem.

EVAP system problems include high emissions and a rich mixture at idle, if the purge valve is stuck open. It should only purge on a warm engine off idle. Sometimes, a novice has attached lines to the wrong locations or a hose has fallen off. The customer might complain of the smell of raw gasoline from inside the car.

If you find liquid fuel in a charcoal canister, check the liquid/vapor separator. Its return line to the tank can become plugged with rust from inside the gas tank. The main vent line can become plugged or kinked. A hole can develop in the vent line from rust. The liquid/vapor separator can become full when a customer keeps topping off the tank during a refill. Gasoline is forced into the liquid/vapor separator. This can cause poor starting on a hot engine along with black smoke. A saturated canister is replaced. When an engine is running poorly, test to see if a saturated canister is the cause by disconnecting or pinching the purge hose and see if the problem changes.

There can be a blockage in the liquid/vapor separator or the vent line between it and the canister. This could cause the fuel tank not to breathe properly. The result is fuel starvation or a collapsed gas tank (on sealed gas cap type).

A loud rush of air entering the tank when the cap is pulled points to a venting problem. To test for this, remove the gas cap and disconnect the gas tank vent line from the charcoal canister. You should be able to blow all the way into the tank through the vent line.

Charcoal canisters rarely cause problems, but a purge valve can fail. Check the purge valve with a vacuum pump. Vacuum should hold for 20 seconds. If the car has an electronically controlled EVAP system,

there could be a DTC if a duty cycle purge solenoid is not operating correctly or if there is a leak in the system. An engine coolant temperature (ECT) sensor can give an erroneous reading that will prevent the solenoid from operating. The winding in the canister purge solenoid can be tested against resistance specifications with an ohmmeter (**Figure 44.9**).

Follow manufacturer specific procedures for testing a tank pressure control valve. A typical procedure is to apply 10 inches of vacuum to the valve while blowing through it from the tank side. With the valve under vacuum, there should be no restriction to airflow.

Testing for EVAP Leaks

EVAP systems on OBD II vehicles self-test for leaks. If there is a DTC, scan for codes to see if an EVAP leak is indicated. There are several ways to check a vacuum purge system for leaks. An exhaust analyzer is one tool that can be used to detect HC around parts of the system.

A flowmeter is another method of checking for leaks (**Figure 44.10**). On a warm engine, after 2 minutes at idle speed, remove the purge hose from the canister and install the flowmeter in series. If the ball is not against the top check for leaks. The purge solenoid is a duty cycle solenoid (see Figure 44.9). Therefore, a properly operating system will pull the ball to the top of the meter, where it will bounce against the seat as the solenoid turns on and off.

Another popular method of EVAP leak testing is to use a smoke tester (see Chapter 48). A very low-pressure smoke tester is available for testing newer EVAP systems.

Check the condition of the gas cap. This is where the majority of the problems are. There are tools available for checking it (**Figure 44.11**). See that it fits snugly and that its gasket is in good condition.

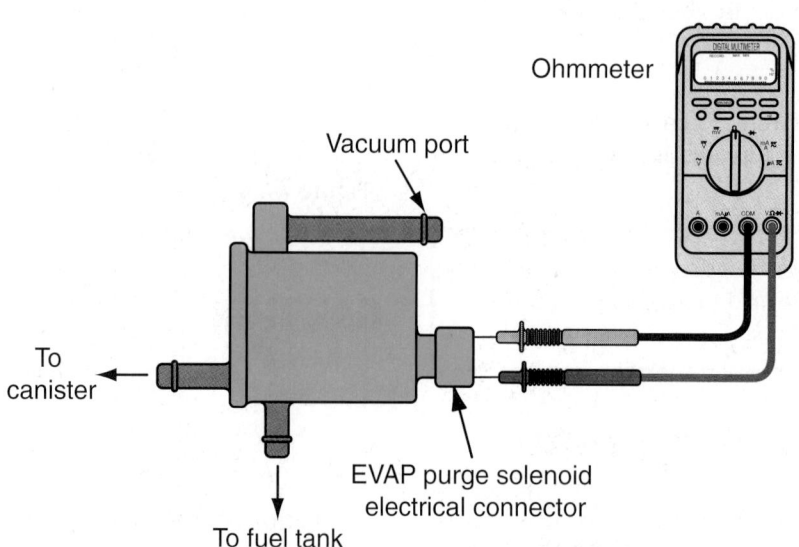

Figure 44.9 Testing an evaporative system purge solenoid.

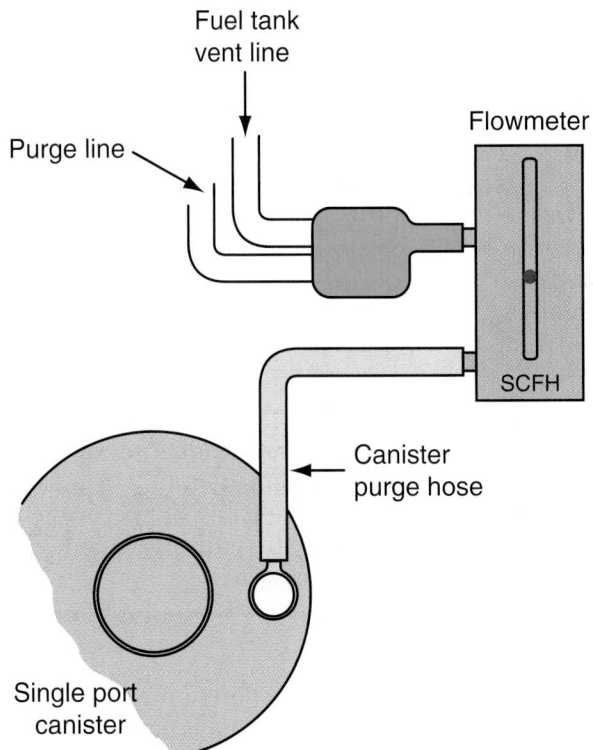

Figure 44.10 A flowmeter is one method of testing for leaks in an evaporative system.

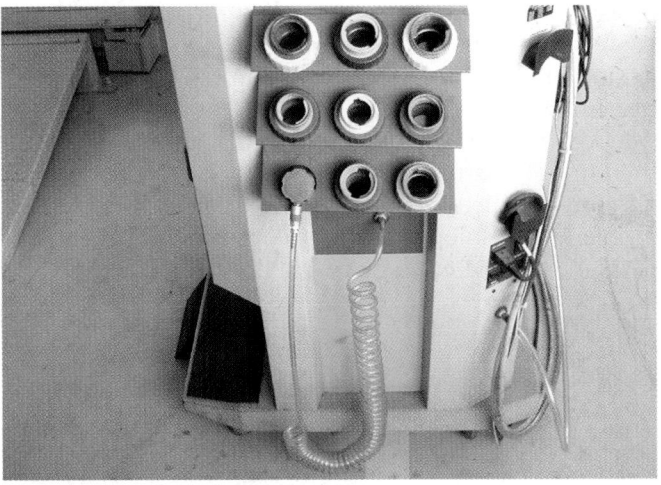

Figure 44.11 A gas cap tester. *(Courtesy of Tim Gilles)*

It is important that the correct gas cap for the vehicle always be used. Be sure to replace a cap with the correct one, sealed or vented.

Older Emission Devices

Older vehicles with carburetors and throttle-body fuel injection had devices for controlling the temperature of incoming air and the temperature in the intake manifold. Some of these older vehicles still undergo vehicle emission inspections in some states, so a very short description of their service is presented here.

A vacuum motor controls the temperature of air entering the air cleaner. Problems with the vacuum motor, its vacuum supply, or the mechanical door can result in driveability and emissions problems. Check service information for the vehicle in question.

Early Fuel Evaporation (EFE). Carbureted vehicles also have devices called manifold heat control valves. These butterfly valves can stick in the open or closed positions, causing hot or cold driveability problems. Some of them are vacuum operated; therefore, a problem with the vacuum supply can result in system failure.

There are also EFE systems that use an electric grid installed beneath a carburetor or TBI.

Current is supplied to the grid when the engine is cold and running. If there is no current at the grid, check the wiring and relay that control it. The grid should have a specified amount of resistance. Check this with an ammeter while there is power to the grid. No current flow indicates an open circuit. Checking this device with an ohmmeter while it is disconnected will also show an open circuit, but if the grid is barely connected and opens under load, the test will not detect a problem.

■ EGR SYSTEM SERVICE

Problems with the EGR system are a common occurrence. A computer-operated EGR valve can experience the same mechanical symptoms as a vacuum switch–operated valve other than vacuum supply problems. Failures in electronic EGR systems often result in zero or limited EGR flow. The computer uses information from the coolant temperature sensor, MAP or MAF sensor, and throttle position sensor to manage EGR valve operation. A problem with one of these systems can affect EGR operation.

Poor EGR performance can also affect speed density (MAP) fuel injection systems. Low EGR flow results in undercalculation of the fuel and a lean air-fuel mixture, and high EGR flow drives the mixture rich. EGR displaces oxygen in the exhaust stream, which confuses the computer. MAF systems are not affected by EGR flow. The computer system will set a fault code if it experiences an electronic failure or sees a sensor or solenoid reading that is out of specification.

Typical symptoms for EGR system problems include stalling or rough idle, hesitation when cold, detonation, and increased cooling system temperature.

If the EGR valve is inoperative, combustion temperatures become too high. On older cars with no knock sensor, the result is often abnormal combustion (pinging or spark knock). Higher NO_x emissions also result. On newer vehicles, a knock sensor will respond to detonation by retarding the ignition timing until the detonation goes away. This will result in poor fuel economy and can cause engine temperature to rise from running with later ignition timing than designed.

The following are other possible causes of spark knock:

- If the engine knocks under load, check the base spark timing first.
- Check the cooling system temperature. A failed cooling fan switch, a stuck thermostat, or a restricted radiator can also cause pinging.
- Spark plugs of too high a heat range will cause preignition.
- Fuel of too low an octane for a high-compression engine can cause spark knock.

Diagnosing EGR Problems

Testing of all types of valves requires these three things:

- Test valve operation
- Clear passageways
- Check sources of vacuum

Check the service library for individual service procedures. There are many different procedures for the different types of valves and controls that have been used over the years.

Testing EGR Valve Operation. If the EGR valve is stuck open, the engine will stumble and have a rough idle. An open EGR valve causes hard starting because it acts like a vacuum leak. If it is open far enough, the engine might not start.

If the EGR valve is of the type where its stem can be seen, the engine can be run while watching the valve stem move with a mirror. The stem can be hot, so do not touch it. If the stem does not move in response to engine revving to about 2,500 rpm when there is a good source of vacuum present, something is wrong with the valve. You can check for vacuum by pulling the hose and feeling the vacuum at the end of it as the engine is revved. An accurate reading can be taken by teeing a vacuum gauge into the hose to the EGR valve.

On some types of EGR valves, you can apply vacuum to the valve with a pump while the engine idles. This causes a vacuum leak that will make the engine stumble. This test will not work with a positive backpressure valve.

Clear Passageways. It is very common to have an EGR valve that appears to operate properly but does not control NO$_x$ or prevent detonation under load. This can be due to an exhaust passageway that has become plugged with carbon.

> **SHOP TIP** A plugged passageway can sometimes be cleaned with a homemade tool. Use a short length of parking brake cable or speedometer cable on a drill. Spread the end of it open, and spin it through the passageway.

OBD II vehicles are required to determine whether EGR flow is sufficient. A passageway that is partially plugged or a valve that is not operating correctly will not control emissions and provide acceptable driveability. All of the major manufacturers use different ways of monitoring EGR flow. Check the service information before proceeding with testing.

When a scan tool is connected to the data link connector to check for DTCs, you can use the EGR control function of the scan tool to operate digital EGR solenoid control valves. Opening a solenoid can cause the engine to stumble momentarily until the computer corrects idle speed. If there is a DTC for a solenoid problem, the solenoids are disconnected and checked with an ohmmeter (**Figure 44.12**).

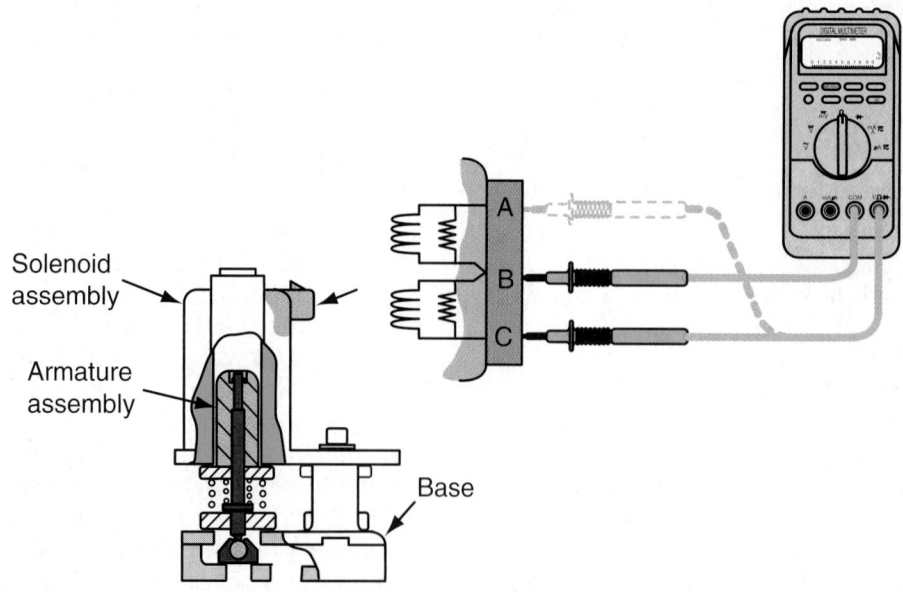

Figure 44.12 Checking the resistance of solenoid windings on a digital EGR valve.

A linear EGR valve has a position sensor on top that tells how far the EGR valve is open. When you use the scan tool to command EGR valve operation, the amount of valve opening is displayed on the scan tool and should be within 10% of the specified amount.

Check the fuse when there is an EGR power supply problem. Then refer to a wiring diagram and test for power and ground and a computer reference signal as noted.

Check Vacuum Source Problems. Check for a vacuum source problem. Run the engine at above 2,000 rpm at normal operating temperature and check to see that there is a good source of vacuum to the valve. Vacuum problems can be caused by thermal valves, delay valves, or improper hose routing.

If the EGR valve operates before the engine is warmed up, the engine can run rough or stumble. A defective thermal vacuum valve could cause this if it sticks in the open or closed position. A valve that is stuck closed will keep the EGR valve from operating, causing fuel knock and excess NO_x.

Most new cars have computer-controlled EGR valves (see Chapter 43). On many of these systems, there will be no vacuum to the EGR valve during park or neutral operation.

■ AIR INJECTION SYSTEM SERVICE

Visual Inspection

Look for obvious conditions that could result in problems.

- Look for rusty check valves and air manifolds on air pump and pulse air systems. Hoses used on the air injection system can be exposed to exhaust gas if a part in the system fails. When there is a burned hose near a check valve, the check valve could be allowing exhaust to flow backwards.
- Hoses that are mounted near exhaust parts can be exposed to excessive heat and are often candidates for replacement. Many of them are formed hoses. Smog hose with the wire in it is a good choice for these applications.
- Inspect the condition and tension of the belt that drives the air pump.

Diagnosing Problems

There are several problems that can happen with an air injection system.

Exhaust Noise Under the Hood. This could be from either a leaking check valve or manifold pipe. These can sometimes rust through.

Backfiring on Deceleration. When a car is backfiring on deceleration, check the diverter valve with a vacuum pump. Vacuum should cause the pump air to be diverted to the muffled exhaust port. This is a typical early system. Check the service information to see what kind of system you have.

Air Pump Noisy or Frozen. If the belt is melted or missing, the pump is undoubtedly frozen and the pump must be replaced. Otherwise, remove the belt and spin the pulley by hand. It is normal for the pump to make some noise, but it should not feel hard to turn or squeal loudly. If the pump is bad, check for a leaking check valve that has allowed exhaust to enter the pump.

Emission Test Failure. When an engine does not pass an emission analyzer test, check the pump to see if it is pumping air. Pinch the hose and listen to the pressure relief valve start releasing air. You can pressure test for pump capacity with a pressure gauge.

To test to see that passageways are clear, you can check the O_2 content of the exhaust with the air pump working. Pinch the hose and compare the reading to the amount of O_2 with the system inoperative.

■ CATALYTIC CONVERTER SERVICE

Catalytic converters can become plugged or inoperative or inefficient to the point of not being able to pass an emissions test. A common failure to a converter is overheating caused by a misfiring cylinder. This is anything that allows unburned fuel to enter the exhaust. If a spark plug wire falls off, the cat can get so hot that it actually melts internally.

A plugged converter restricts the engine's exhaust flow. The symptom of this is that the engine might start and accelerate briefly, but then has trouble breathing and will not produce power. A hissing sound can come from the exhaust, accompanied by a roaring sound or spitting from the fuel intake in the engine compartment.

Do Not Just Replace the Converter

A catalytic converter is designed to reduce normal exhaust emissions. It can mask problems that are related to the fuel or ignition systems, or mechanical problems. This will overwork the converter, and it probably will not last as long. An overworked converter will lose efficiency. Converters do not usually just stop working. They just work less and less effectively. Before replacing a converter, determine whether there is another problem that has contributed to its failure. Installing a new converter might fix the problem for a while, but it is sure to come back when the converter fails once again.

■ CONVERTER TESTING

The cat must be heated to at least 600°F to work. As it begins to oxidize HC and CO, it will become even hotter. When the converter begins to oxidize, this is called **converter light off.** An oxidizing converter that is working will have a hotter exhaust at its outlet than at its inlet. A temperature probe can be used to take temperature measurements of the pipe at the inlet and

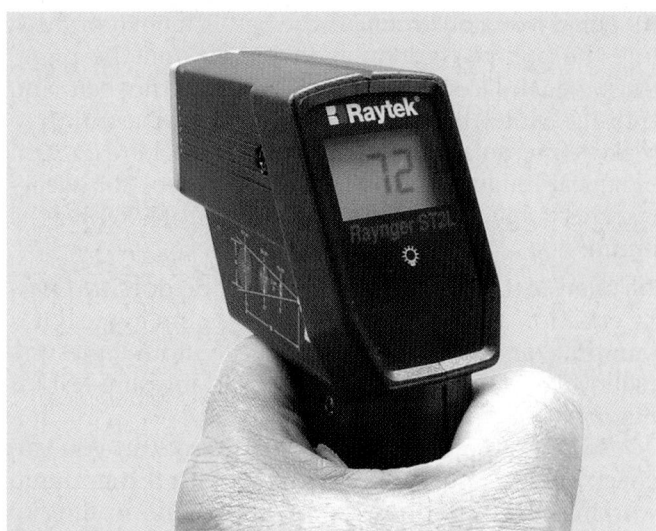

Figure 44.13 An infrared thermometer. *(Courtesy of Tim Gilles)*

the outlet. This is one of the easiest tests to see if the converter is operating:

■ The engine is run at 2,000 rpm until the converter is at normal operating temperature. A converter will cool off at idle speed and might stop working.

■ A good converter should measure 5%–10% hotter at its outlet than its inlet. This is only true when the converter is working. If the engine is running with very clean exhaust, the converter is dormant.

■ Shut off the engine. Create a rich mixture or pull a spark plug wire and ground it. This will result in higher HC and O_2 to the converter.

NOTE: *Do this test quickly (30 seconds should be sufficient). A converter will overheat in a very short time (2–3 minutes).*

The temperature is measured with a **high-temperature pyrometer** that is touched against the surface of the pipe. Another tool is an **infrared thermometer** (**Figure 44.13**). It is not held against the surface, just near it. The surface pyrometer is slower than the infrared thermometer.

A problem with the temperature test for three-way cats is that this test does not tell whether they are working well in both halves.

The best test for a converter is to test its operation with an emission analyzer. This and other emission analyzer tests of the converter are covered later in this chapter.

■ CONVERTER REPLACEMENT

A replacement converter is often bolted on with new gaskets. Some of them require welding for installation. Be sure to get an EPA-approved equivalent replacement. Installation of a used catalytic converter is illegal for an emission technician unless it has been tested and certified (a process that costs as much as a new one).

■ ANALYZING EXHAUST EMISSIONS

Air is mostly nitrogen (N) and oxygen (O_2). Perfect combustion would leave only water (H_2O) and carbon dioxide (CO_2) in the exhaust (**Figure 44.14**). Nitrogen would be in the exhaust stream, but it would not have changed throughout the burning process. If it did, this was caused by high temperatures. Certain things vary in the combustion process that make it less than perfect. These include variations in ignition, air and fuel, combustion temperatures, and physical limitations such as the skin effect and combustion chamber shape.

Fuel Burning

To understand what the readings in the exhaust mean, you will need to understand what happens when fuel burns. Hydrocarbons (HCs) are what gasoline is made up of. In analyzing exhaust readings, the lower the HC's the better because HC is unburned fuel. Burning of the fuel is called **oxidizing**. To be able to burn, fuel requires oxygen and heat. About 21% of the air we breathe is O_2.

Atomized fuel in the correct proportion with air is compressed in the combustion chamber. Compressing the air-fuel mixture pushes the O_2 molecules and fuel molecules closer together. When some of the molecules are ignited by the heat of the spark from the spark plug, they burn. During combustion, heat is given off. The heat ignites molecules next to the burning molecules and a chain reaction takes place. This is called the *flame front*. Any of the mixture that does not burn must go out the tailpipe. That is what is measured in the exhaust as HC.

Some HC in the exhaust is normal. If the reading is excessive, the cause could be anything that interferes with the normal compression, ignition, or burning of the fuel.

Compression. Compression must be enough to push the molecules close together. If not, burning could start but not continue because the chain reaction is broken. HC levels will be higher.

Ignition. With an ignition misfire, the entire charge of raw fuel in the cylinder will go out the tailpipe and be measured as HC.

Air-Fuel. If the air-fuel mixture is too rich, there will not be enough O_2 in the mixture to combine with all of the fuel so it can be burned. As an air-fuel mixture becomes leaner, HC emissions begin to reach their lowest point at about 14.7:1, called *stoichiometric* (**Figure 44.15**). Depending on engine design, fuel systems are run at 14.7:1 or as lean as 16:1. Mixtures leaner than about 16:1 tend to misfire because there is not enough fuel to continue the chain reaction.

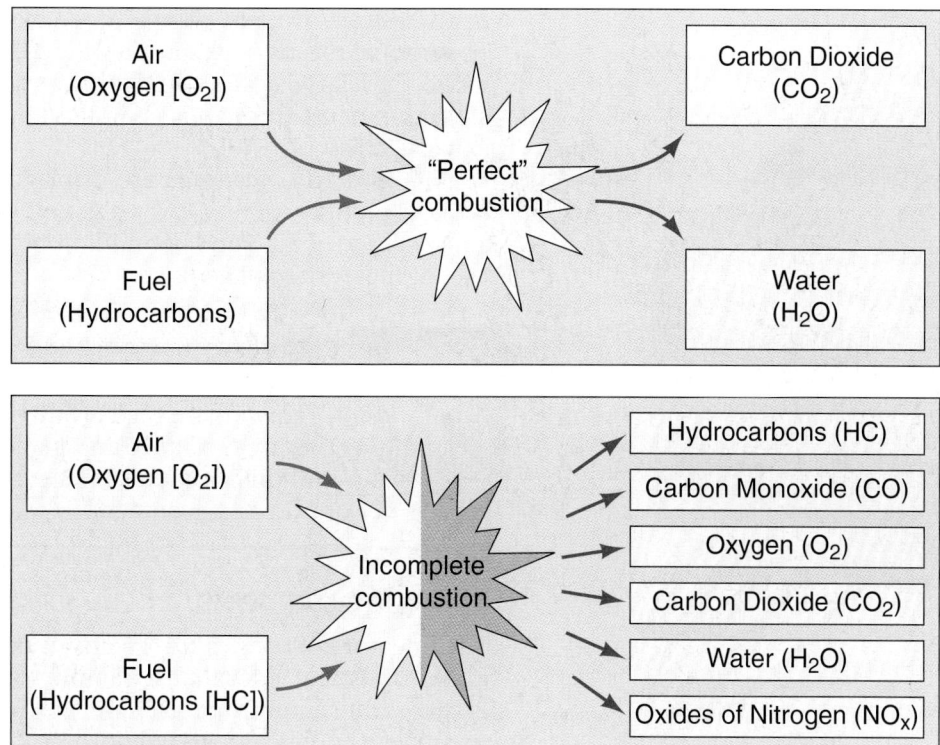

Figure 44.14 By-products of combustion. *(Courtesy of Environmental Systems Products, Inc.)*

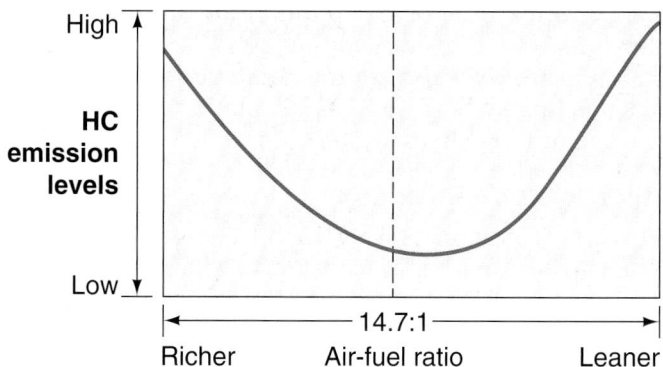

Figure 44.15 HC emissions are lowest at an air-fuel mixture of about 14.7:1. *(Courtesy of NAPA Institute of Automotive Technology)*

With either a rich or lean misfire, HC in the exhaust rises.

EMISSION ANALYZERS

The exhaust emission analyzer measures the content of gases in the exhaust through a probe installed in the exhaust pipe of a running engine.

An exhaust analyzer is helpful in diagnosing problems with the fuel system, catalytic converter, EVAP emission system, and air injection system.

SAFETY NOTE CO is colorless, odorless, and dangerous. It deprives the body of O_2. Headaches, coordination problems, unconsciousness, and death can result from exposure to too much CO. Be sure to use adequate ventilation when running an engine during testing.

The emission analyzer has a pump that pulls an exhaust sample through a hose connected to the exhaust pipe probe. The gas sample is moved through filters to trap solids and remove moisture. There are sampling tubes inside the tester that the exhaust sample flows through. An infrared light beam is focused through the exhaust gas onto a detector. The energy required to penetrate the exhaust sample varies with the amount of gas in the exhaust.

A **two-gas analyzer** is the older type that measures HC and CO. Some of these machines required a long warm-up period. Later model computerized

Figure 44.16 A portable five-gas analyzer. *(Courtesy of Vetronix Corp.)*

machines have automatic features for warmup and calibration. These machines also read four gases or five gases rather than two. A **four-gas analyzer** measures HC, CO, CO_2, and O_2. A **five-gas analyzer** (the most modern) measures five gases: HC, CO, CO_2, O_2, and NO_X (**Figure 44.16**).

▨ SCIENCE NOTE ▨

Wavelenghts of light

When white light is passed through a prism, we see that it is made of many colors. Each color is due to a photon of light of a particular wavelength.

■ *White light contains wavelengths ranging to 400 nanometers (1 nanometer = 10^{-9} meter)*

■ *Violet light = short wavelengths to 700 nanometers*

■ *Red light = long wavelengths*

Visible light is only a small part of the electromagnetic spectrum. At very long wavelengths are x-rays and gamma rays.

NOTE: *When testing for emissions in the exhaust, remember that the air pump, catalytic converter and its temperature, and exhaust leaks will all affect test results.*

Measuring Exhaust Gases

Exhaust gases are measured in different ways. Manufacturer and government emission tests use *constant volume sampling*, which is measured in grams per mile.

In these tests, exhaust samples are collected in a bag. The normal means of exhaust testing in the automotive service industry is called *partial stream sampling*, which collects exhaust through a small hose inserted into the exhaust pipe.

HC and NO_X readings are displayed in parts per million (ppm), whereas CO, CO_2, and O_2 are given as a percentage of air. Parts per million provides a precise way of measuring percentage.

SHOP TIP Think of percentages as "parts per hundred." This will help you realize the relationship between percentage and parts per million. Parts per million is a much smaller amount than parts per hundred (percentage).

▰ HYDROCARBONS

The measurement 1 ppm means that out of every million parts of an exhaust sample, one part is HC. Because the amount of HC we are dealing with is so small, the percentage would be too minute to be read on a percent scale. A reading of 500 ppm is only 0.05%—five-hundredths of 1%. Specifications can range from a high of 700 ppm for an early-model car down to 100 or less for later models.

HCs are oxidized by the catalytic converter and air injection system. The HC sample can be diluted by leakage in the exhaust system.

Hydrocarbon Readings

The most common cause of high HC is unburned fuel. This can be caused by any misfire, mechanical problem, or advanced ignition timing. HCs indicate unburned gasoline resulting from a misfire (**Figure 44.17**). High

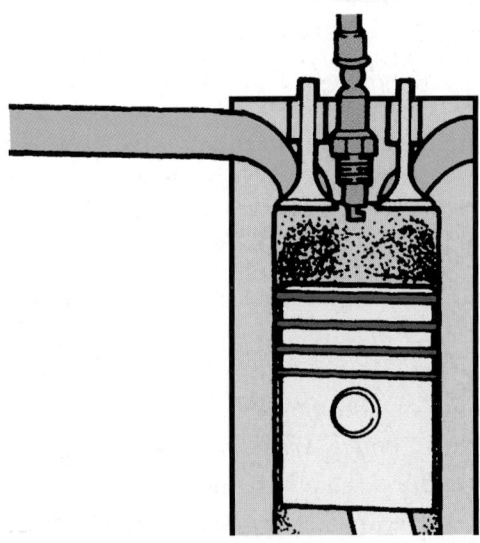

Figure 44.17 Weak or broken piston rings can cause a misfire, higher HC emissions from the exhaust, and increased crankcase pressure. *(Courtesy of DaimlerChrysler)*

HC could be because of too much fuel or too little or too much air.

Too Much Fuel. This can result from anything that allows excessive amounts of fuel to enter the engine's combustion chambers, causing an excessively rich air-fuel mixture. Possible causes include high fuel pressure, bad fuel injectors, or evaporative emissions problems like vapor canister purge or oil saturated with fuel. When any of these result in an air-fuel mixture that is too rich, misfiring can occur.

Not Enough Air. Restricted airflow is usually caused by a mechanical malfunction. These include a burned or misadjusted valve, a broken piston ring, a flat cam lobe, cam timing off, a restricted air cleaner, or a restricted exhaust system. These problems all result in less air flowing through the engine. If intake or exhaust is restricted, the engine cannot bring in enough fresh air and fuel. The distance between the HC molecules is decreased and burning is hampered.

Too Much Air. A manifold vacuum leak is one cause of an air-fuel mixture that is too lean to burn. Excessive air increases the space between the HC molecules, making burning more difficult.

Other Causes and Results of a Misfire. After a misfire, whatever remains in the cylinder ends up in the exhaust stream, increasing the HC level. There are other causes of high HC readings besides problems with the air-fuel mixture. A defective spark plug or problems with ignition wiring are obvious misfire culprits. Advanced ignition timing is another possible cause.

■ When the timing is advanced, the flame starts to burn the mixture before the molecules are fully compressed. This results in a break in the chain reaction and the flame front goes out before all of the fuel is burned.

SHOP TIP The exhaust analyzer can be used to test to see if a fuel-injected engine with a no-start condition is getting fuel. Crank the engine over and watch the tester. If HCs are coming out of the exhaust, there is fuel.

■ An air-fuel mixture that is too rich or too lean can cause a misfire. This causes high HC.
■ A manifold vacuum leak that results in a mixture that is too lean to burn.
■ An excessively rich or lean air-fuel mixture.

Analyzer Readings

A lean mixture can cause fluctuating HC readings (CO remains mostly fixed). With a working catalytic converter, a reading of over 100 ppm can indicate a rich or lean condition or a misfire due to advanced timing, low compression, a fouled spark plug, inoperative air injection, and so on.

Excess carbon on a valve stem can restrict intake flow. The carbon can also absorb fuel like a sponge. This promotes a misfire when the mixture cannot stay stable because of the fuel going into or leaving the carbon at uncontrolled times.

■ CARBON DIOXIDE

Carbon dioxide (CO_2) is a by-product of complete combustion, giving an indication of how thoroughly fuel is burning in the cylinder. The closer combustion is to being complete, the higher the CO_2 reading will be. CO_2 is a good indicator of the efficiency of the engine.

The air-fuel ratio affects CO_2. CO_2 is highest at an air-fuel ratio of 14.7:1. You can begin to see why the stoichiometric ratio (14.7:1) is used for emission control. CO_2 levels drop when lower or higher than this level (**Figure 44.18**).

Unlike HC, CO_2 only results from combustion of the fuel:
■ If there is no flame, no CO_2 is in the exhaust.
■ When the combustion process is incomplete, the CO_2 level will be lower. A rich or lean mixture will cause a *drop* in the CO_2 reading.
■ A high CO_2 level means that the air-fuel mixture must be near stoichiometric.
■ CO_2 is not a good rich/lean indicator. If CO_2 is low, the air-fuel ratio could still be correct, but it can also be rich or lean.
■ If CO_2 is low and O_2 is low, the air-fuel mixture is rich.
■ If CO_2 is low and O_2 is high, the air-fuel mixture is lean (see Figure 44.18).
■ A typical good running engine with a catalyst will have from 13% to 16% CO_2. The amount of CO_2 produced between different vehicles is not consistent and rises as the catalytic converter heats up.

Reading CO_2 is different from reading HC. It is read as a percent of the total rather than parts per million. If the reading is 13.2%, then 13.2% of the total volume of the exhaust is CO_2.

If the air pump is running during a test reading, the volume will be changed and the reading will not be accurate. Also, leaks in the exhaust system can change CO_2 readings.

Figure 44.19 shows the effect extra air in the exhaust has on each of the five gases. Notice that the percentage of CO_2 is lower because the volume of the exhaust gas sample has become larger. O_2 is an exception to this because adding outside air adds O_2 to the exhaust.

NOTES:
■ *Be sure to disable the smog pump or pulse air system during exhaust gas analysis.*
■ *Readings are taken after the catalytic converter.*

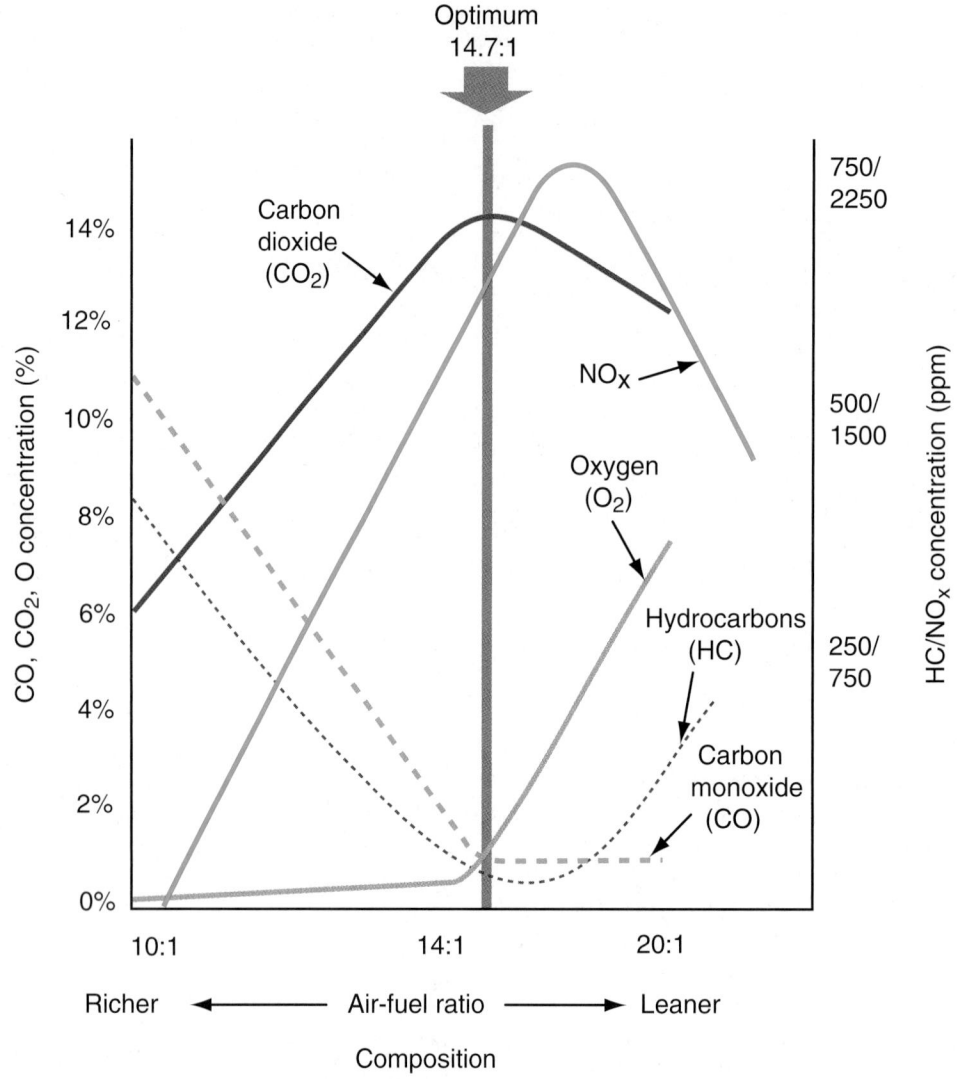

Figure 44.18 If CO_2 is low and O_2 is high, the air-fuel mixture is lean.

EXHAUST DILUTION	
HC ppm	Lower than actual
CO%	Lower than actual
CO_2%	Lower than actual
O_2%	Higher than actual
NO_x ppm	Lower than actual

Figure 44.19 The effect extra air in the exhaust has on each of the five gases. (*Courtesy of NAPA Institute of Automotive Technology*)

An engine that is running will always have at least 4% CO_2. Some state emission tests use a combination of CO and CO_2 to check the condition of the exhaust system. A combined reading of *less* than 7%–8% (depending on the car) would fail the test. When a reading is below that level, look for a source of air entering the exhaust.

A test of catalytic converter efficiency using CO_2 is covered later in this chapter.

■ CARBON MONOXIDE (CO)

CO should be CO_2 but there is not enough O_2. A rich air-fuel mixture (not enough O_2) causes incomplete combustion.

 If you start to feel dizzy or get a headache, get some fresh air. This could be caused by CO poisoning.

Without the effect of the catalytic converter, CO gives a good indication of the air-fuel ratio (**Figure 44.20**). With a rich mixture, levels of CO are high. They are low with a lean mixture (plenty of air). When there is high CO, you know you do not have a lean mixture.

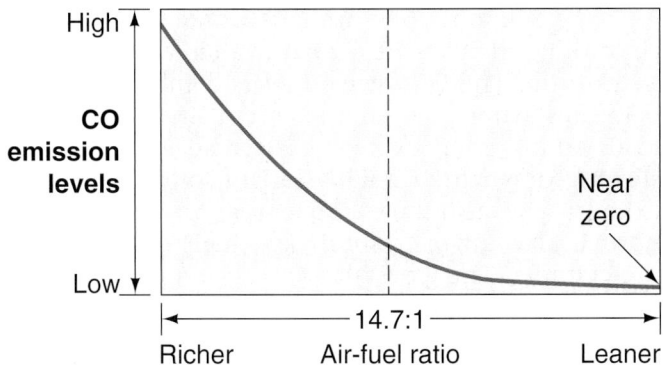

Figure 44.20 Carbon monoxide readings give a good indication of the air-fuel ratio. O_2 levels should follow CO. *(Courtesy of NAPA Institute of Automotive Technology)*

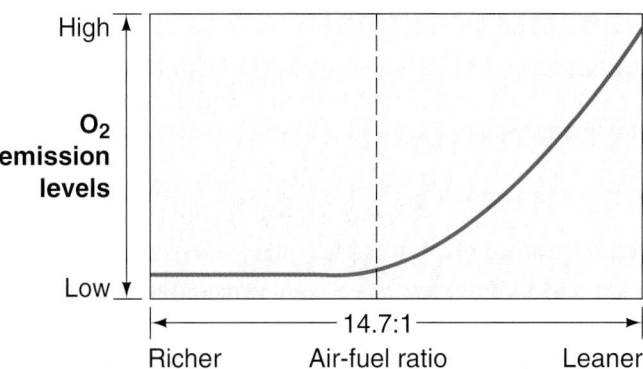

Figure 44.21 When the air-fuel mixture is richer than stoichiometric, the level of oxygen will be low. *(Courtesy of NAPA Institute of Automotive Technology)*

CO is measured as a percentage of the total exhaust. The ideal reading would be 0%. Before the cat, you could expect to read between 0.6% and 2% CO with a 14.7:1 mixture. CO is not lowest at stoichiometric but gets lower beyond that point.

SHOP TIP CO and O_2 should be close to the same (see Figure 44.20).

High readings can be caused by a rich mixture, retarded ignition timing, or a worn engine. CO is oxidized into CO_2 by the catalytic converter and air injection system. It is diluted by exhaust leaks.

VINTAGE ENGINES

When working on older cars with carburetors, a two-gas analyzer was handy for reading and adjusting air-fuel ratios using the CO reading.

CO, like CO_2, is a by-product of combustion. If there is no flame, there is no CO. If a spark plug is disabled, HC readings will be high but CO will not register for that cylinder.

SHOP TIP You can get a good idea of what a problem is by paying attention to your senses. For instance, if you smell exhaust, it is unburned HCs, so fuel is not burning. It cannot be CO, because that has no odor.

■ OXYGEN (O_2)

When the air-fuel mixture is richer than stoichiometric, the level of O_2 will be low (**Figure 44.21**). This is because a rich mixture provides enough fuel to use up all of the O_2 in the cylinder.

As the mixture becomes leaner, O_2 levels increase quickly. When the mixture is lean, O_2 rises because there is not enough fuel to use up all of the O_2.

SHOP TIP When checking O_2 emissions in the exhaust, it is extremely important that there be no supplemental air or exhaust system leak.

Diagnosis with O_2

Like HC, O_2 is not formed during combustion, so it should be consumed during each combustion cycle. Before combustion, there is 21% O_2 in the air. The amount left over after combustion subtracted from 21% leaves the amount of O_2 used during combustion. Without air injection, a normal reading going *into* the catalytic converter is 1%–2% (14.7:1 air-fuel ratio).

■ Lean mixture—Higher O_2 means less O_2 was used during combustion.
■ Rich mixture—Lower O_2 means more O_2 was used during combustion.
■ Over 5% O_2 usually indicates a misfire.
■ O_2 can be higher because of an air leak (vacuum leak).

Other possible causes include a restricted fuel filter or a bad fuel pump.

O_2 is consumed by a catalytic converter. On a late-model car running at 2,000 rpm with a hot converter, normal readings from the exhaust pipe should be below 0.1% with the air injection system disabled. A higher reading can be because the smog pump has not been disabled or because of dilution from an exhaust leak.

O_2 is used in combustion, so anything that ends combustion or prevents it from occurring will raise O_2 levels. Look for an ignition, fuel, or mechanical problem.

■ OXIDES OF NITROGEN

Air is made up of 78% N and 21% O_2. Oxides of nitrogen (NO_x) form when the two bond under high-combustion heat (above 2,500°F) and pressure in the engine.

> ### ■ SCIENCE NOTE ■
>
> *Nitrogen oxide (NO) is one part of nitrogen and one part of oxygen. Nitrogen dioxide (NO_2) means that there are two oxygen atoms. It is five times as toxic as NO. Most of the NO that leaves the exhaust will become NO_2 eventually.*

NO_x include the entire family of nitrogen oxides. The "x" in NO_x means a variable number, such as NO_1, NO_2, NO_3, etc.

Controlling NO_x

- Rich mixture—A rich mixture will cool combustion and has less O_2, so NO_x emissions are lower (**Figure 44.22**).
- Lean mixture—NO_x emissions are also less when the mixture is leaner than 16:1. This is because combustion is slower or fuel fails to burn, which keeps temperatures lower.

NO_x emissions are highest at a point near stoichiometric. This is obviously not the best air-fuel ratio if you want to control NO_x. The catalytic converter is more efficient at controlling HC and CO when the air-fuel mixture is stoichiometric.

Lower NO_x production is achieved in several ways:

- Reducing the compression ratio. This is achieved in newer engines when variable valve timing closes the intake valve later.
- NO_x emissions are prevented from forming by the EGR valve.
- NO_x emissions are reduced after they are formed by the reduction catalytic converter.

NO_x will not form under normal atmospheric conditions. It takes heat **and** available O_2 to make them form. The EGR valve lowers combustion temperatures when the engine is cruising and under load. Adding inert exhaust gas spreads out the HCs, slowing down their chain reaction burn. The inert exhaust gas also takes up space, preventing the normal amount of fresh air and fuel from entering the cylinder. The gases also absorb some of the heat formed during combustion.

Diagnosing High NO_x

Older emission analyzers had provisions for testing only four gases. Newer analyzers test a fifth gas, NO_x, which only forms under load and heat conditions, so a dynamometer or a portable tester is required. **Figure 44.23** shows a comparison of NO_x emissions from the same engine run at 40 mph with different horsepower loads on it.

EGR problems are the biggest contributor to high NO_x emissions. During one dynamometer test done at an emission testing lab, a 1988 car with a stoichiometric air-fuel ratio was run at 50 mph under light load. The five gas readings are shown in **Figure 44.24**. When the vacuum hose to the EGR valve was disconnected and plugged, the NO_x reading increased dramatically (**Figure 44.25**). Driveability did not change.

If you do not have a dyno but have a portable five-gas analyzer, you can still test EGR function. Have an assistant drive the car while you watch the emission levels on the tester. Disable the EGR valve and try the same road test to see if the EGR valve is functioning. The NO_x on the second reading after the EGR valve has been disabled should be much higher than the first. If the shop's five-gas analyzer

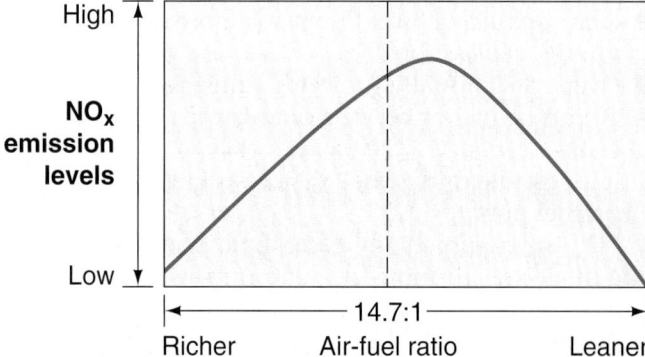

Figure 44.22 A rich mixture will cool combustion and it has less O_2, so NO_x emissions are lower. Slower burning of lean mixtures results in less NO_x, too. *(Courtesy of NAPA Institute of Automotive Technology)*

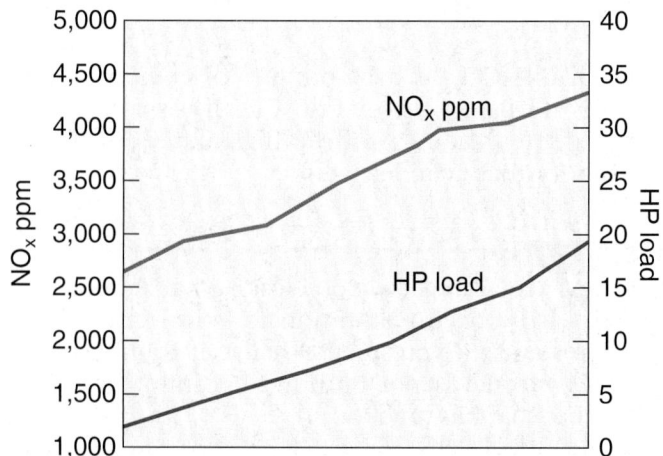

Figure 44.23 The line on the top represents NO_x emissions with the same engine under a higher load. *(Courtesy of NAPA Institute of Automotive Technology)*

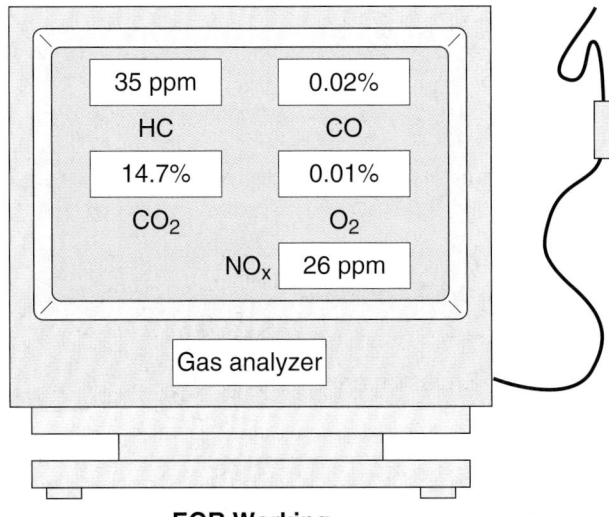

EGR Working

Figure 44.24 Results of a dynamometer test with all emission systems operating. *(Courtesy of NAPA Institute of Automotive Technology)*

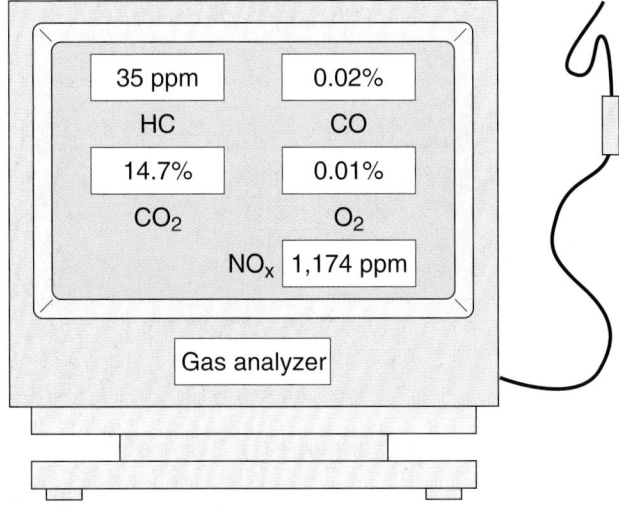

EGR Disabled

Figure 44.25 Results of the same test with the EGR system not working. Notice the level of NO_x emissions. *(Courtesy of NAPA Institute of Automotive Technology)*

is not portable, a stall test (see Chapter 74) will create NO_x.

Anything that increases temperatures can be a possible cause of NO_x levels that are too high. Increases in the temperature of the cooling system or incoming air and compression raised by carbon deposits can all cause NO_x to increase. Ignition timing also affects NO_x levels. Advanced timing raises pressure, causing increased combustion temperatures. Always check the base timing and spark advance when there is high NO_x.

SCIENCE NOTE

NO_x will make your eyes burn. Lean mixtures cause NO_x. Misfiring causes HC, which you can smell, but you cannot smell NO_x. When NO_x is combined with water (tears), a mild nitric oxide burns your eyes. When your eyes are burning at the drag races, NO_x is the cause.

Five-Gas Analyzer Tips

The following are generic five-gas emissions specifications that you can use for easy reference. The specifications are for a computer-controlled engine with a catalytic converter:

- HCs — 30–50 ppm or less
- CO_2 — 0.3%–0.5% or less
- O_2 — 0%–2% (a higher number signifies a leaner mixture)
- CO_2 — 12%–17%
- NO_x — Less than 100 ppm at idle and 1,000 ppm at WOT

SHOP TIP When analyzing exhaust readings there are some constants:
- CO and O_2 are rich/lean indicators.
- CO and O_2 should be equal. If they are the same, the mixture is correct.
- CO, O_2, and HC should be low.
- CO_2 should be high. (This is efficient combustion.)

CATALYTIC CONVERTER TESTS

Tap on a suspected bad cat lightly with a rubber mallet to see if it rattles. If it does, the substrate is broken and the converter will need to be replaced. The cause could be physical damage, overheating, or a defect in manufacturing. The converter can be tested for backpressure by removing the O_2 sensor in front of it and installing a sensitive pressure gauge (**Figure 44.26**). Desirable pressure is less than 1.25 psi.

An oxidizing converter produces CO_2 and H_2O. One of the reasons a converter must have O_2 is to be able to catalyze CO; $CO + O_2 = CO_2$. O_2 levels in the exhaust vary, so air injection is sometimes used. Single bed three-way converters work on only one catalyst and do not have an air pipe. They are popular on newer vehicles without air injection. The air-fuel ratio is important to these catalysts. They depend on the air-fuel mixture fluctuating to both sides of 14.7:1 to be efficient.

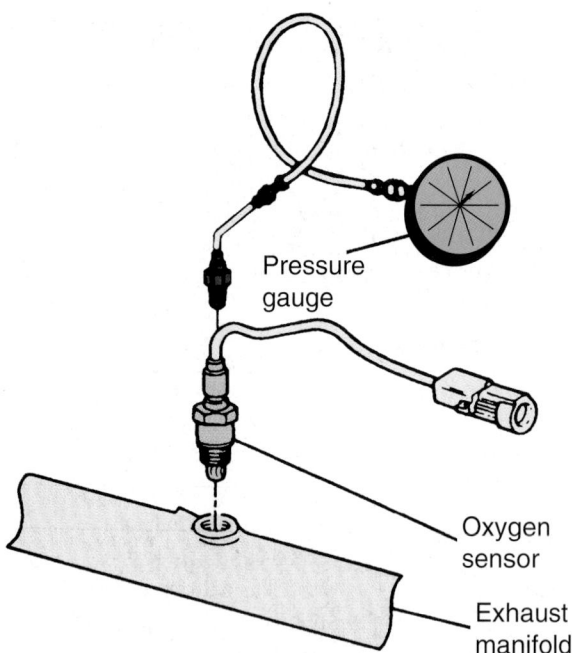

Figure 44.26 To test catalytic converter backpressure, install a sensitive pressure gauge in place of the oxygen sensor.

GAS	BEFORE	AFTER
HC	69 ppm	4 ppm
CO	0.77%	0.06%
CO_2	13.6%	14.5%
O_2	0.6%	0.01%
NO_x	2502 ppm	814 ppm

Figure 44.27 Dynamometer test results showing exhaust analysis done before and after the converter. *(Courtesy of NAPA Institute of Automotive Technology)*

■ A rich mixture will oxidize HC and CO but will not reduce NO_x.

■ A lean mixture will lower reduction but will not do oxidation efficiently.

 **CASE HISTORY** *A converter was tested while operating the car on a dynamometer. The speed was 50 mph with a slight uphill load. Test readings were taken before and after the converter. **Figure 44.27** shows that the converter was working with 94.2% efficiency for HC oxidation and 92.2% efficiency for CO reduction. These tests could have been done without the dynamometer, but NO_x results would not have been significant without the load. Results of the converter efficiency test at idle are shown in **Figure 44.28**.*

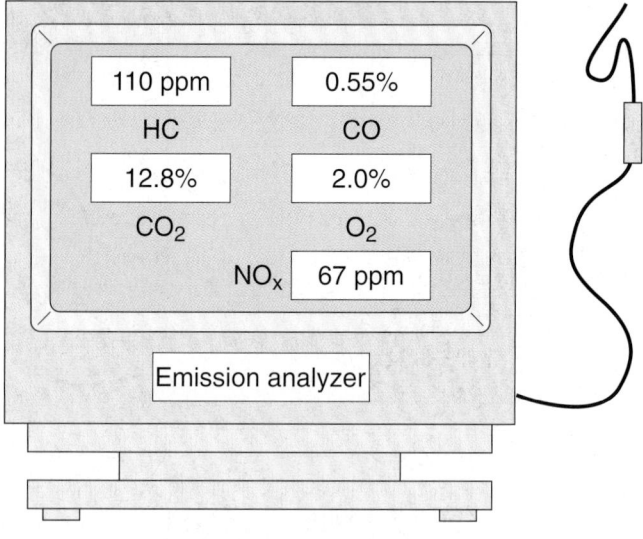

Before Catalyst

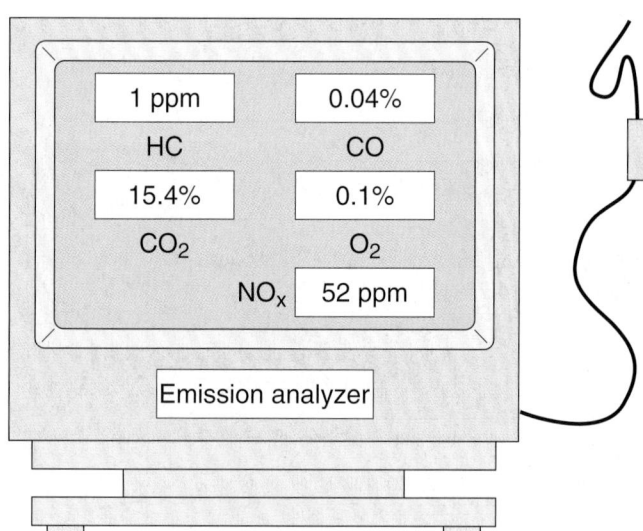

After Catalyst

Figure 44.28 When readings are taken at idle, NO_x does not change much. *(Courtesy of NAPA Institute of Automotive Technology)*

NOTE:

■ *When testing a cat, be sure it is hot. Driving the car on the road or on a dyno is the best way to heat up the cat. Running the engine in a stall is not as effective.*

CO_2 Converter Test

A cranking test for CO_2 can be used to tell the condition of a catalytic converter. With a hot catalytic converter, crank the engine for 10 seconds with the secondary ignition system disabled. Observe the CO_2 reading. A reading below 12% indicates that the converter is weak or defective. Be sure to disable the secondary ignition system and not the primary. Otherwise you will disable the fuel injection system.

OBD II Converter Tests

Catalytic converters on OBD II cars (1996 and later) are tested automatically. The computer uses signals from two O_2 sensors, one located in front of the converter and the other located behind it. Catalytic converters use O_2 in the conversion process, so very little O_2 should remain in the exhaust stream after the converter. When a converter problem is pinpointed by the computer, further tests can be done to it.

Catalytic Converter Testing

A functioning cat stores O_2 when lean and uses O_2 when rich. The function of a catalytic converter can be compared to a sponge, sucking up O_2 and squeezing it back out. New cats are about 90% efficient. A new cat can be compared to sandpaper: a fresh cat is like new sandpaper, whereas an older one is like worn sandpaper.

NOTE: *Approximately 90% of misfires are typically HC misfires; less than 10% are CO misfires. O_2 should not come out of the cat if CO is present. Its O_2 should be used up by the cat during the conversion process.*

During a converter test:
- There should be no exhaust leaks.
- The air injection is off.
- The cat is hot.
- Engine speed is 2,000 rpm.

As the engine runs, watch the exhaust levels on the emission analyzer. Emissions should drop. Observe the O_2 reading. If it drops to zero, the cat is storing O_2 while the mixture is lean.

Snap Test. A snap test can be done only when there is fuel control—closed loop and O_2 sensor switching between rich and lean. To perform the snap test:
- Snap the throttle open and let it drop back down to idle. This loads up the cat with fuel.
- Watch the O_2 and CO. Check the rise in O_2 while the CO is rising. O_2 will continue to rise after the CO peaks, but this is not important. What *is* important is the O_2 level at the point of peak CO.

A cat is 80% efficient at 1.2% O_2. Less than 1.2% is good, but higher than 1.2% is bad. The following are possible results of the snap test:
- If O_2 rises past 1.2%, the cat is not working properly. Replace it, then retest.
- If O_2 rises to about 1.2%, the cat is getting a little weak. It would probably pass an I/M inspection under old cut points but not under the new ones.
- If O_2 remains below 1.2%, the cat is operating as designed. A cat in excellent condition is better than 90% efficient (0.6%, for instance). A marginally good cat will read 1.1% O_2.

Cat Testing Using CO_2 Conversion. When the results of an O_2 storage test are marginal, a second testing method, called the CO_2 conversion method, can be used. The test must be done when there is no fuel control, and it cannot be done on carbureted cars. A large, specified amount of propane is introduced in the intake manifold through a large vacuum port (**Figure 44.29a**) and pumped to a preheated cat. Converter efficiency is measured by looking at the amount of CO_2 produced. First, heat the cat at 2,500 rpm for 3 minutes with a monolithic cat and 5 minutes with a bead style. If you do not know the type of cat, use 5 minutes as a guide.

After heating up the cat, shut the engine off, then do the following:
- Quickly disable the ignition and fuel systems.

NOTE: *This must happen quickly!*

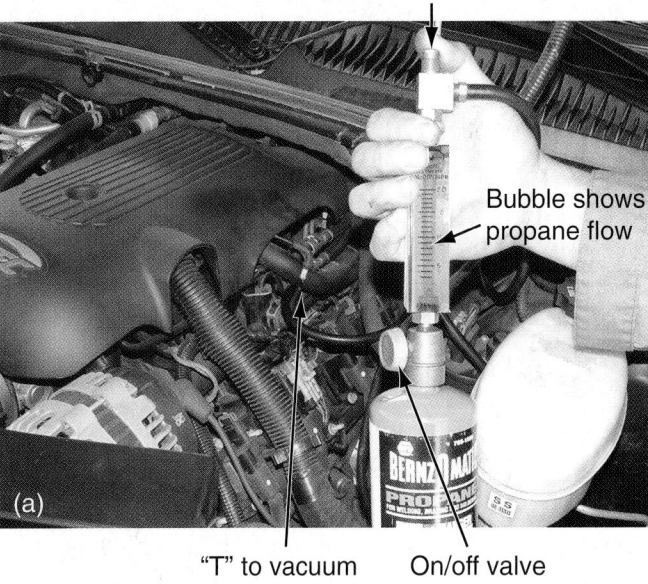

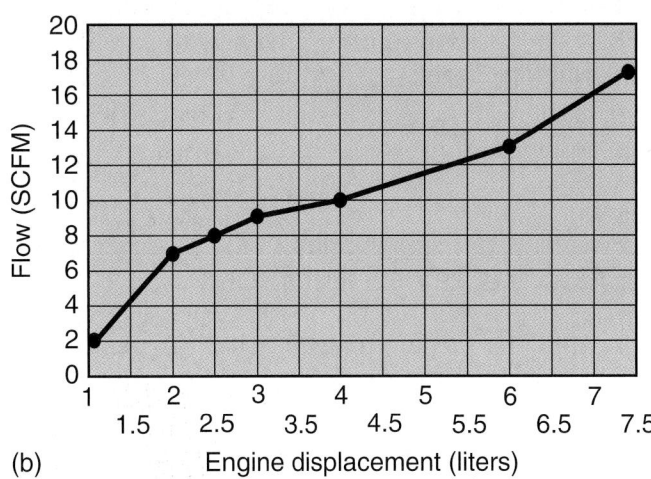

Figure 44.29 (a) A propane enrichment tool bleeds propane into the intake manifold while the engine is running. (b) Adjust the flow rate of the propane enrichment tool based on engine size.

- Crank the engine for 15 seconds at WOT to clear the cat.
- Adjust the flow rate of the propane enrichment tool based on engine size (**Figure 44.29b**). Use less propane rather than more. Using too much propane will result in too much CO_2 and the test will have to be done again. Different rates are used for single or dual cats.

Close the throttle, then do as follows:

- Crank the engine over while propane is flowing.
- After 15 seconds, shut the propane off.
- Stop cranking and watch the CO_2. When it peaks, write down the HC reading.
- Refer to the service information chart that shows the pass/fail line at 80% efficiency (**Figure 44.30**).

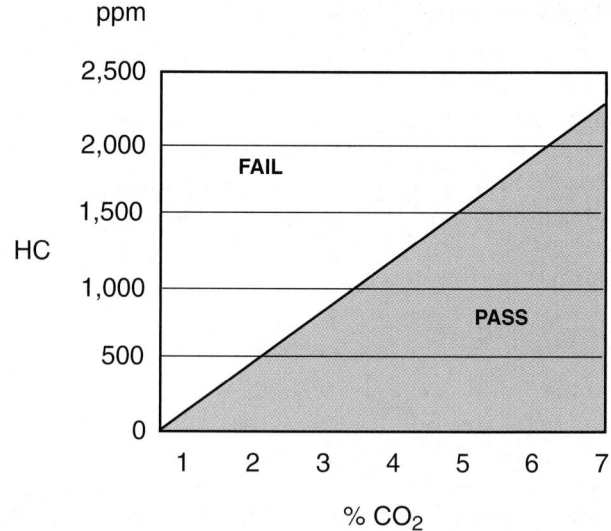

Figure 44.30 A pass/fail line for a catalytic converter.

REVIEW QUESTIONS

1. What will happen to engine idle when idle speed management is disabled and the PCV valve is plugged off?

2. When a knock sensor detects detonation, what does it do to ignition timing?

3. What can an EGR passageway become plugged with?

4. The catalytic converter must be heated to at least _____ °F before it starts to work.

5. What does oxidizing mean?

6. Hydrocarbons are measured in _____ per million.

7. CO_2 is a good indicator of the _____ of the engine.

8. A typical good running engine with a catalyst will have from _____ to 16% CO_2.

9. Is CO read as a percentage or as parts per million?

10. Which of the five exhaust gas readings becomes higher when the exhaust is diluted with air?

11. Which smog device must be disabled during emission analyzer tests?

12. What is the lowest amount of CO_2 a running engine can have without any outside air entering the exhaust system?

13. If an engine that will not start is being cranked with an emission analyzer connected, the presence of what gas tells you that there is a fuel supply?

14. When an O_2 reading is from 2% to 5%, what is suspected?

15. During a cranking CO_2 test, a reading below _____% indicates that the converter is weak or defective.

ASE-STYLE REVIEW QUESTIONS

1. Applying vacuum to an EGR valve while the engine idles can cause:

   a. The engine to stumble
   b. Idle speed to rise
   c. Both A and B
   d. Neither A nor B

2. Technician A says that a valve cover gasket leak can hamper crankcase ventilation system operation. Technician B says that on an idling engine without electronic idle control the rpm should remain constant when the PCV valve is plugged off. Who is right?

   a. Technician A   c. Both A and B
   b. Technician B   d. Neither A nor B

3. Which of the following can be caused by a faulty EGR valve?

   a. Spark knock   **c.** Hard starting

   b. Rough idle   **d.** All of the above

4. Technician A says that a charcoal canister purge valve should open when the engine is cold. Technician B says that a purge valve should open at idle. Who is right?

   a. Technician A   **c.** Both A and B

   b. Technician B   **d.** Neither A nor B

5. Technician A says that a catalytic converter can melt if a spark plug wire falls off. Technician B says that a catalytic converter can be tested for backpressure by removing the O_2 sensor behind the catalytic converter and installing a pressure gauge. Who is right?

   a. Technician A   **c.** Both A and B

   b. Technician B   **d.** Neither A nor B

6. Technician A says that an engine that has more than 14.5% CO_2 probably has a functioning converter. Technician B says that when there is a misfire, there is no CO_2 in the exhaust. Who is right?

   a. Technician A   **c.** Both A and B

   b. Technician B   **d.** Neither A nor B

7. Which of the following indicate(s) rich and lean fuel conditions?

   a. CO_2 **c.** CO

   b. HC   **d.** All of the above

8. Technician A says that high combustion temperatures and load cause high NO_X. Technician B says that anything that increases combustion temperatures can be a possible cause of HC levels that are too high. Who is right?

   a. Technician A   **c.** Both A and B

   b. Technician B   **d.** Neither A nor B

9. Which of the following is/are true about NO_X emissions?

   a. Retarded timing raises NO_X.

   b. Advanced timing raises NO_X.

   c. Both A and B.

   d. Neither A nor B.

10. Technician A says that a good catalytic converter will use up most of its oxygen, leaving very little coming out the exhaust. Technician B says that a catalytic converter's efficiency declines with use. Who is right?

   a. Technician A   **c.** Both A and B

   b. Technician B   **d.** Neither A nor B

CHAPTER 45

Electronics and Computer Systems Fundamentals

■ **OBJECTIVES**

Upon completion of this chapter, you should be able to:

✔ Describe the operation of various semiconductors.

✔ Understand how computers operate.

✔ Explain the operation of various types of sensors and actuators.

✔ Compare the different types of computer memory.

✔ Summarize the various guidelines of on-board diagnostics.

■ **KEY TERMS**

forward bias	ROM	OBD
reverse bias	baud rate	DTC
microprocessor	thermistor	intermittent fault code
bit	piezoelectric crystals	wide area network (WAN)
byte	keep alive memory	local area network (LAN)
RAM	adaptive fuel trim	controller area network (CAN)

■ **INTRODUCTION**

Electronics is the science of using very small amounts of electricity to control larger amounts of electricity. All of the laws of basic electricity still apply. Those basic laws are covered in Chapter 25. The emphasis of this chapter is on the operation of semiconductors, computers, sensors, and actuators.

■ **SEMICONDUCTORS**

Semiconductors, used to make diodes and transistors, are materials capable of being either a conductor or an insulator. Silicon and germanium are common semiconductor materials. These materials have a crystalline structure. This means they share outer electrons with each other (**Figure 45.1**), so they do not gain and lose electrons like conductors do.

Valence rings are discussed in Chapter 25. Refer to that chapter to refresh your knowledge. When a material has fewer than four electrons in its outer

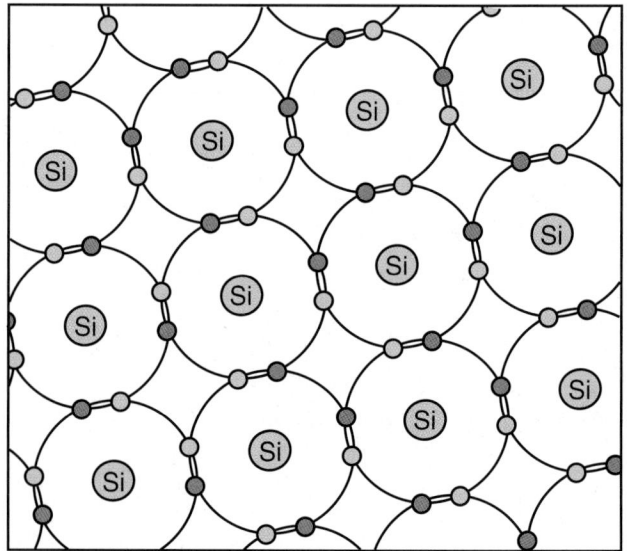

Figure 45.1 Semiconductor materials share outer electrons with each other.

valence ring, it is a conductor. If it has more than four, it is an insulator. Crystals are made of materials that have four electrons in their outer valence rings. A pure crystal has its atoms linked in a way that leaves no *holes* that would allow electron movement. A pure crystal is therefore a good insulator.

The insulating property of a crystal can be changed by *doping* the crystal to make a semiconductor. This process adds a very small amount of *impurity* (one atom in 10 million). Yet, enough holes are added so that the material will conduct electricity when a voltage is applied to its base. This acts on the semiconductor material to change it from an insulator to a conductor. Current in the circuit can then pass through it.

Doping a semiconductor creates one of two types of semiconductor materials, depending on the type of impurity added. The negative type is called an *N-type*. These crystals are doped with an atom like phosphorous, with extra electrons. N-type semiconductors have five or more electrons in their outer valence ring. The extra electron is free, which gives the material a negative charge (**Figure 45.2**).

A *P-type* semiconductor is positively charged and can carry electrical current. An impurity with a three-electron outer ring, such as aluminum or boron, is added to the crystal. Wherever this element fits into the crystal, there is a hole for a fourth electron to fit into (**Figure 45.3**). The hole is a positively charged space. It carries the current in a P-type semiconductor when a voltage is applied. Because the holes are positively charged, they attract electrons. The electrons cannot become free of their atoms, but the crystals can rearrange their patterns to fill a nearby hole. This leaves a hole where the electron came from. That hole is filled by another atom's electron and so forth, resulting in electron flow (**Figure 45.4**). Hole movement only occurs in the semiconductor, while electron movement occurs in the entire circuit.

Electron Flow

Conventional electrical theory is described as electron flow from positive to negative. In the field of electronics,

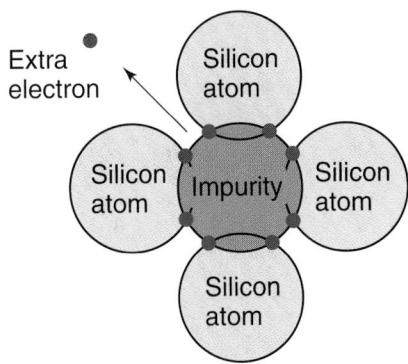

Figure 45.2 N-type semiconductors have extra electrons in their outer ring, which gives the material a negative charge.

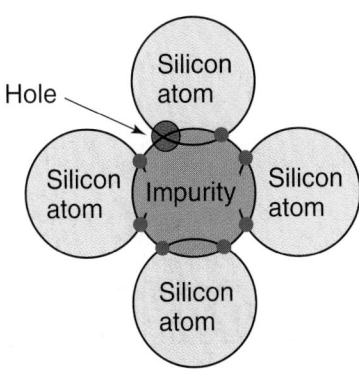

Figure 45.3 A P-type semiconductor, which can carry electrical current, is doped with a material that leaves a hole for a fourth electron to fit into.

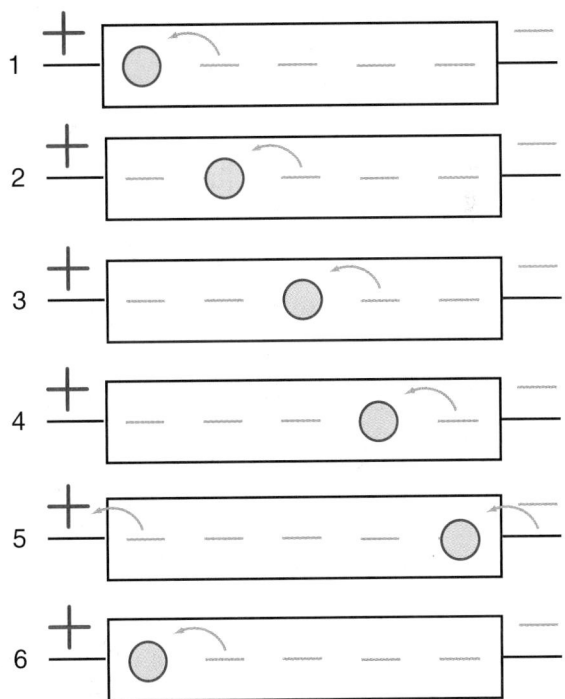

Figure 45.4 Hole movement in a semiconductor.

electron theory describes electrons as flowing from (−) to (+). Compare this to automobiles moving in traffic. As cars move forward after a stoplight to fill holes left from previous cars, did the cars move forward or did the holes move backward? Holes flow from (+) to (−); electrons flow (−) to (+). In a diode, holes flow in the direction of the arrow, while electrons go against the arrow.

Semiconductors are designed to handle a limited amount of current. If too much current is applied in a reverse direction, it can ruin the diode or transistor as it is forced through it.

◼ DIODES

A diode (**Figure 45.5**) is a one-way electrical check valve. It is made by placing P-type and N-type crystals back to back. This *P-N junction* will only allow electrons

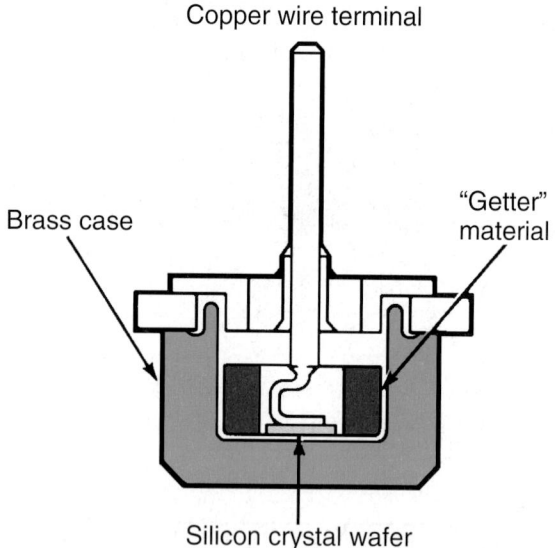

Copper wire terminal

Brass case

"Getter" material

Silicon crystal wafer

Figure 45.5 Parts of a diode. *(Courtesy of Ford Motor Company)*

to pass when a voltage greater than 0.5 to 0.7 volt (in silicon) pushes holes in the positive material toward the extra electrons in the negative material.

When the polarity is such that the P-N junction conducts current, this is called **forward bias** (**Figure 45.6**). Forward bias diodes allow current flow. When current tries to flow in the opposite direction, it cannot. Reverse bias diodes do not allow current to flow. The P and N sides of the diode are connected to opposite charges, which attract (**Figure 45.7**). The P material is drawn toward the negative part of the circuit, and the N material is drawn toward the positive part. With the P-N junction empty, current flow stops.

The electrical symbol for a diode is an arrow with a bar at its point (**Figure 45.8**). The point indicates the direction of conventional current flow. Electron flow is in the opposite direction. The wide end of the arrow indicates the P side. It is often called an *anode* and is like the positive terminal on a battery. The bar end of

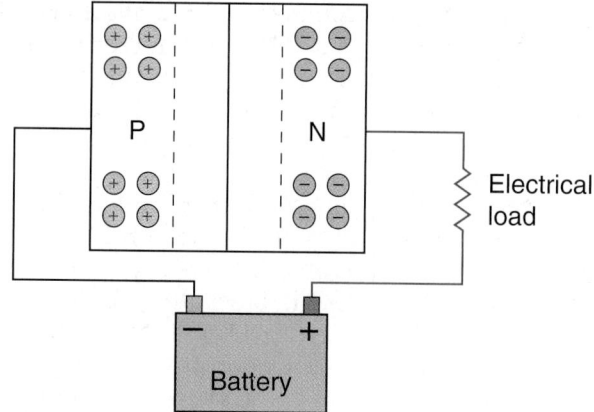

Figure 45.7 A reverse bias diode does not allow current flow.

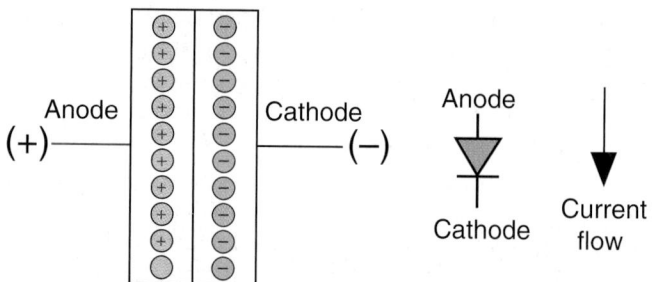

Figure 45.8 An electrical symbol for a diode. The point indicates the direction of conventional current flow.

the anode is the N side. Often called a *cathode*, it is like the negative terminal on a battery.

Clamping Diode. When current flow is halted in a coil winding, the magnetic field around the winding collapses. This results in a voltage surge, which can be seen on a scope as a voltage spike. Voltage spikes can damage electronic components. Older systems used capacitors to control these shocks to the electronic system. Today's vehicles most often use a clamping diode for this purpose. It is installed parallel to the coil (**Figure 45.9**), acting like a shock absorber by

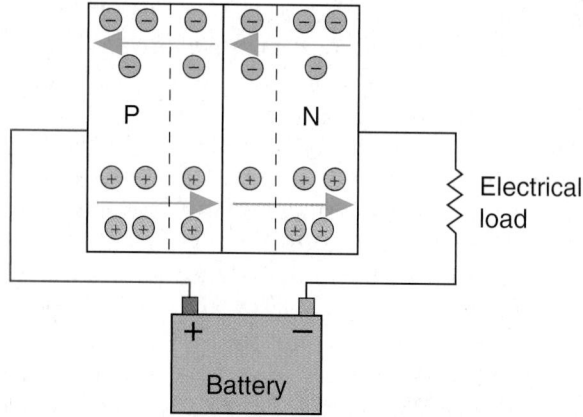

Figure 45.6 A voltage pushes holes in the positive material toward the extra electrons in the negative material.

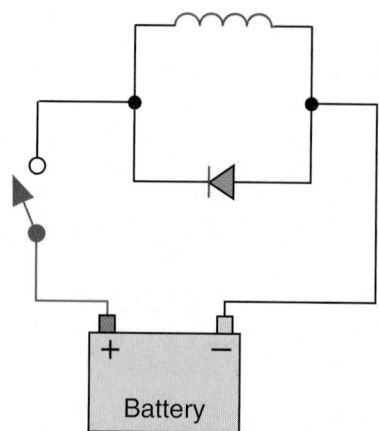

Figure 45.9 A clamping diode is installed parallel to the coil.

providing an alternate electrical path for electron flow when the circuit opens. An air-conditioning clutch is one example of a magnetic coil where a clamping diode is used for circuit protection.

▬ TRANSISTORS

Transistors turn electrical circuits on and off. They are controlled by another electrical circuit. A transistor is an electronically controlled relay or switch. A small amount of current applied to the transistor causes it to relay larger amounts of electricity through it. The difference between a diode and a transistor is that the transistor has two P-N junctions, while the diode has one. It is like a diode with an extra side.

Transistor Operation. There are three semiconductor crystal layers in a transistor, called the *emitter,* the *base,* and the *collector* (**Figure 45.10**). The base is always in between the emitter and the collector. The transistors used in automobiles are called *bipolar transistors*. They have two polarities, *electrons* and *holes*. A transistor is either an *N-P-N* or a *P-N-P* type. Each layer of the transistor has a connection for an electrical lead. The emitter is the input lead; the output lead is the collector. The base layer, in the center, controls the switching function of the transistor.

Electrical current cannot move across the layers of the transistor unless the base (center layer) has a voltage applied to it. Voltage causes the base layer to reverse its charge and become a conductor. Extra electrons or holes are added to the base by the controlling current. The transistor can be controlled by whatever amount of voltage is desired by putting a resistor on its base (input). This controls at which voltage level it opens or closes. It also keeps current flow low through that part of the transistor.

A transistor is *forward biased* when it is allowing current to flow. When the voltage is removed from the base, the circuit shuts off. This is called **reverse bias**.

The schematic for a transistor has the arrow on the emitter pointing in the direction of current flow. The N-P-N transistor is normally off. When the base is forward biased with a more positive voltage, the emitter-to-collector circuit is turned on (**Figure 45.11**). The amount of output current is proportional to the amount of current through the base leg. This type of transistor

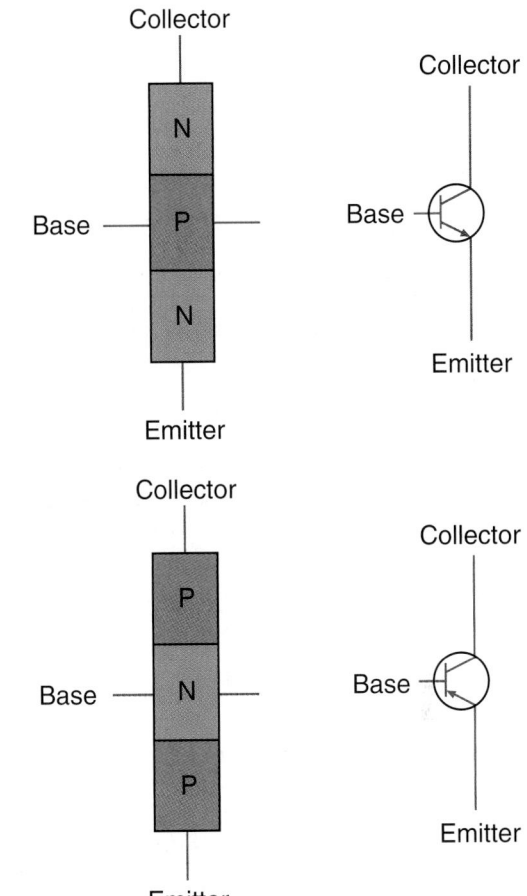

Figure 45.10 P-N-P and N-P-N transmitters.

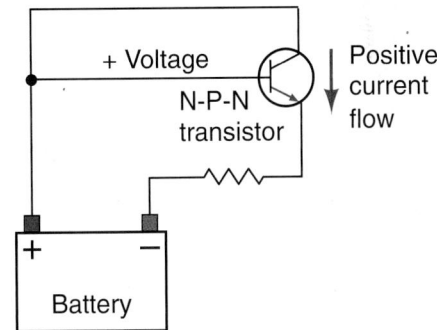

Figure 45.11 When the base is forward biased with a more positive voltage, the emitter-to-collector circuit is turned on.

can also be used as a variable resistor by varying the current applied to the base. More current at the base results in more current passing to the collector. This kind of variable resistance can be found in a power transistor for an automatic climate system blower motor.

A P-N-P transistor is controlled by its ground. A more negative voltage than is in the emitter must be applied to the base of a P-N-P transistor to turn it on. This results in current flow between the emitter and collector. A small amount of current flows from the source of the control voltage through the base and is grounded through the emitter when the base has voltage applied to it.

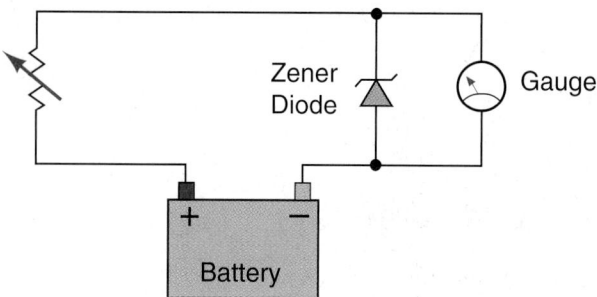

Figure 45.12 A zener diode and its electrical wiring diagram symbol.

A transistor actually never shuts off. Current either flows in from the base or from the collector. It never flows in from the emitter, so it is a "dual diode." When a contact point closes, it takes $\frac{1}{100}$ of a second to accelerate from 0 to 4 amps. A transistor can regulate at 10,000 times a minute because current always continues to flow.

Zener Diodes

A *zener diode* (**Figure 45.12**) is one whose crystals are more heavily doped during manufacturing. Zeners allow current flow at a certain voltage but halt current flow below that. They are used to control transistors and electronic voltage regulators. When a transistor is switched off and on several thousand times a minute, a zener is needed to control the backwash or double bounce of the voltage.

▄▄ ELECTRONIC INSTRUMENT DISPLAYS

There are three kinds of electronic digital displays: light-emitting diodes, liquid crystal displays, and vacuum fluorescent displays. They are used in various forms in instrument panels (see Chapter 32).

Digital and analog displays require an input signal from a sending unit. Sending units for analog and digital displays might be the same type but might not be interchangeable.

Light-Emitting Diodes

Light-emitting diodes (LEDs) (**Figure 45.13**) are often used in lighting applications such as automotive digital displays and test instruments. They are also used in some center, high-mounted brake lights and as a trigger in some ignition and fuel systems. They have a crystal that operates like a light bulb and glows when current flows through it. The most common color is red.

Figure 45.13 An LED and its electrical wiring diagram symbol.

Warning lights on the instrument panel are often LEDs because they have no filament and are less likely to fail than conventional bulbs. They can be a single indicator light or have a seven- or eleven-segment display. They can be arranged to form letters, numbers, or bar graphs. LEDs are more difficult to see in bright light. They use more power than the other types of displays but less power than ordinary light bulbs.

Organic light-emitting diodes (OLEDs) are used in computer displays and lighting systems. They are inexpensive to manufacture and use very little power. They put out a bright light and do not need back lighting like conventional LEDs.

Liquid Crystal Displays

Liquid crystal displays (LCDs) are found in calculators and watches as well as in dash gauges. The LCD does not actually produce light. It is a sandwich of liquid and special glass that has a conductive coating. Behind the LCD is a separate incandescent or halogen light source. When voltage is applied to the glass layers, light can pass through the LCD. Without voltage, light cannot pass.

LCDs produce black-and-white images. Color filters in front of the display can produce colored images. Drawbacks to LCDs include that they operate more slowly in cold weather, and they are fragile.

Vacuum Fluorescent Displays

Vacuum fluorescent displays are miniature fluorescent lights—glass tubes filled with very low-pressure argon or neon gas. Passing current through the tube causes the tube to glow brightly. They can withstand tougher handling and put out a very bright light. Neon lights have been used for some instrument panel displays since the mid-1980s because of their bright light output but are being replaced by OLEDs, which use less power and cost less to manufacture.

▄▄ AUTOMOTIVE COMPUTER SYSTEMS

Integrated Circuits

An integrated circuit (IC) is a complete miniaturized electric circuit. It consumes considerably less electrical current than a large-scale circuit. Transistors, diodes, and resistors are included in the *chip*. The chip consists of tiny sandwiched silicon wafers of P-type or N-type material. One transistor has limited ability when it comes to performing complicated tasks. When many semiconductors are used in a circuit, the functions they can perform are amazing. As many as 30,000 transistors can be placed on a chip that is only $\frac{1}{4}$" square (**Figure 45.14**). The circuit is constructed by photographically reproducing circuit patterns onto a silicon wafer. An electrical circuit that would normally fill a large room can be put in this small area. A chip consists of several layers with underlying electrical connections.

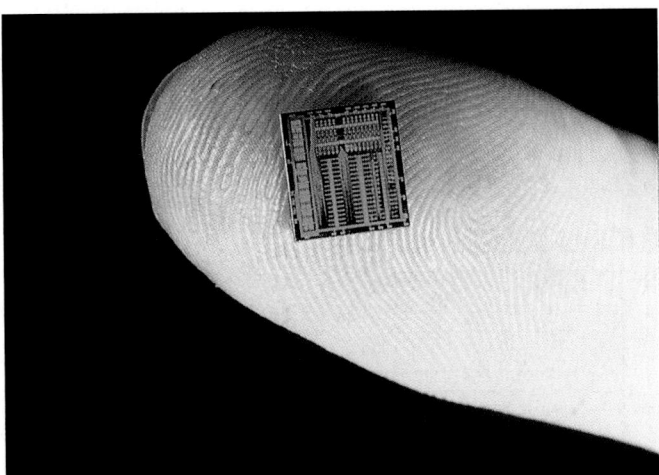

Figure 45.14 The actual size of this integrated circuit (chip) is less than ¼" square. *(Courtesy of Texas Instruments)*

Figure 45.16 Several methods of chip packaging. *(Reprinted with permission from Bosch)*

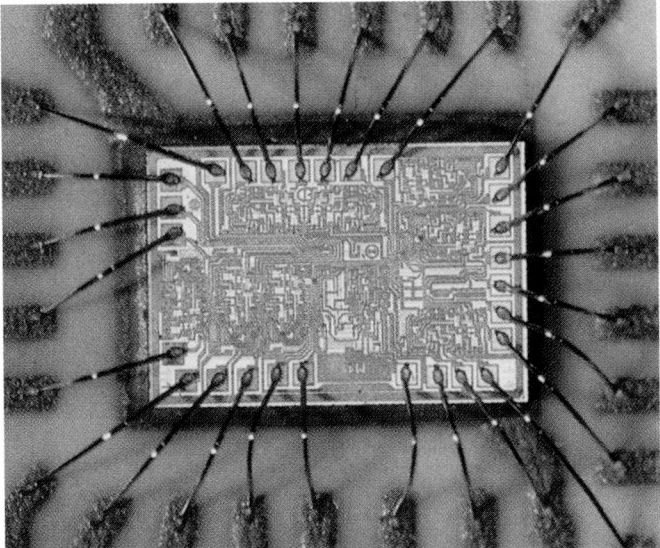

Figure 45.15 This chip from a fuel injection module is enlarged 12 times here. *(Reprinted with permission from Bosch)*

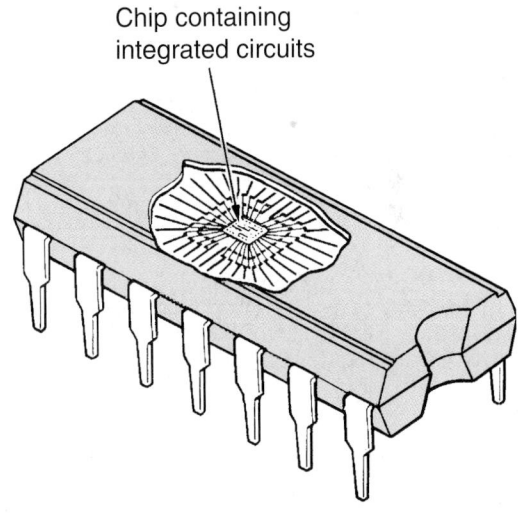

Chip containing integrated circuits

Figure 45.17 Some chips are replaceable and have pins that come out of the bottom that plug into sockets in the circuit board.

External electrical wiring is connected to the chip (**Figure 45.15**). Chips can be mounted on a flat ceramic or metal plate, or they can be surrounded by plastic. Several methods of chip packaging are illustrated in **Figure 45.16**. Some chips are replaceable and have pins that come out of the bottom that plug into sockets in the circuit board (**Figure 45.17**).

ICs are limited in the amount of electrical current they can carry. Too much electrical current causes heat that can damage the circuit.

PARTS OF A COMPUTER SYSTEM

There are three main parts to automotive computer systems: the *computer*, sensors, and actuators. The many sensors and actuators control engine and other vehicle

functions (**Figure 45.18**). Sensors relay information to the computer on air or coolant temperature, airflow, manifold pressure, and barometric pressure. The computer processes the information and sends command signals to actuators.

ON-BOARD COMPUTER

An automotive on-board computer can be called a control assembly, a control module, or a control unit. Electronic part names are set by the Society of

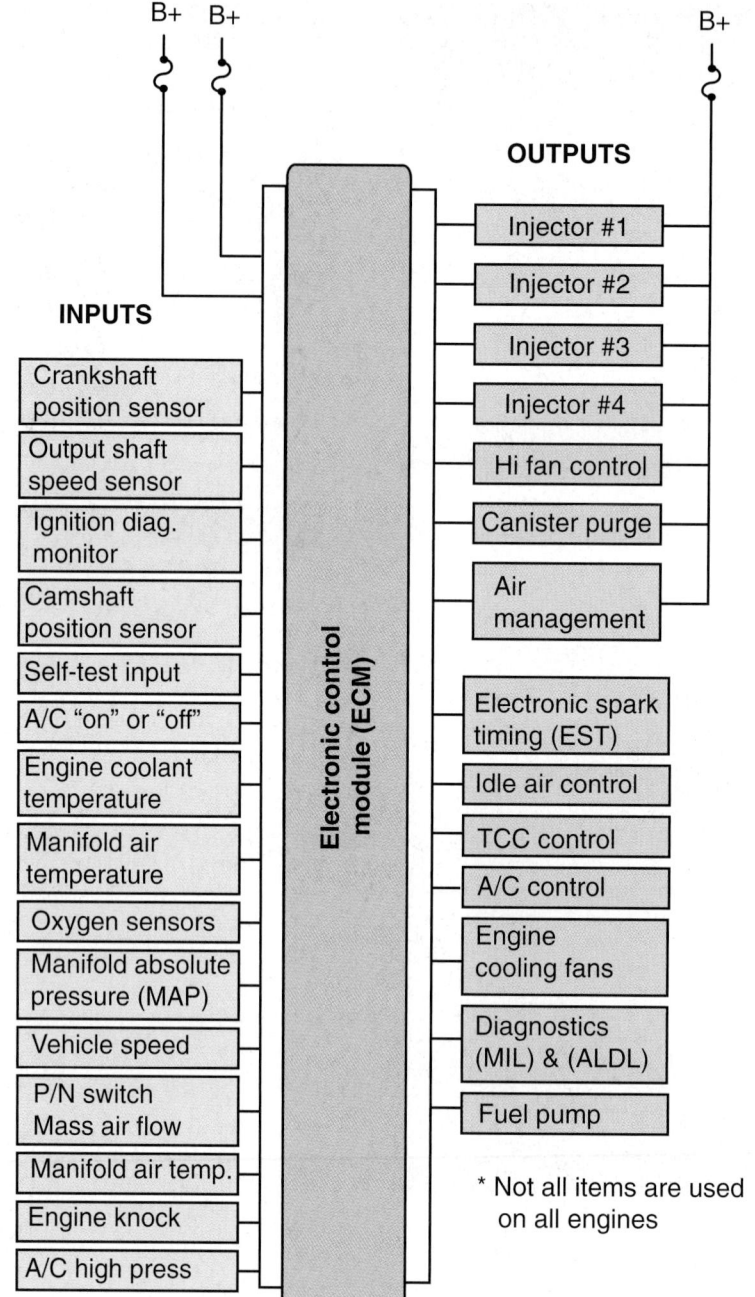

Figure 45.18 The many sensors send signals to the computer, which uses actuators to control engine and other vehicle functions.

Automotive Engineers on-board diagnostics (OBD) guidelines (covered in Chapter 47). According to OBD II guidelines, all new model computers that control things other than engine functions are called *powertrain control modules* (PCMs) (**Figure 45.19**). The PCM has been known in the past as the electronic control unit (ECU), the electronic control module (ECM), the electronic control assembly (ECA), or simply "the controller." A typical system has one central computer and several modules. The modules deal with powertrain, ABS brakes, comfort systems, and body control.

A computer has four functions:

■ To gather input
■ To make decisions and processes information
■ To store information
■ To take action by way of an output command

The **microprocessor** is the calculating and decision-making chip in the computer. It does not actually think, but follows instructions programmed into its memory. There are thousands of miniature diodes and transistors in the microprocessor. Voltage surges in the system are a problem, so diodes are installed throughout the system.

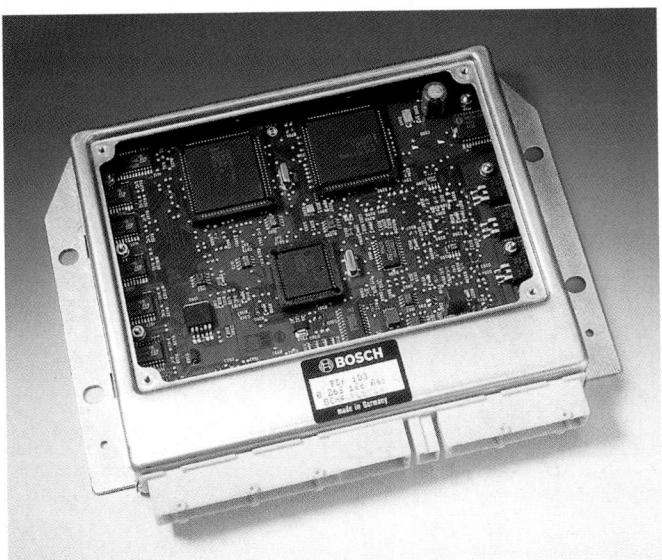

Figure 45.19 A powertrain control module (PCM). *(Courtesy of DaimlerChrysler Corporation)*

COMPUTER NOTE

Transistors act like electronic switches and are either on or off. The computer interprets combinations of zeros and ones to form numbers or words. This binary (zeros and ones) information is processed to determine the meanings of the signals. Then the processed information is stored or delivered from the computer.

*When the computer receives a voltage of at least a certain value, it converts it to a one. When there is no voltage, the computer interprets a zero. Each zero and one represents a **bit** of information. Eight bits makes a **byte**, which is sometimes called a word. Electronic information is exchanged in bytes.*

Computers do the same types of things mechanical systems used to do. But they can do them much faster and more accurately. Newer cars make decisions in 600 thousandths of a second.

Items that make up a computer system include *hardware* and *software*. Hardware consists of the visible mechanical parts. Software, the magic stuff inside, is information stored as electronic signals that can be modified. This is why it is called "soft." Software includes the *programs* for the computer.

Computer Electrical Control

Some computers carry a very light electrical load. They control grounds to output devices rather than power because they cannot turn heavy amp loads off and on. This is called *ground-side switching*. Typical amp loads like turning on the radiator fan relay, vacuum

solenoids, the alternator field, and four or six injectors are loads that are too big for this electronic circuit. Also, if a short circuit were to occur in a powered circuit, the computer would act like a fuse and would be blown. With *power-side switching*, a logic module tells a power module what to do.

NOTE: *Computers in most North American vehicles use ground-side switching with current-limiting devices. This design is more likely to result in galvanic corrosion, which occurs when power goes to ground and moisture is present. Most European and some Japanese manufacturers use power-side switching. Power-side switching minimizes galvanic corrosion because electrons move in their normal direction. The downside is that computers are thought to fail more often because they handle electrical current directly, rather than simply switching electrical devices to ground.*

Automotive computer systems since the mid-eighties have used 5 volts to operate. This voltage is low enough to prevent damage to the circuits in the chip, yet high enough to provide consistent transmission of information. Reference voltage must be less than minimum battery voltage or signals will be inaccurate. Because battery voltage never dips this low, the computer will operate more consistently.

■ INFORMATION PROCESSING

Computer input signals from sensors must be conditioned or processed. The computer's logic circuits either turn these signals into output commands or store them for a short or long time in the computer's memory (**Figure 45.20**). The memory contains values programmed into the computer software that the computer checks incoming information against.

Sensor information can be digital (on/off), but is often *analog* when the sensed information varies gradually. Remember the earlier discussion on binary information. Automotive processors used since the eighties are digital. An analog signal is a sine wave, and a digital processor must see digital input, which means it is something like yes or no, or on/off (**Figure 45.21**).

An analog temperature sensor might have an output voltage range from 0 volt to 5 volts. If the sensor's temperature range is from 0°F to 250°F, then a voltage somewhere between 0 volt to 5 volts will be interpreted by the computer as a specific temperature. Each 1-volt change will be equal to a 50°F change in temperature.

■ COMPUTER MEMORY

A computer's memory contains programs that it refers to as it processes information. The microprocessor can read information from memory or it can write information to it. Memory files are located in many places. Each piece of information has an address. When stored information is needed, the microprocessor looks it up at its address in memory.

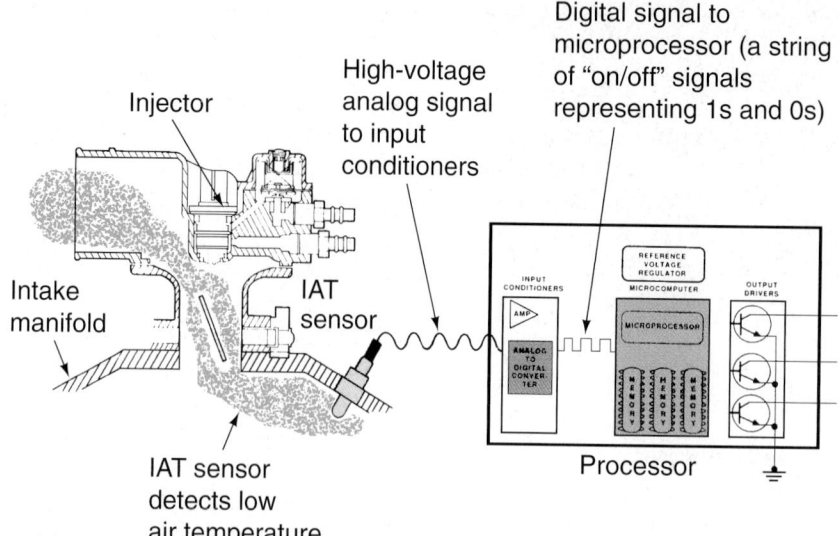

Figure 45.20 The signals from sensors must be conditioned or processed. *(Courtesy of Ford Motor Company)*

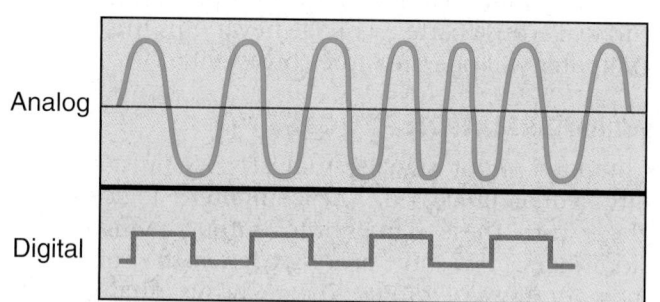

Figure 45.21 Analog signals can be constantly variable. Digital signals are either on/off or low/high.

Random Access Memory

Information that will be stored temporarily is sent from the microprocessor to random access memory (**RAM**). *Random access* means that the information can be retrieved in any order. RAM is like a notepad that you can read from and write to. Input from the sensors changes frequently, so it is stored in RAM (**Figure 45.22**).

Volatile and Non-volatile RAM. When RAM is *volatile*, it is erased each time the ignition is turned off. RAM can also be *non-volatile*. This means that the information is not erased when the ignition switch is turned off. Non-volatile RAM is like the station settings on an electronically tuned radio. When the ignition is off, the radio remembers the station settings, yet the radio

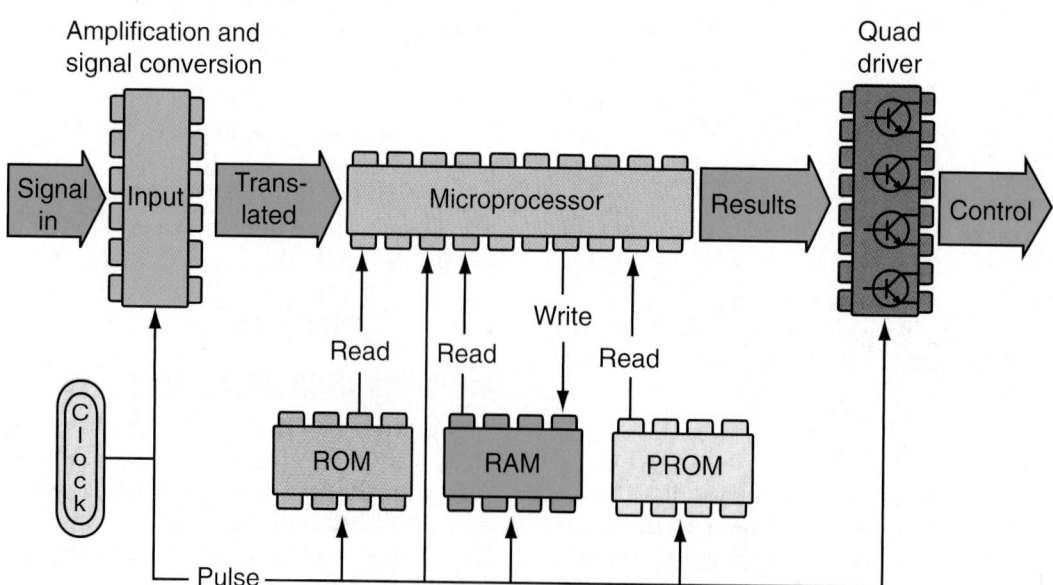

Figure 45.22 A schematic of a typical automotive computer with three kinds of memory.

turns off with the key. If the battery is disconnected, however, the radio will forget the station settings.

Read-Only Memory

Read-only memory (ROM) is permanently programmed information available to the microprocessor (see Figure 45.22). This information is programmed into the chip during manufacturing, so disconnecting the battery does not change it. The computer can read from ROM, but it cannot write to it. ROM contains *lookup tables,* which is program information on how the car is supposed to perform. One system might have 560 programmed instructions for different conditions. These *parameters* are what the engineers feel is the best adjustment for a particular operating condition. The computer compares sensor information to these parameters and makes adjustments as needed.

Programmable Read-Only Memory (PROM)

ROM also contains the *calibration tables.* These are instructions for specific engines and drivetrains. A car with a manual transaxle will have different program requirements than one with an automatic transaxle. Calibration table chips can be either PROMs or EEP-ROMs. A PROM is *programmable read-only memory.* PROMs provide a specific vehicle with a program for its fuel and emission systems, including spark advance. PROMs contain instructions for the EGR valve, vapor canister purge, torque converter clutch, air conditioning, radiator electric fan, and others. There might be 60 specific parameters for that particular car.

On many older GM cars, the computers were the same but the PROM was replaceable. Other manufacturers write the information into ROM and then replace the computer if there is a change; newer vehicles use an EEPROM. An EEPROM is *electronically erasable read-only memory.* When there is an update in the engineering of the computer program, these chips can be reprogrammed in the car, in the service bay, or by telephone modem from the manufacturer's facility. When the information in an EEPROM is erased and reprogrammed, this is called *flashing the PROM.*

OBD II vehicles are required to have soldered PROMS. They cannot be removed. Another kind of memory, keep alive memory, is covered later in this chapter. It is reprogrammed electronically.

■ COMMUNICATION RATE

A quartz crystal uses timed pulses to maintain an orderly flow of information. During the time between each pulse, one bit of binary information is transmitted within the computer. The speed at which this occurs is called **baud rate**. A computer with a baud rate of 56,000 (56K) can transmit 56,000 bits of information

per second. Baud rate is getting faster as technology advances. An early 1980s GM computer had a baud rate of 160. Today's computers have high-speed networks that run at speeds of 500 Kbps (500,000 bits per second) and are even capable of twice that speed.

■ SENSORS AND ACTUATORS

During the input phase, the computer receives signals from sensors. The computer sends a voltage signal called *reference voltage* to many of the sensors. Most computers use 5 volts as a reference signal. Sensors monitor engine functions and modify the voltage signals that return to the computer.

An *actuator* is an electronic or magnetic relay that can perform a desired function. The computer processes information and controls the current flow to these devices (**Figure 45.23**). They act upon the commands when their electrical circuits are opened and closed. This results in adjustments that bring operating conditions back to within the programmed parameters.

Sensors and actuators are all transducers. A *transducer* converts energy from one form to another. Sensors convert energy (like temperature, light, or motion) to voltage signals. Actuators change electricity into a mechanical action (work). Transducers can also be mechanical, such as a vacuum unit that reacts to changing air pressure on its diaphragm to move a control rod. Mechanical transducers like solenoids are often operated by computer systems, too.

Sometimes, an output signal can be an input signal for another device. For instance, the computer might turn on the air-conditioner compressor clutch while the engine is idling. The signal that causes the clutch to turn on also serves as an input signal to increase throttle air flow to compensate for the increased load on the engine. More information on actuators is covered later in this chapter.

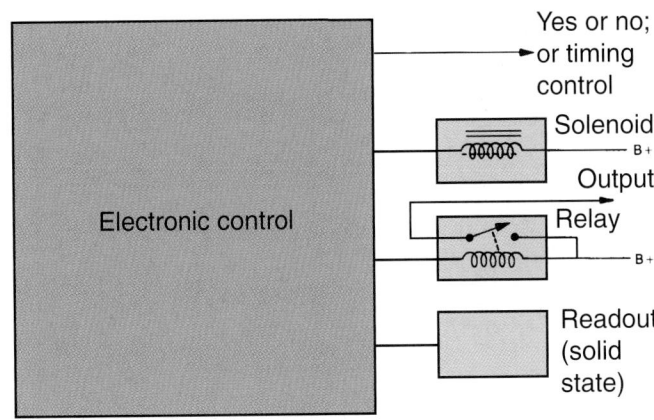

Figure 45.23 The computer processes information and controls the current flow to actuators. (*Courtesy of Ford Motor Company*)

TYPES OF SENSORS

There are five different types of sensors:
- Variable resistors (voltage modifying)
- Variable DC frequency
- Variable voltage generators
- Variable AC voltage/frequency generators
- Switches

Specific operation of various types of sensors is covered in the chapters that deal with their systems. Some of the more popular ones are listed here.

THERMISTORS

A **thermistor** is a variable resistor made from semiconductor material. Its resistance changes predictably as its temperature changes. It is used for measuring air and water temperatures because even a small change in temperature will result in a change in its resistance. Air temperature information is used by the computer to calculate fuel delivery. Thermistors are also used in coolant temperature sensors (**Figure 45.24**). Thermistors are either positive temperature coefficient (PTC) or negative temperature coefficient (NTC). The most common type is the NTC, which provides less resistance as temperature increases.

VOLTAGE DIVIDERS

A *voltage divider* circuit has three wires. Voltage dividers are variable resistors that produce a variable DC voltage signal. The signal varies according to the sensor's mechanical position. A potentiometer is a three-terminal variable resistor. A reference voltage is applied to one of the terminals. The other end of its circuit is ground. A movable center contact or wiper senses voltage between ground and its position on a wire-wound resistor. The ground connection completes the circuit to the wiper (**Figure 45.25**).

The potentiometer measures linear or rotary motion in throttle position, airflow, and EGR valve position sensors. In a throttle position sensor, it is attached to one end of the throttle shaft. Moving the

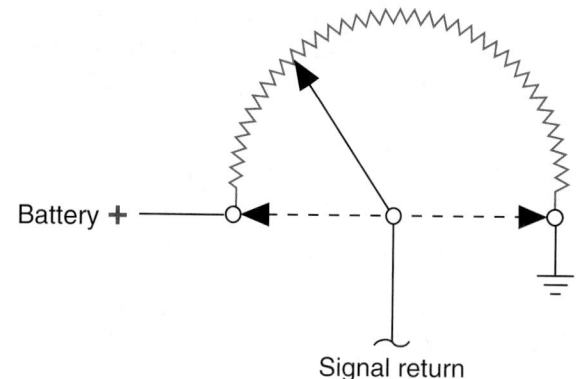

Figure 45.25 In a potentiometer a movable center contact or wiper senses voltage between ground and its position on a wire-wound resistor.

shaft moves the wiper to a different position on the wire-wound resistor coil. A high voltage (4.5 volts) is read as wide-open-throttle, while a lower voltage (0.5 volt) means closed throttle (**Figure 45.26**). Voltages in between give an indication of throttle position. When a potentiometer is used in an airflow meter, it reacts to pivoting of the sensor shaft also.

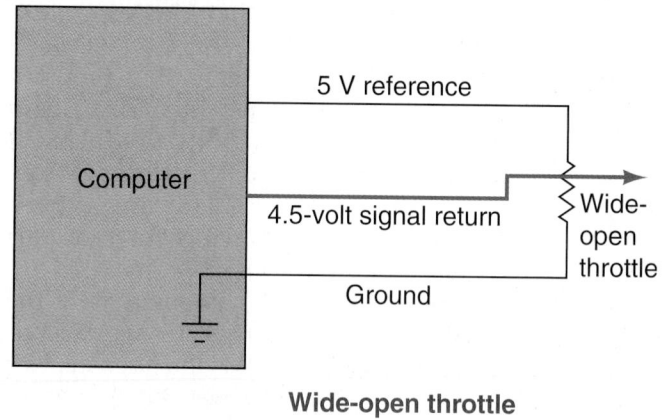

Wide-open throttle

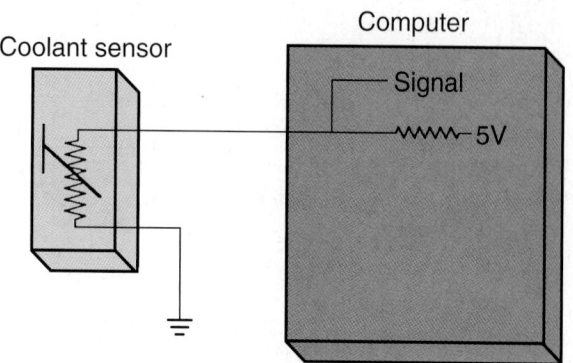

Figure 45.24 Varying current flow through a thermistor to ground in a coolant temperature sensor circuit. (*Courtesy of OTC/SPX Service Solutions*)

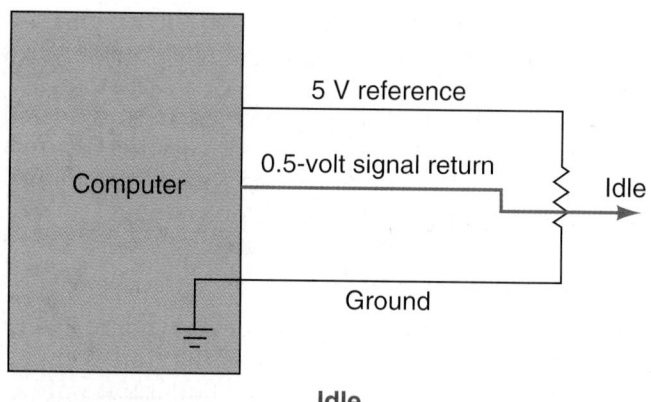

Idle

Figure 45.26 A throttle position sensor will read a high voltage (4.5 volts) as wide-open throttle (WOT). A lower voltage (0.5 volt) means closed throttle.

A rheostat differs from a potentiometer in that it carries current. It also has a movable arm but has only two wires. There is no ground wire, as ground is supplied by the movable arm.

PIEZOELECTRIC AND PIEZORESISTIVE SENSORS

Piezoelectric crystals develop a voltage on their surfaces when pressure is applied to them. In late-model vehicles, these devices are commonly used as switches for measuring pressure in engine oil, power steering, or air-conditioning systems. In air-conditioning and power steering systems, pressure switches give a signal to the computer so that it can control engine idle speed.

A similar crystal is commonly used in MAP sensors for measuring intake manifold pressure. It is called a *piezoresistive* sensor. It is made up of a silicon diaphragm sealed to a quartz plate (**Figure 45.27**). Doping the silicon creates four resistances around the edges of the diaphragm. There is a vacuum chamber between the quartz and the silicon diaphragm. When pressure is applied to the diaphragm, it deflects. This causes a change in the resistance of the resistors. The reference voltage input is changed as it goes through the sensor, resulting in lower output voltage.

HEATED RESISTIVE SENSORS

A heated resistive *sensor* can be used to monitor the amount of air taken into the engine. An electric current is applied to a platinum wire or foil screen. The computer maintains the amount of current necessary for the wire to remain at a constant temperature. The current required to maintain this temperature is interpreted by the computer. From this information, it can calculate the amount of airflow into the engine. This is called a mass airflow sensor (MAF). Refer to Chapter 40 for more on the operation of MAP, BARO, and air density sensors.

VARIABLE DC FREQUENCY SENSORS

Frequency sensors are used for the same things as the previous sensors, but they produce a digital signal. They have three wires like other MAP, BARO, and air density sensors. The signal return line produces a pulsed frequency measured in hertz (cycles per second). Increased airflow causes the sensor to vary how fast it cycles off and on.

VOLTAGE GENERATORS

The sensors covered previously receive a reference voltage that they send a signal response to. Voltage generating sensors have no reference voltage but create their own. There are both AC and DC voltage generators.

In a *variable AC voltage generator,* a magnetic pickup generates an AC analog signal (consisting of a positive and negative sine wave). *Pulse generators* are signal generators like those found in antilock brake wheel sensors, magnetic distributor triggers, or crankshaft or camshaft position sensors.

An oxygen sensor is a *variable DC voltage generator.* It is a galvanic battery that can generate a voltage of from 0.1 volt to 0.9 volt (100 to 900 millivolts) in response to the amount of oxygen in the vehicle's exhaust. A rich mixture (less oxygen) results in a higher voltage (450–900 millivolts). More oxygen results in a lower voltage. The operation of the O_2 sensor is covered in Chapter 40.

A knock sensor has a piezoelectric crystal sensing element that senses vibration and creates a voltage signal of 300 to 500 mV (**Figure 45.28**). When the voltage signal is generated, the computer retards ignition timing until the engine knock goes away.

WHEATSTONE BRIDGES

Variable resistance sensing to measure temperature, pressure, or mechanical strain can be done with a series-parallel circuit in a wheatstone bridge. A wheatstone bridge is two simple series circuits connected in parallel across battery power on its way to ground

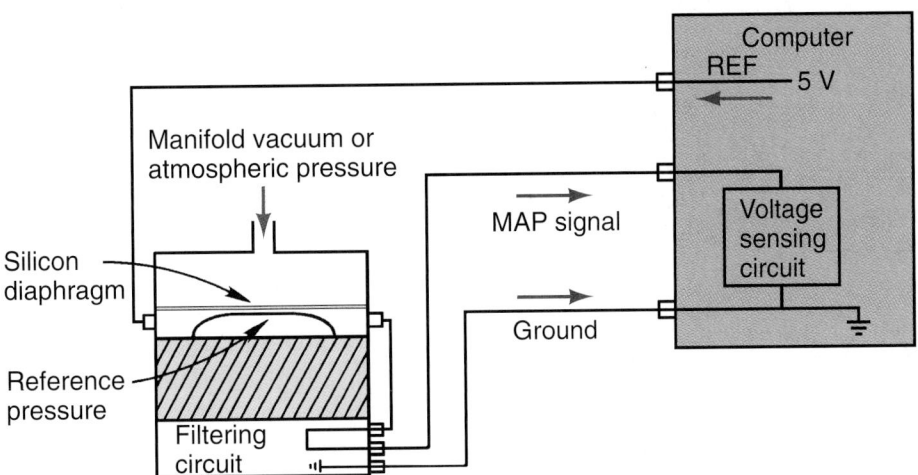

Figure 45.27 A piezoresistive sensor is made up of a silicon diaphragm sealed to a quartz plate.

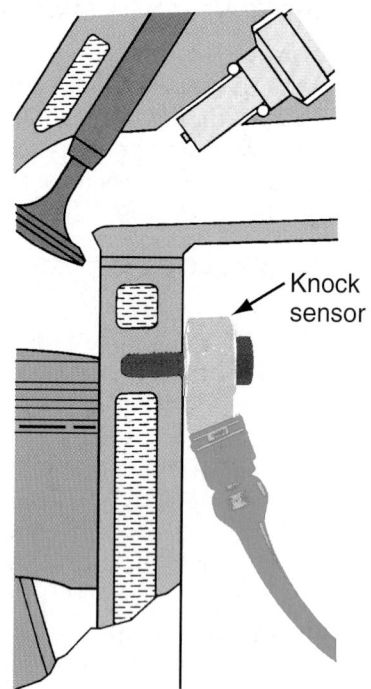

Figure 45.28 A knock sensor has a piezoelectric sensing element that senses vibration and creates a voltage signal of 300 to 500 mV. (*Reprinted with permission from Bosch*)

(**Figure 45.29**). Three of the resistors have the same value. The remaining resistor is a sensing resistor. The bridge is balanced when all four resistors are of the same value, giving 0-volt output from the voltage sensor. When the sensing resistor's resistance changes, the resulting change in circuit balance causes a proportional voltage output to the sensing circuit. A hot wire MAF sensor uses a wheatstone bridge.

▬ SWITCHES

Some of the sensor signals the computer receives are from simple switches. There are three types of switches: switch-to-power, switch-to-ground, and Hall-effect. A switch is used for transmission gear position. A brake on/off switch

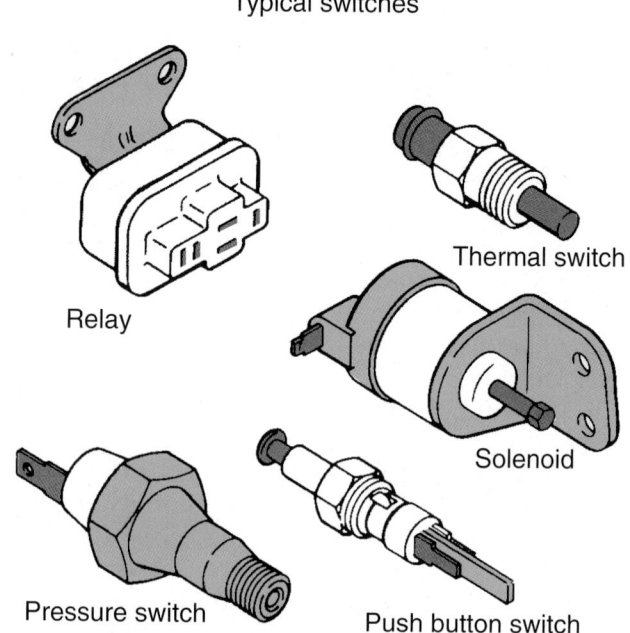

Typical switches

Relay

Thermal switch

Solenoid

Pressure switch

Push button switch

Figure 45.30 Several types of switches. (*Courtesy of Federal-Mogul Corporation*)

gives a signal to the torque converter clutch and cruise control. A power steering switch senses when the wheel is turned against a lock and raises engine idle to compensate for the added load. **Figure 45.30** shows examples of several types of simple switches.

A mechanical pressure switch can sense hydraulic or air pressure. Temperature switches have a bimetal element that opens and closes in response to heat.

On/off switches, Hall-effect switches, and some airflow sensors generate digital signals, so they do not require the computer to convert their output signal from analog.

▬ TYPES OF ACTUATORS

Actuators are the devices that act upon processed signals received from the computer. They can be solenoids,

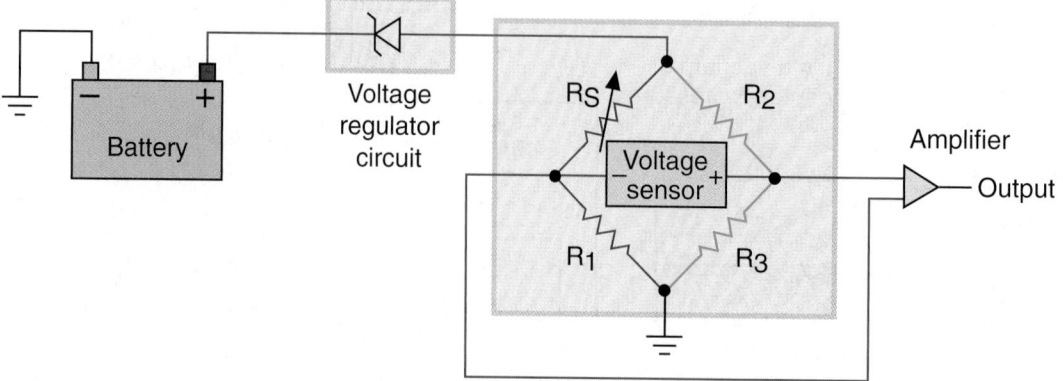

Figure 45.29 A wheatstone bridge is two simple series circuits connected in parallel between battery power and ground.

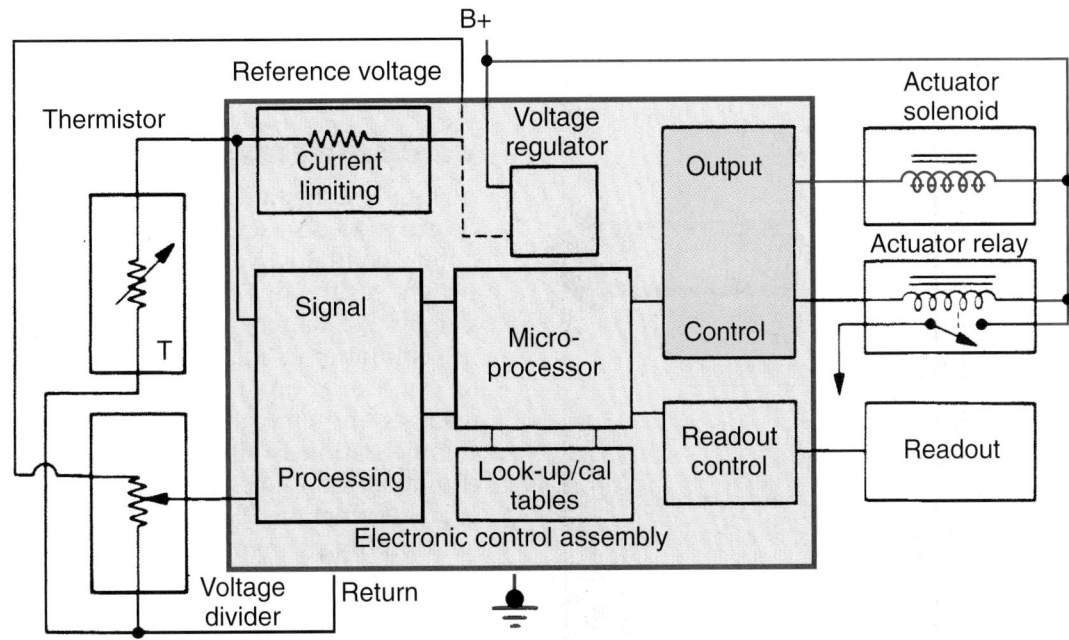

Figure 45.31 Output drivers supply the actuators with a ground to switch them on. This is called ground-side switching. *(Courtesy of Ford Motor Company)*

DC motors, relays, switches, or control modules. With ground-side switching, actuators are always powered. When they are not operating, it is because they do not have a complete circuit to ground. P-N-P transistors in the computer called *output drivers* supply the actuators with a ground to switch them on. **Figure 45.31** shows a schematic of a computer with output drivers. When the output driver does not provide a ground to the actuator, the actuator does not operate.

One single module can have a group of four transistors. This is called a *quad driver*. This space-saving module can control up to four actuators.

■ SOLENOID ACTUATORS

A solenoid is a magnetic switch (see Chapter 28). Solenoids are used for several things:
- An air management system uses solenoids for controlling the flow of air from a smog pump.
- A solenoid opens to allow fumes to be purged from the vapor canister when the engine is warm and above idle speed.
- EGR vacuum control has two solenoids; one is used for vacuum actuation, and the other, for bleeding off vacuum.
- Fuel injectors are solenoids whose pulse width (see Chapter 40) is controlled by the computer.
- An idle speed solenoid controls engine idle.
- Electronic transmissions have shift solenoids and a converter lockup solenoid.

■ RELAY ACTUATORS

Relays trigger the operation of high current load devices such as the fuel pump, cooling fan, air-conditioning

compressor clutch, O_2 sensor heater, computer, alternator, or trunk latch. Most relays are open until they are energized. This is called a *normally open relay*. The computer controls the ground side of the relay, which has low current. When it is grounded, current can pass through the relay.

■ MOTOR ACTUATORS

An idle speed control motor is an example of a motor actuator, as is an electric fuel pump. Engine idle is controlled by the computer. Sensors detect anything that will put a load on the engine at idle, and the computer makes corrections by powering a motor. Some other systems shut off the air conditioning or alternator at idle rather than trying to compensate for the load of these devices.

There are two types of idle motors. One is used on throttle-body injection and computer-controlled carburetors to open the throttle plate. Fuel injection uses an *idle air control (IAC) motor*—a *stepper motor* with two electromagnetic circuits (**Figure 45.32**). It can move in to close off airflow past the throttle plate and lower idle. It can also move outward, opening a passageway to increase idle.

NOTE: *Most computer control devices are controlled by the computer through ground-side switching. An idle air control motor is an exception to this rule.*

A throttle actuator control (TAC) motor is used on drive-by-wire systems (**Figure 45.33**). The throttle plate is opened or closed very rapidly in response to drivers that toggle the motor's polarity back and forth. A spring holds the throttle plate at approximately 7%

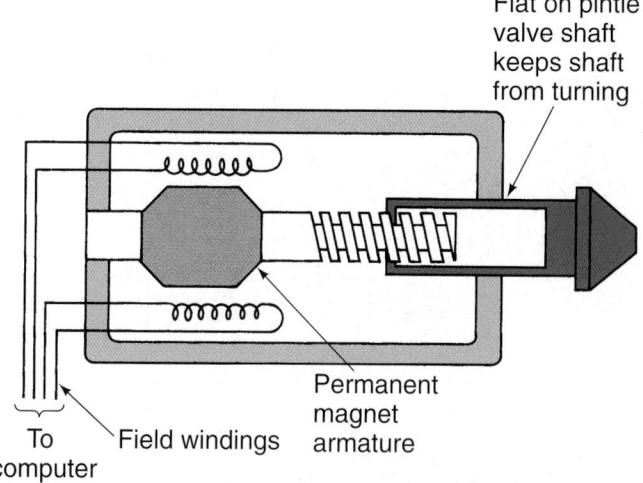

Flat on pintle valve shaft keeps shaft from turning

Permanent magnet armature

To computer   Field windings

Figure 45.32 A stepper motor used to control idle speed.

TAC motor   Throttle body

Figure 45.33 A "drive-by-wire" throttle actuator control motor. *(Courtesy of Tim Gilles)*

throttle opening when no current is flowing to the TAC motor.

SWITCH ACTUATORS/MODULES

The ignition module is one example of an actuator switch. It turns the primary ignition system on and off in response to computer sensor inputs. On modern vehicles, computer spark timing takes the place of vacuum and mechanical spark advance.

Vacuum advance has been replaced by a manifold pressure sensor that provides engine load information to a chip in the ignition module or PCM. The load information is analyzed along with the coolant temperature sensor signal and tachometer signal (for the mechanical advance). A chip provides the correct total spark advance signal to the coil (**Figure 45.34**). The computer also uses information from the throttle position sensor, air temperature sensor, and knock sensor in making its decision.

Other modules control the operation of the air-conditioning compressor. Cruise control is also engaged and disengaged by a module.

ADAPTIVE STRATEGY

Adaptive strategy uses **keep alive memory** (KAM) (**Figure 45.35**). Keep alive memory means the computer maintains power to RAM when the ignition switch is off. This allows it to keep information as long as the battery is not disconnected. KAM allows the computer to adapt to different hardware from car to car. The computer will compensate for these differences and changes resulting from wear and aging. Compensation is also made for variations in vehicle equipment options. Examples of types of adaptive strategies include idle air control, adaptive fuel trim, ignition timing management, and electronic transmission shift schedules.

Adaptive Fuel Trim

As a vehicle ages, wear and mechanical problems can occur. This results in changes in base fuel injector pulse width requirements, which are correctable using oxygen sensor feedback.

Adaptive fuel trim varies the fuel system to operate at the correct air-fuel ratio, which has a large effect on improving emission control. Short-term fuel trim (STFT), also called adaptive memory or block integrator, is a short-term correction in the air-fuel mixture during closed loop. Long-term fuel trim (LTFT) makes long-term corrections.

In OBD I, for short- and long-term fuel trim the computer uses a scale from 0 to 255, with 128 being the preferred value, which is 0% fuel correction. In OBD II, fuel trim is reflected as the percentage of change from the base pulse width calculation (0%). A positive number means the PCM is adding fuel, and a negative number means it is taking it away. STFT and LTFT can make corrections in fuel delivery by about 25% in either direction (rich or lean).

The computer has a short-term fuel chip and a chip that calculates fuel injector pulse width. The signal from the oxygen sensor goes to the short-term chip before it goes to the pulse width chip. One change in the oxygen sensor voltage results in a corresponding change in the pulse width.

A zirconia O_2 sensor cycles repeatedly between rich and lean. STFT is what drives the oxygen sensor signal. If the O_2 signal is rich, STFT adjusts the injector pulse width lean until the O_2 sensor responds by switching lean. Then the STFT drives the sensor rich again, and the cycle repeats.

LTFT changes fuel delivery from normal values. The job of LTFT is to adjust fuel delivery to an amount that will keep STFT from moving too far from 0% correction, which is considered to be normal.

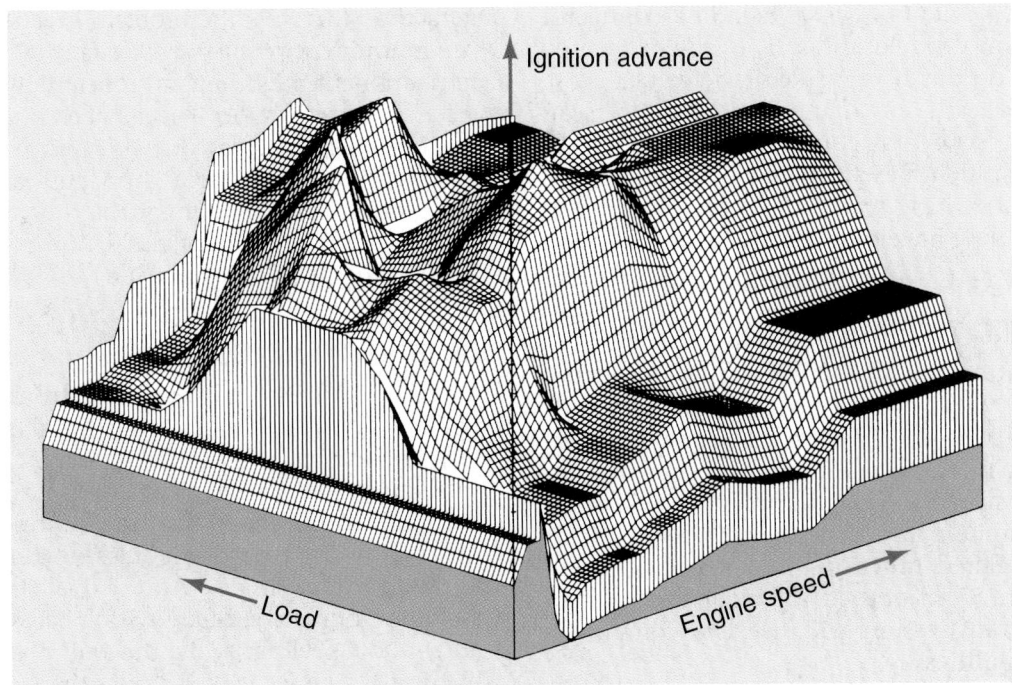

Figure 45.34 An ignition timing map. *(Reprinted with permission from Bosch)*

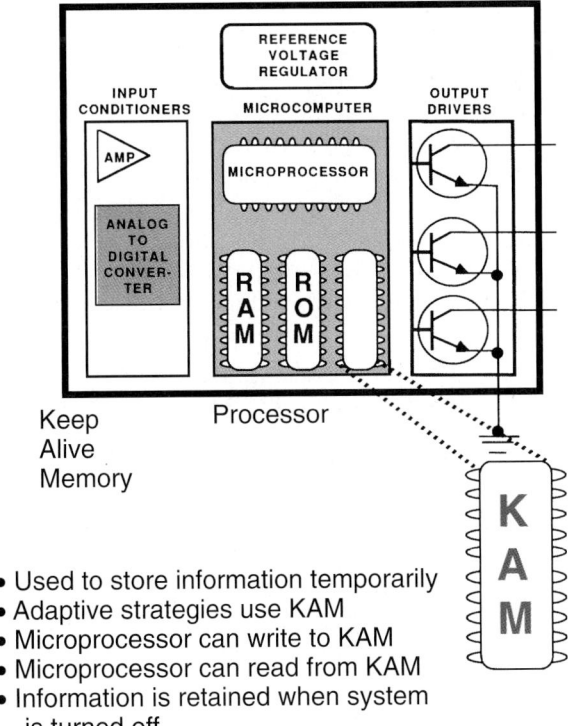

Keep
Alive
Memory

Processor

- Used to store information temporarily
- Adaptive strategies use KAM
- Microprocessor can write to KAM
- Microprocessor can read from KAM
- Information is retained when system
  is turned off

Figure 45.35 Keep alive memory (KAM). *(Courtesy of Ford Motor Company)*

Adaptive fuel trim is covered in more detail in Chapter 46.

KNOCK SENSOR

With a knock sensor, the computer adjusts ignition timing to be as advanced as possible without resulting in engine knock. First, it advances the timing until the engine knocks. Then, it retards the timing until the knock goes away. It will then advance the timing again, searching for the best timing setting for that rpm, load, temperature, and fuel. When it finds the best setting, it stores the information in memory for future reference.

ELECTRONIC THROTTLE CONTROL/ DRIVE-BY-WIRE

Electronic throttle control (ETC), often called drive-by-wire, is used on some late-model vehicles. The ETC module receives an accelerator pedal position signal and controls an electric motor attached to the throttle plate. Other than pedal position, inputs include how fast the pedal is being depressed or released. The aim is to provide better driveability and fuel economy, along with lower exhaust emissions and longer transmission life (due to gentler shifts). Electronic cruise control and traction control systems can also be part of the system. If there is a problem in the system, it reverts to a fail-safe position, allowing part throttle operation only.

ON-BOARD DIAGNOSTICS

On-board diagnostics (OBD) means that the computer has diagnostic capabilities. OBD began with legislation in California that required all 1988 vehicles sold in that state to be able to identify emission-related failures. Later federal legislation called for new OBD II standards. These are covered in detail in Chapter 47. The standards include a standardization of terms among all manufacturers and a universal data link connector (DLC) for reading trouble codes. The DLC is located under the dash on the driver's side of the vehicle.

OBD II also calls for a universal (generic) scan tool. When the scan tool is connected to the DLC, it automatically identifies the vehicle from its VIN. Manufacturers also have scan tools dedicated to their own vehicles. These have more dedicated information on them.

■ DIAGNOSTIC TROUBLE CODES

OBD II legislation mandated universal diagnostic trouble codes (**DTCs**). Interpretation of those codes is covered in Chapter 47. When an electronic problem occurs in a circuit that the computer senses or controls, a DTC is stored in the computer's non-volatile RAM. The computer also sets and stores codes for non-emission-related functions that it controls. For instance, malfunctions of the antilock braking system or electronic transmission will use the computer's diagnostic capabilities.

A malfunction indicator lamp (MIL), formerly called a check engine light, located on the dash display must illuminate if emissions exceed 1.5 times the federal standard (**Figure 45.36**). Emission-related codes are all stored in memory on OBD II systems.

The battery supplies the power to the computer for memory when the engine is off, but this results in very little drain. Computer code memory draws less than 0.005 amp (5 mA).

■ COMPUTER SELF-DIAGNOSTICS

Computer systems have become very complicated. As they have progressed, their self-diagnostic capabilities have improved as well. Computers can diagnose over 90% of the faults that occur in electronic systems. The MIL comes on when the key is turned on and during engine cranking. It is supposed to go out shortly after the engine starts. If it is on when the car is running, there is probably a fault code stored in the computer.

Whenever the key is turned to the "on" position, the system does a *self-check*. Some faults detected during this test cause the light to come on. Others are not serious and are stored for later service. Hard faults are those codes that are present and are stored in memory at the time of the self-test. Most of these codes will result in the MIL coming on and staying on. Others will only

light the MIL while the problem is actually happening. An intermittent problem will set a code that will remain in memory. It might remain indefinitely or for up to 50 restart/warm-up cycles. This is called non-volatile RAM.

An **intermittent fault code** is one that only occurs occasionally for a short period of time and is not present in the system at the time of the fault test. Each system component has a number assigned to it as a fault code.

■ MULTIPLEXING

Modern vehicles have many electronic control units (ECUs) that share information from ECU to ECU. In fact, the average vehicle has over 16 ECUs, and some cars have over 40 modules. Multiplexing, sometimes known as MUX, or in-vehicle networking, allows the many different control modules to communicate on a network using a single circuit or dual circuits. Multiplex systems eliminate separate wiring between each sensor and module, eliminating the bulky wiring harnesses of the past. Older vehicles often had multiple sensors serving different systems. Multiplex systems can share information between modules that employ system-specific sensors. For instance, information from the MAP sensor can be used by both the engine and transmission modules, whereas vehicle speed sensor (VSS) information is shared between different modules controlling the transmission, electronic brake and suspension systems, engine management, and cruise control.

Sensor signals can be analog or digital. Analog signals must be translated to digital signals prior to entering the network. Multiple digital signals can travel through the same channel in alternating time slots. Rapidly changing voltage signals are pulsed from high to low or from low to high. The voltage pulses are streamed together to form a message like Morse code. Data are transmitted through to individual computer modules, which can also be called *nodes*.

Sensor signals go to all of the modules but are only received by those modules with an interest in the information. Information is prioritized, and a chip prevents message overlapping. Digital messages carry an identification code, so something more important supersedes something less important, and only one digital message is transmitted at a time.

Some modules can initiate communication with other modules. These are called *masters*. Other modules, called *slaves*, can only listen for instructions and send confirmation when the task has been completed. In a typical network, the master module can exchange information with other modules. The master module provides reference voltage and a dedicated return line to the slave node.

Twisted Pair Wiring

Many multiplex systems use twisted pair wiring to prevent radio interference. Twisting the wires makes

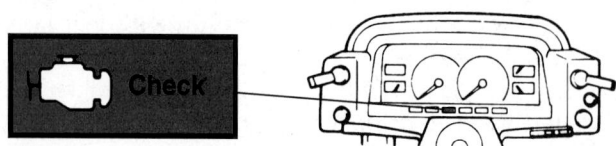

Figure 45.36 A malfunction indicator lamp on the dash display.

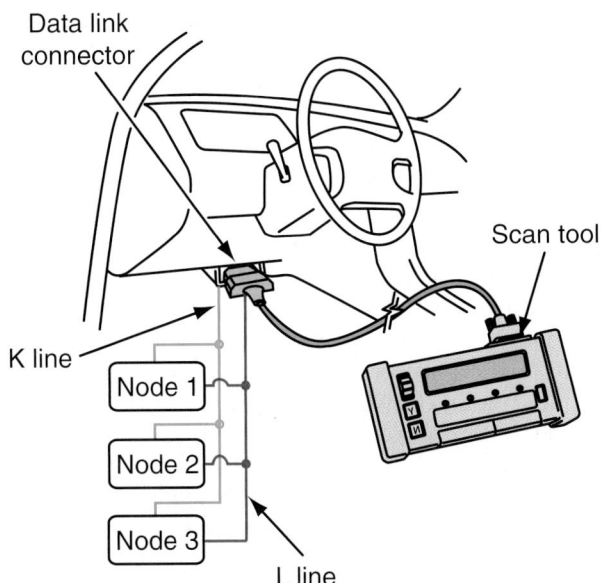

Figure 45.37 A network system with two wires, using one wire for scan tool diagnostics.

it less likely that a magnetic field from another circuit can induce a voltage in the pair of wires, resulting in electrical noise. If there is electrical interference, it will affect both of the wires equally in opposite directions. Twisted pairs must have wires of consistent length. This is especially important in a high-speed CAN bus, which is covered later. The two wire leads are not usually shielded from outside electrical noise, although twisted pairs on some vehicles are shielded.

There are different types of multiplexing. Time division multiplexing (TDM) is the type used on most automobiles with twisted pair wiring. Wavelength division multiplexing (WDM) is used with fiber-optic networks.

Multiplex Communication Protocols

The language used by modules for communication on a network is called a protocol. Protocols are a set of rules that differ according to speed and delivery methods. Protocol communication speed is classified by baud rate, which is defined by the SAE:

- Class A. This is a low-speed protocol with a baud rate of up to 10 kilobytes (10,000 bits) per second (Kbps).
- Class B. This is a medium-speed protocol with a baud rate between 10 and 125 Kbps. Class B is the OBD II emission protocol until 2008. It is also used for audio controls, vehicle speed, temperature sensors, and instrument panel displays. Single- and two-wire versions of Class B are used for fault code storage and scan tool operation (**Figure 45.37**). Single-wire systems, which are the most common, use a variable pulse width signal, and twisted pair wired systems use a pulse width modulated signal.
- Class C. This is a high-speed protocol with a baud rate between 125 and 1,000 Kbps (1 Mb/s). Class C is typically used for high-speed, real-time controls such as ABS, vehicle stability control, and air bag sensing and control.

It is common to find different multiplex classes within the same vehicle's electrical system (**Figure 45.38**) because some systems do not require information to be processed at the highest speeds. A *gateway module* allows slower communicating networks to communicate with faster ones. Sometimes the gateway module is the body control module (BCM), which might also be the instrument panel control module. If all of the serial information were on one bus, traffic would be limited. Performance and safety are the top priorities in network design, so those systems reside on the high-speed bus. Comfort systems like trip computers and electric seats and windows use the low-speed bus, where inexpensive low-speed communications are sufficient. Using different speed networks on the same vehicle also allows the manufacturer greater flexibility when using modules, sensors, and other components.

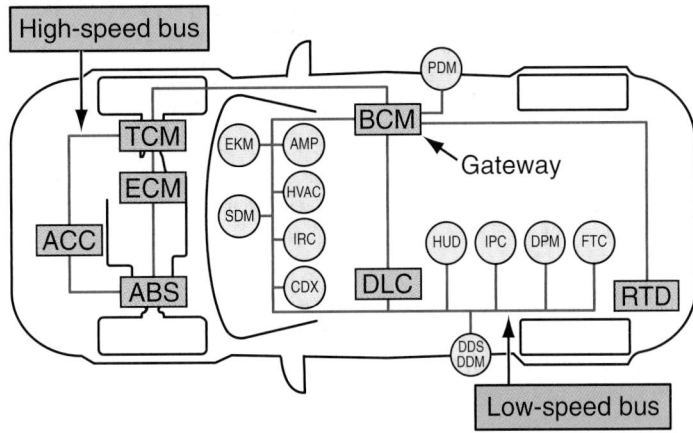

Typical GM LAN Network

Figure 45.38 It is common to find more than one network class within the same vehicle's electrical system. This one has a high- and a low-speed bus connected by a gateway module.

▄ NETWORK SIZES AND TYPES

Networks are classified according to their size. The Internet is an example of a **wide area network (WAN)**. An office, home, or vehicle network is called a **local area network (LAN)**. Two types of LANs are in common use. The most common protocol used for home LANs and the Internet is called *Ethernet*. Automotive LANs use a **controller area network (CAN)**.

CAN is also used in many industries besides the automotive industry. In the computer industry, CAN is sometimes known as small area network (SAN). Ethernet is much faster than CAN because home computers are faster than automotive computers. However, CAN has the advantage of being robust; it can survive within the harsh environment of the automobile. CAN is also capable of fault confinement, which means a defective node can remove itself from the network. If a defective node continued to broadcast, it could slow down the network or even cause the network to stop all communications.

CAN Systems

Early LAN systems used different protocols designed by many of the individual auto manufacturers. A dedicated scan tool was required for each system. Today, most automotive high-speed communication is CAN, which was first used on a mass-produced vehicle in 2001 and has been used on many vehicles since 2003. A form of CAN protocol known as ISO 15765-4 is the legislated protocol for OBD II diagnostics on all 2008 and later vehicles, making it the universal communication method. One scan tool will be able to access OBD II information on all vehicles.

Messages on a network are of two types: *diagnostic mode* and *normal mode* messages. Diagnostic mode messages are when a scan tool communicates with a module on the network, or vice versa. Normal mode messages share information between network modules during normal vehicle operation. An example of normal mode operation is when the instrument cluster sends the fuel level to the main computer.

CAN saves power over some of the earlier network protocols because it allows individual modules to power down when they are not needed. Modules wake up when signaled and power up and perform a needed task before powering back down after a few seconds with no use. CAN systems also have less parasitic loss.

Manufacturers use different CAN system designs. A typical CAN system uses medium- and high-speed CAN for traditional network functions, and high-speed diagnostic CAN C for communicating with a scan tool. The medium-speed CAN B bus remains in operation while the ignition is off, and individual modules can remain active if needed. The high-speed CAN C bus is powered only when the ignition is turned on and uses

twisted pair wiring. **Figure 45.39** shows a schematic of a typical CAN bus system.

Network Topology

The *topology* of a network describes its physical makeup. A network can be wired in several ways:

■ A daisy chain topology, also called series or linear, is one of the earliest wiring methods: it is wired in series like Christmas tree lights.

■ Star (common splice) topology is the most popular topology for home LANs. All modules on the network communicate with one another through a hub, router, or switch (**Figure 45.40**). The star configuration allows for easy addition of modules to the network by plugging a device into the central processor. A fault in a star system will result in failure of only the affected component (**Figure 45.41**).

■ Bus topology uses less cabling than star topology. A single *main run* bus cable has all modules connected to it (**Figure 45.42**). Terminators provide resistance that absorbs frames once they have been transmitted so they are not received again. The terminators can be at each end of the bus or can be inside two of the nodes.

■ Ring, or loop, topologies are wired in series, but messages can travel either way in the network so a single break will not open the circuit (**Figure 45.43**).

One example of an automotive system that combines protocols has a bus used for high-speed communication while a star network is used for low-speed communication. A gateway module manages communication between the two networks.

Input sensors and output actuators are hard wired into the network circuit at the closest module, which puts the sensor information into the system where it can be shared with all other members of the network. This can result in strange things like rear-wheel speed sensors that are not directly wired to the ABS circuit because the information gets into the system in other ways.

CAN Information Transmission and Message Arbitration

The CAN chip manages information transfer between two or more modules according to preset rules. When two modules transmit at the same time, CAN uses nondestructive *bitwise arbitration* so messages do not collide. Collision resolution in Ethernet systems is destructive, which means a collision will jam the network, whereas CAN holds colliding information until it can be sent.

CAN transmits messages using *broadcast technology*, which means messages are broadcast to all computers on the network rather than to one in particular. ECUs use filters to accept or ignore a CAN message. If an ECU is programmed to receive a CAN message, then it will accept it and decode the contents. In CAN, a block of information is defined as a *frame*. A data

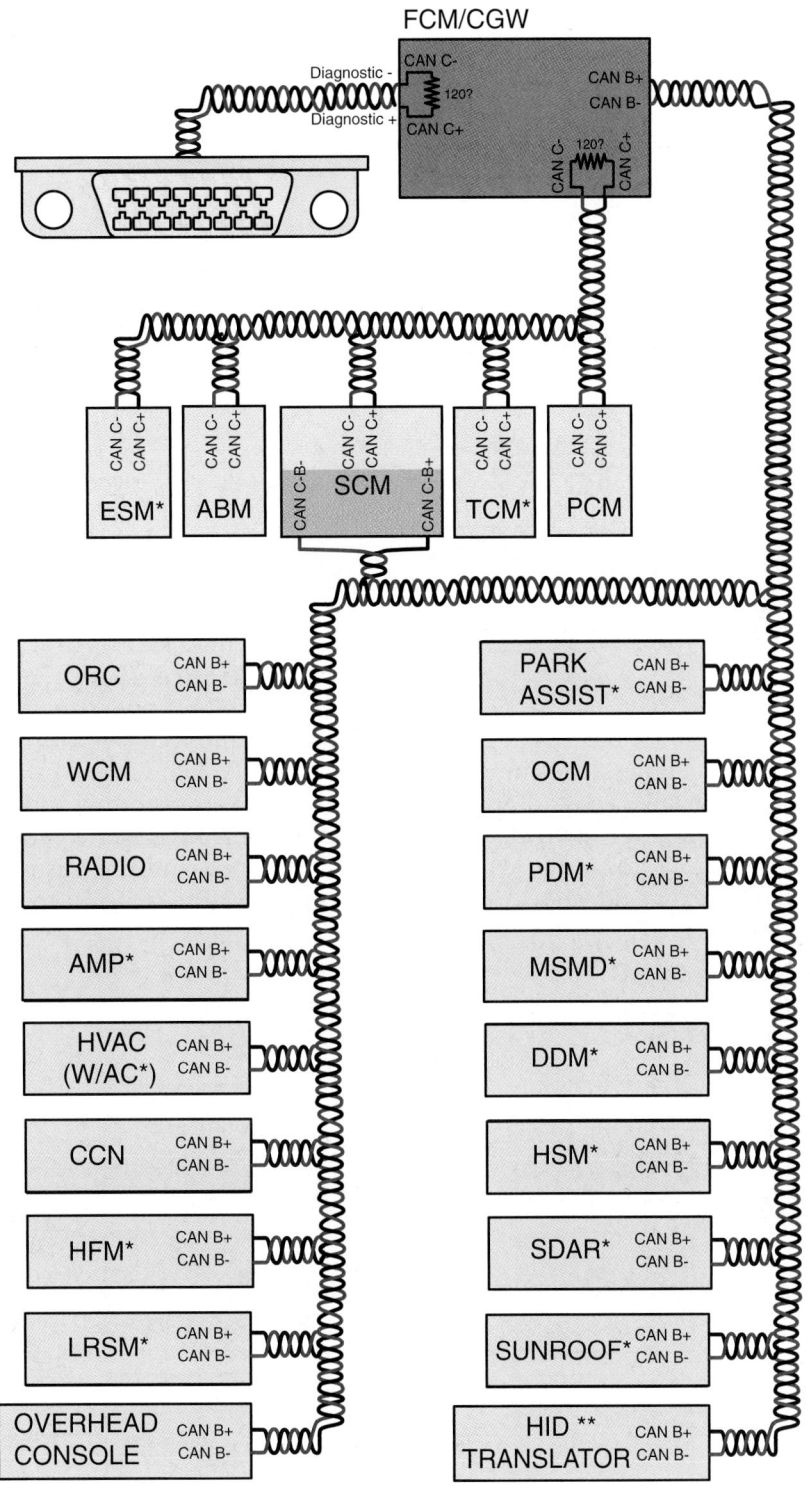

Figure 45.39 On this CAN system, the steering control module with steering angle sensor input has access to both the C and B busses. Note the twisted wires that provide the bus.

frame is one containing the basic information like vehicle speed that a vehicle needs for operation. The beginning of each frame contains a section identifying the vehicle data so each controller can decide whether it wants to process the information in the frame or ignore it.

Each module periodically sends out a message to let the other modules know it is working correctly. If a

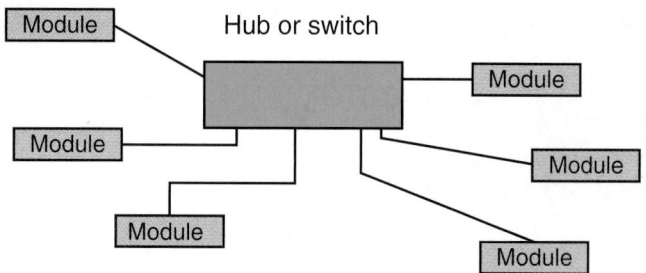

Figure 45.40 In a star topology, all modules communicate through a hub or switch.

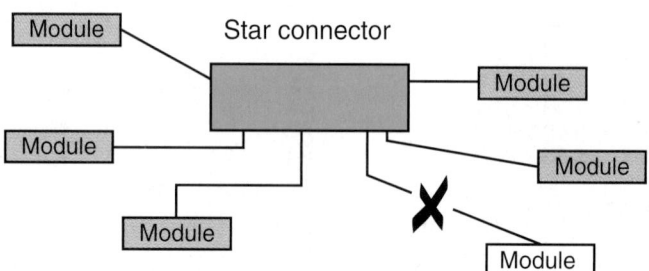

Figure 45.41 A fault in a star network only affects one module.

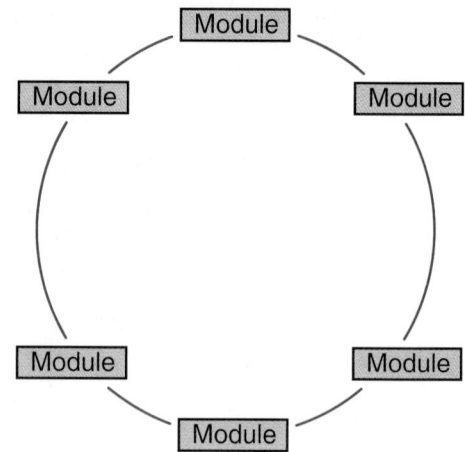

Figure 45.43 Ring, or loop, topology with all modules wired in series. Messages can travel in either direction.

module does not send its expected message, a network DTC may be set. When a module fails, a default setting keeps the entire system from going down with it.

In electronics, a *bias* is when a voltage or current reaches a certain threshold in one direction or the other. In a typical CAN system, when the ignition is on, the CAN C bus becomes active and the bus is biased at about 2.5 volts. The range is always above negative and can vary (**Figure 45.44**). A computer processes information in bits. Nondestructive bitwise arbitration uses dominant and recessive bits to determine the priority of a message. When CAN C (+) and CAN C (−) are equal, the bit "1" is transmitted and the bus is recessive. When CAN C (+) is high while CAN C (−) is low, the bit "0" is transmitted and the bus is dominant. Each controller monitors the bus and the controller with the most

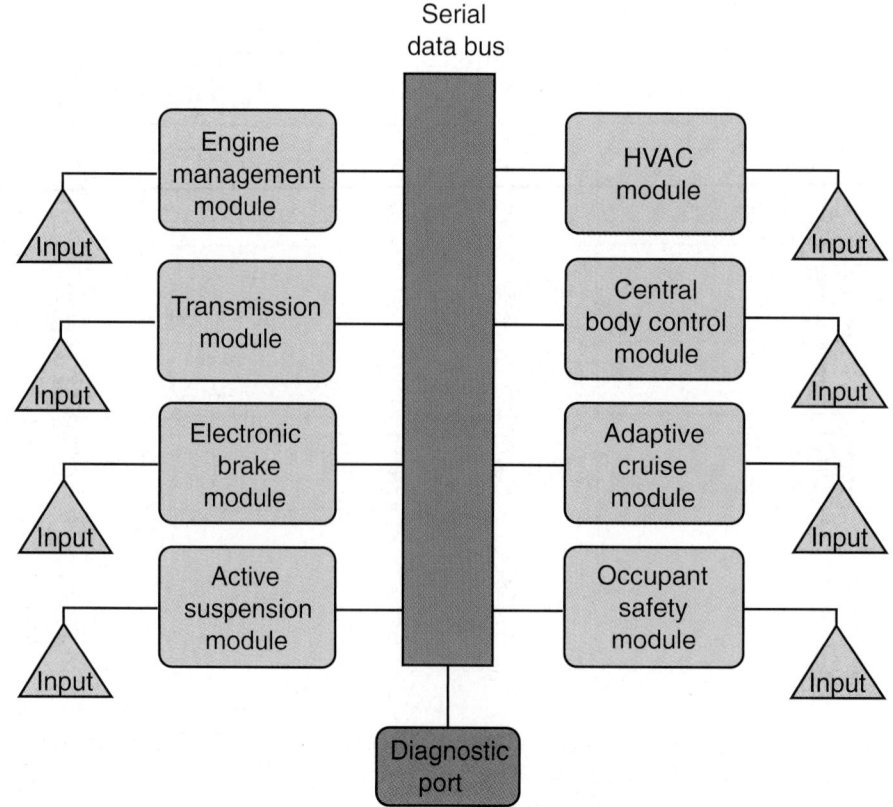

Figure 45.42 Bus topology has a single main-run bus with all modules connected to it.

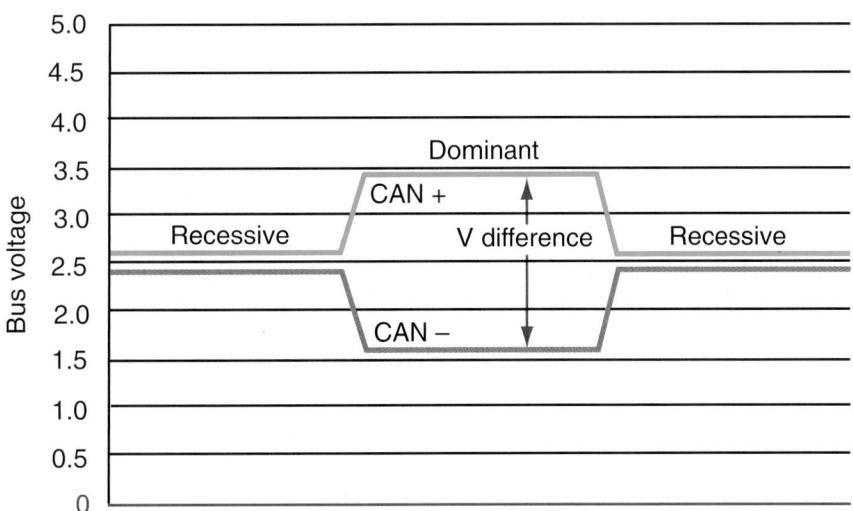

Figure 45.44 CAN C bus voltages are always above negative and can vary.

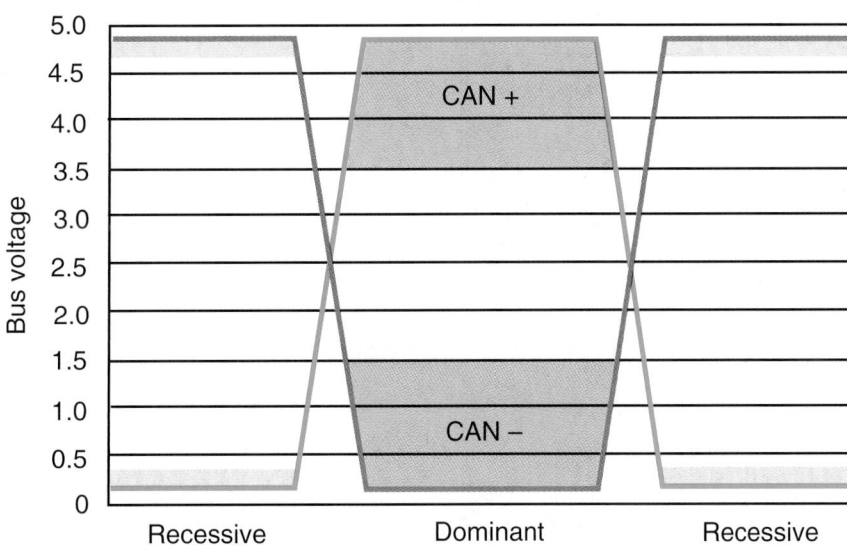

Figure 45.45 Medium-speed CAN uses the same logic as high-speed CAN, but the voltages are different.

dominant bits in a row has the highest priority for messages. Medium-speed CAN uses similar logic, but the voltages are different (**Figure 45.45**).

■ SUPPLEMENTAL DATA BUS NETWORKS

Supplemental bus networks provide additional support to the main bus network. One supplemental system is the early low-speed UART single master module and multiple slave system. The master module controls the speed of data transfer and routes sensor data into the CAN bus (**Figure 45.46**).

Fiber Optics and the Media-Oriented System Transport Data Bus

The media-oriented system transport (MOST) data bus is a networking standard designed to connect audio and video components, providing high performance at a low cost. The protocol was developed and promoted by an alliance of 17 international auto manufacturers and more than 50 component suppliers from the electrical and audio-video industries. A MOST network lowers manufacturing costs and minimizes bulky wiring harnesses. Simple devices like speakers and more complex digital devices with sophisticated controls can be managed by the system.

MOST networks usually use a ring topology, but star configurations and double rings are sometimes used. Networks of up to 64 nodes are possible, using a plug-and-play feature that allows for the easy addition of devices.

Fiber Optics

System information on a MOST network is carried primarily by fiber optics known as plastic optical fibers (POF). Fiber-optics systems are capable of much higher speeds than ordinary networks, on the order of 25 megabits per second. They carry digital

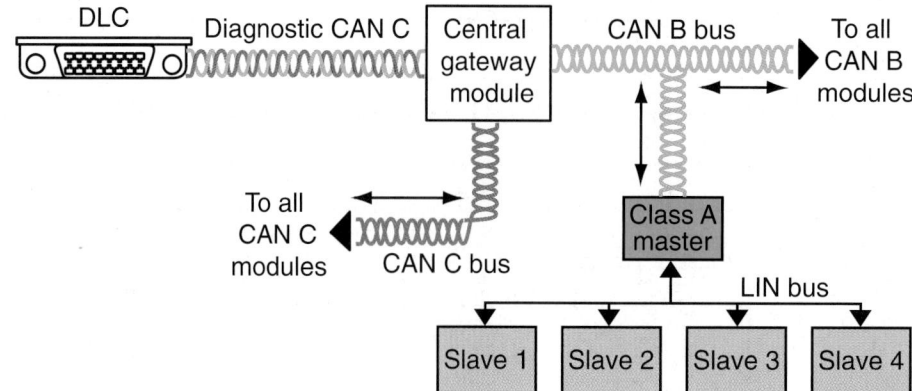

Figure 45.46 A supplemental bus network with master and slaves. The master module controls data transfer speed and routes data from the sensors into the bus.

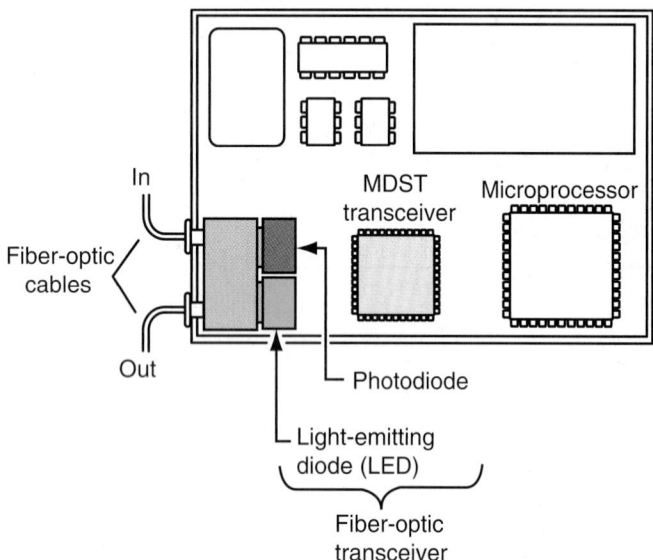

Figure 45.47 Communication on a MOST bus occurs when signals are transmitted between an LED, a photodiode, and a transceiver.

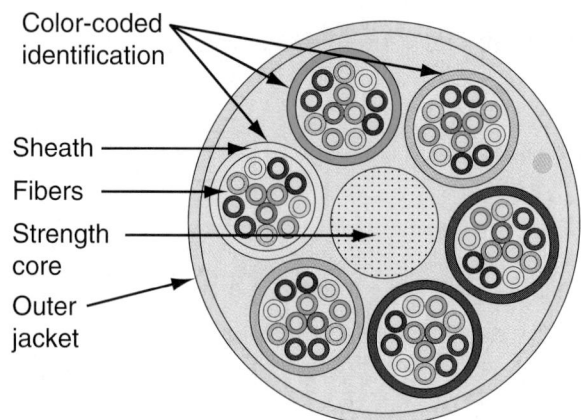

Figure 45.48 A fiber-optic cable is constructed in layers.

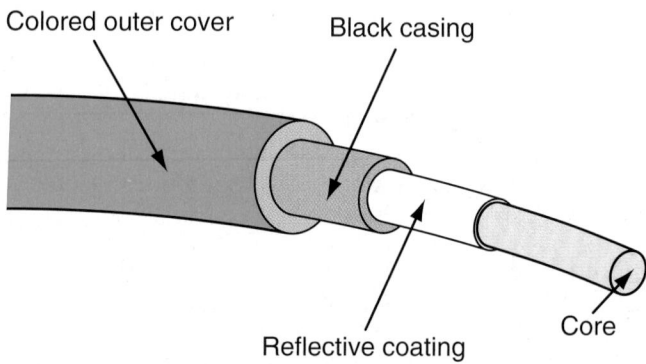

Figure 45.49 The fiber-optic core is protected from outside light by the casing.

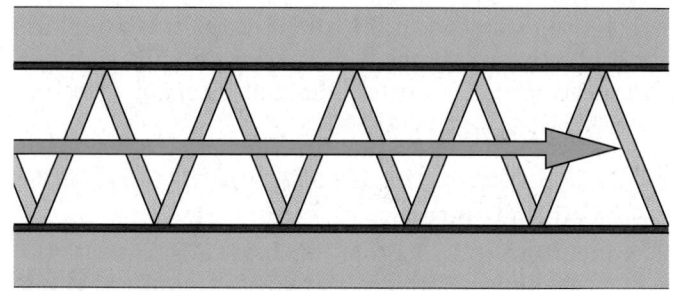

Figure 45.50 Most of the light in a fiber-optic cable travels in a zigzag fashion.

information over plastic or glass fibers using infrared or visible light. Light is reflected inward so it remains within the fiber-optic cable even when there are extreme bends in the conductor. Unlike conventional electrical conductors, fiber-optic information transmits light waves that are relatively free of electrical noise.

Communication between modules on the MOST bus occurs when light signals are transmitted between an LED, a photodiode, and a transceiver (**Figure 45.47**). The light signals are converted to voltage signals by the photodiode and transmitted to the MOST transceiver. From there, the LED converts the voltage signals to light signals for transmission over the fiber-optic cable.

A fiber-optic cable is constructed in layers (**Figure 45.48**). The core protects the casing from outside light (**Figure 45.49**). Most of the light in a fiber-optic cable travels in a zigzag fashion resulting from the total reflection of the light (**Figure 45.50**). When a

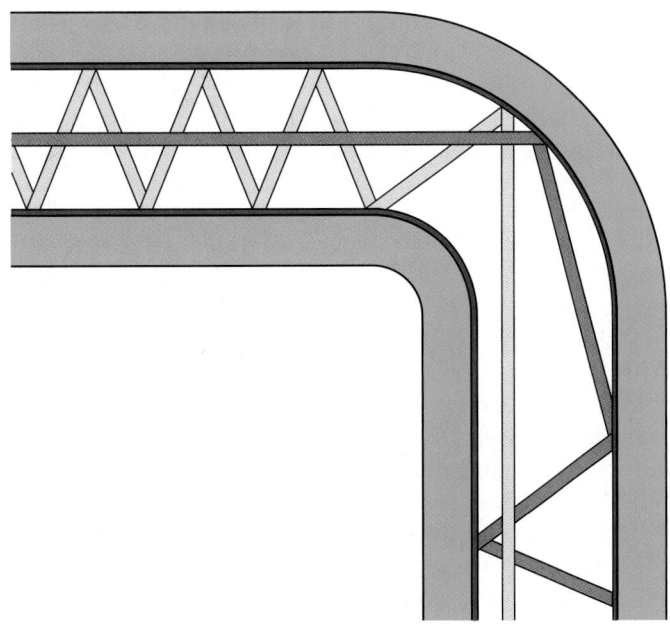

Figure 45.51 When a fiber-optic cable is bent, the light waves are guided through the bend.

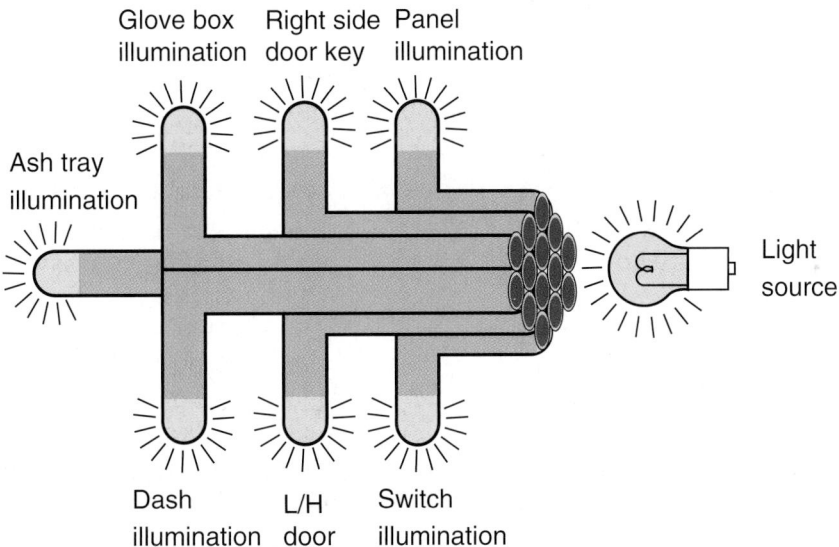

Glove box illumination   Right side door key   Panel illumination

Ash tray illumination

Light source

Dash illumination   L/H door   Switch illumination

Figure 45.52 Fiber optics can illuminate several items from one source.

fiber-optic cable is bent, light waves within the cable are guided through the bend (**Figure 45.51**).

Fiber optics can be used to illuminate several objects from one source, such as indicator lights (**Figure 45.52**). They are often used for side marker lights attached to turn signal indicators or for illumination of a halo around a door lock cylinder.

■ TELEMATICS

New vehicles are being equipped with wireless systems controlled by voice commands that allow access to the Internet and e-mail, downloading of audio and video files, or obtaining navigation information. *Telematics* blends computers and wireless telecommunications, combining global positioning satellite (GPS) tracking and other wireless communications for automatic roadside assistance and remote diagnostics. The General Motors OnStar system is one example of telematics.

■ WIRELESS NETWORKS

Wireless networks transmit information without wires. An example of a wireless network is the tire pressure information monitor, which has a sensor/transmitter inside the tire. In 1997, the Institute of Electrical and Electronics Engineers (IEEE) developed standard 802.11, which describes specifications for wireless LAN technology that interfaces between a wireless master or base station, or between two wireless modules.

Some of the many wireless technologies being evaluated and developed for automotive use include Bluetooth, DSRC, wireless fidelity (Wi-Fi), WiMAX (802.16), ultra wide band (UWB), Zigbee, near field communications (NFC), radio frequency identification (RFID), and satellite. Two of the more popular protocols starting to gain acceptance in automotive networks are Bluetooth and DSRC.

Bluetooth

Bluetooth is a wireless personal area network (PAN) protocol that allows communication between modules from over 2,000 different manufacturers using standard radio transmissions. It can be used to connect devices like laptop computers, PDAs, and hands-free cellular phones to the vehicle electrical system. It is used for streaming music from personal devices into vehicles, and vice versa. Wireless control of rear seat headphones and rear seat gaming ports is also an option. An application of Bluetooth for the automotive service industry is remote vehicle diagnostics (RVD), which eliminates the wired connection to a scan tool.

Radio waves are transmitted on a frequency that does not require a license to use. The normal range is about 33 feet (10 meters). Other wireless devices like garage door openers operate on the same radio frequency, so sophisticated security measures have been designed into the software to protect against interference between signals.

When a device is connected to the system, it is given an identity using a unique personal identification number (PIN) that is 48 bits long. The master module, which has the PIN as well, makes the connection and synchronizes with the devices.

DSRC

Another wireless communication standard based on 802.11 is dedicated short range communications (DSRC). It is used to link vehicles with each other and to link vehicles with roadside data access points. DSRC has initially been focused on entertainment applications.

■ REVIEW QUESTIONS

1. A _____ is a material that can be either a conductor or an insulator.

2. Conventional electrical theory as used in automobiles says that electricity flows from _____ to _____.

3. A small amount of current applied to a _____ causes it to relay larger amounts of electricity through it.

4. The three semiconductor crystal layers in a transistor are called the emitter, the base, and the _____.

5. A _____ diode is used to control a voltage regulator.

6. The three main parts to automotive computer systems are the computer, sensors, and _____.

7. If a temperature sensor's range is 0°F–250°F, each volt of difference will be interpreted by the computer as a _____° change in temperature.

8. The speed at which binary information is transmitted within the computer is called its _____ rate.

9. A _____ is a variable resistor made from semiconductor material whose resistance changes predictably as its temperature changes.

10. _____ is a term for the communication between several computers using a single circuit or dual circuits.

■ ASE-STYLE REVIEW QUESTIONS

1. The layer of a transistor that controls the switching function is called the:
   - **a.** Base
   - **c.** Junction
   - **b.** Emitter
   - **d.** None of the above

2. According to electron theory, electricity flows from:
   - **a.** Positive to negative
   - **c.** Both A and B
   - **b.** Negative to positive
   - **d.** Neither A nor B

3. Technician A says that semiconductors can be either positively or negatively charged. Technician B says that some diodes conduct electricity from positive to negative and others conduct from negative to positive. Who is right?
   - **a.** Technician A
   - **c.** Both A and B
   - **b.** Technician B
   - **d.** Neither A nor B

4. Which of the following is/are true about transistors?

  a. Voltage applied to a transistor causes the base layer to reverse its charge and become a conductor.

  b. A transistor that is forward biased allows current to flow.

  c. Both A and B.

  d. Neither A nor B.

5. Technician A says that temporarily stored information is located in RAM. Technician B says that permanently stored information that the computer refers to is called ROM. Who is right?

  a. Technician A   **c.** Both A and B

  b. Technician B   **d.** Neither A nor B

6. All of the following are true about computer networks *except*:

  a. Class A is a low-speed protocol.

  b. Class C is a medium-speed protocol.

  c. Different multiplex classes can be found in one vehicle's electrical system.

  d. Class B is used for fault code storage.

7. Which of the following is/are true?

  a. Piezoelectric crystals develop voltage in response to pressure applied to them.

  b. When there is no voltage, a computer sees a "1."

  c. Keep alive memory (KAM) means the computer maintains power to ROM when the ignition switch is off.

  d. All of the above.

8. Technician A says that a transistor has two P-N junctions. Technician B says that a diode has one P-N junction. Who is right?

  a. Technician A   **c.** Both A and B

  b. Technician B   **d.** Neither A nor B

9. Which of the following is true about computers?

  a. The programs for the computer are part of its hardware.

  b. Computers usually control the power side of circuits.

  c. Automotive computers use battery voltage for their reference voltage.

  d. None of the above.

10. Technician A says that integrated circuits can carry large amounts of current. Technician B says that actuators change electricity into mechanical work. Who is right?

  a. Technician A   **c.** Both A and B

  b. Technician B   **d.** Neither A nor B

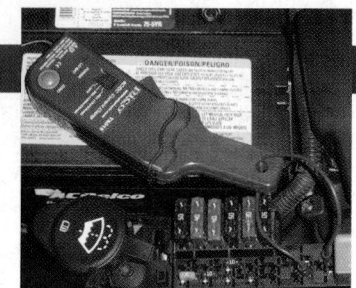
Electronics and Computer Systems Service

■ OBJECTIVES

Upon completion of this chapter, you should be able to:

✔ Diagnose related engine and electrical problems prior to computer repair.

✔ Describe the theory and operation of on-board diagnostics.

✔ Read trouble codes.

✔ Use a scan tool.

✔ Confirm closed loop.

✔ Test sensors and actuators.

✔ Diagnose computer wiring problems.

✔ Diagnose and replace a computer.

■ KEY TERMS

high impedance voltmeter	unidirectional	passive sensors
diagnostic tree	bidirectional	active sensors
DLC	KOEO	negative coefficient
PID	high authority sensors	thermistor

■ INTRODUCTION

Computer systems have become very sophisticated. In order to repair complicated systems, you will need to consult model-specific service information that includes step-by-step procedures for troubleshooting systems. This chapter covers items in a generic way to give you an understanding of the ways various types of sensors and actuators are tested. You should read the fuel and emission systems chapters in this book before studying the information presented here. Study Chapter 33 and be familiar with the use of electrical wiring diagrams, too. Today's technicians must be able to read a wiring schematic in order to diagnose and repair modern computerized vehicles.

■ INSPECTION SEQUENCE

Computers have self-diagnostic ability. The newer the computer, the more accurately it will diagnose problems. A simple problem like a loose or corroded wiring

connection can cause a problem that the computer tries to correct by compensating with other changes. This can make it seem like the computer is at fault.

When diagnosing computer systems, a logical diagnosis sequence must be followed before checking the computer. You cannot overlook the basics and just start replacing expensive electronic parts until a fix is made. Parts stores will usually not accept the return of electronic parts. Also, some models of computer modules are programmed with the vehicle VIN and cannot be swapped into a circuit in place of another module while trying to correct a problem. Be sure to carefully question the customer about the symptoms and when they occur.

On older cars, mechanics could often make successful repairs by jumping to quick conclusions based on past experience. On newer cars, the computer responds to information it receives from the sensor inputs. The sensors are reacting to changes in engine operating conditions. Be careful to eliminate all of the mechanical conditions first before going on to a computer diagnosis. A large majority of cars will not actually have computer problems but can be fixed during the inspection and maintenance process.

Problems in a computer system are isolated using visual and diagnostic checks, a digital voltmeter, an ohmmeter, a DSO, or a scan tool. Most sensors and actuators have specifications for resistance measurement. Voltage to and from devices can also be tested. A mechanical failure can also occur in come cases. Individual devices are not tested until other test procedures point in the direction of that area.

High-Impedance Meters

A digital multimeter (DMM), also called a digital volt-ohmmeter (DVOM) (**Figure 46.1**), is the instrument used To measure electricity in electronic circuits. A **high-impedance voltmeter** must be used to perform tests on these systems. The integrated circuits in computer systems operate on very small amounts of current. An analog meter (one with a needle) (see Chapter 25) operates on magnetism and can load down a computer circuit and actually change what is happening in the circuit. The meter

seems like a short in the circuit, offering an easier path for electrical flow.

A high-impedance meter has very high input resistance (usually about 10 million ohms). It has a digital display instead of a needle. The high input resistance prevents the meter from drawing current while it is connected to a circuit. This protects the circuit and keeps readings accurate.

Other advantages of DMMs:
- They are often self-scaling so adjustments are not required.
- They do not have to be connected in the correct polarity. A plus or minus ID is displayed on the screen with the reading.

SHOP TIP   To check a meter to see if it has high impedance, turn the selector to the DC volts position. Measure the resistance through it with another ohmmeter (**Figure 46.2**). A meter of high enough impedance to use in automotive electronic circuits will read over 10 m/ohms.

Visual Inspection

Electronic systems are quite dependable. The cause of a problem can often be determined during a careful visual inspection. If the engine runs but has poor driveability, perform a visual inspection under the hood for disconnected or broken vacuum lines or electrical connections. While the engine runs, listen for vacuum leaks. Inspect vacuum hoses for horizontal cracks. High vacuum can cause a crack to suck together. Hard neoprene hoses can develop a restriction bubble. The result is poor throttle response due to trapped vacuum. Make sure vacuum hoses are routed correctly.

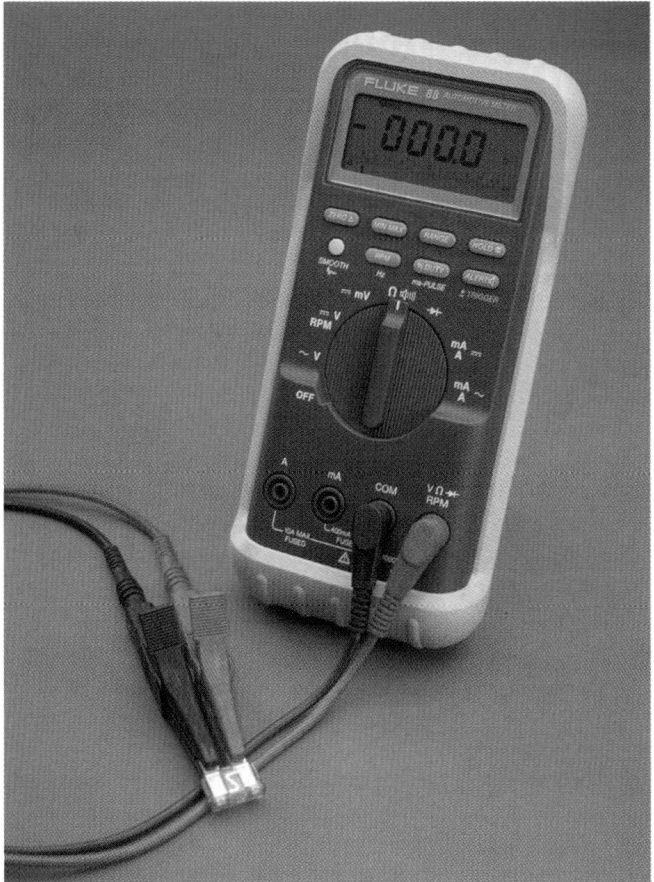

Figure 46.1 A digital volt-ohmmeter (DVOM) is sometimes called a digital multimeter (DMM).

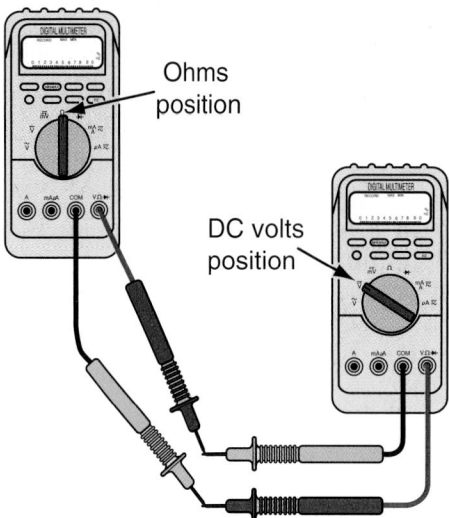

Ohms position

DC volts position

Figure 46.2 Testing a meter's impedance by measuring its internal resistance in the DC volts position.

■ PERFORM DIAGNOSTIC TESTS

Analyze the cause of the problem rather than just fixing the problem's result. If there are no codes stored in memory, check for fuel or ignition problems with a lab scope and exhaust analyzer. If a problem is in one of those areas, check any sensors that might contribute. Remember, a test drive takes time to perform. If you can solve the problem in a few short minutes in the service bay, you will be more productive.

Check Engine Condition

To quickly evaluate compression, listen during cranking for an even rhythm and then for a smooth idle. A vacuum gauge (see Chapter 48) is easy to connect and can be used to qualify engine condition. It should produce between 17 and 21 inches of steady vacuum at idle and 2,500 rpm.

An EGR valve is not open at idle during normal operation. A hand vacuum pump can be used to check that the EGR valve remains closed at idle. When the valve is opened by the vacuum pump, the engine should stumble and run roughly. This is due to the lean condition created by the open EGR valve.

If a cylinder has a regular misfire, the catalytic converter will become very hot as it attempts to catalyze the unburned fuel that results. Run the engine at 1,500 to 2,000 rpm for 30 seconds to run air through the CAT to give it a chance to get rid of extra fuel after a power balance test.

Ignition Checks

On engines with distributor ignition, check the base timing setting. Follow manufacturer's instructions for disconnecting the computer for this test. Otherwise, even in open loop (when the car is cold), the computer will provide extra ignition advance. Base ignition timing that is not within 3 degrees must be adjusted. Use an adjustable timing light to check total advance. Total ignition advance is usually 15 to 25 degrees at idle. At 2,500 rpm, expect to find 30 degrees or more of ignition advance.

When an old cars had ignition points, tune-up work included adjustment. That type of maintenance is gone. With computerized electronic fuel injection, idle speed is more important because it can cause changes in other settings. Most throttle position sensors are no longer adjustable except for the throttle stop that prevents the throttle plate from binding up when all the way closed. Idle speed should remain constant, unless there are deposits on the throttle plate that prevent it from closing all the way. Cleaning the throttle plate is a maintenance procedure on many engines, although some manufacturers recommend against using chemical spray cleaners due to possible damage of the plating.

Charging System Test

A charging system test is one of the core things to do before beginning a diagnostic procedure. A bad diode in an alternator can cause driveability problems and set other trouble codes.

If the battery is not fully charged, the computer system will still work, but related problems can affect engine operation. For electronic fuel injection to work properly, new cars require a minimum of 11 volts. Older fuel injection systems would work on 8 to 9 volts. If voltage drops below about 11.6 volts, the computer will raise the engine's idle to increase the alternator's charging rate.

NOTE: *If the charging system voltage is hovering near 11.6 volts, stepping on the brakes will turn on the brake lights, dropping the voltage further and causing the computer to raise engine idle.*

When voltage drops below 10 volts, the system can go into limp-in mode. Normal computer operation will be lost, resulting in poor driveability.

Testing the insulated and ground circuits for voltage drops is covered in Chapter 29 under the heading of "Circuit Resistance Test."

■ ON-BOARD DIAGNOSTICS

Computer systems today are very complicated. It would be very difficult and time consuming to diagnose problems if the computer did not have its own memory and self-diagnostic capability. Computers can detect incorrect electrical conditions and save diagnostic trouble codes to memory. There are codes for many malfunctions.

Each time the key is turned on, the computer does a *self-check* of its circuits. The computer sends a test voltage signal to the sensors and actuators, checking for continuity and return signal voltage. Sometimes, a value shows up during the self-check that is outside of expectations. When this happens, the computer will either store a code to provide service information to a technician or it will turn on the malfunction indicator light (MIL). The malfunction indicator light is sometimes called a *check engine light,* but MIL is the newer, standard term.

Check to see if the MIL is lit when the engine is running. If it is, check for diagnostic trouble codes.

■ A hard fault is one that is present at the time of the self-test.

■ An intermittent fault is one that occurs for a short period of time and is not present in the system at the time of the fault test. Intermittent faults are the most difficult to diagnose.

When the key is turned on, the MIL should come on for a few seconds and remain on during engine cranking. This provides a check of the bulb and circuit. The MIL should go out shortly after the engine starts. If it comes on when the engine runs, there will be a trouble code stored in the computer's memory.

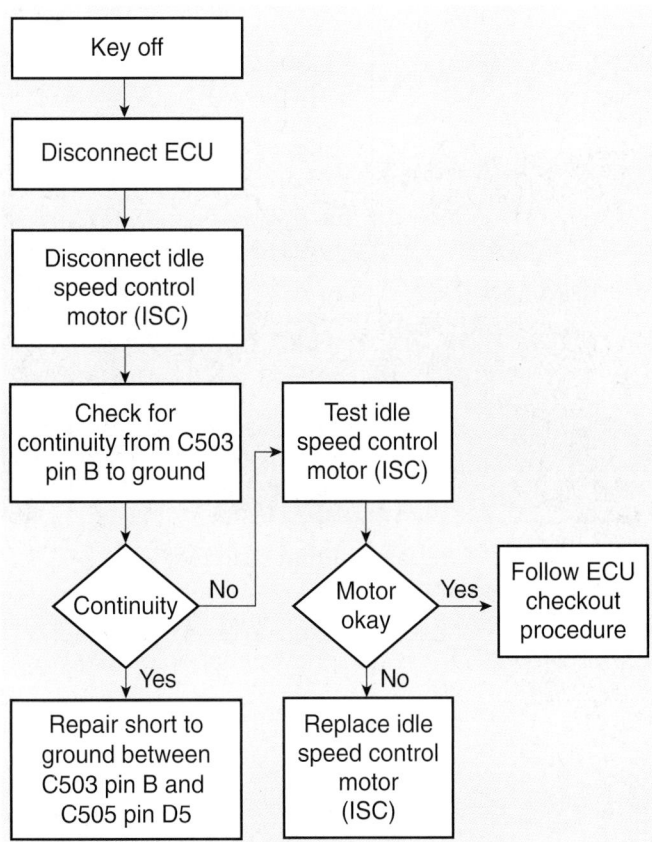

Figure 46.3 A diagnostic tree. *(Courtesy of DaimlerChrysler Corporation)*

NOTE: *The codes are only a starting point. They do not tell you what to replace. They simply inform you that something out of the ordinary has occurred within a certain circuit.*

To isolate a fault, use the correct test instrument and service information. A **diagnostic tree**, or flow chart, in the service Library provides a step-by-step diagnostic procedure to follow when troubleshooting hard-fault problems (**Figure 46.3**).

A common technician error is to replace a component at the bottom of the flow chart. Always follow the tree in order. Going out of sequence can cause you to miss the problem. This is like following a road map. If you miss a turn at a street, you will be lost. If the last step on the tree indicates that the computer is the problem, be sure to thoroughly test the computer's power and grounds before replacing it.

Sensors are more often the cause of electronic control problems than actuators. The logical test sequence would therefore be to check the sensors and wiring before testing actuators and wiring. The computer is tested last.

READING TROUBLE CODES

There are different ways to read trouble codes, depending on the year and make of car. OBD II systems have standardized connectors and procedures. Earlier systems had different procedures for retrieving codes.

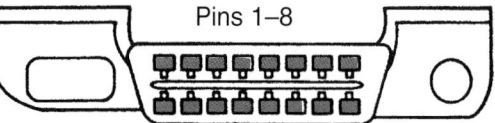

Figure 46.4 A standard OBD II data link connector (DLC).

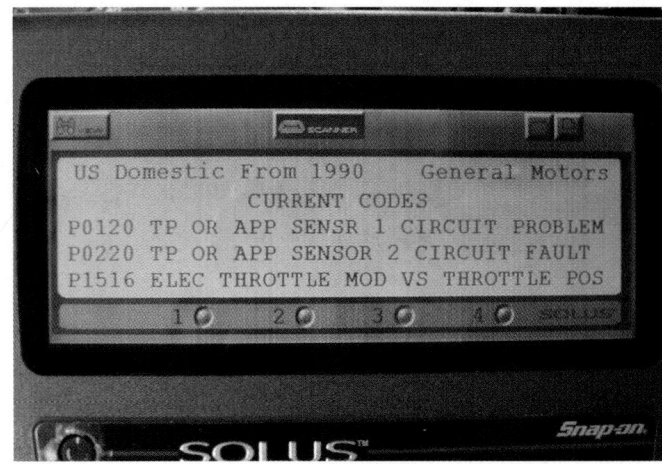

Figure 46.5 Diagnostic trouble codes. *(Courtesy of Tim Gilles)*

Most systems have a diagnostic connector called a data link connector (**DLC**) (**Figure 46.4**). A scan tool can be connected to it to read codes (**Figure 46.5**). The procedure for retrieving fault codes varies. Service literature gives the procedure for each vehicle and describes what each trouble code means. Code numbers on OBD II cars are all the same for a particular emission-related sensor (**Figure 46.6**). Earlier cars had different codes. Older systems call for using a test light, a voltmeter, or a scan tool to read codes. When there is more than one trouble code in memory, the computer will give the lowest number first.

Some imported cars flash on an LED located on the side of the computer. Others flash the check engine light on the instrument panel, and some cars display codes digitally. The easiest way to read codes in all systems is to use a scan tool when possible (**Figure 46.7**).

OBD II codes are covered in detail in Chapter 47.

SCAN TOOLS

A scan tool is a portable computer that reads data from the on-board computer. It is connected to the diagnostic connector (**Figure 46.8**) and provides quick access to diagnostic information.

Late-model vehicles access *live serial data streams* through the DLC.

On-board diagnostics includes parameter identification data (**PID**). These are the processed data used by the engine management program. In OBD II, PID

SAE J2012 standards specify that DTCs have a five-digit numbering and lettering system. The following prefixes indicate the general area to which the DTC belongs

1st Letter
- P — powertrain
- B — body
- C — chassis

1st Number
- 0 — SAE
- 1 — manufacturer

3rd Number (Subgroup)
- 0 — Total system
- 1 — Fuel-air control
- 2 — Fuel-air control
- 3 — Ignition system misfire
- 4 — Auxiliary emission controls
- 5 — Idle speed control
- 6 — PCM and I/O
- 7 — Transmission
- 8 — Non EEC powertrain

The fourth and fifth digits indicate the specific area where the trouble exists. A P1711 code is as follows:

- P — Powertrain DTC
- 1 — Manufacturer-defined code
- 7 — Transmission subgroup
- 11 — Transmission oil temperature (TOT) sensor and related circuit

Figure 46.6 A breakdown of OBD II diagnostic trouble codes.

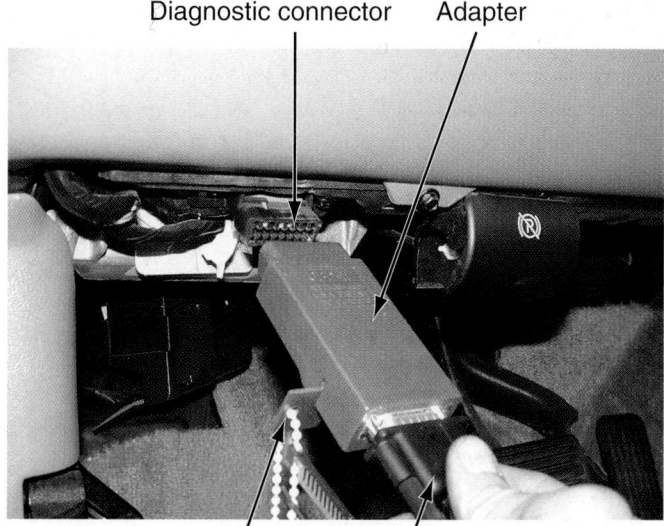

Figure 46.8 An OBD II adapter being used to connect a scan tool to the vehicle's diagnostic connector. (*Courtesy of Tim Gilles*)

Diagnostic connector   Adapter

Program cartridge   From scan tool

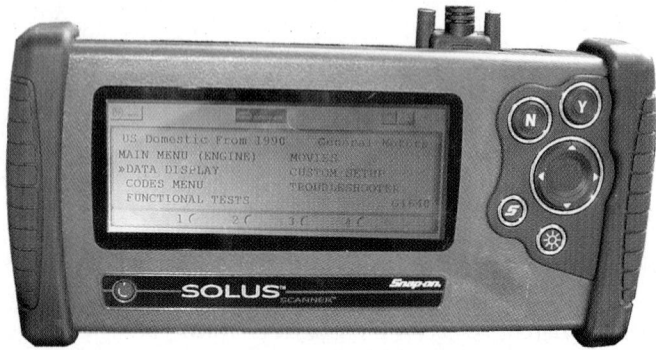

Figure 46.7 A hand-held scan tool. (*Courtesy of Tim Gilles*)

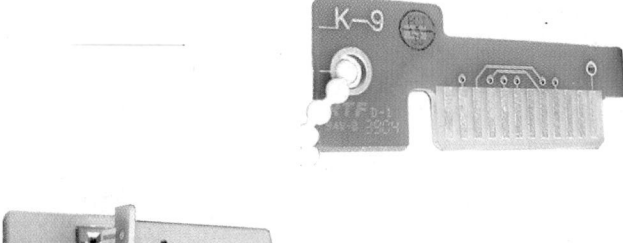
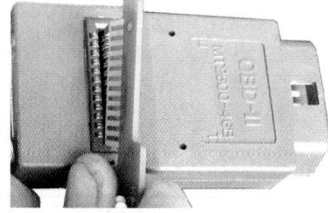

Figure 46.9 An interchangeable program cartridge for the vehicle is inserted into the scan tool. (*Courtesy of Tim Gilles*)

numbers can be displayed on the scan tool. They can be useful in finding a problem when a code has not been set.

Generic vs. Manufacturer Software Cartridges

Service literature includes detailed step-by-step procedures for diagnosis. Scan tools have specific software cartridges (**Figure 46.9**) that sometimes contain service tips and bulletins, as well as specifications for the particular vehicle. Cartridges are regularly updated. This saves the technician time that would have been spent looking up specifications in service information.

Scan Tool Features

The scan tool is hand-held and can be taken on a road test. Live serial data can be displayed on the screen,

or when an intermittent problem occurs, information can be saved for later interpretation. Scan tool data can be saved to a computer; a hard copy can be produced on a printer. This can be used to verify that a problem has been corrected.

The scan tool is limited to diagnosing computer problems. Systems or parts of systems not controlled by the computer will not be diagnosed by the scan tool. When the scan tool lists a problem, that just means to look in that area for a fault. A code could have been mistakenly set during an earlier repair, or there could be a problem with circuit wiring. The computer might also have given a command to an actuator, but a problem with the actuator prevented it from carrying out the command. Symptoms must be followed up to diagnose these problems.

Bidirectional Communications

Communication between the scan tool and the computer can be unidirectional or bidirectional. **Unidirectional** means that the scan tool can read data but cannot give commands to the computer. **Bidirectional** systems can receive commands from the scan tool. Anything that is computer controlled can be operated from the scan tool. For instance, the left window motor can be operated, one headlight can be turned on, or the horn can be honked. Engine actuators can also be operated to check their function. If they function correctly, the problem must be related to the signal on the input side.

■ BREAKOUT BOX

A *breakout box* is sometimes used to diagnose problems. DMM probes are inserted into the pin holes in the breakout box to access various circuits' sensors and actuators through the pin connector to the computer (**Figure 46.10**). A breakout box gives a way of reading what the raw values within the system are. You cannot read processed data.

■ RETRIEVING TROUBLE CODES

When OBD I was on most vehicles, scan tools were not so widely owned by automotive technicians. Manufacturers had different procedures for reading codes at that time. A typical procedure was to follow a series of steps that would trigger the computer to flash codes through the check engine light. The procedure involved cycling the key a number of times or shorting between terminals on the DLC. The code would be flashed on the check engine light. The first code to flash would be the test code 12: one flash, a pause, and then two flashes

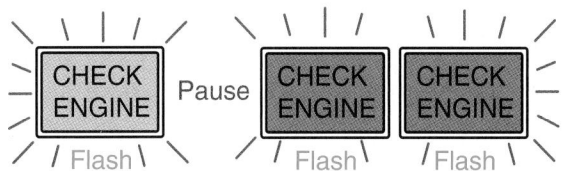

Figure 46.11 Some OBD I vehicles could be directed to display codes through the check engine light. The one shown here is a code 12.

(**Figure 46.11**). Check the service literature for different manufacturers' code checking procedures.

Scan Tool Use

OBD II scan tools are powered through the DLC and do not require an external power source. Do not disconnect or connect the scan tool while the key is on. Before removing connections to electronic components, be certain that the scan tool is removed from the circuit.

■ WORKING WITH CODES

When more than one code is given, fix the lower number code first. A lower code sets a higher code. The lower number code could have caused the higher number code(s) to set. Fix the problem first, then start again working with the lower number codes. Be sure to check power and grounds. After fixing the lower number code first, erase the codes and test drive the car to reset codes from any hard faults that are still present. Then read the codes again.

Reading OBD II Codes

On OBD II cars, a scan tool must be used to read codes. The scan tool can also be used on older models, but it simply reads codes. It cannot read serial data if the system did not have that available. Some late-model vehicles also provide code readout on the instrument panel. OBD II codes are covered in detail in Chapter 47.

■ ERASING TROUBLE CODES

A code can remain in memory even though a problem has been corrected. After repairs have been made in response to codes stored in memory, clear the codes. A procedure that shuts off power to the computer can be followed to erase the codes. Manufacturers' methods for this vary. On some systems, simply pulling the computer fuse from the fuse panel for a short time can do it. A scan tool can erase codes without having to disconnect anything. Disconnecting the battery ground cable will erase the computer's memory, but it will also erase the memory of the clock, radio, power seats, and other things. After erasing the codes, test drive the car to see if it still "sets codes."

Figure 46.10 A breakout box has pin holes for accessing various sensors and actuators.

Test Drive

Before taking a car on a test drive:
■ Write down or save a copy of any codes that are present.
■ On cars with over 50,000 miles, check the backside of the throttle plate for carbon or gum buildup. The buildup results from the cumulative effects of blowby that can cause driveability problems.

On adaptive strategy systems, disconnect the negative battery terminal for 10 minutes to clear computer volatile memory. This will allow the system to reprogram itself. During a failure of part of the system, it might have relearned to compensate for a problem. Erasing the volatile memory is done before fixing computer problems. Be sure to make a note of radio stations and seat positions so you can reprogram them for the customer.

Test drive the car to see if any codes return. These are hard faults. During the test drive, try to duplicate the conditions that the customer highlighted in his or her complaint. If the problem cannot be duplicated, this is an intermittent problem.

■ SCAN TOOL SNAPSHOT

The scan tool has a feature like a flight recorder in an airplane. Information is frozen even if the car has not set a code. This feature is handy for catching glitches and intermittent problems known as soft faults. A *snapshot*, also called an *event* or a *movie*, is like a roll of film. Each picture is a frame. Remember that freeze frame data is not a movie. It is a series of pictures representing the conditions present when the DTC was set. Scan tools vary on the number of pictures that can be recorded. This depends on the computer and the individual scan tool.

A typical snapshot would be as follows. Imagine there are 61 frames on the roll of film. During a test drive, you feel an intermittent problem and press the record button or any number key on the scan tool. The tool freezes (memorizes) the information in the 61 frames that the machine reads, some before and some after you pressed the button.

There is also a setting that automatically records when any fault code occurs during the test drive. The tool can also be programmed to only record when a particular fault code sets. Some tools allow for multiple snapshots, while others will only record one.

Back at the service bay, you can read the information (**Figure 46.12**). Go back ten frames from zero. Zero is when you pushed the button. A throttle position sensor (TPS) might show a value that is off 0.5 volt or so during frame changes in the movie. A TPS

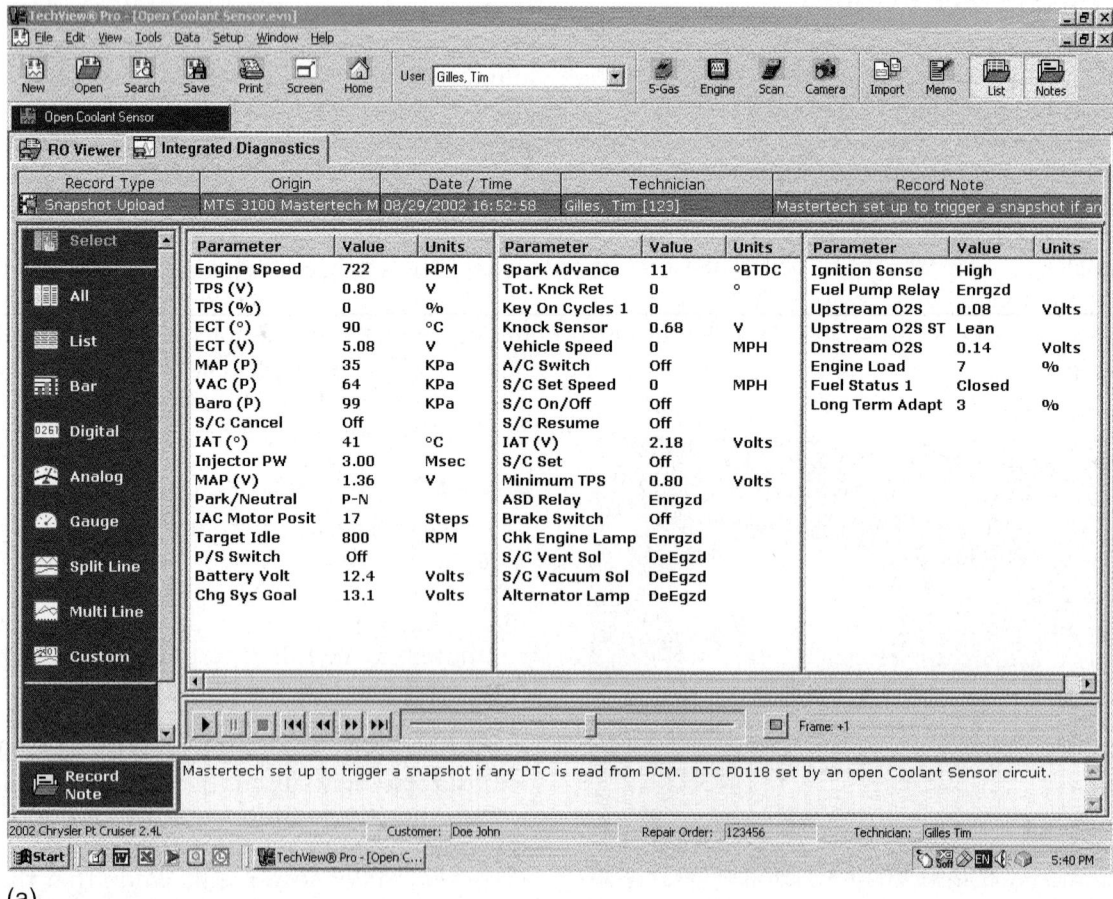

(a)

Figure 46.12 (a) Parameters when a snapshot is triggered. (*Reprinted with permission from Bosch Diagnostics*)

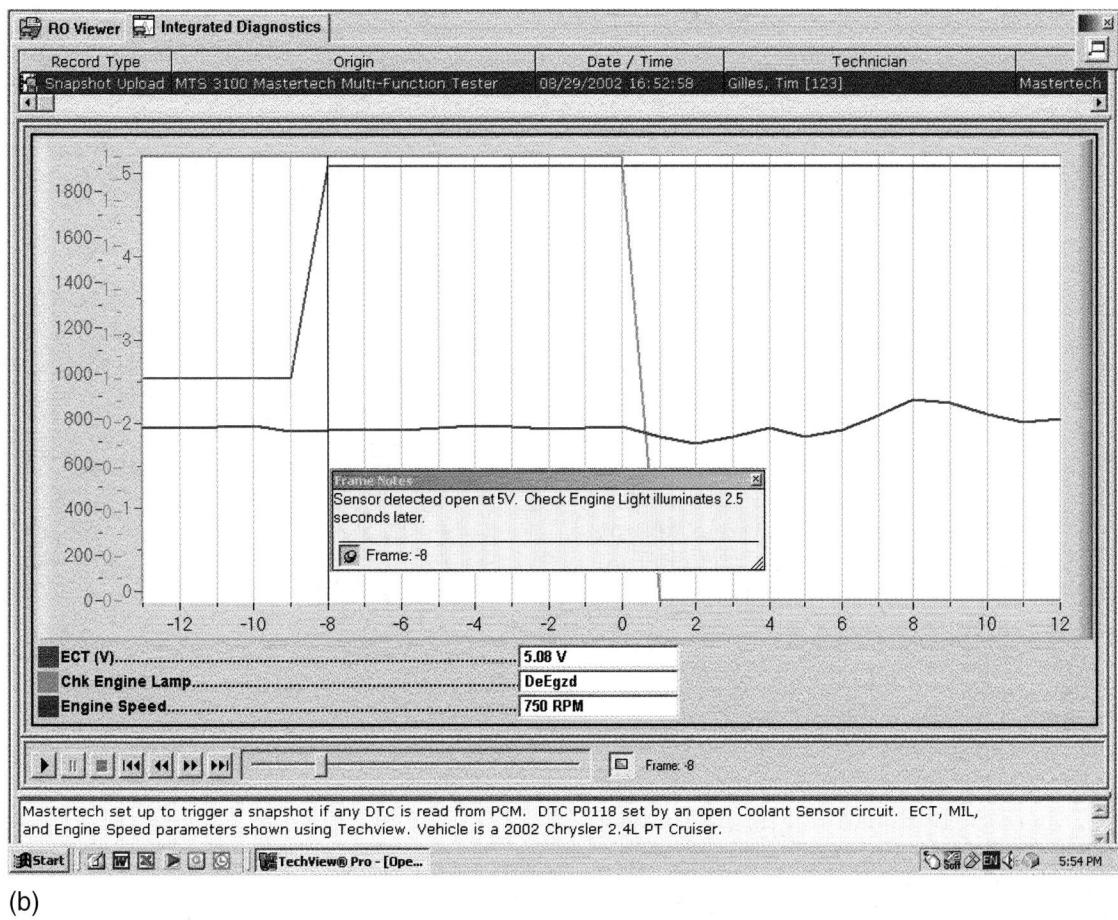

(b)

Figure 46.12 (b) Appearance of a snapshot on a labscope. *(Reprinted with permission from Bosch Diagnostics)*

is a potentiometer with a wiper and a coil. A glitch in the wiper coil could be the cause of the problem. If you do not find a problem, go ten frames ahead of zero. Most problems will be found in this 21-frame span. If a problem is not found, look for an *engine* problem that you might have missed.

There are priorities in the freeze frame data. Fuel problems or misfires are more important to emission control, so they supersede previous freeze frame data. This means the snapshot is covered up with a new one. Once the higher priority snapshot is in memory, nothing can cover it up. If there is a fuel problem and a misfire problem, the one in the snapshot will be whichever one happened first.

Before a test drive with a scan tool attached, be sure to follow the manufacturer's instructions. A GM car that is driven in the wrong test mode can be damaged. Always check the key-on, engine-off (KOEO) data before running the engine. You might find a 4.5-volt signal from a TPS instead of 0.5 volt. This would happen if there was an open in the ground circuit to the TPS. It would have no way to reduce the reference voltage sent by the computer. The computer sees the throttle as being wide open and shuts off fuel while trying to "clear flood." The engine will not start.

Practicing with the scan tool and making comparisons with specifications will give you an understanding of the various acceptable readings to be expected.

Examples of criteria found on the scan tool include:

- engine rpm
- engine load
- state of fuel control
- warm-up status
- O_2 sensor switching ratio
- vehicle speed

When you are looking for glitches, try to duplicate the data. If the car was running at 50 mph under light load with the engine warm, you have something to look for in the freeze frame data.

CLOSED LOOP

The computer requires the correct inputs from sensors and the correct actions from actuators to be able to function properly. When the engine is cold, the system is in open-loop operation (see Chapter 40). During this time, the PCM watches all monitored signals, ignoring the O_2 sensor. It operates the fuel and ignition systems according to preprogrammed parameters.

When the engine has warmed up and goes into closed loop, the computer system receives information from the sensors and decides what action the actuators should take. The result is the best driveability, emission control, and fuel economy for the condition at the time.

Three things must occur for a computer system to go into closed loop. The O_2 sensor must be at operating temperature and start sending signals, engine coolant must reach a specified minimum temperature, and a specified period of time must have passed after the engine is first started. Coolant temperature and time period vary between manufacturers.

When a sensor whose input is needed for the engine to run properly fails to give the correct signal, the computer has a fail-safe or limp-in mode that allows the engine to continue to operate. It will not run at its best, but at least it will run. If the coolant temperature sensor or the O_2 sensor stops working, the system returns to open-loop operation. In open loop, the fuel mixture will be slightly rich and will not vary. Emissions will be higher and fuel economy will be lower.

A faulty thermostat is a common reason for an engine not to operate in closed loop. A thermostat might be stuck so that it does not close all of the way. This can result in too slow a warmup when the car is not driven far enough on a normal trip. Because the engine does not warm up, the computer does not go into closed loop and give feedback to the fuel system. Some people mistakenly install a lower temperature thermostat in an attempt to solve an overheating problem. This causes the same result; no closed-loop operation.

Confirm Closed Loop

If the system is working properly in closed loop, the rest of the computer system is probably working well. Methods for confirming closed loop vary with manufacturers. A DMM, a scan tool, or a lab scope are some common ways.

An artificially rich condition can be created by restricting the air intake. The computer should compensate by reducing the amount of fuel delivery. An artificially lean condition can be created by removing a vacuum hose. Short-term fuel trim should drive the fuel system rich.

Multimeter Test for Closed Loop. Connect a DMM to the oxygen sensor (**Figure 46.13**). Connect the positive lead to the jumper wire and the negative lead to a good ground. Set the meter to DC volts.

NOTE: *Do **not** use an ohmmeter to test an O_2 sensor. To measure resistance in a circuit, an ohmmeter puts voltage through the circuit. Putting voltage into an O_2 sensor can change the reading it responds to.*

Firmly apply the parking brake and run the engine at 2,000 to 2,500 rpm until the cooling fan comes on. When the vehicle goes into closed loop, the voltage will rapidly swing above and below 0.5 volt as it goes from rich to lean. Although you will not be able to see

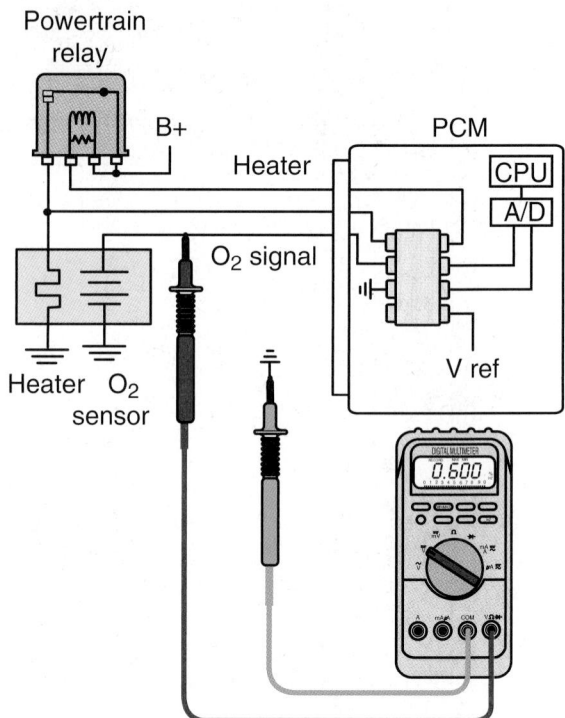

Figure 46.13 Wiring schematic for connecting a DMM to an oxygen sensor.

this with the DMM, a good system will switch between rich and lean several times a second. The speed at which this switching takes place is called cross counts. Late-model computers will pick up an O_2 sensor that is of low quality or has become slower and set a code.

A slow O_2 sensor can cause a stumble during acceleration. After a period of slow operation, the oxygen sensor fails. When this happens, it usually goes rich or stays in open loop. The result is worse fuel economy.

■■■ SCAN TOOL DIAGNOSIS OF O_2 FEEDBACK

With a scan tool, data are displayed showing that feedback fuel control is occurring. Fuel trim allows the computer to change the air-fuel mixture average to either side of stoichiometric to compensate for small problems.

Analyzing scan tool fuel trim is a good diagnostic tool. On an OBD I GM Block Integrator and Block Learn system, a reading of 128 on the scan tool is stoichiometric. Lower numbers mean the computer is attempting to correct for a rich condition. Higher numbers mean that the computer is correcting for a lean condition. The farther away the number is from 128, the worse the condition is.

Short- and long-term fuel trim, the standard used in OBD II vehicles, works the same way but the reading is a percentage of correction. The equivalent to GM's 128 is 0%. A positive percentage indicates that the PCM is adding fuel to compensate for a lean exhaust condition (**Figure 46.14**). A rich condition causes a negative percentage (**Figure 46.15**). If the short-term

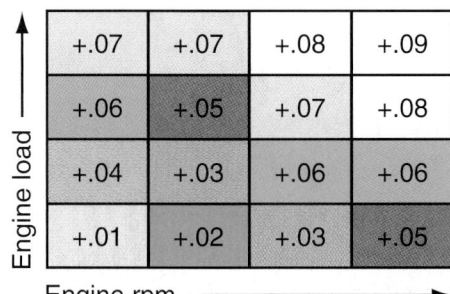

+.07	+.07	+.08	+.09
+.06	+.05	+.07	+.08
+.04	+.03	+.06	+.06
+.01	+.02	+.03	+.05

Engine load →

Engine rpm →

Figure 46.14 A positive percentage on the fuel trim look-up table indicates that the PCM is adding fuel to compensate for a lean exhaust reading.

−.04	−.03	−.03	−.03
−.06	−.05	−.05	−.04
−.08	−.09	−.08	−.06
−.10	−.10	−.09	−.07

Engine load →

Engine rpm →

Figure 46.15 A rich condition causes negative percentages on the look-up tables.

+.04	+.05	+.06	+.09
−.01	+.01	+.05	+.03
−.03	−.02	−.01	+.01
−.05	−.04	−.03	−.01

Engine load →

Engine rpm →

Figure 46.16 A fuel trim table with some cells rich and other cells lean.

trim cannot correct a problem, then the long-term trim will try to bring the mixture back to 0%.

On most vehicles, short-term fuel trim (STFT) and long-term fuel trim (LTFT) data are available on the scan tool. You will need a good understanding of fuel trim in order to be able to use this information for diagnostic purposes. Some newer vehicles use more than the 16 cells in the illustrations shown here. The fuel trim cells or blocks are called electronic look-up tables. They are held in RAM, where the PCM can refer to them as it assigns values to each cell. The cells control injector pulse width depending on driving conditions. This is called load cell mapping. With mass airflow sensor fuel injection systems, increased airflow means engine load has been increased. Therefore, the fuel trim moves into a higher load cell to increase fuel delivery. Load cell maps are more complicated on speed density fuel injection systems, with cells representing specific engine rpm and vacuum (load) levels.

STFT values constantly change according to what the O_2 sensor sees. STFT changes the injector pulse width to drive the O_2 sensor rich when it sees a lean condition and vice versa. This keeps the O_2 sensor switching back and forth.

During closed loop operation, STFT is not stored by the PCM. The problem with STFT is that it takes time to respond and make a correction for what it sees in the exhaust pipe. LTFT is when the PCM learns what STFT is doing so it can make a correction to the base pulse width ahead of time.

LTFT wants to keep the O_2 sensor average as close to 450 mV so the switch rate does not vary too much. Therefore, it makes a change in the cell for any particular driving condition that causes too rich or too lean a condition.

Figure 46.16 shows a fuel trim look-up table for an engine that is running rich in the lower cells and lean in the higher cells. Up to a certain point, LTFT will compensate for the problems that are causing this. However, if the computer memory is erased, the PCM will have to relearn the best LTFT for each driving range. If a problem such as a vacuum leak is bad enough that fuel trim cannot compensate, this is known as the threshold limit.

Diagnosing with Fuel Trim

The following are some things to keep in mind when using a scan tool to diagnose driveability problems while watching fuel trim:

- Drive the vehicle under the same conditions where the problem occurred, and look for a block that is out of range.
- A restricted fuel filter or low fuel pump output will increase fuel trim under load.
- Leaks that allow air into the intake system will result in higher fuel trim values in the idle cells but normal values in the higher speed cells. This is because the air leak is minimal to the engine's total airflow at higher speeds.
- A plugged or sticking fuel injector will affect fuel trim cells equally as rpm increases.

SHOP TIP After erasing codes or replacing a fuel system part, reset the adaptive fuel trim using the scan tool. This provides a much faster base than letting the system relearn on its own.

OBD II systems use dual O_2 sensors. The readings are displayed on the scan tool and can be compared to see if there is a problem in one side of the system or the other. If both sides are the same, a problem is located in something common to them, like fuel pressure, battery voltage, grounds, or the computer. Chapter 41 covers diagnosis, service, and replacement of O_2 sensors.

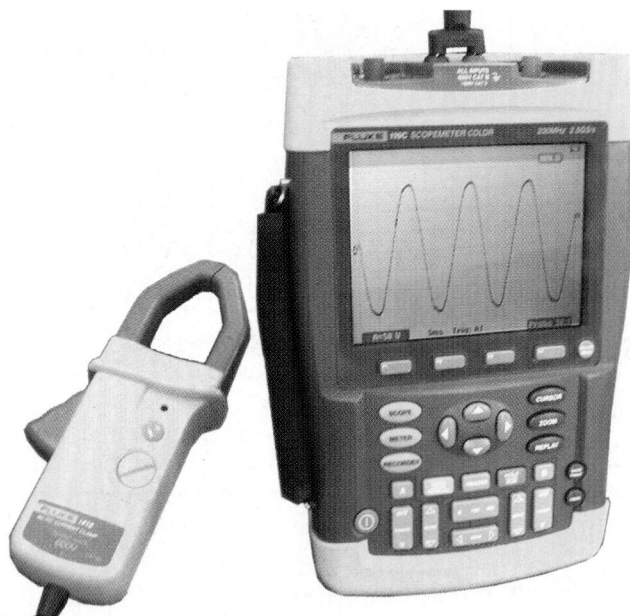

Figure 46.17 A hand-held scope with an inductive clamp. *(Courtesy of Tim Gilles)*

■ DIGITAL WAVEFORMS

A digital volt-ohmmeter does not display in real time; instead, it displays an average reading that "refreshes" periodically. You might miss a glitch while reading a DVOM. Watching O_2 sensor voltages is one example. The meter is not fast enough to keep up and might miss a quick change that would highlight a problem. A waveform would be a better diagnostic tool because its time base can be expanded to catch a problem that occurs in a very short time frame. **Figure 46.17** shows a hand-held scope used with an inductive clamp.

Lab Scopes and Graphing Multimeters

Digital storage oscilloscopes (DSOs) (**Figure 46.18**) and graphing multimeters (**Figure 46.19**) are two

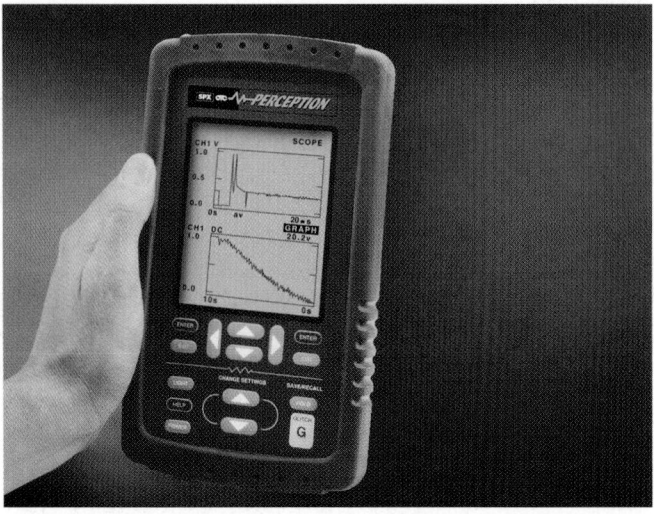

Figure 46.18 A hand-held DSO displaying two waveforms. *(Courtesy of OTC/SPX Service Solutions)*

Figure 46.19 A scan tool with graphing capability. *(Courtesy of Tim Gilles)*

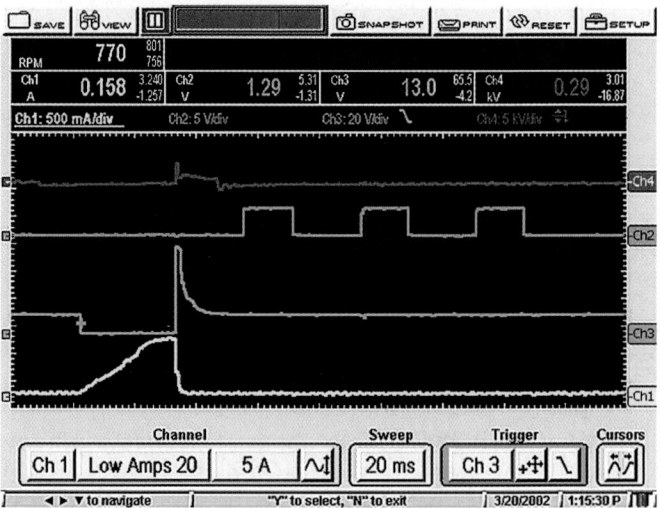

Figure 46.20 A screen showing four waveforms. *(Courtesy of Snap-on Tools Company)*

diagnostic tools capable of displaying voltage or frequency in waveform. Either tool can display multiple waveforms from two to four different sources for comparison purposes (**Figure 46.20**). A scope capable of displaying four waveforms is called a *four-channel scope*. Waveforms can show AC, DC, RFI noise, and frequency all at once. **Figure 46.21** shows how four waveforms or six multimeter functions can be displayed on a four-channel DSO.

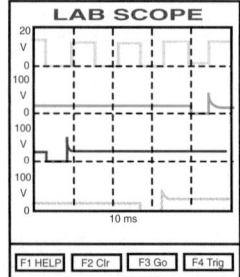

Figure 46.21 On a digital storage oscilloscope four waveforms or six multimeter functions can be displayed at once.

Glitch

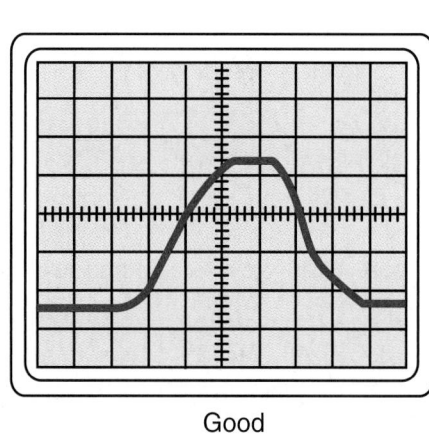 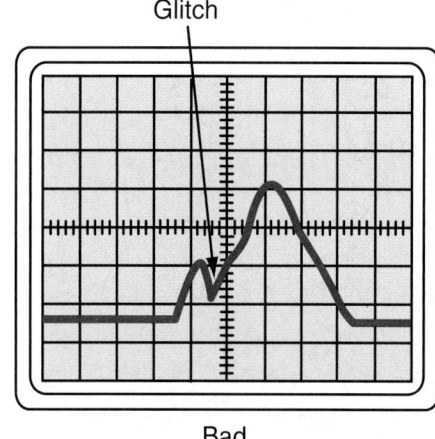

Good   Bad

Figure 46.22 The left waveform shows a good TPS pattern as the throttle is opened and closed. The right pattern shows a defective TPS.

While digital voltmeters can sample information several times a second and scan tools can freeze information that the computer sees, a DSO can do both. A DSO reads in 40 billionths of a second (NS). It measures voltage and time like an ordinary oscilloscope. There are advantages to being able to watch a waveform instead of seeing numbers that are changing very rapidly. For instance, you might want to compare an O_2 sensor waveform with injector pulse width, TPS, and MAP sensor waveforms occurring at the same time. With a faulty TPS, a waveform can show a dropout in the pattern as the throttle is moved from closed to open and the wiper arm crosses the area of greatest wear, which is most likely to occur between idle and part throttle. A scan tool might not be able to see this. **Figure 46.22** shows DSO patterns for good and bad sensors.

These tools do not replace the scan tool, which reads serial data and fault codes, but they are sometimes helpful in finding intermittent problems when there is no DTC to point you in the right direction. Some of the better scan tools also include DSO and graphing multimeter functions.

Some older DSOs require more complicated adjustment, but most automotive scopes are user-friendly. You will need to adjust the voltage scale and time base so you can view the entire waveform or blow it up to see part of the pattern in greater detail. On some scopes, you will also need to adjust the *trigger*, which starts the scope pattern. Other scopes automatically display the signal in a preprogrammed position.

Lab scope patterns for different sensors are shown in the sensor testing section of this chapter, and ignition patterns are covered in Chapter 38.

A DSO is often used with a low current probe. This tool is popular because it is fast, accurate, and easy to use. Many manufacturers still give electrical specifications as resistance values, but current ramping is easier because it is noninvasive. Measuring resistance requires a circuit to be disconnected.

Fuel pumps can be tested using a DSO and low current probe.

SHOP TIP Remember that current flow is the same everywhere in a series circuit, so you do not need to be able to access the fuel pump in order to see its current waveform using your low current probe. You can clamp anywhere in the circuit. You can even clamp around several wires in a loom and look for the right type of pattern for a fuel pump. **Figure 46.23** shows a current probe connected around a jumper wire installed in place of a fuse.

As a fuel pump gets older, it begins to turn slower. If you know the number of bars on the commutator in the pump, you can do the math and determine the rpm of the pump.

Figure 46.23 A low current probe accessing the fuel pump circuit by clamping a jumper wire installed in place of the fuse. *(Courtesy of Tim Gilles)*

The speed of an electric fuel pump can be calculated with a DSO waveform of AC coupled current. An easy way to acquire this pattern is with a current probe connected to the ground circuit of the pump. Commutators have different numbers of bars depending on the manufacturer. You can determine the number of bars by examining the waveform. One pump revolution provides a pattern that joins with the next to form a continuously repeating waveform. **Figure 46.24** *shows a pattern for an electric fuel pump. The dips in the pattern represent the gaps between the commutator bars in the armature. The pattern from a worn pump is not perfect, so there will be a signature you can use to determine where the repeating pattern begins and ends. After you have determined where one rotation begins and ends, you can count the number of AC humps and use the pattern's time-per-division setting to determine the speed of pump rotation. There are approximately eight pattern humps. At a speed of 2 milliseconds (0.002 second) per division on the scope, the complete repeating pattern occurs in 9 milliseconds. Therefore, this is the time it takes to complete one pump revolution.*

To determine the rotation speed of the pump:

1. *Divide the rotation time into 1 second:*
   1 second ÷ 0.009 = 111
2. *Multiply the result by 60 seconds. This is the speed of the pump:*
   111 × 60 seconds = 6,660 rpm

Counterelectromotive force (CEMF). On a scope pattern, voltage reacts very fast because it is not subject to resistance and CEMF, but current has to react to its own

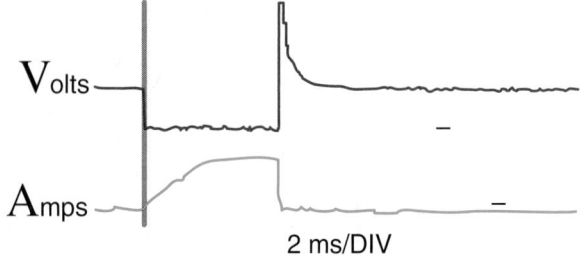

Figure 46.25 Waveforms comparing voltage and amperage for a fuel injector. Current (amps) waveforms tend to be curved due to CEMF, or inductive capacitance.

magnetic field as it flows into a circuit. Scope patterns for amperage tend to be curved (**Figure 46.25**). This is because as current flows into a circuit, a magnetic field builds around the wire. If the wire is straight, the magnetic field will be weak, but when a wire is wound in a coil, a stronger magnetic field builds around the wire. The more coils in the wire, the stronger the magnetic field. Ignition coils, fuel injectors, and solenoids are all magnetic devices. Current can easily begin to flow into them, but as the magnetic field around the coil begins to increase it impedes further current flow. If you look at the current ramp in **Figure 46.26**, you will notice current flowing quickly at the beginning but tapering off as time goes on.

Current reacts in a similar way to magnetic impedance as it does to a resistance. If you wrap a wire in a coil and put it around a nail, you can connect both of its ends across a battery. A magnetic field will build in the nail, but the wire will not burn. However, if you straighten out the wire and connect it across the battery, the wire will get very hot and burn. Impedance is what prevents the coiled wire from flowing too much current.

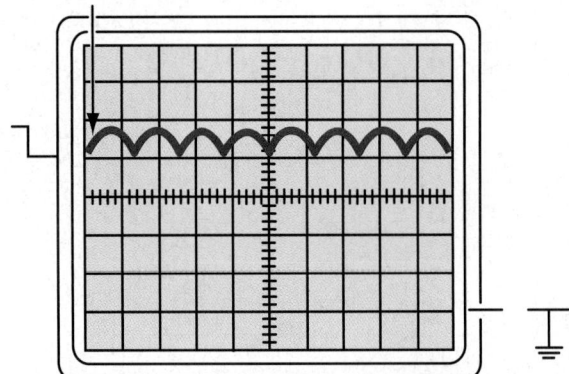

Figure 46.24 A pattern for a motor such as a fuel pump shows dips that represent the gaps between the commutator bars in the armature.

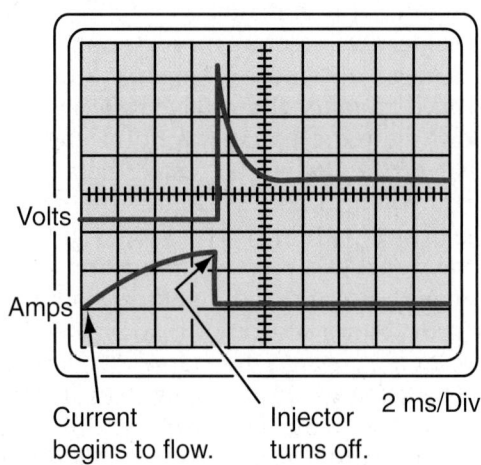

Figure 46.26 In this typical current ramp for a standard fuel injector, current flow builds quickly and then tapers off as the building magnetic field impedes current flow.

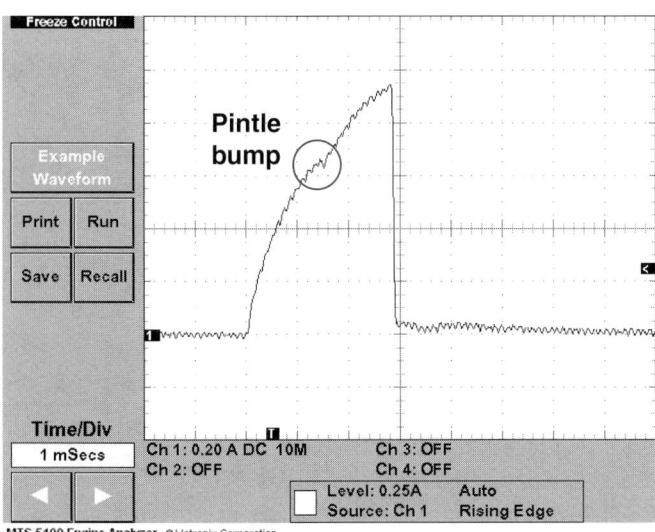

Figure 46.27 A pintle bump in the current ramp occurs when the injector pintle moves open, changing the magnetic field momentarily.

When using a low current probe to read waveforms, zero it while holding it near the circuit you are working on. The pickup is inductive, which means it reacts to magnetism and puts out a voltage equivalent to the meter or scope. Current flow, and therefore magnetism, will vary with temperature and inductive reactance. The zero will tend to creep up or down. If a stable, exact reading is important, you will need to re-zero the meter often.

A DSO provides a good example of inductive reactance when looking at a waveform showing current flow in a fuel injector. With the pattern spread out across time, note the "pintle bump" that forms in the pattern when the fuel injector pintle moves inside the injector's coil (**Figure 46.27**). When the pintle moves open, the strength of the magnetic field surrounding the injector varies momentarily. This causes current flow into the injector coil to change, creating the bump in the current pattern.

> **SHOP TIP** An easy way to spot late-opening injectors on a majority of vehicles is by using a DSO with a current probe to compare fuel injector pintle bumps between cylinders.

▄▄ LOGIC PROBE

A logic probe is like a test light, but instead of one light bulb, it has three colored LEDs (**Figure 46.28**). A logic probe has two leads instead of one. The red lead is connected to the battery positive (B+), which powers the tester. The black lead is attached to ground. This provides the probe with a reference. Touching the probe to ground lights the green LED. The red LED illuminates when touched to a power source. The yellow light comes on when a pulsed voltage, like a fuel injector, is sensed.

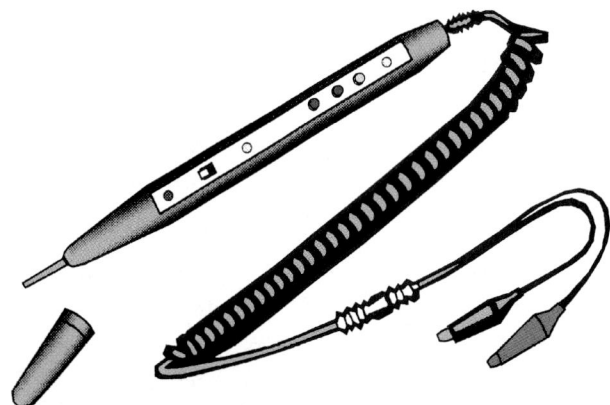

Figure 46.28 A logic probe with three LEDs.

▄▄ SENSOR AND ACTUATOR TESTING

Sensor Testing Strategy

A no-start condition can occur if a distributor reference or crankshaft sensor signal is lost. The computer cannot substitute for the missing information, and the engine will not start. This is also a good place to look for a problem when a vehicle dies intermittently.

A defective or misadjusted TPS can send an excessive voltage to the computer that is interpreted as WOT. The computer thinks the engine is flooded because the throttle is being held to the floor during engine cranking, so it shuts off the fuel and the engine will not start.

NOTE: *With the key on, do not disconnect any electrical components unless instruction specifically says to do this.*

▄▄ DIAGNOSING SENSOR PROBLEMS

When using a scan tool to diagnose sensors and actuators, use the following sequence to perform a quick check.

Properly operating sensors are necessary for an engine to run smoothly during all operating conditions. The signals listed here are listed in the order of probability that they might cause a driveability problem. The higher on the list, the more probability of a problem (**Figure 46.29**).

■ First, check the input sensors.
■ Then, perform a quick check of input switches.
    This includes checking in park, reverse, and neutral

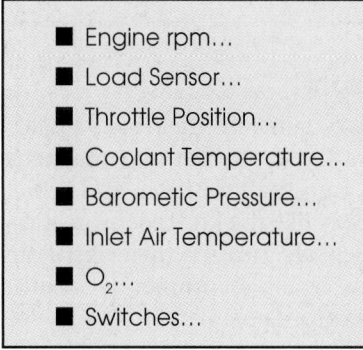

Figure 46.29 Sensors listed in the order of probability of causing a driveability problem.

to see if the neutral start switch is saying the engine is loaded in gear or unloaded in neutral. Work the power steering by turning the steering wheel to see if a load is sensed, raising engine rpm.

- Next, check outputs. Check injector pulse width and rpm with the key on and engine off (KOEO). The idle air control needs a startup position, so it should not be on "O."

Check spark advance to see that it works. You do not need to check the accuracy of the advance curve. Just see that advance is working as the engine speed is raised.

Check to see that the air switching valve (if so equipped) is changing from upstream to downstream.

Some engine sensor signals have a higher priority signal than others. These **high authority sensors** include the following:

- Oxygen sensor (in closed loop only)
- Mass airflow (MAF)
- Manifold absolute pressure (MAP)
- Engine coolant temperature (ECT)
- Throttle position (TP)

Low authority sensors include idle air temperature and barometric (BARO) pressure (when it is not part of the MAP sensor).

If a high authority sensor is disconnected, the computer will substitute a replacement value. If the engine runs better with the sensor disconnected, the sensor is either measuring incorrectly or it is defective.

Passive Sensors

Passive sensors do not generate their own voltage. These include temperature sensors, position sensors, pressure sensors, and airflow sensors. All of these respond to a reference voltage from the computer. Check that the reference voltage signal is the correct amount, which is usually 5 volts.

> **SHOP TIP** When diagnosing a possible coolant temperature sensor problem, keep in mind that the CTS can only add fuel, not take it away. Fuel injection base pulse width is determined before modifications due to battery voltage, coolant temperature, or throttle position.

Active Sensors

Active sensors generate their own signal. Included in this group are the O_2 sensor and magnetic sensors for engine speed, cam shaft position, and vehicle speed; piezoelectric sensors like the knock sensor; and solar sensors. To test these sensors measure their signal output.

The following are recommended methods for diagnosing sensor problems:

- Voltage generating sensors—digital meter
- Switching type sensors—ohmmeter and voltmeter

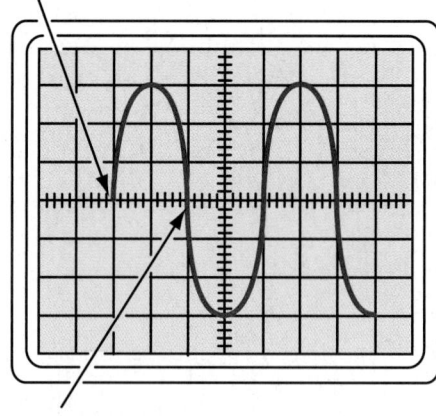

When the pole aligns with the magnet

When the pole is exactly between reluctor poles

Figure 46.30 The effect of magnetism on the waveform of a magnetic sensor.

- Variable resistance sensors—ohmmeter and voltmeter, voltage drop

NOTE: *Remember that the frequency and amplitude of a magnetic sensor change with speed. The signal should continually grow and become closer together with speed. Figure 46.30 shows how magnetism affects the waveform.*

■ SENSOR TESTS

Vehicles have many different sensors. Their descriptions and testing procedures are given in this section.

Vehicle Speed Sensor

A vehicle speed sensor (VSS) supplies input for electronic speedometers and cruise control systems. It is also used to control the torque converter clutch, electronic transmission shift control, some emission controls, variable assist power steering, and electronically adjusted shock absorbers. There are two types of speed sensors. One is photoelectric, and the other is a magnetic AC generator.

A failed sensor or sensor circuit can cause premature or no converter clutch lockup, lack of change in steering assist, or inoperative or inaccurate cruise control and speedometer operation. The system is most easily tested by connecting a scan tool and running the drive wheels with the vehicle off the ground.

NOTE: *Front-wheel-drive cars should not have the front wheels run while they hang unsupported. Place jackstands under the lower control arms before running the drivetrain.*

A speed indication should be read when speed exceeds 3 mph. Codes are set when the sensor does not work properly.

■ OXYGEN SENSOR

The O_2 sensor's job is to enrich the mixture enough so the reduction (NO_x) catalyst can work. It must also provide a lean enough mixture for HC and CO to oxidize. The mixture must go across stoichiometric from rich to lean in order for the catalytic converter to work efficiently. A lab scope or a scan tool is used to look at O_2 sensor operation to tell if it is going rich and lean in the correct amounts and in the right amount of time.

Older O_2 sensors did not work when cold, so a cold driveability problem (such as hard starting, poor idle, or a lack of full power) could not be related to a defective O_2 sensor. Operation and service of the oxygen sensor is covered in Chapter 40 and Chapter 41.

The following are some items for consideration when testing an O_2 sensor:

- An O_2 sensor can only cause fuel trim to compensate by changing the air-fuel mixture by a small amount. It cannot cause a no-start condition.
- A voltage reading that is low could indicate a failing sensor, sending a lean signal, which will cause the computer to drive the fuel system rich.
- A lean condition in the fuel system could also be the cause of a low voltage reading.
- Air pumps on cars with three-way catalytic converters have a switch that pumps air on either side of the converter (*upstream* or *downstream* of the CAT). If an air pump is putting air into the system before the O_2 sensor (upstream) during closed loop, the sensor senses the extra air in the exhaust and registers a lean mixture. The computer responds by increasing the fuel injector pulse width.

 **CASE HISTORY**

A Jeep blew an air pump, and pieces of the fiber vanes blew into the air hose downstream of the pump. A small piece became stuck in the upstream valve on the diverter. Some air was always allowed to pump upstream even in closed loop. This resulted in the computer driving the fuel system rich whenever the engine was in closed loop.

- On a four-cylinder engine, the effect of a vacuum leak is much worse at idle than it would be with a six- or eight-cylinder. The O_2 sensor will stop working at idle, and the system will lose its ability to correct the air-fuel ratio. It returns to operation when on the road.
- With a bad spark plug, the misfire that results causes the engine to use less O_2. When there is more oxygen in the exhaust, there is a lower voltage from the O_2 sensor (**Figure 46.31**). Because it senses that the mixture is lean, the computer drives the system rich.

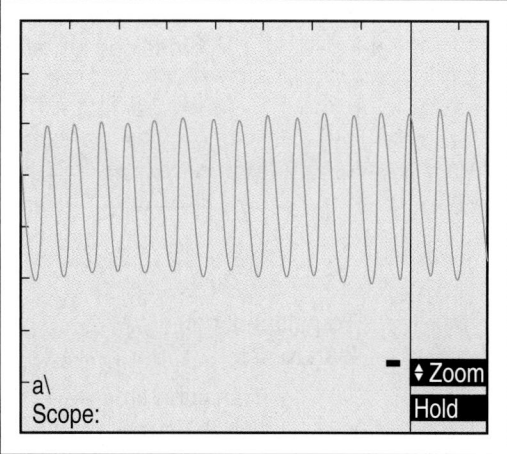

(a)

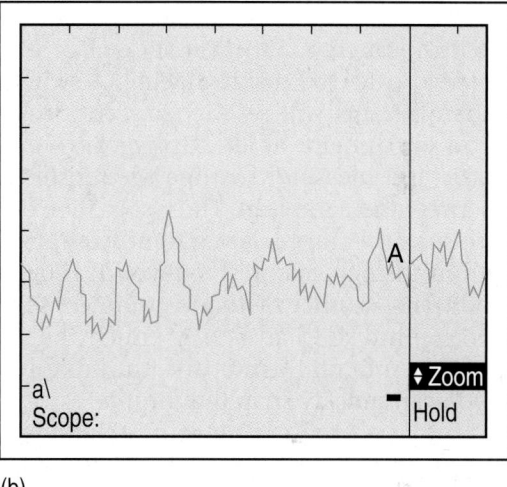

(b)

Figure 46.31 (a) A normal DSO oxygen sensor pattern and (b) one when a plug wire is open, causing the engine to use less O_2.

- An engine running at the stoichiometric ratio (14.7:1) will produce a reading of about 1% to 1.5% O_2 in the exhaust.

Oxygen Sensor Condition

Some newer heated oxygen sensors (HO_2S) become hot and start to cycle almost immediately after a cold engine startup. Other newer sensors can take up to 30 seconds to begin working. At first a typical sensor switches slowly, but as it becomes hotter, it responds more quickly to STFT changes. The PCM monitors the amount of time it takes for the HO_2S to become active and sets a code if it is too slow.

A "lazy" sensor is one that produces voltage slowly and then does not change back and forth between rich and lean signals fast enough. A properly operating sensor will move between 0.2 and 0.8 volt quickly. For an O_2 sensor to be considered to be fast enough on newer systems, it will require a specified cross count speed or switch rate from high to low (**Figure 46.32**) that is

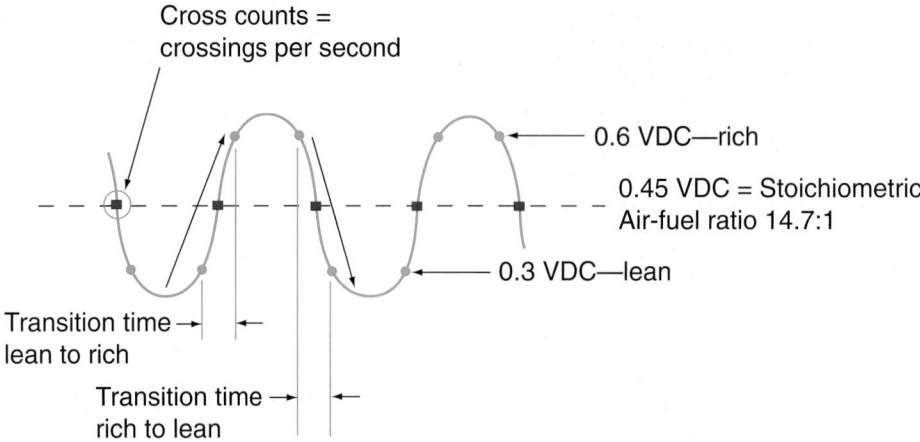

Figure 46.32 Oxygen sensor cross counts or switch rate.

faster than 100 ms (ten times per second). New sensors can switch in less than 50 ms. If the sensor is switching at a speed near 100 ms, it is aging. A switch speed slower than 100 ms will set a code. Perform a switch rate test on a hot engine at idle. Using a lab scope, snap the throttle and measure the time the sensor takes to switch between rich and lean. The test is done on a lean engine, so snap the throttle again right away after deceleration from the first snap test to take advantage of the lean condition. Another method is to inject some propane into the intake to artificially enrich the mixture. As soon as you shut off the propane, the mixture will go lean so you can quickly snap the throttle.

Test an oxygen sensor's range by creating full rich and full lean conditions. When a large vacuum hose is pulled (from the power brake booster, for instance), the voltage should go to at least as low as 0.2 volt. A propane enrichment test uses a propane bottle, hose, and metering device. Propane can be metered slowly through the end of the hose that was removed to create the vacuum leak. Add propane until the maximum voltage value is reached (rich mixture showing on the voltmeter). When the propane is pulled away from the hose, the voltmeter should respond immediately by dropping the voltage.

Replacement O_2 Sensors

Check the vents in the thimble of a replacement O_2 sensor. There should be the same number of holes and they should face clockwise or counterclockwise like the ones on the original sensor. Installing the wrong sensor can result in slower cross counts.

Titanium dioxide O_2 sensors do not produce their own voltages, so they run from 0 to 5 volts. Some of them work backwards of normal, with a higher voltage being a leaner mixture. Check the manufacturer's service information if a voltage higher than 0.9 volt is encountered.

Wide Range Oxygen Sensor Service

Wide range O_2 sensor operation is discussed in Chapter 40. These O_2 sensors are capable of accurately detecting air-fuel ratios over a wider range than an HO_2S and can determine how much leaner or richer the air-fuel ratio is than lambda (1.000). Remember that a wide range O_2 sensor is actually two zirconia sensors, one within the other, and the amount of energy difference between the two sensors is what determines the air-fuel ratio. The computer wants to prevent the sensor from cycling like a normal O_2 sensor, so the PCM cycles current to the pump cell, moving O_2 in and out of the sensor to maintain O_2 sensor output at a constant voltage. The outside part of the sensor is measuring exhaust O_2 content, while the outside of the inner sensor samples outside air. Rather than the outside sensor comparing O_2 content in the exhaust stream to O_2 content in the air outside, a wide range O_2 sensor provides a comparison to O_2 in the sealed diffusion chamber separating the two halves of the sensor controlled by the PCM.

Consult the computer information library before diagnosing a problem with a wide range O_2 sensor. The following provides you with an understanding of the wide range sensor and its interaction with the PCM.

The sensor used in this example has five wires; two are for the heater. Of the remaining three, the first wire (the cell voltage input) is like a simple O_2 sensor output wire, sending a rich or lean output from the outside sensor to the PCM. Instead of reading the difference between the outside air and the exhaust like an ordinary HO_2S, however, the outside sensor compares the exhaust to the oxygen content in the diffusion chamber.

The second wire to the sensor goes from the outside of the inner sensor to the PCM. It is called the pump cell control. The PCM uses this wire to increase or decrease the O_2 level in the diffusion chamber. Applying voltage to this wire causes current to flow through the diffusion chamber as the PCM adjusts O_2 content to provide a 0.45-volt signal output to the outer sensor element.

The third wire is a reference voltage wire that acts as ground between the inner and outer sensors. It is not true chassis ground, but if you were to measure voltage on a wide band O_2 sensor, you would find that the difference between the pump cell voltage input wire and the reference wire is approximately 0.45 volt.

Rather than sensing voltage, the computer monitors the changes it makes in amperage. A rich mixture gives a negative current, and a lean mixture gives a positive current. The computer is able to determine the real air-fuel ratio by monitoring the pump cell control wire to see how much current is required to keep the outer sensor's signal at 0.45 volt.

■ LOAD SENSORS

Load sensors include MAP, vacuum, and MAF. They tell the computer how much air is entering the engine and affect ignition timing and air-fuel ratios. MAF sensors are covered in Chapter 40.

■ MAP SENSOR

Basic fuel delivery to the engine is determined by the MAP sensor and rpm signal. The other sensors support these two. When engine load is high, the fuel injectors remain on longer. The sensor ignores the signal from the O_2 sensor under high load. You can cause the mixture to go very rich by disconnecting the vacuum supply to the MAP sensor. The engine will run extremely rich and possibly die. The computer also retards timing under high load.

When there is a problem in the MAP sensor, its circuit, or its vacuum connection, driveability symptoms can include detonation, power loss, stalling, a rough idle, and poor fuel economy. If there is a problem with

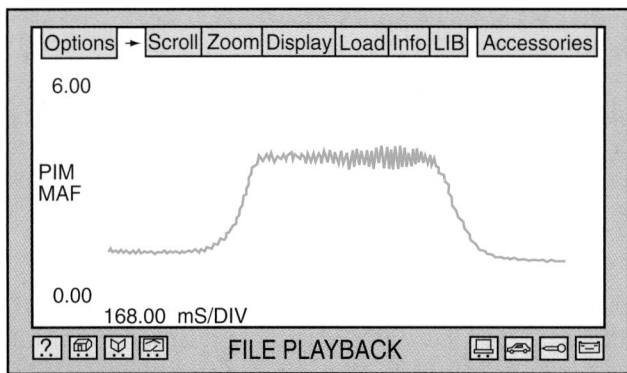

Figure 46.34 A normal digital storage oscilloscope (DSO) MAP sensor pattern.

the MAP sensor vacuum hose or an engine vacuum leak, the computer "sees" high load. It compensates by lengthening the fuel injection pulse width and retards the ignition timing.

A MAP sensor produces a voltage that drops as vacuum becomes higher (**Figure 46.33**). This means the voltage should drop when the engine is quickly accelerated and then returned to normal (**Figure 46.34**). Diagnostic codes will be set when readings are different from expectations. MAP sensor operation can be checked with a scan tool connected to the diagnostic connector. The sensor can also be tested with a digital voltmeter connected to it while applying vacuum to the sensor. With a 5-volt reference signal, a typical MAP sensor will read close to 5 volts with no vacuum signal. Voltage will drop in steps until it reaches about 1 volt with 20 inches of vacuum applied. There is also a digital MAP sensor (Ford) that produces a frequency signal (Hz) that changes based on vacuum conditions.

■ BARO SENSORS

Barometric pressure (BARO) sensors are used on some systems to monitor changes in the weather or altitude. Less air requires less fuel and less timing advance so the computer adjusts to compensate. Some systems combine the BARO sensor with the MAP sensor. When the key is turned on, altitude is read by the computer.

NOTE: *If you drive from low altitude to high altitude, the computer does not change until the key is cycled off and on again. When driving through a big change in altitude, it may be necessary to stop and shut off the engine. After restarting, the computer will compensate for the change in altitude.*

There are different types of BARO sensors. Most work like a MAP sensor, producing a voltage signal that becomes less as altitude increases. Trouble codes can indicate a problem. Testing can be done with a

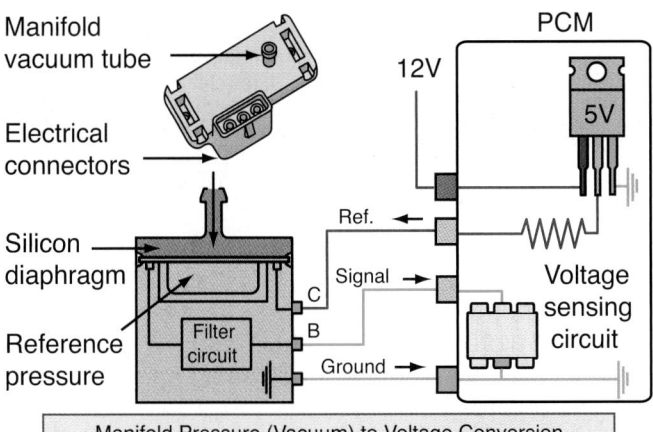

Manifold Pressure (Vacuum) to Voltage Conversion		
Absolute Pressure	Vacuum	Sensor Voltage
14.7 psi	0"	4.5 V
10.5 psi	6.5"	3.75 V
7 psi	13"	2.5 V
3.5 psi	19.5"	1.25 V
0 psi	26"	0.5 V

Figure 46.33 MAP sensor voltages at low and high vacuum.

scan tool. Some scan tools will give actual pressure readings.

A defective sensor or sensor circuit can cause poor high-altitude performance or spark knock because the air-fuel ratio and timing will not be correct. If a sensor reading is in the correct range, apply vacuum to it and watch for a change in voltage output.

■ VACUUM SENSORS

Some systems have vacuum sensors that measure the difference between atmospheric pressure and intake manifold pressure (below the throttle plate). A MAP sensor compares the reading inside the engine to an absolute pressure calibration in the computer. Systems that use vacuum sensors must also use a barometric pressure sensor.

NOTE: *Computer systems that do not have a MAP sensor use a TPS and an airflow sensor to determine the load on the engine.*

If the engine has no BARO sensor, the MAP sensor is a BMAP (combination barometric and MAP) sensor. When the ignition key is on but the engine is off, the BMAP senses barometric pressure. At high altitude, there is not as much air. This calls for less fuel and more spark advance. The BMAP gives information to the computer so it can calculate the correct spark timing and air-fuel mixture. A BMAP works as a MAP sensor when the engine is running. MAP and BARO sensors are not interchangeable.

■ THROTTLE POSITION SENSOR

The TPS is a potentiometer mounted on the throttle shaft (see Chapter 45). It measures the angle of the throttle plate. When the rpm signal is under 600 rpm, the computer senses that the engine is cranking. The TPS also works as an accelerator pump. When the voltage signal changes quickly, the computer gives a signal for more fuel.

A defective or misadjusted TPS will cause a hesitation when accelerating, just like a bad accelerator pump on a car with a carburetor. When the engine is under a heavy load, the air-conditioning compressor is shut off by the computer. A bad TPS can shut off an air-conditioning compressor.

A TPS can be checked with either a voltmeter or an ohmmeter (**Figure 46.35**). To perform ohmmeter tests, the switch must be disconnected, so a voltmeter (or DSO scope, which is actually a voltmeter, too) is the tester of choice. Three tests are made for TPS operation. Reference voltage must be present at the switch with the key on. The base voltage is also compared to specifications. Finally, voltage should change gradually and evenly as the throttle is opened and closed. If the voltage does not rise, or there are skips (glitches) in the voltage measurement (**Figure 46.36**), the sensor is bad.

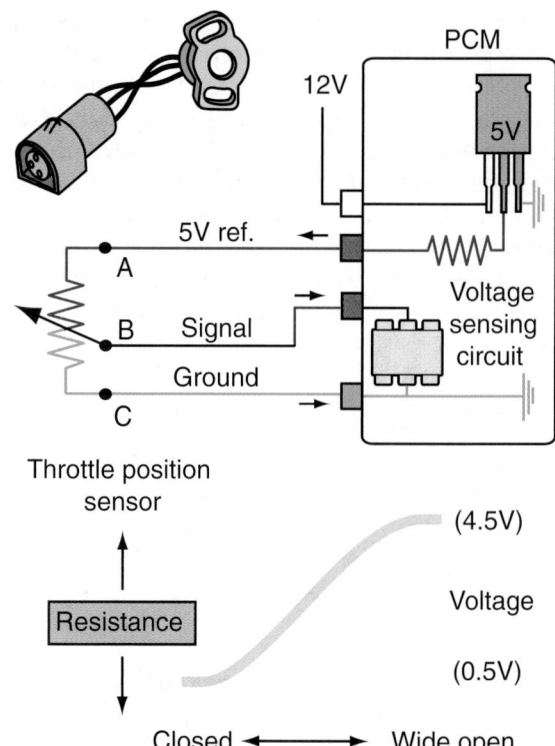

Figure 46.35 A throttle position sensor can be checked with an ohmmeter or a voltmeter. The elongated slots in this TPS allow for adjustment.

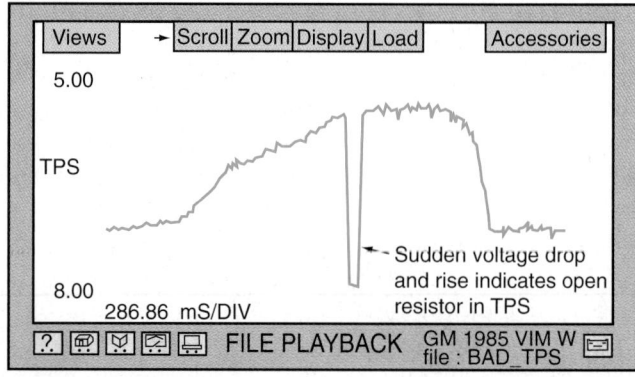

Figure 46.36 A glitch in a DSO pattern on a TPS test.

When checking voltage output of a TPS, typically there would be less than 0.1 volt at idle. Typical voltage at half throttle would be 2.5 volts. Remember, there is a 5-volt reference signal. That is why there is a 2.5-volt signal when the throttle is half open. At WOT, the signal would be 4.5 volts.

Some systems work the opposite way, with high voltage at idle and low voltage at wide-open throttle. Reference voltage can also be different. The principle of operation is the same, however.

Different kinds of problems can happen with a TPS. In cold weather, a TPS gets cold and the feather (wiper arm) sometimes does not wipe. This results in high resistance in the switch. Hard starting, intermittent high idle, or a hesitation can happen when fuel vapors get into the switch. Remember, the switch is mounted on the throttle plate shaft.

Some throttle position sensors are adjustable. Follow the manufacturer's instructions. On older engines, correct adjustment is crucial to proper system operation. On newer engines, this is not as important because the computer uses whatever reading it takes at idle as base voltage (adaptive learn strategy).

■ COOLANT TEMPERATURE SENSOR

The *engine coolant temperature sensor (ECT)* affects how the engine operates in all conditions. The most common problem is when the computer system will not go into closed loop when the engine is warm. This causes poor fuel economy due to the rich mixture with no feedback control. Cold performance problems can also result from a bad ECT sensor. Symptoms can include poor idle, stalling, and hesitation or stumble on acceleration.

When the computer senses an ECT circuit problem on an OBD II car, it sets a code PO115. Problems are most often related to wiring or connectors, rather than failure of the sensor. When there is a failure in the coolant temperature sensor or its circuit, the result is like an automatic choke problem on a car with a carburetor. The computer sends out a small voltage signal (5 volts) to the sensor and monitors the voltage drop in the circuit taking place based on the resistance of the sensor.

If the sensor circuit is open or the sensor has a bad ground, circuit continuity is lost. The voltage will not be able to be reduced. The computer will assume that the coolant temperature is –40°F and set fuel delivery and spark timing accordingly. Spark advance will be less than normal and the mixture will be very rich.

If circuit wiring becomes grounded, the 5-volt reference signal will drop to zero. The computer assumes that the engine is very hot. A cold engine may not start. When the coolant temperature is too high, the PCM shuts off the air-conditioning compressor.

Testing a Coolant Temperature Sensor

An ECT sensor can be tested with an ohmmeter or a voltmeter (**Figure 46.37**). Resistance readings change with temperature. An ECT sensor is a thermistor and will have lower resistance as the temperature of the coolant rises. This is called a **negative coefficient thermistor**.

Check the resistance of the sensor and compare it to the actual temperature of the engine to see if it is working properly. The sensor can also be removed and tested on a hot plate in the same manner a thermostat is tested (**Figure 46.38**).

■ AIR TEMPERATURE SENSORS

Intake air temperature (IAT) sensors have been called a variety of names in the past (ACT, VAT, MCT, MAT,

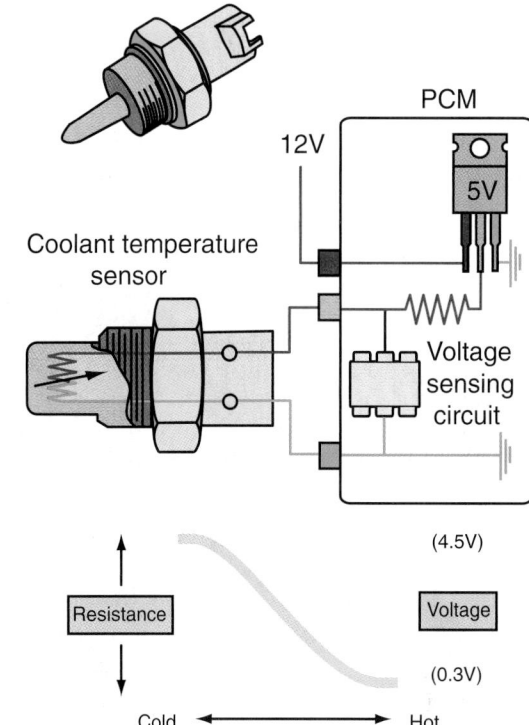

Figure 46.37 The ECT sensor can be tested with an ohmmeter or voltmeter.

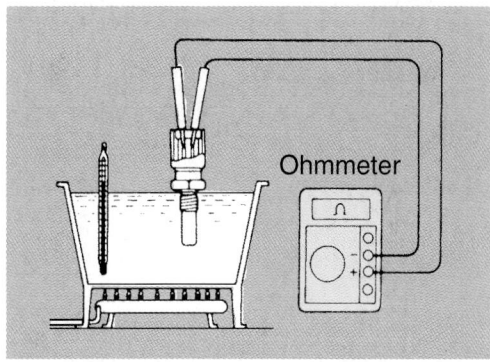

Figure 46.38 An engine coolant temperature sensor can be tested on a hot plate.

and ATS). Under OBD II guidelines, these parts are now called IATs. It is used to fine-tune the air-fuel mixture to compensate for the density of the air. The IAT sensor is mounted in the intake system with its sensor end exposed to the incoming air stream (**Figure 46.39**).

An IAT sensor works like a coolant temperature sensor. Its resistance lowers as temperature rises (**Figure 46.40**). It can be damaged by intake manifold popbacks and its operation can be hampered by contamination with oil and dirt. IAT sensor temperature should be 30°F to 40°F cooler than ECT temperature. Too high a temperature of the air in the intake manifold could be due to EGR or hot air stove problems.

An IAT circuit problem on an OBD II car sets a code PO110. When testing with a scan tool, as the engine

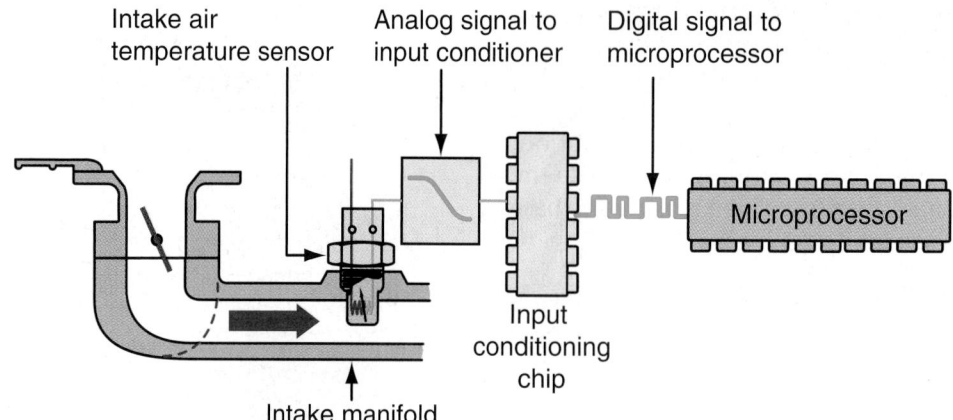

Figure 46.39 The intake air temperature sensor is mounted in the intake manifold with its sensor end exposed to the incoming air stream.

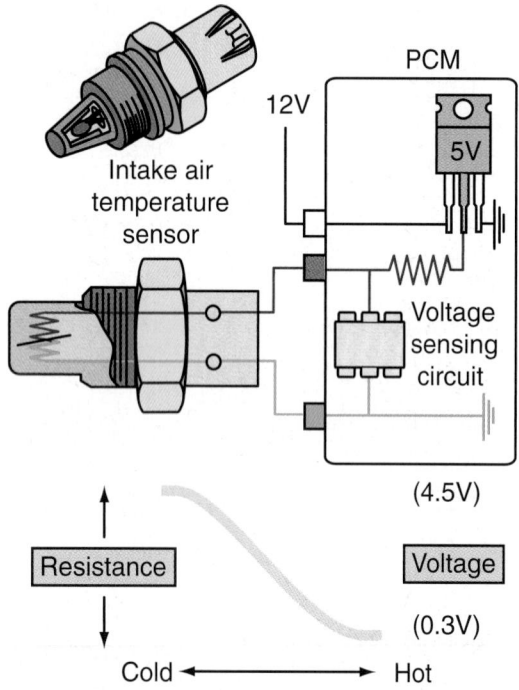

Figure 46.40 An IAT sensor's resistance lowers as temperature rises.

speed is raised, temperature should drop slightly as the sensor is cooled by the increased amount of air moving across it.

If the sensor is removed from the engine, it can be tested with heated water and an ohmmeter the same way an engine coolant temperature sensor is tested. Blow hot air on it with a hair dryer or heat gun. Do *not* use a propane torch. Watch for a drop in resistance as temperature increases.

▬ AIRFLOW SENSOR SERVICE

Fuel systems that are controlled by an airflow sensor react poorly to vacuum leaks. A leak allows unmetered air into the engine, resulting in an excessively lean mixture. Once the engine is warm and is in closed loop, the O_2 sensor can compensate for small amounts of air leaking in. Because the O_2 sensor is in the exhaust, it comes after the leak. Even though it causes the fuel system to compensate for the leak, a hesitation on acceleration could result, especially when cold.

Dirt can cause problems in a vane airflow (VAF) sensor. A dirty air filter, or housing left dirty when an air filter cartridge is changed, can cause the air door pivot shaft to stick. Feel the operation of the door by pushing it open gently. If it is sticky or binding and spray carburetor cleaner does not free it up, it will have to be replaced.

An intake manifold popback can also cause the door to bend or break. A backfire valve is built into some sensor doors to prevent this. If the backfire valve begins to leak, it will cause a rich-running engine.

Trouble codes will set when there is a problem with a VAF sensor. Checking the operation of the electrical portion of the sensor is done with a voltmeter. Follow the service instructions for the car you are working on.

NOTE: *Opening the air door should slowly increase the voltage until reference voltage is reached.*

Unlike a VAF sensor, an MAF sensor has no moving parts that can fail. Trouble codes will set if the sensor or its circuit gives an unexpected reading. Hot wire sensors produce a voltage that varies from about 0.5 volt at idle to reference voltage (5 volts) at WOT. Glitches can be spotted using a DSO (**Figure 46.41**).

Hot film MAF sensors produce a variable frequency instead of a voltage. The frequency varies from about 30 Hz at idle to about 150 Hz at WOT. This can be read on a scan tool as grams per second. A bad MAF frequency sensor DSO pattern is shown in **Figure 46.42**. MAF sensor output can also be checked by monitoring frequency (Hz) on a digital multimeter.

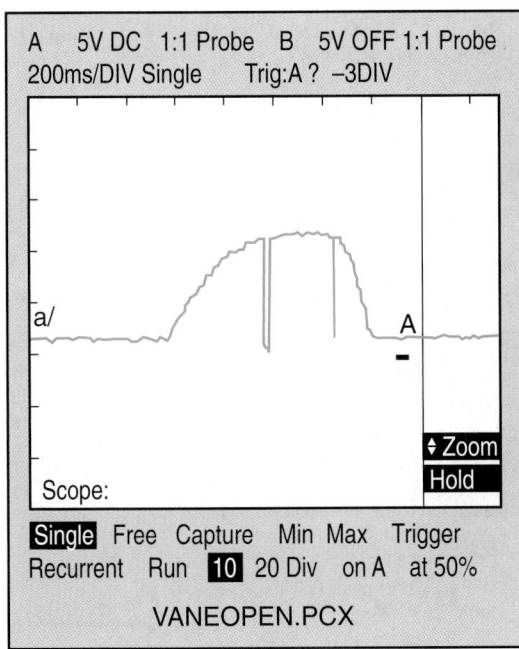

Figure 46.41 Glitches in a vane-type mass airflow sensor pattern.

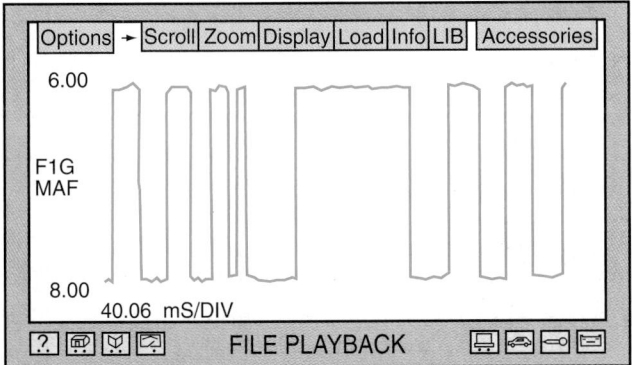

Figure 46.42 A good frequency-type MAF sensor waveform would show proportional increases in the digital signals. This one is bad.

■ KNOCK SENSOR SERVICE

A knock sensor only affects ignition timing. If a knock sensor fails, the computer will not retard the timing to prevent spark knock. A loose bracket or other vibration can also cause the timing to be retarded when it does not need to be. Sometimes, a main bearing knock or piston slap will cause the timing to retard too. Some in-line five- and six-cylinder engines have two knock sensors.

The sensor can be tested by rapping on the engine near the sensor with a metal tool like a socket extension or a wrench. On distributor ignitions, use a timing light with the engine running on fast idle. A decrease in ignition advance should be observed.

A scan tool can be used while rapping on the engine (**Figure 46.43**). Some systems will show the actual amount of spark retard; others will show a yes/no or on/off reading on the scan tool. A normal

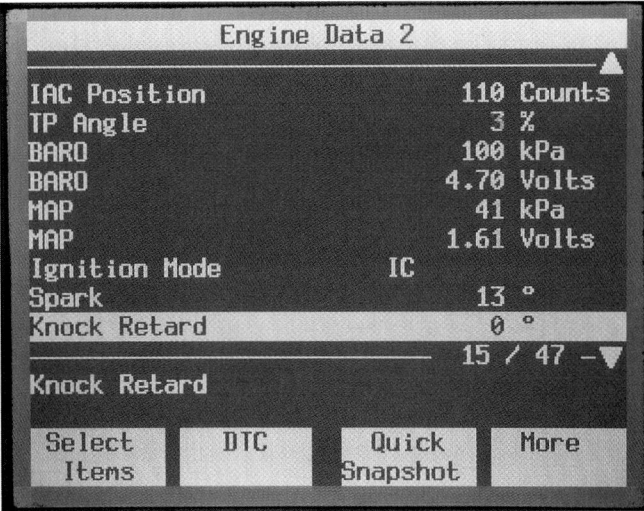

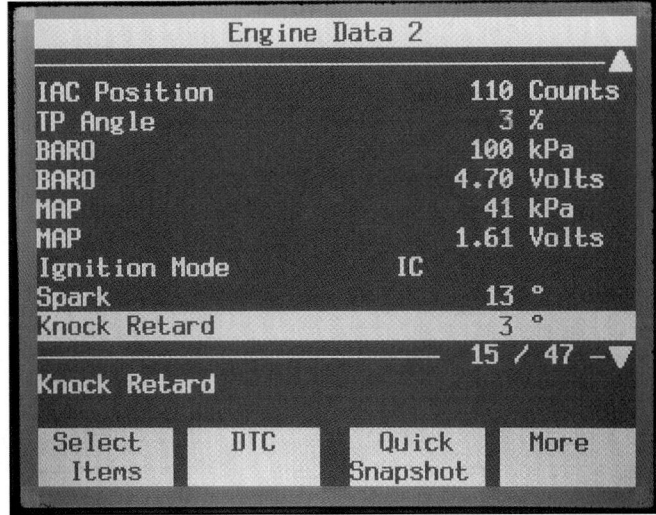

Figure 46.43 Scan tool showing spark retard before and after tapping on the manifold.

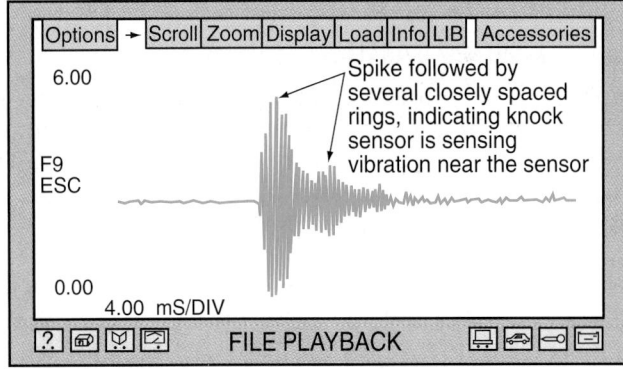

Figure 46.44 A normal knock sensor waveform.

knock sensor waveform on a DSO is a group of closely spaced oscillations with a gradually decreasing voltage (**Figure 46.44**).

■ ACTUATOR SERVICE

Actuators include solenoids, fuel injectors, stepper motors, and motors for electronic suspension hydraulic controls. These actuators are covered in their respective chapters. Testing an actuator involves checking for voltage at the actuator control terminal. If the actuator has voltage to it but it does not work, be sure it has a good ground path. Test individual actuators according to service instructions.

REPAIR THE PROBLEM

After repairing a problem, road test the car again. If the problem caused a trouble code, be sure to erase the computer's memory before testing the car again. When a car has had previous problems, the computer learns to correct for them. The test drive will allow a late-model computer with adaptive strategy to relearn its best adjustments.

NOTE: *Adaptive memory learns faster if the computer's memory is erased first. With old knowledge after a part is changed, it takes longer to adapt.*

SHOP TIP Rather than erasing all memory, it is usually best to use the scan tool to erase only the codes. OBD II systems have monitors, some of which take a long time to reset. An engine might have to be left overnight or driven through a difficult drive cycle in order to run the monitors so the vehicle can pass an emission test.

After replacement of the computer or when a battery has been disconnected, poor driveability and performance can result until the computer relearns the best driveability settings. Hard starting, high idle, stumble, and stalling are possible symptoms. Some scan tools can partially reset previously learned parameters if they were saved before the repair.

COMPUTER WIRING SERVICE

Poor electrical connections are the most common cause of problems in computer systems. Electronic components use very, very small amounts of electricity and are very sensitive to resistance in wiring. As little as 200 ohms resistance can cause a problem.

Wiring problems include loose or corroded connections and grounded wires. An advantage of on-board diagnostics is that it helps keep the wiring connections from becoming corroded or damaged because wires are not removed during diagnosis. This maintains the integrity of the connectors.

SHOP TIP To safely backprobe an electronic connection, probe under the insulation parallel to the wire (**Figure 46.45**). There are special probe kits available from tool suppliers, or you can use an ordinary pin.

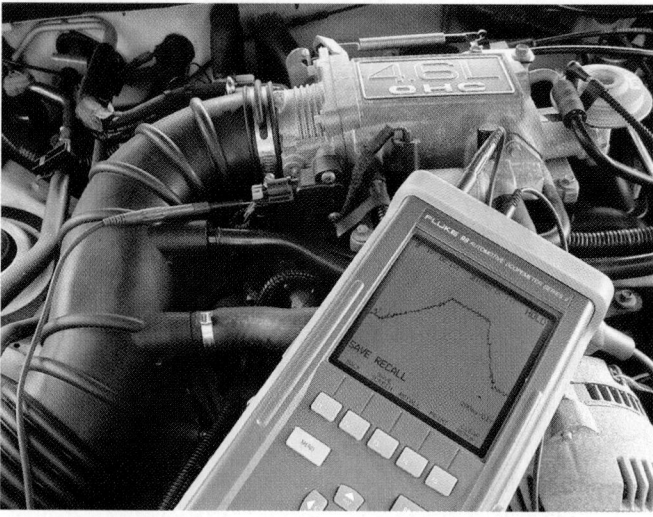

Figure 46.45 Backprobing a connection. *(Courtesy of Tim Gilles)*

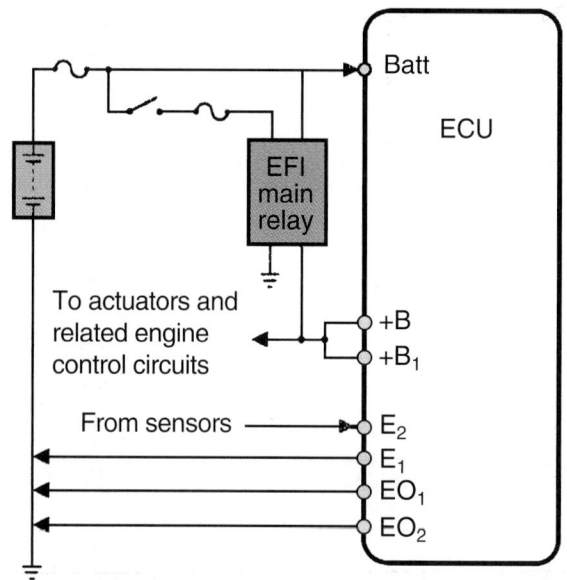

Figure 46.46 One computer lead always has power (BATT). The other two on this diagram are powered from the relay when the key is on.

Always use a wiring diagram when working on computer systems. First, check to see which parts are affected. When several devices are not working, look closer to the power source. When only one or two are not working, look in that part of the circuit for the problem. Be sure to ask the customer if the problem only occurs when hot or cold. Does it always happen or is it intermittent?

The computer must have good power and ground connections. Most computers have two or three power leads (**Figure 46.46**). One lead will always have power. This allows the computer to keep information in volatile memory. The second lead (and sometimes a third lead if used) supplies power when the key is on. Fuse links protect these wires. They are located near the battery, starter, or starter relay.

The number of grounds varies from two to six. They are either power or sensor grounds. Power grounds are used for actuators like motors or solenoids. Sensor grounds return to the computer and then to a ground source.

To test a ground circuit, measure voltage drop across the ground side of the circuit. Connect the meter's positive lead to the ground side of the sensor or actuator. This is the lead that returns to the computer. Connect the negative side of the meter to the negative battery terminal. When the circuit is loaded, voltage drop in a sensor's ground circuit should be less than 0.1 volt. Power ground circuits should not exceed 0.3 volt. Isolate the problem by working your way down the circuit (see Chapter 25). Ground-side resistance that is too high will decrease voltage in the circuit, and the computer will receive too low a return signal from the sensor.

NOTE:

- *If you remove and repair a ground connection, be sure you reconnect it to the same place and do not change the length of the wire.*

Examine wires to see that they are properly routed and are all original in regard to length and shape. *Radio frequency interference (RFI)* is a problem with computer systems. If a computer feed wire is routed near the alternator or secondary ignition wires, signals that the computer receives will be jumbled. This can result in serious driveability problems. If you find a bare wire wrapped around the outside of wires running between a distributor and the computer, these are *shielding wires*. The shielding wire is connected to ground and absorbs RFI that might interfere with the rpm reference signal to the computer.

When a wiring problem is suspected, disconnect and clean the connector using electronic connector cleaner.

Corrosion can be cleaned from the connection with a soft bronze bristle brush, available at electronic supply stores. After cleaning the connection, coat the wire ends with dielectric compound if it was used previously.

When connections are replaced, use solder rather than crimp connectors. Always use shrink tubing or liquid electrical tape to protect the connection from moisture when wires are soldered. This is especially crucial in areas where the roads are salted in the winter. Salt accelerates the corrosion process.

CAUTION ■ Do not use color TV cleaner. It contains acids that will cause further corrosion of the connection.
■ Do not use carburetor cleaner, brake cleaner, alcohol, or electromotive spray. The connector is held in plastic, which has a hydrocarbon base. A strong cleaner can ruin it.
■ Do not touch the connector's metal parts. Acid from your hands can begin the corrosion process.

The quality of ground connections is very important. One of the most prominent causes of ignition module failure is loss of ground. Check ground straps and wires that connect the engine and chassis. The eyelet of a ground wire where it bolts to the cylinder head must be tight and its bolt must not have paint or rust under its head.

Repairing Twisted/Shielded Wire

The twisted pair wiring used in some computer circuits carries very small amounts of current. Therefore, splices in the wire must have no resistance. If the wiring has aluminum/mylar tape shielding, carefully unwrap it. After replacing a damaged section with wiring of *exactly the same length*, re-wrap the shielding tape around the conductors. Shielded wiring also has an uninsulated drain wire (**Figure 46.47a**). The drain wire is wrapped around the outside of the shielding tape (**Figure 46.47b**). If the drain wire was cut, splice it and solder it before sealing the repair with heat shrink tubing or electrical tape.

Computer Reflashing

Electronic updates are a regular occurrence among manufacturers. However, approximately two-thirds of vehicles with updates do not have them done because there has been no customer complaint. General Motors has over 3,000 code possibilities (P3000). There might have been five updates, but only one TSB was generated. Before attempting a repair, always check for TSBs first!

 **CASE HISTORY** *A customer brought a Cadillac in with the MIL illuminated. The technician checked for codes, finding a random misfire code (P0300). He checked for TSBs and found a reference to previous instances of a bad serpentine belt and tensioner causing this problem. He found a bad tensioner bearing, which he replaced. The code was cleared and the customer complaint was solved.*

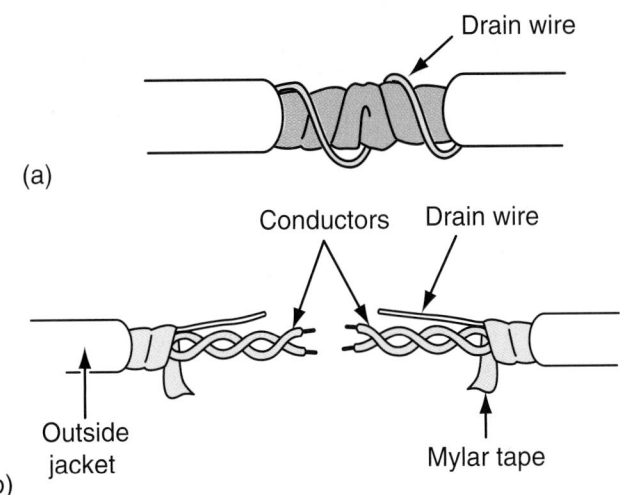

Figure 46.47 (a) Twisted pair shielded wire. (b) The drain wire is wrapped around the outside of the shielding tape.

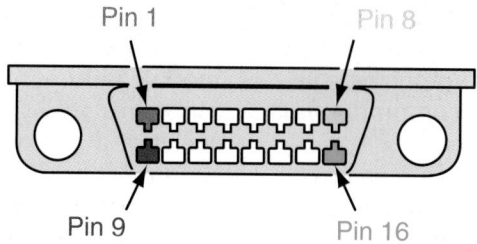

Pin 1: Manufacturer discretionary
Pin 2: J1850 bus positive
Pin 3: Manufacturer discretionary
Pin 4: Chassis ground
Pin 5: Signal ground
Pin 6: ISO 15765-4 CAN – C (+)
Pin 7: ISO 1941-2 "K" line
Pin 8: Manufacturer discretionary

Pin 9: Manufacturer discretionary
Pin 10: J1850 bus negative
Pin 11: Manufacturer discretionary
Pin 12: Manufacturer discretionary
Pin 13: Manufacturer discretionary
Pin 14: ISO 15765-4 CAN – C (–)
Pin 15: ISO 9141-2 "L" line
Pin 16: Battery power

Figure 46.48 CAN is located on pins 6 and 14 on a universal OBD II 16-pin DLC.

Recalls are only covered through the new vehicle warranty period. Some shops routinely reflash vehicles that are beyond warranty whenever an update is found. Volume is a key factor because a subscription service that allows you to be able to reflash can be costly.

When using an earlier scan tool on a CAN network, use the correct adapter or you can ruin the scan tool.

Figure 46.49 Since the 2004 model year, all vehicles allow universal access to reprogramming downloads from the Internet.

CAN is on pins 6 and 14 on the DLC (**Figure 46.48**). To identify a CAN system, check the voltage on pin 6. If 2.5 is measured, the system is CAN. Otherwise it will be toggling between 0 volt and 12 volts or 6 volts (ENC). ENC stands for entertainment and comfort.

All vehicles manufactured since 2004 are governed by SAE Standard J2534, which allows universal access to reprogramming downloads from the Internet (**Figure 46.49**). A generic pass-through device is connected between a personal computer (PC) and the DLC (**Figure 46.50**). The device interfaces with different protocols and comes with drivers that must be loaded onto the PC so it can communicate with the vehicle. Reprogramming in this fashion is also possible on most vehicles back to the beginning of OBD II in 1996 and on some earlier vehicles. Some scan tools can also work as a pass-through device for PCM reprogramming without the need for a PC using both hard-wired and wireless downloads (**Figure 46.51**).

Computer Location

Electronic instruments can be damaged by excessive voltage, heat, moisture, or vibration, so computers are usually mounted in the driver's compartment. Sometimes, they are in the right-hand *kick panel* (in front of the passenger door) (**Figure 46.52**) or near the glove box. Other times, they are located under the passenger

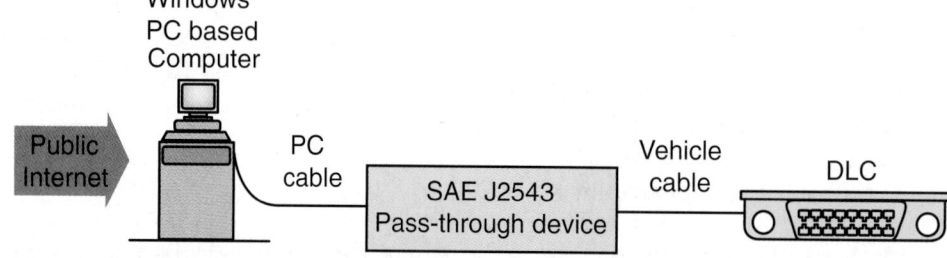

Figure 46.50 A generic pass-through device is connected between a personal computer and the DLC.

Figure 46.51 Some scan tools work as a pass-through device to allow reprogramming with hard-wired or wireless downloads.

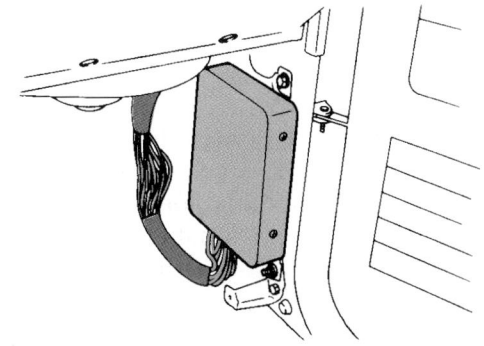

Figure 46.52 A computer located inside the right front kick panel. *(Courtesy of Mazda Motors of America, Inc.)*

seat. Newer computers are designed to dissipate heat more quickly. Some of these are installed in the engine compartment. To be safe, pull the computer fuse before removing its connectors.

■ STATIC ELECTRICITY

Static electricity from sliding around on the front seat is a concern when working around electronic components. People who work around sensitive components sometimes wear a ground strap attached to their wrist. You can eliminate the static electricity by touching ground before you touch a computer. Do not take the computer out of the container until you are already in the front seat. The only time you need to wear a ground strap is if you are backprobing the computer under the hood.

■ ELECTRICAL DAMAGE TO A CIRCUIT

If a circuit never exceeds its voltage or current limits, it is very reliable. Too much electrical current causes heat that can damage an electrical circuit. Voltage is limited to about 20 volts. Damaged connections are usually the reason for failure of electronic components. Connector pins on chips can be damaged or bent during installation.

Semiconductors are designed for only a limited amount of current. A light bulb can draw 2.5 amps,

but transistors in computers draw only 200 milliamps (0.2 A). If too much current is accidentally applied in reverse, it can force through and ruin a diode or transistor. Most computers can tolerate a high current surge for about 5 seconds.

Voltage spikes are the biggest cause of electrical damage to an integrated circuit. If a sensor or wiring connection is disconnected while it is powered up, a spike can occur. Most automotive computer systems operate on 5 volts. A spike while disconnecting a powered connection can be 50 volts or so. The spark occurs because electrons were in motion before the circuit was broken back up at the connection. When they try to push their way across the gap, the spike is created. Manufacturers build safeguards for voltage spikes into their systems, but not enough to protect against disconnecting and reconnecting components.

Voltage spikes can also result from an arc welder. Always disconnect the battery before doing any welding.

Intermittent faults are those that never seem to be present when you are trying to diagnose a problem. They can be very difficult to diagnose. Most of these problems are the result of a faulty electrical connection. Sometimes a problem can be diagnosed by doing a wiggle test with test instruments connected to the circuit (**Figure 46.53**).

Diagnosing Multiplex Systems

Bus diagnosis is similar to other electrical system diagnoses. Verify the complaint, look for related symptoms and analyze them, and troubleshoot in a logical sequence. Cover the basics. If a module is not communicating, check for a loss of sufficient battery voltage or a poor ground.

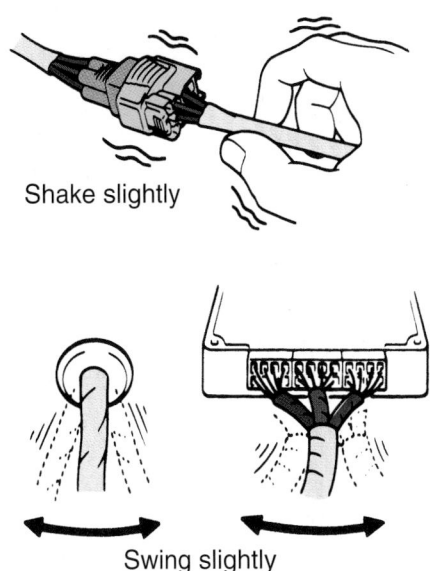

Shake slightly

Swing slightly

Figure 46.53 An intermittent problem can sometimes be diagnosed with a wiggle test while test instruments are connected.

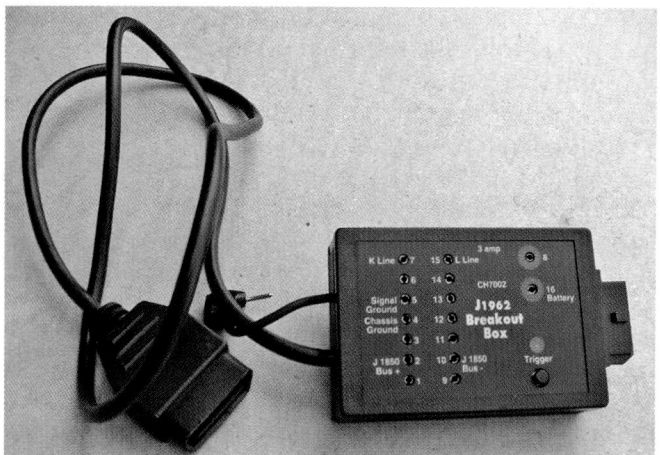

Figure 46.54 A universal breakout box that connects in series between the scan tool and the DLC.

Bus systems use the DLC to communicate with the scan tool. SAE Standard J1962 specifies a DLC connection that will work on any OBD II vehicle. The standard applies not only to scan tools. A universal breakout box that connects in series between the scan tool and the DLC is also available (**Figure 46.54**). As mentioned earlier in this chapter, a breakout box provides a means of testing voltages and resistances in individual circuits accessed by the DLC.

Remember that P-codes are those related to the powertrain. In a similar manner, network communication fault codes are called U-codes and body system faults are called B-codes. Most modules can set a U-code when an abnormal condition is sensed. This can include a failure of the bus or a problem with a module. In addition, a module expecting a message on the bus from another module might set a U-code and a B-code. The scan tool lists the status of a DTC as active if it is a current condition. If the condition no longer exists, the scan tool lists the condition as a stored code.

Many OBD II vehicles use pins 7 (K-line) and 15 (L-line) of the DLC for scan tool communication with the PCM and/or other modules on the bus (**Figure 46.55**). The K-line transmits data to the scan tool from modules, and the L-line carries data from the scan tool to modules. On some vehicles modules have DLC dedicated terminals (**Figure 46.56**).

Some scan tools have a vehicle module scan feature. This queries all of the modules and lists any that did not reply. If a scan tool cannot communicate with the PCM, check the service literature and review the wiring diagram to help analyze the cause of the problem. Look for variations and commonalities in network construction that can help you understand

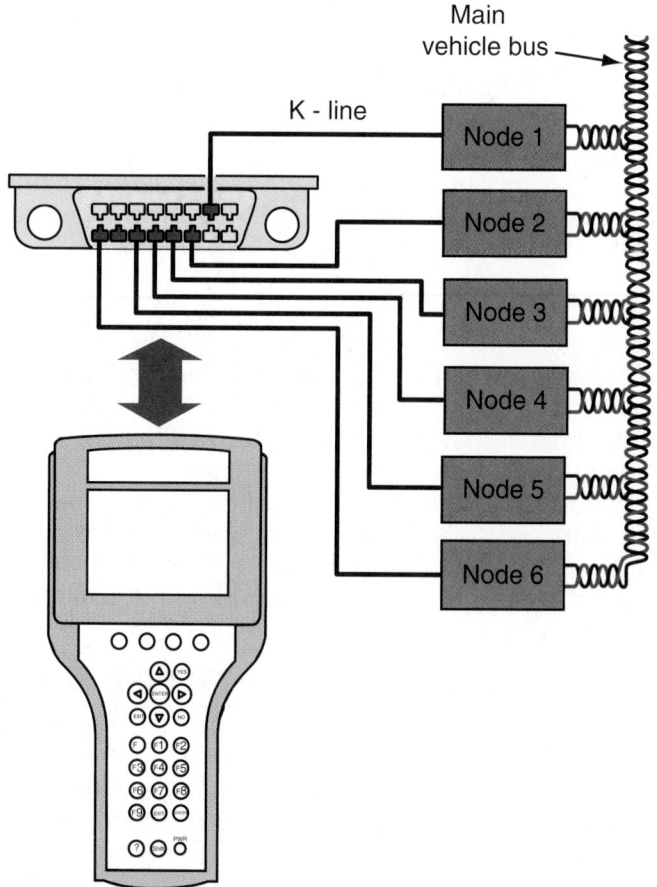

Figure 46.55 Many OBD II vehicles use pins 7 and 15 to link the scan tool with modules.

Figure 46.56 On some vehicles, modules have dedicated terminals on the DLC.

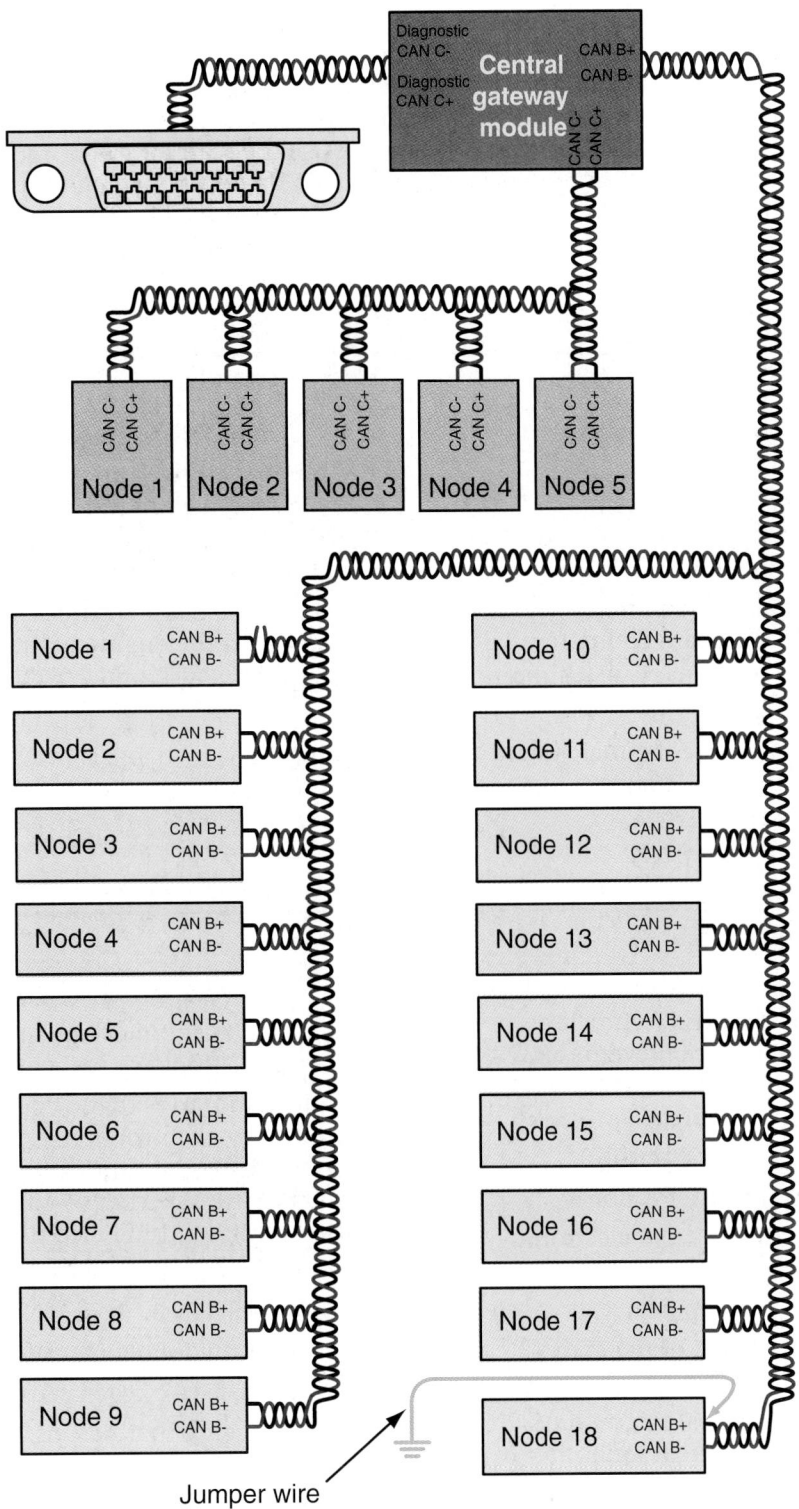

Figure 46.57 To diagnose an open circuit when a CAN B network is in single-wire mode, short the last module to ground with a jumper wire. The problem module will stop communicating because its circuit path has been eliminated.

a simple or widespread fault. Data are shared on the network, so look for something in common among related systems.

CAN systems usually use more than one class of network. Diagnostic CAN C works with the scan tool, which will help you identify which network you are

working on (CAN B, for instance). The central gateway module is the common component to all of the networks and can provide a DTC pointing you to a faulty circuit or component. If you cannot communicate with the central gateway module, the CAN C circuit is the problem.

Determine whether there is a total or partial failure in the bus. In the case of a total bus failure, there will be a short to voltage or ground. The breakout box can be used to test voltages and resistance.

Service information will provide steps for solving a network problem. The following information provides an example of network troubleshooting. CAN B networks can still operate in single-wire mode when there is an open in a circuit, so there might not be a symptom. When the other circuit is shorted to ground, the module will no longer be able to communicate and all other modules on the network will go into single-wire mode. Use the scan tool with a jumper wire shorted to ground to determine which module is not communicating (**Figure 46.57**). The location of the fault will be between the bus and the module.

▬ REVIEW QUESTIONS

1. A high _____ voltmeter is used to measure voltages in computer circuits.

2. If the MIL comes on when the engine is running, there is probably a trouble _____ stored in the computer's memory.

3. A _____ _____ is a portable computer that reads data from the car's on-board computer.

4. A _____ is a scan tool feature that is like a flight recorder.

5. A good oxygen sensor will cause the fuel system to switch between rich and lean about _____ times in 10 seconds.

6. _____ sensors are ones that do not generate their own voltage.

7. Which tool would not require a throttle position sensor to be disconnected for testing, a voltmeter or an ohmmeter?

8. What is the name of the sensor that is tested by rapping on the engine with a tool?

9. Computer grounds are either power or _____ grounds.

10. Voltage _____ happen when electrical connections are pulled apart while there is a load.

▬ ASE-STYLE REVIEW QUESTIONS

1. Technician A says that an analog voltmeter is used to test computer systems. Technician B says that high input resistance prevents an analog meter from drawing current while it is connected to a circuit. Who is right?
   a. Technician A   c.   Both A and B
   b. Technician B   d.   Neither A nor B

2. Which of the following causes more electronic control problems?
   a. Actuators   c.   Computers
   b. Sensors   d.   None of the above

3. Technician A says that an intermittent fault is one that is present at the time of the self-test. Technician B says that intermittent faults are often the result of a faulty electrical connection. Who is right?
   a. Technician A   c.   Both A and B
   b. Technician B   d.   Neither A nor B

4. Each of the following statements about oxygen sensor systems is true *except*:
   a. A slow O_2 sensor will usually cause the system to run lean.
   b. An engine that runs poorly when cold could have a bad O_2 sensor.

   c. Use an ohmmeter to test the O_2 sensor.
   d. A faulty thermostat can cause a lack of O_2 sensor feedback fuel control.

5. Technician A says that a code can still remain in memory even though a problem has been corrected. Technician B says that a scan tool can erase trouble codes without having to disconnect anything electrical. Who is right?
   a. Technician A   c.   Both A and B
   b. Technician B   d.   Neither A nor B

6. Which of the following can cause a no-start?
   a. A defective oxygen sensor
   b. An engine that has lost its crankshaft sensor signal
   c. Both A and B
   d. Neither A nor B

7. With oxygen sensor feedback, a bad spark plug can result in:
   a. A rich air-fuel mixture
   b. A lean air-fuel mixture
   c. Both A and B
   d. Neither A nor B

8. Disconnecting the vacuum supply to a MAP sensor will cause:

   a. The fuel system to go rich

   b. The fuel system to go lean

   c. The engine to die

   d. All of the above

9. Which of the following is/are true about electronic test instruments?

   a. Some scan tools have DSO and graphing multimeter functions.

   b. Current ramping with a DSO and current probe is popular because it is noninvasive.

   c. A low-current probe can be used to determine the rpm of a fuel pump.

   d. All of the above.

10. Which of the following is/are true about twisted pair wiring?

   a. It reduces electrical noise.

   b. It is important that the exact length of replacement wire be used.

   c. Both A and B

   d. Neither A nor B

CHAPTER 47

Advanced Emissions and On-Board Diagnostics (OBD)

◼ OBJECTIVES

Upon completion of this chapter, you should be able to:

✔ Describe the operation of on-board diagnostic systems.

✔ Explain the differences between OBD I and OBD II.

✔ Interpret OBD II scan tool data.

✔ Describe the operation of OBD II monitors.

✔ Use a scan tool to verify the running of various OBD II monitors.

◼ KEY TERMS

OBD II	drive cycle	comprehensive component
non-volatile RAM	pending code	monitor
warm-up cycle	type A code	heat soak
trip	monitor	switch ratio
enabling criteria	readiness indicator	propane enrichment test

◼ INTRODUCTION

Government involvement in vehicle emission controls and fuel economy legislation has been a driving factor in the advanced technology found on motor vehicles. The primary objective of modern on-board diagnostics (*OBD*) is to improve air quality. This chapter is related to earlier chapters on fuel systems and emission controls, and understanding its material is based on reading Chapter 45 and Chapter 46 first, which deal with electronics and computer controls.

There have been vast improvements in technology since the first computers were installed on vehicles. Ninety percent of emissions occur during warmup. Some early computer-controlled cars needed a temperature of 176°F to achieve closed loop. Because of continual improvements in technology through the early 1990s, including the introduction of heated O_2 sensors, by 1996, closed loop could be within achieved at 68°F with OBD II cars, sometimes within seconds of a cold startup.

◼ HISTORY OF ON-BOARD DIAGNOSTICS

Although many earlier vehicles came equipped with OBD, OBD Generation One (OBD I) legislation began with all cars sold in the state of California in 1988. OBD I was regulated by the California Air Resources Board. Its requirements called for vehicles to have a system capable of identifying a vehicle's computer system faults. That 1988 legislation also set requirements for future OBD II diagnostics.

California has the toughest emission control laws in the world, and Los Angeles has the worst air quality in the nation. Los Angeles basin air quality worsened each year until 1992. Since 1994, its air quality has become progressively better. Automotive emission controls have made a large and measurable improvement in air quality.

OBD legislation is constantly evolving as government regulators and industry representatives refine the program. It has evolved from a California-only program to a national program. Although legislation

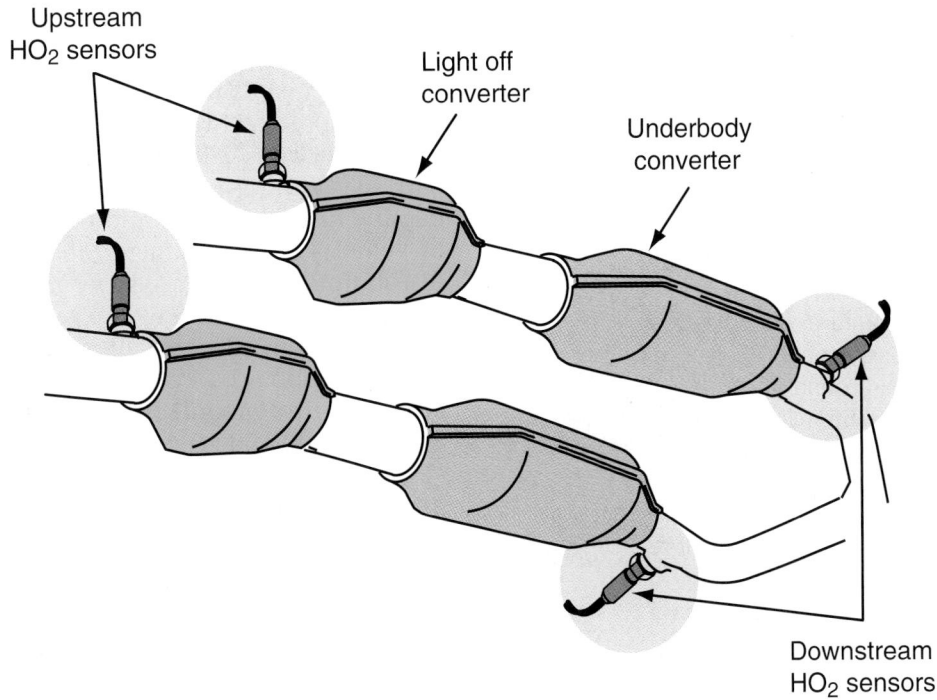

Figure 47.1 The extra O_2 sensor in the rear monitors catalyst efficiency.

covers only the United States, most manufacturers build their cars for the North American continent, so the vehicles are basically the same.

On-board diagnostics II (**OBD II**) guidelines began in 1994 on some cars. The standards were gradually phased in until January 1, 1997, when the law required all cars built in the United States to comply.

NOTE: *Imported cars built overseas had until January 1, 1998, to comply.*

The initial goal of OBD II was for a computer to be able to self-detect a failure when an exhaust emission increased over 50%. Two other goals were to eliminate intermittent illumination of the malfunction indicator light (MIL) and to reduce the time between the occurrence of a malfunction and its detection and repair. Some parts of the emission control system degrade with age. OBD II is capable of monitoring this as it occurs.

OBD has replaced tailpipe emission inspections in some instances.

OBD II OPERATION

The objective of OBD II is to detect exhaust and evaporative emissions that exceed the Federal Test Procedure (FTP) by 1.5 times. The FTP is an emission standard that measures hydrocarbon (HC), carbon monoxide (CO), and oxides of nitrogen (NO_x) emissions by weight, in grams per mile. The FTP allows for engine and emission component wear by listing allowable emission levels for new vehicles, for vehicles with 50,000 miles, and for those with 100,000 miles.

In addition to improving the quality and durability of components so they can last the mandated 100,000 miles or 10 years (whichever comes first), some extra hardware is required for OBD II, including the following:

- An additional heated O_2 sensor located behind the catalytic converter to monitor catalytic converter efficiency (**Figure 47.1**).
- Misfire detection capability with more precise crankshaft and cam position sensors (**Figure 47.2**). A diagnostic trouble code (DTC) identifies the misfiring cylinder when there is only one. There are exemptions to this requirement for certain operating conditions. When there are multiple misfiring cylinders, the code need not identify the cylinders.

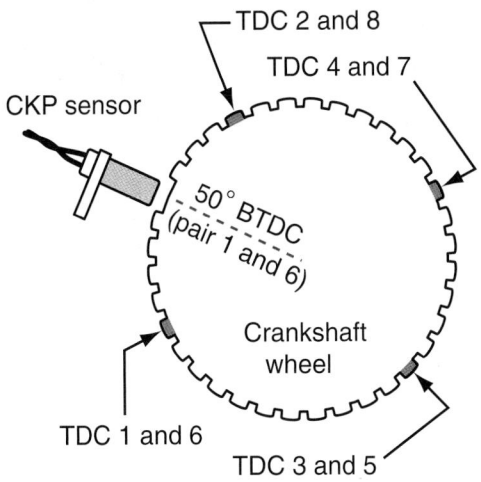

Figure 47.2 A more precise crankshaft sensor for misfire detection.

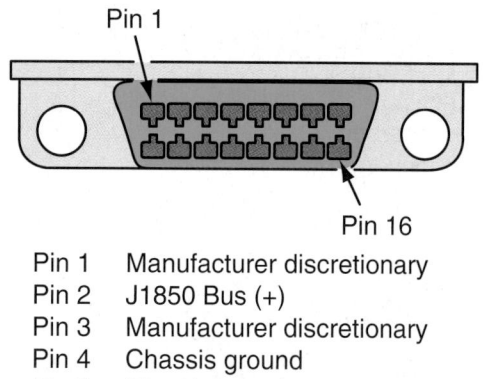

Pin 1

Pin 16

Pin 1	Manufacturer discretionary
Pin 2	J1850 Bus (+)
Pin 3	Manufacturer discretionary
Pin 4	Chassis ground
Pin 5	Signal ground
Pin 6	ISO 15765-4 CAN-C (+)
Pin 7	ISO 9141-2 K-line
	ISO 1423-4 K-line
Pin 8	Manufacturer discretionary
Pin 9	Manufacturer discretionary
Pin 10	J1850 (-)
Pin 11	Manufacturer discretionary
Pin 12	Manufacturer discretionary
Pin 13	Manufacturer discretionary
Pin 14	ISO 1565-4 CAN-C (-)
Pin 15	ISO 9141-2 L-line
	ISO 14230-4 L-line
Pin 16	Unswitched battery power

Figure 47.3 A generic OBD II connector. Discretionary terminals can be used as specified by the manufacturer.

- A new standardized 16-pin data link connector (DLC) (**Figure 47.3**).
- An evaporative system monitor.

■ SOCIETY OF AUTOMOTIVE ENGINEERS (SAE) STANDARDS

OBD II provides common standards to be shared among all makes of vehicles. The Society of Automotive Engineers (SAE) prepared the following standards for several areas:

- Standard communication protocol (SAE J1850)
- Standardization of terms (SAE J1930)
- Standard diagnostic connector (SAE J1962)
- Universal scan tool (SAE J1979)
- Standard diagnostic trouble codes (J2012)
- Common diagnostic test modes (SAE J2190)

These are minimum standards. Manufacturers often include advanced, or enhanced, diagnostic capabilities at their discretion.

Standard Communication Protocol

SAE J1850 specifies the protocol, a previously agreed-upon digital code used to communicate between the computer and the scan tool. OBD II requires each manufacturer to use the same language between the computer, sensors, actuators, and scan tool.

Standardization of Terms

SAE J1930 lists common names for all components that serve a similar purpose. Prior to OBD II, manufacturers provided their own unique names for automotive devices. As high technology became a normal part of automotive service, automotive vocabulary became unmanageable, and repairing multiple makes of vehicles became difficult at best. Under OBD I, a distributorless ignition system might have been called distributorless ignition, integrated direct ignition, direct ignition, DIS, or EDIS. The OBD II term is electronic ignition (EI). A computer, now known as a powertrain control module (PCM), could have been called ECM, ECA, ECU, EEC, MCU, PCM, SBEC, or SMEC.

Standard Diagnostic Connector

Before OBD II, there were no guidelines for the scan tool connector. OBD II specifies SAE Standard J1962, which requires a universal DLC for reading DTCs. The DLC is located under the dash in an accessible location (**Figure 47.4**). It is supposed to be on the left side of the dash. If it is located elsewhere, a sticker on the left side of the dash will indicate its location.

The D-shaped, 16-pin plug can only be connected one way. Some of the 16 pins are assigned according to OBD II standards. Pin #16, located on the lower right, is the unswitched battery positive pin. All OBD II scan tools connect to the vehicle's battery through this connection. Manufacturers can use seven of the remaining pins (1,3,8,9,11,12,13) as they see fit.

Generic Scan Tool

SAE J1979 specifies that a generic scan tool can be used on different makes of vehicles. The scan tool can also

Data link connector

Scan tool adapter

Figure 47.4 Connecting a scan tool to the DLC. (*Courtesy of Tim Gilles*)

clear trouble codes and on some vehicles is necessary to be able to clear trouble codes.

Standard Diagnostic Trouble Codes

SAE J2012 is an SAE-approved list of generic DTCs. The United States government has accepted this list to be used in all automotive service publications in the United States since January 1995. DTCs are covered later in this chapter.

Common Diagnostic Test Modes

Some scan tools have two portions: global and enhanced. The enhanced side is not limited to emission DTCs only. It requires you to input the VIN and includes manufacturer-specific menus and data for the vehicle. The global side of the scan tool is fast because the VIN does not need to be entered. The global side is generic and includes 15 modes. Of these only 1–9 are used.

■ Mode 1 includes parameter information data (PIDs). These are the processed data used by the engine management program. In OBD II, these PID numbers can be displayed on the scan tool. Some of the PID references are on the OBD II generic list. All scan tools must be able to reference these. Others may be manufacturer specific, and a generic scan tool might not be able to access them. PIDs can be useful in finding a problem when a code was not set. Mode 1 also allows the scan tool to display the readiness status of emission-related monitors (covered later).

Figure 47.5 Freeze frame data. *(Courtesy of Tim Gilles)*

> **SHOP TIP** When using a scan tool on a non-CAN network, select a limited number of parameters for viewing rather than selecting "all data." The scan tool must refresh or update in order to display live PIDs. There will be a longer delay in updating if the scan tool must update each PID.

The refresh rate for PIDs on a CAN scan tool can be about ten times as fast as a pre-CAN scan tool, and six PIDs can be listed with no delay.

■ Mode 2 refers to freeze frame data. OBD II requires the computer to store a "freeze frame" of engine conditions present at the time of a malfunction. Freeze frame data includes a number of stored PIDs (**Figure 47.5**). These PIDs are used to identify the operating conditions at the time the fault occurred. This process is useful in identifying intermittent faults. The highest priority DTCs are fuel system and misfire. These can overwrite a lower priority freeze frame.

■ Mode 3 allows scan tools to access stored emission-related codes (DTCs). Complete listings of DTCs from each manufacturer are readily available.

■ Mode 4 allows the scan tool to clear emission-related diagnostic information from the computer's memory. When codes have been cleared, the computer stores a "readiness" code that will remain in memory until all of the OBD II monitors have been run and shown to be without fault. Information on monitors is covered later in this chapter.

■ Mode 5 defines the O_2 sensor monitoring parameters. Specific operating conditions must exist for the test to be completed. The O_2 sensor must operate within defined specifications in order to protect the catalyst.

■ Mode 6 allows the technician to use a scan tool to access test results for noncontinuous monitors and operate system actuators. Monitor operation and more information on Mode 6 are provided later in the chapter.

■ Mode 7 provides test results for continuous monitors. It displays pending DTCs (covered later). Mode 7 is not supported by all global OBD II scan tools.

■ Mode 8 provides control of on-board systems or components, such as sealing the evaporative system before performing a manual pressure test.

■ Mode 9 displays vehicle information, including the VIN and verification of correct software installation and assurance that it has not been altered.

■ TROUBLE CODES AND THE MALFUNCTION INDICATOR LAMP

The MIL located on the dash display must illuminate if emissions exceed 1.5 times the federal standard. Emission-related codes are all stored in memory on OBD II systems.

OBD II only deals with emission codes. A manufacturer's dedicated scan tool will be capable of

MONITOR	MONITOR TYPE (when it completes)	NUMBER OF MALFUNCTIONS (on separate drive cycles to set DTC)	NUMBER OF SEPARATE CONSECUTIVE DRIVE CYCLES (to light MIL and store DLC)	NUMBER AND TYPE OF DRIVE CYCLES (with no malfunction to erase pending DTC)	NUMBER AND TYPE OF DRIVE CYCLES (with no malfunction to turn MIL off)	NUMBER OF WARM-UPS TO ERASE DTC (after MIL is extinguished)
Catalyst efficiency	Once per drive cycle	1	3	1	3 OBD II drive cycles	40
Misfire type A	Continuous	1	1		3 similar conditions	40
Misfire type B/C	Continuous	1	2	1	3 similar conditions	40
Fuel system	Continuous	1	2	1	3 similar conditions	40
Oxygen sensor	Once per trip	1	2	1 trip	3 trips	40
EGR	Once per trip	1	2	1 trip	3 trips	40
Comprehensive component	Continuous when conditions allow	1	2	1 trip	3 trips	40

Figure 47.6 OBD conditions for setting diagnostic trouble codes.

locating more problems because it will use the other pins. A generic scan tool might be able to read 50 fault codes, whereas a manufacturer's tool might do hundreds.

When an electronic problem occurs in a circuit that the computer senses or controls, a DTC is stored in the computer's **non-volatile RAM**. Non-volatile RAM is an OBD II term. It was commonly called keep alive memory (KAM) under OBD I. Other terms used in OBD II include warm-up cycle, trip, and drive cycle.

A **warm-up cycle** occurs every time the engine cools off and temperature rises at least 40°F. Coolant temperature must reach at least 160°F. In OBD II, a warm-up cycle's purpose is to clear codes only. This replaces the key cycles of OBD I. A code is usually erased after 40 warm-up cycles, if it does not recur during that time. Fuel trim and misfire codes take 80 warm-up cycles to clear. **Figure 47.6** lists the various OBD II conditions for setting DTCs and illuminating the MIL.

A **trip** (**Figure 47.7**) requires the ignition switch to be off for a period of time. When the engine is restarted and the vehicle is driven, various emission control monitors on the vehicle operate to complete one trip. These **enabling criteria** can include monitors for engine misfire, catalyst efficiency, fuel system, O_2 sensor, EGR, evaporative system, and air injection. An example of one

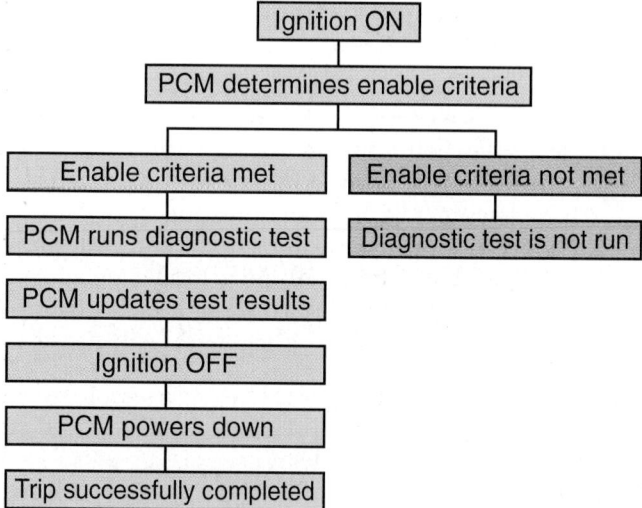

Figure 47.7 OBD trip cycle.

manufacturer's trip would be that five of the monitors—not including the catalyst monitor—would be required to operate to define one trip.

During a **drive cycle**, the engine must enter closed loop and all of the specified trip monitors must operate (**Figure 47.8**). The catalyst monitor must also operate. Drive cycles are not all the same. In 1998, all manufacturer drive cycles changed due to changes in OBD regulations.

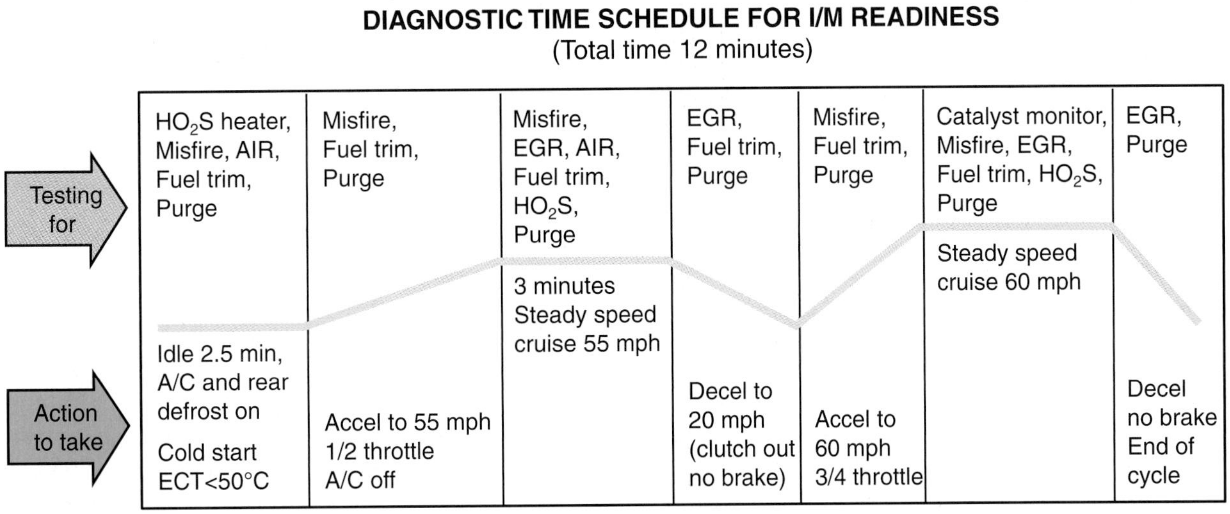

DIAGNOSTIC TIME SCHEDULE FOR I/M READINESS
(Total time 12 minutes)

Figure 47.8 An OBD II drive cycle.

In most cases, the same fault must be detected during two drive cycles to light the MIL. The code will be stored in the computer's non-volatile RAM. The first time DTC enable criteria are met, a **pending code**—or available code—is set. A second consecutive occurrence of the fault illuminates the MIL. Some scan tools can read pending DTCs, depending on the manufacturer. A generic scan tool might not see a pending code, although it will be able to read the code that caused the MIL to light.

If the monitors do not detect a fault for three consecutive drive cycles, the MIL will turn off. The DTC is then erased from memory when there have been 40 engine warm-up cycles after the MIL went out. There are some conditions in which a DTC is not set unless the fault has been detected in six drive cycles. These include failures to the evaporative system and catalytic converter.

OBD II is separate from powertrain control. One of its requirements is adaptive strategy. It can learn as the engine wears out. The newer systems are very sophisticated and can store many trouble codes. A misfire will be stored immediately as a code. Vehicles manufactured since 1996 have at least two O_2 sensors, sometimes as many as six, and are capable of monitoring catalytic converter operation. The catalyst efficiency monitor will set a code if the same fault happens on three consecutive drive cycles. A pending DTC is a code that has not occurred enough times to light the MIL.

When the key is turned on, the MIL should come on for a few seconds and remain on during engine cranking. It should go out shortly after the engine starts. If it comes on when the engine runs, there should be a trouble code stored in the computer's memory. If the MIL is lit when the engine is running, connect a scan tool to the DLC to read the code that was generated. Generic code numbers on all OBD II cars are the same for a particular sensor.

Under OBD I, codes could sometimes be read by counting the flashes from the MIL. Manufacturers gave

different instructions for reading their codes. If you are working on an OBD I vehicle, this information is available in the service manual. With OBD II, a scan tool is required to read codes, and reading a flashing MIL is not possible.

■ OBD II CODES

OBD II DTCs have five characters (**Figure 47.9**), a letter followed by four numbers.

■ The first character is a letter identifying the area of the vehicle that the code relates to. For instance, B is for body, C is for chassis, P is for powertrain, and U is for user defined (Ford uses it for multiplexing, for instance).

■ The second character is a 0 or a 1. A 0 is an SAE generic code, whereas a number 1 is a code assigned by the vehicle manufacturer.

■ The third character is a number from 1 to 8 representing a subsystem of the vehicle that the code belongs to. For example, a number 3 designates ignition system/misfire.

■ The last two characters are numbers that represent the actual fault code listing the sensor or actuator circuit where the problem has occurred. In the case of the above misfire designation, this number would be the cylinder that is misfiring.

Expansion of Code Numbers. The number of available codes has diminished as vehicle electronic systems have become more complex. The SAE has expanded the DTC numbers beyond P0. New SAE code numbers include P2XXX and P3400–P3999. Manufacturer codes will also be increased, adding numbers from P3000–P3399.

Types of Diagnostic Trouble Codes

There are four types of DTCs. Two of them are emission related. The other two are not considered emission

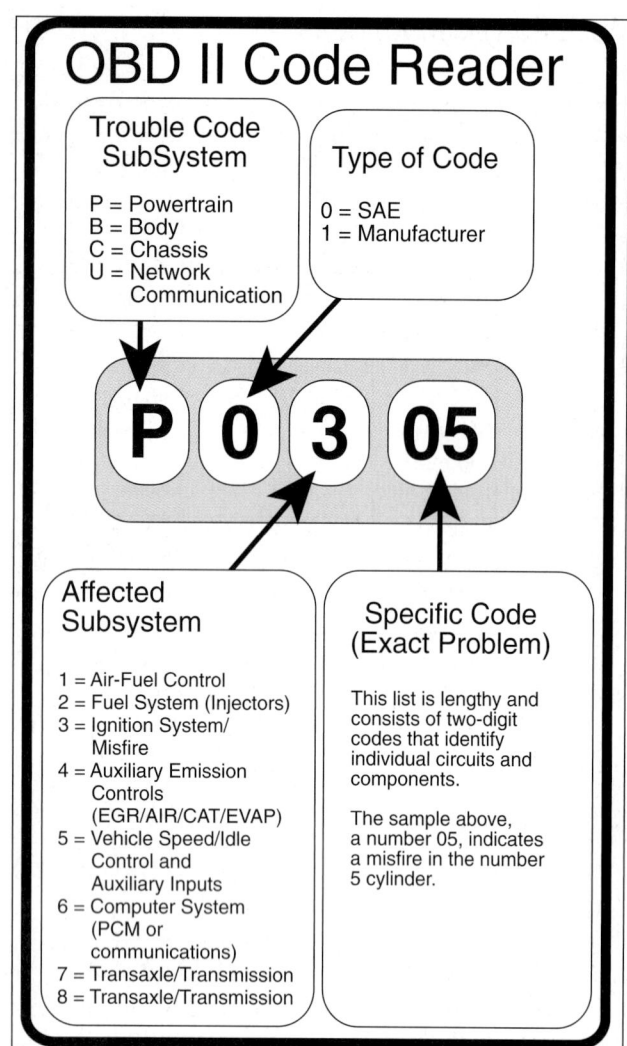

OBD II Code Reader

Trouble Code SubSystem

P = Powertrain
B = Body
C = Chassis
U = Network Communication

Type of Code

0 = SAE
1 = Manufacturer

P 0 3 05

Affected Subsystem

1 = Air-Fuel Control
2 = Fuel System (Injectors)
3 = Ignition System/ Misfire
4 = Auxiliary Emission Controls (EGR/AIR/CAT/EVAP)
5 = Vehicle Speed/Idle Control and Auxiliary Inputs
6 = Computer System (PCM or communications)
7 = Transaxle/Transmission
8 = Transaxle/Transmission

Specific Code (Exact Problem)

This list is lengthy and consists of two-digit codes that identify individual circuits and components.

The sample above, a number 05, indicates a misfire in the number 5 cylinder.

Figure 47.9 How to read OBD II codes. *(Courtesy of NAPA Institute of Automotive Technology)*

related because they do not result in emissions at 1.5 times the federal standard.

A catalyst damaging code, or **type A code**, is always emissions related. It is the most severe type of code that can result in damage to the catalyst. It will light the MIL on the first trip failure. A freeze frame of the event is stored when the code is set and updates every time the diagnostic test fails.

A type A code is something serious, such as a dead cylinder misfire. A type A misfire is one that can cause catalytic converter temperature to rise to a point where it could result in permanent damage (1,832°F [1,000°C]). The MIL will start blinking repeatedly 200 crankshaft revolutions after the first misfire detection occurs. If the misfire continues, the MIL will continue to flash. If the misfire stops, the MIL will remain on until the problem is fixed and the codes have been cleared. Stored codes describe which cylinder(s) misfired. The computers on these vehicles are sophisticated. Some manufacturers disable fuel injectors to the misfiring cylinders.

A type B code also is emission related but is not as serious. A type B misfire might not damage the cat, but it is enough to fail the FTP. Type B is a two-trip monitor. A pending code will be stored, ready to light the MIL with another consecutive trip failure. On the second consecutive trip failure, a DTC is stored, freeze frame data are saved, and the MIL lights up. The freeze frame data are only stored if they have a higher priority than one that is already in memory.

Rationality

OBD are separate programs within the computer. The computer receives multiple input signals, compares them to information in its programs, and decides which ones are rational (make sense). For example, it recognizes inputs from the vehicle speed sensor or the crankshaft position sensor to determine if they are realistic for current operating conditions.

The following is an example of rationality. A throttle position sensor (TPS) has a closed throttle signal, but the manifold absolute pressure (MAP) sensor says there is high engine load. The computer will ignore the closed throttle signal and run the engine with just the MAP signal. A MIL will illuminate to report the problem.

■ OBD II DIAGNOSTIC TESTING

The OBD II system performs diagnostic tests on the emission and engine management systems. Diagnostic tests include passive, active, and intrusive tests. These tests check if a component is operating as designed, according to specifications. Passive tests monitor the system as it operates. During active tests, the computer controls a specific action while monitoring takes place. Intrusive tests are those where the computer will cause an effect on the vehicle's emission output or its performance.

OBD II Monitors

In the OBD II system, **monitors** look for malfunctions. Some devices or systems are monitored all the time. Others are not checked until a component is activated under a predetermined operating condition. Monitors are of two types: continuous or non-continuous. A continuous monitor is operating whenever the engine runs. The engine misfire monitor is an example of a continuous monitor.

NOTE: *"Continuous" means a monitor check is taking place when requested by the computer when the key is on or the engine is running.*

A non-continuous monitor is one that tests a component or system once during each drive cycle. Non-continuous monitors include the catalyst efficiency monitor, fuel system monitor, O_2 sensor monitor, O_2 sensor heater monitor, EGR monitor, evaporative system monitor, and secondary air injection monitor.

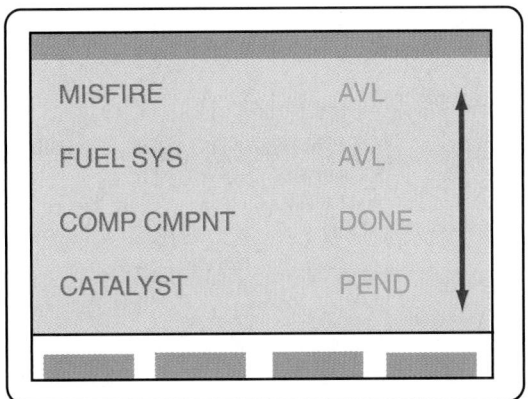

MISFIRE	AVL
FUEL SYS	AVL
COMP CMPNT	DONE
CATALYST	PEND

Readiness status categories

- Misfire
- Comp component
- Heated catalyst
- AIR
- O_2 sensor
- EGR system

- Fuel system
- Catalyst
- EVAP
- A/C refrigerant
- O_2 sensor heater

Figure 47.10 The scan tool tells which monitors have run.

Readiness indicators are readings displayed on the scan tool that tell whether any or all of the OBD II monitors have been completed since the keep alive memory was last cleared (**Figure 47.10**). Scan tools will display "complete," "finished," "ready," or "done" after the test has been performed. Certain enabling criteria must be met before a monitor can run during the OBD II drive cycle, including road speed; coolant temperature; acceleration; and deceleration, sometimes without braking. The key is to get the monitors to run. As mentioned earlier, manufacturers have different drive cycles. Be sure to match enabling criteria. Some technicians use a chassis dynamometer to maximize their ability to get all of the monitors to run quickly. Future emission testing may require technicians to check the status of these monitors. An incomplete monitor will result in a failed test.

The scan tool will indicate if monitors have not run since codes were cleared. The monitor will display "incomplete." The PCM cannot store a DTC unless a monitor has run. There can be several reasons for a monitor not to run, including a component failure or because enabling criteria have not been met. The PCM could also be withholding judgment pending further testing. If the scan tool displays "complete," this does not necessarily mean that the component passed the test.

NOTE: *A DTC tells you that a monitor test has failed. It does not tell you how to fix the vehicle. Other components related to the component listed in the DTC can be the actual cause of the problem. Be certain you understand the system you are working on.*

Monitor Tests

The following sections describe the operation of the various OBD II monitors. Manufacturers use different means of testing all of the required emission components and systems. Their service information gives more detailed descriptions and service procedures. As mentioned previously, certain enabling criteria must be met before non-continuous monitors can be run. A technician must understand what are the enabling criteria for the specific application to be able to do a complete drive cycle in the shortest time possible. In some cases, it might be difficult, if not impossible, to meet all of the enabling criteria to set a monitor.

Comprehensive Component Monitor. A comprehensive component monitor is a continuous monitor that looks at electrically controlled emission devices, sensors, and actuators that are not tested by other OBD II monitors. In most cases, this is done the same way as it was done in OBD I. The monitor not only looks for opens, shorts, and grounds, but also for the correct range of operation. When possible, it also checks sensor inputs for rationality.

The comprehensive component monitor also checks that outputs (actuators) are functioning correctly. Examples of such tests include idle rpm monitoring and transmission shift control solenoid operation. The computer monitors the voltage in an actuator's circuit. Most actuators are activated when they are switched to ground. This reduces the circuit's voltage to almost zero. When the actuator is not energized, voltage in the circuit should be the same as the charging system voltage.

Evaporative Emission Leak Check Monitor. An advanced evaporative system is checked to ensure it has no leaks larger than 0.040, or 0.020, in 2001 and later systems. This is about the size of the end of a ballpoint pen. The system is tested for component function and that it can flow vapors as designed. Besides the normally closed canister purge solenoid, the PCM controls a normally open vent solenoid on the atmospheric side of the charcoal canister (**Figure 47.11**).

Leak testing is done by pressure or by vacuum, depending on the manufacturer.

With some manufacturers, a vacuum is applied to the fuel tank. Then, the entire evaporative system is closed from the atmosphere while it is monitored for a change in pressure. The leak check cannot begin unless inputs from other sensors are received. Other manufacturers use a pump to pressurize the evaporative system. Then they monitor how often the pump cycles to see if there is a leak.

Monitoring the evaporative system has been a difficult task for the automobile manufacturers. In order to run this monitor, the fuel level in the tank must be within a specified range at the correct ambient temperature.

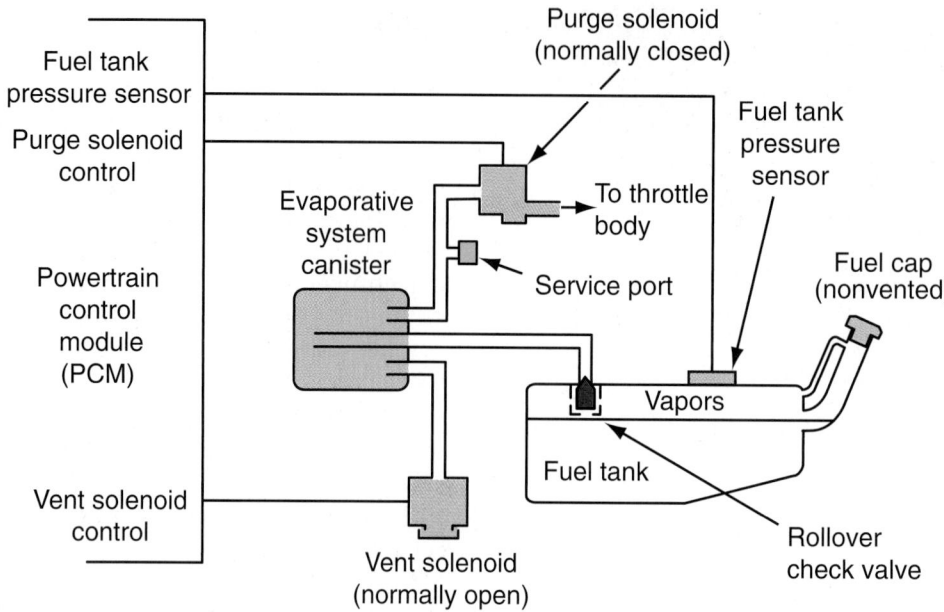

Figure 47.11 An evaporative system is sealed off by the PCM to test for leaks.

On extremely cold days, the test does not occur at all. In some geographic areas, this monitor might not run for several months. Also, the monitor will not run if fuel level in the tank is above 80% or under 15%. This amount varies with different manufacturers. The fuel tank cluster provides the PCM with fuel level information.

A troublesome problem is that of **heat soak**, when gasoline gets hot. Hot fuel causes pressure. The system must be able to accomodate that information in its tests. Some manufacturers use a pressure sensor in the gas tank. Others use a vapor generation test. The system will be tested again once it returns to normal temperature and pressure.

CAUTION Never blow shop air into the evaporative system or the fuel filler neck. This can damage system components—but more importantly, it is a safety hazard. Use an evaporative test smoke machine to locate leaks (see Chapter 44).

One of the common faults found by this monitor has been when a motorist leaves the fuel cap too loose after filling up the fuel tank.

EGR Monitor. The computer checks electronic components that direct vacuum or measure the opening of the EGR valve. They could all be working correctly, but the passage that exhaust gas travels through could be plugged. Manufacturers use different ways of monitoring whether or not the EGR system is working as designed.

Intake manifold pressure is related to the position of the EGR valve. Opening the valve increases mani-

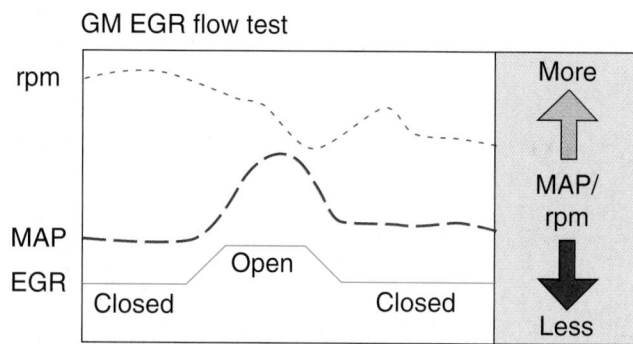

Figure 47.12 The monitor watches the MAP signal as the EGR valve operates.

fold pressure (decreases vacuum). Some manufacturers monitor the MAP signal as the EGR valve is operated to verify system operation (**Figure 47.12**).

NOTE: *GM does not use the MAP sensor for fuel injection. It is used for testing the EGR and only works as a typical MAP sensor when it is called on to serve a role as a backup if the mass airflow sensor (MAF) fails.*

The EGR valve is forced open during a closed throttle deceleration and/or during cruise conditions. This is an example of an intrusive test. A customer might notice this on a vehicle equipped with a tachometer. The manufacturer tries to perform invasive tests in a manner that will not be noticed by the driver.

NOTE: *It is important that service personnel understand the operation of the system, so they can inform the customer when a condition is normal and not attempt to make a repair. Attempting to repair a vehicle that is not defective can result in a lemon law buyback by the manufacturer.*

In another method of testing, a differential (delta) pressure feedback EGR sensor measures changes in

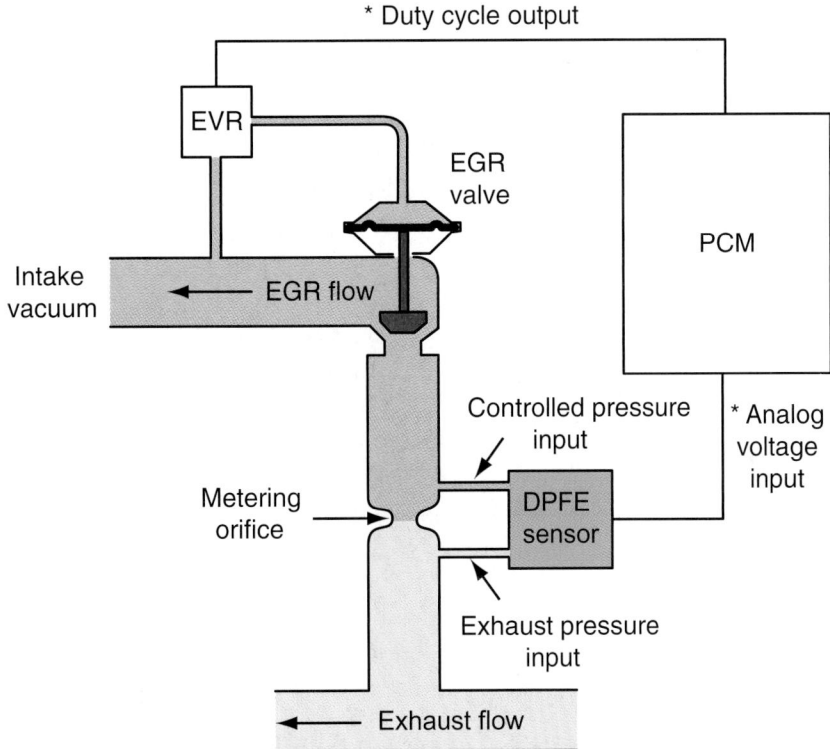

Figure 47.13 An EGR system with a differential pressure feedback sensor.

pressure in the passageway to the EGR valve. Two hoses are connected to the passageway, with an orifice plate between them (**Figure 47.13**). The upstream pressure hose measures exhaust pressure, and the downstream pressure hose measures pressure flowing through the valve. If the valve is closed, pressure should be equal on both sides of the orifice. If exhaust is flowing through the orifice into the EGR valve and onto the intake manifold, pressure will be lower in the upstream hose. To test exhaust gas recirculation, the computer opens the EGR valve and compares the pressure values with specifications. The test is done once every OBD II trip.

> **SHOP TIP** On differential pressure feedback EGR monitoring systems, changing to a larger diameter exhaust system will change backpressure in the exhaust system. This will result in a different EGR flow, which might cause an MIL to light.

Fuel Trim Monitor. The fuel system monitor operates continuously when the fuel system is in closed loop. It looks at short- and long-term fuel trim to check that it is within specified parameters compared to the O_2 sensor signal (**Figure 47.14**). When a problem results in excessive fuel trim corrections for more than the allowable period, emissions increase and the monitor sets a code. If the fault is detected during two consecutive drive cycles, the computer will illuminate the MIL. The maximum allowable correction is usually about 30% plus or minus.

Heated Oxygen Sensor Monitor. The heated oxygen sensor monitor (HO_2S) changes the injector pulse width while it checks the upstream oxygen sensor to see that its oscillating frequency is fast enough. The monitor looks at transition times between rich to lean and lean to rich (**Figure 47.15**). If a response rate is slow due to a lazy sensor (**Figure 47.16**) or if minimum and maximum voltages are insufficient, a code is set. The rear sensor is tested differently to see if its relatively flat voltage line falls outside of specified voltage limits.

O_2 sensors are monitored even before closed loop. OBD II can have a modified closed loop. The O_2 sensor will work, but the system will not have fuel trim yet.

NOTE: *Some later models have fuel trim in open loop, instead of following the traditional totally preprogrammed parameters of open loop. The HO_2S toggles with a bias toward rich. The objective is to try to lower cold engine emissions.*

Sensor Identification. With the multiple O_2 sensors on OBD II exhaust systems, each individual sensor must be identified by a code resulting from the monitor test. A pre-catalyst sensor is called sensor 1 (S1), and a post-cat sensor is called S2. In V-type engines, Bank 1 is the cylinder bank that has cylinder number 1 (**Figure 47.17**).

Oxygen Sensor Heater Monitor. The oxygen sensor's heater is tested electronically. Some systems use the length of time before the O_2 sensor begins operating

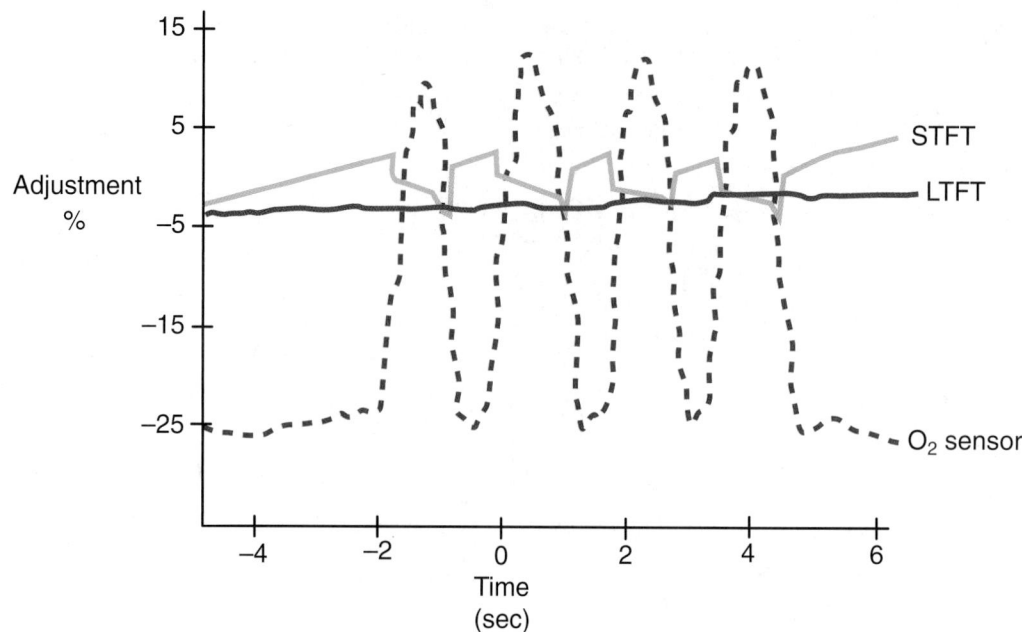

Figure 47.14 The fuel system monitor looks at short- and long-term fuel trim compared to the oxygen sensor signal.

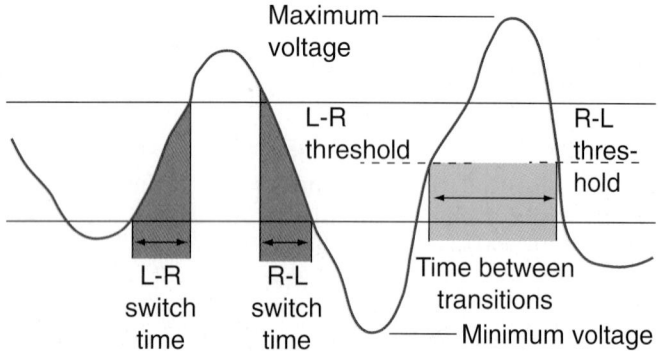

Figure 47.15 The oxygen sensor monitor checks the speed of the switch rates between rich and lean.

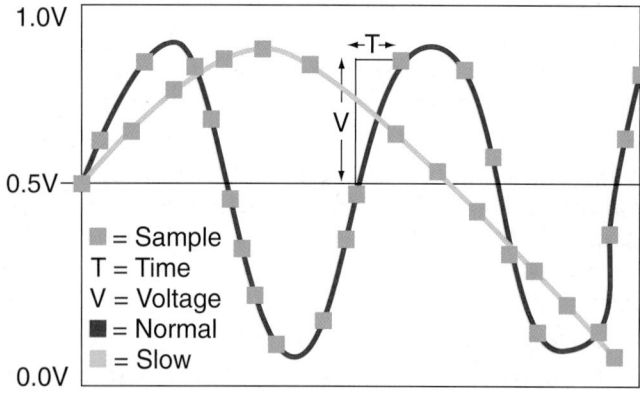

Figure 47.16 A failing oxygen sensor switches at a slower rate.

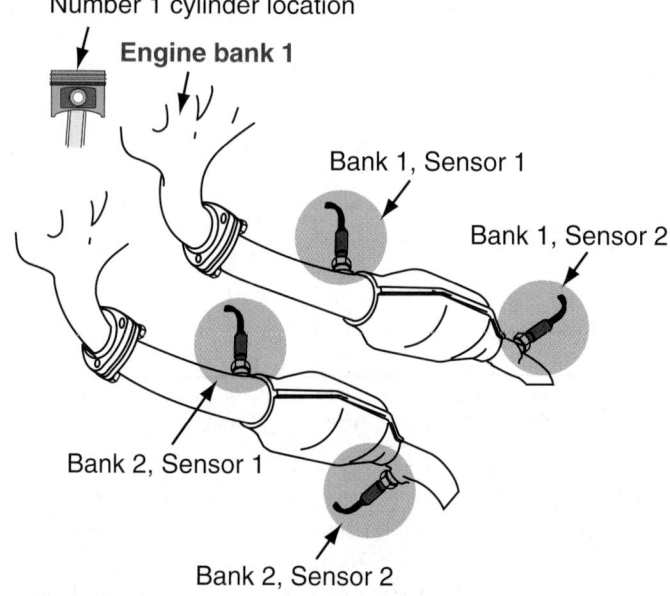

Figure 47.17 OBD II oxygen sensor locations. On V-type engines, Bank 1 is the cylinder bank where the number 1 cylinder is located.

to determine the heater's operation. Others use a measurement of current flow to the heater to determine its operation (**Figure 47.18**). The heater monitor tests right away in the drive cycle.

Misfire Detection Monitor. The misfire monitor is a continuous monitor that detects when the engine slows down momentarily due to a misfire. Some manufacturers use a crankshaft position sensor (CKP) only for these data. Others use both crankshaft and camshaft position sensors (CMP) (**Figure 47.19**). Misfire detection is powerful and uses a good deal of space in the OBD II software.

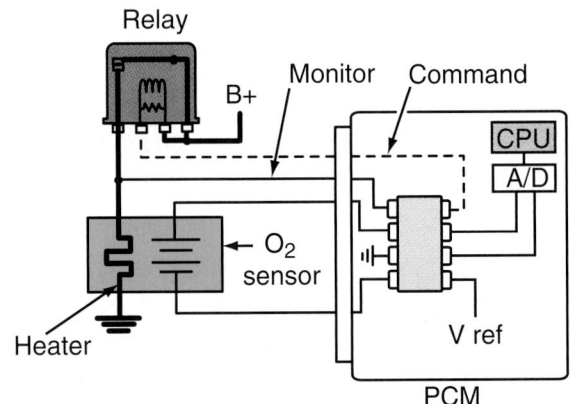

Figure 47.18 An oxygen sensor heater monitor circuit.

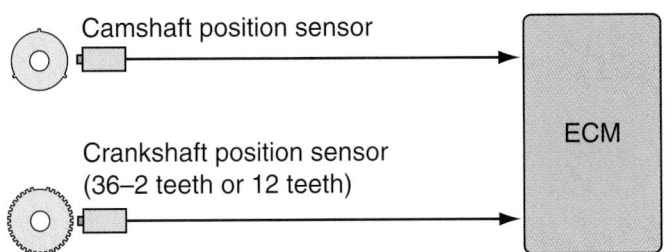

Figure 47.19 For misfire detection, many OBD II systems compare the speed of the crankshaft and camshaft. *(Courtesy of Toyota Motor Sales, U.S.A., Inc.)*

Using misfire counters, the PCM calculates the temperature of the catalyst based on driving conditions and the percentage of misfiring (**Figure 47.20**). The computer watches every 200 crankshaft revolutions,

knowing each cylinder will fire 100 times in that span. The misfire counter is based on the engine's number of cylinders (**Figure 47.21**).

Type A Misfires. For type A misfires where the cat could overheat, the PCM is more likely to start flashing the MIL when there is a heavier engine load and a greater percentage of misfiring (**Figure 47.22**). When the engine operating conditions change from the range where the misfire occurred, the MIL will stop flashing but will remain lit.

During a type A misfire, the fuel system reverts to open loop because the misfire would result in more O_2 in the exhaust, which the O_2 sensor and software would try to correct by adding fuel in error. Some engine management systems shut off fuel to any cylinders that are misfiring. At idle speed, a higher percentage of misfire (up to above 40%) is sometimes allowed. A typical misfire percentage that will set a type A fault is about 4%. **Figure 47.23** shows a diagnostic tester screen with a 5% misfire percentage.

Number of cylinders		Number of firings per cylinder in 200 crankshaft revolutions		Ignition count
4	×	100	=	400
6	×	100	=	600
8	×	100	=	800

Figure 47.21 The misfire counter is based on the engine's number of cylinders.

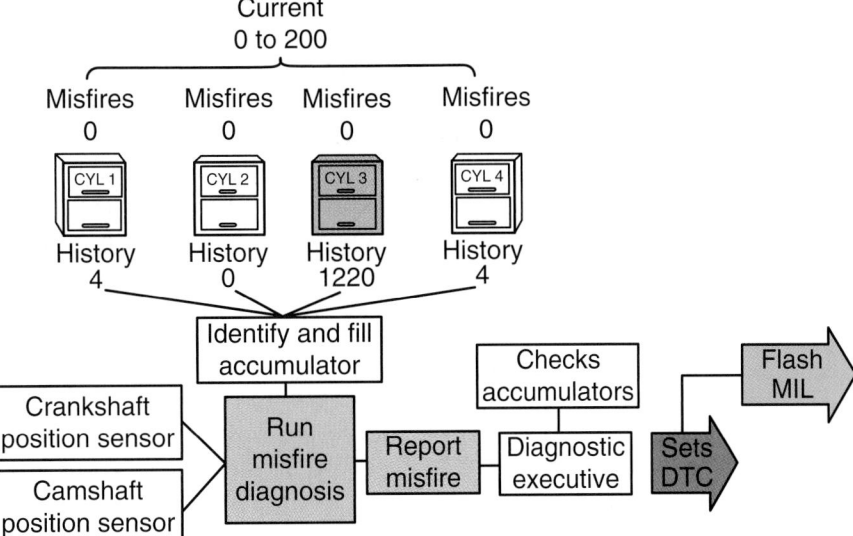

Figure 47.20 The PCM calculates the temperature of the catalyst based on driving conditions and the percentage of misfires.

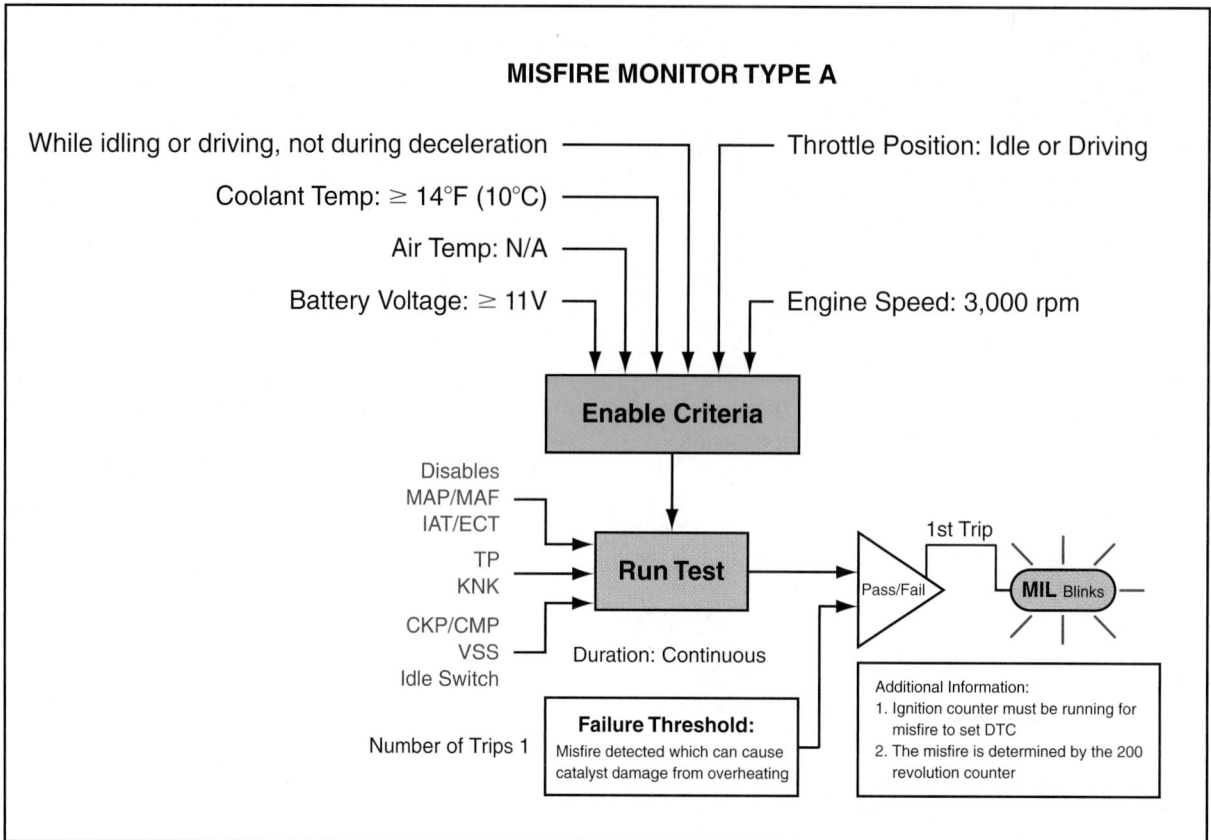

Figure 47.22 A type A misfire will cause the MIL to flash. *(Courtesy of Toyota Motor Sales, U.S.A., Inc.)*

NOTE: *The most misfiring occurs during the first 1,000 crankshaft revolutions after the engine starts, while some of the hydraulic valve lash adjusters are still partially empty.*

Type B Misfires. A typical type B misfire monitor would allow about 2%–3% random misfiring before setting a pending code and recording the operating conditions (PIDs) in a freeze frame. After one more drive cycle, if the same conditions are met, a code is set and the MIL is turned on.

If the misfire does not return in the next trip under similar conditions, the program stores the code and freeze frame data (PIDs) for another 80 trips. If the problem recurs, it is ready to turn on the MIL once again. Misfires above the allowable limit are type A codes and have first priority on freeze frame memory because they can damage a catalytic converter.

Misfire detection is more reliable at lower rpm. Most engines detect misfire by seeing very small variations in crankshaft rpm. Other things can also set a misfire, however. Rough roads are a common cause of false misfire detection. A loose fan belt can also set a code as the crankshaft vibrates excessively.

There are different ways to detect false misfires. Software can detect most false misfires by detecting patterns that have the characteristics of a rough road

(**Figure 47.24**). Some cars use a vertical acceleration sensor. Most cars use the antilock brake system (ABS) to detect a rough road condition. Variations in wheel speed can be detected by ABS sensors the same way a crankshaft position sensor senses engine misfires.

For vehicles with automatic transmissions, some manufacturers disable the torque converter clutch when a misfire indication begins. This isolates the rear wheels and engine, so the monitor can check if there really is a misfire.

Some systems with speed density fuel injection use the MAP sensor for misfire detection.

Secondary Air Injection Monitor. An interesting application of new technology is in the way manufacturers are dealing with air injection. Many newer engines do not have air injection systems. Instead, they use higher engine compression to increase NO_X. The increased NO_X provides more O_2 available for conversion in the catalytic converter. For auxiliary air, electric air pumps are also used (**Figure 47.25**). In addition to electronic monitoring, a functional check can also be performed on the system using the O_2 sensor(s) to test for airflow. The test is usually done at startup when air is needed to help warm the cat and oxidize the extra HC and CO in the exhaust immediately after startup.

Misfire Detection

When the IGNITION counter is cycling, the misfire monitor is operating. A percentage above zero indicates the cylinder(s) that is misfiring.

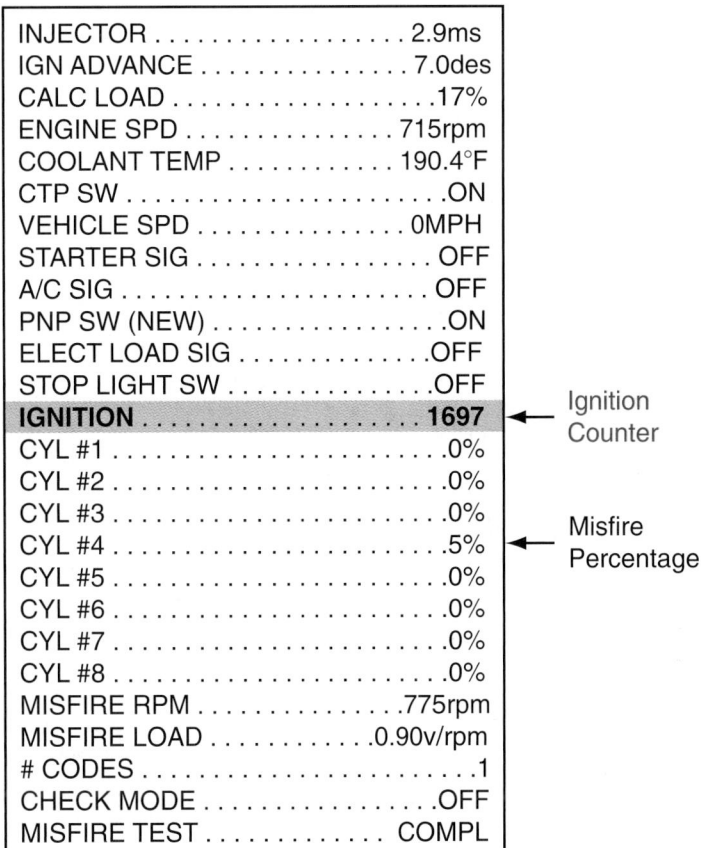

```
INJECTOR . . . . . . . . . . . . . . . . . . 2.9ms
IGN ADVANCE . . . . . . . . . . . . . 7.0des
CALC LOAD . . . . . . . . . . . . . . . . .17%
ENGINE SPD . . . . . . . . . . . . . . 715rpm
COOLANT TEMP . . . . . . . . . 190.4°F
CTP SW . . . . . . . . . . . . . . . . . . . .ON
VEHICLE SPD . . . . . . . . . . . . . 0MPH
STARTER SIG . . . . . . . . . . . . . . OFF
A/C SIG . . . . . . . . . . . . . . . . . . OFF
PNP SW (NEW) . . . . . . . . . . . . . .ON
ELECT LOAD SIG . . . . . . . . . . . .OFF
STOP LIGHT SW . . . . . . . . . . . . .OFF
IGNITION . . . . . . . . . . . . . . . . . 1697        ← Ignition Counter
CYL #1 . . . . . . . . . . . . . . . . . . . .0%
CYL #2 . . . . . . . . . . . . . . . . . . . .0%
CYL #3 . . . . . . . . . . . . . . . . . . . .0%
CYL #4 . . . . . . . . . . . . . . . . . . . .5%        ← Misfire Percentage
CYL #5 . . . . . . . . . . . . . . . . . . . .0%
CYL #6 . . . . . . . . . . . . . . . . . . . .0%
CYL #7 . . . . . . . . . . . . . . . . . . . .0%
CYL #8 . . . . . . . . . . . . . . . . . . . .0%
MISFIRE RPM . . . . . . . . . . . . .775rpm
MISFIRE LOAD . . . . . . . . . . .0.90v/rpm
# CODES . . . . . . . . . . . . . . . . . . .1
CHECK MODE . . . . . . . . . . . . . .OFF
MISFIRE TEST . . . . . . . . . . . . COMPL
```

Figure 47.23 A screen from a scan tool showing 5% misfire in cylinder number 4. *(Courtesy of Toyota Motor Sales, U.S.A., Inc.)*

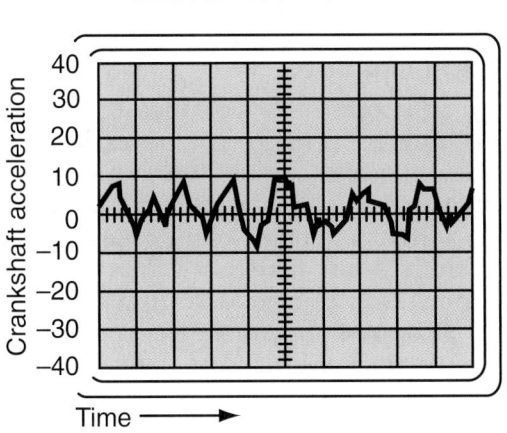
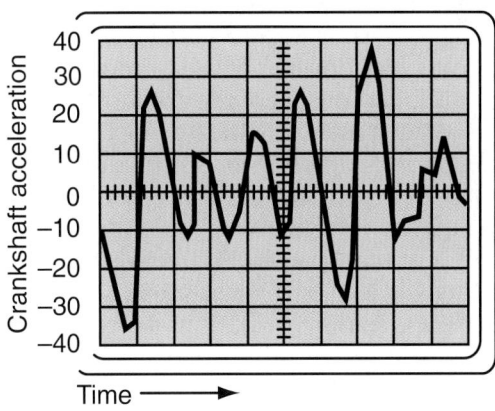

Figure 47.24 Impact of a rough road on misfire monitoring.

Thermostat Monitor. A more recent monitor since the 2000 model year is the thermostat. It is enabled each drive cycle after the engine has been off for at least 2 hours. If the vehicle is driven in a manner that will generate enough heat (sufficient load and speed), engine warmup should be predictable. The monitor considers temperature at startup and has a timer. The target temperature is 20°F less than the rated temperature of the thermostat. A failure will illuminate the MIL.

Positive Crankcase Ventilation (PCV) System Monitor. When a PCV system fails to operate correctly, HC emissions can increase. Since the 2004 model year, all manufacturers have been required to monitor the PCV system as part of OBD II. The PCV valve must be connected to the crankcase source with a positive lock so it cannot become accidentally disconnected. The system uses an extra large diameter hose. If it should become disconnected, the very large vacuum leak that results will

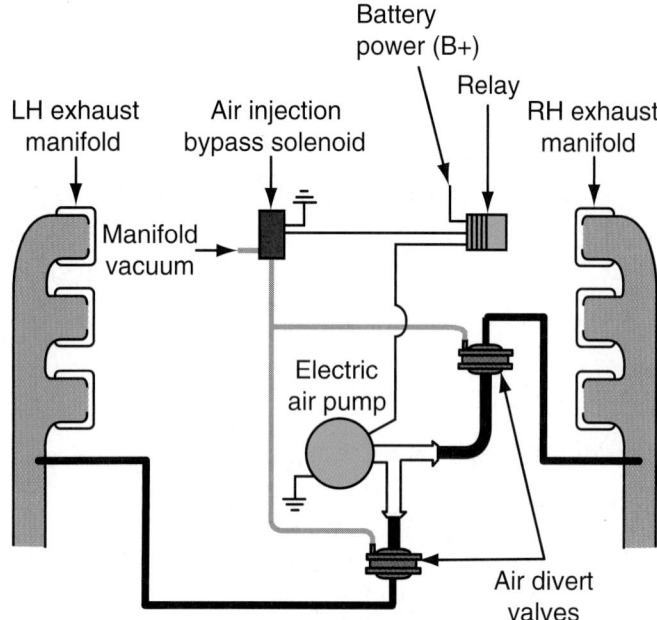

Figure 47.25 An electric air pump is used on some vehicles.

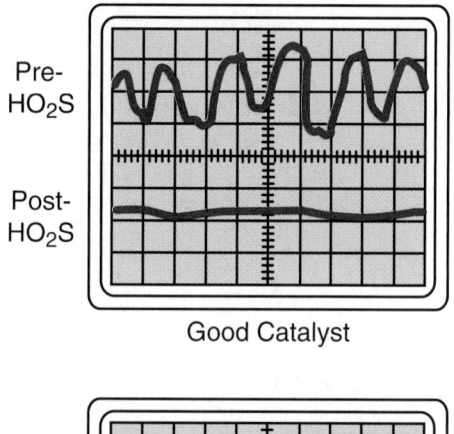

Good Catalyst

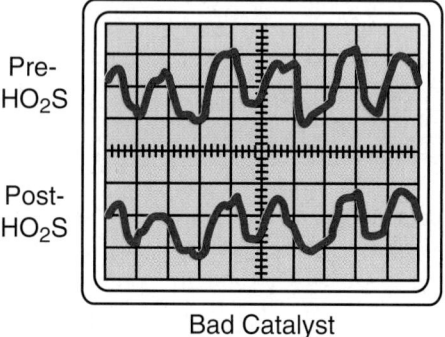

Bad Catalyst

Figure 47.26 Oxygen sensor waveforms showing good and bad catalytic converters.

prevent the engine from restarting. If the engine runs with the hose disconnected, the lean idle that results will cause an illuminated MIL after two drive cycles.

The PCV monitor runs once each drive cycle. It can detect a disconnected or damaged hose or grommet or a restriction in a hose or valve. It also detects if the crankcase vent is restricted or becomes disconnected.

Catalytic Converter Monitor. A good cat stores O_2 when it is working properly. All other monitors must run before the catalytic converter monitor will run. This is because many other things can cause a cat to have difficulty storing O_2. A good catalytic converter will use up most of its O_2, leaving very little coming out of the exhaust—usually less than 1%. Because there is little or no O_2 in the exhaust, the signal from the rear O_2 sensor should not oscillate between rich and lean (100 mV–900 mV). It should remain flat. The voltage level where it remains flat is not important, just that it is flat. **Figure 47.26** shows good and bad catalytic converter waveforms.

The speed of O_2 sensor oscillations between rich and lean is called the **switch ratio**. A correctly operating front O_2 sensor switches at about two times per second at low rpm. With a new converter, the rear sensor will have a switch ratio of 0. As a catalytic converter ages, its efficiency declines. When the efficiency of the catalyst has deteriorated to the point where it will no longer function to OBD II standard, the ratio will be 0.8–0.9 times per second.

NOTE: *Aggressive driving can set a code for the cat. Six successive drive cycles with aggressive driving will light the MIL.*

NOTE: *The rear O_2 sensor provides a means of monitoring catalyst efficiency. To compensate for an inefficient catalyst, it fine-tunes fuel trim. Manufacturer service information says that under normal operating conditions, the rear O_2 sensor will only make a correction of 0.5% to 2%, depending on the manufacturer. Under abnormal conditions, when the front O_2 sensor has reached its point of maximum fuel trim correction, the rear O_2 sensor can add more fuel in an attempt to compensate.*

Mode $06 Data

You will sometimes see Mode 6 referred to as Mode $06. This is because it is a hexadecimal alphanumeric code using 16 characters. The first ten characters are the numbers 0–9. The final six characters are uppercase letters A–F. Interpretation of the code is available in service literature. Many newer scan tools are user friendly and automatically convert Mode 6 data on the screen.

Mode 6 information is not always available on manufacturer-specific scan tools. It is an option on some OBD II global/generic scan tools, although the information available is not consistent among manufacturers. When available, it provides results of sensor and actuator tests. Mode 6 tells you which tests have been run, whether they passed or failed, and if they failed, by how much. You can compare the test results of non-continuous monitors with the designed test limits so you can see if a part is close to failing. Mode 6 data are also helpful in diagnosing intermittent problems.

■ EMISSION TESTING PROGRAMS

Laws governing emission testing differ depending on location. Some government entities specify different

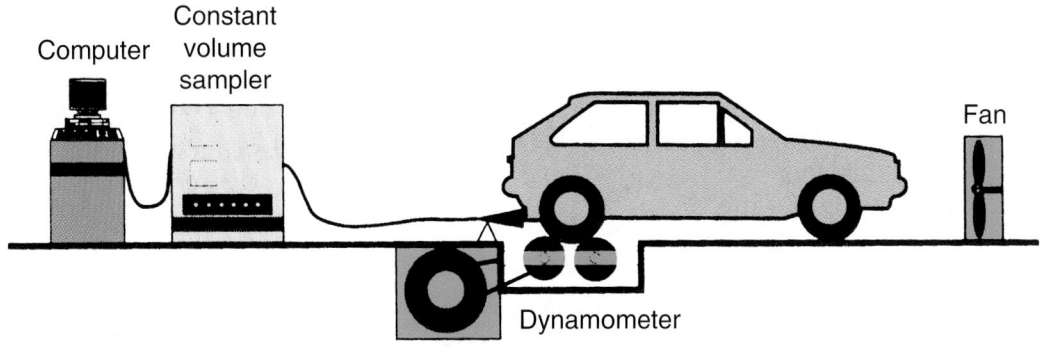

Figure 47.27 A dynamometer is used for enhanced emission testing.

types of "enhanced" testing based on the location. Enhanced testing is done only in areas with higher smog levels. As OBD II has progressed, standards have become tougher to meet. The pass/fail emission standards are called "cut points (see Chapter 47)." Reduced cut points have made it tougher for some vehicles to pass the test.

Some of the different tests include two speed idle (TSI), acceleration simulation mode (ASM), and inspection and maintenance (I/M) 240. ASM and I/M 240 are enhanced tests that use a dynamometer (**Figure 47.27**) and test for NO_x. ASM tests are run at 15 mph under 50% load and at 25 mph under 25% load. The I/M 240 testing covers a variety of driving conditions over a 240-second period.

Causes of Oxides of Nitrogen (NO_x)

NO_x is formed when nitrogen and oxygen combine in the combustion chamber at sustained high temperatures in excess of 2,500°F. When NO_x is combined with HCs in the presence of sunlight, smog is formed. There are several causes of increased NO_x, abnormal combustion (preignition or detonation) being a major one. Abnormal combustion can be caused by several things. Carbon deposits cause abnormal combustion by increasing the compression ratio and creating hot spots. Deposits are more common on engines with high mileage. Advanced ignition timing or using fuel of too low an octane level can also cause abnormal combustion. Hot air and insufficient EGR are also contributing factors.

Rich or lean conditions must be addressed before diagnosing NO_x-related failures. A lean air-fuel ratio can increase NO_x when it causes preignition. Dirty fuel injectors can increase NO_x by causing lean mixtures. A rich air-fuel ratio does not actually increase NO_x, but the resulting misfire will cause increased CO, which can result in an artificially low NO_x reading.

Prior to testing, the MIL must not be illuminated and no codes can be stored in memory. Test the following items in order to diagnose a vehicle that has failed an emissions test:

1. There must be a correctly functioning oxygen sensor (O_2S).
2. The computer must be in fuel control.

3. The cat must be functioning at 80% efficiency or better.

NOTE: *The cat must have a varying rich/lean air-fuel ratio to work. A catalytic converter is only efficient during fuel control. The cat needs O_2, and fewer cars still have air pumps.*

Testing for Fuel Control

The O_2S must be fast enough in its repeated transitions from rich to lean. The measurement of the number of times the O_2S voltage switches over and under the 450 mV midpoint are called cross counts. As they age, O_2 sensors can become slower. To tell if the O_2S is fast enough, use a digital storage oscilloscope (DSO). Digital multimeters and scan tools are too slow for this measurement.

An O_2S must also be within calibration. A typical rich reading is above 800 mV (0.8 volt), and a typical lean reading is below 200 mV (0.2 volt).

To test an O_2S, use a propane enrichment tool to bleed propane into the air intake while the engine is running. This will cause a rich air-fuel mixture. Start with an amount of propane flow that is twice the number of cylinders. For this test, the exact amount is not important. The O_2S pattern should flatline rich at a voltage exceeding 800 mV (0.800 volt). Lower than 800 mV means the sensor does not know what rich is. A reading of 100 mV can be a result of fuel trim, which is able to change the air-fuel mixture by up to 20%. Otherwise, the accuracy of the sensor is questionable.

NOTE: *Some vehicles have been known to have pattern failures from lean misfires due to the O_2 sensor being skewed higher (1.4–1.5 volts). A reading of 1.1 volts is probably not too far out of specification to work.*

Propane Enrichment Test. With propane being injected, the DSO should flatline high. This will cause the PCM to reduce the pulse width, decreasing the amount of fuel from the injectors.

- Run the test at 2,000–2,500 rpm for about 1 second.

■ Shut off the propane while the system is still rich and before the processor reacts. The DSO reading should flatline lean. Voltage should be lower than 175 mV and should not go below 0 volt. If the sensor tests good on rich and lean, it is probably good. However, test it at a higher speed in case there is a crack, which could draw in O_2.

■ To check for quickness of the O_2S, inject a quick burst of propane while the pattern is still lean. Voltage should rise from lean to full rich (approximately 300–600 mV) in less than 100 ms.

NOTE: *During emission testing, a little extra fuel can be handled, but the lack of fuel is what causes misfires and failures.*

■ REVIEW QUESTIONS

1. What percentage of emissions occurs during warmup?

2. In what year did all cars sold in the United States have to comply with OBD II regulations?

3. The OBD II objective is to detect exhaust and evaporative emissions that exceed the Federal Test Procedure (FTP) by how much?

4. What three gases do FTP emission standard measure by weight in grams per mile?

5. A ___ cycle occurs every time the engine cools off and temperature rises at least 40°F and coolant temperature reaches 160°F.

6. What is the purpose of an OBD II warm-up cycle?

7. Not including the catalyst monitor, how many monitors are required to operate in order to define one "trip"?

8. During a ___ cycle, the engine must enter closed loop and all five of the trip monitors, and the catalyst monitor, must operate.

9. A ___ code will light the malfunction indicator lamp (MIL) on the second occurrence.

10. If the monitors do not detect a fault for ___ consecutive drive cycles, the MIL will turn off.

11. A DTC is erased when there have been 40 ___ cycles after the MIL went out.

12. Vehicles produced since the year ___ have had at least two oxygen sensors to monitor catalytic converter operation.

13. If the second character in an OBD II DTC is a 0, is it a generic code or a manufacturer's code?

14. What is a PID?

15. What is the name of the type of diagnostic test in which the computer affects the vehicle's emission output or engine performance?

■ ASE-STYLE REVIEW QUESTIONS

1. Which of the following is/are true about the malfunction indicator light?
   a. An ABS problem can light the MIL.
   b. Whenever the computer detects a fault it will turn on the MIL.
   c. Both A and B.
   d. Neither A nor B.

2. Two technicians are discussing scan tool snapshots. Technician A says that a frame-by-frame recording is taken several seconds before and after a button is pressed. Technician B says that a snapshot is useful in identifying intermittent faults. Who is right?
   a. Technician A   c. Both A and B
   b. Technician B   d. Neither A nor B

3. Which of the following is/are true about OBD II codes?
   a. A type A code is when the computer determines a misfire will raise the emissions above 1.5 times the federal limit.
   b. A type B code is a dead misfire that can damage the catalyst.
   c. Both A and B.
   d. Neither A nor B.

4. Technician A says that a continuous monitor only checks when requested by the computer. Technician B says that a non-continuous monitor tests a component or system once during each drive cycle. Who is right?
   a. Technician A   c. Both A and B
   b. Technician B   d. Neither A nor B

5. Which of the following is/are true about OBD II oxygen sensors?

   a. The signal from the rear oxygen sensor should oscillate between rich and lean.

   b. The signal from the front oxygen sensor should be relatively flat.

   c. Both A and B

   d. Neither A nor B

6. Leaving the fuel cap loose after filling up the fuel tank can affect:

   a. The fuel system monitor

   b. The EGR monitor

   c. Both A and B

   d. Neither A nor B

7. Technician A says that the catalytic converter monitor must run before all of the other monitors can run. Technician B says that a catalytic converter's efficiency declines with use. Who is right?

   a. Technician A   **c.** Both A and B

   b. Technician B   **d.** Neither A nor B

8. Technician A says that if the MIL is on there is a DTC stored in memory. Technician B says that the second time DTC enable criteria are met a pending code is set. Who is right?

   a. Technician A   **c.** Both A and B

   b. Technician B   **d.** Neither A nor B

9. All of the following are true regarding the misfire monitor *except*:

   a. Rough roads are a common cause of false misfire detection.

   b. Most cars use the ABS to detect a rough road condition.

   c. Misfire detection is more reliable at higher rpm.

   d. Some automatic transmission computers disable the torque converter clutch when a misfire indication begins.

10. The computer will turn off the MIL and erase a misfire DTC:

   a. After 40 warm-up cycles.

   b. After 80 wram-up cycles.

   c. After one drive cycle.

   d. If the misfire monitor does not detect a similar misfire in the next three trips.

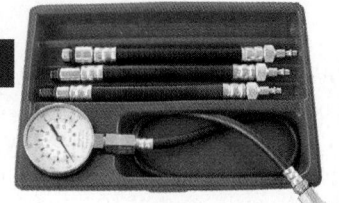

Diagnosing Engine Performance Problems

■ **OBJECTIVES**

Upon completion of this chapter, you should be able to:

✔ Perform compression tests in a correct manner.

✔ Perform cylinder leakage tests in a correct manner.

✔ Describe the procedures for doing various vacuum tests.

✔ Explain the differences in types of carbon deposits.

■ **KEY TERMS**

cranking vacuum test

cylinder power balance test

compression test

wet compression test

cylinder leakage test

■ **INTRODUCTION**

Problems with the engine and ignition system can result in poor fuel economy, high emissions, and poor driveability. Being able to correctly diagnose and repair these problems requires related knowledge from material in other related chapters. Chapter 37 covers ignition system theory. Firing orders and companion cylinders are covered in Chapter 16. Engine mechanical problems are covered in detail in Chapter 49. Refer to those chapters as needed.

For an engine to start and run properly, three things are necessary:

■ Sufficient compression

■ A timed spark

■ Fuel and air mixed in the correct ratio

The spark created by the ignition system must be delivered to the correct spark plug at a specified point before TDC. This is called ignition timing. Vaporized air and fuel in the correct quantity and ratio must be drawn into the cylinder. If the mixture is not compressed enough, it will not burn when ignited.

■ **VISUAL CHECKS**

The first thing to do when an engine does not start or is running poorly is a visual check. A broken vacuum line or an electrical wire that has fallen off is sometimes the cause. Check the fuel gauge to verify there is sufficient fuel. Believe it or not, sometimes the car is just out of

gas. This is comparable to an appliance repair person making a house call only to discover that an appliance does not work because it is not plugged in.

■ **IGNITION SYSTEM CHECKS**

For an engine to run, the ignition system must produce a strong spark at the correct moment. Checking for spark is the quickest thing to do. If an engine will not start, see if there is spark at any of the spark plugs. A tester is available that provides a visible gap (**Figure 48.1**). If there is a strong hot spark, the next tests to make are for fuel and compression.

With fuel injection it is usually easier to test compression than check for fuel. With a spark plug

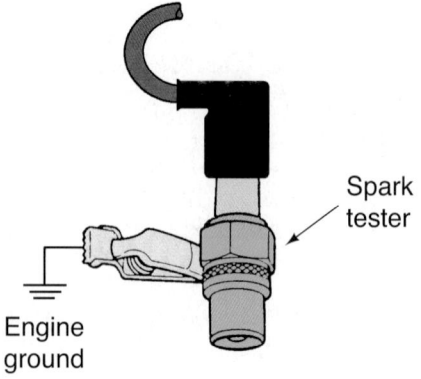

Spark tester

Engine ground

Figure 48.1 A spark tester. *(Courtesy of Ford Motor Company)*

removed, hold your thumb over the spark plug hole while cranking the engine with the ignition system disabled. If the engine has good compression, the piston will blow air hard on your thumb. A compression tester can also be used, but at this point you are just trying to determine that compression is present.

Sometimes an engine starts and runs but one cylinder has a *dead misfire*. This could be due to ignition or mechanical problems. Further testing will be required.

ENGINE PERFORMANCE TESTING

When looking for reasons why an engine runs poorly, first rule out other causes.

Fuel Problems

- Dry black soot at the exhaust pipe indicates an overly rich air-fuel mixture rather than oil consumption.
- Fuel in the crankcase causes a dangerous condition to develop. Leaking fuel injectors or a bad pressure regulator will allow the crankcase to collect fuel.

 If you are working on a vehicle and your clothing should catch fire, fall on the ground and roll around to smother the flames. Do not run!

The air-fuel mixture can also be affected by air leaks in some fuel injection intake systems. Scan tools used in diagnosing computer control problems show whether the oxygen sensor responds quickly to control the air-fuel ratio.

Poor quality fuel, blowby, and EGR gases can cause a gummy deposit buildup behind the throttle plate. After many miles, the deposit gets hard and restricts air travel around the plate, causing driveability problems and erratic changes in idle speed. Use a chemical cleaner to remove the deposits.

Dirty fuel injectors can cause driveability problems. When injectors get dirty, this can cause detonation at full throttle. Use a stethoscope to listen for different sounds between the injectors. Dirty injectors do not always make the clicking sound associated with a normal injector. Following injector cleaning, the normal click can return. Newer ball-type injectors are self-cleaning.

A plugged fuel filter will cause high-speed starvation. Some technicians recommend a fuel injection cleaning and fuel filter change at every 30,000-mile maintenance.

COMPRESSION LOSS

Low compression can be due to two causes: engine *breathing problems* and *compression leaks*. An engine that cannot breathe properly is suffocating and will not be able to develop the necessary compression to function properly. Engine vacuum will drop off, further lowering

compression. Causes of breathing problems include worn camshaft lobes that do not open the valve far enough or incorrect valve timing.

NOTE: *Late valve timing will cause poor low rpm performance. At higher rpm, engine performance might be acceptable.*

Breathing problems can also be traced to carbon buildup around the neck of the valve, intake restrictions such as a dirty air cleaner, or a blocked exhaust. A blocked exhaust will be evident when engine rpm is raised quickly. A roar will be heard through the intake manifold.

NOTE: *A collapsed laminated (two-layer) exhaust pipe (see Figure 42.37) will often cause a stripe of discoloration on the outside of the pipe at the point of restriction.*

Catalytic converters can become plugged when run for a prolonged period of time with an ignition system defect. A rich air-fuel mixture can also result in a plugged converter when it overheats and melts internally. These problems cause a flashing malfunction indicator light (MIL) on OBD II vehicles. A catalytic converter can become so hot it can start a fire.

A car with an exhaust restriction can also experience automatic transmission harsh or late shifts due to the resulting faulty computer input.

Exhaust backpressure can be tested by connecting a fuel pump vacuum/pressure tester to smog pump lines into the exhaust manifold, or an adapter can be substituted in place of the EGR valve. Specifications vary among manufacturers. As a general rule, pressure should not exceed 1.75 psi at wide-open throttle (WOT) under full load.

VACUUM TESTING

A vacuum/pressure gauge (**Figure 48.2**) can be used to measure intake manifold vacuum or fuel pressure. Intake manifold vacuum readings are useful in determining

Figure 48.2 A vacuum/pressure tester used to measure intake manifold vacuum. (*Courtesy of Tim Gilles*)

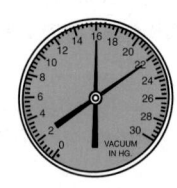

16–22 inches of vacuum is considered normal for altitudes below 1,000 ft

Subtract one (1) inch for every 1,000 ft. higher

Figure 48.3 Normal engine vacuum.

engine problems. Vacuum readings compare pressure in the intake manifold to atmospheric pressure. Connect a vacuum gauge to a manifold vacuum source. Typical vacuum readings should range between 16 and 22 inches of mercury (in. Hg) at idle, and the needle should be steady (**Figure 48.3**).

NOTE: *Vacuum readings will drop approximately 1 inch for each 1,000 feet above sea level.*

Figure 1.10 shows how throttle opening corresponds to engine vacuum.

Rough Idle

A *leaking intake manifold* gasket can cause a rough idle.

NOTE: *At speeds above idle, symptoms of a leaking manifold diminish and do not show up on the vacuum gauge. This is because the size of the leak is proportionally less as the engine breathes more air.*

Sometimes, an engine will idle rough for no apparent reason. Port fuel injected engines use O-rings to seal each individual fuel injector where it enters an intake port (**Figure 48.4**). A leaking O-ring, which causes a vacuum leak, will result in a lean air-fuel mixture for that cylinder. A rough idle is the result.

When there is a leak of unmetered air (a vacuum leak), an engine with an oxygen sensor feedback fuel

Figure 48.4 Fuel injection O-rings can become hard and brittle with age, resulting in a rough idle, especially during open-loop operation when the engine is first started and the computer is not receiving feedback from the oxygen sensor. *(Courtesy of Tim Gilles)*

system can run rough when cold, yet run fine when hot. This is because the oxygen sensor, which does not work until sufficiently heated, tells the computer to compensate with a richer air-fuel mixture (see Chapter 41).

Low Vacuum

When there is low, steady vacuum, check the ignition timing, which can also cause an engine to run hot when retarded. Bad struts bottoming out or a broken A/C bracket can cause a knock sensor to retard the timing.

NOTE: *Vacuum readings for an idling engine with an aftermarket performance camshaft will be much lower. These engines are inefficient at low rpm. If the timing is retarded a great deal, check the cam timing too (see Chapter 52). Excessive timing chain slack can allow the timing chain to skip a tooth on its sprocket. Because the distributor is usually driven by the cam, this also results in late ignition timing.*

A high-performance camshaft will have low vacuum at idle, but the vacuum will go up as rpm is raised. This is caused by valve overlap.

Weak piston rings can also result in low vacuum, but this is unusual.

Leaking

A leaking valve is indicated when the needle drops at regular intervals. A power balance test will pinpoint the low cylinder.

Restricted Exhaust

You can hear cars with restricted exhaust systems if you ever have occasion to be stopped on a hill or grade. A restriction in the exhaust causes a hissing sound from the tailpipe when under load. To test for a restricted exhaust, raise the engine rpm quickly to 2,000 to cause a vacuum reading that is momentarily low. Then, release the throttle quickly. Vacuum should return smoothly and quickly to higher-than-normal levels. A slow, hesitating return usually indicates a restriction.

Cranking Vacuum

A **cranking vacuum test** can be performed to check for internal air leaks. For the test to be effective, the engine must be sealed off with the throttle plate all the way closed and the crankcase ventilation system plugged off.

Plug the PCV hose to the air cleaner and disable the ignition system. During cranking, the needle on the gauge should be steady. This is more important than the amount of vacuum on the gauge. The engine should produce a minimum of 3 inches of vacuum while cranking although most engines will produce far more.

Next, pull the positive crankcase ventilation (PCV) valve and cover its opening with your thumb. Cranking vacuum should rise. If it does not, there is a problem with the PCV system.

OTHER VACUUM TESTS

Sometimes, a V-type engine can have a vacuum leak between the underside of the intake manifold and the crankcase. To test for this, pinch off the PCV valve hose to the manifold and the breather hose to the air cleaner. When the engine is running, if there is vacuum at the oil filler opening, an internal vacuum leak is indicated.

Another test for an internal leak uses propane. All PCV system hoses are closed off for this test. If the engine idle changes when propane is put into the oil filler opening, the gas is getting to the combustion chambers by way of an intake manifold leak.

Vacuum Testing with a DSO

Vacuum can be used with a digital storage oscilloscope (DSO) to test for engine and ignition problems. A vacuum transducer connected to the DSO is synchronized with the number one cylinder. Changes in scope readings are correlated to misfires in particular cylinders. More information on DSO testing is found in Chapter 38 and Chapter 46.

Smoke Testing

A smoke tester is a popular machine for finding vacuum, oil, cooling, and exhaust leaks (**Figure 48.5**). The machine generates thick, white smoke at a pressure of 1–1.5 psi. In the event of an increase in pressure above that amount, the machine vents off the excess smoke. A refillable smoke chamber allows for 200–300 tests. Some machines also use UV dye that can leave a tell-tale trace which can be located using a black light.

Finding Vacuum Leaks. Vacuum leaks are located while the engine is off with the throttle closed. Disconnect the vacuum hose from the PCV valve and pump smoke into the PCV hose. Sometimes it is easier to disconnect the hose from the power brake booster and use that as an access point. An engine can be filled with enough smoke for a thorough check in about a minute to a minute and a half.

Finding Exhaust Leaks. For exhaust leaks, a soft cone adapter is used. An exhaust system takes about 2 minutes to fill with smoke. The smoke machine can even detect smaller exhaust leaks before they can be heard. Manifold leaks can also be tested using smoke and a mirror. Smoke from cracks shows up better when the engine is cold.

Finding Coolant and Oil Leaks. Finding coolant leaks with smoke can be done with the cooling system drained. Potential oil leak locations can be found by putting smoke in through the dipstick tube. First, seal the crankcase and pinch the PCV hose so smoke does not leak back through the intake manifold.

COMPRESSION PROBLEMS

Compression can leak due to several causes such as a blown head gasket (**Figure 48.6**), burned valves (**Figure 48.7**), or worn or broken piston rings and/or a damaged piston (**Figure 48.8**).

When valve clearances are adjusted too tightly, they cannot properly seal the cylinder. This sometimes happens after an engine rebuild.

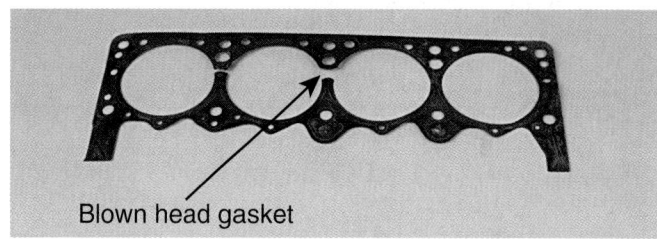

Blown head gasket

Figure 48.6 A blown head gasket. *(Courtesy of Tim Gilles)*

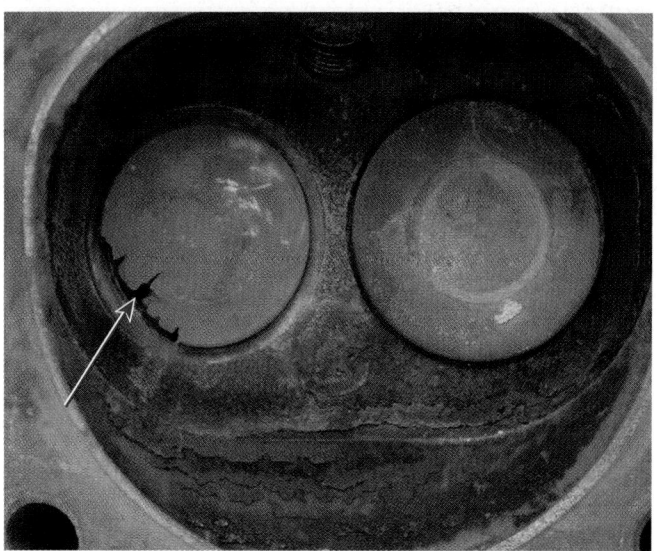

Figure 48.7 A burned valve. *(Courtesy of Tim Gilles)*

Figure 48.5 A smoke tester can be used to pinpoint vacuum, oil, and exhaust leaks. *(Courtesy of Tim Gilles)*

Broken ring land

Figure 48.8 A damaged piston resulted in a broken ring land. *(Courtesy of Tim Gilles)*

Diagnosing Engine Compression Problems

To quickly locate a weak cylinder, perform a **cylinder power balance test**. A **compression test** or leakage test can be performed to pinpoint compression problems located by the power balance test. If any of these tests indicate a possible valve sealing problem, be sure to check the valves for proper adjustment before proceeding with repair.

SHOP TIP A quick judgment about engine compression can often be made by simply cranking the engine with the ignition system disabled and listening for an uneven rhythm. An engine whose cylinders have equal compression will have an even cranking rhythm.

Cylinder Power Balance Test

With computer control of idle disabled, shorting out the spark to a cylinder should result in a drop in engine rpm. A cylinder that does not drop as much as the others is not pulling its share of the load. Variations in rpm drop between cylinders should be less than 5%. The problem could be due to the ignition or fuel system, or the engine could have vacuum leaks or compression problems. A compression test or leakage test (covered later) can be performed to pinpoint problems located by the power balance test.

Hand-held scan tools used with computer-controlled engines also have power balance test capability.

Engine management systems have power balance testing capability that is accessible with a scan tool (**Figure 48.9**). To manually perform a power balance test on a system with computer controls, engine idle

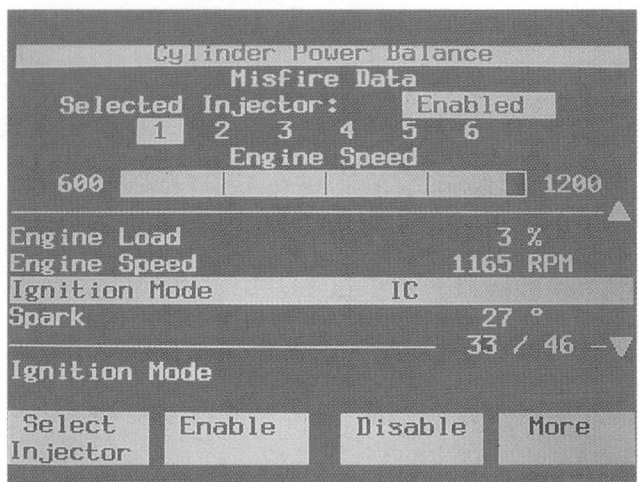

Figure 48.9 A cylinder power balance test disables each cylinder while monitoring rpm drop. *(Courtesy of Tim Gilles)*

control must be disabled. Electronic engine management systems control engine idle. If a cylinder's ignition is momentarily grounded out, the computer will allow more air to enter the intake manifold, raising idle speed to compensate. When engine management is disabled, do not allow a cylinder to be shorted out for longer than 30 seconds to prevent catalytic converter overheating.

Compression Test

One of the most common and least expensive pieces of test equipment is the compression tester (**Figure 48.10**). A compression tester is simply a pressure gauge that is inserted into a spark plug hole and registers a reading as the engine is cranked. Differences in pressure between the cylinders pinpoints problem areas. If all cylinders are performing equally and engine performance is acceptable, the engine passes the test.

NOTE: *A compression test will* **not** *tell the condition of oil control rings, only compression rings.*

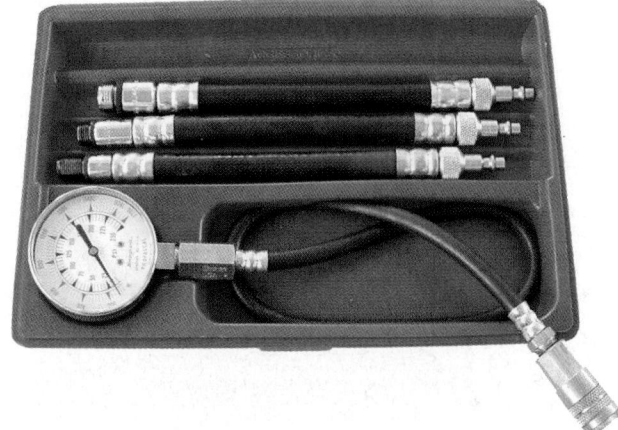

Figure 48.10 A compression tester and adapters. *(Courtesy of Tim Gilles)*

A typical compression tester has two sizes of spark plug threads or multiple adapter hoses. It has a Schrader valve for releasing the pressure that is saved in the gauge so that the technician can read the compression after the engine is cranked.

The compression tester is threaded into the spark plug hole while the engine is cranked. Ideally, the test is done with the engine at normal operating temperature. The battery must be fully charged for the test to be effective.

Compression Test Procedure. Twist the rubber boots on the spark plug cables to loosen them from the spark plugs (**Figure 48.11**). The boots stick to the plugs because of heat radiated from the exhaust manifold. The inside material in the cables can easily be damaged if the cables are handled roughly.

Before removing spark plugs, blow dirt away from their base with compressed air. The specified way to do a compression test calls for the removal of *all* of the spark plugs, so the starter can crank the engine easily. Keep spark plugs in order for comparison. Spark plug diagnosis is covered later in the chapter.

Block the throttle in the wide-open position using a throttle depressor.

Blocking the throttle open during a compression test prevents an engine with a carburetor from sucking fuel into the cylinders and lets the engine breathe air more easily.

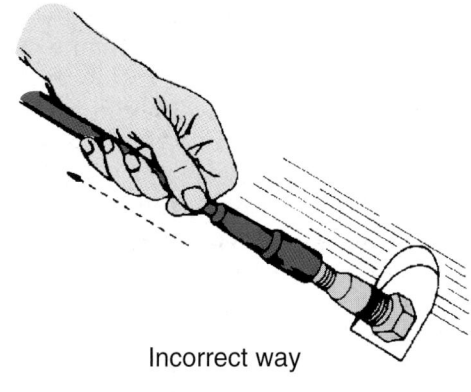

Incorrect way

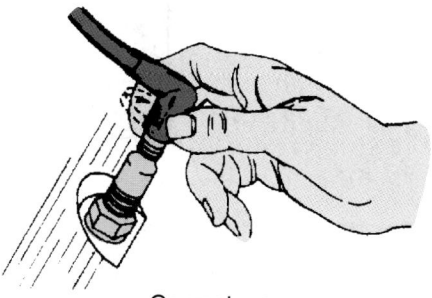

Correct way

Figure 48.11 Twist the rubber boots on the spark plug cables to loosen them from the spark plugs. *(Courtesy of Federal-Mogul Corporation)*

Connect a remote starter switch between the S terminal on the starter solenoid and the ungrounded battery post.

SHOP TIP On some vehicles, when the starter is not easily accessible, a wire in an easy-to-reach location under the hood can often provide access to the starter circuit. Study the wiring diagram for the vehicle.

When cranking with the starter motor, disable the ignition system, following manufacturer's instructions.

Insert the compression gauge into a spark plug hole and crank the engine through at least *four* compression strokes. The gauge will move four times. With all the plugs removed, you can hear each compression stroke as the compression in the cylinder being tested slows the engine. Record each reading on the repair order.

Interpreting Compression Test Results

If all cylinders perform equally and engine performance is acceptable, the engine passes the test.

- Variations in compression between cylinders should be no more than 20%.
- When two cylinders next to each other have low compression, a blown head gasket is usually indicated (see Figure 48.6).
- One or several cylinders with low compression and no apparent pattern of loss often indicates burned exhaust valves. Rough idling is a symptom. At higher rpm, rough running will disappear.

SHOP TIP On an engine at idle, a burned exhaust valve will cause a dollar bill to be sucked against the end of the exhaust pipe every time the bad cylinder's piston has an intake stroke.

■ MATH NOTE ■

When compression test specifications are available, they are only an estimate. If specifications are not available, locate the compression ratio in the specification manual and use the following formula:

$$\frac{Compression}{Ratio} \times \frac{Atmospheric}{Pressure} + \frac{Atmospheric}{Pressure} + \frac{5\ (Volumetric}{Efficiency)}$$

For example, to figure out the approximate compression on an 8:1 engine at sea level (14.7 psi atmospheric pressure): 8.0 × 14.7 + 14.7 + 5 = 137.3 psi.

Wet Compression Test

If any cylinders show poor results, perform a **wet compression test**. Squirt about a tablespoon of oil into each low cylinder. The oil will seal around worn rings, boosting the compression reading. When low cylinder readings increase to normal during a wet test, a piston ring problem is indicated.

After completing the compression test, reinstall the spark plugs. Be especially careful not to strip the threads on an aluminum cylinder head, and use anti-seize compound on the top two or three spark plug threads.

Running Compression Test

A traditional cranking compression test measures how well a cylinder is sealed by the head gasket, rings, and valves. The running compression test, also called the volumetric efficiency test, measures how well the cylinders can draw in air and fuel. It can be used when there is a lack of power or if an engine miss cannot be traced to a cause.

Following a cranking compression test, after all of the readings have been written down, remove one spark plug and install the compression tester. The Schrader valve should remain in the tester. Ground the spark plug cable and start the engine. Bleed pressure off through the Schrader valve and allow the reading to stabilize. A typical test will show the gauge bouncing back and forth around 50–60 psi. The cylinder cannot fill completely as it does during a cranking compression test because the engine is spinning faster and the throttle plate restricts airflow.

Snap the throttle to WOT and let it return to idle. The gauge will hold the peak reading. This reading will be higher than it was at idle because it occurs when the throttle restriction is first removed and the piston is still moving slowly. Record the running compression test reading and repeat the test for the remaining cylinders.

Running compression should be about 80% of cranking compression. If not, the following could be the likely causes:
- If less than 80%, an intake system restriction is a likely cause.
- If more than 80%, the exhaust system is causing a restriction. Look for a restriction in the exhaust, such as a plugged catalytic converter.
- A problem with only one cylinder is a problem specific to that cylinder, such as carbon buildup, a worn cam lobe, or a broken spring or rocker arm.

■ CYLINDER LEAKAGE TEST

The **cylinder leakage test** can be used to accurately pinpoint causes of leakage in a combustion chamber. Regulated compressed air is introduced into a cylinder through the spark plug hole (**Figure 48.12**). If there is a leak, it can be pinpointed by listening:

Figure 48.12 A cylinder leakage tester is connected to the cylinder through the spark plug hole. *(Courtesy of Tim Gilles)*

- Oil filler = leaking rings or piston
- Air cleaner = leaking intake valve
- Exhaust = leaking exhaust valve
- Bubbles in the radiator = blown head gasket, or a crack in the head or block, which allows the regulated air to enter the cooling system

The leakage tester offers several advantages over a compression test.
- The test can be performed on an engine that is removed from a car (such as an engine purchased at a salvage yard).
- A performance camshaft will not affect the results of the test. It would cause lower readings on the compression test because engine vacuum is lower at cranking speeds with a racing cam.
- The exact source of leakage can be pinpointed before engine disassembly.

To perform a leakage test, the piston is positioned at TDC on the compression stroke. This ensures that both valves are completely closed.

Acceptable leakage on the tester's gauge is usually less than 10% to 15%.

NOTE: *Any time there is leakage past the piston rings, the PCV valve can allow air to travel into the intake manifold where it can be mistaken for a leaking intake valve. To avoid this situation, remove the oil filler cap, or unhook the vacuum line to the PCV valve or pinch it with pliers.*

■ CARBON-RELATED PROBLEMS

Carbon can cause performance problems in an engine. It can cause the engine to have increased compression, which can cause fuel to detonate in the cylinder. Fuel can become saturated in carbon deposits on valves. This causes driveability problems when the fuel goes into or comes out of the carbon at the wrong times. The carbon deposits can also block the flow of air and fuel into the cylinder.

There are two kinds of deposits: oil based and carbonaceous. The oil-based deposits are the traditional gummy, black ones like the deposit on the valve in Figure 18.6. They are caused when oil and heat come together.

Carbonaceous deposits are ones that result from fuel. They are called cauliflower deposits because of their resemblance to the vegetable. These deposits are not as thick as oil deposits. They are also hard, dry, and tougher to remove. They cause driveability problems because fuel vapors can be absorbed into them. Driveability problems include rough idling when cold, loss of power, surging, and high exhaust emissions. Valve deposits sometimes cause problems in as little as 5,000 miles. Additives are put into fuels by oil companies to minimize this problem. Removal of carbon deposits is covered in Chapter 41.

REVIEW QUESTIONS

1. An engine will not start. If there is a strong hot spark, the next tests are for _____ and compression.

2. Some technicians recommend a fuel injection cleaning and _____ change at each 30,000-mile service.

3. A roar will be heard through the intake manifold when a(n) _____ system is restricted.

4. What happens to the outside of an exhaust pipe at the point of an internal restriction?

5. Vacuum readings will drop approximately 1 inch for each _____ feet above sea level.

6. An engine should produce a minimum of _____ of vacuum while cranking but most engines will produce far more.

7. When engine idle rises during a cylinder power balance test, what could be the cause?

8. During a compression test, remove all of the _____ so the starter can crank the engine easily.

9. How many compression strokes should the engine crank a minimum of during a compression test?

10. If a dollar bill gets sucked against an exhaust pipe when the engine is idling, what could this indicate?

ASE-STYLE REVIEW QUESTIONS

1. When there is an intake air leak (vacuum leak), an engine with an oxygen sensor feedback fuel system will:
   a. Run fine when cold   c. Both A and B
   b. Run rough when hot   d. Neither A nor B

2. An engine will not start. Technician A says that this could be due to excessive carbon in the combustion chamber. Technician B says that compression could be too weak. Who is right?
   a. Technician A   c. Both A and B
   b. Technician B   d. Neither A nor B

3. Which of the following is the most likely cause of a buildup of dry, black ash or soot in the exhaust pipe?
   a. An overly rich air-fuel mixture
   b. An overly lean air-fuel mixture
   c. Oil consumption
   d. Exhaust pipe tagging

4. Technician A says that dirty fuel injectors can cause an engine to ping at full throttle. Technician B says that a dirty injector might not sound the same as a clean one. Who is right?
   a. Technician A   c. Both A and B
   b. Technician B   d. Neither A nor B

5. Which of the following is a likely vacuum reading for an engine with a performance camshaft?
   a. Lower at idle, rising with rpm increase
   b. Steady and higher than original equipment
   c. Lower in all running conditions
   d. Higher in all running conditions

6. An engine has late valve timing. Technician A says that this will cause poor, low rpm performance. Technician B says that at higher rpm, engine performance might be acceptable. Who is right?
   a. Technician A   c. Both A and B
   b. Technician B   d. Neither A nor B

7. Each of the following is true about compression testing *expect*:
   a. Two neighboring cylinders that are low can indicate a blown head gasket.
   b. Several cylinders with low readings are likely due to burned or tight valves.
   c. The condition of the oil control rings can be determined.
   d. The throttle is held open during the test.

8. If a compression test reading goes up after adding oil to the cylinder:
   a. The compression rings are probably worn.
   b. There is a burned valve.
   c. There is a blown head gasket.
   d. There is a crack in the head or cylinder wall.

9. Technician A says that a cylinder leakage test is better than a compression test for testing the condition of a salvage yard engine before installing it in a car. Technician B says that a compression tester is better than a cylinder leakage tester for testing the condition of an engine with a performance camshaft. Who is right?
   a. Technician A   c. Both A and B
   b. Technician B   d. Neither A nor B

10. Which of the following can be associated with a burned valve?
   a. Rough idle
   b. Smooth running at higher speeds
   c. Uneven exhaust flow
   d. All of the above

Automotive Engine Service and Repair

THEN AND NOW: GASKETS

Man has been presented with the challenge of sealing things up since prehistoric times. Early examples include stuffing mud and leaves into the gaps between the logs of a hut and, according to the Bible, Noah smearing pitch on the hull of the Ark. With the coming of the industrial revolution in the mid-1700s, sealing jobs became more precise. But the basic idea was still to fill up the imperfections in mated surfaces, thus preventing leakage through the seam.

It was not until the beginning of the automotive age, over a century ago, that this challenge began to be faced often by so many people. Engineers and mechanics used all kinds of materials and substances to contain and separate engine oil, coolant, vacuum, and compression. Early vehicles used many different types of sealing materials, including soft metals, leather, rubber, paper, and even beef fat. It was a tough fight, however, and leaks of every kind were a common type of mechanical failure.

The engine gasket with the most difficult job is the head gasket, located between the head and the cylinder block. Sealing this area was a very big problem in early engines. Some makers eliminated this seam altogether by casting the head and block in one piece (not a practical arrangement from a service standpoint).

In the old days, mechanics commonly made gaskets by laying a piece of paper-like material against the casting. Then they would tap along the edges with the rounded end of a ball-peen hammer, which would cut the material.

Early on, bolts were tightened by feel. This method of tightening could only be learned through experience, overstretching bolts and sometimes breaking them off. The tool that solved this problem was the torque wrench. It told the mechanic exactly how many foot-pounds of twisting power was being exerted on the fastener. A recent tightening method, called "torque turn" or "angle torquing," reduces the effect of friction on the torque measurement. After tightening the bolt a small amount with a torque wrench, the bolt is further tightened a specified number of degrees.

Now, gaskets have to face higher temperatures, increased pressures, fewer and thinner bolts, and the differing expansion rates in aluminum head/iron block engines. Gaskets have become highly engineered (and expensive) parts. A head gasket might be constructed of heat-proof graphite on an expanded core or a multilayer steel gasket might be an engineer's choice to stabilize the joint. Cork/rubber is still being used for valve cover or oil pan seams, but often molded silicone rubber is the material of choice.

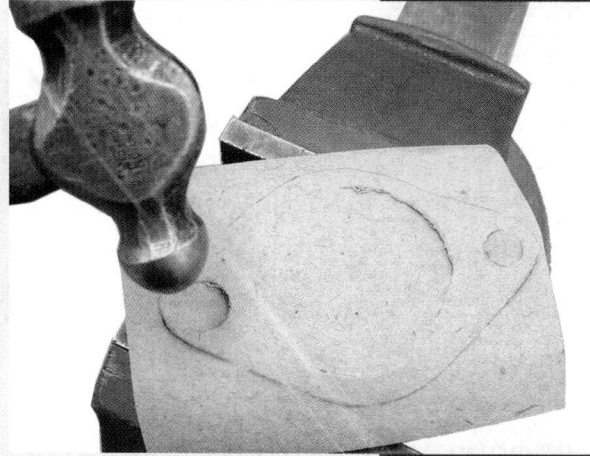

Vintage gasket making. *Cutting a gasket with a ball-peen hammer. (Courtesy of Tim Gilles)*

A modern multilayer steel head gasket. *(Courtesy of Tim Gilles)*

CHAPTER 49

Diagnosing Engine Mechanical Problems

■ OBJECTIVES

Upon completion of this chapter, you should be able to:

✔ Use engine diagnostic tools and equipment safely and properly.

✔ Diagnose engine and related problems prior to repair.

■ KEY TERMS

black light

engine knock

flat cam

oil pressure sending unit

pressure relief valve

oil analysis

cross fluid contamination

seized engine

oil wash

■ INTRODUCTION

This chapter focuses on the diagnosis of problems before engine disassembly. Internal problem diagnosis after disassembly is covered in more detail in later chapters. Those mechanical problems that are related to engine driveability and performance are covered in Chapter 48.

It is very important to diagnose the cause of the problem before performing a repair. It is not unusual for an inexperienced technician to spend many hours of work only to discover that the repair was unnecessary. Four major diagnosis areas are covered:

1. Oil consumption
2. Engine noises
3. Oil pressure problems
4. Cooling system problems

Problems related to engine performance, such as rough idle or low compression, are covered in Chapter 48.

There are many causes of engine problems. Some are the result of normal wear and tear or due to a lack of maintenance. Others might be due to work previously done to the engine. Additionally, problems that appear to be engine related can be due to problems in areas other than the engine like the transmission or emission controls. Sometimes a problem with a system causes an engine to fail. If the problem is not taken care of, the failure will recur.

■ DIAGNOSING PROBLEMS BEFORE A REPAIR

An engine should be properly diagnosed before disassembly for two reasons. First, it should be determined that a repair is really necessary. Second, the exact location of a problem should be determined while the engine is running (before disassembly). A thorough discussion of the problem with the vehicle's owner is also helpful. Many times an owner's driving habits or maintenance procedures can be the cause of the problem.

■ OIL CONSUMPTION

A vehicle owner will usually blame piston rings for oil consumption even though oil can be lost due to a variety of other conditions. Oil can be lost through either *external leakage* or *internal oil consumption*. Internal oil consumption can sometimes be spotted as an oily coating on the inside of the exhaust pipe. Oil consumption of this magnitude will also be accompanied by blue smoke. Black soot at the exhaust pipe often indicates an overly rich air-fuel mixture, not oil consumption. Smoke from an overly rich mixture will be black.

The rate of normal oil consumption depends on the size of the engine, the weight and shape of the vehicle, the viscosity and service rating of the oil, engine rpm during use, engine temperature, and the amount of oxidation and dilution of the oil. Information on oil is found in Chapter 12.

A customer might complain of a unique instance of rapid oil use. Sometimes this happens when the car was driven for 1,000 or more miles of city driving, followed by a highway trip. During the city driving, the engine might have consumed a normal amount of oil, but because of fuel and water dilution from city driving, the dipstick registers "full." When the customer goes on a highway trip, the oil thoroughly heats up and the fuel and water diluting it evaporate. This gives the appearance of rapid oil consumption.

■ CAUSES OF OIL CONSUMPTION

Bad Valve Guides or Seals

The valve guides are one source of oil consumption. When an engine has been rebuilt and has relatively low mileage, oil leakage past the valve guide seals is a more likely cause of oil consumption than leakage past piston rings. When oil leaks through intake valve guides into the combustion chamber, smoke can be visible from the exhaust during deceleration when vacuum in the cylinder is especially high. Smoke at other times might be from exhaust valve guides (see Chapter 52).

When only one side of a spark plug is fouled with carbon, leaking guide seals are indicated. The carbon will be on the side of the spark plug that was facing the intake valve. Leaking valve guide seals also result in carbon deposits in the "neck" area of the intake valves. Watch for carbon deposits when disassembling a cylinder head (**Figure 49.1**).

Different kinds of valve guide seals are described in detail in Chapter 18. Valve guide seals are replaced during a valve job while the heads are disassembled. Chapter 52 includes a procedure for replacing valve guide seals on some engines *without* removing the head(s).

Oil Consumption from Piston Rings

Smoke during deceleration can also be due to worn or broken piston rings. The worn rings allow oil on the cylinder walls to escape into the combustion chamber during the high vacuum conditions associated with engine deceleration.

Piston rings are usually the first thing a customer suspects when a car starts to use oil. When oil is consumed past piston rings, the cause is often a poor maintenance schedule. Engines that suffer from too few oil changes can have plugged oil control rings (**Figure 49.2**). Oil rings scrape oil from the cylinder walls and return it to the crankcase through the inside of the piston (**Figure 49.3**).

NOTE: *According to the Ford Motor Company, there are approximately 36,500 drops of oil in 1 quart. If the engine consumed as little as 1/1100 of a drop of oil on each stroke, it would use 1 quart of oil in 1,000 miles. So you can see the fantastic job the oil rings do and the major problem created when they are plugged up.*

When looking for a cause for oil consumption, technicians will often check an engine's compression (see Chapter 48). While this procedure might locate worn or broken compression rings, it does not check the condition of oil control rings.

Increased Oil Consumption after a Valve Job

The condition of the entire engine must be considered before performing a valve job on a high-mileage engine. A valve job increases *compression*, but it also increases engine *vacuum*, which can cause more oil to be sucked past the worn piston rings into the combustion chamber.

If cylinder heads are removed for a valve job, look at the tops of the pistons to see if oil consumption

Figure 49.1 A lack of carbon formation on the top of the piston, beneath the intake valve, indicates oil leakage through the valve guide. The valve in this photo was found when the head was disassembled. *(Courtesy of Tim Gilles)*

Carbon on valve stem   No carbon on piston

Figure 49.2 Plugged oil control ring.

Figure 49.3 Holes or slots behind the oil control ring allow oil to return to the oil pan. *(Courtesy of Tim Gilles)*

Figure 49.4 A clean area around the outside edge of a piston indicates excessive oil consumption past the rings. *(Courtesy of Tim Gilles)*

caused by worn or stuck piston rings is indicated (**Figure 49.4**).

Oil Consumption from Excessive Rod Bearing Clearance

An engine with high mileage can have worn connecting rod bearings. Results of this are

- Low oil pressure when the engine idles but normal pressure off idle
- Oil consumption when excessive oil leaks out between the rod journals and bearings

The oil is thrown onto the cylinder walls at high speed (**Figure 49.5**). The oil rings do not have enough capacity to return all of this oil to the crankcase. Whatever oil enters the combustion chamber will be burned with the air-fuel mixture.

High-speed driving also causes increased oil consumption because of the extra oil thrown from the rods. In one test, an engine with excessive rod bearing clearance run at 70 mph used seven times the oil that it used at 40 mph.

Vacuum Modulator

Some, mostly older, automatic transmissions have a vacuum modulator that senses engine load by using intake manifold vacuum to tell the transmission when to shift (see Chapter 73). If the diaphragm in the modulator leaks, transmission fluid will be sucked into the intake manifold. This can produce smoke that can be confused with engine oil smoke, even though the engine may be in good condition.

The leaking diaphragm can also cause a rough idle (air leak into the intake manifold) and harsh, late shifts of the transmission. Spark plugs near the vacuum tap on the manifold often become oil fouled, too.

Be sure to question the vehicle owner thoroughly. The combination of any or all of these symptoms can

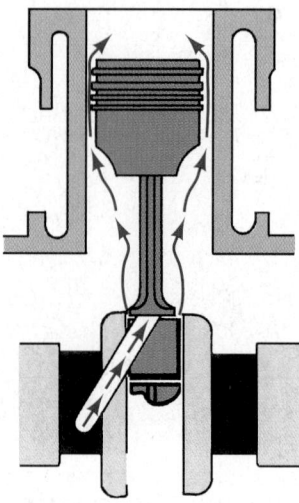

Figure 49.5 Effect of excessive bearing clearance on oil throw-off.

lead an owner to believe an engine rebuild is needed. The key to diagnosing a faulty vacuum modulator is that the transmission is probably using automatic transmission fluid (ATF) without leaking externally.

Incorrect Oil Level

The same engine is often installed in several different model cars. Each engine can be equipped with an oil pan of a different shape, which requires a different length oil dipstick. Sometimes oil consumption is blamed when the only problem is that the car has the wrong dipstick. Every time the owner mistakenly adds oil to the crankcase, the crankshaft whips it up like an egg beater whips cream. Excessive oil is thrown onto the cylinder walls. The oil rings cannot handle this amount of oil, so it migrates into the cylinders and is burned off. It is especially important to check that the correct dipstick is used after an engine change or short-block installation.

Excessive oil can also result in aeration of the oil, noisy lifters, unstable oil pressure, gurgling sounds coming from the crankcase, and oil leakage.

Oil Consumption from Plugged Cylinder Head Drainback Holes

When engine oil is not changed often enough, thick, dirty oil can plug the holes in the cylinder head. These holes are there to allow upper-end oil to drain back to the crankcase (**Figure 49.6**). The problem can be temporarily solved by cleaning out the holes, but it is a symptom of a poorly maintained engine that will soon require major service. The oil stays up in the valve cover area instead of returning to the crankcase. It floods the valve guide, making the valve stem seal ineffective.

Oil Consumption from a Leaking V-type Intake Manifold Gasket

Intake manifold vacuum can draw oil into the intake ports from the lifter valley area under the intake manifold (**Figure 49.7**). This is a tough problem to find. A cranking vacuum check, a smoke test, or a propane

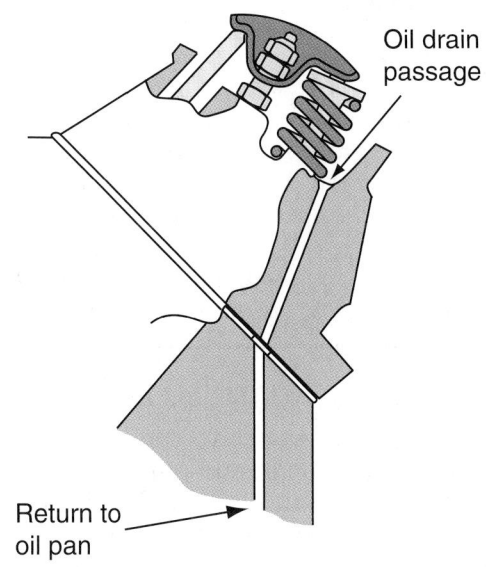

(a)

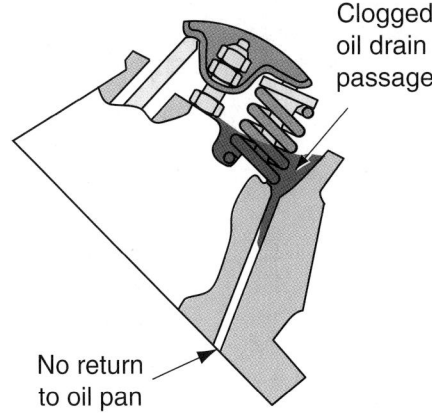

(b)

Figure 49.6 (a) The cylinder head has holes that allow oil to drain back to the oil pan. (b) A plugged oil drainback hole will cause an engine to smoke when oil enters the cylinder through the valve guides.

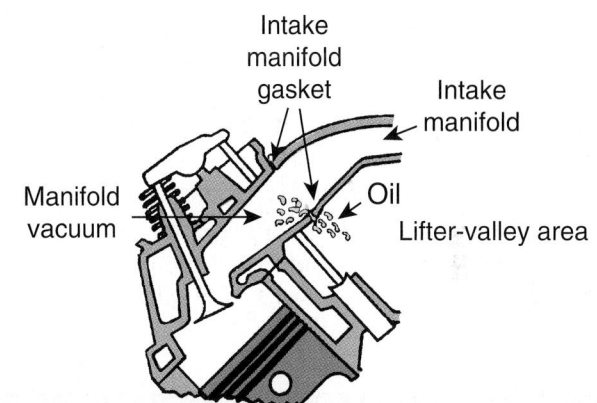

Figure 49.7 Oil can be drawn into the intake manifold past a faulty manifold gasket.

test can be performed to check for internal air leaks before the engine is disassembled. The procedure is covered in Chapter 48. When removing the manifold, always inspect visually for the possibility of previous intake gasket leakage.

V-type engines equipped with an exhaust gas recirculation (EGR) valve on the intake manifold often experience oil fouling of the spark plugs that are closest to the EGR valve. This is caused when the intake manifold warps or the manifold gasket fails. A replacement gasket designed for high-temperature applications is available.

Oil Consumption Due to Crankcase Pressure

One possible reason for excessive oil leakage is a plugged PCV valve, which can cause pressure to build within the crankcase. This pushes on the gaskets and seals, allowing oil to leak out. Operation of the positive crankcase ventilation (PCV) system is covered in Chapter 43. Crankcase pressure can result in increased internal oil consumption, too.

■ OIL LEAK TESTING

Oil leaking past gaskets and seals is a common cause of oil consumption.

NOTE: *A loss of 1 drop of oil every 30 feet results in a loss of about 3 quarts of oil every 1,000 miles.*

A rear main bearing seal leak is indicated when there is oil on the engine side of the flywheel or torque converter. Oil on the transmission side of the torque converter indicates a leakage of a front transmission seal.

NOTE: *If the engine's oil pressure sending unit leaks, a great deal of oil can be lost in a short amount of time. If the sending unit shows any signs of seeping oil, it should be replaced as soon as possible.*

Black light

Figure 49.8 A black light oil leak detector. The leak shows up as yellow-green streaks. *(Courtesy of Tim Gilles)*

Black Light Testing

One effective way to test for hard-to-find oil leaks is to use a **black light**. An ounce of fluorescent liquid is added to the engine oil. After the car is driven to give the oil a chance to heat up and leak out, a black light highlights this source of the leaks in bright yellow-green streaks (**Figure 49.8**). A mirror can be used to bounce the black light into hard-to-see areas. Washing the block first is helpful but not necessary.

Sometimes when a leak is minor, it does not show up after just a short time, so the car might need to be driven for a day or so. After repairing the leak, the engine is cleaned and rechecked with the black light. The fluorescent dye stays in the oil. The dye is not harmful, and the manufacturer says that it dissipates within 300 miles of driving.

CASE HISTORY

A technician was attempting to repair an oil leak on a dual overhead cam six-cylinder engine. He replaced the valve cover gasket but the leak continued and the unhappy customer returned to the shop. After pouring some leak detection fluid into the crankcase and taking the car on a test drive, he aimed the black light at the engine. Bright streaks of yellow oil were visible coming from behind one of the camshaft sprockets. He removed the valve cover, timing belt, and sprocket and replaced the circular rubber seal behind the timing sprocket.

Fluorescent leak detection fluids are also available for automatic transmissions, fuel systems, air-conditioning systems, and power steering leaks.

Spark plugs can be a good indication of what is taking place in a cylinder. **Figure 49.9** shows an abnormal spark plug condition associated with oil consumption.

■ ENGINE PERFORMANCE AND COMPRESSION LOSS

Compression loss can be a result of things like a blown head gasket, burned valves, or broken piston rings. Engine compression testing and mechanical problems

Figure 49.9 These spark plugs are fouled with carbon resulting from plugged oil control rings. *(Courtesy of Tim Gilles)*

that result in compression loss are covered in detail in Chapter 48.

ENGINE NOISES

It is important to try to determine the location of noises before disassembling the engine. There have been many instances where engines have been disassembled, inspected, and rebuilt but when reinstalled in the car still had the same problem. Pinpointing the origin of the noise might have resulted in more careful scrutiny of the offending part while the engine was apart.

Noises are often transmitted from their origins to other locations, and they can be difficult to isolate. Listen through a stethoscope (**Figure 49.10**), or listen at the end of a large screwdriver, a piece of hose, or a long wooden dowel to help pinpoint noises.

CAUTION Both the stethoscope and screwdriver are electrical conductors. They should be used with caution around electrical connections such as those located on the back of the alternator.

Accessory Noises

Accessories often cause noises that can be mistaken for other problems. When there is a noise, inspect alternators, smog pumps, air-conditioning compressors, and coolant pumps carefully.

Belts are a common source of noise. If you suspect a belt of making noise, disconnect the belt and run the engine for a short time, or spray soapy water on the belt while the engine runs.

A fan clutch on the coolant pump can cause a noise that sounds serious and is hard to pinpoint. A stethoscope cannot be used to listen to the noise, because the fan is spinning when the engine runs.

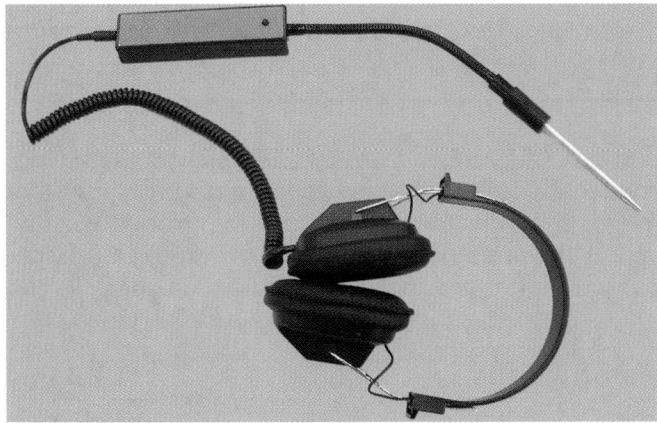

Figure 49.10 A stethoscope. This one amplifies the sound through head phones. *(Courtesy of Tim Gilles)*

Sometimes, the extra looseness (bearing play) in the fan clutch can be felt when the engine is off.

SHOP TIP To isolate noises from defective belt-driven parts or accessories, disconnect the drive belt and run the engine for a short time at idle.

ENGINE KNOCKS

Crank Shaft Noises

Engine knocks can be caused by a variety of things. Crankshaft noises are generally deeper in pitch than other engine noises. It is important to isolate the source of the noise so an accurate diagnosis can be made.

Front Main Bearing Knock. Excessive front main bearing clearance results in a heavy knock after the engine warms up. The knock is most pronounced at about 1,500–2,500 rpm. Loosening an accessory belt often reduces the intensity of the knock. Upper front main bearing wear results from accessory belts adjusted too tightly.

Thrust Bearing Knock. End thrust is movement of the crankshaft in a forward and backward direction. The crankshaft surface that controls end thrust can become worn, allowing the crank to move back and forth. Excessive end thrust will cause a clunk when the vehicle leaves a stop sign. This is more noticeable on vehicles with standard transmissions.

Rod Knock. Too much rod bearing clearance can result in a rod knock. During a cylinder power balance test, the intensity of the knock will diminish or disappear altogether as the offending cylinder's spark plug is shorted out. Sometimes this condition is accompanied by low oil pressure, especially at idle.

Related Noises. A loose flywheel, torque converter, or vibration damper can cause a very serious sounding noise. Torque converter flexplates sometimes crack, causing a serious sounding knock. To test for a cracked flexplate:

■ Run the engine at about 2,000 rpm.
■ Turn the key off and then on, listening for a clunk as the engine restarts.
■ Shut off the engine and use an inspection mirror (**Figure 49.11**) to try to see the crack in the flexplate.

NOTE: *Shining a flashlight into the mirror will bounce the light onto the flexplate so you will be able to look at hard-to-see locations.*

Figure 49.12 shows cracks in a flexplate made visible by magnetic crack inspection. Magnetic crack inspection is covered in Chapter 52.

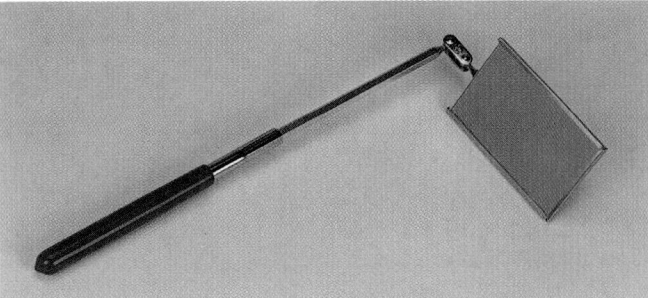

Figure 49.11 An inspection mirror. This one telescopes and has a replaceable mirror on a pivoting head.

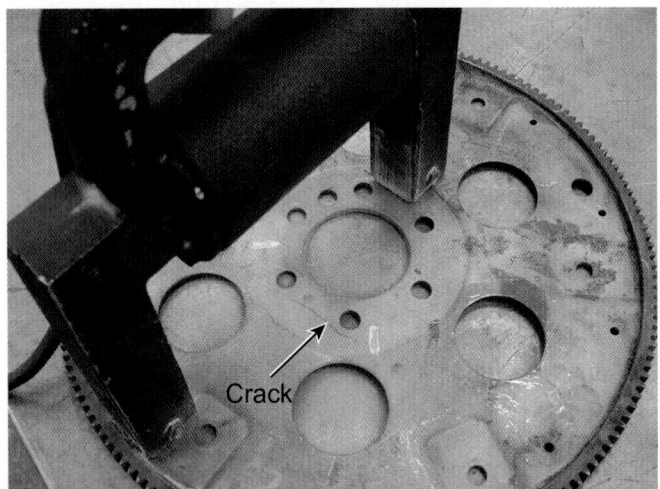

Crack

Figure 49.12 A cracked flexplate. *(Courtesy of Tim Gilles)*

Figure 49.13 This woodruff key was damaged by a loose crankshaft pulley. *(Courtesy of Tim Gilles)*

Bent Oil Pan. A bent sheet metal oil pan can cause a deep knocking sound when a connecting rod hits the dented spot. Complete engines can weigh several hundred pounds. It is not uncommon for engine sheet metal, such as an oil pan or valve cover, to become dented when the engine is removed or replaced in the vehicle.

CASE HISTORY *A student did a complete rebuild of an in-line six-cylinder engine. After installing the engine and priming the lubrication system, he started it up. The engine had a very serious knock originating from the oil pan that could be heard with a stethoscope. The student was very discouraged and wanted to remove the engine from the vehicle to repair the problem. The instructor noticed an indentation in the oil pan and suggested that the student remove the pan to inspect it. Examination of the inside of the oil pan showed a shiny spot where the crankshaft had been hitting it. Knocking out the dent fixed the problem.*

CASE HISTORY *A customer complained of an intermittent knocking sound from the passenger compartment. The technician listened carefully with a stethoscope at several locations along the driveline. The noise seemed to be coming from the torque converter, but replacing the torque converter did not solve the problem. The technician decided to remove the oil pan and inspect the crankshaft. As he loosened the front crankshaft pulley bolt, he noticed that it did not seem to be as tight as it should be. These bolts are usually extremely tight. After removing the damper, he found the damaged woodruff key shown in* **Figure 49.13**.

A new timing belt had been installed 15,000 miles previously. The torque specification for the damper was in the neighborhood of 200 foot-pounds, but the bolt had not been tightened correctly. The result was a vibration that transmitted noise along the crankshaft, where it was heard as a torque converter noise. A perfectly good torque converter was replaced before the true cause of the problem was found. A new crankshaft key and a bolt tightened to the correct specification solved the problem.

Rod Side Clearance. Excessive clearance on the sides of connecting rods sometimes causes a ticking sound that resembles a valvetrain noise but comes from the crankcase. The noise sometimes goes away during a cylinder power balance test. Spark knock (see Chapter 39) can intensify noise caused by excessive rod side clearance.

Piston Noises

There are several different types of piston noise resulting from cracked pistons, piston slap, excess piston pin clearance, and other causes.

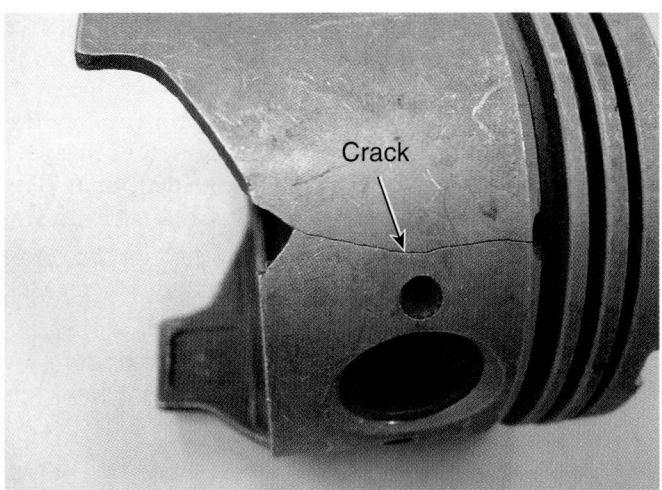

Figure 49.14 A cracked piston. *(Courtesy of Tim Gilles)*

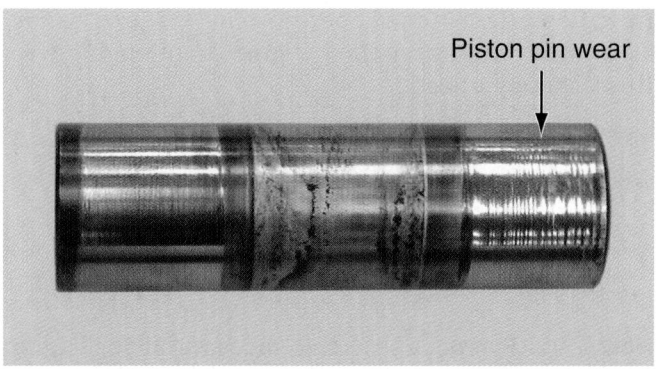

Figure 49.15 This badly worn wrist pin caused an engine knock. *(Courtesy of Tim Gilles)*

Grounding out the plug during the power balance test can *increase* a piston noise. This is the opposite result from the way a bad connecting rod reacts to ignition grounding.

Cracked Piston. A cracked piston is sometimes a source of noise (**Figure 49.14**).

- The noise is sometimes higher pitched than a crank-related noise and could be confused with a valvetrain noise, except that it occurs at a faster rate than a valve noise.
- Cracked pistons are often the result of a broken timing chain or improper valve timing, which can allow a valve to strike a piston on some engines.

Piston Slap. *Piston slap* is caused by excessive clearance between the piston skirt and the cylinder wall.

- Sometimes the noise gets louder during acceleration, often clearing up when the engine warms up.
- Piston slap at TDC and BDC causes oval wear as the piston rings scrub the sides of the cylinder.
- Look for worn or collapsed piston skirts when excess oval cylinder wall wear is found.

Piston Pin Noise. Noise from excessive piston pin clearance makes a "double click" sound at idle or fast idle (**Figure 49.15**).

- Pin noise often becomes more intense after the installation of new piston rings. The noise will gradually become less as the rings wear.
- With the engine running at the speed where the most noise occurs, grounding the plug wire will sometimes increase the noise even more. The noise level might not increase, but it will not become less. Piston inertia causes the noise, which is why the noise does not go away when the spark plug is shorted out.
- The noise usually becomes less or goes away when the engine warms up.

Other Piston Sounds. Another piston noise is caused by a *broken ring*, which can rattle during acceleration.

A cylinder ring ridge that is not removed when new rings are installed can cause a clicking sound because the square edge of the new ring hits against the rounded edge of the ring ridge (see Figure 53.12). This can also force a ring land down on the second ring. The result is a stuck second compression ring or a broken ring or ring land.

Valvetrain Noises

Valvetrain noises make a loud "ticking" sound and are the most common of engine noises. A finger placed on the valve spring retainer while the engine idles will feel the shock each time the loose valve hits its seat.

NOTE: *Valvetrain noises occur at half the speed of engine crankshaft rpm. This is because the cam only turns once for every two turns of the crankshaft.*

Noise from excessive valve guide-to-stem clearance can be pinpointed with the engine running. Squirt oil on the suspected guide to take up the clearance and stop the noise.

Sticking Valve. A sticking intake valve often results in a popping noise through the carburetor as burning gases escape into the intake manifold past the leaking intake valve. To find a sticking valve on a pushrod engine, hook up a timing light to one spark plug wire at a time with the valve cover off. The strobe action of the light will catch the offending valve in the open position.

Worn or Flat Cam Lobe. A similar situation arises when an engine has a smooth idle but runs rough under acceleration, "popping back" through the intake valve. This can often be traced to an exhaust lobe on the cam that has "gone flat," resulting in a **flat cam**. Pressure builds up on the power stroke and cannot escape during the exhaust stroke, so it goes back up the intake port. The lobe can go bad fairly quickly once it starts to wear, so the condition seems to occur overnight.

The pop back can be more severe with port fuel injected engines because each individual injector is near an intake valve port. The drop in suction from the offending cylinder does not cut the flow of fuel as it would in a carbureted engine or an engine with throttle-body injection.

> **SHOP TIP** If you hold your hand over the carburetor of an engine with a bad cam, it might get wet with fuel.

Noisy Fuel Pump. A defective mechanical fuel pump makes a noise resembling a bad hydraulic lifter. The noise is loudest at idle speed, when little fuel is being used by the engine.

Rocker Arms. Lack of lubrication to rocker arms can cause a loud, "squeaky" sound in some older engines. This problem is not as common as it used to be, partly because of the better oils in use today.

Timing Components. Valvetrain noises can also come from inside the timing cover. They can be caused by a bad timing chain or a loose sprocket or gear. The noise usually is a rattle or knock that becomes louder when decelerating. For engines with a timing chain tensioner, a worn chain can become loose enough on the sprocket to rattle whenever the engine floats or cruises between load and coast conditions. In severe cases, a chain can actually wear a hole in the timing cover, resulting in an oil leak (**Figure 49.16**). Severely worn cam bearings can also be the cause of excessive timing chain slack. Depending on the design of the lubrication system, this problem can also be accompanied by low oil pressure at idle.

Lifter Noises

A very common valve noise is caused by a noisy lifter. Sometimes this occurs when the engine is first started because a lifter has lost its oil while the engine was off. Whenever an engine is shut off, there will be some valves held open, putting spring pressure on lifters to bleed them down. When pressurized oil reaches the lifter, the noise goes away. If the noise goes away in less than 15 seconds, this is considered normal. The following are some other lifter noises and their causes:

Intermittent Noise at Idle or Low Speed. This can often be traced to dirt in the lifter check valve or wear (see Chapter 52).

Noise at Idle That Goes Away at Higher Speeds. This usually indicates excessive wear between the lifter body and its plunger. This noise could also be caused by low oil pressure or too thin an oil.

Quiet at Idle but Noisy at High Speed. The oil could be full of air. This occurs when the oil level is so high that the crank whips the oil, filling it with air. It could also happen because there is air leaking into the suction side of the oil pump.

Lifter Noise at All Engine Speeds

Lifter noise can be due to several things:

■ Dirt or varnish buildup inside the lifter

> **SHOP TIP** Lifters that stick because of varnish buildup can sometimes be loosened up by squirting carburetor spray cleaner down the hollow pushrod lubricating channel where it contacts the rocker arm.

■ Worn parts such as worn rocker arms or a cam lobe that is going flat
■ Insufficient oil supply
■ Oil too thin
■ Oil pressure too low

Spark Knock Noise

Abnormal combustion can cause engine noise. Sometimes it can be serious enough that engine damage can result. Spark knock can be a result of carbon buildup, incorrect ignition timing, fuel of too low an octane, an engine with a compression ratio that is too high, or cooling system problems. Chapter 39 deals with abnormal combustion in detail.

Excessive Carbon Buildup. In a car with relatively low mileage, carbon buildup in the combustion chambers can cause an increase in the compression ratio (**Figure 49.17**). This problem is not as common since the introduction of unleaded gas, which leaves fewer deposits, but some late-model engines develop carbon problems in less than 10,000 miles.

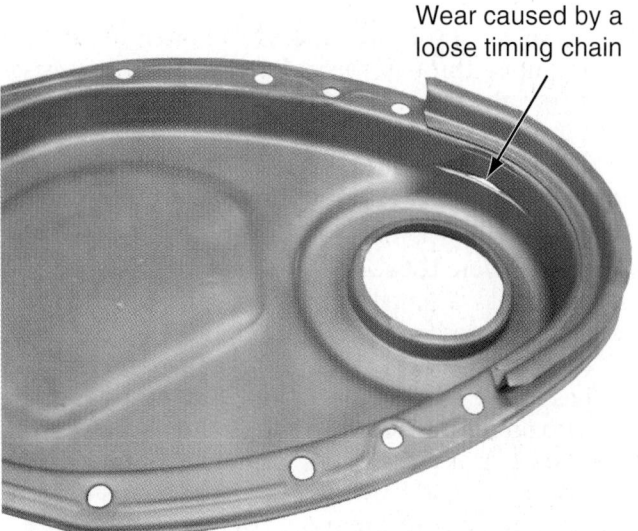

Wear caused by a loose timing chain

Figure 49.16 A worn timing chain rubbed on this timing cover, causing noise and an oil leak. (*Courtesy of Tim Gilles*)

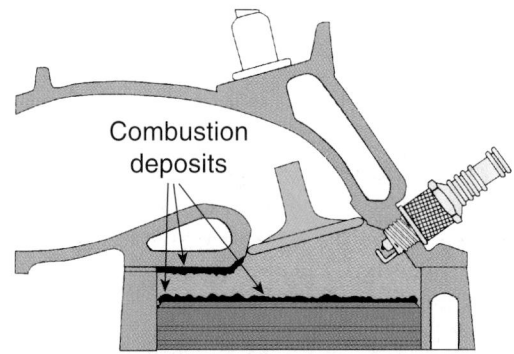

Figure 49.17 Carbon in the combustion chamber can raise the engine's compression ratio. *(Courtesy of DaimlerChrysler Corporation)*

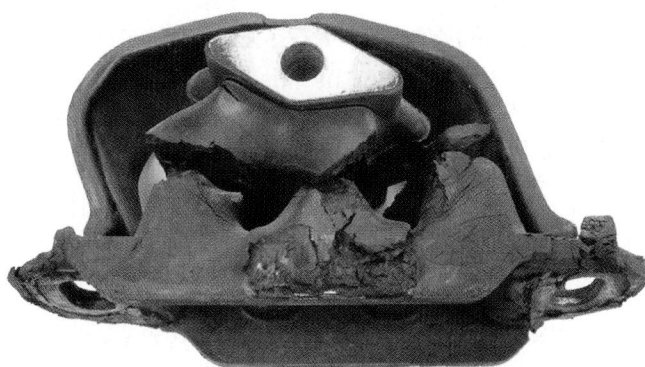

Figure 49.18 A broken engine mount. *(Courtesy of Tim Gilles)*

Broken Motor Mount

The engine is attached to the frame with a rubber and metal motor mount. A broken motor mount (**Figure 49.18**) can cause vibrations, sticking throttle and shift linkages, torn radiator hoses, interference between the fan and radiator, and broken ground straps from the engine to the firewall or frame. Late-model motor mounts are designed with safety provisions so that, although they may fail, the engine is restrained.

To check for a broken motor mount, have another person put the transmission in both forward and reverse ranges while keeping the vehicle braked. This will cause the engine to lift on one side and then on the other.

A broken ground strap (see Figure 50.7) can cause several problems that result as the electricity hunts for its shortest path to ground. These include an etched transmission bushing and drive shaft yoke, burned emergency brake cables, a burned carburetor return spring, flickering headlights, failure of the engine to start, burned front wheel bearings or CV joints (on front-wheel-drive cars), and other seemingly mysterious conditions. Because the problem is not obvious (as it would be if the engine did not run, for instance), a defective ground strap is sometimes ignored.

■ OIL PRESSURE PROBLEMS

Oil pressure can be too low or too high. Low oil pressure is far more common. Engines with high odometer mileage begin to use oil between oil changes. Most people use self-serve gasoline stations, and many of them tend to neglect to check the engine's oil level at each fill-up. Allowing the oil level to drop too low can cause major engine damage. Usually the first thing to happen is lower main bearing wear (**Figure 49.19**). After this happens, oil pressure will remain permanently low at idle speed. Because the pump turns faster when the engine rpm is increased, clearance problems do not usually result in low oil pressure at off-idle speeds.

Low Oil Pressure

Low oil pressure can ruin an otherwise good engine in a short time. Often the cause of a low oil pressure reading on an electric dash gauge or light is a faulty **oil pressure sending unit**.

■ Test the gauge by grounding the wire that leads to the sending unit on the block. When the wire is grounded with the key switch on, the gauge should show maximum oil pressure, or the light should go on. If either happens, the gauge and wiring are good and the sending unit is at fault.

■ Pressure can be tested by temporarily installing an oil pressure gauge in place of the sending unit

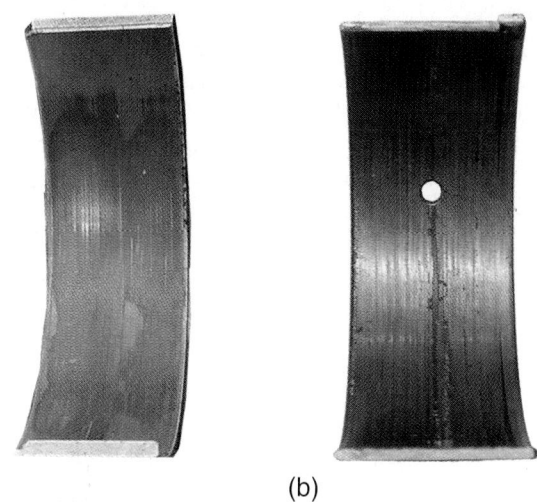

(a)   (b)

Figure 49.19 (a) Lower main bearing wear. (b) More serious bearing wear. *(Courtesy of Tim Gilles)*

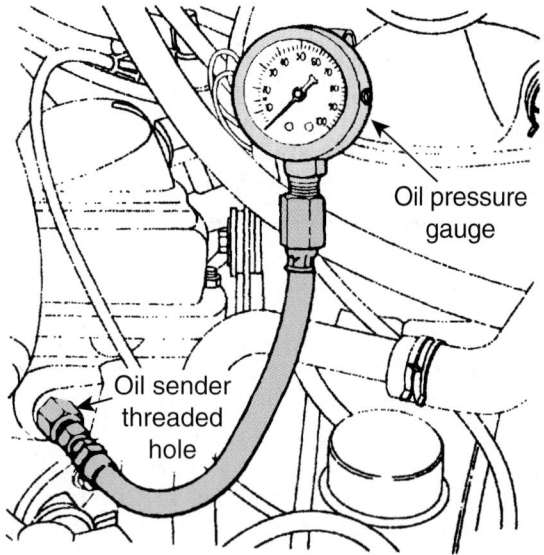

Figure 49.20 An oil pressure gauge is installed in the oil sending unit hole in the block. *(Courtesy of DaimlerChrysler Corporation)*

Figure 49.21 An oil pressure sending unit is removed with a special socket. *(Courtesy of Tim Gilles)*

(**Figure 49.20**). A special *oil sending unit socket* is used to remove the sender (**Figure 49.21**). Using pliers can damage the unit. A static oil clearance test can be performed using a pressure primer with the oil pan removed (see Chapter 56).

The oil pump can also cause low oil pressure. It might be worn excessively or have a sticking relief valve. The intake sump (oil pickup) can have an improperly located or partially plugged screen, or the pump body could be loose on the block (see Chapter 53).

High Oil Pressure

Occasionally, an engine can have oil pressure that is too high. This can result in a bursting oil filter or cause oil consumption or bearing material to be washed from the bearings. The problem can be caused by a stuck **pressure relief valve** in the oil pump or by a severe block-

age in an oil gallery near the cam or crank (close to the oil pump). In either case, the oil pump relief valve would be unable to bypass enough oil at high speeds.

Oil Analysis

A common test done by fleets (large companies with many vehicles) is **oil analysis**. A sample of crankcase oil is sent to a laboratory to be tested. The results of this test can point to mechanical problems before they have become serious. Some of the things that are measured include the following:

- Coolant or moisture, which can indicate internal coolant leaks or insufficient maintenance intervals
- Metals, which can indicate which part of an engine is wearing
- Dirt

■ COOLING SYSTEM PROBLEMS

When a cooling system has been neglected, expensive engine damage often results. When a radiator has become plugged or its fins have become corroded and are flaking away, it cannot conduct enough heat away and the engine will overheat on the highway. Also, the water jackets in the engine can develop a buildup of minerals and scale, which prevents the transfer of heat to the cooling system. Material will continually flake off, resulting in a plugged radiator. Testing for cooling system problems is covered in detail in Chapter 21.

■ INTERNAL ENGINE LEAKAGE

Internal leaks are those that are within the engine. They can be either combustion leaks, coolant leaks, or oil leaks.

Internal Coolant Leaks

When there is an internal coolant leak, the coolant level often drops. Leakage from the outside of the engine is not evident. Internal leaks can happen in the following locations:

- In the water crossover passage of the intake manifold on V-type engines
- In threaded plugs inside of valve covers in cylinder heads
- In combustion areas, such as the head gasket (**Figure 49.22**)
- In a cracked head or block

A leaking thermal vacuum switch can cause a leak into the intake manifold when the engine is warm only. A vacuum-operated heater valve can cause a similar problem.

Internal leaks are diagnosed using the block tester, pressure tester, or infrared analyzer. Water in the exhaust is commonly caused by condensation. Steam in the morning from the exhaust is a normal condition. Excessive water from the exhaust can be caused by a crack in the combustion chamber or a head gasket leak. A coolant leak in a combustion chamber will be evident

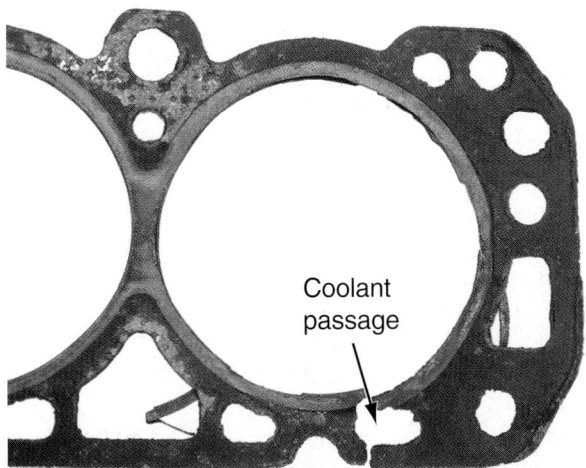

Figure 49.22 This failed head gasket allowed coolant to enter the cylinder. *(Courtesy of Tim Gilles)*

Figure 49.23 A crack in the cylinder head cleaned the carbon off the top of this piston. *(Courtesy of Tim Gilles)*

when the cylinder head is removed from the engine. A normal combustion chamber will be coated with a small amount of carbon. A cylinder with the leak will not have any carbon in it (**Figure 49.23**). Checking for cracks after disassembly is covered in Chapter 52.

Cross Fluid Contamination

Water leaking into the crankcase will contaminate the oil. This is called **cross fluid contamination**. The condition will be evident when a valve cover is removed and the oil is milky and parts are rusted. An internal water leak can occur when a loose timing chain wears a hole in the inside of a timing cover that has a water pump mounted to it. An additional problem that can occur with an internal leak is failure of a computerized fuel system's oxygen (O_2) sensor. This happens when silicates that are a part of the additive package of some coolants coat the sensor. This is much

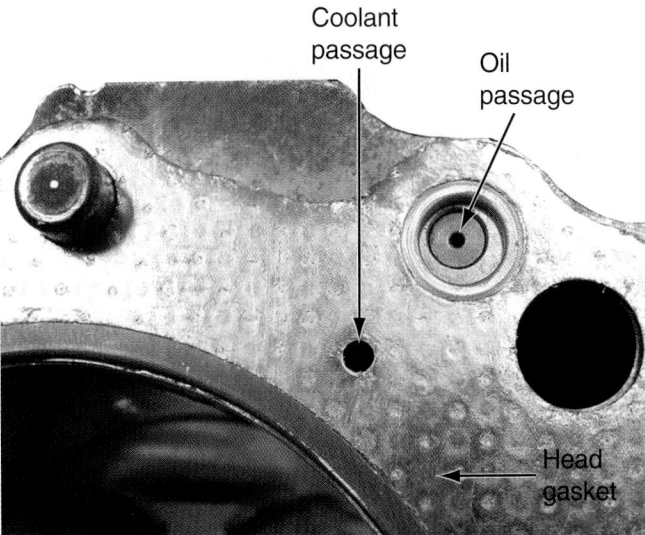

Figure 49.24 A warped head can result in cross fluid contamination when engine oil and coolant mix. *(Courtesy of Tim Gilles)*

the same as O_2 sensor failure caused by some silicone RTV sealants (see Chapter 51).

Internal Oil to Coolant Leaks

When a leak occurs between an oil and water passageway, pressurized oil (approximately 30 psi) will force its way into the cooling system (approximately 15 psi) (**Figure 49.24**). The engine will overheat and pour a messy oil and water mixture from the radiator overflow. This can also be a symptom of a leaking automatic transmission cooler (see Chapter 21). The transmission will also fill with coolant when the engine is turned off.

Internal leakage can be spotted by installing a pressure tester on the radiator filler neck of a warmed-up engine. The pressure tester can also pinpoint the location of external leaks. Experienced technicians will not begin a repair until they are positive of the locations of leaks.

Exhaust Gas in the Coolant

A leaking head gasket will not always show up on a pressure test. A block check tester or an infrared exhaust analyzer can also be used to check to see if there is exhaust gas in the coolant (see Chapter 21). Bubbles in the coolant can also indicate a leak.

■ SEIZED ENGINE

A **seized engine** occurs when a starter motor will not crank the engine over and the engine cannot be turned over by hand. A frozen smog pump, power steering pump, water pump, or other belt-driven accessory can actually keep an engine from turning over. A drive belt on a frozen accessory can become so hot it melts. When the engine is shut off, it vulcanizes

Figure 49.25 This damaged starter motor was the cause of a "frozen" engine. The starter was repaired, and the engine ran fine. (*Courtesy of Tim Gilles*)

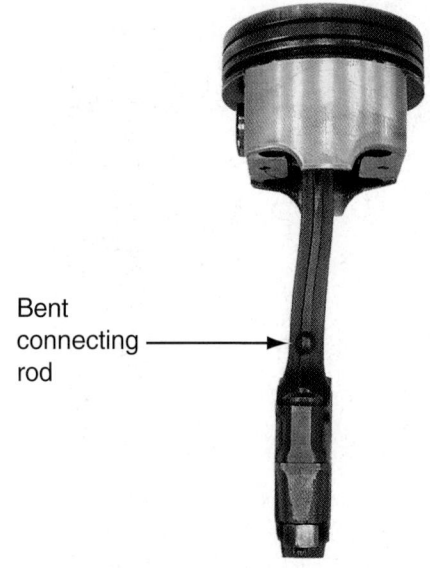

Bent connecting rod

Figure 49.26 This connecting rod was bent when the head gasket failed and water became trapped in the cylinder. (*Courtesy of Tim Gilles*)

to the pulley, preventing the engine from rotating during the next morning's startup attempt. Be sure to loosen the drive belts before condemning an engine (**Figure 49.25**).

If the engine will not turn in either direction with the belts loose, the cause could be coolant thermoplastic seizure that results when coolant mixes with engine oil. Seizure can be a result of other serious engine damage, such as piston seizure, seized bearings, a broken crank, or a seized rod or valve.

Sometimes pressurized coolant from the radiator fills a cylinder after an engine is shut off. If both of that cylinder's valves are closed, the engine will not be able to turn completely over when cranked. This is called hydrolock. If the spark plugs are removed, the engine will be able to crank, and water will pour out of the offending plug hole.

CASE HISTORY *A technician bought an old pickup with the intention of rebuilding the engine. The previous owner said that the engine was seized up (wouldn't turn over). After towing the vehicle to his garage, he began to diagnose the engine before disassembling it. He discovered that the engine wasn't really seized but was hydrolocked. He knew that head gaskets on long in-line six cylinders are more likely to fail in the middle, so he took out the center two spark plugs and cranked the engine. After water came out the spark plug holes, he was able to replace the spark plugs and run the engine. Surfacing the cylinder head, replacing the head gasket, and changing the oil were the only things required to put the truck in good working order.*

Sometimes a bent connecting rod is the result when there is a hydrolock and the piston is forced against the water in the cylinder (**Figure 49.26**).

CAUTION *Be sure the liquid in the cylinder is not gasoline before cranking the engine. A fire could result. See the related case history in Chapter 21.*

◼ ELECTRONIC FAILURES/ENGINE DAMAGE

Engine damage such as burned valves, scuffed pistons, worn bearings, and damaged cylinder heads can sometimes be traced to electronic component failures. Today's fuel and emission control systems are computer controlled. Computers receive input from various engine sensors.

An electronic EGR valve can become inoperative if its input sensor signals are interrupted. When the EGR valve does not operate properly, detonation can result.

An electric cooling fan failure can be due to an inoperative sensor.

An overly rich air-fuel mixture resulting from a failed sensor or electronic part can cause oil dilution, resulting in piston or crankshaft bearing damage. A tip-off here is that the catalytic converter may overheat and melt, causing exhaust backpressure. Always trace a problem to its root cause.

◼ ENGINE PERFORMANCE AND FUEL MIXTURE PROBLEMS

Emission control and fuel system malfunctions sometimes mimic problems related to the engine. Sometimes important items are neglected during an engine repair. A fuel or emission problem that caused an engine problem will probably cause it to happen again if it is not repaired. Larger engine shops will employ a specialist capable of diagnosing these complicated problems.

An air-fuel mixture that is too *lean* (too much air/too little fuel) can increase heat in the combustion chamber, resulting in detonation and/or burned internal engine parts. An overly *rich* mixture (too much fuel/too little air) can cause oil wash. Oil wash is when oil is washed from cylinder walls, resulting in cylinder wall wear. Leaking fuel injectors can be a cause of cylinder wall oil wash. Intake valve deposits, which will affect engine idle and emissions, can also result.

■ REVIEW QUESTIONS

1. Carbon has built up in the neck area of a valve. What is the probable cause?

2. If a spark plug has a carbon deposit on only one side, what is a probable cause?

3. What is one problem that can result from both of the following: worn rod bearings and high-speed driving?

4. What kind of glowing liquid is used to test for oil leaks using a black light?

5. A noise goes away when the spark plug wire for one cylinder is grounded out. What could this noise be?

6. In the above condition, the noise becomes *louder* when the cylinder is grounded out. What could this be?

7. What is a common cause of a cracked piston?

8. When a cylinder is worn excessively oval, what should be looked for?

9. When an engine runs for a long period with an excessively lean air-fuel mixture, what type of engine damage can result?

10. What two different problems can be pinpointed by looking at the tops of the two pistons shown in **Figure 49.27**?

 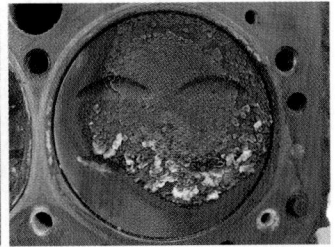

Figure 49.27 What could have caused this type of carbon formation on these piston tops? *(Courtesy of Tim Gilles)*

■ ASE-STYLE REVIEW QUESTIONS

1. What is the most likely cause of oil consumption in an engine with 40,000 miles since a rebuild?

   a. Bad valve guide seals

   b. Worn or plugged piston rings

   c. Both A and B

   d. Neither A nor B

2. An engine has bad valve guide seals. Technician A says that they can be replaced on some engines without removing the cylinder heads. Technician B says that carbon deposits on the necks of the intake valves can result. Who is right?

   a. Technician A   **c.** Both A and B

   b. Technician B   **d.** Neither A nor B

3. Which of the following is/are true about worn main bearings?

   a. The upper main bearing is most likely to have more wear than the lower.

   b. Oil pressure will be low at all engine speeds.

   c. Both A and B.

   d. Neither A nor B.

4. Which of the following could be the cause of a noise that occurs at ½ engine rpm?
   a. A bad lifter
   b. A valve that needs to be adjusted
   c. Both A and B
   d. Neither A nor B

5. An engine has a crack between a water jacket and an oil gallery. Technician A says that oil will go into the cooling system when the engine is running. Technician B says that coolant will go into the oil when the engine is off. Who is right?
   a. Technician A
   b. Technician B
   c. Both A and B
   d. Neither A nor B

6. A car has oil smoke from its exhaust during deceleration. Technician A says that bad valve guide seals could be the cause. Technician B says that worn piston rings could be the cause. Who is right?
   a. Technician A
   b. Technician B
   c. Both A and B
   d. Neither A nor B

7. Technician A says that it is normal for an engine to consume oil during a long freeway trip after an extended period of city driving. Technician B says that adding excessive oil to the crankcase can result in noisy hydraulic valve lifters. Who is right?
   a. Technician A
   b. Technician B
   c. Both A and B
   d. Neither A nor B

8. Any of the following could cause a "seized" engine *except*:
   a. A melted accessory drive belt
   b. A broken starter drive housing
   c. A burned valve
   d. A burned engine bearing

9. Which of the following could cause an intermittent knocking sound from the passenger compartment?
   a. A defective torque converter
   b. A worn crankshaft
   c. A loose vibration damper
   d. Any of the above

10. Which of the following could be caused by a broken electrical ground strap?
    a. A burned transmission bushing
    b. A defective emergency brake cable
    c. Both A and B
    d. Neither A nor B

CHAPTER 50

Engine Removal and Disassembly

■ OBJECTIVES

Upon completion of this chapter, you should be able to:

✔ Label and organize parts prior to engine removal.

✔ Remove an engine from a vehicle in a safe and methodical manner.

✔ Disassemble the engine following the correct procedures.

✔ Keep parts organized for reassembly.

✔ Inspect for and interpret causes of wear to engine parts.

■ KEY TERMS

firewall   ring ridge   rod bolt protector

■ INTRODUCTION

This chapter deals with removal and disassembly of the engine. It is important that procedures be followed carefully. It is very easy to damage the car body or paint when removing the engine. The engine is also very heavy and your personal safety is at stake.

Engine parts must be removed and inspected in an orderly manner. Signs of wear noticed during engine disassembly can be clues to problem areas. The correct repair can prevent the problem from occurring again.

Be sure to consult the applicable repair information before beginning the repair job. Procedures differ slightly between manufacturers.

This section of the chapter describes how to remove an engine from both front- and rear-wheel-drive vehicles. Rear-wheel drive, which was the predominant driveline arrangement in North American vehicles until the 1980s, is still used in many SUVs, pickup trucks, and luxury automobiles. In this chapter, engine removal specifics common to both front- and rear-wheel-drive vehicles are covered first. Rear-wheel-drive engine removal is covered next, followed by front-wheel drive.

NOTE: *Computer information systems like All-Data and Mitchell On Demand, AERA's Prosis, and manufacturer-specific Web sites have technical bulletins. Always refer to these before attempting a large repair.*

■ ENGINE REMOVAL

Install fender covers on both fenders and over the grille to protect the vehicle's paint. Some shops like to protect the windshield with a piece of corrugated cardboard.

Battery Cables

Disconnect the battery cables. The ground cable (usually the negative) should be disconnected first. This is a good habit to follow whenever doing major repairs. Before disconnecting the battery, make a note of which radio stations the customer has set on the radio so they can be reprogrammed after the engine is reinstalled.

The outside of the battery is often covered with a thin film of battery acid. After handling the battery, be sure to clean your hands. Battery acid can ruin clothing that you touch later. It is a good idea to wear gloves when handling a battery.

Hood

Remove the hood before working on the engine. Before loosening the hood bolts, mark the location of the hood to the hood hinges so that it can be properly reinstalled. One effective method is to use a pencil to mark the exact hood positions with an outline.

Be careful not to let it slip back into the windshield or to damage the paint on the hood. Set it down on a fender cover or cardboard.

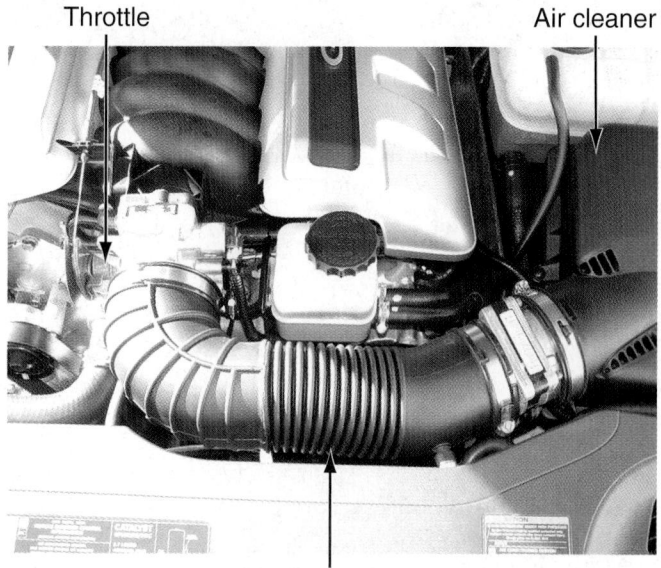

Figure 50.1 Remove the air cleaner and hose. Be careful not to crack the hose. *(Courtesy of Tim Gilles)*

Label Vacuum Lines

Remove the air cleaner and label all wires and vacuum lines. The air cleaner on a typical port fuel injected engine is located off to the side and is connected to the fuel system by a molded hose (**Figure 50.1**). These hoses can become fragile with age and can be expensive to replace. A broken hose can cause driveability complaints on some types of fuel injection systems and can allow dirty air to enter the engine on all systems. Later model vehicles usually have some of the hoses grouped together with male and female connectors on the ends to make it less likely to make an error in a hose connection. Sometimes the hoses are numbered, color coded, or are different sizes.

> ### VINTAGE ENGINES
>
> Most vehicles manufactured since 1985 have had fuel injection systems. Before fuel injection, engines used carburetors. As emission controls became more prevalent, most of them were activated by intake manifold vacuum controlled by thermal vacuum valves. A carburetor often had many hoses connected to it. As engine controls evolved, these functions were managed by computer controls. One result of this is fewer vacuum hoses in the engine compartment.

Late-model vehicles often have a vacuum diagram label under the hood. When there is not, draw a map of the vacuum hoses. Assign a number to each hose on the map. Then use masking tape and a pen to number each line as it is removed. There are several methods used to keep hoses identified so they can be reattached in the correct position. Rolls of numbered

tape are available from automotive tool and parts suppliers (**Figure 50.2**). Affix two pieces of tape with the same number, one on the hose and the other on the connection where the hose was removed (**Figure 50.3**). Some technicians keep several narrow rolls of different color tape for labeling unmarked hoses in this same manner.

NOTE: *Later model cars are usually equipped with dedicated electrical and vacuum connectors that can only be attached to their counterparts. Even if this is true on the vehicle you are working on, label the connection with information as to which accessory it connects to, or draw a map showing electrical and vacuum connections before beginning disassembly.*

Drain Coolant and Oil

Drain all coolant from the radiator and block. If the coolant will not be reused, be sure to comply with local

Figure 50.2 Rolls of numbered tape are available from automotive tool and parts suppliers. *(Courtesy of Tim Gilles)*

Figure 50.3 Tape is numbered to label matching vacuum and electrical lines. *(Courtesy of Tim Gilles)*

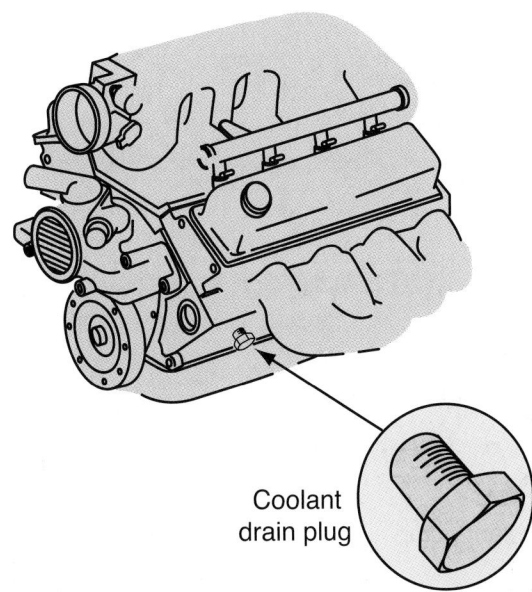

Coolant
drain plug

Figure 50.4 A coolant drain plug in a cylinder block.

regulations for its disposal. If the block is equipped with a coolant drain plug (**Figure 50.4**), the engine block should also be drained.

Drain engine oil and remove the oil filter. The oil filter is made from thin sheet metal that is easily crushed or torn if the filter wrench is not held as close to the filter base as possible.

Radiator

Remove the radiator. Disconnect the radiator hoses from the engine. If the hoses are left attached to the radiator, they do not need to be marked for reassembly. Remove the radiator from the car. It can be stored in the vehicle's trunk with its hoses facing upward and the radiator cap installed so any remaining coolant does not leak out. The radiator is fragile, so be sure that other parts are not stored on it.

NOTE: *Be sure to protect carpets or mats when storing parts in the vehicle.*

If the car has an automatic transmission, it is usually equipped with a heat exchanger (cooler) on the bottom or side of the radiator (see Chapter 20). The two lines leading to the radiator from the transmission must be removed and plugged. Use a flare nut (tubing) and an open-end wrench to disconnect the line. Some newer vehicles require a special release tool to remove the transmission cooler lines.

SHOP TIP A short length of hose plugged at one end with a bolt can be installed over a disconnected line to keep fluid from leaking from it.

SHOP TIP A ruined transmission oil line can be repaired using a union (see Chapter 24). A piece of hose is another alternative, but two considerations must be made:

- Transmission oil cooler (TOC) hose must be used. Fuel hose will be damaged by automatic transmission fluid. A leak and fire could be the result.
- Both ends of the tubing must be double flared or bubble flared to prevent the hose from slipping off. Transmission fluid in this circuit ranges from 35 to 55 psi. A bubble flare works best for this because it will not cut the inside of the hose.

Distributor and Spark Plug Wires

Some engines have distributors. It is a good idea to remove the distributor and spark plug wiring before removing the engine to prevent damage caused by interference with the lifting sling.

Do not remove the spark plug wires from the distributor cap. They are already in the correct firing order and need not be disturbed. Mark the location of the number 1 plug wire on the distributor cap before removing any wires. If the wires are to be replaced, replace them one at a time so they can be measured easily and kept in order.

Alternator, Fan, and Accessory Drives

Remove the alternator. If possible, leave the wires connected to the alternator; simply unbolt it and use wire to fasten it to something out of the way. Use masking tape to label any electrical wiring that must be disconnected. Taking the time to do this will save the time that would be wasted in trying to determine wires during reassembly.

Many of today's engines do not have belt-driven cooling fans or air pumps for the emission control system. To prevent damage to these items if an engine is so equipped, they should be removed, along with the alternator, before attempting to remove the engine (**Figure 50.5**).

If the engine is to be cleaned before removal, protect the alternator. Some soaps can damage alternator bearings. After engine cleaning, be sure to clearly label the electrical connections to the starter and sending units for oil pressure and coolant temperature (**Figure 50.6**).

Heater Hoses and Ground Strap

Remove the heater hoses and ground strap. Some heaters have a control valve in one of the heater hoses. On some engines, it matters which way the heater hoses go. Label the hose that comes from the coolant pump so it can be reattached correctly later. Use care when

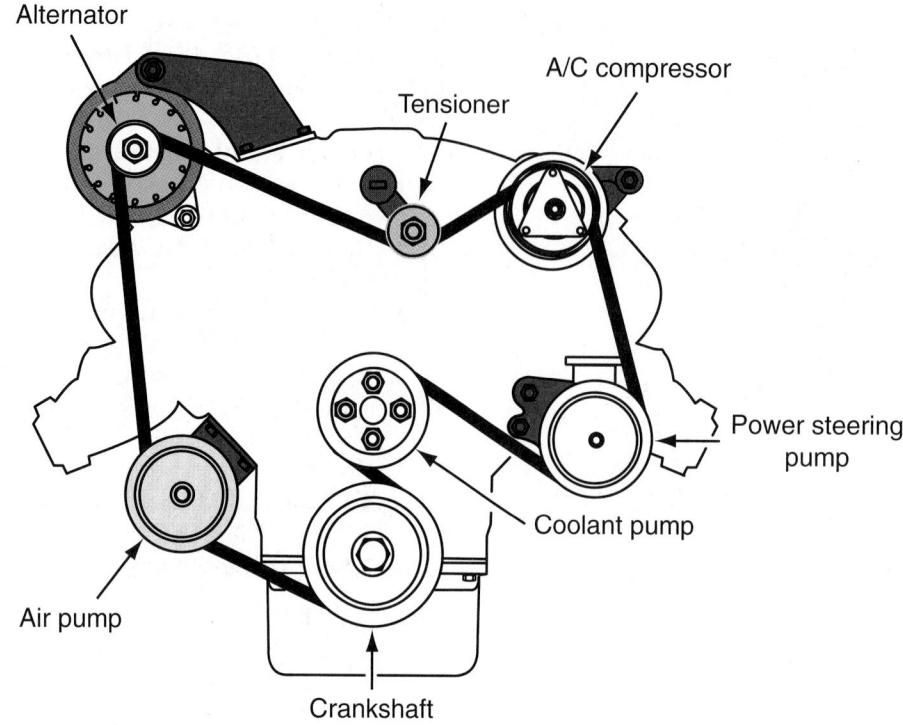

Figure 50.5 Remove the alternator and air pump.

Figure 50.6 Label the wires to the sending units.

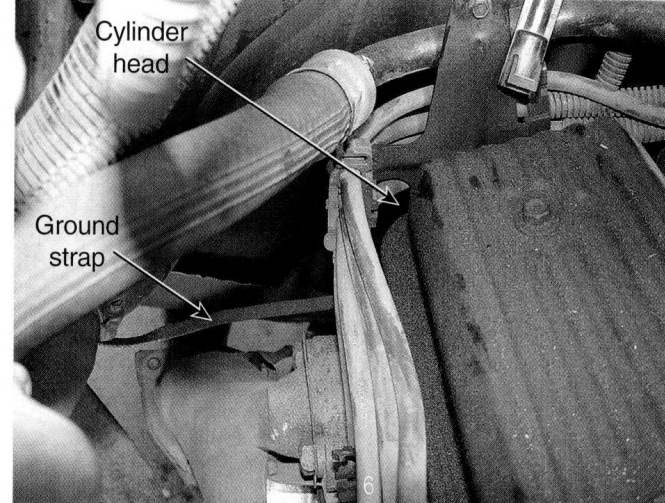

Figure 50.7 Be sure to remove the ground strap.
(Courtesy of Tim Gilles)

removing the heater hose. Cut the hoses and replace them if necessary to avoid damaging the heater core.

> **SHOP TIP** A missing or broken ground strap can cause a transmission bushing, wheel bearing, or emergency brake cable to burn or fail. This is because electricity follows the path of least resistance to ground, causing etching and heat. A bushing, bearing, or cable may become an alternate "ground" circuit if the strap is not attached. A missing ground strap can also interfere with proper operation of electrical circuits.

There is usually a ground strap from the engine to-the **firewall** (bulkhead), or frame (**Figure 50.7**). Be sure to disconnect it before the engine is removed.

Cooling System Switches

Remove switches and sensors. Newer engines have coolant temperature sensors that supply the PCM with information on engine coolant temperature. In response to this input, the PCM commands electrical devices called actuators to switch emission control devices on as engine temperature rises.

Figure 50.8 Remove the crankshaft position sensor. *(Courtesy of Tim Gilles)*

Engines with distributorless ignitions have crankshaft position sensors. Sometimes they have camshaft sensors, too. Carefully remove them and store them where they will not be damaged. **Figure 50.8** shows one of these sensors.

Late-model engines have detonation sensors, also called knock sensors (**Figure 50.9**). Remove this if it is easily accessible and might be broken during engine removal.

After fuel system pressure has been bled (see Chapter 41) the fuel injectors can be removed.

Intake Manifold and Valve Covers

Remove the throttle linkage or cable. Then remove the valve covers and motor mount bolts. Before the engine is removed from the car, the intake manifold can also be removed.

Figure 50.9 A detonation sensor signals the computer to change ignition timing as it responds to vibration. *(Courtesy of Tim Gilles)*

Accessory Brackets and Accessories

Mark accessory brackets and remove accessories. Any accessory brackets, such as air conditioning, that are attached to the head or block may be removed. If the vehicle has many accessories, label the brackets to show their location on the head or block.

> **SHOP TIP** Some accessories have multiple-piece brackets. After removal, reassemble them to the accessory so you will not forget how to reassemble them later. Repair informations do not usually show bracket locations.

Because of the unusual lengths of many accessory bracket bolts, it is a good idea to put them in labeled bags for easier reassembly. An air-conditioning compressor bolt that is too long can damage an expensive compressor housing. When there are bolts of several different lengths, it saves time to assemble each one back into its accessory bracket as it is removed.

> **CAUTION** *Do not disconnect air-conditioning hoses,* as they contain pressurized refrigerant. If a refrigerant line must be disconnected, *use extreme caution and wear eye protection.* Refrigerant vaporizes so fast that it freezes anything it touches, and it can cause blindness. R-12 refrigerant turns to deadly phosgene gas (nerve gas) if allowed to contact an open flame or an extremely hot metal surface. It is unlawful to release refrigerants into the atmosphere. It must be reclaimed and can be recycled (**Figure 50.10**).

Since 1994, most vehicles have used R-134A refrigerant, but there are many older air-conditioning systems that still have R-12 refrigerant. R-12, also known as Freon, has been implicated in the depletion of the Earth's ozone layer. In the Montreal Protocol of 1987, 23 countries, including the United States and Canada, signed an agreement setting limits on the production of ozone-depleting chemicals. Since 1992, the use of refrigerant recycling machines has been required by law. Persons using R-12 equipment must be certified. Certification is available from several industry associations.

The *air-conditioning compressor* can usually be wired up out of the way with the lines still attached (**Figure 50.11**). When air-conditioning lines must be disconnected, be sure to plug all openings immediately. Moisture must not be allowed to enter the system.

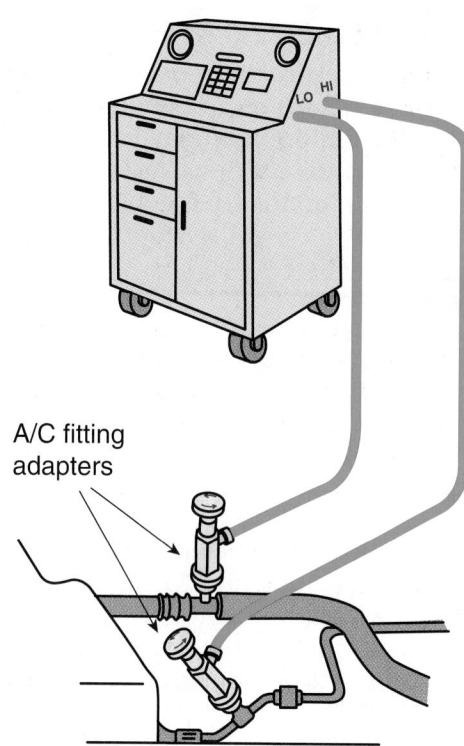

Figure 50.10 An air-conditioning charging and recycling machine.

Figure 50.11 Wire the air-conditioning compressor out of the way with both refrigerant lines still attached. *(Courtesy of Tim Gilles)*

Remove the *power steering pump without* disconnecting the lines and wire it in a position so that fluid cannot leak out.

Exhaust Components

Remove exhaust components. Bolts holding the exhaust manifold and exhaust pipe are often rusted and will be difficult to remove and have a tendency to break. Spray penetrating oil on them and use a

Figure 50.12 Use an impact wrench and impact socket to remove exhaust manifold-to-pipe bolts.

six-point socket. See Chapter 7 for information on how to remove broken fasteners.

> **SHOP TIP**
> ■ A rusted fastener can sometimes be tightened slightly to help penetrating oil get into the threads.
> ■ Rusted manifold bolts are easier to remove with an impact wrench (**Figure 50.12**).
> ■ Use a six-point socket to remove most engine items, especially exhaust bolts and nuts. A 12-point socket will be more likely to slip, rounding off the corners of a nut.

Most cars since the 1980s have computer-controlled fuel systems that use an oxygen sensor to determine the proper air-fuel mixture (**Figure 50.13**). Disconnect the wire to the sensor. Leave the sensor in the exhaust system. Oxygen sensors can be expensive. Use an oxygen sensor socket if it is to be removed from the exhaust manifold (see Chapter 41). Otherwise, leave it in the manifold and be very careful not to damage it.

Fuel Line

Remove and plug the fuel line. The fuel line from the tank must be disconnected. To prevent fuel leakage from the line, the preferred method is to plug it with a bolt and hose clamp (**Figure 50.14**). A popular tool among technicians is pinch pliers, or clamp pliers (**Figure 50.15**). When using pinch pliers on hose or tubing care must be taken not to damage the inner lining of a hose. Also, some fuel injection tubing is designed for high-pressure use. This tubing is rigid and must not be clamped.

Figure 50.13 Oxygen sensors can be expensive. Use an oxygen sensor socket to remove one. It has a slot for the wires to the sensors. *(Courtesy of Tim Gilles)*

Figure 50.14 A fuel hose plugged with a bolt and hose clamp.

Figure 50.15 A fuel line closed off with pinch pliers. Be careful not to damage the hose or tubing. *(Courtesy of Tim Gilles)*

> **SAFETY NOTE**
> - Even though fuel may not pour out of a fuel line when it is first disconnected, a hot day can cause expansion of the fuel in the tank. This can result in the fuel in the tank siphoning out all over the ground. Be sure to plug all fuel lines, including those to the smog-control vapor canister. Also remember to loosen the fuel cap to prevent pressure from building in the fuel tank on a hot day.
> - Fuel-injected engines have residual pressure in the fuel lines when the engine is off. Relieve this pressure before loosening a fuel line connection (see Chapter 41).
> - When working around fuel systems, use a fluorescent trouble light that is cool to touch.

Determine Whether to Remove the Transmission

Before engine removal, locate the recommended procedure in the service literature. On a rear-wheel-drive vehicle, it is generally easier to leave an automatic transmission in the chassis when removing the engine (**Figure 50.16**). Some front-wheel-drive engines are easiest removed without removing the transmission as well.

Separating the Engine and Transmission/Transaxle

Although the first set of instructions here applies to rear-wheel-drive vehicles specifically, the following hints apply to separating the engine from an automatic

Figure 50.16 When an engine is removed from a rear-wheel-drive vehicle, it is usually easier to leave the transmission in the vehicle. *(Courtesy of Tim Gilles)*

transmission or transaxle on both front-wheel-drive and rear-wheel-drive vehicles.

If the transmission is to be removed, on a rear-wheel-drive car, make centerpunch marks on the converter and the flexplate so that they can be correctly aligned on reassembly.

SHOP TIP
- If an impact wrench is used, the crankshaft will not have to be held to keep it from turning.
- Be sure always to turn the engine in its direction of normal rotation. Some OHC engines will skip cam timing if they are turned backwards.

Remove the torque converter attaching nuts and/or bolts from the flywheel flexplate (**Figure 50.17**). To gain access to each bolt, rotate the engine by turning the crankshaft with a special flywheel turning tool or by using a large socket and ratchet on the damper bolt at the front of the engine crankshaft (**Figure 50.18**).

Pry the torque converter toward the transmission, away from the flexplate (**Figure 50.19a**). Leave the converter installed in the front of the transmission during engine removal so transmission fluid will not pour out of it. **Figure 50.19b** and **Figure 50.19c** show the relationship between the torque converter pilot and the mating pilot hole on the rear of the crankshaft.

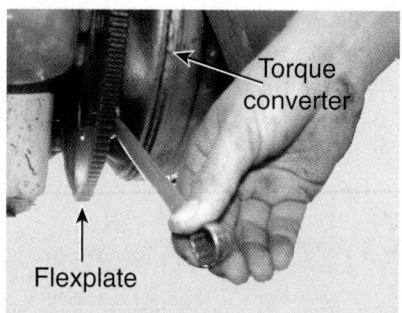

Figure 50.17 Remove bolts that attach the torque converter to the flexplate. *(Courtesy of Tim Gilles)*

Figure 50.18 Turning the crank using the damper bolt. *(Courtesy of Tim Gilles)*

(a)

(b)

(c)

Figure 50.19 (a) Pry the torque converter away from the flexplate. (b) Torque converter and flexplate. (c) A pilot hole in a crankshaft. *(Courtesy of Tim Gilles)*

If the transmission dipstick tube is attached to the cylinder head, unbolt it from the head.

Remove the engine-to-transmission bolts. On rear-wheel-drive vehicles, these bolts are easily loosened from underneath the car using a very long extension and a universal socket (**Figure 50.20**).

NOTE: *When a long extension is used, a larger impact wrench might be needed because some of the impact is absorbed by the long tool.*

Sometimes, it is necessary to unbolt the rear transmission crossmember and allow the rear of the transmission to drop. This provides easier access to the

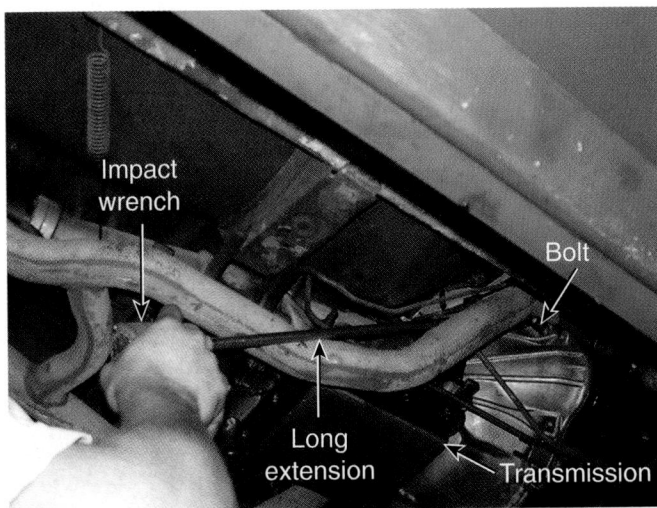

Figure 50.20 This tool setup is good for removing top transmission-to-engine bolts. *(Courtesy of Tim Gilles)*

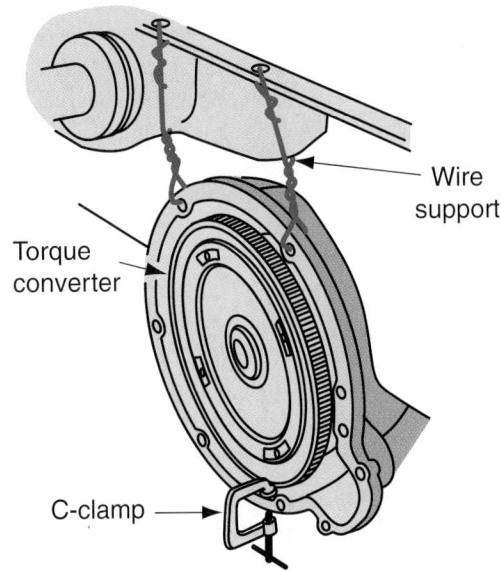

Figure 50.22 A C-clamp installed to keep the torque converter in place when the engine is removed without removing the transmission. Attach wire to the transmission so it does not hang.

Figure 50.21 When removing a torque converter, immediately flip its open end up so fluid does not leak out. *(Courtesy of Tim Gilles)*

If the converter slides too far forward, it will come out of the transmission front pump. It must be realigned with the front-pump drive gear to prevent damage to the pump and flexplate when the engine is reinstalled.

SHOP TIP When the engine and transmission are first separated, measure and record the distance from the edge of the converter housing to the front of the converter (see Chapter 74). Although performing this step is not necessary, it can give peace of mind that the converter is installed all the way during reassembly.

Ideally, converters without studs should be only ⅛" from the flexplate when pushed as far into the transmission as possible. This ensures that the transmission front-pump drive gear is sufficiently engaged by the torque converter drive lug. If the distance is more than ⅛", shims can be installed.

The transmission *must* be supported during and after engine removal. Support the transmission with a jack until it can be wired up so that it will not hang (see Figure 50.22).

It is a good practice to remove the torque converter from the transmission and replace the front transmission seal during an engine rebuild. This is sometimes necessary after an engine rebuild because an old transmission front seal that was not replaced can begin to leak. Front seal replacement is covered in Chapter 74.

top engine-to-transmission attaching bolts and also to the transmission cooler lines located high on some transmissions. Do not allow the transmission to hang without support.

When a torque converter is removed from an automatic transmission or transaxle, quickly rotate it upward so the transmission fluid does not empty out and make a mess (**Figure 50.21**). A typical torque converter can hold 3 or 4 quarts of transmission fluid.

A C-clamp can be clamped on the transmission housing to keep the converter in place in the front pump of the transmission. The converter can also be held into the transmission with safety wire or a bar bolted across the front of the converter housing (**Figure 50.22**).

An engine with an automatic transmission was replaced with a rebuilt unit. While the engine was out of the car, the torque converter, which was no longer supported by the rear of the engine crankshaft, hung on the old seal, which had become brittle with age. When the new engine was started, oil came pouring from the front of the transmission. The transmission had to be removed to make the repair. Removing the engine would have been more difficult. The customer blamed this leak on the shop because the transmission was not leaking when he brought the car in for the engine.

If the transmission is to be removed from a rear-wheel-drive vehicle, disconnect the shift linkages, electrical wires, speedometer cable, and the drive shaft. Tape the rear U-joint cups with masking tape so they will not accidentally fall off the U-joint (**Figure 50.23**).

NOTE: *Some two-piece drive shafts are splined in the middle. The U-joints on both sections must be in "phase" (in the same plane) or serious vibration will occur. If the halves are to be separated, mark them for easier reassembly.*

Plug the end of the transmission after removing the drive shaft so transmission fluid will not leak out (**Figure 50.24**).

Disconnect the speedometer cable or wire from the transmission.

Before removing the transmission from a car with a standard transmission, the clutch activating fork and gearshift linkages must be disconnected.

Engine Mounts

Unbolt the engine mounts. Mark them with a center-punch or marker to show which side of the mount is the front and which side is left or right.

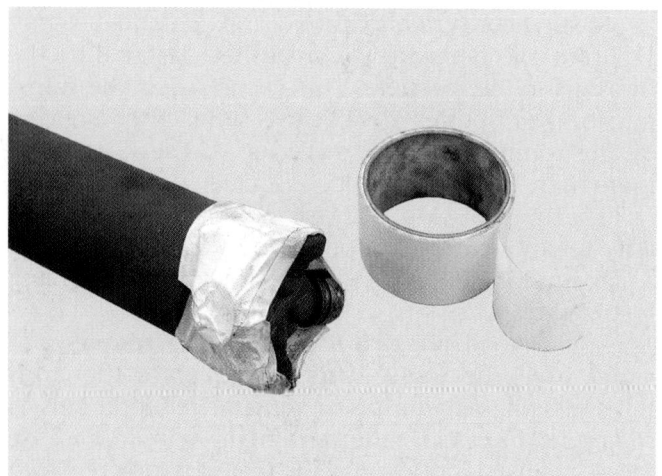

Figure 50.23 Tape wrapped around the U-joint cups to keep them from falling off. *(Courtesy of Tim Gilles)*

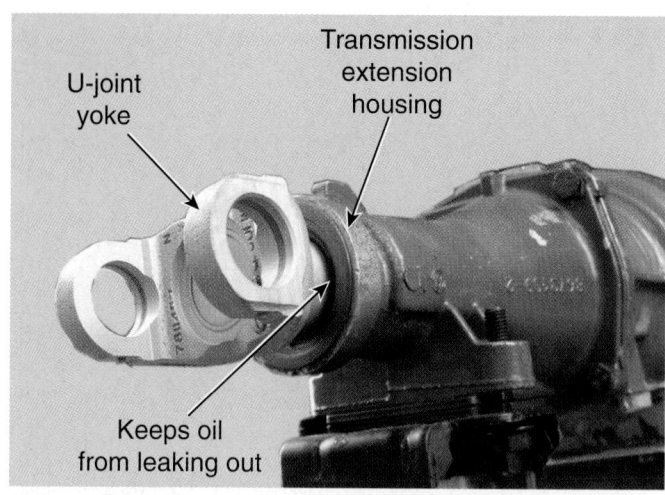

Figure 50.24 The end of the transmission is plugged with an old slip yoke after the drive shaft is removed. *(Courtesy of Tim Gilles)*

An apprentice was installing an engine in an Oldsmobile and was having a very difficult time. Further investigation showed that the engine mounts were installed on the wrong side of the engine, which caused the engine to be at least 1" too far forward in the chassis.

In some vehicles, the engine can actually be installed with the mounts reversed. The rear transmission mount stretches until there is metal-to-metal contact with the mount and the frame. This results in noticeable engine vibration as the vehicle is driven.

Removing the Engine

Remove the engine from the vehicle. Attach a cable sling, a chain, or a special lifting tool to the heads or block (**Figure 50.25**). Some engines are equipped with lifting brackets. Make sure the bolts are tightened all the way up against the sling brackets to protect them from excessive stress that can break them (**Figure 50.26**). Spacers can be made from old piston pins or pieces of pipe cut to different lengths.

A chain hoist, sometimes called a chainfall, is sometimes used for removing an engine from a vehicle. Unlike a hydraulic engine hoist, with a chainfall it is harder to maneuver an engine back into position during reinstallation. Also, raising an engine for access to engine mounts or oil pan removal requires that the hood first be removed.

CAUTION Be careful of the car's paint when using a chain hoist to remove an engine (**Figure 50.27**).

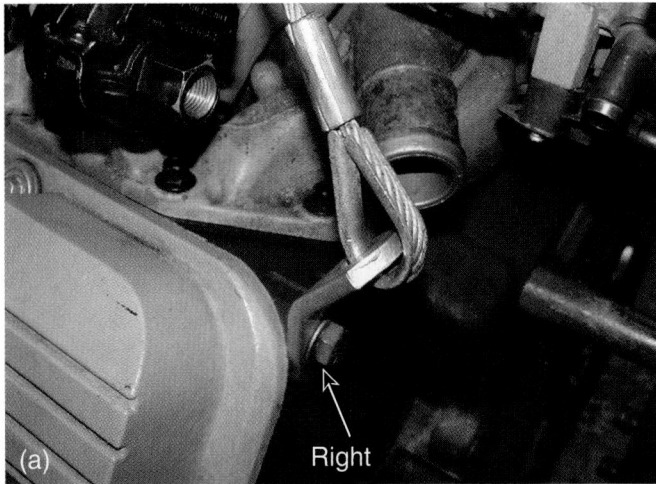

Figure 50.25 (a) Right and (b) wrong ways to install a lifting sling. *(Courtesy of Tim Gilles)*

Figure 50.26 Avoid excessive stress on the sling bolts.

 **CASE HISTORY** *An apprentice technician was using a chain hoist to remove an engine from a car that had just been painted. As the chain was being pulled, it was rubbing against the paint on a fender, damaging it.*

The grille in front of the radiator is also an area where paint can be damaged by a chainfall. Even if the engine rebuild is perfect in every other respect, you can be sure that paint or grille damage is the one thing the customer will notice. The paint can be protected with a fender cover.

Front-Wheel-Drive Engine and Transaxle Removal

The following is a generic procedure for front-wheel-drive engine removal. Check related service information for instructions before attempting to remove the engine. Some engines can be removed leaving the transaxle in

Figure 50.27 The chain hoist can damage paint. *(Courtesy of Tim Gilles)*

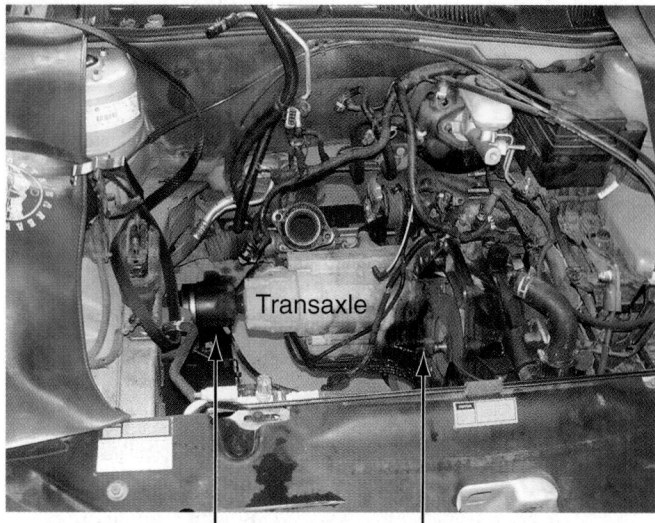

Figure 50.28 Sometimes the engine can be removed while the transaxle is left in the vehicle. *(Courtesy of Tim Gilles)*

Figure 50.29 Some front-wheel-drive engines are easiest to remove with the transaxle. Then the engine and transaxle are separated. *(Courtesy of Tim Gilles)*

Figure 50.30 Some front-wheel-drive engines and transaxle assemblies are more easily removed by lowering them through the bottom. *(Courtesy of Tim Gilles)*

Axles

Figure 50.31 Before removing a front-wheel-drive engine and transaxle, the axles are removed. Do not allow them to hang.

the vehicle (**Figure 50.28**). Other engines are more easily removed with the transaxle (**Figure 50.29**). Some engine/transaxle assemblies are removed through the hood opening, while others are more easily lowered from the bottom of the vehicle (**Figure 50.30**).

Removing the Transaxle

If the transaxle is to be removed with the engine as a unit, the drive axles are removed first (**Figure 50.31**) This will require removal of the lower ball joints from both sides of the vehicle so that the wheel hubs can be moved out far enough to allow the splines on the ends of the axles to be disengaged from the transaxle. Be ready with a drain pan. When the axles are removed,

oil will spill from the transaxle. Remove both axles and store them for later reassembly. Detailed procedures for removing front axles are covered in Chapter 78.

CASE HISTORY *An apprentice removed an engine and transmission from a front-wheel-drive vehicle. After the new engine installation was completed, the customer left with the car. Later that day, the car was towed into the shop with the complaint that it would not go into fifth gear. Further investigation showed that the transmission did not have any oil in it. The apprentice forgot to replenish the oil that had drained out when the axles were pulled. The transmission required extensive and costly repair.*

Figure 50.32 Disconnect the front-wheel-drive transmission shift cable or linkage. *(Courtesy of Tim Gilles)*

Disconnect the speedometer cable, the transmission shift linkage (**Figure 50.32**), and the clutch cable (if this is a manual transaxle). Remove the torque rod, as applicable. Use an impact wrench to disconnect the exhaust pipe from the exhaust manifold and move it out of the way.

Attach a sling to the engine and transaxle assembly at each side (to allow for balance during engine removal). Attach an engine hoist to the sling and apply just enough tension to the sling to unload the engine mounts for easy removal of their bolts. Remove the bolts that hold the front engine mount to the cylinder head and remove the mount from the body to allow room for the engine to clear during removal. Remove the nuts or bolts from the front and rear engine mounts.

Remove the bolts attaching the front and rear torque rods to the engine. Loosen the bolts attaching the torque rods to the vehicle body and swing the torque rods to move them out of the way. Raise the engine a few inches and check to be certain that anything connected between the engine and the vehicle has been disconnected. Then finish removing the engine. Removing the engine is best done with two people. One person raises the engine slowly while the other guides it until it clears parts of the vehicle body and frame.

When the engine and transaxle are free of the vehicle, roll the shop crane until the engine can be lowered safely.

SAFETY NOTE Be careful when using a cherry picker (engine hoist) for moving an engine. Let the engine down as low as possible in order to keep the center of gravity low. If the center of gravity is too high, the cherry picker can tip over.

On some front-wheel drives, the cradle that holds the engine and transaxle is lowered from the bottom of the car.

NOTE: *When removing engines from vans and motor homes, special solutions are often required. Sometimes, it is easier to remove an engine from a van by dropping it out of the bottom. Other times, the cylinder heads might need to be removed first.*

■ ENGINE DISASSEMBLY

A typical engine has many parts (**Figure 50.33**). Rebuilding an engine is not difficult, but a successful outcome depends on being organized. Professional technicians can skip some of the more basic steps described here once they have a track record to build upon. The first time you undertake a large project like this, however, you will want to be more careful not

Figure 50.33 Before and after all the parts are removed from a V6 engine. *(Courtesy of Tim Gilles)*

Figure 50.34 This student was very organized, making reassembly much easier. *(Courtesy of Tim Gilles)*

to forget any important steps. When a repair job is undertaken in a repair shop, it is usually completed within a day or two. After all, the customer is waiting for his or her car. A school or hobby project, on the other hand, often takes weeks or even a month to complete. The professional will not need to separate nuts and bolts into baggies with labels, but this can be a time and error saver on the first-time rebuild. **Figure 50.34** shows an example of one student's organizational efforts. When this engine was reassembled after several delays caused by back-ordered parts, needless to say, everything went back in the same place from which it came.

SHOP TIP The tops of used oil containers can be cut off and used to store nuts and bolts. Plastic baggies are handy for storing small parts that could be easily lost. Make notes during disassembly regarding special parts or service needs.

Before and during an engine disassembly, inspect for problems that might add to the cost or feasibility of the rebuild. Items like broken castings, stripped threads, broken studs, and damaged sending units or vacuum switches should be noted on the work order and included in the estimate to the customer. Noting such items now can eliminate controversy later, should the customer dispute whether such items were damaged when the vehicle or engine entered the shop.

Save all old parts, including gaskets, until the engine is reassembled and running again. Parts might be needed for comparison. On the assembly line sometimes, a damaged engine is salvaged by machining a valve lifter or main bearing bore oversize, or a crankshaft undersize. Watch for these variations.

CAUTION Before beginning engine disassembly, make sure the engine is cold. Tearing down a hot engine can cause warped cylinder heads, especially on engines with aluminum heads.

Pushrod and overhead cam engines have slightly different procedures outlined below.

Clutch Parts

Remove clutch parts. If the engine is equipped with a standard transmission, remove the transmission and clutch before installing the engine stand mounting head. If the clutch pressure plate is the original one and will be reused, mark it so it can be replaced with its original orientation to the flywheel and crankshaft (**Figure 50.35**). The pressure plate and engine were balanced together at the factory, and although this is not crucial, it is the professional approach to mark it to maintain the original balance. Loosen each retaining screw a little at a time (**Figure 50.36**). Then move on to the next (in a star pattern) until all of them are loose enough to remove. The pressure plate is spring loaded against the clutch disc. Leaving one screw tight after all the rest have been removed can bend the pressure plate. It is possible to install some clutch discs backwards. Mark the clutch disc on the flywheel side so that it (or its replacement) can be reinstalled in the same direction (see Figure 50.35). Clutch parts and installation are covered in detail in Chapter 69 and Chapter 70. When the crankshaft is removed from the engine the flywheel or flexplate will make a convenient stand to use while storing the crankshaft on end. If the flywheel must be removed from the crankshaft in order to remove it from

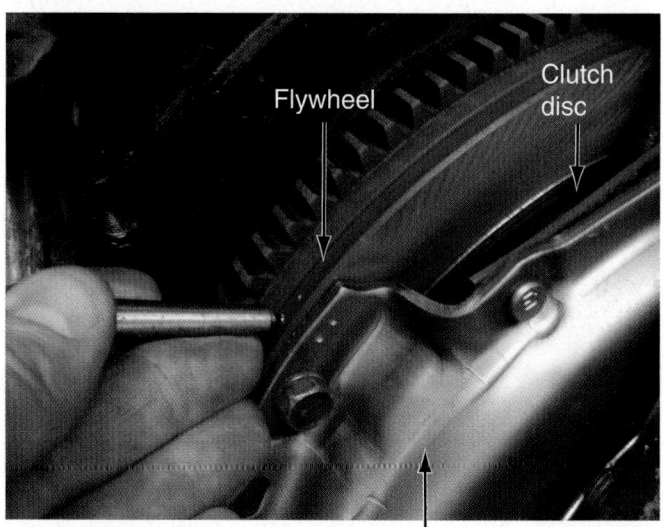

Figure 50.35 Marking the flywheel and pressure plate. *(Courtesy of Tim Gilles)*

Retaining screws

Pressure plate

Figure 50.36 To remove the clutch plate, loosen each retaining screw a little at a time. Do not remove all of them, leaving one tight. The screws on this clutch will require the use of a 12-point socket. *(Courtesy of Tim Gilles)*

the engine, keep the flywheel bolts and washers separated and do not lose them. They are special; their heads are thinner and the washers are thinner than normal bolts and washers.

Hybrid Armature Removal

On hybrid cars, the location of the motor/generator is usually behind the engine. Instead of the flywheel or torque converter on a conventional engine, the motor/generator's permanent magnet rotor is bolted to the back of the crankshaft. The rotor magnet is extremely strong; a puller is required to remove it. **Figure 50.37**

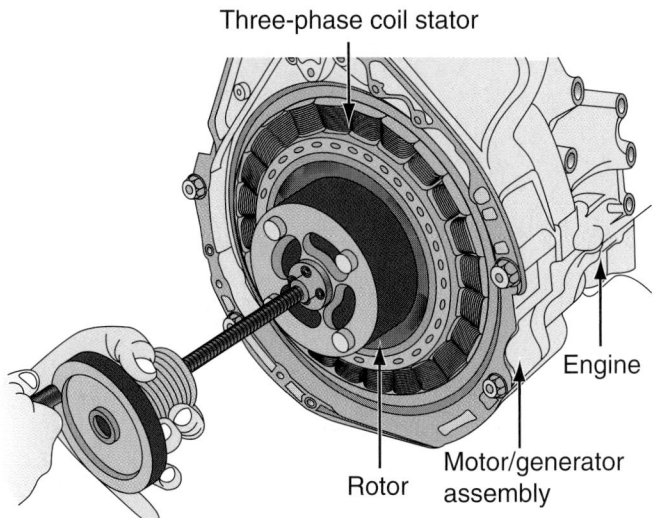

Three-phase coil stator

Engine

Rotor

Motor/generator assembly

Figure 50.37 Removing a hybrid-assist rotor from the rear of the engine's crankshaft.

shows the back of a Honda hybrid engine with the transaxle removed. A large puller is being used to remove it from the center of the three-phase coil stator.

Engine Stand Adapter

Mount the engine on a stand. While the engine is still on the floor, mount the universal mounting head to the engine (see Chapter 9). Then, install the engine and mounting head on the engine stand. Use four bolts of the proper length with washers. The bolts should have a minimum amount of thread contact into the block of at least 1½ times the diameter of the bolt.

 SAFETY NOTE
■ Do not work on the engine while it hangs in the air. It should be mounted on an engine stand or rested on the floor.
■ The mounting head should be mounted so that its center of gravity when it is on the stand will not force the engine to rotate, possibly causing injury to a technician.

NOTE: *Be sure all oil and coolant are out of the engine. Remove the plug from the side of the block and rotate the engine on the stand to eliminate any remaining moisture.*

Coolant Pump

Remove the coolant pump. Inspect the impeller to see that it is undamaged. Also, feel the bearing for roughness and check to see that there is no evidence of leakage from the pump's vent hole. Some shops routinely replace the coolant pump during an engine installation.

Oil Pan

Remove the oil pan. Loosen all bolts to the oil pan. Keep the engine upright and disassemble the top end. It is better to remove the pan before turning the engine over on the stand. This step will help prevent any oil accumulated in the bottom of the pan from pouring back into the engine.

Sometimes the pan will loosen by itself. If not, wedge a gasket scraper or a rolling-head prybar between the pan and block and loosen the pan, being careful not to bend or distort it. If necessary, break the seal by tapping a scraper blade all along the gasket.

Valve Covers

Remove the valve cover(s). On V-type engines, label one of the valve covers "left" or "right" before removing it. Sheet metal parts on engines assembled at the factory are usually vulcanized to the head or block with *RTV sealants* (see Chapter 51). Valve covers and oil pans sealed in this manner can be difficult to remove. Here are two ways:

■ Slip a knife blade between the head and sheet metal valve cover to break the seal.
■ Sheet metal parts can be tapped with a rubber mallet to loosen them. Tap on a curved (strong) area to avoid damaging a part.

■ ENGINES WITH PUSHRODS

Stud-Mounted Rockers

Loosen the nuts on their studs before disassembly and cleaning and turn the rocker arms to the side to remove the pushrods. After the heads are cleaned, they can be removed one at a time to inspect the parts for unusual wear. Keep the pushrods in order so they can be reassembled to their original places.

Shaft-Mounted Rockers

Shaft-mounted rockers should be loosened slowly and evenly. If all the bolts but one are loosened, the pressure of multiple valve springs will be exerted on only one rocker tower, and damage can result. Next, remove the pushrods.

Pushrods can be pushed through holes made in a piece of cardboard. Some engines use pushrods of varying lengths. These *must* be kept in order.

SHOP TIP Keep all parts in order:
■ Inspecting for worn parts will be easier when organized. It is very important to find the root cause of any problem and repair it so it will not happen again.
■ Parts become "wear-mated" to each other; they should be returned to their original positions if they are to be reused.

Valve Lifters

Remove valve lifters. Wipe oil off of the bottoms of valve lifters and label them with a felt marker to keep them in order for reassembly. If they are to be reused, lifters must be used on the cam lobe on which they originally ran. Inspect the lifters for unusual signs of wear. The cam and lifters are usually replaced during a major engine overhaul because of wear factors.

The bottoms of lifters often have varnish built up on them in the area that extends out of the lifter bore (**Figure 50.38**). This makes it difficult to remove them. Varnish can be softened with lacquer thinner or spray carburetor cleaner. If the lifters are hard to remove, wait until the camshaft is removed and try to push them out from the bottom or remove them with a special lifter puller. If the puller is not available, or if the lifters will

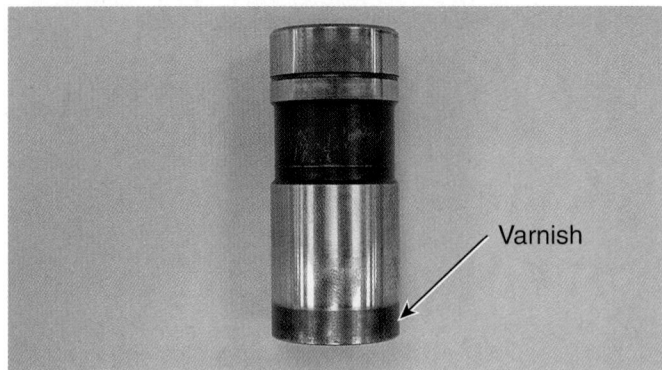

Figure 50.38 Varnish buildup on the bottom of a lifter. *(Courtesy of Tim Gilles)*

not be reused, tap them out from the top with a drift punch after the cam is removed. Be careful not to nick the lifter bores. Excessively dirty lifters that are forced out from the bottom can damage lifter bores.

■ OVERHEAD CAM ENGINES

■ Before the cam belt or chain is removed, position the number 1 piston at TDC and note the location of the timing marks on the cam and crank sprockets (**Figure 50.39**).
■ Locate a sketch in the repair information that shows how the camshaft and crankshaft are timed and compare the sketch to the marks on the engine before disassembling it.
■ Draw a sketch of the cam timing to keep for future reference.
■ Do not remove the OHC camshaft(s) yet. This procedure is described in Chapter 52.

CAUTION When an OHC head is removed, the camshaft will be holding some of the valves open. Be careful not to set the head face down or the open valves can be bent.

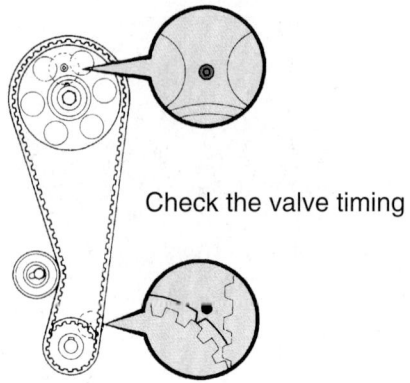

Check the valve timing

Figure 50.39 Align the timing marks for the camshaft and crankshaft sprockets before removing the timing belt or chain.

Vibration Damper

Remove the vibration damper. Some engines have a bolt that holds the vibration damper on the crankshaft. Some vibration dampers will slip off after the bolt is removed. Others are pressed on. A vibration damper puller is required to remove these (**Figure 50.40**). Using the wrong puller can ruin the damper. Grabbing the damper by the outer ring can pull the ring off the damper or pull it off-center (**Figure 50.41**). When using a puller, protect the damper bolt threads in the end of the crankshaft. Sometimes a puller will have a replaceable tip. Select the largest one that fits (see Figure 50.40). Other times, a step plate is used under the puller tip (see Chapter 8).

On some engines, the timing cover seal can be removed before removing the timing cover. Many tim-

ing cover seals are removed and replaced from the inside of the timing cover. These will need to be removed after taking off the timing cover.

SHOP TIP It is easiest to remove the damper using an impact wrench on the puller screw. This will eliminate the need to hold the crankshaft to keep it from turning. Wear safety goggles during this operation. The damper bolt is also easiest removed with an impact wrench.

Timing Cover

Remove the timing cover. Measure timing chain slack as described in Chapter 52.

Cam Drive Assembly

Remove the cam drive assembly.
■ Pushrod engines: Unbolt the cam sprocket and slide or pry it off the cam. Then remove the chain. Reinstall the sprocket and tighten one bolt fingertight. The sprocket will be used to help remove the cam later.
■ Overhead cam: Remove the chain or belt tensioner to remove the cam drive (**Figure 50.42**).

Cylinder Head(s)

Remove the cylinder head(s). Mark one of the cylinder heads (if there are more than one) "left" or "right" and remove them. Remember, left is when viewed from the flywheel end. Most repair informations now give a head bolt *removal* sequence.

Some engines have cast-in pry points. Be careful not to break a casting by using excessive force when prying. *During the entire engine disassembly process, be absolutely certain that all bolts have been removed*

Figure 50.40 This puller has two sizes of replaceable tips. To protect the threads in the crankshaft, use the largest one that will fit. *(Courtesy of Tim Gilles)*

Rubber piece

Figure 50.41 Use of the wrong type of puller ruined this damper. *(Courtesy of Tim Gilles)*

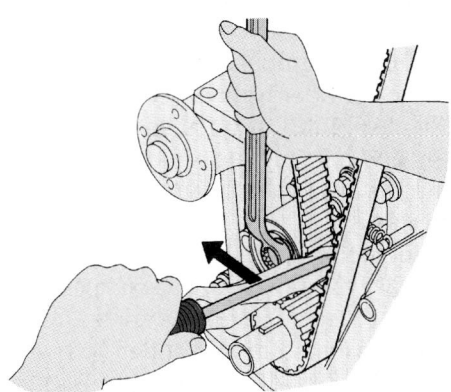

Figure 50.42 This timing belt has a spring-loaded tensioner. Loosen the tensioner bolt, pry it away from the belt, and retighten the bolt.

before using force. If a part cannot be pried loose easily, recheck the repair information.

During removal of the head bolts, look to see if any head bolts are of different lengths or sizes. If they are all of the same lengths and sizes, they do not need to be kept in order. Sometimes, a special bolt is designed with a smaller shank area to allow for the passage of oil to the rocker arm area of the head.

Head Gasket

Inspect the head gasket. After removing the head, look for evidence of coolant or oil leakage. Save the head gasket(s) until the job is completed. It will be handy for diagnostic purposes.

NOTE: *In many areas, laws give the customer the right to inspect all old parts.*

Inspect the head gasket for signs of detonation damage and compression leakage. If the gasket was sealing properly, there will be a well-defined line of thin carbon around the combustion chamber on both the head and block. Carbon deposits on the metal rings of the gasket or a poorly defined combustion chamber seal indicate possible compression leakage.

■ CYLINDER BLOCK DISASSEMBLY

Ring Ridges

Until the 1980s, a pronounced **ring ridge** was a routine occurrence during engine service. Ring ridges are not as common as they used to be, but they are still important.

NOTE: *A good rule of thumb is when a ring ridge is large enough to catch a fingernail moving upward, the engine is a candidate for a rebore and new oversized pistons and rings.*

Modern engines use premium piston rings that will not accommodate a worn cylinder bore.

Figure 50.43 shows a pattern on the cylinder wall where the first and second compression rings wear. The ridge is caused by two things:

■ The pressure of combustion forcing the piston ring against the cylinder wall (especially the top one)
■ A lack of clean lubrication at the top of the cylinder

The pistons can only be removed through the top of the block because main-bearing webs are in the way at the bottom.

Sometimes, the ridge is simply an edge of carbon that can be easily removed with a scraper. Other times, especially on older engines, it can be quite deep. If new rings are installed in a cylinder with a ring ridge, the new square ring will not strike the rounded edge of the ring ridge and possibly break a piston land (**Figure 50.44**).

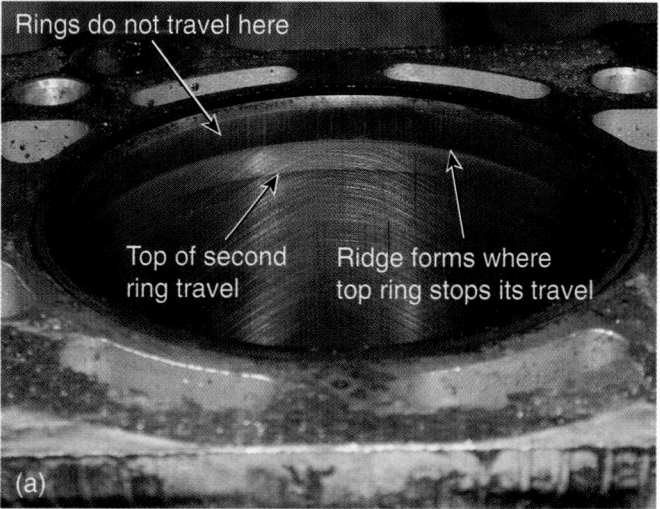

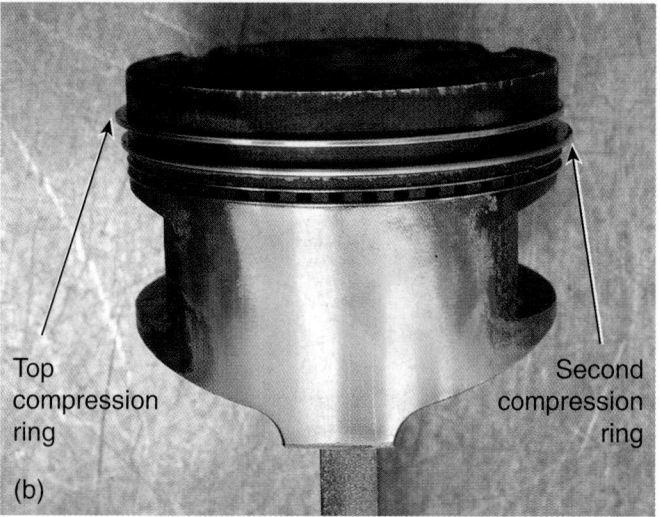

Figure 50.43 The pattern in this cylinder shows wear from the top and second compression rings. *(Courtesy of Tim Gilles)*

SHOP TIP If both the damper and the flywheel have been removed from the engine, turn the crank by adjusting an adjustable-end wrench (crescent wrench) to the size of the crank. When the jaw contacts the woodruff key on the crank, the crank can be turned (**Figure 50.45**). There are also special tools that fit over the end of the crank and engage the woodruff key. Be especially careful not to damage the front of the crankshaft. A damaged snout will make the crankshaft unacceptable as a core return.

Mark Main Bearing and Connecting Rod Caps

Turn the engine over so that the crankshaft is facing up. Mark the main caps and rod caps, if they have not been previously marked. *Main and rod caps are not interchangeable!*

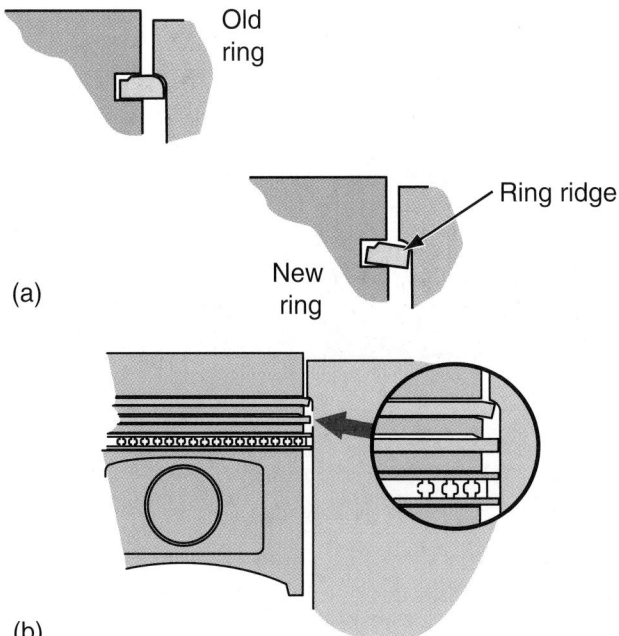

(a)

(b)

Figure 50.44 When new rings are installed in a cylinder with a ridge: (a) A new square ring will strike the old rounded ridge. (b) The ring land above the second ring can be forced down.

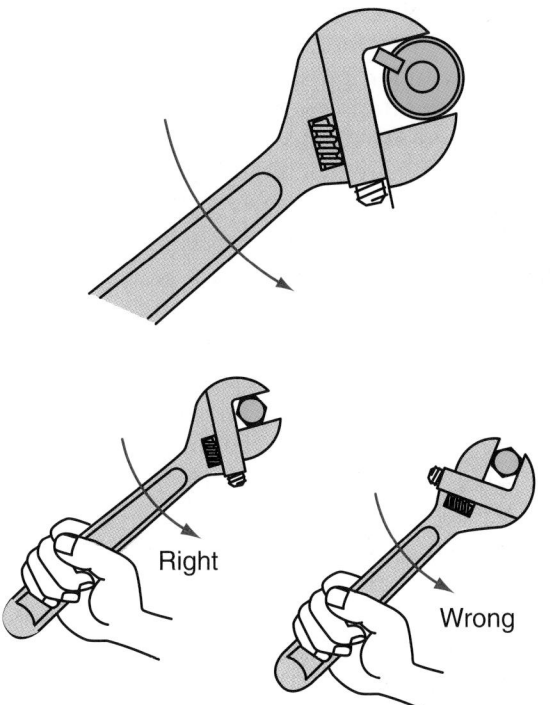

Figure 50.45 An adjustable wrench grasps the woodruff key to turn the crank. Is the wrench in this example being used properly?

Main Bearing Caps

Main caps must be installed in one direction only. Some main caps are labeled with cast numbers. Numbering begins at the damper end of block (#1) and progresses

to the flywheel end. Some main caps have an arrow, labeling the direction the cap should face. Unless you write down whether it faces front or rear, you will need to locate that information when you reassemble the engine.

SHOP TIP Remember, "time is money." Searching the service literature for unnecessary information like the direction that main caps face is a waste of time. Mark main and rod caps the same way when you disassemble any engine. Then you will not need to locate difficult to find service information when reassembling your engine.

Figure 50.46 shows a fast, easy way to mark main caps using a center punch. Mark on both the main cap and the block so you know which direction the cap goes when you put it back.

Connecting Rods and Caps

Connecting rods and caps are mated to one another and must be marked for identification. **Figure 50.47** shows the correct place to stamp the numbers when marking connecting rods. Marking the rod and cap on the side in this manner does not distort the connecting rod and makes it easy for you to reinstall the correct cap in the right direction on the rod. Rod caps should always be marked while they are still installed on the crankshaft.

NOTE: *An average engine has only 0.002" bearing clearance. This amounts to 0.001" on each side of the crankshaft. For reference, the thickness of an average human hair is about 0.0025". If a main or rod cap goes on backward, it could easily be off by 0.001" and have*

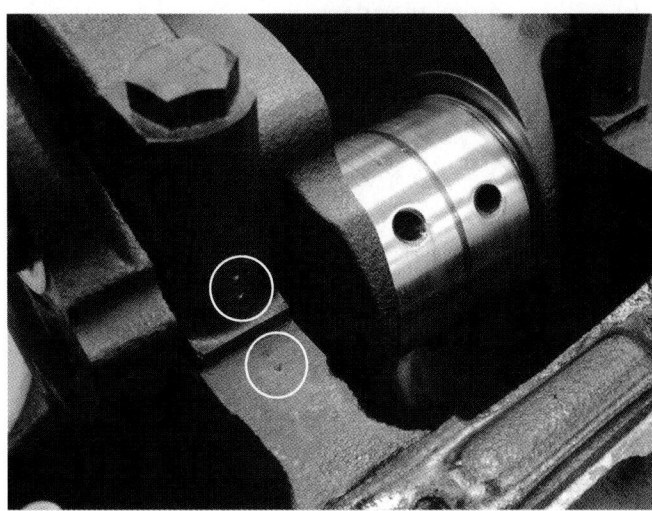

Figure 50.46 A main bearing cap and block marked with a center punch. *(Courtesy of Tim Gilles)*

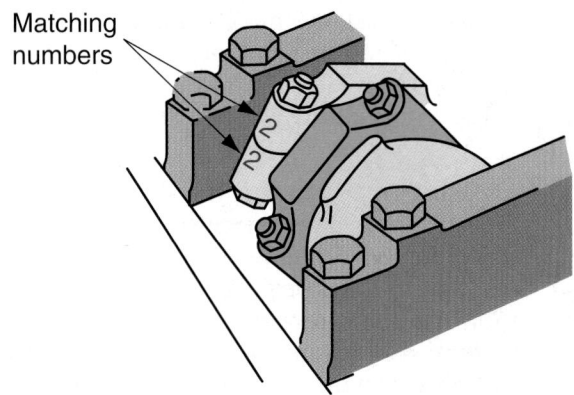

Figure 50.47 Mark rod caps before disassembly.

no oil clearance. Connecting rods and main caps are malleable. This means they can distort when hammering on them with a number stamp or center punch. There is a slight bit of extra clearance at the bearing parting line. That is the reason for marking rods and mains in the positions shown in the previous illustrations. If you mark a main or rod cap at a point 90 degrees to the bolt, you can distort the bearing housing enough to eliminate bearing clearance.

Be certain to inspect connecting rods previously marked (at the factory or during a previous engine rebuild) to see that they are labeled correctly.

NOTES:

■ *Powdered metal connecting rods should not be marked using number stamps.*
■ *Do not file notches on the rod beam. This can cause stress raisers, which weaken the rod.*

When marking rods on a V-type engine, mark them according to the cylinder's number. The only side of the rod that is easily accessible is the side that faces the outside of the engine (away from the cam on a push-rod engine) (**Figure 50.48**). Figure 37.9 shows many

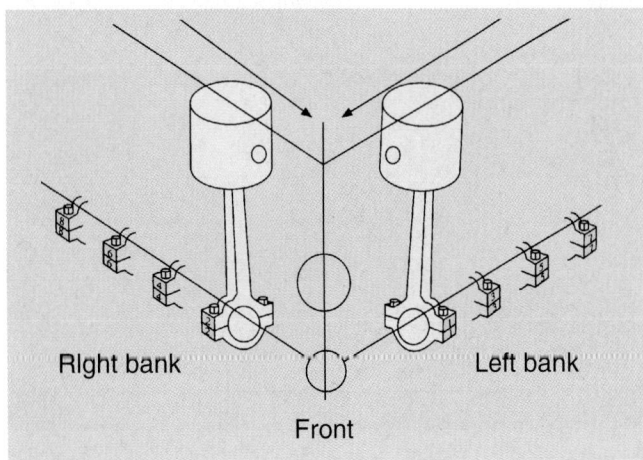

Figure 50.48 Mark the connecting rods on the correct side.

Figure 50.49 Offset connecting rods from a V-type engine. *(Courtesy of Tim Gilles)*

of the firing orders used by various manufacturers. On V-type engines, the number one cylinder will almost always be the one that is the farthest forward on the block. Because cylinders on V-type engines share crankpins with cylinders on the opposite side of the block, the cylinders must be staggered.

Connecting rods must not be reinstalled backwards. **Figure 50.49** shows two V8 connecting rods. Notice that they are machined more on one side than the other.

Piston and Rod Assembly

Remove and inspect the piston and rod assembly.

■ Move each piston to BDC so the rods will clear the crank during piston removal.
■ Loosen each rod nut. The amount of torque required to loosen the bolts can be determined using a dial indicator torque wrench. This will tell you whether nuts were properly torqued before.

NOTE: *The click-type torque wrench should not be used for loosening.*

■ Use a brass hammer to lightly tap on the ends of the rod bolts to loosen the rod caps for easy removal (**Figure 50.50**).
■ The crank is soft and nicks easily. To protect the crank, install **rod bolt protectors** (available from your parts source) or pieces of fuel hose 3" to 4" long on the rod bolts before removing the rods (**Figure 50.51**). If you use pieces of hose, do not cut them too short. Longer ones can be removed easily, even though they become slippery when coated with oil.
■ Use the handle of a soft mallet or use a piece of hickory like a sawed-off baseball bat or shovel

Figure 50.50 Use a brass hammer to tap lightly on each rod bolt. This will loosen the cap from the enlarged area of the bolt that aligns the cap precisely to the rod. *(Courtesy of Tim Gilles)*

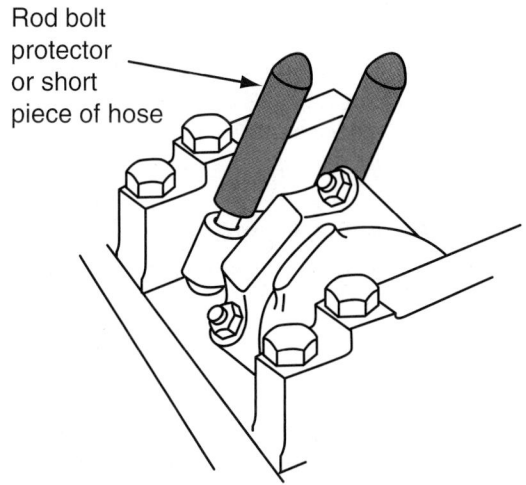

Rod bolt protector or short piece of hose

Figure 50.51 Install hoses or rod bolt protectors on the rod bolts before removing the piston and rod assembly.

handle to push the piston and rod from the bore. The connecting rod is relatively soft and can easily be damaged by contact with hard materials such as a screwdriver or prybar. Immediately reinstall the rod cap on the rod. Remember, connecting rod caps are not interchangeable.

NOTES:

- *Keep the old bearing in place in the rod for future diagnosis.*
- *Be very careful not to drop a piston. When a piston is lying on a bench with the connecting rod attached it is unbalanced and can very easily be knocked off a bench. In fact, this is a common occurrence with apprentice technicians. Lay piston and rod assemblies on their sides to prevent a costly mishap.*

Piston, Rings, Rod, and Bearings

Inspect the piston, rings, rod, and bearings for the following:

- Detonation can affect the rod bearing. If the rod bearing falls out as the piston and rod are removed from the cylinder, this means the bearing has lost its crush. Check for detonation damage, especially on the upper rod bearing.
- Inspect the pistons for obvious wear and breakage.
- This is also a good time to visually inspect each cylinder for corresponding wear.
- Inspect the oil rings to see if they are plugged.
- If compression rings were working properly, the piston land between the top and second rings should be relatively clean. Chapter 54 describes a procedure for checking piston ring wear.

Crankshaft

Remove the crankshaft and inspect for wear.

- Remove the main cap bolts. Use a dial indicator torque wrench to see how tight they were.
- Remove the main caps. They fit tightly in the block (**Figure 50.52**).

NOTE: *Main caps are not interchangeable, and they must be returned to their original positions.*

- Normally, bearings are replaced. If bearings are reused, they *must* be returned to their original positions, as they have been wear-mated to the crank journals.
- Carefully lift out the crankshaft.
- Leave the flywheel or flexplate bolted to the crank to help hold it upright during storage to prevent damage (**Figure 50.53**).
- Check to see if there is a pilot bearing or bushing in the rear of a standard transmission crankshaft. It can be removed with a puller (see Chapter 70).

Pry

Figure 50.52 Pry the main caps loose from the block. *(Courtesy of Tim Gilles)*

Figure 50.53 Store a crankshaft upright to prevent damage. Leave the flywheel installed. *(Courtesy of Tim Gilles)*

- Check the condition of the crankshaft surface where the rear main seal rides.
- Inspect the bearing surfaces of the crankshaft for wear. Measure the main and connecting rod journals with a micrometer and compare them to manufacturer's specifications (see Chapter 53).
- Inspect the thrust bearing surfaces. These surfaces control fore and aft movement of the crankshaft. Wear, which is unusual, is usually greatest on the rear side.
- Inspect the front upper bearing to see if there is any sign of wear resulting from belts that were too tight.
- Label the backs of the main bearings with a felt marker with their positions. The number 1 *upper* main bearing is marked 1U, the number 1 *lower* is marked 1L, and so on. As the main bearings are removed from the block lay them in a row (**Figure 50.54**). If the crankshaft is bent or the crankcase is out of alignment, you will be able to see the wear pattern on the bearings. Bearing wear problems are discussed in Chapter 53.

Normally, bearings are replaced. If bearings are reused, they *must* be returned to their original positions, because they have been wear-mated to the crankshaft bearing journals.

Camshaft

Remove the camshaft.
- Pushrod engines: Some engines use a bolt-on cam thrust plate. Gear-drive cam bolts can be accessed through holes in the cam gear to remove the cam.

Figure 50.54 As main bearings are removed, lay them in a row. If the crankshaft or crankcase is misaligned, you will be able to see the resulting wear pattern on the bearings. *(Courtesy of Tim Gilles)*

An impact screwdriver is sometimes necessary to remove Phillips head screws (see Chapter 8).

Varnish often builds up on the edges of the cam journals, making it difficult to remove the cam (**Figure 50.55**). Squirt some penetrating oil on the varnish. Reinstall the cam sprocket and use it as a "handle" to remove the cam. It is easier to remove the camshaft without damaging the lobes if the engine is stood on end.

NOTE: *If the engine is mounted on an engine stand and the cam will not come out, try supporting the cam gear end of the block. Sometimes the block will sag just enough to bind the cam. This is especially true on in-line six-cylinder engines.*

- Overhead cam engines: The camshaft is usually removed with the cylinder head. Follow manufacturer's instructions for removal of the cam from the head. This varies with the design of the head.

Inspect the cam lobes and journals for visible wear. When several cam lobes are excessively worn, look for a fuel leak or an internal coolant leak causing a loss of critical lubrication.

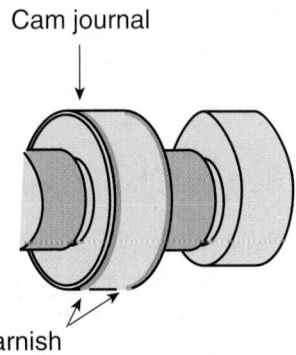

Figure 50.55 Varnish buildup on the edge of the cam journal makes camshaft removal difficult.

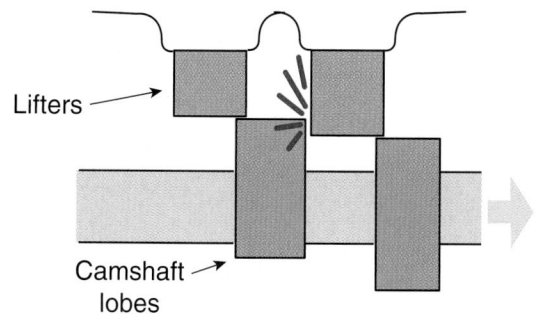

Figure 50.56 The lifters must be moved high enough in their bores to allow the cam to be removed.

If the lifters on a pushrod engine are too worn to be reused, the following procedure will save time in removing the cam:

- With the engine in the upside-down position, move the lifters to the highest positions in their bores by turning the camshaft one complete revolution.
- Use a piece of wooden dowel to finish pushing the lifters into their bores so that they will clear the cam (**Figure 50.56**).
- Now the cam can be carefully removed. This is a *delicate* operation. Be careful not to chip any cam lobes.

Cam Bearings

Remove and label cam bearings. Cam bearings are inexpensive and are usually replaced.

- Remove the cam bearings with a cam bearing tool (see Chapter 53).
- The rear cam plug can be knocked out with the cam bearing tool.
- Label the bearings with a felt marker, or put them in labeled baggies. They may need to be referred to when determining block positions of the new bearings.

Core Plugs

Remove core plugs. Remove core plugs before hot-tanking the block (see Figure 21.32). Use a large punch to knock them sideways. They can then be removed easily with a roll-head punch or a large pair of pliers.

NOTE: *Be careful not to pound them into the side of a cylinder wall (see Case History in Chapter 21).*

Cleaning the Block

Engine parts are cleaned using procedures described in Chapter 10.

Following cleaning, oil galleries *must* be cleaned with a brush after the plugs sealing the ends of the galleries are removed. Over many miles of driving under various maintenance conditions, grime builds up in the oil galleries. This material is loosened during the cleaning process. If it is not physically removed, it will end up ruining new engine bearings and possibly more. Chapter 53 covers the procedure for removing oil gallery core plugs from the block before using a long rifle brush to clean the galleries.

Following cleaning, replace the main caps and torque then bolts to help keep the block properly stressed during any machining.

Crank Sprocket Removal

Remove the crank sprocket. The crank gear, or sprocket, is positioned by a woodruff key or bar key on the front of the crankshaft. Some crank sprockets or gears are pressed-fit. If the crank sprocket is to be replaced, it can be removed as shown in Chapter 55. The woodruff key can be removed from a crank or cam by striking it on its rear end with a brass punch. Then strike the underside of the front to roll it out of its channel. Be certain to install a woodruff key flat in its groove or the crank sprocket may be damaged.

Finish Diagnosis and Repair

Finish diagnosis and repair of the engine assembly. The procedures for cleaning, measuring, and repairing parts are found in subsequent chapters.

NOTE: *Use air pressure and a blowgun to verify that a transmission oil cooler is not restricted. A restricted cooler can cause damage to the engine's crankshaft thrust bearing. Failure to flush the transmission oil cooler and its supply tubing will void an engine remanufacturer's warranty.*

▪▪ REVIEW QUESTIONS

1. What could cause a transmission bushing, wheel bearing, or emergency brake cable to burn or fail?
2. What is the name of the refrigerant that is responsible for depletion of the Earth's ozone layer?
3. Name two belt-driven accessories that can be wired out of the way instead of disconnecting them.
4. What kind of transmission is usually best left in the vehicle during engine removal?
5. Are both engine mounts always interchangeable?
6. When moving an engine that is suspended on an engine hoist, what position should it be in?
7. How long should bolts be that hold the engine to the engine stand?

8. Which parts *must* be kept in order if they are to be reused?

9. Which tool can be used to assist when pulling a damper so that the crankshaft will not need to be held from turning?

10. What is the name of the ledge that forms at the top of a cylinder as the engine wears?

■ ASE-STYLE REVIEW QUESTIONS

1. Which of the following is/are true about refrigerants?
   a. They can instantly freeze anything with which they come into contact.
   b. It is illegal to allow R12 refrigerant to escape into the air.
   c. Both A and B.
   d. Neither A nor B.

2. Technician A says to mark powdered connecting rods using a number stamp. Technician B says to use a click type torque wrench to check torque when loosening rod nuts. Who is right?
   a. Technician A   c. Both A and B
   b. Technician B   d. Neither A nor B

3. Technician A says that a plugged transmission oil cooler can cause damage to a crankshaft thrust bearing. Technician B says that wear on a thrust bearing is usually on the rear side. Who is right?
   a. Technician A   c. Both A and B
   b. Technician B   d. Neither A nor B

4. Which of the following is/are true about impact wrenches?
   a. Using an impact wrench sometimes makes the removal of rusted bolts easier.
   b. When using an impact wrench to remove a crankshaft damper, the crankshaft must be restrained from turning.
   c. Both A and B.
   d. Neither A nor B.

5. Which of the following can result from a broken ground strap?
   a. Transmission bushing failure
   b. Wheel bearing failure
   c. Both A and B
   d. Neither A nor B

Engine Sealing, Gaskets, Fastener Torque

■ KEY TERMS

torque-to-yield
torque turn
preload
non-retorque gasket

surface texture
microinch
rubberized cork
anaerobic chemical

aerobic chemical
Teflon® tape
dynamic seal

■ INTRODUCTION

This chapter deals with the very important area of sealing lubricants and coolants in the engine. Leaks are not only unprofessional, but they run the risk of serious mechanical part failures. Cleanliness is very important when assembling parts. If a leak is to be avoided, there must be *clamping force* on two parts that are held together. Clamping force results from the tension or stretch on bolts securing the two pieces together (**Figure 51.1**). It also results from the springiness of a gasket between parts. Knowledge of the techniques in this chapter will help make you a better technician and avoid costly errors.

■ TORQUE AND FRICTION

Torque is the measurement of the twisting effort required to tighten a fastener. The amount of clamping force actually applied by a fastener changes with the use of lubricants. Heavy lubrication of threads can result in overstretched bolts.

■ Approximately 90% of the torque applied in tightening a fastener is used to overcome friction. About half of that 90% is lost to friction between the head of the bolt and the work surface; the other half is lost between the threads.

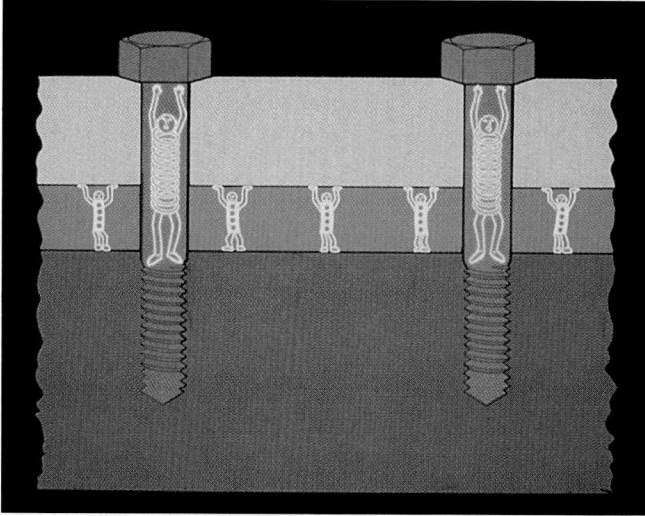

Figure 51.1 Clamping force results from bolt stretch. *(Courtesy of Federal-Mogul Corporation)*

■ The final 10% provides the clamping force necessary to hold the assembly tight.

■ Fastener threads must be sealed and lightly oiled.

Using the wrong torque can distort an engine's cylinder block. Follow the original equipment (OE) manufacturer's torque and fastener recommendations.

In some cases, the manufacturer calls for the use of a sealer or thread lock compound prior to installation and torquing.

■ TORQUE WRENCHES

The amount of torque applied to a fastener is measured with a torque wrench (**Figure 51.2**). There are three styles of torque wrenches:

■ The inexpensive *beam wrench* is very common (Figure 51.2).

■ The *"click" (spring) wrench* is available with a ratchet (**Figure 51.3**). It is the most versatile, because it can be used in hard-to-reach places in the engine compartment. The "click" can be felt, even if the gauge cannot be seen.

NOTE: *Reset this torque wrench to zero after each use; this will unload the spring and maintain its accuracy.*

■ The *dial torque wrench* is the most accurate (**Figure 51.4**). Some are available with a light or a buzzer for use when the dial face cannot be seen.

Torque wrenches used in automobile repair commonly come with ½" or ⅜" drives.

To get an accurate reading with a torque wrench:

■ Hold it at 90 degrees to the fastener being tightened.

■ Torque in three or four steps.

■ Tighten first to about one-third of the torque specification.

■ Then tighten it two-thirds.

■ The third step should be to within 10 foot-pounds of the final specification. When the last step is within 10 foot-pounds, the final torque will be more accurate.

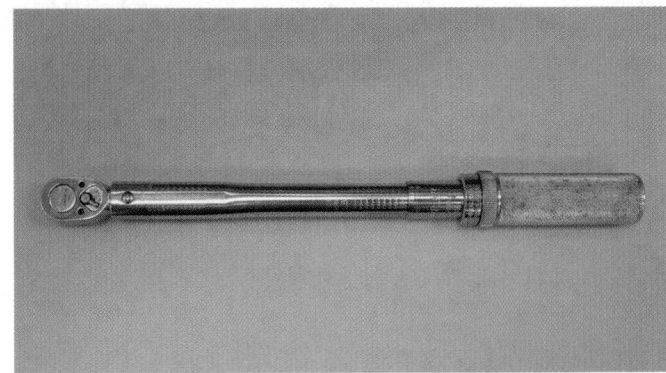

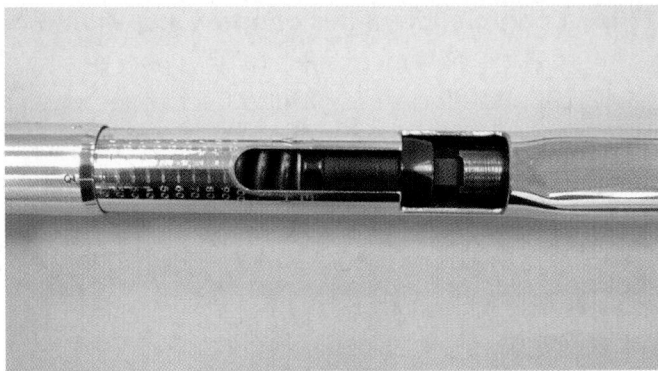

Figure 51.3 A "click" torque wrench. *(Courtesy of Tim Gilles)*

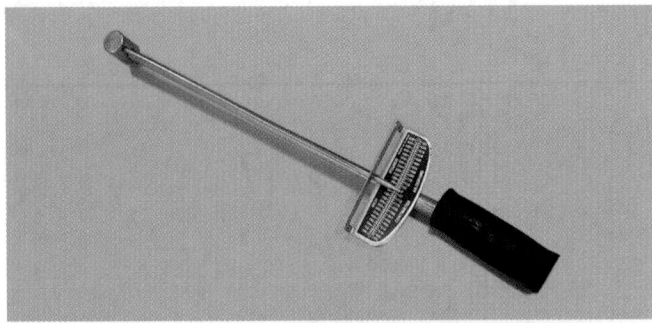

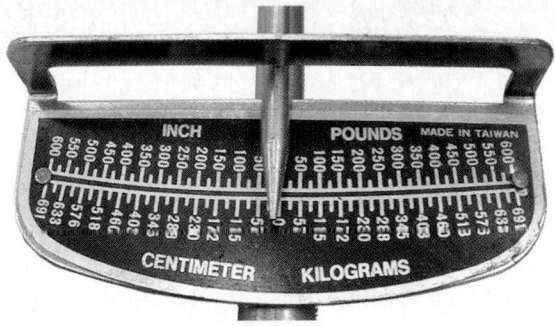

Figure 51.2 A beam scale torque wrench. *(Courtesy of Tim Gilles)*

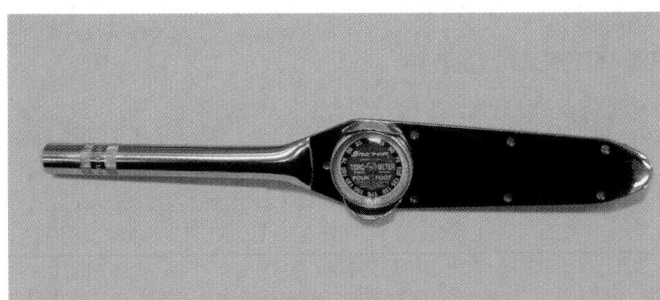

Figure 51.4 A dial indicator torque wrench. Hold the torque wrench at 90 degrees to the fastener being tightened. *(Courtesy of Tim Gilles)*

■ The final step is to double-check all torque readings. During the recheck, if a bolt still moves it should be removed and examined for water or oil in the hole. Liquid can cause initial resistance to torquing and eventually "bleed" up the threads, causing the bolt to loosen.

A bolt can be overstretched because it is too poor a grade of fastener for the torque being applied or because a thread lubricant is used without a corresponding reduction in torque. Some technicians depend on "feel" when torquing. An experiment that tests a special tension gauge against a group of experienced technicians will show how inaccurate this method of torquing really is.

Torque is measured in foot-pounds or inch-pounds. The metric measurement of torque is the Newton-meter.

■ 1 inch-pound equals ¹⁄₁₂ of 1 foot-pound.
■ A foot-pound torque wrench is not accurate below 15 foot-pounds; use an inch-pound torque wrench and multiply the torque figure by 12.

■ TORQUE-TO-YIELD

A common misconception is that when a fastener is torqued to its yield point (see Chapter 7), it will no longer exert the necessary clamping force on the fastened joint. **Figure 51.5** shows the relationship between clamping force, bolt stretch, and yield point. Normal head bolt torque values have a calculated safety factor of about 25% less than the maximum amount a bolt can be torqued without becoming permanently stretched. So clamping force is only 75%

of its potential for the bolt size. Further study of the chart in Figure 51.5 will show that a small amount of installation error in one or more of these head bolts can result in a wide variation in clamping force.

Because torque is a function of friction, head bolts torqued to the same specification can produce variables in clamping force of up to ± 200%. Some of the bolts might be called upon to do more than their share of the clamping, rather than sharing the load equally. The result can include distortion of engine cylinders. Some manufacturers have even used different amounts of torque on head bolts of the same engine in order to stress the block in a desired manner.

In some late-model engines, head bolts are purposely torqued to within 2% of their yield point. This method, known as **torque-to-yield**, provides a more consistent clamping force and the most fatigue resistant joint possible. Notice in Figure 51.5 that the bolt may be elongated considerably at its yield point before it reaches it failure point. Notice also that the clamp load through the entire yield section on the chart is more constant, especially when compared to the large changes that occur in clamping load when the bolts are torqued to different values that are within the elastic range of the fastener. Torque-to-yield bolts work well with aluminum heads, which experience greater expansion with heat.

NOTE: *Bolts torqued into yield have been stretched beyond their elastic limit and must be replaced whenever the head is removed. Gasket sets often come with a new set of head bolts when their replacement is called for.*

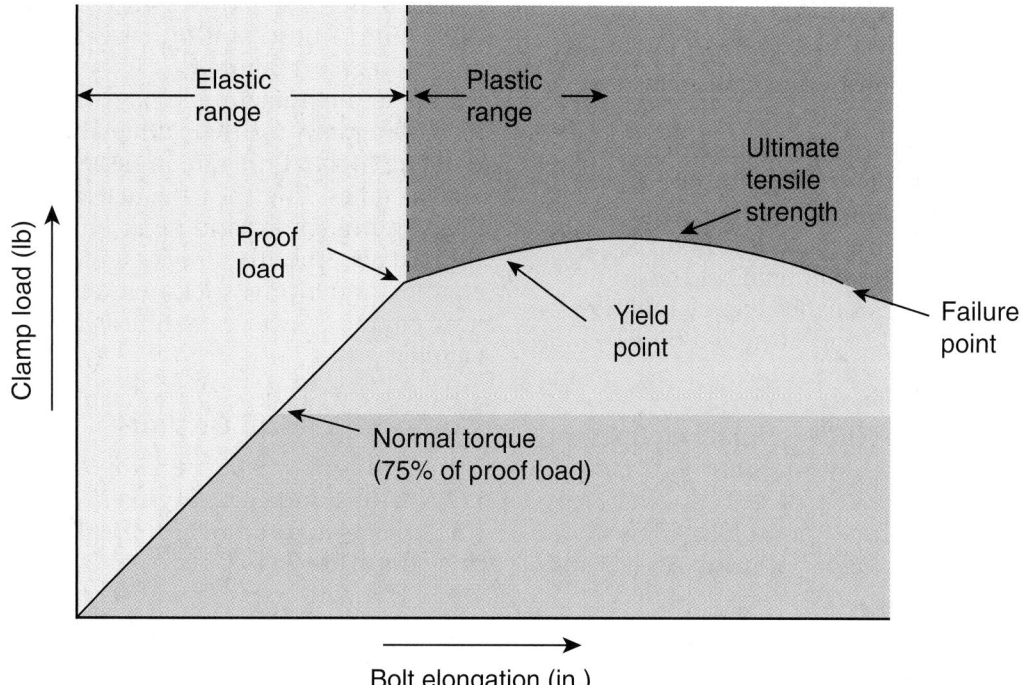

Figure 51.5 Relationship between proper clamp load and bolt failure.

Figure 51.6 A torque angle gauge. *(Courtesy of Tim Gilles)*

TORQUE BY DEGREES

On the assembly line, equipment tightens bolts until they are stretched a prescribed amount. **Torque turn**, or *torque angle*, is one method used by manufacturers to be able to ensure equal tightness of fasteners. To duplicate this method of tightening when a torque turn specification is available:

- First, tighten the bolt to the specified torque.
- Then, turn it an additional 35 to 180 degrees, depending on the specification.

Tightening in this manner lessens the possibility that variations in friction will have an effect on torque readings. A *torque angle gauge* that helps keep track of how far a fastener is turned is available from tool manufacturers (**Figure 51.6**).

FASTENER CLAMPING LOAD

The clamping load on a fastener is known as **preload**. There are three ways of measuring the preload on a fastener:

- The most common is with a torque wrench.
- A torque angle gauge is another method.
- The most accurate method is to measure the actual amount of bolt stretch. This is not practical, however, unless both ends of the bolt are accessible. High-performance engine builders measure bolt stretch on connecting rod bolts.

GASKETS AND SEALS

Gaskets are installed between two surfaces to prevent leakage. Some gaskets seal low-pressure fluids; head gaskets seal the high pressures of combustion. Seals are used to seal in lubricants around a rotating shaft. **Figure 51.7** shows the locations of many gaskets and seals used in an engine.

Gaskets Sets

Gaskets can be purchased individually or in sets. A *full set* (*FS*) contains everything needed for an engine

rebuild. Often, extra unnecessary gaskets are included in a full gasket set. This helps keep to a minimum the number of part numbers stocked by parts suppliers. Although there are more gaskets in the set, the cost is still lower for the manufacturer.

NOTE: *Older cars had carburetors rather than fuel injection. Carburetor gaskets are usually not included in a full gasket set. A 1955–1980 small block General Motors gasket set would have to include 28 different carburetor gaskets to meet all possible needs.*

SHOP TIP
- Be sure to save the old carburetor gasket for comparison.
- Be sure to read the contents on the cover of the box. Sometimes a full set does not include valve guide seals or a crankshaft rear main seal.

Other gasket sets available include the head set, oil pan set, and timing cover set.

Cylinder Head Gaskets

Modern head gaskets have a more difficult job of sealing than those on older engines. Today's cylinder blocks are lighter and more flexible than older blocks. Many engines have an aluminum head and an iron block, or vice versa. These are called *bimetal engines*. A head gasket has an especially difficult job sealing these engines. Aluminum expands two to three times as much as cast iron when hot (**Figure 51.8**). If the engine overheats, extreme stress in an aluminum head can result in head bolt overstretching, head warping, and head gasket failure.

Some manufacturers have had very expensive recall programs for their bimetal engines, after experiencing high failure rate in head gaskets in as little as 50,000 miles. The shearing action resulting from the head and block expanding at different rates destroys the gaskets. Smooth surfaces are the norm for bimetal engines, and biting, sticky gaskets are used on all cast iron engines. Surface texture is covered later in this chapter.

Composite Head Gaskets

A modern **non-retorque** composite **gasket** has a facing, coating, and core (**Figure 51.9**). The core allows the gasket to accommodate imperfections more readily. It is constructed of either solid or perforated clinched steel. The core is faced with either expanded graphite, or Kevlar® fiber bound with nitrile rubber. These materials are dense, so they do not compress much, but they can compress enough to conform to minor irregularities of a sealing surface.

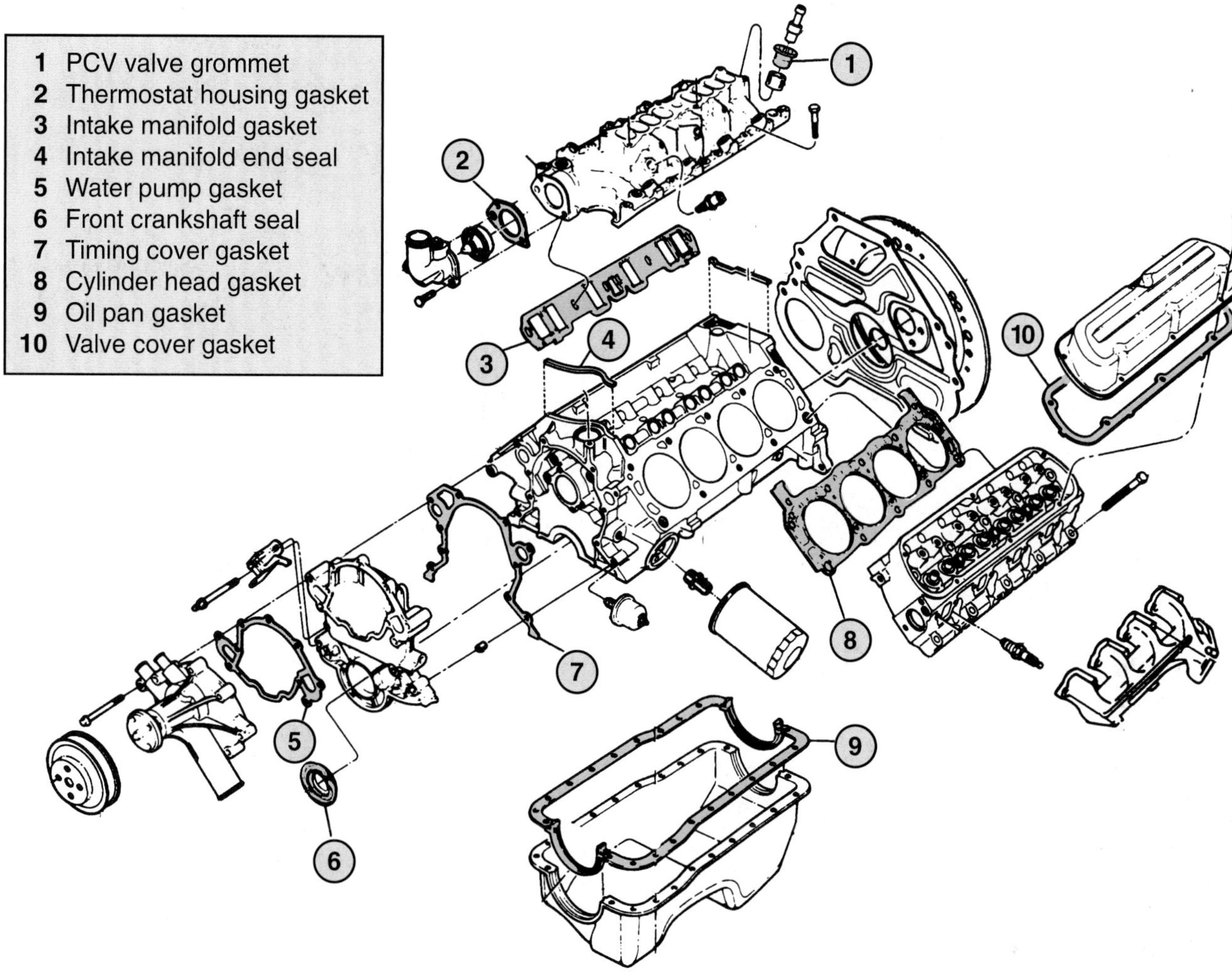

1	PCV valve grommet
2	Thermostat housing gasket
3	Intake manifold gasket
4	Intake manifold end seal
5	Water pump gasket
6	Front crankshaft seal
7	Timing cover gasket
8	Cylinder head gasket
9	Oil pan gasket
10	Valve cover gasket

Figure 51.7 Typical engine gasket and seal locations.

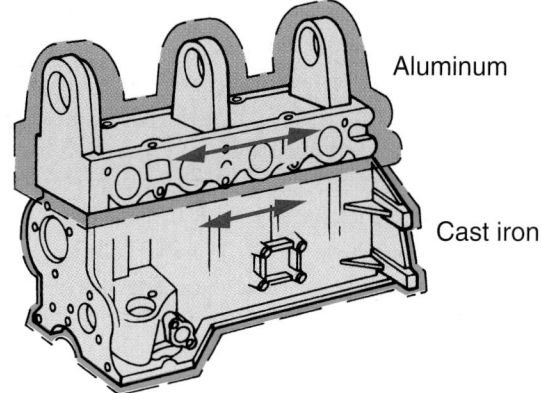

Figure 51.8 Aluminum expands two to three times as much as cast iron when hot.

Graphite Head Gaskets

In the aftermarket, graphite-faced gaskets are considered to be a premium gasket. Graphite is more resistant to heat (2,100°F) than Kevlar (400°F). Graphite also transfers heat from the engine to the coolant more quickly. This is why graphite is the gasket of choice when and engine is prone to detonation.

Because graphite is soft, it does a good job of cold sealing, too. Composite graphite gaskets require a very smooth surface finish. Graphite gaskets can also be coated to lengthen their service life. One of graphite's limitations is its cost. It is more expensive than other non-asbestos gasket materials. Care should be taken when handling these gaskets, as graphite is easily damaged.

Teflon and Silicone-Coated Gaskets

A Teflon, moly, or silicone-based antifriction coating helps seal minor surface imperfections. These gasket coatings do not stick to parts, so they allow much more movement between the head and gasket on bimetal engines (**Figure 51.10**). They also have the added advantage of easy disassembly.

Some premium Teflon gaskets used on all-cast-iron engines have silk-screened silicone beads that help provide a good cold seal around cooling passages (**Figure 51.11**). This is a good feature on engines with cast iron blocks and heads. However, the silicone holds tight to the head and block. Having the head stuck to the block would not be good for bimetal engines.

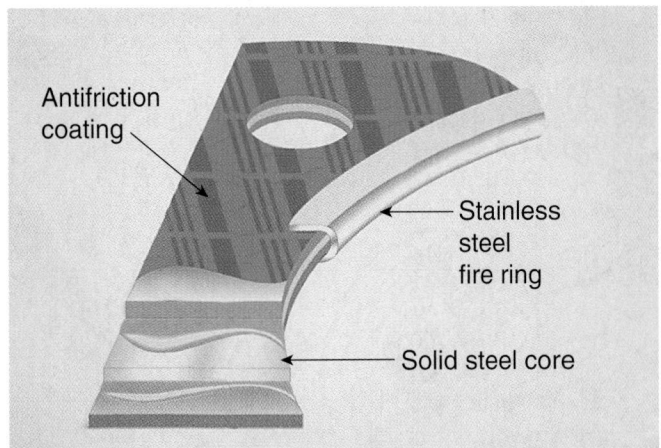

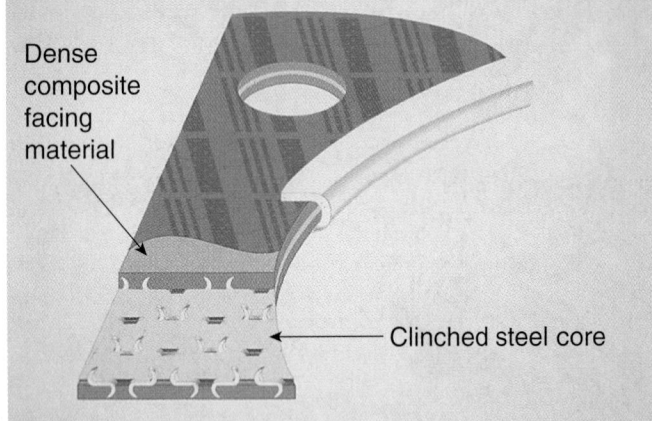

Figure 51.9 Two modern composite gasket designs with a facing, core, and coating. (*Courtesy of Federal-Mogul Corporation*)

NOTES:

■ *Teflon-coated gaskets should be used without gasket sealers.*

■ *Mechanics used to use lacquer as a sealer. Lacquer can eat silicone sealing material, although it will not harm Teflon.*

Metallic Head Gasket

Metallic gaskets are single layer and multilayer. In the 1970s and earlier, an inexpensive embossed steel shim head gasket, called a beaded steel gasket, was used extensively as original equipment on engines with

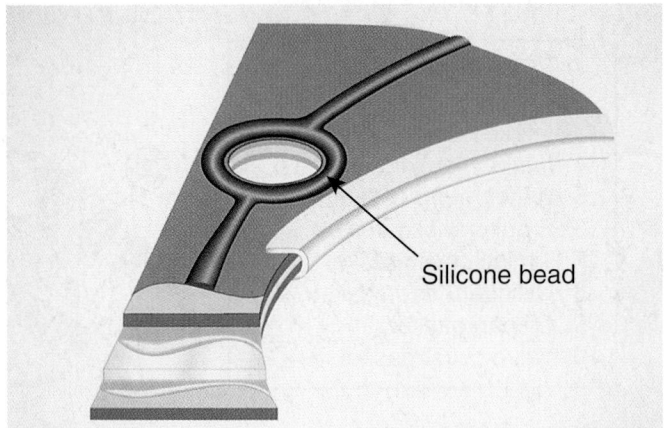

Figure 51.11 A non-retorque gasket with a silicone bead. (*Courtesy of Federal-Mogul Corporation*)

thick iron heads and blocks. Today, most manufacturers do not use these gaskets, and their use is not recommended in the aftermarket. Beaded steel gaskets were used on new engines because mating surfaces between new parts is nearly perfect. This type of gasket will not seal scratches in the head or block, and it will not recover its shape after an engine has been overheated.

The embossed area of a steel shim gasket is created with 100 tons of pressure in a very large (400-ton capacity) press. The embossed area works like a lock washer to help the gasket maintain its seal. A beaded steel gasket was the original equipment gasket on the 350 Chevrolet. Feel the cylinder head surface on one of these heads after it is cleaned. It is sometimes indented (brinelled) from the gasket embossing, which is installed with the raised side toward the head so it will not indent the block. If the surface of the head is indented, the head should be resurfaced.

Replacement gaskets are about twice as thick as the steel shim. Their extra thickness compensates for the amount they will compress and also makes up for metal removed during routine head surfacing.

Beaded steel gaskets and other older steel-faced gaskets are the only gaskets that require sealers.

Multilayer Steel Gaskets. Multilayer steel (MLS) gaskets have been used in other engine locations in the past. Recently, they have become more common

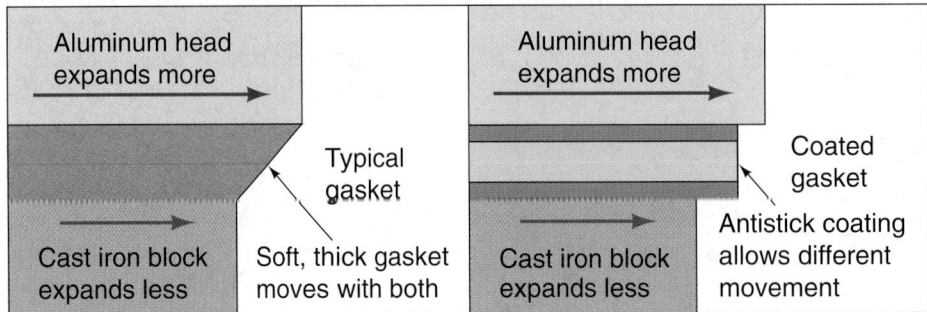

Figure 51.10 Anti-stick gasket coatings allow movement between the head and block on bi-metal engines. (*Courtesy of Federal-Mogul Corporation*)

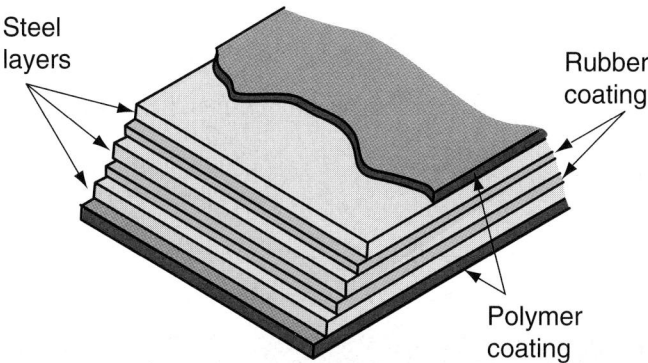

Figure 51.12 A rubber-coated multilayer steel (MLS) head gasket.

as cylinder head gaskets with some manufacturers. Having multiple layers causes the gasket to spring apart, providing a further increase in clamping force. This "springiness" is called recovery. It allows the use of lower torque on fasteners, which results in less cylinder wall distortion.

MLS gaskets have from three to seven layers (**Figure 51.12**). A silicone coating often separates the layers. The outer layers of an MLS gasket are stainless steel, embossed at 200 tons in an 800-ton capacity press. Workers say that the whole building shakes when the press is working!

Most multilayer gaskets are coated with a very thin layer of Viton® or nitrile rubber so they will seal when cold. Multilayered gaskets require more careful surface preparation. The head must be flatter and smoother. Not all shops have equipment capable of producing the fine finish required.

Importance of Correct Surface Texture. Metal surfaces in the engine are designed to have various **surface textures**. If a gasket is used with too smooth or rough a surface, a leak could result. In **Figure 51.13**, the surface left by the machine used to grind or mill the head is quite rough. In fact, it is rougher than what

Figure 51.13 This surface finish on an older cylinder head is rougher than would be used for a modern head gasket. *(Courtesy of Tim Gilles)*

would be used for modern gaskets. For conventional head gaskets used with cast iron heads, a rough surface helps the gasket take a "bite" into the material being sealed. With these surfaces, it should be possible to feel the texture with your fingernail. The surface would have at least the same roughness as a cylinder bore after honing. A fingernail is relatively wide. A profilometer (**Figure 51.14a**) is necessary for reading finer surface textures. It reads with the end of a very small ball, a diamond stylus tip—like a record player needle—with a 0.0004" radius. A visual comparison of surface roughness can be made with a surface roughness gauge (**Figure 51.14b**).

Gasket bite is not recommended on aluminum heads, which require a finer surface finish. An aluminum head must be allowed to expand and contract against an iron block. If the surface is not smooth, it will wear the gasket out and a leak will result.

Measuring Surface Texture. Measurements of surface texture are rated on the American National Standard Institute (ANSI) scale. There are several types of measurements, but roughness measured in **microinches** is the most important.

NOTE: *A microinch is one millionth of an inch.*

Roughness above 63 microinches can be checked visually. Visual scales are available from gasket manufacturers.

Roughness average (RA) is the most widespread surface texture measurement because it is simple and

(a)

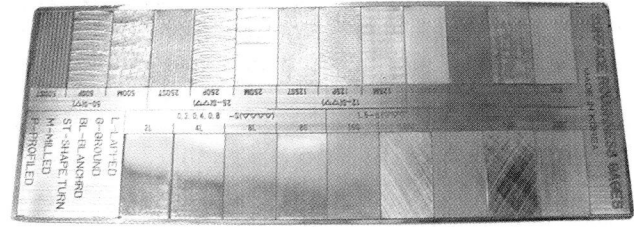

(b)

Figure 51.14 (a) A profilometer measures surface texture. (b) A surface roughness gauge or surface comparitor. *(a, Courtesy of Sunnen Products Company, St. Louis, Missouri; b, Courtesy of Tim Gilles)*

uses inexpensive measuring devices. It is not the best, but it is adequate for automotive reground or polished surfaces. To determine an RMS reading from a meter reading in RA, add 11% to the meter reading.

The following are some manufacturer's recommendations used as guidelines by machinists:

- For cast iron (non-multilayered steel) 80–100 RA (90–110 RMS) is recommended
- For aluminum heads a finer finish is recommended (45 RA). Head flatness is also more critical with aluminum heads
- Viton rubber-coated multilayered steel (MLS) head gaskets: 10–30 RA
- Graphite-coated composite gaskets: less than 60 RA
- Kevlar composite gaskets: 60–125 RA
- Intake or exhaust manifolds (cast iron or aluminum): 45 RA

NOTE: *Using a crosshatch pattern reduces the roughness of the surface finish by half.*

Copper Head Gaskets

Copper head gaskets are available for use in corrosive environments, such as for marine applications. They are also effective in high-vibration uses. Top fuel dragsters and funny cars often use copper head gaskets.

Head Shims. When too much metal has been removed from a head's surface, a shim can be installed to compensate (see Chapter 52). Shims are widely available. They are 0.020" thick and are either copper or steel. Copper conforms better to surface imperfections, whereas steel is more durable. Sealer is used between the head and the shim but not between the shim and head gasket.

No-Retorque Gaskets

Modern head gaskets are called permanent torque or no-retorque. They retain their resistance to compression better than older-style (retorque) gaskets; therefore, bolt stretch during torquing does not diminish their clamping force as much (**Figure 51.15**). A gasket must be able to maintain a good seal over its lifetime. Small changes in the gasket's thickness can result in a substantial loss of clamping force. Continuous elongation of the bolt so that clamping force is maintained is necessary. The greatest torque loss will occur in the first hour of operation of a rebuilt engine.

Although modern head gaskets compress less than older head gaskets, all gaskets will relax some during use. This is called creep relaxation. Older head gaskets required retorquing after about 500 miles. Retorquing cylinder heads is not commonly done with today's gaskets. Many of today's engines have intake manifolds, air conditioning, turbochargers, or other accessories that can make head bolts inaccessible. Non-retorque gaskets can save 20 minutes to 2 hours of the labor time usually required for retorquing.

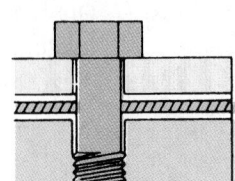

Bolt stretch = 0.015" @ clamp up

Gasket set = 0.005"

Bolt stretch = 0.010" @ end

Therefore $\dfrac{0.010"}{0.015"}$ or 2/3 of clamp up torque retained

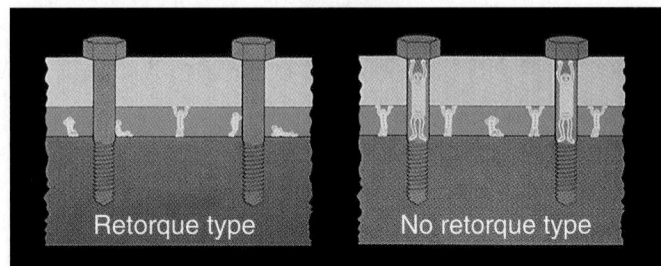

Retorque type   No retorque type

Figure 51.15 Clamping force on retorque and non-retorque gaskets. *(Courtesy of Federal-Mogul Corporation)*

NOTE: *Whether or not the gasket is a permanent-torque type, always follow the gasket manufacturer's torque and retorque recommendations.*

Although retorquing is not required, modern non-retorque gaskets can benefit from retorquing after several heating and cooling cycles. The recommendation is usually for a gasket retorque at 500 miles, because this is when the car should come back for its first oil change and a thorough inspection.

SHOP TIP When cleaning aluminum heads, it is better to use a chemical gasket remover, lacquer thinner, or alcohol with a plastic scraper to prevent damage to the sealing surface. Spray gasket cleaners can be left to penetrate for 5–10 minutes.

■ CLEANING THE HEAD

Clean cylinder head and block surfaces thoroughly before engine assembly. Be careful not to nick the surface when scraping gaskets, especially on aluminum heads, which are softer.

CAUTION Spray cleaners can damage the skin and eyes. Be sure to follow the instructions printed on the can and check the material safety data sheet (MSDS) for hazards.

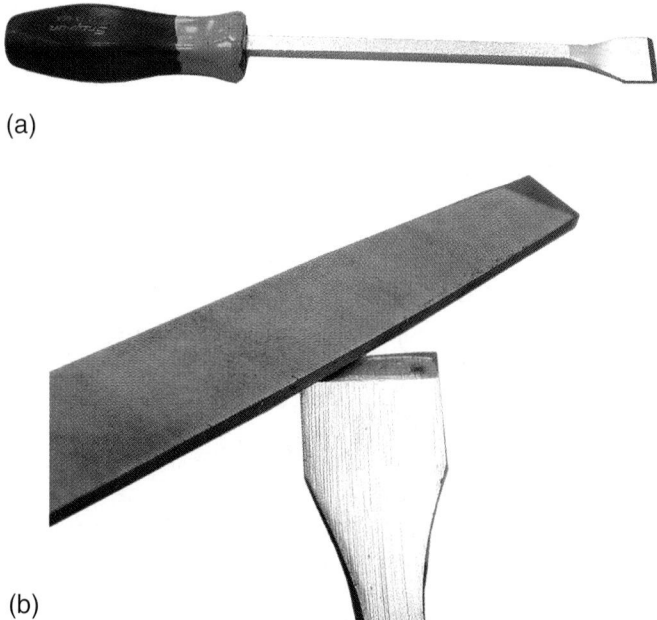

(a)

(b)

Figure 51.16 (a) A gasket scraper. (b) Sharpening a gasket scraper. Use a file on only one edge. *(Courtesy of Tim Gilles)*

The safest way to clean the surfaces of an iron head and block is to *carefully* use a sharp scraper. The scraper should be ground from one side only. Then it is finish-sharpened with a file (**Figure 51.16**). The file removes any high spots (gaskets or nicks), but it does not cut into the original surface, so it does not change the surface texture. Aluminum heads are safely cleaned using a plastic scraper.

Abrasive Disc Cleaning

Some technicians like to use abrasive bristle discs to clean head and block surfaces (**Figure 51.17**). Use of abrasive discs has been the subject of technical service bulletins because they have caused some serious engine damage. Never use them with a die grinder.

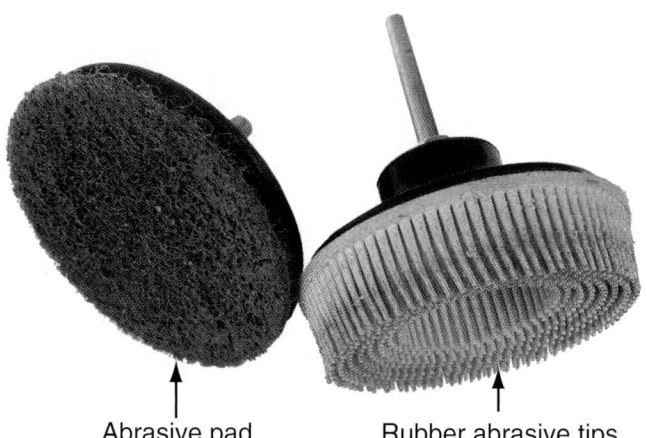

Abrasive pad   Rubber abrasive tips

Figure 51.17 An abrasive disc is sometimes used to clean a gasket sealing surface. On the one shown on the right, as the rubber tips wear, more abrasive material is exposed. *(Courtesy of Tim Gilles)*

Die grinders spin at very high rpm, and damage can quickly be done to the surface of an aluminum head.

Here are some rules when using abrasive discs:

- Do not use them on an aluminum head that uses a multilayer steel gasket.
- Do not use a disc of too coarse a grit. White bristle discs remove material more slowly than yellow ones, so they are safer to use.
- Do not clean the "ghost marks" left when the gasket "stains" the aluminum head surface around water jacket openings. Attempting to do this will result in metal removal.
- Be careful not to clean in one area longer than another.
- Be certain that any material thrown about by the discs is carefully removed so it cannot ruin an engine.

SAFETY NOTE Older gaskets often contain asbestos, which has been phased out as a gasket material. Most manufacturers now use substitute materials. While cleaning a gasket surface, remember that the gasket might contain asbestos, which can be dangerous to your health.

■ HEAD GASKET INSTALLATION

Installing a head gasket correctly will help guard against combustion leakage. Head gaskets perform the most difficult sealing job in the engine. Sealing of combustion is very critical. Cylinder head bolts must hold against the pressure of combustion that is trying to push the head off the block. The clamping pressure on the bolts is greater than the combustion pressure, so the head bolts never feel the stress as long as proper bolt torque is maintained.

NOTE: *About 75% of the clamping load of the head bolts are used to seal combustion. The remaining load is to seal coolant and oil.*

- Chase the bolt holes with a tap.
- Use a wire brush on the bolt threads and the underside of the bolt head. This is where most of the friction during torquing comes from.
- Lubricate each bolt with 10W-30 oil and wipe off the oil with a shop towel.

Head bolts should be torqued in the sequence prescribed in the service literature (**Figure 51.18**). This torque pattern ensures that the fasteners will be pulled evenly against the head and block. Remember to torque the head in three or four stages as described earlier.

Head Gasket Selection

Be certain the correct gasket has been purchased. Carefully match a new gasket to the old one or to the block

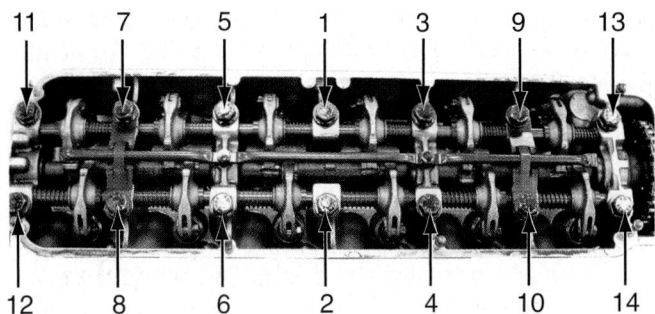

Figure 51.18 Follow the cylinder head bolt tightening sequence. *(Courtesy of Tim Gilles)*

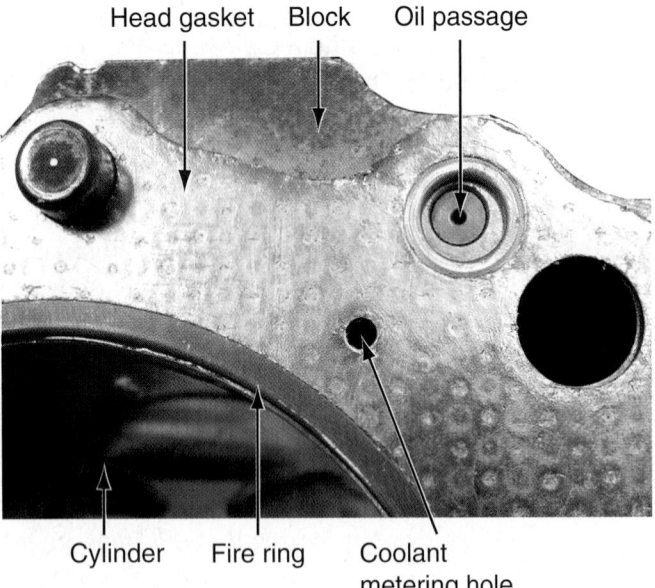

Figure 51.19 Head gaskets have coolant metering holes and oil passages. *(Courtesy of Tim Gilles)*

and head. Head gaskets have small holes that control the flow of coolant in the head (**Figure 51.19**). These should not be cut out to match the old gasket. If the cooling and oil holes do not match up, serious lubrication or cooling problems can result.

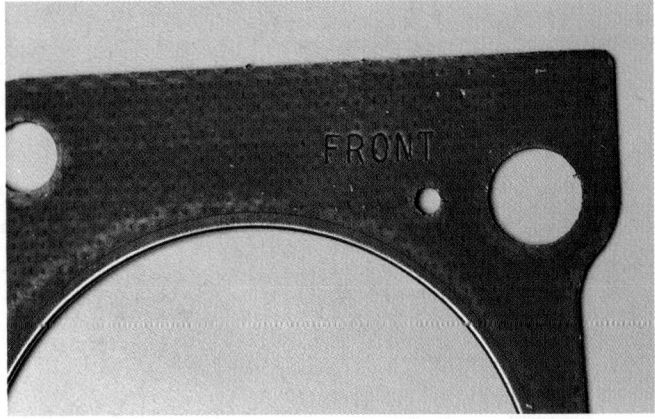

Figure 51.20 Some head gaskets are printed with installation directions.

Some engines have head gaskets that are directional. These have an imprint marked "top" or "front" (**Figure 51.20**). Some V-type engines have head gaskets that interchange from side to side. Other do not. Some manufacturers have used differed head gaskets for the same size engine built in different years. Use of the wrong gasket in one of these engines can result in cooling problems.

■ OTHER ENGINE GASKETS

Other gaskets are used at several locations to seal oil and coolant in the engine. **Figure 51.21** shows several different kinds of gasket materials.

Valve Cover, Timing Cover, and Oil Pan Gaskets

Valve cover and oil pan gaskets can be made of several different materials. Valve cover gaskets are molded to fit the shape of the cover. They often have tabs that fit into slots in the valve cover to hold them in place. Cast aluminum valve covers on overhead cam engines often have rubber semicircular plugs that fill holes in the head left from manufacturing (**Figure 51.22**). These harden with age and begin to leak. They are not included with a valve cover gasket and must be ordered separately. Valve covers used with hemi heads have centrally located spark plugs installed in tubes that intersect the cylinder head and valve cover. Round seals supplied with the valve cover gasket are installed to prevent oil from leaking into the spark plug tubes (**Figure 51.23**). If you find oil in these tubes when changing spark plugs, these seals are the cause of the leak, which can result in a smell of burning oil from under the hood.

Oil pan gaskets used on sheet metal oil pans are sometimes made up of several pieces (**Figure 51.24**). Rubber seals on the ends join with side gasket strips. Sometimes a timing cover gasket is replaced without

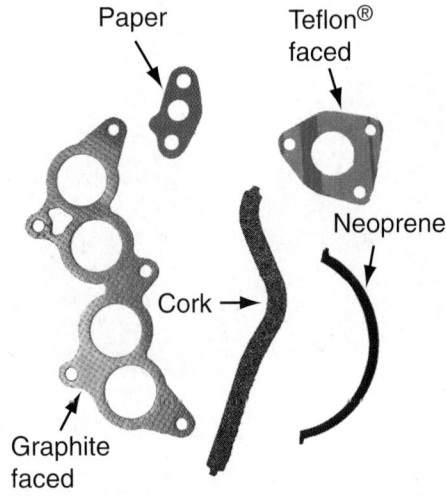

Figure 51.21 Various gasket materials. *(Courtesy of Tim Gilles)*

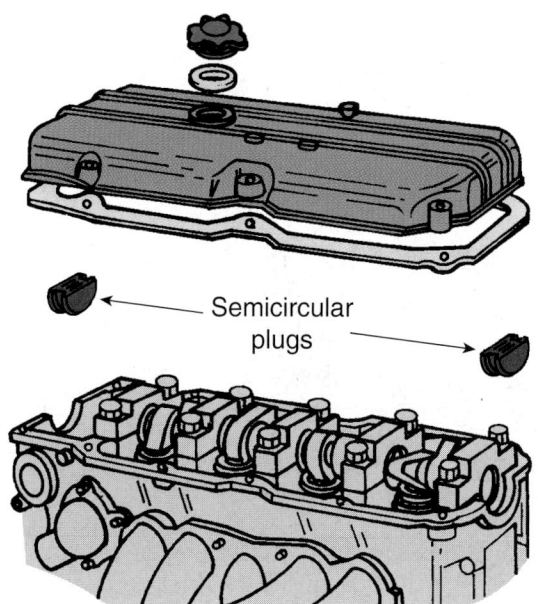

Figure 51.22 Some OHC heads have semicircular plugs at the ends.

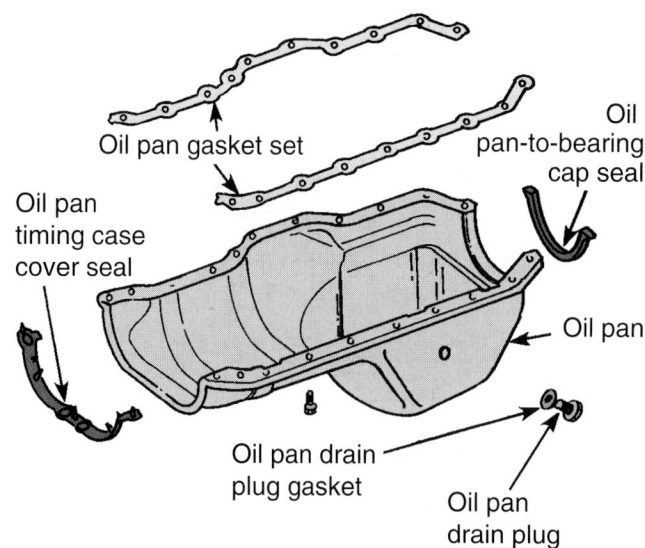

Figure 51.24 Oil pan gaskets and end seals.

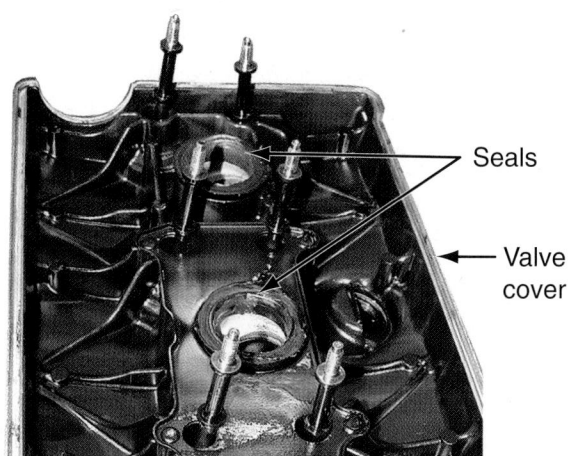

Figure 51.23 Round seals are supplied with the valve cover gasket to seal the spark plug tubes on heads with centrally located spark plugs. *(Courtesy of Tim Gilles)*

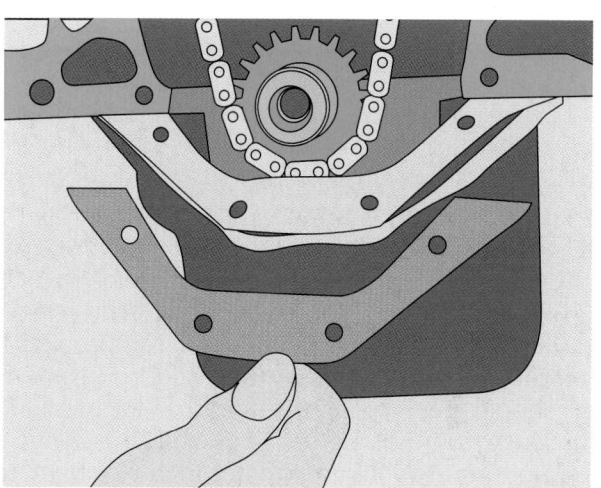

Figure 51.25 When the timing cover is removed without removing the oil pan, the front part of the old pan gasket is removed and replaced.

removing the oil pan. **Figure 51.25** shows how to cut a pan gasket and install a new piece of gasket.

Sheet metal parts often become distorted when bolts are tightened against them during assembly. Always check to see that the valve cover or oil pan is straight before assembly (**Figure 51.26**).

CAUTION Torque applied to sheet metal parts with gaskets is very light. Check the manufacturer's specification. Overtorquing can damage the gasket (**Figure 51.27**).

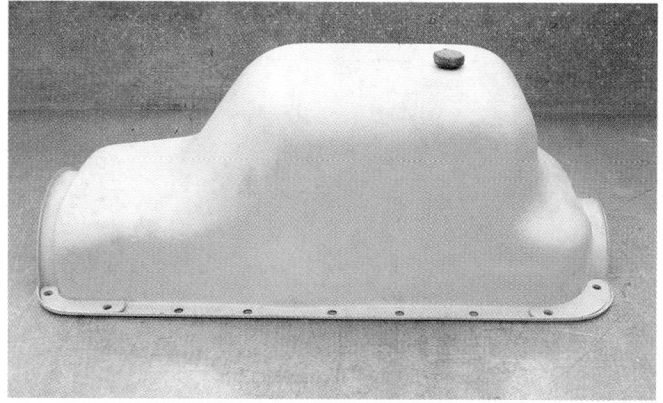

Figure 51.26 Check a sheet metal oil pan to see that it is flat. Place it on a solid, flat surface. Put a flashlight inside to illuminate any high spots. *(Courtesy of Tim Gilles)*

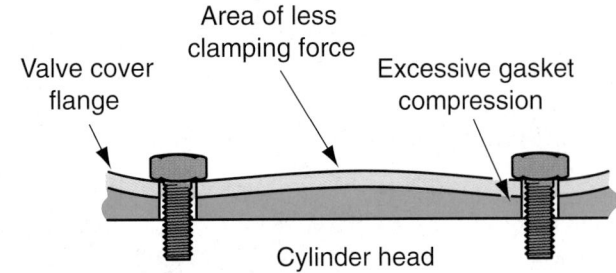

Figure 51.27 Do not overtorque valve cover gaskets.

Cork Gaskets

Cork used to be a common gasket material and is still used in some vintage applications. Because it shrinks in the presence of air, it should come in a vacuum-sealed package.

> **SHOP TIP** A cork that has shrunk can be soaked in water for 5 minutes or so to help it regain its size.

NOTE: *Cork becomes brittle with heat. A brittle gasket can crack if screws are tightened further in an attempt to correct a leak.*

The two most common gasket materials used for valve cover gaskets are **rubberized cork** and synthetic rubber. Modern cork gaskets are rubberized cork (**Figure 51.28**). This is a superior material to regular cork because it does not shrink. All cork gaskets are somewhat resilient—that is, they are "springy"—but they should only be lightly torqued. Recommendations range from 5 to 15 foot-pounds. This is about the same amount the bolts could comfortably be tightened with a screwdriver or nutdriver. Follow the torque

Figure 51.28 When cork gaskets are found today, rubber has been mixed with the cork to improve the gasket characteristics. (*Courtesy of Tim Gilles*)

Figure 51.29 Small fasteners are tightened with an inch-pound torque wrench. (*Courtesy of Tim Gilles*)

recommendations to guard against breaking the oil pan or valve cover gasket screws.

NOTE: *Typical torque on a valve cover or oil pan is about 60 inch-pounds. This is the torque required to correctly compress the gasket, not for stretching a fastener. Use the inch-pound torque wrench because foot-pound torque wrenches are not accurate at less than 15 foot-pounds of torque (**Figure 51.29**). To convert foot-pounds to inch-pounds, multiply by 12.*

Silicone and Neoprene Gaskets

Molded silicone and neoprene (artificial rubber) gaskets are the best, but they are expensive. Because they are reusable sometimes, these gaskets are especially useful for valve covers, which must be periodically removed—to adjust mechanical valve clearance, for instance.

> **SHOP TIP** Silicone and neoprene gaskets are used *without* sealers. If these gaskets must be held in position, use an adhesive on one side of the gasket only—the side that is against the stamped metal part.

Paper Gaskets

Paper gaskets are generally the only gaskets other than retorqueable head gaskets that require the use of a sealer. Paper gaskets are found on some timing covers, water pumps, water outlets, fuel pumps, and at the base of some carburetors. Several manufacturers make tool sets for cutting the holes in paper gaskets (**Figure 51.30**).

> **SHOP TIP** Gasket paper can be purchased in sheets. A gasket can be roughed out by holding the paper against the part and tapping on it with a ball-peen hammer.

Figure 51.30 A gasket cutting set. *(Courtesy of Tim Gilles)*

Figure 51.31 Using RTV to seal where gaskets join. *(Courtesy of Tim Gilles)*

Intake Manifold Gaskets

Intake manifold gaskets must be resistant to air and water leaks. They are made of embossed steel shim, asbestos-faced steel, or the non-retorque materials. On V-type engines, there are two side gaskets and two end gaskets. Silicone RTV (covered later in this chapter) is used to join the gasket pieces and around water passages (**Figure 51.31**). Be sure to read any instructions that come with the gasket set.

Exhaust Manifold Gaskets

Exhaust manifold gaskets are not usually found on new cast iron heads because the joined surfaces seal

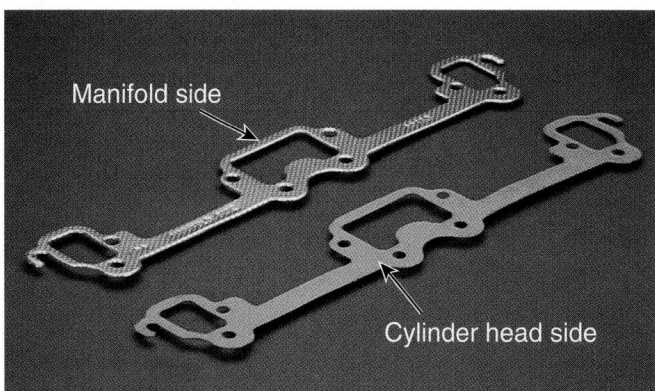

Figure 51.32 One side of an exhaust manifold gasket has a steel facing that goes against the manifold. *(Courtesy of Federal-Mogul Corporation)*

perfectly. Gaskets are used once the parts have been separated during service, however (**Figure 51.32**). Replacement gaskets are made of perforated steel with asbestos or another material. The manifold expands much more than the head.

- The steel side of the gasket faces toward the manifold.
- Do *not* torque the manifold at the outside ends first; it may crack (**Figure 51.33**).
- When there is one hole smaller than the others, its bolt goes in first.

NOTE: *When a manifold cracks or a manifold bolt breaks, it is not unusual to find a burned exhaust valve in a port close to the exhaust leak. This happens because of the shock of the relatively cold air (from the exhaust leak) on the red-hot valve.*

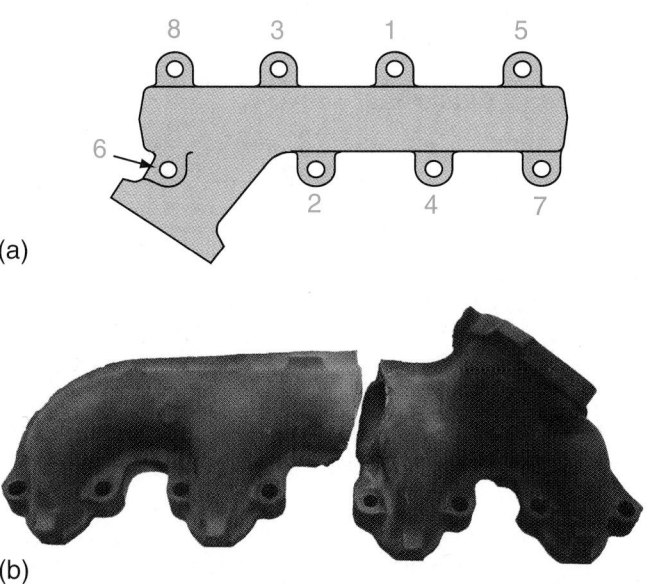

Figure 51.33 Exhaust manifold torque. (a) Exhaust manifold torque sequence. (b) An exhaust manifold that cracked because it was torqued first at the outside ends. *(b, Courtesy of Tim Gilles)*

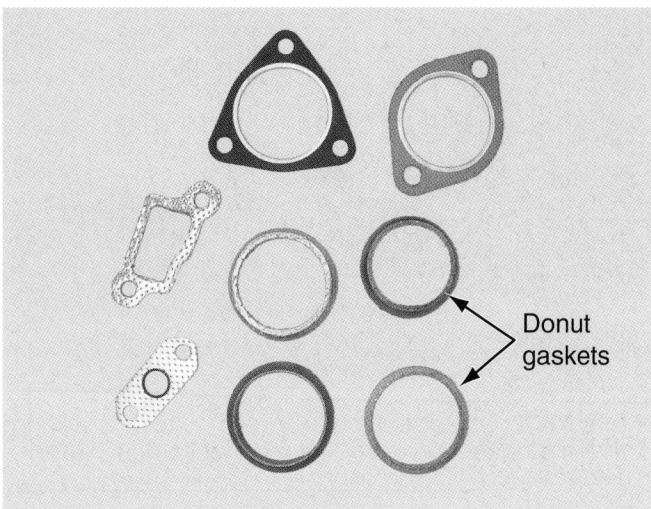

Figure 51.34 *Various exhaust gaskets. (Courtesy of Tim Gilles)*

Figure 51.35 *Some sealers are spread on from a can with a brush. (Courtesy of Federal-Mogul Corporation)*

Some exhaust manifold gaskets use a heat shield to protect the valve cover gasket from excess exhaust heat rising from the manifold.

The bottom of the manifold uses a "donut" O-ring gasket (**Figure 51.34**). In the past, these were made of readily available and inexpensive asbestos. The trend today is toward ceramic O-rings, which are also reusable.

In front-wheel-drive vehicles with transverse (sideways mounted) engines, acceleration and deceleration forces movement of ± 4 degrees (8 degrees total possible) between the manifold and exhaust pipe. Consequently, the seal for this joint must be flexible as well as durable. Stainless steel wire mesh or expanded graphite are used for these gaskets. The gaskets wear out. When they leak, they must be replaced. Simply loosening and retightening will only result in a future failure.

Chemical Gaskets

Silicone and other chemical sealers are sometimes used instead of gaskets. These "formed-in-place" gaskets have been used since the mid-1970s in new cars. Chemical gaskets can be used for everything except the head gasket and fuel systems gasket.

Chemical gaskets are sometimes difficult to use in the repair trade, so aftermarket anaerobic materials are available for these uses.

Anaerobic sealers cure without air and are used only on precision machine parts. The closer the fit, the faster the sealer cures.

■ GASKET SEALERS

Gasket sealers come in several configurations. Some come in tubes that are squeezed to apply the sealer to the part. Others are spread on from a can with a brush (**Figure 51.35**). Sealers also come in spray cans for easy application.

Gasket sealers serve three purposes:
■ They help gaskets to seal.
■ They hold gaskets in place.
■ Some gasket sealers also take the place of gaskets.

SHOP TIP Gasket sealers usually are not necessary. Some major gasket manufacturers suggest that sealers are necessary only on paper gaskets to hold a gasket in position or to seal new core plugs to the core holes in an engines. No sealers should be used on the non-retorque head gaskets.

Neoprene intake manifold end gaskets on V-type engines sometimes "squish" out. Neoprene will usually stay in place if it is completely dry.

Do not use *hardening gasket cements* as sealers in automotive applications; they are inflexible and can begin to leak after vibration or heating and cooling. A small amount of hardening cement can be used on the corners of some gaskets, simply to hold them in place during assembly.

Some technicians use grease to hold gaskets in place for assembly, but the greased parts can shift during assembly. A better choice would be an adhesive or an adhesive sealant (**Figure 51.36**).

Locking Thread Sealers

"Hard glue" sealers, often called thread lockers or *loctite* (after one of several manufacturers of these products), are anaerobic. That is, they harden to their maximum strength when tightened between two metals *without the presence of air* (**Figure 51.37**). Some adhesives have very little shear strength and their effectiveness is affected by elevated temperatures.

NOTE: *Chemical adhesives are not recommended for bolts over ⅜" in diameter.*

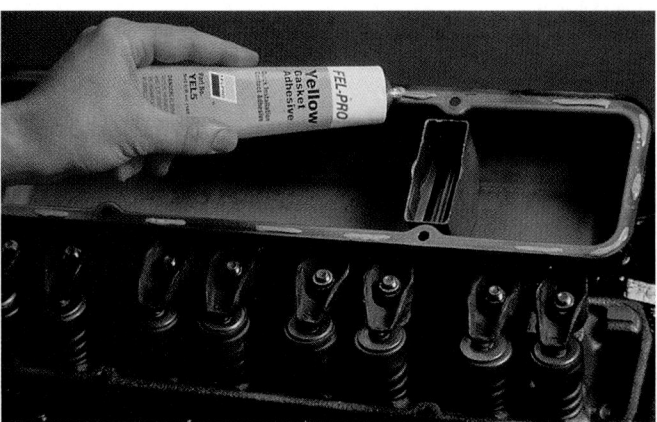

Figure 51.36 An adhesive sealant is sometimes used to hold a gasket in position. *(Courtesy of Federal-Mogul Corporation)*

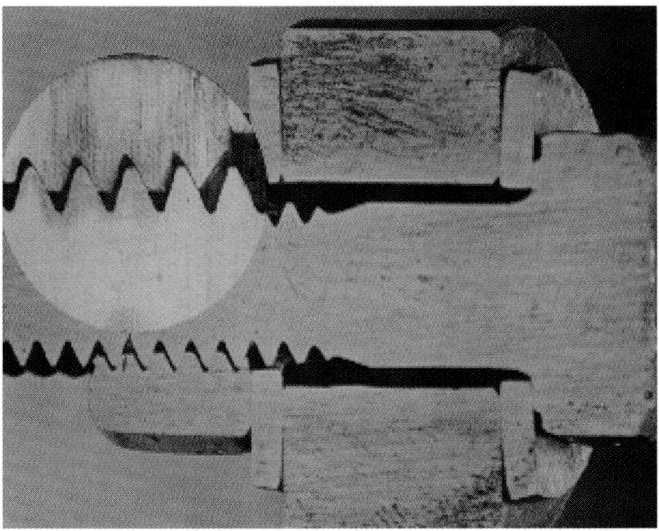

Figure 51.37 Anaerobic sealers harden in the absence of air. *(Courtesy of Permatex Inc.)*

There are several kinds of thread sealers (**Figure 51.38**), which are differentiated by their colors. The *blue* sealer is used when parts may be disassembled again. It is said that the torque required to fasten parts with the blue sealer will have to be doubled for disassembly. This sealer hardens only when torqued.

The other popular color is *red stud and bearing mount*. This sealer is used to fasten parts that are **not** intended to be disassembled in the future, although applying heat will cause it to fail if disassembly is required.

NOTE: *Anaerobic means that the chemical works only in the absence of air. That is why the container it comes in is not full; if all the air were removed, the sealer would harden. Silicone RTV, an aerobic sealer, comes in a full tube because it cures when exposed to air. Although loctite will not harden in its container, it will harden between the container and its cap. Be sure to wipe off the threads on the container before installing the cap.*

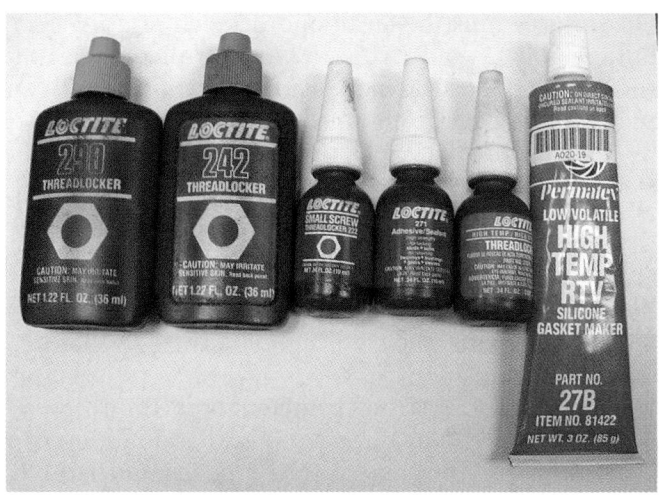

Figure 51.38 Anaerobic and aerobic sealers come in a tube to prevent contact with air. Anaerobic sealers have space in the container for air. *(Courtesy of Tim Gilles)*

■ SILICONE-RTV SEALANT

Room temperature vulcanizing (*RTV*) is a popular sealer commonly known as *silicone rubber*. Because it cures when exposed to air, silicone is an aerobic sealer. It comes in several colors.

Much of the assembly of modern engines is done by robots. With the exception of cylinder head and intake manifold gaskets, many gaskets can be eliminated by substituting a bead of RTV that can be applied by a robot. Aftermarket gaskets are available as a replacement for most factory RTV applications.

Follow manufacturer's directions when making formed-in-place (FIP) gaskets. Excess material could end up plugging an oil pump screen.

SHOP TIP
- Because RTV sealed parts are actually vulcanized together, disassembly can be very difficult. Sometimes a sharp knife can be inserted between the parts to cut the bead of RTV.
- Oil pans and valve covers can be soaked in solvent for an hour or so to help soften the RTV for easier removal.

RTV sets up faster in warm temperatures and high humidity. It usually begins to cure ("skin over") in about 15 minutes, but it does not set completely for about a day. This is a good feature if a part is forgotten, or if parts must be disassembled for any other reason. The gasket will not be ruined.

CAUTION RTV silicone is not used around fuel (it is eaten by gasoline).

Surfaces must be clean and dry before using RTV or anaerobic sealers. Use a non-petroleum cleaner such as alcohol to avoid leaving an oil film that can prevent the sealer from bonding.

When using RTV sealer in place of a gasket:

■ Be sure to apply the bead of sealer to the sheet metal surface on the *inside* of screw holes to prevent leakage.
■ Put the sealer in place and finger-tighten the fasteners.
■ Give the material a chance to vulcanize so that it will not squish out.
■ After it sets, the screws can be tightened further.

If the gasket film is too thin, heat and motion can cause leakage. For that reason, grooves and stopper devices are installed to ensure a thick enough gasket (0.020"–0.060" minimum).

NOTE: *Excessive RTV can squeeze out and get into the oil, where it can clog the oil pump pickup screen and ruin the engine (**Figure 51.39**).*

Silicone is recommended for use where gaskets join and for sealing water passages in intake manifolds.

NOTE: *RTV will not vulcanize or bond to a gasket, so its use as a gasket sealer is not recommended.*

Many new engines use preformed one-piece silicone gaskets that have steel washers around the bolt holes to prevent overtightening.

Low-Volatile RTV

The *acetic acid* in household silicone gives it the smell of vinegar as it sets up. General Motors has reported that acetic acid, which gives off formaldehyde fumes as it sets up, was responsible for coating the oxygen sensors of its computer-controlled fuel system. Low-volatile RTV has been developed by substituting an

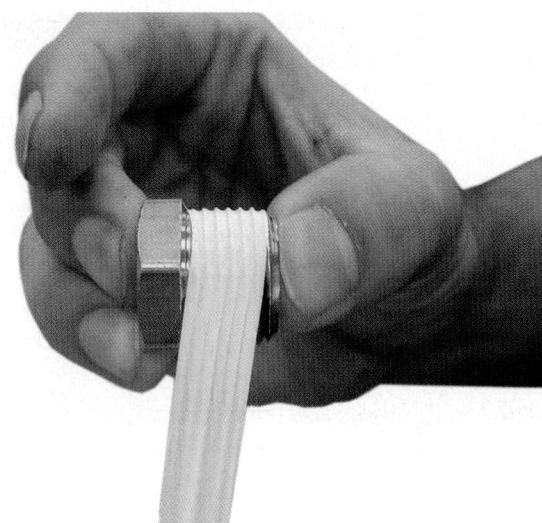

Figure 51.40 Sealing an oil gallery threaded plug with Teflon tape. *(Courtesy of Tim Gilles)*

alcohol-related chemical for acetic acid. Be sure to use a low-volatile silicone on vehicles equipped with oxygen sensors. Low-volatile silicones take somewhat longer to cure than regular silicones.

High-Temperature RTV

Adding iron oxides to RTV can increase its temperature rating by 50°F–100°F. This silicone is *red* in color. Because silicone is not used in high-temperature areas of the engine, it may not be worth the extra cost to buy high temperature RTV.

Teflon® Tape or Liquid Thread Sealer

Threads can be sealed with **Teflon tape** to prevent liquid leakage. **Figure 51.40** shows Teflon tape stretched around a threaded fitting.

SHOP TIP
■ Use caution when Teflon tape is used on a tapered pipe thread. Teflon acts as a lubricant, and overtightening can result in a cracked casting.
■ Apply Teflon tape starting with the second thread to prevent some of it from entering and contaminating the system being sealed.

Rubber Cement

Rubber cement sealers are very popular and easy to use. Be sure to keep the lid on the can as much as possible, because it will thicken when exposed to air.

■ SEALS

Dynamic seals, also called *chevron* seals, are used at the front and the rear of the crankcase. They are known as dynamic seals because they seal moving parts. Gaskets are *static seals* because they seal nonmoving parts.

Figure 51.39 Examples of excess silicone use that resulted in serious engine damage. *(Courtesy of Tim Gilles)*

Seals are generally made of butyl rubber or neoprene. Neoprene seals should always be lubricated during installation. This is to prevent damage due to overheating during initial startup of the engine.

FRONT SEALS

Front seals are usually *lip-type seals* with a garter spring (see Chapter 60).

SHOP TIP The seal lip, or open side of the seal, faces toward the oil.

When a seal is exposed to outside air and dust, such as at the ends of the oil pan, it also has a *dust lip*, which deflects dirt from the outside of the seal.

The front seal is installed in the timing cover and seals against the hub on the vibration damper. Wear on the damper hub is sometimes accelerated by contaminated oil combined with friction of the seal against the hub oil. A *harmonic balancer repair sleeve* kit will repair a worn damper hub at minimal cost. A very thin stainless steel ring, often called a speedi sleeve (**Figure 51.41a**), is pressed over the worn surface of the damper (**Figure 51.41b**) using a small amount of red loctite. The new sealing surface is only slightly larger than the original so the new seal conforms to it. Some replacement seals are designed so that the lip of the

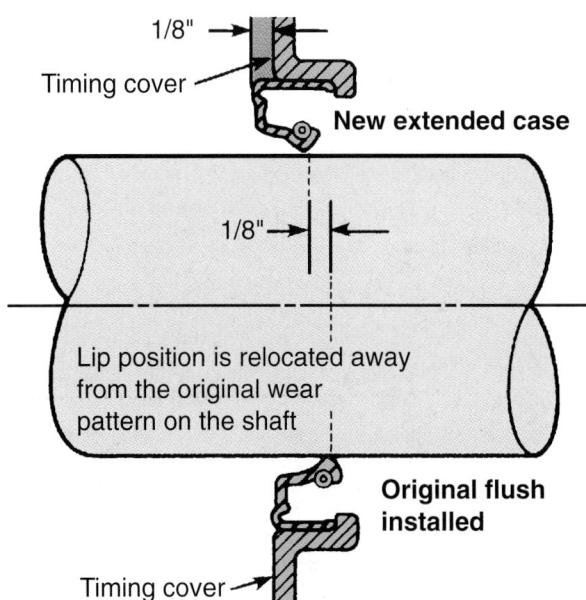

Figure 51.42 A seal with a relocated lip.

seal will meet the damper hub at a different place than the original seal (**Figure 51.42**). Front seals can be replaced with the timing cover on or off the engine. With the cover off the engine, the seal can be removed with a seal puller. Be careful not to ruin a sheet metal timing cover by prying on the sheet metal lip that the seal seats against (**Figure 51.43**). The seal can also be installed without removing the timing cover.

OHC Camshaft Seals

Overhead camshafts with timing belts have seals located behind the camshaft sprocket(s) (**Figure 51.44**). This sometimes presents a difficult diagnosis, and replacing the seals can be a time-consuming job. Be sure to install them carefully.

Figure 51.41 Worn damper. (a) This damper has a groove worn in its hub. (b) A repair sleeve installed on the damper hub. *(Courtesy of Tim Gilles)*

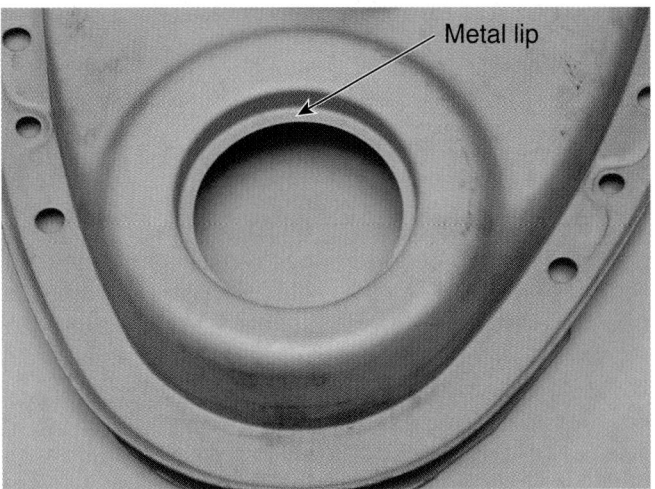

Figure 51.43 With the seal removed, notice the metal lip in the timing cover. *(Courtesy of Tim Gilles)*

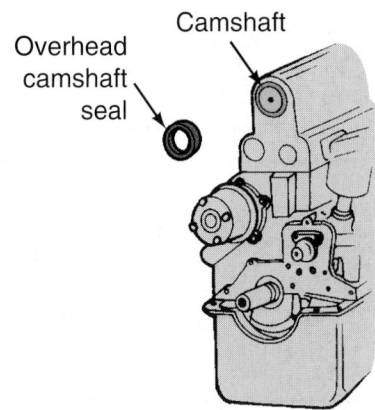

Figure 51.44 A belt-driven OHC engine has a seal at the front of the camshaft.

SHOP TIP The top part of the seal is often held in place by the front camshaft cap. Loosening the cap makes it easier to install the seal.

REAR MAIN SEALS

The crankshaft rear seal is called the rear main seal because it is located behind the rear main bearing. A leaking rear seal can destroy a clutch. There are two types of rear seals: the neoprene lip seal and the rope seal.

NOTE: *Rear main seals are often needlessly replaced when an intake manifold, valve cover, or oil pan gasket is the true source of the leak.*

Neoprene Rear Seal

Newer engines use a neoprene lip seal. Some of these are split in half, and others are a single piece. The split seal should be installed with its parting line offset, as in **Figure 51.45**, and with its lip facing toward the oil (**Figure 51.46**).

NOTE: *Be sure to lubricate neoprene seals with grease during assembly. When the engine is started the first time after a rebuild there is a short interval before the seal gets any oil; greasing the seal will protect it from failure during this interval.*

Prior to installing a crankshaft seal, be certain that the crankshaft rear sealing surface is clean and smooth.

Single-piece neoprene rear seals have become popular in recent years. Manufacturers can reduce engine weight and length by not having a separate flywheel flange at the rear of the crankshaft.

Some full round rear main seals are installed in a small casting that is bolted to the rear of the block with a gasket underneath it. When removing the seal from these, be sure the casting remains fastened to the block to avoid cracking it (**Figure 51.47**).

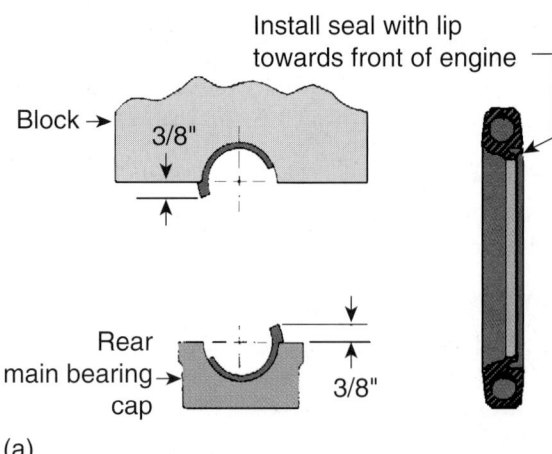

(a)

(b)

Figure 51.45 Neoprene rear seal installation. Offset the parting lines and face the seal lip toward the oil. (*a, Courtesy of Ford Motor Company; b, Courtesy of Tim Gilles*)

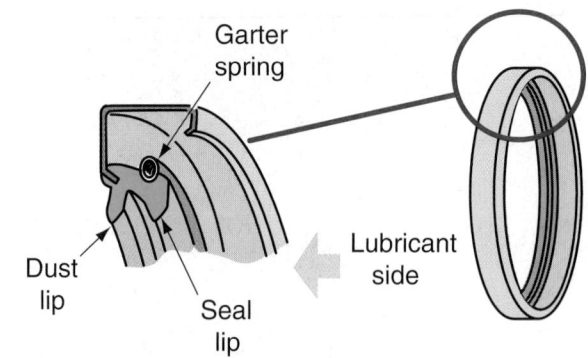

Figure 51.46 The lip of the seal faces the oil.

NOTE: *Care must be taken that flywheel bolts are carefully torqued, or the crankcase rear main seal surface can become distorted, causing a leak.*

Flywheel bolts on some of these engines protrude into the crankcase (**Figure 51.48**). Sealer must be applied to them to prevent oil leakage from the interior of the oil pan to the clutch surface.

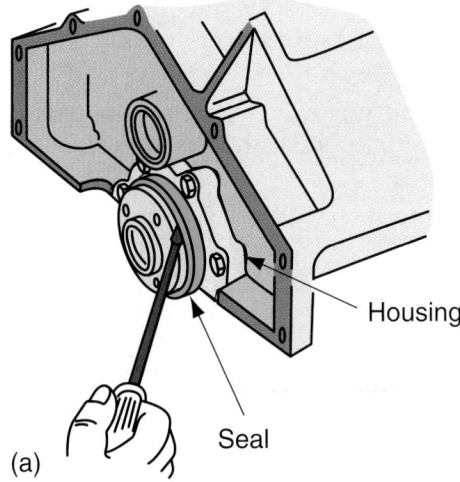

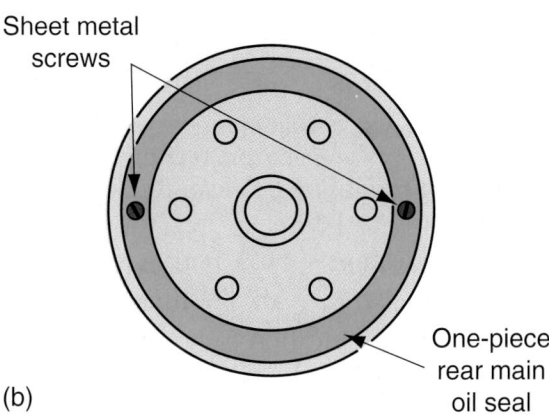

Figure 51.47 (a) When prying a seal out of a cast housing, be sure the casting remains bolted to the block to avoid breaking it. (b) A one-piece seal can be removed by screwing sheet metal screws into it when the crankshaft is in the block.

 *A technician replaced a clutch on an engine with a single-piece rear main seal. When he installed the flywheel, he used an impact wrench. The customer returned later with a rear main bearing seal leak because the crankshaft sealing surface was distorted by the uneven torque.*

Figure 51.48 Oil can leak through the flywheel bolt holes from the crankcase to the clutch. *(Courtesy of Tim Gilles)*

Rope Seal

Almost all vintage engines used another type of rear seal, called the rope, or *wick*, seal. These seals are inexpensive, but their installation is more time consuming than that of the lip seal. The rope seal is formed into place and trimmed to fit.

NOTE: *A conversion from a rope to a lip seal is often possible.*

■ TRANSMISSION FRONT PUMP SEAL

Another seal found on automobiles is the front transmission seal. This seal keeps transmission oil from leaking out between the torque converter and transmission oil pump. It suffers the abuse of severe heat and often becomes brittle with age. If the seal fails, a substantial amount of transmission fluid will leak from between the engine and transmission. As a preventative measure, the seal is often replaced when the engine is out of the car for an overhaul. The procedure for replacing the seal is covered in Chapter 74.

■ REVIEW QUESTIONS

1. Approximately how much of the torque applied to a fastener is used to overcome friction?
2. What is the name of the metric measurement for torque?
3. Refer to a bolt torque chart. What is the proper torque for a grade 5, ⅜" bolt?
4. How much of the clamping load on a head gasket is used for sealing combustion?
5. Does the lip of an oil seal face toward or away from the oil?

6. When installing half of a rear main lip seal in a block, the parting line is positioned offset/even (choose one) with the parting line of the main cap.

7. What color is the locking sealer that is used to glue parts that will *not* be disassembled again?

8. What does RTV stand for?

9. Low _____ RTV silicone sealant that does not use acetic acid to cure is used with oxygen sensor systems.

10. Before putting a click-type torque wrench aside after using it, what should be done?

■ ASE-STYLE REVIEW QUESTIONS

1. Which of the following is/are true about head gaskets?
   a. They have small metering holes to direct coolant flow.
   b. They sometimes have small metering holes to direct oil flow.
   c. Both A and B.
   d. Neither A nor B.

2. Technician A says that bolts that have been torqued to yield have been stretched beyond their elastic limit. Technician B says that bolts that have been torqued to yield must be replaced whenever the head has been removed. Who is right?
   a. Technician A   c. Both A and B
   b. Technician B   d. Neither A nor B

3. Modern head gaskets require retorquing after:
   a. 500 miles   c. 5,000 miles
   b. 1,000 miles   d. None of the above

4. Technician A says that blue loctite is used on parts that are likely to be disassembled in the future. Technician B says that red loctite is used on permanent installations. Who is right?
   a. Technician A   c. Both A and B
   b. Technician B   d. Neither A nor B

5. One microinch is
   a. One thousandth of an inch
   b. One millionth of an inch
   c. One hundredth of an inch
   d. None of the above

6. Which of the following is/are used with a gasket sealer?
   a. A Teflon-coated head gasket
   b. A neoprene gasket
   c. A paper gasket
   d. All of the above

7. Technician A says that a broken exhaust manifold bolt can cause a burned exhaust valve. Technician B says that a broken exhaust manifold bolt can cause a burned intake valve. Who is right?
   a. Technician A   c. Both A and B
   b. Technician B   d. Neither A nor B

8. Technician A says that some types of RTV can damage an oxygen sensor. Technician B says that using too much RTV can result in engine failure. Who is right?
   a. Technician A   c. Both A and B
   b. Technician B   d. Neither A nor B

9. The torque specification for a valve cover gasket is 5 foot-pounds. Which of the following is the best procedure?
   a. Use a ½"-drive foot-pound torque wrench and torque to 5 foot-pounds.
   b. Use a ⅜"-drive foot-pound torque wrench and torque to 5 foot-pounds.
   c. Use an inch-pound torque wrench and torque to 5 foot-pounds.
   d. Use an inch-pound torque wrench and torque to 60 inch-pounds.

10. A multilayer steel head gasket requires a head and block with:
   a. A very smooth surface finish
   b. A rough surface that will bite into the gasket
   c. A wide range of surface finishes (MLS gaskets are very forgiving.)
   d. None of the above

Engine Diagnosis and Service: Cylinder Head and Valvetrain

■ OBJECTIVES

Upon completion of this chapter, you should be able to:

✔ Disassemble a cylinder head in the correct manner.

✔ Clean and inspect a cylinder head for cracks and warpage.

✔ Diagnose cylinder head and valvetrain wear problems and determine the correct repair procedure.

✔ Understand machine shop repair processes for cylinder heads.

✔ Reassemble a cylinder head.

✔ Understand camshaft and cam drive service procedures.

■ KEY TERMS

mushroomed valve tip	pinning or stitching	interference angle
valve stem height	knurling	valve spring installed height
dye penetrant	thinwall	valve clearance

■ INTRODUCTION

This chapter deals with service to the cylinder head and valvetrain. When a cylinder head is removed for valve grinding, this is called a *valve job*. When a head gasket has been leaking, sometimes the head is removed for resurfacing and gasket replacement only. This chapter also deals with timing chain or timing belt service, important maintenance procedures on modern, long-life engines.

■ HEAD DISASSEMBLY

Cylinder heads are easier to work on if they are clean. After cleaning, rinse heads thoroughly and lubricate all machined surfaces immediately to prevent rusting of ferrous parts. Try not to damage the coating on valve springs. Be sure to lubricate valve springs that have been hot-tanked in an alkaline solution. Hot-tanking removes the protective coating from the springs and they can rust very fast. Even the smallest hint of rust on a valve spring will require its replacement.

CAUTION Be careful when handling a cylinder head with an overhead cam. While the cam is still installed in the head, some of the valves will be held open. Do not set the head down on the combustion chamber side or some of the valves can be bent.

Some OHC camshafts act directly on the valve, others use rocker arms. When there are rocker arms, it is necessary to remove them from OHC heads first in order to remove the valve and spring. Pry against the spring retainer and remove the rocker arm. Check the manufacturer's recommendations for camshaft removal. Special tools are often needed. Then the camshaft can be removed.

On OHC heads with removable cam caps, verify that the caps are correctly numbered before removing them (**Figure 52.1**). On DOHC heads, number

Figure 52.1 Verify that the cam caps are correctly numbered before removing them. *(Courtesy of Tim Gilles)*

Figure 52.2 Strike the retainer with a piece of pipe or an old piston pin. *(Courtesy of Tim Gilles)*

the exhaust caps E-1, E-2, and so on, and number the intake caps I-1, I-2, and so on. Varnish buildup often results in the valve retainer being stuck to the valve locks (keepers). Before using a spring compressor, strike each spring retainer with a brass hammer and a short length of pipe, an old piston pin, or a special tool (**Figure 52.2**). If this is not done, the jaws of the spring compressor can be bent or broken.

> **CAUTION** Be especially careful not to bend a valve when striking the retainer. Sometimes the retainer is so firmly stuck that it moves the valve down against the spring when struck. If the cylinder head is face down on the bench, the valve head will strike the bench and be bent. The solution to this problem is to mount the heads on the head stands.

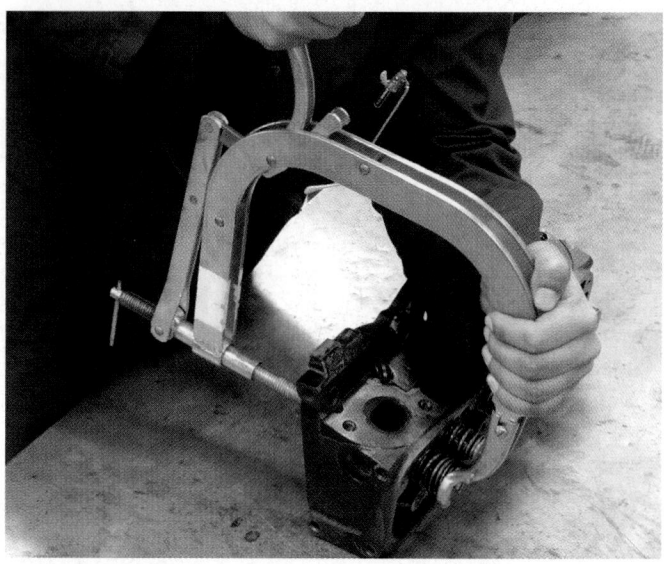

Figure 52.3 A common type of valve spring compressor in use. *(Courtesy of Tim Gilles)*

Spring-Removal Tools

For protection from flying parts, *be sure to wear face protection* when removing valve springs. There are several types of spring-removal tools. Different jaw shapes are available to provide access to the various cylinder head shapes. With a manually operated *spring compressor* (**Figure 52.3**), the jaws must be adjusted to fit the spring retainer (**Figure 52.4**). Adjust the compressor so that the spring will be compressed just enough to allow the keepers to be removed.

Another type of spring compressor is air operated. Be careful that fingers are not accidentally pinched when using this tool, as the plunger moves very fast.

Another type of compressor is the small *screw-type compressor* (**Figure 52.5**). This compressor is especially handy for compressing springs when the heads are

Figure 52.4 Adjust the jaws to fit the spring retainer. *(Courtesy of Tim Gilles)*

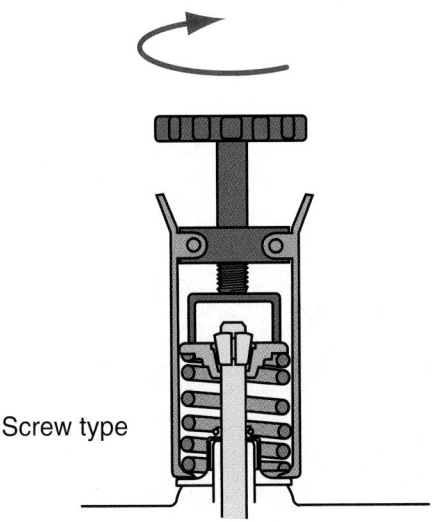

Screw type

Figure 52.5 This valve spring compressor can be used with the heads on or off the car.

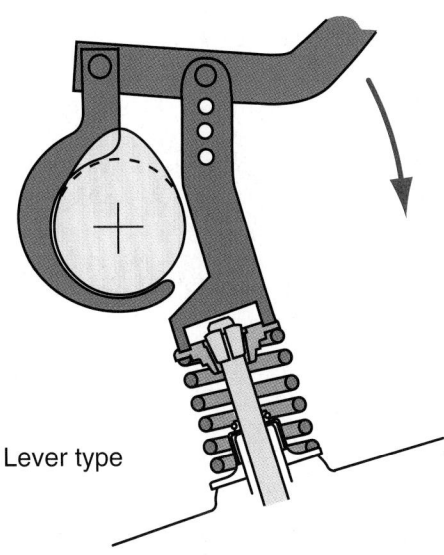

Lever type

Figure 52.6 A valve spring compressor for overhead cam engines.

Figure 52.7 A head with four valves per cylinder. (*Courtesy of Tim Gilles*)

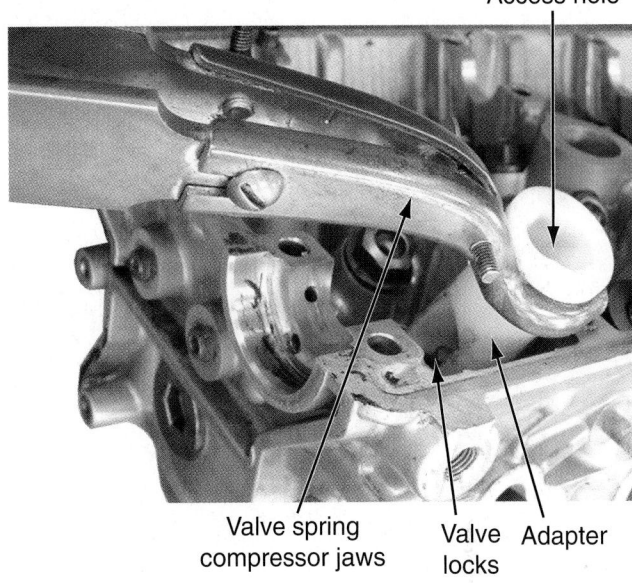

Access hole

Valve spring compressor jaws

Valve locks

Adapter

Figure 52.8 An adapter used with a manual valve spring compressor for removing keepers on heads with bucket-type cam followers. (*Courtesy of Tim Gilles*)

(a)

Figure 52.9 (a) This air-operated bench unit easily compresses valve springs on all types of heads. (*Courtesy of Tim Gilles*)

on the engine. Another type of compressor, shown in **Figure 52.6**, is used on OHC engines. It can be used to replace valve guide seals with the head still installed on the engine.

Heads with four valves per cylinder have become popular in recent years (**Figure 52.7**). These heads typically have bucket-type cam followers (covered later). The valve and spring assembly in these heads is recessed in the head, which makes removal, and especially reinstallation, of the keepers more difficult. **Figure 52.8** shows an adapter that can be used with a standard valve spring compressor. One alternative for shops that do enough of this work to support the investment is an air-operated bench unit (**Figure 52.9a**) that makes disassembly and reassembly of these heads easier and less time consuming. **Figure 52.9b** shows an adapter being used with this bench unit.

Figure 52.9 (b) Different sized adapters are available. They are easily installed in the press ram. This one is being used on a bucket-type OHC head. (*Courtesy of Tim Gilles*)

Figure 52.10 Keep all valves in order. (*Courtesy of Tim Gilles*)

Keep Valves in Order. Keep all valves in the numerical order in which they were removed (**Figure 52.10**).

■ One or more oversized valve stems might have been used in a previous head repair procedure.

■ If the valve guides are not to be serviced, the valves should be removed in numerical sequence and returned to their original guides.

■ Valves can be placed through holes in a piece of wood, metal, or cardboard to ensure their respective order positions.

Be especially careful not to lose any keepers. They are very small and easily lost. Put all small valve components in a sealable storage container, such as a coffee can.

NOTE: *Remember that aluminum heads always have a steel shim beneath the valve spring (**Figure 52.11**). They are very easy to lose, especially during the cleaning process.*

Sometimes the tips of valve stems become mushroomed due to pounding when the engine runs with excessive valve clearance (**Figure 52.12**). A **mushroomed valve tip** area must be dressed with a file or whetstone before the valve can be removed from the guide.

Another reason why a valve might not be easily removable from a valve guide is keeper grooves that have become burred at the edges. A whetstone can be used to

Figure 52.11 Aluminum heads always have a steel shim beneath the valve spring. Do not lose one during the cleaning process. (*Courtesy of Tim Gilles*)

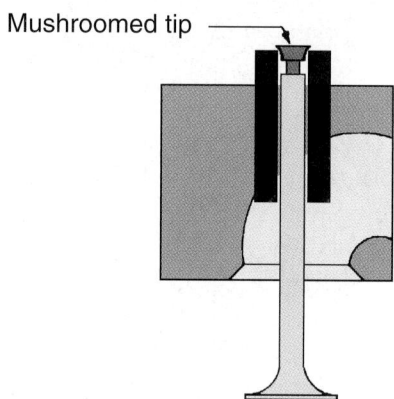

Figure 52.12 A mushroomed valve tip is filed prior to removal.

smooth the burred edges so the valve can be removed and installed without damaging the valve guide.

CAUTION Do *not* try to drive out mushroomed valves with a hammer and punch. The valve or guide can be damaged.

Measuring Stem Height

Measure and record **valve stem height** during head disassembly. Following spring removal, measure the height of the stem tip prior to removal of the valves (**Figure 52.13**) and record these figures on the shop work order. If extensive seat or valve work becomes necessary, obtaining this measurement now will be an important step in the process. Factory specifications are not readily available for stem height. The stem height measurements should be nearly the same from valve to valve. Prior to disassembly, the heights of the valve springs are also checked and recorded (**Figure 52.14**).

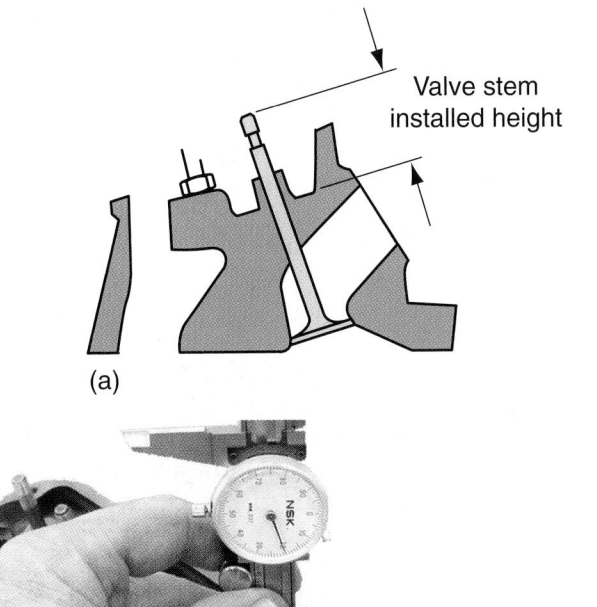

(a)

(b)

Figure 52.13 (a) Valve stem height. (b) Measure valve stem height prior to removing the valves. *(b, Courtesy of Tim Gilles)*

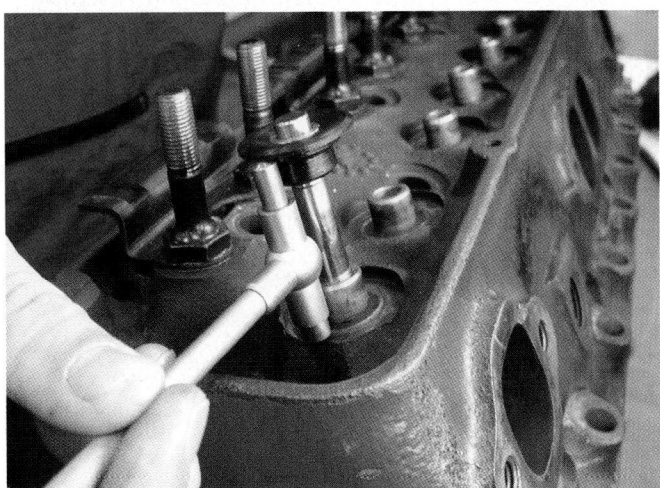

Figure 52.14 Measure valve spring height prior to disassembly of the head. *(Courtesy of Tim Gilles)*

Removing Carbon from Combustion Chambers

Most OHC heads have oil galleries, so they should not be cleaned with a bead blaster (see Chapter 10). Beads will become trapped in the galleries and cause engine failure.

A portable wire wheel with a drill motor, an air drill, or a die grinder are other available carbon removal

tools commonly used, but these methods can damage an aluminum head. Remove carbon and gasket material by hand, using a plastic gasket scraper or carbon removing tool.

Chemical carbon removers can be applied to aluminum heads to help soften the carbon and gasket materials stuck to the surface. When softened, deposits are removed with a plastic scraper.

Removing Carbon from Valves

Carbon can be removed from necks of valves by glass bead blasting, or by using a wire wheel buffer on a grinder (**Figure 52.15**). Clean the keeper grooves as well (**Figure 52.16**), but do not clean the guide rub area of the valve (**Figure 52.17**). This part of the valve stem is typically chrome plated. The wire brush will scratch the chrome, resulting in a rough surface that can cause excessive wear to the valve guide.

Figure 52.15 Using a wire wheel to remove carbon from a valve. *(Courtesy of Tim Gilles)*

Figure 52.16 Cleaning the keeper groove(s) on a wire wheel. *(Courtesy of Tim Gilles)*

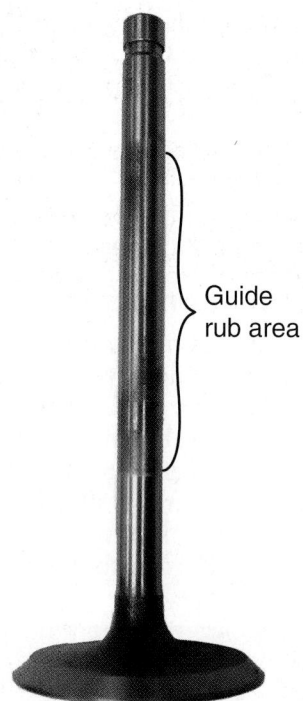

Figure 52.17 The worn area on this valve stem shows where it rides on the valve guide. Do not wire brush this area when cleaning. *(Courtesy of Tim Gilles)*

■ CYLINDER HEAD INSPECTION

Checking for Flatness

Cylinder heads sometimes warp, especially when a head gasket has failed due to engine overheating. Warped heads are resurfaced to ensure head gasket sealing. Clean the head surface before checking for flatness. Use a straightedge and feeler gauge diagonally, vertically, and horizontally on the head (**Figure 52.18**). When checking for warpage on the ends of a head, be sure to rock the straightedge so one edge of it rests against the opposite side of the head. This will leave an opening for the feeler gauge on the other side of the head. If you do not do this, you will only get half of the warp reading.

NOTES:

■ *Warpage on both sides of an OHC head is not always equal. Check the valve cover rail for straightness, too. The spark plug side of the valve cover rail often has more warpage. Maximum warpage on the cam side is 0.002". If it is more, straighten the head as explained*

later. Otherwise the camshaft will not be able to turn without resistance and could break.

■ *Resurface heads warped more than 0.006" on in-line sixes, 0.004" on four or eight cylinders, and 0.003" on three cylinders or V6s.*

■ *Heads should not be warped more than 0.003" in any 6" length.*

■ *Flatness across the width of a head should not vary by more than 0.004".*

When checking head flatness on a four cylinder, try to fit a 0.004" feeler gauge under the straightedge. If it fits, but a 0.005" does not, then it is fine. Also, aluminum heads with corrosion around water passages must be resurfaced. Also check the deck of the block for flatness (see Chapter 51). Sometimes the combined warpage of the block and head exceeds the recommended tolerance.

NOTE: *Always check the manufacturer's specifications. Some newer engines allow only 0.001" warpage on the head or block.*

■ RESURFACING BY GRINDING, CUTTING, OR SANDING

Resurfacing is accomplished either by *"fly-cutting"* the head on a milling machine, grinding the head on a head grinder, or sanding the head on a belt sanding machine. Most heads clean up with less than 0.010" of metal removed. Head resurfacing can increase compression, so remove as little metal as possible. The correct surface finish is very important for correct long-life head gasket operation. In the event more metal is removed than is desirable, gasket companies make 0.020" shims (for some applications only) that can be installed on reinstallation of the head (**Figure 52.19**).

■ STRAIGHTENING CYLINDER HEADS

Warped aluminum OHC heads are commonly straightened. Besides measuring the top surface of the head for

Figure 52.19 A shim used to compensate for metal removed from the head during surfacing. *(Courtesy of Tim Gilles)*

Figure 52.18 Check in several directions for excessive warpage.

Figure 52.20 A head can be straightened using an oven.

Figure 52.21 A portable magnetic crack detector. (*Courtesy of Tim Gilles*)

flatness, check to see if the camshaft turns easily while installed in the head.

NOTE: *If the camshaft does not turn easily and the machine shop only resurfaces the bottom side of the head, a broken camshaft can result.*

There are several methods of straightening cylinder heads. The one considered the best is to use a heating oven (**Figure 52.20**). The head is bolted to a thick plate with shims positioned under the outside edges of the head. The head is heated in the oven for 5 to 6 hours and allowed to cool slowly.

An added advantage to straightening the head prior to surfacing is that combustion chamber volumes will remain equal. Normally during surfacing of the warped head, the outer cylinders' combustion chambers will end up with reduced volume.

NOTE: *Removal of 0.020" from OHC head surfaces results in about 1 degree of retard in valve timing. OHC engines have chain tensioners, but they only control slack on one side of the chain (see Figure 18.45). Slack on the drive side of the chain is taken up by rotation of the crank. This results in valve timing that is retarded after surfacing or chain stretch.*

CRACK INSPECTION

Cracks are sometimes found in combustion chambers, between adjacent combustion chambers, and also, rarely, on the valve spring side of the head.

There are a number of ways to detect cracks. Sometimes after the head is cleaned, cracks will be apparent to the eye. Glass bead blasting is especially helpful in highlighting cracks.

SHOP TIP When water has been entering the combustion chamber from a leaking head gasket or a crack in the head, there will be no carbon on the surfaces of the combustion chamber and piston.

Magnetic Crack Inspection

Crack inspection on *iron* heads can be performed using a simple magnetic crack detector, shown in **Figure 52.21**. The electromagnet sets up a magnetic field. The head is dusted with iron powder and the powder lines up with the magnetic lines of force. A crack interrupts these lines of force, causing the powder to gather around the crack. *Magnaflux* is the common trade name for a more sophisticated type of magnetic crack detection.

Dye Penetrant

Most modern heads are aluminum, so magnetic crack detection will not work. A **dye penetrant** kit can be used to detect cracks in any metal (**Figure 52.22**). It is usually not used on iron because of its expense compared to magnetic detection. After the surface is cleaned, a red dye is sprayed on and allowed to dry for about 3 to 5 minutes. The surface is then wiped clean. Sometimes a dye remover is used to help clean up the surface. A light spray of a white *developer* makes any cracks visible to the eye. Cracks show up as red or pink lines on a white background. A black light crack detector works in a similar fashion. The black light is shined on the surface after spraying a special solution on the head. The fluorescent dye that penetrates the crack is illuminated by the black light.

Pressure Testing

Heads that are especially susceptible to cracks are often inspected by pressure testing (**Figure 52.23**). This test is very effective but can be time consuming. All openings in the head are plugged and the head is filled with water or air. If air is used, soapy water is sprayed over the head surface, or the head is submerged in water to check for air bubbles.

Figure 52.22 Checking for cracks with a dye penetrant. (a) Spray on penetrant. (b) After 5 minutes, clean the surface. (c) Spray on developer to highlight the crack. (*Courtesy of Tim Gilles*)

▬ CRACK REPAIR

Cracks are sometimes repairable, but the repair is only practical if the cost of a bare head is more than twice the cost of the crack repair. If there is any question

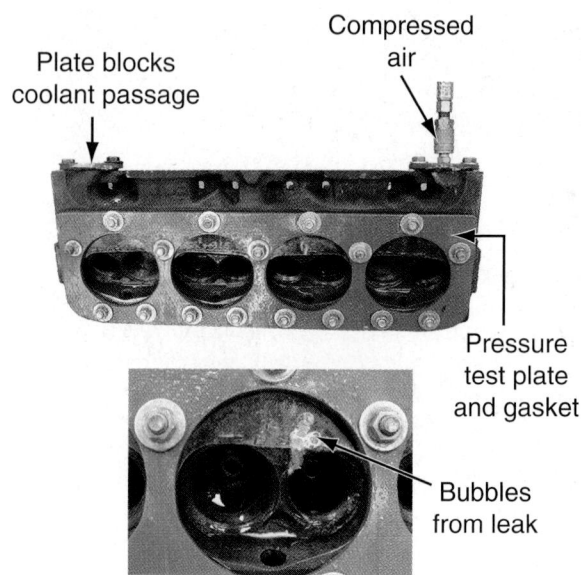

Figure 52.23 A custom-fitted pressure test plate installed on a cylinder head. (*Courtesy of Tim Gilles*)

whatsoever as to the effectiveness of a crack repair, it is not worth taking a chance. A new or used head should be obtained. Also, factory aluminum heads are heat treated. The heat treatment is lost during welding.

Tapered Plugs

Cracks in iron heads are commonly repaired with tapered, threaded plugs. This process is known in the industry as **pinning** or **stitching** a crack. The plugs are usually iron, but sometimes brass plugs are used. Both ends of the crack must be drilled to prevent the crack from spreading any further. If both ends of the crack cannot be seen, the head is usually scrapped.

To pin a crack, holes are drilled and tapped along the crack. The plugs are dipped in ceramic sealer and then threaded into the holes and broken off. More plugs are installed so that they overlap the first plugs (**Figure 52.24a**). Each plug overlaps about one-third

Figure 52.24 Pinning a crack. (a) Install the pins one at a time and then cut them off. Install the plugs so they overlap.

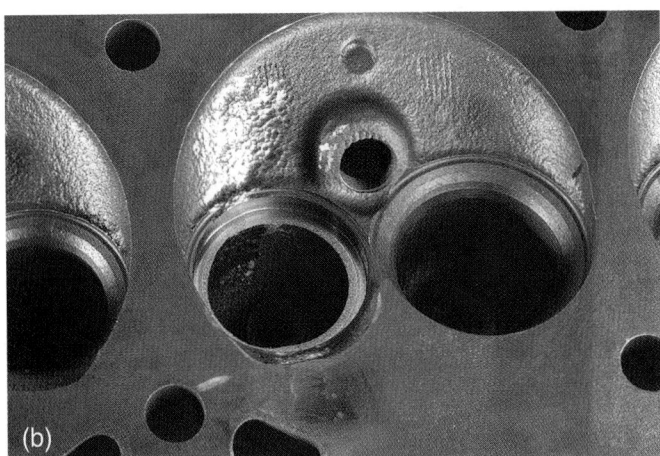

Figure 52.24 (continued) (b) Grind and clean the chamber. Install an insert seat. *(Courtesy of Tim Gilles)*

Figure 52.25 While rotating the spring, there should be no more than ¹⁄₁₆" of variance between the spring and straightedge. *(Courtesy of Tim Gilles)*

of the next plug all along the crack. They are installed at different angles so that they will interlock above and below the surface. To help close the crack, the plugs are peened outward toward the casting and the casting is peened inward toward the plugs. The surface is ground flush to finish the surface.

A combustion chamber crack often extends through the valve seat. After repairing the crack, the valve and seat machine is used to cut a bore in the combustion chamber for a new valve seat (**Figure 52.24b**).

Welding Heads

Welding is a common method of repairing aluminum head cracks. The aluminum is preheated and then welded using an inert gas shield. The process is known in the industry as *heli-arc*, or *tungsten inert gas* (*TIG*), welding.

Iron heads are not welded as often. Welding an iron head is a very specialized job. An iron head must first be heated in an oven to about 700°F, or the head will crack as it is welded. Specialty shops usually weld iron heads using an inert gas welder. The head is resurfaced to complete the repair.

■ CHECKING VALVE SPRINGS

Springs are tested for tension, squareness, and height. Spring tension is tested on a spring tester (see Figure 52.45). A scale on the side of the tester provides a place to check for squareness as well as spring height (**Figure 52.25**). Specifications are available in the service information and in booklets published by spring manufacturers. A spring is allowed to be 10% less than specifications.

■ CHECKING VALVE STEMS

When valve stems wear, the result can be oil consumption. Measure the valve stem with a micrometer at an area of the valve stem above where it normally rides in the valve guide (**Figure 52.26**). Then measure on the

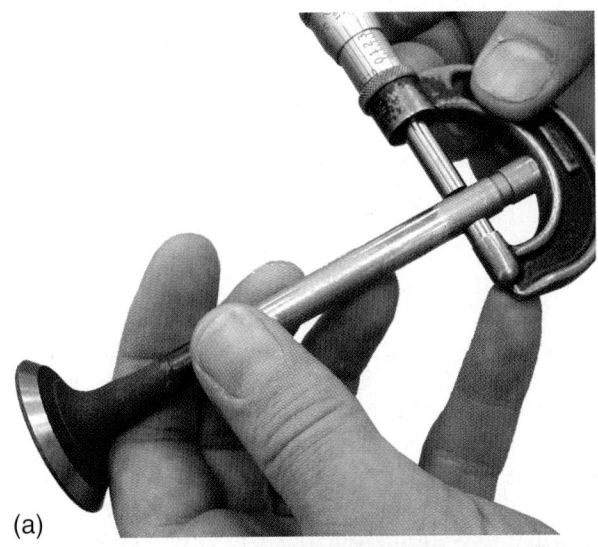

(a)

(b)

Figure 52.26 Measuring the valve stem for wear. (a) Measuring the unworn portion. (b) Measuring the worn portion. *(Courtesy of Tim Gilles)*

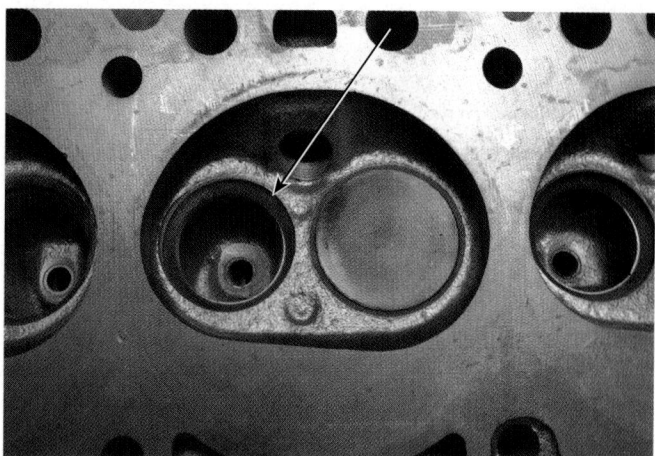

Figure 52.27 This exhaust valve seat is badly worn and is considerably wider on one side than the other. Look for a badly worn valve guide as the cause.

Figure 52.28 Measuring guide wear with a dial indicator. *(Courtesy of Tim Gilles)*

worn area at the top of the stem. Comparing the two will give the amount of wear. Typical wear will be less than 0.001".

VALVE GUIDE SERVICE

Checking Valve Guides

Valve guides wear in a bellmouth fashion, which can result in an increase in oil consumption (see Figure 18.5). When a valve seat has worn and is much wider than usual, look for a worn valve guide as the cause (**Figure 52.27**).

There are two primary ways that *valve stem-to-guide clearance* is checked.

- Use a split ball gauge and a micrometer (see Figure 6.27). Measure at the top, bottom, and center.
- Using a dial indicator, rock the valve back and forth (**Figure 52.28**). When measuring with an indicator, divide the measurement by 2 for the amount of valve-to-guide clearance. Hold the valve in its normally open position during the measurement. A hose can be cut to the correct length to hold each valve open the correct amount.

NOTE: *Many engines have positive valve guide seals (covered later in this chapter). Positive guide seats fit the top of the guide snugly and must be removed prior to checking valve stem-to-guide clearance. On OHC engines, they are sometimes difficult to remove without a special tool like the one shown in* **Figure 52.29**.

GUIDE REPAIR

Guides can be repaired in several ways:
- A worn integral guide can be bored out to accept a pressed-fit insert guide.
- A worn insert guide can be pressed out and replaced with a new one.

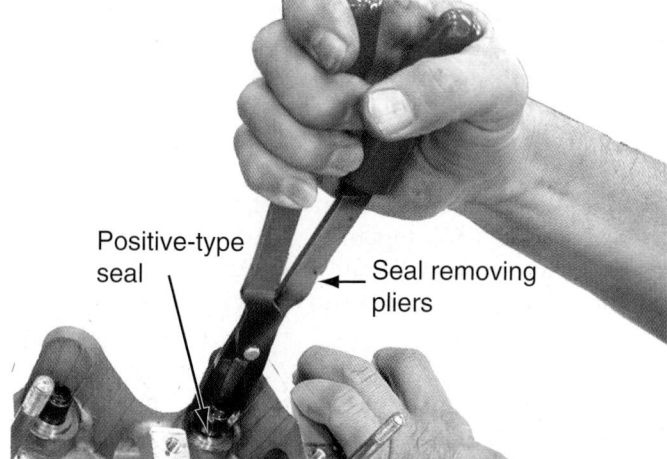

Positive-type seal

Seal removing pliers

Figure 52.29 Removing a positive valve seal using special pliers. *(Courtesy of Tim Gilles)*

- Guides can be repaired using a process called **knurling**. Knurling makes the inside of the guide smaller. Then it is reamed out to fit the valve stem (**Figure 52.30**). Knurling does not usually provide the longevity of other repairs.

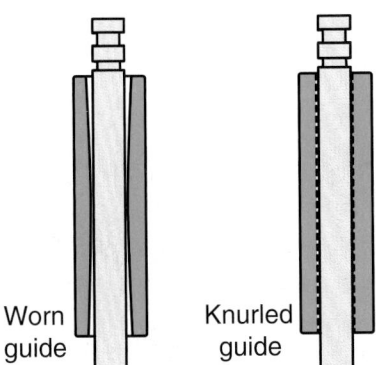

Figure 52.30 A valve guide before and after the knurling process.

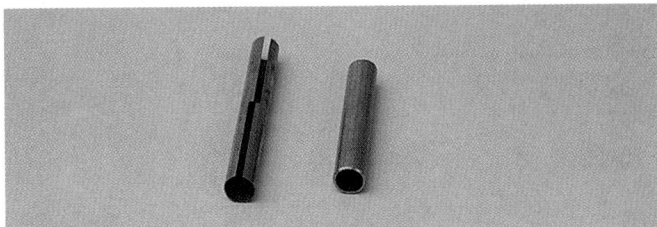

Figure 52.31 A thinwall guide liner (left) and a solid guide liner (right).

■ A **thinwall** insert can be installed in a valve guide (**Figure 52.31**). First, the guide is bored oversize and the insert is pounded in. Next, it is broached to form it to the guide bore. Finally, it is reamed to provide the correct clearance for the valve stem. A pressed-fit guide can also be repaired by knurling (only on cast iron guides) or installing a thinwall insert. This practice is sometimes easier or less expensive than replacing the valve guide.

When reaming guides, be certain to provide sufficient stem-to-guide clearance. **Figure 52.32**

shows valves that burned because they were too tight in their guides.

■■ GRINDING VALVES

Valves are refinished on the face angle using a valve grinder (**Figure 52.33**). The stem tip is reground flat during valve grinding (**Figure 52.34**). It is also ground to a chamfer (**Figure 52.35**). Be sure to wear eye protection when working around grinding wheels. Dust comes off of the grinding wheel, and grinding wheels have been known to explode. The grinding wheel is dressed with an industrial diamond that wears very little as it removes the outer surface of the grinding wheel (**Figure 52.36**). The grinding wheel surface must be squared prior to its use in machining a part. Diamonds are the hardest materials on Earth, so they are very effective in cutting the surface of grinding stones.

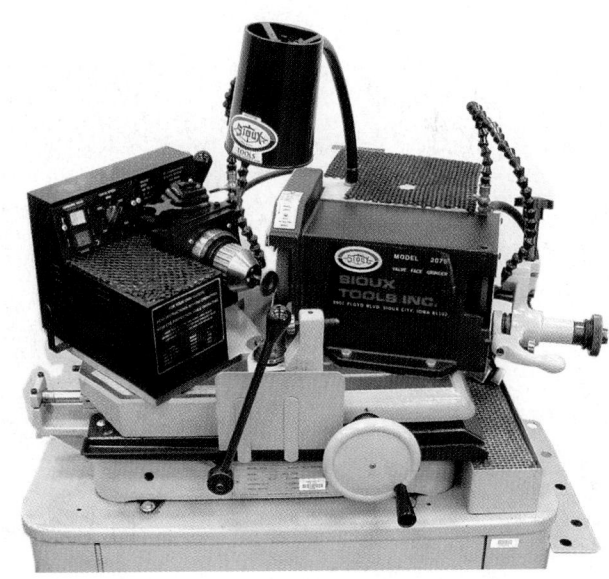

Figure 52.33 A valve grinder. (*Courtesy of Tim Gilles*)

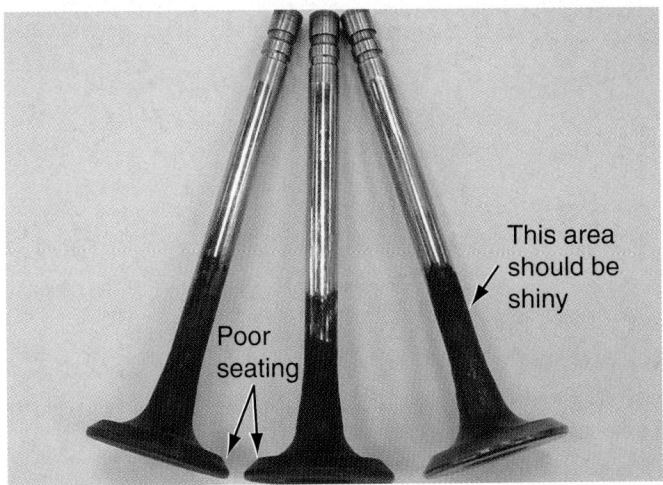

Figure 52.32 Valves that are too tight in the bottom (hottest area) of their guides can stick instead of closing properly. This results in poor valve seating and, ultimately, a burned valve. (*Courtesy of Tim Gilles*)

Figure 52.34 Grinding the valve stem tip. (*Courtesy of Tim Gilles*)

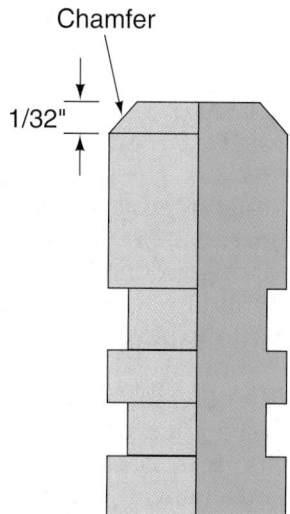

Figure 52.35 The stem tip is ground to a chamfer. *(Courtesy of Federal-Mogul Corporation)*

Figure 52.36 The grinding wheel is dressed with an industrial diamond. *(Courtesy of Tim Gilles)*

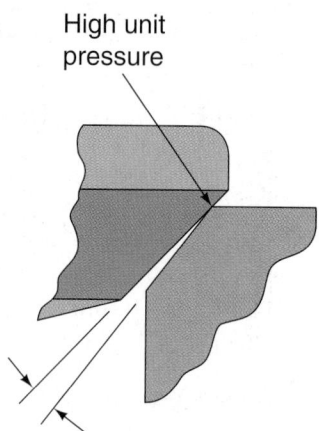

Figure 52.37 An interference angle. *(Courtesy of Federal-Mogul Corporation)*

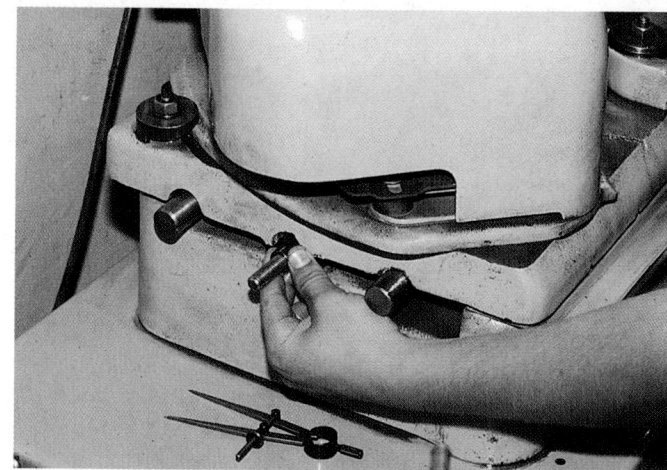

Figure 52.38 Adjust the carriage stop so that the neck of the valve will not come into contact with the stone. *(Courtesy of Tim Gilles)*

Figure 52.39 This valve has been ruined because the neck area was ground. *(Courtesy of Tim Gilles)*

The angle of the chuck is adjustable for grinding valve faces to different angles. The most common angle is 45 degrees. Some machinists grind an **interference angle** between the valve and the seat (**Figure 52.37**). The valve is usually ground to 44 degrees, and the seat is ground to 45 degrees. This helps the newly reground seats and valves to seat into each other.

The valve grinder has an adjustable *stop* (**Figure 52.38**) that prevents the neck of the valve from accidentally contacting the grinding wheel, which would ruin the valve (**Figure 52.39**).

Grinding oil is pumped to a nozzle that flows it onto the face of the valve during grinding (**Figure 52.40**). The valve can be observed as it turns in the chuck. If it wobbles, it is either bent or it is not chucked properly.

Very little metal is removed from the surface of the valve face during grinding. If too much metal is removed from the valve face, the valve face margin will be too thin and the valve could burn. The face is ground until all "pits" are removed. Pits are small, dark indentations in the surface of the metal on the valve face.

Figure 52.40 Adjust the coolant to flow over the valve face during grinding. *(Courtesy of Tim Gilles)*

GRINDING VALVE SEATS

Valve guides must be refinished before attempting to refinish valve seats. Servicing valve guides after seat refinishing will result in an off-center valve seat that must be reground or cut. A valve that is off-center will leak, causing a rough idle when cold. The valve will heat up very fast after startup, becoming red hot and softening to conform to the valve seat, resulting in a smooth idle when warm.

Valve seats are refinished with a grinding stone (**Figure 52.41**) or a seat cutter (**Figure 52.42**). Grinding three angles makes it possible to control the width of the seat (**Figure 52.43**). The intake seat is usually ground to a width of about 1/16", and the exhaust valve is ground to about 3/32" (**Figure 52.44**). Exhaust seats are wider than intake seats because valve cooling is critical. If the seat is too narrow or the valve clearance adjustment is too tight, the valve will not be able to dissipate enough heat to keep it from burning.

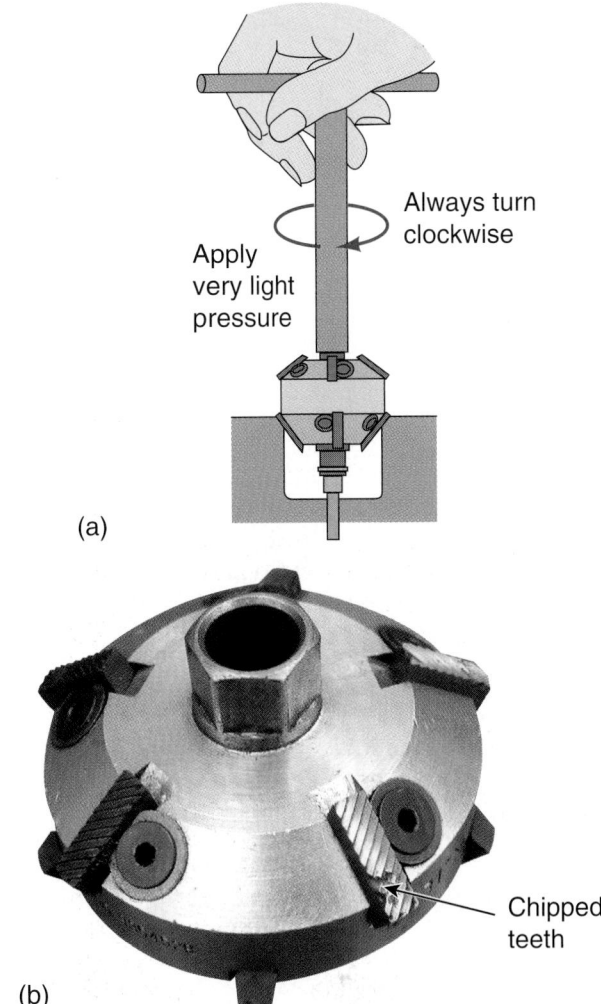

(a)

Apply very light pressure

Always turn clockwise

Chipped teeth

(b)

Figure 52.42 (a) Seats are sometimes refinished with a carbide seat cutter. (b) Be careful not to let the cutters slide down the guide pilot. They are fragile. *(a, Courtesy of Neway Manufacturing, Inc.; b, Courtesy of Tim Gilles)*

Top angle

To remove stock from top of seat, use 30-degree wheel

Throat angle

To remove stock from bottom of seat, use 60-degree wheel

45-degrees

Valve seat width

Figure 52.43 A three-angle valve seat.

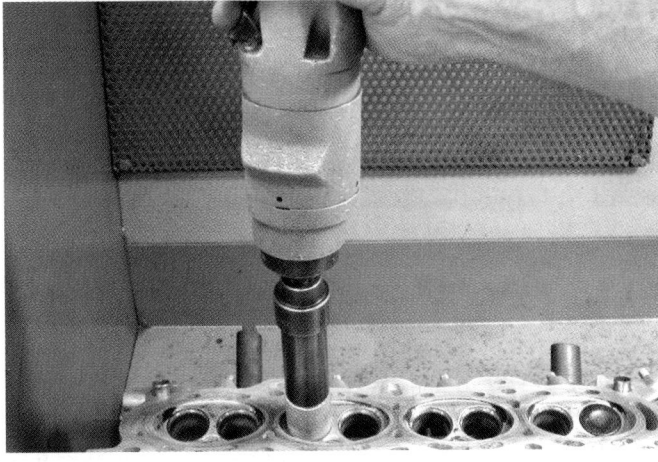

Figure 52.41 Grinding a valve seat. *(Courtesy of Tim Gilles)*

The 45-degree angle that contacts the valve face is ground or cut until the entire seat area is cleaned up and is free of pits. Sometimes the seat will polish up with very

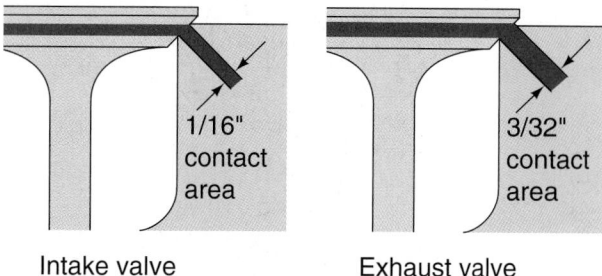

Figure 52.44 Intake and exhaust seat width. *(Courtesy of Federal-Mogul Corporation)*

little metal removed. When this happens, machining of the other two angles may not be necessary.

The 60-degree angle in the bottom of the seat, called the throat angle, is cut *very lightly*. This angle can be done by hand (with no grinding motor). Cut the throat until a ring shows up around the entire bottom of the 45-degree area. Be careful not to encroach into the seat area with this angle.

CAUTION Removing too much of the seat with the 60-degree cutter or grinder will result in a need to replace the seat.

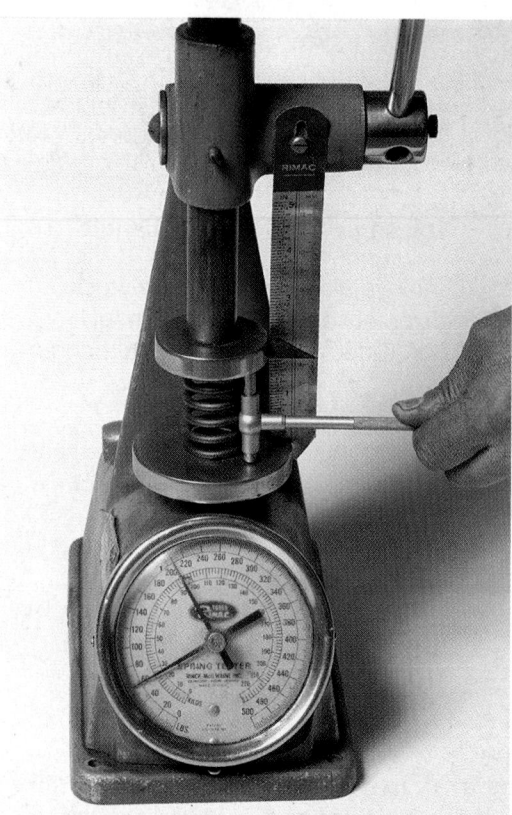

Figure 52.45 Use the installed height measurement to check to see what the actual tension will be.

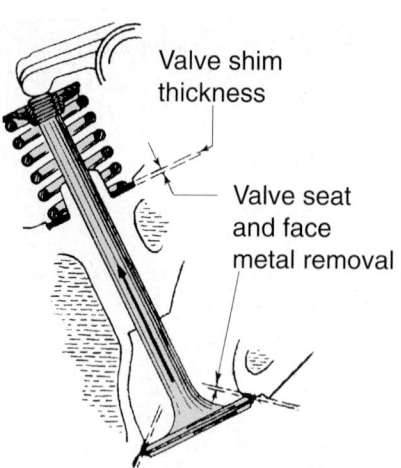

Figure 52.46 The installed height of the spring changes due to wear or metal removal during machining. *(Courtesy of Hastings Manufacturing Company)*

The top angle is cut until the width of the seat is correct. The head must be thoroughly cleaned of all grit before beginning assembly.

■ CHECKING INSTALLED HEIGHT OF THE VALVE STEM

When the seat and valve are reground, the stem moves further into the cylinder head. This results in increased **valve spring installed height**. After grinding the valve and seat, check the installed height. That measurement can be used to check what the tension of the spring will be when installed on the head (**Figure 52.45**). To correct for too excessive valve stem installed height and maintain correct spring tension, shims are installed under the springs when a head is reassembled (**Figure 52.46**).

■ SOLVENT TESTING THE VALVE AND SEAT

After the valve and seat have been ground:
- Turn the head over so that the combustion chambers face up.
- Place the head on head stands and put it on a shelf in the solvent tank.
- Install the valves in their ports.
- Install the spark plugs in their holes.
- Fill the combustion chambers with solvent and check for leaks (**Figure 52.47**).

The solvent test is a very important test. If the valves seal against leakage without springs, they will seal when the engine is started. If any of the valves leak, they can be repaired now while it is easy to do.

■ REASSEMBLING THE HEAD

Clean the head thoroughly before reassembly. Be sure that the guides have been thoroughly cleaned too. Lubricate all valve stems (**Figure 52.48**).

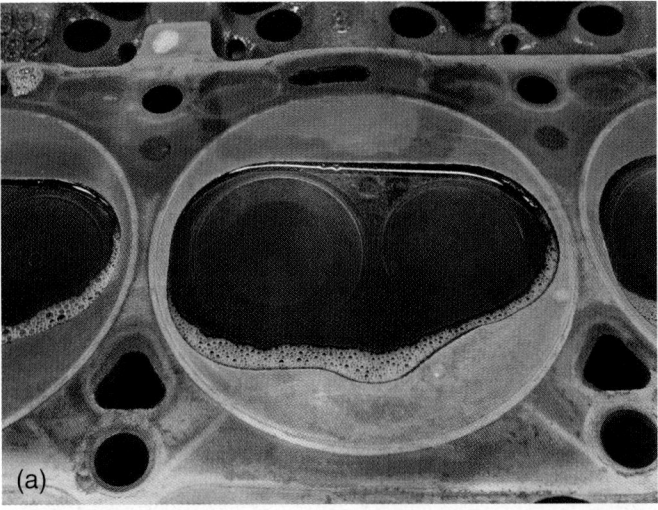

Figure 52.47 Solvent test for valve seating. (a) Fill the combustion chamber. (b) One of these valve seats is leaking. *(Courtesy of Tim Gilles)*

Figure 52.48 It is very important to lubricate the valve stem before assembling the head. Anti-seize lubricant is being applied here. *(Courtesy of Tim Gilles)*

■ VALVE GUIDE SEAL INSTALLATION

Install the guide seals before installing the springs on all but O-ring seals.

NOTE: *Before installing positive seals, remember to install the valve spring shim, if applicable.*

Sometimes intake and exhaust seals are made of different materials or have different shapes. Always check for instructions in the gasket set regarding the placement of these seals. Umbrella exhaust seals sometimes are shorter. To save money, sometimes manufacturers make the intake seals from an artificial rubber material with a lower temperature rating than the exhaust seals.

Be certain to lubricate the seals before installing them (**Figure 52.49**). By the time splash lubrication reaches the seals, they will probably be ruined by excessive heat.

Positive seals are often supplied with a plastic sheath that is installed on the top of the valve stem to protect the sealing surface on the inside of the seal from accidental damage by the valve stem retainer grooves.

CASE HISTORY *A student was restoring a Camaro that had 90,000 miles on it. It did not smoke or use oil, but he wanted to put the engine in as-new condition. After a careful and complete rebuild, he took the car on a trip. It used a quart of oil every 200 miles. All the spark plugs had carbon deposits on one side of the center insulator only. With compressed air injected into the spark plug holes, the valve springs were removed to inspect the valve guide seals. It was discovered that the O-ring seals had been put onto the valve stem before the spring was installed. Each seal was twisted. Excessive oil consumption was the result.*

Figure 52.49 Carefully slip the lubricated valve stem seal onto the valve stem. *(Courtesy of Tim Gilles)*

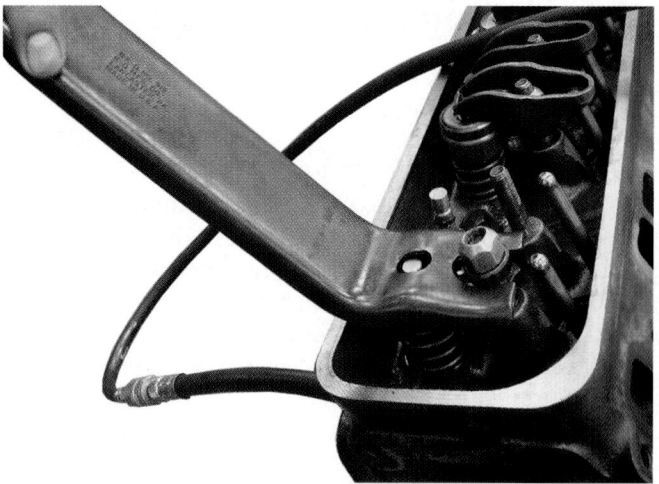

Figure 52.50 The valve is held closed with air pressure while the spring is compressed. *(Courtesy of Tim Gilles)*

On a worn engine, valve guide seals can be replaced with the head on the engine (see Figure 52.5 and Figure 52.6). Air is put into the cylinder to hold the valve while the spring is compressed (**Figure 52.50**). The piston for each cylinder is positioned at TDC with both valves closed. If both rocker arms are removed, piston position is unimportant unless the air is disconnected while the valve keepers and spring are removed from the valve.

▬ INSTALL THE VALVE AND SPRING ASSEMBLY

Some springs have coils more closely spaced at one end than at the other. During assembly, be sure that the end that is more tightly coiled is positioned against the cylinder head. **Figure 52.51** shows different styles of valve springs. To prevent damage to the valve guide seal, compress the spring just enough to install the keepers. Inspect each keeper for wear before installing it. Use grease to help hold keepers in place during reassembly. After assembly, tap the top of each valve tip with a soft face hammer to check that the keepers are seated. Some newer engines use bee hive–shaped springs, with one end of the coils smaller in diameter. The small end goes up, away from the head.

Uniform coil | Coil with damper | Close wound coils toward head

Figure 52.51 Different types of valve springs. *(Courtesy of Tim Gilles)*

Figure 52.52 Lubricate rocker arm pivots with assembly lube. *(Courtesy of Tim Gilles)*

▬ PUSHROD ENGINE ROCKER ARM SERVICE

Engines that do not have hydraulic clearance adjustment sometimes experience wear between the rocker arm and valve stem tip. This makes it difficult to get a proper valve clearance adjustment. Cast rocker arms that are shaft-mounted can be reground. Stud-mounted rocker arms are not serviceable and are replaced when worn.

Sometimes rocker studs pull out of the head, or the rocker wears against the rocker stud. An oversize stud can be installed by a machine shop. A threaded replacement stud, called a *screw-in stud*, is also available.

Be sure to thoroughly lubricate rocker arms before installing them (**Figure 52.52**).

▬ INSPECT PUSHRODS

Inspect the ends of the pushrod and the surface of the socket where it pivots on the rocker arm. Look for pitting or other unusual wear (**Figure 52.53**). Roll the pushrods on a bench to check if they are bent.

Figure 52.53 Inspect both ends of each pushrod for excessive wear. The rocker arm used with this pushrod was also replaced. *(Courtesy of Tim Gilles)*

INSPECT OHC CAMSHAFT

Overhead camshafts often have oil galleries and holes drilled in cam lobes for direct lubrication. On some camshafts, the oil holes are small and are prone to plugging. Check that the oil holes are clear before installing a used camshaft in the head. This is a good practice to follow when installing a new cam, too.

REASSEMBLING OHC HEADS

On overhead cam engines, reinstall the camshaft in the head. Check to see that the camshaft cap alignment bushings are installed and positioned correctly before installing and torquing the cam caps. On bucket-type OHC heads, lubricate the buckets (**Figure 52.54**) and install them in the head prior to installing the cam. Adjust the **valve clearance** (lash) before installing the head on the engine.

Adjust Valve Clearance

Valve lash must be enough to allow heat to dissipate from the valve to the valve seat. As pushrods, rocker arms, and other valvetrain parts wear, lash *increases*. Lash *decreases* as the valve and seat wear, allowing the valve stem to move further into the valve guide (**Figure 52.55**).

NOTE: *If the lash is too tight, the valve will not contact the valve seat for a long enough time to cool. This will cause the valve to burn.*

Overhead cam engine valve lash is usually easy to adjust.

■ On one type of engine, a feeler gauge of the specified thickness is inserted between the rocker or cam follower and the cam while the lobe faces up. Then the clearance is adjusted. Two wrenches are needed: one loosens the locknut while the other makes the adjustment (**Figure 52.56**).

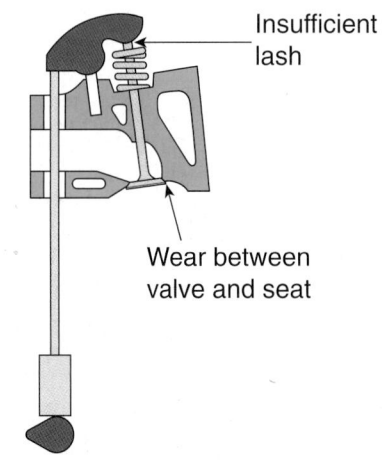

Figure 52.55 Problems related to valve lash. Lash decreases because of head expansion or valve and seat wear. *(Courtesy of Federal-Mogul Corporation)*

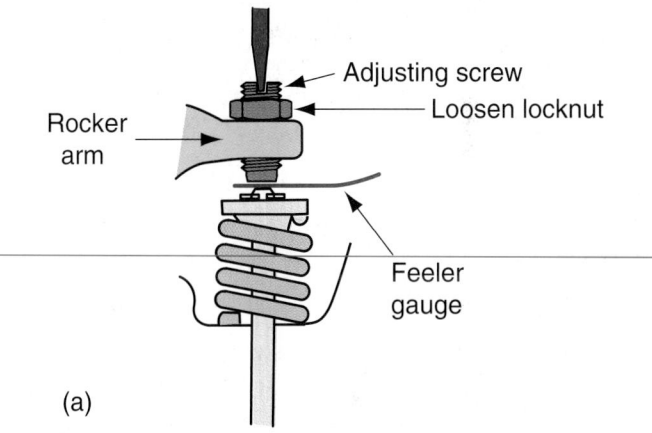

(a)

(b)

Figure 52.56 (a) Mechanical valve lash adjustment. (b) Adjusting OHC valve clearance. *(b, Courtesy of Tim Gilles)*

Figure 52.54 Thoroughly lubricate the OHC buckets before installing them on the head. *(Courtesy of Tim Gilles)*

■ On some OHC engines, the cam pushes on the valve through an adjustable cam follower (often called a "bucket"). One type of head design uses different thicknesses of shims (**Figure 52.57**).

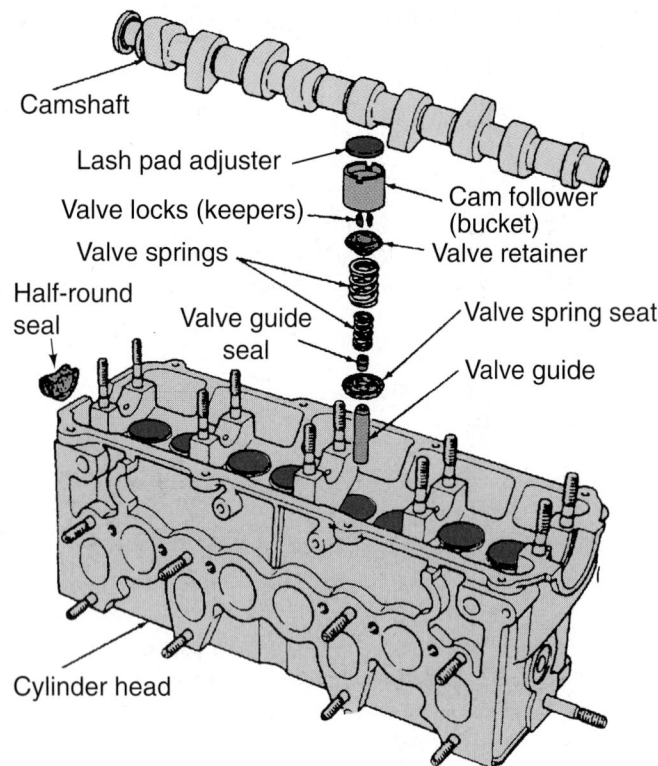

Figure 52.57 Some engines use different thicknesses of disks to adjust valve clearance. *(Courtesy of DaimlerChrysler)*

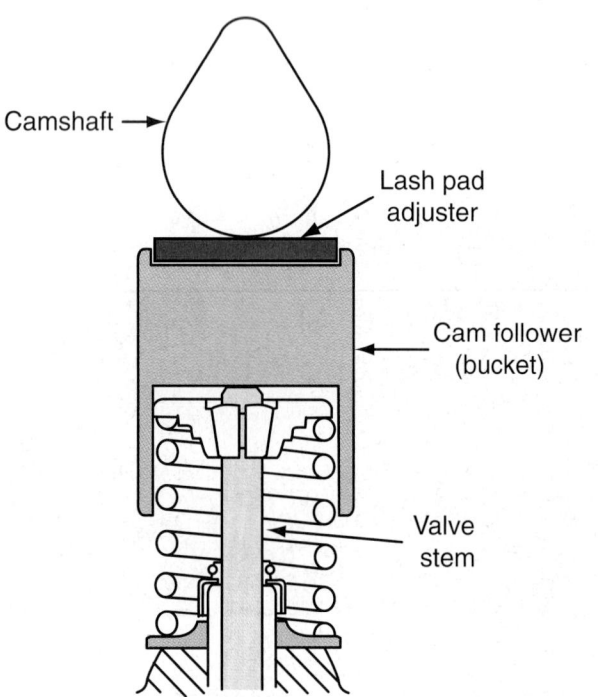

Figure 52.58 Clearance is measured between the shim (lash pad adjuster) and the cam lobe.

Clearance is measured between the shim and the cam lobe (**Figure 52.58**).

On OHC heads with bucket adjusters, the valve clearance can be adjusted using special tools so the

Figure 52.59 Prying the special tool against the camshaft holds the two buckets down against spring pressure while another tool lifts the disc out of its seat in the bucket. *(Courtesy of Tim Gilles)*

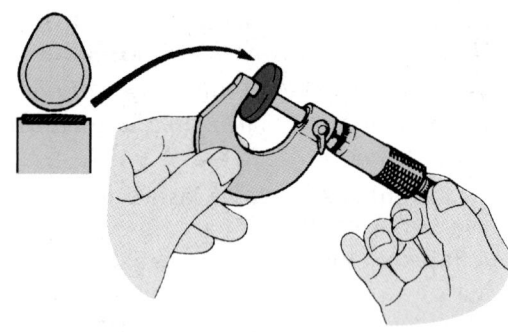

Figure 52.60 The shim is measured and a new one of the correct thickness is installed.

adjustment disks, or pucks, can be removed and replaced as needed. One tool fits on the outside edges of two buckets. Prying the tool against the camshaft holds the two buckets down against spring pressure while the other tool reaches around the camshaft and lifts the disk from the bucket (**Figure 52.59**). Some technicians like to use a rubber-tipped blowgun to release the disk from the oil that tends to hold it to the bucket. The shim is measured (**Figure 52.60**), and then a new one of the proper thickness is installed.

■ Another kind of cam follower has a tapered adjusting screw. Each turn of the screw changes clearance by 0.003". Adjustments must be done in one-turn increments (0.003").

■ CAMSHAFT SERVICE

Measuring the Camshaft for Wear

Specifications are not readily available for the height of cam lobes. A comparison measurement can be made by measuring each lobe and seeing if they are all close to the same. A visual check of the cam lobes is the standard way of looking for wear (see Figure 18.32). The cam and lifters become wear-mated during the first few minutes the engine is run.

NOTE: *If the cam is replaced, replace the lifters also. Used lifters will rapidly wear out a new cam. If a cam is to be reused, it is important that the lifters be kept in order for replacement on their respective lobes.*

Roller Cam Wear

Lobes on roller cams are polished to a fine matte finish. During engine break-in, the lifter burnishes the lobe to a smooth mirror finish. The visual wear pattern is not as important as it is with conventional cam lobes. To inspect for wear, measure the lobe and look for a wear ledge greater than 0.005".

■ LIFTER SERVICE

Hydraulic lifters can fail for several reasons:
- Dirt lodged in the check valve can allow the lifter to leak. Too much wear between the lifter and the body can cause excessive leak-down. This results in a noisy lifter.
- A lifter might be noisy because of an oil pressure problem.

> **SHOP TIP** An oil pressure problem will be evident on hollow pushrod engines if no oil is reaching the rocker arms.

- A lifter can become stuck because of varnish that accumulates between its plunger and body due to a lack of periodic oil changes.

> **SHOP TIP** A stuck lifter can sometimes be freed up by the addition of an additive to the oil, or by spraying fuel system spray cleaner down a hollow pushrod oil channel.

Hydraulic lifters are not rebuilt, because they are relatively inexpensive. It is possible to disassemble and clean lifters, but they *must* be reassembled with their mated parts. Because of the labor time involved, this procedure is usually impractical. Sometimes a piece of foreign material becomes lodged in the lifter check valve. If the lifter is removed and disassembled before serious wear between the lifter and cam lobe results, it may be cleaned.

Worn mechanical lifters can be reground. The base of the lifter is ground to a radius on a special lifter grinder.

Figure 52.61 Camshafts on modern engines are driven by either a belt or a chain. *(Courtesy of Tim Gilles)*

Roller lifters have a longer service life than standard lifters. Wear can occur on the roller pin. Some of these pins are serviceable.

■ CAM AND LIFTER BREAK-IN

Lubrication and break-in are critical to the life of a cam. On a pushrod engine, the cam is usually lubricated by oil that is thrown from the crank, so long periods of idle are hard on the cam. A cam that survives the first half hour of use without wear should last the life of the vehicle with minimal additional wear. Chapter 56 covers the correct way to break in a camshaft and lifters.

■ TIMING CHAIN AND BELT SERVICE

Camshafts on modern engines are driven by either a belt or a chain (**Figure 52.61**). Some older engines used two gears between the crankshaft and camshaft.

■ TIMING BELT SERVICE

Poor alignment, incorrect tension, and worn sprockets will all result in increased timing belt fabric wear. After the fabric begins to wear, cracks will start to appear at the base of the teeth on the belt. This is the most common type of belt failure.

NOTE: *The timing cover serves an important purpose, especially during bad weather. Snow can blow in on the warm engine of a parked vehicle, where it melts and freezes up again on the timing sprocket.*

> **CASE HISTORY** *After a trip, a man parked his vehicle outside for the night. During the night there was a snowstorm with heavy winds. When he attempted to start his car in the morning, the engine would not turn over. His son had replaced the timing belt, and the timing cover had not been properly reinstalled. Snow that blew into the space between the sprocket and belt melted on the hot engine and then froze again during the night.*

To inspect the condition of a timing belt, twist it gently. Do not turn it more than 90 degrees. The tensile cords in timing belts can be damaged. Be certain that all of the small rubber gaskets that separate the parts of a multiple-piece timing cover are reinstalled properly.

Because timing belts are not lubricated, there are no sealing gaskets on the timing cover. A belt's service life can be affected by contact with foreign material: oil, mud, smog, or even ice. Timing covers for belts are designed to *thoroughly* cover the belt. If a foreign object gets into the belt area, damage to the tensile cord can result in belt failure. A damaged flange behind the sprocket can cut the belt.

Timing belts have a very strong fiberglass cord structure, but they are delicate. When overflexed, the cords can break. Do not twist them more than 90 degrees and do not coil them up or hang them for storage.

To inspect a timing belt:
- Look for fraying, cracks at the base of a tooth, or loose fibers.
- Rotate the engine slowly by hand while checking it.
- Do not twist the belt.
- Wear on one side of a tooth indicates a misalignment problem.
- Check for oil leaks, which could damage the belt. An OHC head with a timing belt has a camshaft oil seal, and some have an O-ring behind the cam sprocket. It is a good idea to replace these during a timing belt replacement. If the valve cover has been removed, loosening the front cam cap makes this job easier (**Figure 52.62**).

CASE HISTORY

A young woman had recently purchased a Honda Civic. Although the car was 12 years old at the time of purchase, it had only 40,000 original miles on it. Because of the age of the vehicle, her automotive technician suggested that she have the-timing belt and other rubber parts, such as drive belts and hoses, replaced. She told him to go ahead and replace all of those parts, as well as the timing belt.

*Following the belt replacement, the car ran for about 5,000 miles without incident. Then, while on the freeway on a long trip, her engine suddenly stopped running, and she had the car towed in for repair. The technician discovered that all the teeth were stripped off the timing belt. (**Figure 52.63**). This is unusual. Typically, a few of the teeth are stripped off a belt when it fails. Further investigation showed that her water pump, which was driven by the timing belt, had failed.*

Stripped cogs

(a)

Loosen cam cap

Seal   Socket

Figure 52.62 Overhead cam heads have a seal behind the cam sprocket. Loosen the front cam cap for easy installation of the seal. *(Courtesy of Tim Gilles)*

All cogs stripped from belt

Water pump frozen

(b)

Figure 52.63 (a) A stripped timing belt. (b) This belt was stripped of its teeth when the water pump froze. *(Courtesy of Tim Gilles)*

Experienced technicians realize that it is advisable to replace the water pump whenever a timing belt that drives the pump is replaced. Water pumps rarely last over 100,000 miles.

■ TIMING BELT REPLACEMENT

Follow manufacturer's recommendations for the replacement interval for the belt. This varies from 20,000 to 60,000 miles. Some manufacturers have longer intervals. Goodyear recommends replacing all timing belts at 40,000 to 50,000 miles. When no recommendation is made, timing belts should be changed every 4 years.

Most American cars are free-wheeling (See Chapter 18). According to the Gates Rubber Company, approximately 40% of engines that have timing belts are interference. Japanese imports are more likely to be interference engines. Those American cars that are interference most often have foreign-made engines.

When doing a maintenance replacement of a timing belt, the time requirement can vary widely among vehicles. Most require 3 or 4 hours, but some take 6 or more. On front-wheel-drive cars, remove the front wheel on the side of the vehicle toward which the crankshaft pulley faces. Remove the inner fender panel. Support the engine with a special fixture like the one shown in Figure 56.15. Remove the engine mount bolts. Remove the crankshaft pulley bolts and pulley and remove the upper and lower timing covers (if there is more than one part to the timing cover).

NOTE: *Before removing the belt, verify that the crankshaft timing marks are aligned correctly.*

Loosen the tensioner bolt and remove the timing belt. Before replacing a timing belt, clean the sprockets with a non-petroleum-based solvent like alcohol or brake cleaner, and inspect them for wear that would cause wear to the teeth on the new timing belt. Replace a sprocket that shows any wear.

An OHC head with a timing belt has a camshaft oil seal behind the cam sprocket. Verify that there is no oil leakage into the area surrounding the belt. If camshaft seals or a valve cover gasket are leaking, repair them prior to replacing the timing belt, or premature failure will occur. It is good insurance to replace these during routine replacement of a timing belt.

NOTE: *The spacing of the teeth on a timing belt must exactly match the cogs on the cam and crank sprockets. Belts have different tooth rib profiles. Some have round ribs, and others are square. Be certain of an exact match. Compare an old belt to its replacement for belt width and for tooth shape and spacing.*

Before installing the belt on the sprockets, put the crankshaft at TDC and be sure that the camshaft and crankshaft timing marks are aligned correctly. Check the service information for the correct positions of the marks. They vary among manufacturers.

SHOP TIP If a camshaft must be relocated to set cam timing on a non-free-wheeling, belt-driven OHC engine, turn the crank to about 45 degrees BTDC. Then all pistons will be slightly down in their bores and valve-to-piston interference will not occur.

Install the new belt and adjust the belt tension until snug. Turn the crankshaft manually two revolutions and recheck the timing marks.

Loosen and re-tension the belt. Many engines have spring-loaded belt tensioners. On these engines, make sure that there is tension on the drive side of the belt (**Figure 52.64a**).

Then, loosen the tensioner to allow the spring to force the tensioner roller against the belt (**Figure 52.64b**). Tighten the tensioner bolt and crank the engine over at least ten times. Tension on the belt will drop as it seats into the timing sprocket. Loosen and retighten the tensioner adjusting bolt to complete the job.

NOTE: *Do not adjust timing belt tension on a hot engine. The belt will be too tight, which can result in a broken camshaft.*

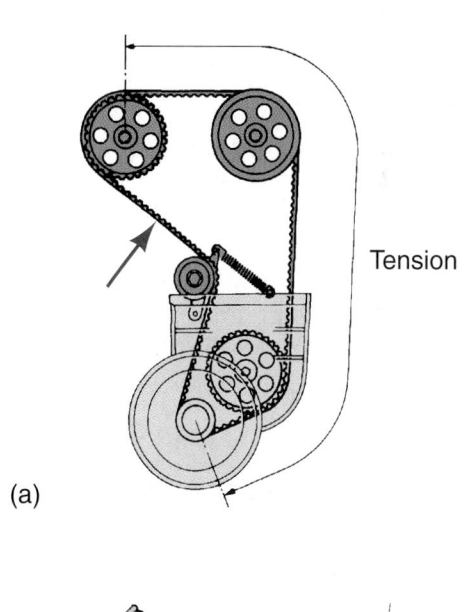

Tension

(a)

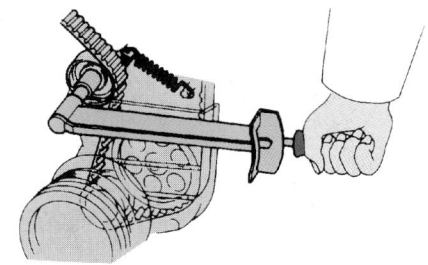

(b)

Figure 52.64 (a) Loosen the tensioner and let the spring pull the roller against the belt. (b) Before tightening the tensioner, see that the belt is tight in the area noted here.

Reassemble all of the parts and affix a sticker to the fender well that tells the mileage when timing belt service was done.

◼ TIMING CHAIN SERVICE

Excessive timing chain stretch can be checked in one of the following ways:

1. On engines without chain tensioners that have distributors driven by the camshaft, watch the timing mark while checking ignition timing at idle. If the mark moves back and forth, the chain is loose.

2. To check a pushrod engine for chain slack while the engine is in the car, remove the distributor cap and turn the vibration damper in one direction until the distributor rotor begins to turn. Then, turn the crank in the opposite direction until the rotor moves again. Observe how far the damper moves before the rotor turns. Because most distributors are driven by the cam, this is a good way to check for chain slack on these engines. The engine may still run acceptably with 10 to 15 degrees of chain slack. But ideally, movement should be less than 5 degrees at the crank.

3. The most obvious way to check chain slack is to measure the slack during engine disassembly after removing the timing cover. Be sure to turn the crank in one direction first, in order to tighten up one side of the chain. Then measure the amount of slack in the chain.

NOTES: *Some timing chain tensioners are spring loaded and are locked by oil pressure when the engine runs. Do not turn these engines over backward by hand. Some types of tensioners will allow the chain to skip off correct valve timing.*

Long chains have chain guides with rubber surfaces that wear. The chain guides are usually replaced when the chain is replaced (**Figure 52.65**).

Figure 52.65 Typical wear to a high-mileage timing chain guide and tensioner. The contact surfaces were smooth when new. *(Courtesy of Tim Gilles)*

Figure 52.66 The chain guide has worn almost all the way through its metal backing. *(Courtesy of Tim Gilles)*

Figure 52.65 shows typical chain guide and tensioner wear. **Figure 52.66** shows severe tensioner wear, with the synthetic rubber surface worn totally away and the aluminum backing almost worn totally through. This can occur due to lack of maintenance or insufficient lubrication. When replacing a damaged chain guide, determine the cause of the failure.

> **CASE HISTORY**
>
> *A customer had a noisy engine in a Toyota pickup with 200,000 miles on the odometer. Some high-mileage engines suffer excess wear and, finally, breakage to a chain guide. With the valve cover removed, wear was evident on the inside of the timing cover (**Figure 52.67**). After removing the timing cover, a broken timing chain guide was found. This resulted in noise and wear to the side of the timing cover as the loose, unrestrained chain pounded against it.*

Before installing a new chain, soak it in oil.

Timing the Cam to the Crank

There are several ways to time the cam to the crank. Be sure to check the service literature before you install the timing chain or belt.

- ◼ Some timing sprockets are properly timed when the marks face each other.
- ◼ On others, there must be a certain number of chain links between the marks.

Timing cover

Wear from loose chain

Figure 52.67 This engine made lots of noise. The cause turned out to be a broken timing chain guide and a loose, worn chain. (*Courtesy of Tim Gilles*)

■ Sometimes chains have colored links that must be aligned with the marks on the sprockets (**Figure 52.68**).

■ Some overhead cams have a mark on the cam gear that lines up with a mark on the cylinder head when the timing mark on the damper is at the TDC mark.

If the old parts are available, carefully compare the new gears or sprockets with the old ones.

Figure 52.68 This timing chain has colored links that align with the timing marks. Check the service literature to verify whether the links go on the left or right side of the sprocket teeth. (*Courtesy of Tim Gilles*)

Figure 52.69 Flatten the bolt holes with a hammer. (*Courtesy of Tim Gilles*)

■ VALVE COVER AND OIL PAN SERVICE

Most valve covers and oil pans are made of sheet metal or cast aluminum. Some engines have been manufactured with plastic valve covers. These are bonded to the cylinder head at the factory with RTV (silicone) sealant. They must be clean of all oil and grease so that the new silicone will bond or stick to the valve cover.

Before installing a sheet metal or cast valve cover or oil pan, check it for straightness. Put it on a flat surface with a flashlight under it. Look for light leaking through.

When bolts are tightened against a gasket, the bolt holes tend to pull down into the gasket. Flatten the bolt holes with a hammer before installing a valve cover (**Figure 52.69**).

A valve cover gasket that is compressed a *small* amount will spring back against the torque of the valve cover screws. If the screw is tightened too much, the gasket is overly flattened and does not hold tension back against the screw. Apprentices often overtighten valve cover screws when a cork or rubber gasket is used. Check bolt tightness with an inch-pound torque wrench. Torque specifications are usually about 50 to 60 inch-pounds. You will be surprised at how little this is.

Besides the cylinder heads and camshaft, other parts of the breathing system include the intake manifold and exhaust manifold. Those topics, as well as turbochargers and superchargers, are covered in Chapter 42.

▨ REVIEW QUESTIONS

1. After removing the valve spring, what measurement is done before removing the valve?

2. What is the maximum amount of warp that a V6 cylinder head can have?

3. If 0.020" is removed from a cylinder head, approximately how much will valve timing change?

4. If a crack in a head is to be repaired with tapered plugs, what is done to both ends of the crack?

5. Three things that a valve spring is checked for are tension, squareness, and _____.

6. The guide repair process that displaces metal inside of the valve guide to make it smaller before reaming it for stem-to-guide clearance is called _____.

7. An _____ angle is when a valve face is ground to an angle that is slightly flatter than the valve seat.

8. A _____ is installed under a valve spring to correct the installed spring height.

9. A _____ test can be done after a valve grind to ensure that the seats do not leak.

10. When _____ are to be reused, they must be kept in order during engine disassembly.

▨ ASE-STYLE REVIEW QUESTIONS

1. Which of the following is/are true about valve seat width?

   a. Seat width is important to valve cooling.

   b. An exhaust valve seat should be wider than an intake seat.

   c. Seat width affects valve spring pressure on the valve seat.

   d. All of the above.

2. A valve has burned. Technician A says that this could be the result of a valve seat that is too wide. Technician B says that this can be the result of a valve margin that is too thin after regrinding a valve. Who is right?

   a. Technician A

   b. Technician B

   c. Both A and B

   d. Neither A nor B

3. Which of the following could be the cause of a warped cylinder head?

   a. Running an engine without the thermostat

   b. Overheating the engine

   c. Both A and B

   d. Neither A nor B

4. Technician A says that OHC cylinder heads often have oil galleries in them. Technician B says that pushrod engine cylinder heads often have oil galleries in them. Who is right?

   a. Technician A

   b. Technician B

   c. Both A and B

   d. Neither A nor B

5. Technician A says that magnaflux is a crack detection method that is commonly used on aluminum heads. Technician B says that bead blasting is a good method for cleaning OHC heads. Who is right?

   a. Technician A

   b. Technician B

   c. Both A and B

   d. Neither A nor B

6. Which of the following engine types can have the valve lash adjusted before installing the head on the engine?

   a. OHC

   b. Cam-in-block (pushrod)

   c. Both A and B

   d. Neither A nor B

7. Two technicians are discussing engine assembly. Technician A says that an acceptable torque recommendation for many valve cover gaskets is 60 foot-pounds. Technician B says to always be sure that timing marks are facing each other when timing the camshaft to the crankshaft. Who is right?

   a. Technician A

   b. Technician B

   c. Both A and B

   d. Neither A nor B

8. Technician A says that if a camshaft does not turn easily, the head probably requires straightening. Technician B says that if the cam side of an OHC head is warped excessively, the head can be straightened. Who is right?

   a. Technician A

   b. Technician B

   c. Both A and B

   d. Neither A nor B

9. A cylinder head has no carbon in only one of its combustion chambers. Which of the following is/are possible causes?

   a. A blown head gasket

   b. A cracked cylinder head

   c. Both A and B

   d. Neither A nor B

10. A cylinder head has required substantial metal removal to make it flat. Technician A says that a shim can be installed to compensate. Technician B says that the valve timing will be retarded. Who is right?

   a. Technician A

   b. Technician B

   c. Both A and B

   d. Neither A nor B

Engine Diagnosis and Service: Block, Crankshaft, Bearings, and Lubrication System

---

■ OBJECTIVES

Upon completion of this chapter, you should be able to:

✔ Analyze wear and damage to the cylinder block.

✔ Select and perform the most appropriate repairs to the block, crankshaft, and bearings.

✔ Analyze wear and damage to the crankshaft and bearings.

✔ Analyze wear and damage to lubrication system parts.

✔ Select and perform the most appropriate repairs to the lubrication system.

■ KEY TERMS

line honing or line boring   crosshatch   tolerance
cylinder glaze   lugging   crank polishing

■ INTRODUCTION

Usually, a cylinder block can be reused after certain service procedures are performed. Some blocks with excessive wear will have to be bored oversize to be used with new, larger pistons; others will need only cleaning and some minor service operations. Others will require major service operations, such as align-boring of main bearing bores, or sleeving of cracked or damaged cylinders. As you read this chapter, you should become familiar with the practice of diagnosing the cause of problems before attempting a repair. This is the skill that separates the successful technicians from the average ones.

■ CLEANING THE BLOCK

First, the block must be thoroughly cleaned. All core plugs, oil gallery plugs, cam bearings, and any other removable parts must be removed.

■ OIL AND WATER PLUG REMOVAL

Oil gallery plugs and threaded heater hose connections have tapered pipe threads. Pipe threads wedge together, making a very tight seal (see Chapter 7).

There are three types of gallery plugs. The female plug is removed with a special plug driver that resembles a ⅜" to ¼" socket adapter, except that it is solid (**Figure 53.1a**).

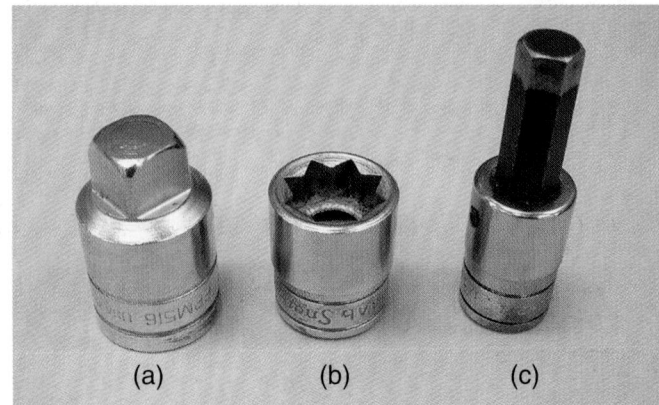

Figure 53.1 Three types of gallery plug tools: (a) Square oil-plug tool. (b) An 8-point socket. (c) An Allen wrench with socket drive. *(Courtesy of Tim Gilles)*

Figure 53.2 Removing stuck oil gallery plugs. (a) Heat the plug with a torch. (b) Apply paraffin to the plug. (c) Remove the plug. (d) This oil gallery is very dirty. (*Courtesy of Tim Gilles*)

SHOP TIP Using a socket adapter to remove these plugs will usually result in a broken tool.

Male plugs can be removed with an 8-point or square drive socket (**Figure 53.1b**). Use an Allen wrench with a socket drive to remove an Allen-type plug (**Figure 53.1c**). An air impact wrench often makes removal of these plugs easier. Following removal, oil passages can be cleaned with a cleaning brush.

The front of the block often has small core plugs that seal the ends of the oil galleries. After the rear threaded plugs are removed, the front core plugs can be knocked out from the rear with a long metal rod.

SHOP TIP Occasionally, plugs will not come out because they are very tightly rusted in place. A rusted plug can usually be removed by heating it with an oxyacetylene torch (**Figure 53.2a**) and then applying paraffin wax (a door-ease wax stick is handy) to the drive hole of the plug (**Figure 53.2b**). The wax acts like a heat sink, shrinking the plug. The plug is then removed (**Figure 53.2c**). If all else fails, an extractor set may be required (see Chapter 7).

Clean the block with one of the methods discussed in Chapter 10. Lubricate all machined areas *immediately* to prevent rusting.

Figure 53.3 A crack in a cylinder wall. *(Courtesy of Tim Gilles)*

Check for Cracks

If the engine was using coolant, check the block for cracks in the cylinder bores after hot-tanking (**Figure 53.3**). Some late-model V-type blocks also have a tendency to crack in the lifter valley area.

Clean Oil Galleries

Deposits in engine oil galleries and supply holes must be removed during block cleaning to prevent loosened particles in the passageways from damaging a newly rebuilt engine. Oil galleries fill with sludge that can trap metal shavings and grinding grit produced during block machining. Use an inexpensive stiff bristle brush (called a "rifle brush") with hot soapy water to clean galleries that run the full length of the block (**Figure 53.4**). Gallery cleaning brush sets are available from engine parts suppliers.

NOTE: *Failure to clean galleries is an invitation to engine failure.*

An apprentice was rebuilding an engine that had burned a rod bearing. He neglected to carefully clean the oil galleries. After the engine was installed and started, metal from the galleries mixed with the engine oil. The result was severe damage to the crankshaft and bearings.

■ OIL AND INSTALLATION

After cleaning the galleries, reinstall the plugs. Threaded oil gallery plugs are coated with liquid sealer or Teflon tape and tightened into the block (see Chapter 7). Be careful not to overtighten threaded plugs. Pipe threads are tapered and wedge tightly to the block as the diameter of the plug increases.

Pressed-fit oil gallery core plugs are installed with red loctite. *Cross-stake* the outside of the core holes with a chisel to prevent high oil pressure from blowing the plugs out (**Figure 53.5**). If the plugs come out, the engine will lose oil pressure.

■ INSPECT AND CLEAN LIFTER BORES

Lifter bores must be clean to ensure that the lifters will spin. The bores can be cleaned with a brake hone turned by hand. Be careful not to enlarge the lifter bore with excessive honing.

■ CHECKING MAIN BEARING BORE ALIGNMENT

Repeated heating and cooling of an engine block can result in misalignment of the main bearing bores. Main bearing bore alignment can be checked with a 0.0015" feeler gauge and a straight bar or a straightedge (**Figure 53.6**).

Figure 53.4 Clean all oil passages with a brush. (a) Main bearing oil passages. (b) Oil galleries.

Figure 53.5 Oil-gallery core holes are cross-staked with a cold chisel. If the plugs come out, the engine will lose oil pressure. *(Courtesy of Tim Gilles)*

Figure 53.6 Checking main bearing bore alignment.

SHOP TIP Do not mount the block on a universal engine stand during the alignment check. Hanging the block from one end could produce a false reading because of the flexibility of the casting (**Figure 53.7**). Set it on the floor or a bench top.

NOTE: *When a main bearing has become hot enough to burn or the block has changed color, the bore will usually shrink (**Figure 53.8**), which means that the main bearing bores must be remachined. When a crankshaft breaks, check for excessive main bearing bore stretch. Any main bearing bore that shows discoloration should also be checked for cracks before doing any repair work on the block.*

Main bearing bores can be checked with a dial bore gauge as shown in **Figure 53.9**. The vertical measurement should not be larger than the horizontal;

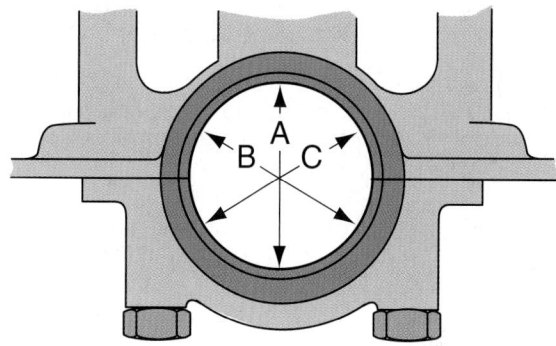

Figure 53.8 If a main bearing has become hot, the measurement at C and B could be less than the measurement at A.

Figure 53.9 Checking main bearing bores with a precision gauge.

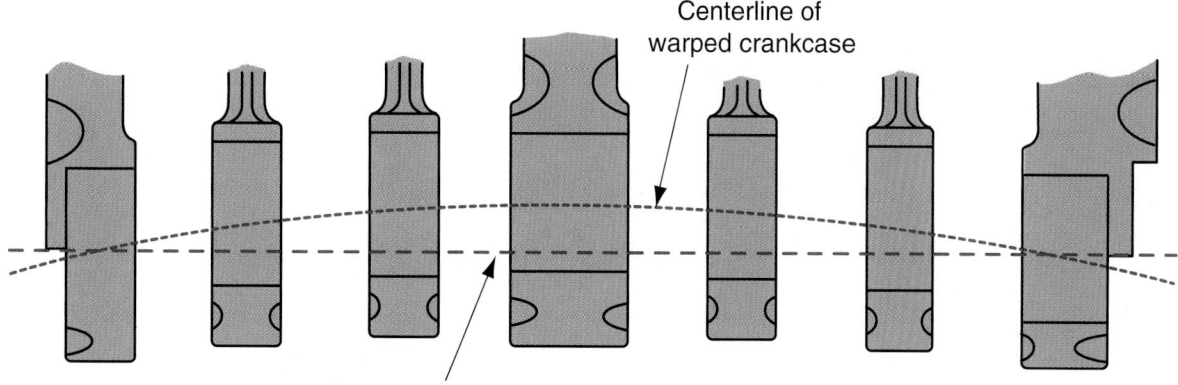

Centerline of warped crankcase

True centerline of crankcase

Figure 53.7 The crankcase can be warped. *(Courtesy of Dana Corporation)*

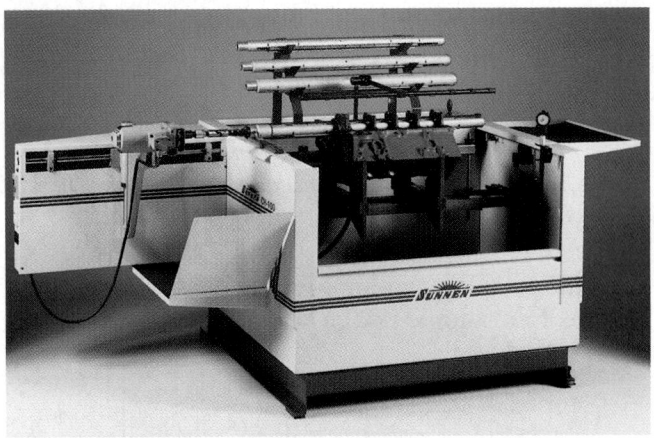

Figure 53.10 Honing main bearing bores with a line honing machine. *(Courtesy of Sunnen Products Company, St. Louis, Missouri)*

if it is, stretch has occurred. An out-of-round measurement of less than 0.001" is acceptable if the horizontal reading is the largest.

Realigning Main Bores

Realignment of the main bores is accomplished either by **line honing** or **line boring** (**Figure 53.10**). First, the main caps are ground or milled on their parting faces, where they meet with the block. Then, the bores are aligned by honing or boring to the original main bearing bore size. The limitation to this procedure is that removing too much metal during this procedure can move the crank shaft up too far into the block. This causes excessive timing chain slack and retarded cam timing. It also means that the piston will come up higher in the cylinder, which causes higher compression.

■ CHECK THE DECK SURFACE FOR FLATNESS

A whetstone or a file can be used to clean the deck surface of the block. Do not make the surface too smooth; head gasket sealing problems may result (see Chapter 51). Just remove any nicks or burrs that might give false readings when checking for surface warpage.

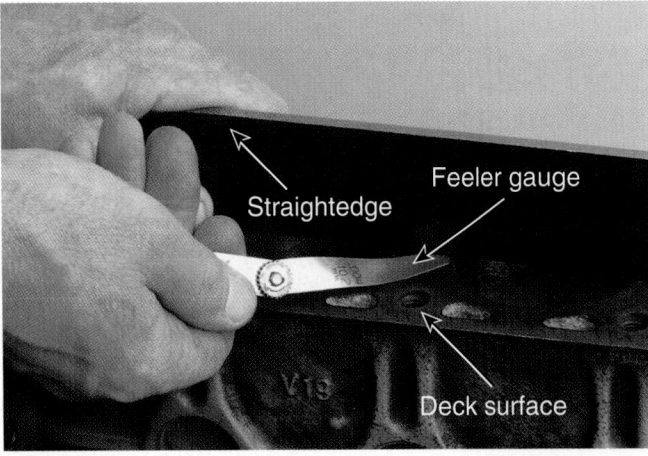

Straightedge
Feeler gauge
Deck surface

Figure 53.11 Check the deck surface for flatness. *(Courtesy of Tim Gilles)*

Check the deck surface for flatness (**Figure 53.11**). Blocks do not warp nearly as often as heads. When the block surfaces are warped, or are not parallel to the main bearing bores, they can be surfaced (*decked*) with a milling machine or grinder in the same manner as a cylinder head.

■ CLEAN ALL BOLT HOLES

Threads in the block must be clean so that correct clamping load is achieved. When a block is hot-tanked, a film develops on the threads. Chase the threads with a tap. Failure to do this can result in improper torque, which can lead to a leaking head gasket. Head bolt holes often run into water jackets, so the threads can rust on the bottom. Rust is very hard, so be very careful not to break a tap while cleaning head bolt threads.

■ INSPECTING CYLINDER BORES

Cylinder Bore Taper

When cylinder bores wear, they do so in a taper and *out-of-round* fashion. Maximum wear is at 90 degrees to the wrist pin, just under the ring ridge. The top of the cylinder receives less lubrication, and it is subjected to the high pressure of the piston rings against the cylinder wall when the air-fuel mixture is ignited. This causes taper wear, which forms the ring ridge at the top of the ring travel. The ridge must be removed or square edges of the new rings can come into contact with the rounded ring ridge (**Figure 53.12**). There is only about 0.002" to 0.004" clearance between the side of the ring and the ring groove. The top ring land can be forced down against the second land, jamming the second ring in its groove (**Figure 53.13**). Ridge removal is covered in Chapter 50.

Older engines were allowed up to 0.010" of taper wear and 0.005" of oval wear. For in-chassis rering jobs, 0.004"–0.006" is now considered to be the maximum amount of upper cylinder taper allowable by most aftermarket piston ring experts. When premium piston rings are to be used during a rebuild, 0.002" is the maximum recommended cylinder bore taper.

By the year 2000, most new engines were equipped with low-friction piston rings. Engine manufacturers

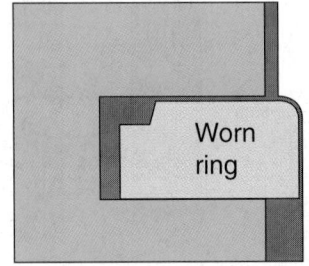
Worn ring

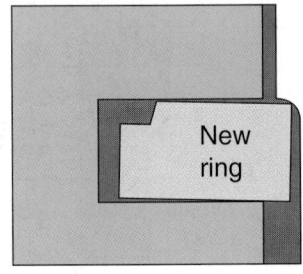
New ring

Figure 53.12 A new square ring will strike the worn round area. *(Courtesy of Dana Corporation)*

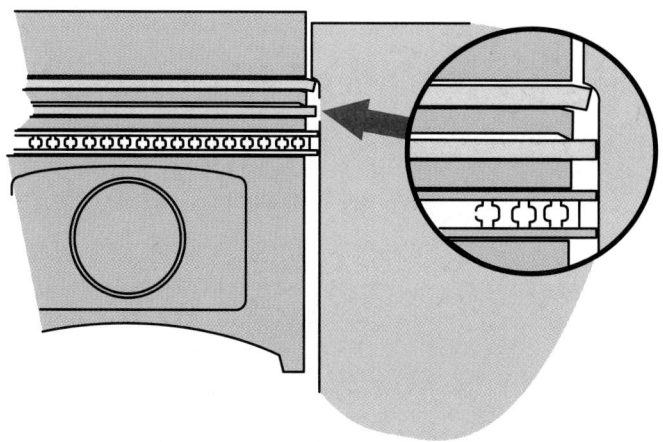

Figure 53.13 The ring land above the second ring can be forced down if the top ring contacts the ring ridge.

are producing engines to far more exact standards than in the past, with very little taper present in the block and life expectancies of 150,000 miles. Newer model engines sometimes have no taper wear evident, even after 150,000 miles of use.

Cylinder Bore Oval Wear. It is more difficult for new rings to seal against an out-of-round cylinder than a tapered one. Out-of-round wear is caused by the rocking of the piston on its wrist pin at TDC and BDC. Oval wear was more common on older engines with carburetors. With a cold engine, gasoline washes the oil from the cylinder walls.

■■■ MEASURING THE BORE

The cylinder bore can be measured in several ways.

■■■ MATH NOTE ■■■

A simple way to check for taper without any special measuring instruments is to use a feeler gauge and an old piston ring:

- *Square up the ring just below the ring ridge and measure the ring end gap with a feeler gauge (**Figure 53.14**).*
- *Compare this measurement to the end gap measurement at the bottom of the cylinder (**Figure 53.15**).*
- *To determine the amount of taper, divide the difference between the two gaps by 3 (**Figure 53.16**). This is done because the change in the gap is really a measurement of change in cylinder circumference, not diameter. So, actually, the number 3.14 (π) would be used.*

The bore can be measured using a telescoping gauge and micrometer, an inside micrometer, or a *cylinder dial bore gauge.* The dial bore gauge is usually

Figure 53.14 Measuring ring end gap at the top of the bore. (a) Insert the ring into the cylinder as shown. (b) Use the piston to square up the ring. (c) Measure the ring gap just below the ring ridge. *(Courtesy of Tim Gilles)*

much more accurate than the other instruments for determining the bore size. To measure the bore using a dial bore gauge, rock the gauge back and forth in the cylinder. The correct size is the smallest diameter that the gauge reads. This is because when the tool moves past the smallest point in either direction, the reading will get larger (**Figure 53.17**).

■■■ DEGLAZING THE CYLINDER BORE

Cylinders become glazed where the piston rings contact the cylinder wall. The **cylinder glaze** can be removed with lacquer thinner or carburetor cleaner or with a glaze breaker that leaves a honed, **crosshatch** appearance

Figure 53.15 (a) Square the ring to the bottom of the bore. (b) Measure the gap at the bottom of the bore. *(Courtesy of Tim Gilles)*

(**Figure 53.18**). The purpose of the crosshatch is to provide channels to hold oil while new piston rings wear into the cylinder walls. If the factory crosshatch is still visible, glaze breaking is not necessary.

The angle of the crosshatch should be between 20 and 60 degrees. Crosshatching that is too steep (more than 60 degrees) will allow oil to run down the cylinder wall; too flat (less than 20 degrees) causes too thick an oil film. The rings will skate over the film, causing oil consumption.

Stroking about once per second will provide the proper crosshatch, depending on the speed of the drill motor. If the crosshatch is flatter than desired, either slow down the drill speed or speed up the stroke speed. A drill with a rotation speed of approximately 450 rpm is recommended for deglazing cylinders.

SHOP TIP Deglaze the ring contact area of the cylinder only (**Figure 53.19**). There is no need to travel through the bottom of the cylinder.

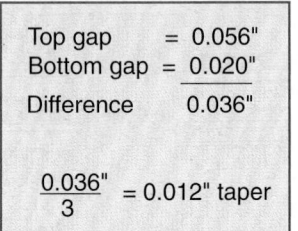

Top gap   = 0.056"
Bottom gap = 0.020"
Difference   0.036"

$$\frac{0.036"}{3} = 0.012" \text{ taper}$$

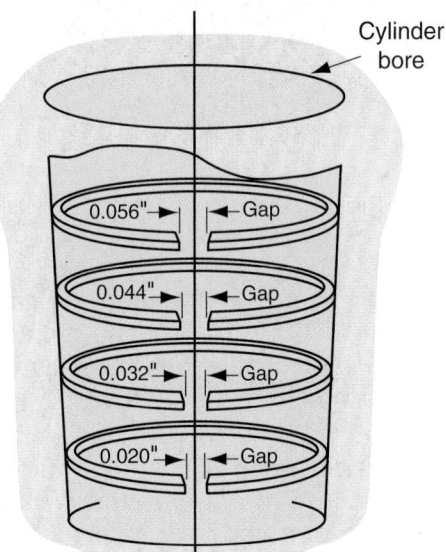

Cylinder bore

0.056" ← Gap
0.044" ← Gap
0.032" ← Gap
0.020" ← Gap

Figure 53.16 Effect of 0.012" cylinder bore taper on the ring gap in a 4" bore.

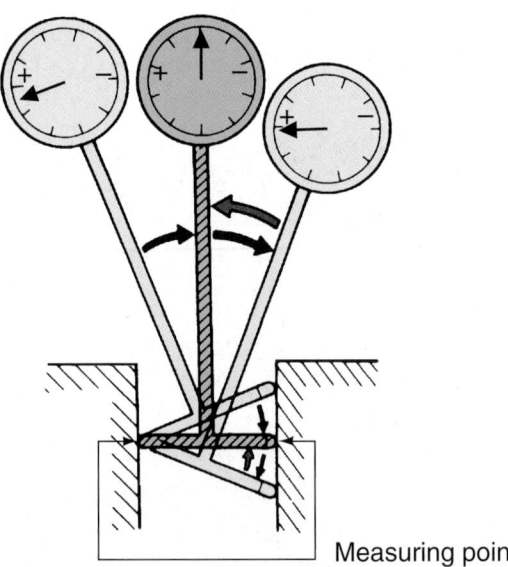

Measuring point

Figure 53.17 Rock the dial bore gauge back and forth. Select the smallest measurement.

NOTE: *Dana Corporation recommends that engines that are rerung while they are still in the vehicle should **not** have the glaze broken with a glaze breaker unless the block is to be thoroughly cleaned afterwards. They feel that glaze breaking is desirable but that the damage done by leaving grit in the engine outweighs the benefits gained by glaze breaking.*

Figure 53.18 Crosshatch appearance of a honed cylinder wall. *(Courtesy of Tim Gilles)*

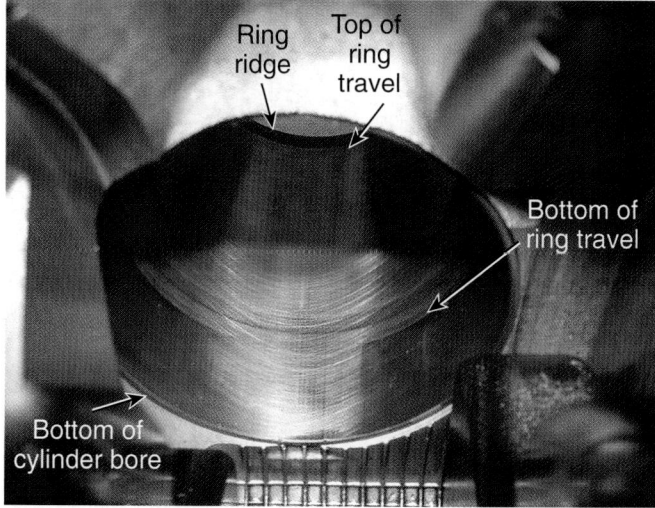

Ring ridge

Top of ring travel

Bottom of ring travel

Bottom of cylinder bore

Figure 53.19 The glazed area contacted by the piston rings. The crosshatch pattern will still be visible in the unworn area below ring travel. *(Courtesy of Tim Gilles)*

Glaze Breakers

There are two principal types of glaze breakers: the spring-loaded glaze breaker and the ball-type glaze breaker, or *flex hone* (**Figure 53.20**). Both are driven by drill motors and cooled with honing oil or cleaning solvent. Glaze breakers remove metal at a much slower rate than the rigid hones used to finish a cylinder to the correct size after boring.

The stones on a spring-loaded glaze breaker can be accidentally broken in two ways:

- This happens when the tool is removed from the cylinder while the drill motor is still spinning.
- Stones also break when they are allowed to spin below the bottom of the cylinder bore and come into contact with main bearing webs.

The flex hone can be rotating as it enters and leaves the cylinder. This will leave a proper crosshatch finish. It is not as susceptible to stone breaking, although the balls

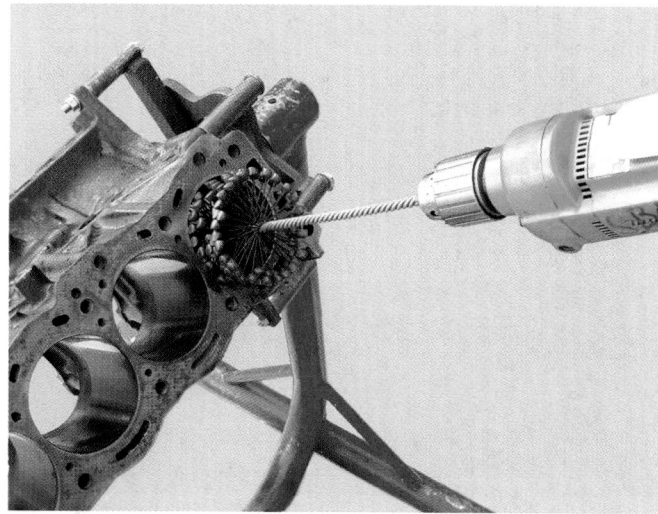

Figure 53.20 A flex hone. *(Courtesy of Tim Gilles)*

tend to wear off on the bottom end of the hone from repeatedly striking the main bearing webs.

■ CLEAN THE BLOCK OF GRIT

A thorough cleanup of the block is necessary after glaze breaking or honing. Any grit left on the block will rapidly wear out new parts. After all block preparation, clean the block with a stiff bristle brush and hot, soapy water (**Figure 53.21**).

- The brush can be used by hand or on the end of an air drill.
- Ordinary cleaning solvent will not lift the grit from the pores of the metal.
- Check cylinder walls and the crankcase for cleanliness with a clean cloth. Grit remaining in the bore will be deposited on the cloth.
- Following cleaning, grit can often be found in the crankcase area, just under the cylinder bores (**Figure 53.22**).

Figure 53.21 Clean the cylinder bores with hot soapy water. *(Courtesy of Federal-Mogul Corporation)*

Figure 53.22 Be certain that this area of the block below the cylinders (seen from the bottom) is cleaned thoroughly of honing grit. *(Courtesy of Tim Gilles)*

Ferrous parts (iron or steel) should be thoroughly coated with oil to prevent rusting, which begins immediately after cleaning.

■ BORING FOR OVERSIZED PISTONS

Cylinders are deglazed for use with new rings *only* if they do not have excessive bore taper. Damaged cylinders or cylinders with too much taper should be rebored and honed by a machine shop to fit new oversized pistons.

Piston Oversizes

Pistons are available in standard 0.020" (0.50 mm), 0.030" (0.75 mm), 0.040" (1.0 mm), and 0.060" (1.5 mm) oversizes. The top of an oversize piston is stamped with the amount of oversize. Some later model blocks do not have enough cylinder wall thickness to accommodate reboring to 0.060" oversize. Pistons of 0.030" oversize are most commonly used. Using a 0.030" cut for each rebore (0.015" of metal removal from each cylinder side) allows for two overhauls (0.030" and 0.060") per block, which is usually more than enough.

MATH NOTE

The compression ratio will change slightly with a rebore. A 0.030" oversize increases compression by about ¹⁄₁₀ of a ratio; 0.060" will result in a change of about ¼ of a ratio.

Example:

4.000" compression ratio	9:1
4.030" compression ratio	9.1:1
4.060" compression ratio	9.25:1

The piston shown in **Figure 53.23** is an inch-standard piston that is 0.040" oversize. With the head off an engine, this tells you that the engine has been bored before and might be at or near its overbore limit.

> **SHOP TIP** North American pistons and bearings and many aftermarket import engine replacement parts still use the inch standard system. To quickly convert metric to inch standard, each 0.25 mm is equal to 0.010". The piston shown in **Figure 53.24** shows a metric oversize piston that is 0.020" oversize.

Figure 53.23 Clean the carbon off the top of the piston to see if it is oversize. This piston is 0.040" oversize. *(Courtesy of Tim Gilles)*

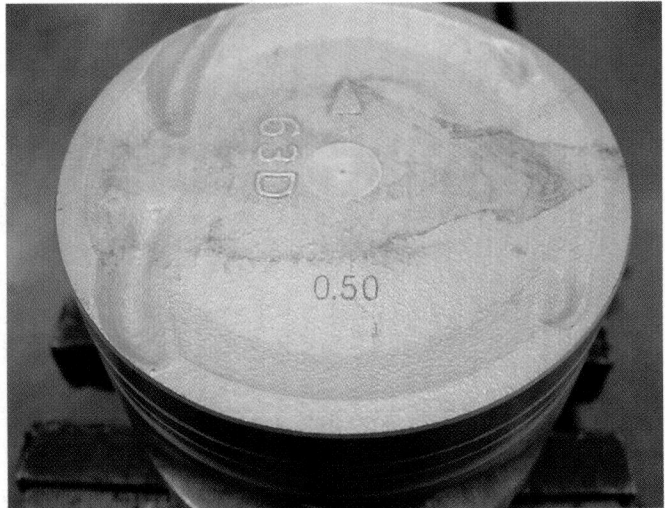

Figure 53.24 To quickly convert metric to inch standard, each 0.25 mm is equal to 0.010". This piston is 0.020" oversize. *(Courtesy of Tim Gilles)*

■ BLOCK DISTORTION

Block castings typically distort when heads and main cap bolts are torqued. Cylinder distortion can be from 0.001" to 0.005". Piston clearance is usually only 0.001" to 0.003".

In extreme cases, this distortion can result in piston scuffing (momentary welding to the cylinder wall). Most often, distortion results in poor piston ring sealing or *piston slap*. Piston slap occurs when excessive piston-to-cylinder wall clearance allows the piston skirt to tap against the cylinder wall.

A boring stand is desirable for reboring and honing blocks (**Figure 53.25**). It supports the block at the main bearing bores.

NOTE: *Cylinders can be up to 0.002" out-of-round when the block is hung from one end on a conventional engine service stand. Boring cylinders with a portable boring bar while the block is mounted to one of these stands can produce oval-shaped cylinders.*

Torque Plates

Because a cylinder block is so flexible when the heads are not holding it in place, a torque plate made of

Figure 53.26 A torque plate. *(Courtesy of Federal-Mogul Corporation)*

1¾"-thick cast iron is sometimes torqued to the top of the block. It stresses the block to simulate assembled conditions when refinishing cylinders (**Figure 53.26**). Main caps should be torqued in place also. A torque plate is also called a *deck plate*, *honing plate*, or *stress plate*.

■ HONING AFTER BORING

The machine shop bores cylinders to within 0.0025" to 0.003" of the desired finished bore size. Then they are finished to the proper size by honing with a *honing machine* (**Figure 53.27**).

Honing after boring provides a much more satisfactory surface for new rings and promotes longer ring life. The ring and bore do not fit exactly to each other, so a crosshatched honed surface provides an area where they can wear into each other.

Manufacturers of honing equipment recommend that cylinders be honed following boring to increase

Figure 53.25 Boring a block with a boring bar and stand. *(Courtesy of Tim Gilles)*

Figure 53.27 An automatic cylinder bore honing machine. *(Courtesy of Tim Gilles)*

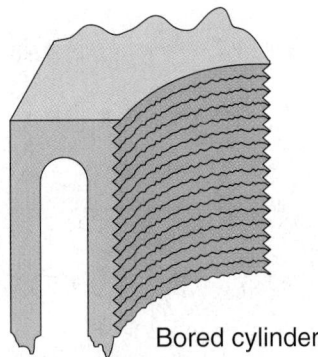

Figure 53.28 A bored cylinder wall is torn and has folded metal, which will cause excessive ring face wear. *(Courtesy of Federal-Mogul Corporation)*

the cylinder's diameter by at least 0.025" to 0.003". The boring cutter leaves a fine thread and microscopic fractures in the cylinder wall (**Figure 53.28**). Small chips break away from the cylinder surface as it is cut. The cavities left behind are about 0.001" deep, depending on the condition of the tool bit.

NOTE: *Honing the cylinder to remove 0.001" of metal from the surface of its wall will result in a cylinder 0.002" larger in diameter.*

Piston Clearance

The machine shop will want to have the new pistons on hand before honing, so that they can be properly fitted to the bores. There is some variation in sizes of pistons within a set. Typical clearance is 0.001" to 0.002" for cast automotive pistons. They come about 0.0015" undersize to the standard bore sizes. For example, a piston for a 4.030" bore will actually measure about 4.0285".

MATH NOTE	
Standard Bore Size	4.000"
Replacement Piston Size	4.0285"
Piston Clearance	+ 0.0015"
Rebore Finished Size	4.030"

The machine shop will measure the bore with a dial bore gauge and compare it to the size of the pistons. The dial bore gauge is set with a very precise micrometer setting fixture (see Figure 6.32a).

Chamfering the Cylinder

After boring and honing, the top of the bore is chamfered about $\frac{1}{16}$". This is so that during installation of the pistons in the block, the new rings can easily enter the cylinder without being chipped.

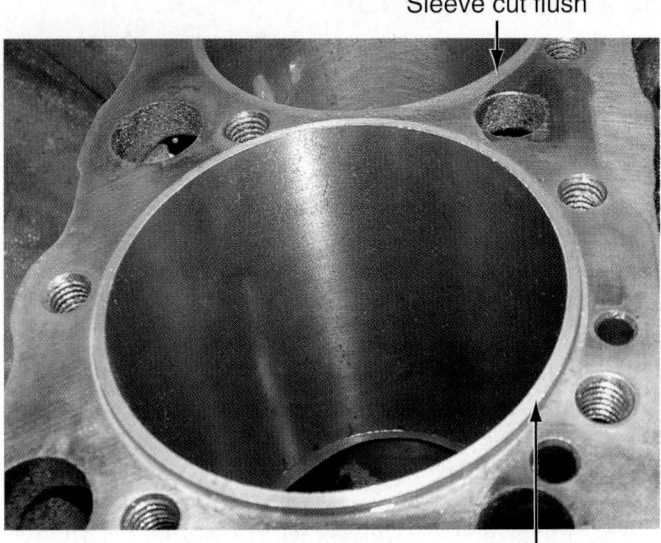

Sleeve cut flush

Sleeve lip

Figure 53.29 The lip of the sleeve will need to be cut. The top of the sleeve on the adjacent cylinder has already been cut. *(Courtesy of Tim Gilles)*

▬ SLEEVES

Sleeves are used to repair a cracked or damaged cylinder (see Figure 53.3). When a sleeve is used to repair a damaged cylinder, it is installed with an interference fit of 0.0005" per inch of cylinder bore. Thus, a 4" bore would require the block to be bored 0.002" smaller than the outside diameter of the sleeve.

After the sleeve is pressed into the bore (**Figure 53.29**), the top of it is finished flush with the block. Then the inside diameter (I.D.) of the sleeve is bored to the finished size.

NOTE: *The I.D. of the sleeve is usually less than a standard bore, so it can be bored to original standard diameter.*

One popular sleeving method is to make a step in the bottom of the bore by cutting the cylinder only to within $\frac{1}{8}$" or $\frac{1}{4}$" of its bottom. The bottom of the sleeve will rest on this step. When the final boring is completed on the sleeve, the line between the sleeve and step in the block will be almost impossible to see. Many machine shops use this procedure.

MATH NOTE	
Measure the top of the sleeve at three places 120 degrees apart.	
4.285"	
4.287"	
4.286"	

Repeat this at the bottom.
4.284"
4.288"
4.286"
Add the six measurements together and divide by 6 to
find the average outside diameter of the above.
4.285"
4.287"
4.286"
4.284"
4.288"
+4.286"
25.716" divided by 6 = 4.286" (average diameter)

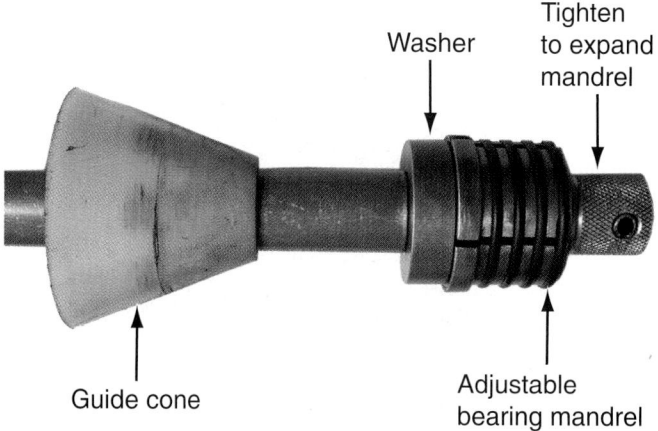

Figure 53.30 A universal cam bearing tool has an expanding mandrel and rubber O-rings. *(Courtesy of Tim Gilles)*

CAM BEARING INSTALLATION (CAM-IN-BLOCK ENGINES)

Cam bearings are pressed-fit into the block. The outside diameter of the bearing is larger than the bearing bores in the block. Before installing cam bearings, emery cloth is used to clean the cam bearing bores. The cam bearings in a pushrod engine are installed before the crankshaft is installed. This makes it easier to align the oil holes in the bearings to the corresponding holes in the block. It will also be easier to shave off any high spots on the bearing that occurred during bearing installation.

Sometimes cam journals are different sizes. The smallest is at the rear of the block. New bearings can be checked for approximate fit by putting them on cam bearing journals before installation.

Cam Bearing Removal and Installation Tools

There are several types of cam bearing removal and installation tools. The most popular one is the universal type (**Figure 53.30**). Its use is described here.

1. A large washer must be used with the driver, or the tool's mandrel pieces may be ruined. Washers of two different sizes are provided with the tool. Use the largest washer possible. Too small a washer can result in broken segments. Line up the segments with the spaces in the mandrel.

2. When starting to press a bearing into a bore, tighten the tool as much as possible by hand. Give the tool one sharp rap with a large hammer, and then loosen the mandrel approximately ⅛ turn before continuing. Be sure to recheck this adjustment occasionally when driving the bearing into place because the tool will sometimes loosen and damage the bearing.

3. Hold the guide cone tightly against the front of the block. Hold the cam bearing driver snugly against the bearing during installation so it does not bounce, or the end of the bearing can be damaged.

Although cam bearings are usually chamfered on the outside edges of both sides, some bearings are chamfered only on one side. Make sure that the chamfer faces the bore before installation. Use a large hammer because of the interference fit of approximately 0.002" to 0.003" that prevents rotation of the bearings in the block while the engine is running. The use of 90- or 140-weight gear lube makes installation much easier. Another helpful tip is to chamfer the inside edge of the bearing with a scraper if it is not already chamfered (**Figure 53.31**).

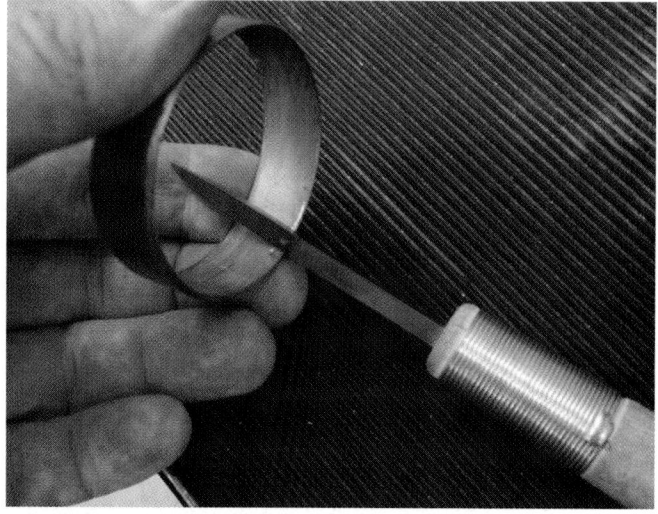

Figure 53.31 Chamfer the inside edge of the cam bearing before installation. *(Courtesy of Tim Gilles)*

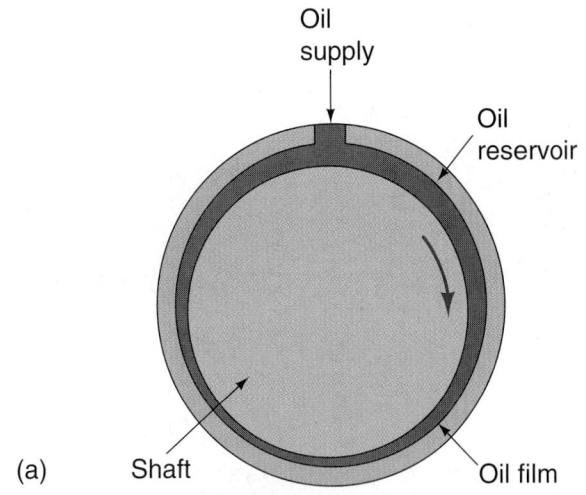

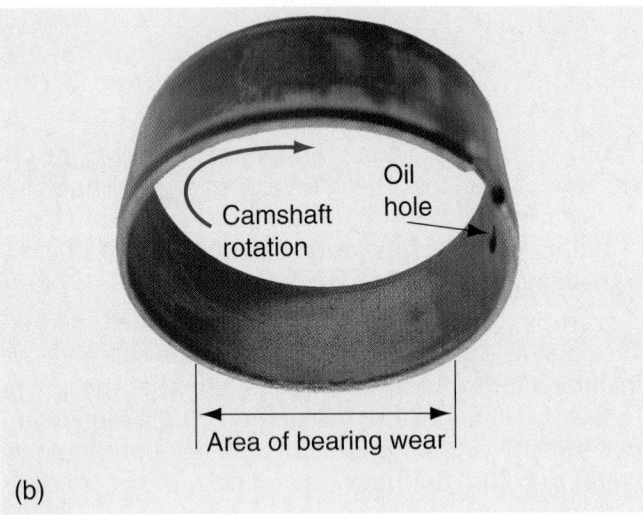

Figure 53.32 (a) Oil is carried under the cam as it rotates. (b) Where was the oil hole on this push rod engine cam bearing positioned on its previous installation? (a, Courtesy of Federal-Mogul Corporation; b, Courtesy of Tim Gilles)

Positioning the Oil Hole. Follow the manufacturer's recommendations when positioning the oil hole. When there is an oil groove around the total cam bore in the block, positioning the oil hole toward the top also allows oil to enter the bearing easily. It will be carried under the cam as it rotates (**Figure 53.32**). The following recommendation from Federal-Mogul Corporation can be followed:

> Install the bearings with their oil holes at approximately the 2 o'clock to 4 o'clock position, viewed from the front, so that oil hole alignment can be observed easily. Avoid positioning the oil hole between the 5 o'clock and 8 o'clock positions.

Double-check oil hole alignment with a piece of bent wire or welding rod.

■ FRONT CAM BEARING INSTALLATION

On pushrod engines, the timing sprockets and chain are lubricated by oil from the front cam bearing. Most front cam bearings are pounded in past the block surface (**Figure 53.33**). This leaves an oil passage in front

Figure 53.33 Install the front cam bearing below the front of the block surface to allow for timing chain oiling. (Courtesy of Tim Gilles)

of the bearing that provides lubrication to the cam timing sprocket and chain. Other bearings have a notch cut into them. The bearing is installed with the notch facing the cam gear.

Check Fit of Bearing

After installing the bearing, install the cam and turn it. If it does not turn easily, remove it and check for high spots. High spots on the bearing will become polished as the cam is rotated after it is first installed. These can be scraped with a bearing scraper.

NOTE: *Bearings should **not** be sanded with emery cloth because pieces of emery grit can become embedded in the bearing surface.*

SHOP TIP A cam bearing reamer can be made by grinding grooves into the journals of an old cam core. As the tool is turned, it shaves off any high spots from the bearing.

■ A special flex hone is available for honing small amounts off of cam bearings.
■ If necessary, Scotch Brite™ can be used to polish cam bearing surfaces. Be sure to clean the bearing surface afterward.

■ CHECKING CRANKSHAFT CONDITION

Checking a Crankshaft for Straightness

Keep bearings in respective position order during disassembly so that the cause of unusual wear conditions can be diagnosed. A bent crank can be indicated when one bearing wears more than the others (**Figure 53.34**). The resulting bearing wear is usually worse at the center.

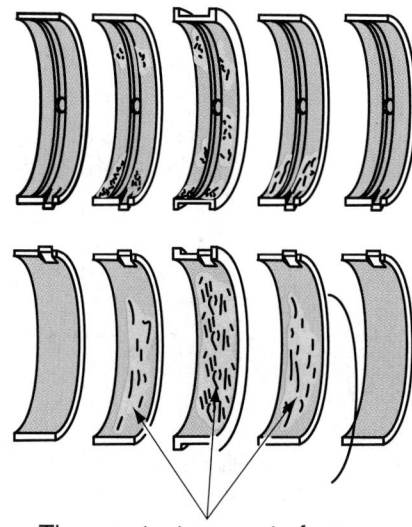

The greatest amount of wear occurs on the middle bearings

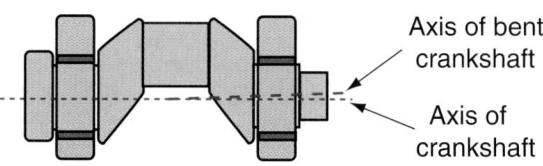

Axis of bent crankshaft

Axis of crankshaft

Figure 53.34 Crankshaft misalignment. Main bearing wear caused by a bent crankshaft or a misaligned crankcase.

Checking for Cracks

The crankshaft can be checked for obvious cracks by "ringing" the counterweights with a light tap of a hammer (**Figure 53.35**). Be careful not to damage the crankshaft. A dull sound indicates the presence of a crack. When ringing a crankshaft, the sprocket

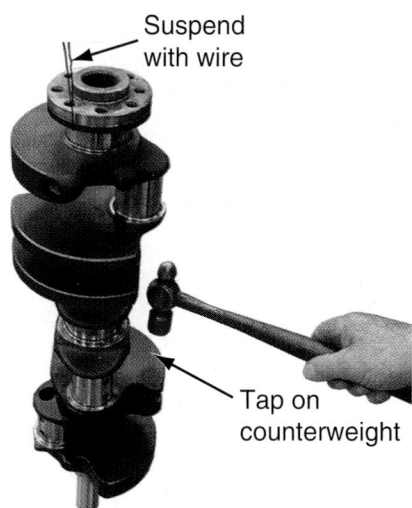

Suspend with wire

Tap on counterweight

Figure 53.35 Ringing a crank to check for cracks. A dull sound indicates a crack. Suspend the crank with wire. Holding it with your hand will invalidate the test. (*Courtesy of Tim Gilles*)

or gear must be removed first, or the crankshaft will sound like it is cracked. Hang the crankshaft by a wire. Holding it by hand will deaden the sound. Machine shops also have more accurate magnetic crack detection methods.

Checking the Vibration Damper

Whenever a crankshaft is broken, be sure to check the vibration damper for obvious signs of damage. The following are some additional vibration damper considerations:

- The outer ring on the damper can slip, causing an out-of-balance condition. With the cylinder head removed and the piston at TDC, make a visual comparison. Do the timing marks for TDC on the timing cover and damper appear to be accurate?
- Be sure the damper is the correct one for the vehicle. The wrong one can be worse than none at all, resulting in a broken crankshaft.
- Be sure all pulleys are straight. A damaged pulley can force a crankshaft to bend during engine operation.
- A very important consideration is to tighten the damper bolt correctly. Crankshafts vibrate; a loose pulley bolt will allow the damper to shake back and forth on the crankshaft snout. Check the keyway for wear that can result from improper damper installation, the wrong size key, or a loose damper hub-to-crank fit.

CASE HISTORY

A customer complained of an intermittent knocking sound from the passenger compartment. The technician listened carefully with a stethoscope at several locations along the driveline. The noise seemed to be coming from the torque converter, but replacing the torque converter did not solve the problem. The technician decided to remove the oil pan and inspect the crankshaft. As he loosened the front crankshaft pulley bolt, he noticed that it didn't seem to be as tight as it should be. These bolts are usually extremely tight. After removing the damper, he found the damaged woodruff key shown in **Figure 53.36**.

A new timing belt had been installed 15,000 miles previously. The torque specification for the damper was in the neighborhood of 200 foot-pounds, but the bolt had not been tightened correctly. The result was a vibration that transmitted noise along the crankshaft, where it was heard as a torque converter noise. A perfectly good torque converter was replaced before the true cause of the problem was found. A new crankshaft key and a bolt tightened to the correct specification solved the problem.

Figure 53.36 A damaged woodruff key. *(Courtesy of Tim Gilles)*

Unworn area

Figure 53.38 The area of a bearing journal that mates with the oil groove in some fully grooved bearings will not wear. *(Courtesy of Tim Gilles)*

A vs. B = vertical taper
C vs. D = horizontal taper
A vs. C and B vs. D = out-of-round

Check for out-of-round
at the end of each journal

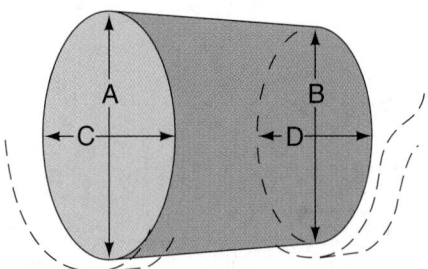

Figure 53.39 Crankshaft journal measurements.

CRANKSHAFT AND BEARING WEAR

Bearings have *loaded* and *unloaded* bearing halves. The upper rod bearing and the lower main bearing are the loaded halves. Upper rod bearing wear (**Figure 53.37**) is often found in combination with lower main bearing wear. A *little* of this kind of wear is considered normal.

It is important to keep all bearings in numerical position order for inspection. The main cause of short bearing life is dirt, including grit, metallic particles, and other abrasives. Sometimes sandy dirt enters the engine through the air cleaner or the dipstick opening. A crankshaft sometimes wears excessively because of abrasives in the oil (**Figure 53.38**).

Journals can wear out-of-round or become tapered. **Figure 53.39** shows how to evaluate crankshaft journal measurements.

Out-of-Round Journal Wear

When the engine is first cranked after it has not been run for a period of time, there is little or no lubrication between the crank and the lower main bearings. The result is that the lower main bearing wears excessively and the main journals wear out-of-round. When the main bearing farthest from the oil pump shows more wear than the other main bearings, a *dry start* condition is indicated. This means that the engine was probably revved before oil had filled the lubricating system. A problem that occurs most often in cold weather. This is one of the reasons why newer vehicles use lower viscosity oil.

Sometimes, all the lower main bearings will be worn except for the front bearing. This bearing usually wears less on the bottom because of the upward tension of the fan belt. Excessive belt tension can cause wear on the *upper* front main bearing.

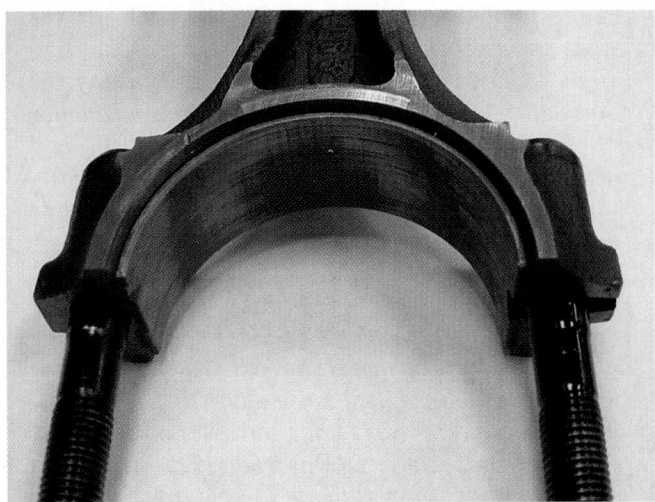

Figure 53.37 Upper rod bearing wear. *(Courtesy of Tim Gilles)*

Connecting rod journals also wear out-of-round, wearing on their top sides because of excessive loads during the power stroke. Crankshaft main bearing journals wear out-of-round. Excessive loads cause the oil film to break down, resulting in wear. Higher loads can be caused by **lugging** the engine or by abnormal combustion.

NOTE: *Lugging occurs when the load on the engine is greater than the rpm needed to develop enough horsepower to pull the load.*

Measure the rod journal in a horizontal and vertical direction to check for out-of-round wear. Crankshaft journals are *miked* (measured with a micrometer) at 90-degree angles to check for out-of-round wear, which should be less than 0.0005".

Tapered Wear

Rod journals sometimes exhibit taper wear due to misalignment of the connecting rod (see Chapter 54). Uneven rod bearing wear, and sometimes piston skirt wear, usually indicates a tapered bearing journal. Connecting rods should be checked for misalignment whenever uneven wear is found.

Thrust Bearing Wear

The thrust bearing surface that faces the rear of the engine sometimes shows excessive wear (**Figure 53.40**). Most thrust bearings have concaved reliefs cut into them to provide lubrication. Under normal conditions, the thrust surface is only under load when the clutch pedal is depressed or if the automatic transmission torque converter is under a load. Thrust bearing wear and failure occur when the load is continuous, such as when there is an improper clutch adjustment (too little free play), when the driver has been "riding the clutch," or when there is excessive automatic transmission oil pressure in the torque converter.

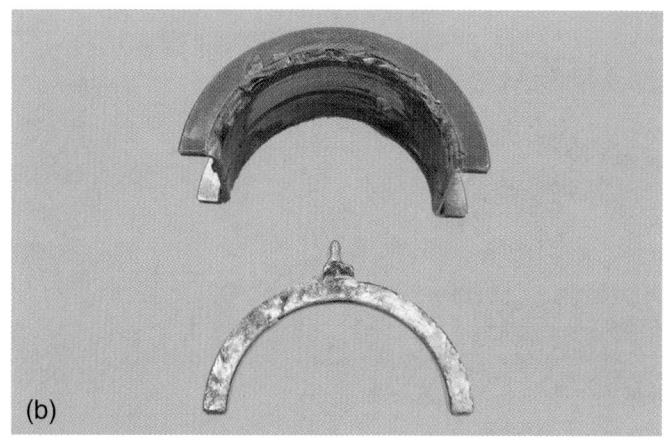

Figure 53.40 (continued) (b) Thrust bearings damaged on one side. *(Courtesy of Tim Gilles)*

▬ CRANKSHAFT JOURNAL TOLERANCE

Wear specifications for crankshaft journals are readily available. An example specification might be 2.2597"–2.2602"; this range is called the **tolerance**. When the crank is worn only slightly and is still within tolerance, it can be reused. **Crank polishing** with an *emery belt* by a machine shop is sometimes desirable. The crankshaft is turned in a lathe or crank grinder while the emery belt is turned against its surface. This is a relatively inexpensive operation.

▬ REGRINDING THE CRANKSHAFT

A machine shop can regrind a crankshaft with a *crankshaft grinder* (**Figure 53.41**). Crankshafts are usually reground to 0.010", 0.020", or 0.030" undersize.

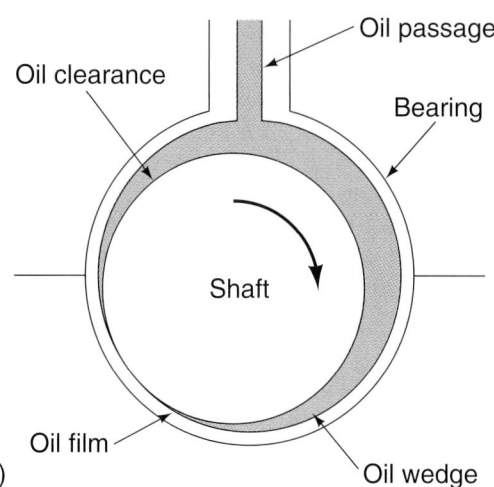

Figure 53.40 (a) A hydrodynamic wedge is formed under a normal bearing journal, but not on a thrust bearing.

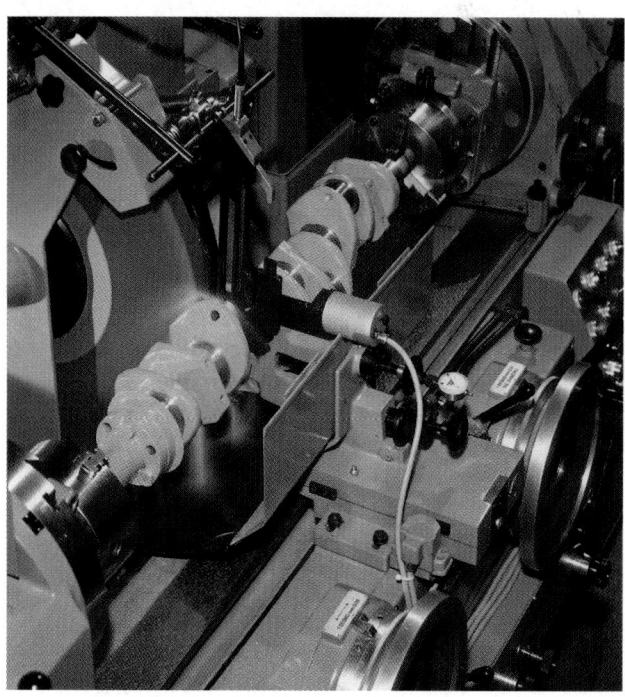

Figure 53.41 A crankshaft grinder. *(Courtesy of Peterson Machine Tools, Inc.)*

Figure 53.42 Oil holes in the crankshaft are chamfered before the journals are polished. *(Courtesy of Tim Gilles)*

NOTE: *A 0.020" undersized crank actually has only 0.010" of metal removed from its surface, but because metal is removed from both sides when the crank is ground, the crank will be 0.020" undersize.*

Sometimes, rod journals and main journals are ground to different undersizes. In this case, the main journal size is listed first—for example, 0.010"–0.020". The main bearing size is 0.010". After grinding, oil holes are chamfered and the crank journals are polished approximately 0.0002" with an emery belt **Figure 53.42** shows a finished journal.

NOTE: *Do not hand polish a crank journal with emery cloth or sandpaper. The shiny, worn crank surface will cause less bearing wear if left alone.*

Burned or severely worn crank journals can be built up on a crank welder before regrinding.

■ MEASURING BEARING CLEARANCE WITH PLASTIGAGE

Bearing clearance can be checked with *plastigage* (**Figure 53.43**) or with a micrometer. Plastigage, a special thickness of plastic string, is applied to the bearing journal between the bearing and the crank. Do **not** rotate the crankshaft while plastigage is in place. When the cap is torqued, the plastic string flattens. The bearing cap is removed to inspect the plastigage. The wider the flattened string, the less clearance there is. Plastigage is oil soluble and is readily cleaned from the bearing or crankshaft.

NOTE: *An engine with double oil clearance will throw off five times as much oil onto the cylinder walls, causing increased oil consumption (**Figure 53.44**).*

Be careful not to allow too much clearance:
■ A front bearing with excessive clearance will knock.
■ Excessive clearance can also cause low oil pressure.

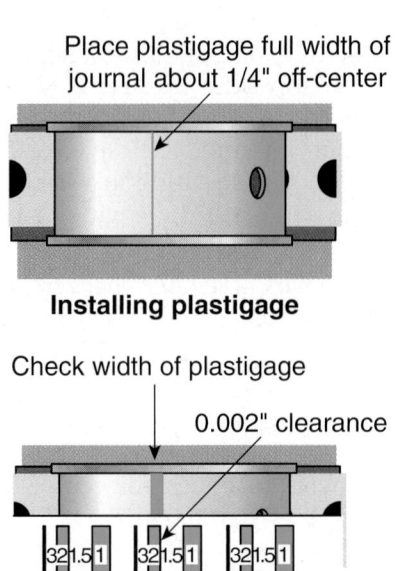

Place plastigage full width of journal about 1/4" off-center

Installing plastigage

Check width of plastigage

0.002" clearance

Measuring plastigage

Figure 53.43 Checking bearing oil clearance with plastigage.

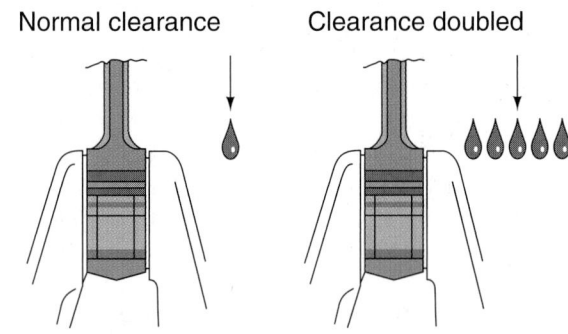

Normal clearance   Clearance doubled

Figure 53.44 Doubling oil clearance results in five times as much oil throw-off from connecting rods. *(Courtesy of Dana Corporation)*

CASE HISTORY

A technician working in a dealership repaired an engine that had a worn crankshaft. The crankshaft was removed and sent to a machine shop for regrinding. After the job was completed, the engine had a minor knock that was diagnosed as a front main bearing. The technician pulled the oil pan and measured all of the crankshaft clearances with plastigage. The front crankshaft bearing had 0.0025" clearance, enough to make noise. He cleaned the main bearing cap and lightly sanded the parting surface with very fine sandpaper. When he rechecked the clearance, it was 0.0015", within specifications. After reassembly the noise was gone. After that, he always checked crankshaft clearances carefully during initial assembly.

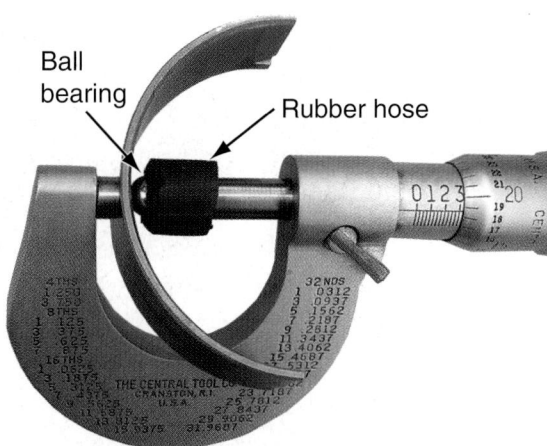

Figure 53.45 Measuring bearing wall thickness. *(Courtesy of Tim Gilles)*

Actual clearance can also be determined by using a micrometer to measure the crankshaft journal, the housing bore, and the bearing insert. When you measure the bearing insert, use a ball bearing and subtract its thickness from your measurement (**Figure 53.45**). Total oil clearance can be determined by subtracting the sizes of the bearing journal and bearing from the housing bore.

MATH NOTE
(Figure 53.46)

Housing Bore	*2.124"*
Bearing Wall Thickness (0.061) × 2	*−0.122"*
Bearing Assembled Inside Diameter	*2.002"*
Shaft diameter	*−2.000"*
Oil Clearance	*0.002"*

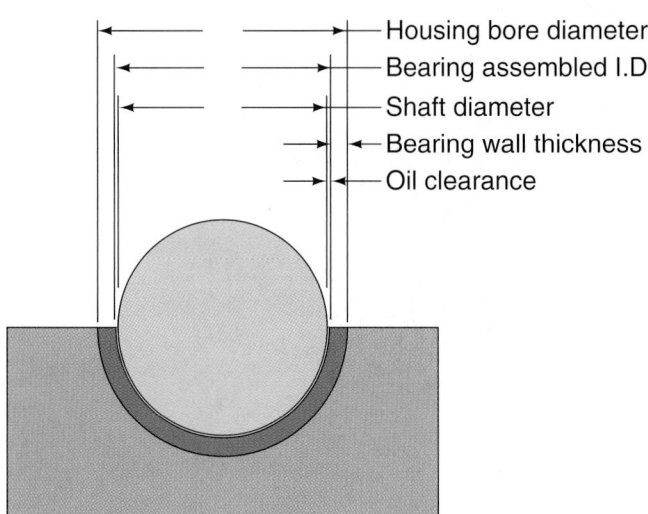

Housing bore diameter
Bearing assembled I.D.
Shaft diameter
Bearing wall thickness
Oil clearance

Figure 53.46 Determining oil clearance. *(Courtesy of Federal-Mogul Corporation)*

Select-Fit Bearings

Some manufacturers use select-fit bearings on their new engines, in other words, those that have not been previously rebuilt. Bearings are matched to the finished crankshaft size, according to the machining tolerance achieved at the factory. With some manufacturers, such as Honda, bearings are selected according to colored marks on the block. Other manufacturers use a numbering system to select bearings.

■ LUBRICATION SYSTEM SERVICE

Oil Pressure

An engine's oil pressure provides a good indication of the condition of its lower end. Low oil pressure at idle can indicate excessive oil clearance at a bearing or a worn oil pump. The engine might take too long to build oil pressure when it is first started. Low oil pressure at idle can also result if the oil pump pressure relief valve is stuck in the "open" position. The relief valve opening is so large that the pump cannot fill the system at low speed, although pressure will usually build to normal levels when the pump produces more volume at higher speeds.

Long periods of idling with low oil pressure can lead to cylinder wall lubrication problems. Throw-off from the connecting rods does not supply the cylinder walls with enough oil for adequate lubrication.

Testing Oil Pressure

To test oil pressure against manufacturer's specifications, remove the indicator light *sending unit* and install an external gauge in its place (see Chapter 49). An engine idling at 800 rpm should have a minimum of 8 psi oil pressure.

An engine with too much oil pressure can have oil consumption problems, and bearing lining material can be washed off the bearing. High oil pressure results when the relief valve sticks in the "closed" position.

■ The oil pressure can be normal at idle but will increase beyond desired levels as engine rpm increases.

■ High pressure can cause oil leaks and can blow out oil gallery core plugs, which will result in no oil pressure at low rpm.

■ Sometimes an oil filter can burst under too much oil pressure.

■ CHECKING OIL PUMPS FOR WEAR

Use a feeler gauge to check an oil pump for wear. The clearances here are approximate and are used only in the absence of service information specifications.

Internal/external gear pumps are checked in three places (**Figure 53.47**). Follow manufacturer specifications. Typical body-to-outer gear clearance is 0.004"–0.008". Inner gear-to-crescent clearance is 0.009"–0.013", and outer gear-to-crescent clearance is 0.008"–0.012".

Figure 53.47 Internal/external gear pumps are checked in three places. (*Courtesy of Tim Gilles*)

Rotor pump clearance tolerances are as follows:
- 0.010" between the inside and outside rotors
- 0.014" between the outside rotor and housing

Gear pump clearance can be checked by inserting plastigage between the cover and gear ends.
- End clearance should be no more than 0.003".
- Side clearance between the gear and housing should be less than 0.005".

Figure 53.48 Examples of restricted oil pickup screens. A plugged screen can ruin a new engine. (*Courtesy of Tim Gilles*)

NOTE: *The pump housing can be easily bruised or distorted during installation of pressed-fit oil pickup screens or because it was carelessly clamped in a vise. This can cause interference between the gears or rotor and the pump housing.*

■ OIL PUMP SCREEN SERVICE

The oil pump sump screen, or pickup, must be checked to be sure it is clean. A new oil pump does not come with a new screen. Some rebuilders replace every screen with a new one. A plugged screen can ruin a new engine in a very short time (**Figure 53.48**).

On a newly cleaned screen, make sure that the bypass is not stuck permanently open (see Figure 19.48). Be certain that a screen does not have any loose or damaged wire mesh that could break loose and cause oil pump failure.

■ OIL PUMP FAILURE

Oil pumps usually wear or seize because of improper engine maintenance or because pieces of a metal part, such as a failed camshaft or bearing, get into the pump (**Figure 53.49**). Sometimes, an oil pump hex drive shaft will twist off when a foreign object becomes wedged in the pump.

Figure 53.49 This gear-type oil pump was damaged when metal particles entered the pump from a failed part elsewhere in the engine. (*Courtesy of Tim Gilles*)

Figure 53.50 These broken umbrella valve guide seals were found in the engine's oil pan. *(Courtesy of Tim Gilles)*

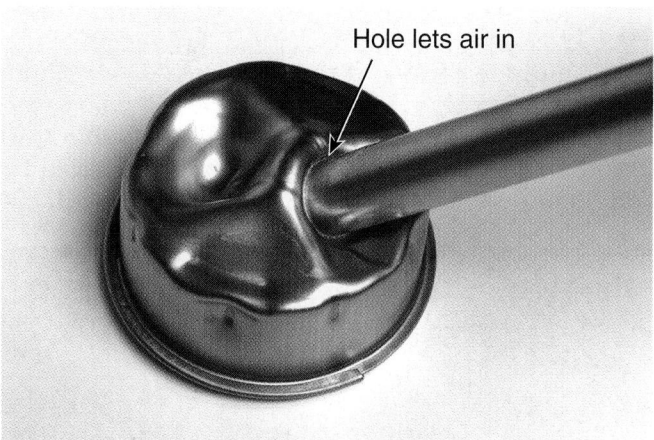

Hole lets air in

Figure 53.51 A leak on the inlet side of the pump will allow air to enter the oil.

Figure 53.52 A bolt-on pump pickup requires a gasket. *(Courtesy of Tim Gilles)*

Another major cause of pump failure is when it draws in plastic material, such as nylon timing sprockets (see Figure 18.41) or deteriorated valve guide seals. Seals become brittle with age and can find their way into the oil system (**Figure 53.50**). Aged or damaged umbrella seals can break up during normal engine operation, but O-ring seals usually enter the system during a careless seal replacement job. Plastic pieces can ruin the oil pump or cause a pump relief valve to stick.

CASE HISTORY

A technician was replacing a timing set in a small-block Chevrolet. The flat-rate manual includes time on this job for removal of the oil pan. From previous experience, the technician knew that he could complete the job without removing the pan. He simply loosened the pan bolts and pried the timing cover out. This cut his time on the job. A short time after the job was completed, the car returned to the shop once again; this time for a failed oil pump and resulting severe engine damage. Pieces of the old timing sprocket were found in the pump screen and oil pump.

Debris that travels beyond the oil pump can also bypass the oil filter (see Figure 12.26) under certain conditions, resulting in serious engine damage.

Air in the System

The bottom of an oil pump does not require a gasket unless it is mounted on the outside of the block. A slight amount of oil leakage on the output side of the pump is harmless. Leakage on the input side of the pump will allow air to be drawn into the oil (**Figure 53.51**).

Check the tube that connects the screen to the oil pump to be sure that there are no holes that can allow the entry of air into the system. Bolt-on oil pickup screens usually require a gasket. Be sure to remove the old gasket carefully. The gasket is only needed on the inlet side of a pump. The outlet side often has no gasket because it is not necessary. The idea is to keep air from entering the pump on the inlet side. **Figure 53.52** shows where a gasket would, or would not, be on a typical crankcase-enclosed oil pump. Some screens are pressed-fit.

NOTE: *Do not pound on the pump when installing the screen. The pump housing can be damaged. Clearances within the pump are so small that if the pump is improperly clamped in a vise or pounded on, interference between the gears or rotors and housing can result.*

- If the screen does not fit tightly, is damaged, or is dirty, it should be replaced.
- Position the screen ¼" to ½" from the bottom of the oil pan. This will prevent it from drawing in sediment that has accumulated in the bottom of the oil pan.

■■ INSTALLING THE OIL PUMP

Oil pumps are filled and tested at the factory before shipment. Before installation, fill the oil pump cavity engine assembly with lubricant (**Figure 53.53**) to ensure that it will prime and will not be damaged by running dry when the engine starts. Turn it by hand to make sure that it has not been damaged by previous mishandling. Priming the lubrication system is covered in detail in Chapter 56. It is also a good idea to check the pressure relief valve to verify that it moves freely (**Figure 53.54**).

Figure 53.53 This pump has been filled with a low melting temperature engine assembly lubricant to allow oil to feed the main oil gallery sooner. *(Courtesy of Tim Gilles)*

Figure 53.54 Checking pressure relief valve operation. Be careful not to scratch the valve bore. *(Courtesy of Tim Gilles)*

■■ REVIEW QUESTIONS

1. What type of threads do oil and heater fitting connections have?
2. When a main bearing bore has changed colors, what should the technician look for?
3. What are two processes that can be used to realign main bearing bores?
4. List two ways that a cylinder bore can wear.
5. What is the crisscross appearance called that results from honing a cylinder?
6. Before measuring a block that is mounted on its end by a universal engine stand, what should be done?
7. What is typical piston-to-cylinder wall clearance?
8. When the main bearing that is farthest from the oil pump shows excessive wear, what is indicated?
9. What is the material called that is used for measuring bearing clearance?
10. Besides metal from worn bearings or a failed camshaft, what are two types of foreign material that can find their way into the oil pump?

■■ ASE-STYLE REVIEW QUESTIONS

1. Technician A says that changing the stroking speed of a drill will change the angle of the crosshatch pattern in the cylinder. Technician B says that it is not necessary to stroke a glaze breaker through the entire cylinder. Who is right?

    a. Technician A   c. Both A and B
    b. Technician B   d. Neither A nor B

2. Technician A says to check a crankshaft for cracks by holding it in the air by hand while ringing the counterweights with a hammer. Technician B says that when a crankshaft is broken, check to see if the vibration damper is defective or loose. Who is right?

    a. Technician A   c. Both A and B
    b. Technician B   d. Neither A nor B

3. Technician A says that excessive belt tension can cause main bearing wear. Technician B says that riding the clutch can cause thrust bearing wear. Who is right?

   a. Technician A   **c.** Both A and B

   b. Technician B   **d.** Neither A nor B

4. Technician A says that a gasket must be installed under an oil pump that is mounted inside of the oil pan. Technician B says that a gasket must be installed under a bolt-on pump screen. Who is right?

   a. Technician A   **c.** Both A and B

   b. Technician B   **d.** Neither A nor B

5. When a crankshaft is ground to 0.030" undersize, how much material is removed from the journal surface?

   a. 0.030"   **c.** 0.060"

   b. 0.015"   **d.** None of the above

6. When measuring cylinder bore taper using a piston ring and feeler gauge:

   a. Subtract the bottom gap measurement from the top gap measurement.

   b. Subtract the top gap measurement from the bottom gap measurement.

   c. Divide the sum of the difference between the top and bottom end gap readings by three.

   d. None of the above.

7. Technician A says to hone at least 0.001" after boring a cylinder. Technician B says to use solvent to clean the block after honing. Who is right?

   a. Technician A   **c.** Both A and B

   b. Technician B   **d.** Neither A nor B

8. A cylinder bore is damaged. Technician A says that it can be repaired with a sleeve. Technician B says that boring it oversize might clean it up. Who is right?

   a. Technician A   **c.** Both A and B

   b. Technician B   **d.** Neither A nor B

9. Technician A says to position the oil hole in a cam bearing directly under the camshaft. Technician B says that the loaded rod bearing is the upper one. Who is right?

   a. Technician A   **c.** Both A and B

   b. Technician B   **d.** Neither A nor B

10. Each of the following statements is true *except*:

   a. Rusted steel is harder than unrusted steel.

   b. Lugging occurs when the engine speed is too low for the load on the engine.

   c. Block deck surfaces warp less often than cylinder head surfaces.

   d. Tapered oil gallery plugs can usually be removed without heating them.

CHAPTER 54

Engine Diagnosis and Service: Piston, Piston Rings, Connecting Rod, Engine Balancing

■ OBJECTIVES

Upon completion of this chapter, you should be able to:

✔ Analyze wear and damage to the piston, piston rings, and connecting rod.

✔ Select and perform the most appropriate repairs to the piston, piston rings, and connecting rod.

✔ Explain the theory of engine balancing.

■ KEY TERMS

stress raiser
scuffing
four-corner scuffing
end gap

connecting rod resizing
internal balancing
bob weights
primary vibration

rocking couple
force, static, or kinetic imbalance
dynamic and couple imbalance

■ INTRODUCTION

This chapter deals with diagnosis and service of the piston, piston rings, and connecting rod. Engine balancing service is also covered. Complete rebuilt engine assemblies are commonly installed in the industry. Information in this chapter deals with actual engine repairs. Repair shops can remove an engine for rebuilding or perform an overhaul while the engine remains in the chassis. Those procedures are covered in Chapter 56.

Occasionally, a piston ring will break or an engine is overheated, which causes rings to lose their spring tension. If a cylinder head is removed for repairs and the oil pan is readily accessible, the customer might opt for the installation of new piston rings. Information learned in this chapter will help you decide on the best repair option.

Piston rings are replaced whenever an engine is disassembled. Pistons are often reused unless the cylinder bores are worn badly enough to require reboring. Connecting rods do not usually require service, but you need to know what to look for to avoid costly engine failures. As in previous service chapters, as you read this chapter you should become familiar with the practice of diagnosing the cause of problems before attempting a repair. This is the skill that separates the most successful technicians from the average ones.

■ PISTON SERVICE

It is easier to service the piston if the piston and rod assembly is held in a vise. Be careful not to damage a slipper piston skirt when clamping the rod in the vise. **Figure 54.1** shows the right way to clamp the rod on a slipper piston in a vise.

NOTE: *The jaws of the vise must be soft metal. Steel-toothed vise jaws can mark the connecting rod and weaken it. Damaging a metal surface raises stress, weakening the part. This is called a "**stress raiser**."*

Removing Piston Rings

Compression rings are removed with a ring expander (**Figure 54.2**). Oil rings are easily removed by rolling off the rails and removing the expander spacer.

Figure 54.1 The correct way to mount a slipper skirt piston and rod assembly in a vise. *(Courtesy of Tim Gilles)*

Figure 54.2 One type of ring expander. *(Courtesy of Tim Gilles)*

Cleaning the Piston

The top of the piston can be cleaned with a scraper or abrasive disc. The *top* (and only the top) of the piston can also be cleaned on a wire wheel, but be especially careful not to round off the edges of the piston head.

CAUTION Do not use the wire wheel to clean the skirt of the piston or the ring groove area.

Supervisors often advise against cleaning any part of the piston with a wire wheel because it requires skill and care. A beginner can accidentally ruin an otherwise good piston by trying to clean the skirt and ring grooves using the wire wheel. When cleaning a piston using a wire wheel, it is a good idea to clean the piston top before removing the piston rings to avoid accidental ring groove damage.

Pistons are never to be chemically cleaned while assembled to the piston pin and connecting rod.

CASE HISTORY *An apprentice cleaned a set of pistons in a chemical cleaner with the connecting rods attached (the wrist pins on most automotive pistons are pressed into the connecting rods). Varnish became trapped between the piston and the wrist pin, freezing them together. To free them up again, it was necessary to press the pistons from the rods. Not only was this an unproductive use of time, but pistons are sometimes damaged during removal with a press.*

Cleaning Piston Ring Grooves

During engine operation, carbon sometimes forms in the back of the compression ring grooves. The carbon must be removed; otherwise, the carbon deposits in the ring grooves might prevent the new rings from compressing enough to enter the cylinder during piston installation. Clean the carbon from the ring grooves with a ring groove cleaner (**Figure 54.3**).

- When using a ring groove cleaner, be careful not to remove aluminum from the rear of the ring groove after all of the carbon has been removed.
- The ring groove cleaning tool works very well on compression ring grooves but can easily nick oil ring grooves, which do not usually get as much hard carbon buildup anyway.
- Oil return holes sometimes become plugged and must be cleaned, or the engine will continue to use oil. Drilled oil holes can be cleaned with a drill bit that correctly fits the oil hole.

Piston rings on most new engines are very narrow; a traditional piston ring groove cleaner might not have a narrow enough cutter. An old ring can be broken off

Figure 54.3 Cleaning carbon from a ring groove with a ring groove cleaner. *(Courtesy of Tim Gilles)*

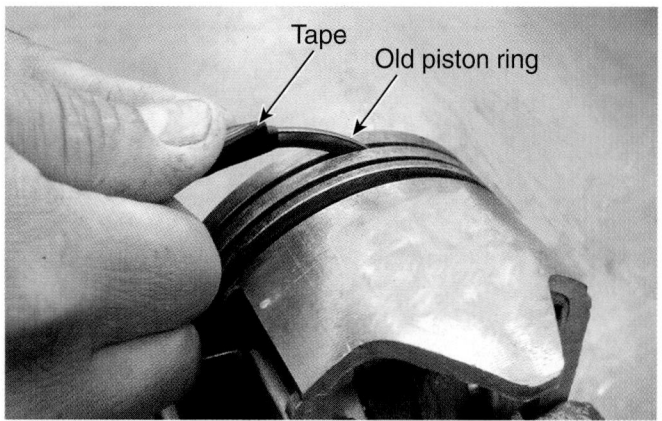

Figure 54.4 Cleaning a piston ring groove using a broken piston ring. *(Courtesy of Tim Gilles)*

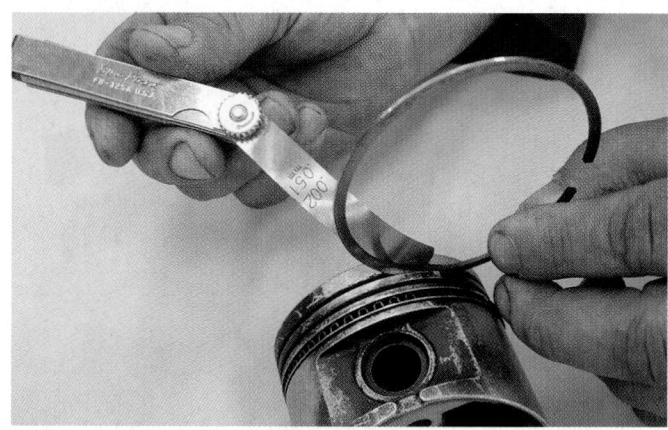

Figure 54.6 Checking a ring groove for wear. *(Courtesy of Tim Gilles)*

and ground sharp to clean out carbon (**Figure 54.4**). Tape the ring so it does not cut you.

Some manufacturers use piston rings of different depths for the same engine. For instance, 5.7L GM engines can have piston rings of four different depths. Pistons designed for use with low-tension rings have ring grooves that are shallower than standard ones. Use a ring to double-check for correct groove depth. Roll the ring around the entire groove to ensure that the ring does not bind up on the edges of the ring land.

Ring Groove Wear/Side Clearance Check

The top ring groove wears the most because the ring is forced against the bottom of the ring groove during combustion. This can cause wear to the piston, and rings with excessive side clearance can break. Modern aluminum pistons have high silicon content that causes them to wear out the piston ring (**Figure 54.5**).

Top ring groove wear was much more common on older pistons. Older cylinder blocks were prone to cylinder bore taper wear. The action of the piston rings moving in and out against the cylinder walls sometimes resulted in severe upper piston ring and groove wear.

Before cleaning the piston, check the top ring groove for excessive wear (**Figure 54.6**). Normal ring-to-groove side clearance is from 0.002" to 0.004". When a new ring is placed in the groove, if a 0.006"

feeler gauge can be inserted under it, the groove is worn excessively. A worn ring can be used for this test if the ring is held in the groove. Only part of the ring wears, so the unworn part is good enough for performing this measurement (see Figure 54.5).

NOTE: *Engines that use low-tension rings usually clean up at 0.020" oversize when boring. They do not wear the cylinder wall as much as standard-tension piston rings, because they do not push as hard against it.*

Measuring a Piston

The place to measure a piston varies among manufacturers, so check the service information. **Figure 54.7** shows where most manufacturers recommend a piston skirt be measured.

Piston Weight

Replacement pistons are designed to weigh the same as originals, even if they are oversize. To maintain engine balance, it is important that a new replacement piston's weight match that of the original piston.

When a replacement piston is heavier than the other pistons it can be lightened at the *balance pads*. Balance pads are usually located on the edge of the skirt under the wrist pin.

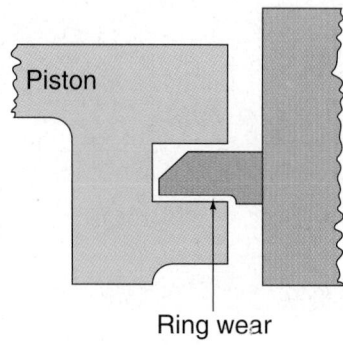

Figure 54.5 When ring wear is evident, it is found on part of the bottom edge.

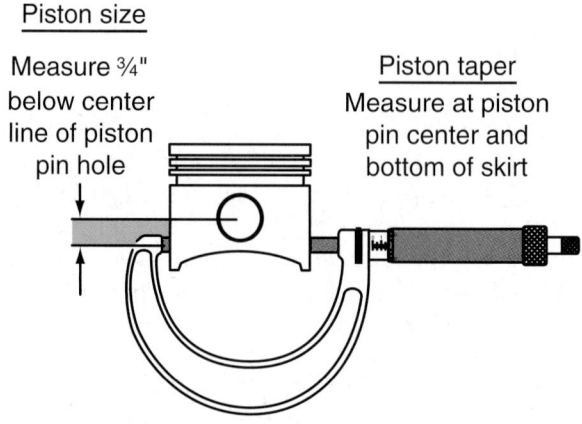

Figure 54.7 Measuring maximum piston skirt diameter.

NOTE: *Replacement pistons for passenger cars and light trucks are destroked 0.020" to compensate for block and head resurfacing. If 0.020" of resurfacing is not done, the compression ratio will be diminished by about 0.25 (9:1 reduced to 8.75:1).*

MATH NOTE

Compression Ratio

Stock block and stock piston	*8:1*
Stock block and new replacement piston	*7.75:1*
Head or block resurfaced 0.020" + new piston	*8:1*

Diagnosing Piston Wear

Scuffing due to excessive heat occurs when the cylinder wall and piston momentarily weld to each other as the piston stops at TDC. The welds are constantly made and broken.

- Scuffing on both skirts is usually caused by insufficient clearance between new pistons and cylinder walls.
- Scuffing on only one skirt can be caused by excessive idling at too low an rpm or by lugging the engine. In either case, there is not enough oil thrown from the rods to provide adequate cylinder wall lubrication (**Figure 54.8**).
- Scuffing can also be caused by cylinder wall hot spots that are the result of poor cooling system maintenance.
- **Four-corner scuffing** occurs when both skirts are scuffed on the edges next to the piston pin (**Figure 54.9**). This is usually a result of an external cause such as too lean an air-fuel mixture, which causes the top of the piston to run too hot, expanding the skirts against the cylinder bore.
- A piston that overheats because of cooling system problems or abnormal combustion will expand excessively near the piston pin. This can cause scuffing of the piston skirt near the pin.

Figure 54.9 This piston skirt it scuffed more toward the outside, toward the wrist pin. *(Courtesy of Tim Gilles)*

■ PISTON RING SERVICE

Ring Wear Diagnosis

The following are some of the things that can cause piston ring wear:

- Leftover honing grit from a careless block cleanup
- Running the engine with a missing or damaged air cleaner or broken vacuum lines
- Using a contaminated oil fill spout or funnel

When inspecting rings for wear, look for the following:

- When wear is due to dirty air getting in, the top ring will show more wear and vertical abrasive lines will be visible. The cylinder wall will show scratches in the area where the rings ride (**Figure 54.10**).
- When ring wear is due to abrasives in the oil, the lower rings and cylinder wall will have more wear and the top ring will have less wear.
- When abrasives cause wear, the bottom side of a ring will wear, leaving a lip on the outside edge (see Figure 54.5). **Figure 54.11** shows how an oil ring can wear.

Figure 54.8 A scuffed piston skirt. *(Courtesy of Tim Gilles)*

Figure 54.10 Contamination by unfiltered air is indicated by vertical scratches in the cylinder wall above the bottom limit of piston ring travel. *(Courtesy of Tim Gilles)*

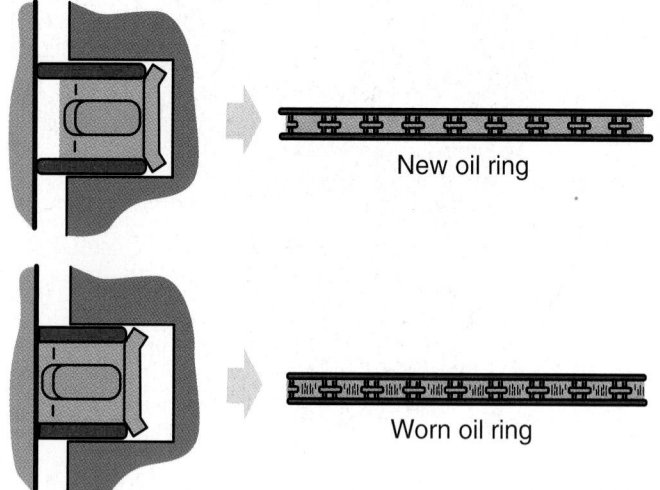

Figure 54.11 New and worn oil ring rails.

Ring Oversizes

When an engine is rebored, oversized pistons and rings are used. Rings are made for standard 0.020" (0.50 mm), 0.030" (0.75 mm), 0.040" (1.0 mm), and 0.060" (1.5 mm) oversizes. North American pistons and bearings and many aftermarket import engine replacement parts still use the inch standard system. To quickly convert metric to inch standard, note that each 0.25 mm is equal to 0.010"

Compression Ring End Gap Clearance

Before installing rings in a cylinder bore, check the ring **end gap** (see Chapter 53). To measure gap clearance, install the ring in the cylinder and square it up with a piston. Then measure the gap with a feeler gauge. The ring must be positioned in the unworn portion of the bore, below ring travel. The end gap is tapered from the outside to inside edge. So be sure to measure at the outside edge of the ring for an accurate measurement.

To give themselves more latitude during production, manufacturers produce rings that have 0.005"–0.010" more gap clearance than the minimum specification. The ring gap should be at least 0.003" to 0.004" for each inch of cylinder bore diameter unless otherwise specified in the service information. The ring end gap can be filed to fit if it is not wide enough. One manufacturer states that maximum gap clearance is not as critical and can actually be as much as 0.030" *more* than the minimum specifications without causing blowby.

The important thing about ring gap is the minimum specification. Too small an end gap can cause the rings to butt together and lock up in the bore as they heat and expand, resulting in scuffing and ring failure. If this happens, the ends of the ring will appear polished.

NOTE: *An increase of 0.002" in the bore size will increase the gap by about 0.006". Installing standard rings in a* 0.030" *oversized cylinder bore will result in an end gap increase of approximately 0.090".*

MATH NOTE

As an interesting experiment, check the gap on a worn ring to see how much metal has been lost from its face surface.

An approximation of wear may be made as follows: If the gap of an old ring is 0.050" and the gap of a new ring is 0.020", divide the difference by 3 (or π) and then by 2 (wear from both sides).

$$\begin{array}{r} 0.050" \text{ Old Ring Gap} \\ -0.020" \text{ New Ring Gap} \\ \hline 0.030" \text{ Difference} \end{array}$$

$$\frac{0.030"}{3} = 0.010" \text{ (Conversion of Circumference to Diameter)}$$

$$\frac{0.010"}{2} = 0.005" \text{ Ring Face Wear}$$

Too much gap clearance can point to a bore too large or the use of rings that are too small.

NOTE: *The gap will change by about 0.030" for each 0.010" error in size.*

A common customer error is to want to install oversized rings in badly worn cylinders. An oversized ring might fit into the top of a tapered cylinder. Because of the tapered cylinder wear, the gap on an oversized ring would lock up as the piston moved down the cylinder wall.

Figure 54.12 shows rings that were too small for the cylinders they were run in. Notice the dark area (carbon deposits) near the gap.

Figure 54.12 Rings that are too small for the cylinder will have the appearance of carbon deposits near the gap.

■ INSTALLATION OF PISTONS AND RINGS

Installing Pins to Connecting Rods

Machine shops separate pressed-fit pins and pistons from their connecting rods using a pin press (**Figure 54.13**). Installation of pressed-fit pins into connecting rods is accomplished using a rod heater to heat the eye of the rod (**Figure 54.14**). While the rod is hot, a new pin will slip into place easily without the risk of ruining a piston.

SHOP TIP

■ When original pistons are to be reused during an overhaul, it is best to simply leave the pistons and rods assembled. Pressing them apart serves no useful purpose and risks ruining an otherwise good piston.

■ Do not glass bead blast pistons while pistons are assembled to the connecting rods.

■ Do not soak the piston and rod assembly in a chemical cleaner. The pin could seize on the piston.

■ Be sure to keep track of the direction the connecting rod faces in relation to the top of the piston. Pistons have a notch on the side of the piston head that faces the front of the engine.

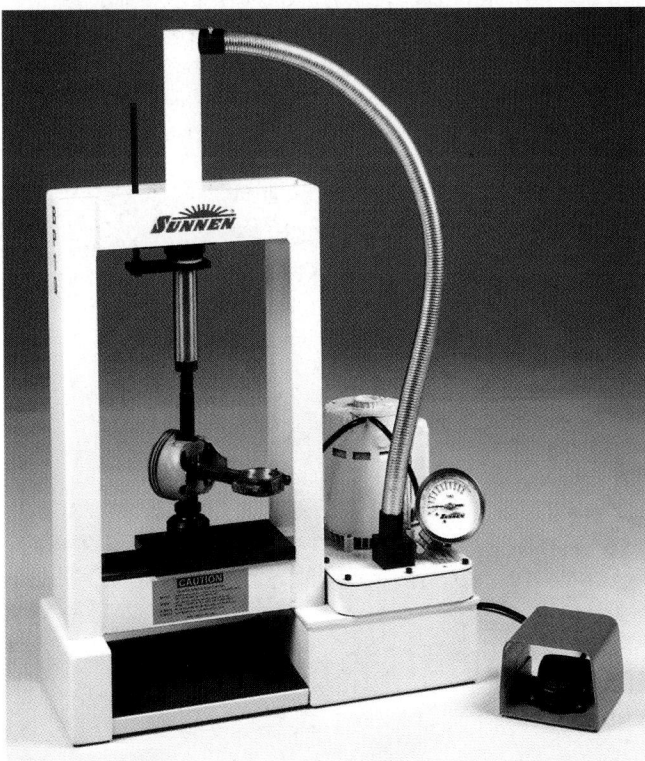

Figure 54.13 Using a piston pin press to remove a pin from a pressed-fit connecting rod. (*Courtesy of Sunnen Products Company, St. Louis, Missouri*)

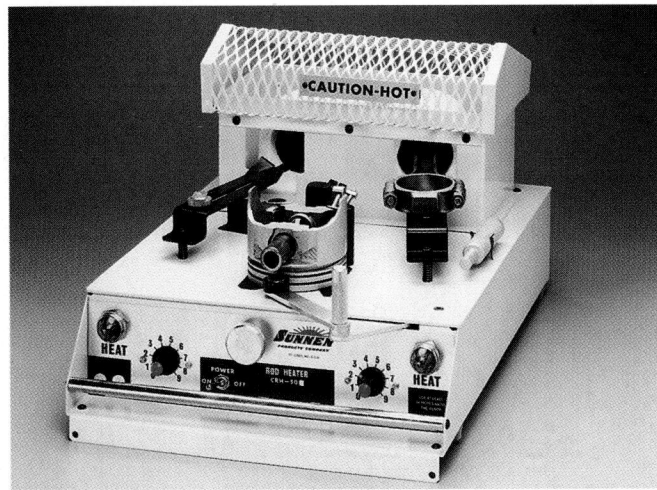

Figure 54.14 A rod furnace. (*Courtesy of Sunnen Products Company, St. Louis, Missouri*)

■ INSTALLING RINGS ON PISTONS

Oil rings are installed first. Then, the second compression ring is installed. Finally, the top ring is installed.

■ OIL RING INSTALLATION

Most automobiles use three-piece oil control rings, consisting of an expander spacer in the center with two outside rails. Install them as shown in **Figure 54.15**.

■ Install the expander spacer, being careful not to overlap its ends. The ends are usually painted different colors to make it obvious to the installer if they are accidentally overlapped. Some expanders are filled with a Teflon button on their ends to prevent improper assembly. In the absence of a recommendation, position the expander gap above one end of the wrist pin.

■ Next, install the rails. Installing the top rail first is easiest. While holding your finger over the butted ends of the expander, install the top rail. Position

Position gap in expander over piston pin

Piston pin

(a)

Figure 54.15 (a) Install the oil ring expander. (*Courtesy of Tim Gilles*)

(b)

Figure 54.15 (continued) (b) Roll the oil ring rails into place. *(Courtesy of Tim Gilles)*

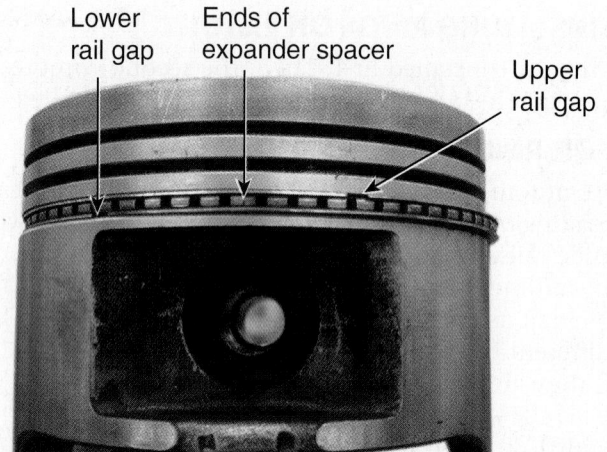

Lower rail gap

Ends of expander spacer

Upper rail gap

Figure 54.16 Correct oil ring installation. *(Courtesy of Tim Gilles)*

its gap above the skirt on one side of the expander spacer gap.

■ Install the lower rail with its gap placed above the skirt on the opposite side of the expander gap (**Figure 54.16**).

Compression Ring Installation

Remember from Chapter 19 that piston rings are often tapered or have chamfers or reliefs to cause them to twist. These piston rings must be installed with their identification marks facing up (**Figure 54.17**). Installing them upside down will result in severe oil consumption. In fact, one compression ring installed upside down can double an engine's oil consumption. The second compression ring actually controls more oil than compression.

Use a ring expander to install the compression rings (see Figure 54.2). If a ring expander is not available, a shop towel can be used as shown in **Figure 54.18**. It is important not to "spiral" or roll the rings on; they can become distorted to resemble a lock

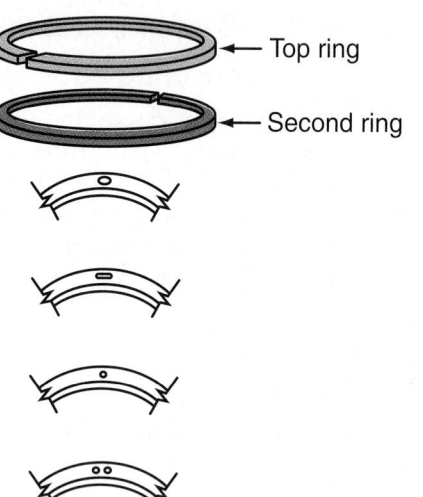

Top ring

Second ring

Figure 54.17 Ring I.D. marks face up. Manufacturers use different markings.

Figure 54.18 Using a shop towel to install a compression ring. *(Courtesy of Tim Gilles)*

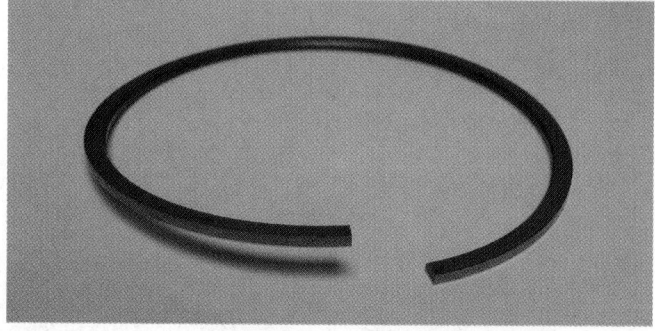

Figure 54.19 Improper installation can ruin a piston ring.

washer (**Figure 54.19**). Overexpanding plain cast iron rings during installation can very easily result in a broken ring.

Compression Ring Gap Position

The gaps are placed at different locations around the piston. Manufacturers specify different gap positions.

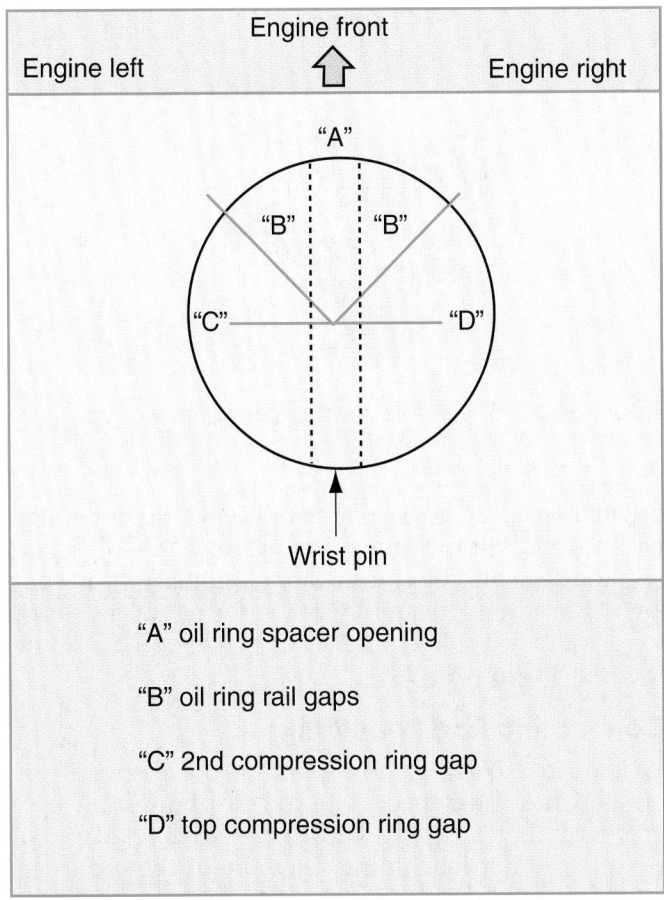

"A" oil ring spacer opening

"B" oil ring rail gaps

"C" 2nd compression ring gap

"D" top compression ring gap

Figure 54.20 A generic chart of ring gap placements popular in the aftermarket. Follow the instructions that come with the rings.

According to information published by engineers in the Perfect Circle Division of Dana Corporation, the reason for the practice of staggering ring end gaps is to guard against scuffing when an engine is started for the first time. As the engine operates, rings will rotate from the position where they were first installed.

■ End gap position is *not* a cause of oil consumption.
■ There are many differing opinions on ring gap placement. The most prudent policy is to follow recommendations of the vehicle manufacturer, when available.

The positions shown in **Figure 54.20** are popular in the aftermarket.

■ CONNECTING ROD SERVICE

During engine disassembly, be sure to keep rod cap rods. Both the upper and lower pieces should be numbered to ensure accurate pairing on reassembly.

CAUTION If the cap is installed backwards, the rod bore will not be round.

Figure 54.21 A twisted connecting rod caused this wear pattern on the piston skirt. *(Courtesy of Tim Gilles)*

Rod Alignment

Closely examine all piston skirts for unusual wear patterns that can indicate a *twisted rod* (**Figure 54.21**). A *bent rod* may show wear on opposite sides of the rod bearings (**Figure 54.22**).

SHOP TIP If an engine has operated for 100,000 miles without an unusual wear problem, no problem should occur if the rod is reused with new bearing inserts. This is one of the advantages of a custom rebuild. Parts can be inspected for alignment and unusual wear problems during disassembly.

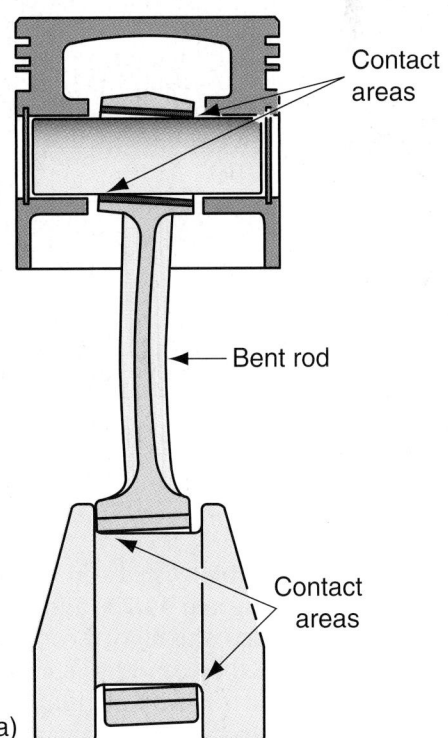

Contact areas

Bent rod

Contact areas

(a)

Figure 54.22 (a) Wear caused by a bent or twisted connecting rod.

Figure 54.22 (continued) (b) Bearing wear from a misaligned rod will show up more on the top bearing half. (*Courtesy of Tim Gilles*)

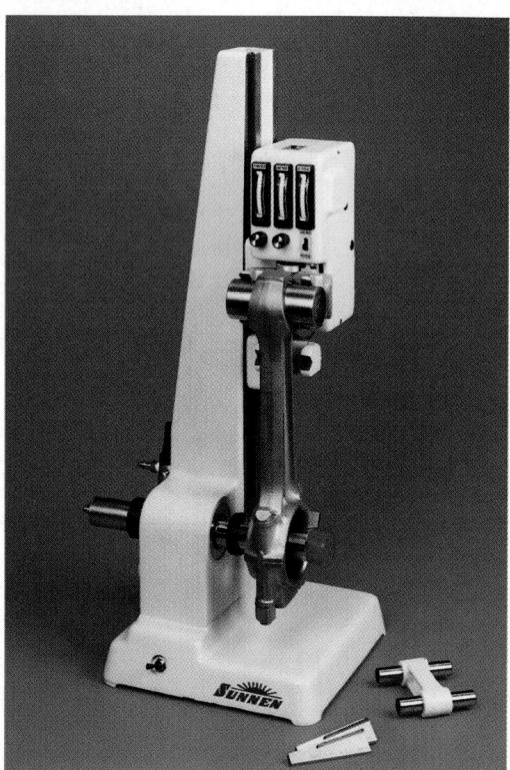

Figure 54.23 Checking connecting rod alignment. (*Courtesy of Sunnen Products Company, St. Louis, Missouri*)

Rods suspected of being twisted can be checked on a rod aligning fixture (**Figure 54.23**) and straightened as needed. Many years ago, misaligned connecting rods were commonly corrected by bending. But metal seems to have "memory," which is why aligning things that are bent is not always successful over the long haul. For instance, when a connecting rod is bent, the molecules in its metal get stretched. When it is bent back into its

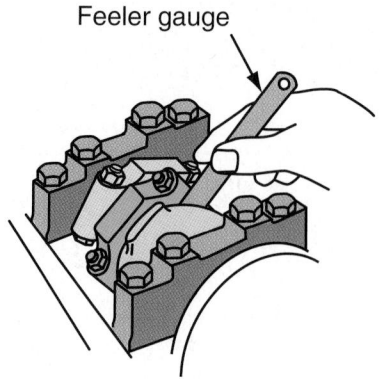

Figure 54.24 Check rod bearing side clearance at several points next to the rod.

original position, however, the molecules do not bunch back together in the same place again.

Rods can also be checked for alignment when they are installed on the crankshaft. A feeler gauge is used to measure rod side clearance at several points around the rod (**Figure 54.24**).

Connecting Rod Rebuilding

When a rod bearing has "spun" or burned, or if the big end has "stretched," the rod can be resized by a machine shop. This is known as **connecting rod resizing**. After machining, the bore of the rod is the same size as original.

Rod stretch sometimes causes the rod to draw closer together at the rod cap parting line. This can cause bearing wear at the ends of the inserts (**Figure 54.25**). Rods are measured for "out-of-round" with a special gauge (**Figure 54.26**). The gauge measures in tenths of thousandths. Usually, rods can be up to 0.001" out-of-round before resizing is necessary.

To resize a rod:

- The pressed-fit rod bolts are pressed or pounded out.
- A small amount of metal (usually less than 0.002") is ground off the rod and cap mating surfaces.
- The rod cap is reinstalled on the rod and the nuts torqued.

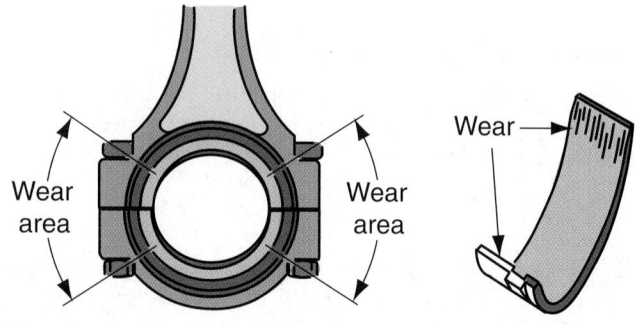

Figure 54.25 Rod stretch causes bearing wear at the parting line.

Figure 54.26 Checking for rod stretch. *(Courtesy of Tim Gilles)*

Figure 54.27 Honing connecting rods to size. *(Courtesy of Sunnen Products Company, St. Louis, Missouri)*

■ The rod bore, which is now smaller as a result of grinding, is honed with a rod hone (**Figure 54.27**) until the original diameter of the rod bore is reached.

NOTE: *Overtorque of rod bolts can cause them to fail during deceleration (when the load is on the bolt instead of on the rod).*

To repair the small end of a full-floating type connecting rod, a bronze bushing is installed and the bore is then honed to the proper size.

■ ENGINE BALANCING SERVICE

Balancing is done by a machine shop or a balancing specialist. Reciprocating parts (the piston assembly, including rings, wrist pins, and the pin end of the rod) are balanced to weigh approximately the same amount. **Figure 54.28** shows a connecting rod being weighed separately on each end. All of the rods are weighed and then lightened to the weight of the lightest one.

Rotating parts are balanced by spinning them on a balancing machine (**Figure 54.29**) to determine the location of heavy spots. Balancers operate at low speed (400 rpm) for safety. This rpm is sufficient to calculate what the amount of imbalance will be at higher speeds.

Heavy counterweights can be lightened by drilling (**Figure 54.30**). **Figure 54.31** shows balance pads on a connecting rod; weight can be removed from them during the balancing procedure. Counterweights that are too light can be drilled for the addition of a heavier metal, like lead or tungsten, sometimes called "mallory metal," which is twice as heavy as steel. Weight can be added to a wrist pin if a piston is too light.

Cast cranks sometimes use a balanced vibration damper and torque converter flexplate. On these cranks,

Figure 54.28 (a) Weighing the big end of the connecting rod. (b) Weighing the small end of the connecting rod. *(Courtesy of Tim Gilles)*

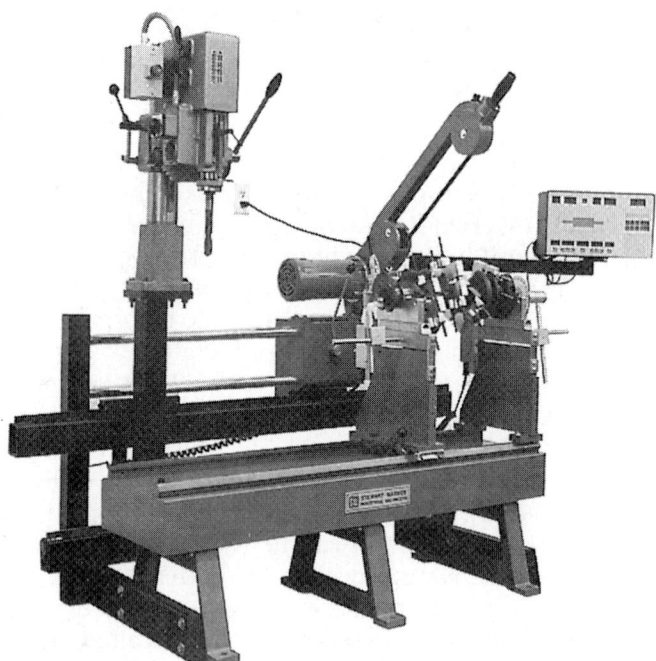

Figure 54.29 A crankshaft in an engine balancer. *(Courtesy of Pro-Bal Industrial Balancers)*

Figure 54.30 Holes are drilled in crankshaft counterweights. *(Courtesy of Pro-Bal Industrial Balancers)*

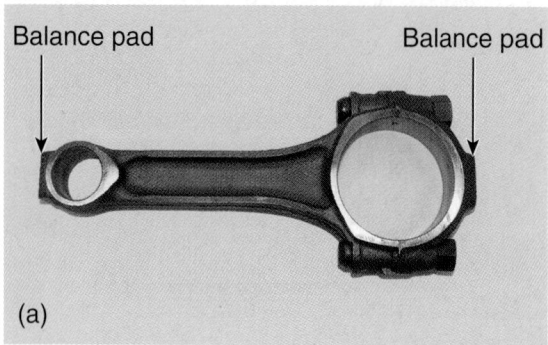

Balance pad   Balance pad

(a)

Figure 54.31 (a) Connecting rod balance pads. *(Courtesy of Tim Gilles)*

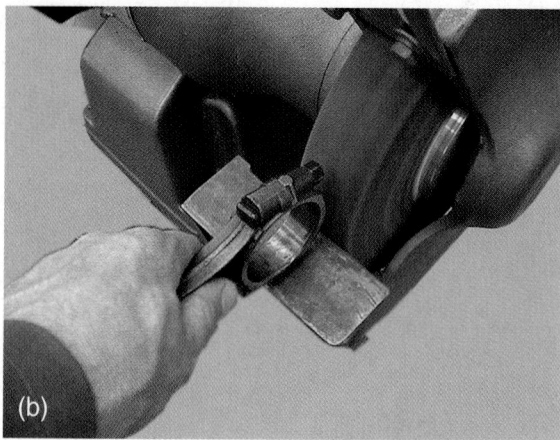

(b)

Figure 54.31 (continued) (b) Removing weight from the crank end of the connecting rod. *(Courtesy of Tim Gilles)*

there might not be any balancing done on the crank counterweights. **Internal balancing** of forged cranks is usually achieved by drilling holes on the counterweights.

When balancing an engine used with a standard transmission, the clutch cover (pressure plate) should be balanced too. It is usually balanced along with the flywheel. Some engines have vibration dampers that are balanced with the belt pulley bolted to them. The relationship of the damper to the pulley should be marked before disassembly.

> **CASE HISTORY**
>
> *A technician rebuilt a 440 Dodge engine for a motor home using spare parts from the shop inventory. He used the existing engine's torque converter and vibration balancer. When the engine was started, a serious vibration was evident. The crankshaft used in the rebuilt engine was forged and the one in the engine that was removed was cast. One of the crankshafts was externally balanced, requiring a special vibration damper and weights on the torque converter.*

To balance the crank, rods, and pistons, the crankshaft must be spun at a specified speed. Pistons and rods would not be able to be spun while attached to the crankshaft unless the parts were assembled in the block. **Bob weights** (**Figure 54.32**) are used when spinning the crankshaft to simulate the correct weight. In-line engines do not use bob weights.

Replacement Piston Balance

Balance is not as important with in-line engines, but it is critical in V-type engines. When replacing a piston on a V-type engine, match the weight of the new piston to the stock piston to maintain correct balance (**Figure 54.33**). Pistons in a new set should not have

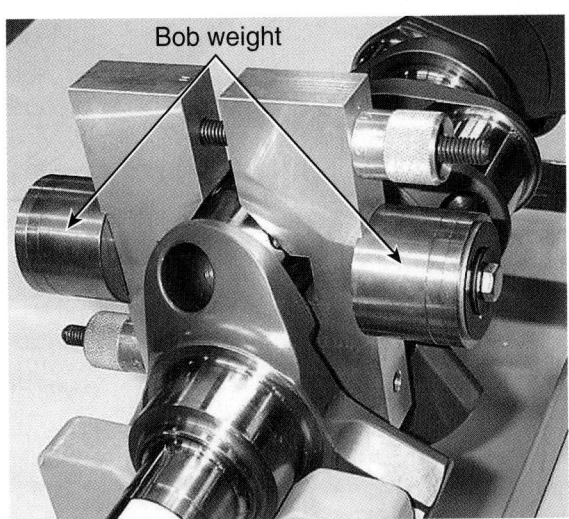

Figure 54.32 A bob weight mounted on a crankshaft throw. *(Courtesy of Pro-Bal Industrial Balancers)*

Figure 54.34 Vibration on a V6 can be reduced by using a single balance shaft. *(Courtesy of Tim Gilles)*

Figure 54.33 One method of removing weight from piston balance pads. *(Courtesy of Tim Gilles)*

CASE HISTORY *After replacing a timing belt on a Mitsubishi 1.8L four cylinder with dual balance shafts, the engine had a serious vibration problem. The timing marks on the front balance shaft and oil pump drive were inspected and found to be aligned properly. After investigation in the service information, the technician discovered that the oil pump drives a rear balance shaft that was actually installed 180 degrees out of position. When putting on the timing belt, installation instructions called for installing a shaft through a hole in the rear of the block to correctly align the rear balance shaft to the oil pump. This corrected the problem.*

a spread of more than 5 grams from the heaviest to the lightest.

NOTE: *A dollar bill weighs about 1 gram.*

SHOP TIP When there is a complaint of engine vibration, start your investigation with the engine mounts. They could be worn out, or stiff ones could have been installed by mistake.

Balance Shafts

Engines with balance shafts (**Figure 54.34**) run very smoothly. The timing of the shafts is critical. They must be replaced in the proper manner to maintain balance.

■ ADVANCED BALANCING INFORMATION

The following section describes the theory of different types of imbalance. It is more advanced information for the serious student. More information on engine balancing theory is included in Chapter 19.

Types of Vibration

When the piston slows down as it approaches TDC, its force pulls the engine up. When it approaches BDC, it pulls the engine down. This is known as **primary vibration**. Engines that have two rod throws 180 degrees apart counterbalance each other to give perfect primary balance.

If there are only two cylinders that fire 180 degrees apart instead of 360 degrees apart, another form of vibration called a **rocking couple** takes place. This results in the engine rocking from end to

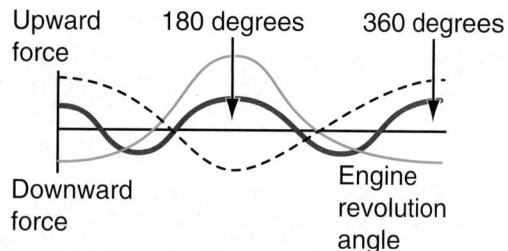

Upward force | 180 degrees | 360 degrees

Downward force | Engine revolution angle

Figure 54.35 The dotted line represents the force generated by cylinders 1 and 4. The fine line represents cylinders 2 and 3.

end. A four-cylinder engine eliminates the rocking couple problem because one pair of couples cancels out the other pair. Cylinders number 1 and 4 move in the opposite direction of the inner two cylinders, numbers 2 and 3 (**Figure 54.35**).

Another form of vibration called *secondary vibration* occurs in in-line four-cylinder engines. Secondary vibrations are only about one-fourth of the strength of primary vibrations but can become quite severe at the higher rpms at which four-cylinder engines quite often operate. Secondary vibration is represented in Figure 54.35 by the solid line. Balance shafts (see Figure 19.42) are used to cancel out secondary vibration. They are driven by the crankshaft in opposite directions at twice crankshaft speed.

Types of Imbalance

Force, also called **static** or **kinetic**, **imbalance** can be compared to the type of imbalance that is corrected when balancing tires with a bubble (level) balancer. As the crankshaft spins, the balancer senses the vibration whenever a heavy area is forcing down. Removing this heavy spot, or counterbalancing it by adding an equal amount of weight to the light side, corrects the force imbalance. Force imbalance is smallest at 90 degrees of rod angle and most at TDC or BDC.

Two other types of imbalance that must be corrected are called **dynamic and couple imbalance**. Both require adding or removing metal at two different places on the part. Correcting dynamic imbalance can result in the correction of force imbalance at the same time. Computer balancers compute the combined amount of dynamic and force imbalance and tell where to remove metal to correct them.

Ninety-degree V6 engines have especially strong primary (up and down) imbalance. These engines are installed transversely in front-wheel-drive cars. The vibration can be felt inside the car on the steering wheel. Soft engine mounts are used that allow the engine to move from side to side. Engine mounts and exhaust connections suffer increased wear and failure because of this, but this is an effective way of smoothing out engine vibrations for front-wheel-drive cars.

REVIEW QUESTIONS

1. What is it called when a metal surface is damaged, weakening it?
2. When the compression ring grooves are cleaned, what substance is removed?
3. What causes scuffing?
4. When both piston skirts are scuffed near the wrist pin, what is this called?
5. What oversizes (in inches) are pistons and rings made for?
6. What minimum size should the ring gap be per inch of cylinder bore?
7. If the bore size is increased by 0.010", how much does the ring gap increase by?
8. If a worn ring's end gap is 0.060" and a new ring's end gap is 0.015", how much is the old ring face worn?
9. What is it called when the big end of a connecting rod is remachined?
10. Where is the extra weight added when an engine is externally balanced?

ASE-STYLE REVIEW QUESTIONS

1. Technician A says that the ring belt area of a piston (just above the piston skirt) is best cleaned with a wire wheel. Technician B says that ring grooves used with low-tension rings are deeper than standard ring grooves. Who is right?
   a. Technician A
   b. Technician B
   c. Both A and B
   d. Neither A nor B

2. Which of the following is/are true about pistons?
   a. Aluminum piston ring grooves are more likely to show wear than piston rings.
   b. Piston and rod assemblies are best cleaned by immersing them in a chemical cleaner.
   c. Both A and B.
   d. Neither A nor B.

3. Technician A says that pressed-fit pins are best installed on connecting rods using a rod heater. Technician B says that pressed-fit piston pins are best removed from connecting rods using a press. Who is right?

   a. Technician A   **c.** Both A and B

   b. Technician B   **d.** Neither A nor B

4. Technician A says that if the piston ring face wears by 0.005", the end gap will change by 0.010". Technician B says that a 0.030" oversize piston should weigh the same as a standard piston. Who is right?

   a. Technician A   **c.** Both A and B

   b. Technician B   **d.** Neither A nor B

5. Compared to older engines, piston rings on late model engines are:

   a. Wider   **c.** The same

   b. Narrower   **d.** Any of the above

6. A cylinder wall has scratches only in the area where the piston rings ride. Technician A says that this can happen if unfiltered air has been entering the engine's air intake. Technician B says that this is due to dirty engine oil. Who is right?

   a. Technician A   **c.** Both A and B

   b. Technician B   **d.** Neither A nor B

7. Technician A says that a connecting rod bolt is under its highest load during deceleration. Technician B says that if the ring gap at the top of the cylinder is larger by 0.010" than the gap at the bottom, the cylinder has 0.010" of bore taper. Who is right?

   a. Technician A   **c.** Both A and B

   b. Technician B   **d.** Neither A nor B

8. Which piston ring wears the most?

   a. The top   **c.** The oil ring

   b. The second   **d.** All wear the same

9. Technician A says that a piston marked with .50 is 0.020" oversize. Technician B says that a piston marked with .75 is 0.030" oversize. Who is right?

   a. Technician A   **c.** Both A and B

   b. Technician B   **d.** Neither A nor B

10. On modern engines, which part tends to wear the most?

    a. The top piston ring

    b. The top piston ring groove

    c. The oil ring

    d. The second compression ring

Ordering Parts, Short and Long Blocks, Engine Assembly

▪ KEY TERMS
engine kit
short block

custom engine rebuild
long block

spin test

▪ INTRODUCTION

Engine Assembly

Before assembly of the engine:
■ Look through the service manual for special instructions.
■ Have all tightening specifications handy.
■ Be sure all parts are thoroughly cleaned.
■ Obtain replacement parts.

▪ ORDERING PARTS

After the block has been stripped, inspect all parts for wear and damage, and compile a list of new parts that are needed. Factory replacement parts are categorized as OE (original equipment). *Stock* means the part is the same as intended by the manufacturer. *Aftermarket* is a broad term that refers to parts that are sold by the non-OE market. Many OE engine parts are manufactured by the same manufacturer as aftermarket parts.

▪ ENGINE KITS

Parts kits are available at wholesale prices for most of the more common foreign and domestic engines. The **engine kit** contains all or most of the parts necessary to completely rebuild an engine. Many of the parts are not individually boxed, and a kit usually costs far less than the individual parts when purchased separately.

There are kits that contain various groups of parts—for example:
■ a crank kit
■ a timing chain set
■ an overhaul kit
■ a master kit

A crankshaft kit includes a reground crankshaft and bearings. A timing chain set includes the chain and sprockets, or chain guides and a tensioner for OHC engines. An overhaul kit includes gaskets, piston rings, crankshaft bearings, and sometimes, a timing set.

The parts normally found in an engine "master kit" are pistons, rings, reground crank and bearings, reground cam and bearings, new or rebuilt oil pump, timing chain set, and a complete gasket set (**Figure 55.1**). Lifters may or may not be included. Many different types and qualities of kits are available. Be sure to compile the parts list carefully. It is not uncommon to spend twice as much on separate engine parts and not get as many new parts as would come in a packaged engine kit.

NOTE: *Sometimes a jobber will promise parts they do not have in stock. If you have a deadline, be certain none of the parts in your order are to be back-ordered.*

Part Cores

When an engine or an engine kit is purchased from a parts source, the crankshaft, camshaft, and oil pump

Figure 55.1 An engine master kit.

are usually returned to them as *cores*. The "core charge" that is paid will be refunded when these parts are returned, provided the core is rebuildable.

Before returning a crankshaft as a core, remove the timing gear or sprocket. If the gear does not have threaded holes, use the puller setup shown in **Figure 55.2**, or a bearing separator and puller like the one shown in Figure 8.42. Save the crankshaft *woodruff key* and pilot bushing. They are not always easy to

(a)

(b)

Figure 55.3 Removing a woodruff key. (a) Pound on the back of the key with a brass punch. (b) Lift the key out of its groove. *(Courtesy of Tim Gilles)*

locate in the correct size. The procedure for removing a woodruff key is shown in **Figure 55.3**.

■ DETERMINING PART SIZES

The engine size can be determined in several ways. Engine rebuilders use books that list casting numbers (**Figure 55.4**). These numbers identify blocks, crankshafts, cylinder heads, and so forth by groups as they were cast at the foundry. The block usually has numbers stamped somewhere on it that can be used to identify the engine before ordering parts.

The cylinder bore diameter and the diameter of the crankshaft journals can be measured and compared to the specifications in the repair manual.

During inspection of engine parts, watch out for unusual oversized or undersized parts. When errors are made at the factory, an engine block is often salvaged. An original engine might have a crankshaft that has been ground undersized on its main and/or rod journals. Cylinder bores, core plug holes, and valve guides and stems can be oversize. Sometimes one or more lifter bores will be machined oversize and fitted with oversize lifters.

Figure 55.2 Puller setup for crankshaft gears and sprockets. *(Courtesy of Tim Gilles)*

Figure 55.4 Engine Builders Association book of block and head casting numbers. *(Courtesy of AERA)*

Manufacturers have codes (usually in their service information) to indicate the use of nonstandard parts. Numbers or letters may be stamped on the parts or on the oil pan rail or crankshaft, or there may be paint on the part (green is a favorite color). Do not leave this to chance. Measure all parts.

TYPES OF ENGINE REBUILDS

A **custom engine rebuild** is when a customer's engine is rebuilt for use in the same vehicle from which it was removed. There are also short blocks and long blocks available from engine rebuilding companies. Some larger companies rebuild engines on an assembly line. Some of these rebuilders are factory authorized by automakers such as DaimlerChrysler or Ford, and they rebuild engines to the authorizing manufacturers' standards.

Short blocks are sometimes used by independent shops and dealerships. Short blocks are completely assembled rebuilt blocks purchased from automotive machine shops and engine rebuilding companies (**Figure 55.5**). They do not include any external parts such as mounting brackets, sheet metal, pumps, or accessories.

Figure 55.5 A short block. *(Courtesy of Jasper Engines and Transmissions)*

Lifters, gaskets, and/or an oil pump may have to be purchased in addition to the short block. An assembly that needs these parts is known as a "short" short block.

Many shops prefer **long blocks** to short blocks. A long block, which includes the cylinder heads assembled on a short block, might be a brand new factory engine or a rebuilt engine. Long blocks are tested in run-in stands or spin testers so more problems can be spotted before installation. An advantage to the shop owner over buying an engine kit is that a long block carries a time and mileage guarantee from the rebuilder.

Engine Cores

When buying a short block or a long block, the old block assembly is returned to the rebuilder as a core. If the block core is unacceptable or if the crankshaft, camshaft, cylinder block, or heads are found to be defective, a core charge is charged. (**Figure 55.6**) Typical practice is for the core charge to be returned only after the rebuilder inspects the parts.

Figure 55.6 This engine ran low on oil and threw a rod. It will not be an acceptable core and one of the heads might even be damaged. *(Courtesy of Tim Gilles)*

SELECTING THE CORRECT REPLACEMENT ENGINE

When you swap one engine for another, be sure you have the correct replacement. Some of the engine parts can be different. Engines can have different mounting points and fittings. If a customer asks for an engine with a different displacement than was originally installed in the vehicle, he or she will probably have problems with emission certification and other concerns.

NOTE: *Federal law prohibits anyone from substituting one engine with another engine of a different year. The engine's emission components must all be the same.*

WARRANTY

Some rebuilt engines have warranties that expire in as little as 90 days or 4,000 miles. A longer guarantee usually means a higher cost. When a rebuilt engine fails while under warranty, some rebuilders pay for repairs according to the time listed as flat rate (see Chapter 5). The percentage of flat rate paid is usually such that the installation shop will not make a profit on the warranty repair, but the technician will be compensated.

Proper diagnosis is very important. If a problem that is not the fault of the rebuilder is causing the concern, under the terms of the warranty the rebuilder is not liable for the cost of the repair. A needlessly replaced long block can be a very costly expense to the installing shop. If an engine failed due to an overheating problem, it is possible the sensors for coolant temperature and the fan relay could be defective. Sometimes a piston will *seize* in the cylinder due to an air leak in the intake manifold or a fuel system problem, resulting in too lean an air-fuel mixture.

Any fuel and cooling system hoses that are not in excellent condition must be replaced. The preliminary estimate for the customer should include these items. If the customer does not agree to their replacement, the shop cannot guarantee the job. Be certain that the water pump and radiator are in good condition. If a careful and methodical engine reassembly is not done, all of your work will be wasted.

REASSEMBLY

Before beginning engine reassembly, inspect and count all new parts. Too many parts could be packaged in a master parts kit. For instance, if there were too many oil gallery plugs or compression rings in a set, the extra one might appear to be a leftover. It would be frustrating to disassemble the engine just to see if a part had been forgotten.

Prepare the cylinder block for reassembly as described in Chapter 53. Clean the block with soap and water,

Figure 55.7 Sludge trapped behind an oil gallery plug. *(Courtesy of Tim Gilles)*

and oil it after it dries. Be sure all oil galleries have been cleaned with a rifle brush before installing new oil gallery plugs. This is extremely important. Failure to do a thorough job can result in a catastrophic engine failure (**Figure 55.7**). Install the oil gallery and coolant core plugs in the block. On pushrod engines, install the cam bearings as described in Chapter 53. Chase all threads in the block with a tap.

BEGIN REASSEMBLY

After the block has been thoroughly cleaned and the cam bearings and core plugs have been installed, you can begin to reassemble the engine.

SHOP TIP

- Once an engine is assembled, the only way for dirt to get in is past the air cleaner, so cleanliness during assembly is a must. If work is stopped at any time during the reassembly, cover the engine completely with a large trash bag to keep dirt out (**Figure 55.8**).
- If a bolt cannot be turned into a hole by finger pressure only, something is wrong. *Never force threads together with hand tools or an impact wrench.*
- Finger-start *all* bolts that fasten a particular part before proceeding to torque any of them down. This will allow a part to be shifted around so that the threads of all the bolts can be started more easily.
- Bolts or studs that are threaded into aluminum should be coated with anti-seize compound to prevent the aluminum from oxidizing to the bolt (see Figure 7.3).

Figure 55.8 Cover the engine with a large trash bag to keep out dirt. *(Courtesy of Tim Gilles)*

Assembly Lubricants

During assembly, lubricate all possible wear areas generously. Assembly lubricants are used on high-load parts such as cam lobes on cam-in-block (**Figure 55.9**).

There are assembly products that have low melting temperatures and are soluble in oil. An assembly lube with the consistency of grease at room temperature is desirable in case the engine is left sitting for a long period of time before it is installed in a vehicle.

- Install the cam on a pushrod engine and be certain that it turns easily in the newly installed bearings. Be careful not to nick a lobe.
- Install the sprocket and use it as a temporary handle to turn the cam as it is installed.

Following installation, make sure that the cam can be turned by hand! Although it does not seem so, cylinder blocks are flexible. If the block is not supported,

it is possible that the camshaft will not be able to enter it easily.

> **CASE HISTORY** *An apprentice was installing cam bearings in an in-line six cylinder mounted on a universal engine stand. After all of the bearings were installed, he attempted to install the camshaft in the block. It would not go in and light was visible between part of the journal and the bearing. The machinist was consulted. He suggested removing the block from the engine stand and resting it on a workbench. After this was done, the camshaft went easily into the block. On the engine stand, the block had been sagging under its own weight.*

Once the head and oil pan are installed on the block, the block will be more rigid.

■ PREPARE THE CRANKSHAFT FOR INSTALLATION

The following items should be observed when preparing the crankshaft for installation:

- Remove the old crankshaft sprocket and install the new one (**Figure 55.10**). Some crank sprockets slide easily into place.
- Make certain that the woodruff key is perfectly flat in its groove in the crank. An improperly installed woodruff key can cause a cracked sprocket (**Figure 55.11**).
- If the crank is already in the block, the sprocket can be heated for easier installation.
- Install the sprocket with the timing mark facing outward. The inside edge of the sprocket has a chamfer that corresponds to the fillet on the crank; if the sprocket is backward, it cannot be installed all the way (**Figure 55.12**).

Figure 55.9 A special assembly lube is used on pushrod engine cam lobes. *(Courtesy of Tim Gilles)*

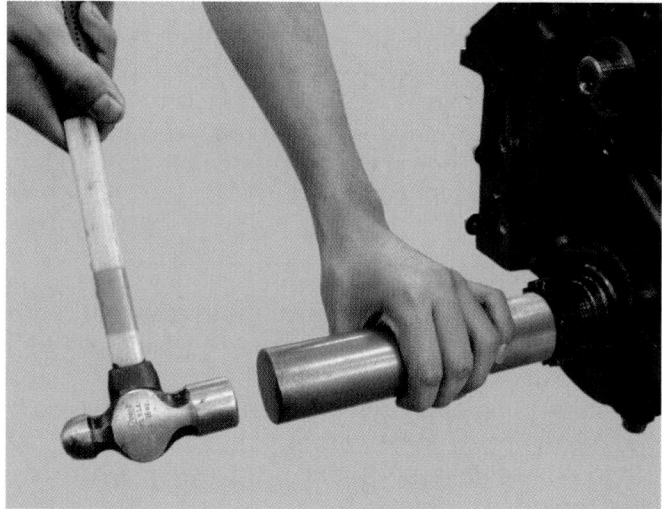

Figure 55.10 Replace the crankshaft sprocket. *(Courtesy of Tim Gilles)*

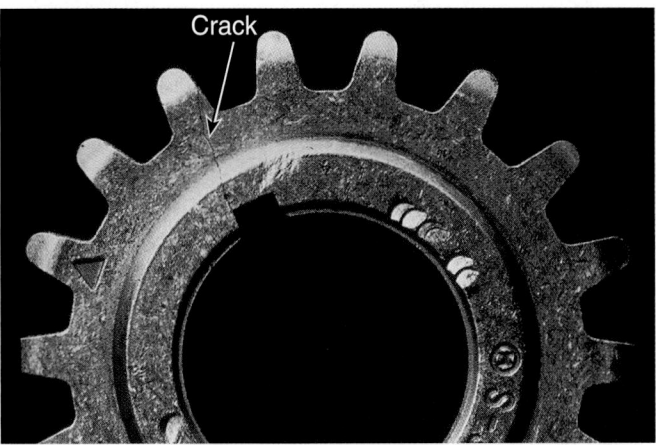

Figure 55.11 Careless installation of the crank sprocket can ruin it. *(Courtesy of Federal-Mogul Corporation)*

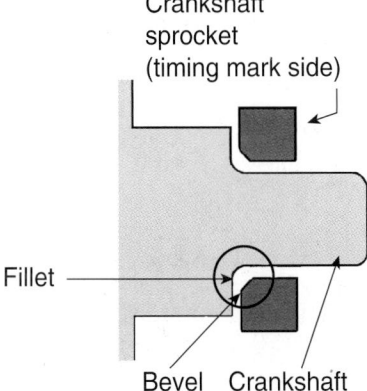

Figure 55.12 The inside edge of a crankshaft sprocket is beveled to clear the fillet on the crankshaft.

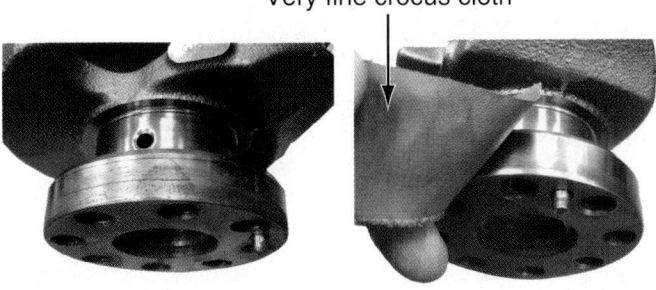

Figure 55.13 Before installing the crankshaft and rear seal, be certain that the sealing surface is clean. *(Courtesy of Tim Gilles)*

■ Be sure that the surface the rear seal rides on is clean. If it is not, clean it with very fine emery or crocus cloth **Figure 55.13**.

SHOP TIP Be sure that all oil passages in the crank are clean (**Figure 55.14**). The machinist is not paid to clean the crankshaft after grinding; this is the assembler's responsibility.

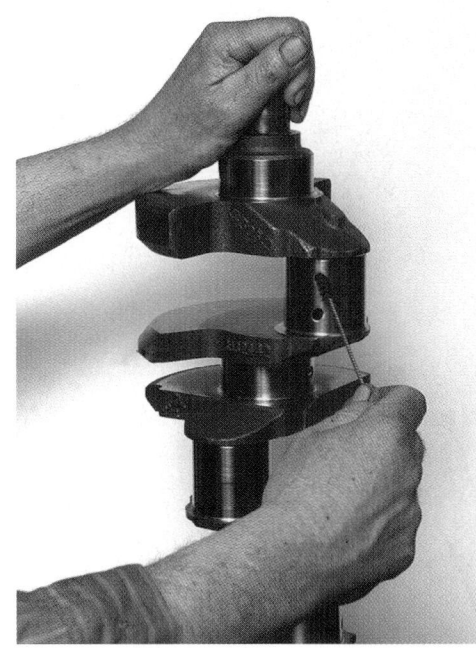

Figure 55.14 Make sure that all crankshaft oil holes are clean.

■ INSTALL THE CRANK

Clean the main bearing bores to prevent oil clearance and heat transfer problems (**Figure 55.15**). If dirt is left on the bearing bore, the bearing can be forced against the crankshaft (**Figure 55.16**).

Be especially careful to clean the recesses where the bearing lock tab will fit (**Figure 55.17**). Failure to clean this area thoroughly is often the cause of a tight crankshaft.

Often, during an oil clearance check on a newly rebuilt engine, excessive clearance is discovered. This could be due to dirt on the block or the main bearing parting halves (**Figure 55.18**). It takes only a small film of dirt left from the hot tank, bake oven, or shot-

Figure 55.15 Clean the main bearing bores. *(Courtesy of Tim Gilles)*

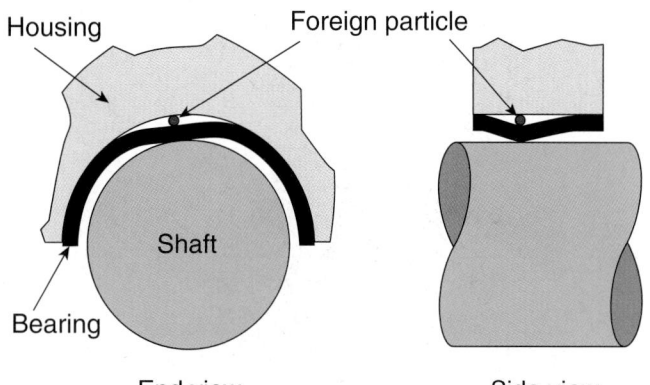

Figure 55.16 Results of careless cleaning. *(Courtesy of Dana Corporation)*

peening process to increase the oil clearance by 0.001" or more.

Main bearings often come with an oil hole and groove in only one half; install these bearings in the upper bearing position (**Figure 55.19**). Lower main bearings (especially for heavy-duty use) are sometimes solid with no oil hole or groove. Other times, the same bearing style with a hole and groove is used in both the upper and lower positions.

■ Install the upper main bearings by pushing them into the main bearing bores (**Figure 55.20**).
■ Be sure that the lubricating holes in the bearings line up properly with the corresponding holes in the block (**Figure 55.21**).

Figure 55.17 Clean the locating lug notch. *(Courtesy of Tim Gilles)*

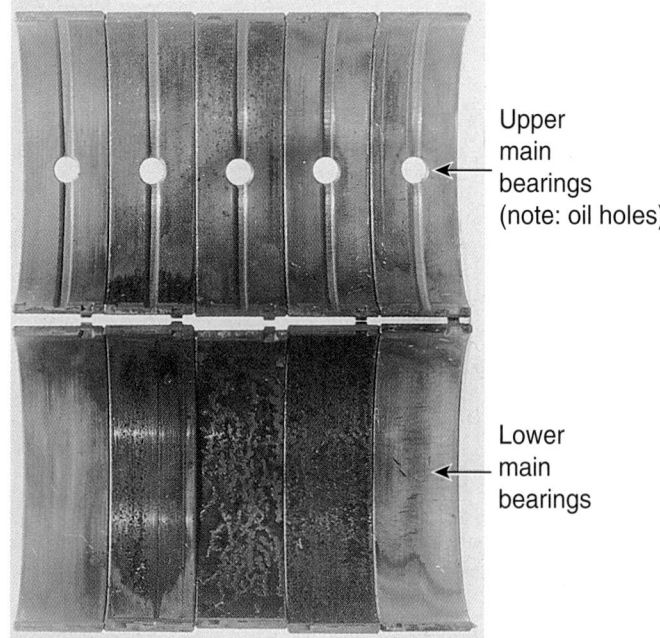

Figure 55.19 These upper main bearings have oil holes and the lower ones do not. The lower bearings are extremely worn. *(Courtesy of Dana Corporation)*

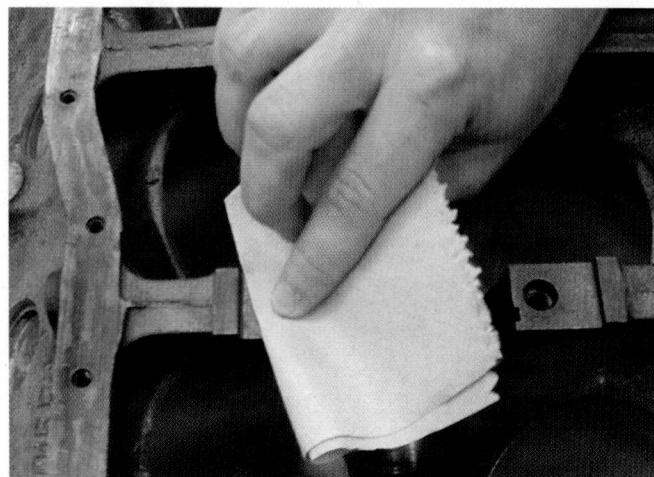

Figure 55.18 Clean the parting surfaces, between the block and main caps, thoroughly. *(Courtesy of Tim Gilles)*

Figure 55.20 Snap the bearing into place.

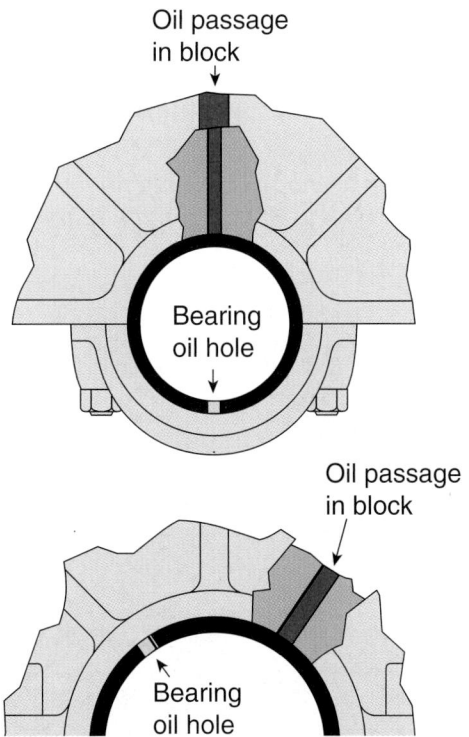

Figure 55.21 Installing the main bearings in the wrong positions can block the passage of oil to the journal. *(Courtesy of Dana Corporation)*

Figure 55.22 Apply assembly lube to the bearings (only on the side that faces the crankshaft). *(Courtesy of Tim Gilles)*

■ Do not touch the bearing surface with your hands.
■ Lubricate the bearings only on the surface that is *toward* the crankshaft (**Figure 55.22**).

NOTE: *Do not oil the bearing backs. They are not a bearing surface.*

Install the Main Bearing Caps

Main bearing caps fit tightly in their recesses in the block. Be certain that they are facing the correct

Figure 55.23 The cap fit is snug enough to require the use of a brass hammer to seat the cap to the block. Be sure the cap is facing the correct direction. *(Courtesy of Tim Gilles)*

Figure 55.24 Apply extra lubrication to the surfaces of the thrust bearings or thrust inserts. *(Courtesy of Tim Gilles)*

direction and then seat them by lightly tapping with a brass hammer (**Figure 55.23**).

NOTE: *Be sure to lubricate the thrust bearing faces (**Figure 55.24**).*

Check Bearing Oil Clearance

Install the crank and check the clearance with plastigage as described in the section on bearing oil clearance in Chapter 53. If the crank and/or the bearings are being reused, it is advisable to check each bearing's clearance.

■ When using a reground crank and new bearings, check the main and rod bearings for proper clearance and to be sure that the right bearings are being installed.
■ Remove the crankshaft and install the rear main seal according to the instructions in Chapter 51. Be sure to offset the parting lines on a two-piece seal.

Figure 55.25 Installing a full-round crankshaft seal without the need of a special tool. *(Courtesy of Tim Gilles)*

> **SHOP TIP** When installing a full-round crankshaft rear seal that fits in the back of the bearing cap, position it on the crankshaft sealing surface before installing the rear main bearing cap (**Figure 55.25**).

Tighten the Main Caps

For a five-main bearing block, the torque sequence is 1-4-3-2-5.

NOTE: *As each main cap is torqued down, check to see that the crank continues to turn easily.*

After the rear cap is installed, check the rear seal drag. Some manufacturers give a torque specification for the amount of effort required to turn the crank with the damper bolt in an assembled engine.

Align the Thrust Bearing Halves

- Torque all bearing caps except the thrust main. Its halves should be aligned before torquing. Misaligned thrust halves could eliminate end play (**Figure 55.26**). This is done by prying on the

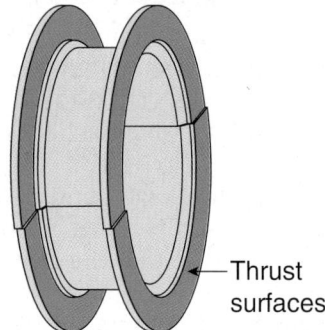

Figure 55.26 Misaligned thrust flanges would prevent end play. This upper thrust bearing is shifted to the left.

crankshaft while the thrust main is still loose (**Figure 55.27**).
- After aligning the thrust halves, check crankshaft end play (**Figure 55.28**) with a feeler gauge or a dial indicator.

NOTE: *End play tolerance is usually from 0.004"–0.006" for a crankshaft with a 2" to 2¾" main bearing diameter.*

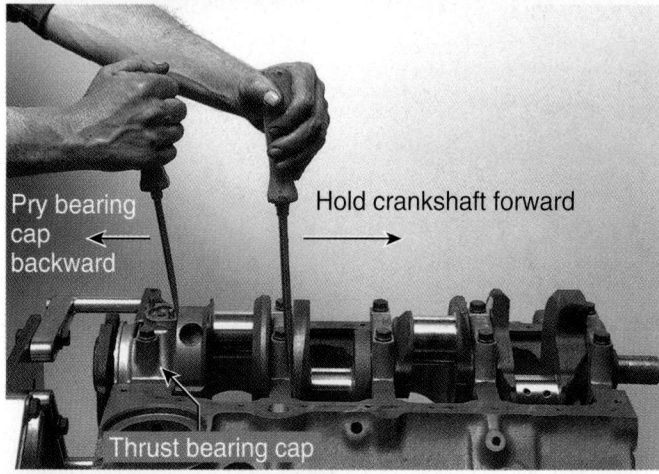

Figure 55.27 Align the thrust surfaces as shown.

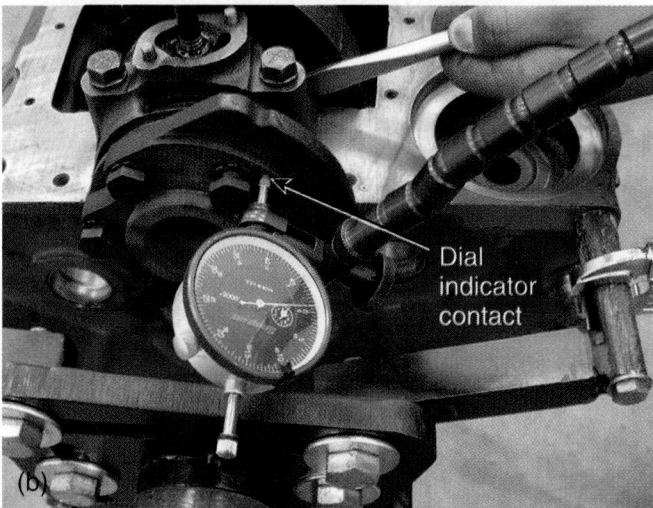

Figure 55.28 Checking crankshaft end play: (a) With a feeler gauge. (b) With a dial indicator. *(Courtesy of Tim Gilles)*

Assemble the Piston and Rings

■ Assemble the new or serviced pistons to the connecting rods. Be sure that the rods have been installed on the pistons in the right direction (see Chapter 54).

NOTE: *Remember to be sure to lubricate the wrist pin. This is often forgotten.*

■ Check the butt gap of the compression rings (see Chapter 54). Then, install them according to the instructions in the ring package with the dots facing up and the gaps properly placed.

Install the Rod Bearings.

■ Install the new bearing inserts in the rod and cap and lubricate them (**Figure 55.29**).

■ Bearing spread keeps the bearings in the rod while the pistons are being installed (see Chapter 19). A good bearing that is going to be reused might have lost its spread. According to Federal-Mogul

Figure 55.29 Use assembly lube on the rod bearing surface before installing the piston assembly into the block. *(Courtesy of Tim Gilles)*

Corporation, bearings can be respread by placing the bearing on a hardwood surface with the parting face down and gently tapping the back with a soft-faced mallet.

NOTE: *When a bearing has lost its spread, inspect the piston for evidence of abnormal combustion.*

■ Make sure that the parting surface of the rod cap does not have any burrs or foreign material that might prevent proper mating when torquing (**Figure 55.30**).

■ Make sure the bearing locating lugs are aligned in the cap (**Figure 55.31**).

Install the Piston and Rod Assembly in the Block.

■ Oil the piston, rings, and piston pin thoroughly (**Figure 55.32**).

■ Install a short length of fuel hose or rod bolt protectors on each rod bolt (**Figure 55.33**). This will guard against nicking the crank, which would result in a damaged bearing insert (**Figure 55.34**).

■ Face the notch on the piston head to the front of the engine (**Figure 55.35a**).

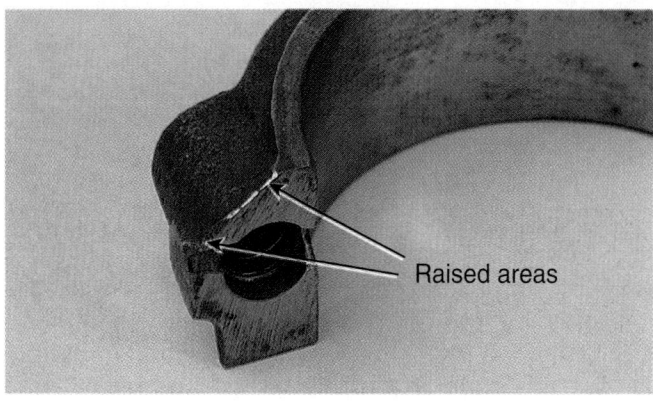

Figure 55.30 A burred connecting rod cap parting surface. *(Courtesy of Tim Gilles)*

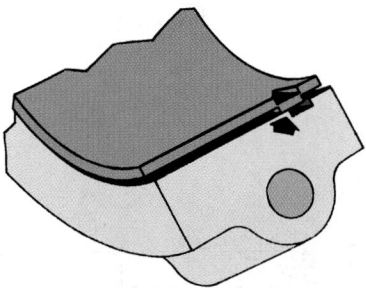

Figure 55.31 Make sure that the bearing tang is properly located before torquing the rod cap. *(Courtesy of Dana Corporation)*

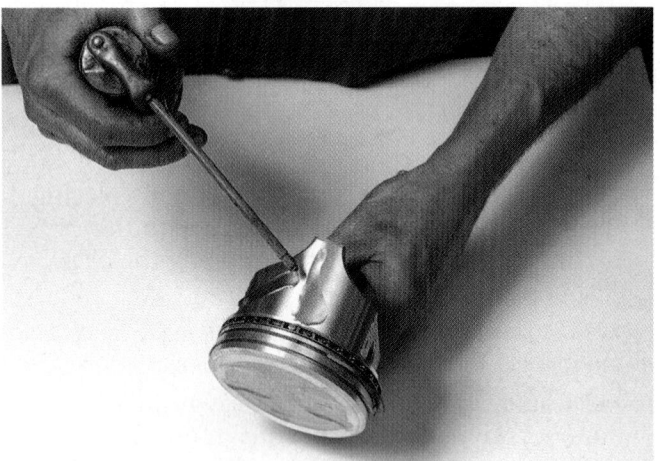

Figure 55.32 Oil the pistons, rings, wrist pins, and rod bearings thoroughly.

Rod bolt protector

Figure 55.33 Install pieces of hose or a special tool on the connecting rod bolts to protect the crankshaft from accidental nicks. *(Courtesy of Tim Gilles)*

■ Using a ring compressor, hold a rubber hammer against the top of the piston and tap the hammer lightly with the palm of your hand while holding the ring compressor firmly against the block until all the rings have entered the cylinder (**Figure 55.35b**).

Figure 55.34 A nick on the rod journal will cause a line all around the bearing insert. *(Courtesy of Tim Gilles)*

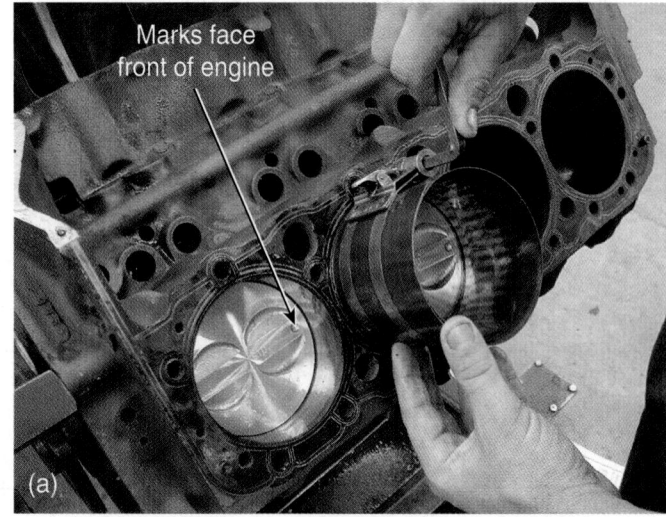

Marks face front of engine

(a)

(b)

Figure 55.35 (a) Face the mark on the piston toward the front of the engine. (b) Using a ring compressor. *(Courtesy of Tim Gilles)*

SHOP TIP Installation of the three-piece oil ring occasionally presents some difficulty. Sometimes, there is a small space between the bottom of the ring compressor and the top of the block. Try to quickly push the oil ring past the end of the ring compressor and into the cylinder bore before one of the narrow oil ring rails can spring out into this space.

- If the piston does not go in easily, something might be wrong. Do not force it! You can damage a ring or a ring land (**Figure 55.36**).
- Once all the rings have entered the cylinder, the rod can be pulled down against the rod journal by hand from the crankcase side of the block. Twist the rod if necessary to align it properly with the journal.
- Be certain that the connecting rod faces in the right direction. On a V-type engine, with the notches on the piston facing forward, the left cylinder bank's rods should face the opposite direction from the right bank's (**Figure 55.37**). If they are facing the wrong way, the rods might have been improperly installed on the pistons. Do not continue with assembly until the rods are correctly installed.

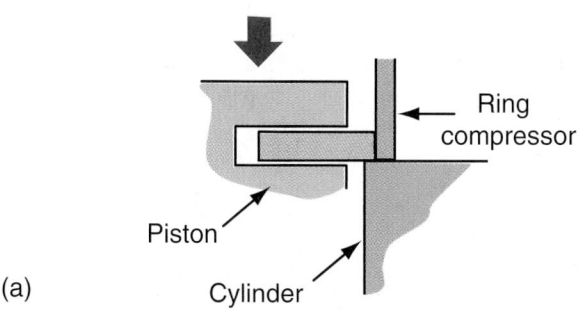

(a)

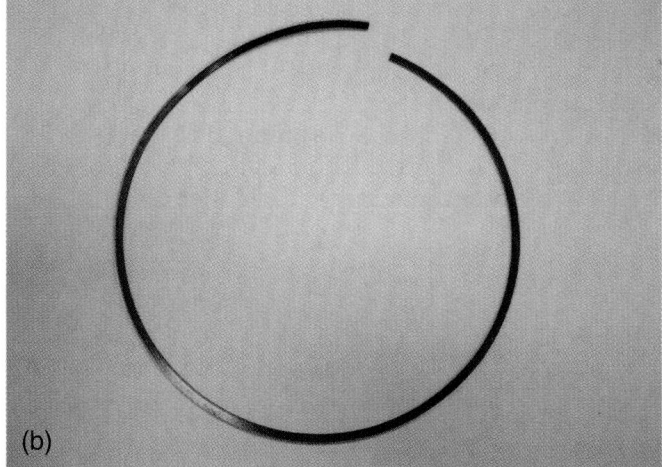

(b)

Figure 55.36 (a) Do not force a piston into the cylinder. (b) This oil ring rail was damaged during piston installation. *(Courtesy of Tim Gilles)*

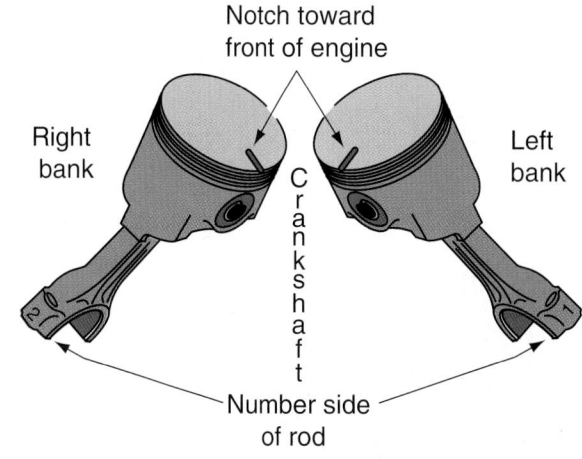

Figure 55.37 On V-type engines, the rods are installed with their numbers facing away from the crankshaft.

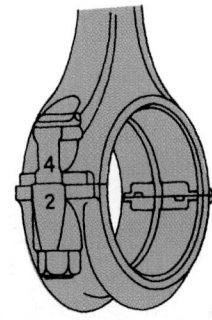

Figure 55.38 Bearing cap on wrong connecting rod. *(Courtesy of Dana Corporation)*

- Remove the hoses or rod bolt protectors from the rod bolts and install the rod caps. Make sure that the numbers on the rod caps correspond to the numbers on the rods (**Figure 55.38**) and that they face in the right direction. The numbers on each rod and cap should be on the same side with the lock tabs facing each other (**Figure 55.39**).
- Rod nuts are square and flat on the bottom side and often are curved on the top. Be sure that the flat side of each rod nut is faced against the cap. Use some loctite, or similar adhesive, when rod nuts are reused (see Chapter 51). Torque the nuts to specifications. Ideally, nuts should not be reused (see Chapter 7). In the industry, rod nuts are commonly reused without adhesives.

Check Installation of Rod and Bearings.
- After installing each piston, rotate the crank one complete turn to check for unacceptable drag (**Figure 55.40**).
- On V-type engines, check the side clearance between the connecting rods with a feeler gauge and compare it to specifications. Use the low side of the tolerance when selecting a feeler gauge.

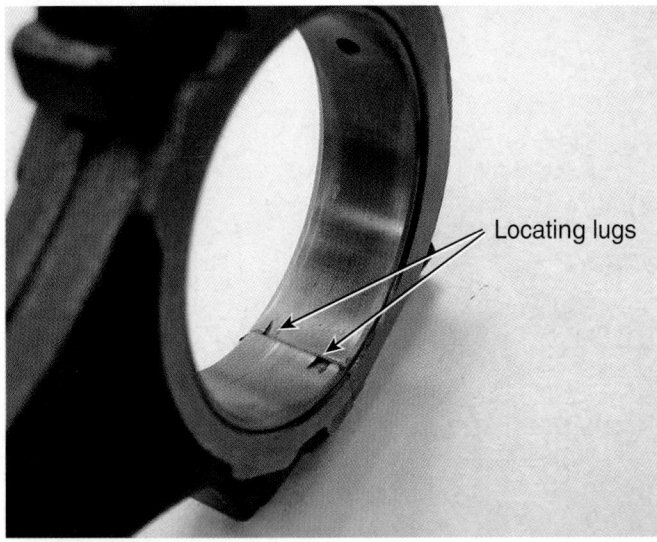

Figure 55.39 Locating lugs usually face each other. *(Courtesy of Tim Gilles)*

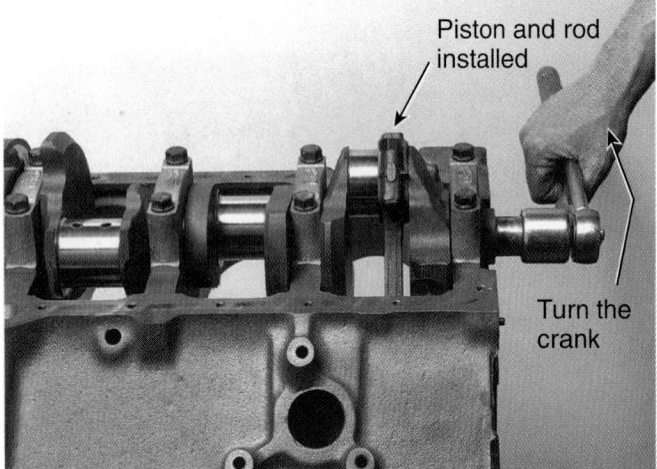

Figure 55.40 Rotate the crankshaft after installation of each piston and rod.

Check at two or three locations around the circumference of the rod. Variations in clearance indicate a bent or twisted rod.

▄▄ INSTALL THE CYLINDER HEADS

- Be sure to install the cylinder head gaskets in the right direction.
- Most engines have dowels to align the heads. If there are no dowels, make alignment pilots by cutting the heads off of two long bolts and hacksawing screwdriver grooves in their tops (**Figure 55.41**). These aligning studs will help protect the head gasket from damage during installation and can be removed easily after the head is installed.
- Be sure to compare the head gasket to the head to see that it is the correct one. Check that the fire

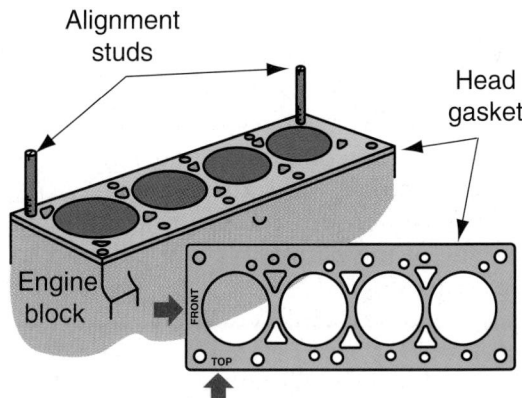

Figure 55.41 Alignment studs can be made using a hacksaw.

rings around the combustion chambers fit. Also see that oil and coolant passageways line up.

▄▄ INSTALL OHC HEADS

Before installing overhead cam heads on the block, the number 1 piston must be at TDC and the camshaft must be turned in the head until its timing mark is properly located. Otherwise, valves held open by the cam might be forced against a piston and become bent as head bolts are tightened.

Install the Head Bolts

After the head bolt threads have been cleaned on the wire wheel, install the head on the block and bolt it on.

- Any bolts with rusted shanks should be replaced.
- Torque the bolts in the proper sequence (see Chapter 51).
- Apply sealer to any bolts that go into holes that go all the way through into water jackets (**Figure 55.42**), so that coolant cannot migrate into the oil. A long, narrow screwdriver can be used to probe the head bolt holes to see if they are "blind" holes (holes that have bottoms).
- Align the TDC timing mark found on the cam or cam sprocket with the mark on the cylinder head (**Figure 55.43**).
- Position the crank-pulley timing mark at TDC. To double-check that the cam is actually positioned at TDC, either the number 1 cylinder or its

Figure 55.42 This head bolt came from a hole that was threaded into a water jacket. *(Courtesy of Tim Gilles)*

Figure 55.43 Position the cam correctly before installing the cylinder head. *(Courtesy of Tim Gilles)*

Cam lobes at TDC

Figure 55.44 When the cam lobes face away from the valves, the cam is positioned at TDC before the power stroke. *(Courtesy of Tim Gilles)*

Timing marks must be in position shown with no. 1 piston at TDC

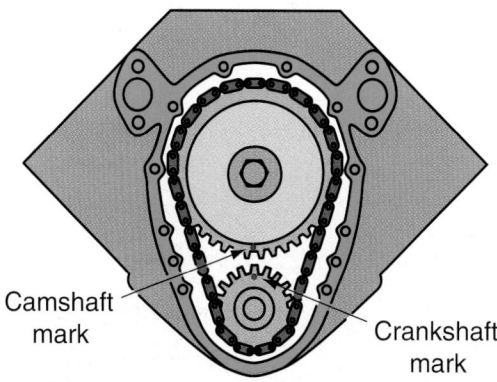

Camshaft mark

Crankshaft mark

Figure 55.45 Install the chain and cam sprocket with the timing marks aligned according to specifications.

- Turn the crankshaft until the cam sprocket aligns with its pin or key.
- Install the cam sprocket on the cam. The marks should now be in perfect alignment (**Figure 55.45**).
- Double-check cam timing by rotating the crankshaft until the marks line up again.

CAUTION Tightening down an OHC camshaft sprocket with an impact wrench can sometimes result in a broken camshaft snout.

■ INSTALL VALVETRAIN PARTS

- Oil and install the lifters or cam followers and make sure they turn freely in their bores.
- On pushrod engines, coat the bottom of each lifter with an assembly lube, an EP 90-weight oil, or an engine oil supplement (EOS) to protect them against wear when the engine is first started.

NOTE: *The Automotive Engine Rebuilders Association recommends against prefilling hydraulic lifters. If they are overfilled, the valves can be held off their seats, which will make starting the engine difficult.*

- Install pushrods and rocker arms if applicable. Lubricate all wear areas thoroughly (**Figure 55.46**).

SHOP TIP If it moves, lube it!

companion cylinder should have its cam lobes facing as shown in **Figure 55.44**. The companion cylinder's cam lobes will face down in the valve overlap position.
- When the cam and crank sprockets are in their proper positions, install the timing chain or belt. For timing belt information, refer to Chapter 52.

Sometimes the chain has plated links that align with marks on the cam and crank sprockets. Be sure to check the service information.

■ INSTALL THE CAM DRIVE (PUSHROD ENGINES)

Install the chain and cam sprocket. Be sure to set the cam timing properly.
- If the sprocket was not previously installed on the crank, lay the timing chain and sprockets on the bench and align their marks carefully.
- Hold both sprockets in the chain while installing the crank gear.

Figure 55.46 Lubricate rocker arm pivots with assembly lube. *(Courtesy of Tim Gilles)*

← Vibration damper

A large nut is turned against this bearing

Figure 55.47 A damper installation tool. *(Courtesy of Tim Gilles)*

■ On shaft-mounted rocker arms, tighten the bolts on the rocker shaft that are closest to the center first. Then, pull all the rest of the bolts slowly tight.

▰ INSTALL THE OIL PUMP

■ Fill the pump with oil so that it will not be run while dry and will "prime" right away.
■ Install the pump as described in Chapter 53. Oil pumps do not usually require a gasket if they are housed inside the oil pan.
■ Make sure that the pickup screen is properly positioned so it will be about ¼" from the bottom of the oil pan.

▰ INSTALL THE TIMING COVER

■ Some older engines with carburetors had a bolt-on fuel pump eccentric. Use loctite when installing it.
■ Install the crankshaft woodruff key and the oil slinger (if the engine uses one) before installing the timing cover.

NOTE: *Be sure to install an oil slinger in the correct direction, or it may rub and make noise. Tearing down a newly rebuilt engine to replace a backwards or forgotten oil slinger is time consuming and frustrating.*

■ Install the timing cover seal as described in Chapter 51. Be sure to grease the lip of the seal. On engines with timing chains, timing covers are usually installed before installing the oil pan because the oil pan and timing cover often share a gasket surface.
■ If the engine does not have timing cover aligning pins, use a special tool to align the damper to the timing cover seal. If this tool is not available, temporarily install the damper. After tightening a

pair of the timing cover screws, the damper can be removed to permit tightening of the remaining screws.

NOTE: *Timing cover screws are small. Torque the screws to specs (usually 50–60 inch-pounds) with an inch-pound torque wrench to prevent breaking the screws.*

▰ INSTALL THE DAMPER

Install the vibration damper. Some technicians use a special installation tool (**Figure 55.47**). Generally, the damper is installed until it bottoms out against the oil slinger and the timing sprocket.

SHOP TIP If you are unsure whether the damper is all the way on, install the coolant pump and pulley. Then, install the damper until its pulley aligns with the coolant pump pulley.

Some vibration dampers are not pressed-fit on the crankshaft. Be sure to install a large washer behind the damper retaining bolt on these engines; otherwise, the damper might fly off, causing damage and a safety hazard.

Gasket Installation

Before installing any gaskets, read the instructions included in the gasket set. They often contain special tips that can be important in assembly of the engine.

▰ INSTALL THE OIL PAN

The installation of the oil pan is especially important. An otherwise perfect engine overhaul will appear amateurish to a car owner if the pan leaks oil. Be careful here. Replacing a pan gasket is a much more difficult repair job when the engine is in the car.

- Be sure that the oil pan is flat.
- Sheet metal oil pan or valve cover bolt holes are often distorted. Straighten them by carefully flattening them with an anvil and hammer.
- Make sure that the pan was not dented when it was removed from the car. A dented pan that is hit by a rotating crank counterweight can sound like a rod knock.

CASE HISTORY *An apprentice technician was installing a rebuilt long block in a vehicle. When the engine was started, a loud knock was heard. The disheartened apprentice sought help from the shop owner. Further investigation showed that the oil pan had been damaged during installation of the engine. The pan was removed from the engine and a shiny spot was visible where the crankshaft had been contacting it. After the pan was straightened and reinstalled, the engine ran fine and the car was returned to the customer.*

- Apply a small amount of RTV where the cork gaskets join the rubber gasket on the front of the pan. Do not use too much; the excess silicone can break off and plug the oil pump screen (**Figure 55.48**).
- When installing the oil pan screws, start each one into its threads before tightening any of them (**Figure 55.49**).

Figure 55.49 Leave all screws loose until each screw has been threaded into its hole. This allows the oil pan to be repositioned until all holes line up. *(Courtesy of Tim Gilles)*

SHOP TIP During assembly of pushrod V-type engines, it is best to adjust *hydraulic lifter* preload before the intake manifold is installed because you will be able to see that the lifters are on the heel of the cam lobes and are positioned low in their bores (**Figure 55.50**). Valve adjustment procedure is covered in Chapter 52.

Be sure nothing has been accidentally left in the engine before the oil pan or intake manifold are installed. **Figure 55.51** shows the result of a careless engine assembly where a shop towel was left inside the engine.

Figure 55.48 When too much silicone RTV is used, it can break off and plug the oil pump intake screen. *(Courtesy of Tim Gilles)*

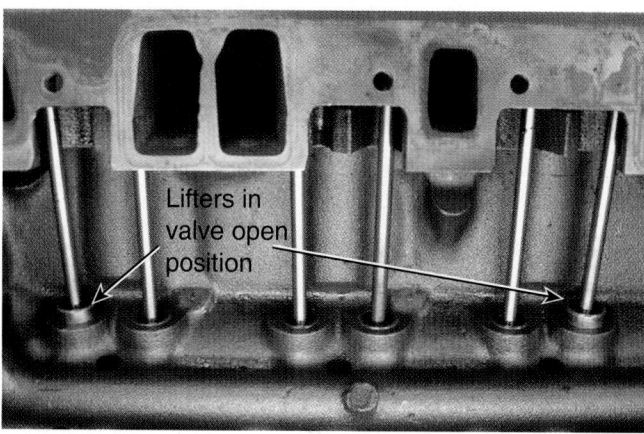

Lifters in valve open position

Figure 55.50 If hydraulic lifter lash is adjusted before installing the intake manifold, it is easy to see which lifters are on the heel of the cam. *(Courtesy of Tim Gilles)*

Figure 55.51 A shop towel was left inside this engine during a careless reassembly. *(Courtesy of Tim Gilles)*

■ INSTALL THE INTAKE MANIFOLD

Chapter 51 describes the precautions to take when sealing an intake manifold. Install the intake manifold and torque according to specifications (**Figure 55.52**).

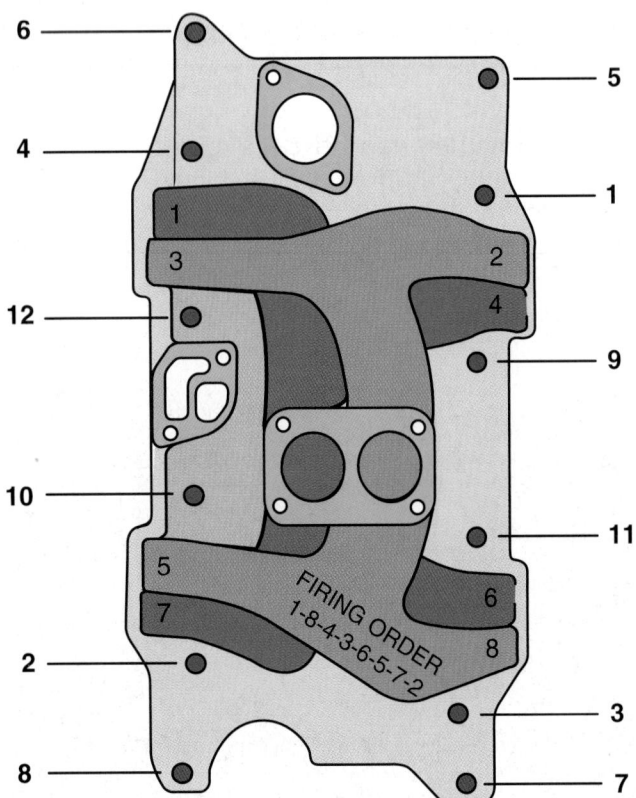

Figure 55.52 Intake manifold torque sequence. Notice the firing order and cylinder numbers cast on the manifold.

■ INSTALL THE THERMOSTAT AND WATER OUTLET HOUSING

Install the thermostat with the temperature sensor facing into the block. If the thermostat is installed upside down, the engine will overheat.

> **CASE HISTORY** *An apprentice was finishing the installation of a rebuilt engine by installing the thermostat and all of the belts and hoses. After the engine was started, he took the car on a test drive. After a very short distance, the engine overheated. Further investigation by the shop foreman showed that the thermostat had been installed backwards. With the temperature sensing bulb facing toward the radiator, the thermostat could not sense the heated coolant in the block. Correctly installing the thermostat corrected the overheating problem.*

■ INSTALL THE FUEL PUMP

Older engines with carburetors had mechanical fuel pumps. Rotate the crankshaft until the fuel pump eccentric is at its lowest position in relation to the pump. Failure to do this makes it difficult to install the pump.

■ COMPLETION OF ASSEMBLY

Paint the Engine

A good paint job is an important part of a professional engine rebuild. It is easier to paint an engine after it is completely assembled. Exhaust manifolds should not be painted, so paint the engine before they are installed.

> **SHOP TIP** Apply a thin layer of grease to exhaust gasket surfaces and carburetor gasket surfaces. This makes it easier to remove paint from these areas.

- Thread some old spark plugs into the plug holes before painting the engine.
- Spray one light coat of paint (*tack coat*). Wait until it becomes tacky to the touch.
- Spray the second coat after the first coat becomes tacky. This will prevent runs in the paint and promote better paint coverage.

Spin Testing

High-volume engine rebuilders use a *run-in stand* to **spin test** rebuilt engines following assembly. The engine is rotated at 600 rpm by the machine while the lubrication system has full oil pressure. Major problems can be spotted using this test before a defective engine leaves the rebuilder. On a long block, certain tests and adjustments are possible during spin testing.

- Compression and oil pressure can be checked.
- Cam, rod, and main bearings can be visually checked for excessive bearing oil leakage.
- Pushrod engine lifter rotation can be verified.

INSTALL EXHAUST MANIFOLD(S)

After the block assembly is painted, install the exhaust manifold(s). Some in-line engines use a combined intake and exhaust manifold gasket.

- Tighten the bolts in the center of the manifold first to prevent cracking it.
- If there are dowel holes in the exhaust manifold that align with dowels in the cylinder head, make sure these holes are clean. If the dowels do not have enough clearance, the manifold will not be able to expand properly and might crack.
- On engines with shared bolts between the intake and exhaust manifolds, tighten the individual manifold-to-engine bolts first. Then, tighten the bolts where the two parts meet.

REVIEW QUESTIONS

1. Use a _____ to clean bolt holes in the block before reassembly.
2. Pry the crank forward and rearward before torquing the thrust bearing cap to align the halves of the _____ bearing.
3. Crankshaft end play can be checked with a dial indicator or with a _____ gauge.
4. Install pieces of _____ or rod bolt protectors on the rod bolts to protect the crank against accidental nicks during piston installation.
5. The crankshaft should be _____ after installing each piston to be sure that nothing is too tight.
6. When installing an OHC cylinder head, be sure that the camshaft and piston are correctly _____ so they do not hit each other.
7. If the block does not have aligning pins for the timing cover, what can be used to align the timing cover?
8. Distorted valve cover or oil pan bolt _____ are flattened with a hammer.
9. When installing the fuel pump, the fuel pump _____ should be in the down or away position so that the pump arm is not activated.
10. Before painting, _____ is applied to areas that are not to be painted.

ASE-STYLE REVIEW QUESTIONS

1. Two technicians are discussing pushrod engine assembly. Technician A says to prefill all hydraulic lifters with oil before installing them. Technician B says to adjust hydraulic lifter preload before installing the intake manifold. Who is right?
   a. Technician A
   b. Technician B
   c. Both A and B
   d. Neither A nor B

2. Which oil pumps require a gasket where the pump bolts to the cylinder block?
   a. Pumps enclosed within the oil pan
   b. Pumps bolted to the outside of the block
   c. Both A and B
   d. Neither A nor B

3. Which of the following is the correct tightening procedure for exhaust manifold bolts?
   a. Start at the outside and work your way toward the center.
   b. Start at the center and work your way toward the outside.
   c. Tighten the second bolt from the outside of the manifold first.
   d. The tightening sequence is unimportant.

4. Two technicians are discussing main bearings. Technician A says that both upper and lower main bearings can have oil holes. Technician B says that upper main bearings often do not have an oil hole. Who is right?
   a. Technician A
   b. Technician B
   c. Both A and B
   d. Neither A nor B

5. Technician A says that crankshaft end play can be checked with a dial indicator. Technician B says that crankshaft end play can be checked with a feeler gauge. Who is right?
   a. Technician A
   b. Technician B
   c. Both A and B
   d. Neither A nor B

6. When installing an oil pan, Technician A says to torque each screw as it is installed. Technician B says that on engines with timing chains, the oil pan is usually installed before the timing cover. Who is right?

   a. Technician A   **c.** Both A and B

   b. Technician B   **d.** Neither A nor B

7. Which of the following is/are true about painting an engine?

   a. It is faster to paint an engine after it is completely assembled except for the exhaust manifold(s).

   b. Use a tack coat to prevent paint from running.

   c. Before painting the engine, put some grease on the exhaust manifold mounting surfaces of the head.

   d. All of the above.

8. Which of the following is/are terms for a factory replacement part?

   a. OE   **c.** Aftermarket

   b. Stock   **d.** All of the above

9. All of the following are true statements *except*:

   a. A custom engine rebuild is when a customer's engine is rebuilt for use in the same vehicle from which it was removed.

   b. A cylinder block can flex until the head and other engine parts are bolted to it.

   c. Lubricate the back side of the bearing insert.

   d. An identifying notch on the crown of a piston faces toward the front of the engine.

10. On which of the following engine designs is the number 1 piston positioned at TDC prior to installing the cylinder head(s)?

   a. Cam-in-block   **c.** Both A and B

   b. Overhead cam   **d.** Neither A nor B

Engine Installation, Break-In, and In-Chassis Repairs

◼ OBJECTIVES

Upon completion of this chapter, you should be able to:

✔ Install an engine in a vehicle.

✔ Pre-lube and make all required adjustments prior to starting an engine.

✔ Inspect and complete the job following engine starting and break-in.

✔ Overhaul an engine while it is in the vehicle.

◼ KEY TERMS

guide pins
pressure primer

WOT
overhaul

steering linkage taper
tempilstick

◼ INTRODUCTION

The content of this chapter deals with installing an engine in the vehicle. After installation, certain procedures are followed to break it in. This chapter also deals with some of the repairs that can be done to an engine while it is in the vehicle.

◼ ENGINE INSTALLATION

Installing a completely assembled engine in a chassis is an important part of engine service. Before installing the engine in the vehicle, be sure that fender covers are installed on the fenders. Be especially careful when using a chainfall. The chain can nick the car's paint.

◼ INSTALL ENGINE MOUNTS

Install the engine mount bolts *loosely* on the block. Leave them loose during engine installation so that the mounts can be more easily aligned with the frame mount brackets.

◼ INSTALL THE ENGINE

Raise the engine and position it in the engine compartment. A *rolling head prybar* (**Figure 56.1**), also called a heel bar, is handy when aligning mounts. Use the pointed end to line up the mount bolt holes before installing the mount bolts.

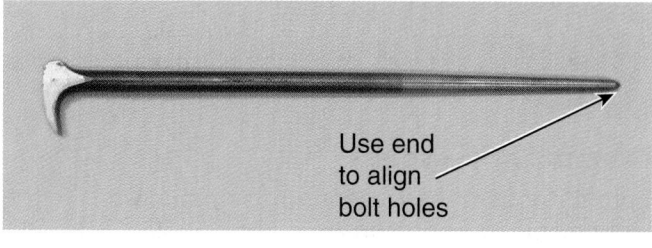

Use end to align bolt holes

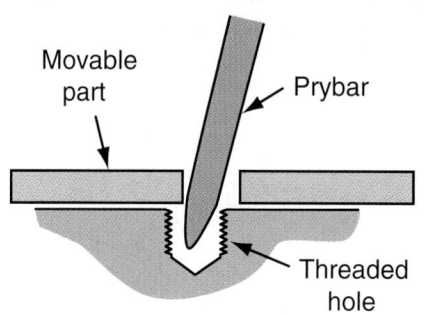

Movable part

Prybar

Threaded hole

Figure 56.1 A rolling head prybar is helpful when aligning motor mounts. *(Photograph, courtesy of Tim Gilles)*

NOTE: *Do not use the engine mount bolts to pull a V-type engine into place. Use shims if necessary to fill any gaps between the mount and block. The Automotive Engine Rebuilders Association reports numerous cases of block distortion, resulting in scuffed pistons in cylinders, near mounts that were forced tight.*

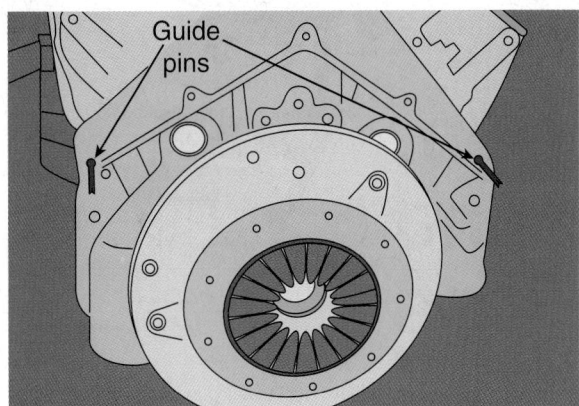

Figure 56.2 Guide pins.

If the transmission was left installed in the car, a floor jack or a transmission jack under the transmission will help to align it with the engine. **Guide pins** will be especially helpful when a standard transmission is still in the vehicle (**Figure 56.2**). They can be made by cutting the heads off some bolts and using a hacksaw to cut a screwdriver slot in them. A standard transmission is often left in the vehicle when a heavy truck transmission is involved.

NOTE: *Engines also have dowel pins attached to the engine block. These are important for engine-to-flywheel alignment and must not be left off the vehicle. Engine-to-transmission housing screws do not provide alignment. Without dowels, damage, especially to an automatic transmission flexplate, can occur in a short period of time.*

Connect Accessories

Bolt the engine to the transmission housing, if the engine and transmission were not installed as a unit. On automatic transmissions, slide the torque converter forward and bolt it to the flexplate (**Figure 56.3**). Be

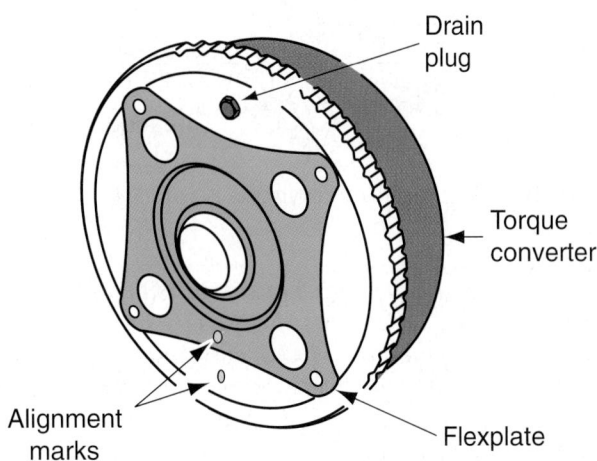

Figure 56.3 Make sure that the flexplate and torque converter are properly aligned.

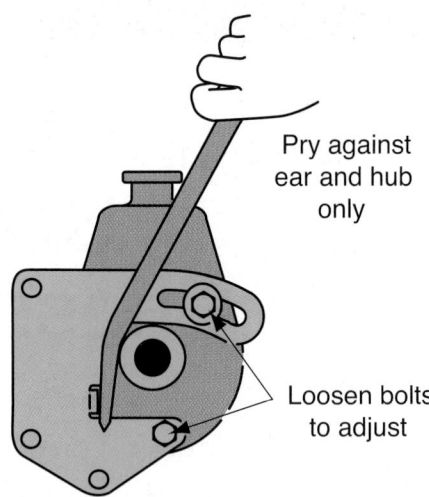

Figure 56.4 When tightening the power steering belt, pry only in the designated areas.

sure that it is in the right position. Some converters have drain plugs that must align with holes in the flexplate. Align the converter with the factory marks or with marks made during disassembly.

Install all previously disconnected parts:
- Bolt on the exhaust pipes. Use new nuts (brass nuts are preferable, if they are available).
- Install all pulleys, accessories, and belts. On non-serpentine belt drives, use a belt tension gauge to adjust belt tension (see Chapter 22). When adjusting power steering belt tension, be careful not to pry against the sheet metal or plastic portion of the pump; the pump may be damaged. Manufacturers usually provide a suitable place to pry on the pump housing (**Figure 56.4**).
- Install the temperature and oil pressure sending units.
- Install the radiator and hoses and fill the cooling system with coolant. Make sure that heater hoses are installed correctly.
- All electrical wires and vacuum lines should be reattached as labeled.

Install the fuel system components, including the fuel injection rails, hoses, and electrical connectors. Be sure that any hose clamps that are replaced are of the full-round type used with high-pressure fuel injection systems, not the kind used with carburetors or cooling systems.

Until the mid-1980s, many North American vehicles still had carburetors. A carburetor must be installed with the correct gasket. Some engines came with several different carburetors, so always save the old gasket to compare with the new one.

Install the oil filter and add oil to the crankcase. Prime the system before installing the distributor (**Figure 56.5**).

Figure 56.5 Priming the lubrication system with a drill. When the distributor for an engine rotates counterclockwise, a reversible drill is used. *(Courtesy of Tim Gilles)*

■ PRIMING THE LUBRICATION SYSTEM

When an engine has a distributor ignition system, the lubrication system can usually be primed by removing the distributor and driving the oil pump with an electric drill. This procedure is easier to do before engine installation while the engine is mounted on the engine stand. A priming tool can be made from an old distributor with the gear removed. Commercially made pump-priming tools are also available. To prime the system by driving the oil pump, note the following items:

■ Drive the tool with a slow drill in the normal direction of distributor rotation.

NOTE: *The oil pump must be turned in the same direction that the distributor rotates if it is to suck oil. The pump will not fill with oil if it is turned in the wrong direction. Chapter 38 explains how to determine the direction of the distributor's rotation.*

■ On engines that have hydraulic lifters, turn the pump until pressure builds in the system, then rotate the crankshaft one complete revolution by hand.

■ Turn the pump once more. This will fill all hydraulic lifters with oil.

The distributor must have a gear on the bottom of its shaft, or it will not be possible to prime the system by driving the oil pump (**Figure 56.6**). If there is no gear on the bottom of the distributor, the oil pump has a gear that is always in mesh with the cam gear, making oil pump priming impossible.

■ PRESSURE PRIMING

When it is not possible to prime the system by removing the distributor, either use a **pressure primer** or remove all spark plugs and crank the engine. Be sure to disable the fuel and ignition systems before cranking the engine. Continue cranking until oil is distributed throughout the engine or until the gauge has registered oil pressure for 30 seconds. It is much easier on the engine to prime

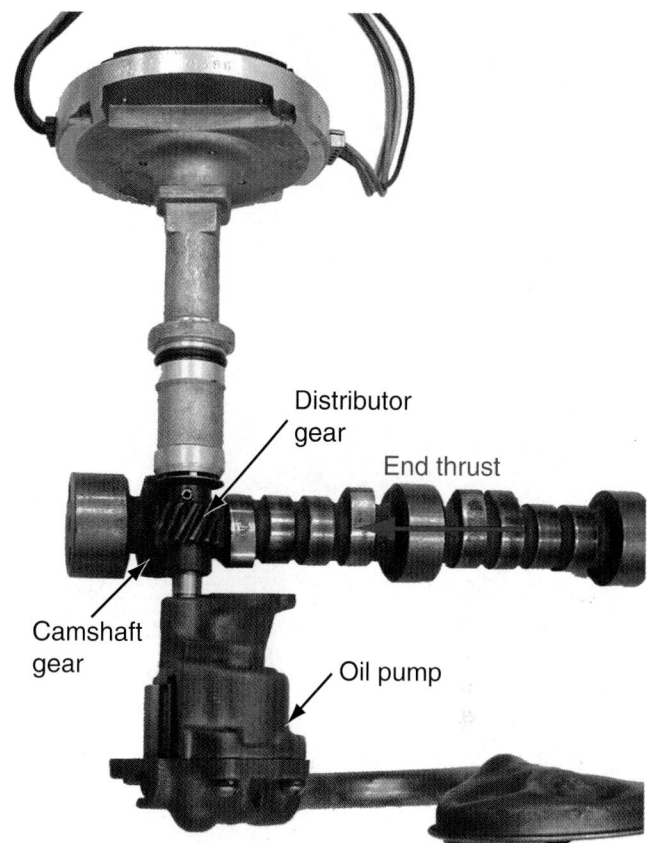

Figure 56.6 The camshaft helical gear drives the distributor and oil pump. Camshaft end thrust results. *(Courtesy of Tim Gilles)*

the system first than to start the engine and wait for the pump to prime itself while the engine runs without oil.

CASE HISTORY

*A group of students rebuilt a small block Chevrolet engine and installed it in a boat. When the engine was started, they noticed that no oil pressure was displayed on the dash gauge. They removed the oil sending unit and replaced it with an oil pressure gauge. They started the engine. The gauge showed low oil pressure that did not increase with engine rpm. They pulled the engine and removed the oil pan. An oil pressure primer (**Figure 56.7**) was attached to the hole where the sending unit had previously been threaded. With pressurized oil introduced to the system, oil could be seen gushing out near the front camshaft bearing. When the timing cover and cam sprocket were removed, it was apparent that the oil gallery plugs had been left out of the front of the engine. Replacing them solved the problem.*

Another group of students was not so lucky. They continued to run their newly rebuilt engine while it did not have oil pressure and burned all of the pistons and cylinder bores. Pressure priming would have saved both of these groups of students a good deal of disappointment.

Figure 56.7 A pressure primer is attached to the oil sending unit hole in the block. *(Courtesy of Tim Gilles)*

A pressure primer can be used to prime the system, check for excessive bearing clearance, check for sufficient oil pressure, and flush oil galleries during an in-car engine repair job.

The supply of oil in the primer tank should be enough to fill the new oil filter as well as the oil lines in the block.

With the oil pan removed, block off the oil pump inlet to perform the test. Be sure to remove the gasket when finished.

■ Acceptable leakage from bearings can be from 20 to 150 drops in a minute. A more rapid flow indicates excessive clearance (**Figure 56.8**).

■ Rotate the crank ½ turn and test again before condemning the bearing. If the oil holes in the crank and block were indexed, the appearance of excessive leakage could result.

■ An absence of oil flow indicates a blocked oil line or insufficient clearance.

Some pumps (the internal/external in particular) are not self-priming and *must* be filled with assembly lube before installation. Follow the manufacturer's directions.

■ INSTALL VALVE COVERS

When oil is apparent at some of the rocker arms, the lubrication system is primed and the valve covers can be installed. If oil is not reaching the valve area during priming, double-check to see that an internal oil leak does not exist. It is easier to position the engine at TDC on #1 now, before installing the valve covers.

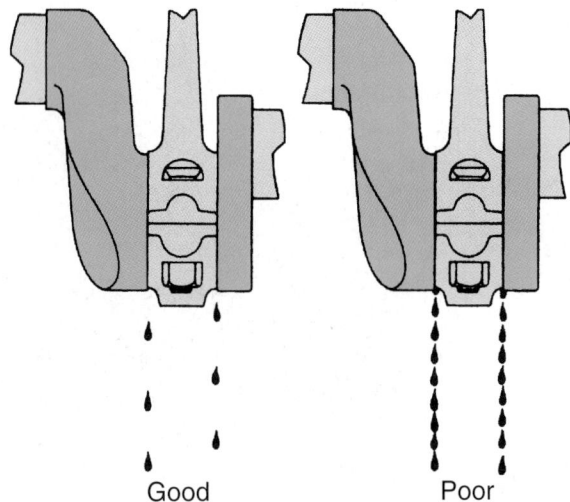

Good   Poor

Figure 56.8 The amount of oil leakage depends on the amount of oil clearance. *(Courtesy of Dana Corporation)*

NOTE: *On V-type engines, be sure to install the valve covers on their proper sides to ensure that breather and oil filler openings are in the right positions.*

■ IGNITION SYSTEM INSTALLATION AND TIMING

There are two predominant ignition system designs: distributor and distributorless. Some engines with an ignition distributor must be timed correctly after installation.

Installing and Timing an Ignition Distributor

Align the timing mark on the damper with the pointer on the timing cover (**Figure 56.9**). Some engines have their timing mark located on the flywheel. It is visible

Figure 56.9 A typical ignition timing mark. *(Courtesy of Tim Gilles)*

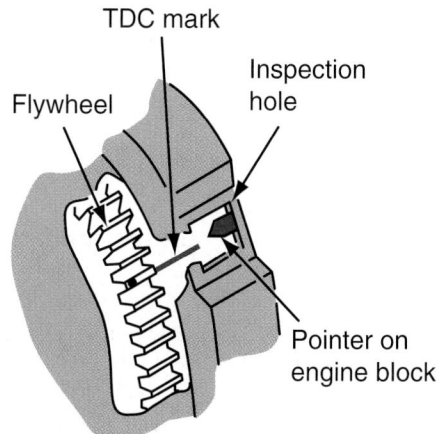

Figure 56.10 Some engines have the timing mark located on the flywheel.

through a small opening in the top of the transmission housing (**Figure 56.10**). When the marks are aligned, either the number 1 cylinder or its companion cylinder is at the top of its compression stroke. The spark plug should be ready to fire. If the distributor is installed 180 degrees off, backfiring will occur and the engine will not run. Directions for distributor installation are included in Chapter 38. Tighten the distributor hold-down. Following installation of the distributor, the ignition timing should be set closely enough so that the engine will start right up.

> **SHOP TIP** On older engines, the idle speed is not controlled by the computer. The engine idle will be off if the timing is wrong. Remember this before changing the idle adjustment screw.
> - If the timing is retarded, the idle will be too low.
> - If the timing is advanced, the idle will be too high.

Before starting the engine, fill the coolant overflow reservoir at least half full of coolant.

ENGINE STARTING AND INITIAL BREAK-IN OF THE CAMSHAFT

The first few minutes of an engine's operation are critical to its long life. A new part might look quite smooth to the naked eye, but compared to a used, worn-in part, it is quite rough. Moving parts are continuously separated by a film of oil; only the peaks on rough surfaces touch. If this contact occurs without *scuffing* (welding), the parts will wear into each other. Temperatures must be controlled so that the oil film will not become too thin or the clearances become too little. Heavy engine loads should also be avoided.

> **SAFETY NOTE** Make sure that everything has been reinstalled and connected properly. Block the front wheels and apply the parking brake before starting the engine. If the transmission linkage was disturbed during the repair, it is possible that the car might start in gear.

Camshaft Break-In—Pushrod Engines

It is especially important that the engine starts up immediately to avoid excessive loading between the cam lobes and the lifters in pushrod engines.
- Start the engine and run it at fast idle (1,500–2,000 rpm) for about 20 minutes to allow the cam and lifters to begin to wear in to each other.
- Idling should be avoided during this period to prevent cam and lifter failure and because, during idle, less oil is thrown off the connecting rod journals onto the cylinder walls, cam, and other parts that need this critical lubrication.

Engine Startup Checks

When the engine is first run, check the following items:
- Make sure the engine has oil pressure.
- Watch the coolant temperature to see that it does not climb too high.
- Check for oil leaks.
- If any adjustments are needed, the engine should be shut off immediately.

Reinstall the hood. Squeeze the top radiator hose to see if it is hard (indicating a full cooling system).

> **CAUTION** Do not open the radiator cap unless the hose collapses when squeezed. Escaping coolant will boil when the pressure is released. Someone could be burned, and coolant will be wasted.

VALVE CLEARANCE ADJUSTMENT

Very few newer engines have mechanical valve clearance adjustment; most are hydraulically compensated. When an engine requires valve adjustment, the clearance is set during engine assembly.

This procedure can be messy because of the oil that is being supplied to the rocker arms. **Figure 56.11** shows a tool that will cut down on the oil mess. Attempting an *OHC adjustment* with the engine running results in oil being thrown from the timing chain and cam.

NOTE: *Try to minimize oil spillage. When running an engine with the valve cover(s) removed, engine oil can run onto a hot exhaust manifold where it might catch fire. Oil*

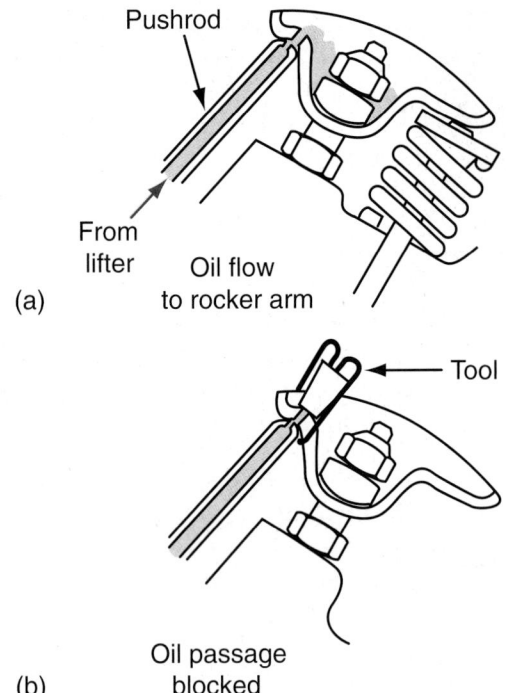

Figure 56.11 (a) Oil is supplied to the rocker arm and valve through the pushrod. (b) An inexpensive way to keep oil from spraying all over during a valve adjustment with the engine running.

is not flammable at room temperature but will ignite when it reaches its flash point, in the same way that diesel oil self-ignites when heated under high compression.

Adjusting Hydraulic Valve Lash—Engine Running

Hydraulic lifter lash on pushrod engines is best done during engine assembly but can also be done with the engine running, if necessary.

With the engine running (idling), loosen the rocker arm adjusting nut until a clicking sound is heard. Then, tighten the nut just until the noise stops. This point is called zero lash. Then, shut the engine off and tighten each adjuster the specified amount. Wait 20 minutes for the lifters to bleed down before starting the engine.

▬ ROAD TEST AND BREAK-IN

Road Test

Take safety precautions before taking a test drive:
- Double-check all hose connections and fluid levels.
- With the key *off*, push the accelerator to wide-open throttle (**WOT**). Be certain there are no binds or obstructions in the accelerator linkage. When released, the throttle pedal should return quickly to the idle position.

NOTE: *Be sure there are no loose parts that could cause a rattle. Many late-model cars have knock (detonation) sensors that will react to the frequency of the vibration*

of spark knock, retarding spark timing. A knock sensor can be triggered by the rattle of a bad fuel pump, a loose valve adjustment, or a muffler vibrating against the frame. When the sensor retards timing, the engine idle will drop and performance will be erratic.

Piston Ring Seating

Because they are pre-lapped during manufacture, most piston ring companies now say that their rings will seat in a very short period of time. If in doubt about whether ring seating is necessary, use the following procedure to help seat the piston rings:
- Drive the car on the freeway.
- In high gear, accelerate from 35 mph to 50 mph and coast back to 35 mph several times. This helps to seat the rings under pressure and vacuum conditions. The deceleration (vacuum) phase also helps in sucking extra oil up into the cylinder walls to prevent scuffing.

It may take 2,000 to 3,000 miles to completely seat the rings. Drive easily (no high rpm) during this period. Engine bearings are soft and will also conform to irregularities in the bearing surfaces during this break-in time.

▬ FINAL INSPECTION

After the road test and break-in procedures have been completed:
- Double-check the engine for oil leaks.
- Make certain that all wires and lines have been correctly reinstalled.
- Check that all warning lights or gauges are operating properly.
- If the malfunction indicator comes on, determine the cause of the DTC and repair it before returning the car to the customer.

Vapor lock occurs when fuel vaporizes in a line and a bubble blocks fuel flow. Pump pressure will then cause the bubble to contract each time the pump pulses. When the line cools and the bubble condenses, the problem disappears. To avoid vapor lock, be sure that all fuel lines are positioned at least ½" from any heat source and that fuel filters are positioned where they can be cooled by air. A common cause of vapor lock is when a fuel line runs too close to an exhaust manifold or pipe.

▬ RETURNING THE CAR TO THE CUSTOMER

The vehicle should be returned to the owner in a clean condition. Grease on the fenders, steering wheel, seats, or carpet will contribute to a poor first impression of the job. An owner who is favorably impressed will be the best source of free advertisement.

When returning the car to customers, it is good practice to raise the hood to give them an opportunity to inspect the professionalism of your work. Explain what was done and give them an opportunity to

500-Mile Inspection

Following an engine rebuild, check the following items:

- ☐ Check oil level.
- ☐ Check for oil leaks.
- ☐ Change oil and filter.
- ☐ Check coolant level.
- ☐ Pressure test cooling system.
- ☐ Inspect hoses and retighten hose clamps.
- ☐ Retorque all exposed fasteners.
- ☐ Check tension of belt(s).
- ☐ Adjust valves as required.
- ☐ Retorque all head bolts as required.

Figure 56.12 A 500-mile inspection checklist.

ask questions. The customer should use the following break-in procedures suggested by Clevite Engine Parts:

1. Do not allow excessive engine idle for the first three hours of rebuilt engine operation.
2. Keep in the normal rpm range at about 75% load for the first 2 to 3 hours.
3. Engine speed should be varied as much as possible.
4. Full load or high-speed operation should be limited to less than 2 to 3 minutes at a time.
5. After high-load operation, allow the engine to return to a stable operating temperature by running at light load before shutting it off.

500-Mile Checkup

The customer should return to the shop after 500 miles for an oil and filter change. A phone call reminder is a good public relations measure and will offer a chance to check up on any complaints. A 500-mile inspection checklist is shown in **Figure 56.12**.

Oil consumption may not stabilize until all parts have broken in (seated). The customer should not be concerned until the vehicle has been driven at least 4,000 miles.

■ ENGINE REPAIR—ENGINE IN THE VEHICLE

Technicians often perform major repair to an engine while it is still in the vehicle. In addition to rebuilding the cylinder heads (valve job), they might do lower end work. This could be the replacement of a single defective piston, or it might be a piston-ring-and-crank-bearing-replacement, called an engine **overhaul**. This job often

includes replacing the timing chain and sprockets. Many OHC engines require a new chain tensioner (see Chapter 18). In-the-car repair is often less expensive for the vehicle owner. An engine that is to be completely rebuilt must be removed from the car.

■ VALVE JOB OR HEAD GASKET REPAIRS

The engine should be cold before removing the head. When the head is bolted to the block, it forms a rigid unit. Unbolting a hot cylinder head (especially aluminum) can cause parts to warp.

- It is a good habit to unbolt the head in a direction opposite to the normal tightening sequence.
- Check the cleanliness of head bolt threads as each one is removed. Chase all bolt holes with a tap.
- When reinstalling the head, be careful that nothing is accidentally pinched between the head and the block (**Figure 56.13**).

CAUTION Before reinstalling a cylinder head, use a suction gun or air nozzle to remove any oil or water from blind head bolt holes. Failure to do this can result in a cracked block next to the bolt hole. The water or oil will not compress during torquing.

Maintaining Valve Timing

During a valve job, it is essential to keep the timing chain or belt in place to maintain correct valve timing. Position the number 1 cylinder at TDC. Some OHC engines use a single long chain for a cam drive. The chain can be wedged against its guides with a tapered block of wood (**Figure 56.14**). If the crankshaft is turned, an unsecured chain tensioner can fall out, causing much extra work. Some engines have a lower and upper chain. These engines do not require special attention to wedging the chain. Be sure to look for hidden head bolts and to check the service information before removing an OHC head.

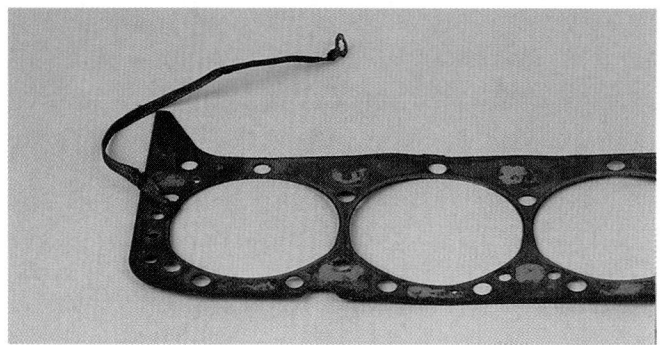

Figure 56.13 This ground strap was clamped between the head and the block during a careless installation. *(Courtesy of Tim Gilles)*

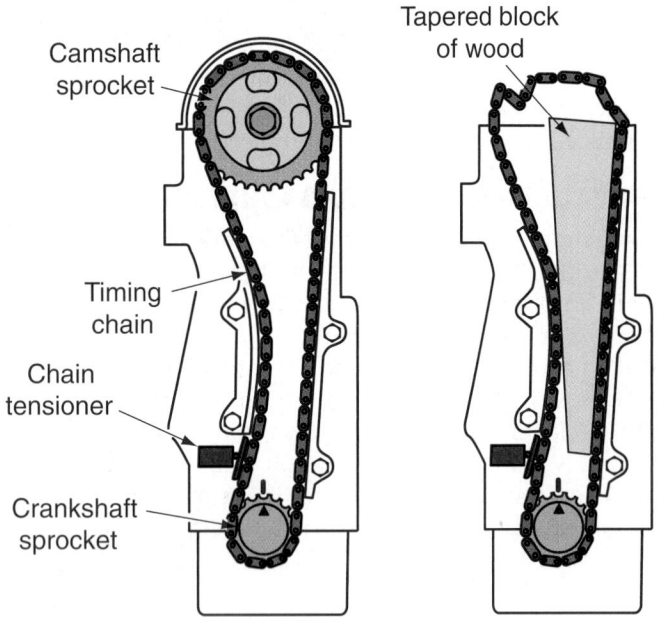

Figure 56.14 The chain tensioner on some OHC engines can be wedged to keep the chain in position during cylinder head removal.

▬ VALVE JOB OR COMPLETE ENGINE OVERHAUL

The question of whether to do a valve job only, or an engine overhaul, is an important one. Valve guide seals are responsible for some oil consumption complaints. Figure 49.4 shows what pistons look like when rings have not been controlling oil. There is a band around the top of each piston where no carbon is present. A valve job on this engine would not solve the oil consumption problem.

▬ HEAD GASKET PROBLEMS

Inspect the head gasket for damage. A faulty head gasket often points to a need for head surfacing, cooling system service, or repair to the engine's fuel or ignition system. Excessive temperatures can turn a metal head gasket blue or black. A Teflon® head gasket can turn brown. Look for signs of coolant leaks and damage from abnormal combustion.

▬ IN-CHASSIS LOWER END REPAIRS

An engine that is to be completely rebuilt must be removed from the car. A low-mileage engine is often repaired without removing it from the car.

An engine hoist can be used to raise the engine in order to remove the engine's oil pan. The engine is blocked up under the engine mounts or hung from a fixture from above during repair (**Figure 56.15**).

▬ REMOVING THE OIL PAN

Removing the oil pan could require the removal of some steering linkage. Often, simply unbolting the

Figure 56.15 The engine can be lifted and supported from the top. *(Courtesy of Tim Gilles)*

idler arm bracket (**Figure 56.16**) will provide enough clearance for the oil pan to be removed. If not, one or more of the **steering-linkage tapers** (**Figure 56.17**) might have to be broken loose. Many technicians use a *"pickle fork"* with an air chisel, but this procedure often tears an otherwise good tie-rod seal. A better way to break a taper is to use a special tie-rod puller

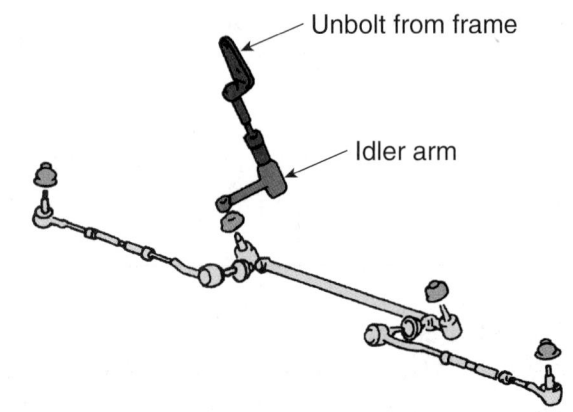

Figure 56.16 Unbolt the idler arm bracket to allow the steering linkage to drop. *(Courtesy of DaimlerChrysler Corporation)*

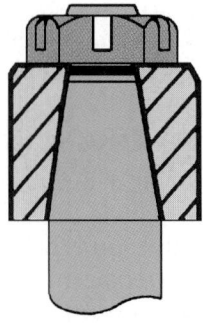

Figure 56.17 A steering-linkage taper.

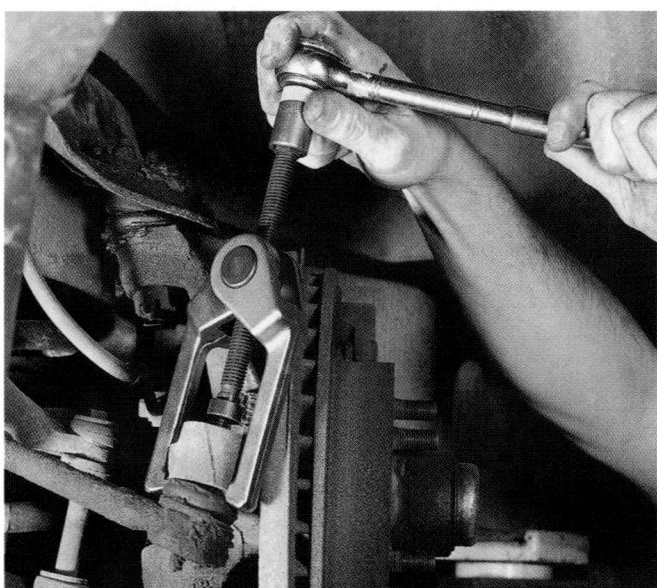

Figure 56.18 Using a tie-rod-end puller. *(Courtesy of Federal-Mogul Corporation)*

(**Figure 56.18**) or two large hammers as shown in **Figure 56.19**. Sometimes the engine mounts must be loosened and the engine raised a few inches in order to remove the oil pan.

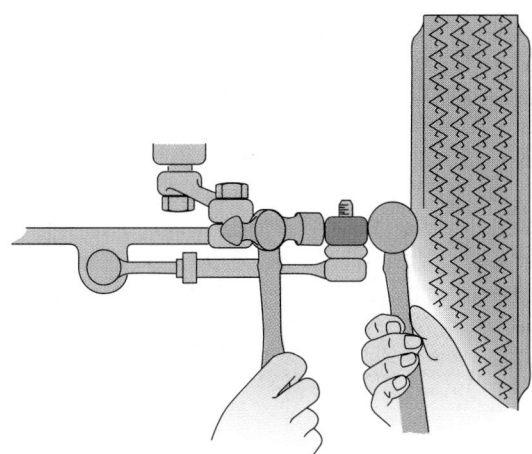

Figure 56.19 Using two hammers to loosen a steering taper connection.

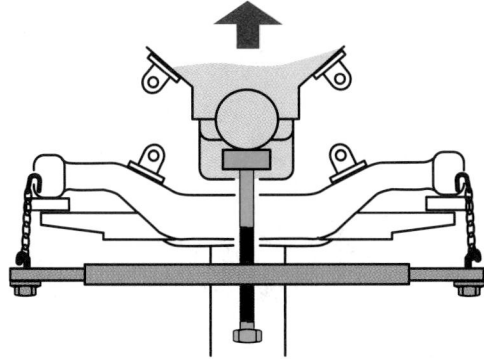

Figure 56.20 The engine can be raised from the bottom.

NOTE: *Be careful not to damage the radiator or the fan shroud when raising an engine off its mounts. Sometimes the radiator hoses must be removed or the radiator unbolted before the engine is raised. Use a piece of plywood to protect the oil pan, and jack the engine up with a hydraulic jack. The engine can then be blocked up at the engine mounts.*

SAFETY NOTE
- Be sure to use plywood instead of ordinary lumber when jacking or supporting something. The many layers of wood that make up plywood are laminated with the grain running in different directions. This makes plywood unlikely to split under pressure like ordinary lumber would.
- If the oil pan is to be removed with the car on a lift, the safest choice is a drive-on lift that raises the car by the wheels (see Chapter 11). Raising the engine from the bottom can upset the balance of the car on a frame-contact hoist.

If a frame-contact hoist is to be used, there are special fixtures that can be used to raise the engine from its top side (see Figure 56.5), or from the bottom (**Figure 56.20**).

SHOP TIP Sometimes it is necessary to unbolt the oil pump after the pan is loosened in order to get enough clearance to remove the pan. The pump is dropped into the pan and removed along with it. When reinstalling the pump, wrap a rubber band around the screws to prevent them from falling out of the holes in the pump while you reinstall it on the engine through the small opening between the pan and the block.

■ REMOVE THE PISTON AND ROD ASSEMBLY

If the lower end is to be repaired, remove the cylinder head(s). If there is a ring ridge in the cylinder, remove it. First, move the piston to BDC and place a rag in the cylinder to catch the metal chips. With the pan removed, the connecting rods can be unbolted and the pistons removed according to the procedure described earlier in this chapter.

Rod Bearing Replacement

Although bearings are usually replaced, an in-car repair might call for the replacement of only one defective

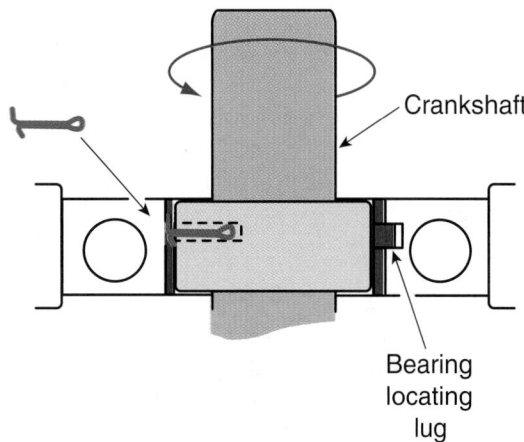

Figure 56.21 Main bearings can be rolled out and new bearings rolled in using a cotter pin.

piston. In this case, an unworn rod bearing might be reused. If a rod bearing is to be reused, mark the back of the bearing with a felt marker so it can be returned to its original position. If a bearing has lost its spread (see Chapter 19), it may have to be spread slightly before reinstallation.

Main Bearing Replacement

Main bearings are replaced with the crankshaft in the engine using a tool installed in the oil feed hole in the journal (**Figure 56.21**). The bearings must be rolled out on the side opposite the bearing locating lug, or tang.

■ REMOVE THE TIMING COVER

To remove the timing cover, first remove the radiator, accessory drive belts, and vibration damper. The damper bolt is loosened using a large socket and impact wrench.

> **SHOP TIP** An angle attachment is available for use with an impact wrench in case an air-conditioning condenser (in front of the radiator) is too difficult to move. However, using the angle attachment lowers the amount of torque available to the damper bolt.

Some OHC engines use a chain drive; others use a belt drive. Service on belt drives is relatively simple because no oil is sealed by the timing cover, as with chain drives. Removing the timing cover on some OHC engines with timing chains is more difficult because the cover often fits between the oil pan and the cylinder head (see Figure 56.16). There are special procedures for replacing cam timing components in these engines. Sometimes the head and pan might have to be

be loosened, which can result in leaks after the repair. When the cover intersects the pan, the front part of the pan gasket needs to be cut. Next, part of a new pan gasket is cut and installed to match the intersection of the oil pan and timing cover (see Chapter 51).

■ FREE-WHEELING AND INTERFERENCE ENGINES

An engine that has enough piston-to-valve clearance to prevent contact is known as a free-wheeling engine. Whenever timing chain service is performed because the chain has broken or skipped, it is possible that valves have come into contact with pistons. Cracked pistons or bent valves can result. Before a chain repair job, perform a leakage test on non-free-wheeling (interference) engines to check for bent valves (see Chapter 48). This step will help ensure that an accurate repair estimate can be made.

■ REPLACE THE TIMING COMPONENTS

When reinstalling the head on an OHC engine, it is very important that the number 1 piston be at TDC and that the cam in the head be timed properly. Otherwise, it is possible that a valve could be open and be forced against a piston. If the head is tightened in that position, a valve will be bent. Valve clearance is adjusted as needed before installing the head.

If the camshaft and crankshaft drive sprockets require replacement, this can be done with the engine in the car. **Figure 56.22** shows a crank sprocket being removed while the crankshaft is in the block.

NOTE: *If the cam sprocket is the nylon type and chunks of gear teeth are missing, the oil pan must be removed in order to extract teeth from the oil pump screen and oil pan (see Figure 53.48). Failure to do this will result in oil pump failure later when the pieces get sucked into the pump and block oil flow.*

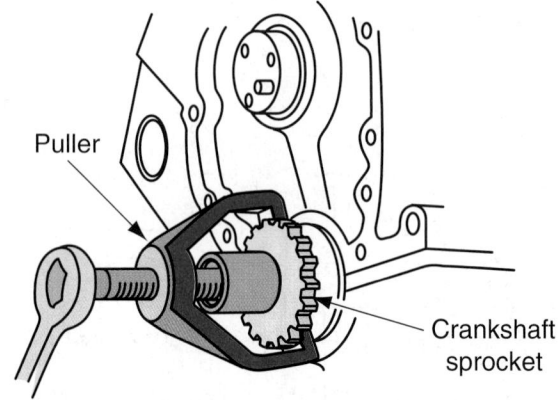

Figure 56.22 The crankshaft sprocket can be removed with the crankshaft still in the engine.

CRANKSHAFT SEAL REPLACEMENT

Crankshaft front and rear seal replacements can be performed with the engine in the car. The procedures are covered in Chapter 51.

> **SHOP TIP** For single-piece (full-round) seals that are in a casting bolted to the rear of the engine block, remove the seal by prying between the crankshaft and seal. Do this before removing the casting. The casting is fragile and is easily broken during seal removal unless it is bolted to the block.

FLYWHEEL RING GEAR SERVICE

Most vehicles with standard transmissions have a replaceable ring gear on the flywheel (**Figure 56.23**). Sometimes, the flywheel ring gear has been worn by a defective starter motor drive (**Figure 56.24**). To remove the worn ring gear from the flywheel, drill a hole between the teeth and break the ring with a chisel.

Heating a Ring Gear

Heat the new ring gear evenly around its circumference during installation. It should not be heated to more than about 400°F; too much heat can remove the hardness from the gear.

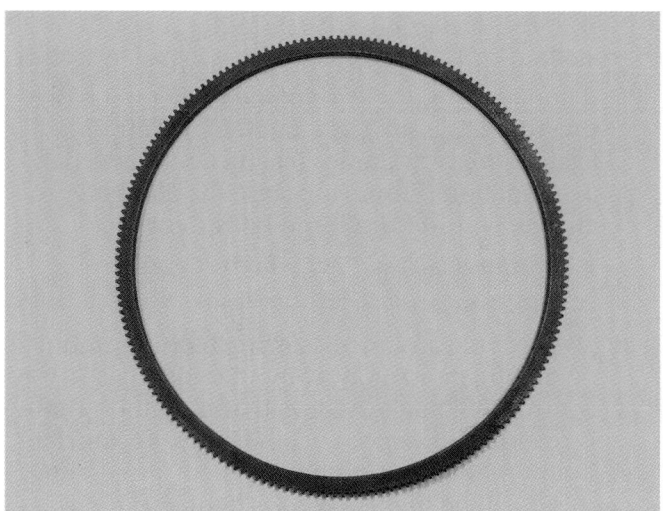

Figure 56.23 A replaceable ring gear. *(Courtesy of Tim Gilles)*

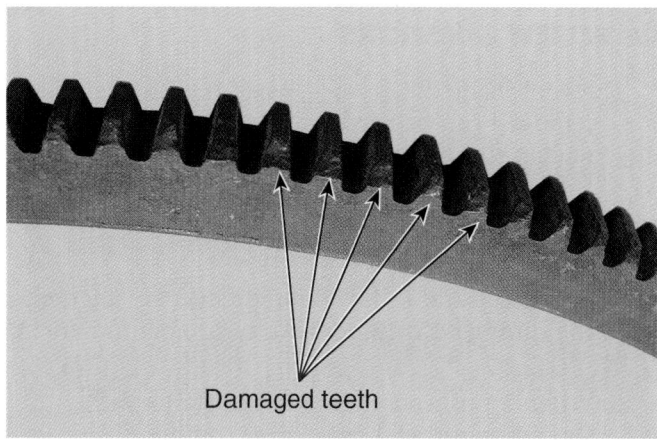

Figure 56.24 Damaged ring gear teeth. *(Courtesy of Tim Gilles)*

> ### SCIENCE NOTE
> *Pure iron is a silvery-white soft metal that rusts rapidly. The addition of small amounts of carbon strengthens it greatly. The hardness of iron is also related to cooling it at different rates. Cooling iron slowly allows carbon to separate from iron almost completely, giving large smooth crystals of iron. Cooling it rapidly gives small jagged crystals in which the carbon does not separate but remains combined to iron as iron carbide (Fe_3C).*

The temperature of the ring gear can be checked with a **tempilstick** (**Figure 56.25**). Tempilsticks are pencil-shaped sticks that melt at different temperatures. A 400°F tempilstick will leave a melted film on the ring gear when it is stroked across a surface hotter than 400°F.

Temperature can also be checked by polishing several spots on the ring gear using emery cloth or sandpaper. Heat the ring gear until the spots turn blue.

Solder can also be used to check the temperature of a ring gear. When the solder melts, the ring gear is hot enough and can be positioned onto the flywheel.

NOTE: *The chamfered side of the teeth should be on the same side as they were on the old ring gear.*

Figure 56.25 A temperature indicating stick melts when it contacts metal heated to its rated temperature. This one melts at 250°F. *(Courtesy of Tim Gilles)*

◼◼◼ REVIEW QUESTIONS

1. A tool made of a _____ shaft with the gear removed can be used for driving the oil pump when priming the lubrication system.

2. During lubrication system priming, the tool is turned in the same direction the _____ rotates.

3. When the crankshaft timing marks are aligned, the number 1 cylinder and its companion are at _____.

4. When a pushrod engine is first started, it should be run at 1,500–2,000 rpm for _____ minutes.

5. The point where valve clearance is eliminated is called zero _____.

6. What does WOT mean?

7. _____ is when fuel boils in a fuel line.

8. At what mileage should the first service and checkup be performed after an engine rebuild?

9. If a head gasket needs to be replaced, what machine work is probably necessary?

10. Because it is less likely to break than standard lumber, what kind of wood should be used when jacking or supporting something.

◼◼◼ ASE-STYLE REVIEW QUESTION

1. Forcing an engine into alignment with its mounts can result in:
   a. A distorted cylinder bore
   b. A bad engine ground
   c. Both A and B
   d. Neither A nor B

2. When adjusting power steering pump belt tension:
   a. Pry on the sheet metal part of the pump.
   b. Pry on the plastic part of the pump.
   c. Both A and B.
   d. Neither A nor B.

3. On which engines can the lubrication system be primed by turning the oil pump?
   a. Those with a gear on the bottom of the distributor
   b. Those without a gear on the bottom of the distributor
   c. Both A and B.
   d. Neither A nor B.

4. If the timing mark on the crankshaft damper on a GM V8 with the firing order 1-8-4-3-6-5-7-2 is aligned with the mark on the timing cover:
   a. The number 6 cylinder is at TDC.
   b. The number 1 cylinder is at TDC.
   c. Both A and B.
   d. Neither A nor B.

5. Immediately after initial engine startup, a pushrod engine should be run
   a. At varying engine speeds
   b. At fast idle
   c. On the freeway at a constant speed
   d. None of the above

6. Two technicians are discussing valve adjustment. Technician A says that valves are sometimes adjusted with the engine running. Technician B says that valves are sometimes adjusted with the engine off. Who is correct?
   a. Technician A   c. Both A and B
   b. Technician B   d. Neither A nor B

7. Technician A says that on an engine with a detonation sensor, the engine's idle can change due to a rattle such as a muffler vibrating against the car's frame. Technician B says that a detonation sensor can cause the computer to change ignition timing. Who is right?
   a. Technician A   c. Both A and B
   b. Technician B   d. Neither A nor B

8. Two technicians are discussing ignition timing on an engine without computer idle speed control. Technician A says that if ignition timing is retarded, idle speed will be lower. Technician B says that if ignition timing is advanced, idle speed will be higher. Who is right?
   a. Technician A   c. Both A and B
   b. Technician B   d. Neither A nor B

9. When installing an engine in a chassis:
   a. Torque the engine mount bolts on one side before installing the mount on the other side.
   b. Leave all mount bolts loose until all engine mount bolts have been installed.
   c. Bolt the engine mounts tightly to the engine before installing the engine in the vehicle.
   d. None of the above.

10. Two technicians are discussing the use of a lubrication system pressure primer. Technician A says that if oil leakage from a bearing appears to be excessive, turn the crankshaft ½ turn to see if the problem resolves itself. Technician B says to block off the oil pump inlet when performing the test. Who is right?
   a. Technician A   **c.** Both A and B
   b. Technician B   **d.** Neither A nor B

10

Brakes and Tires

THEN AND NOW: BRAKING SYSTEMS

As self-propelled vehicles came on the scene, it became apparent that an effective means of stopping them would be required. Borrowing from wagons, the first cars used *"spoon" brakes*. This was simply a block of wood pressed against the wheel by a lever. Next came the *contracting brake,* a steel strap or cable wrapped around the outside of the wheel's hub that could be tightened by the driver.

In 1903, a Frenchman named Louis Renault dramatically improved stopping power with the *internal expanding drum brake*. This brake design has curved friction material-lined "shoes" that are pushed against the inside of a drum to cause drag. Internal expanding drum brakes are still in wide use today.

A drawback to a brake drum is that it is a lot like a cast iron cooking pot. Getting rid of heat has always been a big problem with this brake design. The development of the disc brake did much to help solve that problem. A disc brake uses friction elements that "pinch" a flat rotor that is exposed to outside air. One of the earliest disc brake versions appeared on the front wheels of an electric car designed by Elmer Ambrose Sperry in 1898. Then came the 1949 Crosley with its *"spot" brakes* (round pads). Front-wheel disc brakes became popular on domestic cars in the mid-1960s.

Another early challenge was how to transmit the driver's signal to stop from his foot to the brake. The pedal replaced the hand lever early on, but mechanical means of applying the brakes (cams, cables, and levers) remained for some time. These systems were dangerous and could cause skidding, however, because they were impossible to equalize perfectly. They also required constant adjustment. The idea of using an enclosed liquid to do the job was proposed back in 1897. This would solve the unequal adjustment problem because hydraulic systems have equal pressure throughout the area in which the brake fluid is enclosed.

The 1910 Buick had the pedal for its transmission brake on the right. *(Courtesy of Bob Freudenberger)*

A modern disc brake rotor and caliper. *(Courtesy of Tim Gilles)*

It took many years to develop dependable hydraulic designs, however. The first American car with hydraulically actuated brakes was the 1921 Dusenberg, followed by Chrysler in 1924. In 1967, a federal law stated that all cars sold in the United States must have two separate hydraulic circuits.

Antilock braking systems (ABS) have become popular today. ABS pulses the brakes on hard stops or slippery surfaces, preventing a wheel from locking up and causing a skid. The idea is far from new. Patent applications were made for mechanical versions in the mid-1920s. Electronic systems were offered for a while in the early 1970s. Neither of these systems was dependable or affordable enough to be acceptable. Today, most vehicles come with ABS as standard equipment.

Brake Fundamentals

■ **KEY TERMS**

kinetic energy	hygroscopic	servo action
coefficient of friction	primary cup	brake fade
bonded lining	secondary cup	unsprung weight
riveted lining	vent port	metering valve
semimetallic lining	replenishing port	proportioning valve
metallic lining	tandem master cylinder	combination valve
ceramic lining	residual check valve	bulkhead
hydraulics	residual pressure	fire wall
master cylinder	longitudinally split	drum-in-hat
force	diagonally split	ABS
Pascal's law	self-energizing	

■ INTRODUCTION

When a car is traveling at freeway speed, a large amount of **kinetic energy** is stored. This is energy that wants to remain in motion (*inertia*). Energy cannot be consumed, but its form can be changed. When you apply the brakes to stop the car, *dry friction* is used to change energy of motion (kinetic energy) to *heat energy*.

The amount of horsepower changed to heat during stopping can amount to several times the power developed by the engine during acceleration. The amount of kinetic energy to be changed into heat depends on the weight and speed of the vehicle. The temperature in the brake linings during a stop can approach 600°F. The amount of dry friction varies, depending on the force applied, the material that the friction surfaces are made of, and the roughness or finish of the friction surface.

■ Friction is the force that resists movement between any two contacting surfaces.

■ The ratio of the force holding two surfaces in contact to the force required to slide one over the other is known as the **coefficient of friction**.

If it takes 70 pounds of force to drag one piece of material across another, the coefficient of friction is 70 divided by 100, or 0.70.

■ The coefficient of friction varies with the temperature of the surfaces in contact, the rubbing speed, and the condition of the surfaces.

Weight Transfer

During a stop, the weight of the vehicle shifts onto the front brakes. Because of this weight transfer, rear brakes do not usually wear out as fast as front brakes. Front brakes must be able to absorb more heat than rear brakes, so linings with more surface area are required. Heavier vehicles require wider linings with more surface area to carry off the increased heat.

The ratio between the front and the rear brakes is about 60/40 for rear-wheel-drive vehicles. When the vehicle has front-wheel drive, the braking ratio is about 80% for the front and 20% for the rear. This is because of the added weight of the powertrain components in the front.

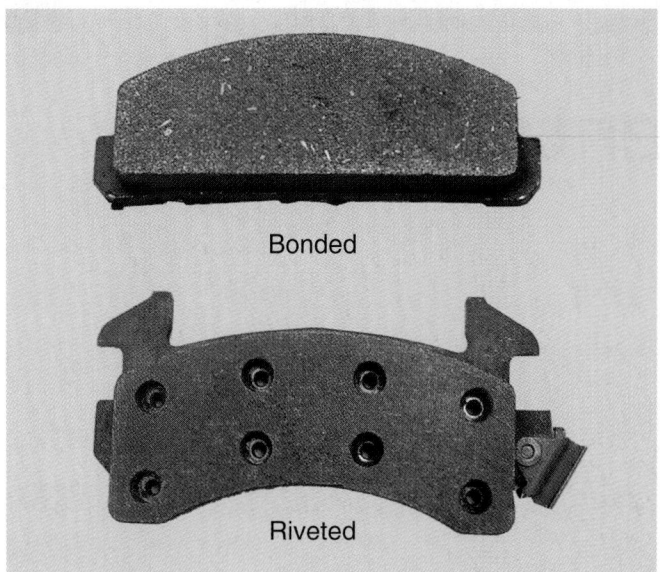

Figure 57.1 Bonded and riveted brake pads. (*Courtesy of Tim Gilles*)

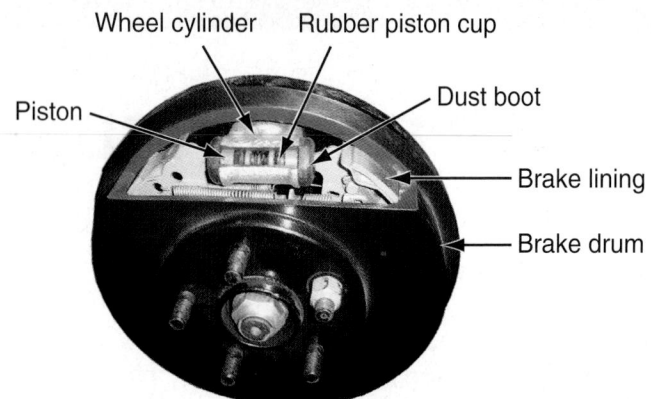

Figure 57.2 Parts of a drum brake. (*Courtesy of Tim Gilles*)

BRAKE LININGS

The friction materials used in cars and trucks are called brake linings. Linings are either **bonded** (glued) or **riveted** to the disc backing (**Figure 57.1**) or shoe (on drum brakes). Some newer pads are *integrally molded*. From the back of the pad, you observe holes that are either full or partially full of lining material. The linings are actually molded to the metal disc backing plate to become one unit.

Linings are either asbestos, nonmetallic organic, semimetallic, metallic, or ceramic. *Asbestos linings* are a health hazard and have been mostly phased out as a brake material. The materials used to replace asbestos are organic materials bonded together with a resin binder.

Semimetallic linings are organic linings with sponge iron and steel fibers mixed into them to add strength and temperature resistance. They are fast at taking heat away from the rotor and putting it into the lining. This heat transfer does *not* affect the service life of the lining. The hotter they get, the better they work.

Metallic linings are used in very heavy-duty and racing conditions. They work poorly when cold.

Ceramic linings are original equipment on about half of all new vehicles. They use ceramic and copper fibers to control heat. The ceramic material dampens some of the noise, too. They also produce lighter colored brake dust.

More complete information on brake lining materials is found in Chapter 58.

DRUM AND DISC BRAKES

In the past, almost all vehicles were equipped with drum brakes on all four wheels. Today, they are only found in some rear brake applications. Late-model

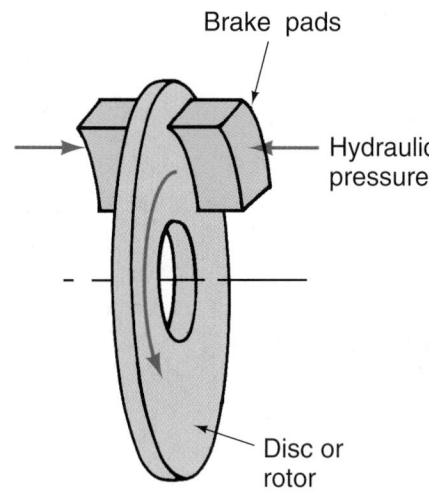

Figure 57.3 Disc brake systems have a rotor and caliper similar to those used on a bicycle.

vehicles use disc brakes in the front and drum or disc brakes in the rear.

There are two main types of brakes: *drum* and *disc*. Drum brake systems have metal brake drums that are bolted to the wheels (**Figure 57.2**). Drums are made in several materials and styles. The linings and braking components are mounted on a fixed backing plate. Disc brake systems have a rotor and caliper, similar to that used on a bicycle (**Figure 57.3**). On drum brakes, the friction materials are called linings or *shoes*. On disc brakes, they are called linings or *pads*.

HYDRAULIC BRAKE SYSTEM OPERATION

Pressurizing liquid to transfer motion or multiply and apply force is called **hydraulics**. When the brake pedal is depressed, it moves a piston in the **master cylinder** (**Figure 57.4**). This pushes fluid under pressure through brake lines and hoses to a "slave cylinder" at each wheel, where the hydraulic pressure acts on pistons to produce **force** (**Figure 57.5**).

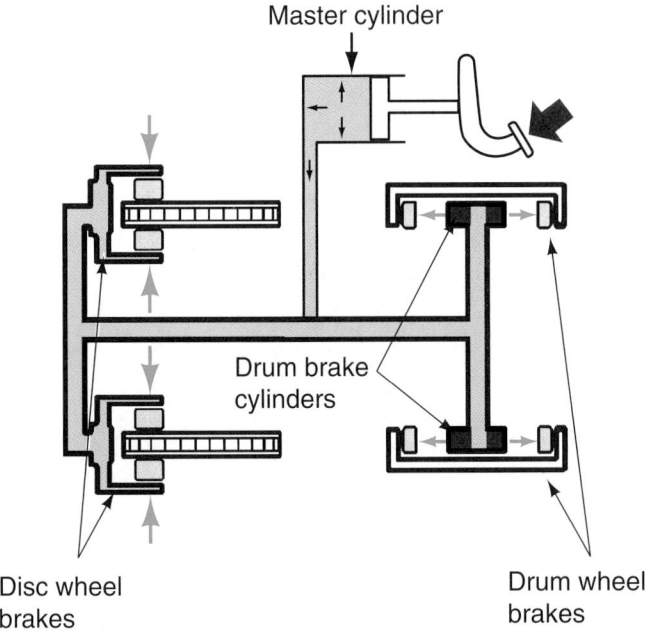
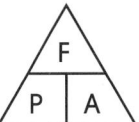

Figure 57.4 Depressing the brake pedal moves a piston in a master cylinder.

Pascal's Law

Pascal's law is fundamental to the development of your automotive diagnostic ability. It applies to such items as the hydraulic brake system, suspension and steering, the lubrication system, engine bearing clearances, the pressurized cooling system, engine compression, and air conditioning. Laws of hydraulic pressure can even be applied to electrical fundamentals. One of the most important applications of Pascal's law is in the hydraulic brake system.

Pascal's law states that "pressure in an enclosed system is equal and undiminished in all directions."

"Enclosed" means that the fluid is not moving. Hydraulic results can be determined using the following formulas:

Pressure = force ÷ area
Force = pressure × area

The following facts apply to the hydraulic braking system:

■ The force applied to the brake linings when the master cylinder operates is increased with a larger diameter wheel cylinder.
■ When larger wheel cylinders are used, the distance that the pedal has to travel before building up pressure increases.
■ In order to increase the pressure coming from the master cylinder, the diameter of its bore must be *smaller*.

In the example shown in **Figure 57.6**, the input piston is smaller than the output piston. Using the correct formula (Pressure = force ÷ area), you can determine that with 200 pounds of force applied to a 2-square-inch piston, the resulting pressure is 100 psi.

$$\frac{200 \text{ pounds}}{2 \text{ square inches}} = 100 \text{ psi (pounds per square inch)}$$

When the 100 psi developed by Piston A is applied to the 5-square-inch piston on the right-hand side of the sketch, the resulting force is 500 pounds. The correct formula for determining this is Force = pressure × area.

100 psi × 5 square inches

In **Figure 57.7** you can see the effects of pressure on pistons of two different sizes. Notice that the smaller

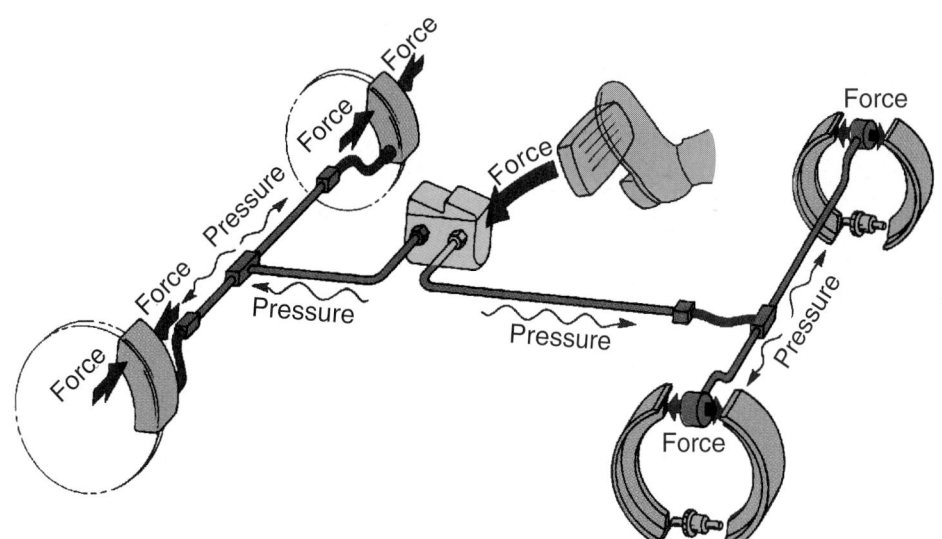

Figure 57.5 Pressure acts on hydraulic cylinders at each wheel to produce force. *(Courtesy of Federal-Mogul Corporation)*

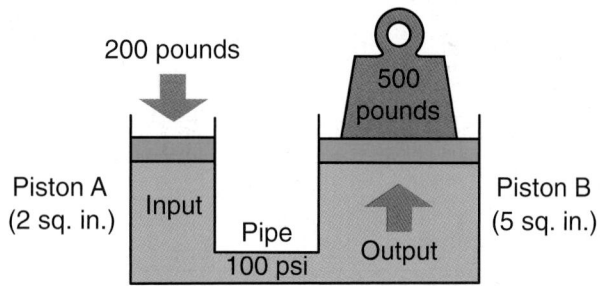

Figure 57.6 The force applied when the master cylinder operates is increased with a larger diameter wheel cylinder.

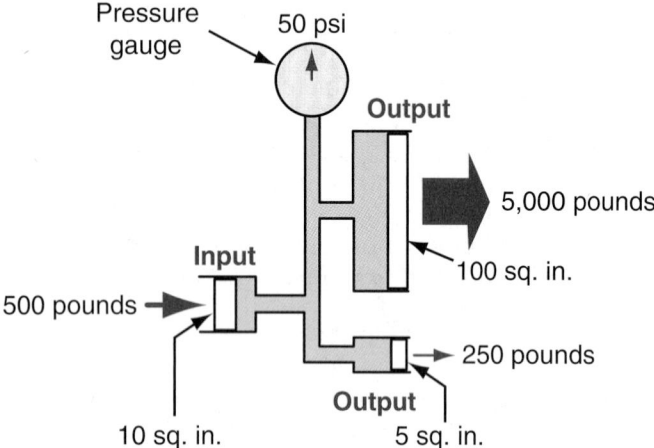

Figure 57.7 Application of formulae for hydraulic pressure.

piston travels much farther in response to movement of the input piston than the larger piston.

■ HYDRAULIC BRAKE FLUID

Hydraulic brake fluid meets standards set by the *Society of Automotive Engineers* (SAE) Standards J1703 and FMVSS 116. Brake fluid number classifications have

Figure 57.8 The DOT number is always on the label of a brake fluid container. *(Courtesy of Tim Gilles)*

been assigned by the Department of Transportation (DOT) and include DOT 1, 2, 3, 4, 5, and 5.1. The DOT number is always listed on the fluid container (**Figure 57.8**). DOT 5 is *synthetic brake fluid.* DOT 3, 4, and 5.1 are fluids made from a *polyglycol* base, similar to that of engine coolant.

Glycol-based fluids are **hygroscopic**, which means that they absorb water. Being hygroscopic is an advantage for brake fluid because it can absorb moisture that enters the system. This prevents the formation of water drops that could boil or freeze. Dispersing the moisture throughout the fluid also prevents localized corrosion that could cause holes to form in brake lines.

NOTE: *Brake fluid can absorb a large enough quantity of water to be ruined in as little as 1 hour if left uncovered. According to the EIS Brake Company, a typical vehicle's brake fluid that has been changed in the last 18 months will have accumulated 2% to 3% water. A water content of 3% lowers the boiling point of DOT 3 brake fluid by 25%. DOT 4 fluid absorbs moisture at a slower rate but is more affected by moisture contamination. With a 3% water concentration, its boiling point can be as much as 50% lower. When there is a choice between using fluid from a large, partly empty container or two unopened smaller containers, the two smaller containers are the best choice.*

The friction produced by the brakes causes the brake fluid to become hot. Consequently, brake fluid is designed to have a high boiling temperature. Water has a far lower boiling point (212°F) than brake fluid (over 400°F). If brake fluid absorbs enough water, it can boil in the lines. This can result in a loss of braking efficiency, a brake pedal that feels "spongy" during application, and even a total loss of brakes.

STANDARDS			
	DOT 3	**DOT 4**	**DOT 5.1**
Boiling Point	min. 205°C/401°F	min. 230°C/446°F	min. 260°C/500°F
Wet Boiling Point*	min. 140°C/284°F	min. 155°C/401°F	min. 180°C/356°F
* Indicates decline in boiling point caused by increasing water content.			

Figure 57.9 DOT specifications for both dry and wet boiling points of polyglycol brake fluids.

DOT specifications list both *dry* and *wet boiling points* (**Figure 57.9**). The dry specification is for new fluid, and the wet specification is for fluid that has absorbed 3.5% water. Brake fluid can absorb enough moisture to reach its wet boiling point within approximately 2 years. Many manufacturers recommend brake fluid replacement at 2- or 3-year intervals.

■ DOT 3 fluid has a dry boiling point of 401°F and a wet boiling point of 284°F.
■ DOT 4 fluid has a higher boiling point (446°F dry and 401°F wet).

Fluid that meets *DOT 3* or *DOT 4* specifications can be used in drum and disc systems. Most domestic manufacturers specify DOT 3. Some DOT 3 fluids are heavy duty and have an equally high boiling point as DOT 4. Under perfect conditions where no moisture is allowed to contact the fluid, DOT 3 is said to have a longer life than DOT 4.

Glycol-based fluids are *not generic* (not all the same). They are a mixture of various substances. Up to ten different ingredients can be blended to make up the fluid. Some fluids perform better than others. All brake fluids are made up of four ingredients:

■ A lubricant to keep parts sliding freely
■ A solvent-diluent, which determines the fluid's viscosity and boiling point
■ A modifier-coupler, which changes the amount of swelling of rubber parts exposed to it
■ Inhibitors to prevent corrosion and oxidation

The lubricant in brake fluid makes up 20%–40% of its content. It is a synthetic polyglycol, such as polyethylene or polypropylene. Most of the content of the fluid (50%–80%) is the solvent-diluent glycol ether. In some fluids, ethylene glycol (the same compound as automotive coolant) is also used for this purpose because it also controls rubber swelling. A wide range of chemicals make up the small remainder of the fluid (0.5%–3%).

NOTE: *Brake fluid prices often vary with the price of engine coolant, due to the amount of glycol used.*

Synthetic Brake Fluid

Synthetic silicone-based brake fluid was developed in the early 1970s. A new DOT category, DOT 5, was created for it. The DOT 5 standard requires a minimum dry boiling point of 500°F and a wet boiling point of 356°F. The wet boiling point does not come into play with silicone fluid, however, because it is *not* hygroscopic. Because it does not absorb moisture, it is called a lifetime fluid.

Silicone fluid has a "spongy" pedal feel because it is more compressible than ordinary brake fluid. The pedal will also have more travel before the brakes are fully applied. This is because silicone fluid contains about three times as much dissolved air as glycol-based fluid. Glycol has 5% dissolved air, whereas silicone has 15%.

NOTE: *Silicone fluid tends to develop air bubbles when it is cycled rapidly. Most manufacturers recommend it **not** be used in ABS applications. Check the manufacturer's recommendation for the correct fluid. Antilock brake systems (ABS) usually call for DOT 3 fluid.*

A newer brake fluid, first used by manufacturers in 1999, is called DOT 5.1. Like DOT 3 and DOT 4, it is glycol based, not silicone based like DOT 5. It has a lower viscosity and higher boiling point than DOT 3 and 4 fluids, making it popular in race cars. Because it is hygroscopic, it must be replaced often when used for racing.

Mineral Oil Brake Fluid

Mineral oil brake fluid, called hydraulic system mineral oil (HSMO), is green rather than the light gold color of ordinary glycol-based brake fluid. It is a good lubricant, is not hygroscopic, and has a high boiling point, but it will cause rubber parts in conventional brake systems to swell. It is not compatible with ordinary brake systems and is not covered by a DOT classification. Some European manufacturers (Rolls Royce, Citroën, and Audi) have used it in their brake systems, although no manufacturers use it today.

■ BRAKE HOSE AND TUBING

Steel hydraulic brake tubing runs the length of the vehicle frame. Rubber hoses on both front wheels and above the rear axle provide a flexible connection from the steel tubing to brake system components (**Figure 57.10**). A flexible rubber connection is needed because the front wheels pivot during steering and the rear axle moves up and down with the suspension. When the car has an independent rear suspension, each rear wheel is served by its own flexible hose. Some vehicles with four-wheel disc brakes use hoses at each wheel.

SCIENCE NOTE

Gore-Tex® is a membrane that is laminated to a variety of fabrics. Its expanded PTFE (similar to Teflon) membrane is hydrophobic (water hating) and contains 9 billion pores per square inch. Each pore is 20,000 times smaller than a drop of water, yet 700 times larger than a molecule of water vapor. Therefore, it allows

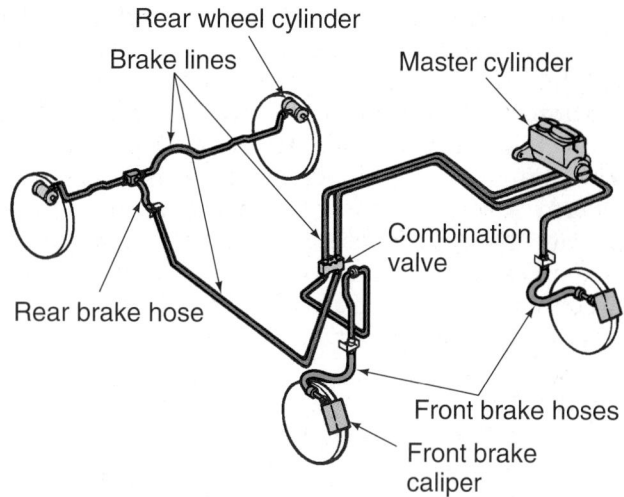

Figure 57.10 Hydraulic lines and hoses connect the master cylinder to the wheel brake units.

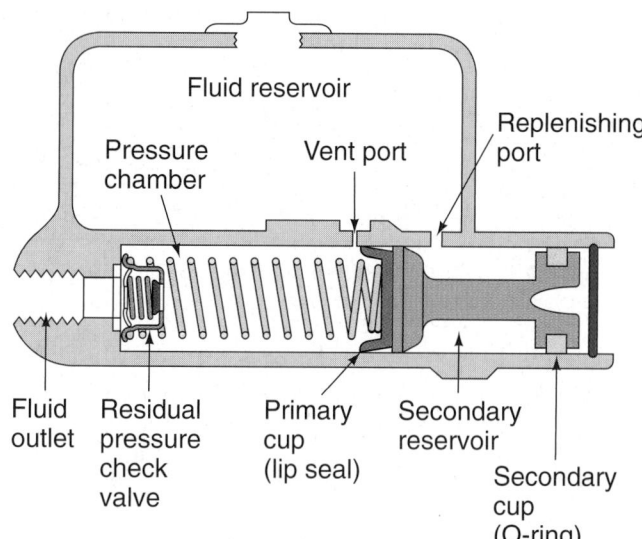

Figure 57.11 A single-piston master cylinder.

moisture from perspiration to escape during exercise, yet liquid water cannot penetrate the pores of the material. An oleophobic (oil hating) substance is also integrated into the membrane to prevent the entry of contaminants (such as bug spray, body oil, suntan lotions, and so forth).

NOTE: *Water molecules are smaller than brake fluid molecules. Over a long period of time, since brake fluid is hydrophilic (water seeking), water vapor from the outside migrates to the brake fluid. This can happen even though brake fluid does not leak through the woven fibers and outer rubber layer of a brake hose.*

Brake lines are usually made of double-walled steel tubing coated with a rust-preventative material. Replacement brake lines are available in several precut lengths with flared ends already on. Sometimes a loop or bend can be put into a line to shorten it to the desired length without having to cut it and reflare one of its ends. When replacing a brake line, be careful to copy the original line as closely as possible. Complete information on brake tubing theory and service is covered in Chapter 24.

■ MASTER CYLINDER OPERATION

When a driver depresses the brake pedal, the linkage applies force to the back of the master cylinder. The master cylinder pressurizes brake fluid and routes it to the wheel cylinders and calipers in the vehicle's brakes. Most master cylinders have two chambers, but the operation of a simple, single-piston master cylinder is discussed first. Service technicians who do not understand the operation of the hydraulic system often perform brake repairs simply by replacing parts. This is unfortunate because it often results in repairs that are not necessary while searching for the solution to a problem. To be able to efficiently

diagnose brake system problems, an understanding of the operation of the master cylinder and the complete hydraulic system is necessary. Pay close attention to the following explanation of master cylinder operation.

The master cylinder, connected to the foot pedal, supplies hydraulic pressure to operate the wheel cylinders during braking. **Figure 57.11** shows a simple single-piston master cylinder with one seal called the **primary cup**. The primary cup compresses fluid when the pedal is depressed. Another seal, called the **secondary cup**, keeps fluid from leaking out of the back of the master cylinder bore. This seal does not seal against pressure.

NOTE: *A master cylinder has two ports for each piston. These ports have had many names over the years. SAE Standard J1153 defines the front port as the* **vent port** *and the rear port as the* **replenishing port**. *The front port is also commonly called the compensating port. The rear port is also called the inlet port.*

Fluid from the reservoir fills the master cylinder bore through the vent (compensating) port, which also allows for expansion of the fluid as it absorbs heat (**Figure 57.12**). When the pedal is applied, the primary cup moves forward in the master cylinder bore, past the vent port. With the vent port isolated, fluid movement is restricted. As the pedal is forced further down, fluid pressure builds in the hydraulic system.

Seal Lips Are Directional

In the simple drum brake system shown in **Figure 57.13**, the lips of the cup seals face inward against the pressurized fluid created by the master cylinder. The lip on the master cylinder primary cup faces toward the fluid. The wheel cylinder also has a pair of rubber cup lip seals. Their lips face in an opposing direction to master cylinder fluid pressure. When a seal is installed in this direction, it *seals*. A seal installed backward will leak.

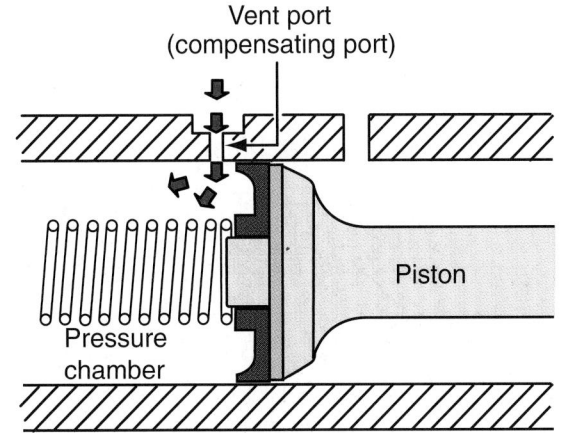

Figure 57.12 Fluid from the reservoir fills the master cylinder bore through the vent port.

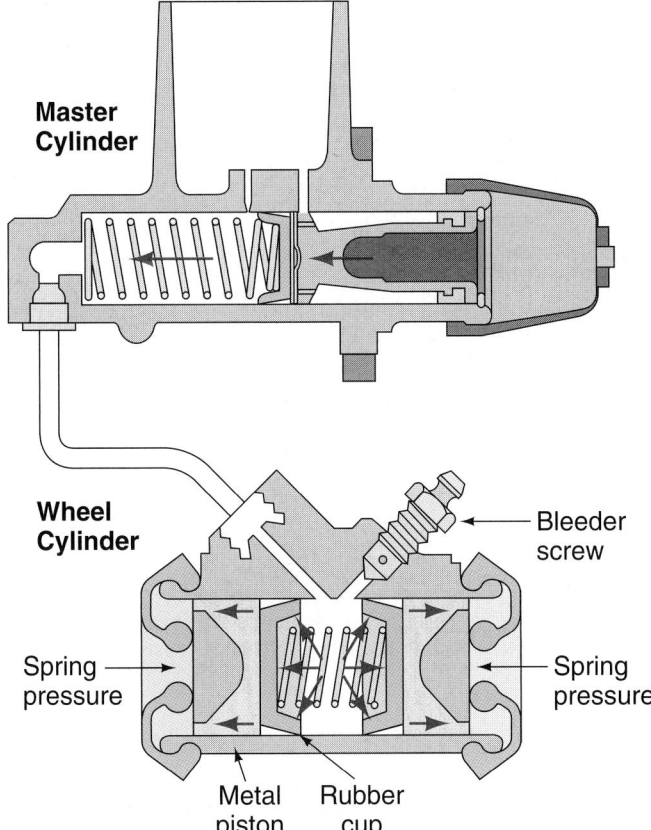

Figure 57.13 The lips of the rubber cups must face toward the brake fluid.

◾ LOW BRAKE PEDAL

As brake linings wear, clearance between the friction surfaces becomes greater. Disc brakes and most drum brakes are *self-adjusting*. When there are no self-adjusters or if a self-adjusting mechanism fails to work, a drum brake will develop extra clearance. The result is that the brake pedal moves closer to the floor before the brakes are applied. This is called a *low pedal*.

When excessive wear of the friction materials occurs, the pedal might be able to travel all the way to the floor. When this situation develops, a second application of the pedal will result in a higher pedal that will stop the car. What makes this happen?

When the brakes are adjusted correctly, only a small volume of fluid in the master cylinder is moved to operate the brakes. When a low pedal is released and then quickly reapplied, fluid becomes momentarily trapped in the cylinders at the wheels. The fluid cannot return quickly to the master cylinder, because the inside diameter of the brake line is very small. This small size prevents the quick movement of a large volume of fluid. Large diameter tubing would allow fluid to move more quickly.

When the pedal is released, the primary cup moves back quickly in the master cylinder bore. An area in front of the primary cup, called the *pressure chamber*, is momentarily empty of fluid. Stored just behind the primary cup is a reservoir of fluid (at atmospheric pressure). The sealing lips on the primary cup lip face away from the fluid pressure, so fluid can leak past the cup from the back to the front. This refills the temporarily empty pressure chamber in the master cylinder bore (**Figure 57.14**). The secondary area is refilled through the replenishing (inlet) port (**Figure 57.15**).

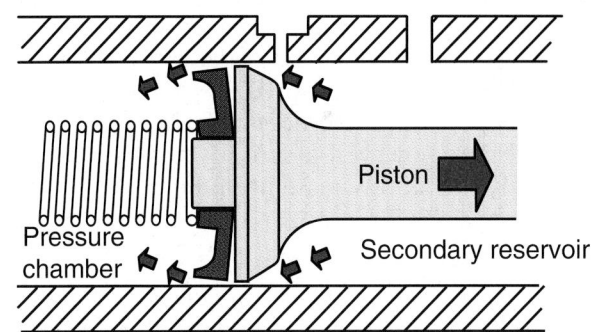

Figure 57.14 Fluid moves from the secondary reservoir to the pressure chamber when the brakes need adjusting.

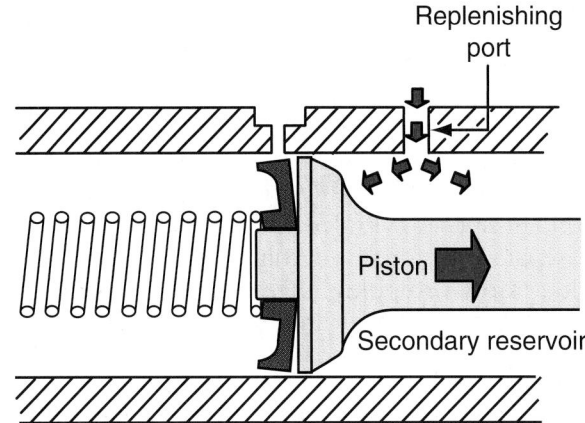

Figure 57.15 Fluid is replenished to the secondary area through the inlet port.

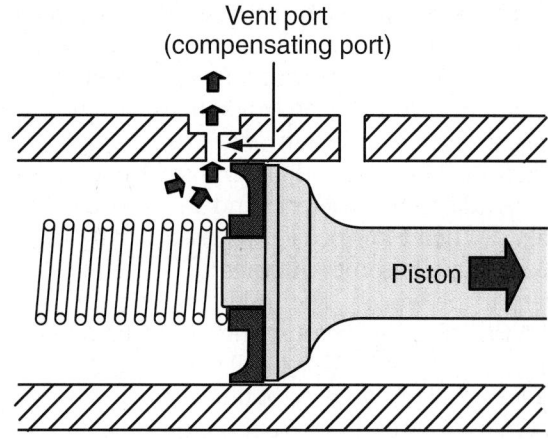

Figure 57.16 Excess fluid bleeds back to the reservoir through the vent port.

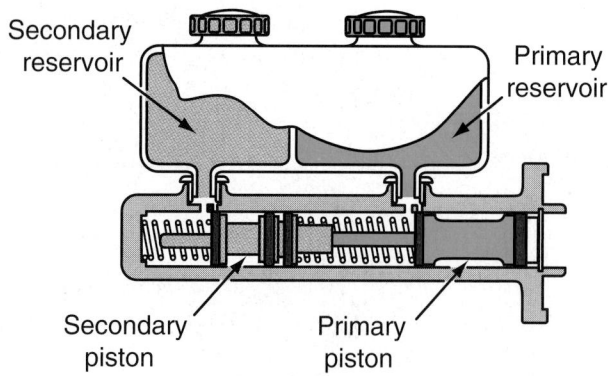

Figure 57.17 A tandem master cylinder.

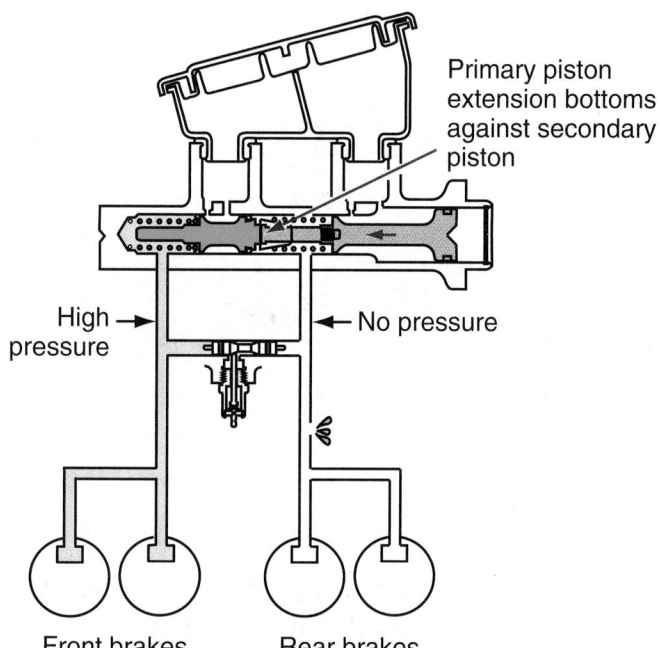

Figure 57.18 The rear piston bottoms against the front piston when the rear half of the hydraulic system fails.

When the pedal is quickly reapplied for the second time, there will now be enough fluid in the system to cause the linings to contact the drum, and pressure will build on the fluid in the wheel cylinders. When the pedal is released again, the brake shoe return springs *slowly* force the excess fluid back through the brake line from the wheel cylinder. The excess fluid bleeds off to the reservoir through the now-uncovered vent (compensating) port (**Figure 57.16**). Until the brakes are correctly adjusted, the brake pedal height will increase on the second pedal application. Low pedal height can also be caused by an internal leak in one part of the master cylinder or by an ABS problem. In these cases, however, a second application of the brake pedal will not increase pedal height.

Tandem Master Cylinder

Older cars were equipped with a simple single-piston master cylinder like the one in the previous description. These cars were more dangerous than cars with modern brake systems. When a car with a single-piston master cylinder had a hydraulic system failure, it would experience a total brake system failure.

Modern cars have master cylinders with two separate hydraulic systems. Since 1967, these **tandem master cylinders** have been mandated by law on all cars.

NOTE: *During the following discussion of tandem master cylinder operation, remember Pascal's law:* **Pressure is equal everywhere in an enclosed system.**

A tandem master cylinder has one cylinder bore with two separate pistons and chambers (**Figure 57.17**). During normal stopping, the primary cup on the rear (primary) piston pushes fluid to the brakes it serves. It also pushes fluid forward against a rearward facing lip seal on the front (secondary) piston. When the friction materials begin to apply force at the wheels, pressure begins to build in both front and rear systems at the same time.

When one-half of this system fails, the pedal will be lower but the remaining half of the hydraulic system should have enough braking capacity to stop the car. During a failure in the part of the brakes served by the rear piston, a stub on the front of that piston bottoms out against the back of the front piston (**Figure 57.18**). Thus, the primary piston is applied mechanically instead of hydraulically.

When the hydraulic system served by the front part of the master cylinder experiences a failure, it cannot build up pressure. A stub on the end of the piston bottoms out at the end of the bore (**Figure 57.19**). This allows the rear piston to build up pressure as it forces fluid against the rearward-facing seal on the back of the front piston.

and wear. At about the same time that drum brakes disappeared from use in vehicle front brakes, residual check valves also disappeared from master cylinders serving rear drum brakes. A residual check valve would be a problem in split diagonal hybrid systems, and *cup expanders* on the ends of the center spring in the wheel cylinder make the valve unnecessary (**Figure 57.20**).

Master Cylinder Reservoir

Many of today's master cylinders are aluminum, with a plastic reservoir sealed to the cylinder with rubber grommets (**Figure 57.21**). They are called *composite* master cylinders because they are made of two materials. Aluminum is lighter than cast iron and less expensive to manufacture. Plastic reservoirs are transparent to allow the fluid level to be observed without removing the reservoir cover, which could lead to dirt or moisture contamination.

Master cylinder fluid level fluctuates through the vent port as the brakes are used because the fluid heats up or cools down (see Figure 57.16). Therefore, the cover to the master cylinder must include a feature that prevents a vacuum lock as the expanded brake fluid cools down. If a vacuum lock were to occur, air would be drawn into the brake fluid past the secondary cup at the rear of the master cylinder.

Master cylinder reservoirs are prevented from vacuum locking in one of two ways. There is either a flexible rubber diaphragm in the cover (**Figure 57.22**) or screw cap, or there is a plastic float. These devices allow atmospheric pressure to act on the fluid in the reservoir without the fluid becoming contaminated with moisture from the outside air.

Master Cylinder Location

Master cylinders on almost all cars and light trucks since the 1950s have been mounted on the bulkhead, also called the fire wall (**Figure 57.23**). Older vehicles had the master cylinder mounted under the floorboard.

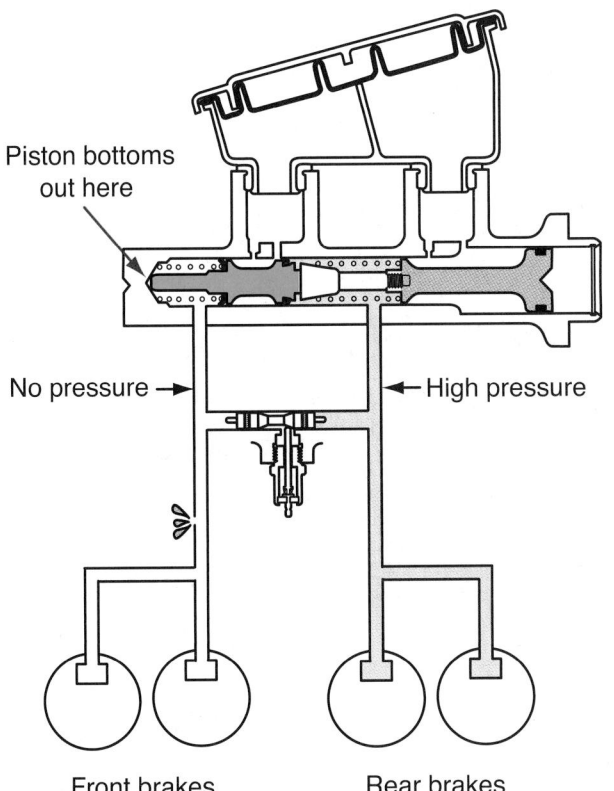

Figure 57.19 The front piston bottoms against the front of the bore, allowing pressure to build in the rear half of the system.

Master Cylinder Check Valve

Older master cylinders used with drum brakes have a **residual check valve** at the fluid outlet. The purpose of this check valve is to keep a small amount of **residual pressure** (6–25psi) in the brake system when the brakes are not applied. This helps keep the wheel cylinder rubber cup sealed against the wall of the wheel cylinder to prevent air from entering.

Disc brake systems do *not* have a check valve. If they did, the pressure would overcome the return action of the disc brake seal, causing brakes to drag

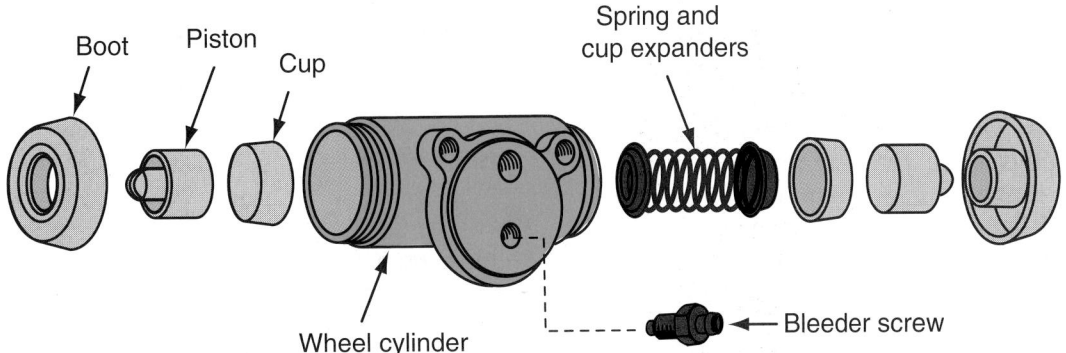

Figure 57.20 Cup expanders on the ends of the center spring take the place of a residual check valve.

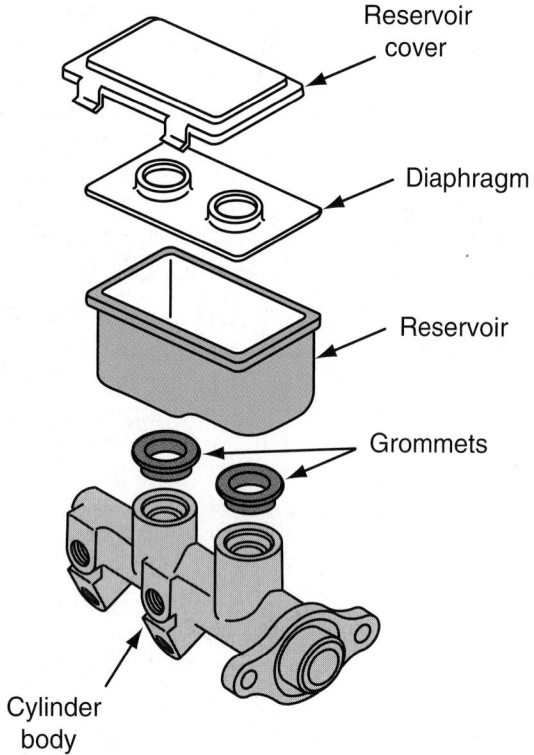

Figure 57.21 A master cylinder with a removable plastic reservoir.

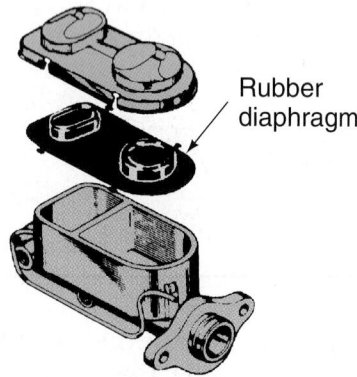

Figure 57.22 The reservoir cover has a rubber diaphragm or a plastic float.

■ SPLIT HYDRAULIC SYSTEM

Tandem systems are either **longitudinally split** (front/rear) or **diagonally split** (**Figure 57.24**). A longitudinally split system operates the front and rear brakes as separate hydraulic systems. Most rear-wheel-drive vehicles use this system.

Chrysler was the first manufacturer to mass-produce a diagonally split system in 1978. The diagonally split system operates the brakes on opposite corners of the vehicle. This system is used on most front-wheel-drive vehicles, which can experience weight transfer of about 80% at the front during a hard stop. If the front brakes were longitudinally split and

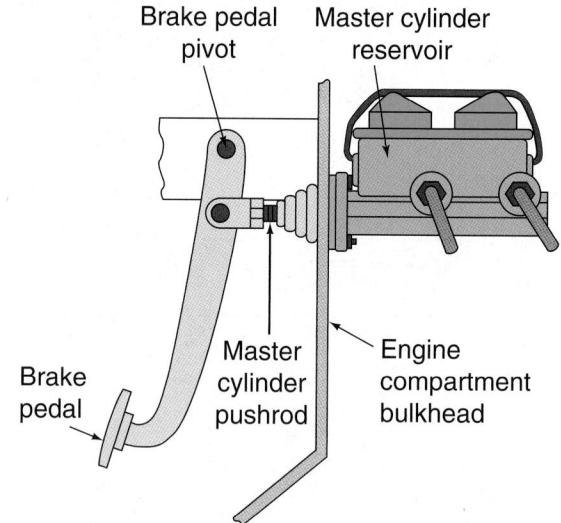

Figure 57.23 A master cylinder with a suspended pedal is mounted on the bulkhead.

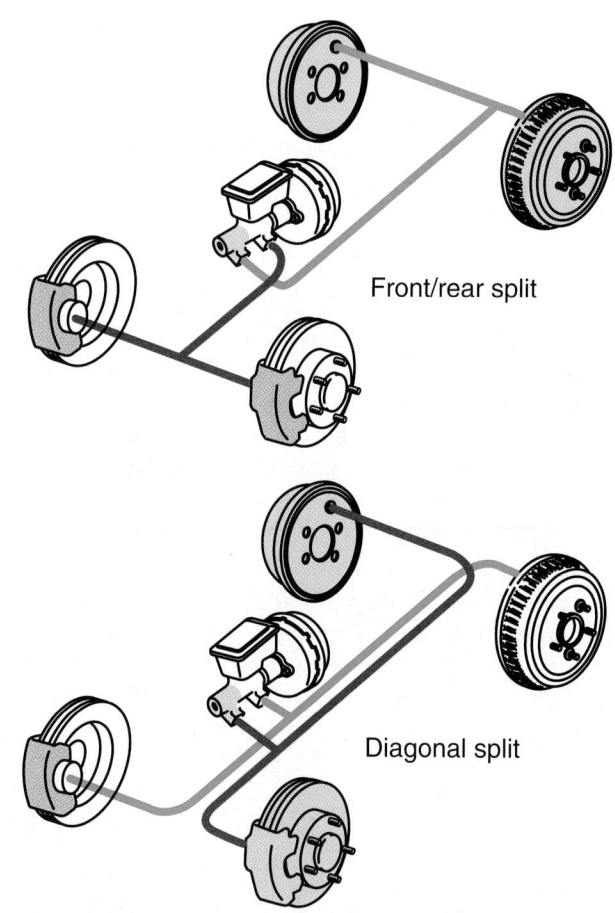

Figure 57.24 Tandem brake cylinders are hydraulically split in different ways.

the front hydraulic system failed, the 20% of braking done by the rear brakes would contribute very little to stopping the vehicle.

The solution to the weight shift problem is to split the hydraulic system diagonally rather than front to rear. With a split diagonal system, 50% of the braking will still be available in the event of a hydraulic failure in one-half of the system. The circuit that includes the left front and right rear brakes comprises one-half of the brake hydraulic system. The other circuit includes the right front and left rear brakes.

Front suspension geometry helps negate the brakes' tendency to pull to one side or the other in the event of a partial hydraulic system failure.

■ QUICK TAKE-UP MASTER CYLINDER

To increase fuel economy, some disc brake calipers are designed to cause less drag when the brakes are not applied. They have specially designed piston O-ring seals and grooves that cause the piston and pads to retract farther from the rotor than normal. Retracting the piston farther means more fluid is needed in order to take up the clearance before pressure can build when the pedal is applied.

One master cylinder design moves a larger amount of fluid when the pedal is first applied. These master cylinders are called *quick take-up, step-bore, fast-fill,* or *dual diameter bore* master cylinders. The bore has two different diameters (**Figure 57.25**). The rear part of the primary piston is larger in diameter than the front part. A larger bore in the master cylinder would normally result in very low braking force at the wheel cylinders. Remember, smaller master cylinder bores create more braking force but move less fluid.

The larger part of the bore, called the low-pressure chamber, allows its piston to move a large volume of fluid more quickly when the pedal is first applied. When the friction materials contact the braking surfaces, the smaller part of the bore comes into play to give a large pressure boost during stopping. The result is less pedal travel.

■ DRUM BRAKES

In the past, almost all vehicles were equipped with drum brakes. Today, drum brakes are found in some rear brake applications. Disc brakes are used mostly in the front. Drum brakes provide good initial stopping ability and also provide a means for a good, inexpensive, mechanical parking brake.

Several different drum brake designs have been used on motor vehicles during their long history, with two of them still used extensively today. The choice of design used depends on the application. Some brakes work better with the heavier loads of rear-wheel-drive vehicles. Others are used primarily on front-wheel-drive cars, which have less weight in the back of the car. The lack of load on the rear wheels creates more of a tendency for rear wheel lockup during a stop.

Dual-Servo Drum Brake

During stopping, a drum brake's leading shoe digs into the brake drum. This is called **self-energizing**. Originally called Bendix brakes, brakes of this design are also called *dual-servo* or *duo-servo* brakes because *both* brake shoes are self-energizing. They use **servo action**, which is when a small force is applied to make a larger force. With no anchor at the bottom, between the front and rear shoes, the primary (front) shoe floats during braking and applies additional pressure against the secondary (rear) shoe (**Figure 57.26**). When stopping in reverse, the secondary shoe is energized and pushes on the primary shoe.

The primary shoe acting on the secondary shoe causes the rear shoe to do a higher percentage of the braking. If you compare the linings on primary and secondary shoes, you will find that the front lining has a shorter length of friction material than the rear (**Figure 57.27**). This is so they will wear at similar

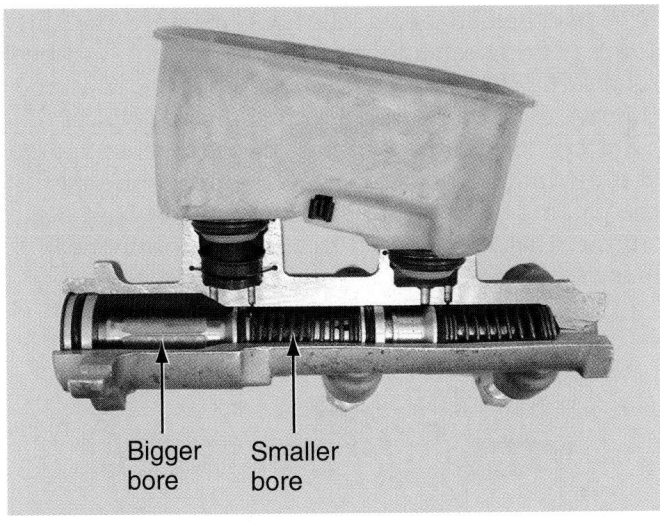

Figure 57.25 A quick take-up master cylinder has a bulge in the rear of the casting. (*Courtesy of Tim Gilles*)

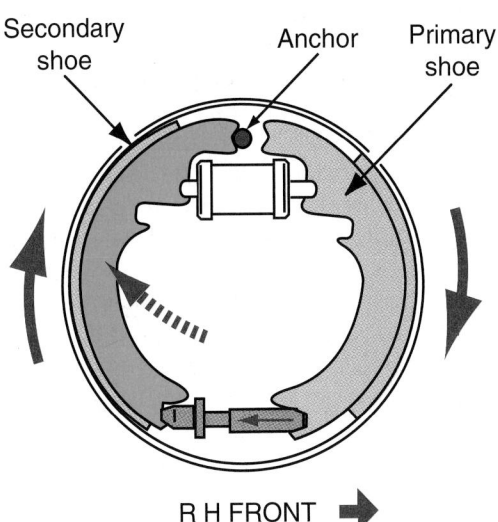

R H FRONT ➡

Figure 57.26 In a self-energizing brake, the energized front shoe pushes against the rear shoe.

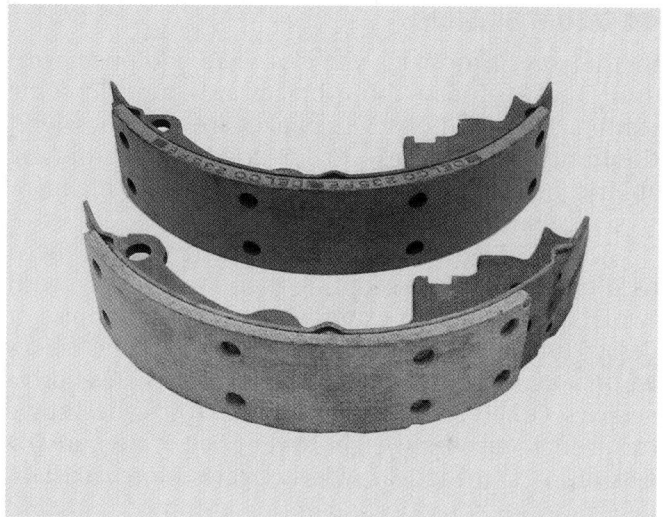

Figure 57.27 A dual-servo brake's front lining has a shorter length of friction material than its rear lining. (*Courtesy of Tim Gilles*)

rates. Sometimes different friction materials are used on primary and secondary linings as well.

Leading-Trailing Drum Brake

Some vehicles use a *leading-trailing* brake (**Figure 57.28**), also called a *simplex* brake. A leading-trailing brake is a *non-servo* brake with an anchor at the bottom end of each shoe. The *leading (front)* shoe is self-energized, while the *trailing* shoe is not. Because the front shoe is self-energized, it wears more than the rear.

Weight distribution in front-wheel-drive cars can be 80% or more toward the front. Leading-trailing brakes are used because they are less effective in stopping than servo-type brakes. Front-wheel-drive cars with split diagonal hydraulic systems and light-duty trucks without ABS use leading-trailing rear brakes because they minimize the tendency for the relatively underweighted rear wheels to lockup. On older vehicles without antilock brakes, this usually eliminates the need for a proportioning valve.

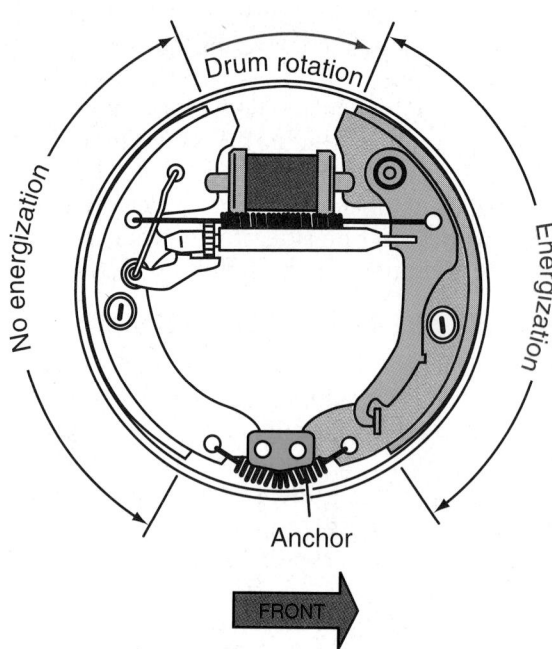

Figure 57.28 In a leading-trailing brake, or simplex, only the front lining is energized.

■ DRUM BRAKE ADJUSTMENT

Disc brakes automatically adjust themselves, but drum brakes require periodic adjustment. As drum brake linings wear, increased clearance between the lining and the drum results in more brake pedal travel before pressure can build and the brakes start to apply. The disc brakes in the front and the drum brakes in the rear are connected hydraulically in the master cylinder. The tandem master cylinder design prevents the front brakes from applying before the rear brakes contact the brake drum. When drum brakes need to be adjusted, pedal travel drops for the entire system.

A typical drum brake adjuster has a threaded shaft attached to an integral starwheel. Drum brakes have had self-adjusting mechanisms since the early 1960s. During self-adjustment, the integral starwheel rotates the threaded shaft, taking up excess clearance (**Figure 57.29**). **Figure 57.30** compares typical self-adjusters for leading-trailing and duo-servo brakes. Leading-trailing brake starwheel adjusters are part of the parking brake strut. Dual-servo brakes have a star adjuster that fits between the bottoms of the floating brake shoes.

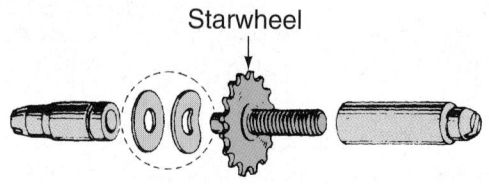

Figure 57.29 A starwheel is turned to adjust the brake clearance. (*Courtesy of Federal-Mogul Corporation*)

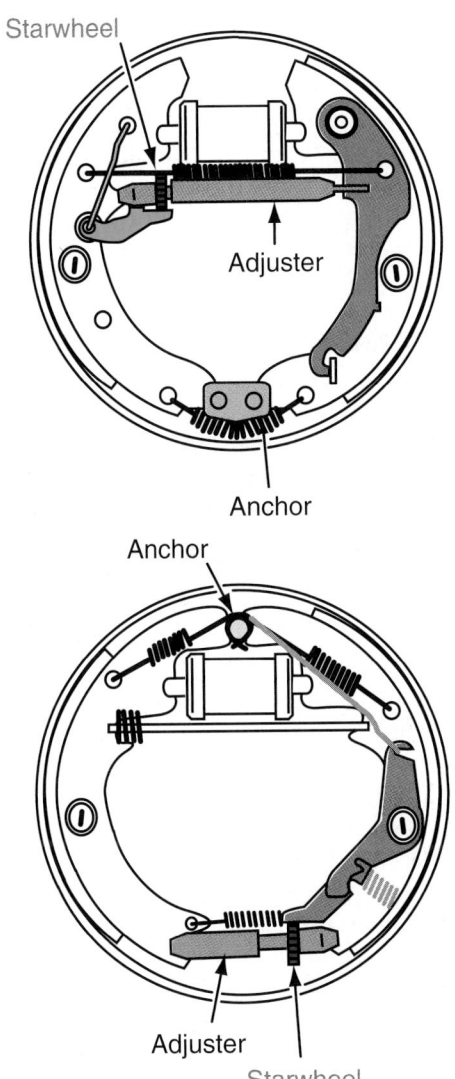

Figure 57.30 Typical starwheel adjuster locations for leading-trailing and dual-servo brakes.

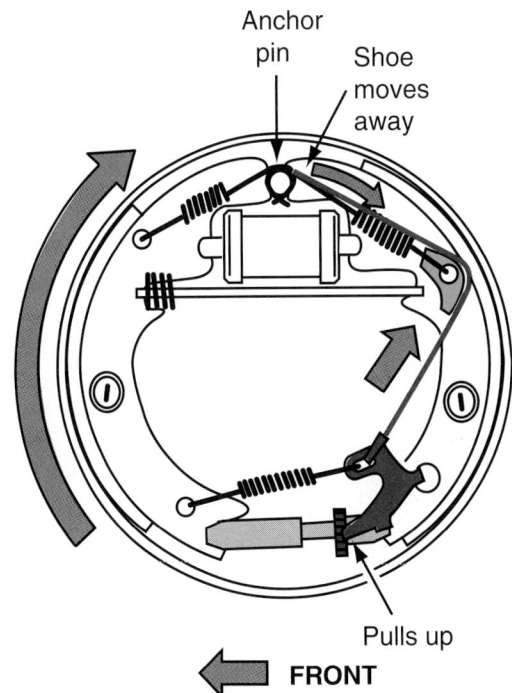

Figure 57.31 When braking in reverse, the top of the shoe moves away from the anchor when there is clearance between the shoe and the drum.

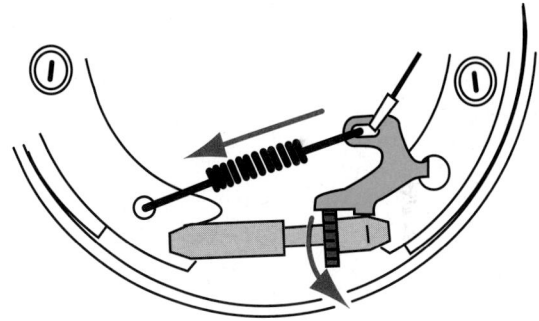

Figure 57.32 When the brake is released, the spring pulls the adjusting lever, turning the starwheel.

Dual-servo self-adjusters operate only when the brakes are applied during a stop when backing up. Wear to the friction material causes excessive lining-to-drum clearance. When this happens, the top of the rear brake shoe moves further away from the anchor pin when stopping. In one common type of cable-type self-adjusting system, the adjusting cable lifts the adjusting lever against its spring (**Figure 57.31**). When the brake is released, the spring pulls the adjusting lever back down. This turns the starwheel to adjust the brakes (**Figure 57.32**).

Many dual-servo self-adjusters have the adjuster lever located above the starwheel teeth. Some self-adjusters, however, have the lever positioned below the starwheel (**Figure 57.33**). Positioning the adjusting lever below the starwheel results in a more positive adjustment.

There are several leading-trailing brake self-adjuster designs that operate as part of the parking brake strut assembly that fits between the two brake shoes. The

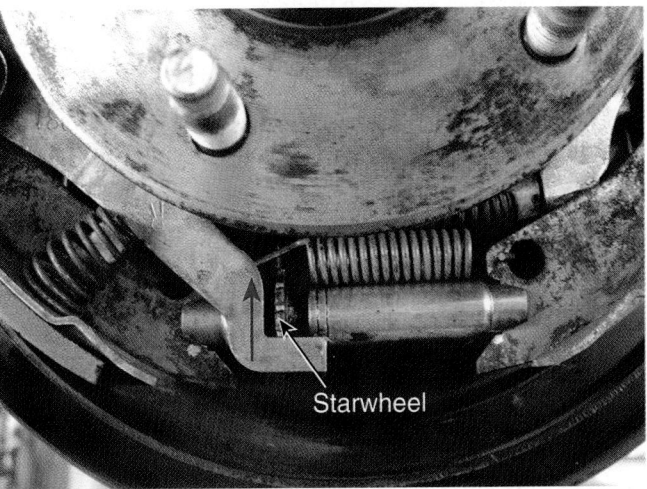

Figure 57.33 This adjusting lever pulls up against the starwheel to adjust the brakes. (*Courtesy of Tim Gilles*)

most common type is called an *incremental adjuster* because during a self-adjustment its starwheel moves only one tooth at a time. The starwheel is part of the parking brake strut and operates in either forward or reverse, whenever too much clearance develops between the lining and drum.

An *expanding strut* self-adjuster also works off the parking brake rod and strut assembly (**Figure 57.34**). If there is too much lining-to-drum clearance when the parking brake is applied, the strut and telescoping rod assembly expands to readjust the clearance when the parking brake is released.

Another parking brake strut adjuster has a rotating ratchet that repositions in response to increasing clearance. Other self-adjusters include ones that use movement of a cam to adjust clearance.

Brake Fade

Brake fade results from excessive brake heat. One drawback to drum brakes is that they do not dissipate heat as well as disc brakes. Disc brakes are exposed to and cooled by abundant air. In drum brakes, heat is trapped within the drum. With increased heat, the drum expands, causing the pedal to move closer to the floor before the brakes are applied. The coefficient of friction of the lining also drops off (**Figure 57.35**). This means that more effort is required to stop the car. As fade becomes worse with the addition of more heat, stopping ability will drop off and may disappear altogether. Brake drums often have fins to help dissipate the heat.

▮▮ DISC BRAKES

Many American cars since the mid-1960s have been equipped with front disc brakes. Today, many cars have four-wheel disc brake systems. A disc brake system has a *rotor* (disc) and a *caliper* that clamps friction pads against the rotor like a bicycle brake caliper squeezes pads against the wheel rim (**Figure 57.36**). **Figure 57.37** shows the parts of a typical disc brake system.

Brake Rotors

Disc brake rotors are either *solid* or *ventilated*. Disc brake parts represent **unsprung weight** (weight that is not supported by the springs) for good vehicle handling. It is desirable for unsprung weight to be kept as low as possible. Lightweight solid rotors are used in lighter cars. Ventilated rotors are heavier but have abundant surface area, so they can dissipate the extra heat generated by the brakes of heavier vehicles.

Brake Calipers

There are two types of disc brake calipers: the *fixed caliper* and the *floating (sliding) caliper* (**Figure 57.38**). Fixed calipers have pistons on both sides of the caliper. A floating caliper has one or two pistons on only one side. The caliper slides as the piston moves out of its bore (**Figure 57.39**). This results in equal force being

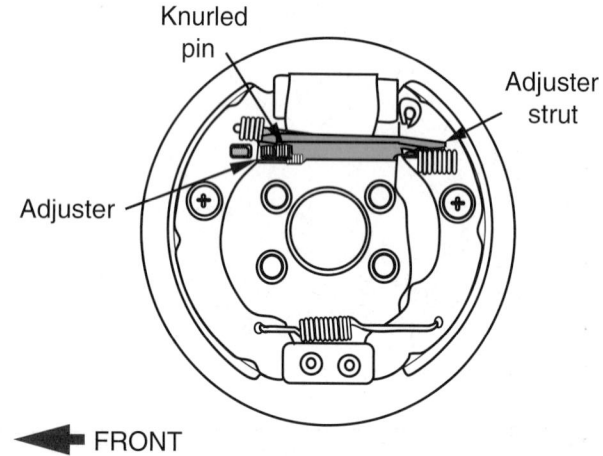

Figure 57.34 An expanding strut self-adjuster is part of the parking brake strut assembly.

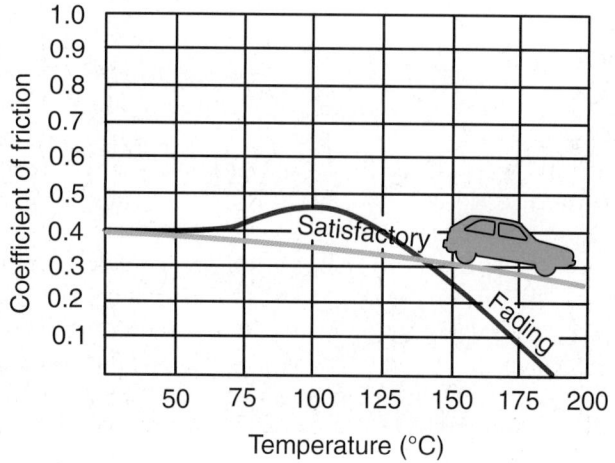

Figure 57.35 Coefficient of friction drops off as brakes fade.

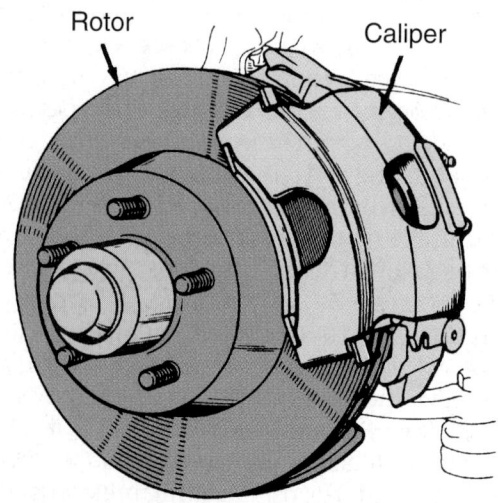

Figure 57.36 A disc brake rotor and caliper resembles a bicycle brake.

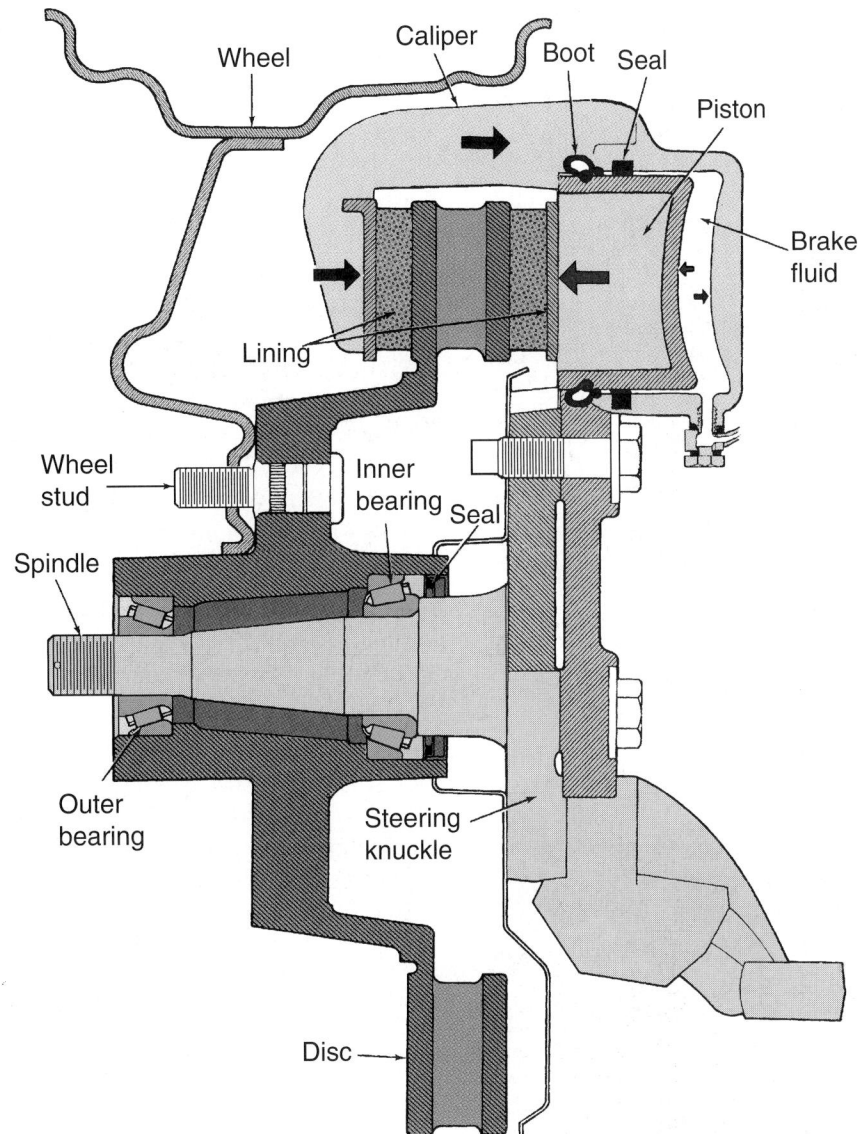

Figure 57.37 A typical disc brake system shown here with the wheel bearing assembly. (*Courtesy of DaimlerChrysler Corporation*)

applied to linings on both sides of the caliper from the hydraulic action on only one side.

One of the drawbacks to floating caliper disc brakes is that a power assist is usually required. Early American disc brakes used the fixed caliper design because more piston area could be used, so a power brake would not be necessary. Most later designs use a floating caliper, which has fewer parts and is easy to repair.

A rubber, square-cut O-ring seals the disc brake piston to its bore. Disc brakes do not require a return spring like drum brakes do. When the brakes are applied, the seal distorts (**Figure 57.40**). When the brakes are released, the seal retracts to its original position, pulling the piston back and allowing the linings to release the rotor.

One advantage to disc brakes is that they are *self-adjusting*. As linings wear, the piston slides forward in its bore to take up the slack (**Figure 57.41**). As this occurs, the caliper requires more fluid. On a master cylinder with reservoirs of different sizes, the larger one is for the disc brakes and the smaller one is for the drum brakes.

Different floating caliper designs include the pin slider, the center abutment, and the swing caliper. The majority of cars use a pin slider caliper. A pin slider caliper slides on guide pins during and after brake application (**Figure 57.42**). The guide pins allow the caliper to twist slightly. This allows for minor changes in pad alignment but still allows the caliper to return to its original position after the brakes are released.

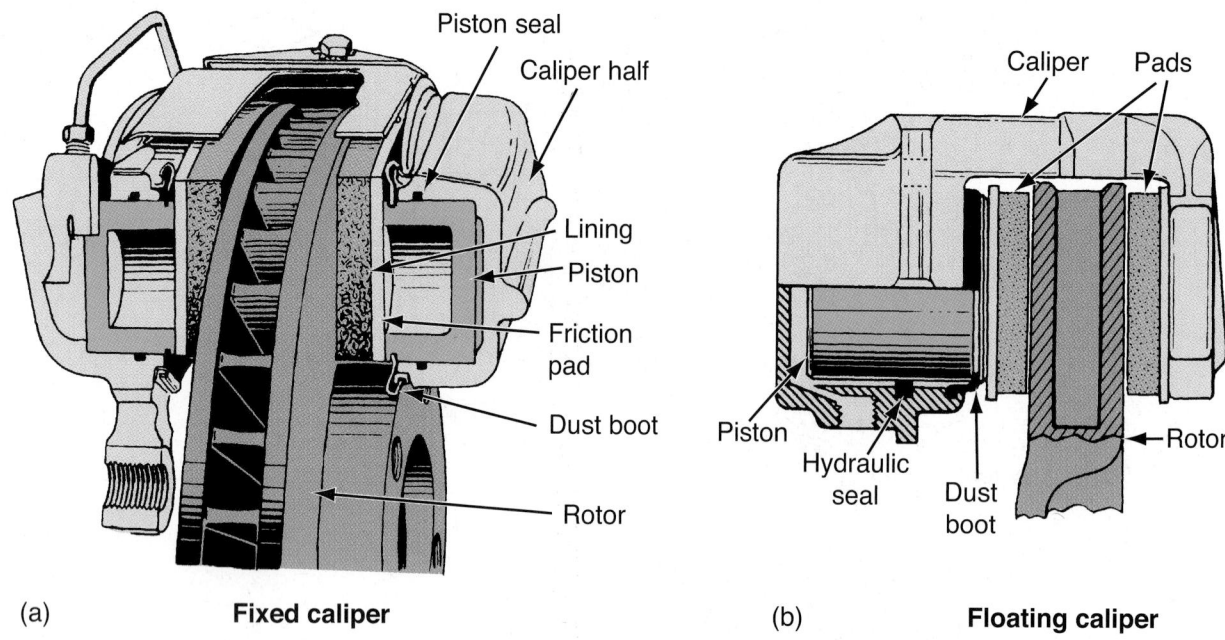

Figure 57.38 (a) A fixed caliper. (b) A floating caliper. (*Courtesy of Federal-Mogul Corporation*)

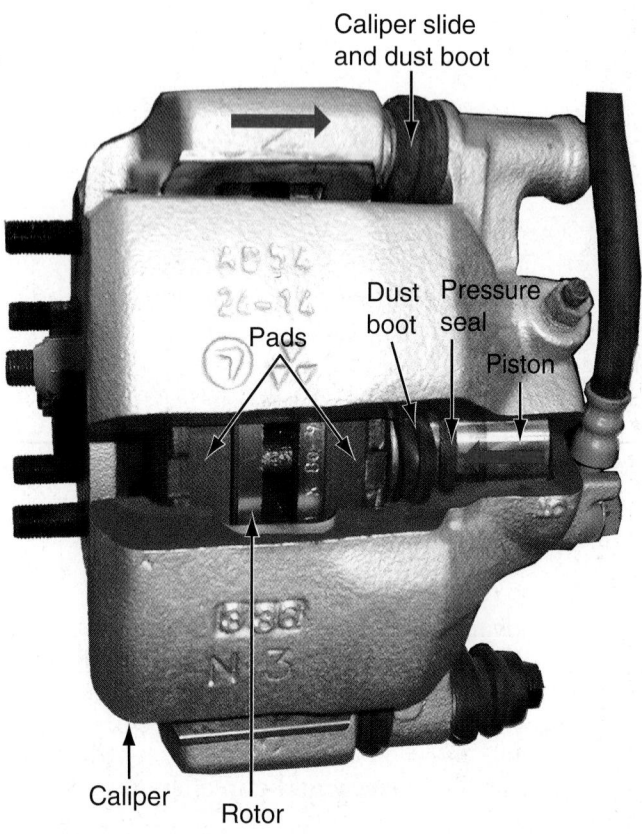

Figure 57.39 A floating caliper moves as the brakes are applied. (*Courtesy of Tim Gilles*)

Repeated sliding would result in excessive wear to the guide pins or the holes in the caliper, but the bushings or O-rings (also called insulators or sleeves), which are made of rubber or Teflon, prevent metal-to-metal contact between the guide pins and the caliper.

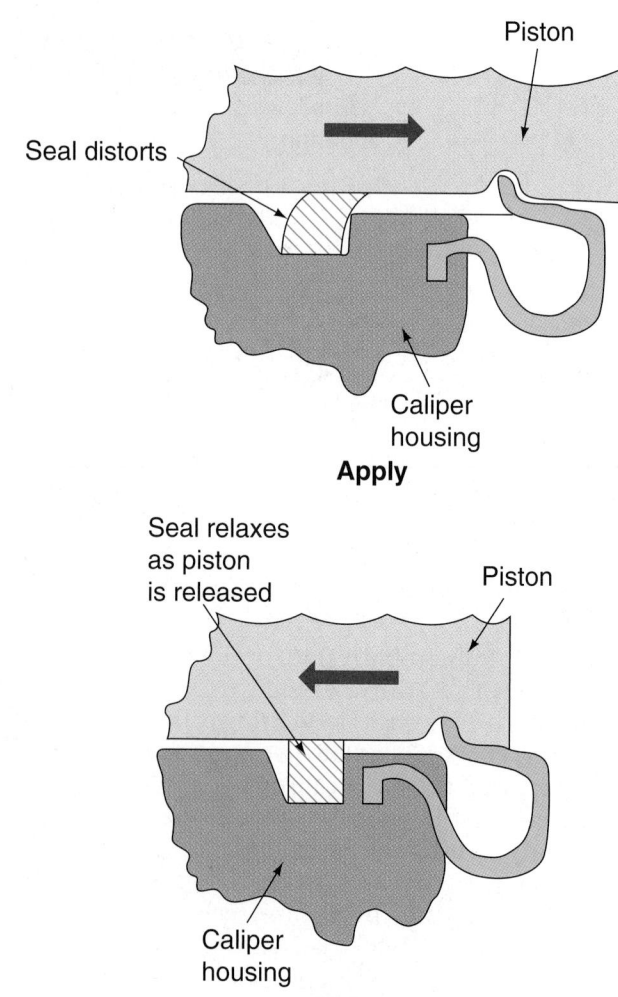

Figure 57.40 When the brakes are applied, the seal distorts. When the brakes are released, the seal retracts, pulling back the piston.

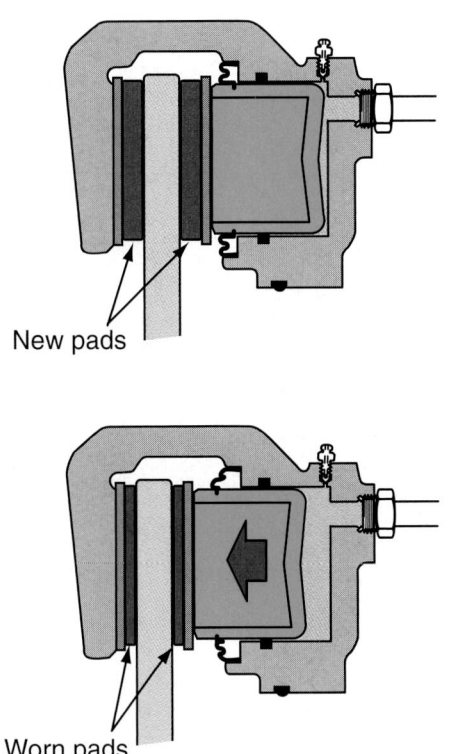

Figure 57.41 As the disc linings wear, the piston moves out to self-adjust.

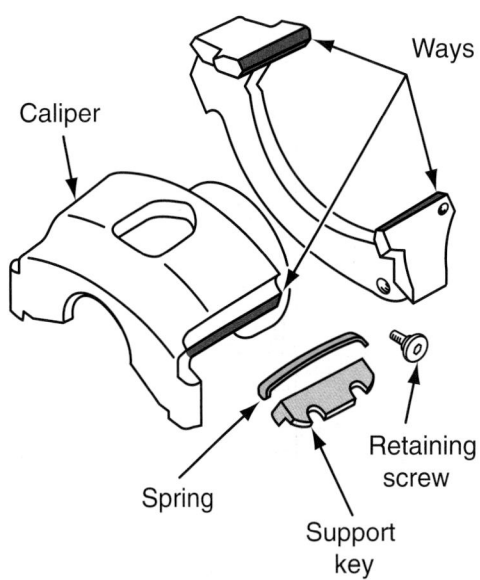

Figure 57.43 A center abutment caliper slides between two wedged sliding surfaces called "ways".

Some, mostly older, American vehicles use a sliding caliper design called a *center abutment caliper*. It has two wedged sliding surfaces called "ways" (**Figure 57.43**). The caliper slides between the ways. A small amount of clearance between the abutment and ways is normal because it allows the caliper to slide. Excessive clearance would result in noise as the caliper rocked against the ways when the brakes were applied.

NOTE: *"Ways" is a term from the machine tool industry describing the guides that a machine tool slides against when it moves on its mount.*

Some import cars use a caliper called a *swing caliper*. The caliper is attached to the spindle support arm by a pivot pin. As the brakes are applied and released, the caliper pivots back and forth on the pin. Unlike the other floating caliper designs, the piston does not remain parallel to the rotor, so the linings on these cars are wedge shaped.

Caliper Pistons

Caliper pistons are hollow and shaped like a cup. They are installed with the open side against the back of the friction pad. This provides very little surface area for heat to be able to transfer back into the brake fluid. Most pistons are made of steel, some of which are chrome plated. Fiberglass-reinforced phenolic resin pistons are also used in some calipers. Plastic pistons

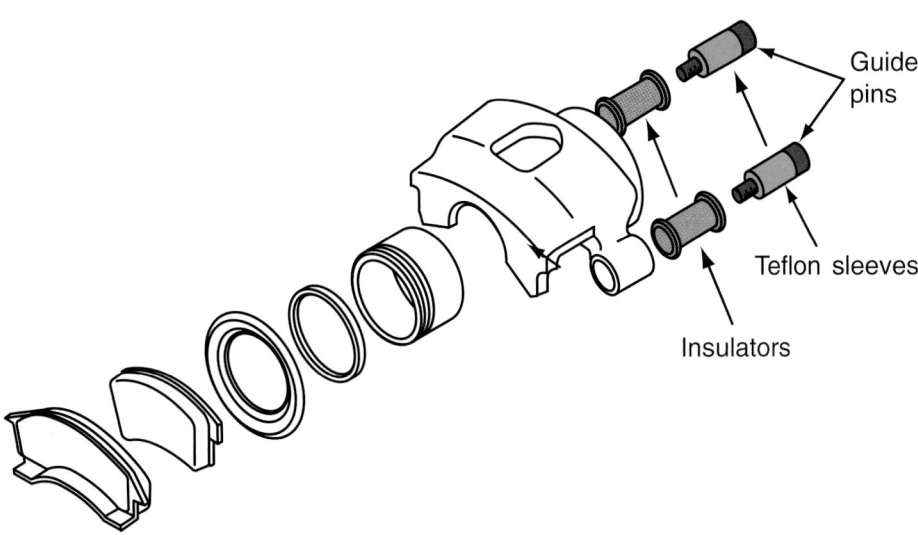

Figure 57.42 A pin slider caliper slides on guide pins that allow it to twist slightly.

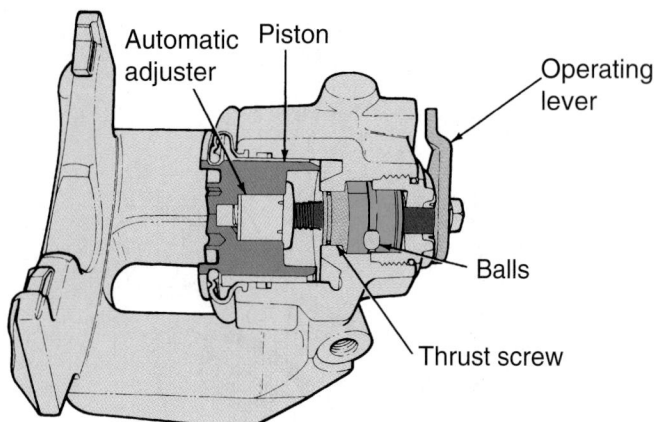

Figure 57.44 A rear disc brake caliper with a self-adjuster operated by the parking brake. (*Courtesy of Ford Motor Company*)

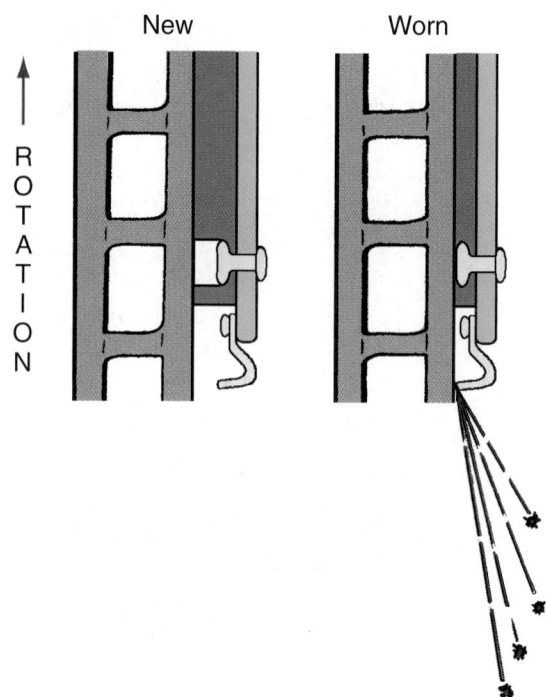

Figure 57.45 When the lining is worn too much, the metal warning sensor contacts the brake rotor.

like these are lightweight, do not conduct heat readily, and do not corrode like metal ones do.

Rear Disc Brakes

Four-wheel disc brakes, used on most luxury and performance vehicles since the end of the 1970s, are more costly than hybrid disc/drum brakes. Rear disc brake systems can have either fixed or floating calipers. They are typically the same as front calipers, although the diameter of the rear pistons is smaller and they have some sort of arrangement for a parking brake.

When a rear disc brake includes a mechanism for the parking brake, there will be an automatic adjuster (**Figure 57.44**). Operating the parking brake causes the piston to thread outward to adjust the clearance.

Disc Linings

Disc brake linings are fastened to a metal back. Some have tabs on the back of the pad that need to be bent during installation to hold it snugly to the caliper. If the pad vibrates, there will be noise.

Some disc brake linings include a *wear sensor* (**Figure 57.45**), which is a small metal tab that rubs against the rotor when the lining wears thin. The resulting noise alerts the driver that brake work is needed before serious damage to the rotor occurs. If installed on the wrong side, rotor damage may result. The wear sensor should pull away from the rotor as it turns.

■ HYDRAULIC SYSTEM VALVES AND SWITCHES

Valves and switches are found in several places in the brake hydraulic system.

Pressure Differential Switch/Brake Warning Light

Tandem systems are equipped with a brake warning light that alerts the driver when half of the hydraulic system has failed. When pressure on one side of the

system drops, a piston inside the *hydraulic safety switch* (**Figure 57.46**) moves off-center. This completes an electrical circuit, which illuminates a warning lamp on the driver's compartment instrument panel. On some cars, the same lamp illuminates when the parking brake is applied. This system also illuminates the brake warning lamp. Some vehicles have wear sensors in the disc brake pads that also turn on the same lamp when the pads wear beyond their limit.

Master Cylinder Fluid Level Switch

Instead of a hydraulic safety switch, some master cylinders have a fluid level switch located in the master cylinder reservoir. There are several designs. A typical fluid level switch has a float (**Figure 57.47**) with contacts above it. When the float level drops beyond the allowable limit, the contacts close and complete the circuit to the warning light. This system can replace the pressure differential switch because a leak in the system will be evident by a drop in master cylinder fluid level.

Metering and Proportioning Valves

Additional hydraulic control valves were required on older vehicles with hybrid (disc/drum) brake systems. Called **metering** and **proportioning valves**, they were required to balance braking force between the front and rear brakes (**Figure 57.48**).

Metering Valve. A metering valve (**Figure 57.49**) is used on front disc brakes when the car has rear drum

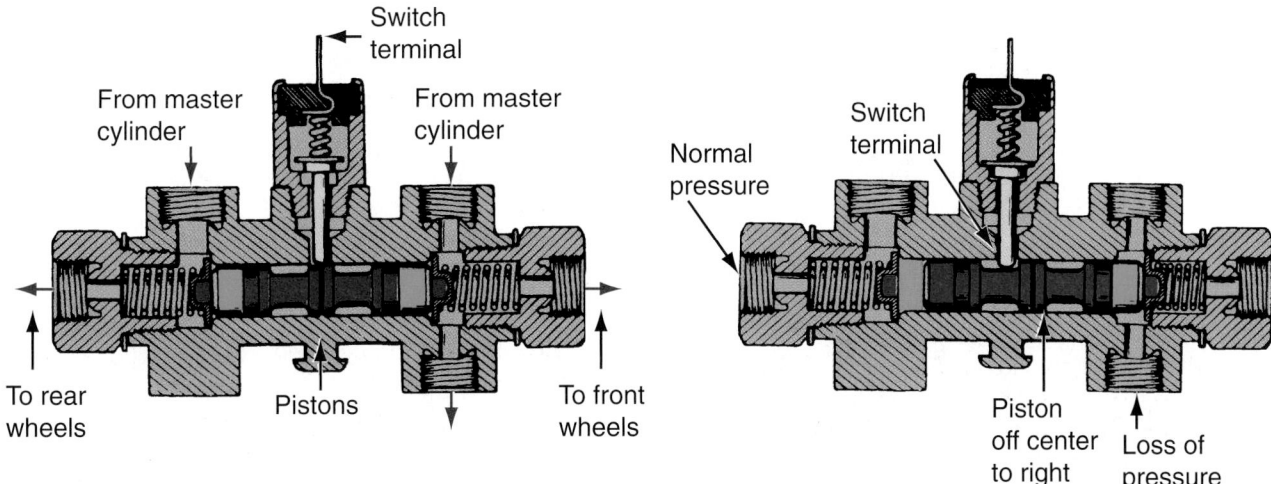

Figure 57.46 During a hydraulic system failure, the piston moves off-center to complete an electrical circuit to the dash light.

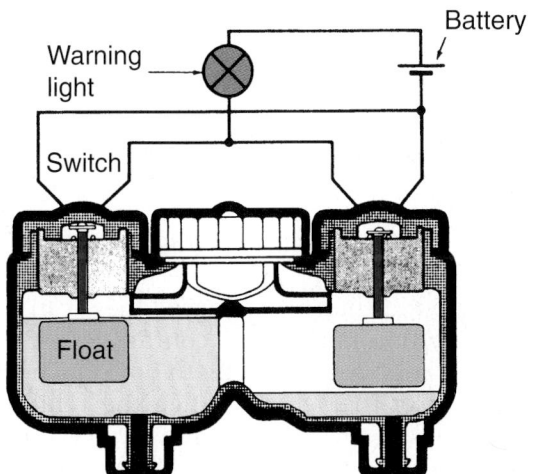

Figure 57.47 Some master cylinders use a float switch instead of a hydraulic switch.

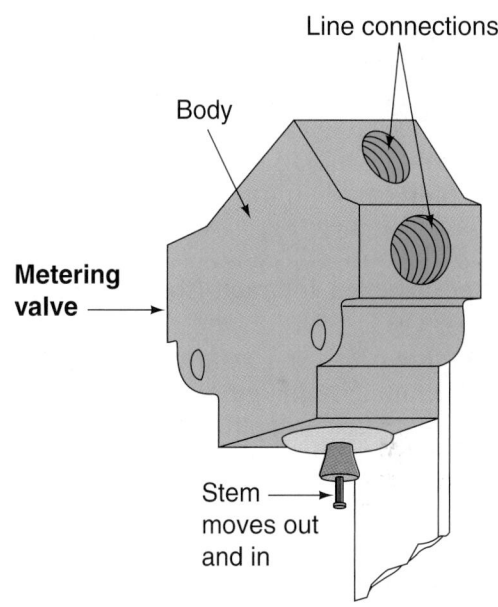

Figure 57.49 A metering valve keeps the front disc brakes from operating until rear drum linings are in contact with the brake drum. (*Courtesy of DaimlerChrysler Corporation*)

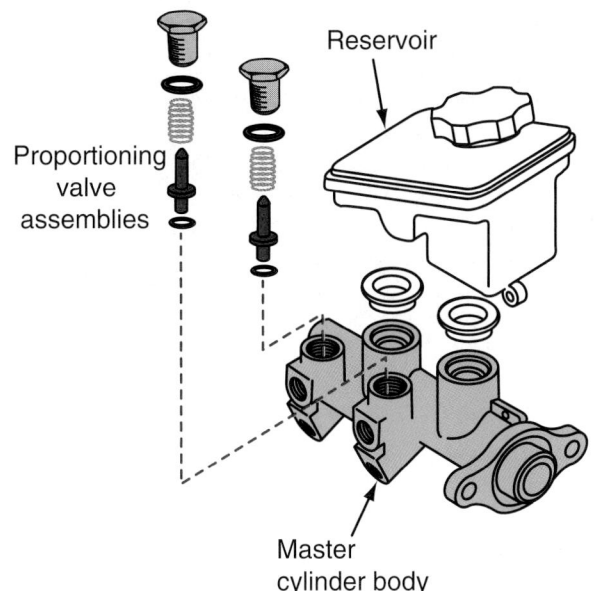

Figure 57.48 These proportioning valves are built into the master cylinder.

brakes. Its purpose is to prevent the front brakes from applying until the rear drum brake shoes overcome return spring pressure and contact the drums. The metering valve does not operate until system pressure reaches about 10 psi. Keeping the valve open below this pressure allows the fluid to expand and contract normally when the brakes are not applied. Once pressure builds to 10 psi, the metering valve shuts off pressure to the front brakes until the rear brake hydraulic system has developed pressure of 75 to 300 psi. The amount of pressure depends on the vehicle. When the metering valve operates, the stem can be seen moving out, away from the body of the valve.

The movement of the valve can be felt very slightly at the brake pedal during very light pedal application.

Preventing the front brakes from applying until the rear brakes start to build pressure serves two purposes:

■ It keeps front pads from doing too much of the light braking, preventing them from wearing out too soon.

■ It helps prevent dangerous skids that could result on slick surfaces if the front brakes were to apply before the rears.

Without a metering valve, the rear brake cylinders would have to overcome the drum brake return spring pressure before the rear brakes could begin to operate. Front disc brakes do not have return springs, so they would be able to start to apply before the rears, even though both sides of the system had equal pressures. With four-wheel disc brakes, a metering valve is unnecessary. Also, front-wheel-drive vehicles with split diagonal systems do not use a metering valve. With front-wheel drive, the front brakes do the vast majority of the braking, so you want them to apply as soon as possible to overcome the torque of the front driving wheels. Because the weight is forward on these cars, the problem of front wheel lockup is greatly reduced and the metering valve is unnecessary. This is fortunate because a separate metering valve would need to be installed for each front wheel in a split diagonal system.

Proportioning Valve. The ability of the brakes to do their job is limited by the grip of the tires to the road surface. During hard stopping, weight shifts forward and away from the rear wheels. The proportioning valve was introduced in the late 1960s to help prevent the rear wheels from locking up during a panic stop or when brakes are applied hard.

Disc brakes require greater force to apply than drum brakes. Remember that dual-servo drum brakes are self-energized, which causes rear wheel lockup to be a problem on some vehicles. Drum and disc hybrid systems are not the only ones to use a proportioning valve. Four-wheel-disc-equipped vehicles also use them.

During a low-pressure stop, the proportioning valve does nothing, but when pressure reaches a predetermined level the proportioning valve prevents a further increase in pressure to the rear brakes. The pressure at which this occurs is called the *split point* because greater pressure will now be applied to the front wheels while the pressure going to the rear wheels remains the same (**Figure 57.50**). As master cylinder pressure continues to increase, the proportioning valve opens once again. It will cycle open and closed, allowing pressure to the rear brakes to increase but not as much as pressure to the front brakes. When the brakes are released, a spring opens the valve and fluid returns to the master cylinder.

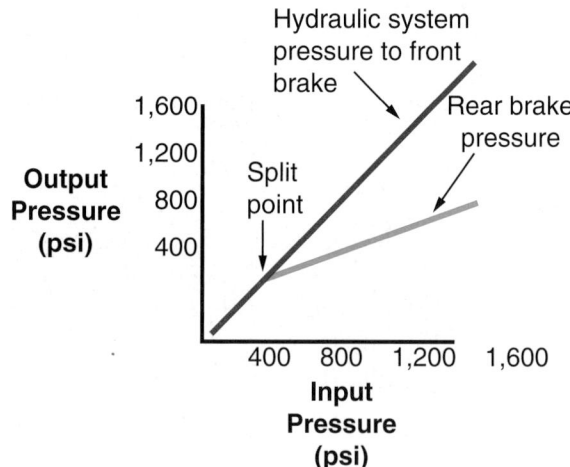

Figure 57.50 The split point happens when pressure reaches a predetermined level and the proportioning valve prevents a further increase in pressure.

Early systems had a single proportioning valve in the line to the rear brakes. With split diagonal systems, two proportioning valves are required. A dual valve can be located near the master cylinder, or the valves can be built into the master cylinder or threaded into the master cylinder outlet ports (**Figure 57.51**).

Some proportioning valves are load sensitive. As chassis-to-ground height changes, they can either bypass or function normally to control stopping pressure to the rear brakes according to how much weight is on the rear of the vehicle. When the vehicle does not have a load, the proportioning valve functions normally. Height-sensing proportioning valves were originally used on pickup trucks, but now they are often found on front-wheel-drive passenger cars, too. A four-door front-wheel-drive car loaded with a family

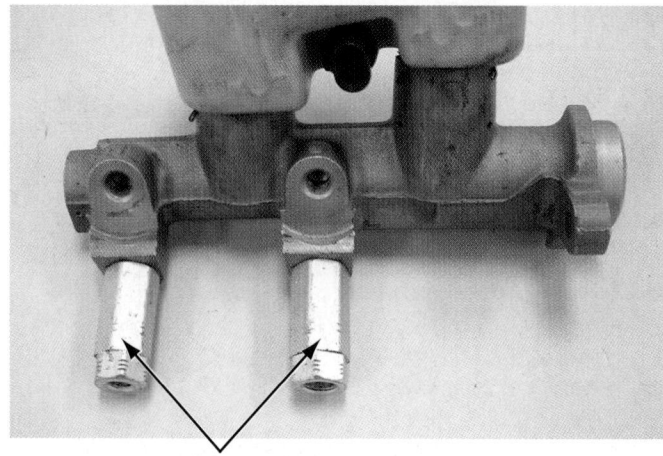

Proportioning valves

Figure 57.51 These proportioning valves are installed at the master cylinder outlets. (*Courtesy of Tim Gilles*)

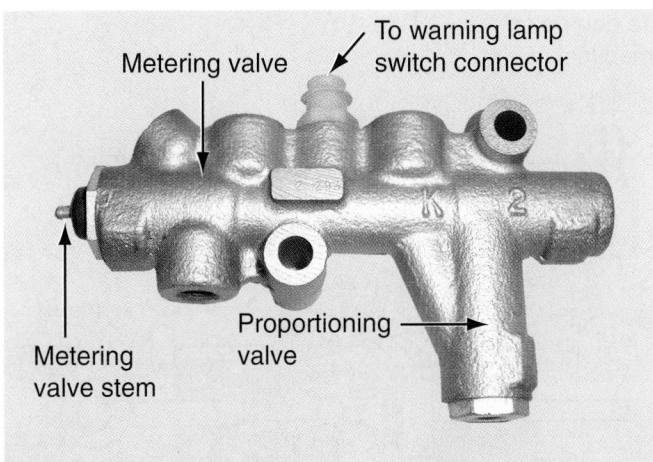

Figure 57.52 A three-function combination valve with a metering and proportioning valve and a safety switch. *(Courtesy of Tim Gilles)*

and luggage can have a considerable change in rear braking requirements.

ABS Brake Pressure Control. Most newer cars are equipped with antilock brakes as standard equipment. Sensors at the wheels provide input to a computer, which monitors for unacceptable differences in speed between the wheels and modulates the brakes to compensate. ABS computer control is more accurate than earlier mechanical devices like the metering and proportioning valves. With ABS, they are no longer needed. Earlier, mechanical proportioning valves were called *fixed* proportioning valves because the pressures at which they would activate were always constant. *Dynamic* proportioning is the term used in newer ABS that modulate rear brake pressures based on the speed of the wheels at the front and rear axles.

Combination Valve. Since the 1970s, metering and proportioning valves have usually been combined with a hydraulic safety switch in one valve called a **combination valve** (**Figure 57.52**). The three-function combination valve is the most common, but there can also be just a metering and proportioning valve or a safety switch and metering valve.

■ POWER BRAKES

Power brakes reduce brake pedal effort but still allow a reasonable feedback feel to the driver during braking. Disc brakes have been common on the fronts of vehicles since the 1970s. Disc brakes have the ability to apply more stopping energy at the wheels, but they also require more application effort and do not self-energize like drum brakes. Gripping the smooth, shiny rotor between two relatively smooth brake pads requires a substantial increase in braking effort. As a result, power brakes have been installed on most vehicles as original equipment since the late 1970s.

For modern brake systems, the most effective and economical means of boosting brake pressure is to use a vacuum-assist brake device called a *booster*, which is installed behind the master cylinder. A brake booster often allows the master cylinder to have a larger bore, allowing the brakes to apply with less pedal travel. A larger bore master cylinder will exert less pressure on the brake fluid but will move a larger volume of fluid in the same amount of pedal movement.

The type of power brake used on most vehicles is mounted behind the master cylinder (**Figure 57.53**). It uses the difference between atmospheric pressure and vacuum to apply force (**Figure 57.54**) from a brake booster powered by vacuum supplied from the intake manifold of the engine. The booster and master cylinder are mounted on the **bulkhead** (also

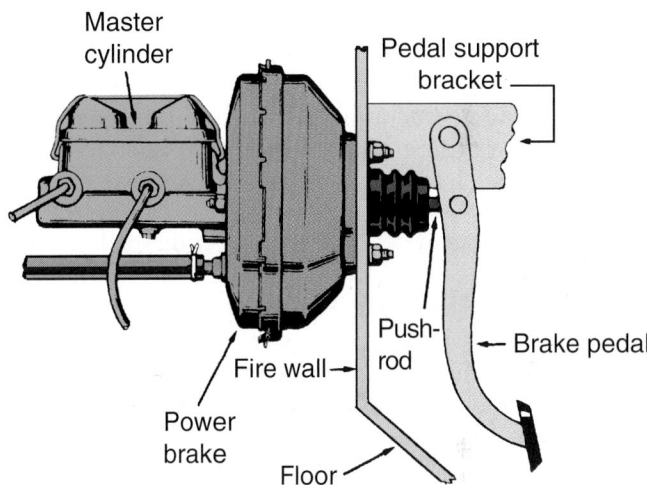

Figure 57.53 A typical power brake combined with a master cylinder.

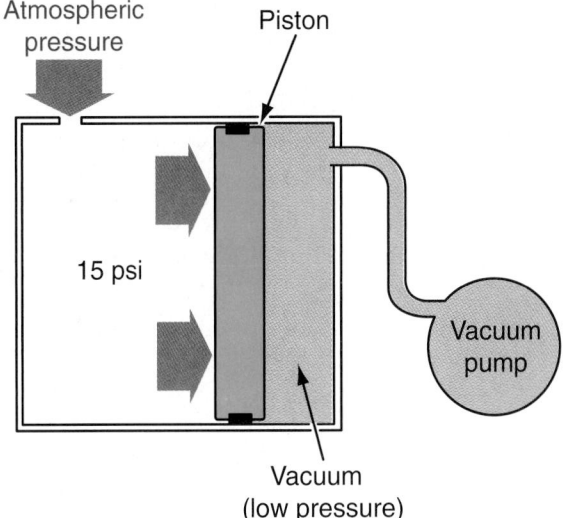

Figure 57.54 The difference between atmospheric pressure and vacuum can be used to apply force.

called the **fire wall**), which is the metal wall between the engine and passenger compartments. The brake booster provides an increase in the braking effort supplied by the driver. In the event of a power brake failure, the service brakes will still operate.

Brake Booster Check Valve

Most power brake systems depend on engine vacuum, but in the event of engine stall, the power brake will still work. Federal standards call for at least one power-assisted stop if engine power is lost. Pedal effort will increase substantially if power is lost, but the first stop will be normal. The second stop will have less assist than the first stop; the third stop will have less assist than the second stop, and so on until there is no further assist.

The booster is able to provide reserve braking because it has a check valve that is typically mounted on the front of the brake booster at the end of the hose that supplies vacuum from the intake manifold (**Figure 57.55**). The check valve is a one-way valve that prevents manifold vacuum from leaving the power booster if engine vacuum drops below the tension of its spring or if the engine is shut off. When vacuum drops, the check valve moves against its seat to keep vacuum in the booster and prevent air from entering the reservoir.

Vacuum Booster Operation

The most common kind of power brake booster in use today is called the vacuum-suspended power brake. The booster is a metal chamber divided in half by a rubber diaphragm. Each side is isolated from the other by a valve at the end of the brake pedal apply rod in the back of the power booster. When the brakes are released while the engine is running, vacuum is present on both sides of a diaphragm inside the booster (**Figure 57.56**). Valves that are part of the pedal pushrod assembly control air movement in the power booster.

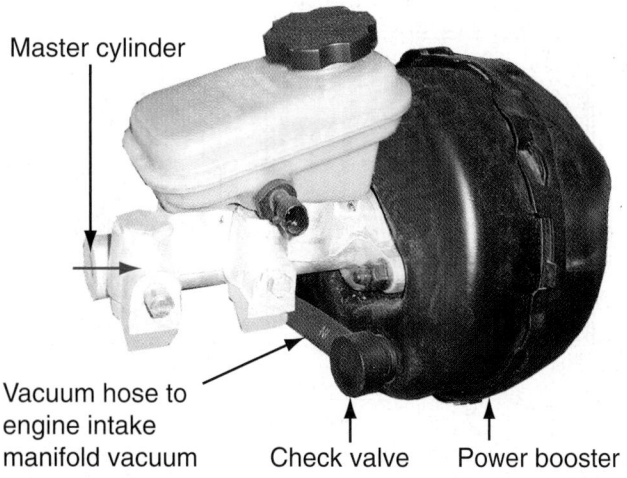

Figure 57.55 A check valve located on the vacuum booster. *(Courtesy of tim Gilles)*

Master cylinder

Vacuum hose to engine intake manifold vacuum   Check valve   Power booster

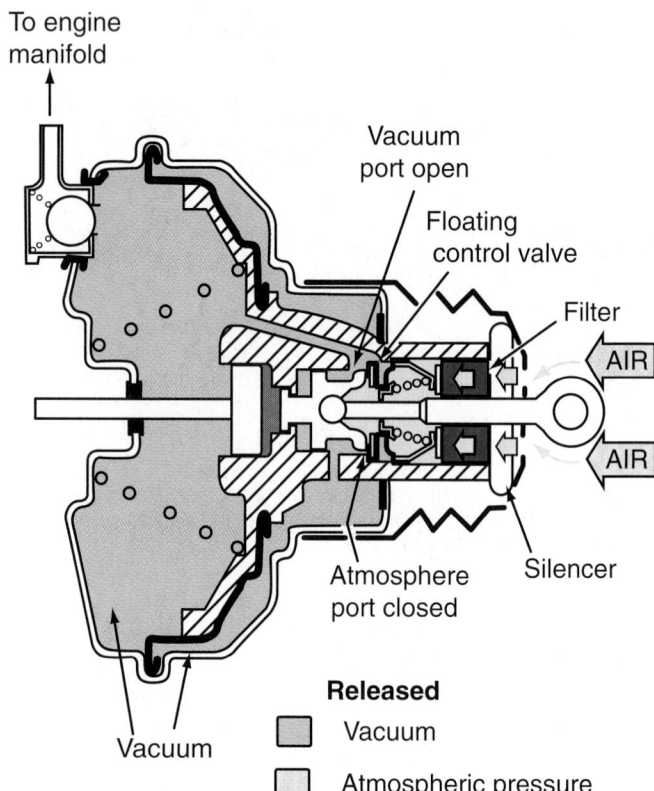

To engine manifold

Vacuum port open

Floating control valve

Filter

AIR

AIR

Silencer

Atmosphere port closed

Vacuum

Released
Vacuum
Atmospheric pressure

Figure 57.56 A power brake in the released position.

During a stop, the vacuum valve is closed and the air valve is opened in response to pedal movement. This uncovers a passageway, allowing air to be drawn from the passenger compartment into the rear chamber of the booster (**Figure 57.57**). Another passageway is blocked at the same time, preventing the incoming air from entering the front chamber of the power booster. The result is a difference in pressure between the two chambers. The pressure is low in the front chamber compared to atmospheric pressure in the rear chamber.

With a pressure differential between the two chambers, there is a forward assist against the master cylinder apply piston. The amount of assist depends on the difference between the pressures in the front and rear chambers of the brake booster. When the pedal is lightly held, a spring-loaded valve prevents more pedal assist from occurring. During holding, the vacuum and atmospheric valves are both closed.

The term "vacuum suspended" comes from the fact that vacuum is present on both sides of the diaphragm when no braking is taking place. During a stop, atmospheric pressure is directed to the back side of the diaphragm (**Figure 57.58**), resulting in an increase in braking force.

Power Booster Air Filter. During application of the brakes, air enters the power booster through a filter behind the pedal pushrod boot (see Figure 57.58).

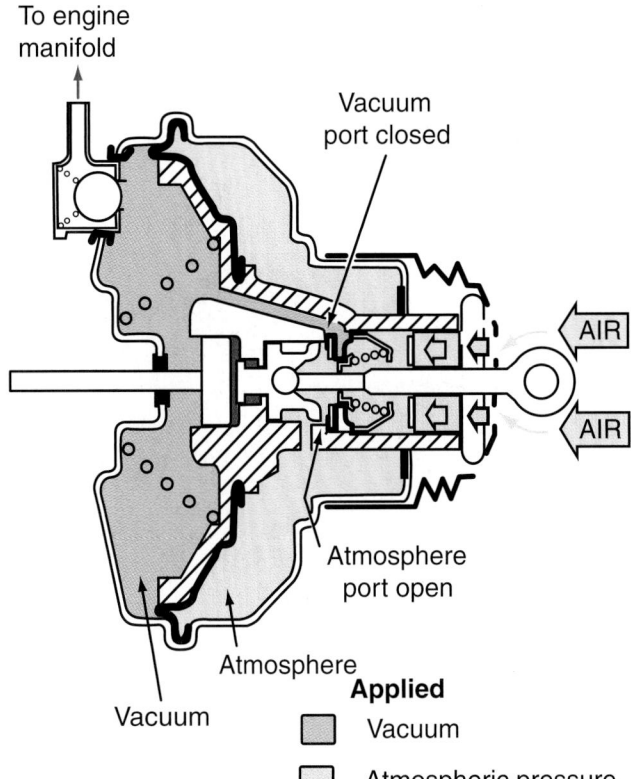

Figure 57.57 A power brake in the applying position.

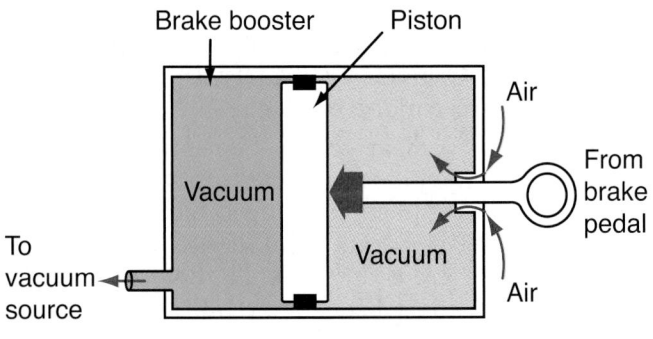

Vacuum Suspended

Figure 57.58 A vacuum-suspended power brake.

The filter cleans the air as it is drawn into the booster from the driver's compartment and reduces the hissing sound of the air entering the power brake booster from the passenger compartment. The driver would hear these noises. In fact, if you listen carefully, they can be heard if the engine is off and you apply the brakes. This can be used to diagnose problems with the power brake system.

Other Power Brake Types

Other power brake types include hydraulic power assist and electric power assist. Some power brake systems use fluid pressure instead of vacuum to produce

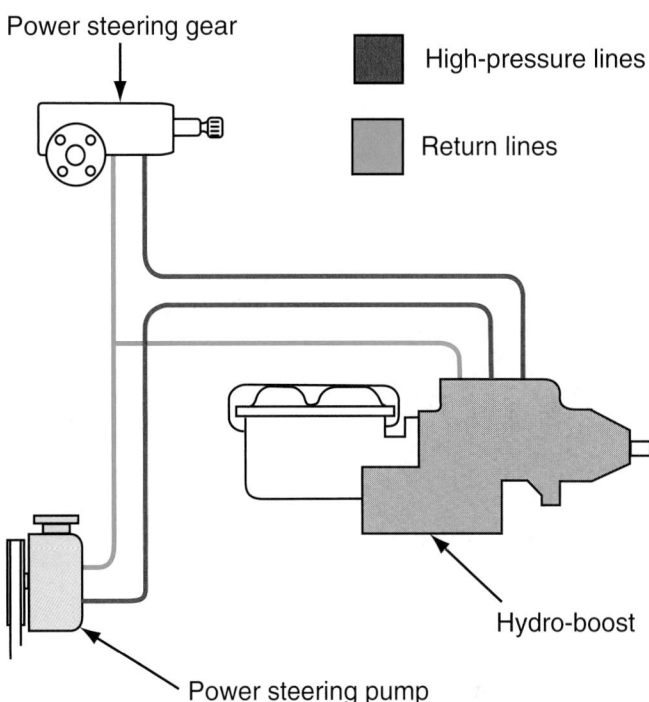

Figure 57.59 A hydro-boost power brake operates off pressure from the power steering pump.

the power assist. A power brake system powered off the power steering pump is called a hydro-boost brake (**Figure 57.59**). The power booster is located in the same place and fits into a similar amount of area as the conventional vacuum power booster (**Figure 57.60**).

Hydro-boost systems are found on some turbocharged gasoline engines, which have problems with vacuum supply because they have pressure in the intake manifold when the turbocharger is working. Hydro-boost is also found on vehicles with diesel engines. Diesel engines do not have a throttle plate and do not produce sufficient vacuum to operate a power booster.

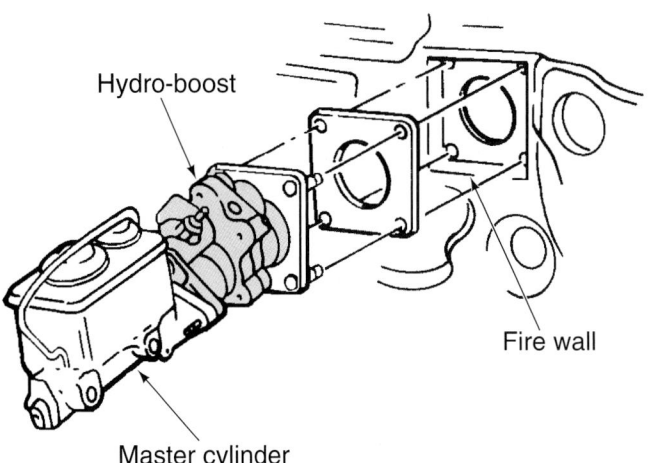

Figure 57.60 The hydro-boost unit is located behind the master cylinder.

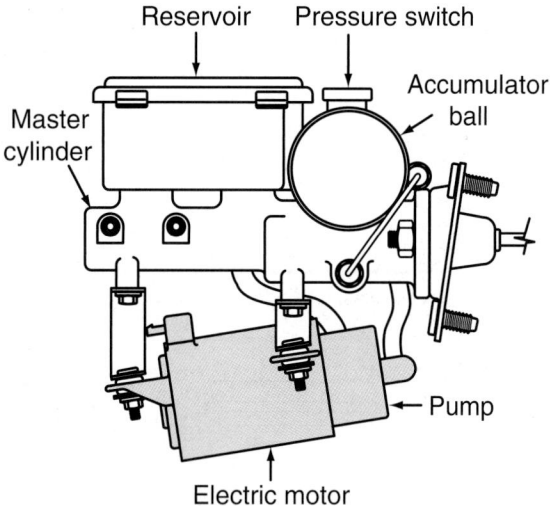

Figure 57.61 A Powermaster brake booster.

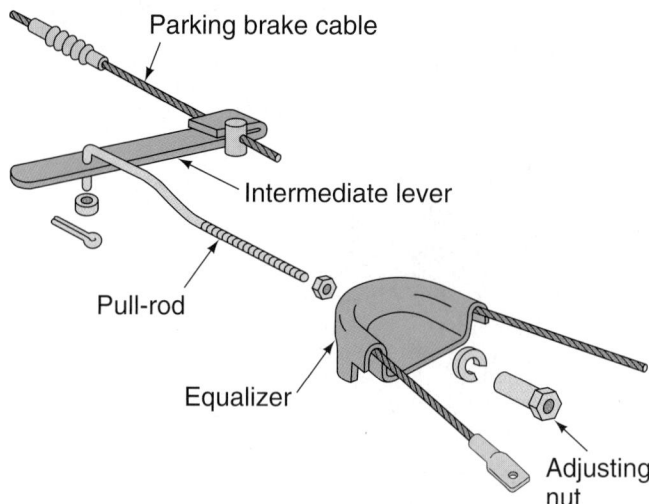

Figure 57.63 An equalizer applies pressure to parking brake cables from each side of the vehicle.

One electric hydraulic power brake type called Powermaster was developed by General Motors. It uses an auxiliary electric motor–driven pump to develop hydraulic pressure for the brake system (**Figure 57.61**). It has a self-contained power booster built into the master cylinder and uses a pump driven by an electric motor rather than one driven by the power steering pump.

▪ PARKING BRAKE

Vehicles are required by federal law to have a *parking brake*. The parking brake, sometimes called the *emergency brake*, must operate independently of the service brakes (the normal brake system). The parking brake can share the same parts as the service brakes, but it must be able to be applied independently. For instance, service brakes are typically applied hydraulically by applying the foot brake pedal, and parking brakes are applied mechanically through a cable operated by a hand lever or foot pedal (**Figure 57.62**).

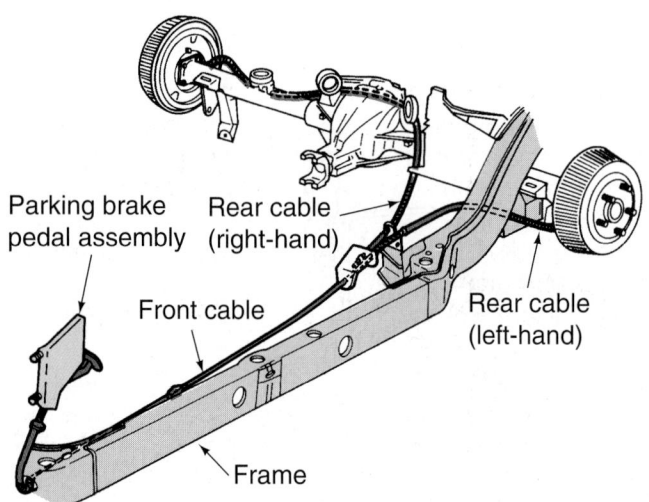

Figure 57.62 Typical parking brake with a cable attached to a lever at each of the rear wheels.

NOTE: *The name "emergency brake" is not really an accurate name. Although a parking brake might help stop a moving car, most are not very effective at that. Federal requirements simply call for the parking brake to "hold a parked vehicle," not stop a vehicle in motion.*

On *parking brakes*, a cable is connected to either a hand brake or foot pedal in the driver's compartment and to an *equalizer* (**Figure 57.63**). The equalizer has a cable from each rear wheel attached to both of its sides. Because it pivots in the center, the equalizer applies each rear parking brake equally.

Parking Brake Warning Lamp. A red warning light illuminates on the instrument panel to warn the driver that the brake is applied. It is hoped that this will prevent damage to the parking and service brakes, which will occur if the car is driven while the parking brake linings are in contact with the friction surfaces of the drum or rotor.

▪ TYPES OF PARKING BRAKES

Almost all cars have the parking brake located at the rear wheels. Parking brakes are of three main types:
- Cars with front disc brakes and rear drum brakes typically use a parking brake that is part of the rear drum brakes.
- Cars with four-wheel disc brakes use one of several types of parking brakes on the rear brakes.
- Some vehicles use a parking brake that operates on the driveline, not the rear brakes.

Drum brakes use an *integral-type parking brake* wherein a cable-actuated bar applies the drum-type emergency brake (**Figure 57.64**). The bar, called the *parking brake strut*, separates the brake linings, wedging them into the brake drum with equal force.

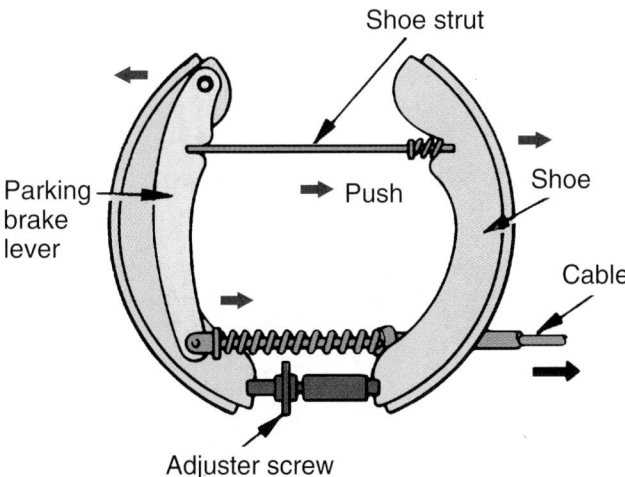

Figure 57.64 A parking brake used with a duo-servo brake.

NOTE: *When parking on a steep hill with the vehicle facing upward, a driver can maximize his or her parking brake effort by applying the service brakes while allowing the car to slowly sink back downhill until the turned front tire comes into contact with the curb (as prescribed by law). Using the service brakes in this manner while applying the parking brake does two things. First, it increases the effort that would normally be applied by the parking brake lever. Second, it takes advantage of servo action to wedge the linings tightly into the drum for greater holding ability. This is especially true with dual-servo brakes.*

There are two main types of parking brakes used on cars with four-wheel-disc brakes. One uses a miniature drum and shoes housed within the inside of the rotor (**Figure 57.65**). This parking brake is often referred to as a **drum-in-hat** brake. The disc is the hat, and the drum is the parking brake section found inside it. Drum-in-hat parking brake shoes wear very little, unless the car is driven with the parking brake partly applied.

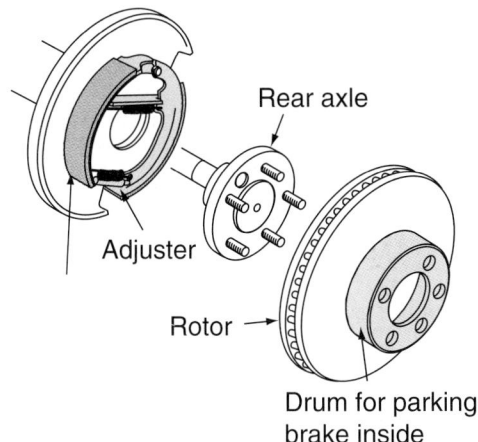

Figure 57.65 Parts of an internal drum-type parking brake.

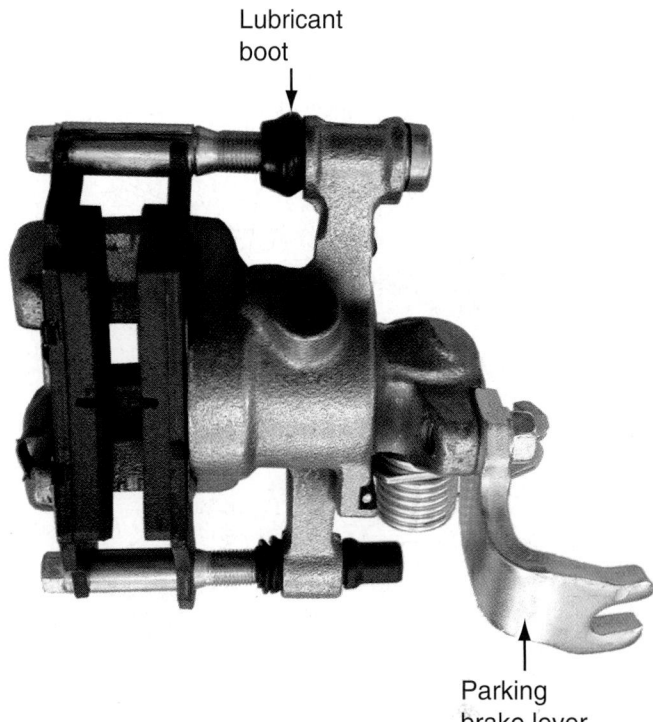

Figure 57.66 A simple cam and lever disc parking brake. *(Courtesy of Tim Gilles)*

Many cars have a parking brake that is an integral part of the rear disc service brakes. Most cars use a conventional parking brake cable arrangement to rotate a lever on the back of the caliper (**Figure 57.66**). Self-adjustment of the service brakes occurs during normal operation, and self-adjustment of the parking brake occurs when the piston moves past a predetermined point in relation to the parking brake self-adjusting mechanism.

NOTE: *Some older integral disc brakes self-adjusted when using the parking brake rather than the service brake.*

Auxiliary Transmission Parking Brake

An independent-type emergency brake mounted behind the transmission on the front of the drive shaft was used on some older cars. This design is still found today on medium-duty trucks and motorhomes. It can be either an internal-expanding or an external-contracting type of brake. The internal-expanding-type brake resembles the brake used with the drum-in-hat type found on some four-wheel disc brake cars (**Figure 57.67**).

■ STOPLIGHT SWITCHES

The stoplights are turned on by a stoplight switch, usually mounted on the brake pedal arm (**Figure 57.68**). When the pedal is depressed, contacts in the switch are closed to complete the circuit to the lights. Some older cars' stoplight switches are operated by hydraulic pressure.

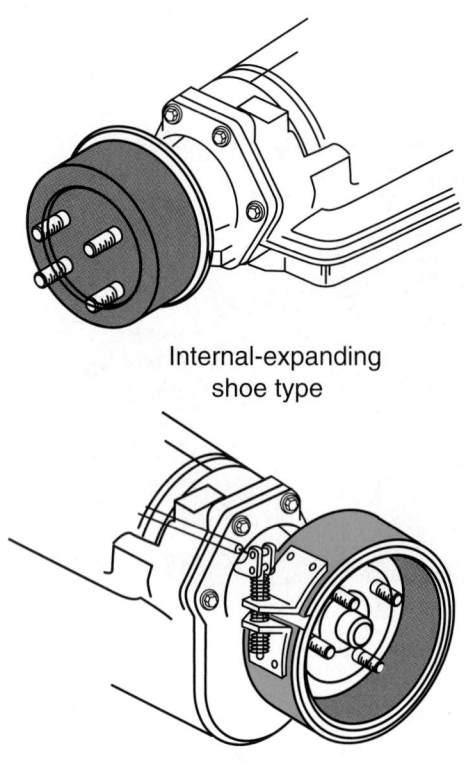

Internal-expanding
shoe type

External-contracting
band type

Figure 57.67 Two types of driveline or transmission parking brakes found on motorhomes and trucks.

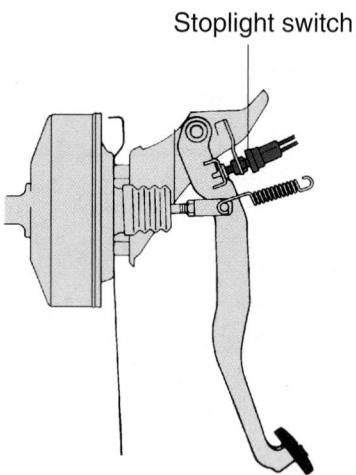

Stoplight switch

Figure 57.68 A stoplight switch.

■ ANTILOCK BRAKES

The ability of the brakes to do their job is limited by the grip of the tires to the road surface. Even with the best quality tires, if the car skids, there will be a loss of stopping ability and control. If the driver can release pressure on the brake pedal just before the wheel locks up, the skid can be avoided. Most newer cars are equipped with computerized antilock brake systems (**ABS**) to keep the wheels from locking up.

The ABS uses sensors and a computer to monitor wheel speed. Wheel speed sensors measure the rotational speed of the wheel. If a wheel locks up, an ABS controller pulsates (modulates) the pressure to that wheel in much the same manner as a race car driver pumps the brakes during a turn to avoid a skid. ABS can do this pulsation much faster than a human, however. A typical ABS can pulsate the pressure to the brake system from 10–20 times per second.

Theory and service of ABS is covered in detail in Chapter 59.

Hybrid Vehicle Brakes

The braking system on a hybrid vehicle has the same parts as a conventional braking system. The system is different, however, in that it has regenerative braking and computer controls to operate the hydraulic brake system rather than a direct link to the master cylinder by the driver. **Figure 57.69** shows a diagram of a complete brake system used on a late-model Prius. Computers control the conventional braking system as well as all other braking systems, including traction control, skid control, and regenerative braking control. These systems are covered in Chapter 59.

Because the regenerative and hydraulic brakes are working together, hybrid brakes feel slightly different than conventional brakes. When the driver releases the accelerator pedal regenerative braking begins. The hydraulic brakes are not used at this time. At low vehicle speeds and when rapid stopping power is required, the hydraulic brakes come into play and provide most of the stopping force. If there is a failure in either the electronic or hydraulic brake system, the brakes will operate normally but with increased pedal pressure and stopping distance, and a warning light on the instrument panel will come on.

Operation of the motor(s) during regenerative braking is covered in more detail in Chapter 73.

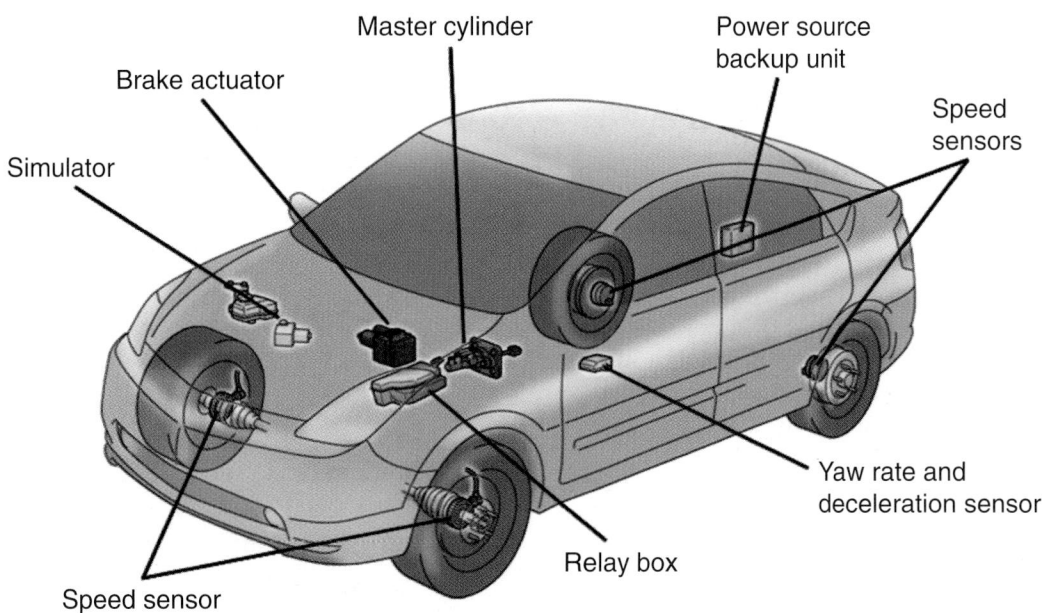

Master cylinder

Brake actuator

Simulator

Power source
backup unit

Speed
sensors

Yaw rate and
deceleration sensor

Relay box

Speed sensor

Figure 57.69 The braking system for a late-model hybrid. Braking is controlled electronically. *(Courtesy of Toyota Motor Sales, U.S.A., Inc.)*

■ REVIEW QUESTIONS

1. If it takes 50 pounds of force to drag one material across another one, what is the coefficient of friction?

2. On disc brakes, the friction linings are called _____.

3. _____'s law states that "pressure in an enclosed system is equal and undiminished in all directions."

4. What is the DOT number for synthetic brake fluid?

5. What is the term for a material that absorbs water?

6. What material is part of brake fluid that is also part of automotive coolant?

7. When the leading shoe on a drum brake is forced into the brake drum as it rotates, this is called _____ action or self-energization.

8. The two main kinds of drum brake designs are the duo-servo and the leading-_____ shoe.

9. What do most power brake booster diaphragms use to provide an assist to pedal effort?

10. The parking brake must operate independently of the _____ brakes.

■ ASE-STYLE REVIEW QUESTIONS

1. All of the following statements about drum brake fade are true *except*:
   a. Brake pedal travel increases.
   b. Lining coefficient of friction becomes higher.
   c. Stopping time is increased.
   d. More pedal effort is required.

2. Which of the following brake designs use(s) return springs?
   a. Disc brakes
   b. Drum brakes
   c. Both A and B
   d. Neither A nor B

3. Which of the following statements is/are true about front disc brakes?
   a. They are self-adjusting.
   b. The brake fluid level drops as linings wear.
   c. When brakes are released the seal retracts, pulling the piston back.
   d. All of the above.

4. The brake pedal on a disc/drum car becomes higher on the second application. Technician A says that the rear brakes might be in need of adjustment. Technician B says that the front brakes might be in need of adjustment. Who is right?

   a. Technician A   **c.** Both A and B

   b. Technician B   **d.** Neither A nor B

5. Technician A says that when the master cylinder bore size is increased beyond original size, stopping effort becomes less. Technician B says that when wheel cylinder bores are increased beyond original size, stopping effort becomes greater. Who is right?

   a. Technician A   **c.** Both A and B

   b. Technician B   **d.** Neither A nor B

6. Which of the following statements is/are true?

   a. Pressure equals force divided by area.

   b. Force equals pressure times area.

   c. Both A and B.

   d. Neither A nor B.

7. All of the following statements about the parking brake are true *except*:

   a. Vehicles are required by federal law to have a parking brake.

   b. Federal law requires that the parking brake be able to stop a vehicle in motion.

   c. The parking brake must operate independently of the service brakes.

   d. Service brakes are applied hydraulically; the parking brake is applied mechanically.

8. Technician A says that the fixed caliper design has a piston or pistons on only one side. Technician B says that the floating caliper design must be able to slide during and after the brakes are applied. Who is right?

   a. Technician A   **c.** Both A and B

   b. Technician B   **d.** Neither A nor B

9. Technician A says that tandem brake systems can be split front to rear. Technician B says that tandem brake systems can be split diagonally. Who is right?

   a. Technician A   **c.** Both A and B

   b. Technician B   **d.** Neither A nor B

10. Technician A says that sometimes linings are bonded to the metal brake backing. Technician B says that sometimes linings are riveted to the metal brake backing. Who is right?

   a. Technician A   **c.** Both A and B

   b. Technician B   **d.** Neither A nor B

CHAPTER 58

Brake Service

■ OBJECTIVES

Upon completion of this chapter, you should be able to:

✔ Inspect brake systems and recommend needed repairs.

✔ Diagnose brake system problems.

✔ Perform brake repairs and adjustments using the correct materials and procedures.

■ KEY TERMS

spongy brake pedal	vacuum bleeding	swaged lug studs
brake drag	reverse fluid injection (RFI)	carbide
brake pedal pulsation	vulcanized	hard spots
brake grab	wheel cylinder kit	unloaded caliper
brake pull	anodized aluminum	loaded caliper
bleeding brakes	brake spoon	service brakes
pressure brake bleeding	discard diameter	

■ BRAKE INSPECTION

A thorough inspection of the braking system should be performed before attempting any repairs. Start with the pedal and master cylinder.

■ CHECK BRAKE PEDAL FEEL

Apply the foot brakes and check the travel of the brake pedal. There should be an ample amount of *pedal reserve* (above the floor) after the brakes begin to stop the car. When there is no pedal reserve, this is a serious problem indicating failure in both halves of the brake system (see Chapter 57). A low pedal can indicate a leak in one half of the system. Check the fluid level first. A leak can be either external or internal (no fluid leaking out).

The pedal should feel firm, not "spongy." A **spongy pedal** indicates air or moisture in the system, calling for a brake bleed.

NOTE: *Cars with ABS brakes will have a feeling of pedal pulsation during hard stops. Some of these systems will also experience a rise in pedal height during hard stops.*

■ MASTER CYLINDER INSPECTION

Hold the pedal down very lightly. It should not fade away toward the floor. If the pedal drifts toward the floor and there is no visible sign of fluid leakage anywhere in the system, the master cylinder is leaking internally past the primary cup. Usually, pushing harder on the pedal will result in the brakes working properly as the seal distorts against the wall of the cylinder, cutting off the internal leak.

Check the level of the fluid in both halves of the master cylinder reservoir. Disc brake lining wear will lower the fluid level. This is because the brake caliper pistons move out in their bores to compensate for the lining wear. If one of the chambers is larger and it is lower in fluid than the other one, the disc linings served by that chamber are probably worn.

If a secondary piston seal is defective, fluid can be pumped from one of the fluid chambers into the other. The result is that one chamber is low, while the other is overflowing.

Look for external leaks from the master cylinder. Leakage can occur past the master cylinder cover when the metal bails are bent and loose. Fluid can also escape past the secondary cup on the rear piston. If there is no power booster, fluid will be visible on the carpet or on the inside of the fire wall or bulkhead. If there **is** a power booster, fluid will leak from its front. A small amount of seepage is acceptable.

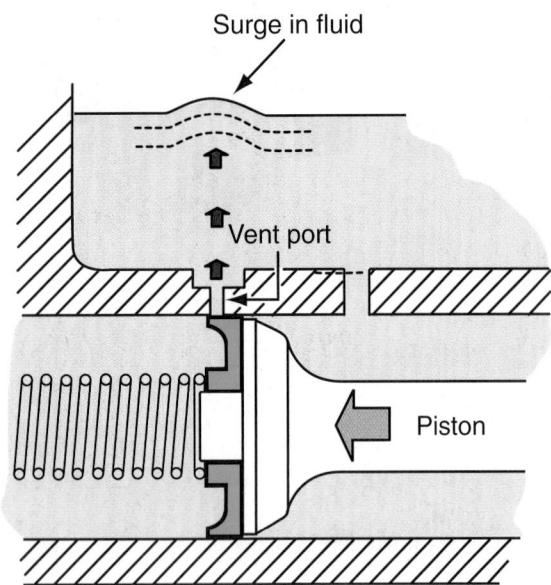

Figure 58.1 Fluid movement past the vent port should always be visible when the pedal is applied.

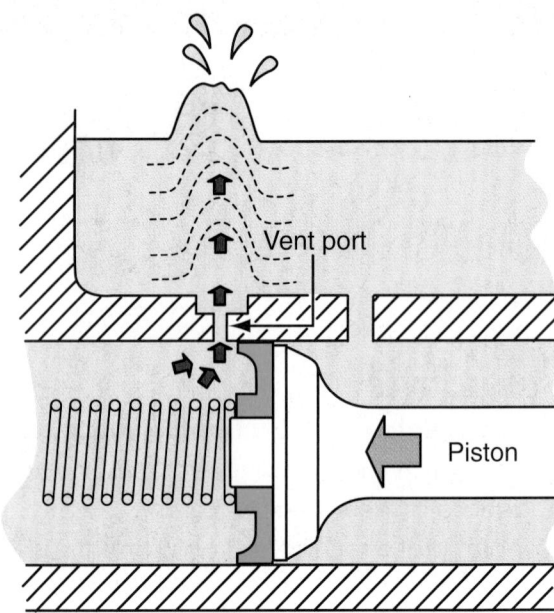

Figure 58.2 A heavy surge from the vent (compensating) port indicates air in the system.

Check the Vent Port

Fluid movement past the vent port should always be visible when the pedal is applied (**Figure 58.1**). The vent port is a very small opening and can become plugged. The result is brakes that do not release all the way after the fluid absorbs heat resulting from braking action. This can cause a high, very hard pedal. As the driver attempts to get the vehicle to move after a stop, the brakes might continue to hold.

Sometimes, brake pedal travel is incorrectly adjusted. Too little pedal free play can cause the primary cup to be positioned past the vent port even when the brakes are released. This will cause the same symptoms as a plugged vent port.

Brake drag is the term given when the brakes stay on after a stop.

Checking for Air in the System

To check for air in the system, have a helper apply the brakes about ten times, holding it the last time. With the cover removed from the master cylinder, watch the movement of fluid while the pedal is released. A heavy surge from the compensating port (**Figure 58.2**) indicates air in the system. Note which chamber has the high fluid surge. This is the side of the system that has the trapped air.

NOTE: *Be sure fender covers are installed on the fender. Splashed brake fluid can ruin paint. Wear eye protection.*

Power Brake Checks

A power brake booster can experience problems such as a hole in a diaphragm or a sticking valve. To test the operation of a power brake booster, exhaust all vacuum reserve from the power booster by applying the foot pedal several times with the engine off. The sound of the air rushing into the booster from the passenger compartment should be heard. Next, hold your foot on the pedal while starting the engine. If the power booster is operating correctly, the pedal will move about an inch closer to the floor after the engine starts.

NOTE: *With the vacuum valve open and the air valve closed, vacuum will equalize on each side of the diaphragm. Unless the booster is defective, this is the only stage of booster operation where air will be able to enter the vacuum side. As the brakes are released, a minor increase in engine rpm might be noticed when the air from the back side of the diaphragm is drawn into the vacuum chamber.*

Stoplight Switch

Apply the brakes and test the stoplights. If they do not come on, check to see if the key must be on first. Then, check the fuse. There are two wires to the stoplight switch. Remove the connector from the switch and use a jumper wire to connect them together. If the stoplights work with the wires connected, the stoplight switch is faulty.

NOTE: *Many stoplight problems are caused by a defective turn signal switch.*

Hydraulic Safety Switch Service

In some early systems from the 1960s, the bulb indicating a hydraulic system failure would light and remain on until it was recentered by a technician after repair. On modern systems, the light only operates during braking. Some systems share the same bulb to indicate when the parking brake is applied. Testing the bulb in these systems can be done with the parking brake applied and the key turned on. Low fluid level is also indicated by the same bulb.

Chassis Problems

Problems related to the chassis can cause symptoms felt in the brakes, such as brake pull. Chassis parts involved include loose wheel bearings, worn ball joints or bushings, and worn steering linkage parts.

Road Test

Before a road test, be sure there is fluid in both reservoir sections of the master cylinder. Check to see that pedal height is normal. If not, do not road test the car. During the road test, stop the car on a deserted, straight section of road. Notice if the car pulls to one side or the other when stopping. Also look for pedal pulsation, a wheel grabbing or locking up, noises, or other unusual conditions.

■ BRAKE DIAGNOSIS

Brake pedal pulsation results when hydraulic pistons are moving during a stop. This can be caused from rear drum brakes that have become out-of-round. This happens when overheated brakes have the parking brake set firmly against them.

Pedal pulsation also results when disc brake rotors have become overheated or when wheel lugs were improperly torqued. A primary cause of pedal pulsation is dirty conditions when installing a rotor. This is covered later in the chapter.

Brakes sometimes **grab**. This means that they apply quickly and tend to stick on. This happens when there is oil or grease on a lining. It can also happen when metallic linings are used. Their coefficient of friction increases as they become hot (**Figure 58.3**).

Extra braking effort can be required when a power booster is not operating properly or when the wrong linings have been installed.

Brake pull happens due to several reasons.
■ When there is a difference in friction between sides of a vehicle, it will pull.
■ Linings can have different friction due to oil or grease.
■ A hose can be restricted or a caliper can be frozen.
■ Changes in wheel alignment resulting from loose parts can also cause pull.
■ A tire of the wrong size can cause pull.

NOTE: *A low tire will probably pull all of the time, not just during stopping.*

When the *warning light* is on, there is either a problem in the brake hydraulic system or a problem ground in the switch's electrical circuit. Also, brake pads on some vehicles have electrical sensors that illuminate the light when the pad wears thin.

Noises during stopping can include squeaks, metal-to-metal sounds from excessive wear, rattles due to loose parts, and rubbing from a distorted backing plate. Noise usually results from vibration. This is covered in more detail later in this chapter.

■ UNDERCAR CHECKS

With the car in the air, visible checks can be made of the brake system. After a complete brake inspection, the customer should be informed of the need for brake work. Write on the repair order that the inspection was performed. A service writer will usually contact the customer in case the customer wants to have the work done while the car is already in the shop for the day.

■ DISC BRAKE INSPECTION

Disc brakes are inspected to see that a sufficient amount of friction material remains on the pads (**Figure 58.4**). When the friction material wears completely away, rotor damage will result. Brake pads are typically replaced before the thickness of the remaining friction material on the steel backing plate reaches 1/8" to 3/16". Some shops

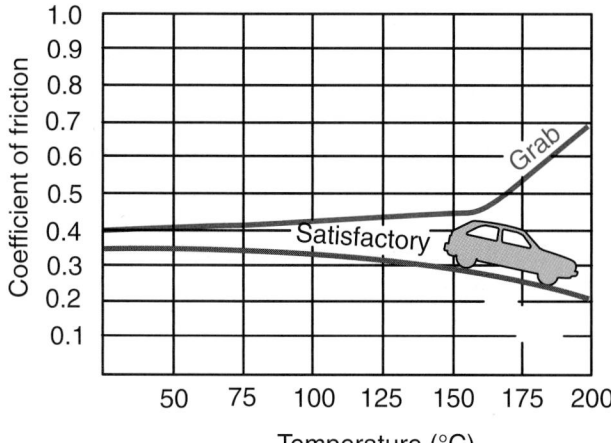

Figure 58.3 Linings constructed of various materials perform differently with changes in temperature.

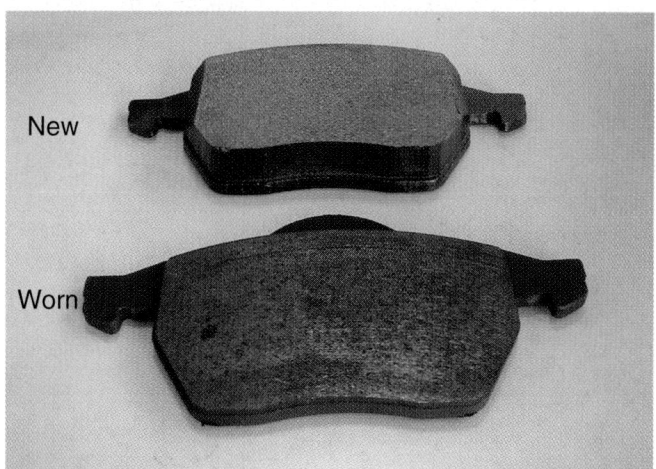

Figure 58.4 New and worn disc pads. *(Courtesy of Tim Gilles)*

Figure 58.5 These rotors are worn so badly that one of them actually separated into two pieces. *(Courtesy of Tim Gilles)*

recommend pad replacement when the remaining pad material is the same thickness as the steel backing.

NOTE: *As friction material wears, heat transfer to the caliper increases.*

Some customers use their brakes well beyond the point where metal is wearing on metal. This most often results in much more costly repairs due to damage to the rotors and sometimes the calipers. **Figure 58.5** shows two such examples.

On most vehicles, you can remove a front wheel and visually inspect the pad thickness without removing the brake caliper. Sometimes the thickness of the friction material is visible through a hole in the caliper. On other calipers, a flashlight can be used to look down the side of the caliper (**Figure 58.6**).

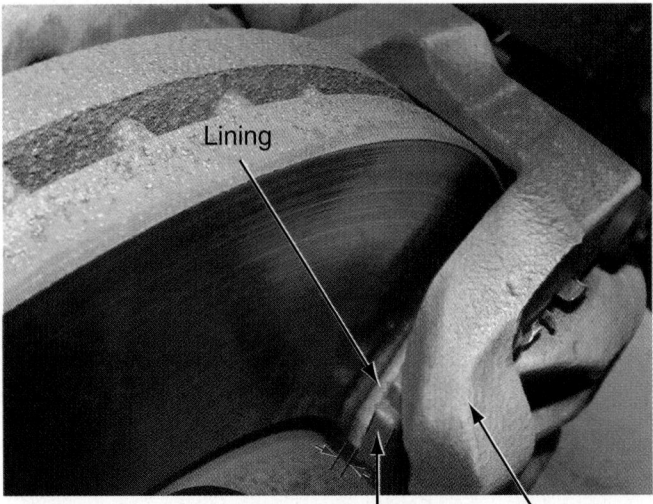

Lining

Metal back   Brake caliper

Figure 58.6 Inspect the thickness of the disc lining. This lining is worn to a thickness of less than its metal back and should be replaced. *(Courtesy of Tim Gilles)*

Disc pads occasionally wear unevenly. If one lining is worn more than the other on a floating caliper disc brake, a caliper slide is probably rusted or dirty. This prevents the caliper from floating freely. Some tapered wear on disc pads results from the tendency of the caliper to twist during braking. This is considered to be normal as long as the taper is less than 1/8".

When a symptom such as noise or pull is present, the caliper must be removed for a more thorough visual inspection of the pad. Sometimes disc pads must be replaced because they are worn unevenly or are heat damaged and/or cracked. The rotor will often suffer damage as a result, so inspect it carefully. When reinstalling calipers, be sure to lube the slides. Check all sliding surfaces to see that they are not rusted and are in good condition. Inspect sliding caliper guide pins. Worn guide pins can cause a brake to drag or squeak after it is released. If they are not clean and free of rust, they must be replaced.

Bushings and sleeves often become gummed up to the point where the caliper cannot slide easily. This can result in one lining that wears more than the other. Another possible result is a brake that will not release all the way after a stop. This results in an intermittent annoying squeal that goes away whenever the brakes are applied.

Inspect the Rotor

Inspect the rotor. Scoring and unusual wear will be visible as the rotor is turned. Feel it with your hand.

CAUTION If a car has been recently driven, the rotor will hold heat for some time. You could be burned when you touch a hot rotor.

A rotor is discarded when it is too thin for service. Measure the thickness of the rotor using a micrometer. Minimum thickness specifications are readily available in manufacturers' pamphlets or in the service information library. Most rotors have a casting or stamping listing the minimum thickness. Depending on the manufacturer, at least 0.015" must remain above this specification to allow for wear. Some manufacturers specify 0.030".

While rotating the rotor, rest a screwdriver or pencil on the caliper mount and hold it near the rotor as a reference while watching for visible runout. Runout that can be seen visually is already excessive, and correction will be required. If you cannot see visual runout, check for runout with a dial indicator.

Checking Rotor Runout with a Dial Indicator

Rotor runout causes the rotor to swing from side to side as it rotates (**Figure 58.7**). Ideally runout should be

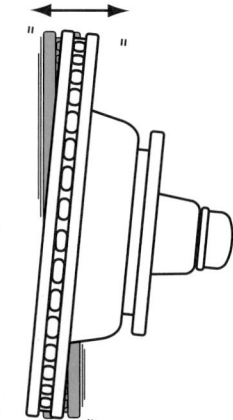

Figure 58.7 Disc brake "rotor runout."

(a)

less than 0.003", although some manufacturers allow as much a 0.008". Some late-model cars specify as little as 0.0005" (½ thousandth) of runout.

NOTE: *Runout can be a result of hub distortion caused when lug nuts are tightened with an impact wrench. It can also be caused by dirt or rust on the mating surfaces of the rotor and hub. Be certain these areas are clean (**Figure 58.8**).*

A runout measurement is taken off the outboard surface of the disc brake rotor using a dial indicator (**Figure 58.9**). Attach the indicator to a rigid part. If the rotor has a serviceable wheel bearing, temporarily remove end play at the adjusting nut. You do not need to tighten the bearing. Simply remove any looseness so this test can be accurate. A roller on the end of the dial indicator provides for a more stable measurement as the rotor is turned against it (**Figure 58.10**). Lug nuts might need to be installed with washers to secure the rotor against the hub.

Rotor Thickness Variation

Excessive runout can result in thickness variations of the rotor. Measure the rotor's thickness in four or five equally spaced places. Measurements should be within 0.005" (one-half of one thousandth) of each other. Thickness variations result in brake pedal or steering wheel vibration.

Drum Inspection

Brake drums must be of a minimum thickness in order to be reused (with or without machining). Drum inspection is covered later in this chapter.

Inspect the Caliper

Look the caliper over carefully for signs of problems. Check the caliper piston dust boots to see if they are torn. Damaged dust boots will allow moisture in, which will ruin the caliper. There should be no sign of fluid leakage from the caliper piston seals.

(b)

(c)

Figure 58.8 Be certain that the mating surfaces of the rotor and hub are clean. *(Courtesy of Tim Gilles)*

Figure 58.9 A dial indicator is mounted on a rigid part to measure rotor runout. Lug nuts hold the rotor tightly to the hub. *(Courtesy of Tim Gilles)*

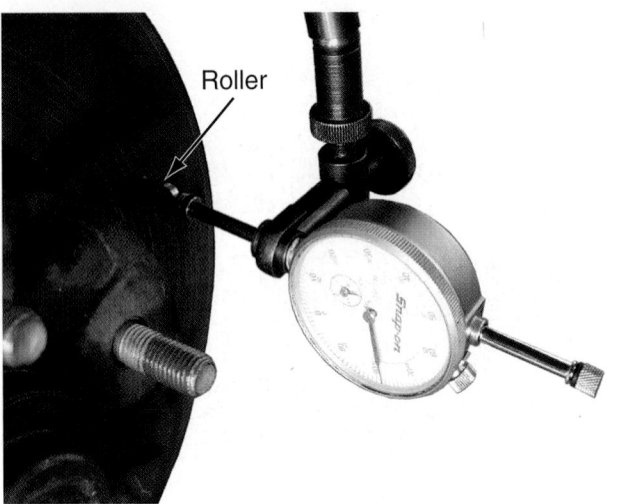

Figure 58.10 A roller on the end of a dial indicator reduces friction and provides a more accurate measurement. *(Courtesy of Tim Gilles)*

Inspect Wheel Seals

Look at backing plates to see that there is no sign of a fluid leak at the axles or grease leaking from the wheel bearings.

Parking Brake Inspection

Apply the parking brake, checking to see that it engages before it is at half travel. Check for damage or rust on the cable.

Inspect Tubing and Hoses

Look for physical damage to hoses, such as cracking, swelling (**Figure 58.11a**), or softening from exposure to oil. A hose that was installed improperly can rub until it fails (**Figure 58.11b**). Woven hoses sometimes experience internal failure, especially when a careless

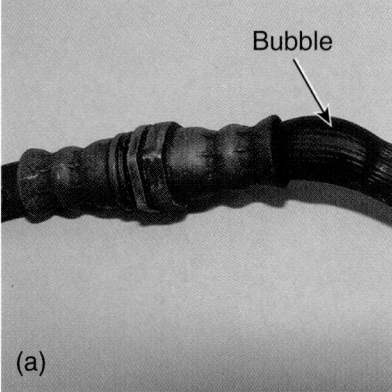

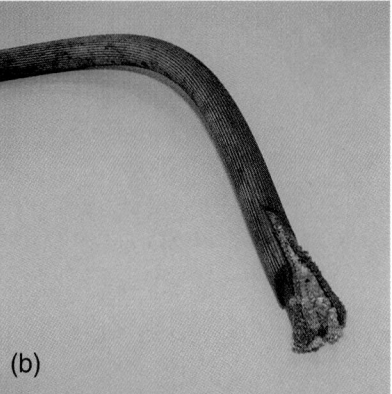

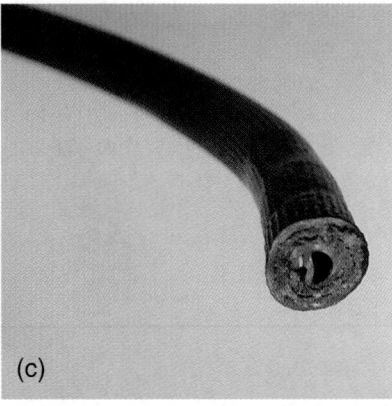

Figure 58.11 (a) This brake hose has a bubble. (b) This hose was damaged when it rubbed on a tire. (c) This hose collapsed on the inside. *(Courtesy of Tim Gilles)*

technician allows a disc brake caliper to hang from a hose during service to brakes, suspension, or wheel bearings. Another cause of internal failure is clamping the hose with locking pliers. There are special tools designed to clamp a brake hose without damaging them, but many technicians avoid clamping a hose altogether. Even though a hose appears to be good from the outside, the inside of the hose can be collapsed, trapping fluid inside the front brake and causing excessive wear or brake pull to one side. Internal damage is difficult to diagnose. Depending on which way the internal rubber flap is torn, the hose could cause delayed engagement of the brake or act as a one-way valve, preventing fluid pressure from releasing the brake after application.

With the brakes released, trapped brake fluid is indicated when a bleed screw is opened and fluid spurts out.

Excessive brake pad wear results when the brakes do not fully release.

A partially plugged brake hose (**Figure 58.11c**) will cause the car to pull at the beginning of a stop. When the pressures equalize, the pull goes away. If a pad on one side is worn more than on the other, this could be due to a bad hose.

Checking for Fluid Leaks

When the master cylinder fluid level is low, look for an external leak. If the seal at the rear of the master cylinder bore (secondary cup) leaks, fluid can escape. When a wheel cylinder is leaking, the insides of the tires will often be streaked with fluid and dirt. Occasionally the outside of the tire can even show signs of fluid leakage. A wet rear tire can be caused from a leaking axle seal too.

To tell the difference between an axle leak (gear oil) and brake fluid leak, try washing the fluid off with water. Brake fluid is water soluble and gear oil is not. Also, brake fluid causes paint damage. The paint on the brake backing plate can become bubbled, indicating a brake fluid leak.

■ BRAKE LINING INSPECTION

Inspecting Drum Brake Assemblies

Removing the Brake Drum. Before inspection, a wheel and drum must be removed. First, mark the drum so it can be installed in the same position (for instance, "LR" for left rear). Some cars have hubs with removable wheel bearings. These can be repacked with grease during this service (see Chapter 60). These are found on the front wheel drums of older cars and on the rear wheel drums of many front-wheel-drive cars. The center hole on the drums of these style brakes pilots off the bearing hub. This is also true of rear-wheel-drive cars, where the drum centers off a machined area on the axle flange. Removal of these drums is often made difficult due to rusting that occurs between the center hole of the drum and the pilot flange (**Figure 58.12**).

There are two ways to remove tight drums. First, spray the flange area with penetrating oil. Many front-wheel-drive rear drums have two threaded holes. Screws are tightened into these holes (**Figure 58.13**). The end of the screw pushes against the flange to force the drum off. Tighten the screws ½ turn at a time, first on one side and then on the other.

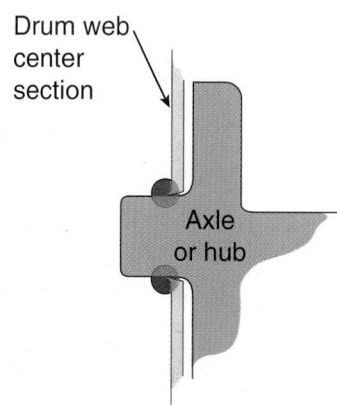

Figure 58.12 The drum sometimes rusts to the axle flange.

Figure 58.13 Turning screws into threaded holes in the drum forces the drum off the axle flange. Do not use an air impact tool to tighten one of these bolts; it will strip the thread. *(Courtesy of Tim Gilles)*

When there are no threaded holes, the drum will usually loosen following a couple of sharp raps with a hammer between the wheel studs (**Figure 58.14**). Be careful not to damage a wheel stud. Pounding here is relatively safe because the drum is supported by the axle flange, which is very sturdy. However, **Figure 58.15** shows a drum that was ruined by someone who did not understand the relationship between the axle and drum.

NOTE: *Pounding on the rear lip of a brake drum is likely to damage the drum.*

Some rear drum assemblies on front-wheel-drive cars have sealed permanent wheel bearings that are tight on the axle. These are removed using a slide hammer (**Figure 58.16**).

NOTE: *When a car owner continues to drive a car after the linings have worn down to the shoe, the shoe will wear a channel in the drum. Then the self-adjusters operate, adjusting the shoe into the worn out area of the drum.*

Figure 58.14 Carefully give a sharp, firm rap with a hammer to break loose rust between the axle flange and drum. (*Courtesy of Tim Gilles*)

Figure 58.15 A drum ruined by someone who did not understand the relationship between the axle and the drum. (*Courtesy of Tim Gilles*)

Figure 58.16 Removing a pressed-fit hub and drum assembly with a slide hammer. (*Courtesy of Tim Gilles*)

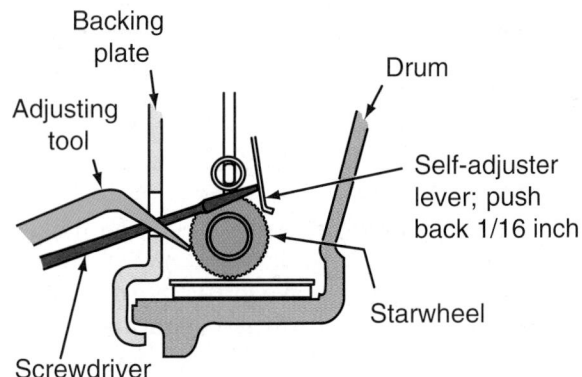

Figure 58.17 To increase the lining to drum clearance, move the self-adjusting lever out of the way so you can turn the starwheel.

*When this has happened, the drum cannot be removed unless the brake adjustment is backed off. The self-adjuster lever must be moved out of the way while performing the adjustment (**Figure 58.17**).*

SHOP TIP Difficult brake drums can sometimes be removed by cutting the holddown pins where they protrude from the backing plate. The drum can then be forced off.

Inspecting Drum Brake Cylinders

Pull back the rubber boots on the wheel cylinders to check for leakage and rust (**Figure 58.18**). A small amount of dampness from brake fluid is acceptable, but fluid should not be dripping. A cast iron wheel cylinder bore should not have rust, and an aluminum piston should not be oxidized.

Figure 58.18 Inspect under the wheel cylinder boot for fluid leakage and rust. (*Courtesy of Tim Gilles*)

Rust (oxidation of iron or steel) is a serious problem. About one-seventh of the iron produced each year by industry is used to replace iron that has rusted. The exact process that occurs during rusting is not known. What is known is that oxygen and water are necessary for rusting to take place. Rusting occurs more rapidly in the presence of acid.

A sacrificial metal can be used to protect iron or steel from rusting. Oxidation requires the removal of one or more electrons from an atom. The oxidation potential of one metal is higher than for another if it requires less energy to remove an electron from the former than the latter. For example, the oxidation potential of magnesium is higher than the oxidation potential of iron or steel. This is because it is easier to remove an electron from magnesium than from steel. This is why blocks of magnesium are strapped to steel ships to prevent rusting. Not only does the magnesium oxidize instead of the iron, but the electrons stripped from the magnesium migrate to the steel. This happens because electrons associated with steel are held more strongly, and if electrons are given a choice, they prefer to be held more strongly. The transferring of electrons from magnesium to steel means that the steel cannot rust until it loses all of its excess electrons. The magnesium will prevent this electron loss by continually supplying the steel with electrons, giving the steel even more protection. Application of this principle can also be found in the magnesium rod screwed into your household water heater to prevent rusting of the tank.

Inspect Drum Brake Linings

Examine the condition of the friction lining surfaces. The front brakes wear more than the rears, so check one of them first.

- A bonded lining should be at least the same thickness as its metal backing. **Figure 58.19** shows the lining (friction material) and the brake shoe.
- Riveted linings should be above the rivet heads.

Some vehicles have a small hole in the brake backing plate through which the thickness of the linings can be verified without removing the drum. A protective plug is removed to uncover the hole. Shine a light into the hole and observe the lining thickness.

Check Self-Adjuster Operation

Self-adjusters can become rusty, requiring disassembly so that their threads can be cleaned. Operate the self-adjusters to see that they move freely.

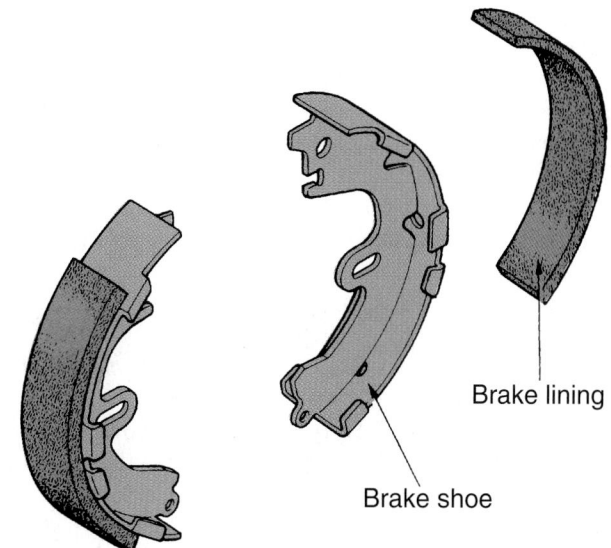

Brake lining

Brake shoe

Figure 58.19 The brake lining is attached to the shoe.

■ BRAKE FLUID SERVICE

The correct fluid will be listed on the reservoir. It is important not to fill the reservoir all the way to the top. Space must be left to allow for fluid expansion as the brakes heat up. The maximum fill level is about ¼" from the top of the reservoir chamber. Be careful not to allow brake fluid to contact the car's finish because it can damage paint. Rinse thoroughly with soap and water in case of a spill. Wiping spilled brake fluid from a fender using a shop towel is more likely to damage the paint.

■ ROUTINE BRAKE FLUID REPLACEMENT

Some brake manufacturers estimate that half of all operating vehicles have never had a brake fluid change. When water is absorbed by brake fluid, the inside of the brake system corrodes, so periodic changing of the brake fluid is necessary. To avoid moisture contamination, brake system flushing is done whenever a brake job is performed, and sometimes more often. Manufacturers are recommending more frequent fluid flushes because most cars are now equipped with ABS brakes, which have very small passages that are more sensitive to contamination than older brake systems.

Most of today's brake systems are "bimetal." This means that system parts are made up of more than one type of metal. Common hydraulic system materials include aluminum, iron, steel, and copper. Master and wheel cylinder housings and pistons are often made of aluminum. Housings are also made of cast iron, and pistons are made of steel. Steel hydraulic tubing is lined with copper.

You probably remember from science lessons that two dissimilar metals in an electrolyte solution can produce a voltage. New brake fluid includes additives to prevent it from becoming an "electrolyte." These additives become depleted, however. Moisture accumulation in the hygroscopic brake fluid also causes battery (galvanic) action. The corrosion that results plugs valves in the ABS and causes pitting in the wheel cylinders and master cylinder.

One of the reasons glycol is the material of choice for brake fluids is that it absorbs water so it will not be able to cause problems in the hydraulic system. In humid (lots of moisture in the air) areas of the country, brake fluid can absorb an average of 3% of its volume in water during 18 months. If the fluid is not changed, moisture content can reach 8% or more in another year. Once the fluid reaches its saturation point, it cannot absorb any more water.

DOT 3 fluid has a minimum dry boiling point of 401°F. Its saturated boiling point must be above 284°F. The following are DOT 3 boiling points with different volumes of moisture accumulation:

% Water Volume	Boiling Point
1%	369°F
2%	320°F
3%	293°F

Fluid Change Interval

During average driving, brake linings usually wear out fast enough for brake fluid to remain in good condition between brake jobs. A general rule of thumb is to change fluid every 2 years or 30,000 miles, but check the manufacturer's recommendations. Antilock brake systems sometimes call for more frequent fluid changes. Service bulletins for Ford, Mercedes, BMW, Volvo, and others call for a brake fluid change every 24 months or less. Some manufacturers do not list a fluid change requirement, but others specify intervals as low as 18,000–24,000 miles.

As you learned in Chapter 57, there are differences between types of brake fluids. To avoid doubt, always use quality brake fluid of a type specified by the manufacturer.

NOTE: *No manufacturers currently recommend DOT 5 silicone brake fluid for their cars. If you encounter an older car with DOT 5 fluid, do not mix it with a glycol-based fluid (DOT 3 or DOT 4). Silicone brake fluid looks different from glycol-based fluid. Because dye is added to the clear silicon base, it is purple in color. Silicone fluid costs several times as much as glycol fluid. Also, rubber brake seals are made of EPDM or SBR rubber. SBR seals can swell when exposed to silicone, causing them to soften or to leak.*

Brake Fluid Testing

Several methods of testing brake fluid are used in the industry.

Figure 58.20 A moisture content tester. *(Courtesy of Tim Gilles)*

Moisture Content Testers. Testers are available to check the moisture content of hydraulic brake fluid (**Figure 58.20**). Over time, moisture gets into the system with air that permeates rubber brake hoses. Newer hoses have EPDM inner liners that prevent this. Moisture can also enter the hydraulic system past rubber seals that are hardened with age. An aging seal might prevent brake fluid from leaking out, but the smaller molecules of air and water can still seep in.

Refractometer Testing. Refractometer testing is another method of testing brake fluid (see Chapter 21). A drop of fluid is placed on the face of the tester, and the result is read by viewing a gauge inside the instrument. Refractometers are designed for testing the specific gravity of specific fluids, including brake fluids, coolant, and battery electrolyte.

Voltmeter Testing. A voltmeter can be used to measure the conductivity of the brake fluid in the master cylinder housing. One probe is put into the fluid, and the other is connected to ground. A reading of 0.3 volt DC is the maximum.

Fluid Test Strips. As brake fluid wears out, it becomes thicker. Test strips can be used to check brake fluid for wear (**Figure 58.21**). They indicate the amount of breakdown in the anticorrosive additives in the brake fluid as they change in color from white to purple.

NOTE: *Test strips do not give an indication of the amount of moisture in the brake fluid.*

Hydraulic System Flushing

Occasionally a brake hydraulic system is accidentally contaminated with a petroleum-based product and must be flushed. Alcohol is used to clean the hydraulic system because it removes petroleum without leaving

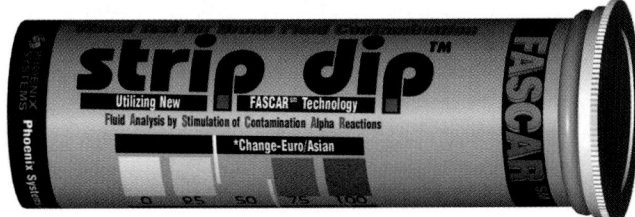

Figure 58.21 FASCAR test strips. *(Courtesy of Phoenix Systems, L.L.C.)*

a residue. Alcohol also absorbs any water that might be present and carries it out during the flush. Following a system flush, use traditional bleeding methods as fluid is restored to the hydraulic system.

■ BLEEDING BRAKES

To operate correctly, the hydraulic system must be able to push a solid column of fluid. Air or moisture in the hydraulic system can result in a soft, "spongy" pedal. Air is compressible, brake fluid is not. Water in the brake fluid can vaporize when it becomes hot, adding gas pockets to the fluid.

When brakes are bled, air and fluid are removed from the hydraulic system through a bleed screw located at the highest point on the back of the wheel cylinder or caliper (**Figure 58.22**). When the bleed screw is loosened, a hollow opening is uncovered (**Figure 58.23**) and fluid escapes through the center of the screw.

> **SHOP TIP** Sometimes a bleed screw is rusted, making it impossible to remove. A trick that works well for loosening stuck threads is to heat the screw and quench it with a wax stick, then remove it quickly.

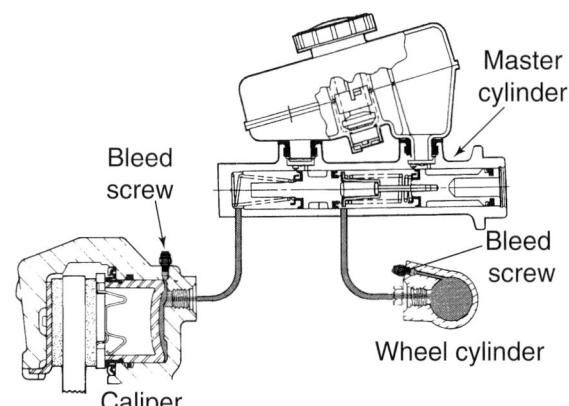

Figure 58.22 Bleed screws are located at the highest point on the back of the wheel cylinder or caliper.

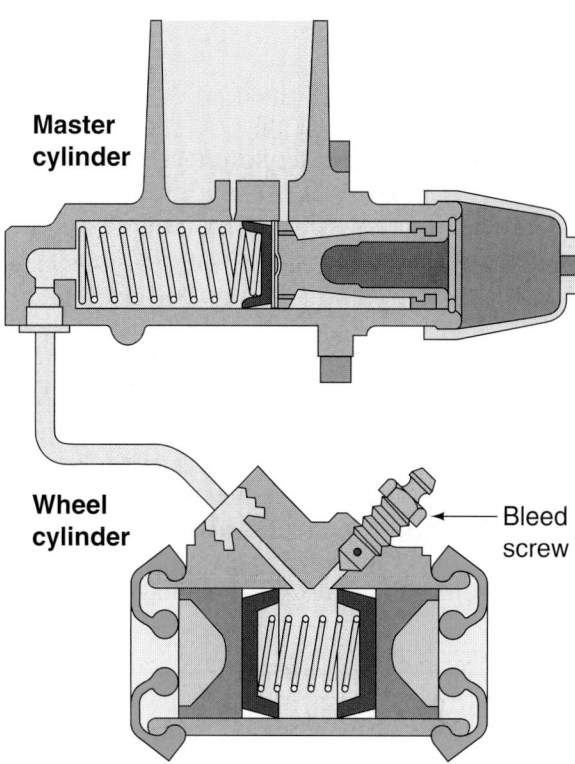

Figure 58.23 Loosening the bleed screw uncovers a passage for fluid flow.

Brake Bleeding Sequence

Manufacturers specify different bleeding sequences, depending on the design of the brake system. Correct brake bleeding procedures are very important. Check the service literature for the procedure. Most front-wheel-drive vehicles have diagonally split hydraulic systems. These require a special bleeding sequence that is described in the service literature. If the brakes are not bled in the correct sequence, a solid pedal might never be achieved. Some import calipers have two or three bleed screws per caliper. Follow the manufacturer's specified procedures for bleeding these calipers.

NOTE: *Brake manufacturers report that perfectly good power brake boosters are often returned under warranty because of incorrect brake bleeding procedures.*

■ BRAKE BLEEDING METHODS

Brakes can be bled in several ways, including manually, with a pressure bleeder, by vacuum, by reverse injection, or by gravity. The different procedures are covered next.

Manual Brake Bleeding

Bleeding brakes manually is best done with two people. The pedal is depressed and released slowly. Here is a typical manual bleeding procedure:

■ Loosen the bleed screw approximately one turn.
■ Ask an assistant to push the brake pedal to the floor.

- Close the bleed screw.
- Release the brake pedal.
- Repeat the process until the fluid coming out of the bleed screw is free of air.
- Check the master cylinder fluid level often and refill it before the fluid level drops too low to prevent air from entering the hydraulic system.

NOTE: *When manually bleeding brakes, pumping the brake pedal is not recommended because it tends to aerate the fluid.*

One problem with manual brake bleeding is that the master cylinder pistons are moved deeper into their bore than they usually travel. If the master cylinder is dirty, the seals will move past the area of the bore normally contacted by the seals, which is constantly wiped clean by the seals. When the seals are pushed across the dirty area beyond the normal seal wipe area during manual bleeding, the seals can be damaged.

SHOP TIP It is best if your helper does not depress the pedal further than ¾ of full travel. Tell them to position their left foot under the pedal to prevent excess travel. If your helper releases the pedal before you have tightened the bleed screw completely, air will be drawn back into the hydraulic system as the piston in the master cylinder retracts.

Manual Bleeding with a Hose

Another manual bleeding method uses a hose installed from the open bleed screw into a container with brake fluid in it (**Figure 58.24**). The pedal is depressed and released several times. Bleeding is complete when no more bubbles show in the jar.

This method can be done without a helper, but you will not be able to observe the bubbles in the container while you are applying the brake pedal. If you continually refill the master cylinder reservoir and use

a sufficient amount of brake fluid, this should not be a problem.

Pressure Brake Bleeding

Pressure bleeding is another method of bleeding brakes. Its advantage over manual bleeding is that the master cylinder seal is not dragged through the dirt in the bottom of the master cylinder bore.

A pressure bleeder is a canister with two chambers separated by a rubber diaphragm: an air chamber and a fluid chamber (**Figure 58.25**). The air chamber is filled with compressed air, which pushes the diaphragm against the fluid. Compressed air contains substantial moisture, which would be absorbed by the brake fluid if the diaphragm were not there to keep the air and fluid separated. Pressurized brake fluid is pumped through a hose to an adapter on the top of the master cylinder (**Figure 58.26**). When a

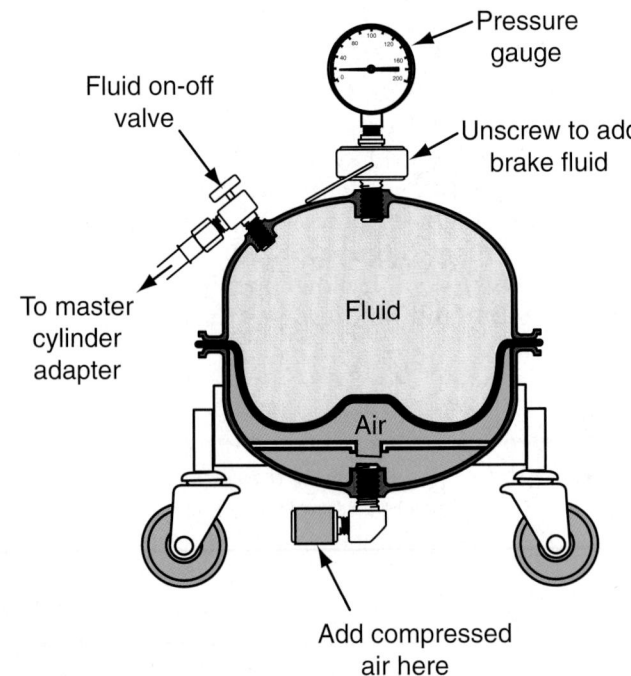

Figure 58.25 A pressure brake bleeder.

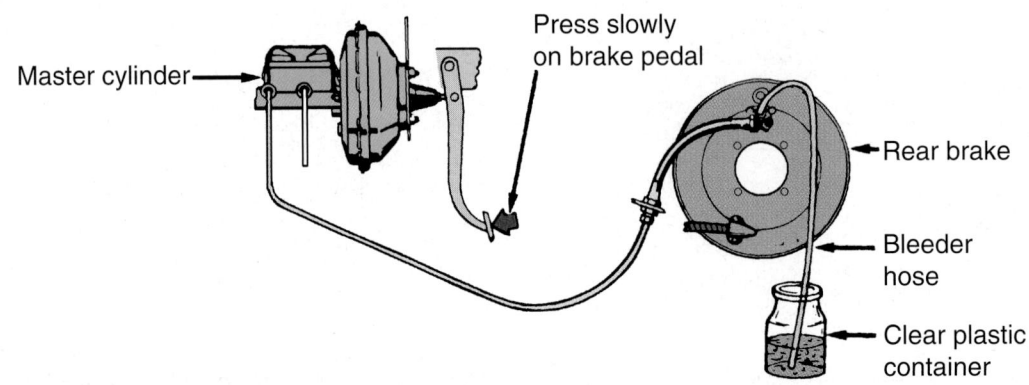

Figure 58.24 Manual brake bleeding procedure.

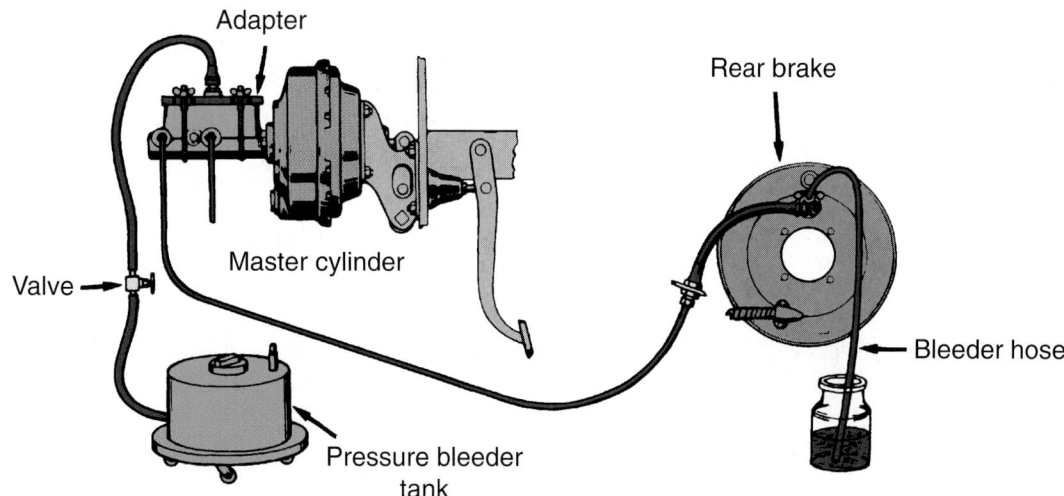

Figure 58.26 Pressure bleeder setup.

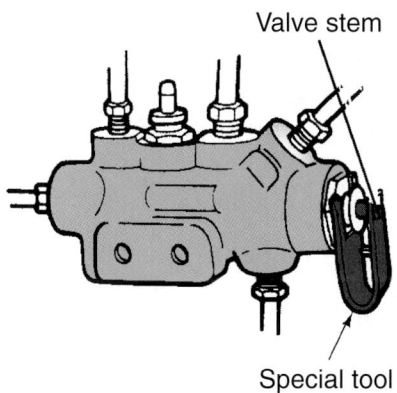

Figure 58.27 Installing this tool on the metering valve allows fluid to reach the wheel cylinders when pressure bleeding.

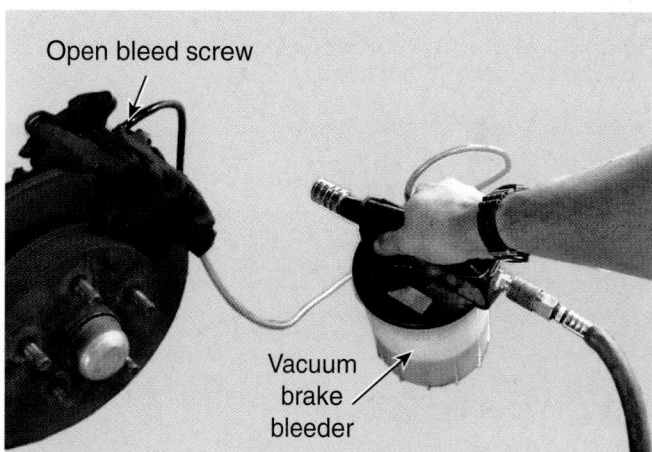

Figure 58.28 A vacuum brake bleeder. *(Courtesy of Tim Gilles)*

bleed screw at the wheel is opened, fluid and air are pumped out.

NOTE: *When bleeding brakes with a pressure bleeder, the metering valve must be disabled using a special tool. Otherwise, the metering valve will prevent fluid from reaching the wheel cylinders (**Figure 58.27**). This is not a problem when manually bleeding brakes using the brake pedal, because the metering valve can be opened with pedal pressure. Do not forget to remove the metering valve tool following bleeding.*

Vacuum Brake Bleeding

Vacuum bleeding, also known as *Vacula™* or *suction bleeding*, is very popular in the brake service industry. This bleeding method does not require adapters for the master cylinder. Setup is fast, and the shop floor remains clean because fluid is collected into a small hand-held vacuum container as bleeding is completed.

A vacuum bleeder pulls fluid from the master cylinder through the bleed screw by suction. A hose from the bleeder is connected to the open bleed screw. The vacuum bleeder is connected to a compressed air supply hose (**Figure 58.28**). Compressed air creates a vacuum as it moves past an opening in the brake bleeder's empty fluid container.

During bleeding, air can be observed in the clear fluid hose. Small air bubbles are normal to see as air leaks past the threads of the bleed screw (**Figure 58.29**). The air does not cause harm, because it is not entering the hydraulic system but going into the container with the spent fluid. Using silicone grease on the bleeder threads can minimize this condition.

> **SHOP TIP** Use the vacuum brake bleeder to empty a master cylinder of dirty brake fluid prior to bleeding the brakes (**Figure 58.30**).

Fluid Injection

The newest brake bleeding method is **reverse fluid injection (RFI)**, or Phoenix Injection™ (**Figure 58.31**).

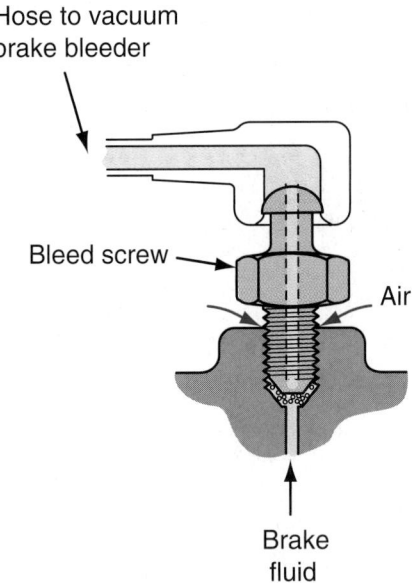

Figure 58.29 Air drawn past the bleeder screw threads may create foam in the fluid.

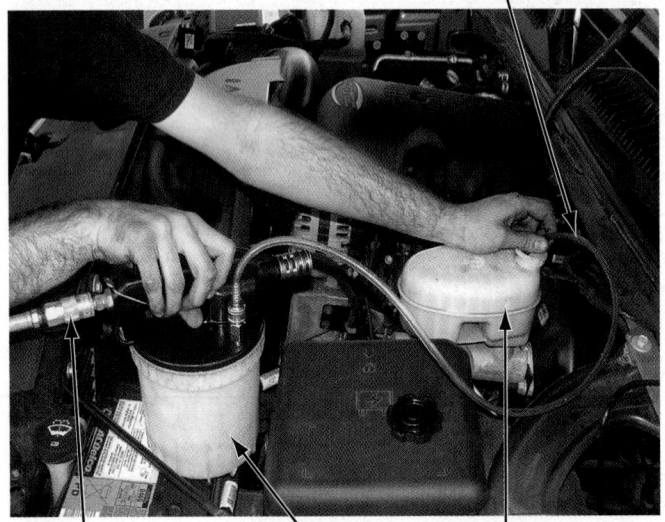

Figure 58.30 The vacuum bleeder can be used to empty the master cylinder reservoir. *(Courtesy of Tim Gilles)*

Fluid is injected through the bleed screws from the underside of the vehicle (**Figure 58.32**). The master cylinder reservoir is emptied at the start of the procedure. The fluid injector is also capable of vacuum and can be used to pump fluid from the reservoir into a container.

Air tends to rise to higher places in the hydraulic system, where it often becomes trapped. Other bleeding methods are sometimes not successful in removing trapped air bubbles. With RFI, the upward flow of fluid finds trapped air bubbles and carries them out of the

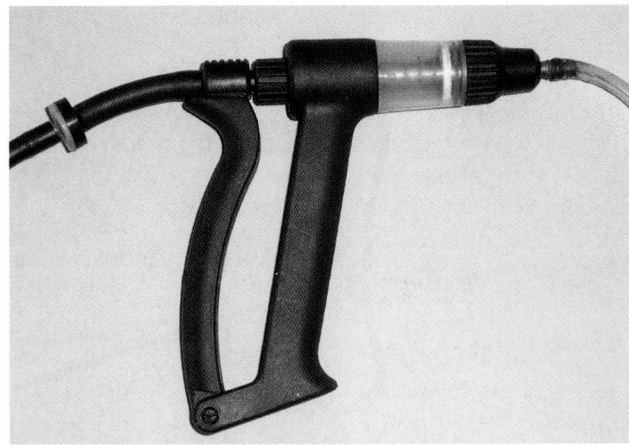

Figure 58.31 A Phoenix injector. *(Courtesy of Tim Gilles)*

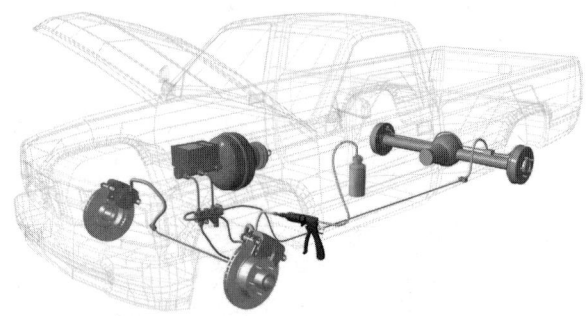

Figure 58.32 Reverse fluid injection (RFI). *(Courtesy of Phoenix Systems, L.L.C.)*

master cylinder into the reservoir. The reservoir refills during the injection process.

RFI is good with ABS, which often has small fluid passageways. Dirt tends to settle out of the brake fluid at the bottom of the master cylinder reservoir. With other bleeding methods, this dirt can be forced into the ABS, where it might plug a small passage. Injecting fluid from the bottom of the brake system to the top eliminates this potential problem.

Gravity Bleeding

Gravity bleeding is a simple and safe but time consuming way of bleeding all brake and clutch systems. To gravity bleed a system, open the bleed screws and allow the system to drain. Be sure to keep the master cylinder fluid level above the bottom of the reservoir.

 **SHOP TIP** In most cases, gravity bleeding will work on ABS that require special procedures.

Brake Bleeding Problems

Several problems can occur during brake bleeding. One problem is trapped air, which results in a chronic low, spongy pedal. Pockets of air form in high spots in the hydraulic system, such as the bends of brake lines and hoses. Some master cylinders are mounted on a slant, which also results in trapped air. With the nose of the master cylinder higher, air cannot escape from the vent port and the pressure chamber remains filled with air. This problem is solved by removing the master cylinder and bench bleeding it.

SHOP TIP Some technicians lift one end of the vehicle to make the master cylinder level during brake bleeding.

The bleed screw is located at the top of the caliper to allow air to escape during bleeding. Disc brake calipers are different on the left and right sides of the vehicle. When a caliper is installed on the wrong side of the car, the bleed screw will be on the bottom. Purging air from the caliper will be difficult, if not impossible.

SHOP TIP Sometimes a brake pedal is consistently low and soft, even after thorough and correct brake-bleeding practices. Air bubbles can become trapped within calipers. Air trapped by the piston seals, between the piston and bore, can be difficult to remove. Sometimes trapped air can be released by tapping lightly on the caliper with a hammer during brake bleeding.

Bleeding Antilock Brake Systems (ABS)

Some vehicles require a scan tool to cycle the ABS during bleeding. Check the manufacturers' bleeding procedures before bleeding brakes. The gravity method will work on many, but not all, of these vehicles.

NOTE: *Avoid accidental contamination of the brake hydraulic system with petroleum products. Rubber parts are used in the brake system. Rubber is not resistant to attack by petroleum.*

As an experiment, put an old rubber dust boot and wheel cylinder cup into a jar of engine oil. Contact with oil will cause rubber parts to swell. When this happens, the following are necessary:

- Completely disassemble the entire hydraulic system.
- Thoroughly flush the system with alcohol.
- Install new rubber parts.

Alcohol is the solvent of choice in cleaning the brake system. It can be used to clean petroleum, leaving no residue. It also absorbs water. Bleeding ABS brakes is covered later in this chapter.

SCIENCE NOTE

Rubber is a polymer (a molecule formed by joining many smaller molecules together) made up of smaller units of isoprene molecules. Natural rubber is tacky because the isoprene polymer chains are not connected to one another. When rubber is **vulcanized**, *sulfur bridges, or cross links, are formed between different polymer chains. These cross-links allow the polymer chains to be stretched to several times their original length without breaking. This elastic behavior makes synthetic rubber an elastomer. Synthetic rubbers are made by joining different molecules together, resulting in rubbers with different characteristics. Oil is soluble in rubber. It dissolves in and around the polymer chains, unlinking them and stretching them out. The result is that the rubber swells when it comes into contact with oil.*

■ ADJUSTING BRAKES

Brake fluid is not compressible. That is why it works to transmit pedal pressure to the wheel cylinders. What happens when air is in the fluid of the hydraulic system? Air *is* compressible, so a "spongy pedal" results. Air in the system will *not* cause a low, firm pedal, but improper brake adjustment will.

To make the correct diagnosis, pump the pedal twice quickly. If the pedal height rises higher on the second application, the brakes need adjustment. To understand why, a familiarity with of the operation of a simple master cylinder is needed (see Chapter 57).

Most brakes today have self-adjusting mechanisms, but some small cars and light trucks still require manual adjustment. Brake adjustment is covered later in this chapter.

SCIENCE NOTE

An understanding of hydraulics is valuable when it is applied to the theory of brake adjustment. Pascal's law states that pressure in an enclosed system is equal in all parts of that system. The key to this is the word "enclosed," which also means "not moving." When fluids are moving, different laws apply.

Consider this example. Both drum brakes on a rear axle are served by one of the master cylinder's chambers. They are adjusted to different lining-to-drum clearances. Will the car pull to one side? If both wheel cylinders have the same pressure applied to their pistons when the fluid is not moving, then the drum with the excessive clearance will be contacted by its linings before the cylinder on the other side of the car will allow pressure to build. In wet weather, the correctly adjusted wheel might grab before the loose one. But as soon as fluid movement in the system stops, both sides will begin to receive equal pressure.

■ MASTER CYLINDER SERVICE

Check Master Cylinder Reservoir Vent

When a master cylinder cover vent is obstructed, air can be drawn in at the back of the master cylinder, aerating the fluid. When there is air in the fluid, it becomes compressible and causes a spongy pedal rather than a firm one. When checking the fluid level, the vent can be easily checked for an obstruction.

■ MASTER CYLINDER REMOVAL

When removing the master cylinder from a vehicle, be sure to use fender covers to protect the paint from accidental spills. Loosen and remove the flared metal lines using a flare-nut wrench. When there is no vacuum power brake booster, disconnect the pedal pushrod from under the dash. This is not necessary on power brake cars. Remove the nuts that hold the master cylinder in place (**Figure 58.33**). Remove the master cylinder and dump the brake fluid in a suitable container for disposal.

■ MASTER CYLINDER DISASSEMBLY

When a master cylinder is defective, take it apart and use what you know about master cylinder operation to locate the source of the problem. Master cylinders can be purchased as new or rebuilt units. Rebuild kits are sometimes available for use when the cylinder bore is not damaged by corrosion. Liability and shop labor

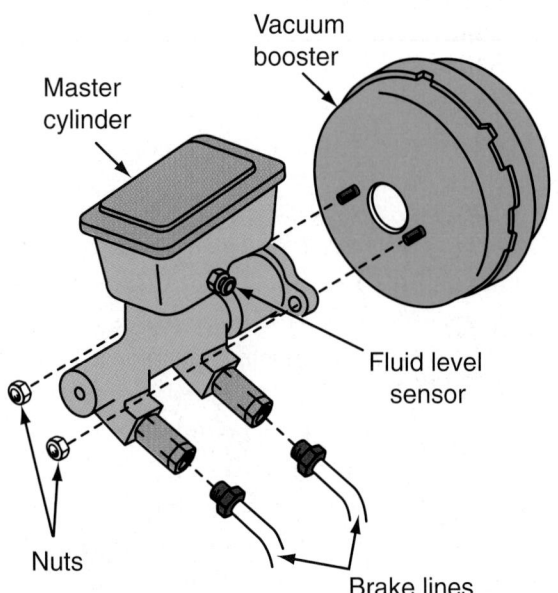

Figure 58.33 Disconnect the brake fluid lines, electrical connections, and attaching bolts when removing a master cylinder.

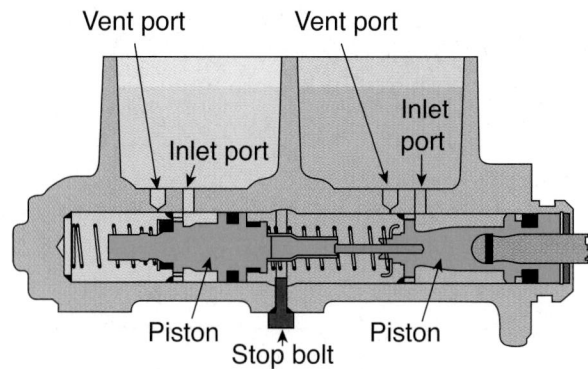

Figure 58.34 This master cylinder has a stop bolt at the front of the rear master cylinder piston.

cost issues have made the practice of a shop rebuilding a unit less popular than it was in the past.

To disassemble a master cylinder, push the piston deeper into the bore using a large Phillips head screwdriver or metal rod. While holding the piston in against piston return spring pressure, remove the circlip at the outside of the bore. Remove the piston from the bore. Sometimes, there is a stop screw in the bottom of the center of the bore that must be removed before the front piston can be removed (**Figure 58.34**).

Clean the inside of the cylinder bore and use a flashlight to inspect it. If the bore of a cylinder is corroded or pitted, the cylinder must be replaced.

Quick take-up master cylinders are serviced in the same manner as other master cylinders. They are usually made of aluminum, and aluminum cylinders cannot be cleaned with abrasives such as hone stones. They are anodized to provide a protective coating to the aluminum.

■ SCIENCE NOTE ■

Anodizing is a chemical process that accelerates and controls the formation of an oxide coating on aluminum. The oxide coating provides corrosion protection, abrasion resistance, and improves the metal's appearance. During the anodizing process, the part is first dipped in a series of cleaners to prepare its surface. Next, it is dipped in an anodizing tank full of an electrolyte (acid and water). When electrical current is applied to the electrolyte, oxygen atoms are attracted to the aluminum. This happens because the electrolyte is negatively charged (cathode) and the aluminum is positively charged (anode). An aluminum oxide begins to grow on the part like roots on a human hair. The oxide grows until the coating rises above the surface of the part.

Anodizing prevents the conduction of electricity. This is important when making such things as aluminum ladders.

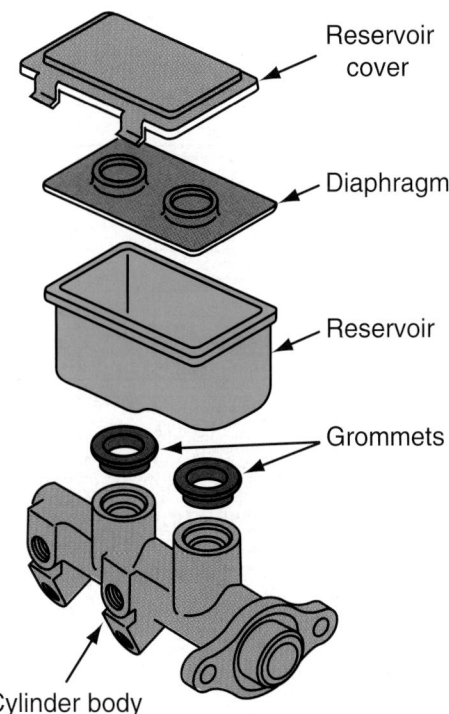

Figure 58.35 A master cylinder with a removable plastic reservoir.

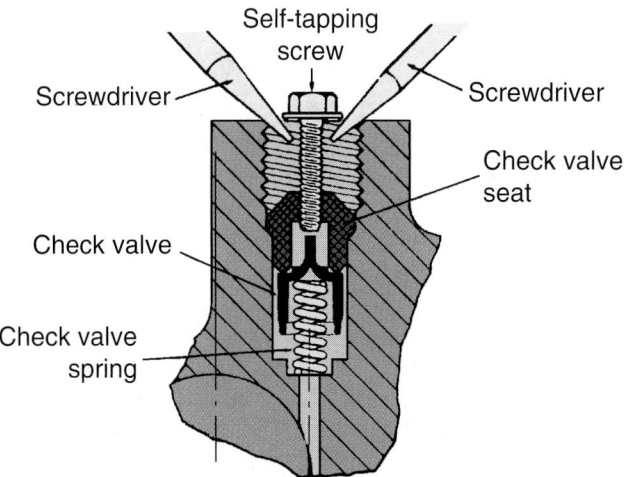

Figure 58.36 Removing a residual pressure check valve. (*Courtesy of Dana Corporation*)

If the reservoir is plastic (**Figure 58.35**), remove it and clean it with hot water. After drying it thoroughly, install it on the new master cylinder. Lube the grommets with brake fluid to ease installation. Be careful not to install it backwards.

> **SHOP TIP**
> To avoid cross-threading when reinstalling flared lines, turn the fitting counterclockwise first to help seat the threads. Then, turn the fitting carefully clockwise to tighten it. Jiggle the metal tubing while turning the fitting so you can turn it all the way against its seat using only your fingers. Then use a flare-nut wrench to tighten it securely. Be sure to use another wrench to hold the flare fitting from turning.

Some master cylinders for drum brake systems have a residual pressure check valve in the fluid outlet under the flare nut. If a check valve is used in a disc brake system, rapid lining wear will occur. The check valve is located under a brass insert that fits tightly in the fluid outlet. To remove the check valve, install a self-tapping screw in the hole in the middle of the insert. Then, pry it out with a pair of screwdrivers (**Figure 58.36**).

> **CASE HISTORY**
> *A student in an automotive brakes class installed disc brakes in place of drum brakes on a 1968 Camaro. He used the original master cylinder. About a year later, his girlfriend enrolled in the same brake class. She brought his car in for installation of new disc brake linings (for the third time since the disc brakes were installed). From what she had learned in class, she had a hunch that the residual pressure check valve might still be in the master cylinder. She removed it and solved her boyfriend's problem.*

A restricted check valve can cause excessive pedal travel before the brakes apply. To test the operation of the check valve, apply and release the brakes. Then open a wheel cylinder bleed screw. A brief spurt of fluid is normal as the residual pressure is released.

■ BENCH BLEEDING THE MASTER CYLINDER

Before installing a master cylinder, it can be filled with fluid and bled of air. **Figure 58.37** shows a master cylinder in a vise with two fabricated tubes returning

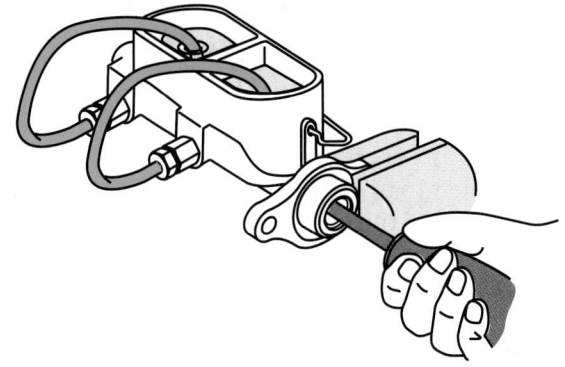

Figure 58.37 Procedure for bench bleeding a master cylinder.

fluid to the reservoir. Pumping the piston using a large screwdriver forces air and fluid out of the piston chambers. New, airless fluid enters the master cylinder through the reservoir. A variation of this process is to hold your fingers over the fluid outlets while releasing the pedal rod. This prevents air from being sucked back into the piston chamber. When the pedal rod is applied, fluid and air escape into a drain pan. The process is repeated slowly until all air is expelled from the master cylinder.

Quick take-up master cylinders have a step bore in which one piston is larger than the other. The large piston provides low pressure but high volume to take up clearance between the linings and rotor with less pedal travel. Feel the bottom of the master cylinder. If the casting is larger near the rear, it is a quick take-up cylinder. These are bench bled with a suction device. Without that, use a large Phillips screwdriver or wooden dowel to slowly depress the rear piston ¾ of the way into the bore. Do not bottom out the piston.

Adjusting Brake Pedal Free Travel

Adjustment of pedal free travel is not usually required, but it should be checked when the master cylinder is replaced. There should be less than ⅛" of free play between the brake pushrod and the back of the primary master cylinder piston or power booster. This translates to free pedal travel of about ¼" to ½".

■ BRAKE JOB

Front or rear linings are replaced in pairs. Parts are purchased in *axle sets* (both wheels on the front or the rear of the car).

> **SHOP TIP** It is a good idea to disassemble only one side of an axle at a time. The other side can then be used for comparison until the job has been repeated often enough so that a reference is no longer needed.

Be sure to consult the appropriate service manual for the vehicle being worked on. The major suppliers of brake parts have brake service manuals available to users of their products.

A complete brake job includes all of the internal hydraulic parts of the wheel cylinders and disc calipers, as well as new brake fluid throughout. New hardware and springs are often included. Drums or rotors may be turned on a lathe. A master cylinder and new hoses are usually *not* included in a brake job. **Figure 58.38** shows what one manufacturer recommends for a complete brake job.

THE CONDITION OF THE BRAKE SYSTEM ON THIS VEHICLE IS AS FOLLOWS:

BRAKE SYSTEM COMPONENTS	REQUIRED SERVICES	$ ESTIMATE PARTS	$ ESTIMATE LABOR
Diagnosis ABS	❏ Recondition ❏ Replace		
Disc Pads*	❏ Replace Front ❏ Replace Rear		
Disc Caliper*	❏ Recondition ❏ Replace		
Disc Hardware*	❏ Replace		
Disc Rotor*	❏ Resurface ❏ Replace		
Grease Seals*	❏ Replace		
Front Wheel Bearings*	❏ Repack ❏ Replace		
Brake Shoes*	❏ Replace Front ❏ Replace Rear		
Brake Drums*	❏ Resurface ❏ Replace		
Wheel Cylinders*	❏ Recondition ❏ Replace		
Brake Hardware*	❏ Replace		
Parking Brakes	❏ Adjust/Lubricate ❏ Replace		
Power Brake Booster	❏ Service ❏ Replace		
Master Cylinder	❏ Recondition ❏ Replace		
Brake Fluid*	❏ Flush System, Add New Fluid		
Lines, Hoses, Combination Valve	❏ Replace		
Stop Light	❏ Replace Bulb ❏ Replace Switch		
Other			
	SUB-TOTAL		

*Total brake service includes reconditioning or replacement of these items.

Figure 58.38 Items in a complete brake job. *(Courtesy of Federal-Mogul Corporation)*

Ethics in Brake Work

A common practice in recent years was for large merchandisers to advertise incomplete brake jobs at very low prices. Once the car was on the lift, the cost of the job often rose rapidly as needed items were sold to the customer. When the job was finished, the price was often much higher than it would have been if the customer had patronized his or her regular repair shop. Consumer affairs divisions in some states have determined this type of selling to be unfair advertising. When an incomplete brake job is advertised in those states, a disclaimer must be included that says something to the effect that the job may cost substantially more when other necessary parts are added.

■ DRUM BRAKE LINING REMOVAL

Before disassembling brakes, clean the entire assembly using either a high-efficiency particulate arresting (HEPA) vacuum or a brake parts washer (see Chapter 3).

Brake linings are made of asbestos or an asbestos substitute. Brake dust is dangerous to breathe. A HEPA vacuum is used for safely removing asbestos dust from brakes. The vacuum will trap the small asbestos fiber, rather than allowing it to escape through the vacuum bag. The HEPA vacuum fits over the brake assembly, and the technician's hands go into gloves in the side. There is a window for viewing and a blow gun for blowing off the parts while the vacuum sucks the dust away.

CAUTION Never blow off brake assemblies unless a HEPA vacuum is installed.

Low-pressure wet brake washers are the most popular way of cleaning brake assemblies. They are fast, convenient, and do a good job cleaning the brakes in preparation for service. There are companies that provide regular service for these cleaners.

SAFETY NOTE Brake dust is hazardous to breathe. Repair shops are required by law to have a brake parts washer or a HEPA vacuum when servicing brakes.

CASE HISTORY *An apprentice was cleaning brakes with chemical brake cleaner dripping into a drain pan positioned beneath the axle. An incandescent shop light (trouble light) was hanging above the brakes, illuminating the work area. He accidentally knocked the trouble light into the pan of brake cleaner, instantly sparking a fire.*

The removal of brake linings is accomplished using special tools. **Figure 58.39** shows a typical tool used to remove and reinstall drum brake springs. Before disassembly, be sure to carefully inspect the linings being replaced. Sometimes there are differences in materials or lengths of the primary and secondary linings.

Figure 58.39 A special tool is used to remove (Left) and replace (Right) brake springs. *(Courtesy of Tim Gilles)*

■ REBUILDING HYDRAULIC CYLINDERS

Hydraulic wheel cylinders and disc brake calipers can be rebuilt if they are made of cast iron and are not corroded. Cylinders can often be rebuilt on the car while they are still bolted to the backing plate. After removing the brake linings, remove the dust covers from the wheel cylinders. Push on the piston on one side of the cylinder to force the parts from the cylinder (**Figure 58.40**). Fluid will leak out.

Two kinds of hone are available: one has two or three stones, and the other is a flex hone similar to those used in engine rebuilding (**Figure 58.41**). A two-stone hone can go into smaller bores. It is used for rebuilding master cylinders and smaller import car cylinders.

After the wheel cylinder is lightly honed to clean it up (**Figure 58.42**), new rubber parts are installed from a **wheel cylinder kit** (**Figure 58.43**).

NOTES:

■ *If the size of the wheel cylinder is increased during honing by more than 0.005", the cylinder must be replaced.*

■ ***Anodized aluminum*** *cylinders, found on many late-model cars, should not be honed because their hard surface layer will be removed and they will corrode.*

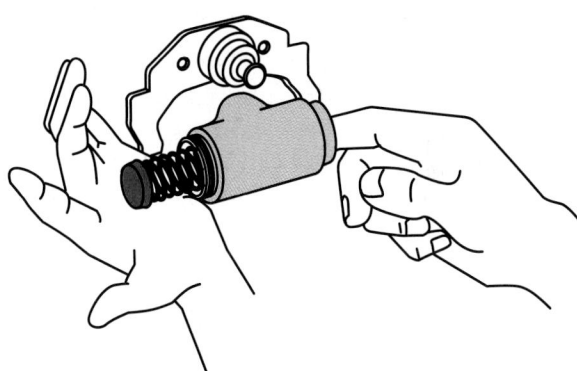

Figure 58.40 Push the piston on one side of the cylinder to force the parts from the cylinder.

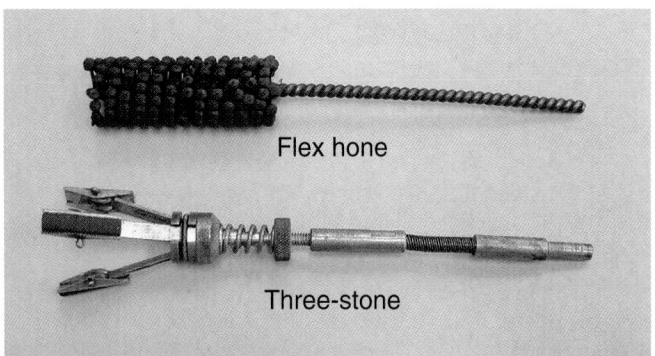

Figure 58.41 Two types of brake cylinder hones. *(Courtesy of Tim Gilles)*

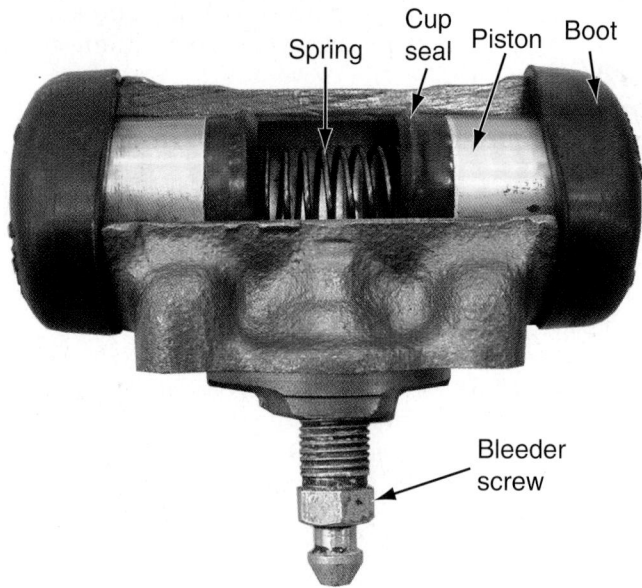

Figure 58.42 Honing a wheel cylinder. *(Courtesy of Federal-Mogul Corporation)*

Figure 58.44 Location of parts in a wheel cylinder. *(Courtesy of Tim Gilles)*

- *If in doubt as to whether a wheel cylinder is cast iron or aluminum, check with a magnet. A magnet will not be attracted to an aluminum cylinder.*
- *If an aluminum cylinder's bore has pit marks, it must be replaced rather than rebuilt.*

REASSEMBLING A WHEEL CYLINDER

Reassemble the wheel cylinder as shown in **Figure 58.44**. The lips on the wheel cylinder cups must face toward the fluid. A seal lip installed backwards will result in fluid leaking from the wheel cylinder when the brakes are applied (**Figure 58.45**).

When reassembling brake hydraulic parts, use brake fluid liberally as an assembly lubricant. After assembling a wheel cylinder, use a clamp to hold the assembly together while installing the linings (**Figure 58.46**).

REMOVING WHEEL CYLINDERS

When a wheel cylinder bore is pitted or corroded, it must be replaced. Use a flare-nut wrench to remove the brake tubing fitting where it threads into the back of the cylinder. The cylinder is held to the backing

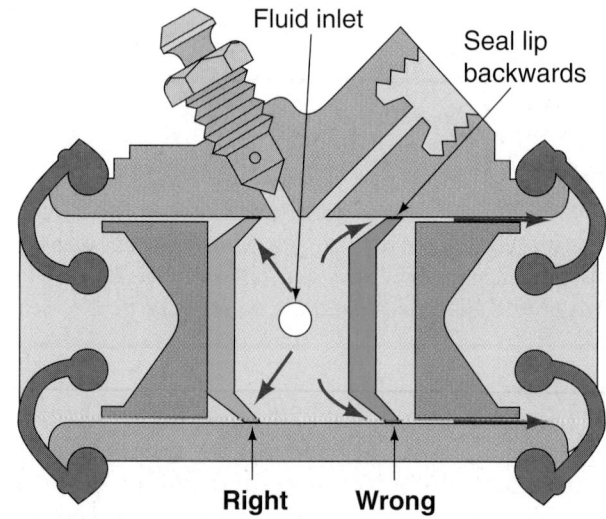

Figure 58.45 A lip seal installed backwards will leak.

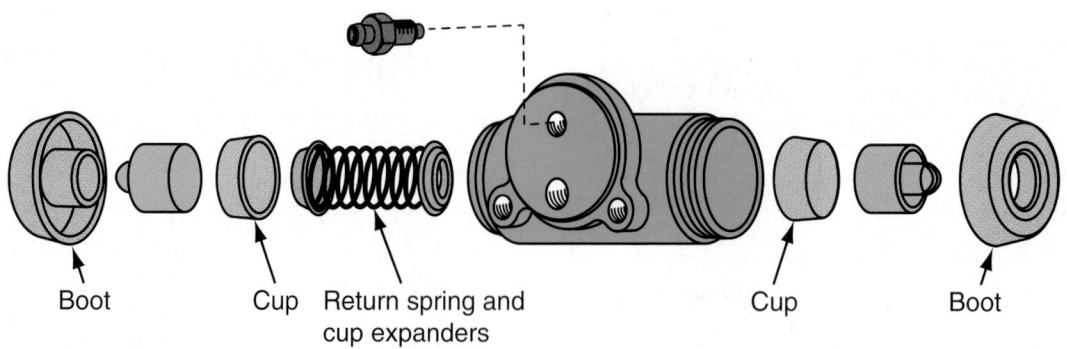

Figure 58.43 Parts of a wheel cylinder kit.

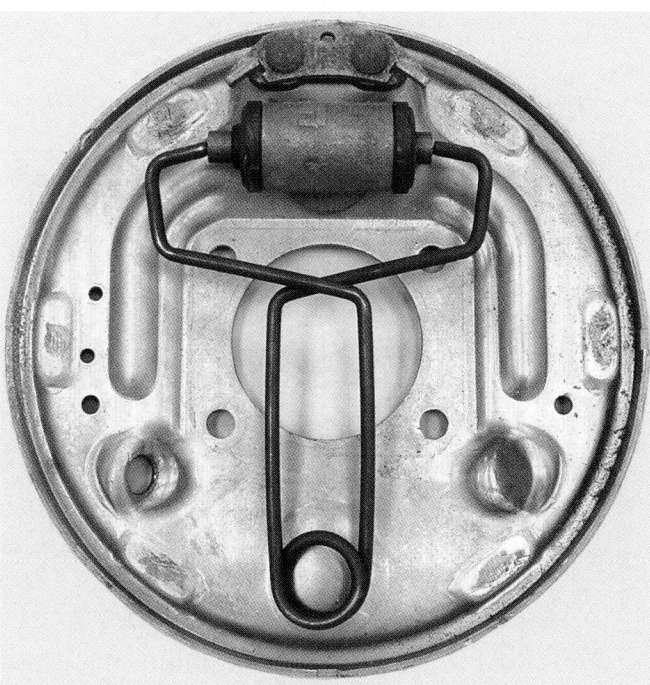

Figure 58.46 A brake cylinder clamp. *(Courtesy of Tim Gilles)*

plate by two screws or a clip that requires a special service tool.

◼ REPLACING DRUM BRAKE SHOES

Clean the backing plates. Use high-temperature lubricant to lubricate the pads on the backing plates that the linings slide on after checking them for grooves (**Figure 58.47**). When reinstalling brake shoes on the backing plates, it is sometimes easier to assemble the springs to the brake linings first. Then, the linings can be folded into each other making installation easier (**Figure 58.48**).

Figure 58.47 First check the backing plate pads where the linings slide for wear. Then lubricate them with a small amount of high temperature lubricant. *(Courtesy of Tim Gilles)*

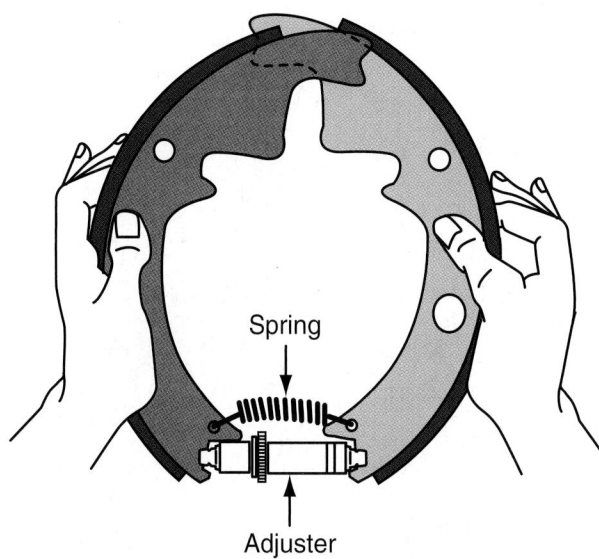

Figure 58.48 The linings can be folded into each other to make installation easier.

Check the return springs visually. Shoe return springs should not be loose or broken. The paint on the springs should be in good condition, indicating that they have not been overheated. When a return spring has been removed from the brake assembly, see if a piece of paper can be inserted between its coils. If so, replace the spring. Another test is to hold the spring up to light. If light is visible between the coils of the spring, replace it.

 **SHOP TIP** If a spring is dropped on a hard surface, it should not ring or bounce.

Self-adjusters must be reinstalled on the correct side of the car because they have either right- or left-hand threads and cannot be interchanged (**Figure 58.49**). Lube the threads of the self-adjusting screw with high-temperature lubricant.

◼ ADJUSTING DRUM BRAKE CLEARANCE

After brake linings are installed on the backing plates, an initial clearance adjustment is made before the drums are installed. A brake adjusting gauge is adjusted to the size of the drum (**Figure 58.50**). Then, the star-wheel of the brake adjuster is turned until the shoes expand to the size of the adjusting gauge. This will provide about 0.010" of clearance between the lining and drum. If additional adjustment is necessary after the brake job is completed, drive the car slowly in reverse while applying the brakes repeatedly. This completes the adjustment.

Figure 58.49 (a) Self-adjuster levers are not interchangeable from side to side. (b) The end of the adjusting screw has a left (L) or right (R) designation. *(Courtesy of Tim Gilles)*

To do an adjustment while the drum is installed requires accessing the starwheel through a hole in the backing plate or front of drum. While turning the wheel by hand, an adjusting tool called a **brake**

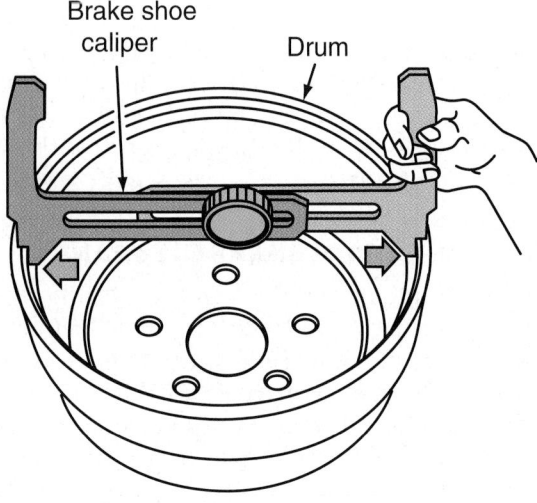

(a)   Setting tool to drum

Figure 58.50 A brake adjusting gauge. (a) Adjust the tool to fit the drum.

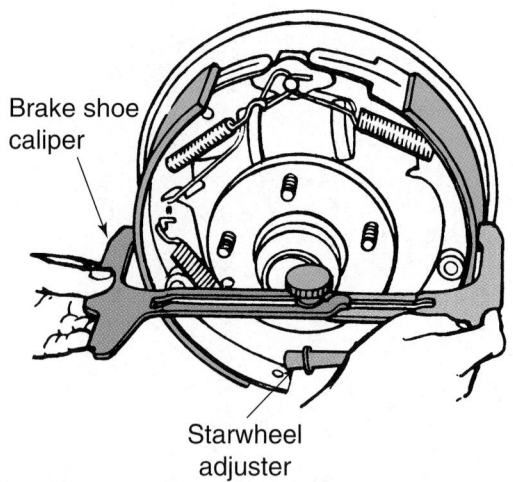

(b)   Setting brake shoes to tool

Figure 58.50 (continued) (b) Adjusting the brake shoes to fit the tool will provide a good initial drum-to-lining clearance.

spoon is used to turn the starwheel until the wheel will no longer turn (**Figure 58.51**). Then, the self-adjusting mechanism is held out away from the starwheel (see Figure 58.21) while loosening it about five to ten teeth until the wheel turns freely once

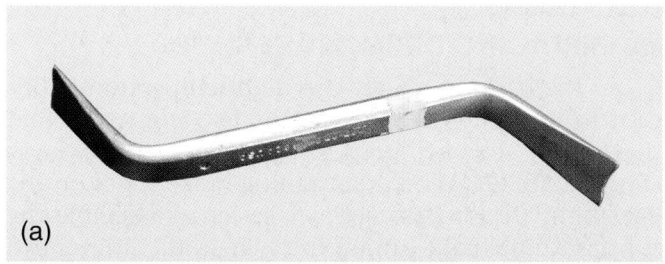

Figure 58.51 (a) A brake spoon used for adjusting drum brakes. (b) Inserting a brake spoon through the adjuster access hole in the backing plate. *(Courtesy of Tim Gilles)*

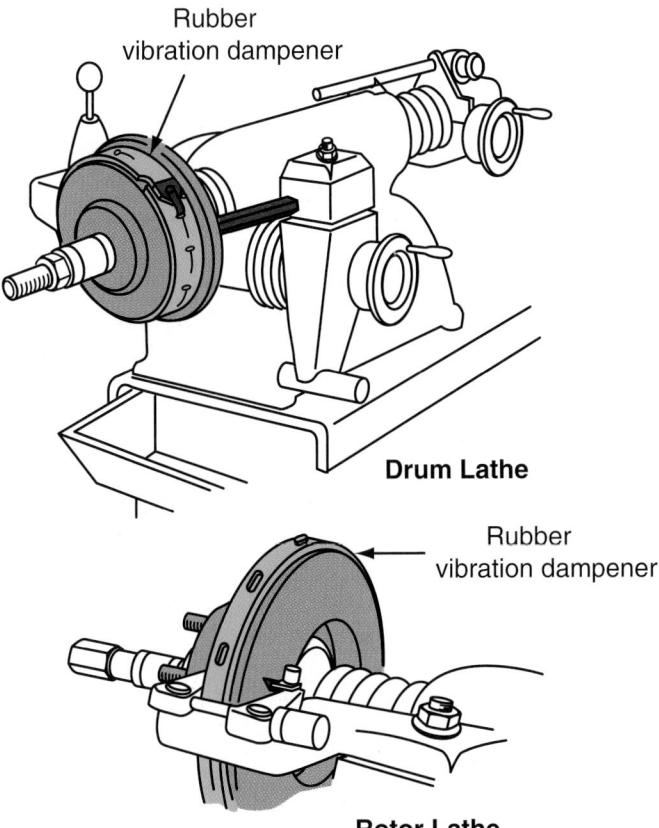

Figure 58.52 Typical multiuse brake lathe setups for machining drums and discs.

again. During tightening, the self-adjuster does not have to be held out of the way. The starwheel will move easily in that direction. Older cars and trucks without self-adjusters use the same procedure, but the self-adjusting mechanism does not have to be held out of the way during loosening.

■ DRUM AND ROTOR SERVICE

When drum or disc linings are replaced, it is a common practice to resurface (*turn*) the drums or rotors using a drum or rotor lathe (**Figure 58.52**). If too much metal needs to be removed to clean up its surface, the drum or rotor must be replaced. Service manuals (available from brake parts manufacturers) give the maximum discard diameter for brake drums and the minimum thickness of disc rotors.

CAUTION When using a brake lathe to machine a disc or a drum, do mot wear loose clothing. Hanging jewelry or long hair should be secured ot of the way.

■ DRUM SERVICE

A drum is measured with a special caliper. The one in **Figure 58.53** has a dial indicator. Since 1972, the maxi-

Figure 58.53 The flat areas on the drum mike rest on the edges of the drum. *(Courtesy of Tim Gilles)*

Figure 58.54 The maximum diameter is cast into a brake drum. *(Courtesy of Tim Gilles)*

mum allowable size for a brake drum has been cast into the outside or inside of brake drums (**Figure 58.54**). Each brake job usually results in a drum turned to 0.030" oversize. The maximum amount that can be cut from most drums is 0.060". A typical drum will list a **discard diameter** that is 0.090" larger than stock. This means that the drum can be machined 0.060", leaving 0.030" for future wear. The size listed on a 10" drum will be 10.090", leaving 0.030" for wear after turning to 0.060".

NOTE: *A 0.030" increase in the diameter means that 0.015" of surface wear has occurred (**Figure 58.55**).*

Inspect drums to see that they are not out-of-round or scored. Be sure there is no grease or oil on the drum. Cast iron absorbs oil. It is very difficult to remove all of the oil from the pores of the metal when an axle seal has leaked. Soaking the drum in a caustic tank is sometimes sufficient to remove the oil from the iron.

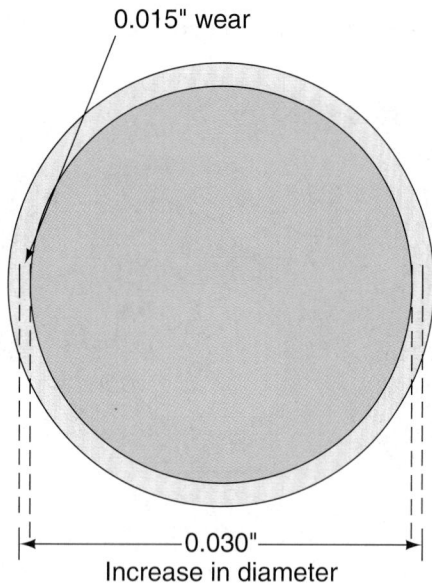

Figure 58.55 Surface wear of 0.015" results in a 0.030" change in diameter.

Turn on the motor and watch to see that the drum or rotor turns without any visible runout (wobble). If the part is not mounted properly, disassemble it and remount it.

NOTE: *Noise is the result of vibration. Vibration can cause a very rough cut when turning parts on a lathe. To prevent this, a silencing band is used (see Figure 58.52).*

The end of the cutter bit has a **carbide** insert. Carbide is a harder metal than ordinary tool steel. It has a longer life if vibration is kept to a minimum. When cutting with carbide, a minimum cut of over 0.005" is recommended. A typical cut will increase the diameter of a drum by 0.015".

Sometimes, **hard spots** appear in the surface of a drum or rotor. This condition happens as a result of excessive heat. The spots, which are blue in color, are places where the metallurgical composition has been changed from cast iron to steel. Attempting to cut them on a lathe results in high spots in the finish. The

Some drums have lug studs that are swaged to hold the drum to the wheel hub (**Figure 58.56**). If a drum with **swaged lugs** must be replaced, it must first be removed from the wheel hub. A cutter is used to assist in removing the studs. When the new stud is installed, a tool is used to deform it to hold it tightly to the hub.

Turning Drums and Rotors

Drums and rotors can be mounted on a lathe using adapters that fit into the wheel bearing races (**Figure 58.57**). Rear drums and FWD rotors are usually mounted on tapered cones and clamped between adapter cups.

NOTE: *Before turning any drums or rotors, any rust needs to be removed from mounting surfaces to eliminate runout.*

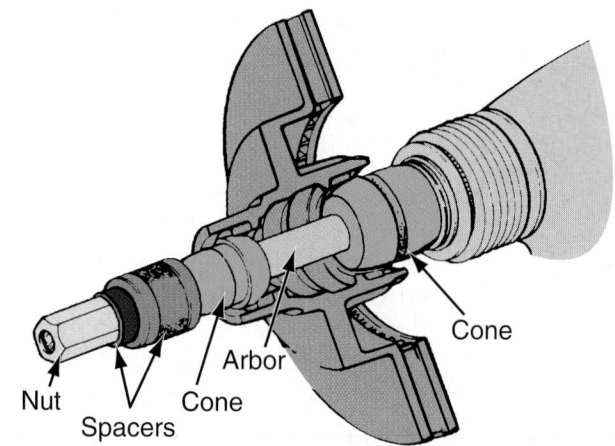

Figure 58.57 Cone adapters fit into the wheel bearing races. (*Courtesy of Federal-Mogul Corporation*)

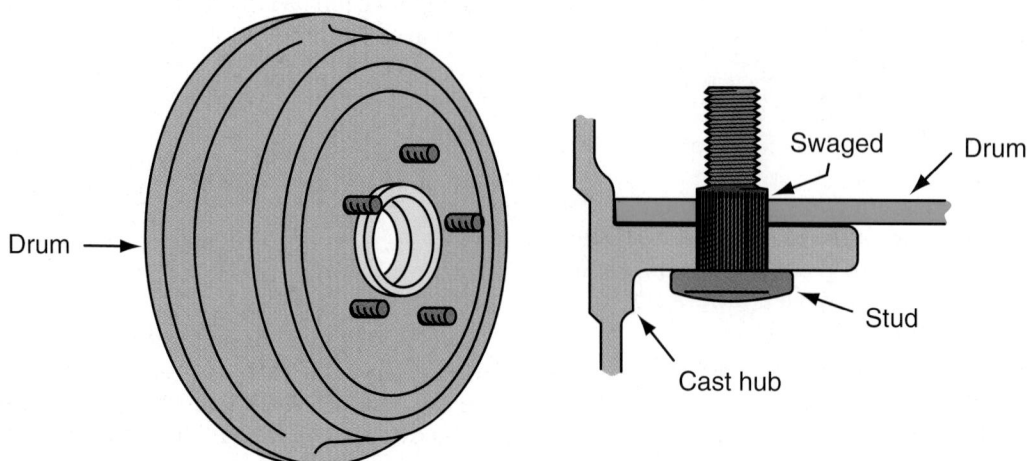

Figure 58.56 Some wheel studs are swaged, deforming the stud to hold it tight against the drum.

carbide cutter cuts the surrounding material but skips over the hard spots. Grinding can be used to remove hard spots, but they may return.

ROTOR SERVICE

A rotor in good condition will most likely provide a better surface than a freshly machined rotor. A used rotor surface on properly operating brakes will be smooth and have some friction material embedded within the pores of the metal. A used rotor can provide better stopping performance than a new one.

When a used rotor is not machined, it is important that the corresponding rotor on the other side of the vehicle be not machined as well. Otherwise a brake pull can result. Rotors are always machined in pairs. If there is any doubt as to the quality of the rotor's surface finish and straightness, remachine or replace it.

The thickness of a rotor must be measured to see if it is still usable. There are special micrometers available that have a pointed tip so the depth of a groove can be measured. This will save machining a rotor that will turn out to be too thin.

NOTE: *According to Wagner Brake, turning the rotors is only necessary if there is pedal pulsation due to warpage. Score marks up to 0.050" deep are said not to affect brake operation.*

Resurfacing drums or rotors any time the linings are replaced is a good idea because it provides a smooth surface. This results in less break-in time for the new lining material. Check rotors for parallelism (**Figure 58.58**) and runout (see Figure 58.11).

When machining a rotor, it is best to cut both sides at once, maintaining equal force on both sides. This reduces the rotor's tendency to distort during machining. Sometimes one side of the rotor is worn more than the other. In this case, remove less metal from the unworn side. Rotors with fixed calipers, however, should have the same amount of metal removed from each side.

A lathe cuts a thread in the surface of the rotor as it is machined (**Figure 58.59**). The thread will be very fine if the lathe cross-feed speed is slow, or it can be coarse if the cross-feed speed is fast. The better the quality of the rotor surface finish, the less tendency there will be for the pads to make noise. The final surface finish should be quite smooth. Feel the finish of a resurfaced rotor with your fingernail. It should be at least as smooth as the finish on a new rotor. **Figure 58.60** lists recommended speeds and feeds.

A *nondirectional finish* (**Figure 58.61**) can be applied to the face of the finished rotor using a rotating sander (**Figure 58.62**) Use 120- to 150-grit sandpaper for about 60 seconds, or until the surface is smooth.

Figure 58.58 Checking a rotor for parallelism. *(Courtesy of Federal-Mogul Corporation)*

Silencing band

Figure 58.59 Threaded appearance as a rotor is refinished. *(Courtesy of Tim Gilles)*

SHOP TIP To get the best possible finish that is equal on the inside and outside surfaces of the left and right side rotors, use a fresh piece of sandpaper on each side every time.

NOTE: *When cleaning a disc or drum lathe, do not use compressed air. Metal chips can get inside of the machine, causing damage.*

Many new rotors allow for only 0.030" of machining. Some allow as little as 0.020". Specification charts are available from all of the brake manufacturers. It is not necessary to refinish new rotors.

Rotor Refinishing Guide

	Rough Cut	Finish Cut
Spindle Speed		
10" & under	150–170 RPM	150–170 RPM
11"–16"	100 RPM	100 RPM
17" & larger	60 RPM	60 RPM
Depth of Cut (Per Side)	0.005"–0.010"	0.002"
Tool Cross Feed (Per Rev.)	0.006"–0.010"	0.002" max
Vibration Dampener	Yes	Yes
Sand Rotors Final Finish	No	Yes
Sanding Instructions	60 seconds per side with 150-grit sandpaper	
Cleaning Instructions	Use brake parts cleaner and dry with paper towels	

NOTE: Some resources say to wash the rotor in hot, soapy water, rinse thoroughly and dry. This process pulls metal shavings out of the casting pores.

Instructions for breaking in the new disc brake pads:
Make ten stops from 30 mph to 5 mph under moderate braking pressure. Allow the brakes to cool between stops.

Figure 58.60 Recommendations for rotor refinishing. *(Courtesy of Federal-Mogul Corporation)*

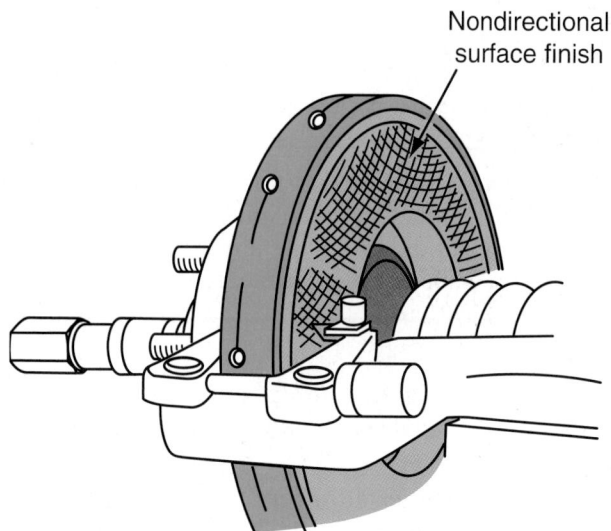

Figure 58.61 A nondirectional crosshatch finish smoothes the threads in the rotor surface.

Nondirectional surface finish

Figure 58.62 Use an orbital sander to sand off the threaded finish while the lathe turns. *(Courtesy of Tim Gilles)*

Wash the Machined Surfaces

It is very important to wash freshly machined surfaces. If you fail to do this, loose metal dust left in the pores of the machined surfaces will become embedded in the friction material, resulting in noise and wear. The only solution will be to replace the new friction materials for a second time. Use soap and hot water or a nonpetroleum solvent like brake parts cleaner or denatured alcohol. Do not dry with compressed air, because it contains small amounts of oil. Paper towels work well. Hot water will evaporate quickly on its own.

SHOP TIP When machined surfaces have been freshly cleaned, do not touch the friction surfaces. Pick up a drum or rotor from the outer edge or from the center hole. It is a good idea to wear vinyl gloves when assembling brake parts.

ON-VEHICLE ROTOR MACHINING

Some earlier disc brakes had rotors that were difficult to remove from the vehicle, sometimes resulting in damaged front wheel bearings. In response to this, on-the-car lathes were developed (**Figure 58.63**). In recent years, runout tolerances have become tighter and on-the-car lathes are becoming more popular once again. On-car machining is also a good choice when machining composite rotors.

Because the rotor does not need to be removed, there will not be a chance for foreign material to become lodged between the rotor and hub. Also, machining the rotor while it is on the hub ensures a true rotor and hub assembly.

Figure 58.63 An on-the-car lathe. *(Courtesy of Tim Gilles)*

CAUTION One extra caution is needed when turning rotors on the vehicle. Rotors are made of iron, which is magnetic. Metal chips removed from the rotor as it is machined will stick to antilock brake wheel speed sensors. Be sure to clean any chips from the sensor before reassembly.

Installing a Rotor

Before reinstalling a brake rotor, clean any rust preventive material from its surface. Be sure that any rust or dirt is removed from the rotor and hub mating surfaces. Failure to do this can result in lateral runout of the rotor. Some rotors are designed to be installed on only one side of the car (**Figure 58.64**). The fins on these rotors slant in one direction.

When tightening lug nuts, be sure to use a torque wrench and torque in a star pattern to avoid warping the rotor.

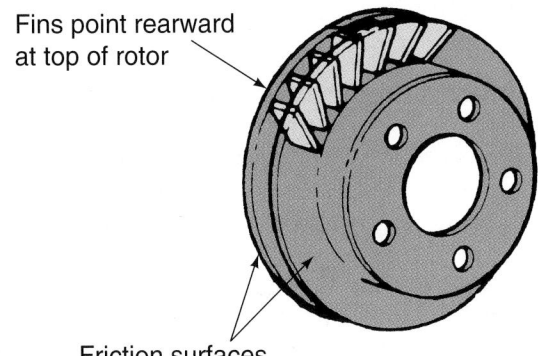

Fins point rearward at top of rotor

Friction surfaces

Figure 58.64 Some rotors are designed to be installed on only one side of the car. *(Courtesy of Federal-Mogul Corporation)*

■ DISC BRAKE SERVICE

Replacing Disc Linings

Disc linings are usually easy to replace. Occasionally, the linings on fixed-type calipers can be replaced without removing the caliper from its mount (**Figure 58.65**). This type of caliper will have an access hole in the top through which the pads can be removed and installed.

Most caliper designs require removal of the caliper to remove the pads. **Figure 58.66** shows typical pin slider and center abutment caliper removal.

Some calipers are easily removed by loosening two bolts on the back of the spindle support. Sometimes there are other bolts that hold the caliper together. Removing these when doing a simple pad replacement or rotor turning will result in needless disassembly of the caliper (**Figure 58.67**).

When the caliper is unbolted, use wire or a hook to hold it so its weight is not allowed to hang on the hose (**Figure 58.68**). The hose can be damaged internally.

Before replacing linings, spin the rotor and inspect it for roughness on the front *and back* that could result from a worn-out brake pad.

NOTE: *Be sure to check both sides of the rotor. Sometimes one pad wears out and not the other.*

Prior to installing replacement pads in a floating caliper, check the condition of its slides. Before installing the pads, slide the caliper back and forth to verify that it can slide freely on its pins or ways. Lube the slides with high-temperature lubricant.

When replacing linings on floating calipers, a C-clamp or a large pair of pliers can be used to move the piston back in its bore (**Figure 58.69**). This is so that the caliper can be easily removed and the new

Figure 58.65 Installing pads on a fixed caliper with access through the top of the caliper. *(Courtesy of Tim Gilles)*

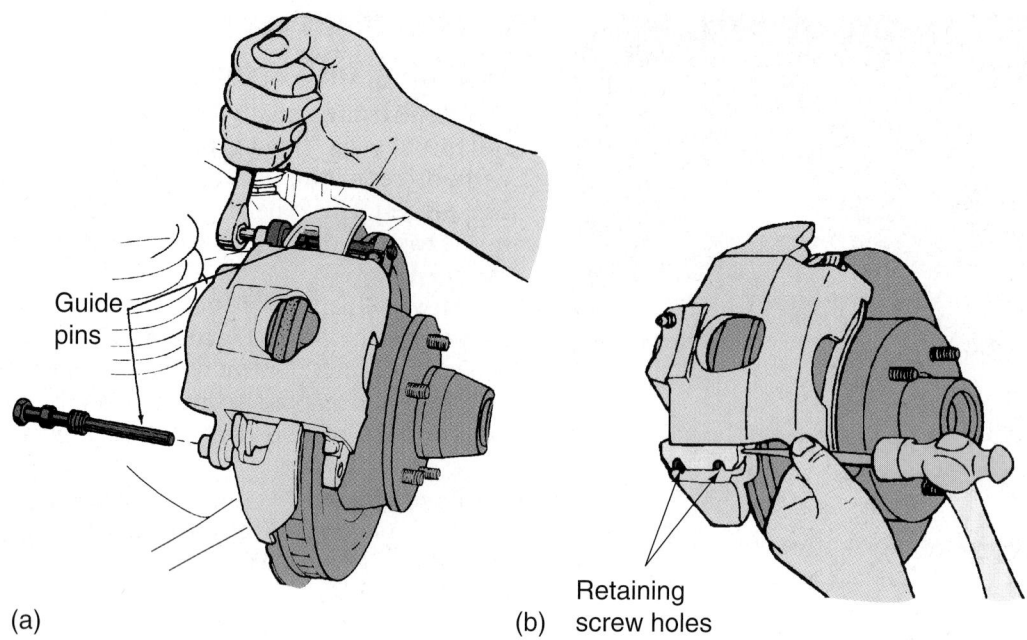

Guide pins

Retaining screw holes

(a) (b)

Figure 58.66 Removing a caliper. (a) Pin slider. (b) Center abutment. *(Courtesy of Federal-Mogul Corporation)*

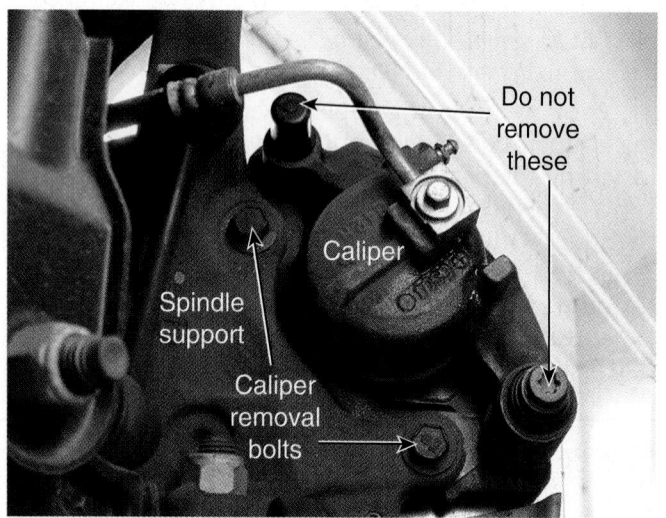

Do not remove these

Caliper

Spindle support

Caliper removal bolts

Figure 58.67 When removing a caliper, be sure to remove the correct bolts. *(Courtesy of Tim Gilles)*

Wire

Figure 58.68 Do not allow the caliper to hang from the brake hose. *(Courtesy of Tim Gilles)*

linings, which are considerably thicker than the worn ones, will fit during reinstallation of the caliper.

NOTE: *Before attempting to push the piston back in its bore, open the bleed screw on the back of the caliper. Then, move the piston all of the way back into its bore. Tighten the bleed screw immediately so it is not accidentally left loose. If left loose, the master cylinder will empty of fluid.*

Opening the bleed screw before retracting the piston is very important. Rust and sediment result as moisture accumulates in the brake fluid. Disc brake calipers and wheel cylinders are the lowest points in the hydraulic system, and they tend to be the dirtiest

areas. When performing a simple brake pad replacement, some unknowing technicians will bottom out the pistons in the caliper bores without first opening a bleeder screw. This forces the sediment from the low parts of the system back into the ABS and master cylinder reservoir. It can also push fluid out the top of the master cylinder if a previous service shop has topped off the fluid level. This is one of the reasons topping off fluid is not recommended.

When the caliper has already been removed from the vehicle, a floating caliper piston can be pushed into its bore as shown in **Figure 58.70**.

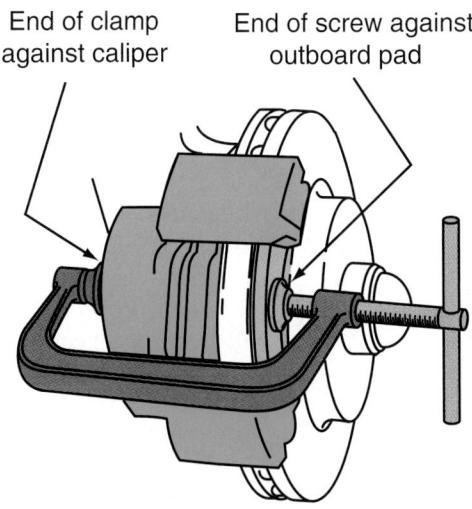

Figure 58.69 Use a C-clamp or large pliers to move the piston back in its bore.

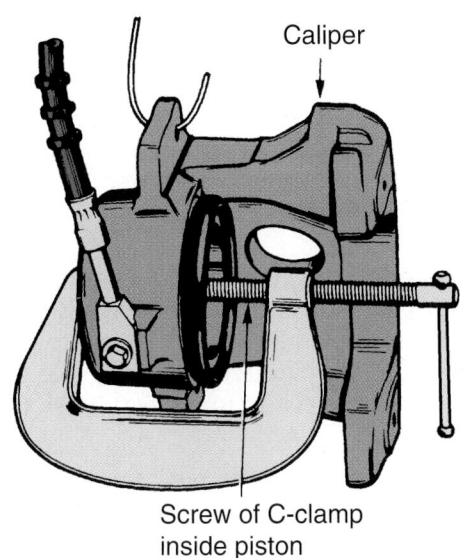

Figure 58.70 Push the piston to the bottom of its bore.

Figure 58.71 (a) This tool spreads the pads to compress pistons in fixed calipers and two-piston floating calipers. (b) A piston compressor used with the old pad to push the piston back into the bore. *(Courtesy of Tim Gilles)*

Attempting to compress pistons in a fixed caliper or a two-piston floating caliper presents a problem. Pushing in on one piston results in the other piston coming out of its bore. All pistons must be compressed at the same time. A tool for that purpose is shown in **Figure 58.71a**. The old pad is used to assist in this operation. Another tool for fixed calipers and multiple piston calipers is shown in **Figure 58.71b**.

■ REAR DISC PAD INSTALLATION

Servicing and repairing rear-wheel disc brakes calls for a few special considerations. As with any repair that is new to you, be certain to consult service literature before beginning your repair.

Rear Disc Parking Brake Caution. Rear-wheel disc brakes have a parking brake built in. There is either a miniature drum brake that is inside of the disc rotor or a mechanism to clamp the disc brake pads to the rotor when the emergency brake is set. The latter will have either a screw or a cam attached to a lever. The lever on the back of the caliper is the clue. If there is a lever, do not try to force the piston into the bore. When installing pads on rear disc brakes with an emergency brake that is not a separate drum within the rotor, the piston will have notches that must be aligned with the pegs protruding from the back of the pad (**Figure 58.72**). These keep the piston from rotating when the parking brake is applied. If only the pads are being replaced and the caliper is not being rebuilt, the piston must be screwed back into the caliper before the pads can fit. You cannot force them using a C-clamp as you did with front brakes. **Figure 58.73** shows a tool that is used to turn the piston into the caliper.

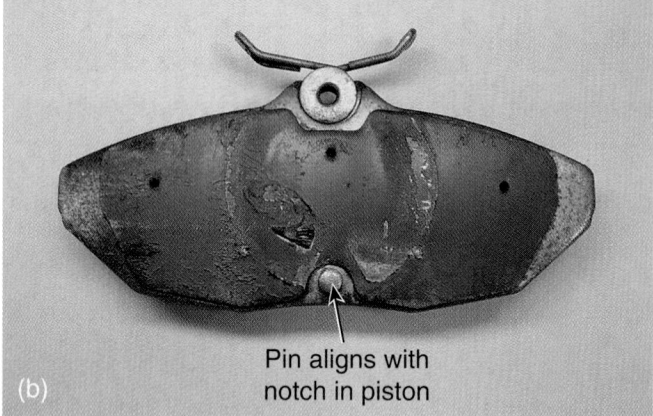

Figure 58.72 (a) A threaded-type rear disc brake piston. (b) This is the back of a pad used with a threaded caliper piston. (*Courtesy of Tim Gilles*)

Figure 58.73 A tool used to retract the piston. (*Courtesy of Tim Gilles*)

Sometimes rust forms in the self-adjusting mechanism when the parking brake has not been used regularly. This freezes up the caliper. Like other self-adjusting mechanisms, left and right sides are not inter-changeable. Because of the complexity of these calipers and the chance that parts are not reusable, many shops install rebuilt units instead of servicing them themselves. If you use a rebuilt caliper, be sure to keep all of the old parts on the old caliper so it will be acceptable as a core return.

Pedal travel on four-wheel disc brakes tends to be further than with disc/drum combinations. This is because the rear pads have to travel further during brake application. When there is a problem with the self-adjuster, pedal travel will become excessive more quickly. Also, after replacing a caliper, the initial clearance must be adjusted to within 1/16" or less. Adjustment can usually be done by repeatedly applying the parking brake.

The drum-in-hat rear disc parking brake design (with a miniature drum brake) should last the life of the vehicle. The thickness of the parking brake lining needs only to be sufficient to hold the car. It does not have to dissipate heat or resist wear, because it is only used to hold the vehicle when parked.

Disc Caliper Rebuilding

A caliper rubber parts kit (**Figure 58.74**) contains a new boot, a piston seal, and sometimes rubber O-rings (for sliding caliper bolts). A square-cut piston seal is held in position by a groove in the bore. A rubber boot keeps contaminants away from the sealing area.

To disassemble a caliper using compressed air and a rubber-tipped blowgun, position a piece of wood or a folded shop towel between the piston and caliper (**Figure 58.75**). Some technicians like to place a shop towel over the piston to prevent brake fluid from spraying onto them.

NOTE: *A typical blowgun for general shop use is OSHA approved and will only provide 35 psi of air pressure. A rubber-tipped blowgun will provide full shop air to the caliper.*

Figure 58.74 A caliper rubber parts kit. (*Courtesy of Tim Gilles*)

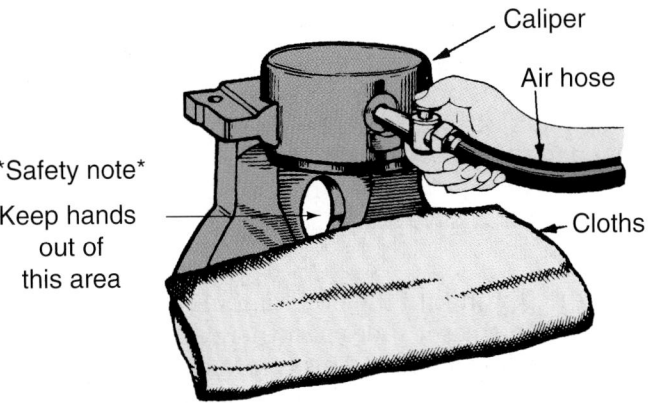

Figure 58.75 Use compressed air and a rubber-tipped blowgun to remove a caliper piston.

CAUTION Before applying air pressure, be sure that your fingers are out of the way. Apply pressure in short bursts to gently push the piston from the bore. Full shop air pressure from the rubber-tipped blowgun can pop the piston out with dangerous force. Pressure behind the piston will become equal to shop air pressure and output force will be multiplied by the area of the piston (remember Pascal's law).

If a caliper has more than one piston, position a piece of wood against the loose piston(s) to prevent the piston from coming all the way out of its bore before the stuck one does. If one piston is out of its bore, the air pressure needed to push the stuck piston from its bore will not be able to build but will leak out through the other piston's open bore (Pascal's law once again).

Rebuilt Calipers

Many shops find it more cost-effective to install rebuilt calipers rather than rebuilding them themselves. Loaded and unloaded calipers are available from remanufacturers. **Unloaded calipers** are also called bare calipers. The **loaded caliper** comes assembled with new friction pads, hardware, and shims (**Figure 58.76**). When a piston is stuck in a caliper bore, there is a good chance its surface finish is corroded. In this case, most shops will elect to purchase a rebuilt caliper rather than attempting to rebuild it.

Clean and Inspect Caliper Parts

Remove the piston and dust boot from the caliper. Some dust boots need to be pried from the caliper, while others come off easily. Remove the old piston seal from the bore and thoroughly clean the bore (**Figure 58.77**). Remember the bore is not a sealing surface, so it does not need to be honed like a drum brake wheel cylinder bore. In drum brake cylinders, the seal rides on the surface of the cylinder bore. With disc brakes, the caliper pistons are the

Figure 58.76 A loaded caliper. *(Courtesy of Tim Gilles)*

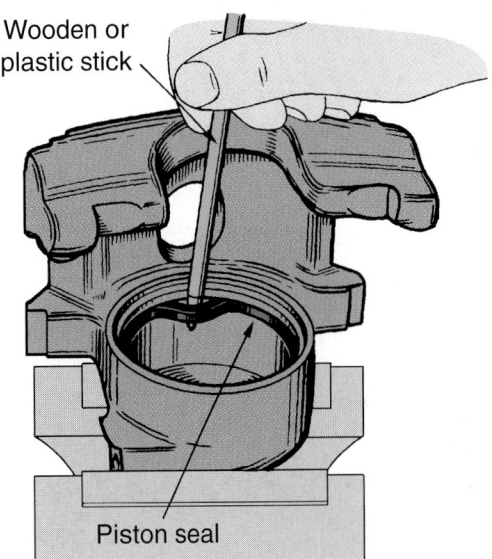

Figure 58.77 Removing the old seal from the caliper bore.

sealing surface, rather than the bore in the caliper. The caliper bore does need to be cleaned, however, and there are special honing tools available to use for this.

Inspect the Pistons

Inspect the pistons for scratches, rust, and corrosion. The outer surface of the piston slides against the square O-ring seal in the caliper bore. If there are flaws in the surface finish of the piston, it must be replaced. Varnish can be cleaned from the piston using spray brake parts cleaner.

SHOP TIP If you use emery cloth to clean a piston, its polished sealing surface will be ruined. Stubborn varnish buildup can be cleaned with crocus cloth. Crocus cloth, unlike abrasive emery cloth, is a very fine polishing cloth that does not leave scratches or grit.

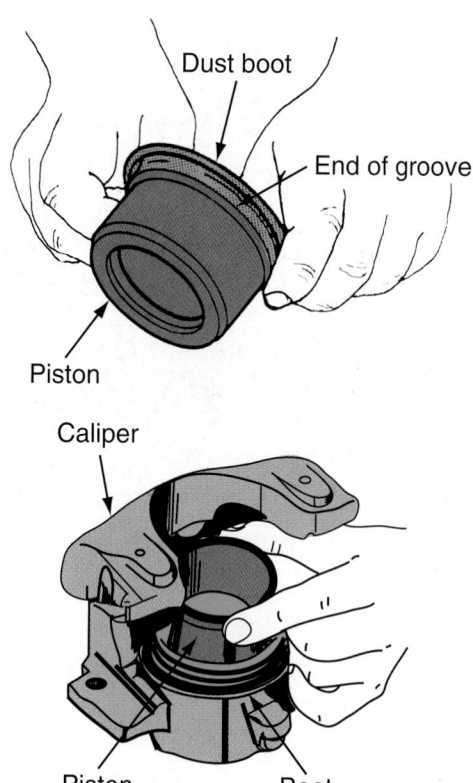

Figure 58.78 Caliper reassembly.

Caliper Reassembly

There are different procedures for caliper reassembly. **Figure 58.78** shows typical parts for a floating caliper assembly. Commercial brake assembly lubricants make reassembly of disc brake calipers easier, but brake fluid can also be used. Apply a liberal amount of lubricant to the piston seal, install it in the channel in the bore, and push on it to seat it.

Dust boot installation procedures vary. On some calipers, the seal is installed into position on the caliper prior to installing the piston in its bore. Hold the piston above the bore as you fit the outside of the dust boot to the caliper. Then push the piston into the bore (**Figure 58.79**).

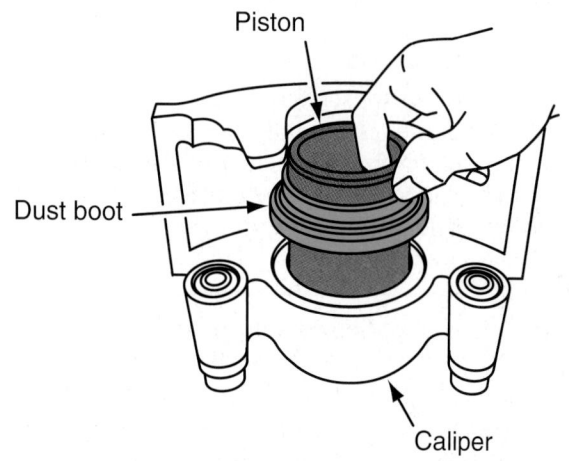

Figure 58.79 Installing the piston in the caliper.

SHOP TIP When reassembling some large-bore, single-piston calipers, it is helpful to apply short bursts of air into the caliper's fluid inlet to help square up the piston and inflate the dust boot while working the piston down into the bore.

Push the piston into the bore carefully, gently work it back and forth as it slides past the seal until it reaches the bottom of the bore. This is to make enough room for the pads when the caliper is reinstalled.

NOTE: *If force is required, something is wrong and you will ruin the seal if you continue trying to install the piston. Once a caliper has been disassembled, a new seal must be used. The old seal will usually expand, making reassembly impossible.*

Install Disc Pads in the Caliper

Most *fixed* calipers use interchangeable inner and outer pads that can usually be slid in from the top after caliper installation. They are positioned by one or two steel pins that are held in place by a retaining clip or small cotter pin.

When the brakes are released, the pads are free to move. In some calipers, the inside pad floats, and the outside pad is fastened tightly to the caliper.

Selecting Linings

Various lining materials are available depending on the application and the friction characteristics desired. Some brakes stop well when they are cold, but they do not work well when hot. Others work very well when hot but will not stop the car until after a few stops (after they have had a chance to become hot (see Figure 58.3). Both of the previous situations can be very dangerous. The best policy is to use only the type of lining material that came on the vehicle as original equipment and use premium quality materials. Besides taking a chance with someone's safety, there could be major liability for a repair shop if an accident results from the use of inferior linings.

Due to weight transfer, front linings are expected to wear out about twice as often as rear linings. When softer linings are used for noise reduction, a shorter service life can be expected.

Brake Lining Materials

Older brake lining materials were classified as organic or inorganic. Asbestos linings were called *organic*, and semimetallic linings were called *inorganic*. Asbestos has not been used for original equipment brakes since the early 1990s, although it is still available as an imported aftermarket lining material. Newer original equipment

Figure 58.80 Dark brown dust on the wheel is a result of semimetallic pad wear. *(Courtesy of Tim Gilles)*

Figure 58.81 SAE friction codes are stamped on the edge of brake linings. *(Courtesy of Tim Gilles)*

friction material classifications include *semimetallic*, *non-asbestos organic* (*NAO*), and *ceramic*.

Semimetallic linings, original equipment on some new front-wheel-drive cars, are very good for quickly removing and absorbing heat from the rotor or drum. The hotter they get, the better they work. For noise reduction, low-metallic linings have become popular, especially in European cars. Drawbacks to semimetallic linings include noise, increased rotor wear, and dark brown dust that develops on front wheels (**Figure 58.80**).

NAO linings, made without iron or steel, usually use aramid fibers or fiberglass. These linings are found most often in low-temperature disc pads and drum brake linings.

About half of vehicles have ceramic linings as original equipment. Instead of steel, ceramic and copper fibers are used for heat control. The ceramic material dampens some of the noise, with any remaining vibrations at a frequency above the range of human hearing. An additional benefit is ceramic brake dust is light in color and does not rust like steel. A disadvantage is that their production is more costly. Sometimes other types of linings work as well or better.

Lining Edge Codes

The side edge of a new brake lining is stamped with a code number established by the Society of Automotive Engineers (SAE) (**Figure 58.81**). The SAE *edge code* provides a uniform means of identification and rates the friction characteristics of different linings based on tests performed in a laboratory. The edge code does not indicate quality. A lining with good stopping characteristics could wear out quickly or cause excessive wear to a rotor or drum. The three groups of numbers or letters in the edge code identify the brake manufacturer, materials used in the lining, and the lining's coefficient of friction.

Aftermarket Friction Material Certifications

Unfortunately, the federal motor vehicle safety standard for brakes only applies to new car production. Industry-based certification programs have been developed to assure customers that aftermarket linings have performance characteristics equal to or better than as the new car (OEM) standards. Two industry brake test programs were developed to give technicians and consumers a means of evaluating aftermarket friction materials. With a vehicle in a special dynamometer test bay, the wheels can be spun and braking is tested under controlled conditions to determine whether the replacement friction materials are equal to or better than OEM standards. The two programs are the D3EA (dual dynamometer differential effectiveness analysis) certification (**Figure 58.82**) and the Brake Manufacturers Council (BMC) program called BEEP (brake effectiveness evaluation procedure) (**Figure 58.83**).

Figure 58.82 D3EA testing uses front and rear brake dynamometer tests. *(Courtesy of Dr. Thomas A. Flaim, Brake Technology, LLP)*

Figure 58.83 The Brake Manufacturers Council single-end brake dynamometer evaluation program, called BEEP for brake effectiveness evaluation procedure, tests brakes against OEM standards.

DISC BRAKE NOISE

Brake noise and vibration are the most likely complaints from a customer. During brake application, the lining sticks to the rotor momentarily before slipping and then sticking again. This causes a high-frequency vibration that results in annoying squeaks and squeals.

Noise during a stop can include squeaks, metal-to-metal sounds from excessive wear, rattles due to loose parts, and rubbing from a distorted backing plate. A very common cause of brake noise occurs when the metal brake pad back vibrates against the metal caliper piston. Harder lining materials have a tendency to make noise, especially when cold, but they last longer and provide better hot stopping than softer linings. Many newer vehicles come from the factory with semimetallic linings, which are more prone to low-frequency vibration and noise. Some consider a small amount of noise from these linings to be normal.

It is important that disc linings be firmly attached to the apply piston or caliper so that vibration is avoided. Some floating calipers have tabs on the pads that hold them tightly to the caliper. These are adjusted to fit, using either a hammer (**Figure 58.84**) or pliers (**Figure 58.85**).

Most pads have small parts called anti-rattle clips, which position the pad and keep it from moving in the caliper (**Figure 58.86**). Some are attached to the back of the pad; others work off guide pins. This hardware is serviced when brake pads are replaced. Sometimes replacement pads come with new hardware, especially with original equipment parts. Some aftermarket brake manufacturers do not provide new hardware with their brakes. The old hardware is reused or must be separately ordered if damaged or broken. Damaged or missing anti-rattle hardware can cause a brake to squeak or drag.

Aftermarket glue/insulator materials are available that can be spread on the metal back of the pad

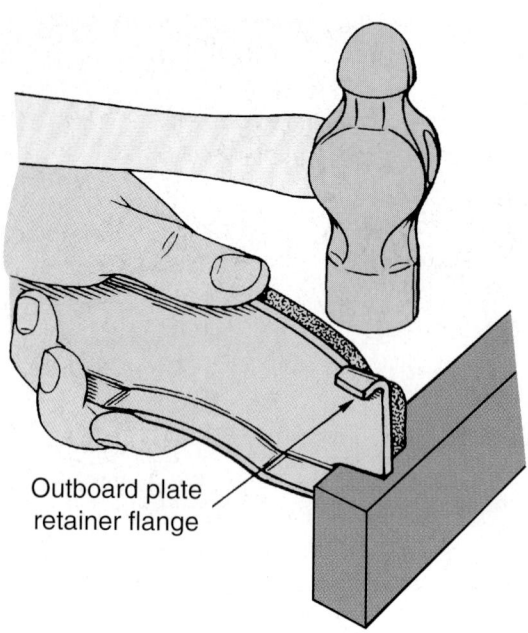

Figure 58.84 Using a hammer to fit the disc pad to the caliper.

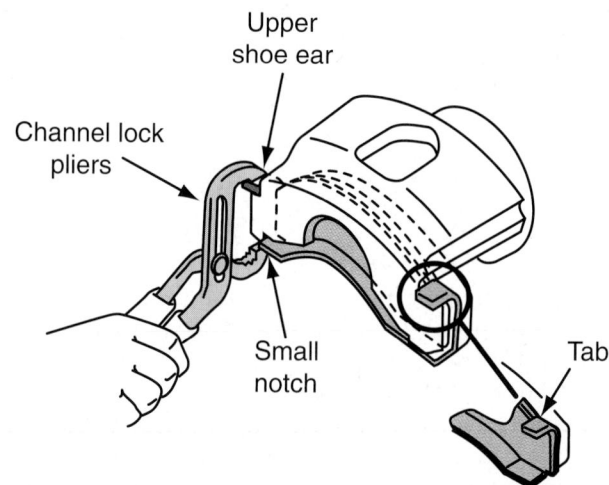

Figure 58.85 Crimping the outer shoe ears tightly to the caliper.

before it is installed in the caliper. After it gets hot and vulcanizes the parts to one another, it helps prevent vibration and dampens noise. With high heat, however, these materials can sometimes be short-lived. Some newer calipers come with dampening materials sandwiched between stainless steel shims.

The manufacturer often provides lubricants and/or shims. Some shims have a rubber insulation coating. Others use thick grease or anti-seize compound applied to the mating sides of the steel brake pad backing and shim. This dampens vibration that could result in brake squeal. Careful installation of the shims and lubricant is important. Be sure to coat both sides of the shims with lubricant and follow specific instructions that are supplied with them.

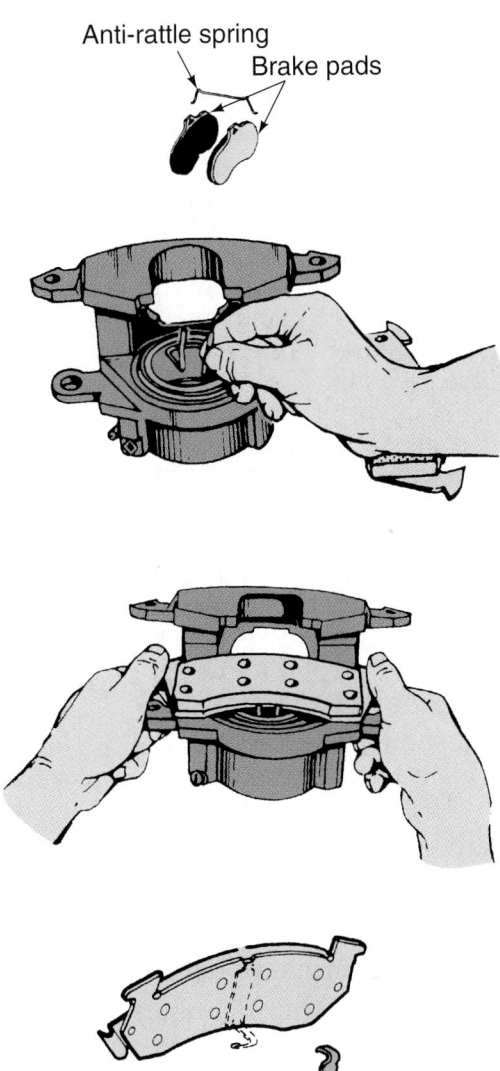

Anti-rattle spring
Brake pads

Figure 58.86 Various disc brake clips and springs. *(Courtesy of Federal-Mogul Corporation)*

Metal-to-metal contacts must be lubricated to prevent noise. Lubricant is always applied to caliper sliding surfaces. Various brake hardware lubricants include synthetic brake grease, silicone, anti-seize compound, or molybdenum-disulfide (called moly lube). A satisfactory brake lubricant must be suitable for use in a high-temperature environment.

NOTE: *Do not use wheel bearing grease on brake parts, and be careful not to allow lubricant to contaminate the friction surface of the pad or rotor.*

SHOP TIP Anti-seize compound has a high boiling point and works well for lubricating brake system sliding surfaces.

Floating calipers must be able to slide freely on their mounts. Caliper kits usually include O-ring bushings along with the other rubber parts. These are important for noise reduction. The sliding caliper mounting bolts must also be clean and free of rust, or they must be replaced. Bushings and sleeves can become gummed up to the point that the caliper cannot slide easily. A result of this can be that one lining wears more than the other. One other result is that a brake does not release all of the way after a stop. This can result in an intermittent annoying squeal that goes away whenever the brakes are applied. Be attentive to the caliper design and be sure that all sliding surfaces are rust-free and in good shape.

Caliper Installation

Install the caliper and verify that it moves freely. Use a torque wrench to tighten the bolts that hold it to the steering knuckle. Brake calipers vibrate and are prone to coming loose if not tightened properly.

Floating calipers must be able to slide freely on their mounts. Guide pins, O-ring bushings, and other rubber parts sometimes require lubrication. Use the correct lubricant supplied with the brake kit. Do *not* use a petroleum-based lubricant that might damage a rubber bushing.

CAUTION Immediately after installing the caliper, be sure to apply the brake pedal. Only hand pressure is necessary, so if the car is on the lift, this can still be done. Sometimes more than one application of the pedal is required to push the linings against the rotor. If the car is backed off the lift before checking the brake pedal, an accident can occur.

Breaking in New Linings

In the past, linings were not fully cured and had to be broken in easily as they finished curing. Today's linings are more fully cured but still require some break-in. To help them finish curing and to seat them to the rotors or drums, accelerate to 30 mph and make 20 to 30 stops with medium to firm pressure. Do not overheat the brakes.

■ SERVICE THE PARKING BRAKE CABLE

During a rear brake job, the emergency brake cables are disconnected from the brake linings. This provides a good opportunity to lubricate the cables. From the underside of the vehicle, pull the cables as far as possible. Wipe them off and inspect them. Apply clean grease to them and push them back into their sheaths. Another way of lubricating a cable is to apply penetrating oil to its sheath (the metal cover that surrounds it).

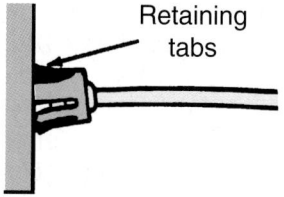

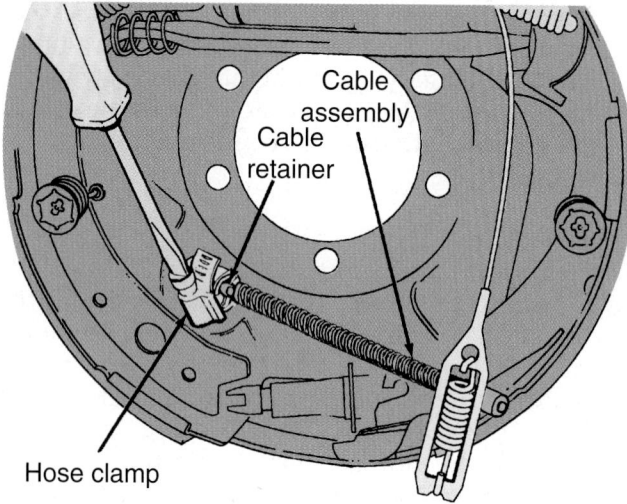

Figure 58.87 Use a small hose clamp to depress the retaining tabs. *(Courtesy of DaimlerChrysler)*

Usually, you can only do this on the ends where the cable has no plastic insulating cover.

The parking brake cable usually has a spring-loaded clip that fits into the brake backing plate. If a parking brake cable requires replacement, use a small hose clamp to help remove it from the backing plate (**Figure 58.87**).

Parking Brake Travel

Service information often describes parking brake adjustment as the number of clicks that the brake handle or pedal makes before being fully applied. The specification might say "less than nine clicks." A good rule of thumb is that the brake should be fully applied at half travel. **Figure 58.88** shows a typical undercar adjustment.

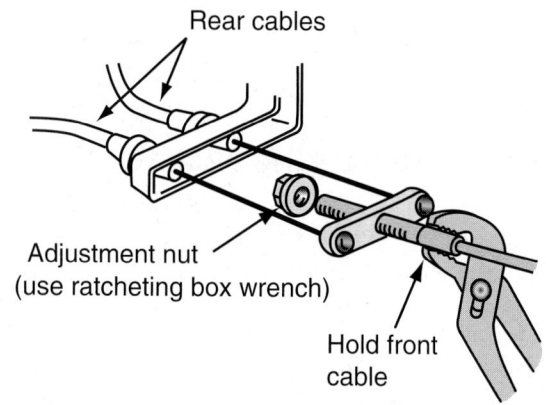

Figure 58.88 Typical parking brake adjustment.

NOTE: *Changing the service brake adjustment will affect the parking brake adjustment (the* **service brakes** *are the brakes that stop the car). Do not adjust the parking brake unless the service brakes have been adjusted first.*

■ VACUUM POWER BRAKE SERVICE

Repairs to power boosters include replacement of the hose, filter, or check valve. New and rebuilt units are available for most cars and trucks. Very few shops rebuild power brake boosters. Power brake units can be purchased as a unit or with a rebuilt master cylinder attached.

Most complaints regarding power brakes have to do with increased pedal effort. This could be due to several things, sometimes not brake related. According to brake manufacturers, a high percentage of power brake boosters returned during warranty are not defective, but misdiagnosed, with many of the complaints related to incorrect brake bleeding techniques.

Wheels and Tires Too Large

Hard pedal effort can also result from the installation of larger diameter tires and wheels on pickup trucks and SUVs. This results in more leverage being exerted by the wheel against the brake system. A smaller bore master cylinder or larger bore wheel cylinders can help correct the problem.

Defective Power Brake Booster

A defective power booster will result in a hard brake pedal. Low brake pedal height or a soft brake pedal are **not** power brake–related complaints. A soft pedal is typically due to air in the hydraulic system, and a low pedal can be related to brake adjustment or a failure in one side of the master cylinder's hydraulic system.

Vacuum Booster Operation Test

To test the operation of a power brake booster, exhaust all vacuum reserve from the power booster by applying the foot pedal several times with the engine off. The sound of the air rushing into the reservoir should be heard. Next, hold your foot on the pedal while starting the engine. If the power booster is operating correctly, the pedal will move about an inch closer to the floor after the engine starts.

Vacuum Booster Leak

Power brake booster problems can be due to a hole in the booster diaphragm or a stuck valve that is leaking. To test for a booster internal leak, shut off the engine and apply steady pressure to the brakes. The pedal height should remain constant for at least 30 seconds. If the booster has an internal leak, the pedal will slowly rise during this test.

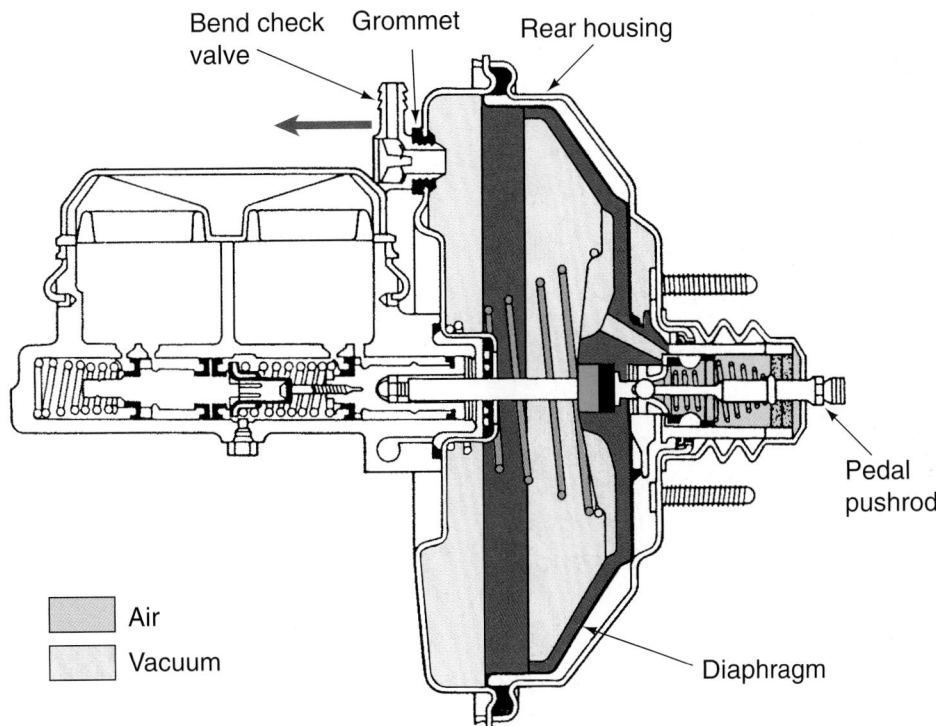

Figure 58.89 Testing the check valve.

Brake Drag

A defective power booster can sometimes cause brakes to drag, resulting in rapid brake lining wear. To verify this, apply the brakes several times with the engine off. With the wheels off the ground, rotate the tires. They should turn freely. Start the engine and recheck to see that the wheels still turn freely. If not, the booster is attempting to apply the brakes. In the passenger compartment, verify that there is free travel between the brake pedal apply rod and the back of the power booster. If not, adjust the pedal apply rod. If there is free travel but the brakes still drag, replace the booster.

Vacuum Supply Checks

Check the hose that supplies vacuum to the power booster from the intake manifold. It should not be damaged, hard, or swollen. The brake booster has a one-way check valve that traps vacuum when the engine is off. If the check valve is bad, the brake effort will vary according to the load on the engine. Checking the power brake check valve is done with the engine off. Carefully bend the valve against its rubber grommet (**Figure 58.89**). If the check valve is holding, air will rush into the front part of the booster. With the valve removed, you should be able to blow through it in one direction and it should seal from the other side.

When you replace a check valve, use water or brake fluid to lubricate the grommet during installation. Check valves are plastic and can be damaged with careless handling.

If there is a problem with braking effort, install a vacuum gauge into the line at the power booster using a "T" (**Figure 58.90**). A typical *minimum* vacuum specification at idle is 15 inches of mercury.

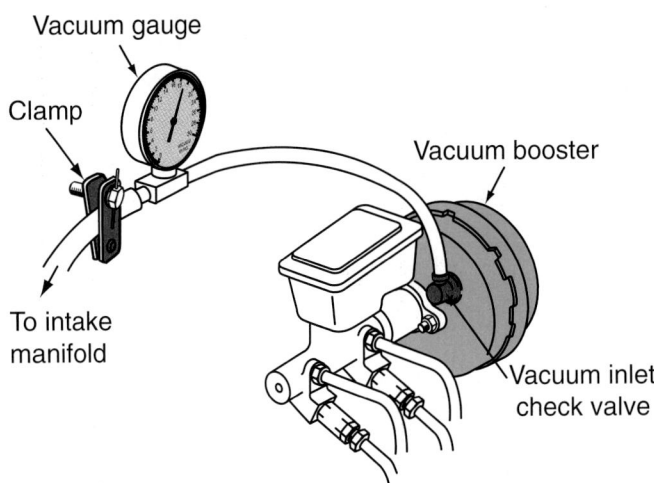

Figure 58.90 Use a vacuum gauge to check vacuum supply. You can check to see that it holds vacuum by clamping off the line.

SCIENCE NOTE

*Low pressure is commonly referred to as vacuum. True vacuum is the complete absence of air, but this condition is only possible in space or in a laboratory. In the United States, pressure is measured in psi, or pounds per square inch. Vacuum (low pressure) is measured in inches of mercury (in. Hg). When the temperature is 68°F, atmospheric pressure at sea level is 14.7 psi (**Figure 58.91**).*

An understanding of engine vacuum can be important to your ability to diagnose power brake booster problems. The vehicle's engine is an air pump. Each intake stroke results in suction in the intake manifold that results in low pressure of about 16–20 in. Hg at idle. When the brakes are not in use, this intake manifold vacuum is applied to both sides of the diaphragm in the vacuum booster.

*Atmospheric pressure (14.7 psi at sea level) is what is in the passenger compartment, at the back side of the power brake booster. Remember, the power booster is mounted on the fire wall, and its back side extends into the passenger compartment. This air pressure is waiting to be used to assist in applying the brakes (**Figure 58.92**).*

NOTE: *A complete vacuum is either 29.9 in. Hg below atmospheric pressure or 14.7 psi below atmospheric pressure. Rounding off 29.9 to 30 and 14.7 to 15 makes it easier to compare between the two.* **Figure 58.93** *compares vacuum and pressure equivalents. You can see that the ratio is approximately 2:1. This is handy for conversions:*

1. *To convert pounds per square inch to inches of mercury, multiply by 2.*

2. *To convert inches of mercury to pounds per square inch, divide by 2.*

For proper operation of the power brake, manufacturers typically specify a minimum vacuum level of 15 in. Hg at idle.

If intake manifold vacuum is 15 inches and atmospheric pressure is 15 psi, what is the pressure differential?

Vacuum side pressure = 15 inches ÷ 2 = 7.5 psi

so,

Atmospheric side pressure = 15 psi – 7.5 psi = 7.5 psi (pressure differential)

Answer : 7.5 psi (pressure differential)

In the preceding example, a power booster makes about 7.5 pounds of force for every square inch of diaphragm area. If the diaphragm has 50 square inches of area, it will develop 375 pounds of extra force to add to the effort of the driver's foot (7.5 psi × 50 square inches = 375 pounds). **Figure 58.94** *explains how this works.*

At higher altitudes, brake assist will be diminished. Engine vacuum is reduced by about 1 inch for every 1,000 feet in elevation above sea level.

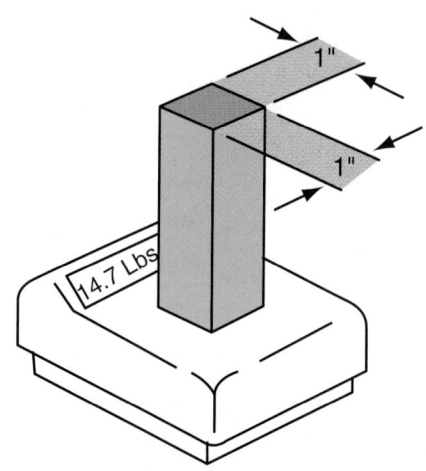

Figure 58.91 If the temperature is 68°F, atmospheric pressure at sea level is 14.7 psi.

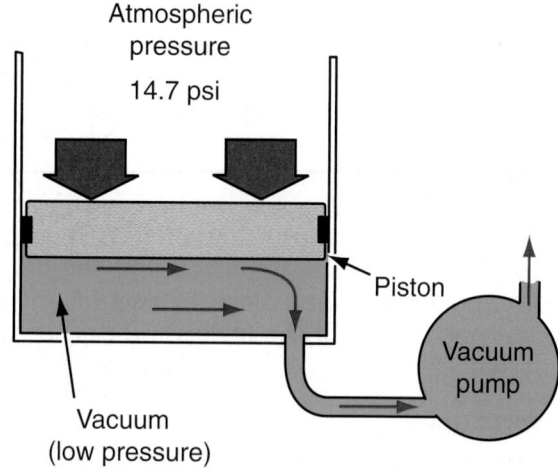

Figure 58.92 A difference between atmospheric pressure and vacuum is used to assist in applying the brakes.

Leaking Power Booster Front Seal

A leak in the seal in the front of a power booster can allow fluid to be drawn into the power booster and burned in the intake manifold, and the rear chamber of the master cylinder will continually require fluid. The key to diagnosing this is that there is no evidence of external fluid leakage.

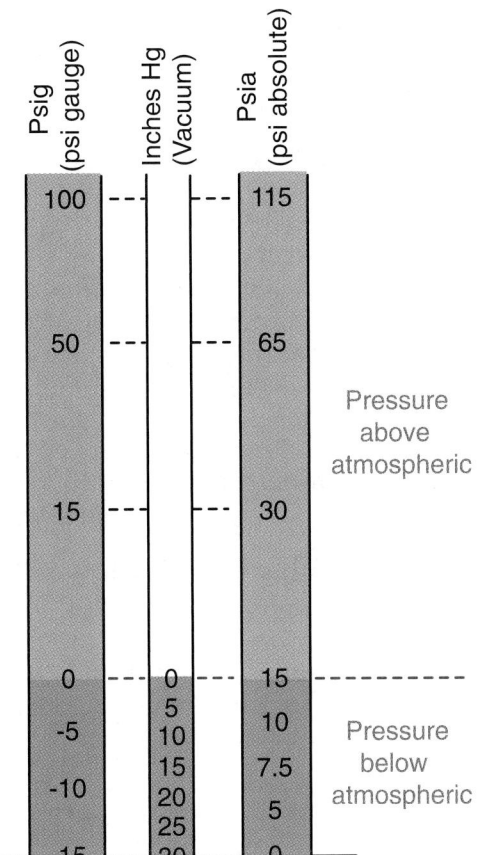

Figure 58.93 When pressure is measured from a theoretical perfect vacuum, which is 29.2 in. Hg, the pressure reading is expressed in pounds per square inch absolute (psia). Pressure measured on a gauge that measures vacuum (pressure less than atmospheric) is expressed in pounds per square inch gauge (psig).

Power Brake Booster Replacement

A typical power booster has four studs that protrude through the bulkhead into the passenger compartment (**Figure 58.95**). When these studs are easily accessible, replacement of the booster is not a difficult job.

When replacing a power booster, it is important that the pushrod depth is correct. If the pushrod is too high, the master cylinder compensating port will be covered, resulting in brake drag after the fluid warms. A pushrod too low will cause excessive brake pedal travel. There are special tools and different methods specified for making this adjustment (**Figure 58.96**). Complete rebuild units are available that include a rebuilt master cylinder assembled to them. These will already be correctly adjusted.

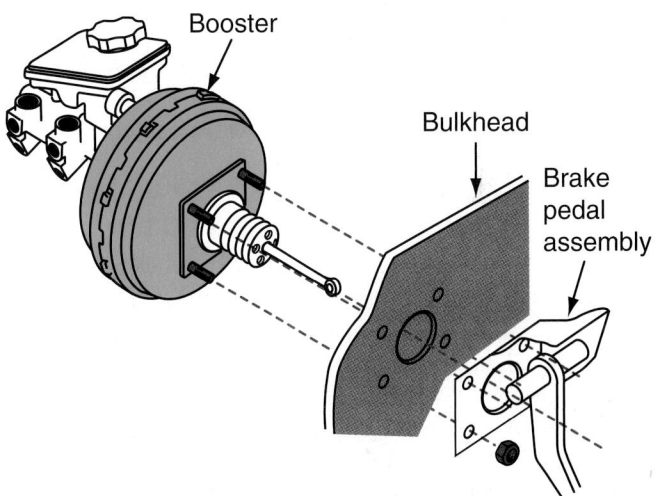

Figure 58.95 A typical power booster has four studs that protrude through the bulkhead into the passenger compartment.

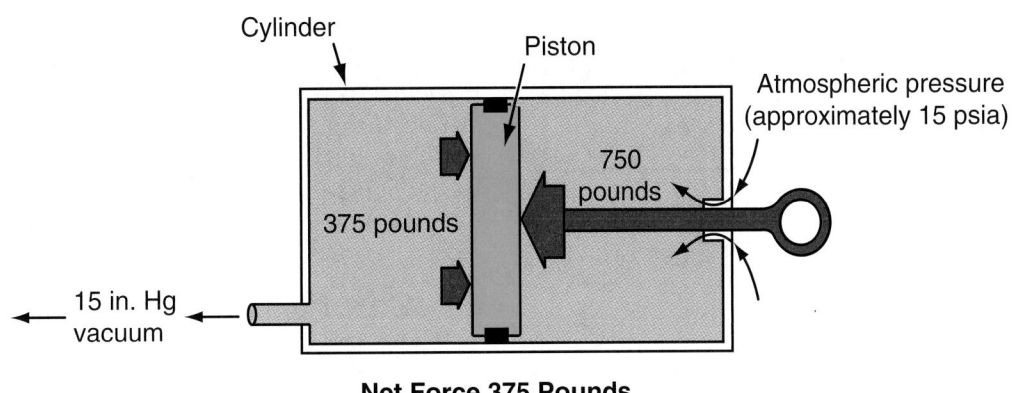

Net Force 375 Pounds

15" vacuum = 7.5 psi absolute

15 psi X 50 square inches = 750 pounds force
7.5 psi X 50 square inches = 375 pounds force

375 pounds pressure differential

Figure 58.94 Explanation of vacuum assist.

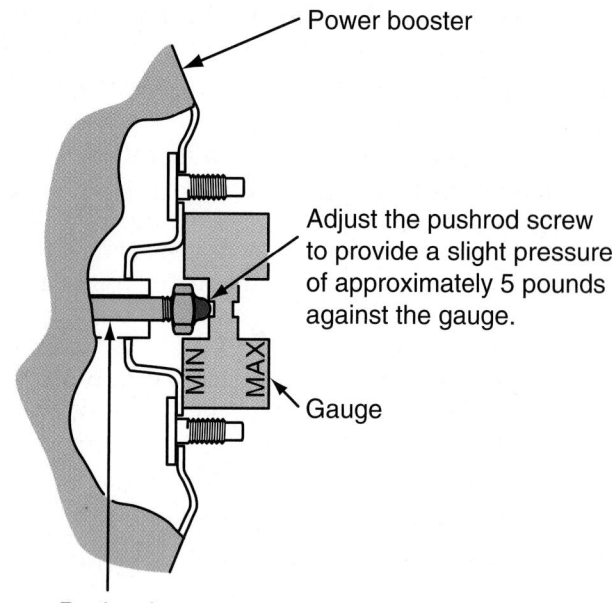

Figure 58.96 A tool for checking power booster pushrod depth.

■ BRAKE WARNING LAMP DIAGNOSIS

Hydraulic brake systems have either a low brake fluid switch or a pressure differential sensing device called a hydraulic safety switch. If either of these devices triggers a signal saying that brake fluid is low or there is low hydraulic pressure in one half of the system, a *red* brake light on the instrument panel should illuminate. When the key is on, the light will also come on whenever the parking brake is on (even a little bit). On some cars, a bulb check feature illuminates the light during engine cranking.

SHOP TIP When a warning lamp is illuminated, disconnecting the switches one at a time until the light goes out will usually identify the cause of the problem.

The amber light is assigned to the ABS. See Chapter 59 for more information on amber light diagnosis. The red light should illuminate when the engine cranks, and the amber ABS light should illuminate some time during key-on, engine crank, or engine start. Both lights should come on and go back off before driving

begins. If a light does not come on or if it does not go off *after* coming on, there is a problem in the brake system or the electrical circuit for the lamp.

NOTE: *If a stoplight is out or if the two lights are of different intensity, the ABS can be disabled.*

Hydraulic Safety Switch Service

Hydraulic safety switches are found in vehicles that do not have fluid level sensing systems. Most of these valves are self-centering and spring loaded. When there is a drop in system pressure, the dash light *only* comes on during a stop and does *not* remain on.

Most safety switches share the same light that indicates when the parking brake is applied. To test the light in these systems, simply turn on the key with the parking brake applied. The light should illuminate.

Master Cylinder Fluid Level Switch Service

Fluid level switches are found on vehicles that do not have a hydraulic safety switch. To test the lamp on the instrument panel, turn on the key. It should light for a few seconds. The float in the master cylinder reservoir can be checked to see that it moves freely. It should turn the lamp on and off as it goes from full length (fluid empty) to its depressed position.

Stoplight Switch Service

To check the operation of the stoplight switch, apply the brakes and verify that the stoplights come on. If not, check to see if the ignition key must be turned to the "on" position first. Then check the fuse. There are at least two wires to the stoplight switch. Remove the connector from the switch and use a jumper wire to connect them together. If the stoplights work with the wires connected together, the stoplight switch is faulty.

Use a wiring diagram to determine how the circuit operates. On North American vehicles, the switch is normally open and power is often supplied to the switch through the turn signal switch. Some stoplight switches have extra contacts for cruise control, the torque convertor clutch, or antilock brake control. According to the SAE term list, a brake on-off switch is called a *BOO switch*.

■ ANTILOCK BRAKE SYSTEM (ABS) SERVICE

Antilock brake systems (ABS) service is covered in Chapter 59.

■ REVIEW QUESTIONS

1. A soft brake pedal due to air in the system is called a _____ pedal.
2. With a vacuum power brake booster, when you apply the brake pedal with the engine off you should hear the sound of _____ entering the booster.
3. What color is silicone brake fluid?
4. Does alcohol absorb water?

5. What do you do to a master cylinder before installing it on a car?

6. What is the name of the type of vacuum cleaner filter that is used to clean asbestos?

7. How much lining-to-drum clearance will there be after brakes are adjusted with a brake adjustment gauge?

8. What can you use to remove the threaded finish left after machining a brake rotor?

9. A _____ is used to push a piston to the bottom of the caliper bore before installing new brake linings.

10. A rebuilt brake caliper that includes all of the parts is called a _____ caliper.

■■■ ASE-STYLE REVIEW QUESTIONS

1. Which of the following is/are true about brakes that need bleeding?

   a. The pedal will rise on the second application of the brakes.

   b. The pedal will feel soft and spongy.

   c. The pedal will be higher than normal.

   d. All of the above.

2. A vehicle has a high, hard pedal and sometimes the brakes remain applied. Technician A says that the vent port in the master cylinder could be plugged. Technician B says that the brake pedal rod might be adjusted too long. Who is right?

   a. Technician A   **c.** Both A and B

   b. Technician B   **d.** Neither A nor B

3. A brake pedal drops slowly toward the floor under very light pressure. Technician A says that a leaking master cylinder piston seal could cause this. Technician B says that a leaking rear wheel cylinder could be the cause. Who is right?

   a. Technician A   **c.** Both A and B

   b. Technician B   **d.** Neither A nor B

4. A car has a low, firm brake pedal. Technician A says that this can be due to brakes that need to be bled. Technician B says that this could be due to brakes in need of adjustment. Who is right?

   a. Technician A   **c.** Both A and B

   b. Technician B   **d.** Neither A nor B

5. Which of the following could cause a low fluid level in the master cylinder?

   a. Worn drum brake linings

   b. Worn disc pads

   c. Both A and B

   d. Neither A nor B

6. Technician A says that wheel cylinder cups must face toward fluid pressure. Technician B says that the master cylinder primary piston cup must face away from fluid pressure. Who is right?

   a. Technician A   **c.** Both A and B

   b. Technician B   **d.** Neither A nor B

7. Two technicians are discussing checks that are made on a disc brake rotor. Technician A says to check it for parallelism. Technician B says to check it for runout. Who is right?

   a. Technician A   **c.** Both A and B

   b. Technician B   **d.** Neither A nor B

8. Which of the following is/are true about brakes?

   a. The master cylinder is normally replaced as part of a brake job.

   b. Brake linings are purchased separately for each wheel.

   c. A good brake return spring will ring if dropped on a hard surface.

   d. None of the above.

9. A master cylinder is leaking from the rear chamber. Technician A says that this could be due to a leaking power brake booster seal. Technician B says that this could be due to a leaking primary cup on the rear piston of the master cylinder. Who is right?

   a. Technician A   **c.** Both A and B

   b. Technician B   **d.** Neither A nor B

10. Technician A says that it is not necessary to hone a disc brake caliper bore during a rebuild because the bore is not the sealing surface. Technician B says to hone an aluminum wheel cylinder before installing a wheel cylinder kit. Who is right?

   a. Technician A   **c.** Both A and B

   b. Technician B   **d.** Neither A nor B

CHAPTER 59

Antilock Brakes, Traction, and Stability Control

■ OBJECTIVES

Upon completion of this chapter, you should be able to:

✔ Describe the reason for an antilock brake system (ABS).

✔ Explain the theory of operation of ABS.

✔ Describe the parts of two-, three-, and four-wheel ABS.

✔ Explain the differences between integral and nonintegral ABS.

✔ Explain how ABS provides traction control and stability enhancement.

✔ Explain ABS and normal brake warning light operation.

✔ Describe how to bleed ABS brakes.

✔ Describe service procedures for ABS brakes.

■ KEY TERMS

antilock brake system (ABS)
antilock brake controller
electronic brake control
   module (EBCM)
controller antilock brake (CAB)
electronic brake and traction
   control module (EBTCM)

wheel speed sensor
lateral acceleration sensor
electromechanical
   hydraulic (EH) unit
integral ABS
nonintegral ABS
rear-wheel antilock (RWAL)

rear antilock brake
   system (RABS)
dump
traction control system (TCS)
acceleration slip
   regulation (ASR)

■ INTRODUCTION

The ability of brakes to do their job is limited by the grip of a vehicle's tires to the road surface. If the tires do not slip, the vehicle will go in the direction it is steered. Once it loses traction, however, steering control is lost. Even with the best quality tires, when a car skids, there is a loss of stopping ability and control. If a driver could release pressure on the brake pedal just before a wheel locked up, the skid could be avoided. When a wheel stops turning, friction between the tire and road generates heat. This softens the tire, causing it to lose traction.

Total traction loss is referred to as 100% slip. A slip rate of 50% means that the wheel is rolling at 50% slower speed than a freely rolling tire at the same vehicle speed. Maximum traction occurs at about 10% to 20% slip (**Figure 59.1**)

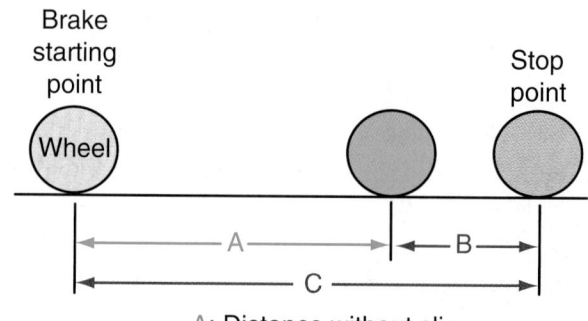

A: Distance without slip
B: Slipped distance
C: Actual distance to stop

$$\text{Slip rate} = \frac{B}{C} = \frac{\text{Vehicle speed} - \text{Wheel speed}}{\text{Vehicle speed}}$$

Figure 59.1 Slip rate.

▄▄ ANTILOCK BRAKES

Most newer cars are equipped with computerized **antilock brake systems** (ABS) to prevent the wheels from locking up (**Figure 59.2**). The ABS uses sensors and a computer to monitor wheel speed. During normal operation an antilock system works like a conventional brake system. It only functions differently when a wheel locks up. Wheel speed sensors measure the rotational speed of the wheel. If a wheel locks up, an **antilock brake controller** pulsates (modulates) the pressure to that wheel in much the same manner as a race car driver pumps the brakes during a turn to avoid a skid. ABS can do this pulsation much faster than a human, however. A typical ABS can pulsate the pressure to the brake system from 10 to 20 times per second.

NOTE: *At some types of auto races, you can hear a pulsating chirp from ABS-equipped cars as the tires lock and unlock during turns. This noise results as the ABS dump solenoid ratchets.*

An antilock system helps avoid loss of control when a wheel loses traction. It does not necessarily result in a shorter stopping distance. One saying is "ABS does not mean you will avoid an accident. It just means you can pick your target."

When the road is bumpy, there is loose gravel, or the road is slick, less pedal force is needed to activate the ABS. The ABS is disabled below a certain speed. If the ABS starts working at low speed on an icy road, the vehicle will skid on the ice, straight, but still without traction. Stopping on snow is faster if the wheel locks, because the snow builds a mound in front of the tire as it skids.

When an ABS senses a failure, the brake system reverts to conventional-only braking. The ABS light on the dash comes on and a code is set in the computer, but braking is normal (except for the integral high-pressure systems, which can lose their rear brakes during a failure).

Pedal Feel

Cars with ABS use the conventional braking system during normal stops. During an ABS stop, the pedal feels different than a conventional braking system. When the ABS becomes active, a small bump followed by rapid pulsation is felt in the brake pedal. The bump is caused by the pump returning fluid to the master cylinder reservoir. The pulsation is noticed in some systems more than in others.

During normal stops there is no pedal pulsation. A pulsating pedal during a non-ABS stop could be due to a warped brake drum or rotor. Some early antilock systems had an increase in pedal height during an ABS stop.

▄▄ ANTILOCK BRAKE SYSTEM COMPONENTS

Various ABS designs are covered later in this chapter. Some components are common to all types of ABS, whether they are integral, add-on, or one, three, or four channel.

Electronic Control Unit

The ABS computer is known by different names and can be found in several places. Often referred to as "the controller," its official names include the **electronic brake control module** (EBCM) or **controller antilock brake** (CAB). If the system includes traction

① **hydraulic modulator with attached ECU**
② **wheel speed sensors**

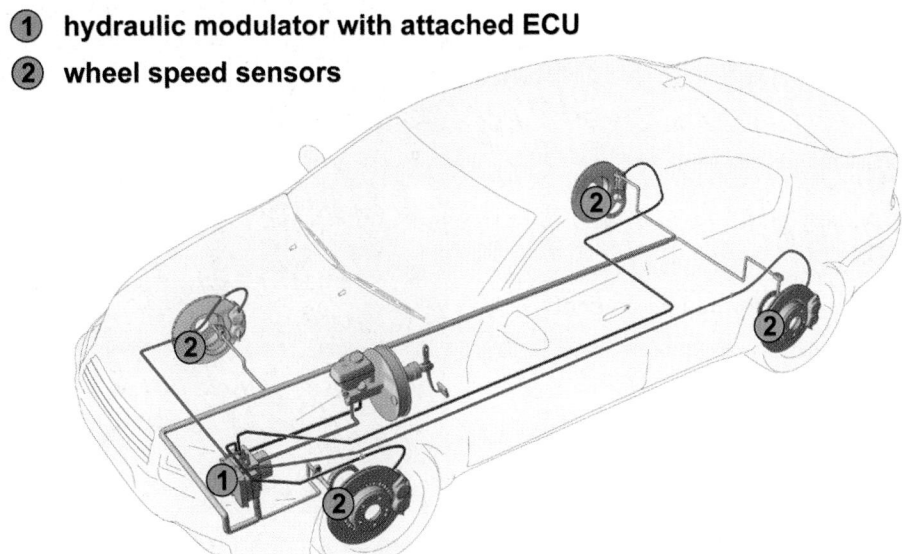

Figure 59.2 The basic electrical and hydraulic components of a four-wheel antilock brake system. *(Reprinted with permission from Bosch)*

control (covered later), the acronym is **EBTCM**, or **electronic brake and traction control module**. The controller can be located inside the trunk, in the passenger compartment, on the master cylinder, or attached to the hydraulic control unit.

Inputs such as those from the wheel speed sensors, brake pedal sensor, and fluid level sensor provide data to the computer. The computer acts on those inputs to correct differences in wheel speed during a loss of traction. Brake fluid output from the master cylinder is interrupted by solenoid-operated valves called pressure modulator valves (PMV). When the computer senses a wheel locking up, electrical current is directed to the solenoid. This energizes a magnetic field to operate the valve. The computer also monitors electronic operation of the system with a self-test every time the ignition system is cycled and the first time the vehicle is driven after a key cycle.

Wheel Speed Sensors

Each wheel with skid control has a sensor to detect the speed of wheel rotation. **Wheel speed sensors** operate in the same way that the magnetic trigger in the distributors of some electronic ignition systems operates. Most sensors are *permanent magnet (PM) generators*, with a coil of wire wound around a permanent magnet core (**Figure 59.3**). The sensor is positioned

near a toothed ring called a *tone ring* or *exciter ring*. The toothed ring spins with the wheel.

On front wheels, the tone ring is mounted to the inside hub of a rotor (**Figure 59.4a**) or on the outer CV joint housing (**Figure 59.4b**). As each tooth moves past the magnet, the magnetic field around the coil increases. As the tooth moves away, the strength of the magnetic field weakens. The space between the tooth and the magnet is called the air gap. Movement of the teeth causes a constantly changing air gap. These changes in the magnetic lines of flux around the sensor cause an alternating current (AC) and voltage. **Figure 59.5a** shows the points where maximum positive and negative voltages are reached. **Figure 59.5b** shows conditions resulting in no voltage induction.

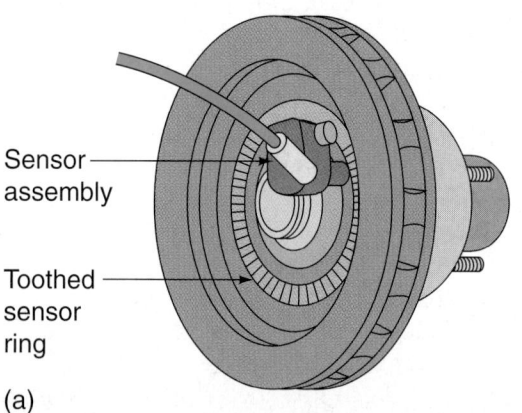

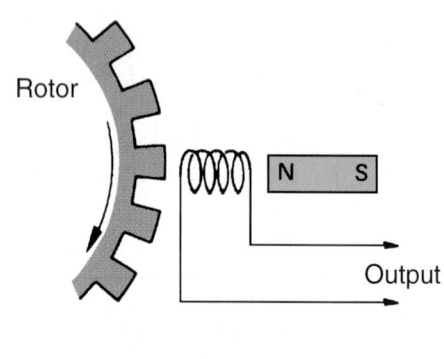

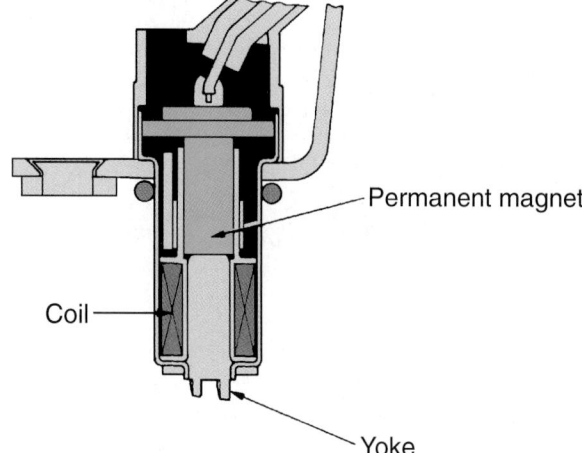

Figure 59.3 Most sensors are a coil of wire wrapped around a permanent magnet core.

Figure 59.4 (a) An ABS sensor and sensor ring mounted inside the rotor. *(Courtesy of Ford Motor Company)* (b) A wheel speed sensor with the tone ring mounted on the outer CV joint. *(Courtesy of DaimlerChrysler)*

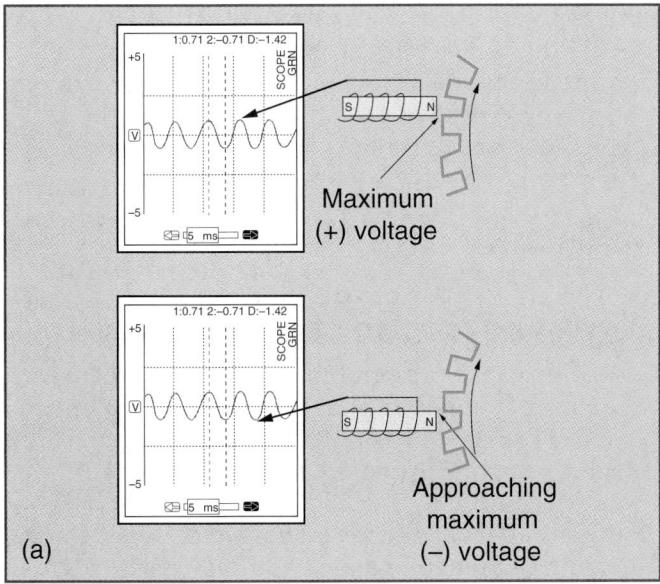

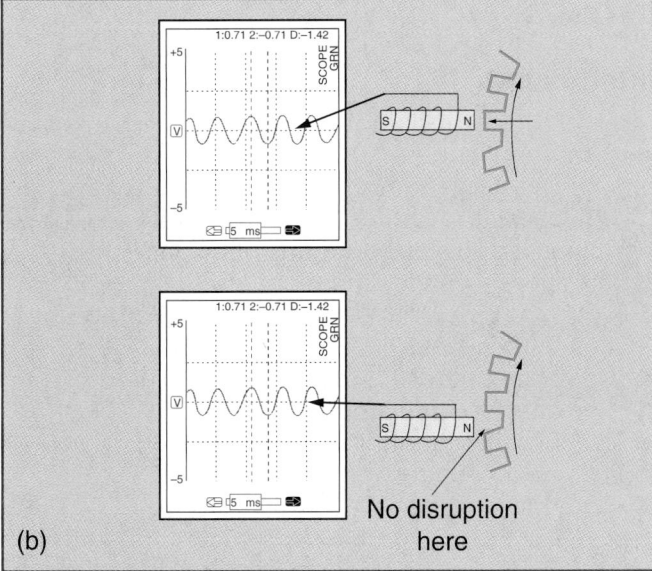

Figure 59.5 (a) Conditions that cause maximum voltage induction. (b) Both of these conditions result in zero voltage because the magnetic field is not being interrupted.

The sensor's frequency changes with the speed of wheel rotation (**Figure 59.6**). The faster the teeth pass the magnet, the higher the voltage output of the sensor and the higher the frequency of the voltage oscillations. When the brakes are applied, the electronic control unit "wakes up" and looks at this information

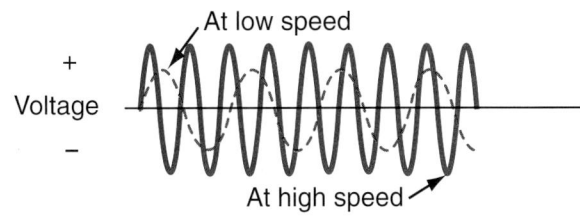

Figure 59.6 Wheel speed sensor frequency changes with wheel speed.

to calculate the speed of the wheel and compare it to the speed of the other wheels.

A **lateral acceleration sensor** is found on some antilock systems. This sensor measures "yaw" forces encountered while turning. The controller uses this information to change ABS control during hard cornering.

Wheel Speed Sensor Installations

There are many different ABS sensor installations. A rear drum brake ABS for front-wheel-drive cars can have the tone ring located inside the brake drum, with the sensor installed through a hole in the backing plate (**Figure 59.7a**). Some have sensors in the hubs or on the back of the hubs (**Figure 59.7b**). Many rear-wheel-drive ABS vehicles have a sensor mounted in the differential (**Figure 59.8**). A tone ring on the differential ring gear provides the necessary speed information (**Figure 59.9**). Some systems have a sensor on the transmission tailshaft or on the transfer

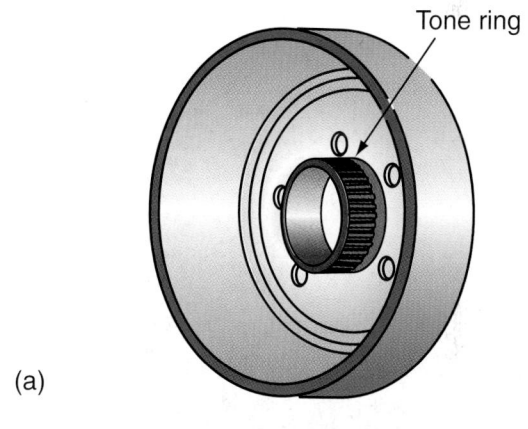

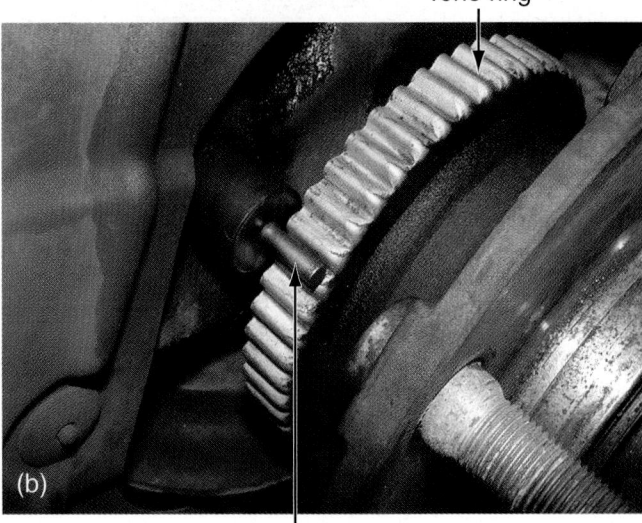

Figure 59.7 (a) Rear wheel ABS with the tone ring inside the drum. (b) Rear wheel ABS sensor and tone ring on the back of the hub. (b, Courtesy of Tim Gilles)

Figure 59.8 An ABS sensor mounted on the differential. *(Courtesy of Tim Gilles)*

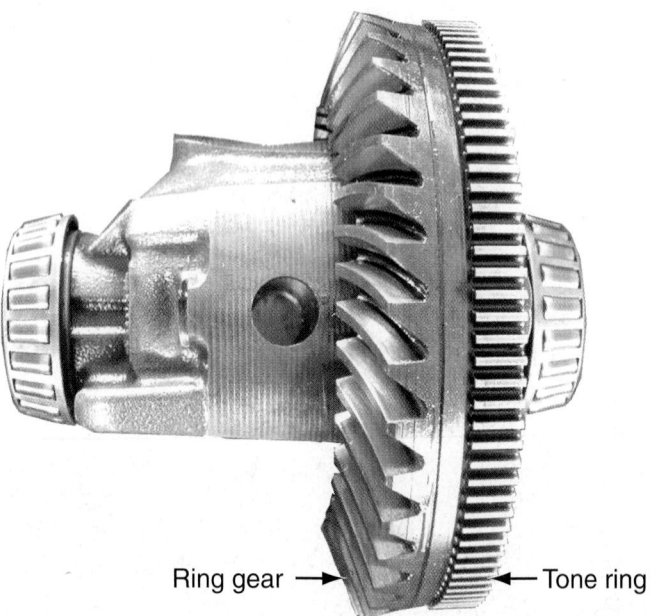

Figure 59.9 An ABS tone ring on a differential ring gear. *(Courtesy of Tim Gilles)*

case in four-wheel drives. AC waveforms are prone to radio frequency interference (RFI) with their signals. To avoid RFI, wheel speed sensor wiring is enclosed in shielded (braided metal) tubing or is wound in twisted pairs. The latter are wires that are wound around each other about five to ten times per foot. Twisted pairs self-cancel RFI.

Hydraulic Control Valve Assembly

A **hydraulic control valve assembly**, sometimes called an **electromechanical hydraulic (EH) unit**

or **electrohydraulic control unit (EHCU)**, has mechanical and electrical parts that cause hydraulic pressure to pulsate or modulate. It operates when one wheel's speed drops a certain amount below the speed of the other wheels on the vehicle.

NOTE: *When the wheels are rotating normally during braking, the ABS is bypassed and the brake system operates in its normal base mode.*

■ TYPES OF ANTILOCK BRAKE SYSTEMS

The control valve assembly can be combined with the master cylinder and brake booster in an integral system (**Figure 59.10**), or it can be separate in a non-integral system (**Figure 59.11**). Several manufacturers produce ABS. Bendix, Bosch, Delco/Delphi (GM), Continental/Teves, Kelsey-Hayes, and Lucas Girling are some of the companies. You will need to know which system you are working on before starting any repairs.

Integral ABS

Most of the earlier antilock systems were **integral** systems (see Figure 59.10). This means that they combine the master cylinder, power brake booster, and ABS hydraulic circuitry in one single hydraulic assembly. Hydraulic circuitry includes the following:
■ The master cylinder and reservoir
■ A brake booster that operates hydraulically, rather than with vacuum
■ A pressure pump and motor
■ A pressure accumulator
■ Switches that monitor pressure
■ Pressure modulator valves
■ A fluid level sensor

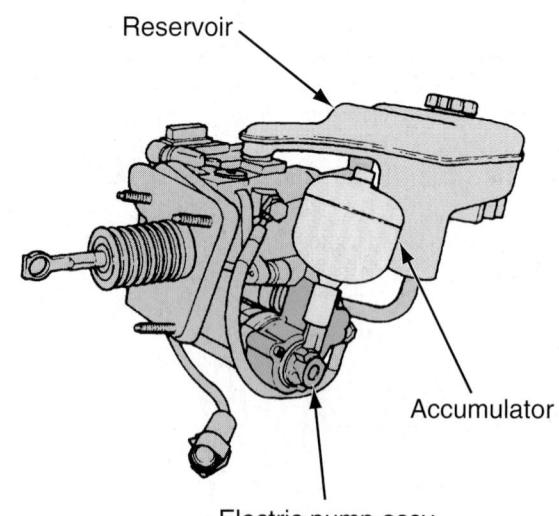

Figure 59.10 An integral ABS. *(Courtesy of Ford Motor Company)*

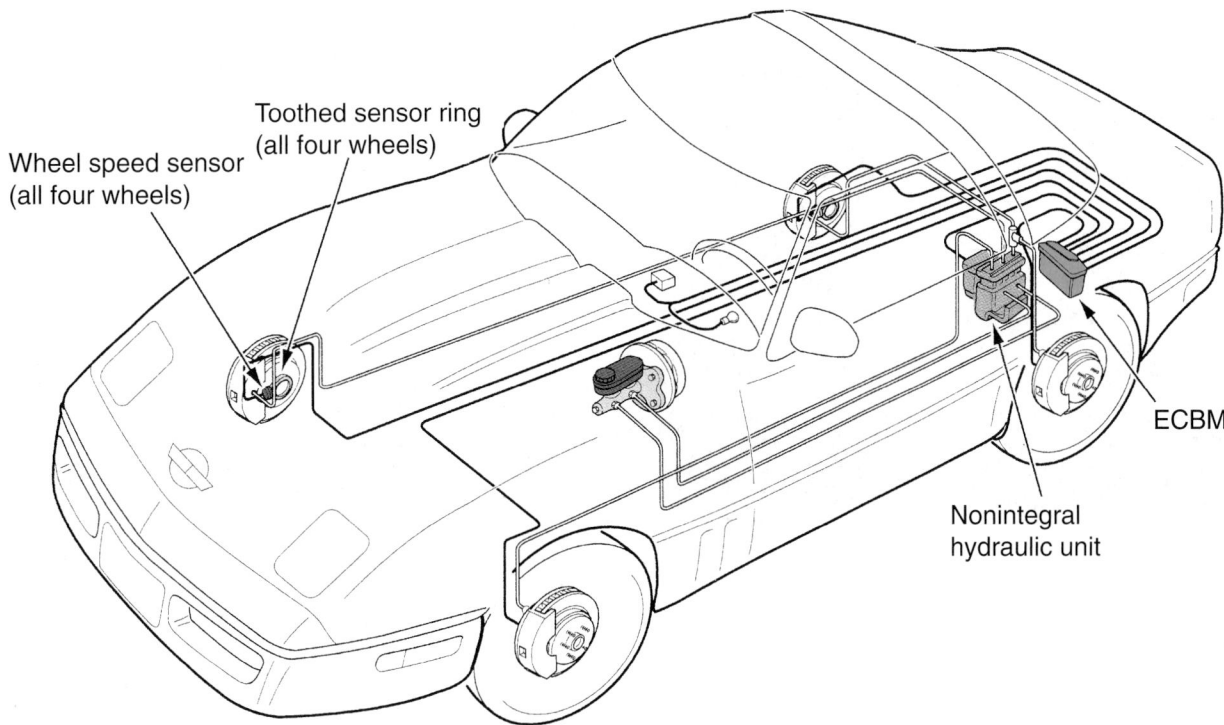

Figure 59.11 A nonintegral ABS.

Notice that this system includes an integral hydraulic brake pressure booster. The booster works in both ordinary and ABS stops. It has a pump that provides the boost pressure. Hydraulic pressure to the rear brakes comes from the pump only, not from the master cylinder. If the pump fails, the pedal will be hard. The rear brakes will not work either. These systems are found often on front-wheel-drive cars, so this is not as severe a problem as it would seem at first. Rear brakes on front-wheel-drive cars only provide about 20% of the braking power during a stop.

Integral systems were used from 1985 until the early 1990s. Some of the integral antilock systems include Bendix 4, 9, and 10; Bosch III; Teves Mark II; and Delco Moraine III. Bendix 4, the last integral system used, stopped production with the 1995 Jeep.

Pumps and Motors. Earlier integral systems used the pump for a source of hydraulic pressure. They also have accumulators as part of the modulator (see Figure 59.10). An accumulator is filled with a charge of high-pressure nitrogen gas (**Figure 59.12**). It stores brake fluid under very high pressure to keep a ready source of constant pressure to operate the ABS when needed. Safety precautions regarding the service of these systems are covered later.

Reservoir. The brake system reservoir is usually much larger than a normal brake system reservoir. Some systems use a second reservoir. This reservoir, which acts like a holding tank, collects pressurized fluid as it returns from the brakes at the wheels. Its location in the hydraulic circuit is before the pressure pump.

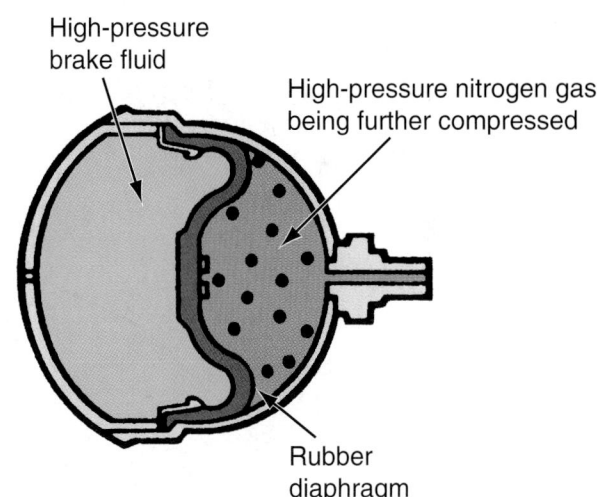

Figure 59.12 Pressure in an accumulator. (*Courtesy of Federal-Mogul Corporation*)

Pump Control Switches. Some systems have a pressure-sensitive switch that monitors accumulator pressure. It turns on the pump when pressure drops to a certain level. If pressure drops below a certain level, an amber ABS warning light on the dash will come on. This light is in addition to the red brake hydraulic system warning light.

Nonintegral ABS

Some of the earlier and all of the later model antilock systems are **nonintegral**, also called remote or add-on ABS. They have gained in popularity because of their

Hydraulic brake lines

ABS unit   ABS module   ABS motor

Figure 59.13 Typical nonintegral ABS. *(Courtesy of Tim Gilles)*

lower cost and relative simplicity. They have a conventional power brake and master cylinder. The ABS unit is separate from the master cylinder, in series with its brake lines (**Figure 59.13**). Since the early 1990s nonintegral three- and four-wheel ABS use a hydraulic pump to circulate fluid. They do not use the pump for power assist and do not have to be depressurized prior to repairing the brakes like an integral system.

Nonintegral ABS include:
- Bendix III, 4, 6, and Mecatronic II
- Bosch 2, 2S, 2U, Micro, and ABSR
- Delco Moraine VI with and without traction control
- Kelsey-Hayes RWAL, RABS, and 4WAL
- Nippondenso
- Sumitomo 1 and 2
- Teves Mark IV
- Toyota Rear Wheel

The method by which fluid pressure is controlled depends on the design of the system. ABS can be either *two wheel* or *four wheel*. They can be *one-*, *three-*, or *four-channel ABS*. Their components and operation are typical of more complicated systems as well. A four-wheel system is basically a single channel times four.

■ TWO-WHEEL ABS

Two-wheel ABS only works on the rear wheels (**Figure 59.14**). Called *single channel*, they are found on sport utility vehicles (SUVs) and light trucks. ABS is especially needed on rear wheels in these vehicles. The brakes are designed to be able to stop a fully loaded truck. When the truck is empty, the rear brakes are prone to locking up. Single-channel systems offer a big improvement to an earlier addition to rear brakes—the proportioning valve. Proportioning valves are still included in the rear brake system along with ABS.

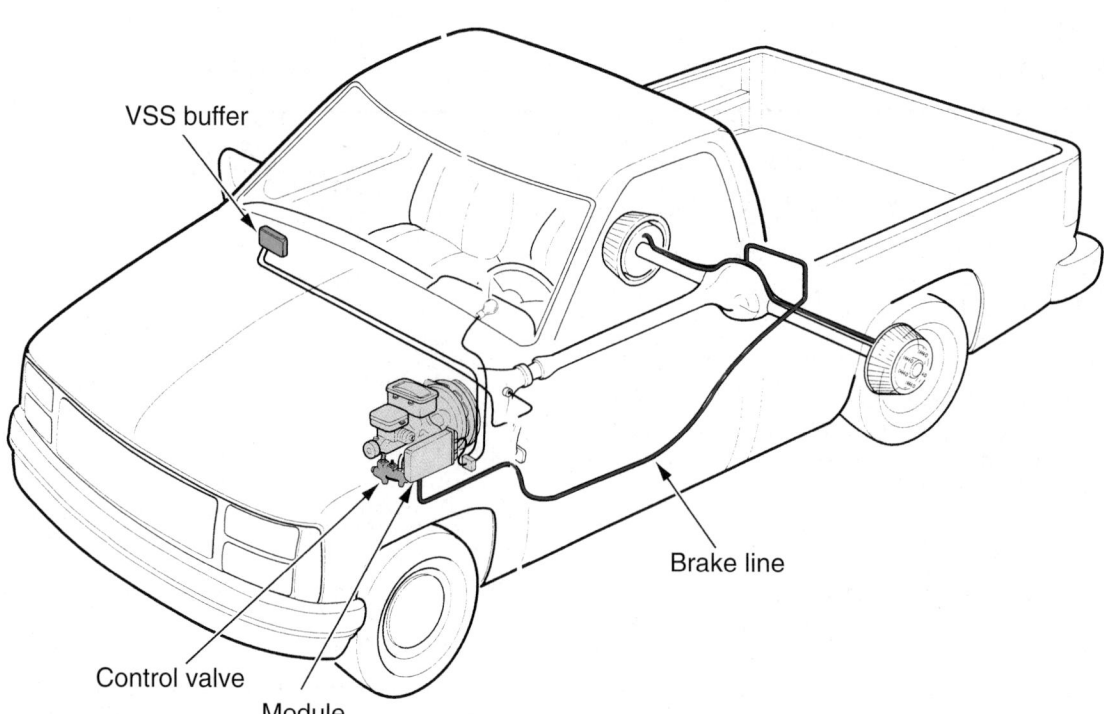

VSS buffer

Brake line

Control valve

Module

Figure 59.14 This Kelsey-Hayes two-wheel RWAL system provides anti-lock control on only the rear wheels.

Two common names of two-wheel Kelsey-Hayes ABS are **rear-wheel antilock (RWAL)** and **rear antilock brake system (RABS)**. RWAL is the name used by GM, and RABS is the name Ford and Chrysler use. The difference in the two systems is that GM uses a single warning light. Ford and Chrysler use two lights. Toyota has a two-wheel ABS for its light trucks, too.

To control skids, the rear brakes are modulated at the same time. A centrally located speed sensor is normally found on the top of the differential, in the transmission, or in the four-wheel-drive transfer case. The front brakes are normal brakes without any extra control.

The system only needs one sensor. With a standard differential, when one rear wheel locks, the other wheel will try to speed up and the drive shaft changes speed rapidly. The signal goes to the ABS computer, which shuts off flow to the rear brakes. There is no need for the vehicle speed sensor to compare the speed of the differential to vehicle speed. Speed simply changes, and the sensor reacts to it. The system logic knows how fast it is possible for the wheels to slow down during braking. This is called *deceleration factor.*

Some manufacturers offer an off/on switch for ABS because sometimes it may be desirable to shut it off. When driving in snow, for instance, a locked-up wheel builds a wedge of snow in front of it, which aids in stopping. The single-channel system is disabled on four-wheel-drive vehicles when in four-wheel drive. Stopping power is more equalized in four-wheel drive. Also, the sensor has too much trouble making good decisions when it is getting input from all four wheels through the drive shaft and transfer case.

Four-Wheel ABS

Four-wheel ABS can be either three channel or four channel. Front wheels in both types are controlled separately. Most rear-wheel drives and some front-wheel drives use *three-channel* systems. A three-channel system has a sensor at each front wheel. The rear works like the single-channel, two-wheel system. One sensor controls both rear wheels together. The rear sensor can be mounted in the differential housing, receiving signals from a notched ring mounted to the outside of the differential ring gear (see Figure 59.8). This sensor often doubles as the vehicle speed sensor (VSS).

The most effective ABS is the *four-channel* system with a sensor monitoring each wheel. It is the most expensive system because a wheel speed sensor is needed at all four wheels.

Some of the newer systems use dynamic proportioning, which eliminates the traditional proportioning valve to the rear brakes (see Chapter 57). The brake pressure modulator valve's (BPMV's) rear inlet valves are cycled during braking to maintain correct front-to-rear brake balance. If the rear brakes decelerate more than the front brakes, the system activates.

Unlike two-wheel ABS, four-channel systems continue operating when in four-wheel drive. On some systems, a G-sensor (an electronic device that measures inertia) tells the controller the rate of stopping while in four-wheel drive.

NOTE: *Having a sensor at each wheel does not necessarily mean that the ABS is four channel. Some three-channel systems have four-channel braking. They have four-wheel sensors but control the rear brakes together hydraulically.*

▬▬ ANTILOCK BRAKE SYSTEM OPERATION

Signals from each of the wheel sensors are pulsed to the ABS control module. When a wheel begins to lock up, the signal from its sensor drops off rapidly. To prevent lockup, the control module blocks further hydraulic fluid pressure to that wheel. If the frequency of the signal continues to drop off, fluid pressure will be *released* to that wheel. This is called pressure **dump** or *decay*. When the wheel begins to rotate freely once again, pressure is reapplied and braking can continue. If the wheel locks again, the cycle can repeat for up to 10–20 times a second.

In most ABS, solenoid valves control the holding and releasing of hydraulic system pressure (**Figure 59.15**). Solenoid valves are small and weigh very little, allowing them to move fast. A typical ABS has two valves for each channel, an isolation valve and a dump valve. An *isolation* valve, also called a *block* valve or a *hold* valve, is normally open, allowing brake fluid to flow. The isolation valve is always the first valve to operate during an ABS stop.

Single-Channel Antilock Brake Operation

During a two-wheel ABS stop, the isolation (hold) valve closes to prevent further hydraulic pressure from reaching the rear brakes. When action by the isolation solenoid is not sufficient to prevent wheel lockup, the normally closed dump (release) valve cycles open and closed rapidly to bleed system pressure. Excess brake fluid flows into a spring-loaded accumulator chamber (**Figure 59.16**). A drop in pedal height can occur that is limited to the size of the accumulator reservoir. Once the accumulator is full, pedal height cannot drop any further. The accumulator is very small because very little fluid needs to flow to achieve a drop in fluid pressure.

On some systems, the pedal can drop to the floor as the accumulator fills. The driver will have to pump the pedal to get a fresh charge of fluid from the master cylinder to raise pedal height. Rear-wheel ABS does not have a pump to increase pressure or raise pedal height.

Once pressure to the rear brakes has been relieved and the computer senses the rear wheels turning again, the dump valve closes. The isolation valve remains closed, and pressure to the rear is resumed

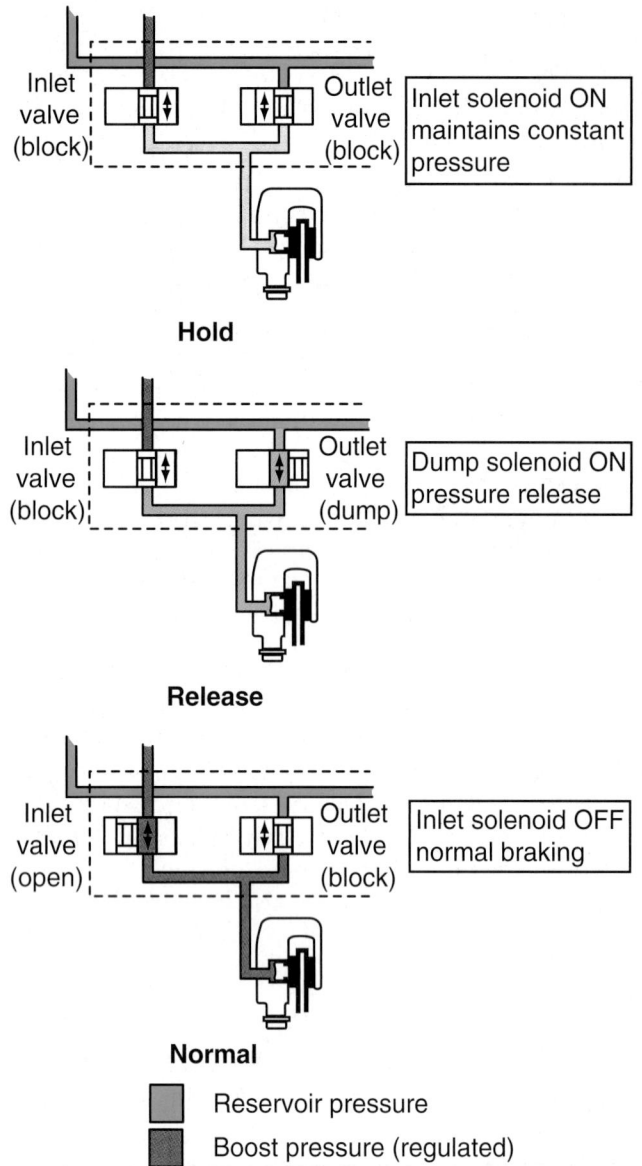

Hold

Inlet valve (block)   Outlet valve (block)

Inlet solenoid ON maintains constant pressure

Release

Inlet valve (block)   Outlet valve (dump)

Dump solenoid ON pressure release

Normal

Inlet valve (open)   Outlet valve (block)

Inlet solenoid OFF normal braking

■ Reservoir pressure
■ Boost pressure (regulated)

Figure 59.15 Solenoid valves control the holding and releasing of hydraulic system pressure.

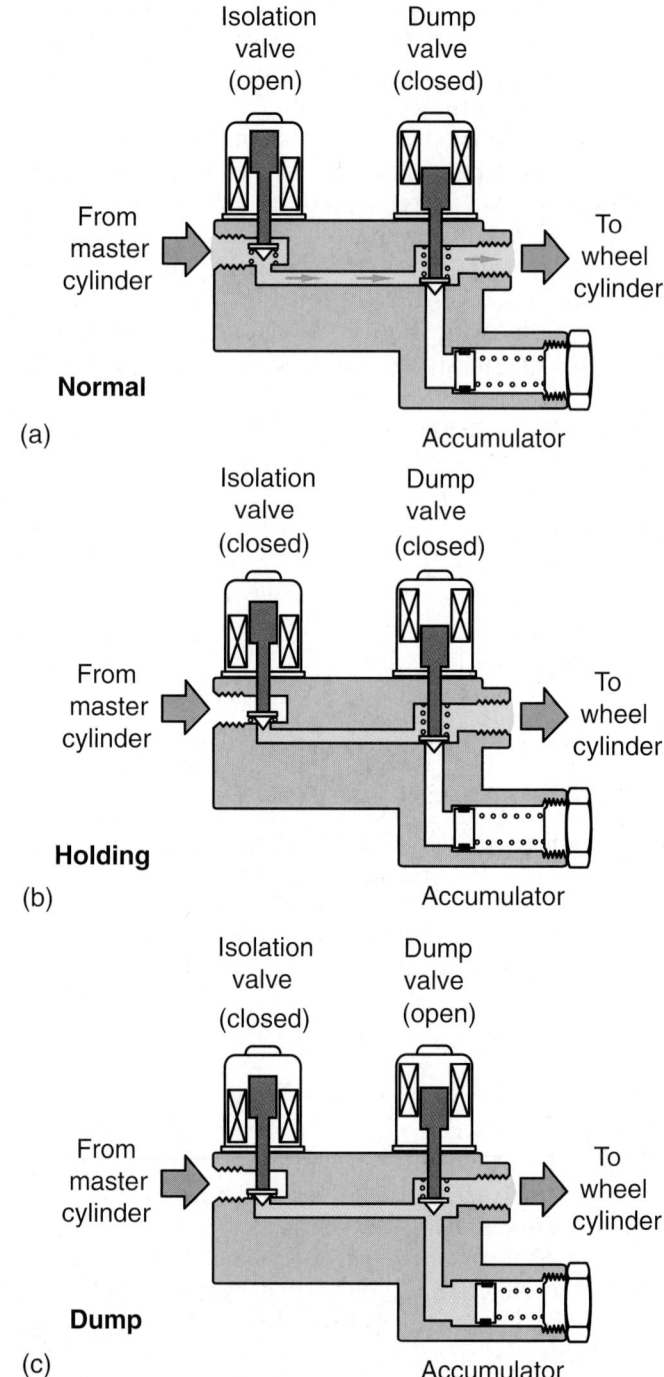

Normal
(a)
Isolation valve (open)   Dump valve (closed)
From master cylinder   To wheel cylinder   Accumulator

Holding
(b)
Isolation valve (closed)   Dump valve (closed)
From master cylinder   To wheel cylinder   Accumulator

Dump
(c)
Isolation valve (closed)   Dump valve (open)
From master cylinder   To wheel cylinder   Accumulator

Figure 59.16 (a) During normal braking of this RWAL system, the isolation valve is open and the dump valve is closed so fluid flows between the master cylinder and the rear brakes. (b) During the pressure-holding stage, the isolation valve closes. (c) During the pressure-reduction stage, the dump valve opens to bleed pressure to the accumulator.

as fluid pressure returns to the rear brakes from the accumulator. When the wheels are turning normally again, the isolation valve reopens to allow normal braking to resume. Remember, all of this happens very fast.

Three- and Four-Channel ABS Operation

In some three- and four-channel systems, a single combination valve is used. It has a three-position solenoid that controls normal fluid flow, hold, and release. These are the three stages of ABS solenoid operation:

1. Pressure *buildup*, also called *increase*. This is *normal braking*. Neither the inlet nor outlet solenoid valves are energized by the computer. With both valves open, pressure from the master cylinder flows normally to the brakes. This mode is called buildup, or increase, because the

driver can still *increase* pressure to the brakes by increasing pedal pressure.

2. Pressure *hold*. The computer detects a wheel slowing rapidly during a stop. It shuts off further flow from the master cylinder, blocking

a further increase in pressure at the wheel. System pressure remains constant at that point.

3. Pressure *reduce*. The computer reduces pressure to a wheel by opening a solenoid valve. Pressure escapes to a low-pressure area in the system. This can result in a drop in pedal height. When pedal travel reaches about 40%, the electric pump increases pressure in the system. The computer energizes a relay, turning on the pump until the pedal height rises enough to close the switch. The pump is capable of more pressure than that required by the ABS. Relief valves return the excess fluid pressure to the master cylinder reservoir.

Electromagnetic Antilock Brakes

Some nonintegral systems, such as the Delco ABS-VI, do not use an accumulator or electric pressure pump. These systems use a motor pack instead (**Figure 59.17**). Three small, high-speed, bidirectional screw-drive electric motors increase and decrease fluid pressure in each wheel circuit. They are positioned quickly and accurately in their bores.

The following describes the systems' operation:
■ One of the motors controls both rear brakes.
■ A separate motor controls each of the front brakes (**Figure 59.18**).

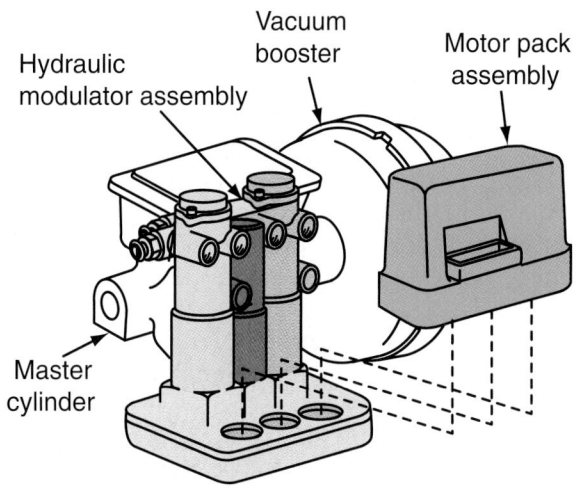

Figure 59.17 An ABS motor pack.

■ An electromagnetic brake in the pump provides precise control. It stops the pump immediately as it alternately raises or bleeds pressure. This brake is important because it must hold the piston's position against hydraulic pressure applied by the foot brake. Some of these systems have nine brake pads in the ABS unit, three on each piston.

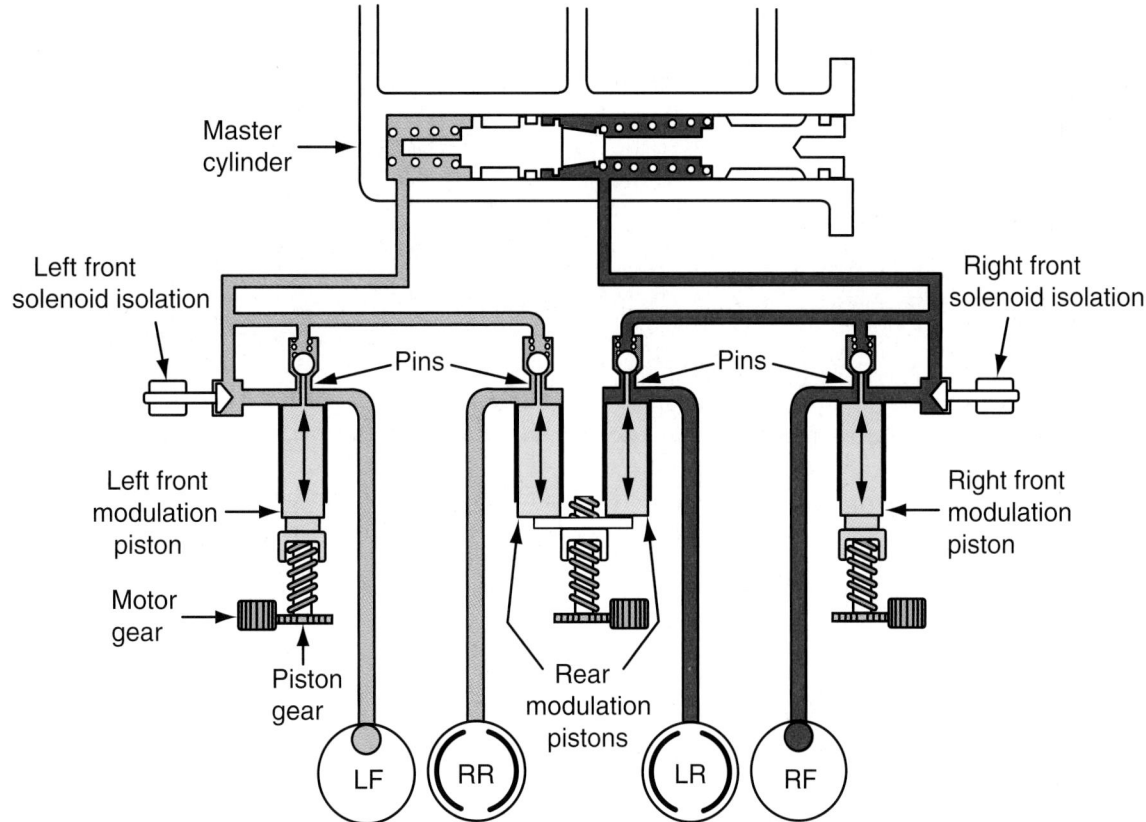

Figure 59.18 An electromagnetic ABS. This is a split diagonal system. Pins at the tops of the pistons hold the check valves open during normal braking.

Conventional wheel speed sensors and a computer are used with this system. A conventional master cylinder and power booster are used as well. The system does not have a high-pressure pump, so system pressure is close to normal brake system pressure.

During normal brake operation, the modulator pistons are in their highest position, called "home." A pin extends from the end of each piston to hold its check valve open (see Figure 59.18). This allows normal braking. The system is a split diagonal system. However, during an ABS stop, the rear pistons are isolated from the master cylinder and the system is no longer split diagonally.

During an ABS stop, one or more of the pistons are driven downward by the electric motor(s). If a front brake piston is moving, the solenoid must be closed. When the piston moves down, the pin moves away from the check valve, isolating the brake cylinder that it controls from the master cylinder. Pressure remains the same as long as the piston does not move farther. There is no dump valve as in other antilock systems. If additional pressure reduction is needed, the computer commands the motor to move the piston farther downward. This drops, or dumps, the pressure. When the wheel starts turning, the pistons are moved back up in their bores to resume normal braking.

Electrically operated pistons can cycle 7 times per second, compared to 10–20 times for a typical solenoid-type ABS. The two solenoids in an electromagnetic system are a safety backup provided on the front brakes only. They are normally open. During an ABS stop, they close to allow the system to be isolated from the master cylinder. If ABS fails while one or both of the front pistons are down in their bores, the solenoids will provide an alternate circuit so the front brakes can still operate. There will be no rear brakes in this instance.

ABS Safeguards

If a malfunction occurs in ABS operation, the computer shuts the system off. The control module has a diagnostic procedure that starts when the vehicle starts and finishes at a speed of from 4 to 10 mph. The ABS remains asleep until it receives a signal from the brake switch. When the ABS is shut off, the brake system operates as a normal system.

NOTE: *If an emergency spare is used, for instance, the difference in wheel rotation speed on that axle is sensed. The ABS will not function, and the ABS light will come on.*

Brake Performance during an ABS Stop

When testing ABS in a hard stop, you should feel the solenoids or motors pulsating in the brake pedal. The amount varies by system.

During an ABS stop due to a loss of traction, the correct response by a driver should be to press even harder on the brake pedal. Even if only one wheel is locking, you will feel the pulses of the controller. The ABS should allow you to steer away from a hazard, but harsh moves of the steering wheel during an ABS stop can result in spinout. ABS braking distances can be longer on some surfaces because a locked tire digs into a rough surface with greater adhesion.

CASE HISTORY *Emergency vehicles are equipped with a switch to turn the ABS on and off. A skid is sometimes desired when attempting to get a car to spin. A highway patrol officer received notice of a burglary in progress. As he sped toward the business address, the suspect's speeding vehicle passed him going in the opposite direction on the street. From habit, the officer slammed on the brakes while attempting to throw his vehicle into a skid in order to make a quick 180-degree turn. The antilock brakes would not allow the inside wheel to stop turning, and he lost control of the vehicle.*

Traction Control

The ABS is sometimes used to limit the amount the wheels can spin during acceleration. This is called a **traction control system (TCS)** or **acceleration slip regulation (ASR)**. If a driving wheel starts to spin during acceleration, the controller applies hydraulic pressure to its brakes.

Electronic engine controls work with the ABS to make traction control even more precise. Most systems limit slip by reducing the effect of accelerator pedal movement and engine power. Bosch, the company who developed the first ASR system in 1986, refers to the application of traction control as "throttle relaxation."

The computer's objective is to match traction with engine power. An engine's power output can be controlled by:

- Disabling fuel injectors (**Figure 59.19**)
- Retarding ignition timing
- Upshifting the transmission
- Closing the throttle (**Figure 59.20**)

Sensors provide the computer with information on vehicle speed and cornering forces. The computer makes a slip threshold calculation. Traction (brake intervention) has a higher priority at low speeds, and directional control has priority at speeds above 50 mph. The system can engage engine control or brake intervention, either separately or at the same time. It is possible to implement engine controls very quickly, while brake intervention is relatively slow. At lower speeds, brake intervention is quick enough to prevent wheel spin, however.

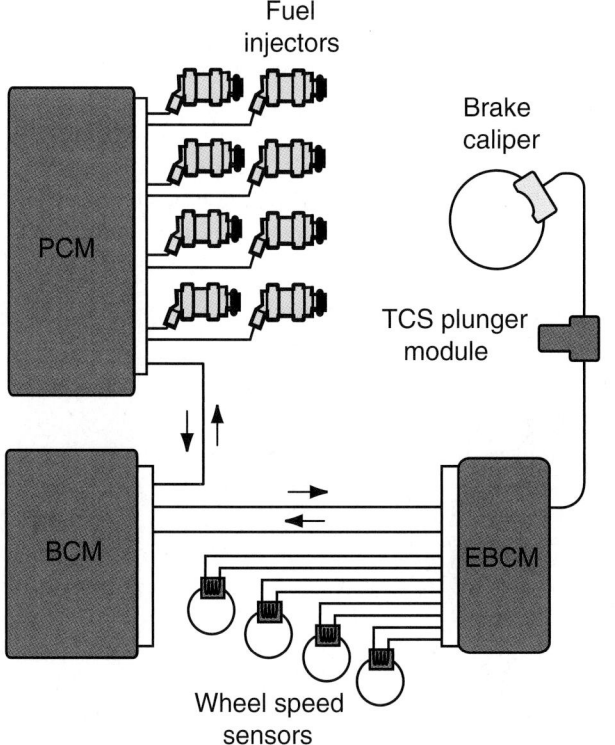

Fuel
injectors

PCM

Brake
caliper

TCS plunger
module

BCM

EBCM

Wheel speed
sensors

Figure 59.19 TCS controls for the Bosch III system. The PCM can cut off one to four fuel injectors on the V8 engine for torque management.

Electronic Stability Control

Stability programs use engine and ABS brake computers to stabilize the vehicle in the event a sudden evasive maneuver results in an unstable condition (**Figure 59.21**). Using a combination of the features from antilock brakes (ABS), TCS, electronic brake-force distribution (EBD), and active yaw control (AYC), the computers determine if the car is actually traveling in the direction that its driver steered. If understeer is sensed, hydraulic pressure is increased to the inside rear brakes to help correct it (**Figure 59.22**). Oversteer is corrected by applying the outside front brake. Previously on luxury vehicles only, this system can now be found even on compact cars. The GM Delphi Traxxar vehicle stability enhancement and the Continental Teves electronic stability program (ESP) are examples of these systems. Electronic stability control (ESC) has now been mandated by the National Highway and Traffic Safety Administration (NHSTA) for all cars and light trucks by 2012. Many SUVs already have ESC as standard equipment to reduce rollover.

The program can apply brakes at any one of the vehicle's wheels. The computers monitor the following:
- Wheel speed sensors
- A steering angle sensor
- A yaw rate sensor
- A lateral acceleration sensor

■ ANTILOCK BRAKE (ABS) SERVICE

Antilock systems have proven to be very trouble-free. In fact, less than 1% of problems in the brake system come from the ABS. If there is a brake problem and the ABS light on the dash is not illuminated, the likelihood is that the conventional brake system is the cause of the problem. Be sure to check out the normal brakes first.

The ABS controller disables traction control if it decides that the brakes are likely to overheat.

Another feature made possible by traction control is *electronic limited slip*. Limited slip differentials work mechanically. Electronic limited slip applies the brakes to a wheel that is spinning in mud or on ice, sending the torque to the wheel on the other side of the vehicle.

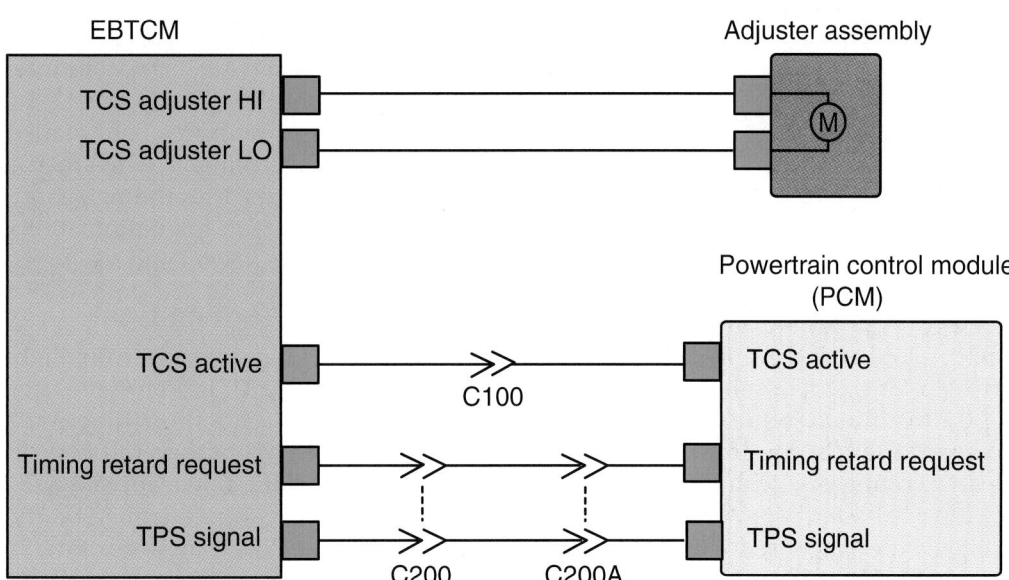

EBTCM

TCS adjuster HI
TCS adjuster LO

Adjuster assembly

M

TCS active

C100

Timing retard request

TPS signal

C200   C200A

Powertrain control module
(PCM)

TCS active

Timing retard request

TPS signal

Figure 59.20 Diagram of Delphi TCS system with throttle adjuster.

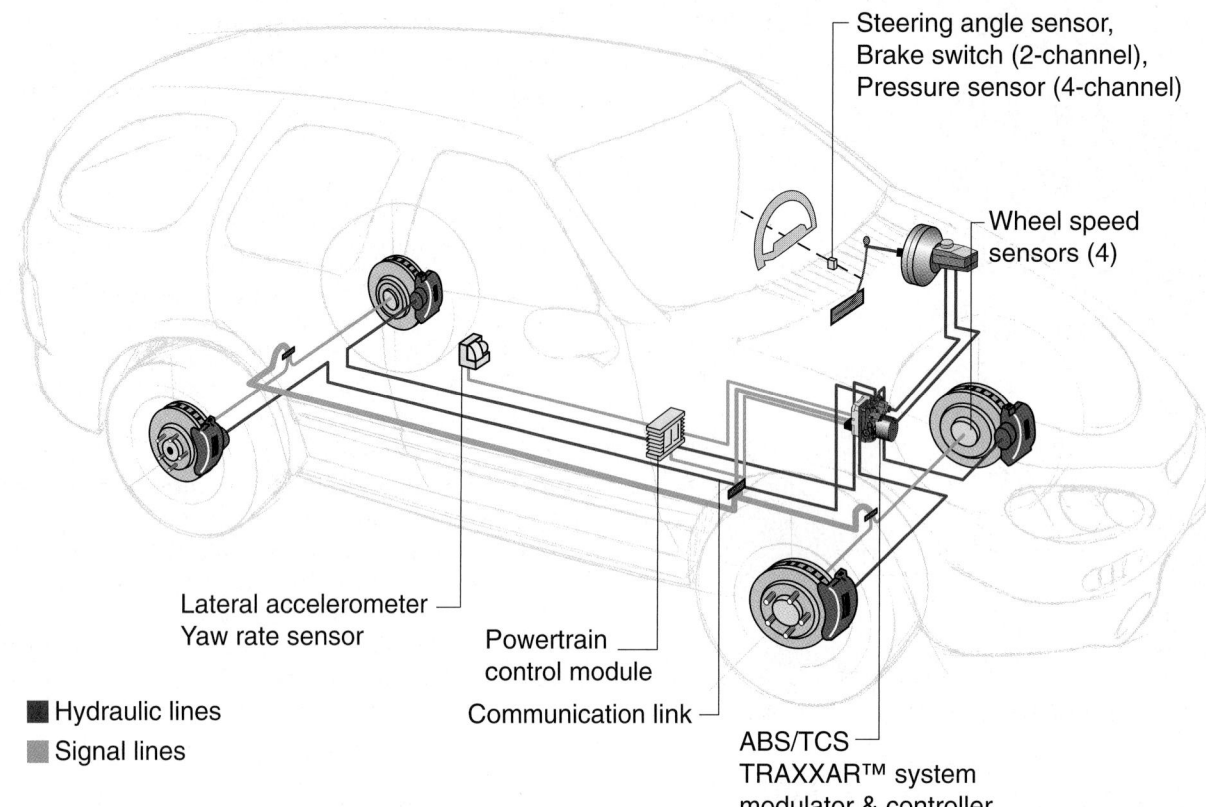

Figure 59.21 Parts of an electronic stability control system. *(Courtesy of Delphi)*

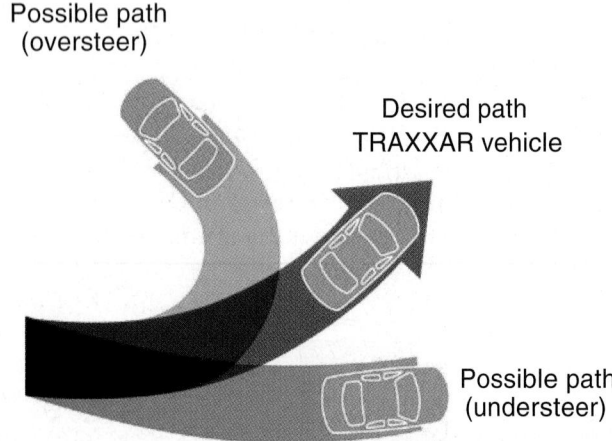

Figure 59.22 Understeer and oversteer are corrected by the electronic stability control program. *(Courtesy of Delphi)*

One way to test for ABS operation is to test drive the car in an empty parking lot. Apply full pressure to the brakes during a hard stop at about 25 mph. If the ABS is functioning, the system should prevent a skid. Pulsation can be felt in the brake pedal, and sometimes there are buzzing or clicking sounds from the modulator.

Electrical problems are diagnosed using the vehicle's on-board diagnostics. Detailed ABS service manuals are available from brake manufacturers.

Red Brake Warning Light

There are two kinds of brake warning lights. An amber light is for ABS problems only. A *red light* is the typical hydraulic system warning light. A low fluid level or low pressure in half of the hydraulic system will illuminate this lamp. It also serves the dual purpose of providing a light that warns the driver that the parking brake is applied.

On integral ABS, the red light can come on when there is a failure. This results in the rear part of the system not operating. Some ABS have three intensities of brightness to the light. The parking brake is the brightest, a hydraulic failure results in a light of medium brightness, and the red light is dimly lit when there is an ABS failure. The warning signal passes through a resistance. To verify whether the light is at full illumination, apply the parking brake with the key on and see if it is brighter.

Amber Brake Warning Light

The amber (yellow) brake warning light is for ABS failures (**Figure 59.23**). It comes on after engine startup and remains on for a short time during the self-test. The amber light operates as follows:

■ With the parking brake applied and key on/engine off, both lights should come on. The amber light will go out in less than 4 seconds.

■ When the engine starts, the amber light will come on for about 4 seconds before going out (if there

ABS light

Figure 59.23 An amber warning light is used for ABS failures. *(Courtesy of Tim Gilles)*

are no problems). The self-test is taking place during this time.

■ After the engine is running, if both lights are on at the same time, the master cylinder has run low on fluid.

■ The pressure modulator valve (PMV) has its own fluid reservoir. Low fluid in that system would light the amber light but not the red light. A code will not set with this condition.

Diagnosing ABS Problems

When diagnosing a problem with the ABS, the first step is to inspect and repair the brake system as if it were non-ABS controlled. An example of a problem with the normal brake system that can mimic an ABS problem is false modulation. *False modulation* is when the ABS operates when it is not supposed to. Some early cars with ABS had this problem during their warranty period because rear drum brake shoes had the wrong coefficient of friction, which caused the ABS to operate during moderate stops. Replacing the rear linings was the recommended repair that solved the problem. Glazed or oil-soaked linings can result in a similar situation. When a pulsating pedal is the symptom, check for grabbing brakes.

SHOP TIP To check for a grabbing brake, drive the vehicle and operate the parking brake.

Other things that can affect ABS operation include damaged or incorrect drum brake return springs or having two primary linings installed on one side of the car with two secondaries on the other side.

A customer might complain that sometimes his car feels like it is accelerating during a stop. This can result when a problem with a wheel speed sensor causes the ABS to restrict hydraulic pressure to one wheel at low speeds.

Distortion of the disc brake caliper seal is what normally retracts its piston. If a rotor is warped or a wheel bearing is loose, the piston can be moved deeper into its bore, causing the ABS to operate when stopping.

NOTE: *Tires that are not the same size can cause variations in the speed sensor input to the control module, causing an ABS light to illuminate and store a trouble code.*

Check to see that all of the tires are the correct size. Next check all fuses and connectors. ABS electrical problems share those common to all computer systems. Power and ground connections must all be tested prior to replacing components. The ground connection for the computer is especially important because the computer controls the valves by switching them to ground. Connectors are susceptible to resistance, and there are often several connections. The computer will not allow much deviation from its expected measurement. Poor connections can cause multiple trouble codes. Follow these steps:

1. Use a scan tool to retrieve trouble codes (DTCs); write them down and then erase them. Some older systems use a flashing ABS light to signal trouble codes. Check the service literature for information on these.

2. Test drive to see if a DTC resets.

3. Follow the repair procedure recommended in the service information. Always follow the detailed diagnostic troubleshooting charts in order. Look over the entire chart first and note resistance specifications for electrical components. Experience with a particular repair will sometimes lead you to go right to the test of one part.

4. Erase stored DTCs and test drive again to verify the repair.

Review the wiring diagram before testing an ABS electrical circuit. When measuring voltage and resistance, use a high-impedance multimeter. An analog meter can cause damage to an electronic circuit. Always turn off the ignition before disconnecting components.

▬ ABS BRAKE FLUID SERVICE

There are many types of ABS, but they share a good deal of service similarities. Those items are covered here first, followed by some more specific service items. Always remember to locate the manufacturer's information specific to the vehicle you are working on before attempting to service an antilock system.

Inspecting Brake Fluid Level

Follow the correct procedure for inspecting brake fluid level. Check the service manual for the car or the underhood label. Integral systems appear to be low because fluid is in the accumulator. Some systems must be depressurized before checking. Others require the key to be on with the accumulator pressurized.

Depressurizing ABS

Some integral ABS operate under extremely high pressure. These systems have an accumulator that must be bled of pressure before attempting to service the system.

> **CAUTION** Extra care is necessary when servicing high-pressure ABS. Up to 2,700 psi can be available at the wheels even when the engine is off. Pressure this high is about 20 times as high as shop air pressure. Failing to bleed off pressure before opening a brake line or bleed fitting could result in serious injury to the technician. **It can force brake fluid to go right through the skin on your hand!**

NOTE: *When bleeding the brakes, accumulator pressure will not be in the brake lines unless the brake pedal is depressed. Do not attempt to bleed brakes by depressing the brake pedal while opening and closing bleed screws.*

Before removing a caliper or master cylinder, be sure the system is depressurized. Most systems are depressurized by applying the brake pedal between 20 and 40 times while the key is off. The pedal should feel like a normal power brake with no vacuum left in the booster. After restarting the vehicle, the ABS indicator light will go out when the system is repressurized.

Some Hondas and Acuras require special equipment for depressurizing.

Flushing and Bleeding ABS

Clean brake fluid is especially important to ABS. Contaminated brake fluid can result in costly repairs. Fluid can be checked for moisture contamination using an electronic moisture tester. The test strips change color when there is moisture contamination.

An annual fluid flush is a good recommendation with ABS. Fluid should be changed at least once every 2 years and, certainly, any time the brake pads are replaced. Some manufacturers have longer change interval recommendations because their factory fill is a premium, longer life brake fluid.

NOTE: *Do not use DOT 5 (synthetic) brake fluid with ABS. Silicone brake fluids are not hygroscopic (do not accumulate water) like ordinary brake fluid. Water can accumulate in some areas of the system, causing problems. Use the recommended fluid, which is usually DOT 3 because it can flow easier. If the fluid does not flow easily, air bubbles can*

be introduced, which can cause spongy brakes. The correct fluid type is usually specified on the master cylinder cap or reservoir body.

Before bleeding an ABS, always check the service information. Brake part manufacturers provide brake bleeding sequence manuals. A typical bleed sequence for a front-to-rear hydraulic split system is shown in **Figure 59.24**.

There are many different procedures. Some integral systems require bleeding of accumulator pressure and disconnecting the power to the ABS controller. Then the system is bled like conventional brakes.

Sometimes there is a special procedure for bleeding a brake pressure modulator valve. A scan tool is necessary to run the autobleed sequence (**Figure 59.25**). The ABS solenoids are cycled, and the pump runs to purge air from the modulator. These circuits are

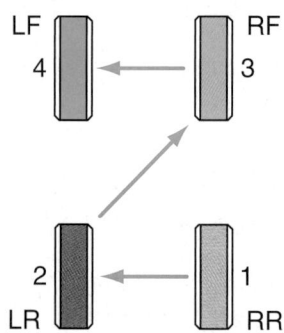

Figure 59.24 Typical bleeding sequence for a split hydraulic system.

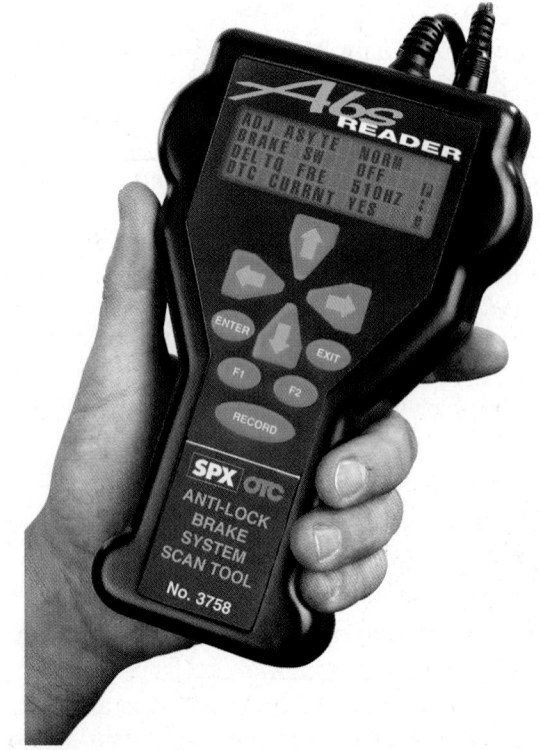

Figure 59.25 A hand-held antilock brake system scan tool. *(Courtesy of OTC/SPX Service Solutions)*

normally closed during normal braking, so the scanner is necessary to put the system into ABS mode.

If you do not need a scan tool to bleed the brakes, the system should be easily bled following procedures in the ABS repair manual. A good rule to follow is not to let the ABS system or master cylinder run dry. Then special bleeding procedures should not be required.

After bleeding the system, put the ABS into operation with a couple of hard stops during a test drive. This will help get any remaining air out of the hydraulic control unit (HCU). It will not bleed the air from the device completely, but it can help after the correct procedures are followed.

Bleed antilock brakes with the key off, or a false code can be set.

Dirty fluid can be forced into the ABS by careless repair practices. Always open a caliper bleed screw before pushing a piston into its bore during a brake pad replacement. Pushing the caliper pistons back in their bores without opening the bleed screws can result in trouble codes due to debris entering the HCU.

Remove fluid from the master cylinder and look for sediment at the bottom. If there is sediment here, it is probably in the control unit also. It might also be possible to clean the control unit as follows:

1. Flush the entire system with a pressure bleeder at 20–25 psi.
2. Be sure to clean the sediment from the bottom of the master cylinder first.
3. Use 2–3 quarts of DOT 3 fluid for the flush.
4. Follow the correct bleeding instructions.

SHOP TIP A safe and easy way to flush all brake systems, especially when you do not have a dedicated scan tool for the vehicle, is to use the gravity method. Simply place containers at all four wheels and open the bleed screws one turn each. Let the system drain slowly while you replenish the fluid supply. The fluid gets replaced but no air gets into the system.

Wheel Speed Sensor Service

The majority of ABS problems result from failure of a wheel speed sensor. Some of them fail because of exposure to harsh operating conditions, and others fail due to abuse. Wheel speed sensors can become demagnetized or polarized by physical impact. This is caused most often by hammering during the removal or installation of front-end components during CV joint service or a Macpherson strut repair.

The computer runs a self-test every time the ignition is cycled. It checks for low or high resistance. When the value is out of range, the computer turns on the dash light and disables the ABS. The computer senses the resistance at the sensor and at the computer and makes a judgment as to which code to set. Verify the resistance of a suspected bad sensor:

1. Infinite resistance is usually due to an open in the sensor's coil winding.
2. High resistance in the sensor is often caused by corrosion in the harness connector due to exposure to water and salt.
3. Too little resistance could be due to a sensor coil that has shorted against another winding. A short like this lowers the total resistance of the coil.
4. If there is almost no resistance, look for sensor wires contacting each other.

Testing a Sensor

A diagnostic flow chart will ask you to check a wheel speed sensor for AC voltage output. To perform the test, spin a wheel while reading its AC voltage output with a voltmeter or an oscilloscope. The flow chart will ask for a voltage reading at a specified speed. Typical voltage output ranges upward with speed from a minimum of at least 650 millivolts (0.65 volt). A faster turning wheel gives higher amplitude (voltage) and frequency. Notice in **Figure 59.26a** how far apart

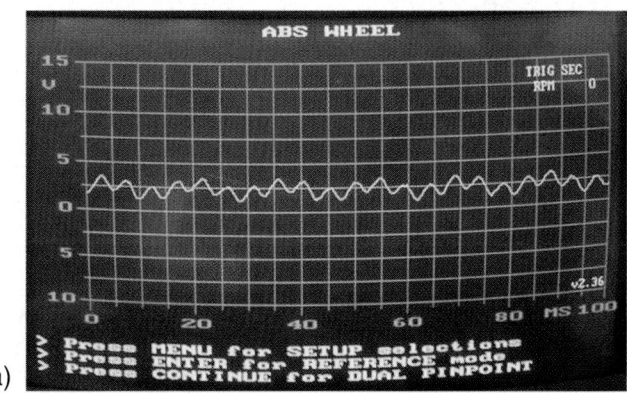

(a)

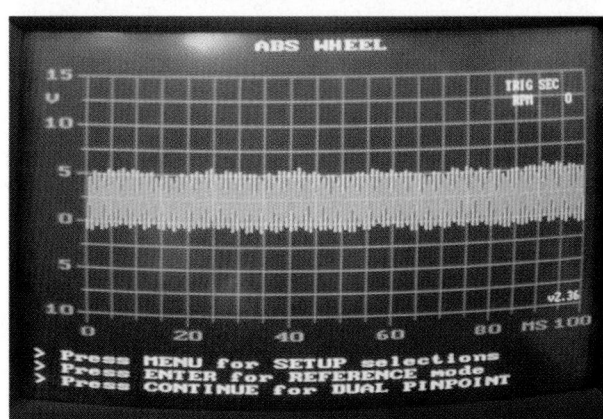

(b)

Figure 59.29 (a) A wheel speed sensor pattern from a slow turning wheel. (b) A wheel speed sensor pattern from a faster turning wheel. *(Courtesy of Tim Gilles)*

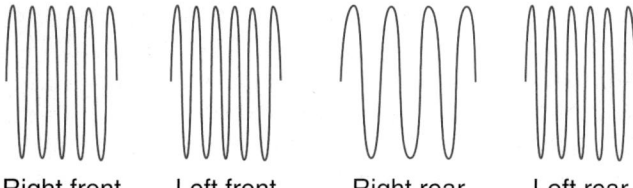

Right front   Left front   Right rear   Left rear

Figure 59.27 When one wheel begins to lock, the sensor sends a signal to the computer.

the oscillations are. The wheel is turning slowly and voltage is low. **Figure 59.26b** shows a faster turning wheel with higher voltage and frequency. If one wheel slows up (**Figure 59.27**), the computer will spot this and act to prevent it from locking. The computer will set a code and turn on the ABS light.

NOTE: *Twisted pair wiring should not be pierced for testing. The wires are purposely long with few connections. Do not use butt connectors to repair them. Do not use a sensor with damage to a twisted wire pair.*

To connect a scope or multimeter, back probe the connector at the wheel. Do not attempt to pierce the insulation or shielding on the wiring loom. Sometimes a rear sensor can be hard to access. You can use the wiring diagram to go to the ABS computer. The sensor generates its own current, so you need both of its wires to make the test. The signal is AC so polarity does not matter as long as you have the correct two wires.

If there is no signal at a wheel, disconnect the connector and measure directly off the pins for the sensor. If there now is a signal, check for a short in the vehicle's wiring harness. If there is no signal with the wheel spinning, check the sensor for continuity with an ohmmeter. **Figure 59.28** shows a differential mounted sensor being tested for continuity.

Metal Shavings

Wheel speed sensors are magnetic, so they attract metal shavings. An erratic sensor signal is the typical result.

The following are conditions where metal can be attracted to a sensor:
- When drum brake linings or disc brake pads wear down to their metal backings, shavings can be picked up by the sensor.
- A drum brake sensor is most apt to collect metal because it is located in close proximity to the brake linings and drum.
- Sensors mounted on the transfer case or differential are also prone to picking up metallic debris from normal wear of the internal parts. **Figure 59.29** shows where metal collects on a differential mounted sensor.
- Turning the rotors with an on-the-car lathe can also result in metal shavings being attracted to sensors located near the rotor.

NOTE: *Some sensors have plastic housings. In addition to being fragile, they could be damaged by spray cleaners.*

Replacing a Wheel Speed Sensor

Damage to the harness is the most common wheel sensor service problem. The wiring on a sensor is not serviceable, and it is sensitive to abuse. Changing the resistance or configuration of the wiring could change the signal sent to the computer by the speed sensor.

Replacing a sensor is not difficult. Simply disconnect its wiring and unbolt it. Be sure to route wires where intended. Interference from the ignition, the alternator, wiper motors, or blower motors can result in electrical "noise" that the computer tries to interpret.

The sensor and harness are sold together as a unit. Sensors can be quite expensive, sometimes costing hundreds of dollars *each*. The wiring is routed like a brake hose so it has enough slack to allow for suspension deflection.

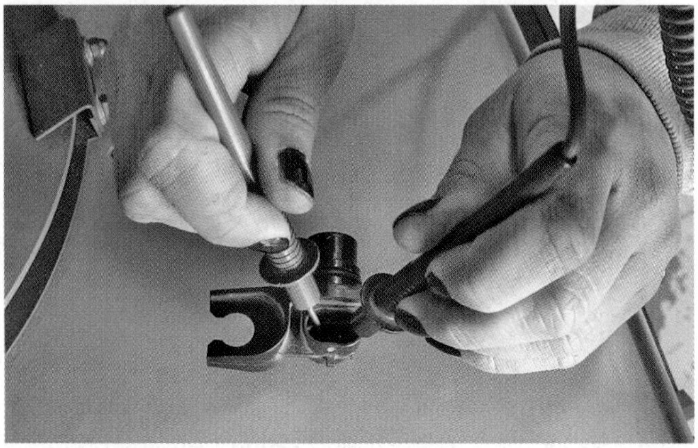

Figure 59.28 Testing a speed sensor for resistance. *(Courtesy of Tim Gilles)*

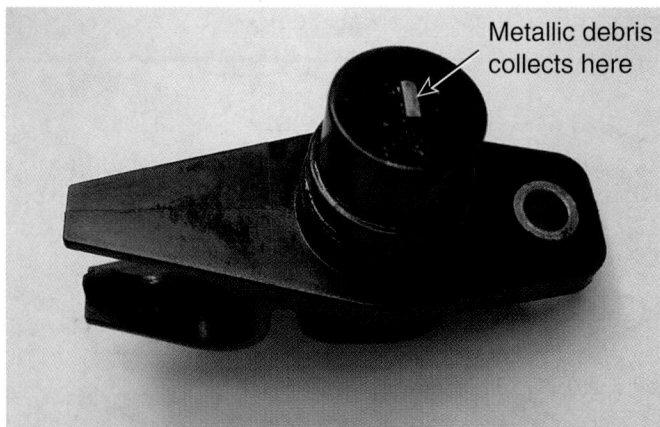

Metallic debris collects here

Figure 59.29 Wheel speed sensors are magnetic. *(Courtesy of Tim Gilles)*

When transmission or suspension work is being done, a safe practice is to disconnect the wiring from its connection at the chassis to prevent the chance of damaging it.

Sensor Air Gap

The gap affects the voltage produced by the sensor. Older systems had adjustable air gaps (**Figure 59.30**). Most newer ones do not. Air gaps range from 0.005" to 0.050". Check the gap all around the sensor ring using a brass feeler gauge (it will not stick to the sensor's magnet). If the gap is too large, the steering knuckle could be damaged, requiring replacement. A gap that is too small can result in contact between the tone ring and sensor if the wheel bearing is loose or when the suspension flexes. This can damage the tone ring or the sensor.

Wheel bearings can change operation of the ABS. When the sensor is on the rotor or axle, looseness in the wheel bearing will affect the air gap. Be sure wheel bearings are not loose or worn. Tighten spindle nuts on front-wheel-drive axles to the correct specification. A wheel bearing that is bad can also result in a trouble code.

Tone Ring Inspection

Some types of tone rings are susceptible to being damaged. Look for damaged teeth, which can cause the ABS to stop functioning. Externally mounted (on the CV joint) tone rings live in the harshest environment. They are exposed to all sorts of road hazards in all kinds of weather. Look for tone ring damage if a new problem has surfaced after recent front-end service work, or after replacing a half shaft with a rebuilt one lacking a tone ring.

The diameter and number of teeth on the tone ring vary with the application. This can result in a problem if an axle was replaced that had the wrong number of teeth on a tone ring mounted on its outer CV joint.

Use a press when replacing a tone ring. Do not hammer it into place. Besides the chance of bending or warping the ring, this can cause magnetism or change the polarity of the ring. A tone ring that has suffered an impact that causes it to be out-of-round can confuse the computer with variations in air gap. It sees this wheel going at different speeds and causes the system to go into antilock mode.

Some rear-wheel sensors are integral with the wheel bearing. These cannot be replaced separately. An erratic sensor signal can also result from a bad wheel bearing. In either event, the entire wheel bearing assembly must be replaced.

Additional Precautions with ABS

There are several precautions common to all antilock systems.

- Do not fast charge the battery with the computer connected. A bad battery will allow voltage to rise too high.
- Do not use a battery charger on the fast charge/boost setting to jump start a vehicle with ABS. Slow charge first or disconnect the negative battery cable before fast charging.
- Do not arc weld on the frame with the computer connected.
- Do not install an antenna (CB radio, cellular phone) near the ABS controller.
- Do not change the tire size other than width. Taller or shorter tires have different rotational speeds. Using the mini spare can disable the ABS.
- Do not disconnect or reconnect electrical ABS parts while the ignition switch is on. This can cause an electrical current surge, which can damage electronic parts.

Integral ABS Service

An integral ABS uses the same inputs as nonintegral. Sensor and fluid concerns are the same. Follow system depressurization instructions provided by the manufacturer.

Most manufacturers of integral ABS service their hydraulic assemblies as a unit. This can be very costly to the customer. This means that the motor, pump, and other parts are not readily available. Other manufacturers sell the pump, motor, and valve body assemblies as separate parts.

If the pump fails in an integral ABS, power-assisted braking will not work. Additional pedal effort will also be necessary because only the front brakes will operate.

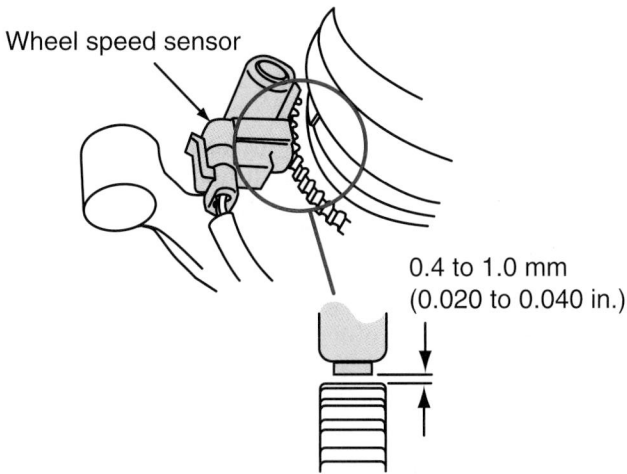

Wheel speed sensor

0.4 to 1.0 mm
(0.020 to 0.040 in.)

Figure 59.30 Wheel speed sensor gap measurement.

Rear-Wheel Antilock Brake Service

Rear-wheel ABS is found on many light trucks and SUVs. Since 1990, RWAL has recorded soft codes but will only store one code at a time. If a code is not a hard fault, the controller shuts off the MIL and the codes remain in memory.

The following also apply to rear-wheel ABS:

- ABS is disabled in four-wheel drive.
- The red light will illuminate if there is no battery or ignition voltage to the ABS controller.
- Codes can be cleared by pulling a fuse with the key in the off position.

CAUTION With some rear-wheel systems, a permanent code can be set if the fuse is pulled to clear codes in key-on position. The controller must be replaced to regain ABS function.

ABS activation at low speeds can be caused when the sensor does not give enough of a signal for the processor. It can be caused by a defective sensor, metal shavings, or a sensor not fully seated after it was replaced.

One common problem in two-wheel ABS is that the dump valve can get some dirt in it. This results in a drop in pedal height but with a firm pedal. At first the pedal feels okay, but then with continued foot pressure it will drift lower toward the floor.

In a rear-wheel-only system, the brake pedal switch is a key input. The computer monitors vehicle speed all the time, but ABS only functions when the brakes are applied and voltage from the switch drops from system voltage to less than 1 volt. If the computer receives system voltage during a stop, it disables the ABS and sets a code.

A DTC can be activated by a two-footed driver. A code sets if the controller senses input from the brake light circuit while accelerating to about 40 mph. If the left foot rests lightly on the brake pedal while accelerating, the brake switch is activated. Although the brakes are not being used, the ABS computer does not know this and can set a code for a bad brake switch. Check the switch for correct adjustment.

Electromagnetic ABS Service

The Delco VI ABS sometimes suffers a failure that results in a low pedal. Follow the diagnostic chart in order to repair the problem. Part of the procedure calls for bleeding the modulator at various bleed screw locations.

One problem that sometimes occurs with this system is when a brake on a piston fails. When this happens, the motor pack must be replaced. If the brake

Figure 59.31 Valves in this ABS are driven by gears. *(Courtesy of Tim Gilles)*

does not hold the piston from turning, it can move down in its bore, resulting in a drop in pedal height. These antilock systems are driven by gears (**Figure 59.31**). Occasionally one of these gears falls off. The gears must all be turned counterclockwise as far as they can be turned. This is called "rehoming" the gears. They must remain in this position while installing the motor pack and cover.

The Delco VI ABS sometimes looks like an integral system because the master cylinder is mounted to the hydraulic modulator. They are connected by fluid transfer tubes and O-rings. The tubes are not reusable, and replacement master cylinders often are supplied with new tubes and O-rings. Be sure to bench bleed the master cylinder before installing it.

Common ABS Electrical Problems

Speed sensor buffers are known to fail quite often. A blown fuse can lead you to this. A buffer is a solid-state (electronic) device. Check inputs and outputs first before condemning the buffer. See if its input wire has power or is open.

Check the resistance of the VSS against specifications. A typical resistance would range from 900 to 2,000 ohms. Check for an AC output voltage from the VSS. The best way is to use an oscilloscope. When replacing a buffer, it will have to be calibrated for the tire size by an authorized parts supplier.

A failed ABS/TCS relay with bad electrical contacts is another common problem. This can be hard to pinpoint because it is often intermittent. This switch is inexpensive and is often replaced when no other test has pinpointed the problem.

On some cars, the use of an incorrect brake lamp or dash lamp can cause the ABS to shut off and the dash warning to illuminate. Some systems are so prone to voltage changes that a bad alternator can sometimes cause an ABS problem.

■■■ REVIEW QUESTIONS

1. At what percentage of slip does maximum traction occur?

2. How fast can a typical ABS pulsate the pressure to the brake system?

3. What is the name of the type of ABS with a conventional power brake and master cylinder that is also called remote or add-on ABS?

4. Which ABS can have up to 2,700 psi available at the wheels even when the engine is off?

5. Changes in the magnetic lines of flux around the wheel speed sensor cause a(n) _____ current voltage.

6. Why do manufacturers enclose wheel speed sensor wiring in braided metal tubing or wind it in twisted pairs?

7. What are two names for the wheel speed sensor's toothed ring that spins with the wheel?

8. What is the name of the type of high-pressure ABS that can lose its rear brakes and power assist during an ABS failure?

9. How many channels does an antilock brake system have when it has separate hydraulic controls for the front brakes but the rear brakes share a common sensor and solenoids?

10. What are the three stages of ABS solenoid operation in three- and four-channel systems?

11. When fluid pressure is released to a wheel, this is called pressure dump or _____.

12. In most ABS, _____ control the holding and releasing of hydraulic system pressure.

13. Which type of ABS uses a motor pack instead of solenoids?

14. In a traction control system, which can be implemented faster, engine controls or application of the brakes?

15. What is the name of the system that can control understeer and oversteer?

■■■ ASE-STYLE REVIEW QUESTIONS

1. Technician A says that when the ABS senses a failure, the brake system reverts to conventional braking. Technician B says that during normal operation, the ABS works as a conventional brake system. Who is right?
   - **a.** Technician A
   - **b.** Technician B
   - **c.** Both A and B
   - **d.** Neither A nor B

2. Technician A says that most wheel speed sensors are permanent magnet types. Technician B says that some ABS use a Hall-effect wheel speed sensor. Who is right?
   - **a.** Technician A
   - **b.** Technician B
   - **c.** Both A and B
   - **d.** Neither A nor B

3. Technician A says that a wheel speed sensor's voltage output does not change with the speed of wheel rotation. Technician B says that a wheel speed sensor's frequency changes with the speed of wheel rotation. Who is right?
   - **a.** Technician A
   - **b.** Technician B
   - **c.** Both A and B
   - **d.** Neither A nor B

4. Which of the following is/are locations for a wheel speed sensor tone ring?
   - **a.** Inside the brake drum
   - **b.** In the differential or on the transmission tailshaft
   - **c.** On a CV joint
   - **d.** All of the above

5. Technician A says that the purpose of ABS is to help avoid loss of steering control when a wheel loses traction. Technician B says that an ABS stop will result in a shorter stopping distance than a stop with normal braking. Who is right?
   - **a.** Technician A
   - **b.** Technician B
   - **c.** Both A and B
   - **d.** Neither A nor B

6. The amber (yellow) brake warning light is on. Technician A says that there is an ABS failure and the ABS is deactivated. Technician B says that this can be due to low fluid level in one-half of the master cylinder. Who is right?
   - **a.** Technician A
   - **b.** Technician B
   - **c.** Both A and B
   - **d.** Neither A nor B

7. A left front wheel locks up during an ABS stop. Technician A says that if the blocking solenoid for the left front brake is energized, the master cylinder is isolated from the left front wheel cylinder. Technician B says that if the dump solenoid opens, normal braking will be reestablished. Who is right?
   - **a.** Technician A
   - **b.** Technician B
   - **c.** Both A and B
   - **d.** Neither A nor B

8. Which of the following could cause a code and illumination of the ABS warning light?

   a. An ABS solenoid problem

   b. A faulty wheel speed sensor signal

   c. A faulty ABS control module

   d. All of the above

9. Technician A says that a scan tool is needed to bleed some types of hydraulic control units. Technician B says to use DOT 5 brake fluid with ABS. Who is right?

   a. Technician A   c. Both A and B

   b. Technician B   d. Neither A nor B

10. In a traction control system, the engine's power output can be controlled by:

   a. Disabling fuel injectors

   b. Retarding ignition timing

   c. Upshifting the transmission

   d. All of the above

Bearings, Seals, and Greases

■ OBJECTIVES

Upon completion of this chapter, you should be able to:

✔ Understand terms that relate to wheel bearings.

✔ Select the correct grease to use for a particular application.

✔ Describe the various wheel and axle bearing arrangements.

✔ Service wheel bearings on front and rear axles.

■ KEY TERMS

frictionless bearing	live axle	copolymer
needle bearing	axle bearing	spalling
race	semi-floating axle	brinelling
bearing cage	full-floating axle	pressed-fit
radial load	grease	push fit
thrust load	NLGI	
wheel bearing	dropping point	

■ INTRODUCTION

Bearings of different types are found throughout the automobile. This chapter first deals with fundamentals of bearings, seals, and lubricants. The last part of the chapter covers bearing service. Bearing diagnosis is covered in more detail in Chapter 76.

■ PLAIN BEARINGS

Plain bearings are the kind used as engine crankshaft bearings. They do not use rolling parts, and they provide sliding contact between two mating surfaces. Frictionless, or anti-friction, bearings provide a rolling contact and allow easy rotation when any part moves.

■ FRICTIONLESS BEARINGS

Frictionless bearings are *ball, roller,* or needle bearings (**Figure 60.1**). Bearings are made of hardened steel alloys. They are ground to a precise finish and size. Frictionless bearings must be lubricated. Some bearings are sealed, having a seal on either or both sides of the bearing. The function of the seal is to keep grease in the bearing and dirt or oil out.

In a ball or roller bearing, the balls or rollers ride between an inner **race** (*cone*), or raceway, and an outer race (*cup*) (**Figure 60.2**). The balls or rollers are usually held in position in a **bearing cage** or *separator*. The cage, usually made of stamped steel or plastic, keeps them from bunching up and rubbing against each other. It also serves to keep the bearings properly located so that the load on them is evenly distributed. When the bearing is the type that can be disassembled, the cage prevents the loss of the parts.

■ DIRECTION OF BEARING LOAD

Bearing manufacturers make bearings to handle different types of loads. A bearing load that is in an up-and-down direction is called a **radial load** (**Figure 60.3**). When the load is in a front-to-rear direction, this is called a **thrust load** or *axial thrust* (**Figure 60.4**).

■ BALL BEARINGS

Ball bearings ride in grooves that are ground into the surfaces of the inner and outer races (see Figure 60.2). A big disadvantage to a ball bearing is that most of the load of the vehicle is exerted on the bottom of the bearing, on a very small surface area of the ball and race.

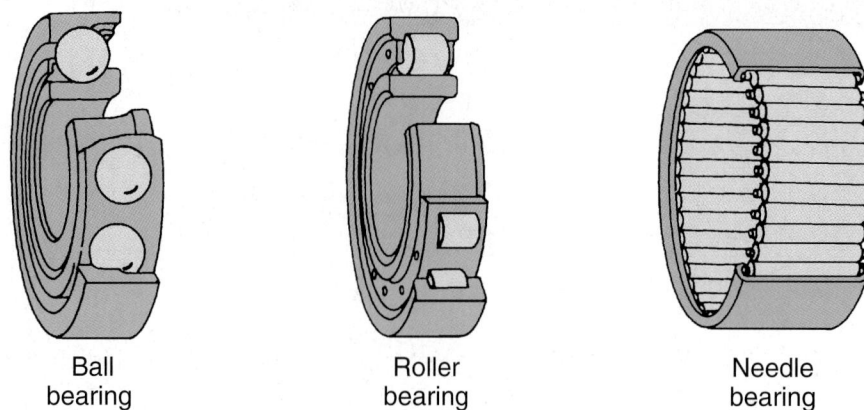

Ball
bearing

Roller
bearing

Needle
bearing

Figure 60.1 Types of frictionless bearings. *(Reproduced by permission of Deere & Company, John Deere Publishing, Moline, IL. All rights reserved.)*

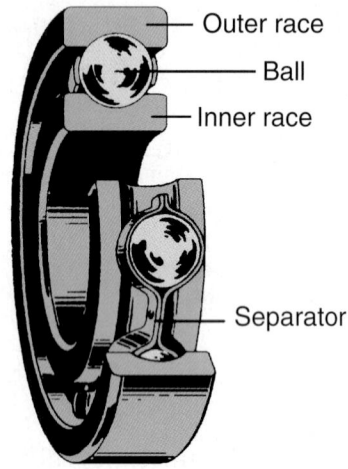

Outer race

Ball

Inner race

Separator

Figure 60.2 Parts of a ball bearing. *(Courtesy of Chicago Rawhide)*

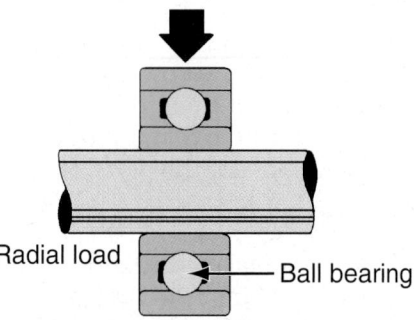

Radial load

Ball bearing

Figure 60.3 Direction of radial loads on a bearing. *(Courtesy of Chicago Rawhide)*

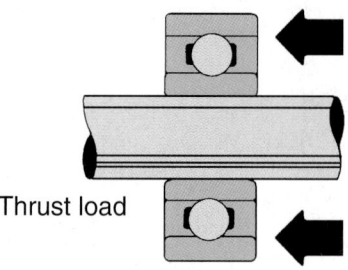

Thrust load

Figure 60.4 Direction of thrust loads on a bearing. *(Courtesy of Chicago Rawhide)*

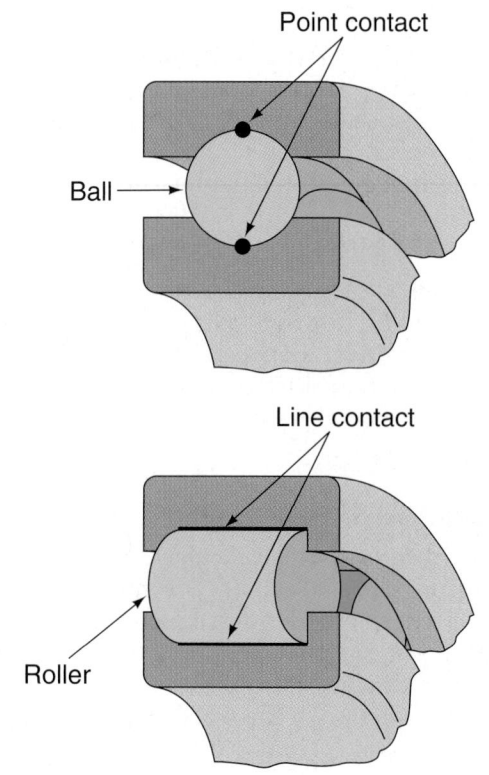

Point contact

Ball

Line contact

Roller

Figure 60.5 Ball and roller bearings have different contact areas. *(Courtesy of Ford Motor Company)*

The rest of the balls do not do much in supporting the load. This can be an advantage, though. Ball bearings have less friction than roller bearings and can operate at higher speeds. **Figure 60.5** shows the difference between ball and roller contact areas.

When a ball bearing is at rest, its load is distributed equally wherever the balls and races are in contact with each other. When there is motion and the ball begins to roll, material in the race actually bulges out

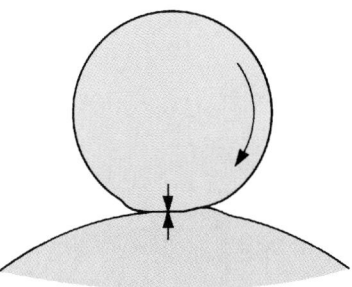

Figure 60.6 Metal bulges out in front of the ball when a load is applied to the bearing. *(Courtesy of Chicago Rawhide)*

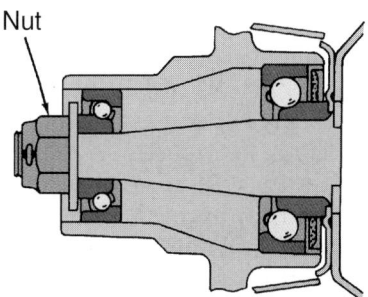

Figure 60.8 A radial thrust, or angular contact, ball wheel bearing set controls thrust in both directions. *(Reproduced by permission of Deere & Company, John Deere Publishing, Moline, IL. All rights reserved.)*

in front of the ball (**Figure 60.6**). Then, it flattens out behind the ball. Metal-to-metal contact cannot be allowed, so lubrication is critical.

Ball bearings control both end thrust (side movement) and radial movement (up and down) (**Figure 60.7**). Single-row ball bearings are popular in transmissions, alternators, steering gears, and differentials. They are designed primarily for radial loads but can withstand thrust loads also. When more load capacity is needed, extra balls are added to the bearing. This reduces the bearing's thrust capacity, however.

When a ball bearing must control thrust, the groove in the bearing race will be offset to one side. **Figure 60.8** shows a typical *radial thrust*, or *angular contact*, ball wheel bearing setup where two radial thrust control bearings face each other. This controls thrust in both directions.

Single-row bearings are susceptible to damage when a shaft is misaligned. When a ball bearing is required to have more thrust capacity or when misalignment is likely to be a problem, a *double-row bearing* is used.

This bearing is typically found in air-conditioning clutches and front-wheel-drive (FWD) front bearing hubs (**Figure 60.9**).

Ball bearings can also be designed primarily to control thrust. This is done by facing the races sideways, rather than up and down (**Figure 60.10**).

Figure 60.9 A non-serviceable sealed double row ball bearing shown with and without the seal. *(Courtesy of Tim Gilles)*

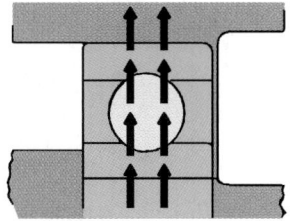

Radial load

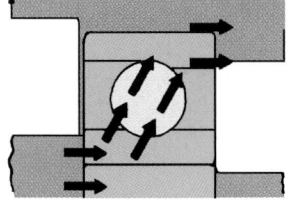

Thrust load

Figure 60.7 Ball bearings control both end thrust and radial movement. *(Courtesy of Chicago Rawhide)*

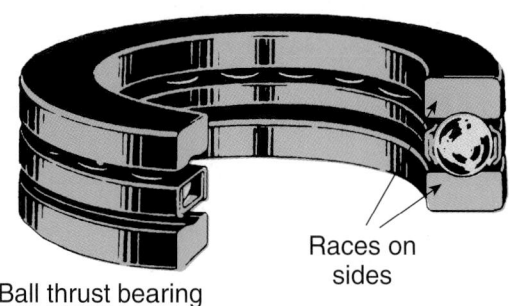

Ball thrust bearing

Races on sides

Figure 60.10 In a ball thrust bearing, the races are faced sideways, rather than up and down. *(Courtesy of Chicago Rawhide)*

■ ROLLER BEARINGS

When a greater load-carrying capacity is needed, roller bearings are used instead of ball bearings. They provide more surface area of contact with the race and can carry a greater load (see Figure 60.5). There are several types of roller bearings. Some roller bearings have only an outer or inner bearing race. The shaft or bore will be precision ground and hardened to serve as the other race. This is often the case with rear-wheel-drive (RWD) rear axles in light vehicles (**Figure 60.11**). RWD cars have straight roller bearings on the rear axle. These bearings support radial loads well but cannot control thrust. Axles must be retained or keyed so that end thrust cannot be applied to the bearing.

Roller bearings do not control end thrust. A thrust bearing (or a pair of tapered roller bearings) must be used when end thrust is to be controlled. A thrust bearing or bushing is one that is mounted at 90 degrees to the wheel bearing's load.

Tapered Roller Bearings

The most popular type of roller bearing for use in automobiles is the tapered roller bearing. Tapered roller bearings are used for front wheel bearings because they can control end thrust when two of them are installed with their tapers facing in opposite directions. The small diameters of the tapers face toward each other (**Figure 60.12**). In front wheel bearings, the larger of the two bearings is installed on the inside (where it supports the weight of the vehicle). The outer (smaller) bearing simply keeps the wheel aligned and provides end thrust control. Two outside bearing cups (races) are pressed into the *hub*.

Sometimes, two tapered roller bearings are used in a single bearing assembly. These are found where thrust loads are greater, such as on front wheel hubs on FWD cars (**Figure 60.13**).

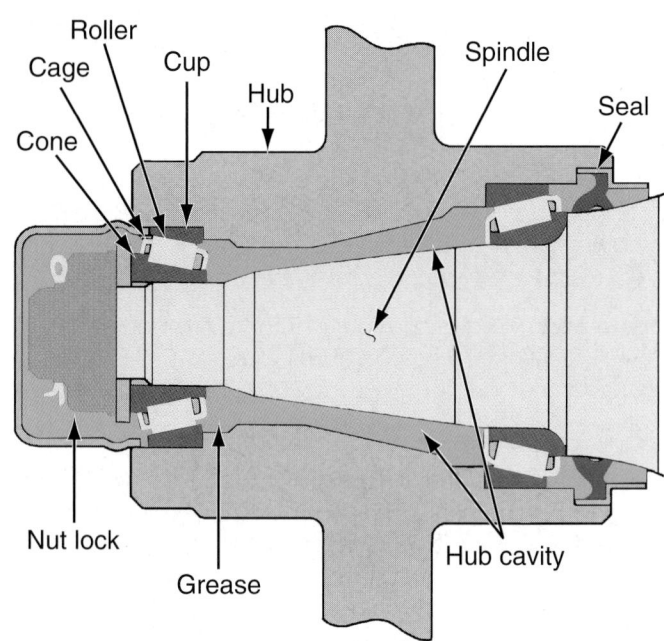

Figure 60.12 Tapered front wheel bearings are installed with their tapers facing in opposite directions. *(Courtesy of DaimlerChrysler Corporation)*

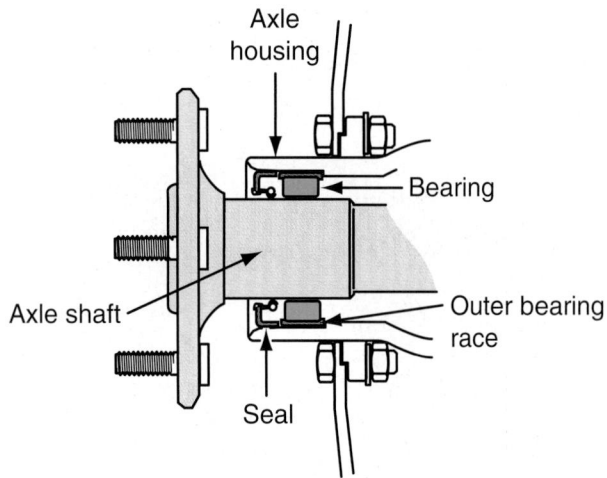

Figure 60.11 A straight roller rear axle bearing for a rear-wheel-drive car.

Figure 60.13 This FWD drive axle bearing has two tapered roller bearings. *(Courtesy of DaimlerChrysler Corporation)*

Tapered roller bearings tend to be self-aligning. The angles of the tapers of the bearing rollers are such that all of the angles meet at a common point (**Figure 60.14**). This makes each bearing roller align itself with the shaft.

Tapered roller bearings assembled with the inner race, bearings, and cage as a single unit (**Figure 60.15a**) have lips on the inside and outside edges of the inner bearing race (**Figure 60.15b**). This prevents the rollers from being able to slide out of the bearing.

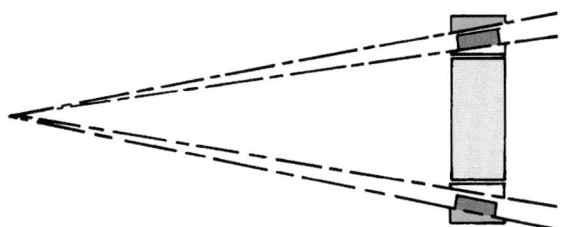

Figure 60.14 The angles of the tapers of the bearing rollers meet at a common point. *(Courtesy of Chicago Rawhide)*

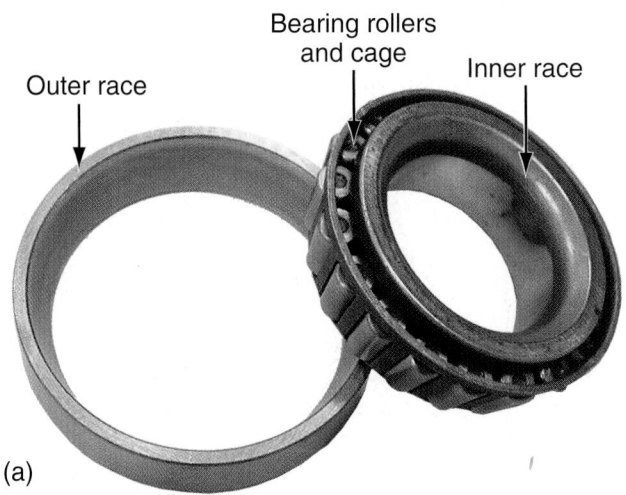

(a)

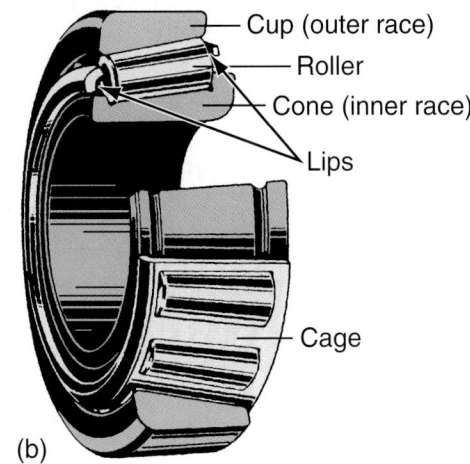

(b)

Figure 60.15 (a) Tapered roller bearing construction. (b) A tapered roller bearing has lips on the inside and outside edges of the inner bearing race. *(a, Courtesy of Tim Gilles; b, Courtesy of Chicago Rawhide)*

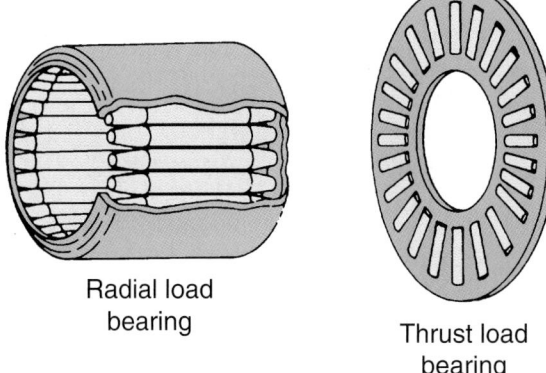

Radial load bearing

Thrust load bearing

Figure 60.16 Needle bearings can be used to control thrust or radial loads. *(Reproduced by permission of Deere & Company, John Deere Publishing, Moline, IL. All rights reserved.)*

Needle Bearings

When a roller bearing is very small, it is called a needle bearing. Needle bearings can be used to control thrust or radial loads (**Figure 60.16**). They are most often used where space is limited and are not as good for high speeds as roller or ball bearings. Needle bearings do not tolerate misalignment. One example of a common use of needle bearings is in universal joints (see Chapter 75).

■ WHEEL BEARINGS

Wheel bearings or axle bearings are found on all of the wheels of a vehicle. Some of them require service, and others are simply replaced when they fail. The term, **wheel bearing**, refers to non-drive front- and rear-wheel bearings. Bearings on **live axles** (those that drive wheels) are referred to as **axle bearings**.

Drive Axle Bearings

Drive axle bearings are located at the ends of the rear axle housing on a RWD car or on the hub on an FWD car. Passenger car rear axles use a **semi-floating axle** design (**Figure 60.17**). This design has a bearing that rides on the axle. This design is unsafe for heavy loads. If the axle breaks, the wheel can fall off. This axle design

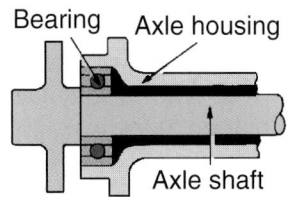

Bearing   Axle housing

Axle shaft

Semi-floating type

Figure 60.17 A semi-floating axle has the bearing mounted on the axle.

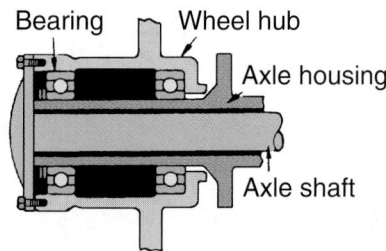

Full-floating type

Figure 60.18 A full-floating axle has the bearing mounted on the axle housing.

is found on vehicles up to ½ ton pickup trucks. That is why they are not often used with large campers.

Full-Floating Axles

Full-floating axles are found on some light-duty and larger trucks (usually ¾ ton and larger). In this axle design, the bearings do not touch the axle. They are located on the outside of the axle housing (**Figure 60.18**). If the axle breaks, the rear brake drum and wheel will still be supported. This axle design is for heavy loads.

Front-Wheel-Drive Bearings

FWD bearings are compact because they must fit into a tight space. There are either a pair of ball or tapered roller bearings housed in the steering knuckle (see Figure 60.13). The small stub axle that the bearing supports is usually the end of the CV joint. It has splines that attach it to the front hub. A nut holds the hub to the CV joint.

▰ GREASES

Lubricants have several purposes. They reduce friction and wear. They must also dissipate heat and protect metal surfaces from rusting. Seals are lubricated by lubricants too. This helps them keep the lubricant in and dirt out.

Greases are used in wheel bearings, chassis joints, universal joints, and gearboxes. Some components are equipped with grease fittings for replenishing grease. Others must be disassembled to replace the grease.

Grease is a combination of oil and a thickening agent. In some lubricating situations, grease has certain advantages over oil. Oil circulation systems are not needed, the tendency for leakage is reduced, and the lubricant remains in place after a long shutdown.

Grease does not leak or flow out of a bearing like an oil would. It acts like a liquid lubricant when a shearing force is applied to it, but the direction of fluid flow is only in the direction that the bearing turns. It does not flow out of the bearing, because it is not fluid at right angles to the direction that the bearing is turning.

Soap Type Comparison				
Soap Type	Calcium Sulfonate	Polyurea	Lithium	Lithium Complex
Texture	Smooth	Smooth	Smooth	Smooth
Dropping Point,°F	550	530	500+	340/375
Mechanical Stability	Excellent	Good	Excellent	Good
Shock Resistance	Excellent	Good	Good	Good
Water Resistance	Excellent	Good	Good	Good
Rust Protection	Excellent	Good	Good	Good
Adhesion	Good	Good	Good	Good
Temperature Resistance	Excellent	Excellent	Excellent	Good

Figure 60.19 Characteristics of different types of greases. *(Courtesy of Kendall Motor Oil)*

How Grease Is Made

The properties of a grease are limited by the quality of oil that it is made of. Greases are usually 70%–90% liquid petroleum oils made semi-fluid or solid by adding a thickening agent. The thickening agent, at a concentration of from 5%–25%, is usually a soap such as *aluminum, lithium, sodium,* or *calcium.* Calcium soap is often used for pressure gun lubricant. Sodium is found in wheel bearing grease, and lithium is used for multipurpose grease. Additives make up 0%–10% of a grease.

Soaps used in greases are similar to soaps made for household uses except they are soluble in oil and not soluble in water. Soaps have more ability to stick to metals than oils do. The lubricating oil in grease is held tightly against a metal surface where it can do a better job of lubricating and of protecting against rust than oil does. Lubricating greases are made to suit a particular application. **Figure 60.19** shows some of the different characteristics of greases.

> ### ▰ SCIENCE NOTE ▰
>
> *Aluminum, lithium, sodium, and calcium are not actually soaps, but soaps can be made from them by incorporating these atoms in place of the acidic hydrogen in an organic acid. Soaps used in greases are similar to soaps made for household uses. Household soaps, which are metallic salts of long-chain organic acids, appear to dissolve in water. They do not actually dissolve in water but form micelles in which the long-chain organic parts of the soap congregate together, leaving the metal salt on the outside. It is the metal salt that is dissolved in water.*

Greases are usually long-chain organic compounds. "Like dissolves like" so they dissolve and thus are held together by the long-chain portion of the soap.

Consistency of Grease

Greases are fibrous. Different sizes of fibers are available. Some are so large that they are visible to the naked eye, while others appear to be smooth.

The National Lubricating Grease Institute (NLGI) defines grease consistency in *NLGI grade numbers*. Numbers begin with grade #000 and increase in whole numbers through grade #6. The grade is determined during a penetration test in which a standard cone is penetrated into the grease under controlled circumstances of weight, time, and temperature. The depth in millimeters that the cone penetrates the grease is what determines the NLGI number of the grease. This standard denotes the consistency of a grease only and does *not* indicate quality or characteristics. A lower NLGI number relates to a softer grease (**Figure 60.20**).

When a grease is softer (lower NLGI number), it has a greater tendency to leak out. A grease that is too thick can cause friction losses, become channelled during use, and not perform satisfactorily. An *all-season lubricant* is used in most climates, but thicker and thinner greases are available for severe situations.

Temperature Characteristics

A grease does not have a sharp (definite) melting point. As it is heated, it will become softer until it becomes fluid. A general idea of the safe temperature range of a grease can be gotten by referring to its **dropping**

Thickener	Structure	Dropping Point (°F)	Max Service Temperature (°F)	Other Properties
Sodium Soap	Fibrous	350	200-275	Natural rust resistance, high water resistance, poor low-temperature properties
Calcium Soap Simple	Smooth	270-290	250	Excellent water resistance
Complex	Smooth, buttery	>450	300	Inherent extreme pressure load-carrying properties, good water resistance
Lithium Soap Simple	Smooth	390	325	Good water resistance, good mechanical stability
Complex	Smooth, slightly stringy	>450	350	Same as above
Aluminum Complex	Smooth gel	>450	300	Excellent water resistance, shear stability and pumpability
Clay	Smooth	>500	350	Nonmelting, very water resistant, marginal shear stability
Polyurea	Opaque, Slightly mealy	>450	350	Good oxidation resistance and water resistance

Figure 60.21 Characteristics of various greases. *(Courtesy of The Lubrizol Corporation)*

point (melting temperature). This is the temperature at which the grease turns into a liquid.

To test for its dropping point, a sample of grease is heated in a special test cup that has a small orifice (hole) of a specific size. The dropping point is the temperature when the first drop of material falls from the orifice. This is *not* the temperature that it would be safe to use the grease. According to Quaker State, greases should generally be used at temperatures no higher than 100°F below their dropping points. **Figure 60.21** shows dropping points of various thickeners.

Viscosity

The viscosity of the oil used in making a grease is important to the grease's *apparent viscosity*. A lubricant must remain fluid in cold weather while still having enough body to protect parts. The ability of a lubricant to maintain its viscosity across a wide variation in temperatures is called its viscosity index.

Lubricant manufacturers make up their greases to fit many applications. **Figure 60.22** is a photo of two samples of the same base oil taken at a very low temperature. The oil on the left is frozen on a popsicle stick. At the right is the same oil with a thickening agent added to make a grease. The grease is still able to flow at that low temperature, while the oil is solidified.

NLGI Consistency Grades	
NLGI Grade	**Worked Penetration at 25°C (77°F) mm/10**
000	445 to 475
00	400 to 430
0	355 to 385
1	310 to 340
2	265 to 295
3	220 to 250
4	175 to 205
5	130 to 160
6	85 to 115

Figure 60.20 A lower NLGI number is a softer grease. *(Courtesy of Shell Lubricants)*

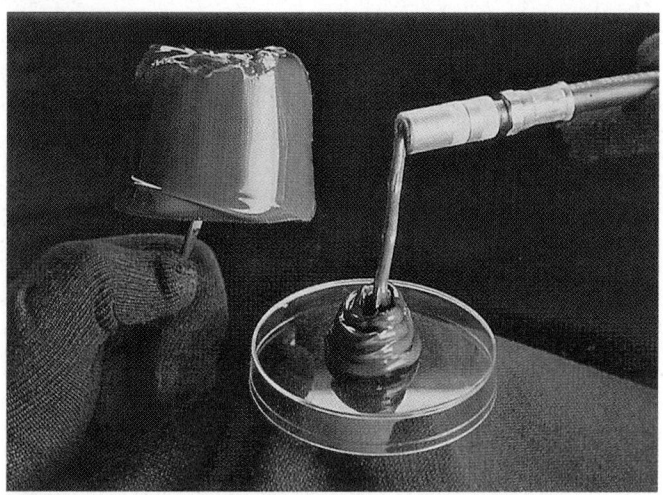

Figure 60.22 The frozen sample of oil on the left is the same viscosity of oil as the grease on the right. The grease has had a thickening agent added and is still able to flow. *(Courtesy of Imperial Oil Limited)*

Because the rate of oil oxidation doubles with every 20°F increase in temperature, when a bearing runs hot its grease will require more frequent changing. A grease with a life expectancy of 1,000 hours at 100°F would have a 500-hour life expectancy if used at 120°F. Grease made from petroleum should not be used at temperatures in excess of 325°F unless it is being changed every hour or two.

Grease Performance Classification

Automotive grease is classified by NLGI for performance. This standard is based on an ASTM standard grease specification approved in May 1990. Automotive grease is classified in two general groups. If the standard has an "L" prefix, the grease is for the lubrication of suspension components such as ball joints and steering pivots. If the prefix is "G," the grease is intended for wheel bearings. **Figure 60.23** shows a further breakdown of the classifications.

Some greases might overlap and be suitable for both chassis and wheel bearings. The NLGI has a logo, which can be displayed on the product (**Figure 60.24**). Only the highest of each category of a combination of the two can be included in the logo.

Extreme Pressure Lubricants

Extreme pressure (EP) lubricants, which are added to some greases, are the same as those found in gear lubricants. Metallic soap thickening agents add a small amount of EP effect. Graphite or molybdenum do not add EP properties, but they increase anti-wear and frictional modification characteristics.

Chassis Lubricants

A chassis lubricant is a grease of a consistency that allows it to be applied through a zerk fitting with a grease gun. It must adhere to the bearing surface and seal out dirt and water. Chassis lubricant is highly resistant to being washed away with water.

Chassis parts move up and down repeatedly, which can break down the structure of the grease. Grease that does not have adequate *shear resistance* will break down, become like oil, and flow away from the surfaces where it is needed.

LA — Service typical of chassis components and universal joints in passenger cars, trucks, and other vehicles under mild duty only. Mild duty will be encountered in vehicles operated with frequent relubrication in noncritical applications.

LB — Service typical of chassis components and universal joints in passenger cars, trucks, and other vehicles under mild to severe duty. Severe duty will be encountered in vehicles operated under conditions which may include prolonged relubrication intervals, or high loads, severe vibration, exposure to water or other contaminants, etc.

GA — Service typical of wheel bearings operating in passenger cars, trucks, and other vehicles under mild duty. Mild duty will be encountered in vehicles operated with frequent relubrication in noncritical applications.

GB — Service typical of wheel bearings operating in passenger cars, trucks, and other vehicles under mild to moderate duty. Moderate duty will be encountered in most vehicles operated under normal urban, highway, and off-highway service.

GC — Service typical of wheel bearings operating in passenger cars, trucks, and other vehicles under mild to severe duty. Severe duty will be encountered in certain vehicles operated under conditions resulting in high bearing temperatures (disc brakes). This includes vehicles operated under frequent stop-and-go service (buses, taxis, urban police cars, etc.), or under severe braking service (trailer towing, heavy loading, mountain driving, etc.).

Reference: ASTM Designations D 4950

Figure 60.23 Automotive grease classifications. *(Reprinted with permission from the National Lubricating Grease Institute)*

Figure 60.24 NLGI logos. *(Reprinted with permission from the National Lubricating Grease Institute)*

Wheel Bearing Grease

Wheel bearing lubricant is grease with a high resistance to heat. It is usually *thicker* and specially formulated for use with various types of bearings. Because the grease is on a spinning part, it must resist the tendency to be thrown off. A poor-quality bearing grease can cause a safety hazard by leaking onto the brake linings. It can also cause bearing failure due to lack of lubrication.

Universal Joint Grease

Universal joint grease is made specifically for universal joints. Some universal joint designs require special lubricants. Be sure to follow manufacturer's recommendations.

Multipurpose Grease

A multipurpose grease satisfies the requirements of chassis, wheel bearing, and universal joint lubricants. It is the most common type of grease used in service shops. But multipurpose does *not* mean all-purpose. It meets only certain requirements. The lubricant most often used in chassis lubrication is a multipurpose *lithium-based grease.*

Solid Lubricant Greases

Some greases contain solid lubricant materials such as *molybdenum* (*moly*) or *graphite*. These are often used to lubricate speedometer cables, emergency brake cables, splines, and leaf springs.

■ WHEEL BEARING SEALS

Automobiles and equipment can use seals for the following reasons:
■ To seal in lubricants
■ To keep different lubricants separated
■ To keep out dirt
■ To maintain vacuum or pressure

Seals are either *dynamic* (for sealing moving parts) or *static* (for sealing fixed parts) (**Figure 60.25**).

Wheel seals are usually lip seals (**Figure 60.26**). There is usually a garter spring behind the seal lip to further increase the pressure of the seal on the shaft. The spring can be separate or cast into the seal itself. The open side of the sealing lip always faces the lubricant (**Figure 60.27**). This is so that any pressure from the lubricant will increase the wiping pressure of the seal on its shaft.

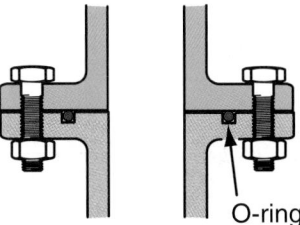

Static seal

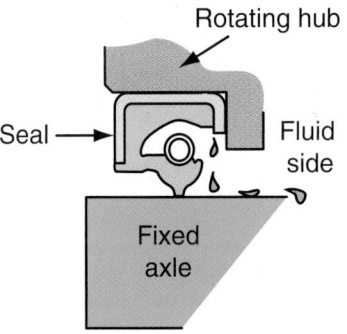

Dynamic seal

Figure 60.25 Static seals do not move. Dynamic seals have parts moving against them.

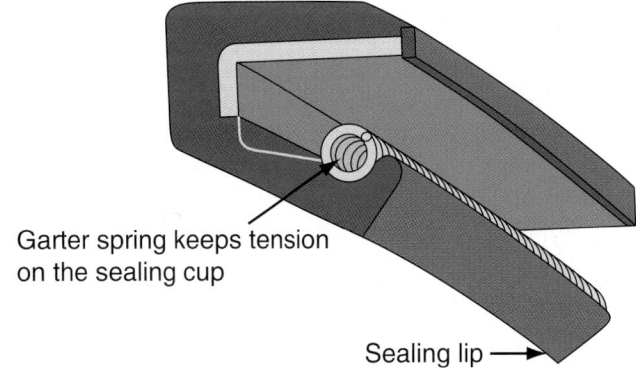

Figure 60.26 A lip seal.

Some seals have more than one lip; one is the *sealing lip,* and the other is the *auxiliary lip* for sealing dust (**Figure 60.28**). Auxiliary seals are only used when there is a large amount of dirt because the auxiliary lip causes increased frictional heat.

Some bearings are sealed on one or both sides. When there is a seal on both sides, the bearing is not serviceable, and must be replaced if there is a problem.

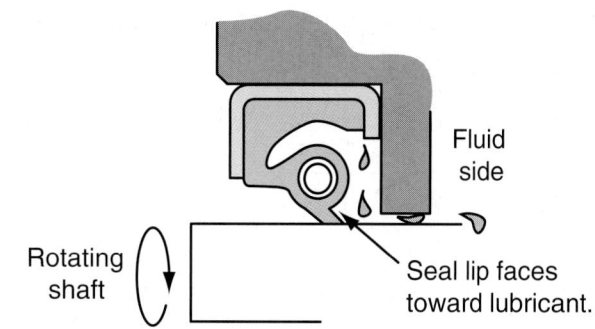

Figure 60.27 The open side of the sealing lip faces the lubricant.

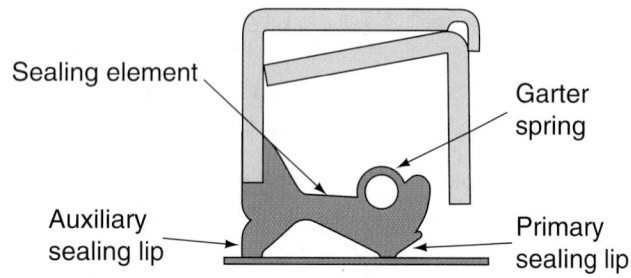

Figure 60.28 This seal has an auxiliary dust sealing lip. (*Courtesy of Chicago Rawhide*)

If lubricant leaks from one of the seals, replace the bearing. Be sure not to immerse the bearing in solvent. If solvent gets into the bearing, it will dilute the lubricant. This can lead to bearing failure.

Often, there is a seal on only one side of a bearing. This type of bearing is commonly found on the rear axle of RWD vehicles. These bearings are lubricated by oil from the differential.

■ SEAL MATERIALS AND DESIGN

Seals are made of different types of materials depending on their intended use. The most popular seals are made of synthetic materials. These materials include nitriles, polyacrylates, silicones, and fluoroelastomers.

Most lip seals are made of *nitrile*. Nitrile is a grey-black or shiny black mixture of two synthetic rubbers: the polymers Buna and acrylonitrile. The combination of the two polymers is called a **copolymer**.

■ SCIENCE NOTE ■

A polymer is defined as a substance consisting of giant molecules formed from smaller molecules of the same kind. A continuous chain results from the repeated addition of small molecules. This chain is the polymer. Polymers are generally known as plastics.

The disadvantage to nitrile seals is that they are not resistant to some synthetic oils and greases. They are good with most mineral oils and greases. Their operating range is between –65°F to 225°F.

Polyacrylate seals are *elastomers* designed for higher temperatures (to 300°F–350°F depending on the lubricant). They are also resistant to extreme pressure lubricants. Polyacrylate is often used for differential pinion bearing seals. They look the same as nitrile seals.

Silicone seals are often used in engines and transmissions. They are usually colored gray, red, orange, or blue. Silicone is a polymer that is very flexible and absorbs lubricant. Silicone seals are resistant to friction and wear. They can operate at temperatures of from 325°F to 350°F depending on the lubricant used.

Fluoroelastomers are used with special lubricants and chemicals, which the other seal materials cannot handle. Viton® is one popular fluoroelastomer. An example of its use is for carburetor inlet needle valves and other fuel system applications because of its resistance to methanol fuels. The color can be brown, black, blue, or green.

Non-Synthetic Seals

Leather used to be the main seal material used. It is expensive, though, and has been mostly replaced by synthetics. Some truck seals are still made of leather, however. Leather does not do a good job around moisture or higher temperatures (above 200°F). To tell if a used seal is synthetic rubber or leather, scratch its surface. It will turn to a rougher, lighter color like brown or beige if it is leather. Rubber will appear smooth and black.

Felt is a material that is sometimes used to keep out dirt. It is made of wool, which absorbs oil well and keeps out dirt. It does not do a good job of keeping oil confined and can also absorb water, contributing to rusting.

Other Seals. When a lubricant is to be directed back to its source, the lip of the seal is sometimes fluted (**Figure 60.29**). A newer seal design is the *wave seal*.

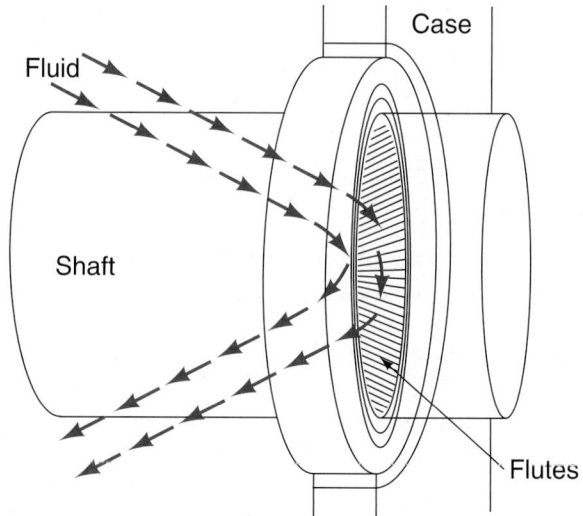

Figure 60.29 This fluted lip seal directs oil back to its source. (*Courtesy of DaimlerChrysler Corporation*)

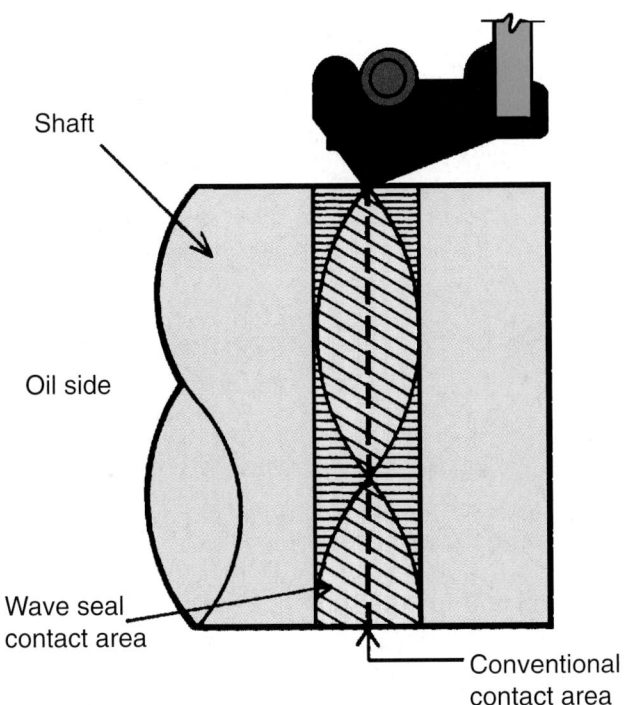

Shaft

Oil side

Wave seal contact area

Conventional contact area

Figure 60.30 A wave seal oscillates, pumping oil back toward its source. *(Courtesy of Chicago Rawhide)*

This seal does not ride on the shaft in a straight line like a conventional seal. It oscillates, pumping oil back toward its source (**Figure 60.30**). It has less friction, lasts longer, and works equally well no matter which direction the shaft turns.

SEAL TOLERANCE

Seals can usually accommodate a shaft that is undersized up to $1/64''$ (0.016") but only if all parts are in perfect alignment. The Rubber Manufacturers Association (RMA) recommends that runout tolerance be held to + or –0.003" for shafts up to 4 inches in diameter. The surface finish should be smooth (10–20 microinches).

WHEEL BEARING DIAGNOSIS AND SERVICE

Seal and Bearing Failure

If the lubricant can leak out, moisture can leak in. Also, once grease has leaked out, the bearing will soon suffer shock loads due to the lack of lubrication. Pieces will come off the bearing and races. The pieces will circulate around the bearing, resulting in heat buildup and failure of the bearing.

Periodic maintenance to bearings on non-drive axles consists of cleaning the bearings, repacking them with grease, and adjusting the clearance or preload after reinstallation. Parts of the wheel bearing assembly are shown in **Figure 60.31**. On drive axles, the only service done to a bearing is to press it off and replace it if it is bad. That service is covered in Chapter 60.

Boat Trailer Bearing Failures. Boat trailer bearings are well sealed and often have abundant high-quality grease. Boat trailers for smaller boats use tires that are of a smaller diameter than normal too. This causes the bearings to rotate at a higher rpm.

One common problem happens when the boat trailer has just been towed for a long distance. Bearings are at operating temperature when the boat trailer is backed into the water. As the warm air inside of the bearing shrinks, cold lake or ocean water is drawn into the bearing. If the trailer is parked for a long period of time before it is used again, the balls or rollers in the bearings can rust.

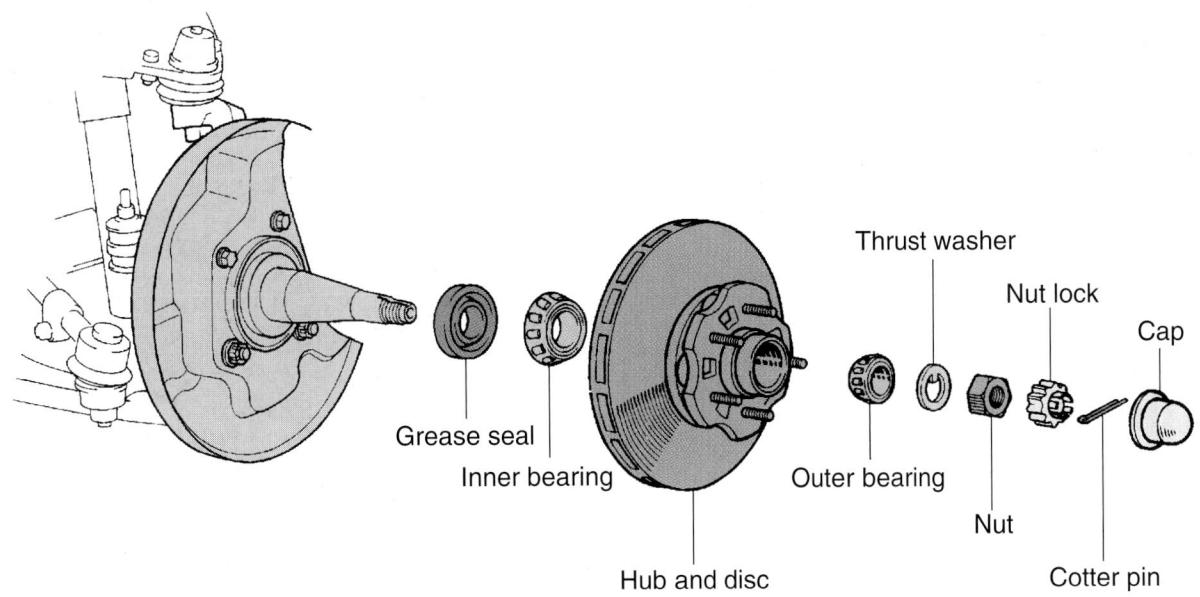

Thrust washer

Nut lock

Cap

Grease seal

Inner bearing

Outer bearing

Nut

Cotter pin

Hub and disc

Figure 60.31 Parts of the wheel bearing assembly.

■ WHEEL BEARING ADJUSTMENT

Wheel bearings must be properly adjusted. If they are too loose, the wheel will be able to move. This can mimic problems that result from worn parts. Some of these problems are shake, noise, wander, steering wheel play, cupped tire wear, and an intermittent low brake pedal on disc brake cars.

The bearing is designed to operate with very little clearance. As a bearing rotates under a load, some heat is developed and the bearing expands. Manufacturers list methods of adjusting their bearings (**Figure 60.32**). A common adjustment will result in the bearing having from about 0.001"–0.005" of running clearance. This is somewhere around the thickness of a hair. A bearing adjusted too tightly will develop more friction, use more power, and may ultimately fail.

An easy method of adjusting a loose bearing can be done with the tire raised off the ground. There is a *dust cap* that is pressed into the hub at the end of the spindle. The dust cap is removed with a large slip-joint pliers or a special dust cover tool (**Figure 60.33**).

Some spindle nuts have lock tabs or locknuts, but most are kept in place with a cotter pin. Remove the cotter pin so the spindle nut can be tightened. Using both hands, grasp the top and bottom of the tire and try to rock it back and forth. As the spindle nut is tightened further, less and less movement will be felt.

The washer under the spindle nut has a tab that fits into a groove on the spindle (see Figure 60.31). Its function is to keep the bearing from trying to turn the

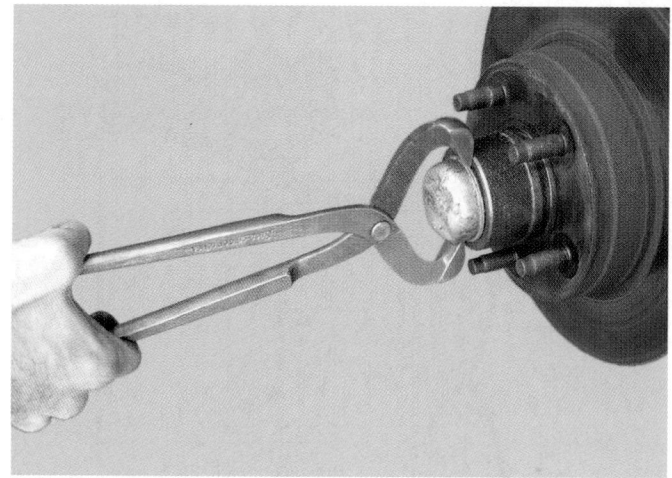

Figure 60.33 A special tool used for removing a wheel bearing dust cap. *(Courtesy of Tim Gilles)*

spindle nut. It fits the spindle loosely, so it should be easy to move.

> **SHOP TIP**
> As a test to see that the bearing is not too tight, insert a screwdriver under the tabbed washer that is under the spindle nut and pry the washer up and down. There is enough slack between the washer and the spindle that it should move freely. If the washer is hard to move, the adjustment is too tight.

NOTE: *The above shop tip works on most cars; however, be certain to always check the service information for the correct procedure when you are working on a make of automobile for the first time.*

Wheel bearing nuts are typically hexagonal (six sided). A popular feature used with wheel bearing adjusting nuts is a bearing retainer, a sheet metal cup that has 12 notches (**Figure 60.34**). This is used for fine-tuning the adjustment so that the cotter pin that locates the nut can be installed closer to its intended position.

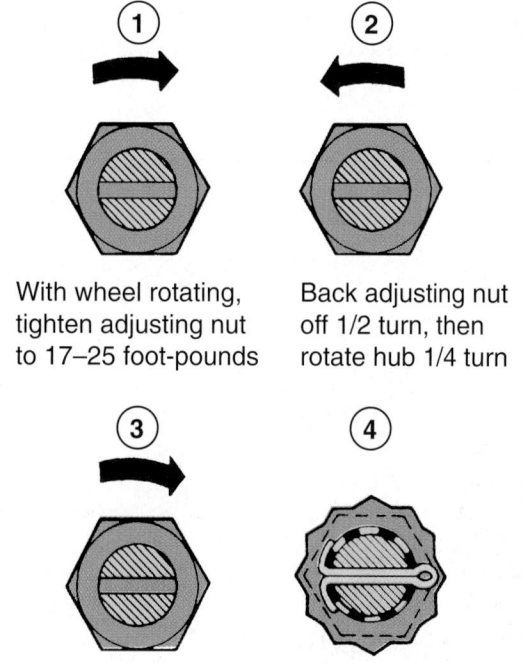

① With wheel rotating, tighten adjusting nut to 17–25 foot-pounds

② Back adjusting nut off 1/2 turn, then rotate hub 1/4 turn

③ Tighten adjusting nut to 24–28 inch-pounds

④ Install the retainer and a new cotter pin

Figure 60.32 Typical wheel bearing adjustment directions. *(Courtesy of Ford Motor Company)*

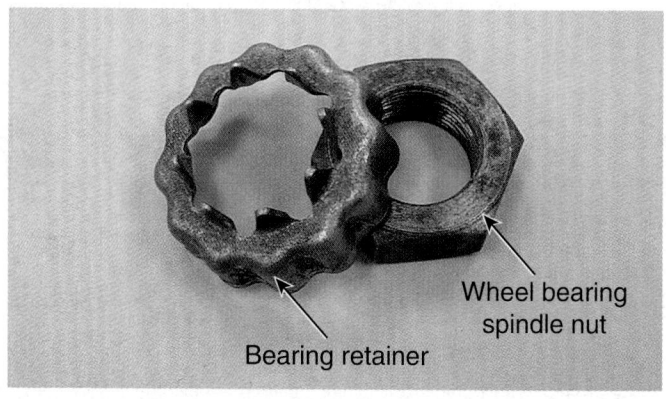

Wheel bearing spindle nut

Bearing retainer

Figure 60.34 A bearing retainer used to "fine-tune" the spindle nut adjustment. *(Courtesy of Tim Gilles)*

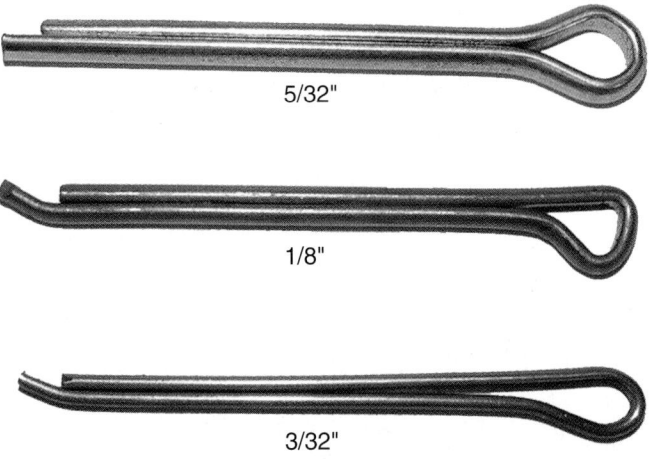

Figure 60.35 Select the largest diameter cotter pin that fits into the hole. *(Courtesy of Tim Gilles)*

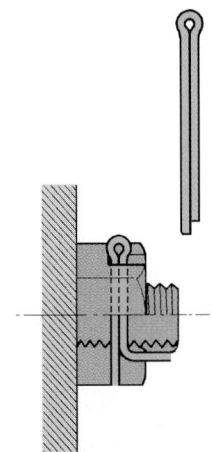

Figure 60.36 Cotter pin installation.

Right

Wrong

Figure 60.37 The right (a) and wrong (b) ways to install a cotter pin. *(Courtesy of Tim Gilles)*

■ SELECTING AND INSTALLING A COTTER PIN

When selecting a new cotter pin, use the largest diameter one that will fit into the hole (**Figure 60.35**). To keep less inventory in stock, repair shops commonly purchase cotter pins of the same length (2" for instance) and cut them with diagonal cutters during installation.

One of the ends of the cotter pin will be longer than the other. Pull on this end with the diagonal cutters to seat the cotter pin fully in its hole. Then, pull the long end out and over the end of the spindle. Cut it off. Then, cut off the remaining end of the cotter pin flush with the spindle (**Figure 60.36**). This provides a securely fastened cotter pin that will be easy to remove during the next service. **Figure 60.37** shows the right and wrong ways to install a cotter pin.

■ REPACKING WHEEL BEARINGS

Tapered wheel bearings are lubricated with grease, which also protects the metal from corrosion and helps carry away heat. Front wheel bearings are sealed from the elements, but over a period of time, moisture, brake,

and road dust can accumulate. It is customary to clean and repack the wheel bearings with grease whenever the front brakes are relined or at 30,000-mile intervals.

On FWD cars, the procedure for repacking the rear axle bearings (if they are not the sealed type) is similar to the procedure for repacking the front axle bearings on RWD cars. Do one side at a time so that parts are not accidentally interchanged from side to side.

After removing the dust cap and cotter pin, remove the spindle nut and the tabbed washer that is under it. To easily remove the outer bearing, rock the tire back and forth at the top. The bearing will usually pop out on the spindle (**Figure 60.38**).

Remove the Seal

Pull the wheel and hub from the spindle. When working with the wheel bearings, be sure to keep grease and solvent off braking surfaces.

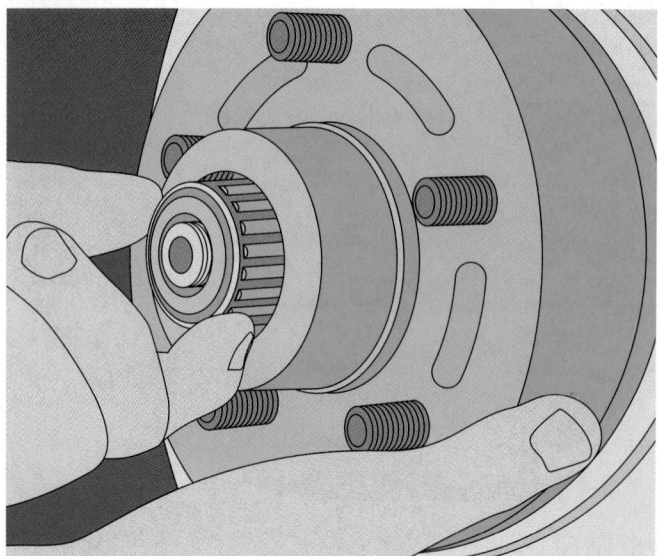

Figure 60.38 Remove the outer bearing.

Figure 60.39 Use a long dowel or drift punch to pound on the bearing and seal from the inside.

Seals are usually replaced during a bearing repack. Using the old seal over is taking a chance on grease leaking out and contaminants leaking into the bearing. To remove the seal, use a long dowel or drift to pound on the bearing from the inside (**Figure 60.39**). The seal can also be removed with a screwdriver or a special seal removal tool (**Figure 60.40**).

A favorite trick used by many undercar technicians is to *carefully* remove the bearing seal using the spindle nut. First remove the outer bearing. Then reinstall the spindle nut (**Figure 60.41**) and pull the rotor gently and firmly against the back of the inner wheel bearing (**Figure 60.42**) to remove the seal (**Figure 60.43**). Be careful! Rough handling can damage the bearing cage. Also, some seals are too tight in the hub to be removed by this method.

Clean out the old bearing; do not just add grease. Wipe all of the old bearing grease from the spindle, the bearing, the bearing cups, and the hub. Check the grease on the shop towel for metal flakes, which would indicate bearing failure.

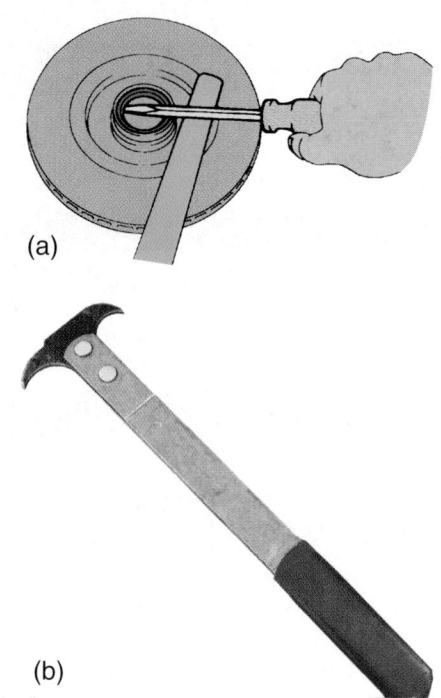

(a)

(b)

Figure 60.40 (a) Using a screwdriver to remove a bearing. (b) A seal removal tool. *(Courtesy of Tim Gilles)*

Figure 60.41 After removing the outer bearing, re-install the spindle nut. *(Courtesy of Tim Gilles)*

CAUTION ■ Do not spin the bearing with air. Spinning a dry bearing can damage it and can also be dangerous if bearings come loose from the bearing cage.
■ When blowing off parts, blow into the solvent tank to avoid making a mess.

NOTE: *It is important to remove all grease from the bearings and hub because the ability of the new grease to lubricate can be compromised if it is mixed with a grease that includes an incompatible soap.*

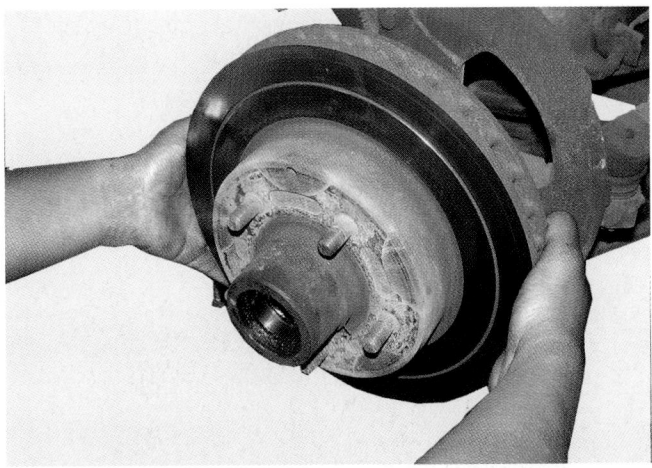

Figure 60.42 Use the rotor to *gently* pull against the inner bearing and seal. *(Courtesy of Tim Gilles)*

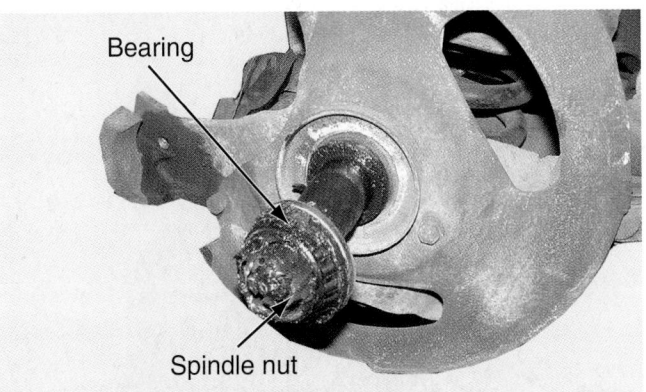

Figure 60.43 The nut, bearing, and seal after seal removal. *(Courtesy of Tim Gilles)*

Figure 60.44 Clean the wheel bearing. *(Courtesy of Tim Gilles)*

Clean the bearing (**Figure 60.44**) and let it air dry on a paper towel or blow it dry from the ends, parallel to the rollers, using compressed air. Be careful not to blow the old, dirty grease into the bearing. Rinse the bearing off with alcohol or brake cleaner afterwards.

NOTE: *Keep the bearing parts together so they can be reassembled in their original bearing races. Bearing parts become wear-mated to each other.*

■ BEARING INSPECTION AND DIAGNOSIS

After the bearings are cleaned, inspect them for damage. If a bearing is damaged, the cause should be determined so the problem does not happen again. Check the bearing and bearing race for pitting and other signs of damage. If any damage is apparent, replace the bearing and its race. Save the old bearing to compare it with the new one. Bearing replacement is covered later in this chapter.

Bearings usually fail slowly. Noise is the clue that something is wrong. As metal is deformed or comes off the bearing races, the noise will become more pronounced. There are two main kinds of damage to a bearing, spalling and brinelling.

Spalling is when pieces break off the bearing metal (**Figure 60.45**). The noise from spalling is a random, high-pitched sound. Brinelling is when the bearing or race has indentations from shock loads (**Figure 60.46**). The noise that results from brinelling damage is a regular, low-pitched sound.

The biggest shock load on a bearing is often during installation. Press directly on the pressed-fit bearing race only. Another cause of bearing damage can be due to an improperly grounded arc welder.

Figure 60.45 A spalled bearing. *(Courtesy of Tim Gilles)*

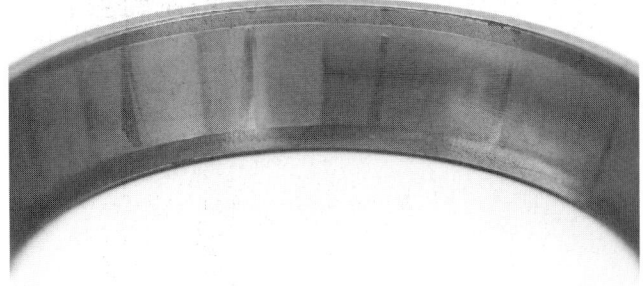

Figure 60.46 Brinelling is the name for dents in the bearing race that result from the rollers hammering against the race.

*A vehicle had a trailer hitch welded to its frame using an electric arc welder. The owner went on a vacation pulling his new trailer. Shortly into the trip, a front wheel bearing failed. This happened because the negative cable to the welder was not clamped directly to the frame but to the energy absorbing bumper mount. The vehicle had a broken ground strap between the engine and chassis. Electricity always takes the shortest path to ground. This time it went through the wheel bearing (**Figure 60.47**).*

Adding Grease to Bearings

To pack bearings by hand, put a small amount of grease in the palm of your hand. Stroke the large open end of the bearing cage against the grease until ribbons of grease start to appear at the opposite ends of the bearing (**Figure 60.48**). Smear a fresh layer of grease all around the bearing and on the clean bearing races.

A *pressure bearing packer* is owned by most shops. A grease gun is applied to a grease fitting to pack the bearing (**Figure 60.49**).

After the bearings are packed with fresh grease, place them on a paper towel. Remember to reinstall them into the same bearing races. Leave a ring of

Figure 60.47 This bearing and race were damaged by electrical current. *(Courtesy of Chicago Rawhide)*

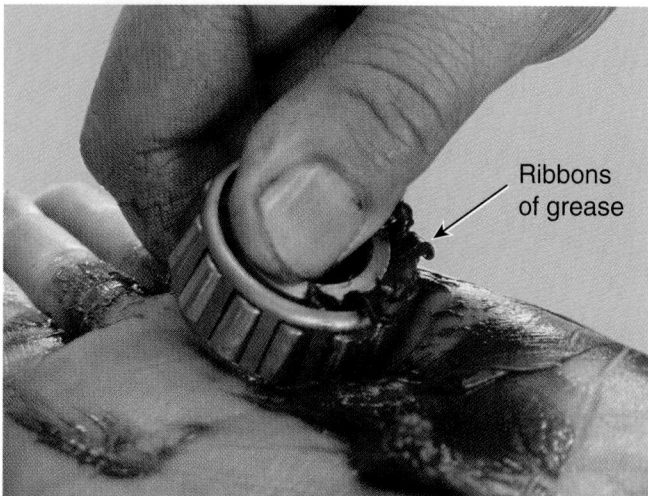

Figure 60.48 Apply grease until ribbons of grease start to appear at the opposite ends of the bearing. *(Courtesy of Tim Gilles)*

Figure 60.49 A pressure bearing packer. *(Courtesy of Tim Gilles)*

grease below the bearing race to help keep the fresh grease inside the bearing area after it heats up and begins to flow.

Grease the Inside of the Hub

Put a small amount of grease in the cavity of the hub. Do not fill it up. According to bearing engineers, the amount of grease that is packed into a bearing determines how well it will work. If the cavity is full, there is no place for the excess grease or pressurized air to go when the bearing heats up. If the hub is full, there is no place for excess grease in the bearing to go. This results in excess heat and fluid friction, which can cause the bearing or seal to fail.

High-speed bearings, like those on race cars, are supposed to be filled to only 25% of their free space. When a bearing is totally full, it should be used at low

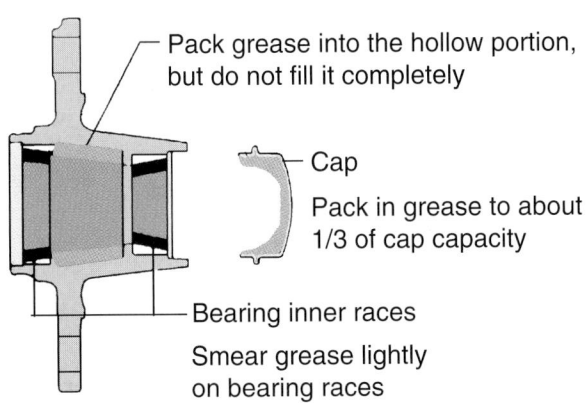

Figure **60.50** Partly fill the hub with grease.

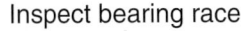

Figure **60.52** Inspect the bearing races in the hub. *(Courtesy of Tim Gilles)*

speeds only. **Figure 60.50** shows the recommended fill for passenger cars.

Inspect the Spindle

Just behind the area where the bearing rides on the spindle is a raised area that the seal rides on (**Figure 60.51**). Clean this area so that the new seal is not accidentally ruined. Inspect the condition of the spindle and check the fit of the large bearing on it.

NOTE: *A worn spindle is usually caused by a bearing seizure that forces the inner race of the bearing to spin on the spindle. The resulting heat softens the hardened spindle, which requires its replacement.*

A bearing is designed to *creep* on the spindle when loaded, so sometimes there are marks on the spindle. This is normal as long as the bearing feels snug and yet moves freely on the spindle. Inspect the bearing cups in the hub (**Figure 60.52**).

After the inner bearing is installed in the hub, the seal can be installed. Be careful to install it straight. A special tool is helpful for this (**Figure 60.53**). Install

(a)

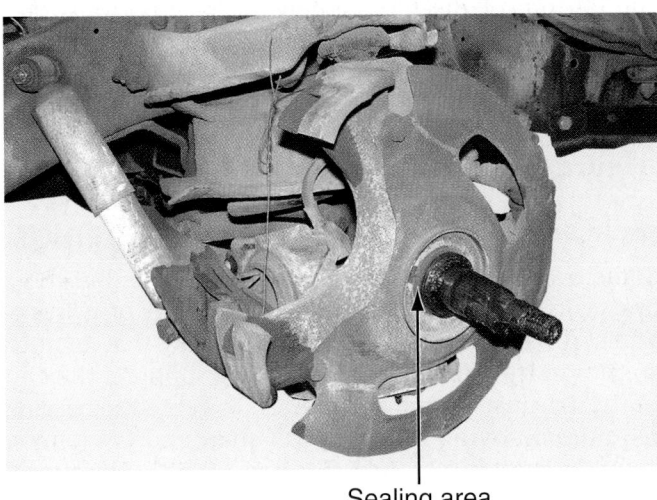

Sealing area

Figure **60.51** Clean the sealing area on the spindle so that the new seal is not accidentally ruined. *(Courtesy of Tim Gilles)*

(b)

Figure **60.53** (a) Select the correct adapter and position the seal over its bore. (b) Pound the seal flush with the top of the bore. *(Courtesy of Tim Gilles)*

Figure 60.54 Lubricate the seal lip before installing the bearings on the spindle.

the seal with its open end, or lip, facing in toward the bearing (see Figure 60.27). Be sure to lubricate the seal lip so it does not burn up (**Figure 60.54**). Adjust the wheel bearing and install the cotter pin according to the instructions provided earlier.

Pack some grease into the dust cap. Do not fill it all the way though. This keeps out contaminants and provides a reservoir for fresh oil. As the grease in the bearing oxidizes, it can dry out. Oil in the fresh grease in the cap can replenish the old grease through *capillary action*.

REPACKING DISC BRAKE WHEEL BEARINGS

The procedure for repacking disc brake wheel bearings is the same as that followed for drum brakes, except that the disc *caliper* must be removed in order to gain access to the inside wheel bearing. The caliper must be supported or wired to the steering knuckle support. Do not let calipers hang on brake hoses. The inside of the brake hose can be damaged. When reinstalling the caliper, torque the caliper bolt to specifications. When the caliper bolts have an Allen head, a socket drive Allen head tool should be used with a torque wrench.

DIAGNOSING WHEEL BEARING NOISE

If a possible wheel bearing groan is heard, driving the car can sometimes help pinpoint the problem. First, check the tires for damage and be sure they are properly inflated. Find an empty parking lot or deserted road and make slow left and right turns. This shifts the weight of the vehicle from one side to the other. When the weight is increased on the bearing, the noise increases. The inside tire always turns at a lower rpm than the outside tire.

- The noise from a wheel bearing that is bad will change pitch as the wheel speeds up and slows down when turning from one side to the other.
- When the outside wheel has the bad bearing, the noise will become worse because that wheel is turning faster.

- Applying the brakes can also cause the noise level from a bad bearing to become less as they contact the drum or rotor with the bad bearing.
- Spinning the wheels can be done using an on-the-car wheel balancer for non-drive wheels.

REPLACING BEARING RACES

Anti-friction bearings usually have one race that is **pressed-fit**. The other race is a **push fit**. A push-fit race slides into place by hand. The pressed-fit race is usually pressed onto or into the rotating part (outer race in brake hub). The push-fit race is usually pushed onto or into the stationary part (inner race on spindle).

When a damaged wheel bearing is replaced, the pressed-fit race must be removed from the bearing hub. The race might still look good to the eye but has been subjected to loads for just as long as the failed bearing. Leaving the old race to be used with a new bearing is an invitation for failure. Bearings should last about 150,000 miles.

The old bearing race is removed by pounding it with a drift punch or a special tool. When using a drift punch, there are recesses in at least two places in the side of the hub that allow for hammering on the back of the race (**Figure 60.55a**). Be sure to hammer a little bit from one side then the other so the race is not distorted during removal, damaging the hub. A special tool or a punch can be used to remove the race (**Figure 60.55b**).

The new race must fit the hub tightly. If not, replacement of the hub is usually required. A loose bearing race is often the result of a bearing seizure, which can be caused by a bent spindle. A special tool is handy for installing a bearing race (**Figure 60.56**). Drive the race into the hub, being careful not to cock it off to one side. If using a soft punch, hit on one side of the cup and then on the other. Drive it all of the way into the bore until it seats solidly against the ridge at the bottom of the hub. When the race bottoms out, the sound from the pounding will change.

A new race can be chilled in a refrigerator to make it easier to install. If a bearing is allowed to run in a misaligned position, it will be overloaded and fail (**Figure 60.57**).

SERVICING FRONT-WHEEL-DRIVE BEARINGS

Some bearings on these cars are serviceable, but most are sealed, requiring no service. The end of a front-wheel-drive axle shaft has drive splines that fit into splines in the rotor hub. The bearing supports the end of the axle just behind the splines. To get to the bearing requires removing the axle (half-shaft and CV joints). A puller is often required (**Figure 60.58**). The procedure is covered in more detail in Chapter 78.

The front wheel bearing is either pressed on or bolted onto the steering knuckle. Special tools are available for removing pressed-on bearings without having

Figure 60.55 (a) This recess leaves room for a punch to remove the race. (b) Removing the bearing with a brass punch. *(Courtesy of Tim Gilles)*

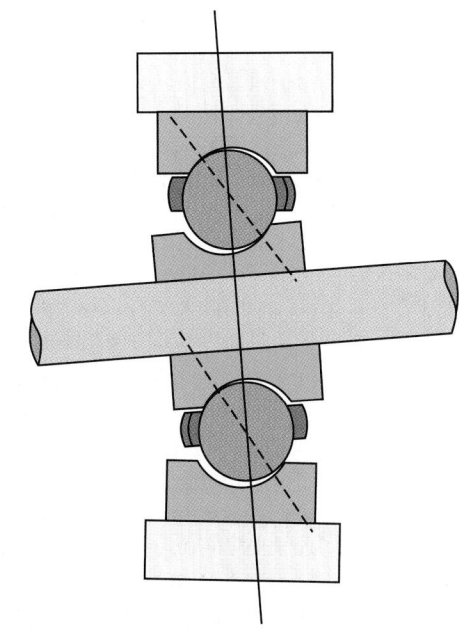

Figure 60.57 Misaligned bearing race wear. *(Courtesy of Tim Gilles)*

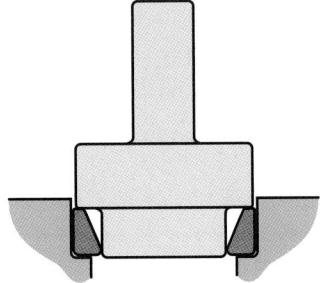

Figure 60.56 A special tool for installing a bearing race.

Figure 60.58 A puller used for removing a hub from a front-wheel-drive car. *(Courtesy of Tim Gilles)*

to remove the steering knuckle from the vehicle. Check the service manual for the correct procedure.

Some manufacturers recommend that the bearing be replaced any time a front-wheel-drive assembly is disassembled (when the front hub is removed from the spindle). End play is controlled by the size of the parts. When they are tightened together, the end play should be correct.

Special care is required during reassembly. When a puller is required to disassemble an axle from the hub, a certain degree of force will be required to reinstall it in the hub. Be certain that the parts are aligned before using any force. There are usually two rows of ball bearings with their races tapered toward each other to avoid end thrust. Be sure that the bearing assembly is held together as a unit before forcing it onto the axle shaft. The bearing can be ruined during this process.

NOTE: *The vehicle should not be rolled on its wheels with the axle removed, or the bearings can be damaged.*

On ball bearings, which is what most front-wheel-drive cars are now equipped with, the bearings are pre-loaded. Torque specifications range up to 200 pounds. Do not torque with the vehicle on the ground. Have an assistant hold pressure on the brakes while torquing. Then, stake the nut into the groove in the spindle (**Figure 60.59**). Do not overtighten and then back off.

Figure 60.59 A front-wheel-drive bearing retaining nut is torqued and staked to keep it in place. *(Courtesy of Tim Gilles)*

■ REVIEW QUESTIONS

1. What kind of bearing is a ball or roller, anti-friction or plain?

2. The separator that holds roller or ball bearings properly spaced in the bearing assembly is called a _____.

3. An up-and-down load on a bearing is called a _____ load.

4. A front-to-rear or side-to-side load on a bearing is called a _____ load.

5. Which bearing design has greater load-carrying capacity, roller or ball?

6. A _____ bearing is a very small roller bearing.

7. What kind of soap would probably be used as a thickening agent for wheel bearing grease?

8. If a grease has a lower NLGI number, is it thicker or softer?

9. What is the term for the temperature at which the grease turns into a liquid?

10. Which grease would be intended for use in wheel bearings, L or G?

11. Which kind of thickener is most often used for multipurpose grease?

12. What kind of oil can be damaging to a nitrile seal?

13. How much running clearance would a correctly adjusted tapered roller front wheel bearing have?

14. What is it called when the bearing or race has indentations from shock loads?

15. On a race car, how far should the wheel bearing hub cavity be filled with grease?

■ ASE-STYLE REVIEW QUESTIONS

1. Two technicians are discussing RWD tapered front wheel bearings. Technician A says that the outside wheel bearing supports the load of the vehicle. Technician B says that the inside bearing holds the wheel in alignment. Who is right?

   a. Technician A   **c.** Both A and B

   b. Technician B   **d.** Neither A nor B

2. Technician A says that roller bearings can support a greater load than ball bearings. Technician B says that most of a vehicle's load is exerted on the bottom of a roller bearing. Who is right?

   a. Technician A   **c.** Both A and B

   b. Technician B   **d.** Neither A nor B

3. Which of the following is/are true about tapered roller bearings?

   a. They have lips on the inner and outer edges on the inner race only.

   b. They can control end thrust.

   c. They are self aligning.

   d. All of the above.

4. Technician A says that axle bearing is the term for those bearings found on non-drive front and rear wheels. Technician B says that wheel bearing is the term for those bearings found on live axles. Who is right?

   a. Technician A   **c.** Both A and B

   b. Technician B   **d.** Neither A nor B

5. All of the following statements are true about axles *except*:

   a. Semi-floating axles are used on passenger cars.

   b. Full-floating axles are used on heavy-duty trucks.

   c. If an axle breaks on a semi-floating axle, the rear brake drum and wheel will still be supported.

   d. Semi-floating axles are used on some light duty trucks.

6. Technician A says that grease is made up mostly of lubricating oil. Technician B says that soaps are what make grease thick. Who is right?

   a. Technician A   **c.** Both A and B

   b. Technician B   **d.** Neither A nor B

7. Which of the following is/are true about lip seals?

   a. The open side of a lip seal always faces toward the lubricant.

   b. A wave seal pumps air toward the outside, to keep dirt out.

   c. Both A and B.

   d. Neither A nor B.

8. Technician A says that during a bearing repack, the hub cavity should be completely filled with grease. Technician B says that a wheel bearing fits the spindle tightly and should be tapped into place with a brass punch. Who is right?

   a. Technician A   **c.** Both A and B

   b. Technician B   **d.** Neither A nor B

9. Which of the following is/are true about cotter pins?

   a. When selecting a new cotter pin, use the largest diameter one that will fit into the hole.

   b. A cotter pin must be cut to the correct length using a special cotter pin cutter.

   c. Both A and B.

   d. Neither A nor B.

10. Technician A says that the rate of oil or grease oxidation doubles with every 20°F increase in temperature. Technician B says that the dropping point is the highest temperature at which a grease can be safely used. Who is right?

   a. Technician A   **c.** Both A and B

   b. Technician B   **d.** Neither A nor B

CHAPTER 61

Tire and Wheel Theory

■ OBJECTIVES

Upon completion of this chapter, you should be able to:

✔ Describe how a tire is constructed.

✔ Understand the various size designations of tires.

✔ Tell the design differences between radial and bias tires.

✔ Be able to select the best replacement tire for a car.

■ KEY TERMS

belt	speed rating	drop center
traction	load index	hub-centric
hydroplaning	GVW	stud-centric
radial-ply	GVWR	SEMA
bias-ply	M+S, MS, M&S, M/S	alloy
footprint	UTQG	negative wheel offset
placard	traction grade	positive wheel offset
profile	size equivalent	
aspect ratio	plus sizing	

■ INTRODUCTION

A service technician should be able to advise customers about tires, discuss aspects of tire design, and help the customer to make the safest (and best) choice when purchasing new tires and/or wheels. Tires and wheels are an important automotive safety and service specialty area. In-depth information about them is presented in this chapter and Chapter 62.

■ TIRE CONSTRUCTION

Tires are constructed of several layers of rubber materials, cords, and two rings of wire, called beads (**Figure 61.1**). The *casing* or *carcass* is the internal structure of the tire. A *ply* is metal or fabric *cord* that is rubberized (covered with a layer of rubber). The plies provide strength to the tire to support the load of the vehicle.

The ends of the plies wrap around the steel *bead* before being bonded to the side of the tire. The beads

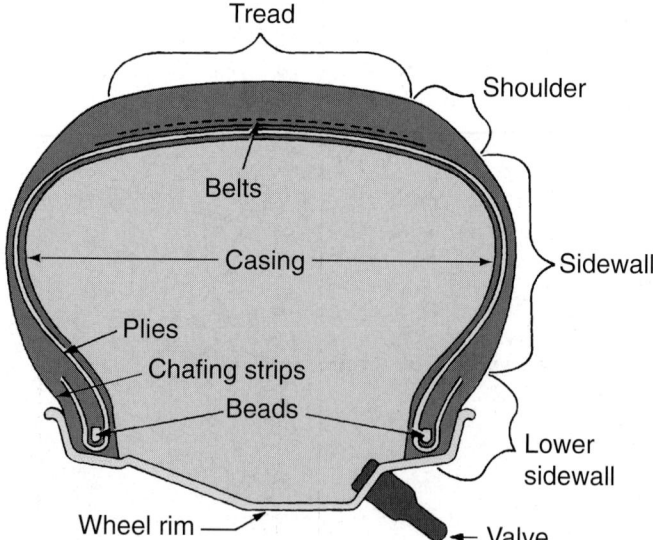

Figure 61.1 Construction of a tire. *(Courtesy of Goodyear Tire & Rubber Company)*

are coils of wires at the side edges of the tire. These give the tire the strength to stay firmly attached to the wheel. *Chafing strips* are hard strips of rubber that protect the beads from damage that could result from chafing against the rim.

A **belt** is a cord structure made up of plies. It is only in the area of a tire under the tread and does not extend under the sidewalls.

The *tread* is the section of the tire that rides on the road. A sidewall covering of rubber protects the casing plies between the tire tread and the tire bead.

◼◼ TUBELESS TIRES

Because of safety considerations and their ease of servicing, car manufacturers in the 1950s began to put tubeless tires on all of their cars. Almost all passenger car tires sold since the early 1960s are of the tubeless design. Some imported cars still had tube-type tires until the mid-1970s. Wire wheels have tubes to prevent leakage from the spoke holes.

The inside of a tubeless tire found on a passenger car has an *inner liner* bonded to it that seals air into the tire. The liner is thicker than the liner on a *tube-type tire*. Tubeless tires are actually safer than tube tires. When a tubeless tire is punctured, it will usually not go flat immediately. A nail tends to be held in the tire by the inner liner, allowing air to escape more slowly. A tube-type tire tends to go flat instantly when punctured because the walls of the inner tube tend to tear.

◼◼ TRACTION

A tire's **traction** is defined as how well it grips the road. Traction is affected by the road surface and contaminants such as water, ice, or debris. It is also affected by the tire's tread, the tread material, inflation pressure, width of the tread, cord ply design, wheel alignment, and other things.

◼◼ TIRE TREAD

The tread is a band made of a rubber compound designed to have various traction and wear characteristics. A federal grading standard (discussed later in this chapter) that is cast into the sidewall of the tire describes a tire's traction and wear characteristics. Grooves in the tread allow traction on wet surfaces, giving the water a place to go. They also allow the tire to flex without squirming, which would cause wear.

Treads are designed for specific types of weather and conditions. The design selected is always a compromise. The best traction on a dry paved road would be with a racing *slick*, or a bald tire. That same tire would be dangerous in the rain. Water forms a wedge under a tire that can actually float the car. This is called **hydroplaning** or *aquaplaning*. A deep tread pattern will break through a water film and grip the road at low speeds, but at high speed the tire can hydroplane (**Figure 61.2**).

Tires with large grooves are designed for use in mud and snow. But the large tread pattern can result in noise on the highway. Treads are often spaced at random intervals to minimize noise.

Sipes are small grooves in the tire tread that look like knife cuts (**Figure 61.3**). They allow extra gripping as the tire flexes. Sipes also clear water off of the road, wiping the contact area to provide a better grip. *Ribs* in the tire tread are designed to pump water from the road through the grooves to the back of the tire, where it is thrown out onto the road.

Tread Pattern Designs

Tires have tread patterns for differing driving conditions. Symmetric tires have treads of the same design on both sides (**Figure 61.4**). They can be installed on either side of the vehicle.

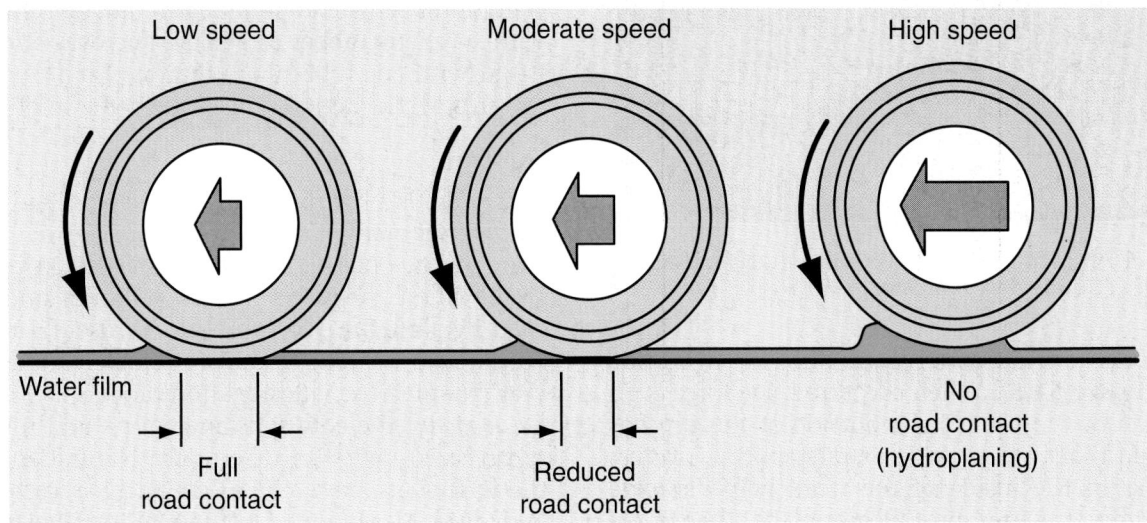

Figure 61.2 Hydroplaning.

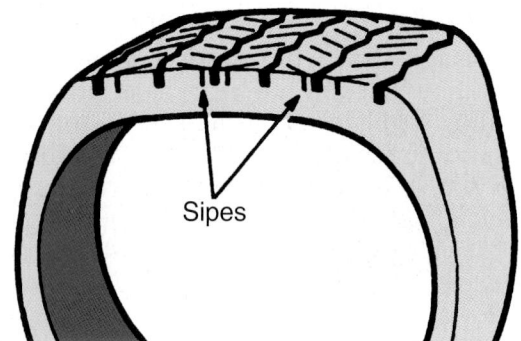

Figure 61.3 Tread sipes.

Figure 61.4 Symmetric tread pattern. (*Courtesy of Tim Gilles*)

Figure 61.5 An asymmetric tread pattern is different from side to side. (*Courtesy of Tim Gilles*)

Asymmetric tread patterns are different from side to side (**Figure 61.5**). When cornering, the force is on the outside of the tire, so the outside of the tire has larger tread blocks to provide extra stability. Grooves and blocks in the inside of the tread help dissipate water. Asymmetric tires must be mounted in one direction only, which makes them more expensive, and

Figure 61.6 A unidirectional tread. (*Courtesy of Tim Gilles*)

their position on the vehicle cannot be rotated except from front to rear.

Tread grooves for water diversion are located on the inside, and a harder sidewall is located on the outside. Directional tread patterns move water out to the sides of the tire, but asymmetric tread designs improve wet performance only. Performance is unchanged in dry conditions.

Some tire treads are unidirectional (**Figure 61.6**). They allow a vehicle to accelerate faster because they have less rolling resistance. They also allow faster stopping. When they are rotated, they must remain on the same side of the vehicle.

■ TIRE TREAD MATERIAL

The tread material calls for compromise. Hard materials might wear longer but not provide sufficient traction. Materials for mud and snow tires must remain soft in cold weather. Soft materials must provide sufficient wear. Natural rubber is compounded in different proportions with synthetic rubber to achieve the desired characteristics. Synthetic rubbers are more resistant to heat and solvents, while natural rubbers are better in other areas.

Rubber

Pure rubber is a hydrocarbon derived from the latex of a tree grown in all of the subtropical areas of the world. It freezes at only 40°F and becomes sticky at 86°F. It swells when contacted by many liquids and is damaged by sunlight. For rubber to be useful, it must be vulcanized (heated) to make it stable. Charles Goodyear patented the vulcanization process in 1842.

Isoprene, the core substance of natural rubber, was synthesized in 1910 in Germany. Natural and/or synthetic rubbers have chemicals such as carbon black and antioxidants added to them to improve grip, abrasion resistance, flexibility, and oxidation resistance.

Hysteresis is a term used by chemical engineers to describe a rubber's energy absorption characteristics. A high hysteresis compound results in quiet running, a comfortable ride, and better wet and dry grip. A low hysteresis compound has good lateral stability, low rolling resistance, and minimized tread wear.

TIRE CORD

Because rubber is elastic and not very strong, it must be reinforced with material such as fabric, fiber, or steel cords. Without these materials, a tire would blow up like a balloon. The most common cord material until World War II was cotton. Today, cord material in the casing is made of either rayon, nylon, or polyester. Cord material for belts can be steel, rayon, nylon, fiberglass, or aramid (Kevlar®), which was developed specifically for the tire industry.

TIRE PLY DESIGN

Most of today's tires are **radial-ply** tires, although some trucks and RVs still use **bias-ply** tires. **Figure 61.7** shows the difference between radial and bias tire construction. A bias tire is made in a *full-circle mold* that has two halves, like a bagel cut in half. The parting line for a tire made in this type of mold will run down the center of the tire's tread. A segmented mold is often used for constructing radial tires. A radial tire that has been made in a segmented mold will have several radial parting lines running from bead to bead across its tire tread.

Bias-Ply Tires

Bias-ply, diagonal, or cross-ply tires have casing plies that cross each other at angles of 35–45 degrees. They ride softer than radials, but their tread tends to squirm when rolling. This results in tire wear. Belts beneath the tread give the tire stability. Bias tires with belts under the tread last longer than unbelted bias tires because the belts keep the tread from squirming (**Figure 61.8**).

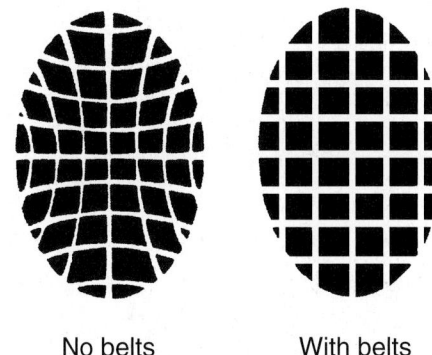

No belts   With belts

Figure 61.8 Belts stabilize the tread in the tread contact area.

Radial-Ply Tires

Radial tires have casing plies that run across the tire from bead seat to bead seat in the "radial" direction of the wheel. The outside circumference of the tire is held together by reinforcing belt rings of slightly angled cord material (**Figure 61.9**).

Radial tires give longer tread life, better grip to the road surface, and improved fuel economy. They ride rougher at low speeds than bias tires but can actually be smoother at faster speeds while travelling over highway expansion joints. A radial is more expensive to construct than a bias tire because during manufacturing it requires more labor.

A tire acts like a part of the suspension system as it supports the load of the car, isolating the passengers from road shock as its sidewall deflects. *Sidewall deflection* allows more of the tread to actually be in contact with the road surface. The larger area of contact, called the tire's **footprint**, allows the load on the tire to be spread across a wider area of the tire. A larger footprint also causes the tire to grip better so it can transmit forces of the engine and brakes to the road surface.

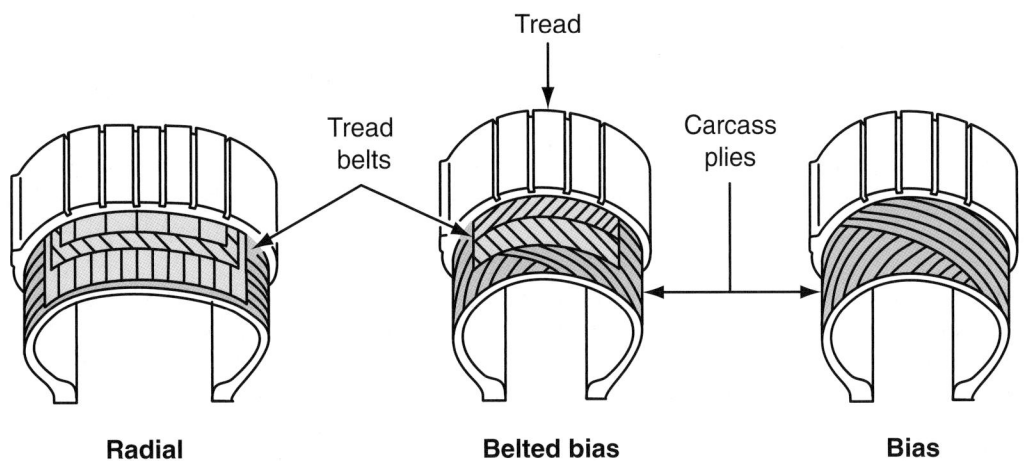

Figure 61.7 Comparison of radial and bias tire construction.

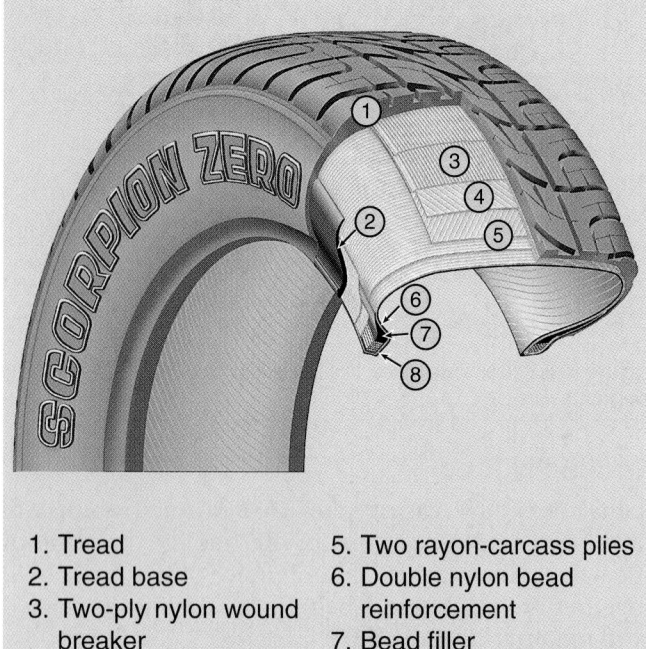

1. Tread
2. Tread base
3. Two-ply nylon wound breaker
4. Two steel-cord plies
5. Two rayon-carcass plies
6. Double nylon bead reinforcement
7. Bead filler
8. Bead core

Figure 61.9 Typical tubeless tire. *(Courtesy of Pirelli Tire of North America LLC)*

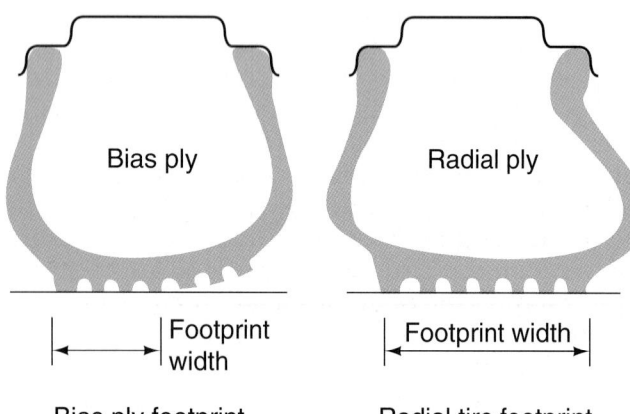

Bias ply   Radial ply

Footprint width   Footprint width

Bias ply footprint   Radial tire footprint

Figure 61.10 The flexible radial sidewall allows more of the tread to remain in contact with the road.

A radial tire flexes on its sidewall and is more resistant to wear because its tread surface stays flat on the ground (**Figure 61.10**). Because of the bulging sidewall, a properly inflated radial will appear to be low on air.

Radial tires have less resistance to rolling, which improves a car's gas mileage. Fuel economy standards have been mandated by the United States Congress for several years. Each manufacturer must meet a *corporate average fuel economy* (CAFE) standard or pay a *"gas-guzzler"* penalty to the government. Because radial tires help the vehicle to achieve better fuel economy, they are included on all new cars as original equipment.

VINTAGE TIRES

Switching to Radials

Customers with bias tires often wish to upgrade to radial tires. Cars built prior to 1972 usually do not have suspension systems designed for radial tires. Installing radials on these cars can result in a somewhat harsher ride at slow speeds. Installing radials on wheels that were designed only for bias tires can result in a dangerous wheel failure because radials exert more pressure against the sides of the rim. Cars produced after 1975 have numbers on the rims that designate their use with radial tires. If the number includes an R, the rim is designed to be used with radials.

The difference in handling characteristics between radials and diagonal bias tires makes it best not to mix them on the same vehicle. Sometimes, a customer's car will have two belted-bias tires that are in good condition. The customer might want to begin to make the switch to radials without buying all four new tires. The *Rubber Manufacturers Association* (*RMA*) recommends that the two new radial tires be installed on the rear.

See Chapter 62 for more information on radial tire service.

■■■ TIRE SIDEWALL MARKINGS

The U.S. Department of Transportation (DOT) requires the listing of certain information on the tire (**Figure 61.11**). Included on the tire sidewall for a typical passenger car are the following:

■ The tire size
■ Maximum permissible cold air pressure
■ Load rating—an indication of the load limit for each of the vehicle's tires under cold inflation
■ The name of the material that the cords of the tire are made of
■ The number of plies in the tread and sidewall areas
■ If the tire is a radial tire
■ Whether the tire is tube-type or tubeless
■ The DOT manufacturing code
■ M+S—indicating that the tire meets the RMA definition for a mud and snow tire
■ Uniform Tire Quality Grade Standard (UTQG)—DOT grading for traction, treadwear, and temperature

Tire Size

A tire information sticker called a **placard** (see Figure 13.32) has been required on cars sold in the United States since 1968. It is located on the door post, the edge of the door, the gas filler door, or on the glove box door. The placard indicates the correct original equipment (OE) tire size, the cold inflation pressure, and the gross axle weight (for commercial vehicles). If there is no placard, check the owner's manual for the information.

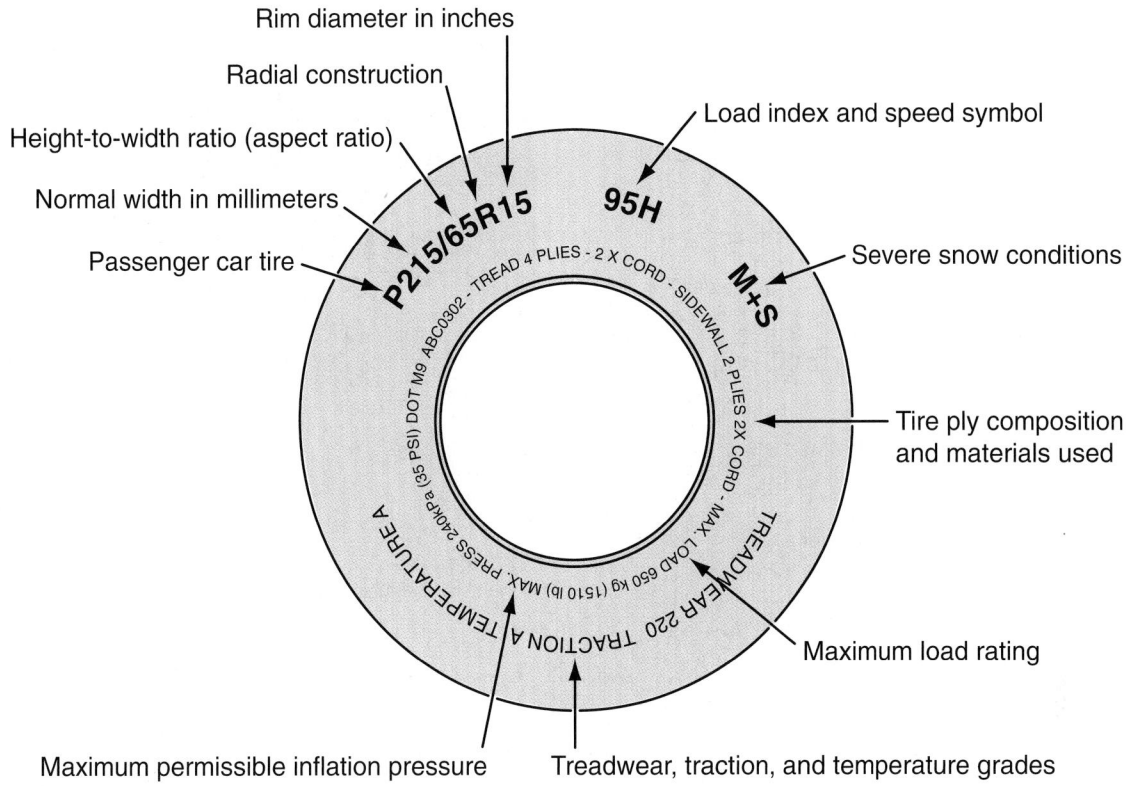

Rim diameter in inches

Radial construction

Height-to-width ratio (aspect ratio)

Normal width in millimeters

Passenger car tire

Load index and speed symbol

Severe snow conditions

Tire ply composition and materials used

Maximum load rating

Maximum permissible inflation pressure

Treadwear, traction, and temperature grades

Figure 61.11 Information found on a typical passenger tire sidewall.

The tire's size is listed on the sidewall, using one of several ratings (**Figure 61.12**).

■ P-Metric (P205-75 R15)
■ European Metric (185/70 R14)
■ 4WD Tires Numeric (6.70-15) (36 × 10.5 × 15)
■ Alphanumeric (FR78-15)
■ Light trucks (LT 235/75R-15)(**Figure 61.13**)
■ European Metric (205/40 R17). Note that European tires do not list a "P" or "LT" at the front of the rating (**Figure 61.14**).

The alphanumeric rating was commonly used until the early 1970s, but metric cross-section measurement is universal today. **Figure 61.15** shows how the tire size designation is interpreted for the P-metric radial tire, the most common tire in use today. The first letter tells the type of tire it is:

■ P means passenger car.
■ LT means light trucks, or C means commercial.
■ T—temporary spare (covered in more detail later).

The tire's cross-section width (215 mm) is listed next. With each size increase (from 215 to 225, for instance), the width of the tire increases by 10 millimeters.

There is usually a letter in the size designation. Some of the possible letters are

■ R (radial)
■ B (belted bias construction)—sometimes left blank
■ D (diagonal bias construction)

A tire's height is called its **profile**. A low-profile tire is shorter than a normal tire. The number that comes after the cross-sectional width of the tire is the **aspect ratio**, which is a measurement of the height-to-width ratio (**Figure 61.16**). In Figure 61.15, aspect ratio is expressed as the number 65. A 60 would have a lower profile. When a designation does not include a designation for the aspect ratio, the tire is an 80 series (P215 R15).

The last part of the sequence (R15) tells that the tire is a radial tire to be mounted on a 15"-diameter rim. This is sometimes followed by an optional load/speed index.

Speed Rating

Sometimes a **speed rating** (**Figure 61.17**) is listed as part of the size designation. This rating was originally developed in Europe, where higher freeway speeds are legal. Tires made for use in the United States have speed ratings based on tests that meet SAE J1561 standards.

Newer tires have the speed rating and load index listed separately after the size designation (**Figure 61.18**). The new designation is a two- or three-digit **load index** followed by the speed symbol. The load index, developed by the International Standards Organization (ISO), provides an industry standard for a tire's maximum load at the designated speed rating. A tire's load-carrying ability is related to the strength

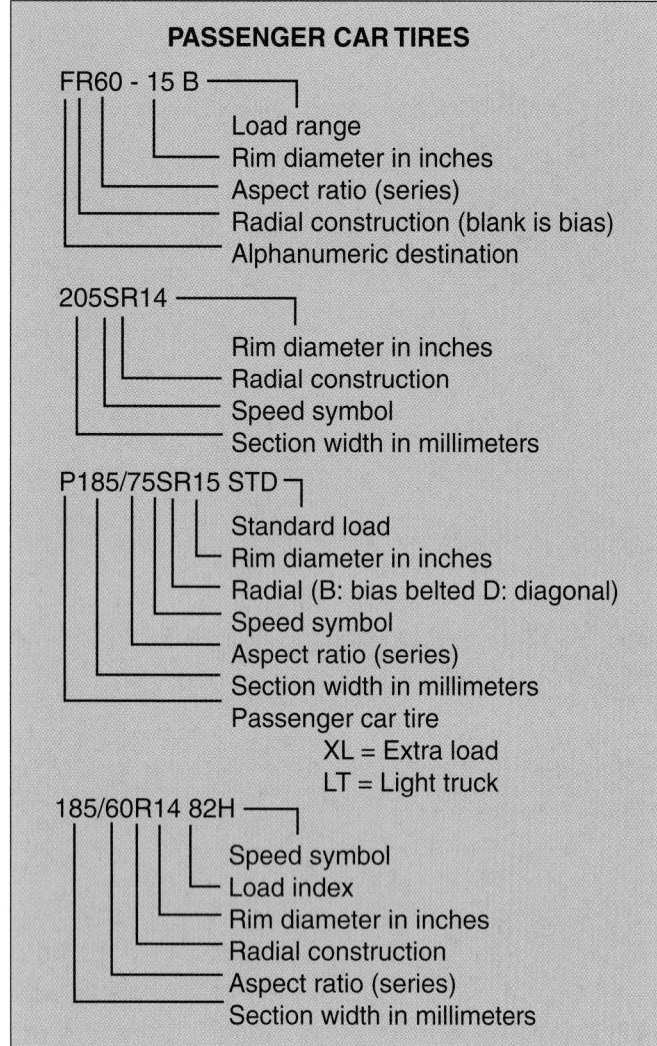

PASSENGER CAR TIRES

FR60 - 15 B
— Load range
— Rim diameter in inches
— Aspect ratio (series)
— Radial construction (blank is bias)
— Alphanumeric destination

205SR14
— Rim diameter in inches
— Radial construction
— Speed symbol
— Section width in millimeters

P185/75SR15 STD
— Standard load
— Rim diameter in inches
— Radial (B: bias belted D: diagonal)
— Speed symbol
— Aspect ratio (series)
— Section width in millimeters
— Passenger car tire
    XL = Extra load
    LT = Light truck

185/60R14 82H
— Speed symbol
— Load index
— Rim diameter in inches
— Radial construction
— Aspect ratio (series)
— Section width in millimeters

Figure 61.12 Different ways of measuring tire size.

LT label for light truck

Figure 61.13 A tire size label for a light truck tire. *(Courtesy of Tim Gilles)*

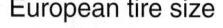
European tire size

Figure 61.14 A European tire size label. Notice that there is no P in the size designation. *(Courtesy of Tim Gilles)*

of its sidewall plies. A tire with a higher load capacity will also have a higher inflation pressure.

The speed rating also serves as an indicator of better handling characteristics that result from improvements to the tire. At high speeds, non-speed-rated tires distort in the sidewall and tread areas (**Figure 61.19**). Speed-rated tires require extra reinforcement in the sidewall, including sidewall bead stiffeners and nylon cap plies or belt edge strips (**Figure 61.20**). Bead stiffeners are made of extra-hard rubber that prevents the sidewall from bulging. Tires with these additions provide a quicker steering response time. You can feel the difference at as little as 35 mph.

Low-profile tires often have sections of sidewall that extend beyond the flange of the rim to protect the rim from damage when it rubs against a curb.

Nylon cap plies are extra plies on the sides of the normal tire plies. Additional centrifugal force results from higher speed, but nylon shrinks when heated, pulling back on the sidewall and flattening the tire footprint. This helps to keep the rear end from breaking loose. Tires used for extreme high speeds have a sidewall reinforced with a band of steel.

Speed symbols for passenger cars range from the L rating (74.5 mph/120 km/h), to ZR (over 149 mph/240 km/h). The letters that denote changes in speed ratings change in 20-kilometer per hour (km/h) increments. H-rated tires were the first speed-rated tires, so that is why this letter is out of order.

These are some of the most common ratings:
- Q—99 mph (160 km/h) - winter tires
- R—106 mph (170 km/h) - heavy-duty light truck tires
- S—112 mph (180 km/h) - family cars and vans
- T—118 mph (190 km/h) - family cars and vans

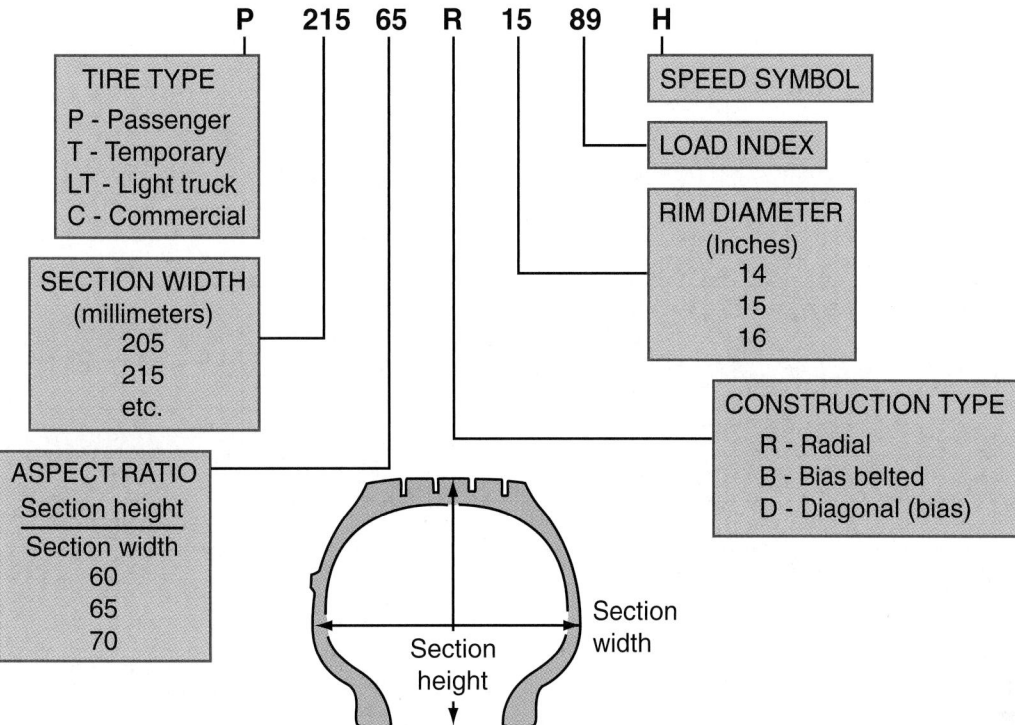

Figure 61.15 Tire size designation on a P-metric radial.

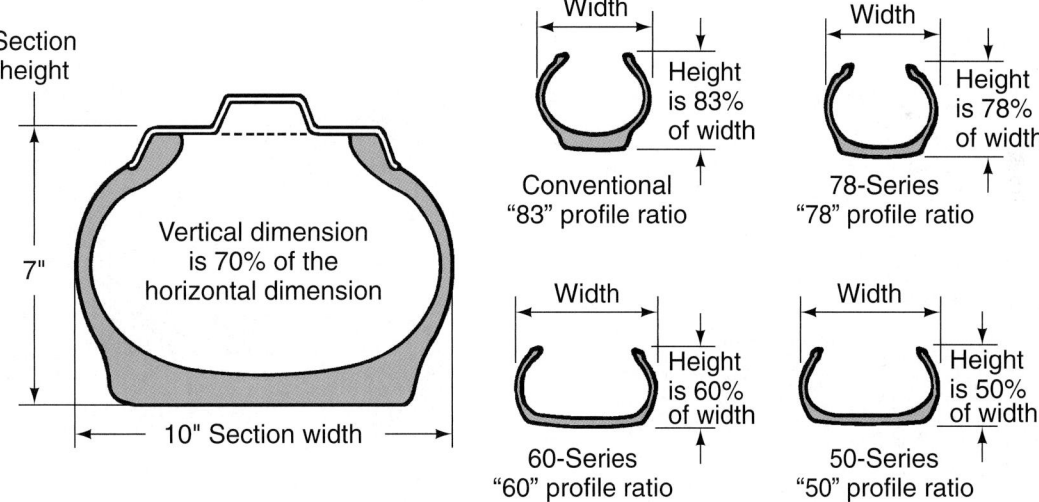

Figure 61.16 Aspect ratio compares the tire's height and width.

- H—130 mph (210 km/h) - high-performance passenger cars
- V—149 mph (240 km/h) - high-performance sports cars

NOTE: *Good winter tires tend to only have a Q-rating because they generate a great deal more heat than conventional tires.*

Light truck tires run hotter and do not dissipate heat as well as passenger car tires. Some of the newer truck tires carry an R or S speed rating.

VINTAGE TIRES

Under the original speed rating system, V-rated tires (130 mph/210 km/h) were the highest achievable speed rating. This category was originally defined as "unlimited." As tire manufacturing abilities improved and safety considerations increased, the V-rating became a "limited" rating. The Z-rating was added as the top speed rating, for speeds in excess of 149 mph, with the exact speed rating determined by the manufacturer and varying with size.

SPEED SYMBOLS/RATINGS

SPEED SYMBOL	MAXIMUM SPEED	APPLIES TO PASSENGER CAR TIRES	APPLIES TO LIGHT TRUCK TIRES
ZR	Above 149 mph (240 km/h)	Yes	No
Y	186 mph (300 km/h)	Yes	No
W	168 mph (270 km/h)	Yes	No
V	(with service description) 149 mph/(240 km/h)	Yes	Yes
H	130 mph (210 km/h)	Yes	Yes
U	124 mph (200 km/h)	Yes	Yes
T	118 mph (190 km/h)	Yes	Yes
S	112 mph (180 km/h)	Yes	Yes
R	106 mph (170 km/h)	No	Yes
Q	99 mph (160 km/h)	(Winter tires only)	Yes
P	93 mph (150 km/h)	No	Yes
N	87 mph (140 km/h)	No	Yes
M	81 mph (130 km/h)	Temporary spare tires	No

Figure 61.17 Tire speed ratings.

Load rating   Speed rating

Figure 61.18 This tire has a Y speed rating and a 95 load rating. It is rated for 186 mph and can carry a load of 1,521 pounds. The ZR is required in the size designation on "Y" rated tires. *(Courtesy of Tim Gilles)*

Figure 61.19 This tire does not have a speed rating. Under high speed it distorts in the tread and sidewall areas. *(Courtesy of Bridgestone Firestone North American Tire, LLC)*

Very High-Speed Ratings

Because some new vehicles are capable of very high speeds, new tire speed ratings have been developed. W-rated tires are rated at 168 mph (270 km/h), and a Y-rated tire is rated at 186 mph (300 km/H). These tires will still carry a Z in their tire number designation, with the W or the Y listed after the load rating, such as *275/35/ZR19 (99Y)*. When the load and Y speed rating are enclosed in parenthesies, the speed rating is in excess of 186 mph (300 km/h).

■ LOAD RATING

A tire's load rating tells how much weight it can safely support at a specified air pressure. It is very important *not* to use a tire that has too low a load rating for the weight of the vehicle.

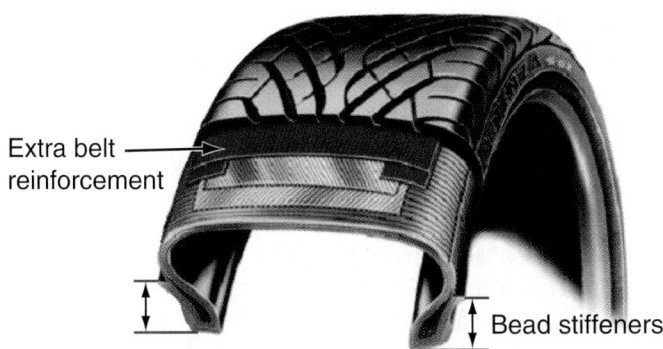

Extra belt reinforcement

Bead stiffeners

Figure 61.20 Speed-rated tires have stiffer sidewalls and more belts. *(Courtesy of Bridgestone Firestone North American Tire, LLC)*

Figure 61.21 A door jamb decal showing gross vehicle weight rating and tire pressures. *(Courtesy of Tim Gilles)*

P-metric radial tires found on today's passenger cars are all of a uniform standard-load rating or an extra-load rating. The amount of load one of these tires can support is determined by the area of the tire and the amount of air pressure in it. Standard-load tires reach their maximum load-carrying capacity when inflated to 35 psi. Extra-load tires achieve maximum load at 41 psi. An extra-load tire is labeled with "XL," as in P205 70R15**XL**.

Although a P-metric standard-load tire has a normal maximum inflation of 35 psi, its tire sidewall might be branded with a maximum pressure of 44 psi. Although the tire can be safely inflated to this pressure, its maximum load-carrying capacity is not increased.

Tire companies recommend that heavy RVs be weighed separately at each wheel. This is to determine the load rating and the air pressure that should be used to safely support the load at each corner of the vehicle.

Gross Vehicle Weight

The vehicle's gross vehicle weight rating (**GVW** or **GVWR**) includes the weight of the vehicle, the weight of the passengers it has seats for (estimated at 150 lb each), and the maximum amount of luggage load. The GVW can be found on a plate or sticker on the door jamb (**Figure 61.21**). It is sometimes listed on the vehicle's registration as well.

Curb weight is the weight of the vehicle without passengers or luggage but includes a full tank of fuel and all fluids filled in the vehicle.

Vehicles, especially pickup trucks, are often overloaded. It is important that tires, brakes, and axles be of sufficient size or capacity to support the load and that a vehicle not be loaded beyond its weight rating.

When towing, be sure that the weight of the trailer is within the maximum capacity of the vehicle. The best way to prevent overloading is to weigh each axle of the vehicle on platform scales. The Rubber Manufacturers Association (P.O. Box 3147, Medina,

OH 44258) provides free information on vehicle weighing procedures.

DOT Codes

The DOT symbol (**Figure 61.22**) signifies that the tire meets DOT safety standards. Before the year 2000, there were 10 characters (a combination of numbers and letters) in the DOT code. Today there are up to 12, but usually 11. All of the characters except for the fifth, sixth, and seventh are regulated by the DOT:

- The first two characters list the plant and manufacturer where the tire was made. There is a separate code for the same manufacturer in different countries so that the country of origin can be determined.
- The second set of characters in the code tells the size and type of the tire.
- The third set group consists of three characters that are not regulated and can be used by manufacturers as they choose.

Figure 61.22 The Department of Transportation (DOT) identification number on a tire sidewall. *(Courtesy of Tim Gilles)*

DOT 8X72 WL1 1603	
8X	The location of the plant where the tire was manufactured
72	The tire size (regulated by the U.S. government)
WL1	The manufacturer's option code, used to tell the difference between tires
16	The week of the year the tire was manufactured
03	The year the tire was manufactured

Figure 61.23 A typical DOT code and the meaning of its numbers and letters.

■ The final four digits tell the week and year the tire was made. The first two digits are the week and the last two are the year. Prior to the year 2000, the date of manufacture was listed as three digits, but since then four digits have been used so it is possible to tell the decade when the tire was manufactured. **Figure 61.23** shows a typical DOT code and the meaning of its numbers and letters.

ALL-SEASON TIRES

When radial tires became common on vehicles, they were found to have more traction on snow than the bias-ply tires they replaced. When a tire has specially designed pockets and slots in at least one tread edge, it can be labeled with a mud and snow designation. This can be any combination of the letters M and S (M+S, MS, M&S, M/S) on the tire sidewall and means that the tire meets definitions set by the Rubber Manufacturers Association (**Figure 61.24**).

SNOW TIRES

M&S tires do not guarantee winter driving performance or safety, however. A snow, or winter, tire is specially designed for winter performance. In 1999, the U.S. Rubber Manufacturers Association and the Rubber Association of Canada decided on a performance-based standard for snow tires that would allow consumers to identify tires designed to enhance traction in harsh winter conditions. Winter tires must meet traction tests on packed snow specified by the American Society for Testing and Materials (ASTM). Tires that meet this standard are labeled with a "snowflake on the mountain" symbol on the tire sidewall next to the M&S symbol (see Figure 61.24).

Summer tires become harder in the winter. Winter tires are made with a tire rubber compound that remains soft in the winter. If these tires are used during the summer, they will be excessively soft and will experience rapid tire wear and generate a good deal of heat. As mentioned earlier in the chapter, they have the low Q-speed rating.

Snow tires have deeper tread grooves designed to provide a better grip when driving on snow-covered roads (**Figure 61.25**). Tread depth for snow tires is $\frac{13}{32}$"–$\frac{15}{32}$" deep compared to new passenger car tire depth of $\frac{10}{32}$".

Most manufacturers recommend that snow tires be installed on all four wheels to prevent handling problems. Wide tires do not cut through snow as easily as narrow tires. Snow tires can be fitted in the OE size, but when the vehicle has wide tires, a tire dealer will often recommend narrower tires of the same load capacity. Changing tire sizes is covered later in this chapter.

Some states allow tires with the severe snow symbol to be used in place of studded tires or snow chains.

Snowflake on mountain symbol

Mud and snow

Figure 61.24 This winter-rated tire has a snowflake symbol showing it is designed for driving in snow. *(Courtesy of Tim Gilles)*

Figure 61.25 A snow tire has an aggressive tread pattern. *(Courtesy of Tim Gilles)*

Winter tires can provide better handling in snow than is afforded by four-wheel drive without winter tires.

Snow Chains

Snow chains are used in some mountain areas during severe winter weather when roads have become covered with ice and snow. Most manufacturers recommend against the use of chains. In fact, some new vehicle sales procedures include having the buyer sign a paper saying that only cable chains can be used. When a vehicle has sufficient clearance between the tires, fenders, and suspension components, tire chains provide a viable means of achieving traction on snow-covered roads.

Tire chains can be of either the cable or chain type. Cable chains (**Figure 61.26**) are not as effective as conventional chains, but they work well in low-clearance applications and are not as apt to cause damage due to incorrect installation. **Figure 61.27** shows the different conventional types of tire chain. Heavy-duty conventional chains have reinforced lugs for a better bite into ice and snow.

Run-Flat Tires

Some new cars do not carry a spare but use one of several methods that allow them to be run with little or no tire pressure. Because an underinflated tire will develop heat that will damage the tire, a run-flat tire must be used with a low-pressure detection system.

A typical run-flat tire has a stiffer sidewall and a tighter tire bead (**Figure 61.28**). The stiff reinforced sidewalls are four to six times as thick as a normal

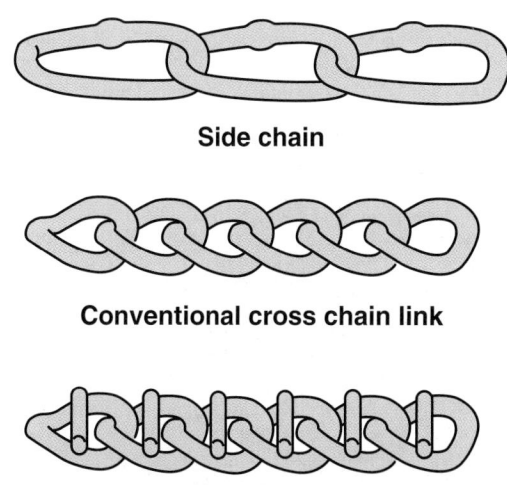

Side chain

Conventional cross chain link

Reinforced cross chain

Figure 61.27 Conventional and heavy-duty chains.

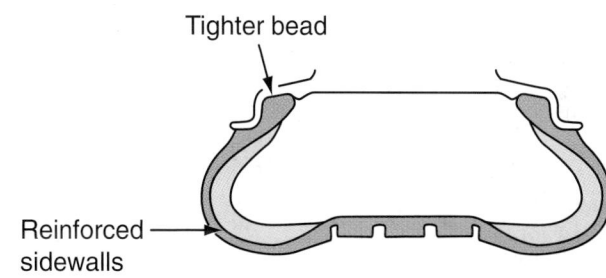

Tighter bead

Reinforced sidewalls

Figure 61.28 A typical run-flat tire has a stiffer sidewall and a tighter tire bead.

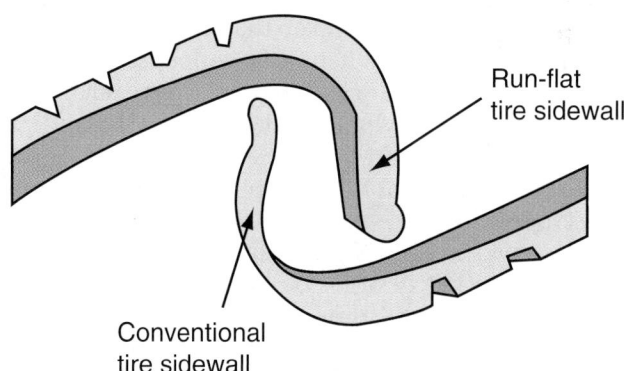

Run-flat tire sidewall

Conventional tire sidewall

Figure 61.29 The stiff reinforced sidewalls of a run-flat tire are four to six times as thick as a normal tire's sidewalls.

tire's sidewalls (**Figure 61.29**). When a conventional tire is driven without air, its beads tend to fall out of the rim's safety bead areas into the drop center. The vehicle can lose control and cause a serious accident. Run-flat tires have a special bead design to prevent this from happening. A run-flat tire can partly support the vehicle even when the tire is completely empty of air and can be driven up to 70 miles without air before suffering damage. When driving over expansion joints on the freeway, a driver should expect the reinforced sidewalls to contribute to a rougher ride.

Figure 61.26 Cable snow "chains" are installed when the clearance between the tire and fender well is minimal. *(Courtesy of Tim Gilles)*

Figure 61.30 A cutaway of a run-flat tire with an insert to support the tire in case of air loss. *(Courtesy of Michelin)*

Many modern vehicles have high-torque engines and high-performance brakes that can cause the tire to slip on the rim. When this happens, the wheel weights will move to a different place in relation to the tire, resulting in vibration due to imbalance. High-performance and run-flat tires with tighter fitting beads prevent the tire from slipping on the rim. Some luxury performance cars and SUVs have "bead lock" rims designed to hold the tire tightly to the rim.

Other run-flat designs use bead retention systems with special tires and rims. One method uses a foam insert in the drop center of the wheel (**Figure 61.30**). The insert maintains the original shape of the tire by preventing the beads from moving into that area during an air loss.

Low-Pressure Warning

The Tread Act, legislated by the United States Congress in 2000, requires new vehicle manufacturers to install a tire low-pressure warning system on all cars. Tire pressure monitoring was phased into cars until all vehicles had it by the year 2006. Sensors transmit a radio frequency (RF) to a receiver (**Figure 61.31**). In some pressure monitor systems, when the pressure in a tire drops below a predetermined point, 25 psi for instance, a warning light on the instrument panel illuminates. Other systems monitor tire pressure continuously.

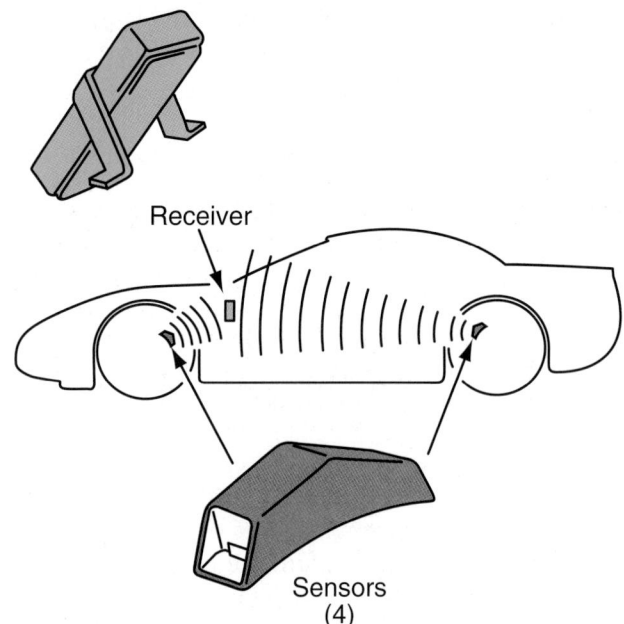

Figure 61.31 Direct tire pressure sensing uses pressure sensors and a receiver.

Direct and Indirect Pressure Monitors. Low tire pressure can be monitored using a direct or indirect method. Direct tire pressure monitoring, which uses individual wheel sensors and a computer, is more costly. Early sensors (prior to 1997) were strapped in the drop center of the rim, opposite the valve stem (**Figure 61.32**). Later sensors, called *integral sensors*, are part of the valve stem (**Figure 61.33**). Each wheel's sensor sends a different signal to the monitor so it can determine which wheel has a pressure problem. A typical sensor is powered by a lithium battery with a 10-year service life. To save the battery, it operates only at speeds over 20 mph (32 kmph) and sends a signal once an hour when parked. Pressure below

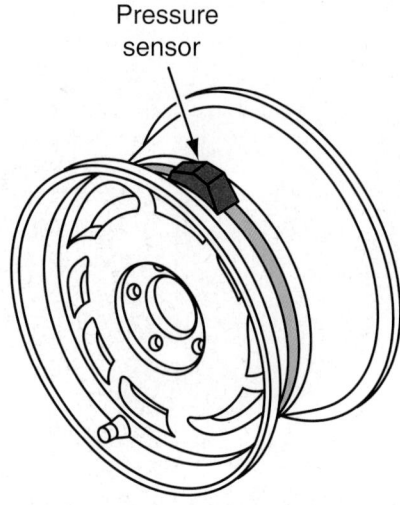

Figure 61.32 Early sensors were strapped in the dropwell of the rim, opposite the valve stem.

Figure 61.33 Late-model direct tire sensors are part of the valve stem. *(Courtesy of Siemens VDO Automotive)*

24 psi (168 kPa) or above 39 psi (272 kPa) illuminates a warning light on the instrument panel.

An indirect tire pressure monitor system uses the antilock brake system to compare the speed of all of the vehicle's wheels, allowing a 10 psi difference in pressure. But if all four tires are low, it does not detect a problem. This presents a new concern for tire makers: Consumers might have a false sense of security if they believe their tires are correctly inflated because no instrument panel light is illuminated.

■ TIRE QUALITY GRADING

American manufacturers use the uniform tire quality grade (UTQG) system, which rates treadwear, traction, and *temperature* dissipation ability. It is printed on the sidewall of the tire. The rating of the tire shown in **Figure 61.34** is 420AA.

Figure 61.34 Uniform Tire Quality Grade (UTQG) markings on a tire sidewall. *(Courtesy of Tim Gilles)*

■ TREAD WEAR

Tire manufacturers test their tires following government specified procedures. The amount a tire's tread wears will vary with wheel alignment, road surface texture, tire rotation maintenance, vehicle speed and braking practices, the weight of the vehicle, and the size of the tire. The government prescribes a test procedure and course. A convoy of less than four test vehicles drives the same 400-mile test course on public roads in Texas. The test sequence lasts 7,200 miles, with tread depth being measured every 800 miles. Tire wear results are compared with those done on a control group of tires. Tread wear ratings are simply an overall indicator of performance, and ratings are not identical between manufacturers.

Tread wear ratings range from under 100 to over 500, increasing in increments of 20. The number 100 represents a standard tire. A 200 would be expected to last twice as long on the government test course, and a 150 would last about 1½ times as long. The actual life of a tire can vary due to road conditions, climate, air pressure, alignment, driving habits, vehicle loading, and other factors.

■ TRACTION GRADE

With regard to the actual ratings, AAA is the highest rating, whereas C is the lowest. Prior to 1997, A was the highest rating for wet braking traction. The first letter is the **traction grade**, which indicates stopping ability on wet asphalt pavement and concrete. This rating, done on specified government test surfaces, covers braking only in a straight ahead direction, not cornering.

Temperature Grade

The second letter is the temperature grade. It indicates a properly inflated tire's resistance to generating heat and its ability to dissipate heat at highway speeds. Temperature ratings are determined using specified government tests in a laboratory on a test wheel. Grade C is the minimum standard required by law. Standards B and A exceed this standard. Continuous high-speed driving can result in deterioration of the tire's material. This and excessive temperatures can lead to sudden tire failure.

Compact Spare Tire

A compact spare tire is what most new cars come equipped with. The compact tire is considerably smaller than a regular tire and is to be used only temporarily. Many have a limited speed of 50 m/hr (31 mph) and a distance of 50 kilometers (31 miles). The speed and distance warning will be printed on the sidewall of the tire.

■■ CHANGING TIRE SIZE

Customers often want to make a change in the size of an OE tire. Selecting a replacement tire that is the exact same size as shown on the placard is not always possible. If tire size is changed, be sure to substitute a tire that has an equal or greater load-carrying capacity. Tire companies provide charts that give the maximum load that various sizes of tires will provide at listed cold pressures.

Tire Size Equivalents

A change in tire size can usually be accomplished without sacrificing safety or design considerations when the correct **size equivalent** is used. At first, this can be confusing. Tires with different bead-to-bead diameters (13", 14", 15", etc.) can all have the same outside diameter. For instance, the following five tire sizes all have the same diameter *and* load capacity:

- 175/70SR13
- 205/60R13 H
- 185/60R14 H
- 205/55VR14
- 195/50VR15

When changing tire sizes, the following five things need to be considered in the replacement tire:

- The intended use of the tire
- The width of the wheel rim
- The overall diameter of the replacement tire assembly
- The speed rating, which must be equal to or greater than that of the original tire
- The overall load-carrying capacity, which must be equal to or greater than the load index number listed on the original tire

Usually, as the diameter of the tire increases, the load capacity of the tire increases also.

 **SAFETY NOTE** Changing from a P185/65R14 (85 load index/1,124-pound maximum carrying capacity) to a P185/60R14 (82 load index/1,047-pound capacity) will result in almost 7% less load capacity.

Tire manufacturers publish application handbooks that give information such as tire dimensions, revolutions per mile, diameter, acceptable rim sizes for each tire size, and the cross section of the tire. When unsure about a possible change in tire size, technical assistance is available through all of the major tire manufacturers.

Since the early 1970s, tires with lower profiles have become increasingly popular. When a tire with a lower profile is being installed, a wider tire and a wheel with a larger diameter will be used to make up the difference in overall tire assembly height. This method, called "**plus sizing**," produces the same overall diameter as the OE tire (**Figure 61.35**). The new combination has less sidewall flexibility but has the same load capacity. It also has a larger footprint.

Lower profile tires grip better and they are more responsive. A high-profile tire provides better loose snow or mud traction because it does a better job of cutting through these materials. Narrow wheels and tires with the proper load capacity can be a better choice for customers who find their snow traction to be inadequate.

Wheel rim width must also be considered when changing tire size. A wider tire provides more support

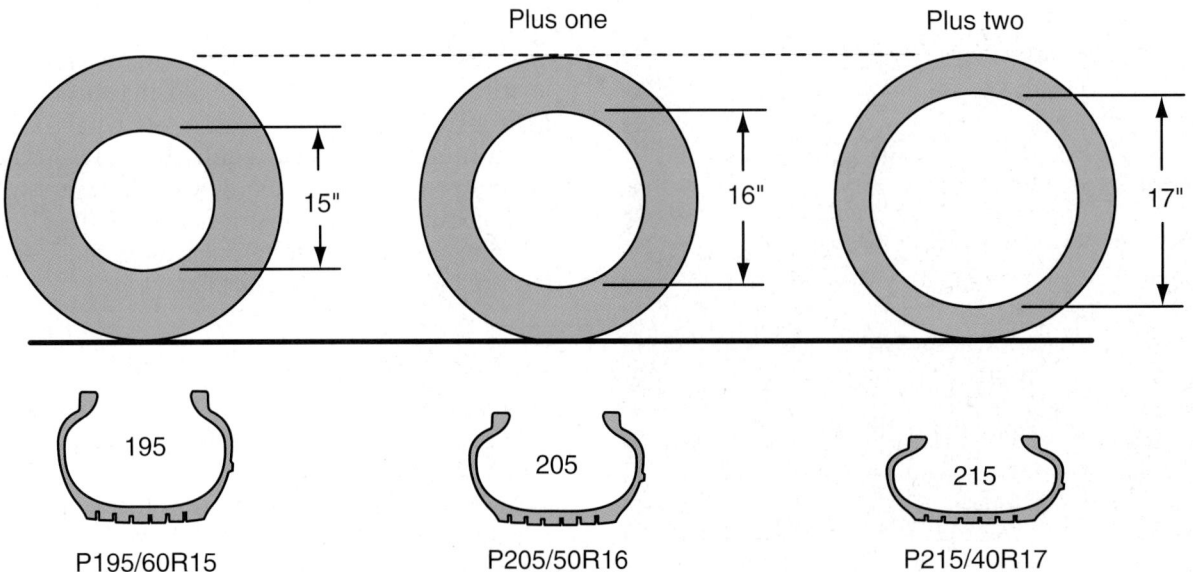

Figure 61.35 Plus sizing maintains the same diameter tire assembly by using a larger wheel and lower tire profile.

to the tire sidewall. A wider wheel provides more support to the tire sidewall. A narrower wheel allows the sidewall to flex more easily, providing a softer ride.

> **SAFETY NOTE** A tire should never be installed on a wheel that is narrower or wider than the manufacturer's recommended wheel width. Installing wider tires on a vehicle will probably mean that wider wheels will have to be installed.

When changing tire size, tire load capacity and diameter can be maintained by using the following formula: When changing to a 5% lower profile (reducing the aspect ratio by 5), choose a tire with a 10 mm wider cross section. For instance, change from 215/75 R15 to 225/70 R15.

■ Tire manufacturers recommend against changing from a lower profile to a higher profile tire, which results in reduced vehicle performance. When making a change in the size of tires on a vehicle, the recommendation is to replace tires in sets of four.

Overall Tire Diameter

DOT standards require the overall diameter of a replacement tire to be within +2% to −3% of what the tire was as OE. Changing the tire diameter can affect antilock brakes, the vehicle speed sensor signals to the computer, the speedometer, gear ratios, and four wheel drive.

Many newer vehicles are equipped with antilock brake systems (ABS) (see Chapter 59). A small pickup coil at each wheel measures wheel rotation speed and generates a signal that is sent to the ABS computer. Tires of a different height than original can cause excessive tire chirping or erratic system operation during a panic stop. When the size difference is large enough, the ABS warning light can come on. A small amount of tire height difference between tires on different sides of the vehicle is enough to cause the computer to turn on the ABS warning light and set a trouble code. In this case, the brakes will continue to operate normally, but ABS function will be disabled. On some computer controlled vehicles, the technician is able to access the computer using a scan tool to change the tire values.

When the diameter of a tire is changed, front-end geometry is altered and the speedometer will need to be recalibrated. When higher diameter tires are installed to provide increased load, this is called "oversizing." Higher diameter tires raise the vehicle's center of gravity. This reduces a vehicle's ability to hold the road and maneuver quickly in an emergency.

"Undersizing" is when a smaller sized tire is installed, often because they are less expensive. Never install undersized tires on a vehicle. They will wear faster and have less load-carrying ability. The vehicle will be lower, the speedometer will no longer be accurate, and the increase in engine rpm for a given speed will result in a decrease in fuel economy.

■ WHEELS

Wheels are made of many different materials, including steel, aluminum, or an alloy of aluminum. OE wheels on less-expensive passenger cars are made of steel. Wheels have two parts: the *center* or *flange* and the *rim* (**Figure 61.36**). The center flange of steel wheels is stamped because this is the least expensive production method. A strip of steel is rolled and butt welded at the ends to form the rim, which is then spot welded to the center flange.

A **drop center** or *rim well* provides a means of removing and installing a tire from the wheel. The tire bead is reinforced with wire and it will not stretch. When a tire is installed on the wheel, one side of the bead is pushed into the drop center so that the other side of the bead can be pulled over the edge of the rim (see Chapter 62).

The raised sections on either side of the drop center area are called *bead seats*. This is where the tire actually seals. There are raised sections on the inside edges of the bead seats called *safety beads* (**Figure 61.37**). These help to keep the tire bead on the bead seat in

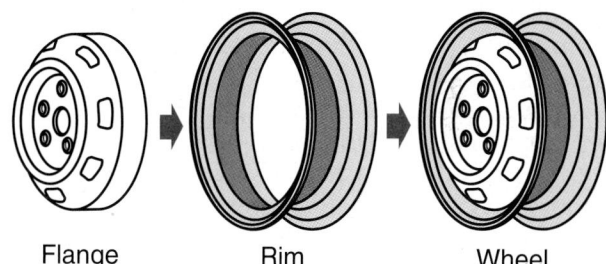

Flange   Rim   Wheel

Figure 61.36 Parts of a wheel rim.

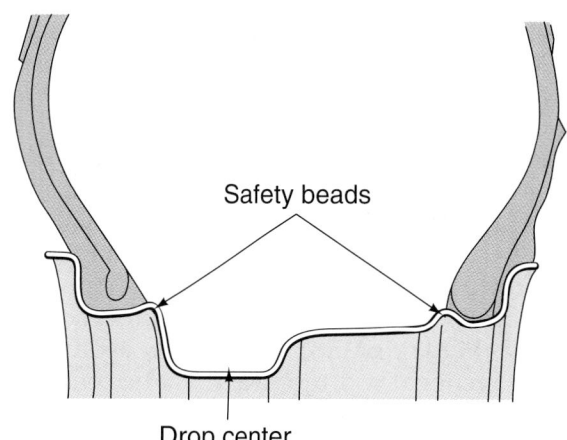

Safety beads

Drop center

Figure 61.37 Safety beads keep the tire on the wheel in case of a flat tire.

case of an "air out," until the car can be safely stopped. Rims for tubeless tires must have safety beads in order to be DOT approved.

Wheels are centered on the hub by one of two methods:

- **Hub-centric**—This means that the center of the wheel has a machined counterbore that pilots on a machined area of the hub. This is the most precise method of centering the wheel to the axle. Most OE wheels are hub-centric.
- **Stud-centric**, or *lug-centric*, wheels locate on the wheel studs. Most aftermarket or custom wheels are stud-centric. Custom wheels that are made for a specific model of vehicle can be hub-centric, but they are more expensive because they are not universal to many makes of vehicle. When stud-centric wheels, which have a bigger center hole, are installed on hub-centric vehicles, the result can be an improperly centered wheel, which causes vibration.

Service information on hub and stud centering is found in Chapter 62, under the heading of wheel balancing.

CUSTOM WHEELS

Customers sometimes purchase custom wheels when they want to have a cosmetic change in the appearance of the vehicle or when different sized tires are installed. Aftermarket wheel quality is rated by the Specialty Equipment Manufacturers Association (**SEMA**) by their affiliate, The SEMA Foundation (*SFI*). Wheels carrying their certification are manufactured to SFI standards.

Aluminum wheels can be cast, forged, or rolled. They can be either a single-piece casting, or they can have lighter rolled rim halves bolted to a cast center section. Race cars use alloy wheels. An **alloy** occurs when two or more metals are combined to make one metal. In the fifties, sixties, and seventies, aluminum wheels were commonly called "mags," due to the magnesium part of the alloy.

The alloy used for mags is a combination of magnesium and silicon. These wheels are strong and light but are not practical for passenger cars, because they are expensive and do not resist corrosion.

Custom wheels for street use can be single-piece castings of light alloy aluminum with a weather resistant coating. The more costly custom wheels fit a single application only. Less expensive wheels are made of weaker materials. They are stud-centric (center off of the wheel lugs) and are made to fit a variety of applications.

Rim width is the measurement from bead seat to bead seat (**Figure 61.38**). It is usually about 80% of the cross-sectional width of the tire. The centerline is at one-half of the rim width.

Wheel offset is the difference between the rim centerline and the mounting surface of the wheel. Sometimes a certain amount of offset is designed into

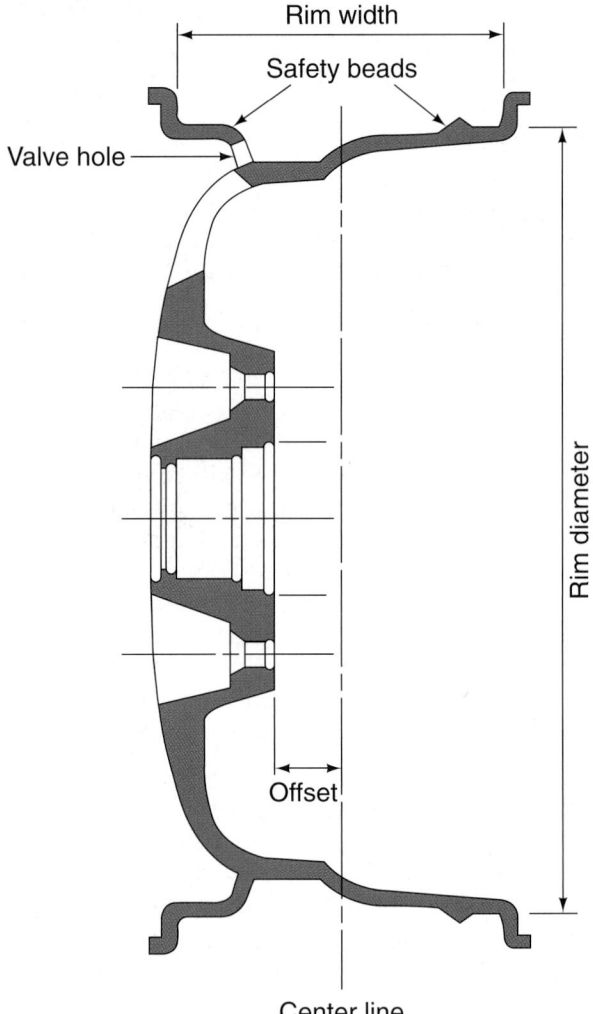

Figure 61.38 Wheel terminology. *(Courtesy of Pirelli Tire of North America LLC)*

a wheel to allow it to clear the fender well. The offset of the wheel is also important to proper brake cooling because it affects the distance between the brake caliper and the wheel. Improper offset will also cause wheel bearing wear.

Wheel clearance is *not* included in application handbooks. This is something that must be carefully checked on the vehicle. The use of wider tires and wheels, or wheels that are offset a different amount than stock, can result in interference between the tire and fender well or suspension components.

Wheel offset is described in various ways. The most common offset classification is as follows (**Figure 61.39**):

- **Negative wheel offset** increases the track width of the tires.
- **Positive offset** (the opposite of negative offset) is found often on front-wheel-drive cars.

When a wheel is replaced, the new wheel should be of the same offset to maintain the proper scrub radius.

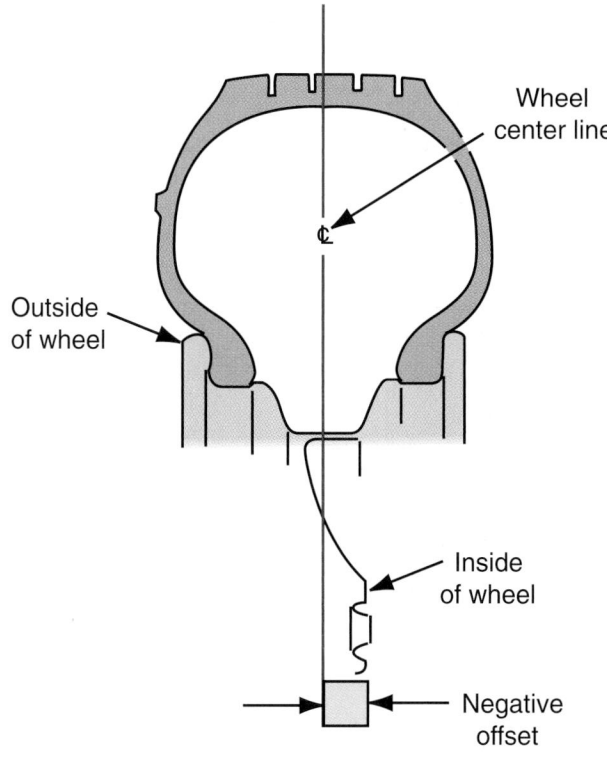

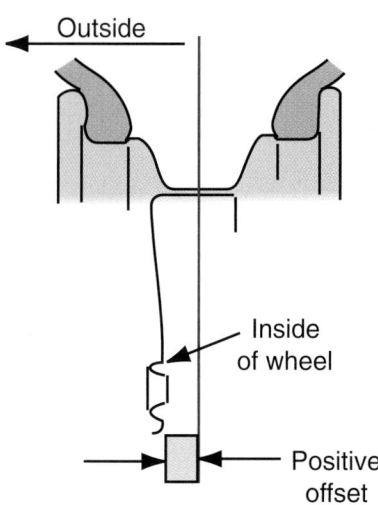

Figure 61.39 Positive and negative wheel offset.

LUG STUDS

Wheels have different numbers of lug studs (between three and eight), depending on the load on the vehicle. Most passenger cars use four or five lug studs, while light trucks usually use six or eight. Heavier trucks and some RVs sometimes use fewer lug bolts or studs, but they are larger in diameter and are tightened to a much higher torque.

There are also various bolt patterns used. A *bolt pattern* that is listed in a catalog as being 6–5½, is a six-bolt pattern spaced around a 5½-inch circle. Bolt patterns with an even number of bolt holes are easy to measure. Simply measure the distance from the center of one bolt to the center of the one across from it. Five bolt patterns are more difficult to measure. Templates are available to help determine the size of a bolt pattern.

LUG NUTS

Lug nuts can use either metric or standard screw threads. Always check stud threads when installing a new nut.

Lug studs have a serrated shank so that they will remain tight in the hole in the hub during tightening. Lug nuts for cast wheels (mags) are long and thick and must fit into a large, deep hole. These lug nuts must be used with a washer to avoid damaging the wheel. Lug nuts for cast wheels (mags) are long and thick and must fit into a large, deep hole. These lug nuts must be used with a washer to avoid damaging the wheel. Most lug nuts are made with the washer permanently installed on them.

TIRE VALVE STEMS

Passenger car tire valve stems are usually rubber and are designed to be used at pressures of less than 4.2 bar (62 psi). A spring-loaded valve core is screwed into the valve stem (**Figure 61.40**). Valve stems have a screw-on dust cap, some of which have a gasket that prevents air loss past the valve core.

NOTE: *According to the Pirelli Armstrong Tire Corporation, an imperceptible leak of one bubble per minute can result in the loss of 0.1 bar (1.5 psi) per month.*

CAUTION Scrub radius (see Chapter 67) has an effect on the handling and steering effort of the vehicle. Changing the height or centerline of a wheel from that designed by the manufacturer can result in a change in scrub radius from positive to negative, or vice versa. The result can seriously affect vehicle handling. This is just one reason why replacing stock tires and wheels with ones of a different size or offset is something that many shops avoid.

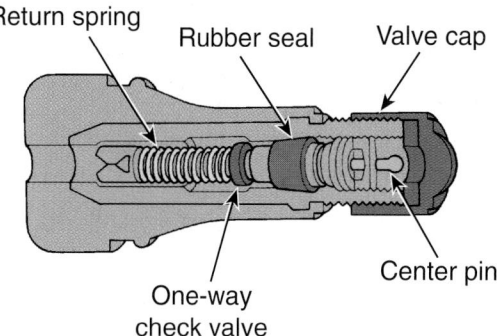

Figure 61.40 Parts of a rubber valve stem.

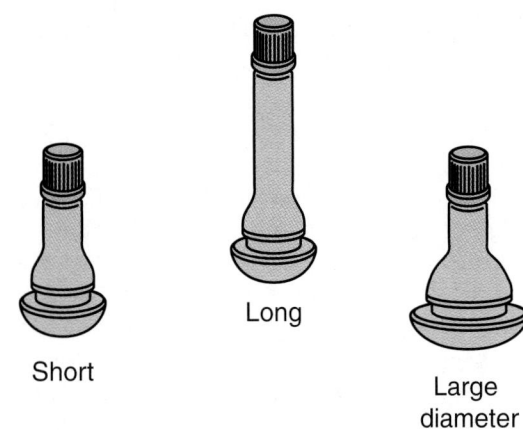

Short

Long

Large diameter

Figure 61.41 Different sizes of rubber valve stems.

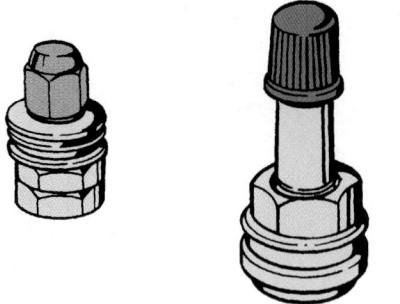

Figure 61.42 Metal valve stems. *(Courtesy of Plews/ Edelmann Division. A Gates Group Company)*

There are two common valve stem lengths available. A short stem is used when there is a hubcap, and a long stem accommodates the use of full wheel covers (**Figure 61.41**). Rubber valve stems come in common diameters of 11.3 mm or 15.7 mm to match the two common wheel rim hole sizes. When a longer stem is required, tire valve extensions can be screwed onto an existing valve stem. Extensions have a spring-loaded check valve so that air pressure can be checked and adjusted as needed.

Light trucks and custom wheels are usually equipped with threaded metal stems that have a rubber bushing and are fastened to the wheel with a nut (**Figure 61.42**). The common wheel hole size for metal stems is 11.5 mm.

◼◼ REVIEW QUESTIONS

1. What is it called when water forms a wedge under the tire, causing it to lose traction?
2. Which tire allows better fuel economy, radial or bias?
3. What is the area of tread contact called?
4. What is another name for a tire's height?
5. What is the miles-per-hour limit for a speed rating of H?
6. What are the two letters that are found on snow or all-season tires?
7. When changing a tire sized 195/75 R15 to a 5% lower profile, the correct tire size would be _____.
8. What is the lower area between the bead seats of a wheel rim called?
9. When a wheel is offset to increase the track width of the front tires, this is called _____ offset.
10. A small leak of one bubble per minute can result in the loss of _____ psi per month.

◼◼ ASE-STYLE REVIEW QUESTIONS

1. Technician A says that a tire with a tread wear rating of 200 would be expected to last twice as long under the same conditions as one with a 100 rating. Technician B says that when the diameter of a tire is changed, front-end geometry is altered. Who is right?
   a. Technician A   **c.** Both A and B
   b. Technician B   **d.** Neither A nor B

2. Which of the following is/are true about custom wheels?
   a. Custom wheels are usually hub-centric.
   b. An alloy wheel is made of two or more metals combined to make one.
   c. Both A and B.
   d. Neither A nor B.

3. Technician A says that the DOT code identifies the manufacturer and country where the tire was made. Technician B says that the DOT code provides the week and year the tire was made. Who is right?

   a. Technician A   **c.** Both A and B

   b. Technician B   **d.** Neither A nor B

4. All of the following are true about winter tires *except*:

   a. A winter tire must have a *snowflake on the mountain* symbol.

   b. Winter tires are softer and wear more rapidly in the summer.

   c. Winter tires have a low speed rating.

   d. M&S tires guarantee winter driving performance or safety.

5. Which of the following is/are true about direct tire pressure monitoring?

   a. It uses individual wheel sensors.

   b. Integral sensors are a part of the valve stem.

   c. Pressure that is too low or too high will illuminate a warning light.

   d. All of the above.

6. Two technicians are discussing *plus sizing*. Technician A says that when a lower profile tire is installed, a wider tire and a larger diameter wheel are used. Technician B says that the same overall diameter of the tire and wheel assembly is achieved. Who is right?

   a. Technician A   **c.** Both A and B

   b. Technician B   **d.** Neither A nor B

7. Tires of a different size than OE have been installed on a vehicle. This can cause:

   a. The ABS light to come on

   b. The speedometer to be off

   c. Both A and B

   d. Neither A nor B

8. Technician A says that a 70 is a lower profile tire than a 75. Technician B says that an H rating is for higher speeds than a Z rating. Who is right?

   a. Technician A   **c.** Both A and B

   b. Technician B   **d.** Neither A nor B

9. Which of the following is/are true about unidirectional tires?

   a. They allow faster vehicle acceleration.

   b. They allow faster stopping.

   c. Both A and B.

   d. Neither A nor B.

10. Technician A says that the P in a P-metric radial's size means *performance*. Technician B says that European tires do not display a P at the front of their tire rating. Who is right?

   a. Technician A   **c.** Both A and B

   b. Technician B   **d.** Neither A nor B

Tire and Wheel Service

■ OBJECTIVES

Upon completion of this chapter, you should be able to:

✔ Adjust tire pressures correctly for all cars.

✔ Rotate tires.

✔ Repair tire punctures in the correct manner.

✔ Determine causes of tire-related vibration.

✔ Understand and perform tire balancing.

■ KEY TERMS

tire rotation

swaged

radial runout

lateral runout

tire plug

patch

■ INTRODUCTION

Tire service is a large area of automobile repair. The average owner can expect to replace at least one set of tires on his or her car. Tire life depends on tire quality, air pressure, vehicle weight, driving conditions, suspension condition, and wheel alignment.

■ TIRE INFLATION

Tire wear can be caused by incorrect inflation pressure (**Figure 62.1**). Tires typically lose about 1–2 psi of pressure a month through permeation of the sidewall, and maybe more during hot weather. This is a normal occurrence. Think about a balloon you have found a few days after a birthday party. Like tires, balloons hold air—they leak slowly.

Maintaining correct air pressure is the most important factor in the safety, performance, and life expectancy of a tire. Underinflation is the most common cause of radial tire failures. Low tire pressure in a radial tire changes the normal deflection of its sidewall. This raises the amount of heat generated within the tire's carcass.

The following are results of low tire pressure:
■ The temperature of the tire rises.
■ The load-carrying capacity of the tire is lowered.
■ Tire tread life is reduced.
■ Fuel consumption increases.
■ The outside edges of the tire wear excessively.

A car will usually pull to the side that has a low tire, especially when it is a front tire. When an older car with bias tires had low air pressure, the driver would usually notice it in the handling of the vehicle because bias type tires experience a strong pull to the side of a tire with lower inflation. However, a radial tire can often be run at a tire pressure that is below specifications without exhibiting handling symptoms.

The following are results of high tire pressure:
■ The center of the tread can wear excessively.
■ A rough ride can result. The tire is actually the first part of the suspension and spring system.

■ CHECKING AIR PRESSURE

Vehicle owners should be encouraged to check tire air pressure at least once a month and prior to a long trip. The following items refer to the effect of temperature on the tire's pressure:
■ As the air in the tire expands due to heat, pressure normally increases. Air pressure should be checked when the tires are cold.
■ It takes less than 3 minutes, or 1 mile of driving at moderate speeds, to make tires too hot to check accurately. According to Michelin Tires, this can increase air pressure by 4 psi or more. Therefore, when adjusting pressure in a hot tire, add 4 psi to the maximum gauge reading desired. For instance,

Center tread wear

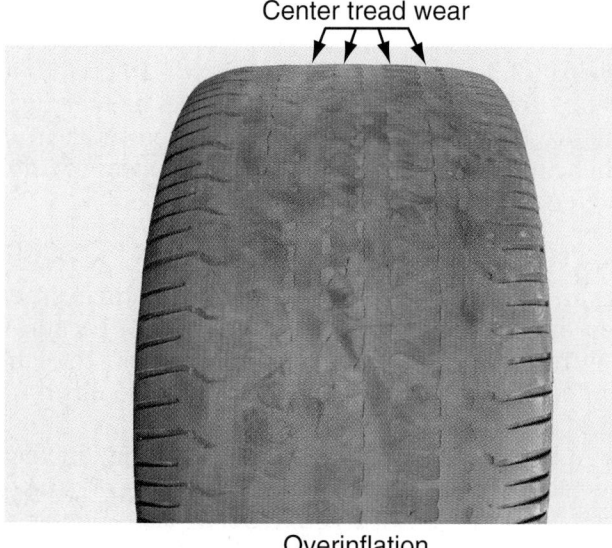

Overinflation

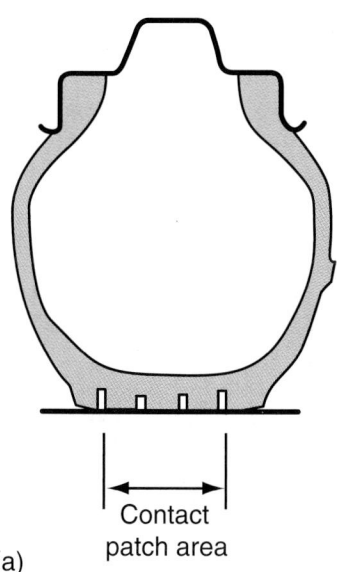

Contact
patch area

(a)

Figure 62.1 (a) Overinflation wears the center of the tread. *(Photograph courtesy of Tim Gilles)*

Outside tread wear

Underinflation

Contact
patch area

(b)

Figure 62.1 (b) Underinflation results in excessive outer tread wear. *(Photograph courtesy of Tim Gilles)*

if the recommended cold pressure is 24 psi and the gauge reads 26 psi, fill the tire to 28 psi. Be sure to recheck it when cold the next day.

NOTE: *According to the Rubber Manufacturers Association (RMA), customers should be advised to check their tire pressures prior to driving while the tires are still cold. In the event they find a low tire, after a short drive to a service facility that has an air compressor, air can be added in the amount that the tire was underinflated.*

- Air should not be let out of a tire when it is hot.
- Each change in outside temperature of 10°F will result in about 1 psi change in tire pressure.
- A hot tire that has lower pressure than the recommended cold pressure is seriously underinflated.
- Cold tire pressure should never be higher than the maximum pressure molded into the tire sidewall.

Read pressure
here

Figure 62.2 A tire pressure gauge showing 34 psi. *(Courtesy of Tim Gilles)*

Tire inflation pressures should always be the same for both tires on one axle so ride and handling are not affected.

An accurate air pressure gauge must be used to check tire pressures (**Figure 62.2**). A normally inflated

30 psi

20 psi

Figure 62.3 The appearance of a radial tire sidewall changes very little with inflation pressure. *(Courtesy of Tim Gilles)*

radial tire has a bulging sidewall, and tire pressure must be dangerously low before there is a visible difference. **Figure 62.3** compares two tires with a 10-pound difference in pressure. Older, bias-type tires had relatively stiff tire sidewalls, compared to modern radial tires. When they were low on pressure, there was a visible difference in the sidewall appearance.

■ CHECKING AND ADJUSTING TIRE PRESSURE

All valve stems should have screw caps on them. These keep out dirt and moisture and provide a backup in case the valve core leaks. Before adding air to a tire, blow air through the air chuck to clear it so dirt is not forced into the valve core.

A high-quality tire gauge should be used. Inexpensive gauges are often inaccurate. One type of gauge is a part of the air chuck (**Figure 62.4**). These air gauges are often abused and become inaccurate when left installed on an air hose.

Tire pressure gauges are usually equipped with a pin that can be used to release air (**Figure 62.5**). Some tire air chucks and gauges have two sides that can be used to inflate and check tire pressures (**Figure 62.6**). The top side is handy to use when checking and

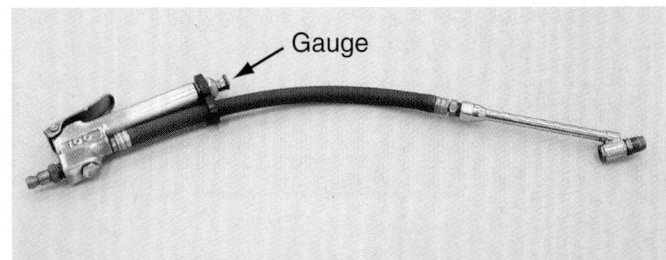

Gauge

Figure 62.4 This type of tire pressure gauge is rarely accurate. *(Courtesy of Tim Gilles)*

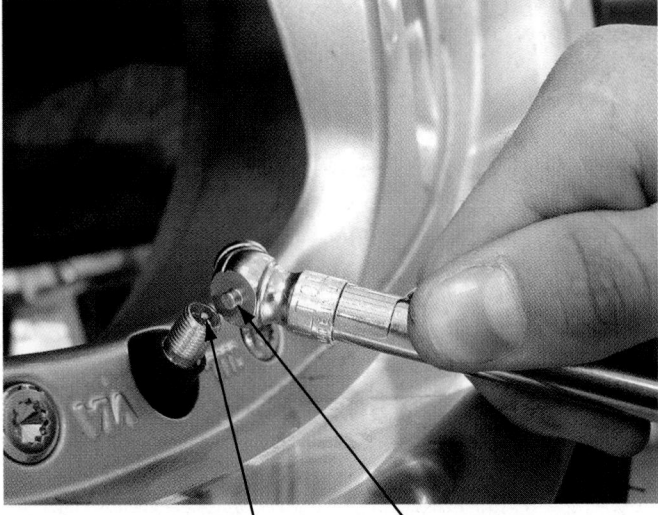

Valve core   To release air

Figure 62.5 A pin on a typical tire pressure gauge used to deflate the tire after too much pressure is added. *(Courtesy of Tim Gilles)*

This side is used for inner duals

Figure 62.6 Either side of this tool can be used for easier access to the valve stem. *(Courtesy of Tim Gilles)*

inflating inner tires on trucks and motorhomes with dual wheels.

Let us examine the effects of not checking tire pressures regularly. If you fill your tires during the summer when the outside air temperature is 90°F and do not check them again for 6 months, the pressure will be considerably lower. A typical tire loses 1 psi in pressure each month. When combined with a 60°F temperature change in seasonal climates, the pressure change in the tire can be substantial. **Figure 62.7** shows typical pressure changes under three conditions.

■ TIRE WEAR

According to Goodyear, a 4 psi decrease in pressure below the recommended amount can result in a 10% loss of tread life. Additional losses of air pressure can result in even more wear and the possibility of serious damage to the tire. Underinflation can also cause the edges of the tire to wear (see Figure 14.2).

The fastest tire wear occurs during hard cornering, braking, and acceleration. Rough pavement also contributes to accelerated tire wear. Slow-speed sharp cornering wears the front tires, while high-speed cornering will remove tread from the tires on the side of the vehicle to which the weight is transferred.

When a tire wears to within ¹⁄₁₆" (²⁄₃₂") of the bottom of its tread, wear bars begin to become more obvious at regularly spaced areas around the tread circumference (see Figure 14.3). The wear bars are raised areas cast into the bottom of the tire tread area to indicate when the tread has become worn beyond its safe limit. The RMA recommends that tires with ¹⁄₁₆" (²⁄₃₂") of remaining tread depth be replaced. These tires are unsafe in wet weather and more apt to be damaged by road hazards. Federal regulations require vehicles in excess

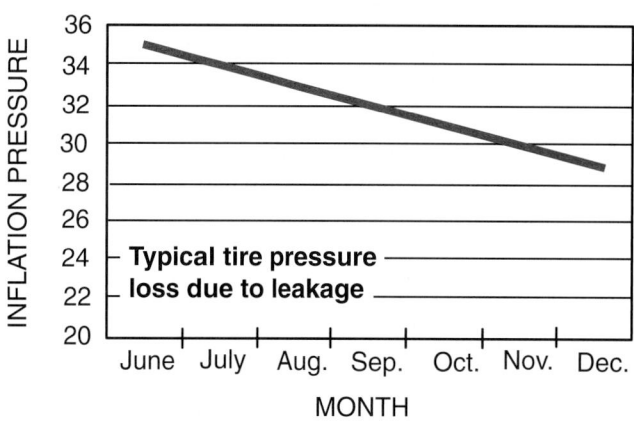

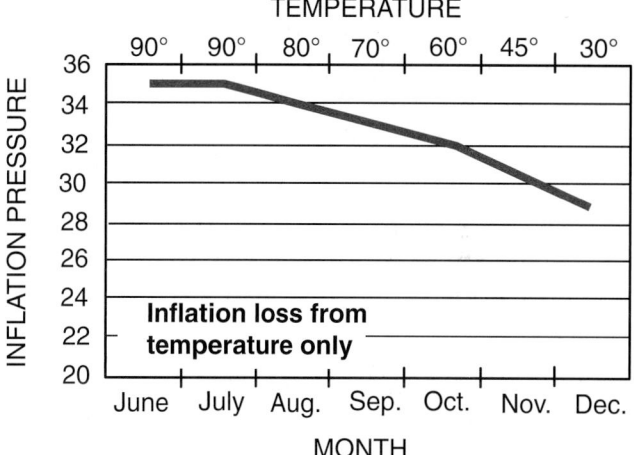

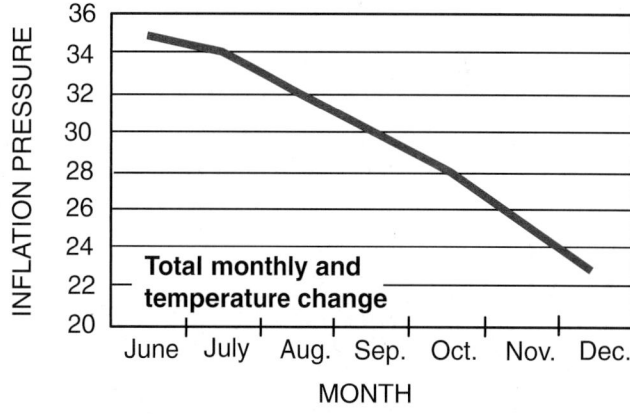

Figure 62.7 Changes in tire inflation due to time and variations in temperature.

of 10,000 pounds GVW to have ⁴⁄₃₂" minimum tread depth on front tires.

When a tire exhibits *scalloped* or *cupped* wear, the cause is usually that the tire has been hopping up and down on the road. This movement, known as *wheel tramp* (see Figure 62.59), results from bad shock absorbers, worn parts (such as ball joints or control arm bushings), out-of-balance tires, or too much runout of a tire.

Front tires on some front-wheel-drive cars develop abnormal tire wear. Some manufacturers specify tire rotation at very low mileage intervals to compensate for this. Tire rotation is covered in more detail later in this chapter.

Inspect the tire for physical damage. When a car is driven with a tire that is flat or underinflated, the tire can become damaged. Look for evidence of tread or sidewall separation. This might show up as an out-of-round tread or visible deformities on the outside of the tire. However, sometimes damage from underinflation might not be visible from the outside of the tire and can only be determined when the tire is removed from the wheel.

■ SIDEWALL CHECKS

Sidewall cracks are usually caused by age and years of exposure to the sun and ozone. Cracks on the outside of sidewalls often develop on recreational vehicles (RV) because RV driving is seasonal. Motorhomes are used mostly on the highway (very low tire wear). Tires can be many years old before the tread wears out. When an RV is parked in the same spot for long periods of time and the tires have not been regularly rotated, only the tires on one side of the vehicle that have been exposed to more sunlight might be cracked.

According to the RMA, a slight sidewall indentation (also known as sidewall undulation) is a common characteristic of radial tire construction (**Figure 62.8**). These indentations are due to the construction of the tire and are purely a visual characteristic and will not affect the performance of the tire. The indentations are due to overlaps in the cord material of about ⅜". Indentations are normal, but if the tire bulges, there is no supporting cord and the tire must be replaced. If there is any question concerning the sidewall appearance, the tire should be removed from service and inspected by a knowledgeable tire dealer, or the tire manufacturer's representative should be contacted. Cuts or cracks in the sidewall that allow cords to be exposed are cause for replacement of the tire.

RVs and light trucks with dual rear tires often experience uneven wear on the rear tires. The diameters of dual wheel tires must be within ¼" of each other.

■ TIRE ROTATION

Front wheels on all cars experience the most wear because they are used for steering and also because weight transfers forward during a stop. Moving tires to different locations on the vehicle is called **tire rotation**. Regular tire rotation allows the tires to wear more evenly, allowing them to be replaced in complete sets. Most manufacturers specify regular rotation intervals. When there is no recommendation, rotation is recommended at 6,000 to 8,000 miles, or before that if tire wear is evident.

In the past, moving radial tires to the other side of the car was not recommended, but this is no longer true as long as there is not a specified rotation pattern to the tire. If one front tire shows uneven wear, it can be switched with the other front tire.

Front-wheel-drive cars experience far more wear on the front tires. Rear tires on these cars last much longer. A typical rotation pattern for front-wheel-drive cars with radial tires involves moving the front tires to the rear on the same side of the car. The rear tires are then moved to the front, but to opposite sides of the vehicle. This is said to even out and minimize "drive cornering scrub" wear that occurs as the footprint of a radial tire squirms against the pavement (**Figure 62.9**). This results in a sawtooth heel-and-toe wear pattern on the tread (**Figure 62.10**). Rotating the tires to another side of the vehicle smoothes this wear. If it is allowed to develop, noise and vibration can result, along with accelerated tread wear.

Figure 62.11 shows typical rotation patterns for front- and rear-wheel-drive vehicles. Four-wheel-drive

Figure 62.8 Slight indentations in the sidewall of a radial tire are normal. *(Courtesy of Tim Gilles)*

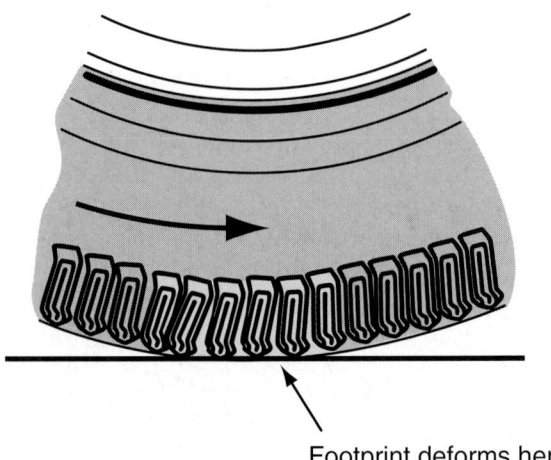

Footprint deforms here

Figure 62.9 Scrub wear that occurs as the footprint of a radial tire squirms against the pavement during cornering.

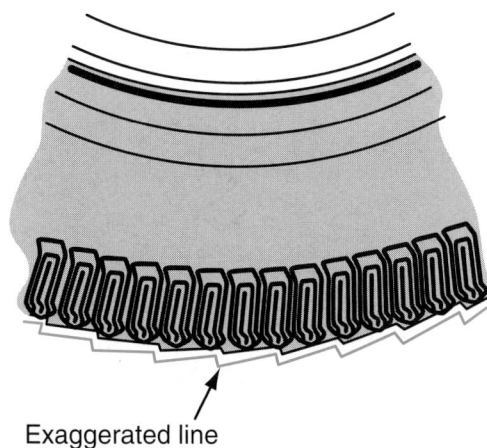

Exaggerated line

Figure 62.10 Sawtooth heel-and-toe wear pattern on radial tire tread.

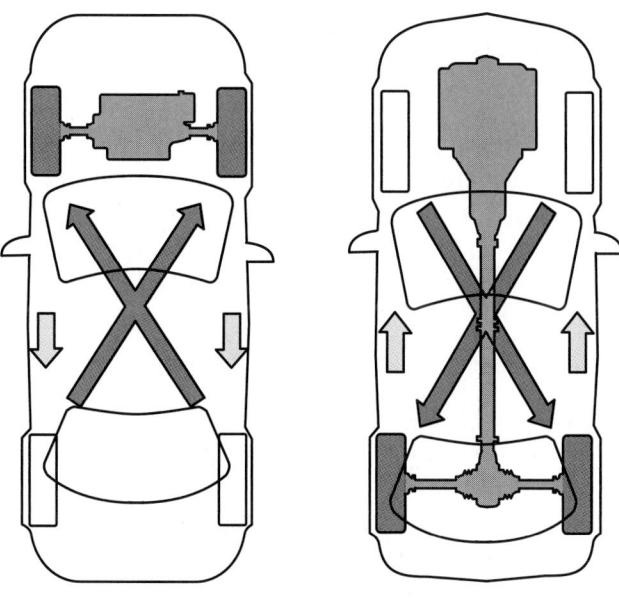

Front-wheel drive   **Rear-wheel drive**

Figure 62.11 Typical tire rotation patterns. With front-wheel drive, "X-to-the-front"; with rear-wheel drive, "X-to-the-rear."

vehicles use the same pattern as rear-wheel drive. Notice how the front-wheel-drive rotation pattern has rear tires crisscrossed to the front? The front tires are then moved directly to the rear on the same side. The opposite is true with rear-wheel-drive and four-wheel-drive rotation patterns.

> **SHOP TIP** Cross the non-driving wheels to the drive wheels. With front-wheel drive, "X-to-the-front." With rear wheel drive, "X-to-the-rear."

Follow the manufacturer's recommended tire rotation pattern.

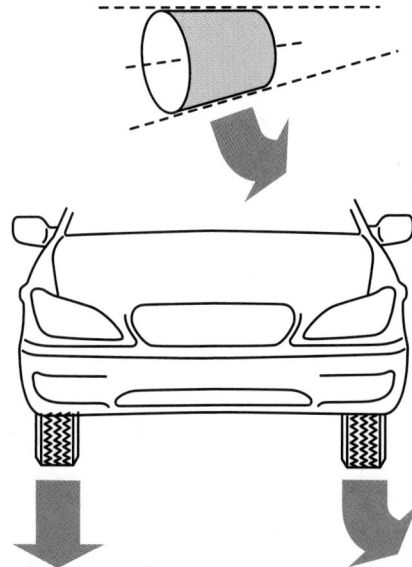

Figure 62.12 Tire conicity results in a pull.

- Some rotation patterns include rotation of the spare tire. In this case, insert the spare into the rotation pattern at the right rear and reposition the tire that would normally have been installed on the right rear as the new spare. Other cars are equipped with only an emergency compact spare, which is not rotated onto the vehicle.
- Some vehicles have tires of a smaller size on the front. These should not be rotated to the rear.
- Some tires are designed to be mounted and run only in a specified direction of rotation. These tires are rotated in the front-to-rear pattern and are cross-rotated only if they are removed from the rim and remounted so that they will still rotate in the designated direction.

Paired tires should be of the same size designation, construction, and tread design. According to the RMA, if radial and non-radial tires are used on the same vehicle, put the radials on the rear.

Sometimes after a tire rotation the car can exhibit a pull to one side or the other. This can be because of an inherent pull within a radial tire caused by tire conicity (when the tread is tapered like a cone), or off-center belts (**Figure 62.12**). If the offending tire was previously installed on the rear of the car (non-steering wheels), it can cause a problem when installed on the front of the car. **Figure 62.13** shows some recommendations for isolating the problem. Off-center belts can also result in outside shoulder wear on the tire.

■ REMOVING AND TIGHTENING LUG NUTS

Most lug nuts have right-hand threads and are loosened by turning counterclockwise. A few vehicles have left-hand threads on the lug nuts on one side of the car. These are labeled with an "L" on the end

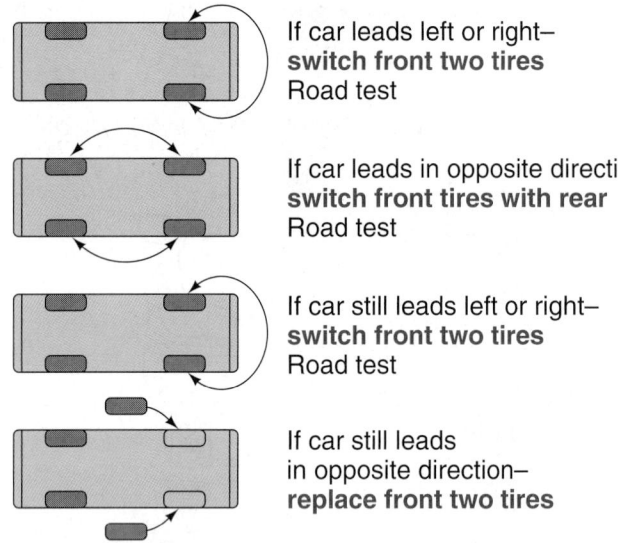

If car leads left or right–
switch front two tires
Road test

If car leads in opposite direction–
switch front tires with rear
Road test

If car still leads left or right–
switch front two tires
Road test

If car still leads
in opposite direction–
replace front two tires

Figure 62.13 Diagnosing tire pull caused by conicity. *(Courtesy of DaimlerChrysler Corporation)*

of the lug stud. Loosening and removing lug nuts is easiest when done with an impact wrench. When the wheel is in the air, it will not have to be held from turning.

CAUTION Do not loosen the lug nuts when the wheel is on the ground. The car can fall down. Have someone hold the brakes applied while you loosen them with the wheel off the ground.

SHOP TIP Remove lug nuts using the tightening pattern to avoid warping a hot rotor.

Tighten lug nuts evenly in a crisscross pattern. Use a torque wrench to avoid warping a disc brake rotor (**Figure 62.14a**). Specifications are available for the different makes of cars. Just as with other fasteners, lug bolts are stretched when they are properly installed so that they maintain clamping force. Lug nuts that are loose allow the wheel to exert all of the weight of the vehicle on the lug bolts, rather than on the wheel. Be sure to install a socket extension between the socket

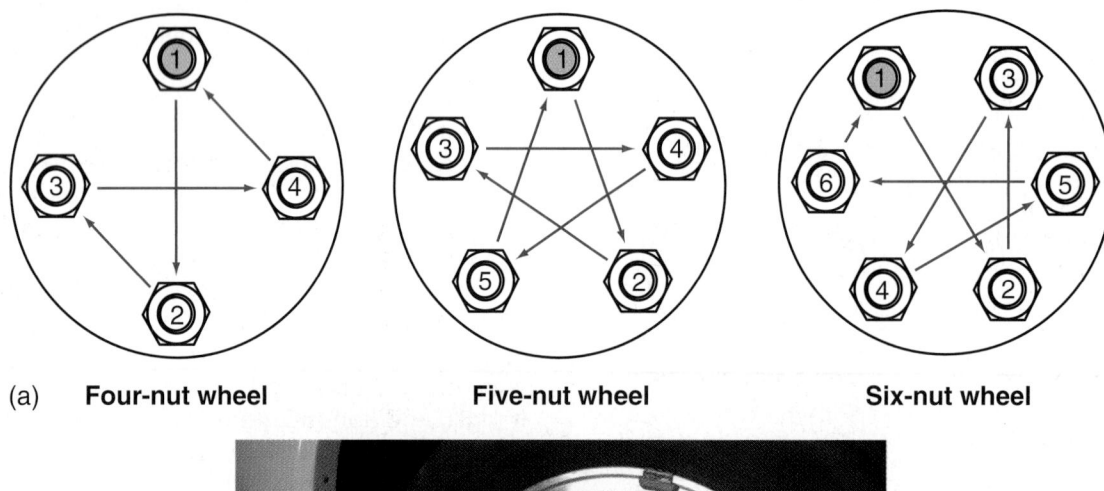

(a)   **Four-nut wheel**   **Five-nut wheel**   **Six-nut wheel**

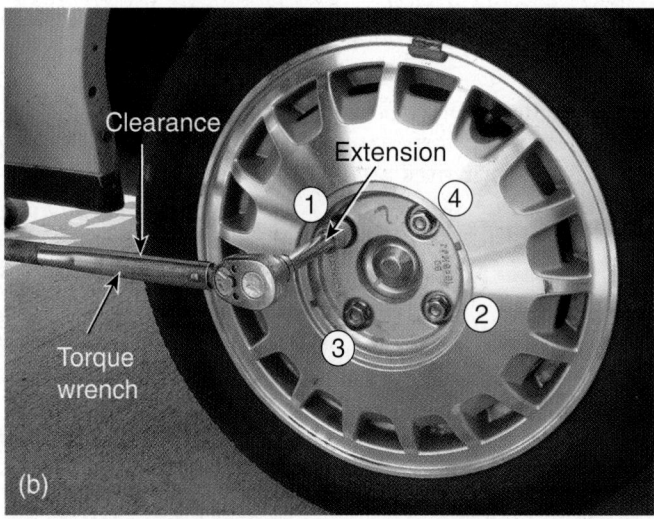

Clearance

Extension

Torque wrench

(b)

Figure 62.14 (a) The torque sequence for wheel lugs. (b) Use a socket extension to provide clearance to the tire sidewall. *(b, Courtesy of Tim Gilles)*

and the torque wrench as needed so your hand and the handle of the torque wrench will be able to clear the tire sidewall (**Figure 62.14b**). Use a ½" drive extension only. Inaccurate torque readings can result from the use of excessively long or ⅜" drive extensions.

Lug nuts for steel wheels and some aluminum wheels are tapered on the side that faces the wheel (**Figure 62.15a**). Some wheels have special lug nuts that cover the ends of the wheel lugs, so these can only be installed in one direction.

> **CASE HISTORY**
>
> *A student complained of a clunking sound and poor vehicle handling. He asked the teacher to inspect the vehicle. The teacher was amazed to find all of the lug nuts on inside out. All of the holes in the rims were damaged so badly that new wheels were required.*

Anti-theft lug nuts are popular on custom wheels. They require a special key (**Figure 62.15b**). Sometimes, the customer loses the key or forgets to bring it to the shop for a tire or brake repair. Most shops that do tire repair have a special tool kit for removing these without the key. It has several adapters that are wedged against the outside of the lug nut. The inside of each adapter is tapered (**Figure 62.16a**). A hammer is used to force the adapter onto the outside of the wheel lock (**Figure 62.16b**). To loosen the wheel lock, the removal tool is driven rapidly by an impact wrench lock while holding it firmly against the wheel lock (**Figure 62.16c**).

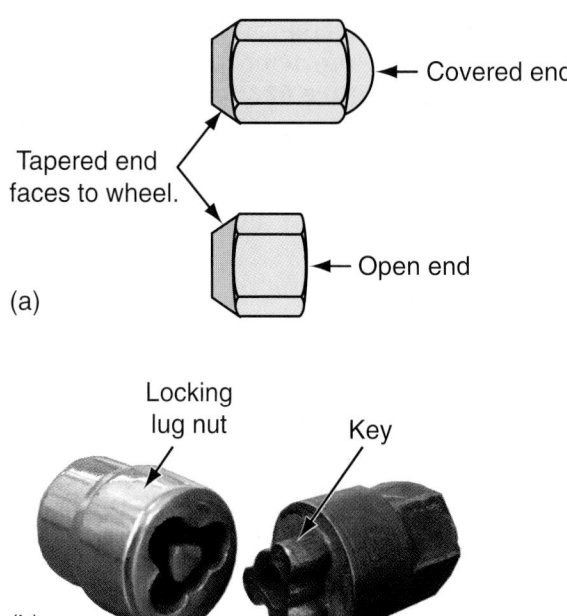

Figure 62.15 (a) Lug nuts for steel wheels are tapered on the side that faces the wheel. (b) A wheel lock nut and key. *(b, Courtesy of Tim Gilles)*

Figure 62.16 (a) The inside of the adapter is tapered. (b) A hammer is used to force the adapter over the lug nut. (c) The lug nut is removed using an impact wrench. *(Courtesy of Tim Gilles)*

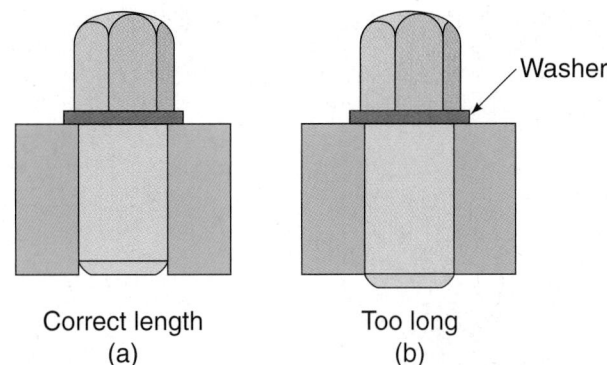

Figure 62.17 Lug nuts in cast aluminum wheels.
(a) is correct, and (b) would allow the wheel to be loose.

A special type of lug nut is used with aluminum wheels. It is usually equipped with a washer that cannot be removed. The shank on the nut must not be too long or it can bottom out, leaving the wheel loose (**Figure 62.17**). Also, the lug nut must have a small amount of clearance so it is free to turn inside of the hole in the wheel.

> **SHOP TIP** Put an anti-seize lubricant on the outside of aluminum wheel lug nuts before installation. This will help avoid the electrolysis that can occur between the steel lug nuts and the aluminum wheel.

■ REPAIRING WHEEL STUDS

Occasionally, a lug bolt will be stripped or broken. If only a couple of threads are damaged, they can be cleaned up with a thread chaser. Broken lug bolts must be replaced. On drum brake cars, when the drum is removed with the hub, the studs are swaged (see Figure 58.56), which means that they are deformed to keep them tight. The swaged area can be cut off with a special tool to make them easier to remove. Drum brakes are no longer found on the fronts of newer vehicles.

If the rotor or drum separates easily from the hub, the lugs can usually be driven from the hub with a brass hammer or punch. If much effort is needed, a tie-rod press can be used to force the lug bolt out of its hole. The new lug stud can be installed with an inverted lug nut and washers as shown in **Figure 62.18**.

■ REMOVING AND MOUNTING TIRES ON RIMS

There are some important points to be aware of before attempting to remove and install tires on wheel rims. *Tires can explode and fingers can be cut off if proper caution is not observed.*
■ Be sure to use the proper size and construction of tire to match the wheel rating.
■ Be sure that the rim diameter matches the diameter molded on the tire sidewall. Be especially careful

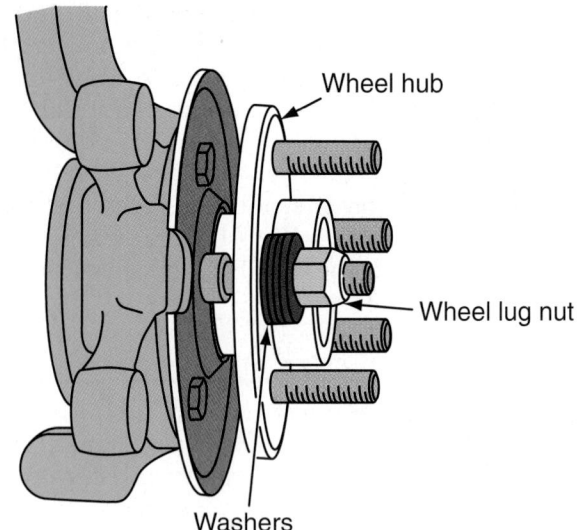

Figure 62.18 Installing a new lug stud.

that a tire or wheel is not a metric size that was manufactured for the overseas market. Although you might be able to mount a metric tire on a standard wheel, it will not fit properly and can come off.
Tire problems such as leaks and vibrations can sometimes be traced to improperly mounted tires.

The following information applies to either of the two most common tire changing machines.

Deflate the Tire

Before removing the tire from the wheel, remove the valve core to deflate the tire completely. **Figure 62.19** shows a tool used to remove the valve core. If the tire is to be patched and reinstalled without rebalancing, be sure to mark the locations of the valve stem and any wheel weights with a marking crayon. If another tire is to be installed on the rim, remove any balance weights from the rim first (**Figure 62.20**).

Tire Changers

There are three types of tire changers in common use. Two of them, the rim clamp and the center post, have

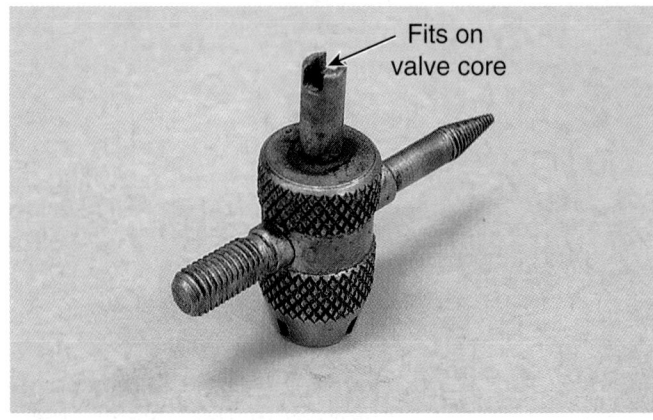

Figure 62.19 A tool for removing and replacing valve cores. *(Courtesy of Tim Gilles)*

Figure 62.20 Remove wheel weights from the rim.
(Courtesy of Tim Gilles)

been used for many years (**Figure 62.21**). The rim clamp changer was originally developed in Europe and has become the predominant tire changer in North America in recent years. Tire changing requirements for high-performance and specialty aftermarket wheels and tires have presented new challenges. A new tire changer design has emerged in response to these needs. **Figure 62.22** shows this new tire changer.

The procedures for mounting and dismounting conventional tires are similar and are covered first. High-performance tire service, which is very similar, is covered later.

Unseating the Beads

Before the tire can be removed from the wheel, the beads must be unseated from the bead seats (**Figure 62.23**). This is called "breaking the beads." Lubricate the tire bead with rubber lube. Rubber lube reduces friction between the tire and wheel, preventing damage to the tire and making removal easier.

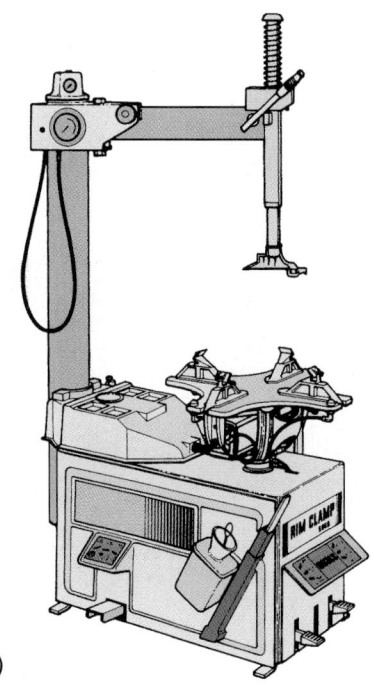

(b)

Figure 62.21 *(continued)* (b) A rim clamp tire changer.

Figure 62.22 A new tire changer design that can be used with low profile, speciality, and performance tires and wheels. *(Courtesy of Tim Gilles)*

(a)

Figure 62.21 (a) A center post tire changer. *(Courtesy of Tim Gilles) (continued)*

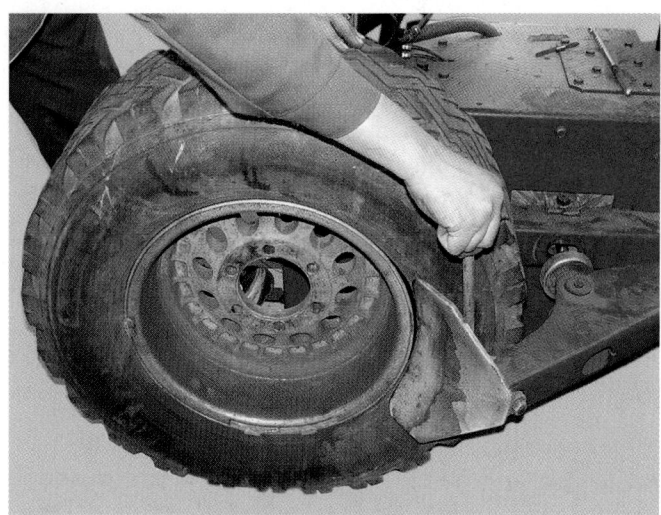

Figure 62.23 Breaking the bead using a tire changer. *(Courtesy of Tim Gilles)*

NOTE: *Be certain that a valve stem tire pressure monitor is rotated 90 degrees away from the bead breaker.*

 ■ Do not attempt to unseat the beads of an inflated tire.
■ Do not hit the tire or rim with a hammer. A hammer should not be necessary when removing a passenger car tire.
■ Never let go of the tire iron when in use. It can flip up and hit you.

Removing the Tire from the Wheel

The bead on the bottom side of the tire is forced into the drop center of the wheel so that the upper bead can be pulled over the edge of the rim (**Figure 62.24**). On either type of tire machine, removal of the tire from the wheel requires the use of a bar called a *tire iron* (**Figure 62.25**). First, the upper bead is pulled over the outside of the rim. Then, the lower one is. Do not try to remove both beads at the same time.

NOTE: *Be careful not to tear the bead. This will ruin the tire. If the bead does not come off the rim without binding:*
■ *Be sure there is enough rubber lube on the bead.*
■ *Double-check to see that the lower bead is totally into the drop center of the wheel.*

■ INSPECTING THE TIRE AND WHEEL

After the tire is removed from the rim, inspect its inside for cuts, carcass damage, penetrating objects, loose cords, dirt, and liquid. Inspect the condition of the bead by pulling out on it in several places around its circumference. If there is a sharp bend in the bead,

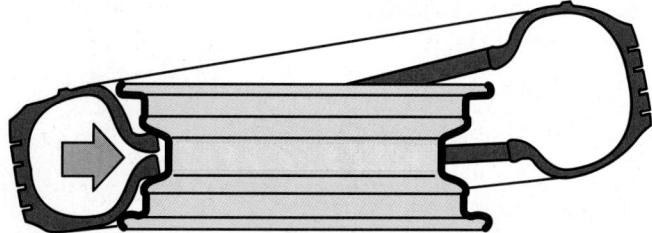

Figure 62.24 Both beads are in the drop center of the wheel while the bead is pulled over the edge of the tire.

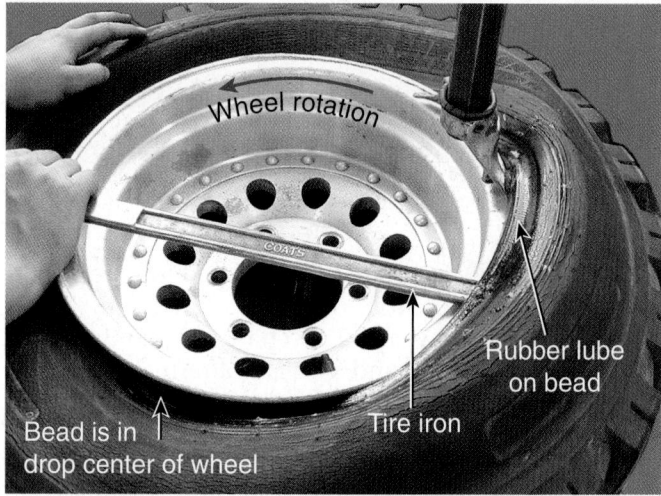

Figure 62.25 A tire iron is used to pull the top bead over the edge of the rim. Be sure that the other side of the tire bead is in the drop center of the wheel. *(Courtesy of Tim Gilles)*

do *not* mount the tire. The bead seat could be broken. It is not worth taking a chance with someone's safety.

Never mount a tire or wheel that is damaged. This could result in injury or death to the occupant(s) of the vehicle; in such a case, the shop may be held liable in court for negligence, and the technician can face criminal charges.

Inspect the condition of the wheel rim for sharp edges, dents, cracks, and other damage. Small dents in the rim flange can be straightened. When there is a larger dent in the wheel, it must be checked for **radial** and **lateral runout** before it can be used. Excessive runout can cause the car to shake at highway speeds. Checking runout is best done with a dial indicator, but spinning the wheel while holding something (a screwdriver or a pencil) near it for reference will give a good indication of whether a wheel is bent.

Rust can damage the bead seat on a wheel. If the rust is on the surface only, it can be removed with a wire brush. If the bead seat is not smooth, the tire will have a slow leak and the wheel will need to be replaced.

■ VALVE STEM SERVICE

Rubber valve stems are customarily replaced when new tires are installed. The valve stem can also be replaced without removing the tire from the rim. This is commonly done when an old valve stem becomes old and cracked or starts to leak. Leaks can be found by prying the stem to side and observing any air leakage.

■ To remove a valve stem, cut it off with a knife (**Figure 62.26a**), or force it through the hole in the rim using the valve stem installing tool.

■ To install the new stem, thread it into the installing tool, lubricate it with rubber lube, and pull it into the hole (**Figure 62.26b**). Be sure that it is pulled all of the way into place and is properly seated in the hole in the rim.

SHOP TIP A new valve stem comes with a new valve core. Remove the valve core before attempting to inflate the tire.

■ RUBBER LUBRICANT

Lubricate both tire beads with an appropriate rubber lube. (**Figure 62.27**) Good-quality rubber lube is slippery and fast drying. Lubricate the bead seats on the rim with rubber lube also. Using rubber lube provides the following advantages:

■ Rubber lube reduces friction between the tire beads and the edge of the rim during mounting.

■ Rubber lube helps to seal around the bead during initial inflation of the tire.

■ Friction between the bead seats and the tire bead will be reduced when inflating the tire. This is important so that the beads will be all of the way seated and the tire tread will not be distorted.

Safety glasses should always be worn when inflating the tire.

Radial tire sidewalls are flexible and their tire beads are designed with a close tolerance for a tight fit to the rim. During inflation, the beads are not easily forced into position. If the beads are not seated properly, the tire can have an out-of-round condition that will make it difficult to balance. The following are some cautions to observe regarding rubber lubricants:

■ Do not use petroleum products, which will damage the rubber in the tire.

■ Rubber lube should not be diluted with water, which can rust the rim.

■ Do not use silicone lubricants or liquid soaps, which will allow the tire to spin on the rim.

Figure 62.26 (a) Cut the bottom of the old valve stem. (b) The new valve stem is installed with a special tool. *(Courtesy of Tim Gilles)*

Figure 62.27 Lubricate both tire beads with an appropriate rubber lubricant. *(Courtesy of Tim Gilles)*

Figure 62.28 A tire's serial number is usually found on the back of the tire. *(Courtesy of Tim Gilles)*

Figure 62.29 Some tires are directional and must be kept on the same side of the vehicle during tire rotation. *(Courtesy of Tim Gilles)*

Directional Tires

Some tires are designed to be run in only one direction. Check the sidewall of the tire to see if there are direction arrows. Be certain the correct side of the tire is facing out. Some tires have whitewalls or special lettering. All tires have a serial number, which is on the back side of the tire (**Figure 62.28**). Other tires are designed to rotate in one direction only. These will have an arrow indicating the direction of rotation (**Figure 62.29**).

■ INSTALL THE TIRE

Clamp the wheel to the tire machine with the narrow bead ledge up. Install the inside bead of the tire over the flange of the rim (**Figure 62.30**). As more and more of the bead is passed over the edge of the rim

Narrow bead ledge

Wide bead ledge

Figure 62.30 The narrow bead ledge is up and the lower tire bead is in the drop center of the wheel. *(Courtesy of Tim Gilles)*

flange, force the bead down into the drop center well of the wheel. This is important! If one side of the bead is not in the drop center, the diagonal mounting distance will be excessive and the other side will not be able to be stretched over the flange (**Figure 62.31**). Be careful not to damage a tire pressure monitor during bead installation. When the first bead is totally installed and positioned in the drop center of the wheel, install the other bead over the flange of the rim (**Figure 62.32**).

Before inflating the tire, rotate it on the rim to align the colored dot with dot with value stem.

Inflating the Tire

Most tire machines include a provision for safely inflating the tire.

> **SAFETY NOTE**
> ■ Inflation can be the most dangerous part of tire installation, as tires can explode during inflation.
> ■ If a machine is not equipped with a lock-down clamp to hold the wheel, inflate the tire in a safety cage. Use an extension hose with a clip-on air valve. This is so the technician can stand away from the tire during inflation.
> ■ Do not leave tools on the tire sidewall when inflating the tire.

Seating the Beads

Tubeless tires require a substantial volume of airflow for the beads to start to seat on the rim. Occasionally, getting the tire to take on air can be quite difficult.

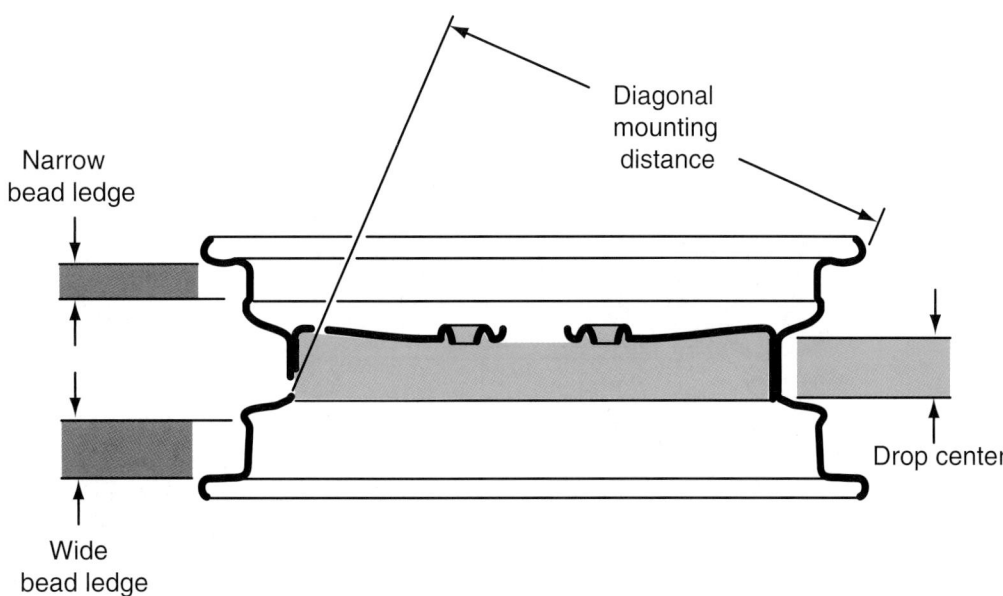

Figure 62.31 The bead must be all the way past the edge of the narrow bead ledge or the diagonal mounting distance will not be sufficient to allow the bead to pass over the edge of the wheel.

Figure 62.32 As the bead is drawn over the edge of the wheel rim, it is important that the other side of the bead stays in the drop center. *(Courtesy of Tim Gilles)*

SHOP TIP Removing the valve core from the valve stem during filling allows a greater volume of air to enter the tire.

Most tire machines have an inflation chamber with a sealing ring. The sealing ring forces a good deal of air from the air chamber into the area surrounding the lower bead of the tire (**Figure 62.33**). The air chamber is required so that there will be a tremendous amount of airflow at once. If air was simply pumped through the normal shop air lines, the volume of air could not be maintained.

CASE HISTORY *A shop purchased a new glass bead blaster, and a technician was attempting to use it. The pressure gauge showed 120 psi. When the pedal was first applied, the glass beads and air from the blast nozzle cleaned effectively for only about 1 second. Then, the pressure would drop to about 60 psi. Further investigation showed that the machine had been fitted with a small regulator with ⅜" pipe nipples on its inlet and outlet sides. The air supply line from the air compressor was ¾" pipe. Installing a larger ¾" regulator solved the problem. The smaller regulator was acting as a restriction or orifice in the line. This is an application of Pascal's Law.*

Pressure after an orifice will be lower than the pressure before it. When trying to fill a tire with air from a remote air compressor, the air line acts as a similar orifice. That's why an inflation chamber is used. If the tire is pulled against the upper bead seat while inflating it, both beads will usually seat.

If the tire is pulled against the upper bead seat while inflating it, both beads will usually seat. It is more difficult to seat the beads on wide rims. If the beads will not seat:

■ Sometimes, the outer bead can be pushed against the inner bead using the bead breaking attachment to force it down over the safety hump.

■ Another trick is to bounce the tire vertically on the floor so that the air in it can force the beads outward.

When inflating a tire on a center post tire machine, the holddown clamp should be in place. When the tire begins to inflate, loosen the holddown clamp one turn. Otherwise as the tire expands, the lower sidewall can wedge against the tire machine, making it difficult to loosen the holddown clamp.

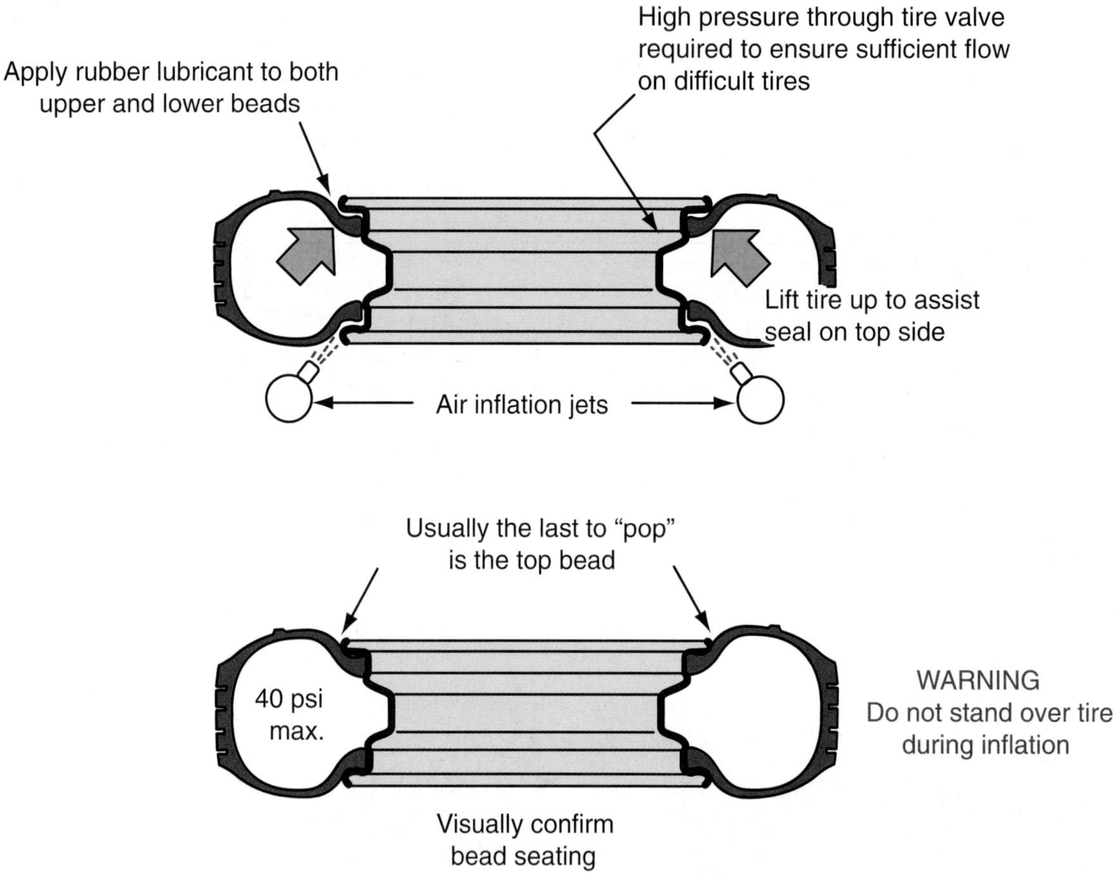

Apply rubber lubricant to both
upper and lower beads

High pressure through tire valve
required to ensure sufficient flow
on difficult tires

Lift tire up to assist
seal on top side

Air inflation jets

Usually the last to "pop"
is the top bead

40 psi
max.

WARNING
Do not stand over tire
during inflation

Visually confirm
bead seating

Figure 62.33 Hold the tire against the top bead seat and use the air jets to seal the bottom bead.

The last bead to seat is usually the top one. As the tire beads slip over the safety humps on the bead seats of the rim, a loud pop is usually heard. This is normal.

Many tire machines include an in-line dial indicator–type tire gauge. Do not put more pressure than 40 psi (or the maximum pressure listed on the tire sidewall) when attempting to seat the beads. If the beads will not seat with 40 psi, break the beads again and thoroughly lubricate the tire and rim again. Reposition the tire on the rim and reinflate it. When a tire is inflated in a safety cage, it can be inflated to a maximum of about 50 psi (3.5 bar) to seat the beads. Run-flat tires (covered later) often require higher pressures to seat the beads.

For especially difficult tire inflations, such as with wider wheels, some tire specialty shops have special bead seating bladders. A different size bead seater is needed for each rim diameter.

SAFETY NOTE
■ When inflating a conventional tire to a pressure higher than 40 psi, be sure it is in a safety cage.
■ Be careful that the direction the valve stem is pointing is away from you when inflating a tire. It can be shot out of the wheel.

There are several potential problems that can occur with conventional tire changers. Center post tire changers were originally designed to be used with steel wheels. The pilot hole in the wheel is clamped against the tire changer, and the wheel can be overly stressed at the center hole during bead breaking, resulting in a bent rim.

Rim clamp tire changers are sometimes referred to as *table top "steel jaw"* tire changers. There are some potential cautions to be aware of when using these machines.

Custom rims can be damaged on the inside by the teeth of the "steel jaw" clamps. Some of today's rim designs allow much of the inside of the rim to be visible. Also, the anticorrosion coating that was applied during manufacture can be damaged.

Custom rims should be clamped from the outside. If the rim does not remain tight on the clamping table it can spin, resulting in scuffing to the outside of the inner rim edge. This can be caused by difficult installations or a problem with the tire changer.
■ If shop air pressure drops due to high air demand, the clamps can slip.
■ With repeated use, jaws wear out and no longer grip tightly.
■ The jaws on the clamping table are not all applied in the same manner and as parts of the apply assembly wear, clamping force can become uneven.

CAUTION Inflating a tire while it is clamped from the outside is dangerous.

Low-profile tires often have sections of sidewall that extend beyond the flange of the rim to protect the rim from damage when it rubs against a curb (**Figure 62.34**). When using a shovel-type bead breaker, this can cause problems. This area is where the shovel would normally be applied. If the shovel slips toward

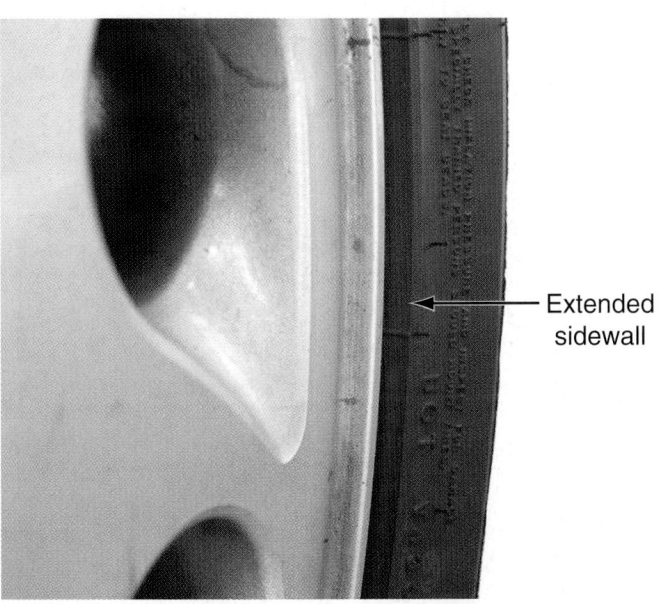

Figure 62.34 Low-profile tires often have an extended tire sidewall that protects the rim from curb scrapes. *(Courtesy of Tim Gilles)*

Extended
sidewall

Figure 62.35 The sidewall can be damaged if the bead breaker is positioned too far away from the edge of the wheel rim as shown here. *(Courtesy of Tim Gilles)*

the rim, it can scratch or chip the rim. If the shovel is applied to the sidewall closer to the tread area, it can damage the tire sidewall (**Figure 62.35**). There are other types of tire changers designed to prevent this.

■ MOUNTING HIGH-PERFORMANCE TIRES

A vehicle can have many thousands of dollars invested in the tires and wheels. Aftermarket wheels can cost between $100 and $1,500 each, and speed-rated tires can cost up to $600 or more each. An incorrectly serviced tire might cost the shop more than $1,000.

Servicing high-performance tires with tighter tire beads can result in a tear to the bead or damage to the tire sidewall if done incorrectly. Removing these tires is done with a tire changer that breaks the bead loose with rollers, instead of the typical "shovel-type" bead breaker used on conventional tire changers.

NOTE: *Tire service should not be performed without proper equipment. Tires with aspect ratios less than 60 will typically call for more sophisticated tire equipment.*

CAUTION Tires that require more than 40 psi to inflate must always be installed in an inflation cage. When reinflating these tires, 60–80 psi could be required to seat the beads, so these tires must be installed in a safety cage during inflation.

Low-profile tires are also more apt to suffer bead damage during removal from the rim. Always apply rubber lubricant to both beads during removal and installation of tires.

■ BEAD ROLLER AND TULIP CLAMP TIRE CHANGER

The newest tire changer design (see Figure 62.22) was developed for use with high-performance wheels and tires. It uses hydraulic bead rollers (**Figure 62.36**) on the top and bottom beads to loosen the beads with a force of up to 4,000 lb. The tulip clamping system uses pads protected by rubber, which allows the wheel to shift without damage to its surface. A spring-loaded device centers the wheel by its hub mount. Practically any size wheel can be mounted in this changer, from 5" to 23" in diameter and up to 19" wide.

In the event the tire is incorrectly mounted and binding occurs, the mounting head will break. This saves expensive tires and wheels from damage. Mounting heads for this tire changer are easy to replace and are inexpensive.

Removing a tire is similar to the rim clamp type, except that the bead is automatically pushed into the

Figure 62.36 Hydraulic bead rollers break the bead seat on the top (a) and bottom (b). *(Courtesy of Tim Gilles)*

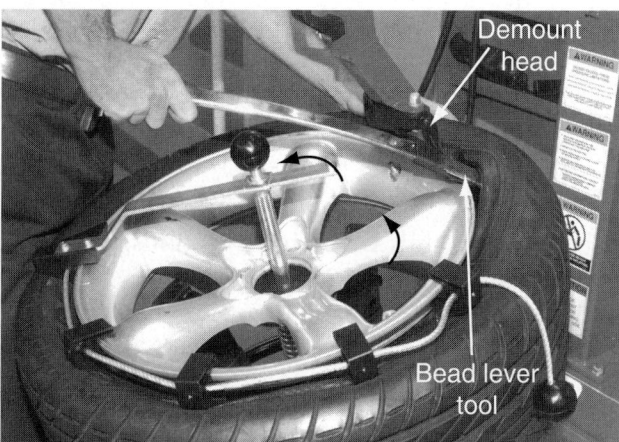

Figure 62.37 As the wheel rotates counterclockwise, the demount head rolls the bead into the drop center while the bead lever pulls the bead over the top of the rim. *(Courtesy of Hunter Engineering Company)*

Figure 62.38 When mounting the bottom bead, be sure the tire pressure sensor is in this position. *(Courtesy of Hunter Engineering Company)*

Figure 62.39 The tire is held in position using blocks. *(Courtesy of Tim Gilles)*

drop center while the bead lever tool pulls the bead over the edge of the rim (**Figure 62.37**).

When installing a tire on a rim with an integral tire pressure sensor, the sensor must be positioned correctly in relation to the tire bead (**Figure 62.38**) or it can be damaged. The tire is held in position on the rim using blocks (**Figure 62.39**).

What to do when the beads will not seat:
- Sometimes, the outer bead can be pushed against the inner bead using the bead breaking attachment to force it down over the safety hump.
- Another trick is to bounce the tire vertically on the floor so that the air in it can force the beads outward.

Radial tires are more difficult to seat than bias tires. An inflatable rubber tube can be used when filling a bias tire to distort the tire tread, which forces the beads outward. The inflatable tube does not work on steel-belted tires such as radials, because the tire is too stiff in the tread area.

When inflating a tire on a center post tire machine, the holddown clamp should be in place. When the tire begins to inflate, loosen the holddown clamp one turn. Otherwise as the tire expands, the lower sidewall can wedge against the tire machine, making it difficult to loosen the holddown clamp.

The last bead to seat is usually the top one. As the tire beads slip over the safety humps on the bead seats of the rim, a loud pop is usually heard. This is normal.

Many tire machines include an in-line dial indicator-type tire gauge. Do not put more pressure than 40 psi (or the maximum pressure listed on the tire sidewall) when attempting to seat the beads. If the beads will not seat with 40 psi, break the beads again and thoroughly lubricate the tire and rim again. Reposition the tire on the rim and reinflate it. When a tire is inflated in a safety cage, it can be inflated to a maximum of about 50 psi (3.5 bar) to seat the beads.

> **SAFETY NOTE**
> ■ Do not inflate a tire higher than 50 psi. The metal bead wire might break.
> ■ Be careful in which direction the valve stem is pointing when inflating a tire. It can be shot out of the wheel.

For especially difficult tire inflations, many tire specialty shops have special *bead seaters*. These represent an extra cost to the shop because a different size bead seater is needed for each rim diameter. On wider wheels, a bead seater might be necessary.

Install the Valve Core

Install the valve core. The valve core should not extend above the top of the valve stem. If it does, the valve cap could contact the top of the valve core and let air out. A shorter valve core should be installed. After the valve core is installed, inflate the tire to the amount recommended on the vehicle placard. Check to see that the valve core is sealing. Then, install the valve stem cap.

▬ TIRE RUNOUT

When a tire is not correctly mounted on a rim or is out-of-round, it will have runout. Runout is when the tire is not round as it spins. If the tire is higher on one side than the other, handling problems will occur.

Pirelli Tires recommends that tires be inflated initially to 40 psi to thoroughly seat the beads. Sometimes, a tire is not properly seated on the rim. The sidewall has *centering ribs* or *locating rings* (**Figure 62.40**) that can be inspected to see that they are even with the edge of the rim flange all the way around the tire. If not, the beads need to be broken down again to

Figure 62.40 Following inflation, check the locating ring to see that the tire is evenly mounted.

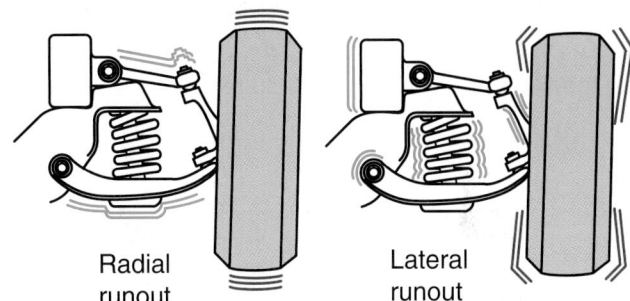

Figure 62.41 Tire runout can be either radial or lateral.

center the tire. Dismounting and remounting the tire on the rim 180 degrees from its original position can sometimes result in improvement.

Another possible cause of runout is that a second-quality or blemished tire might be out-of-round. Runout can be either lateral (sidewall wobble) or radial (tread up and down) (**Figure 62.41**). Runout can be checked with a dial indicator. Some shops have *tire-truers* for correcting radial runout.

▬ TIRE REPAIR

Checking for Leaks

To check for leaks, a water tank is commonly used. Roll the inflated tire slowly around in the tank while looking at the area where the tread meets the water (**Figure 62.42**). Small bubbles from a slow leak can usually be found in this manner. If a leak is not evident, push the valve stem from side to side. Also, check carefully where the tire bead meets the rim. If a soak tank is not available, apply soapy water to the outside of the tire.

Mark the location of the leak and remove the nail or screw if it is still there (**Figure 62.43**). Notice whether it comes out at an angle or not. During the repair, the hole will be reamed at the same angle as the injury to the tire.

Inspecting the Tire

When inspecting a tire before repairing it, remove it from the wheel. A special tool called a bead spreader

Figure 62.42 Check for bubbles at the water line as you spin the tire slowly in the tank. *(Courtesy of Tim Gilles)*

Figure 62.43 Mark the location of the injury and remove the nail or screw. *(Courtesy of Tim Gilles)*

can be used to hold the beads apart during the tire inspection and repair. Check the entire outside and inside of the tire carefully to see that it is not damaged from having been run flat. Use a bright light. The tire must be dry prior to inspection. Inspect the outside of the tire for anything that can allow moisture in like weather checking, cracks, or tread separation. Tires that should *not* be repaired include those that:

■ Smell of burned rubber or have bluish discolored rubber on the sidewall flex area
■ Have evidence of loose cords
■ Have an inner liner that has blisters, bubbles, cracks, or the casing cord pattern showing through
■ Are worn beyond the wear bars

■ REPAIRING A TIRE

Tire puncture repairs can be done with a rubber **tire plug**, a **patch**, or both. Pinhole repairs can be made using a patch only. A pinhole is a very small hole that closes up visually when the nail is removed.

Tire repairs can be made when they are within the tread area (crown) (**Figure 62.44**) as long as the puncture is less than ¼" in diameter. Most shops will *not* perform repairs to tire sidewalls, and repairs to the bead area can *never* be safely done.

NOTE: *With expensive off-road or farm tires, repairs to holes in the sidewall area or holes larger than ¼" in the tread area can sometimes be made but only by a special full-service tire facility (not traditional retail tire outlets).*

The RMA publishes guidelines for the proper repair of all types of tires. Tires are sometimes repaired improperly. Rubber plugs are often installed without removing the tire from the wheel. This is a substandard repair because the tire should be removed from the wheel and carefully inspected to see that it is in good shape. Damage can happen when a repair is done to a tire that has been run flat, has a hole that is too large to be repaired, or has suffered damage from an impact or a nail. **Figure 62.45** shows types of tire damage that are not seen from the outside.

There are differing opinions as to the repairability of some tires. When the cost of a quality repair is considered against the cost of a new tire, the correct decision is often to replace the tire with a new one. Sometimes the puncture appears to be in an acceptable area of the tread. Here is a good way to tell if a tire can be repaired: During the tire repair process, a puncture is reamed. If the resistance of the steel belts is not felt while reaming, the passenger car tire should be replaced. The puncture shown in **Figure 62.46** turned out to be outside the area supported by the tire's belts. Although the tire was in very good condition, it was replaced with a new one.

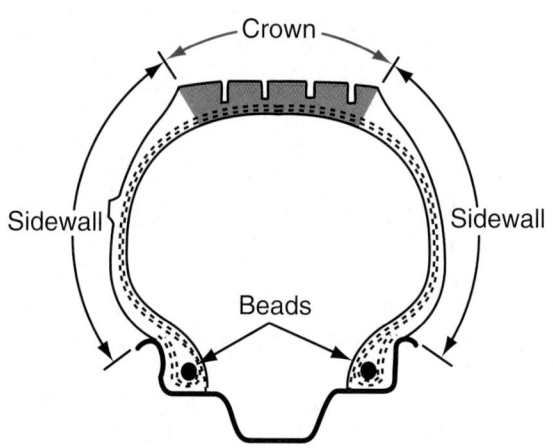

Figure 62.44 Repairs are usually made only to the crown area.

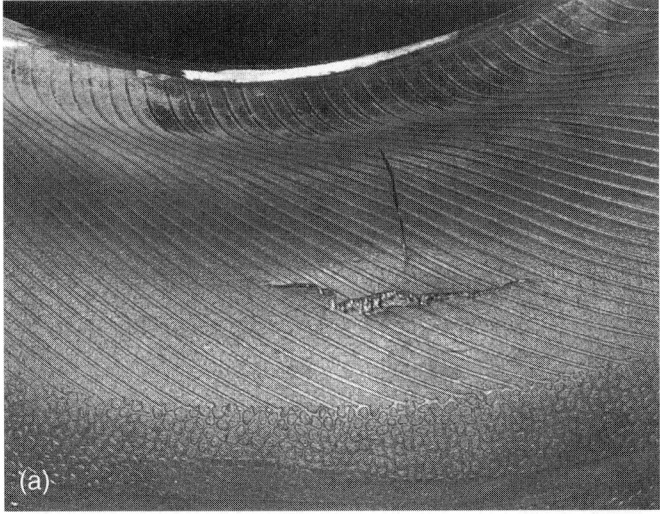

Figure 62.45 (a) A rim bruise break is caused by impact against something like a curb or pothole. (b) This damage was caused by driving for a short distance on a severely underinflated tire. *(Courtesy of Rubber Manufacturers Association)*

Each tire manufacturer publishes repairability standards that they distribute to their dealers. Some manufacturers say that high-speed tires are not repairable, because these tires are used on powerful cars at high rates of speed. Others say that a high-speed tire will lose its speed rating if repaired. Still others give repair procedures for these tires. In any event, the tire should be removed from the rim if a repair is to be made.

A proper repair consists of:
- Removing the tire from the wheel
- Inspecting the tire carefully
- Running a reamer or drill through the hole
- Installing a tire plug
- Smoothing off the inside of the plug
- Installing a patch on the inside of the tire

■ PREPARING THE TIRE FOR REPAIR

Before attempting to repair a tire, the inside of the tire liner must be cleaned. The rough inner surface of a tire is covered with a material that will prevent the adhesion of a patch unless the surface is properly prepared. Using rubber-cleaning solvent, clean the area where the new patch will be installed. A scraper is used to clean the liner because it is more effective in removing oils, grease, or silicone (**Figure 62.47**).

Plugging the Hole

A steel-belted radial tire that has a puncture through a steel belt probably has steel exposed. Moisture can damage the inside of the belt area of the tire. It is advisable that this tire be filled with a rubber plug as well as patched.

Ream the Hole

The hole must be drilled or reamed first so that the plug will fit into it (**Figure 62.48**). Holes that go through steel belts beneath the tread can have sharp pieces of

Figure 62.46 This puncture is outside the crown area. It should not be repaired. *(Courtesy of Tim Gilles)*

Figure 62.47 Use rubber cleaning solvent and a scraper to clean the tire liner. *(Courtesy of Tim Gilles)*

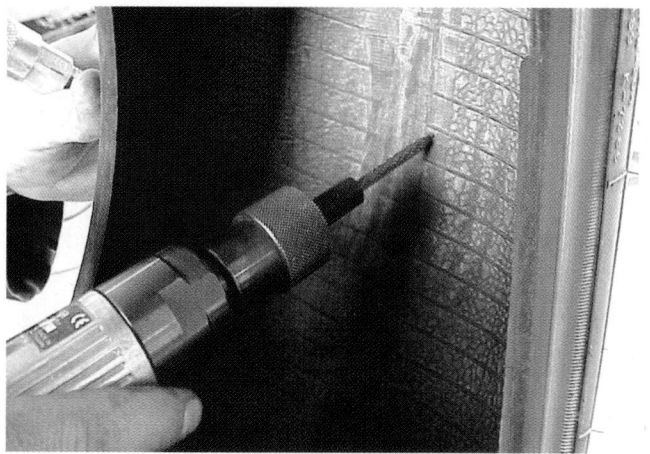

Figure 62.48 Ream the hole to remove metal burrs from the belts. *(Courtesy of Tim Gilles)*

Figure 62.50 Cut the plug off flush with the tread surface. *(Courtesy of Tim Gilles)*

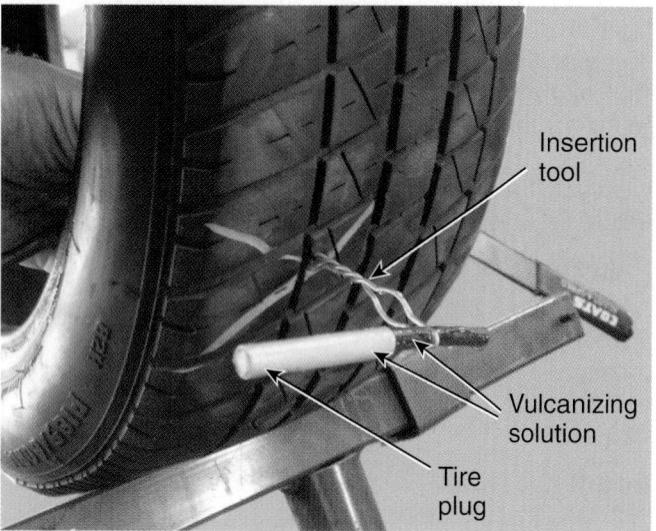

Insertion tool

Vulcanizing solution

Tire plug

Figure 62.49 Apply vulcanizing solution to the plug and install it in the hole. *(Courtesy of Tim Gilles)*

Figure 62.51 Buff the area where the patch will be installed. *(Courtesy of Tim Gilles)*

steel sticking into them. The result can be a cut plug. This is another reason that the hole is drilled or reamed before installing a plug. If a drill is used, it should have a speed of less than 5,000 rpm to avoid excess heat. Before drilling, probe the hole with an awl to determine the direction of the puncture.

A rubber plug is inserted into the hole in the tire. First, vulcanizing cement is applied to the plug (**Figure 62.49**). The cement is the same as that which is used to vulcanize a tire patch to the tire liner. Next, the plug is forced into the hole until ½" of the plug extends above the tread surface of the tire. Last, the end of the plug is cut off with a flexible knife to ⅛" to ¹⁄₁₆" above the inner liner. Cut the plug flush with the tread surface on the outside of the tire (**Figure 62.50**). Be careful not to stretch the plug while cutting it.

The liner is lightly buffed with a fine buffing stone and a low-speed buffer motor (less than 5,000 rpm) to complete the preparation of the sur-

face (**Figure 62.51**). First, outline an area ½" around the outside of the patch with a marking crayon. Be careful not to buff through the liner into the casing plies. Just clean the surface of the rubber. If a portion of steel belt extends into the inside of the tire, it is carefully ground away. Because compressed air contains oils and moisture, it should not be used to remove the dust. The dust from the buffing operation is vacuumed away.

■ PATCHING THE TIRE

After the plug is cut flush with the inside of the tire liner, clean and scrape the area with rubber cleaning solvent and a scraper. After the cleaned area dries, apply an even amount of the vulcanizing cement to the buffed area where the patch will be installed (**Figure 62.52**). Vulcanizing cement is liquid when it is in good condition. If the cement has solidified or is like jelly, it should be discarded. It is a good idea that the

Figure 62.52 Apply vulcanizing cement to the buffed area. *(Courtesy of Tim Gilles)*

Figure 62.53 Use the stitcher to seat the patch to the tire casing. *(Courtesy of Tim Gilles)*

cement be the same brand as the patch so that the two are compatible.

Allow the vulcanizing cement to *completely* dry before applying the patch. Drying time varies with temperature and humidity.

> **SHOP TIP** Moving the puncture to the up position will allow the cement to dry quicker and more evenly. Solvents are heavier than air.

Install the Patch

Carefully remove the backing from the adhesive side of the patch. Be careful not to touch the sticky adhesive.

NOTE: *The patch should be installed while the tire is in its natural shape. A bead spreading tool is commonly used by many shops when affixing a patch to the inside of the tire. This is not recommended for radial tires when applying the patch.*

Center the patch over the hole, making sure that the bead arrows on the patch point to the beads of the tire. Roll the patch into place with a corrugated tire stitcher (**Figure 62.53**). Roll firmly from the center outward, using as much hand pressure as possible. After stitching is complete, remove the thin plastic covering from the top of the patch.

Combination Plug/Patches

Some plugs are a combination of plug and patch (**Figure 62.54**). These can only be used when the hole goes straight into the tire. If the hole is slanted, use a plug and a patch. Some combination plug/patches are mushroom-shaped and must be installed using a special plug installation tool. Be sure to dip a blunt probe in vulcanizing cement and push it through

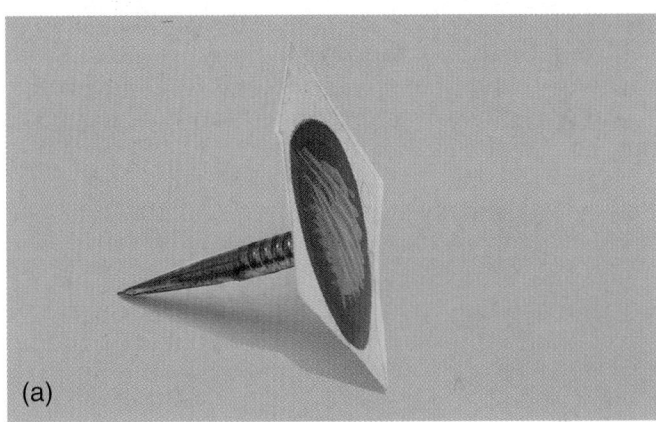

(a)

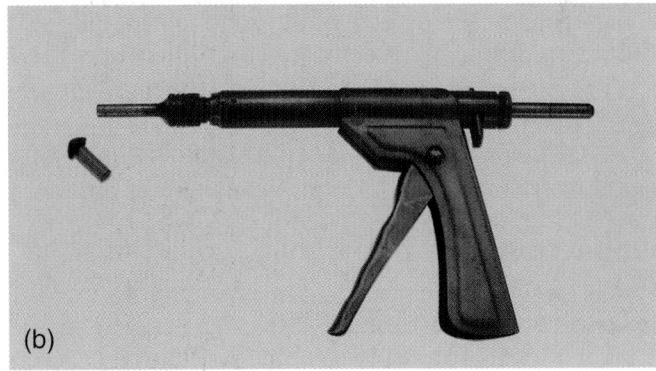

(b)

Figure 62.54 (a) A combination plug/patch. (b) One type of combination plug/patch installed with an insertion tool. *(Courtesy of Tim Gilles)*

the hole to thoroughly coat the walls of the hole. After reinstalling the tire on the wheel and inflating it, double-check with water to see that the leak has been fixed.

Liquid Puncture Sealants. Liquid puncture sealants are commonly available in automotive stores. While these products may be handy when a flat tire occurs, they are not recommended by tire manufacturers.

SAFETY NOTE Some puncture sealants are flammable and can result in an explosion from a spark caused during an external plug repair. The spark results when drilling through the steel belts of the tire. Removing the tire from the wheel lessens the danger of an explosion.

TIRE AND WHEEL BALANCE

Tire imbalance is one of several possible causes of vehicle vibration. Some other possibilities are a bent axle or wheel rim, an out-of-phase drive shaft, bad universal joints, or misaligned drivetrain components. Tire imbalance can result in a cupped tire, a loss of traction, and premature wear to steering and suspension parts.

While tire imbalance is not always the cause of vehicle vibration, it is a major cause. Front tires are the most prone to exhibit symptoms from imbalance. Vibration from rear tire imbalance may not be as obvious, but tire wear can still result.

Imbalance results when the weight of a tire's materials is not equally distributed around the tire. The result is that one side of the tire is heavier than the other. This can be due to errors in manufacturing, or it can result from tire wear. Sometimes new tires can have a minor imbalance that is correctable by adding specific amounts of weight to the wheel rim at specified points.

When an excessive amount of weight is required to achieve balance, a defective tire could be the cause, or the tire might be incorrectly seated on the wheel rim.

NOTE: *Out-of-round tires can be balanced but will still cause vibration from wheel tramp. Runout of less than 0.030" will not usually result in noticeable vibration. Runout of more than 0.125" calls for replacement of the tire.*

Wheel Weights

Wheel weights are attached to the rim to correct imbalance (**Figure 62.55**). Clip-on wheel weights are attached to the rim using a special wheel weight hammer (see Figure 62.20).

Several styles of lead clip-on weights are available in increments of ¼ ounce. There are different types of clips available. Some have an extra long clip for fitting under wheel covers. Another style has a wider, larger clip to fit on pickup truck wheels. Aluminum wheels require alloy or coated clips if they have a flange that will accept a clip-on weight. The coated weight will not corrode the wheel.

Lead strips that are cut to length are also available for aluminum wheels (**Figure 62.56**). These are attached to the wheel with double-sided tape after cleaning a place on the inside of the wheel with

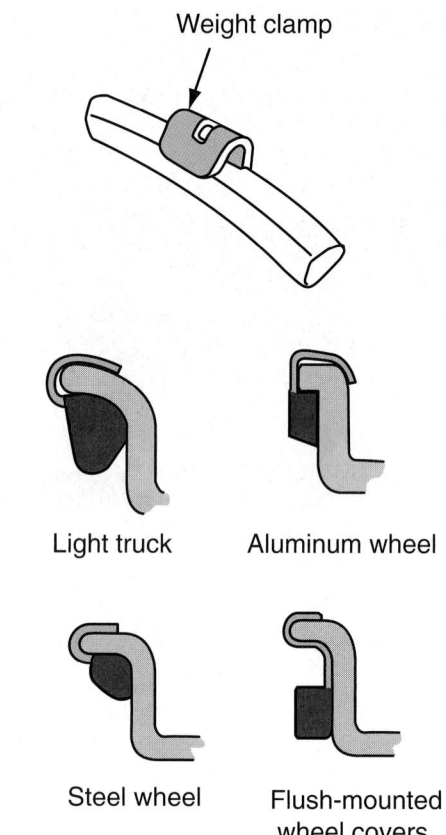

Figure 62.55 Lead wheel weights are installed on the wheel to counterbalance a tire. There are several styles for different applications.

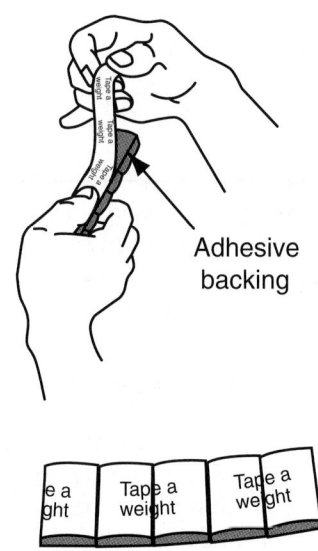

Figure 62.56 Lead tape weights.

sandpaper. Sometimes they are installed on the inside and outside of the wheel, but this is unsightly. Chrome weight strips are available also.

NOTE: *Be careful when working around aluminum wheels. They are easily scratched. Plastic shields are available that cover tire tools to prevent damage.*

■ TYPES OF WHEEL BALANCE

The two types of wheel-balancing methods used on cars are static and dynamic. Static, which means an object is stationary, imbalance is measured with the wheel at rest. If a wheel with static imbalance were mounted on a spindle with the heavy spot at the top, the heavy spot would rotate to the lowest possible position on its own. If the wheel was statically balanced, it would *not* have a tendency to rotate by itself.

Static imbalance subjects the wheel to vertical impacts that become worse with higher speed. The impacts occur because the tire has a heavy spot on one end of its tread (**Figure 62.57**). A small amount of imbalance when the wheel is at rest can amount to a great deal of pounding force when the wheel is spinning at highway speeds (**Figure 62.58**). For instance, if a 15" wheel is 1 ounce out of balance, when the wheel rotates at 60 mph, the pounding force will be 4.6 pounds.

Static imbalance causes wear to mechanical parts, vibration, and gouged tire tread wear, called *cupping* (see Figure 14.4). In severe cases, especially when accompanied by a bad shock absorber, a tire can actually hop so badly that it leaves the road surface. This is called wheel tramp (**Figure 62.59**).

The part of the tire that is contacting the ground is actually traveling at 0 mph. This is called *static friction* or *static contact*. During wheel tramp when the tire recontacts the road, a small amount of the tire's rubber is scrubbed off. This is because the tire was traveling at the same speed as the vehicle when it left the road surface. This scrubbing happens all around the tire, accounting for the cupped wear all around the tire's tread surface.

Older, narrower bias tires were static-balanced. Static means an object is stationary. Although today's modern wheel balancers are capable of measuring the static imbalance of a spinning wheel, the name remains from when imbalance was measured with the wheel at rest on a bubble balancer. Static balancing, also called single-plane balancing, is done in a single plane where compensating weight is added on the opposite side of the wheel (**Figure 62.60**). If it were possible to put the compensating weight at exactly the same place on the opposite side of the tire, the amount of the weight would be exactly the same as the imbalance. Unfortunately, tires are sometimes heavier on one side of the tread than the other. Modern wheel balancers correct this imbalance. Static

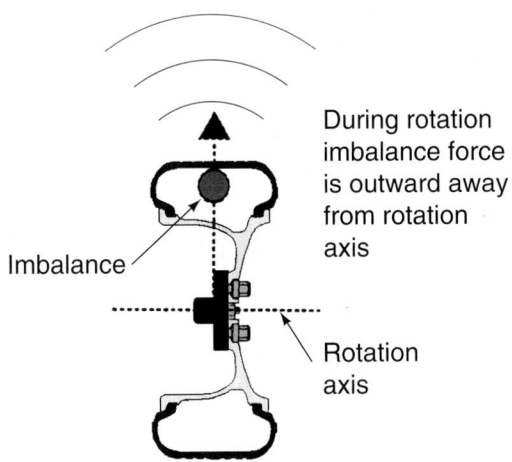

Figure 62.57 When the heavy spot is rotated, force is directed out away from the center of the wheel. *(Courtesy of John Bean Company)*

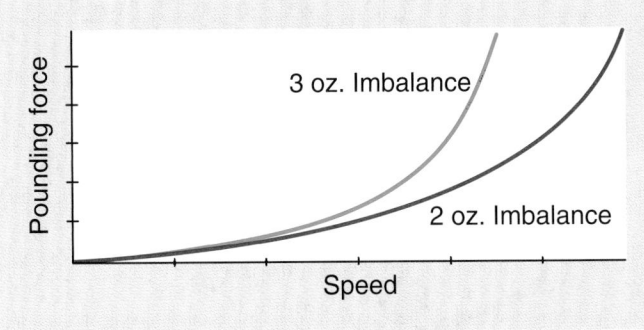

Figure 62.58 Pounding force increases as speed and the amount of imbalance increase. *(Courtesy of John Bean Company)*

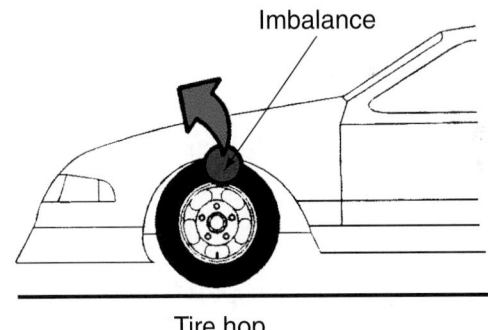

Tire hop

Figure 62.59 Wheel tramp happens when speed and force to the imbalance become large enough to cause the tire to leave the road surface. *(Courtesy of John Bean Company)*

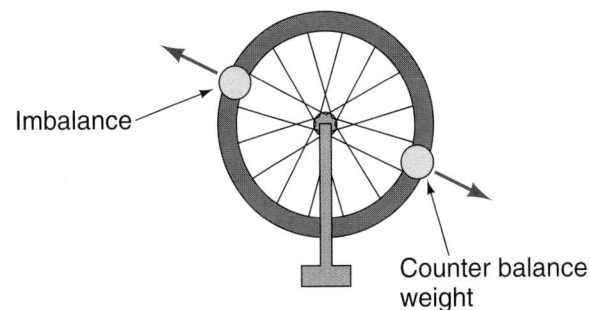

Figure 62.60 To balance a tire in a single plane, compensating weight is added on the opposite side of the wheel. *(Courtesy of John Bean Company)*

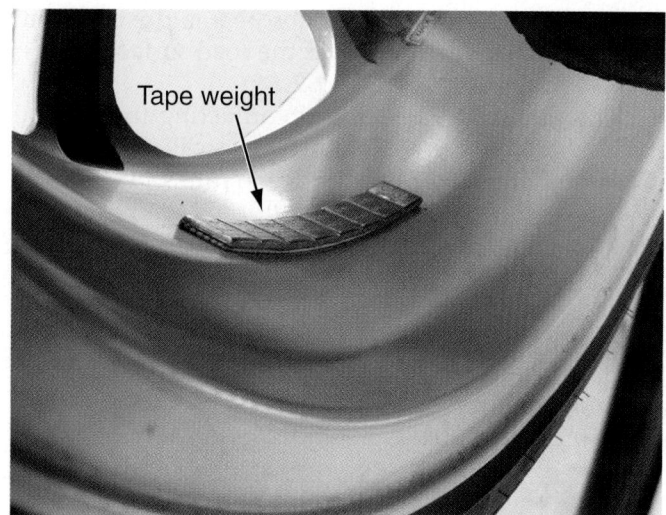

Figure 62.61 Location of weights for static balance. *(Courtesy of Tim Gilles)*

balancing is an option when using a computerized spin balancer. This option might be used on an aluminum wheel when unsightly weights might not be desired on the outside of the wheel, although there are better ways of balancing these wheels. When static-balancing an aluminum wheel, weights with tape on their back sides are fixed to the inside of the wheel at the center plane (**Figure 62.61**).

■ COUPLE IMBALANCE

When a tire tread is lopsided, with more of its weight on the outside or the inside of its tread, this is known as couple imbalance. Couple imbalance only shows up when the wheel is spinning. The wheel will have a tendency to shimmy, which also results in faster tire wear.

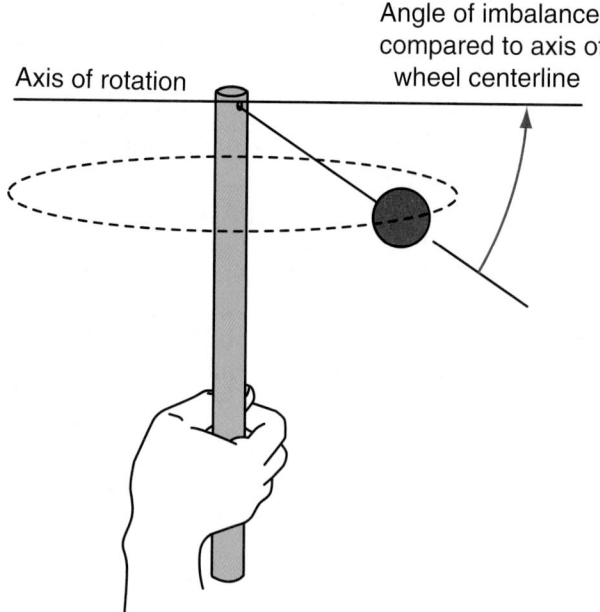

Figure 62.62 As the speed of the rotation increases, the imbalance moves toward the axis of rotation.

Figure 62.62 shows how a spinning heavy spot will seek the centerline of the tire. This is why the tire shimmies (wobbles from side to side). When the heavy spot is in the front, it causes the wheel to move one way. After the tire rotates one-half revolution, the heavy spot causes the wheel to try to turn in the opposite direction (**Figure 62.63**).

The previous example is of an imbalance on only one side of the tire. Usually there are imbalances on both sides of the tire, requiring counterbalancing weights to be installed on both sides of the wheel.

A wheel may be in static balance but not couple balance. If a tire has been statically balanced by adding

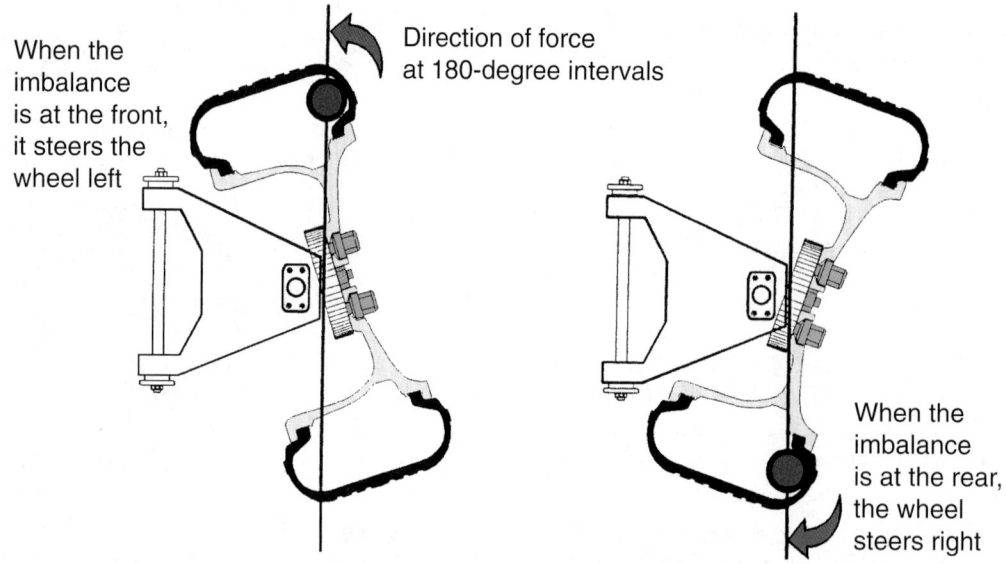

Figure 62.63 Action of dynamic or couple imbalance. *(Courtesy of John Bean Company)*

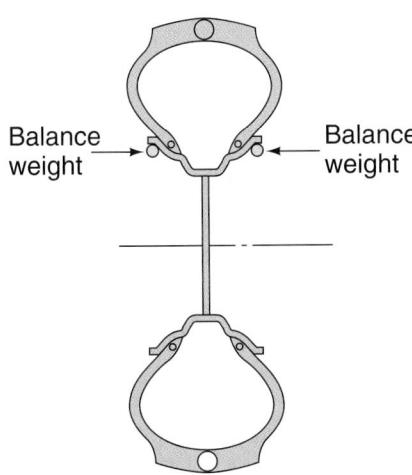

Figure 62.64 The amount of weight is split in half and put on both sides of the rim to avoid creating a couple imbalance.

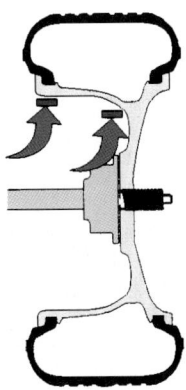

Figure 62.65 Tape weights added to the inside of the wheel for cosmetic purposes. (*Courtesy of John Bean Company*)

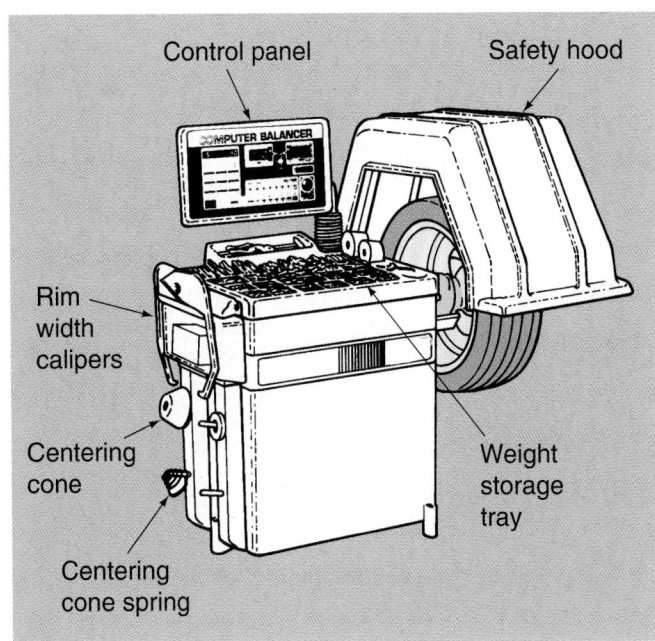

Figure 62.66 A computer wheel balancer. (*Courtesy of Hennessy Industries, Inc.*)

all of the weight to one side of the offset wheel rim, that weight will be thrown side-to-side in a different direction as the tire spins, resulting in shimmy. The proper way to balance a wheel statically is to split the amount of weight to be used in half, placing an equal amount of weight on both sides of the rim (**Figure 62.64**). Pirelli Tires recommends this procedure anytime the amount of weight to be added is in excess of 20 grams (0.71 oz.). With tape weights, weights are often added in two places on the inside of the wheel (**Figure 62.65**).

■ DYNAMIC BALANCE

It is not likely that a tire and wheel will have only static imbalance, and not couple imbalance, too. Dynamic imbalance is the combination of both static and couple imbalance. Dynamic balance means balance in motion. It is also called *two-plane balance* because it measures side-to-side (lateral) force as well as up-and-down (axial) force. Lateral forces are felt when a steering wheel moves back and forth.

Dynamic wheel balancers spin the wheel and locate vibration in it. The computer splits the tire into two halves and measures lateral and radial (axial) forces on each side of the tire's center. Weights are added to the proper side of the rim to correct the imbalance. A tire that is dynamically balanced will be statically balanced, too. Dynamic balancing can be done with a computer balancer or with an on-the-car spin balancer.

■ COMPUTER BALANCERS

Off-the-car computerized wheel balancers (**Figure 62.66**) are very popular and easy to use. Computer balancers balance in both the static and dynamic planes. When the wheel is off the car for tire replacement, computer balancing is usually performed on the new tire. The wheel is mounted on either a horizontal or a vertical threaded shaft using adapters that are supplied with the machine.

■ CENTERING THE WHEEL ON THE BALANCER

Another cause of imbalance is that the wheel was not properly installed on the wheel balancer. According to the FMC Corporation, a 36-pound tire and wheel assembly that is off-center by only 0.006" will result in a ½-ounce imbalance error.

When mounting a wheel on the wheel balancer, the best method is to center the wheel by the same method as designed by the manufacturer. There are two ways that wheels are centered (see Chapter 61):
■ Hub-centric—this is when the hole at the center of the wheel locates the wheel (**Figure 62.67**).
■ Lug-centric—this is when the lug nuts center the wheel.

Figure 62.67 The center hole of a hub-centric wheel fits on the hub. This is a rear wheel and hub. *(Courtesy of Tim Gilles)*

To determine whether or not a wheel is hub-centric, see if the wheel fits the hub snugly with the lug nuts removed.

Preparation for Mounting the Wheel

The backing plate and wheel lug flange must be clean and undamaged. Be sure that the wheel and tire have all foreign objects removed from them before attempting tire balance. When the wheel and tire assembly is first spun, look for signs of obvious runout and double check the mounting.

Mounting Hub-Centric Wheels

When mounting a hub-centric stamped-steel wheel, use a centering cone. The most accurate centering system is to install it from the back side of the wheel. This is because the center hole was originally stamped from

the back. **Figure 62.68** shows the recommended mounting arrangement on the wheel balancer.

It is important that the centering cone fit the shaft snugly. When new, it has about 0.001" clearance. Expandable collets are available that provide the most accuracy in centering. This is because they expand when tightened, eliminating all clearance between the collet and the shaft.

Mounting Lug-Centric Wheels

To mount lug-centric wheels, use a special lug-centering adapter (**Figure 62.69**). The arms on the adapter are mounted in different places, depending on the wheel's number of lug holes. There are only five adapter arms. With six lug wheels, use three arms. With eight lug wheels, use four arms. After the arms are installed and tightened on the adapter, the adapter is held against the mounting flange on the wheel balancer while

Figure 62.69 A lug-centric adapter installed on a wheel balancer. It can be configured with three, four, or five adjustable arms. *(Courtesy of Tim Gilles)*

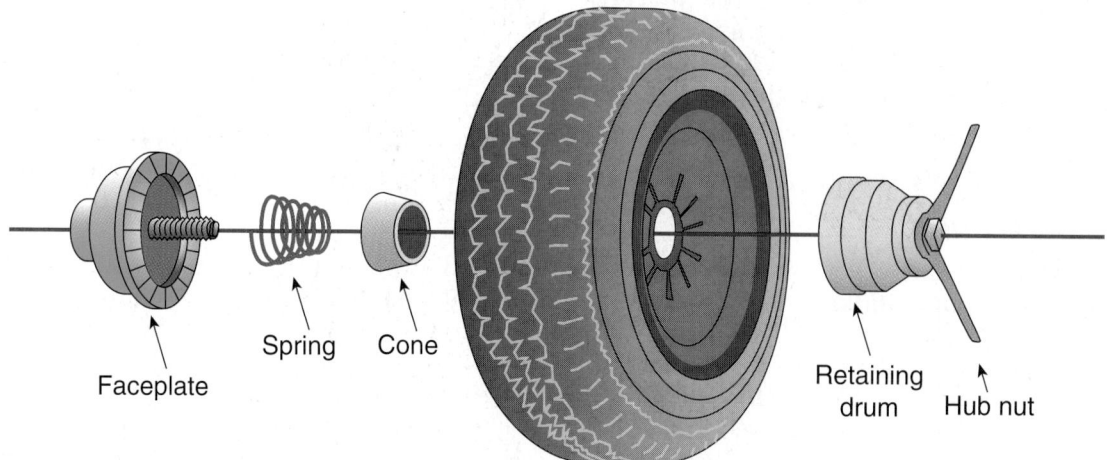

Figure 62.68 Recommended computer balancer mounting arrangement for hub-centric wheels.

Figure 62.70 The lug-centric adapter is held against the balancer mounting flange while retaining screws are installed and tightened. *(Courtesy of Tim Gilles)*

screws are installed and tightened (**Figure 62.70**). Special lug nuts hold the wheel against the adapter (**Figure 62.71**).

> **CASE HISTORY**
> *A man purchased a used Suburban that had aftermarket aluminum wheels and oversized tires. The tires appeared to be in good condition, with less than half of the tread worn. Unfortunately, they had been balanced incorrectly using adapters in the center hole instead of a lug-centric adapter. The tires had worn unevenly, developing cupped tread wear. They were now too far out of balance and had to be replaced.*

Adapter lug nuts

Figure 62.71 Special lug nuts hold the wheel against the lug-centric adapter. *(Courtesy of Tim Gilles)*

Program the Wheel Balancer

After the wheel is mounted, the technician needs to program three references into the computer:
- The width of the rim, measured by a rim caliper (**Figure 62.72**)
- The location of the flange on the wheel, measured by a gauge (**Figure 62.73**)
- The diameter of the wheel (for instance: 13", 14", 15")

To balance a tire, the safety hood is lowered over the wheel. The wheel is spun for a short time and then stops. The machine gives a readout telling how much weight to use and where to put it. After installing the weights, the wheel is spun once again to check for the accuracy of the balance job. A reading of "zero" on both sides of the wheel means that the wheel is ready for installation on the car.

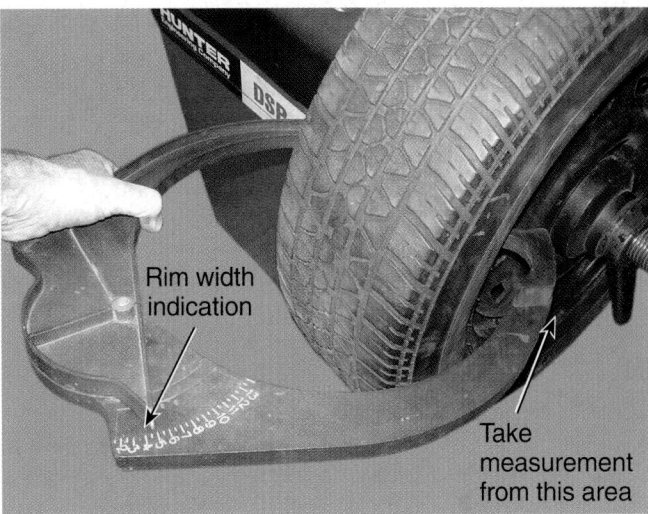

Rim width indication

Take measurement from this area

Figure 62.72 Use a special caliper to measure the width of the rim. Apply its ends to the places on the inside and outside of the rim where wheel weights would be applied. *(Courtesy of Tim Gilles)*

Measure to wheel weight location

Distance number indication

Figure 62.73 This measurement tells the computer where the wheel is located on the balancer shaft. *(Courtesy of Tim Gilles)*

SHOP TIP Weights for aluminum wheels have double-sided tape attached to their back side. When balancing aluminum wheels with tape weights, use duct tape to temporarily attach the weights to the wheel. Install one weight on the wheel as near to the outside as possible. The other weights are installed on the inside edge of the wheel.

NOTE: *Make certain there is enough clearance between the wheel weights and the disc brake caliper.*

■ MATCH MOUNTING

Some computer balancers have an extra feature that matches a tire's imbalance to a wheel's imbalance. This procedure is used when a weight of more than 2 ounces is required on one side of a wheel. The tire is given an initial spin on the balancer. Then, it is deflated and installed on a tire machine so that the beads can be broken down. After rotating the tire on the rim 180 degrees, the tire is reinflated and rebalanced. The computer can now tell how much the wheel is out of balance and how much the tire is out of balance. If neither is unacceptable, the tire is once again rotated on the rim to complete the match mount.

■ FORCE VARIATION

Today's lighter vehicle designs are more sensitive to road feel. When a load is placed against a tire, there is a change in the stiffness of the sidewall and the tire's footprint. There can be stiffer or weaker areas of the tire when under load (**Figure 62.74**). Manufacturers have specifications and acceptable limits for this change, which is called force variation. A wheel and tire might be free of runout when measured with a dial

indicator. Yet, the tire might vibrate under load due to excessive force variation.

Some tires come with a mark or a paint dot on their sidewall to indicate the high or low side of force variation. Unfortunately, this marking is not uniform among manufacturers, so simply lining the valve stem up with the mark could result in an exaggeration of the problem. Also, some manufacturers use the mark to label the high point, while others use it to label the low point. There are tire balancers that use a special device to measure and correct force variation (**Figure 62.75**). This type of wheel balancer identifies and locates tire and wheel problems caused by force variation, runout in the wheel (**Figure 62.76**), and the random position between the force variation and runout when the tire is mounted on the rim. The balancer detects variations in wheel runout, and the computer

Figure 62.75 A tire balancer with a force roller. The force roller presses against the treads as the wheel spins. *(Courtesy of Tim Gilles)*

Sensing roller

Figure 62.76 This wheel balancer detects wheel runout. *(Courtesy of Tim Gilles)*

Figure 62.74 There can be stiffer or weaker areas of the tire when under load. *(Courtesy of Hunter Engineering Company)*

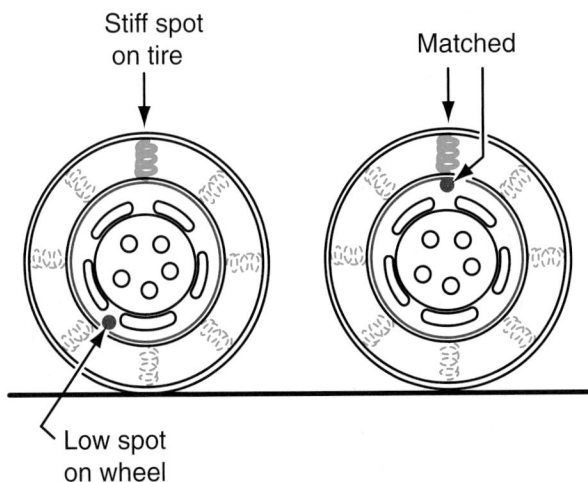

Stiff spot on tire

Matched

Low spot on wheel

Figure 62.77 Matching the high or stiff spot on the tire with the low spot on the wheel reduces vibration.

recommends a change in the tire and wheel positions to match the stiff or high spot on the tire with the low spot on the rim (**Figure 62.77**). This minimizes problems caused by force variation.

MOUNTING THE WHEEL ON THE CAR

Before installing a wheel on a car, double-check to see that the bolt holes in the wheel center are in good condition. It is not uncommon to find steel wheels where the bolt holes are worn from a wheel that was loose, lug nuts that were installed upside down, or lug nuts that were overtightened. Finger-tighten the lug nuts. Shake the wheel to center it over the centering area on the hub. Cross-tighten the lug nuts with a torque wrench.

REVIEW QUESTIONS

1. What kind of tire has a bulging sidewall when properly inflated, radial or bias?

2. About how much does pressure in a tire increase as the tire warms up?

3. Wear bars show up around the tire tread at what tread depth?

4. What is the name for the up-and-down action of the tire that results in scalloped tire wear?

5. What is applied to the tire bead before installing a tire on a rim?

6. Should the valve core be in the valve stem or out of it when filling the tire with air to seat the bead?

7. What kind of hole can be repaired using only a patch?

8. A hole in the tire tread area of a steel-belted tire must be _____ or _____ before installing a plug in it.

9. What kind of tool is used to clean the inside of a tire liner before installing a patch?

10. What type of wheel balance is measured with the wheel stationary?

11. If a 15" wheel is 1 ounce out of balance, when the wheel rotates at 60 mph, the pounding force will be _____ pounds.

12. How fast is the part of the tire that is contacting the ground traveling?

13. When a tire is lopsided, with more weight on one side than the other, what kind of imbalance results?

14. What is the combination of both static and couple imbalance called?

15. To mount _____ centric wheels, use a special lug-centering adapter.

ASE-STYLE REVIEW QUESTIONS

1. Which of the following is/are true about wheel balancing?
   a. When static-balancing a wheel, it is better to place the weight on the inside of the wheel.
   b. Tire balance problems become less as speed increases.
   c. Both A and B.
   d. Neither A nor B.

2. Technician A says that a computer tire balancer balances the tire statically. Technician B says that the computer tire balancer balances a tire dynamically. Who is right?
   a. Technician A   c. Both A and B
   b. Technician B   d. Neither A nor B

3. Technician A says that wheel lugs on most cars are loosened by turning them clockwise. Technician B says that lug nuts are installed

with the tapered side away from the rim. Who is right?

a. Technician A   c. Both A and B

b. Technician B   d. Neither A nor B

4. Technician A says that a tire's inflation pressure drops as the car is driven. Technician B says that tire pressure should be checked when the tire is hot. Who is right?

a. Technician A   c. Both A and B

b. Technician B   d. Neither A nor B

5. Which of the following is/are true about tire pressure?

a. A tire typically loses pressure of about 1 psi per month.

b. Changes in outside air temperature can have a dangerous effect on tire pressure.

c. Both A and B.

d. Neither A nor B.

6. When rotating tires:

a. Cross the non-drive wheels to the drive wheels.

b. Cross the front tires to the rear on rear-wheel-drive vehicles.

c. Both A and B.

d. Neither A nor B.

7. When mounting tires on a tire changer, which of the following is/are true?

a. Be sure that a valve stem tire pressure monitor is rotated 90 degrees from the bead breaker.

b. If the tire has match mount marking, rotate on the rim to align the colored dot with the valve stem before inflating the tire.

c. Both A and B.

d. Neither A nor B.

8. Technician A says that the part of the tire that is in contact with the ground is traveling at the same speed as the car. Technician B says that the combination of static and couple balance is called dynamic balance. Who is right?

a. Technician A   c. Both A and B

b. Technician B   d. Neither A nor B

9. Technician A says that not to inflate a tire over 25 pounds when seating a bead. Technician B says that if a hot tire has a lower pressure than the recommended cold pressure, it is seriously underinflated. Who is right?

a. Technician A   c. Both A and B

b. Technician B   d. Neither A nor B

10. Technician A says that the valve core should be installed before inflating a tire to seat its beads. Technician B says that to unseat the tire beads while the tire is inflated. Who is right?

a. Technician A   c. Both A and B

b. Technician B   d. Neither A nor B

Suspension, Steering, Alignment

THEN AND NOW: STEERING

Some of the early automobiles could reach surprisingly high speeds. From a safety standpoint, the steering systems used in these vehicles were very primitive. Direct linkage from a tiller to the drag link could result in a broken arm when you hit a bump. The first domestic vehicle equipped with a steering wheel was the 1901 Packard.

Even after the tiller evolved to the less lethal steering wheel, automobile steering systems were not satisfactory. Various types of steering mechanisms were designed to reduce kickback and improve mechanical advantage. These included a *chain drive* and the *worm-and-sector steering gear* found on some cars still on the road today. The best idea was the *recirculating ball steering gear,* which made its debut on V-16 Cadillacs. Migrating ball bearings on the steering shaft reduced the high friction that occurred between the internal gears of the earlier gear boxes. The recirculating ball steering gear also resists the shock of running over a pothole.

The 1901 Mercedes had a steering wheel instead of a tiller. *(© Courtesy of Mercedes-Benz USA, LLC)*

There were earlier attempts at *power steering,* but quality systems did not appear until after World War II. The war effort resulted in much research on hydraulics for aircraft servos and tank controls. In 1951, Chrysler applied this new technology to the production of the first automotive power steering system.

The *rack-and-pinion* steering gear used on many vehicles today uses a round gear on the steering wheel shaft. It meshes with a long, flat gear cut into the top of a long shaft that connects between the front wheels. Rack-and-pinion steering is not a new idea. In fact, it was used on huge steam tractors in the mid-19th century. The 1886 Daimler use a curved-rack version to control its center-pivoted axle. The planetary steering used on Ford's Model T was based on the same principle.

Rack-and-pinion steering was abandoned on early vehicles due to severe kickback and its limited mechanical advantage. It was not until the 1950s that European engineers began to rediscover its potential and improve its

The Lexus RX 400 H has electrical power steering. *(Courtesy of Lexus)*

design to eliminate these problems. In 1971, Ford showed its European connection by producing the Pinto, the first volume domestic car to come with rack-and-pinion steering.

An electrically powered, electronically controlled rack and pinion steering gear is found on some vehicles. They are especially found on hybrid electric and fuel cell electric vehicles. When a hybrid's piston engine shuts off at idle and low speed, the electric steering gear can still provide steering assist using battery power. An electric motor assists the rotation of the pinion, giving fast response and eliminating the pump and hoses. This system has not come into widespread general use in gasoline engine vehicles due to the excessive amount of electrical energy it consumes.

Suspension Fundamentals

■ KEY TERMS

suspension	monoleaf spring	aeration
sub-frame	composite springs	gas shock
bounce	rigid axle	Macpherson strut
compression	independent suspensions	load carrier ball joint
jounce	control arms	follower ball joint
rebound	A-arm	passive suspension system
coil spring	wishbone	adaptive suspension
torsion bar	ball joints	system
leaf spring	shock absorber	dive
shackle	double-acting shock	squat
center bolt	absorber	active suspension system
overload leaves	body roll	

■ INTRODUCTION

The vehicle chassis includes the frame, shocks and springs, steering parts, tires, brakes, and wheels. Part of the chassis is the **suspension** system (**Figure 63.1**). There are several suspension designs in use today. Over the years, there have been many non-standardized part names to go with these designs. The names used in this chapter are the ones that have become standard. As you read the chapter, become familiar with the names of the parts.

■ SUSPENSION

The suspension supports the vehicle, cushioning the ride while holding the tire and wheel correctly positioned in relation to the road. Suspension system parts include the springs, shock absorbers, control arms, ball joints, steering knuckle, and spindle or axle.

Sprung and Unsprung Weight

Sprung weight is the weight that is supported by the car springs. Vehicle control increases with the reduction of

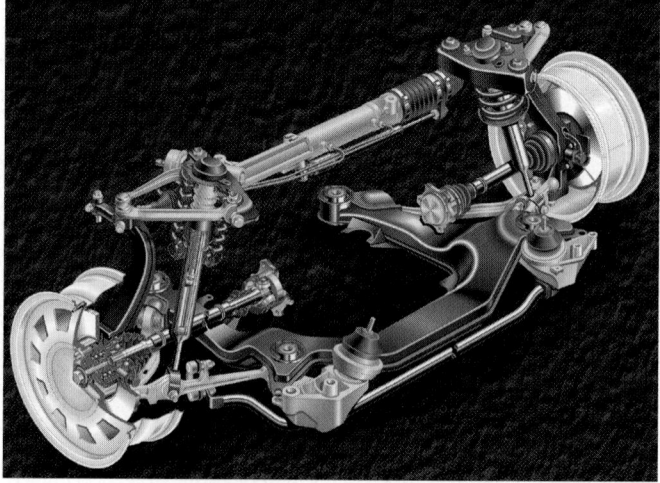

Figure 63.1 Suspension and steering systems.

unsprung weight. Anything *not* supported by the springs is unsprung weight. This includes the tires and wheels, brakes (on most vehicles), bearings, axles, and differential.

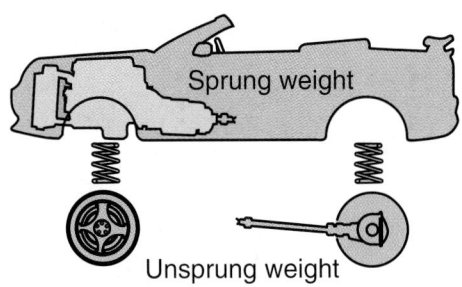
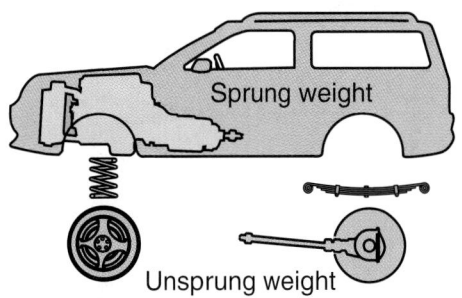

Figure 63.2 A comparison of sprung and unsprung weight.

Sprung weight includes the powertrain, body and frame, and anything else carried by the weight of the springs. The tire and wheel assemblies react to irregularities in the surface of the road, while the sprung components are relatively insulated from those effects (**Figure 63.2**). Having low unsprung weight means that inertia has less of an effect on those parts as they react to potholes and bumps in the road. This means the springs will have less work to do.

■ FRAME AND SUSPENSION DESIGNS

There are various types of frame designs used in cars and trucks. Cars are designed to be as lightweight as possible to improve fuel economy. Many newer cars have front-wheel drive. Almost all older, heavier cars had rear-wheel drive.

A front-wheel-drive car is made with a **sub-frame** in the front that includes the engine, transaxle, and steering/suspension system (see Chapter 65). These cars, as well as modern rear-wheel-drive cars, rarely have a frame running the entire length of the car. Instead, they have a sheet metal floor pan with small sections of frame at the front and rear.

■ SPRINGS

Springs support the load of the car, absorbing the up-and-down motion of the wheels that would normally be transmitted directly to the frame and the car body. Up or down motion is referred to as **bounce**. When the wheel moves up as the spring compresses, this is called **compression** or **jounce**. When the wheel moves back down, it is called **rebound**.

A tire can absorb some of the shock caused when it hits a bump in the road, but the tire passes over the bump so fast that the spring needs to absorb the rest of the shock. There are four types of springs used in vehicles: the coil spring, the torsion bar, the leaf spring, and the air spring.

Coil Spring

The **coil spring** is the most common type of spring used in the front and rear of passenger cars (**Figure 63.3**). It is constructed of a spring steel rod wound into a coil. Springs are painted or coated with vinyl or epoxy to reduce noise and prevent rust or nicks, which

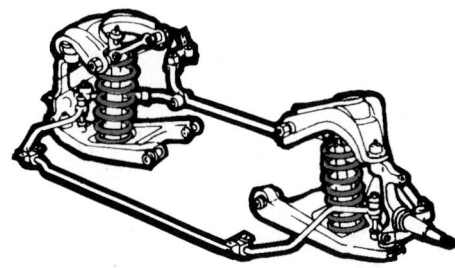

Figure 63.3 A suspension with coil springs.

could raise stress in the spring and result in breakage. The coil spring is dependable and relatively inexpensive. It can carry a heavy load, but it is relatively light in weight, especially when compared to leaf springs.

> ### ■ SCIENCE NOTE ■
>
> *An alloy is a metal mixed with other metals. Coil springs are an alloy of different types of steel, usually mixed with silicon or chromium. Springs are tempered, which hardens the steel alloy. Tempering is a very precise process where the spring is heated to a specific temperature and then cooled at a precise rate. If the spring is cooled too slowly, it will anneal, or become soft. Cooling it too fast will make it brittle.*

There are different ways to make variable-rate springs (also called progressive-rate springs). One way is to use a tapered rod. The ends of the spring do not work as hard as the center coils. As the spring is being compressed, it becomes stiffer. This makes for a smoother ride when going over smaller bumps and still allows for heavier carrying capacity.

The most commonly used variable-rate spring is made of metal of a consistent diameter, with unequally spaced coils in a cylindrical shape wound with the coils more closely spaced on one end than the other (**Figure 63.4**). The more closely spaced coils at one end of the spring do not function until the spring is compressed sufficiently at the other end and in the middle to cause them to produce force. The spring's *active* coils work throughout the complete range of spring compression. The spring's *transitional* coils (more widely spaced) progressively bottom out, becoming inactive once they are compressed to their maximum capacity.

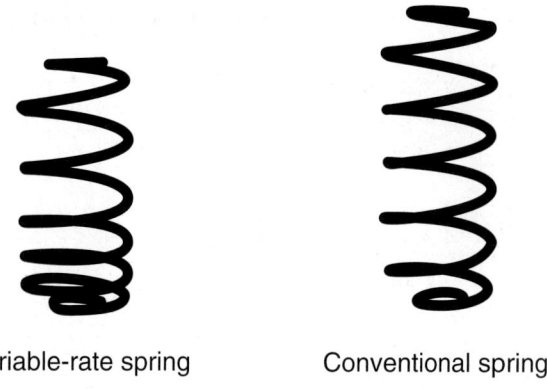

Variable-rate spring   Conventional spring

Figure 63.4 The most common variable-rate coil springs have a consistent diameter and unequally spaced coils.

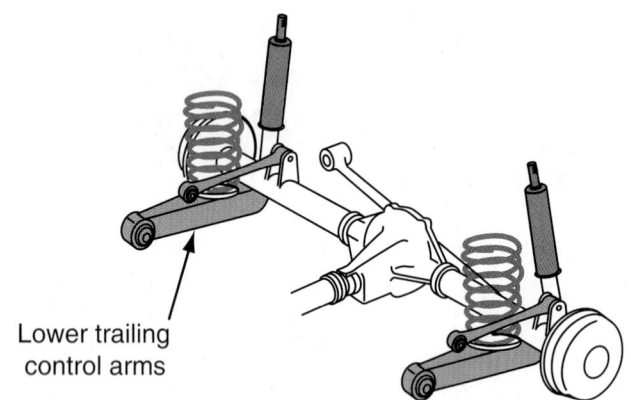

Lower trailing control arms

Figure 63.5 A coil spring used in a rear suspension has control arms or trailing arms to hold the axle in alignment.

When a car has leaf springs, the leaf spring maintains the position of the axle. But when a coil spring is used with a rigid rear axle, the axle wants to move out of its correct position, so upper and lower control arms are used to control fore and aft movement (**Figure 63.5**). A reinforcement bar, or track bar, is sometimes needed to keep the coils in a stable side-to-side position.

Torsion Bar Spring

A **torsion bar** is a straight rod that twists when working as a spring (**Figure 63.6**). When the wheel moves up during jounce, the torsion bar twists in one direction. When the wheel rebounds, the torsion bar unwinds. Torsion bars are made of heat-treated alloy steel with a hex head or splines at each end. One end of the torsion bar fits into a mating surface at the frame (see Figure 63.4). The other end attaches to the movable lower control arm of the vehicle's suspension system.

A torsion bar can be mounted in the chassis to run either front-to-rear or side-to-side. Light-duty trucks and sport utility vehicles (SUVs) use longitudinal torsion bars. Transversely mounted torsion bars were used on some older cars. Almost all torsion bar installations are on the front.

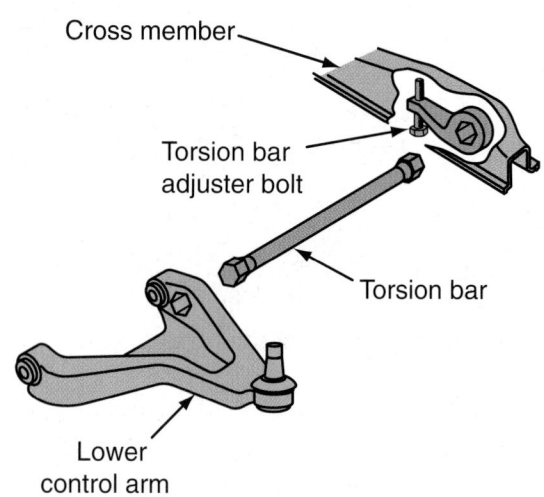

Cross member

Torsion bar adjuster bolt

Torsion bar

Lower control arm

Figure 63.6 A typical torsion bar suspension.

Some vehicles use torsion bars because they do not require much vertical space and the car can be designed to be lower on the front. Compared to coil and leaf springs, torsion bars can store a higher amount of maximum energy. A shorter, thicker torsion bar can carry more load than a longer, thinner one. Spring tension can be adjusted by turning a screw against a bracket mounted on one of the torsion bar ends. This allows restoration of the vehicle's correct ride height prior to an alignment.

Leaf Spring

Most leaf springs are mounted at a right angle to the axle (**Figure 63.7**). They are very resistant to lateral movement, so control arms or struts are not needed. A **leaf spring** is made of a long, flat strip of spring steel or composite fiber rolled at both ends to accept a pressed-fit rubber insulating bushing. The front end of the spring is attached directly to the frame, and the rear of the spring is connected to the frame with a spring **shackle** (**Figure 63.8**). The spring shortens and lengthens as it compresses and rebounds, and the shackle compensates for these changes.

As a leaf spring is deflected, it becomes progressively stiffer. To provide a variable spring rate, extra springs of varying lengths, called *leaves*, are added to the *master* leaf. Each leaf is curved more than the one above it. Only the master leaf is rolled at the ends. A **center bolt** extends through a hole in the center of all of the leaves to maintain their position in the spring, and metal clips keep the leaves centered. The main leaf

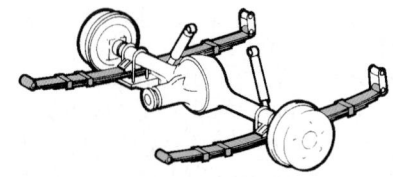

Figure 63.7 A rear suspension with multi-leaf springs.

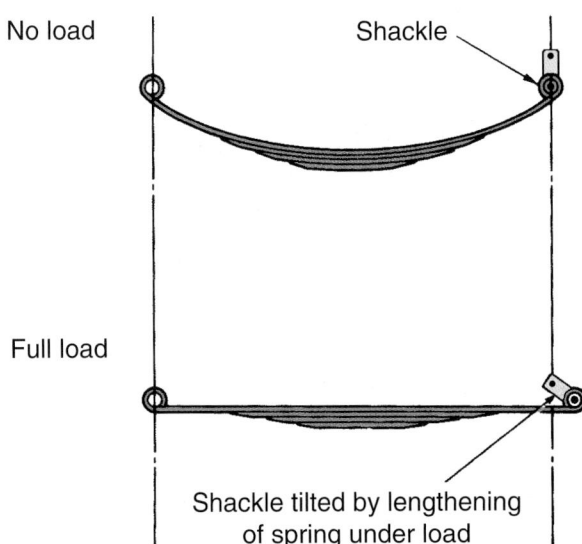

Figure 63.8 Action of a leaf spring.

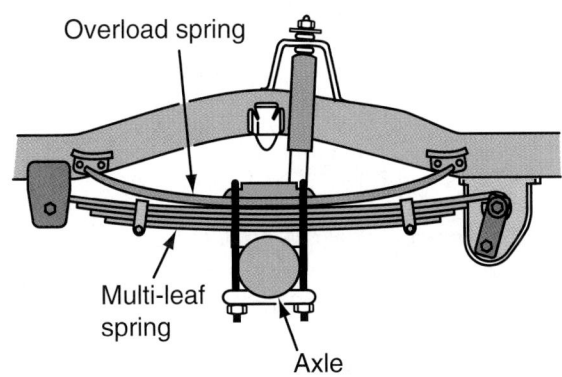

Figure 63.9 An overload spring does not work until the other leaves have deflected enough under load to contact it.

is the strongest leaf in the spring pack. The rest of the leaves are progressively shorter, as they are positioned farther from the main leaf. Some trucks have one or more **overload leaves** that do not work until the other leaves have deflected enough under load to allow them to come into contact (**Figure 63.9**).

When a multiple-leaf spring deflects, the lengths of the leaves change. Because each leaf slides on the one next to it, friction and noise can result. The friction between the leaves dampens spring oscillations, but it also causes a rougher ride than a coil spring. A **monoleaf spring** is a single strip of steel, thicker in the center and gradually tapering thinner toward the outside ends. Manufacturing a leaf to be tapered gives it a variable spring rate, providing for a better ride. A single-leaf spring also has no problem with friction and noise. Monoleaf springs can be mounted longitudinally or transversely and can be used in front or rear suspensions.

Some vehicles made since the early 1980s have **composite springs** made of reinforced fiberglass or graphite reinforced plastic. Corvettes, for instance, use a composite leaf spring mounted transversely in the

rear. Weight is about 30 pounds less per spring, and composite materials do not rust. They are expensive, but they can be easily manufactured to a taper across their length. Because they are a single leaf, they have no center bolt.

Air Springs

An air spring has a rubber air chamber attached by tubing to an air compressor. Air springs are found on suspension systems that control ride height. Some systems can control spring rate as well. Some vehicles use air springs as the *only* springs. Others use coil springs to support the weight of the vehicle, while an auxiliary air spring enclosed within the coil spring adjusts ride height. Some light trucks and SUVs have air suspension systems used in conjunction with the regular leaf springs that act as adjustable overload springs. The different air spring designs are covered in more detail later in this chapter.

■ SUSPENSION CONSTRUCTION

There are different suspension designs. Some parts are common to different suspension types, while others are unique to one suspension design. Different suspension designs and parts are covered in the next section.

Independent and Solid Axle Suspensions

A **rigid axle**, called a *straight axle* or *solid axle*, is found on the rear of most RWD vehicles, some FWD vehicles, and some heavy truck front ends. On truck front ends, a rigid axle is called an *I-beam* (**Figure 63.10**). I-beam

(a)   Rigid axle suspension

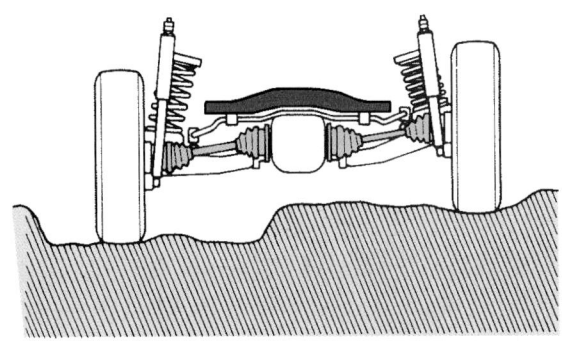

(b)   Independent suspension

Figure 63.10 When an independent suspension goes over a bump, only one wheel deflects.

axles are very strong and can support a great deal of weight. When a wheel attached to a rigid axle goes over a bump, the wheel on the other side is affected by that movement, too (**Figure 63.10a**). Because the rigid axles on I-beam front ends are heavy, they increase the vehicle's unsprung weight, which results in a rougher ride.

Independent suspensions are found on most passenger car front ends, and on some rear ends. When a wheel on a car with independent suspension goes over a bump, only that wheel will move up and down (**Figure 63.10b**). Independent suspension systems have less unsprung weight than rigid axle suspensions, which provides improved ride quality as well.

Control Arms

Control arms are used on independent suspensions to allow the springs to deflect (move up or down). When they are A-shaped like the ones shown in **Figure 63.11**, they are called an **A-arm** or a **wishbone**. Another type of control arm commonly found on lighter cars has a single bushing with a *strut rod* (also called *radius rod*) for stability (**Figure 63.12**).

Bushings

Rubber bushings are used to keep many suspension parts separated. The bushing has an outer and inner metal shell (**Figure 63.13**). The outer part of the bushing is pressed into the control arm and the inner part fits against a pivot shaft (**Figure 63.14**).

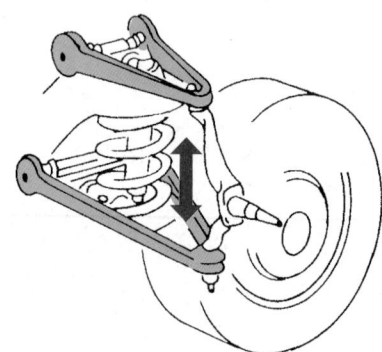

Figure 63.11 Control arms allow the springs to deflect. (*Courtesy of Federal-Mogul Corporation*)

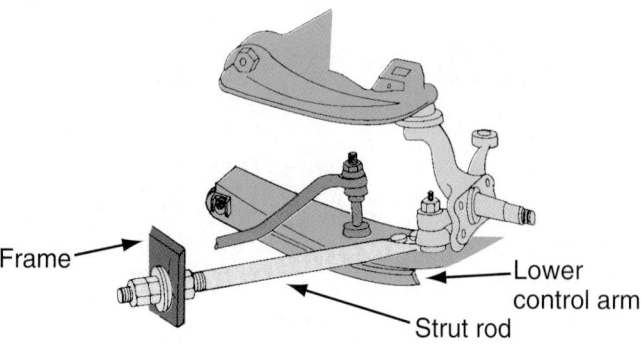

Frame — Lower control arm — Strut rod

Figure 63.12 The strut rod provides stability to a narrow control arm. (*Courtesy of Federal-Mogul Corporation*)

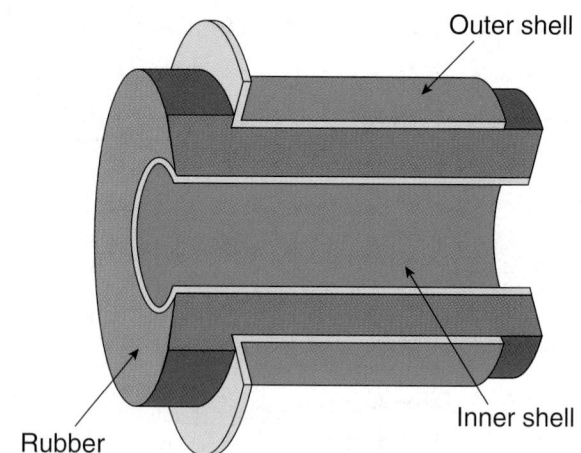

Outer shell — Inner shell — Rubber

Figure 63.13 Rubber bushings keep suspension parts separated.

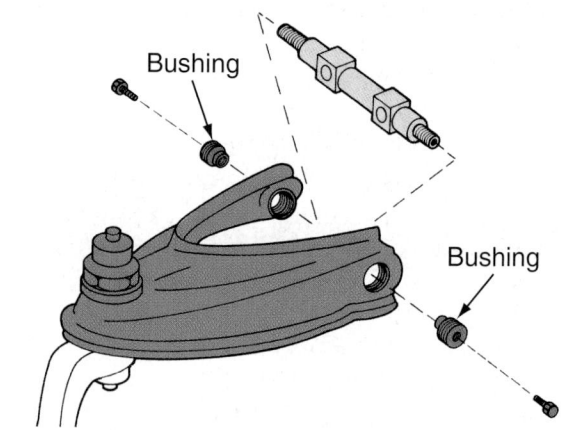

Bushing — Bushing

Figure 63.14 Control arm bushings.

During compression or rebound, the control arm moves up or down, twisting the rubber bushing and allowing parts to pivot without any metal-to-metal contact (somewhat like a vulcanized motor mount). Some of the suspension's resistance to body roll comes from the resistance of the bushings to twisting.

Ball Joints

Ball joints (**Figure 63.15**) attach the control arm to the spindle. A ball joint allows motion in two directions, moving with the same up-and-down motion as the bushings on the other end of the control arm. A ball joint also allows the spindle to pivot for steering. Depending on the design of the suspension, a ball joint will either be pulled apart or compressed together as it supports its load. More information on types of ball joints is found later in this chapter.

■ SUSPENSION TYPES

Several front suspension designs have been used on vehicles. Two of the most popular ones are the short-and-long arm suspension (SLA) and the Macpherson strut

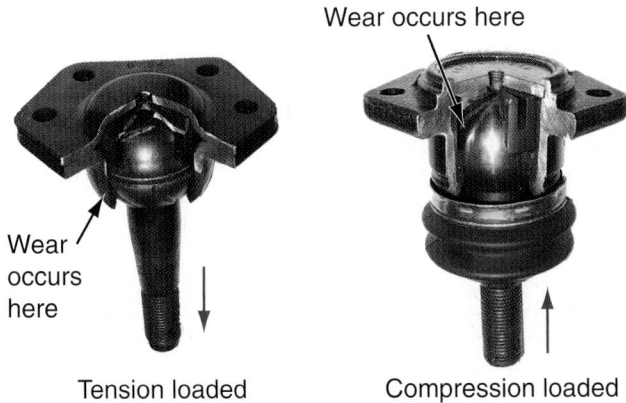

Wear occurs here

Wear occurs here

Tension loaded   Compression loaded

Figure 63.15 Two ball joint designs. A tension-loaded joint pulls apart when loaded. A compression-loaded joint is pushed together. *(Courtesy of Tim Gilles)*

(**Figure 63.16**). Rear suspensions on cars are sometimes independent, but most SUVs have rigid axles. Some are used with leaf springs, but most have coil springs.

Short-and-Long Arm Suspension

SLAs are used on both front and rear wheels. Because both upper and lower control arms resemble the shape of a wishbone, this suspension system is called a "double wishbone." The SLA double wishbone suspension uses two control arms of unequal length that are not parallel to one another. The top control arm, the shorter one, slants downward toward its outer end. As the vehicle travels over a bump, the spring compresses and the outer end of the control arm moves upward in its arc of travel. This causes the top of the wheel and tire to tilt slightly *outward*. The tread width of the front tires remains nearly constant as the spring deflects, however. Although the short upper control arm travels in an arc, the outer end of the longer lower control arm remains in relatively the same plane (**Figure 63.17**).

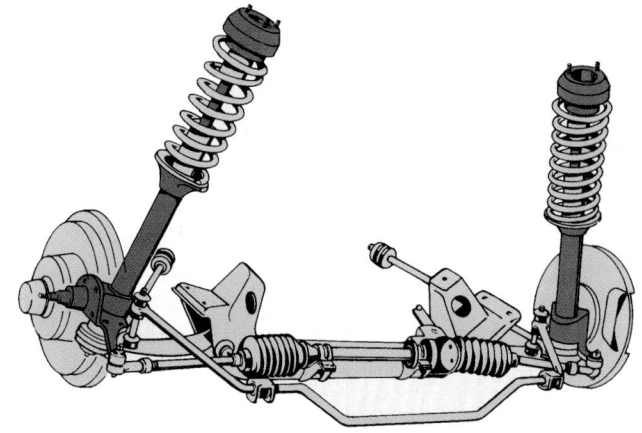

Macpherson strut

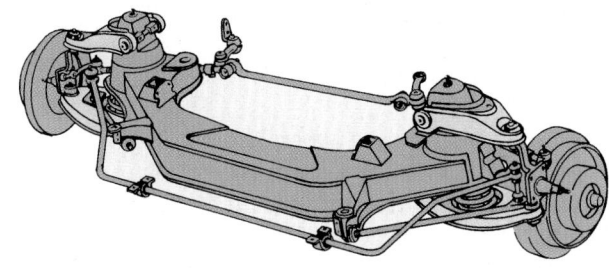

Short/long arm

Figure 63.16 Two popular suspension designs. *(Courtesy of Federal-Mogul Corporation)*

If the control arms were both the same length, the tire would slide from side to side as it goes over bumps (**Figure 63.18**).

Some SLA suspensions on unibody cars have the coil spring above the upper control arm. On vehicles with frames, the spring is situated below the upper control arm.

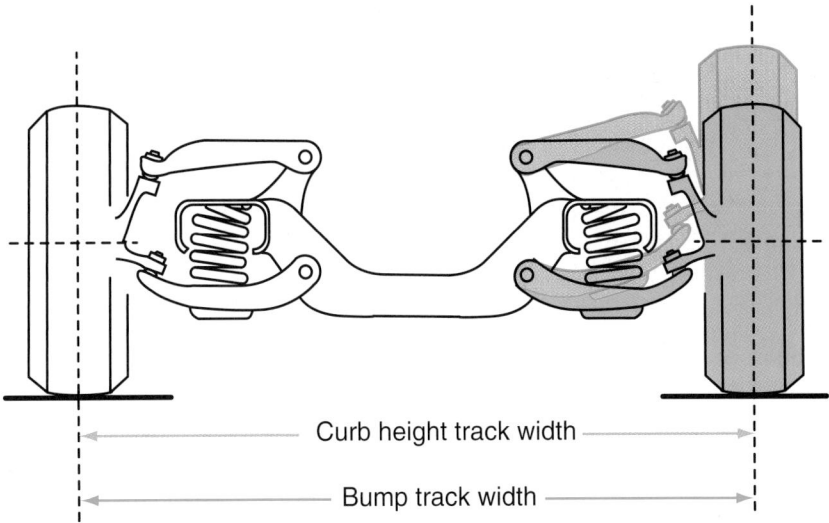

Curb height track width

Bump track width

Figure 63.17 SLA suspensions allow the track width to remain constant, reducing tire scrub.

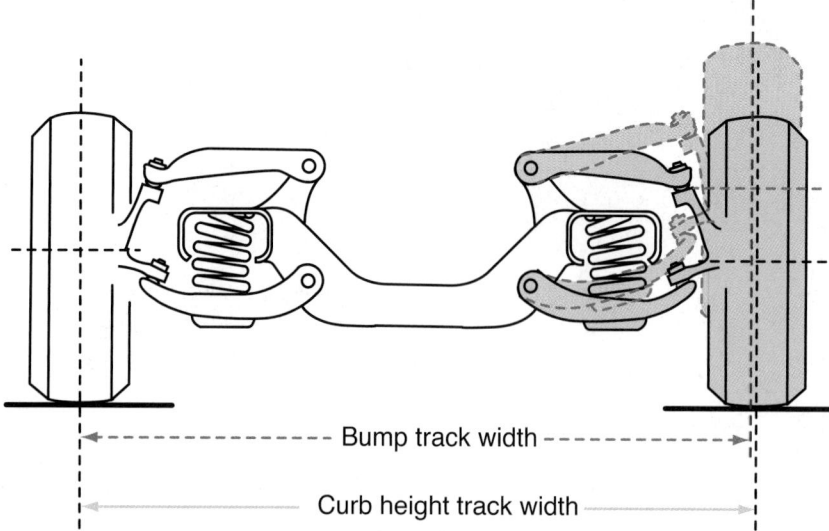

Bump track width

Curb height track width

Figure 63.18 With control arms of equal length, the tire slides as it goes over a bump, scrubbing the tires and wearing them out.

Double wishbone suspensions are popular for several reasons. They have improved directional stability and steering control and react to body roll less than any other type of independent suspension. SLA suspensions also maintain precise wheel position under all driving conditions and can have a large amount of vertical travel without scrub, a change in the distance between the treads of the two front tires. These suspensions are good for absorbing bumps, and the tires maintain more surface area on the road.

Sometimes a double wishbone suspension is not an option because it requires more space than Macpherson struts or some multilink designs.

Macpherson Strut Suspensions

Many smaller cars use the Macpherson strut design. Most consider the SLA suspension to be a better suspension, but weight, space savings, and cost considerations account for the Macpherson strut's popularity.

> ### ■ HISTORY NOTE ■
>
> *Earl Steele MacPherson designed the Macpherson strut suspension during the late 1940s. MacPherson, who was born in Britain, became an engineer, first working for Chevrolet and later with the European division of Ford. The Macpherson strut was first used on the 1949 French Ford Vedette.*

A Macpherson strut incorporates the coil spring and shock absorber into its front suspension (**Figure 63.19**) using only a single control arm on the bottom. It does not have an upper control arm and upper ball joint. The spindle is attached to the strut housing, and a strut bearing at the top allows the entire unit to rotate when steering. Shock absorbers are covered later in this chapter.

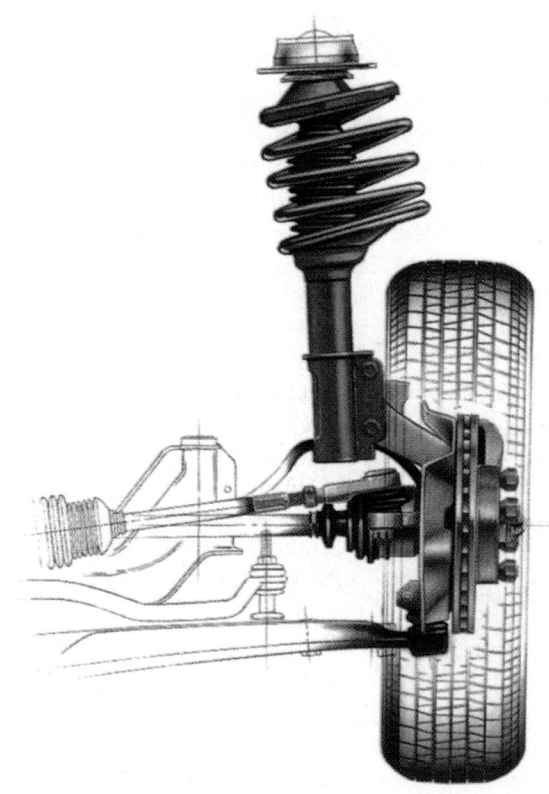

Figure 63.19 A Macpherson strut assembly. *(Courtesy of Ford Motor Company)*

■ HIGH-PERFORMANCE SUSPENSIONS

Several suspension types have been used in sport and racing vehicles. The multilink and the double wishbone are the most popular ones.

Multilink Suspensions

Multilink suspensions give the designer more options in tuning the suspension. When an independent suspension

Figure 63.20 A more complicated five-link suspension.

has more than two control arms, it is called a multilink. Extra links are used to keep the wheel in a more precise position during cornering and on bumps. Steering control is improved and tire wear is minimized. A multilink suspension is basically a double wishbone suspension that has each arm of the wishbone as a separate part. **Figure 63.20** shows a complicated five-link suspension. This suspension is a double wishbone with a fifth control arm. Several variations of the multilink suspension have been used on different luxury and sport vehicles beginning in the late 1980s.

Rear suspensions can be independent, but often they have rigid axles. Some are used with leaf springs, but most have coil springs.

■ SHOCK ABSORBERS

There are four **shock absorbers**, commonly called shocks or dampers, one at each corner of the vehicle. Shock absorbers do not actually absorb shock; their function is to *dampen spring oscillations* by converting the energy from spring movement into heat energy. Normal motion of the springs, which can be controlled using shock absorbers, is uncomfortable to passengers, causes cupped tire wear, and is unsafe.

Shock absorbers are designed to resist or damp out excess and unwanted motion in the suspension. When a spring is compressed, it absorbs energy. During spring rebound, this energy is released. The first cycle is followed by more compression/rebound cycles called oscillations (**Figure 63.21**). Shock absorbers damp out the excess motion. Poor shock absorbers are known to aggravate potential SUV rollovers, especially of top-heavy vehicles like SUVs and vans. Shock absorbers are also critical for proper tire-to-road

Figure 63.21 When a compressed spring rebounds, it begins to oscillate.

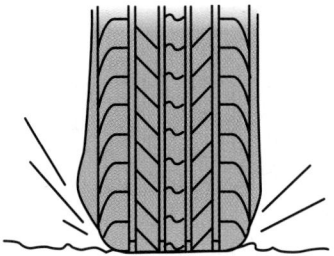

Figure 63.22 Under load, the tire forms a mechanical interlock with the road surface.

contact (**Figure 63.22**). This affects braking, steering, cornering, and overall stability. The effectiveness of antilock brakes is also determined in part by the good tire-to-road contact provided by shock absorbers.

NOTE: *Although shock absorbers absorb much of the road shock, the vehicle's tires are actually the primary shock absorbers.*

■■■ HISTORY NOTE ■■■

*Early shock absorbers were mechanical friction devices. Friction material similar to that found on a clutch or brake was fitted between two levers. One of the levers was attached to the frame and the other to the spring mount or spring (**Figure 63.23**). They required frequent adjustment because they were prone to wear.*

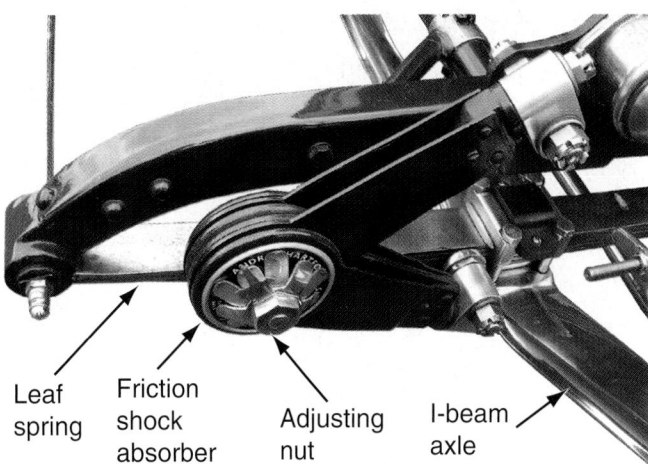

Leaf spring   Friction shock absorber   Adjusting nut   I-beam axle

Figure 63.23 An early friction shock absorber. (*Courtesy of Tim Gilles*)

▓▓ HYDRAULIC SHOCK ABSORBER OPERATION

In a modern hydraulic shock absorber, one end is attached to the suspension; the other is attached to the car body or frame (**Figure 63.24**). The shock is mounted on rubber bushings to allow for slight changes in its angle of installation as it is compressed and extended.

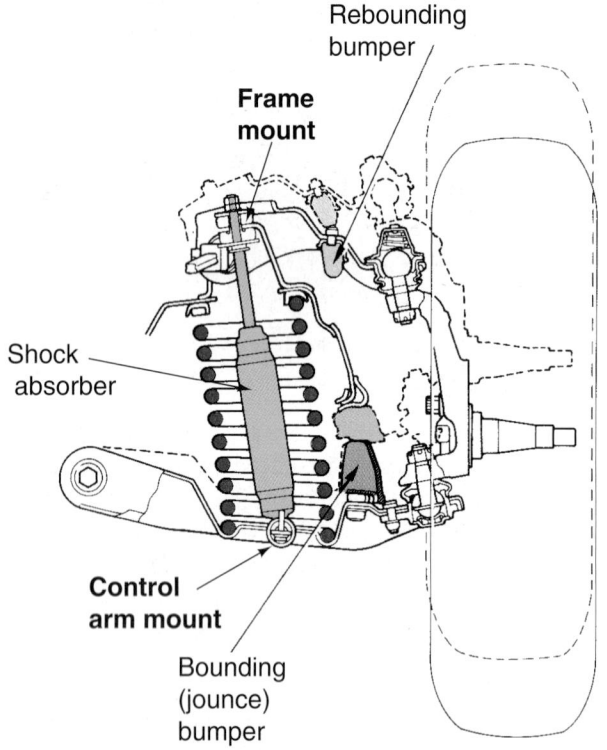

Figure 63.24 One end of the shock is attached to the control arm, and the other end is attached to the frame.

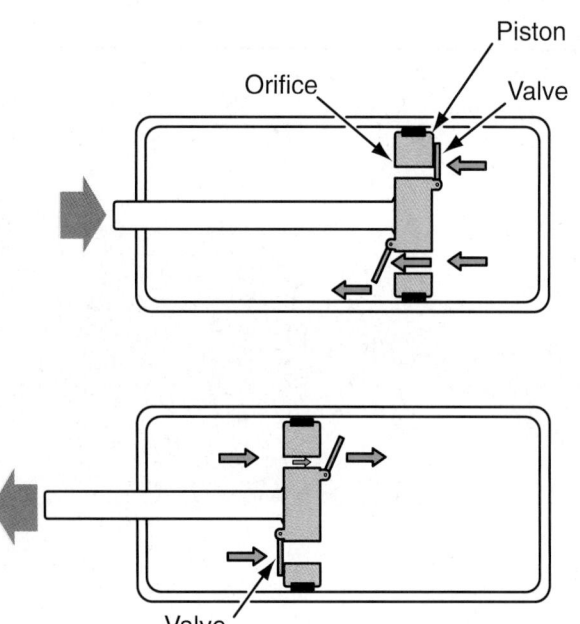

Figure 63.25 Fluid is forced through an orifice to dampen spring action.

To damp the action of a spring, a hydraulic shock absorber forces oil through small openings, called *valves*, like water through a squirt gun (**Figure 63.25**). This action generates hydraulic friction, which converts motion energy to heat energy as it reduces unwanted motion. Using energy in this fashion lessens the oscillations of the springs.

NOTE: *The shock only converts unwanted motion to heat. If it were to remove all of a spring's energy, all motion would stop.*

A shock absorber has two chambers, with a piston that forces fluid through the valves from one chamber to the other. The faster the piston moves, the more resistance it encounters **Figure 63.26** shows the parts of a conventional shock absorber. **Figure 63.27** shows a cutaway Macpherson strut shock absorber.

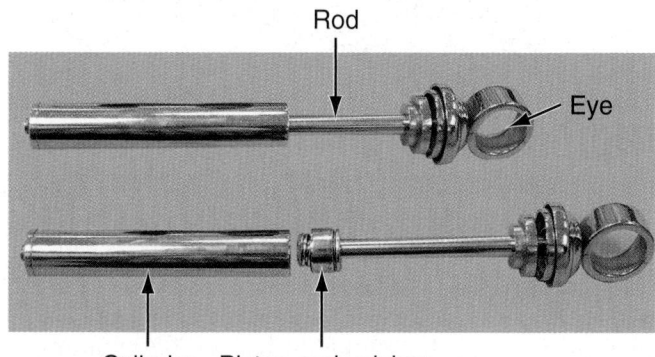

Figure 63.26 Parts of a conventional shock absorber. *(Courtesy of Tim Gilles)*

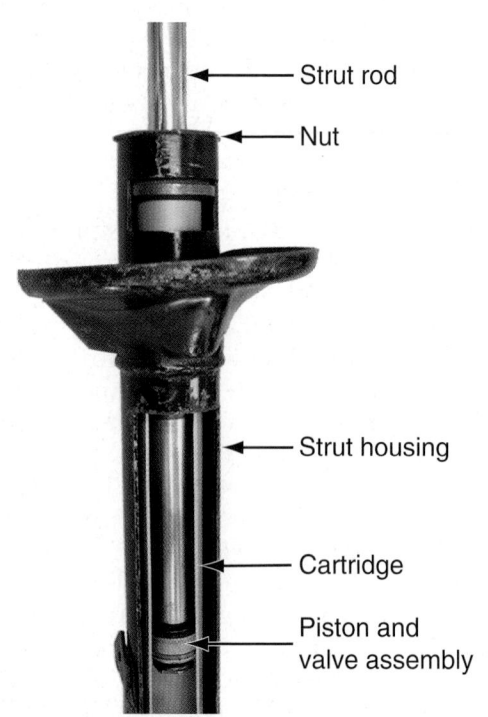

Figure 63.27 Cutaway of a Macpherson strut shock absorber. *(Courtesy of Tim Gilles)*

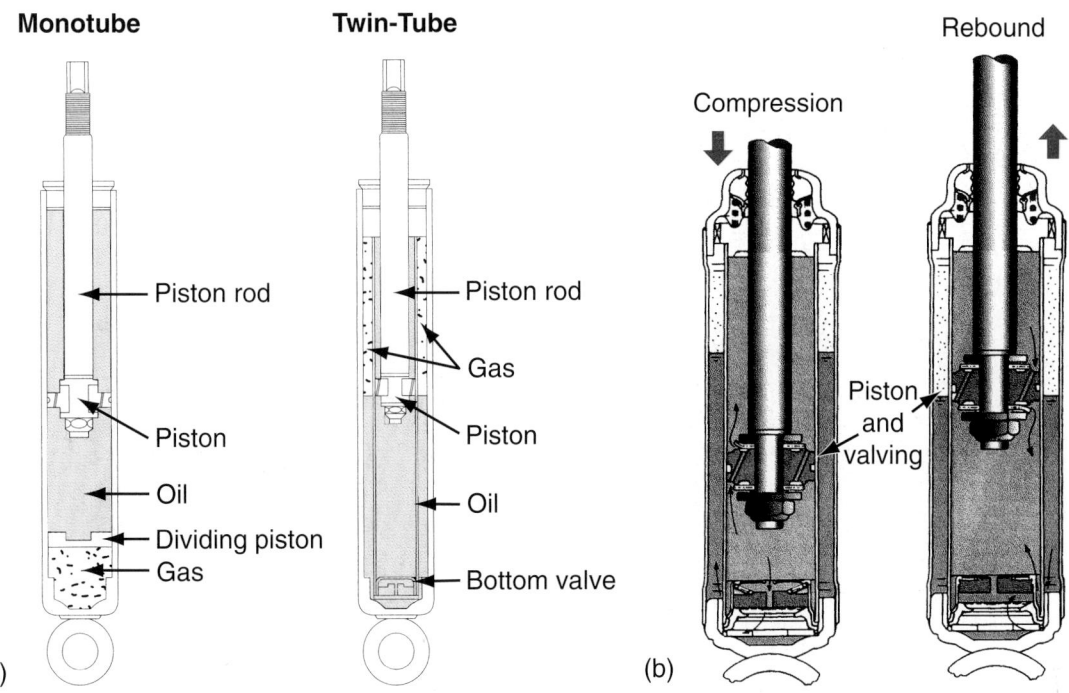

Figure 63.28 (a) Monotube and twin-tube shock absorbers. (b) Shock action during the compression and rebound cycles. *(Courtesy of Tokico (USA), Inc., Performance Division)*

Today's shock absorbers are either twin-tube or monotube (single-tube) designs (**Figure 63.28a**). Monotube shocks are not as common as the twin-tube design but are still found in many applications. As the shock is compressed, the piston rod displaces oil. There must be a reserve space to accept the extra oil (**Figure 63.28b**). In twin-tube designs, the outer tube provides the reserve chamber. A monotube shock usually has the reserve chamber in the same column as the working piston. When the shock extends, oil returns once again to the pressure chamber.

Front and rear shocks are not the same, usually having chambers and valves of different sizes to control the differences in weight between the front and rear of the car. Also, the center of gravity shifts when the vehicle stops or is thrown into a turn, causing the front shocks to do more of the work.

The up-and-down movements of a shock are called compression (or jounce) and *extension* (or *rebound*). Shock absorbers are called **double-acting** because they control motion when moving both up and down. Early friction shocks had a *ratio of* 50–50, controlling the motion of the springs equally on compression and rebound. A typical modern shock provides more resistance on compression than rebound (70/30, for instance).

■ COMPRESSION AND REBOUND RESISTANCE

Unlike springs, which are sensitive to loads, hydraulic shocks are sensitive to velocity. The faster the piston moves through the oil in the shock, the higher the resistance.

> ### ■ SCIENCE NOTE ■
>
> *A basic formula says that the resistance to flow (oil through the piston) goes up as a square of the velocity. Aerodynamics engineers try to reduce resistance, while shock engineers use it to control unwanted suspension motion. Air is also a fluid. Four times as much power is required to push a car through the air at 100 mph as is needed at 50 mph.*

Shocks must deal with a wide variety of suspension motions and velocities. Resistance can be very high, so a series of orifices and valves is necessary to manage the flow properly. Either too little or too much resistance can result in poor adhesion to the road. Too much resistance can cause a harsh ride. Too little resistance allows excess body motion, poor control of unsprung weight, and wheel bounce. A defective shock will have too little resistance.

A chart of resistances provided at various piston velocities is called a damping force curve (**Figure 63.29**).

NOTE *Most modern shocks compress more easily than they extend.*

■ Compression damping is used to control the relatively light unsprung weight of the tires, wheels, and brakes. It works with the spring to keep the tire in contact with the road surface. A car with a wheel that hops off the pavement has a compression control problem with its shock absorber.

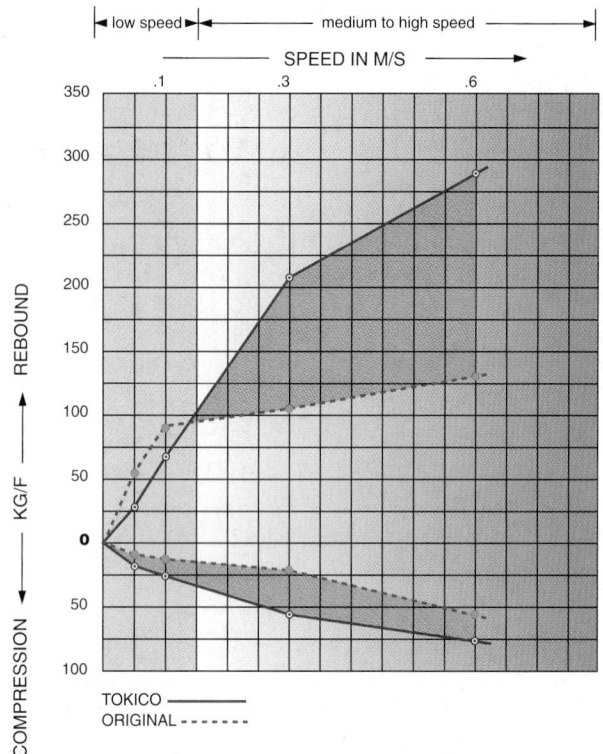

Figure 63.29 A dampening force graph comparing two different shocks for the rear of a pickup truck. Notice how rebound damping increases dramatically at medium and high fluid speed. *(Courtesy of Tokico (USA), Inc., Performance Division)*

■ Rebound damping controls excess chassis motion as the shock extends, when the heavier weight of the car body is in motion. A car that floats as it travels down the road has a rebound control problem with its shock absorber.

Because piston velocities on the rebound side (body motion) tend to be much lower than on the compression side (hitting a chuck-hole), compression damping force is usually much lower than rebound resistance. Fluid movement changes depending on whether the car hits a hole (fast fluid movement), or whether the car leans over (called **body roll**) as it goes through a turn (slow fluid movement). Virtually all shock absorbers use three stages of valving. The initial valve (first damping stage) is a small hole because little force is generated by low piston velocities. As the piston moves faster, an additional (second stage) valve opens at a level of pressure determined by the manufacturer. The final stage of valving (third stage) is called the high-speed restriction.

■ BUMP STOPS AND LIMITERS

Shock absorber movement follows the travel of the rest of the vehicle's suspension. If the shocks or struts reach their extreme travel limit by topping or bottoming, damage can result. Some vehicles use rubber bump stops (sometimes called snubbers) on control arms to

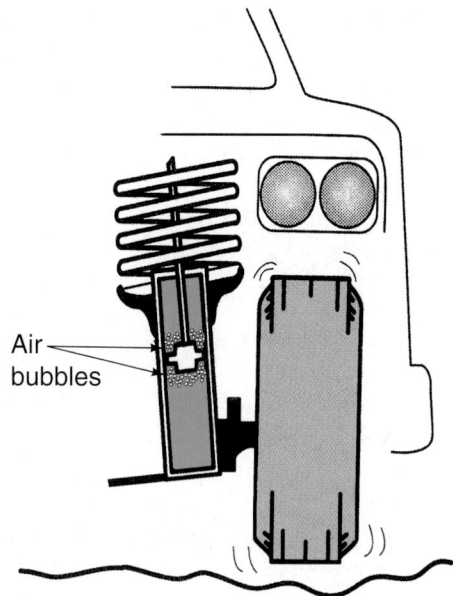

Figure 63.30 Air bubbles form around the shock absorber valves during bumpy rides. *(Courtesy of Tokico (USA), Inc., Performance Division)*

limit excessive travel. Internal bump stops are generally found only in monotube struts or strut cartridges.

Normal full shock absorber travel is considered to be a good thing. Altered vehicle height, however, is a typical cause of *reduced* total travel that allows a shock to top out or bottom out. Raising or lowering a vehicle excessively in either direction can force shocks and struts to their extreme travel points. This can result in damage to the shock as well as other suspension parts.

Aeration of Fluid

The shock must be installed in a nearly vertical position so when it extends, air will not be drawn in place of fluid from the outer reservoir. This would result in a "skip" as the shock moves through its range of motion. **Aeration**, or cavitation, is when hydraulic fluid becomes mixed with air.

As the shock piston moves, pressure builds up in the fluid in front of it. A drop in pressure happens behind the piston, causing air bubbles and making the fluid foamy. When driving on rough roads, fluid is rapidly forced through the check valves. The entire area of fluid around the piston becomes aerated, and shock operation suffers (**Figure 63.30**). During normal operation, the air will usually work its way back to the air chamber.

There are two ways that designers avoid the tendency toward aeration. One of them is to use spiral grooves or a flat spiral ring around the outside of the reservoir tube. Another way to avoid aeration of the shock fluid is to use gas shocks.

■ GAS SHOCKS

Gas shocks were invented to control cavitation or foaming of the oil. All oil has some air or gas bound up in solution. Pressurizing the oil keeps the air in solution so the piston works in clear oil and can provide consistent damping.

■ HISTORY NOTE ■

In 1953, French physicist Christian Bourcier de Carbon designed the monotube high-pressure gas shock absorber and founded the De Carbon Company the same year. A short time later, a license was sold to the German company Bilstein. The shocks became original equipment on the 1957 Mercedes-Benz. The patent has now expired, and many top companies use monotube technology.

Pressurizing the oil column in the shock absorber keeps the bubbles in the solution. Compare this to what happens when a carbonated beverage is first opened. When the top is removed, the pressure drops and the bubbles come out of the soft drink. Pressurizing the shock keeps the shock oil clear and eliminates the skips that occur in the action of a regular shock absorber.

Some gas shocks have a pressurized gas-filled cell. This is a plastic bag that takes the place of the free air in a normal shock. These bags usually have low pressure (10–20 psi of refrigerant gas) in them. True gas shocks have a reservoir pressurized with nitrogen gas at 100–200 psi (**Figure 63.31**).

Nitrogen is used because it is dry (no water vapor), it is chemically inert at normal shock temperatures, and it is fairly inexpensive.

Twin-tube gas shocks are pressurized with nitrogen gas at 90–140 psi, depending on the manufacturer and the damping force desired. They are usually less expensive to manufacture than a monotube type.

Monotube gas shocks have pressures between 250–400 psi. Because the monotube has no base or bottom valve, it needs the high pressure to support the compression valving as well as control cavitation.

NOTE: *Gas shocks are packaged with a strap to hold them in the compressed position. They extend on their own when the strap is removed for installation.*

Monotube gas shock absorbers, also called DeCarbon shocks after their inventor, offer better cooling of the shock fluid and can be mounted upside down. Monotube gas shocks are more susceptible to physical damage than twin-tube shocks.

Rear Shock Mounts

To minimize vibration and improve ride quality, rear shock absorbers on RWD vehicles are mounted in two ways. Some of them have the shocks slanted inward and toward the rear at the top (**Figure 63.32**). Others have one shock mounted in front of the axle, with the other one mounted behind it (**Figure 63.33**).

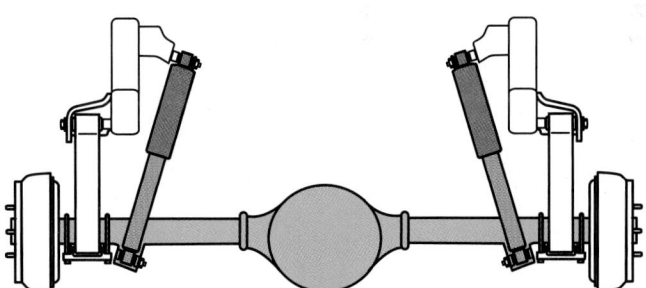

Figure 63.32 Shock absorbers mounted to the rear of the axle.

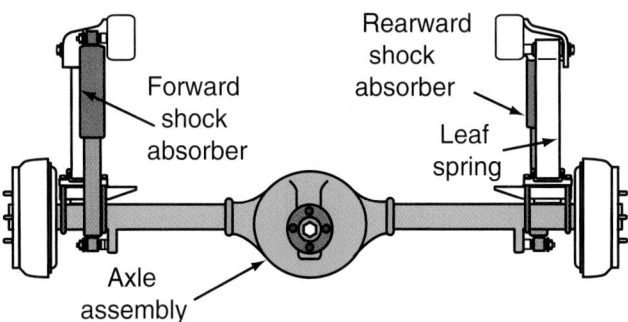

Figure 63.33 Shock absorbers staggered on each side of the rear axle.

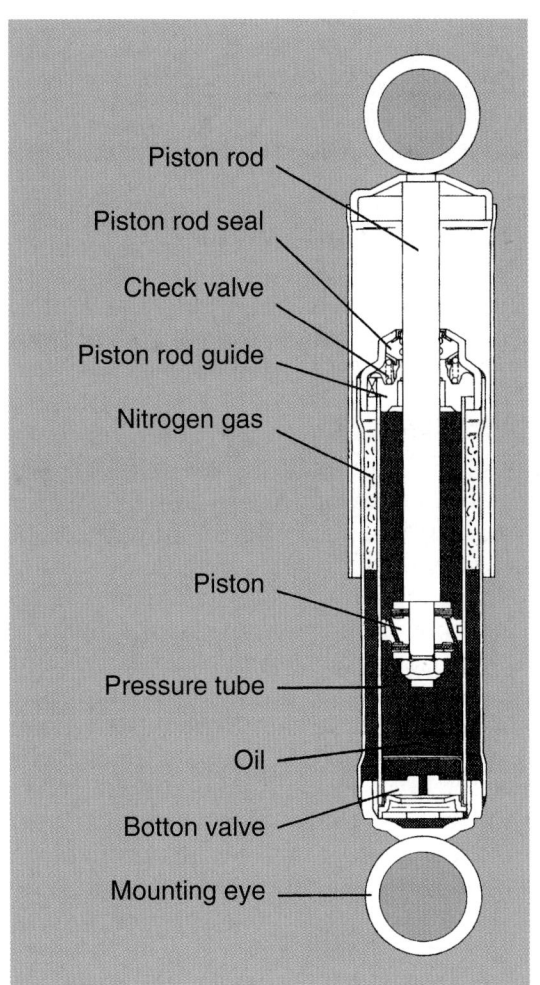

Figure 63.31 Parts of a pressurized gas shock. *(Courtesy of Tokico (USA), Inc., Performance Division)*

Labels on Figure 63.31:
Piston rod
Piston rod seal
Check valve
Piston rod guide
Nitrogen gas
Piston
Pressure tube
Oil
Bottom valve
Mounting eye

Labels on Figure 63.33:
Forward shock absorber
Rearward shock absorber
Leaf spring
Axle assembly

■ AIR SHOCKS/LEVELING DEVICES

Shock absorbers are not normally designed to carry the weight of the vehicle. If they were, the height of the vehicle would be affected when they were removed or when they wore out.

There are aftermarket devices that use the shock absorber as a means of correcting or adjusting the height of the vehicle. The two common ones are air shocks and coil springs that are mounted on the outside of the shock body. There is a disadvantage to leveling the vehicle in either of these ways. When shocks support the weight of the vehicle, the shocks and shock mounts are prone to breakage. Air shocks (**Figure 63.34**) have a rubber bladder that, when filled with varying amounts of air, raises the rear of the vehicle to compensate for a heavier-than-normal load.

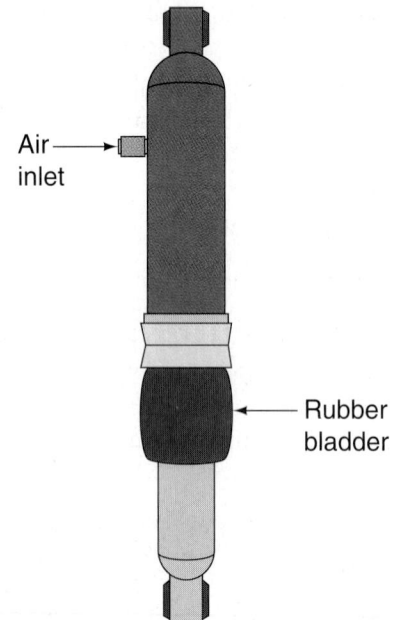

Figure 63.34 An air shock. *(Courtesy of Tenneco Automotive Operating, Inc.)*

Figure 63.35 A coil spring shock.

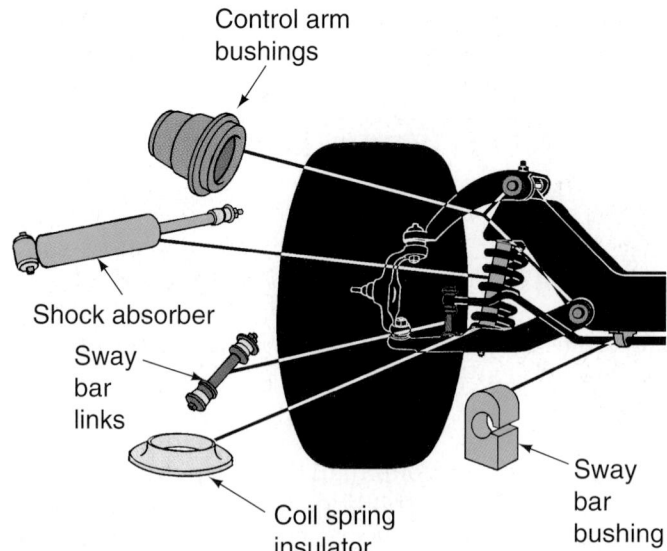

Figure 63.36 Parts are insulated from each other with rubber bushings. *(Courtesy of Federal-Mogul Corporation)*

NOTE: *A small amount of air must be kept in the bladder at all times. If the vehicle is driven while the bladder is empty, the bladder can be folded into the shock and become torn.*

A coil spring shock (**Figure 63.35**) has a constant-rate spring that works the same during both extension and compression. Its action in preventing body roll is similar to that of a stabilizer bar (covered later in the chapter).

■ OTHER FRONT END PARTS

Other parts are attached to the suspension to help control the ride. Parts like stabilizers and strut rods are insulated from front suspension parts and the frame with rubber bushings (**Figure 63.36**).

■ STABILIZER BAR

A stabilizer bar, also called a *sway bar* or an *anti-roll bar,* is used on the front or rear of many suspensions (**Figure 63.37**). It connects the lower control arms on both sides of the vehicle together, reducing sway and functioning as a spring when the car leans to one side. When both tires move up or down an equal amount, the stabilizer simply rotates in its bushings (**Figure 63.38**). If one of the wheels moves up, the bar twists as it tries to move the other wheel along with it.

Stabilizer links and bushings provide some flexibility and softness to the sway control so the suspension can still operate somewhat independently during minor bumps.

Spindle and Ball Joints

The *spindle support arm* (also called the *steering knuckle*) includes the axle that the wheel bearing is mounted on. Ball joints (see Figure 63.15) attach the control arm

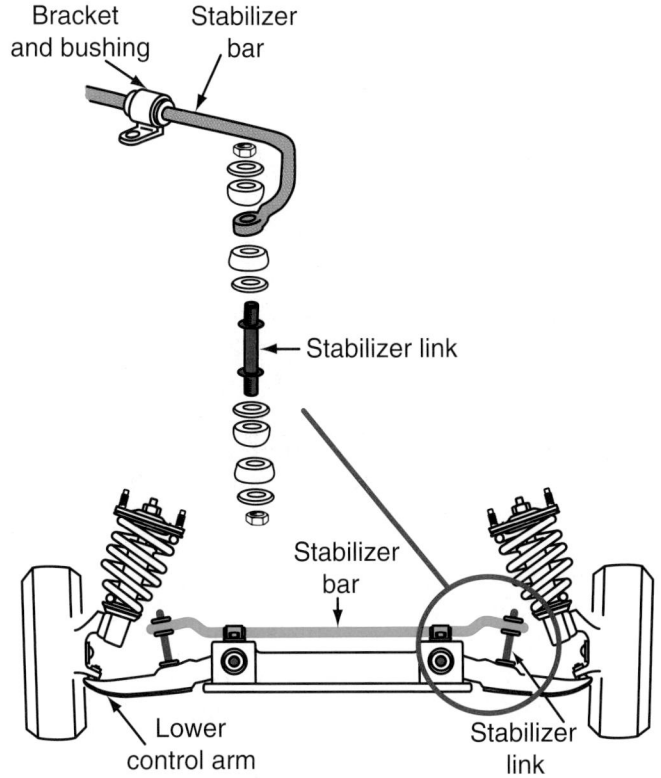

Figure 63.37 A stabilizer bar and links.

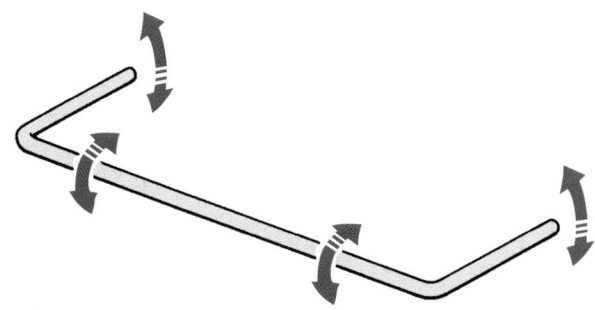

Figure 63.38 Action of a stabilizer.

to the spindle. Depending on the design, ball joints will be located on top of or under the control arm.

On suspensions with two control arms, ball joints function as either a **load carrier** or a **follower ball joint** (**Figure 63.39**). The ball joint on the control arm that has the spring mounted on it is the load carrier. The function of the follower ball joint is to maintain parts in the proper position.

There are two styles of ball joints. They are either compressed all the time *(compression type)* or always pulling apart *(tension type)* (**Figure 63.40**).

The Macpherson strut suspension system uses only one ball joint because it has only one control arm. It has a pivot bearing at the top of the strut that allows the strut to rotate for steering (**Figure 63.41**). On a

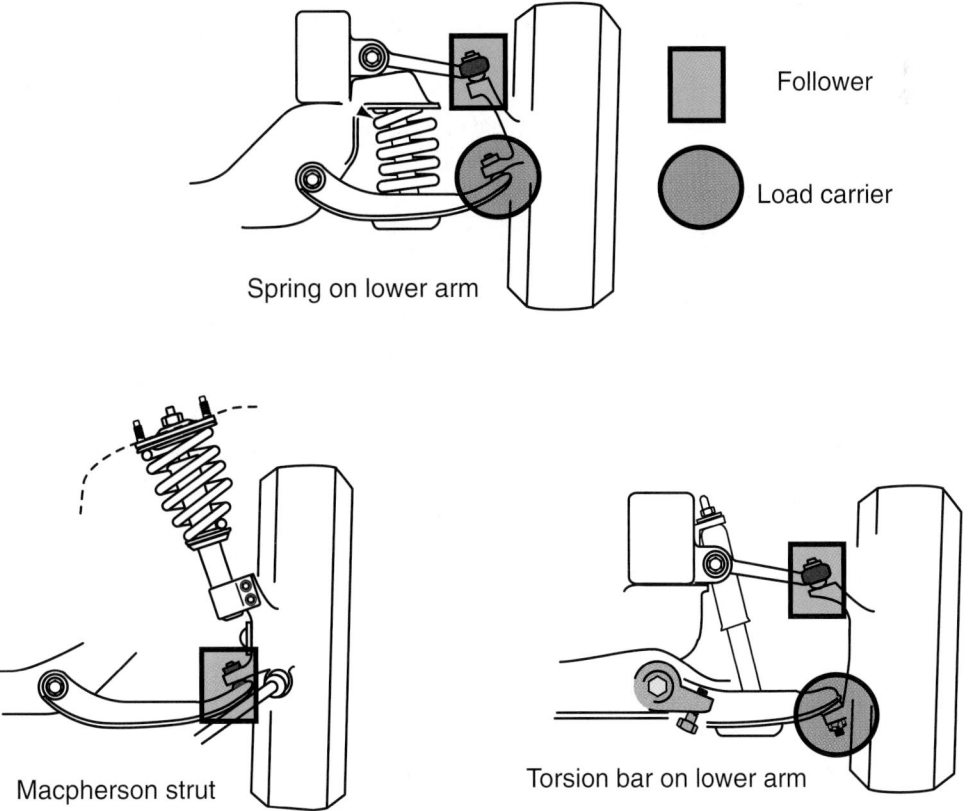

Figure 63.39 The location of load carrier and follower ball joints.

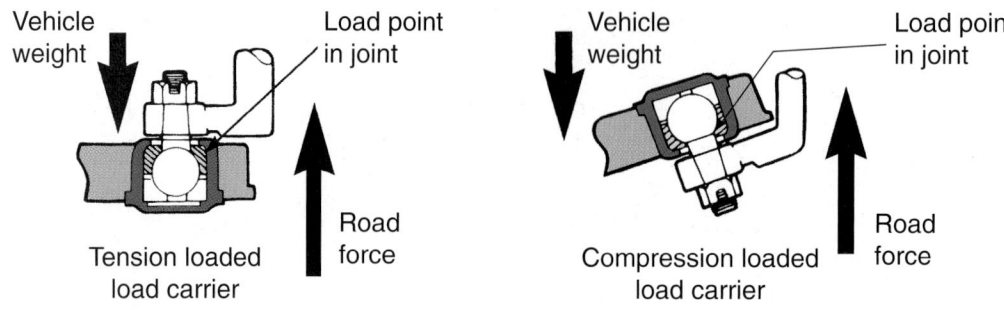

Figure 63.40 Tension and compression loaded ball joints. *(Courtesy of Federal-Mogul Corporation)*

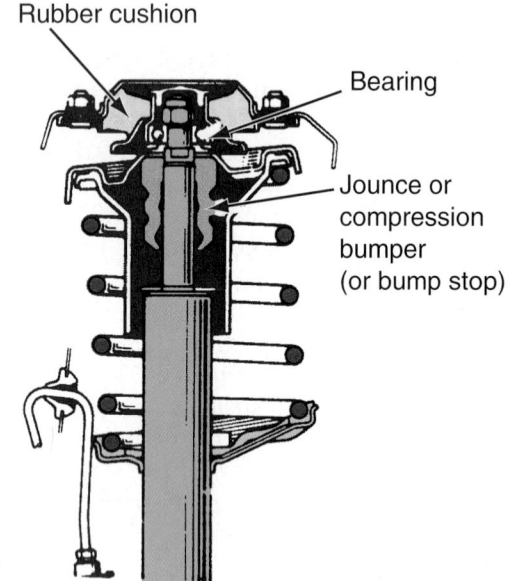

Figure 63.41 A pivot bearing allows the strut to rotate.

strut suspension, the pivot bearing carries the load and the ball joint is a follower.

■ SUSPENSION LEVELING SYSTEMS

Normal suspension systems are called **passive** systems. Passive systems have either a firm or soft ride. Their height varies according to mechanical forces on the suspension, and they do not adjust to these changes. Early leveling systems were manual, using air shocks and a compressor. A manual switch was used to change the height of the car body. Today's systems are automatic, and they are called **adaptive suspension systems**.

Electronically controlled suspension systems are used by many manufacturers on some of their luxury vehicles. They keep the vehicle at the same height when weight is added to different parts of the car. Some of the advanced systems can vary the damping capability of the shock absorbers as well. Most of the systems have air or conventional springs.

Automatic Suspension Leveling

There are two-wheel and four-wheel automatic leveling systems. The simplest leveling systems use air

shocks or air springs filled by air from a compressor (**Figure 63.42**). An air dryer is attached to the pump to condition the air before it enters the shocks. A height sensor connected to the frame and axle housing is used for vehicle height input (**Figure 63.43**). It can turn on the compressor or bleed air to correct changes in height.

With electronically controlled systems, a computer reacts to signals from sensors at all four wheels to change the amount of air in air springs at each wheel (**Figure 63.44**). The aim is to keep the vehicle level to the road from side to side and front to rear.

Several sensors are found with different systems. Some of them and their functions include:

■ Three or four height sensors are located at the wheels. They are often rotary Hall-effect sensors. When there are three sensors, one is for the solid rear axle. Height sensors electronically measure the distance between the control arm or axle and the frame. The computer can also use their signal to prevent the vehicle from bottoming out when going over major variations in the road surface, like railroad tracks.

■ Signals from brake and door sensors help the computer to decide to disable automatic height adjustment when the car is stopping or when passengers are getting in or out.

■ Speed sensors are used in some systems to lower the vehicle in the front or both front and rear for high-speed aerodynamics. In some systems, spring rate is increased in response to higher speed, too.

■ A photo diode and shutter in the steering columns of some advanced systems cause the spring rate to change when the vehicle is turning.

■ A G-sensor on some cars helps the system accommodate severe maneuvers with a change in spring rate.

■ An acceleration sensor to sense the rate of acceleration uses the throttle position sensor or mercury switches.

A mode switch is installed in the dash of some cars. It lets the driver select the degree of ride harshness desired.

Some light trucks and SUVs have air suspension systems used in conjunction with the regular leaf springs. The air spring, positioned between the truck

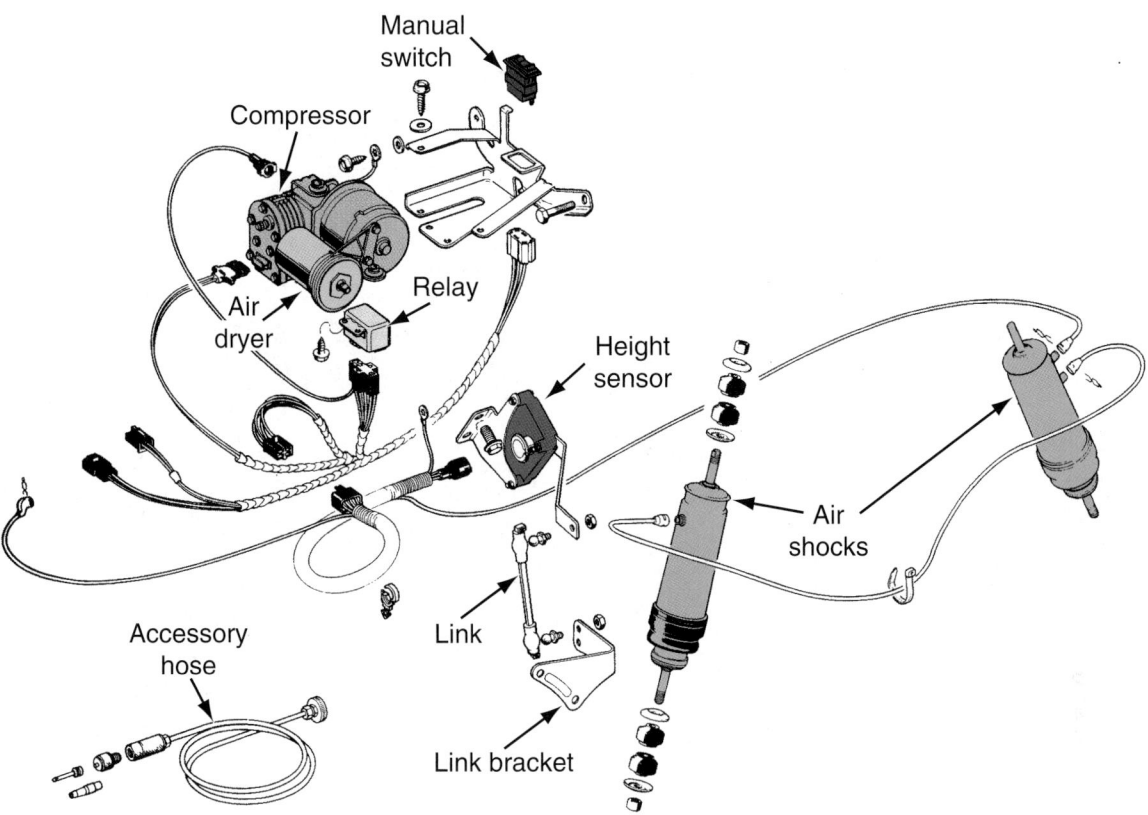

Figure 63.42 An automatic leveling system that uses air shocks. *(Courtesy of DaimlerChrysler Corporation)*

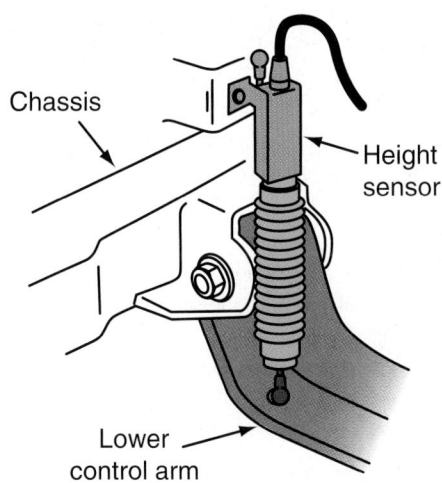

Figure 63.43 Height sensors are connected to the body and a suspension member. It sends a signal to the control module if a height change occurs.

frame and the leaf spring, acts as an adjustable and overload spring (**Figure 63.45**).

Electronically Controlled Shock Absorbers

Electronically controlled shock absorbers have variable valving. The amount of shock damping changes as the size of a metered orifice within the shock is manipulated by the computer. An actuating motor on the top of the shock (**Figure 63.46**) turns a control rod that changes the size of the orifice (**Figure 63.47**).

The newest adaptive systems use solenoid actuated shock valves, which allow almost instantaneous changes in the size of a shock's damping orifice. The suspension can react to changes in body height in 0.010 second (10 milliseconds). These devices can be found on shocks and struts.

Some electronically controlled shocks use a variable orifice controlled by gas or air. Normal shock absorber valves are used.

Shock absorbers deflect in one direction to control fluid flow:

- At lower speeds, they operate normally (the valve is fully open). This is the comfort mode, and fluid flows unhampered through the orifice. The valves work as normal.
- The orifice is partly restricted between 40 and 60 mph. Fluid flow is balanced between the small orifice and the normal shock valve. This provides a medium degree of control.
- At speeds above 60 mph, or during acceleration or braking, the orifice shifts to a firm ride position. Flow through the orifice is restricted to increase effort required to move oil through the shock absorber.

Magneto-Rheological Fluid Shock Absorbers

Another type of electronically controlled shock absorber that does not use electromechanical solenoids or valves is the magneto-rheological, or MR, shock absorber (**Figure 63.48**). It uses a fluid that rapidly changes its

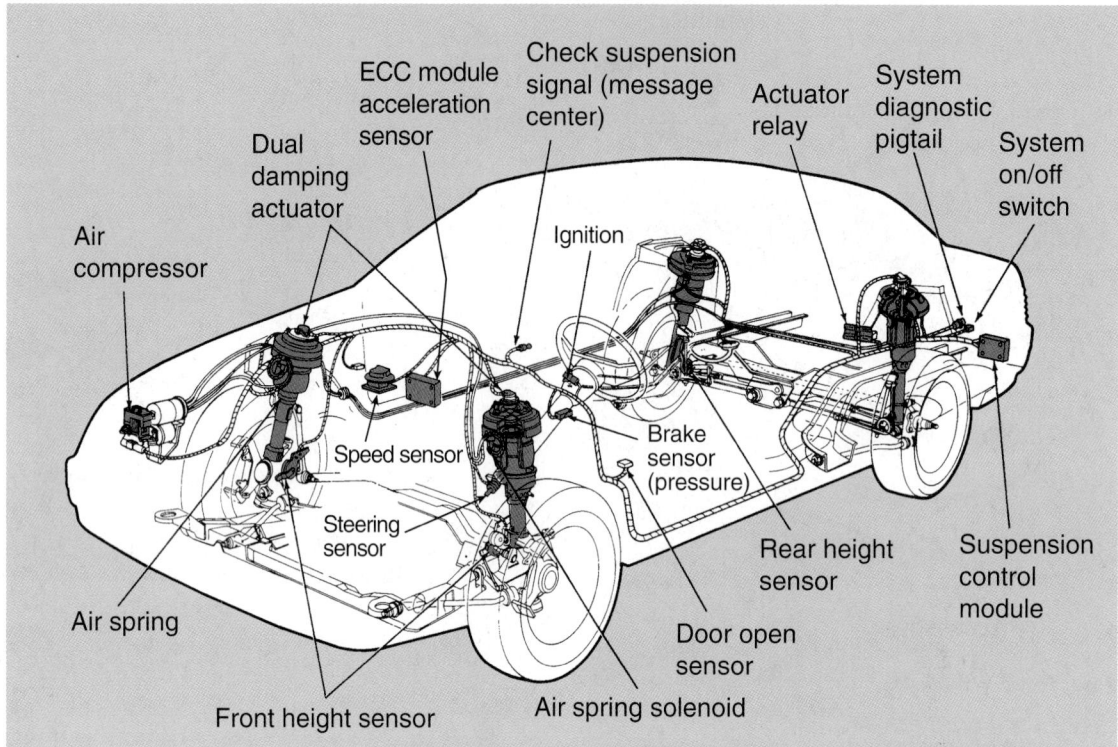

Figure 63.44 A computer system controls the air springs at each wheel. *(Courtesy of Ford Motor Company)*

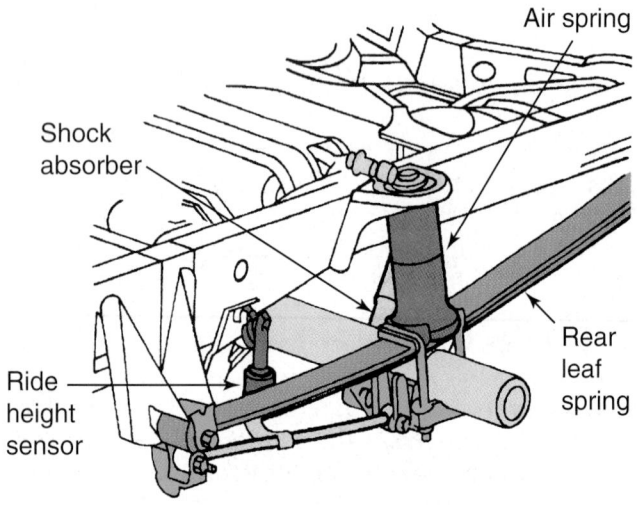

Figure 63.45 An air spring used with a conventional spring.

Figure 63.46 An electronically controlled shock absorber. *(Courtesy of Tim Gilles)*

viscosity in response to computer-controlled signals. The fluid is a synthetic oil with iron particles in suspension. The iron is dispersed throughout the fluid, allowing the shock to operate as a normal shock under ordinary driving conditions. Each shock has an electrical winding that can be energized by the control module in response to a signal from a wheel position sensor when a large bump in the road is encountered. This causes the iron particles in the shock fluid to align themselves (**Figure 63.49**), turning the fluid into a very viscous gooey mass. The current is supplied up to 1,000 times

per second so it can very quickly vary the damping characteristics of the shock from firm to normal.

During hard braking, a vehicle wants to **dive**. The front of the car is pushed down and the rear of the car slides up. During hard acceleration the front of the vehicle lifts and the rear lowers. This is called **squat**. Advanced active suspensions help control some of the forces normally encountered in driving. They can reduce pitch and body roll as well as helping to control squat during acceleration and dive during braking.

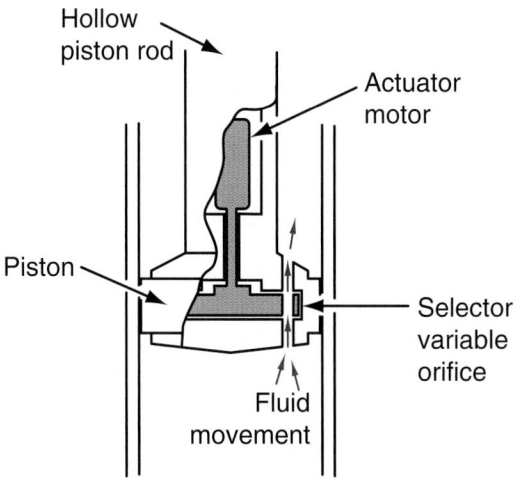

Figure 63.47 The control rod changes the size of the orifice.

Figure 63.48 The viscosity of the magnetic fluid in this MR shock absorber changes very quickly in response to computer commands. *(Courtesy of Delphi)*

Active Suspensions

An **active suspension system** works with sensors, a computer, and activators to solve these problems. Active suspensions were first developed by Lotus in England. A few luxury cars have active suspensions.

An active suspension does not require conventional shock absorbers or springs. Each wheel has a double-acting hydraulic cylinder (high-speed actuator) to keep the car body level during all driving conditions. Shock absorbers and springs are not necessary on these vehicles, because the actuators replace them. They provide a smooth ride and yet give the handling characteristics of a vehicle with a stiff suspension. The suspension can be programmed by the driver as to the type of ride desired.

The active suspension is powered by a hydraulic pump driven by the engine. It requires about the same amount of power as a typical power steering pump during a turn (3–5 horsepower). The high-speed actuators can raise or lower the vehicle in 0.003 second. As soon as a bump has been absorbed by an actuator, pressure is re-established to keep the wheel in contact with the road and maintain ride height. During hard stops, active suspension reduces the tendency for the vehicle to dive or lean. Power consumed by the system is least when the vehicle is riding flat.

The computer uses signals received from sensors to track the position of each actuator. It can sense whether a wheel is in jounce or rebound. It also senses how heavily each wheel is loaded and whether the wheel is turning or pointing straight ahead. The computer sees a bump and immediately releases pressure from a control valve. It can release pressure instantly or relatively slowly, depending on what the computer program specifies. After the suspension absorbs the shock, pressure is forced back into the actuator to keep the tire contacting the road and maintain ride height.

NOTE: *In case of a flat tire, the system can be told to raise the tire so a jack is not needed.*

Sensors provide the computer with necessary information regarding extension and compression of each actuator and how heavily the vehicle is loaded. Different sensors are covered in earlier sections in this chapter.

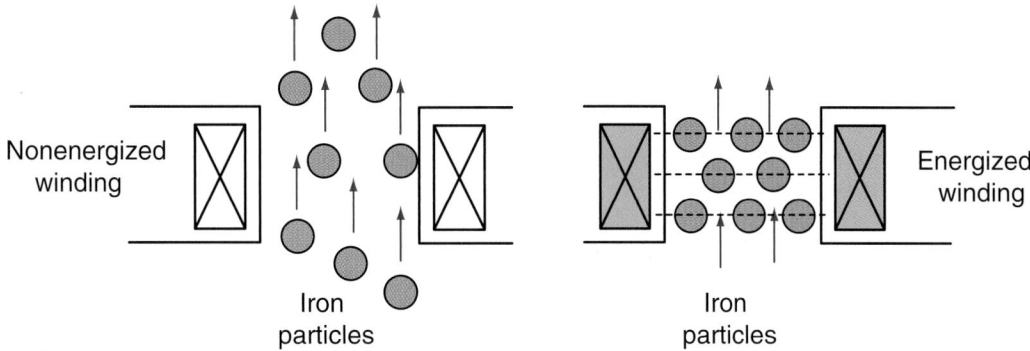

Figure 63.49 The magneto-rheological fluid changes its viscosity, resulting in a firm shock absorber.

REVIEW QUESTIONS

1. Which kind of spring is adjustable?

2. What is the name of the suspension type that uses two different length control arms?

3. What is the name of the part that is used to control oscillations of a spring?

4. Hydraulic shock absorbers are sensitive to velocity/load (circle one).

5. Which shock absorber motion controls the least amount of weight, compression or rebound?

6. What is the term that describes the softness or harshness of the ride?

7. What kind of shock can be mounted upside down?

8. If a small amount of air is not kept in an air shock, what can happen?

9. What is the name of the kind of front suspension that has a built-in shock absorber and only one ball joint?

10. What is the name of the ball joint design that is always trying to pull apart?

ASE-STYLE REVIEW QUESTIONS

1. When one wheel of a vehicle goes into a pothole and the wheel on the other side of the vehicle moves too, the suspension system is:
   a. A rigid axle suspension
   b. An independent suspension
   c. Both A and B
   d. Neither A nor B

2. Which of the following is/are true about Macpherson strut suspensions?
   a. The ball joint is a follower.
   b. It has only one control arm.
   c. Both A and B.
   d. Neither A nor B.

3. A typical shock absorber will:
   a. Compress easier than it extends
   b. Extend easier than it compresses
   c. Only restrict movement as it is compressed
   d. Have extension and compression that are equal

4. Two technicians are discussing gas shocks. Technician A says that gas under pressure prevents oil from foaming. Technician B says that gas under pressure prevents cavitation of the fluid. Who is right?
   a. Technician A   c. Both A and B
   b. Technician B   d. Neither A nor B

5. Technician A says that shock absorbers are designed to carry the weight of the vehicle. Technician B says that shock absorbers dampen spring oscillations. Who is right?
   a. Technician A   c. Both A and B
   b. Technician B   d. Neither A nor B

6. Technician A says that the Macpherson strut provides a better ride and handling than the SLA. Technician B says that a Macpherson strut suspension is often called a double wishbone suspension. Who is right?
   a. Technician A   c. Both A and B
   b. Technician B   d. Neither A nor B

7. Technician A says that some types of ball joints carry a load and others do not. Technician B says that a load carrying ball joint on an SLA suspension is on the control arm that has the spring attached to it. Who is right?
   a. Technician A   c. Both A and B
   b. Technician B   d. Neither A nor B

8. All of the following are true about unsprung weight *except*:
   a. Lower unsprung weight is desirable for better handling.
   b. Tires are unsprung weight.
   c. The engine is unsprung weight.
   d. Independent suspension systems have less unsprung weight than rigid axle suspensions.

9. Technician A says that independent suspensions with more than two control arms are called multilink suspensions. Technician B says that "dive" is when the front of the vehicle lifts and the rear lowers during hard acceleration. Who is right?
   a. Technician A   c. Both A and B
   b. Technician B   d. Neither A nor B

10. Technician A says that shock absorber rebound damping controls the lighter weight of unsprung weight. Technician B says that a car that floats as it travels down the road has a shock absorber rebound control problem. Who is right?
    a. Technician A   c. Both A and B
    b. Technician B   d. Neither A nor B

CHAPTER 64

Suspension System Service

◼ OBJECTIVES

Upon completion of this chapter, you should be able to:

✔ Diagnose suspension system problems.

✔ Service suspension system components.

✔ Describe suspension system repairs.

✔ Replace Macpherson strut cartridges.

✔ Replace suspension bushings.

◼ KEY TERMS

axial movement
radial movement

wear indicator ball joint
dominant end

◼ INTRODUCTION

Emphasis in this chapter is on the diagnosis and service of suspension system parts. The reader will become familiar with commonly performed chassis diagnosis and repair procedures. When suspension parts are in good condition and properly aligned, they are subjected to two forces. These forces are the weight of the car on the springs, and the force of the road on the tires (**Figure 64.1**).

Suspension parts take a large amount of abuse as the car is regularly driven over potholes, speed bumps, and concrete expansion joints. Forces are transmitted through the suspension parts, causing wear over time (**Figure 64.2**).

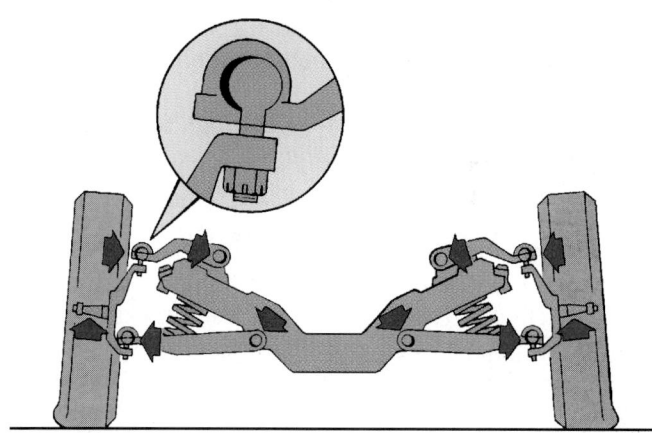

Figure 64.2 Locations of wear. *(Courtesy of McQuay Norris)*

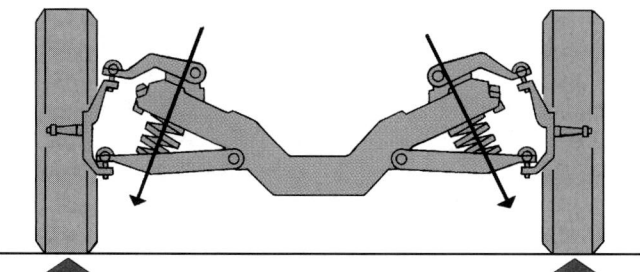

Figure 64.1 The weight of the car and the force of the road act on the suspension system. *(Courtesy of McQuay Norris)*

◼ DIAGNOSING SUSPENSION SYSTEM PROBLEMS

When diagnosing suspension problems, carefully question a customer about the symptoms. Problems with the suspension system usually come to light when the customer complains of noises such as a clunk or squeak, vibration, or a handling problem. All of the suspension system parts depend on the others for their performance. The rest of the system is affected when one part becomes worn.

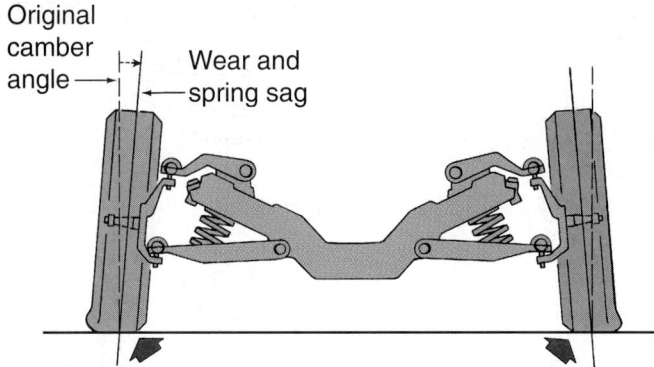

Figure 64.3 The camber angle will go toward negative on most cars when the ride height drops. *(Courtesy of McQuay Norris)*

Wheel alignment settings can change when a suspension component wears. Bushing wear can cause suspension parts to change position (see Figure 64.15). This can cause a car to pull to one side either all the time or just when braking.

A normal condition that occurs with the passage of time is spring sag. This results as the springs lose their ability to keep the vehicle at the correct ride height. As the springs weaken and the vehicle sits closer to the ground, wheel alignment angles change and abnormal tire wear can result. This is because the camber angle will change on most cars when the ride height drops (**Figure 64.3**). Ride height measurement is covered in Chapter 68.

Check all ball joints and bushings for looseness. Check rubber bushings and the sway bar links for wear or cracking.

SHOCK ABSORBER SERVICE

Shock absorbers can be functioning poorly before the appearance of obvious signs of failure, such as tire cupping or excessive front-end float. Signs of inadequate shock absorbers include excessive body motion, road wander (especially in wind), poor adhesion of the tire to the surface of rough roads, a harsh ride, suspension bottoming, and poor braking. A bad shock absorber on one corner of the vehicle can cause the entire vehicle to feel strange.

A defective shock absorber can cause a tire to "hop." Feel the tire tread around the total circumference of the tire. When a scalloped or gouged wear pattern develops on a tire look for a bad shock absorber. Tires that are unbalanced or out-of-round can also be a contributing factor.

A shock can be defective because it is leaking or physically damaged. When one damaged shock is found, both shocks on that end of the vehicle (front or rear) are replaced. A leaking shock can result from a defective seal, an improper dust cover, or lowering a vehicle too far. A damaged piston rod on a Macpherson strut can result from holding the rod with

a pliers during installation of the top nut. This can cause a leak as well.

Some shocks still work well after 50,000 miles. Others may be weak after 4,000 miles. Original equipment shock absorbers are designed to give a satisfactory (soft) ride when used with tight, new suspension bushings. By the time a vehicle has accumulated over 20,000 miles and the suspension has loosened up, some owners choose to replace the shocks simply to stiffen the ride on the highway. Internal parts can also experience wear. Shocks should be checked to see that they still provide good control.

TESTING A SHOCK

A preliminary shock test with the car on the ground is the *bounce test.* Push down hard two or three times on the fender at each corner of the car. After the fender is pushed down, it should oscillate only about 1.5 cycles and then settle. Hand-testing shocks tests only the very lowest piston speed, so it does not fully indicate the condition of the shock. When the operation of the shock is in question, it can be unbolted from its lower mount. Then the shock can be moved to see that it moves slowly and with equal resistance through its normal range of motion. A skip of lag in a shock as its direction of motion is changed indicates a defective shock.

NOTE: *The bounce test only checks the first stage of shock absorber operation. A failure in the second or third stage valves would not be evident from this test. Therefore, if the shock passes the bounce test, it still might be defective.*

The bounce test and looking for cupped tires are worst case, last resort types of evaluation. Evaluating worn shocks is best done by driving the vehicle over a variety of roads.

Shocks do not often fail at the same rate. Usually one shock has a problem, requiring the replacement of both.

Sometimes there is a complaint of noise, the location of which can be determined during the bounce test. The sound of the fluid being forced through the valves in the shock is normal.

Perform a visual inspection of the shock:

- Inspect the condition of the shock mounts and rubber cushions.
- Look to see if any fluid has leaked out of the shock, indicating a bad seal. Fluid cannot be replenished and the shock will have to be replaced. It is normal for a slight amount of moisture to be on the seal (**Figure 64.4**).
- If the outside of the shock body is damaged, replace both shocks.
- If rubber bump stops show signs of contact, a shock could be damaged or the vehicle might have been lowered or raised excessively.

Shocks sometimes limit spring travel. In fact, on some coil spring cars, if the shocks are removed,

A slight amount of moisture is ok, but this shock is wet

Figure 64.4 A slight amount of oil on the seal is normal. However, the body of the shock should not be wet. *(Courtesy of Tim Gilles)*

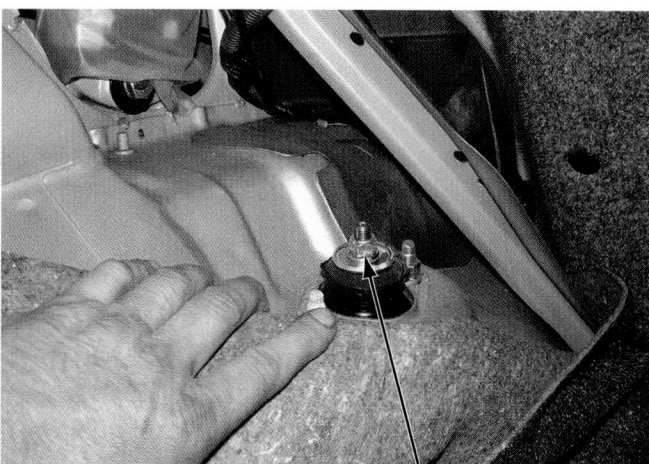

Shock absorber

Figure 64.6 The top of this shock absorber is located under the floor mat in the trunk. *(Courtesy of Tim Gilles)*

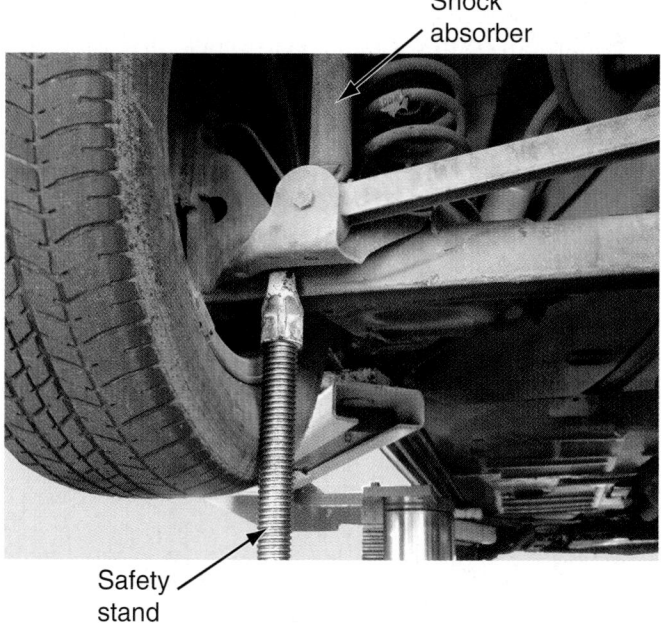

Shock absorber

Safety stand

Figure 64.5 Be sure to support the axle before removing a shock. *(Courtesy of Tim Gilles)*

Figure 64.7 A special shock tool. This end of the tool is used to hold the shock rod while loosening or tightening the retaining nut. *(Courtesy of Tim Gilles)*

the spring can actually fall out. When the shock is removed, the control arm can drop quickly downward in response to spring pressure. If the shocks are to be replaced, the car should be raised on a hoist that supports the wheels. If this is not possible, support the axle with a high lift stand while removing the shock (**Figure 64.5**).

The top of a shock absorber is often hidden within the trunk (**Figure 64.6**) or under a rear seat. A special tool is available to help with shock absorber removal and installation. One end is used to hold the shock rod while loosening or tightening the retaining nut

(**Figure 64.7**). The other end of the tool is installed through the hole in the trunk and threaded onto the end of the shock rod to assist in pulling it into its socket in the trunk (**Figure 64.8**).

Shock Mounts

Check shock mounts to see that the shock is secured to the vehicle. A common problem with a shock absorber is for it to become loose. Shocks are mounted to the chassis with rubber cushions. Shock mounts for passenger cars are usually the *single stud* (*bayonet*) type or the *ring mounting type*. Some of the ring mountings have a bar (cross-pin), stud, or bolt pressed into them (**Figure 64.9**).

Figure 64.8 This end of the tool is installed through the hole in the trunk and threaded onto the end of the shock rod to assist in pulling it through the hole in the trunk. *(Courtesy of Tim Gilles)*

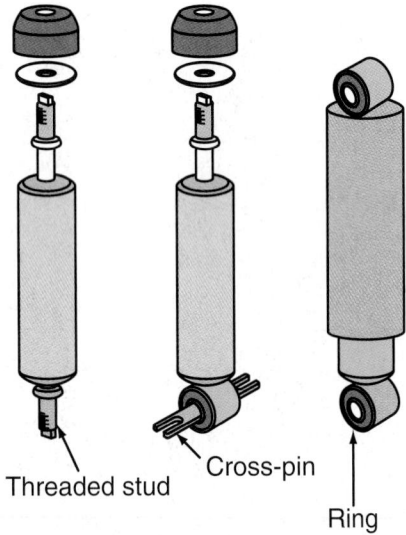

Threaded stud

Cross-pin

Ring

Figure 64.9 Different shock mounts.

Rubber cushions allow the shock to have some flexibility in the mount. On some single stud-type shocks, the tightness of the nut can determine how well the shock cushion functions. Some stud mount nuts are properly tightened when torqued against a shoulder on the stud. Others are tightened only until the rubber bulges out even with the end of its metal retainer (**Figure 64.10**).

Air and Gas Shocks

An air shock can sometimes fail because of a hole in its rubber bladder. Sometimes the rubber will rot if oil comes into contact with it. If the shock is installed in a tight location, the rubber can rub on suspension parts when the air shock is inflated. Air shocks can be checked for leaks using a soapy water solution. Rub the solution over the lines, fittings, and bladder while

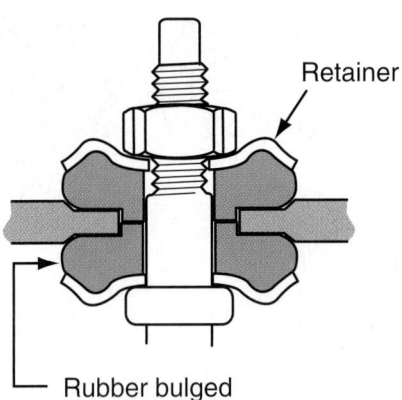

Retainer

Rubber bulged

Figure 64.10 Tighten the nut until the cushions bulge almost to the outer edge of the retainers.

the shocks are inflated. Look for bubbles, which would indicate a leak.

Gas Shocks

A gas-filled shock absorber will expand to its fully extended position if not restrained. When gas shocks are removed from the box, there is a band around them to hold them compressed. If a gas shock has lost its gas charge, it will no longer expand on its own. It may still have oil in it, but the gas charge will no longer be there to prevent foaming when the oil becomes heated.

■ MACPHERSON STRUT SERVICE

A large majority of vehicles have Macpherson struts. When a Macpherson strut shock fails, there are two common repair procedures:
■ Replace the entire strut assembly.
■ Install a strut cartridge into the original shock housing (**Figure 64.11**).

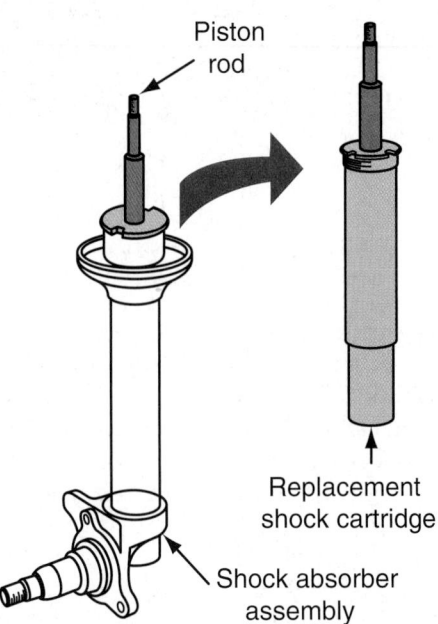

Piston rod

Replacement shock cartridge

Shock absorber assembly

Figure 64.11 A replacement cartridge is sometimes installed in the old strut.

The entire strut assembly is removed from the car (**Figure 64.12**). First, mark one of the bolts and its location at the top of the strut tower (**Figure 64.13**). The strut is easily removed from the top. Removing the bottom part of the strut sometimes requires that the lower ball joint be disconnected from the spindle support (steering knuckle). This will allow the lower control arm

CAUTION Hammering on a pinch bolt can cause damage to the bolt and the ball joint. If the pinch bolt does not come out easily, pry on the lower control arm to release any tension on it. To minimize the danger of the compressed spring, unload it prior to removing the strut. Raise the vehicle by the frame and allow the spring to extend as far as possible.

to be pried down so the strut can be removed. Some lower ball joints have a pinch bolt (**Figure 64.14**) that holds the ball joint. Others will have a cotter pin and castle nut that will need to be removed.

Some struts are easily removed at the bottom by removing two bolts that fasten the strut to the spindle support (**Figure 64.15**). On some imports, removing a brake line from a caliper is often necessary on this style. Brake bleeding will be required following the strut repair or replacement. Sometimes one of the two bolts has a slot that allows the camber adjustment during a wheel alignment. If the vehicle has this adjustment feature, be sure to mark the location of the bolt so it can be replaced in the same position.

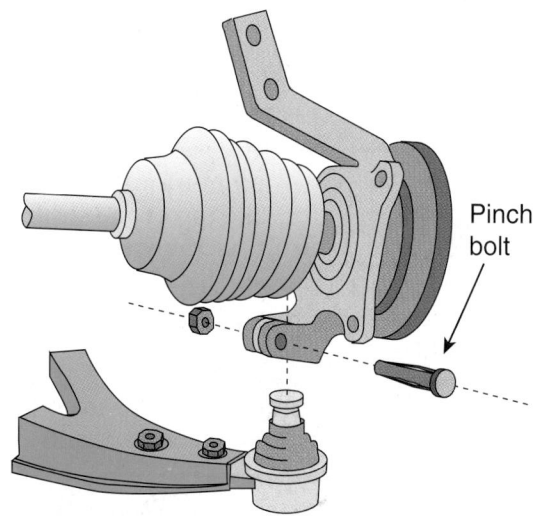

Figure 64.14 A pinch bolt holds the steering knuckle to this ball joint. (*Courtesy of Federal-Mogul Corporation*)

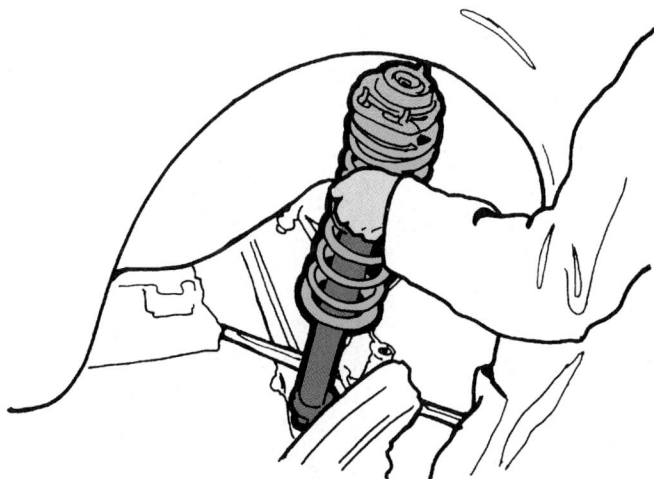

Figure 64.12 The entire strut assembly is removed from the car. (*Courtesy of Tenneco Automotive Operating, Inc.*)

Figure 64.13 Prior to loosening the strut to chassis bolts, make an alignment mark on one strut bolt and the chassis.

Figure 64.15 Remove these bolts to remove the spindle support and caliper from the strut. Remove the brake hose only if it is in the way. (*Courtesy of Tim Gilles*)

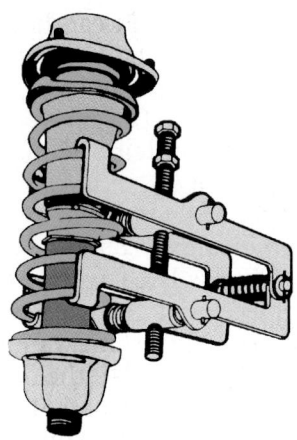

Figure 64.16 A Macpherson strut compressor, sometimes called a "clamshell." *(Courtesy of Federal-Mogul Corporation)*

Figure 64.17 Using a spanner wrench to remove a nut from the strut. *(Courtesy of Tim Gilles)*

SAFETY NOTE
- The center nut must not be removed from the top of the strut before the strut assembly is removed from the vehicle. This would allow the spring to be decompressed during strut removal.
- Be careful with the compressed spring. Dropping it can cause it to dislodge from the spring compressor, with dangerous results.

A spring compressor is used to compress the coil spring (**Figure 64.16**). With the spring compressed, the nut at the top of the strut can be removed from the strut rod. Once the nut has been removed, the parts can be disassembled from the strut.

SHOP TIP
A tight nut can be removed using an impact wrench, squeezing the trigger rapidly to produce short bursts of power. Avoid spinning the shaft.

Replacing a Strut Cartridge

Struts that are serviceable with a shock cartridge have a nut at the top. Sometimes a special spanner wrench is required to remove it (**Figure 64.17**). Other times a conventional hex nut is used. One of these is used in **Figure 64.18**, which shows an exploded view of the parts being removed from a strut housing.

Add Oil to the Strut Body

Unlike a replacement cartridge, an original equipment strut is not enclosed in a cartridge. The old shock fluid will need to be poured out of the strut. Some lightweight oil is poured into the strut body to help conduct heat between the new strut cartridge and the outside of the

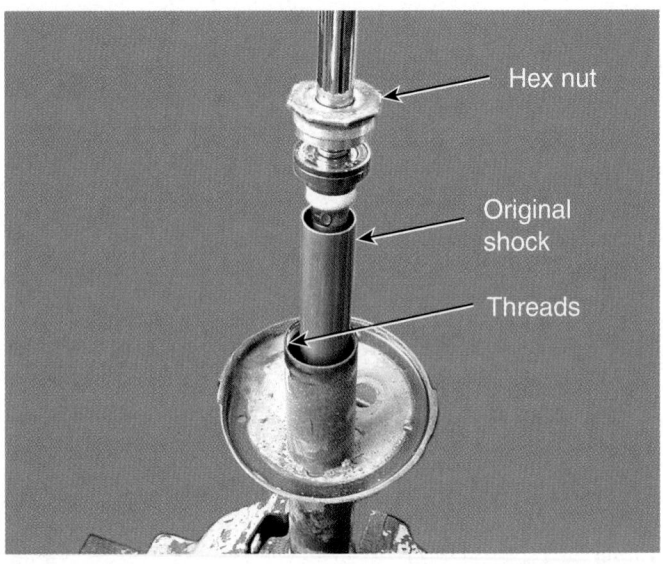

Figure 64.18 Removing parts from a strut housing. *(Courtesy of Tim Gilles)*

strut housing. The old shock oil can be used for this. Two or 3 ounces of oil is usually enough. The amount is not important, but it should almost reach the top of the strut tube with the cartridge replaced.

SHOP TIP
ATF is a good choice of oil to add to the strut because it is red in color. If the shock leaks later on, you will be able to determine if the oil is from the strut or is ATF.

Center the strut cartridge in the strut body before installing the locknut.

■ INSPECT THE UPPER STRUT BEARING

Inspect the condition of the upper strut bearing while the strut assembly is disassembled. A questionable bearing can easily be replaced at this time.

Figure 64.19 The end of the spring is aligned with its seat. *(Courtesy of Tim Gilles)*

INSTALL THE COIL SPRING

Install the coil spring and tighten the locknut. Be sure both ends of the spring are correctly seated before removing the compressor (**Figure 64.19**).

> **CAUTION** Do not use an impact wrench to tighten the locknut. The piston rod can spin rapidly, damaging the new seal.

REINSTALL THE STRUT ASSEMBLY

Reinstall the strut assembly on the car in the same position it was in before. A wheel alignment should always be performed after a strut replacement. If the brake caliper hose was disconnected, brakes will need to be bled as well.

BUSHING SERVICE

Bushings are made of synthetic rubber. They insulate suspension parts from noise and road shock. Rubber bushings deteriorate with age. They are also susceptible to heat damage. Upper control arm bushings are especially prone to heat damage because of their close proximity to engine exhaust manifolds. Driving on bumpy roads can also result in heat and fatigue of the bushing's rubber.

Bushings should not be lubricated, because petroleum attacks rubber and can ruin it. Rubber lubricant (the kind used to aid in mounting tires) can be used to help quiet a squeaky, hardened bushing, but the fix is only temporary.

CONTROL ARM BUSHINGS

Damage or distortion to a control arm bushing can result in changes in wheel alignment settings (**Figure 64.20**). Inspect the control arm bushings for

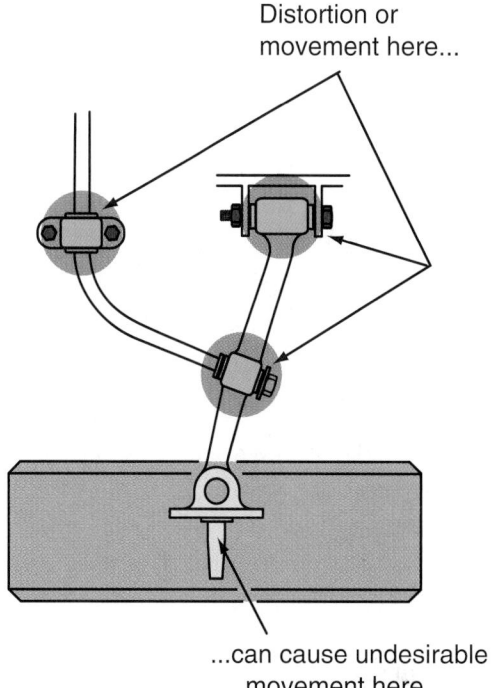

Figure 64.20 Defective bushings can change alignment settings.

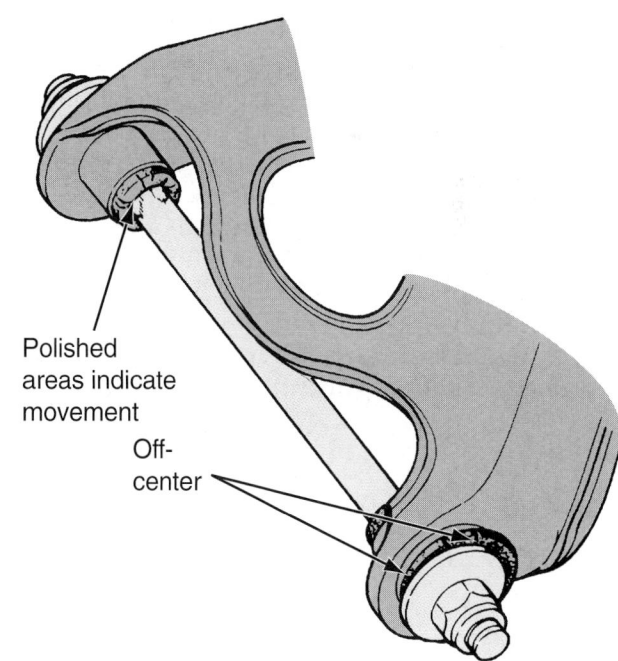

Figure 64.21 Inspect control arm bushings. *(Courtesy of Federal-Mogul Corporation)*

deterioration and splits in the rubber (**Figure 64.21**). See if the bushing is off-center. Sometimes bushings are in a position where visual inspection is difficult. Using a flashlight and mirror will sometimes help.

Push on the fenders of the vehicle while listening for noise. To inspect bushings for looseness, use a prybar to see if the control arm can be moved.

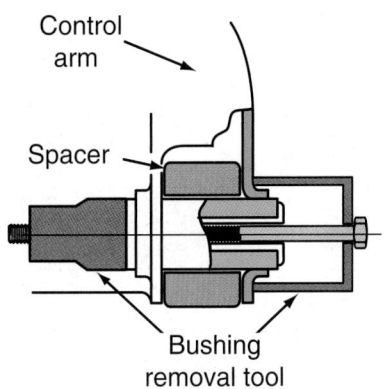

Figure 64.22 Control arm bushing removal with a puller.

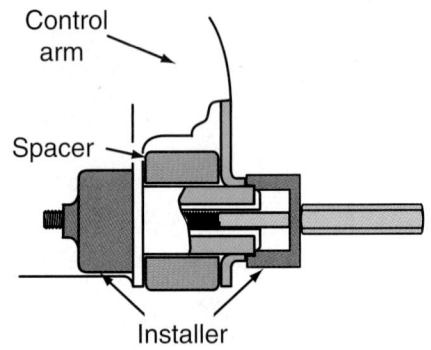

Figure 64.23 Control arm bushing installation with a puller.

Special offset replacement control arm bushing sets are available that compensate for sagged frame components (such as strut towers leaning together) that can cause front-end alignment problems.

Control arm bushings are pressed or driven out. It is important that the control arm holes not be damaged by the process. If an air chisel is used, be sure that the chisel bit is wide and dull. If it is sharp, it will cut the bushing rather than push it out. The new bushing is pressed or pounded in with a special driving tool.

Figure 64.22 shows removal of a bushing using a bushing puller. The puller kit also includes adapters for bushing installation (**Figure 64.23**).

NOTE: *Tighten bolts or nuts that hold bushings in place only when the suspension is in its normal ride position.*

■ STRUT ROD BUSHING SERVICE

Refer to Figure 63.12 for the relationship between the strut rod, lower control arm, and bushing. When replacing a strut rod bushing, the nut is removed. This will allow the lower control arm to move fore or aft. Next remove the fasteners that hold the strut rod to the control arm and slide the strut rod out of the bushing. Most bushings have a center spacer that will only allow the bushing to be compressed a certain amount during reinstallation when the strut rod nut is tightened against it.

■ STABILIZER BAR SERVICE

Inspect the bushings at both ends of the stabilizer bar (**Figure 64.24**). Stabilizer links do not have to be removed to replace the inner bushings. Replacement stabilizer bushings have a split in them. Some original bushings do not have a split. These can be cut off with a razor blade. Then install the new bushing.

CAUTION When replacing stabilizer bar bushings, be sure that *both* front wheels are either on the ground or in the air. Otherwise, the sway bar will be spring loaded. With one wheel jacked up, removal of the nut on the top of the stabilizer link could result in a serious injury.

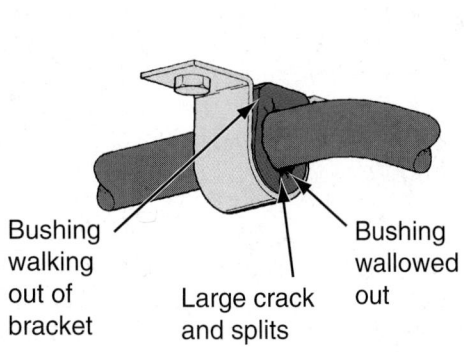

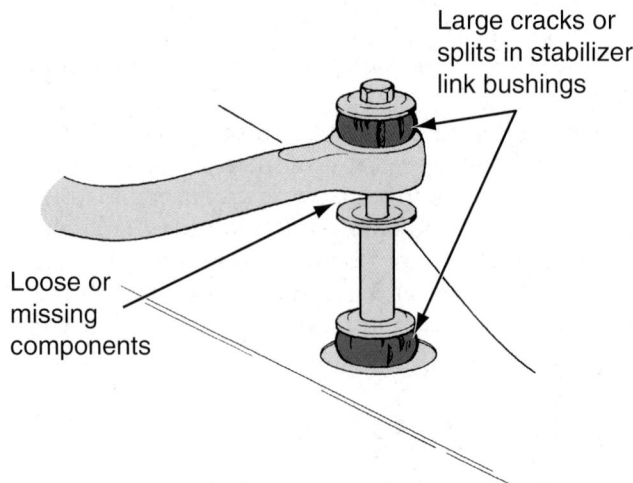

Figure 64.24 Inspect stabilizer bushings. *(Courtesy of Federal-Mogul Corporation)*

Figure 64.25 This spindle broke after a burned wheel bearing was replaced. *(Courtesy of Tim Gilles)*

SPINDLE SERVICE

The steering knuckle and spindle are not serviced. If the spindle is damaged, it is replaced. Damage from a collision will be noticed during the SAI alignment angle check. If a wheel bearing fails, it can wear or heat the spindle. This changes the metallurgy of the spindle and it must be replaced.

 **CASE HISTORY** *A front-wheel bearing failed on a 1965 Mustang. There was wear on the spindle that the technician failed to notice. He replaced the wheel bearing and returned the car to the customer. Several months later, the customer's daughter was driving the car when she drove over a small pothole in the road. The spindle broke off just behind the area where the old bearing had burned (**Figure 64.25**). She was not traveling fast and, luckily, was not hurt when the front wheel fell off the car.*

BALL JOINT SERVICE

Ball joints are relatively trouble-free, but occasionally they wear out. The ball joint is sealed within a rubber boot filled with grease. A small bleed hole in the ball joint boot allows for grease movement during lubrication. Other cuts or tears in the boot will allow water and dirt into the joint, causing it to wear.

Feel around the outside of the boot, looking for tears. If the boot is torn, the joint will probably fail soon and should be replaced. Inspect also for signs of rust or cracks on the control arm near the joint.

Manufacturers list specifications for movement of ball joints. Always check the specifications and be sure that the proper procedure is followed. Vertical or **axial movement** is usually specified, although some

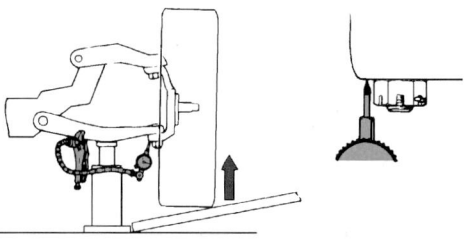

Axial check

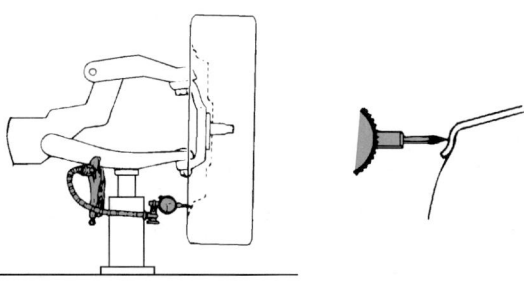

Radial check

Figure 64.26 Ball joint checks. *(Courtesy of Federal-Mogul Corporation)*

manufacturers specify horizontal or **radial movement** (**Figure 64.26**).

A load-carrying ball joint usually has some movement when unloaded. In the past, the unnecessary replacement of good ball joints was a common occurrence. As a result, some states now require measurement and documentation of the amount of clearance found on a worn ball joint before it can be replaced. Most load-carrying ball joints have a wear limit of 0.060" of vertical movement. But some joints can have as much as 0.200" movement and still be within specified limits.

Before checking ball joint wear, determine whether the ball joint is a load carrier or follower (see Figure 63.39). It is important to know this when testing ball joints for wear because the load-carrying joint must be unloaded to test it. Load-carrying ball joints are either tension or compression types (see Figure 63.40). Tension-loaded ball joints are more common.

- The ball joint on the control arm that has the spring mounted on it is the load carrier.
- The ball joint on a strut suspension is the follower joint and therefore carries no load.

MEASURING BALL JOINT WEAR

Check the manufacturer's recommended procedures before checking ball joints. For an accurate wear check, the ball joint must not support the weight of the vehicle. Some SLA suspension systems have the spring above the upper control arm. **Figure 64.27** shows how to unload the joint for both the spring-above and

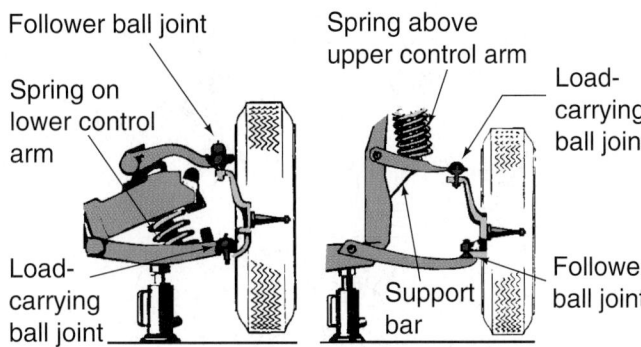

Figure 64.27 Methods of checking ball joints. *(Courtesy of Federal-Mogul Corporation)*

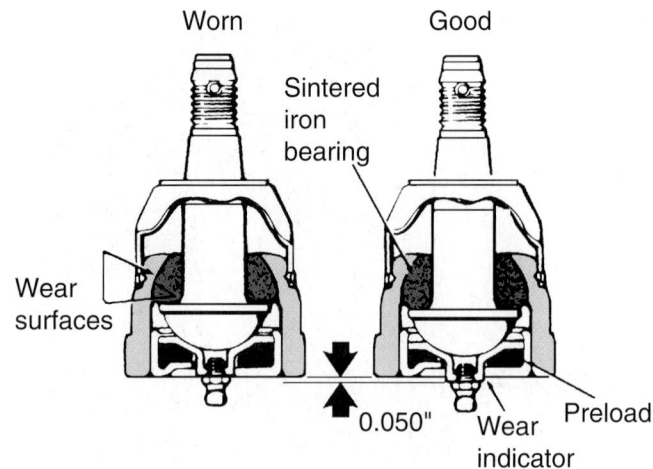

Figure 64.28 Checking a wear indicator ball joint. *(Courtesy of Federal-Mogul Corporation)*

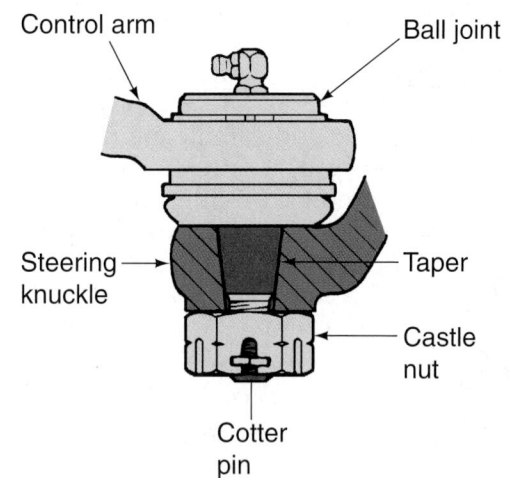

Figure 64.29 A ball joint taper. *(Courtesy of Federal-Mogul Corporation)*

the spring-below types. Prying between the tire and ground will show any clearance.

The follower joint should be checked at the same time that the loaded ball joint is unloaded for testing. Follower ball joints hold the steering knuckle in the correct position and allow it to pivot during bumps and turns. Recommendations for most follower ball joints call for replacement if there is "any perceptible movement" in the joint. Always check specifications.

To check a follower ball joint for movement, unload the joint and try to move the tire back and forth while looking for movement. For this test to be accurate, wheel bearing clearance cannot allow the wheel to move.

SHOP TIP
- A quick way to eliminate wheel bearing clearance from the measurement is to apply the brakes during the test. When a helper is not available, brakes can be applied using a pedal depressor.
- To measure ball joint clearance without a dial indicator, measure the length of the ball joint with a caliper and then measure it again when the joint has been unloaded.

Wear Indicator Ball Joints

Some ball joints have a **wear indicator** built into them. The most common type of wear indicator has a shoulder that sticks out of the bottom of the joint about 0.050" when it is new (**Figure 64.28**). If the ball joint has worn, this shoulder will recede into the ball joint housing. When it is flush, the ball joint should be replaced. A wear indicator ball joint must be loaded and at normal ride height to read the indicator.

▬ SEPARATING BALL JOINT TAPERS

The ball joint is connected to the steering knuckle with a tapered connection, called a *steering taper* (**Figure 64.29**), which is also used in other steering connections. Removing a ball joint can be done with a special tool. Another way of "breaking" the taper is to use a large hammer.

- First, remove the cotter pin from the ball joint nut and loosen the nut several turns.
- Position the vehicle so that the coil spring is pushing on the ball joint. This could require lifting the vehicle or allowing its weight to rest on the wheels.
- Use a hammer to pound sharply on the steering knuckle on the outside of the taper. This will deform the taper and spring pressure will separate the ball joint from the steering knuckle.

NOTE: *A pickle fork (see Figure 66.9) will probably ruin the rubber seal on the ball joint. If the ball joints are not going to be replaced, this will be a costly subtraction from the profit of the job.*

▬ REPLACING THE BALL JOINT

A ball joint can be retained in one of several ways. **Figure 64.30** shows three of the most common methods.

Some original equipment ball joints are fastened to the control arm with rivets. These rivets must be removed

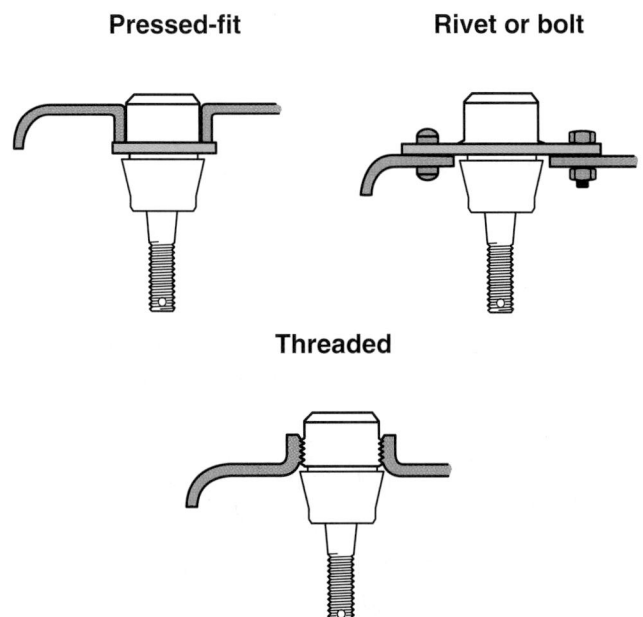

Figure 64.30 Three common ball joint retaining methods.

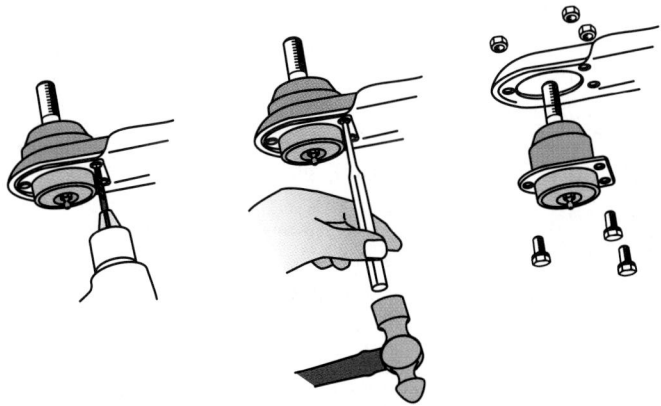

Figure 64.31 Replacing a riveted ball joint.

in order to extract the ball joint (**Figure 64.31**). First, a small drill bit is used to bore a hole about halfway into the rivet. Then, the rivet head is drilled using a large bit until the rivet head falls off. Finally, bolts and nuts are used to hold the replacement joint in place.

NOTE: *Many technicians use an air chisel to cut the heads off the rivets, but this calls for ear protection for the technician and others in the shop.*

Ball joints can also be pressed or threaded into the hole in the control arm. If the control arm has been removed from the vehicle, a pressed-fit ball joint can be removed using a standard hydraulic press. When the control arm is still on the car, a special press set allows the removal and replacement of these ball joints. **Figure 64.32** shows this process.

Some pressed-fit ball joints have a spot weld that holds them in place. This must be carefully removed. When a replacement ball joint is installed, a snap ring often takes the place of the weld.

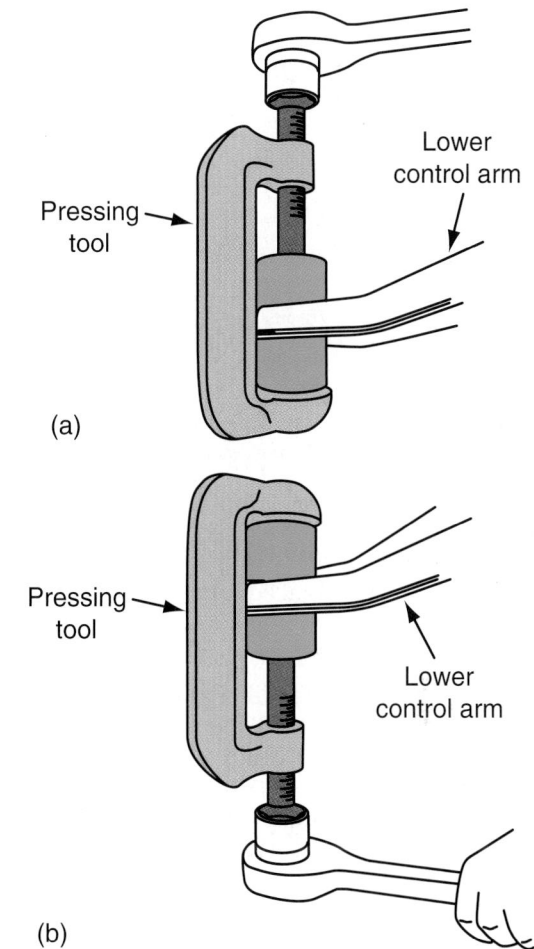

Figure 64.32 (a) Ball joint removal. (b) Installing a pressed-in ball joint in the lower control arm.

NOTE: *Be sure the grease release hole in the new ball joint's rubber boot is aimed away from the brakes.*

■ COIL SPRING SERVICE

A coil spring will rarely break unless it has been constantly overloaded or its surface has a stress raiser from a nick, corrosion, or defect. Over the years, the weight of the car causes the coil spring to lose some of its tension. This results in a lowered ride height. A ride height check can be performed to see if the height of the car conforms to specifications.

Incorrect ride height affects the camber and toe alignment angles. With an SLA suspension system, the car is designed to ride at a height that minimizes changes in tread width. When the suspension system drops due to sagged springs, the upper control arm is in a different place in its arc of travel. This means that there will be even more camber change during bumps (**Figure 64.33**). Tires will wear faster and handling is affected.

A vehicle that is too low cannot be aligned properly, so this test should be done before attempting a wheel alignment. The ride height check is covered in Chapter 68.

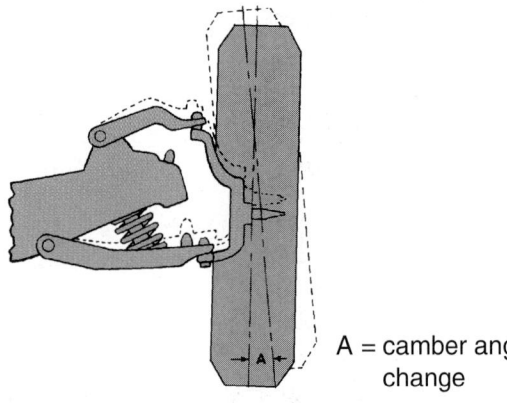

Figure 64.33 The position of the control arm changes in its arc due to spring sag. *(Courtesy of McQuay Norris)*

ADJUSTING SPRING HEIGHT

Coil springs must be replaced when they have sagged beyond specifications. An advantage of the torsion bar spring used on a few vehicles is that spring height can be adjusted. A screw is turned against a bracket mounted on one of the torsion bar ends. The vehicle's correct ride height can be restored prior to an alignment.

Air shocks or shock absorbers with coil springs around them are designed to be used only for temporary overload conditions. The weight of the vehicle will rest on the shock mounts instead of the spring seat. Shock mounts are not designed to continually support the vehicle.

COIL SPRING REPLACEMENT

Replacement springs are purchased from a dealer or aftermarket parts supplier. They are replaced in front or rear pairs. Some aftermarket springs have a part number stamped on the end of the coil. Original equipment springs are often tagged with a part number wrapped to one of the coils. The tag might be missing or unreadable. Replacement springs must be of the same kind as the one in the vehicle. Coil springs have different shaped ends. Square-ended coils can be tapered or untapered. Full wire spring ends can also be cut off squarely, called *tangential*. **Figure 64.34** shows different spring ends.

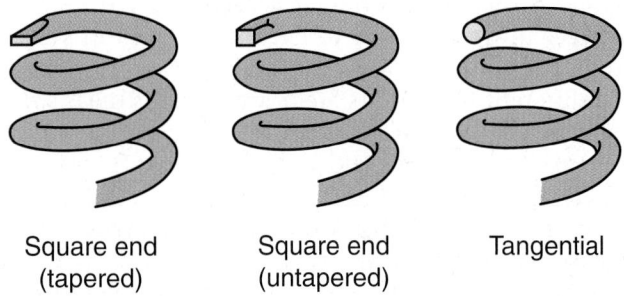

Square end   Square end   Tangential
(tapered)   (untapered)

Figure 64.34 Different types of spring ends.

Aftermarket regular duty coil springs are similar to original equipment springs but are sometimes designed to replace different springs. See Chapter 63 for more information on springs.

SLA COIL SPRING REPLACEMENT

Most passenger cars require the use of a coil spring compressor for coil spring removal and replacement. When a spring is installed in a vehicle, it is held in a compressed position. When it is fully extended, it is much longer. Before removing a spring, the wheel, shock absorber, and stabilizer links are removed. The outer tie rod is disconnected from the steering arm.

During a coil spring replacement, only the lower ball joint needs to be removed. The upper ball joint can remain in place. The upper control arm and steering knuckle will be moved out of the way during the coil spring replacement. Sometimes, both upper and lower ball joint tapers must be broken to be able to get the disc brake splash shield and spindle to clear the lower control arm. Remove the cotter pin from the ball joint nut and back the nut off several turns.

When using the special ball joint press to loosen a taper connection, unload the ball joint with a floor jack. For maximum leverage, position the jack so it is as close to the outer ball joint as possible (**Figure 64.35**). A piece of wood will help get more leverage and keep the jack level.

> **SAFETY NOTE** Reach across to jack up the control arm from the opposite side of the vehicle (see Figure 64.35). This will allow the jack to roll toward the inside of the vehicle as the control arm is lowered, releasing spring pressure.

Figure 64.35 The correct way to use a jack when doing SLA coil spring work. *(Courtesy of Tim Gilles)*

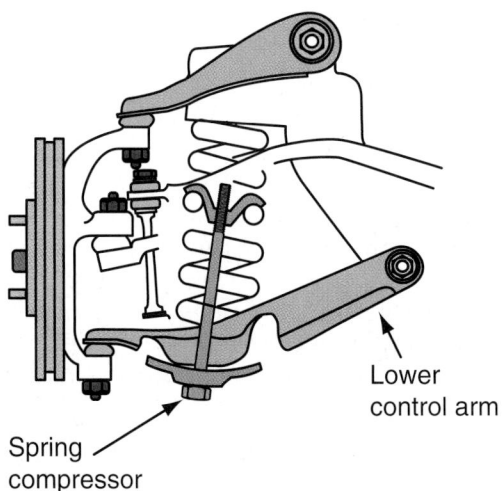

Figure 64.36 A spring compressing tool installed on a spring in a short/long arm suspension system.

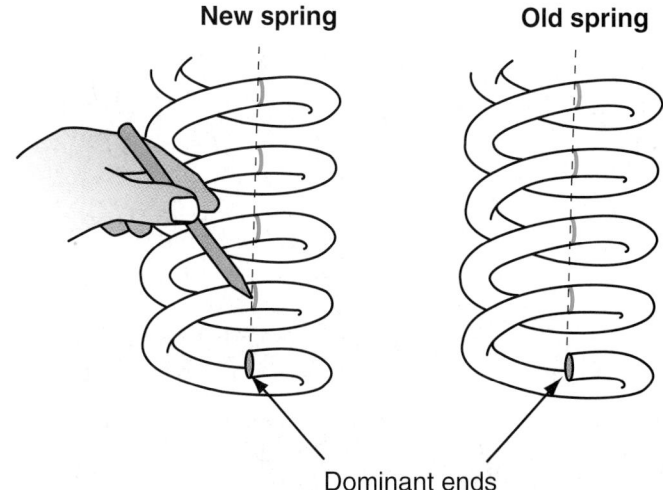

Figure 64.37 Align the dominant ends of the old and new springs.

Raise the jack to compress the spring until the vehicle begins to lift. Remove the lower ball joint nut and lift the steering knuckle assembly away from the ball joint stud. Support it out of the way. A coil spring compressor can also be used to keep the spring compressed while repairs are performed to the spindle or upper control arm (**Figure 64.36**).

Use two clips to hold the spring compressed. The clips are either short or long. Use the one that fits best. Short clips are used to hold four rungs of the spring. Long clips are used to clip five rungs. Use two clips positioned side by side on the same spring rungs. Position the clips as low as possible on the inboard side of the spring.

Pry down on the lower control arm and pry the coil spring from its place. Mark the locations of the clips. Use a coil spring compressor to compress the spring and remove the clips.

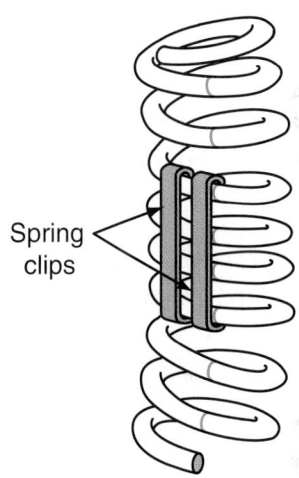

Figure 64.38 Compress the spring and install the spring clips.

 **SAFETY NOTE** Never compress the spring far enough for the coils to stack up.

Lay the old spring next to the new spring. Align the dominant ends (**Figure 64.37**). The **dominant end** is the end that aligns in the coil spring seat in the frame or lower control arm. Mark the new spring in the same location of the old spring. Compress the spring and install the spring clips (**Figure 64.38**).

Because of the spring clips, the spring will be bowed. This will make aligning the upper and lower spring seats easier during installation. If there are insulators that fit into the spring seats (**Figure 64.39**), tape them onto the spring to make installation easier.

The spring seats must be accurately aligned. During installation, align the upper end first. Then, push

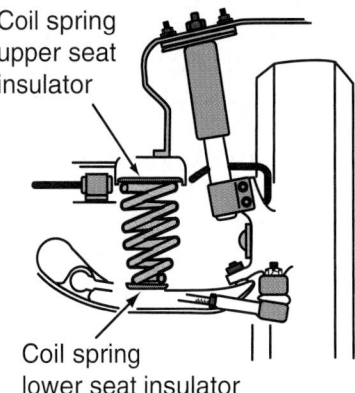

Figure 64.39 Insulators sometimes fit into the spring seats.

the bottom of the spring into place. Be sure the lower spring end fits into the pocket in the spring seat. Jack the control arm to compress the spring and install the ball joint stud in the tapered hole (**Figure 64.40**). Be sure to align the cotter pin hole so it is parallel to the

Figure 64.40 Align the cotter pin hole and install the ball joint in the steering knuckle. *(Courtesy of Federal-Mogul Corporation)*

brake backing plate. Otherwise, it will be difficult to install the new cotter pin in the castle nut. Torque the castle nut to specifications.

NOTE: *If the cotter pin hole does not line up, tighten until the next hole does. Do not back off on the nut to align the hole.*

After connecting the ball joint(s), remove the spring clips from the spring. A small prybar may be needed for this.

Complete the assembly of all of the components. Before lowering the wheels onto the ground, loosen the bolts that compress the control arm bushings. Drive the vehicle a short distance and jounce the suspension several times. Retorque all bushings to specifications before doing a wheel alignment.

Torsion Bar Removal and Replacement

The torsion bar adjusting bolt must be loosened before removing a torsion bar (**Figure 64.41**). First, measure how far its adjusting bolt extends above the surface of the bracket. This will save time when readjusting ride height after reinstallation. Be sure the wheels are

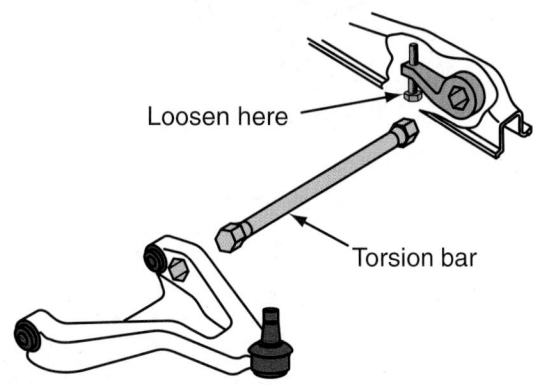

Figure 64.41 Remove tension from the torsion bar adjusting bolt.

lifted off the ground before attempting to loosen the adjusting bolt.

If the same torsion bar is to be reinstalled, mark its end and its adjusting arm prior to removal so it can be reinstalled in the same position.

CAUTION Some SUVs with torsion bars have air suspension systems. To avoid injury, disable the air suspension system before working on the system.

Before reinstalling a torsion bar, be sure it is going into the correct side of the vehicle. Torsion bars are directional. They are marked left or right, as viewed from the driver's seat, and must be installed on the proper side of vehicle.

After the torsion bar is installed, check and adjust ride height as needed (see Chapter 68).

Leaf Spring Service

Leaf spring problems include broken leaves, spring sag, differences in ride height from side to side, squeaks, worn bushings, and broken center bolts. Bushings and shackles can be replaced without removing the spring from the vehicle.

When leaf springs have sagged, they are removed and sent to a facility where they can be "re-arced." This reestablishes the original ride height.

Multiple leaf spring leaves have holes through their centers. A *center bolt* runs through the holes of all the springs, with the head of the bolt fitting into an opening in the spring mount on the axle. The center bolt holds the position of the axle in relationship to the springs so the rear wheels can track behind the front wheels. Sometimes the center bolt breaks. This can alter the position of the rear axle, changing vehicle tracking and alignment.

When insulators need replacing or there is a broken spring leaf, secure the spring with a C-clamp or in a large vise. Then remove the center bolt and disassemble the spring pack. Spring leaves are different lengths. Before disassembly, use a crayon to number the springs so they can be easily replaced in their original positions.

Removing leaf springs is easiest done on a wheel contact lift with the vehicle frame supported on jack stands or the lift's air jack. If a wheel contact lift is not available, the job is performed with the vehicle on the ground. Raise the body of the vehicle up until the wheels just start to leave the wheel contact surface. Then disconnect the spring from one side at a time. When possible, leave the vehicle in the same position so the spring can easily be reinstalled. When positioning the spring be sure that the head of the center bolt fits into the hole or recess in the spring seat. This correctly positions the spring fore and aft.

WHEEL ALIGNMENT

Following completion of suspension repair jobs, a wheel alignment will be required. This will reposition suspension components so that the car will be safe to drive and will go straight without unusual tire wear.

Remember to torque bolts that go through bushings only after the vehicle is resting on its tires.

ELECTRONIC SUSPENSION SERVICE

Electronic suspension system problems are often related to mechanical failures, including leaks in the system. A leak can occur in a rubber spring bladder, or nylon tubing can be damaged.

Electrical failures can also occur in the air system. The length of time a compressor can run is limited to a few minutes in case of a solenoid malfunction. Compressor failures do, however, occur.

If the battery is being charged, the ignition switch must be in the "off" position if the air suspension switch is on. Damage to the compressor relay or motor could result.

Wiring connections are also a common source of problems. Sometimes simply disconnecting and reconnecting wiring connections can correct a problem.

Electronic failures will set a computer code and cause an instrument panel light to flash or illuminate continuously. The source of electronic problems can usually be traced using a scan tool to read diagnostic trouble codes (DTCs) stored in computer memory. Manufacturer's scan tools are best for diagnosing the most troublesome problems on specific makes of vehicles. Most aftermarket scan tools also have the ability to diagnose the most probable electronic failures.

Scan tools were originally designed to diagnose emission-related problems. There are many other computer-managed systems on automobiles today. Scan tool cartridges are available to help in the diagnosis of electronic system problems in these systems.

There are several sensors for electronic suspension systems (see Chapter 63). Service literature contains detailed charts and diagrams for each manufacturer's system. **Figure 64.42** shows a typical list of DTCs for a programmed ride control (PRC) system.

CAUTION Be sure the leveling system is disabled before raising the vehicle on a lift, jacking it up, or towing. Some systems stay on all the time (not activated by the ignition switch). The ignition switch must be off, and the leveling system switch—often located in the trunk—must be turned off before lifting or towing the vehicle.

When raising a vehicle with an electronically controlled air suspension, most manufacturers recommend

PRC Diagnostic Trouble Codes (DTCs)

Code	Defect
6	No system problem
1	Left rear actuator circuit
2	Right rear actuator circuit
3	Right front actuator circuit
4	Left front actuator circuit
5	Soft relay control circuit shorted
7	PRC control module
13	Firm relay control circuit shorted
14	Relay control circuit

Figure 64.42 PRC system diagnostic trouble codes (DTCs).

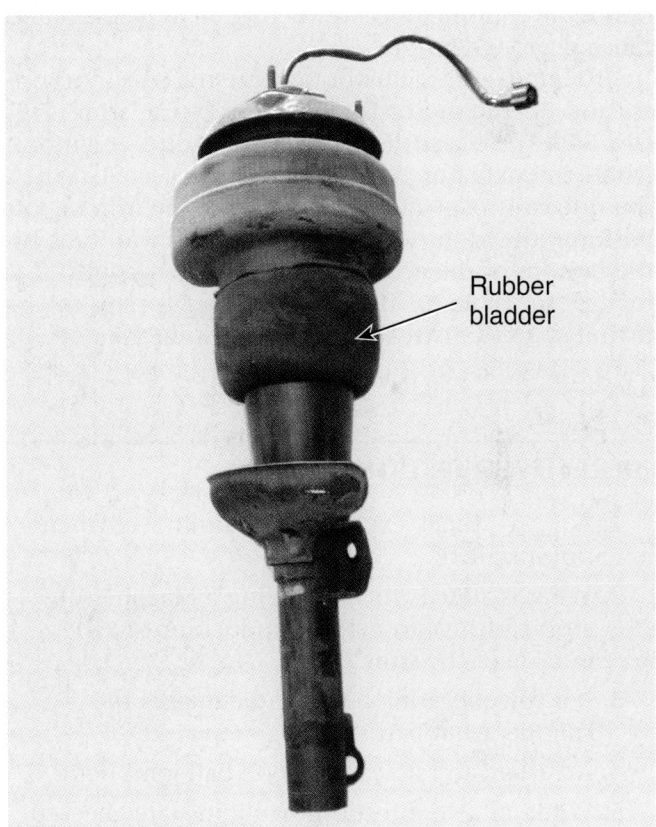

Rubber bladder

Figure 64.43 An air spring strut assembly from an electronic suspension system. *(Courtesy of Tim Gilles)*

a frame contact hoist (sometimes called a body hoist). Lifting by the frame allows air springs to be in their extended position, where the rubber bladder (**Figure 64.43**) will not be damaged. With the suspension hanging free, height sensors lengthen to their maximum position. If the system is on, the computer sees this as a "high car" condition and responds by bleeding pressure from the air springs. If the vehicle is lowered to the ground while the springs are in this position, the rubber could be damaged if the air spring bags happen to fold inward.

When lifting a vehicle with a hydraulic jack, lift by the front crossmember and manufacturer's specified rear lift points in front of the rear tires. Before towing a vehicle with an automatic leveling system, be sure to check the service manual for precautions.

Prior to doing a wheel alignment, the vehicle must be set at regular curb height. Failing to do this will result in a faulty wheel alignment.

ELECTRONICALLY CONTROLLED SHOCK ABSORBERS

Some shock absorbers are electronically controlled to provide a ride that is more firm or more soft than normal. When a firm ride is selected, an actuator motor rotates in response to an electronic signal to reposition a valve in the shock, restricting oil movement and causing a rougher ride.

To remove the actuator, squeeze the plastic retainers and lift the motor from the top of the strut (**Figure 64.44**). To test the actuator, turn on the ignition while the actuator wiring is still connected. When the ride control switch is activated to the firm or soft position, the motor should rotate the control tube on the bottom of the actuator within a few seconds. If it rotates, the actuator is operating. If not, refer to the manufacturer's instructions before testing the unit.

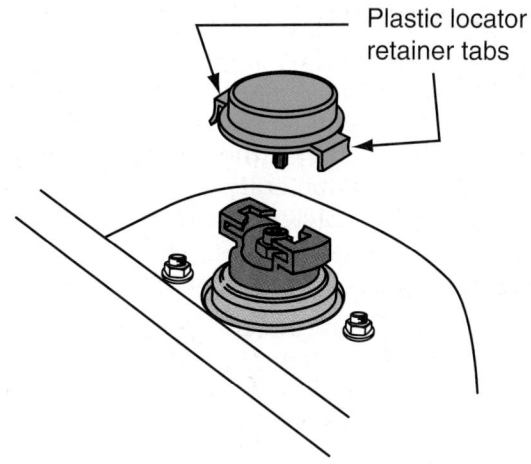

Plastic locator retainer tabs

Figure 64.44 Squeeze the plastic retainers and lift the motor from the top of the strut.

To diagnose the actuator system, you will need wiring color codes and specific instructions. A typical diagnosis procedure will have you rotate the actuator motor to where it is normally positioned under different commands and make identifying marks on the actuator and the control tube. With the electrical wiring disconnected, the actuator control tube can be turned with a small screwdriver to put it into the desired position. In different positions, electrical resistance is tested using an ohmmeter and the results are compared with specifications.

REVIEW QUESTIONS

1. A _____ shock expands to its full travel when not installed.

2. What is added when installing a Macpherson strut cartridge to help it conduct heat to the outside of the strut?

3. If a wheel bearing burns and damages the spindle, what part should be replaced?

4. Are there any holes in a good ball joint boot?

5. What are two directions ball joints are checked for wear?

6. Which kind of ball joint is more common, tension or compression?

7. When a wear indicator ball joint wears, does the shoulder move into or out of the ball joint?

8. When a replacement ball joint is held in place with nuts and bolts, what held the previous ball joint in position?

9. What is used with a spring compressor to aid in the removal and replacement of coil springs?

10. What is the name of the end of a coil spring that aligns in the coil spring seat in the frame or lower control arm?

■ ASE-STYLE REVIEW QUESTIONS

1. Technician A says that it is normal for a small amount of moisture to be on the outside of a shock absorber. Technician B says that if the outside of the shock body is damaged, replace it along with the one on the other side of the car. Who is right?

   a. Technician A   **c.** Both A and B

   b. Technician B   **d.** Neither A nor B

2. Which of the following is/are true about shock absorber rubber mounts?

   a. A shock mount should be tightened until the rubber bumper is totally compressed.

   b. Shock absorber bushings should be lubricated with light oil.

   c. When tightening a nut on the top of a shock rod, the rod must be held from turning.

   d. None of the above.

3. When replacing stabilizer link bushings, which of the following statements is/are true?

   a. Both front wheels can be on the ground.

   b. Both front wheels can be off the ground.

   c. Both A and B.

   d. Neither A nor B.

4. Technician A says that a load-carrying ball joint must be unloaded to test it. Technician B says that load carrying ball joints must be replaced if there is any perceptible movement. Who is right?

   a. Technician A   **c.** Both A and B

   b. Technician B   **d.** Neither A nor B

5. Which of the following is/are true statements?

   a. When checking ball joints for wear, applying the brakes will temporarily eliminate wheel bearing clearance.

   b. Spring sag can result in abnormal tire wear.

   c. Both A and B.

   d. Neither A nor B.

6. When jacking on a control arm to compress a coil spring, the jack should be positioned:

   a. Perpendicular (90°) to the tires

   b. Parallel to the tire tread, in line with the vehicle

   c. Both A and B

   d. Neither A nor B

7. Technician A says to align the cotter pin hole in a ball joint so it is parallel to the brake backing plate. Technician B says that after tightening to the correct torque specification, loosen the ball joint locknut to align the cotter pin hole. Who is right?

   a. Technician A   **c.** Both A and B

   b. Technician B   **d.** Neither A nor B

8. Which of the following is/are true about shock absorbers?

   a. Cupped tire wear could result from a bad shock absorber.

   b. The sound of the fluid being forced through the valves in the shock is normal.

   c. Air shocks are designed to raise a vehicle's height.

   d. All of the above.

9. A Macpherson strut shock absorber is leaking badly. Technician A says that sometimes an entire strut assembly must be replaced. Technician B says that some vehicles use a strut cartridge installed into the original shock housing. Who is right?

   a. Technician A   **c.** Both A and B

   b. Technician B   **d.** Neither A nor B

10. Which of the following is/are true about worn suspension bushings?

   a. Worn bushings can change alignment settings.

   b. The car can pull to one side when braking.

   c. The car can pull to one side all of the time.

   d. All of the above.

Steering Fundamentals

Upon completion of this chapter, you should be able to:

✔ List the parts of steering systems.

✔ Describe the principles of operation of steering systems.

✔ Compare linkage systems to rack and pinion.

✔ Describe how power steering systems operate.

✔ Understand the operation of four wheel steering systems.

■■ **KEY TERMS**

recirculating ball and nut
  steering gear
rack-and-pinion steering
steering ratio

lock to lock
parallelogram steering
ball socket
turnbuckle

toe-out-on-turns
flow control valve
spool valve

■■ **STEERING SYSTEMS**

The steering system works with the suspension system. It allows the driver to steer the car while providing a comfortable amount of steering effort. Steering system parts include the *steering gear,* the *steering linkage,* the *steering wheel,* and the *steering column.* There are two styles of steering. One has a gear box and parallelogram linkage (**Figure 65.1**). The other is a simple long rack with linkage extending from its ends (**Figure 65.2**).

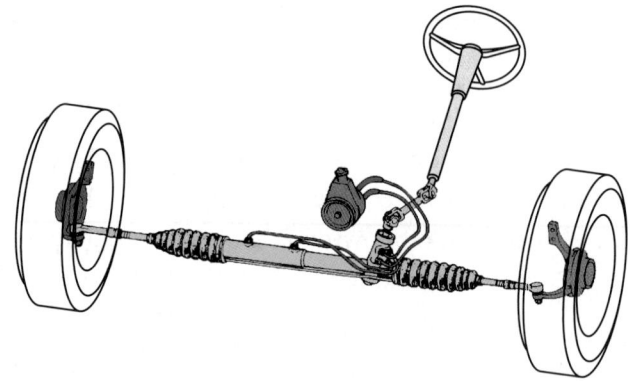

Figure 65.2 Rack-and-pinion steering with linkage. *(Courtesy of Federal-Mogul Corporation)*

■■ **STEERING GEARS**

The two common types of automotive steering gears are the conventional **recirculating ball and nut steering gear** and the **rack-and-pinion steering**. There are other types of steering gear box designs, but these are the primary ones used in automobiles and light trucks.

    The number of teeth on the driving gear compared to the number of teeth on the driven gear help determine the **steering ratio**. The length of the

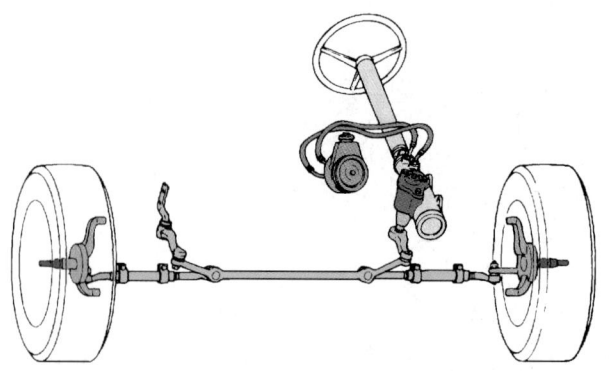

Figure 65.1 A steering gearbox and parallelogram linkage. *(Courtesy of Federal-Mogul Corporation)*

steering arms, pitman arms, and idler arms (parts of the conventional steering linkage) also play a part in determining steering ratio.

When the steering wheel is turned all the way in one direction, it stops against a *lock*. Turning the wheel all the way from one lock to the other is called **lock to lock**. The steering ratio refers to the amount of space it takes for a vehicle to be able to turn around. A "fast" steering ratio is about three turns lock to lock. Slower ratios require the wheel to be turned about four times. The ratio of a steering gear varies, depending on whether or not the car has a power assist. Power steering cars usually have faster ratios. A 15:1 ratio means

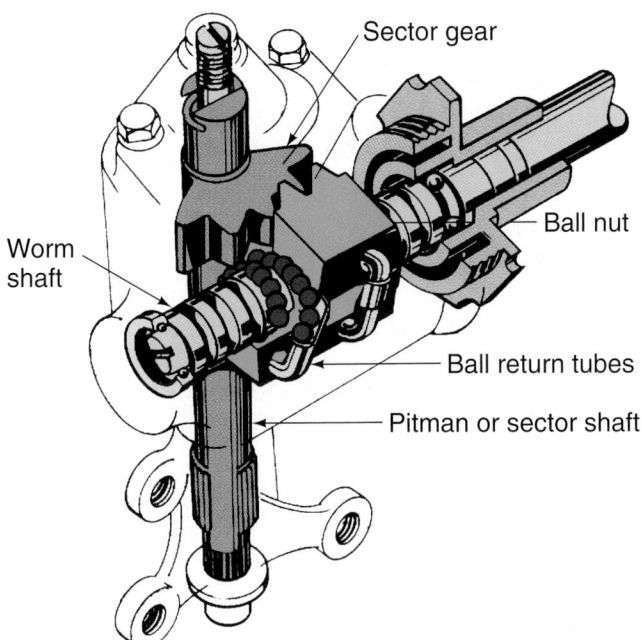

Figure 65.3 A recirculating ball and nut steering gear. *(Courtesy of Federal-Mogul Corporation)*

that when the steering wheel is turned 15 degrees the front wheels will turn 1 degree.

RECIRCULATING BALL AND NUT STEERING GEAR

In a recirculating ball and nut steering gear (**Figure 65.3**), a *sector gear* (part of the *pitman* or *sector shaft*) meshes with a *ball nut* that rides on bearings on the *worm (steering) shaft* to provide a smooth steering feel. The ball nut has curved channels for ball bearings to ride in. The steering shaft also has bearing channels. The balls rotate and recirculate through tubes (ball returns).

RACK AND PINION STEERING

On a rack-and-pinion steering system (**Figure 65.4**), the end of the steering shaft has a *pinion gear* that meshes with the *rack gear* (**Figure 65.5**). They are used on many cars because they are lighter than standard steering gears and are easier to assemble to the vehicle at the factory. A typical rack and pinion often has a faster ratio.

A spring-loaded damper preloads the rack gear to prevent the rack from flexing, which would cause gear backlash (**Figure 65.6**). An adjustment screw or shims are used to adjust the amount of tension against the rack. The pinion shaft is usually supported by needle bearings. The rack is most often supported by plain bushings.

Rack-and-pinion steering systems are more easily damaged when the front wheels hit a curb or rock. They transmit more road shock, which can be felt through the steering wheel. The rack is mounted on rubber bushings to help cushion shocks (**Figure 65.7**). The rack-and-pinion unit can be mounted in several locations. Some are on the sub-frame and others are on the firewall.

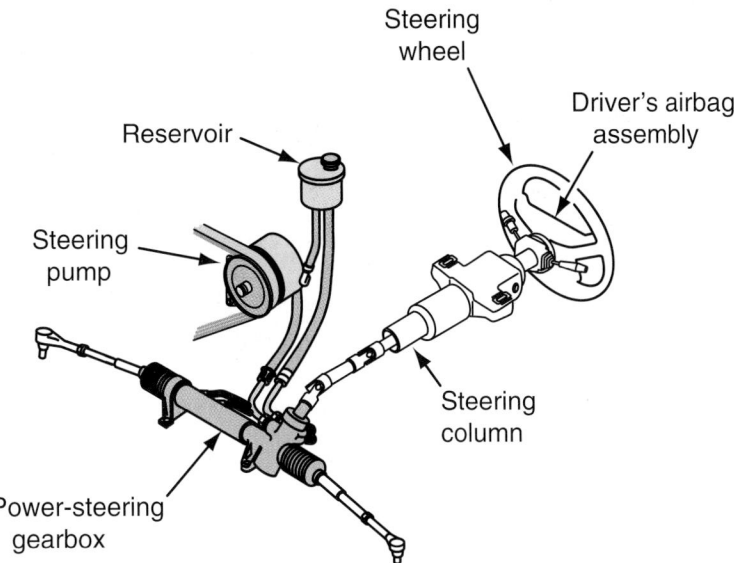

Figure 65.4 A power assisted rack-and-pinion steering system.

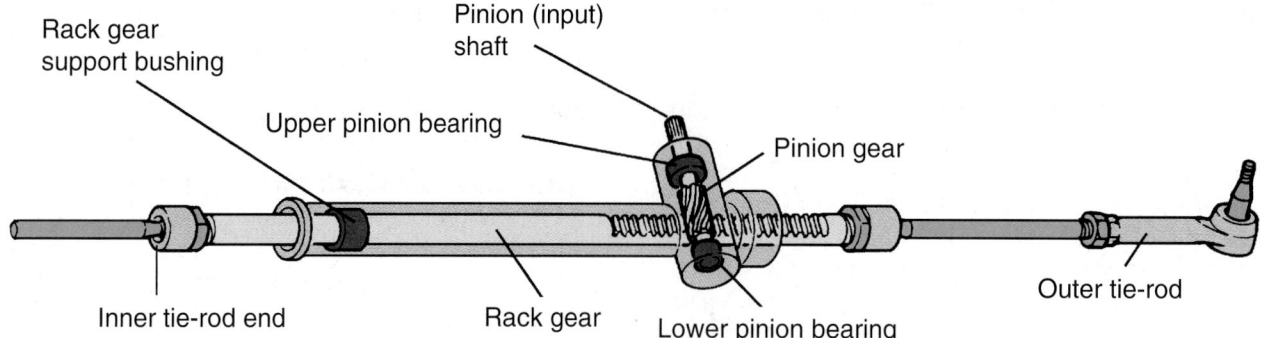

Figure 65.5 Parts of a rack-and-pinion steering gear. *(Courtesy of Federal-Mogul Corporation)*

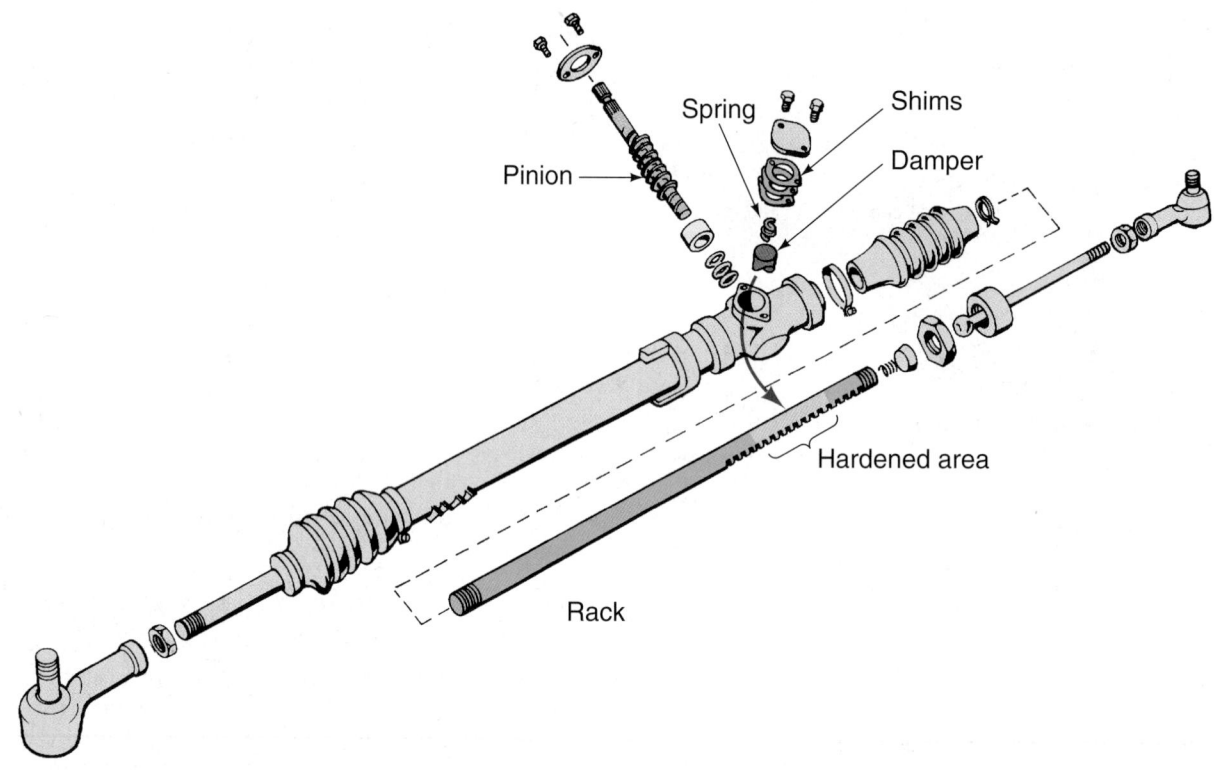

Figure 65.6 A damper holds tension against the rack. *(Courtesy of Federal-Mogul Corporation)*

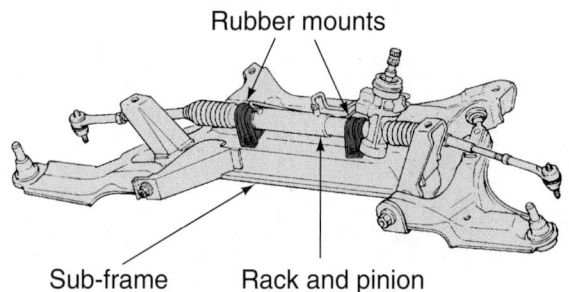

Figure 65.7 The rack is mounted on rubber bushings. *(Courtesy of Federal-Mogul Corporation)*

STEERING LINKAGE

The steering gear is connected to the wheels by the steering linkage. Steering linkage parts vary depending on the design used, but all designs include *tie-rods,* *steering arms,* and a *steering-knuckle.* A comparison of the steering linkage parts of the two major steering systems is shown in **Figure 65.8**.

When a recirculating ball and nut steering gear box is used, there can be a number of different linkage designs used, depending on the suspension design. The most popular steering design in use with the long and short arm suspension is the parallelogram.

PARALLELOGRAM STEERING LINKAGE

On passenger cars, a recirculating ball gear usually uses a **parallelogram steering** system (**Figure 65.9**). The name comes from the parallelogram shape made by the steering linkage during a turn. Parallelogram steering parts are shown in **Figure 65.10**. *Tie-rods* on each side are connected by the *center link.* The *pitman arm* is the part that connects the steering box to the center link. An *idler arm* supports the center link on the passenger side (**Figure 65.11**).

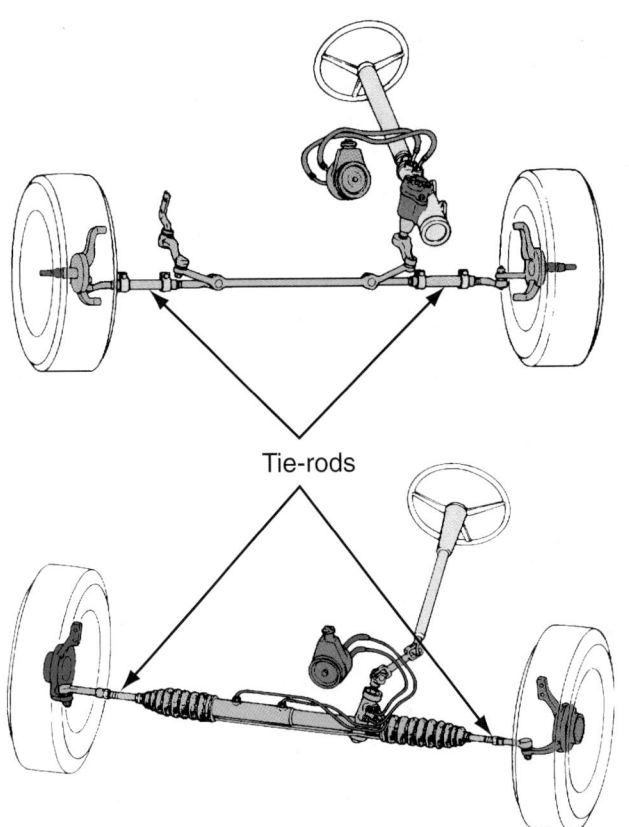

Figure 65.8 Both types of steering have tie-rods, steering arms, and a steering knuckle. *(Courtesy of Federal-Mogul Corporation)*

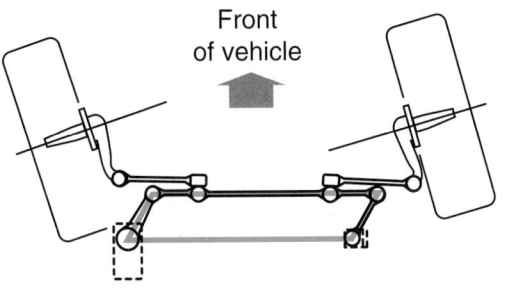

Figure 65.9 The shape of a parallelogram is made by the steering linkage during a turn.

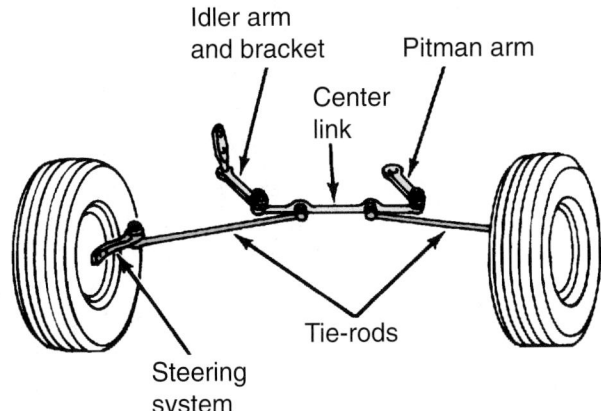

Figure 65.10 Parts of a parallelogram steering system during a turn. *(Courtesy of DaimlerChrysler Corporation)*

Figure 65.11 Idler arm cutaways. *(Courtesy of Tim Gilles)*

■ BALL SOCKETS

Ball sockets connect the steering linkage parts. They allow parts to rotate during a turn and pivot as the steering deflects during a bump.

■ TIE-RODS

Tie-rod (**Figure 65.12**) ends are attached to pivot points at the front wheels. Not only do they transmit motion from the steering wheel to the front wheels, but they maintain the correct front wheel toe (the amount that the front tires are aimed inward or outward at the front). There is a threaded adjusting sleeve that connects the inner and outer tie-rods (**Figure 65.13**). It has a right-hand thread on one end and a left-hand thread on the other end. During an adjustment, when it is turned it acts like a **turnbuckle**. Turning it one way shortens the tie-rod assembly. Turning it the other way lengthens it.

Figure 65.12 A cutaway of a tie rod end. *(Courtesy of Tim Gilles)*

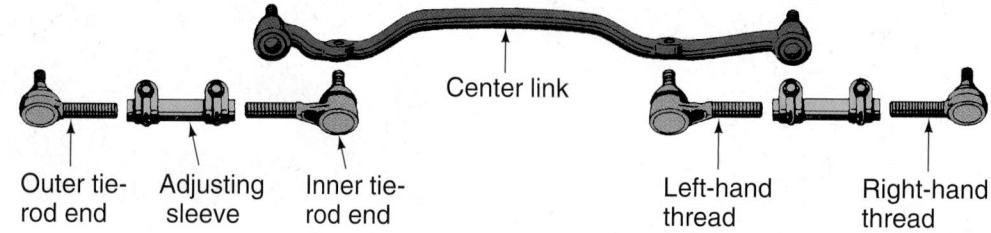

Outer tie-rod end — Adjusting sleeve — Inner tie-rod end — Center link — Left-hand thread — Right-hand thread

Figure 65.13 A threaded sleeve connects the inner and outer tie-rods. *(Courtesy of Federal-Mogul Corporation)*

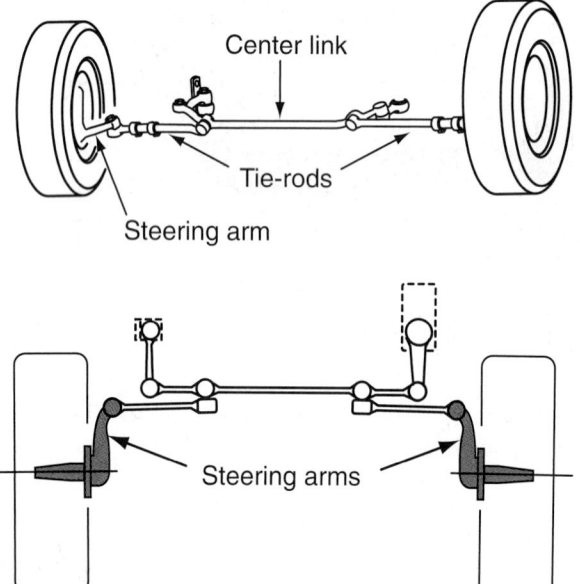

Figure 65.14 The tie rods attach to the front wheels at the steering arms.

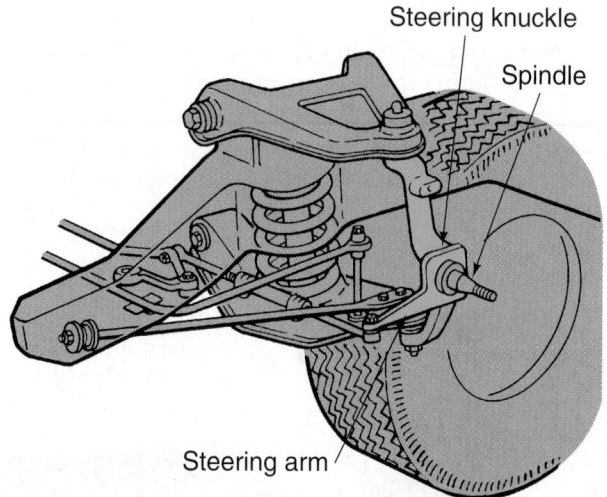

Figure 65.15 Steering arm and knuckle. *(Courtesy of McQuay Norris)*

STEERING ARM

The tie-rods attach to the front wheels at the steering arms (**Figure 65.14**). The steering arm is attached to the *steering knuckle,* which includes the *spindle* (**Figure 65.15**).

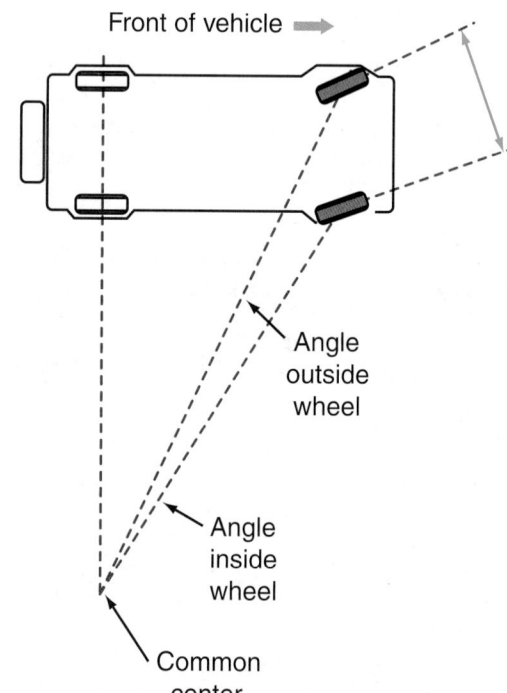

Figure 65.16 The inside wheel turns sharper than the outside wheel because they follow different circular paths. This is called *toe out on turns* or *turning radius.*

During a turn, the inside wheel must turn sharper than the outside wheel because they follow different circular paths (**Figure 65.16**). The steering arms are angled inward, rather than being parallel to the frame. This angle, called the Ackerman angle, provides an important steering angle, **toe-out-on-turns** (**Figure 65.17**).

RACK-AND-PINION STEERING LINKAGE

Rack and pinion does not have the complicated steering linkage used with a recirculating ball and nut steering box. In most systems, two tie-rods come out of the steering rack (see Figure 65.7). They have conventional tie-rod end ball sockets only on the outer ends. These attach to the steering knuckles. The inner tie-rod ends are ball sockets, enclosed within *rubber bellows* or *boots* on the rack (**Figure 65.18**). In another type of rack-and-pinion system, the inner tie-rods attach to the center of the rack gear (**Figure 65.19**).

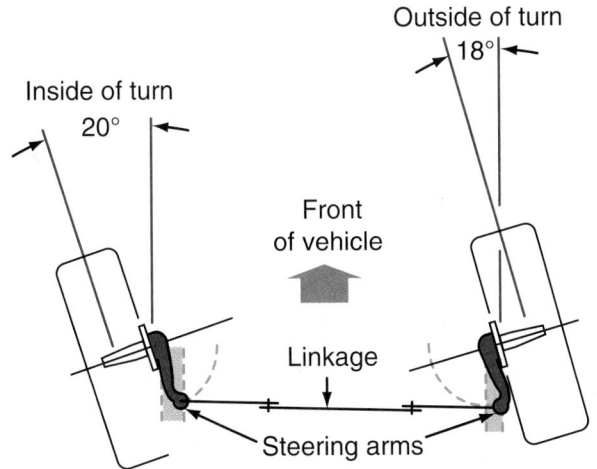

Figure 65.17 The steering arms are bent at an angle to allow the fronts of the tires to toe out during a turn.

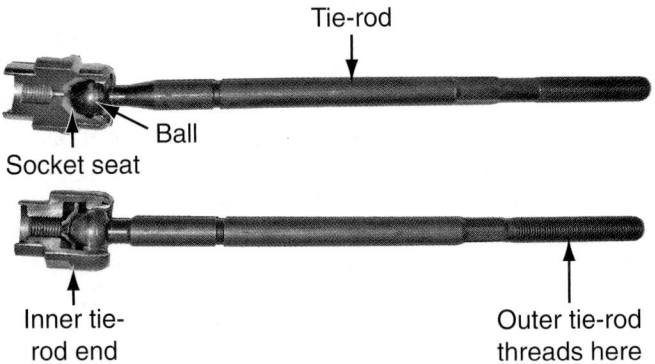

Figure 65.18 Two types of inner tie rod sockets. *(Courtesy of Tim Gilles)*

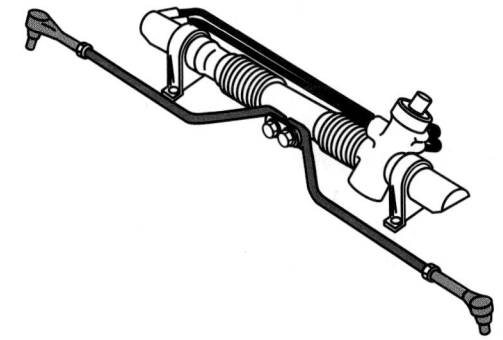

Figure 65.19 Inner tie-rods attach at the center of this rack.

Steering Damper

Some steering linkages use a *steering damper*, a horizontally mounted shock absorber, to minimize the effect of road shocks to the steering wheel (**Figure 65.20**).

■ STEERING COLUMN

The steering wheel is splined to a steering shaft located in the center of the steering column. A locknut retains the steering wheel to the shaft. The shaft is supported by bearings at the top and bottom of the steering column.

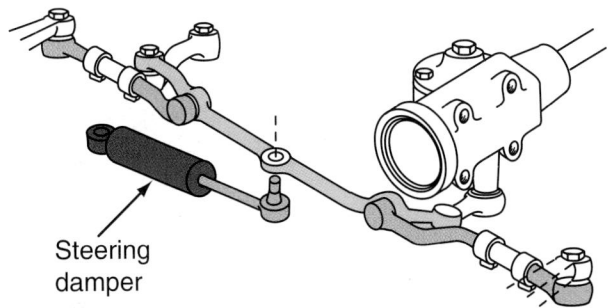

Figure 65.20 A steering damper absorbs road shocks to the steering linkage.

Included in the steering column is the turn signal switch and horn control. Sometimes a headlight and dimmer switch, windshield wiper and washer controls, transmission shift selector, cruise control, and ignition switch and lock (between the shift selector and the steering wheel) are built into the steering column. This is called the *European style* of controls.

Tilt and Collapsible Steering Columns

Tilt steering wheels and collapsible columns are used on many vehicles (**Figure 65.21**). Tilt columns provide

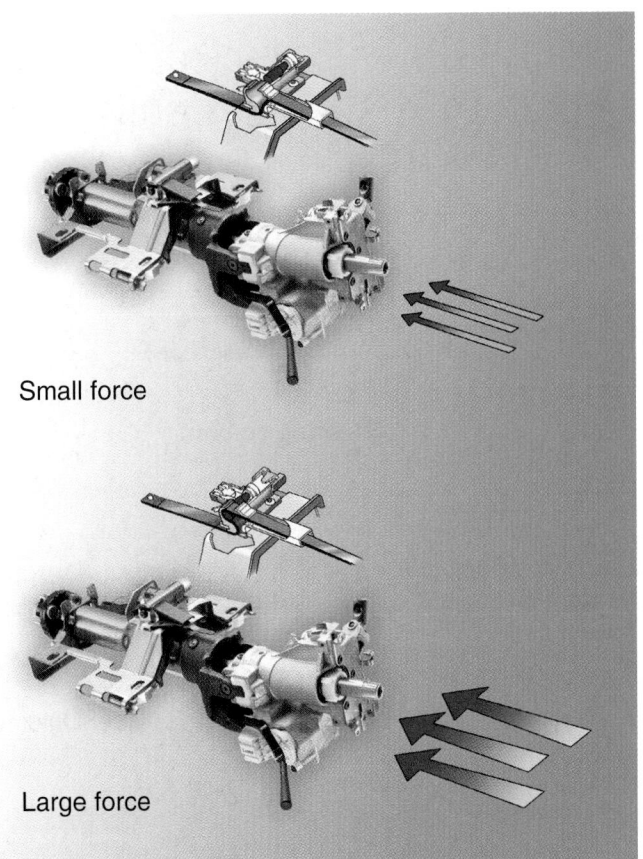

Small force

Large force

Figure 65.21 This tilting steering column can also collapse during an impact. The insets show how the collapsing mechanism operates. *(Courtesy of Delphi)*

the driver with a means of adjusting the angle of the steering wheel. A tilt column includes a short section of shaft beneath the steering wheel connected to the steering column using gears or a universal joint. A typical tilt column uses a spring-loaded ratcheting mechanism to hold the steering wheel in position. A tilt release lever compresses the spring to release tension on the ratchet when steering wheel position adjustment is desired. Some columns are telescopic, allowing the height of the steering wheel in relation to the instrument panel to be adjusted as well.

Air bags are installed on the steering wheel in most vehicles, and, by law, steering columns and shafts are required to be collapsible in case of an accident. There are different designs of collapsible steering columns and shafts. One type uses a two-piece outer section retained with plastic shear pins (**Figure 65.22**). It is combined with a two-piece inner steering shaft, also retained with plastic shear pins.

Service to the steering column and air bags, called supplemental inflatable restraints (SIR), is covered in Chapter 66.

The steering shaft connects the steering wheel to the steering gear. Because the steering column is

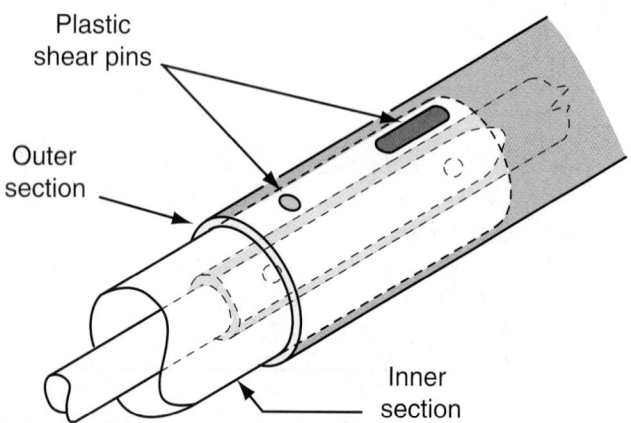

Steering column

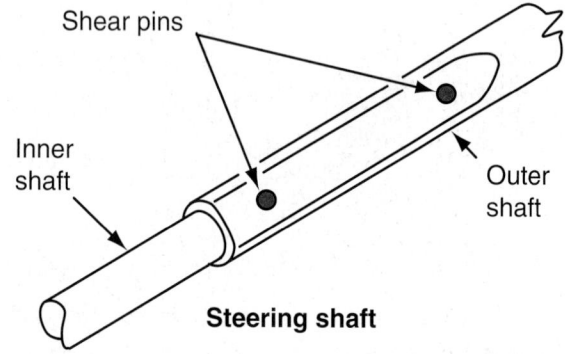

Steering shaft

Figure 65.22 A collapsible steering column and shaft.

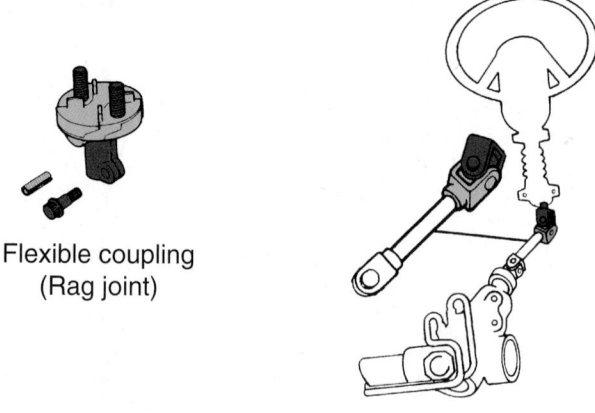

Flexible coupling (Rag joint)

Universal joint

Figure 65.23 A flexible coupling or a universal joint provides the attachment between the steering shaft and the splined input shaft of the steering gear. *(Courtesy of Federal-Mogul Corporation)*

mounted to the body of the car and the steering gear box is mounted to the frame, there is movement or flex between the two. There is a *flexible coupling* or a *universal joint* between the steering shaft and the splined input shaft of the steering gear (**Figure 65.23**). It allows a small amount of misalignment between the steering column and the steering gear. The flex coupling also keeps shock from transferring from the road to the steering wheel. When there is a greater angle, a universal joint is used.

When the transmission shift selector is located in the steering column, a slotted lock plate attached to the upper steering shaft is engaged with a lever to lock the steering wheel when the shift lever is in park and the ignition switch is off.

■ POWER STEERING

Steering systems found on most cars today are power assisted, although there are still some manual units made. Most power steering is hydraulic, with pressure supplied from the crankshaft by a belt-driven pump (**Figure 65.24**).

■ POWER STEERING PUMP

A steering pump driven by a crankshaft belt supplies hydraulic power to assist steering effort. Three main types of power steering pumps have been used on cars (**Figure 65.25**). They are the *roller, vane,* and *slipper* types. Of these, the vane design is the most used. There are several types of power steering pumps. All of them work in the same way. As the pump shaft is turned, oil is drawn into the pump. The oil is squished into a smaller area, which traps it and pressurizes it for delivery to the steering gear.

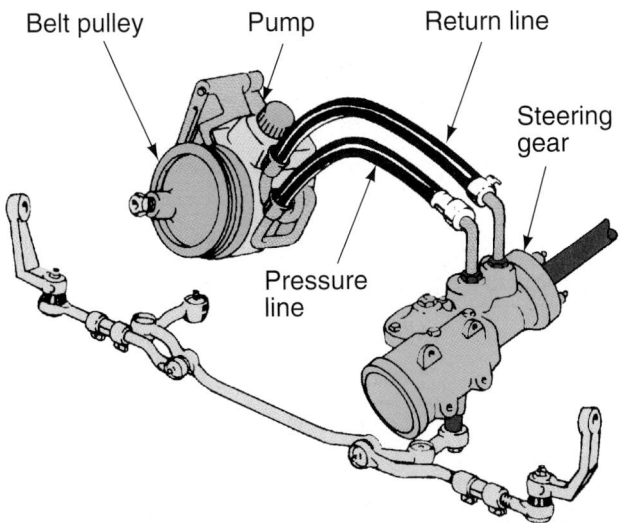

Figure 65.24 Parts of a hydraulic power steering system. *(Courtesy of Federal-Mogul Corporation)*

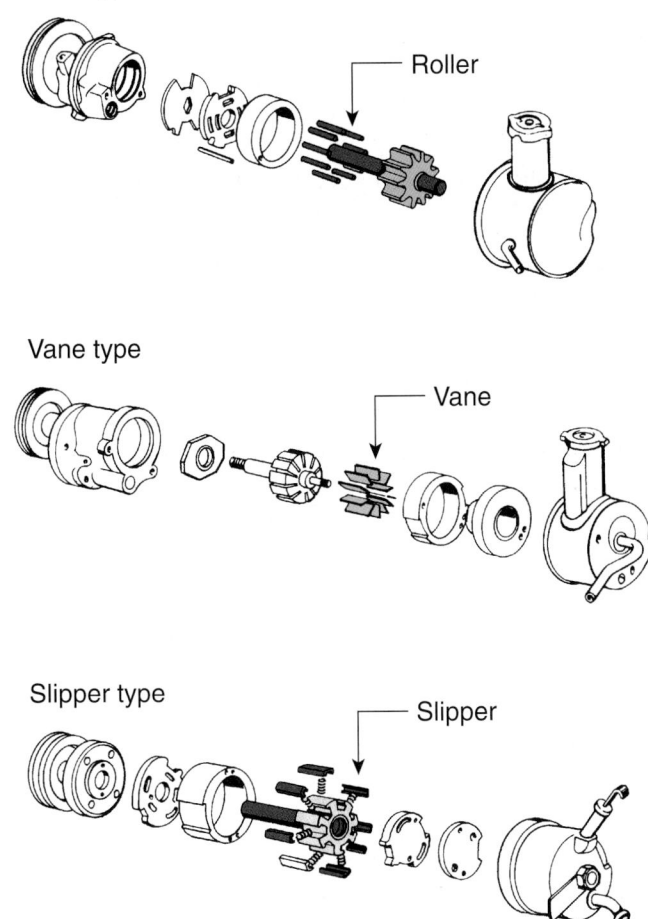

Figure 65.25 Types of power steering pumps. *(Courtesy of Federal-Mogul Corporation)*

Power Steering Pump Operation

The power steering pump works under difficult conditions. Steering assist is needed most when the car is stopped or nearly stopped. So a steering pump must deliver sufficient flow to be able to provide steering assist at low engine rpm and idle. A pump develops more flow as it is driven at higher speeds. When the pump causes excess fluid to flow at cruising rpm, it must divert this fluid back to the inlet side of the pump so it can be returned to the reservoir without creating pressure.

There is a two-stage relief valve in the pump. A *control valve* monitors the turning effort on the wheel to provide the correct amount of assist (**Figure 65.26**). Its two functions are to control flow and limit maximum pressure. The **flow control valve** is almost always working. The pressure relief part of the valve hardly ever opens. It opens when the steering wheel is held all the way to one side, against a steering lock. This causes noise as pressure is released.

When the pump turns at low speed, fluid action is as shown in Figure 65.26. At higher speeds, the pump can flow too much fluid. The by-pass port opens, allowing the excess fluid to return to the pump intake. When pressure in the steering system becomes too high, the pressure relief valve opens to allow the pressure to bleed off to the fluid intake side of the pump.

Steering Power Consumption

Power steering pumps require a considerable amount of horsepower to operate. Sometimes on small vehicles, when steering effort is high, the air conditioning compressor will be shut off to compensate for the draw of the power steering system. Many late-model cars have computer-controlled charging systems that also shut off the alternator at idle when the power steering system is under load.

Power assist on some late-model rack-and-pinion steering units is supplied by an electric motor. These

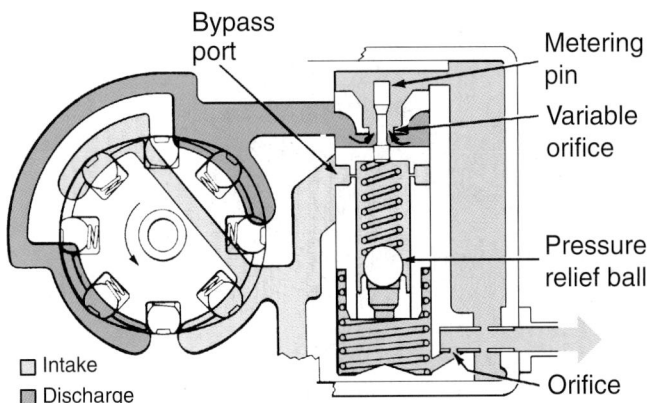

Figure 65.26 A control valve regulates flow in the power steering pump. *(Courtesy of Ford Motor Company)*

vehicles are mostly hybrid-electric, requiring electrical power assist when the engine is shut off during idle stop and off idle operation.

■ TYPES OF POWER STEERING

Power steering systems are either rack-and-pinion or conventional recirculating ball and nut units with a hydraulic control system added. Most are integral power steering systems (see Figure 65.24). Integral means "part of." This means the power steering components are within the steering gear. Integral units can be either recirculating ball or rack and pinion (**Figure 65.27**). Integral recirculating ball gear boxes have a ball nut and sector gear, just like a manual gear box. The ball nut is housed within a *power piston*. Pressurized oil enters a chamber on either side of the power piston to provide steering assist (**Figure 65.28**).

To sense and control power steering assist, gear boxes use either a pivot lever or a torsion bar acting on a **spool valve**. On the *pivot lever type,* turning effort causes a spool valve to be moved by a pivot lever (**Figure 65.29**). When the spool valve moves, fluid is directed to one side or the other of the power piston to provide the assist.

On the *torsion bar type,* also called a *rotary valve type,* a small, sensitive torsion bar twists in response to steering effort. This turns a rotary spool valve to direct pressure to the correct side of the power piston (**Figure 65.30**).

A *linkage-type* gear box, found on older cars and light trucks, uses a standard gear box with a piston attached to the steering linkage. **Figure 65.31** shows fluid assist during left and right turns.

In a rack-and-pinion system, fluid is directed to a chamber on either side of the rack, in a manner similar to the linkage-type power steering. **Figure 65.32** shows how a power rack-and-pinion steering gear works during left and right turns. To sense and control the amount of assist, some racks use a torsion bar attached to the input shaft. Other units use a spool valve that

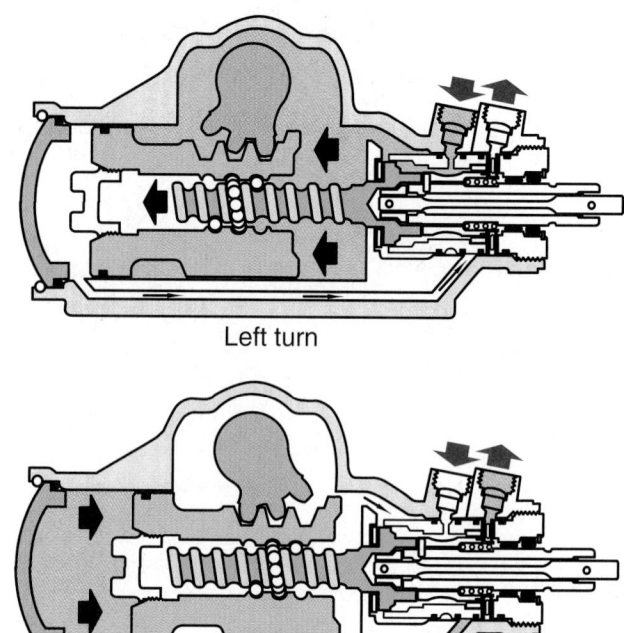

Left turn

Right turn

Figure 65.28 Pressurized oil enters a chamber on either side of the power piston to provide steering assist.

moves in response to up-and-down movement of the pinion gear as it tries to drive the rack.

■ ELECTRONICALLY CONTROLLED VARIABLE EFFORT POWER STEERING

Power steering is installed on vehicles so that a car can be easily steered at low speeds. Once the vehicle achieves a reasonable speed, a fixed level of power assist is no longer necessary and can interfere with a driver's "feel" for the road.

On some late-model vehicles, the speed of the vehicle determines how much power assist is given. One

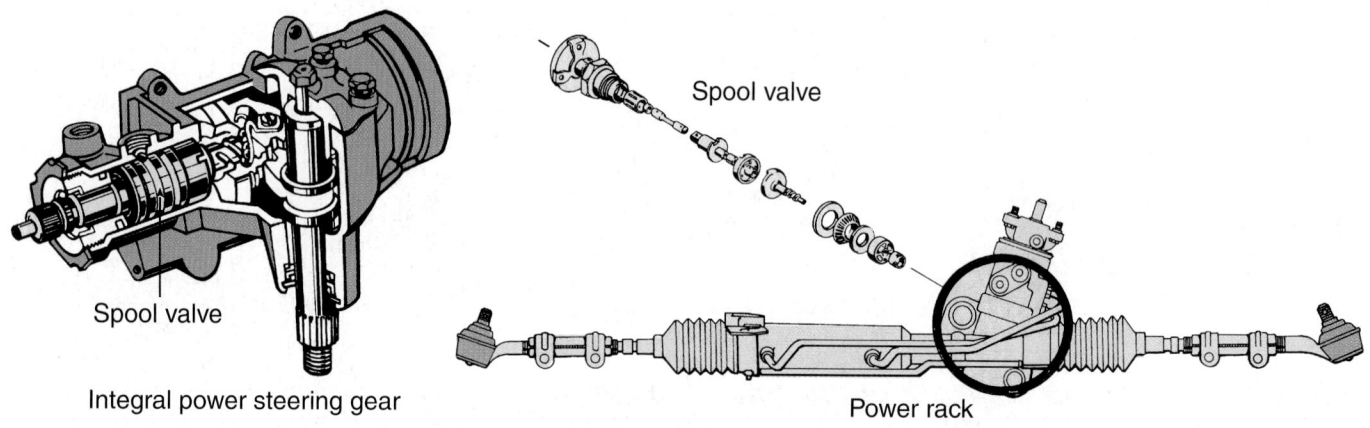

Spool valve

Integral power steering gear

Spool valve

Power rack

Figure 65.27 Types of integral power steering. (*Courtesy of Federal-Mogul Corporation*)

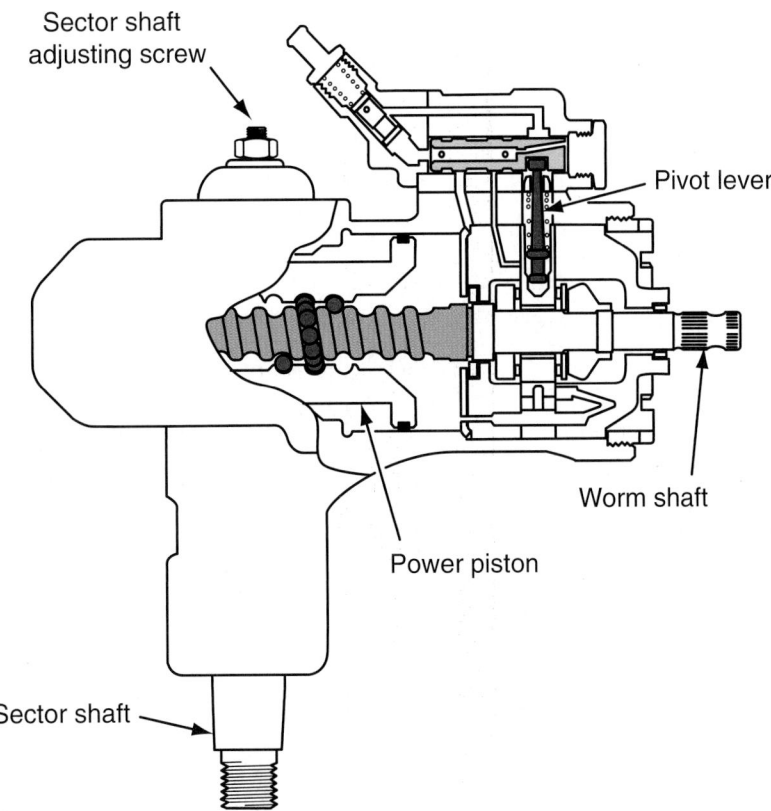

Figure 65.29 A pivot lever-type power steering gear.

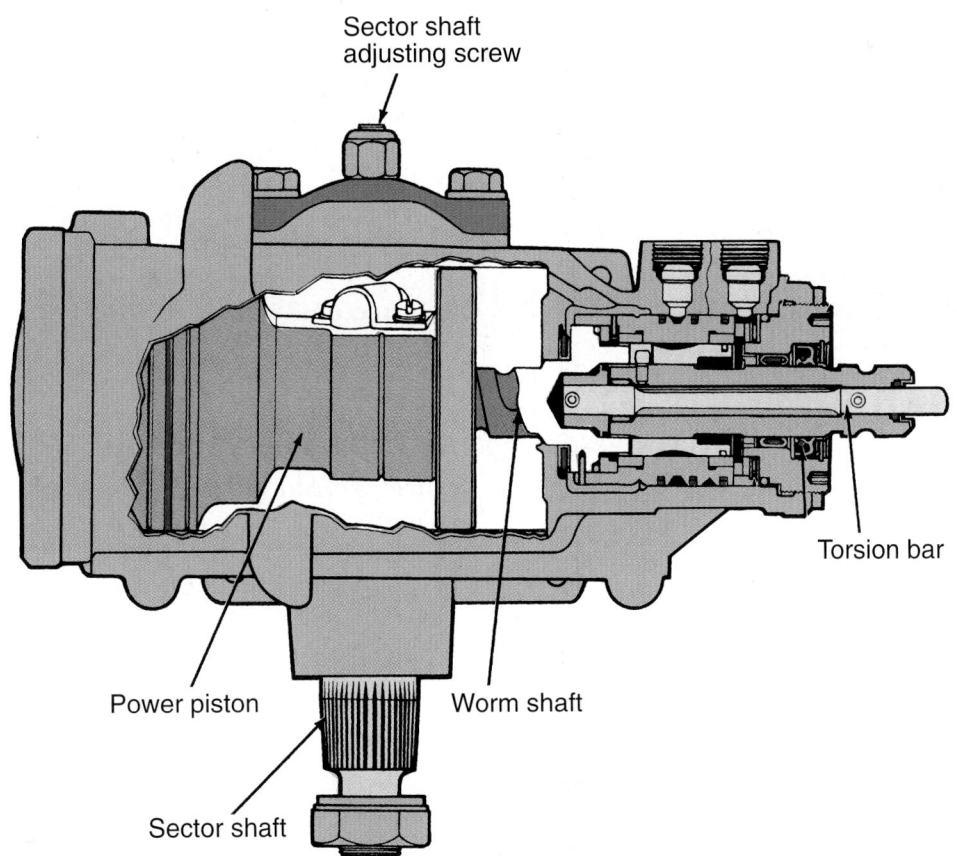

Figure 65.30 A torsion bar-type power steering gear. *(Courtesy of DaimlerChrysler Corporation)*

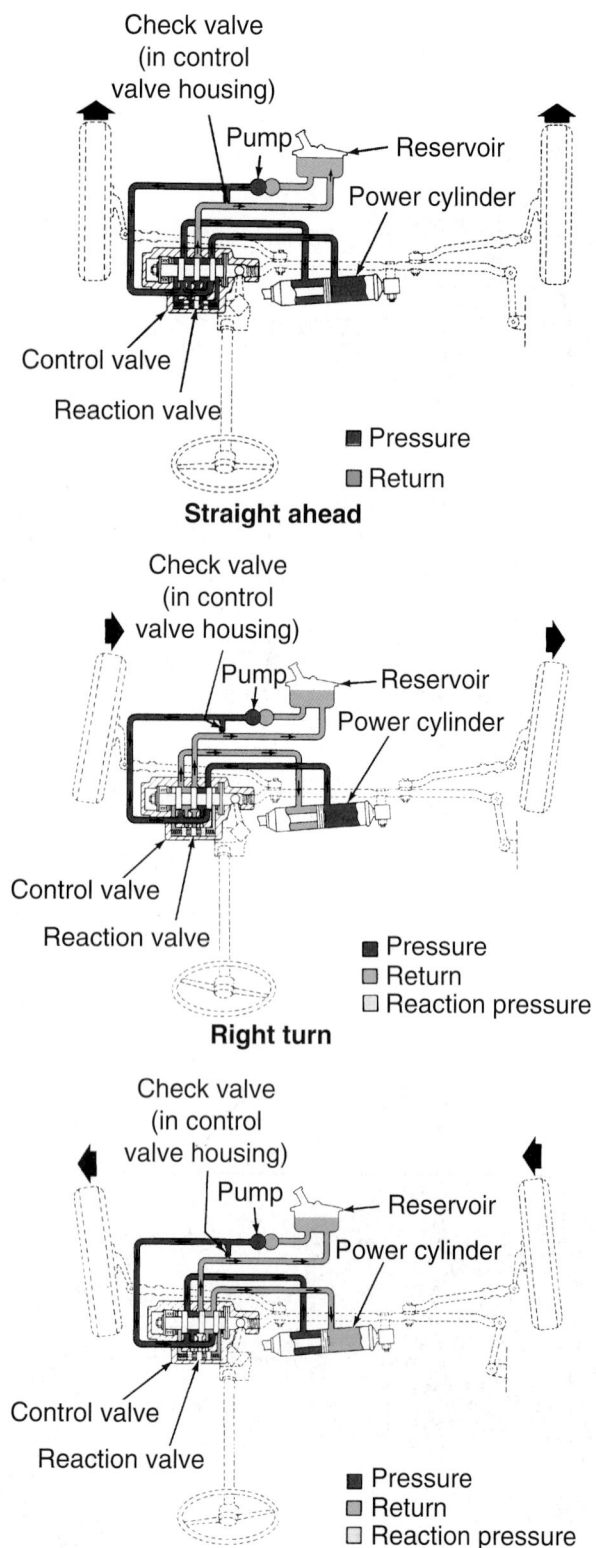

Check valve
(in control
valve housing)

Pump — Reservoir

Power cylinder

Control valve

Reaction valve

■ Pressure
□ Return

Straight ahead

Check valve
(in control
valve housing)

Pump — Reservoir

Power cylinder

Control valve

Reaction valve

■ Pressure
□ Return
□ Reaction pressure

Right turn

Check valve
(in control
valve housing)

Pump — Reservoir

Power cylinder

Control valve

Reaction valve

■ Pressure
□ Return
□ Reaction pressure

Left turn

Figure 65.31 Fluid flow in a linkage power steering system. (*Courtesy of Ford Motor Company*)

type controls fluid output from the pump. Another type controls the amount of fluid pressure available in the power steering gear.

In pump-controlled units, an electronically managed actuator solenoid changes fluid flow in the power steering pump control valve. This is done with a pulse width modulated signal, like in fuel injection systems. Maximum power assist occurs at 1,500 rpm (fast idle) while the vehicle is at rest. The computer varies the solenoid on-time, which allows pump pressure to be higher. As the speed of the vehicle increases, the amount of pump flow is decreased. This increases the steering effort and gives the driver a better feel for the road.

In the steering gear–controlled units **Figure 65.33**, the amount of boost available at the power steering gear is sensed and a module responds by changing fluid flow in the pump control valve to provide the correct amount of assist.

General Motors' electronic variable assist steering system, Magnasteer, has a rotary actuator attached to the input shaft of the hydraulically powered rack-and-pinion steering gear (**Figure 65.34**). A control module varies a supply of electrical current to the actuator, which uses electromagnetic force on the steering gear input shaft to increase or decrease steering effort.

Turning the steering wheel rotates the input shaft/spool valve with its 16 permanent magnet segments. The rotary actuator solenoid that surrounds the input shaft has 16 matching segments, electromagnetically powered by a coil. At high speeds, the control module causes the electromagnetic segments to attract the permanent magnet segments, increasing steering effort to improve road feel. At low speeds, the module reverses the polarity on the actuator's magnetic segments, reducing steering effort.

More advanced variable-effort steering systems can adjust steering effort in response to lateral forces. A computer senses G-forces from a lateral accelerometer. It compares this signal with vehicle speed and a signal from the digital steering angle sensor that tells how fast the steering wheel is being turned (**Figure 65.35**). The steering effort is adjusted to fit the condition.

■ FOUR-WHEEL STEERING

Since the late 1980s, some automotive and truck manufacturers have produced four-wheel steering systems as an extra-cost item in a few select vehicles. All four wheels steer, improving handling and helping the vehicle make tighter turns. The turning radius of a truck with four-wheel steering is similar to that of a compact car.

To reduce turning radius and provide assistance during parking, at low to medium speeds the rear wheels steer in a direction opposite to the front wheels. At high speeds, they turn in the same direction as the front wheels to improve maneuverability during

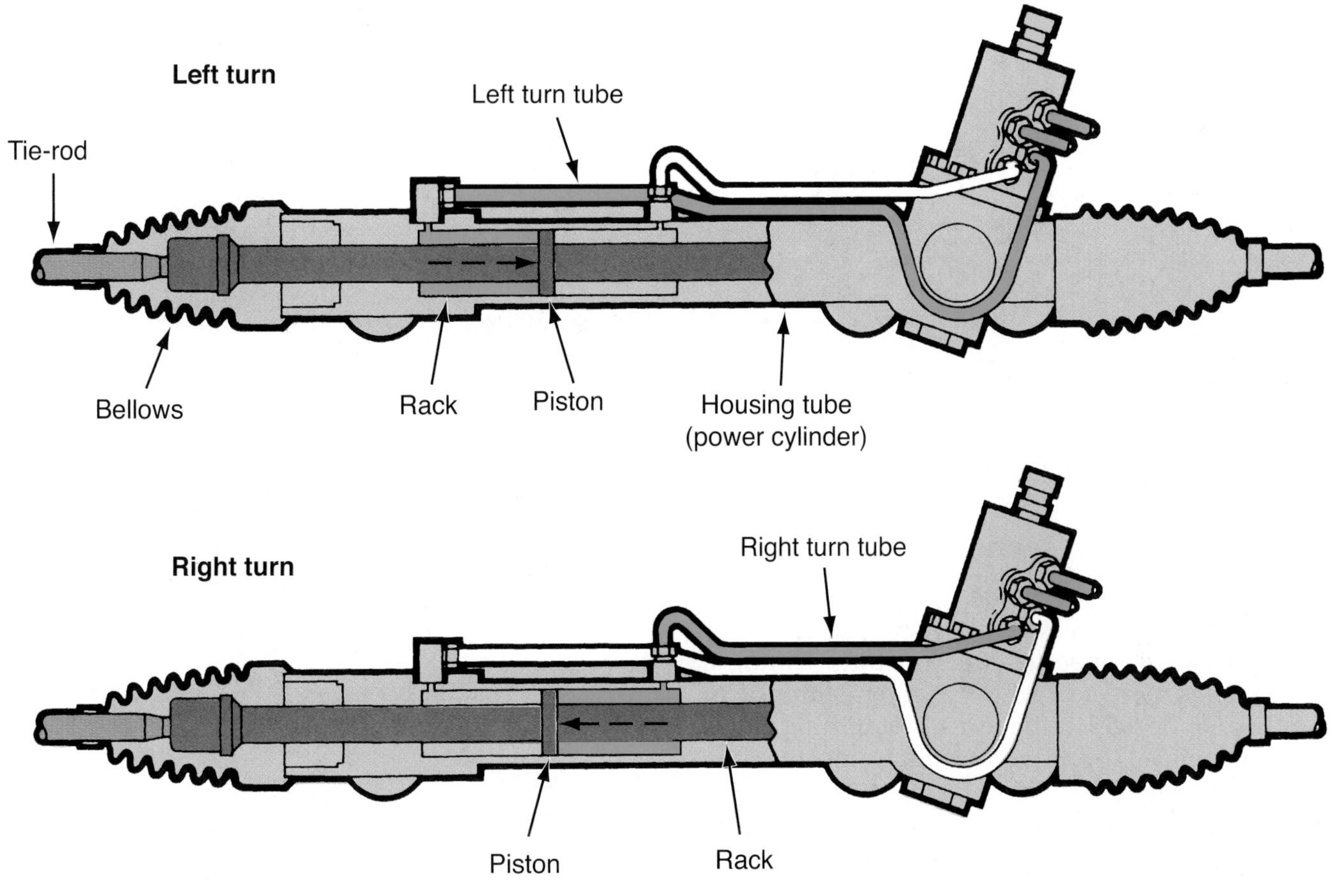

Left turn

Tie-rod

Left turn tube

Bellows

Rack

Piston

Housing tube
(power cylinder)

Right turn

Right turn tube

Piston

Rack

Figure 65.32 Fluid power assist in a rack-and-pinion unit. *(Courtesy of Hunter Engineering Company)*

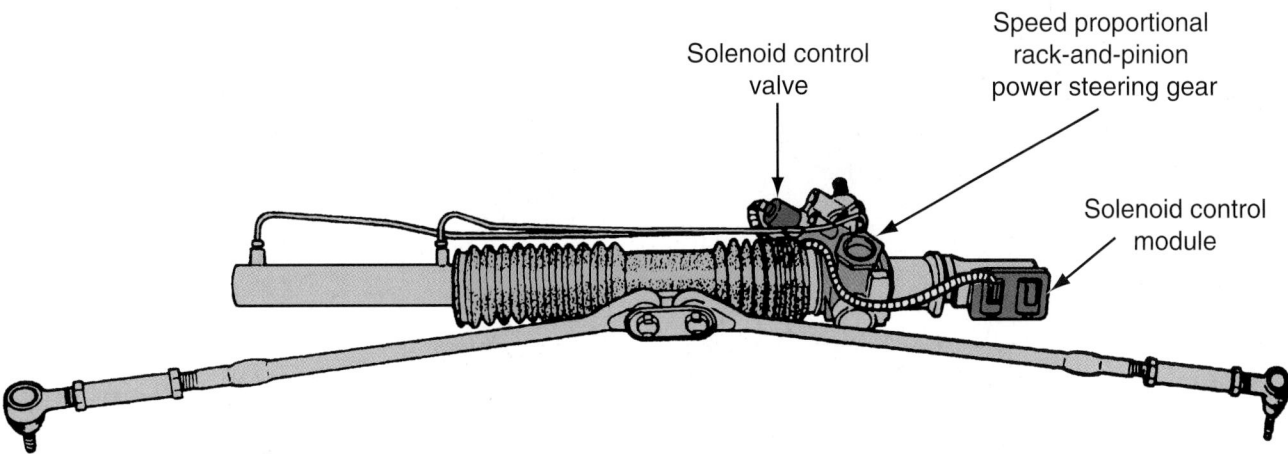

Solenoid control
valve

Speed proportional
rack-and-pinion
power steering gear

Solenoid control
module

Figure 65.33 A speed proportional power steering unit. *(Courtesy of DaimlerChrysler Corporation)*

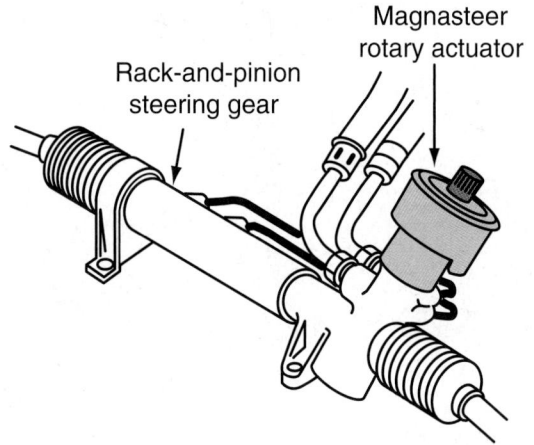

Figure 65.34 The Magnasteer system uses a magnetic actuator.

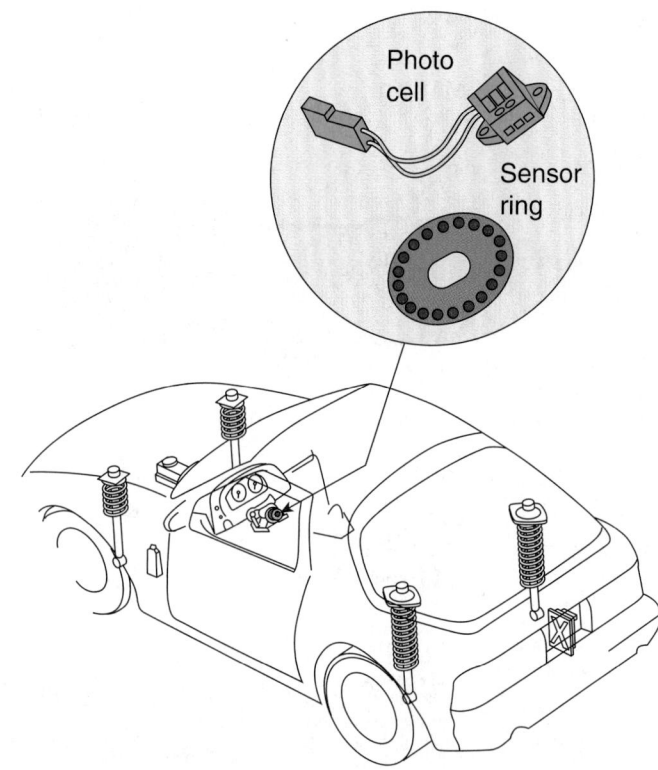

Figure 65.35 A digital steering angle sensor tells the computer how fast and how far the steering wheel is turned.

lane changes. Front wheels do most of the steering and rear-wheel turning is generally limited to 5 to 6 degrees during an opposite-direction turn. During a same-direction turn, rear-wheel steering is limited to about 1 to 1.5 degrees. If the rear wheel turning radius were not limited, the wheels would bump into the curb as the vehicle attempted to maneuver out of a parallel parking spot (**Figure 65.36**). There are different types of four-wheel steering, including mechanical, hydraulic, and electric/hydraulic.

■ ELECTRONICALLY CONTROLLED STEERING SYSTEMS

Electronically controlled steering gears can be "drive-by-wire" (**Figure 65.37**), with no mechanical connection, or they can be mechanically connected using an electro-hydraulic steering gear (**Figure 65.38**). Some active front-steering (AFS) systems are capable of electronically changing the steering gear ratio so the electronic control unit (ECU) steers the front wheels at a different rate than the steering wheel is turned by the driver. The assist requirements of highway and in-town steering are different, and the AFS system changes between

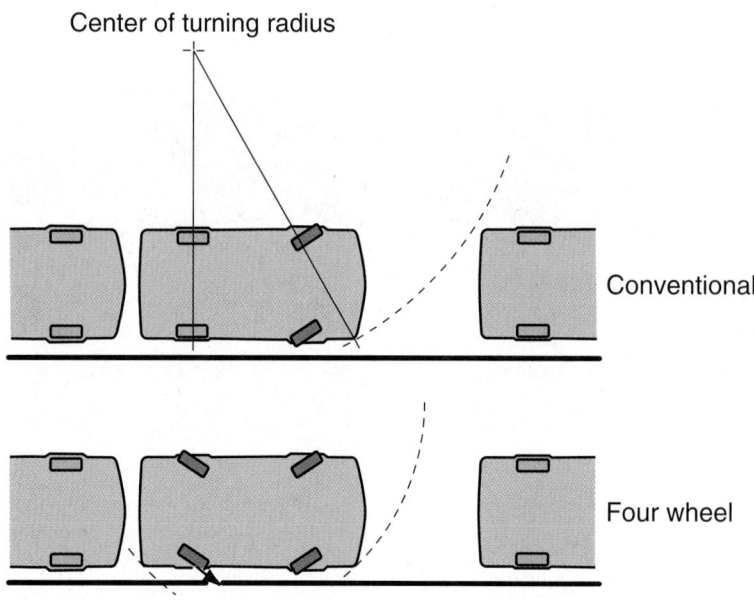

Figure 65.36 The action of conventional front-wheel steering and four-wheel steering when parallel parking.

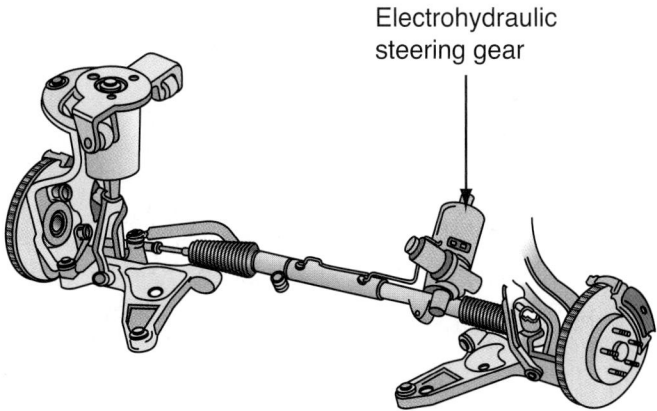

65.38 An electro-hydraulic steering gear provides active front steering control mechanically, without drive-by-wire.

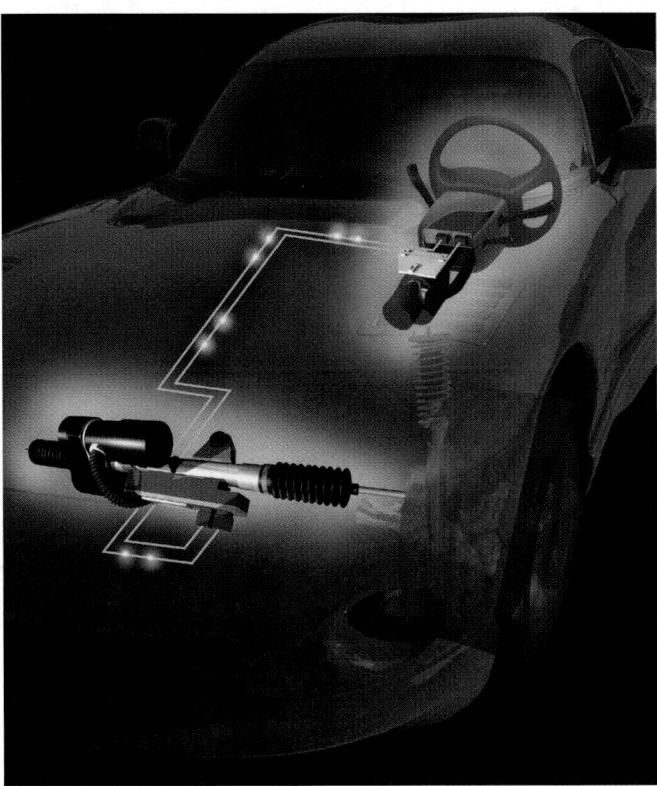

Figure 65.37 An electronically controlled "drive-by-wire" active front steering system. *(Courtesy of Delphi)*

low-and high-speed steering assist seamlessly. At lower vehicle speeds, the amount of assist is increased. At higher speeds, when power assist is not necessary, the amount of assist is lowered.

Electric Steering Gears

Electric motor assists for automobiles are rack-and-pinion steering gears. Some motors operate directly on the steering gear at the pinion shaft (**Figure 65.39**). Others act on the top of the steering shaft, with the motor located in the passenger compartment (**Figure 65.40**).

Because an electric motor can operate when the engine is not running, electric power steering is the design of choice for hybrid and fuel cell vehicles. Electric steering improves fuel economy because it only operates when needed, but electric steering

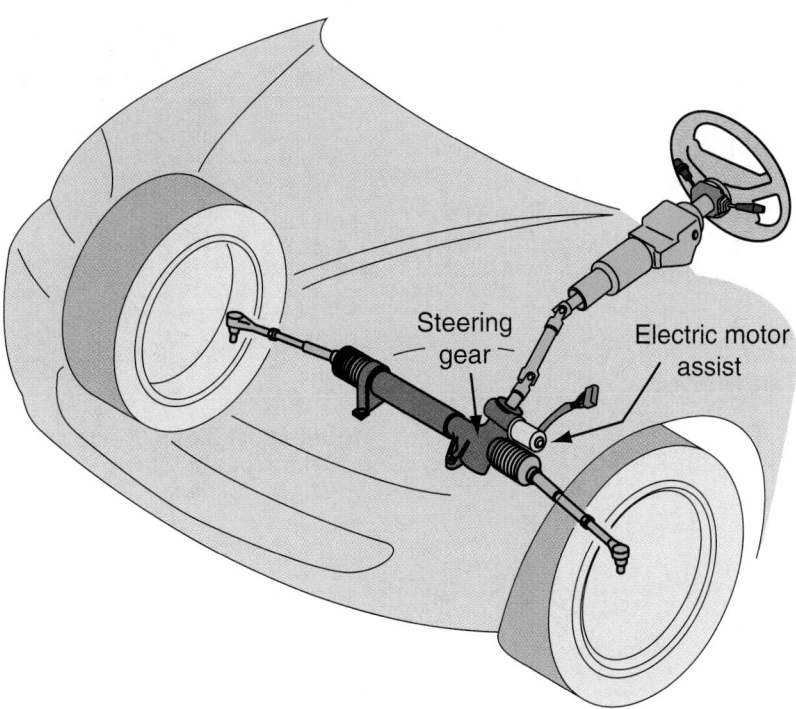

Figure 65.39 An electric steering assist located on the steering gear.

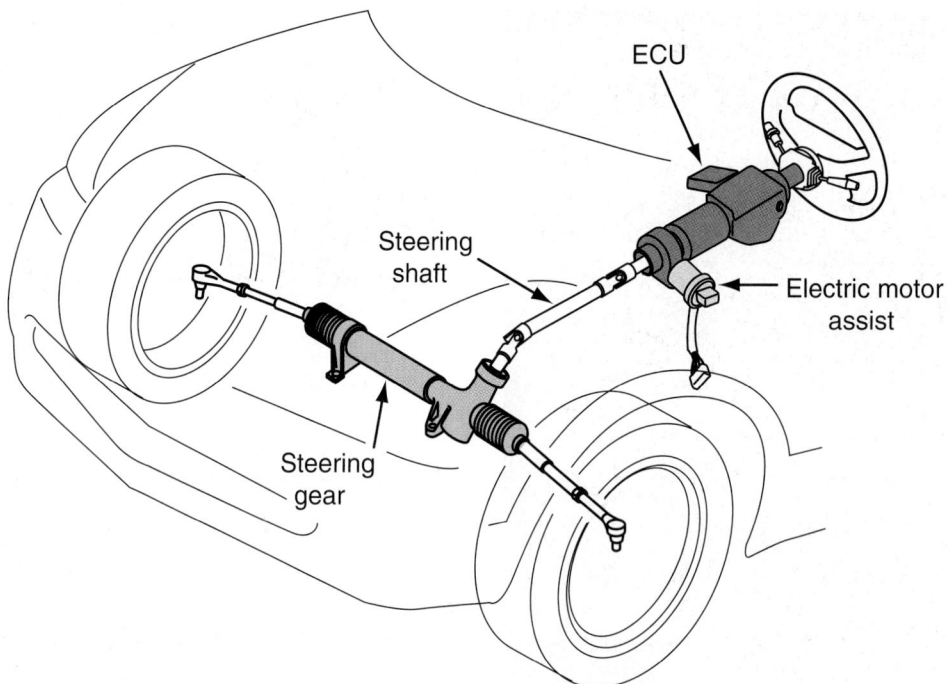

Figure 65.40 An electric steering assist located at the top of the steering shaft, in the passenger compartment.

requires at least a 42-volt electrical system, so there are cost and safety considerations associated with it.

The ECU controls the amount of steering assist by regulating the current applied to the steering assist motor. When the driver turns the steering wheel, the steering gear input shaft twists a torsion bar until the twisting effort equalizes with the opposing, or reaction, force. The amount the torsion bar twists is monitored by a torque sensor. The torque sensor compares magnetic inductance between an input and an output coil. The coils are called resolvers (**Figure 65.41**). The output signal from the torque sensor to the ECU reflects the turning effort of the driver. Other vehicle conditions, like vehicle speed, steering angle, and the skid control system's yaw and deceleration rate are also inputs to the computer decision-making process.

Planetary Gear Active Steering

BMW active steering uses a planetary gearset between the steering wheel and the steering gear. The input is the sun gear and the output is the planetary carrier.

The ring rear is "held," but its speed is regulated by a computer-controlled electric motor. The computer controls the speed the electric motor turns the ring gear, providing a variable steering gear ratio.

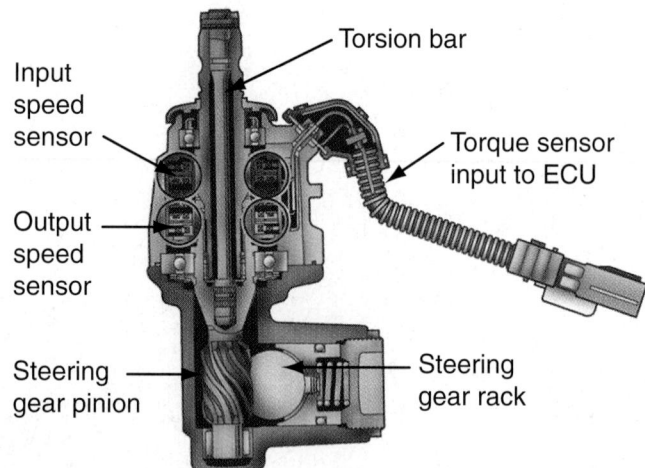

Figure 65.41 When the torsion bar twists, sensors at each end monitor the turning effort of the driver. *(Courtesy of Toyota Motor Sales, U.S.A., Inc.)*

■ REVIEW QUESTIONS

1. One of the two most common types of automotive steering gears is the recirculating ball and nut. What is the name of the other one?

2. What is the term that compares the number of teeth on the driving gear to the number of teeth on the driven gear?

3. When the steering wheel is turned all the way in one direction, it stops against a _____.

4. What is the name of the steering part that is angled so that the front wheels toe out during a turn?

5. What is another type of steering shaft coupling besides a flex coupling?

6. What are the names of the three main types of steering pumps that have been used on automobiles?

7. Which is the most common type of power steering pump design?

8. What is the name of the power steering design that has the power steering components within the steering gear?

9. What is the name of the power steering design that uses a standard gear box with a piston attached to the steering linkage?

10. What type of power steering is used on hybrid vehicles?

■ ASE-STYLE REVIEW QUESTIONS

1. Which of the following is/are true about turning radius?

  a. The outer wheel turns sharper than the inner wheel.

  b. Turning radius is an adjustable angle.

  c. Both A and B.

  d. Neither A nor B.

2. Front tires are supposed to toe _____ during a turn.

  a. Outward   **c.** Both A and B

  b. Inward   **d.** Neither A nor B

3. Technician A says that steering assist is needed most when the car is stopped or nearly stopped. Technician B says that the power steering must be able to provide more steering assist at low engine rpm and idle. Who is right?

  a. Technician A   **c.** Both A and B

  b. Technician B   **d.** Neither A nor B

4. Two technicians are comparing rack-and-pinion steering to recirculating ball steering. Technician A says that recirculating ball steering is more easily damaged when a wheel hits a rock. Technician B says that recirculating ball steering gears are mounted on rubber bushings to help cushion shocks. Who is right?

  a. Technician A   **c.** Both A and B

  b. Technician B   **d.** Neither A nor B

5. All of the following are true about steering columns *except*:

  a. Some steering columns can tilt, allowing the angle of the steering wheel to be changed.

  b. Some steering columns can telescope, allowing the length of the steering column to be changed.

  c. Steering columns can collapse during an accident.

  d. The steering shaft is held in position in the steering column by a bearing at the top and a rag joint or universal joint at the bottom.

6. Which of the following is/are true about electric motor power steering?

  a. They are rack-and-pinion steering gears.

  b. The motor can be located in the passenger compartment.

  c. Both A and B.

  d. Neither A nor B.

7. The steering design used with short and long arm suspensions is the:

  a. Rack and pinion

  b. Recirculating ball and nut

  c. Both A and B

  d. Neither A nor B

8. A "fast" steering ratio is about _____ turns lock to lock.

    a. Three   **c.** Five

    b. Four   **d.** Ten

9. In the figure shown below, the part indicated by the arrow is a(n):

    a. Idler arm   **c.** Center link

    b. Tie rod   **d.** Pitman arm

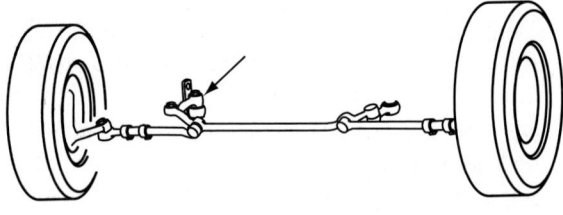

10. An electronically controlled power steering system

    a. Increases assist at lower vehicle speeds.

    b. Decreases assist as vehicle speed increases.

    c. Both A and B.

    d. Neither A nor B.

Steering Service

Upon completion of this chapter, you should be able to:

✔ Describe problems with steering system components.

✔ Inspect the condition of the steering system.

✔ Explain different types of repairs that can be made to the steering system.

✔ Describe repair procedures for rack-and-pinion and parallelogram steering systems.

✔ Perform steering system repairs in a safe and professional manner.

■ **KEY TERMS**

bump steer
jam nut
SIR

pyrotechnics
clock spring
preload adjustment

high point
morning sickness

■ **INTRODUCTION**

In normal use, with good lubricants and seals, steering parts remain separated by lubricants. When seals are damaged and a lubricant leaks out, parts can rub against each other. Also, road hazards such as potholes can cause hard impacts to steering parts. This can result in immediate damage or damage that will result in later wear. Belts and hoses can fail, resulting in operating problems with the steering system.

■ **FLUID LEVEL CHECKS**

Power steering fluid should be hot when checking its level. With the engine idling, turn the wheel several times in each direction to raise the temperature. Before shutting off the engine to check the fluid level. Sometimes a system has a reservoir located in a remote location from the pump.

■ **TYPE OF FLUID**

Power steering fluid is recommended in power steering systems (**Figure 66.1**). Earlier systems traditionally used automatic transmission fluid, which is red in color. Late-model vehicles usually use power steering fluid colored either yellow or yellow/green and of a low viscosity. When manufacturers specify a special type of power steering fluid, using the wrong type can sometimes damage seals within the system.

Figure 66.1 Power steering fluid. *(Courtesy of Tim Gilles)*

■ **DIAGNOSING STEERING PROBLEMS**

Steering and suspension problems can be caused by play, looseness, incorrect height, and wheels that are not aligned to specifications. Wheel alignment theory is discussed in a later chapter, but some of the terms will be discussed here when they apply to a particular problem.

NOISE DIAGNOSIS

Noise from the steering system can be due to several things. A loose power steering belt can result in a jerky feeling during a turn. It can also squeal when the wheel is turned. Belt squeal will be especially evident when the wheel is turned all the way to the end of its travel against a steering lock. If allowed to continue slipping without a belt tension adjustment, the belt will become glazed. The pulley can also wear out.

When the steering wheel is turned against a lock, the pump must bypass the extra pressure that develops. A "whirring" sound is normal as bypass occurs.

Noise can also occur due to a lack of fluid in the power steering reservoir. This will cause a whine when the steering wheel is turned, especially under load.

Loose parts will cause a clunk when going over bumps or when turning the steering wheel from side to side.

HARD STEERING

Hard steering can be caused by binding in the steering linkage, steering box problems, low tires, or incorrect wheel alignment settings. Always check power steering belt tension and fluid level. These are common causes of hard steering.

TIRE WEAR

Steering and suspension problems often result in unusual tire wear. Worn parts can cause scalloped or gouged tire wear. Tire wear can also be due to improper alignment adjustments (see Chapter 68).

STEERING PART INSPECTION

The condition of the steering system must be inspected before doing a wheel alignment. Excessive looseness of front-end parts can allow the wheels to move on their own. This can cause vibration, noise, tire wear, and unsafe driving conditions.

Remember to inspect the wheel bearings to see that they are not too loose or too tight. Also, check to see if they feel rough before performing other tests on the front end. If a wheel bearing is loose, the wheel alignment equipment will not be able to make accurate readings.

STEERING LINKAGE INSPECTION

The fastest way to discover obvious looseness is by performing a "dry park check." Have an assistant turn the wheel from side to side with the tires on the ground or supported by a wheel contact lift. Look for loose parts while observing the steering and suspension as the wheels turn. Sometimes you can feel looseness by holding steering linkage parts during this test.

NOTE: *A vehicle with power steering must have its engine running for this test to prevent looseness that might be present in the steering gear when there is no fluid pressure.*

If there is looseness at one steering position more so than at others, perform the check with the steering wheel in that position. When looseness is difficult to diagnose, some technicians prefer to jack up only one wheel at a time and check for looseness on the opposite side. This puts more effort against the loaded side, making loose parts more apparent. Switch sides to test the other wheel.

While the vehicle is in the air during a lube inspection, the steering linkage can be inspected. With the tires raised and hanging free, turn the steering wheel through its full range of travel. Check to see that there is no binding in the steering linkage or gear box.

STEERING GEAR LOOSENESS

Excessive steering wheel free play can be caused by a worn steering column flexible coupling, worn steering linkage, or a worn or misadjusted steering gear. Steering gear box looseness is usually adjustable.

Check steering wheel free play with the engine off. Power steering cars should have the engine on. Move the wheel back while checking to see how far it moves before the wheels start to move.

NOTE: *Check the steering gear box to see if it feels smooth throughout its entire travel. Check it in the straight-ahead position for binding at the center position and from lock to lock to see that there is no looseness or rough feel.*

PARALLELOGRAM INSPECTION

Tie-rods are held in position by the pitman arm, idler arm, and center link. Parts should allow pivoting and turning, but prevent movement. They are tested by attempting to move them.

Pivot sockets are filled with lubricant and sealed with rubber seals. Check sockets for looseness or damage to the seal.

NOTE: *When checking for steering looseness, be sure that the steering wheel is in the straight-ahead position. This is where the most wear would tend to be.*

Idler Arm Inspection

To check the idler arm, grasp the center link at a point that is as near to the idler arm as possible. Push it firmly up and down while looking for movement (**Figure 66.2**). Movement here can result in a change in toe-in of the front wheels, causing tire wear.

Pitman Arm Inspection

The pitman arm is the lever that attaches the steering linkage to the steering gear box. A small amount of movement where the pitman arm attaches to the

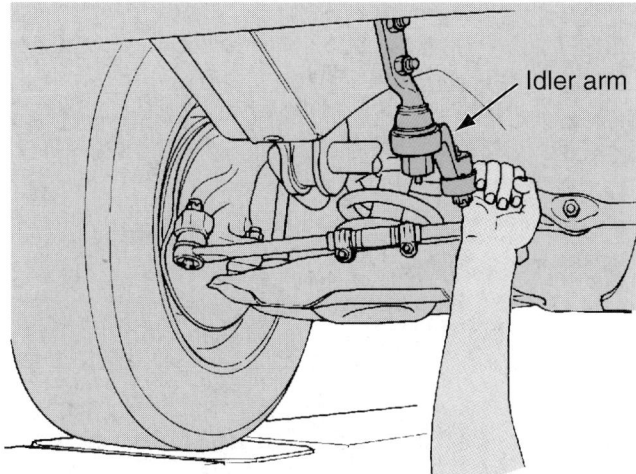

Figure 66.2 Checking an idler arm for looseness. *(Courtesy of Federal-Mogul Corporation)*

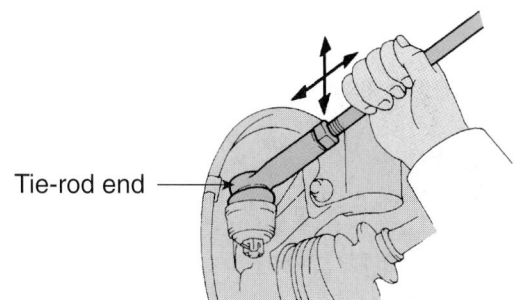

Tie-rod end

Figure 66.3 Checking for tie-rod end wear.

steering linkage is sometimes considered normal. Check manufacturer's specifications.

Tie-Rod Inspection

Inspect tie-rods for wear by moving them by hand to see if there is any perceptible looseness (**Figure 66.3**). There should be none. The tie-rod should be free to pivot, but when worn the resulting looseness can cause tire wear. There are two types of tie-rod sockets. One is preloaded, with no clearance. The other is spring loaded and can be compressed vertically. One test for a tie-rod is to compress it (**Figure 66.4**). Spring-loaded sockets should have firm spring pressure when compressed all the way. Preloaded sockets should not give at all.

■ RACK-AND-PINION STEERING LINKAGE INSPECTION

The inside tie-rod is housed within the bellows boot (**Figure 66.5**). Feel for looseness by grasping the tie-rod through the bellows boot. There should not be any looseness between the stud and the housing. The steering linkage must be in its usual position for this test. If the tie-rod is hanging down in its socket, it could be bound up, preventing an accurate check.

■ STEERING LINKAGE REPAIRS

Steering linkage parts are locked by wedging them against each other with beveled connections called "tapers"

Figure 66.4 Testing a compressible tie-rod end. *(Courtesy of Tim Gilles)*

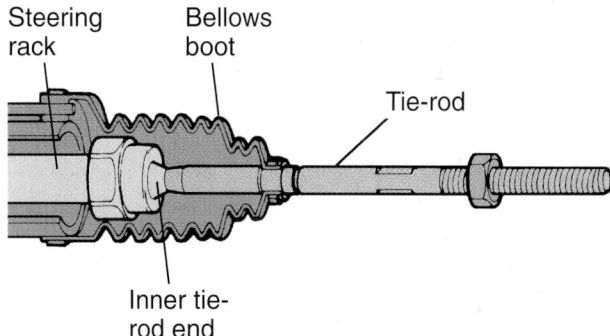

Steering rack   Bellows boot   Tie-rod

Inner tie-rod end

Figure 66.5 Location of the inner tie-rod end. *(Courtesy of Federal-Mogul Corporation)*

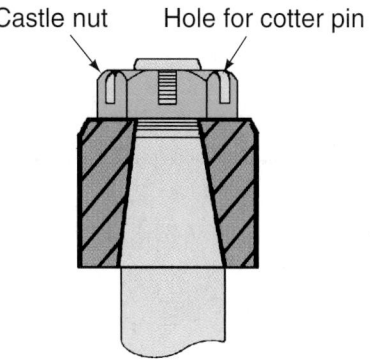

Castle nut   Hole for cotter pin

Figure 66.6 A steering linkage taper connection.

(**Figure 66.6**). Locknuts, usually castle nuts with cotter pins, provide additional protection. When disassembling tapers to service or replace parts, there are several ways to get them apart, listed here in order of preference:

- Tapers can usually be separated using a puller (**Figure 66.7**).
- Linkage parts can also be separated by deforming the outside of the taper using two hammers (**Figure 66.8**).
- A tie-rod separator, sometimes called a *pickle fork*, can be used if the part will not be reused

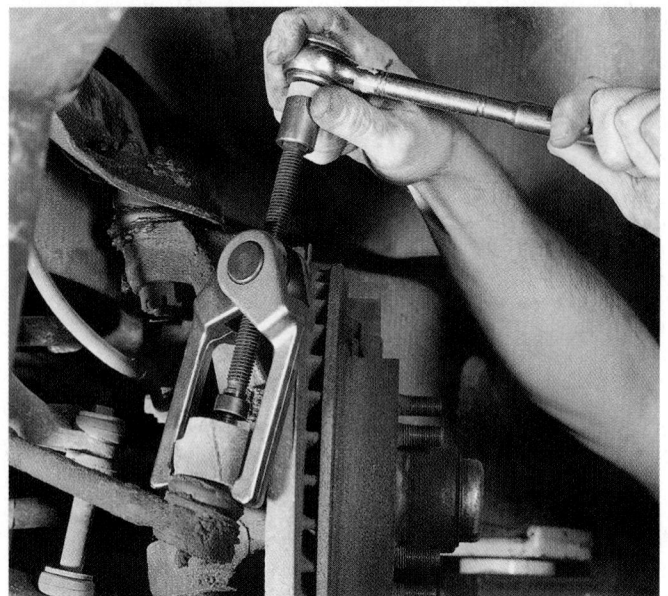

Figure 66.7 Using a tie-rod puller. *(Courtesy of Federal-Mogul Corporation)*

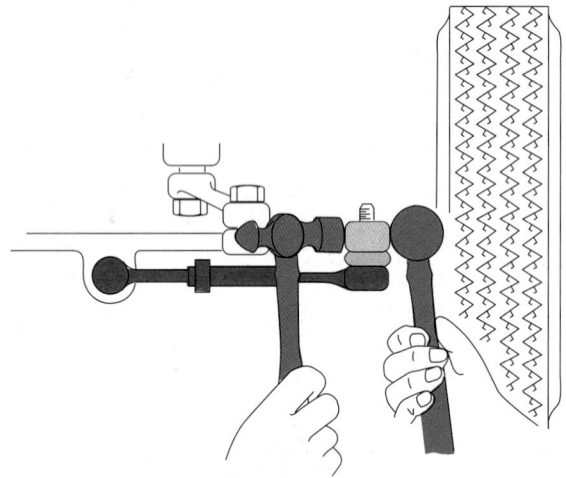

Figure 66.8 Using two hammers to loosen a steering taper connection.

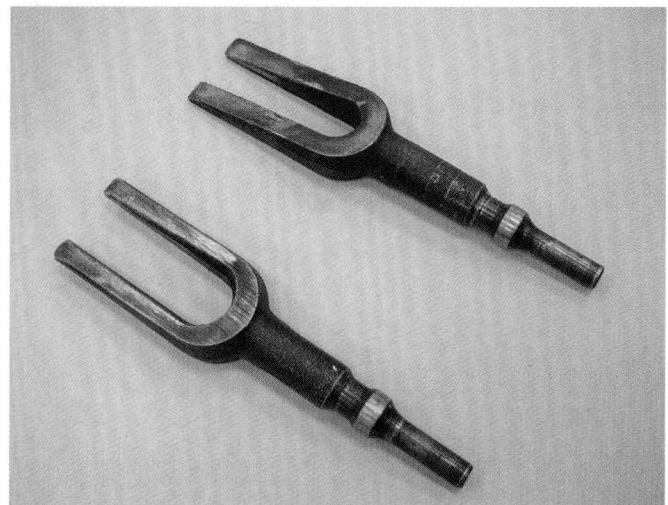

Figure 66.9 Two sizes of pickle forks for use with an air chisel. *(Courtesy of Tim Gilles)*

Figure 66.10 These slotted holes are for mounting an adjustable idler arm. *(Courtesy of Federal-Mogul Corporation)*

(**Figure 66.9**). The pickle fork often cuts the seal and ruins the part, so be very careful when performing this procedure.

■ IDLER ARM REPLACEMENT

During replacement of an idler arm, check to see if it is one of the adjustable types (**Figure 66.10**). If the idler arm and pitman arm are not level to each other, the car will experience **bump steer**. Bump steer causes toe to change when going over bumps, which will cause tire wear (see Chapter 68).

■ PITMAN ARM REPLACEMENT

Removal of the pitman arm requires a special puller (**Figure 66.11**). It must be replaced in the same position on the splines of the steering gear. Be sure to check its position before removing it. Pitman arms often have a "blind spline." One wide spline ensures that the pitman arm can only be installed in one position.

■ TIE-ROD END REPLACEMENT

Tie-rod ends are replaceable, whether the steering system is a rack-and-pinion or parallelogram steering system. Tie-rod ends are threaded to provide a means of adjusting toe. When a tie-rod is replaced, measure the old tie-rod assembly before disassembling it. An approximate toe-in adjustment can be made to the new one prior to installation (**Figure 66.12**). When removing the old tie-rod end, count the number of turns it takes to remove it. Then, turn the new tie-rod end onto its threads the same number of turns.

Figure 66.11 A pitman arm puller. *(Courtesy of Federal-Mogul Corporation)*

Figure 66.12 Measure the tie-rod before disassembling it. *(Courtesy of Tim Gilles)*

Some vehicles have tie-rods that appear to be the same on both ends. The difference is in the threads. One end has left-hand threads; the other has right-hand threads. It is possible to install the entire shaft backwards (which will work). Mark the tie-rod to identify the inner or outer end before removing it.

Before tightening tie-rod clamps, check to see that they are in good condition and are positioned properly so they can be clamped tightly (see Chapter 68). Before doing a front-end adjustment, spray penetrating oil on the threads of the tie-rods. Do this during the steering linkage inspection so the lubricant has time to soak in.

CASE HISTORY

A technician ordered a tie-rod end for a small pickup truck. He specified that the tie-rod end was for the outside left. When the new tie-rod was delivered from the parts store, he attempted to install it. The threads did not match up. Further investigation showed that the tie-rod end was for the inside. The parts store double-checked and found that they had sent the correct part. The tie-rod had been previously installed backwards. This had no effect on the operation of the steering but resulted in some confusion during service.

■ RACK-AND-PINION TIE-RODS

Outer rack and pinion tie-rod ends are serviced the same way as other tie-rod ends. There are several ways that the inner tie-rod socket is held to the ends of the rack. Be sure to follow the instructions that come with the replacement part.

- One method uses a *jam nut*. Some of these types use a thread adhesive. The rack must be held from turning while tightening the jam nut (**Figure 66.13**).
- Another method uses a special washer that is staked into a flat area in the end of the rack.
- When a roll pin is used to hold the tie-rod to the rack, it must be drilled to remove it. After installing the new tie-rod end, a new hole is drilled for a new pin 90 degrees to the original one.

New bellows boot kits are available. Urethane boots are used by many domestic manufacturers. These are superior to neoprene boots.

■ STEERING WHEEL, COLUMN, AND AIR BAG SERVICE

Removal of the steering wheel is sometimes necessary. This could be because a turn signal switch requires replacement or a horn does not work.

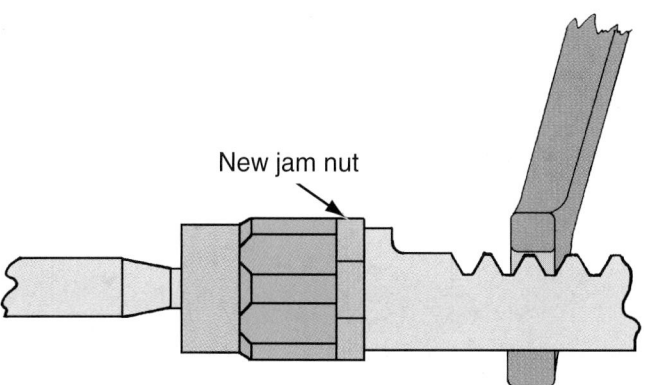

New jam nut

Figure 66.13 Hold the rack with a wrench while tightening the jam nut.

NOTE: *On most steering systems, when the steering wheel is off-center, adjustment is usually made by turning the tie-rods. It is* **not** *done by removing the steering wheel.*

■ AIR BAGS

Many newer vehicles come equipped with air bags, also called supplemental inflatable restraints (**SIR**) to protect occupants of a vehicle during a crash.

CAUTION Accidentally deploying (activating) an air bag can result in an injury or death.

- Air bags contain rocket propellant. They are shipped in cartons that say "**pyrotechnics**" on their label.
- When an air bag deploys, it opens so fast that only a very good video camera can freeze the action enough to slow it down for frame-by-frame investigation (**Figure 66.14**).
- It is not unusual for a windshield to be broken by the passenger-side air bag.

The driver-side air bag is located in the steering wheel (**Figure 66.15**). It is usually easily removable by loosening two or three screws beneath the steering wheel. Before attempting to remove a steering wheel on one of these vehicles, consult a service manual.

Air Bag Clock Spring

The driver-side air bag has a **clock spring** or spiral cable (**Figure 66.16**). It delivers the electrical signal from the air bag module to the steering wheel to deploy the air bag in case of an accident. The spring is called a clock spring because of its resemblance to the spring in a mechanical clock.

NOTE: *The clock spring must be wound correctly and must be "timed" to the steering column.*

The clock spring has a certain amount of travel from fully wound to fully unwound (usually 4–5 turns of the wheel). It is in its midway position when the steering wheel is straight ahead. When the wheel is turned one way against a steering lock, it unwinds. Turning the wheel the other way winds it up.

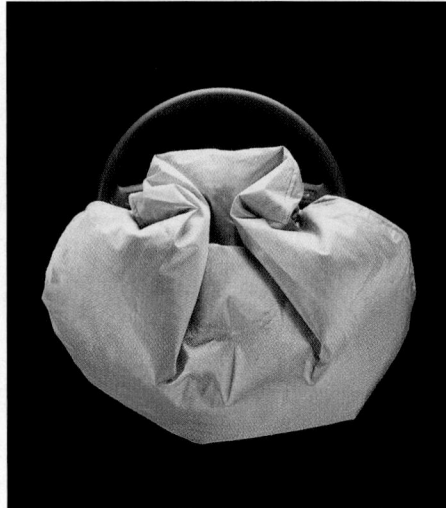
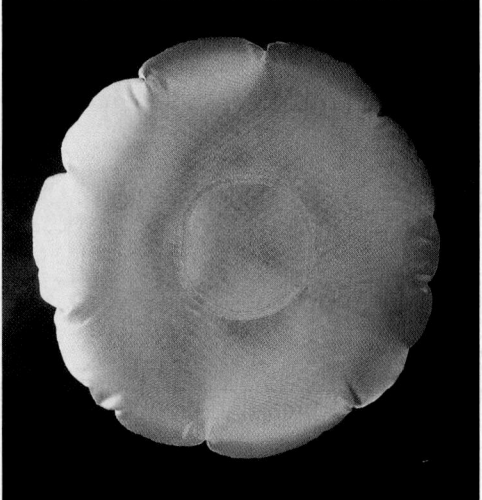

Figure 66.14 Various stages of air bag inflation. *(Courtesy of TRW, Inc.)*

Steering column

Clock spring

Air bag

Figure 66.15 Steering column parts of an air bag. *(Courtesy of Federal-Mogul Corporation)*

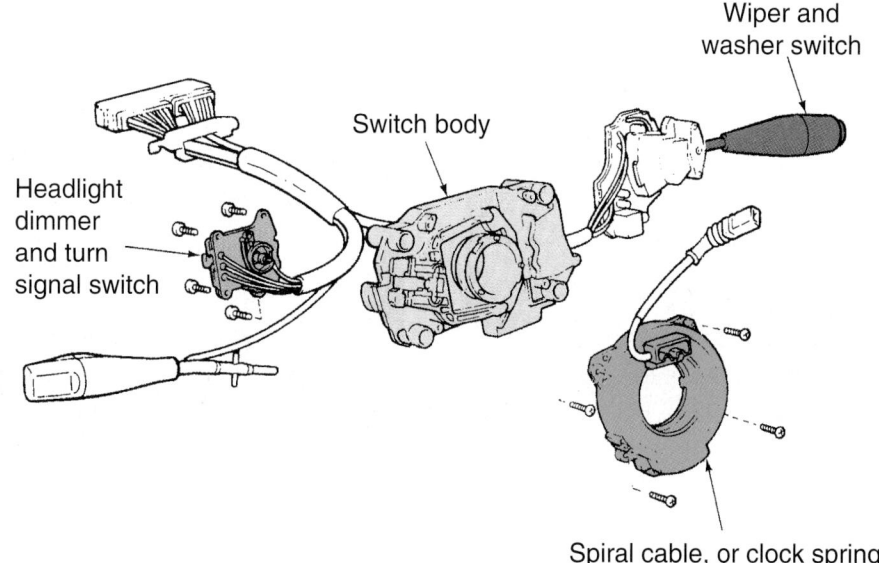

Figure 66.16 The spiral cable is commonly called a clock spring.

Figure 66.17 shows a clock spring after the steering wheel has been removed. If the clock spring is not installed in the correct position, it will break when the wheel is turned in both directions. The air bag will not operate, the SRS dash light will come on, and the clock spring will have to be replaced. A new clock spring comes with an aligning mark so it can be installed properly. Put the steering wheel in the straight-ahead position before removing a steering column. The arrows should be lined up on the clock spring.

 When removing an entire steering gear or column assembly, turn the wheel all the way against the steering lock and remove the ignition key, locking the steering column. During reinstallation, be sure the wheels are positioned in the same full travel position.

 When removing a steering wheel on a vehicle with an air bag, always check the manufacturer's service information for any specific procedures that might apply.

Simply disconnecting the battery will not ensure that the bag will not deploy. There is a backup capacitor that will deploy the air bag in the event the battery becomes disconnected during a crash. A typical precautionary procedure before disconnecting an air bag would be to short the battery terminal ends together or depress the brake pedal after the battery has been disconnected.

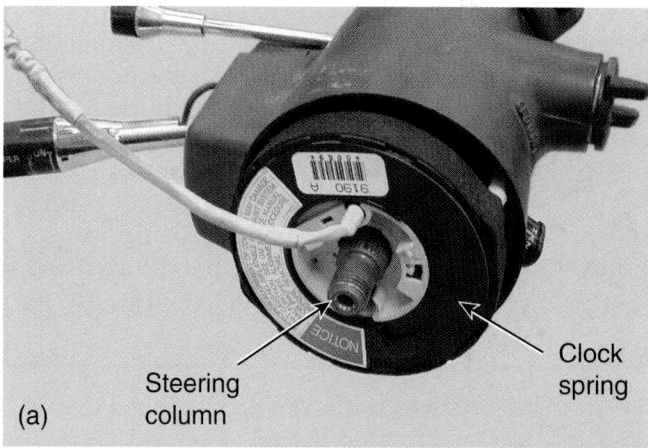

(a)

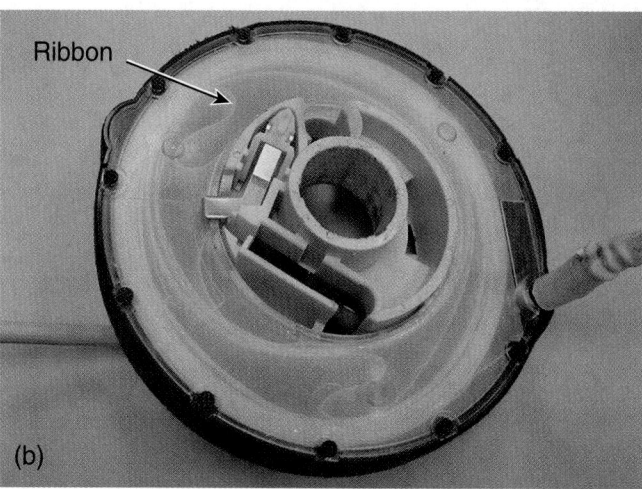

(b)

Figure 66.17 (a) The clock spring is located under the steering wheel. (b) A clock spring viewed from the back side. The ribbon is made of nylon with a gold filament. *(Courtesy of Tim Gilles)*

Figure 66.18 Air bag electrical connections are yellow. (*Courtesy of Tim Gilles*)

Stepping on the brake pedal is an OBDII procedure driveability technicians use to clear computer memory.

With the battery disconnected and the key off, step on the brakes. Since the taillights are normally energized without the key on, they try to light. This works to discharge the air bag capacitor.

■ Be certain the electrical connection is the correct one before disconnecting it. Air bag connectors are colored yellow (**Figure 66.18**). Always store an air bag facing up on the workbench. If it accidentally deploys, an entire steering column could be shot into the ceiling. If the air bag is facing upward, the chance of injury to bystanders is less.

Air bags are covered in detail in Chapter 34.

▬ STEERING WHEEL SERVICE

Pulling the Steering Wheel

Do not hammer on the end of the steering shaft when removing the steering wheel. This can damage the column. Install a puller in the threaded holes in the steering wheel (**Figure 66.19**). If there are no alignment marks, make some with a centerpunch and hammer. When reinstalling a steering wheel, double-check to see that it is in the neutral position (half-way between the steering locks).

▬ STEERING COLUMN SERVICE

Steering column service can be complicated, and the proper service manual should be consulted before repairs are attempted. In addition to disassembly information, a troubleshooting chart is typically included in this service material.

Noises in the steering column can be due to a loose or damaged steering coupling, misalignment, or a bearing or horn ring that lacks lubricant. Shaft

Figure 66.19 Removing a steering wheel with a steering wheel puller.

bearings or bushings are typically held in place with snaprings. When bearings are not permanently lubricated, an O-ring retains lubricant.

Mechanical problems with the ignition lock are also included in steering column service. These can include failure to lock or unlock and high effort needed to turn the switch, which can be due to a defective lock cylinder or bent or misaligned parts.

Flexible Joint Replacement

A worn flex coupling can cause looseness in the steering wheel. To replace a flex coupling, loosen and remove the bolts attaching it to the steering gear and lower steering shaft. **Figure 66.20** shows the locations of the parts. Remove the steering column mounting bolts where the column mounts to the instrument panel inside the passenger compartment. Move the column far enough toward the rear of the car for the coupling to be removed from the steering gear.

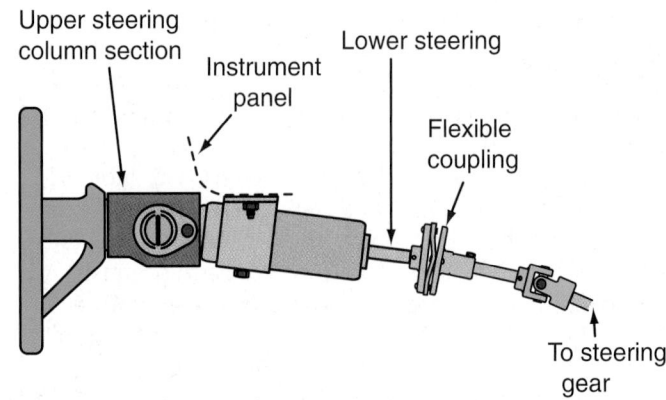

Figure 66.20 Parts of a steering column.

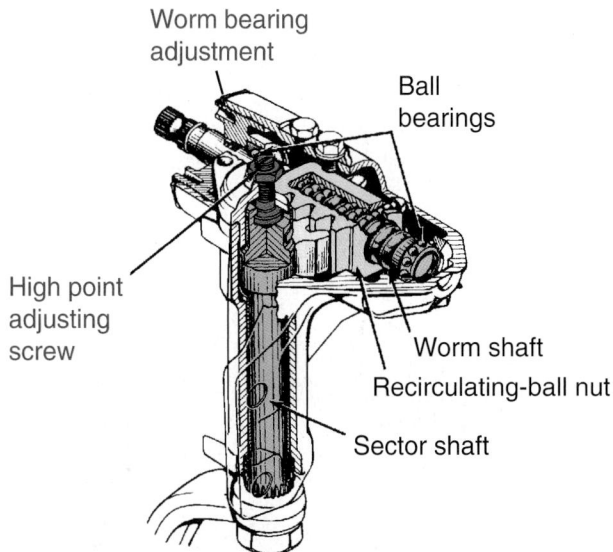

Figure 66.21 High point and worm bearing pre-load adjustment locations on a recirculating ball and nut steering gear. *(Courtesy of DaimlerChrysler Corporation)*

■ STEERING GEAR SERVICE

Steering gears are generally reliable. The most common service is the repair of fluid leaks. Steering gear damage can also be caused by collisions or deep potholes. Steering gears are occasionally rebuilt by the technician in the repair shop. When rebuilt steering gears are available at a reasonable price, they are most often the repair of choice.

Recirculating Ball and Nut Steering Gear Service

Recirculating ball and nut manual steering gears have two adjustments to make (**Figure 66.21**). One of them is a **preload adjustment** to the bearings at the ends of the steering shaft (worm shaft). These are ball bearings, which call for a small amount of load to be on them. The load adjustment is measured with an inch-pound torque wrench (**Figure 66.22**).

The second adjustment is to the **high point**. The sector gear teeth in a steering gear have a high point where the gear teeth come closer together when the car is traveling straight ahead (**Figure 66.23**). The high point must be centered when the vehicle travels straight ahead so the steering response is not faster in one direction than the other.

■ MANUAL RACK SERVICE

It is important that the steering wheel on a rack and pinion be properly centered. This is for two reasons:

■ The center of the rack is hardened.
■ If the tie-rods are of different lengths, the car can steer to one side farther than it can to the other.

Figure 66.22 The bearing preload adjustment is measured with an inch-pound torque wrench. *(Courtesy of DaimlerChrysler Corporation)*

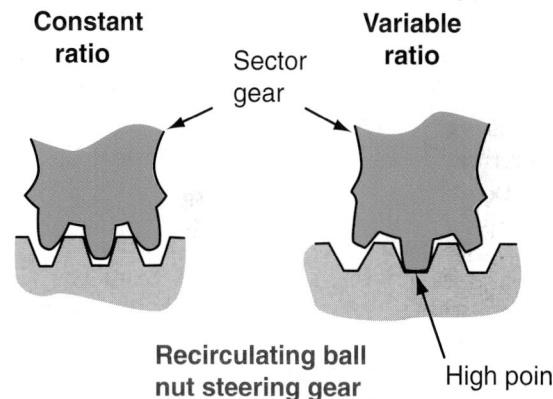

Figure 66.23 The high point is the position where the gear teeth are at their closest mesh when the steering wheel is centered.

■ RACK-AND-PINION LOOSENESS

A steering damper holds tension against the steering rack. A rack gear preload adjustment can be made (see Chapter 65), but be sure to check the service manual for the correct procedure and specification. If the rack is worn in the center, tightening this adjustment can result in binding steering at the outsides during turns.

■ POWER STEERING SYSTEM SERVICE

Power steering system service includes repair and replacement of the pump, flushing of the hydraulic system, seal replacement, belt and hose service, and service and repairs to the steering gear.

Reservoir Service

The fluid reservoir is made of sheet metal or plastic and is either attached to the pump or remotely located. An attached unit is called an *integral reservoir*. Some vehicles have a stand-alone pump that is attached by hoses to a reservoir mounted in a remote location.

◼ POWER STEERING SYSTEM FLUSHING

Be sure to check the condition of the hoses. Power steering pumps or gears often fail because the inside of the power steering hose has deteriorated.

In the past, power steering fluid was considered to be a long-life item. Today, rack-and-pinion steering is common on modern vehicles, and the housing and valve that control steering are susceptible to wear. When the fluid becomes contaminated with metal particles, more wear results. The service life of today's rack-and-pinion steering systems can be lengthened considerably by changing the steering fluid at regular intervals. The most popular vehicles—that is, those vehicles produced in the largest quantities—will have rebuilt steering units available at competitive prices. Low-volume vehicles do not enjoy the same availability of rebuilt parts, and the replacement cost for a steering unit can be substantially higher.

When replacing a rack-and-pinion steering unit, the system must be thoroughly flushed. A common power steering problem is **morning sickness**—a loss of power assist on cold startups, often in one direction only. Failures from morning sickness result in contaminants being circulated throughout the fluid system. These contaminants can ruin a new rack if they are not flushed from the system. Some manufacturers recommend the installation of a filter when a rack is replaced. Service to power rack-and-pinion steering gears is covered later in this chapter.

Power Steering Flush Methods

Commercial fluid exchange units are available (**Figure 66.24**). These machines provide a convenience to the fast lube service market. Fluid can easily and cleanly be exchanged without disconnecting hoses.

When the shop does not have a steering flusher, a technician can flush a steering system using the following procedure (**Figure 66.25**):
- Raise the vehicle and remove the return hose from the reservoir. When the pump has an integral reservoir the return hose often has an ordinary hose clamp rather than a crimped connection.
- Plug the reservoir outlet.
- Put the end of the hose in a drain pan.
- Run the engine at idle and turn the steering wheel from lock to lock. Then, shut the engine off.
- Refill the reservoir with clean fluid.

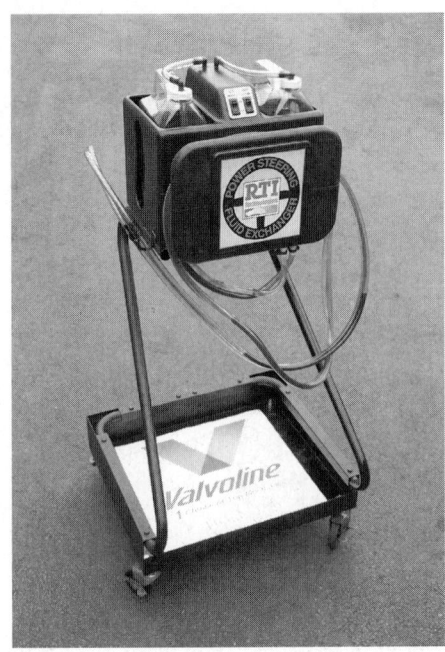

Figure 66.24 A power steering fluid exchange machine.

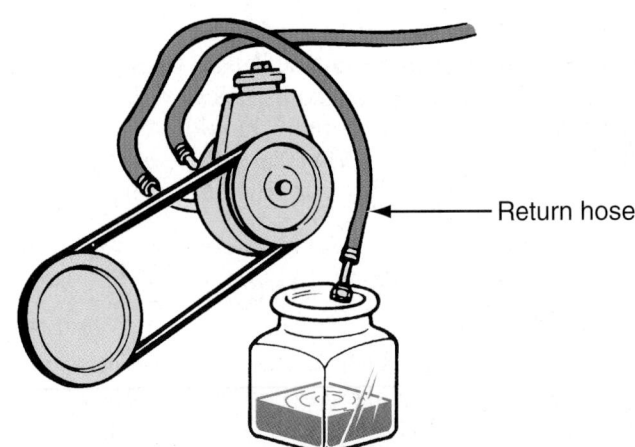

Return hose

Figure 66.25 Run the engine at idle and turn the steering wheel from lock to lock to pump the old power steering fluid from the return hose. *(Courtesy of Federal-Mogul Corporation)*

- Start the engine and wait until fluid begins to flow from the hose. Then, shut the engine off.
- Repeat the cycle of refilling the reservoir and running the engine until the fluid coming from the return hose is clean and free of air bubbles.

◼ BLEEDING THE SYSTEM OF AIR

Air in the power steering system will cause erratic operation and/or a growling sound, especially when the steering wheel is turned.

The best practice when replacing a power steering component is to avoid allowing air to enter the system. Do not let the pump reservoir become empty. Allowing the fluid level to drop too low allows air to be pumped into the system. When refilling the system after a steering gear or hose replacement, fill the reservoir, start the engine, then shut it off immediately. When the engine is started, the reservoir will be sucked empty in as little as 1 second as the pump fills empty voids in the hose or steering gear with fluid. Refill the reservoir and repeat the procedure until the fluid level no longer drops.

Check that the reservoir is full and remove air from the system as follows.

- Be sure the tires are lifted off the ground so you don't cause a flat spot on the tire tread as you cycle the steering wheel back and forth.
- When the fluid is warm, run the engine at idle and cycle the steering wheel from lock to lock several times, holding it against the steering lock at each side for 2 or 3 seconds. Then, inspect the fluid in the reservoir for bubbles. If the fluid appears to be foamy after bleeding is attempted, allow it to sit for several minutes until the foam disappears.

NOTE: *Ford recommends turning the wheel from lock to lock 20 to 25 times.*

Test-drive the vehicle. If there is still a groan from the power steering system, there is probably air remaining in the fluid. Do the following to check for air:

- With the engine running at 1,000 rpm, verify that the fluid level is at the hot/full mark.
- Shut off the engine and check the fluid level again. If the fluid is bled of air, the fluid level should not increase by more than 0.020" (5mm) (**Figure 66.26**)

Some technicians like to bleed power steering air with the engine off. This prevents larger air bubbles from turning into many more smaller ones (which will be more difficult to expel). Sometimes an external vacuum source is needed to remove air from the fluid. A rubber stopper is installed in the reservoir filler neck. Vacuum is applied through the stopper for about 15 minutes.

NOTE: *A few vehicles have hydraulically assisted power brakes, known in the industry as hydro-boost, which are operated off the power steering pump. On these systems, depress the brakes several times while turning the steering wheel to help rid the system of air.*

■ POWER STEERING PUMP REPLACEMENT

Prior to removing a power steering pump, disconnect the return hose to drain the fluid. Loosen and remove the belt. On some vehicles, access to power steering pump bracket fasteners is easier with the vehicle raised on a lift. Sometimes a pump has more than one bracket. If bracket removal from the pump is required, make a careful observation of the bracket locations.

■ REPAIRING POWER STEERING PUMP OIL LEAKS

An integral reservoir is sealed to the pump with a large O-ring (**Figure 66.27**). Seals are also located at the pump shaft and the fittings for the hoses (**Figure 66.28**). Occasionally there are seals at the mounting bolts.

Pulley and Seal Removal and Replacement

Replacement of the pump shaft seal requires removal of the pulley. The pulley or pump can be easily damaged with a conventional puller, so special pullers and installers are used for this purpose. **Figure 66.29** shows a popular puller used to remove the pulley from the steering pump. Another tool used to reinstall the pulley is usually included with a replacement pump.

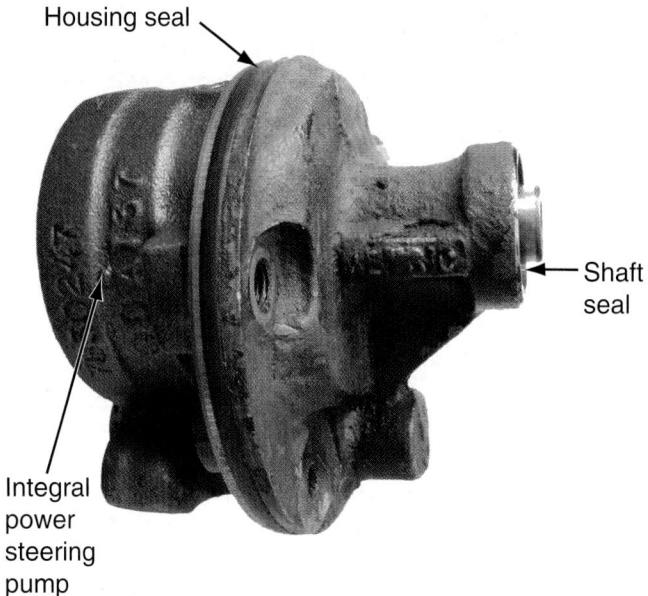

Figure 66.27 Power steering pump seals. (*Courtesy of Tim Gilles*)

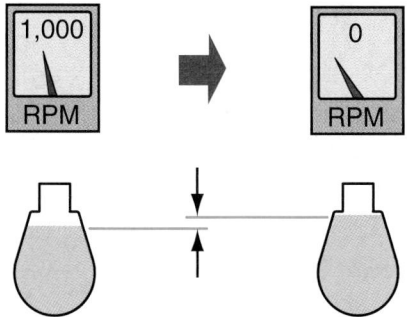

Figure 66.26 If the fluid is bled of air, when the engine is shut off the level of the fluid should rise less than 0.020 inch (5 mm). "0.020" is less than ¹⁄₃₂", barely visible.

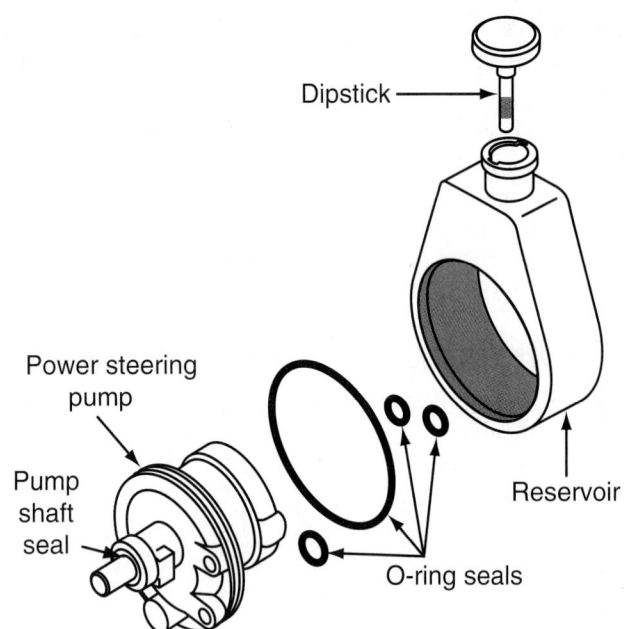

Figure 66.28 Seals are located at the pump shaft, at the host fittings, and on the reservoir housing, when applicable.

Figure 66.29 Removing a power steering pump pulley using a special tool. *(Courtesy of Tim Gilles)*

It is turned into internal threads on the pump shaft (**Figure 66.30**). When there is a leaking front seal and space permits, the pulley can be removed without removing the pump.

■ POWER STEERING PRESSURE DIAGNOSIS

When power steering system hydraulic pressure is low, there will probably be noise, and it will be harder to steer the vehicle. When pressure varies, steering effort will be erratic, causing the steering wheel to jerk during a turn. Belt tension should be checked first. This is the most common cause of erratic steering effort. Steering system pressure can be tested using a pressure gauge. Pressure testing varies among manufacturers. Check the service information for the vehicle.

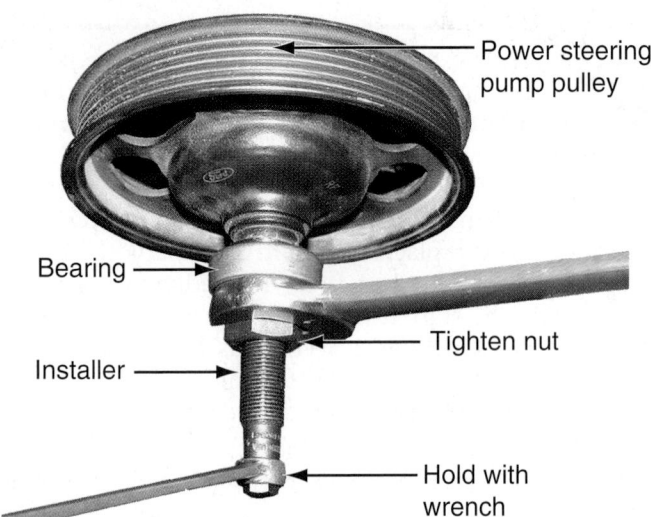

Figure 66.30 Installing a power steering pump pulley using a special tool. *(Courtesy of Tim Gilles)*

■ POWER STEERING PUMP SERVICE

When a power steering pump has a problem, it is typically replaced with a rebuilt pump. A few shops still perform power steering pump repairs.

Flow Control Valve Service

The high-pressure hose connects to a fitting that also serves as a part of the flow control and pressure relief valve assembly (**Figure 66.31**). The flow control valve piston and spring will fall out after the fitting is removed. The seat for the pressure relief valve ball is threaded into the control valve assembly (**Figure 66.32**).

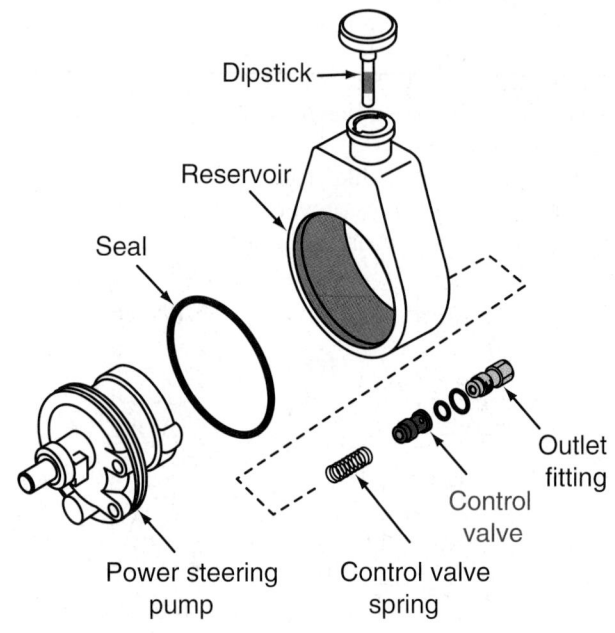

Figure 66.31 The control valve is located behind the power steering pump outlet fitting that connects to the pressure hose.

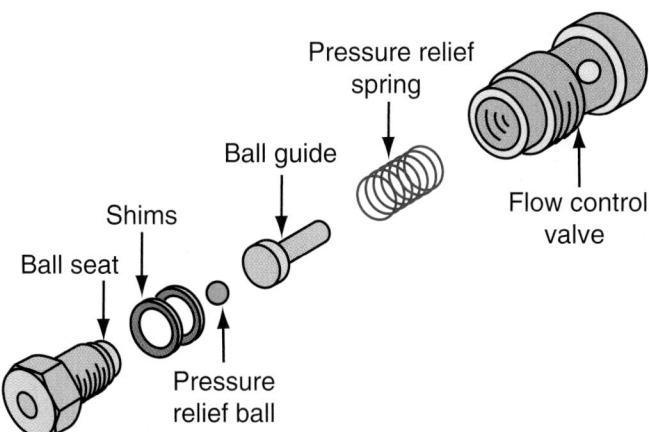

Figure 66.32 A typical flow control and pressure relieve valve assembly.

Disassemble and inspect all parts. Crocus cloth can be used to remove minor scratches from the control valve. Clean the housing and valve, and lubricate them with power steering fluid before reassembly. When the relief valve is mounted on the outside of the pump housing, be sure to use a new O-ring during reinstallation to prevent leaks.

■ POWER STEERING HOSES

The system has two hoses: a pressure hose and a return hose. The pressure hose is made of reinforced oil-resistant synthetic rubber that is crimped to metal tubing. It must be able to withstand temperatures of 300°F and handle pressures up to 1,500 psi.

Sometimes the pressure hose has two different ends. The larger end fits on the pump side and acts as an "accumulator," which absorbs pulsations in the hose from variations in pressure. The ends of the pressure hose have flared connections. Others have captured seals that allow the tubing to be able to turn, even when tight. The return hose is usually attached only with hose clamps because it is not under high pressure.

When checking a hose, look for signs of leakage or dampness at the connections. Also, look for signs of deterioration such as cracks, signs of rubbing, or swelling. A failed pressure hose can leak, blowing oil onto an exhaust manifold where it can cause a fire. More information on hoses can be found in Chapter 23.

SAFETY NOTE Power steering fluid, which is not flammable at engine operating temperature, is flammable at temperatures a little above 300°F (lower than the temperature of an exhaust manifold).
Get the correct replacement hose and replace the old one. Compare the new hose to the old hose to be sure that the length and fittings are identical.

SHOP TIP Sometimes a replacement hose is not available. Some businesses have special equipment for making hydraulic hoses. Check with a business that deals with heavy truck, industrial, or farm implement accounts.

■ REFILLING THE POWER STEERING SYSTEM

Be sure to use the correct fluid when refilling the system. After refilling the system, run the engine and check for leaks. On hoses that use O-rings, install new O-rings.

SHOP TIP If a flared fitting leaks, try loosening the fitting and twisting the tubing to seat it against the flare seat. Then retighten the fitting.

Bleed the system of air as discussed earlier in the chapter.

■ POWER STEERING GEAR SERVICE

Power steering failures include a lack of power assist in one direction and leakage from the steering gear. Power steering seal kits are available. Some shops rebuild power steering gears, but most buy rebuilt ones.

A majority of today's new cars are equipped with power rack-and-pinion steering units. Rack-and-pinion steering is more prone to problems because its weight has been cut over the years. Earlier power rack assemblies weighed as much as 54 pounds. Some of today's modern racks weigh as little as 8–10 pounds. Weight reductions of this magnitude have had an effect on the durability of the steering gears. Some of these lighter units have not been very durable, requiring replacement while the vehicle still has relatively low mileage.

A rack-and-pinion steering gear can have an external leak at the pinion shaft (attached to the steering column) or from the rubber bellows at either end of the rack (**Figure 66.33**). Rack failures can be due to a torsion bar bent from a hard impact or serious internal or external leaks.

■ REPLACING RACK AND PINION UNITS

Rack-and-pinion gears are often replaced with rebuilt units. Some units have a good record of success when rebuilt by technicians, and others do not. Sometimes grooves wear in the control valve housing (**Figure 66.34**). The symptom of this problem is usually a loss of power assist on cold startups (morning sickness). Rebuilt units usually have a nickel-plated sleeve installed to correct this and prevent it from happening again. When deciding whether to rebuild a unit or buy a replacement, an auto parts supplier is a good reference source.

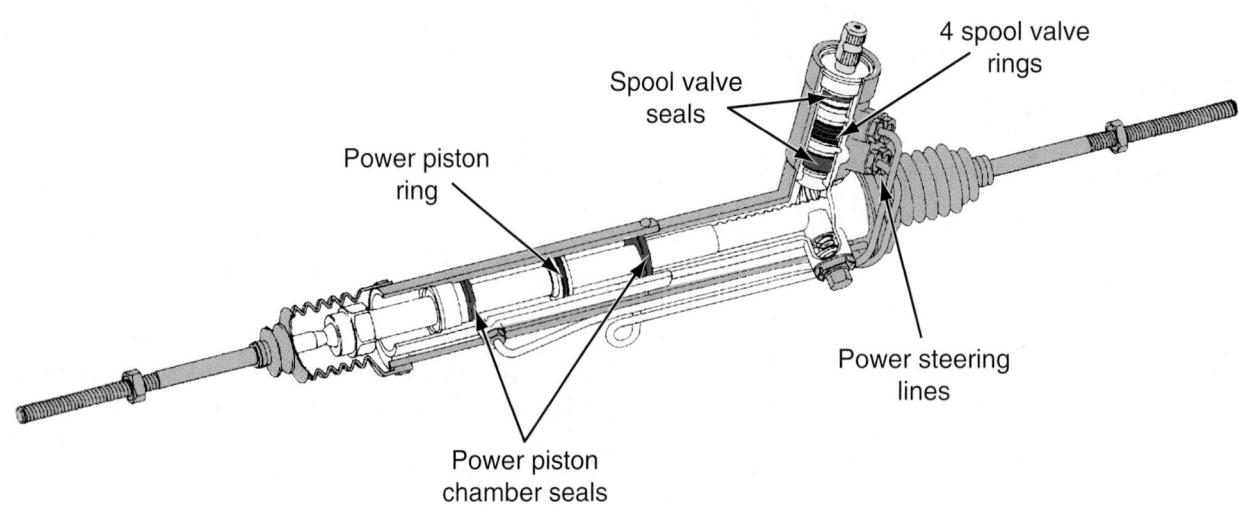

Figure 66.33 Rack-and-pinion sealing points. *(Courtesy of Federal-Mogul Corporation)*

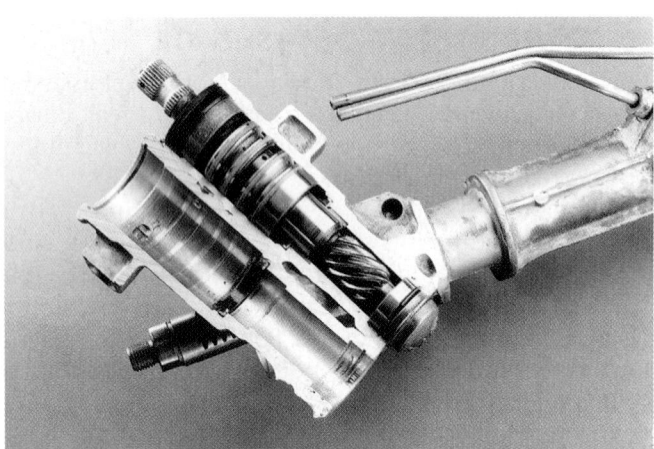

Figure 66.34 Grooves worn in the control valve housing. *(Courtesy of Federal-Mogul Corporation)*

NOTE: *To prevent damage to an air bag clock spring, lock the steering column and remove the key, or wrap the seat belt through the steering wheel to prevent it from turning while the steering unit is removed.*

Before removing a rack-and-pinion unit from the car, the steering shaft coupler must first be removed. There are two main types of coupler designs. Both types use a roll pin or threaded shoulder bolt that fits into a groove on the pinion stub shaft (**Figure 66.35**). One type uses a roll pin or bolt at the top, too. The other uses a D-shaped spring-loaded slip fit coupler. The spring clip can fall out on the interior carpet of the car as the steering gear is lowered for removal. It must be reused.

Roll pins must be removed with a punch and chisel. On the type that has two roll pins, the lower roll pin is usually corroded and difficult to remove while

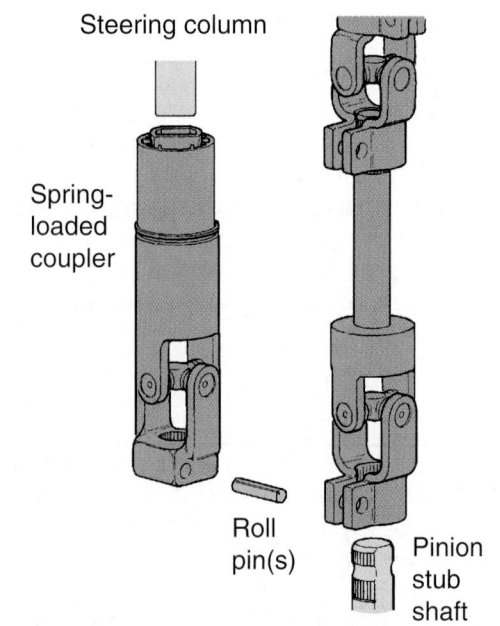

Figure 66.35 Two designs of steering shaft couplers. *(Courtesy of Federal-Mogul Corporation)*

the rack is on the car. Once the steering rack is on the workbench, this pin can more easily be removed.

There are usually four tubing connections into the top of the rack-and-pinion unit. The two that are from the hoses to the power steering pump must be removed. Disconnecting and reconnecting the hoses is often not possible with ordinary flare-nut wrenches. A special crowfoot flare socket can be used. Some power steering lines have swivel ends. When these are fully tightened, they can still rotate. Be sure to tighten them to the correct torque specification and replace the two seals that seal them to the tube fitting. Sometimes, it is necessary to lower the crossmember under the engine to gain access to the rack and pinion.

When replacing an entire rack and pinion unit, be sure that the brackets are installed properly. If the rack is not level, bump steer can result.

More information on power steering fluid and bleeding is included in Chapter 23.

VARIABLE POWER STEERING SERVICE

Problems with variable power steering found in some newer cars can be either mechanical or electrical. When there is an electrical system failure, the power steering system goes back to operating on full-time power assist. In other words, steering becomes easier at higher speeds. If this is the case, refer to the electrical troubleshooting procedure in the vehicle service information. When there is a mechanical problem, there will almost always be noise or a loss of power assist.

ELECTRONICALLY CONTROLLED POWER STEERING SYSTEM SERVICE

A typical complaint with electronically controlled power steering is steering effort that is too easy or too hard. Before performing any other tests, perform visual inspections for correct fluid level and belt tension, and check electrical connections at the steering gear actuator solenoid. If the visual inspection does not locate any possible causes for the problem, connect a scan tool to the diagnostic connector under the dash near the steering column.

CAUTION Be certain that the ignition key is off when connecting or disconnecting a scan tool.

Turn the ignition to the "on" position and select the correct item from the scan tool menu.

If there is an electrical defect, a diagnostic trouble code (DTC) will be displayed. Follow instructions in the service literature and perform required tests with a high-impedance digital multi-meter. Service literature will include a detailed diagnostic chart.

REVIEW QUESTIONS

1. What happens to power steering temperature when the wheel is turned several times?

2. What is the name of the test for looseness where an assistant turns the wheel from side to side with the tires on the ground?

3. How much looseness should there be between a rack-and-pinion inner tie-stud and housing?

4. What is used to remove a pitman arm from a steering gear?

5. What measuring tool is used to make an approximate toe setting when replacing a tie-rod end?

6. What is another name for the spiral cable that sends an electrical signal from the steering column to the steering wheel?

7. What can happen if the clock spring is not centered before steering linkage is reconnected to a steering gear?

8. What is the name of the point in a steering gear where the gear teeth come closer together when the car is traveling straight ahead?

9. Which kind of power steering system is more prone to problems, rack and pinion or conventional?

10. What is it called when there is a loss of steering power assist on cold startups?

ASE-STYLE REVIEW QUESTIONS

1. Technician A says that automatic transmission fluid can be used in all power steering systems. Technician B says that the engine must be running when checking steering looseness with a dry park check on a power steering vehicle. Who is right?

   a. Technician A   **c.** Both A and B
   b. Technician B   **d.** Neither A nor B

2. Technician A says that a loose belt can cause the steering wheel to jerk during a turn. Technician B says that a loose steering pump belt is more likely to squeal when driving straight. Who is right?

   a. Technician A   **c.** Both A and B
   b. Technician B   **d.** Neither A nor B

3. Which of the following is/are true about a vehicle that has been driven for a long time with a loose belt?
   a. The belt has probably become glazed.
   b. The pulley groove could be worn out.
   c. Both A and B.
   d. Neither A nor B.

4. Technician A says that when testing looseness on a rack and pinion tie-rod, the wheels should be hanging free. Technician B says that inner and outer parallelogram tie-rod ends are identical. Who is right?
   a. Technician A   c. Both A and B
   b. Technician B   d. Neither A nor B

5. Technician A says that if the steering wheel is off-center when driving on the highway, remove the steering wheel using a special puller and replace it in the straight-ahead position. Technician B says that when removing a steering wheel, loosen the nut and then pound on the end of the shaft until the wheel pops loose. Who is right?
   a. Technician A   c. Both A and B
   b. Technician B   d. Neither A nor B

6. Technician A says to store an air bag face down on a workbench. Technician B says that the air bag connector wiring must be disconnected before removing an air bag. Who is right?
   a. Technician A   c. Both A and B
   b. Technician B   d. Neither A nor B

7. Which of the following is/are true about rack-and-pinion steering?
   a. The center of a rack-and-pinion steering rack is hardened.
   b. If a rack is not centered, the vehicle can steer further to one side than to the other.
   c. Both A and B.
   d. Neither A nor B.

8. Which type steering gear must be centered so the steering response is not faster in one direction than the other?
   a. Recirculating ball and nut
   b. Rack and pinion
   c. Both A and B
   d. Neither A nor B

9. If the ends of the idler arm and pitman arm are not at equal height:
   a. Bump steer will result.
   b. The vehicle will lean to one side.
   c. Both A and B.
   d. Neither A nor B.

10. Which power steering hose has a hose clamp on its connection to the pump?
    a. The pressure hose
    b. The return hose
    c. They both have hose clamps
    d. None of the above

Wheel Alignment Fundamentals

■ KEY TERMS

toe alignment
scuff
camber
camber roll
caster
SAI

included angle
crossmember
cradle
turning radius
Ackerman angle
wheel base

tracking
dog tracking
set-back
slip angle
understeer
oversteer

■ INTRODUCTION

Correct alignment of the wheels provides a vehicle with the ability to run straight on the highway with very little steering effort. This chapter deals with the principles of the different alignment angles.

■ ALIGNMENT ANGLES

There are five wheel alignment angles:
■ Toe
■ Camber
■ Caster
■ Steering axis inclination (SAI)
■ Turning radius

■ TOE

The alignment angle most responsible for tire wear is **toe**. Toe is a comparison of the distances between the fronts and the rears of a pair of tires (**Figure 67.1**). When the tires are closer together at the front, this is called *toe-in*. When the tires are further apart at the front, this is *toe-out*.

Every 1/16" of toe-in results in 11 feet per mile of **scuff**. This means that the tires actually move sideways for 11 feet out of every mile traveled. If a car has 1" of toe, the tire is dragged 182 feet sideways every mile. The result is severe tire wear and decreased fuel economy.

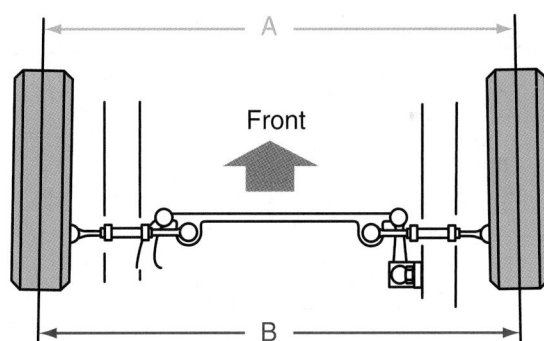

Figure 67.1 Toe is a comparison of the distances between the fronts and the rears of a pair of tires.

Several things can cause incorrect toe:
■ An improper adjustment
■ Looseness in the steering linkage due to wear
■ A collision with a curb
■ A change in either the caster or camber adjustment

Tie-rods and other steering linkage parts are built to be flexible if there is an impact. If steering linkage parts were brittle, they could break during an impact, causing a dangerous loss of vehicle control. Because they are somewhat flexible, steering linkage parts will bend rather than break.

NOTE: *Bent steering linkage will result in a change in the toe setting, and the steering wheel will usually be off-center when traveling straight down the road.*

Front toe is adjustable on all vehicles and rear toe is adjustable on some. When a car is driven, the toe adjustment changes.

Whether the tires deflect inward or outward when rolling depends on if the car has front- or rear-wheel drive. On rear-wheel-drive cars, rolling tires tend to toe out when driving. This happens as the steering linkage deflects due to the rolling resistance of the tires, taking up a clearance that exists.

During an alignment, compensation is made so that front tire toe will be as close as possible to zero while rolling. The specification for adjustment for rear-wheel-drive cars usually calls for the tires to be slightly toed-in (closer together at the front) while on the alignment rack.

Front-wheel-drive cars react in an opposite fashion. Wheels on these cars tend to toe inward as they are pushed by engine torque. Specifications for front-wheel-drive cars usually call for a slight toe-out setting or zero.

NOTE: *A change in toe affects the position of the steering wheel.*

Information on front and rear toe adjustment is found in Chapter 68.

■ CAMBER

Camber, an adjustable angle on most vehicles, is the inward or outward tilt of a tire at the top. A tire that tilts out at the top has *positive camber* (**Figure 67.2**). A tire tilted in at the top has *negative camber*.

Camber is a *tire wearing* angle. The inside and outside edges of the tread on a cambered tire actually have two different radii (**Figure 67.3**), so they rotate at different speeds. When the tire leans to one side, the tread will wear on that side as the smaller diameter of the tire squirms against the road surface.

The camber angle is controlled by the position of the control arms or struts. The control arms are quite strong and are not usually affected by an impact, like running into a curb. On long- and short-arm suspen-

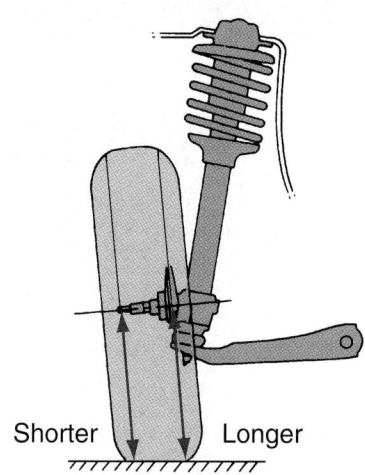

Figure 67.3 The tread will wear on the shorter side as the smaller diameter of the tire squirms against the road surface.

sion systems, negative camber typically develops as springs sag with age. Sagged springs are often accompanied by wear on the inside edges of the front tires from the resulting negative camber.

Camber is a *directional control* angle also. A cambered tire tends to roll in a circle, as if it were at the large end of a cone (**Figure 67.4**). Think about what happens if you put an ice cream cone on a table and attempt to roll it. This is called **camber roll**. Unequal camber between tires on opposite sides of the vehicle can cause a steering pull (to the side with the most camber).

The inner wheel bearing is larger because it is designed to support the vehicle's load. Camber is usually positive, which loads the inner wheel bearing (**Figure 67.5**).

■ CASTER

Caster is the angle that describes the forward or rearward tilt of the spindle support arm (**Figure 67.6**). When the top is tilted to the rear as shown in **Figure 67.7**, the wheel has *positive caster*. Its *lead point*, or *point of load*, is in front of true vertical. Negative caster is the forward tilt of the steering axis. **Figure 67.8** compares positive and negative caster.

Caster is a *directional control* angle but does not cause tire wear when the car is going straight. When riding a bicycle (which has positive caster), it tends to continue in a straight-ahead direction without holding the handle bars. If the handle bars are aimed in the wrong direction (negative caster), the bicycle loses its stability.

When the front wheels on a car have different caster settings, the car will pull toward the side with the most negative caster. That wheel will have its point of load behind the wheel on the other side of the car. Caster causes the spindle to move either toward the ground or away from the ground during a turn depending on whether it is negative or positive. Positive caster causes the spindle to move toward the

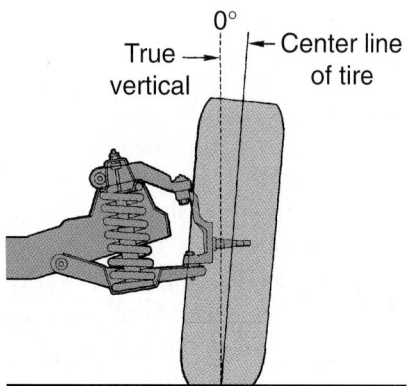

Figure 67.2 This wheel has positive camber. *(Courtesy of FederalMogul Corporation)*

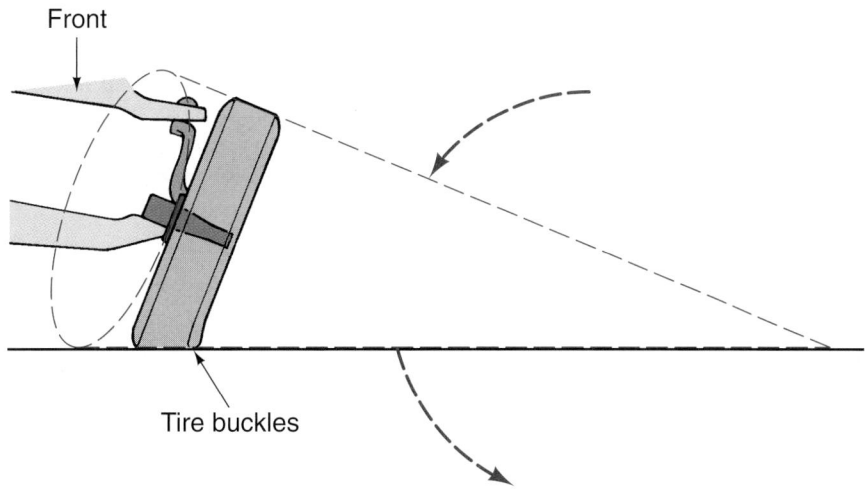

Figure 67.4 A cambered tire tends to roll in a circle.

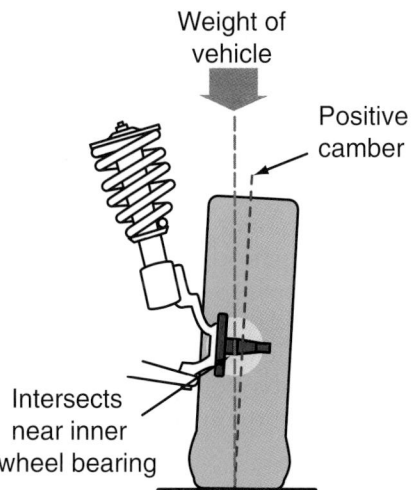

Figure 67.5 Positive camber loads the inner wheel bearing.

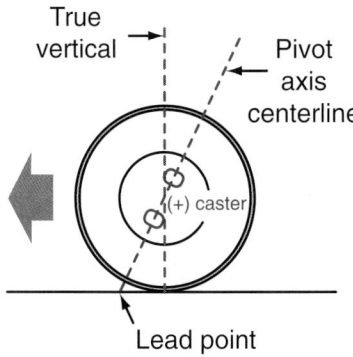

Figure 67.6 Caster.

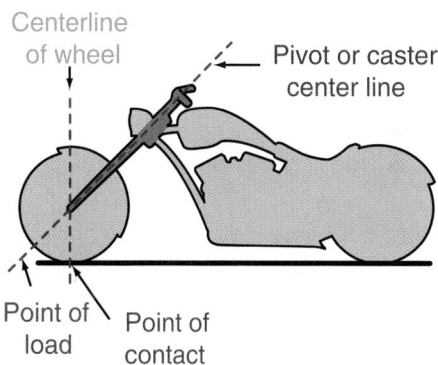

Figure 67.7 Bicycles and motorcycles have positive caster.

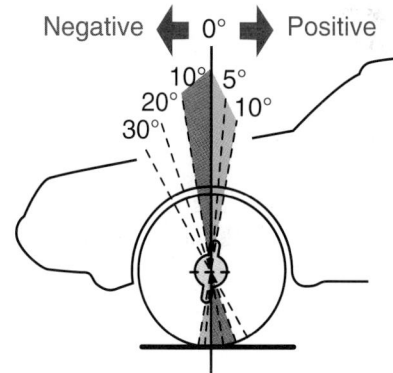

Figure 67.8 Comparison between positive and negative caster.

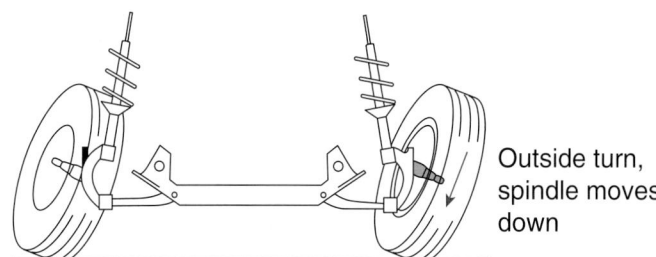

Figure 67.9 Because the tire cannot move closer to the ground, the vehicle must lift during a turn. *(Courtesy of Hunter Engineering Company)*

ground when the wheel is steered to the left. Because the tire cannot move closer to the ground, the vehicle must lift (**Figure 67.9**). When turning effort is removed from the steering wheel after the turn, the tire attempts to move back toward the center. Steering a positive-castered wheel to the right causes the spindle to move up, with the opposite results.

Wheels with equal caster will cancel each other out as a return to straight-ahead force.

NOTE: *Steering axis inclination (SAI) exerts a similar force in terms of returnability to straight ahead, but in a different and consistent direction. SAI is covered later in the chapter.*

Positive caster aligns the steering axis with bumps encountered in the road, so road shock is more likely to be felt in the steering wheel (**Figure 67.10**). Too much positive caster can result in a front wheel shimmy. Some vehicles that have a large amount of positive caster use a steering damper to prevent this. When the damper is worn out, the front wheels will shimmy uncontrollably when a front tire hits a small bump.

NOTE: *Excessive positive caster can cause cupped tire wear if the resulting shimmy is allowed to continue.*

Negative caster makes a vehicle easier to steer. It can also cause a car to wander and weave on the highway. Old cars had narrow tires. Their alignment specification called for positive caster to keep the car going straight without holding the steering wheel. Newer cars tend to have wider tires, which tend to keep rolling straight (**Figure 67.11**).

NOTE: *Light trucks have different caster settings depending on their rear axle ride height. Changing the caster setting can have a drastic effect on the driveability of the vehicle.*

Caster and Tire Wear

Caster is actually a measurement of the camber angle as it changes during a turn. A high amount of caster can result in tire wear if the car is driven excessively on winding roads or in the city. This is because a high amount of caster influences camber roll. Although caster is known as primarily a directional control angle, it can cause tire wear too. Alignment equipment manufacturers teach this differently. Some say that tire wear is not a result of caster.

◼ STEERING AXIS INCLINATION

Steering axis inclination (**SAI**) is the amount that the spindle support arm leans in at the top (**Figure 67.12**). SAI is also known as *ball joint inclination* (*BJI*) or *king pin inclination* (*KPI*). It is *not* a tire wearing angle. SAI basically helps the vehicle steer in the straight-ahead direction.

How SAI Works

Although the spindle is horizontal to the ground (except for the camber angle) when the wheels are pointed straight ahead, it attempts to move closer to the ground during a turn (see Figure 67.9). The spindle cannot move any closer to the ground because of the tire, so the car must lift. During a turn, the weight of the car puts pressure on the wheel, causing it to want to return it to the straight-ahead position.

SAI has three functions:
- ◼ After a turn, SAI helps the vehicle return to straight ahead (**Figure 67.13**).
- ◼ SAI keeps the vehicle going straight down the road. (Positive caster does this, too. But the more positive caster, the harder it is to steer the car.)

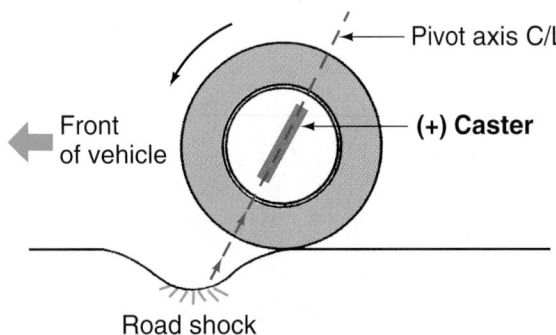

Figure 67.10 Positive caster results in road shock being transmitted through the steering column.

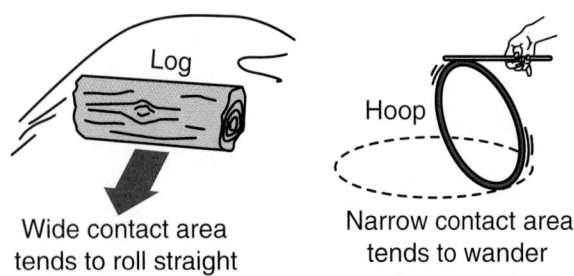

Figure 67.11 Wider tires tend to roll straight.

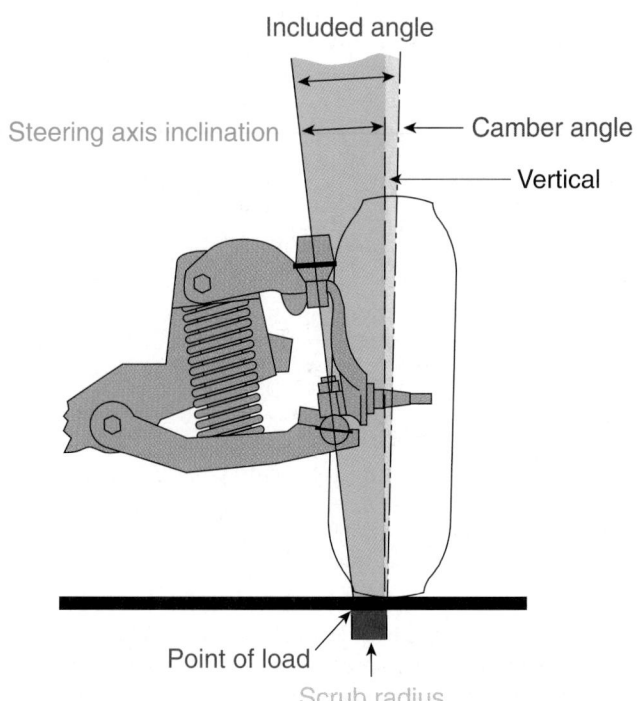

Figure 67.12 Steering axis inclination puts the pivot point under the tire.

View from side

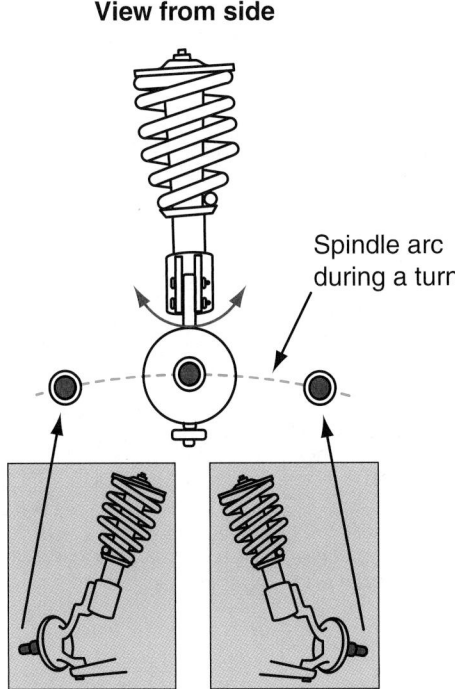

Spindle arc during a turn

Figure 67.13 SAI aids in returning to a straight-ahead position after a turn when vehicle weight causes the spindle to aim straight ahead.

■ SAI allows the car to have less positive caster (for easier steering) while still having good directional stability.

The combination of SAI and camber is called the **included angle**.

Some cars with a large amount of SAI will wear the outsides of the tires (especially on non-radial tires) because of the excessive amount that the wheel cambers during turns (see Figure 67.9). These cars do not require as much positive caster to keep the car from wandering.

In city driving, right-hand turns are made about 80% of the time. Delivery trucks are especially hard on right-hand springs, bearings, and tires (which must turn sharper).

NOTE: *SAI can be used as a diagnostic angle in determining whether parts are bent.*

▬ SCRUB RADIUS

Scrub radius is a factor of steering axis inclination. It is the pivot point for the front tire's footprint. Scrub radius is the distance at the road surface between the centerline of true vertical at the center of the tire tread and the steering axis pivot centerline.

The junction of the steering axis and centerline pivot point is normally below the surface of the road (see Figure 67.12). How far below the surface of the road determines how much scrub radius there is.

More scrub radius makes it harder to steer the car. Positive camber reduces scrub radius, but tire life is best when the running camber angle is zero.

■ Scrub radius on rear-wheel-drive cars is called *positive scrub radius*. The front wheels toe out when rolling (**Figure 67.14**).
■ Front-wheel-drive cars usually have *negative scrub radius*. The tires toe in when rolling (**Figure 67.15**).

NOTE: *If a left front tire on a front-wheel-drive vehicle with positive scrub radius blows out, the tire will pull hard to the left. The still-inflated right tire will pull inward. To prevent this dangerous situation, SAI has been increased on front-wheel-drive cars. This results in negative scrub radius, which causes the blown tire to pull inward. The car will continue to go straight because this direction of motion counteracts the motion of the right front tire.*

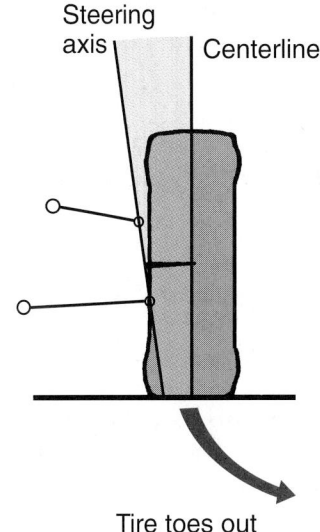

Steering axis | Centerline

Tire toes out

Figure 67.14 With positive scrub radius, the tire toes out when rolling.

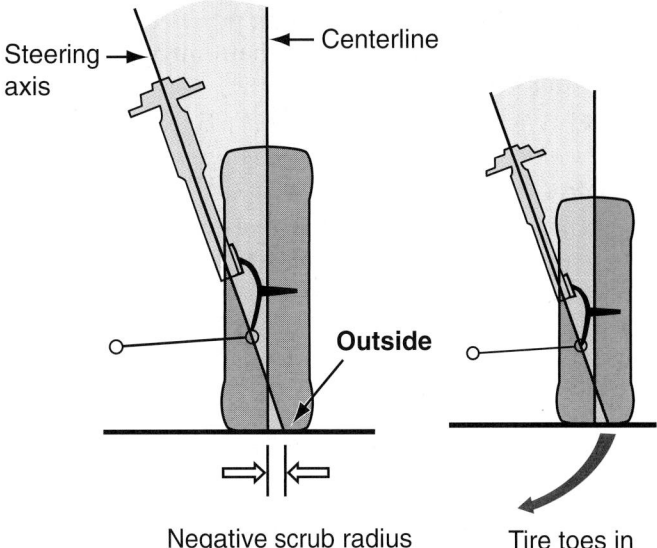

Steering axis → | ←— Centerline

Outside

Negative scrub radius | Tire toes in

Figure 67.15 With negative scrub radius, the tire toes in when rolling.

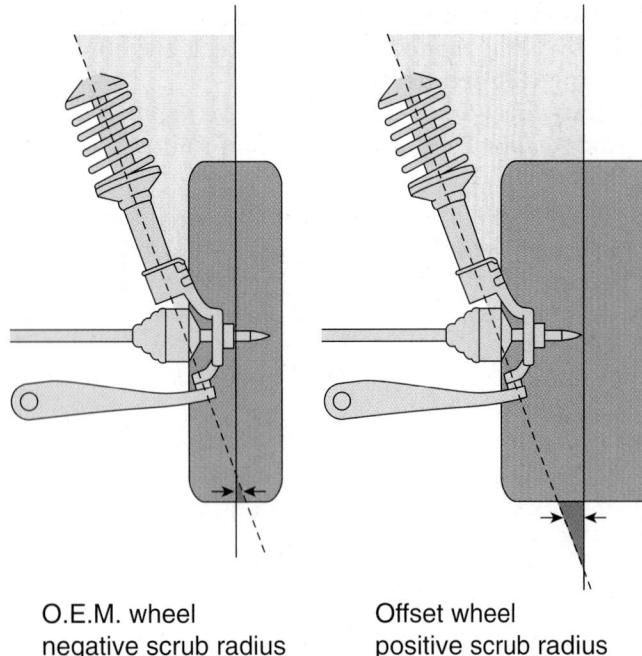

O.E.M. wheel
negative scrub radius

Offset wheel
positive scrub radius

Figure 67.16 Installing lower profile tires and offset wheel rims on a vehicle will change scrub radius. *(Courtesy of SPX Corporation)*

Incorrect Scrub Radius

Installing lower profile tires and offset wheel rims on a vehicle will result in a big increase in scrub radius (**Figure 67.16**). The pivot point can actually move outside of the tires' footprint area. The result is much harder steering, wheel shimmy, and a tendency to wander.

Negative scrub radius can result when tires and wheels that are too tall are installed on a rear-wheel-drive vehicle. The result can be instability. When a tire on an SLA suspension system goes over a bump, the scrub radius can change from positive to negative as the camber changes. Vehicle handling can become dangerous.

NOTE: *When changing to tires and wheels of a different size, modifications to the suspension must be made. Wheel alignment cannot compensate for this.*

Other causes of incorrect scrub radius are a bent front suspension part or damage to the frame at the **crossmember** or *strut tower*. The crossmember, or **cradle**, is the large steel part of the frame beneath the engine and between the front wheels. The strut tower is the area inside of the front fenders that supports a MacPherson strut.

■ TURNING RADIUS

When a car makes a turn, the outside wheel must travel in a wider arc than the inside wheel (**Figure 67.17**). The alignment angle that controls this is called

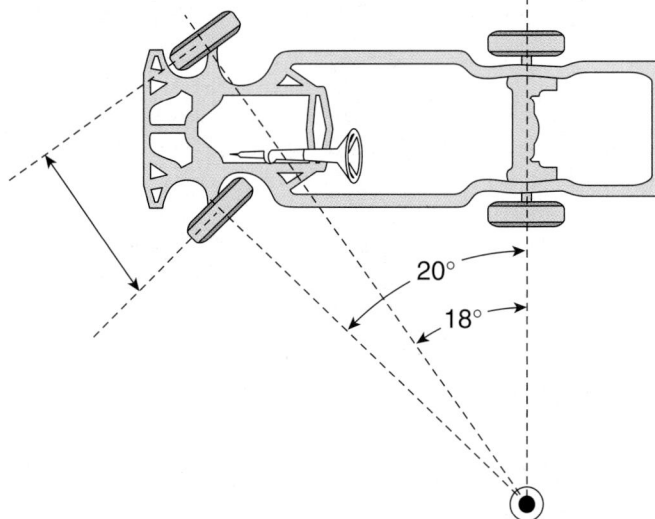

Figure 67.17 Front tires must toe out during a turn. *(Courtesy of Hunter Engineering Company)*

turning radius, toe-out-on-turns, or the **Ackerman angle**. The tires actually toe out during a turn because the steering arms are angled inward or outward (see Figure 65.14). When the wheels are turned, they move at different amounts as they move through their arc of travel (**Figure 67.18**).

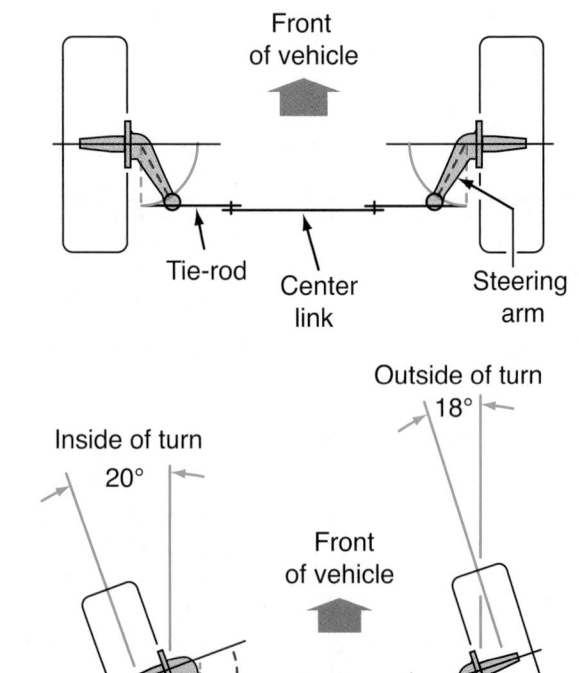

Figure 67.18 The tires turn different amounts as the steering arms move through their arcs of travel.

▬ TRACKING

The distance between the front and rear tires is called the **wheel base**. The side-to-side distance between an axle's tires is called the *track*. For minimum tire wear and good fuel economy, the wheels must run on track. The rear tires are supposed to follow in the tracks of the front tires. All four wheels should form an exact rectangle.

Tracking is a term that refers to the relationship between the average direction that the rear tires point when compared to the front tires. When tracking is off, the front wheels do not follow the rear wheels (**Figure 67.19**). This causes the car to try to steer to the side, increasing tire wear and upsetting handling.

If the rear axle is out of line to the right, it will cause the steering wheel to be aimed to the right. This is often referred to as **dog tracking** due to its resemblance to the way that some dogs walk, with their rear feet not in alignment with their front feet.

This situation will cause right front toe-out and left front toe-in. The result will be inside or outside toe wear. Toe wear is described in the pre-alignment section of this chapter.

▬ SET-BACK

Set-back is the amount that one front wheel is behind the front wheel on the other side of the car (**Figure 67.20**). It is measured in degrees. A negative angle indicates that the wheel on the left side is set back. This will cause the vehicle to steer to the left. It

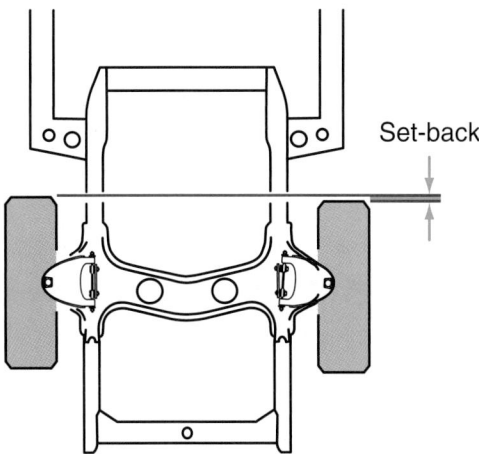

Figure 67.20 Set-back is the amount that one front wheel is behind the one on the other side of the car.

can also cause a brake pull. A collision can cause incorrect setback.

▬ SPECIAL HANDLING CHARACTERISTICS

The handling characteristics of a vehicle are described by terms such as slip angle, understeer, and oversteer.

Slip Angle

Tire manufacturers sometimes recommend different inflation pressures or alignment angles to change a vehicle's handling characteristics. A tire with lower inflation has a greater **slip angle**. Slip angle is the tendency during a turn for a tire to continue to go in the direction it was going, even though the rim has turned in response to steering wheel movement (**Figure 67.21**). Besides tire pressure, the amount of slip angle depends on the weight exerted vertically on the tire, its the structure and type, and the wheel alignment setting on the wheel. Positive camber causes a tire to have a greater slip angle.

Understeer and Oversteer

When a car does not seem to respond to movement of the steering wheel during a hard turn, this is called

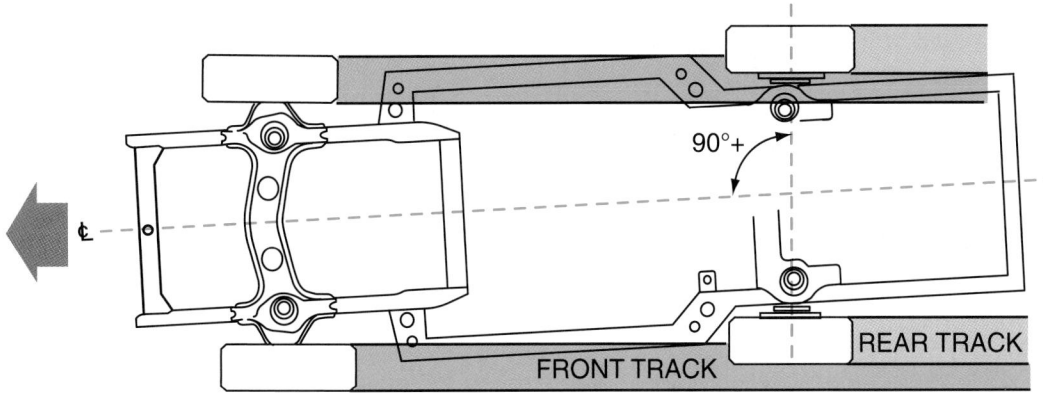

Figure 67.19 When tracking is off, the front wheels do not follow the rear wheels.

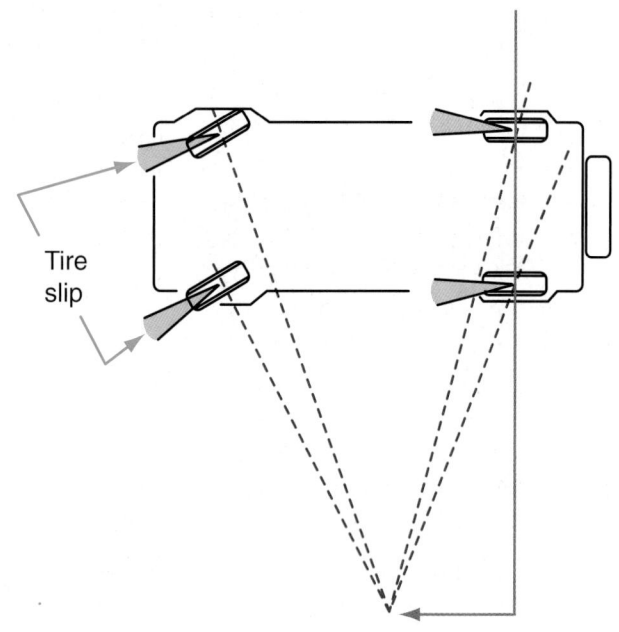

Figure 67.21 Slip angle.

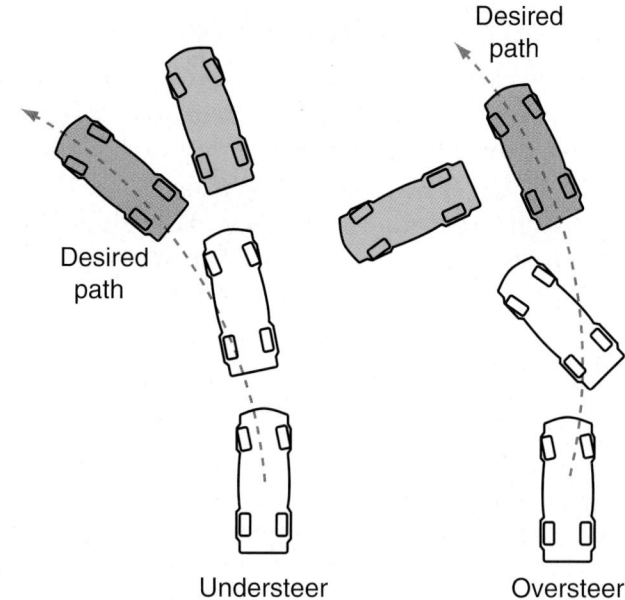

Figure 67.22 Understeer and oversteer.

understeer. When a car turns too far in response to steering wheel movement, this is called **oversteer** (**Figure 67.22**). Desirable steering is said to be *neutral*.

Lower inflation pressures are sometimes specified for the rear tires of rear-wheel-drive cars to prevent the car from understeering. That way, when a car begins to slide and the driver lets up on the accelerator, the driver regains control. If the car had oversteer, easing up on the gas would result in a spin-out.

When a car has four-wheel alignment capability, setting rear-wheel camber more negative than the front tires results in less tendency to oversteer. When tires of different profiles are installed on the same vehicle, the lower profile tires are installed on the rear to prevent oversteer.

▬▬ REVIEW QUESTIONS

1. List the five wheel alignment angles.

2. Which alignment angle is most responsible for excessive tire wear?

3. If a tire has 1" of toe-in, how far is it dragged sideways in one mile of driving?

4. Which kind of car has its toe set inward when not rolling, front-wheel drive or rear-wheel drive?

5. When the top of a tire tilts inward (toward the center of the vehicle), what positive or negative angle is this?

6. The angle that keeps a bicycle going straight when your hands are off the handle bars is called _____.

7. What is the name of the alignment angle that keeps the car going straight while allowing easy steering?

8. When tires of a different height or wheel offset are installed on a vehicle, what alignment factor changes?

9. During a turn, which front wheel turns sharper, the outside or the inside?

10. When a car does not seem to respond to movement of the steering wheel during a hard turn, what is this called?

◼◼◼ ASE-STYLE REVIEW QUESTIONS

1. Technician A says that camber can cause a car to pull to one side. Technician B says that camber can cause a tire to wear on one side of its tread. Who is right?

   a. Technician A **c.** Both A and B
   b. Technician B **d.** Neither A nor B

2. Which of the following is/are true about caster?

   a. When springs sag with age, caster changes.
   b. Caster is a measurement of camber change as the wheel is turned.
   c. Caster is a directional control angle.
   d. All of the above.

3. Technician A says that scrub radius is what causes the tires to toe in or out when rolling. Technician B says that front-wheel-drive cars usually have positive scrub radius. Who is right?

   a. Technician A **c.** Both A and B
   b. Technician B **d.** Neither A nor B

4. Technician A says that front tires toe out during a turn. Technician B says that turning radius is caused by the toe setting. Who is right?

   a. Technician A **c.** Both A and B
   b. Technician B **d.** Neither A nor B

5. All of the following will cause a vehicle to pull to one side *except*:

   a. Set-back **c.** A low tire
   b. Tracking **d.** Toe-in

6. Two technicians are discussing set-back. Technician A says that set-back can cause a brake pull. Technician B says that incorrect setback can be due to a collision. Who is right?

   a. Technician A **c.** Both A and B
   b. Technician B **d.** Neither A nor B

7. Which of the following is/are true about front-wheel toe?

   a. Wheels on front-wheel-drive cars tend to toe inward when driving.
   b. Wheels on rear-wheel-drive cars tend to toe outward when driving.
   c. Both A and B.
   d. Neither A nor B.

8. Technician A says that uneven caster can cause a car to pull to the side. Technician B says that excessive caster will cause a tire to wear on one side of its tread. Who is right?

   a. Technician A **c.** Both A and B
   b. Technician B **d.** Neither A nor B

9. Which of the following is/are true about camber?

   a. Camber is affected by vehicle ride height.
   b. Negative camber loads the inner wheel bearing.
   c. Both A and B.
   d. Neither A nor B.

10. When a car makes a turn, the outside wheel must travel in a wider arc than the inside wheel. The alignment angle that controls this is called:

    a. Turning radius
    b. Toe-out-on-turns
    c. The Ackerman angle
    d. All of the above

Wheel Alignment Service

■ KEY TERMS

toe change	slip plates	toe angle
bump steer	four-wheel alignment	spread
torque steer	crowned road	geometric centerline
steering pull	shims	thrustline
radius plates	eccentric cam	thrust angle

■ INTRODUCTION

Before performing a wheel alignment, a thorough inspection of the steering and suspension systems must be performed. Loose parts will prevent an accurate and lasting adjustment. Looseness of suspension or steering parts can result in slack in the steering wheel, shimmy, or an intermittent pull to one side or the other. For an alignment to be successful, suspension and steering components must be in good working order.

NOTE:

■ *A vehicle with worn or loose parts cannot be aligned.*

■ *Incorrect alignment settings can result in pull, instability, and tire wear.*

Unusual tire wear or vehicle handling problems are usually what cause a driver to bring a vehicle in to a shop for a wheel alignment. Front axles experience far more stress than rear axles because they support the weight of the engine and provide for steering. Although the rear axle is usually not the cause of problems, the technician must consider the entire frame and the steering and suspension systems to be able to properly align a vehicle.

This chapter first deals with suspension and steering system wear and inspection procedures. The latter part of the chapter deals with wheel alignment measuring and adjusting procedures.

■ PRE-ALIGNMENT INSPECTION

Prior to a wheel alignment, the suspension and steering systems must be inspected. If parts are loose or worn, an alignment will not be successful:

■ Tire pressures must be adjusted to specifications.

■ The frame must be at the correct height.

■ Worn bushings and pivots parts must not allow movement of suspension and steering parts.

■ Steering gear and linkage coupling points must not have excessive clearance.

■ A car's tires must be new or be worn evenly for the vehicle to be level during alignment measurement.

■ TIRE WEAR INSPECTION

Unusual tire wear can be caused by worn parts, incorrect inflation, hard cornering, or incorrect wheel alignment. Camber and toe are the two adjustable wheel alignment angles that often cause wear. It is better if the alignment technician can look at worn tires before they are replaced. If tires are evenly worn and there is no pull or wandering, caster and camber should not require adjustment.

Tire Wear from Camber

Wear from incorrect camber shows up on either the outside or inside of the tire tread (**Figure 68.1**).

Figure 68.1 Excess camber results in more wear on one side of the tire tread. *(Courtesy of Tim Gilles)*

Camber causes the inside and outside of the tire to have different diameters (see Figure 67.3). Outside wear is due to positive camber. It usually results from incorrect settings or a vehicle with a high amount of steering axis inclination driven with a high amount of cornering.

NOTE: *Tire wear from camber results as springs sag over time, changing the height of suspension components.*

Tire Wear from Toe

Driving a vehicle with excessive toe is dangerous because the front tires are sliding. When toe is incorrect, the tire will develop a feathered edge, also called a sawtooth pattern (**Figure 68.2**). The feathered edge can be especially severe on bias tires, where it is often apparent to the eye. Feathered wear also occurs on radial tires and can often be felt by hand, although it is more difficult to detect visually unless the toe is severely out of spec. Tire wear resulting from incorrect toe on radial tires often appears as wear to one side of the tire, similar to camber wear.

Before moving your hand across the tread surface, be sure the tire is not worn to the point where steel wire is exposed (**Figure 68.3**).

To diagnose excessive toe on a worn tire:
- Move your hand across the tire's footprint area from the outside to the inside. If you feel a feathered edge, the car has excessive toe-out (see Figure 68.2).
- If you feel a feathered edge when moving from the inside to the outside, the car has excessive toe-in (**Figure 68.4**).

Figure 68.3 Be careful not to cut your hand on exposed steel belts. *(Courtesy of Tim Gilles)*

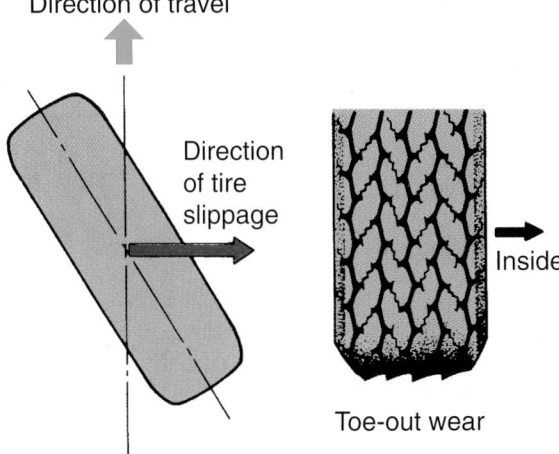

Figure 68.2 Toe wear results in a sawtooth pattern or feathered edge. The illustration depicts toe-out wear.

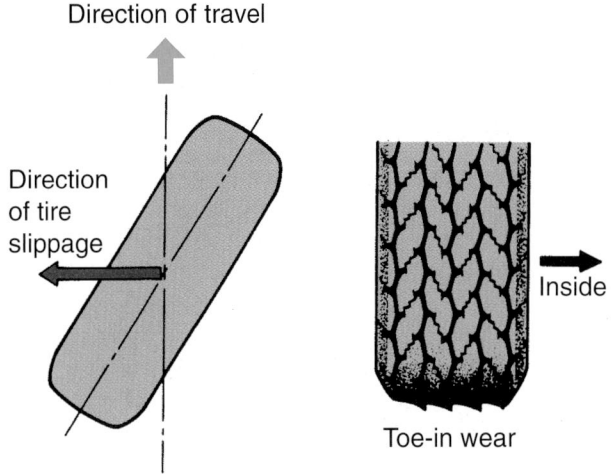

Figure 68.4 Toe-in wear.

Figure 68.5 A common wear pattern resulting from incorrect toe on a rear wheel. *(Courtesy of Tim Gilles)*

SHOP TIP One saying often used by technicians as a memory aid is "smooth in, toe-in."

Other Toe Wear Factors

On a rear-wheel-drive vehicle with toe-in, the right front tire tends to toe in more than the left and its outside edge wears because the tire is rolling under at that edge. With toe-out, the inside edge of the left tire will roll under, resulting in wear on the inside edge of the left front tire.

■ Radial tires with toe-in will both roll under, resulting in wear that looks like positive camber.

■ Toe on the rear of a car does not equalize like toe on the front does. If the tires are not rotated, they will develop diagonal wear (**Figure 68.5**).

▬ RIDE HEIGHT CHECK

Because the weight of the car is always resting on them, springs tend to sag as a car ages. Alignment specifications are based on the assumption that the ride height, or *curb riding height*, of the vehicle is correct. Although it is possible to adjust alignment when the springs have sagged beyond specifications, tire wear and unusual handling can result. Prior to making any wheel alignment adjustments, ride height must be measured and compared to specifications (**Figure 68.6**).

NOTE: *A vehicle that is either too high or too low will probably have a camber setting that is outside of specified limits.*

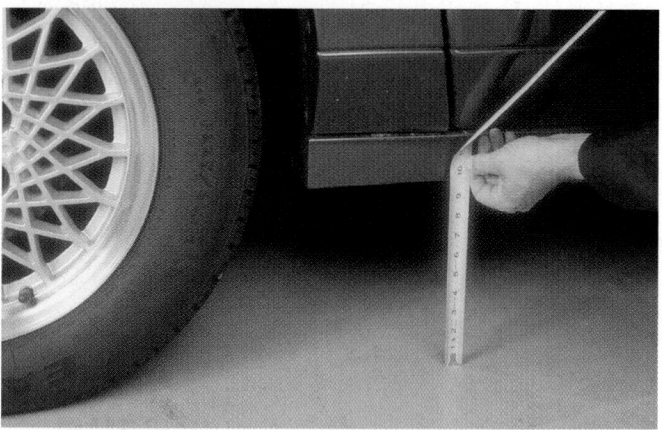

Figure 68.6 Measure and record ride height prior to aligning wheels.

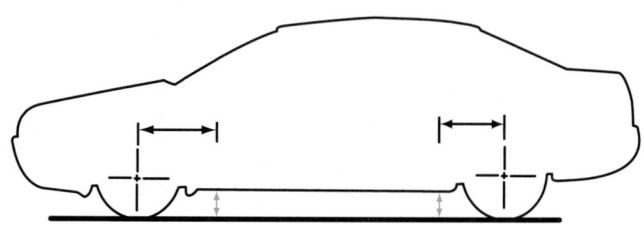

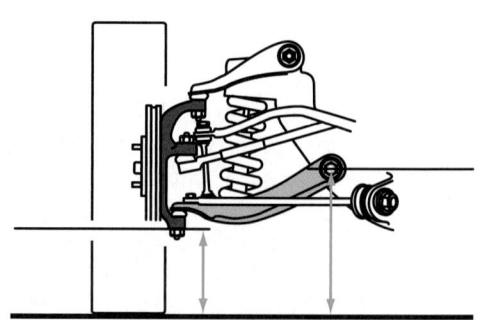

Figure 68.7 Typical ride height measurements. The lower sketch shows the typical measurement locations on an SLA suspension.

A car with a short/long arm (SLA) suspension with springs that have sagged will not function within its desired alignment range as the springs deflect. This can cause changes in the way the car reacts to bumps. **Figure 68.7** shows where typical ride height measurements would be checked on an SLA suspension.

▬ TOE CHANGE

As suspension height changes, toe measurement can change with it (**Figure 68.8**). The tie-rods are designed to remain parallel to the lower control arms when they pivot (**Figure 68.9**). When springs have sagged, **toe change** can result. Toe change causes the tire to move on the road surface, scrubbing away tread. A small amount of toe change can be absorbed by the flexing of a properly inflated tire. Loose parts or incorrect

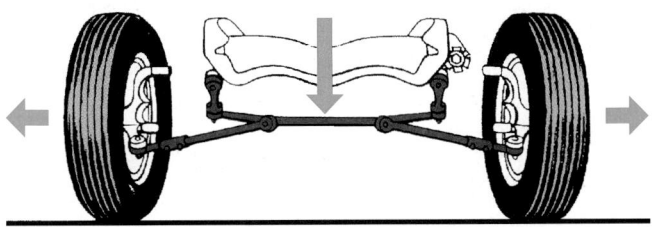

Figure 68.8 Toe change resulting from changing ride height. *(Courtesy of Hunter Engineering Company)*

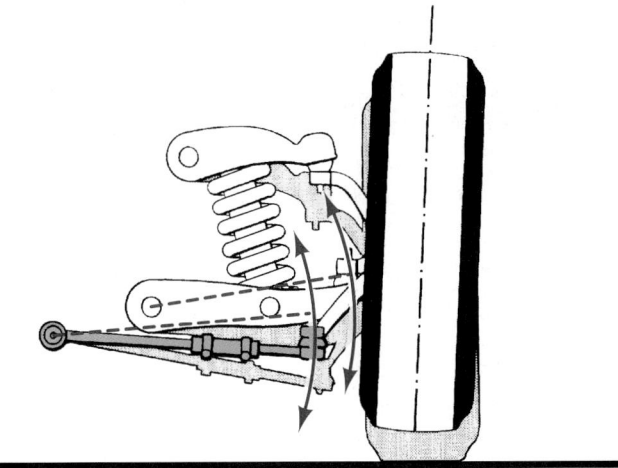

Figure 68.9 The tie-rods are designed to remain parallel to the lower control arms when they pivot. *(Courtesy of Hunter Engineering Company)*

wheel alignment can move the amount of allowable toe change out of normal limits.

When toe change is confined to one side of the vehicle, **bump steer**, or *orbital steer*, can result. Bump steer happens when a wheel with tie-rods at unequal heights goes over a bump. The car momentarily steers in the direction that the wheel turns as the toe changes (**Figure 68.10**). Toe change measurement and correction is covered later in this chapter.

NOTE: *The height of the steering arm changes as caster changes. Toe change can be caused by having different caster angles from side to side.*

TORQUE STEER

Sometimes a vehicle will turn abruptly to the side during initial acceleration. This action, called **torque steer**, is usually found on a front-wheel-drive car with axles of unequal lengths. It can also be caused by anything that causes the axles to be at different heights. This results in unequal CV joint angles. The height difference could be due to a loose sub-frame or a problem with unequal spring height.

HISTORY NOTE

On cars with small engines, torque steer wasn't a problem. When front-wheel drive became popular and horsepower increased, torque steer was more noticeable. Older front-wheel-drive cars sometimes had front driving axles of different lengths. During acceleration, the long axle would twist, delaying power transfer to its front wheels. During quick deceleration, it would recoil. Some of the solutions attempted by auto makers were to use a thicker axle, change the scrub radius, or use an intermediate axle shaft so that drive axles would both be the same length.

SUSPENSION LOOSENESS

A motto of one aftermarket suspension and steering part manufacturer is "you can't align looseness." Steering and suspension components are designed to pivot, without allowing any change in the positions of parts.

Perform a dry park check for steering and suspension looseness as described in Chapter 13. Always check the adjustment of the wheel bearings to see that they are not loose before attempting a wheel alignment.

TEST DRIVE

Unless a vehicle is unsafe to drive, a test drive should be done before performing repairs. Before driving the

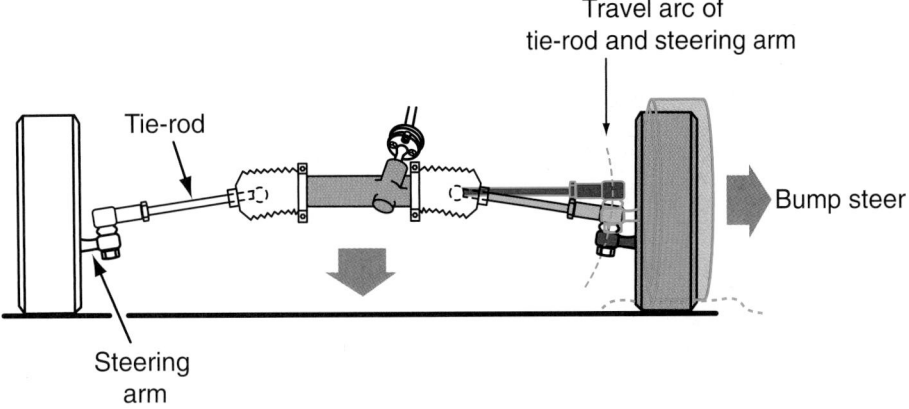

Figure 68.10 Bump steer.

vehicle, a visual inspection must be performed to check the following items:

■ Suspension bushings, visually and with a prybar
■ Steering linkage pivot connections—firmly grasping the part and rocking it to check for looseness
■ Rubber grease boots on tie-rod ends and ball joints
■ Shock absorbers

Also check to see if the vehicle has any signs of collision damage.

▬ TIRE CHECKS

To perform a tire check, follow these steps:

■ Adjust tire pressures to specifications. Check the condition of the valve stems and look for signs of impact damage that might have resulted in a bent rim (**Figure 68.11**).
■ Look for unusual wear to the tire tread.
■ Check for signs of damage to the sidewalls or tread area (**Figure 68.12**).
■ Be sure that tires of the correct size are used.
■ Radials and bias tires should not be mixed. Front tires should be of the same brand and tread pattern.

Power Steering Checks

To perform a power steering check, follow these steps:

■ Check power steering fluid level. Look for evidence of fluid leaking.
■ Check drive belt tension.
■ Check the power assist to see that it works with equal ease in both directions.

During the test drive, the vehicle is checked for several conditions:

■ Hard steering can be caused by binding parts, incorrect alignment, low tires, or a failure in the power steering system. Be sure to note whether the vehicle is easier to steer at higher speeds.

Figure 68.11 Inspect wheels and tires for impact damage. *(Courtesy of Tim Gilles)*

Figure 68.12 Distorted tread like this will affect a wheel alignment. *(Courtesy of Tim Gilles)*

■ Do tires squeal on turns? This could be due to a bent steering arm or low tire pressures.
■ Are there squeaks and clunks? This could be the result of bad bushings, which can also cause changes in camber, resulting in brake pull.
■ Is there a shimmy or tramp (the steering wheel shakes from side to side)? This could indicate a bent or out-of-balance wheel, or excessive caster.
■ A vehicle that wanders might have an incorrect caster angle setting.
■ Does the vehicle wheel pull to one side? When the steering wheel wants to go to the side by itself, especially when you let go of the wheel, this is called **steering pull**.

NOTE: *Sometimes pull can be the result of the crown of the road. Be sure to perform the test on a flat surface.*

■ Does the vehicle pull to one side or the other? Does it pull always, or only during braking? This could be due either to alignment or brakes.
■ If the vehicle pulls, does the direction or amount of pull change if the tires are rotated? A defective or damaged tire can cause pull (usually to the side with the bad tire).
■ Is it possible that a power steering problem could be causing the pull? Some vehicles have adjustable spool valves. With others, the steering box might require service. With the wheels off the ground, start the engine and verify that the wheels do not self-steer in either direction.
■ Does the owner complain of a rough ride? This could be due to tire pressures that are too high, a bent or frozen shock absorber, or, on an older vehicle, the installation of radial tires.
■ Does the vehicle have body roll, where it leans excessively to one side or the other during fast turns? The shock absorbers could be worn out.
■ Is there a noise during turns that changes in pitch as the vehicle weaves to the left and then to the right? The wheel bearings on the outside wheels spin faster

during a turn. Due to this, and weight shift (which loads bearings), a bad bearing will make more noise when turning to one side than to the other.

■ A wheel bearing noise will often be accompanied by looseness.

Before attempting a wheel alignment, check for looseness in any related parts. With the vehicle raised, grasp the tires with your hands at the six o'clock and twelve o'clock positions and check for movement. If there is movement, readjust the wheel bearing. If there is still movement, locate its source in the suspension system.

NOTE: *A check for similar movement at the three o'clock and six o'clock positions verifies steering linkage condition.*

Test ball joints for looseness as described in Chapter 64. It is best to consult the appropriate service information because procedures vary. Some ball joints are designed to have a specified clearance. Several part manufacturers provide specification tables.

■ INSPECTION CHECKLIST

A checklist that can be used by technician to make sure that no steps are accidentally forgotten during

a suspension inspection and test drive is shown in **Figure 68.13**.

■ WHEEL ALIGNMENT PROCEDURES

The front suspension is designed to keep the wheels in the best possible position when rolling. Wheels must roll freely with as little tire scuff (side-to-side wear) as possible. Alignment settings can change according to vehicle speed, roughness of the road surface, acceleration, braking, weight distribution, or cornering. Specifications are developed by manufacturers so that alignment can be adjusted with the vehicle at rest on a level alignment rack (**Figure 68.14**). If tire wear is experienced when the settings are within specifications, the technician can make adjustments to compensate.

Adjustments to original alignment settings might be needed because of wear on the inside of pivoting parts, bad road conditions, collision damage, spring sag, or unusual loads. A new car might need an alignment because it came from the factory with adjustments outside of the normal range of specifications. Design engineers list a range of adjustment limits for

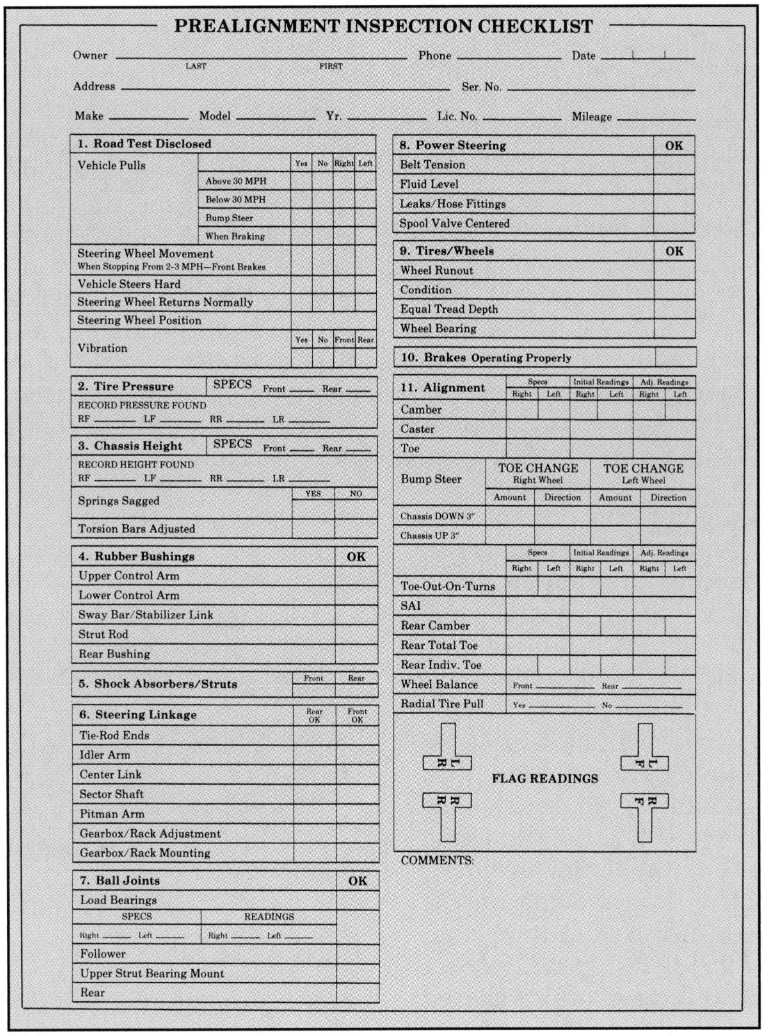

Figure 68.13 A pre-alignment inspection checklist.

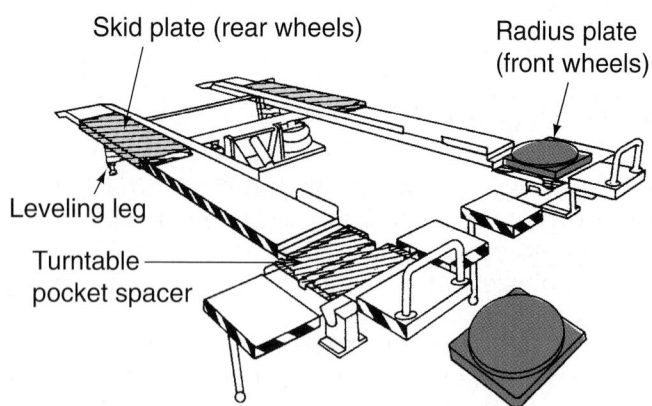

Figure 68.14 A wheel alignment rack with ball bearing supported plates for the front and rear wheels. *(Courtesy of Snap-on Tools Company, www.snapon.com)*

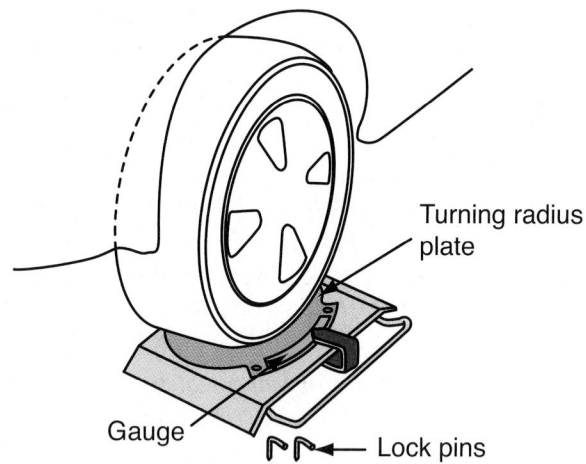

Figure 68.15 A radius plate.

wheel alignment angles. If the adjustments fall within those limits and the car "tracks" straight (goes straight on a level road), little tire wear should occur.

Only three of the five alignment angles are normally adjustable:

■ Caster
■ Camber
■ Toe

The other two angles are measured to check for damage to suspension and steering parts. Of the three adjustable angles, tire wear when traveling straight only results from incorrect camber and toe. Directional pull can be caused by camber and caster.

NOTE: *On some MacPherson strut systems, only toe is adjustable.*

■ MEASURING ALIGNMENT

Most alignment measurements are read in degrees and parts of degrees. Degrees represent part of a 360-degree circle in both the metric and inch systems. In the United States, remaining portions of degrees are either fractional or decimal. In the metric system, as in the world of science, portions of degrees are read in minutes (') and seconds ("). Like a clock, there are 60 minutes in a degree and 60 seconds in a minute.

When taking alignment measurements, ball bearing–supported plates are placed beneath the tires. These allow the tires to assume a relaxed position (see Figure 68.14).

On most alignment racks, the front wheels are positioned on **radius plates** (**Figure 68.15**). A radius plate has a gauge that measures in degrees how far a wheel is turned to the right or left. On four-wheel alignment racks, **slip plates** are under the rear tires.

Pins hold the upper plate from moving on the bearings. After the vehicle is driven onto the plates, the pins are pulled out of the plates. Pushing down on the bumpers lets the wheels creep into a relaxed position, as they would be in when rolling on the road. Air or hydraulic jacks are used to raise the car off the lift during alignment adjustments or to reposition the radius plates.

Computerized alignment machines are used to do **four-wheel alignment** inspection (**Figure 68.16**). Many newer vehicles are equipped with rear-wheel adjustment, which requires four-wheel alignment capability. Mechanical measuring systems are covered here as well because it is important to understand what you are actually measuring when taking alignment readings.

To get accurate measurements, the vehicle must be level. Toe is measured in either inches, millimeters, or angle of toe. The remaining angles are measured in degrees of a circle. When adjustment is possible, caster and camber are usually adjusted together since adjusting caster affects the camber reading. Toe is adjusted last, after caster and camber. When other alignment angles are changed or when parts are replaced, toe changes.

Figure 68.16 A four-wheel alignment machine with sensors attached to each wheel. All four wheel heads send alignment position information to a computer. *(Courtesy of Hunter Engineering Company)*

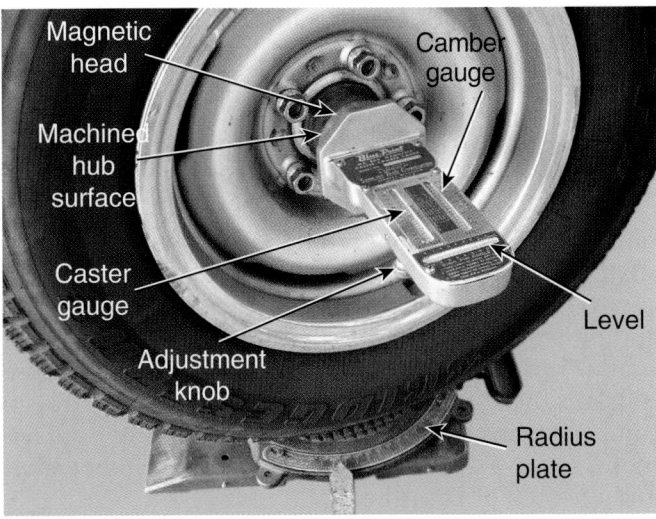

Figure 68.17 A portable alignment gauge attached to the hub with a magnet. *(Courtesy of Tim Gilles)*

NOTE: *If tires are to be replaced, the new tires should be installed before attempting an alignment. An alignment rack provides a level location for checking suspension angles. The heights of worn tires will affect alignment measurements.*

The tools used for measuring caster and camber in older and portable mechanical alignment measuring systems have bubble levels. These are used for making comparisons to an exact level position. To attach the gauges to the wheels, a wheel rim adapter clamp is used or a large magnet holds the gauge to the wheel hub (**Figure 68.17**). The tool is mounted against a machined area of the hub. The wheel bearing dust cap is removed so that a pilot can center the gauge on the center hole in the end of the spindle.

■ MEASURING CAMBER

Camber is simply a comparison measurement to true vertical, using a level. To measure camber, position the wheels straight ahead while reading the gauge. If the wheels are not straight ahead, the caster angle can cause an incorrect camber reading.

SHOP TIP If the toe adjustment is very far off, as it could be after parts are replaced, an inaccurate camber reading will result. When parts are replaced, make a preliminary toe adjustment first. Then read the camber setting.

■ MEASURING CASTER

Caster causes the wheel's camber angle to change during a turn.

NOTE: *The caster measurement is actually a reading of camber change while turning.*

To measure the caster setting, the wheel is first turned either inward or outward a specified amount. The amount and direction of the turn depend on the manufacturer. They are not all the same.

The older, portable bubble alignment gauges measure both caster and camber. There are two levels on the gauge. One of the levels, used to measure caster, is adjustable using a thumbscrew. When a front wheel has been turned to 20 degrees on the radius plate, set the level to zero. Then turn the wheel 40 degrees until the wheel is 20 degrees in the opposite direction. The reading on the gauge at this point is the caster reading.

Alignment settings for caster and camber are usually within a range. However, adjusting alignment to within that range will not always ensure that the vehicle will go straight.

■ ROAD CROWN AND PULL

Roads are **crowned** (higher at the center than the outside) so that rain will run off (**Figure 68.18**). A vehicle aligned with equal settings from side to side will drift to the outer edge of a crowned road.

To compensate for road crown, two methods can be used:

■ *Camber* can be set slightly more positive on the driver's side.

■ *Caster* can be set so that it is slightly more negative on the driver's side of the vehicle.

Many alignment technicians like to correct for road crown using caster because camber that is too far positive will result in wear to the outer edge of the tire.

NOTE: *When correcting for road crown, caster creates less of a pull than camber. Caster must be within ½ degree of the caster setting on the other side of the vehicle.*

Adjusting Caster and Camber

There are many methods of caster and camber adjustment. Alignment equipment manufacturers provide detailed charts and catalogs that describe specific

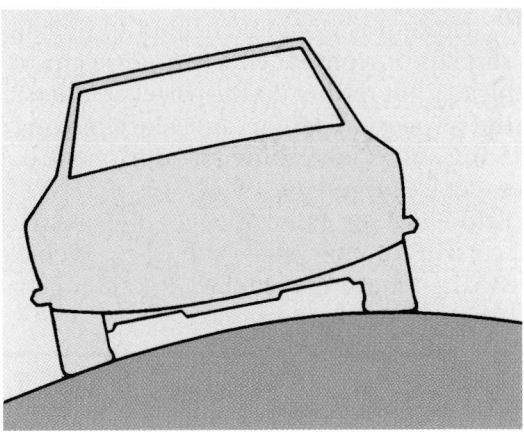

Figure 68.18 Roads are crowned so water can run off them.

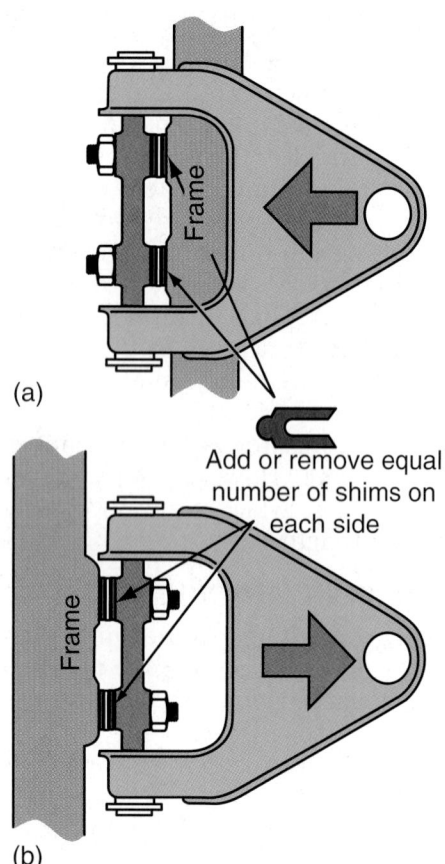

(a)

Add or remove equal
number of shims on
each side

(b)

Figure 68.19 Changing camber with shims on inboard and outboard pivot shafts. (a) With this suspension design, it is easier to remove shims with the weight of the vehicle on the tire. (b) On this suspension, removing shims is easiest with the weight off the tire.

adjustment methods. On SLA suspensions, the camber adjustment is done with shims or eccentrics. **Figure 68.19** shows a common SLA adjustment in which **shims** are removed or installed to reposition the upper control arm.

> **SHOP TIP** Before making an alignment adjustment, decide if it would be helpful to unload the suspension components so the weight of the vehicle is not resting on the wheels. Understanding the suspension design can make removing and replacing shims or turning an adjustment eccentric easier (see Figure 68.19).
>
> Before making an alignment adjustment, note the current camber angle while the vehicle is raised. Then make the required amount of change to the "new" camber reading.

■ PLAN AHEAD

When there are shims, caster and camber are changed together. Removing and replacing shims is sometimes a

fair amount of work. Plan ahead. Think about the effect the shim will have. Look at the control arm inner shaft to see what the effect of adding or removing shims will be. Some vehicles (usually light trucks) have the pivot shaft located outboard of the frame. Shims have the opposite effect as the normal control arm with the pivot shaft inboard of the frame. Figure 68.19 shows how to change camber on inboard and outboard pivot shaft locations.

To change caster with shims requires removing or adding shims at either end of the control arm pivot shaft (**Figure 68.20**). This will change camber because it moves the control arm out or in on one side. To keep camber from changing much during a caster adjustment, remove a shim from one side of the pivot shaft and install it at the other end.

Service literature sometimes states the effect of a particular shim thickness on both the caster and camber readings. More often this information is not available. When a technician works repeatedly on one make of vehicle, experimenting with one size of shim will reveal the effect that shim has on alignment settings.

An example of the amount a shim would change a vehicle's alignment is as follows:

$\frac{1}{16}$" shim on one side only = $\frac{1}{2}$ degree caster change
or
Remove a $\frac{1}{32}$" shim from one side and add it to the other for $\frac{1}{2}$-degree caster change.

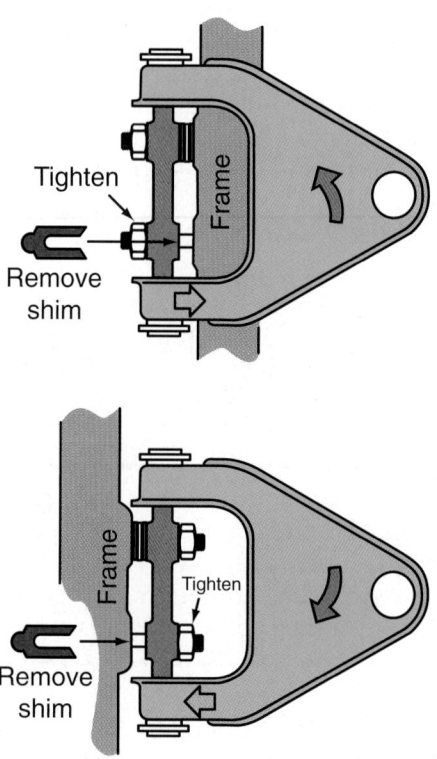

Tighten

Remove shim

Tighten

Remove shim

Figure 68.20 Changing caster. Removing shims from one side or adding them to the other moves the position of the upper ball joint toward the front or rear, depending on the position of the frame.

Service literature sometimes includes a chart showing the amount of shim change needed to get a desired setting. Some computer four-wheel alignment machines do the calculations for you.

Some vehicles use an **eccentric cam** adjustment on the upper or lower control arm, or strut (**Figure 68.21**). Turning the adjustment repositions the camber and caster angle. Less common alignment adjustment methods include slotted holes in the upper control arm shaft and an eccentric bushing installed under a ball joint. Some computer alignment machines have a feature that calculates the position of the bushing.

When MacPherson struts are adjustable (not all of them are), there are slots in the upper bearing bracket for adjusting caster (**Figure 68.22**) or an eccentric between the shock tube and steering knuckle (**Figure 68.23**).

When a vehicle has a narrow lower control arm, there is a strut rod running from it to the frame for strength. There are often threads on one end of the strut. Repositioning the nuts on the threaded area moves the control arm in an arc (**Figure 68.24**). The

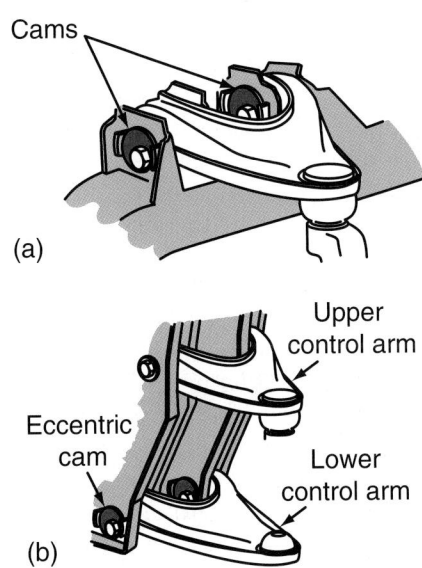

Figure 68.21 (a) An eccentric adjustment on an upper control arm. (b) An eccentric adjustment on a lower control arm.

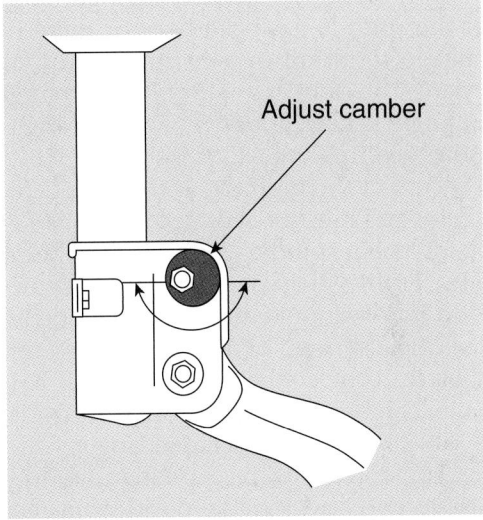

Figure 68.23 Camber can be adjusted at the bottom of this strut by turning the eccentric. (*Courtesy of Hunter Engineering Company*)

Figure 68.22 Slots allow movement of the top of the strut to adjust alignment. (*Courtesy of Tim Gilles*)

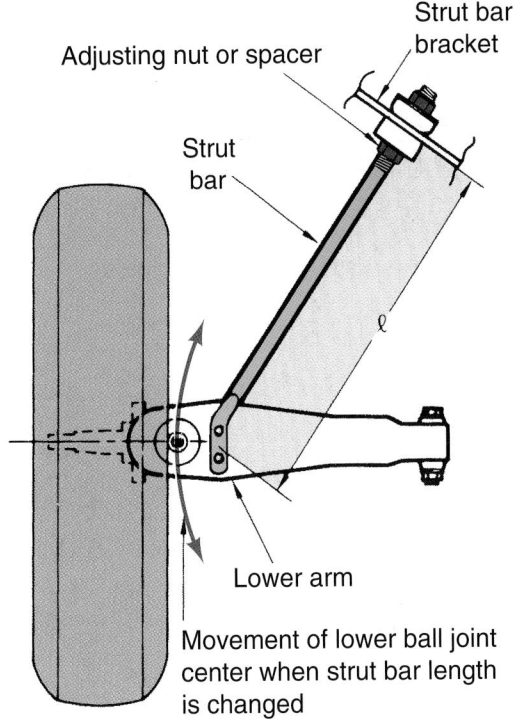

Figure 68.24 Adjusting the length of the strut bar changes the caster.

result is that caster changes. Moving the lower control arm toward the rear makes caster more negative.

■ MEASURING STEERING AXIS INCLINATION

SAI does not change, it is not adjustable, and some equipment does not measure it. On front-wheel-drive MacPherson strut vehicles, SAI is the primary directional control angle and adjustable angles can be used to correct vehicle aim toward straight ahead. Front-wheel-drive MacPherson strut vehicles have lighter suspensions that are more easily damaged than heavier rear-wheel-drives vehicles. A change in SAI only occurs if the spindle has been bent or there has been body damage resulting in a bent strut tower. When this happens, the camber angle will also change. If the included angle is correct but the camber angle is not, the spindle/steering knuckle is not bent, so replacing it will not correct the problem. If the included angle is not correct, parts must be replaced.

Look for a bent spindle when a front wheel bearing wears out. Front wheel bearings on rear-wheel-drive vehicles do not wear out very often, but a bent spindle can cause the bearings to be misaligned, increasing the load on it. If the cradle (the sub-frame) has shifted to one side, camber will change on both front wheels (**Figure 68.25**) but the included angle will remain the same. Some vehicles have an alignment hole that is used to verify correct cradle position in relation to the body.

Unless there is a symptom pointing to the possibility of incorrect SAI as a contributing cause, most technicians do not go through the extra procedures required to check SAI. Before checking SAI, camber must be set correctly. Comparing SAI from side to side is a good indicator of correct SAI angles.

Measuring Included Angle

The included angle is the amount of SAI minus camber. If camber is negative, subtract it from the SAI reading. If camber is positive, add it to the SAI reading. Included angle is usually within ½ degree from side to side. When camber cannot be adjusted, check the included angle. If the included angle is off, the spindle, strut, or steering knuckle is bent, which requires replacement of the part.

■ MEASURING TOE

Checking and adjusting toe after replacing a tie-rod end or other steering linkage component is an important part of the job. When measuring toe, the distances between the fronts and the rears of the front tires are compared (see Figure 67.1).

Before computer alignment machines, a measuring tape, trammel bar, or optical toe measuring gauge was used. When making a mechanical measurement with a trammel bar (also called a tram gauge), a technician would sight down the side of the front tire to the rear tire and make an adjustment so that each front wheel looked to be in line with the rear wheel on the same side of the vehicle. Measuring toe without an optical device requires an accurate line to be scribed on the tread's footprint. With the tire raised off the ground, the tire was spun while applying chalk to the center of its tread. Next, a narrow line was scribed in the middle of the chalked area using a special scribing tool or a board with a nail pounded through it. After the vehicle was lowered to the ground, the distances between the scribed lines on the tread surfaces at the front and rear of the tire were compared (**Figure 68.26**). The trammel bar was adjusted to measure at the same height as the spindles at the centers of the wheel hubs, which resulted in toe being measured at the largest circumference of the tire tread.

On earlier wheel alignment equipment, an optical toe device projects an image to a gauge on the opposite side of the vehicle.

Methods of Calculating Toe

Toe has traditionally been measured as a distance in inches or millimeters, called total toe or toe distance. A more recent trend is to measure the **toe angle**, which is the angle between the two front tires. Toe angle is equal between the two front tires when the tires are

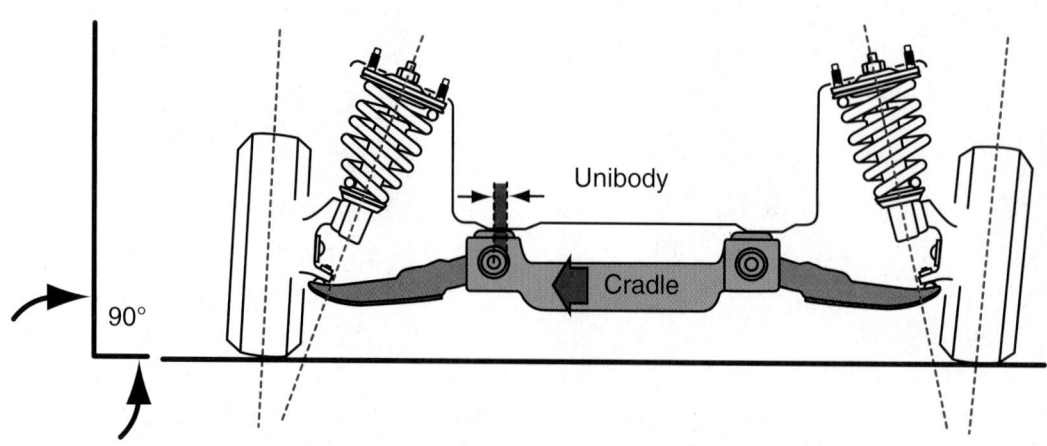

Figure 68.25 When the cradle shifts to one side, camber will change but the included angle will remain the same.

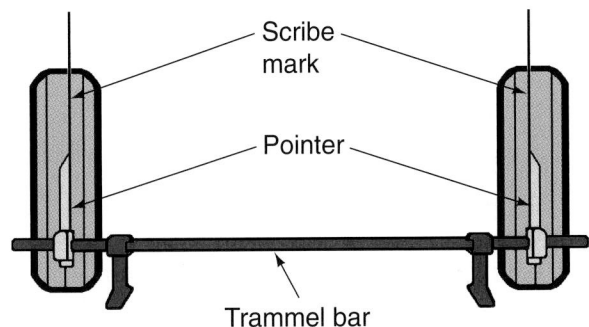

Figure 68.26 A trammel bar for measuring toe.

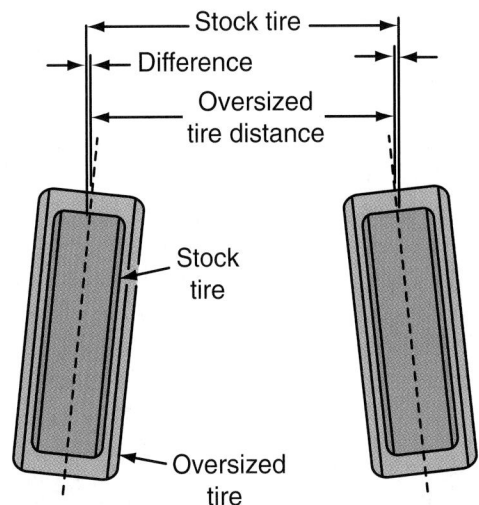

Figure 68.27 Oversized tires will affect toe distance measurement but not toe angle.

rolling forward. The diameter of the tire does not affect the toe angle reading as it does with traditional toe distance measurement (**Figure 68.27**). This is an advantage, especially when oversized tires are used. When toe angle is measured in degrees, toe-in is a positive (+) angle and toe-out is a negative (−) angle. Zero toe is when the wheels are parallel. **Figure 68.28** lists conversions of lengths and angles used in wheel alignment.

Toe in FWD and RWD Vehicles

Front- and rear-wheel-drive vehicles have different toe specifications because they react differently when their tires are in motion. The objective is to have zero toe when the vehicle is in motion.

■ Rear-wheel-drive vehicles are usually set slightly toed-in (about 1⁄16" on new vehicles). This is because the front tires tend to push outward at the

front when rolling due to the rolling resistance of the tires and slack (also called compliance) in steering linkages and bushings (**Figure 68.29**).

■ Front-wheel-drive vehicles are usually set with zero toe or a slight amount of toe-out. This is because the driving front tires tend to push inward at the front as they pull the vehicle forward on the road. The running toe setting can undergo more of a change in response to scrub radius and to increased rolling resistance caused by low tire pressures or wider tires. Scrub radius changes when wheels with more negative offset than original equipment are installed.

Length and Angle Conversions Used in Wheel Alignments					
Inch (Fractional)	Inch (Decimal)	Metric (mm)	Degrees (Decimal)	Degrees (Fractional)	Degrees (Minutes)
1⁄32	0.0312	0.793	0.0625	1⁄16	3.75
1⁄16	0.0625	1.587	0.125	1⁄8	7.5
3⁄32	0.0937	2.381	0.1875	3⁄16	11.25
1⁄8	0.125	3.175	0.25	1⁄4	15
5⁄32	0.1562	3.968	0.3125	5⁄16	18.75
3⁄16	0.1875	4.762	0.375	3⁄8	22.5
7⁄32	0.2187	5.556	0.4375	7⁄16	26.25
1⁄4	0.250	6.35	0.5	1⁄2	30
9⁄32	0.2812	7.143	0.56625	9⁄16	33.75
5⁄16	0.3125	7.937	0.625	5⁄8	37.5
11⁄32	0.343	8.7317	0.6875	11⁄16	41.25
3⁄8	0.375	9.525	0.75	3⁄4	45
13⁄32	0.4062	10.318	0.8125	13⁄16	48.75
7⁄16	0.4375	11.112	0.875	7⁄8	52.5
15⁄32	0.4687	11.906	0.9375	15⁄16	56.25
1⁄2	0.500	12.7	1.0	1	60

Figure 68.28 Length and angle conversions used in wheel alignments.

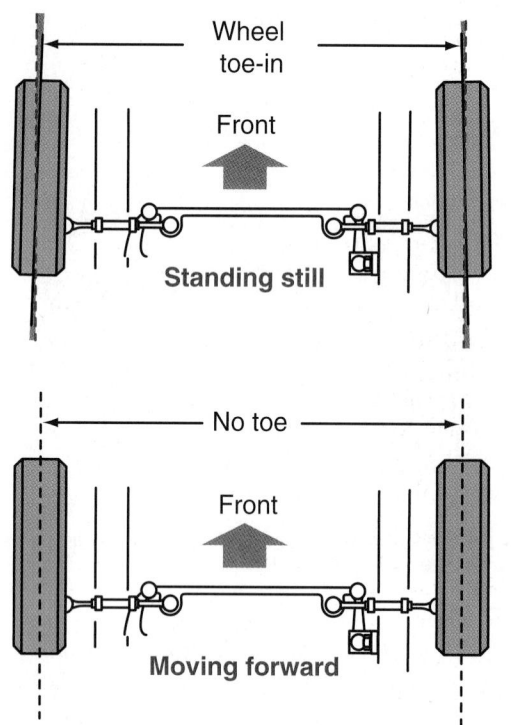

Figure 68.29 On a rear-wheel-drive car, tires tend to toe out when rolling.

ADJUSTING TOE

Steering linkages on most vehicles have either two or four tie-rod ends. On conventional parallelogram steering, the toe adjustment is made by turning the threaded "turnbuckle" sleeves between the tie-rod ends (**Figure 68.30**). Because one of the tie-rod ends has a left-hand thread and the other has a right-hand thread, as the sleeve is turned the tie-rod assembly will become either longer or shorter. This changes the toe setting.

First, center the steering wheel and hold it in place with a steering wheel clamp (**Figure 68.31**). Then make the adjustment.

After turning the adjusting sleeves, the clamp must be properly positioned before tightening it. It must not come into contact with anything when the wheels are turned from side to side. Be sure that the opening of the clamp is not positioned over the split in the adjusting sleeve (**Figure 68.32**). On parallelogram systems, if one tie-rod happens to be tilted one way and the other is tilted the other way, the tie-rod will not be able to pivot (**Figure 68.33**). The tie-rods are positioned so that they are not binding up. This is most easily done by turning both tie-rods in the same direction before tightening the clamp.

Figure 68.31 A steering wheel holder and a brake pedal depressor. (*Courtesy of Tim Gilles*)

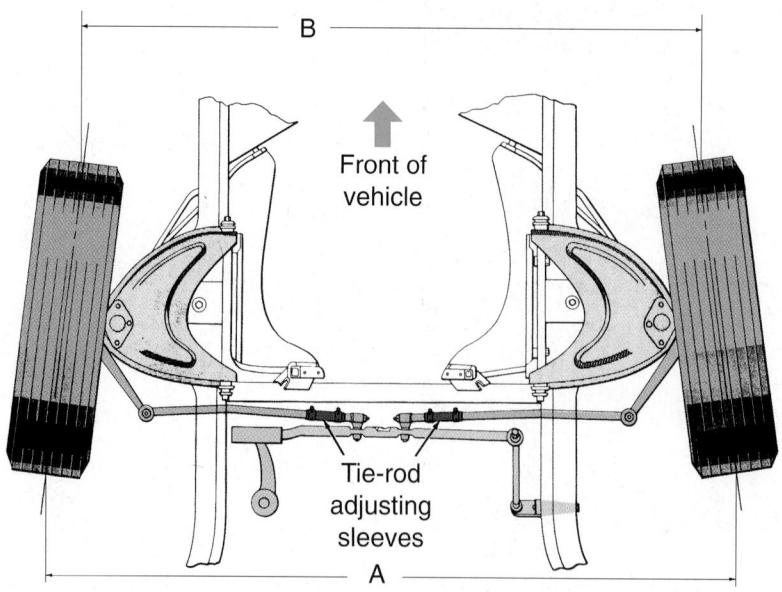

Figure 68.30 Turn the adjusting sleeves to adjust toe. (*Courtesy of SPX Corporation*)

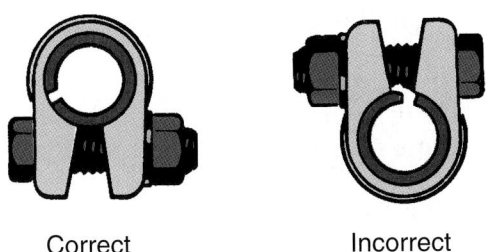

Correct   Incorrect

Figure 68.32 Position the clamp correctly.

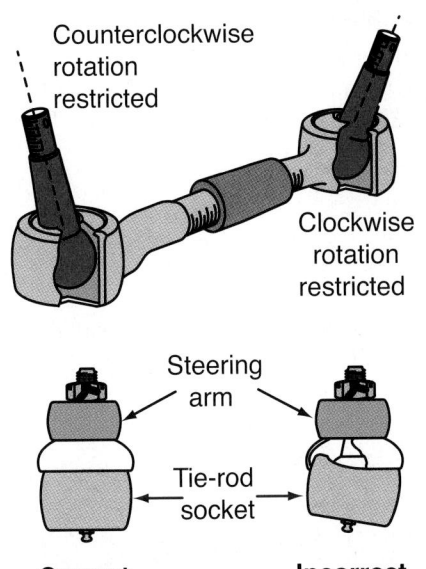

Counterclockwise rotation restricted

Clockwise rotation restricted

Steering arm

Tie-rod socket

Correct   **Incorrect**

Figure 68.33 Position the tie-rods so that they will not bind.

Rack-and-pinion steering systems have a conventional outer tie-rod and an inner tie-rod end with a jam nut on each side (**Figure 68.34**). Hold the tie-rod end with a wrench or pliers while loosening and tightening the jam nut (there is usually a provision on the tie-rod to accommodate a wrench). The inner end of the tie-rod is a ball socket that can be rotated to adjust toe. It will not bind if a tie-rod is turned off-center, like the parallelogram type shown previously in Figure 68.33.

NOTE: *Loosen the clamp on the boot as needed to be sure the boot does not twist while you adjust toe.*

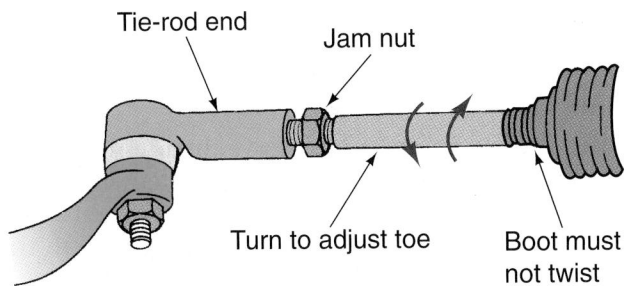

Tie-rod end   Jam nut

Turn to adjust toe   Boot must not twist

Figure 68.34 Rack-and-pinion tie-rod end. *(Courtesy of Federal-Mogul Corporation)*

■ CENTERING THE STEERING WHEEL

Adjusting one of the tie-rods more than the other will cause the steering wheel to be off-center.

NOTE: *In most cars it is important that the steering wheel be centered using the tie-rods, not by removing the steering wheel and putting it back on straight.*

If the steering wheel is not straight ahead when driving, follow this procedure to straighten it:
- Count the number of turns of the steering wheel while turning it from lock to lock.
- Position the steering wheel so that it is half-way between the locks. It should be straight. If not, remove it and put it back on straight.
- Use a steering wheel clamp to hold the steering wheel in the centered position (see Figure 68.31).

This procedure will allow turn signals to cancel correctly during turns and minimize looseness in recirculating ball and nut steering gears.

When making a rough adjustment, turn each tie-rod an equal amount in opposite directions. This will approximately maintain the current toe setting. A small amount of change in the toe setting can occur when tightening the tie-rod clamps or jam nuts because of slack between the threads.

> **SHOP TIP** Correct adjustment will often require that the initial setting be slightly off, in anticipation of the change that will occur when the jam nut is completely tight.

> **SHOP TIP** From the front of the car, sight down the tires on each side of the car to see if the front one aligns with the rear one. This will give a rough estimate of what direction the tie-rods need to be turned.

Test drive the car on a straight, level road. If the steering wheel is off-center and points to the left, adjust the tie-rods so that the tires also point to the left (**Figure 68.35**). Adjust the wheels to face the opposite direction if the wheel points to the right.

NOTE: *People often perceive things differently from each other. A customer might describe a steering pull because the steering wheel is off-center. The customer sees the wheel off-center and straightens it, causing the vehicle to steer to the side. Manufacturers' specifications for maximum allowable steering wheel angle variation are typically +/− 3 degrees.*

■ TOE CHANGE CHECK

Because toe can change with bumps, toe is accurate when the vehicle is at the correct ride height only. Use

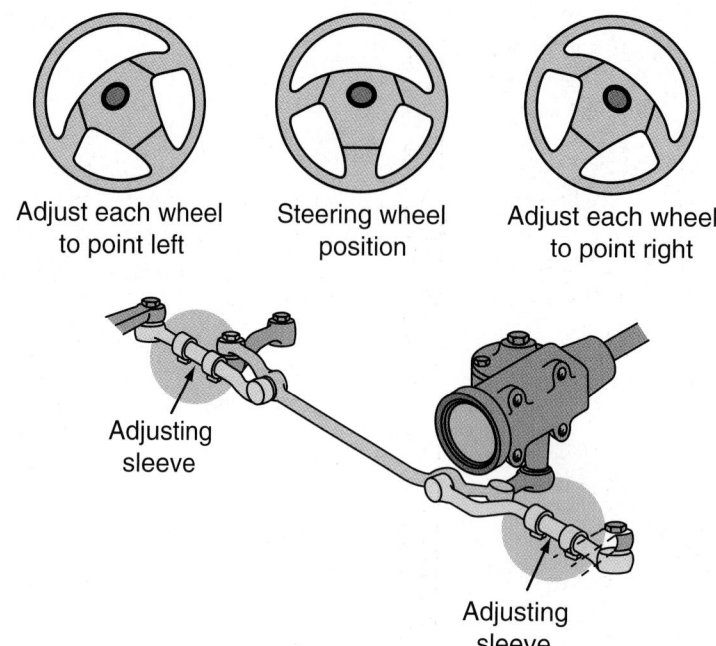

Adjust each wheel to point left

Steering wheel position

Adjust each wheel to point right

Adjusting sleeve

Adjusting sleeve

Figure 68.35 Adjusting steering wheel center using the tie-rods. Adjust tie-rod sleeves equally in opposite directions.

the jack on the alignment rack to raise the vehicle an equal amount on each side (**Figure 68.36**). Check to see that toe changes equally on each wheel. If not, one end of the steering linkage is at an incorrect height.

When a rack-and-pinion steering gear is mounted in a non-level position, its tie-rods will be at unequal angles. First, check the mounting of the rack-and-pinion steering gear. Sometimes one or both of its holding brackets are loose or a bushing has become damaged. Measure and compare the height of the steering gear and linkage at several locations. Sometimes a long straightedge is helpful in spotting irregularities (**Figure 68.37**). Some vehicles use shims to adjust rack-and-pinion height to correct for toe change (**Figure 68.38**).

Sometimes an idler arm will have slotted mounting holes so that its height can be adjusted. The vehicle might have had an accident and require the services of a frame straightening shop.

NOTE: *Aftermarket suspension parts sometimes relocate steering linkage connections, which can cause extreme*

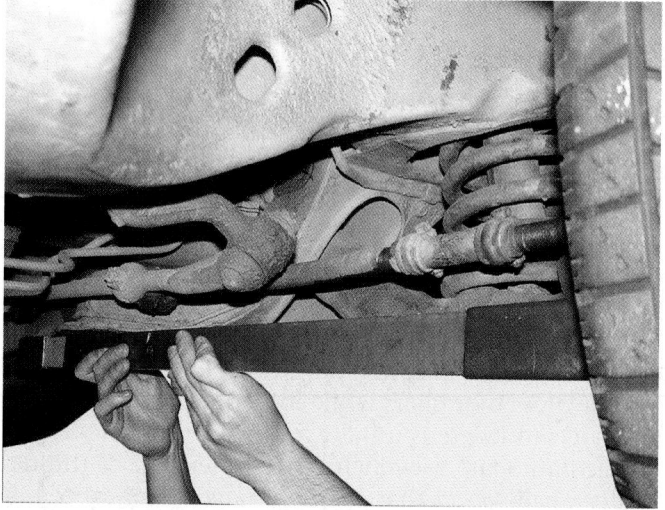

Figure 68.37 Sometimes a long straightedge is helpful in spotting irregularities in steering linkage angle. *(Courtesy of Tim Gilles)*

changes in toe as the suspension system moves. After installation of new parts, it is always a good idea to check for toe change during suspension system travel.

■ MEASURING TURNING RADIUS

To measure turning radius, observe the pointer on the radius plate while making a caster measurement. As the wheels are turned from side to side, the outer wheel should make a turn that is 2 or 3 degrees less than the inside wheel. An example: When the outside wheel is turned 18 degrees, the inside wheel's radius plate gauge should read about 20 degrees (**Figure 68.39**). The

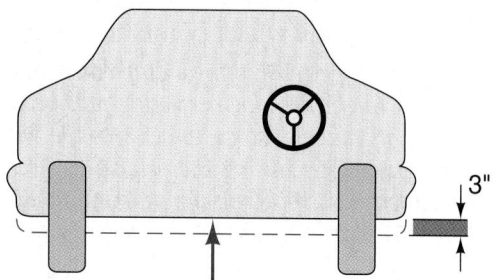

3"

Figure 68.36 Raise the vehicle equally from side to side to check for toe change. *(Courtesy of Hennessy Industries, Inc.)*

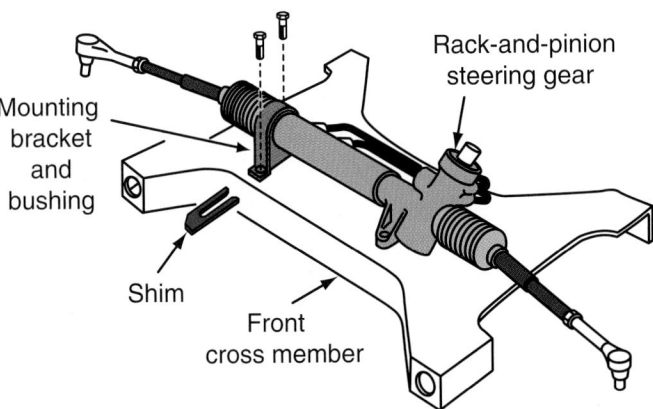

Figure 68.38 Some vehicles use shims to adjust rack-and-pinion height to correct for toe change.

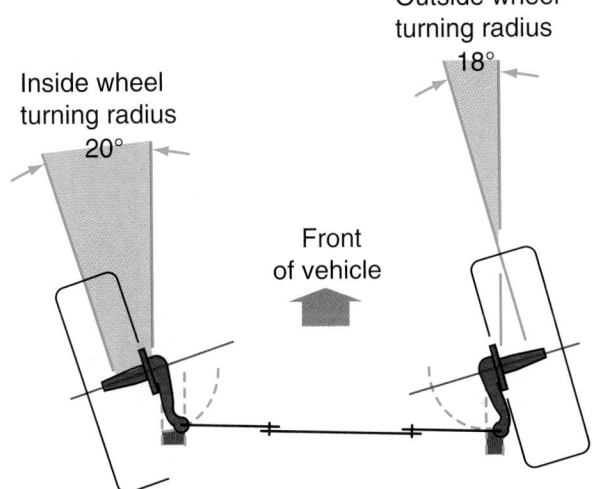

Figure 68.39 The inner wheel turns at a sharper angle than the outer wheel.

reading should be within 1½ degrees of the specifications found in the service manual.

The steering arms are angled to point to the center of the rear axle (**Figure 68.40**). This is the Ackerman angle.

NOTE: *Turning radius is not an adjustable angle, but if a steering arm becomes bent the turning radius is affected and the tires can squeal on turns.*

■ GENERAL WHEEL ALIGNMENT RULES

A suspension/steering specialist will experiment with different vehicles to see what changes result from various adjustments. The following general rules are provided as a starting point. Due to variations in design, they will not always be true.

Caster/Camber

■ The vehicle will pull to the side with the least (most negative) caster.
■ The vehicle will pull to the side with the most (most positive) camber (**Figure 68.41**).
■ Adjusting for more negative caster results in easier steering.

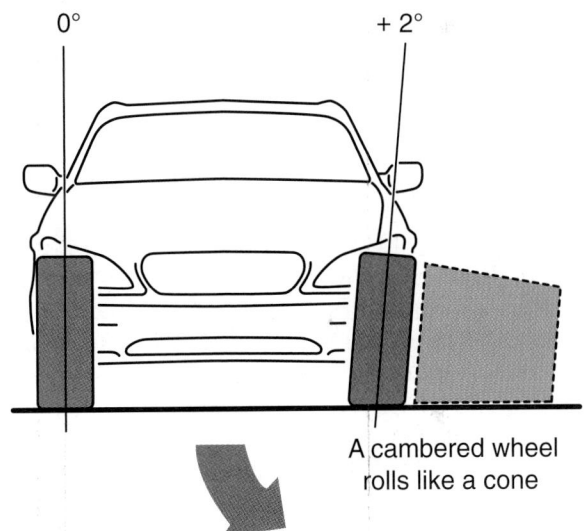

Figure 68.41 The vehicle will pull to the side with the most positive camber.

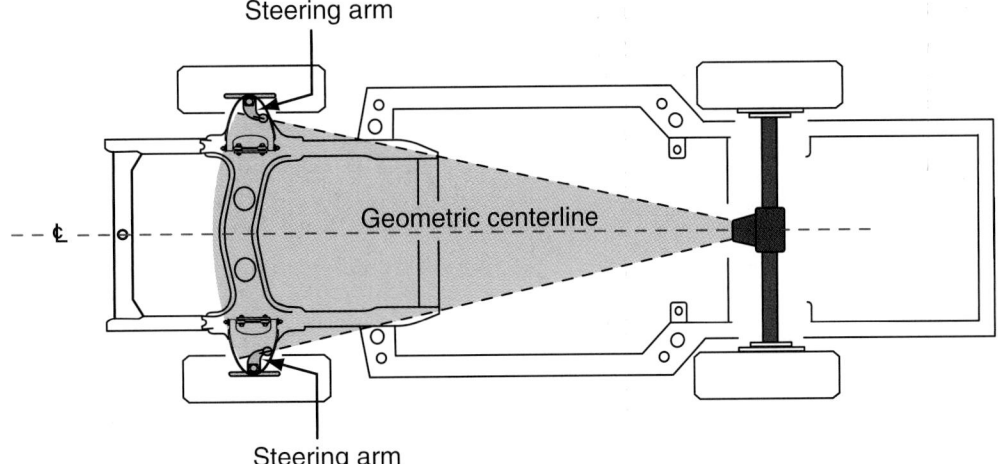

Figure 68.40 The Ackerman angle has the steering arms point to the center of the rear axle.

- The driver's weight will usually cause camber to increase on the left front wheel and decrease on the right front wheel.
- On shim-type vehicles, changing camber will not affect caster, but changing caster can affect camber.
- Caster for both wheels should be set either positive or negative, not one positive and one negative.
- Caster **spread** between the front wheel settings should not be more than ½ degree.

NOTE: *Spread is the difference between alignment settings from side to side. It is also called cross caster or cross camber.*

Camber is sometimes specified with more than a ½-degree of spread. Always follow manufacturer's specifications.

- Make caster equal from side to side. Use camber to compensate for road crown. Set the left side ¼ degree more positive than the right (if caster is equal).
- Vehicles with manual steering are uncommon today, but on these vehicles caster is generally 0 to 1 degree negative. Power steering vehicles can have caster as high as 10 degrees (Mercedes).
- On a MacPherson strut vehicle, jounce the vehicle while measuring camber. If camber changes drastically on either front wheel, the strut is bent.

Toe

- Before driving on the alignment rack, be sure the pins are in the radius plates.
- Every ¼" turn of the adjusting sleeve results in about ¹⁄₁₆" of change in toe.
- Changes in caster and camber affect toe, so toe is adjusted last.
- Toe is measured in inches, millimeters, or degrees.

■ FOUR-WHEEL ALIGNMENT

Modern wheel alignment equipment measures wheel alignment of all four wheels. Rear-wheel toe and camber are often adjustable. When they are not, the front wheels are adjusted, with reference to the settings in the rear. That concept will be covered later.

The **geometric centerline** of the vehicle is a line drawn between the center of the front axle and the center of the rear axle. When individual toe is measured at all four wheels, the geometric centerline is used. Several measurement factors are considered in relation to the geometric centerline. The following describes some of these factors.

Thrustline is the direction in which the rear wheels are pointing (**Figure 68.42**). When the rear wheels are held in a fixed position, the thrustline defines the wheels' true straight-ahead position. If the rear wheels are aimed to the right, the thrustline is to the right.

Thrust angle is the angle formed by the thrustline and the geometric centerline (**Figure 68.43**).

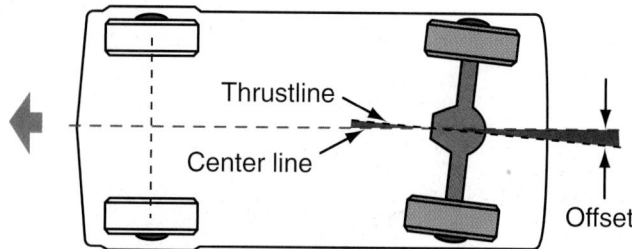

Figure 68.42 Thrustline is the direction that the rear wheels are pointing.

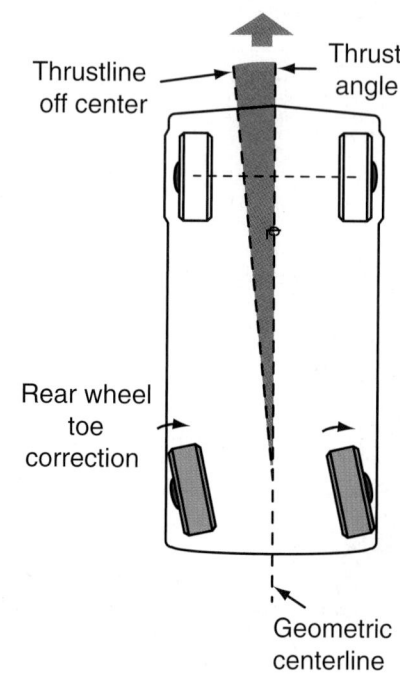

Figure 68.43 Correct rear-wheel alignment adjusts the rear wheel toe to specifications with the thrust angle near zero.

A thrust angle is created when the rear wheels are not parallel to the vehicle centerline. This is most often the result of an accident or impact that moves the position of the axle away from being perpendicular to the centerline. The thrust angle is positive when off to the right and negative when off to the left. Full four-wheel alignment involves adjusting rear-wheel toe to factory specifications with the thrust angle at or near zero. Independent rear suspensions typically have a means of toe adjustment. When rear toe is adjustable, adjusting individual rear toe will compensate for incorrect thrust angle.

NOTE: *Individual toe is measured at each wheel and is in reference to the geometric centerline of the vehicle. In order to center the steering wheel, it will be necessary to adjust the front individual toe for each wheel.*

When the wheels are closer together on one side of the vehicle than the other, this is called wheelbase difference (**Figure 68.44**). *Setback* is the amount that one front wheel is behind the one on the other side of the

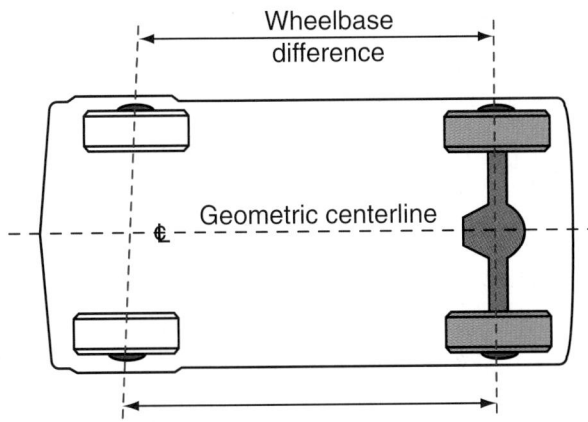

Figure 68.44 The wheels on the right side are closer to each other than the wheels on the left side.

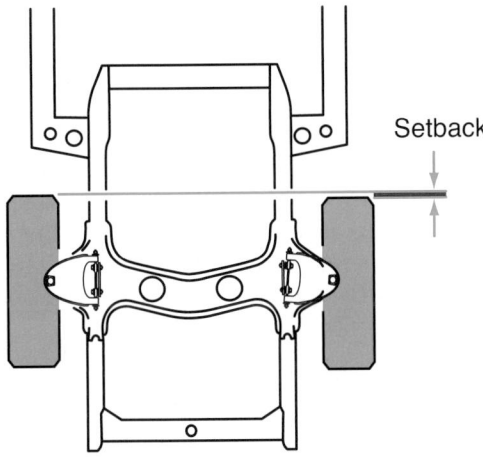

Figure 68.45 Setback is the amount that one front wheel is behind the one on the other side of the car.

car (**Figure 68.45**). Though not adjustable, it is measured during a four-wheel alignment. A collision can cause incorrect setback. It can cause a vehicle to steer to one side or the other and can also cause brake pull.

Track width difference exists when the distance between the two front wheels is different than the distance between the two rear wheels (**Figure 68.46**). Front wheels on some vehicles have a narrower track width than the rear wheels. In winter weather, this can pose a problem when the rear tires try to follow tracks made in snow or mud by the front tires.

SHOP TIP Sight across the sidewall bulges of the front tires as you look at the rear tires. If the rear wheels have a wider track width, you will most likely see a small amount of tread on each side. When the rear wheel thrustline is straight, the steering wheel will be centered if you can observe an equal amount of tread on each side.

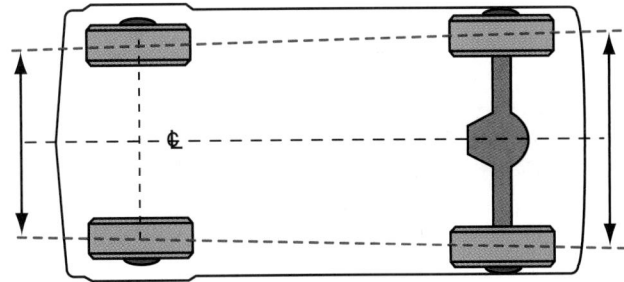

Figure 68.46 The front wheels often have narrower track width than the rear wheels.

PERFORMING A FOUR-WHEEL ALIGNMENT

During a computer wheel alignment, sensors are installed on all four wheels (**Figure 68.47**). Some of the newest alignment machines use targets (**Figure 68.48**) rather than gauges on the wheels. Four digital cameras monitor the position of the targets. When all four wheels are checked at the same time,

Figure 68.47 Mount the wheel alignment head on the wheel. *(Courtesy of Tim Gilles)*

Figure 68.48 Some newer alignment machines use a target on each wheel. Digital cameras monitor the location of each target. *(Courtesy of Tim Gilles)*

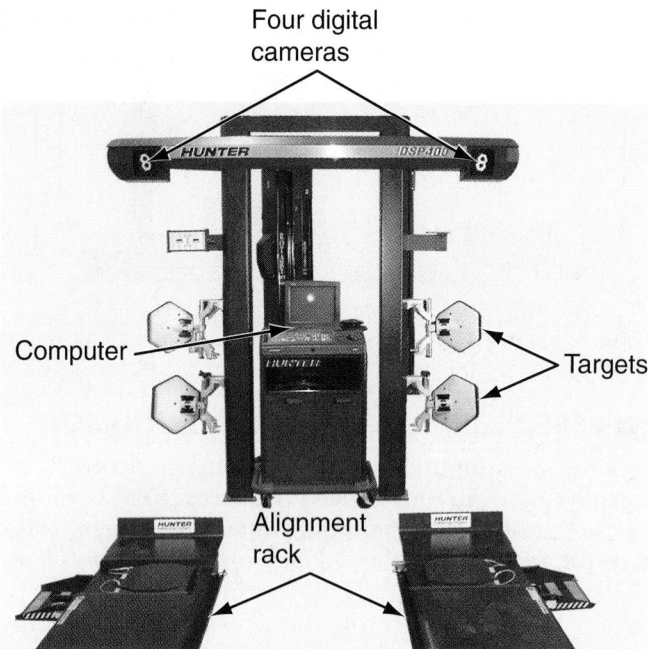

Figure 68.49 A modern alignment machine that automatically compensates the wheels by simply rolling the vehicle on the alignment rack. *(Courtesy of Tim Gilles)*

the thrust angle is calculated. If the thrust angle is the same as the geometric centerline, the steering wheel will be correctly centered.

COMPENSATING THE ALIGNMENT HEADS

Alignment machines that use targets automatically compensate the alignment heads by simply rolling the vehicle on the alignment rack (**Figure 68.49**). Many older wheel alignment machines required a higher degree of skill to adjust alignment heads that were mounted on the wheels. Modern alignment machines are much quicker and easier.

MEASURING CASTER AND CAMBER

Caster and camber are measured in the same manner as with manual alignment equipment. The amount of wheel sweep during a caster check is determined by the alignment program. The technician watches the monitor (**Figure 68.50**) while slowly turning the wheels a few degrees in one direction on the radius plates. Then, a turn in the opposite direction is required. When the wheels have been positioned correctly and the computer has the information needed to make the calculations, alignment readings are displayed on the monitor. One popular feature on some computer alignment equipment is the availability of still photo and video tutorials on various procedures.

ADJUSTING REAR-WHEEL ALIGNMENT

Camber and toe adjustments are possible on some vehicles.

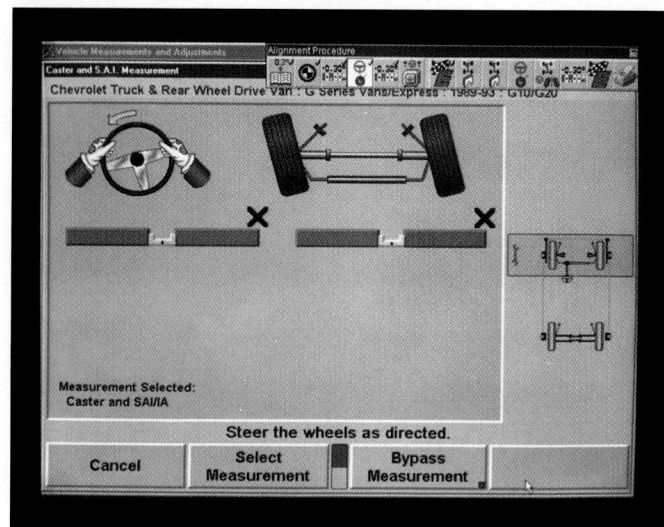

Figure 68.50 Using a computer alignment machine to make a caster reading.

Adjusting Rear Camber

Camber adjustment on a vehicle with a double wishbone rear suspension is usually done by turning an eccentric adjuster. Another camber adjustment used on rear strut suspensions calls for the installation of a tapered wedge between the top of the rear knuckle and the strut (**Figure 68.51**).

Adjusting Rear Toe

Rear-wheel toe is adjusted in several ways, depending on the manufacturer. One common method involves moving the lower control arm with an adjustable linkage attached to the knuckle (**Figure 68.52**). Some manufacturers have a toe link with a slotted hole that allows the wheel to be moved to adjust toe when its hold-down screw is loosened (**Figure 68.53**).

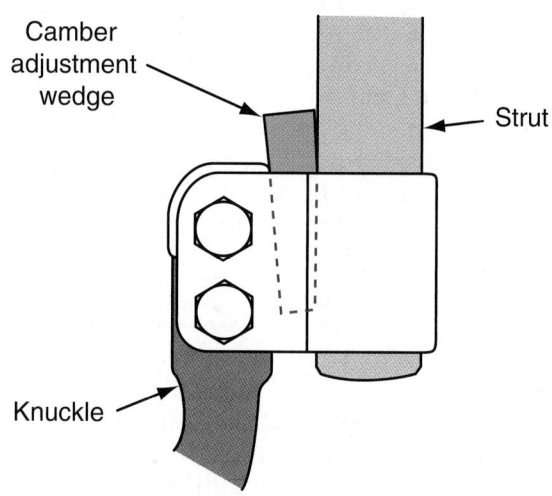

Figure 68.51 A rear strut suspension camber adjustment using a tapered wedge between the top of the rear knuckle and the strut.

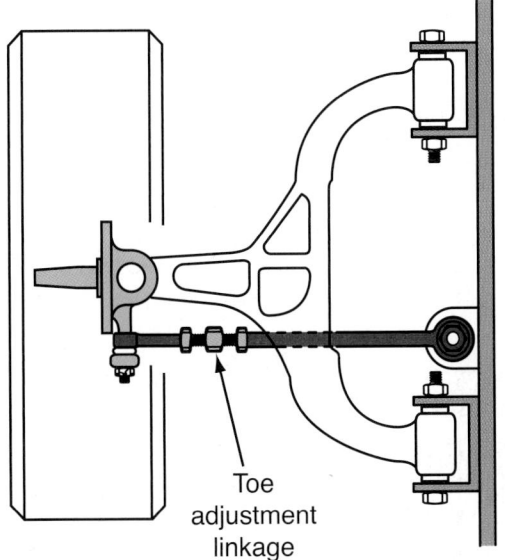

Figure 68.52 Top view of rear toe adjustment linkage.

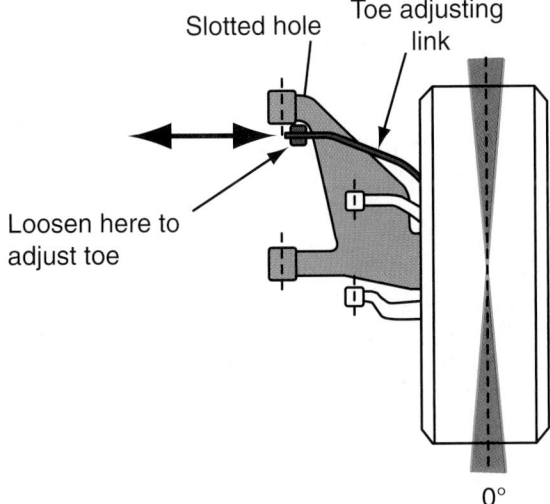

Figure 68.53 A slotted hole in the toe adjusting link allows the wheel to be moved to adjust toe.

Another rear-wheel alignment adjustment method requires the installation of a tapered shim typical of that used to adjust front-wheel alignment on four-wheel-drive vehicles (**Figure 68.54**). Camber and/or toe can be changed, depending on the installation position of the shim.

More Wheel Alignment Rules

- Be sure there are no heavy loads in the vehicle when reading alignment angles. The rear axles of smaller front-wheel-drive cars are so light that they can flex under load, resulting in camber change.
- Ideally, the fuel tank should be full.
- The vehicle should be aligned in the condition it is normally driven—for example, a light truck normally operated with a camper, or a car with a trunk full of sales literature.

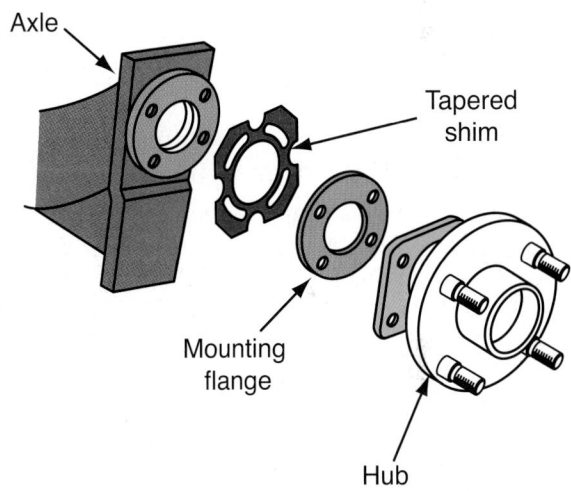

Figure 68.54 A tapered shim installed to correct rear-wheel alignment.

▬▬ REVIEW QUESTIONS

1. When a front-wheel-drive car has two drive axles of different lengths, what can happen during acceleration?

2. If you move your hand across the tire's footprint area from the outside to the inside and feel a _____ edge, there is excessive toe-in.

3. On a rear-wheel-drive car with excessive toe-in, the _____ edge of the right front tire tends to wear.

4. What is the name of the test where an assistant turns the steering wheel back and forth a short distance while you look for looseness in the steering linkage?

5. Which of the alignment angles are normally adjustable?

6. Which alignment angle is usually measured in inches or millimeters, rather than degrees of a circle?

7. _____ caster results in easier steering.

8. When the road is higher in the center than on the outside edges, what is this called?

9. Is SAI adjustable?

10. The included angle is the amount of SAI _____ camber.

■■■ ASE-STYLE REVIEW QUESTIONS

1. Several thousand miles after a wheel alignment, a vehicle's steering wheel has shifted off-center when driving on a straight highway. The required repair is to:
   a. Remove the steering wheel and put it back on in the correct position.
   b. Adjust the tie-rod ends during a wheel alignment.
   c. Rotate the tires.
   d. Adjust the steering gear box.

2. Which of the following will cause a change in turning radius?
   a. A bent steering arm
   b. A change in camber
   c. A change in caster
   d. Adjusting the steering axis inclination

3. Technician A says that a change in camber can cause a change in toe. Technician B says that a change in toe will cause a change in camber. Who is right?
   a. Technician A   c. Both A and B
   b. Technician B   d. Neither A nor B

4. Technician A says that toe can change when springs sag. Technician B says that centering the steering wheel is done by adjusting tie-rod length. Who is right?
   a. Technician A   c. Both A and B
   b. Technician B   d. Neither A nor B

5. Technician A says that caster is read with the wheels facing straight ahead. Technician B says that caster spread between the front wheel settings can be up to 2 degrees. Who is right?
   a. Technician A   c. Both A and B
   b. Technician B   d. Neither A nor B

6. Which of the following is *not* true regarding camber?
   a. The car will pull to the side with the most positive camber.
   b. The car will pull to the side with the most negative camber.
   c. Excessive camber can cause a tire to wear on one side.
   d. Sagged springs can cause negative camber.

7. Which of the following is true regarding caster?
   a. The car will pull to the side with the most positive caster.
   b. The car will pull to the side with the most negative caster.
   c. Excessive caster can cause a tire to wear on one side.
   d. Caster does not cause a tire to pull.

8. Which of the following is/are true about toe change?
   a. It can cause bump steer.
   b. It can happen when tie-rods are at unequal heights and a wheel goes over a bump.
   c. It can be caused by different caster angles from side to side.
   d. All of the above.

9. A vehicle turns abruptly to the side during initial acceleration. Technician A says that this problem can be caused by unequal spring height. Technician B says that this could be due to a loose sub-frame. Who is right?
   a. Technician A   c. Both A and B
   b. Technician B   d. Neither A nor B

10. Two technicians are discussing how to compensate for the crown designed into a typical road. Technician A says that camber can be set slightly more positive on the driver's side of the vehicle. Technician B says that caster can be set so that it is slightly more negative on the driver's side of the vehicle. Who is right?
    a. Technician A   c. Both A and B
    b. Technician B   d. Neither A nor B

DriveTrain

THEN AND NOW: FRONT-WHEEL DRIVE

Although *front-wheel drive* (FWD) did not appear as a regular design in American cars until the mid-1980s, it dates back to 1877, when New York attorney George B. Selden filed a patent on a "road engine" with a transversely mounted engine driving its front wheels. The car, which was not actually built until 1905, could not really be called a usable vehicle.

The advantages of pulling with the front wheels instead of pushing with the rears have long been an attraction to vehicle designers. French automotive pioneer Andre Citroën is credited with saying, "Driving the front wheels is a natural idea. After all, the horse doesn't push the cart." The idea was fine in theory, but front wheels are required to steer while rising and falling with bumps in the road. The ordinary universal joint used in rear-wheel-drive cars simply could not handle the sharp angles required for directing torque to front-wheel-drive wheels. Front-wheel-drive cars required a component that had not been invented yet.

In 1928, Alfred Rzeppa invented a new type of joint capable of transmitting twisting power through very sharp bends. This device was the *constant velocity,* or *CV, joint.* There were serious durability problems with early CV joints. They needed more development and better lubricants before they could be considered dependable, so most car makers stayed with RWD. There were some notable exceptions, however, such as the beautiful Cord of the 1930s.

Improved technology and the genius of English engineer Sir Alec Issigonis came together in 1960 with the introduction of the first truly successful FWD automobile. The Austin-Morris Mini stands as a landmark in automotive history. In the United States, the 1966 Oldsmobile Toronado was the first modern-day FWD car.

Today, well over half of the automobiles on our roads are equipped with front-wheel drive. It is popular because it makes more efficient use of space, improves traction, and helps avoid spinouts. But FWD also provides more complicated and expensive repair and service opportunities.

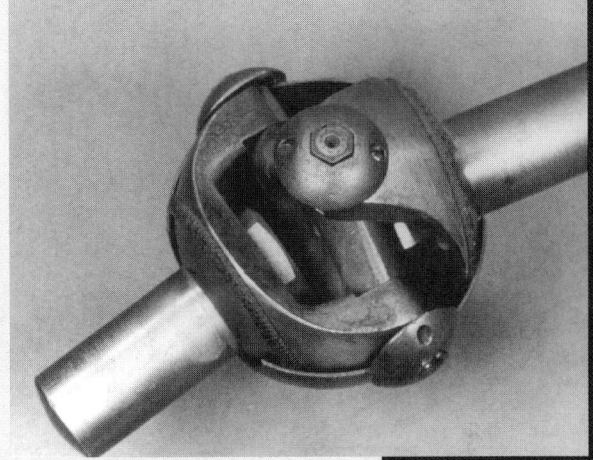

Plain universal joints, such as Clarence Spicer's original 1901 version, cannot cope with the angles FWD requires. *(Courtesy of Dana Corporation)*

Inside a modern CV joint. *(Courtesy of Bob Freudenberger)*

Clutch Fundamentals

■ **KEY TERMS**

clutch hub	release bearing	overcenter assist spring
dampened hub	diaphragm spring	slave cylinder
pressure plate assembly	quill	clutch free play
release lever	clutch fork	

■ **INTRODUCTION**

A clutch is found on cars with manually shifted transmissions. The clutch disengages the engine from the transmission, allowing it to continue running while the car is at rest (**Figure 69.1**). It also allows for releasing the engine from the transmission during gear shifts. The driver controls the application of a clutch from inside the vehicle, using the clutch pedal.

Driver control allows the clutch to gradually be applied. This is important because the internal combustion engine does not make sufficient torque at lower engine rpm to be able to move the car. The clutch must be "slipped" so it gradually connects the rear wheels to the engine. There are two *driving* members of the clutch: the pressure plate and the flywheel. An additional part—a friction disc that acts as a *driven* member—is positioned between the driving parts.

■ **CLUTCH PARTS AND OPERATION**

The clutch has several parts: a flywheel, pressure plate, disc, and release mechanism (**Figure 69.2**). If the clutch disc is pushed against the flywheel with enough force, the disc will rotate with the flywheel.

■ **CLUTCH DISC**

The clutch disc has splines that fit over splines on the transmission input shaft. The splines allow the disc to slide back and forth on the input shaft as it is first clamped to the flywheel and then released. It has several main parts: a hub and damper, facings, and plates.

Hub and Damper

The **clutch hub** is the inner part of the disc. It is the part that has the splines that slide over the input shaft.

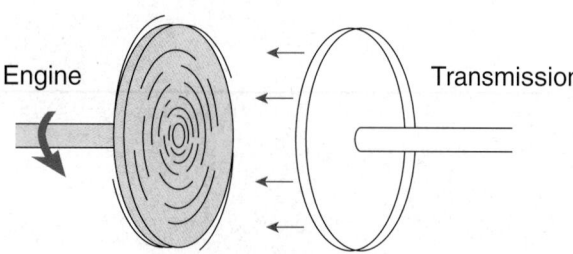

Only one disc is turning

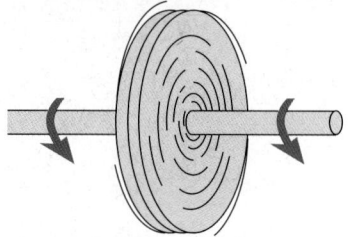

Both discs are turning

Figure 69.1 The clutch disengages and engages the engine to the transmission. *(Reproduced by permission of Deere & Company, John Deere Publishing, Moline, IL. All rights reserved)*

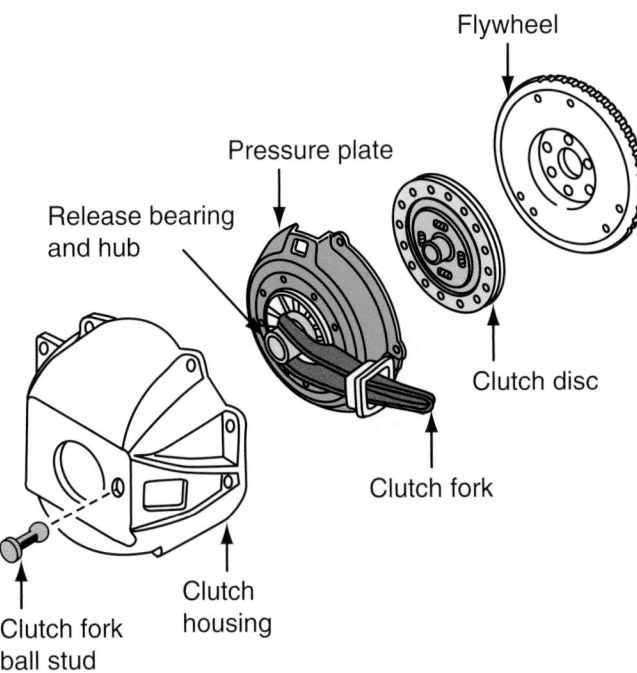

Figure 69.2 Parts of the clutch assembly.

The number of splines differs from make to make. Some hubs have as many as 26 splines. Others have as few as 10.

Most clutch disc hubs are **dampened hubs** (**Figure 69.3**). *Torsional dampers* that are either coil springs or rubber are positioned between the disc plate and the clutch hub to absorb shock during engagement. Some hubs used on smaller cars are rigid, with no dampening.

A series of large rivets, called *stop pins,* prevent the hub from excessively compressing the springs or rubber. The springs cushion the application. Engineers can tune the design of the springs and windows to provide different characteristics to the clutch. Torsional dampers keep engine vibration from causing noise and wear in the transmission.

Clutch Disc Facing

The clutch disc has facings made of *friction material,* traditionally containing *molded* or *woven asbestos.* Facing material must be able to withstand the extreme heat that is generated as the clutch is slipped during engagement. Fibers are visible in a woven disc. Woven asbestos transfers heat better than does molded asbestos.

Asbestos is an excellent friction material for brakes and clutches, but it is regulated because it is dangerous to breathe. It can cause lung problems and a cancer of the chest cavity. Non-asbestos materials using fiberglass or Kevlar® are being substituted for asbestos.

NOTE: *Semi-ceramic clutch discs, also called cerametallic, are made of bronze, iron, or steel matrix materials, which contain ceramic friction particles. These discs are used for racing and heavy-duty applications. There are wedge-shaped paddle wheels spaced around the facing of the disc* **Figure 69.4**. *Semi-ceramic discs provide increased clutch efficiency but also engage abruptly and tend to chatter. Some semi-ceramics that have a solid disc are for severe racing only and are not intended for street use.*

Clutch Disc Plate

Facings are riveted to both sides of a cushion plate. The *cushion plate* is riveted to the *disc plate* (**Figure 69.5**). As the clutch disc is compressed between the flywheel and pressure plate, the *cushion plate* lets the facings compress. This results in smoother engagement of the clutch.

The clutch facings have grooves so they do not stick to the flywheel. Air is trapped in the grooves when the clutch is engaged. When the clutch is released, the centrifugal force of the trapped air pushes the disc away from the pressure plate and flywheel.

Figure 69.4 A semi-ceramic clutch disc. *(Courtesy of Tim Gilles)*

Figure 69.3 A dampened clutch hub. *(Courtesy of LuK Automotive Systems)*

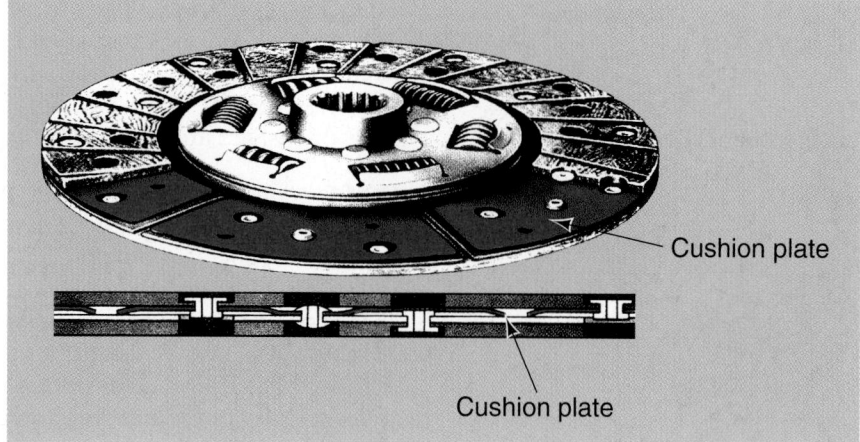

Cushion plate

Cushion plate

Figure 69.5 Between the two clutch facings is a cushion plate. *(Courtesy of LuK Automotive Systems)*

PRESSURE PLATE

The **pressure plate assembly** is also called the *clutch cover assembly.* The pressure plate is a cast iron plate that is actually part of the cover assembly. The cover assembly is bolted to (and rotates with) the flywheel.

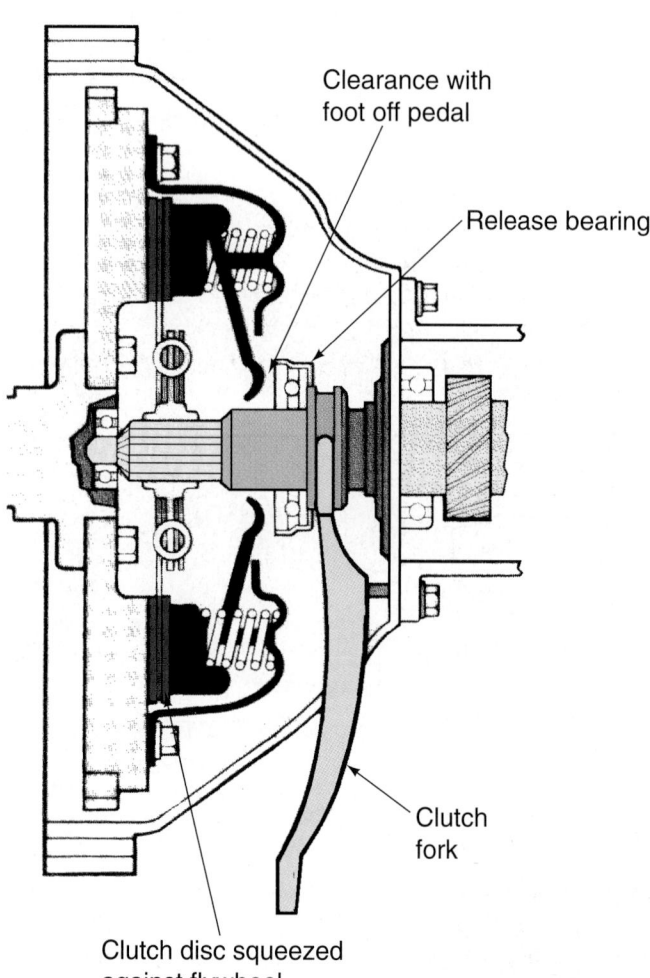

Clearance with foot off pedal

Release bearing

Clutch fork

Clutch disc squeezed against flywheel

Figure 69.6 The clutch disc is squeezed against the flywheel by spring pressure when the pedal is up. *(Courtesy of Federal-Mogul Corporation)*

The clutch disc is wedged between the pressure plate and the flywheel. There is spring pressure on the disc during normal operation (**Figure 69.6**). This is because the space between the pressure plate and the flywheel is less than the thickness of the clutch disc. Remember, the inside of the clutch disc hub has splines that connect it to the input shaft on the transmission. This means that the engine and transmission are physically connected when the clutch pedal is released.

TYPES OF CLUTCH COVERS

There are two main types of clutch covers: coil spring and diaphragm (**Figure 69.7**). There are other designs of clutches, but these are most prevalent.

COIL SPRING CLUTCH

A coil spring clutch cover is shown in Figure 69.6. Pressure plate springs are preloaded when the clutch cover is assembled at the factory. More compression of the springs takes place when the cover assembly is bolted to the flywheel. When the clutch is engaged (foot off of the pedal), the pressure plate exerts a force of between 1,000 and 3,000 pounds on the disc. Heavy-duty clutches exert force around 3,000 pounds and light duty is around 1,000 pounds. The pressure plate is designed so that when the disc is totally worn out, there will be about 10% more torque carrying capacity left in the clutch than the engine can deliver.

NOTE: *There is often confusion regarding the terms that apply to clutch application and release.*
- *When you* **apply** *the pedal, you* **release** *the clutch.*
- *When you* **release** *the pedal, you* **apply** *the clutch.*

RELEASE LEVERS

Release levers, also called *fingers*, are attached to the cover assembly at pivot points. One end of the release lever contacts the **release bearing**. The other end pulls or pushes on the pressure plate. Pushing on the clutch pedal moves the pivot lever, which pulls the pressure plate away from the flywheel (**Figure 69.8**).

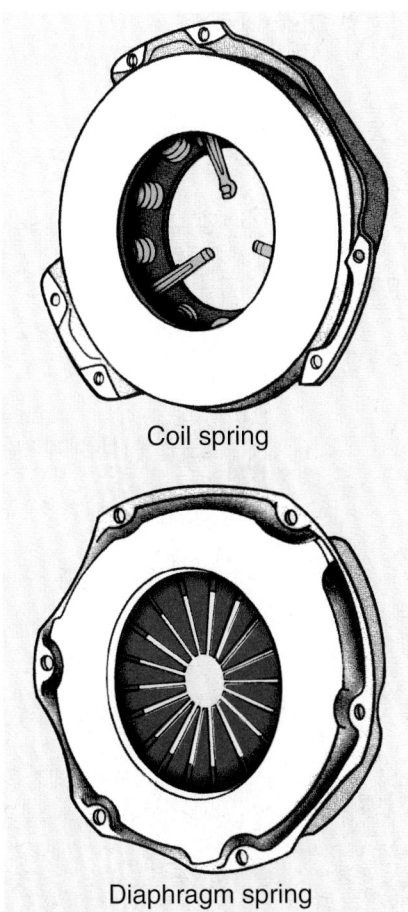

Coil spring

Diaphragm spring

Figure 69.7 Coil and diaphragm spring clutch covers. *(Courtesy of Federal-Mogul Corporation)*

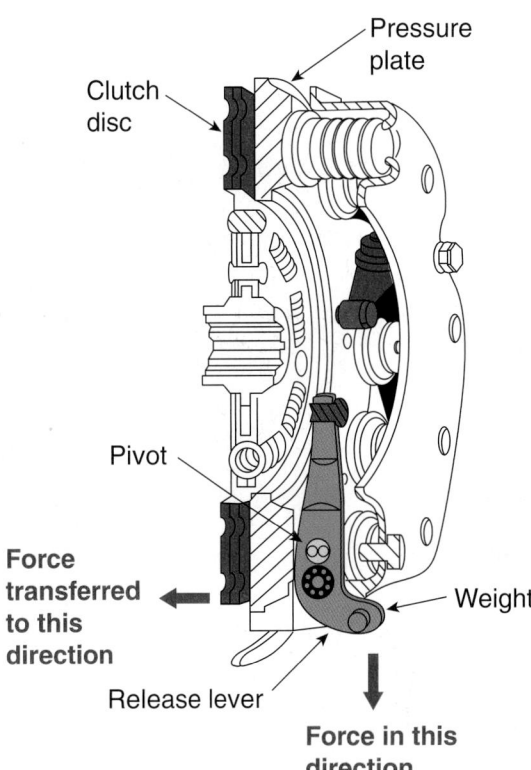

Figure 69.9 A weight on the release lever causes centrifugal force to apply more force against the disc at higher speeds. *(Courtesy of Ford Motor Company)*

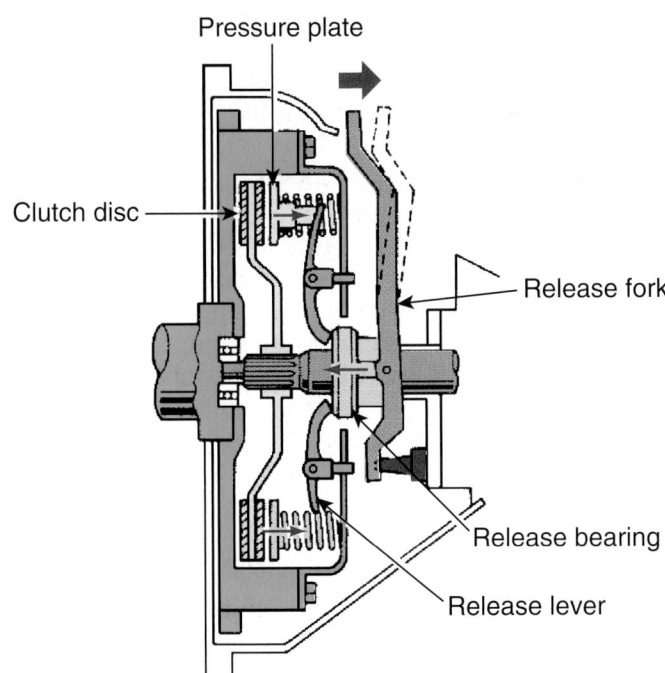

Figure 69.8 Pushing on the pedal moves the pivot lever to pull the pressure plate away from the disc.

This action releases the disc's link between the engine and the transmission.

There are some advantages to a coil spring over a diaphragm spring:
■ It is better for heavy-duty uses because more coil springs can be installed to make a clutch apply with more force.
■ Putting a weight at the end of the release lever results in centrifugal force applying the clutch more tightly at higher speeds (**Figure 69.9**).

Coil spring disadvantages include:
■ More pedal pressure is required from the driver to disengage it.
■ It does not apply the clutch as heavily as the disc wears.
■ Coil spring clutch covers must be precisely balanced after assembly.

■ DIAPHRAGM CLUTCH

A **diaphragm spring** (see Figure 69.7), also called a Belleville spring, replaces the release levers and coil springs in a diaphragm clutch. The spring works much like an older compression-style oil can (**Figure 69.10**). The diaphragm pivots off of pivot rings when the clutch pedal is depressed.
■ A diaphragm clutch requires lower pedal operating pressure.

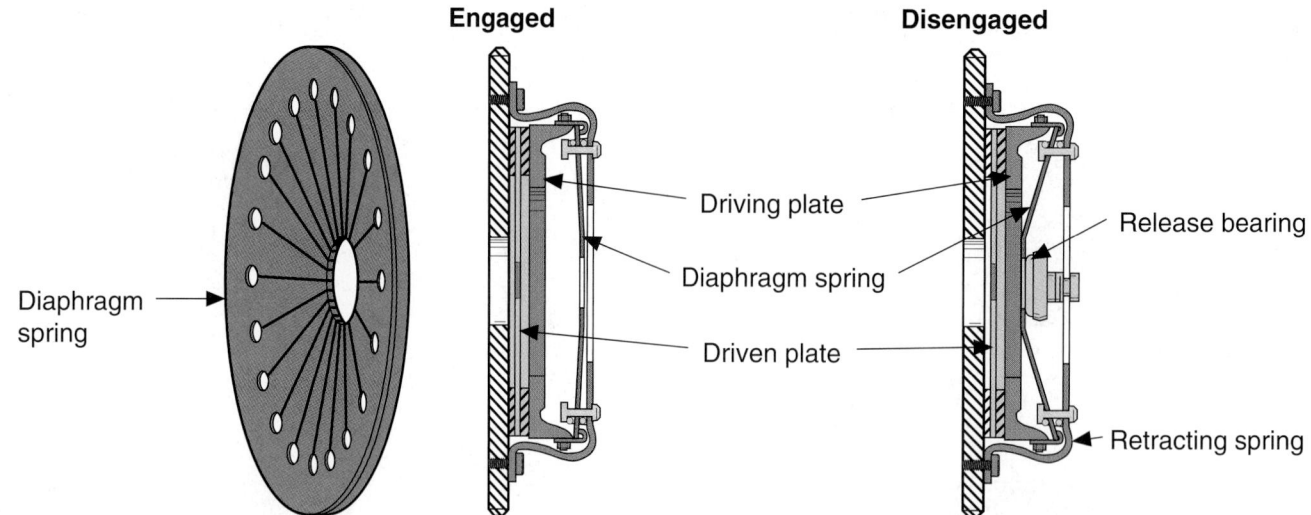

Figure 69.10 A clutch pressure plate assembly with a diaphragm spring.

- It also takes up less space. This can be helpful when a clutch is being installed in a tight spot in a front-wheel-drive car.
- As the clutch disc wears, spring pressure exerted on the disc actually becomes greater for the first half of disc wear) due to the overcenter design of the diaphragm.
- Diaphragm clutches are well balanced when compared to coil spring clutches because of their type of construction.

Flywheel

The clutch disc is applied against the flywheel, which is a heavy mass (**Figure 69.11**). Some high-performance flywheels are aluminum, which weighs less but allows the engine to rev very quickly between shifts,

Figure 69.11 The flywheel provides one of the surfaces against which the clutch disc applies.

Figure 69.12 An aluminum flywheel with a steel insert friction surface. (*Courtesy of Tim Gilles*)

sometimes dangerously so. Aluminum flywheels have a steel insert friction surface (**Figure 69.12**).

A *dual mass* flywheel (**Figure 69.13**) is used on a few vehicles to reduce noise and vibration and allow smoother gear shifting. It is made up of two flywheel plates connected by a damper and spring. The forward flywheel plate is bolted to the rear of the crankshaft like an ordinary flywheel. The clutch pressure plate bolts to the back half. The two plates can turn at different rates to absorb shocks, isolating the engine and transmission from one another.

▬ PILOT BEARING OR BUSHING

The engine side of the transmission input shaft is supported by a sealed pilot bearing or sintered bronze bushing pressed into the end of the crankshaft (see Figure 69.2). **Figure 69.14** shows three styles of pilots. Some front-wheel-drive transaxles do not use a pilot

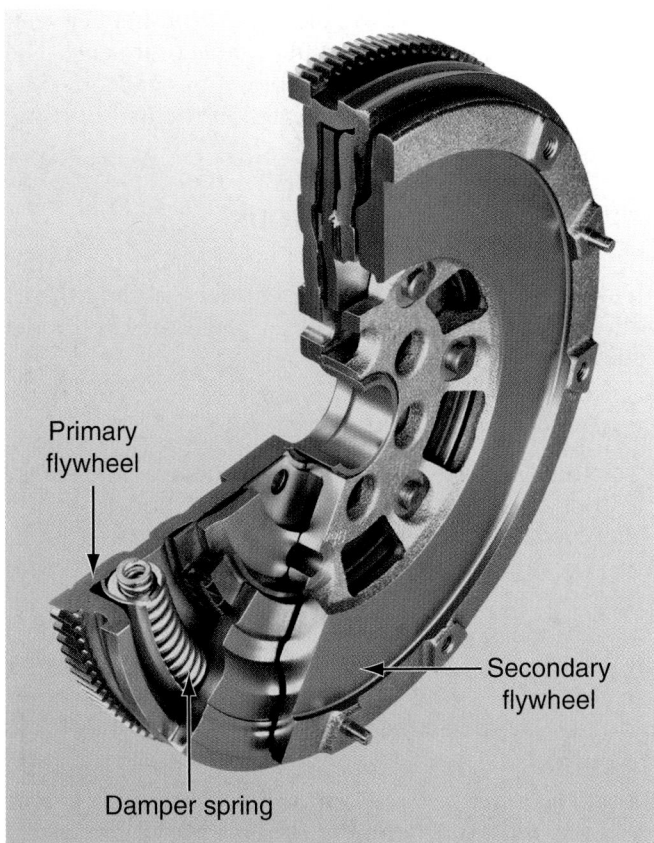

Figure 69.13 A dual mass flywheel redues vibration and noise. *(Courtesy of Luk Automotive Systems)*

Primary flywheel

Secondary flywheel

Damper spring

Figure 69.14 A pilot bearing and a pilot bushing. *(Courtesy of Tim Gilles)*

bearing, because a roller or two ball bearings support the front of the transmission shaft (**Figure 69.15**).

■ RELEASE BEARING

The release bearing, also called a *throwout bearing* (see Figure 69.6), allows the pressure plate release mechanism to operate as the crankshaft rotates. The release bearing slides on the front transmission bearing retainer, often called a quill (**Figure 69.16**). Conventional release bearings are used in rear-wheel-drive

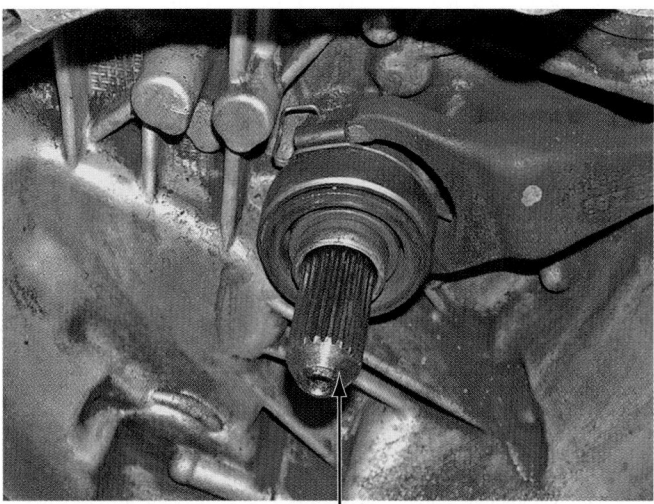

No pilot bearing needed

Figure 69.15 This front-wheel-drive transaxle does not use a pilot bearing. *(Courtesy of Tim Gilles)*

Clutch release lever

Front transmission bearing retainer (Quill)

Transmission input shaft

Release bearing

Figure 69.16 The release bearing slides on the transmission front bearing retainer. *(Courtesy of Tim Gilles)*

cars. A special kind of release bearing is used in some front-wheel-drive cars.

The release bearing is lubricated and then sealed at the factory. A *hub*, or *collar*, is sometimes part of the bearing. Other times, the hub is pressed-fit and replaceable.

Some release bearings are flat on their face and others are curved. This depends on whether the release levers are flat or curved. Curved release levers go against flat bearings, and vice versa.

Self-Centering Release Bearing. Self-centering release bearings are used on front-wheel-drive cars because they do not use a pilot bearing in the crankshaft.

Without a pilot bearing, the transmission input shaft might not be correctly aligned with the rear of the crankshaft. If there is misalignment during clutch application, the bearing can apply the clutch while it is not correctly aligned, resulting in noise.

Other Release Bearings. Some release bearings found on older European cars use a round carbon ring that slowly wears away with each use. There are also special release bearings found on cars that have pressure plates that pull to release.

CLUTCH FORK

The release bearing hub has a provision to attach it to the **clutch fork** (**Figure 69.17**). The clutch fork,

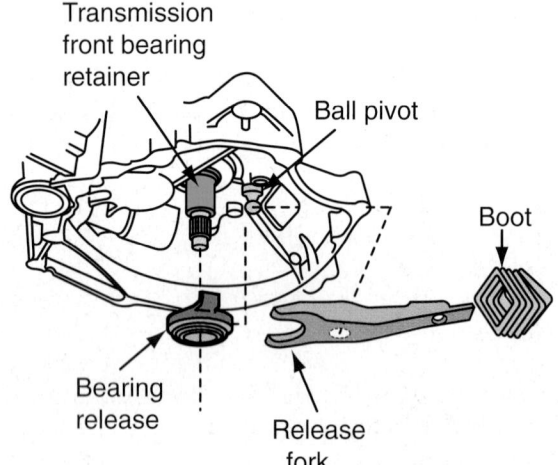

Figure 69.17 The clutch fork fits between the release bearing and the clutch linkage.

also called a *throwout lever* or *release arm,* fits between the release bearing and the clutch cable or linkage. The clutch fork has a pivot shaft (sometimes called a *cross shaft*) or a pivot ball or raised area in the bell housing off which it pivots.

CLUTCH RELEASE METHODS

The clutch pedal operates the clutch fork using either linkage, a cable, or hydraulic cylinders. A clutch start switch is included on the clutch pedal on late-model vehicles.

MECHANICAL LINKAGE

Mechanical linkage found on some older cars and trucks has rods and pivot arms that carry motion from the clutch pedal to the release fork (**Figure 69.18**). Clutch free play adjustment can be made on a threaded rod. An **overcenter assist spring** pulls the pedal upward during the first half of travel and toward the floor during the second half of travel. This makes releasing the clutch (applying the pedal) easier. The overcenter spring is the same concept used on hood springs.

NOTE: *To see how an overcenter spring works, carefully operate the clutch pedal with the linkage disconnected. The pedal will fall strongly toward the floor during the second half of travel. Keep your fingers out of the way.*

A drawback to mechanical linkage systems is that when an engine or transmission moves on its rubber mounts, clutch adjustment and amount of application changes.

CLUTCH CABLE

Some newer cars use a less expensive cable to operate the clutch (**Figure 69.19**). A cable is flexible so

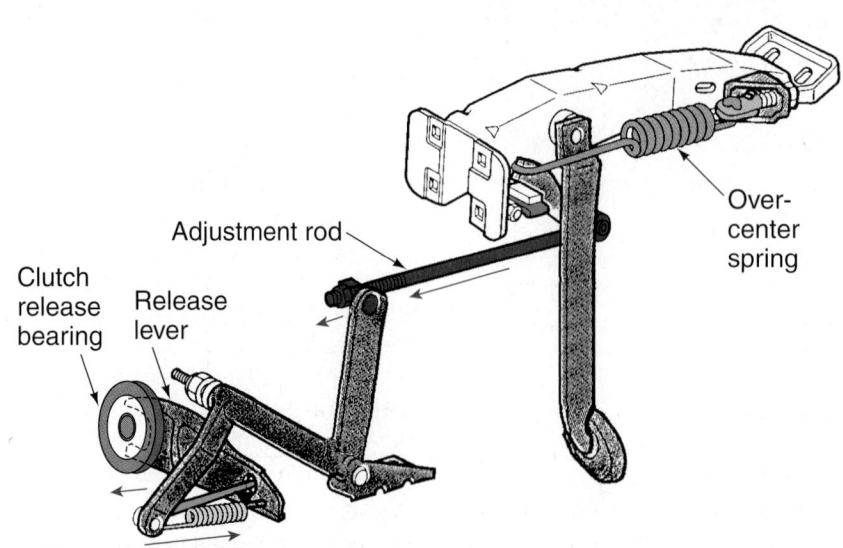

Figure 69.18 A mechanical clutch linkage.

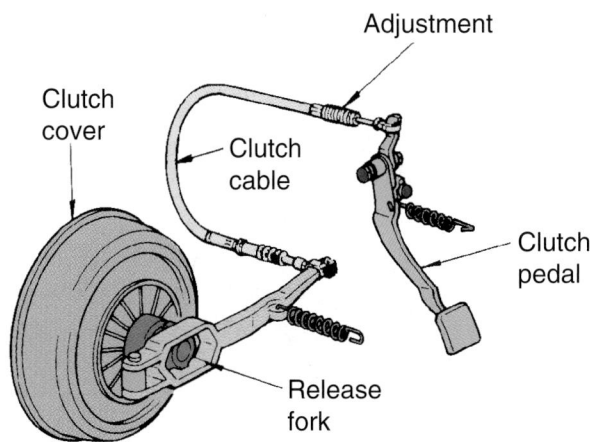

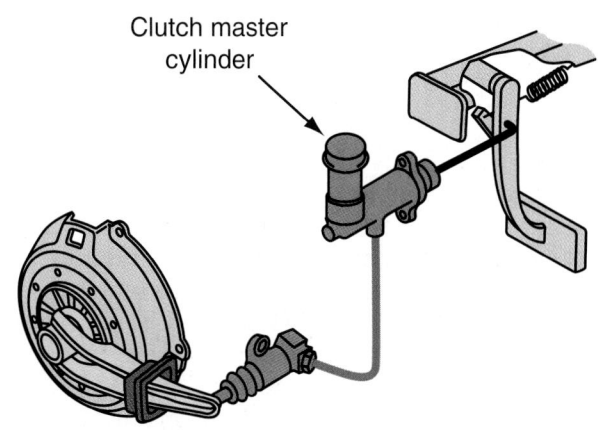

Figure 69.19 A cable-operated clutch release.

Figure 69.20 A hydraulically operated clutch.

its adjustment remains the same as the engine moves. This is especially important on four-cylinder, front-wheel-drive cars, in which the engine the moves quite a bit. A drawback to cables is that they develop friction and wear with repeated use.

Linkage can push on a clutch arm, while a cable can only pull on it. With a cable, the pivot point of the fork must be on the outside of the input shaft, away from the cable end. The amount of available adjustment to a cable system is less than with linkage.

■ HYDRAULIC CLUTCH OPERATION

Hydraulic clutches are found on many vehicles. Hydraulic operation of clutch linkage is similar to the way brakes operate. Liquids cannot be compressed, so they can transmit motion and increase force. Brake master cylinder operation is covered in detail in Chapter 57.

The input piston is located in the *master cylinder* (action cylinder) connected to the clutch pedal (**Figure 69.20**). It is located next to the brake master cylinder. The output piston is located in the reaction or actuator cylinder, commonly called a slave cylinder. It is attached to the release lever at the clutch. The two cylinders are attached hydraulically by tubing and hose, just like in the brake system. Hydraulic systems are popular on custom vehicles because adaptation of the clutch release mechanism is easy.

The only difference between a clutch master cylinder and a single brake master cylinder is that the brake master cylinder has a residual check valve if it is for drum brakes. A clutch master cylinder does not have one because the clutch would stay applied as if a foot were always resting on the pedal. This would result in failure of a standard release bearing.

Some newer hydraulic systems have the slave cylinder connected to the release bearing. This eliminates the need for a release fork or cross shaft.

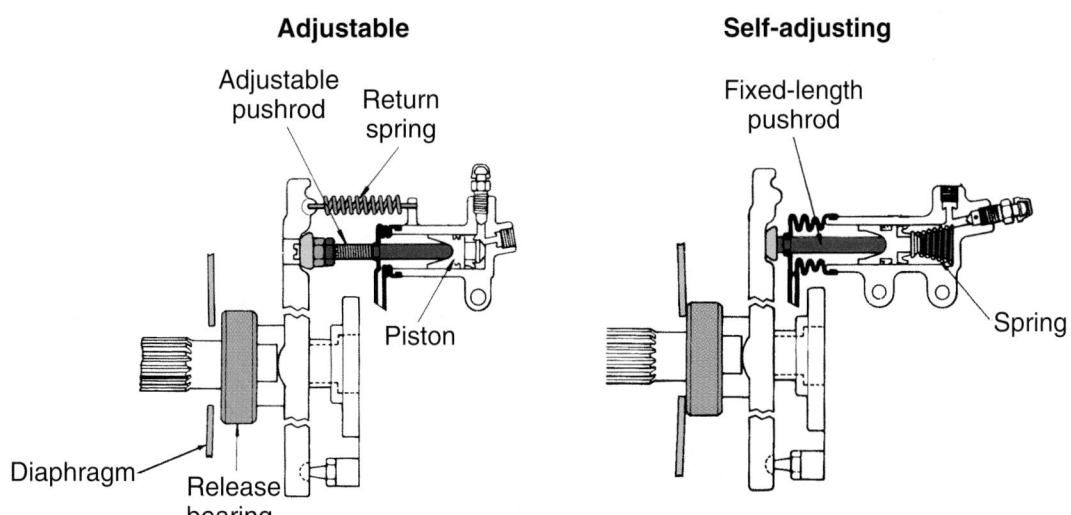

Figure 69.21 Comparison of adjustable and self-adjusting slave cylinders. Note the spring at the right side of the self-adjusting slave cylinder.

CLUTCH FREE PLAY

Often, the clutch linkage has a spring that pulls the release bearing away from the pressure plate after the clutch pedal is released. **Clutch free play**, or free travel, at the pedal is the result. Free travel is usually adjusted to about 1" at the pedal, which translates to about ⅛" at the release lever on the clutch.

Newer cars that have self-adjusting clutches maintain contact between the release levers and release bearing. **Figure 69.21** shows a clutch slave cylinder with a spring to keep the release bearing in contact with the clutch cover. This requires a different style of release bearing. Standard release bearings do not remain in constant contact with the clutch cover like the ones on self-adjusting systems do.

Many newer cars have self-adjusting cables. A spring-loaded toothed sector gear is pinned to the pedal arm (**Figure 69.22**). Whenever the clutch is released, a *pawl* is lifted and released. If there is any slack in the adjustment, the gear's spring causes it to rotate. This takes up the slack and the pawl slips into its new position on the gear to complete the adjustment.

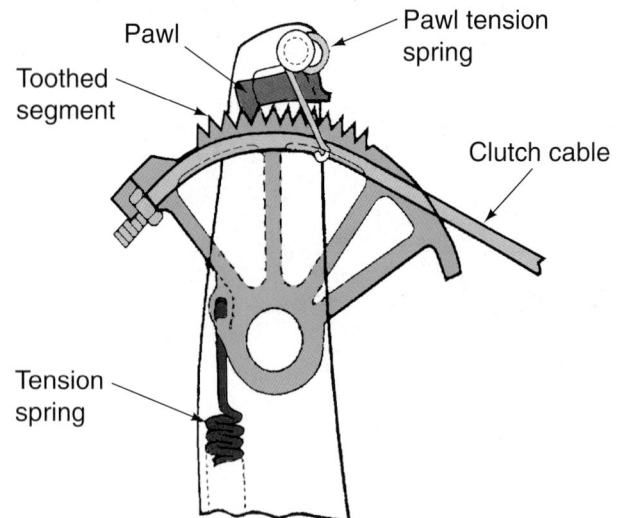

Figure 69.22 A clutch cable self-adjuster. (*Courtesy of Ford Motor Company*)

REVIEW QUESTIONS

1. Where do the clutch disc splines ride?
2. Clutch torsional dampers are either rubber or _____ springs.
3. What is the name of the part of the clutch disc that is between the two facings and lets them compress?
4. What is another name for the pressure plate assembly?
5. When the clutch disc is totally worn out, about how much more torque carrying capacity is left in the clutch than the engine can deliver?
6. What is another name for clutch release levers?
7. The front of the transmission input shaft is supported in the crankshaft by a _____ bearing or bushing.
8. What is another name for a release bearing?
9. Which clutch release system changes its adjustment as the engine moves on its mounts: linkage, cable, or hydraulic?
10. What is the name of the hydraulic cylinder to which the clutch master cylinder sends fluid pressure?

ASE-STYLE REVIEW QUESTIONS

1. Which of the following springs is/are not used in clutches?
   a. Coil springs.
   b. Diaphragm springs.
   c. Torsion bar springs.
   d. All of the above are used in clutches.
2. Which of the following is/are true?
   a. Depressing the clutch pedal releases the clutch.
   b. Releasing the pedal applies the clutch.
   c. Both A and B.
   d. Neither A nor B.
3. Technician A says that diaphragm clutches are easier to apply than coil spring clutches. Technician B says that coil spring clutches are more common for heavy duty uses. Who is right?
   a. Technician A
   b. Technician B
   c. Both A and B
   d. Neither A nor B

4. Which of the following must have a pilot bearing or bushing to support the transmission input shaft?

   a. Any transmission without support built into the transmission case and input shaft

   b. Front-wheel-drive transaxles

   c. Both A and B

   d. Neither A nor B

5. Technician A says that grooves in the clutch disc facing help the clutch to release. Technician B says that semi-ceramic clutch disks are found in many passenger cars. Who is right?

   a. Technician A   **c.** Both A and B

   b. Technician B   **d.** Neither A nor B

6. Two technicians are discussing hydraulic clutch operation. Technician A says that a clutch master cylinder has a residual check valve. Technician B says that the output piston is located in the slave cylinder. Who is right?

   a. Technician A   **c.** Both A and B

   b. Technician B   **d.** Neither A nor B

7. Which of the following clutch release mechanisms can be self-adjusting?

   a. A clutch cable

   b. A hydraulic clutch

   c. Both A and B

   d. Neither A nor B

8. Technician A says that some clutch designs maintain constant contact between the release levers and release bearing. Technician B says that some clutch designs keep the release bearing away from the release levers except during shifting. Who is right?

   a. Technician A   **c.** Both A and B

   b. Technician B   **d.** Neither A nor B

9. A clutch can be released

   a. By linkage   **c.** Hydraulically

   b. By a cable   **d.** All of the above

10. Technician A says that liquids can be compressed. Technician B says that liquids can be used to increase force. Who is right?

   a. Technician A   **c.** Both A and B

   b. Technician B   **d.** Neither A nor B

70

Clutch Diagnosis and Service

■ OBJECTIVES

Upon completion of this chapter, you should be able to:

✔ Diagnose clutch problems before disassembly.

✔ Adjust a clutch.

✔ Install a replacement clutch.

✔ Inspect worn or damaged clutch parts and determine the cause of the problem.

■ KEY TERMS

dragging clutch

chattering clutch

grabbing clutch

gravity bleeding

riding the clutch

■ INTRODUCTION

Early automobiles were all produced with manually shifted transmissions, which require a clutch for engagement and disengagement from the engine. Some new model cars and light trucks have manual transmissions that utilize a clutch as well. Because the clutch wears with use, it is not unusual for a car to have at least one new clutch installed during its lifetime. Clutch work is not that difficult to learn and can be a lucrative service area for technicians. As in other automotive service areas, the important thing a true professional learns is to be able to diagnose problems. Problems are diagnosed before disassembly and after disassembly.

SAFETY NOTE Many clutches contain asbestos. Breathing asbestos can be dangerous to your health. Raw asbestos fibers are very small in diameter and resemble long, curly hairs. Your lungs cannot rid themselves of these fibers as they normally do with other dusts. Before working on a clutch assembly, wash it with a brake parts cleaner or vacuum it with a HEPA vacuum.

■ DIAGNOSIS OF CLUTCH PROBLEMS

A clutch will normally last in excess of 100,000 miles. City driving in heavy traffic reduces clutch life. Extra heavy-duty uses can actually damage a clutch before its normal life expectancy is reached. Such abuse includes excessive slipping of the clutch in stop-and-go hilly traffic. Pulling a boat up a launching ramp or carrying a heavy load also represent potential abuse. When pulling a heavy load, the objective is to have the clutch fully engaged (pedal up) as soon as possible. Pulling heavy loads calls for a transmission with a much lower gear ratio in first gear.

Diagnosing a problem before disassembly is very important. You must first remove a transmission or transaxle before you can remove a clutch. Be sure to verify the problem. The labor in a clutch job is time consuming and costly.

■ CLUTCH NOISES

Listen for unusual noises and try to isolate them to the clutch. Sometimes transmission noise can be mistaken for clutch noise. When the noise only happens when the pedal is moved, try moving the pedal while the engine is off.

With the engine running, use your knowledge of clutch operation while determining the cause.

■ A noise when the clutch is first engaged (pedal let out and disc being wedged against the flywheel and pressure plate) is due to a problem with the friction lining (**Figure 70.1**).

■ The release bearing on cars without self-adjusting features is not designed to contact the clutch

Figure 70.1 The disc and pressure plate will wear into one another when the friction material is worn too thin. The rivets will be the first thing to touch. *(Courtesy of LuK Automotive Systems)*

release levers *except* when the pedal is applied. If the noise only happens when your foot is resting lightly on the pedal, the release bearing is probably at fault.

- The pilot bearing is pressed into the rear of the crankshaft. For it to make noise, the input shaft must be held still. Put the transmission in gear, depress the clutch pedal, and start the engine. The input shaft will be held still while the crankshaft and pilot bearing rotate.

◼ TRANSMISSION NOISE

Sometimes noise occurs only when the engine is idling in neutral with the clutch engaged (pedal all the way up). The transmission input shaft is turning at this time. The front transmission bearing on the input shaft and the mainshaft pilot on the back of the input shaft are the only bearings that are rotating. One of these bearings is probably the source of the noise.

◼ PEDAL PROBLEMS

Some diagnosis can be done on the basis of pedal feel:
- A pulsating pedal is due to something internal in the clutch. The clutch will probably have to be disassembled, but looseness or misalignment elsewhere could also be the culprit.
- Sometimes a pedal is hard to depress. A binding linkage or cable could be the problem.

Further exploration of these problems is covered in the next section. Problems are often accompanied by photos of the resulting worn parts.

◼ SLIPPING CLUTCH

A clutch is tested for slipping by putting it in the highest gear range (this would be fifth gear on a five-speed). Set the parking brake firmly and attempt to slip the clutch as if trying to make the vehicle move. If the clutch is slipping, engine rpm will rise. If the clutch holds, the engine will die.

NOTE: *This is a test of short duration in order to avoid damage to clutch components.*

There are several possible reasons for a clutch to slip.

Partial Engagement

Partial engagement is a common cause of a slipping clutch. In a manually adjusted system, as a clutch disc wears, free play becomes *less*. Because of the multiplication effect of the clutch linkage parts, a small amount of wear on the clutch disc can result in a total loss of original free travel. If the wear progresses beyond this point, it will be just as if the driver is always "riding the clutch" (pushing slightly on the clutch pedal). Slipping is the result. This ruins the clutch disc (if it is not worn out already).

When a clutch slips, temperatures inside the clutch housing can reach 500°F in just a few seconds. Burn marks on the pressure plate are one of the results.

Partial Disengagement

One cause of partial disengagement is a problem with a hydraulic clutch release system. Look for a low fluid level first, then check for a failed master or slave cylinder. Replacement of the leaking part fixes the problem. A leaking clutch or brake master cylinder might be leaking into the passenger compartment. Check the carpet or floormat for evidence of leakage.

There might be no evidence of leakage because a master cylinder might have an *internal* leak. Test for an internal leak by pumping the clutch pedal and holding it to the floor while starting the engine. If the clutch slowly begins to engage, there is an internal leak.

Air in the lines can cause a spongy pedal on hydraulically released cars. Bleed the system to solve this problem.

◼ DRAGGING CLUTCH

When the clutch *drags,* it does not release properly. The disc stays attached to the flywheel. Too much pedal free travel can result some or all of the time. The transmission is often difficult to shift between gears or to put in low or reverse after starting.

To test for a **dragging clutch**:
- Start the engine and run it until the engine and transmission are at normal operating temperatures.
- With the engine at idle and the transmission in neutral, depress the clutch pedal to the floor.
- Wait 10 seconds.
- Shift the transmission into reverse.

Ten seconds should be enough time for the clutch to come to a complete stop, allowing a shift into reverse without gear clashing. If there is a gear clash, the clutch is dragging (not releasing completely). Most cars will take only three or four seconds before allowing a smooth, quiet shift into reverse.

A dragging clutch can be due to a bent disc or clutch bearing retainer or to a rusted input shaft. Discs are checked at the factory for runout. They can be damaged in shipping. A more likely cause of a bent clutch disc is that the transmission was allowed to hang on the disc during installation.

Figure 70.2 shows a disc that failed after the transmission sagged during installation. The symptom of the problem was that it was impossible to disengage the clutch. The thin piece of sheet metal that makes up the center portion of the disc is called the *retainer plate*. This thin sheet metal is very strong in a radial direction (perpendicular to the input shaft). But sideways, the disc is very lightweight and flimsy. It can be easily bent, resulting in later damage.

Sometimes, there is too much friction between the splines in the clutch hub and input shaft. The splines can be rusty because they were never lubricated. Rust can also result from a leaky core plug on the back of the engine, driving through stream beds, or when the transmission fills with water through its vent hole when driving during a flood.

Clutch linkage can become stiff. This is an easy repair. Simply lube the linkage and free it up. Any worn or bent linkage part should be replaced or straightened.

OILY CLUTCH FACINGS

Oil leaking onto the clutch disc can result in clutch slipping. The disc will often overheat from the slipping, resulting in a burned disc. The disc and pressure plate must be replaced and the other parts thoroughly cleaned. If any oil remains in the pores of the metal surfaces of the flywheel, the clutch will probably begin to chatter soon after the disc is replaced.

The source of the oil leak must be determined and repaired or the problem will return. Oil leakage that is evident on the front side of the flywheel is from either a crankshaft rear seal or external engine leaks such as a valve cover or oil pan gasket. Oil on the rear side of the flywheel is often due to a bad front transmission seal or gasket or an overfilled transmission.

DAMAGED FRICTION SURFACES

Damage to friction surfaces of the clutch cover or flywheel can cause a clutch to slip or give inconsistent performance. When a clutch slips, heat is generated. A flywheel or pressure plate can be warped from the heat. This can cause uneven or incomplete contact with the disc. The flywheel, pressure plate, or disc can all become glazed from excess heat. A poor friction coefficient between the disc and metal friction surfaces results. The flywheel is commonly surfaced to ensure proper operation as part of the clutch job.

CHATTERING OR GRABBING CLUTCH

A **chattering clutch** occurs when the pedal shakes as the clutch is engaged (pedal raised from the floor). **Figure 70.3** shows a pressure plate with marks on it

Figure 70.2 This disc failed because it was bent during installation. *(Courtesy of LuK Automotive Systems)*

Figure 70.3 The marks on this pressure plate are from chattering. *(Courtesy of LuK Automotive Systems)*

that resulted from clutch chatter. A **grabbing clutch** occurs when the friction disc does not slip normally but grabs all at once. There are several common causes for these conditions. One is oil on the friction lining. This calls for replacing the disc and repairing the oil leak that caused the problem.

Worn or broken motor mounts often cause clutch chatter. Worn motor mounts are especially noticeable on four-cylinder engines. When there are only four cylinders, each firing impulse is more clearly defined. These engines will vibrate noticeably when resting at idle at a signal light. Coupled with clutch chattering, this points to mounts that, while not broken, might have become too soft. Oil leaking from a longtime leak onto a motor mount will soften the rubber too.

A worn or bent clutch fork can cause clutch chatter because it does not allow the release bearing to contact the pressure plate evenly (**Figure 70.4**). The clutch does not fully disengage when stepping on the pedal. This can also cause hard shifting, especially into first or reverse.

Uneven release fingers will cause the same problems as a worn or bent clutch fork. The force on the clutch disc will be diminished when the clutch pedal is released (clutch engaged). One release lever remaining in contact with the release bearing will diminish clutch clamping force by one-third. This problem can be caused by shipping damage, improper installation, uneven bolting to the flywheel, or a poor machine job on the flywheel. Be sure to resurface the flywheel when replacing the pressure plate.

Misalignment of parts or a bent friction disc can also cause uneven clutch operation. Misaligned parts can happen when pieces of dirt or debris become trapped between the transmission and its housing or the housing and the engine block. A bent friction disc or even a bent transmission input shaft can happen during a careless installation.

CLUTCH SERVICE

Clutch Adjustment

Some clutches require adjustment. As the disc wears, it becomes thinner. This results in the release levers moving *closer* to the release bearing, which means less clearance (**Figure 70.5**). If the disc wears too far, the release bearing would release the clutch disc at all times. Adjust the clearance to provide about ½"–1" of free travel at the pedal.

When a car has a cable, adjustment is made by adjusting the length of the outside of the cable housing (**Figure 70.6**). Adjusting a clutch linkage requires a similar adjustment to a threaded rod. Hydraulic systems are often self-adjusting or self-compensating and have a full-time contact release bearing. On the systems that require adjustment, there is an adjustment on the slave cylinder pushrod.

SERVICING HYDRAULIC COMPONENTS

Leaks in the hydraulic system can be in hoses or lines but usually result from aging of internal rubber sealing

Figure 70.4 The release bearing was not contacting this diaphragm spring evenly. *(Courtesy of LuK Automotive Systems)*

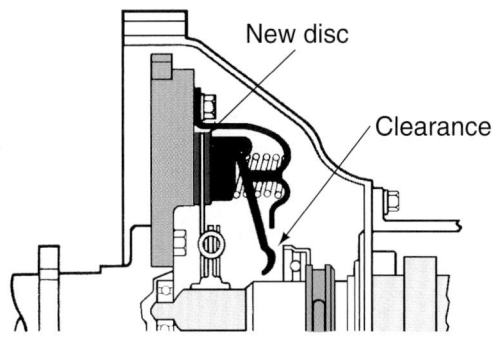
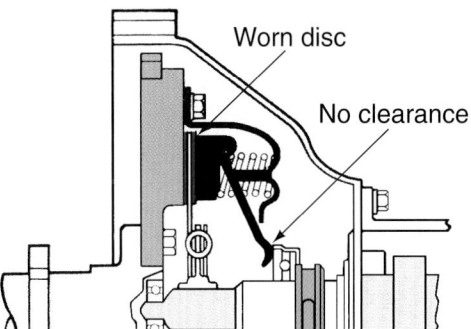

Figure 70.5 As the clutch disc wears, clearance becomes less. *(Courtesy of Federal-Mogul Corporation)*

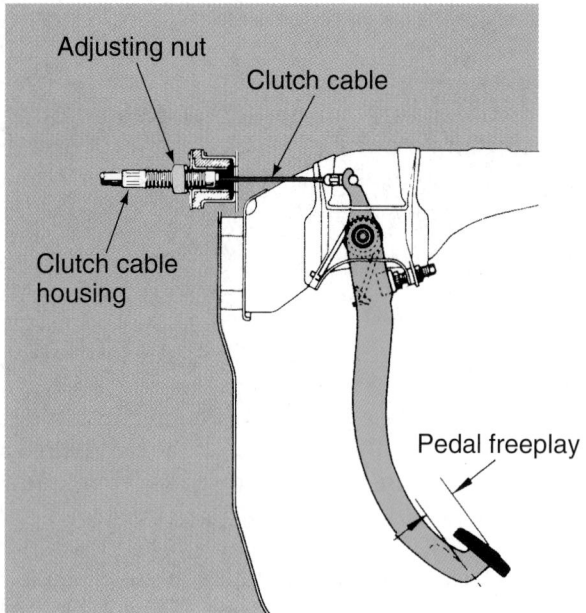

Figure 70.6 Cable adjustment.

parts. The system has a master cylinder and a slave cylinder, either of which can develop leaks. The master cylinder (**Figure 70.7**) has a primary cup that can cause internal leaks. The secondary cup can cause external leaks onto the floor in the passenger compartment.

The slave cylinder has a dust boot and one or two sealing rings that fluid pressure from the master cylinder acts upon (**Figure 70.8**). A slave cylinder will usually leak fluid onto the ground when it fails. Sometimes it leaks slowly, so adding fluid can put off the repair.

Bleeding the Hydraulic System

Air is removed from the hydraulic system by using a **gravity bleeding** procedure:

■ Open the bleeder screw on the slave cylinder.

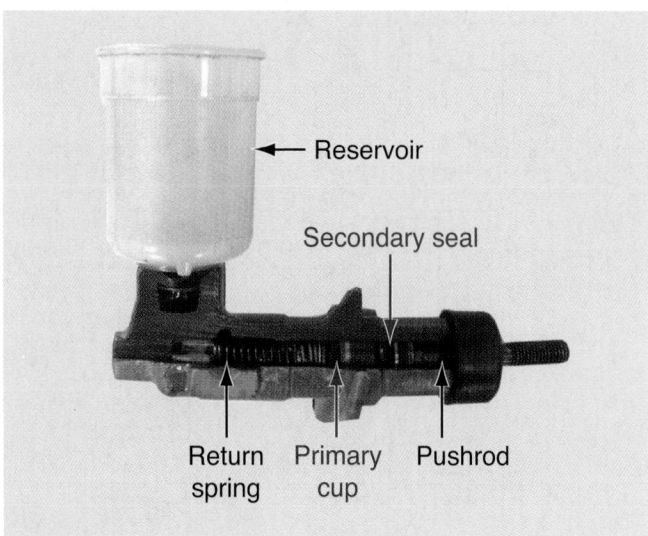

Figure 70.7 A clutch master cylinder has two rubber parts that can fail. (*Courtesy of Tim Gilles*)

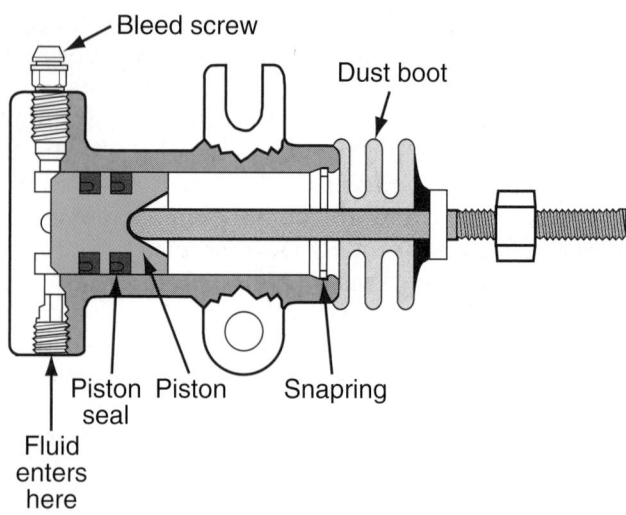

Figure 70.8 Parts of a slave cylinder. Replaceable rubber parts are highlighted.

■ Remove the lid from the master cylinder reservoir and let the cylinder drain. Refill it with fresh fluid.
■ Occasionally close the bleeder screw and refill the master cylinder with fresh fluid.
■ Repeat until all of the air is out of the system.
■ Sometimes it is necessary to have an assistant push on the pedal while the bleeder screw is open (**Figure 70.9**) and let up on the pedal when the screw is closed.

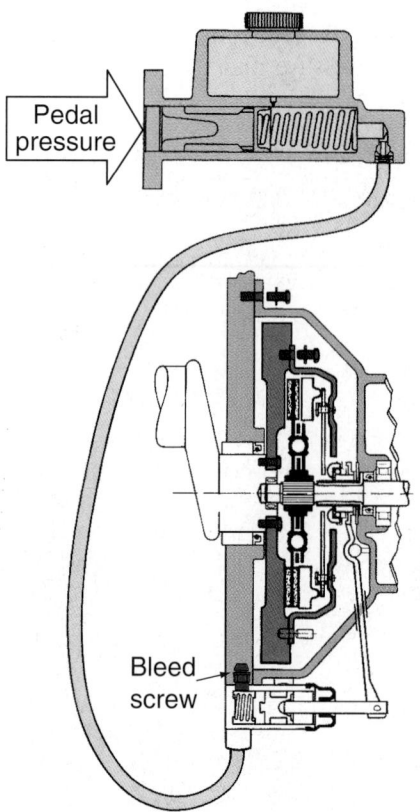

Figure 70.9 Fluid is moved from the master cylinder to the slave cylinder and exits from the bleed screw. (*Courtesy of LuK Automotive Systems*)

The system can also be bled using a pressure bleeder. Use fender covers. Be careful that fluid does not spill on paint.

Some hydraulic systems can be difficult to bleed. Possible solutions to this are to use the gravity or vacuum bleeding procedures described in Chapter 58. A newer type of bleeder, called a Phoenix Injector™, is also popular for bleeding difficult systems. Fluid is injected toward the master cylinder from the bleed screw on the slave cylinder.

■ CLUTCH REPLACEMENT

When a clutch job is done, it is common to replace the disc, release bearing, clutch cover, and pilot bearing. These often come in a clutch parts kit. The pressure plate and disc are often damaged and usually require replacement. The release bearing and pilot bushing are usually replaced on a preventive maintenance basis. With all of the parts replaced, a comeback from the failure of an old part two months later will not occur. If a new part fails, it is customary for the supplier to pay labor on the warranty replacement.

One advantage to buying an entire kit is that the shop will get a price break on volume, making the kit less expensive than its individual parts. Another advantage is that the shop can make a firm price quote to the customer.

■ REMOVE THE TRANSMISSION OR TRANSAXLE

Remove the transmission or transaxle. Follow the instructions in the service information.

CAUTION Before attempting to remove a transmission or transaxle, disconnect the battery ground cable.

On rear-wheel-drive vehicles, mark and remove the drive shaft, release mechanism, speedometer drive gear, crossmember, and transmission. On front-wheel-drive vehicles, remove the halfshafts (drive axles) and transaxle. Sometimes it is necessary to remove the engine. A transmission jack is used to lift the transmission in and out of the vehicle. A more complete description of transmission removal is given in Chapter 72. Transaxle removal is covered in Chapter 78.

■ CLUTCH REMOVAL

The parts of the clutch assembly that will be removed are shown in Figure 69.2. Unbolt the clutch housing from the engine. Some engines have integral clutch housings (part of the transmission housing). Note the dowels

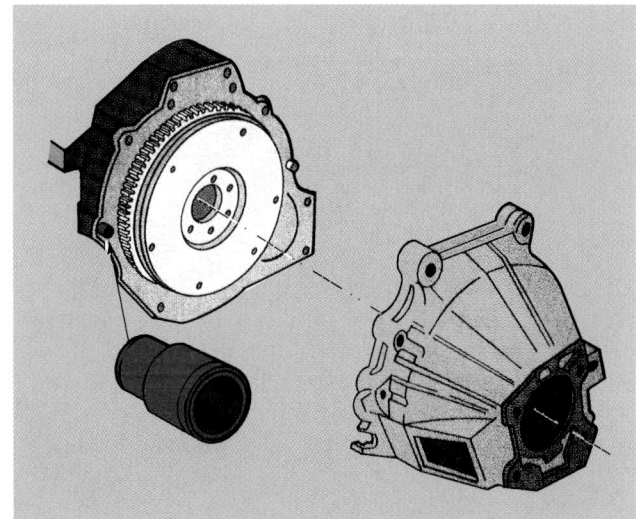

Figure 70.10 Dowels align the clutch housing with the engine. *(Courtesy of Federal-Mogul Corporation)*

that align the housing to the engine (**Figure 70.10**). Be sure that they are not lost when the transmission is removed.

SHOP TIP Pay close attention to the positioning of components during disassembly, and keep all of the parts until the job is completed.

Mark Parts Before Disassembly

It always makes good sense to mark clutch and drive shaft parts before disassembling them. Clutch parts are balanced at the factory while on the engine. If the clutch cover is to be used again, it should be replaced in its original position. Make a centerpunch mark on the flywheel and on the clutch.

SAFETY NOTE Use a HEPA vacuum to clean any asbestos dust from the assembly before disassembling it. Asbestos fibers can pass through a regular vacuum bag.

Use an aligning arbor to hold the disc while removing the clutch cover (**Figure 70.11**). This will prevent it from falling accidentally. Unbolt the pressure plate bolts evenly to prevent bending the plate when the last bolt is removed. Carefully remove the assembly. Be careful not to drop the parts; they are heavy.

Inspect the clutch parts for damage. You will want to determine the cause before replacing the parts. A

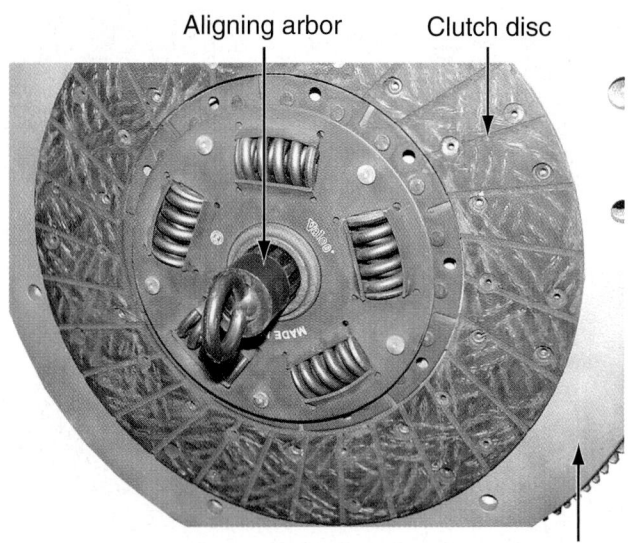

Aligning arbor   Clutch disc

Flywheel

Figure 70.11 Use an aligning arbor to hold the disc while removing the clutch cover. *(Courtesy of Tim Gilles)*

repeat failure must be avoided. The disc shown in **Figure 70.12** was damaged because the engine was pulling a load at too low an engine rpm. This situation, called *engine lugging,* is caused by driver error.

Clean Parts

Use hot soapy water, brake parts cleaner, or alcohol to clean off any grease from clutch parts. Metal surfaces can be deglazed with sandpaper.

Figure 70.12 This clutch hub was damaged by a driver lugging the engine. *(Courtesy of LuK Automotive Systems)*

Figure 70.13 A flywheel damaged from heat and slippage. *(Courtesy of Tim Gilles)*

CAUTION ■ Do not wash the release bearing in solvent. It is packed with grease and will be ruined if solvent enters it.
■ Do not wash clutch parts in mineral-based Stoddard solvent. Solvent contains oil that will be absorbed into the friction surfaces of the metal.
■ Do not blow off parts with compressed air because of the danger of blowing asbestos around.

■ FLYWHEEL REMOVAL

Before removing the flywheel, mark it to make realignment with the crankshaft easier. Usually the bolt pattern on the flywheel is such that the flywheel will only bolt on one way.

Inspect the flywheel for damage. If the surface is flat, a sanding disc can be used to deglaze the surface. Heat from a slipping clutch causes cracks and warpage (**Figure 70.13**). A flywheel with this type of damage can be resurfaced by a machine shop. This is necessary to validate most clutch disc warranties.

■ FLYWHEEL STARTER RING GEAR REPLACEMENT

Check the starter ring gear teeth to see that they are not damaged. Standard transmission flywheels have a pressed-fit ring gear. Sometimes a ring gear will be damaged due to starter motor problems (**Figure 70.14**). It can be removed and replaced with a new one. To remove a worn ring gear, drill a hole in it between two teeth and break it with a chisel.

To install the new ring gear, it must first be heated to expand it. It must be heated evenly all around its circumference. If it gets hotter on one side than on the

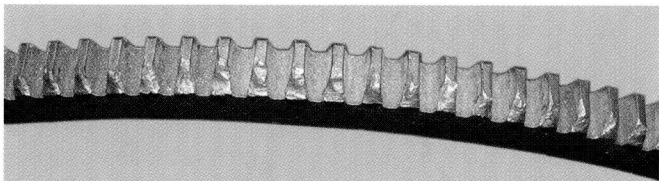

Figure 70.14 Damaged ring gear teeth due to starter motor problems. *(Courtesy of Tim Gilles)*

other, it will become smaller in diameter on the cold side. Do not heat it to more than 400°F.

The chamfered side of the teeth should be aimed to the same side of the flywheel as the old teeth were.

■ FLYWHEEL INSTALLATION

Use the correct quality screws on the flywheel and the clutch cover. These screws are usually shoulder screws (screws that do not have thread all the way to the top). When lock washers are used, they are usually serrated (especially on the flywheel) (**Figure 70.15**).

Figure 70.15 Flywheel lock washers are usually serrated. *(Courtesy of Tim Gilles)*

When installing the flywheel, use the correct grade of screw (usually SAE Grade 8 or metric property class 10.9 or higher, see Chapter 7). Torque the screws to their correct specification. This is especially important on crankshafts that have a full round rear seal. The sealing area can be distorted if not evenly torqued.

Inspect the rear main engine seal and the front transmission seal for leaks while the flywheel is removed.

■ INSPECT NEW PARTS

Before attempting to replace parts, always compare the new ones with the old ones. If the clutch was disassembled before the new parts were ordered, take the old parts to the parts store with you for comparison. There are often different sizes of release bearings and different diameters of clutches available within the same make of car.

■ PILOT BUSHING SERVICE

The pilot bushing or bearing in the crankshaft is often replaced as part of a clutch job. When the bearing is not to be replaced, inspect it by pushing on it while attempting to rotate it (**Figure 70.16**). It should spin smoothly without binding. **Figure 70.17** shows a badly worn pilot bushing.

If the bushing must be removed, use a puller. A slide hammer (**Figure 70.18**) or a bridge yoke bearing puller can be used (**Figure 70.19**). A screw is turned to expand a puller head inside of the pilot bushing.

> **SHOP TIP** A trick that sometimes works is to pack the cavity behind the pilot bearing with grease. Then, insert the largest bolt that will fit into the ID of the bearing. Pound on the bolt to force out the bearing.

Check the fit of the new bearing or bushing on the transmission input shaft before installing it in the crankshaft. The new bushing or bearing is installed

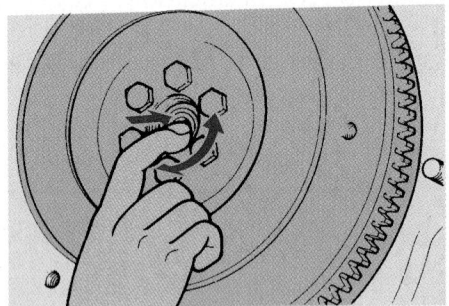

Figure 70.16 Checking the pilot bearing.

Figure 70.17 A badly worn pilot bushing. *(Courtesy of Tim Gilles)*

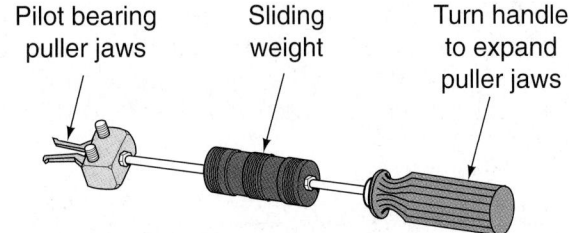

Pilot bearing puller jaws   Sliding weight   Turn handle to expand puller jaws

Figure 70.18 Turning the handle on the slide hammer expands the jaws of the puller. *(Courtesy of Kent-Moore Division, SPX Corporation)*

Figure 70.19 Sometimes the puller must be disassembled for the jaws to fit into the part. After reassembling the puller, expand its jaws against the bearing or bushing and turn the nut to pull it from the bore in the crankshaft. *(Courtesy of Tim Gilles)*

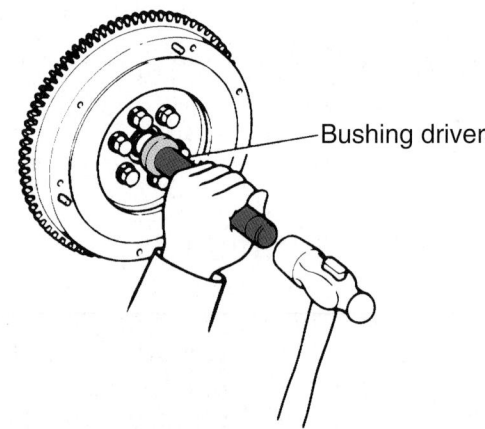

Bushing driver

Figure 70.20 Use a bushing driver to install the new bearing or bushing.

with a bushing driver (**Figure 70.20**) or an old input shaft with a washer installed over the end of it.

■ CLUTCH DISC SERVICE

The clutch disc is normally replaced any time the clutch is apart. A disc is relatively inexpensive and is the one part that must wear with use. If the pressure plate and flywheel have not been overheated, replacing the disc will often result in like-new clutch performance.

■ CLUTCH COVER/PRESSURE PLATE SERVICE

A clutch cover assembly is usually replaced when a disc is. If it is unworn, it may be reused. If you choose to reuse a clutch cover that appears to be in good

Figure 70.21 Check the pressure plate for warpage.

condition, inspect the pressure plate to see that it is flat (**Figure 70.21**).

■ CLUTCH INSTALLATION

Before installing a clutch disc, be sure to clean your hands. When handling the disc, hold it on its edges like you would hold a photograph. Clutch discs often get a small amount of grease on them from careless handling.

A thin film of grease or anti-seize compound is applied to the input shaft splines. If too much grease is applied, it can be thrown outward until it is trapped on the friction lining (**Figure 70.22**). Even this small amount of grease on the friction lining results in clutch chatter.

Figure 70.22 Excess grease on the input shaft was thrown from the center of this clutch hub to the friction material. *(Courtesy of LuK Automotive Systems)*

Figure 70.23 This clutch disc was installed backwards. Notice the marks on the flywheel bolts and clutch hub. *(Courtesy of Tim Gilles)*

Be careful to install the disc in the right direction. There should be a marking on it that says *"flywheel side."* If in doubt, check the service information. **Figure 70.23** shows a clutch disc that was installed backwards. The flywheel screws rubbed against the clutch disc, and the transmission had to be removed to correct the problem.

Attach the disc and clutch cover to the flywheel. Align the punch marks on the flywheel and clutch cover if the parts are being reused.

Using a clutch-aligning tool or an old transmission input shaft, align the clutch disc using the pilot. Injection molded plastic clutch pilots are available for each make of vehicle (see Figure 70.11). There are different sizes of splines and pilot ends to match the different sizes of clutch pilots available. One company makes a kit that includes 35 different sized pilots. A universal aligning tool is also available. Check the pilot shaft to see that it fits the pilot and the disc splines.

With the pilot holding the disc, tighten all of the clutch cover screws finger tight. To avoid any chance of warping the pressure plate, tighten the screws in a crisscross pattern. Tighten each one no more than one-half turn at a time. Then turn the next one one-half turn. After all of the screws are tight, torque them in gradual increments in a star pattern to specifications (**Figure 70.24**).

SHOP TIP If you do not have a pilot shaft and the clutch housing is separate from the transmission, a helper can depress the clutch pedal to activate the release bearing. This will leave the clutch disc loose so the transmission can be moved around until the pilot can be aligned and slipped into the bearing or bushing.

Figure 70.24 Torque in a star pattern. *(Courtesy of Tim Gilles)*

Sometimes, technicians are tempted to try to lift a manual transmission into place without the aid of a jack. While this is often a necessity, especially when a job is being done at home, the weight of the transmission should never be allowed to hang on the clutch disc. This can easily bend the clutch disc, ruining the job (**Figure 70.25**).

▬ RELEASE BEARING SERVICE

The release bearing sometimes fails. This happens often on clearance-type release bearings when the driver has a habit of **riding the clutch** (resting a foot on the clutch pedal when driving). With the engine running and the transmission in neutral, if resting your foot on the pedal produces a noise the release bearing is probably bad.

A release bearing is usually replaced on a preventive maintenance basis during a clutch job. If the rest of the clutch parts are not being replaced, check the bearing for roughness or excessive looseness.

Sometimes when a new release bearing is purchased, the collar (hub) is reused. The fit between the bearing and collar is snug but not very tight. Tap it off with a hammer or use a press. Install the collar in the new bearing in the same manner (**Figure 70.26**).

Inspect the end of the input shaft. The splines must be clean, unworn, and free of rust. Inspect the pilot area.

Check the *front bearing retainer,* often called a quill. It fits around the outside of the transmission input shaft (see Figure 69.17). It must be centered exactly and be smooth and undamaged, with no dents or worn spots. If free movement of the release bearing is prevented, the result will be clutch grab or chatter. If there are signs of off-center wear on the clutch cover release levers or diaphragm fingertips, the release bearing is misaligned (see Figure 70.4).

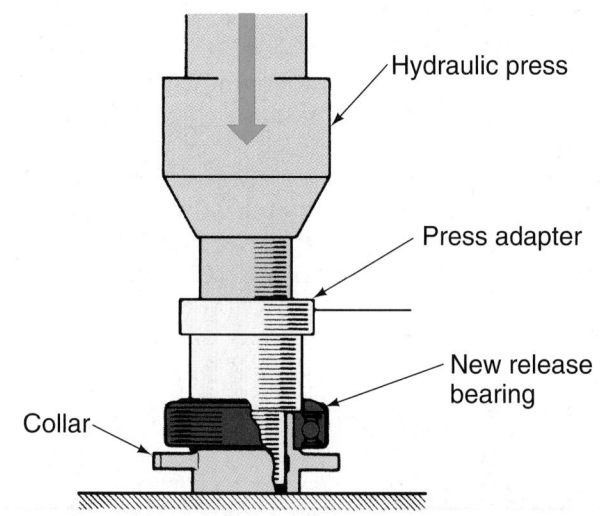

Figure 70.26 Installing a used release bearing collar in a new bearing.

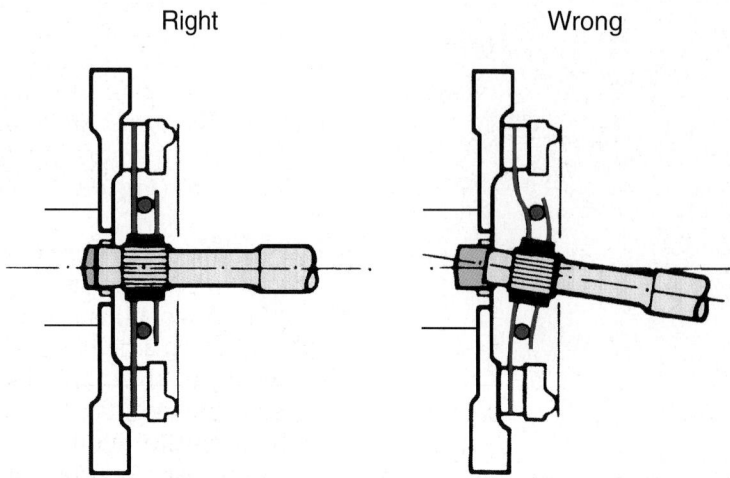

Figure 70.25 Allowing the transmission to hang on the clutch disc can bend the disc. *(Courtesy of Luk Automotive Systems)*

NOTE: *Be sure that the oil return hole on the front bearing retainer is correctly lined up with the hole in the transmission case. Otherwise, the oil that lubricates the front bearing will not be able to return to the transmission case. Clutch failure will be the likely result when the oil runs out of the input shaft and onto the clutch disc.*

Lubricate Release Mechanism Contacts

Various parts of the release mechanism have sliding contacts that require application of a small amount of grease (**Figure 70.27**). Install the release bearing. Be sure that the clutch fork is not damaged. Locations of parts that attach to the clutch fork are shown in **Figure 70.28**. Attach the clutch fork to the release bearing and clutch housing (**Figure 70.29**).

■ CLUTCH HOUSING INSTALLATION

Be sure to install any sheet metal pieces that might fit between the engine block and the clutch housing (**Figure 70.30**). These are commonly left out by mistake. Install the clutch housing, or the entire transmission if there is no separate clutch housing. Be sure all of the dowels are in place. Missing dowels can result in misalignment problems (see Figure 70.10).

Be careful that nothing becomes accidentally trapped between the surfaces of the clutch housing

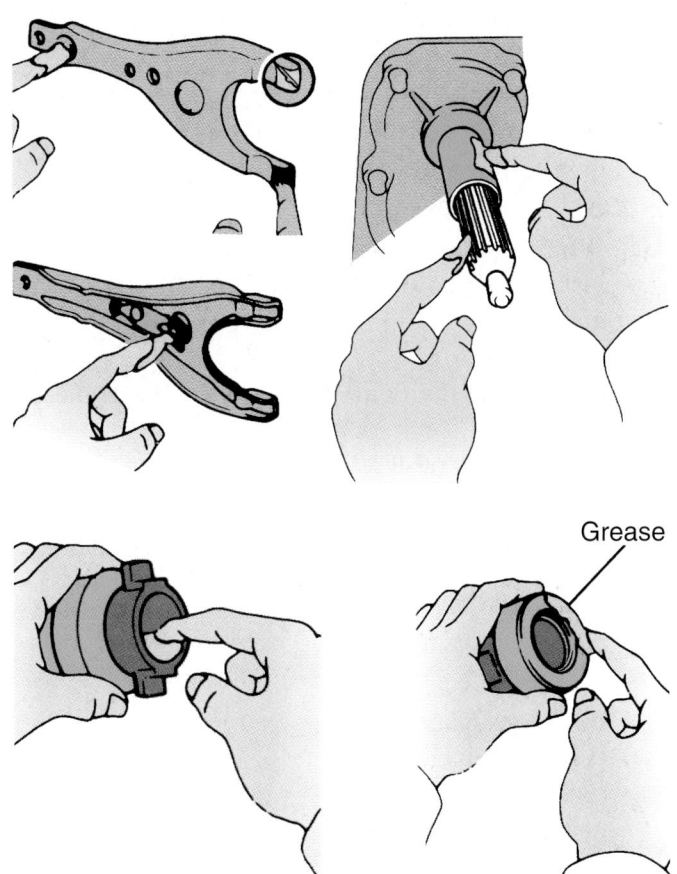

Figure 70.27 Lightly lubricate the various contacts in the release mechanism.

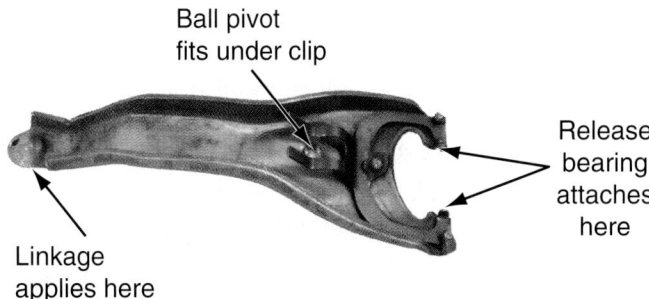

Figure 70.28 Locations of parts that attach to the clutch fork. *(Courtesy of Tim Gilles)*

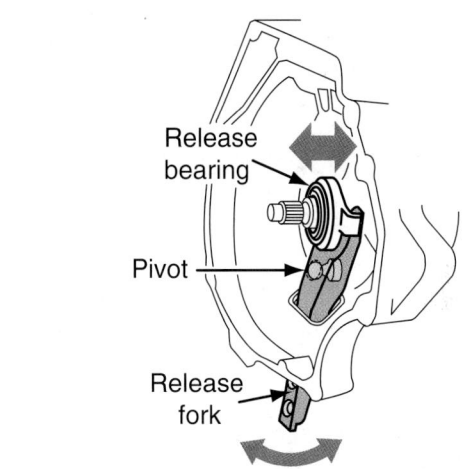

Figure 70.29 Attach the clutch fork to the release bearing and clutch housing.

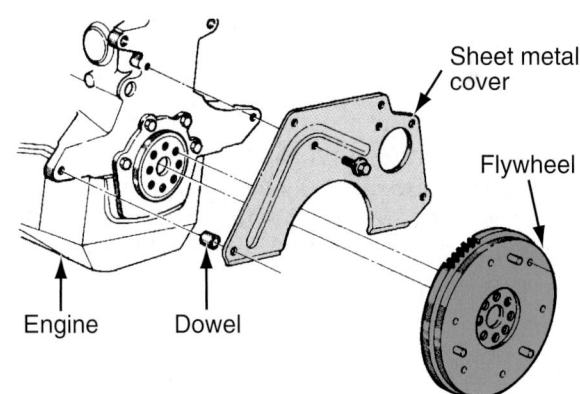

Figure 70.30 Be sure to install any sheet metal pieces that might fit between the engine block and the clutch housing. *(Courtesy of DaimlerChrysler Corporation)*

and the engine block. Check to see that there are no nicks or pieces of debris on the engine and clutch housing mating surfaces. Misalignment will cause clutch problems. If the clutch assembly was removed because of chattering and a previous misalignment is suspected, inspect the clutch housing surfaces.

To complete the job, install the transmission and remaining parts. Perform any needed clutch adjustment and fluid level checks, and test drive the car.

■ REVIEW QUESTIONS

1. When the clutch pedal is let up from the floor, is the clutch being engaged or disengaged?

2. Should the transmission be in gear or in neutral when checking for pilot bearing noise?

3. If there is no sign of a fluid leak but the clutch slowly engages with your foot on the pedal, what kind of leak does a clutch master cylinder have?

4. If the transmission grinds when shifting into reverse after 10 seconds in neutral with the clutch in, what is the problem?

5. When the pedal shakes as the clutch is engaged, what is this called?

6. When something becomes trapped between the engine and the clutch housing during installation, what condition results?

7. How much free travel at the pedal should you adjust a clutch to?

8. The names of the cylinders in a hydraulic clutch system are the master cylinder and the _____ cylinder.

9. What is the name of the puller that has a weight on it that is slid against a stop to pull a pilot bearing?

10. What is another name for the front transmission bearing retainer?

■ ASE-STYLE REVIEW QUESTIONS

1. Technician A says to test for a slipping clutch by shifting the transmission into low gear. Technician B says to test for a slipping clutch with the parking brake firmly engaged. Who is right?
   a. Technician A   c.   Both A and B
   b. Technician B   d.   Neither A nor B

2. As a manually adjusted clutch disc wears, free play:
   a. Becomes less.
   b. Becomes more.
   c. Remains the same.
   d. The result depends on the design of the clutch pressure plate.

3. Technician A says that if a clutch noise only happens when the pedal is all the way to the floor, the release bearing is probably at fault. Technician B says that noise that only happens when the engine is idling in neutral with the clutch engaged is probably an input shaft bearing. Who is right?
   a. Technician A   c.   Both A and B
   b. Technician B   d.   Neither A nor B

4. Which of the following is is/are true when installing a clutch?
   a. Tighten clutch cover screws in a circular pattern.
   b. Use a torque wrench to tighten to the correct specification.
   c. Both A and B.
   d. Neither A nor B.

5. Which of the following can result in clutch chatter?
   a. Defective engine mounts
   b. A small amount of grease on the clutch disc
   c. Misaligned parts
   d. All of the above

6. Technician A says that clutch parts should be cleaned thoroughly in mineral-based cleaning solvent. Technician B says that clutch parts can be cleaned using hot water or brake cleaner. Who is right?
   a. Technician A   c.   Both A and B
   b. Technician B   d.   Neither A nor B

7. Which of the following clutches are self-adjusting?
   a. Some hydraulic clutches
   b. Some mechanical clutches
   c. Both A and B
   d. Neither A nor B

8. Which of the following is/are methods of bleeding the clutch hydraulic system of air?
   a. Putting fluid into the slave cylinder bleed screw
   b. Letting fluid out of the slave cylinder bleed screw
   c. Both A and B
   d. Neither A nor B

9. Technician A says that using an impact wrench to tighten a flywheel can result in a leaking rear engine seal. Technician B says that metric flywheel bolts are property class 10.9 or higher. Who is right?
   a. Technician A   c.   Both A and B
   b. Technician B   d.   Neither A nor B

10. Which of the following is/are usually replaced during a clutch job?
    a. The release bearing
    b. The pilot bushing or bearing
    c. The clutch disc and pressure plate
    d. All of the above

CHAPTER 71

Manual Transmission Fundamentals

■ OBJECTIVES

Upon completion of this chapter, you should be able to:

✔ Describe the relationship between gears and torque.

✔ Understand the basic types of gears.

✔ Calculate gear ratios.

✔ Trace the power flow through three-, four-, and five-speed transmissions.

✔ Name all of the transmission parts.

■ KEY TERMS

manual transmission	close ratio transmission	helical gear
lower gear	granny gear	idler gear
gear radius	final drive ratio	synchronizer
gear ratio	mesh	blocker ring synchronizer
overdrive	pitch diameter	countergear
wide ratio transmission	spur gear	dog teeth

■ INTRODUCTION

A **manual transmission** is used with a clutch. It must be shifted between gears manually (**Figure 71.1**). This kind of transmission is also called a *stick shift* or a *standard transmission*. Standard transmission is no longer an accurate name, because more cars are built with automatic transmissions. The name became popular when automatic transmissions first appeared as an option on cars.

A transmission is used in rear-wheel-drive cars. A similar unit, called a transaxle, is used in front-wheel-drive cars. It is covered in Chapter 77, along with drive axles and constant velocity (CV) joints.

■ PURPOSE OF A TRANSMISSION

A four-stroke engine does not produce equal torque at all times. It has lower torque when starting from a stop. That is why you have to slip the clutch during take-off to keep the engine from stalling. A transmission provides a means of changing torque to fit an engine's operating requirements.

When the engine is operated in low gear, the engine's crankshaft turns approximately three times to one turn of the transmission output shaft. This is called a 3:1 *ratio*. Without the extra leverage that this gear reduction provides, the engine would stall or lug during takeoff. **Figure 71.2** compares the torque an engine produces in various gear ranges. Compare this to riding a ten-speed bicycle. When you shift into a lower gear, the pedal crank revolves faster and you climb the hill easier. In a **lower gear**, a small gear

Figure 71.1 Cutaway of a manual transmission. *(Courtesy of DaimlerChrysler Corporation)*

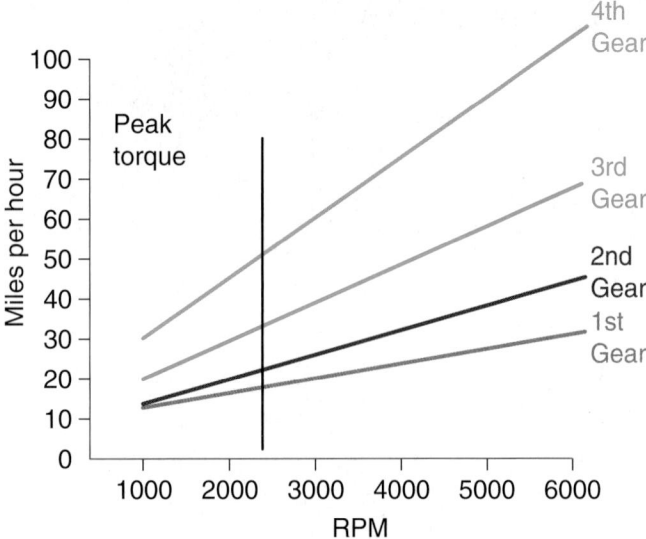

Figure 71.2 The vehicle must be traveling faster before the transmission is shifted into fourth gear.

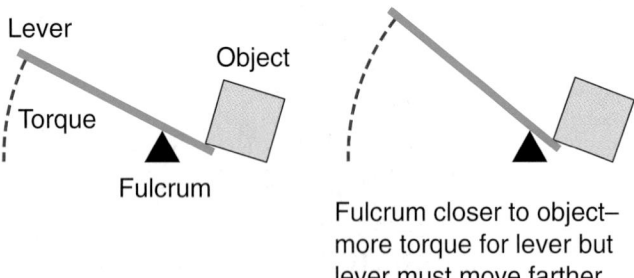

Figure 71.3 Gears provide leverage like that of a light person lifting a heavy person on a teeter-totter. *(Reproduced by permission of Deere & Company, John Deere Publishing, Moline, IL. All rights reserved)*

drives a larger gear. The gears provide leverage like that of a light person lifting a heavy person on a teeter-totter (**Figure 71.3**).

NOTE: *If you ever drive an electric golf cart, notice how much torque the electric motor makes when you first take off. Electric motors produce excellent torque from a standing start.*

■ USING GEARS TO INCREASE TORQUE

To measure torque, the force applied is multiplied by the distance from the centerline of rotation. Twenty pounds of force applied to the end of a 1-foot rod produces 20 foot-pounds of torque. Ten pounds of force applied to a 2-foot rod also produces 20 foot-pounds of torque.

When the driving gear is smaller than the driven gear, its output speed decreases and the output torque increases. The distance from the center of a gear to its outside edge is called the **gear radius**. The radius is where the torque is measured. In **Figure 71.4**, the radius of the smaller gear is 1 foot, and 25 pounds of force is applied to the gear. The output of the gear is

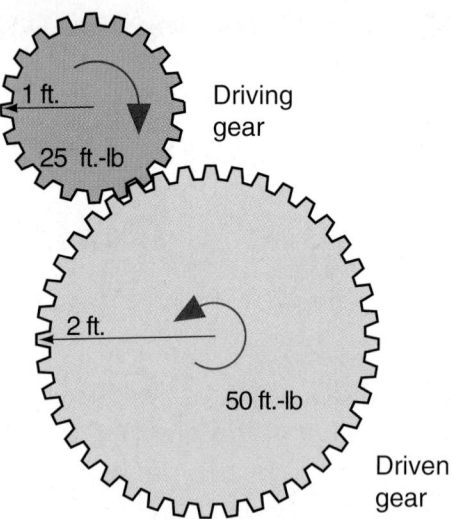

Figure 71.4 A driving gear of half the size results in double the torque output.

25 foot-pounds of torque. That torque is applied to the second gear, which has a 2-foot radius. The resulting output is 50 foot-pounds of torque.

NOTE: *The amount of torque increase is proportional to the speed decrease. The output speed is half, so torque is doubled.*

■ GEAR RATIO

A **gear ratio** can be calculated by dividing the number of teeth on the driven gear by the number of teeth on the driving gear. The input gear is called the *driving* gear. The output gear is called the *driven* gear. If there are 12 teeth on the driving gear and 24 teeth on the driven gear, the resulting gear ratio is 2:1 (**Figure 71.5**). The driving gear must turn two times to one turn of the driven (larger) gear. The torque on the larger gear would be twice that on the input shaft (smaller) gear.

When figuring a gear ratio when there are more than two gears, number the gears first and be certain which is driving and which is driven. Then, multiply

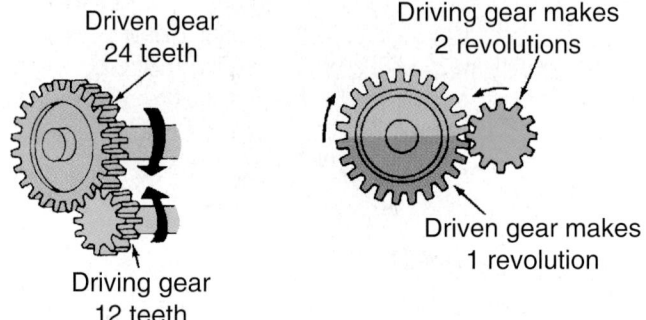

Figure 71.5 A 2:1 gear ratio. *(Reproduced by permission of Deere & Company, John Deere Publishing, Moline, IL. All rights reserved)*

driving gears together and driven gears together before dividing. When figuring reverse gear ratio, ignore the idler gear. Divide the output gear count by the input gear count.

TRANSMISSION GEAR RANGES

Transmissions in cars and light trucks can have either three, four, five, or six forward gear ranges, called speeds. A reverse gear range allows the vehicle to back up. A neutral gear range is useful for long periods of idle and when servicing a vehicle. Lower power engines require more gear ranges to keep the engine at maximum torque.

Approximate transmission gear ratios for a three-speed transmission are
- 3:1 first
- 2:1 second
- 1:1 high
- 3:1 reverse

Typical ratios with a four-speed transmission are
- 3:1 first
- 2.5:1 second
- 1.5:1 third
- 1:1 fourth
- 3:1 reverse

When there is a fifth gear, its ratios are usually the same as for a four-speed but with an added overdrive gear:
- 0.75:1 fifth gear (overdrive)

OVERDRIVE

Overdrive is the opposite of gear reduction. The output shaft turns *faster* than the input shaft. It provides a ratio that is another step beyond the 1:1 ratio of high gear—0.75:1, for instance (a 25% change in speed). With a modern five-speed transmission, it is when the large gear drives a smaller gear. There are also other types of overdrives.

Besides the common overdrive found in five-speed transmissions, there are planetary gear types. Overdrive is achieved using planetary gears in an automatic transmission with a lock-up torque converter (except for Honda and Acura transmissions). Planetary overdrive manual transmissions were popular in the 1950s and 1960s, before automatic transmissions became popular. A planetary gearset was attached to the rear of a manual transmission output shaft. The overdrive unit was located in the extension housing.

CLOSE AND WIDE RATIO

In the muscle car days of the 1960s and early 1970s, a four-speed manual transmission could sometimes be ordered with a close ratio or a **wide ratio transmission**, when there is a larger difference between

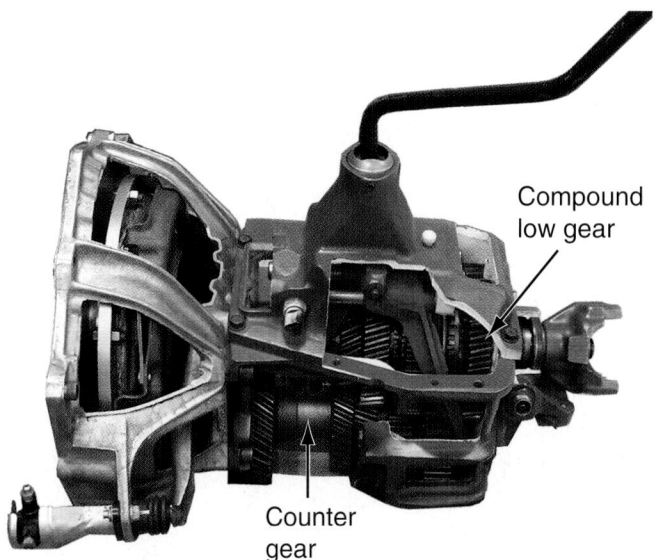

Figure 71.6 This cutaway truck transmission has an extra-low compound first gear. During normal driving, second gear is selected for taking off from a stop. *(Courtesy of Tim Gilles)*

the ratios of the gears. This transmission might have a first gear ratio close to 3:1, while a **close ratio transmission** would have a first gear ratio closer to 2.2:1. The wide ratio transmission provides more torque increase in first gear.

Shifting between gears in a close ratio transmission would result in less drop of engine rpm, keeping the engine within its torque band. It would be more applicable to racing. Shifting while driving in traffic could feel like staying in the same gear and letting out the clutch. The close ratio transmission was most commonly used with a low (i.e., 4.11:1) final drive gear ratio. This made up for the 2.1 first gear ratio, which would make accelerating from a standing stop more difficult.

Some light trucks come equipped with a four-speed transmission that has a very low first gear, sometimes called a **granny gear**. These transmissions usually have a very long floor shift lever (**Figure 71.6**). When driving these trucks, it is customary to start from a stop in second gear. This is because the low gear ratio is in the 7:1 range, handy for idling a boat up a boat ramp or pulling a very heavy load. The second gear ratio is comparable to first gear in a three-speed transmission.

FINAL DRIVE RATIO

The **final drive ratio** is the ratio between the transmission output shaft and the differential ring gear. Say a differential has a ratio between its pinion gear and ring gear of 3:1. If the transmission was shifted into a 3:1 first gear, the final drive ratio to the rear wheels would be 9:1.

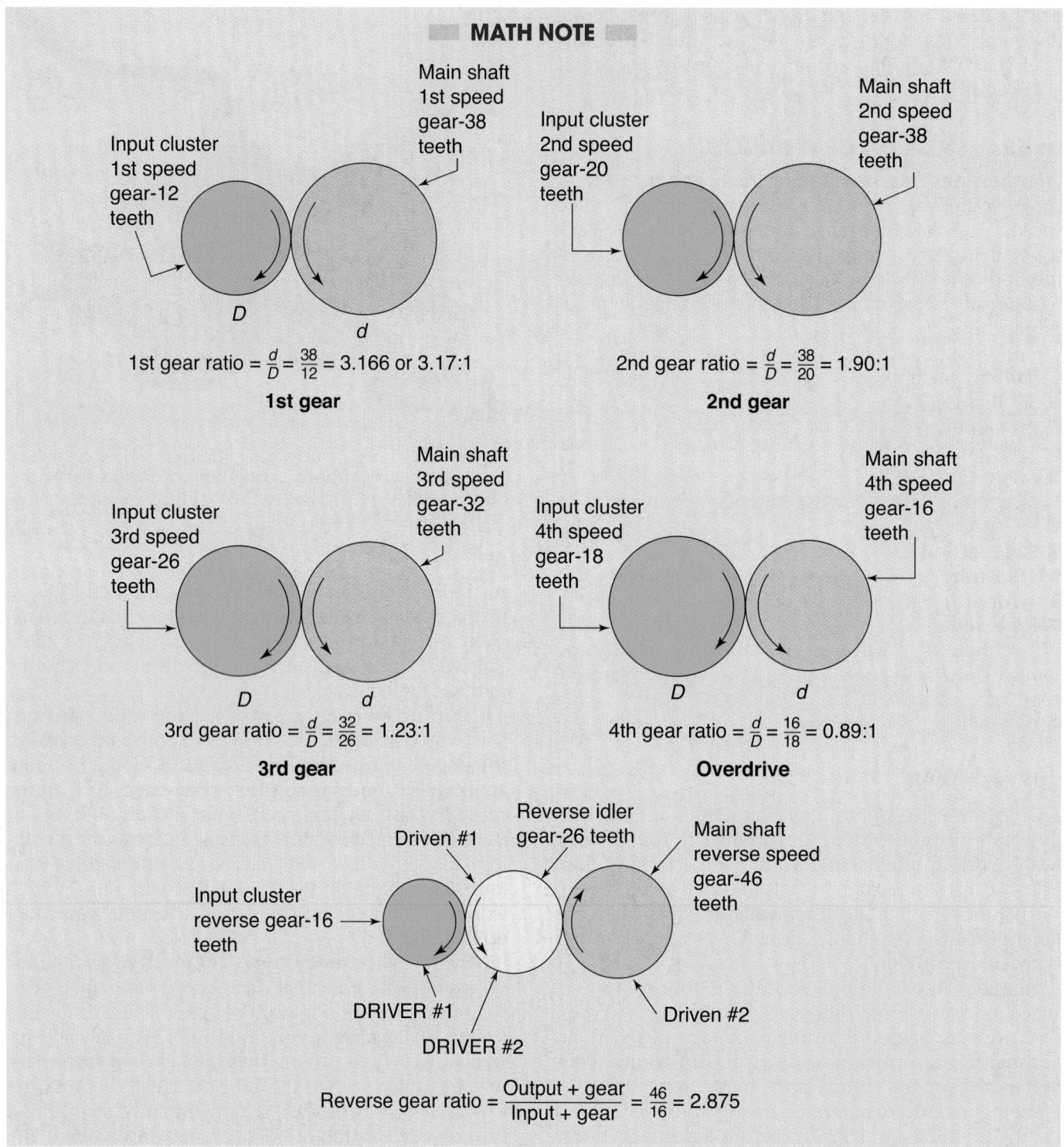

Input cluster
1st speed
gear-12
teeth

Main shaft
1st speed
gear-38
teeth

D d

1st gear ratio $= \frac{d}{D} = \frac{38}{12} = 3.166$ or $3.17{:}1$

1st gear

Input cluster
2nd speed
gear-20
teeth

Main shaft
2nd speed
gear-38
teeth

2nd gear ratio $= \frac{d}{D} = \frac{38}{20} = 1.90{:}1$

2nd gear

Input cluster
3rd speed
gear-26
teeth

Main shaft
3rd speed
gear-32
teeth

D d

3rd gear ratio $= \frac{d}{D} = \frac{32}{26} = 1.23{:}1$

3rd gear

Input cluster
4th speed
gear-18
teeth

Main shaft
4th speed
gear-16
teeth

D d

4th gear ratio $= \frac{d}{D} = \frac{16}{18} = 0.89{:}1$

Overdrive

Driven #1

Reverse idler
gear-26 teeth

Main shaft
reverse speed
gear-46
teeth

Input cluster
reverse gear-16
teeth

DRIVER #1

DRIVER #2

Driven #2

Reverse gear ratio $= \frac{\text{Output} + \text{gear}}{\text{Input} + \text{gear}} = \frac{46}{16} = 2.875$

■ GEAR TYPES AND OPERATION

The shape of a gear tooth is designed to allow the teeth to roll into and out of **mesh** with a minimum of friction. As gears contact each other, the load rolls across the gear teeth. The contact pattern (patch) is where the teeth of the two gears meet. The effective diameter of the meshed gear (used for determining its circumference) is called the **pitch diameter** (**Figure 71.7**).

Gear teeth can also slide against each other, depending on the design. Manual transmissions use two kinds of gears, a spur gear and a helical gear.

■ SPUR GEARS

Spur gears are simple gears with straight-cut teeth (**Figure 71.8**). Spur gears have only one tooth in contact at a time. With this design, there is no end thrust and the transmission will not attempt to pop out of

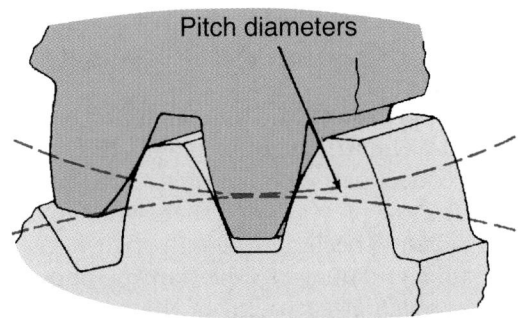

Figure 71.7 The point where the teeth of the two gears meet is called the pitch diameter. *(Reproduced by permission of Deere & Company, John Deere Publishing, Moline, IL. All rights reserved)*

Spur gear   Helical gear

Figure 71.8 This photo compares a spur gear with a helical gear. Spur gears have square-cut gear teeth. *(Courtesy of Tim Gilles)*

gear during acceleration or deceleration. That is one reason spur gears are often used in reverse. Reverse often has no synchronizer clutch (covered later), as forward gears do. Spur gears can also slide together easily.

Backlash is the clearance between meshing gear teeth (**Figure 71.9**). It allows for expansion and lubrication oil to get between the gear teeth. Because there is only one tooth in contact at a time, a clicking sound

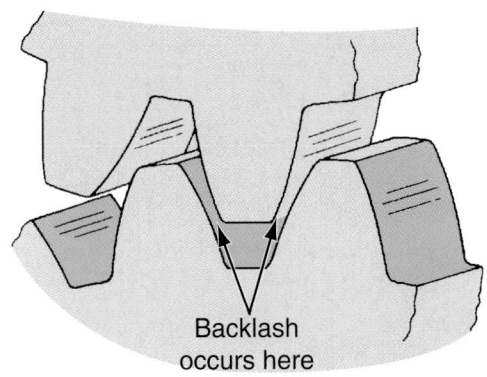

Backlash
occurs here

Figure 71.9 Backlash is the clearance between meshing gear teeth. *(Reproduced by permission of Deere & Company, John Deere Publishing, Moline, IL. All rights reserved)*

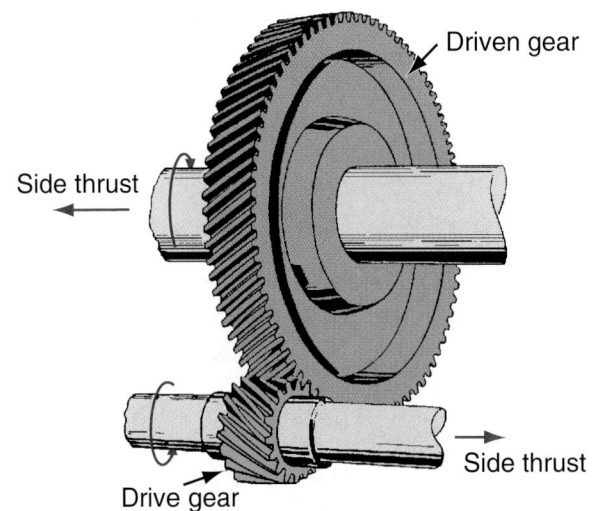

Figure 71.10 Helical gears have slanted gear teeth and move away from each other under load (thrust).

results as a gear tooth rolls out of contact and a new one takes its place. As this noise gains speed, it turns into *gear whine*.

■ HELICAL GEARS

Helical gears have replaced spur gears in transmissions because they are much quieter. They are machined at an angle which gives them a continuous flow of power across the gear teeth (**Figure 71.10**). Engagement starts at the tip of one end of the gear tooth and rolls down the tooth. This keeps backlash to a minimum and makes these gears quieter. It also results in more gear strength because there is more area of tooth contact. A problem with helical gears is that they cause end thrust under load.

■ IDLER GEARS

When gears rotate against each other, the driven gear will be turned in the opposite direction. An **idler gear** is used between two other gears. Its purpose is to change the direction of rotation (**Figure 71.11**).

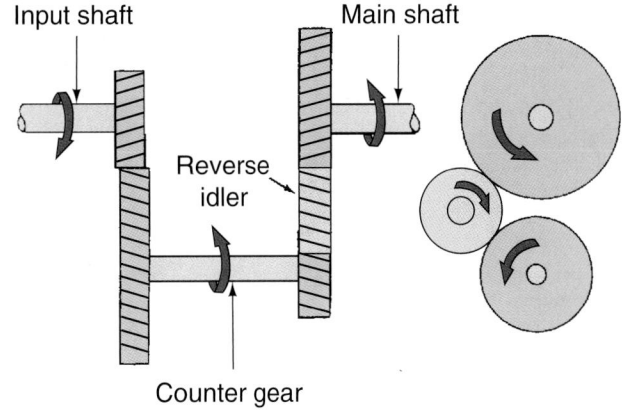

Figure 71.11 A reverse idler gear changes the direction of rotation. *(Reproduced by permission of Deere & Company, John Deere Publishing, Moline, IL. All rights reserved)*

■ TRANSMISSION PARTS

Figure 71.12 shows the basic parts of a four-speed manual transmission. Gear flow is from the clutch disc to the input shaft. The various gears provide a means of changing torque and speed of the output shaft. Each forward gear has a **synchronizer**, sometimes called a synchro, that keeps two meshing gears from clashing during a shift. Shift linkage acts on shift forks within the transmission to select a gear. Power flows from the input shaft to a countergear and then to the mainshaft or output shaft.

Parts are housed in a transmission *gear case* made of iron or aluminum (**Figure 71.13**). The case is usually made of aluminum because of its lighter weight. Bolted to the back part of the case is the *extension housing*, or tailshaft housing. Sometimes a case is iron and the extension housing is aluminum. The extension housing or case often has threaded flange holes for the transmission-to-crossmember mount.

A *front bearing retainer* holds the input shaft bearing against the case. It also acts as a sleeve for the throwout bearing.

The case has drain and fill plugs for adding and draining oil. The drain plug often has a magnet attached to collect particles that might break or wear off gear teeth. If these pieces attach themselves to the magnet, they will not become lodged between gear teeth and cause further damage to the transmission.

There is a seal at each end of the transmission, at the input and output shafts. Gaskets seal between the parts of the case. A breather is provided at the top of the case.

■ TRANSMISSION LUBRICATION

Transmission parts are separated by oil at all times. Oil is moved throughout the case by the rotating gears. This is called *splash lubrication*. The lubricant varies by manufacturer. Some use SAE 80 or 90 gear oil, the same

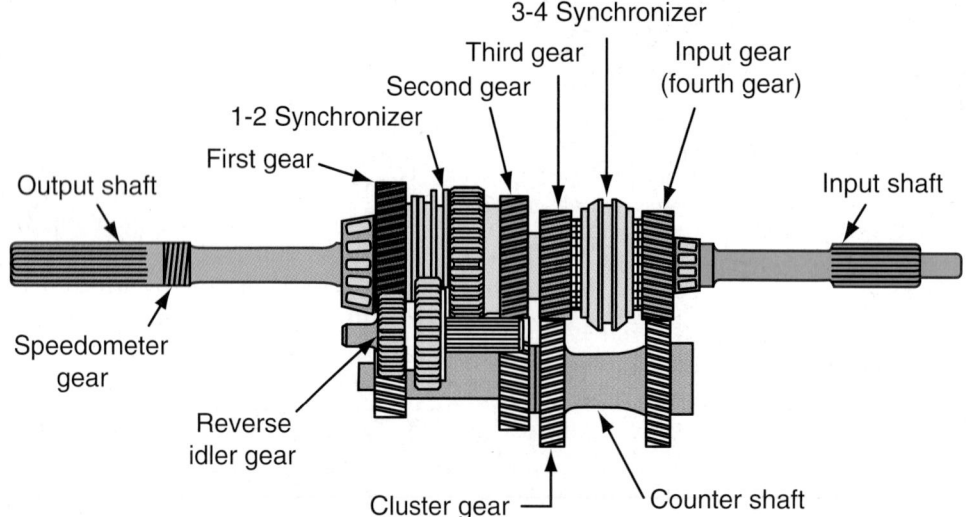

Figure 71.12 Parts of a four-speed manual transmission.

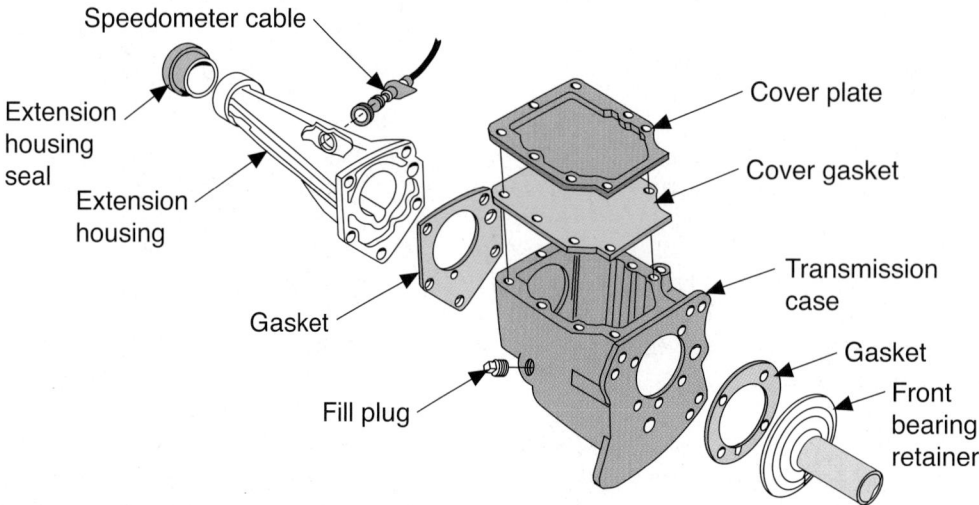

Figure 71.13 Parts attached to a transmission case. *(Courtesy of Ford Motor Company)*

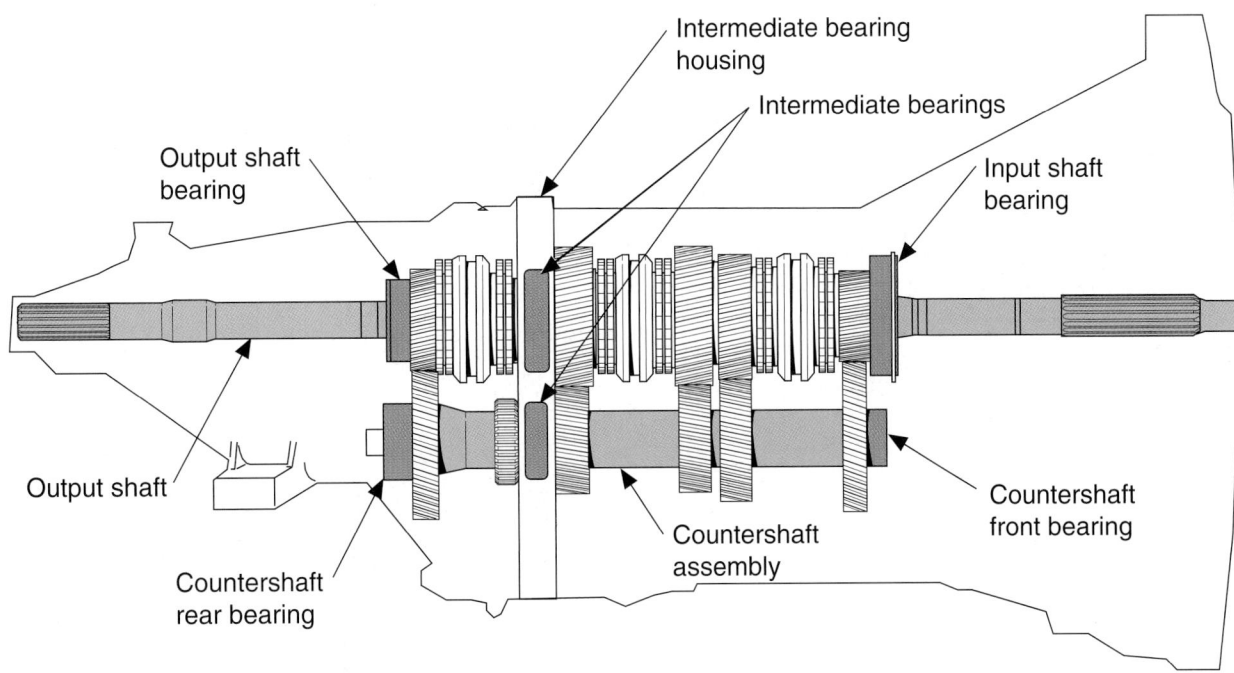

Figure 71.14 Locations of various bearings in a manual transmission. *(Courtesy of Ford Motor Company)*

as is used in differentials. Others use SAE 30 motor oil, automatic transmission fluid (ATF), or a specialized synthetic lubricant. Using the incorrect oil can drastically affect the transmission's operation. Be sure to follow the manufacturer's recommendations.

■ TRANSMISSION BEARINGS

Bearings support the ends of almost all rotating parts within a transmission. They allow parts to rotate with very little friction. Several types of bearings are found in manual transmissions (**Figure 71.14**). Reverse idler shafts and gears are sometimes supported by bushings, but the bearings are ball, roller, or needle.

■ TRANSMISSION GEARS AND SHAFTS

Different parts of the geartrain are commonly called "shafts." These include the input shaft, countershaft, and output shaft or mainshaft.

The *input shaft* is often called a *clutch shaft* (**Figure 71.15**). Its bearing is called an input shaft bearing or *clutch bearing*. (You may hear a front bearing retainer called a clutch bearing retainer.)

The *countershaft* is usually one gear made up of a series of gears that mesh with the various gears on the mainshaft (**Figure 71.16**). It is often called a *cluster gear*. Older transmission designs have short, plain steel shafts that ride inside of the countergear and reverse idler gear. Rows of small needle bearings separate the shaft and the inside of the gear to support the load. The ends of the shafts are pressed fit in the case. Some heavy load transmissions have two rows of bearings on each end. Later model transmissions often have the countergear supported by a large, tapered roller bearing at the ends.

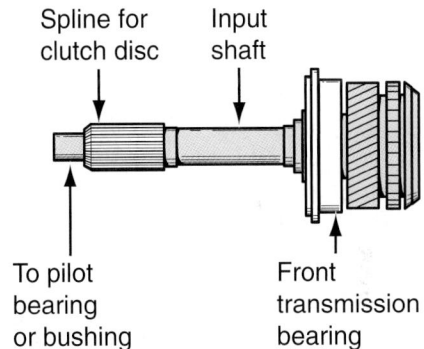

Figure 71.15 Parts of an input shaft.

Figure 71.16 A counter gear. *(Courtesy of Tim Gilles)*

The *output shaft* or *mainshaft* has all gears and synchronizers mounted on it (**Figure 71.17**). On manual transmissions made since the mid-1960s, all forward gears are in *constant mesh*. This means that only shift collars move to engage each gear to the mainshaft.

The *reverse idler gear* is the only gear that moves into mesh with another gear (**Figure 71.18**). The reverse gearset is often made up of spur gears, rather than helical gears. That is why a transmission will often howl when in reverse gear.

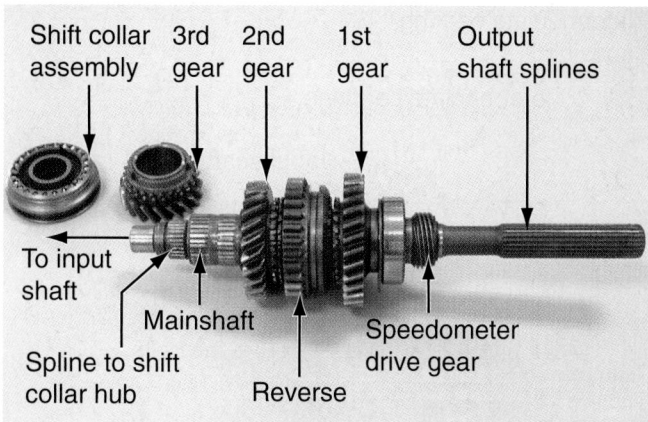

Shift collar assembly | 3rd gear | 2nd gear | 1st gear | Output shaft splines

To input shaft

Spline to shift collar hub | Mainshaft | Reverse | Speedometer drive gear

Figure 71.17 Parts on the mainshaft of a four-speed transmission. *(Courtesy of Tim Gilles)*

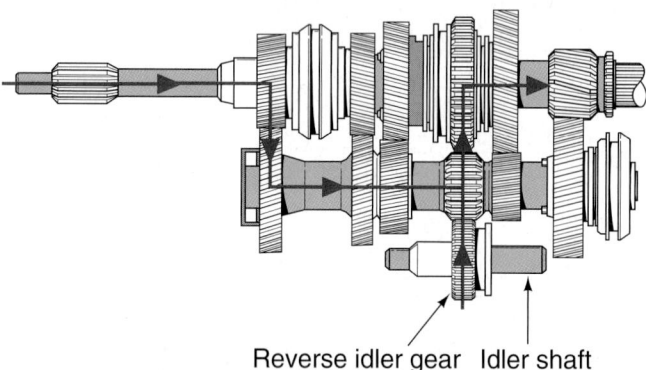

Reverse idler gear | Idler shaft

Figure 71.18 A reverse idler gear slides into contact with another gear. The one in this transmission is a spur gear. *(Courtesy of Ford Motor Company)*

■ SYNCHRONIZER ASSEMBLY

The synchronizer helps two gears spinning at different speeds mesh without clashing. The synchro blocks the shift and brakes the two parts together using a cone clutch type of action.

Blocker ring synchronizers are commonly used in automobiles. There are other kinds, but they are rare. Parts of a typical synchro are shown in **Figure 71.19**. Locate the *hub* in the sketch. It is splined to the mainshaft (**Figure 71.20**). A *shift collar,* or *synchronizer sleeve,* fits around the outside of the hub (**Figure 71.21**). The gears are in constant mesh with their counterparts on

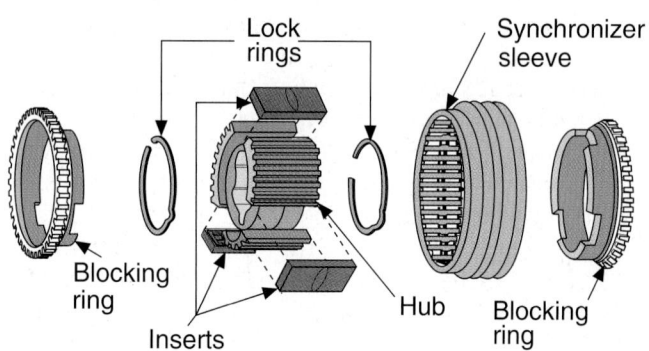

Lock rings | Synchronizer sleeve

Blocking ring | Inserts | Hub | Blocking ring

Figure 71.19 Parts of a synchronizer assembly.

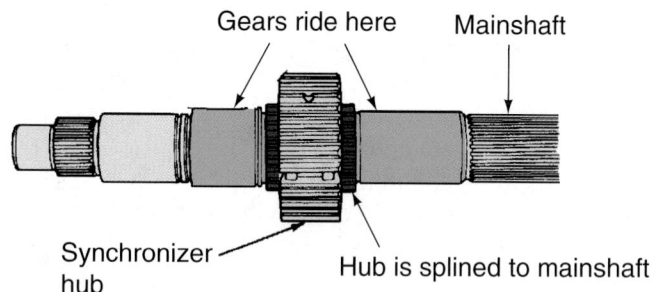

Gears ride here | Mainshaft

Synchronizer hub | Hub is splined to mainshaft

Figure 71.20 The hub is splined to the mainshaft and the gears are free to rotate. *(Courtesy of Ford Motor Company)*

Key | Key

Teeth | Spline to mainshaft | Spring

Hub

Figure 71.21 A sleeve is splined to the outside of the hub. *(Courtesy of Tim Gilles)*

the **countergear** and rotate freely on bearing areas on the mainshaft (**Figure 71.22**). To make a gear shift, the splines on the outside of the hub attach to a gear. The assembly is sometimes called a *dog clutch.* Each gear has **dog teeth,** or clutch teeth, little teeth around the circumference of the edge of the gear. When a gear shifts,

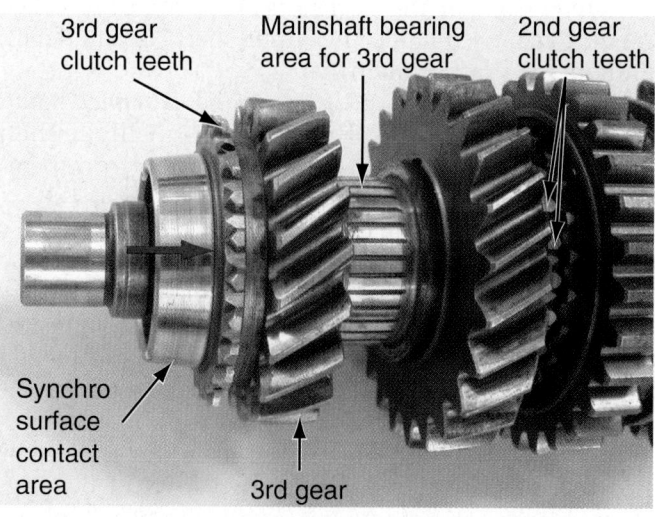

3rd gear clutch teeth | Mainshaft bearing area for 3rd gear | 2nd gear clutch teeth

Synchro surface contact area | 3rd gear

Figure 71.22 Each constant mesh gear has clutch teeth. The gear rotates freely on a bearing area on the mainshaft. *(Courtesy of Tim Gilles)*

the sleeve is moved by shift linkage to select the desired gear. The synchro sleeve has splines around its inside diameter. As it is moved toward the gear, it engages its clutch teeth.

Within each shift collar is a blocking ring synchronizer called a *synchro ring* or blocker ring (**Figure 71.23**). Blocker rings are usually made of brass. Some newer blocker rings are paper lined. Automatic transmission fluid must be used with these.

Besides preventing the clashing of gears, the synchro's purpose is to lock the input shaft gear to the output shaft gear. Keys fit into notches in the synchro rings to tie the hub to the synchro ring and keep the two spinning together (**Figure 71.24**). The keys prevent the blocker ring teeth from moving more than half of a tooth's width in either direction. Figure 71.24 shows a shift collar engaging the third-gear clutch teeth. To shift to fourth gear, the collar is moved to the left by the shift fork. As the shift pro-

gresses, the tapered ends of the clutch teeth on the blocker ring attempt to mesh with the tapered ends of the sleeve teeth (**Figure 71.25a**). As the shift is completed, the teeth on the sleeve overlap the clutch teeth on the gear (**Figure 71.25b**). **Figure 71.26** shows how the blocker ring wedges against the cone of the gear to stop it from spinning.

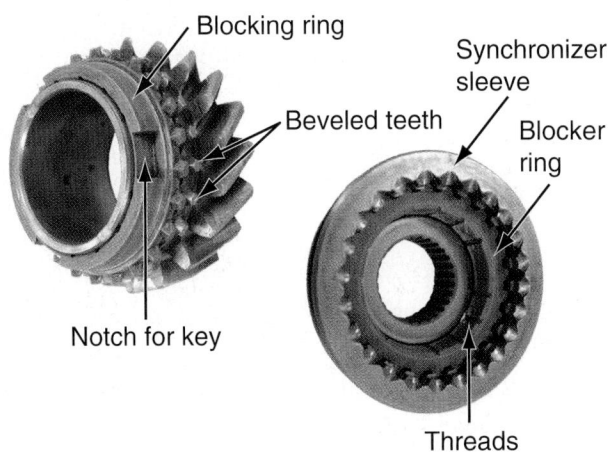

Figure 71.23 A blocking ring wedges against the shoulder of the gear. *(Courtesy of Tim Gilles)*

Figure 71.24 A shift collar engaging the third-gear clutch teeth. *(Courtesy of Tim Gilles)*

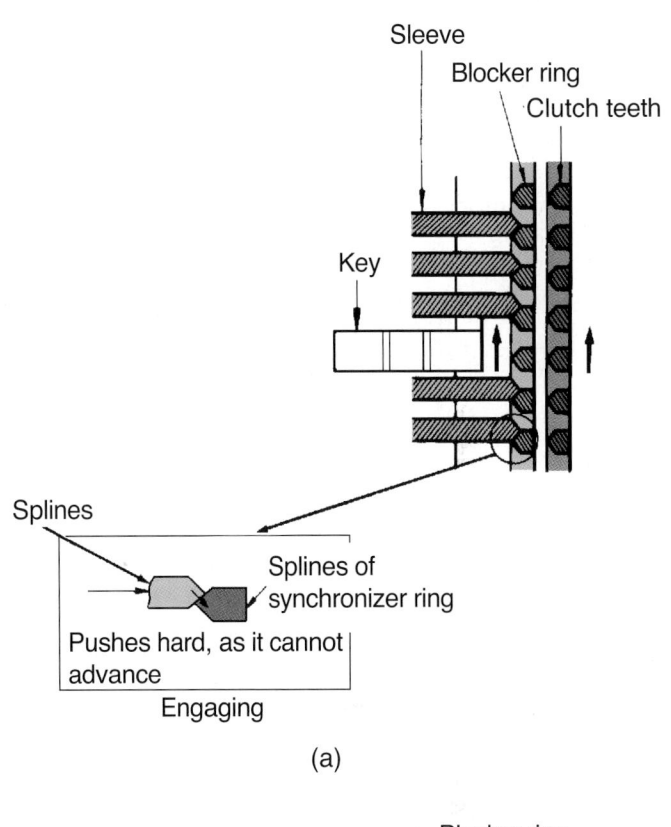

(a)

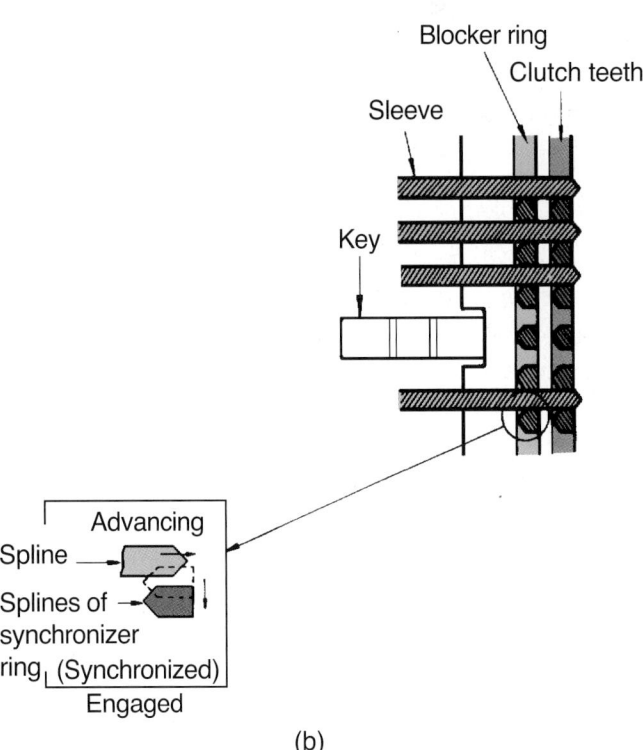

(b)

Figure 71.25 Action of a synchronizer assembly.

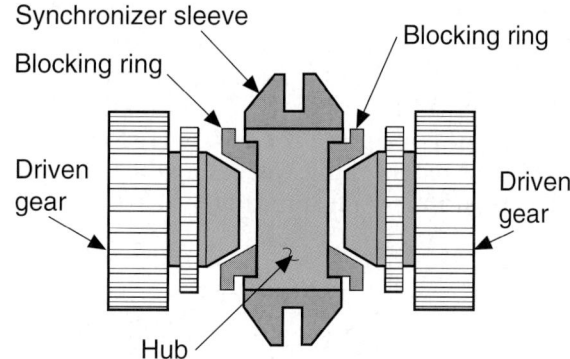

Synchronizer sleeve

Blocking ring

Blocking ring

Driven gear

Driven gear

Hub

Synchronizer in neutral position before shift

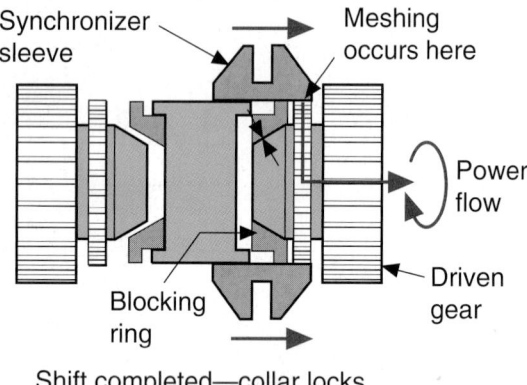

Synchronizer sleeve

Meshing occurs here

Power flow

Driven gear

Blocking ring

Shift completed—collar locks driven gear to hub and shaft

Figure 71.26 The blocker ring in neutral position (above) and wedging against the cone of the gear (bottom).

■ GEAR SHIFT MECHANISMS

Shift forks fit into grooves cut in the outside of the synchro collar. Shift linkage can be either the internal shift rail type (**Figure 71.27**) or the external rod type (**Figure 71.28**). There are also mechanisms that use cables instead of linkage. Each type has features for keeping the transmission in gear and for keeping it out of two gears at once. The internal linkage type will be used here for illustration purposes.

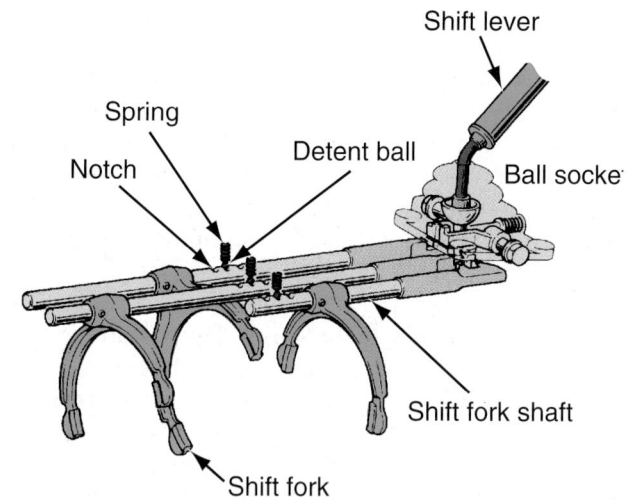

Shift lever

Spring

Detent ball

Notch

Ball socket

Shift fork shaft

Shift fork

Figure 71.27 A typical internal shift mechanism.

Figure 71.28 An external shift mechanism on a four-speed manual transmission.

A *detent mechanism* holds the transmission in gear (provided there are no worn parts in the synchronizer shift collar). Spring tension holds the detent balls into the detent notches in the shift rail (see Figure 71.27). This keeps the shift rails in place, preventing the transmission from popping out of gear.

An interlock mechanism prevents the selection of two gears at once (which would destroy the transmission). When one of the shift shafts is moved during a shift, a set of interlock pins holds the other shafts in their neutral positions (**Figure 71.29**). There is also a shift restricting pin so that a shift cannot be made from fifth to reverse during a downshift.

■ SHIFT PATTERN

Various shift patterns used for different transmissions, but most are use a standard one (**Figure 71.30**). With four or five speeds, sometimes reverse is in a different position. Other times, the handle is pushed down to locate the reverse gate.

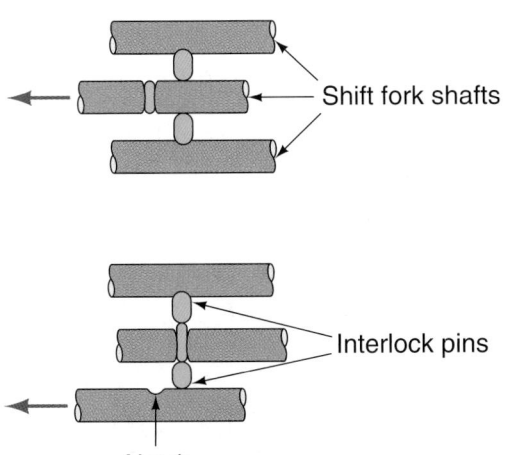

Figure 71.29 Interlock pins prevent shifting into two gears at the same time.

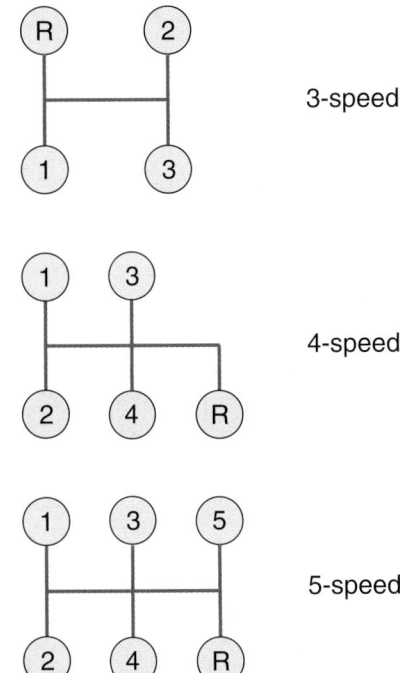

Figure 71.30 Typical gear shift patterns.

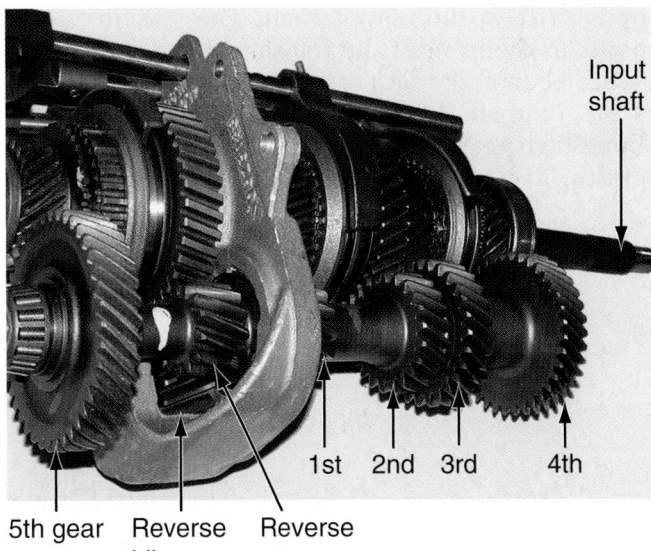

Figure 71.31 This five-speed transmission has reverse and fifth gear in the rear section. *(Courtesy of Tim Gilles)*

■ TRANSMISSION POWER FLOW

Old transmissions had sliding spur gears. Today's transmissions are constant mesh. The only thing that moves is the synchro collar, which selects the gear. Until the mid-1960s, most transmissions were three-speed manuals. Engines of the day were large and powerful. Three-gear ranges were plenty to pull these cars. Second and third speeds had synchronized constant mesh gears and worked in the same manner as a modern transmission.

In three-speeds produced before the mid-1960s, first gear (called low) and reverse shared one sliding gear. The sliding gear came into contact with an idler gear during a shift to reverse. There was no synchronizer for first gear. The vehicle had to come to a complete stop before shifting into first gear, or grinding would occur. When these transmissions were disassembled, the first gear side of the sliding gear teeth was commonly found to be chipped.

All manual transmissions operate in a similar fashion, whether there are three speeds or five speeds. In fact, a three-speed transmission is not much smaller than a five-speed. Because three-speeds are from older cars, a three-speed will probably even be heavier.

In most five-speed transmissions, power in fourth gear is direct, for a 1:1 ratio, and fifth gear is an overdrive. The fifth gear is located in the extension housing (**Figure 71.31**). A four-speed is simply this kind of five-speed without the overdrive. It has the same gears in the case, but no extra gear in the extension housing.

■ FOUR-SPEED TRANSMISSION POWER FLOW (NON-OVERDRIVE)

Neutral

In neutral, the synchro sleeves are all centered and do not mesh with the clutch teeth of any gear (**Figure 71.32**). In the illustration, the spur gear on the outside of the synchronizer sleeve is the reverse gear, a typical setup in transmissions. Notice that there is a blocking ring on each side of the synchro assembly. If this is a three or four synchronizer, one ring is for stopping third gear on the mainshaft, while the other is for stopping fourth gear. The other synchro assembly is for one to two shifts.

High Gear

The easiest gear is high gear, with a 1:1 ratio. Its power flow runs straight through the transmission from the input shaft to the output shaft. The synchro sleeve moves to the front of the transmission to engage the clutch teeth on the back of the input shaft. This attaches the input shaft to the output shaft. **Figure 71.33** shows high gear in a three-, four-, or five-speed transmission.

Third Gear

In third gear, power comes in through the input shaft, which is engaged to the countergear. The synchro sleeve has moved to the right, and the clutch teeth of third gear are engaged (**Figure 71.34**). This locks third gear to the mainshaft.

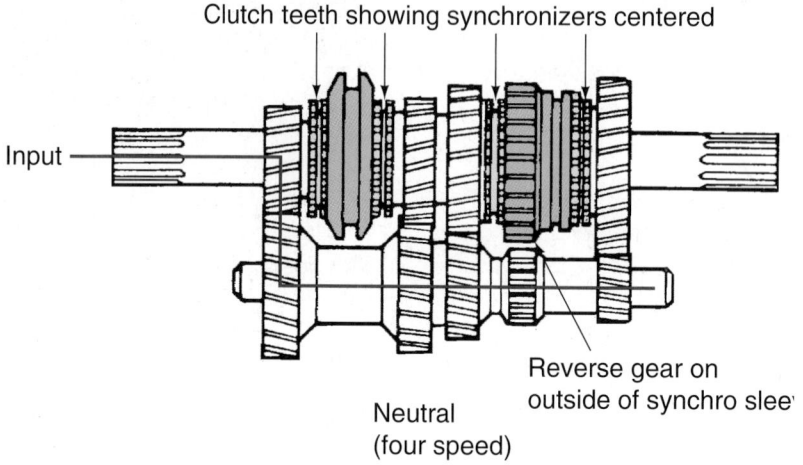

Figure 71.32 In neutral, none of the clutch teeth are engaged. (*Courtesy of Volvo Cars of North America*)

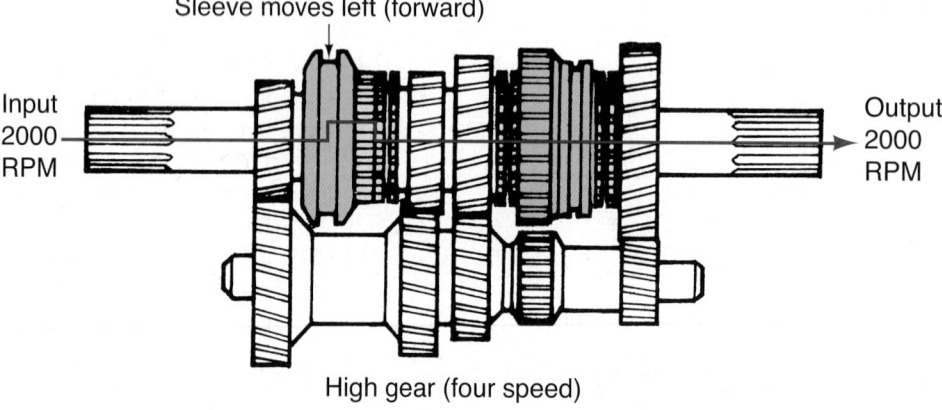

Figure 71.33 In high gear, the synchro sleeve engages the clutch teeth of the input shaft (clutch gear). (*Courtesy of Volvo Cars of North America*)

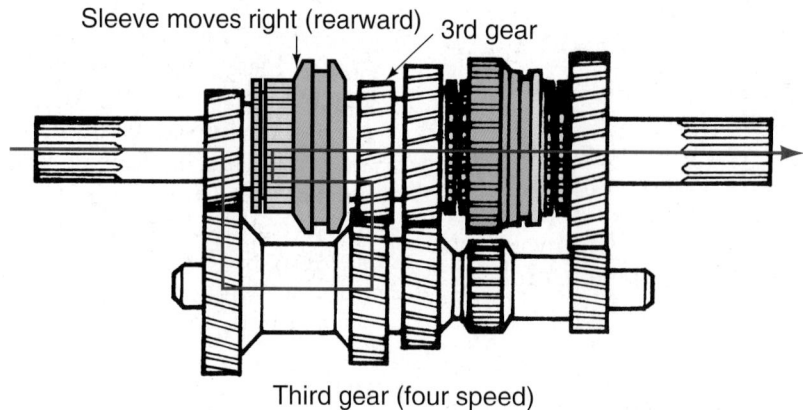

Figure 71.34 Power flow in third gear. The synchro sleeve has moved to the right to engage the third-gear clutch teeth. (*Courtesy of Volvo Cars of North America*)

Second Gear

In second gear, the synchro sleeve in the rear of the transmission is moved toward the left to engage the second-gear clutch teeth (**Figure 71.35**). This attaches second gear to the output shaft.

First Gear

In first or low gear, the synchro sleeve in the rear of the transmission is moved toward the right to engage the first-gear clutch teeth. This attaches first gear to the output shaft (**Figure 71.36**).

Reverse

In reverse, both synchro sleeves are in the neutral position. The reverse idler gear is slid into mesh between the spur gear on the outside of the rear synchro sleeve and the reverse gear on the countershaft (**Figure 71.37**). The hub inside of the synchro sleeve is splined to the mainshaft. Because the idler gear is

one more gear than is normally used, the direction of rotation is in reverse.

■ FIVE-SPEED TRANSMISSION

Gear flow in a typical five-speed transmission is the same in the first four speeds and reverse as that described for a four-speed transmission. As previously noted, a five-speed has an extra set of gears in the extension housing.

In fifth gear, both synchro sleeves in the transmission case are in the neutral position. Power flow is through the end of the countergear to a gear at its end. The gear has clutch teeth and a synchro sleeve that meshes with it during fifth-gear operation (**Figure 71.38**). In some five-speeds, both reverse and fifth gear are in the extension housing or rear section of the case. These transmissions have a dog clutch that is between the fifth and reverse gears. **Figure 71.39** shows power flow in fifth and reverse in one of these transmissions.

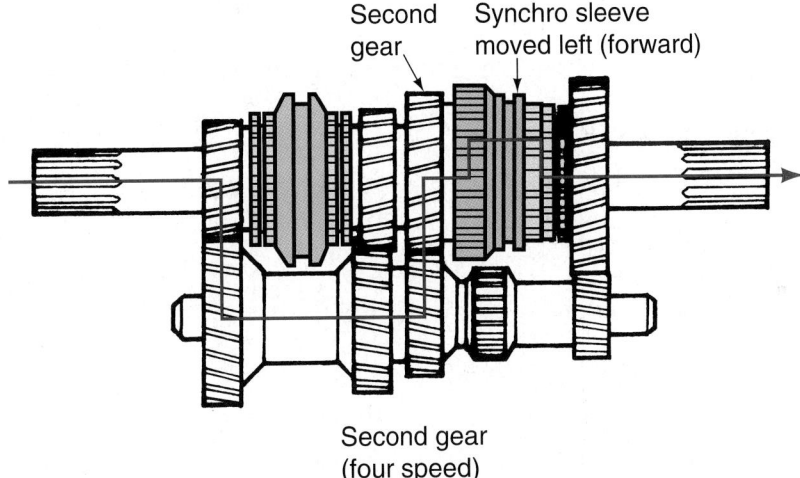

Second gear
(four speed)

Figure 71.35 Power flow in second gear. The synchro sleeve in the rear of the transmission has moved toward the left to engage the second-gear clutch teeth. *(Courtesy of Volvo Cars of North America)*

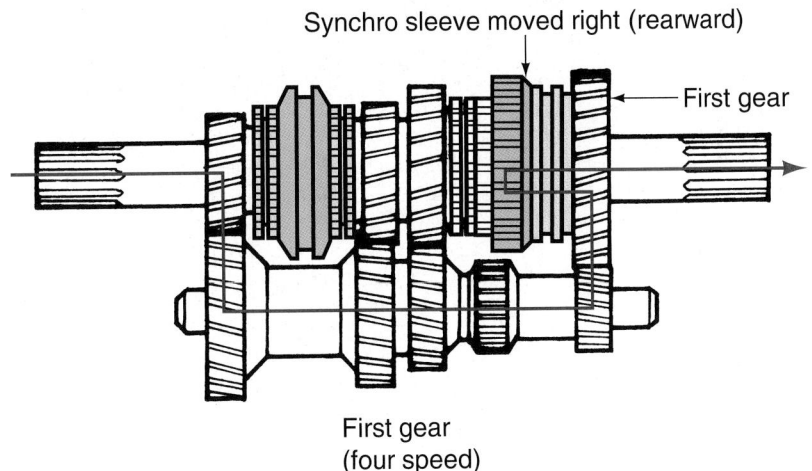

First gear
(four speed)

Figure 71.36 Power flow in first gear. The synchro sleeve in the rear of the transmission is moved toward the right to engage the first-gear clutch teeth. *(Courtesy of Volvo Cars of North America)*

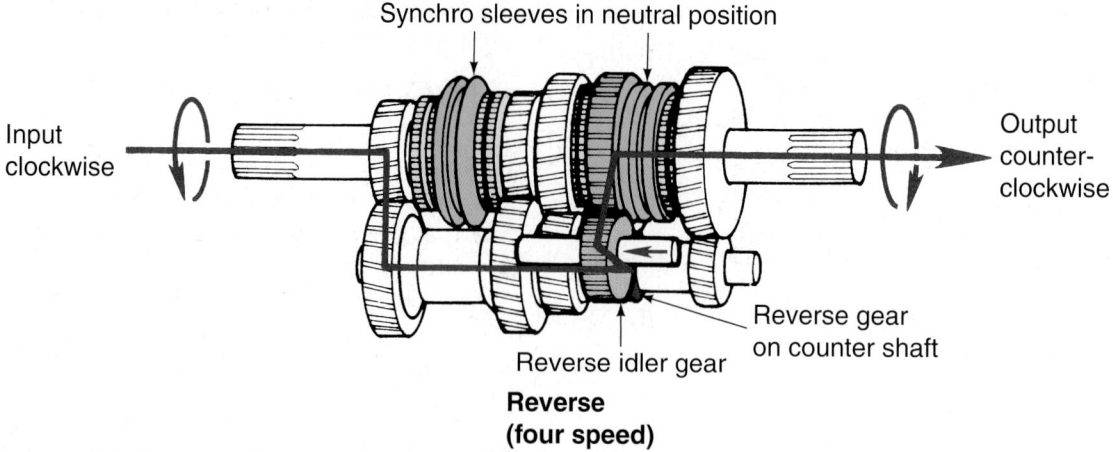

Synchro sleeves in neutral position

Input
clockwise

Output
counter-
clockwise

Reverse gear
on counter shaft

Reverse idler gear

**Reverse
(four speed)**

Figure 71.37 Power flow in reverse. The reverse idler gear is slid into mesh between the spur gear on the outside of the rear synchro sleeve and the reverse gear on the countershaft. *(Courtesy of Volvo Cars of North America)*

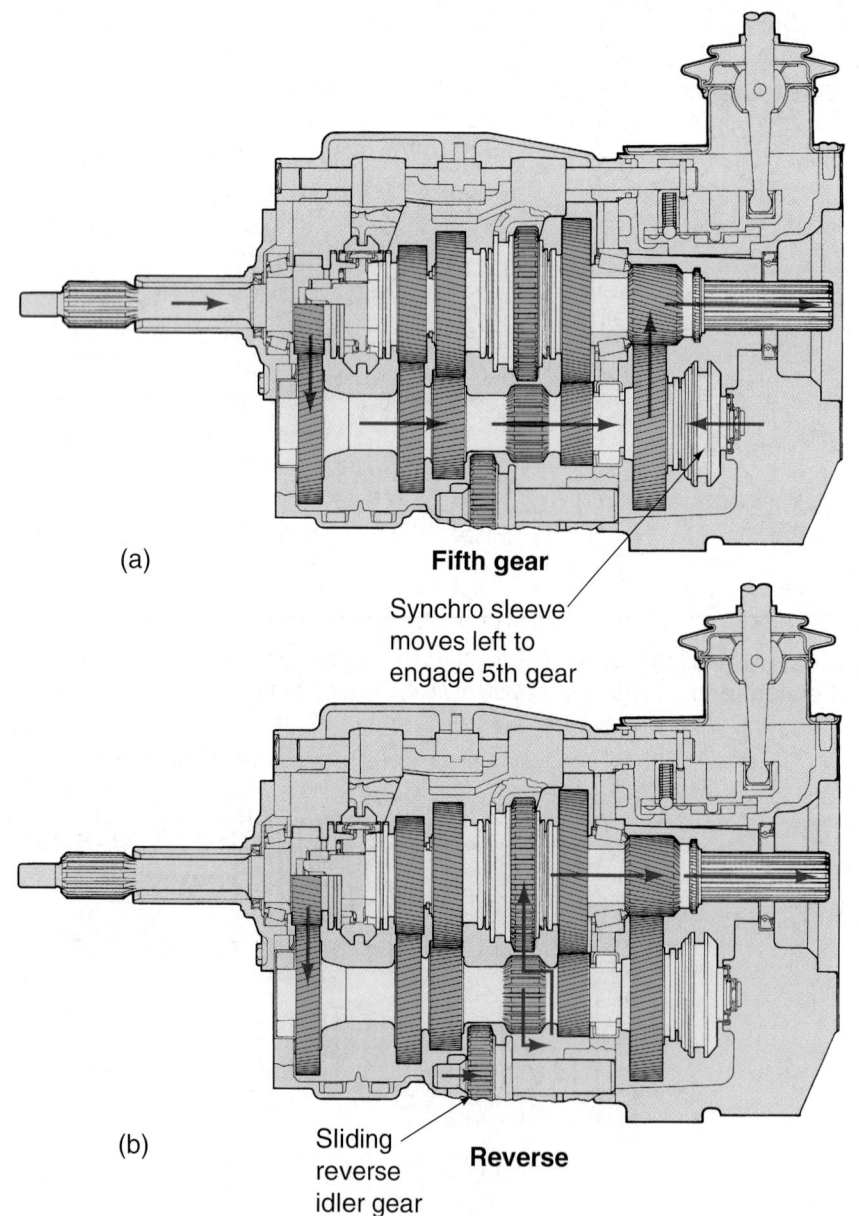

(a)   **Fifth gear**

Synchro sleeve
moves left to
engage 5th gear

(b)   Sliding   **Reverse**
reverse
idler gear

Figure 71.38 (a) Power flow in fifth gear in some five-speed transmissions. (b) In reverse, a sliding idler gear moves into mesh. *(Courtesy of Borg Warner)*

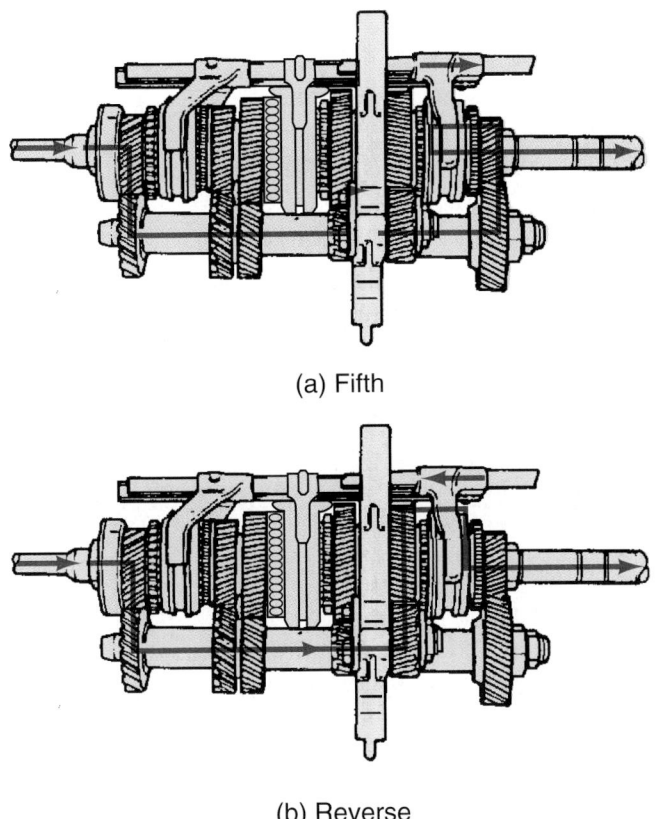

(a) Fifth

(b) Reverse

Figure 71.39 (a) Power flow in fifth and reverse in some five-speeds. (b) In reverse a sliding idler gear moves into mesh. *(Reprinted with permission by American Isuzu Motors, Inc.)*

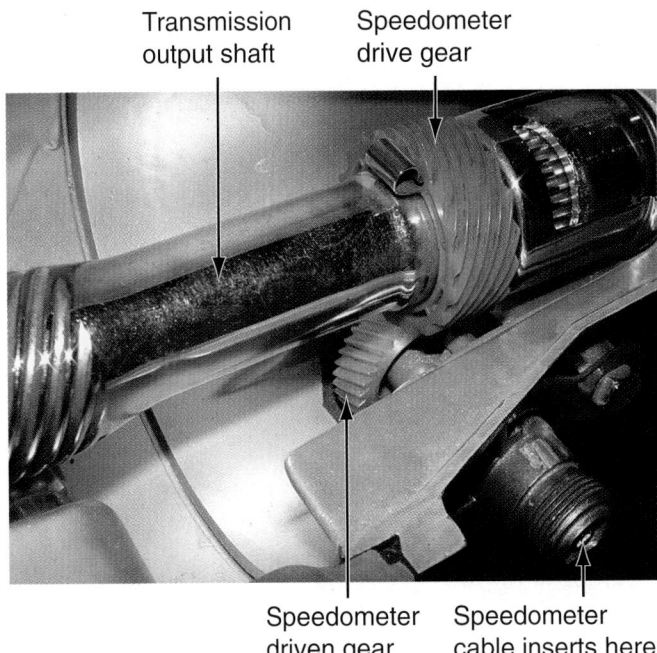

Transmission output shaft   Speedometer drive gear

Speedometer driven gear   Speedometer cable inserts here

Figure 71.40 A speedometer gear driven by a gear on the transmission output shaft. *(Courtesy of Tim Gilles)*

SPEEDOMETER DRIVE

Some cars use electric speedometers that receive a signal from a vehicle speed sensor. A mechanically operated speedometer has a cable driven by a gear on the transmission output shaft (**Figure 71.40**). The gear on the output shaft is either part of the shaft or a replaceable metal or plastic part. The gear that meshes with it is usually plastic. Speedometer gears come with different numbers of teeth so that they can be replaced to adjust for differences in tire diameter. Some makes of cars come with different wheel sizes and use the same transmission.

SWITCHES AND SENSORS

A transmission will often have switches that tell the computer or a mechanical system what gear range it is in. Switches are used for backup lights or ignition advance control. Shifting into reverse will close a switch to light the backup lights. Shifting into a particular gear can result in a change in ignition timing to assist in controlling emissions.

Computer technology allows the addition of several electronic features to transmissions. A vehicle speed sensor (VSS) installed on late-model transmissions takes the place of the mechanical speedometer gear and cable.

Shift lamps on the dash respond to signals from the engine computer as it determines the load on the engine and helps the driver decide to shift for better fuel economy. The light is disabled in high gear.

Some six-speed transmissions have a shift blocking solenoid that makes the driver achieve high enough rpm before shifting when the engine is cold.

Reverse lockout, another feature on some transmissions, prevents the transmission from being shifted accidentally into reverse while moving forward.

■ REVIEW QUESTIONS

1. A manual or standard transmission can also be called a _____ shift.

2. If there are 16 teeth on the driving gear and 48 teeth on the driven gear, the resulting gear ratio is _____ to one.

3. What is the approximate gear ratio in low gear in a transmission with a granny gear?

4. What is the name for the ratio between the transmission output shaft and the differential ring gear?

5. A _____ gear is a simple gear design with straight-cut teeth.

6. The clearance between meshing gear teeth is called _____.

7. What is another name for the input shaft?

8. A countergear can also be called a _____ gear.

9. What is the term used to describe the method of clutching between shifts when there are no synchronizers in a transmission?

10. What is a common gear ratio for high gear in a manual transmission?

■ ASE-STYLE REVIEW QUESTIONS

1. Which of the following is/are true about transmission speeds?
   a. When speed goes down, torque goes up.
   b. The input shaft turns faster than the output shaft.
   c. Both A and B.
   d. Neither A nor B.

2. Two transmissions have different ratios. One has a 3:1 gear ratio in first gear and the other has a 2.2:1 gear ratio in first gear. Technician A says that the 2.2:1 transmission is a close ratio. Technician B says that the 3:1 ratio is a wide ratio. Who is right?
   a. Technician A   c.   Both A and B
   b. Technician B   d.   Neither A nor B

3. Which of the following is/are true about spur gears?
   a. Reverse gears are often spur gears.
   b. Spur gears have replaced helical gears in transmissions because they are quieter.
   c. Both A and B.
   d. Neither A nor B.

4. Technician A says that some manual transmissions use SAE 80 or 90 gear oil. Technician B says that some manual transmissions use automatic transmission fluid. Who is right?
   a. Technician A   c.   Both A and B
   b. Technician B   d.   Neither A nor B

5. Which of the following is/are true about synchronizers?
   a. The synchronizer hub is splined to the mainshaft.
   b. The synchronizer wedges against the gear teeth to stop the gear from spinning.
   c. The synchronizer is needed for shifting into reverse.
   d. None of the above.

6. Technician A says that a typical five-speed transmission has an extra set of gears located in the extension housing. Technician B says that some five-speed transmissions have a dog clutch that is between the fifth and reverse gears. Who is right?
   a. Technician A   c.   Both A and B
   b. Technician B   d.   Neither A nor B

7. Each of the following statements is true about gear design *except*:
   a. A spur gear has more backlash than a helical gear.
   b. A spur gear is stronger than a helical gear.
   c. A spur gear is noisier than a helical gear.
   d. A spur gear has straight-cut gear teeth.

8. Which of the following is/are true about transmission gearing?
   a. In a higher gear range, a small gear drives a larger gear.
   b. In a lower gear, more torque is produced.
   c. Both A and B.
   d. Neither A nor B.

9. Technician A says that an idler is a gear that changes the direction of rotation. Technician B says that a synchro keeps two meshing gears from clashing during a shift. Who is right?
   a. Technician A   c.   Both A and B
   b. Technician B   d.   Neither A nor B

10. Technician A says that a transmission changes torque to fit an engine's operating requirements. Technician B says that in low gear the crankshaft turns once while the transmission output shaft turns approximately three times. Who is right?
   a. Technician A   c.   Both A and B
   b. Technician B   d.   Neither A nor B

Manual Transmission Diagnosis and Repair

■ **KEY TERMS**

dummy shaft   fibrous grease

■ INTRODUCTION

This chapter deals with the diagnosis and repair of manual transmissions. Emphasis is on generic diagnosis and repair procedures. Knowing these procedures will make diagnosing and repairing transmissions an orderly process.

Learning how to use a press and pullers properly is a very important part of transmission repair. There are many different ways that transmissions are disassembled and reassembled. Be sure to locate service information for the transmission being worked on. Disassembling transmissions usually requires the use of controlled force. Do not use a press, puller, or hammer until you are sure that what you are doing is correct.

NOTE: *Unless you are an apprentice or are working on the transmission with someone who has repaired that particular transmission before, service information will be necessary to avoid damage to the transmission.*

Instructions specific to a particular transmission will not be covered in this chapter. Each transmission has its own peculiarities when it comes to disassembly, but the basic operation and diagnostic procedures are similar. Gears require a certain amount of clearance between each other. Synchronizers are mostly of the same design, and procedures for testing them are similar.

■ TRANSMISSION DIAGNOSIS

Quite often complaints of hard shifting or a transmission that does not go into gear are due to a clutch that is out of adjustment or defective. Always check clutch adjustment and operation of linkage or the hydraulic assist. See that the clutch fork moves the expected distance before removing the transmission from the car.

Transmission failure can result from abuse or misapplication of shocks to the transmission. This can be because of hot rodding or due to a wet foot slipping off of the clutch during a downshift. Worn linkage, driving with a defective clutch, or wear resulting from extra-high mileage can also cause transmission problems.

Before removing the transmission, try to determine the cause of the failure. Ask the driver about the symptoms. A test drive will often help confirm your diagnosis if the car is driveable.

Some typical symptoms a transmission can have include gear clashing during a shift, hard shifting, jumping out of gear, or unusual noise. Noises can be due to a failed front or rear bearing, or defective needle bearings in the countergear or input shaft. Bad linkage often contributes to a transmission becoming locked in gear. This can be because of the internal failure that results from someone forcing a faulty linkage to shift gears.

NOTE: *Driver error can result in low-mileage manual transmission damage. This occurs when the overdrive ratio*

is selected at speeds below approximately 50 miles per hour. The result is that too much load without enough leverage is exerted on the helical gear teeth. Helical gear teeth wipe against each other, creating end thrust. This can cause the transmission to begin to jump out of gear, something that will become worse and worse each time it happens again. Selecting overdrive at too low a speed can also result in engine damage due to lugging.

■ LUBRICANT CHECKS

Oil leaks often contribute to transmission failure. If the transmission runs low on lubricant, check for leaks at the front and rear seals, any switches, plugs or vents, and gaskets. Lubricant level is checked by removing the fill plug and inserting your little finger inside the hole. You should feel oil within ½" of the bottom of the hole (see Figure 14.21).

■ TRANSMISSION REMOVAL

This section is a generic description of a rear-wheel-drive transmission removal. Removing a transmission can be dangerous if you do not understand exactly what you are doing. Seek assistance before pulling your first transmission. Use a hoist and transmission jack or jack stands and a hydraulic jack if doing the job on the ground.

Before installing the jack under the transmission, drain the gear oil. Be certain that the bolt being removed is the correct one for draining the fluid.

> **CASE HISTORY** *An apprentice at a lube and oil service outlet was told to drain the transmission oil out of a Ford Mustang with a five-speed transmission. After he drained the transmission, he loosened and removed a bolt in the side of the transmission that he figured was the fill plug. Unfortunately, this was not the fill plug. The transmission had to be removed and disassembled for repair (**Figure 72.1**).*

Sometimes, it may be necessary to remove exhaust pipes before the transmission can be removed. Be sure to use penetrating oil on the rusty studs. Use an impact wrench with a deep socket to remove the nuts.

Disconnect the battery. Before the transmission can be removed, the drive shaft must be removed. Mark the rear universal joint on the drive shaft and its corresponding yoke for proper balance upon reinstallation. With a two-piece drive shaft, mark each half so they can be reassembled with the correct phasing. When the gear oil is not to be drained ahead of time, a yoke or a seal installation tool can be installed on the output shaft splines to keep oil from running out (**Figure 72.2**).

Remove the speedometer cable from the extension housing. Remove and label any wires to electrical

Figure 72.1 An apprentice thought this bolt was the transmission fill plug when he mistakenly removed it. This resulted in having to remove and partially disassemble the transmission.

Figure 72.2 A slip yoke installed on the output shaft will prevent lubricant leakage. *(Courtesy of Tim Gilles)*

switches (backup lights, transmission controlled spark switches, and so forth).

Disconnect external shift linkage at the transmission. This can involve pulling the handle from the top of a shift rail type of transmission, or it can mean removing the shift linkage from the shift levers on the side of the transmission.

> **SHOP TIP** To gain easier access to the nuts that hold a shifter handle in place, remove the cross-member bolts (**Figure 72.3**) and slowly lower the jack a small amount. This will usually let the transmission hang low enough to gain access to the shifter nuts. Be careful not to put too much pressure on the front motor mounts when you lower the jack.

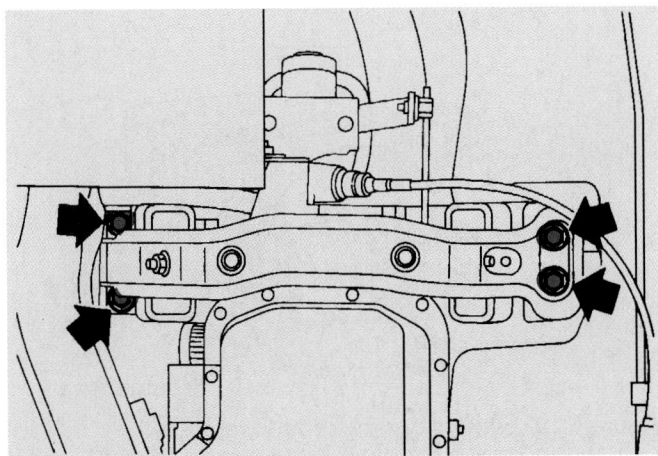

Figure 72.3 Support the transmission and remove the crossmember bolts. *(Courtesy of Nissan North America, Inc.)*

If the vehicle is equipped with one, remove the crossmember. Support the rear of the engine.

 **SHOP TIP** If lifting beneath the engine from the oil pan with a jack, make sure to use a piece of plywood to protect the oil pan.

Use a jack under the transmission to support it. Remove the transmission. Slide it straight back away from the flywheel until the input shaft is clear of the clutch. Do not let the engine hang only by its two mounts. Be aware of the cooling fan-to-radiator shroud clearance.

SAFETY NOTE Use a jack to remove the transmission or be sure to have someone help you so you do not hurt your back.

Remove the transmission and clean it. After cleaning the transmission, put it on a workbench for disassembly.

Identify the Transmission

When ordering parts from a dealer or parts store, you will need to identify the transmission. There will be either a transmission I.D. tag under one of the bolts or numbers stamped on the transmission case. The VIN can also be helpful.

■ TRANSMISSION DISASSEMBLY

Before disassembling a transmission, always check the service information. Procedures vary from transmission to transmission. Sometimes a case can be damaged by pressing a bearing in the wrong direction.

NOTE: *Sometimes heat is necessary for expanding a gear during disassembly or reassembly. A heat gun or an oven are the tools of choice rather than a torch, which can damage a gear if it is overheated.*

With external shift rail type linkage, remove the shift levers. Remove the extension housing. If it does not come off easily after unbolting it, use a soft mallet or brass hammer to knock it loose (**Figure 72.4**). On some transmissions with internal shift rails, the linkage will have to be disconnected before the extension housing can be removed. The extension housing on these types must be removed before the cover can be removed from the transmission. Then, remove the shift fork assembly and cover (**Figure 72.5**).

After removing the side cover or top cover, try to shift the gears and rotate the input shaft. Inspect the

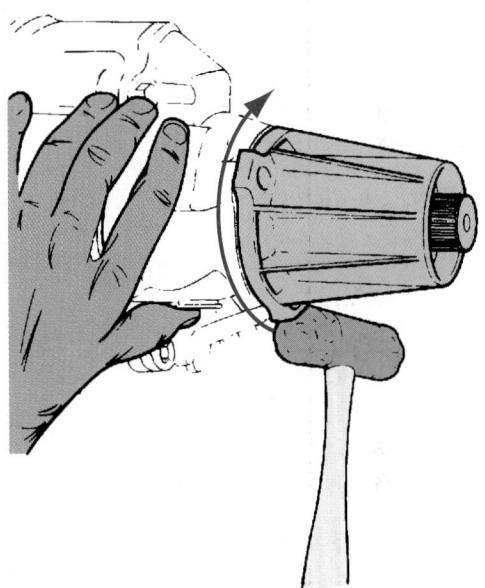

Figure 72.4 If the extension housing does not come loose easily, tap it lightly with a brass hammer. *(Courtesy of DaimlerChrysler Corporation)*

Figure 72.5 Remove the cover and shift forks.

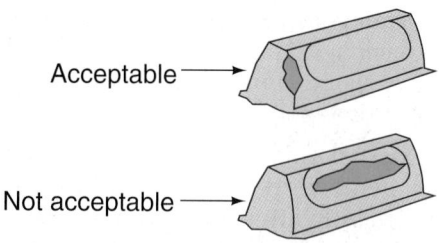

Figure 72.6 Acceptable and unacceptable gear tooth damage. *(Courtesy of Ford Motor Company)*

gears for any obvious signs of damage. As you disassemble the transmission, look for unusual signs of wear on parts that rub against each other. Gears will develop a normal wear pattern with use. They should not have any damage on the face of the gear in the tooth contact area. Damage to the edges of the gear tooth can be ground down if there is a high spot (**Figure 72.6**).

Remove the front bearing retainer (see Figure 71.13). Note the position of its oil return hole so it can be correctly reinstalled. Save the gasket. There are different thicknesses available, and when it is replaced, one of the same size must be used. They control how tightly the bearing is held by the bearing retainer. Next, remove the snap rings on the outside and inside of the bearing.

On many transmissions, before the mainshaft assembly can be removed from the case, the countershaft must be removed to allow the countergear to drop away from the gears on the mainshaft.

> ### SHOP TIP
> ■ The end of a shaft can easily be damaged by pounding on it with a steel hammer. Develop the habit of using brass hammers or drifts on hard steel parts.
> ■ A short shaft can be used to keep the needle bearings in the countershaft. This shaft, called a **dummy shaft** (**Figure 72.7**), is slightly shorter than the case. This will allow the gears to drop free of the mainshaft gears.

Knock the countershaft and reverse idler shafts out of the case. Check the service information to see which direction to drive the shafts. Sometimes they can only go one way or the case will be damaged. A countershaft might be tapered or might have a pin through it on one side.

The mainshaft is removed from the rear or the top of the transmission, depending on transmission design. The countergear can be removed next. On some transmissions, the countergear bearing must be removed with a puller before the countergear can be removed (**Figure 72.8**).

Remove the input shaft from the case. Be careful not to damage the gears. Depending on the type of transmission, sometimes the input shaft is removed

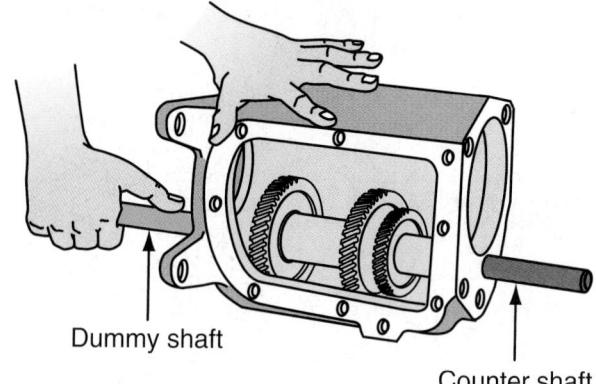

Figure 72.7 Using a dummy shaft to remove the countershaft when there are loose needle bearings.

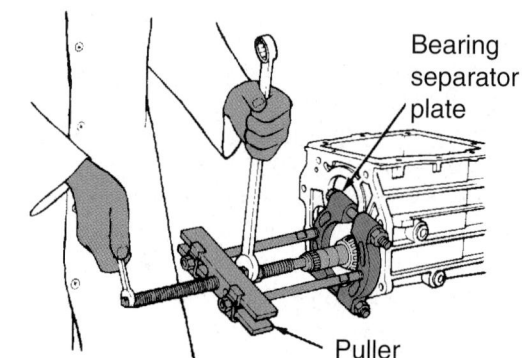

Figure 72.8 Removing the countergear bearing. *(Courtesy of Ford Motor Company)*

before removing the mainshaft. Other times, the countershaft must be removed first. Check the recommendation in the service information.

■ DISASSEMBLE THE MAINSHAFT ASSEMBLY

When disassembling the parts, remember to lay them out as they are disassembled and keep them in order. The high-gear synchro clutch hub is usually held in place on the mainshaft with a snap ring. It is removed with snap ring pliers (**Figure 72.9**). The synchronizer clutch hub fits snugly on splines on the mainshaft. A press or puller will probably be required to pull it

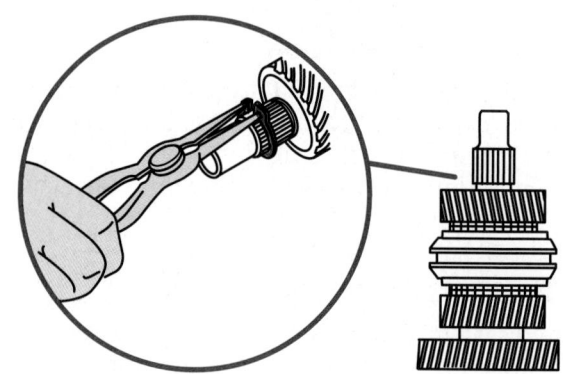

Figure 72.9 Removing a snapring on the front of the mainshaft.

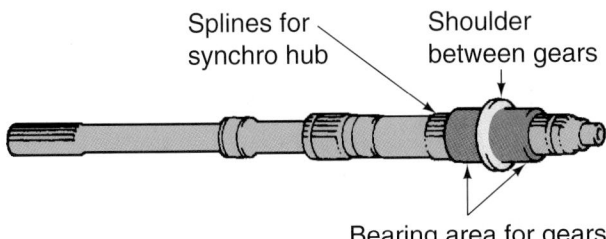

Figure 72.10 Be aware of the location of the shoulder in the middle of the mainshaft when pressing or pulling gears.

from the mainshaft. It must be removed to release the gear behind it. Most synchronizer hubs have a specific "front" and "back" side.

CAUTION There is often a shoulder on the mainshaft that separates the gears (**Figure 72.10**). Be aware of this shoulder when pressing or using a puller. Sometimes, a removable collar has the shoulder on it.

■ SYNCHRONIZER SERVICE

If synchronizers are to be reused, scribe a line on the outside so that they can be reassembled in the same position (**Figure 72.11**). Some synchronizers are marked from the factory.

CAUTION Do not use a punch to mark a synchro sleeve. It can cause distortion. An electric engraver is a good alternative.

After the gears and parts are disassembled from the shaft and cleaned, inspect them. Check the bottom of

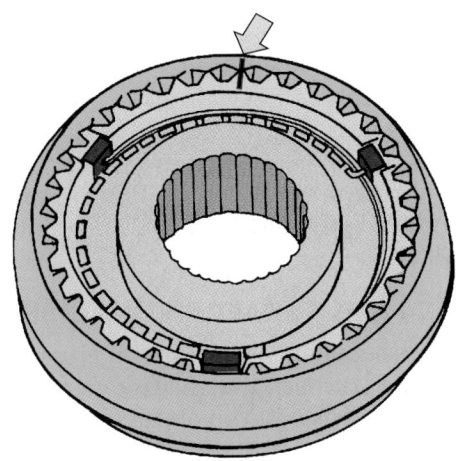

Figure 72.11 Mark the synchro before disassembly.
(*Courtesy of Volkswagen of America, Inc.*)

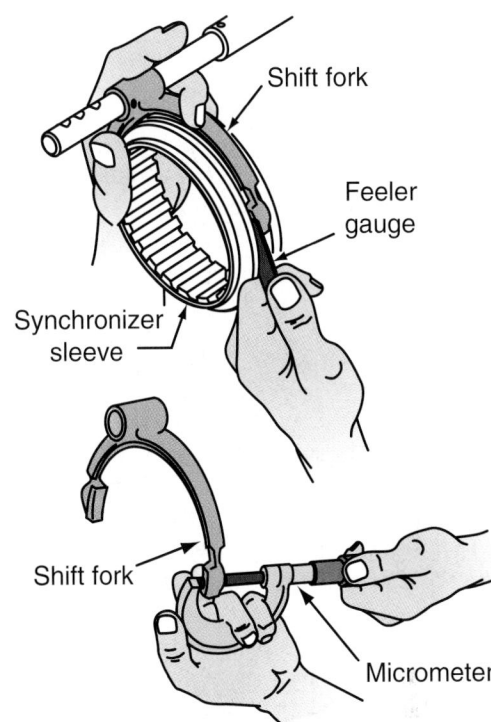

Figure 72.12 Checking for wear on the synchro sleeve and shift fork.

the transmission case for metal. You might find brass dust in the oil due to synchronizer or thrust washer wear. You can check for gear wear by dragging a magnet through any remaining oil in the case. This will not work for brass synchros or thrust washers, because brass is not magnetic. Thoroughly clean the oil from the case with solvent and clean all the transmission parts.

CAUTION When blowing bearings dry, do not allow them to spin. They can fly apart, possibly causing an injury.

Look for wear on the gears and shift sleeves. Check the shift mechanism for wear and warpage. Wear can occur between the shift fork and synchro sleeve (**Figure 72.12**) or between parts of the linkage.

■ SYNCHRONIZER INSPECTION

The best way to tell if a synchro is not working is to test drive the car before you remove the transmission. A worn synchro will cause gear clashing, especially on downshifts (from third to second, for instance).

Inspect the synchronized gears (**Figure 72.13**). Gears are in constant mesh and rarely experience excessive wear unless the transmission has been operated without oil or with the wrong lubricant. The place that often wears is where teeth on the synchronizer hub

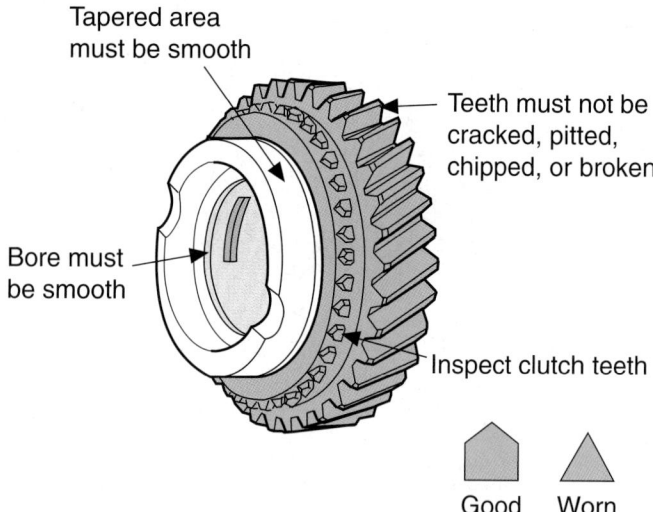

Tapered area must be smooth

Teeth must not be cracked, pitted, chipped, or broken

Bore must be smooth

Inspect clutch teeth

Good   Worn

Figure 72.13 Inspect the gears. *(Courtesy of DaimlerChrysler Corporation)*

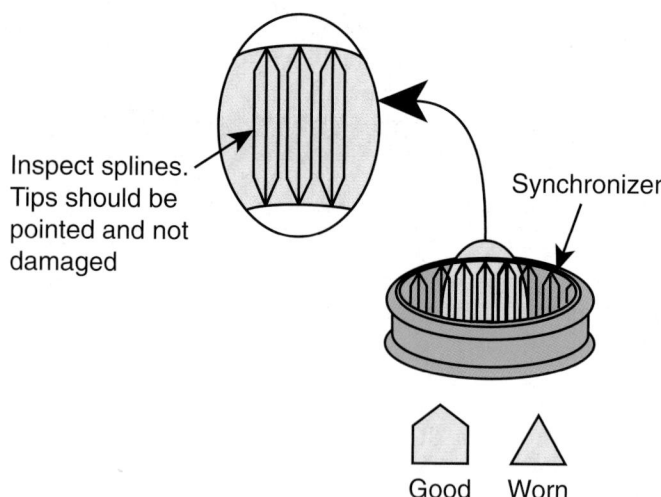

Inspect splines. Tips should be pointed and not damaged

Synchronizer

Good   Worn

Figure 72.14 Inspect the synchro clutch sleeve.

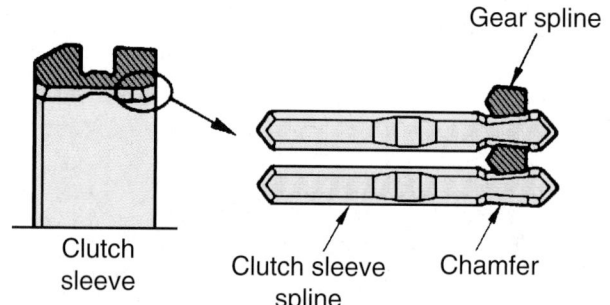

Gear spline

Clutch sleeve

Clutch sleeve spline

Chamfer

Figure 72.15 Shape of the synchro clutch sleeve teeth.

rub against the teeth on the gear. Wear in this area can result in a transmission that slips out of gear or sticks in gear. When synchros have experienced wear, the teeth on the corresponding gear could also be worn.

Inspect the synchro sleeve (**Figure 72.14**). The teeth are sometimes smaller toward the inside (**Figure 72.15**). Their width should be even across their length, not worn more on one side. Synchro sleeves wear either on the outside edge of the tooth or on the inside, partway into the sleeve. If you replace a clutch sleeve, look for problems with the shift linkage that could have caused the problem.

CASE HISTORY

A customer had her Jeep towed into a repair shop because it jumped and made terrible grinding sounds when driven. She had loaned it to a friend and it had "stopped working." The customer related that she had been having difficulty shifting the transmission into fifth gear for some time and that sometimes it would get stuck in fifth.

When the transmission was disassembled, a stripped fifth gear was discovered. The synchro hub was also severely worn. Part of the shift mechanism that keeps the transmission from shifting into two gears at once was also broken. Third gear was damaged badly on the countergear and mainshaft. Before reading the next paragraph see if you can figure out why these parts were broken.

Diagnosis:

The transmission became stuck in fifth gear and the friend forced it with enough pressure to break the shift rail. With the shift lockout mechanism broken, the transmission was able to shift into third gear while it remained locked in fifth. The result was a very costly transmission overhaul. Gears, especially the countergear, are very expensive.

■ INSPECT BLOCKER RINGS

Carefully inspect the brass blocker (synchro) rings if they are to be reused. The inside surface of the blocker rings should have sharp edges to the thread cut into it, and the teeth should be unworn and not damaged (**Figure 72.16**). Because removing and disassembling a transmission is difficult and time consuming, synchros are often replaced when a transmission is apart. This will depend on the cost of the new parts. The test drive is especially important for this part of transmission diagnosis.

Push the blocker ring against the polished tapered surface on the gear that it rides on. It should grab the chrome surface when pushed against it. If it is too worn, it will slip. Manufacturers often specify a minimum clearance between the blocker ring and the gear (**Figure 72.17**).

■ INSPECT INPUT SHAFT AND MAINSHAFT

Inspect the input shaft. It has several areas that could require service. Inspect the mainshaft splines. They occasionally get twisted from abusive driving. A twisted spline will prevent the drive shaft slip yoke from sliding in and out of the transmission when the car goes over bumps.

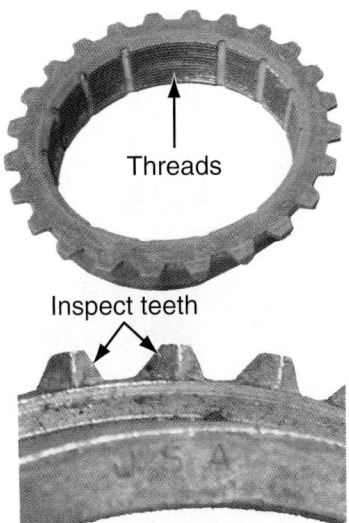

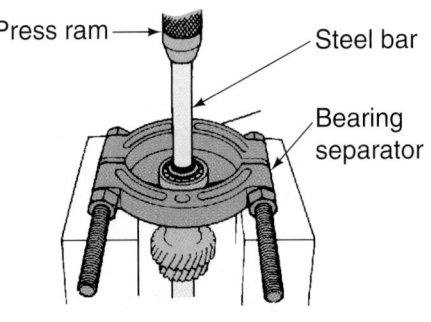

Removing a bearing

Figure 72.18 Using a press to remove a bearing from a shaft.

Figure 72.16 Inspect synchronizer blocker ring threads and teeth. The threads should be sharp. *(Courtesy of Tim Gilles)*

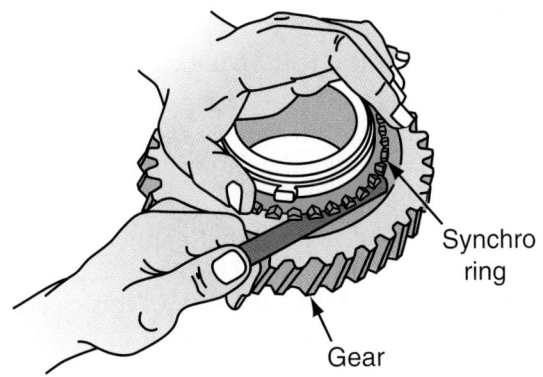

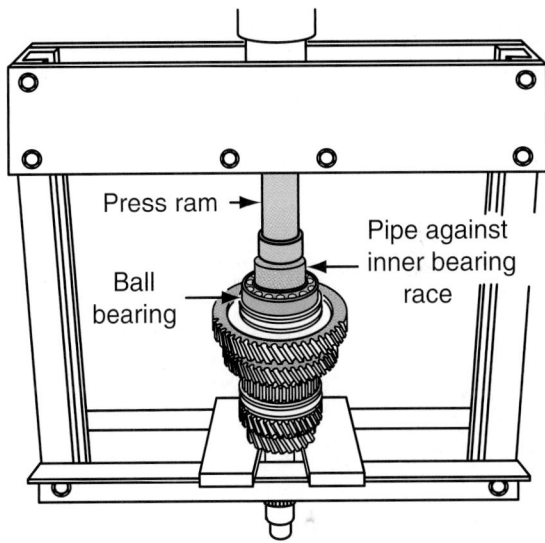

Figure 72.19 When installing a bearing on a shaft, press against the inner bearing race only.

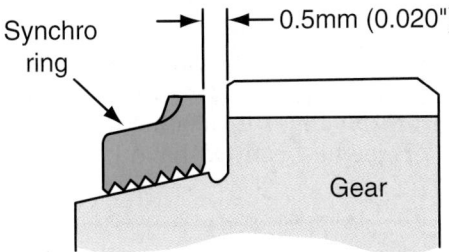

Figure 72.17 Checking synchro wear.

The speedometer gear or vehicle speed sensor (VSS) gear is also located on the output shaft. Inspect it for damage. This gear rarely suffers damage, because the gear from the speedometer that meshes with it is usually made of a softer material. Sometimes, the speedometer gear on the mainshaft is replaceable. A clip can hold it in place or it can be a pressed fit. A special puller is available to pull these gears on the car.

■ REPLACE WORN BEARINGS

Replace any worn bearings. Use a press and support the inside of the bearing with a bearing separator

(**Figure 72.18**). When reassembling the new bearing to the shaft, always put pressure on the inside of the bearing only (**Figure 72.19**). Pressing on the outside of the bearing can damage it. It is important that the bearing be installed all of the way onto the shaft or clearance problems can result.

■ REASSEMBLE THE TRANSMISSION

Before assembling the transmission, locate and assemble any new replacement parts. Two gears operating in mesh create a wear pattern between them. When a gear is replaced, the corresponding gear that rides on it is customarily replaced too. Otherwise, using a new gear with an old gear can result in unacceptable gear noise. The cost of replacement gears often makes it advantageous to consider the cost of a guaranteed rebuilt unit.

■ REASSEMBLE THE SYNCHRONIZERS

Reassemble the synchronizer hub parts (**Figure 72.20**). There are usually three inserts held in place with a round spring. Install the springs so that they rotate away from the same insert but in opposite directions.

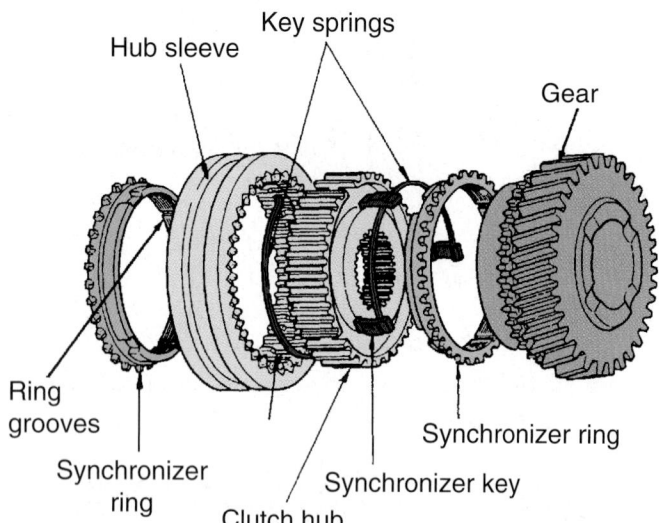

Figure 72.20 Parts of a synchronizer hub.

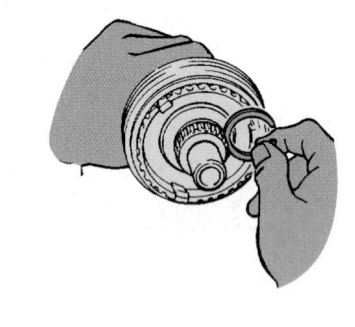

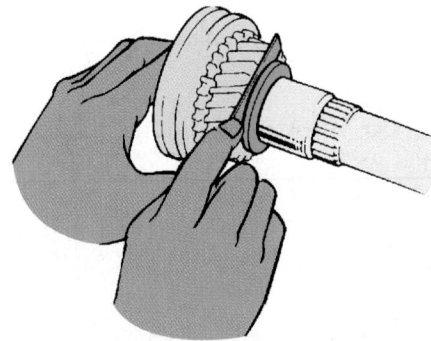

Figure 72.21 The thickness of the snapring determines the clearance between the gears.

■ REASSEMBLE THE MAINSHAFT

Install the gears on the mainshaft. Lubricate all parts liberally. Install the synchro hub with a press or pipe and hammer. Check end play of assembled parts as described in the service manual. On some transmissions, selective snap rings are available. The thickness of the snap ring determines the clearance between the gears (**Figure 72.21**).

■ END PLAY

The thrust washers at the ends of the countergear determine where the gear will be positioned in the case and also the amount of clearance. There might be a

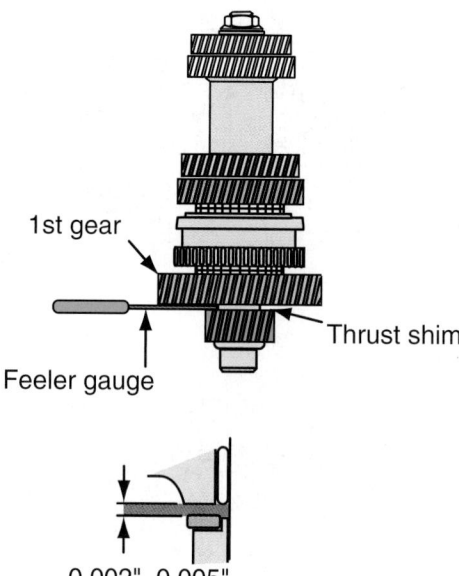

Figure 72.22 Measuring clearance between the gears on the mainshaft.

thicker washer on the rear of the countergear, or vice versa.

NOTE: *Positioning the washers on the wrong ends of the gear can result in interference between the countergear and other gear(s) on the mainshaft.*

Clearance is checked on the mainshaft after all of the gears are assembled on it (**Figure 72.22**). Normal clearance is usually 0.002" to 0.005". This clearance is adjusted by changing the thickness of the snap ring or by installing a thrust washer at the center of the shaft. Clearance is checked at each of the gears.

■ NEEDLE BEARING INSTALLATION

Needle bearings are sometimes *caged* (they are held in place by a cage). They can also be loose and are installed one at a time (**Figure 72.23**). If this is the case, use very heavy grease to hold them in place during assembly. Be sure that all needle bearings have been installed.

SHOP TIP
- A non-fibrous grease works best (a **fibrous grease** is sticky). It leaves no strings of grease when you put a small amount between two fingers and then spread them apart.
- A dummy shaft can be used during transmission assembly to make reinstallation of the countershaft much easier. The short shaft is installed and the bearings are inserted around it. The countergear is installed in the bottom of the transmission case. When the countershaft is pounded into the case and into the end of the countergear, it displaces the short shaft. The short shaft comes out the other end of the transmission case during assembly.

Figure 72.23 Installing loose needle bearings in the countergear and input shaft.

The service manual often specifies the number of needle bearings. Experienced technicians will also count the needle bearings during disassembly.

■ INSTALL NEW GASKETS

Purchase a new gasket set for the transmission. Gaskets are not reusable. A gasket that was working fine previously could leak after reassembly. Replacing this gasket would be a complicated repair that would be done at no charge to the customer.

NOTE: *Silicone RTV should only be used when recommended. Some gaskets are a specified thickness to provide the correct bearing end play or preload.*

Install the shift mechanism.

NOTE: *Check to see if the transmission shifts properly before completing the assembly. On some transmissions, if the shift forks are installed backwards, the transmission will not shift properly. Fixing the problem could require removal of the transmission once again.*

CAUTION Be sure that metric and English fasteners are not interchanged by accident.

On rear-wheel-drive cars, replace the extension housing bushing and seal. The bushing rides against the slip yoke on the front of the drive shaft. It can be replaced with a hydraulic press or bearing driver before the extension housing is installed. After the extension housing is installed, the bushing can be removed and replaced with special tools (**Figure 72.24**). A seal

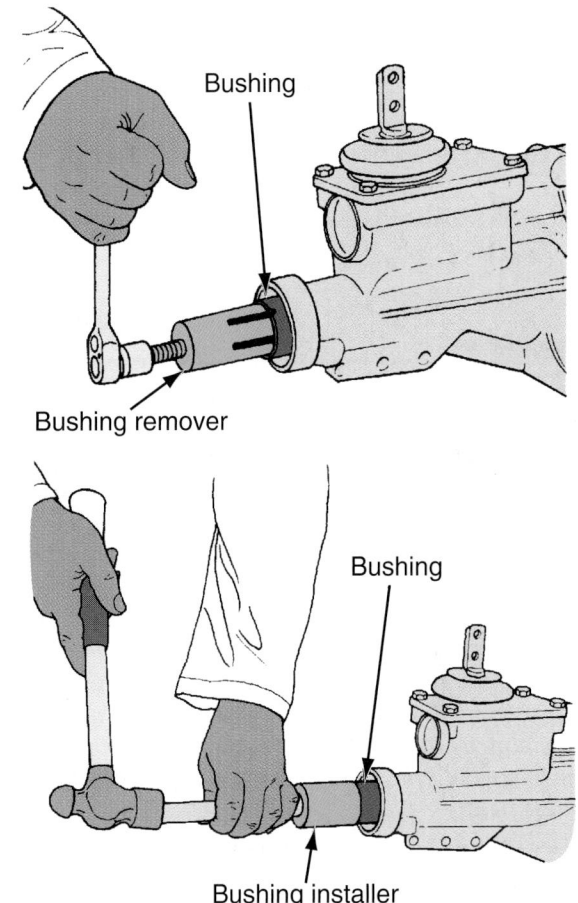

Figure 72.24 Removing and installing a transmission bushing. *(Courtesy of Ford Motor Company)*

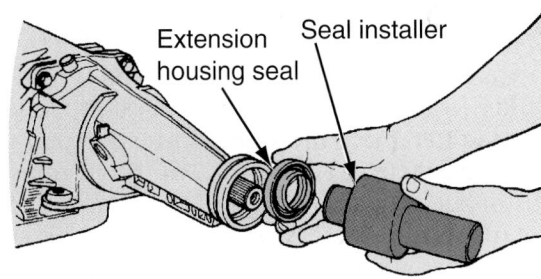

Figure 72.25 Installing a seal with a seal installer. *(Courtesy of Ford Motor Company)*

installer is used to install a seal without distorting it (**Figure 72.25**).

NOTE: *When installing rubber seals and O-rings, be sure to oil them. Failure to do this can result in damage to the seal.*

■ INSTALL THE TRANSMISSION

Before attempting to install the transmission, be sure that the clutch fork is properly seated on its pivot. Align the transmission with the engine. The transmission must be at the same angle so it is able to slip into the splines in the clutch disc. If the clutch has been disassembled, see the instructions on clutch disc alignment in Chapter 70. Shake the transmission while rotating it slightly back and forth to help the input shaft to slip into the clutch disc and pilot bearing or bushing.

> **CAUTION** Do not use the transmission bolts to draw the transmission into the clutch. The transmission should slide all the way in by hand.

CASE HISTORY *An apprentice technician was installing a clutch in a rear-wheel-drive car. During reinstallation of the transmission, he found that it would only slide in to about ½" of the housing. He installed the bolts and tightened them lightly and evenly until the transmission was flush with the housing. After completion of the job, he started the car in neutral. When he attempted to shift it into gear, it would not go. He shut off the engine and put the transmission in gear. When he attempted to start it this time it lurched forward. The clutch would not release because the pilot bushing was deformed and held the transmission input shaft tightly.*

Install the bolts in the transmission. Install the crossmember and its vulcanized mount, if applicable. Reinstall any clutch or transmission linkage that was removed previously.

■ TRANSMISSION LINKAGE ADJUSTMENT

Most external linkages include a provision for aligning the shifter to neutral. With the linkages disconnected from the shifter, place the shifter in neutral and put a pin through the hole (**Figure 72.26**). Then, adjust the lengths of the rods until they can be attached to the arms of the shifter. There is usually an adjustable swivel that is threaded up or down the shift rod to adjust it.

■ ADD LUBRICANT

Remember to add the recommended lubricant before driving the vehicle. The lubricant can be added prior to installation in the vehicle. This can result in oil leaking out of the extension housing as the transmission is installed if a yoke is not installed on the output shaft. The lubricant level should come to the bottom of the fill hole (see Figure 14.21).

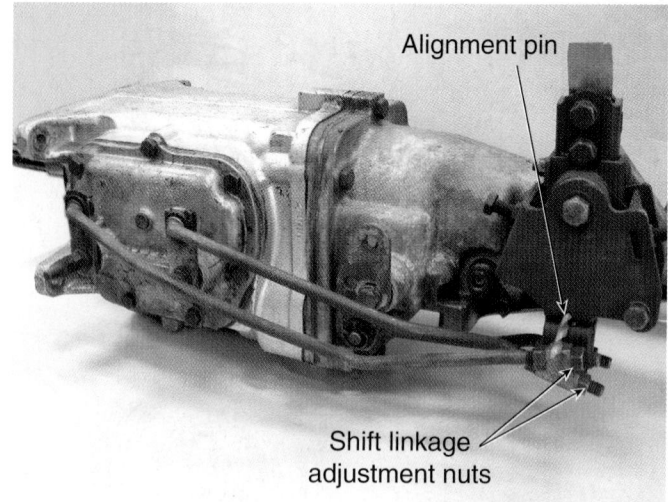

Figure 72.26 A pin is used to align the shift gate while adjusting the transmission linkage pieces to the correct length. *(Courtesy of Tim Gilles)*

CASE HISTORY *An apprentice finished installing a recently rebuilt transmission in a car. He asked the master technician how much gear lube to put in the transmission. The technician said to just fill it until the gear lube begins to come out of the fill hole in the side of the transmission. The drain plug on this transmission happened to be on the side, rather than the bottom. Not seeing the fill plug, he mistakenly pumped some oil through the drain opening. Of course, it soon began to leak out. Had he checked the service information for the system's capacity, he would not have destroyed the second gear synchronizer early in the test drive.*

NOTE: *Be sure to check the service information for the correct type of lubricant to use. In some manual transmissions, ATF or engine oil is specified. Using the incorrect oil can result in transmission damage.*

■ TEST DRIVE

During the test drive, check for proper operation of the transmission. Be sure that shifts are smooth. The shifter should feel firm and the neutral gate should be accurately aligned. After returning to the shop, check the outside of the transmission for leaks.

■■■ REVIEW QUESTIONS

1. Before condemning a transmission that shifts hard, check the operation of the _____.
2. A yoke or a _____ installation tool can be installed on the output shaft splines to keep oil from running out.
3. To gain easier access to the nuts that hold a shifter handle in place, remove the _____ bolts and slowly lower the jack a small amount.
4. What is the name of the short shaft used to drive out a countershaft?
5. There is often a _____ on the mainshaft that separates the gears.
6. What should you do to synchronizer rings before disassembling them from the hub and shift collar?
7. What is the best way to determine if synchronizers are working?
8. How many inserts are there usually on a synchronizer assembly?
9. What will an experienced technician count during transmission disassembly?
10. What gear position is the transmission shifter placed in to perform a transmission linkage adjustment?

■■■ ASE-STYLE REVIEW QUESTIONS

1. Which of the following is/are true about transmission gear tooth wear?
   a. Damage to the edges of the gear tooth can be ground down if there is a high spot.
   b. Gears should have no damage on the face of the gear in the tooth contact area.
   c. Damage to a gear's dog teeth can result in a transmission that jumps out of gear.
   d. All of the above.
2. Technician A says that the speedometer or VSS drive gear on the mainshaft is sometimes replaceable. Technician B says that the gear on the transmission output shaft is softer than the meshing gear from the VSS or speedometer. Who is right?
   a. Technician A   c. Both A and B
   b. Technician B   d. Neither A nor B
3. Which of the following could result from a worn synchronizer hub?
   a. A transmission that jumps out of gear
   b. A transmission that sticks in gear
   c. Both of the above
   d. None of the above
4. Technician A says that blocker rings should have sharp threads. Technician B says that there is often a specification for blocker ring to gear clearance. Who is right?
   a. Technician A   c. Both A and B
   b. Technician B   d. Neither A nor B
5. When reassembling a new bearing onto a shaft:
   a. Apply pressure on the outside of the bearing.
   b. Apply pressure on the inside of the bearing.
   c. Apply pressure to the entire bearing.
   d. None of the above.
6. When assembling a manual transmission, all of the following are true *except*:
   a. On some transmissions the thickness of the snapring determines the clearance between the gears.
   b. Clearance is checked between the gears after they are assembled on the countergear.
   c. Installing a gasket of the wrong thickness can change input shaft end play.
   d. It is possible to add lubricant to a transmission prior to installing it in the vehicle.
7. Which of the following can be used on loose needle bearings to keep them in place during installation of a countershaft?
   a. Oil   c. Both A and B
   b. Grease   d. Neither A nor B
8. Technician A says that the extension housing bushing can be replaced when the extension housing is off the transmission. Technician B says that the extension housing bushing can be replaced when the extension housing is on the transmission. Who is right?
   a. Technician A   c. Both A and B
   b. Technician B   d. Neither A nor B

9. Which of the following is/are true?

  a. Bearings should be spun dry with compressed air before lubricating them.

  b. Synchro rings can be checked for damage by dragging a magnet through the transmission oil.

  c. Both of the above.

  d. None of the above.

10. Technician A says to use the transmission bolts to draw the transmission into the clutch disc. Technician B says to rotate the transmission while the input shaft slides into the pilot bearing. Who is right?

  a. Technician A   **c.** Both A and B

  b. Technician B   **d.** Neither A nor B

Automatic Transmission Fundamentals

■ **KEY TERMS**

fluid coupling	Simpson geartrain	shift quadrant
vortex	Ravigneaux geartrain	upshift
one-way clutch	rotor-type pump	downshift
coupling speed	internal/external gear	throttle pressure
stall speed	crescent-type pump	governor pressure
lock-up torque converter	vane-type pump	detent
compound planetary	orifice	park pawl
gearset	valve body	CVT

■ **INTRODUCTION**

An automatic transmission shifts gears automatically and does not require a manual clutch. It has the same purpose as a standard transmission, to change (or multiply) the amount of torque produced by the engine to match driving conditions. Automatic transmissions can be found in both front- and rear-wheel-drive vehicles. Front-wheel-drive vehicles combine the transmission with a differential in a single unit called an automatic *transaxle* (**Figure 73.1**). The operation of the automatic transmission transaxle is the same as that of a rear-wheel drive. Transaxles are covered in Chapter 77.

Today's automatic transmission is a *torque converter automatic*. Several designs of automatic transmissions existed before the first torque converter automatic was introduced by General Motors in the 1948 Buick. The first automatic transmission was actually developed in Boston in 1904 by the Sturdevant brothers, but its design was not the same as today's reliable unit.

Figure 73.1 An automatic transaxle. (© *Courtesy of Mercedes-Benz USA, LLC*)

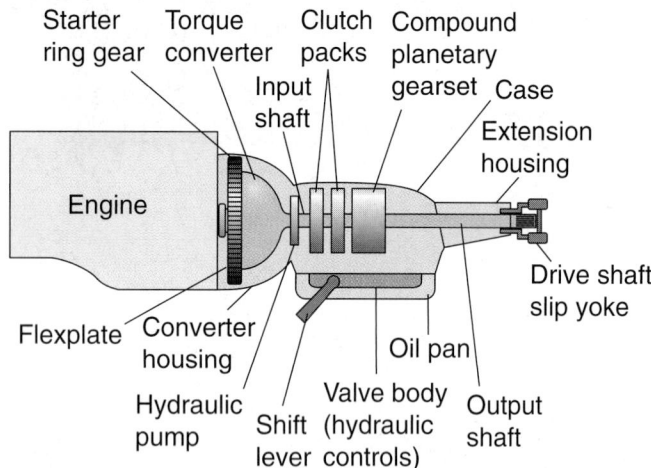

Figure 73.2 Basic parts of an automatic transmission.

AUTOMATIC TRANSMISSION PARTS

The automatic transmission consists of several parts. Refer to **Figure 73.2** for the names and locations of the basic parts.

- A torque converter takes the place of the clutch used with a manual transmission.
- An *input shaft* connects the torque converter to the transmission.
- The transmission *pump* provides hydraulic fluid to the torque convertor, valve body, and lubrication circuits.
- A valve body makes shift decisions and acts on them.
- *Bands* and clutches, called *planetary holding members*, are applied by hydraulic pressure to drive or hold parts of a compound *planetary gear system* to make the gear shifts.
- The *transmission case* includes the torque converter housing, fluid pan, oil passages, and extension housing.

POWER TRANSMISSION

Automatic transmissions use three methods of transmitting power: fluid, friction, and gears. The torque converter transmits power using fluid. The planetary holding members use fluid and friction. Gears in the transmission transmit power and also change speed and torque.

The basic areas of the transmission are the torque converter, planetary gears, and hydraulic controls (**Figure 73.3**). Automatic transmissions sense road speed and engine load to determine when to make a correct shift. In conventional automatic transmissions, gear shifts are made by hydraulic pressure. Electronically controlled transmissions are shifted by electric solenoids triggered by the onboard computer.

FLEXPLATE

A standard transmission has a flywheel. A *flexplate* and torque converter are used with automatic transmissions instead of a flywheel (**Figure 73.4**). It is a thin

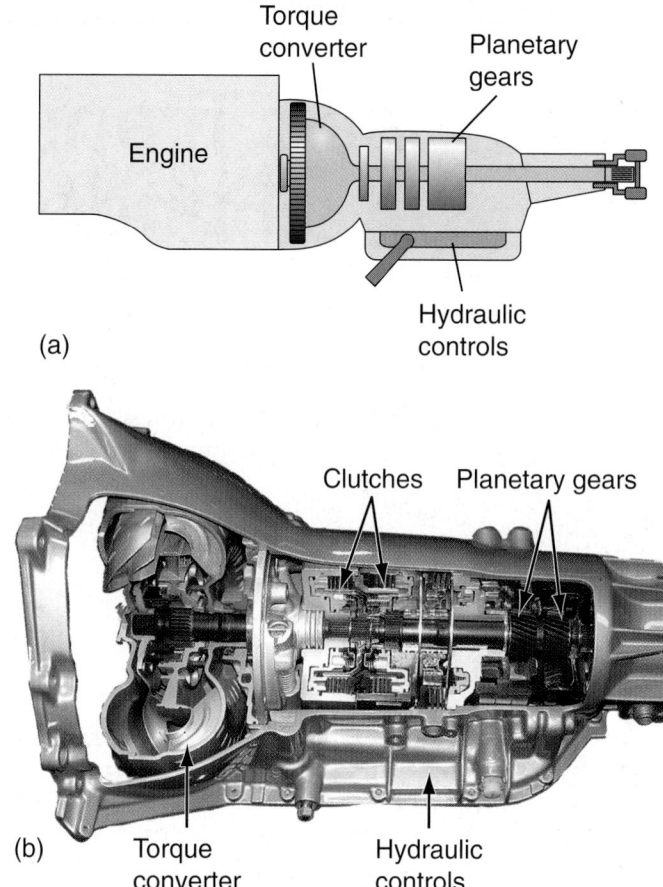

Figure 73.3 Basic areas of an automatic transmission. (b, Courtesy of Tim Gilles)

stamped version of the heavy steel flywheel, usually with a starter ring gear on its outside circumference. Some torque converters have their starter ring gears on the outside of the housing.

The flexplate is bolted to the engine's crankshaft. Bolts or studs on the outside of the torque converter housing are used to fasten the converter to the flexplate. A torque converter is large and holds 3 or 4 quarts of fluid. The torque converter's weight takes the place of the flywheel's function of smoothing the engine's power impulses.

TORQUE CONVERTER

The torque converter takes the place of the clutch. It allows the vehicle to idle at a stop sign. It also slips during initial acceleration to prevent engine stalling.

HISTORY NOTE

Chrysler developed the predecessor to the torque converter in 1937. Called a fluid coupling, it was used with a manual transmission that also had a clutch. With the clutch pedal all of the way up and the vehicle stopped, the engine would not stall.

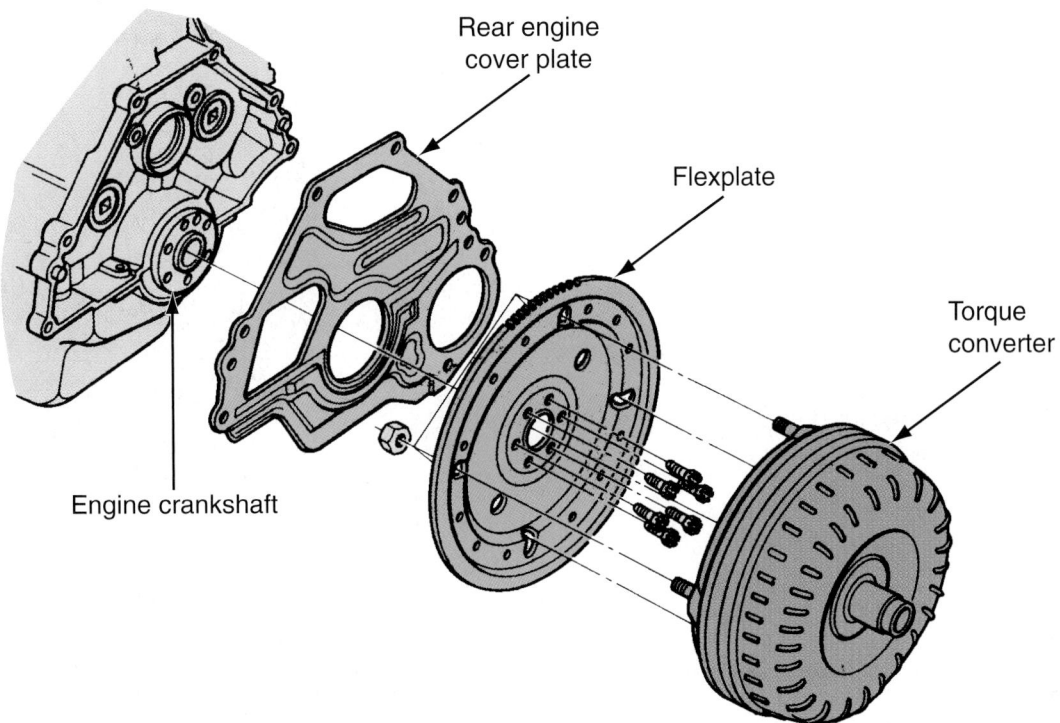

Figure 73.4 A flexplate instead of a flywheel is used with automatic transmissions. *(Courtesy of Ford Motor Company)*

Operation of the **fluid coupling** can be compared to two fans (**Figure 73.5**) with fluid used to transmit motion instead of air. As the first fan turns faster, the second fan picks up energy and turns. There are two parts contained within the welded fluid coupling housing. The driving member is called the *impeller* or *pump* (not to be confused with the hydraulic pump). The driven member is called the *turbine*. There is another part of a transmission that is also called a pump. That pump produces fluid flow to develop the pressure necessary to operate the transmission. To avoid confusion when referring to the names of transmission parts, impeller will be used in the following discussion when referring to the converter pump.

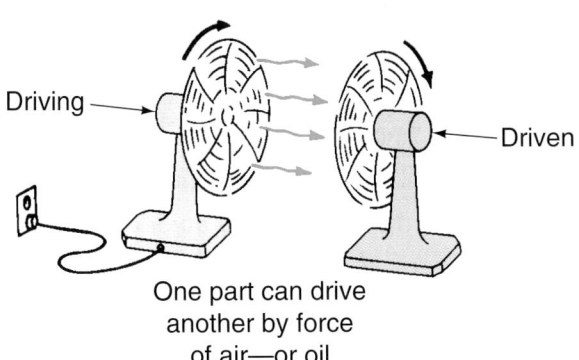

Figure 73.5 Operation of the fluid coupling can be compared to two fans. *(Reproduced by permission of Deere & Company, John Deere Publishing, Moline, IL. All rights reserved)*

■■ TORQUE MULTIPLICATION

A torque converter has the same parts as a fluid coupling but also has an additional member, a *stator*. The difference between a fluid coupling and a torque converter is that the torque converter can actually *increase* torque. Some industrial torque converters can increase torque to the equivalent of 9:1 gear reduction when the pump first starts to drive the turbine. The amount of torque increase in an automotive torque converter when starting from a stop is about 2:1.

Torque is multiplied whenever the impeller spins faster than the turbine. When an automatic transmission equipped vehicle is used to pull a boat up a launching ramp, the following ratios are combined to increase torque available to the drive wheels:
- Transmission low-gear ratio   = about 3:1
- Torque converter ratio   = about 2:1
- Differential ratio   = about 2.5:1

Total ratio = about 15:1

■■ TORQUE CONVERTER OPERATION

The turbine has splines in its center that mesh with splines on the end of the transmission input shaft (**Figure 73.6**). The converter impeller (pump) is actually part of the torque converter housing. It is the part that is closest to the transmission. When the converter housing (impeller) rotates at idle speed, fluid is thrown from the impeller toward the turbine (the driven member of the torque converter). The centrifugal force of

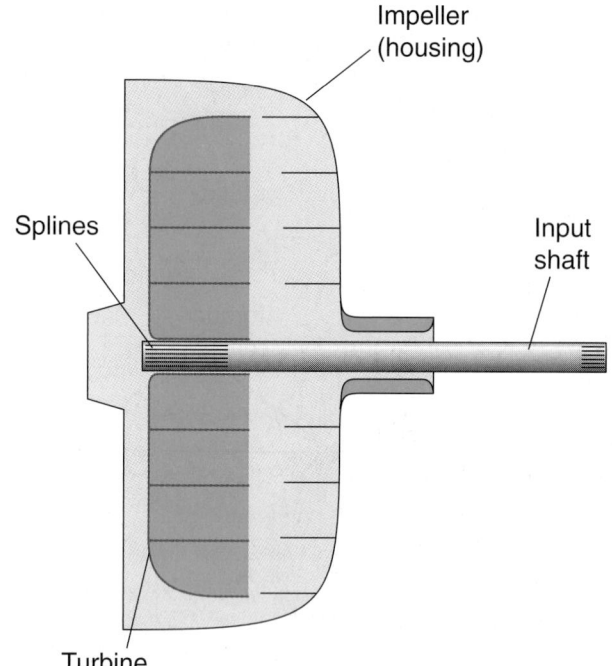

Figure 73.6 Parts of a fluid coupling.

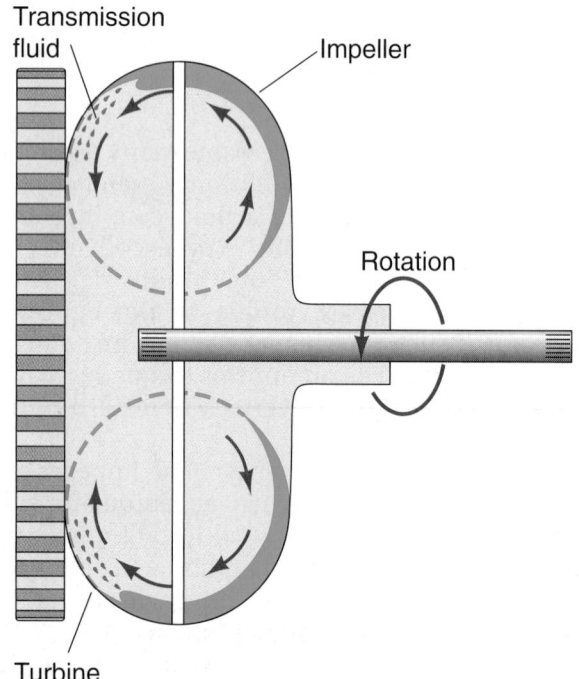

Figure 73.7 The centrifugal force of the rotating torque converter throws oil to the outside of its housing.

the rotating torque converter also throws fluid to the outside of its housing (**Figure 73.7**).

■ TORQUE CONVERTER STATOR

A fluid coupling has no stator and *straight blades* (**Figure 73.8a**). A torque converter has a stator and *curved blades* (**Figure 73.8b**). The stator is installed between the impeller and the turbine to redirect fluid flow. This makes torque increase possible. The stator is mounted

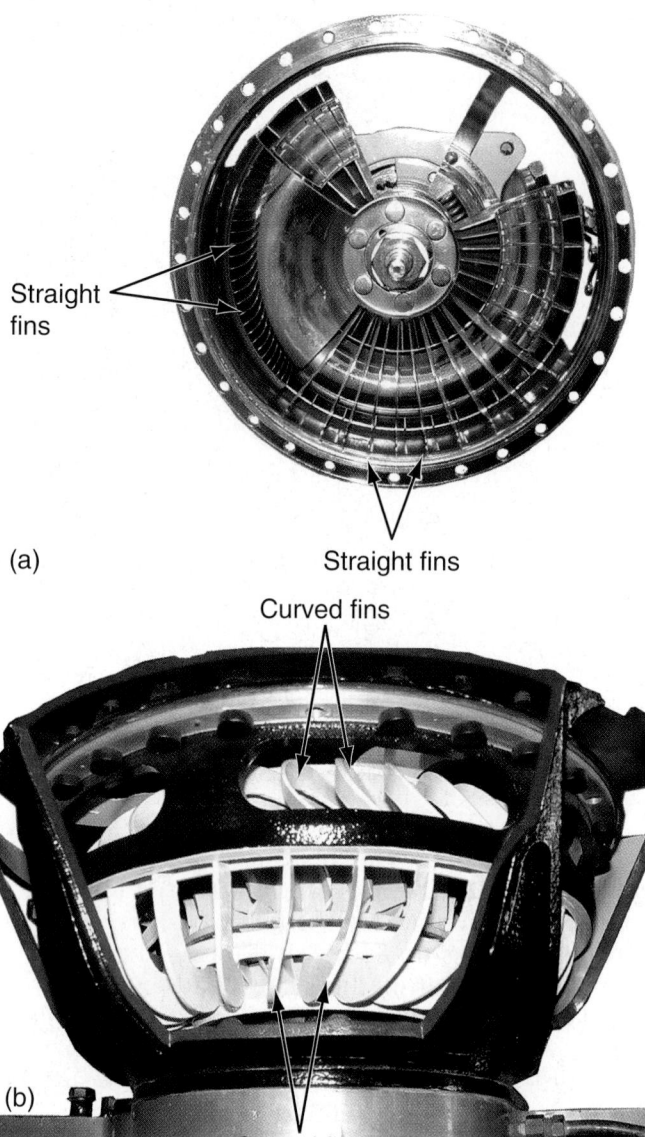

Figure 73.8 (a) A fluid coupling has straight blades. (b) A torque converter has curved blades and a stator in the center. *(Courtesy of Tim Gilles)*

on a one-way clutch supported by the outside of the stationary front pump housing.

When the engine is idling, the converter housing (impeller) spins slowly at engine speed. A small amount of fluid is thrown from the impeller to the turbine, but not enough to lock the engine to the transmission. Unless the engine's idle is too high, the car will remain at rest when the brakes are released.

When the engine is accelerated, more fluid is thrown from the impeller to the turbine. Fluid flows toward the outside, away from the center of the impeller (**Figure 73.9**). When the fluid hits the turbine, it is redirected inside toward the center of the converter (**Figure 73.10**).

As the turbine is driven by the impeller, the transmission input shaft turns and the vehicle begins to move. There is still some slippage between the

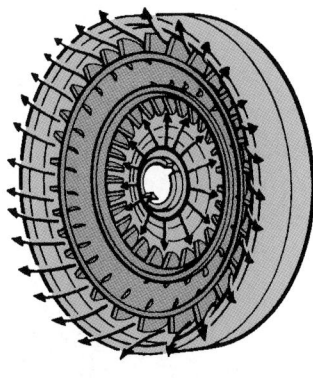

Figure 73.9 Oil flows toward the outside, away from the center of the impeller. *(Courtesy of Ford Motor Company)*

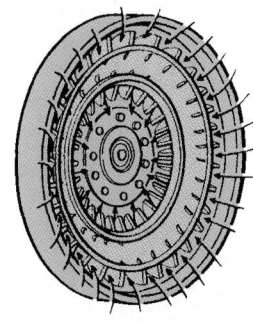

Figure 73.10 When the oil hits the turbine, it is directed inside toward the center of the converter. *(Courtesy of Ford Motor Company)*

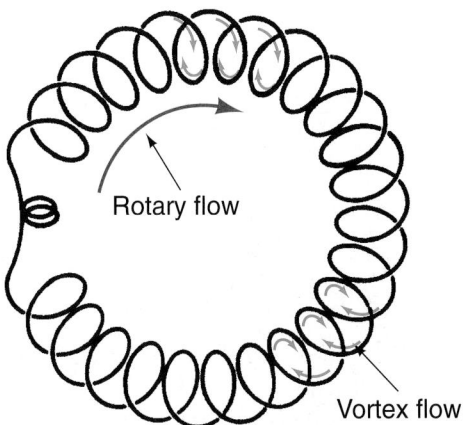

Figure 73.11 The spiral pattern of oil flow is called vortex.

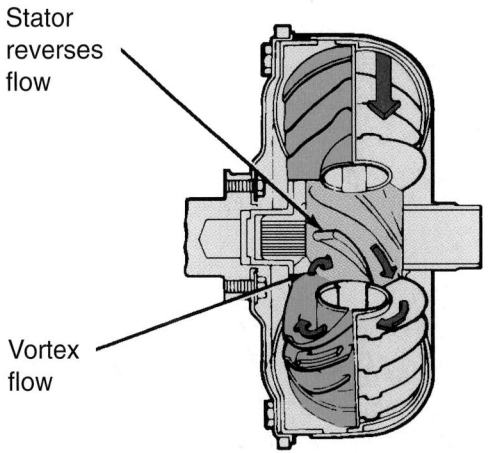

Figure 73.12 The stator catches the fluid coming out of the turbine and redirects fluid flow. *(Courtesy of Ford Motor Company)*

impeller and turbine. Because they are not turning at almost the same speed, torque is being increased.

When the vehicle reaches cruising speed, the impeller and turbine are turning at almost the same speed with very little slippage. There is no further torque multiplication because the parts are turning at the same speed.

■ STATOR OPERATION

The stator is installed between the impeller and the turbine to redirect fluid flow. In a torque converter, the blades are curved so that transmission fluid is thrown at an angle against the turbine. This increases torque. A fluid coupling has straight blades and does not provide torque increase.

As it rotates, the engine's flywheel turns the transmission pump, creating flow in the transmission and converter. Fluid is *circulated* in a pattern throughout the torque converter. This spiral pattern of fluid flow is called vortex flow (**Figure 73.11**).

From a standing start, the fluid is thrown against the turbine at a sharp angle. The turbine can only be driven clockwise but fluid exits from it moving in a counterclockwise direction. If this fluid were allowed to hit the impeller, it would stop it from turning. The stator catches the fluid coming out of the turbine and redirects fluid flow (**Figure 73.12**). When fluid leaves

the stator it strikes the impeller, flowing in the same direction. This actually accelerates the impeller.

The torque converter has split rings attached to the blades of the turbine and impeller (**Figure 73.13**). These direct the fluid in a smooth pattern. Without the split rings, flow in the center of the converter would be turbulent during periods of the most torque multiplication.

■ STATOR CLUTCH OPERATION

The stator has a one-way clutch, also called an overrunning clutch, that locks in one direction and freewheels in the other (**Figure 73.14**). When the fluid strikes the stator at a high angle, the stator's one-way clutch locks up, preventing rotation of the stator (**Figure 73.15**). When the speed of the turbine starts to catch up with the speed of the impeller, the stator does not need to redirect fluid flow any longer. In fact, redirecting the fluid at this point in time would actually work against the turbine. This would cause the stator to act as a brake to the engine. To prevent this, the stator's one-way clutch freewheels. This allows the stator, impeller, and turbine to turn freely with the mass of

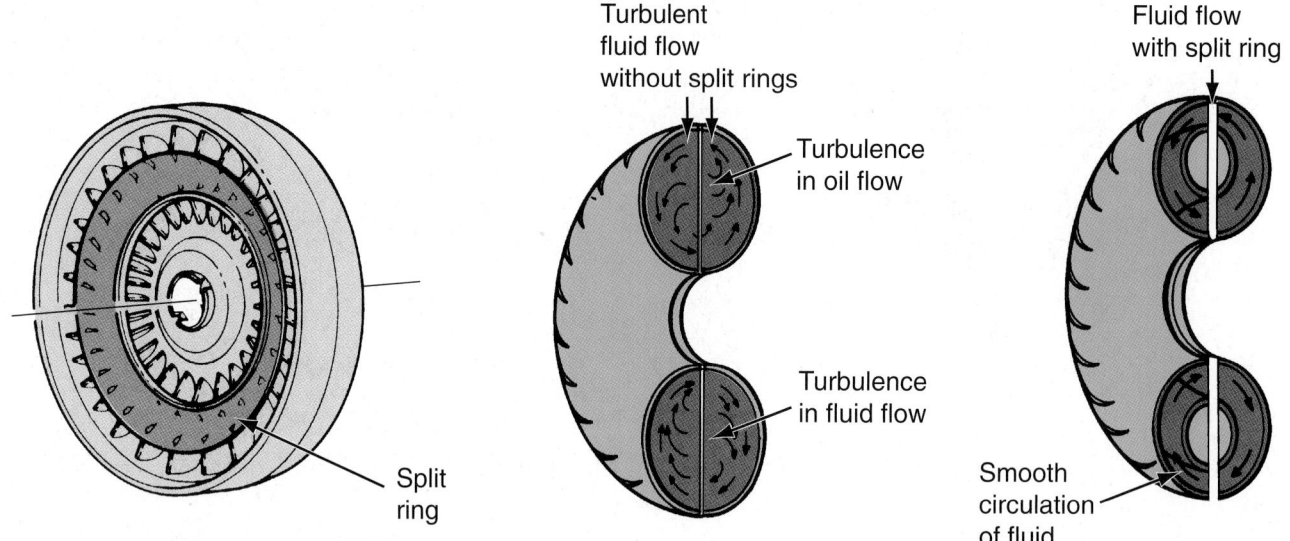

Turbulent
fluid flow
without split rings

Turbulence
in oil flow

Turbulence
in fluid flow

Split
ring

Fluid flow
with split ring

Smooth
circulation
of fluid

Figure 73.13 The torque converter has split rings attached to the blades of the turbine and impeller. *(Courtesy of Ford Motor Company)*

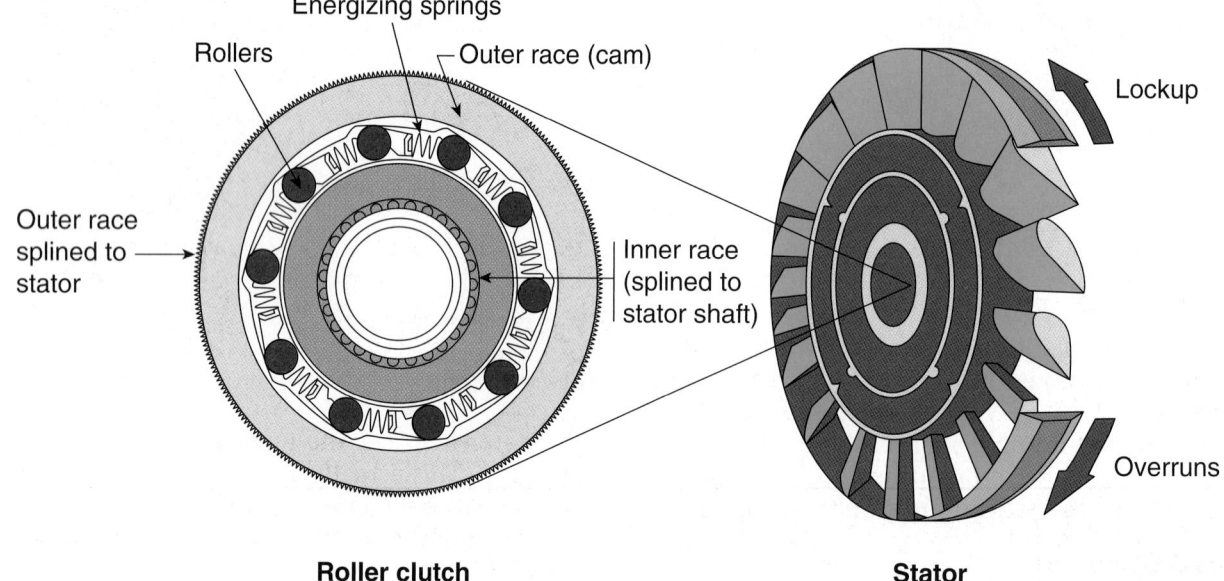

Energizing springs

Rollers

Outer race (cam)

Outer race
splined to
stator

Inner race
(splined to
stator shaft)

Lockup

Overruns

Roller clutch

Stator

Figure 73.14 The stator clutch locks in one direction and freewheels in the other.

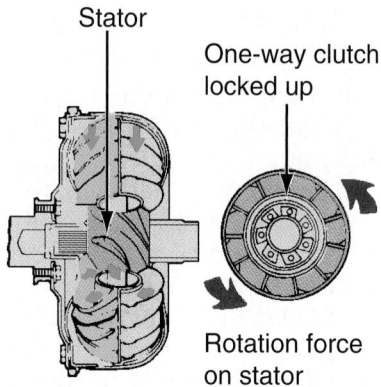

Stator

One-way clutch
locked up

Rotation force
on stator

Figure 73.15 Stator clutch locked up. *(Courtesy of Ford Motor Company)*

fluid that is now turning with rotary flow (without vortex). There is little vortex flow because the impeller and turbine have reached **coupling speed**. At this point, all of the converter parts and the ATF turn as a unit (**Figure 73.16**).

The converter does not become efficient at transferring power from the engine until coupling speed is reached at about 2300 rpm. When the turbine rpm reaches about $\%_0$ that of the impeller, there is no more torque multiplication.

■ STALL SPEED

Stall speed occurs when the vehicle is prevented from moving while the engine is accelerated. The torque converter slips, allowing engine rpm to increase. When

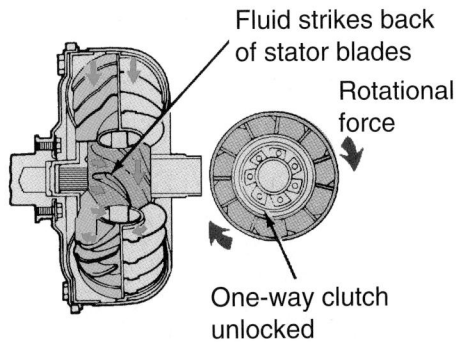

Fluid strikes back of stator blades

Rotational force

One-way clutch unlocked

Figure 73.16 When the stator clutch unlocks at coupling speed, all of the converter parts and the ATF turn as a unit. *(Courtesy of Ford Motor Company)*

the engine rpm can no longer rise, this is referred to as stall speed. Stall speed is the point of maximum torque multiplication.

NOTE: *A high stall speed converter is used in drag racing. This is a torque converter that is too small for a normal vehicle. It allows the engine to accelerate to a higher rpm before reaching stall speed. The camshaft in the engine is such that its maximum torque is not reached until higher engine speeds. A higher stall speed converter allows the engine to be at a higher rpm during engine braking at the start of the drag race for a faster takeoff.*

Lower stall speed converters are used for motor homes and for pulling trailers because they produce less heat and are more efficient.

LOCK-UP TORQUE CONVERTERS

Modern vehicles have **lock-up torque converters**. With a standard torque converter, actual coupling speed only happens when the vehicle crests a hill. Coupling occurs at that point when the engine stops driving

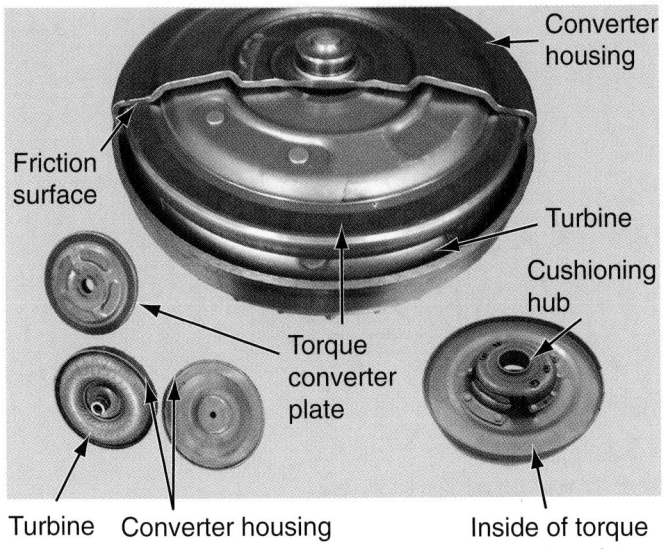

Converter housing

Friction surface

Turbine

Cushioning hub

Torque converter plate

Turbine   Converter housing

Inside of torque converter plate

Figure 73.17 When oil is applied to the inside of this torque converter plate and exhausted from the other, the converter is locked. *(Courtesy of Tim Gilles)*

the wheels and the wheels start to drive the engine. The rest of the time, there is up to about 10% slippage between the impeller and the turbine. Slippage causes a cutting action on the transmission fluid. The result is worse fuel economy and heat buildup.

Some transmissions do not lock up the converter until high gear or overdrive. In overdrive the gear ratio is 0.7:1. With this ratio and lower engine speed, a standard converter would have more tendency to slip. Newer transmissions often have lockup in all gears except first and reverse.

In a lock-up converter, a pressure plate behind the turbine locks it to the back part of the converter housing (**Figure 73.17**). This provides a mechanical link between the engine's crankshaft and the transmission input shaft to prevent slippage within the converter. To lock up the impeller and turbine, fluid is directed to one side of the pressure plate and exhausted from the other side.

AUTOMATIC TRANSMISSION COMPONENTS

Like a manual transmission, an automatic transmission has a *mainshaft* or *output shaft*. Output of the transmission is directed to the differential. An *input shaft* or *turbine shaft* fits between the torque converter and front of the transmission.

The *stator support* is a large splined support that extends out of the transmission's front pump body (**Figure 73.18a**). It surrounds the input shaft. The splines mesh with splines on the inside of the stator's overrunning clutch. The transmission input shaft extends through it and is supported by bushings. **Figure 73.18b** shows how the stator is positioned within the torque converter.

PLANETARY GEARS

The planetary gearset provides a means of changing gear ratios by coupling and releasing members to get different results.

> ### HISTORY NOTE
>
> *James Watt improved the steam engine in many ways. The reciprocating motion of the beam attached to the piston needed to be changed into rotary motion in order to be able to harness the engine's energy. Watt didn't want to pay royalties to the inventor who had patented one device that could be used to attach the connecting beam to a crankshaft, so he devised the planetary gear system consisting of a sun and planets to do the conversion.*

One major advantage of a planetary gearset is that all gears are in *constant mesh*. This means no gear clashing or wear as gears turning at different speeds engage and disengage. Another advantage is that the load is distributed over several gears instead of only two.

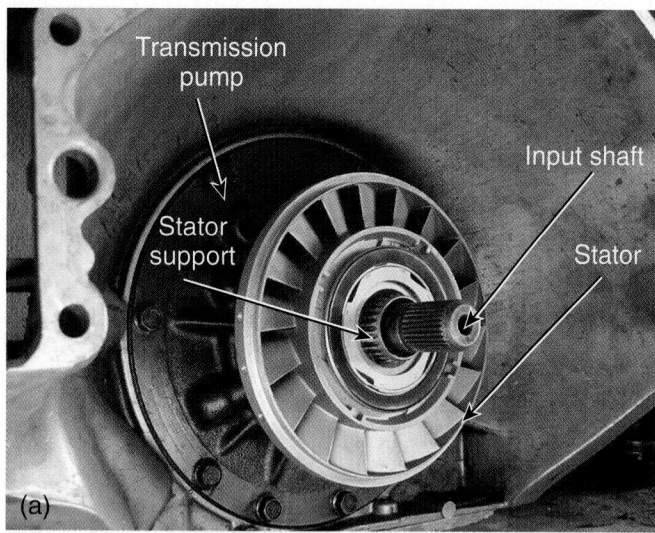

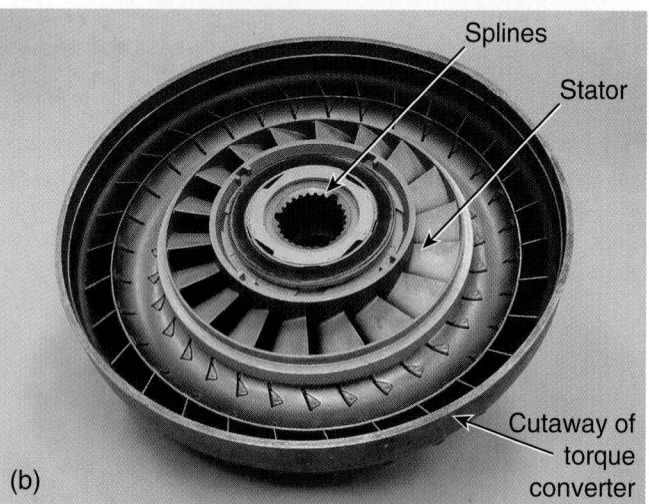

Figure 73.18 (a) The torque converter stator fits on splines on the pump's stator support. (b) Cutaway parts of a torque converter showing the relationship between the stator and the stator support splines. *(Courtesy of Tim Gilles)*

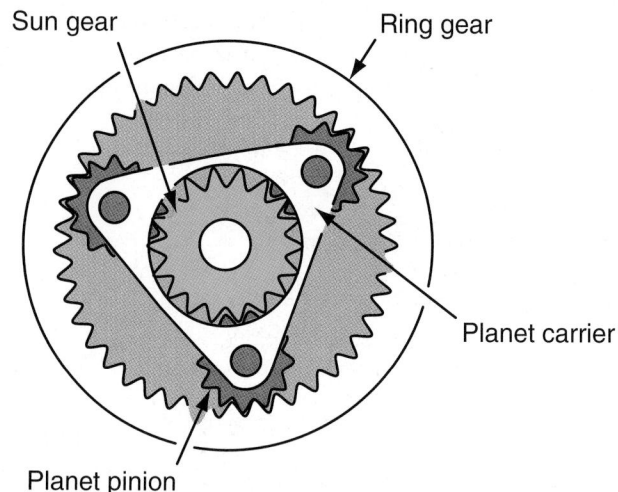

Figure 73.19 Parts of a simple planetary gear.

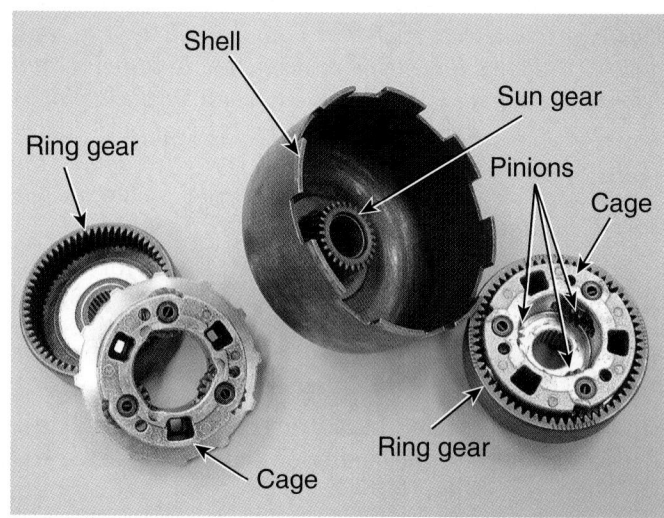

Figure 73.20 Parts of a Simpson gearset. *(Courtesy of Tim Gilles)*

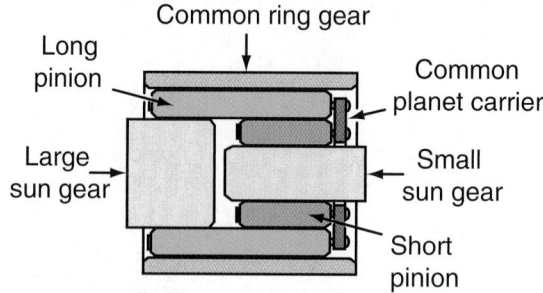

Figure 73.21 Parts of a Ravigneaux gearset.

▬▬ SIMPLE PLANETARY GEARSET

In a simple planetary gearset there is a *sun gear, planetary pinions*, a *carrier*, and a *ring gear* (**Figure 73.19**). The ring gear is also called an *internal gear* because of its internal teeth. The planetary carrier is also called a *cage*. A **compound planetary gearset** combines two planetary gearsets to provide more gear ratio possibilities.

One popular design is the **Simpson geartrain**. It has one long sun gear that operates two cages with pinions and two ring gears. Invented by a retired transmission designer, it uses interchangeable gears in each gearset. This lowers manufacturing costs. The Simpson geartrain is often called a "salad bowl" because of the shape of its outside metal shell (**Figure 73.20**).

Another transmission design is the **Ravigneaux geartrain** or *compound gear design*. It has long and short planetary pinions and two sun gears of different sizes. Only one ring gear is used (**Figure 73.21**).

Tandem Planetary Gears

Some automatic transmissions use two simple planetary gearsets, connected in series with each other (**Figure 73.22**). They operate in the same manner as a Simpson compound planetary gearset, but without sharing a common sun gear.

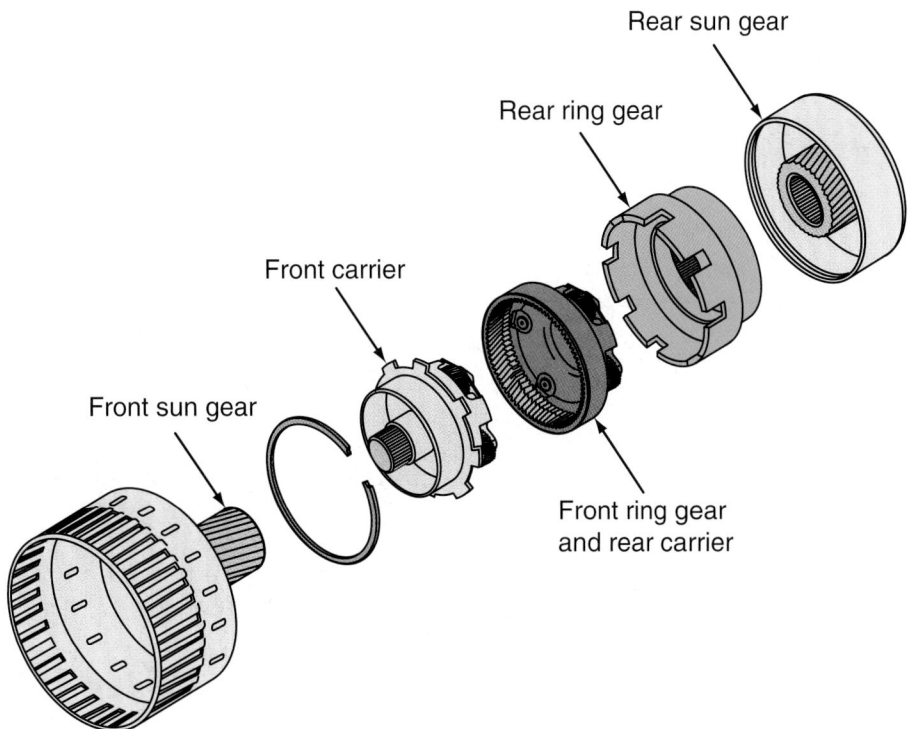

Figure 73.22 Two simple planetary gearsets connected together in series.

Lepelletier Geartrain

Some newer automatic transmissions use the *Lepelletier geartrain*, which combines different planetary arrangements. Five-speed transmissions use three simple planetary gearsets, with six clutches.

Six-speed transmissions use one simple and one Ravigneaux gearset with five clutches. In the simple gearset, the ring gear is the input. The sun gear is fixed to the housing so it cannot rotate. Clutches connect the carrier to the Ravigneaux's large and small sun gears. The input ring gear can be connected to the Ravigneaux carrier with a clutch and the output shaft is connected to the Ravigneaux ring gear.

Figure 73.24 This transmission is produced with four, five, or six speeds. *(Courtesy of Tim Gilles)*

Some luxury transmissions have more speed ranges and use as many as 10 to 15 clutches. Many overdrive transmissions use simple and Simpson planetary gearsets in series (**Figure 73.23**). The automatic transmission shown in **Figure 73.24** is produced with four, five, or six speeds, depending on the cost of the vehicle and the intended use. Due to fuel economy and emission concerns, more five- and six-speed transmissions are being produced.

■ SIMPLE PLANETARY OPERATION

Several shift options are available from one simple gearset. The chart in **Figure 73.25** shows the six possible conditions with a simple planetary gearset.

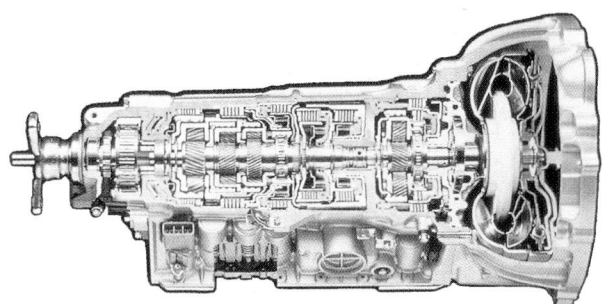

Figure 73.23 A five-speed automatic with an extra planetary gearset.

Conditions	1	2	3	4	5	6
Ring	D	T	H	H	T	D
Carrier	T	D	D	T	H	H
Sun	H	H	T	D	D	T
Speed	I	L	L	I	IR	LR

D = driven (output)	L	= reduction
H = hold	R	= reverse
I = increase (overdrive)	T	= turn (input)

Figure 73.25 The six possible conditions, other than direct drive, with a simple planetary gearset.

Basic gear rules are:
- Two gears with external teeth in mesh will rotate in opposite directions.
- Two gears in mesh, one with internal and one with external teeth (ring and planet pinions), will rotate in the same direction.

When planetary pinion gears move in an automatic transmission, this is referred to as *"walking"* or *"idling."*
- When walking inside of the ring gear, pinions are moving in a direction that is opposite to their own rotation.
- When the carrier is held from turning, the pinions idle. This produces a change in direction (reverse).

Forward gear reduction can be accomplished in a simple planetary set in two ways. Maximum gear reduction is accomplished by turning the sun gear while holding the ring gear (**Figure 73.26**). Output is through the planet carrier. This is condition number 3 on the chart:

Sun	= input	(T)
Ring	= held	(H)
Carrier	= output	(D)
Output	= normal rotation, gear reduction	

If the sun gear is held while turning the ring gear, output is through the carrier for a lesser gear reduction (**Figure 73.27**). This condition (number 2 on the chart) is used in the Simpson geartrain for second-gear operation. The Simpson geartrain will be used here to illustrate how an automatic transmission operates.

Ring	= input	(T)
Sun	= held	(H)
Carrier	= output	(D)
Output	= normal rotation, gear reduction	

Reverse is accomplished by using only the rear gearset (**Figure 73.28**). Input power is through the sun gear shell to the sun gear. Holding the carrier turns the pinions into idler gears. This results in reverse operation of the ring gear for output.

Sun	= input	(T)
Carrier	= held	(H)
Ring	= output	(D)
Output	= reverse	

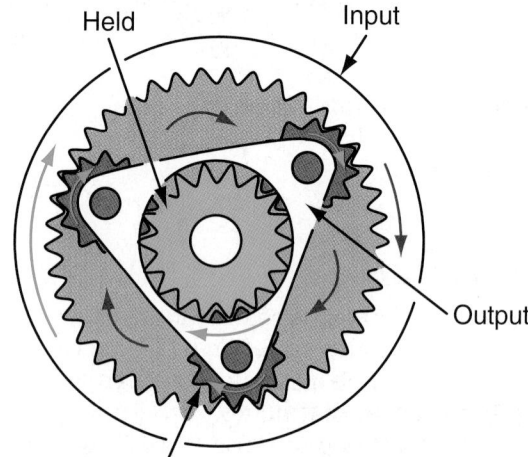

Pinions walking around the sun gear

Figure 73.27 If the sun gear is held while turning the ring gear, output is through the carrier for a lesser gear reduction.

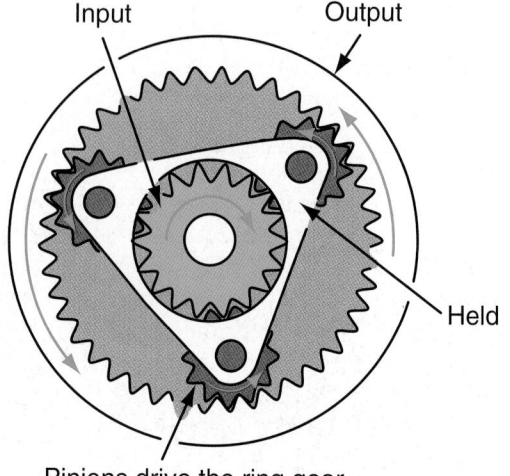

Pinions drive the ring gear

Figure 73.28 In reverse, the pinions act as idlers.

Input   Held

Output

Pinion walking inside ring gear

Figure 73.26 Maximum gear reduction is accomplished by turning the sun gear while holding the ring gear.

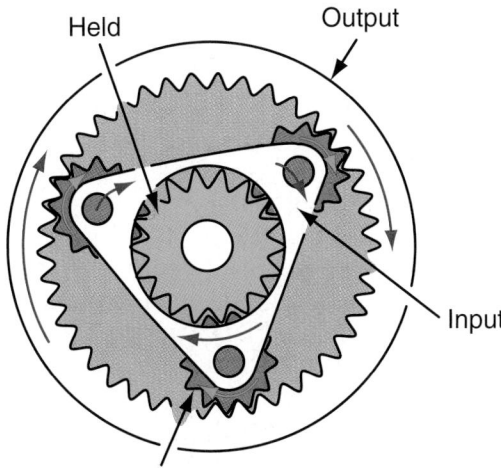

Pinions walk around the sun gear

Figure 73.29 Overdrive is when the output shaft turns faster than the input shaft.

Overdrive is when the output shaft turns faster than the input shaft. It results when the carrier is turned. The sun gear is held and output is through the ring gear. This is condition number 1 in the chart (**Figure 73.29**).

Carrier = input (T)
Sun = held (H)
Ring = output (D)
Output = normal rotation, overdrive

■ COMPOUND PLANETARY OPERATION

In compound planetary gearsets, two sets of planets are used together. In non-overdrive transmissions, high gear (third or drive) is direct drive, giving a 1:1 ratio. This is easily accomplished by turning on two gear units at once to lock up the geartrain.

Low-gear operation in a Simpson geartrain is a "double reverse," which results in forward operation (a combination of conditions number 5 and 6).

Here is what happens in the front gearset (condition number 5) (**Figure 73.30a**):

Ring = input (T)
Carrier = held (H)
Sun = output (D)
Output = reverse

■ The input shaft is turned clockwise, driving the forward ring gear clockwise.
■ The planetary carrier is held by the rear wheels to turn the pinions into idlers. This is because the carrier is splined to the output shaft. Although the carrier does rotate in a clockwise direction, it can only turn at the speed of the output shaft.
■ The ring gear drives the planetary pinions clockwise.
■ The pinions act as idlers to drive the sun gear counterclockwise.
■ Output from the front gearset is in a reverse direction.

Here is what happens in the rear gearset (**Figure 73.30b**):

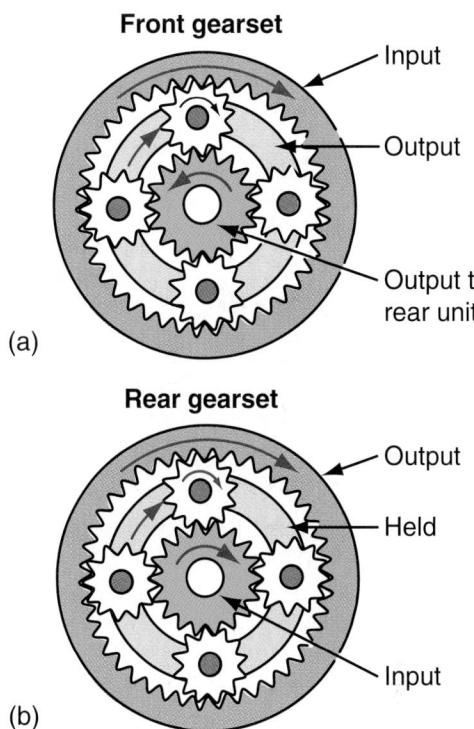

Figure 73.30 Low gear uses both planetary gearsets.

Sun = input (T)
Carrier = held (H)
Ring = output (D)
Output = reverse

■ Input is through the sun gear in a counterclockwise direction.
■ The rear carrier is held by a one-way clutch or band. This causes the pinions to be driven clockwise.
■ The ring gear is driven clockwise. It is splined to the output shaft.

Combining the two reverse actions results in forward operation.

In neutral, all holding devices are released, allowing engine freewheeling when the transmission gear shift selector is in neutral or park.

To achieve overdrive, some transmissions have an overdrive planetary set added to either the front or the rear of the existing Simpson geartrain.

Ravigneaux Operation

A Ravigneaux gearset uses two sun gears, two sets of planetary pinions connected to a single carrier, and a ring gear. The part layout is shown in Figure 73.21. There is a large and small sun gear and six planetary pinions, three long and three short. The planetary pinions rotate on shafts in the planetary carrier. The ring gear meshes with the long planetary pinions.

In first gear, the input shaft drives the small sun gear clockwise (**Figure 73.31**). The carrier is held by a one-way clutch that causes the short pinions to rotate

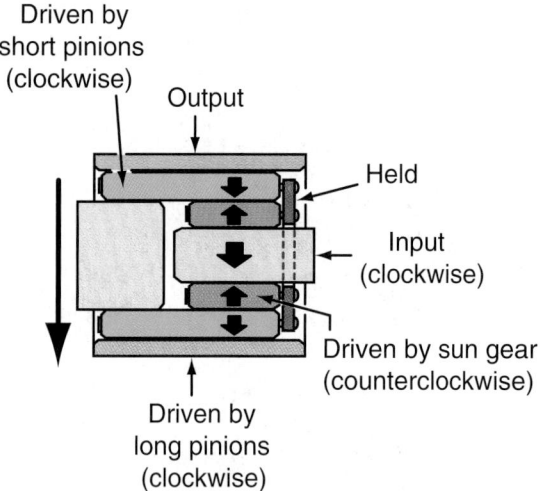

Figure 73.31 In first gear, the input shaft from the torque converter drives the small sun gear clockwise.

counterclockwise. The short planets turn the long planets clockwise at a lower speed. The long pinions turn the output ring gear clockwise, resulting in a gear ratio of about 2.5:1.

In second gear, a clutch locks the outer race of the one-way clutch (**Figure 73.32**) to hold the large

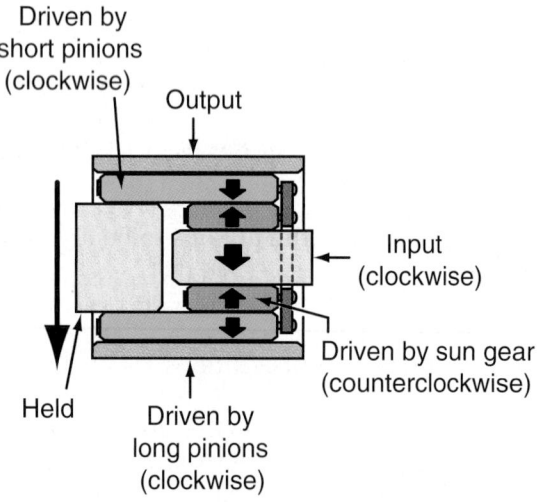

Figure 73.32 In second gear, a clutch locks the outer race of the one-way clutch.

sun gear from turning. Input from the small sun gear drives the short pinions counterclockwise. They drive the long planets, which walk around the held large sun gear to drive the ring gear clockwise.

In third gear two members are driven at the same time, which locks all the gears together for a 1:1 output. The direct clutch drives the planetary carrier and the forward clutch drives the small sun gear.

In overdrive, the direct clutch drives the planetary carrier clockwise, causing the long pinions to walk clockwise around the held large sun gear. The pinions drive the output ring gear clockwise.

In reverse, input is from the reverse clutch to the large sun gear, which is driven clockwise with engine rotation. The planetary carrier is held from turning, which causes the long pinions to be driven clockwise, which turns the output ring gear counterclockwise at about a 2:1 gear reduction.

■ DRIVING AND HOLDING DEVICES

To operate a planetary gearset, one member must be held while another is driven. A driving device, a *fluid clutch*, is used to lock a rotating planetary member to the input shaft. *Bands* and locking one-way clutches are used to hold a part of the planetary set. Clutches and bands are friction devices. They operate any time fluid pressure is applied to them.

■ CLUTCHES

Multiple disc clutches can be used for *holding* or *driving* parts of the planetary gearset. They contain steel plates (called *"steels"*) and friction discs (**Figure 73.33**). The steels are held against one element of the clutch pack, while the friction discs are splined to the corresponding part. Half of the clutch plates in a clutch pack can be splined to the case of the transmission or to another component of a planetary gearset. When clutches have half of their plates splined to the transmission case, they are used as holding devices. **Figure 73.34** shows a transmission cutaway where one of the clutch packs is connected to a clutch drum and the other is connected to the transmission case. Notched tabs, or tangs, on

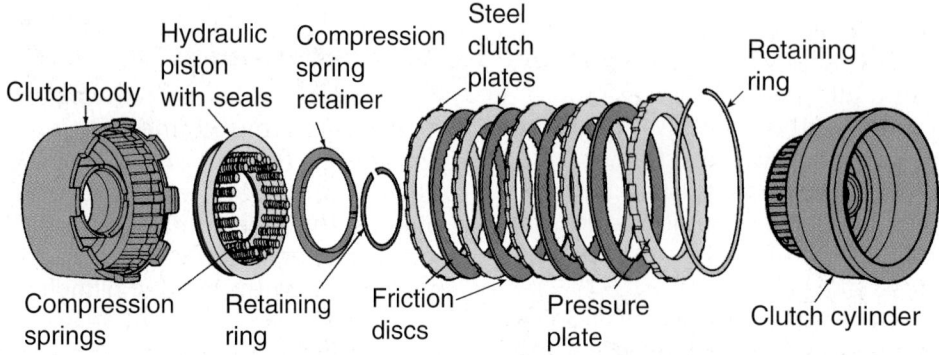

Figure 73.33 Parts of a multiple-disc clutch pack. *(Courtesy of Ford Motor Company)*

Figure 73.34 Two clutch packs. Steels in the front pack are notched to its drum. Steels in the rear pack are notched to the transmission case. *(Courtesy of Tim Gilles)*

the outside of the steel plates fit into corresponding notches in the clutch drum or case.

■■■ CLUTCH OPERATION

There is an apply piston in the clutch pack that is applied by hydraulic pressure and released by spring pressure (**Figure 73.35**). This means that whenever hydraulic pressure is not directed at the clutch, the clutch will release and the friction discs and steels will be free to turn independently of each other.

When a driving clutch is engaged, fluid is directed into the clutch drum. The fluid passageway is drilled

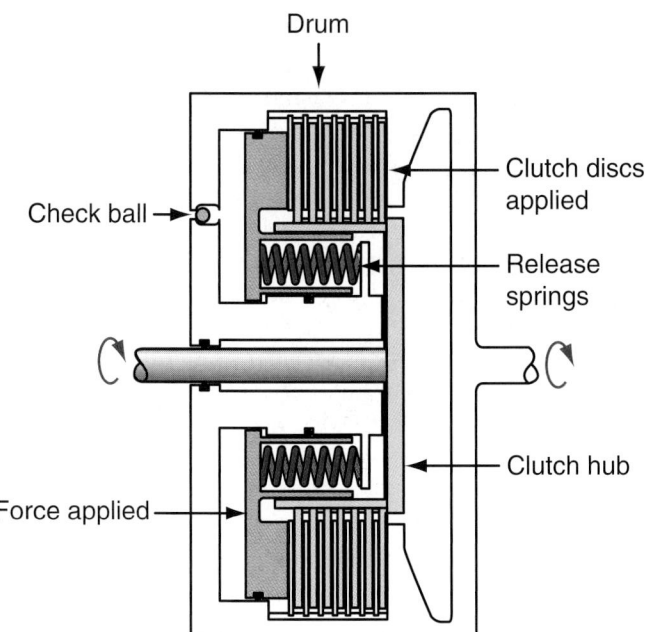

Figure 73.35 This clutch pack is applied by hydraulic pressure and released by spring pressure.

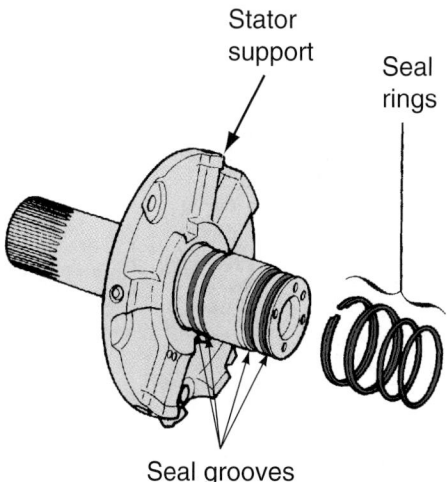

Figure 73.36 Seal rings. *(Courtesy of Ford Motor Company)*

into the mainshaft. Round seals (**Figure 73.36**) isolate the fluid and keep it under pressure. The fluid travels through the passages between the seals into a drilled fluid passage on the inside of the clutch drum. Fluid pressure is applied to a large piston on the inside of the drum. The piston has rubber seals on its inside and outside diameters (**Figure 73.37**).

The piston is applied against the discs to compress the return springs and lock up the clutch through the pressure plate. When the pressure is released, the piston is pushed away by the return springs. It is important that this occur quickly because another action will be taking place in another clutch pack during a gear shift. The piston will often have a check ball that allows fluid to be released from the clutch when pressure is released (**Figure 73.38**). The check ball also releases air when the clutch refills with fluid on the next application.

NOTE: *A major advantage of the hydraulic clutch is that it does not require periodic adjustment like a brake band does.*

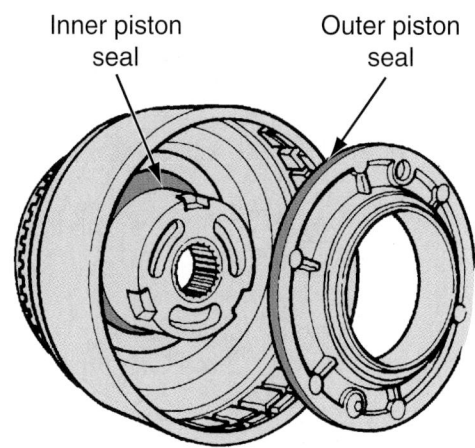

Figure 73.37 The piston has rubber seals on its inside and outside diameters. *(Courtesy of Ford Motor Company)*

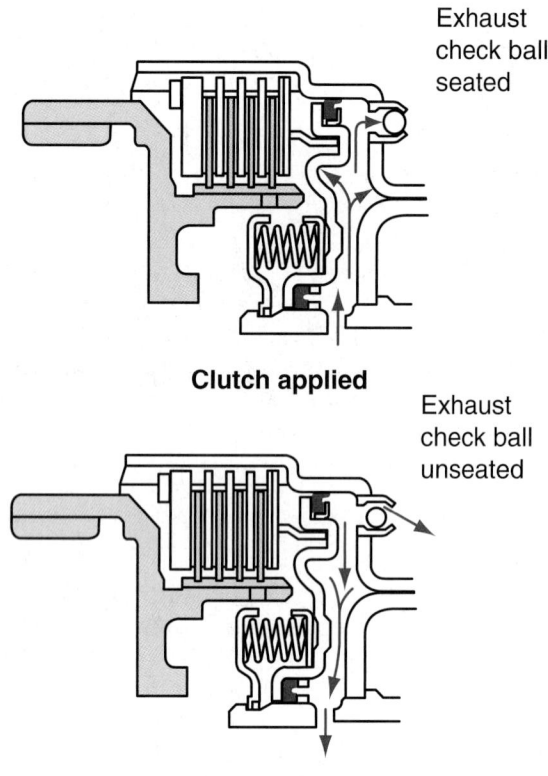

Clutch applied

Clutch released

Figure 73.38 A check ball allows fluid to be released from the clutch when pressure is released.

■ ONE-WAY CLUTCHES

In order for a planetary gearset to operate, various holding devices must be used. An overrunning, or one-way, clutch can be used to hold a part of a planetary gearset from turning. It locks in one direction and freewheels in the other. It locks instantly when an attached planetary member tries to reverse itself. An overrunning clutch is commonly used in *drive low* gear (when the gear shift is positioned in drive). The car freewheels when coasting.

NOTE: *If a driver wants the engine to slow the car down during deceleration, he or she must select the manual low position. In that position, a band or clutch holds the planetary device from turning.*

Overrunning clutches have an inner and outer race and a set of springs and rollers, just like the overrunning clutch used in a starter motor drive (**Figure 73.39**). Another type of overrunning clutch is called a *sprag* clutch (**Figure 73.40**). It works in the same manner but has a different shaped locking device between the inner and outer races. Sprag clutches are strong because more sprags are possible in a smaller area. A drawback to a sprag is that machining tolerances must be held quite close. Otherwise, a sprag can "flip" under heavy torque load, such as when shifting from reverse to low while the vehicle is rolling.

■ BANDS

A *brake band*, commonly called a band, is clamped around the outside of a drum to hold it from turning (**Figure 73.41**). An external band can only hold a component of the transmission stationary to its case. A clutch pack can hold two rotating planetary components together. A band is a steel strap with friction lining material on its inside diameter. A band can be either flexible or rigid. Some bands are *single wrap* and others are *double wrap* (**Figure 73.42**). Torque in low and reverse is higher than in other gears so more holding power is necessary. Double wrap bands are used for low and reverse gears because they can grip tighter with a lesser amount of force applied to them. Double

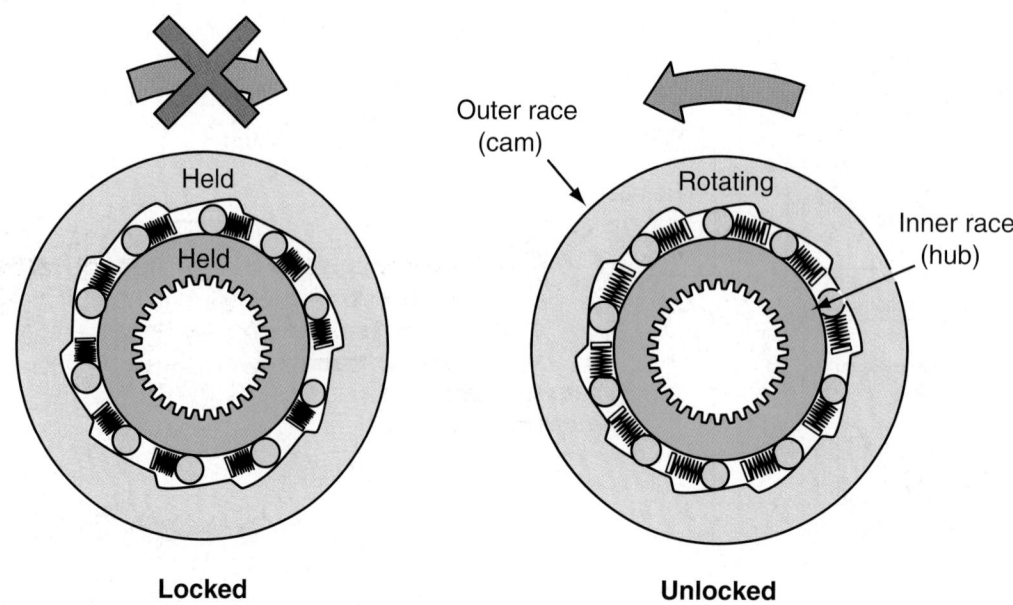

Locked   **Unlocked**

Figure 73.39 One-way roller clutch operation.

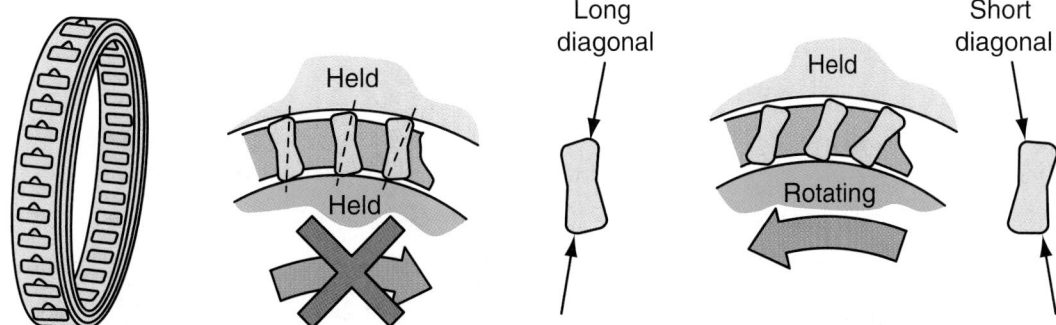

Figure 73.40 Operation of a sprag clutch.

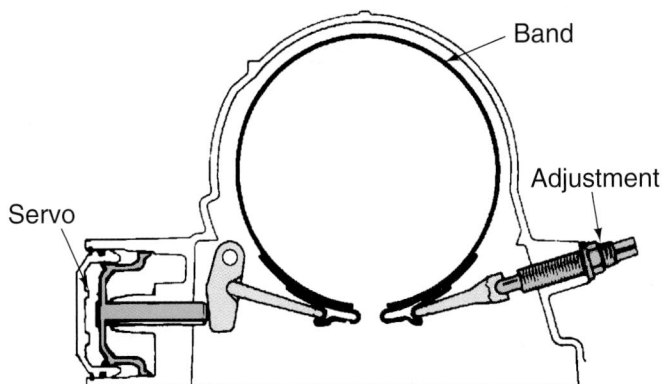

Figure 73.41 A band and servo. (*Courtesy of Ford Motor Company*)

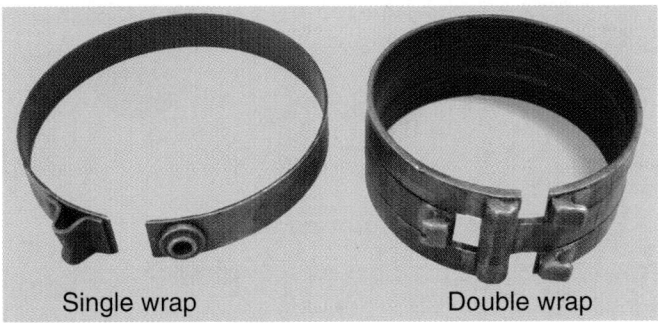

Figure 73.42 Single wrap and double wrap bands. (*Courtesy of Tim Gilles*)

wrap bands are also more flexible for better friction material contact.

Servos

A servo is used to operate a band. It has an apply piston with a seal on its outside diameter. The servo pushes on a rod that applies to one side of the band. There is a band adjustment screw or selective servo pin on the other end of the band (see Figure 73.41). Bands require periodic adjustment as they wear, so they sometimes not found in late-model transmissions. Some bands are not adjustable.

■ BAND OPERATION

When the band is applied, fluid pressure is directed into the servo's cylinder. The piston applies pressure

to one end of the band. The other end of the band is fixed in place, so the band tightens around the outside of the drum. The friction material on the inside of the band keeps the drum from turning. A servo includes a spring or springs to return the piston after fluid pressure is exhausted. One disadvantage of bands is that they require periodic adjustment to compensate for wear of the friction material.

NOTE: *Some transmissions have bands that operate only when the shift selector is placed in the D1 or D2 position for grade retard or engine braking. In normal drive range, a one-way clutch is the holding member in lower gear ranges. During deceleration the clutch freewheels, so engine braking is not possible without a band.*

■ ACCUMULATOR

During shifts, some parts are held from turning while other parts are driven. Shifts must occur at a specified time. Shuddering or damage can result as the transmission is momentarily locked up if two components are applied at the same time. An accumulator has a piston and reservoir that must fill before pressure can be applied to the driving or holding device (**Figure 73.43**).

NOTE: *Pressure results from a resistance to flow. If a reservoir (accumulator) is empty, it must fill before pressure can be used to apply force to a clutch or band.*

■ HYDRAULIC SYSTEM

The hydraulic system makes fluid pressure that transmits power through the torque converter. It also senses vehicle speed and engine load, directing fluid to reaction members at the correct time to cause shifts. Parts of the hydraulic system include a pump, pressure regulator, manual valve, vacuum modulator, governor, shift valves, servos, pistons, and control valve body. A transmission technician will use a service manual chart of the hydraulic system to solve transmission problems.

■ FLUID PUMP

Fluid is distributed throughout the transmission to control shifts and provide lubrication to critical parts.

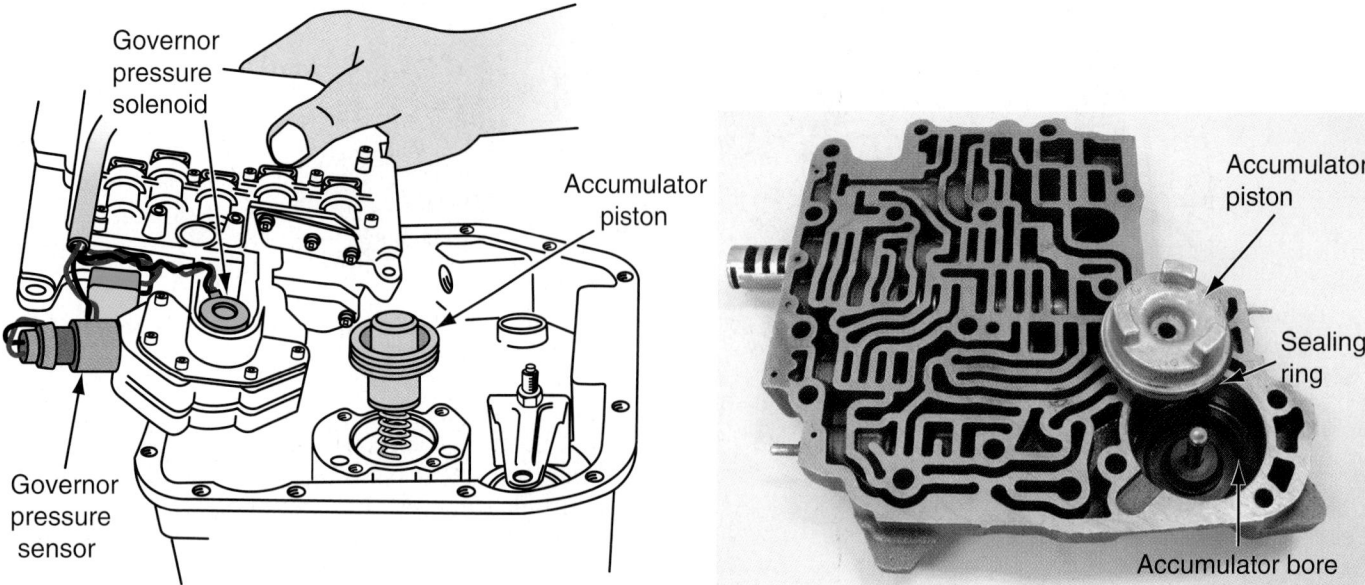

Figure 73.43 An accumulator has a reservoir that must fill before pressure can be applied to a driving or holding device. *(right, Courtesy of Tim Gilles)*

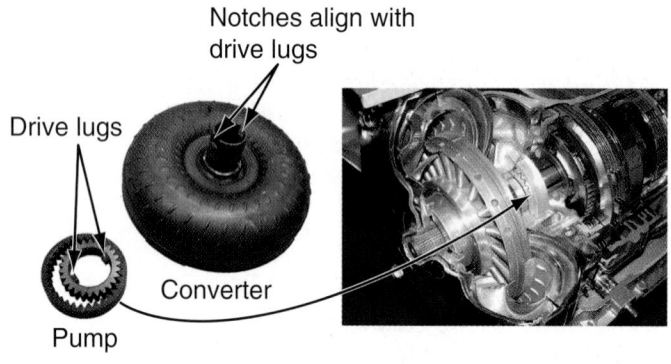

Figure 73.44 The pump is driven by lugs on the torque converter. *(Courtesy of Tim Gilles)*

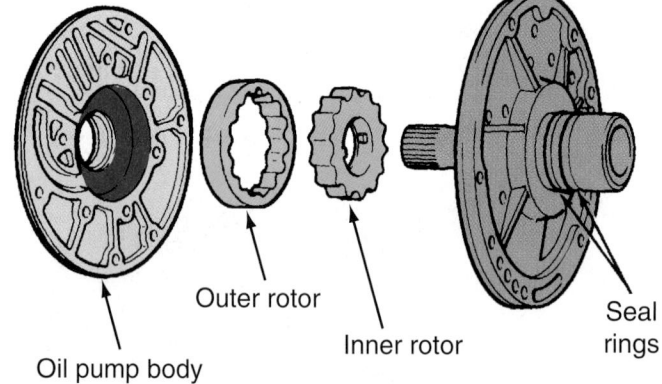

Rotor pump

Figure 73.45 A rotor-type pump. *(Courtesy of DaimlerChrysler Corporation)*

The transmission fluid pump, located just behind the torque converter at the front of the transmission, is driven by lugs on the outside of the snout of the torque converter (**Figure 73.44**). It is often called the *front pump* because old transmissions often had a front and rear pump.

The fluid pump does several things:

- It creates the fluid flow and hydraulic pressure to apply to the clutches and bands.
- It provides lubrication to the parts in the transmission.
- It fills the torque converter.
- It circulates fluid throughout the transmission and to the heat exchanger in the radiator in the front of the vehicle to cool the transmission.
- Its pressure operates valves in the hydraulic valve body.

■ TYPES OF PUMPS

There are three types of pumps: the **rotor type** (**Figure 73.45**), the **internal/external gear crescent** type (see Figure 73.48), which is the most common, and the **vane type** (**Figure 73.46**). Gear-type pumps are fixed displacement pumps. Any unused oil returns to the oil pan through the relief valve. Vane-type pumps are variable displacement pumps. The amount of oil output varies based on the transmission's needs. A spring holds the pump housing at its full output position (**Figure 73.46a**). When a decrease in oil pressure is desired, oil is directed to the back of the sliding housing to cause it to rotate against spring pressure (**Figure 73.46b**).

Fluid is drawn through a sump screen or *filter* in the bottom of the fluid pan (**Figure 73.47**). The filter can be a simple fine mesh metal screen or it can be made of paper or felt. A filter only keeps the larger particles out because it must allow sufficient fluid flow for transmission operation. Paper filters are made of cellulose or Dacron® fabric. Cellulose is the resin-based

 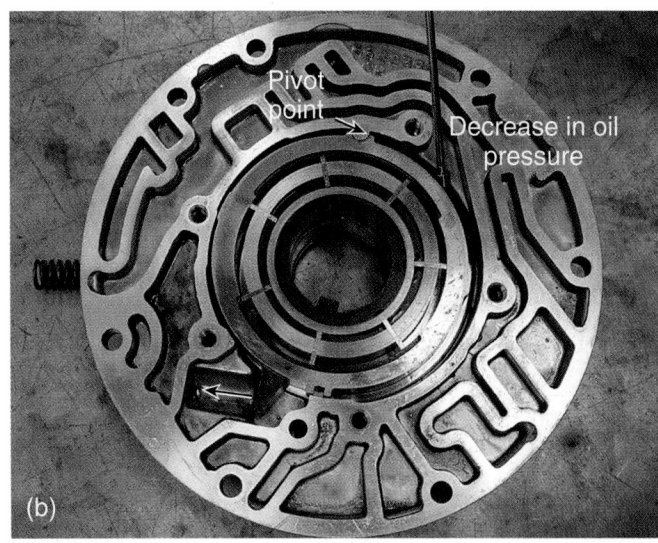

Figure 73.46 A variable displacement vane-type pump. (a) A spring holds the pump in its high-output position. (b) Oil pressure overcomes spring pressure to decrease pump output. (*Courtesy of Tim Gilles*)

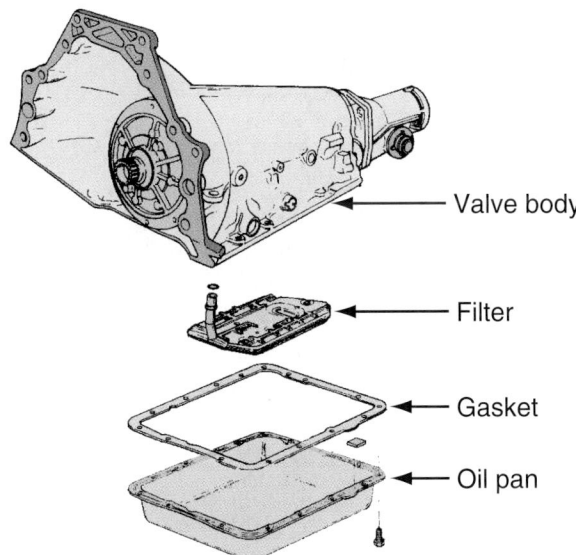

Figure 73.47 The filter is located within the oil pan.

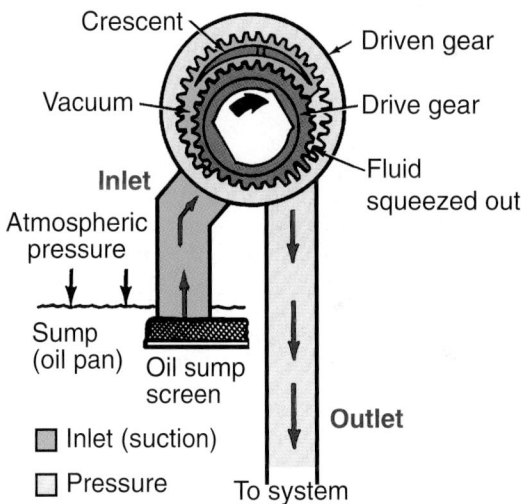

Figure 73.48 Operation of an oil pump. (*Courtesy of Ford Motor Company*)

material that paper is made from. These filters keep out smaller particles and can plug more easily. Some transmissions have a by-pass circuit in case the filter becomes plugged.

Felt-type filters are the most common. They trap contaminants throughout the filter and not just on its surface. Felt filters are made from polyester material that is randomly spaced to trap smaller particles and yet allow sufficient fluid flow.

Secondary filters are found in transmissions, too. They are simply screens located in fluid passageways.

The fluid, called automatic transmission fluid (ATF), is sent from the pump (**Figure 73.48**) to the pressure regulator before being distributed by the valve body. The pressure in this circuit is called *line pressure*. In the converter fluid circuit, it is pumped to the torque converter and cooler before returning to

the transmission. This fluid is under pressure of about 35-psi. Fluid returning to the transmission from the cooler is used to lubricate transmission parts.

■ TRANSMISSION VALVES

Spool valves control fluid flow in an automatic transmission. They have *lands* and *valleys* to control fluid flow. **Figure 73.49** shows a basic hydraulic system.

Valves can be moved in three ways:
■ By spring
■ By a lever or rod
■ By hydraulic pressure

When all of a valve's faces are the same diameter, this is called a balanced valve. If one face is larger than the other, fluid directed between them will result in movement of the valve in the direction of the larger face (**Figure 73.50**).

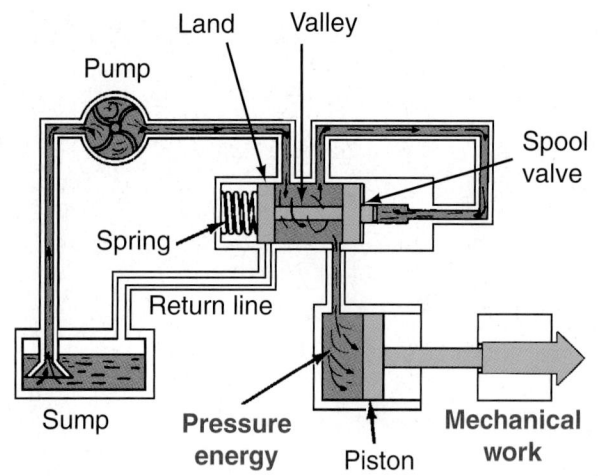

Figure 73.49 A basic hydraulic system. *(Courtesy of Ford Motor Company)*

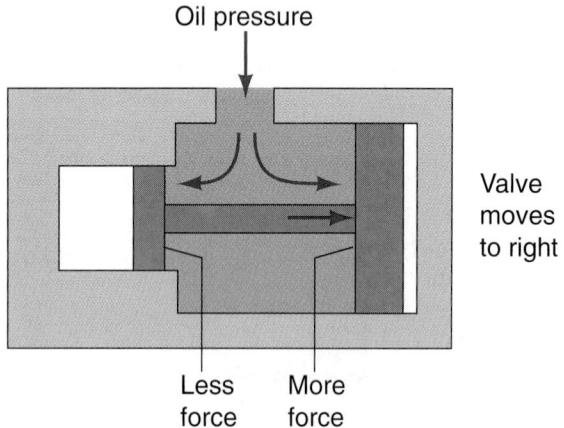

Figure 73.50 Fluid pressure will move the piston in the direction of the larger land.

■ PRESSURE REGULATOR

The pressure regulator valve determines the pressure in the system. As engine rpm increases, the pump turns faster and puts out more volume, which can result in excess pressure. The excess pressure is relieved by allowing fluid to flow back to the fluid pan (**Figure 73.51**) instead of forcing it through galleries, which would increase pressure (remember Pascal's law).

The pressure regulator is a variable-pressure regulator. When higher pressures are required, fluid is diverted to the back side of the valve. This works with the relief valve spring to increase the pressures throughout the transmission.

An **orifice** restricts fluid flow. As fluid tries to move through the orifice, pressure on the other side of it drops (**Figure 73.52a**). When fluid movement is restricted, pressure equalizes on both sides of the orifice (**Figure 73.52b**). Orifices can be used to reduce pressure of moving fluid, such as when a delay in a shift is desired.

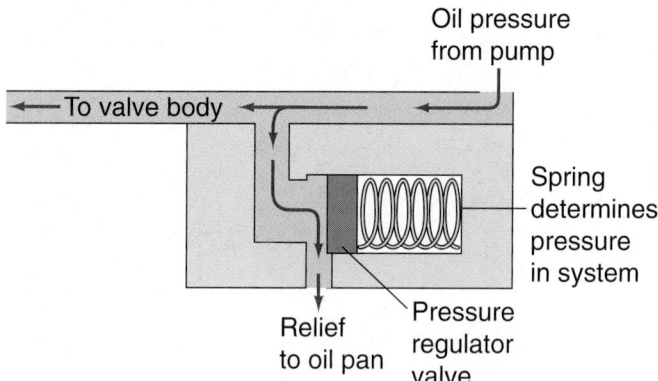

Figure 73.51 Excess pressure is relieved by allowing oil to flow back to the oil pan.

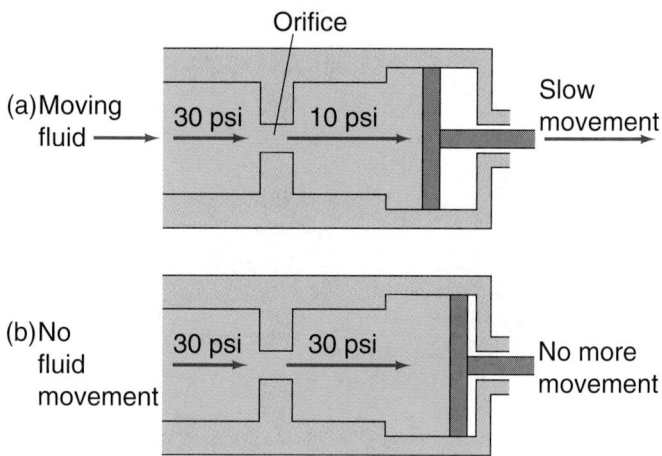

Figure 73.52 As fluid tries to move through the orifice, pressure on the other side of it drops.

■ HYDRAULIC VALVE BODY

The **valve body** is the brains of the transmission (**Figure 73.53**). It senses engine load and vehicle speed and adjusts the transmission's shift points and fluid pressures to match. The valve body contains many valves, springs, and orifices to control the shifting of the transmission. It is usually bolted to the bottom of the transmission, inside of the pan. The pump inlet filter or screen is usually attached to the bottom of the valve body.

A *valve body separator plate*, also called a spacer plate or channel plate (**Figure 73.54**) fits between the valve body and the transmission. It has holes of many different sizes to regulate fluid flow.

The *manual control valve* is the longest valve in the valve body. It is attached by linkage to the shift lever. When the driver moves the shift lever, the manual valve moves to direct fluid through the valve body to the correct driving or holding devices in the transmission.

The **shift quadrant** is the readout on the gear selector that tells what gear the transmission is in. For safety reasons, the positions of automatic

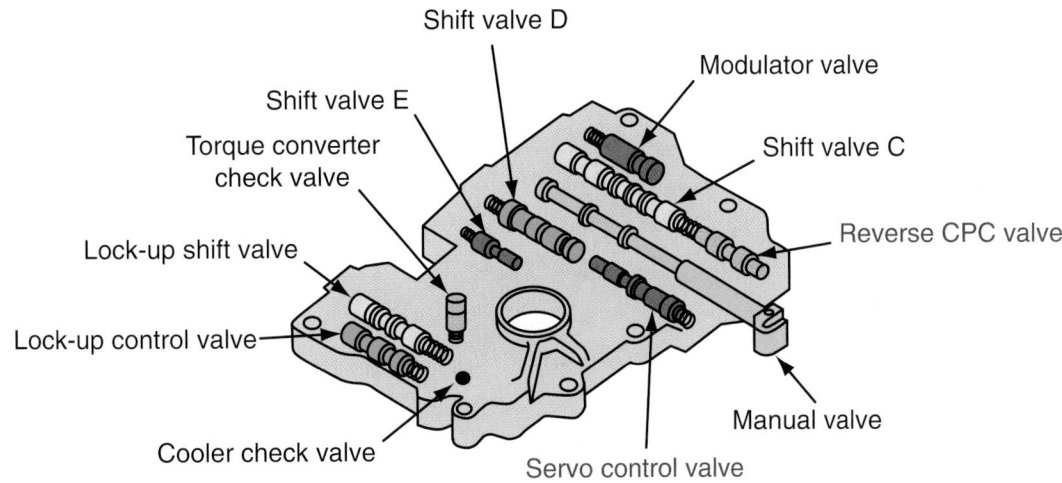

Figure 73.53 Parts of a valve body.

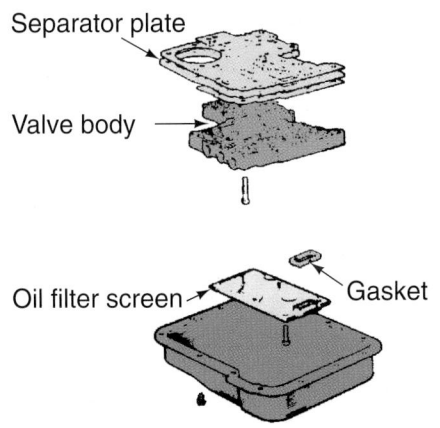

Figure 73.54 Location of a separator plate. *(Courtesy of Ford Motor Company)*

transmission gears are controlled by federal law. The shift order is always PRNDL, which sounds like "prindle" when spoken. Later model transmissions have several lower gear ranges (PRNDD2L). On some vehicles (notably those with Hydramatic transmissions) up until the mid-fifties, reverse was positioned on the opposite end of the selector from normal (NDLR). An owner of two vehicles with different shift positions could accidentally back up instead of going forward, or vice versa.

■ TRANSMISSION AUTOMATIC SHIFT SELECTION

The transmission selects the correct gear range based on the load on the engine and the speed of the vehicle. For instance, as a vehicle on the highway approaches a hill, the engine starts to work harder and the vehicle begins to slow down. Pressing harder on the accelerator pedal results in lower engine vacuum because the engine cannot breathe enough air to be able to produce maximum vacuum, as it would when going downhill or idling.

An understanding of the following terminology is necessary prior to a discussion of transmission operation:

- **Upshift** is when the transmission shifts from a low gear to a higher gear; second to third, for instance.
- **Downshift** is when the transmission shifts to a lower gear; second to first, for instance.

Shift valves respond to engine speed and load by moving back and forth in their *shift cylinders* to create gear shifts. There are shift valves for each gear shift position. For instance, in a three-speed automatic transmission, there is a *1–2 shift valve* and a *2–3 shift valve*. The shift valve has two forces working on it: throttle pressure and governor pressure.

Throttle pressure results when engine vacuum changes. Low vacuum, due to an open throttle or heavy load, results in high throttle pressure. High throttle pressure will force the shift valve to move into a lower gear position. For instance, a 2–3 shift valve will attempt to move to the second gear position due to the increased load on the engine (**Figure 73.55a**).

Governor pressure is results from increases in vehicle speed. When a car is traveling at 75 miles per hour and the throttle is moved to the wide-open position (called WOT or wide-open throttle), the 3–2 shift valve wants to move to the second gear position. High governor pressure prevents this shift until the speed drops (**Figure 73.55b**). The amount of drop in speed would be whatever the engineers who designed the valve body determined would be acceptable for such a shift.

■ GOVERNOR

The governor is located on the output shaft of the transmission. Some governors are gear-driven by the speedometer gear. Other governors are pinned to the output shaft (**Figure 73.56**).

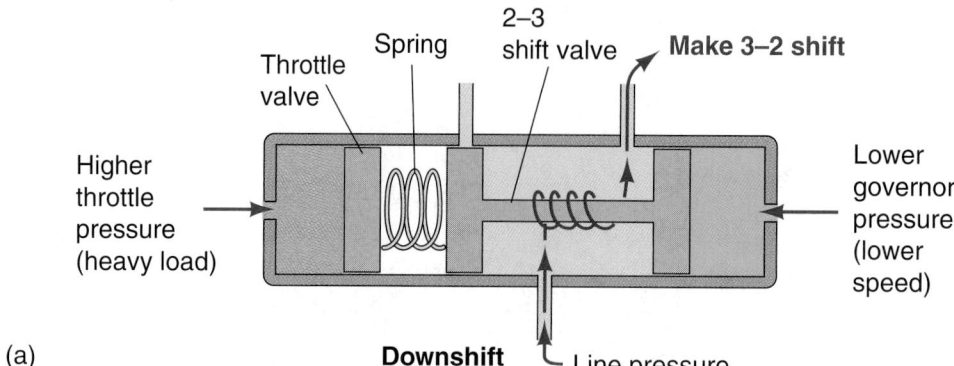

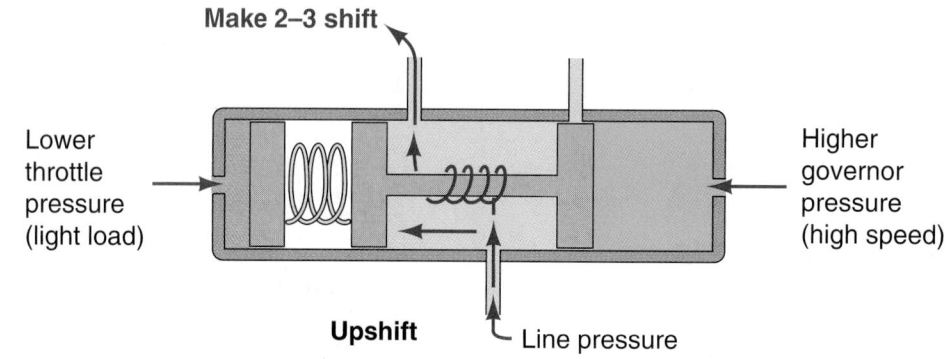

Figure 73.55 (a) The 2-3 shift valve will attempt to move to the second gear position due to the increased load on the engine. (b) Under light load, governor pressure causes a 2-3 shift.

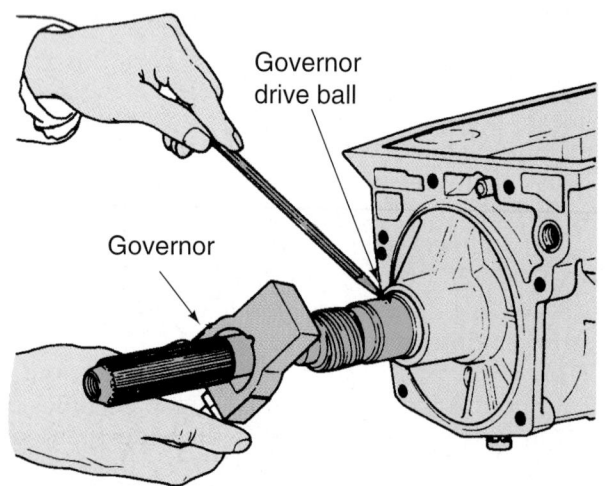

Figure 73.56 A governor that is pinned to the output shaft. *(Courtesy of Ford Motor Company)*

The governor is a variable-pressure relief valve. Governor pressure can be equal to its source (line pressure) but no greater. When the transmission's output shaft is moving at high speeds, weights in the governor are thrown outward. They are attached to a spool valve that closes off a fluid passageway, causing governor fluid pressure to climb. When the speed is slow, the weights are in and the fluid passage is open. This results in lower governor fluid pressure.

■ VACUUM MODULATOR

To control throttle pressure, some transmissions have a vacuum modulator valve (**Figure 73.57**). Others have a cable that moves the *throttle valve* based on the position of the accelerator (gas pedal). Throttle pressure times upshifts and also increases pump pressure under load. It aids the spring on the shift valve to work against governor pressure.

A vacuum modulator is similar in appearance to a distributor vacuum advance. It has a diaphragm inside

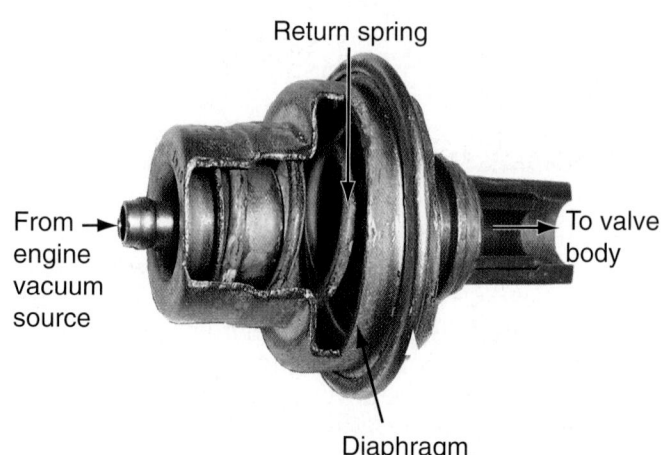

Figure 73.57 A vacuum modulator. *(Courtesy of Tim Gilles)*

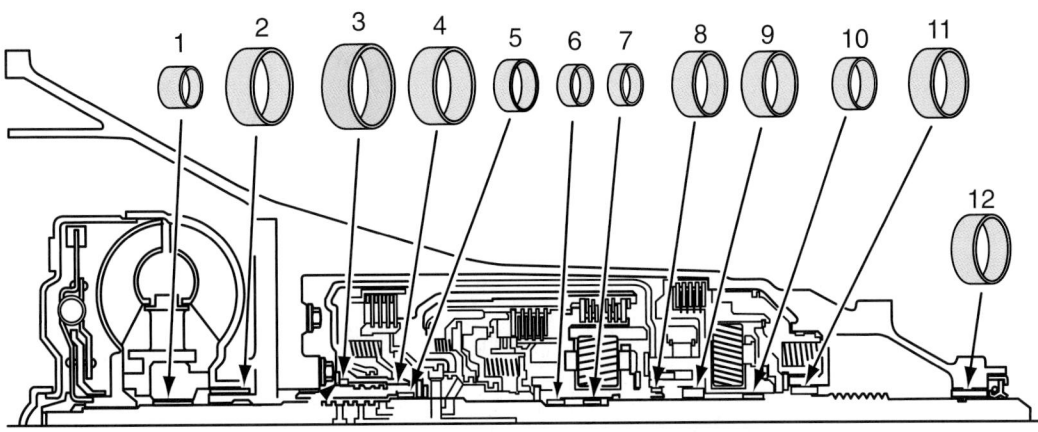

Figure 73.58 Bushings are located in several places in an automatic transmission.

it and a hose fitting that is attached to a manifold vacuum source (a hose to the intake manifold).

■ KICKDOWN VALVE

At wide-open throttle, a kickdown (also called downshift or **detent**) valve comes into operation. The kickdown valve is either manually operated by a cable or linkage, or it is an electrically operated solenoid. Operating the kickdown valve causes throttle pressure to go to its highest point. Attempting a *forced downshift* like this will still not result in a kickdown until governor pressure is low enough to allow it.

It is easy to find a manual kickdown valve when inspecting a valve body. It is the one that is spring loaded. It extends out of the side of the valve body where it contacts the lever that applies it from outside of the transmission.

Bushings, Thrust Washers, and Snaprings

Bushings are located at several places in an automatic transmission (**Figure 73.58**). They are round and are made of soft bronze alloy, or steel with a soft bearing surface, like a camshaft bearing.

Thrust washers or bearings are used to control fore and aft movement within the transmission (**Figure 73.59**). Some of the transmission's thrust washers are available in different thicknesses to allow for end play adjustment during transmission assembly.

Torrington bearings are thrust devices with small roller bearings (**Figure 73.60**). They are sometimes used because they reduce friction between parts moving at different speeds.

Snaprings are used to keep parts positioned on a shaft and can be either internal or external. Some snaprings are available in different thicknesses for adjusting clearances.

■ AUTOMATIC TRANSMISSION FLUID (ATF)

Automatic transmission fluid is oil that is specially formulated for use in automatic transmissions. Facts about the different types of fluids and their correct uses are covered in Chapter 13.

■ AUTOMATIC TRANSMISSION COOLING

The transmission develops a great deal of heat during operation. This is especially true during starting and stopping, in hot weather, and when pulling a load. Whenever there is a difference in the speeds of the torque converter impeller and turbine, the fluid is heated as it is cut.

Heat damages transmission fluid. Under normal conditions, transmission fluid will last for over 50,000 miles without being damaged. When fluid is heated to excessive temperatures, however, it oxidizes. Ruined fluid can damage a transmission. Oxidized fluid gets gummy or varnishy. This causes clutch pistons and valves to stick throughout the transmission. When this happens the transmission will need to be disassembled and rebuilt.

Most transmissions have a *fluid cooler*, called a heat exchanger. It is located in the bottom or side tank of the radiator. If your transmission has a heat exchanger, there will be two lines between the transmission and radiator (see Figure 20.10). It is called a heat exchanger because in cold weather, the fluid must reach operating temperature as quickly as possible. The engine's cooling system has a thermostat that results in quick warming of the engine's coolant. The warm coolant heats the transmission fluid by way of the heat exchanger in the radiator.

One cause of transmission failure is leakage of the cooler. If a radiator heat exchanger leaks, the results are:

■ When the engine is operating: Fluid pressure in the transmission is higher than that in the cooling system so ATF migrates into the radiator. This results in a lower cooling ability for the radiator because fluid does not cool as well as coolant.

■ When the engine is turned off: Pressure remains in the radiator (usually about 14 to 17 pounds). There is no pressure on the ATF unless the engine

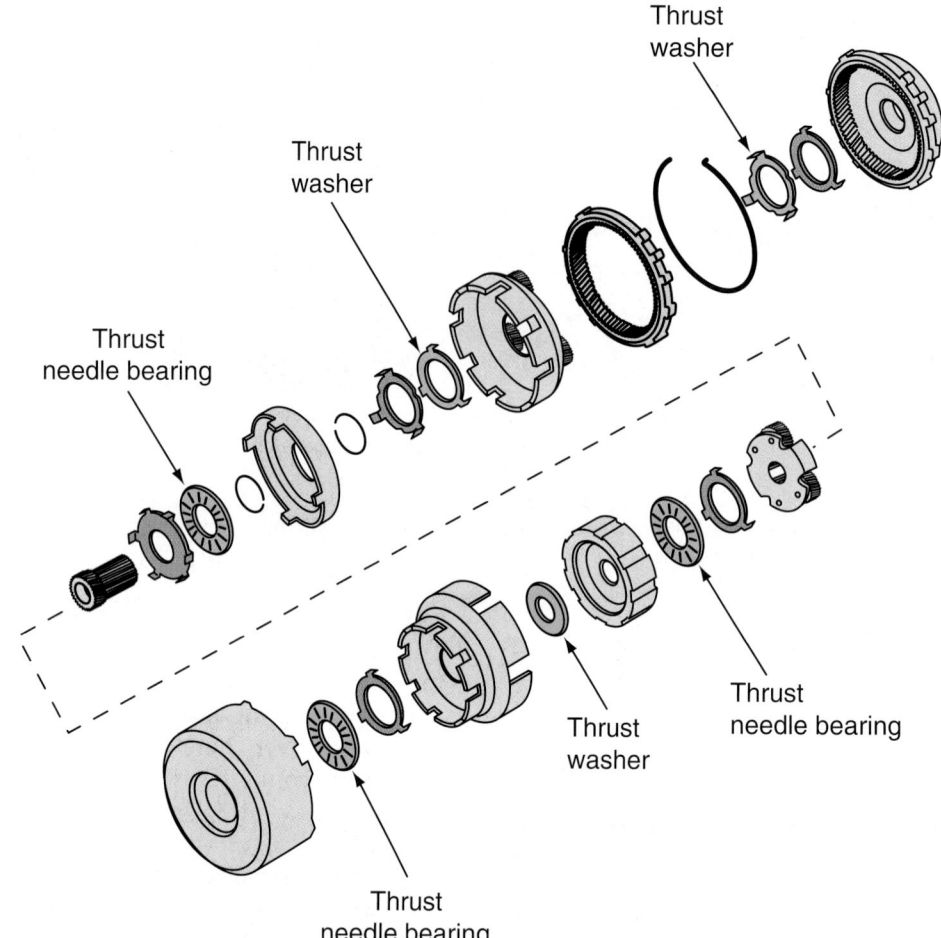

Figure 73.59 Thrust washers or bearings control the fore and aft motion of the gear train and reduce friction between rotating parts.

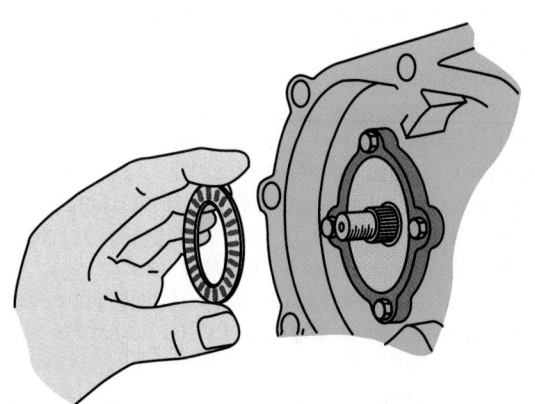

Figure 73.60 A Torrington-type thrust bearing.

is running, so coolant migrates into the transmission. Coolant that is mixed with fluid becomes very gummy. If the leak is serious, the transmission will require rebuilding after the internal coolant leak is fixed.

■ AUXILIARY COOLER/HEAT EXCHANGER

An auxiliary cooler/heat exchanger is often added to motor homes and vehicles that pull trailers. It resembles a small radiator and is hooked into the cooler line in series (**Figure 73.61**). Some manufacturers of auxiliary coolers recommend that they be installed in the line *before* the radiator cooler so that flash heat is removed by the cooler before transmission heat is exchanged with the engine's coolant. If it is installed after the radiator, engine coolant can actually become hotter during heavy-duty operation.

■ PARK PAWL

The **park pawl** is a lever that locks the output shaft of the transmission when the shift lever is placed in park (**Figure 73.62**).

SAFETY NOTE The park pawl is not supposed to be used in place of an emergency brake. In fact, it is dangerous to trust this device. If you study Figure 73.62, you will see that there are high spots between the notches in the park pawl gear.

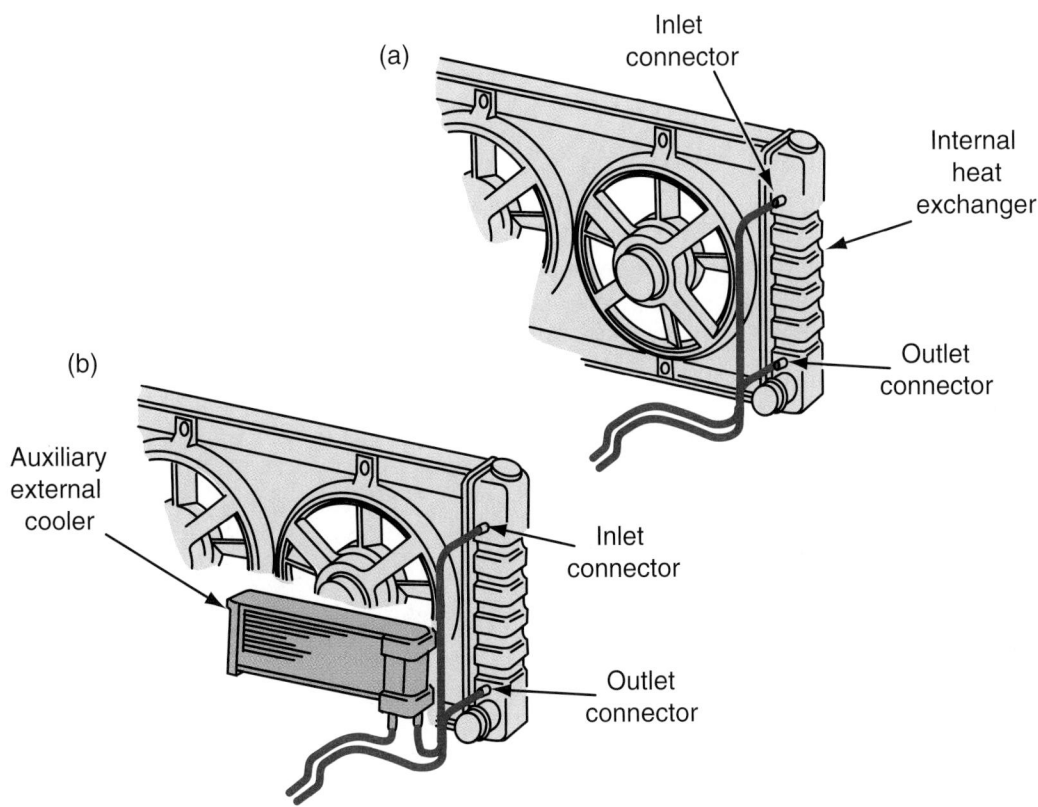

Figure 73.61 An auxiliary cooler is installed in series before the radiator transmission cooler.

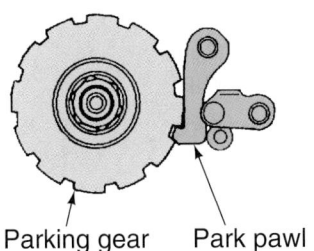

Figure 73.62 A park pawl.

CASE HISTORY *A driver left his car idling with the gear shift in the park position when he went back into his house for something he had forgotten. The park pawl was not fully engaged and engine vibration caused the transmission to slip into reverse. The car backed down his driveway and ran over a chicken.*

■ ELECTRONIC AUTOMATIC TRANSMISSIONS

Electronic controls have been used on automatic transmissions since the mid 1980s. Transmission shifts are controlled by the onboard computer using engine load, vehicle speed, and other inputs. Electronic controls provide more precise shift control and replace part of the relatively costly valve body with less expensive solenoids.

In a conventional hydraulically controlled transmission, shift valves are moved by hydraulic pressure from the governor. In an electronically controlled transmission, upshifts are controlled by solenoids (**Figure 73.63**).

Electronic transmissions vary by manufacturer, but they all operate in a similar fashion using basically the same parts. The computer is the heart of the electronic transmission shift control (**Figure 73.64**). It receives signals from various sensors and acts on them through one or more fluid control solenoids.

Shift solenoids used in an automatic transmission are similar to solenoids once used in earlier automatic transmissions for forced downshifts and for engaging the torque converter clutch. A solenoid can open or close a passageway and is similar to a fuel injector used

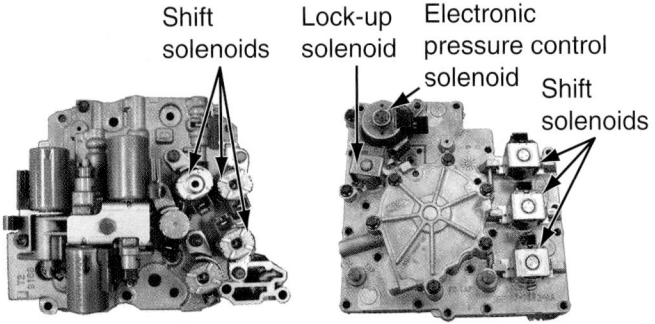

Figure 73.63 Electronic transmission shift solenoids. Note each one's electrical connection. *(Courtesy of Tim Gilles)*

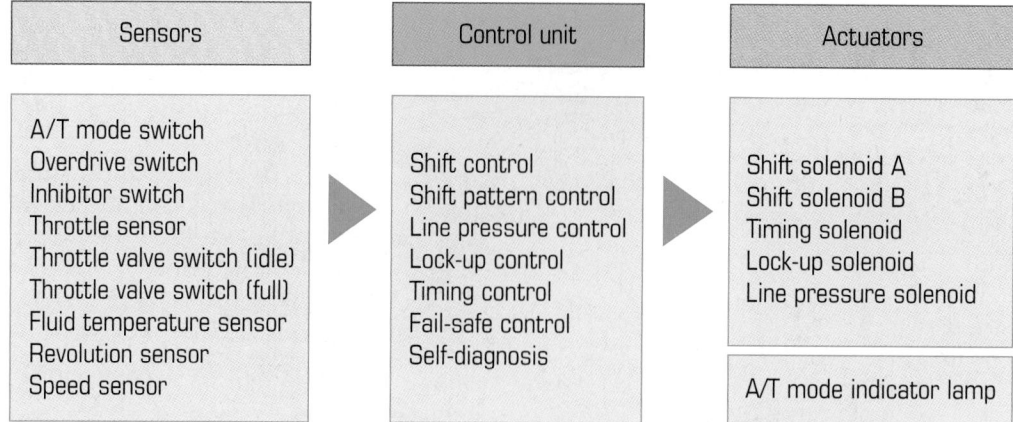

Sensors	Control unit	Actuators
A/T mode switch Overdrive switch Inhibitor switch Throttle sensor Throttle valve switch (idle) Throttle valve switch (full) Fluid temperature sensor Revolution sensor Speed sensor	Shift control Shift pattern control Line pressure control Lock-up control Timing control Fail-safe control Self-diagnosis	Shift solenoid A Shift solenoid B Timing solenoid Lock-up solenoid Line pressure solenoid A/T mode indicator lamp

Figure 73.64 The control unit (computer) receives signals from various sensors and acts on them through fluid control solenoids. *(Courtesy of Nissan North America, Inc.)*

with electronic fuel injection. The solenoid operates when its coil of wire is energized. It is returned to its original position by a spring when current flow stops and the magnetic field in its coil breaks down. Solenoids can be either normally open or normally closed when no electricity flows through them.

In direct control transmissions, fluid flows to a clutch or servo. Cycling the solenoid rapidly on and off (like pulse width in a fuel injector) controls how fast the clutch applies and thus the shift feel. In other transmissions, the solenoid controls a typical spool-type shift valve that sends fluid to the clutch or servo.

Depending on the manufacturer, the computer controls either the ground or the power side of the circuit. Some manufacturers have transmission controls within a powertrain control module (PCM). Others have a separate transmission control module (TCM) that gives input to the PCM. Shifts are controlled according to programming data stored within the computer. Gearshift position, throttle angle, and vehicle speed are primary inputs, although the computer program also considers other inputs, such as engine temperature and load.

Conventional transmissions use road speed and engine load information to make shift decisions. These same inputs, along with others, are used to make electronic transmissions shift.

- A vehicle speed sensor (**Figure 73.65**) takes the place of the governor. It frequently sends to the computer an output voltage that varies with road speed. This information is also used for the electronic speedometer. Speed sensors are usually permanent magnet generators, although some manufacturers use a modified Hall-effect sensor. A magnetic speed sensor is called a variable reluctance sensor.
- A throttle position sensor sends the computer information previously provided by a throttle cable. A voltage signal changes in response to throttle opening.

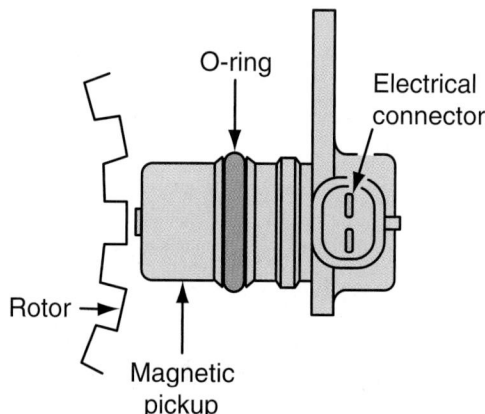

Figure 73.65 A speed sensor takes the place of the governor.

- Engine load signals are already known to the computer through the existing devices, like the mass airflow sensor, used to operate the engine's fuel and emission systems.
- The automatic transmission shift lock control solenoid prevents shifting into gear without the brake applied.
- A neutral start switch mounted on the transmission prevents the engine from starting in any gear but park or neutral. On electronically controlled transmissions, it is not just for park and neutral. Known as the MLPS (manual lever position switch), it also operates the back-up lights and gives information to the computer regarding other gear range positions when selected by the driver.
- A power or sport range selection switch on some cars is controlled by the driver. It allows the computer to make transmission shifts at a higher rpm.
- Torque converter lockup is controlled by the computer.
- Engine coolant temperature monitored by the computer prevents the torque converter clutch from operating when the engine is cold. Some

Figure 73.66 Steering wheel controls for shifting an electronic transmission.

transmissions also have later upshifts when the engine is cold.

- A brake on/off (Boo) switch functions as part of the cruise control and stoplight switch. It tells the computer to release the torque converter clutch during a stop to prevent the engine from stalling.
- An overdrive cancel switch shuts overdrive off in response to computer inputs.
- A transmission fluid temperature (TFT) sensor monitors fluid temperature. When the transmission fluid becomes too hot, the transmission operating characteristics change to help it cool down. The program typically calls for applying the torque converter clutch in lower gear ranges to prevent excessive converter heat. If the temperature continues to rise beyond a predetermined temperature, a diagnostic trouble code is set. When the fluid is too cold, the computer prevents overdrive operation and torque converter lockup.

Manual Range Shifting

Electronic transmission shifting allows the possibility of the driver shifting gears, without needing a clutch. This is accomplished by moving the shift selector lever or pushing a button on the shifter or steering wheel (**Figure 73.66**). If engine speed becomes too high for conditions, the computer will override the driver and make a shift. There are many such systems available, depending on the manufacturer and the vehicle's intended use.

ELECTRONIC AUTOMATIC TRANSMISSION OPERATION

On late-model transmissions, the computer decides on shift points based on the power output from the engine. It makes changes in shift timing, shift feel, and torque converter clutch operation, which result in increased fuel economy, lower emissions, and better driveability.

As a transmission wears, more clearance occurs between the parts. In a conventional transmission, more clearance would change the way the transmission shifts. In electronic transmissions, shift timing is altered by the computer as it "learns" the transmission's best shift points. A late-model transmission can discern a driver's habits and alter the shift conditions to match. This is called adaptive learning. Some manufacturers improve shift feel by reducing engine output during a shift, as fuel injectors are shut off. Also, during a shift the converter clutch goes off and then comes back on.

NOTE: *Since On-Board Diagnostics (OBD II) requirements were established in 1996, transmission problems are designated an emission-related component. This is because if a transmission slips, emissions can rise. A transmission slip will activate a malfunction indicator lamp (MIL) on the dashboard. Computers need .070 second (70 milliseconds) to find a problem by comparing signals from sensors that give input and output shaft speed.*

ELECTRONIC TORQUE CONVERTER CONTROL

The computer controls the torque converter clutch (TCC) solenoid in response to signals from the computer. The TCC cannot come on unless the engine is warm—at least 150°F (65°C). The speed required for lockup is about 40 mph. If the brake switch is closed (no foot is on the brake) and the throttle position sensor signal does not show a closed throttle, the computer will send a signal for the TCC to engage.

ELECTRONIC PRESSURE CONTROL

There are two types of electronic pressure control (EPC):
- On/off variable force solenoids (PWM), also called force motors
- Pulse width modulated (also called duty cycle)

Variable force solenoids are electronic modulators. A variable hydraulic pressure from a solenoid is exerted on a valve. The solenoid is either on or off.

Pulse width modulation does not operate a valve like a variable force solenoid does. It slides back and forth, opening or closing a passage. This bleeds off pressure or allows pressure to rebuild. Changing a solenoid's on time or off time results in differences in fluid flow. A duty cycle—just as with fuel injection—reflects how long it is on. This is like ignition dwell. If it changes with rpm and changes its length, it is "pulse width modulated." PWM solenoids are subject to increased wear because they move back and forth so often during their lives.
- Solenoid power   1%   = high pressure
- Solenoid power   100% = low pressure

Electronic transmission control provides a measure of protection to the transmission. When electronics fail, pressure should go to its maximum. This will cause very hard shifts and bring the customer in for service, but the transmission will not burn up.

The solenoids that control shift valves are on/off solenoids. The pressure relief valve is controlled by a PWM solenoid. The TCC solenoid can be either an on/off solenoid or PWM.

■ TRANSMISSION SHIFT CONTROL

Electronic shift control is different from older mechanical controls. It has line pressure only. There is no governor or modulator pressure. Forward gears are controlled by the computer. Reverse only works when all of the solenoids are off.

In a typical electronic system, there are two solenoids with four possible forward gears. Here is how a simple three-speed electronic automatic transmission would operate:

- First gear: Solenoid A is on and solenoid B is off. Only the manual valve is used in first gear. First gear has the forward clutch and a one-way clutch only. It has no shift valve. All of the shift valves are seated by spring pressure. Line pressure changes in response to the pressure regulator valve and also in response to throttle pressure.
- Second gear: Solenoids A and B are on (B went from off to on). The brake band is on and the 1-2 shift valve provides the fluid. Shift solenoid B moves the 1-2 shift valve by dumping pressure off the end of the shift valve.
- Third gear: To apply third gear, solenoid A is off and solenoid B remains on.

The following are examples of front- and rear-wheel-drive transmission solenoid functions for two General Motors transmissions:

1. Rear-Wheel Drive (4L80-E)

Gear	Solenoid A	Solenoid B
1st	on	off
2nd	off	off
3rd	off	on
4th	on	on

Notice that second gear has both solenoids off. This would be the default or limp-in mode gear for this transmission.

2. Transaxle (front-wheel drive) 4T60-E

Gear	Solenoid A	Solenoid B
1st	on	on
2nd	off	on
3rd	off	off
4th	on	off

Both solenoids are off in third gear, so this would be the limp-in mode for this transmission.

Constant Mesh Helical Gear Automatic Transmissions

Transmissions used in some Honda and Saturn vehicles use multiple-disc hydraulic clutches and shift solenoids but have gearing similar to a manual transmission (**Figure 73.67**). Four hydraulic clutches and a

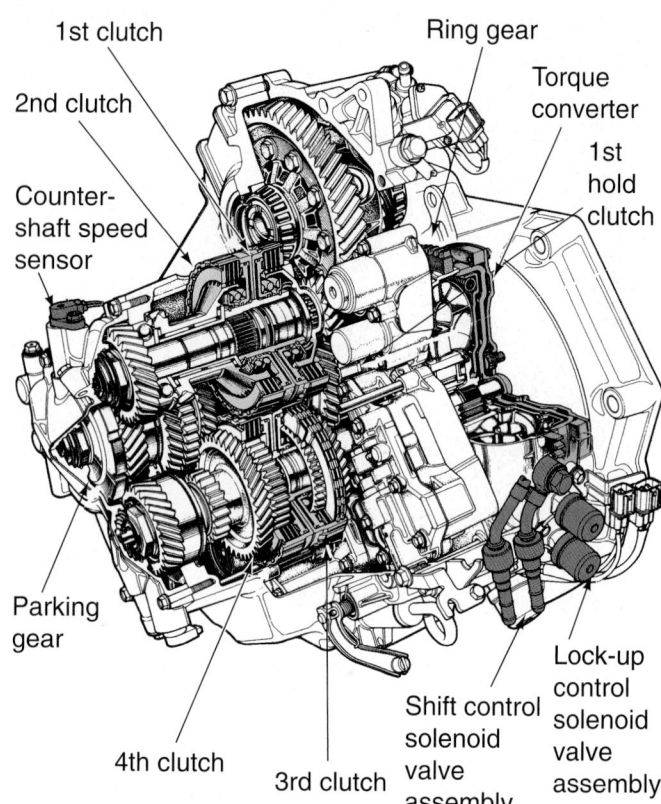

Figure 73.67 Multiple-disc hydraulic clutches and shift solenoids are used in some transmissions. *(Courtesy of Automatic Transmission Rebuilders Association (ATRA))*

one-way clutch are used to control forward shifting between gears. Reverse shifts are made using a manual transmission-style shift fork.

■ CONTINUOUSLY VARIABLE TRANSMISSION

Continuously variable transmissions (**CVT**) (**Figure 73.68**) are being used by many manufacturers. They work in a manner similar to a variable-speed drill press (**Figure 73.69**) and are similar to the transmissions used in snowmobiles and some motorscooters and small motorcycles. Where a conventional transmission has two, three, four, or more speeds, the CVT has infinite driving ratios. The aim of these transmissions is to increase fuel economy in the range of 25%. This is possible because the engine can be run with a relatively constant rpm where it is at its best efficiency. Because the engine does not have to accelerate through each gear, a smooth increase in vehicle speed is also possible. CVTs do not handle torque as well as conventional transmissions so they are found primarily in smaller vehicles intent on high fuel economy.

Because torque within the transmission is transferred between the steel cones and a steel chain, a special lubricant is used that changes phase to a glassy solid. The chain is a pushing (compressive) chain, which uses less surface area than a flat link pulling (traction) chain. According to Nissan, there is 3000 psi of pressure on the driving cones.

Figure 73.68 Drive assembly from a continuously variable transmission. *(Courtesy of Tim Gilles)*

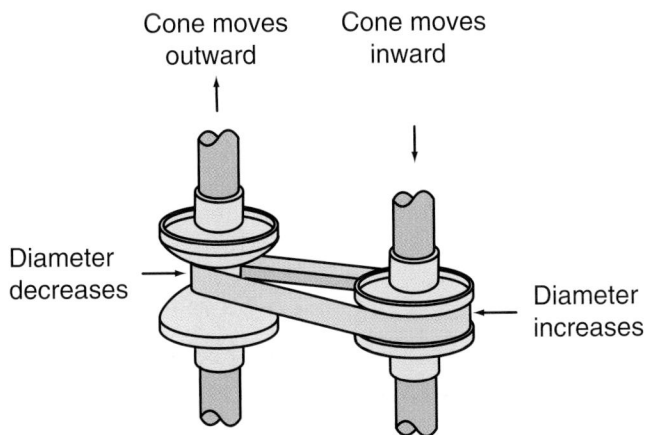

Figure 73.69 The gear ratio changes as the cones move.

Cone moves outward

Cone moves inward

Diameter decreases

Diameter increases

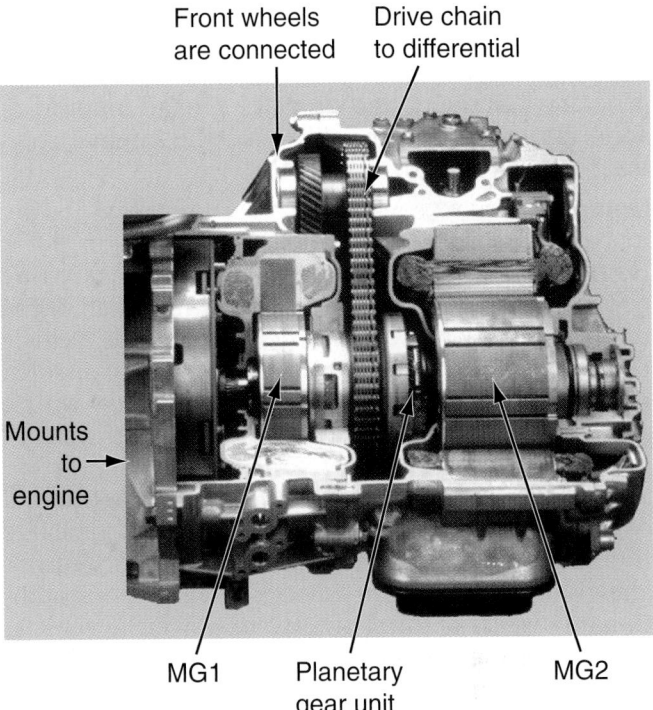

Front wheels are connected

Drive chain to differential

Mounts to engine

MG1 Planetary gear unit MG2

Figure 73.70 This Toyota system uses a planetary gear unit to transmit power between two motors and the engine. MG1 is motor/generator1 and MG2 is motor/generator 2. *(Courtesy of Toyota Motor Sales, U.S.A., Inc.)*

■ HYBRID PLANETARY TRANSMISSION OPERATION

Many hybrids use constantly variable (CVT) or conventional automatic transmissions, but many hybrids use a unique transmission. The hybrid drive system used in Toyota, Lexus, and Ford Escape vehicles uses a large planetary gearset to transmit power between one or two motors and the engine (**Figure 73.70**). This arrangement provides a constantly variable transmission. Depending on the manufacturer or vehicle, it is called a hybrid transaxle, a final drive, or a power split device. A torque converter is unnecessary with these vehicles. Basic hybrid electric vehicle information is covered in

Chapter 16. Engine speed is controlled using drive-by-wire, rather than a mechanical throttle connection. Computers calculate the best conditions for emissions and economy and control the transaxle, mixing input from the engine and motor(s) to balance the demands of electrical generation and power.

A conventional automatic transmission planetary gearset has one input and three possible outputs. A hybrid planetary transaxle has three inputs and one output. Low speed power is provided by an electric motor, which can supply maximum torque from a standing start. As torque from the electric motor declines, an internal combustion engine provides supplemental power.

Electric Motor/Generators

There are two motor/generators (MG) in the Toyota hybrid system (THS). Both motor/generator 1 (MG1) and motor/generator 2 (MG2) are part of the transaxle assembly. MG2 is the main power plant, supplying about 50 kilowatts of power. In comparison, MG1 provides about 10 kilowatts, about $\frac{1}{5}$ the power of MG2.

Hybrid Component Interaction

Hybrid motor/generators operate as motors when they power a vehicle. They can also generate electricity to recharge the battery pack.

Motor/Generator1. MG1 is splined to the sun gear. It controls the transaxle gear ratio and is a starter for the internal combustion engine (**Figure 73.71**). It also generates electricity and provides a small amount of acceleration assist, especially when the internal combustion engine is off.

Motor/Generator2. When starting from a stop, power is supplied solely by the main electric motor, MG2 (**Figure 73.72**). MG2 is splined to the planetary ring gear and is used to propel the vehicle nearly all of the time. It also generates electricity when the vehicle decelerates (**Figure 73.73**). During this process, called regenerative braking, MG2 is the main brake source and each wheel is modulated by the computer system's brake-by-wire system.

At speeds above 15 mph, MG1 starts the internal combustion engine, which drives MG1 to recharge the battery pack (**Figure 73.74**). If the battery has already reached its intended charge, MG1's additional power is applied to MG2 to move the vehicle. **Figure 73.75** shows planetary gear operation as the engine and MG2 moving the vehicle.

Starting engine

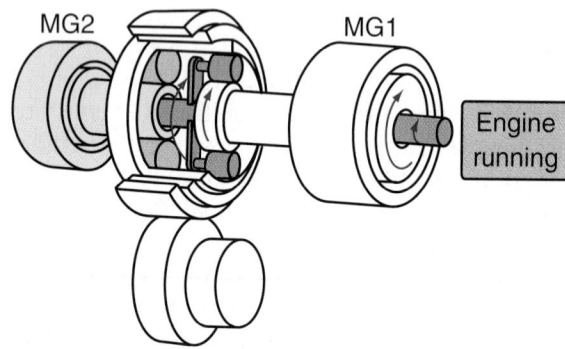

Figure 73.71 Planetary operation when starting the engine.

MG2 drives vehicle

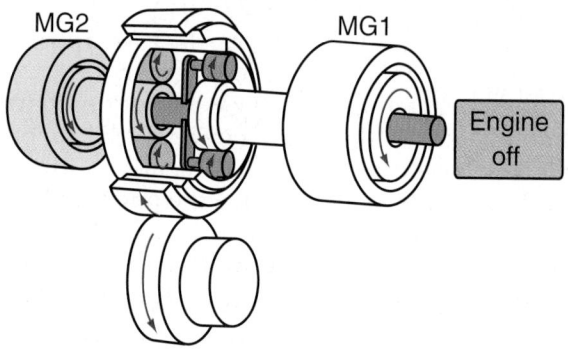

Figure 73.72 When starting from a stop, power comes only from MG2.

Deceleration D Range

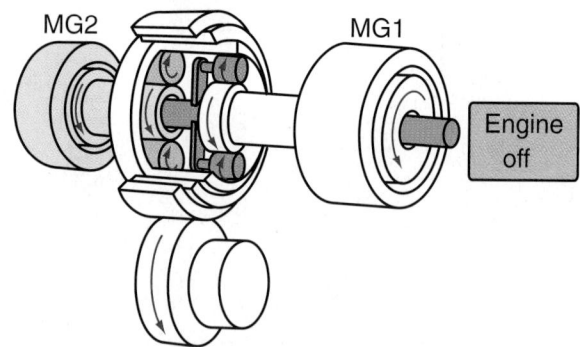

Figure 73.73 During regenerative braking, MG2 is the main brake source.

MG1 as generator

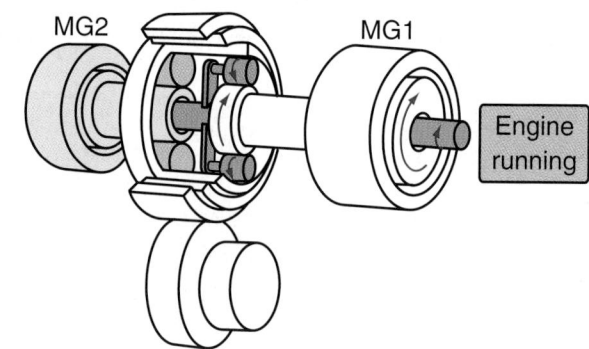

Figure 73.74 At speeds above 15 mph, the engine runs to power MG1 and recharge the battery pack.

Engine and MG2 drive vehicle

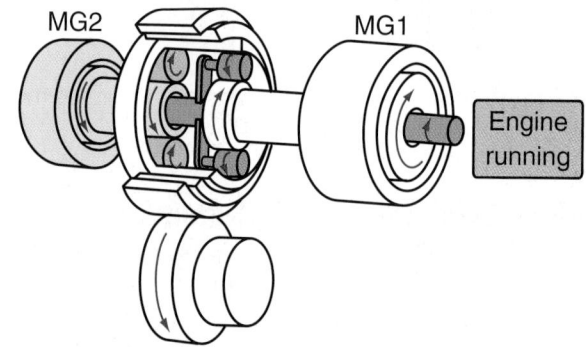

Figure 73.75 When the battery pack has been recharged, both MG1 and MG2 drive the vehicle.

Internal Combustion Engine. The engine (ICE) is splined to the planetary carrier. Any time the planet carrier spins, the engine must be rotating.

Reverse. In reverse the engine does not supply power to the wheels. It only starts and runs when needed to power MG1 when battery voltage drops below a

Reverse

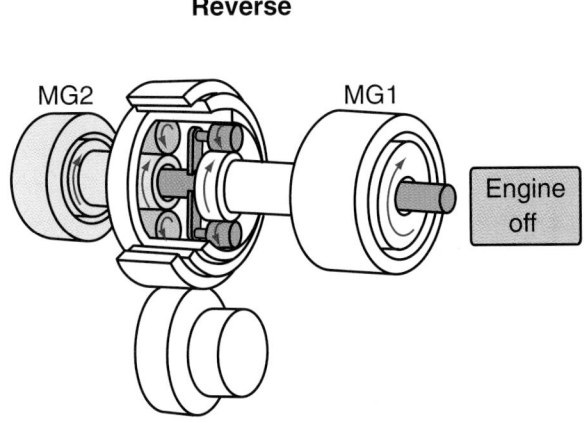

Figure 73.76 In reverse, the engine is off and MG2 supplies power to drive the vehicle.

certain level. MG2 supplies power in reverse, as shown in **Figure 73.76**.

Hybrid Computer Controls. The hybrid computer determines when to turn the engine on and off. It also controls the interaction between the engine and both motors, which determines the transaxle gear ratio. During regenerative braking, if the engine begins to overspeed, the MG1 electrical field can be energized to put a load on the engine. The current that is generated can either go to the battery or be dissipated as heat.

Hybrid Differential. Output from the planetary assembly provides a final drive ratio using a chain to drive a normal transaxle differential. Because output from the planetary is counterclockwise, an idler gear is used to change direction of rotation to clockwise.

> **SHOP TIP** You cannot power brake the engine. It has drive-by-wire, which will only allow 800 rpm or so if you floor it in neutral.

Hybrid Operation

The larger Toyota and Lexus hybrids are powerful vehicles that provide excellent acceleration. These models use an additional third electric motor on the rear axle to provide extra acceleration, all-wheel drive, and traction control. Double regenerative braking is an added benefit. The extra motor/generator in the rear is called motor/generator rear (MGR). The following figures illustrate some of the operations of a complete hybrid system with an MGR. In the illustrations, the planetary assembly is called a planetary gear unit (PGU).

Intial Start from a Stop. When starting from a stop (**Figure 73.77**), MG2 spins the ring gear. The engine is off (unless needed), holding the planet carrier stationary. This turns the planetary pinions into idler gears, driving the sun gear counterclockwise and spinning MG1 under no load in a reverse direction. The generator does not necessarily conduct electricity

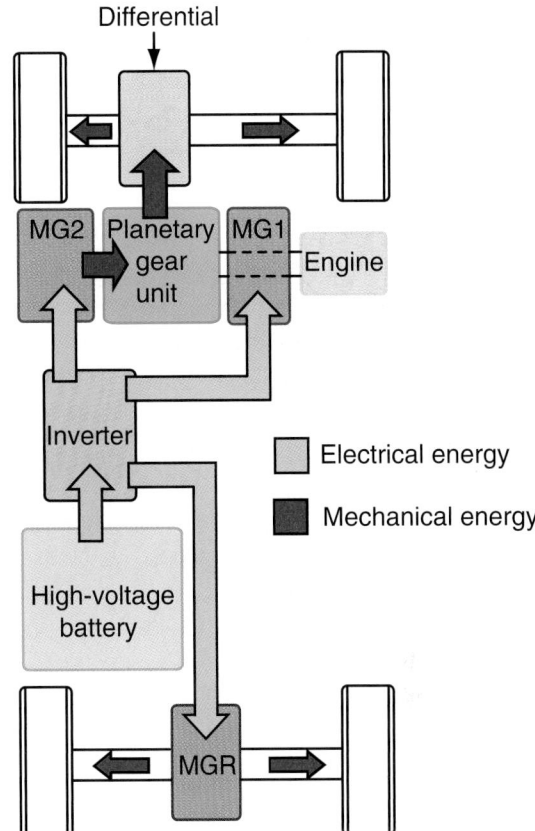

Figure 73.77 Electrical and mechanical power flow when starting from a stop. The engine does not normally run.

just because it is spinning. Like the alternator in a conventional charging system, it generates current only when the computer controls its field.

Starting the Engine. At speeds above 15 mph, the engine is started by MG1 (**Figure 73.78**). Remember that MG1 is spinning in reverse because the engine is holding the planetary carrier stationary. The computer energizes the field in MG1 enough so that the planetary carrier attached to the engine starts to spin, turning the engine's crankshaft. When the engine is running, it does not control vehicle speed. Rather, motor, generator, and engine output are controlled through the power split device.

Light Acceleration. During light acceleration, MG2 and the engine drive the vehicle and MG1 recharges the battery.

Light Load Cruise. When cruising under light load, the engine supplies power while MG2 works as a generator to power MG1.

Heavy Acceleration. When full power is called for, MG2, the engine, and the high-voltage battery all supply additional power. In all-wheel-drive hybrids, the rear motor supplies power as well (**Figure 73.79**).

Shift Control. Shift control is provided by three driver inputs—shift position, accelerator pedal position, and braking input from master cylinder pressure.

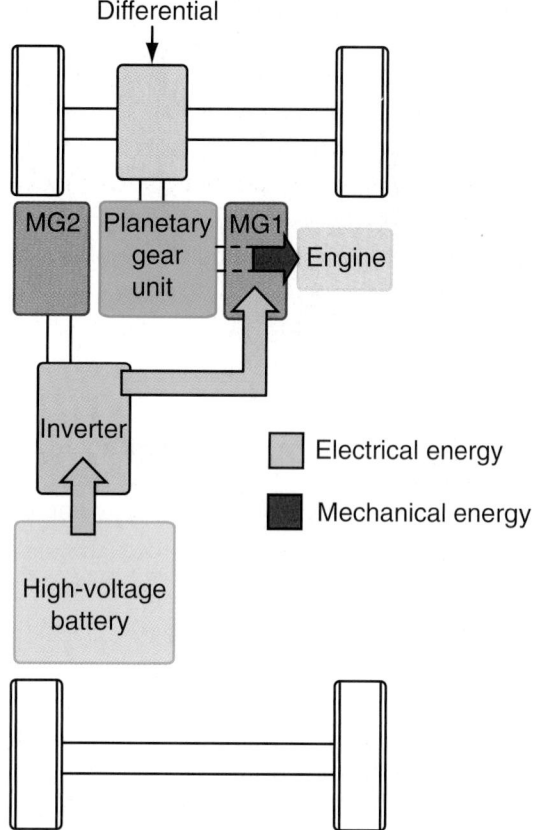

Figure 73.78 When starting the engine, electrical power flows from the high-voltage battery to MG1.

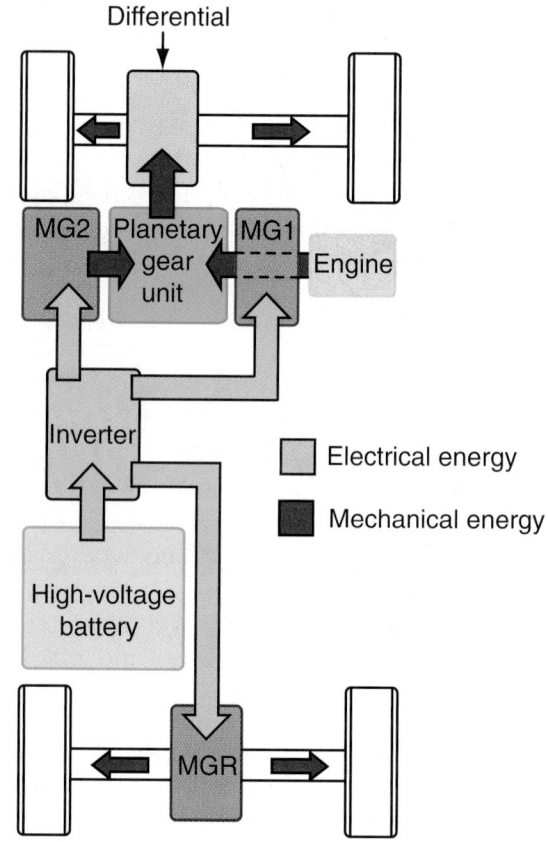

Figure 73.79 Full acceleration in an all-wheel-drive hybrid. MG2, the engine, MG1, and the rear motor (MGR) all work together to supply additional power.

■■ REVIEW QUESTIONS

1. What is the name of the part that is actually part of the converter housing, the impeller or the turbine?

2. What is it called when the impeller and turbine turn at the same speed?

3. What is the name of the part that redirects fluid flow at the center of the converter?

4. In a simple planetary gearset there is a sun gear, planets, a carrier, and a _____ gear.

5. In a planetary gearset with input in a clockwise direction, if the planet carrier is held from turning, what is the direction of output?

6. _____ pressure is in the main line of the transmission.

7. What is the order of gear shifts on the shift quadrant?

8. A vacuum _____ is a vacuum-operated diaphragm that controls shift points on some automatic transmissions.

9. What is a detent valve for?

10. What is another name for a transmission cooler?

ASE-STYLE REVIEW QUESTIONS

1. Which of the following is/are true about an automatic transmission clutch?

   a. It can drive planetary gears.

   b. It can hold planetary parts from turning.

   c. It does not require periodic adjustment.

   d. All of the above.

2. Technician A says that the device that controls shift points on an automatic transmission is the valve body. Technician B says that the device that controls shift points on an automatic transmission is the computer. Who is right?

   a. Technician A **c.** Both A and B

   b. Technician B **d.** Neither A nor B

3. Some cars must be manually shifted to a specified gear range in order to get engine braking during

   a. Deceleration **c.** Both A and B

   b. Acceleration **d.** Neither A nor B

4. As a vehicle climbs a hill, what happens to engine vacuum?

   a. Vacuum decreases.

   b. Vacuum increases.

   c. Vacuum remains the same.

   d. Vacuum disappears altogether.

5. Technician A says that the torque converter stator clutch freewheels when coupling speed is reached. Technician B says that there is less than 10% slippage in a lock-up converter when its clutch is not applied. Who is right?

   a. Technician A **c.** Both A and B

   b. Technician B **d.** Neither A nor B

6. As a vehicle slows down, what happens to governor pressure?

   a. It decreases. **c.** It remains the same.

   b. It increases. **d.** It disappears altogether.

7. Which of the following is/are true about an automotive torque converter?

   a. When starting from a stop, the approximate torque increase in the torque converter is 2:1.

   b. The stator redirects fluid flow from the turbine to the impeller.

   c. The least heat in the converter is created during lockup.

   d. All of the above.

8. Which of the following is/are true?

   a. The type of one-way clutch that uses rollers is called a sprag.

   b. A band can be used to hold or drive a planetary gearset.

   c. Bands and roller clutches are holding members.

   d. None of the above.

9. Hydraulic pressure is directed to the inside of a spool valve with one face smaller than the other. Technician A says that the valve will move toward the larger side. Technician B says that the valve will move toward the smaller side. Who is right?

   a. Technician A **c.** Both A and B

   b. Technician B **d.** Neither A nor B

10. Technician A says that a car with an automatic transmission has a flexplate. Technician B says that a car with an automatic transmission has a torque converter. Who is right?

    a. Technician A **c.** Both A and B

    b. Technician B **d.** Neither A nor B

74

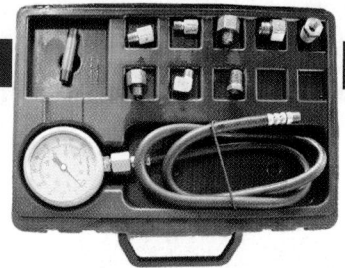

Automatic Transmission Diagnosis and Service

■ **KEY TERMS**

Dexron III/Mercon
severe service
varnish
vacuum leaks
mushy shift

harsh shift
stall test
pressure test
air test

ground side switching
power side switching
soft parts
hard parts

■ **INTRODUCTION**

Major automatic transmission repairs are typically done by an automatic transmission specialty shop. But a knowledge of transmission service and diagnosis is important for all technicians. Problems with throttle pressure, a leaking vacuum modulator, or a locked-up torque converter stator can all cause problems that can mimic engine problems. A technician lacking basic knowledge of transmission operation could do extensive repairs to the fuel system or replace an engine because he or she lacks an understanding of transmission operation.

When a transmission has been tested and proves to need major repairs, it must be removed from the vehicle. Manufacturers' service information gives detailed instructions on re-building procedures. Although this kind of work is no more difficult than other types of automotive work, it is a complicated specialty area. Special schooling is recommended before attempting to perform major repairs. Some special tools are also required to perform transmission work.

CASE HISTORY

A customer called a repair shop to make an appointment to have a rebuilt engine installed. Before ordering an engine, the shop owner suggested that the car be properly diagnosed. The technician test drove the car and found that it had a rough idle, white smoke from the exhaust, and harsh, late shifts. When he returned to the repair shop, he pulled the hose from the vacuum modulator and found fluid in it. The transmission fluid level was low, too. The spark plugs closest to the vacuum tap on the intake manifold were oil-fouled. He replaced the modulator, cleaned the spark plugs, and returned the car to the very happy customer.

In a specialty automatic transmission repair shop, there are two levels of employees: a remove and replace (R&R) specialist and a *rebuilder*. The R&R specialist removes the transmission and reinstalls it after it is

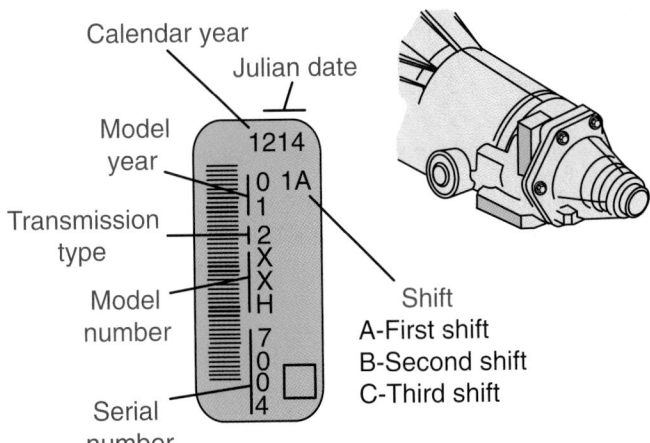

Figure 74.1 An OEM identification label for a GM transmission.

Figure 74.2 Typical automatic transmission fluid (ATF).

rebuilt. The rebuilder typically performs rebuilds at the workbench.

Because most transmissions operate in a similar manner, maintenance and diagnosis procedures are similar. References in this chapter to transmissions also include transaxles.

■ AUTOMATIC TRANSMISSION IDENTIFICATION

The best way to identify a transmission is by locating its identification label (**Figure 74.1**). This can be either a metal tag or a number stamped on the case. Service literature will identify the transmission from its number.

Another popular way of identifying transmissions is via a wall chart provided by one of the transmission part manufacturers. The chart shows various oil pan shapes and notes what vehicles use them, the manufacturer of the transmission, and the transmission's model number.

■ AUTOMATIC TRANSMISSION MAINTENANCE

Transmission service is most often preventive maintenance, rather than correction of a problem. Automatic transmissions generally require very little maintenance. A transmission that has not been run under severe service conditions should easily last for 100,000 miles or more, especially with today's extended-life fluid.

■ TRANSMISSION FLUID SERVICE

The oil in an automatic transmission is called automatic transmission fluid (ATF). Most types of ATF are dyed red in color to help differentiate them from engine oil. ATF has a low viscosity, similar to an SAE 10 engine oil. Additives include viscosity index improvers, oxidation and corrosion inhibitors, extreme pressure lubricants, anti-foam additives, detergents and dispersants, friction modifiers, and pour point depressants.

There are several types of automatic transmission fluids. **Dexron III/Mercon** is the latest and most used (**Figure 74.2**). It was designed with electronic transmissions in mind. These transmissions require fluid of a low viscosity, especially when outside temperatures are very low. It is used in many late model domestic and imported transmissions and can be used as a replacement fluid in many earlier transmissions.

ATF is longlasting and does not collect by products of combustion like engine oil does. Manufacturers have different recommendations on the length of service before a fluid change is necessary. Automatic transmission fluid additives can be depleted under severe service.

Anything that causes heat is **severe service**. This includes trailer towing and stop-and-go city driving in hot weather. Repeated starts from a stop cause the torque converter to shear the fluid. A transmission in a vehicle pulling a trailer can easily develop 300°F of fluid temperature. Fluid starts oxidizing rapidly at about 250°F, and the rate of oxidation doubles every 20°F after that point. Fluid in older non-electronic transmissions generally operated at lower temperatures and had a longer life expectancy. At 195°F, fluid lasts about 50,000 miles and at 215°F about 25,000 miles. It is not unusual for a modern transmission to operate at 215°F.

■ FLUID LEVEL

The dipstick is located in a tube that extends from the transmission pan. The top end of the dipstick is in the engine compartment. The fluid level should be between the add and full marks on the dipstick (see Figure 13.33). Correct fluid level is very important to proper transmission operation. Too low a fluid level can result in transmission damage.

A low fluid level is usually caused by a leak. A *leaking vacuum modulator* will also result in an internal leak when ATF is pulled into the intake manifold where it is burned. When burned, ATF causes white smoke out of the exhaust.

A high fluid level can result in foamed fluid when the transmission gears churn up the fluid. With a normal fluid level the gears do not dip into the fluid sump. A high fluid level can result when too much fluid is added after a fluid change and the excess is not removed.

High fluid level can also result because a customer adds fluid mistakenly after improperly checking the fluid level. When checking engine oil level, the add line means that the crankcase is one quart low. It only takes one *pint* of fluid to raise the level from add to full on a hot transmission.

Checking Fluid Level

The following conditions must be met when checking ATF level:
- Car on level ground
- Engine and transmission warm (dipstick should be uncomfortable to touch)
- Parking brake set
- Gear selector in park position

NOTE: *Look for special fluid checking information stamped on the dipstick. For instance, many DaimlerChrysler rear-wheel-drive vehicles specify neutral as the correct position for checking fluid level because they do not circulate fluid through the torque converter in park. Some General Motors automatic transmissions use a thermostatic device to raise ATF level when the transmission is cold. The level actually goes down as the transmission warms up.*

■ CHECK FLUID CONDITION

Fluid condition can tell a good deal about the overall condition of a transmission. The fluid should be clean and red, not pink and foamy or milky from contamination with water or coolant. Fluid should be easy to wipe from the dipstick and should not smell burnt.

Following are some conditions that might be found:
- *Milky fluid* is usually found when coolant and fluid mix together when the transmission cooler in the radiator leaks.
- **Varnish** is indicated when the dipstick will not wipe clean. Internal parts in the transmission cannot operate properly when the fluid has become oxidized like this. The transmission will need to be removed, disassembled, and rebuilt.
- *Burned fluid* will be dark and have a burnt smell.

NOTE: *Some fluids have a burnt smell when they are in good condition, so be sure this is not the normal condition. Burned fluid is usually caused by failed friction parts in the clutch packs or bands. Friction material might be evident on the dipstick. The transmission will probably require rebuilding soon.*

When adding fluid, be sure to use the correct type. Add fluid through the dipstick hole with a funnel with a small opening that will fit into the dipstick tube. Be sure the funnel is clean.

SHOP TIP A funnel that has had oil in it will collect shop dust. After use, rinse it and keep it in a plastic bag.

■ CHANGING TRANSMISSION FLUID

When changing transmission fluid, check the service manual for special instructions. Some manufacturers call for draining the converter also. In this case, the converter has drain plugs. Most transmissions do not have drain plugs in the converter, so they do not require a complete change of the fluid. Only the ATF in the pan is removed. Adding new fluid after the pan is reinstalled replenishes any additives that have been depleted during service.

Automatic Transmission Flushing

Electronic transmissions are more sensitive to problems with fluid condition. PWM valves tend to wear, and solenoids can stick or become sluggish. One recommendation calls for transmission service at every third oil change. At today's extended intervals, this would be at about 22,250 miles.

Transmission flush machines are a common sight in many shops (**Figure 74.3**). Many of them are easily

Figure 74.3 A transmission flush machine. *(Courtesy of Tim Gilles)*

connected to one transmission cooler line. One of the flushing machine's lines goes to the cooler, and the other line goes to the cooler line from the transmission.

A specified amount of ATF of the correct type is poured into the machine's reservoir. A typical amount is 16 quarts. The reservoir has a rubber bladder that keeps the fresh fluid separate from the old fluid. The new fluid pushes the old fluid out and continues to flush the transmission with new fluid until all 16 quarts have been run through the transmission. The transmission will have the correct amount of fluid remaining in its sump because only the same amount is installed as is removed.

TRANSMISSION FILTER SERVICE

The pan must be removed to clean or replace the filter (**Figure 74.4**). Newer transmissions work harder, so they leave more clutch material in the fluid. While ATF might not need to be changed, the filter should be.

Some filters are replaceable. Some older cars had reusable brass screens. It is customary to replace either the filter or the screen. They usually come in a filter kit with a new gasket. Some filters have a drainback check valve to prevent fluid from draining from the valve body when the engine is off.

Transmission filters are *unlike* engine oil filters, which have a by-pass valve in case they become plugged. Polyester felt filters are the most popular filters in late-model transmissions. Felt will trap materials as small as 60 microns (0.0025"), about the size of a human hair. A clogged transmission filter can cause shifting problems, pressure loss, and possibly transmission failure. A plugged filter will sometimes cause a whining sound like an empty power steering pump.

Removing the Pan

Most transmission pans do not have drain plugs. To drain the fluid requires removal of the pan.

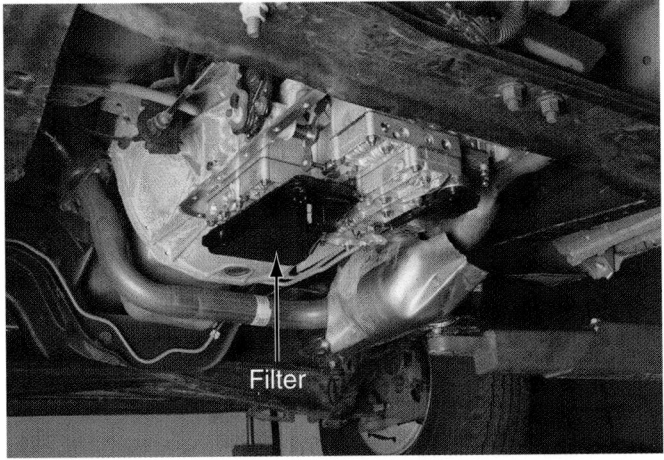

Figure 74.4 With the oil pan removed, the transmission fluid filter is visible. It is typically fastened to the bottom of the valve body.

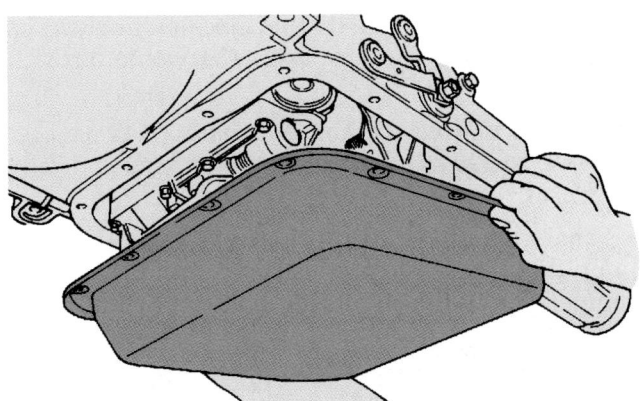

Figure 74.5 Slightly loosen the screws in the front of the pan and let the oil flow out of the back. (*Courtesy of Ford Motor Company*)

CAUTION There are usually 3 to 4 quarts of fluid in the pan. This weighs about 6 to 8 pounds. On the first transmission fluid change, beginners often make a large mess. Be careful! The fluid is often quite hot and if the pan is not carefully removed, the fluid will spill out fast.

SHOP TIP When removing the pan, leave a row of screws in the front of the pan. Remove all other screws. Then, loosen the remaining screws in the front of the pan (**Figure 74.5**). Be sure to have a large fluid catch tray ready. Let the fluid spill into the catch tray and then carefully remove the pan.

Inspect the Pan

Inspect the inside of the pan for pieces of friction material or metal particles. These might indicate serious transmission problems. A small amount of debris from friction material is normal. Clean the pan with a lint-free rag.

NOTE: *Occasionally, lint from shop rags is found plugging up the transmission screen or filter.*

Check the screw holes in the pan to see that they are not raised up from previous overtightening. Use a hammer to flatten them against an anvil.

REMOVE AND REPLACE THE FILTER

It is important that a pump suck only fluid and not air. Some filters have a gasket. Others plug into the transmission case or pump and are sealed with only a neoprene O-ring or sleeve. Be sure to remove the old O-ring or gasket and replace it with a new one. Lubricate the new O-ring with ATF. Do not use sealer. Install the filter and check to see that its tube fits snugly in its bore. Be sure to replace any parts that come off with the filter.

NOTE: *Most transmissions are aluminum. They are easy to strip. Be sure to use a torque wrench on filter and valve body screws.*

Some filters have nylon housings and others are aluminum or steel. There is usually a protective metal grommet where the mounting bolt goes through to the valve body.

If air gets into the torque converter, it will not be mixed with the fluid but will gather around the input shaft. Too much air can result in a burned up torque converter. The converter acts like a centrifuge, separating the fluid and air. Fluid is thrown to the outside of the converter. The air that is separated is sent on to the cooler.

NOTE: *The bottom of anything spinning is always the outside.*

SHOP TIP To check for aeration from the converter:
- Run a hose from the cooler line to the transmission dipstick tube.
- Start the engine.
- If there is an air leak, for a few seconds air will be visible in the fluid.

Replace the Pan

Gasket sealer is not normally used on pan gaskets except to hold a gasket in place. Rubber cement or gasket positioning sealer can be used if necessary.

NOTES:
- *Usually, the screw holes in the gasket are smaller than the screw diameter. When the screws are installed through the pan and the gasket, they will remain there during installation to position the gasket.*
- *Do not use silicone on the pan. ATF attacks it. It is not uncommon to find silicone plugging a transmission filter or screen.*

Tighten the screws evenly. Do not tighten any screw all the way until all screws are started in their threads. Use an inch-pound torque wrench to torque the screws to specifications. Look this up in the service manual. Specifications usually range from 60 to 150 inch-pounds. This is not very much, but it compresses the gasket and allows it to work properly. Overtightening a gasket will take it past its elastic limit and cause it to lose its sealing ability. Gaskets often split in the middle when they are overtightened.

■ REFILL THE TRANSMISSION

Check the service manual for the capacity of the pan.

SHOP TIP
- The torque converter contains about 3 to 4 quarts of fluid. If only the pan was drained, be sure that a smaller amount of fluid than transmission capacity is added.
- If in doubt about the fluid capacity, add 3 quarts (the approximate amount contained in the pan). Then, run the engine, step on the brake, put the transmission in each gear, and recheck the fluid level. Putting the transmission in each gear range fills the cavities in each circuit. If you have a helper, you can run the engine with the transmission in low gear while you fill the transmission.
- If the fluid does not touch the bottom of the dipstick, add one quart. Then, start the engine and recheck the fluid level.
- When fluid is on the dipstick, remember that only ½ quart will move the fluid level from *add* to *full*.
- Cold fluid will expand by about 1 pint when heated to operating temperature, so leave the fluid level at the add mark until the car can be test driven to heat the fluid.

Do not overfill the transmission. Foaming and siphoning can result. Some older Ford C6 and GM TH400 transmissions had shallow pans and were prone to problems from overfilling. The fluid would siphon out the dipstick tube. Later models of those transmissions had deeper pans to solve that problem.

SHOP TIP Excess fluid can be removed by one of the following methods:
- Loosen and remove one of the transmission cooler lines from the radiator. Run the engine to pump ATF out into a pan. Be sure to use two wrenches (one of them a flare-nut wrench) to prevent damage to the radiator.
- Through the transmission dipstick hole:
- Use a hand suction gun.
- Use a solvent siphon gun hooked to shop air.

■ DIAGNOSIS AND REPAIR OF LEAKS

Leaks can happen from the front pump seal, the rear seal to the drive shaft yoke (rear-wheel-drive cars), the shift lever shaft seal, the pan gasket, or the extension housing gasket (**Figure 74.6**). Sometimes, one of the transmission cooler lines or the dipstick tube leaks.

External leaks are obvious. Locating the source of a leak can be as easy as cleaning the outside of the

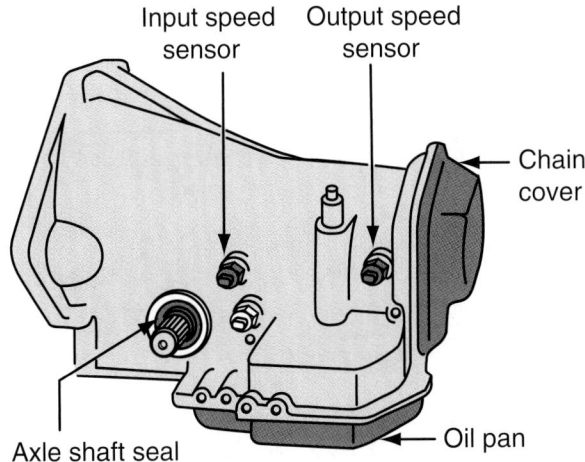

Figure 74.6 Some sources of leaks on a transaxle.

transmission and watching for new fluid. ATF is usually red in color, which helps differentiate it from engine oil. A fresh leak will usually clean the dirt off of the outside of the transmission or leave a washed area that looks like water running through dirt.

If a leak is particularly stubborn, fluorescent dye and a black light can be used. The leaking fluid will glow under the black light. Most leaks (except for the front pump seal) can be repaired without removing the transmission.

■ LEAKS FROM THE CONVERTER HOUSING

A leak from the converter housing can be due to several causes (**Figure 74.7**). To check the converter for leaks, remove the access cover on the housing. Some housings do not have an access cover. In this case,

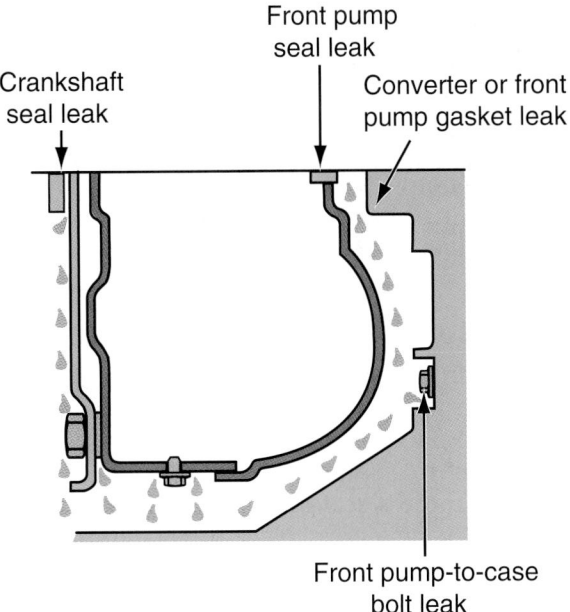

Figure 74.7 Determine the direction of fluid travel to identify the cause of a fluid leak around the torque converter.

disconnect the battery ground cable and remove the starter. An oil leak from the engine crankshaft seal could be the cause. This will usually leave a film on the front side of the flexplate. Engine oil is not red, so it is usually easy to tell the difference.

■ TRANSMISSION COOLER LINE LEAK

Pressure in the transmission cooler lines is usually about 35 psi. Some technicians attempt to repair leaks in these lines by cutting out the bad section of line and replacing it with a length of fuel hose. Ordinary fuel hose is not recommended for use with transmission fluid. Use transmission oil cooler (TOC) hose (see Chapter 23). The hose can easily blow off, even though there are hose clamps on the line, so both ends of the line must be flared.

> **SAFETY NOTE** Leaking fluid can cause a fire. Although fluid is not flammable at room temperature, it ignites just like gasoline does when heated in excess of about 300°F. A leak from a pressurized line can vaporize more easily. If transmission fluid is sprayed onto a hot exhaust manifold, it could cause a fire.

NOTE: *The ends of the lines must be double-flared (see Chapter 24) so that the flare is not so sharp that it cuts its way through the line. The body and engine move and vibrate on different mounts. In time, the hose could suffer a cut.*

A better repair is to cut out the old section of line and replace it with a *union* (see Figure 24.15) or install flare fittings and install a short piece of reinforced hose that has threaded ends crimped onto it.

■ SPEEDOMETER DRIVE GEAR LEAK

Many late-model vehicles have an electronic speedometer pickup, which has no mechanical connection. Mechanical speedometers have an O-ring seal that can leak from the extension housing.

■ SHIFT LEVER SEAL REPLACEMENT

The shift lever to the manual valve is sealed with a lip seal or O-ring. It is sometimes necessary to remove the pan and valve body in order to remove the shift lever shaft. Check the service manual for procedures.

■ PUMP SEAL REPLACEMENT

The transmission has a full round seal that rides on the front of the torque converter where it enters the transmission pump (**Figure 74.8**). This seal suffers from a good deal of heat because the torque converter operates at a temperature higher than the rest of the transmission. When this seal fails, transmission fluid *pours* out of it. A front pump seal often fails after an engine R&R.

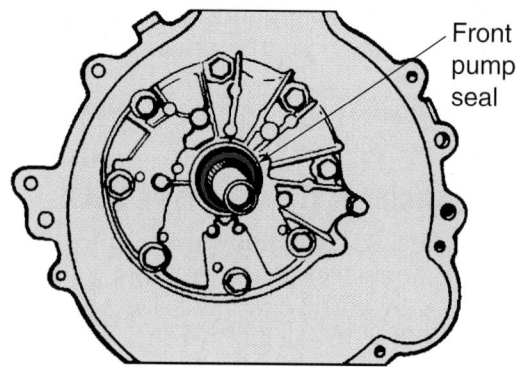

Figure 74.8 A front pump seal inside the converter housing. *(Courtesy of Ford Motor Company)*

An engine in a van was worn out. The engine was pulled and sent to the machine shop to be rebuilt. While the engine was out of the vehicle, the torque converter was left hanging in the transmission, unsupported on its front end.

NOTE: *When installed between the engine and the transmission, the front end of the converter is supported in the rear of the engine's crankshaft.*

The old transmission seal, which had lost some of its elasticity, became distorted and began to leak after the engine was reinstalled and started. The transmission had to be removed to replace the seal.

A *front pump seal* replacement requires that the transmission be removed from the vehicle. Follow the manufacturer's instructions. Front-wheel-drive procedures vary. Be sure to support the engine from the top (see Figure 50.17). Sometimes, engine cradles, suspension parts, or complete engine/transmission assemblies must be removed. If axles are removed, fluid drains from the transaxle. Save the fluid in a clean can for refilling later. If the container is not completely clean or is not kept covered, the fluid must be replaced to prevent contamination. Before reinstalling a transaxle, examine the CV joint boots. If they are torn, replace them.

Rear-wheel-drive procedures are more straightforward.

These general instructions apply to all transmissions.

- Remove the inspection cover or starter motor to gain access to the engine side of the flexplate.
- Mark the converter and flexplate so they can be reassembled in the same position.
- Unbolt the torque converter from the flexplate and push the converter all the way into the transmission pump.
- Install a transmission jack under the transmission.

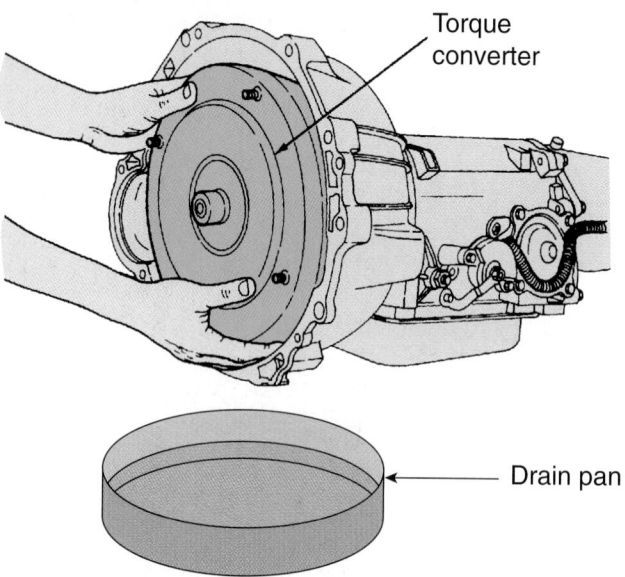

Figure 74.9 Remove the torque converter from the transmission. *(Courtesy of Ford Motor Company)*

- Remove any crossmembers or braces supporting the engine.
- Mark and remove any linkage.
- Remove the speedometer cable or wire.
- Unbolt the transmission from the engine.
- Check to see that any dowel pins that align the engine to the transmission are in place (see Figure 70.10).

■ FRONT SEAL REPLACEMENT

To replace the seal:
- Place a drain pan under the torque converter and remove it from the transmission. (**Figure 74.9**). If you quickly rotate it so the transmission end faces up, very little fluid will leak out.
- Remove the old seal with a chisel, hook puller, or slide hammer seal puller (**Figure 74.10a**).
- Install the seal on the installer and drive the seal into the pump (**Figure 74.10b**).

Before installing a new seal, be certain the oil return hole to the pan is unrestricted. A restricted oil return can cause the seal to pop out due to excessive pressure buildup.

■ PUMP BUSHING REPLACEMENT

A bushing in the transmission pump supports the snout of the torque converter. When a pump seal is replaced, the pump is often removed and the bushing replaced too. Excessive pump bushing clearance (over 0.003") can cause a seal to pop out. Be sure to install the bushing using a press. Never use a hammer.

Check the service manual for the procedure on bushing replacement. This often requires removal and disassembly of the pump. Removing the pump sometimes requires the use of a pair of slide hammers threaded into two of the bolt holes in the pump (**Figure 74.11**).

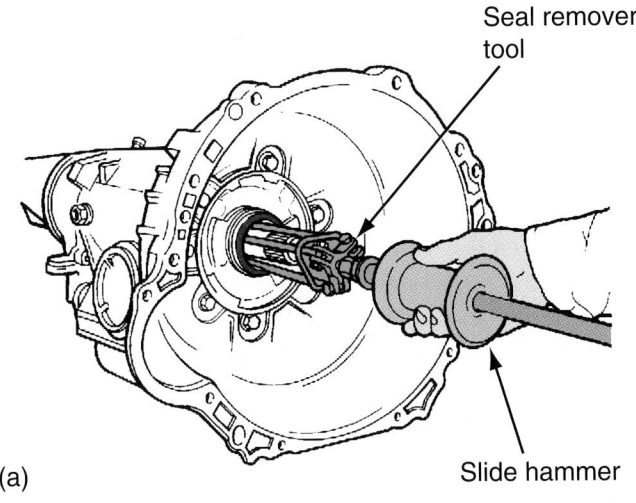

(a)

Seal remover tool

Slide hammer

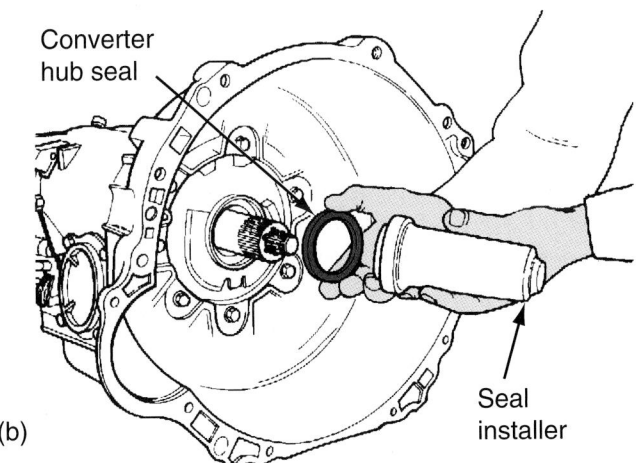

Converter hub seal

(b)

Seal installer

Figure 74.10 (a) Remove the old seal from the pump housing. (b) Install the seal on the installer and drive the seal into the pump. *(Courtesy of Ford Motor Company)*

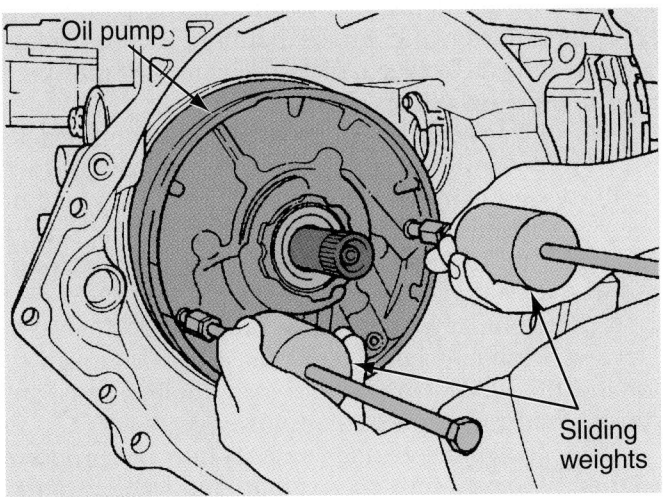

Oil pump

Sliding weights

Figure 74.11 Slide hammer pullers for removing the pump from the transmission body. *(Courtesy of DaimlerChrysler Corporation)*

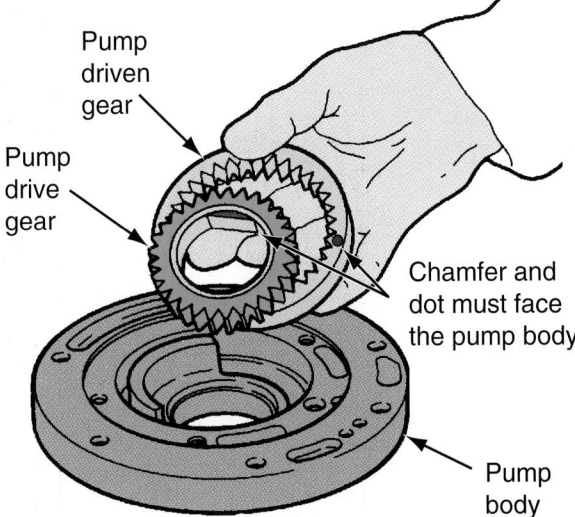

Pump driven gear

Pump drive gear

Chamfer and dot must face the pump body

Pump body

Figure 74.12 Install the pump gear with the chamfer down so the torque converter can fit easily into it. *(Courtesy of Ford Motor Company)*

Mark the gears or rotors so they can be put back in the same position in relation to each other. The pump gear must be reinstalled with the chamfer down so that the converter can be installed into it (**Figure 74.12**). Some pumps require the use of a special clamp that fits around the outside of the pump body to align the halves.

Before reassembling the pump, put a straightedge across the pump body and check the clearance to the faces of the pump gears. Gear-type pumps should have 0.001" to 0.003" clearance, and vane-type pumps should have 0.001" to 0.0015" clearance.

■ REINSTALLING THE TRANSMISSION

Install the transmission and bolt it to the engine. Be sure the engine and transmission are touching each other completely before installing any bolts. The torque converter must be correctly engaged with the transmission front pump gear (see Figure 73.43). It will drop deeper into the bell housing (toward the transmission) and the drive lugs on the converter should be felt engaging the transmission front pump gear (**Figure 74.13**). Failure to install the converter correctly can result in damage to parts, including the transmission pump.

CASE HISTORY *A student was bolting an automatic transmission to an engine. There was a gap of about ⅜" between the engine and transmission. He installed two bolts through the transmission into the back of the engine and began to tighten them, forcing the engine and transmission together. The result was a cracked transmission housing.*

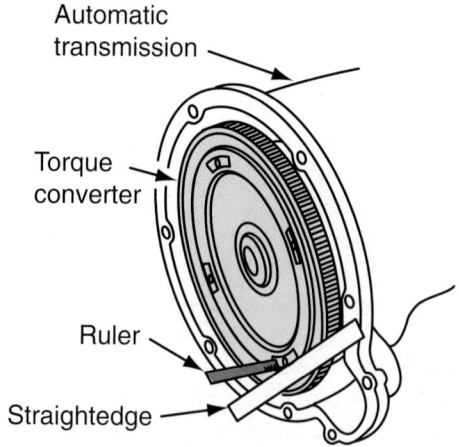

Figure 74.13 Checking to be sure that the converter is installed all the way into the transmission housing.

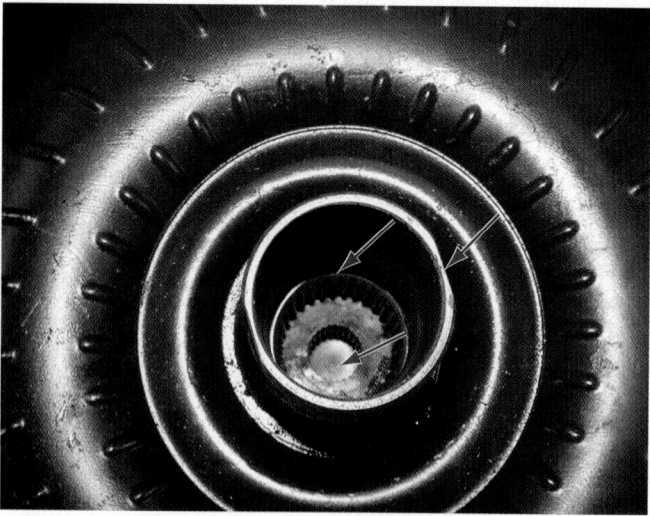

Figure 74.14 Three areas that must slide onto or into other members during torque converter installation into an automatic transmission transaxle. *(Courtesy of Tim Gilles)*

Figure 74.14 shows a view looking into a torque converter. When a torque converter is installed into an automatic transmission, at least three areas must align with other members: the pump drive, the stator support splines, and the input shaft splines. Some lock-up converters have an additional splined area for another small input shaft.

When the converter is installed correctly, there should be at least ⅛" between the end of the converter studs and the flexplate. There will be enough room to rotate it and align it with the bolt holes in the flexplate. Then, slide the converter toward the engine and bolt it to the flexplate.

NOTE: *When installing the torque converter to the flexplate, be sure to use the correct bolts. Bolts that are too long can be forced through the converter housing or they can leave the flexplate loose. This can result in elongated holes and a cracked flexplate due to excessive vibration.*

The result is a clunk when shifting from neutral into gear. Bolts too long can also deform the converter housing and lock up the converter clutch.

■ REAR OIL SEAL AND BUSHING REPLACEMENT

Replacement of the extension housing seal and bushing on rear-wheel-drive cars requires removal of the drive shaft (see Chapter 70). The old seal can be removed with a chisel or a special seal puller. Special drivers are available for installing the new seal (see Figure 72.25).

Using a special puller, the old bushing can be removed without removing the extension housing (see Figure 72.24). The bushing sometimes has an oil return hole that must be aligned with a hole in the extension housing during installation. The puller set includes an installation tool, too.

■ AUTOMATIC TRANSMISSION PROBLEM DIAGNOSIS

This section relates to both electronic and older, mechanically operated automatic transmissions. Diagnosing automatic transmission mechanical problems is the same whether the transmission is mechanically or electronically controlled. From a diagnostic point of view, the hydraulic control system of an electronic transmission operates in the same manner as a mechanical transmission. The primary difference is that electronically controlled units do not have a mechanical governor or a vacuum modulator. Instead, they use computer inputs from the vehicle speed sensor and the engine fuel management system's intake manifold pressure sensor. Electronic transmission tests are covered later in this chapter.

Shop Tests

Before condemning a transmission, check for diagnostic trouble codes and check the condition of the engine. It should accelerate normally and idle smoothly. On an unmodified engine, intake manifold vacuum below 15" can indicate a mechanical problem. Be sure there are no **vacuum leaks**.

On older cars, a problem in the vacuum feed to the modulator would result in harsh, late shifts. Fluid in the vacuum line is a pretty good indication that the diaphragm is leaking. The modulator diaphragm should be able to hold at least 24" applied by a vacuum pump.

Throttle linkage cannot be binding. If there is a kickdown cable, check to see that it is not broken or forcing the kickdown valve to stay applied. This can cause a transmission to stay in lower gears.

The *quadrant indicator* might not be properly aligned. As you move the shifter, count the number of clicks from park to drive. Perform the test drive with the shifter in the correct position and make any needed adjustment to the indicator later.

Perform a visual check of *electrical connections*. Look for corrosion, bad insulation, and disconnections.

A hot-running transmission could be caused by several things. If the transmission is running hot, the torque converter clutch (TCC) might not be operating correctly. TCCs were added to overdrive transmissions because the torque converter would generate so much heat from slippage that the oil cooler could not eliminate all of it. An added benefit of TCCs is increased fuel economy.

Another cause of high heat could be a restricted oil cooler. To test cooler flow, remove the return line from the oil cooler. Run the engine with the line over a container to catch the ATF that pumps out. A good amount of cooler oil is a quart in 20 seconds. If less than that flows through, clean the cooler with cooler flush, which is thinner than ATF and can run through the cooler. Always test cooler flow rate after a flush job is completed. A plugged cooler will have to be replaced.

Test Drive

Before taking a test drive, talk to the customer and get a clear idea of the complaint. Road testing is usually done on a warm vehicle. Ask the customer how the transmission operates when it is cold.

Be sure there is at least some fluid on the dipstick before driving the car. Operation without fluid can damage a transmission.

Some technicians suggest that the car be driven before adjusting the fluid level. This way, if adding fluid solves the problem, that can be noted during a retest.

NOTE: *Be sure to inspect the fluid on the dipstick before wiping it clean.*

Use a *road test checklist* to help diagnose a problem. Be sure to turn off the radio during the test drive so unusual noises can be heard.

Start your test drive by checking *minimum throttle upshifts*. The specified speed varies between manufacturers. Minimum throttle is at a speed just above idle. A typical 1–2 shift should happen no later than about 22 to 24 mph (this would be under heavy load). The rest of the shifts should happen a few miles per hour apart and not be stacked on top of each other. Note shift quality to see that shifts are not too soft and the rpm does not slip up. Try forced detent downshifts at WOT.

■ SLIPPAGE

Slippage can be caused by a low fluid level, incorrect linkage adjustment, leakage in a clutch pack or servo, a plugged pump inlet screen or filter, or a problem with the valve body.

■ MORNING SICKNESS

As a transmission ages, heat takes a toll on the synthetic rubber seals within it. As seals age, they harden. This was especially true with the lower-quality seals used on older vehicles. A hard seal will sometimes leak when the transmission is first put in gear on a cold morning. The transmission will hesitate for anywhere from a few seconds to a minute, leaving the vehicle standing still, as if in neutral, while the engine races. Usually the seals soften sufficiently to quit leaking and the vehicle begins to move. The transmission might operate perfectly fine after this initial problem. A transmission such as this is said to be "sleepy" or to have "morning sickness."

A similar condition results all of the time when a transmission is slightly low on fluid. Once the fluid expands, the transmission will work in a normal manner.

■ TRANSMISSION DRAINBACK

Another problem resembling morning sickness could be worn input shaft bushings. The complaint might be that the transmission slips on initial takeoff for about 5 to 10 seconds. When the engine is shut off, the torque converter is supposed to stay full and ready for the next engine start. A one-way check valve in the line provides fluid to the converter circuit. The bushing inside the stator supports the input shaft. Worn input shaft bushings can allow fluid to run out of the converter and overfill the pan (**Figure 74.15**). The owner might complain of an occasional puddle of ATF under the vehicle. This is because of fluid escaping from the vent hole high on the side of the transmission case.

A key to diagnosing converter drainback is that the problem goes away as the engine runs and the converter

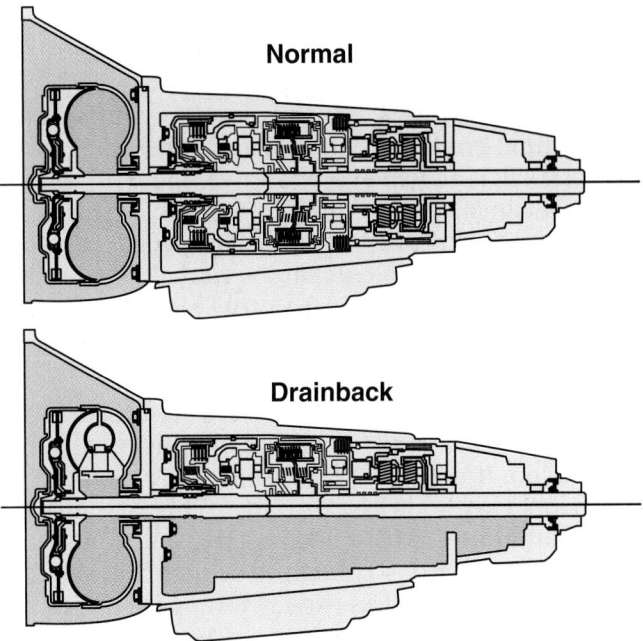

Figure 74.15 The torque converter can empty past worn input shaft bushings.

refills. If you suspect a drainback condition in a running vehicle, check the fluid level on the dipstick at night after the transmission has cooled down. Check it again in the morning to see how much higher it is.

Shift Feel

One symptom of leaking seals occurs as the transmission shifts into a higher gear. During the shift, the engine rpm will increase in speed briefly until the clutch pack or servo piston finally seals and pressure builds up. This is called a **mushy shift**. Using speed sensors on the input and output shafts, the computer on an electronically controlled transmission can sense slippage and will increase pressure within the leaking transmission circuit to compensate.

A **harsh shift**, the opposite of a mushy shift, occurs when the transmission makes a gear change that is too fast. It might shudder due to actually having clutches applied for two planetary operations at once.

Mechanical Transmission Problems

The following are related to mechanically shifted transmissions.

Shift Feel. From the factory, shift feel is usually so soft that the customer cannot notice it. A properly timed harsh shift is better for transmission life because aggressive shifts are easier on clutch facing materials, but they allow shocks to damage transmission hard parts in response to driver abuse. A shift kit can be installed to make a mechanical transmission shift more aggressively. The kit includes springs for the valve body and instructions for drilling enlarged holes in the valve body separator plate to allow fluid to reach a clutch sooner.

Incorrect Shift Timing. Shifts can be either too late, with the engine rpm too high, or too early, with the engine lugging. A transmission that shifts at the wrong time could be the result of a pressure problem or a sticking valve in the valve body. An engine problem can also cause the transmission to shift later because more throttle opening is needed to move the car. Pressure problems can be caused by a leaking modulator, a misadjusted throttle valve cable, or a sticking governor. Adjustments are covered later in the chapter.

Shift Chatter. Several problems can cause a clutch to chatter during a shift. Two factors influence clutch application: how fast it applies and how much torque it can hold without breaking free. This is similar to a dry clutch used with a manual transmission. If the springs are not strong enough the clutch will slip, but it will not chatter unless excessive slippage has glazed the surface of the clutch. When chatter continues after initial clutch application, something is causing the clutch to slip.

One common cause of chattering is using the wrong type of ATF. The clutch friction material or the finish on the steel plates or clutch drum can also change the shift feel. These problems are indicated when the clutch chatters only during clutch application.

Torque Converter Clutch Chatter. TCC chatter is caused by the same factors as other clutch chatter. To isolate this problem, add friction modifier to the transmission fluid. If this does not solve the problem, it could be related to the engine. If you artificially increase line pressure and chatter ceases, the torque converter is the probable cause.

Sticking Governor. When a governor sticks in the "out" position, the valve body is receiving erroneous information that the vehicle is travelling at high speed. In response, the shift valve will try to move toward high gear. The 3–2 shift valve will be in the third gear position during a start, so the car will take off in high gear. If the governor sticks in the "in" position, the vehicle will stay in a lower gear range.

Vacuum Modulator Diagnosis and Adjustment. If the vacuum modulator has a hole in its diaphragm, the valve body acts as if the throttle is wide open (WOT) and the shift valve moves into the lower gear position. This causes harsh, late shifts, if the transmission upshifts at all. When there is a hole in the diaphragm, several symptoms can result. One is a vacuum leak, which can result in a rough, slightly higher idle and a flat spot (hesitation) during take off. Also, the hole in the diaphragm means that the transmission acts as if the engine is under full load, even though the car might be barely cruising. The result is that the transmission shifts really late, if at all. One side effect of high modulator pressure is that pressures increase within the transmission to provide a firm shift under the heavier load of full throttle. Thus, the transmission also shifts harshly. Check the vacuum supply to see that it is sufficient.

No Upshifts

When a vehicle remains in low gear, you can do several things to diagnose. Shift the vehicle into each manual gear range during your test drive. If the vehicle has a vacuum modulator that controls line pressure only, remove the vacuum hose. On mechanically shifted transmissions, see if wide open throttle makes a difference. On electronic transmissions, pulling the shift solenoid connector will put the transmission into the failsafe gear range. This will be a gear range higher than first. A scan tool or transmission tester can be used to command each gear shift.

Wrong Gear Starts

Sometimes a transmission will start in the wrong gear and then proceed with its shifts. On both mechanical and electronic transmissions, this can be due to a stuck shift valve. In a mechanically controlled transmission, the governor circuit pressure could be too high. Remember that only the governor can cause a valve to shift from hydraulic pressure and it takes more pressure to move the shift valve than is required to hold it in its shifted position. Therefore, only a small amount of extra governor pressure can prevent a downshift.

On electronic transmissions, wrong gear starts can be caused by outside wiring problems or by a shift solenoid that is plugged or leaking. Check for electrical problems by verifying that the computer is signaling a shift.

Whenever you begin a repair, be sure to consult the service information. Late-model computer systems can experience very strange electrical problems. For instance, a problem with a center high-mounted brake light can sometimes cause a transmission problem. For example, on some sophisticated vehicles, the vehicle speed sensor prompts the computer to raise the volume on the radio in response to increased road noise. If the radio is exchanged, a transmission problem can result.

Engine Dies in Gear

When the engine runs in neutral but dies when put in gear, a lean-running engine could be the cause. When the cause is transmission-related, it is probably due to torque converter clutch staying applied. Before removing the transmission, verify the problem by running the engine in gear while the vehicle is raised on the lift. As you slowly apply the brakes, engine performance will begin to suffer if the converter clutch remains applied.

No Vehicle Movement

When the vehicle will not move in forward or reverse ranges, the cause could be several things, including no hydraulic pressure or a mechanical problem. Sometimes the converter does not fill with fluid, even though line pressure is sufficient. In this case, disconnect the cooler line and check it for flow. Converter drainback does not usually cause vehicle immobility because the converter will start to fill when the engine starts and performance will improve as the converter is purged of air.

If there is a problem with fluid intake, like a plugged filter, the transmission will start having problems after driving for a while. After sitting for a time, it might start to work fine again.

Sometimes the car will move in one direction but not the other. In this case you can eliminate several possibilities, including hydraulic pressure. If it works in reverse, but not in forward ranges, pressure must be good enough. The torque converter is also not to blame. This type of problem is usually of a mechanical nature or results from an internal component leak.

Backward One-way Clutch

A backward one-way roller or sprag clutch can cause the car to not engage in drive low, but to work in manual low (because the band is holding, instead of the roller clutch). During the 1–2 shift, binding can occur.

> **SHOP TIP** On Simpson or Ravigneaux geartrains, to test for a backwards one-way clutch, raise the vehicle's drive wheels. If you can rotate both drive wheels at the same time in a forward direction with less resistance than when you turn them backwards, the one-way clutch is correctly installed. On some later transmissions there will be no resistance difference between forward and backward.

Internal Hydraulic System Cross Leaks

A *cross leak* can cause hydraulic binding when *foreign oil* enters the wrong circuit. Sometimes a cross leak can occur when the gasket between the valve body and its separator plate leaks. Sometimes the wrong gasket is installed, the old gasket surface was not thoroughly cleaned, or the valve body bolts were improperly torqued. Cross leaks can also be caused by sealing rings that are too small or by missing valve body check balls.

A cross leak can be difficult to diagnose. Higher line pressure will often make the problem more intense as the leak worsens. Methods of raising line pressure vary. On some transmissions, this is done by removing the vacuum hose to the modulator. On others, pulling on the throttle cable raises pressure. If the test is positive and the transmission has pressure tap points, pressure test each component to isolate the problem.

▬ NOISES

Transmission noises can result from a high or low fluid level, gear or bearing wear or damage, or a bad torque converter bushing or one-way clutch. Noises can be transmitted through other parts.

Engine accessories such as the power steering pump, smog pump, timing belt idler bearing, or bad fan belts can also mimic transmission noise. A bad universal joint or axle bearing can also be mistaken for transmission noise.

Sometimes a torque converter will make noise. In park or neutral, the stator does not hold and all of the converter pieces turn together. If the converter noise goes away in park and neutral, the converter is not the source of the noise. Noise that seems to come from a torque converter can be due to a loose crankshaft pulley (see related case history in Chapter 53).

■ Use a stethoscope to pinpoint the source of a noise.
■ Check to see if the noise occurs in all gear ranges. This could be a torque converter or pump; something that operates in all gears.

- A defective pump will cause humming and buzzing that gets louder with increased engine rpm. It will occur in any gear position.
- A grinding noise when the vehicle is moving that increases under load could be a failure in a planetary gearset, a needle bearing, or a bushing.

▬ FLEXPLATE

The flexplate is so named because it flexes inward toward the engine. The converter normally slides forward during deceleration. When the engine is under load, the converter is driving its turbine and there is no forward load on the crankshaft. During deceleration the rear wheels drive the converter, reversing the direction of thrust. When the converter moves forward, the pilot hub on the nose of the torque converter (**Figure 74.16**) moves into the bore in the rear of the crankshaft. The flexplate will allow movement of up to 0.080" to 0.100". Rust, paint, or damage to the crankshaft pilot bore can result in damaging loads to the engine's crankshaft thrust surface.

SHOP TIP
- One test for an intermittent flexplate sound is to run the engine at about 2000 rpm (fast idle). Shut off the ignition and quickly turn it back on. A loud rattle will be heard when the engine is restarted.
- Another test is to shift the transmission from neutral into gear while listening for a clunk.

A cracked flexplate (see Figure 49.12) can cause a knocking sound. A badly cracked flexplate will cause a sound similar to a fan hitting the radiator. The noise changes with speed and load.

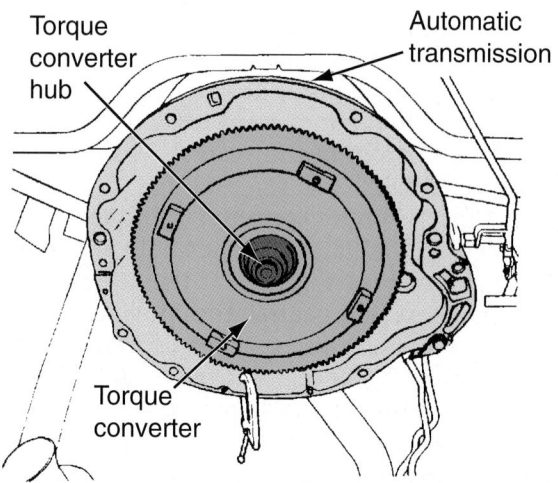

Figure 74.16 The torque converter hub moves in and out of the crankshaft hub pilot hole. *(Courtesy of DaimlerChrysler Corporation)*

▬ TRANSMISSION TESTS

Stall Test

A **stall test** is used to determine if a transmission is slipping or a torque converter is defective.

To perform a stall test:
- Connect a tachometer to the engine's ignition to measure engine rpm.
- Set the emergency brake firmly.
- Apply the *service brakes* (foot brake).
- Start the engine.
- Put the gear selector in low gear.
- Accelerate the engine until speed no longer increases. Note engine rpm at this point.

NOTE: *Be sure that you do not do the stall test for longer than 5 seconds because of the excess heat that the test causes. Also, if rpm climbs beyond specifications, stop the test immediately. Always check the service manual for manufacturer's specifications and special instructions prior to performing the test.*

Stall Test Results.
- If the rpm is lower than specifications, the torque converter stator clutch could be slipping.
- If the rpm is higher than specifications, the transmission is slipping. A band adjustment might correct this (if the transmission has bands), but the transmission will probably require major repair.

Pressure Test

A **pressure test** will tell if the transmission is experiencing internal leakage or requires adjustment to linkages or the vacuum modulator. Pipe plugs are installed in the outside of the transmission case and tap into various circuits in the transmission. The service manual will provide each pressure tap point's location as well as the specifications.

The pressure gauge that is used should have at least a 300 psi capacity. It should have a long hose so it can be stretched through the window into the driver's compartment.

To perform a pressure test:
- Connect a pressure gauge to the correct pressure tap (**Figure 74.17**).

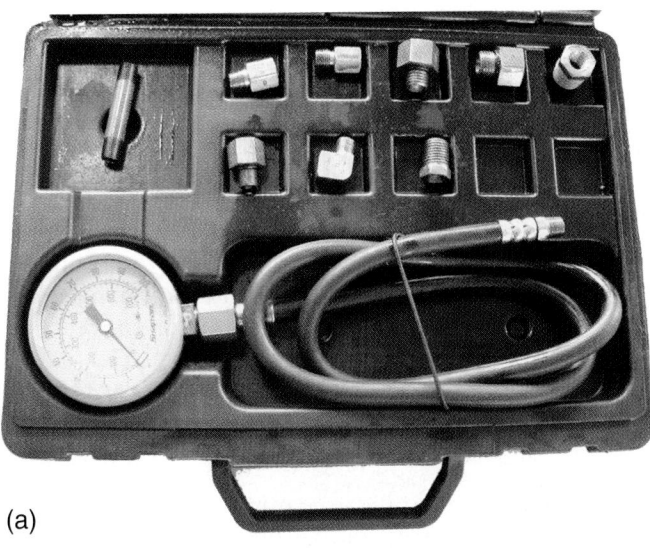

(a)

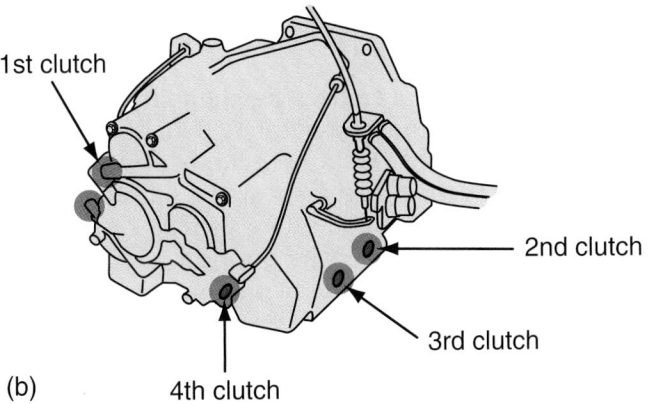

1st clutch

2nd clutch

3rd clutch

(b)   4th clutch

Figure 74.17 (a) A pressure gauge set. (b) Install a pressure gauge in the correct pressure tap. *(Photograph, courtesy of Tim Gilles)*

- Run the engine until it reaches operating temperature.
- Shift through all of the gear ranges and note the pressures that result.
- Test drive the car to note shift points and pressures.

SHOP TIP
- Governor pressure and miles per hour are approximately equivalent on most valve bodies.
- If the transmission has good pressure in reverse, the hydraulic system must be good.

Air Test

An **air test** can check the operation of various clutch packs and servos before disassembling a transmission fully. The pan and valve body are removed and air is directed into the fluid passages that supply the clutch or servo. The service manual tells where each passage leads.

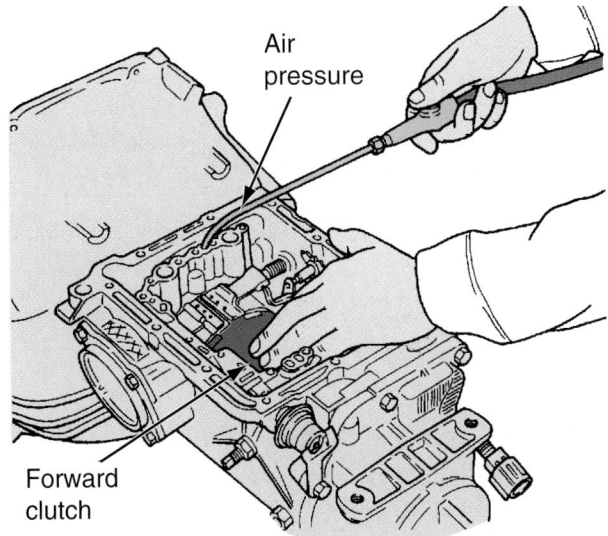

Air pressure

Forward clutch

Figure 74.18 When the valve body is removed, you can use compressed air to feel a clutch apply or watch a band apply. *(Courtesy of Ford Motor Company)*

A rubber-tipped blowgun is used for performing an air test. When pressure is introduced into the passageway, you will hear a clunk as the clutch applies. If it is leaking, you will hear air leaking out of the leaking seal.

NOTE: *An air test will not always pinpoint a small leak in a seal. Transmission fluid must first pass through an orifice in the separator plate. A higher volume of air can pass through this orifice, enough to compensate for the leak.*

CAUTION A rubber-tipped blowgun does not have a pressure relief feature like a normal blowgun. Pressure at the tip will equal shop air pressure. Use caution with this high air pressure. If you have an air pressure regulator, set it for no more than 35 psi. If the clutch or servo seals work at 35 psi, they will work at normal transmission pressure.

On some transmissions, the geartrain is visible when the valve body is removed. You can feel a clutch apply or watch a band apply (**Figure 74.18**). When air testing a governor, sometimes a vibrating or whistling sound is heard, depending on the style of governor.

VALVE BODY REMOVAL

If the valve body is being removed while the transmission is in the car, be sure it is cold.

NOTE: *If valve body is removed when hot, it can warp.*

When reinstalling the valve body, the torque sequence is important to avoid warping the valve body.

When removing a valve body, there are often *check balls* that must be replaced in the same position on

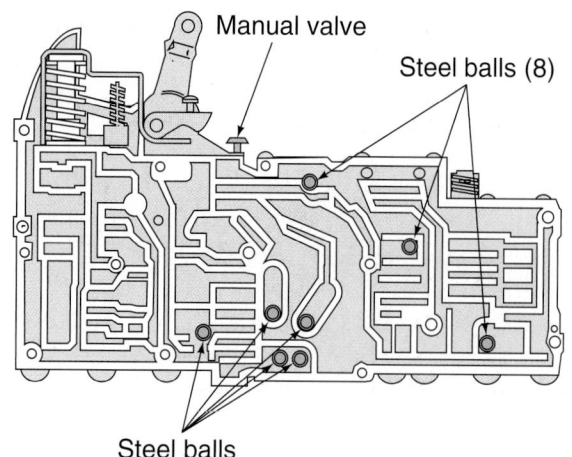

Figure 74.19 Check balls must be installed in the correct location in the valve body. *(Courtesy of DaimlerChrysler Corporation)*

reassembly (**Figure 74.19**). Failure to put these in the correct position will result in faulty transmission operation. With the transmission in the car, reassembly can be difficult. A good trick is to put the check balls in their correct positions on top of the separator plate and put it into place first. Petroleum jelly can also be used to hold the check balls in place.

▰ TRANSMISSION ADJUSTMENTS

Adjustments that can be made on a transmission include those to the transmission linkage, the kickdown lever or switch, the neutral safety switch, and band clearance.

Linkage Adjustment

Check to see if the shift linkage has any slack or looseness in it. This can cause a transmission manual valve to be in the wrong position (**Figure 74.20**). Adjustment

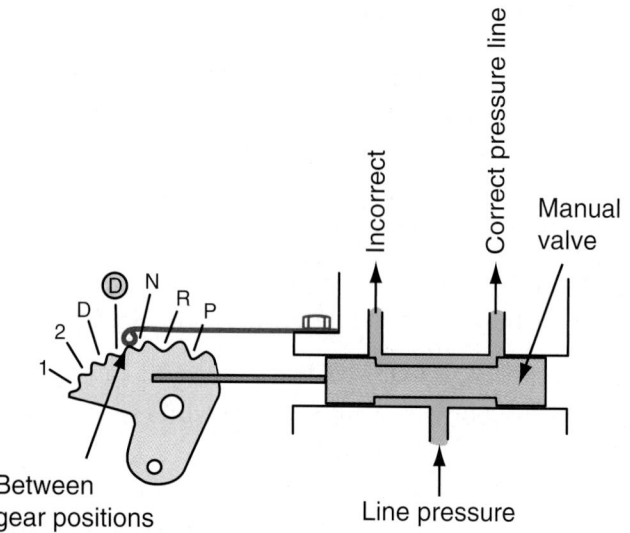

Figure 74.20 If the shift linkage adjustment is off, fluid can be diverted into the wrong passage.

procedures vary between manufacturers. Check the service manual for the exact procedure. On most shift linkages, there is a locknut on the shift rod. Loosening it allows the adjustment to be shortened or lengthened as needed.

> **SHOP TIP** There is usually no reason for an adjustment to change. Whenever a linkage part is removed, it should be marked or removed in such a manner that it can be easily replaced in its original position.

The quadrant indicator might not be properly aligned. To check this, move the shifter as you count the number of clicks from park to drive.

Neutral Safety Switch

If the engine does not crank when the shift selector is in the park position, the neutral safety switch might need adjustment. Moving the shift selector while turning the key often results in engagement of the starter motor.

To readjust the neutral safety switch:

■ Put the selector in the park position.
■ Loosen the screws on the neutral safety switch.
■ With the ignition switch held in the start position, move the safety switch until the engine cranks.
■ Tighten the switch.
■ Double-check the adjustment by attempting to start the engine in the park and neutral positions.
■ The switch sometimes controls the operation of the back-up lights also. With the key on and the engine off put the selector in the reverse position. Check the back-up lights to see that they are on. Then, check in the drive position to see that they are out.

Sometimes, there is a pin hole to align while adjusting the neutral safety switch.

To check a neutral safety switch to see if it is bad:

■ Connect an ohmmeter across the terminals of the switch.
■ With the switch in park and neutral, the reading should be zero ohms. This means the switch is "closed."
■ With the switch in the other gear positions, the reading should be infinite. This means the switch is "open."

Throttle Cable Adjustment

Some transmissions use a throttle cable to give the transmission valve body information on how far the throttle is open. A throttle cable is usually adjustable. Incorrect adjustment of a throttle cable can result in incorrect shift points and transmission damage from pressures that are too low for the condition at hand. Check the repair manual for the correct procedure.

Band Clearance Adjustment

A brake band is used on some transmissions as a holding device to keep a planetary gearset drum from turning. The friction material on a band is on its inside surface. When this surface wears, the timing of a shift can be affected. Slippage might occur also.

Some transmissions have adjustable bands. Others have bands that only operate occasionally when the shift selector is placed in manual second or low for engine braking. These bands are not usually adjustable.

To adjust a band:

- Locate the specification in the service manual.
- Locate the band adjustment bolt. It has a square head and a large locknut (**Figure 74.21**). Use an 8-point socket and an inch-pound torque wrench.
- Tighten the band adjustment bolt (turn clockwise) by the specified amount of torque to seat the friction material against the drum.
- Loosen the bolt by the specified amount to provide the correct clearance.
- Tighten the locknut while holding the adjustment bolt from turning.

The locknut has a sealing surface to keep transmission fluid from leaking around the adjusting screw. These nuts are usually replaced.

Some transmissions have more than one band. Adjustment specifications on these can differ because the bands are often of different designs. One band adjustment specification might call for loosening the adjustment screw by 1¼ turns and the other might call for loosening 2¾ turns.

Speedometer Adjustment

A conventional speedometer can be adjusted to read lower or higher by changing either the driving gear on the transmission mainshaft or the speedometer-driven gear installed through the side of the tailshaft housing. To reduce the speedometer reading, increase the number of teeth on the driven gear or decrease the number of teeth on the driving gear. To increase the speedometer reading, do the opposite.

Electronic speedometers can be reprogrammed for accuracy or when tire sizes are changed.

■ ELECTRONIC TRANSMISSION SERVICE

Most automatic transmissions manufactured since the early 1990s have computer-controlled shifting. Problems can be either mechanical or electronic in nature. Use a scan tool to check for diagnostic trouble codes (DTCs). It is good practice to always check for electronic failures first because they can result in hydraulic/mechanical problems.

Scan Tools

Some generic scan tools can work on different makes of vehicles, but they must have the applicable data cartridge (**Figure 74.22**). The scan tool is connected to the under-dash data link connector. Bi-directional scan tools are capable of displaying DTCs and solenoid signals from the TCM. They can also read other TCM inputs and allow the technician to send commands to the TCM to cause shifts and alter other transmission functions. Snapshots can also be recorded during a test drive (see Chapter 45).

NOTE: *On electronic transmissions, battery voltage is especially important. A slipping belt can cause voltage to drop to the point where computer performance is affected. Always check the battery connections and state of charge, especially when the scan tool displays a solenoid code.*

Transmission Diagnostic Codes

Electronic systems can self-diagnose electrical components and circuits (see Chapter 45) and display the diagnosis on the scan tool. Transmission diagnostic trouble codes begin with the letter "P" (for powertrain), followed by four numbers, for example, PO731. The "0" following the "P" indicates an SAE generic code. When the third character is a 7 or 8, this denotes a transmission code. The last two characters describe the type of problem. The PO731 code indicates an incorrect gear 1 ratio. This

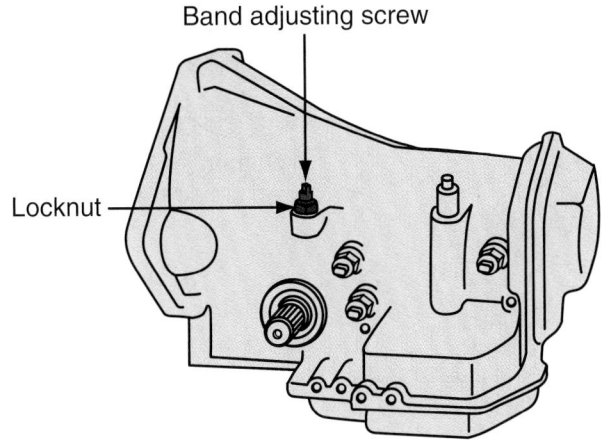

Figure 74.21 A typical band adjusting screw and locknut.

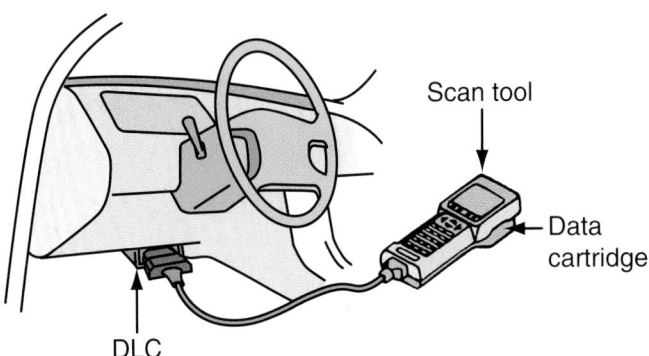

Figure 74.22 A data cartridge for automatic transmissions is used with a scan tool connected to the vehicle's data link connector.

code could be the result of several unusual factors, like the wrong chain drive in FWD vehicles or a slipping clutch.

Electronic Transmission Diagnosis

Before removing an electronic transmission from the vehicle, be certain of the cause of transmission failure. Carefully verify the complaint with the customer, during a test drive if possible. Test the vehicle in all shift ranges.

Typical electronic components are shown in **Figure 74.23**. Electronic transmission operation was covered in Chapter 73. Be sure to read the information related to the basic theory found in that chapter, as it will apply to the following.

Switches and Sensors

Computers in North American vehicles use **ground side switching** with current limiting devices. European vehicles use **power side switching**. More information on this is covered in Chapter 39. Switches in electronic transmissions either connect two wires together, or they provide a ground. Prior to electronic controls, TCC switches were grounded by governor

pressure when the specified speed was reached. Today's transmissions ground their switches through the computer in response to various inputs. Electrical problem diagnosis includes loads (solenoids) and switches (TCM). On the ground side, when the circuit is under load, the voltage drop reading should be zero (see Chapter 29). If it is higher, a switch problem is indicated. Here are some switches and sensors that can cause problems.

Brake On/Off Switch. The brake on/off switch (BOO) disconnects the TCC. The brake switch is important. Many transmission failures are caused by someone adding trailer brakes. If the converter is not locked up in overdrive, it has a .78:1 ratio. The burn the transmission feels is like the burn in your legs when you have not shifted your 10-speed bicycle into the correct, lower gear ratio. Check the brake switch for battery voltage on one side or on the other. At 0.4–0.6 volt (400–600 mV) the computer sees that the brakes are on.

Vehicle Speed Sensors. Speed sensors are usually permanent magnet generators that can be tested with a digital multimeter using its AC voltage scale. A lab scope can also be used. Raise the vehicle on a wheel-free lift

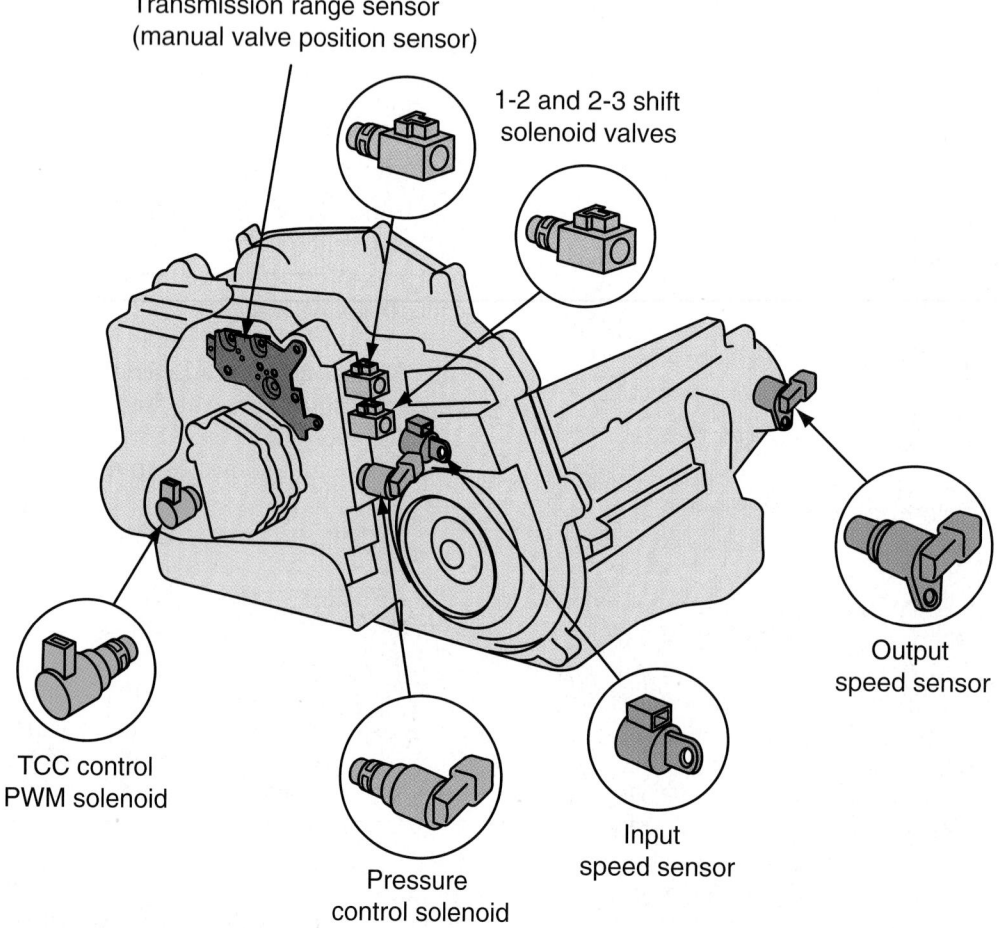

Figure 74.23 Typical electrical components on a transaxle.

and connect the meter to the sensor. Start the engine, shift into gear, and increase wheel speed slowly until you reach approximately 20 mph. Then read the voltage. It should increase evenly and smoothly as speed increases.

In older vehicles, installing bigger tires would cause a late shift due to the increased governor input. A late shift on electronic transmissions is due to the vehicle speed sensor (VSS). Tire size is important. Big tires can fool the computer and cause a lack of power. The air-fuel ratio determined by the computer includes information regarding the transmission, differential ratio, and tire circumference. Taller tires will tend to overheat the torque converter because it does not get into overdrive, where the converter will lock up. With taller tires, a lower axle ratio is called for.

Transmission Range Sensor. A part that frequently fails is the transmission range (TR) sensor. The sensor is part of the park/neutral safety switch. When there is wear on the TR, the engine races and then drops into a gear. Sensor wear between the park and drive positions is most common. To test this sensor, check voltage as the shift lever is moved. Wiggle the lever while in drive to see if the voltage remains constant. The computer might not set a code because it does not know that you did not move the shifter.

Solenoids. When diagnosing a solenoid problem, one of three conditions must be present.

1. There is an electrical feed problem.
2. There is a problem with ground.
3. A solenoid is defective.

If the transmission does not smell like burned asbestos, there is a good possibility of an electronic cause for the problem, so get out your scan tool. The problem could be related to a solenoid.

Solenoids are called *force motors*. A force motor can be flow-checked by attempting to blow air through it. You need to know if the solenoid is normally open or normally closed. If the solenoid is open, it flows easily. If it is closed, it should not leak at all. If it is open and restricted, attempt to clean it out. If it is still restricted, replace it. Also replace solenoids when there is metal contamination from a hard part failure or when a transmission has been overheated severely.

To verify that a shift solenoid is operating, energize it and listen for a click. To check a solenoid electrically, disconnect the electrical connector and measure its resistance between ground and the connector.

Rebuilders use a flow test on solenoids. They must have a 1.4-psi tolerance or they cannot be used. When oil goes through solenoid coils, they become magnetized over time and collect debris. If a transmission has 50,000 miles and is being rebuilt, the solenoids should probably be replaced, or, at the very least, tested and flushed.

The electronic pressure control solenoid (EPC) is pulse-width modulated, so it moves back and forth all of the time (some move at frequencies of 6,000 times per second). Therefore, it is always replaced during a rebuild because a bad EPC can cause a transmission failure, but a bad shift solenoid will not.

Electronic Transmission Shift Control

Transmission pressures that are quickly varied by a computer provide a benefit in terms of efficiency, driveability, and control, but problems can develop if tolerances are not met during manufacturing and maintained during the rebuilding process. In an electronic automatic transmission, line pressure can change to accommodate variations in shifts as the computer reacts to input shaft and output shaft speed differences.

NOTE: *Late-model automatic transmission shift diagnosis requires a bi-directional scan tool. Generic scan tools will not work on all transmissions.*

Transmission pressure is almost 500 psi in reverse or at WOT, and pressure is high when the engine initially starts. The extra pressure is needed to help flush the solenoids. With variable oil pressure, line pressure is severely reduced at idle. This makes it vitally important that an oil pressure increase of from 5 to 50 psi is quickly achieved. Pressure testing can eliminate many inaccurate electronic diagnoses.

Transmissions are designed to go into a "limp-in" mode if the computer detects a failure. The transmission can put itself into a default gear, and that will be the only driving condition available to the driver. Hydraulic pressure drops or severely worn clutches can cause a default.

NOTE: *Limp-in mode will be either second, third, or fourth gear, depending on the vehicle manufacturer. Limp-in is whatever gear operates without solenoids. Reverse also operates because it doesn't require a solenoid.*

If a transmission slips during limp-in mode, it has a mechanical/hydraulic problem and will require a rebuild.

 **SHOP TIP** Pulling a fuse can cause limp-in mode.

All solenoids are two-wire connections. When there are four wires, one powers each solenoid, one is for the TCC, and the other is for ground. Disconnect all wires and control A and B switches and drive the car. You can make a test box or buy one from the tool suppliers.

Testing Shift Solenoid Current Draw

If the B+ wire into the transmission is accessible, it can be used with an amp probe. Current flow will vary depending on the number of solenoids energized at one time. In a typical transmission with third-gear limp-in mode, during a second gear shift, current should drop. When the transmission shifts to third gear, another drop will take place when both solenoids turn off. On electronic transmissions, this method is quicker than a pressure test.

> **SHOP TIP** A handy testing tool can be made by soldering wires to a minifuse so current flow can be measured in series with the circuit.

■ MORE TESTS BEFORE TRANSMISSION REMOVAL

Some tests can be done in the service bay or on the road or dyno. Other tests are done on the lift. If the problem could be due to the hydraulic system, perform a pressure test and verify flow through the oil cooler. Try some tests that can cause the problem to change. On mechanical transmissions, this can include adjusting the throttle cable or varying vacuum to the modulator and driving under differing loads. On electronic transmissions, this can include energizing shift solenoids or driving the vehicle with shift solenoids disconnected.

On the lift, some things can help pinpoint your diagnosis. If the transmission has bands, like the A4LD OD, you can try tightening a band all the way or pry on a servo. If the car moves while in neutral, you can try running it without ATF (for a few seconds only). If the vehicle still moves, a mechanical condition is the cause. An air test of various hydraulic circuits can also be done.

Remember, do not remove a computer-controlled transmission until you have determined that the failure is within the transmission and not related to a DTC or wiring to the transmission.

■ HYBRID AUTOMATIC TRANSMISSION SERVICE

Many hybrids use conventional automatic transmissions and are serviced in traditional ways. If the transmission is removed, however, special considerations are required. The torque converter is bolted to the rotor of the high-voltage motor. Before removing the transmission from the vehicle, the high-voltage electrical system must be disabled. Also, the rotor is a very strong permanent magnet. Do not allow anything magnetic to come into contact with the rotor.

Hybrids with unique automatic transmissions can have fluid service, but if they require repair, they are replaced with a complete replacement unit. When a vehicle speed sensor fails, the speedometer will not work.

■ REMANUFACTURED AUTOMATIC TRANSMISSIONS AND TRANSAXLES

Many dealer and independent repair shops do transmission diagnosis and removal, but do not do rebuilds. Parts for some transmissions are very expensive, and remanufactured transmissions have become readily available at a reasonable cost and with a warranty. A shop that does not rebuild a particular type of transmission on a regular basis will often opt for a remanufactured transmission.

When replacing one transmission with another, it is important that the correct one be acquired, especially with some imports, which can have year-to-year changes. Installing the wrong transmission can result in a permanently illuminated MIL on the instrument panel.

■ TRANSMISSION REBUILDING

Procedures vary for rebuilding various transmissions. Some things are typical to almost all of them. Parts can be purchased separately or in kits (**Figure 74.24**). Parts contained in a rebuild kit are called **soft parts**. These include all rubber seals and gaskets, bands, clutch friction discs, modulator, thrust washers, and bushings. A rebuilt converter will be installed too.

When internal damage has occurred, **hard parts** will be necessary. This raises the cost of the rebuild considerably. Hard parts include metal parts such as gears, valves, pump bodies, clutch drums, and other items that would not normally be replaced in a rebuild.

Whenever a clutch or band has failed, look for one of two causes.

1. Hydraulic pressure has dropped too low due to a seal leak or line pressure problem.
2. There could be a shift timing problem where a clutch or band applies before it should. This can be caused by a cross leak, which sometimes occurs when the gasket between the valve body and its separator plate leaks. To help prevent

Figure 74.24 A transmission rebuild kit. (*Printed with permission of Transtar® Industries, Inc.*)

this, always carefully clean gasket sealing surfaces and torque the valve to specifications.

NOTE: *When soft or hard parts have failed in a transmission, be sure to flush the cooler thoroughly with compressed air from a rubber-tipped blowgun. A new rebuild can be ruined when the particles in the cooler enter the transmission. Be careful of the transmission fluid exiting the cooler when blowing it out.*

Always check manufacturer's technical service bulletins before rebuilding a transmission.

Seal Kit Overhaul

Sometimes, a transmission is overhauled simply to replace hard seals that are causing morning sickness. If all of the transmission soft parts are still in good condition, a very inexpensive seal kit (under $20) might be all that is needed to put the transmission back in good order.

Valve Body Service

If a valve body is disassembled, be sure to lay out each piece on a towel as it comes from the valve body. The correct locations of valves and springs are important to the proper operation of the transmission. Service manuals show a layout of a valve body, but changes are often made in the middle of model years to some of the valves.

When there has been no internal failure, there is a safer means of servicing the valve body. Use carburetor cleaner spray on each valve. Carefully move each valve with a plastic screwdriver to avoid accidentally scratching it. Valves are pressure applied and spring released. See that each valve moves and returns easily. If valve movement seems to be sluggish, disassemble that valve and clean its bore. Crocus cloth can be used to clean the valve bore. Unroll it in the bore and twist it to polish the bore.

Clutch Pack Service

Remove the large snapring from the clutch pack and remove the clutch discs (**Figure 74.25**). Some clutch pistons require a spring compressor to disassemble them. The compressor pushes the piston return spring or springs down while the snapring is removed (**Figure 74.26**). The piston can be removed with shop air (**Figure 74.27**).

Replace the old seals on the piston and inside the clutch pack (**Figure 74.28**). Lubricate the seals and install the piston in the clutch pack. Sometimes, there is a special tool that is installed over the inside of the clutch drum during piston installation (**Figure 74.29**). Some piston seals are O-ring seals and some are lip seals. Lip seals are more delicate and require more care during installation so they are not cut. A special "piano wire" type tool has rounded edges to help install the seals (**Figure 74.30**).

Figure 74.25 Remove the large snapring from the clutch pack and remove the clutch discs.

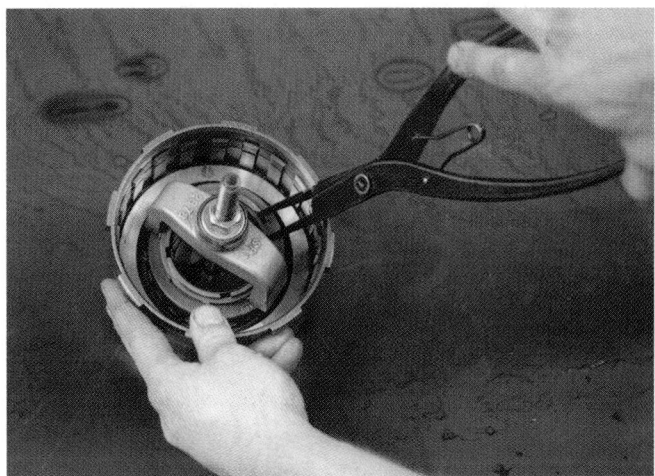

Figure 74.26 Compress the piston return spring and remove the snapring.

Figure 74.27 The piston can be removed with shop air.

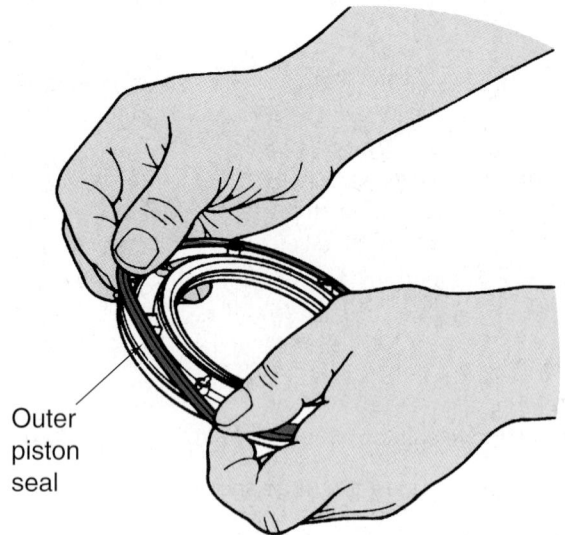

Figure 74.28 Replace the piston seals. *(Courtesy of Ford Motor Company)*

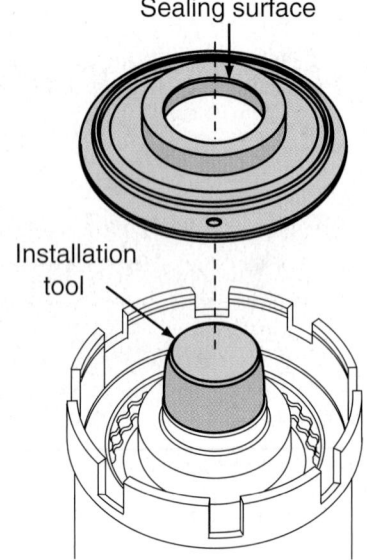

Figure 74.29 A tool that helps install the inside piston seal over the drum.

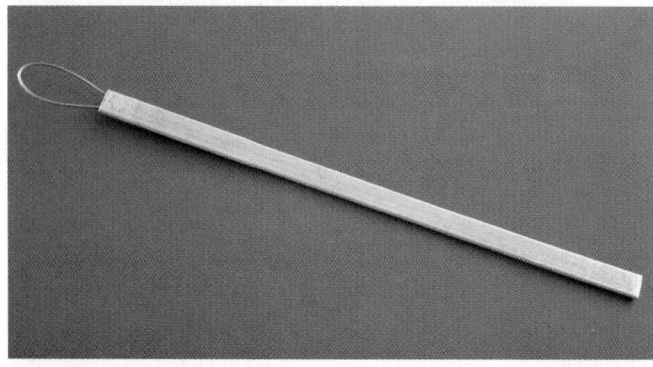

Figure 74.30 This tool is used to help install lip piston seals. *(Courtesy of Tim Gilles)*

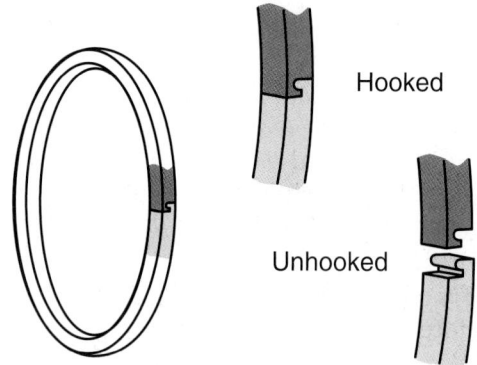

Figure 74.31 Some seal rings must be hooked together at the ends during installation.

After assembling the clutch piston, air check it to see that it does not leak excessively. A leakage of 15 to 20% is designed into clutch packs to lube and cool Torrington® bearings and other transmission parts.

Assemble the clutch pack without seals or oil and check its end play. Sometimes a clutch pack is too loose. This can affect shift timing by delaying the shift. To eliminate excessive clutch end play, put an extra steel at the bottom (but never at the top). Before clutch discs are installed be sure to soak them in automatic transmission fluid.

Install Seal Rings

Seal rings fit into grooves on the shaft that supports the center of the clutch drums. There are different types of rings and they must be installed carefully. Some seal rings must be hooked together at the ends during installation (**Figure 74.31**). Teflon® rings that have a butt gap should have no more than 0.003" clearance at the gap. Scarf cut Teflon® rings should butt squarely (**Figure 74.32**). There should not be an overlap or a gap. Solid Teflon® rings are stretched carefully to fit them in place, then they are compressed with a special tool.

Bushing Service

Check all bushings for visible signs of wear. Bearings often have a soft grey metal surface when new. A worn bushing will appear shiny. Bushings are pressed fit and sometimes require alignment of lubricant openings during installation.

▬ TRANSMISSION REASSEMBLY

When reassembling the transmission parts in the case, be sure not to force anything. It is common practice to replace low gear sprags or roller clutches during a rebuild.

SHOP TIP When installing one-way clutches, remember that they always turn in the same direction the engine rotates.

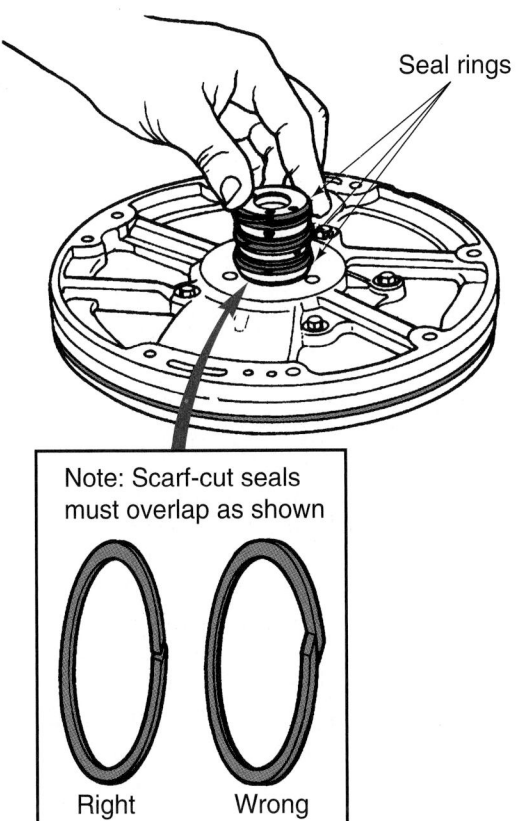

Figure 74.32 Scarf-cut seals. *(Courtesy of Ford Motor Company)*

Do not use moly assembly lube on a one-way clutch because it will cause the clutch to slip. It needs to be able to take a bite into its race. Many technicians like to rough up the surface of the races slightly with very fine sandpaper before assembly.

The spring in the roller clutch is bimetal, and heat is an enemy. It does not take much pressure to get it to work (10 inch-pounds is the specification). If rollers fall out of the clutch, the springs have become hot and weak. The clutch should be replaced.

End Play Adjustment

End play is checked with a dial indicator (**Figure 74.33**). Selective washers or shims are used to adjust end play. Too little end play during this test will let you know something is wrong.

Many transmissions have more than one end play adjustment. Always perform the rear end play adjustment first. It will affect the front end play reading. Adjust to the minimum end of the specification. As wear of thrust bushings occurs, clearance will increase. When end play is too loose, the clutch drum can move back and forth, causing wear on the seal rings. Transmissions with Torrington bearings instead of thrust washers have a tighter end play tolerance.

After the final end play check, torque the bolts to the transmission pump. Install the valve body and torque its bolts. Overtorquing valve body bolts can cause valves to stick in their bores, preventing shifts.

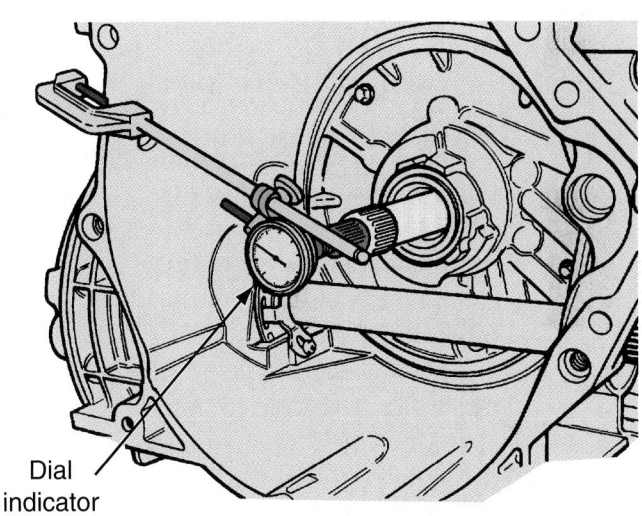

Figure 74.33 End play is checked with a dial indicator. *(Courtesy of DaimlerChrysler Corporation)*

■ REVIEW QUESTIONS

1. _____ service is the name for the condition during stop-and-go hot weather driving and trailer towing.

2. What color smoke does ATF cause when burned in the engine?

3. What is indicated if the dipstick will not easily wipe clean?

4. If there is fluid in a vacuum line to the modulator, what could be wrong?

5. What is better for transmission life, a harsh shift or a mushy shift?

6. If the governor sticks in the "out" position, what gear would the transmission want to be in, second or third?

7. If the vacuum modulator had a hole in its diaphragm, what gear would the transmission want to be in, second or third?

8. How much is a flexplate designed to be able to move?

9. How much pressure capacity should a transmission fluid pressure gauge have?

10. What is the name of the test that is done by pressurizing hydraulic passages to see if clutches are working?

▬ ASE-STYLE REVIEW QUESTIONS

1. All of the following can describe limp-in mode *except*:
   a. It will be a gear that does not require any solenoids to be on.
   b. It can be first gear.
   c. It can be second gear.
   d. It can be third gear.

2. Technician A says that all automatic transmission vehicles have their fluid checked with the engine running in park. Technician B says that when checking ATF level, the fluid on the transmission dipstick should be uncomfortable to touch. Who is right?
   a. Technician A   c. Both A and B
   b. Technician B   d. Neither A nor B

3. What happens to ATF level as the engine warms up?
   a. The fluid level will *climb* on the dipstick about 1 quart from cold to hot.
   b. The fluid level will *drop* on the dipstick about 1 quart from cold to hot.
   c. The fluid level will *drop* on the dipstick about 1 pint from cold to hot.
   d. The fluid level will *climb* on the dipstick about 1 pint from cold to hot.

4. Technician A says that automatic transmission filters are designed to last the life of the car. Technician B says that transmission fluid filters have a bypass valve in case they become plugged. Who is right?
   a. Technician A   c. Both A and B
   b. Technician B   d. Neither A nor B

5. What happens if the governor sticks in the "in" position?
   a. The transmission will shift late.
   b. The transmission will stay in low gear longer.
   c. Both A and B.
   d. Neither A nor B.

6. If the hose to a vacuum modulator falls-off:
   a. The transmission will probably shift late.
   b. Transmission shifts will be soft or mushy.
   c. Both A and B.
   d. Neither A nor B.

7. Technician A says that a stall test is when the brakes are applied and the engine is accelerated to its highest possible rpm. Technician B says that a stall test should not last longer than 30 seconds. Who is right?
   a. Technician A   c. Both A and B
   b. Technician B   d. Neither A nor B

8. Which tool is used when adjusting a band?
   a. An inch-pound torque wrench
   b. A foot-pound torque wrench
   c. A band-adjusting wrench
   d. None of the above

9. Which of the following is/are true about automatic transmissions?
   a. When a vehicle speed sensor fails, the speedometer will not work.
   b. Most torque converters have a drain plug.
   c. Noise that seems to come from the torque converter can be from a loose crankshaft vibration damper.
   d. If a converter noise goes away in park or neutral, the converter is not the source of the noise.

10. Which of the following is/are true about the electronic pressure control solenoid?
    a. It moves back and forth at a high frequency.
    b. It is replaced during a rebuild.
    c. A bad EPC can cause a transmission failure.
    d. All of the above.

CHAPTER 75
Driveline Operation

■ OBJECTIVES

Upon completion of this chapter, you should be able to:

✔ Describe the operation of universal joints.

✔ Explain how a differential works.

✔ Understand the differences between types of limited slip differentials.

✔ Describe the different types of rear axles and bearings.

✔ Select the correct gear oil for different applications.

✔ Identify differences between four-wheel-drive types.

✔ Explain the parts and operation of a transfer case.

■ KEY TERMS

driveline	differential	non-hunting gearset
live axles	third member	partial non-hunting gearset
halfshaft	pumpkin	limited slip differential
Hotchkiss drive	banjo housing	swing axle
cross and yoke universal joint	Salisbury axle	EP additive
phase	spider gear	AWD
constant velocity universal joint	hunting gearset	locking hubs

■ INTRODUCTION

Most vehicles until the early 1980s were equipped with rear-wheel drive (RWD). Today, pickup trucks, sport utility vehicles (SUVs) and some higher-end vehicles still use rear-wheel drive. On rear-wheel-drive vehicles, **driveline** is a term that describes the parts that transfer power from the transmission to the rear wheels. The driveline includes the driveshaft and universal joints (U-joints), axles and axle bearings, and differential. Rear-wheel-drive vehicles have a long driveshaft between the transmission and differential (**Figure 75.1**).

This chapter will cover parts of the driveshaft, and also **live axles**. Live axles are ones that turn with the wheels. Be aware that differentials are sometimes also referred to as "axles." Front-wheel-drive vehicles have a transaxle with two **halfshafts** that deliver power to the front wheels. Information related to FWD theory and service is found in Chapters 77 and 78.

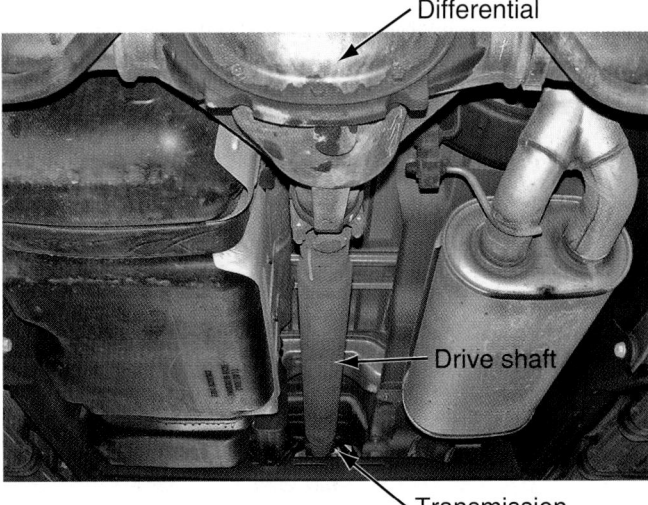

Figure 75.1 On rear-wheel-drive vehicles, a drive shaft connects the transmission to the differential. *(Courtesy of Tim Gilles)*

As you study the driveline, you will learn more than just driveline service and repair procedures. Pay attention to the methods used in pressing bearings and other parts. Press skills are used in many other areas of automotive repair.

■ DRIVE SHAFT (RWD)

> ### ▰▰▰ HISTORY NOTE ▰▰▰
>
> *The open drive shaft, or* Hotchkiss drive *design, has been in use since the 1950s and is the only one found on most vehicles since that time. The older design that it replaced was an enclosed type called the* torque tube. *The drive shaft not only transfers power, but it allows for changes in driveline length as a car goes over bumps.*

The drive shaft, or *propeller shaft,* is usually made of steel tubing, although some late-model drive shafts are aluminum or carbon fiber. Drive shafts are strong and light. They must be balanced and straight. Universal joints at both ends attach the drive shaft to other components. Yokes to accept the universal joints are welded onto the shaft at both ends. A typical drive shaft (**Figure 75.2**) includes two universal joints and a slip yoke. Sometimes a rear yoke bolts to a flange on the differential.

■ SLIP YOKE

As the vehicle goes over bumps, the rear springs allow the rear axle assembly to go up and down (**Figure 75.3**). The distance between the differential and the transmission changes, so the drive shaft must be able to move in and out of the transmission. A slip yoke is attached to a universal joint on the front end

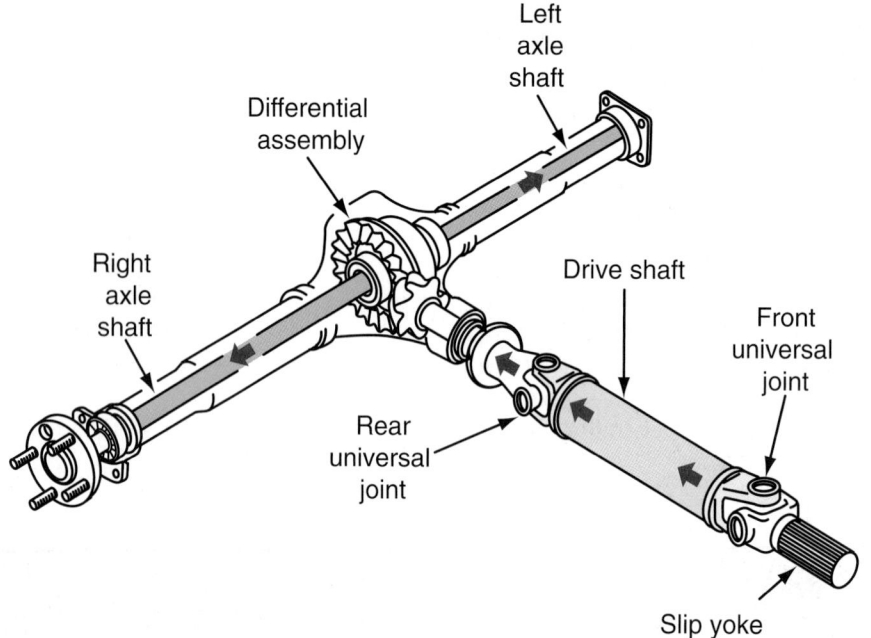

Figure 75.2 A typical drive shaft.

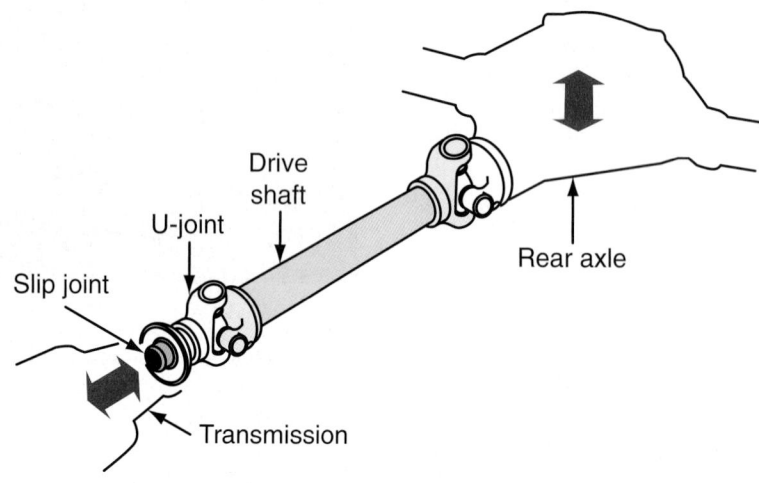

Figure 75.3 A slip yoke allows the drive shaft length to change as the car goes over bumps.

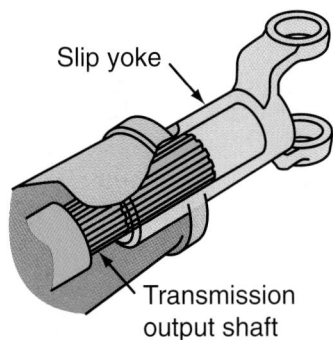

Figure 75.4 The slip yoke fits over splines on the output shaft.

of the drive shaft. The other end of the slip yoke fits over splines on the output shaft or mainshaft of the transmission (**Figure 75.4**). It slides in and out of the transmission as the distance between the transmission and differential changes.

The slip yoke is machined smooth on its outside diameter. This provides a sealing surface for the extension housing seal. It also provides a bearing surface for the extension housing bushing to act upon.

When a slip yoke is used with an automatic transmission, there is sometimes a seal that goes over the output shaft. It rides against the inside of the slip yoke (**Figure 75.5**). Its purpose is to keep ATF, which is as thin as SAE 10 engine oil, from leaking out of the slip yoke through its vent hole (**Figure 75.6**). The splines on the output shaft are lubricated by the ATF. There

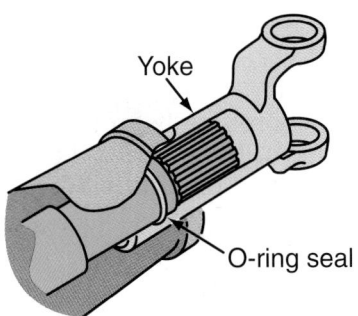

Figure 75.5 Automatic transmission slip yokes sometimes have seals.

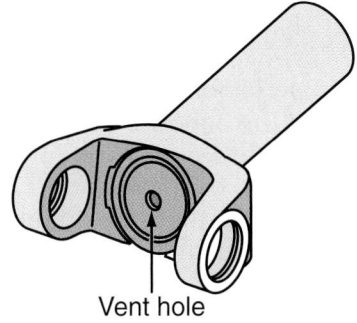

Figure 75.6 A slip yoke with a vent hole.

must be a vent hole to allow the slip yoke to move in and out. Some yokes are greased and have a grease fitting. This design must be sealed at the end of the yoke.

If the rear yoke is not attached to the drive shaft, there will be a *flange* that is bolted to the front of the *differential pinion shaft,* the splined shaft that comes out of the front of the differential (covered later).

■■ UNIVERSAL JOINTS

Universal joints, called U-joints, are located at both ends of the drive shaft. They transmit power at an angle (**Figure 75.7**). When the axle moves up or down, the universal joint allows for the necessary changes in angle at the ends of the drive shaft.

The most popular universal joint design is called a **cross and yoke**, or *Cardan.* It is two Y-shaped yokes connected by a cross, called a *spider* (**Figure 75.8**). Most U-joints are made of forged, carburized steel. At the

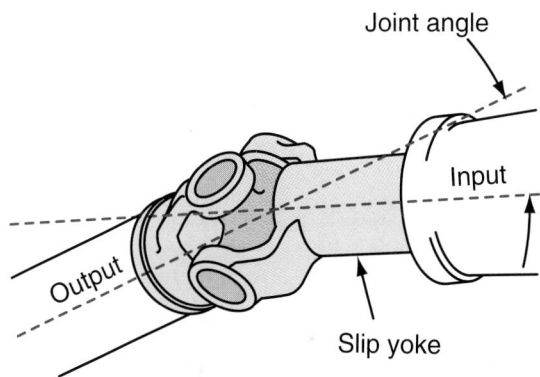

Figure 75.7 Universal joints allow power to be transmitted at an angle.

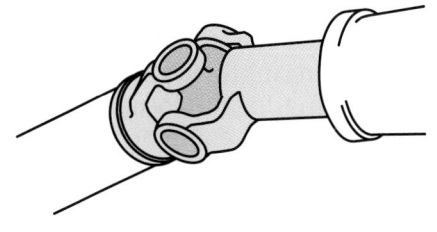

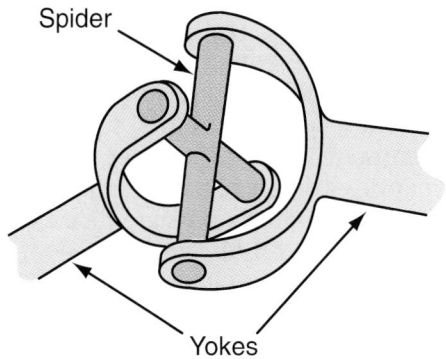

Figure 75.8 A universal joint is two Y-shaped yokes connected by a cross called a spider.

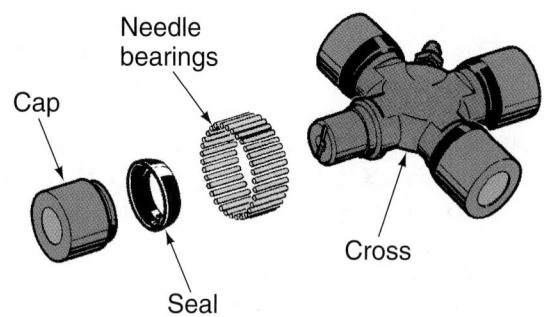

Figure 75.9 Parts of a universal joint. *(Courtesy of Federal-Mogul Corporation)*

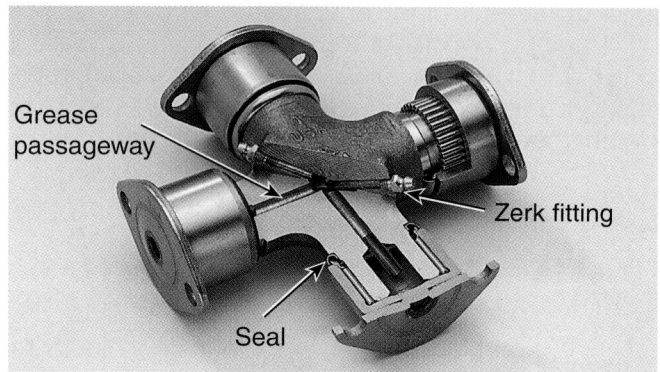

Figure 75.10 Replacement joints usually have a zerk fitting for lubrication during service. *(Courtesy of Federal-Mogul Corporation)*

ends of the universal joints are four bearing caps with needle bearings (**Figure 75.9**). The needle bearings ride on *trunnions,* which are bearing areas ground on the ends of the cross or spider. The caps and bearings allow the joint to swivel as its angle changes. The caps and the ends of the trunnions must have a groove for grease.

Grease seals fit onto the ends of the bearing caps to keep dirt out and lubricant in. Replacement joints usually have a zerk fitting for lubrication during service (**Figure 75.10**). Original equipment joints do not have this feature.

There are several methods of holding the U-joints to the yokes (**Figure 75.11**). Universal joints are usually pressed fit into the yokes on the drive shaft. Snaprings fit into grooves in the yoke to keep the joint centered and retained. The most popular mounting found on cars is one where U-bolts hold the U-joint in place between tabs in the rear flange. On the rear flange there must be a feature that aligns the joint on center. Alignment is done by either tabs on the outsides or snaprings on the ends. Instead of a snapring, some original equipment joints are held in place with injected molded plastic. If the drive shaft is not installed exactly on center, serious vibration will result.

■ TWO-PIECE DRIVE SHAFT

Balance is very important on a drive shaft. During balancing done at the factory, weights are spot-welded to

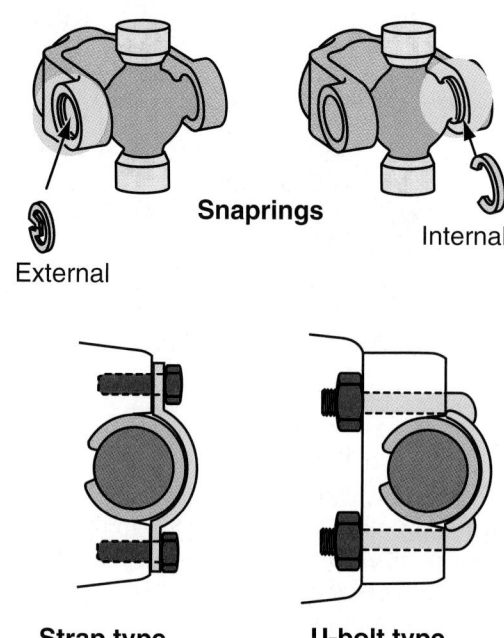

Figure 75.11 Several methods of holding universal joints to yokes.

the drive shaft. When a drive shaft is too long, it can flex, upsetting balance. Most cars have a single drive shaft, but a vehicle with a longer wheel base will have a two-piece drive shaft. These are often found on large RWD cars, SUVs, and light trucks.

A properly assembled two-piece driveshaft will have its U-joint cups in alignment with one another. An improperly assembled two-piece driveshaft will be severely out of balance, causing major vibration. This condition, called out-of-phase, is discussed later in the chapter.

On a two-piece drive shaft, a *center support bearing* holds the center of the shaft where the two shafts attach to each other. It bolts to the frame, crossmember, or underside of the vehicle. The sealed bearing is supported in a rubber mount that dampens noise and vibration (see Figure 14.27).

■ DRIVE SHAFT ANGLE

Cardan universal joints have a problem that limits their use. When they are operated at an angle, the speed of the driven shaft varies as it revolves. The front of the drive shaft is attached to the yoke splined to the transmission output shaft, so they both turn at the same average speed. The yoke and drive shaft are coupled together by a universal joint, however, so the speed of the driven yoke at the opposite end of the drive shaft increases twice and decreases twice during each drive shaft revolution (**Figure 75.12**).

The elliptical path that the input and output yokes take can be compared to the face of a clock (**Figure 75.13**). At 12 and 6 o'clock, the speeds of the input and output yokes are the same. In between, the output yoke turns at a speed that is not constant.

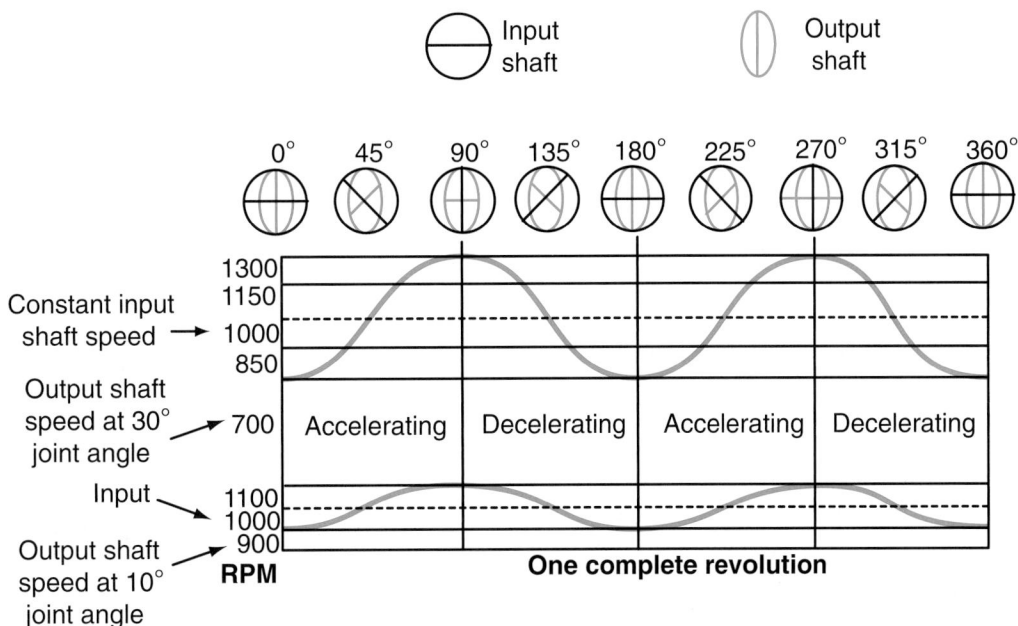

Figure 75.12 A comparison of the speed change of universal joints with the shaft at both 10-degree and 30-degree angles.

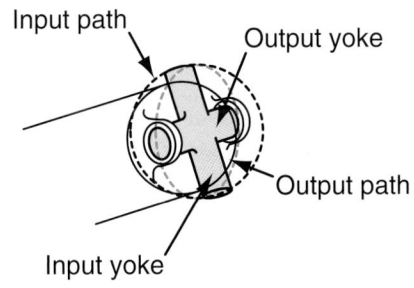

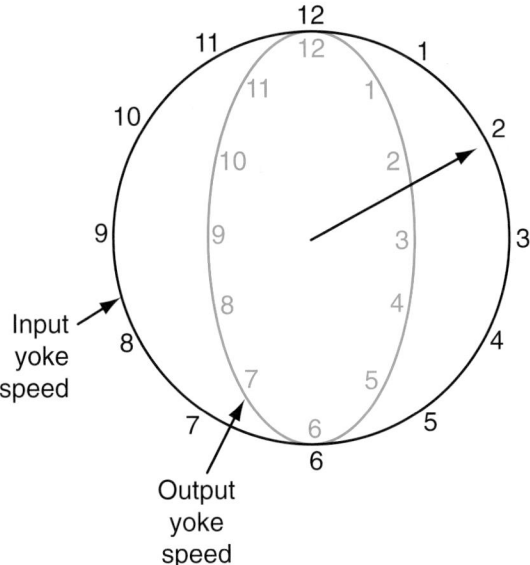

Figure 75.13 The elliptical path that the input and output yokes take can be compared to the face of a clock.

The arrow in the sketch shows that the output yoke has traveled to approximately the 2.2 o'clock position compared to the 2 o'clock position of the input yoke.

If the angle is increased, the change in *velocity* during each revolution increases. This can cause the drive shaft to whip like a jump rope. Most vehicle manufacturers do not use angles greater than 3 to 4 degrees with single U-joints because they cause excessive vibration. Figure 75.13 compares the speed change of the universal joints with the shaft at both 10-degree and 30-degree angles. Study the illustration. If the drive shaft angle were set at 30 degrees, at an input speed of 1000 rpm the speed of the driven yoke would vary from 700 RPM to 1300 RPM during 90 degrees of drive shaft movement. Then it would change back. Imagine the vibration that would result from this situation.

With a one-piece drive shaft, arranging the universal joints so that they can cancel out each other's angle solves the problem (**Figure 75.14**). The angle is canceled out by the universal joint at the opposite end of the drive shaft. Both ends of the drive shaft must be in **phase** (**Figure 75.15**) for the canceling action to take place. This means that the trunnions of the front and rear U-joints are in the same plane (parallel). On a two-piece drive shaft, the front and rear halves can be assembled out of phase, resulting in vibration.

It is important that the angles at each end of the drive shaft be almost equal, or vibration will result. When an angle is steeper, a CV joint is used. The type used in front-wheel-drive axles is covered in Chapter 77.

■ CONSTANT VELOCITY JOINTS

The vibration caused by the speeding up and slowing down of the drive shaft can be canceled by putting two Cardan U-joints next to each other. These **constant velocity universal joints** are called *double Cardan* universal joints (**Figure 75.16**). The two joints

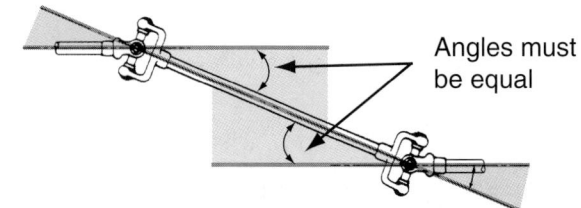

Angles must
be equal

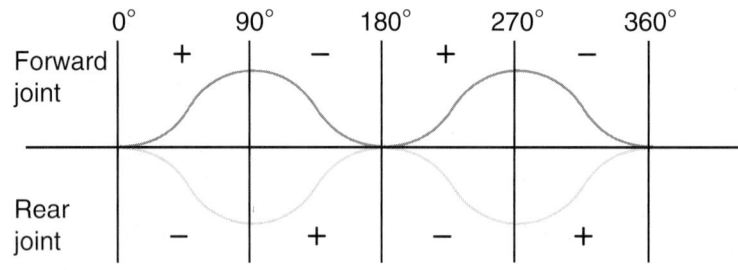

Figure 75.14 U-joints can be positioned to cancel each other's angle.

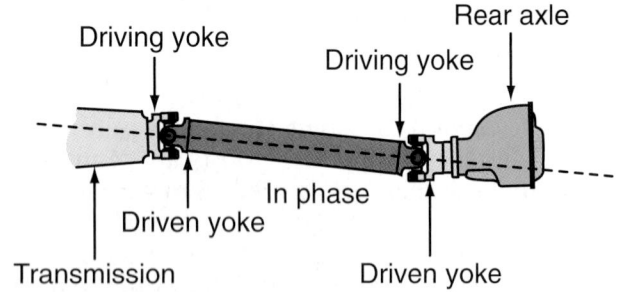

Figure 75.15 Both U-joints are in phase.

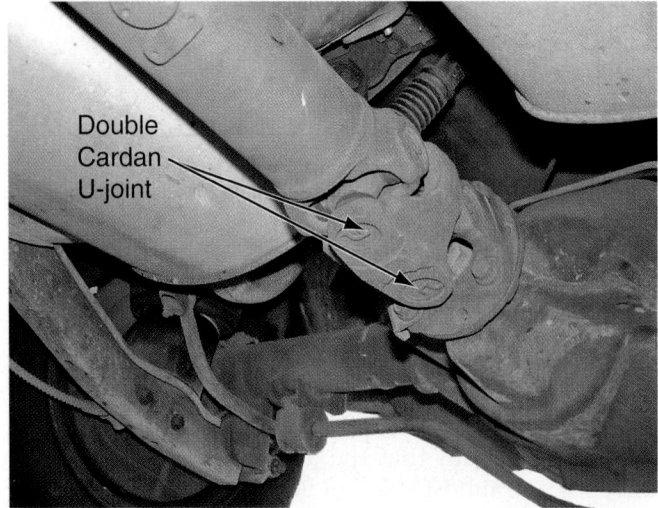

Figure 75.16 A double Cardan U-joint assembly.
(Courtesy of Tim Gilles)

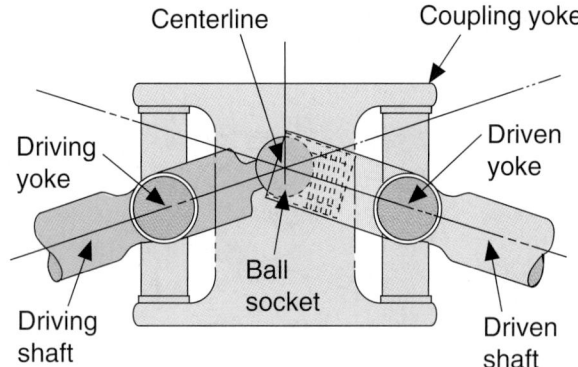

Figure 75.17 A double Cardan U-joint has two single
Cardan U-joints connected by a centering socket and
yoke.

are connected by a centering socket and yoke (**Figure 75.17**). The two joints are phased to cancel out each other's angle before the change in speed can go to the drive shaft. This is not a true constant velocity joint, but the speed change never leaves the joint. Its output is at a constant speed, so the drive shaft does not get any of the vibrations it would get from a single joint. Because of their lack of vibration, constant velocity joints are found on larger luxury cars and pickup trucks.

Another type of constant velocity universal joint is the *ball and trunnion*, also called tripod tulip (see Figure 77.25). Though not common on rear-wheel-drive cars, it is used in front-wheel drives because it is a true constant velocity joint.

■ DIFFERENTIAL

The purpose of a **differential** is to send power to the wheels. It also increases engine torque by providing a final drive gear reduction, usually of about 2.5 to 3.5 to 1. After power flows from the transmission to the differential, it must be turned 90 degrees and exit through the axles (**Figure 75.18**). During a turn, the inner and outer wheels rotate at different speeds (**Figure 75.19**). The differential must also accommodate this.

NOTE: *The differential is called a **third member**. The engine is the first member; the transmission is the second member.*

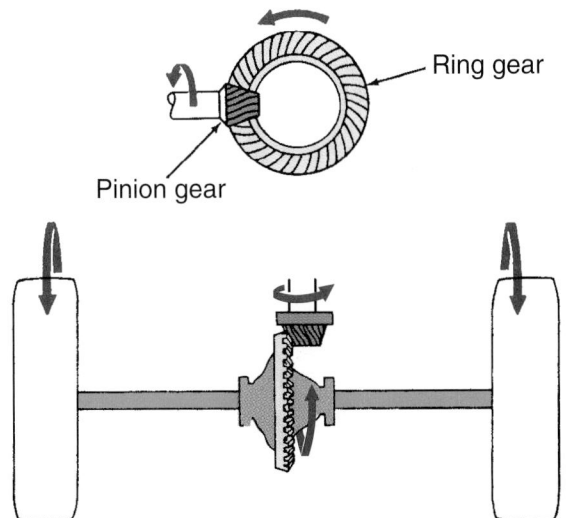

Figure 75.18 Power is turned 90 degrees and exits through the axles. *(Reproduced by permission of Deere & Company, John Deere Publishing, Moline, IL, all rights reserved)*

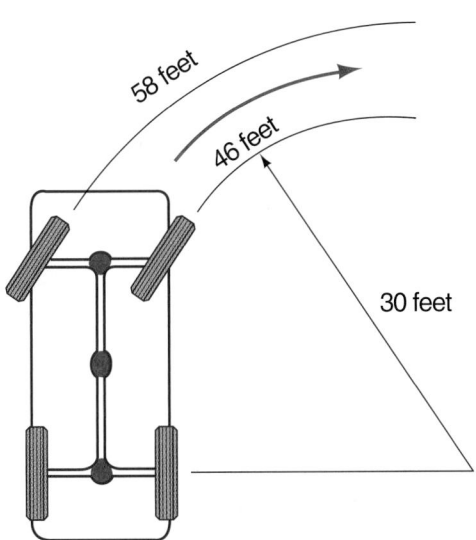

Figure 75.19 During a turn the inner wheel does not turn as far as the outer wheel.

◼ DIFFERENTIAL CONSTRUCTION

Parts of the differential include the differential pinion gear, the ring gear, the case, two drive axles, and rear bearings, all housed within the axle housing (**Figure 75.20**). The pinion gear is splined to a *flange*, or *yoke*, that attaches to the drive shaft. When the flange is flat (**Figure 75.21**) it is called a *companion flange*. Compare the two flanges in Figures 75.20 and 75.21.

The *pinion gear* drives the *ring gear*. The ring has more teeth than the pinion. The *differential case* to which the ring gear is bolted is mounted between two tapered roller bearings. The tapers of these bearings face each other. There is either a *crush sleeve*, sometimes called

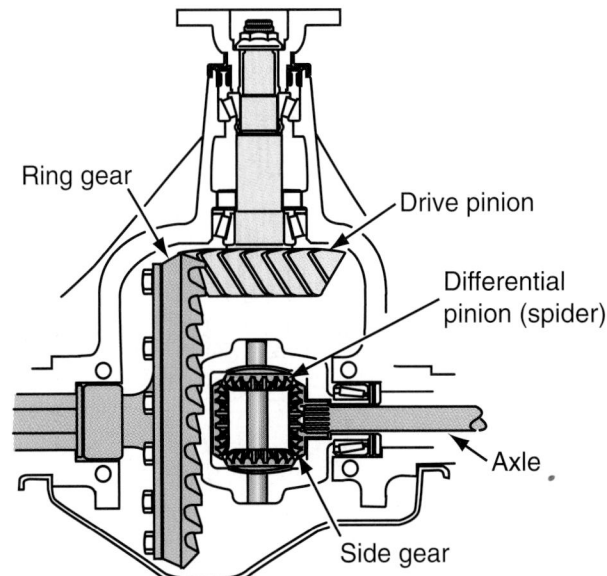

Figure 75.20 Parts of a differential assembly.

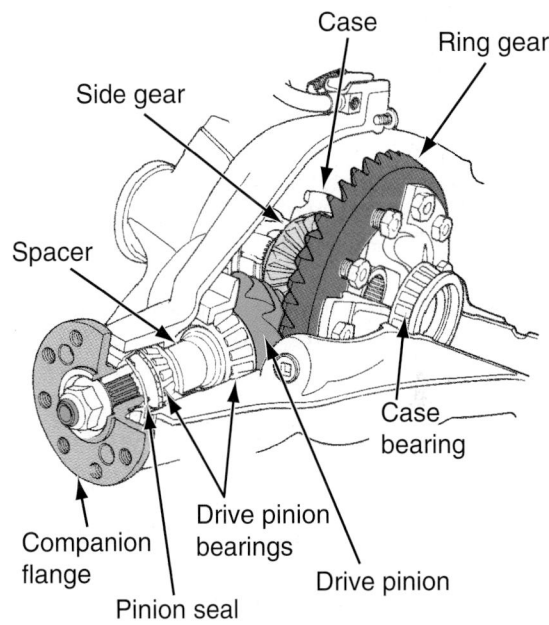

Figure 75.21 The flat flange on the front of this differential is called a companion flange. *(Courtesy of Ford Motor Company)*

a *collapsible spacer*, or shims between the two bearings (**Figure 75.22**). The crush sleeve has three functions:

◼ It keeps the bearings separated from their bearing races.

◼ It maintains the preload (tension) on the bearings. The bearing must be held close against the bearing races to prevent the pinion from moving back and forth on the ring gear. This would cause damage, looseness, or noise.

◼ It keeps the front bearing race from spinning with the bearing.

A shim controls how deeply the pinion is positioned into the ring gear.

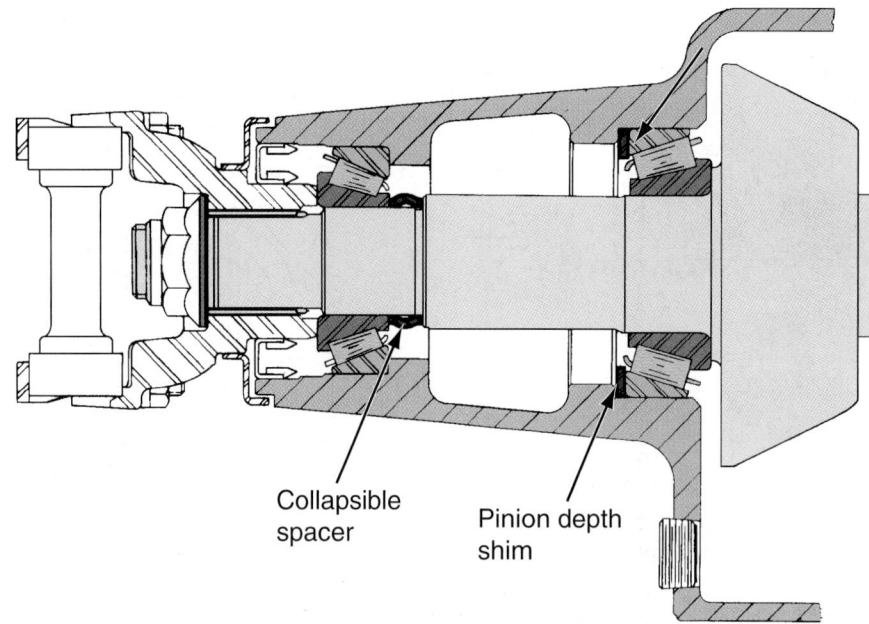

Figure 75.22 A collapsible spacer separates the two pinion bearings. (*Courtesy of DaimlerChrysler Corporation*)

■ DIFFERENTIAL HOUSING

The differential housing, or carrier, holds the drive pinion and case. There are two types. One has a removable third member, called a **pumpkin**. The housing is called a **banjo housing** because it resembles a two-sided banjo when the third member is removed (**Figure 75.23**). The other is called an *integral* or **Salisbury axle**. It has a cover of either stamped steel or cast aluminum (**Figure 75.24**). The third member is not removable from this axle as a unit.

There is a vent and a breather tube in the top of the axle housing. As lubricant warms up during use it expands. The breather makes room for this. The tube usually runs up into a fenderwell. It prevents water from entering the axle when the car is driven through puddles or deep water. The vent often has a one-way check valve in it that allows only pressurized air to escape from the valve. Air can leak back in past the wheel seals as the lubricant cools.

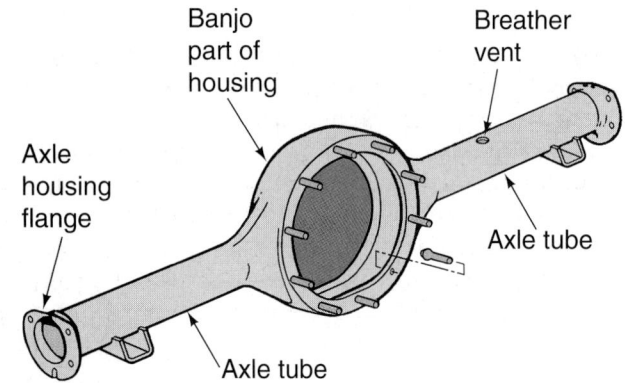

Figure 75.23 A banjo housing. (*Courtesy of Ford Motor Company*)

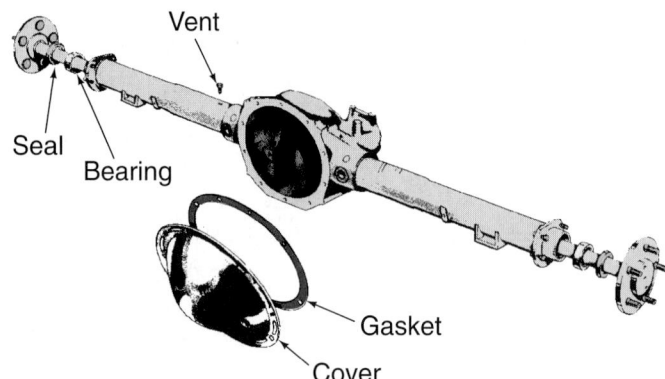

Figure 75.24 A Salisbury or integral axle has a cover on the back. (*Courtesy of DaimlerChrysler Corporation*)

■ DIFFERENTIAL OPERATION

During a turn, the outside wheel must be able to roll farther than the inside one. On the end of each axle is a set of splines. The *side gears* have splines that mesh with the axle splines (**Figure 75.25**). *Differential pinions* mesh with the side gears (**Figure 75.26**).

NOTE: *These are called differential pinions, not to be confused with the drive pinion, which meshes with the ring gear.*

The differential pinions are mounted on a *pinion shaft*. It must be removed to get the axles out of some types of cars. This procedure is covered in Chapter 76.

The side gears and differential pinions are called **spider gears**. The differential case has the ring gear mounted on it. It has the spur bevel spider gears inside of it. There are always two side gears, but sometimes there are more than two differential pinion gears that mesh between the side gears.

Figure 75.25 The side gears have splines that mesh with the axle splines. *(Courtesy of Eaton Corporation)*

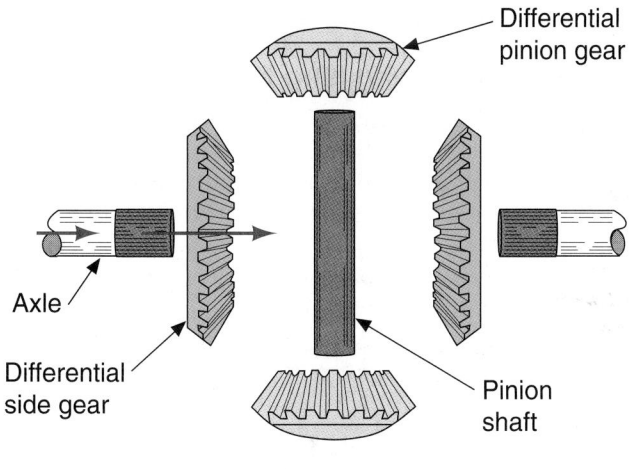

Spider gears

Figure 75.26 Differential pinions and side gears mesh with each other.

Power Flow

When the car is traveling straight ahead, the pinion gears do not rotate against each other. They turn with the case. The pinion shaft is mounted in the differential case and rotates with it (**Figure 75.27**). During a corner, when one of the wheels turns faster than the other, the side gears (splined to the axles) rotate against the differential pinions (**Figure 75.28**).

■ DIFFERENTIAL GEARS

Differential gears have progressed from plain bevel gears to spiral bevel, and finally to hypoid gears

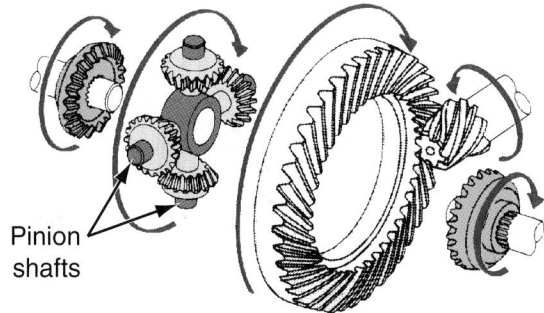

Figure 75.27 When traveling straight ahead, the pinion shaft is mounted in the differential case and rotates with it. *(Reproduced by permission of Deere & Company, John Deere Publishing, Moline, IL, all rights reserved)*

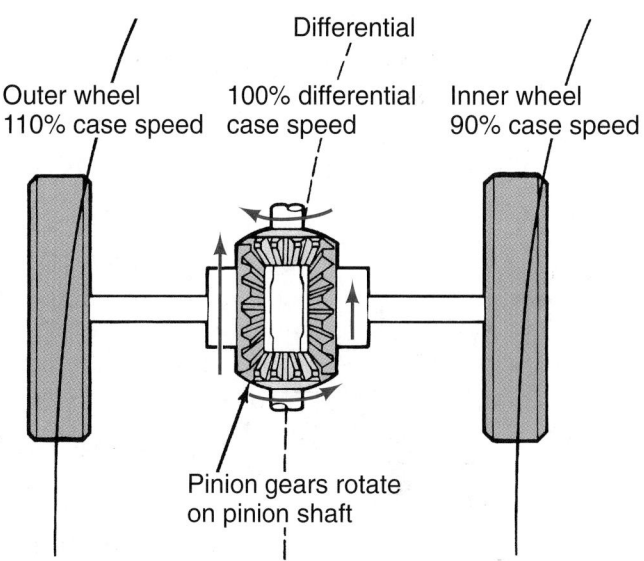

Figure 75.28 When one of the wheels turns faster than the other, the side gears rotate against the differential pinions. *(Courtesy of DaimlerChrysler Corporation)*

(**Figure 75.29**). In a hypoid gearset, the pinion gear is lower than the centerline of the ring gear. This is so that the transmission hump can be lower. The teeth of a hypoid gear are curved in a spiral shape. Sometimes, a drive pinion gear has a pilot bearing that supports the front of it (**Figure 75.30**).

NOTE: *When gears are machined, they are made by hobbing, a rough machine process that leaves heavy machine marks on the surface of the gear tooth (**Figure 75.31a**). Next, the gear is finish-machined by a shaving process. The root of the gear tooth remains as hobbed without the finish-machining (**Figure 75.31b**).*

Because hypoid gears are curved, each tooth has a concave and a convex side. The convex side is the drive side and the concave side is the coast side (**Figure 75.32**).

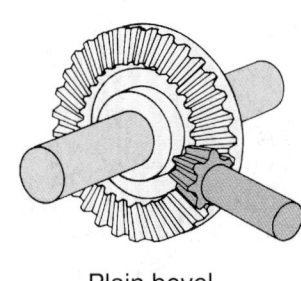

Plain bevel

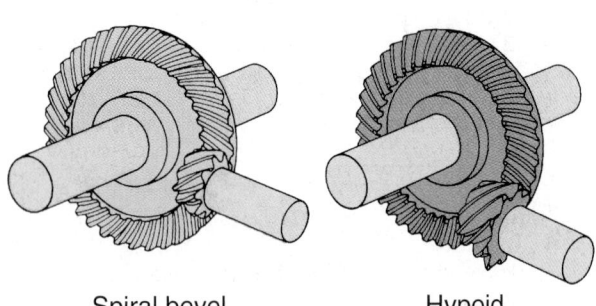

Spiral bevel   Hypoid

Figure 75.29 Types of differential gears. *(Reproduced by permission of Deere & Company, John Deere Publishing, Moline, IL, all rights reserved)*

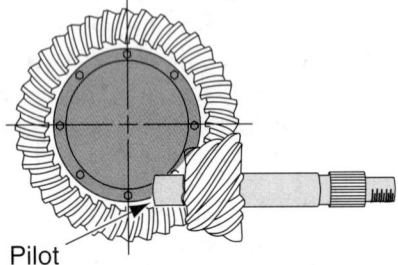

Pilot

Figure 75.30 Sometimes the drive pinion has a pilot supporting its front.

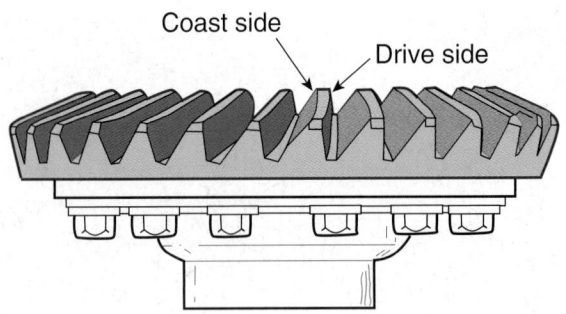

Coast side
Drive side

Figure 75.32 The convex side is the drive side and the concave side is the coast side.

■ GEARSETS

Ring and pinion gears are produced as a *matched set*. They are like the parts of a hydraulic lifter. They are either matched electronically or lapped together. A ring and pinion set must be mounted together in exactly the right position, or noise and wear will take place. The pinion gear must be mounted at a correct depth in the ring gear. Differential setup is critical. That subject is covered in detail in Chapter 76. In a matched gearset, the ring and pinion gears are marked (**Figure 75.33**) so that they can be assembled with the marks facing each other.

Hunting and Non-hunting Gearsets

Gearsets are either hunting or non-hunting. Most ring and pinion sets are hunting. With a **hunting gearset**, a pinion gear tooth will move around until it contacts all of the ring gear teeth. This gearset will not have to be timed.

With a **non-hunting** gearset, one tooth on the pinion gear will mesh with the same tooth on the ring gear again and again, rather than "hunting" for all of the rest of the teeth. This will have an even gear ratio,

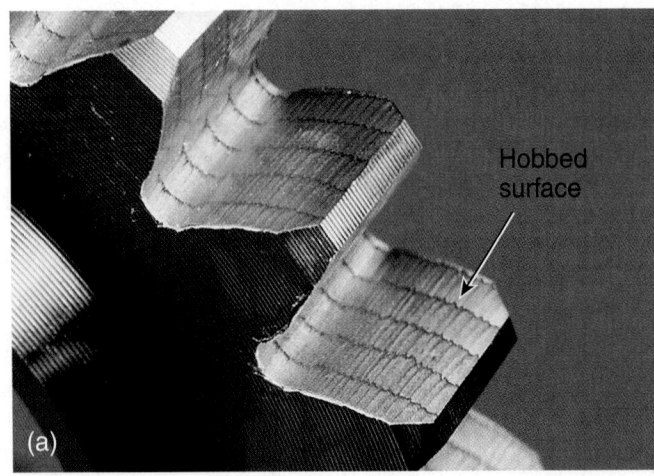

(a)

Hobbed surface

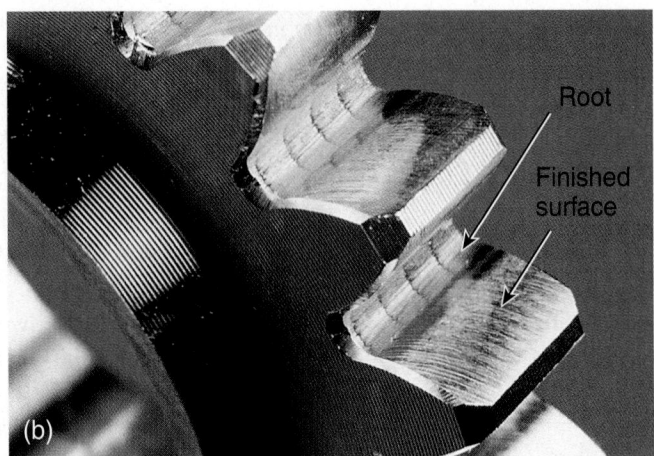

(b)

Root

Finished surface

Figure 75.31 (a) Gears are made by a process called hobbing, a rough machine process that leaves heavy machine marks on the surface of the gear tooth. (b) When the tooth is finish-machined (shaved), the root of the tooth remains rough. *(Courtesy of Caterpillar, Inc.)*

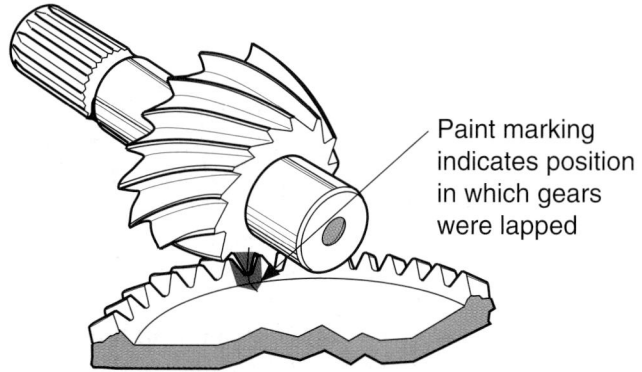

Figure 75.33 In a matched gearset, the ring and pinion gears are marked. *(Courtesy of Ford Motor Company)*

such as 4:1. One pinion tooth would contact four different ring gear teeth every revolution. A non-hunting gearset has a timing mark that needs to be aligned.

A **partial non-hunting gearset** is one with a ratio like 2.5:1 that allows one pinion tooth to contact two or three different ring gear teeth during one revolution and two or three different teeth the next revolution. At the third revolution it will be back to the same teeth on the ring gear.

■ AXLE RATIO

To find the ratio in the differential, divide the number of teeth on the pinion gear into the number of teeth on the ring gear. If there are 10 teeth on the pinion gear and 33 teeth on the ring gear, this would be a 3.30:1 ratio.

NOTE: *A higher ratio means a lower number; 3.5 is a lower ratio than 3.3. For racing, differential ratios can go as low as 5:1 or 6:1.*

A car with a manual transmission will have a lower ratio, i.e., 3.50:1. With an automatic transmission, the ratio will be higher; 3.0:1, for instance. For better fuel economy, axle ratios tend to be higher (lower number).

■ LIMITED SLIP DIFFERENTIAL

A standard differential allows one wheel to rotate if it begins to slip. The other wheel will remain at rest and the vehicle will be stuck. A **limited slip differential** prevents this from happening. A limited slip differential locks up the spider gears when one wheel starts to lose traction. This puts traction to the still wheel. The spider gears are not locked up during normal operation.

There are other names for locking differentials— *Positraction* (Chevrolet), *Equa-lok* or *Traction-lok* (Ford), *Sure-Grip* (DaimlerChrysler), *Power-Loc* (Toyota), *Trac-loc* (Dana), and *Super Brute* (GM).

■ TYPES OF LIMITED SLIP DIFFERENTIALS

There are several designs of locking differentials. The most popular one has clutch packs similar to an automatic transmission (**Figure 75.34**). When torque is applied to the clutch pack, the side gear locks to the case, causing the wheel on that side to be driven (**Figure 75.35**).

The clutch pack is applied to the side gear in different ways. Usually a spring (either a coil, multiple coils, an S-shaped spring, or a Belleville spring) pushes the

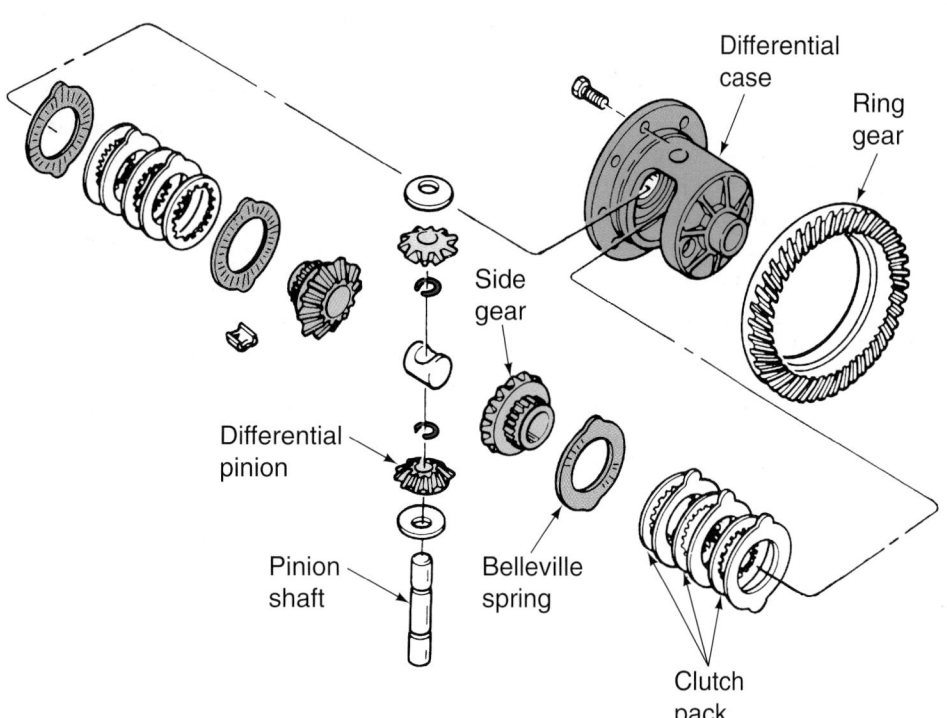

Figure 75.34 A limited slip differential with clutch packs. *(Courtesy of DaimlerChrysler Corporation)*

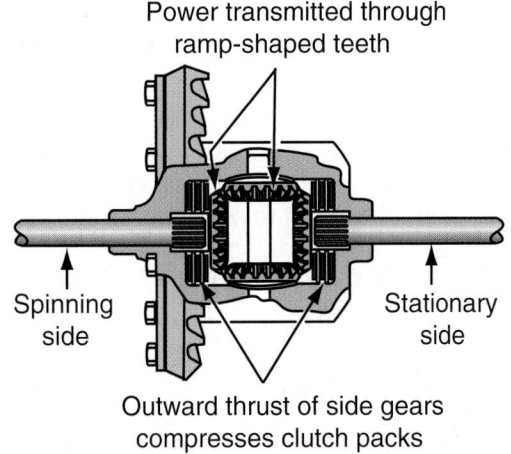

Power transmitted through ramp-shaped teeth

Spinning side

Stationary side

Outward thrust of side gears compresses clutch packs

Figure 75.35 When torque is applied to the pinion gears, pressure on the side gears compresses the clutch packs.

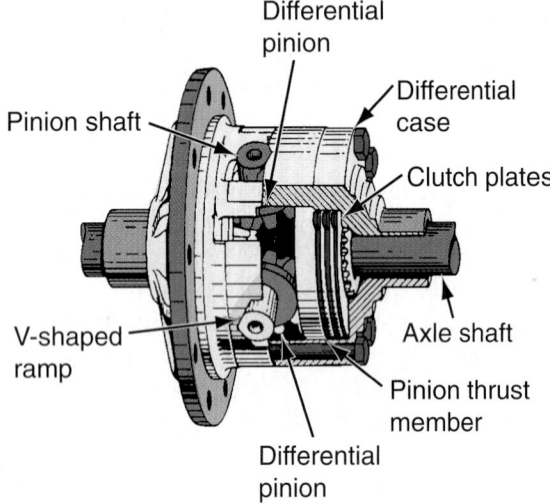

Differential pinion

Pinion shaft

Differential case

Clutch plates

V-shaped ramp

Axle shaft

Pinion thrust member

Differential pinion

Figure 75.36 These pinion shafts are positioned in V-shaped ramps instead of a round hole. *(Courtesy of DaimlerChrysler Corporation)*

side gear against the clutch pack. When the side gear on the side that is spinning tries to climb up out of the pinion gears with which it is meshed, the pinions push against the clutch pack splined to the side gear on the opposite side. This is called torque loading.

Some older units have four pinion gears and two pinion shafts. The pinion shafts are positioned in V-shaped ramps, instead of a round hole (**Figure 75.36**). This allows the pinion gears to move to one side or the other and apply pressure to the clutch packs.

The other main type of limited slip differential is the *cone-type* (**Figure 75.37**). Cones are forced against the case by springs. Because the axles are splined to the cones, they will want to rotate with the differential case.

One other design used in some applications is the *viscous coupling* differential. In the past it was used in the Jenson Interceptor. It had an expensive patent that is now expired, so it is expected to be seen in more and more vehicles. **Figure 75.38** shows a viscous coupling used in an all-wheel drive. It is located in the center

of the drive shaft that runs between the front and rear differentials.

A viscous coupling is a sealed unit that contains silicone fluid. The clutch plates in the coupling are all steel with many holes drilled in them to dissipate heat. There is no spring to apply the clutches. Silicone does not change in viscosity as it heats up. The viscous coupling is not totally full of fluid. When there is slippage between the spider gears, air bubbles from in the silicone. The fluid is sticky, and when it gets hot the bubbles create pressure that locks the plates. When they cool down, they release once again. This cycling can take place several times a minute. This is called hysteresis. The viscous coupling is not serviceable and is expensive if it fails. Be careful not to mismatch tires or the unit will work all of the time.

NOTE: *It is important that tires be the same size on all limited slip differentials. Otherwise, the clutches will always be working against each other and cause premature wear or failure.*

Other differential designs are used only for racing or off-road applications. One popular design is the Detroit Locker. It has a ratcheting pair of clutch packs that physically lock up the unit. A drawback is that it makes a ratcheting noise when it releases. Off-road enthusiasts recommend that this differential not be used on the front of a four-wheel drive because it affects steering.

▪ DRIVE AXLES AND BEARINGS

Drive axles usually support the weight of the car. Bearings are covered in Chapter 60. Refer to that chapter for more information about particular types of bearings. The coverage here is limited to specific rear-wheel bearing designs.

▪ SEMI-FLOATING AXLE BEARING TYPES

Semi-floating bearings (see Figure 60.17) are found on passenger cars. There are two semi-floating designs in common use.

Bearing Retaining Axle

On one design, called a *bearing retained axle,* a bearing with an inner race is pressed onto the axle shaft. A bearing *retainer ring* (**Figure 75.39**) is pressed onto the axle shaft after the bearing is installed. The outside of the bearing fits tightly into the axle housing so oil cannot leak out of the differential. The axle bearing used with this axle is usually packed with grease and sealed.

A bearing retainer that is installed on the axle before the bearing is pressed on provides a means of bolting the assembly to the brake backing plate. The bolts holding the bearing retainer also clamp the brake backing plate to the axle housing.

C-Lock Axle

The other style of axle is called a *C-lock* or *C-clip axle.* On most axles of this type, the bearing rides on a hardened

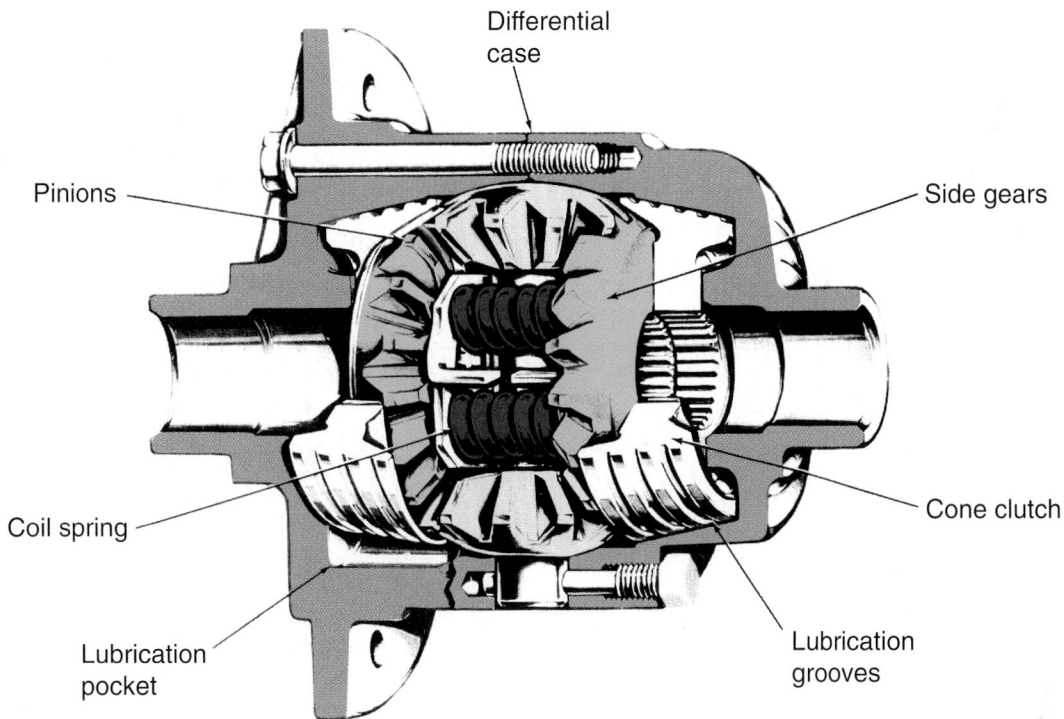

Figure 75.37 A cone-type limited slip differential. *(Courtesy of DaimlerChrysler Corporation)*

Labels on figure: Differential case, Pinions, Side gears, Coil spring, Cone clutch, Lubrication pocket, Lubrication grooves

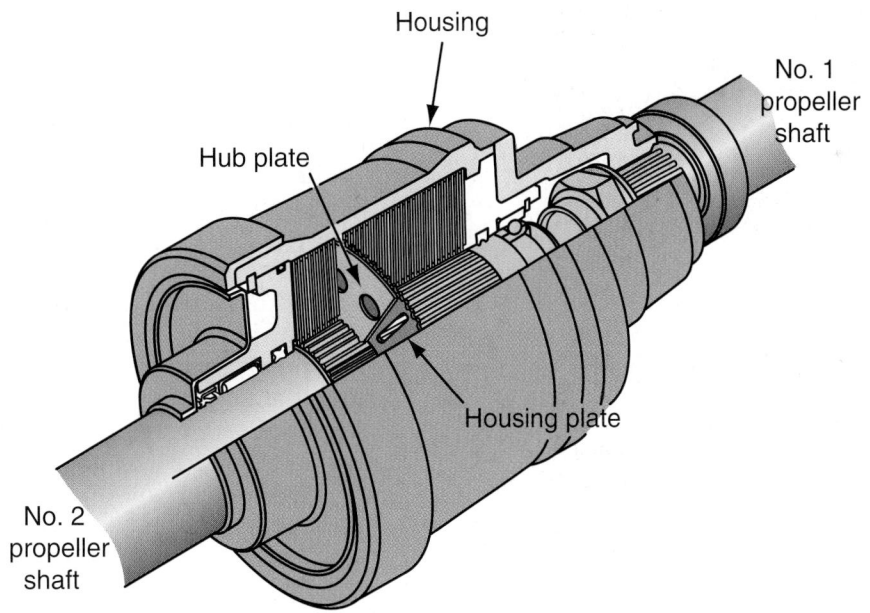

Figure 75.38 A viscous coupling has multiple disc clutches, is sealed and is almost full of silicone.

Labels on figure: Housing, Hub plate, No. 1 propeller shaft, No. 2 propeller shaft, Housing plate

area of the axle rather than having an inner bearing race (**Figure 75.40**). The outer bearing race and its bearing rollers fit tightly into the end of the axle housing. The axle bearing is lubricated by a mist of oil from the differential. A lip seal keeps the oil from leaking out onto the brake shoes.

This retainer type is found only on Salisbury axles. Salisbury axles are those with a cover on the rear that do not have a removable center differential section (third member). A C-lock at the inside end of the axle holds the axle into the differential (see Figure 76.17). The outside of the C-lock fits into a recess in the differential side gear (the gear that is splined to the rear axle and drives it). When the differential pinion shaft is installed, the C-locks are trapped within the side gears.

NOTE: *With either of the semi-floating axle designs, a broken axle can result in the wheel moving out away from the axle housing. With the C-lock axle, there will be nothing to*

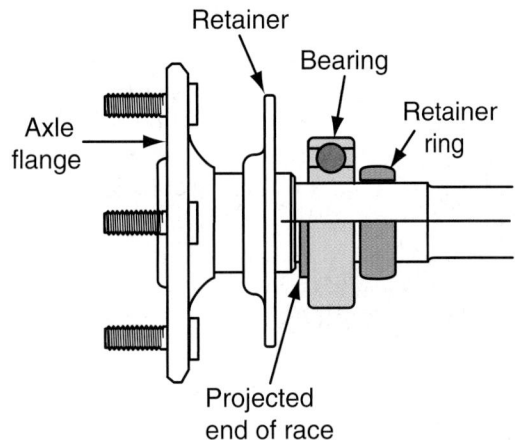

Figure 75.39 A retaining ring is pressed onto the shaft after the bearing.

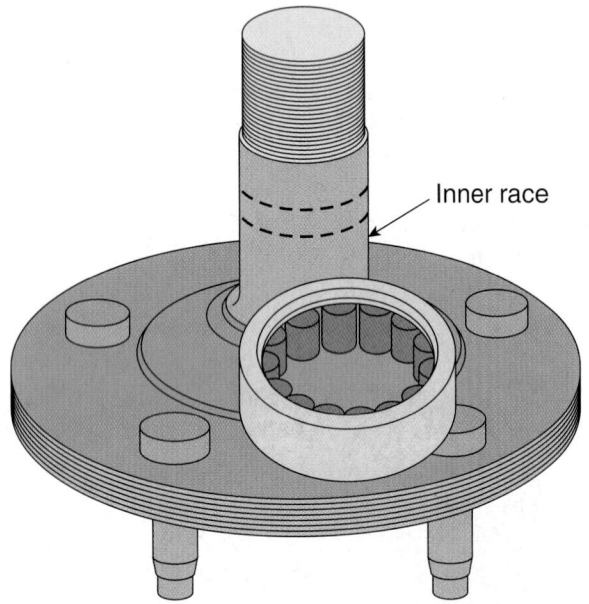

Figure 75.40 On a C-lock axle, the bearing rides on a hardened area of the axle. (*Courtesy of Ford Motor Company*)

retain it. *With the bearing retained axle, the bearing can wear through the retainer if the car is driven in this condition. That is why light-duty trucks and heavier use full floating axle bearings (see Figure 60.18). The full floating axle can break and the wheel and hub will still be supported by the bearings.*

■ INDEPENDENT REAR SUSPENSION

When a car has independent rear suspension, the axles must be able to pivot independently as each wheel goes over a bump. Instead of solid axles, there are two **swing axles** (**Figure 75.41**). Each axle has a universal joint at one or both ends. These are similar to the ones used for front axles in front-wheel-drive cars (see Chapter 77).

■ GEAR OILS

Gear oils are special heavy liquid lubricants used to lubricate gears and bearings in manual transmissions

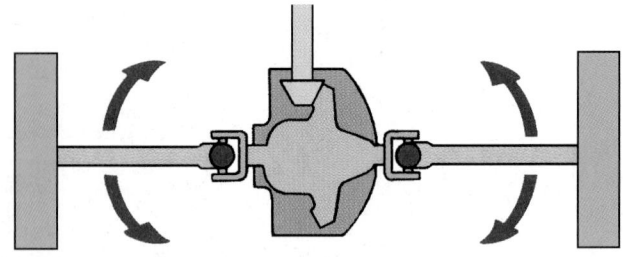

Flexible axle shaft

Figure 75.41 An independent rear suspension has two swing axles. (*Reproduced by permission of Deere & Company, John Deere Publishing, Moline, IL, all rights reserved*)

and differentials. A good oil prevents high temperatures and scoring of the parts. A hypoid axle set is the most difficult of gear applications to lubricate. The gear teeth can come under pressures as high as 400,000 psi. Because of their shape, a high amount of sliding occurs between the mating surfaces of the gear teeth. Gears must remain separated by a film of oil, even when under heavy loads. A good oil prevents high temperatures and scoring of parts. As one gear tooth moves against another, oil is forced from between them. This oil must be continuously replenished.

When the lubricant film fails, wear results. Gear oils have additives similar to those found in engine oils. Additives prevent oxidation and corrosion, reduce friction, limit wear, and prevent foaming. **EP additives** (extreme pressure) are chemicals that cause the formation of compounds on the gear teeth that prevent welding of the gear teeth under extreme load conditions.

There are many varieties of gear oils. Some gear oils are thin, like those in automatic transmissions. Others are thick, like those in differentials. Oil used in manual transmissions must remain fluid enough to allow easy shifting in cold weather. Automatic transmission fluids must remain fluid to as low as –40°F and remain at almost the same viscosity at over 300°F.

Viscosities for gear oil range from 75 to 250. There are also multigrades like 80W–140. The W is for winter grade, like in engine oil.

NOTE: *Gear oil viscosity is* not *the same as engine oil. It is rated in centiStokes. An SAE 90 gear lube is comparable in viscosity to an SAE 40 or 50 engine oil.*

The chart in **Figure 75.42** compares viscosities of SAE gear and engine lubricants.

API Gear Oil Ratings

The American Petroleum Institute (API) and *Coordinating Research Council* (CRC) have established a system for classifying gear lubricants. The Society of Automotive Engineers (SAE) lists these as shown in **Figure 75.43**. The GL in the number stands for gear lube. GL1 is the lowest grade, only for light duty and low speed. Automotive gear lubes are GL4, GL5, and GL6. GL6 is a synthetic classification.

Viscosity equivalent chart

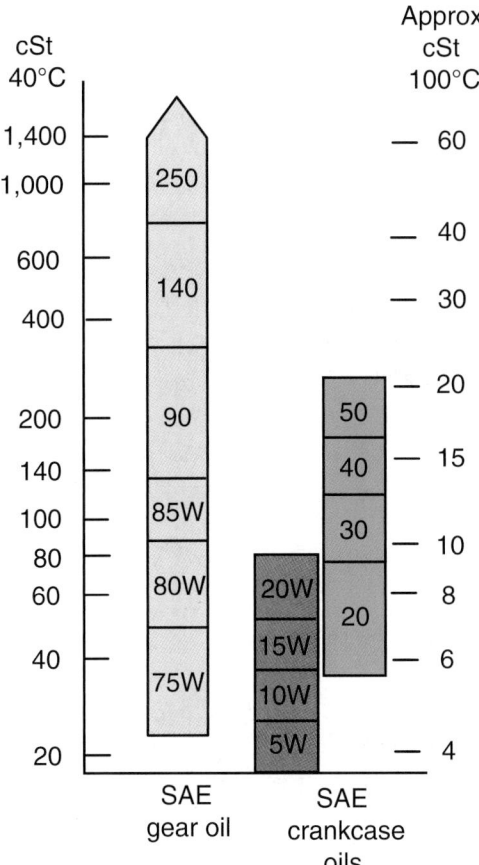

Figure 75.42 A comparison of viscosities of SAE gear and engine lubricants.

API classification	Amount of EP additives	Type of service
GL-1	None	Light duty only. Spiral bevel and worm gear axles. Some manual transmissions.
GL-2	Small amount	Worm gear axles under more severe conditions than GL-1.
GL-3	Small amount	Spiral bevel axles and manual transmissions under more severe conditions than GL-1.
GL-4	Medium amount	Hypoid gears—light service. Spur and bevel gears—medium service.
GL-5	High amount	Hypoid gears—medium service. Spur and bevel gears—heavy-duty service.
GL-6	Highest amount	Most gears—extremely heavy-duty service.

Figure 75.43 Gear lubricant ratings.

Limited Slip Gear Oils

Limited slip gear oils require a special friction modifier additive. The fill plug is usually marked to indicate this. When there is not enough friction modifier, the clutch discs stick together and grab rather than slipping. As the car makes a turn, a chirping or clunking sound can sometimes be heard. Be sure the lubricant you install in a limited slip axle is suited for limited slip uses. Friction modifier additives can be added to gear lube to make it compatible with limited slip axles.

NOTE: *Limited slip gear oil will work in standard axles.*

■ FOUR-WHEEL DRIVE

There are different types of four-wheel drives. One type is the conventional four-wheel drive that operates like an RWD vehicle with an extra differential in the front (**Figure 75.44**). Another type is like an FWD vehicle with an extra differential in the rear (**Figure 75.45**). Still another is called all-wheel drive (**AWD**). It remains in four-wheel drive all of the time.

■ FOUR-WHEEL-DRIVE AXLE ASSEMBLY

An axle assembly for a four-wheel drive is very similar to the two-wheel-drive rear axle. A drive shaft and universal joints transfer power from the transmission by way of a transfer case. The ends of the axle housing on front axles are different because they have to pivot to allow the front wheels to be turned during steering (**Figure 75.46**).

■ TRANSFER CASE

On four-wheel-drive vehicles there is a transfer case (**Figure 75.47**) on the rear of the transmission that provides for power transmission to the front axle assembly (see Figure 75.44). Two drive shafts come out of the transfer case; one goes to the front and the other goes to the rear.

Most four-wheel drives provide the options of high and low ranges for steep climbing or highway driving. The transfer case allows for shifting between low range and high range. Low range usually provides an additional

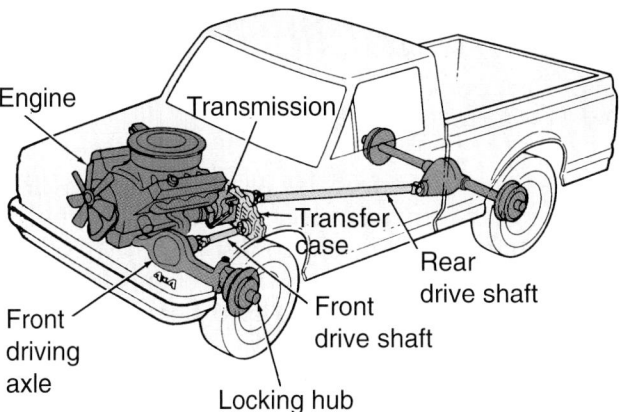

Figure 75.44 Parts of a four-wheel drive. (*Courtesy of Ford Motor Company*)

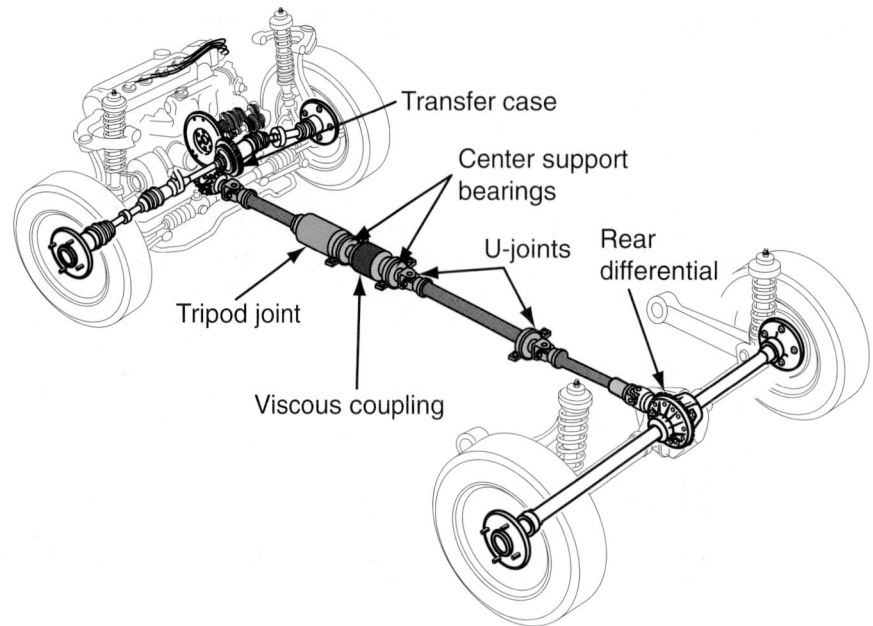

Figure 75.45 A four-wheel drivetrain with a front transaxle.

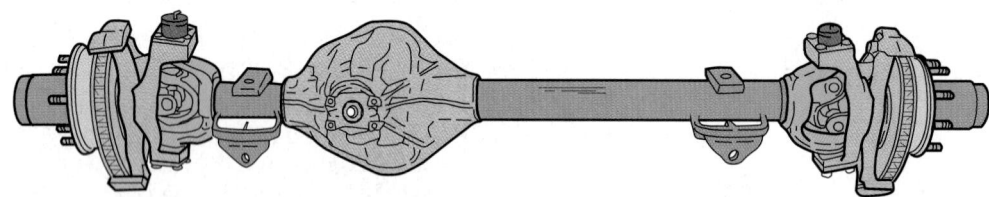

Figure 75.46 The ends of the four-wheel-drive front axle must pivot for steering. *(Courtesy of Ford Motor Company)*

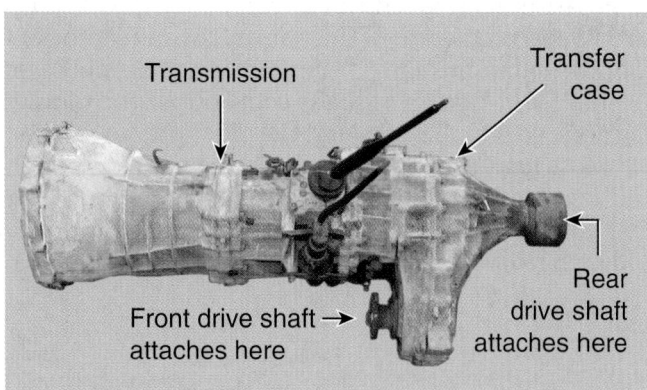

Figure 75.47 A four-wheel-drive transmission and transfer case. *(Courtesy of Tim Gilles)*

2:1 reduction. High range is 1:1. A transfer case also allows for shifting between two-wheel and four-wheel drive.

The transfer case can consist of either a gear drive or a planetary gearset with a chain drive (**Figure 75.48**). Power flow is covered here using a gear drive transfer case.

- In *two-wheel-drive high range (2H)*, power flows through the locked planetary gearset to the mainshaft and to the rear differential.
- In *four-wheel-drive high range (4H)*, power flows from

the input shaft to the locked planetary gearset in the same way it does in 2WD (**Figure 75.49**). A *sliding clutch*, called a *range clutch*, shifts into the mainshaft clutch gear. Power flow is through the drive chain and front output yoke to the front drive shaft. Power also flows through the rear drive shaft so that both axles drive the vehicle.

- In *four-wheel-drive low range (4L)*, everything is the same as in high range except that the shifter moves the range clutch rearward to engage low gear (**Figure 75.50**).

Planetary Low Range

In a planetary transfer case, the ring gear is pressed into the housing, so it is held stationary. Power flows into the sun gear and out through the cage or carrier (**Figure 75.51**). A gear reduction is produced that is transferred to both the front and rear output shafts.

■ LOCKING HUBS

Many axles also have **locking hubs**. There are three types of locking hubs. Their purpose is to allow a four-wheel-drive vehicle to be used in two-wheel drive without the axles and differential being attached to the front wheels.

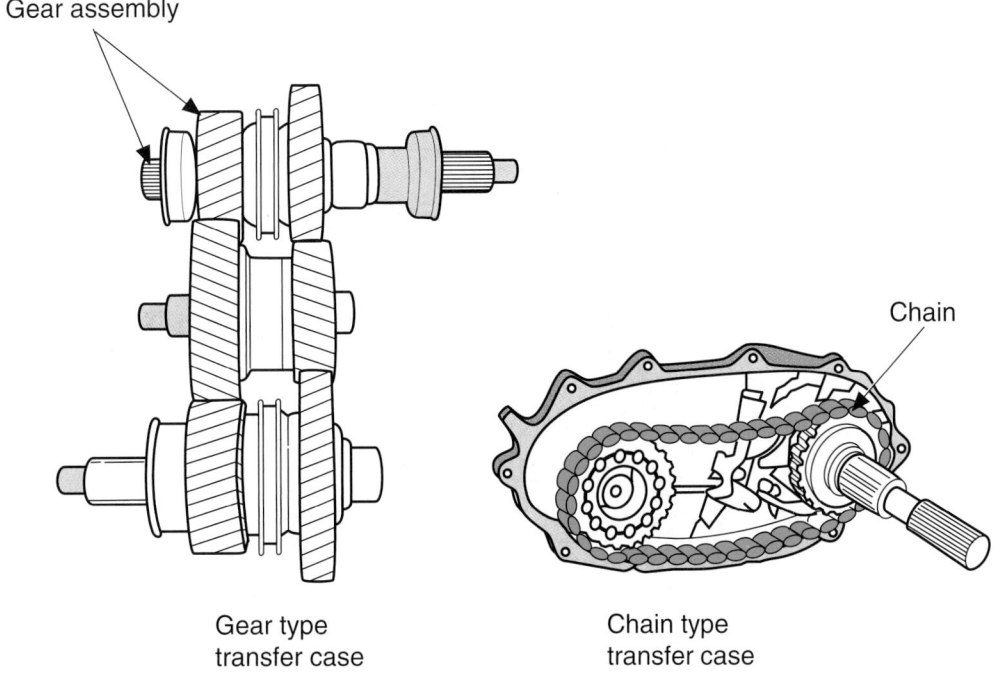

Gear type
transfer case

Chain type
transfer case

Figure 75.48 Gear drive and chain drive transfer cases. *(Courtesy of Ford Motor Company)*

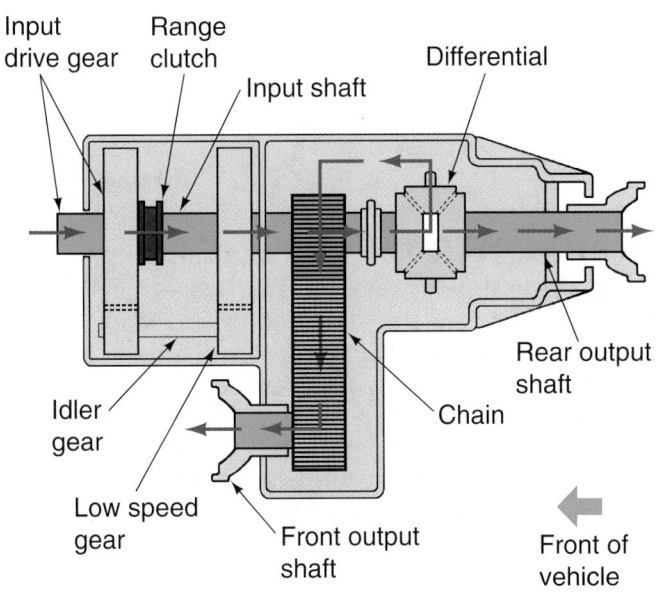

High range

Figure 75.49 Power flow through the transfer case in four-wheel-drive high range. *(Courtesy of Ford Motor Company)*

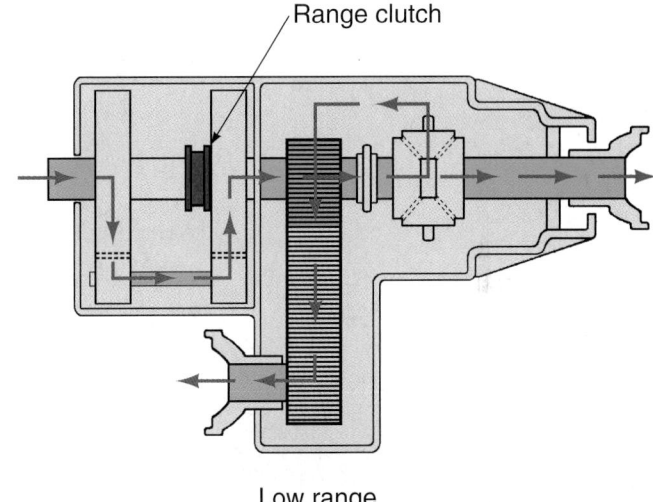

Low range

Figure 75.50 Power flow through the transfer case in four-wheel-drive low range. *(Courtesy of Ford Motor Company)*

With a *manual locking hub,* the hubs have on and off locks on the front spindles (**Figure 75.52**). These must be turned to lock and unlock the axles from the wheels. When the hubs are unlocked, the front wheels are not driven.

■ The vehicle should not be operated in four-wheel drive with the hubs disengaged or driveline wear will occur.

■ When shifting the hubs back out of locking range, the transfer case must be shifted to two-wheel drive. This releases the pressure from the hubs and makes them easy to turn.

Automatic locking hubs come on whenever the vehicle is shifted into four-wheel drive. *Full-time hubs* are always locked to the front axles and differential.

NOTE: *Automatic hubs do not always disengage when four-wheel drive is shut off. Occasionally, a vehicle must be stopped and backed up for about 3 feet to disengage the hubs. A disadvantage to these hubs over the manual ones is that they do not work in reverse.*

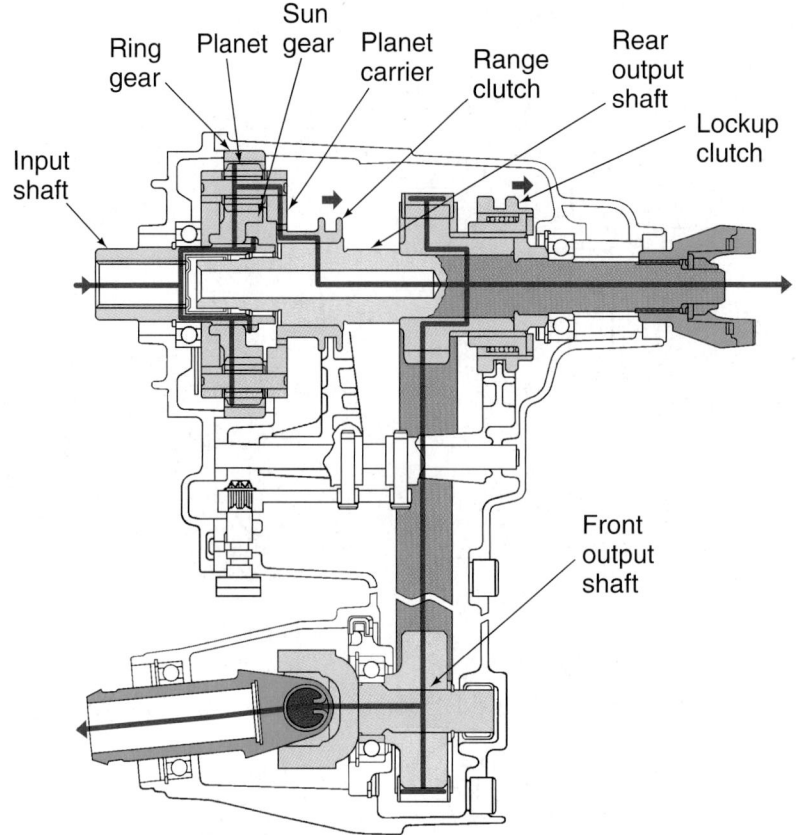

Figure 75.51 Power flow through a planetary gear transfer case in low range. *(Courtesy of Ford Motor Company)*

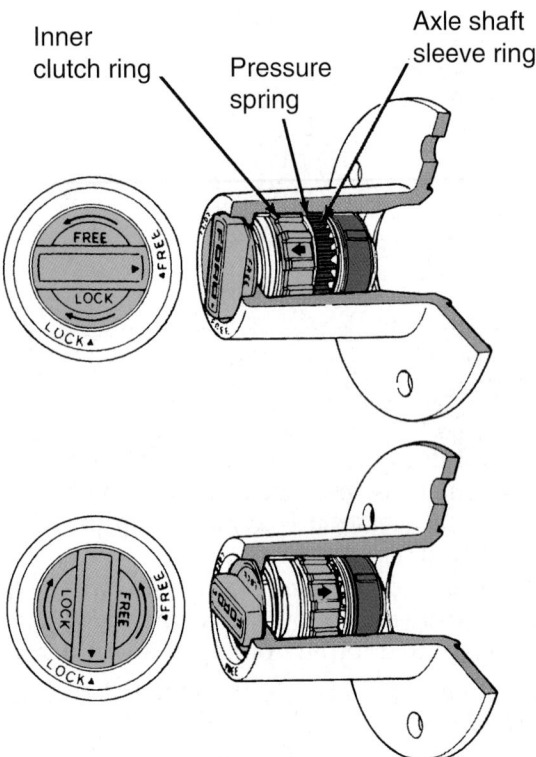

Figure 75.52 Locking hubs. *(Courtesy of Ford Motor Company)*

■ ALL-WHEEL DRIVE

In all-wheel drive (AWD), all four wheels are driven. These vehicles are not designed for off-road use but are to improve poor traction in icy or snowy driving. Control is improved by transferring engine power to the wheel with the most traction. Most AWD systems have a center differential to split the power between the front and rear axles (**Figure 75.53**). The transfer case shown in **Figure 75.54** has the differential within the transfer case. Four-wheel-drive transfer cases do not have this differential.

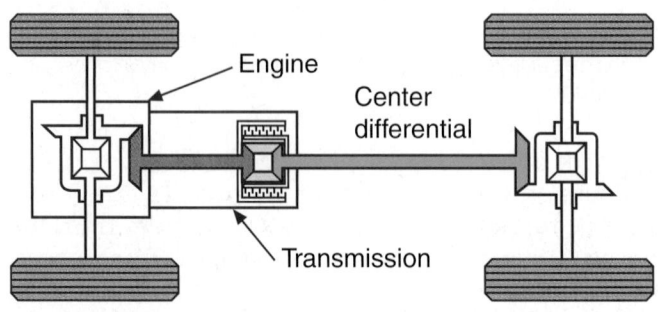

Figure 75.53 A center differential splits the power between the front and rear axles. *(Courtesy of DaimlerChrysler Corporation)*

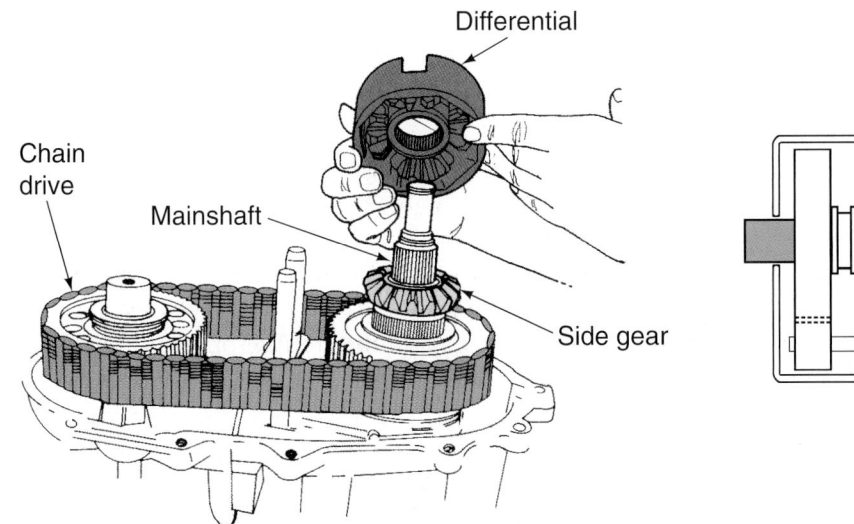

Figure 75.54 This transfer case includes a differential to provide four-wheel drive. *(Courtesy of DaimlerChrysler Corporation)*

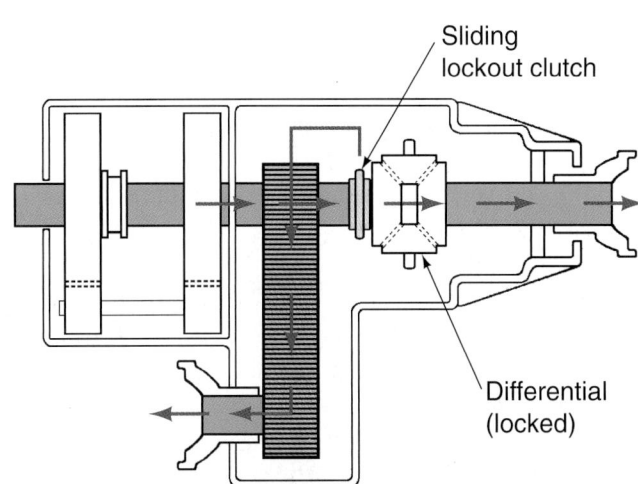

Figure 75.55 A lockout for the center differential. *(Courtesy of Ford Motor Company)*

When there is a center differential, if one wheel is lifted free of the ground, it is possible for all of the torque to go to that wheel. This means no power goes to the other wheels. A lockout for the center differential must be included in these types (**Figure 75.55**). A popular new way of connecting the front and rear axles is the viscous coupling, which allows the front and rear wheels to revolve at different speeds (**Figure 75.56**).

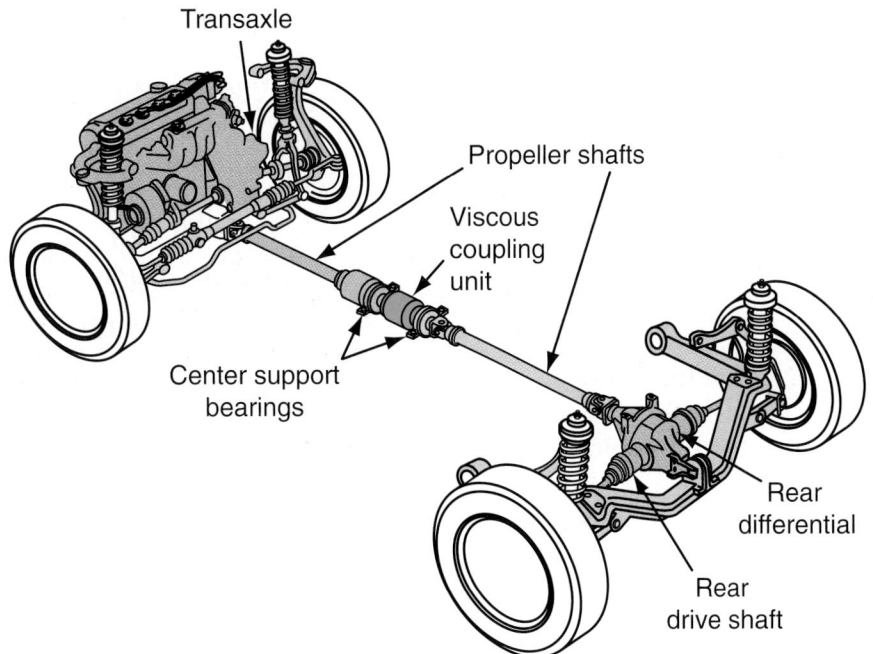

Figure 75.56 All-wheel drive with a viscous coupling.

■ REVIEW QUESTIONS

1. What is the name of the open drive shaft system?

2. What is the name of the part that slides in and out of the transmission as the distance between the transmission and differential changes?

3. How many times in one revolution does speed increase and decrease when a U-joint is operated at an angle?

4. Angles of greater than _____ degrees are not used with single U-joints because they cause excessive vibration.

5. At 30 degrees of drive shaft angle, when a shaft is turned at 1000 rpm, the speed of the driven yoke will change from _____ rpm to 1155 rpm in one-quarter of a revolution.

6. What is the name of the CV joint that uses two U-joints?

7. The differential is called the _____ member.

8. Which has more teeth, the ring gear or the pinion gear?

9. The name of the part that separates the two pinion bearings is called a crush sleeve or collapsible _____.

10. During a turn, which wheel must move farther than the other, the inside or the outside?

11. The side gears and differential pinions are called _____ gears.

12. To find the ratio in the differential, divide the number of teeth on the _____ gear by the number of teeth on the _____ gear.

13. Which car will have a higher numerical axle ratio, one with an automatic or a manual transmission?

14. An SAE 90 gear lube is comparable in viscosity to an SAE 40 or 50 _____ oil.

15. What does the GL stand for in the gear lube rating?

■ ASE-STYLE REVIEW QUESTIONS

1. Which of the following is/are true about universal joints?

   a. Original equipment universal joints have a grease fitting.

   b. When a Cardan universal joint is operated in a straight line, the speed of the output member is different than the speed of the input member.

   c. Both A and B.

   d. Neither A nor B.

2. Which of the following is/are true about driveshaft vibration?

   a. Universal joints are angled to cancel each other out.

   b. On a one-piece driveshaft, the front and rear yokes can be out of phase, resulting in vibration.

   c. Both A and B.

   d. Neither A nor B.

3. Technician A says that constant velocity joints are found on smaller rear-wheel-drive cars. Technician B says that the speed of the input yoke is constantly changing as it revolves. Who is right?

   a. Technician A   **c.** Both A and B

   b. Technician B   **d.** Neither A nor B

4. Which of the following is/are true about differentials?

   a. A differential allows its two wheels to turn at different speeds.

   b. A differential increases the torque from the engine.

   c. Both A and B.

   d. Neither A nor B.

5. Technician A says that the differential is removed from a Salisbury axle as a complete unit. Technician B says that a removable third member is called a pumpkin. Who is right?

   a. Technician A   **c.** Both A and B

   b. Technician B   **d.** Neither A nor B

6. Which of the following is/are true about differential gears?

   a. Differential pinion gears have splines that mesh with the axle splines.

   b. When the car is traveling straight ahead, the differential pinion gears rotate.

   c. Both A and B.

   d. Neither A nor B.

7. Technician A says that the convex side of a gear tooth is the drive side. Technician B says that with a hunting gear set, one pinion gear tooth will contact all of the ring gear teeth. Who is right?

   a. Technician A   **c.** Both A and B
   b. Technician B   **d.** Neither A nor B

8. Which of the following is/are true about differentials?

   a. A standard differential will not allow one wheel to spin without putting power to the other.
   b. A limited slip differential requires a different lubricant than a standard differential.
   c. Both A and B.
   d. Neither A nor B.

9. Technician A says that silicone in a viscous coupling changes its viscosity to apply the clutches. Technician B says that a four-wheel-drive vehicle requires a differential or viscous coupling between the front and rear axles. Who is right?

   a. Technician A   **c.** Both A and B
   b. Technician B   **d.** Neither A nor B

10. Which of the following statements is/are true?

   a. Gear lube viscosity is different than motor oil viscosity.
   b. C-lock axles are sometimes installed on differentials with removable third members.
   c. Both A and B.
   d. Neither A nor B.

Driveline Diagnosis and Service

■ OBJECTIVES

Upon completion of this chapter, you should be able to:

✔ Diagnose and repair universal joints.

✔ Remove and replace axle bearings and seals.

✔ Disassemble and reassemble a differential.

✔ Run a gear pattern and adjust a differential.

■ KEY TERMS

galling	howl	heel
drive	breakaway torque	toe
cruise	gear pattern	pitch line
coast	drive side	face
float	coast side	flank

■ INTRODUCTION

In this chapter, service to the driveline is covered. The driveline includes the drive shaft and U-joints, axles and axle bearings, and differential. Live axles are covered in this chapter. They turn with the wheels. Transaxles, a later chapter, have similar service to their differentials. Your understanding of differential service procedures will apply later. Differentials are also referred to as front or rear *axles*.

In this chapter, you will learn more than just driveline service and repair procedures. Pay attention to the methods used in pressing bearings and other parts. Press skills are used in many other areas of automotive repair.

■ DRIVE SHAFT DIAGNOSIS

Drive shaft problems can result in noise or vibration from worn or rusted U-joints, a worn slip yoke, or a bad center support bearing. Worn U-joints can cause *squeaking*, *clunking*, *knocking*, or *grinding* sounds.

SHOP TIP Before disassembling a universal joint, check to see that a wheel cover or hub cap is not making the noise. Remove all of them and drive the car to see if the noise goes away.

Sometimes a car will make a *clunking* sound when changing from acceleration to deceleration. This can be due to worn slip yoke splines or a bad extension housing bushing. It can also be caused by problems in the differential or a very worn U-joint. Sometimes leaf springs can be loose at the differential, allowing the housing to "wind up." This allows the input (pinion) shaft to the differential to move higher or lower in relation to its normal position, changing the drive shaft angle.

A worn center support bearing can cause a *whining* sound that varies with vehicle speed. The noise is constant in pitch, rather than changing or intermittent like U-joint noise. A U-joint noise changes pitch because of the changing angle of the U-joint.

NOTE: *The drive shaft is not always the source of noises that seem to be coming from it. A ringing sound, heard in the driveshaft, sometimes results from a bad clutch disc damper. Replacing the clutch disc usually solves the problem. If the vehicle has an automatic transmission, the problem can be due to a bad lock-up converter.*

Drive Shaft Balance

Drive shafts are a possible source of vibration. In high gear, the drive shaft spins at engine rpm. If the shaft is bent or a universal joint is worn, vibration can occur. Drive shaft balance is covered in Chapter 79.

Some drive shafts are built in two pieces with rubber dampening rings inside of them. Another drive shaft style has a damper like a crankshaft vibration damper mounted on its outside. This absorbs torsional vibration.

Drive Shaft Inspection

Several checks are made when inspecting a drive shaft.
- Look for undercoating, missing balance weights, or obvious physical damage.
- Move the drive shaft up and down while watching the U-joints for looseness.
- Look for a rusty appearance around U-joint cups.
- Move the slip yoke up and down against the extension housing bushing while looking for excessive movement.
- Check to see that the motor mount on the transmission crossmember is in good condition.

A dent in the drive shaft tubing will weaken it, which can cause it to kink easily under load. The strength of a drive shaft is longitudinal. Try stepping on a soda can and quickly touch both sides of the can. It will immediately collapse when you touch it. Cracks in a drive shaft result from physical damage. They always start on the surface, never at the inside.

■ UNIVERSAL JOINT DIAGNOSIS AND SERVICE

When a universal joint begins to fail, a squeaking sound is often noticed just when the car begins to go forward. The most common cause of U-joint failure is that its grease has dried out. This often happens because the seal on the U-joint has failed, allowing moisture in. A vibration can also occur when a U-joint starts to fail. With a worn U-joint, a sharp, one-time click sound often occurs when the vehicle direction is changed from forward to reverse or when the vehicle first takes off.

■ DRIVE SHAFT SERVICE

To remove the drive shaft:
- Mark the drive shaft so that it can be replaced in the same position. Use a crayon to mark the rear differential yoke and the companion flange (**Figure 76.1**).
- Unbolt the rear U-joint from the differential companion flange.
- Pry the rear U-joint forward away from the differential.
- Wrap tape around the U-joint cups so that they cannot fall off of the U-joint cross.

NOTE: *If one of the cups falls off the U-joint, one or more of the small needle bearings might fall out. If one gets lost, the entire U-joint must be replaced.*

- On a two-piece drive shaft, unbolt the center support bearing.

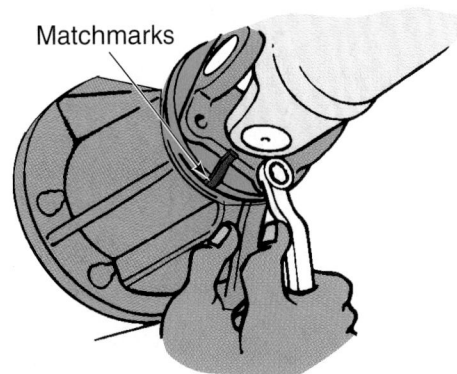

Matchmarks

Figure 76.1 Mark the rear differential yoke and the companion flange before removing the driveshaft.

NOTE: *Be sure to mark both halves of the shaft in the center where the splines are. The shaft will be out of phase if it comes apart and is not reassembled correctly (see Figure 75.15). Serious vibration will result.*

- The drive shaft will now slip out of the transmission. When it is removed, oil will probably come out of the transmission.

SHOP TIP To prevent oil from leaking out of the transmission, install an old slip yoke onto the mainshaft splines. If an old yoke is not available, use a plastic plug or a bushing installation tool.

■ UNIVERSAL JOINT DISASSEMBLY

The procedure described here is for single cross and yoke (Cardan) universal joints. If the U-joint has any snaprings, remove them (**Figure 76.2**). Some snaprings are on the inside of the yoke. Others are on the outside. When the snapring is on the outside, a sturdy pair of pliers can be used.

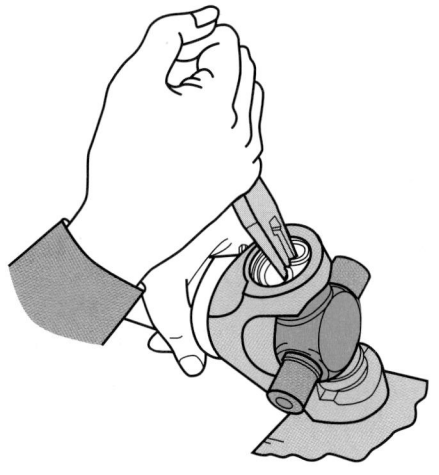

Figure 76.2 Removing a snapring from a U-joint.
(Courtesy of Ford Motor Company)

Figure 76.3 Sometimes snaprings go on the inside of the yoke.

NOTE: *Small needle-nose pliers can become damaged because they are not sturdy enough.*

When snaprings are on the inside (**Figure 76.3**), a punch or a special tool can be used to remove them. If the U-joint is retained by plastic resin, follow the manufacturer's service instructions. A small tube of resin is usually used.

There are three ways U-joints are commonly disassembled. The most common method is to use a bench vise. When pressing a universal joint out of its yoke, the bearing cup on one side must be pushed into a socket or pipe that is slightly larger in diameter (**Figure 76.4**). After the cup is removed from one side, the cross is forced against the other cup to force it out of the other side. The cross can be removed at this time. Then the cup is pounded out with a punch (**Figure 76.5**). The process is repeated on the other side of the cross to complete the removal of the U-joint.

NOTE: *One problem with using a vise is that it is weak when opened as far as it takes to accommodate the U-joint. A vise can break if excessive force is used when its jaws are far apart.*

Many shops have a special universal joint tool. It is used in the same manner (**Figure 76.6**). The hole in the opposite side of the tool is larger than the U-joint bearing cup that will be pressed into it during removal. A third method of universal joint removal is to use a press with a special tool.

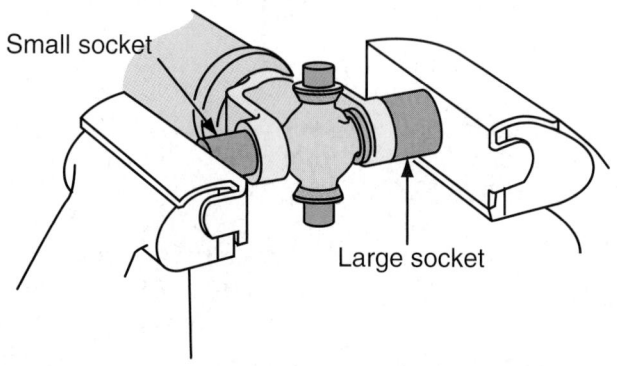

Figure 76.4 Removing a U-joint with sockets and a vise.

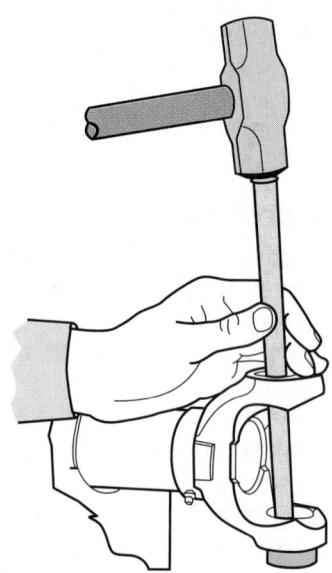

Figure 76.5 Pound out the bearing cup with a punch.

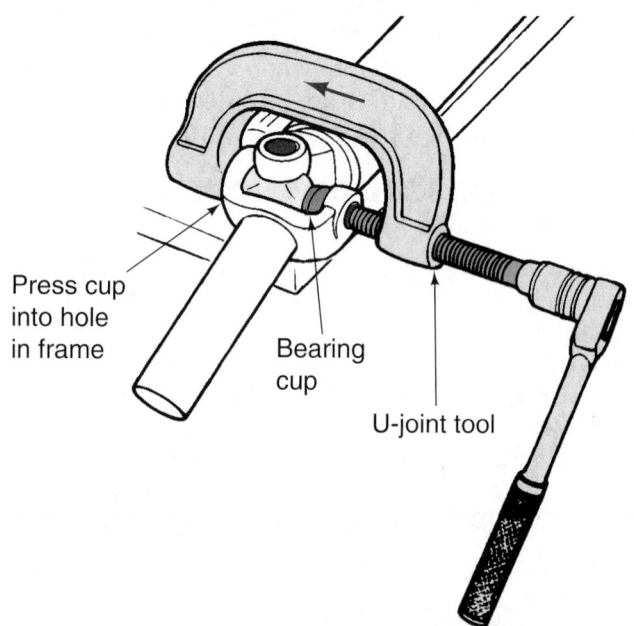

Figure 76.6 A special universal joint tool. *(Courtesy of Ford Motor Company)*

Some shops pound the U-joint out with a hammer. This process works, but you must be careful not to bend a yoke. Even a small amount of misalignment is enough to cause a vibration. Also, if you install the tube part of the drive shaft in a vise during installation, be careful not to damage the tube.

One problem caused during installation in a vise is that drive shaft yoke ears commonly become sprung inward. This results in brinelling. Brinelling is when small indentations wear into the bearing surface (**Figure 76.7**). Brinelling is often the result of a faulty U-joint installation. The joint should always feel loose and relaxed (not binding up) after a correct installation.

Figure 76.7 Brinelling on the trunnion. *(Courtesy of Tim Gilles)*

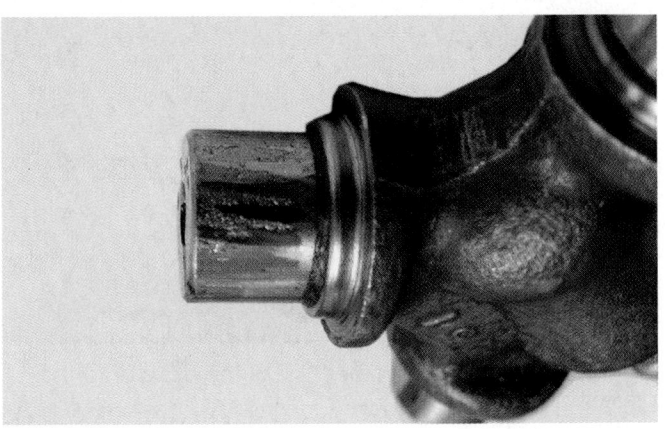

Figure 76.8 Galling on the trunnion. *(Courtesy of Tim Gilles)*

A problem with drive shaft angles can cause **galling**, which often happens on the ends of the trunnions. Usually galling will be found on a trunnion 180 degrees to brinelling (**Figure 76.8**).

■ UNIVERSAL JOINT REASSEMBLY

Replacement U-joints usually have a zerk fitting. If so, it will probably be slanted in one direction (**Figure 76.9**).

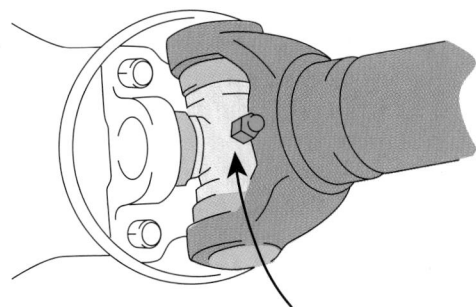

Figure 76.9 A correctly installed U-joint has the zerk fitting slanted toward the drive shaft.

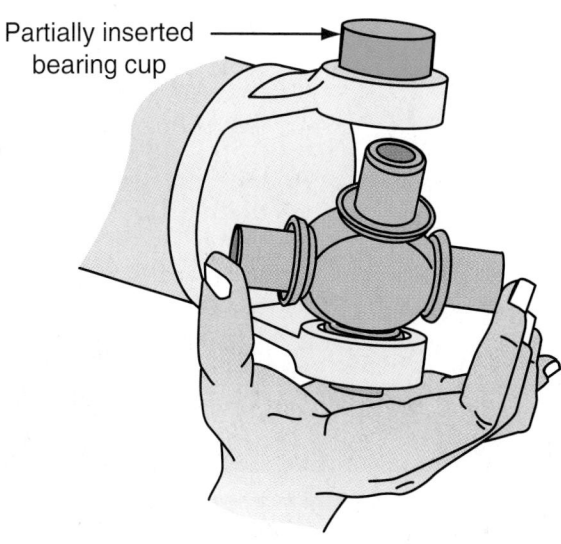

Partially inserted bearing cup

Figure 76.10 Put a bearing cup partially into the yoke.

SHOP TIP When installing the cross into the yoke be sure that the zerk fitting is angled inward toward the drive shaft. Installing the cross with the zerk fitting backwards will make it difficult, if not impossible, to lube the U-joint after the drive shaft is installed on the vehicle.

The grease used in universal joints is an NLGI #1, with an extreme pressure (EP) additive. If the U-joint doesn't have a zerk fitting for lubrication, be sure to check to see that it has been packed with lubricant before you install it in the drive shaft. Too much lubricant can bind it up, making installation difficult.

Put a bearing cup partially into the yoke (**Figure 76.10**). Put the U-joint into the cup and compress the cup into the yoke (**Figure 76.11a**). *Carefully* install the other cup in the opposite side of the yoke. Be sure that you do not accidentally knock one of the needle bearings out of position. A very common occurrence is for one bearing to be knocked into the bottom of the cup. When the cups are pressed together all the way, the joint locks up. When you disassemble it to fix the problem, you might find a damaged needle bearing. This would require the purchase of a new U-joint.

Move the joint back and forth to see that it is free and nonbinding in both cups (**Figure 76.11b**). Place the yoke between the jaws of the vise and compress the remaining cup into the yoke (**Figure 76.11c**). While you are pressing it in, move the joint back and forth and watch for binding. Do not force anything.

NOTE: *If the bearing cup turns in the hole in the yoke, the yoke is defective. The cup is hardened and the ear of the yoke is not. A new yoke will need to be installed on the drive shaft by a machine shop.*

Figure 76.11 (a) Put the U-joint into the cup and compress the cup into the yoke. (b) Move the U-joint to see that it is not binding up. (c) Compress the remaining cap into the yoke.

Install one of the snaprings before completing the pressing procedure in the vise. Use a socket that is smaller than the hole that the bearing cup goes into and press the cup until it comes up against the snapring. Then, install the other cup until it is deep enough to install the remaining snapring. Move the joint in each direction to see that it moves freely

Figure 76.12 Striking the yoke realigns the needle bearings in the bearing cup.

Figure 76.13 Install the other half of the cross onto the drive shaft.

without binding. If it is slightly tight, strike the drive shaft yoke with a punch to free it up (**Figure 76.12**). A tight joint will cause severe vibration.

NOTE: *Sometimes, universal joint cups are held in place with plastic material that is injected into a hole in the yoke. Follow manufacturers' instructions when dealing with one of these U-joints.*

Some replacement U-joints use internal snaprings when replacing injection molded plastic retainers (see Figure 76.3).

After two bearing cups are installed in the slip yoke, install the other half of the cross onto the drive shaft (**Figure 76.13**).

■ DRIVE SHAFT INSTALLATION

When installing a drive shaft that bolts to a yoked flange, be sure that the universal joints fit exactly between the tabs on the flange. If the drive shaft is not installed exactly on center, there will be serious vibration. Be sure that all of the contact surfaces are clean before installing the drive shaft. Dirt or a burr on the companion flange where the U-joint cap will fit can cause the shaft to vibrate after installation. With the transmission in neutral, slide the slip yoke into the transmission. Align the marks that you made during removal and install the rear U-joint onto the companion flange. Install the retaining bolts to complete the job.

NOTE: *If a vehicle experiences vibration after the installation of new universal joints, sometimes it can be corrected by removing the drive shaft and reinstalling it, turning it 180 degrees in the companion flange.*

■ TWO-PIECE DRIVE SHAFT SERVICE

On a two-piece drive shaft, the center support bearing sometimes fails. The seized bearing can tear away the rubber mount that supports the outside of it, allowing the bearing to rotate with the drive shaft. This allows the drive shaft to wobble up and down, causing a vibration. The bearing is pressed off and a new one is pressed on. Be sure to press on the inside bearing race so you do not damage the bearing. Some bearings are designed to be installed in one direction only. Installing them backwards will result in damage to the rubber ring on the bearing.

If the two pieces of the drive shaft are separated, be sure that it is reassembled in phase. Assembling the front and rear halves out of phase can cause extreme vibration. If the shafts were not marked before disassembly, use a tape measure or a steel rod to check the alignment of the halves.

■ DIFFERENTIAL AND AXLE DIAGNOSIS AND SERVICE

Differential gears rarely wear out. In fact, they often last the life of the vehicle on the original lubricant. Causes of damage to the gears include moisture and dirt getting into the differential and lubricant leaking out. Abuse by a driver is another prime cause of damage to a differential. Sometimes, a gear ratio change is desired and this is the reason why a differential is disassembled.

■ PROBLEM DIAGNOSIS

When diagnosing a noise or vibration, isolate the problem. Be sure it is not from the transmission. The differential and transmission are connected by the drive shaft. Noises can transmit from one to the other. When two gears mesh, some noise is normal. Also, a van or station wagon, can act like a big, speaker box, amplifying noises.

Road Test

Does the noise change under different driving conditions? There are several driving conditions to be aware of when listening for noises or feeling for vibration. Become familiar with the following terminology:

■ **Drive**—under acceleration, power is on the convex side of the gear tooth.
■ **Cruise**—the car is maintaining its speed.
■ **Coast**—deceleration, power is on the concave side of the gear tooth.
■ **Float**—car speed is slowly dropping.

Noise that resembles a **howl** or whine can be due to adjustment of the ring and pinion or due to worn bearings. Incorrect differential gear adjustment can result in a howl that occurs only under drive or only under coast conditions. Worn bearings will make a constant sound that changes in relation to road speed. Clunking noises can be due to damaged gears or bearings. When a gear is badly damaged, a shudder can sometimes be felt along with the noise.

Noise that happens only during a turn is probably due to a problem with the spider gears. They can become damaged when a wheel is allowed to spin in a puddle and then gets traction. The differential pinions are very small gears. They cannot withstand the punishment of a heavy load that the large ring and pinion can. Remember that the pinion gears (spiders) are only moving during a turn. When they come to an abrupt halt, they can easily lose teeth. Damaged side gears are usually on the side that received the stress.

Other problems related to spider gears include:

■ Pinion gears too tight on the shaft
■ Side gears too tight or too loose in the differential case
■ Excessive backlash between the spider gears

Lubricant Leaks

Gear oil can leak out of the differential at the pinion seals, from the cover, from the banjo housing, or from the axle seals. When axle seals leak, a rear brake could become covered with oil and not work properly. Because rear wheels do not steer, this is not usually an obvious problem to the driver. Because the oil is thrown by the spinning brake drum, the entire brake backing plate might be stained. Oil should fill the axle housing, as shown in Figure 14.21.

SHOP TIP To tell the difference between a brake cylinder leak and an axle seal leak, spray the area with water. Brake fluid will wash off. Gear oil will require solvent to remove.

■ AXLE BEARING DIAGNOSIS

A worn or damaged axle bearing can result in a fluid leak or bearing noise. If a groan is heard that could be due to a wheel bearing, driving the car can sometimes help pinpoint the problem. First, check the tires for damage and be sure they are properly inflated. Find an empty parking lot or deserted road and make slow left and right turns. The inside tire always turns at a lower rpm than the outside tire. The noise from a bad wheel bearing will change pitch as the wheel speeds up and slows down when turning from one side to the other. When the outside wheel has the bad bearing, the noise will become worse because that wheel is turning faster and is loaded more heavily due to weight shift of the vehicle to the outside.

Applying the brakes can also cause the noise level from a bad bearing to become less as the brakes contact the drum or rotor on the axle that has the bad bearing.

After the road test, raise the car on a hoist and try to move the tires and wheels by hand. The only kind of bearing that has any normal movement is the one used with a C-lock axle, which sometimes has end play that can be felt. The other kinds of wheel and axle bearings should have less than 0.005" end play. Up and down movement is not normal. When pushing in and out on the top and bottom of a tire on a wheel with a tapered roller bearing, a small amount of movement is normal. Testing an axle bearing in this manner should result in no movement.

Listening for Noises

With the car on the lift, have a helper run the engine with the transmission engaged. Sometimes the vibration from a bad bearing can be felt by placing your hand on a part close to it. Listen for noise with a stethoscope. Put the end of the tool against the brake backing plates to locate bad axle bearings. Listen below the pinion gear for a bad pinion bearing.

CAUTION Be careful of spinning wheels during this test. Be especially careful that tires do not accidentally touch parts of the lift. This could result in a very serious accident!

If the noise is related to tires, it will stop when the car is lifted. Axle bearing noise will also diminish because the load has been removed.

Bearing Problems

A differential can experience bad carrier bearings or a bad pinion bearing. Carrier bearings and a pinion bearing will make a constant sound because the carrier is always driven at the same speed as the ring and pinion. The sound will usually become louder with an increase in speed.

Limited Slip Problems

Limited slip differentials can make a chattering sound during turns. The recommended fix is to drain the differential and refill it with fluid with a friction modifier. There are three kinds of differential lubricants. Experienced technicians like to put limited slip fluid in all differentials to avoid problems. The lubricant can be tested by mixing some alcohol with it. If it turns blue it is the correct lubricant. If it turns yellow it is the wrong lubricant.

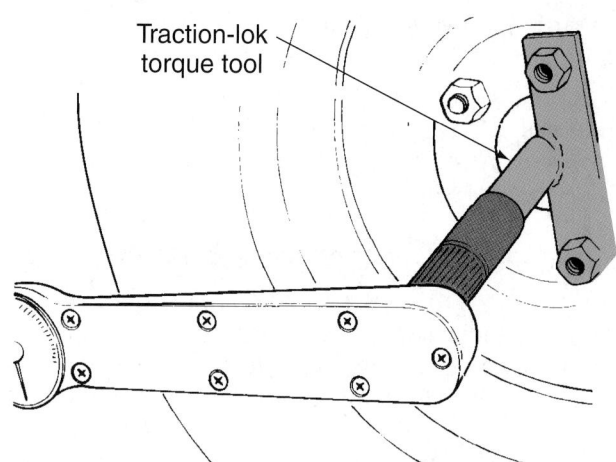

Figure 76.14 Checking a limited slip axle. (*Courtesy of Ford Motor Company*)

A limited slip axle can be tested for **breakaway torque** (**Figure 76.14**). This is the amount of torque needed to make a side gear rotate the clutches. A special tool is used to center the torque wrench on the axle. Typical turning torque on a clutch disc type axle with used clutch plates is 75 foot-pounds. With new discs, torque can range from 155 to 195 foot-pounds. If breakaway torque is too low or too high, there is a problem in the limited slip part of the axle.

▬ AXLE BEARING SERVICE

Non-serviceable rear axle bearings have no service interval. When they fail, replacement is required. The axle must be removed from the housing to replace the bearings. Axle removal is commonly done by service establishments, but the presswork required to install the new bearing on the axle is usually done by parts stores, many of which have machine shops. Axle seals are sometimes part of the axle bearing; at other times they are separate from the axle bearing. Axles are supported on the inside end by the side gears in the differential. Repairing an axle seal requires removal of the axle from the differential.

▬ REMOVING A BEARING-RETAINED AXLE

Recall that axle bearings on vehicles with removable third member differentials are called *bearing-retained axles*.

NOTE: *These differentials are often called "pumpkins" because the outer ribbed structure of their casting resembles a pumpkin.*

The bearing is pressed fit on the axle shaft and fits tightly in the axle housing. There is a retaining plate that must be removed before pulling the axle. Its bolts are accessed through holes in the axle flange (**Figure 76.15**). Use a socket and extension (usually with an air impact wrench) to remove each nut and bolt, then turn the axle to allow access to the next bolt. Some axles have four bolt holes, but most have

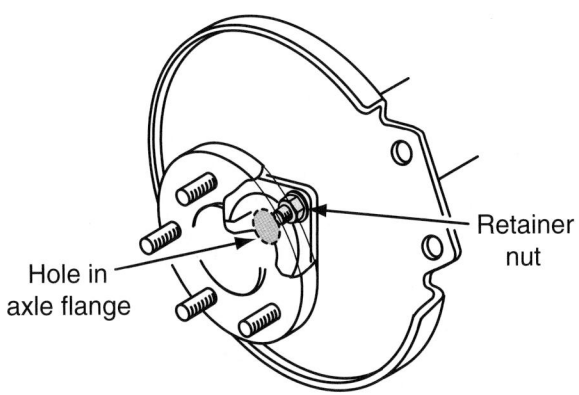

Figure 76.15 A hole in the axle flange allows for removal of the retainer nuts.

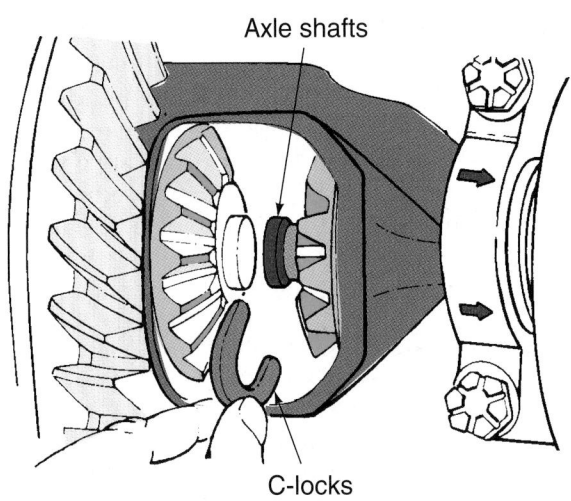

Figure 76.17 When the axle is pushed inward the C-lock becomes free of the inside of the side gear. *(Courtesy of Ford Motor Company)*

Figure 76.16 Removing an axle with a slide hammer.

only one. The outside of the bearing is often tight in the hole in the axle housing. A slide hammer is often required to remove it (**Figure 76.16**).

Removing a C-Lock Axle

The cover on the back of the differential must be removed prior to removal of a C-lock axle. The end of a C-lock axle has a machined groove into which a C-lock fits. The differential pinions must be removed to allow access to the C-locks. The pinion shaft has a lock bolt. After it is removed, the shaft is pulled out. After the pinion lock bolt, shaft, and pinions are removed, the axles can be pushed inward, allowing the C-locks to be removed (**Figure 76.17**).

SHOP TIP ■ If you slide the pinion shaft out and do not turn the axle, the gears will stay in place. ■ Occasionally a pinion shaft lock bolt will break off. There are special tools available for removing these.

On most C-lock axles, the bearing rides on a hardened area of the axle, rather than having an inner bearing race (see Figure 75.40). The outer race and bearing rollers are held in the axle housing. The hardened area of the axle can be damaged where the seal rides on it. Some replacement seals are designed so that the lip will ride in a different area of the axle, providing a new sealing surface.

Axle Seal Service

After the axle is removed, pry out the old seal (**Figure 76.18**). The seal can be pried out with a prybar, a seal puller, or the end of the removed axle. Check the surface of the axle shaft to see that it is clean and undamaged. Apply lubricant to the lip of the new seal before installing it. Use a seal installing tool to avoid cocking it to the side and damaging it. During reinstallation, support the axle and do not let it drag on the surface of the new seal.

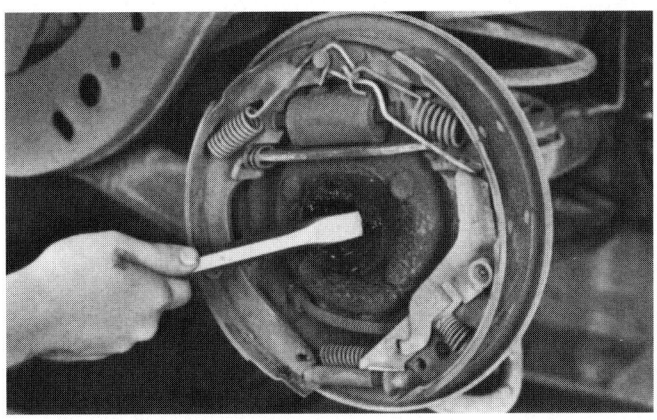

Figure 76.18 Removing the old axle seal.

■ AXLE BEARING REPLACEMENT

Some older imported cars had bearings that were pressed onto the axle after the brake backing plate. When removing the axle on these vehicles, the brake line is first disconnected. The backing plate is unbolted and the still-assembled brakes are removed along with the axle (**Figure 76.19**). A press or a special puller (**Figure 76.20**) can be used to remove the bearing. Other types of axles with pressed-fit bearings come out independent of the backing plate (**Figure 76.21**).

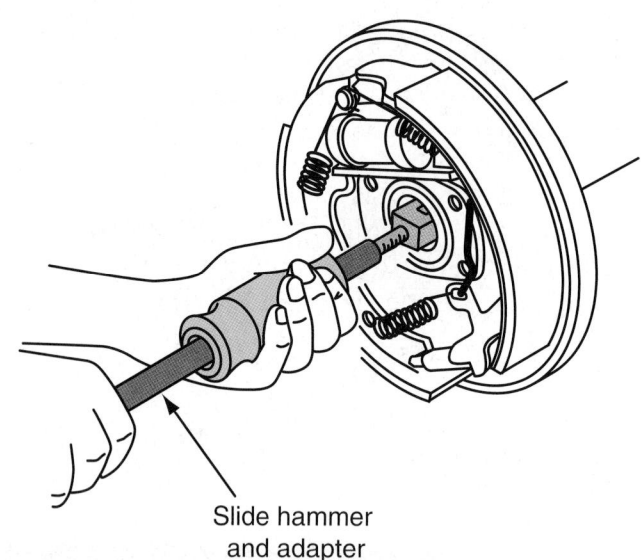

Slide hammer
and adapter

Figure 76.22 Removing an axle bearing with a slide hammer.

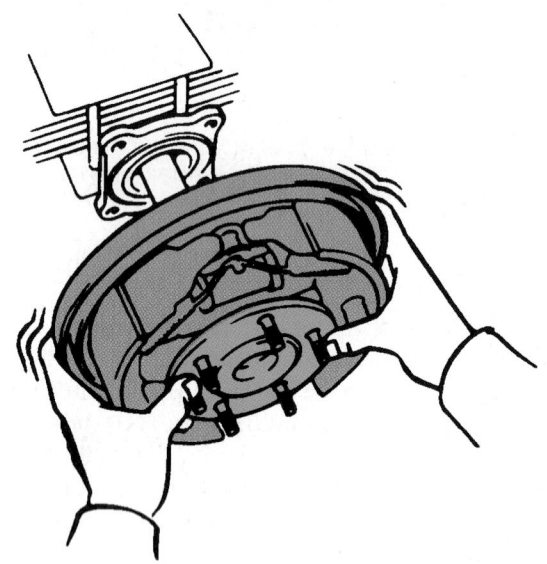

Figure 76.19 Some axles must be pulled with the brake backing plate.

On C-lock type axles, the bearing remains in the housing after the axle is removed. The bearing is pulled with a slide hammer and special attachment (**Figure 76.22**).

Pressed-Fit Bearing Replacement

Do not try to press the bearing and retaining ring off at the same time. Grind a notch on the retaining ring and strike it with a chisel to weaken it prior to pressing it off (**Figure 76.23**). Presses usually have a large

Puller

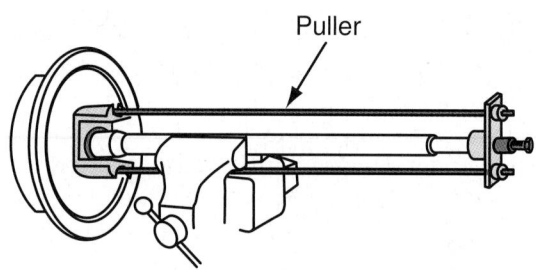

Figure 76.20 A puller setup for removing an axle bearing.

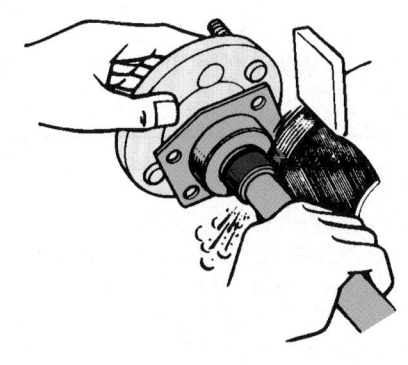

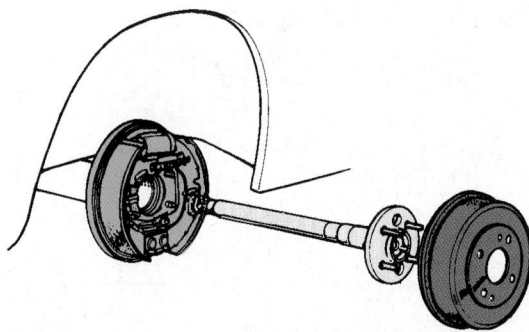

Figure 76.21 Some axles are removed without the backing plate.

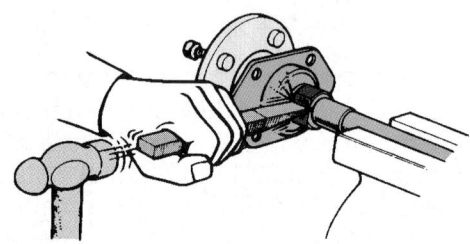

Figure 76.23 Removing a bearing retaining ring.

safe cage made of large-diameter steel pipe that can be installed around the bearing when pressing it off. It is not unusual for a bearing to require in excess of 20 tons of pressure to remove it.

CAUTION If a bearing explodes during removal (and they quite often do), this can be comparable to a grenade exploding!

If the gauge on the press reads over 15 tons, something could be wrong. Be certain that you are pressing parts apart and not pressing against something solid.

■ AXLE BEARING INSTALLATION

When pressing a bearing onto the axle it is supported by the inner race (**Figure 76.24**). The cage and rollers should always be able to be turned during the installation process. This ensures that there is no load on them. First, the bearing is installed on the axle. The retaining ring is often heated (150°C/300°F) for easier assembly. Then, the retaining ring is pressed into place against the axle bearing (**Figure 76.25**).

CAUTION Do not attempt to press the bearing and the retaining ring onto the axle at the same time.

■ REINSTALL THE AXLE

Reinstall the axle in the housing and bolt on the retainer. Be sure that the oil return slot in the retainer

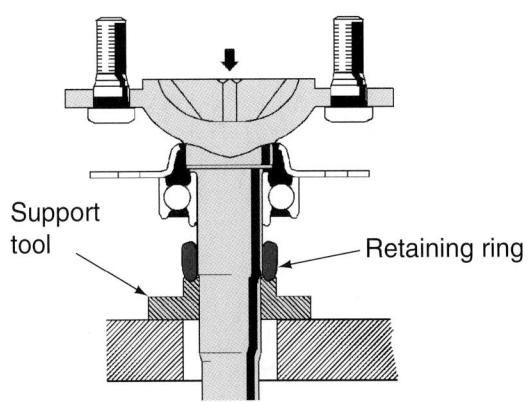

Figure 76.25 Pressing the retaining ring into place against the axle bearing.

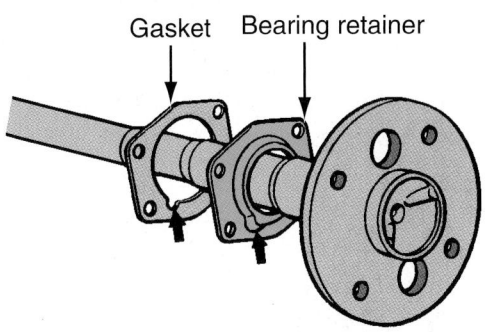

Figure 76.26 Be sure that the oil return slot in the retainer lines up with the return hole in the gasket and the axle housing.

lines up with the return hole in the axle housing (**Figure 76.26**). This is similar to the oil return on the clutch bearing retainer on a manual transmission.

■ FULL FLOATING AXLE SERVICE

Full floating axles are found on trucks and vans that are ¾ ton and larger. Axle bearings are located in the hub

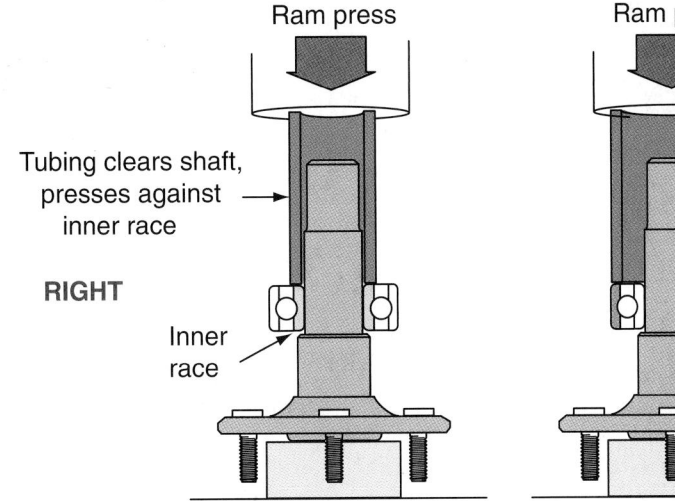
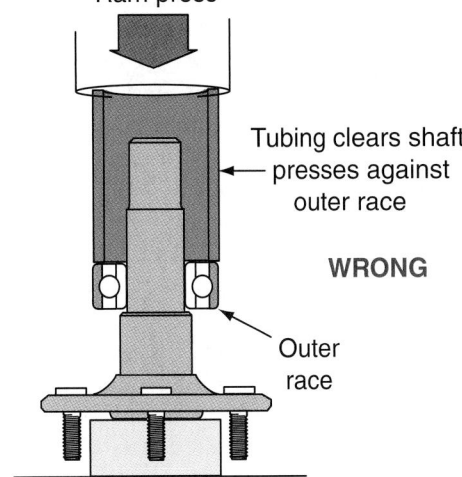

Figure 76.24 Press bearings off by putting pressure on the inner race.

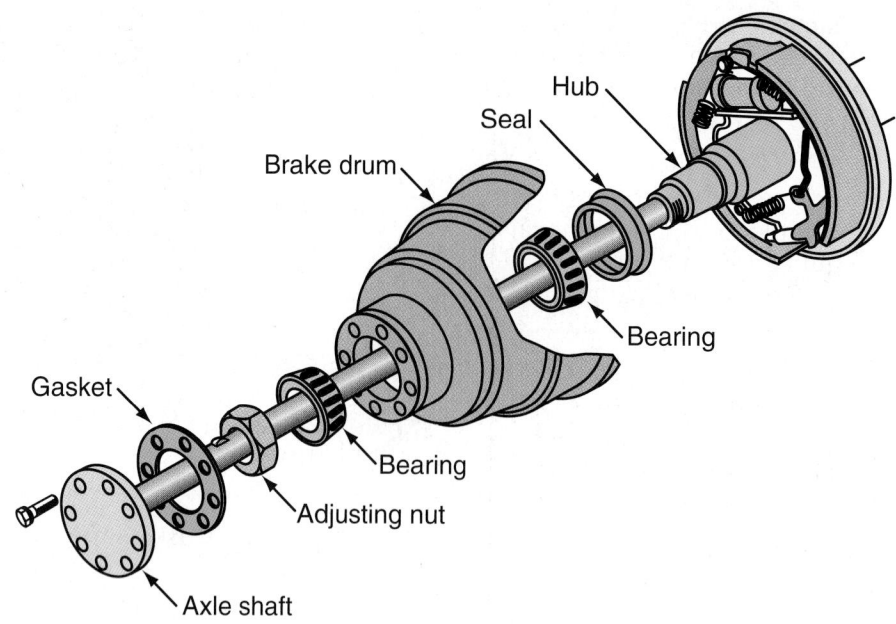

Figure 76.27 Typical full floating axle and hub assembly. The bearings ride on the hub, not on the axle.

and brake drum assembly (**Figure 76.27**). To remove the drum on a full floating system, the axle must be removed first (**Figure 76.28**). Often, there are soft metal beveled washers that surround the studs that hold the axle to the hub. Remove them by striking the outside of the axle flange with a brass hammer. Do not force a chisel in between the axle and hub or you could create a future seal leak.

After the axle is removed, the locknut that holds the brake drum can be removed. There are different types of locknut arrangements. A typical locking arrangement has two nuts, the outer nut being a locknut. Lift off the brake drum. Catch the outer bearing so it does not fall on the ground.

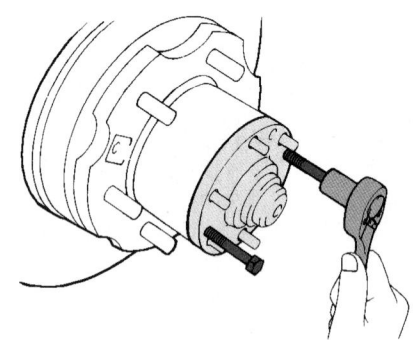

Figure 76.28 Removing a full floating axle.

 *An apprentice did a brake job on a ¾-ton pickup truck. When he reinstalled the rear drum and hub, he didn't pay attention to how the locking nut was supposed to work and assembled it improperly. The owner was driving his truck later that day and the axle, hub, and wheel came off during a turn. The result was that expensive repairs were required on the truck. Luckily, there was no accident and no one was hurt.*

Install a new inner bearing seal. During reassembly, the hub and drum are reinstalled on the outside of the axle housing. The retaining nut is tightened until it has little or no clearance.

After the hub is correctly reinstalled, install the axle and beveled washers. Pound the beveled washers in until the whole axle is seated (**Figure 76.29**).

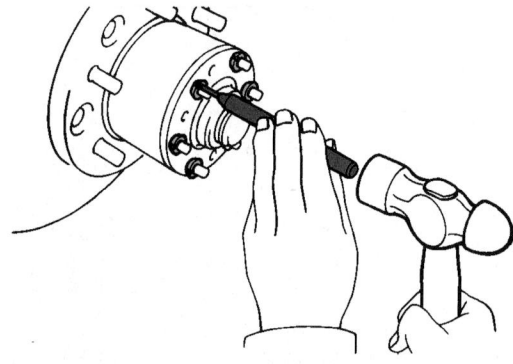

Figure 76.29 Installing the beveled washers.

■ DIFFERENTIAL PINION SEAL REPLACEMENT

A common repair to a differential is the replacement of the pinion seal. Parts involved in a typical pinion seal replacement are shown in **Figure 76.30**. A very important part of this job is to maintain tension on the pinion bearing crush sleeve. One way of maintaining

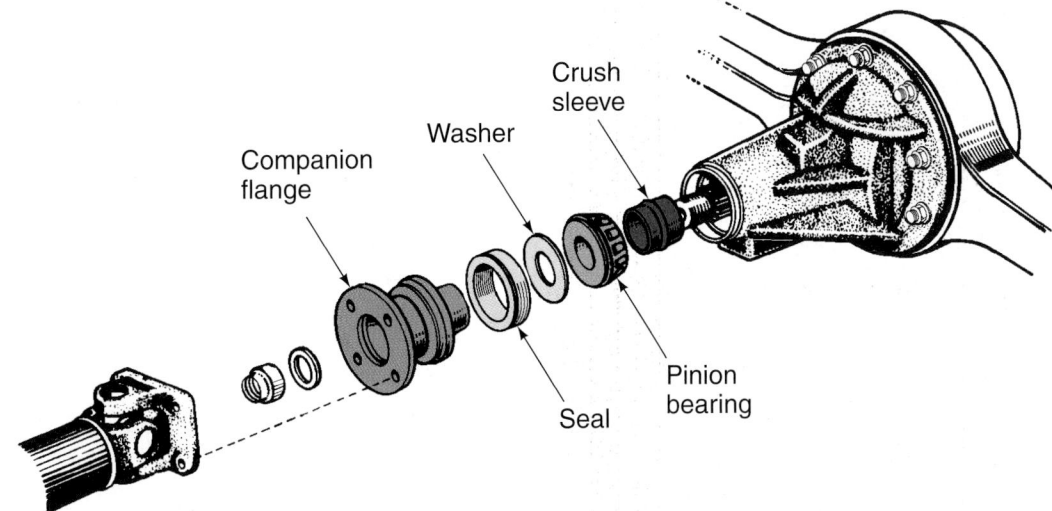

Figure 76.30 Parts involved in a pinion seal replacement.

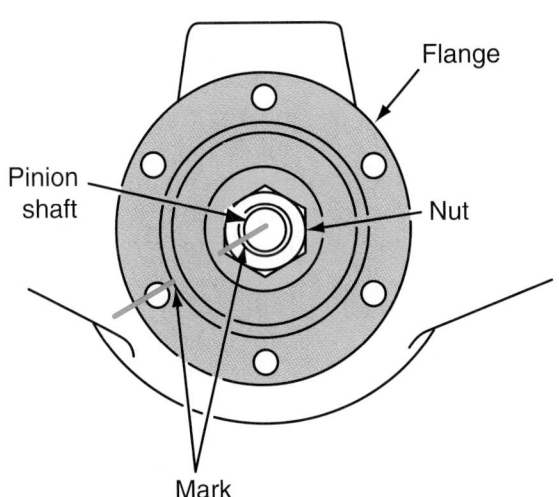

Figure 76.31 Mark the flange, nut, and pinion shaft before disassembly.

crush sleeve tension is to mark the nut and pinion shaft before disassembly (**Figure 76.31**). After reassembly, tighten an additional ⅛ turn past the mark.

Another method of maintaining pinion bearing crush sleeve tension is to use a dial indicator torque wrench when removing the nut. Note the torque

required to loosen it. During reinstallation of the nut, add 5 pounds to the former torque to recompress the old crush sleeve.

NOTE: *When a bearing fails because it was too tight, the small end of the bearing will experience wear first.*

CASE HISTORY

*A student's mother's car had a leaking pinion seal. The student removed the drive shaft and pinion flange. He used a hammer to remove the flange. Then he replaced the seal and reassembled all of the parts. He tightened the pinion nut securely, without a torque wrench. Within 3000 miles, the pinion bearing failed because the bearing adjustment was too tight (**Figure 76.32**).*

Always use a long bar to hold the yoke from turning while loosening the pinion nut (**Figure 76.33**).

Figure 76.32 This pinion bearing failed because it was too tight. The small ends of the rollers failed first. *(Courtesy of Quaker State Corporation)*

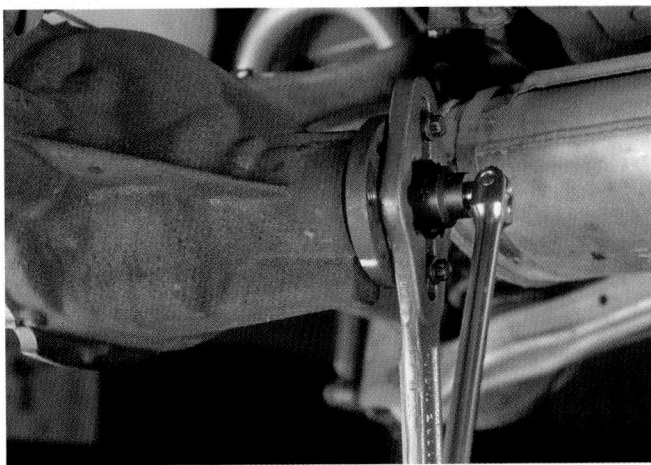

Figure 76.33 Use a bar to restrain the flange while loosening the pinion nut.

Figure 76.34 Use a puller to remove the flange.

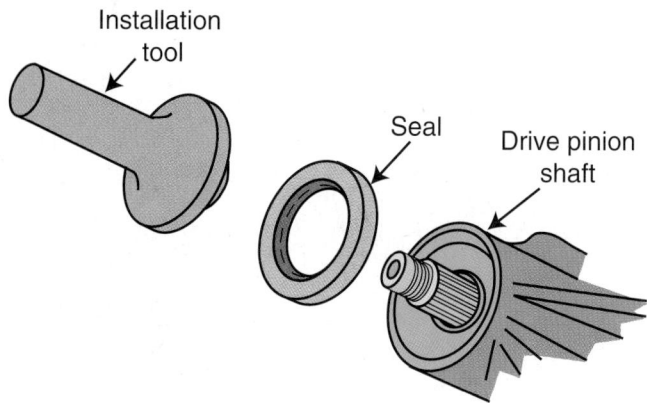

Figure 76.36 To avoid distorting the seal, install it using a seal installer.

Using an impact wrench can damage the ring and pinion or the pinion bearings.

Use a puller to remove the flange (**Figure 76.34**). Do not use a hammer. A flange can be easily bent, which will cause a driveline vibration. The student's car in the case history always had a driveline vibration after the pinion seal job because he bent the flange.

Remove the seal with a hammer and chisel (**Figure 76.35**). Install the new seal with a seal installer to avoid cocking or damaging it (**Figure 76.36**). Lube the lip of the new seal and the mating surface on the companion flange before reinstalling it.

■ DIFFERENTIAL REPAIR

Identify the Differential

Look for an identification tag somewhere on the differential. It is usually located under one of the nuts holding the third member or one of the bolts holding the cover.

■ REMOVING A THIRD MEMBER

To remove a separable third member from a banjo housing, first pull the axles. If there is a drain plug, drain the housing of gear oil. Then remove all of the nuts from the studs around the outside of the third member. Each stud will usually have a copper washer under its nut. These are sometimes hard to see because of an accumulation of grease. Use a sharp scraper to remove each of these washers. If they are not removed, the third member will not come out of the housing after the nuts are removed.

A gasket will occasionally hold the third member to the housing if a previous technician has glued it in place on both of its sides. Usually the third member can be easily removed. Be careful not to drop it. It is heavier on one side than the other and must be carefully supported when lifting it away from the housing. Use a hydraulic jack if the vehicle is raised on jackstands (**Figure 76.37**).

■ DISASSEMBLING A SALISBURY AXLE

To remove a carrier from a Salisbury axle requires removal of the cover. Be sure to have a drain pan under it to catch the gear oil as it flows from the housing. Remove

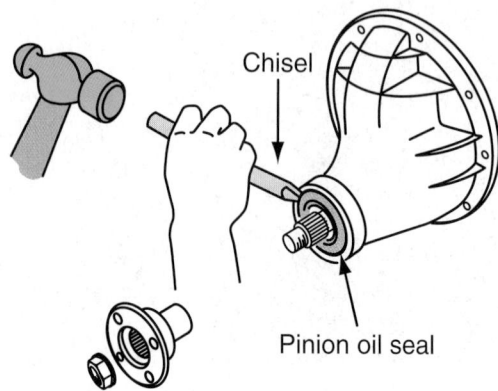

Figure 76.35 Removing a pinion seal.

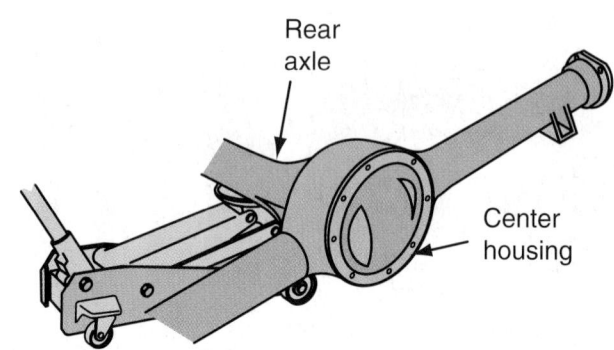

Figure 76.37 Use a jack to help lift out a differential.

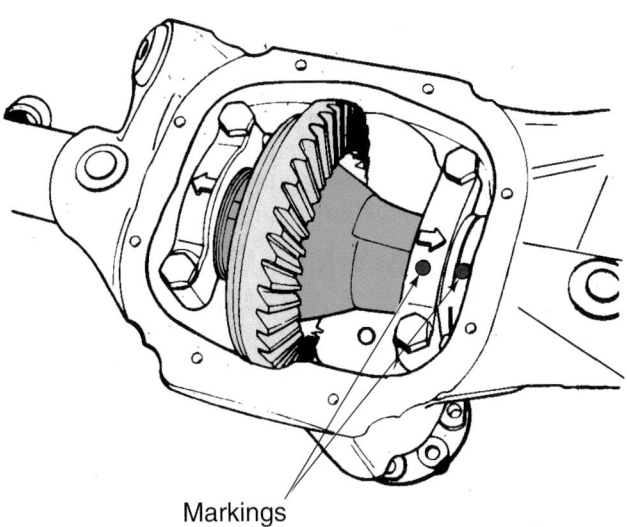

Markings

Figure 76.38 Mark the side bearing caps before removing them. *(Courtesy of Ford Motor Company)*

the C-locks and the axles as described earlier. Mark the carrier housing caps if they are not already marked.

Mark the side bearing caps before removing them (**Figure 76.38**). After removing the bolts, use the

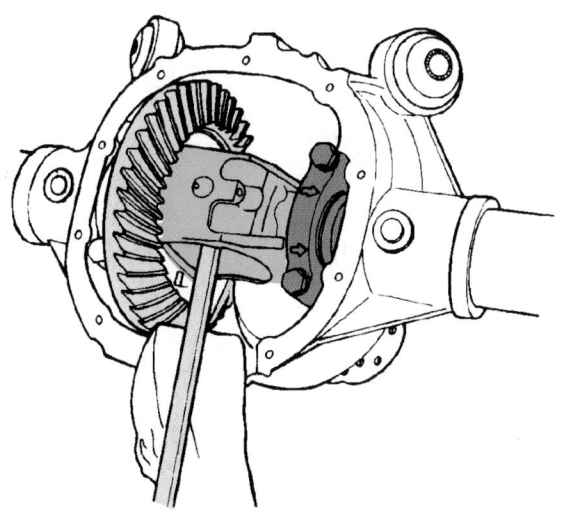

Figure 76.39 Removing an integral carrier from the housing. *(Courtesy of Ford Motor Company)*

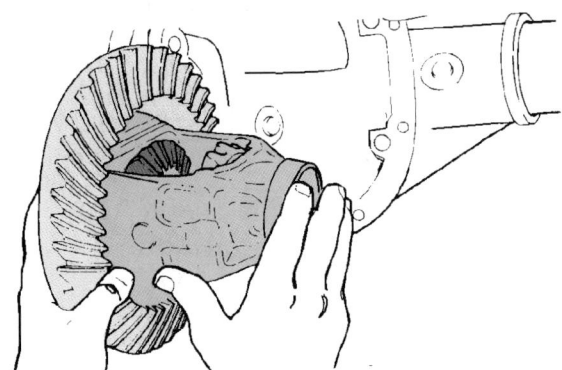

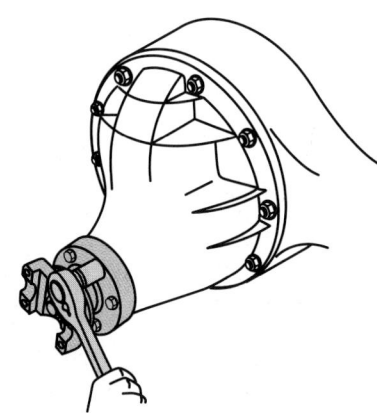

Figure 76.40 This differential has a removable pinion housing.

pointed end of a roll head prybar in the bolt holes to pry the caps off. Pry back and forth, first on one side and then the other. Once the caps are loose, pry against a solid part of the carrier to remove it from the axle housing (**Figure 76.39**). Use wire to attach any shims to the caps so they can be returned to their original positions.

On some types of differentials, the drive pinion gear can be removed by unbolting it from the housing (**Figure 76.40**). On other types, the pinion nut and flange are removed, as described earlier. To loosen a pinion nut, use a holding bar, as described earlier, or clamp the yoke in a vise if working on a replaceable third member or a replaceable pinion.

■ CLEAN AND INSPECT PARTS

As you disassemble a differential, look for signs of wear or misalignment on bearing caps. Bearings can also fail due to lack of lubricant, use of the wrong lubricant, or an incorrect bearing adjustment.

Be sure to keep bearing caps with their bearings in case they will be reused. Use only a puller to remove bearings that are worn or damaged. Good bearings can remain in place on the case or pinion gear. When removing a carrier bearing from the case, be sure to use the correct pullers to avoid damage. During reinstallation, be sure the case bearings are seated all the way against the shoulder on the case.

Inspect the ring and pinion gears for damaged teeth. Check the case to see that the lubricant passageway to the pinion bearings is clear (**Figure 76.41**). If gears are worn, adjusting the ring and pinion clearances will not solve the problem. The gearset will have to be replaced.

■ DIFFERENTIAL REASSEMBLY

A ring and pinion must be replaced as a matched set. If one gear is damaged, they must both be replaced. Before reassembling any components in the differential, clean all parts in solvent and blow them dry. Remember to blow solvent back into the solvent tank instead of onto

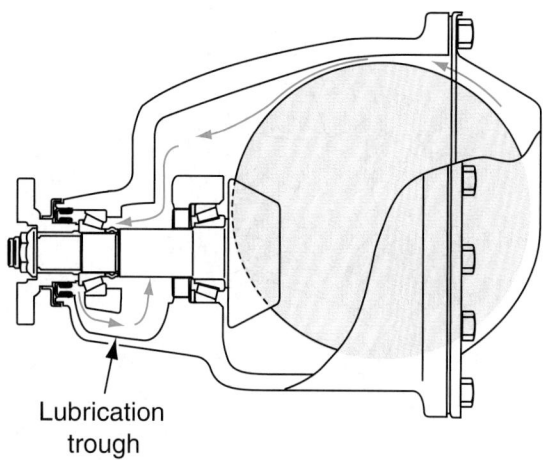

Figure 76.41 Be sure that the passageway that provides lubrication to the pinion bearings is clear.

the floor or a workbench. Coat all parts with differential oil prior to reassembly. Install the side gears and pinions in the case and torque all of the case bolts.

If a ring gear is being replaced, heat it in oil to make it easier to install on the case. It must be installed perfectly flat. Lightly run a file across the top of the bolt holes to remove any burred edges. Pilot studs made from cutoff screws with a hacksaw slot for a screwdriver are a good idea (**Figure 76.42**). Be sure to torque the ring gear retaining screws after installation (**Figure 76.43**).

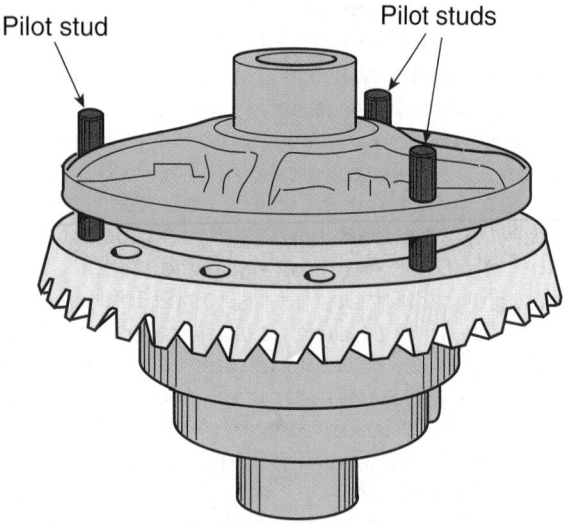

Figure 76.42 Pilot studs assist in installation of a ring gear.

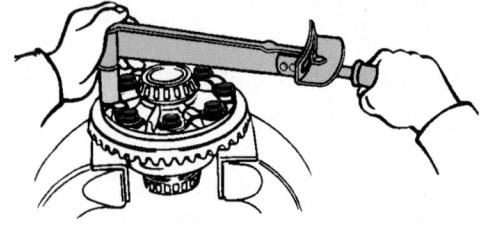

Figure 76.43 Torque the ring gear screws.

ADJUSTING A DIFFERENTIAL

Ring and Pinion Adjustments

A ring and pinion must mesh with each other in the correct position or noise and rapid wear can result. The two adjustments that are made are *pinion gear depth* and *ring and pinion backlash* clearance.

PINION GEAR DEPTH

Pinion gear depth is a measurement of how far the pinion gear extends into the differential housing. The measurement can vary according to where the pinion bearing race is positioned in the case. During manufacture, if more was machined from the case or if the ledge on the pinion gear had more or less metal on it, pinion depth will vary. This adjustment requires a special tool when starting from scratch without existing gears (**Figure 76.44**).

If a gearset is produced that will result in a depth that is different than stock, there will be a number on it telling how much compensation to make when setting up the clearance between the two (**Figure 76.45**). A +2 on the pinion gear means that the shoulder for

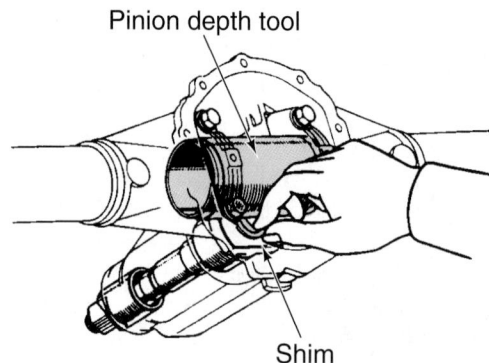

Figure 76.44 A special tool is used to adjust pinion depth.

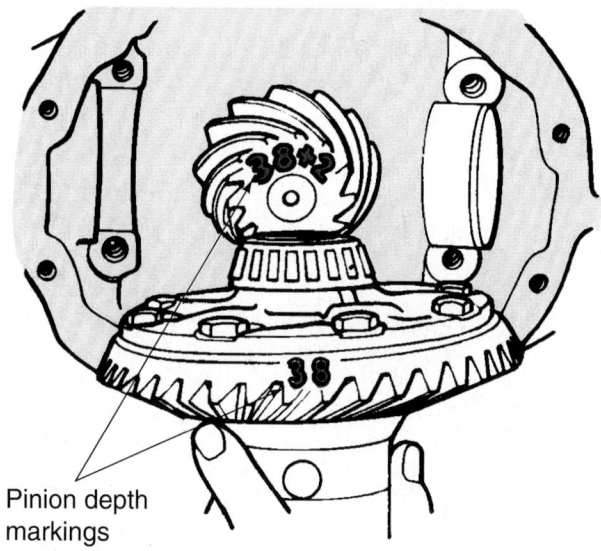

Figure 76.45 Pinion depth shim mark. *(Courtesy of DaimlerChrysler Corporation)*

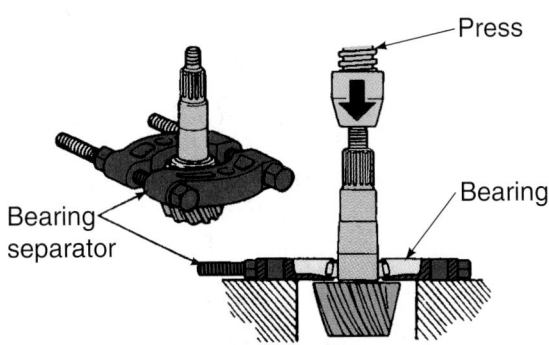

Figure 76.46 Removing a pinion bearing.

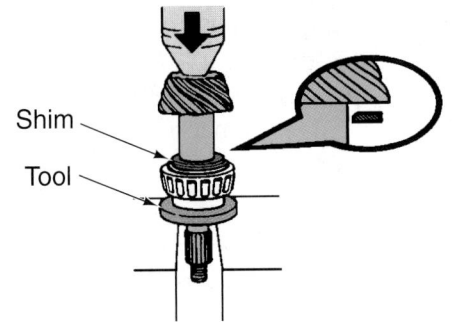

Figure 76.47 Installing a pinion bearing.

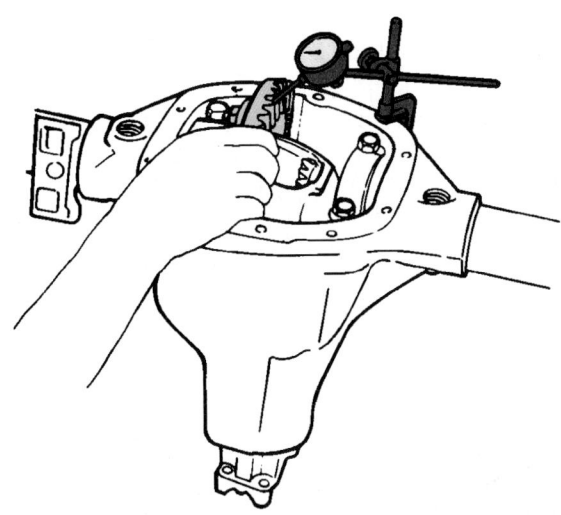

Figure 76.48 Checking ring gear backlash. *(Courtesy of DaimlerChrysler Corporation)*

the shims should be 0.002" too large for stock. This calls for a shim that is 0.002" less than the shim that was on the old gear (if that gear had a "0" mark).

Shims vary in thickness. Use a micrometer to double-check the size of labeled shims. If a pinion depth gauge is not available, try checking the gear tooth pattern (covered later) while using a shim of half the thickness of the maximum shim specification. Tighten the pinion nut to 25 foot-pounds for the preliminary test so you do not ruin the crush sleeve.

To remove a pinion bearing, use a bearing separator and press (**Figure 76.46**). When pressing the shaft back into the bearing, use a tool or piece of pipe that is larger than the diameter of the pinion shaft (**Figure 76.47**). Do not forget to install the pinion depth shim under the bearing.

RING GEAR BACKLASH

Check gear backlash with a dial indicator (**Figure 76.48**). Backlash specifications vary by manufacturer. Try to adjust backlash to the middle of the specifications. A setting that will work on any hypoid gearset is 0.007". Too little backlash can result in a broken gearset after expansion takes place. Also, the lubricant will not have enough room to be able to separate the metal of the gear teeth, so they might become burned.

To check runout of the ring and pinion, read the backlash at several points around the gear. The maximum variance should be no more than 0.002" to 0.003".

SIDE BEARING PRELOAD

Case side bearings are adjusted to a specified preload. If bearings are too tight, they can fail. If they are too loose, the ring gear can move in the case, causing noise and wear. A good indication of preload is that your little finger should strain when trying to turn the ring gear.

On a Salisbury axle, preload and backlash adjustments are made with shims (**Figure 76.49**). Hold the ring gear away from the pinion and measure the clearance between the bearing and the case with feeler gauges (**Figure 76.50**). Adding a 0.010" shim will change the backlash by about 0.005". Add an additional 0.004" thickness to the shim pack on each side of the gear (0.008" total) to preload the bearings.

Shims are installed by pounding them in, either with a punch or with a special tool. Some shops have a case spreader that can be used instead (**Figure 76.51**). A dial indicator is used with most spreaders. Be sure not to spread the case more than 0.020".

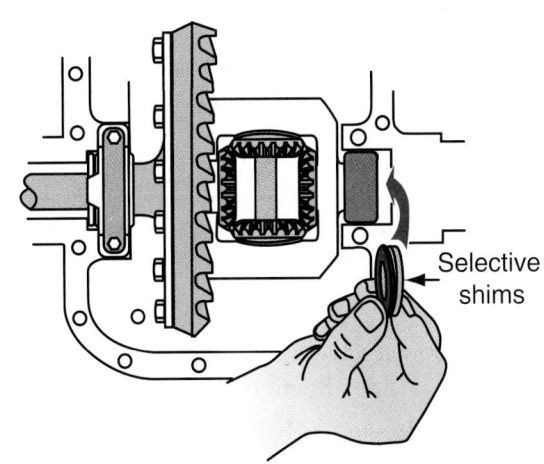

Figure 76.49 Preload and backlash settings made with shims.

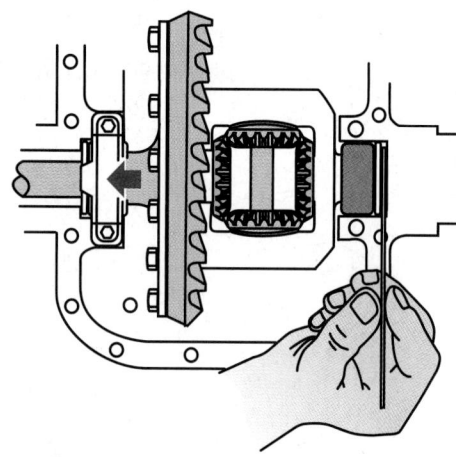

Figure 76.50 Measure clearance before determining the shim size.

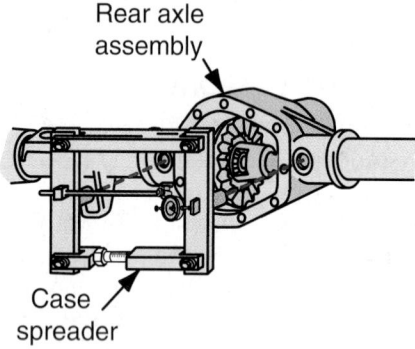

Rear axle assembly

Case spreader

Figure 76.51 A case spreader.

NOTE: *A case spreader is one of the more expensive tools that a service manual will list for repair of a component. In the real world, many dealerships and independent shops do not purchase many of the more costly tools. The job can be done without this tool, but it is more time consuming to perform.*

In a separable third member, the side bearing adjustment is made in combination with the backlash adjustment (**Figure 76.52**). Adjusting nuts with holes in them are turned with a spanner wrench (**Figure 76.53**). With the caps torqued to 25 foot-pounds, the adjusting nuts should be easy to turn in their threads. Be sure to keep the adjusting nut in its original bearing cup.

■ Tighten the adjusters to move the ring gear against the pinion until all backlash clearance is removed.
■ Tighten the adjusting nut on the side facing the ring gear teeth until the correct amount of backlash is attained.
■ Tighten each side an additional one to two notches to preload the side bearings. Check the repair manual for the specification.

■ **CONTACT PATTERN**

When a differential is adjusted, a **gear pattern** is taken off of the ring gear teeth. A coating of colored

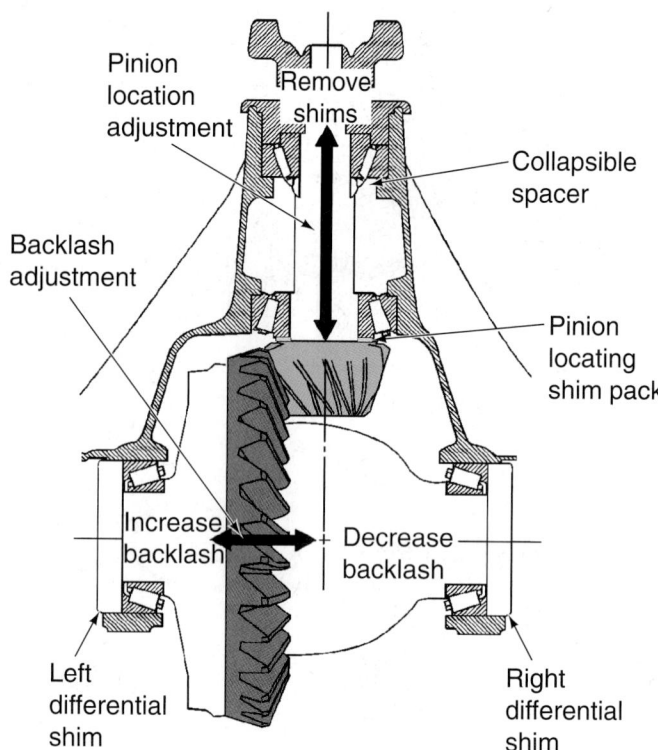

Figure 76.52 Adjustments to the ring and pinion gear positions. *(Courtesy of Ford Motor Company)*

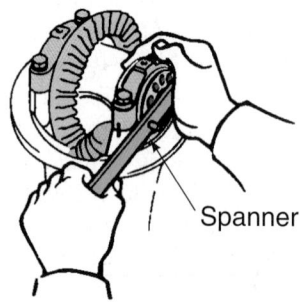

Spanner

Figure 76.53 Adjusting the side bearing preload with a spanner.

paste is painted onto the gear teeth. Manufacturers sell gear marking compound in different colors. Yellow is popular because the pattern is easier to see. A desirable pattern is located near the center of the gear teeth on both the drive and coast sides.

Gear Tooth Terms

The **drive side** of the gear teeth is convex and the **coast side** is concave (see Figure 75.32). The **heel** of a gear tooth is the outer end and the **toe** is the inner end (**Figure 76.54**). A **pitch line** runs through the center of the tooth from heel to toe. Other terms used when analyzing gear tooth patterns are **face** and **flank**. The face, or *top*, of the tooth is the area above the pitch line. The flank, or *root*, is the area below the pitch line.

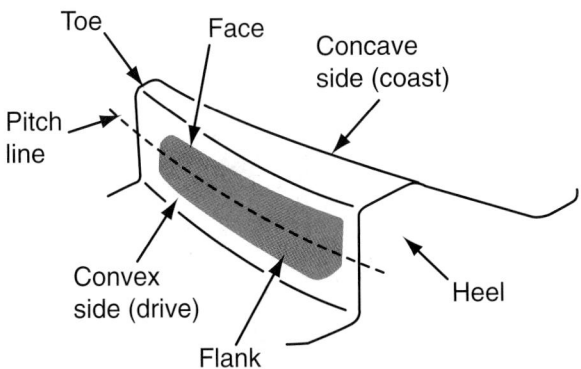

Figure 76.54 Gear tooth terms.

Making a Pattern

A good pattern will tend to be toward the toe of the tooth when tested under light load conditions (**Figure 76.55**). Under heavier load a more complete pattern can be read across the gear tooth. The pattern is made on the desired side of the ring gear teeth by rotating the ring gear while holding the pinion gear lightly (**Figure 76.56**). Wrapping a rag around the yoke is a good way to put resistance on the pinion gear. Another method is to put a prybar between the side of the ring gear and the differential housing. Install a ½" ratchet and socket on the pinion gear nut to drive it.

Changing backlash or pinion depth changes both patterns.

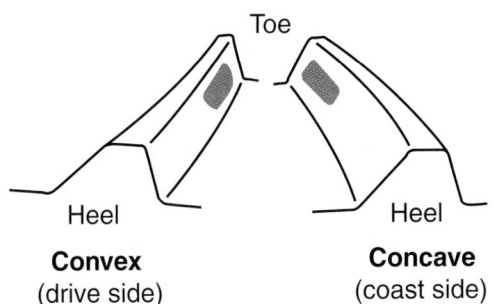

Figure 76.55 A pattern will be toward the toe of the gear tooth under light load conditions.

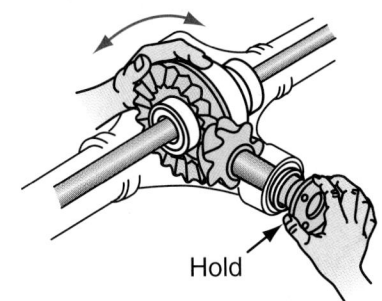

Figure 76.56 Hold resistance against one of the gears while rotating the other in both directions.

■ BACKLASH PATTERN CHANGE

When the drive pinion is correctly positioned, an increase in backlash causes the pattern on both sides of the gear tooth to move higher on the face of the tooth (**Figure 76.57a**). Moving the ring gear closer to the pinion will reduce backlash clearance and move the pattern toward the toe and flank on both sides of the gear tooth.

■ PINION DEPTH PATTERN CHANGE

Altering the pinion depth results in a different pattern change than changing the backlash clearance. This is

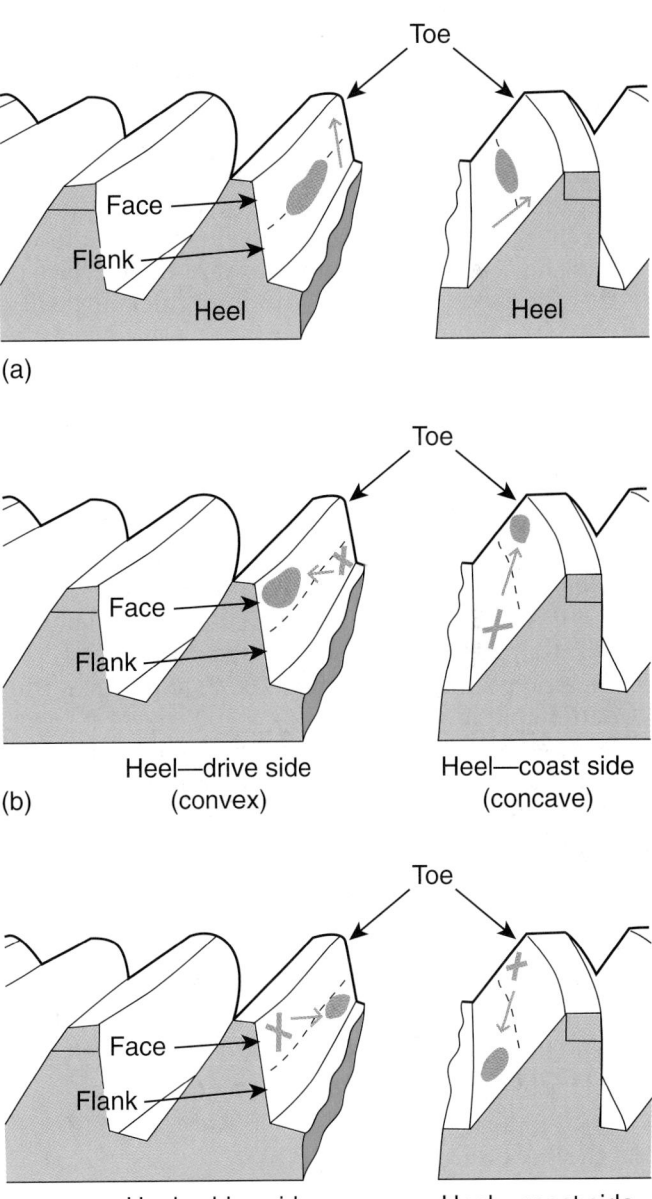

Figure 76.57 (a) Increasing backlash causes the pattern to rise on the gear tooth. (b) Moving the pinion out of mesh with the ring gear (too thin a shim) results in this pattern change. (c) Moving the pinion into mesh with the ring gear (too thick a shim) results in this pattern change.

due to the shape of the helical gear tooth. When the patterns are at opposite ends of the tooth on the drive and coast sides, this is a pinion depth problem.

Here is a typical scenario when moving the pinion gear on today's passenger cars and trucks, most of which have integral axles:

■ On the drive side of the gear tooth, pulling the pinion gear away from the ring gear (toward the heel of the gear tooth) moves the pattern from the toe flank to the heel face (**Figure 76.57b**). The coast side pattern moves the opposite way.

■ When the pinion is moved out of mesh, backlash increases and the pattern moves up the gear tooth.

With the backlash set in the correct range of 0.006–0.008", the following would be conditions and corrections to make with pinion depth:

■ When the shim is too thin, the drive pinion will be moved out of mesh (too far away) from the centerline of the ring gear (see Figure 76.52). The pattern on the drive side (convex side of the gear tooth) will move toward the heel/face of the tooth and the coast side will move toward the toe/face of the gear tooth. Increasing the shim thickness will move the pinion gear deeper into mesh toward the toe of the ring gear. This will cause the pattern to move back toward the center of the gear faces (see Figure 76.57b).

■ When the shim is too thick, the drive pinion is too close to the center of the ring gear tooth. The resulting pattern will have moved from heel/face to toe/flank on the drive side of the tooth and from toe/face to heel/flank on the coast side of the tooth (**Figure 76.57c**). Decreasing the thickness of the shim moves the drive pinion away from the ring gear (refer to Figure 76.52) and moves the pattern closer to center.

When changing pinion depth, leave the pinion seal out until the correct depth has been established. Then install the seal and the crush sleeve, and complete the assembly procedure.

NOTE: *Some larger passenger cars and trucks have a removable housing in which the pinion gear and its supporting bearings are mounted. When changing pinion depth in these axles, adding and removing shims results in a change in the pattern opposite of the norm. Pulling the pinion gear away from the ring gear moves the drive side pattern reading toward the toe instead of the heel.*

■ PINION BEARING PRELOAD

The pinion bearings must be *preloaded* because of the tremendous amount of force on the gears during acceleration and deceleration. If the drive pinion gear were allowed to move, clearance between it and the ring gear would change. Use a collapsible spacer to preload the bearings (see Figure 75.20). Check the preload by seeing how difficult it is to turn the drive pinion in its

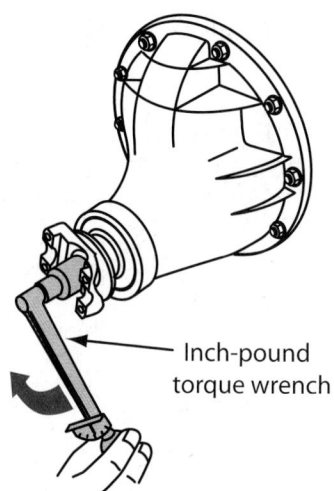

Figure 76.58 Checking pinion bearing preload.

bearings. The more the nut on the pinion shaft is tightened, the more preload there will be on the bearings. The nut is tightened in small increments while checking repeatedly for drag with an inch-pound torque wrench (**Figure 76.58**). A dial-type indicator is best. Do not overtighten, because you cannot back off the adjustment. Once the crush sleeve is compressed, you will have to replace it.

Check for the amount of effort required to turn the pinion nut. A typical reading is 15 to 25 inch-pounds—not very much. Do not take the reading right at the start of turning, but during turning. Be sure the bearings have been lubricated first if you want an accurate reading. Pound lightly on both ends of the pinion shaft to seat the bearings.

When reusing pinion bearings, the preload is set to one-half the new preload specification. Always replace the crush sleeve when a differential is disassembled.

NOTE: *A new crush sleeve is difficult to collapse at first. Start to collapse it with an impact wrench and then use the bar and a breaker bar (see Figure 76.33).*

Some differentials have solid spacers rather than collapsible ones. These use shims to adjust the preload. The torque is adjusted to a specified amount.

■ RING AND PINION NOISE

If the pattern is good but there is a gear howl, replace the ring and pinion. Incorrect backlash can be isolated during a test drive. Noise during acceleration points to heavy heel contact. The ring gear needs to be moved closer to the pinion. The pattern is usually set slightly toward the toe because during acceleration it will move toward the heel.

If the noise happens when coasting in gear, the pattern will be too heavy on the toe of the gear tooth. Move the ring gear away from the pinion. Noise does not usually occur during float.

■ FOUR-WHEEL-DRIVE SERVICE AND REPAIR

Four-wheel-drive (4WD) components are the same as two-wheel-drive components, with the exception of the transfer case and front hubs. Drive shafts and differentials are the same, and they are serviced and repaired in the same manner as covered earlier in this chapter.

Transfer Case

Before attempting to service a transfer case, check service literature for procedures. Prior to removing a transfer case, mark both drive shafts at each end so they can be reassembled in the same positions. These parts are often balanced to correct for runout that occurs during manufacturing. Installing them 180 degrees from their originally installed position can result in a vibration.

Disassembling a transfer case is very similar to working on a manual transmission. Chain drive transfer cases are the most popular style. The Warner 13-56, pictured in Figures 76.59 to 76.63, is representative of a typical chain drive transfer case and will give you an idea of the parts. Remove the case bolts (**Figure 76.59**) and separate the housing (**Figure 76.60**). After removing the clutch coil (**Figure 76.61**), the chain assembly is removed (**Figure 76.62**). This allows access to the planetary gear assembly (**Figure 76.63**). Be sure to follow all manufacturer instructions.

Locking Front Hubs

Axles and locking hubs must be removed in order to remove a differential from the axle housing. Locate

Figure 76.59 Remove the bolts that attach the case halves.

Figure 76.61 Remove the clutch coil.

Figure 76.60 Separate the case halves.

Figure 76.62 Remove the chain assembly.

Figure 76.63 Unbolt the planetary gear plate and remove it from the case.

service information for each manufacturer's four-wheel-drive system before beginning disassembly. Some hubs can be removed easily by removing the bolts on the hub and sliding it off the axle. Another type has a snapring that is removed after partially taking the hub apart. Hub replacement is the reverse of the disassembly procedure. A typical disassembly procedure for a manual locking hub is as follows:

■ Position the hub to its unlocked position.
■ Remove the cover bolts.
■ Remove the outer snapring and the bolts that hold the hub in place (**Figure 76.64**).

A disassembled manual locking hub is shown in **Figure 76.65**.

Figure 76.64 Remove the snapring and the bolts that hold the hub in place. (*Courtesy of DaimlerChrysler Corporation*)

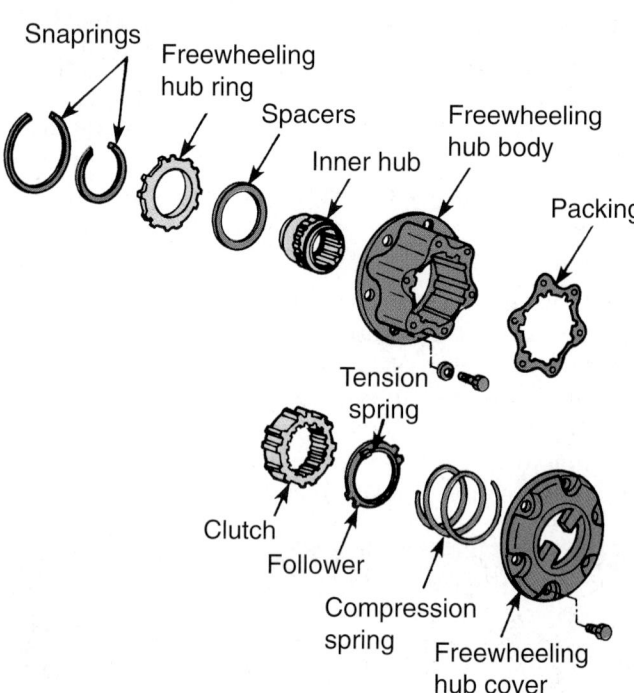

Figure 76.65 Parts of a manual locking hub. (*Courtesy of Mitsubishi Motor Sales of America, Inc.*)

NOTES:
■ *Locking hubs are not serviceable. If they are defective, they are replaced as a unit.*
■ *For long life, locking hubs should be disassembled for cleaning and new lubrication if the unit has been exposed to high water.*

Removal of an automatic locking hub is similar. Refer to applicable service information. An exploded view of an automatic locking hub is shown in **Figure 76.66**.

Wheel Hub and Rotor Removal

A four-wheel-drive wheel hub has a pair of tapered roller wheel bearings similar to those found on two-wheel-drive vehicles. They are larger than 2WD bearings because a 4WD axle must be able to pass through the inside of the bearings. Two nuts are required to keep the bearing adjustment from changing. This is similar

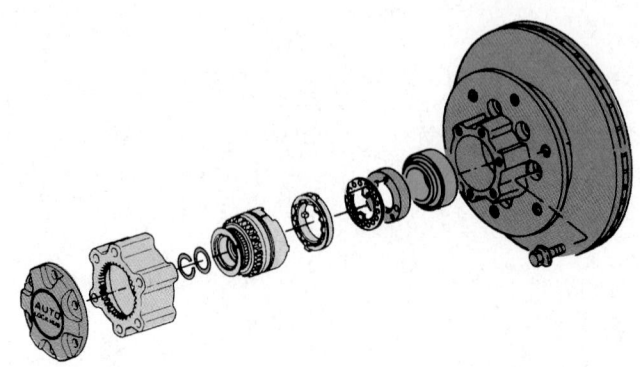

Figure 76.66 Parts of an automatic locking hub. (*Reprinted with permission by American Isuzu Motors, Inc.*)

Figure 76.67 A special socket is usually needed to loosen or adjust 4WD wheel bearings. *(Courtesy of Tim Gilles)*

to bearings found on floating axles (covered earlier in this chapter). Various types of locking arrangements exist, depending on the manufacturer. Removing these nuts usually requires a special socket (**Figure 76.67**).

Axle Shaft Removal

A four-wheel-drive axle is the same on the inside as a two-wheel-drive axle. It is splined into the differential

side gear. The outer end is different. It has a U-joint or CV joint. The wheel hub and spindle keeps it in place. **Figure 76.68** shows a cutaway of a four-wheel-drive outer U-joint. After the brakes and backing plate are removed, the spindle is removed from its support in the steering knuckle (**Figure 76.69**). Once the spindle is removed, the axle shaft assembly can be pulled through the steering knuckle (**Figure 76.70**).

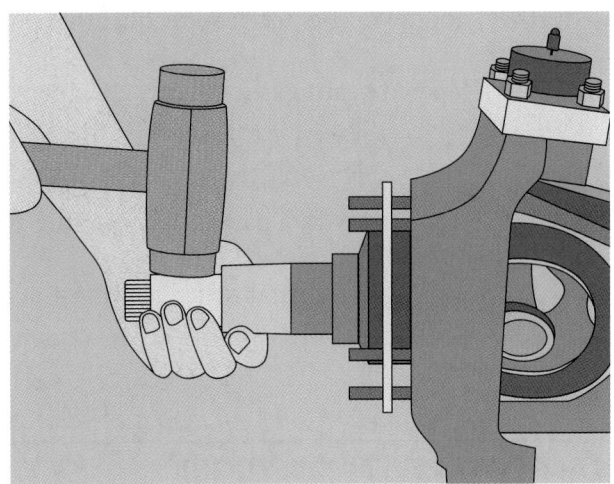

Figure 76.69 Removing the spindle from its support in the steering knuckle.

Figure 76.68 A four-wheel drive outer universal joint at a front wheel. *(Courtesy of Tim Gilles)*

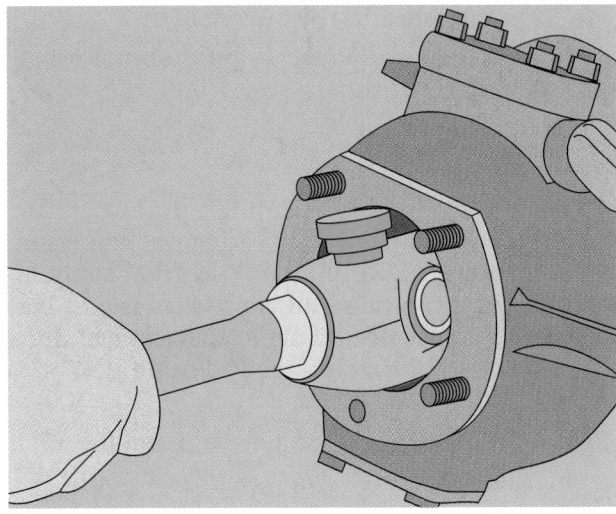

Figure 76.70 After the spindle is removed, pull the axle shaft through the steering knuckle.

■■■ REVIEW QUESTIONS

Drive Shaft and Axle

1. What is the term for an axle that turns with the wheel?

2. What can you wrap around the U-joint caps to keep them from failling off?

3. What is it called when a drive shaft is assembled in a position where the front and rear U-joints are not parallel?

4. What is it called when small indentations wear into the bearing surface?

5. Which part is hardened, the U-joint cap or the ear of the drive shaft yoke?

6. The four test-drive conditions for the driveline are drive, cruise, coast, and _____.

7. Which gears in a differential are the ones that tend to break when a spinning rear wheel gets traction?

8. A limited slip axle can be tested for _____ torque.

9. After a pinion nut is retightened, how much further should you tighten it to recrush the sleeve?

10. When a tapered bearing is too tight, which end will fail?

Differential

1. What part is held with a bar while loosening the drive pinion nut?

2. If a ring gear is worn, what other gear must also be replaced?

3. The two adjustments that are made are pinion gear depth and ring and pinion _____ clearance.

4. Which setting does the number found on a ring and pinion affect?

5. When testing pinion depth setting, tighten the pinion nut to 25 foot-pounds for the preliminary test so you do not ruin the _____ sleeve.

6. On a Salisbury axle, adding a 0._____ " shim will change the backlash by about 0.005".

7. What is the name of the outer end of the gear tooth, heel or toe?

8. Does pulling the pinion away from the ring move the pattern reading toward the toe or heel of the tooth?

9. When changing pinion depth, leave the ____ ____ out until the correct depth has been established.

10. A typical pinion bearing preload reading is _____ inch-pounds.

▦ ASE-STYLE REVIEW QUESTIONS

1. Which of the following is/are true about universal joint zerk fittings?
- **a.** Install the U-joint with the zerk fitting facing towards the driveshaft.
- **b.** Original equipment U-joints do not usually have zerk fittings.
- **c.** Both A and B.
- **d.** Neither A nor B.

2. Technician A says that during a turn when the outside wheel has a bad bearing, the noise can become worse. Technician B says that applying the brakes can cause the noise level from a bad bearing to become higher as they contact the drum or rotor that has the bad bearing. Who is right?
- **a.** Technician A
- **b.** Technician B
- **c.** Both A and B
- **d.** Neither A nor B

3. Which of the following is/are true about axle seals?
- **a.** They are sometimes part of the axle bearing.
- **b.** They are sometimes separate from the axle bearing.
- **c.** Both A and B.
- **d.** Neither A nor B.

4. Technician A says that a limited slip axle that chatters during turns could have the wrong lubricant. Technician B says that an axle with C-locks must have the cover removed from the back of the differential housing in order to remove the axles. Who is right?
- **a.** Technician A
- **b.** Technician B
- **c.** Both A and B
- **d.** Neither A nor B

5. Technician A says that a dent in the driveshaft can weaken it. Technician B says that if a driveshaft is very slightly off center, this is enough to cause vibration. Who is right?
- **a.** Technician A
- **b.** Technician B
- **c.** Both A and B
- **d.** Neither A nor B

6. Technician A says that a backlash setting of 0.007" will work on any differential. Technician B says that a pattern taken under light load will tend to be toward the heel of the gear tooth. Who is right?
- **a.** Technician A
- **b.** Technician B
- **c.** Both A and B
- **d.** Neither A nor B

7. On which type of differential are the side bearing preload and gear backlash settings made with shims?
- **a.** Salisbury
- **b.** Pumpkin
- **c.** Both A and B
- **d.** Neither A nor B

8. Technician A says to be sure there is no resistance against the ring or pinion gears when rolling a contact pattern. Technician B says that changing backlash or pinion depth changes both patterns. Who is right?

   a. Technician A   **c.** Both A and B

   b. Technician B   **d.** Neither A nor B

9. When the patterns are at opposite ends of the tooth on the drive and coast sides, _____ is the problem.

   a. Backlash   **c.** Both A and B

   b. Pinion depth   **D.** Neither A nor B

10. Ring gear backlash is checked with:

    a. A feeler gauge

    b. A dial indicator

    c. A special backlash checking fixture

    d. None of the above

CHAPTER 77

Front-Wheel-Drive (Transaxle and CV Joint) Fundamentals

■ OBJECTIVES

Upon completion of this chapter, you should be able to:

✔ Describe differences between front- and rear-wheel drivetrains.

✔ Tell the names of parts of a transaxle.

✔ Trace the power flow through four- and five-speed transaxles.

■ KEY TERMS

stub shaft	plunge joint	tulip
CV joint	fixed joint	double offset plunge joint
inboard side	Rzeppa CV joint	cross groove plunge joint

■ INTRODUCTION

Today, most automobiles are sold with front-wheel drive. However, until the early 1980s, most vehicles came equipped with rear-wheel drive. This chapter covers the theory of front-wheel-drive axles, CV joints, and manual transaxles.

A front-wheel-drive car has a transaxle, which combines a transmission and differential into one unit. Drive axles extend to the front wheels out of each side of the transaxle. At each end of the drive axle is a constant velocity universal joint, called a CV joint.

A transaxle can be either manual (**Figure 77.1**) or automatic (**Figure 77.2**). As you will see, the theory of operation of the clutch, manual or automatic transmission, and differential sections of a transaxle is almost identical to those components on rear-wheel-drive vehicles. Those fundamentals are covered in previous chapters. This chapter will emphasize aspects that are unique to transaxles. Prior to reading this chapter, you should have read the previous driveline chapters.

■ FRONT-WHEEL DRIVE

Until the last twenty years, most cars came equipped with rear-wheel drive. Front-wheel drive (FWD) first appeared on an American car in 1930, on the Auburn. The Cord, built later in the 1930s, also used front-wheel drive. The next American-made car to be built with front-wheel drive was the 1966 Oldsmobile Toronado. Front-wheel drive showed up on a few more vehicles but remained limited until the mid-1980s. Today, a majority of cars come with front-wheel drive.

Advantages of front-wheel drive include reduced weight and a more efficient drivetrain, resulting in better fuel economy. When combined with MacPherson struts there is less unsprung weight for better handling. The transmission hump necessary in rear-wheel drives is also eliminated.

A few front-wheel-drive engines have been mounted in the engine compartment longitudinally, from front to rear with the transaxle (**Figure 77.3**). Most transaxles are mounted in a sideways (transverse) fashion (**Figure 77.4**). The transaxle is bolted to the engine. A clutch or torque converter connects it to the crankshaft. The transaxle is bolted to the engine.

■ MANUAL TRANSAXLE

Manual transaxles and transmissions use the same kind of clutch. Some clutch release mechanisms operate in

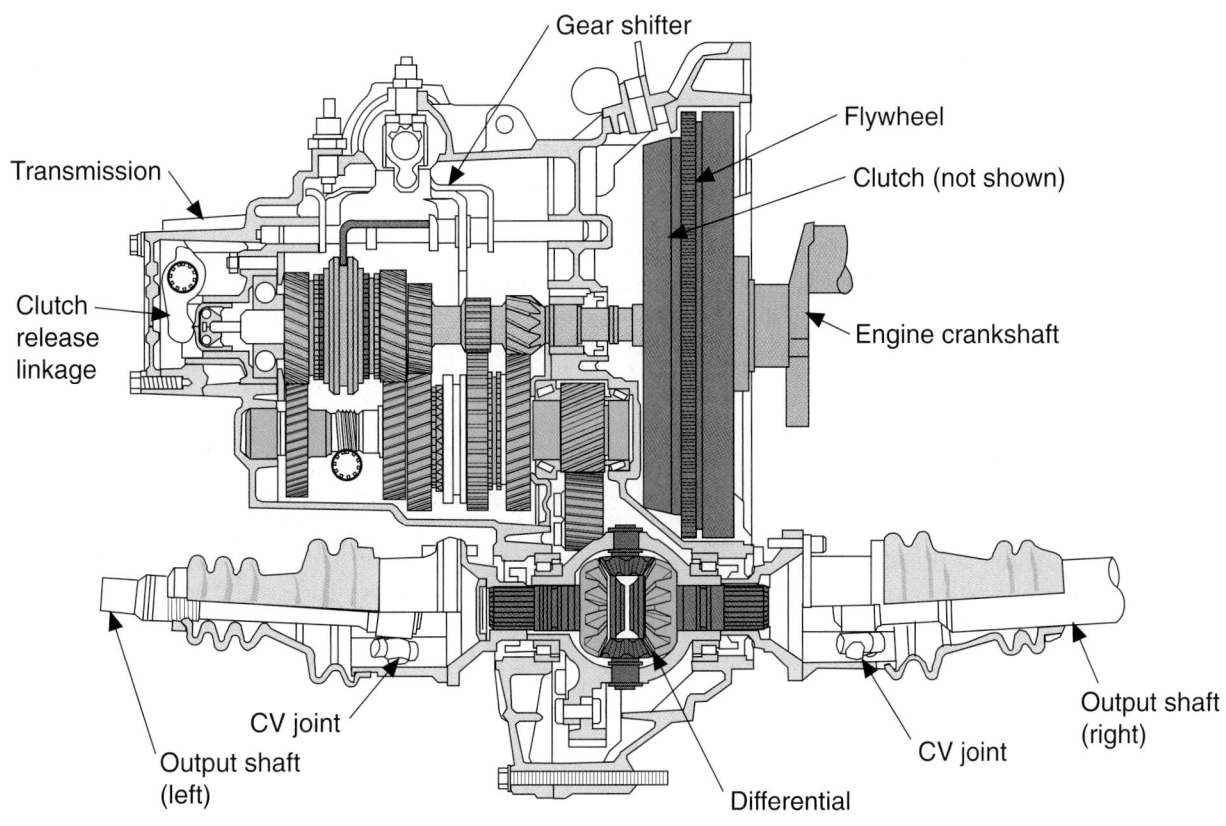

Figure 77.1 A manual transaxle. *(Courtesy of DaimlerChrysler Corporation)*

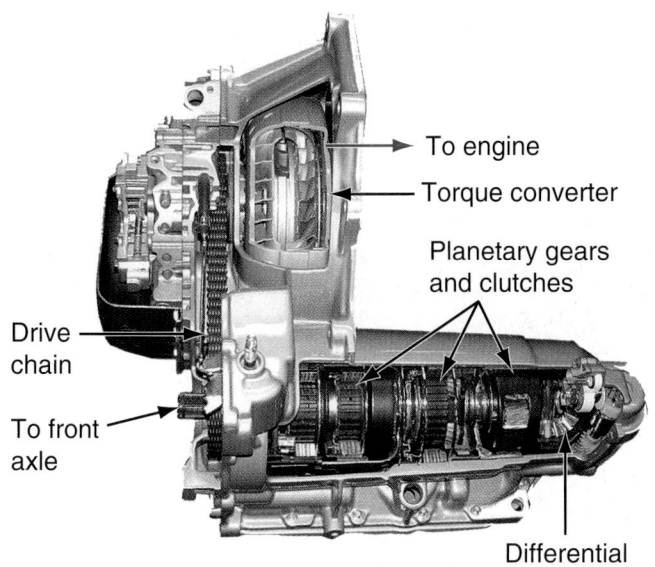

Figure 77.2 An automatic transaxle. *(Courtesy of Tim Gilles)*

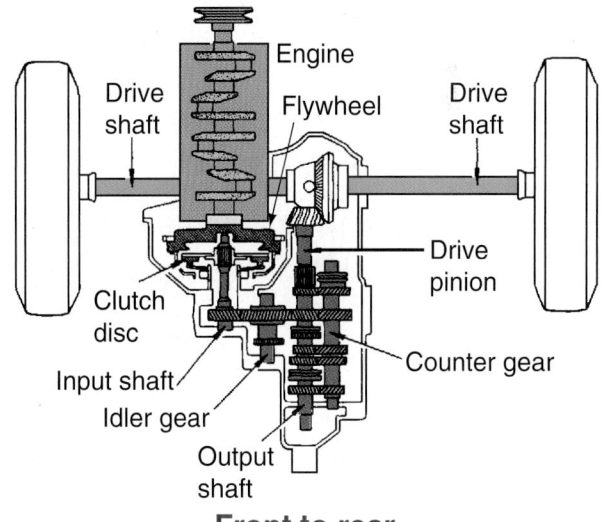

Figure 77.3 A front-to-rear engine and transaxle.

the normal manner described in Chapter 69. Other clutches are released by a pushrod running through the center of the input shaft (**Figure 77.5**).

Transaxle part names vary with different manufacturers. There is an input shaft and an output shaft. The output shaft is commonly called an *intermediate*

shaft. It can also be called a mainshaft, countershaft, or pinion shaft because it drives the pinion gear that turns the differential ring gear. In this chapter it will be called an intermediate shaft.

The gears and synchronizers operate similarly to a manual transmission (see Figure 77.1). All gears are in constant mesh except for a sliding reverse idler. To save

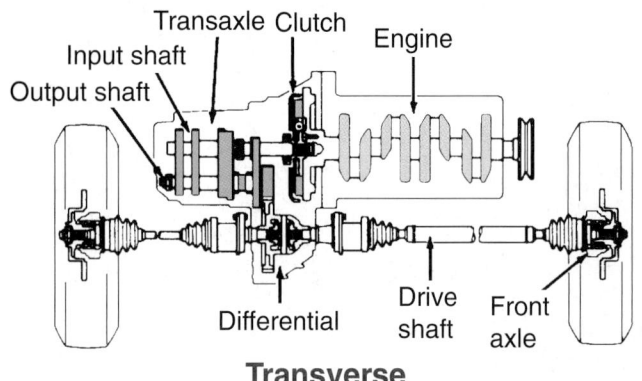

Transverse

Figure 77.4 A side-to-side engine and transaxle.

space, narrower gears and pressed-fit synchro hubs are sometimes used.

Transaxles have three parallel shafts. The input shaft is usually located above the intermediate shaft (so called because it is in the middle of power flow). The third parallel shaft is the differential assembly. Input shaft gears directly drive output shaft gears. Remember,

in rear-wheel-drive transmissions, the input shaft drives a countergear that powers gears on the mainshaft. In the transaxle, one of the shafts has been eliminated. Some transaxles have an additional shaft when the design calls for even more space saving, but most have the kind of geartrain described here.

The gear shafts are primarily supported by larger ball, roller, or tapered roller bearings rather than shafts and needle bearings. End play between the gears is controlled by thrust washers or spacers between the bearing end plate and case.

■ SHIFT LINKAGE

Transverse transaxles are sometimes shifted by cables (**Figure 77.6**). Other times, shift linkage is used (**Figure 77.7**) There are two shift cables or rods. One of them moves a selector on the transaxle that determines which of the shift forks will move. The other one moves the shift fork back and forth. An advantage to cables is that engine shake is not transmitted back to the driver's hand on the shift lever.

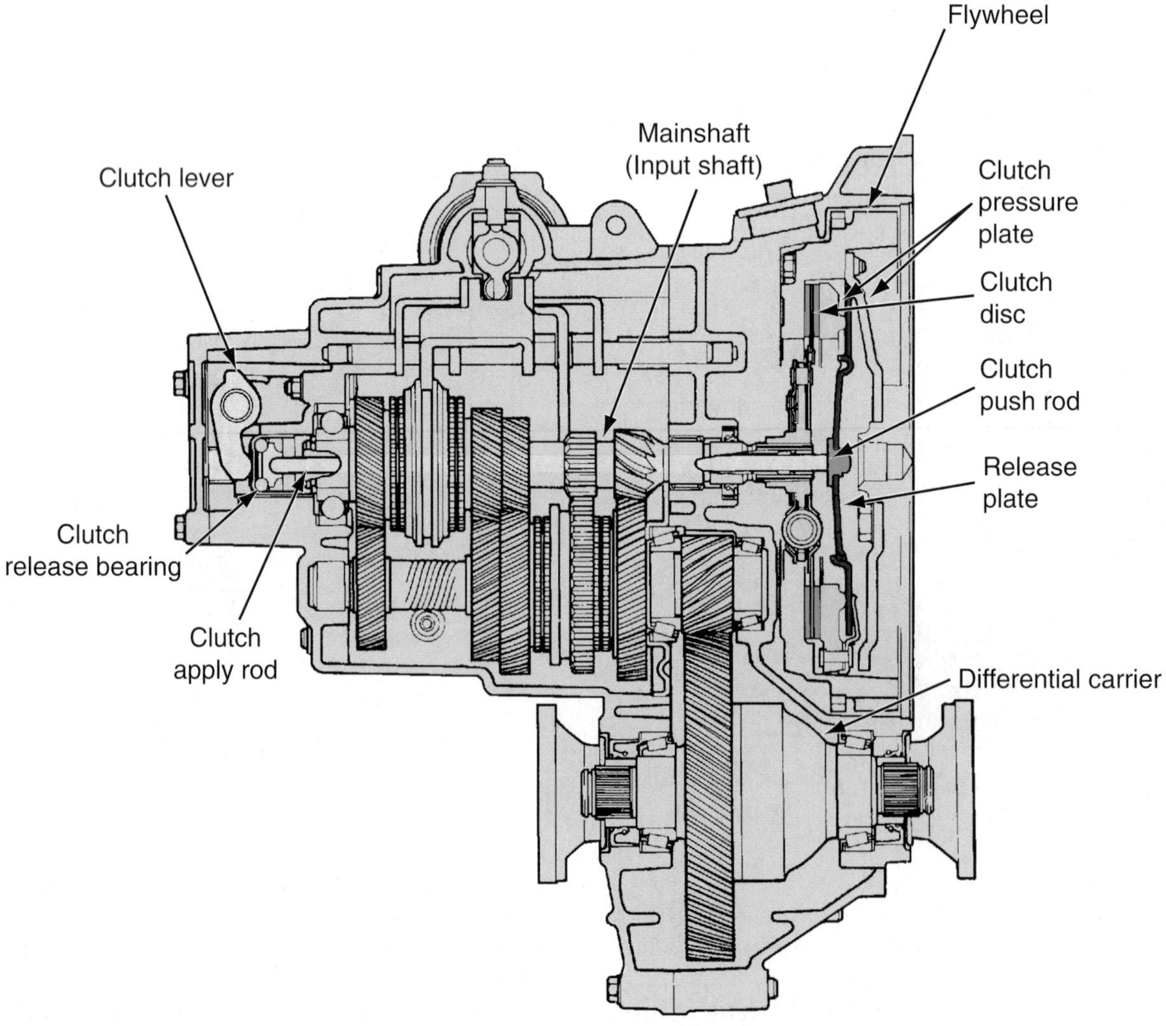

Figure 77.5 This transaxle has a clutch apply rod that runs through the inside of the input shaft. (*Courtesy of DaimlerChrysler Corporation*)

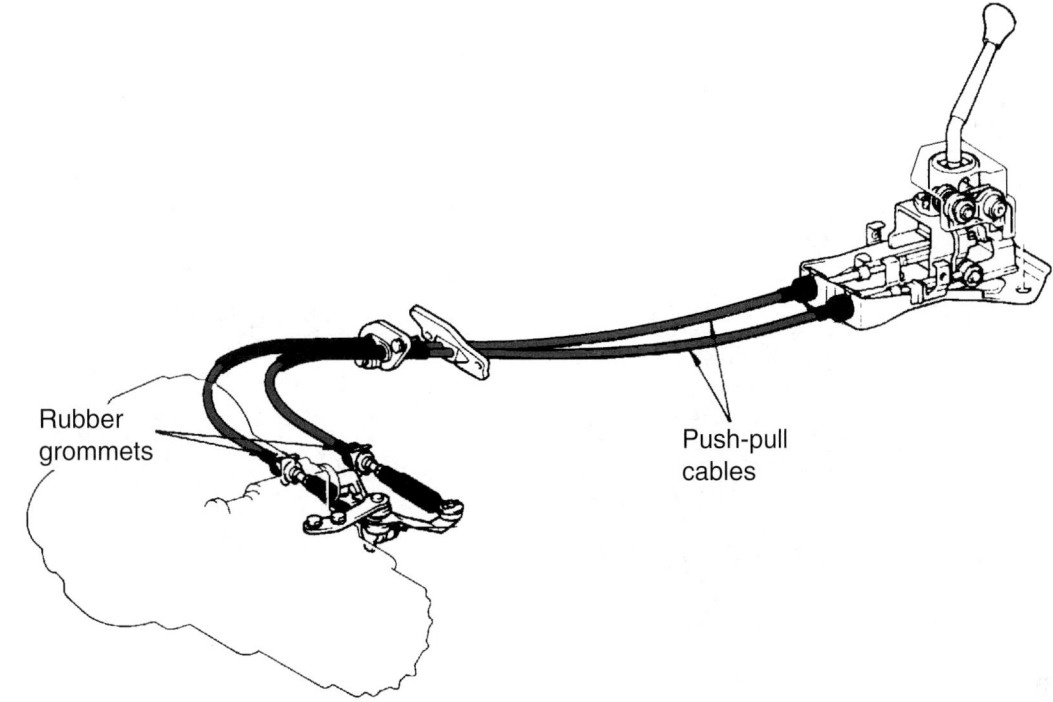

Figure 77.6 A transaxle shifted by cables.

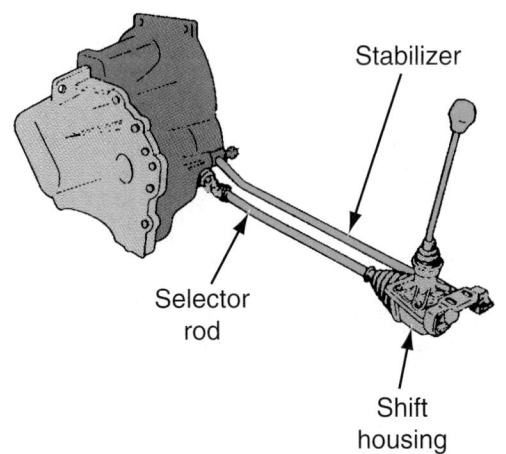

Figure 77.7 A transaxle shifted by linkage. *(Courtesy of Ford Motor Company)*

■ TRANSAXLE DIFFERENTIAL

A differential allows the wheels to turn at different speeds when rounding corners. This is accomplished the same way as in the rear-wheel-drive differential, with pinion gears and side gears (**Figure 77.8**). There is no need to turn power 90 degrees as in rear-wheel drive, so an ordinary helical gearset (like the kind used in the transmission section) is used instead of bevel gears. Power from the differential side gears is transmitted to the front drive axles through axle shafts (**Figure 77.9**).

■ TRANSAXLE POWER FLOW

There are different designs in transaxles. Some have gears clustered (two or more fixed gears) on the intermediate shaft, and others have them clustered on the input shaft. The four-speed transaxle in **Figure 77.10** has an input shaft with all of the fixed gears on it. This is a true cluster gear. Compare it to the four-speed transaxle in **Figure 77.11**. It has two fixed gears and two synchronized gears on the input shaft.

In the five-speed transmission used in the following power flow explanation, the fixed gears for first, second, and reverse are on the input shaft. The fixed gears for third, fourth, and fifth are on the intermediate shaft. In all gear ranges, power flow leaves the transmission intermediate shaft and continues through the drive pinion to the axles. Because the engine is mounted sideways, the axles run parallel to the input shaft.

■ Low gear—power comes in through the input shaft first gear. The first and second synchro clutch selects first gear on the intermediate shaft, providing a gear reduction (**Figure 77.12**).

■ Second gear—power comes in through the input shaft second gear. The first and second synchro clutch selects second gear on the intermediate shaft, providing a gear reduction (**Figure 77.13**).

■ Third gear—power comes in through the input shaft third gear, which is connected to the input shaft by the third and fourth synchro clutch. The fixed third gear on the intermediate shaft receives the power, providing a gear reduction (**Figure 77.14**).

■ Fourth gear—power comes in through the input shaft fourth gear, which is connected to the input shaft by the third and fourth synchro clutch. The fixed fourth gear on the intermediate shaft receives the power, providing a gear reduction (**Figure 77.15**).

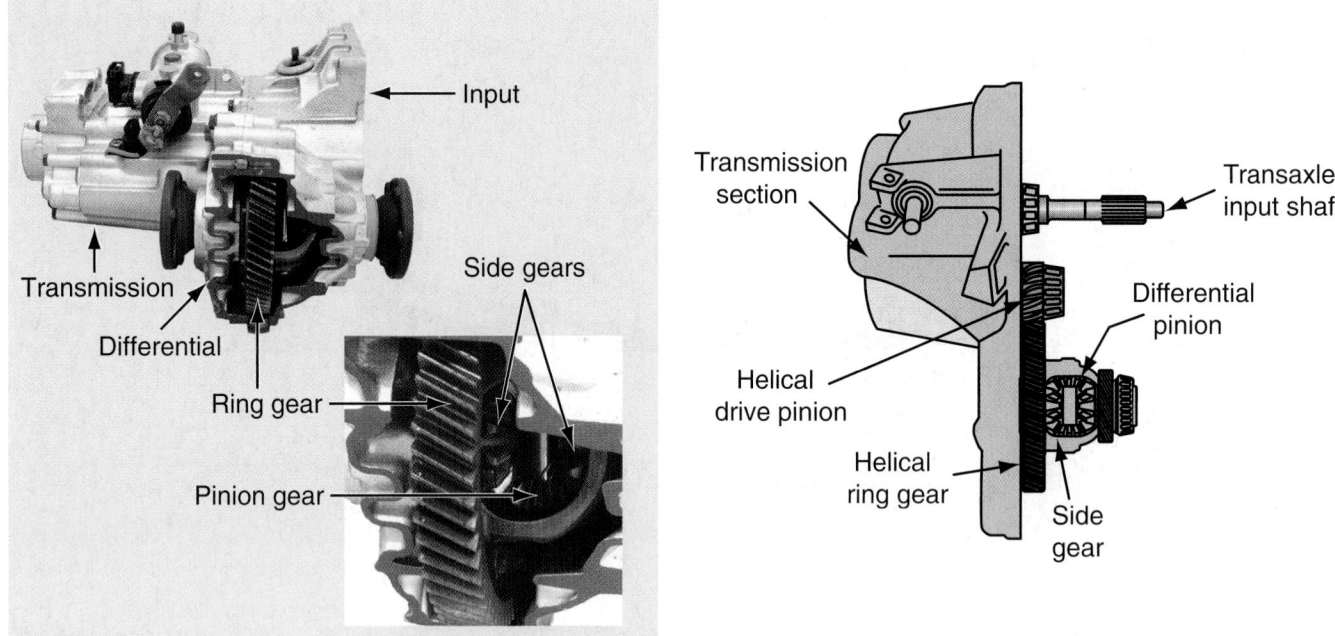

Figure 77.8 A transaxle has a differential section and helical drive gears. *(Photographs courtesy of Tim Gilles)*

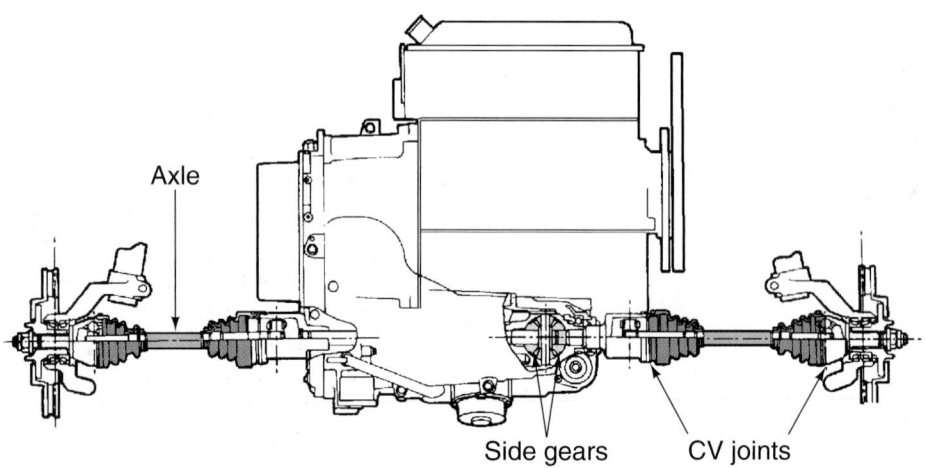

Figure 77.9 Power from the differential side gears is transmitted to the front drive axles through axle shafts. *(Courtesy of Ford Motor Company)*

NOTE: *In a typical rear-wheel-drive transmission, fourth gear would be accomplished by locking the input shaft to the output shaft for a 1:1 ratio. Notice that in a transaxle, power flow is always through gears in each gear range.*

- Fifth gear—power comes in through the input shaft fifth gear, which is connected to the input shaft by the fifth gear synchro clutch. The fixed fifth gear on the intermediate shaft receives the power (**Figure 77.16**). It is smaller than the gear on the input shaft, so it provides an overdrive.
- Reverse gear—in reverse, power comes in through the input shaft where it leaves through a fixed spur gear (**Figure 77.17**). With all of the other synchro clutches in the neutral range, a sliding spur idler

gear is moved into contact with a spur gear on the outside of the 1–2 synchro clutch, which is splined to the intermediate shaft. The direction of rotation of the differential pinion gear is reversed.

■ AUTOMATIC TRANSAXLE

An automatic transaxle is a combination of an automatic transmission and a differential. The same parts and operation apply, with the differential being the difference. With a transverse engine, power flow is either through gears or a sprocket and chain. **Figure 77.18** shows a typical automatic transaxle with a gear drive. A chain drive (**Figure 77.19**) allows the transaxle to be mounted slightly below and to the side of the engine.

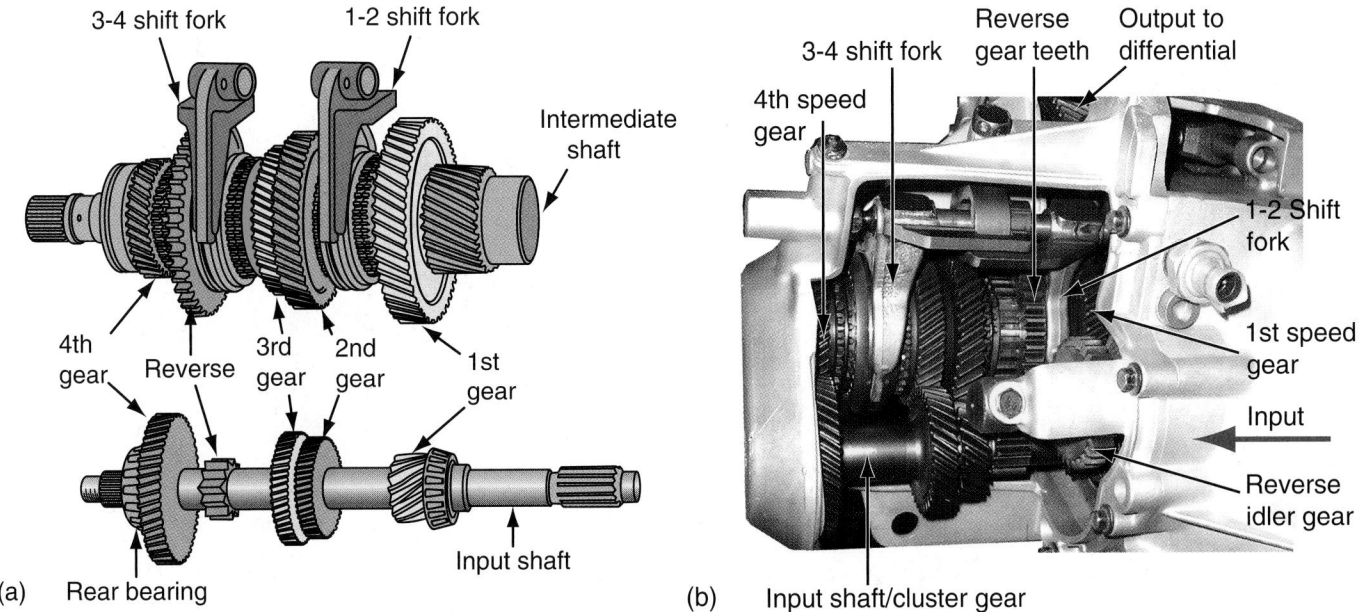

Figure 77.10 (a) A transaxle with all of the fixed gears on the input shaft. (b) The input shaft on this four-speed transaxle includes all of the fixed gears, like a cluster gear in a conventional transmission. *(Photograph courtesy of Tim Gilles)*

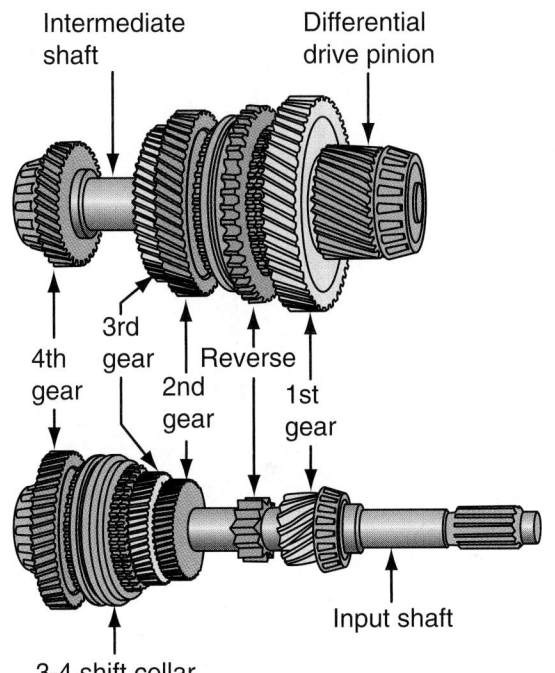

Figure 77.11 The input shaft on this transaxle has two fixed gears, reverse, and two synchronized gears.

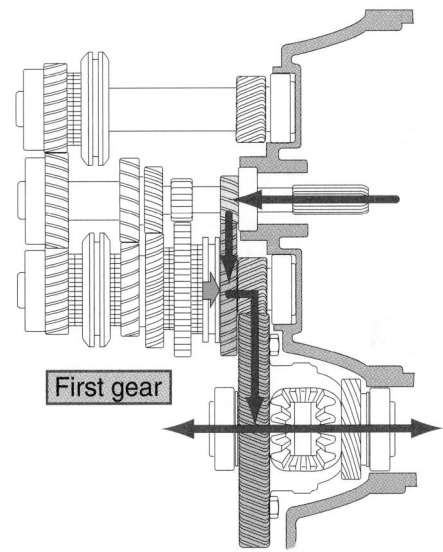

Figure 77.12 Power flow in first gear.

▣ FRONT DRIVE AXLES

There is a major difference between rear-wheel-drive axles and front-wheel-drive axles. Front-wheel-drive axles have constant velocity (CV) joints at each end. CV joints are necessary because the axles are driven at sharper angles than can be tolerated by ordinary universal joints (see Chapter 75). They also allow steering of the front wheels to take place during power transmission (**Figure 77.20**).

An ordinary universal joint changes its output speed twice in every revolution when run at an angle. A true constant velocity joint provides a constant output speed no matter what the shaft angle.

In a rear-wheel-drive car, the drive shaft turns very fast because it is positioned before the gear reduction of the differential. Front-wheel-drive axles turn slower than a rear-wheel drive shaft by the difference of the axle ratio (about one-third as slow).

▣ AXLE SHAFT PARTS

The drive axle is called a half shaft or an *axle shaft*. There is a short shaft at the outside end called a **stub shaft** or *stub axle*. It is splined to the front hub so it

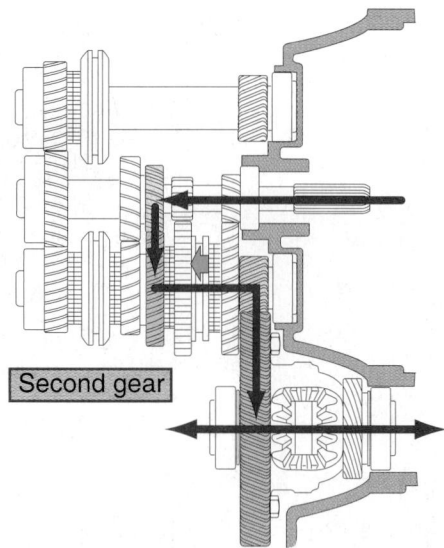

Figure 77.13 Power flow in second gear.

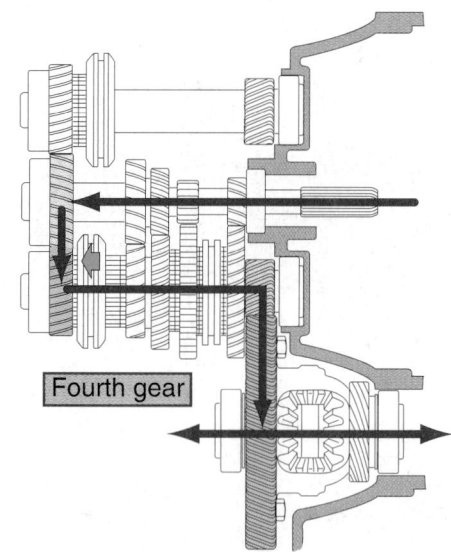

Figure 77.15 Power flow in fourth gear.

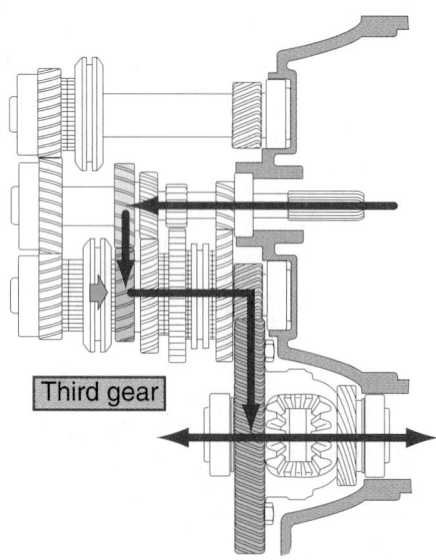

Figure 77.14 Power flow in third gear.

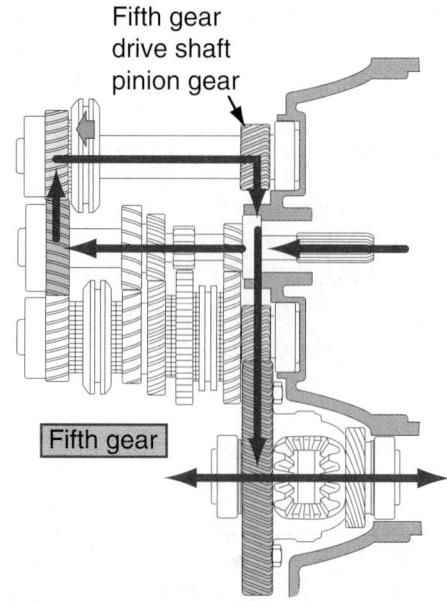

Fifth gear
drive shaft
pinion gear

Figure 77.16 Power flow in fifth gear.

can drive the front wheels. Some axle shafts have a stub shaft on the inside, too. Others have a female-type splined yoke that slides over male splines—similar to a rear-wheel-drive transmission output shaft.

CV Joints

At each end of a front-wheel-drive axle is a constant velocity joint, called a **CV joint** (see Figure 77.9). There are three ways of classifying CV joints.

- Inboard and outboard
- Fixed and plunge
- Ball and tripod

The inside end of the axle shaft is commonly called the **inboard side** and the outside end nearest the wheel is called the *outboard side*.

Plunge CV Joints

The inboard CV joint is called a **plunge joint**. When the suspension system deflects, the axle's length must change (**Figure 77.21**). The plunge joint takes care of this change in length. It acts like a slip yoke does in a rear-wheel-drive car. As its name implies, it plunges in and out to compensate for changes in axle length. Because of its design, it can also transmit motion through an angle.

Fixed CV Joints

The outboard joint is called a **fixed joint**. This means that it does not allow for a change in length like the inboard joint. It allows for a change in the angle of the axle as the suspension moves over bumps in the road. A fixed joint also allows for severe pivoting during

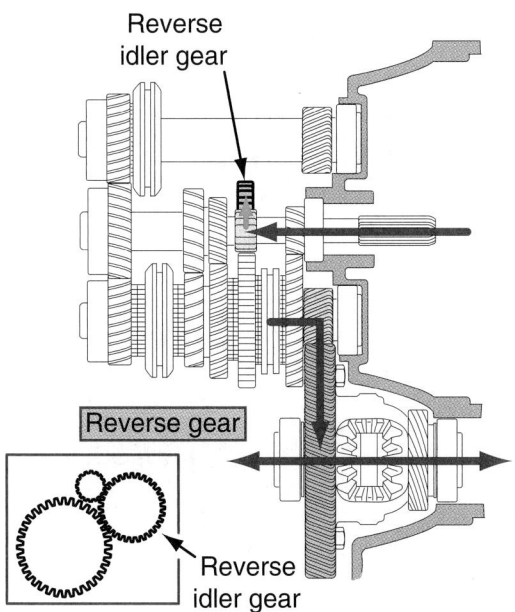

Figure 77.17 Power flow in reverse gear.

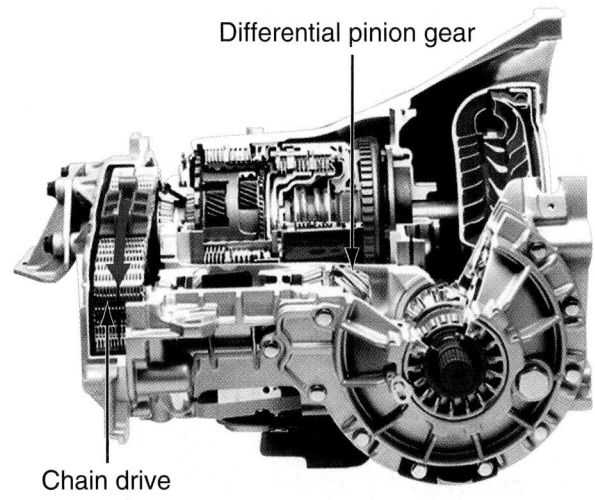

Figure 77.19 Cutaway of an automatic transaxle with a chain drive. (© *Courtesy of Mercedes-Benz USA, LLC*)

steering. It can accommodate angles up to 50 degrees, although the maximum amount of travel is usually limited to 46 degrees. Outer joints wear out more often than inner joints because of the large angle to which they are subjected. Plunge joints cannot change angles as effectively as fixed joints (**Figure 77.22**). They are limited to about 20 degrees.

■ CV JOINT CONSTRUCTION

There are different types of plunge and fixed joints. Different combinations of joints are found on axles. The most common combination uses a **Rzeppa CV joint** for the outside fixed joint and a *tripod* for the inside plunge joint. The tripod has become more popular because it costs less to manufacture.

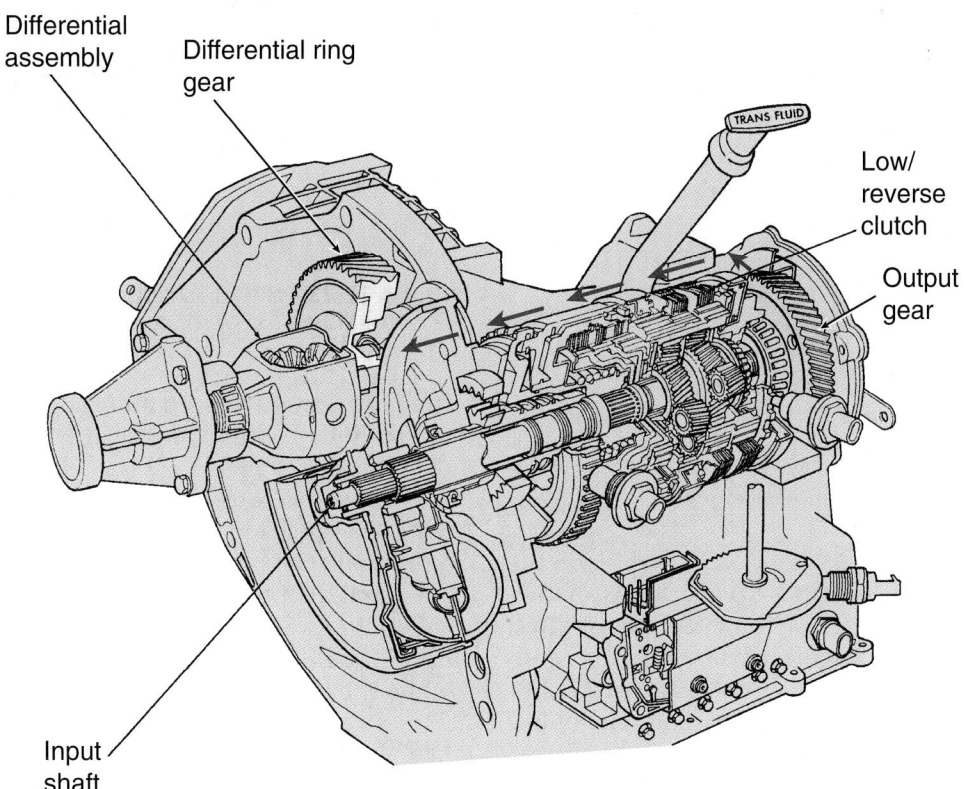

Figure 77.18 An automatic transaxle with a gear drive. The output gear drives a gear on the opposite end of the shaft that turns the differential pinion gear. (*Courtesy of DaimlerChrysler Corporation*)

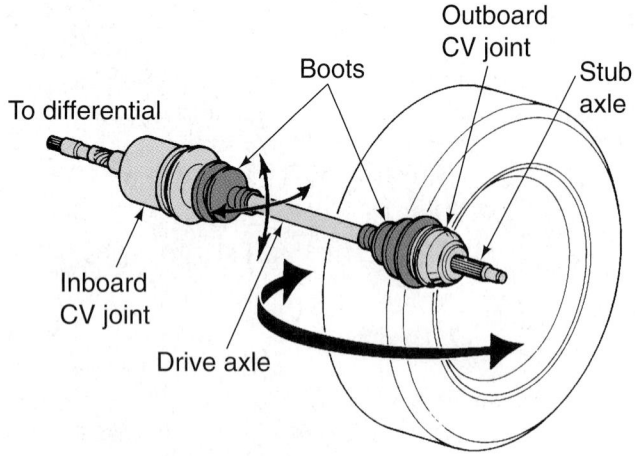

Figure 77.20 CV joints pivot during steering. *(Courtesy of Federal-Mogul Corporation)*

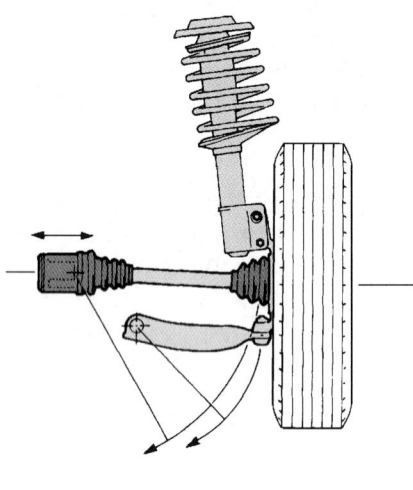

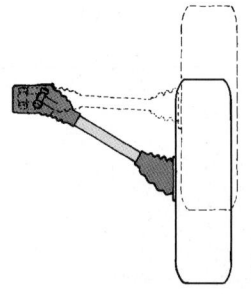

Figure 77.21 A plunge joint allows the axle to change length. *(Courtesy of Federal-Mogul Corporation)*

Fixed Joints

Almost all newer vehicles use a Rzeppa joint for the fixed outer joint. The Rzeppa joint is named after Alfred Rzeppa, the Ford engineer who invented it in 1929. It is a ball-and-socket type joint (**Figure 77.23**) with the inner race splined to the axle shaft. The outer race is either part of the stub shaft or splined to the stub shaft. Splined parts are held to their shafts with snaprings. Six ball bearings fit in between the inner and outer races, housed in a cage. The joint flexes in

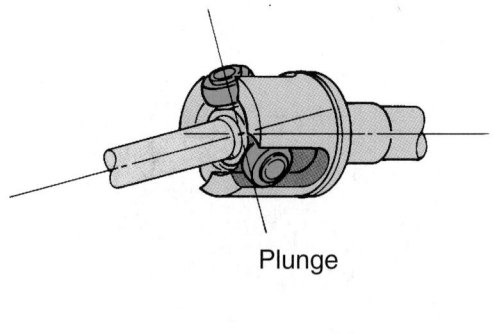

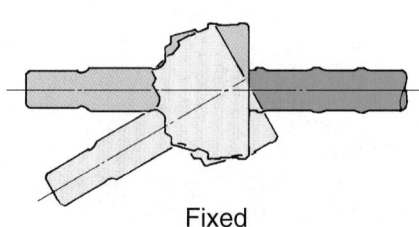

Figure 77.22 A plunge joint does not allow as much angle change as a fixed joint. *(Courtesy of Federal-Mogul Corporation)*

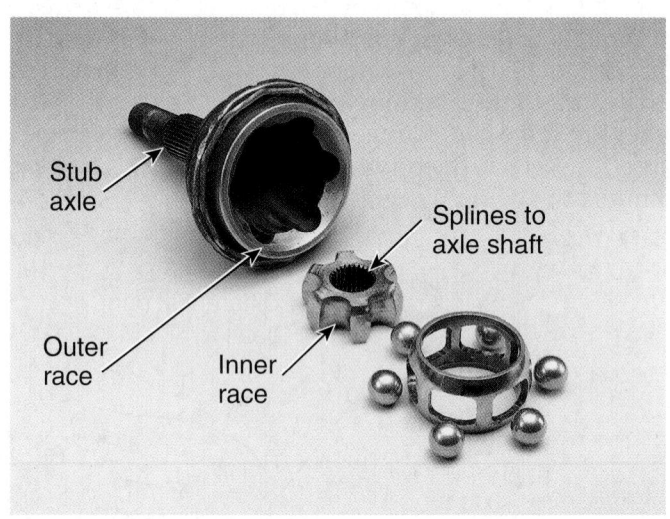

Figure 77.23 Parts of a Rzeppa CV joint. *(Courtesy of Federal-Mogul Corporation)*

tracks in the inner race and outer housing. The balls act as both bearings and a medium for transferring power to the wheels.

A Rzeppa joint is like a bevel gear with balls instead of teeth. It provides even power transmission at all angles. As the inner and outer races move back and forth compensating for the angle change, the balls remain in a constant perpendicular plane (**Figure 77.24**). The cage holds the balls in this position while the inner and outer races move.

Types of Plunge Joints

There are three types of plunge joints. A *tripod tulip plunge joint* (**Figure 77.25a**) has a spider assembly with three spherical rollers that ride on needle bearings,

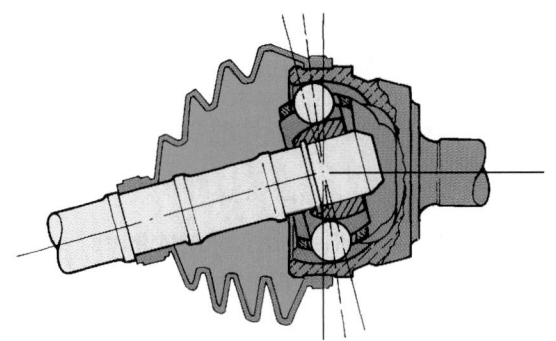

Figure 77.24 As the inner and outer races move back and forth compensating for the angle change, the balls remain in a constant perpendicular plane. *(Courtesy of Federal-Mogul Corporation)*

similar to a cross and yoke universal joint. The yoke or housing, often called a **tulip**, has three grooves on which the rollers can roll in and out as the axle length changes. The housing is usually spline-mounted, but can also be bolted to a flange.

A **double offset plunge joint** (**Figure 77.25b**) uses six balls evenly spaced by a cage. Most double off-set joints are spline-mounted to the transaxle. The outer housing is considerably deeper than a cross groove so it can plunge further. The races have straight grooves.

A **cross groove plunge joint** (**Figure 77.25c**), also called a *pancake joint,* has grooves in the bearing races that would cross each other if they extended far enough (**Figure 77.26**). There are six balls and a cage, in addition to the inner and outer races.

▬ AXLE SHAFTS

Axle shafts can be either solid or hollow. Some have damper weights to absorb vibration (**Figure 77.27**). Because the axle shafts turn much slower than a rear-wheel-drive drive shaft, balance is not as important.

Axle shafts can be the same length or they can be different lengths (**Figure 77.28**). Earlier cars had unequal length axle shafts because the transaxle was mounted off to one side of the engine. When there are different length shafts, the long one wraps up and there

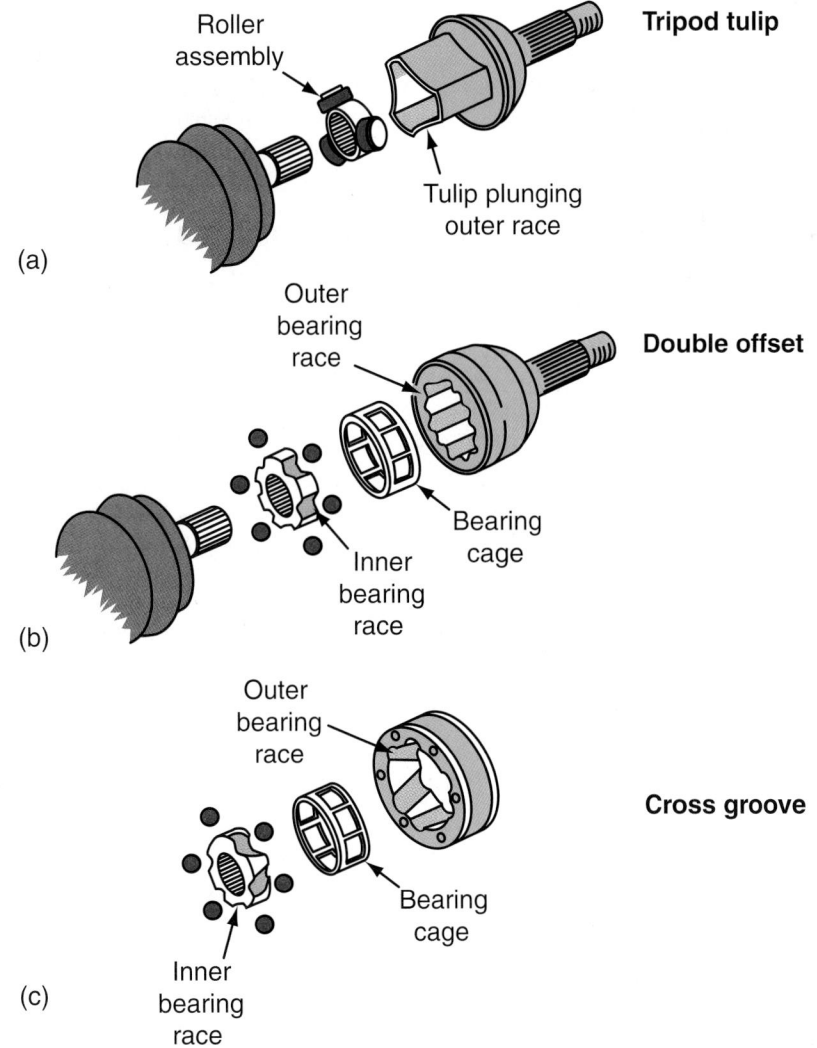

Figure 77.25 Plunge joints. (a) A tripod tulip joint. (b) A double offset joint. (c) a cross groove joint.

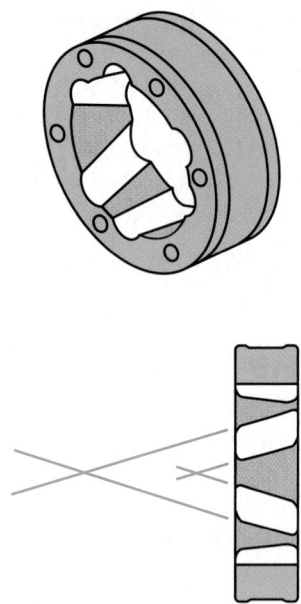

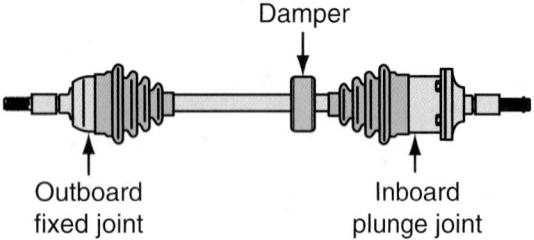

Figure 77.26 In a cross groove joint, the upper and lower sides of the grooves in the bearing races would cross each other if they extended far enough.

Damper

Outboard
fixed joint

Inboard
plunge joint

Figure 77.27 On some axles a damper absorbs vibration.

is a slight lag before it puts its torque to the wheel. The more horsepower the engine has, the greater the effect of

Typical front-wheel-drive drivelines

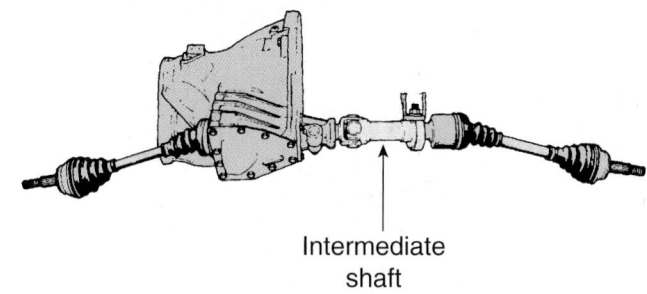

Intermediate
shaft

Figure 77.28 When axle shafts are the same length there is an intermediate shaft. (*Courtesy of Federal-Mogul Corporation*)

this torque steer. One way of preventing torque steer is to make the longer axle shaft of larger diameter tubing.

The trend is to have halfshafts of equal length on both sides of the car. An intermediate shaft is used on the equal length system so that the axle shafts can be the same length. The Ford Taurus actually has the intermediate shaft as part of the transaxle.

■ CV JOINT BOOTS

Boots at each end of the axle contain grease and protect the joint from the elements. The CV joint boot is attached to the axle and stub shafts with plastic or steel bands or straps. Boots are made of natural rubber, neoprene, silicone, or urethane.

■ REVIEW QUESTIONS

1. How are most cars produced today, with front-wheel drive or rear-wheel drive?

2. When a transaxle is connected to the shifter by cables, how many cables will there be?

3. The parts of the transaxle that allow the front wheels to turn at different speeds during a turn are the differential pinion gears and _____ gears.

4. A FWD axle is also called a _____ shaft.

5. What is the name of the short splined shaft at the outside end of a FWD axle?

6. The two kinds of CV joints are fixed and _____.

7. A _____ joint is like a bevel gear with balls instead of teeth.

8. What kind of sound does a worn Rzeppa joint make if its cage becomes worn, allowing the races to move back and forth in their planes?

9. What is it called when a car with unequal length halfshafts steers one way during acceleration?

10. If one halfshaft is thicker than the other, will it probably be the longer one or the shorter one?

■■■ ASE-STYLE REVIEW QUESTIONS

1. Technician A says that transaxles have all of the fixed gears on the input shaft. Technician B says that FWD axles run parallel to the input shaft. Who is right?

 a. Technician A   **c.** Both A and B

 b. Technician B   **d.** Neither A nor B

2. Which of the following is/are true about five-speed transmissions/transaxles?

 a. In fourth gear in a five-speed transaxle, the input shaft and output shaft are locked together.

 b. In fifth gear the input shaft turns faster than the output shaft.

 c. Both A and B.

 d. Neither A nor B.

3. Technician A says that ordinary universal joints cannot handle as steep an angle as CV joints. Technician B says that a RWD drive shaft turns at a faster rpm than a FWD axle. Who is right?

 a. Technician A   **c.** Both A and B

 b. Technician B   **d.** Neither A nor B

4. Technician A says that a plunge joint cannot change angles as effectively as a fixed joint. Technician B says that the outboard joint is a fixed joint. Who is right?

 a. Technician A   **c.** Both A and B

 b. Technician B   **d.** Neither A nor B

5. Which kind of outboard CV joint is used on almost all front-wheel-drive vehicles?

 a. Rzeppa   **c.** Double offset

 b. Cross groove   **d.** Tripod tulip

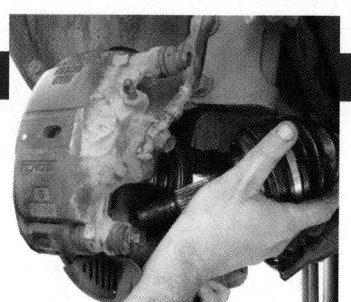

CHAPTER 78

Front-Wheel-Drive (Transaxle and CV Joint) Service

■ OBJECTIVES

Upon completion of this chapter, you should be able to:

✔ Diagnose CV joint problems.

✔ Service CV joints.

✔ Replace CV joint boots.

✔ Disassemble and repair transaxles.

■ KEY TERMS

pinch bolt
knock-off type CV joint

bridge type clamp
earless clamp

rebuilt halfshaft

■ INTRODUCTION

The emphasis in this chapter is the removal, service, and replacement of front drive axles and CV joints. Not only is this a lucrative service area, but transaxle removal is relatively straightforward, once axle removal is accomplished. A related area, vibration analysis, is covered in Chapter 79.

■ TRANSAXLE AND FRONT-WHEEL-DRIVE SERVICE AND REPAIR

Inspecting Axle Assemblies

Some checks of the axle assemblies can be made with the axle on the car.

■ LEAKING CV JOINT BOOT

Front-wheel-drive axles often require service as a vehicle ages. Some checks and repairs of axle assemblies can be made with the axle on the vehicle. Others require removal and disassembly of the components.

■ CV JOINT BOOT SERVICE

Over a period of time, boots age and suffer the effects of the elements and repeated flexing. When a boot fails, grease is thrown from it. If the boot is not replaced before contaminants enter it, the entire joint will fail.

Failure of the boot is the most common problem with CV joints. They are designed to last about 75,000 miles. Boot replacement is an often-required service. The outboard boot is the one that fails most of the time.

NOTE: *CV joints are lubed for life. If a boot tears or the clamp falls off, the joint will fail if not serviced immediately. The estimated life of a joint is eight hours without lubricant.*

The action of an operating CV joint is like a vacuum cleaner, causing suction of water into the joint. Storm runoff water in a gutter carries a good deal of mud and grit that will destroy the bearing areas inside of a CV joint. When a boot has failed, the correct repair is to disassemble and inspect the CV joint before replacing the boot and grease. Be sure to check the seal at the transaxle and plunge joint for leakage, too.

NOTE: *A noisy joint will not get quieter with the installation of a new boot and grease.*

■ BOOT KITS

Replacement boot kits come with new clamps and the correct amount of high-temperature grease required to refill the boot and CV joint (**Figure 78.1**).

The halfshaft is removed to replace the boot. A new nut is required for the stub shaft, but it does not come

1370

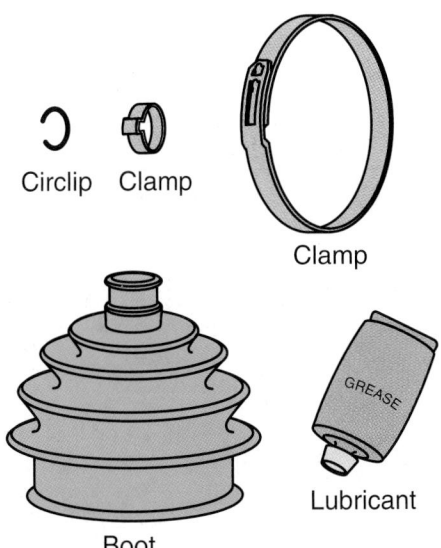

Figure 78.1 A CV joint boot kit.

in the boot kit because there are too many different nuts used. The halfshaft is removed to replace the boot. Removing the stub shaft from the hub is all that is required to replace the outer boot.

There are a few axle shafts where the fixed joint cannot be disassembled. These can have the boot installed from the plunge joint end.

 **SHOP TIP** Spray the clamp with penetrating oil to reduce friction during installation.

AXLE INSPECTION AND DIAGNOSIS

Check the halfshaft for obvious looseness. Noises coming from the front axle can be detected during a test drive. To test for noise, weight must be on the suspension. Otherwise the joint is not in the same position where all of the wear occurs. Just feeling the joint does not work for diagnosing noise unless the joint is extremely damaged. Questions to ask the customer include:

■ What does the noise sound like?
■ Does it occur during acceleration, deceleration, or when turning?

CV JOINT DIAGNOSIS

The following are diagnostic procedures for fixed and plunge joints.

Fixed Joint Problems

Test the car during turns, as you did when diagnosing axle bearings. A clicking sound during a turn could indicate a bad outboard joint. The clicking will increase and decrease with wheel speed. If the cage becomes worn, the result is a clicking sound when turning as

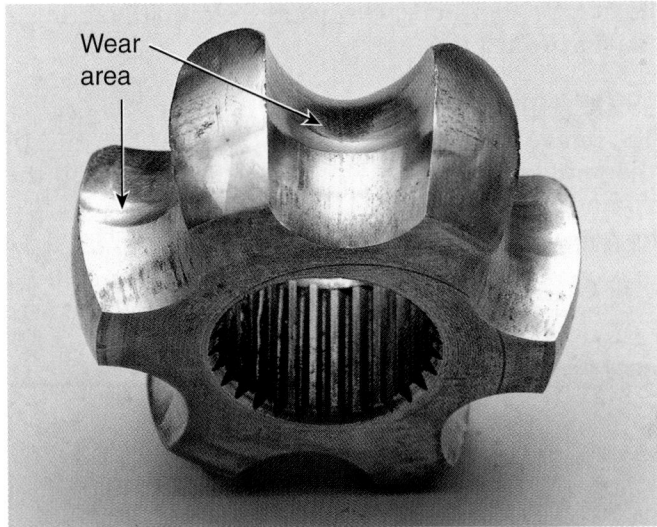

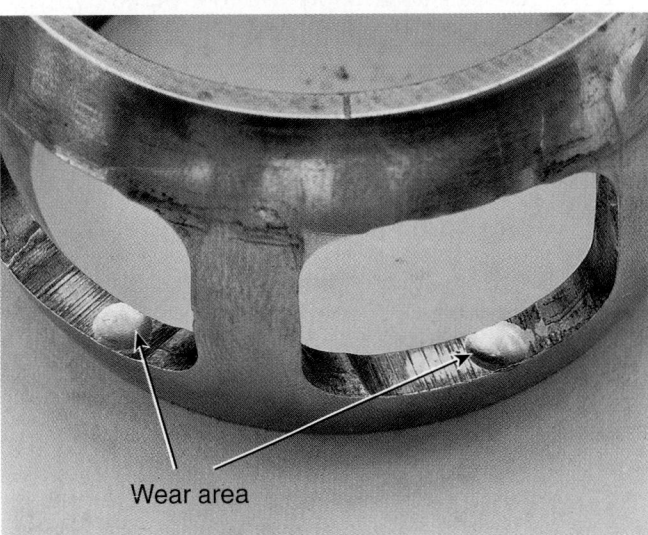

Figure 78.2 A worn cage or race can cause a clicking sound during a turn. (*Courtesy of Federal-Mogul Corporation*)

the races move back and forth in their planes (**Figure 78.2**). Opposing balls in a Rzeppa joint always work together in a pair. When one ball wears its track, the opposite side will have identical wear.

Always check the splines on a shaft. They could cause a clicking sound, too.

Plunge Joint Problems

A bad plunge joint will make a clunking sound when starting from a stop, during deceleration, or when braking. If an accurate diagnosis cannot be made with the axle on the car, remove it and disassemble it to find the cause of the problem. Excess wear is hard to visually detect, and slight, (almost imperceptible) wear can result in a problem.

AXLE SHAFT REMOVAL

There are some differences in the methods whereby axle shafts are removed and serviced. Check the service library for the proper procedure before attempting a job for the first time.

Loosen the Axle Nut

To remove the axle, first loosen the stub shaft nut in the center of the wheel while the wheel is still on the ground.

NOTE: *Most service information advises against using an impact wrench on the nut with the vehicle raised on a lift because of possible damage to parts, including the transaxle.*

> **SHOP TIP** When the vehicle has custom wheels, or if the wheel is already off the vehicle, you can block a ventilated rotor from turning. Wedge a large screwdriver in one of the vent holes. Have a helper apply the foot brake tightly while you loosen the nut.

Some axle nuts are staked to lock them in place (**Figure 78.3**). A new nut will be required when these are disassembled.

Remove the Stub Shaft from the Hub

On some cars it is easier to move the hub away from the stub shaft if the tie rod is removed first (see Figures 59.7 and 59.8).

Front-wheel-drive vehicles typically have MacPherson strut suspensions. A stub shaft is located at the end of the fixed-type outer CV joint. The inner plunge-type joint cannot move far enough inward to allow removal of the stub shaft from the hub (see Figure 77.22 for the difference between fixed and plunge joints). First, the ball joint or the strut must be disconnected so the hub can be moved away from the stub shaft.

Some vehicles have a conventional ball joint with a tapered connection (**Figure 78.4**). Others use a **pinch bolt**-type ball joint connection. A pinch bolt fits through a groove in the ball joint (**Figure 78.5**).

Figure 78.3 Most axle nuts are staked to lock them in place. *(Courtesy of Federal-Mogul Corporation)*

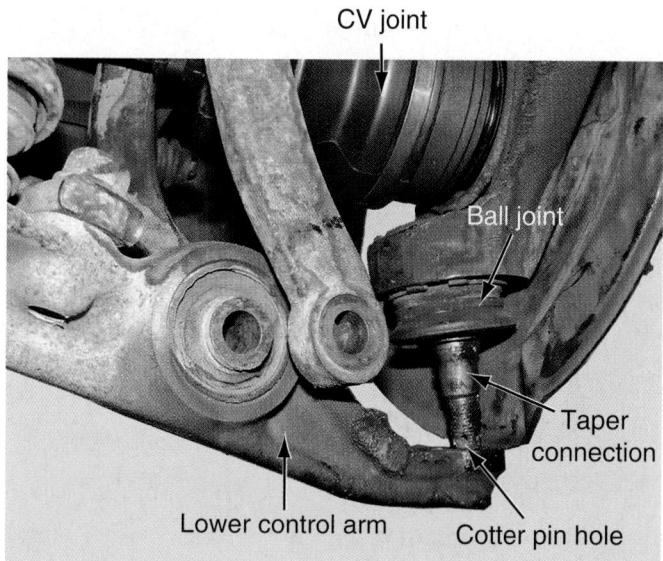

Figure 78.4 The control arm is pried away from the ball joint to allow removal of the CV joint and stub shaft from the wheel hub. *(Courtesy of Tim Gilles)*

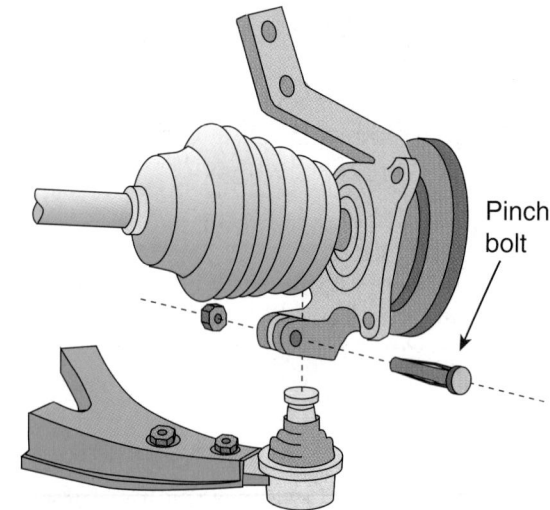

Figure 78.5 This ball joint design uses a pinch bolt.

When it is tightened, it "pinches" the slotted connection to secure it to the ball joint.

After the ball joint is disconnected, the control arm can be pried down to allow removal of the stub shaft from the hub (**Figure 78.6**). On some suspensions, removal of a link bolt might also be required (see Figure 78.4).

> **SHOP TIP**
> ■ Disconnecting the sway bar at both ends will make it easier to pry down on the control arm to separate it from the ball joint.
> ■ If you remove a strut, mark it either at the top or the bottom so that it can be reinstalled in the same position. Otherwise the wheel alignment will be off after the job is completed.

Figure 78.6 Prying against the strut to force the control arm down allows the ball joint to be separated from its connection.

Remove the Stub Shaft

Removing the stub shaft from the hub can sometimes be done by tapping it with a brass hammer. Other times a puller is required (**Figure 78.7**). When using the puller, the ball joint needs to be removed after the shaft is partway loose or the plunge joint will bottom out.

When removing the stub shaft, be careful not to tear the front wheel bearing seal on the back of the hub. After the outer joint is removed from the hub, be careful not to tear it on the suspension parts (**Figure 78.8**). Watch that the plunge joint does not come out of its housing. The boot can stretch enough to allow this.

NOTE: *Be careful not to nick the splines on the outboard axle stub shaft. Doing so will make it difficult to reinstall the splines in the hub.*

Remove the Axle

Be sure to note which kind of axle retention method is used at the transaxle. On a few vehicles, the inner

Figure 78.7 Removing a stub shaft. On some cars a puller is needed. *(Courtesy of Federal-Mogul Corporation)*

Figure 78.8 Move the stub shaft away from the hub and be careful not to tear the boot. *(Courtesy of Tim Gilles)*

end of the axle is simply unscrewed from a flange (**Figure 78.9**). A slide hammer is used to remove some types of splined inner joints (**Figure 78.10**). Other types might simply be pried away with a screwdriver (**Figure 78.11**). If you are not certain how the inner end of the axle is restrained, check the service information. You might damage an aluminum transaxle if you use excessive force.

If the plunge joint has a spline, oil will leak from the differential when the axle shafts are pulled from the housing of a manual transaxle. If you did not drain the transaxle first, place a drain pan under the axle before removing it. Replace the axle seal in the transaxle if it is hard or was leaking.

SHOP TIP Use a clean pan to collect all of the fluid that spills from the transmission. The fluid can be reinstalled. This way you do not have to locate the correct type of transmission fluid or measure the quantity that you install. Be absolutely certain not to allow the fluid to drain over parts contaminated with road grime.

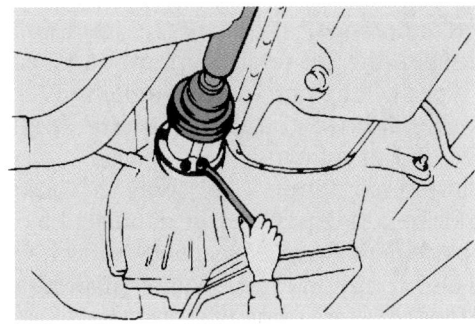

Figure 78.9 Sometimes a plunge joint is unbolted from a flange.

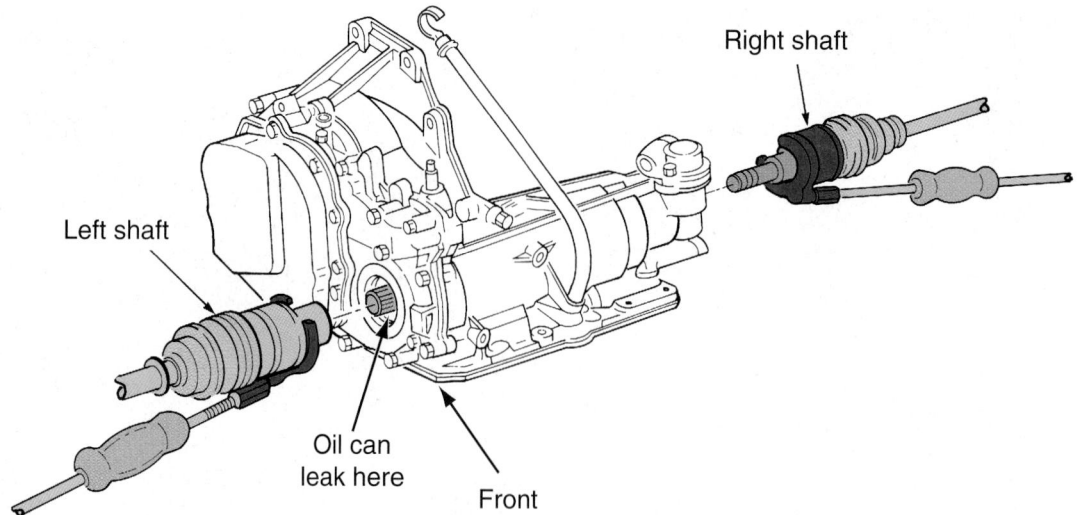

Figure 78.10 A slide hammer is used to remove some types of splined inner joints.

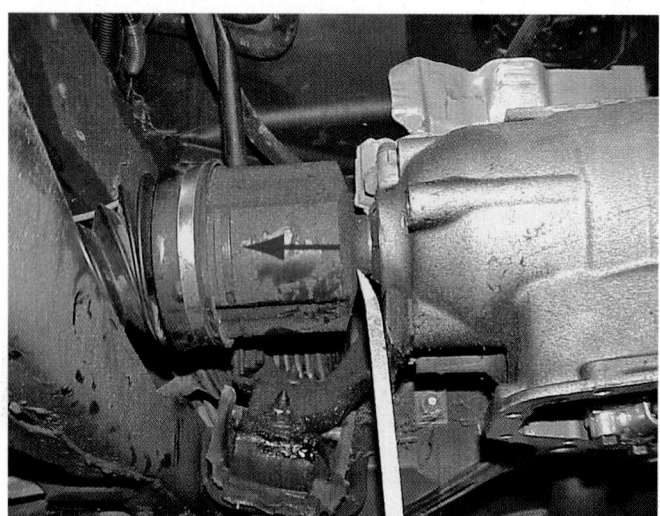

Figure 78.11 Prying a plunge joint out of a transaxle. *(Courtesy of Tim Gilles)*

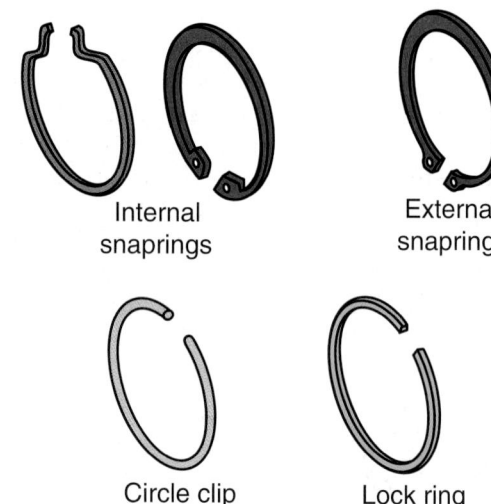

Figure 78.12 Several types of retaining rings.

Boot Removal

Position the axle in a vise. Cut off the boot clamps and remove the boot.

NOTE: *Always mark the location of the end of the old boot on the axle with tape so the new one will have the clamp installed in the same position.*

Clips and SnapRings. Wipe off the joint and inspect it to see what kind of retaining method is used on it. Several types of clips are used to retain CV joints on the axle shafts or the transaxle (**Figure 78.12**). Clips are not reused so they must be replaced.

Circlips are common. They come in various thicknesses and sizes. Measure the wire diameter before selecting a replacement. *Internal snaprings* and *external snaprings* are also common. They require snaping pliers for removal. *Lock rings* are sometimes installed in a groove to locate or limit movement of a part on a shaft.

■ CV JOINT REPLACEMENT

An internal circlip is the common way of retaining a fixed joint to the shaft. The clip rides in a groove in the shaft. Its outer diameter goes against the inner race (**Figure 78.13**). At the other side of the inner race there is often a lock ring or a thrust washer. It will be the only thing visible when this type of joint is wiped off during an inspection. This joint design is called the **knock-off-type CV joint**. It is removed using a brass punch and hammer. There is a chamfer on the inner race that helps compress the clip during removal.

Some Japanese vehicles use another version of the knock-off-type joint. A circlip rides in a groove in the axle and fits into a corresponding groove inside the inner race (**Figure 78.14**). To remove this joint, pressure must be applied to the inner race. There is a special tool for doing this (**Figure 78.15**). If you use this tool to knock off another type of fixed joint that

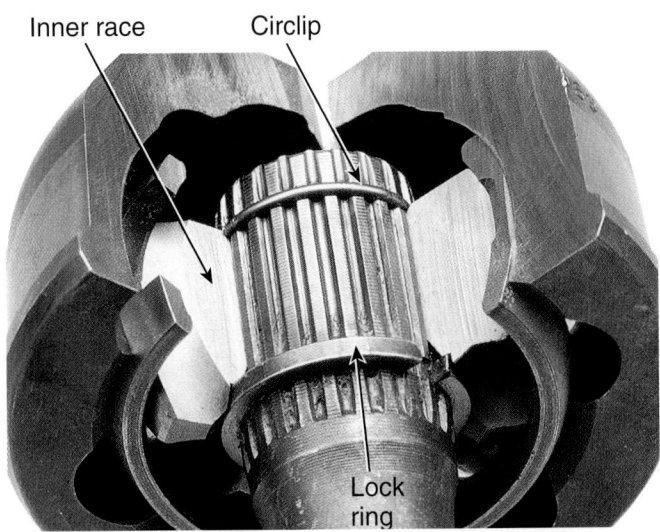

Inner race   Circlip

Lock ring

Figure 78.13 This inner race has a lock ring on one side and a circlip on the other. *(Courtesy of Federal-Mogul Corporation)*

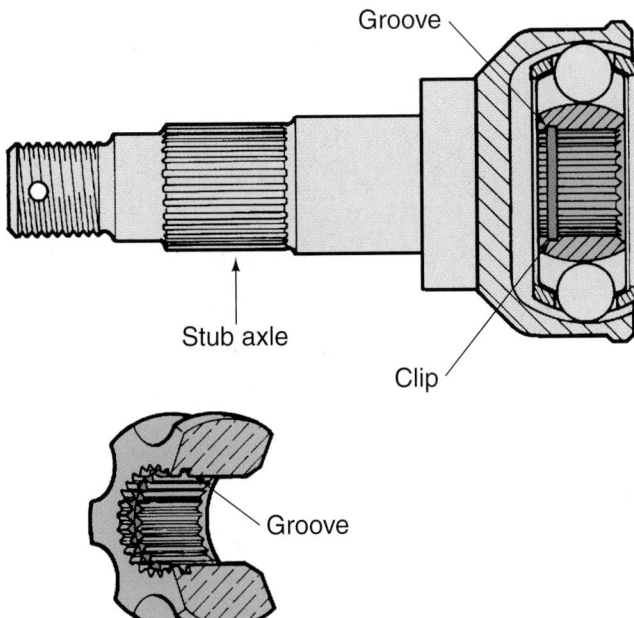

Groove

Stub axle

Clip

Groove

Figure 78.14 Some CV joints have a circlip that fits into a groove inside the inner race. *(Courtesy of Federal-Mogul Corporation)*

has a lock ring, be sure you're not driving against the lock ring.

When the CV joint has been removed from the axle, the boot is removed.

Fixed Rzeppa joints are held to their shafts by one of two methods. For one type, when the joint is wiped off during an inspection, an external snapring will be visible. There is a recess in the inner race where the tabs on the snapring fit (**Figure 78.16**). After removing the snapring with snapring pliers, the CV joint can be removed from the axle shaft. The other type of retaining method uses an internal clip that fits into a

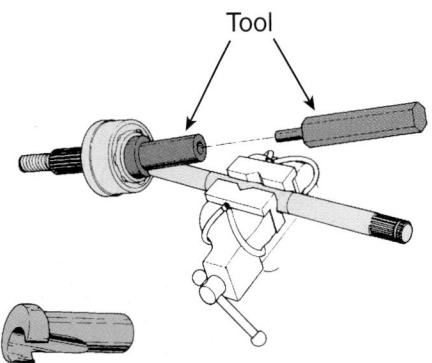

Tool

Figure 78.15 A special tool used to remove knock-off-type CV joints. *(Courtesy of Federal-Mogul Corporation)*

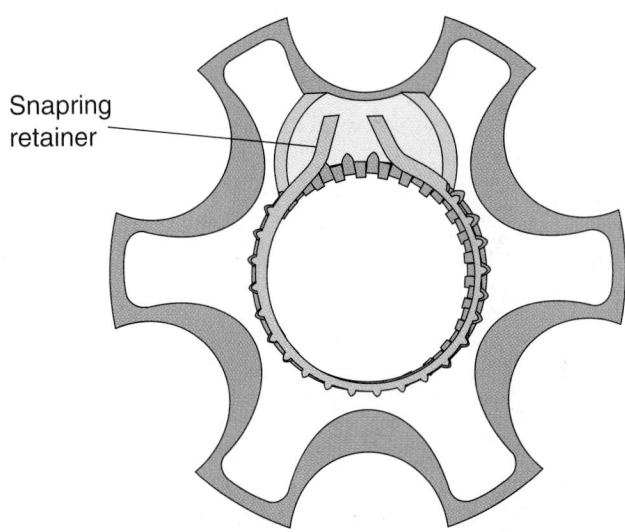

Snapring retainer

Figure 78.16 The tabs of the snapring fit into a recess in the inner race. *(Courtesy of DaimlerChrysler Corporation)*

groove in the shaft. It expands into the joint's inner race. Rzeppa joints retained in this manner are sometimes called knock-off joints because they are driven off using a soft hammer.

▨ FIXED JOINT DISASSEMBLY AND INSPECTION

After the boot is removed, use a drift punch or a special tool to move the inner race to the side for a more complete inspection (**Figure 78.17**). Wipe the parts off and look for evidence of wear (**Figure 78.18**). If the metal is colored blue, this is acceptable if the bearing surfaces are all smooth and unworn.

To disassemble a Rzeppa joint, remove the balls (**Figure 78.19**). First, mark all major components with a felt marker so they can be reassembled in their original positions. To remove the balls, the inner race must be turned sideways. This can be done using a brass punch or a special tool. After removing the balls, the inner race can be turned so that it can be removed from the stub shaft (**Figure 78.20**). All parts are cleaned and inspected once again for wear.

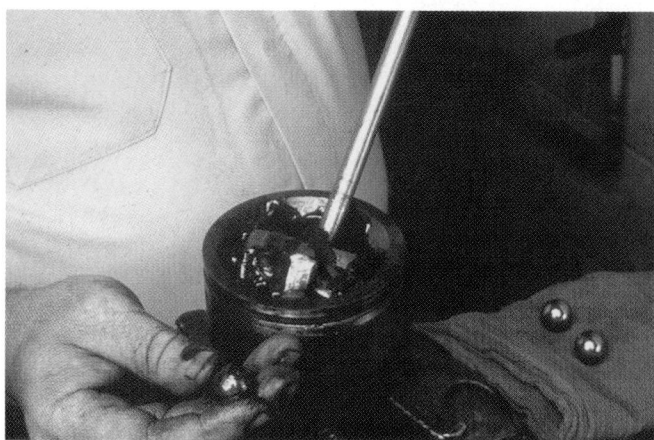

Figure 78.17 Move the CV joint to the side for cleaning and inspection. (*Courtesy of Federal-Mogul Corporation*)

Figure 78.18 These CV joint balls are pitted. The joint must be replaced. (*Courtesy of Federal-Mogul Corporation*)

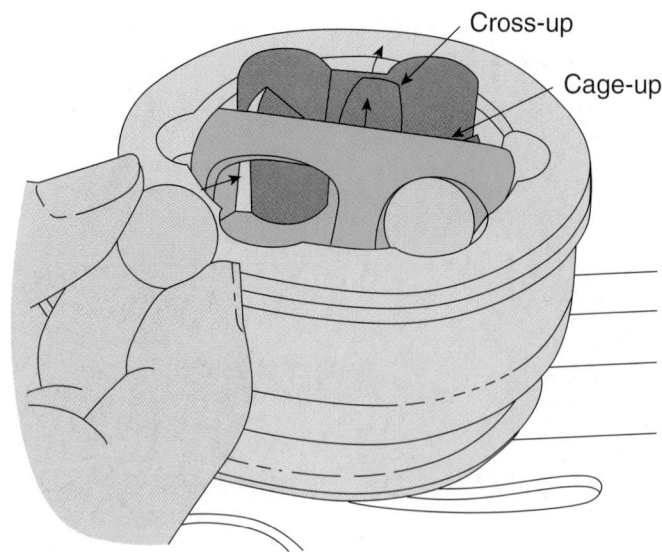

Figure 78.19 With the cage turned to the side, the balls can be removed. (*Courtesy of DaimlerChrysler Corporation*)

Figure 78.20 After removing the balls, the inner race can be turned so that it can be removed from the stub shaft.

Grease in a CV joint has a very difficult job to do. Parts of the CV joint should *not* be cleaned in solvent. The solvent film left behind can ruin the grease. Some technicians like to remove the balls one at a time, clean and inspect them, and put them back. This makes reassembly easier. After wiping the joint clean, lightly grease all parts. Reassemble the joint and install it on the axle using a new clip.

Install the Boot and Joint

The boot kit includes one or more tubes with the correct amount of CV joint grease. Be sure to follow any instructions that come with the kit. Some of the grease is applied to the CV joint itself and the rest goes into the boot.

The lock ring can be temporarily removed from the shaft to make installation of the boot easier. After the boot is slipped over the end of the axle, install a new circlip. Be careful not to overstretch the circlip. After the joint is pounded onto the shaft, pull on it to see that its snapring or circlip is seated in its groove.

Install the end of the boot and its clamp. A large screwdriver with a dull tip is handy for installing the large end of the boot over the CV joint.

■ CV JOINT BOOT CLAMPS

There are several types of clamps (**Figure 78.21**).
- A *universal clamp* is a continuous strap made of stainless steel. It comes in two sizes. A special tool pulls it tight around the boot. A universal clamp will work on any boot except a urethane boot. Spray the clamp with penetrating oil to reduce friction during installation.
- The **bridge type clamp**, the most common, requires a special crimping tool.

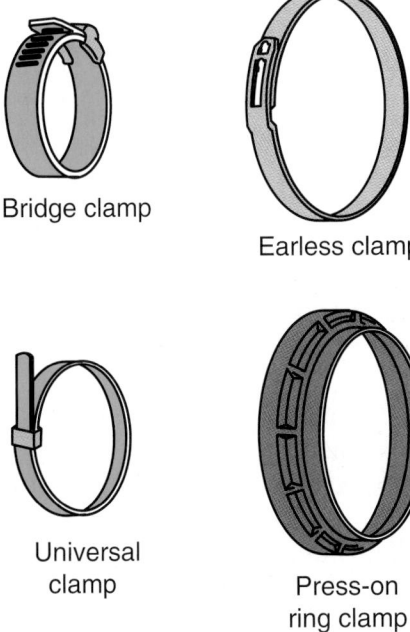

Figure 78.21 Types of CV joint boot clamps.

Universal strap- or bridge-type retaining clamps (**Figure 78.22**) can be installed in either of two directions. Position the clamp so that its tail (the portion that sticks out) will not be able to wrap itself up on road debris. If it faces so the tail follows the band on the clamp as it rotates toward the front, debris will not catch on it easily.

■ Urethane boots are stiffer and a tighter crimp is necessary, so a heavy-duty crimping tool is needed. The tool is used with a torque wrench to ensure the correct clamping load on the boot (**Figure 78.23**). Bridge-type universal clamps that can be cut to length are available.

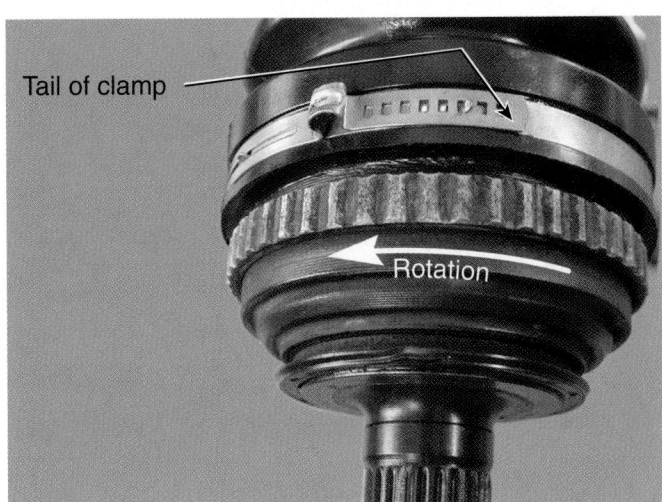

Figure 78.22 Install the clamp with its tail facing away from the axle's direction of rotation. *(Courtesy of Tim Gilles)*

Figure 78.23 Crimping a bridge clamp on a urethane boot using a torque wrench and a special tool. *(Courtesy of Federal-Mogul Corporation)*

■ An **earless clamp**, also called a *low profile* or *locking clamp*, is used when there is a problem with the axle clearing other parts. Another type of clamp might accidentally get knocked off.
■ Another type of clamp is a pressed fit ring. It must be installed in perfect alignment or it will become overexpanded. These clamps are not reusable.

NOTE: *Do not use a regular hose clamp to replace a boot clamp. It will probably ruin the boot. In a pinch, a nylon tie strap can be used for a day or two until a clamp can be purchased.*

Installing the Small-end Clamp

After pulling the big end of the boot over the CV joint and install its retaining clamp, vent the small end of the boot on the axle (**Figure 78.24**). Slide the small end of the boot until it is even with the mark you made on the shaft before disassembly and install the clamp in that position.

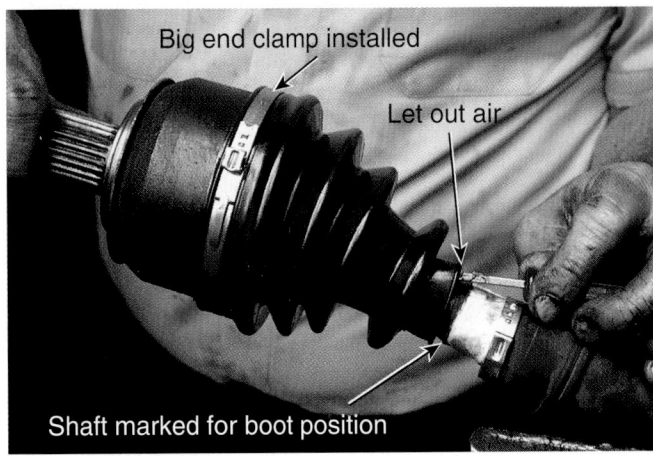

Figure 78.24 After installing the large clamp on the boot, let the air out of the small end before installing the clamp. *(Courtesy of Federal-Mogul Corporation)*

■ SERVICING AN INNER TRIPOD JOINT

Before removing the inner boot, tape the shaft and joint to show where the clamps should be reinstalled (**Figure 78.25**). This is especially important on plunge joints because they must be able to move in and out. Mark the location of the housing to the shaft so parts can be reassembled in their original locations. Cut away the clamps and boot. Some tripods will simply come out of the tulip. Others will have retaining tabs, which must be bent out of the way, or a wire ring retainer.

Check to see if the tripod rollers can come off of the spider. Put tape around a tripod so the balls and rollers do not come out (**Figure 78.26**). It can be difficult to get all of the rollers back in. Mark the tape to locate the tripod to the shaft on reassembly. It is not necessary to disassemble these. Simply roll them to feel that they are free and smooth.

Inspect the condition of the tulip. Feel its grooves for depressions in the metal. A polished appearance in the center of travel is normal. The tripod will have to be removed to install the boot.

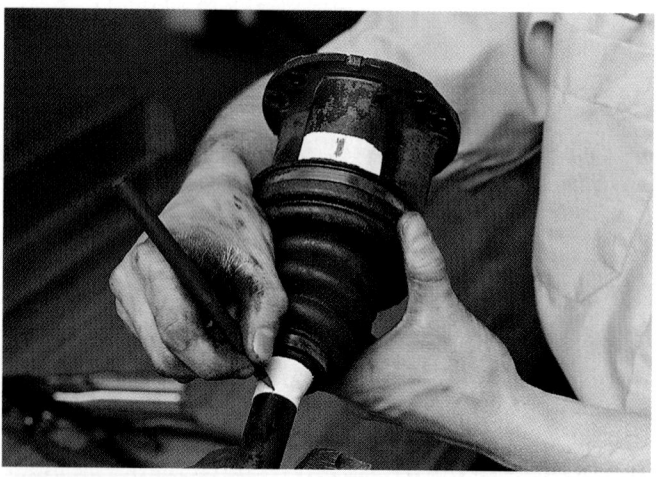

Figure 78.25 Put tape next to the ends of the boot and make marks to make reassembly easier. *(Courtesy of Federal-Mogul Corporation)*

Figure 78.26 Put tape around the rollers so they cannot come off, spilling out their needle bearings. *(Courtesy of Federal-Mogul Corporation)*

When reinstalling the joint onto the axle shaft, be sure to align any marks made during disassembly. Use a soft-faced hammer to tap the joint back onto the shaft. If it has a circlip, you will hear it click when it reaches the correct position.

> **SHOP TIP**
>
> On some axles, the boot can be installed from the other side while the fixed joint is off.

After the boot and tripod are reinstalled, grease the rollers and fill the boot with the rest of the grease. Put the tulip back over the tripod and bend any retaining tabs as needed. Position the joint at mid-travel and crimp the clamps.

Remove the retaining ring that holds the tripod to the shaft and remove the tripod. Some have different types of circlips and others have to be driven off.

■ DOUBLE OFFSET PLUNGE JOINTS

Double offset plunge joints vary in how they are connected to the axle shaft. Most are held by an internal circlip. The joint must be disassembled before it can be removed. This kind of joint is identified by a large round retaining ring in a groove in the joint's outer housing (**Figure 78.27**). The other kind of double offset joint has a circlip that can be removed by pounding on the outer housing. A metal retaining ring holds the parts in the housing.

■ CROSS GROOVE JOINT SERVICE

A cross groove joint is serviced in a similar manner to Rzeppa joints. It is disassembled by turning the inner race and cage perpendicular to the outer cage and removing them.

Figure 78.27 This double offset plunge joint is disassembled by removing a retaining ring. *(Courtesy of Federal-Mogul Corporation)*

REBUILT HALFSHAFTS

Installing a complete **rebuilt halfshaft** has become a popular repair. Although both the outer and inner joints are replaced, an entire rebuilt shaft is sometimes less expensive than just one joint kit, especially with some imports. The determining factor in whether a technician will rebuild a shaft or buy a complete rebuilt shaft is often whether there is other work to be done. With the reduced cost of parts and labor, installing rebuilt halfshafts is often in the best interest of the customer.

INSTALLING THE AXLE

To install a spline-type joint into the transaxle, push in on the axle while turning it to align the splines. Push the splines all the way in. If there is a circlip, it will click when it is in its groove. Pull on the housing of the joint to see if the snapring is seated. Do not pull on the axle shaft. The plunge joint could come apart.

After the inner end is installed, position the stub shaft into the hub as far as it can go. Install the washer and a new axle nut and tighten the nut until it is snug. Reinstall the front end components. Do not forget to torque the ball joint stud bolt if that was reassembled. A new pinch bolt is recommended. Leaving these loose can cause a serious accident. Some pinch bolts have a knurled area under the head to keep them from turning. Be sure to turn only the nut.

A staff member brought her car into a school shop to have the students replace her CV joint boots. After the job was completed, she drove the car home with no problem. But when she backed down her driveway the next morning the wheel fell off to one side and turned 90 degrees to the other. She left a big skid mark on her driveway and ruined her tire before she realized there was a problem. The car was towed to the shop, where it was discovered that the pinch bolt had come loose and the ball joint had fallen out of its hole. Luckily no one was killed in an accident that surely would have happened if the wheel had fallen off while on the highway.

When the car is back on the ground, torque the new stub shaft nut. The torque on the spindle nut can be substantial. A planetary torque multiplier works well for this (see Figure 7.46). Use a dull chisel to *stake* the nut into the slot in the spindle (see Figure 78.3). Always check the clearance of the boot clamps before test driving the car.

TRANSAXLE REPAIR

Transaxle repair is similar to transmission repair on rear-wheel-drive cars. On transaxles used with transverse engines, the differential gears are helical gears,

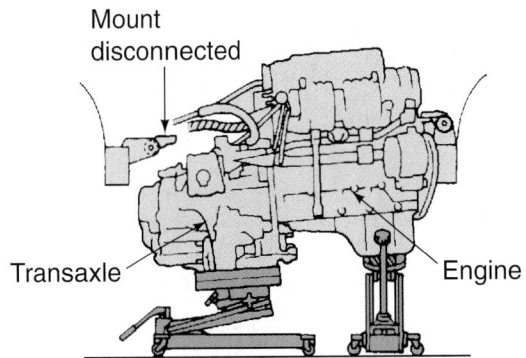

Figure 78.28 Removing the transaxle using a pair of jacks.

Figure 78.29 Lowering the transaxle to the ground.

like the rest of the gears in the transmission section (see Figure 77.8). Because these are not hypoid gears, there are no pinion depth or backlash adjustments. The only adjustment to the differential is side bearing preload. Manual transaxle clutch adjustment is the same as for rear-wheel drive.

TRANSAXLE REMOVAL

On front-wheel-drive cars, it is often easier to pull the engine with the transaxle. If it is easier to leave the engine in the car, be sure to support the engine. The engine can sometimes be supported from the top with a special fixture (see Figure 56.19). Most shops that do this type of work will have one.

A pair of jacks can be used to perform the job with the car supported on jackstands (**Figure 78.28**). The transaxle is unbolted from the engine and lowered to the ground (**Figure 78.29**).

The axles are pulled in the manner described previously. After the axles are removed, removing the transaxle is the same as with rear-wheel-drive cars.

MANUAL TRANSAXLE REPAIR

Separate the halves of the case on a transaxle (**Figure 78.30**). As you did with a RWD transmission, check the service manual for any specifics prior to

Figure 78.30 Separating the halves of the transaxle.

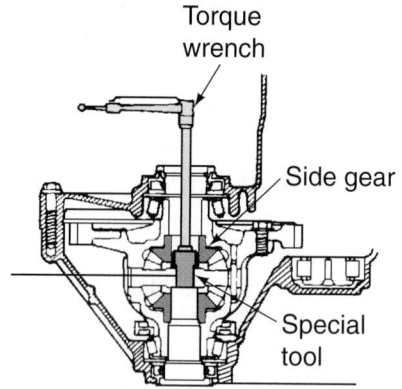

Figure 78.31 A torque wrench and special tool are used to check how difficult it is to turn the side bearings.

disassembly. Service and assembly of the components are similar to that of RWD transmissions. Those items and press and puller procedures are covered in detail in Chapter 72.

End play of the gears is checked with a feeler gauge or dial indicator. Selective thrust washers and snaprings can be used to make adjustments. If you are disassembling a transmission, be sure to keep all of the snaprings and shims wired to their respective parts.

■ AUTOMATIC TRANSAXLE REPAIR

Automatic transaxle repair is the same as that for conventional rear-wheel-drive automatic transmissions. Refer to Chapter 73 and Chapter 74 for automatic transmission theory and repair information. One difference with automatic transaxles is that either a gear or a chain is driven by the transmission output shaft to turn the differential section of the transaxle. See Figures 70.18 and 70.19, which describe the different styles.

The main difference between conventional automatic transmissions and automatic transaxles is that a transaxle has a differential. Differential service procedures are the same for manual and automatic transaxles.

■ TRANSAXLE DIFFERENTIAL SIDE BEARING ADJUSTMENT

When the case halves are reassembled, the differential side bearing preload is adjusted. Check the manufacturer's instructions. Some of them use special tools. **Figure 78.31** shows a torque wrench and special tool used to check to see how difficult it is to turn the differential.

■ INSTALL THE TRANSAXLE

Installation of the transaxle is the reverse of the removal sequence. After the transaxle is reinstalled be sure to check the transaxle lubricant level. Some transaxles have a fill plug on the side like an ordinary RWD transmission. Checking the lubricant on others is sometimes different. Some transaxles have a dipstick. Other transmissions require the removal of the speedometer gear to check the fluid level.

■ REVIEW QUESTIONS

1. What is the only adjustment to a transaxle differential?
2. How many hours is the estimated life of a CV joint that has lost its lubricant?
3. What is the most common repair to CV joints?
4. What material is a universal clamp made out of?
5. Before pulling a splined axle from a manual transaxle, what do you position under it?
6. Before removing a boot from a CV joint, what do you do to the axle?
7. When there is no snapring visible when a CV joint is wiped off for inspection, this kind of joint is called a _____- off joint.
8. Which end of a CV joint boot is installed first, the big end or the little end?
9. What do you do to a plunge joint's tripod rollers to keep them from coming apart?
10. What do you do to the stub shaft nut after it is torqued to specification?

ASE-STYLE REVIEW QUESTIONS

1. Technician A says that transverse engine transaxles use helical differential gears. Technician B says that most transaxles do not have pinion depth and backlash adjustments. Who is right?
   - **a.** Technician A
   - **b.** Technician B
   - **c.** Both A and B
   - **d.** Neither A nor B

2. Which of the following is/are true about CV joint boots?
   - **a.** Some boots tend to fail more often than others.
   - **b.** The outboard boot on the passenger side fails more often than the one on the drivers side.
   - **c.** Both A and B.
   - **d.** Neither A nor B.

3. Technician A says that it is not necessary to remove an axle and disassemble a CV joint just because the boot has failed. Technician B says to spray the CV joint boot with penetrating oil to reduce friction during installation. Who is right?
   - **a.** Technician A
   - **b.** Technician B
   - **c.** Both A and B
   - **d.** Neither A nor B

4. Which of the following is/are true when checking for axle shaft noise?
   - **a.** The weight must be off of it.
   - **b.** A clicking sound during a turn could be due to a bad plunge joint.
   - **c.** Both A and B.
   - **d.** Neither A nor B.

5. A bad plunge joint will:
   - **a.** Make a clunking sound when starting from a stop
   - **b.** Make a clunking sound when braking
   - **c.** Both A and B
   - **d.** Neither A nor B

6. All of the following are true statements about the stub shaft nut *except*:
   - **a.** Use an impact wrench to remove the stub shaft nut with the car in the air.
   - **b.** A new stub shaft nut is required when replacing CV joint boots.
   - **c.** Torque the stub shaft nut with the wheel on the ground.
   - **d.** Some stub shaft nuts must be staked with a punch after tightening.

7. Technician A says that removing a stub shaft from a hub always requires a puller. Technician B says that sometimes a stub shaft can be removed from the hub with a brass hammer. Who is right?
   - **a.** Technician A
   - **b.** Technician B
   - **c.** Both A and B
   - **d.** Neither A nor B

8. Technician A says to clean all CV joint metal parts in solvent. Technician B says to bottom out the tripod in its bore before clamping the boot to the axle shaft. Who is right?
   - **a.** Technician A
   - **b.** Technician B
   - **c.** Both A and B
   - **d.** Neither A nor B

9. Which of the following is/are true about CV joint repair?
   - **a.** An impact wrench is the best tool for installing the stub shaft nut with the vehicle in the air.
   - **b.** Installing a rebuilt halfshaft is often less expensive for the customer.
   - **c.** Both A and B.
   - **d.** Neither A nor B.

10. Technician A says that a transaxle must have the correct drive pinion depth and backlash. Technician B says to roll a gear contact pattern before assembling a transaxle. Who is right?
    - **a.** Technician A
    - **b.** Technician B
    - **c.** Both A and B
    - **d.** Neither A nor B

Driveline Vibration and Service

■ **KEY TERMS**

vibration	first order vibration	whine
frequency	second order vibration	match mount
hertz	beat/boom vibration	transducer
amplitude	EVA	launch shudder
velocity	snapshot	
natural frequency	moan	

■ **VIBRATION ANALYSIS**

Tire and wheel imbalance is the most common cause of vibration complaints. Those procedures were covered in Chapter 55. Broken engine and transmission mounts are another frequent cause. After tires, wheels, and mounts have been eliminated as a cause of vibration, the items dealt with in this chapter come into play.

Vibration does not usually shut the car down. It does annoy the driver and can result in his or her unhappiness with the repair job or the car in general.

A process of elimination is used to analyze vibration problems. The following terms are used in the analysis:

■ **Vibration** describes a part that is in motion in waves, or *cycles*. **Figure 79.1** is an example of vibration as it cycles through a material.

■ **Frequency** is the number of cycles that take place in a period of time. The number of cycles in one second is measured in **hertz** (**Figure 79.2**).

■ **Amplitude** measures vibration intensity (**Figure 79.3**). An intense vibration has more amplitude.

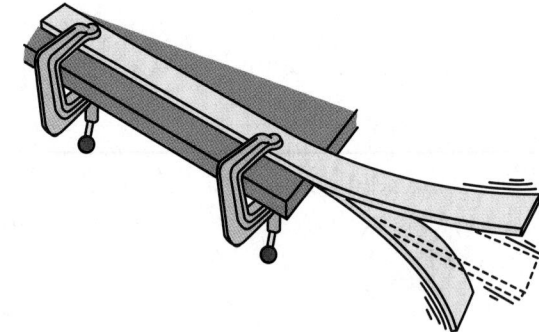

Figure 79.1 Vibration happens in cycles.

■ **Velocity** is a combination of amplitude and frequency (**Figure 79.4**). Velocity is not constant. It is an average, measured in inches per second. If amplitude is 4" and frequency is 10 Hz, this is 40"/sec velocity.

Resonant or **natural frequency** is the frequency at which a body vibrates. A tuning fork can be used to illustrate this. Unibody vehicles often have inherent vibration characteristics that cause them to vibrate

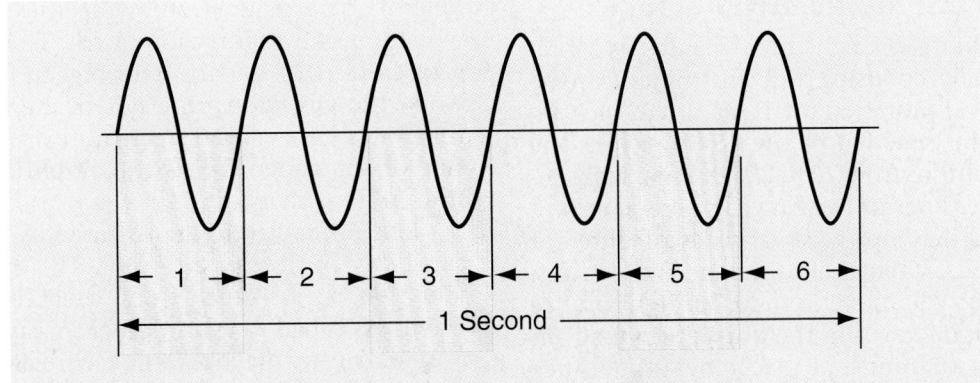

Figure 79.2 Frequency is the number of cycles taking place in a period of time.

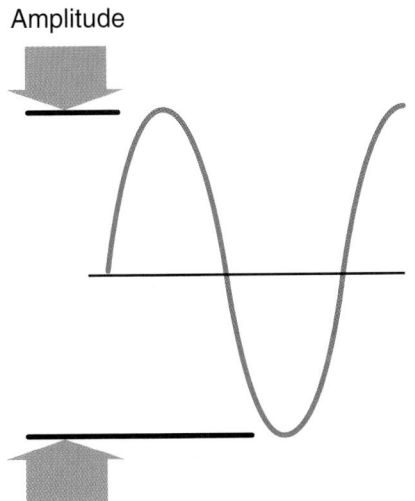

Figure 79.3 Amplitude is a measurement of the intensity of a vibration.

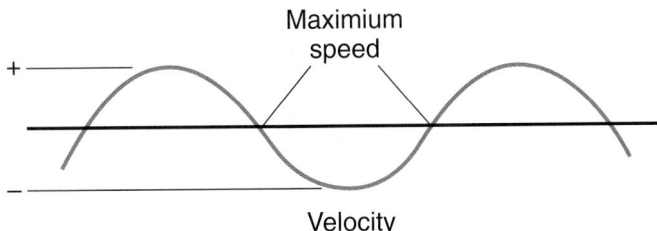

Figure 79.4 Velocity is a combination of amplitude and frequency. Maximum speed (velocity) is reached when the vibrating wave is at half travel.

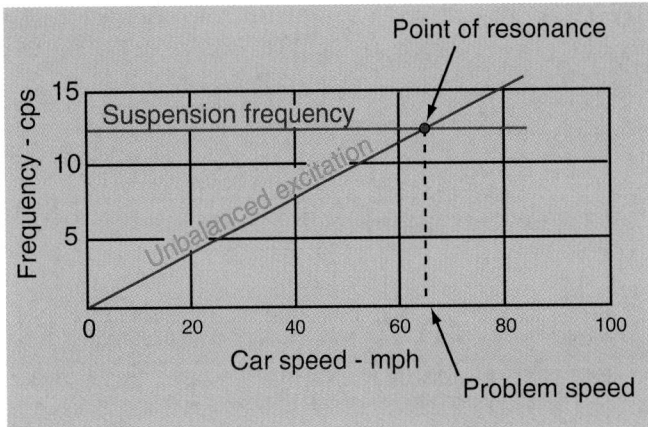

Figure 79.5 A drive shaft that is out of balance will have a frequency that can become the same as the suspension frequency at a particular speed.

at a predictable road speed. This was a problem especially with older unibody vehicles. More recently engineers have designed these vehicles so that their vibration range is outside of the normal driving range. Engines, suspension systems, and drive shafts all vibrate at different frequencies. Stiffer materials tend to have higher natural vibrating frequencies. For example, when you tighten a string on a guitar, it becomes stiffer and the sound from its vibration occurs at a higher frequency.

The severity (amplitude) of a vibration will be greatest at its *point of resonance*. An automotive suspension system has a natural frequency in the range of 10 to 15 Hz. The suspension's frequency is always the same, but a drive shaft that is out of balance will have a frequency that can become equal to the suspension frequency at a particular speed (**Figure 79.5**). This is the point of resonance, the speed where the vibration will be greatest. Front and rear suspension systems are designed with frequencies that are different by 3 Hz to avoid a combination of their vibrations.

■ TYPES OF VIBRATIONS

First order vibration is anything that spins at drive shaft speed and vibrates only once every revolution.

Second order vibration is a universal joint. Remember, as the drive shaft rotates, the U-joints speed up and slow down twice during each revolution. Two vibrations per revolution is second order vibration. This vibration is always related to road speed and is usually worse under torque.

A **beat/boom vibration** occurs when one vibration interacts with another. Canceling one of the vibrations can sometimes solve the problem.

■ VIBRATION TEST INSTRUMENTS

Because every vibration has its own frequency, vibrations can be analyzed using test instruments. One type of mechanical vibration analyzer, a *reed tachometer* (**Figure 79.6**) invented in the 1940s, senses the frequency of vibrations from 10 to 80 Hz. It is a tachometer because it measures in hertz (cycles per second).

The reed tach has two rows of reeds of different lengths. The reeds vibrate at different frequencies, lower or higher, as you move from left to right or vice versa. The vibration shown in the illustration is strongest at 16 Hz. The tool must be placed near the source of the vibration or it cannot react.

A newer and better version of the reed tachometer is the electronic vibration analyzer (**EVA**) (**Figure 79.7**). It has a 20-foot cord and an electronic pickup. The pickup is held in place with putty,

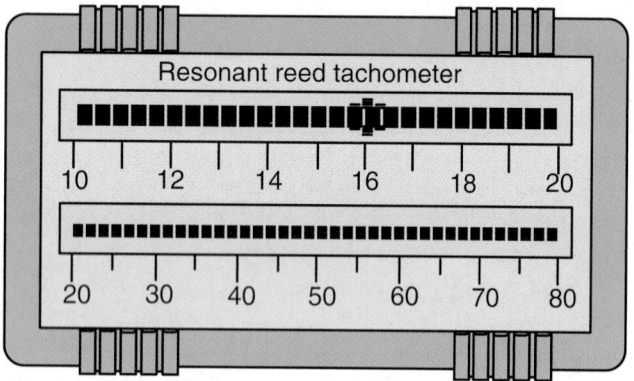

Figure 79.6 A reed tachometer.

Figure 79.7 An electronic vibration analyzer (EVA). *(Courtesy of Kent-Moore Division, SPX Corp.)*

Velcro®, or a magnet, allowing the car to be driven while checking for vibrations. The EVA measures acceleration of a part. The bigger the reading, the worse the vibration. One side of the gauge shows the frequency of a vibration in rpm or hertz. The other side of the gauge shows the amplitude of the vibration in *Gs (gravity)*.

Like other electronic instruments, the EVA has the capability of *freezing* an event. This is handy when a vibration occurs quickly and then ceases. A recorded event is called a **snapshot**. A snapshot has ten frames. During playback, the technician can freeze and unfreeze the display to isolate the event.

The EVA is handy for balancing drive shafts, too. It has a coil of wire that can hook to the inductive pickup on an ignition timing light.

■ VIBRATION AND FREQUENCY

General Motors has done extensive vibration testing. Earlier front-wheel-drive cars had unibodies with resonant frequencies that came into play in the normal driving range. This has been controlled by stiffening the floor pans until their frequency is beyond the normal range of use. Different types of vibrations and their frequencies are shown in **Figure 79.8**.

Some vibrations can be felt; others can be heard. Those that are heard go into higher frequencies, above 100 Hz. When dealing with vibrations that cannot be heard, work on the problem with the highest amplitude and ignore the rest until the worst are fixed. The following definitions can be used to isolate vibrations that can be felt, even if you do not have access to a vibration analyzer.

■ *Shake* is a low-frequency vibration of 5 to 20 Hz. It is commonly the feeling an unbalanced tire produces, felt through the steering wheel or the seat. Shake is also called shimmy, wobble, shudder, or hop. When the vibration is road-speed-related, it can be caused by tires, wheels, or brake drums. When it is engine-rpm-related, check the engine and accessories.

■ *Roughness* is a higher frequency vibration than shake (20–50 Hz). It is usually related to driveline parts and is similar to the feeling one gets when cutting wood with a jigsaw.

■ *Buzz* comes in at a frequency of 50 to 100 Hz, about the frequency of an electric razor. It is often caused by a vibration in the exhaust system or an engine accessory like the air conditioning compressor. If the buzz is road-speed-sensitive, a driveline problem can be the cause.

■ *Tingling* is vibration above 100 Hz. It feels like pins and needles in the steering wheel or through the floor. The reed tachometer does not measure above 100 Hz. The EVA can pick up tingling problems. Again, road-speed-related is drivetrain, and engine rpm-related is engine.

Engine speeds and frequencies

First order any engine			Firing frequency		
	RPM	Hz	4-Cylinder second order	6-Cylinder third order	8-Cylinder fourth order
Shake	500	8.3	16.8	24.9	33.2
	750	12.5	25	37.5	50
	1000	16.5	33.3	49.8	66.6
	1500	25	50	75	100
Roughness	2000	33.3	66.6	99.9	133.2
	2500	41.6	83.2	124.8	166.4
Buzz	3000	50	100	150	200
	3500	55.3	110.6	174.9	233.2
	4000	66.6	132.4	199.8	266.4

Figure 79.8 Different types of vibrations and frequencies.

The following vibrations can be heard:
- *Boom* is a low-frequency noise in the range of 20 to 60 Hz. It is described as droning, moaning, or humming. Pressure can be felt in the ears. There can be a feeling of roughness with it. Look at the chart. Boom is normally caused by the driveline, like shake.
- **Moan** or *drone* is a noise that is compared to the hum of a bumblebee. Its range is from 60 to 120 Hz. Powertrain mounts or exhaust system problems are possible causes.
- *Howl* is a noise like the wind across the top of a soda bottle. Its range is from 120 to 300 Hz.
- **Whine** sounds like a turbine or mosquitos. The range is from 300 to 500 Hz. It is usually related to gear noise.

Road Test

During the road test, use a process of elimination. Driveline vibration is slightly different than the feeling that results from tire imbalance. It is not normally identified as a possible problem until wheel balance has been done first and a vibration still remains. The vibration usually occurs at a specific rpm and goes away at other speeds, just like wheel balance.

When test driving for drive shaft vibration, drive in high gear at the engine rpm where the vibration is worst. Then, shift into a different gear or put the transmission into neutral to see if the vibration changes or stops. When there is no change in the vibration, the driveline is a possible source of the problem. If the problem does change, one of the components whose rpm changed with the gear change is responsible. These include the engine, clutch or torque converter, or the transmission.

Be sure you understand the difference between amplitude and frequency when attempting to diagnose vibration problems. When road testing, note the speed at which a vibration occurs. A vibration should always occur at the same speed or frequency. If a second road test results is less intensity but occurs at the same speed, the amplitude has changed. Drive the vehicle and note the speed at which vibration amplitude is worst. Then, repair the suspected problem and retest at the same speed.

Modern test instruments for vibration are so sensitive that you could be attempting to fix a problem that cannot be felt. If you cannot feel the vibration, do not try to fix the problem.

Testing in the Service Bay

To test for a first order vibration, raise the tires in the air and run the engine in gear to check the driveline. Move the pickup of the vibration analyzer around and try to determine where the energy from the vibration is located. If there is no vibration without a load on the tires, this is not a balance problem and is probably related to misalignment of parts. Angles are covered later in this chapter.

While the drive shaft is spinning, check for vibrations at its front and rear. A torque-sensitive problem is usually in the differential. This is often caused by a bearing cap that is not seated squarely in its bore.

■ DRIVE SHAFT RUNOUT

After checking for vibration, check the drive shaft for runout. When there is excessive runout in a drive shaft, suspect the drive shaft or the pinion flange. First, measure the runout of the drive shaft (**Figure 79.9**).

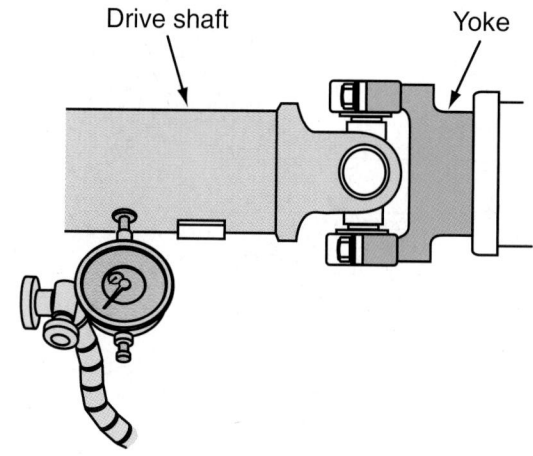

Figure 79.9 Measuring drive shaft runout.

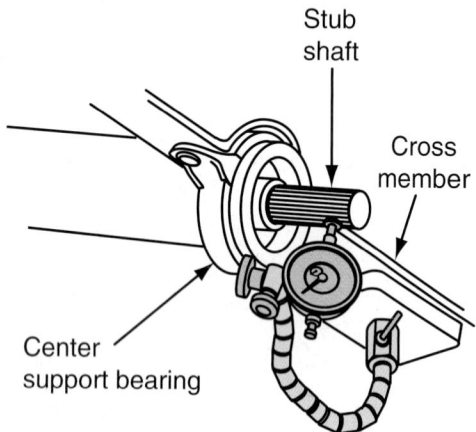

Figure 79.10 Measuring stub shaft runout.

Mark the high spot on the shaft. Then remove it and remount it at a spot 180 degrees away on the flange. If the runout remains the same, the shaft is at fault. Otherwise the flange is the problem.

One part of a two-piece shaft is called a stub shaft (**Figure 79.10**). These are allowed 0.003" runout. Excessive runout here is usually the cause of vibrations felt at the center support bearing.

Some manufacturers **match mount** instead of balancing to correct runout. This means that runout of one part and runout of the other are considered during assembly. This is especially prevalent on front-wheel-drive axles, which are often marked for reassembly. If a drive shaft was off by 0.010", and a flange was also off by 0.010", you could have either a perfectly straight combination or one that is 0.020" off.

NOTE: *A part with runout cannot be balanced if it causes parts to bind up once per revolution. Consider that drive shaft parts that are square can be balanced. This is not true with tires—because they touch against the road surface, they must not have runout. You cannot balance a box and expect it to run smoothly on the road.*

Repairing or Replacing a Drive Shaft

Drive shafts can be aluminum or steel. Most steel drive shafts are shorter than 60" to 65". A 3" to 3.5" diameter passenger car drive shaft should be shorter than 60" to 70". Aluminum drive shafts can be 85" to 90" because aluminum has a higher "critical speed." Longer steel shafts use a center support bearing. Do not swap an aluminum drive shaft for a steel one unless the length is less than 60" to 65".

Drive shafts can be straightened by a specialty machine shop. They heat and quench it until it is straight within 0.010". Longer shafts are allowed less runout in the center.

■ OTHER CAUSES OF VIBRATION

Other things besides the driveline can cause vibrations—for example, engine accessories. Test to eliminate outside causes of vibrations in the repair bay. If a vibration is still evident when the car is not moving, it cannot be a driveline problem. Some engines are especially prone to vibrations. Engine accessory brackets have been designed to *excite* at 400 Hz or higher. This is equivalent to 6000 engine rpm on a V8, a driving condition that the average driver should not encounter.

The engine, torque converter, and drive shaft change in frequency as rpm changes. A rebuilt torque converter is often the cause of a vibration that is rpm related.

Check to see that a muffler is not vibrating against the frame. All muffler hangers should be in place and undamaged.

■ DRIVE SHAFT BALANCE

Drive shafts are balanced on the ends, not in the middle. The drive shaft can be compared to a very wide, low-profile tire. At the factory, weights are welded onto the ends of the drive shaft when the shaft is balanced. The balance weights should be at least 1" down the shaft from the weld so the integrity of the weld is not disturbed. Power takeoffs (PTOs) and shafts that spin at under 1000 rpm need not be balanced. Specialty shops in the aftermarket shorten or lengthen drive shafts and do balancing.

When the drive shaft is out of balance, a technician can correct it using an on-the-car strobe-type wheel balancer. Heavy spots are counterbalanced by installing hose clamps with the screw positioned on the opposite side of the heavy spot (**Figure 79.11**). The pinion nose end of the drive shaft is where most of the first order driveline vibrations come from. Drive shaft flanges are sometimes balanced. This is to compensate for runout created during manufacturing. Weight will either be on the front of a flange or the outside depending on whether the correction was for flange runout or differential balance (**Figure 79.12**). Runout of the flange is checked with a dial indicator

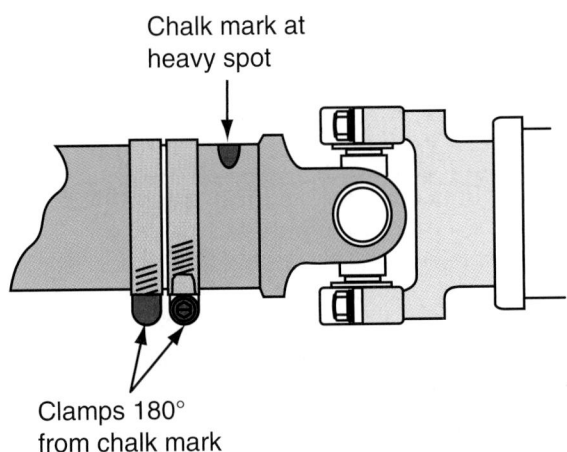

Chalk mark at heavy spot

Clamps 180° from chalk mark

Figure 79.11 Counterbalancing a heavy spot with hose clamps.

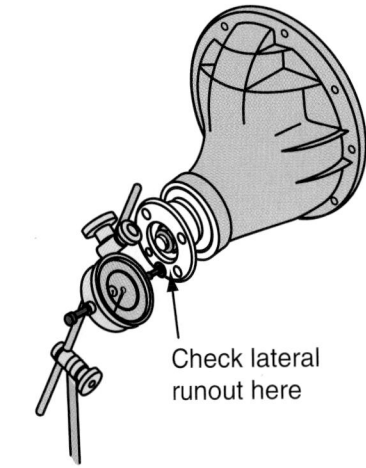

Check lateral runout here

Figure 79.13 Checking flange runout.

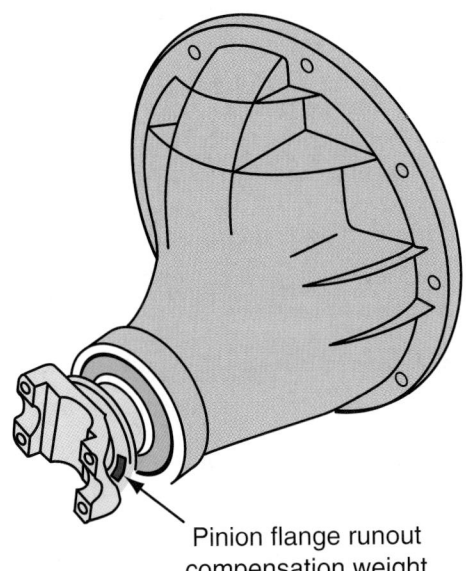

Pinion flange runout compensation weight

Figure 79.12 When there is weight on a pinion flange, it was put there to correct for runout.

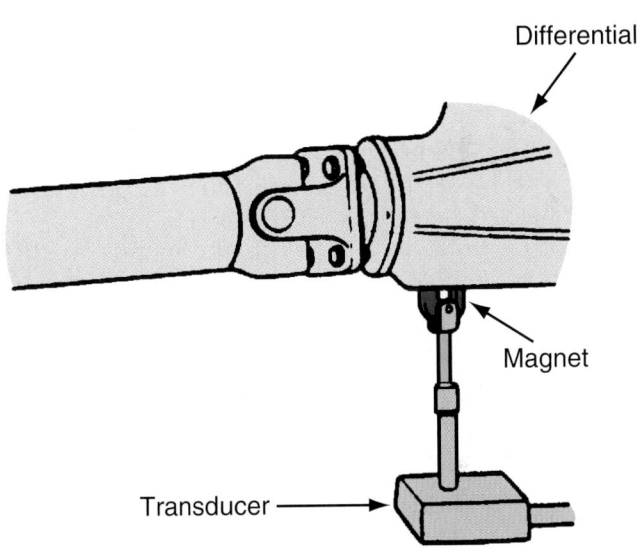

Differential

Magnet

Transducer

Figure 79.14 A transducer is used to pick up vibrations.

(**Figure 79.13**). It is limited to 0.006" when there is no weight on the flange.

Checking Drive Shaft Balance

One end of the drive shaft is balanced first. After the other end is balanced, the first end is rechecked. A **transducer** is a good tool for testing for drive shaft imbalance. This is part of an on-the-car wheel balancer. It is mounted on the differential case, under the pinion gear (**Figure 79.14**).

A transducer is a magnet that is spring loaded on both sides. It moves back and forth in a coil of wire, sending a signal to a meter in response to vibration. The drive shaft is marked anywhere with a piece of chalk. When the drive shaft is spun at the speed where the vibration occurs, the transducer triggers a strobe light. When the light illuminates the mark on the drive shaft, make a mental note of its position. When the

shaft is stopped, position the mark at the same place (6 o'clock, for instance). At this position, the heavy spot on the drive shaft is on the bottom of the shaft.

Large hose clamps are positioned on the shaft opposite the heavy spot. Then, the shaft is spun again to see whether more clamps are needed to correct the problem. You can move the hose clamp to one side to see if the vibration improves. If it gets worse, try it the other direction. The EVA has a pickup wire that can be used with an induction timing light that will determine the heavy spot.

Drive shaft balance can also be checked without special instruments. First, put four numbers around the circumference of the drive shaft (**Figure 79.15**). Because imbalance can be related to runout, put the #1 mark at a point opposite to the amount of greatest runout. Put the clamp with the weight at the #1 position and run the shaft. Check to see if it feels better

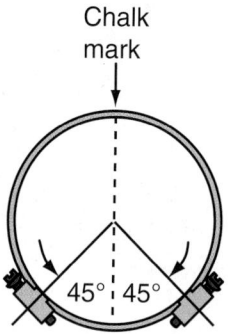

Figure 79.15 Two hose clamps can be separated or brought closer together to change the amount of weight.

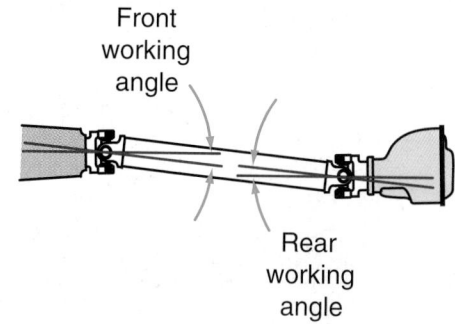

Figure 79.16 The working angle is the difference between the angle of the drive shaft and the angles of the transmission or differential.

or worse. Then, move it to the #2 position and try it again. Try all four positions. You are trying to find out where to put the weight.

When you find the best position, move the hose clamp halfway between the numbers to one side to see if the vibration improves. If it gets worse, try it the other direction. When you find the best position, add another hose clamp and repeat the procedure. The clamps can be separated to fine tune the amount of weight needed (see Figure 79.15).

Another procedure is to run the engine to turn the shaft while holding a piece of chalk near the shaft. The heavy spot will tend to be marked by the chalk. The hose clamp will go opposite to the chalk mark.

◼ DRIVE SHAFT ANGLE

To extend U-joint service life, the angle of the transmission output shaft and the front of the differential should be within ½ degree of each other. If not, either the transmission or differential should be shimmed with tapered wheel alignment shims. A ¼" shim will result in a change of about ¾ degrees. Leaf springs can be shimmed on the rear axle housing or the transmission can be shimmed on the crossmember mount. A faster shaft rpm requires a lower angle.

NOTE: *Check the condition of engine and transmission mounts. They are a frequent cause of drive shaft vibration.*

The *working angle* is the difference between the drive shaft angle and the angle of the transmission or differential (**Figure 79.16**). For passenger cars, the working angle should not be more than 4 degrees. There must always be at least a small angle (at least ½ degrees) to keep the needle bearings rolling, or brinelling will occur.

The trim height (ride height) measurement (see Figure 68.7) is important to universal joint angle, and also to front-wheel-drive tripod joints (see Chapter 77).

Four-wheel drives have a front shaft that is short and has a sharp angle. Riser kits that make room for tall tires cause an even sharper angle. Slip yokes and splines tend to wear out in these drive shafts.

Launch shudder is a common second order vibration complaint in raised trucks with a high drive shaft angle. It occurs on acceleration from a standing start to about 25 mph before disappearing.

Several methods can be used to measure the drive shaft angle. These include a digital measuring tool (**Figure 79.17**), a protractor, or a magnetically attached inclinometer. The angle is measured in both the front and the rear.

Suggested maximum drive shaft angles are as follows:

5000 rpm–3¼ degrees	3000 rpm–5⅝ degrees
4500 rpm–3⅔ degrees	2500 rpm–7 degrees
4000 rpm–4¼ degrees	2000 rpm–8⅔ degrees
3500 rpm–5 degrees	1500 rpm–11½ degrees

You can see from the chart that a four-wheel-drive truck with high drive shaft angles is better operated at low shaft rpm. If the transmission and differential angles are not the same a double Cardan U-joint will be needed.

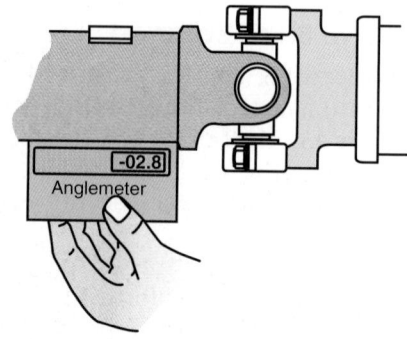

Figure 79.17 A digital drive shaft alignment tool.

REVIEW QUESTIONS

1. The number of cycles in one second is measured in _____.

2. A _____ fork can be used to illustrate resonant frequency.

3. A vibration that interacts with another is called a _____ boom vibration.

4. What is the name for the vibration that is about the same frequency as an electric razor?

5. When road testing, note the _____ at which a vibration occurs.

6. A differential vibration is often caused by a _____ cup that is not seated squarely in its bore.

7. Excessive runout of the stub shaft is usually the cause of vibrations felt at the _____ support bearing.

8. When the runout of one part and the runout of another part are considered during assembly, this is called _____ mounting.

9. The angle of the transmission output shaft and the front of the differential should be within _____ degrees of each other.

10. The difference between the drive shaft angle and the angle of the transmission or differential is called the _____ angle.

ASE-STYLE REVIEW QUESTIONS

1. Technician A says that front and rear suspension systems are designed to have different resonant frequencies to avoid wheel shimmy. Technician B says that a vibration that occurs twice every revolution is related to a universal joint. Who is right?

   a. Technician A   c. Both A and B
   b. Technician B   d. Neither A nor B

2. Which of the following is/are true about electronic vibration analyzers?

   a. They can measure acceleration of a part.
   b. They have a memory function.
   c. Both A and B.
   d. Neither A nor B.

3. All of the following are true about vibration problems *except*:

   a. An electronic vibration analyzer can be used with a timing light to balance a drive shaft.
   b. Vibrations above 100 hertz are usually felt, rather than heard.
   c. When dealing with a vibration that cannot be heard, work on the problem with the highest amplitude first.
   d. Vibrations that can be felt often occur at speeds of less than 100 times per second.

4. Technician A says that a vibration should always occur at the same speed or frequency. Technician B says that if you cannot feel the vibration, do not try to fix the problem. Who is right?

   a. Technician A   c. Both A and B
   b. Technician B   d. Neither A nor B

5. Which of the following could be a cause of excessive runout in a drive shaft?

   a. The pinion flange
   b. A bent driveshaft
   c. Both A and B
   d. Neither A nor B

6. Technician A says that a square drive shaft could theoretically be balanced. Technician B says that a square tire could theoretically be balanced. Who is right?

   a. Technician A   c. Both A and B
   b. Technician B   d. Neither A nor B

7. Technician A says that a rebuilt torque converter is often the cause of driveline vibrations. Technician B says that drive shafts are balanced in the middle. Who is right?

   a. Technician A   c. Both A and B
   b. Technician B   d. Neither A nor B

8. Technician A says that heavy spots on a drive shaft can be counterbalanced by installing hose clamps with the screw positioned even with the heavy spot. Technician B says that most of the first order driveline vibrations come from the transmission end of the drive shaft. Who is right?

   a. Technician A   c. Both A and B
   b. Technician B   d. Neither A nor B

9. Which of the following is/are true about altered vehicles?

   a. Four-wheel-drive vehicles with tall tires are more likely to have vibration problems from the drive shaft.

   b. Shuddering in a raised truck will occur at speeds above 25 mph.

   c. Both A and B.

   d. Neither A nor B.

10. Technician A says that ride height does not affect the universal joint angle. Technician B says that shafts that spin at under 1,000 rpm do not need to be balanced. Who is right?

   a. Technician A   **c.** Both A and B

   b. Technician B   **d.** Neither A nor B

English-Metric Conversion Chart

CONVERSION FACTORS

Unit	To	Unit	Multiply By
LENGTH			
Millimeters		Inches	.03937
Inches		Millimeters	25.4
Meters		Feet	3.28084
Feet		Meters	.3048
Kilometers		Miles	.62137
Miles		Kilometers	1.60935
AREA			
Square Centimeters		Square Inches	.155
Square Inches		Square Centimeters	6.45159
VOLUME			
Cubic Centimeters		Cubic Inches	.06103
Cubic Inches		Cubic Centimeters	16.38703
Liters		Cubic Inches	61.025
Cubic Inches		Liters	.01639
Liters		Quarts	1.05672
Quarts		Liters	.94633
Liters		Pints	2.11344
Pints		Liters	.47317
Liters		Ounces	33.81497
Ounces		Liters	.02957

Unit	To	Unit	Multiply By
WEIGHT			
Grams		Ounces	.03527
Ounces		Grams	28.34953
Kilograms		Pounds	2.20462
Pounds		Kilograms	.45359
WORK			
Centimeter Kilograms		Inch Pounds	.8676
Inch Pounds		Centimeter Kilograms	1.15262
Meter Kilograms		Foot Pounds	7.23301
Foot Pounds		Newton Meters	1.3558
PRESSURE			
Kilograms/ Sq. Centimeter		Pounds/Sq. Inch	14.22334
Pounds/Sq. Inch		Kilograms/Sq. Centimeter	.07031
Bar		Pounds/Sq. Inch	14.504
Pounds/Sq. Inch		Bar	.06895
Atmosphere		Pounds/Sq. Inch	14.696
Pounds/Sq. Inch		Atmosphere	.06805
TEMPERATURE			
Centigrade Degrees		Fahrenheit Degrees	$(C° \times 9/5) + 32$
Fahrenheit Degrees		Centrigrade Degrees	$(F° - 32) \times 5/9$

Inches	Decimals	MM
1/64	.016	.397
1/32	.031	.794
3/64	.047	1.191
1/16	.063	1.588
5/64	.078	1.984
3/32	.094	2.381
7/64	.109	2.778
1/8	.125	3.175
9/64	.141	3.572
5/32	.156	3.969
11/64	.172	4.366
3/16	.188	4.763
13/64	.203	5.159
7/32	.219	5.556
15/64	.234	5.953
1/4	.250	6.350
17/64	.266	6.747
9/32	.281	7.144
19/64	.297	7.541
5/16	.313	7.938
21/64	.328	8.334
11/32	.344	8.731
23/64	.359	9.128
3/8	.375	9.525
25/64	.391	9.922
13/32	.406	10.319
27/64	.422	10.716
7/16	.438	11.113
29/64	.453	11.509
15/32	.469	11.906
31/64	.484	12.303
1/2	.500	12.700

Inches	Decimals	MM
33/64	.516	13.097
17/32	.531	13.494
35/64	.547	13.891
9/16	.563	14.288
37/64	.578	14.684
19/32	.594	15.081
39/64	.609	15.478
5/8	.625	15.875
41/64	.641	16.272
21/32	.656	16.669
43/64	.672	17.066
11/16	.687	17.463
45/64	.703	17.859
23/32	.719	18.256
47/64	.734	18.653
3/4	.750	19.050
49/64	.766	19.447
25/32	.781	19.844
51/64	.797	20.241
13/16	.813	20.638
53/64	.828	21.034
27/32	.844	21.431
55/64	.859	21.828
7/8	.875	22.225
57/64	.891	22.622
29/32	.906	23.019
59/64	.922	23.416
15/16	.938	23.813
61/64	.953	24.209
31/32	.969	24.606
63/64	.984	25.003

Appendix A.1 English-metric conversion chart. (*Courtesy of Mitchell 1*)

Size Conversion Chart

Fractions	Decimal In.	Metric MM.	Fractions	Decimal In.	Metric MM.
1/64	.015625	.39688	33/64	.515625	13.09687
1/32	.03125	.79375	17/32	.53125	13.49375
3/64	.046875	1.19062	35/64	.546875	13.89062
1/16	.0625	1.58750	9/16	.5625	14.28750
5/64	.078125	1.98437	37/64	.578125	14.68437
3/32	.09375	2.38125	19/32	.59375	15.08125
7/64	.109375	2.77812	39/64	.609375	15.47812
1/8	.125	3.1750	5/8	.625	15.87500
9/64	.140625	3.57187	41/64	.640625	16.27187
5/32	.15625	3.96875	21/32	.65625	16.66875
11/64	.171875	4.36562	43/64	.671875	17.06562
3/16	.1875	4.76250	11/16	.6875	17.46250
13/64	.203125	5.15937	45/64	.703125	17.85937
7/32	.21875	5.55625	23/32	.71875	18.25625
15/64	.234375	5.95312	47/64	.734375	18.65312
1/4	.250	6.35000	3/4	.750	19.05000
17/64	.265625	6.74687	49/64	.765625	19.44687
9/32	.28125	7.14375	25/32	.78125	19.84375
19/64	.296875	7.54062	51/64	.796875	20.24062
5/16	.3125	7.93750	13/16	.8125	20.63750
21/64	.328125	8.33437	53/64	.828125	21.03437
11/32	.34375	8.73125	27/32	.84375	21.43125
23/64	.359375	9.12812	55/64	.859375	21.82812
3/8	.375	9.52500	7/8	.875	22.22500
25/64	.390625	9.92187	57/64	.890625	22.62187
13/32	.40625	10.31875	29/32	.90625	23.01875
27/64	.421875	10.71562	59/64	.921875	23.41562
7/16	.4375	11.11250	15/16	.9375	23.81250
29/64	.453125	11.50937	61/64	.953125	24.20937
15/32	.46875	11.90625	31/32	.96875	24.60625
31/64	.484375	12.30312	63/64	.984375	25.00312
1/2	.500	12.70000	1	1.00	25.40000

Appendix A.2 Size conversion chart. *(Courtesy of General Motors Corporation, Service Technology Group)*

General Torque Specifications Chart
(When SAE 10 oil is used as a lubricant)

Material	SAE 2 Mild Steel		SAE 5		SAE 8	Socket Head Cap Screws
Minimum Tensile P.S.I. Strength	74,000	60,000	120,000	105,000	150,000	160,000
Proof P.S.I. Load	55,000	33,000	85,000	74,000	120,000	136,000
Steel Grade Symbols	⬡		⬡		✳	⊙
Bolt Diameter Inches	Torque: pound foot					
¼	7	—	10	—	14	16
⁵⁄₁₆	14	—	21	—	30	33
³⁄₈	24	—	37	—	52	59
⁷⁄₁₆	39	—	60	—	84	95
½	59	—	90	—	128	145
⁹⁄₁₆	85	—	130	—	184	210
⅝	117	—	180	—	255	290
¾	205	—	320	—	450	510
⅞	—	200	515	—	730	825
1	—	300	775	—	1090	1235
1⅛	—	425	—	955	1545	1750
1¼	—	600	—	1345	2180	2000

**NOTE: Use only when manufacturer's specifications are not available. These values are for stiff metal-to-metal joints and are based on 90% of proof load. Do not use for gasketed joints or joints of soft materials.

Appendix A.3 General torque specifications chart (when SAE 10 oil is used as a lubricant). *(Courtesy of Snap-on Tools Company, Copyright Owner)*

General Torque Specifications Chart for I.S.O.* Metric Fasteners**
(When SAE 10 oil is used as a lubricant)

Minimum Tensile Strength	kg/mm²	40		50		60			80	100	120
	P.S.I.	56,900		71,100		85,340			113,800	142,200	170,700
Proof Load	kg/mm²	22.6	29.1	28.2	36.4	33.9	43.7	47.5	58.2	79.2	95.0
	P.S.I.	32,150	41,390	40,110	51,770	48,220	62,160	67,560	82,780	112,650	135,130
Property Class		4.6	4.8	5.6	5.8	6.6	6.8	6.9	8.8	10.9	12.9

☐ kilogram centimetre
☐ kilogram metre

Bolt Diameter metric	inch	Torque									
6mm	.236	49	63	61	79	74	95	103	126	172	206
8mm	.315	119	153	148	191	178	230	250	306	417	500
10mm	.394	235	303	294	379	353	455	495	606	8.2	10
12mm	.472	411	529	427	662	616	7.9	8.6	10.5	14	17
14mm	.551	654	8.4	8.2	10.5	10	12	13	17	23	27
16mm	.630	10	13	12	16	15	20	21	26	36	43
18mm	.709	14	18	17	23	21	27	30	36	49	59
22mm	.866	27	35	34	44	41	52	57	70	95	114

*I.S.O. = International Standardization Organization.
**NOTE: Use only when manufacturer's specifications are not available. These values are for stiff metal-to-metal joints and are based on 90% of proof load. Do not use for gasketed joints or joints of soft materials.

Appendix A.4 General torque specifications chart for I.S.O.* metric fasteners** (when SAE 10 oil is used as a lubricant). *(Courtesy of Snap-on Tools Company, Copyright Owner)*

Metric Conversion Chart

METRIC CONVERSION: lb. ft. to N•m

The chart below can be used to convert pound foot to newton metre. The left hand column lists pound foot in multiples of 10 and the numbers at the top of the columns list the second digit. Thus 36 pound foot is found by following the 30 pound foot line to the right to "6" and the conversion is 49 N•m

lb. ft.	0	1	2	3	4	5	6	7	8	9
	N•m	N•m	N•m	N•m	N•m	N•m	N•m	N•m	N•m	N•m
0	0	1.36	2.7	4.1	5.4	6.8	8.1	9.5	10.9	12.2
10	13.6	14.9	16.3	17.6	19.0	20.3	21.7	23.1	24.4	25.8
20*	27	28	30	31	33	34	35	37	38	39
30	41	42	43	45	46	47	49	50	52	53
40	54	56	57	58	60	61	62	64	65	66
50	68	69	71	72	73	75	76	77	79	80
60	81	83	84	85	87	88	90	91	92	94
70	95	96	98	99	100	102	103	104	106	107
80	109	110	111	113	114	115	117	118	119	121
90	122	123	125	126	127	129	130	132	133	134
100	136	137	138	140	141	142	144	145	146	148

* Above 20 lb. ft. the converted N•m readings are rounded to the nearest N•m.

METRIC CONVERSION: kg•cm to N•m

The chart below can be used to convert kilogram centimetre to newton metre. The left hand column lists kg•cm in multiples of 10 and the numbers at the top of the columns list the second digit. Thus 72 kg•cm is found by following the 70 kg•cm line to the right to "2" and the conversion is 7.1 N•m.

kg. cm.	0	1	2	3	4	5	6	7	8	9
	N•m	N•m	N•m	N•m	N•m	N•m	N•m	N•m	N•m	N•m
0	0	.098	.20	.29	.39	.49	.59	.69	.78	.88
10	.98	1.08	1.18	1.27	1.37	1.47	1.57	1.67	1.76	1.86
20	2.0	2.1	2.2	2.3	2.4	2.5	2.6	2.7	2.8	2.8
30	2.9	3.0	3.1	3.2	3.3	3.4	3.5	3.6	3.7	3.8
40	3.9	4.0	4.1	4.2	4.3	4.4	4.5	4.6	4.7	4.8
50	4.9	5.0	5.1	5.2	5.3	5.4	5.5	5.6	5.7	5.8
60	5.9	6.0	6.1	6.2	6.3	6.4	6.5	6.6	6.7	6.8
70	6.9	7.0	7.1	7.2	7.3	7.4	7.5	7.6	7.7	7.8
80	7.9	7.9	8.0	8.1	8.2	8.3	8.4	8.5	8.6	8.7
90	8.8	8.9	9.0	9.1	9.2	9.3	9.4	9.5	9.6	9.7
100	9.8	9.9	10.0	10.1	10.2	10.3	10.4	10.5	10.6	10.7

One oz. in. = 28.35 gms. in.
One lb. in. = 1.152 kg•cm
One lb. ft. = .138 kg•m

One kg•cm = .8679 lb. in.
One kg•cm = 7.233 lb. ft.
One N•cm = .0885 lb. in.
One N•m = .7375 lb. ft.

Courtesy of Snap-on Tools

Appendix A.5 Metric conversion chart. *(Courtesy of Snap-on Tools Company, Copyright Owner)*

Size	Threads per inch			Outside Diameter Inches	Tap Drill Approx. 75% Full Thread	Decimal Equivalent of Tap Drill	Size	Threads per inch			Outside Diameter Inches	Tap Drill Approx. 75% Full Thread	Decimal Equivalent of Tap Drill
	NC	NF	NS					NC	NF	NS			
0	...	80	...	.0600	3/64	.0469	10	...	...	28	.1900	23	.1540
1	...	...	56	.0730	54	.0550	10	...	...	30	.1900	22	.1570
1	64	...	...	.0730	53	.0595	10	...	32	...	.1900	21	.1590
1	...	72	...	.0730	53	.0595	12	24	...	...	.2160	16	.1770
2	56	...	...	.0860	50	.0700	12	...	28	...	.2160	14	.1820
2	...	64	...	.0860	50	.0700	12	...	...	32	.2160	13	.1850
3	48	...	...	.0990	47	.0785	1/4	20	...	...	.2500	7	.2010
3	...	56	...	.0990	45	.0820	1/4	...	28	...	.2500	3	.2130
4	...	...	32	.1120	45	.0820	5/16	18	...	...	.3125	F	.2570
4	...	...	36	.1120	44	.0860	5/16	...	24	...	.3125	I	.2720
4	40	...	...	.1120	43	.0890	3/8	16	...	...	.3750	5/16	.3125
4	...	48	...	.1120	42	.0935	3/8	...	24	...	.3750	Q	.3320
5	...	...	36	.1250	40	.0980	7/16	14	...	...	.4375	U	.3680
5	40	...	...	.1250	38	.1015	7/16	...	20	...	.4375	25/64	.3906
5	...	44	...	.1250	37	.1040	1/2	13	...	...	.5000	27/64	.4219
6	32	...	...	.1380	36	.1065	1/2	...	20	...	.5000	29/64	.4531
6	...	...	36	.1380	34	.1110	9/16	12	...	...	.5625	31/64	.4844
6	...	40	...	.1380	33	.1130	9/16	...	18	...	.5625	33/64	.5156
8	...	...	30	.1640	30	.1285	5/8	11	...	...	.6250	17/32	.5312
8	32	...	...	.1640	29	.1360	5/8	...	18	...	.6250	37/64	.5781
8	...	36	...	.1640	29	.1360	3/4	10	...	...	.7500	21/32	.6562
8	...	...	40	.1640	28	.1405	3/4	...	16	...	.7500	11/16	.6875
10	24	...	...	.1900	25	.1495	7/8	9	...	...	.8750	49/64	.7656

Appendix B.1 A tap drill chart for inch-sized drills. *(Courtesy of the L.S. Starrett Company)*

Bore/stroke engine displacement...

STROKE

BORE \ STROKE	3.000	.062	.125	.187	.250	.312	.375	.437	.500	.562	.625	.687	.750	.812	.875	.937	4.000	.062	.125	.187	.250	.312	.375	.437	.500
3.000	170	173	177	180	184	187	191	194	198	201	205	209	212	216	219	223	226	230	233	237	240	244	247	251	254
.062	177	180	184	188	192	195	199	203	206	210	214	217	221	225	228	232	236	239	243	247	250	254	258	261	265
.125	184	188	192	196	199	203	207	211	215	219	222	226	230	234	238	242	245	249	253	257	261	265	268	272	276
.187	192	195	199	203	207	211	215	219	223	227	231	235	239	243	247	251	255	259	263	267	271	275	279	283	287
.250	199	203	207	211	216	220	224	228	232	236	241	245	249	253	257	261	265	270	274	278	282	286	290	294	299
.312	207	211	215	220	224	228	233	237	241	246	250	254	258	263	267	271	276	280	284	289	293	297	302	306	310
.375	215	219	224	228	233	237	242	246	250	255	259	264	268	273	277	282	286	291	295	300	304	309	313	318	322
.437	223	227	232	237	241	246	250	255	260	264	269	274	278	283	288	292	297	302	306	311	315	320	325	329	334
.500	231	236	241	245	250	255	260	265	269	274	279	284	289	293	298	303	308	313	318	322	327	332	337	342	346
.562	239	244	249	254	259	264	269	274	279	284	289	294	299	304	309	314	319	324	329	334	339	344	349	354	359
.625	248	253	258	263	268	273	279	284	289	294	299	304	310	315	320	325	330	335	341	346	351	356	361	366	372
.687	256	262	267	272	278	283	288	294	299	304	310	315	320	326	331	336	342	347	352	358	363	368	374	379	385
.750	265	271	276	282	287	293	298	304	309	315	320	326	331	337	342	348	353	359	364	370	376	381	387	392	398
.812	274	280	285	291	297	302	308	314	320	325	331	337	342	348	354	359	365	371	377	382	388	394	399	405	411
.875	283	289	295	301	307	312	318	324	330	336	342	348	354	360	366	371	377	383	389	395	401	407	413	419	425
.937	292	298	304	310	317	323	329	335	341	347	353	359	365	371	377	383	390	396	402	408	414	420	426	432	438
4.000	301	308	314	320	327	333	339	346	352	358	364	371	377	383	390	396	402	408	415	421	427	433	440	446	452
.062	311	318	324	331	337	344	350	356	363	369	376	382	389	395	402	408	415	421	428	434	441	447	454	460	467
.125	321	328	334	341	348	354	361	367	374	381	388	394	401	408	414	421	428	434	441	448	454	461	468	475	481
.187	330	337	344	351	358	365	372	379	385	392	399	406	413	420	427	434	441	447	454	461	468	475	482	489	496
.250	341	348	355	362	369	376	383	390	397	404	411	419	426	433	440	447	454	461	468	475	482	489	497	504	511
.312	351	358	365	372	380	387	394	402	409	416	424	431	438	446	453	460	467	475	482	489	497	504	511	519	526
.375	361	368	376	383	391	398	406	414	421	429	436	444	451	459	466	474	481	489	496	504	511	519	526	534	541
.437	371	379	387	394	402	410	418	425	433	441	449	456	464	472	479	487	495	503	510	518	526	534	541	549	557
.500	382	390	398	406	414	421	429	437	445	453	461	469	477	485	493	501	509	517	526	533	541	549	557	565	573

The above chart is for 8-cylinder engines. For the displacement of a 6-cylinder engine multiply the 8-cylinder figure by .75. For the displacement of a 4-cylinder engine multiply the 8-cylinder figure by .5 or divide by 2

Appendix C.1 Bore/stroke engine displacement. (*Courtesy of Crane Cams, Inc.*)

AMERICAN WIRE GAUGE SIZES

Gauge Size	Conductor Diameter (Inch)	Cross Section Area (Circular Mils)
20	.032"	1,020
16	.051"	2,580
12	.081"	6,530
8	.128"	16,500
2	.258"	66,400
0	.325"	106,000
2/0	.365"	133,000

AWG Size	Metric Size
20	0.5
18	0.8
16	1.0
14	2.0
12	3.0
10	5.0
8	8.0
6	13.0
4	19.0

Appendix D.1 American wire gauge sizes.

Total Approx. Circuit Amperes	Total Circuit Watts	Total Candle Power	Wire Gage (For Length in Feet)											
12V	12V	12V	3'	5'	7'	10'	15'	20'	25'	30'	40'	50'	75'	100'
1.0	12	6	18	18	18	18	18	18	18	18	18	18	18	18
1.5		10	18	18	18	18	18	18	18	18	18	18	18	18
2	24	16	18	18	18	18	18	18	18	18	18	18	16	16
3		24	18	18	18	18	18	18	18	18	18	18	14	14
4	48	30	18	18	18	18	18	18	18	18	16	16	12	12
5		40	18	18	18	18	18	18	18	18	16	14	12	12
6	72	50	18	18	18	18	18	18	16	16	16	14	12	10
7		60	18	18	18	18	18	18	16	16	14	14	10	10
8	96	70	18	18	18	18	18	16	16	16	14	12	10	10
10	120	80	18	18	18	18	16	16	16	14	12	12	10	10
11		90	18	18	18	18	16	16	14	14	12	12	10	8
12	144	100	18	18	18	18	16	16	14	14	12	12	10	8
15		120	18	18	18	18	14	14	12	12	12	10	8	8
18	216	140	18	18	16	16	14	14	12	12	10	10	8	8
20	240	160	18	18	16	16	14	12	10	10	10	10	8	6
22	264	180	18	18	16	16	12	12	10	10	10	8	6	6
24	288	200	*18	18	16	16	12	12	10	10	10	8	6	6
30			18	16	16	14	10	10	10	10	10	6	4	4
40			18	16	14	12	10	10	8	8	6	6	4	2
50			16	14	12	12	10	10	8	8	6	6	2	2
100			12	12	10	10	6	6	4	4	4	2	1	1/0
150			10	10	8	8	4	4	2	2	2	1	2/0	2/0
200			10	8	8	6	4	4	2	2	1	1/0	4/0	4/0

HOW TO USE CHART

1
Measure Length of Wire in Circuit — Chart applies to ground return. Two-wire circuits will be total of both wire lengths. Be sure to include both vehicles on auto and trailer applications.

2
Find the total amperes, watts or candlepower and choose nearest value in proper column.

3
Move horizontally to proper footage column and find nearest wire gage.

Based on maximum of 10% voltage drop

*18 gage indicated above this line could be 20 gage electrically — 18 gage is recommended for mechanical strength.

Appendix E.1 Wire gauge changes with increases in wire length. *(Courtesy of Federal-Mogul Corporation)*

NOTE: Be sure to save all old engine parts.

Engine year and size _____

Piston and ring size: STD _____ .020" over _____ .030" over _____ .040" over _____ .060" over _____
(Check one)

Crank and bearing sizes: Mains STD _____ .010" under _____ .020" under _____
Rods STD _____ .010" under _____ .020" under _____
(Check one)

Core plug size _____

✓ Those items needed	Quantity	Price
*1. Reground crankshaft and bearings (Crankshaft kit) (*be sure to save old woodruff key*)		
**2. Piston, wrist pins		
*3. Piston rings		
**4. Camshaft		
*5. Cam bearings		
**6. Valve lifters or cam followers (as needed)		
**7. Cam and crank sprockets		
*8. Timing chain or belt		
**9. Oil pump		
**10. Core plugs and oil gallery plugs		
**11. Engine gasket set (includes the following)		
A. Head gaskets		
B. Valve guide seals		
C. Valve cover gasket(s)		
D. Intake manifold gasket(s)		
E. Timing cover gasket		
F. Oil pan gasket		
G. Rear main oil seal		
H. Front crankshaft seal		
I. Exhaust manifold gasket(s)		
J. Coolant outlet gasket or seal		
12. Oil. *NOTE:* Rating and viscosity		
13. Large garbage bag. (to keep engine clean during assembly)		

NOTE: If a kit is purchased, the old crank, cam, and oil pump must be returned to the parts supplier or a core charge must be paid.

Engine Master Kit (parts marked * or ** are included in a high-quality kit)
Engine overhaul kit (parts marked * only are included in this kit)

The following items are inspected during engine repair and may require replacement at extra cost to the vehicle owner.

	Quantity	Price
15. Hoses		
A. Radiator hose		
B. Heater hoses		
C. Thermostat bypass		
D. Vacuum hose		
E. Fuel hose		
F. Other hose		
16. Hose clamps		
A. Heater		
B. Fuel		
C. Radiator		
17. Fuel filter		
18. Thermostat		
19. Coolant outlet housing		
20. Radiator cap		
21. Coolant		
22. Engine mounts		
23. Ignition parts		
A. Spark plugs		
B. Rotor and distributor cap (if applicable)		
C. Plug wires (if applicable)		
24. Air filter		
25. PCV valve		
26. Accessory drive		
27. Ground straps or battery cables		
28. Valves		
A. Intake		
B. Exhaust		
29. Valve springs		
30. Pushrods (if applicable)		
31. Clutch parts		
32. Automatic transmission front seal		
33. Automatic transmission fluid. Type: _____		

Appendix F.1 An engine rebuilding parts checklist.

PREALIGNMENT INSPECTION CHECKLIST

Owner _____ Phone _____ Date ___|___|___
          LAST   FIRST

Address _____ Ser. No. _____

Make _____ Model _____ Yr. _____ Lic. No. _____ Mileage _____

1. Road Test Disclosed

Vehicle Pulls		Yes	No	Right	Left
	Above 30 mph				
	Below 30 mph				
to	Bump Steer				
	When Braking				
Steering Wheel Movement — When Stopping From 2 to 3 mph—Front Brakes					
Vehicle Steers Hard					
Steering Wheel Returns Normally					
Steering Wheel Position					

Vibration		Yes	No	Front	Rear

2. Tire Pressure SPECS Front ____ Rear ____

RECORD PRESSURE FOUND
RF ____ LF ____ RR ____ LR ____

3. Chassis Height SPECS Front ____ Rear ____

RECORD HEIGHT FOUND
RF ____ LF ____ RR ____ LR ____

	YES	NO
Springs Sagged		
Torsion Bars Adjusted		

4. Rubber Bushings OK

	OK
Upper Control Arm	
Lower Control Arm	
Sway Bar/Stabilizer Link	
Strut Rod	
Rear Bushing	

5. Shock Absorbers/Struts

	Front	Rear

6. Steering Linkage

	Rear OK	Front OK
Tie-Rod Ends		
Idler Arm		
Center Link		
Sector Shaft		
Pitman Arm		
Gearbox/Rack Adjustment		
Gearbox/Rack Mounting		

7. Ball Joints OK

Load Bearings		OK

SPECS	READINGS
Right ____ Left ____	Right ____ Left ____

Follower	
Upper Strut Bearing Mount	
Rear	

8. Power Steering OK

	OK
Belt Tension	
Fluid Level	
Leaks/Hose Fittings	
Spool Valve Centered	

9. Tires/Wheels OK

	OK
Wheel Runout	
Condition	
Equal Tread Depth	
Wheel Bearing	

10. Brakes Operating Properly

11. Alignment

	Specs Right	Specs Left	Initial Readings Right	Initial Readings Left	Adj. Readings Right	Adj. Readings Left
Camber						
Caster						
Toe						

Bump Steer	TOE CHANGE Right Wheel Amount	Direction	TOE CHANGE Left Wheel Amount	Direction
Chassis DOWN 3″				
Chassis UP 3″				

	Specs Right	Specs Left	Initial Readings Right	Initial Readings Left	Adj. Readings Right	Adj. Readings Left
Toe-Out On Turns						
SAI						
Rear Camber						
Rear Total Toe						
Rear Indiv. Toe						
Wheel Balance	Front ____		Rear ____			
Radial Tire Pull	Yes ____		No ____			

FLAG READINGS

COMMENTS:

Appendix G.1 Prealignment inspection checklist.

Pressure vs. Vacuum			
PSIG	PSI (abs)	BAR	IN.Hg
2	16	1.14	–
1	15	1.07	–
0	14.2	1.00	0
−1	13	0.94	2
−2	12	0.87	4
−3	11	0.80	6
−4	10	0.72	9
−5	9	0.64	11
−6	8	0.57	13
−7	7	0.50	15
−8	6	0.44	18
−9	5	0.37	20
−10	4	0.30	22
−11	3	0.22	24
−12	2	0.14	26
−13	1	0.07	28
−14	0	0.00	30

Appendix H.1

°C																						
−30	−25	−20	−17.8	−15	−10	−5	0	5	10	15	20	25	30	35	40	50	60	70	80	90	100	
−22	−13	−4	−0	−5	−14	−23	32	41	50	59	68	77	86	95	104	122	140	158	176	194	212	
°F																						

Appendix I.1

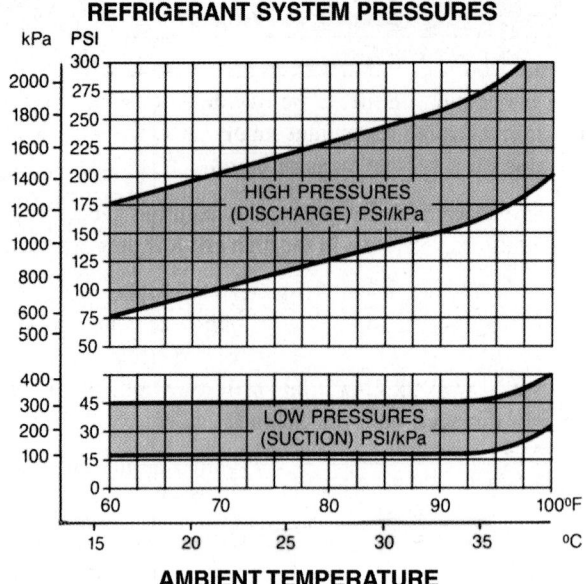

REFRIGERANT SYSTEM PRESSURES

AMBIENT TEMPERATURE

Ambient Temp °F	High Side PSIG, R-12	Low Side PSIG, R-12	High Side PSIG, R-134a	Low Side PSIG, R-134a
60	120–150	5–15	120–170	7–15
70	140–180	8–16	150–250	8–16
80	160–250	10–18	190–280	10–20
90	200–280	12–25	220–330	15–25
100	220–300	15–30	250–350	20–30
110	250–320	20–35	280–400	25–40

Appendix J.1 *(Top, Courtesy of Ford Motor Company)*

A

abnormal combustion—also called spark knock, it is the term for preignition and detonation, which can cause spark knock and engine damage.

combustión anormal—tambien llamado detonación de las bujías, es el término para el encendido anticipado y la detonación, que puede causar la detonación de las bujías y los daños al motor.

abrasion—rubbing.

abrasión—frotamiento.

ABS—antilock brake systems; ABS comes from the German word *Antiblockiersystem*. It is a computerized safety device installed in newer cars to prevent the wheels from locking up.

ABS—sistema de frenos antibloqueantes.

AC—alternating current; oscillating electrical current that surges from positive to negative and back again.

AC—corriente alterna; un corriente eléctrico oscilante que surge del positivo al negativo y luego regresa.

accumulator (air conditioning)—a refrigerant reservoir and dryer that supplies refrigerant vapor to the compressor.

acumulador (aire acondicionado)—un suministro de refrigerante y un secador que proporciona el vapor de refrigerante al comprimidor.

accumulator (transmission)—a reservoir used in timing and cushioning gear shifts.

acumulador (transmisión)—un suministro usado en sincronizar y en suavizar los cambios de velocidades.

acid—solutions that are below 7 on the pH scale are acidic. A pH rating of 1 is a strong acid.

ácido—las soluciones que son menos del 7 en la escama de pH se consideran acídicos. Un rango pH de 1 es un ácido fuerte.

A-circuit—a charging system with the regulator in the ground circuit.

circuito A—un sistema de cargar que tiene el regulador en el circuito a tierra.

Ackerman angle—also called turning radius or toe-out-on-turns. The tires toe out during a turn because the steering arms are bent at an angle.

ángulo Ackerman—tambien se llama radio de viraje o divergencia en las vueltas. Los neumáticos se inclinan hacia afuera en una vuelta porque los brazos de dirección son torcidos en un ángulo.

active alarm system—an alarm system that must be turned on manually.

sistema de alarma activo—un sistema de alarma que debe encenderse manualmente.

active restraint—a restraint that passengers must activate.

cinturón de seguridad activo—un cinturón de seguridad que los pasajeros deben activar.

active sensors—those that generate their own signals.

sensores activos—los que generan sus propios señales.

active suspension—an automatic suspension where each wheel has a hydraulic cylinder to keep the car body level during all driving conditions.

suspension activa—una suspensión en la cual cada rueda tiene un cilindro hidráulico para mantener nivel al coche en cualquier condición de manejo.

actuator—an electronic or magnetic relay that can perform a desired function.

actuador—un relé electrónico o magnético que puede cumplir una función deseada.

actuators—the devices that act upon processed signals received from the computer. They can be solenoids, DC motors, relays, switches, or control modules.

impulsores—los dispositivos que actuan sobre los señales recibidos de la computadora. Pueden ser solenoides, los motores DC, los relés, los interruptores o los módulos de control.

AC voltage generator—a magnetic pickup that generates an AC analog signal.

generador de voltaje AC—un captador magnético que genera una señal AC análoga.

adaptive fuel trim—it allows the fuel system to operate at the correct air-fuel ratio, no matter what the differences in fuel systems are.

compensador adaptivo de combustible—permite operar al sistema de combustible en la tasa correcta del aire-combustible, sin que importa las diferencias entre los sistemas de combustible.

adaptive suspension system—a suspension leveling system that keeps the vehicle at the same height when weight is added to parts of the car.

sistemas adaptivas de suspensión—una sistema anivelador de suspensión que mantiene al vehículo en la misma altura cuando se añade el peso en las áreas del coche.

AERA—Automotive Engine Rebuilders Association.

AERA—la Asociación de Reconstructores de Motores Automotrices.

aeration—also called cavitation, it is when hydraulic fluid becomes mixed with air.

aeración—tambien llamado cavitación, es cuando el fluido hidráulico se mezcla con el aire.

aerobic chemical—a chemical that hardens only when exposed to air.

química aeróbica—una química que se endurece solamente al ser expuesta al aire.

aftercooler—also called an intercooler or charge air cooler, it is a sophisticated air cooler for blowers.

post refrigerador—tambien llamado un enfriador intermedio o un enfriador de aire cargado, es un enfriador de aire sofisticado para los ventiladores.

aftermarket—the distribution system that goes from parts manufacturers to customers; not original equipment (OE).

(posventa) piezas de reposición—sistema de distribución del fabricante de piezas al cliente; no es equipo original (OE).

air bag module—includes the air bag and its inflator.

módulo de bolsa de aire—incluye la bolsa de aire y su inflador.

air conditioning—the process in which air inside the passenger compartment is cooled, dried, cleaned, and circulated.

aire acondicionado—el proceso por el cual el aire dentro del compartimento de pasajeros se enfría, se seca, se limpia y se circula.

air density system—when an airflow sensor measures the volume of air entering the engine.

sistema de densidad del aire—cuando el sensor del flujo del aire mide el volumen del aire entrando en el motor.

air filter capacity—the amount of dirt an air filter can hold before it becomes restricted.

capacidad del filtro de aire—la cantidad de partículos que puede contener un filtro de aire antes de obstruirse.

air injection system—a smog pump or aspirator valve that lowers HC and CO in the exhaust. It also provides air to the catalytic converter to help heat it up when it is cold.

sistema de inyección de aire—una bomba de esmog o una válvula aspiradora que disminuye la cantidad del HC y CO en el escape. Tambien proporciona el aire al convertidor catalítico para ayudar en calentarlo cuando hace frío.

air mix damper—also called a blend air door, it controls heating and air-conditioning action.

compuerta de mezcla de aire—tambien llamado una compuerta combinadora, controla la acción de calentamiento y aire acondicionado.

air test—a test of the operation of various clutch packs and servos before disassembling a transmission fully.

prueba de aire—una prueba de la operación de varios equipos de embrague y los servos antes de completamente desarmar una transmisión.

alkaline—a strong alkaline cleaner has a pH rating above 10.

alcalino—una limpiadora alcalina fuerte tiene la clasificación pH más alta de las 10.

alloy—when two or more metals are combined to make one.

aleación—cuando se combinan dos o más metales para hacer uno.

alternator—an alternating current (AC) generator that produces electrical current and forces it into the battery to recharge it.

alternador—un generador de corriente alterna (AC) que produce un corriente eléctrico y lo dirige a una batería para cargarla.

ambient air—surrounding air.

aire ambiente—el aire alrededor.

ambient temperature switch—keeps an air-conditioning compressor from working when outside temperatures are cold.

interruptor de la temperatura del ambiente—previene de que funciona el comprimidor del aire acondicionado cuando las temperaturas externas son frías.

amp—a unit of electrical current flow, 1 coulomb per second of electron flow.

amperio—una unidad del flujo de un corriente eléctrico, 1 culombio por segundo de flujo de los electrones.

amplitude—the intensity of a vibration.

amplitúd—la intensidad de una vibración.

amplitude modulation—AM radio; varies the strength of the broadcast signal.

modulación de amplitud—radio AM; varía la potencia de la señal de transmisión.

anaerobic chemical—a chemical that hardens only in the absence of air.

química anaeróbica—una química que se endurece solamente con la ausencia del aire.

analog meter—a meter with a dial whose needle is moved by a magnet.

medidor análogo—un medidor con un cuadrante cuyo aguja indicador se mueve por un imán.

anneal—to soften a metal, it is heated and allowed to cool slowly.

recocer—para ablandar un metal, se calienta y se permite enfriar lentamente.

anode—like the positive terminal on a battery.

ánodo—como el terminal positivo de una batería.

anodized aluminum—aluminum with a protective plating.

aluminio anodizado—el aluminio que tiene un blindaje protectivo.

ANSI—American National Standards Institute.

ANSI—Instituto Nacional de Normas Americanas.

antiblockiersystem—a computerized safety device installed in newer cars to prevent the wheels from locking up.

sistema antibloqueo—un dispositivo de seguridad computerizado instalado en automóviles más nuevos para evitar el bloqueo de las ruedas.

anti-drainback valve—a valve in a horizontally mounted oil filter that prevents it from emptying when the engine is off.

válvula que evita el drenaje de regreso del aceite—una válvula en un filtro de aceite montado horizontalmente que previene que se vacía cuando está apagado el motor.

anti-friction bearings—also called frictionless bearings, provide a rolling contact with balls, rollers, or needles.

balero antifricción—tambien llamado balero sin fricción, provee un contacto rodante utilizando bolas, rodillos o agujas.

API—American Petroleum Institute.

API—Instituto Americano de Petroleo.

arcing—burning.

formación del arco—quemando.

arc welding—welding using electricity.

soldadura con arco—la soldadura que usa la electricidad.

ASA—Automotive Service Association.

ASA—Asociación de Servicios Automotores.

ASC—Automotive Service Council.

ASC—Consejo de Servicios Automotores.

ASE—National Institute for Automotive Service Excellence.

ASE—Instituto Nacional de la Excelencia en el Servicio Automotor.

ASME—American Society of Mechanical Engineers - 345 E. 47th Street, New York, NY 10017.

ASME—La Sociedad Americana de Ingenieros Mecánicos - 345 E. 47th Street, New York, NY 10017.

aspect ratio—a measurement of the height-to-width ratio of a tire.

relación de aspecto—una medida de la relación entre altura a anchura de un neumático.

aspirator valve air system—also called a pulse-air system, it uses the pulses created by the exhaust gases to blow fresh air into the exhaust.

sistema de aire de válvula aspirador—tambien llamado un sistema de aire de impulso, usa los impulsos creados por los gases del escape para introducir el aire fresco dentro del escape.

ATF—automatic transmission fluid.

ATF—fluido para transmisión automática.

atomization—the name of the process by which fuel is suspended in air in a mist of tiny droplets like fog.

atomización—el nombre del proceso por el cual el combustible se suspenda en el aire en un vapor de gotitas pequeñas como una niebla.

atoms—the building blocks of life composed of protons, neutrons, and electrons.

átomos—los elementos esenciales de toda vida compuesta de los protones, los neutrones y los electrones.

ATRA—Automatic Transmission Rebuilders Association.

ATRA—Asociación de Reconstructores de Transmisiones Automáticas.

AWD—all-wheel drive; when all four wheels are driven.

AWD—tracción a cuatro ruedas; cuando impulsan todas cuatro ruedas.

AWG—American wire gauge.

AWG—calibre americano de los alambres.

axial compressor—an air-conditioning compressor that has pistons that move lengthwise in the compressor body.

comprimidor axial—un comprimidor de aire acondicionado que tiene pistones que muevan por la longitud del cuerpo del comprimidor.

axial movement—vertical movement.

movimiento axial—un movimiento vertical.

axle bearing—the term for a bearing that is on live axles (those that drive wheels).

cojinete del eje—el término para un cojinete que se encuentra en los ejes vivos (los que impulsan las ruedas).

B

backfire—a term for combustion that occurs in the vehicle's exhaust (after it leaves the cylinder).

petardear—un término para la combustión que ocurre en el escape del vehículo (después de que sale del cilindro).

backlash—the clearance between meshing gear teeth.

juego libre—la holgura entre los dientes engrenados de un engranaje.

backpressure transducer valve—part of an EGR valve that senses engine load and changes the amount of exhaust flow into the intake manifold.

válvula transductor de contrapresión—parte de una válvula EGR que percibe la carga del motor y cambia la cantidad del flujo del escape en el múltiple de admisión.

ball and trunnion—a type of constant velocity universal joint.

bola y muñón—un tipo de unión universal de velocidad fija.

ball socket—part that connects steering linkage parts.

rótula—una parte que conecta las partes de enlace de dirección.

banjo fitting—a fitting, resembling a banjo, that allows fluid to turn 90 degrees in a short area.

montaje banjo—un montaje, parecido a un banjo, que permite que el fluido gira el 90 grados en una área muy corta.

banjo housing—a differential housing without the third member, called this because it resembles a two-sided banjo.

cárter de la caja del diferencial (banjo)—una caja de un diferencial sin tercer miembro, llamado "banjo" porque parece un banjo de dos lados.

barrel—a term for the number of throttle passages found in a carburetor.

cañón—un término para el número de los pasos del acelerador que se encuentra en el carburador.

barrier creams—creams that are rubbed on the skin prior to working in a greasy environment to aid in cleanup.

cremas decapantes—las cremas que se aplican a la piel antes de trabajar en un ambiente con mucha grasa para ayudar en desengrasar.

base—a pH rating of above 7 is an alkaline cleaner, caustic or base.

base—una clasificación pH más del 7 es una limpiador alcalino, cáustico o un base.

base circle—what the cam would be if there were no lobe.

círculo base—lo que sería la leva excéntrica si no tuviera lóbulo.

base timing—also called initial timing, it is the timing setting before the computer has a chance to make changes.

sincronización de referencia—tambien llamado el encendido inicial, es la colocación del encendido antes de que la computadora haya tenido la oportunidad de hacer los cambios.

battery terminal—the connection on the side or top of a battery that the clamp connects to.

terminal de la batería—la conexión en el lado o en la parte superior de una batería a la cual se conecta la abrazadera de la batería.

baud rate—the speed that timed pulses (bits) of information are transmitted within the computer. A computer with a baud rate of 28,800 (28.8K) can transmit 28,800 bits of information per second.

coeficiente baúd—la velocidad con que los impulsos (bits) cronometizados se transmiten con una computadora. Una computadora con una coeficiente baúd de 28,800 (28.8K) puede transmitir 28,800 bits de información por segundo.

BCA—Battery Council International.

BCA—Concilio Internacional de la Batería.

B-circuit—a charging system with the regulator between the positive feed (insulated side) and the field coil (which is grounded inside of the alternator).

circuito B—un sistema de cargar que tiene el regulador entre la alimentación positiva (lado aislado) y la bobina inductora (que va a tierra dentro del alternador).

bearing cage—also called separator, it is a stamped steel or plastic insert that keeps bearing balls or rollers properly spaced around the bearing assembly.

jaula del cojinete—tambien llamado un espaciador, es una inserción hecha del acero embutido o del plástico que mantiene un espacio adecuado entre las bolas o las rodillas del cojinete y la asamblea del cojinete.

bearing clearance—clearance between a bearing and bearing journal.

holgura del cojinete—la holgura entre un cojinete y un muñón del cojinete.

bearing crush—the bearing insert extends out of the bearing bore. When the two halves are tightened against each other, the bearing is held from turning in its bore.

compresión del cojinete—la inserción del cojinete se extiende más allá del agujero del cojinete. Cuando las dos mitades se aprietan la una contra la otra, el cojinete no puede girar en su agujero.

bearing retained axle—a pressed-fit axle with a bearing retainer ring.

eje sujetado por cojinete—un eje de ajuste a presión con un anillo de retén para el cojinete.

bearing spread—the spread of the bearing insert makes it possible for it to stay in place in its bore during engine assembly.

extensión del cojinete—la extensión de la inserción del cojinete hace posible quedarse en su posición en el agujero durante la asamblea del motor.

beat/boom vibration—one vibration interacting with another.

vibración latido/zumbido—una vibración que reciproca con otra.

belt—a cord structure made up of plies.

correa/banda—una estructura de cordón compuesto de telas.

BHP—brake horsepower; the usable horsepower at the crankshaft.

BHP—caballos de fuerza del freno; los caballos de fuerza disponibles en el cigüeñal.

bias-ply—a tire structure, also called diagonal or cross-ply, that has casing plies that cross each other at an angle of 35–45 degrees.

doble capa—una estructura de neumático, tambien llamada bandas diagonales, o telas al sesgo que tiene las capas que se cruzan en un ángulo de 35 a 45 grados.

bidirectional—when the computer can receive commands from the scan tool and anything that is computer controlled can be operated from the scan tool.

bidireccional—cuando la computadora puede recibir órdenes de un aparato explorador y cualquiera cosa que se controla por medio de la computadora se puede operar desde el aparato explorador.

bimetal coil spring—a thermostatic coil consisting of two types of metal wound together. When the coil is heated, it expands. When it cools, it shrinks.

resorte helicoidal bimetal—un arrollamiento termostático que consiste de dos tipos de metal enredados. Cuando el arrollamiento se calienta, expande. Cuando se enfría, encoge.

bimetal engine—an engine made of iron and aluminum.

motor bimetal—un metal hecho de hierro y aluminio.

bimetal strip—two metal strips with different expansion rates.

banda bimetálica—dos bandas de metal con una coeficiente de expansión distintas.

binary information—transistors are electronic switches that are either on or off (zeros and ones). The computer interprets combinations of zeros and ones to form numbers or words, processing the information to determine the meanings of the signals.

información binaria—los transistores son los interruptores electrónicos que están o prendidas o apagadas (ceros o unos). La computadora interprete la combinación de ceros y unos para formar los números o las palabras, procesando la información para determinar el significado de las señales.

bit—each zero and one represents a bit of information.

bit—cada cero y uno representa un bit de información.

black light test—a test for oil leakage that uses fluorescent dye and an ultraviolet light source.

prueba de luz negra—una prueba para fuga de aceite que usa un tinte fluorescente y una fuente de luz ultravioleta.

bleeding brakes—process by which air and fluid are released from the brake system.

purga de frenos—proceso a través del cual se elimina el aire y líquido del sistema de frenos.

bleeding shocks—a process of removing air from shocks that have been stored on a shelf. Also called purging shocks.

purgar los amortiguadores—el proceso de remover el aire de los amortiguadores que se ha almacenado en un estante.

blend air door—also called an air mix damper, it controls heating and air-conditioning action.

puerta para mezclar el aire—tambien llamado una compuerta para mezclar aire, controla la acción de calentamiento y aire acondicionado.

block check test—a test for an internal coolant leak in which a dye changes color when exposed to exhaust gas in the radiator.

prueba del bloqueo—una prueba para encontrar una fuga interior de fluido refrigerante en la cual un tinte cambia de color al exponerse al gas del escape en el radiador.

blocker ring synchronizer—the most popular style of synchronizer.

sincronizador tipo anillo—el estilo más popular de sincronizador.

blowby gases—gases that leak past the piston rings into the crankcase during the power stroke. They contain acids as well as unburned and burned fuel.

fuga de gases a través de los anillos de los cilindros—la fuga de los gases desde los anillos de los pistones al cárter durante la carrera de fuerza. Contienen los ácidos tanto como el combustible no consumido y consumido.

blower—a belt-driven pump, also called a supercharger.

ventilador—una bomba impulsada por banda, tambien llamado un sobrealimentador.

blow-through—when a supercharger or turbocharger pressurizes the air cleaner above the carburetor.

inyección superior—cuando un sobrealimentador o un turbocompresor presuriza el limpiador de aire en la parte superior del carburador.

BMEP—brake mean effective pressure; a term relating to the pressure in the cylinder calculated from the horsepower reading on a dynamometer.

BMEP—presión efectivo mediano del freno; un término relacionado a la presión en el cilindro que se calcula por la lectura de la potencia de fuerza en un dinamómetro.

bob weights—are used when spinning the crankshaft to simulate the correct weight of the piston and related parts during machine shop testing.

pesos de contra balance—se usan al girar el cigüeñal para simular el peso correcto del pistón y sus partes relacionadas durante la prueba en el taller de máquinas.

body roll—when the car leans excessively to one side or the other during a fast turn.

balanceo de la carrocería—cuando el automóvil se inclina hacia un lado u otro durante un giro rápido.

bolt grade—an SAE or metric classification system that rates the minimum tensile strength of a fastener.

grado de perno/clase de propriedad—un sistema de clasificación SAE o métrica que clasifica la fuerza mínima de tensión de un sujetador.

bonded lining—a lining that is held to its backing by a binder (glue).

forro aglomerado—un forro que se pega al respaldo con un ligante (pegamento).

boost pressure—the amount of air density a turbo provides.

presión auxiliar—la cantidad de densidad del aire que provee un turbo.

bore—the diameter of the cylinder.

diámetro interior—el diámetro de un cilindro.

boundary lubrication—when an oil film becomes too thin or starts to break down under load.

límite de lubricación—cuando una película de aceite se espesa demasiado y comienza a deteriorarse bajo una carga.

brake band—an external brake planetary holding device.

banda de freno—un dispositivo planetario externo de sujeción.

brake drag—when the brakes stay on after a stop.

arrastre de freno—cuando los frenos siguen funcionando después de parar.

brake fade—resulting from excessive brake heat, the drum expands and the pedal moves closer to the floor. The coefficient of friction of the lining also drops off.

amortiguamiento del freno—resulta de un calor excesivo del freno, el tambor expande y el pedal se acerca al piso. La coeficiente de fricción del forro tambien se disminuya.

brake grab—when brakes apply quickly and tend to stick on.

amarro del freno—cuando los frenos se aplican rápidamente y suelen quedarse aplicados.

brake pads—disc brake friction linings.

almohadilla de freno—los forros de fricción en un freno de disco.

brake pedal pulsation—when the brake pedal moves up and down during a stop due to out-of-round drum brakes or overheated disc brake rotors.

pulsación del pedal de frenos—cuando el pedal de freno brinca durante un paro debido a la eccentricidad de los frenos de tambor o a los rotores sobrecalentados de un freno de disco.

brake pedal reserve—pedal travel distance before brakes begin to apply.

reserva del pedal de frenos—la distancia que avanza el pedal antes de que comienzan a aplicar los frenos.

brake pull—when a vehicle pulls to one side during a stop.

jalón del freno—cuando un vehículo jala hacia un lado durante una parada.

brake shoes—the friction material backing on drum brake systems.

zapatos de freno—el material de forro de fricción en los sistemas de freno de tambor.

brake spoon—a tool used to adjust drum brake clearance.

cuchara de freno—una herramienta que sirve para ajustar la holgura del tambor de freno.

brake thermal efficiency—brake horsepower converted to British thermal units (Btu), divided by the fuel's heat input in British thermal units, with the result multiplied by 100.

eficiencia termal del freno—los caballos de fuerza convertido a los Btus, dividido por los Btus de calor del combustible con el resultado multiplicado por 100.

breakaway torque—the amount of torque needed to make a side gear rotate the clutches in a limited-slip differential.

par máximo de salida—la cantidad del par requirido para hacer que un engranaje lateral gira los embragues en un diferencial autoblocante.

breakout box—a tool in which the probes of a DVOM are inserted into its pin holes to access various sensors and actuators through the pin connector to the computer.

caja electróncia indicadora de códigos de problemas—una herramienta en la cual las sondas de un DVOM se insertan en los agujeros para pasador obteniendo aceso a varios sensores e impulsores por el conectador de púas en la computadora.

bridge-type clamp—a type of clamp used to fasten CV joint boots to the axle.

abrazadera tipo puente—un tipo de abrazadera usado para sujetar fundas de conexión de CV al eje.

brinelling—when the bearing or race has indentations from shock loads.

efecto brinell—cuando el cojinete o la pista tiene muescas debido al choque de las cargas.

British Imperial (U.S.) System—a measuring system that uses fractions and decimals, based on inches, feet, and yards.

Sistema Imperial Inglesa (U.S.)—un sistema de medida que usa las fracciones y los decimales, basado en las pulgadas, los pies y las yardas.

brush—a part, usually carbon, that provides electricity to rotating parts.

escobilla—un parte, tipicamente del carbón, que proveé la electricidad a las partes giratorias.

Btu—British thermal unit; the amount of energy required to raise the temperature of one pound of water by 1°F.

Btu—unidad térmica inglesa; la cantidad de la energía requerida para aumentar el calor una libra de agua por 1°F.

buildup—also called increase or normal braking. This is the first stage of the ABS solenoid operation. Neither the inlet nor outlet solenoid valves are energized by the computer.

aumento—también llamado frenado en aumento o normal. Esta es la primera etapa de la operación de solenoides del ABS. Ni las válvulas de entrada ni las de la salida son energizadas por la computadora.

bulb trade number—the number that identifies a bulb for all manufacturers.

marca de fabricante de bombilla—el número que identifica una bombilla para todos los fabricantes.

bulkhead or fire wall—the metal wall between the engine and passenger compartments.

tabique o mamparo de encendios—el muro de metal entre el motor y los compartimentos de los pasajeros.

bump steer—also called orbital steer, happens when a wheel with tie-rods at unequal heights goes over a bump. The car momentarily steers the direction that the wheel turns as the toe on that one wheel changes.

dirección de topes—tambien llamado la dirección orbital, ocurre cuando una rueda cuyos tirantes de tracción son de niveles desiguales pasa sobre un tope. El coche se desvía momentariamente en la dirección en que gira la rueda mientras que cambia la divergencia de esa rueda.

butterfly valve—a valve that controls airflow.

válvula de mariposa—una válvula que controla el flujo del aire.

by-pass oil filter—an oil filtering system that only filters about 10% of the oil at once, trapping very fine materials and allowing very little oil to flow through it. Also called a partial-flow oil filter.

filtro del aceite de desvío—un sistema para purificar el aceite que sólo filtra aproximadamente el 10% del aceite a la vez, atrapando a las partículas muy finas y permitiendo que pase muy poco aceite a través del filtro. También se llama un filtro del aceite de flujo limitado.

by-pass valve—a valve in the oil filter that opens to let unfiltered oil flow to the bearings when the oil is cold and thick or the filtering material is plugged due to a lack of proper maintenance.

válvula de desvío—una válvula en el filtro de aceite que abre para permitir que el aceite no filtrado pase a los cojinetes cuando el aceite está frío y espeso o cuando el material de filtro está atascado por falta del mantenimiento adecuado.

byte—8 bits, sometimes called a word.

byte—8 bits, a veces llamado una palabra.

C

cable—a large wire that allows more electrical current flow.

cable—un alambre grande que permite más flujo del corriente.

CAFE—Corporate Average Fuel Economy; each manufacturer must meet a standard or pay a "gas-guzzler" penalty to the government.

CAFE—Normas de la Economía del Combustible indicadas por la Corporación; cada fabricante debe adherir a una norma o pagar una multa de "consumidor de combustible" al gobierno.

cam ground piston—when a piston is cold, its skirt is oval shaped. As it warms up, it takes on a round shape.

pistón excéntrico—cuando está fría su faldilla es de forma ovalada. Al calentarse toma una forma redonda.

camber—the inward or outward tilt of a tire at the top.

curvatura—la inclinación hacia adentro o hacia afuera en la parte superior de un neumático.

camber roll—when a cambered tire rolls in a circle, as if it were at the large end of a cone.

rodar en comba—cuando un neumático rueda en un círculo, como si estuviera en la extremidad de un cono.

candlepower—a rating for the intensity of a headlamp.

unidad de intensidad luminosa—una potencia indicada de la intensidad de un faro.

capacitor—also called a condenser, it stores electricity.

capacitor—tambien llamado un condensador, almacene la electricidad.

capscrew—an externally threaded fastener that is used without a nut in a threaded or blind hole application to hold components together.

casquete fileteado—un sujetador enroscado exteriormente que se usa sin tuerca en una aplicación de orificio enroscado o ciego para sujetar los componentes.

carbide—a harder metal than ordinary tool steel.

carburo—un metal más duro que el acero común de herramientas.

carbon blaster—a tool that uses crushed walnut shells blasted by compressed air to remove the carbon deposits.

chorro de carbón—una herramienta que usa las cáscaras molidas de nueces chorreadas con el aire comprimido para despegar los depósitos del carbón.

carbon pile—alternating positive and negative layers of carbon that are compressed against each other when a large knob on the face of a volt amp tester is tightened to cause a rapid discharge of the battery.

pila de carbón—las capas alternas de carbón positivo y negativo que se unen bajo presión al apretar un gran botón en la cara de un comprobador de voltamperios para causar una descarga rápida de la batería.

carbon trail—a line of electrically conducting carbon that forms in a cracked distributor cap.

rastro de carbón—una línea de carbón que conduce la electricidad que forma en una tapa agrietada de un distribuidor.

carbonaceous deposits—also called cauliflower deposits, they are thin, hard carbon deposits that result from fuel.

depósitos carbonosos—tambien llamados depósitos de coliflor, son los depósitos de carbón delgados y duros que resultan del combustible.

carburetor—a mechanical device that mixes fuel and air in response to how much air is flowing through it.

carburador—un dispositivo mecánico que mezcla el combustible con el aire como una respuesta de cuánto aire lo atraviesa.

case-hardened washer—a washer that is hard on the surface only; the core of the washer is soft and will compress.

arandela con endurecido de superficie—una arandela que es dura sólo en su superficie; el centro de la arandela es blando y se comprime.

casewashers—they are hard on the surface only; the core of the washer is soft and will compress.

arandela cementada—son duras solamente en sus superficies; el núcleo de la arandela es blando y comprimirá.

caster—the forward or rearward tilt of the spindle support arm.

ángulo del caster—la inclinación hacia afrente o hacia atrás del brazo de soporte del husillo.

catalyst—a substance that causes a chemical reaction to occur faster without going under any measurable change to itself.

catalizador—una sustancia que causa ocurrir más rapidamente una reacción química sin efectuar un cambio apreciable a si mismo.

catalytic converter—an emission device located in front of the muffler in the exhaust system. It looks very much like a heavy muffler and contains catalysts to clean up an engine's emissions before they leave the end of the exhaust pipe.

convertidor catalítico—un dispositivo de emisiones ubicado en la parte delantera del silenciador en el sistema del escape. Se parece mucho a un silenciador de trabajo pesado y contiene los catalizadores para limpiar las emisiones del motor antes de que salgan del tubo del escape.

cathode—like the negative terminal on a battery.

cátodo—parecido al terminal negativo de una batería.

caustic—a pH rating of above 7 is an alkaline cleaner, caustic or base.

cáustico—una clasificación pH de más de 7 es un limpiador alcalino, cáustico o un base.

cavitation—also called aeration, it is when hydraulic fluid becomes mixed with air.

cavitación—tambien llamado aeración, es cuando el fluido hidráulico se mezcla con el aire.

CCA—cold cranking amps.

CCA—amperios de arranque en frío.

CCOT—cycling clutch orifice tube.

CCOT—tubo del orificio del embrague en ciclo.

CD-ROM—compact disc read-only memory (CDs); CDs store a great deal of information that can be accessed by a personal computer with a CD player.

CD-ROM—disco compacto de memoria-solo-lectura (CDs); los CDs almacenen una gran cantidad de información que se puede acesar con una computadora personal con un tocadisco de discos compactos.

center of gravity—the point between the front and the rear wheels where the weight will be distributed evenly.

centro de gravedad—el punto entre las ruedas delanteras y traseras en donde se distribuirá el peso igualmente.

centrifugal advance—also called mechanical advance, it senses the speed of the engine.

avance centrífuga—tambien llamado un avance mecánico, percibe la velocidad del motor.

cetane number—describes how easily a diesel fuel will ignite.

índice de ceteno—describe qué tal facilmente un combustible de diesel encenderá.

CFC—chlorofluorocarbon; CFCs are depleting the protective ozone layer through a chemical reaction.

CFC—clorofluorocarburo; los CFCs están deteriorando la capa protectiva de ozono por medio de una reacción química.

chainfall—a chain hoist.

caída de cadenas—una montacargas con cadena.

charcoal canister—a small plastic or steel container filled with activated charcoal that can store gasoline vapors until the right time for them to be drawn into the engine and burned.

bote de carbón—un pequeño receptáculo de plástico o de acero llenado con el carbón activo que puede almacenar los vapores de la gasolina hasta el tiempo apropiado de que entran al motor para quemarse.

charge air cooler—also called an intercooler or an aftercooler, it is a sophisticated air cooler for blowers.

enfriador de aire cargado—tambien llamado un enfriador intermedio o postrefrigerador, es un enfriador del aire sofisticado para los ventiladores.

charging system output test—a test that loads the charging system and measures its current output.

prueba de salida del sistema de carga—una prueba que pone una carga en el sistema de carga y mide su salida de corriente.

chassis—the group of parts that includes the frame, shocks and springs, steering parts, tires, brakes, and wheels. It supports the engine and the car body.

chasis—el conjunto de partes que incluyen el armazón, los amortiguadores y resortes, las partes de dirección, los neumáticos, y las ruedas. Soporta el motor y la carrocería del coche.

chassis lubricant—a grease of a consistency that allows it to be applied through a zert fitting with a grease gun.

lubricante del chasis—una grasa de una consistencia que permite que se aplica por un niple de grasa con una pistola engrasadora.

chattering clutch—when the pedal shakes as the clutch pedal is raised from the floor.

castañateo del embrague—cuando el pedal tiembla al subir el pedal del embrague del piso.

cherry picker—an engine hoist.

burro—un aparato para izar el motor.

chip—tiny sandwiched silicon wafers constructed by photographically reproducing circuit patterns onto them make up a chip. As many as 30,000 transistors are placed in an area that is only ¼" square.

ficha—las obleas diminutas de silicio intercaladas construidas con los patrones de circuitos reproducidos fotográficamente en ellos. Tantos como 30,000 transistores se colocan en una área que mide un cuadro de sólo un ¼" de una pulgada.

CI engine—a diesel compression ignition engine.

motor CI—un motor de diesel con encendido de compresión.

circuit—when a complete circle is provided for electrical flow.

circuito—cuando un círculo completo se provee para el flujo de la electricidad.

circuit breakers—electrical circuit protection devices that open in the event of an electrical overload to prevent damage from occurring elsewhere in an electrical circuit. They reset automatically or can be manually reset after they trip.

rompedor de circuito—los dispositivos de protección de los circuitos eléctricos que se abren en caso de una sobrecarga eléctrica para prevenir que ocurren daños en otras áreas del circuito eléctrico. Se vuelven a cerrar automáticamente o pueden ser cerrados a mano después de que se hayan desconectado.

circuit resistance test—a test that checks the resistance in the ground and insulated circuits by measuring voltage drop.

prueba de resistencia del circuito—una prueba que verifica la resistencia en la tierra y los circuitos aislados al medir la caída del voltaje.

clamping force—results from the tension or stretch on bolts. It also results from the springiness of a gasket between parts.

fuerza de apriete—resulta de la tensión o la estiración en los pernos. Tambien resulta de la elasticidad de un empaque entre las partes.

Class A fire—one that can be put out with water.

encendio Grado A—uno que se puede apagar con el agua.

Class B fire—one in which there are flammable liquids such as grease, oil, gasoline, or paint.

encendio Grado B—uno en el cual hay líquidos inflamables tal como la grasa, el aceite, la gasolina o la pintura.

Class C fire—an electrical fire.

encendio Grado C—un encendio eléctrico.

Class D fire—a fire involving a flammable metal such as magnesium or potassium.

encendio Grado D—un encendio que involucra un metal inflamable tal como el magnesio o el potasio.

C-lock or C-clip axle—an axle with a groove on the inside that a clip fits into to keep it from coming out.

eje de cerradura C o grapa C—un eje que tiene una muesca en el interior en la cual queda una grapa que previene que salga.

clock spring—also called spiral cable, it connects the steering column to the air bag in the steering wheel.

resorte espiral—tambien llamado un cable espiral, conecta el tubo de dirección a la bolsa de aire en el volante de dirección.

closed loop—when a computer-controlled engine reaches operating temperature and the computer is receiving feedback from the oxygen sensor and making adjustments to the air-fuel mixture.

bucle cerrado—cuando un motor controlado por computadora llega a la temperatura de operación y la computadora recibe la información del detector de oxígeno y hace los ajustes a la mezcla de combustible-aire.

closed ventilation system—a crankcase ventilation system that does not allow blowby to escape into the atmosphere.

sistema de ventilación cerrado—un sistema de ventilación del cárter que no permite que la fuga de gases a través de los anillos de los cilindros salgan a la atmósfera.

close nipple—a small section of pipe that has male tapered threads on each end that join in the middle.

niple cerrado—una pequeña sección de tubo que tiene una rosca cónica macho en cada extremidad que se unen el en medio.

close ratio transmission—when there is a small difference between the ratios of a transmission's forward gears.

transmisión de relación aproximada—cuando hay una pequeña diferencia entre las relaciones de las velocidades de marcha en adelante de una transmisión.

cloud point—when diesel fuel appears cloudy when the wax separates out in very cold weather.

punto de turbiedad—cuando el combustible de diesel tiene una apariencia nebulosa al separarse la cera en el clima muy frío.

clutch—a device that uncouples the powertrain from the engine.

embrague—un dispositivo que desenganche el tren de potencia del motor.

clutch cushion plate—a metal cushion in between the clutch facings that lets the facings compress.

placa amortiguadora del embrague—un amortiguador de metal entre los forros del embrague que deja que se compriman los forros.

clutch facings—the friction material part of the clutch disc.

forros del embrague—la parte de materia de fricción del disco del embrague.

clutch fork—also called a throwout or release lever, it fits between the release bearing and the clutch cable or linkage.

horquilla del embrague—tambien llamado una palanca de desenganche o descarga, queda entre el cojinete de desenganche y el cable o la biela del embrague.

clutch free play—also called free travel, it is movement measured at the clutch pedal.

juego del embrague—tambien llamado carrera libre, es el movimiento medido en el pedal del embrague.

clutch hub—the inner part of a clutch disc.

cubo del embrague—la parte interior de un disco de embrague.

CNG—compressed natural gas.

CNG—gas natural comprimido.

CO—carbon monoxide, emissions that result when gasoline is not completely burned.

CO—las emisiones del monóxido de carbono que resultan cuando la gasolina no se consume completamente.

CO_2—carbon dioxide.

CO_2—anhídrido carbónico.

coast—deceleration, power is on the concave side of the gear tooth.

desembrague (vuelo)—la deceleración, la potencia está en el lado cóncavo del diente del engranaje.

coast side—the concave side of a gear tooth.

lado del desembrague (vuelo)—el lado cóncavo del diente del engranaje.

coefficient of friction—the ratio of the force holding two surfaces in contact to the force required to slide one over the other.

coeficiente de la fricción—la relación de la fuerza que sostiene el contacto entre dos superficies con la fuerza que se requiere en deslizar una sobre la otra.

coil saturation—when the magnetic field has finished its buildup inside of the coil.

saturación de la bobina—cuando se ha terminado la acumulación del campo magnético dentro de la bobina.

coil spring—a spring steel rod wound into a coil.

resorte helicoidal—una barra del acero para resortes que se ha enrollado en la forma de espiral.

cold solder connection—a defective electrical connection made when the solder is melted by the soldering iron, but the wire is too cold to bond to it.

conexión de soldadura fría—una conexión eléctrica defectuosa hecho cuando la soldadura se derrite con el hierro de soldadura, pero el alambre es demasiado frío para adherir con ella.

collet—a device that fits around the outside of a round part. When clamped around the outside, it holds that round part tightly.

collar de apriete—un dispositivo que queda alrededor de una parte redonda. Al apretarse al exterior, sujete esa parte fuertemente.

companion cylinders—two cylinders whose pistons come to TDC and BDC at the same time.

cilindros acoplados—dos cilindros cuyos pistones llegan a PMS y PMI a la misma vez.

composite headlamp—a headlamp housing with a glass balloon that the halogen lamp fits inside of.

faro compuesto—una caja del faro que tiene un receptáculo de vidrio en el cual queda una lámpara halógena.

compound gauge—a gauge that reads in either pressure or vacuum.

indicador combinado—un medidor que lee bajo presión o en vacío.

compound planetary gearset—two planetary gearsets combined to provide more gear ratio possibilities.

conjunto compuesto de engranajes planetarios—dos conjuntos de engranajes planetarios combinados para proveer más posibilidades de relaciones entre velocidades.

comprehensive component monitor—looks at electronically controlled emission devices, sensors, and actuators that are not tested by other OBD II monitors.

monitor de componentes completo—observa los dispositivos de emisión, sensores e impulsores controlados electrónicamente, en los que los demás monitores OBD II no realizan pruebas.

compression—when the wheel moves up as the spring compresses. Also called a jounce.

compresión—cuando la rueda se mueva hacia arriba al comprimir el resorte. También se llama un sacudo.

compression fittings—fittings that compress the end of a nut or a sleeve to provide a seal.

guarniciones de compresión—los accesorios que comprimen en la extremidad de una tuerca o una manguilla para proveer un sello.

compression pressure—the amount of pressure made by the piston moving up in the cylinder.

presión de compresión—la cantidad de presión creado por el pistón subiendo en el cilindro.

compression ratio—the difference between a cylinder's volumes with the piston at TDC and BDC.

relación de compresión—la diferencia entre los volúmenes de un cilindro cuando el pistón está en PMS y PMI.

compression test—a test of engine condition where a pressure gauge is inserted into a spark plug hole and registers a reading as the engine is cranked.

prueba de compresión—una prueba de la condición del motor en la cual un indicador de presión se introduce en un orificio de bujía y registra una lectura al arrancar el motor.

condensation—when a vapor changes to a liquid.

condensación—cuando el vapor se convierta en líquido.

condenser (air conditioning)—part of an air-conditioning system, it is a radiator for refrigerant.

condensador (aire acondicionado)—parte del sistema de aire acondicionado, es el radiador para el fluido refrigerante.

condenser (electrical)—also called a capacitor, it stores electricity.

condensador (eléctrica)—tambien se llama un capacitor, almacene la electricidad.

conductor—a material with atoms that allow electricity to flow freely.

conductor—una materia cuyos átomos permiten fluir fácilmente la electricidad.

connecting rod resizing—a machine shop process that restores the big end of the connecting rod to original size and roundness.

recalibración de la biela—un proceso del taller de máquinas que restaura la extremidad grande de la biela a su tamaño y redondez original.

constant velocity universal joint—also called a CV joint, a universal joint whose output speed is constant like its input speed.

junta universal de velocidad constante—una junta universal cuyo salida de velocidad es constante como su entrada de velocidad.

control force—the amount of damping control designed into a shock absorber by engineers. This is sometimes adjustable.

fuerza de control—la cantidad de control de amortiguación que los ingenieros incorporan en un amortiguador. Suele ser ajustable.

convection—when a body gives off heat, the surrounding air becomes warmer and moves upward.

convección—cuando un cuerpo despide el calor, el aire del ambiente se calienta y suba.

converter light off—when the catalytic converter reaches a temperature at which it begins operating.

luz del conversor apagada—cuando el conversor catalítico alcanza la temperatura en la que comienza a funcionar.

coolant—also called antifreeze, it is a combination of ethylene glycol and water used instead of pure water in the engine's cooling system.

fluido refrigerante—tambien llamado anticongelante, es una combinación del glicol etileno y el agua que se usa en vez del agua puro en el sistema de enfriamiento del motor.

coolant hydrometer—a tool that compares the weight of pure coolant to water to give an indication of coolant strength.

hidrómetro del fluido refrigerante—una herramienta que compara el peso del fluido refrigerante puro al del agua para dar una indicación de la concentración del fluido refrigerante.

copolymer—a combination of two polymers.

copolímeros—una combinación de dos polímeros.

core—a part that is returned for rebuilding.

núcleo—una parte que se entrega para reconstrucción.

cost leader—service items, such as oil, that are priced with little to no markup; also called loss leader.

líder de costos—ítems de servicio, tales como el aceite, con precio con poco o ningún margen de ganancias; también llamado líder de pérdidas.

countergear—one gear made up of a series of gears that mesh with the various gears on the mainshaft. It is often called a cluster gear.

engranaje auxiliar—un engranaje compuesto de una serie de engranajes que se endientan con varios engrenajes en el árbol principal. Muchas veces se refiere como el conjunto de ruedas dentadas.

coupling speed—the point at which all of the converter parts and ATF all turn as a unit.

velocidad del acoplamiento—el punto en el cual todas las partes del convertidor y el ATF giran como si fueran una unidad.

CPU—central processing unit; a computer.

CPU—unidad de proceso central; una computadora.

cradle—also called the crossmember, it is the large steel part of the frame beneath the engine and between the front wheels.

cuna—tambien llamado una traviesa, es una parte grande del armazón debajo del motor y entre las dos ruedas delanteras.

crank angle sensor—a primary trigger that senses the position of the piston for the computer.

sensor del ángulo de arranque—un disparador primario que detecta la posición del pistón para la computadora.

crank polishing—a motor-driven emery belt is used to polish the crankshaft, usually after grinding.

pulido del cigüeñal—una banda de esmeril impulsado por motor se usa en pulir el cigüeñal, normalmente después del esmerilado.

crank throw—one-half of the total piston stroke.

cigüeña—mitad del recorrido total del pistón.

crankcase—the area surrounding the crankshaft.

cárter—el área alrededor del cigüeñal.

cranking vacuum test—a test to see how much vacuum an engine can produce when cranking.

prueba de arranque en vacío—una prueba para ver cuánto vacío puede producir un motor al arrancarse.

cross and yoke universal joint—also called Cardan, it is the most popular universal joint design.

junta de horquilla y cruceta—tambien llamada Cardán, es el diseño más popular de junta universal.

cross counts—the speed at which the fuel signal from an O_2 sensor fluctuates from rich to lean.

salidas reversas—la velocidad con la que un señal de combustible de un sensor de oxígeno varía de rico a pobre.

crossfire induction—when one spark plug firing induces a spark in the one next to it, causing it to fire before its time.

inducción de encendido cruzado—cuando el disparo de una bujía induce una chispa en el que está a lado, causando que dispare prematuramente.

cross-flow head—a cylinder head that has the intake and exhaust ports on opposite sides of an in-line engine.

cabeza de flujo transversal—una cabeza del cilindro que tiene las puertas de entrada y salida en lados opuestos de un motor en línea.

cross-flow radiator—a radiator design where coolant flows from side to side.

radiador de flujo transversal—un diseño de radiador en el cual el fluido refrigerante fluye de lado a lado.

cross fluid contamination—when oil and coolant are interchanged with each other.

contaminación entre los fluidos—cuando el aceite y el fluido refrigerante se intercambian.

cross groove plunge joint—also called a pancake joint, it has balls and grooves in the bearing races that would cross each other if they extended far enough.

junta profundizada de muesca en cruz—tambien llamado una junta achatada, tiene las bolas y las muescas en las pistas de los cojinetes que se cruzarían si se extendieran lo bastante.

crosshatch—the criss-cross finish left when the cylinder is honed.

marcas cruzadas—el acabado de razguños cruzados que resulta cuando se rectifica un cilindro.

crossmember—also called the cradle, it is the large steel part of the frame beneath the engine and between the front wheels. A crossmember is sometimes used to support a rear-wheel-drive transmission.

traviesa—tambien llamado una cuna, es la parte grande de acero de un armazón debajo del motor y entre las dos ruedas delanteras. Una traviesa a veces se usa para soportar una transmisión de tracción trasera.

crossover—an amplifying device that blocks certain frequencies to a speaker.

divisor de frecuencias—un dispositivo de ampliación que bloquea ciertas frecuencias a un altavoz.

crowned road—when the road is higher at the center than the outside.

carretera corcovada—cuando la carretera es más alta en el centro que en las orillas.

cruise—the car is maintaining its speed.

crucero—el coche mantiene su velocidad.

current—the amount of electrons flowing in a circuit.

corriente—la cantidad de los electrones que fluyen en un circuito.

current draw—the amount of current required to operate a load.

consumo de corriente—la cantidad del corriente que se requiere en operar una carga.

current limiting system—when the ignition module shuts back on current flow as soon as the coil primary winding is saturated.

sistema limitador de corriente—cuando un módulo de encendido disminuya el flujo de corriente tan pronto que se satura la bobina del encendido.

custom rebuild—when an engine is rebuilt as a unit, rather than assembled from a variety of parts.

reconstrucción especializada—cuando un motor se reconstruye como una unidad, en contraste de asemblarse de una variedad de partes.

CV joint—constant velocity universal joint used on both ends of front-wheel-drive axles, also used on some rear-wheel drives.

junta CV—junta universal de velocidad constante usado en ambas extremidades de los ejes de tracción delantera, tambien usada en algunas de tracción trasera.

CVT—continuously variable transmission.

CVT—transmisión variable continua.

cylinder bank—a row of cylinders.

banco de los cilindros—una fila de cilindros.

cylinder bore taper wear—the tapered ring that is worn into a cylinder by the pressure of the piston rings against the cylinder wall.

aro cónico del calibre del cilindro—el aro cónico metido dentro del cilindro por desgaste, por la presión de los aros de pistón contra la pared del cilindro.

cylinder glaze—the area where the rings ride in the cylinder develops a glazed appearance.

porcelana (barníz) del cilindro—el área en donde los anillos viajan en el cilindro toma una apariencia barnizado.

cylinder leakage test—a test where regulated compressed air is introduced into a cylinder through the spark plug hole.

prueba de fuga del cilindro—una prueba en que el aire comprimido regulado se introduce al cilindro por el orificio de la bujía.

cylinder power balance test—a test where the spark to a cylinder is shorted out, resulting in a drop in engine rpm.

prueba del balance de fuerza en el cilindro—una prueba en la cual se efectúa un corto en la chispa del cilindro, resultando en una caída del rpm en el motor.

D

dampened hub—a clutch hub with torsional dampers that are either coil springs or rubber positioned between the disc plate and the clutch hub to absorb shock during engagement.

cubo amortiguado—un cubo del embrague que tiene amortiguadores de torsión que son de resortes helicoidales o de caucho entre la placa disco y el cubo del embrague para absorber los choques durante el enganchamiento.

DC—direct current; when electrons flow in only one direction.

DC—corriente directo; cuando los electrones fluyen en una sóla dirección.

deep-cycle—when the battery is allowed to run almost completely dead, and then is recharged.

ciclo prolongado—cuando se permite que se descarga casi completamente una batería, y luego se recarga.

delta winding—an alternator stator winding that resembles the Greek letter delta (Δ).

devanado triangular (en delta)—un devanado del estátor del alternador que parece la letra griega delta (Δ).

dermatitis—irritated skin.

dermatitis—la piel irritada.

desiccant—a material that removes moisture from the system.

desicante—una materia que remueva la humedad del sistema.

detent—forced kickdown.

detención—un disminuyo forzado.

detonation—self-ignition due to excessive pressure in the cylinder.

detonación—el autoencendido debido a una presión excesiva en el cilindro.

detonation sensor—also called a knock sensor, it is a piezoelectric crystal that detects the frequency of the vibration caused by detonation and tells the computer to retard the spark timing until the vibration goes away.

sensor de detonación—tambien llamado un sensor de golpeteos, es un cristal piezoeléctrico que percibe la frecuencia de la vibración causada por la detonación y mande que la computadora retarda la chispa al encendido hasta que se desaparece la vibración.

Dexron III/Mercon—the latest and most popular ATF.

Dexron III/Mercon—el ATF más nuevo y más popular.

diagnostic tree—part of the service information that provides a step-by-step diagnostic procedure to follow when troubleshooting hard fault electrical problems.

esquema diagnóstica—parte de la información de servicio que provee un procedimiento paso a paso diagnóstico que uno puede seguir en diagnosticar los problemas de fallos eléctricos difíciles.

diagonal braking system—operates the brakes on opposite corners of the vehicle.

sistema de enfrenamiento diagonal—opera los frenos en las esquinas opuestas del vehículo.

diaphragm spring—also called a Belleville spring, replaces the release levers and coil springs in a diaphragm clutch.

resorte de diafragma—tambien llamado un resorte Belleville, reemplaza las palancas de desconexión y los resortes helicoidales en un embrague de diafragma.

dielectric—a nonconductive grease or coupling.

dielétrico—una grasa o acoplamiento aislante.

diesel-cycle—an engine that ignites its air and fuel using compression.

ciclo diesel—un motor que encienda su aire y combustible utilizando la compresión.

dieseling—also called run-on, is when an engine continues to run even after the ignition key is turned off.

autoencendido—tambien llamado marcha continua, es cuando un motor continua a marchar aún después de que se haya apagado la llave del encendido.

differential—a device that allows the rear wheels to be able to rotate at different speeds as the vehicle goes around corners.

diferencial—un dispositivo que permite girar las ruedas traseras en velocidades diferentes cuando un vehículo da la vuelta en una esquina.

digital processed information—signals that are yes or no, or on/off.

información procesado digitalmente—los señales que son de sí o no, o de prendido/apagado.

diode—a one-way electrical check valve made by placing P-type and N-type crystals back to back.

diodo—una válvula de detención eléctrica de una vía que se hace poniendo los cristales de tipo P y de tipo N espalda a espalda.

DIS—distributorless ignition system or direct ignition system; it has multiple ignition coils that fire spark plugs based on a signal received from the computer.

DIS—sistema de encendido sin distribuidor o sistema del encendido directo; tiene múltiples bobinas del encendido que disparan las bujías según un señal recibido de la computadora.

discard diameter—the maximum allowable size for a brake drum.

diámetro de rechazo—el tamaño máximo permisible de un tambor de freno.

discharge service valve—the high side service valve in an air-conditioning system.

válvula de descarga de servicio—la válvula del lado alto de servicio en un sistema de aire acondicionado.

displacement—the volume that the piston displaces in the cylinder as it moves from TDC to BDC.

desplazamiento—el volumen que desplaza el pistón en el cilindro al moverse del PMS al PMI.

display pattern—also called parade pattern, it displays all of the cylinders next to each other, side by side so that the heights of the voltage spikes can be compared.

modelo indicativo—tambien llamado modelo desfile, demuestra todos los cilindros juntos, lado a lado para que se puede comparar las alturas de los impulsos de tensión.

dive—during hard braking, the front of the car is pushed down and the rear of the car raises up.

empinado—al frenar de repente, la delantera del automóvil baja y la trasera se eleva.

diverter valve—a device that prevents a backfire during deceleration on carbureted cars.

válvula derivador—un dispositivo que previene un retorno de encendido durante la deceleración en los coches con carburadores.

DIY—do-it-yourself.

DIY—siglas en inglés para 'hágalo usted mismo'.

DLC—data link connector; a diagnostic connector that a test instrument can be connected to for reading computer serial data.

DLC—conector de enlace de datos; un conector diagnóstico al cual se puede conectar un probador para leer los datos seriales de la computadora.

DMM—digital multimeter.

DMM—multímetro digital.

dog teeth—little teeth around the circumference of the edge of the gear.

dientes pontiagudas—los pequeños dientes alrededor de la circunferencia en la orilla del engranaje.

DOHC—dual overhead cam.

DOHC—doble árbol leva sobre la cabeza.

dominant end—the end of a coil spring that aligns in the coil spring seat in the frame or lower control arm.

extremidad dominante—la extremidad de un resorte helicoidal que se alinea en el asiento del resorte helicoidal que se encuentra en el armazón o en el brazo de mando inferior.

doping—a process of adding an impurity to a crystal to make a semiconductor.

preparación del semiconductor—el proceso de añadir una impureza a un cristal para fabricar un semiconductor.

DOT—Department of Transportation.

DOT—Departamento de Transportación.

DOT dry specification—for new fluid.

DOT especificación seca—para un fluido nuevo.

DOT wet specification—for fluid that has absorbed 2% water.

DOT especificación en húmedo—para el fluido que ha absorbido el 2% del agua.

double-acting shock absorber—controls motion when moving both up and down.

amortiguador de choque de doble acción—controla ambos movimientos de hacia arriba y de hacia abajo.

double flare—one way in which the ends of tubing are formed.

extremidad de bocinado doble—una manera en la cual se forman las extremidades de los tubos.

double offset plunge joint—has six balls evenly spaced by a cage. The outer housing is considerably deeper than a cross groove so it can plunge further. The races have straight grooves.

junta profundizada de desviación doble—tiene seis bolas separadas uniformemente por una pista. La caja exterior es de una profundidad considerable más de una muesca en cruz para que puede profundizar aún más. Las pistas tienen las muescas rectas.

double-walled tubing—a type of tubing that has two layers of metal and can bend easier.

tubo de doble muro—un tipo de tubo que tiene dos capas de metal y que puede doblarse más facilmente.

down-flow radiator—a radiator design where coolant flows from top to bottom.

radiador de flujo vertical—un diseño del radiador en el cual el fluido fluye desde arriba hacia abajo.

downshift—when the transmission shifts to a lower gear; second to first, for instance.

cambio descendente—cuando la transmisión cambia de velocidad a un engranaje más bajo, del segundo al primero, por ejemplo.

dragging clutch—when a clutch does not release easily.

embrague arrastrante—cuando un embrague no quiere desconectarse facilmente.

draw-through—when a supercharger or turbocharger pressurizes the intake manifold after the carburetor and air cleaner.

aspirar—cuando un sobrealimentador o un turbocompresor presuriza el múltiple de admisión después del carburador y el filtro de aire.

drive—when under acceleration, power is on the convex side of the gear tooth.

impulso—al estar en aceleración, la potencia está en el lado convexo del diente del embrague.

drive cycle—a cycle in which the engine must enter closed loop and all five of the trip monitors and the catalyst monitor must operate.

ciclo de transmisión—el ciclo en el que el motor debe ingresar en anillo cerrado y todos los cinco monitores de viaje y el monitor catalizador deben funcionar.

driveline—a term that describes the parts that transfer power from the transmission to the rear wheels.

flecha motríz—un término que describe las partes que transfieren la potencia de la transmisión a las ruedas traseras.

drive shaft—a hollow metal tube that has a universal joint at each end.

árbol de transmisión—un tubo metálico hueco que tiene una junta universal en ambas extremidades.

drive shaft phase—when trunnions of the front and rear U-joints are in the same plane (parallel).

fase del árbol de transmisión—cuando los muñones de las juntas en U delanteras y traseras están en el mismo plano (son paralelos).

drive side—the convex side of a gear tooth.

lado del impulso—el lado convexo de un diente del engranaje.

drop center—also called rim well, provides a means of removing and installing a tire from the wheel.

llanta de centro cóncavo—tambien llamado llanta cóncava, provee una manera de remover e instalar un neumático de la rueda.

dropping point—the temperature at which a grease turns into a liquid.

punto de caída—la temperatura en la cual una grasa se convierte en líquido.

dry park check—with the tires on the ground, an assistant turns the steering wheel back and forth a short distance. Looseness in the steering linkage or suspension is detected.

verificación de park en seco—con los neumáticos en contacto con el suelo, un asistente gira el volante de dirección de un lado al otro por una corta distancia. Se detecta así cualquier juego libre en las bielas de dirección o en la suspensión.

dry start—when the crankshaft rubs against the bearings before oil has been pumped to them.

arranque en seco—cuando el cigüeñal frota contra los cojinetes antes de que se les ha proporcionado el aceite.

DTC—diagnostic trouble code.

DTC—código diagnóstico de errores.

dual bed converter—a catalytic converter that has separate reduction and oxidation parts.

convertidor de doble alojamiento—un convertidor catalítico que tiene partes separadas para reducción y oxidación.

dual exhaust system—when the exhaust system has two pipes, two mufflers, and two catalytic converters.

sistema de doble escape—cuando el sistema de escape tiene dos tubos, dos silenciadores y dos convertidores catalíticos.

dual function electronic ignition system—spark timing by the computer takes the place of vacuum and mechanical spark advance.

sistema de arranque electrónico de doble función—la sincronización de la chispa reemplaza un avance de la chispa por medio del vacío o por avance mecánico.

dual pass machine—an air-conditioning station that requires two procedures; first recover, then recycle.

máquina de dos pasos—una estación de aire acondicionado que requiere dos procedimientos; el primero de recobrar, luego de reciclar.

dual-plane manifold—when each barrel supplies half of the engine's cylinders.

múltiple de doble plano—cuando cada cañón provee para la mitad de los cilindros de un motor.

dummy shaft—a short shaft that is used to keep the bearings in place while removing a transmission shaft.

árbol falso—un eje corto que sirve para mantener en posición los cojinetes mientras que se remueva el árbol de la transmisión.

dump—also called decay. Fluid pressure that will be released to the wheel if the frequency of the signal continues to drop off.

vaciado—también llamado descarga; presión de líquido que será liberada a la rueda si la frecuencia de la señal sigue cayendo.

dump valves—release valves that open and close rapidly to bleed system pressure.

válvulas de vaciado—válvulas de descarga que se abren y se cierran rápidamente para purgar la presión del sistema.

duration—the number of degrees of crankshaft travel while the valve is off its seat.

duración—el número de grados que avance un cigüeñal mientras que la válvula está levantada de su asiento.

duty cycle—a measurement of the maximum practical capacity of a piece of equipment. If an air compressor's maximum capacity is when it runs for 7 out of every 10 minutes, this is referred to as a 70% duty cycle.

ciclo de trabajo—una media de la capacidad máxima práctica de una pieza de equipo. Si la capacidad máxima de un comprimidor de aire es cuando trabaja 7 de cada 10 minutos, esto se llama un ciclo de trabajo de 70%.

DVOM—digital volt-ohmmeter.

DVOM—voltío-ohmiometro digital.

dwell—the length of time in degrees of distributor rotation that primary current is flowing in the coil primary winding.

ángulo de cierre—la cantidad del tiempo en grados de la rotación del distribuidor en que el corriente primario fluye en las bobinas primarias.

dye penetrant—a crack detection process that uses a dye and developer.

tinte penetrante—un proceso de detectar las grietas que usa un tinte y un revelador.

dynamic—moving.

dinámico—en movimiento.

dynamic and couple imbalance—types of imbalance that require adding or removing metal at two different places on the part. Computer balancers compute the combined amount of dynamic and force imbalance and tell where to remove metal to correct them.

desequilibrado de emparejado dinámico—los tipos de desequilibrado que requieren el añadir o remover el metal en dos lugares distintos de la parte. Los equilibradores computerizados calculan la cantidad combinada del desequilibrio dinámico y forzado e indica en dónde hay que remover el metal para corregirlos.

dynamic seal—a seal that works against moving parts.

sello dinámico—un sello que funciona contra las partes en movimiento.

dynamometer—a power absorption device for measuring engine power.

dinamómetro—un dispositivo de absorción de fuerza para medir la fuerza de un motor.

E

earless clamp—a type of clamp used to fasten CV joint boots to the axle.

abrazadera sin orejilla—tipo de abrazadera usado para sujetar fundas de conexión de CV al eje.

ECA—electronic control assembly, a computer.

ECA—asamblea de control electrónica, una computadora.

ECU—electronic control unit, a computer.

ECU—unidad electrónica de control, una computadora.

eddy current—an electric current resulting from a magnetic field.

EDM—electrical discharge machining.

EDM—mecanizado por descarga eléctrica.

EEPROM—electronically erasable read-only memory.

EEPROM—memoria solo lectura borrable electrónicamente.

EFE—early fuel evaporation.

EFE—evaporación temprana del combustible.

EGR—the exhaust gas recirculation system that allows a small amount of exhaust gas to be routed into the incoming air-fuel mixture to reduce NO_x emissions.

EGR—el sistema de recirculación de los gases de escape que permite que una pequeña cantidad del gas del escape se desvía a la mezcla del combustible-aire para disminuir las emisiones NO_x.

elastic limit—this point is reached when a fastener will no longer return to its original shape when loosened.

límite elástico—este punto se alcanza cuando un sujetador no volverá a su forma original al ser aflojado.

electricity—the flow of electrons from one atom to another.

electricidad—el flujo de los electrones desde un átomo a otro.

electrolysis—using electricity to break down water into hydrogen and oxygen.

electrólisis—usando la electricidad para descomponer el agua al hidrógeno y el oxígeno.

electrolyte—a mixture of sulfuric acid and water used in automotive batteries.

electrólito—una mezcla del ácido sulfúrico y el agua que se usa en las baterías automotivas.

electromagnetic induction—when electricity is produced by moving a magnetic field over a conductor.

inducción electromagnética—cuando la electricidad se produce por medio de pasar un campo magnético sobre un conductor.

electromechanical voltage regulator—a regulator that operates using coils of wire and electrical contact points.

regulador de voltaje electromecánico—un regulador que opera utilizando las bobinas de alambre y los puntos de contacto eléctricos.

electron theory—electron flow is from negative to positive.

teoría de electrones—el flujo de los electrones es del negativo al positivo.

electronic computer advance—uses information available to the computer for the fuel and emission systems to control spark advance.

avance electrónico computerizado—usa la información disponible a la computadora de los sistemas de combustible y emisiones para controlar el avance de la chispa.

electronic ignition system—when a transistor triggers the buildup and collapse of the magnetic field in the coil primary.

sistema de arranque electrónico—cuando un transistor dispara la acumulación y el colapso del campo magnético en la bobina primaria.

element—a group of battery plates connected together in parallel.

elemento—un grupo de placas de batería conectado juntos en paralelo.

EMF—electromotive force, also called voltage, which pushes or pulls an electron out of its orbit.

EMF—fuerza electromotríz, tambien llamado voltaje, que empuja o jala un electrón fuera de su órbito.

end gap—the gap between the ends of the piston ring when it is installed in the cylinder.

holgura en la extremidad—la holgura entre las extremidades del anillo de pistón al instalarse en el cilindro.

end play—back-and-forth clearance.

juego en la extremidad—holgura lateral.

end thrust—side-to-side or front-to-rear force against a shaft.

empuje de la extremidad—la fuerza de un lado a otro o de frente a atrás contra un eje.

energy—the ability to do work.

energía—la habilidad a trabajar.

engine displacement—determined by multiplying the cylinder displacement by the number of cylinders.

desplazamiento del motor—se determine al multiplicar el desplazamiento por el número de cilindros.

engine kit—parts sold together as a group for rebuilding an engine.

piezas del motor sueltas—las partes vendidas juntas en grupo para reconstruir un motor.

engine knock—a noise that results from excessive clearance or abnormal combustion.

golpeteo del motor—un ruido que resulta de una holgura excesiva o de la combustión anormal.

engine sling—a cable or chain used with an engine hoist.

eslinga del motor—un cable o una cadena que se usa con una grúa para el motor.

E.P.—extreme pressure.

E.P.—presión extrema.

EPA—Environmental Protection Agency.

EPA—Agencia de Protección del Medio Ambiente.

EP additive—extreme pressure additive added to lubricants to prevent welding between metal surfaces.

aditivo EP—un aditivo de presión extrema que se agrega a los lubricantes para prevenir la soldadura entre los superficies metálicos.

EPR—evaporator pressure regulator valve.

EPR—válvula reguladora de la presión de evaporación.

EPROM—an older PROM that had to be removed from the computer and put under an ultraviolet light for a specified period of time to reprogram it.

EPROM—un modelo PROM más antiguo se tenía que remover de la computadora y ser puesta bajo una luz ultravioleta por un período específico para reprogramarla.

ester—polyalester oil for R-134A air-conditioning systems.

ester—el aceite polialaster para los sistemas de aire acondicionado R-134A.

ether starting fluid—a spray containing ether, used to start a gasoline engine that has run out of fuel.

líquido de arranque de éter—un spray que contiene éter, usado para dar arranque a un motor a gasolina al que se le ha acabado la gasolina.

EVA—electronic vibration analyzer.

EVA—analizador electrónico de vibraciones.

evaporation—a method of heat transfer where moisture is vaporized as it absorbs heat.

evaporación—un metodo de transferencia del calor en el cual la humedad se vaporiza al absorber el calor.

evaporator—part of an air-conditioning system. A small, radiator-like device used to remove heat from the passenger compartment.

evaporador—parte de un sistema de aire acondicionado. Un dispositivo pequeño parecido a un radiador que sirve para remover el calor del compartimento de pasajeros.

exhaust backpressure—resistance to an engine's airflow caused by restrictions in the exhaust system.

contrapresión del escape—la resistencia al flujo del aire del motor debido a las restricciones en el sistema del escape.

expansion tank—a small tank sometimes located inside of the main gas tank that allows for expansion of fuel in a full tank on a hot day.

tanque de expansión—un tanque pequeño que a veces se encuentra dentro del tanque de gas principal que permite expandir el combustible dentro de un tanque lleno en un día caliente.

external combustion engine—an engine where the fuel is burned outside of the engine, a steam engine.

motor de combustión externa—un motor en el cual el combustible se consume fuera del motor, un motor de vapor.

F

face—also called top, the part of a gear tooth that is the area above the pitch line.

cara—tambien llamado parte superior, la parte de un diente de engranaje que está en el área superior a la línea cero.

fan clutch—a temperature or torque sensitive clutch attached to a belt-driven cooling fan.

embrague ventilador—un embrague sensitivo a la temperatura o a la torsión que está conectada a un ventilador impulsado por correa.

feedback carburetor—a computer-controlled carburetor that meters the fuel according to how much oxygen is found in the engine's exhaust.

carburadores de retroalimentación—un carburador computerizado que mide el combustible de acuerdo con la cantidad de oxígeno encontrada en el escape del motor.

feedback system—a computer fuel system that monitors the oxygen content in the exhaust.

sistema de retroalimentación—un sistema de combustible computerizado que amonesta la cantidad del oxígeno en el escape.

ferrous—iron and steel.

férreo—hierro y acero.

ferrule—a sleeve used in a compression fitting.

férula—un manguito que se usa en un guarnición de compresión.

fhp—frictional horsepower; the power lost due to friction.

fhp—caballos de fuerza friccional; la fuerza perdida debido a la fricción.

fibrous grease—a grease that is sticky and therefore not good for use in transmission assembly.

grasa fibrosa—una grasa que es pegajosa y, por lo tanto, su uso en el conjunto de la transmisión no es bueno.

field coil—a heavy copper ribbon wound around soft iron cores to form an electromagnet.

bobina inductora—una gruesa cinta del cobre enredada alrededor de los núcleos de hierro blando para formar un imán.

filament—a wire in a light bulb that provides a resistance to electron flow. When it heats up, it causes light.

filamento—un alambre en un foco que provee una resistencia al flujo de electrones. Cuando se calienta, alumbra.

filings—the pieces of metal that come off the workpiece during filing.

limaduras—los pedazos del metal que se desprenden de la pieza al limar.

filter sock—a filter inside of the gas tank.

filtro con colador—un filtro dentro del tanque de gas.

final drive ratio—the ratio between the transmission output shaft and the differential ring gear.

relación de mando final—la relación entre la salida de la transmisión y la corona.

fire wall—the metal wall between the engine and passenger compartments.

mamparo de encendios—el muro de metal entre los compartimentos del motor y de los pasajeros.

firing line—the upward line that starts the scope pattern.

indicación del encendido—la línea ascendente con la cual comienza el patrón en el osciloscopio.

firing order—the order in which the spark plugs in a multicylinder engine fire.

orden del encendido—el orden en el cual disparan las bujías en un motor con múltiples cilindros.

first order vibration—anything that spins at drive shaft speed that vibrates only once every revolution.

vibración del primer orden—cualquier cosa girando con la velocidad del árbol de la transmisión que vibra solamente una vez por cada revolución.

five-gas analyzer—an infrared exhaust analyzer that measures HC, CO, CO_2, O_2, and NO_x.

analizador de cinco gases—un analizador de escape infrarrojo que mide HC, CO, CO_2, O_2 y NO_x.

fixed joint—the outboard CV joint that allows for a change in the angle of the axle in response to bumps and for steering.

junta fija—la junta CV fuera de borda que permite un cambio en el ángulo de un eje como respuesta a los choques y para la dirección.

flame front—when heat ignites molecules next to already burning molecules and a chain reaction takes place, which results in a flame expanding evenly across the cylinder.

frente de llama—cuando el calor encienda las moléculas juntas a las moléculas ya quemando y resulta una reacción en cadena, lo cual causa una llama que se expande uniformemente por el cilindro.

flank—also called root, the part of a gear tooth that is the area below the pitch line.

flanco—tambien llamado el raíz, la parte de un diente de un engranaje que está en el área inferior a la línea cero.

flapper valve—also called a duckbill valve, a rubber valve that allows fluid to flow through its center in one direction only.

válvula de charnela—tambien llamado una válvula de aleta, una válvula de caucho que permite que el fluido fluye por su centro en solamente una dirección.

flare nuts—hollow fittings on fuel, brake, or hydraulic lines.

tuercas bocinadas—los guarniciones huecos en las líneas de combustible, de frenos o hidraúlicas.

flashing the PROM—when the information in an EEPROM is erased and reprogrammed.

limpiar el PROM—cuando la información en un EEPROM es borrada y reprogramada.

flash point—the temperature at which the vapors of a flammable liquid will ignite when brought into contact with an open flame.

temperatura de inflamabilidad—la temperatura en que los vapores de un líquido inflamable encenderán al ponerse en contacto con una llama expuesta.

flat cam—when a cam lobe wears out.

leva aplanada—cuando se desgaste el lóbulo excéntrico de una leva.

flat rate/commission—method of payment in which the technician is paid a flat hourly rate for the amount of time a job is supposed to take, regardless of how much time is actually spent completing the job.

tarifa fija/comisión—método de pago en el que se le paga al técnico una tarifa por hora fija por el tiempo que se supone que llevará un trabajo, independiente del tiempo que efectivamente haya llevado el trabajo.

Flat-Rate Manual—also called a Parts and Time Guide, it lists the estimated cost of parts. A service writer or a business owner will use it to give a fairly accurate estimate to a customer.

Manual de Precio Fijo—tambien llamado una Guía de Partes y Labor, cataloga los precios estimados de las partes. Un girador de servicios o un negociante lo usará para fijar un presupuesto bastante acertado para una clientela.

fleet shop—a shop that exclusively services a fleet of vehicles, such as a taxicab company.

taller de flotas—taller exclusivo para servicio de flotas de vehículos, tales como una compañía de taxis.

float—car speed is slowly dropping.

flotar—la velocidad del coche se disminuya poco a poco.

flooding—when an engine gets too much fuel to support combustion.

ahogar—cuando un motor recibe demasiado combustible para efectuar la combustión.

fluid coupling—a fluid clutch.

acoplador flúido—un embrague de fluido.

flux—rosin or acid used to clean metal so that solder will stick to it.

flujo eléctrico—la resina o el ácido que se usa para limpiar el metal para que la soldadura se adherirá.

follower ball joint—the nonweight carrying ball joint that keeps the steering knuckle in alignment.

junta de casquillo de rótula—la articulación esférica de no soportar peso que mantiene alineado el muñón de dirección.

foot-pound—a measurement of work where 1 pound is moved for a distance of 1 foot.

libra-pie—una medida del trabajo en la cual 1 libra se mueve la distancia de 1 pie.

foot-pounds/Newtons—torque readings are expressed in foot-pounds in the English system, Newtons in the metric system.

libras-pie/Newtons—las lecturas de torsión se describen en libras-pie en el sistema inglés, en Newtons en el sistema métrico.

footprint—the area of the tire tread that contacts the road.

huella—el área de la banda de rodamiento que toca la carretera.

force—any action that changes, or tends to change, the position of something.

fuerza—cualquier acción que cambia, o tiene tendencias a cambiar, la posición de algo.

force, static or kinetic imbalance—different names for the same thing. A type of imbalance that can be compared to balancing tires with a bubble (level) balancer.

fuerza, estático o desequilibrio cinético—los nombres distintos para la misma cosa. Un tipo de desequilibrio que se puede comparar al equilibrar los neumáticos con un compensador anivelador (de burbuja).

forward bias—causes a P-N junction to conduct current.

polarización directa—causa que un enlace P-N conduzca el corriente.

fouled spark plug—one with a buildup of carbon that shorts it out.

bujía ensuciado—uno que tiene una acumulación del carbón que causa un cortocircuito.

four-channel system—the most effective ABS because it has sensors that monitor each wheel.

sistema de cuatro canales—el freno antibloqueo (ABS) más eficaz, pues tiene sensores para el monitoreo de cada rueda.

four corners scuffing—damage to the piston that usually occurs as a result of an external cause such as too lean an air-fuel mixture, which causes the top of the piston to run too hot.

rozamiento en cuatro puntos—el daño al pistón que suele ocurrir como resultado de una circunstancia externa, tal como una mezcla demasiada pobre del combustible-aire, lo cual causa que la parte superior del pistón marcha con excesivo calor.

four-gas analyzer—an infrared exhaust analyzer that measures HC, CO, CO_2 and O_2.

analizador de cuatro gases—un analizador de escape infrarrojo que mide HC, CO, CO_2 y O_2.

four-stroke cycle—also called the Otto-cycle, intake, compression, power, exhaust.

ciclo de cuatro carreras—tambien llamado el ciclo Otto, entrada, compresión, fuerza, escape.

four-wheel alignment—a method of wheel alignment done on many newer cars that are equipped with rear wheel adjustment capability.

alineación a cuatro ruedas—un método de alineación en las ruedas que se efectúa en muchos de los coches de último modelo que se equipan con la capacidad de ajustar las ruedas traseras.

free-air delivery—the best measurement of an air compressor's useful capacity measured in standard cubic feet per minute (SCFM).

entregada de aire libre—la mejor medida de la capacidad útil de un comprimidor de aire que se mide en pies cúbicos calibrados por minuto (SFCM).

free-wheeling engine—an engine that will not experience piston-to-valve interference if the timing chain or belt skips or breaks.

motor de sobremarcha—un motor que no experiencia una interferencia de pistón a válvula si brinca o se rompa la cadena o correa de tiempo.

frequency—how many cycles take place in a period of time.

frecuencia—cuántos ciclos ocurren en un período del tiempo.

frequency modulation—FM radio; varies the frequency of the broadcast signal.

modulación de amplitud—radio FM; varía la potencia de la señal de transmisión.

friction disc—the driven member of a clutch that is positioned between the driving parts.

disco de fricción—el miembro arrastrado de un embrague colocado entre dos partes impulsores.

frictionless bearing—also called antifriction bearing, it provides a rolling contact with balls, rollers, or needles.

cojinete sin fricción—tambien llamado cojinete antifricción, proveé un contacto rodante por medio de bolas, rodamientos, o las agujas.

full-field test—a test that takes the regulator out of the circuit and causes the alternator to give full output.

prueba completa de campo—una prueba que remueva el regulador del circuito y causa que el alternador dé su salida completa.

full-floating axle—an axle design in which the bearings do not touch the axle but are located on the outside of the axle housing.

eje flotante—un diseño de eje en el cual los cojinetes no tocan al eje pero se encuentran al exterior de la caja del eje.

full-floating plain bearings—bearings that have oil clearance on both sides, they spin at about one-third shaft rpm.

cojinetes lisos flotantes—los cojinetes que tienen holgura para el aceite en ambos lados, giran una tercera parte de la velocidad rpm.

full-flow oil filter—a filtering system designed to flow all of the oil supplied by the pump to the oil filter on its way to the engine bearings.

filtro de aceite de pleno flujo—un sistema de filtración diseñado para dirigir todo el aceite suministrado por la bomba en su rumbo a los cojinetes del motor.

furnace brazing—a process used to make seamless tubing that fuses copper to steel. When the copper melts, the seam disappears.

cobresoldadura en horno—un proceso usado en la fabricación de los tubos sin cordón que funde el cobre en el hierro. Cuando se derrite el cobre, el cordón se desaparece.

fuse—a circuit protection device designed to melt when the flow of current becomes too high for the wires or loads in the circuit.

fusible—un dispositivo para la protección de un circuito diseñado a fundir cuando el flujo de corriente exceda lo que puede tolerar los alambres o las cargas de un circuito.

fusible link—a circuit protection device that is a small length of wire, smaller in diameter than the wire it is connected to.

cartucho de fusible—un dispositivo para la protección de un circuito que consiste de un trozo pequeño del alambre, de un diámetro más pequeño del alambre al cual está conectado.

FWD—front-wheel drive.

FWD—tracción delantera.

G

galling—wear caused by metal-to-metal contact in the absence of adequate lubrication. Metal is transferred from one surface to another.

desgaste por fricción—el desgaste causado por el contacto entre un metal y otro en la ausencia de la lubricación adecuada. El metal se transfiere de una superficie a otra.

gas shock—a shock with the oil column pressurized to keep the bubbles from forming in the fluid.

amortiguador de gas—un amortiguador con una columna de aceite bajo presión que previene que forman las burbujas en el fluido.

gasohol—a mixture of gasoline and alcohol.

gasohol—una mezcla de la gasolina y el alcohol.

Gauss meter or gauge—a meter or gauge that detects magnetism.

medidor o calibrador Gauss—un medidor o calibrador que detecta el magnetismo.

gear pattern—a pattern taken off of the ring gear teeth from a coating of colored paste painted on the gear teeth.

patrón de engranajes—una impresión que se toma de los dientes de la corona por medio de una pasta de color pintado en los dientes del engranaje.

gear radius—the distance from the center of a gear to its outside edge. It is where the torque is measured from.

radio de engranaje—la distancia del centro de un engranaje a su orilla exterior. Es en donde se mide la torsión.

gear ratio—the difference in speed between two gears calculated by dividing the number of teeth on the driven gear by number of teeth on the driving gear.

relación de engranajes—la diferencia en velocidad entre dos engranajes que se calcula al dividir el número de los dientes en el engranaje arrastrado por el número de los dientes en el engranaje propulsor.

generator—an electrical device that produces alternating current that is rectified to DC current by the brushes and commutator.

generador—un dispositivo eléctrico que produce el corriente alterna que se rectifica en un corriente DC por medio de las escobillas y el conmutador.

geometric centerline—a line drawn between the center of the front axle and the center of the rear axle.

línea central geométrico—una línea dibujada entre el centro del eje delantero y el centro del eje trasero.

ghp—gross horsepower; engine power available with only the water pump and alternator using power.

ghp—caballo de fuerza brutola fuerza del motor disponible cuando sólo usan potencia la bomba de agua y el alternador.

glitch—a momentary interruption of an electrical signal.

glitch—una interrupción momentánea de una señal eléctrica.

governor pressure—the kind of transmission pressure that results from increases in vehicle speed.

presión del regulador—el tipo de presión de transmisión que resulta de aumentar la velocidad del vehículo.

grabbing clutch—when the friction disc does not slip normally, but grabs all at once.

embrague agarrador—cuando el disco de fricción no se desliza normalmente, sino que se agarra de una vez.

granny gear—when a transmission has a very low first gear, it is sometimes called a granny gear.

engranaje abuelita—cuando una transmisión tiene la primera velocidad muy lenta, a veces se llama un engranaje abuelita.

gravity bleeding—bleeding a hydraulic clutch by opening the slave cylinder bleed screw and letting fluid flow out.

purgar con gravedad—purgar un embrague hidráulico abriendo el tornillo de purgar del cilindro secundario y dejando que salga el fluido.

grease—a combination of oil and a thickening agent.

grasa—una combinación del aceite y un agente espesativo.

greasesweep—an absorbent material, such as rice hull ash or kitty litter used for cleaning spills.

limpiagrasa—un material absorbente, tal como la ceniza de casco de arroz o la arena usado para limpiar los derrames.

ground side switching—when the ground sides of circuits to output devices are controlled, rather than the power sides.

conmutación a tierra—cuando los lados a tierra de los circuitos a los dispositivos de salida son controlados, en vez de los lados de potencia.

ground spark plug electrode—the electrode on the end of the metal case, it can be bent toward or away from the center electrode to make the correct spark plug gap.

electrodo de bujía a tierra—el electrodo en la extremidad de una caja metálica, se puede torcer hacia o en dirección contraria del electrodo central para hacer una abertura de chispa correcta.

grounded circuit—when current goes directly to ground.

circuito a tierra—cuando el corriente va directamente a tierra.

guide pins—capscrews with the heads cut off. They have a hacksaw slot for a screwdriver.

clavijas de guía—los tornillos de casquete cuyos casquetes se han quitado cortándolos. Tienen una muesca cortada con una sierra en donde se puede meter un destornillador.

GVWR—gross vehicle weight rating; includes the weight of the vehicle, the weight of the passengers it has seats for (estimated at 150 lb each), and the maximum amount of luggage load.

GVWR—siglas en inglés para 'clasificación de peso bruto del vehículo'; incluye el peso del vehículo, el peso de los pasajeros para los que cuente con asientos (estimado en 150 lb cada uno) y la cantidad máxima de carga de equipaje.

H

half shaft—also called axle shaft, it is a drive axle on a front-wheel-drive vehicle.

semi-eje—tambien llamado una árbol motor, es un eje propulsor en un vehículo de tracción delantera.

Hall effect—current is passed through a thin semiconductor material while a magnetic field passes through it. This produces a small voltage (about 0.4 volt) in the semiconductor.

efecto Hall—el corriente pasa por un material delgado semiconductor mientras que lo atraviesa un campo magnético. Esto produce un pequeño voltaje (aproximadamente 0.4 voltio) en el semiconductor.

Hall-effect switch—a stationary sensor and rotating trigger wheel that uses the Hall effect to control the ignition primary.

interruptor efecto Hall—un sensor estático y rueda disparadora rotativa que utiliza el efecto Hall para controlar el encendido primario.

halogen headlamp—a brighter headlamp used on newer cars.

faro halógeno—un faro más luminoso que se usan en los coches de último modelo.

hand tools—a term for such tools as sockets, wrenches, and screwdrivers.

herramientas manuales—un término para las herramientas tales como las llaves de caja, las llaves de tuerca, y los destornilladores.

hard faults—those codes that are present and are stored in memory at the time of the self-test.

fallos fijos—esos códigos que están presentes y se almacenan en la memoria en el tiempo de autoprueba.

hard parts—metal parts such as gears, valves, pump bodies, clutch drums, and other things that would not normally be replaced in an automatic transmission rebuild.

partes duras—las partes metálicas tales como los engranajes, las válvulas, los cárteres de las bombas, los tambores de los embragues y otras cosas que normalmente no se reemplazan durante una reconstrucción de la transmisión.

hard spots—where the metallurgical composition of cast iron has been changed to steel by heat.

zonas duras—en dónde la composición metalúrgica del hierro colado se ha convertido en el acero por medio del calor.

hardware—the mechanical parts of an electronic system.

hardware—las partes mecánicas de un sistema electrónico.

harsh shift—the opposite of a mushy shift, it is when the transmission makes a gear change that is too fast.

cambio brusco—la opuesta de un cambio blando, es cuando la transmisión cambia de velocidades demasiado rápidamente.

Hazard Communication Standards—In the United States, the Environmental Protection Agency (EPA) regulations that outline disposal requirements for cleaning chemicals, used oil, heavy metals, antifreeze/coolant, asbestos, and other hazardous materials.

reglas de la Comunicación de Riesgos—En los Estados Unidos, las regulaciones de la Agencia de Protección del Medio Ambiente (EPA) que detallan los requerimientos de la disposición de las químicas de limpieza, el aceite usado, los metales pesados, los fluidos de anti-congelante o refrigerante, el amianto, u otros materiales peligrosos.

HC—hydrocarbon.

HC—hidrocarburo.

header—an aftermarket manifold made of tube steel.

cabezal de tubo—un múltiple no original fabricado del acero tubular.

heater core—a small heat exchanger in the passenger compartment that engine coolant is circulated through.

núcleo del calentador—un pequeño intercambiador de calor en el compartimento de pasajeros por el cual se circula el fluido refrigerante del motor.

heat exchanger—another name for a cooler.

cambiador de calor—otro nombre para un enfriador.

heat range—an indication of how fast heat can travel away from the spark plug center electrode to the cooling system's water jackets in the cylinder head.

gama de calor—una indicación de qué tal rápido puede viajar el calor del electrodo central de bujía a las camisas de agua del sistema enfriador de la cabeza del cilindro.

heat sink—a part that helps to dissipate heat caused by electrical flow.

fuente fría—una parte que ayuda en dispersar el calor causado por un flujo eléctrico.

heat soak—when gasoline gets hot and evaporates.

saturación de calor—cuando la gasolina se calienta y se evapora.

heat transfer—the movement of heat that occurs whenever there is a difference in temperatures between two objects.

transferencia del calor—el movimiento del calor que ocurre cuando hay una diferencia de temperatura entre dos objetos.

heat-shrink tubing—tubing that shrinks to seal an electrical joint when heated.

tubo que se contrae con el calor—el tubo que se contrae para sellar una junta eléctrica al calentarse.

heel—the outer end of the gear tooth.

talón—la extremidad exterior de un diente de engranaje.

helical gear—a gear that is machined at an angle that gives it a continuous flow of power across the gear teeth and makes them quieter in operation.

engranaje helicoidal—un engranaje que se maquina en un ángulo lo cual les proporciona un flujo de fuerza a través de los dientes de engranaje y los hace más silenciosos durante la operación.

hemispherical combustion chamber—a nonturbulent combustion chamber design shaped like half a globe.

cámara de combustión hemiesférica—un diseño de cámara de combustión sin turbulencia que tiene una forma media esférica.

HEPA vacuum—a vacuum cleaner with a special filter that is used for asbestos.

apiradora HEPA—una aspirador con un filtro especial que se usa para el amianto.

hertz—cycles per second.

hertz—ciclos por segundo.

high cordline belt—a higher quality belt with the tensile cord above center.

correa de alta tensión—una correa de alta calidad que tiene la cuerda de tensión superior al centro.

high-impedance voltmeter—a voltmeter with a very high input resistance (usually about 10 million ohms). The high-input

resistance prevents the meter from drawing current while it is connected to a circuit.

medidor de alta impedancia—un voltímetro con una resistencia de entrada muy alta (normalmente aproximadamente 10 millónes de ohmios). La resistencia muy alta de entrada previene que el calibrador jala el corriente mientras que esté conectado a un circuito.

high point—the point where the teeth of a conventional steering gear come closer together when the car is traveling straight ahead.

punto alto—el punto en el cual los dientes un engranaje de dirección convencional se acercan mientras que el coche viaja en una línea recta.

high side—the side of an air-conditioning system that comes after the compressor.

lado alto—el lado del sistema de aire acondicionado después de que pase por el comprimidor.

high-temperature pyrometer—a temperature measuring device that is touched against a surface to obtain a reading.

pirómetro de alta temperatura—un dispositivo de medir la temperatura que se toca contra una superficie para obtener una lectura.

hobbing—a rough machine process that leaves heavy machine marks on the surface of the gear tooth.

fresar—un proceso desbastador por máquina que deja las huellas de maquinar profundas en la superficie del diente de engranaje.

hold—the computer detects a wheel slowing rapidly during a stop. It is the second stage of the ABS solenoid operation.

interrupción—la computadora detecta que una rueda reduce rápidamente su movimiento durante una parada. Es la segunda etapa de la operación de solenoide del ABS.

horseless carriage—an early vehicle built on the principle of the horse and wagon.

carruaje sin caballo—un vehículo primitivo fabricado sobre los fundamentales del caballo y carro.

horsepower—the measurement of an engine's ability to perform work.

potencia de caballo—la medida de la habilidad de un motor en efectuar el trabajo.

Hotchkiss drive—an open drive shaft.

propulsión tipo Hotchkiss—un árbol motor descubierto.

hot soak—when the engine is shut off on hot days and gasoline boils out of the float bowl into the engine.

evaporación por calor—cuando se apaga un motor en tiempo cálido y la gasolina escapa hirviendo de la taza flotante al motor.

howl—a noise like the wind across the top of a soda bottle. Its range is 120–300 Hz.

aullido—un ruido parecido al viento pasando por la boca de una botella. Su gama es 120 al 300 Hz.

hub-centric—means that the center of the wheel has a machined counterbore that pilots on a machined area of the hub.

cubo-céntrico—quiere decir que el centro de la rueda tiene un agujero abocardado que sirve de guía en un área maquinado del cubo.

humidity—the moisture content of the air. When humidity is 50%, the air is holding half the amount of moisture that it is capable of holding at a given temperature.

humedad—el contenido de humedad en el aire. Cuando la humedad es del 50%, el aire contiene la mitad de la cantidad de la humedad que puede contener en una temperatura dada.

hunting gearset—when a pinion gear tooth will move around until it contacts all of the ring gear teeth.

conjunto de engranaje con diente suplementario—cuando un diente del engranaje piñón dará la vuelta hasta que se ha puesto en contacto con todos los dientes de la corona.

hybrid vehicle—a vehicle that uses more than one type of energy for its power.

vehículo híbrido—un vehículo que usa más de un tipo de energía para su propulsión.

hydraulic control valve assembly—also called electromechanical hydraulic unit or electrohydraulic control unit; contains mechanical and electrical parts that cause hydraulic pressure to pulsate or modulate.

conjunto de válvulas de control hidráulico—también llamada unidad hidráulica electromecánica o unidad de control electrohidráulico; contiene piezas mecánicas y eléctricas que hacen que la presión hidráulica pulse o module.

hydraulics—when liquid under pressure is used to transfer motion or apply force.

hidráulica—cuando un líquido bajo presión se usa para transferir el movimiento o para aplicar la fuerza.

hydrolocked engine—when coolant or fuel in a cylinder prevents an engine crankshaft from turning over.

motor bloqueado hidráulicamente—cuando el líquido refrigerante o el combustible en un cilindro previene que capota el cigüeñal del motor.

hydrometer—a device that compares the weight of electrolyte to the weight of pure water.

hidrómetro—un dispositivo que compara el peso del electrólito con el peso del agua puro.

hydroplaning—when water forms a wedge under a tire that can actually float the car.

efecto hidroavión—cuando el agua forma una cuña debajo de un neumático que puede en efecto hacer flotar a un coche.

hydrostatic lock—when a liquid is trapped in a blind hole when the fastener contacts the oil, it cannot compress it so the fastener cannot be properly tightened.

cerradura hidrostática—cuando un líquido se atrapa en un orificio ciego al ponerse en contacto con el aceite el sujetador, no puede comprimir con que el sujetador no se puede apretar correctamente.

hygroscopic—when a material can absorb water.

higroscópico—cuando una materia puede absorber el agua.

hypoid gear—a gear design with teeth that are spiraled and curved used when a ring gear and a pinion gear intersect below the centerline of the ring gear.

engranaje hipoide—un diseño de engranaje cuyos dientes son en espiral y curvados usado cuando la corona y el engranaje de piñón cruzan debajo de la línea central de la corona.

hysteresis—a term used by chemical engineers to describe a rubber's energy absorption characteristics.

histéresis—un término usado por los ingenieros químicos para describir las características de absorción de la energía del caucho.

I

IAC motor—idle air control motor; a stepper motor with two electromagnetic circuits that open and close an air passage to control idle speed.

motor IAC—motor de control del aire de marcha en vacío; un motor paso a paso con dos circuitos electromagnéticos que abren y cierran el paso del aire para controlar la velocidad de la marcha en vacío.

IC—integrated circuit; a complete miniaturized electric circuit.

IC—circuito integrado; un circuito eléctrico en miniatura completo.

I.D.—inside diameter.

I.D.—diámetro interior.

idler gear—a gear used between two other gears, the purpose of which is to change direction of rotation.

engranaje de piñón loco—un engranaje que se usa entre dos otros engranajes, su propósito es cambiar la dirección de rotación.

ignition system—creates and distributes a timed spark to the cylinder.

sistema del encendido—crea y distribuya una chispa sincronizada al cilindro.

ignition timing—the point at which ignition occurs in a cylinder.

tiempo del encendido—el punto en el cual ocurre el encendido en el cilindro.

ignition timing setting—the point at which the secondary ignition circuit fires the spark plugs.

ajuste de temporización de ignición—el punto en el que el circuito de ignición secundario dispara las bujías

I-head—valve placement in modern engines is in the cylinder head, above the piston.

cabeza I—una colocación de las válvulas en los motores más modernos es en la cabeza del cilindro, superior al pistón.

ihp—indicated horsepower; the amount of pressure made in the combustion chambers.

ihp—potencia de caballo indicado; la cantidad de presión creada en las cámaras de combustión.

impedance—a meter's internal resistance.

impedancia—la resistencia interna de un medidor.

inboard side—the inside end of an axle shaft. The opposite side is called the outboard side.

lado abordo—la extremidad interior de una flecha del eje. El lado opuesto se llama el lado fuera de bordo.

inches of mercury (in. Hg)—the measurement increment for vacuum or low pressure.

pulgadas de mercurio (in. Hg)—el incremento de dimensión de la presión en vacío o baja presión.

included angle—the combination of SAI and camber.

ángulo incluído—la combinación de SAI y combadura.

independent suspension—a type of suspension system that when it goes over a bump, only that wheel will deflect.

suspensión independiente—un tipo de sistema de suspensión que al pasar por un tope, sólo se desvía esa rueda.

induction-hardened valve seat—an integral valve seat that is harder than the surrounding area of a cylinder head.

asiento de válvula endurecido por inducción—un asiento de válvula íntegro que es más duro que el área al su alrededor en la cabeza del cilindro.

inductive pickup—the pickup for an electrical meter that wraps around the wire being sensed and measures current flowing through it using principles of magnetism.

captador inductivo—el captador de un medidor eléctrico que se enreda alrededor del alambre que se tiene que detectar y mide el corriente que lo atraviesa usando los fundamentales del magnetismo.

inert gas—a gas that does not burn and takes up space in the combustion chamber.

gas inerta—un gas que no se quema y ocupe el espacio en la cámara de combustión.

inert gas shield—an inert gas shield (argon, helium, or CO_2) is applied over the arc area during welding to prevent oxidation of the metal.

blindaje de gas inerta—un blindaje de gas inerta (argón, helio o el CO_2) se aplica al arco durante la soldadura para prevenir la oxidación en el metal.

inertia—the tendency of a body to keep its state of rest or motion.

inercia—la tendencia de un cuerpo de mantener su estado de descanso o movimiento.

inertia starter drive—also called Bendix drives, they operate like a heavy nut on a screw.

acoplamiento del arrancador de inercia—tambien llamado los acoplamientos Bendix, opera como una tuerca muy gruesa en un tornillo.

inertia wheel—a device that allows the seat belt to be more comfortable for the passengers and eliminates the need to adjust the belt for each passenger.

rueda de inercia—un dispositivo que permite que el cinturón de seguridad les resulte más cómodo a los pasajeros y que elimina la necesidad de realizar ajustes en el cinturón para cada pasajero.

infrared thermometer—a thermometer that takes a temperature reading when it is aimed toward a surface.

termómetro infrarrojo—un termómetro que toma una lectura de temperatura cuando se apunta a una superficie.

initial timing—also called base timing, it is the timing setting before the computer has a chance to make changes.

tiempo inicial—tambien llamado tiempo base, es la regulación del tiempo antes de que la computadora haya tenido una oportunidad de efectuar los cambios.

insulator—the opposite of a conductor; they have no, or few, free electrons.

aislador—lo opuesto de un conductor; no tiene ningunos, o muy pocos, electrones libres.

intake manifold vacuum—the pressure inside of the intake manifold when the engine is running.

vacío del múltiple de entrada—la presión en el interior de la entrada del múltiple mientras que esté en marcha el motor.

integral seat—a part that is one with the head.

asiento íntegro—un parte que es parte de la cabeza.

intercooler—also called a charge air cooler or an aftercooler, it is a sophisticated air cooler for blowers.

enfriador intermedio—tambien llamado enfriador de aire cargado o un postrefrigerador, es un enfriador sofisticado para los ventiladores.

interference angle—a difference between the valve and seat face angles that results in increased pressure to aid valve seating.

ángulo de interferencia—la diferencia entre los ángulos de la válvula y de la cara del asiento que resulta en una presión aumentada para ayudar en asentar las válvulas.

interference engine—also called non-free-wheeling, an engine that will experience piston-to-valve interference if the timing chain or belt skips or breaks.

motor de interferencia—tambien llamado de no rodamiento libre, un motor que experiencerá la interferencia entre pistón y válvula si se brinca o quiebra la cadena o correa de tiempo.

interference fit—when two parts have a pressed fit.

ajuste de interferencia—cuando dos partes tienen un ajuste embutido.

intermittent fault—a problem that occurs occasionally and without a pattern; more difficult to diagnose than a hard fault.

defecto intermitente—un problema que ocurre ocasionalmente y sin un padrón; es más difícil de diagnosticar que un defecto constante.

intermittent fault code—one that only occurs occasionally for a short period of time and is not present in the system at the time of the fault test.

código de fallos intermitentes—uno que sólo ocurre a veces por muy corto tiempo y no se presenta en el sistema durante el tiempo de la prueba de fallos.

internal balancing—balancing that is done by drilling holes on the crankshaft counterweights.

equilibración interna—la equilibración que se efectúa taladrando los orificios en los contrapesos del cigüeñal.

internal/external gear crescent-type pump—a type of oil pump with an internal gear, an external gear, and a crescent.

bomba semilunar de engranajes internos/externos—un tipo de bomba de aceite con un engranaje interno, engranaje externo y una estructura semilunar.

inversion layer—when warm air becomes trapped within 1,000 feet of the ground. The trapped air produces smog.

capa de inversión—cuando el aire tibio se atrapa a una altura de menos de 1000 pies de la tierra. El aire atrapado produce el esmog.

inverted flare nut—the common type of SAE flare used on automobiles.

tuerca bocinada invertida—un tipo de abocinamiento SAE usado en los automóviles.

ISC motor—idle speed control motor; opens the throttle plate to control idle speed.

motor ISC—motor de control de velocidad en marcha en vacío; abre la placa del acelerador para controlar la velocidad de marcha en vacío.

ISO—International Standards Organization.

ISO—International Standards Organization (Organización Internacional de Estándares)

ISO flare—one style of forming the ends of tubing, also called a bubble flare.

ISO abocinamiento—un estilo de formar las extremidades de los tubos, tambien llamado un abocinamiento de burbuja.

isolated field—an alternator with two field leads that is energized and grounded externally.

campo aislado—un alternador con dos líneas del campo que se energetiza y va a tierra al exterior.

J

jackscrew—a screw for adjusting belt tension.

gato—una tuerca para ajustar la tensión de la correa.

jam nut—a locknut.

tuerca de seguridad—una contra tuerca.

jobber—auto parts wholesaler.

jobber—palabra en inglés para vendedor mayorista de piezas para automóviles.

joule—an equivalent value that compares heat energy (Btu) to mechanical energy (ft.-lb.). 1 Btu = 778 ft.-lb.

julio—un valor equivalente que compara la energía calorífica (Btu) a la energía mecánica (ft.-lb.). 1 Btu = 778 ft.-lb.

K

keep-alive memory—KAM; means the computer maintains power to random access memory (RAM) when the ignition switch is off. This allows it to keep information as long as the battery is not disconnected.

memoria de retención—KAM; quiere decir que la computadora mantiene la potencia a la memoria de acceso aleatoria (RAM) cuando está apagado el interruptor del encendido. Este lo permite mantener la información mientras que no se disconecte la batería.

kinetic energy—energy of motion.

energía cinética—la energía del movimiento.

Kirchoff's law—the total voltage drop in an electrical circuit will always be equal to the available voltage at the source.

ley de Kirchoff—la caída total del voltaje en un circuíto eléctrico siempre será igual al voltage disponible en la fuente.

knock-off type CV joint—an internal circlip CV joint retaining design.

junta CV de fácil desenganche—un diseño de retención de junta CV con grapa circular.

knock sensor—also called a detonation sensor, it is a piezoelectric crystal that detects the frequency of the vibration caused by detonation and tells the computer to retard the spark timing until the vibration goes away.

sensor de golpeteo—tambien llamado un sensor de detonación, es un cristal eléctrico que detecta la frequencia de la vibración causado por la detonación y manda que la computadora retrasa la chispa del tiempo hasta que desaparezca la vibración.

knurling—a metal displacement process that results in the enlargement of the surface of a metal.

moletear—un proceso de desplazamiento de metal que resulta en el ensanchamiento de la superficie de un metal.

KOEO—key-on, engine-off.
KOEO—llave prendida, motor apagado.

L

labor intensive—a process that requires a human's labor.
labor de mano intensivo—un proceso que requiere el labor de un ser humano.

laminated—more than one layer.
laminado—de más de una capa.

latent heat—the extra heat required before matter can change its state.
calor latente—el calor extra que se requiere antes de que la materia puede cambiar su estado.

latent heat of condensation—the heat released during condensation. When steam condenses back to water, it releases 970 Btu of heat per pound.
condensación latente del calor—el calor soltado durante la condensación. Cuando el vapor se condensa y se convierta al agua, suelta 970 Btu del calor por libra.

latent heat of vaporization—the heat required to heat water at 212°F to turn it into steam. It will require an additional 970 Btu of heat to make it boil.
calor latente de la vaporización—el calor requerido a calentar el agua de 212°F para convertirlo al vapor. Requerirá unos 970 Btu adicionales del calor para hacerlo hervir.

lateral acceleration sensor—a sensor that measures the force encountered while turning.
sensor de aceleración lateral—un sensor que mide la fuerza encontrada al girar.

lateral runout—wobble of a part in a side-to-side direction.
corrimiento lateral—el bamboleo de una parte en una dirección de lado a lado.

launch shudder—a common second order vibration complaint in raised trucks with a high drive shaft angle.
temblor inicial—una queja común de vibración de segunda orden en los camiones elevados con un ángulo muy alto del árbol motor.

LCD—liquid crystal display.
LCD—presentación de cristal líquido.

lead oxidation—when lead is exposed to air it forms a black oxidized coating.
oxidación de plomo—cuando el plomo se expone al aire forma una capa negra de oxidación.

LED—light-emitting diodes found in digital displays on the dashboard and in test instruments.
LED—los diodos emisores de luz que se encuentran en las presentaciones digitales en el tablero de instrumentos y en los intrumentos para efectuar pruebas.

L-head—also called flat head, this older engine design has the valves in the cylinder block, next to the cylinder.
cabeza L—tambien llamado una cabeza plana, este diseño más antiguo tiene las válvulas en el bloque del cilindro, junto al cilindro.

licensing—when the government requires a certain testing or skill level before allowing specialized work to be performed.
emisión de licencia—cuando el gobierno requiere ciertas pruebas o nivel de destreza para permitir la realización de trabajos especializados.

lift—the height to which the lobe raises the lifter.
altura de izado—la altura a la cual el lóbulo levanta el aparato de izar.

light off—when the catalytic converter becomes hot enough and begins to oxidize pollutants.
luces apagados—cuando el convertidor catalítico se calienta lo suficiente y comienza a oxidar los contaminantes.

limited slip differential—it locks up the spider gears when one wheel starts to lose traction.
diferencial autoblocante—enclava los satélites del diferencial cuando una rueda comienza a perder la tracción.

line honing or line boring—methods of refinishing the main bearing bores to assure correct alignment.
barrenado en línea o rectificado en línea—los métodos de acabado de los diámetros interiores de los cojinetes principales.

liquid/vapor separator—a part of the fuel tank or the expansion tank that keeps liquid fuel from being drawn into the charcoal canister.
separador de vapor líquido—una parte del tanque de combustible o el tanque de expansión que previene que el combustible en líquido se aspira al bote de carbono.

list price—the suggested price of an item, paid by the customer who is paying for a car repair; also called retail price.
precio de lista—el precio sugerido de un artículo, pagado por el cliente que esté pagando por una reparación de automóvil; también llamado precio minorista.

live axle—an axle that turns with the wheels.
eje vivo—un eje que se gira con las ruedas.

load carrier ball joint—the ball joint in the control arm with the spring attached to it.
articulación esférica de carga—la articulación esférica en el brazo de mando que tiene un resorte conectado.

loaded caliper—a rebuilt caliper complete with new friction pads, hardware, and shims.
calibre cargado—un calibre reconstruído completo con las almohadillas, la herramienta, y las chapas nuevas.

load index—the maximum load at the designated speed rating.
índice de carga—la carga máxima en la tasa de velocidad designada.

locking hubs—allow a four-wheel-drive vehicle to be used in two-wheel drive without the axles and differential being attached to the front wheels.
cubos enclavadores—permite que un vehículo de tracción en cuatro ruedas utiliza la tracción de dos ruedas sin que se conectan los ejes y el diferencial con las ruedas delanteras.

lock to lock—when the steering wheel is turned all the way from one direction to the other.
cierre a cierre—cuando el volante de dirección se da la vuelta completamente de una dirección a la otra.

lock-up torque converter—a converter with a friction disc that locks the impeller and turbine together.
convertidor de par enclavado—un convertidor con un disco de fricción que enclava al impelador con la turbina.

LOF—lube, oil, and filter.
LOF—la lubricación, el aceite, y el filtro.

long block—a complete engine assembly with heads.
bloque largo—una asamblea completa del motor con las cabezas.

longitudinal braking system—operates the front and rear brakes separately.
sistema de enfrenamiento longitudinal—opera los frenos delanteros y traseros por separado.

long nipple—a nipple with a section of plain pipe separating the threads.
niple larga—un niple con una sección de tubo simple que separa las roscas.

look-up tables—program information on how the car is supposed to perform.
tablas de verificación—información del programa de cómo debe operar el coche.

lower end—the parts that make up a short block.
extremidad baja—las partes que constituyen un bloque corto.

lower gear—when a small gear drives a larger gear.

velocidad baja—cuando un engranaje pequeño propulsa un engranaje más grande.

low-maintenance battery—a battery that does not normally require water to be added.

batería de bajo mantenimiento—una batería que no suele requerir que se añade el agua.

low-pressure cutout switch—shuts off the compressor clutch when air conditioning pressure drops too low, usually because the system is low on refrigerant.

interruptor de corto-circuito en baja presión—apaga al embrague del comprimidor cuando la presión del aire acondicionado cae demasiado, normalmente debido a que le falta refrigerante al sistema.

low side—the side of an air conditioning system that comes after the flow control device before the evaporator.

lado bajo—el lado de un sistema de aire acondicionado después del dispositivo de control de flujo y antes del evaporador.

LPG—liquified petroleum gas.

LPG—gas de petroleo liquado.

lugging—lugging occurs when the load on the engine is greater than the rpm needed to develop enough horsepower to pull the load.

arrastre—el arraste ocurre cuando la carga en el motor supera los rpm que se requieren para desarrollar la potencia de caballo para tirar de la carga.

M

Macpherson strut—a suspension design that incorporates the shock absorber into the front suspension.

poste Macpherson—un diseño de suspensión que incorpora el amortiguador de choque en la suspensión delantera.

MAF—mass airflow sensor.

MAF—sensor del flujo del aire en masa.

maintenance-free battery—a battery with no provision for adding water.

batería de no mantenimiento—una batería que no tiene provisiones para añadir el agua.

malfunction indicator light—an OBD II term for a light on the dash display, also called a check engine light, that tells when a hard code has been detected.

luz indicador de defecto—un término OBD II para una luz en el indicador del tablero de instrumentos, tambien llamado la luz revisa el motor, que informa de que se ha descubierto un código fijo.

manifest—an EPA form for tracking hazardous wastes.

manifestación—una forma del EPA para rastrear los desechos tóxicos.

manifold heat control valve—a device also called a heat riser, it is a butterfly valve that fits between the exhaust manifold and exhaust pipe and routes exhaust gas under the floor of the intake manifold when the engine is cold to improve vaporization of the cold fuel.

válvula de control del calor del múltiple—un dispositivo que tambien se llama una columna de calor, es una válvula de mariposa que queda entre el múltiple del escape y el tubo del escape que desvía el gas debajo del piso del múltiple de admisión cuando el motor está frio para mejorar la vaporización del combustible frío.

manual transmission—a transmission that is manually shifted and used with a clutch. It is also called a stick shift or a standard transmission.

transmisión manual—una transmisión que cambia de velocidades manualmente que se usa con un embrague. Tambien se llama una palanca de cambiar velocidades o una transmisión estándar.

master automobile technician—a journey level professional who is certified in all eight of the ASE areas of specialization.

técnico maestro en automóviles—un profesional por jornada, certificado en todas las ocho áreas ASE de especialización.

match mount—when runout of one part and runout of the other are considered during assembly to cancel each other out.

montaje adaptativo—cuando el corrimiento de una parte y el corrimiento de otra parte se consideran durante la asamblea para cancelarse mutuamente.

mechanical efficiency—describes all of the ways friction is lost in an engine. Brake horsepower (Bhp) divided by the indicated horsepower.

rendimiento mecánico—describe todas las maneras en que se puede perder la fricción en un motor. La potencia de caballo del enfrenamiento (Bhp) se divide por la potencia indicada en caballo.

mesh—when two gears run against each other.

endentado—cuando dos engranajes funcionan en contacto.

metallic lining—a lining made of metal that is used in very heavy-duty and racing conditions.

forro metálico—un forro hecho del metal que se usa bajo condiciones de trabajo pesado y en coche de carreras.

metering valve—shuts off pressure to the front brakes until about 125 psi builds up in the rear brakes. This keeps front pads from doing too much of the light braking and helps prevent dangerous skids that could result on slick surfaces if the front brakes were to apply before the rears.

válvula de medida—apaga la presión en los frenos delanteros hasta que acumula aproximadamente 125 psi en los frenos traseros. Esto previene que las almohadillas delanteras efectúan demasiado del enfrenamiento ligero y ayuda en prevenir los deslizamientos peligrosos que podrían resultar en las superficies resbaladizas si se aplicaran los frenos delanteros antes de los traseros.

metric system—the international system (S.I.) of measurement based on the meter, which is 39.37" long, slightly longer than a yard.

sistema métrica—el sistema internacional (S.I.) de medida basado en el metro, que mide 39.37 pulgadas de longitud, un poquito más largo que una yarda.

microfiche—a small plastic film card that is magnified by a microfiche reader.

microfiche—una pequeña tarjeta de película de plástico que se magnifica con un lector de microfiche.

microinch—a microinch is one millionth of an inch.

micropulgada—una micropulgada es la millonésima parte de una pulgada.

microprocessor—the calculating and decision-making chip in the computer.

microprocesador—la ficha que calcula y de forma las decisiones en una computadora.

MIG welding—metal inert gas welding.

soldadura MIG—soldadura de metal con blindaje de gas inerta.

MIL—malfunction indicator lamp located on the dash display.

MIL—luz indicador de defectos situado en el indicador del tablero de instrumentos.

min/max—a feature of some voltmeters that compares the maximum and minimum readings during a dynamic test.

min/max—una característica de algunos voltímetros que compara las lecturas máximas y mínimas durante una prueba dinámica.

mini-fuse—also called blade fuse, it has a fuse element cast into a clear plastic outer body.

fusible miniatura—tambien llamado un fusible de hoja tiene el elemento de fusible incorporado en un cuerpo exterior de plástico transparente.

moan—also called drone, it is a noise that is compared to the hum of a bumble bee. Its range is 60–120 Hz.

quejido—tambien llamado zumbido, es un ruido que se compara al zumbido de una abeja. Su gama es del 60 a 120 Hz.

molecular sieve—a material that will not crumble that has been used for desiccants since the early 1980s.

criba molecular—una materia que no se desmenuza que ha sido usado como desecante desde los principios de la decada ochenta.

momentum—a body going in a straight line will keep going the same direction at the same speed if no other forces act on it.

impulso mecánico—un cuerpo viajando en una línea recta continuará viajando en la misma direccíon en la misma velocidad si ninguna otra fuerza actúa en su contra.

monitor—a device used to look for malfunctions in the OBD II system.

monitor—un dispositivo usado para buscar defectos de funcionamiento en el sistema OBD II.

monolithic catalyst—the inside, honeycomb-shaped part of a catalytic converter.

catalizador monolítico—la parte interior, con forma de nido de abejas, de un conversor catalítico.

monolithic converter—a converter with honeycomb structure.

convertidor monolítico—un convertidor que tiene una estructura de nido de abeja.

Montreal Protocol—an agreement that set limits on the production of ozone-depleting chemicals, with a total phase out by the year 2000.

Protocol de Montreal—un acuerdo que definó los límites en la producción de las químicas que deterioran la capa de ozono, con su prohibición completa por el año 2000.

morning sickness—a term applying to power steering or automatic transmissions that describes a lack of hydraulic pressure buildup when seals are cold and hard. A cold automatic transmission will hesitate from a few seconds to a minute, leaving the vehicle standing still as if in neutral while the engine races.

achaques—un término que aplica a la dirección hidráulica o a las transmisiones automáticas que describe la carencia de acumulación de presión hidráulica cuando los sellos son duros y fríos. Una transmisión automatica fría vacilará de unos segundos a un minuto, dejando parado al vehículo como si estuviera en neutral mientras que el motor acelera.

motor mounts—engine mounts; they attach the engine to the frame.

soportes de motor—sujetan el motor a la estructura.

MPI—multiport fuel injection; a fuel injection system that opens its injectors in pairs or groups a sufficient amount of time prior to intake valve opening so the intake port is filled with fuel before the valve opens.

MPI—siglas en inglés para inyección de combustible multipuerta; un sistema de inyección de combustible que abre sus inyectores en pares o grupos una cantidad de tiempo suficiente antes de la abertura de la válvula de entrada, de modo que la puerta de entrada se llene de combustible antes de que la válvula se abra.

M+S, MS, M&S, M/S—any combination of the letters M and S on the tire sidewall means that the tire meets snow tire definitions set by the Rubber Manufacturers Association.

M+S, MS, M&S, M/S—cualquier combinación de las letras M y S en el refuerzo lateral de un neumático quiere decir que el neumático cumple con los requerimientos definidos por la Asociación de Fabricantes de Caucho.

MSDS—material safety data sheet; provides details of the composition of a chemical, lists possible health and safety problems, and gives precautions for its safe use.

MSDS—hoja de datos de seguridad de las materias; provee los detalles de la composición de las químicas, detalla los problemas de salud y seguridad y avisa de precauciones en el uso prudente.

multiple viscosity—also called multi-vis, a lubricant with a rating for cold and hot viscosity.

viscosidad múltiple—tambien llamado multi-vis, un lubricante con una clasificación de viscosidad fría y caliente.

multiplexing—a term used when several computers are linked together by one pair of wires allowing one sensor to provide information to several computers.

unir con multiplex—un término que se usa cuando varias computadoras se eslabonan juntos con un par de alambres permitiendo que un sensor provee la información a varias computadoras.

multistage ABS—a system in which air bags can have more than one squib and can inflate an air bag in stages.

sistema de bolsa de aire multietapa—un sistema en el que las bolsas de aire pueden tener más de un detonador; puede inflar las bolsas de aire en etapas.

mushroomed valve tip—a description of the shape of the tip of a valve stem that results from the repeated impacts due to excessive clearance.

boquilla de la válvula aplanada—una descripción de la forma de la boquilla del vástago de la válvula que resulta de choques repetidos debido a una holgura excesiva.

mushy shift—the engine increases in speed briefly during a shift until the clutch pack or servo piston finally seals and pressure builds up.

cambio de velocidades blanda—la velocidad del motor aumenta brevemente durante un cambio de velocidades hasta que el equipo de embragues o el servo del pistón sella finalmente y aumenta la presión.

N

NA—a natural amber light bulb.

NA—un foco de luz de ámbar natural.

NADA—National Automobile Dealers Association.

NADA—Asociación Nacional de Concesionarios de Automóviles.

NATEF—National Automotive Technicians Education Foundation.

NATEF—Fundación Nacional para la Educación de Técnicos en Automóviles.

natural frequency—also called resonant frequency, it is the frequency at which a body vibrates. A tuning fork is an example.

frecuencia natural—tambien llamado una frecuencia resonante, es la frecuencia en la cual vibra un cuerpo. Un diapasón es un ejemplo.

NCFR—no cause for removal.

NCFR—no hay causa para remover.

needle bearing—a very small roller bearing used to control thrust or radial loads.

cojinete de agujas—un cojinete de rodamientos muy pequeño que se usa para controlar las cargas de empuje o radiales.

negative camber—when a tire is tilted in at the top.

comba negativa—cuando un neumático está inclinado en la parte superior.

negative caster—the forward tilt of the steering axis.

ángulo de caster negativo—la inclinación hacia adelante del eje de dirección.

negative coefficient thermistor—will have lower resistance as the temperature of the coolant rises.

termistor de coeficiente negativo—tendrá una resistencia más baja al subir la temperatura del fluido refrigerante.

negative offset—when the wheel is offset in such a way as to increase the track width of the tires.

escentrado negativo—cuando la rueda está descentrado en una manera que aumenta la anchura de la banda de los neumáticos.

neoprene—a kind of oil-resistant artificial rubber.

neopreno—un tipo de caucho artificial resistente al aceite.

net horsepower—the maximum power available from the engine when all the accessories are turned on.

potencia neta de caballo—la cantidad máxima de la fuerza disponible del motor cuando se han prendido todos los accesorios.

net price—the price that the wholesale auto repair shop pays.

precio neto—el precio que paga el taller de reparación de automóviles mayorista.

NLGI—National Lubricating Grease Institute.

NLGI—Instituto Nacional de Grasa Lubricante.

non-hunting gearset—when one tooth on the pinion gear will mesh with the same tooth on the ring gear again and again, rather than "hunting" for all of the rest of the teeth.

conjunto de engranaje sin diente suplementario—cuando un diente del engranaje del piñón se endenta con el mismo diente en la corona repetidamente, en vez de "buscar" todos los otros dientes.

nonintegral—remote or add-on ABS. It contains a conventional power brake and master cylinder.

no integral—ABS remoto o agregado; contiene un freno eléctrico convencional y un cilindro maestro.

nonretorque gasket—a non-retorque gasket compresses less than a conventional gasket so the head does not require retorquing.

junta de tipo no retorsión—una junta de no retorsión se comprime menos que una junta convencional con que no se tiene que retorcer la cabeza.

non-volatile RAM—when the information in memory is not erased when the ignition switch is turned off; commonly called keep-alive memory (KAM).

RAM no volátil—cuando la información en la memoria no se borra al apagar el interruptor del encendido.

normally aspirated—an engine that is not supercharged.

aspirado normalmente—un motor que no es supercargado.

N-type—a negative-type crystal doped with an atom-like phosphorous with extra electrons.

tipo N—un cristal de tipo negativo al cual se ha agregado un átomo como el fósforo que tiene electrones extra.

O

O$_2$—oxygen.

O$_2$—oxígeno.

O$_2$ sensor safe—an RTU that does not emit vapors that can damage an O$_2$ sensor.

seguro para sensor de O$_2$—un vehículo para todo terreno que no emite vapores que pueden dañar al sensor de O$_2$.

OBD—on-board diagnostics; regulations that require the computer to monitor the engine's oxygen sensor, EGR valve, and charcoal canister purge solenoid to see that all of these systems continue operating properly.

OBD—diagnósticos a bordo; las regulaciones que requieren que la computadora amonesta el sensor de oxígeno, la válvula EGR, y el solenoide de purga del bote de carbono del motor para verificar que todos estos sistemas continuan a operar correctamente.

OBD II—guidelines that began in 1994 on some cars and are on all gasoline-powered new cars built since 1996. These standards came about to provide standardization of terms and connections between makes.

OBD II—póliza que comenzó en el 1994 en algunos coches y en todos los coches nuevos de combustión de gasolina desde el 1996. Estas normas se crearon para proveer una normalización de términos y conexiones entre los fabricantes.

octane—a measurement of a fuel's ability to resist explosion during combustion.

octano—una medida de la habilidad de un combustible en resistir la explosión durante la combustión.

O.D.—outside diameter.

O.D.—diámetro exterior.

OE—original equipment.

OE—siglas en inglés para equipo original.

OEM—original equipment manufacturer.

OEM—siglas en inglés para fabricante de equipo original.

OHC—overhead cam valvetrain design, also called cam-in-head.

OHC—diseño de tren de válvulas con leva en cabeza, tambien llamado leva-en-cabeza.

ohm—the unit of electrical resistance measurement. One ohm is the resistance that will allow 1 ampere to flow when pushed by 1 volt.

ohmio—la unidad de medida de la resistencia eléctrica. Un ohmio es la resistencia que permitirá 1 ampere a fluir al ser impulsado por 1 voltio.

Ohm's law—the law governing the relationship between volts, ohms, and amps.

ley de Ohm—la ley que gobierna la relación entre los voltios, los ohmios y los amperes.

oil analysis—a process for determining contaminants in engine oil.

analisis de aceite—un proceso para determinar los contaminantes presente en el aceite del motor.

oil-based carbon deposits—the traditional gummy, black carbon deposits like those found on valves. They result when oil and heat come together.

depósitos de aceite debasados en carbón—los depósitos pegajosos tradicionales de carbón negro parecidos a los que se encuentran en las válvulas. Resulten cuando el aceite y el calor se juntan.

oil cooler—also called a heat exchanger, it is usually in the bottom or side of the radiator that has automatic transmission fluid running through it.

enfriador de aceite—tambien llamado un intercambiador de calor, suele estar en la parte inferior o en un lado del radiador por el cual fluye el fluido de transmisión automática.

oil pressure sending unit—the part that is threaded into the cylinder block oil passageway. It sends an electrical signal to the gauge or light on the dashboard.

unidad emisor de presión del aceite—una parte enroscada en el paso de aceite del bloque del cilindro. Manda una señal al indicador o a la luz en el tablero de instrumentos.

oil wash—when excessive fuel washes oil from the cylinder walls.

lavado del aceite—cuando el combustible excesivo enjuaga el aceite de los muros del cilindro.

one-way clutch—an overrunning clutch or the roller or sprag design that locks in one direction and freewheels in the other.

embrague de una vía—un diseño de embrague de sobremarcha o la rodilla o puntal que se enclava de una dirección y rueda libremente en la otra.

open circuit—when there is a break in the path of electrical flow in a circuit.

circuito abierto—cuando hay un corto en la vía del flujo eléctrico en un circuito.

open loop—when a computer-controlled engine is cold and the computer is not receiving feedback from the oxygen sensor.

bucle abierto—cuando un motor controlado por computadora está frío y la computadora no está recibiendo datos del sensor de oxígeno.

orifice—a restriction in a passage to slow down the flow of fluid.

orificio—una restricción en un paso para retardar el flujo de un fluido.

OSS—occupant sensor system; a smart air bag system feature that determines the size and position of a passenger so that the airbag will deploy efficiently.

OSS—siglas in inglés para sistema de sensor de ocupantes; una característica de sistema de bolsas de aire inteligente que determina el tamaño y posición del pasajero para que la bolsa de aire se abra con eficiencia.

Otto-cycle—the four-stroke gasoline spark ignition engine design.

ciclo Otto—el diseño de motor de gasolina de cuatro carreras encendido por chispa.

overcenter assist spring—a spring that pulls one way during the first half of travel and the other way during the second half of travel.

resorte sobrecentro—un resorte que jala en una dirección durante la primera parte de la carrera y de la otra dirección en la segunda parte de la carrera.

overdrive—the opposite of gear reduction, it is when the output shaft turns faster than the input shaft.

sobremarcha—lo opuesto de reducción de velocidad, es cuando el eje de salida gira más rapidamente que el eje de entrada.

overhaul—when an engine is disassembled, usually in the chassis, and new rings and bearings are installed; not a complete engine rebuild.

reacondicionamiento—cuando se desarma un motor, generalmente en el chasis, y se le instalan anillos y cojinetes nuevos; no es una reconstrucción completa del motor.

overhead cam engine—an engine with the cam above the cylinder head.

motor de levas sobre la cabeza—un motor que tiene la leva superior a la cabeza del cilindro.

overload spring—an extra spring leaf that does not work until the other leaves have deflected enough under load to allow them to come into contact with the overload spring.

resorte de sobrecarga—un resorte adicional de lámina que no trabaja hasta que las otras láminas hayan desviado lo bastante de la carga para permitir que se ponga en contacto con el resorte de sobrecarga.

overrunning clutch—a one-way clutch.

embrague de sobremarcha—un embrague de una vía.

oversquare—when an engine has a cylinder bore that is larger than its stroke.

sobrecuadrada—cuando un motor tiene un diámetro interior del cilindro que es más grande que su carrera.

oversteer—when a car turns too far in response to steering wheel movement.

sobregirar—cuando un coche da la vuelta demasiado en respuesta al movimiento del volante de dirección.

owner's manual—a booklet that comes with a new car.

manual del dueño—un librito que viene con un coche nuevo.

oxidation catalyst—a two-way catalyst that controls HC and CO.

catalizador de oxidación—un catalizador de dos vías que controla los HC y el CO.

oxides of nitrogen (NO_x)—the entire family of nitrogen oxides that form when oxygen and nitrogen bond under high heat and pressure in the engine. The "x" in NO_x means a variable number.

óxidos de nitrógeno (NO_x)—toda clase de óxidos de nitrógeno que forman cuando el oxígeno y el nitrógeno se unen bajo el alto calor y presión en el motor. El "x" en NO_x quiere decir un número variable.

oxidizing—burning of the fuel.

oxidando—combustión del combustible.

oxyacetylene welding—gas welding.

soldadura oxiacetileno—soldadura con gas.

P

PAG—polyalkylene glycol refrigerant oil for R-134A.

PAG—el aceite refrigerante polialquílico glicol para el R-134A.

parallel circuit—a circuit with different branches that current can flow through, starting from a common point.

circuito paralelo—un circuito con derivaciones diferentes por los cuales puede fluir el corriente, comenzando de un punto común.

parallelogram steering—linkage that makes the shape of a parallelogram during a turn.

dirección en paralelogramo—la biela que hace la forma de un paralelogramo al efectuar un viraje.

parasitic load—an electrical load when the key is off.

carga parásita—una carga eléctrica cuando la llave está en posición apagada.

park pawl—a lever that locks the output shaft of the transmission when the shift lever is placed in park.

trinquete de park (estacionamiento)—una palanca que enclava el eje de la salida de la transmisión cuando la palanca selectora está en la posición de park.

part cores—used parts that are returned to the rebuilder for use in another rebuild.

almas—las partes usadas se regresan al reconstruidor para ser utilizadas en otra reconstrucción.

partial non-hunting gearset—when one pinion tooth contacts two or three different ring gear teeth during one revolution and two or three different ones the next revolution.

conjunto de engranaje con diente suplementario parcial—cuando un diente del engranaje del piñón se endenta con dos o tres dientes de la corona durante una revolución y con dos o tres distintos en la próxima revolución.

particulate—an airborne microscopic particle or dust or soot that contains lead and carbon.

partícula—una partícula microscópica transmitida por el aire o el polvo o el hollín que contiene el plomo y el carbón.

Pascal's law—the hydraulic law that states that pressure in an enclosed system is equal in all parts of that system.

ley de Pascal—la ley de las hidráulicas que dice que la presión en un sistema cerrado es igual en cada parte de ese sistema.

passive alarm system—the system is automatically "armed" when the key is removed from the engine, all the doors are locked, and the trunk and hood are closed.

sistema de alarma pasiva—el sistema se "arma" automáticamente cuando se retira la llave del motor, con todas las puertas trabadas y el maletero y el capó cerrados.

passive restraint—a restraint that takes place automatically to protect the occupant of the vehicle.

sujeción pasiva—un sistema de sujeción que ocurre automáticamente para proteger al ocupante del vehículo.

passive sensors—those that do not generate their own voltage.

sensores pasivos—aquellos que no generan su propio voltaje.

passive suspension system—a normal suspension whose height varies according to mechanical forces on the suspension and they do not adjust to these changes.

sistema de suspensión pasiva—una suspensión normal en la que la altura varía de acuerdo con las fuerzas mecánicas ejercidas sobre la suspensión y que no se ajusta a estos cambios.

patch—a piece of rubber vulcanized to the inner liner of a tire to repair a leak.

parche—un pedazo del caucho vulcanizado al forro interior de un neumático para reparar una fuga.

PCV system—positive crankcase ventilation. It prevents the emission of blowby gases from the crankcase.

sistema PCV—ventilación positivo del cárter del cigüeñal. Previene la emisión de los gases de fuga del cárter del cigüeñal.

PDI—pre-delivery inspection.

PDI—siglas en inglés para inspección previa a la entrega.

pending code—also known as available code; a code that is set the first time DTC criteria are met.

código pendiente—también conocido como código disponible; un código colocado al ajustarse por primera vez los criterios para DCT.

pH scale—a scale that gives the strength or alkalinity of a solution. The scale ranges from 1 through 14. Pure water is rated at 7.

escama pH—una escama que da la fuerza o lo alcalino de una solución. La escama varía del 1 al 14. El agua pura tiene una clasificación de 7.

photochemical smog—the name given to visible air pollution.

esmog fotoquímica—el nombre dado a la contaminación visible del aire.

PID—parameter identification data; processed data used by the engine management program useful in finding a problem when a code was not set. In OBD II, PID numbers can be displayed on the scan tool.

PID—siglas en inglés para datos de identificación de parámetros; datos procesados utilizados por el programa de administración de motores, útiles para encontrar un problema cuando no se ha fijado un código. En el OBD II, los números PID pueden exhibirse en la herramienta de escaneo.

piezoelectric crystals—crystals that develop a voltage on their surfaces when pressure is applied to them.

cristales piezoeléctricos—los cristales que desarrollan un voltaje en sus superficies cuando se les aplica la presión.

pinion bearing preload—tension between the pinion bearings to keep them at the correct clearance to their bearing races.

precarga del cojinete de piñón—la tensión entre los cojinetes del piñón para mantenerlas en la holgura correcta con respeto a las pistas del cojinete.

pinning or stitching—a crack repair process that uses threaded tapered plugs.

chavetear o ribetear—un proceso de reparación de grietas que usa los espárragos cónicos.

pipe dies—tools used to form the threads on the outside of pipe.

macho para tubo—las herramientas que sirven para cortar las roscas al exterior del tubo.

piston ring—a part that seals the piston to the cylinder.

anillo del pistón—la parte que sella el pistón al cilindro.

piston slap—when a piston skirt fits the cylinder too loosely, the resulting noise is called piston slap.

palmada del pistón—cuando la falda del pistón queda muy floja en el cilindro, el ruido que resulta se llama una palmada del pistón.

piston stroke—is the distance that the piston moves from TDC to BDC.

carrera del pistón—es la distancia que mueva el pistón del PMS al PMI.

pitch diameter—the point where the teeth of the two gears meet. This is the effective diameter of the gear.

diámetro del paso—el punto en que se juntan los dientes de dos engranajes. Este es el diámetro efectivo del engranaje.

pitch line—a line that runs through the center of the tooth from heel to toe.

línea cero—una línea que corre por el centro del diente del talón al dedo (la punta).

placard—a tire information sticker.

cartel—una etiqueta de información del neumático.

plain bearings—a bushing type of bearing that uses no rolling parts and provides sliding contact between two mating surfaces.

cojinetes sencillos—un cojinete de tipo buje que no usa las partes rodantes y provee un contacto deslizante entre dos superficies apareadas.

plastigage—a thin strip of plastic that deforms when crushed, it is used to measure bearing oil clearance.

plasticímetro—una lámina delgada del plástico que se deforma al ser aplastada, se usa para medir la holgura del aceite del cojinete.

plenum—the air space in the intake manifold below the carburetor.

pleno—el espacio del aire en el múltiple de admisión debajo del carburador.

plunge joint—the inboard CV joint that changes in length like a slip yoke does in a rear-wheel-drive car.

junta profundizada—la junta CV abordo que cambia en su longitud parecido a como lo hace una horquilla deslizante en un coche de tracción trasera.

POA—pilot operated absolute valve.

POA—válvula absoluta operada por piloto.

points—called contact points or breaker points, they are an electromechanical trigger for the ignition primary circuit.

puntos—llamados los puntos de contacto o contactos platinados, son un disparador electromecánico para el circuito primario del encendido.

pole shoe—the soft iron core of an electromagnet.

pieza polar—el núcleo de hierro blando en un electroimán.

polymer—a substance, called a plastic, consisting of giant molecules, formed from smaller molecules of the same kind.

polímero—una sustancia, llamado un plástico, consistiendo de las moléculas gigantescas, formadas de las moléculas más pequeñas del mismo tipo.

pop-back—a term for combustion occurring in the fuel induction system (before it enters the cylinder).

detonación—un término por la combustión ocurriendo en el sistema de inducción del combustible (antes de que entra al cilindro).

poppet valve—the style of valve used by four-stroke cycle internal combustion engines.

válvula elevación—el estilo de la válvula usado en los motores de combustión interna de cuatro carreras.

ported vacuum—when the vacuum port is above the throttle plate.

vacío con lumbrera—cuando la lumbrera del vacío queda arriba de la placa del acelerador.

port fuel injection—a fuel injection system that uses a fuel rail with individual fuel injectors at each intake port.

inyección del combustible por lumbrera—un sistema de inyección del combustible que usa un carríl de combustible con los inyectores individuales en cada lumbrera de admisión.

positive camber—when a tire is tilted out at the top.

comba positiva—cuando un neumático está inclinado hacia afuera en la parte superior.

positive caster—the rearward tilt of the steering axis at the top.

ángulo del caster positivo—la inclinación hacia atrás del eje de dirección.

positive offset—when the wheel is offset in such a way as to decrease the track width of the tires, it is often found on FWD cars.

descentrado positivo—cuando la rueda está descentrado en una manera que disminuya la anchura de la banda de los neumáticos, se suele encontrar en los coches FWD (tracción delantera).

positive spark plug electrode—the end of the conductor.

electrodo positivo de la bujía—en la extremidad del conductor.

positive stop—an automatic adjustment feature used with hydraulic valve lifters.

paro positivo—una característica de ajuste automático usado con las levantaválvulas hidráulicas.

potentiometer—a variable resistor that measures linear or rotary motion.

potenciómetro—un resistor variable que mide el movimiento lineal o giratorio.

power—how fast work is done or how fast motion is produced against a resistance.

fuerza—el rapidez con el cual se efectúa el trabajo o el rapidez con el cual se produce un movimiento contra una resistencia.

power side switching—a logic module tells a power module to control power to output devices.

conmutación lateral eléctrica—un módulo lógico le indica a un módulo eléctrico que controle la electricidad a los dispositivos de salida.

preignition—also called ping, occurs when the air-fuel mixture ignites before the regular spark occurs.

autoencendido—tambien llamado "ping", ocurre cuando la mezcla de aire-combustible encienda antes de que ocurre la chispa normal.

preload—the clamping load on a fastener.

precargar—la carga de apriete en un sujetador.

preload adjustment—an adjustment where a bearing has an additional load placed against it after it is snug (zero-lash).

ajuste de la carga previa—un ajuste en el cual se impone una carga adicional en el cojinete después de que esté apretada (juego de cero).

pressed-fit—when two parts with an interference fit are forced together.

ajuste con presión (embutido)—cuando dos partes con un ajuste de interferencia se unen bajo fuerza.

pressure bleeder—a canister that uses air pressure against a diaphragm to pressurize brake fluid.

purga de presión—un bote que usa la presión del aire contra un diafragma para presurizar el fluido de freno.

pressure cycling switch—a device that turns a magnetic air-conditioning clutch on and off.

interruptor del ciclo de presión—un dispositivo que encienda y apaga un embrague magnético del aire acondicionador.

pressure plate assembly—also called the clutch cover assembly, it is bolted to (and rotates with) the flywheel. It compresses the clutch disc against the flywheel.

asamblea de placa de presión—tambien llamado la asamblea de la tapadera del embrague, es empernada al (y gira con) volante. Oprime el disco del embrague contra el volante.

pressure primer—used to prime the lubrication system.

cebador a presión—usado para cebar el sistema de lubricación.

pressure priming—a method of priming the lubrication system with a pressure tank.

cebado de presión—un método de cebar el sistema de lubricación con un tanque de presión.

pressure relief valve—a valve in a hydraulic system that opens to allow excess pressure to bleed off to the fluid intake side of the pump.

válvula de descompresión—una válvula en un sistema hidráulico que abre para permitir que la presión excesiva se purga hacia el lado de admisión de fluidos de la bomba.

pressure test—a test to tell if the transmission is experiencing internal leakage or requires adjustment to linkages or the vacuum modulator.

prueba de presión—una prueba para verificar si la transmisión tiene una fuga interior o si requiere un ajuste en las bielas o en el modulador del vacío.

primary trigger—the electronic or electromechanical device that controls the switching of current flow in the primary winding.

disparador primario—el dispositivo electrónico o electromecánico que controla el cambio del flujo del corriente en el devanado primario.

primary vibration—when the piston slows down as it approaches TDC, its force pulls the engine up. When it approaches BDC, it pulls the engine down.

vibración primaria—cuando el pistón retarda al acercarse al PMS, su fuerza hace subir al motor. Cuando se acerca al PMI, hace jale al motor hacia abajo.

primary winding—the large winding in the ignition coil made up of 150 to 250 turns of wires.

devanado primario—el devanado grande de la bobina del encendido que consiste de 150 a 250 vueltas de los alambres.

primary wiring—low-voltage wiring.

alambre primario—el alambre de bajo voltaje.

production engine—an engine produced at the factory.

motor fabricado—un motor producido en una fábrica.

profile—a tire's height.

perfíl—la altura de un neumático.

PROM—programmable read-only memory.

PROM—memoria programable de solo lectura.

prony brake—a simple friction brake dynamometer.

freno prony—un dinamómetro para los frenos de fricción simple.

propane enrichment test—a test where propane or acetylene gas is put into the oil filler opening to see if the engine idle changes in response to an intake manifold gasket leak.

prueba de enriquecimiento de propano—una prueba en la cual se introduce el propano o el gas acetileno en la apertura de agregar aceite para ver si se cambia la marcha en vacío como resultado de una fuga en el empaque del múltiple de admisión.

psi—pounds per square inch.

psi—libras por pulgada cuadrada.

P-type—a crystal like aluminum or boron that is positively charged and can carry electrical current.

tipo P—un cristal como el aluminio o el boro que tiene una carga positiva y puede transportar el corriente eléctrico.

pull—when the steering wheel is held straight ahead and the car goes to the side.

desviación—cuando el volante de dirección se mantiene recta y el coche va a un lado.

pulse width—the length of time that an injector remains open.

impulso en anchura—la cantidad del tiempo en que queda abierto un inyector.

pulse width modulation—turning the alternator rapidly on and off to regulate voltage.

modulación al impulso en anchura—prender y apagar rápidamente al alternador para regular el voltaje.

pumpkin—a removable differential third member.

calabaza—un tercer miembro móvil del diferencial.

purge—when a vapor canister empties itself of vapor storage when the engine is running.

purga—cuando un bote de vapor se vacía si mismo del vapor almacenado mientras que marcha el motor.

push fit—when a part slides into place by hand.

ajuste a mano—cuando una parte se puede instalar a mano.

pushrod engine—this valvetrain design uses pushrods and has the cam located in the cylinder block.

motor de varillas empujadoras—este diseño de tren de válvulas usa las varillas empujadoras y tiene la leva situada en el bloque del cilindro.

pyrotechnics—a term that describes fireworks.

pirotecnia—un término que describe los fuegos artificiales.

Q

quad driver—when one single space-saving module has a group of four transistors and can control up to four actuators.

impulsor de cuatro—cuando un módulo eficiente en espacio tiene un grupo de cuatro transistores que pueden controlar hasta cuatros actuadores.

quill—the front transmission bearing retainer.

árbol hueco—el retenador del cojinete delantero de la transmisión.

R

R&R—remove and replace job; the estimated time it should take a professional technician to perform a specified job.

R&R—un trabajo remover y reemplazar; el tiempo estimado que debe tomar un técnico profesional en efectuar un trabajo especificado.

R-12 refrigerant—also called Freon®, it is a CFC (chlorofluorocarbon).

fluido refrigerante R-12—tambien llamado el Freón®, es un CFC (clorofluorocarbono).

R-134A refrigerant—a refrigerant that is used in newer vehicles. It is hydrofluorocarbon.

fluido refrigerante R-134A—un fluido refrigerante que se usa en los vehículos de ultima moda. Es un hidrofluorocarbono.

race—a bearing cup.

pista—un anillo exterior de un cojinete.

rack-and-pinion steering—the end of the steering shaft has a pinion gear that meshes with a rack gear.

dirección de piñón y cremallera—la extremidad de un eje de dirección tiene un engranaje de piñón que endenta con un engranaje de cremallera.

radial compressor—an air-conditioning compressor that resembles a radial aircraft engine. It has multiple cylinders with double ended pistons and one eccentric crankshaft throw.

comprimidor radial—un comprimidor del aire acondicionado que parece un motor de aeroplano. Tiene múltiple cilindros con pistones de doble cara y un radio del cigüeñal excéntrico.

radial load—a load that is in an up-and-down direction.

carga radial—una carga que mueve en la dirección de arriba a abajo.

radial movement—horizontal movement.

movimiento radial—un movimiento horizontal.

radial ply—a tire structure that has casing plies that run across the tire from bead seat to bead seat in the "radial" direction of the wheel.

carcasa radial—una estructura de neumático que tiene cordón de las telas atravesando el neumático desde el asiento de la ceja al asiento de la ceja en una dirección "radial" de la rueda.

radial runout—wobble of a part in an up-and-down direction.

corrimiento radial—un bamboleo de una parte en una dirección de arriba a abajo.

radiation—a method of heat transfer where heat bounces off of a surface. Dark car colors absorb and radiate heat better than light colors.

radiación—un método de transferencia del calor en el cual el calor bota de una superficie. Los coches de colores oscuros absorben y radian el calor mejor que los colores claros.

radius plate—an alignment gauge that fits under the front wheels and measures in degrees how far a wheel is turned to the right or left.

placa radial—un medidor de alineación que queda debajo de las ruedas delanteras y mide en grados cuánto gira una rueda a la derecha o a la izquierda.

RAM—random access memory; like a notepad that you can read from and write to.

RAM—memoria de acceso aleatorio; como un cuadernito en que puede leer y escribir.

raster pattern—also called stacked pattern, it displays all of the cylinders vertically, one above the next.

patrón de cuadro—tambien llamado un patrón amontonado, presenta todos los cilindros verticalmente, el uno sobre el otro.

Ravigneaux geartrain—a compound planetary gear design with long and short pinions.

tren de embragues Ravigneaux—un diseño de conjunto de embrague planetarios con los piñones largos y cortos.

reach—the length of the threaded area of a spark plug.

alcance—la longitud del área enroscado de una bujía.

readiness indicators—readings found on the scan tool that tell whether any or all of the OBD II monitors have been completed since the keep-alive memory was last cleared.

indicadores de prontitud—lecturas que se encuentran en la herramienta de escaneo e indican si alguno o todos los monitores OBD II han sido completados desde la última vez en que se borró la memoria 'keep-alive' (mantener vivo).

rebound—when the wheel moves back down.

rebote—cuando la rueda desciende de nuevo.

rebound bumper—a rubber bumper that is bolted to the lower control arm or chassis that comes into use when the suspension reaches the full limit of its travel.

parachoque del rebote—un parachoque de caucho empernado en el brazo de control inferior o en el chasis que funciona cuando la suspensión llega al límite total de su extensión.

rebuilt halfshaft—a front-wheel-drive axle that includes rebuilt inner and outer CV joints.

medio eje reconstruido—un eje de transmisión de la rueda delantera que incluye juntas de CV interiores y exteriores.

receiver/dryer—a refrigerant reservoir and dryer that supplies liquid refrigerant to the expansion valve.

colector desecador—un recipiente de fluido refrigerante que proporciona el fluido refrigerante a la válvula de expansión.

reciprocating engine—an engine with a piston that goes up and down and a crankshaft that rotates.

motor reciprocante—un motor con un pistón que suba y baja y un cigüeñal que gira.

recirculating ball and nut steering gear—a popular low friction steering gear box. A sector gear meshes with a ball nut that rides on ball bearings on the worm shaft to provide a smooth steering feel.

embrague de dirección de bola y tuerca recirculatoria—un cárter de engranajes de dirección de baja fricción popular. Un engranaje sectorial endenta con una tuerca esférica montado sobre los rodamientos de bolas en un eje fileteado para proveer una sensación de dirección uniforme.

red light—a typical hydraulic system warning light.

luz roja—una luz de advertencia típica de los sistemas hidráulicos.

reduce—the third stage of the ABS solenoid operation. The computer reduces pressure to a wheel by opening a solenoid valve.

reducir—la tercera etapa de la operación de solenoides del ABS. La computadora reduce la presión a la rueda al abrir la válvula solenoide.

reduction catalyst—one part of the reaction in a catalytic converter. During the reduction phase, diatomic oxygen molecules are removed from oxides of nitrogen (NO_x). This happens during reactions with water (H_2O), CO, and HC.

catalizador de reducción—una parte de la reacción en un conversor catalítico. Durante la fase de reducción, se eliminan moléculas de oxígeno diatómicas de los óxidos de nitrógeno (NO_x). Esto sucede durante reacciones con el agua (H_2O), CO y HC.

reduction converter—a catalytic converter that catalyzes NO_x.

convertidor reductor—un convertidor catalítico que cataliza el NO_x.

regenerative braking—when an EV decelerates, the motor becomes a generator and produces electricity to recharge the batteries as it slows the vehicle down.

frenado regenerativo—al desacelerarse el EV, el motor se convierte en un generador y produce electricidad para recargar las baterías a medida que reduce la velocidad del vehículo.

regulator maximum voltage test—a test that checks to see that the voltage regulator is energizing the rotor field and is not allowing the alternator to overcharge.

prueba del voltaje máximo del regulador—una prueba para verificar que el regulador de voltaje proporciona la energía al campo del rotor y no permite que sobrecarga el alternador.

relay—a magnetically controlled switch used when a large load must be controlled by a small wire.

relé—un interruptor controlado por magnetismo que se usa cuando se tiene que controlar una carga grande con un alambre pequeño.

release bearing—also called a throwout bearing, it contacts the rotating clutch to release its disc.

cojinete de empuje del embrague—tambien llamado el cojinete de desembrague, se pone en contacto con el embrague giratorio para desengranar el disco.

release lever—also called finger, a part of the coil spring clutch that when compressed pulls the pressure plate away from the flywheel.

palanca de desembrague—tambien llamado patilla, una parte del embrague de resorte helicoidal que al ser comprimido retira la placa de presión del volante.

reserve capacity—a measurement of the battery's ability to provide current when there is no electricity from the charging system.

capacidad de reserva—una medida de la habilidad de la batería en proveer el corriente cuando no viene electricidad del sistema de cargar.

resistance—when there is an obstruction to electrical flow.

resistencia—cuando hay una obstrucción al flujo eléctrico.

resistance key—a normal key with a resistance pellet imbedded in it.

llave de resistencia—una llave normal con un gránulo de resistencia incrustado.

resistor—an electrical device used to make heat or to control the intensity of a load.

resistor—un dispositivo eléctrico que se usa para producir el calor o para controlar la intensidad de una carga.

resonator—a second muffler in line with the other muffler.

resonador—un silenciador secundario en línea con el otro silenciador.

returnless fuel system—a fuel system used on some newer vehicles with its fuel pressure regulator and filter mounted in a gauge and pump cluster mounted in the top of the fuel tank. There is no return line from the engine to the fuel tank, and the regulator exhausts excess fuel pressure directly to the tank.

sistema de combustible sin retorno—un sistema de combustible utilizado en algunos vehículos más recientes con su regulador de presión de combustible y filtro montados en un grupo de bomba y medidor, montado a su vez en la parte superior del tanque de combustible. No existe una línea de retorno del motor al tanque de combustible y el regulador descarga la presión excesiva del combustible directamente al tanque.

reverse bias—not allowing current to flow.

polarización reversa—no dejando fluir el corriente.

reverse bleeding—moving fluid through a hydraulic system from the slave cylinder to the master cylinder.

purga invertida—mudando el fluido a través del sistema hidráulico del cilindro secundario al cilindro maestro.

rheostat—varies current flow through the circuit as a movable wiper arm runs along a resistor.

reóstato—modifica el flujo del corriente atravesando un circuito cuando el brazo de contacto móvil desliza por un resistor.

riding the clutch—resting a foot on the clutch when driving.

desgaste continuo del engrane—descansando el pie en el pedal del embrague mientras que uno maneja.

rigid axle—a non-independent axle, also called a straight axle or I-beam, found on most rear ends and some heavy truck front ends.

eje rígido—un eje no independiente, tambien llamado un eje recto o un hierro en T, encontrado en la mayoría de las asambleas traseras y en algunas asambleas delanteras de camiones de trabajo pesado.

ring ridge—the ledge that forms at the top of the cylinder as the cylinder wears.

nervadura en el pistón—la arruga formada en la parte superior de un cilindro al desgastarse el cilindro.

riveted lining—a lining that is held to its backing by rivets.

forro con remaches—un forro que se adhiere al respaldo por medio de los remaches.

RMA—Rubber Manufacturers Association.

RMA—Asociación de Fabricantes de Caucho.

road horsepower—measures horsepower available at the car's drive wheels.

potencia de caballo en carretera—mide la potencia en caballos disponible en las ruedas de tracción del coche.

rocking couple—another form of vibration that occurs when there are only two cylinders that fire 180 degrees apart instead of 360 degrees apart.

oscilación con dos—otra forma de vibración que ocurre cuando sólo hay dos cilíndros que encienden en la posición de 180° en vez de en la posición de 360°.

rod bolt protector—a piece of soft plastic or rubber used to protect the crankshaft from nicks during engine disassembly and assembly.

protector del vástago—una pieza flexible de plástico o caucho que se usa para evitar los daños al cigüeñal cuando se desmonta o se arma un motor.

rod out—when a radiator becomes plugged, this is done by a radiator shop to clean its cooling passageways.

escariar—cuando un radiador está atascado, un taller de radiadores efectúa esto para limpiar los pasos de enfriar.

rolled edge clamp—hose clamp used for fuel injection and other high pressure uses. It is designed so that it will not cut into a hose.

grapa con bordes redondeados—la grapa para tubos que se usan en la inyección del combustible y en otras aplicaciones de alta presión. Se ha diseñado para no cortar un tubo.

rollover valve—a valve in the vent line from the fuel tank that blocks the line if the car is in an accident and rolls over, preventing liquid fuel from being able to leak out.

válvula de seguridad—una válvula en la línea de ventilación del tanque de combustible que cierra la línea en caso de un accidente en el cual el coche se voltea, previniendo una fuga del combustible en forma líquido.

ROM—read-only memory; permanent programmed information available to the microprocessor programmed into the chip during manufacturing.

ROM—memoria de solo lectura; la información programada permanente disponible al microprocesador programado en la ficha durante la fabricación.

Roots-type supercharger—also called a lobe-type supercharger, it is the most popular positive displacement supercharger.

sobrealimentador tipo Roots—tambien llamado un sobrealimentador con lóbulo, es el tipo más popular de sobrealimentador de desplazamiento positivo.

rotary vane pump—a pump with blades used for power steering pump, air conditioning, or a smog pump.

bomba giratoria de aletas—una bomba con palas que se usa en la bomba de dirección hidráulica, el aire acondicionador, o una bomba de esmog.

rotor-type pump—a type of oil pump that has an internal and an external rotor.

bomba del tipo rotor—un tipo de bomba de aceite que tiene un rotor interno y un rotor externo

RTV—room temperature vulcanizing silicone rubber.

RTV—el caucho de silicona que vulcaniza en temperatura ambiente.

rubberized cork—a combination of cork and artificial rubber that is a superior material to regular cork because it does not shrink.

corcho cauchotado—una combinación del corcho y el caucho sintético que es una materia superior al corcho natural porque no se encoje.

runners—the passages in an intake manifold.

conductos—los pasos en un múltiple de admisión.

runout—the amount that a part wobbles up and down or side to side when rotated.

corrimiento—la cantidad que oscila una parte de arriba a abajo o de un lado a otro al ser girado.

RVP—Reid vapor pressure; a standard by which volatility is measured.

RVP—presión de vapor Reid; una norma por la cual se mide la volatilidad.

RWD—rear-wheel drive.

RWD—tracción trasera.

Rzeppa CV joint—a popular ball and socket type joint used in most outboard fixed CV joints.

junta CV Rzeppa—una junta de articulación esférica popular que se usa en la mayoría de las juntas CV fuera de bordo.

S

SAE—Society of Automotive Engineers.

SAE—Asociación de Ingenieros Automotivos.

safing sensors—determine whether or not there has been a crash. Also called arming sensors.

sensores de choque—también llamados sensores de armado, determinan si ha habido un choque.

SAI—steering axis inclination; the amount that the spindle support arm leans in at the top. SAI is also known as ball joint inclination (BJI) or king pin inclination (KPI).

SAI—inclinación del eje director; la cantidad en que inclina hacia adentro el husillo del brazo de apoyo en la parte superior. El SAI tambien se conoce como la inclinación de la articulación esférica (BJI) o la inclinación de la clavija maestra (KPI).

Salisbury axle—also called integral, its third member is not removable as a unit.

eje Salisbury—tambien llamado un integral, su tercer miembro no se puede remover como una unidad.

saturated vapor—a liquid that is in contact with its vapor within an enclosed space.

vapor saturado—un líquido que está en contacto con su vapor en un espacio encerrado.

scale—this forms as minerals in the local water supply settle out of the water during heating and become deposited on heated parts.

incrustación—esto forma cuando los minerales del suministro del agua local sedimentan al ser calentados y se depositan en las partes calentadas.

scan tool—a portable computer connected to the vehicle's diagnostic connector that reads data from the car's on-board computer.

aparato de exploración—una computadora portátil conectada al conector diagnóstico que lee los datos de la computadora abordo del coche.

SCFM—standard cubic feet per minute.

SCFM—estándar de pies cúbicos por minuto.

schematic—a wiring diagram.

esquemático—una diagrama del cableado.

scroll compressor—a compressor with a fixed and a moveable scroll.

comprimidor voluta—una comprimidora que tiene un espiral fijo y uno móvil.

scrub radius—the distance at the road surface between the centerline of true vertical at the center of the tire tread and the steering axis pivot centerline.

radio viraje en rueda—la distancia en la superficie de la carretera entre la línea central vertical verdadera del centro de la banda de rodamiento y la línea central de pivote del eje de dirección.

scuff—due to incorrect toe, when the tires move sideways for a certain distance out of every mile traveled.

desgaste—debido a la divergencia incorrecta, cuando los neumáticos muevan lateralmente una cierta distancia por cada milla de viaje.

scuffing—the damage caused when parts momentarily weld to one another. It occurs due to excessive heat.

rozamiento—el daño causado cuando las partes se sueldan momentariamente la una a la otra. Ocurre debido al calor excesivo.

secondary winding—the smaller winding within the center of the coil primary winding made up of several thousand turns of very small, hairlike insulated wire.

devanado secundario—el arrollamiento más pequeño dentro del centro del devanado de la bobina primaria compuesta de varias miles de vueltas de alambre aislada pequeña y fina.

secondary wiring—high-voltage ignition wiring.

cableado secundario—el cableado de encendido de alto voltaje.

second order vibration—a vibration from a universal joint.

vibración del segundo orden—una vibración de una junta universal.

seized engine—when a crankshaft will not turn.

motor agarrotado—cuando un cigüeñal no dará la vuelta.

Selden patent—a patent covering all gasoline-powered self-propelled road vehicles.

patente Selden—un patente que se aplica a todos los vehículos de autopropulsión de combustión de gasolina.

self-energization—when the leading shoe on a drum brake is forced into the brake drum.

autoenergetización—cuando la zapata de guía de un freno de tambor se empuja contra el tambor del freno.

SEMA—Specialty Equipment Manufacturers Association.

SEMA—Asociación de Fabricante de Aparatos Especiales.

semiconductor—a solid-state material that acts as both an insulator or a conductor.

semiconductor—un material de estado sólido que se comporta como un aislador o tambien un conductor.

semi-floating axle—an axle with a bearing that rides on the axle.

eje semi-flotante—un eje con un cojinete que va montado sobre el eje.

semimetallic lining—organic lining with sponge iron and steel fibers mixed into it to add strength and temperature resistance.

forro semi-metálico—un forro orgánico que tiene incorporado el hierro esponjoso y las fibras del acero para proporcionarles más fuerza y resistencia a la temperatura.

sending unit—a device that sends a signal to a gauge regarding oil pressure or coolant temperature.

unidad emisor—un dispositivo que envía una señal a un indicador referente a la presión del aceite o la temperatura del fluido refrigerante.

sensible heat—the heat that goes into matter as its temperature increases.

calor sensible—el calor que entra en la materia en cuanto aumenta su temperatura.

sensor—a device that relays to the computer such information on throttle position, air and coolant temperature, airflow, manifold pressure, and barometric pressure.

sensor—un dispositivo que releva a la computadora la información referente a la posición del acelerador, la temperatura del aire y del fluido refrigerante, el flujo del aire, la presión del múltiple y la presión barométrica.

sensor safe RTV—low-volatile silicone that does not give off vapors that can damage an oxygen sensor.

RTV para sensores—un silicio resistente a la volatilidad que no emite los vapores que pueden dañar un sensor de oxígeno.

series circuit—an electrical circuit where current flowing equally through all parts of a circuit flows first to one load and then on to the next one.

circuito en serie—un circuito eléctrico en el cual el corriente fluyendo igualmente por todas partes del circuito fluye primero a una carga y luego a la próxima.

serpentine belt—a belt that is called serpentine because of the snake-like path that it follows.

correa serpentín—una correa que se llama serpentín debido a que la senda que sigue parece la de una culebra.

service dispatcher—the person who organizes the repair orders and dispatches them to the technicians in the service bays.

despachante de servicio—la persona que organiza las órdenes de reparación y las despacha a los técnicos en las áreas de servicio.

service manager—the person responsible for the operation of the service department.

gerente de servicio—la persona responsable por las operaciones del departamento de servicio.

service manual—manual containing all of the most used specifications for a vehicle.

manual de servicio—manual que contiene todas las especificaciones más usadas del vehículo.

service record—the written record of the service performed on a vehicle; also called a repair order (R.O.) or work order (W.O.).

registro de servicio—el registro escrito del servicio realizado en un vehículo; también llamado orden de reparación (R.O., según sus siglas en inglés) u orden de trabajo (W.O., según sus siglas en inglés).

service writer/service advisor—the person who greets the customer, listens to the complaint, interprets it, and then writes the repair order.

redactor de servicio/consejero de servicio—la persona que recibe al cliente, escucha su queja, la interpreta y luego redacta la orden de reparación.

servo action—when the leading shoe on a drum brake is forced into the brake drum.

acción servo—cuando la zapata de guía de un freno de tambor se empuja contra el tambor del freno.

set-back—the amount that one front wheel is behind the one on the other side of the car.

retroceso—la cantidad de que una rueda delantera queda atrás de la del otro lado del coche.

severe service—includes trailer towing and stop-and-go city driving in hot weather.

servicio severo—incluye usando remolque y viajando en tránsito pesado en clima cálida.

SFE—Society of Fuse Engineers.

SFE—Sociedad de Ingenieros de Fusibles.

SFI—sequential fuel injection; a more refined fuel injection system where each injector is opened just before its intake valve opens, allowing adjustments to the fuel mixture to be made very quickly.

SFI—siglas en inglés para inyección secuencial de combustible; un sistema de inyección de combustible más refinado, en el que se abre cada inyector justo antes de que su válvula de entrada se abra, permitiendo la realización rápida de ajustes en la mezcla de combustible.

shift quadrant—the readout on the gear selector that selects what gear the transmission is in.

sector de cambios—el indicador en el selector de velocidades que selecciona en cuál velocidad va la transmisión.

shock absorber—a device that dampens spring oscillations by converting the energy from spring movement into heat energy.

amortiguador—un dispositivo que amortigua las oscilaciones de los resortes al convertir la energía del movimiento de los resortes a la energía calorífica.

shock absorber control force—determines the softness or harshness of the ride.

fuerza del control del amortiguador—determina la uniformidad o la brusquedad del viaje.

shock ratio—the difference between the amount of control on compression and extension.

relación de choques—la diferencia entre la cantidad del control en la compresión y la extensión.

shop foreman—the person responsible for keeping repair work on track.

supervisor de taller—la persona responsable por mantener el control de los trabajos de reparación.

short block—an engine without heads.

bloque corto—un motor sin las cabezas.

short circuit—when the electrical path has been shortened by a wire contacting another.

corto circuito—cuando una trayectoria eléctrica se ha cortado debido a una alambre tocando a otra.

shorting bar—a safety device that automatically shorts the circuit when an airbag module connector is disconnected.

barra de corto—un dispositivo de seguridad que provoca automáticamente un corto en el circuito cuando se desconecta un conector del módulo de la bolsa de aire.

siamese runner—when one runner feeds two neighboring cylinders.

conducto siamés—cuando un conducto alimenta dos cilindros que son juntos.

SI engine—an engine that ignites its air and fuel with a spark.

motor SI—un motor que encienda su aire y combustible con una chispa.

signal flasher—a device activated by the heat of the electricity traveling through it to cause the turn signal bulbs to flash.

pulsador indicador—un dispositivo activado por el calor de la electricidad que lo atraviesa para causar que destellan los focos indicatores de vueltas.

Simpson geartrain—a compound planetary gear design that shares the same sun gear between the gearsets.

tren de engranajes Simpson—un diseño de engranajes planetarios combinados que comparten un engranaje solar entre los conjuntos de engranajes.

sine wave voltage—an oscilloscope pattern showing positive and negative voltage.

voltaje de onda senoidal—un patrón del osciloscopio que muestra el voltaje positivo y negativo.

single channel—a two-wheel antilock brake system found on sport utility vehicles and light trucks. Two-wheel antilock brake systems only work on the rear wheels.

canal único—un sistema de frenos antibloqueo en dos ruedas que se encuentra en vehículos utilitarios deportivos y camionetas; sólo funciona en las ruedas traseras.

single exhaust system—when there is only one pipe to the rear of the car.

sistema simple del escape—cuando hay sólo un tubo en la parte trasera del coche.

single pass machine—a machine for air conditioning that removes the refrigerant and recycles it during the 15-minute minimum vacuum cycle.

máquina de un solo paso—una máquina para el aire acondicionado que remueva el refrigerante y lo recicla durante el ciclo de vacío mínimo de 15 minutos.

single-plane manifold—when each of a manifold's barrels serve all of the engine's cylinders.

múltiple monoplano—cuando cada uno de los cañones del múltiple sirve para todos los cilindros del motor.

SIR—supplemental inflatable restraints; air bags.

SIR—resguardos suplementales inflables; las bolsas de aire.

size equivalents—when different tire sizes have the same diameter and load capacity.

equivalentes de tamaño—cuando los tamaños distintos de los neumáticos tienen el mismo diámetro y capacidad de carga.

skin effect—the name for the thin area of fuel that does not burn when the surface area of the combustion chamber is cold so fuel burning is quenched.

efecto de película—el nombre de la superficie delgada del combustible que no quema cuando la superficie de la cámara de combustíon está fría lo cual apaga al combustible que está quemando.

SLA—short/long arm; a suspension design that uses two control arms of unequal length.

SLA—brazo corto/largo; un diseño de suspensión que usa dos brazos de control de longitud disparejos.

slave cylinder—also called the reaction or actuator cylinder, it is the output piston in a hydraulic clutch that is attached to the release lever at the clutch.

cilindro secundario—tambien llamado un cilindro de reacción o actuador, es el pistón de salida de un embrague hidráulico conectado a la palanca de desenganche en el embrague.

slip angle—the tendency during a turn for a tire to continue to go in the direction it was going, even though the rim has turned in response to steering wheel movement.

ángulo de desviación—la tendencia durante un viraje a que un neumático continua a viajar en la dirección en la cual iba, aunque la llanta se ha girado respondiendo al movimiento al volante de dirección.

sludge—a mixture of moisture, oil, and contaminants that occurs when engine temperatures are too low.

cieno—una mezcla de la humedad, el aceite y los contaminantes que ocurre cuando las temperaturas del motor son demasiadas bajas.

smog pump—an air injection system air pump.

bomba del esmog—una bomba de sistema de aire de aire inyección.

snapshot—during a test drive when a driveability symptom is felt, this records frame by frame a period of several seconds before and after the button on the scan tool is pressed. This helps diagnose intermittent faults.

instantáneo—durante un viaje de prueba cuando se siente una síntoma, esto registra cuadro a cuadro un período de varios segundos antes y después de que se haya oprimido el botón en el aparato explorador. Esto ayuda en el diagnósis de los fallos intermitentes.

soft parts—these are part of a rebuild kit, including all rubber seals and gaskets, bands, clutch friction discs, modulator, thrust washers, and bushings.

partes flexibles de repuesto—estas son parte de un juego de piezas de reconstrucción, incluye todos los sellos y empaques de caucho, las correas, los discos de fricción del embrague, el modulador, las arandelas de empuje y los bujes.

software—information stored as electronic signals that can be modified.

software—la información almacenado en forma de señales electrónicas que se pueden modificar.

SOHC—single overhead cam.

SOHC—árbol leva única sobre la cabeza.

solenoid—a combination magnetic switch and mechanical device.

solenoide—una combinación de interruptor magnético y dispositivo mecánico.

sound—vibration in the air.

sonido—una vibración en el aire.

spalling—when pieces break off of the bearing metal.

escamación—cuando se despegan pedazos del metal del cojinete.

spark line—a horizontal line that begins at the voltage level where electrons start to flow across the spark plug gap.

línea de chispa—una línea horizontal que comienza en el nivel de voltaje en donde comienzan a fluir los electrones a través del entrehierro de la bujía.

spark plug deposit test—a test in which spark plugs are examined to see if an engine is using oil.

prueba de depósitos en la bujía—una prueba en la cual se examinan las bujías para verificar si el motor está usando el aceite.

specific gravity—the weight (strength) of electrolyte.

gravedad específica—el peso (la fuerza) de un electrólito.

speed density system—when the computer uses a manifold absolute pressure (MAP) sensor and engine rpm (tach) signal to calculate the amount of air entering the engine.

sistema de densidad en velocidad—cuando una computadora usa un sensor MAP (presión absoluta del múltiple) y el señal del rpm del motor (taquímetro) para calcular la cantidad del aire entrando al motor.

speed rating—a tire rating that means that a properly inflated tire has been designed to operate at up to a certain mph for short periods of time.

capacidad de la velocidad—una clasificación del neumático quiere decir que un neumático inflado correctamente ha sido diseñado a operar en ciertas mph por cortos períodos de tiempo.

spider gear—the side gear and differential pinion.

satélite del diferencial—el engranaje lateral y piñón del diferencial.

spin test—when an engine is spun with a spin tester to check such things as oil pressure and compression.

prueba rotativa—cuando se gira un motor con un probador giratorio para verificar las cosas como la presión del aceite y la compresión.

spiral cable—also called clock spring, it connects the steering column to the air bag in the steering wheel.

cable espiral—tambien llamado un resorte espiral, conecta la columna de dirección a la bolsa de aire en el volante de dirección.

spongy brake pedal—a soft pedal feel due to air in the brake hydraulic system.

pedal de freno esponjoso—una sensación blanda en el pedal debido al aire en el sistema hidráulico de frenos.

spool valve—a valve that has lands, valleys, and faces.

válvula de carrete—una válvula que tiene partes planas, las acanaladuras y las caras.

spread—the difference between alignment measurements from side to side.

aplastamiento—la diferencia entre las medidas de alineación de un lado a otro lado.

sprung weight—the weight supported by the car springs.

peso del resorte—el peso soportado por los resortes de un coche.

spur gear—a simple gear design with straight cut teeth.

engranaje recto—un diseño simple de engranaje que tiene dientes rectos.

squat—during hard acceleration, the front of the vehicle lifts and the rear lowers.

agacharse—al acelerar de golpe, la parte delantera del vehículo se eleva y la trasera baja.

squibs—one or more fuse-like ignition devices.

buscapiés—uno o más dispositivos de ignición similares a fusibles.

squirrel cage fan—a fan used with a blower motor that resembles a squirrel cage.

ventilador caja de ardilla—un ventilador con un motor soplador que parece una caja de ardilla.

stall speed—the level of engine rpm when the engine cannot continue to increase speed with the brake and accelerator applied at the same time.

velocidad de atascado—el nivel de rpm del motor cuando el mismo no logra seguir aumentando la velocidad aplicando el freno y el acelerador al mismo tiempo.

stall test—with brakes applied, the engine is accelerated until speed no longer increases. Note engine rpm at this point.

prueba de paro—con los frenos puestos, se acelera el motor hasta que no aumenta en velocidad. Nota el rpm en este punto.

static—at rest.

estático—en descanso.

static pressure—the pressure reading in the system when it is not operating.

presión estático—la lectura de la presión en el sistema mientras que no está operando.

static timing—timing the distributor with the engine stopped.

regulación estática—regulando el distribuidor con el motor parado.

steering damper—a sideways shock absorber for road shocks.

amortiguador de la dirección—un amortiguador que absorba los choques de la carretera lateralmente.

steering linkage taper—a part that connects the steering gear to the wheels.

biela ahusamiento de dirección—una parte que conecta el engranaje de dirección a las ruedas.

steering ratio—the number of teeth on the driving gear compared to the number of teeth on the driven gear.

relación de dirección—el número de dientes en el engranaje impulsor comparado al número de dientes en el engranaje arrastrado.

stock—original equipment.

fondo—aparatos originales.

stoichiometric—the ideal air-fuel ratio, 14.7:1 by weight, in which all of the oxygen is consumed in the burning of the fuel.

estoquiométrico—la relación ideal del aire-combustible, 14.7:1 por peso, en la cual se consume todo el oxígeno al quemarse el combustible.

stratified charge—the spark is ignited in a richer air-fuel mixture. The flame from this mixture burns a leaner mixture in its path.

carga estratificada—la chispa se enciende en una mezcla más rica de aire-combustible. La llama de esta mezcla enciende la mezcla más pobre que está en su ruta.

street elbow—a 90 degree connection with a male thread on one end and a female thread on the other.

codo roscado macho y hembra—un conector de 90° con una rosca macho en una extremidad y una rosca hembra en la otra.

stress raiser—when a metal surface is damaged it raises stress, weakening the part.

deformación de tensión—cuando la superficie del metal se daña crea la tensión, debilitando la parte.

stress test—when ignition parts like a module or pickup coil are cooled or heated while watching the scope pattern.

prueba de esfuerzo—cuando las partes del encendido, como un módulo o un devanado detector se enfrían o se calientan mientras que se observa el patrón en el osciloscopio.

stroke—the movement of the piston from the top of its travel to the bottom of its travel.

carrera—el movimiento del pistón desde el punto superior de su viaje al punto inferior de su viaje.

strut cartridge—a replacement shock absorber for Macpherson struts.

cartucho del poste—un amortiguador de reemplazo de los postes Macpherson.

stub shaft—the live axle end at the outside of the outboard CV joint on a front-wheel-drive axle shaft.

mangueta del eje—la extremidad viva del eje en la parte exterior de la junta CV fuera de bordo en un árbol motor de tracción delantera.

stud-centric—wheels that center using the wheel studs.

centrado en el husillo—las ruedas cuyos centros son los husillos de las ruedas.

STV—suction throttling valve.

STV—válvula de aceleración por aspiración.

subcooling—a liquid with a temperature considerably beneath its boiling point.

subenfriar—un líquido con una temperatura bastante más baja que su punto de ebullición.

sub-frame—a group of parts in the front of a front-wheel-drive vehicle that includes the engine, transaxle, and steering/suspension system.

bastidor—un grupo de partes en la parte delantera de un vehículo de tracción delantera que incluye al motor, el transeje, y el sistema de dirección/suspensión.

suction service valve—the low side service valve on the inlet to the air-conditioning compressor (low side).

válvula de aspiración de servicio—la válvula de servicio del lado bajo en la admisión al comprimidor del aire acondicionador (lado bajo).

sulfation—when the battery is allowed to remain in a discharged state, the lead sulfate in the battery's plates becomes hard and resistant to recharging.

sulfatación—cuando se permite que la batería se queda en un estado de descarga, el sulfato de plomo en las placas de la batería se endurece y resiste la carga.

supercharger—a belt-driven pump, also called a blower.

sobrealimentador—una bomba conducida por correa, tambien llamado un ventilador.

superheat—temperatures above the liquid's boiling point.

supercalentar—las temperaturas más altas que el punto de ebullición.

superimposed pattern—a scope pattern used to compare all of the cylinders while their patterns are displayed one on top of the other.

patrón superimpuesta—un patrón de osciloscopio usado para comparar todos los cilindros mientras que sus patrones su muestran uno sobre el otro.

surface charge—the charge on the surface of a battery plate.

carga superficial—la carga en la superficie de una placa de batería.

surface texture—measurements of the roughness or smoothness of a surface are rated on the American National Standard Institute (ANSI) scale.

textura de la superficie—las medidas de lo áspero o lo liso de una superficie se clasifican en la escala Instituto Americano de Normas Nacionales (ANSI).

surface-to-volume ratio—a term for the amount of surface area exposed in the combustion chamber.

relación de superficie a volumen—un término para la cantidad del área de la superficie expuesta en la cámara de combustión.

surge—a driveability condition that results from a lean air-fuel mixture.

sacudida—una condición del manejo que resulta de una mezcla de aire-combustible pobre.

suspension—a group of parts that supports the vehicle, cushioning the ride while holding the tire and wheel correctly positioned in relation to the road. Suspension system parts include the springs, shock absorbers, control arms, ball joints, steering knuckle and spindle, or axle.

suspensión—un grupo de partes que soportan al vehículo, suavizando el viaje mientras que mantiene correctamente posicionado la rueda y el neumático en su relación a la carretera. Las partes del sistema de suspensión incluyen los resortes, los amortiguadores, los brazos de mando, las articulaciones esféricas, la articulación y muñón de dirección, o el eje.

swaged—a means of permanently locking a part to another part by deforming part of it or the material that surrounds it.

estampada—una manera de enclavar permanentemente una parte a otra deformando una porción de ella o a la materia que la rodea.

swaged lug studs—a method of holding a lug stud to a wheel hub or axle.

agarradera estampada—un método de fijar un pasador saliente en el cubo o el eje.

swept volume—another means of determining a cylinder's displacement.

cilindrada—otra manera de determinar el desplazamiento de un cilindro.

swing axle—an axle with a universal joint at one or both ends.

eje articulado—un eje que tiene una junta universal en una o ambas extremidades.

switch—a device used to interrupt a circuit, providing a means of turning it on and off.

interruptor—un dispositivo usado para interrumpir un circuito, proveyendo una manera de encenderlo y apagarlo.

switch ratio—the speed of oxygen sensor oscillations between rich and lean.

relación de conmutación—la velocidad de las oscilaciones del sensor de oxígeno entre rico y pobre.

synchronizer—also called a synchro, it keeps two meshing gears from clashing during a shift.

sincronizador—tambien llamado un sincro, previene que chocan dos engranajes endentados durante un cambio de velocidades.

T

TAC—thermostatic air cleaner.

TAC—limpiador de aire termostático.

tap drill—a tap drill provides the correct hole size prior to tapping a thread.

tamaño del barreno—un barreno provee el tamaño correcto del agujero antes de que se cortan las roscas.

taper—a steering linkage connection.

transición ahusada—una conexión de la biela de dirección.

taper cylinder wear—wear occurring in the top inch of a cylinder.

desgaste del cilindro cónico—el desgaste que ocurre in la primera pulgada de la parte superior de un cilindro.

TBA—tires, batteries, and accessories.

TBA—los neumáticos, las baterías y los accesorios.

Teflon® tape—tape used to seal between male and female threads.

cinta de teflón—cinta adhesiva utilizada para sellar entre roscas macho y hembra.

temper—to toughen a metal, it is heated to a specific temperature and then quenched.

templar—para endurecer un metal, se calienta a una temperatura específica y luego se enfría rapidamente.

tempilstick—a stick that melts at a predetermined temperature, giving an indication of a metal's temperature during heating.

varilla de templar—una varilla que se derrite en una temperatura predeterminada, proporcionando una indicación de la temperatura de un metal al calentarlo.

tensile cords—cords in the belt that provide strength.

cordón de tensión—los cordones en la correa que proveen la fuerza.

tensile strength—the maximum stress a material can withstand without breaking.

fuerza de tensión—el esfuerzo máximo que puede aguantar una materia sin romper.

thermal efficiency—the ratio of how effectively an engine converts a fuel's heat energy into usable work.

rendimiento termal—la relación de qué tal eficaz es un motor en convertir la energía calorífica del combustible al trabajo útil.

thermistor—a variable resistor made from semiconductor material. Its resistance changes predictably as its temperature changes. Although voltage varies in response to changes in resistance, current flow remains constant.

termistor—un resistor variable hecho de la material de semiconductor. Su resistencia cambia según cambia su temperatura. Aunque varía su voltaje en respuesta a los cambios en resistencia, el flujo del corriente queda constante.

thermoplastic seizure—when coolant has migrated into the engine oil, it can seize the engine.

agarrotamiento termoplástico—cuando el fluido refrigerante ha migrado al aceite del motor, puede agarrotar al motor.

thermostat bypass—a passageway that allows coolant to circulate when the thermostat is closed.

termostato de derivación—un paso que permite que el fluido refrigerante circula cuando está cerrado el termostato.

thickening agent—a constituent of grease at a concentration of from 5% to 25%, it is usually a soap such as aluminum, lithium, sodium, or calcium.

agente espesativo—un constituyente de la grasa en una concentración del 5% a 25%, suele ser un jabón tal como el aluminio, el litio, el sodio o el calcio.

thinwall—a valve guide repair process that uses a bronze bushing.

reparación aislada—un proceso de reparar las guías de las válvulas usando un buje de bronce.

third member—the differential assembly is called this. The engine is the first member and the transmission is the second member.

tercer miembro—así se llama la asamblea del diferencial. El motor es el primer miembro, y la transmisión es el segundo miembro.

three-channel system—contains a sensor at each front wheel and the rear works like the single-channel two-wheel system.

sistema de tres canales—contiene un sensor en cada rueda delantera; el trasero funciona como un sistema de dos ruedas y un único canal.

three-phase electrical output—when the three windings of the stator produce their currents in phases.

salida eléctrica trifásica—cuando los tres devanados del estator producen sus corrientes en fases.

three-way converter—converts HC and CO, and also changes harmful NO_x into harmless nitrogen and oxygen.

convertidor tridireccional—convierta el HC y el CO, y tambien cambia el NO_x nocivo al nitrógeno y oxígeno inocuo.

three-way oxidation converter—used with computer cars, it has a tube in the center to provide air from the smog pump to the reduction part of the catalytic converter.

convertidor de la oxidación tridireccional—usado en los coches de computadora, tiene un tubo en el centro para proveer el aire de la bomba del esmog a la parte de reducción del convertidor catalítico.

throttle pressure—the kind of pressure that results when engine vacuum changes that gives an indication of engine load.

presión del acelerador—el tipo de presión que resulta cuando cambia el vacío del motor que da una indicación de la carga en el motor.

throttle-body injection (TBI) system—also called central fuel injection (CFI) system, it uses an intake manifold similar to a carbureted system.

sistema inyección al cuerpo del accelerador—tambien llamado el sistema inyección central del combustible, usa un múltiple de admisión parecido a un sistema carburado.

through-hardened washer—a washer that is hardened throughout, not just on the surface.

arandela de temple profundo—una arandela que es totalmente templados, no sólo en la superficie.

throw—the amount that the rod journal moves out from the main bearing center line or one-half of the total stroke.

recorrido—la cantidad que mueva el muñón de la biela de la línea central del cojinete principal o la mitad de la carrera total.

thrust angle—the angle formed by the thrustline and the geometric centerline.

ángulo de empuje—el ángulo formado por el eje de empuje y la línea central geométrica.

thrust bearing—a bearing mounted at 90 degrees to the load.

cojinete de empuje de bolas—un cojinete montado en 90 grados a la carga.

thrustline—when the rear wheels are held in a fixed position, this defines the wheels' true straight ahead position.

eje de empuje—cuando las ruedas traseras se sujeten en una posición fija, esto define la posición verdadera de las ruedas directamente hacia afrente.

thrust load—a load in a front-to-rear or side-to-side direction.

carga de empuje—una carga en una dirección delantera a lado o de lado a lado.

timing light—a strobe light that is triggered by the voltage going through the number one spark plug cable.

luz de encendido—una lámpara estroboscópica disparada por el voltaje atravesando el cable de la bujía número uno.

tinning a wire—when soldering, a wire is preheated and has solder applied to it.

preparación al estañado—al soldar, se precalienta un alambre y se le aplica la soldadura.

tire plug—a piece of rubber vulcanized into a hole in a tire.

tapón de neumático—un pedazo de caucho vulcanizado en un hoyo de un neumático.

toe—the inner end of a gear tooth.

interior del engranaje—la extremidad interior de un diente de un engranaje.

toe alignment—a comparison of the distances between the fronts and the rears of a pair of tires.

alineación convergencia o divergencia—una comparación de las distancias entre las partes delanteras y las traseras de un par de neumáticos.

toe change—when toe changes in response to changes in ride height.

cambio de divergencia—cuando la divergencia cambia como respuesta a los cambios de la altura del viaje.

toe-in—when the tires are closer together at the front.

convergencia—cuando los neumáticos son más cercas en la parte delantera.

toe-out—when the tires are further apart at the front.

divergencia—cuando los neumáticos quedan más lejos en la parte delantera.

toe-out-on-turns—also called Ackerman angle or turning radius. The tires toe out during a turn because the steering arms are bent at an angle.

divergencia-en-viraje—tambien llamado ángulo Ackerman o el radio de viraje. Los neumáticos divergen durante una viraje porque los brazos de dirección son doblados en un ángulo.

tolerance—the range of measurement that is allowable for a part.

tolerancia—el límite de medida que se permite en una parte.

ton—12,000 Btu/hr.

tonelada—12,000 Btu/hr.

tone ring—a toothed ring that spins with the wheel. Also called exciter ring.

anillo de tono—también llamado anillo de excitador; un anillo con dientes que gira con la rueda.

tons rating—how much heat needs to be added in 24 hours to turn 1 ton of ice at 32°F into water.

potencia indicada de tonelada—cuánto calor se requiere añadir en 24 horas para convertir una tonelada del hielo en 32° al agua.

torque—the tendency of force to rotate a body on which it acts.

torsión—la tendencia de la fuerza en girar un cuerpo en el cual obra.

torque converter—a fluid coupling that multiplies torque.

convertidor del par—un coplamiento fluido que multiplica la torsión.

torque steer—when there are different length shafts, the long one wraps up and there is a slight lag before it puts its torque to the wheel, causing the car to pull to the passenger side.

torsión de dirección—cuando hay ejes de distintas longitudes, el eje largo se tuerza y hay una pequeña pausa antes de que aplica su torsión a la rueda, causando que el coche jale al lado del pasajero.

torque-to-yield—when head bolts are purposely torqued to within 2% of their yield point.

torsión-al-límite—cuando los pernos de la cabeza se tuerzan a propósito dentro del 2% de su límite de proporcionalidad.

torque tube—an enclosed drive shaft.

tubo de torsión—un eje encerrado.

torque turn—when a bolt is tightened to a specified torque and then turned an additional 35 to 180 degrees.

viraje de torsión—cuando un perno se aprieta a un par específico y luego se le da una vuelta adicional de 35 a 80 grados.

torsional vibration—vibrations occurring in the crankshaft when force on the pistons is imparted to the crankshaft of a V-type engine.

vibración torsional—las vibraciones ocurriendo en el cigüeñal cuando la fuerza en los pistones se transfiere al cigüeñal de un motor tipo V.

torsion bar—a straight rod that twists when working as a spring.

barra de torsión—una varilla recta que se retuerza cuando funciona como un resorte.

track—the side-to-side distance between an axle's tires.

pista—la distancia de lado a lado entre los neumáticos de un eje.

tracking—a term that refers to the relationship between the average direction that the rear tires point, when compared to the front tires.

seguimiento—un término que serefiere a la relación entre la dirección normal hacia cual apuntan los neumáticos traseros, al ser comparado con los neumáticos delanteros.

traction—how well a tire grips the road.

tracción—la habilidad con la cual un neumático agarra la carretera.

transaxle—a unit that combines a transmission and differential.

flecha de transmisión—una unidad que combina una transmisión y un diferencial.

transducer—a device that converts energy from one form to another. It is a magnet that is spring loaded on both sides. It moves back and forth in a coil of wire to send a signal to a meter in response to vibration.

transductor—un dispositivo que convierta la energía de una forma a otra. Es un imán cargado por resorte en ambos lados. Oscila en una bobina de alambre para mandar una señal a un aparato de medida como respuesta a la vibración.

transistor—an electronic relay.

transistor—un relé electrónico.

transmission oil cooler—also called a heat exchanger, it is usually in the bottom or side of the radiator that has automatic transmission fluid running through it.

enfriador del aceite de la transmisión—tambien llamado un intercambiador de calor, suele estar en la parte inferior o en un lado del radiador por el cual atraviesa el fluido de transmisión.

transponder key—used by some late-model systems as a theft deterrent, it receives a radio signal each time the engine is started and disables the starting, fuel, and ignition systems if the signal is not as expected.

llave de transpondedor—utilizado en ciertos sistemas de modelos recientes para disuadir el robo; recibe una señal de radio cada vez que se da arranque al motor y bloquea los sistemas de arranque, combustible e ignición si la señal no es la esperada.

tread—the section of the tire that rides on the road.

banda de rodamiento—la sección de un neumático que viaja en la carretera.

tread wear indicator—also called a wear bar, this is molded into the tire's tread and indicates when 1⁄16" of tread is remaining.

indicador del desgaste de la banda—tambien llamado barra del desgaste, esto está moldeado dentro de la banda de rodamiento de un neumático e indica cuando le queda un 1⁄16" de uso.

trigger—switching device.

gatillo—un dispositivo de conmutación.

trip—when the ignition switch has been off for a period of time, when the engine is restarted and the vehicle is driven, various emission control monitors on the vehicle operate to complete one trip.

disparo—cuando el interruptor de la ignición ha estado apagado por un período largo; cuando se vuelve a dar arranque al motor y se maneja el vehículo, funcionan varios monitores de control de emisiones en el vehículo para completar un disparo.

TSB—technical service bulletin.

TSB—boletín de servicio técnico.

tulip—the yoke or housing of a tripod CV plunge joint. It has three grooves that the rollers can roll in and out as the axle length changes.

tulipán—la brida o el cárter de una junta trípoda tipo CV. Tiene tres muescas en las cuales pueden rodar los rodillos adentro y afuera al alargar el eje.

turbo lag—the time required to bring the turbo up to a speed where it can function effectively.

retraso del turbo—el tiempo requerido en accionar el turbo a una velocidad en la cual puede funcionar eficazmente.

turbocharger—a small radial fan pump driven by the energy of the exhaust flow.

turbocompresor—un pequeño ventilador radial impulsado por la energía del flujo del escape.

turbocharger cartridge—a new or rebuilt replacement unit.

bote del turbocompresor—una unidad de reemplazo nueva o reconstruida.

turnbuckle—a part between two rods. Turning it one way shortens a rod assembly. Turning it the other way lengthens it.

tensor de tornillo—una parte entre dos varillas. Dándole vuelta en una dirección acorta la asamblea de varilla. Volteándolo en la otra dirección lo alarga.

turning radius—also called Ackerman angle or toe-out-on-turns. The tires toe out during a turn because the steering arms are bent at an angle.

radio de viraje—tambien llamado el ángula Ackerman o divergencia-en-vueltas. Los neumáticos divergen en las vueltas porque los brazos de dirección son torcidos en un ángulo.

TVRS—television/radio suppression resistor spark plug cables made of braided aramid fiber impregnated with graphite and latex.

TVRS—los cables de bujía con resistores de supresión de la televisión/radio hechos de fibra aramid impregnada con el grafito y el látex.

TVS—thermal vacuum switch; also called a ported vacuum switch (PVS), it is found in a cooling passage on precomputer cars. It controls the flow of vacuum according to engine temperature.

TVS—interruptor de vacío termal; tambien llamado un interruptor de vacío con lumbrera, se encuentra en los pasos de enfriamiento en los coches precomputerizados. Controla el flujo del vacío conforme a la temperatura del motor.

two-gas analyzer—an older infrared exhaust analyzer that measures HC and CO.

analizador de dos gases—un analizador de escape infrarrojo más antiguo que mide HC y CO.

two-stroke engine—an engine that completes its entire operation cycle in one turn of the crankshaft.

motor de dos carreras—un motor que completa su ciclo de operación completo con una vuelta del cigüeñal.

TXV—thermostatic expansion valve; a device that controls the amount of refrigerant allowed to flow to the evaporator.

TXV—válvula de expansión del termostato; un dispositivo que controla la cantidad del fluido refrigerante que fluye al evaporador.

type A code—a type of diagnostic trouble code that can result in damage to the catalyst; it is always emissions related.

código del tipo A—un tipo de código de problema de diagnóstico que puede provocar daños en el catalizador; siempre está asociado a las emisiones.

type I headlamp—has high beam only.

faro tipo I—sólo tiene la luz larga.

type II headlamp—has both low and high beams.

faro tipo II—tiene ambas luces cortas y largas.

U

underhood emission control label—a label found on cars since 1972 that describes the size of the engine, ignition timing specifications, idle speed, valve lash clearance adjustment, and the emission devices that are included on the engine.

etiqueta de control de emisión bajo el capót—una etiqueta encontrada en los coches desde el 1972 que describe el tamaño del motor, las especificaciones del encendido, la velocidad de marcha en vacío, los ajustes de holgura del juego de las válvulas, y los dispositivos de emisión que se incluyen en el motor.

undersquare—when the bore is less than the stroke.

subcuadrado—cuando el diámetro interior es menos que la carrera.

understeer—when a car does not seem to respond to movement of the steering wheel during a hard turn.

subdirección—cuando un coche no parece responder al movimiento del volante de dirección durante un viraje abrupto.

unibody—a chassis design that has a floorpan and a small subframe section in the front and rear.

monocasco—un diseño del chasis que tiene un piso y una pequeña sección del armazón en la parte delantera y trasera.

unidirectional—when the scan tool can read data but cannot give commands to the computer.

unidireccional—cuando un aparato explorador puede leer los datos, pero no puede enviar los órdenes a la computadora.

union—a connection that joins two pieces of tubing together often used on vacuum or air pressure lines.

unión—una conexión que une dos piezas de tubo que suele usarse en las líneas de vacío o de presión de aire.

unloaded caliper—a bare rebuilt caliper with no pads or hardware.

calibre descargado—un calibre reconstruido desprovisto de las almohadillas o la herramienta.

unsprung weight—the weight of parts of the suspension that are not supported by springs. It affects vehicle handling.

peso desprovisto de muelles—el peso de las partes de suspensión que no se soportan con muelles. Afecta el manejo del vehículo.

upper end—parts of the upper end of the engine include the cylinder head(s) and valvetrain.

extremidad superior—las partes de la extremidad superior del motor incluye la(s) cabeza(s) y el tren de válvulas.

upshift—when the transmission shifts from a low gear to a higher gear; second to third, for instance.

cambiar a una velocidad más alta—cuando la transmisión cambia desde una velocidad baja a una más alta; de la segunda a la tercera, por ejemplo.

UTQG—uniform tire quality grade system that rates tread wear, traction, and temperature resistance.

UTQG—sistema uniforme de calificar la calidad de los neumáticos que valúa el gasto de la banda, la tracción y la resistencia a la temperatura.

V

V-ribbed belt—a belt that has multiple ribs on one side and is flat on the other side.

banda con refuerzas en V—una banda que tiene varias nervaduras en un lado y está plana del otro lado.

vacuum advance—part of the distributor that senses engine load and changes the timing to compensate.

avance de vacío—una parte del distribuidor que detecta la carga del motor y cambia el encendido para compensar.

vacuum motor—a vacuum-operated device that opens and closes doors or valves.

motor de vacío—un dispositivo operado por vacío que abre y cierra las puertas o las válvulas.

valve body—the hydraulic control assembly of the transmission.

cuerpo de la válvula—la asamblea del control hidráulico de la transmisión.

valve clearance—also called valve lash; usually adjustable clearance in the valvetrain.

holgura de las válvulas—también llamada juedo de las válvulas; la holgura que normalmente se puede ajustar en el tren de válvulas.

valve overlap—the number of degrees of crankshaft rotation that occurs during the time that both the intake and exhaust valves are open at the same time.

solape de las válvulas—el número de grados de rotación del cigüeñal que ocurre durante el tiempo en que ambas válvulas de admisión y escape están abiertas al mismo tiempo.

valve port—a passage that allows the flow of intake or exhaust into or out of the cylinder.

puerto de válvula—un paso que permite el flujo de admisión o del escape entrar or salir del cilindro.

valve spring installed height—a measurement that increases with machining of the valve and valve seat. Shims can be used to restore the correct height and spring tension.

altura instalada del resorte de la válvula—una medida que aumenta cuando la válvula y el asiento de la válvula se han mecanizado. Las chapas se pueden usar para restaurar la altura y la tensión del resorte correcta.

valve stem height—a measurement that reflects how much a valve and valve seat have been ground.

altura del vástago de la válvula—una medida que refleja cuanto se han rectificado la válvula y el asiento de la válvula.

valvetrain—the parts that open and close the valves.

tren de válvula—las partes que abren y cierran las válvulas.

vane-type pump—a type of oil pump that has vane drives.

bomba de excéntricas—un tipo de bomba de aceite que tiene paletas.

vapor lock—a condition that occurs to a carbureted car when fuel boils in the fuel line or filter.

cierre de vapor—una condición que ocurre en un coche con carburador cuando hierve el combustible en la línea del combustible o en el filtro.

vaporization—when atomized fuel turns into a gas.

vaporización—cuando un combustible atomizado se convierta en gas.

variable DC voltage generator—a zirconia oxygen sensor is a galvanic battery that can generate a voltage of 0.1 volt to 0.9 volt (100 to 900 millivolts) in response to the amount of oxygen in the vehicle's exhaust.

generador de voltaje DC variable—un sensor de oxígeno zirconía es una batería galvánico que puede generar un voltaje de 0.1 voltio a 0.9 voltio (100 a 900 milivoltios) como respuesta a la cantidad del oxígeno en el escape del vehículo.

variable displacement turbo—a turbocharger that changes its nozzle opening in response to changes in engine load.

turbo de desplazamiento variable—un turbocargador que cambia la apertura de la boquilla como respuesta a los cambios en la carga en el motor.

variable dwell—in an electronic ignition system, when at low engine rpm, dwell is shorter and at higher rpm it is longer.

parada variable de movimiento—en un sistema de encendido electrónico cuando en bajas rpm del motor, la parada es más corta y en rpm más altas es más larga.

variable rate spring—a tapered rod spring that gives a smoother ride when going over smaller bumps and still allows for heavier carrying capacity.

muelle de capacidad variable—un muelle de varilla cónico que proporciona un viaje más uniforme al pasar por los topes más chiquitos y que todavía permite una capacidad de cargas pesadas.

variable resistors—called rheostats or potentiometers, used to control speed and intensity of electrical loads.

resistores variables—llamados reostatos o potenciómetros, usados en controlar la velocidad y la intensidad de las cargas eléctricas.

varnish—forms in a transmission when the fluid has become oxidized from heat.

vidriado—forma en una transmisión cuando el fluido se ha oxidado debido al calor.

VE—volumetric efficiency; the measurement comparing the volume of airflow actually entering the engine with the maximum that theoretically could enter it.

VE—rendimiento volumétrico; la medida que compara el volumen del flujo del aire actualmente entrando al motor con lo máximo que podría entrar teoréticamente.

velocity—a combination of amplitude and frequency.

velocidad—una combinación del amplitud y frequencia.

venturi—a smaller area in the carburetor that restricts airflow resulting in lower pressure to draw fuel from the float bowl. Air also speeds up due to the restriction.

venturi—una área pequeña en el carburador que restringe el flujo del aire resultando en que la presión más baja aspira el combustible de la taza del flotador. El aire tambien se acelera debido a la restricción.

vibration—a part that is in motion in waves, or cycles.

vibración—una parte en movimiento en forma de ondas, o ciclos.

VIN—vehicle identification number.

VIN—siglas en inglés para número de identificación del vehículo.

viscosity—a measurement of how easily a liquid can flow.

viscosidad—una medida de qué tal facilmente puede fluir un líquido.

viscosity index—an oil's ability to resist change in viscosity as it heats up.

índice de la viscosidad—la habilidad del aceite de resistir un cambio en viscosidad al calentarse.

viscous coupling—a sealed clutch unit that contains silicone fluid.

acoplador viscoso—una unidad hermética del embrague que contiene el silicio líquido.

volatile RAM—it is erased each time the ignition is turned off.

RAM volátil—se borra cada vez que se apaga el interruptor del encendido.

volatility—a measure of how easy a fuel evaporates.

volatilidad—una medida de qué tal facilmente se evapora un combustible.

volt—the measurement of electrical pressure.

voltio—la medida de presión eléctrica.

voltage dividers—variable resistors that produce a variable DC voltage signal.

divisor de tensión—los resistores variables que producen una señal de voltaje DC variable.

voltage drop—a loss of voltage caused by current flow through a resistance.

caída de voltaje—la pérdida del voltaje causada por el flujo del corriente a través de una resistencia.

voltage drop testing—a test using a voltmeter to check for resistance in a loaded circuit.

prueba de caída de voltaje—una prueba usando un voltímetro para verificar la resistencia en un circuito cargado.

voltage regulator—a device that controls the strength of the electromagnetic field in the rotor of an alternator.

regulador del voltaje—un dispositivo que controla la fuerza del campo electromagnético en el rotor de un alternador.

voltage spike—a high voltage that occurs while disconnecting a powered circuit lead that happens because electrons that were in motion before the circuit was broken back up at the connection. When they try to push their way across the gap, the spike is created.

impulsos de tensión—un voltaje alto que ocurre al desconectar un hilo de circuito bajo tensión que pasa porque los electrones estaban en movimiento antes de que se abriera el circuito, en la conexión. Cuando tratan a traversar la abertura, se crea el impulso.

vortex—a spiral pattern of fluid flow from impeller to turbine.

torbellino—un patrón espiral del flujo del fluido del impulsor a la turbina.

vulcanized—heating rubber to make it stable.

vulcanizado—calentando al caucho para estabilizarlo.

vulcanizing—a process that cures rubber by heating it.

vulcanizando—un proceso que cura el caucho calentándolo.

W

Wankel engine—a rotary engine design.

motor Wankel—un diseño de motor rotativo.

warm-up cycle—a cycle that occurs every time the engine cools off and temperature rises at least 40°F.

ciclo de recalentamiento—un ciclo que ocurre cada vez que un motor se enfría y la temperatura llega al menos a los 40°F.

waste gate—a relief valve for excessive turbocharger boost.

compuerta de residuos—una válvula de relieve para soplante excesivo del turbocompresor.

waste spark method—when a distributorless ignition system has both ends of each coil's secondary winding attached to one of a pair of companion cylinders spark plugs and both cylinders' spark plugs fire every revolution of the crankshaft.

método de encendido perdido—cuando un sistema de encendido sin distribuidor tiene ambas extremidades del devanado secundaria de cada bobina conectada a un par de bujías en cilindros acoplados y la bujía de ambos cilindros disparan cada revolución del cigüeñal.

watt—the horsepower equivalent in the metric system. One horsepower equals 0.746 kw.

vatio—la potencia de caballo equivalente en el sistema métrico. Una potencia de caballo igual a 0.746 kw.

WD—warehouse distributor.

WD—siglas en inglés para distribuidor de depósito.

wear indicator ball joint—a ball joint with a shoulder that indicates when it is worn out.

articulación esférica indicador del desgaste—una articulación esférica con un collar que indica cuando está gastado.

wedge combustion chamber—a turbulent combustion chamber design shaped like a wedge.

cámara de combustión en forma de cuña—un diseño de cámara de combustión turbulente en forma de cuña.

wet compression test—a compression test where oil is squirted into the cylinder to test for worn rings.

prueba de compresión en húmedo—una prueba de compresión en la cual se chorrea el aceite dentro del cilindro para comprobar el gasto en los anillos.

wheel base—the distance between the front and rear tires.

batalla (empate)—la distancia entre los neumáticos delanteros y traseros.

wheel bearing—the term for a non-drive front and rear wheel bearing.

cojinete de las ruedas—el término para un cojinete no impulsor de las ruedas delanteras y traseras.

wheel cylinder kit—a parts kit for rebuilding wheel cylinders.

piezas de reconstrucción del cilindro de la rueda—el conjunto de las partes sueltas para reconstruir los cilindros de las ruedas.

whine—a sound like a turbine or mosquitoes. The range is 300–500 Hz. It is usually related to gear noise.

gemido—un ruido como una turbina o los mosquitos. La gama es del 300 a 500 Hz. Suele ser relacionado al ruido de los engranajes.

wide ratio transmission—when there is a large difference between the ratios of a transmission's forward gears.

transmisión de relación extensiva—cuando hay una gran diferencia entre las relaciones de las velocidades de marcha en adelante de una transmisión.

work—when an object is moved against a resistance or opposing force.

trabajo—cuando un objeto se mueva contra una resistencia o una fuerza opuesta.

work harden—when a material is flexed or pounded on and it becomes harder.

endurecido por acritúd—cuando una materia es doblada o molida y se endurece.

worm gear clamp—one of the most popular hose clamps for all applications except fuel injection.

abrazadera graduable—una de las abrazaderas de tubo flexible más populares para todas aplicaciones menos las de inyección del combustible.

WOT—wide-open throttle.

WOT—acelerador completamente abierto.

wye (Y) winding—an alternator stator winding in which three leads are branched out in a "Y" pattern.

devanado estrella (Y)—un devanado del estator del alternador en el cual tres hilos se extienden para formar una "Y".

Z

zener diode—a semiconductor used to control transistors and electronic voltage regulators that only conducts electricity when a certain voltage is reached.

diodo zener—un semiconductor usado para controlar los transistores y los reguladores de voltaje electrónicos que sólo conduce la electricidad llegando a un cierto voltaje.

zerk fitting—a grease fitting threaded to fit into holes in the part to be lubricated.

niple de grasa—una válvula de lubricación enroscada que se ajusta en un agujero de una parte para que se puede lubricar.

zero-lash—when there is no clearance and no interference between parts of the valvetrain.

juego de cero—cuando no hay holgura ni interferencia entre las partes del tren de válvulas.

ZEV—zero-emission vehicle.

ZEV—vehículo de cero emisión.

INDEX

Page numbers followed by a "f" indicate that the entry is included in a figure.

IMPORTANT-READ CAREFULLY: This End User License Agreement ("Agreement") sets forth the conditions by which Delmar Learning, a division of Thomson Learning Inc. ("Thomson") will make electronic access to the Thomson Delmar Learning-owned licensed content and associated media, software, documentation, printed materials and electronic documentation contained in this package and/or made available to you via this product (the "Licensed Content"), available to you (the "End User"). BY CLICKING THE "I ACCEPT" BUTTON AND/OR OPENING THIS PACKAGE, YOU ACKNOWLEDGE THAT YOU HAVE READ ALL OF THE TERMS AND CONDITIONS, AND THAT YOU AGREE TO BE BOUND BY ITS TERMS CONDITIONS AND ALL APPLICABLE LAWS AND REGULATIONS GOVERNING THE USE OF THE LICENSED CONTENT.

1.0 SCOPE OF LICENSE

1.1 <u>Licensed Content.</u> The Licensed Content may contain portions of modifiable content ("Modifiable Content") and content which may not be modified or otherwise altered by the End User ("Non-Modifiable Content"). For purposes of this Agreement, Modifiable Content and Non-Modifiable Content may be collectively referred to herein as the "Licensed Content." All Licensed Content shall be considered Non-Modifiable Content, unless such Licensed Content is presented to the End User in a modifiable format and it is clearly indicated that modification of the Licensed Content is permitted.

1.2 Subject to the End User's compliance with the terms and conditions of this Agreement, Thomson Delmar Learning hereby grants the End User, a nontransferable, non-exclusive, limited right to access and view a single copy of the Licensed Content on a single personal computer system for noncommercial, internal, personal use only. The End User shall not (i) reproduce, copy, modify (except in the case of Modifiable Content), distribute, display, transfer, sublicense, prepare derivative work(s) based on, sell, exchange, barter or transfer, rent, lease, loan, resell, or in any other manner exploit the Licensed Content; (ii) remove, obscure or alter any notice of Thomson Delmar Learning's intellectual property rights present on or in the License Content, including, but not limited to, copyright, trademark and/or patent notices; or (iii) disassemble, decompile, translate, reverse engineer or otherwise reduce the Licensed Content.

2.0 TERMINATION

2.1 Thomson Delmar Learning may at any time (without prejudice to its other rights or remedies) immediately terminate this Agreement and/or suspend access to some or all of the Licensed Content, in the event that the End User does not comply with any of the terms and conditions of this Agreement. In the event of such termination by Thomson Delmar Learning, the End User shall immediately return any and all copies of the Licensed Content to Thomson Delmar Learning.

3.0 PROPRIETARY RIGHTS

3.1 The End User acknowledges that Thomson Delmar Learning owns all right, title and interest, including, but not limited to all copyright rights therein, in and to the Licensed Content, and that the End User shall not take any action inconsistent with such ownership. The Licensed Content is protected by U.S., Canadian and other applicable copyright laws and by international treaties, including the Berne Convention and the Universal Copyright Convention. Nothing contained in this Agreement shall be construed as granting the End User any ownership rights in or to the Licensed Content.

3.2 Thomson Delmar Learning reserves the right at any time to withdraw from the Licensed Content any item or part of an item for which it no longer retains the right to publish, or which it has reasonable grounds to believe infringes copyright or is defamatory, unlawful or otherwise objectionable.

4.0 PROTECTION AND SECURITY

4.1 The End User shall use its best efforts and take all reasonable steps to safeguard its copy of the Licensed Content to ensure that no unauthorized reproduction, publication, disclosure, modification or distribution of the Licensed Content, in whole or in part, is made. To the extent that the End User becomes aware of any such unauthorized use of the Licensed Content, the End User shall immediately notify Delmar Learning. Notification of such violations may be made by sending an Email to delmarhelp@thomson.com.

5.0 MISUSE OF THE LICENSED PRODUCT

5.1 In the event that the End User uses the Licensed Content in violation of this Agreement, Thomson Delmar Learning shall have the option of electing liquidated damages, which shall include all profits generated by the End User's use of the Licensed Content plus interest computed at the maximum rate permitted by law and all legal fees and other expenses incurred by Thomson Delmar Learning in enforcing its rights, plus penalties.

6.0 FEDERAL GOVERNMENT CLIENTS

6.1 Except as expressly authorized by Delmar Learning, Federal Government clients obtain only the rights specified in this Agreement and no other rights. The Government acknowledges that (i) all software and related documentation incorporated in the Licensed Content is existing commercial computer software within the meaning of FAR 27.405(b)(2); and (2) all other data delivered in whatever form, is limited rights data within the meaning of FAR 27.401. The restrictions in this section are acceptable as consistent with the Government's need for software and other data under this Agreement.

7.0 DISCLAIMER OF WARRANTIES AND LIABILITIES

7.1 Although Thomson Delmar Learning believes the Licensed Content to be reliable, Thomson Delmar Learning does not guarantee or warrant (i) any information or materials contained in or produced by the Licensed Content, (ii) the accuracy, completeness or reliability of the Licensed Content, or (iii) that the Licensed Content is free from errors or other material defects. THE LICENSED PRODUCT IS PROVIDED "AS IS," WITHOUT ANY WARRANTY OF ANY KIND AND THOMSON DELMAR LEARNING DISCLAIMS ANY AND ALL WARRANTIES, EXPRESSED OR IMPLIED, INCLUDING, WITHOUT LIMITATION, WARRANTIES OF MERCHANTABILITY OR FITNESS OR A PARTICULAR PURPOSE. IN NO EVENT SHALL THOMSON DELMAR LEARNING BE LIABLE FOR: INDIRECT, SPECIAL, PUNITIVE OR CONSEQUENTIAL DAMAGES INCLUDING FOR LOST PROFITS, LOST DATA, OR OTHERWISE. IN NO EVENT SHALL DELMAR LEARNING'S AGGREGATE LIABILITY HEREUNDER, WHETHER ARISING IN CONTRACT, TORT, STRICT LIABILITY OR OTHERWISE, EXCEED THE AMOUNT OF FEES PAID BY THE END USER HEREUNDER FOR THE LICENSE OF THE LICENSED CONTENT.

8.0 GENERAL

8.1 <u>Entire Agreement</u>. This Agreement shall constitute the entire Agreement between the Parties and supercedes all prior Agreements and understandings oral or written relating to the subject matter hereof.

8.2 <u>Enhancements/Modifications of Licensed Content</u>. From time to time, and in Delmar Learning's sole discretion, Thomson Thomson Delmar Learning may advise the End User of updates, upgrades, enhancements and/or improvements to the Licensed Content, and may permit the End User to access and use, subject to the terms and conditions of this Agreement, such modifications, upon payment of prices as may be established by Delmar Learning.

8.3 <u>No Export</u>. The End User shall use the Licensed Content solely in the United States and shall not transfer or export, directly or indirectly, the Licensed Content outside the United States.

8.4 <u>Severability</u>. If any provision of this Agreement is invalid, illegal, or unenforceable under any applicable statute or rule of law, the provision shall be deemed omitted to the extent that it is invalid, illegal, or unenforceable. In such a case, the remainder of the Agreement shall be construed in a manner as to give greatest effect to the original intention of the parties hereto.

8.5 <u>Waiver</u>. The waiver of any right or failure of either party to exercise in any respect any right provided in this Agreement in any instance shall not be deemed to be a waiver of such right in the future or a waiver of any other right under this Agreement.

8.6 <u>Choice of Law/Venue</u>. This Agreement shall be interpreted, construed, and governed by and in accordance with the laws of the State of New York, applicable to contracts executed and to be wholly preformed therein, without regard to its principles governing conflicts of law. Each party agrees that any proceeding arising out of or relating to this Agreement or the breach or threatened breach of this Agreement may be commenced and prosecuted in a court in the State and County of New York. Each party consents and submits to the non-exclusive personal jurisdiction of any court in the State and County of New York in respect of any such proceeding.

8.7 <u>Acknowledgment</u>. By opening this package and/or by accessing the Licensed Content on this Website, THE END USER ACKNOWLEDGES THAT IT HAS READ THIS AGREEMENT, UNDERSTANDS IT, AND AGREES TO BE BOUND BY ITS TERMS AND CONDITIONS. IF YOU DO NOT ACCEPT THESE TERMS AND CONDITIONS, YOU MUST NOT ACCESS THE LICENSED CONTENT AND RETURN THE LICENSED PRODUCT TO THOMSON DELMAR LEARNING (WITHIN 30 CALENDAR DAYS OF THE END USER'S PURCHASE) WITH PROOF OF PAYMENT ACCEPTABLE TO DELMAR LEARNING, FOR A CREDIT OR A REFUND. Should the End User have any questions/comments regarding this Agreement, please contact Thomson Delmar Learning at <u>delmarhelp@thomson.com</u>.